# Advanced Modern Algebra

## Second Edition

Joseph J. Rotman

Graduate Studies
in Mathematics

Volume 114

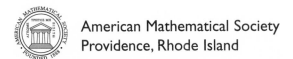

American Mathematical Society
Providence, Rhode Island

First edition © 2002 held by the American Mathematical Society

The 2002 edition of this book was previously published by Pearson Education, Inc.

2000 *Mathematics Subject Classification.* Primary 11–XX, 12–XX, 13–XX, 15–XX, 16–XX, 18–XX, 19–XX, 20–XX.

For additional information and updates on this book, visit
**www.ams.org/bookpages/gsm-114**

**Library of Congress Cataloging-in-Publication Data**

Rotman, Joseph J., 1934–
    Advanced modern algebra / Joseph J. Rotman. — 2nd ed.
        p. cm. — (Graduate studies in mathematics ; v. 114)
    Includes bibliographical references and index.
    ISBN 978-0-8218-4741-1 (alk. paper)
    1. Algebra.  I. Title.  II. Series.

QA154.3.R68   2010
512—dc22

2009052217

To my wife

Marganit

and our two wonderful kids

Danny and Ella,

whom I love very much

# Contents

# Preface to Second Edition

Algebra is used by virtually all mathematicians, be they analysts, combinatorists, computer scientists, geometers, logicians, number theorists, or topologists. Nowadays, everyone agrees that some knowledge of Linear Algebra, Group Theory, and Commutative Algebra is necessary, and these topics are introduced in undergraduate courses. We continue their study.

This book can be used as a text for the first year of graduate Algebra, but it is much more than that. It can also serve more advanced graduate students wishing to learn topics on their own. While not reaching the frontiers, the book does provide a sense of the successes and methods arising in an area. In addition, this is a reference containing many of the standard theorems and definitions that users of Algebra need to know. Thus, this book is not merely an appetizer; it is a hearty meal as well.

When I was a student, Birkhoff–Mac Lane, *A Survey of Modern Algebra*, was the text for my first Algebra course, and van der Waerden, *Modern Algebra*, was the text for my second course. Both are excellent books (I have called this book *Advanced Modern Algebra* in homage to them), but times have changed since their first publication: Birkhoff and Mac Lane's book appeared in 1941; van der Waerden's book appeared in 1930. There are today major directions that either did not exist 75 years ago, or were not then recognized as being so important, or were not so well developed. These new areas involve Algebraic Geometry, Category Theory,[1] Computer Science, Homological Algebra, and Representation Theory.

Let me now address readers and instructors for whom this book is a text in a beginning graduate course. Instead of devoting the first chapters to a review of more elementary material (as I did in the first edition), here I usually refer to FCAA (my book, *A First Course in Abstract Algebra*, 3rd ed.). I have reorganized and rewritten this text, but here are some other major differences with the first edition.

---

[1] *A Survey of Modern Algebra* was rewritten, introducing categories, as Mac Lane–Birkhoff, *Algebra*, Macmillan, New York, 1967.

Noncommutative rings are now discussed earlier, so that left and right modules can appear at the same time. The existence of free groups is shown more simply (using only deletions instead of deletions and insertions). In fact, much of the section on presentations has been rewritten. Division rings and Brauer groups are introduced in the same chapter as the Wedderburn–Artin Theorems; their discussion then continues after cohomology has been studied. The section on Grothendieck groups has been completely rewritten. Some other items appearing here that were not treated in the first edition are Galois Theory for infinite extensions, the Normal Basis Theorem, abelian categories, and module categories.

The Table of Contents enumerates the highlights of each chapter; here are more details. Chapters 1 and 2 present the elements of Group Theory and of Commutative Algebra, along with Linear Algebra over arbitrary fields.

Chapter 3 discusses Galois Theory for finite field extensions (Galois Theory for infinite field extensions is discussed in Chapter 6, after inverse limits have been introduced). We prove the insolvability of the general polynomial of degree 5 and the Fundamental Theorem of Galois Theory. Among the applications are the Fundamental Theorem of Algebra and Galois's Theorem that a polynomial over a field of characteristic 0 is solvable by radicals if and only if its Galois group is a solvable group. The chapter ends by showing how to compute Galois groups of polynomials of degree $\leq 4$.

Chapter 4 continues the study of groups, beginning with the Basis Theorem and the Fundamental Theorem for finite abelian groups (finitely generated abelian groups are discussed in Chapter 8). We then prove the Sylow Theorems (which generalize the Primary Decomposition to nonabelian groups), discuss solvable groups, simplicity of the linear groups $PSL(2, k)$, unitriangular groups, free groups, presentations, and the Nielsen–Schreier Theorem (subgroups of free groups are free).

Chapter 5 continues the study of commutative rings, with an eye to discussing polynomial rings in several variables; unique factorization domains; Hilbert's Basis Theorem; applications of Zorn's Lemma (including existence and uniqueness of algebraic closures, transcendence bases, and inseparability); Lüroth's Theorem; affine varieties; Nullstellensatz over $\mathbb{C}$ (the full Nullstellensatz, for varieties over arbitrary algebraically closed fields, is proved in Chapter 10); primary decomposition of ideals; the Division Algorithm for polynomials in several variables; Buchberger's algorithm and Gröbner bases.

Chapter 6 introduces noncommutative rings, left and right $R$-modules; categories, functors, natural transformations, and categorical constructions; free modules, projectives, and injectives; tensor products; adjoint functors; flat modules; inverse and direct limits; infinite Galois Theory.

Chapter 7 continues the study of noncommutative rings, aiming toward Representation Theory of finite groups: chain conditions; Wedderburn's Theorem on finite division rings; Jacobson radical; Wedderburn–Artin Theorems classifying semisimple rings. These results are applied, using character theory, to prove Burnside's Theorem (finite groups of order $p^m q^n$ are solvable). After discussing multiply transitive groups, we prove Frobenius's Theorem that Frobenius kernels are normal subgroups of Frobenius groups. Since division rings have arisen naturally, we introduce Brauer groups. Chapter 9 views Brauer groups as cohomology groups, allowing us to prove existence of division rings of positive characteristic. This chapter ends with abelian categories and the characterization of categories of modules, enabling us to see why a ring $R$ is intimately related to the matrix rings $\text{Mat}_n(R)$ for $n \geq 1$.

Chapter 8 considers finitely generated modules over principal ideal domains (generalizing earlier theorems about finite abelian groups), and then goes on to apply these results to rational, Jordan, and Smith canonical forms for matrices over a field (the Smith normal form yields algorithms that compute elementary divisors of matrices). We also classify projective, injective, and flat modules over PIDs. Bilinear forms are introduced, along with orthogonal and symplectic groups. We then discuss some multilinear algebra, ending with an introduction to Lie Algebra.

Chapter 9 introduces homological methods, beginning with abstract simplicial complexes and group extensions, which motivate homology and cohomology groups, and Tor and Ext (existence of these functors is proved with derived functors). Applications are made to modules, cohomology of groups, and division rings. A descriptive account of spectral sequences is given, sufficient to indicate why Ext and Tor are independent of the variable resolved. We then pass from Homological Algebra to Homotopical Algebra: Algebraic $K$-Theory is introduced with a discussion of Grothendieck groups.

Chapter 10 returns to Commutative Algebra: localization, the general Nullstellensatz (using Jacobson rings), Dedekind rings and some Algebraic Number Theory, and Krull's Principal Ideal Theorem. We end with more Homological Algebra, proving the Serre–Auslander–Buchsbaum Theorem characterizing regular local rings as those noetherian local rings of finite global dimension and the Auslander–Buchsbaum Theorem that regular local rings are UFDs.

Each generation should survey Algebra to make it serve the present time.

It is a pleasure to thank the following mathematicians whose suggestions have greatly improved my original manuscript: Robin Chapman, Daniel R. Grayson, Ilya Kapovich, T.-Y. Lam, David Leep, Nick Loehr, Randy McCarthy, Patrick Szuta, and Stephen Ullom; and I give special thanks to Vincenzo Acciaro for his many comments, both mathematical and pedagogical, which are incorporated throughout the text.

Joseph Rotman

Urbana, IL, 2009

# Special Notation

| | | | |
|---|---|---|---|
| $\lvert X\rvert$ | cardinal number of set $X$ | $\mathbb{C}$ | complex numbers |
| $\mathbb{N}$ | natural numbers | $\mathbb{Q}$ | rational numbers |
| $\mathbb{R}$ | real numbers | $\mathbb{Z}$ | integers |
| $1_X$ | identity function on set $X$ | $A^{\top}$ | transpose of matrix $A$ |

# Chapter 1

# Groups I

### Section 1.1. Classical Formulas

As Europe emerged from the Dark Ages, a major open problem in Mathematics was finding roots of polynomials. The quadratic formula had been known for over two thousand years and, arising from a tradition of public mathematical contests in Venice and Tuscany, formulas for the roots of cubics and quartics had been found in the early 1500s.

Consider the cubic $F(X) = X^3 + bX^2 + cX + d$.[1] The change of variable $X = x - \frac{1}{3}b$ yields a simpler polynomial $f(x) = x^3 + qx + r$ whose roots give the roots of $F(x)$: if $u$ is a root of $f(x)$, then $u - \frac{1}{3}b$ is a root of $F(x)$. Special cases of the cubic formula were discovered by Scipio del Ferro around 1515, and the remaining cases were completed by Niccolò Fontana (Tartaglia) in 1535 and by Girolamo Cardano in 1545. The roots of $f(x)$ are

$$g + h, \quad \omega g + \omega^2 h, \quad \text{and} \quad \omega^2 g + \omega h,$$

where $g^3 = \frac{1}{2}\left(-r + \sqrt{R}\right)$, $h = -q/3g$, $R = r^2 + \frac{4}{27}q^3$, and $\omega = -\frac{1}{2} + i\frac{\sqrt{3}}{2}$ is a primitive cube root of unity. This formula is derived as follows. If $u$ is a root of $f(x) = x^3 + qx + r$, write

$$u = g + h,$$

and substitute:

(1) $$0 = f(u) = f(g + h) = g^3 + h^3 + (3gh + q)u + r.$$

Now the quadratic formula can be rephrased to say, given any pair of numbers $M$ and $N$, that there are (possibly complex) numbers $g$ and $h$ with $g + h = M$ and

---

[1]We must mention that modern notation was not introduced until the late 1500s, and was generally agreed upon only after the influential book of Descartes in 1637. For example, letters for variables were invented (by Viète) in 1591, the equality sign $=$ was invented (by Recorde) in 1557, and exponents were invented (by Hume) in 1636. The symbols $+$ and $-$ were introduced (by Widman) in 1486 (see Cajori, *A History of Mathematical Notation*).

$gh = N$ (for $g^2 - Mg + N = 0$). In particular, we may assume that $g + h = u$ and $gh = -\frac{1}{3}q$. This forces $3gh + q = 0$, and Equation (1) now gives $g^3 + h^3 = -r$. After cubing $gh = -\frac{1}{3}q$, we obtain the pair of equations

$$g^3 + h^3 = -r$$
$$g^3 h^3 = -\tfrac{1}{27}q^3.$$

Thus, there is a quadratic equation in $g^3$:

$$g^6 + rg^3 - \tfrac{1}{27}q^3 = 0.$$

The quadratic formula gives

$$g^3 = \tfrac{1}{2}\left(-r + \sqrt{r^2 + \tfrac{4}{27}q^3}\right) = \tfrac{1}{2}\left(-r + \sqrt{R}\right)$$

[note that $h^3$ is also a root of this quadratic, so that $h^3 = \frac{1}{2}(-r - \sqrt{R})$]. There are three cube roots of $g^3$, namely, $g$, $\omega g$, and $\omega^2 g$. Because of the constraint $gh = -q/3$, each of these has a "mate:" $g$ and $h = -q/(3g)$; $\omega g$ and $\omega^2 h = -q/(3\omega g)$: $\omega^2 g$ and $\omega h = -q/(3\omega^2 g)$ (since $\omega^3 = 1$).

**Example 1.1.** If $f(x) = x^3 - 15x - 126$, then $q = -15$, $r = -126$, $R = 15376$, and $\sqrt{R} = 124$. Hence, $g^3 = 125$, so that $g = 5$. Thus, $h = -q/(3g) = 1$. Therefore, the roots of $f(x)$ are

$$6, \quad 5\omega + \omega^2 = -3 + 2i\sqrt{3}, \quad 5\omega^2 + \omega = -3 - 2i\sqrt{3}.$$

Alternatively, having found one root to be 6, the other two roots can be found as the roots of the quadratic $f(x)/(x-6) = x^2 + 6x + 21$. ◀

**Example 1.2.** The cubic formula is not very useful because it often gives the roots in unrecognizable form. For example, let

$$f(x) = (x-1)(x-2)(x+3) = x^3 - 7x + 6;$$

the roots of $f(x)$ are, obviously, $1, 2$, and $-3$. The cubic formula gives

$$g + h = \sqrt[3]{\tfrac{1}{2}\left(-6 + \sqrt{\tfrac{-400}{27}}\right)} + \sqrt[3]{\tfrac{1}{2}\left(-6 - \sqrt{\tfrac{-400}{27}}\right)}.$$

It is not at all obvious that $g + h$ is a real number, let alone an integer. There is another version of the cubic formula, due to Viète, which gives the roots in terms of trigonometric functions instead of radicals (FCAA, pp. 360–362). ◀

Before the cubic formula, mathematicians had no difficulty in ignoring negative numbers or square roots of negative numbers when dealing with quadratic equations. For example, consider the problem of finding the sides $x$ and $y$ of a rectangle having area $A$ and perimeter $p$. The equations $xy = A$ and $2x + 2y = p$ give the quadratic $2x^2 - px + 2A$. The quadratic formula gives

$$x = \tfrac{1}{4}\left(p \pm \sqrt{p^2 - 16A}\right)$$

and $y = A/x$. If $p^2 - 16A \geq 0$, the problem is solved. If $p^2 - 16A < 0$, one did not invent fantastic rectangles whose sides involve square roots of negative numbers; instead, one merely said that there is no rectangle whose area and perimeter are so related. But the cubic formula does not allow us to discard "imaginary" roots, for we have just seen, in Example 1.2, that an "honest" real and positive root can appear in terms of such radicals: $\sqrt[3]{\frac{1}{2}\left(-6 + \sqrt{\frac{-400}{27}}\right)} + \sqrt[3]{\frac{1}{2}\left(-6 - \sqrt{\frac{-400}{27}}\right)}$ is an integer! Thus, the cubic formula was revolutionary. For the next 100 years, mathematicians reconsidered the meaning of *number*, for understanding the cubic formula raises the questions whether negative numbers and complex numbers are legitimate entities.

Consider the quartic $F(X) = X^4 + bX^3 + cX^2 + dX + e$. The change of variable $X = x - \frac{1}{4}b$ yields a simpler polynomial $f(x) = x^4 + qx^2 + rx + s$ whose roots give the roots of $F(x)$: if $u$ is a root of $f(x)$, then $u - \frac{1}{4}b$ is a root of $F(x)$. The quartic formula was found by Luigi Ferrari in the 1540s, but we present the version given by Descartes in 1637. Factor $f(x)$,

$$f(x) = x^4 + qx^2 + rx + s = (x^2 + jx + \ell)(x^2 - jx + m),$$

and determine $j$, $\ell$ and $m$ [note that the coefficients of the linear terms in the quadratic factors are $j$ and $-j$ because $f(x)$ has no cubic term]. Expanding and equating like coefficients gives the equations

$$\ell + m - j^2 = q,$$
$$j(m - \ell) = r,$$
$$\ell m = s.$$

The first two equations give

$$2m = j^2 + q + r/j,$$
$$2\ell = j^2 + q - r/j.$$

Substituting these values for $m$ and $\ell$ into the third equation yields the **resolvent cubic**:

$$(j^2)^3 + 2q(j^2)^2 + (q^2 - 4s)j^2 - r^2.$$

The cubic formula gives $j^2$, from which we can determine $m$ and $\ell$, and hence the roots of the quartic. The quartic formula has the same disadvantage as the cubic formula: even though it gives a correct answer, the values of the roots are usually unrecognizable.

Note that the quadratic formula can be derived in a way similar to the derivation of the cubic and quartic formulas. The change of variable $X = x - \frac{1}{2}b$ replaces the quadratic polynomial $F(X) = X^2 + bX + c$ with the simpler polynomial $f(x) = x^2 + q$ whose roots give the roots of $F(x)$: if $u$ is a root of $f(x)$, then $u - \frac{1}{2}b$ is a root of $F(x)$. An explicit formula for $q$ is $c - \frac{1}{4}b^2$, so that the roots of $f(x)$ are, obviously, $u = \pm\frac{1}{2}\sqrt{b^2 - 4c}$; thus, the roots of $F(X)$ are $\frac{1}{2}\left(-b \pm \sqrt{b^2 - 4c}\right)$.

It is now very tempting, as it was for our ancestors, to seek the roots of a quintic $F(X) = X^5 + bX^4 + cX^3 + dX^2 + eX + f$ (of course, they hoped to find roots of polynomials of any degree). Begin by changing variable $X = x - \frac{1}{5}b$ to eliminate the $X^4$ term. It was natural to expect that some further ingenious substitution together with the formulas for roots of polynomials of lower degree, analogous to the resolvent cubic, would yield the roots of $F(X)$. For almost 300 years, no such formula was found. In 1770, Lagrange showed that reasonable substitutions lead to a polynomial of degree six, not to a polynomial of smaller degree. Informally, let us say that a polynomial $f(x)$ is **solvable by radicals** if there is a formula for its roots which has the same form as the quadratic, cubic, and quartic formulas; that is, which uses only arithmetic operations and roots of numbers involving the coefficients of $f(x)$. In 1799, Ruffini claimed that the general quintic formula is not solvable by radicals, but his contemporaries did not accept his proof; his ideas were, in fact, correct, but his proof had gaps. In 1815, Cauchy introduced the multiplication of permutations, and he proved basic properties of what we call the *symmetric group* $S_n$; for example, he introduced the cycle notation and proved unique factorization of permutations into disjoint cycles. In 1824, Abel gave an acceptable proof that there is no quintic formula; in his proof, Abel constructed permutations of the roots of a quintic, using certain rational functions introduced by Lagrange. Galois, the young wizard who was killed before his 21st birthday, modified Lagrange's rational functions but, more important, he saw that the key to understanding which polynomials are solvable by radicals involved what he called *groups*: subsets of the symmetric group $S_n$ that are closed under composition—in our language, *subgroups* of $S_n$. To each polynomial $f(x)$, he associated such a group, nowadays called the *Galois group* of $f(x)$. He recognized conjugation, normal subgroups, quotient groups, and simple groups, and he proved, in our language, that a polynomial (over a field of characteristic 0) is solvable by radicals if and only if its Galois group is a *solvable group* (solvability being a property generalizing commutativity). A good case can be made that Galois was one of the most important founders of modern Algebra. For an excellent account of this history, we recommend the book, Tignol, *Galois' Theory of Algebraic Equations*.

## Exercises

**1.1.** Given $M, N \in \mathbb{C}$, prove that there exist $g, h \in \mathbb{C}$ with $g + h = M$ and $gh = N$.

\* **1.2.** The following problem, from an old Chinese text, was solved by Ch'in Chiu-shao (Qin Jiushao) in 1247. There is a circular castle, whose diameter is unknown; it is provided with four gates, and two *li* out of the north gate there is a large tree, which is visible from a point six *li* east of the south gate. What is the length of the diameter?

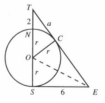

**Figure 1.1.** Castle problem.

**Hint.** The answer is a root of a cubic polynomial.

**1.3.** (i) Find the complex roots of $f(x) = x^3 - 3x + 1$.

(ii) Find the complex roots of $f(x) = x^4 - 2x^2 + 8x - 3$.

**1.4.** Show that the quadratic formula does not hold for $f(x) = ax^2 + bx + c$ if we view the coefficients $a, b, c$ as lying in the integers mod 2.

## Section 1.2. Permutations

For Galois, groups consisted of certain permutations (of the roots of a polynomial), and groups of permutations remain important today.

**Definition.** A **permutation** of a set $X$ is a bijection[2] from $X$ to itself.

A **rearrangement** of a set $X$ is a list, with no repetitions, of all the elements of $X$. For example, there are six rearrangements of $X = \{1, 2, 3\}$:

$$123; \quad 132; \quad 213; \quad 231; \quad 312; \quad 321.$$

Now let $X = \{1, 2, \ldots, n\}$. All we can do with such lists is count the number of them; there are exactly $n!$ rearrangements of the $n$-element set $X$.

Now a rearrangement $i_1, i_2, \ldots, i_n$ of $X$ determines a function $\alpha \colon X \to X$, namely, $\alpha(1) = i_1, \alpha(2) = i_2, \ldots, \alpha(n) = i_n$. For example, the rearrangement 213 determines the function $\alpha$ with $\alpha(1) = 2$, $\alpha(2) = 1$, and $\alpha(3) = 3$. We use a two-rowed notation to denote the function corresponding to a rearrangement; if $\alpha(j)$ is the $j$th item on the list, then

$$\alpha = \begin{pmatrix} 1 & 2 & \cdots & j & \cdots & n \\ \alpha(1) & \alpha(2) & \cdots & \alpha(j) & \cdots & \alpha(n) \end{pmatrix}.$$

That a list contains *all* the elements of $X$ says that the corresponding function $\alpha$ is surjective, for the bottom row is $\operatorname{im} \alpha$;[3] that there are no repetitions on the list says that distinct points have distinct values; that is, $\alpha$ is injective. Thus, each list determines a bijection $\alpha \colon X \to X$; that is, each rearrangement determines a permutation. Conversely, every permutation $\alpha$ determines a rearrangement, namely, the list $\alpha(1), \alpha(2), \ldots, \alpha(n)$ displayed as the bottom row. Therefore, rearrangement and permutation are simply different ways of describing the same thing. The advantage of viewing permutations as functions, however, is that they can now be composed and, by Exercise 1.7 on page 14, their composite is also a permutation.

**Definition.** The family of all the permutations of a set $X$, denoted by $S_X$, is called the **symmetric group** on $X$. We denote $S_X$ by $S_n$ when $X = \{1, 2, \ldots, n\}$, and we call it the **symmetric group on $n$ letters**. The identity permutation $1_X$ is usually denoted by $(1)$.

---

[2]A function $f \colon X \to Y$ is an **injection** (or is *one-to-one*) if $x, x' \in X$ and $x \neq x'$ implies $f(x) \neq f(x')$; equivalently, $f(x) = f(x')$ implies $x = x'$. A function $g \colon Y \to Z$ is a **surjection** (or is *onto*) if, for each $z \in Z$, there exists $y \in Y$ with $z = g(y)$. A function is a **bijection** (or is a *one-one correspondence*) if it is both an injection and a surjection. The **identity function** on a set $X$, denoted by $1_X \colon X \to X$, is given by $1_X \colon x \mapsto x$ for all $x \in X$; it is a bijection.

[3]If $f \colon X \to Y$ is a function, then $X$ is called its **domain**, $Y$ is called its **target** and, if $x \in X$, then $f(x)$ is called the **value**. The **image** of $f$, denoted by $\operatorname{im} f$, is the subset of the target $Y$ consisting of all the values. Thus, $\operatorname{im} f \subseteq Y$, and $\operatorname{im} f = Y$ if and only if $f$ is surjective.

Notice that composition in $S_3$ is not *commutative*; it is easy to find permutations $\alpha, \beta$ of $\{1, 2, 3\}$ with $\alpha\beta \neq \beta\alpha$.

We now introduce some special permutations.

**Definition.** Let $i_1, i_2, \ldots, i_r$ be distinct integers in $X = \{1, 2, \ldots, n\}$. If $\alpha \in S_n$ fixes[4] the other integers in $X$ (if any) and if

$$\alpha(i_1) = i_2, \quad \alpha(i_2) = i_3, \ldots, \alpha(i_{r-1}) = i_r, \quad \alpha(i_r) = i_1,$$

then $\alpha$ is called an **r-cycle**. We also say that $\alpha$ is a cycle of **length** $r$, and we denote it by $\alpha = (i_1 \ i_2 \ \ldots \ i_r)$.

The term *cycle* comes from the Greek word for circle. The cycle $\alpha = (i_1 \ i_2 \ \ldots \ i_r)$ can be pictured as a clockwise rotation of the circle, as in Figure 1.2.

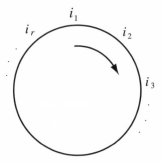

**Figure 1.2.** Cycle $\alpha = (i_1 \ i_2 \ \ldots \ i_r)$.

A 2-cycle interchanges $i_1$ and $i_2$ and fixes everything else; 2-cycles are also called **transpositions**. A 1-cycle is the identity, for it fixes every $i$; thus, all 1-cycles are equal. We extend the cycle notation to 1-cycles, writing $(i) = (1)$ for all $i$ [after all, $(i)$ sends $i$ into $i$ and fixes everything else].

There are $r$ different cycle notations for any $r$-cycle $\alpha$, since any $i_j$ can be taken as its "starting point":

$$\alpha = (i_1 \ i_2 \ \ldots \ i_r) = (i_2 \ i_3 \ \ldots \ i_r \ i_1) = \cdots = (i_r \ i_1 \ i_2 \ \ldots \ i_{r-1}).$$

**Definition.** Two permutations $\alpha, \beta \in S_n$ are **disjoint** if every $i$ moved by one is fixed by the other: if $\alpha(i) \neq i$, then $\beta(i) = i$, and if $\beta(j) \neq j$, then $\alpha(j) = j$. A family $\beta_1, \ldots, \beta_t$ of permutations is **disjoint** if each pair of them is disjoint.

For example, two cycles $(i_1 \ldots i_r)$ and $(j_1 \ldots j_s)$ are disjoint if and only if $\{i_1, \ldots, i_r\} \cap \{j_1, \ldots, j_s\} = \varnothing$.

---

[4]Let $f \colon X \to X$ be a function. If $x \in X$, then $f$ **fixes** $x$ if $f(x) = x$, and $f$ **moves** $x$ if $f(x) \neq x$.

**Proposition 1.3.** *Disjoint permutations* $\alpha$, $\beta \in S_n$ *commute.*

**Proof.** It suffices to prove that if $1 \leq i \leq n$, then $\alpha\beta(i) = \beta\alpha(i)$. If $\beta$ moves $i$, say, $\beta(i) = j \neq i$, then $\beta$ also moves $j$ [otherwise, $\beta(j) = j$ and $\beta(i) = j$ contradict $\beta$'s being an injection]; since $\alpha$ and $\beta$ are disjoint, $\alpha(i) = i$ and $\alpha(j) = j$. Hence $\beta\alpha(i) = j = \alpha\beta(i)$. The same conclusion holds if $\alpha$ moves $i$. Finally, it is clear that $\alpha\beta(i) = i = \beta\alpha(i)$ if both $\alpha$ and $\beta$ fix $i$. •

Aside from being cumbersome, there is a major problem with the two-rowed notation for permutations. It hides the answers to elementary questions such as: Is a permutation a cycle? Is the square of a permutation the identity? The **algorithm** we now introduce, which factors a permutation into a product of disjoint cycles, will remedy this defect. Let

$$\alpha = \begin{pmatrix} 1 & 2 & 3 & 4 & 5 & 6 & 7 & 8 & 9 \\ 6 & 4 & 7 & 2 & 5 & 1 & 8 & 9 & 3 \end{pmatrix}.$$

Begin by writing "(1." Now $\alpha\colon 1 \mapsto 6$; write "(1 6." Next, $\alpha\colon 6 \mapsto 1$, and the parentheses close: $\alpha$ begins "(1 6)." The first number not having appeared is 2, and we write "(1 6)(2." Now $\alpha\colon 2 \mapsto 4$; write "(1 6)(2 4." Since $\alpha\colon 4 \mapsto 2$, the parentheses close once again, and we write "(1 6)(2 4)." The smallest remaining number is 3; now $3 \mapsto 7$, $7 \mapsto 8$, $8 \mapsto 9$, and $9 \mapsto 3$; this gives the 4-cycle (3 7 8 9). Finally, $\alpha(5) = 5$; we claim that

$$\alpha = (1\ 6)(2\ 4)(3\ 7\ 8\ 9)(5).$$

Since multiplication in $S_n$ is composition of functions, our claim is that

$$\alpha(i) = [(1\ 6)(2\ 4)(3\ 7\ 8\ 9)(5)](i)$$

for every $i$ between 1 and 9 [after all, two functions $f$ and $g$ are **equal** if and only if they have the same domain, the same target,[5] and $f(i) = g(i)$ for every $i$ in their domain]. The right side is the value of the composite $\beta\gamma\delta$, where $\beta = (1\ 6)$, $\gamma = (2\ 4)$, and $\delta = (3\ 7\ 8\ 9)$ [we may ignore the 1-cycle (5) when we are evaluating, for it is the identity function]. Now $\alpha(1) = 6$; let us evaluate the composite on the right when $i = 1$:

$$\begin{aligned} \beta\gamma\delta(1) &= \beta(\gamma(\delta(1))) \\ &= \beta(\gamma(1)) &&\text{because } \delta = (3\ 7\ 8\ 9) \text{ fixes } 1 \\ &= \beta(1) &&\text{because } \gamma = (2\ 4) \text{ fixes } 1 \\ &= 6 &&\text{because } \beta = (1\ 6). \end{aligned}$$

Similarly, we can show that $\alpha(i) = \beta\gamma\delta(i)$ for every $i$, proving the claim.

We multiply permutations from right to left, because multiplication here is composition of functions;[6] that is, to evaluate $\alpha\beta(1)$, we compute $\alpha(\beta(1))$.

---

[5] If $X \subsetneq Y$ is a proper subset, then this definition shows that the inclusion $i\colon X \to Y$ is not equal to the identity $1_X$, for they have different targets.

[6] There are authors who multiply permutations differently, so that their $\alpha\beta$ is our $\beta\alpha$. This is a consequence of their putting "functions on the right": instead of writing $\alpha(i)$ as we do, they write $(i)\alpha$. Consider the composite of permutations $\alpha$ and $\beta$ in which we first apply $\beta$ and then apply $\alpha$. We write $i \mapsto \beta(i) \mapsto \alpha(\beta(i))$. In the right-sided notation, $i \mapsto (i)\beta \mapsto ((i)\beta)\alpha$. Thus, the notational switch causes a switch in the order of multiplication; we write $\alpha\beta$; they write $\beta\alpha$.

Here is another example: let us write $\sigma = (1\ 2)(1\ 3\ 4\ 2\ 5)(2\ 5\ 1\ 3)$ as a product of disjoint cycles in $S_5$. To find the two-rowed notation for $\sigma$, evaluate, starting with the cycle on the right:

$$\sigma: 1 \mapsto 3 \mapsto 4 \mapsto 4;$$
$$\sigma: 4 \mapsto 4 \mapsto 2 \mapsto 1;$$
$$\sigma: 2 \mapsto 5 \mapsto 1 \mapsto 2;$$
$$\sigma: 3 \mapsto 2 \mapsto 5 \mapsto 5;$$
$$\sigma: 5 \mapsto 1 \mapsto 3 \mapsto 3.$$

Thus,

$$\sigma = (1\ 4)(2)(3\ 5).$$

**Proposition 1.4.** *Every permutation $\alpha \in S_n$ is either a cycle or a product of disjoint cycles.*

**Proof.** The proof is by induction on the number $k$ of points moved by $\alpha$. The base step $k = 0$ is true, for now $\alpha$ is the identity, which is a 1-cycle.

If $k > 0$, let $i_1$ be a point moved by $\alpha$. Define $i_2 = \alpha(i_1)$, $i_3 = \alpha(i_2)$, ..., $i_{r+1} = \alpha(i_r)$, where $r$ is the smallest integer for which $i_{r+1} \in \{i_1, i_2, \ldots, i_r\}$ (since there are only $n$ possible values, the list $i_1, i_2, i_3, \ldots, i_k, \ldots$ must eventually have a repetition). We claim that $\alpha(i_r) = i_1$. Otherwise, $\alpha(i_r) = i_j$ for some $j \geq 2$. But $\alpha(i_{j-1}) = i_j$; since $r > j - 1$, this contradicts the hypothesis that $\alpha$ is an injection. Let $\sigma$ be the $r$-cycle $(i_1\ i_2\ i_3\ \ldots\ i_r)$. If $r = n$, then $\alpha = \sigma$. If $r < n$, then $\sigma$ fixes each point in $Y$, where $Y$ consists of the remaining $n - r$ points, while $\alpha(Y) = Y$. Define $\alpha'$ to be the permutation with $\alpha'(i) = \alpha(i)$ for $i \in Y$ that fixes all $i \notin Y$, and note that

$$\alpha = \sigma\alpha'.$$

The inductive hypothesis gives $\alpha' = \beta_1 \cdots \beta_t$, where $\beta_1, \ldots, \beta_t$ are disjoint cycles. Since $\sigma$ and $\alpha'$ are disjoint, $\alpha = \sigma\beta_1 \cdots \beta_t$ is a product of disjoint cycles.  •

The **inverse** of a function $f\colon X \to Y$ is a function $g\colon Y \to X$ with $gf = 1_X$ and $fg = 1_Y$. A function has an inverse if and only if it is a bijection (FCAA, p. 95), and inverses are unique when they exist. Every permutation is a bijection; how do we find its inverse? In the pictorial representation on page 6 of a cycle $\alpha$ as a clockwise rotation of a circle, its inverse $\alpha^{-1}$ is just the counterclockwise rotation.

**Proposition 1.5.**

(i) *The inverse of the cycle $\alpha = (i_1\ i_2\ \ldots\ i_r)$ is the cycle $(i_r\ i_{r-1}\ \ldots\ i_1)$:*

$$(i_1\ i_2\ \ldots\ i_r)^{-1} = (i_r\ i_{r-1}\ \ldots\ i_1).$$

(ii) *If $\gamma \in S_n$ and $\gamma = \beta_1 \cdots \beta_k$, then*

$$\gamma^{-1} = \beta_k^{-1} \cdots \beta_1^{-1}.$$

**Proof.** FCAA, p. 115.  •

Usually we suppress the 1-cycles in the factorization of a permutation in Proposition 1.4 [for 1-cycles equal the identity function]. However, a factorization of $\alpha$ in which we display one 1-cycle for each $i$ fixed by $\alpha$, if any, will arise several times.

**Definition.** A ***complete factorization*** of a permutation $\alpha$ is a factorization of $\alpha$ into disjoint cycles that contains exactly one 1-cycle $(i)$ for every $i$ fixed by $\alpha$.

For example, a complete factorization of the 3-cycle $\alpha = (1\ 3\ 5)$ in $S_5$ is $\alpha = (1\ 3\ 5)(2)(4)$.

There is a relation between the notation for an $r$-cycle $\beta = (i_1\ i_2\ \ldots\ i_r)$ and its ***powers*** $\beta^k$, where $\beta^k$ denotes the composite of $\beta$ with itself $k$ times. Note that $i_2 = \beta(i_1)$, $i_3 = \beta(i_2) = \beta(\beta(i_1)) = \beta^2(i_1)$, $i_4 = \beta(i_3) = \beta(\beta^2(i_1)) = \beta^3(i_1)$, and, more generally,

$$i_{k+1} = \beta^k(i_1)$$

for all positive $k < r$.

**Theorem 1.6.** *Let $\alpha \in S_n$ and let $\alpha = \beta_1 \cdots \beta_t$ be a complete factorization into disjoint cycles. This factorization is unique except for the order in which the cycles occur.*

**Proof.** Since every complete factorization of $\alpha$ has exactly one 1-cycle for each $i$ fixed by $\alpha$, it suffices to consider (not complete) factorizations into disjoint cycles of lengths $\geq 2$. Let $\alpha = \gamma_1 \cdots \gamma_s$ be a second factorization of $\alpha$ into disjoint cycles of lengths $\geq 2$.

The theorem is proved by induction on $\ell$, the larger of $t$ and $s$. The inductive step begins by noting that if $\beta_t$ moves $i_1$, then $\beta_t^k(i_1) = \alpha^k(i_1)$ for all $k \geq 1$. Some $\gamma_j$ must also move $i_1$ and, since disjoint cycles commute, we may assume that $\gamma_s$ moves $i_1$. It follows that $\beta_t = \gamma_s$ (Exercise 1.15 on page 15); right multiplying by $\beta_t^{-1}$ gives $\beta_1 \cdots \beta_{t-1} = \gamma_1 \cdots \gamma_{s-1}$, and the inductive hypothesis applies. •

**Definition.** Two permutations $\alpha, \beta \in S_n$ have the ***same cycle structure*** if, for each $r \geq 1$, their complete factorizations have the same number of $r$-cycles.

According to Exercise 1.12 on page 15, there are

$$\frac{1}{r}[n(n-1)\cdots(n-r+1)]$$

$r$-cycles in $S_n$. This formula can be used to count the number of permutations having any given cycle structure if we are careful about factorizations having several cycles of the same length. For example, the number of permutations in $S_4$ of the form $(a\ b)(c\ d)$ is $\frac{1}{2}\left[\frac{1}{2}(4 \times 3)\right] \times \left[\frac{1}{2}(2 \times 1)\right] = 3$, the "extra" factor $\frac{1}{2}$ occurring so that we do not count $(a\ b)(c\ d) = (c\ d)(a\ b)$ twice.

The types of permutations in $S_4$ and in $S_5$ are counted in Tables 1 and 2.

Here is a computational aid.

**Lemma 1.7.** *If $\gamma, \alpha \in S_n$, then $\alpha\gamma\alpha^{-1}$ has the same cycle structure as $\gamma$. In more detail, if the complete factorization of $\gamma$ is*

$$\gamma = \beta_1\beta_2 \cdots (i_1\ i_2\ \ldots) \cdots \beta_t,$$

| Cycle Structure | Number |
|---|---|
| (1) | 1 |
| (1 2) | 10 |
| (1 2 3) | 20 |
| (1 2 3 4) | 30 |
| (1 2 3 4 5) | 24 |
| (1 2)(3 4 5) | 20 |
| (1 2)(3 4) | 15 |
|  | $\overline{120}$ |

| Cycle Structure | Number |
|---|---|
| (1) | 1 |
| (1 2) | 6 |
| (1 2 3) | 8 |
| (1 2 3 4) | 6 |
| (1 2)(3 4) | 3 |
|  | $\overline{24}$ |

**Table 1.** Permutations in $S_4$.              **Table 2.** Permutations in $S_5$.

*then $\alpha\gamma\alpha^{-1}$ is the permutation obtained from $\gamma$ by applying $\alpha$ to the symbols in the cycles of $\gamma$.*

**Remark.** For example, if $\gamma = (1\ 3)(2\ 4\ 7)(5)(6)$ and $\alpha = (2\ 5\ 6)(1\ 4\ 3)$, then

$$\alpha\gamma\alpha^{-1} = (\alpha 1\ \alpha 3)(\alpha 2\ \alpha 4\ \alpha 7)(\alpha 5)(\alpha 6) = (4\ 1)(5\ 3\ 7)(6)(2). \quad \blacktriangleleft$$

**Proof.** Observe that

(1) $$\alpha\gamma\alpha^{-1} \colon \alpha(i_1) \mapsto i_1 \mapsto i_2 \mapsto \alpha(i_2).$$

Let $\sigma$ denote the permutation defined in the statement.

If $\gamma$ fixes $i$, then $\sigma$ fixes $\alpha(i)$, for the definition of $\sigma$ says that $\alpha(i)$ lives in a 1-cycle in the factorization of $\sigma$. Assume that $\gamma$ moves a symbol $i$; say, $\gamma(i) = j$, so that one of the cycles in the complete factorization of $\gamma$ is

$$(i\ j\ \ldots).$$

By definition, one of the cycles in the complete factorization of $\sigma$ is

$$\big(\alpha(i)\ \alpha(j)\ \ldots\big);$$

that is, $\sigma \colon \alpha(i) \mapsto \alpha(j)$. Now Equation (1) says that $\alpha\gamma\alpha^{-1} \colon \alpha(i) \mapsto \alpha(j)$, so that $\sigma$ and $\alpha\gamma\alpha^{-1}$ agree on all numbers of the form $\alpha(i)$. But every $k \in X = \{1, \ldots, n\}$ lies in $\operatorname{im}\alpha$, because the permutation $\alpha$ is surjective, and so $\sigma = \alpha\gamma\alpha^{-1}$. $\quad \bullet$

**Example 1.8.** We illustrate the converse of Lemma 1.7; the next theorem will prove that this converse holds in general. In $S_5$, place the complete factorization of a 3-cycle $\beta$ over that of a 3-cycle $\gamma$, and define $\alpha$ to be the downward function. For example, if

$$\beta = (1\ 2\ 3)(4)(5)$$
$$\gamma = (5\ 2\ 4)(1)(3),$$

then

$$\alpha = \begin{pmatrix} 1 & 2 & 3 & 4 & 5 \\ 5 & 2 & 4 & 1 & 3 \end{pmatrix},$$

and the algorithm gives $\alpha = (1\ 5\ 3\ 4)$. Now $\alpha \in S_5$ and

$$\gamma = (\alpha 1\ \alpha 2\ \alpha 3),$$

so that $\gamma = \alpha\beta\alpha^{-1}$, by Lemma 1.7. Note that rewriting the cycles of $\beta$, for example, as $\beta = (1\ 2\ 3)(5)(4)$, gives another choice for $\alpha$.  ◄

**Theorem 1.9.** *Permutations $\gamma$ and $\sigma$ in $S_n$ have the same cycle structure if and only if there exists $\alpha \in S_n$ with $\sigma = \alpha\gamma\alpha^{-1}$.*

**Proof.** Sufficiency was proved in Lemma 1.7. For the converse, place one complete factorization over the other so that each cycle below lies under a cycle of the same length:

$$\gamma = \delta_1\delta_2 \cdots (i_1\ i_2 \dots) \cdots \delta_t$$
$$\sigma = \eta_1\eta_2 \cdots (k\ \ell\ \dots) \cdots \eta_t.$$

Now define $\alpha$ to be the "downward" function, as in the example; hence, $\alpha(i_1) = k$, $\alpha(i_2) = \ell$, and so forth. Note that $\alpha$ is a permutation, for there are no repetitions of symbols in the factorization of $\gamma$ (the cycles $\eta$ are disjoint). It now follows from Lemma 1.7 that $\sigma = \alpha\gamma\alpha^{-1}$.  •

Here is another useful factorization of a permutation.

**Proposition 1.10.** *If $n \geq 2$, then every $\alpha \in S_n$ is a transposition or a product[7] of transpositions.*

**Proof.** In light of Proposition 1.4, it suffices to factor an $r$-cycle $\beta$ into a product of transpositions, and this is done as follows:

$$\beta = (1\ 2\ \dots\ r) = (1\ r)(1\ r-1) \cdots (1\ 3)(1\ 2).  \bullet$$

Every permutation can thus be realized as a sequence of interchanges, but such a factorization is not as nice as the factorization into disjoint cycles. First, the transpositions occurring need not commute: $(1\ 2\ 3) = (1\ 3)(1\ 2) \neq (1\ 2)(1\ 3)$; second, neither the factors themselves nor the number of factors are uniquely determined. For example, here are some factorizations of $(1\ 2\ 3)$ in $S_4$:

$$(1\ 2\ 3) = (1\ 3)(1\ 2)$$
$$= (2\ 3)(1\ 3)$$
$$= (1\ 2)(2\ 3)$$
$$= (1\ 3)(4\ 2)(1\ 2)(1\ 4)$$
$$= (1\ 3)(4\ 2)(1\ 2)(1\ 4)(2\ 3)(2\ 3).$$

Is there any uniqueness at all in such a factorization? We will prove that the parity of the number of factors is the same for all factorizations of a permutation $\alpha$; that is, the number of transpositions is always even or always odd [as suggested by the factorizations of $\alpha = (1\ 2\ 3)$ displayed above].

---

[7]It is convenient to generalize the term *product* to apply when there is only one factor. We can now rephrase the statement of Proposition 1.10 to say that every permutation in $S_n$ is a product of transpositions. Similarly, we can rephrase Proposition 1.4 to say that every permutation in $S_n$ is a product of disjoint cycles.

**Example 1.11.** The *15-puzzle* has a *starting position* that is a $4 \times 4$ array of the numbers between 1 and 15 and a symbol $\square$, which we interpret as "blank." For example, consider the following starting position:

| 12 | 15 | 14 | 8 |
|----|----|----|---|
| 10 | 11 | 1  | 4 |
| 9  | 5  | 13 | 3 |
| 6  | 7  | 2  |   |

A *move* interchanges the blank with a symbol adjacent to it; for example, there are two beginning moves for this starting position: either interchange $\square$ and 2 or interchange $\square$ and 3. We win the game if, after a sequence of moves, the starting position is transformed into the standard array $1, 2, 3, \ldots, 15, \square$.

To analyze this game, note that the given array is really a permutation $\alpha \in S_{16}$ (if we now call the blank 16 instead of $\square$). More precisely, if the spaces are labeled 1 through 16, then $\alpha(i)$ is the symbol occupying the $i$th square. For example, the given starting position is

$$\begin{pmatrix} 1 & 2 & 3 & 4 & 5 & 6 & 7 & 8 & 9 & 10 & 11 & 12 & 13 & 14 & 15 & 16 \\ 12 & 15 & 14 & 8 & 10 & 11 & 1 & 4 & 9 & 5 & 13 & 3 & 6 & 7 & 2 & 16 \end{pmatrix}.$$

Each move is a *special* kind of transposition, namely, one that moves 16 (remember that the blank $\square = 16$). Moreover, performing a move (corresponding to a special transposition $\tau$) from a given position (corresponding to a permutation $\beta$) yields a new position corresponding to the permutation $\tau\beta$. For example, if $\alpha$ is the position above and $\tau$ is the transposition interchanging 2 and $\square$, then $\tau\alpha(\square) = \tau(\square) = 2$ and $\tau\alpha(2) = \tau(15) = 15$, while $\tau\alpha(i) = \alpha(i)$ for all other $i$. That is, the new configuration has all the numbers in their original positions except for 2 and $\square$ being interchanged. To win the game, we need special transpositions $\tau_1, \tau_2, \ldots, \tau_m$ such that

$$\tau_m \cdots \tau_2 \tau_1 \alpha = (1).$$

It turns out that there are some choices of $\alpha$ for which the game can be won, but there are others for which it cannot be won, as we shall see in Example 1.15. ◀

**Definition.** A permutation $\alpha \in S_n$ is *even* if it is a product of an even number of transpositions; $\alpha$ is *odd* if it is not even. The *parity* of a permutation is whether it is even or odd.

It is easy to see that $(1\ 2\ 3)$ and $(1)$ are even permutations, for there are factorizations $(1\ 2\ 3) = (1\ 3)(1\ 2)$ and $(1) = (1\ 2)(1\ 2)$ as products of two transpositions. On the other hand, we do not yet have any examples of odd permutations! It is clear that if $\alpha$ is odd, then it is a product of an odd number of transpositions. The converse is not so obvious: if a permutation is a product of an odd number of transpositions, it might have another factorization as a product of an even number of transpositions. After all, the definition of an odd permutation says that there does not exist a factorization of it as a product of an even number of transpositions.

**Proposition 1.12.** *Let $\alpha$, $\beta \in S_n$. If $\alpha$ and $\beta$ have the same parity, then $\alpha\beta$ is even, while if $\alpha$ and $\beta$ have distinct parity, then $\alpha\beta$ is odd.*

**Proof.** Let $\alpha = \tau_1 \cdots \tau_m$ and $\beta = \sigma_1 \cdots \sigma_n$, where the $\tau$ and $\sigma$ are transpositions, so that $\alpha\beta = \tau_1 \cdots \tau_m \sigma_1 \cdots \sigma_n$ has $m + n$ factors. If $\alpha$ is even, then $m$ is even; if $\alpha$ is odd, then $m$ is odd. Hence, $m + n$ is even when $m, n$ have the same parity and $\alpha\beta$ is even. Suppose that $\alpha$ is even and $\beta$ is odd. If $\alpha\beta$ were even, then $\beta = \alpha^{-1}(\alpha\beta)$ is even, being a product of evenly many transpositions, and this is a contradiction. Therefore, $\alpha\beta$ is odd. Similarly, $\alpha\beta$ is odd when $\alpha$ is odd and $\beta$ is even. $\bullet$

**Definition.** If $\alpha \in S_n$ and $\alpha = \beta_1 \cdots \beta_t$ is a complete factorization into disjoint cycles, then *signum* $\alpha$ is defined by

$$\mathrm{sgn}(\alpha) = (-1)^{n-t}.$$

Theorem 1.6 shows that sgn is well-defined, for the number $t$ is uniquely determined by $\alpha$. Notice that $\mathrm{sgn}(\varepsilon) = 1$ for every 1-cycle $\varepsilon$ because $t = n$. If $\tau$ is a transposition, then it moves two numbers, and it fixes each of the $n - 2$ other numbers; therefore, $t = (n - 2) + 1 = n - 1$, and so $\mathrm{sgn}(\tau) = (-1)^{n-(n-1)} = -1$. Lastly, observe that if $\alpha \in S_n$, then $\mathrm{sgn}(\alpha)$ does not change when $\alpha$ is viewed in $S_{n+1}$ by letting it fix $n + 1$. If the complete factorization of $\alpha$ in $S_n$ is $\alpha = \beta_1 \cdots \beta_t$, then its complete factorization in $S_{n+1}$ has one more factor, namely, the 1-cycle $(n + 1)$. Thus, the formula for $\mathrm{sgn}(\alpha)$ in $S_{n+1}$ is $(-1)^{(n+1)-(t+1)} = (-1)^{n-t}$.

**Theorem 1.13.** *For all $\alpha$, $\beta \in S_n$,*

$$\mathrm{sgn}(\alpha\beta) = \mathrm{sgn}(\alpha)\,\mathrm{sgn}(\beta).$$

**Proof.** If $k$, $\ell \geq 0$ and the letters $a$, $b$, $c_i$, $d_j$ are all distinct, then it is easy to check (FCAA, p. 120) that

$$(a\ b)(a\ c_1\ \dots\ c_k\ b\ d_1\ \dots\ d_\ell) = (a\ c_1\ \dots\ c_k)(b\ d_1\ \dots\ d_\ell);$$

multiplying this equation on the left by $(a\ b)$ gives

$$(a\ b)(a\ c_1\ \dots\ c_k)(b\ d_1\ \dots\ d_\ell) = (a\ c_1\ \dots\ c_k\ b\ d_1\ \dots\ d_\ell).$$

These equations show that $\mathrm{sgn}(\tau\alpha) = -\mathrm{sgn}(\alpha)$ for every $\alpha \in S_n$, where $\tau$ is the transposition $(a\ b)$. If $\alpha \in S_n$ has a factorization $\alpha = \tau_1 \cdots \tau_m$, where each $\tau_i$ is a transposition, then $\mathrm{sgn}(\alpha\beta) = \mathrm{sgn}(\alpha)\,\mathrm{sgn}(\beta)$ for every $\beta \in S_n$ is proved by induction on $m$ (the base step has just been proved). $\bullet$

**Theorem 1.14.**

(i) *Let $\alpha \in S_n$; if $\mathrm{sgn}(\alpha) = 1$, then $\alpha$ is even, and if $\mathrm{sgn}(\alpha) = -1$, then $\alpha$ is odd.*

(ii) *A permutation $\alpha$ is odd if and only if it is a product of an odd number of transpositions.*

**Proof.**

(i) If $\alpha = \tau_1 \cdots \tau_q$ is a factorization of $\alpha$ into transpositions, then Theorem 1.13 gives $\mathrm{sgn}(\alpha) = \mathrm{sgn}(\tau_1) \cdots \mathrm{sgn}(\tau_q) = (-1)^q$. Thus, if $\mathrm{sgn}(\alpha) = 1$, then $q$ must be even, and if $\mathrm{sgn}(\alpha) = -1$, then $q$ must be odd.

(ii) If $\alpha$ is odd, then it is a product of an odd number of transpositions (for it is not a product of an even number of such). Conversely, if $\alpha = \tau_1 \cdots \tau_q$, where the $\tau_i$ are transpositions and $q$ is odd, then $\mathrm{sgn}(\alpha) = (-1)^q = -1$; hence, $q$ is odd. Therefore, $\alpha$ is not even, by part (i), and so it is odd.    •

**Example 1.15.** An analysis of the 15-puzzle, as in Example 1.11, shows that a game with starting position $\alpha \in S_{16}$ can be won if and only if $\alpha$ is an even permutation that fixes $\square = 16$. For a proof of this, we refer the reader to McCoy–Janusz, *Introduction to Modern Algebra*, pp. 229–234 (see Exercise 1.17 on page 15). The proof in one direction is fairly clear, however. Now $\square$ starts in position 16, and each move takes $\square$ up, down, left, or right. Thus, the total number $m$ of moves is $u + d + l + r$, where $u$ is the number of up moves, and so on. If $\square$ is to return home, each one of these must be undone: there must be the same number of up moves as down moves (i.e., $u = d$) and the same number of left moves as right moves (i.e., $r = l$). Thus, the total number of moves is even: $m = 2u + 2r$. That is, if $\tau_m \cdots \tau_1 \alpha = (1)$, then $m$ is even; hence, $\alpha = \tau_1 \cdots \tau_m$ (because $\tau^{-1} = \tau$ for every transposition $\tau$), and so $\alpha$ is an even permutation. Armed with this theorem, we see that if the starting position $\alpha$ is odd, the game starting with $\alpha$ cannot be won. In Example 1.11,

$$\alpha = (1\ 12\ 3\ 14\ 7)(2\ 15)(4\ 8)(5\ 10)(6\ 11\ 13)(9)(\square).$$

Now $\mathrm{sgn}(\alpha) = (-1)^{16-7} = -1$, so that $\alpha$ is an odd permutation. Therefore, it is impossible to win this game.    ◄

# Exercises

**1.5.** Give an example of functions $f \colon X \to Y$ and $g \colon Y \to X$ such that $gf = 1_X$ and $fg \neq 1_Y$.

**1.6.** Prove that composition of functions is associative: if $X \xrightarrow{f} Y \xrightarrow{g} Z \xrightarrow{h} W$, then

$$h(gf) = (hg)f.$$

∗ **1.7.** Prove that the composite of two injections is an injection, and that the composite of two surjections is a surjections. Conclude that the composite of two bijections is a bijection.

∗ **1.8. (Pigeonhole Principle)** (i) Let $f \colon X \to X$ be a function, where $X$ is a finite set. Prove equivalence of the following statements.

(a) $f$ is an injection.

(b) $f$ is a bijection.

(c) $f$ is a surjection.

(ii) Prove that no two of the statements in (i) are equivalent when $X$ is an infinite set.

(iii) Suppose there are 501 pigeons, each sitting in some pigeonhole. If there are only 500 pigeonholes, prove that there is a hole containing more than one pigeon.

* **1.9.** Let $Y$ be a subset of a finite set $X$, and let $f \colon Y \to X$ be an injection. Prove that there is a permutation $\alpha \in S_X$ with $\alpha | Y = f$.

* **1.10.** Find $\operatorname{sgn}(\alpha)$ and $\alpha^{-1}$, where

$$\alpha = \begin{pmatrix} 1 & 2 & 3 & 4 & 5 & 6 & 7 & 8 & 9 \\ 9 & 8 & 7 & 6 & 5 & 4 & 3 & 2 & 1 \end{pmatrix}.$$

**1.11.** If $\alpha \in S_n$, prove that $\operatorname{sgn}(\alpha^{-1}) = \operatorname{sgn}(\alpha)$.

* **1.12.** If $1 \le r \le n$, show that there are

$$\frac{1}{r}[n(n-1)\cdots(n-r+1)]$$

$r$-cycles in $S_n$.

**Hint.** There are exactly $r$ cycle notations for any $r$-cycle.

* **1.13.** (i) If $\alpha$ is an $r$-cycle, show that $\alpha^r = (1)$.

**Hint.** If $\alpha = (i_0 \ldots i_{r-1})$, show that $\alpha^k(i_0) = i_j$, where $k = qr + j$ and $0 \le j < r$.

(ii) If $\alpha$ is an $r$-cycle, show that $r$ is the smallest positive integer $k$ such that $\alpha^k = (1)$.

**1.14.** Show that an $r$-cycle is an even permutation if and only if $r$ is odd.

* **1.15.** (i) Let $\alpha = \beta\delta$ be a factorization of a permutation $\alpha$ into disjoint permutations. If $\beta$ moves $i$, prove that $\alpha^k(i) = \beta^k(i)$ for all $k \ge 1$.

(ii) Let $\beta$ and $\gamma$ be cycles both of which move $i$. If $\beta^k(i) = \gamma^k(i)$ for all $k \ge 1$, prove that $\beta = \gamma$.

* **1.16.** Given $X = \{1, 2, \ldots, n\}$, let us call a permutation $\tau$ of $X$ an **adjacency** if it is a transposition of the form $(i\ i+1)$ for $i < n$.

(i) Prove that every permutation in $S_n$, for $n \ge 2$, is a product of adjacencies.

(ii) If $i < j$, prove that $(i\ j)$ is a product of an odd number of adjacencies.
**Hint.** Use induction on $j - i$.

* **1.17.** (i) Prove, for $n \ge 2$, that every $\alpha \in S_n$ is a product of transpositions each of whose factors moves $n$.

**Hint.** If $i < j < n$, then $(j\ n)(i\ j)(j\ n) = (i\ n)$, by Lemma 1.7, so that $(i\ j) = (j\ n)(i\ n)(j\ n)$.

(ii) Why doesn't part (i) prove that a 15-puzzle with even starting position $\alpha$ which fixes $\square$ can be solved?

* **1.18.** Define $f \colon \{0, 1, 2, \ldots, 10\} \to \{0, 1, 2, \ldots, 10\}$ by

$$f(n) = \text{the remainder after dividing } 4n^2 - 3n^7 \text{ by } 11.$$

(i) Show that $f$ is a permutation.[8]

---

[8] If $k$ is a finite field, then a polynomial $f(x)$ with coefficients in $k$ is called a **permutation polynomial** if the evaluation function $f \colon k \to k$, defined by $a \mapsto f(a)$, is a permutation of $k$. A theorem of Hermite–Dickson characterizes permutation polynomials (Small, *Arithmetic of Finite Fields*, p. 40).

(ii) Compute the parity of $f$.

(iii) Compute the inverse of $f$.

**1.19.** If $\alpha$ is an $r$-cycle and $1 < k < r$, is $\alpha^k$ an $r$-cycle?

\* **1.20.** (i) Prove that if $\alpha$ and $\beta$ are (not necessarily disjoint) permutations that commute, then $(\alpha\beta)^k = \alpha^k\beta^k$ for all $k \geq 1$.

**Hint.** First show that $\beta\alpha^k = \alpha^k\beta$ by induction on $k$.

(ii) Give an example of two permutations $\alpha$ and $\beta$ for which $(\alpha\beta)^2 \neq \alpha^2\beta^2$.

\* **1.21.** (i) Prove, for all $i$, that $\alpha \in S_n$ moves $i$ if and only if $\alpha^{-1}$ moves $i$.

(ii) Prove that if $\alpha, \beta \in S_n$ are disjoint and if $\alpha\beta = (1)$, then $\alpha = (1)$ and $\beta = (1)$.

\* **1.22.** Prove that the number of even permutations in $S_n$ is $\frac{1}{2}n!$.

**Hint.** Let $\tau = (1\ 2)$, and define $f\colon A_n \to O_n$, where $A_n$ is the set of all even permutations in $S_n$ and $O_n$ is the set of all odd permutations, by $f\colon \alpha \mapsto \tau\alpha$. Show that $f$ is a bijection, so that $|A_n| = |O_n|$ and, hence, $|A_n| = \frac{1}{2}n!$.

\* **1.23.** (i) How many permutations in $S_5$ commute with $\alpha = (1\ 2\ 3)$, and how many *even* permutations in $S_5$ commute with $\alpha$?

**Hint.** Of the six permutations in $S_5$ commuting with $\alpha$, only three are even.

(ii) Same questions for $(1\ 2)(3\ 4)$.

**Hint.** Of the eight permutations in $S_4$ commuting with $(1\ 2)(3\ 4)$, only four are even.

**1.24.** Give an example of $\alpha, \beta, \gamma \in S_5$, with $\alpha \neq (1)$, such that $\alpha\beta = \beta\alpha$, $\alpha\gamma = \gamma\alpha$, and $\beta\gamma \neq \gamma\beta$.

\* **1.25.** If $n \geq 3$, prove that if $\alpha \in S_n$ commutes with every $\beta \in S_n$, then $\alpha = (1)$.

**1.26.** If $\alpha = \beta_1 \cdots \beta_m$ is a product of disjoint cycles and $\delta$ is disjoint from $\alpha$, show that $\beta_1^{e_1} \cdots \beta_m^{e_m}\delta$ commutes with $\alpha$, where $e_j \geq 0$ for all $j$.

---

## Section 1.3. Groups

Since Galois's time, groups have arisen in many areas of Mathematics other than the study of roots of polynomials, for they are the precise way to describe the notion of symmetry, as we shall see.

The essence of a "product" is that two things are combined to form a third thing of the same kind. For example, ordinary multiplication, addition, and subtraction combine two numbers to give another number, while composition combines two permutations to give another permutation.

**Definition.** A ***binary operation*** on a set $G$ is a function

$$* : G \times G \to G.$$

In more detail, a binary operation assigns an element $*(x, y)$ in $G$ to each ordered pair $(x, y)$ of elements in $G$. It is more natural to write $x * y$ instead of $*(x, y)$; thus, composition of functions is the function $(f, g) \mapsto g \circ f$; multiplication, addition, and subtraction are, respectively, the functions $(x, y) \mapsto xy$, $(x, y) \mapsto x+y$, and $(x, y) \mapsto x - y$. The examples of composition and subtraction show why we want ordered pairs, for $x * y$ and $y * x$ may be distinct. In constructing a binary operation on a set $G$, we must check, of course, that $x, y \in G$ implies $x * y \in G$; if this is so, we say that $G$ is **closed** under $*$. For example, if $G$ is the set of all odd integers, then $G$ is closed under multiplication (for $xy$ is odd when both $x$ and $y$ are), but $G$ is not closed under addition.

As any function, a binary operation is well-defined; when stated explicitly, this is usually called the **law of substitution**:

$$\text{if } x = x' \text{ and } y = y', \text{then } x * y = x' * y'.$$

**Definition.** A **group** is a set $G$ equipped with a binary operation $*$ such that

(i) the **associative law**[9] holds: for every $x$, $y$, $z \in G$,

$$x * (y * z) = (x * y) * z;$$

(ii) there is an element $e \in G$, called the **identity**, with $e * x = x = x * e$ for all $x \in G$;

(iii) every $x \in G$ has an **inverse**: there is $x' \in G$ with $x * x' = e = x' * x$.

The set $S_X$ of all permutations of a set $X$, with composition as binary operation and $1_X = (1)$ as the identity, is a group (the **symmetric group** on $X$).

Some of the equations in the definition of group are redundant. When verifying that a set with a binary operation is actually a group, it is obviously more economical to check fewer equations (see Exercise 1.31 on page 27: a set $G$ containing an element $e$ and having an associative binary operation $*$ is a group if $e * x = x$ for all $x \in G$ and, for every $x \in G$, there is $x' \in G$ with $x' * x = e$).

We are now at the precise point where Algebra becomes *abstract* Algebra. In contrast to the concrete group $S_n$ consisting of all the permutations of $\{1, 2, \ldots, n\}$, we have passed to abstract groups whose elements are unspecified and whose products are not explicitly computable (instead, multiplication is merely subject to certain rules). It will be seen that this approach is quite fruitful, for theorems now apply to many different groups, and it is more efficient to prove theorems once for all instead of proving them anew for each group encountered. In addition to this obvious economy, it is often simpler to work with the "abstract" viewpoint even when dealing with a particular concrete group. Indeed, when we deal with abstract groups, we can more easily focus on the essential facts we need without being distracted by "noise." For example, we will see that certain properties of $S_n$ are simpler to treat without recognizing that the elements in question are permutations and the binary operation is composition (see Example 1.28).

---

[9]Not all binary operations are associative. For example, subtraction is not associative: if $c \neq 0$, then $a - (b - c) \neq (a - b) - c$, and so the notation $a - b - c$ is ambiguous. The cross product of two vectors in $\mathbb{R}^3$ is another example of a nonassociative operation.

**Definition.** A group $G$ is called **abelian**[10] if it satisfies the **commutative law**:

$$x * y = y * x$$

for every $x, y \in G$.

The groups $S_n$, for $n \geq 3$, are not abelian because $(1\ 2)$ and $(1\ 3)$ are elements of $S_n$ that do not commute: $(1\ 2)(1\ 3) = (1\ 3\ 2)$ and $(1\ 3)(1\ 2) = (1\ 2\ 3)$.

**Lemma 1.16.** *Let $G$ be a group.*

(i) *The **cancellation laws** hold: if either $x * a = x * b$ or $a * x = b * x$, then $a = b$.[11]*

(ii) *The element $e$ is the unique element in $G$ with $e * x = x = x * e$ for all $x \in G$.*

(iii) *Each $x \in G$ has a unique inverse: there is only one element $x' \in G$ with $x * x' = e = x' * x$ (henceforth, this element will be denoted by $x^{-1}$).*

(iv) *$(x^{-1})^{-1} = x$ for all $x \in G$.*

**Proof.**

(i) Choose $x'$ with $x' * x = e = x * x'$; then

$$a = e * a = (x' * x) * a = x' * (x * a)$$
$$= x' * (x * b) = (x' * x) * b = e * b = b.$$

A similar proof works when $x$ is on the right.

(ii) Let $e_0 \in G$ satisfy $e_0 * x = x = x * e_0$ for all $x \in G$. In particular, setting $x = e$ in the second equation gives $e = e * e_0$; on the other hand, the defining property of $e$ gives $e * e_0 = e_0$, so that $e = e_0$.

(iii) Assume that $x'' \in G$ satisfies $x * x'' = e = x'' * x$. Multiply the equation $e = x * x'$ on the left by $x''$ to obtain

$$x'' = x'' * e = x'' * (x * x') = (x'' * x) * x' = e * x' = x'.$$

(iv) By definition, $(x^{-1})^{-1} * x^{-1} = e = x^{-1} * (x^{-1})^{-1}$. But $x * x^{-1} = e = x^{-1} * x$, so that $(x^{-1})^{-1} = x$, by (iii).   •

From now on, we will usually denote the product $x * y$ in a group by $xy$, and we will denote the identity by $1$ instead of by $e$. When a group is abelian, however, we usually use the **additive notation** $x + y$; in this case, the identity is denoted by $0$, and the inverse of an element $x$ is denoted by $-x$ instead of by $x^{-1}$.

**Example 1.17.**

(i) The set $\mathbb{Q}^\times$ of all nonzero rationals is an abelian group, where $*$ is ordinary multiplication: the number $1$ is the identity, and the inverse of $r \in \mathbb{Q}^\times$ is $1/r$. Similarly, $\mathbb{R}^\times$ and $\mathbb{C}^\times$ are multiplicative abelian groups.

---

[10]Commutative groups are called *abelian* because Abel proved (in modern language) that if the Galois group of a polynomial $f(x)$ is commutative, then $f(x)$ is solvable by radicals.

[11]We cannot cancel $x$ if $x * a = b * x$. For example, we have $(1\ 2)(1\ 2\ 3) = (2\ 1\ 3)(1\ 2)$ in $S_3$, but $(1\ 2\ 3) \neq (2\ 1\ 3)$.

Note that the set $\mathbb{Z}^\times$ of all nonzero integers is *not* a multiplicative group, for none of its elements (aside from $\pm 1$) has a multiplicative inverse in $\mathbb{Z}^\times$.

(ii) The set $\mathbb{Z}$ of all integers is an additive abelian group with $a * b = a + b$, with identity 0, and with the inverse of an integer $n$ being $-n$. Similarly, we can see that $\mathbb{Q}$, $\mathbb{R}$, and $\mathbb{C}$ are additive abelian groups.

(iii) The *circle group*,
$$S^1 = \{z \in \mathbb{C} : |z| = 1\},$$
is the group of all complex numbers of modulus 1 (the *modulus* of $z = a + ib \in \mathbb{C}$ is $|z| = \sqrt{a^2 + b^2}$) with binary operation multiplication of complex numbers. The set $S^1$ is closed, for if $|z| = 1 = |w|$, then $|zw| = 1$. Complex multiplication is associative, the identity is 1 (which has modulus 1), and the inverse of any complex number $z = a + ib$ of modulus 1 is its complex conjugate $\overline{z} = a - ib$ (which also has modulus 1). Thus, $S^1$ is a group.

(iv) For any positive integer $n$, let
$$\mu_n = \{z \in \mathbb{C} : z^n = 1\}$$
be the set of all the $n$th *roots of unity* with binary operation multiplication of complex numbers. Now $\mu_n$ is an abelian group: the set $\mu_n$ is closed [if $z^n = 1 = w^n$, then $(zw)^n = z^n w^n = 1$]; $1^n = 1$; multiplication is associative and commutative; the inverse of any $n$th root of unity is its complex conjugate, which is also an $n$th root of unity.

(v) The plane $\mathbb{R}^2$ is a group with operation vector addition; that is, if $\alpha = (x, y)$ and $\alpha' = (x', y')$, then $\alpha + \alpha' = (x + x', y + y')$. The identity is the origin $O = (0, 0)$, and the inverse of $(x, y)$ is $(-x, -y)$. ◀

**Example 1.18.** Let $X$ be a set. The *Boolean group* $\mathcal{B}(X)$ (named after the logician Boole) is the family of all the subsets of $X$ equipped with addition given by *symmetric difference* $A + B$, where
$$A + B = (A - B) \cup (B - A)$$
(recall that $A - B = \{x \in A : x \notin B\}$). Symmetric difference is pictured in Figure 1.3. It is plain that $A + B = B + A$, so that symmetric difference is

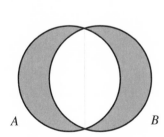

**Figure 1.3.** Symmetric difference.

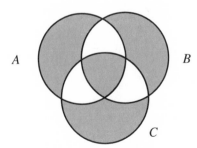

**Figure 1.4.** Associativity.

commutative. The identity is $\varnothing$, the empty set, and the inverse of $A$ is $A$ itself, for

$A + A = \varnothing$. The reader may verify associativity by showing that both $(A + B) + C$ and $A + (B + C)$ are described by Figure 1.4. ◀

**Example 1.19.** An $n \times n$ matrix $A$ with real entries is called **nonsingular** if it has an inverse; that is, there is a matrix $B$ with $AB = I = BA$, where $I$ is the $n \times n$ identity matrix. Since $(AB)^{-1} = B^{-1}A^{-1}$, the product of nonsingular matrices is itself nonsingular. The set $\mathrm{GL}(n, \mathbb{R})$ of all $n \times n$ nonsingular matrices having real entries, with binary operation matrix multiplication, is a (nonabelian) group, called the **general linear group**. [The proof of associativity is routine, though tedious; a "clean" proof of associativity can be given (Corollary 2.128) once the relation between matrices and linear transformations is known.] ◀

**Definition.** If $G$ is a group and $a \in G$, define the **powers**[12] $a^k$, for $k \geq 0$, inductively:
$$a^0 = 1 \quad \text{and} \quad a^{n+1} = aa^n.$$
If $k$ is a positive integer, define
$$a^{-k} = (a^{-1})^k.$$

A binary operation allows us to multiply two elements at a time; how do we multiply three elements? There is a choice. Given the expression $2 \times 3 \times 4$, for example, we can first multiply $2 \times 3 = 6$ and then multiply $6 \times 4 = 24$; or, we can first multiply $3 \times 4 = 12$ and then multiply $2 \times 12 = 24$; of course, the answers agree, for ordinary multiplication of numbers is associative. If a binary operation is associative, the notation $abc$ is not ambiguous. Let us now consider the definition of powers. The first and second powers are fine: $a^1 = aa^0 = a1 = a$ and $a^2 = aa$. There are two possible cubes: we have defined $a^3 = aa^2 = a(aa)$, but there is another reasonable contender: $(aa)a = a^2a$. If we assume associativity, then these are equal:
$$a^3 = aa^2 = a(aa) = (aa)a = a^2a.$$
There are several possible products of $a$ with itself four times; is it obvious that $a^4 = a^3a = a^2a^2$? And what about higher powers?

Let $G$ be a set with a binary operation; an **expression** in $G$ is an $n$-tuple $(a_1, a_2, \ldots, a_n) \in G \times \cdots \times G$ ($n$ factors) which is rewritten as $a_1a_2 \cdots a_n$. An expression yields many elements of $G$ by the following procedure. Choose two adjacent $a$'s, multiply them, and obtain an expression with $n - 1$ factors: the new product just formed and $n - 2$ original factors. In this shorter new expression, choose two adjacent factors (either an original pair or an original one together with the new product from the first step) and multiply them. Repeat this procedure until there is a *penultimate* expression with only two factors; multiply them and

---

[12]The terminology $x$ *square* and $x$ *cube* for $x^2$ and $x^3$ is, of course, geometric in origin. Usage of the word *power* in this context arises from a mistranslation of the Greek *dunamis* (from which dynamo derives) used by Euclid. *Power* was the standard European rendition of *dunamis*; for example, the first English translation of Euclid, in 1570, by H. Billingsley, renders a sentence of Euclid as, "The power of a line is the square of the same line." However, contemporaries of Euclid (e.g., Aristotle and Plato) often used *dunamis* to mean *amplification*, and this seems to be a more appropriate translation, for Euclid was probably thinking of a one-dimensional line segment sweeping out a two-dimensional square. (I thank Donna Shalev for informing me of the classical usage of *dunamis*.)

obtain an element of $G$ which we call an ***ultimate product***. For example, consider the expression $abcd$. We may first multiply $ab$, obtaining $(ab)cd$, an expression with three factors, namely, $ab$, $c$, $d$. We may now choose either the pair $c$, $d$ or the pair $ab$, $c$; in either case, multiply these, obtaining expressions with two factors: $(ab)(cd)$ having factors $ab$ and $cd$ or $((ab)c)d$ having factors $(ab)c$ and $d$. The two factors in either of these last expressions can now be multiplied to give an ultimate product from $abcd$. Other ultimate products derived from the expression $abcd$ arise from multiplying $bc$ or $cd$ as the first step. It is not obvious whether the ultimate products from a given expression are all equal.

**Definition.** Let $G$ be a set with a binary operation. An expression $a_1 a_2 \cdots a_n$ in $G$ ***needs no parentheses*** if all its ultimate products are equal elements of $G$.

**Theorem 1.20 (Generalized Associativity).** *If $G$ is a group having elements $a_1, a_2, \ldots, a_n$, then the expression $a_1 a_2 \cdots a_n$ needs no parentheses.*

**Proof.** The proof is by induction on $n \geq 3$. The base step holds because the operation is associative. For the inductive step, consider two ultimate products $U$ and $V$ obtained from a given expression $a_1 a_2 \cdots a_n$ after two series of choices:

$$U = (a_1 \cdots a_i)(a_{i+1} \cdots a_n) \quad \text{and} \quad V = (a_1 \cdots a_j)(a_{j+1} \cdots a_n);$$

the parentheses indicate the penultimate products displaying the last two factors that multiply to give $U$ and $V$, respectively; there are many parentheses inside each of these shorter expressions. We may assume that $i \leq j$. Since each of the four expressions in parentheses has fewer than $n$ factors, the inductive hypothesis says that each of them needs no parentheses. It follows that $U = V$ if $i = j$. If $i < j$, then the inductive hypothesis allows the first expression to be rewritten as

$$U = (a_1 \cdots a_i)\left([a_{i+1} \cdots a_j][a_{j+1} \cdots a_n]\right)$$

and the second to be rewritten as

$$V = \left([a_1 \cdots a_i][a_{i+1} \cdots a_j]\right)(a_{j+1} \cdots a_n),$$

where each of the expressions $a_1 \cdots a_i$, $a_{i+1} \cdots a_j$, and $a_{j+1} \cdots a_n$ needs no parentheses. Thus, these expressions yield unique elements $A$, $B$, and $C$ in $G$, respectively. The first expression gives $U = A(BC)$ in $G$, the second gives $V = (AB)C$ in $G$, and so $U = V$ in $G$, by associativity. $\bullet$

**Corollary 1.21.**

   (i) *If $a_1, a_2, \ldots, a_{k-1}, a_k$ are elements in a group $G$, then*

$$(a_1 a_2 \cdots a_{k-1} a_k)^{-1} = a_k^{-1} a_{k-1}^{-1} \cdots a_2^{-1} a_1^{-1}.$$

  (ii) *If $a \in G$ and $k \geq 1$, then $(a^k)^{-1} = a^{-k} = (a^{-1})^k$.*

**Proof.**

   (i) The proof is by induction on $k \geq 2$. Using generalized associativity,

$$(ab)(b^{-1}a^{-1}) = [a(bb^{-1})]a^{-1} = (a1)a^{-1} = aa^{-1} = 1;$$

a similar argument shows that $(b^{-1}a^{-1})(ab) = 1$. The base step $(ab)^{-1} = b^{-1}a^{-1}$ now follows from Lemma 1.16(iii). The proof of the inductive step is left to the reader.

(ii) Let every factor in part (i) be equal to $a$. Note that we have defined $a^{-k} = (a^{-1})^k$, and we now see that it coincides with the other worthy candidate for $a^{-k}$, namely, $(a^k)^{-1}$.  •

**Corollary 1.22.** *If $G$ is a group, $a \in G$, and $m$, $n \geq 1$, then*

$$a^{m+n} = a^m a^n \quad and \quad (a^m)^n = a^{mn}.$$

**Proof.** In the first case, both elements arise from the expression having $m + n$ factors each equal to $a$; in the second case, both elements arise from the expression having $mn$ factors each equal to $a$.  •

It follows that any two powers of an element $a$ in a group commute:

$$a^m a^n = a^{m+n} = a^{n+m} = a^n a^m.$$

**Proposition 1.23 (Laws of Exponents).** *Let $G$ be a group, let $a$, $b \in G$, and let $m$ and $n$ be (not necessarily positive) integers.*

(i) *If $a$ and $b$ commute, then $(ab)^n = a^n b^n$.*

(ii) $(a^m)^n = a^{mn}$.

(iii) $a^m a^n = a^{m+n}$.

**Proof.** The proofs, while routine, are lengthy double inductions.  •

The notation $a^n$ is the natural way to denote $a * a * \cdots * a$, where $a$ appears $n$ times. However, if the operation is $+$, then it is more natural to denote $a + a + \cdots + a$ by $na$. Let $G$ be a group written additively; if $a, b \in G$ and $m$ and $n$ are (not necessarily positive) integers, then Proposition 1.23 is usually rewritten as

(i) $n(a + b) = na + nb$.

(ii) $m(na) = (mn)a$.

(iii) $ma + na = (m + n)a$.

Theorem 1.20 holds in much greater generality.

**Definition.** A *semigroup* is a set having an associative operation; a *monoid* is a semigroup $S$ having a (two-sided) identity element 1; that is, $1s = s = s1$ for all $s \in S$.

Of course, every group is a monoid.

**Example 1.24.**

(i) The set of natural numbers $\mathbb{N}$ is a commutative monoid under addition (it is also a commutative monoid under multiplication). The set of all even integers under addition is a monoid; it is a semigroup under multiplication, but it is not a monoid.

(ii) A direct product of semigroups (or monoids) with cooordinatewise operation is again a semigroup (or monoid). In particular, the set $\mathbb{N}^n$ of all $n$-tuples of natural numbers is a commutative additive monoid.

(iii) The set of integers $\mathbb{Z}$ is a monoid under multiplication, as are $\mathbb{Q}$, $\mathbb{R}$, and $\mathbb{C}$. There are noncommutative monoids; for example, $\mathrm{Mat}_n(\mathbb{R})$, the set of all $n \times n$ matrices with real entries, is a multiplicative monoid. ◄

**Corollary 1.25 (Generalized Associativity).** *If $S$ is a semigroup and $a_1, a_2, \ldots, a_n \in S$, then the expression $a_1 a_2 \cdots a_n$ needs no parentheses.*

**Proof.** The proof of Theorem 1.20 assumes neither the existence of an identity element nor the existence of inverses. •

Can two powers of an element $a$ in a group coincide? Can $a^m = a^n$ for $m \neq n$? If so, then $a^m a^{-n} = a^{m-n} = 1$.

**Definition.** Let $G$ be a group and let $a \in G$. If $a^k = 1$ for some $k \geq 1$, then the smallest such exponent $k \geq 1$ is called the **order** of $a$; if no such power exists, then we say that $a$ has **infinite order**.

In any group $G$, the identity has order 1, and it is the only element of order 1. An element has order 2 if and only if it is equal to its own inverse; for example, $(1\ 2)$ has order 2 in $S_n$. The additive group of integers, $\mathbb{Z}$, is a group, and 3 is an element in it having infinite order (because $3 + 3 + \cdots + 3 = 3n \neq 0$ if $n > 0$). In fact, every nonzero element in $\mathbb{Z}$ has infinite order.

The definition of order says that if $x$ has order $n$ and $x^m = 1$ for some positive integer $m$, then $n \leq m$. The next theorem says that $n$ must be a divisor of $m$.

**Proposition 1.26.** *If $a \in G$ is an element of order $n$, then $a^m = 1$ if and only if $n \mid m$.*

**Proof.** If $m = nk$, then $a^m = a^{nk} = (a^n)^k = 1^k = 1$. Conversely, assume that $a^m = 1$. The Division Algorithm provides integers $q$ and $r$ with $m = nq + r$, where $0 \leq r < n$. It follows that $a^r = a^{m-nq} = a^m a^{-nq} = 1$. If $r > 0$, then we contradict $n$ being the smallest positive integer with $a^n = 1$. Hence, $r = 0$ and $n \mid m$. •

What is the order of a permutation in $S_n$?

**Proposition 1.27.** *Let $\alpha \in S_n$.*

(i) *If $\alpha$ is an $r$-cycle, then $\alpha$ has order $r$.*

(ii) *If $\alpha = \beta_1 \cdots \beta_t$ is a product of disjoint $r_i$-cycles $\beta_i$, then the order of $\alpha$ is $\mathrm{lcm}\{r_1, \ldots, r_t\}$.*[13]

(iii) *If $p$ is prime, then $\alpha$ has order $p$ if and only if it is a $p$-cycle or a product of disjoint $p$-cycles.*

**Proof.**

(i) This is Exercise 1.13 on page 15.

---

[13]The **least common multiple** is abbreviated to lcm.

(ii) Each $\beta_i$ has order $r_i$, by (i). Suppose that $\alpha^M = (1)$. Since the $\beta_i$ commute, $(1) = \alpha^M = (\beta_1 \cdots \beta_t)^M = \beta_1^M \cdots \beta_t^M$. By Exercise 1.21 on page 16, disjointness of the $\beta$'s implies that $\beta_i^M = (1)$ for each $i$, so that Proposition 1.26 gives $r_i \mid M$ for all $i$; that is, $M$ is a common multiple of $r_1, \ldots, r_t$. On the other hand, if $m = \mathrm{lcm}\{r_1, \ldots, r_t\}$, then it is easy to see that $\alpha^m = (1)$. Therefore, $\alpha$ has order $m$.

(iii) Write $\alpha$ as a product of disjoint cycles and use (ii).  •

For example, a permutation in $S_n$ has order 2 if and only if it is a product of disjoint transpositions.

**Example 1.28.** Suppose a deck of cards is shuffled, so that the order of the cards has changed from $1, 2, 3, 4, \ldots, 52$ to $2, 1, 4, 3, \ldots, 52, 51$. If we shuffle again in the same way, then the cards return to their original order. But a similar thing happens for any permutation $\alpha$ of the 52 cards: if one repeats $\alpha$ sufficiently often, the deck is eventually restored to its original order. One way to see this uses our knowledge of permutations. Write $\alpha$ as a product of disjoint cycles, say, $\alpha = \beta_1 \beta_2 \cdots \beta_t$, where $\beta_i$ is an $r_i$-cycle (our original shuffle is a product of disjoint transpositions). By Proposition 1.27, $\alpha$ has order $k$, where $k$ is the least common multiple of the $r_i$. Therefore, $\alpha^k = (1)$.

Here is a more general result with a simpler proof (abstract Algebra can be easier than Algebra): we show that if $G$ is a finite group and $a \in G$, then $a^k = 1$ for some $k \geq 1$. Consider the list $1, a, a^2, \ldots, a^n, \ldots$. Since $G$ is finite, there must be a repetition occurring on this infinite list: there are integers $m > n$ with $a^m = a^n$, and hence $1 = a^m a^{-n} = a^{m-n}$. We have shown that there is some positive power of $a$ equal to 1. [Our original argument that $\alpha^k = (1)$ for a permutation $\alpha$ of 52 cards is still worthwhile, because it gives an algorithm computing $k$.]  ◀

Let us state formally what we proved in Example 1.28.

**Proposition 1.29.** *If $G$ is a finite group, then every $x \in G$ has finite order.*

Table 3 for $S_5$ augments Table 2 on page 10.

| Cycle Structure | Number | Order | Parity |
|---|---|---|---|
| (1) | 1 | 1 | Even |
| (1 2) | 10 | 2 | Odd |
| (1 2 3) | 20 | 3 | Even |
| (1 2 3 4) | 30 | 4 | Odd |
| (1 2 3 4 5) | 24 | 5 | Even |
| (1 2)(3 4 5) | 20 | 6 | Odd |
| (1 2)(3 4) | 15 | 2 | Even |
| | $\overline{120}$ | | |

**Table 3.** Permutations in $S_5$.

**Definition.** If $G$ is a finite group, then the number of elements in $G$, denoted by $|G|$, is called the ***order*** of $G$.

The word *order* in Group Theory has two meanings: the order of an *element* $a \in G$; the order $|G|$ of a *group* $G$. Proposition 1.38 will explain this by relating the order of a group element $a$ with the order of a group determined by it.

Here are some geometric examples of groups.

**Definition.** An ***isometry*** is a distance preserving bijection[14] $\varphi \colon \mathbb{R}^2 \to \mathbb{R}^2$; that is, if $\|v - u\|$ is the distance from $v$ to $u$, then $\|\varphi(v) - \varphi(u)\| = \|v - u\|$. If $\pi$ is a polygon in the plane, then its ***symmetry group*** $\Sigma(\pi)$ consists of all the isometries $\varphi$ for which $\varphi(\pi) = \pi$. The elements of $\Sigma(\pi)$ are called ***symmetries*** of $\pi$.

**Example 1.30.** Let $\pi_4$ be a square having vertices $\{v_1, v_2, v_3, v_4\}$ and sides of length 1; draw $\pi_4$ in the plane so that its center is at the origin $O$ and its sides are parallel to the axes. It can be shown that every $\varphi \in \Sigma(\pi_4)$ permutes the

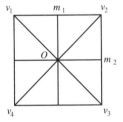

**Figure 1.5.** Square.

vertices (Exercise 1.65 on page 46); indeed, a symmetry $\varphi$ of $\pi_4$ is determined by $\{\varphi(v_i) : 1 \leq i \leq 4\}$, and so there are at most $24 = 4!$ possible symmetries. Not every permutation in $S_4$ arises from a symmetry of $\pi_4$, however. If $v_i$ and $v_j$ are adjacent, then $\|v_i - v_j\| = 1$, but $\|v_1 - v_3\| = \sqrt{2} = \|v_2 - v_4\|$; it follows that $\varphi$ must preserve adjacency (for isometries preserve distance). The reader may now check that there are only eight symmetries of $\pi_4$. Aside from the identity and the three rotations about $O$ by $90°$, $180°$, and $270°$, there are four reflections, respectively, in the lines $v_1v_3$, $v_2v_4$, the $x$-axis, and the $y$-axis (for a generalization to come, note that the $y$-axis is $Om_1$, where $m_1$ is the midpoint of $v_1v_2$, and the $x$-axis is $Om_2$, where $m_2$ is the midpoint of $v_2v_3$). The group $\Sigma(\pi_4)$ is called the ***dihedral group***[15] of order 8, and it is denoted by $D_8$. ◄

---

[14]It can be shown that $\varphi$ is a linear transformation if $\varphi(0) = 0$ (FCAA, Proposition 2.59). A distance preserving function $f \colon \mathbb{R}^2 \to \mathbb{R}^2$ is easily seen to be an injection. It is not so obvious (though it is true) that $f$ must also be a surjection (FCAA, Corollary 2.60).

[15]Klein was investigating those finite groups occurring as subgroups of the group of isometries of $\mathbb{R}^3$. Some of these occur as symmetry groups of regular polyhedra (from the Greek *poly* meaning "many" and *hedron* meaning "two-dimensional side"). He invented a degenerate polyhedron that he called a *dihedron*, from the Greek *di* meaning "two" and *hedron*, which consists of two congruent regular polygons of zero thickness pasted together. The symmetry group of a dihedron is thus called a *dihedral group*. It is more natural for us to describe these groups as in the text.

**Example 1.31.** The symmetry group $\Sigma(\pi_5)$ of a regular pentagon $\pi_5$ with vertices $v_1, \ldots, v_5$ and center $O$ (Figure 1.6) has 10 elements: the rotations about the origin by $(72j)^\circ$, where $0 \leq j \leq 4$, as well as the reflections in the lines $Ov_k$ for $1 \leq k \leq 5$. The symmetry group $\Sigma(\pi_5)$ is called the **dihedral group** of order 10, and it is denoted by $D_{10}$.   ◄

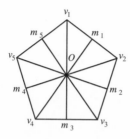

**Figure 1.6.** Pentagon.                                    **Figure 1.7.** Hexagon.

**Definition.** If $\pi_n$ is a regular polygon with $n \geq 3$ vertices $v_1, v_2, \ldots, v_n$ and center $O$, then the symmetry group $\Sigma(\pi_n)$ is called the **dihedral group** of order $2n$, and it is denoted[16] by $D_{2n}$. We define the **dihedral group** $D_4 = \mathbf{V}$, the **four-group**, to be the group of order 4

$$\mathbf{V} = \big\{(1), (1\ 2)(3\ 4), (1\ 3)(2\ 4), (1\ 4)(2\ 3)\big\} \subseteq S_4$$

[see Example 1.33(i) on page 28].

**Remark.** Some authors define the dihedral group $D_{2n}$ as a group of order $2n$ generated by elements $a, b$ such that $a^n = 1$, $b^2 = 1$, and $bab = a^{-1}$. Of course, one is obliged to prove existence of such a group; this is done in Proposition 4.86.   ◄

The dihedral group $D_{2n}$ of order $2n$ contains the $n$ rotations $\rho^j$ about the center by $(360j/n)^\circ$, where $0 \leq j \leq n-1$. The description of the other $n$ elements depends on the parity of $n$. If $n$ is odd (as in the case of the pentagon; see Figure 1.6), then the other $n$ symmetries are reflections in the distinct lines $Ov_i$, for $i = 1, 2, \ldots, n$. If $n = 2q$ is even (the square in Figure 1.5 or the regular hexagon in Figure 1.7), then each line $Ov_i$ coincides with the line $Ov_{q+i}$, giving only $q$ such reflections; the remaining $q$ symmetries are reflections in the lines $Om_i$ for $i = 1, 2, \ldots, q$, where $m_i$ is the midpoint of the edge $v_i v_{i+1}$. For example, the six lines of symmetry of $\pi_6$ are $Ov_1$, $Ov_2$, and $Ov_3$, and $Om_1$, $Om_2$, and $Om_3$.

## Exercises

**1.27.** Let $G$ be a semigroup. Prove directly, without using generalized associativity, that $(ab)(cd) = a[(bc)d]$ in $G$.

---

[16]Some authors denote $D_{2n}$ by $D_n$.

**1.28.** (i) Compute the order, inverse, and parity of

$$\alpha = (1\ 2)(4\ 3)(1\ 3\ 5\ 4\ 2)(1\ 5)(1\ 3)(2\ 3).$$

(ii) What are the respective orders of the permutations in Exercises 1.10 and 1.18 on page 15?

**1.29.** (i) How many elements of order 2 are there in $S_5$ and in $S_6$?

(ii) Make a table for $S_6$ (as the Table 3 on page 24).

(iii) How many elements of order 2 are there in $S_n$?
**Hint.** You may express your answer as a sum.

* **1.30.** If $G$ is a group, prove that the only element $g \in G$ with $g^2 = g$ is 1.

* **1.31.** This exercise gives a shorter list of axioms defining a group. Let $H$ be a semigroup containing an element $e$ such that $e * x = x$ for all $x \in H$ and, for every $x \in H$, there is $x' \in H$ with $x' * x = e$.

(i) Prove that if $h \in H$ satisfies $h * h = h$, then $h = e$.
**Hint.** If $h' * h = e$, evaluate $h' * h * h$ in two ways.

(ii) For all $x \in H$, prove that $x * x' = e$.
**Hint.** Consider $(x * x')^2$.

(iii) For all $x \in H$, prove that $x * e = x$.
**Hint.** Evaluate $x * x' * x$ in two ways.

(iv) Prove that if $e' \in H$ satisfies $e' * x = x$ for all $x \in H$, then $e' = e$.
**Hint.** Show that $(e')^2 = e'$.

(v) Let $x \in H$. Prove that if $x'' \in H$ satisfies $x'' * x = e$, then $x'' = x'$.
**Hint.** Evaluate $x' * x * x''$ in two ways.

(vi) Prove that $H$ is a group.

* **1.32.** Let $y$ be a group element of order $n$; if $n = mt$ for some divisor $m$, prove that $y^t$ has order $m$.
**Hint.** Clearly, $(y^t)^m = 1$. Use Proposition 1.26 to show that no smaller power of $y^t$ is equal to 1.

* **1.33.** Let $G$ be a group and let $a \in G$ have order $k$. If $p$ is a prime divisor of $k$ and there is $x \in G$ with $x^p = a$, prove that $x$ has order $pk$.

* **1.34.** Let $G = \mathrm{GL}(2, \mathbb{Q})$, let $A = \left[\begin{smallmatrix} 0 & -1 \\ 1 & 0 \end{smallmatrix}\right]$, and let $B = \left[\begin{smallmatrix} 0 & 1 \\ -1 & 1 \end{smallmatrix}\right]$. Show that $A^4 = I = B^6$, but that $(AB)^n \neq I$ for all $n > 0$, where $I = \left[\begin{smallmatrix} 1 & 0 \\ 0 & 1 \end{smallmatrix}\right]$. Conclude that $AB$ can have infinite order even though both factors $A$ and $B$ have finite order (of course, this cannot happen in a finite group).

* **1.35.** If $G$ is a group in which $x^2 = 1$ for every $x \in G$, prove that $G$ must be abelian. [The Boolean groups $\mathcal{B}(X)$ in Example 1.18 are such groups.]

* **1.36.** If $G$ is a group of even order, prove that the number of elements in $G$ of order 2 is odd. In particular, $G$ must contain an element of order 2.

**Hint.** Pair each element with its inverse.

**1.37.** Let $L(n)$ denote the largest order of an element in $S_n$. Find $L(n)$ for $n = 1, 2, \ldots, 10$.

The function $L(n)$ is called **Landau's function**. No general formula for $L(n)$ is known, although Landau, in 1903, found its asymptotic behavior:

$$\lim_{n \to \infty} \frac{\log L(n)}{\sqrt{n \log n}} = 1.$$

See Miller, The maximal order of an element of a finite symmetric group, *Amer. Math. Monthly* 94 (1987), 315–322.

## Section 1.4. Lagrange's Theorem

A *subgroup* $H$ of a group $G$ is a group contained in $G$ such that $h, h' \in H$ implies that the product $hh'$ in $H$ is the same as the product $hh'$ in $G$. Note that the multiplicative group $H = \{\pm 1\}$ is not a subgroup of the additive group $\mathbb{Z}$, for the product of 1 and $-1$ in $H$ is $-1$ while the "product" in $\mathbb{Z}$ is their sum, 0. The formal definition of subgroup is more convenient to use.

**Definition.** A subset $H$ of a group $G$ is a **subgroup** if

(i) $1 \in H$,

(ii) $H$ is **closed**; that is, if $x, y \in H$, then $xy \in H$,

(iii) if $x \in H$, then $x^{-1} \in H$.

Observe that $G$ and $\{1\}$ are always subgroups of a group $G$, where $\{1\}$ denotes the subset consisting of the single element 1. A subgroup $H \subsetneq G$ is called a **proper subgroup**; a subgroup $H \neq \{1\}$ is called a **nontrivial subgroup**.

**Proposition 1.32.** *Every subgroup $H$ of a group $G$ is itself a group.*

**Proof.** Property (ii) shows that $H$ is closed, for $x, y \in H$ implies $xy \in H$. Associativity $(xy)z = x(yz)$ holds for all $x, y, z \in G$, and it holds, in particular, for all $x, y, z \in H$. Finally, (i) gives the identity, and (iii) gives inverses.  •

For Galois, groups were subgroups of symmetric groups. Cayley, in 1854, was the first to define an "abstract" group, mentioning associativity, inverses, and identity explicitly. He then proved Theorem 1.95: every abstract group with $n$ elements is isomorphic to a subgroup of $S_n$ (we discuss *isomorphism* in the next section).

It is easier to check that a subset $H$ of a group $G$ is a subgroup (and hence that it is a group in its own right) than to verify the group axioms for $H$: associativity is inherited from $G$, and hence it need not be verified again.

**Example 1.33.**

(i) The set of four permutations,

$$\mathbf{V} = \big\{ (1), (1\ 2)(3\ 4), (1\ 3)(2\ 4), (1\ 4)(2\ 3) \big\},$$

is a subgroup of $S_4 : (1) \in \mathbf{V}$; $\alpha^2 = (1)$ for each $\alpha \in \mathbf{V}$, and so $\alpha^{-1} = \alpha \in \mathbf{V}$; the product of any two distinct permutations in $\mathbf{V} - \{(1)\}$ is the third one. It follows from Proposition 1.32 that $\mathbf{V}$ is a group; it is called the **four-group** ($\mathbf{V}$ abbreviates the original German term *Vierergruppe*).

Consider what verifying associativity $a(bc) = (ab)c$ would involve: there are four choices for each of $a$, $b$, and $c$, and so there are $4^3 = 64$ equations to be checked.

(ii) If $\mathbb{R}^2$ is the plane considered as an (additive) abelian group [see Example 1.17(v)], then any line $L$ through the origin is a subgroup. The easiest way to see this is to choose a point $(a, b) \neq (0, 0)$ on $L$ and then note that $L$ consists of all the scalar multiples $(ra, rb)$. The reader may now verify that the axioms in the definition of subgroup do hold for $L$.

(iii) The group $\mu_n$ of $n$th roots of unity [see Example 1.17(iv)] is a subgroup of the circle group $S^1$, but it is not a subgroup of the plane $\mathbb{R}^2$. ◄

We can shorten the list of items needed to verify that a subset is, in fact, a subgroup.

**Proposition 1.34.** *A subset $H$ of a group $G$ is a subgroup if and only if $H$ is nonempty and $xy^{-1} \in H$ whenever $x, y \in H$.*

**Proof.** Necessity is clear. For sufficiency, take $x \in H$ (which exists because $H \neq \varnothing$); by hypothesis, $1 = xx^{-1} \in H$. If $y \in H$, then $y^{-1} = 1y^{-1} \in H$, and if $x, y \in H$, then $xy = x(y^{-1})^{-1} \in H$. •

Note that if the binary operation on $G$ is addition, then the condition in the proposition is that $H$ is a nonempty subset such that $x, y \in H$ implies $x - y \in H$. Of course, the simplest way to check that a candidate $H$ for a subgroup is nonempty is to check whether $1 \in H$.

**Corollary 1.35.** *A nonempty subset $H$ of a finite group $G$ is a subgroup if and only if $H$ is closed; that is, $x, y \in H$ implies $xy \in H$.*

**Proof.** Since $G$ is finite, Proposition 1.29 says that each $x \in G$ has finite order. Hence, if $x^n = 1$, then $1 \in H$ and $x^{-1} = x^{n-1} \in H$. •

This corollary can be false when $G$ is an infinite group. For example, let $G$ be the additive group $\mathbb{Z}$; the set $\mathbb{N} = \{0, 1, 2, \dots\}$ of **natural numbers** is closed under addition, but $\mathbb{N}$ is not a subgroup of $\mathbb{Z}$.

**Example 1.36.** The subset $A_n = \{\alpha \in S_n : \alpha \text{ is even}\} \subseteq S_n$ is a subgroup, by Proposition 1.12, for it is closed under multiplication: even $\circ$ even = even. The group $A_n$ is called the **alternating group**[17] on $n$ letters. ◄

**Definition.** If $G$ is a group and $a \in G$, then the **cyclic subgroup** of $G$ **generated** by $a$, denoted by $\langle a \rangle$, is

$$\langle a \rangle = \{a^n : n \in \mathbb{Z}\} = \{\text{all powers of } a\}.$$

---

[17]The alternating group first arose in studying polynomials. If

$$f(x) = (x - u_1)(x - u_2) \cdots (x - u_n),$$

where $u_1, \dots, u_n$ are distinct, then the number $D = \prod_{i<j}(u_i - u_j)$ can change sign when the roots are permuted: if $\alpha$ is a permutation of $\{u_1, u_2, \dots, u_n\}$, then $\prod_{i<j}[\alpha(u_i) - \alpha(u_j)] = \pm D$. Thus, the sign of the product alternates as various permutations $\alpha$ are applied to its factors. The sign does not change for those $\alpha$ in the alternating group.

A group $G$ is called **cyclic** if there exists $a \in G$ with $G = \langle a \rangle$, in which case $a$ is called a **generator** of $G$.

The Laws of Exponents, Proposition 1.23, show that $\langle a \rangle$ is, in fact, a subgroup: $1 = a^0 \in \langle a \rangle$; $a^n a^m = a^{n+m} \in \langle a \rangle$; $a^{-1} \in \langle a \rangle$.

**Example 1.37.**

(i) The multiplicative group $\mu_n$ of all $n$th roots of unity [see Example 1.17(iv)] is a cyclic group; a generator is the primitive $n$th root of unity $\zeta = e^{2\pi i/n}$ [by De Moivre's Theorem, $e^{2\pi i k/n} = \cos\left(\frac{2\pi k}{n}\right) + i \sin\left(\frac{2\pi k}{n}\right) = \zeta^k$].

(ii) The (additive) group $\mathbb{Z}$ is an infinite cyclic group with generator 1.

(iii) We recall the definition of the integers modulo $m$. Given $m \geq 0$ and $a \in \mathbb{Z}$, the **congruence class** of $a$ mod $m$, denoted by $[a]$, is

$$[a] = \{b \in \mathbb{Z} : b \equiv a \bmod m\} = \{a + km : k \in \mathbb{Z}\}.$$

**Definition.** The **integers mod $m$**, denoted by $\mathbb{I}_m$,[18] is the family of all congruence classes mod $m$ with binary operation

$$[a] + [b] = [a + b].$$

It is easy to see that $\mathbb{I}_m$ is a group; it is a cyclic group, for $[1]$ is a generator. Note that if $m \geq 1$, then $\mathbb{I}_m$ has exactly $m$ elements, namely, $[0], [1], \ldots, [m-1]$.

Even though the definition of $\mathbb{I}_m$ makes sense for all $m \geq 0$, one usually assumes that $m \geq 2$ because the cases $m = 0$ and $m = 1$ are not very interesting. If $m = 0$, then $\mathbb{I}_m = \mathbb{I}_0 = \mathbb{Z}$, for $a \equiv b \bmod 0$ means $0 \mid (a - b)$; that is, $a = b$. If $m = 1$, then $\mathbb{I}_m = \mathbb{I}_1 = \{[0]\}$, for $a \equiv b \bmod 1$ means $1 \mid (a - b)$; that is, $a$ and $b$ are always congruent. ◄

The next proposition relates the two usages of the word *order* in Group Theory.

**Proposition 1.38.** *Let $G$ be a group. If $a \in G$, then the order of $a$ is equal to $|\langle a \rangle|$, the order of the cyclic subgroup generated by $a$.*

**Proof.** The result is obviously true when $a$ has infinite order, and so we may assume that $a$ has finite order $n$. We claim that $A = \{1, a, a^2, \ldots, a^{n-1}\}$ has exactly $n$ elements; that is, the displayed elements are distinct. If $a^i = a^j$ for $0 \leq i < j \leq n-1$, then $a^{j-i} = 1$; as $0 < j - i < n$, this contradicts $n$ being the smallest positive integer with $a^n = 1$.

It suffices to show that $A = \langle a \rangle$. Clearly, $A \subseteq \langle a \rangle$. For the reverse inclusion, take $a^k \in \langle a \rangle$. By the Division Algorithm, $k = qn + r$, where $0 \leq r < n$; hence, $a^k = a^{qn+r} = a^{qn} a^r = (a^n)^q a^r = a^r$. Thus, $a^k = a^r \in A$, and $\langle a \rangle = A$. •

A cyclic group can have several different generators; for example, $\langle a \rangle = \langle a^{-1} \rangle$.

---

[18] We introduce this new notation because there is no commonly agreed one; the most popular contenders are $\mathbb{Z}/m\mathbb{Z}$ and $\mathbb{Z}_m$. The former notation is too complicated to use many times in a proof; the latter is ambiguous because, when $p$ is prime, $\mathbb{Z}_p$ often denotes the ring of $p$-adic integers and not the integers mod $p$.

**Definition.** If $n \geq 1$, then the **Euler $\phi$-function** $\phi(n)$ is defined by

$$\phi(n) = |\{k \in \mathbb{Z} : 1 \leq k \leq n \text{ and } (k, n) = 1\}|.$$

**Theorem 1.39.**

(i) If $G = \langle a \rangle$ is a cyclic group of order $n$, then $a^k$ is a generator of $G$ if and only if $(k, n) = 1$.[19]

(ii) If $G$ is a cyclic group of order $n$ and $\text{gen}(G) = \{\text{all generators of } G\}$, then

$$|\text{gen}(G)| = \phi(n),$$

where $\phi(n)$ is the Euler $\phi$-function.

**Proof.**

(i) If $a^k$ generates $G$, then $a \in \langle a^k \rangle$, so that $a = a^{kt}$ for some $t \in \mathbb{Z}$. Hence, $a^{kt-1} = 1$; by Proposition 1.26, $n \mid (kt-1)$, so there is $v \in \mathbb{Z}$ with $nv = kt-1$. Therefore, 1 is a linear combination of $k$ and $n$, and so $(k, n) = 1$.

Conversely, if $(k, n) = 1$, then $ns + kt = 1$ for $s, t \in \mathbb{Z}$; hence

$$a = a^{ns+kt} = a^{ns}a^{kt} = a^{kt} \in \langle a^k \rangle.$$

Therefore, $a$, hence every power of $a$, also lies in $\langle a^k \rangle$, and so $G = \langle a^k \rangle$.

(ii) Since $G = \{1, a, \ldots, a^{n-1}\}$, this result follows from Proposition 1.38.  •

**Proposition 1.40.** *The intersection $\bigcap_{i \in I} H_i$ of any family of subgroups of a group $G$ is again a subgroup of $G$. In particular, if $H$ and $K$ are subgroups of $G$, then $H \cap K$ is a subgroup of $G$.*

**Proof.** This follows easily from the definitions.  •

**Corollary 1.41.** *If $X$ is a subset of a group $G$, then there is a subgroup $\langle X \rangle$ of $G$ containing $X$ that is **smallest** in the sense that $\langle X \rangle \subseteq H$ for every subgroup $H$ of $G$ that contains $X$.*

**Proof.** There do exist subgroups of $G$ that contain $X$; for example, $G$ contains $X$. Define $\langle X \rangle = \bigcap_{X \subseteq H} H$, the intersection of all the subgroups $H$ of $G$ containing $X$. By Proposition 1.40, $\langle X \rangle$ is a subgroup of $G$; of course, $\langle X \rangle$ contains $X$ because every $H$ contains $X$. Finally, if $H_0$ is any subgroup containing $X$, then $H_0$ is one of the subgroups whose intersection is $\langle X \rangle$; that is, $\langle X \rangle = \bigcap H \subseteq H_0$.  •

There is no restriction on the subset $X$ in the last corollary; in particular, $X = \varnothing$ is allowed. Since the empty set is a subset of every set, we have $\langle \varnothing \rangle \subseteq H$ for every subgroup $H$ of $G$. In particular, $\langle \varnothing \rangle \subseteq \{1\}$, and so $\langle \varnothing \rangle = \{1\}$.

**Definition.** If $X$ is a subset of a group $G$, then $\langle X \rangle$ is called the **subgroup generated by** $X$.

---

[19]The gcd (greatest common divisor) of $k$ and $n$ is denoted by $(k, n)$. If $(k, n) = 1$, we say that $k$ and $n$ are **relatively prime**.

If $X$ is a nonempty subset of a group $G$, a **word**[20] on $X$ is an element $g \in G$ of the form $g = x_1^{e_1} \cdots x_n^{e_n}$, where $x_i \in X$ and $e_i = \pm 1$ for all $i$.

**Proposition 1.42.** *If $X$ is a nonempty subset of a group $G$, then $\langle X \rangle$ is the set of all the words on $X$.*

**Proof.** We claim that $W(X)$, the set of all the words on $X$, is a subgroup. If $x \in X$, then $1 = xx^{-1} \in W(X)$; the product of two words on $X$ is also a word on $X$; the inverse of a word on $X$ is a word on $X$. It now follows that $\langle X \rangle \subseteq W(X)$, for $W(X)$ is a subgroup containing $X$. The reverse inclusion is clear, for any subgroup of $G$ containing $X$ must contain every word on $X$. Therefore, $\langle X \rangle = W(X)$.   •

**Definition.** If $H$ and $K$ are subgroups of a group $G$, then

$$H \vee K = \langle H \cup K \rangle$$

is the **subgroup generated by $H$ and $K$.**

It is easy to check that $H \vee K$ is the smallest subgroup of $G$ that contains both $H$ and $K$.

**Corollary 1.43.** *If $H$ and $K$ are subgroups of an abelian group $G$, then*

$$H \vee K = H + K = \{sh + tk : h \in H, k \in K, s, t \in \mathbb{Z}\}.$$

**Proof.** Words $x_1^{e_1} \cdots x_n^{e_n} \in \langle H \cup K \rangle$ are written $e_1 x_1 + \cdots + e_n x_n$ in additive notation, and they can be written in the displayed form because $G$'s being abelian allows us to collect terms.   •

**Example 1.44.**

(i) If $G = \langle a \rangle$ is a cyclic group with generator $a$, then $G$ is generated by the subset $X = \{a\}$.

(ii) Let $a$ and $b$ be integers, and let $A = \langle a \rangle$ and $B = \langle b \rangle$ be the cyclic subgroups of $\mathbb{Z}$ they generate. Then $A \cap B = \langle m \rangle$, where $m = \mathrm{lcm}\{a, b\}$, and $A + B = \langle d \rangle$, where $d$ is the gcd $(a, b)$.

(iii) The dihedral group $D_{2n}$ (the symmetry group of a regular $n$-gon, where $n \geq 3$) is generated by $\rho, \sigma$, where $\rho$ is a rotation by $(360/n)°$ and $\sigma$ is a reflection. Note that these generators satisfy the equations $\rho^n = 1$, $\sigma^2 = 1$, and $\sigma\rho\sigma = \rho^{-1}$. We defined the dihedral group $D_4 = \mathbf{V}$, the four-group, in Example 1.33(i); note that $\mathbf{V}$ is generated by elements $\rho$ and $\sigma$ satisfying the equations $\rho^2 = 1$, $\sigma^2 = 1$, and $\sigma\rho\sigma = \rho^{-1} = \rho$.   ◄

Perhaps the most fundamental fact about subgroups $H$ of a finite group $G$ is that their orders are constrained. Certainly, we have $|H| \leq |G|$, but it turns out that $|H|$ must be a divisor of $|G|$. To prove this, we introduce the notion of *coset*.

**Definition.** If $H$ is a subgroup of a group $G$ and $a \in G$, then the **coset** $aH$ is the subset $aH$ of $G$, where

$$aH = \{ah : h \in H\}.$$

---

[20]This term will be modified a bit when we discuss free groups in Chapter 4.

Each element of a coset $aH$ (e.g., $a$) is called a **representative**.

The cosets just defined are often called **left cosets**; there are also **right cosets** of $H$, namely, subsets of the form $Ha = \{ha : h \in H\}$. In general, left cosets and right cosets may be different, as we shall soon see.

If we use the $*$ notation for the binary operation on a group $G$, then we denote the coset $aH$ by $a * H$, where $a * H = \{a * h : h \in H\}$. In particular, if the operation is addition, then this coset is denoted by

$$a + H = \{a + h : h \in H\}.$$

Of course, $a = a1 \in aH$. Cosets are usually not subgroups. For example, if $a \notin H$, then $1 \notin aH$ (otherwise $1 = ah$ for some $h \in H$, and this gives the contradiction $a = h^{-1} \in H$).

**Example 1.45.**

(i) If $[a]$ is the congruence class of $a \bmod m$, then $[a] = a + H$, where $H = \langle m \rangle$ is the cyclic subgroup of $\mathbb{Z}$ generated by $m$.

(ii) Consider the plane $\mathbb{R}^2$ as an (additive) abelian group and let $L$ be a line through the origin (see Figure 1.8); as in Example 1.33(ii), the line $L$ is a subgroup of $\mathbb{R}^2$. If $\beta \in \mathbb{R}^2$, then the coset $\beta + L$ is the line $L'$ containing $\beta$ that is parallel to $L$, for if $r\alpha \in L$, then the parallelogram law gives $\beta + r\alpha \in L'$.

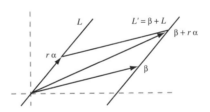

**Figure 1.8.** The coset $\beta + L$.

(iii) Let $A$ be an $m \times n$ matrix with real entries. If the linear system of equations $A\mathbf{x} = \mathbf{b}$ is *consistent*; that is, the **solution set** $\{\mathbf{x} \in \mathbb{R}^n : A\mathbf{x} = \mathbf{b}\}$ is nonempty, then there is a column vector $\mathbf{s} \in \mathbb{R}^n$ with $A\mathbf{s} = \mathbf{b}$. Define the **solution space** $S$ of the homogeneous system $A\mathbf{x} = \mathbf{0}$ to be $\{\mathbf{x} \in \mathbb{R}^n : A\mathbf{x} = \mathbf{0}\}$; it is an additive subgroup of $\mathbb{R}^n$. The solution set of the original inhomogeneous system is the coset $\mathbf{s} + S$.

(iv) Let $A_n$ be the alternating group, and let $\tau \in S_n$ be a transposition [so that $\tau^2 = (1)$]. We claim that $S_n = A_n \cup \tau A_n$. Let $\alpha \in S_n$. If $\alpha$ is even, then $\alpha \in A_n$; if $\alpha$ is odd, then $\alpha = \tau(\tau\alpha) \in \tau A_n$, for $\tau\alpha$, being the product of two odd permutations, is even. Note that $A_n \cap \tau A_n = \varnothing$, for no permutation is simultaneously even and odd.

(v) If $G = S_3$ and $H = \langle (1\,2) \rangle$, there are exactly three left cosets of $H$, namely

$$\begin{aligned}
H &= \quad \{(1), (1\,2)\} \ = (1\,2)H, \\
(1\,3)H &= \{(1\,3), (1\,2\,3)\} = (1\,2\,3)H, \\
(2\,3)H &= \{(2\,3), (1\,3\,2)\} = (1\,3\,2)H,
\end{aligned}$$

each of which has size two. Note that these cosets are also "parallel"; that is, distinct cosets are disjoint.

Consider the right cosets of $H = \langle (1\,2) \rangle$ in $S_3$:

$$\begin{aligned}
H &= \quad \{(1), (1\,2)\} \ \ = H(1\,2), \\
H(1\,3) &= \{(1\,3), (1\,3\,2)\} = H(1\,3\,2), \\
H(2\,3) &= \{(2\,3), (1\,2\,3)\} = H(1\,2\,3).
\end{aligned}$$

Again, we see that there are exactly 3 (right) cosets, each of which has size two. Note that these cosets are "parallel"; that is, distinct (right) cosets are disjoint.

Finally, observe that the left coset $(1\,3)H$ is not a right coset of $H$; in particular, $(1\,3)H \neq H(1\,3)$.  ◄

**Lemma 1.46.** *Let $H$ be a subgroup of a group $G$, and let $a, b \in G$.*

(i) $aH = bH$ *if and only if* $b^{-1}a \in H$. *In particular,* $aH = H$ *if and only if* $a \in H$.

(ii) *If* $aH \cap bH \neq \varnothing$, *then* $aH = bH$.

(iii) $|aH| = |H|$ *for all* $a \in G$.

**Remark.** Exercise 1.38 on page 37 has the version of (i) for right cosets: $Ha = Hb$ if and only if $ab^{-1} \in H$, and hence $Ha = H$ if and only if $a \in H$.  ◄

**Proof.** The first statement follows from observing that the relation on $G$, defined by $a \equiv b$ if $b^{-1}a \in H$, is an equivalence relation[21] whose equivalence classes are the left cosets. Since the equivalence classes of an equivalence relation form a partition, the left cosets of $H$ partition $G$ (which is the second statement). The third statement is true because $h \mapsto ah$ is a bijection $H \to aH$ [its inverse is $ah \mapsto a^{-1}(ah)$].  •

For example, if $H = \langle m \rangle \subseteq \mathbb{Z}$, then $a + H = b + H$ if and only if $a - b \in \langle m \rangle$; that is, $a \equiv b \bmod m$.

The next theorem is named after Lagrange who asserted, in 1770, that the orders of certain subgroups of $S_n$ are divisors of $n!$. The notion of group was invented by Galois 60 years later, and it was probably Galois who first proved the theorem in full.

---

[21]An *equivalence relation* on a set $X$ is a binary relation $\equiv$ which is reflexive, symmetric, and transitive. If $a \in X$, then its *equivalence class* is $[a] = \{x \in X : x \equiv a\}$. If $\equiv$ is an equivalence relation on $X$, then the family of all equivalence classes forms a *partition* of $X$; that is, they are pairwise disjoint nonempty subsets of $X$ whose union is all of $X$. Conversely, given a partition $(A_i)_{i \in I}$ of $X$, there exists an equivalence relation on $X$ whose equivalence classes are the $A_i$ (FCAA, p. 102).

**Theorem 1.47 (Lagrange's Theorem).** *If $H$ is a subgroup of a finite group $G$, then $|H|$ is a divisor of $|G|$.*

**Proof.** Let $\{a_1H, \ldots, a_tH\}$ be the family of all the distinct left cosets of $H$ in $G$. We claim that
$$G = a_1H \cup a_2H \cup \cdots \cup a_tH.$$
If $g \in G$, then $g = g1 \in gH$; but $gH = a_iH$ for some $i$, because $a_1H, \ldots, a_tH$ is a list of *all* the left cosets of $H$. Now Lemma 1.46(ii) shows that the cosets partition $G$ into pairwise disjoint subsets, and so
$$|G| = |a_1H| + |a_2H| + \cdots + |a_tH|.$$
But $|a_iH| = |H|$ for all $i$, by Lemma 1.46(iii); hence, $|G| = t|H|$, as desired. •

**Definition.** The *index* of a subgroup $H$ in $G$, denoted by $[G : H]$, is the number of left[22] cosets of $H$ in $G$.

The index $[G : H]$ is the number $t$ in the formula $|G| = t|H|$ in the proof of Lagrange's Theorem, so that
$$|G| = [G : H]|H|;$$
this formula shows that the index $[G : H]$ is also a divisor of $|G|$; moreover,
$$[G : H] = |G|/|H|.$$

**Example 1.48.**

(i) Here is another solution of Exercise 1.22 on page 16. In Example 1.45(iv), we saw that $S_n = A_n \cup \tau A_n$, where $\tau$ is a transposition. Thus, there are exactly two cosets of $A_n$ in $S_n$; that is, $[S_n : A_n] = 2$. It follows that $|A_n| = \frac{1}{2}n!$.

(ii) Recall that the dihedral group $D_{2n} = \Sigma(\pi_n)$, the symmetries of the regular $n$-gon $\pi_n$, has order $2n$, and it contains the cyclic subgroup $\langle \rho \rangle$ of order $n$ generated by the clockwise rotation $\rho$ by $(360/n)°$. Thus, $\langle \rho \rangle$ has index $[D_{2n} : \langle \rho \rangle] = 2n/n = 2$, and there are only two cosets: $\langle \rho \rangle$ and $\sigma\langle \rho \rangle$, where $\sigma$ is any reflection outside of $\langle \rho \rangle$. It follows that $D_{2n} = \langle \rho \rangle \cup \sigma\langle \rho \rangle$; every element $\alpha \in D_{2n}$ has a unique factorization $\alpha = \sigma^i\rho^j$, where $i = 0, 1$ and $0 \le j < n$. ◄

**Corollary 1.49.** *If $G$ is a finite group and $a \in G$, then the order of $a$ is a divisor of $|G|$.*

**Proof.** Immediate from Lagrange's Theorem, for the order of $a$ is $|\langle a \rangle|$. •

**Corollary 1.50.** *If $G$ is a finite group, then $a^{|G|} = 1$ for all $a \in G$.*

**Proof.** If $a$ has order $d$, then $|G| = dm$ for some integer $m$, by the previous corollary, and so $a^{|G|} = a^{dm} = (a^d)^m = 1$. •

**Corollary 1.51.** *If $p$ is prime, then every group $G$ of order $p$ is cyclic.*

---

[22]Exercise 1.44 on page 38 shows that the number of left cosets of a subgroup $H$ is equal to the number of right cosets of $H$.

**Proof.** If $a \in G$ and $a \neq 1$, then $a$ has order $d > 1$, and $d$ is a divisor of $p$. Since $p$ is prime, $d = p$, and so $G = \langle a \rangle$. •

In Example 1.45(iii), we saw that the additive group $\mathbb{I}_m$ is cyclic of order $m$. Now multiplication $\mathbb{I}_m \times \mathbb{I}_m \to \mathbb{I}_m$, given by

$$[a][b] = [ab],$$

is also a binary operation on $\mathbb{I}_m$ (which is well-defined, by FCAA, Proposition 1.60); it is associative, commutative, and $[1]$ is an identity element. However, $\mathbb{I}_m$ is not a group under this operation because inverses may not exist; for example, $[0]$ has no multiplicative inverse.

**Proposition 1.52.** *The set $U(\mathbb{I}_m)$, defined by*

$$U(\mathbb{I}_m) = \{[r] \in \mathbb{I}_m : (r, m) = 1\},$$

*is a multiplicative group of order $\phi(m)$, where $\phi$ is the Euler $\phi$-function. In particular, if $p$ is prime, then $U(\mathbb{I}_p)$ is a multiplicative group of order $p - 1$.*

**Remark.** Theorem 2.46 says that $U(\mathbb{I}_p)$ is a cyclic group for every prime $p$. ◀

**Proof.** If $(r, m) = 1 = (r', m)$, then $(rr', m) = 1$: if $sr + tm = 1$ and $s'r' + t'm = 1$, then

$$(sr + tm)(s'r' + t'm) = 1 = (ss')rr' + (st'r + ts'r + tt'm)m;$$

hence $U(\mathbb{I}_m)$ is closed under multiplication. We have already mentioned that multiplication is associative and that $[1]$ is the identity. If $(a, m) = 1$, then $[a][x] = [1]$ can be solved for $[x]$ in $\mathbb{I}_m$. Now $(x, m) = 1$, for $rx + sm = 1$ for some integer $s$, and so $(x, m) = 1$. Hence, $[x] \in U(\mathbb{I}_m)$, and so each $[r] \in U(\mathbb{I}_m)$ has an inverse in $U(\mathbb{I}_m)$. Therefore, $U(\mathbb{I}_m)$ is a group, and the definition of the Euler $\phi$-function shows that $|U(\mathbb{I}_m)| = \phi(m)$. The last statement follows because $\phi(p) = p - 1$ when $p$ is prime. •

Here is a group-theoretic proof of Fermat's Theorem.

**Corollary 1.53 (Fermat).** *If $p$ is prime and $a \in \mathbb{Z}$, then*

$$a^p \equiv a \bmod p.$$

**Proof.** It suffices to show that $[a^p] = [a]$ in $\mathbb{I}_p$. If $[a] = [0]$, then $[a^p] = [a]^p = [0]^p = [0] = [a]$. If $[a] \neq [0]$, then $[a] \in \mathbb{I}_p^\times$, the multiplicative group of nonzero elements in $\mathbb{I}_p$. By Corollary 1.50 to Lagrange's Theorem, $[a]^{p-1} = [1]$, because $|\mathbb{I}_p^\times| = p-1$. Multiplying by $[a]$ gives the desired result: $[a^p] = [a]^p = [a]$. Therefore, $a^p \equiv a \bmod p$. •

Euler generalized Fermat's Theorem.

**Theorem 1.54 (Euler).** *If $(r, m) = 1$, then*

$$r^{\phi(m)} \equiv 1 \bmod m.$$

**Proof.** Since $|U(\mathbb{I}_m)| = \phi(m)$, Corollary 1.50 gives $[r]^{\phi(m)} = [1]$ for all $[r] \in U(\mathbb{I}_m)$. In congruence notation, this says that if $(r, m) = 1$, then $r^{\phi(m)} \equiv 1 \bmod m$. •

**Example 1.55.** It is easy to see that the square of each element in the group

$$U(\mathbb{I}_8) = \big\{[1], [3], [5], [7]\big\}$$

is $[1]$ [thus, $U(\mathbb{I}_8)$ resembles the four-group **V**], while

$$U(\mathbb{I}_{10}) = \big\{[1], [3], [7], [9]\big\}$$

is a cyclic group of order 4 with generator $[3]$ [were the term *isomorphism* available (we introduce it in the next section), we would say that $U(\mathbb{I}_8)$ is isomorphic to **V** and $U(\mathbb{I}_{10})$ is isomorphic to $\mathbb{I}_4$]. See Example 1.60. ◄

**Theorem 1.56 (Wilson's Theorem).** *An integer $p$ is prime if and only if*

$$(p-1)! \equiv -1 \bmod p.$$

**Proof.** Assume that $p$ is prime. If $a_1, a_2, \ldots, a_n$ is a list of all the elements of a finite abelian group $G$, then the product $a_1 a_2 \cdots a_n$ is the same as the product of all elements $a$ with $a^2 = 1$, for any other element cancels against its inverse. Since $p$ is prime, $\mathbb{I}_p^{\times}$ has only one element of order 2, namely, $[-1]$ (if $p$ is prime and $x^2 \equiv 1 \bmod p$, then $x = [\pm 1]$). It follows that the product of all the elements in $\mathbb{I}_p^{\times}$, namely, $[(p-1)!]$, is equal to $[-1]$; therefore, $(p-1)! \equiv -1 \bmod p$.

Conversely, assume that $m$ is composite: there are integers $a$ and $b$ with $m = ab$ and $1 < a \leq b < m$. If $a < b$, then $m = ab$ is a divisor of $(m-1)!$, and so $(m-1)! \equiv 0 \bmod m$. If $a = b$, then $m = a^2$. If $a = 2$, then $(a^2 - 1)! = 3! = 6 \equiv 2 \bmod 4$ and, of course, $2 \not\equiv -1 \bmod 4$. If $2 < a$, then $2a < a^2$, and so $a$ and $2a$ are factors of $(a^2 - 1)!$; therefore, $(a^2 - 1)! \equiv 0 \bmod a^2$. Thus, $(a^2 - 1)! \not\equiv -1 \bmod a^2$, and the proof is complete. •

**Remark.** We can generalize Wilson's Theorem in the same way that Euler's Theorem generalizes Fermat's Theorem: replace $U(\mathbb{I}_p)$ by $U(\mathbb{I}_m)$. For example, if $m \geq 3$, we can prove that $U(\mathbb{I}_{2^m})$ has exactly 3 elements of order 2, namely, $[-1]$, $[1+2^{m-1}]$, and $[-(1+2^{m-1})]$ (Rotman, *An Introduction to the Theory of Groups*, p. 121). It follows that the product of all the odd numbers $r$, where $1 \leq r < 2^m$, is congruent to $1 \bmod 2^m$, because

$$(-1)(1+2^{m-1})(-1-2^{m-1}) = (1+2^{m-1})^2 = 1 + 2^m + 2^{2m-2} \equiv 1 \bmod 2^m. \quad ◄$$

---

## Exercises

∗ **1.38.** Let $H$ be a subgroup of a group $G$.

   (i) Prove that right cosets $Ha$ and $Hb$ are equal if and only if $ab^{-1} \in H$.

   (ii) Prove that the relation $a \equiv b$ if $ab^{-1} \in H$ is an equivalence relation on $G$ whose equivalence classes are the right cosets of $H$.

**1.39.**   (i) Define the **special linear group** by

$$\mathrm{SL}(2, \mathbb{R}) = \{A \in \mathrm{GL}(2, \mathbb{R}) : \det(A) = 1\}.$$

  Prove that $\mathrm{SL}(2, \mathbb{R})$ is a subgroup of $\mathrm{GL}(2, \mathbb{R})$.

  (ii) Prove that $\mathrm{GL}(2, \mathbb{Q})$ is a subgroup of $\mathrm{GL}(2, \mathbb{R})$.

∗ **1.40.** (i) Give an example of two subgroups $H$ and $K$ of a group $G$ whose union $H \cup K$ is not a subgroup of $G$.

   **Hint.** Let $G$ be the four-group **V**.

   (ii) Prove that the union $H \cup K$ of two subgroups is itself a subgroup if and only if $H$ is a subset of $K$ or $K$ is a subset of $H$.

∗ **1.41.** Let $G$ be a finite group with subgroups $H$ and $K$. If $H \subseteq K \subseteq G$, prove that

$$[G : H] = [G : K][K : H].$$

**1.42.** If $H$ and $K$ are subgroups of a group $G$ and $|H|$ and $|K|$ are relatively prime, prove that $H \cap K = \{1\}$.

   **Hint.** If $x \in H \cap K$, then $x^{|H|} = 1 = x^{|K|}$.

∗ **1.43.** Let $G$ be a group of order 4. Prove that either $G$ is cyclic or $x^2 = 1$ for every $x \in G$. Conclude, using Exercise 1.35 on page 27, that $G$ must be abelian.

∗ **1.44.** If $H$ is a subgroup of a group $G$, prove that the number of left cosets of $H$ in $G$ is equal to the number of right cosets of $H$ in $G$.

   **Hint.** The function $\varphi \colon aH \mapsto Ha^{-1}$ is a bijection from the family of all left cosets of $H$ to the family of all right cosets of $H$.

**1.45.** If $p$ is an odd prime and $a_1, \ldots, a_{p-1}$ is a permutation of $\{1, 2, \ldots, p - 1\}$, prove that there exist $i \neq j$ with $ia_i \equiv ja_j \bmod p$.

   **Hint.** Use Wilson's Theorem.

## Section 1.5. Homomorphisms

An important problem is determining whether two given groups $G$ and $H$ are somehow the same. For example, we have investigated $S_3$, the group of all permutations of $\{1, 2, 3\}$. The group $S_Y$ of all the permutations of $Y = \{a, b, c\}$ is a group different from $S_3$, because permutations of $\{1, 2, 3\}$ are not permutations of $\{a, b, c\}$. But even though $S_3$ and $S_Y$ are different, they surely bear a strong resemblance to each other (see Example 1.57). More interesting is the strong resemblance of $S_3$ to $D_6$, the symmetries of an equilateral triangle. The notions of homomorphism and isomorphism will allow us to compare different groups.

**Definition.** Let $(G, *)$ and $(H, \circ)$ be groups (we have displayed the binary operations on each). A **homomorphism**[23] is a function satisfying

$$f(x * y) = f(x) \circ f(y)$$

for all $x, y \in G$. If $f$ is also a bijection, then $f$ is called an **isomorphism**. Two groups $G$ and $H$ are called **isomorphic**, denoted by $G \cong H$, if there exists an isomorphism $f : G \to H$ between them.

---

[23]The word *homomorphism* comes from the Greek *homo* meaning "same" and *morph* meaning "shape" or "form." Thus, a homomorphism carries a group to another group (its image) of similar form. The word *isomorphism* involves the Greek *iso* meaning "equal," and isomorphic groups have identical form.

**Definition.** Let $a_1, a_2, \ldots, a_n$ be a list with no repetitions of all the elements in a group $G$. A **multiplication table** for $G$ is the $n \times n$ matrix whose $ij$ entry is $a_i a_j$.

| $G$ | $a_1$ | $a_2$ | $\cdots$ | $a_j$ | $\cdots$ | $a_n$ |
|---|---|---|---|---|---|---|
| $a_1$ | $a_1 a_1$ | $a_1 a_2$ | $\cdots$ | $a_1 a_j$ | $\cdots$ | $a_1 a_n$ |
| $a_2$ | $a_2 a_1$ | $a_2 a_2$ | $\cdots$ | $a_2 a_j$ | $\cdots$ | $a_2 a_n$ |
| $a_i$ | $a_i a_1$ | $a_i a_2$ | $\cdots$ | $a_i a_j$ | $\cdots$ | $a_i a_n$ |
| $a_n$ | $a_n a_1$ | $a_n a_2$ | $\cdots$ | $a_n a_j$ | $\cdots$ | $a_n a_n$ |

A multiplication table for a group $G$ of order $n$ depends on the listing of the elements of $G$, and so $G$ has $n!$ different multiplication tables. (Thus, the task of determining whether a multiplication table for a group $G$ is the same as a multiplication table for another group $H$ is a daunting one, involving $(n!)^2$ comparisons, each of which involves checking $n^2$ entries.) If $a_1, a_2, \ldots, a_n$ is a list of all the elements of $G$ with no repetitions, and if $f \colon G \to H$ is a bijection, then $f(a_1), f(a_2), \ldots, f(a_n)$ is a list of all the elements of $H$ with no repetitions, and so this latter list determines a multiplication table for $H$. That $f$ is an isomorphism says that if we superimpose the given multiplication table for $G$ (determined by $a_1, a_2, \ldots, a_n$) upon the multiplication table for $H$ [determined by $f(a_1), f(a_2), \ldots, f(a_n)$], then the tables match: if $a_i a_j$ is the $ij$ entry in the multiplication table of $G$, then $f(a_i a_j) = f(a_i) f(a_j)$ is the $ij$ entry of the multiplication table for $H$. In this sense, isomorphic groups have the **same multiplication table**. Thus, isomorphic groups are essentially the same, differing only in the notation for the elements and the binary operations.

**Example 1.57.** Let us show that $G = S_3$, the symmetric group permuting $\{1, 2, 3\}$, and $H = S_Y$, the symmetric group permuting $Y = \{a, b, c\}$, are isomorphic. First, list $G$:

$$(1), \quad (1\ 2), \quad (1\ 3), \quad (2\ 3), \quad (1\ 2\ 3), \quad (1\ 3\ 2).$$

We define the obvious function $\varphi \colon S_3 \to S_Y$ that replaces numbers by letters:

$$(1), \quad (a\ b), \quad (a\ c), \quad (b\ c), \quad (a\ b\ c), \quad (a\ c\ b).$$

Compare the multiplication table for $S_3$ arising from this list of its elements with the multiplication table for $S_Y$ arising from the corresponding list of its elements. The reader should write out the complete tables of each and superimpose one on the other to see that they do match. We will check only one entry. The $4, 5$ position in the table for $S_3$ is the product $(2\ 3)(1\ 2\ 3) = (1\ 3)$, while the $4, 5$ position in the table for $S_Y$ is the product $(b\ c)(a\ b\ c) = (a\ c)$. The same idea shows that $S_3 \cong D_6$, for symmetries of an equilateral triangle correspond to permutations of its vertices. This result is generalized in Exercise 1.46 on page 44. ◄

**Lemma 1.58.** *Let $f \colon G \to H$ be a homomorphism of groups.*

   (i) $f(1) = 1$.

   (ii) $f(x^{-1}) = f(x)^{-1}$.

   (iii) $f(x^n) = f(x)^n$ *for all $n \in \mathbb{Z}$.*

**Proof.**

(i) $1 \cdot 1 = 1$ implies $f(1)f(1) = f(1)$. Now use Exercise 1.30 on page 27.

(ii) $1 = x^{-1}x$ implies $1 = f(1) = f(x^{-1})f(x)$.

(iii) Use induction to show that $f(x^n) = f(x)^n$ for all $n \geq 0$. Then observe that $x^{-n} = (x^{-1})^n$, and use part (ii).  •

**Example 1.59.** If $G$ and $H$ are cyclic groups of the same order $m$, then $G$ and $H$ are isomorphic. (It follows from Corollary 1.51 that any two groups of prime order $p$ are isomorphic.) Although this is not difficult, it requires some care. We have $G = \{1, a, a^2, \ldots, a^{m-1}\}$ and $H = \{1, b, b^2, \ldots, b^{m-1}\}$, and the obvious choice for an isomorphism is the bijection $f \colon G \to H$ given by $f(a^i) = b^i$. Checking that $f$ is a homomorphism, that is, $f(a^i a^j) = b^i b^j = b^{i+j}$, involves two cases: $i + j \leq m - 1$, so that $a^i a^j = a^{i+j}$, and $i + j \geq m$, so that $a^i a^j = a^{i+j-m}$. We give a less computational proof in Example 1.77.  ◄

A property of a group $G$ that is shared by all other groups isomorphic to it is called an ***invariant*** of $G$. For example, the order $|G|$ is an invariant of $G$, for isomorphic groups have the same order. Being abelian is an invariant. In fact, if $f$ is an isomorphism and $a$ and $b$ commute, then $ab = ba$ and

$$f(a)f(b) = f(ab) = f(ba) = f(b)f(a);$$

that is, $f(a)$ and $f(b)$ commute. The groups $\mathbb{I}_6$ and $S_3$ have the same order, yet are not isomorphic ($\mathbb{I}_6$ is abelian and $S_3$ is not). In general, it is a challenge to decide whether two given groups are isomorphic. See Exercise 1.49 on page 44 for more examples of invariants.

**Example 1.60.** We present two nonisomorphic *abelian* groups of the same order. Let $\mathbf{V} = \{(1), (1\ 2)(3\ 4), (1\ 3)(2\ 4), (1\ 4)(2\ 3)\}$ be the four-group, and let $\mu_4 = \langle i \rangle = \{1, i, -1, -i\}$ be the multiplicative cyclic group of fourth roots of unity, where $i^2 = -1$. If there were an isomorphism $f : \mathbf{V} \to \mu_4$, then surjectivity of $f$ would provide some $x \in \mathbf{V}$ with $i = f(x)$. But $x^2 = (1)$ for all $x \in \mathbf{V}$, so that $i^2 = f(x)^2 = f(x^2) = f((1)) = 1$, contradicting $i^2 = -1$. Therefore, $\mathbf{V}$ and $\mu_4$ are not isomorphic.

There are other ways to prove this result. For example, $\mu_4$ is cyclic and $\mathbf{V}$ is not; $\mu_4$ has an element of order 4 and $\mathbf{V}$ does not; $\mu_4$ has a unique element of order 2, but $\mathbf{V}$ has 3 elements of order 2. At this stage, you should really believe that $\mu_4$ and $\mathbf{V}$ are not isomorphic!  ◄

**Definition.** If $f \colon G \to H$ is a homomorphism, define

$$\boldsymbol{kernel}^{24} f = \{x \in G : f(x) = 1\}$$

and

$$\boldsymbol{image}\ f = \{h \in H : h = f(x) \text{ for some } x \in G\}.$$

We usually abbreviate kernel $f$ to ker $f$ and image $f$ to im $f$.

---

[24] *Kernel* comes from the German word meaning "grain" or "seed" (*corn* comes from the same word). Its usage here indicates an important ingredient of a homomorphism.

**Example 1.61.**

(i) If $\mu_2$ is the multiplicative group $\mu_2 = \{\pm 1\}$, then sgn: $S_n \to \mu_2$ is a homomorphism, by Theorem 1.13. The kernel of sgn is the alternating group $A_n$, the set of all even permutations, and its image is $\mu_2$.

(ii) Determinant is a surjective homomorphism det: $\mathrm{GL}(n, \mathbb{R}) \to \mathbb{R}^\times$, the multiplicative group of nonzero real numbers, whose kernel is the special linear group $\mathrm{SL}(n, \mathbb{R})$ of all $n \times n$ matrices of determinant 1, and whose image is $\mathbb{R}^\times$ (det is surjective: if $a \in \mathbb{R}^\times$, then det: $\left[\begin{smallmatrix} a & 0 \\ 0 & 1 \end{smallmatrix}\right] \mapsto a$).

(iii) Let $H = \langle a \rangle$ be a cyclic group of order $n$, and define $f \colon \mathbb{Z} \to H$ by $f(k) = a^k$. Then $f$ is a homomorphism with $\ker f = \langle n \rangle$.  ◀

**Proposition 1.62.** *Let $f \colon G \to H$ be a homomorphism.*

(i) *$\ker f$ is a subgroup of $G$ and $\operatorname{im} f$ is a subgroup of $H$.*

(ii) *If $x \in \ker f$ and $a \in G$, then $axa^{-1} \in \ker f$.*

(iii) *$f$ is an injection if and only if $\ker f = \{1\}$.*

**Proof.**

(i) Routine.

(ii) $f(axa^{-1}) = f(a)1f(a)^{-1} = 1$.

(iii) $f(a) = f(b)$ if and only if $f(b^{-1}a) = 1$.  •

**Definition.** A subgroup $K$ of a group $G$ is called a ***normal subgroup*** if $k \in K$ and $g \in G$ imply $gkg^{-1} \in K$. If $K$ is a normal subgroup of $G$, we write

$$K \lhd G.$$

Proposition 1.62(ii) says that the kernel of a homomorphism is always a normal subgroup (the converse is Corollary 1.75). If $G$ is an abelian group, then every subgroup $K$ is normal, for if $k \in K$ and $g \in G$, then $gkg^{-1} = kgg^{-1} = k \in K$. The converse of this last statement is false: in Proposition 1.69, we shall see that there is a nonabelian group of order 8 (the *quaternions*), each of whose subgroups is normal.

The cyclic subgroup $H = \langle (1\,2) \rangle$ of $S_3$, consisting of the two elements $(1)$ and $(1\,2)$, is *not* a normal subgroup of $S_3$: if $\alpha = (1\,2\,3)$, then

$$\alpha(1\,2)\alpha^{-1} = (1\,2\,3)(1\,2)(3\,2\,1) = (2\,3) \notin H$$

[alternatively, Theorem 1.9 gives $\alpha(1\,2)\alpha^{-1} = (\alpha 1\ \alpha 2) = (2\,3)$]. On the other hand, the cyclic subgroup $K = \langle (1\,2\,3) \rangle$ of $S_3$ is a normal subgroup, as the reader should verify.

It follows from Examples 1.61(i) and (ii) that $A_n$ is a normal subgroup of $S_n$ and $\mathrm{SL}(n, \mathbb{R})$ is a normal subgroup of $\mathrm{GL}(n, \mathbb{R})$ (it is also easy to prove these facts directly).

**Definition.** Let $G$ be a group. A ***conjugate*** of $a \in G$ is an element in $G$ of the form $gag^{-1}$ for some $g \in G$.

It is clear that a subgroup $K \subseteq G$ is a normal subgroup if and only if $K$ contains all the conjugates of its elements: if $k \in K$, then $gkg^{-1} \in K$ for all $g \in G$.

**Example 1.63.**

(i) Theorem 1.9 states that two permutations in $S_n$ are conjugate if and only if they have the same cycle structure.

(ii) In Linear Algebra, two matrices $A, B \in \mathrm{GL}(n, \mathbb{R})$ are called **similar** if they are conjugate; that is, if there is a nonsingular matrix $P$ with $B = PAP^{-1}$. We shall see that $A$ and $B$ are conjugate if and only if they have the same rational canonical form (Theorem 8.36).  ◄

**Definition.** If $G$ is a group and $g \in G$, define **conjugation** $\gamma_g \colon G \to G$ by

$$\gamma_g(a) = gag^{-1}$$

for all $a \in G$.

**Proposition 1.64.**

(i) *If $G$ is a group and $g \in G$, then conjugation $\gamma_g \colon G \to G$ is an isomorphism.*

(ii) *Conjugate elements have the same order.*

**Proof.**

(i) If $g$, $h \in G$, then $(\gamma_g \gamma_h)(a) = \gamma_g(hah^{-1}) = g(hah^{-1})g^{-1} = (gh)a(gh)^{-1} = \gamma_{gh}(a)$; that is,

$$\gamma_g \gamma_h = \gamma_{gh}.$$

It follows that each $\gamma_g$ is a bijection, for $\gamma_g \gamma_{g^{-1}} = \gamma_1 = 1 = \gamma_{g^{-1}} \gamma_g$. We now show that $\gamma_g$ is an isomorphism: if $a$, $b \in G$,

$$\gamma_g(ab) = g(ab)g^{-1} = ga(g^{-1}g)bg^{-1} = \gamma_g(a)\gamma_g(b).$$

(ii) If $a$ and $b$ are conjugate, there is $g \in G$ with $b = gag^{-1}$; that is, $b = \gamma_g(a)$. But $\gamma_g$ is an isomorphism, and so Exercise 1.49 on page 44 shows that $a$ and $b = \gamma_g(a)$ have the same order.   •

**Example 1.65.** The **center** of a group $G$, denoted by $Z(G)$, is

$$Z(G) = \{z \in G : zg = gz \text{ for all } g \in G\}.$$

Thus, $Z(G)$ consists of all elements commuting with everything in $G$.

It is easy to see that $Z(G)$ is a subgroup of $G$; it is a normal subgroup because if $z \in Z(G)$ and $g \in G$, then $gzg^{-1} = zgg^{-1} = z \in Z(G)$.

A group $G$ is abelian if and only if $Z(G) = G$. At the other extreme are groups $G$ with $Z(G) = \{1\}$; such groups are called **centerless**. For example, $Z(S_3) = \{(1)\}$; indeed, all large symmetric groups are centerless, for Exercise 1.25 on page 16 shows that $Z(S_n) = \{(1)\}$ for all $n \geq 3$.  ◄

**Example 1.66.** If $G$ is a group, then an ***automorphism***[25] of $G$ is an isomorphism $f \colon G \to G$. For example, every conjugation $\gamma_g$ is an automorphism of $G$ (it is called an ***inner automorphism***); its inverse is conjugation by $g^{-1}$. An automorphism is called ***outer*** if it is not inner. The set $\mathrm{Aut}(G)$ of all the automorphisms of $G$ is itself a group under composition, called the ***automorphism group***, and the set of all conjugations,

$$\mathrm{Inn}(G) = \{\gamma_g : g \in G\},$$

is a subgroup of $\mathrm{Aut}(G)$ (see Proposition 1.105). ◀

**Example 1.67.** The four-group $\mathbf{V} = \big\{(1), (1\ 2)(3\ 4), (1\ 3)(2\ 4), (1\ 4)(2\ 3)\big\}$ is a normal subgroup of $S_4$. By Theorem 1.9, every conjugate of a product of two transpositions is another such; Table 1 on page 10 shows that only three permutations in $S_4$ have this cycle structure, and so $\mathbf{V}$ is a normal subgroup of $S_4$. ◀

**Proposition 1.68.** *Let $H$ be a subgroup of index $2$ in a group $G$.*

(i) *$g^2 \in H$ for every $g \in G$.*

(ii) *$H$ is a normal subgroup of $G$.*

**Proof.**

(i) Since $H$ has index 2, there are exactly two cosets, namely, $H$ and $aH$, where $a \notin H$. Thus, $G$ is the disjoint union $G = H \cup aH$. Take $g \in G$ with $g \notin H$, so that $g = ah$ for some $h \in H$. If $g^2 \notin H$, then $g^2 = ah'$, where $h' \in H$. Hence,

$$g = g^{-1}g^2 = h^{-1}a^{-1}ah' = h^{-1}h' \in H,$$

and this is a contradiction.

(ii) [26] It suffices to prove that if $h \in H$, then the conjugate $ghg^{-1} \in H$ for every $g \in G$. If $g \in H$, then $ghg^{-1} \in H$, because $H$ is a subgroup. If $g \notin H$, then $g = ah_0$, where $h_0 \in H$ (for $G = H \cup aH$). If $ghg^{-1} \in H$, we are done. Otherwise, $ghg^{-1} = ah_1$ for some $h_1 \in H$. But $ah_1 = ghg^{-1} = ah_0hh_0^{-1}a^{-1}$. Cancel $a$ to obtain $h_1 = h_0hh_0^{-1}a^{-1}$, contradicting $a \notin H$. •

**Definition.** The group of ***quaternions***[27] is the group $\mathbf{Q}$ of order 8 consisting of the following matrices in $\mathrm{GL}(2, \mathbb{C})$:

$$\mathbf{Q} = \{\, I, A, A^2, A^3, B, BA, BA^2, BA^3 \,\},$$

where $I$ is the identity matrix, $A = \begin{bmatrix} 0 & 1 \\ -1 & 0 \end{bmatrix}$, and $B = \begin{bmatrix} 0 & i \\ i & 0 \end{bmatrix}$.

The element $A \in \mathbf{Q}$ has order 4, so that $\langle A \rangle$ is a subgroup of order 4 and, hence, of index 2; the other coset is $B\langle A \rangle = \{B, BA, BA^2, BA^3 \,\}$. Note that $B^2 = A^2$ and $BAB^{-1} = A^{-1}$.

---

[25] The word *automorphism* is made up of two Greek roots: *auto*, meaning "self," and *morph*, meaning "shape" or "form." Just as an isomorphism carries one group onto a faithful replica, an automorphism carries a group onto itself.

[26] Another proof of this is given in Exercise 1.57 on page 45.

[27] Hamilton invented an $\mathbb{R}$-*algebra* (a vector space over $\mathbb{R}$ which is also a ring) that he called *quaternions*, for it was four-dimensional. The group of quaternions consists of eight special elements in that system; see Exercise 1.68 on page 47.

**Proposition 1.69.** *The group* **Q** *of quaternions is not abelian, yet every subgroup of* **Q** *is normal.*

**Proof.** By Exercise 1.67 on page 46, **Q** is a nonabelian group of order 8 having exactly one subgroup of order 2, namely, the center $Z(\mathbf{Q}) = \langle -I \rangle$, which is normal. Lagrange's Theorem says that the only possible orders of subgroups are 1, 2, 4, or 8. Clearly, the subgroups $\{I\}$ and **Q** itself are normal subgroups and, by Proposition 1.68(ii), any subgroup of order 4 is normal, for it has index 2.   •

A nonabelian finite group is called ***hamiltonian*** if every subgroup is normal. The group **Q** of quaternions is essentially the only hamiltonian group, for every hamiltonian group has the form $\mathbf{Q} \times A \times B$, where $a^2 = 1$ for all $a \in A$ (Exercise 1.35 on page 27 says that $A$ is necessarily abelian) and $B$ is an abelian group of odd order (*direct products* will be introduced in the next section) (Robinson, *A Course in the Theory of Groups*, p. 143).

Lagrange's Theorem states that the order of a subgroup of a finite group $G$ must be a divisor of $|G|$. This suggests the question, given a divisor $d$ of $|G|$, whether $G$ must contain a subgroup of order $d$. The next result shows that there need not be such a subgroup.

**Proposition 1.70.** *The alternating group* $A_4$ *is a group of order* 12 *having no subgroup of order* 6.

**Proof.** First, $|A_4| = 12$, by Example 1.48(i). If $A_4$ contains a subgroup $H$ of order 6, then $H$ has index 2, and so $\alpha^2 \in H$ for every $\alpha \in A_4$, by Proposition 1.68(i). But if $\alpha$ is a 3-cycle, then $\alpha$ has order 3, so that $\alpha = \alpha^4 = (\alpha^2)^2$. Thus, $H$ contains every 3-cycle. This is a contradiction, for there are eight 3-cycles in $A_4$.   •

## Exercises

* **1.46.** Show that if there is a bijection $f: X \to Y$ (that is, if $X$ and $Y$ have the same number of elements), then there is an isomorphism $\varphi: S_X \to S_Y$.

**Hint.** If $\alpha \in S_X$, define $\varphi(\alpha) = f\alpha f^{-1}$. In particular, show that if $|X| = 3$, then $\varphi$ takes a cycle involving symbols 1, 2, 3 into a cycle involving $a$, $b$, $c$, as in Example 1.57.

**1.47.**   (i) Show that the composite of homomorphisms is itself a homomorphism.

  (ii) Show that the inverse of an isomorphism is an isomorphism.

  (iii) Show that two groups that are isomorphic to a third group are isomorphic to each other.

  (iv) Prove that isomorphism is an equivalence relation on any set of groups.

**1.48.** Prove that a group $G$ is abelian if and only if the function $f: G \to G$, given by $f(a) = a^{-1}$, is a homomorphism.

* **1.49.** This exercise gives some invariants of a group $G$. Let $f: G \to H$ be an isomorphism.

(i) Prove that if $a \in G$ has infinite order, then so does $f(a)$, and if $a$ has finite order $n$, then so does $f(a)$. Conclude that if $G$ has an element of some order $n$ and $H$ does not, then $G \not\cong H$.

(ii) Prove that if $G \cong H$, then, for every divisor $d$ of $|G|$, both $G$ and $H$ have the same number of elements of order $d$.

(iii) If $a \in G$, then its **conjugacy class** is $\{gag^{-1} : g \in G\}$. If $G$ and $H$ are isomorphic groups, prove that they have the same number of conjugacy classes. Indeed, if $G$ has exactly $c$ conjugacy classes of size $s$, then so does $H$.

**1.50.** Prove that $A_4$ and $D_{12}$ are nonisomorphic groups of order 12.

**1.51.** (i) Find a subgroup $H$ of $S_4$ with $H \neq \mathbf{V}$ and $H \cong \mathbf{V}$.

(ii) Prove that the subgroup $H$ in part (i) is not a normal subgroup.

**1.52.** Let $G = \{x_1, \ldots, x_n\}$ be a monoid, and let $A = [a_{ij}]$ be a multiplication table of $G$; that is, $a_{ij} = a_i a_j$. Prove that $G$ is a group if and only if $A$ is a **Latin square**, that is, each row and column of $A$ is a permutation of $G$.

$*$ **1.53.** Let $G = \{f : \mathbb{R} \to \mathbb{R} : f(x) = ax + b, \text{where } a \neq 0\}$. Prove that $G$ is a group under composition that is isomorphic to the subgroup of $\mathrm{GL}(2, \mathbb{R})$ consisting of all matrices of the form $\left[\begin{smallmatrix} a & b \\ 0 & 1 \end{smallmatrix}\right]$.

$*$ **1.54.** (i) If $f : G \to H$ is a homomorphism and $x \in G$ has order $k$, prove that $f(x) \in H$ has order $m$, where $m \mid k$.

(ii) If $f : G \to H$ is a homomorphism and $(|G|, |H|) = 1$, prove that $f(x) = 1$ for all $x \in G$.

**1.55.** (i) Prove that $\begin{bmatrix} \cos\theta & -\sin\theta \\ \sin\theta & \cos\theta \end{bmatrix}^k = \begin{bmatrix} \cos k\theta & -\sin k\theta \\ \sin k\theta & \cos k\theta \end{bmatrix}$.

**Hint.** Use induction on $k \geq 1$.

(ii) Prove that the **special orthogonal group** $SO(2, \mathbb{R})$, consisting of all $2 \times 2$ orthogonal matrices of determinant 1, is isomorphic to the circle group $S^1$. (Denote the transpose of a matrix $A$ by $A^{\top}$; if $A^{\top} = A^{-1}$, then $A$ is **orthogonal**.)

**Hint.** Consider $\varphi : \begin{bmatrix} \cos\alpha & -\sin\alpha \\ \sin\alpha & \cos\alpha \end{bmatrix} \mapsto (\cos\alpha, \sin\alpha)$.

**1.56.** Let $G$ be the additive group of all polynomials in $x$ with coefficients in $\mathbb{Z}$, and let $H$ be the multiplicative group of all positive rationals. Prove that $G \cong H$.

**Hint.** List the prime numbers $p_0 = 2, p_1 = 3, p_2 = 5, \ldots$, and define

$$\varphi(e_0 + e_1 x + e_2 x^2 + \cdots + e_n x^n) = p_0^{e_0} \cdots p_n^{e_n}.$$

$*$ **1.57.** (i) Show that if $H$ is a subgroup with $bH = Hb = \{hb : h \in H\}$ for every $b \in G$, then $H$ must be a normal subgroup.

(ii) Use part (i) to give a second proof of Proposition 1.68(ii): if $H \subseteq G$ has index 2, then $H \triangleleft G$.

**1.58.** (i) Prove that if $\alpha \in S_n$, then $\alpha$ and $\alpha^{-1}$ are conjugate.

(ii) Give an example of a group $G$ containing an element $x$ for which $x$ and $x^{-1}$ are not conjugate.

* **1.59.** (i) Prove that the intersection of any family of normal subgroups of a group $G$ is itself a normal subgroup of $G$.

   (ii) If $X$ is a subset of a group $G$, let $N$ be the intersection of all the normal subgroups of $G$ containing $X$. Prove that $X \subseteq N \lhd G$, and that if $S$ is any normal subgroup of $G$ containing $X$, then $N \subseteq S$. We call $N$ the **normal subgroup of** $G$ **generated by** $X$.

   (iii) If $X$ is a subset of a group $G$ and $N$ is the normal subgroup generated by $X$, prove that $N$ is the subgroup generated by all the conjugates of elements in $X$.

* **1.60.** If $K \lhd G$ and $K \subseteq H \subseteq G$, prove that $K \lhd H$.

* **1.61.** Define $W = \langle (1\ 2)(3\ 4) \rangle$, the cyclic subgroup of $S_4$ generated by $(1\ 2)(3\ 4)$. Show that $W$ is a normal subgroup of $\mathbf{V}$, but that $W$ is not a normal subgroup of $S_4$. Conclude that normality is not transitive: $W \lhd \mathbf{V}$ and $\mathbf{V} \lhd G$ do not imply $W \lhd G$.

* **1.62.** Let $G$ be a finite abelian group written multiplicatively. Prove that if $|G|$ is odd, then every $x \in G$ has a unique square root; that is, there exists exactly one $g \in G$ with $g^2 = x$.

   **Hint.** Show that squaring is an injective function $G \to G$.

**1.63.** Give an example of a group $G$, a subgroup $H \subseteq G$, and an element $g \in G$ with $[G : H] = 3$ and $g^3 \notin H$. Compare with Proposition 1.68(i).

   **Hint.** Take $G = S_3$, $H = \langle (1\ 2) \rangle$, and $g = (2\ 3)$.

* **1.64.** Show that the center of $\mathrm{GL}(2, \mathbb{R})$ is the set of all **scalar matrices** $aI$ with $a \neq 0$.

   **Hint.** Show that if $A$ is a matrix that is not a scalar matrix, then there is some nonsingular matrix that does not commute with $A$. [The generalization of this to $n \times n$ matrices is true; see Corollary 2.133(ii)].

* **1.65.** Prove that every isometry in the symmetry group $\Sigma(\pi_n)$ permutes the vertices $\{v_1, \ldots, v_n\}$ of $\pi_n$. (See FCAA, Theorem 2.65.)

* **1.66.** Define $A = \begin{bmatrix} \zeta & 0 \\ 0 & \zeta^{-1} \end{bmatrix}$ and $B = \begin{bmatrix} 0 & 1 \\ i & 0 \end{bmatrix}$, where $\zeta = e^{2\pi i/n}$ is a primitive $n$th root of unity.

   (i) Prove that $A$ has order $n$ and $B$ has order 2.

   (ii) Prove that $BAB = A^{-1}$.

   (iii) Prove that the matrices of the form $A^i$ and $BA^i$, for $0 \leq i < n$, form a multiplicative subgroup $G \subseteq \mathrm{GL}(2, \mathbb{C})$.
   **Hint.** Consider cases $A^i A^j$, $A^i B A^j$, $BA^i A^j$, and $(BA^i)(BA^j)$.

   (iv) Prove that each matrix in $G$ has a unique expression of the form $B^i A^j$, where $i = 0, 1$ and $0 \leq j < n$. Conclude that $|G| = 2n$.

   (v) Prove that $G \cong D_{2n}$.
   **Hint.** Define a function $G \to D_{2n}$ using the unique expression of elements in $G$ in the form $B^i A^j$.

* **1.67.** Let $\mathbf{Q} = \{\, I, A, A^2, A^3, B, BA, BA^2, BA^3 \,\}$, where $A = \begin{bmatrix} 0 & 1 \\ -1 & 0 \end{bmatrix}$ and $B = \begin{bmatrix} 0 & i \\ i & 0 \end{bmatrix}$.

   (i) Prove that $\mathbf{Q}$ is a nonabelian group with binary operation matrix multiplication.

   (ii) Prove that $A^4 = I, B^2 = A^2$, and $BAB^{-1} = A^{-1}$.

(iii) Prove that $-I$ is the only element in **Q** of order 2, and that all other elements $M \neq I$ satisfy $M^2 = -I$. Conclude that **Q** has a unique subgroup of order 2, namely, $\langle -I \rangle$, and it is the center of **Q**.

∗ **1.68.** Prove that the elements of **Q** can be relabeled as $\pm 1$, $\pm \mathbf{i}$, $\pm \mathbf{j}$, $\pm \mathbf{k}$, where

$$\mathbf{i}^2 = \mathbf{j}^2 = \mathbf{k}^2 = -1, \quad \mathbf{ij} = \mathbf{k}, \quad \mathbf{jk} = \mathbf{i}, \quad \mathbf{ki} = \mathbf{j},$$
$$\mathbf{ij} = -\mathbf{ji}, \quad \mathbf{ik} = -\mathbf{ki}, \quad \mathbf{jk} = -\mathbf{kj}.$$

∗ **1.69.** Prove that the quaternions **Q** and the dihedral group $D_8$ are nonisomorphic groups of order 8.

∗ **1.70.** Prove that $A_4$ is the only subgroup of $S_4$ of order 12.

∗ **1.71.** (i) For every group $G$, show that the function $\Gamma \colon G \to \mathrm{Aut}(G)$, given by $g \mapsto \gamma_g$ (where $\gamma_x$ is conjugation by $g$), is a homomorphism.

(ii) Prove that $\ker \Gamma = Z(G)$ and $\mathrm{im}\,\Gamma = \mathrm{Inn}(G)$; conclude that $\mathrm{Inn}(G)$ is a subgroup of $\mathrm{Aut}(G)$.

(iii) Prove that $\mathrm{Inn}(G) \lhd \mathrm{Aut}(G)$.

## Section 1.6. Quotient Groups

The construction of the additive group of integers modulo $m$ is the prototype of a more general way of building new groups, called *quotient groups*, from given groups. The homomorphism $\pi \colon \mathbb{Z} \to \mathbb{I}_m$, defined by $\pi \colon a \mapsto [a]$, is surjective, so that $\mathbb{I}_m$ is equal to im $\pi$. Thus, every element of $\mathbb{I}_m$ has the form $\pi(a)$ for some $a \in \mathbb{Z}$, and $\pi(a) + \pi(b) = \pi(a + b)$. This description of the additive group $\mathbb{I}_m$ in terms of the additive group $\mathbb{Z}$ can be generalized to arbitrary, not necessarily abelian, groups. Suppose that $f \colon G \to H$ is a surjective homomorphism between groups $G$ and $H$. Since $f$ is surjective, each element of $H$ has the form $f(a)$ for some $a \in G$, and the operation in $H$ is given by $f(a)f(b) = f(ab)$, where $a, b \in G$. Now $\ker f$ is a normal subgroup of $G$, and the *First Isomorphism Theorem* will reconstruct $H = \mathrm{im}\, f$ and the surjective homomorphism $f$ from $G$ and $\ker f$ alone.

We begin by introducing a binary operation on the set

$$\mathcal{S}(G)$$

of all nonempty subsets of a group $G$. If $X, Y \in \mathcal{S}(G)$, define

$$XY = \{xy : x \in X \text{ and } y \in Y\}.$$

This multiplication is associative: $X(YZ)$ is the set of all $x(yz)$, where $x \in X$, $y \in Y$, and $z \in Z$, $(XY)Z$ is the set of all such $(xy)z$, and these are the same because $(xy)z = x(yz)$ for all $x, y, z \in G$. Thus, $\mathcal{S}(G)$ is a semigroup; in fact, $\mathcal{S}(G)$ is a monoid, for $\{1\}Y = \{1 \cdot y : y \in Y\} = Y = Y\{1\}$.

An instance of this multiplication is the product of a one-point subset $\{a\}$ and a subgroup $K \subseteq G$, which is the coset $aK$.

As a second example, we show that if $H$ is any subgroup of $G$, then

$$HH = H.$$

If $h$, $h' \in H$, then $hh' \in H$, because subgroups are closed under multiplication, and so $HH \subseteq H$. For the reverse inclusion, if $h \in H$, then $h = h1 \in HH$ (because $1 \in H$), and so $H \subseteq HH$.

It is possible for two subsets $X$ and $Y$ in $\mathcal{S}(G)$ to commute even though their constituent elements do not commute. For example, if $H$ is a nonabelian subgroup of $G$, then we have just seen that $HH = H$. Here is another example: let $G = S_3$, let $X$ be the cyclic subgroup generated by $(1\ 2\ 3)$, and let $Y$ be the one-point subset $\{(1\ 2)\}$. Now $(1\ 2)$ does not commute with $(1\ 2\ 3) \in X$, but $(1\ 2)X = X(1\ 2)$. In fact, here is the converse of Exercise 1.57 on page 45.

**Lemma 1.71.** *A subgroup $K$ of a group $G$ is a normal subgroup if and only if*

$$gK = Kg$$

*for every $g \in G$. Thus, every right coset of a normal subgroup is also a left coset.*

**Proof.** Let $gk \in gK$. Since $K$ is normal, $gkg^{-1} \in K$, say $gkg^{-1} = k' \in K$, so that $gk = (gkg^{-1})g = k'g \in Kg$, and so $gK \subseteq Kg$. For the reverse inclusion, let $kg \in Kg$. Since $K$ is normal, $(g^{-1})k(g^{-1})^{-1} = g^{-1}kg \in K$, say $g^{-1}kg = k'' \in K$. Hence, $kg = g(g^{-1}kg) = gk'' \in gK$ and $Kg \subseteq gK$. Therefore, $gK = Kg$ when $K \lhd G$.

Conversely, if $gK = Kg$ for every $g \in G$, then for each $k \in K$, there is $k' \in K$ with $gk = k'g$; that is, $gkg^{-1} \in K$ for all $g \in G$, and so $K \lhd G$.    •

A natural question is whether $HK$ is a subgroup when both $H$ and $K$ are subgroups. In general, $HK$ need not be a subgroup. For example, let $G = S_3$, let $H = \langle (1\ 2) \rangle$, and let $K = \langle (1\ 3) \rangle$. Then

$$HK = \{(1), (1\ 2), (1\ 3), (1\ 3\ 2)\}$$

is not a subgroup because it is not closed: $(1\ 3)(1\ 2) = (1\ 2\ 3) \notin HK$. Alternatively, $HK$ cannot be a subgroup because $|HK| = 4$ is not a divisor of $6 = |S_3|$.

**Proposition 1.72.**

(i) *If $H$ and $K$ are subgroups of a group $G$, at least one of which is normal, then $HK$ is a subgroup of $G$; moreover, $HK = KH$ in this case.*

(ii) *If both $H$ and $K$ are normal subgroups, then $HK$ is a normal subgroup.*

**Remark.** Exercise 1.80 on page 59 shows that if $H$ and $K$ are subgroups of a group $G$, then $HK$ is a subgroup if and only if $HK = KH$.    ◄

**Proof.**

(i) Assume first that $K \lhd G$. We claim that $HK = KH$. If $hk \in HK$, then $k' = hkh^{-1} \in K$, because $K \lhd G$, and

$$hk = hkh^{-1}h = k'h \in KH.$$

Hence, $HK \subseteq KH$. For the reverse inclusion, write $kh = hh^{-1}kh = hk'' \in HK$. (Note that the same argument shows that $HK = KH$ if $H \lhd G$.)

We now show that $HK$ is a subgroup. Since $1 \in H$ and $1 \in K$, we have $1 = 1 \cdot 1 \in HK$; if $hk \in HK$, then $(hk)^{-1} = k^{-1}h^{-1} \in KH = HK$; if $hk, h_1k_1 \in HK$, then $hkh_1k_1 \in HKHK = HHKK = HK$.

(ii) If $g \in G$, then Lemma 1.71 gives $gHK = HgK = HKg$, and the same lemma now gives $HK \lhd G$. •

Here is a fundamental construction of a new group from a given group.

**Theorem 1.73.** *Let $G/K$ denote the family of all the left cosets of a subgroup $K$ of $G$. If $K$ is a normal subgroup, then*

$$aKbK = abK$$

*for all $a$, $b \in G$, and $G/K$ is a group under this operation.*

**Proof.** Generalized associativity holds in $\mathcal{S}(G)$, by Corollary 1.25, because it is a semigroup. Thus, we may view the product of two cosets $(aK)(bK)$ as the product $\{a\}K\{b\}K$ of four elements in $\mathcal{S}(G)$:

$$(aK)(bK) = a(Kb)K = a(bK)K = abKK = abK;$$

normality of $K$ gives $Kb = bK$ for all $b \in K$ (Lemma 1.71), while $KK = K$ (because $K$ is a subgroup). Hence, the product of two cosets of $K$ is again a coset of $K$, and so a binary operation on $G/K$ has been defined. As multiplication in $\mathcal{S}(G)$ is associative, so, in particular, is the multiplication of cosets in $G/K$. The identity is the coset $K = 1K$, for $(1K)(bK) = 1bK = bK = b1K = (bK)(1K)$, and the inverse of $aK$ is $a^{-1}K$, for $(a^{-1}K)(aK) = a^{-1}aK = K = aa^{-1}K = (aK)(a^{-1}K)$. Therefore, $G/K$ is a group. •

It is important to remember what we have just proved: the product $aKbK = abK$ in $G/K$ does not depend on the particular representatives of the cosets. Thus, the law of substitution holds: if $aK = a'K$ and $bK = b'K$, then

$$abK = aKbK = a'Kb'K = a'b'K.$$

**Definition.** The group $G/K$ is called the ***quotient group*** $G \bmod K$. When $G$ is finite, its order $|G/K|$ is the index $[G : K] = |G|/|K|$ (presumably, this is the reason why *quotient groups* are so called).

**Example 1.74.** We show that the quotient group $G/K$ is precisely $\mathbb{I}_m$ when $G$ is the additive group $\mathbb{Z}$ and $K = \langle m \rangle$, the (cyclic) subgroup of all the multiples of a positive integer $m$. Since $\mathbb{Z}$ is abelian, $\langle m \rangle$ is necessarily a normal subgroup. The sets $\mathbb{Z}/\langle m \rangle$ and $\mathbb{I}_m$ coincide because they are comprised of the same elements; the coset $a + \langle m \rangle$ is the congruence class $[a]$:

$$a + \langle m \rangle = \{a + km : k \in \mathbb{Z}\} = [a].$$

The binary operations also coincide: addition in $\mathbb{Z}/\langle m \rangle$ is given by

$$(a + \langle m \rangle) + (b + \langle m \rangle) = (a + b) + \langle m \rangle;$$

since $a + \langle m \rangle = [a]$, this last equation is just $[a] + [b] = [a + b]$, which is the sum in $\mathbb{I}_m$. Therefore, $\mathbb{I}_m$ and the quotient group $\mathbb{Z}/\langle m \rangle$ are equal (and not merely isomorphic). ◄

There is another way to regard quotient groups. After all, we saw, in the proof of Lemma 1.46, that the relation $\equiv$ on $G$, defined by $a \equiv b$ if $b^{-1}a \in K$, is an equivalence relation whose equivalence classes are the cosets of $K$. Thus, we can view the elements of $G/K$ as equivalence classes, with the multiplication $aKbK = abK$ being independent of the choices of representative.

We remind the reader of Lemma 1.46(i): two cosets $aK$ and $bK$ of a subgroup $K$ are equal if and only if $b^{-1}a \in K$. In particular, when $b = 1$, then $aK = K$ if and only if $a \in K$.

We can now prove the converse of Proposition 1.62(ii).

**Corollary 1.75.** *Every normal subgroup $K \lhd G$ is the kernel of some homomorphism.*

**Proof.** Define the **natural map** $\pi \colon G \to G/K$ by $\pi(a) = aK$. With this notation, the formula $aKbK = abK$ can be rewritten as $\pi(a)\pi(b) = \pi(ab)$; thus, $\pi$ is a (surjective) homomorphism. Since $K$ is the identity element in $G/K$,

$$\ker \pi = \{a \in G : \pi(a) = K\} = \{a \in G : aK = K\} = K,$$

by Lemma 1.46(i).  •

The next theorem shows that every homomorphism gives rise to an isomorphism and that quotient groups are merely constructions of homomorphic images. Noether emphasized the fundamental importance of this fact, and this theorem is often named after her.

**Theorem 1.76 (First Isomorphism Theorem).** *If $f \colon G \to H$ is a homomorphism, then*

$$\ker f \lhd G \quad \text{and} \quad G/\ker f \cong \operatorname{im} f.$$

*In more detail, if $\ker f = K$, then $\varphi \colon G/K \to \operatorname{im} f \subseteq H$, given by $\varphi \colon aK \mapsto f(a)$, is an isomorphism.*

**Remark.** The following diagram describes the proof of the First Isomorphism Theorem, where $\pi \colon G \to G/K$ is the natural map $a \mapsto aK$ and $i \colon \operatorname{im} f \to H$ is the inclusion.

**Proof.** We have already seen, in Proposition 1.62(ii), that $K = \ker f$ is a normal subgroup of $G$. Now $\varphi$ is a well-defined function: if $aK = bK$, then $a = bk$ for some $k \in K$, and so $f(a) = f(bk) = f(b)f(k) = f(b)$, because $f(k) = 1$.

Let us now see that $\varphi$ is a homomorphism. Since $f$ is a homomorphism and $\varphi(aK) = f(a)$,

$$\varphi(aKbK) = \varphi(abK) = f(ab) = f(a)f(b) = \varphi(aK)\varphi(bK).$$

It is clear that $\operatorname{im}\varphi \subseteq \operatorname{im}f$. For the reverse inclusion, note that if $y \in \operatorname{im}f$, then $y = f(a)$ for some $a \in G$, and so $y = f(a) = \varphi(aK)$. Thus, $\varphi$ is surjective.

Finally, we show that $\varphi$ is injective. If $\varphi(aK) = \varphi(bK)$, then $f(a) = f(b)$. Hence, $1 = f(b)^{-1}f(a) = f(b^{-1}a)$, so that $b^{-1}a \in \ker f = K$. Therefore, $aK = bK$, by Lemma 1.46(i), and so $\varphi$ is injective. We have proved that $\varphi\colon G/K \to \operatorname{im}f$ is an isomorphism. $\bullet$

Note that $i\varphi\pi = f$, where $\pi\colon G \to G/K$ is the natural map and $i\colon \operatorname{im}f \to H$ is the inclusion, so that $f$ can be reconstructed from $G$ and $K = \ker f$.

Given any homomorphism $f\colon G \to H$, we should immediately ask for its kernel and image; the First Isomorphism Theorem will then provide an isomorphism $G/\ker f \cong \operatorname{im}f$. Since there is no significant difference between isomorphic groups, the First Isomorphism Theorem also says that there is no significant difference between quotient groups and homomorphic images.

**Example 1.77.** Let us revisit Example 1.59, which showed that any two cyclic groups of order $m$ are isomorphic. If $G = \langle a \rangle$ is a cyclic group of order $m$, define a function $f\colon \mathbb{Z} \to G$ by $f(n) = a^n$ for all $n \in \mathbb{Z}$. Now $f$ is easily seen to be a homomorphism; it is surjective (because $a$ is a generator of $G$), while $\ker f = \{n \in \mathbb{Z} : a^n = 1\} = \langle m \rangle$, by Proposition 1.26. The First Isomorphism Theorem gives an isomorphism $\mathbb{Z}/\langle m \rangle \cong G$. We have shown that every cyclic group of order $m$ is isomorphic to $\mathbb{Z}/\langle m \rangle$, and hence that any two cyclic groups of order $m$ are isomorphic to each other. Of course, Example 1.74 shows that $\mathbb{Z}/\langle m \rangle = \mathbb{I}_m$, so that every finite cyclic group of order $m$ is isomorphic to $\mathbb{I}_m$.

The reader should have no difficulty proving that any two infinite cyclic groups are isomorphic to $\mathbb{Z}$. $\blacktriangleleft$

**Example 1.78.** What is the quotient group $\mathbb{R}/\mathbb{Z}$? Take the real line and identify integer points, which amounts to taking the unit interval $[0, 1]$ and identifying its endpoints, yielding the circle. Define $f\colon \mathbb{R} \to S^1$, where $S^1$ is the circle group, by

$$f\colon x \mapsto e^{2\pi i x}.$$

Now $f$ is a homomorphism; that is, $f(x + y) = f(x)f(y)$. The map $f$ is surjective, and $\ker f$ consists of all $x \in \mathbb{R}$ for which $e^{2\pi i x} = \cos 2\pi x + i\sin 2\pi x = 1$; that is, $\cos 2\pi x = 1$ and $\sin 2\pi x = 0$. But $\cos 2\pi x = 1$ forces $x$ to be an integer; since $1 \in \ker f$, we have $\ker f = \mathbb{Z}$. The First Isomorphism Theorem now gives

$$\mathbb{R}/\mathbb{Z} \cong S^1. \quad \blacktriangleleft$$

Here is a counting result.

**Proposition 1.79 (Product Formula).** *If $H$ and $K$ are subgroups of a finite group $G$, then*

$$|HK||H \cap K| = |H||K|.$$

**Remark.** The subset $HK = \{hk : h \in H \text{ and } k \in K\}$ need not be a subgroup of $G$, but see Proposition 1.72 and Exercise 1.80 on page 59. $\blacktriangleleft$

**Proof.** Define a function $f\colon H \times K \to HK$ by $f\colon (h, k) \mapsto hk$. Clearly, $f$ is a surjection. It suffices to show, for every $x \in HK$, that $|f^{-1}(x)| = |H \cap K|$, where $f^{-1}(x) = \{(h, k) \in H \times K : hk = x\}$ [because $H \times K$ is the disjoint union $\bigcup_{x \in HK} f^{-1}(x)$]. We claim that if $x = hk$, then

$$f^{-1}(x) = \{(hd, d^{-1}k) : d \in H \cap K\}.$$

Each $(hd, d^{-1}k) \in f^{-1}(x)$, for $f(hd, d^{-1}k) = hdd^{-1}k = hk = x$. For the reverse inclusion, let $(h', k') \in f^{-1}(x)$, so that $h'k' = hk$. Then $h^{-1}h' = kk'^{-1} \in H \cap K$; call this element $d$. Then $h' = hd$ and $k' = d^{-1}k$, and so $(h', k')$ lies in the right side. Therefore, $|f^{-1}(x)| = |\{(hd, d^{-1}k) : d \in H \cap K\}| = |H \cap K|$, because $d \mapsto (hd, d^{-1}k)$ is a bijection.  •

The next two results are consequences of the First Isomorphism Theorem.

**Theorem 1.80 (Second Isomorphism Theorem).** *If $H$ and $K$ are subgroups of a group $G$ with $H \lhd G$, then $HK$ is a subgroup, $H \cap K \lhd K$, and*

$$K/(H \cap K) \cong HK/H.$$

**Proof.** Since $H \lhd G$, Proposition 1.72 shows that $HK$ is a subgroup. Normality of $H$ in $HK$ follows from a more general fact: if $H \subseteq S \subseteq G$ and $H$ is normal in $G$, then $H$ is normal in $S$ (if $ghg^{-1} \in H$ for every $g \in G$, then, in particular, $ghg^{-1} \in H$ for every $g \in S$).

We now show that every coset $xH \in HK/H$ has the form $kH$ for some $k \in K$. Since $x \in HK$, we have $x = hk$, where $h \in H$ and $k \in K$; hence, $xH = hkH$. But $hk = k(k^{-1}hk) = kh'$ for some $h' \in H$, so that $hkH = kh'H = kH$. It follows that the function $f\colon K \to HK/H$, given by $f\colon k \mapsto kH$, is surjective. Moreover, $f$ is a homomorphism, for it is the restriction of the natural map $\pi\colon G \to G/H$. Since $\ker \pi = H$, it follows that $\ker f = H \cap K$, and so $H \cap K$ is a normal subgroup of $K$. The First Isomorphism Theorem now gives $K/(H \cap K) \cong HK/H$.  •

The Second Isomorphism Theorem gives the product formula in the special case when one of the subgroups is normal: if $K/(H \cap K) \cong HK/H$, then $|K/(H \cap K)| = |HK/H|$, and so $|HK||H \cap K| = |H||K|$.

**Theorem 1.81 (Third Isomorphism Theorem).** *If $H$ and $K$ are normal subgroups of a group $G$ with $K \subseteq H$, then $H/K \lhd G/K$ and*

$$(G/K)/(H/K) \cong G/H.$$

**Proof.** Define $f\colon G/K \to G/H$ by $f\colon aK \mapsto aH$. Note that $f$ is a (well-defined) function (called *enlargement of coset*), for if $a' \in G$ and $a'K = aK$, then $a^{-1}a' \in K \subseteq H$, and so $aH = a'H$. It is easy to see that $f$ is a surjective homomorphism.

Now $\ker f = H/K$, for $aH = H$ if and only if $a \in H$, and so $H/K$ is a normal subgroup of $G/K$. Since $f$ is surjective, the First Isomorphism Theorem gives

$$(G/K)/(H/K) \cong G/H.  •$$

The Third Isomorphism Theorem is easy to remember: the $K$s can be canceled in the fraction $(G/K)/(H/K)$. We can better appreciate the First Isomorphism Theorem after having proved the third one. The quotient group $(G/K)/(H/K)$ consists of cosets (of $H/K$) whose representatives are themselves cosets (of $K$). A direct proof of the Third Isomorphism Theorem could be nasty.

The next result, which can be regarded as a fourth isomorphism theorem, describes the subgroups of a quotient group $G/K$. It says that every subgroup of $G/K$ is of the form $S/K$ for a unique subgroup $S \subseteq G$ containing $K$.

**Proposition 1.82 (Correspondence Theorem).** *Let $G$ be a group, let $K \lhd G$, and let $\pi\colon G \to G/K$ be the natural map. Then*

$$S \mapsto \pi(S) = S/K$$

*is a bijection between* $\mathrm{Sub}(G; K)$, *the family of all those subgroups $S$ of $G$ that contain $K$, and* $\mathrm{Sub}(G/K)$, *the family of all the subgroups of $G/K$. Moreover, $T \subseteq S \subseteq G$ if and only if $T/K \subseteq S/K$, in which case $[S:T] = [S/K : T/K]$, and $T \lhd S$ if and only if $T/K \lhd S/K$, in which case $S/T \cong (S/K)/(T/K)$.*

**Remark.** The following diagram is a way to remember this theorem:

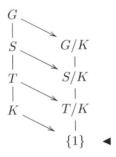

**Proof.** Define $\Phi\colon \mathrm{Sub}(G; K) \to \mathrm{Sub}(G/K)$ by $\Phi\colon S \mapsto S/K$ (it is routine to check that if $S$ is a subgroup of $G$ containing $K$, then $S/K$ is a subgroup of $G/K$).

To see that $\Phi$ is injective, we begin by showing that if $K \subseteq S \subseteq G$, then $\pi^{-1}\pi(S) = S$. As always, $S \subseteq \pi^{-1}\pi(S)$. For the reverse inclusion, let $a \in \pi^{-1}\pi(S)$, so that $\pi(a) = \pi(s)$ for some $s \in S$. It follows that $as^{-1} \in \ker \pi = K$, so that $a = sk$ for some $k \in K$. But $K \subseteq S$, and so $a = sk \in S$. Assume now that $\pi(S) = \pi(S')$, where $S$ and $S'$ are subgroups of $G$ containing $K$. Then $\pi^{-1}\pi(S) = \pi^{-1}\pi(S')$, and so $S = S'$ as we have just proved in the preceding paragraph; hence, $\Phi$ is injective.

To see that $\Phi$ is surjective, let $U$ be a subgroup of $G/K$. Now $\pi^{-1}(U)$ is a subgroup of $G$ containing $K = \pi^{-1}(\{1\})$, and $\pi(\pi^{-1}(U)) = U$.

Now $T \subseteq S \subseteq G$ implies $T/K = \pi(T) \subseteq \pi(S) = S/K$. Conversely, assume that $T/K \subseteq S/K$. If $t \in T$, then $tK \in T/K \subseteq S/K$ and so $tK = sK$ for some $s \in S$. Hence, $t = sk$ for some $k \in K \subseteq S$, and so $t \in S$.

Let us denote $S/K$ by $S^*$. To prove that $[S:T] = [S^* : T^*]$, it suffices to show that there is a bijection from the family of all cosets of the form $sT$, where $s \in S$, and the family of all cosets of the form $s^*T^*$, where $s^* \in S^*$, and the reader may check that $sT \mapsto \pi(s)T^*$ is such a bijection.[28] If $T \lhd S$, then $T/K \lhd S/K$ and $(S/K)/(T/K) \cong S/T$, by the Third Isomorphism Theorem; that is, $S^*/T^* \cong S/T$. It remains to show that if $T^* \lhd S^*$, then $T \lhd S$; that is, if $t \in T$ and $s \in S$, then $sts^{-1} \in T$. Now $\pi(sts^{-1}) = \pi(s)\pi(t)\pi(s)^{-1} \in \pi(s)T^*\pi(s)^{-1} = T^*$, so that $sts^{-1} \in \pi^{-1}(T^*) = T$.  •

**Example 1.83.** Let $G = \langle a \rangle$ be a (multiplicative) cyclic group of order 30. If $\pi \colon \mathbb{Z} \to G$ is defined by $\pi(n) = a^n$, then $\ker \pi = \langle 30 \rangle$. The subgroups $\langle 30 \rangle \subseteq \langle 10 \rangle \subseteq \langle 2 \rangle \subseteq \mathbb{Z}$ correspond to the subgroups

$$\{1\} = \langle a^{30} \rangle \subseteq \langle a^{10} \rangle \subseteq \langle a^2 \rangle \subseteq \langle a \rangle.$$

Moreover, the quotient groups are

$$\frac{\langle a^{10} \rangle}{\langle a^{30} \rangle} \cong \frac{\langle 10 \rangle}{\langle 30 \rangle} \cong \mathbb{I}_3, \quad \frac{\langle a^2 \rangle}{\langle a^{10} \rangle} \cong \frac{\langle 2 \rangle}{\langle 10 \rangle} \cong \mathbb{I}_5, \quad \frac{\langle a \rangle}{\langle a^2 \rangle} \cong \frac{\mathbb{Z}}{\langle 2 \rangle} \cong \mathbb{I}_2. \quad \blacktriangleleft$$

Here are some applications of the Isomorphism Theorems.

**Proposition 1.84.** *If $G$ is a finite abelian group and $d$ is a divisor of $|G|$, then $G$ contains a subgroup of order $d$.*

**Remark.** We have already seen, in Proposition 1.70, that this proposition can be false for nonabelian groups.  ◀

**Proof.** We first prove the result, by induction on $|G|$, for prime divisors $p$ of $|G|$. The base step $|G| = 1$ is true, for there are no prime divisors of 1. For the inductive step, choose $a \in G$ of order $k > 1$. If $p \mid k$, say $k = p\ell$, then Exercise 1.32 on page 27 says that $a^\ell$ has order $p$. If $p \nmid k$, consider the cyclic subgroup $H = \langle a \rangle$. Now $H \lhd G$, because $G$ is abelian, and so the quotient group $G/H$ exists. Note that $|G/H| = |G|/k$ is divisible by $p$, and so the inductive hypothesis gives an element $bH \in G/H$ of order $p$. If $b$ has order $m$, then Exercise 1.54 on page 45 gives $p \mid m$. We have returned to the first case.

Let $d$ be any divisor of $|G|$, and let $p$ be a prime divisor of $d$. We have just seen that there is a subgroup $S \subseteq G$ of order $p$. Now $S \lhd G$, because $G$ is abelian, and $G/S$ is a group of order $n/p$. By induction on $|G|$, $G/S$ has a subgroup $H^*$ of order $d/p$. The Correspondence Theorem gives $H^* = H/S$ for some subgroup $H$ of $G$ containing $S$, and $|H| = |H^*||S| = d$.  •

We now construct a new group from two given groups.

**Definition.** If $H$ and $K$ are groups, then their ***direct product***, denoted by $H \times K$, is the set of all ordered pairs $(h, k)$, with $h \in H$ and $k \in K$, equipped with the operation

$$(h, k)(h', k') = (hh', kk').$$

---

[28] When $G$ is finite, we may prove that $[S : T] = [S^* : T^*]$ as follows:

$$[S^* : T^*] = |S^*|/|T^*| = |S/K|/|T/K| = (|S|/|K|) / (|T|/|K|) = |S|/|T| = [S : T].$$

It is easy to check that the direct product $H \times K$ is a group [the identity is $(1,1)$ and $(h,k)^{-1} = (h^{-1}, k^{-1})$].

We now apply the First Isomorphism Theorem to direct products.

**Proposition 1.85.** *Let $G$ and $G'$ be groups, and let $K \triangleleft G$ and $K' \triangleleft G'$ be normal subgroups. Then $(K \times K') \triangleleft (G \times G')$, and there is an isomorphism*

$$(G \times G')/(K \times K') \cong (G/K) \times (G'/K').$$

**Proof.** Let $\pi \colon G \to G/K$ and $\pi' \colon G' \to G'/K'$ be the natural maps. It is easy to check that $f \colon G \times G' \to (G/K) \times (G'/K')$, given by

$$f \colon (g, g') \mapsto (\pi(g), \pi'(g')) = (gK, g'K'),$$

is a surjective homomorphism with $\ker f = K \times K'$. The First Isomorphism Theorem now gives the desired isomorphism. $\bullet$

**Proposition 1.86.** *If $G$ is a group containing normal subgroups $H$ and $K$ with $H \cap K = \{1\}$ and $HK = G$, then $G \cong H \times K$.*

**Proof.** We show first that if $g \in G$, then the factorization $g = hk$, where $h \in H$ and $k \in K$, is unique. If $hk = h'k'$, then $h'^{-1}h = k'k^{-1} \in H \cap K = \{1\}$. Therefore, $h' = h$ and $k' = k$. We may now define a function $\varphi \colon G \to H \times K$ by $\varphi(g) = (h, k)$, where $g = hk$, $h \in H$, and $k \in K$. To see whether $\varphi$ is a homomorphism, let $g' = h'k'$, so that $gg' = hkh'k'$. Hence, $\varphi(gg') = \varphi(hkh'k')$, which is not in the proper form for evaluation. If we knew that $hk = kh$ for $h \in H$ and $k \in K$, then we could continue:

$$\varphi(hkh'k') = \varphi(hh'kk') = (hh', kk') = (h, k)(h', k') = \varphi(g)\varphi(g').$$

Let $h \in H$ and $k \in K$. Since $K$ is a normal subgroup, $(hkh^{-1})k^{-1} \in K$; since $H$ is a normal subgroup, $h(kh^{-1}k^{-1}) \in H$. But $H \cap K = \{1\}$, so that $hkh^{-1}k^{-1} = 1$ and $hk = kh$. Finally, we show that the homomorphism $\varphi$ is an isomorphism. If $(h, k) \in H \times K$, then the element $g \in G$, defined by $g = hk$, satisfies $\varphi(g) = (h, k)$; hence $\varphi$ is surjective. If $\varphi(g) = (1, 1)$, then $g = 1$, so that $\ker \varphi = 1$ and $\varphi$ is injective. Therefore, $\varphi$ is an isomorphism. $\bullet$

**Remark.** We must assume that both subgroups $H$ and $K$ are normal. For example, $S_3$ has subgroups $H = \langle (1\,2\,3) \rangle$ and $K = \langle (1\,2) \rangle$. Now $H \triangleleft S_3$, $H \cap K = \{1\}$, and $HK = S_3$, but $S_3 \not\cong H \times K$ (because the direct product is abelian). Of course, $K$ is not a normal subgroup of $S_3$. ◄

**Theorem 1.87.** *If $m$ and $n$ are relatively prime, then*

$$\mathbb{I}_{mn} \cong \mathbb{I}_m \times \mathbb{I}_n.$$

**Proof.** If $a \in \mathbb{Z}$, denote its congruence class in $\mathbb{I}_m$ by $[a]_m$. The reader can show that the function $f \colon \mathbb{Z} \to \mathbb{I}_m \times \mathbb{I}_n$, given by $a \mapsto ([a]_m, [a]_n)$, is a homomorphism. We claim that $\ker f = \langle mn \rangle$. Clearly, $\langle mn \rangle \subseteq \ker f$. For the reverse inclusion, if $a \in \ker f$, then $[a]_m = [0]_m$ and $[a]_n = [0]_n$; that is, $a \equiv 0 \bmod m$ and $a \equiv 0 \bmod n$; that is, $m \mid a$ and $n \mid a$. Since $m$ and $n$ are relatively prime, $mn \mid a$ (FCAA, Exercise 1.60), and so $a \in \langle mn \rangle$, that is, $\ker f \subseteq \langle mn \rangle$ and $\ker f = \langle mn \rangle$. The First

Isomorphism Theorem now gives $\mathbb{Z}/\langle mn \rangle \cong \operatorname{im} f \subseteq \mathbb{I}_m \times \mathbb{I}_n$. But $\mathbb{Z}/\langle mn \rangle \cong \mathbb{I}_{mn}$ has $mn$ elements, as does $\mathbb{I}_m \times \mathbb{I}_n$. We conclude that $f$ is surjective. •

For example, it follows that $\mathbb{I}_6 \cong \mathbb{I}_2 \times \mathbb{I}_3$. Note that there is no isomorphism if $m$ and $n$ are not relatively prime. For example, $\mathbb{I}_4 \not\cong \mathbb{I}_2 \times \mathbb{I}_2$, for $\mathbb{I}_4$ has an element of order 4 and the direct product (which is isomorphic to the four-group $\mathbf{V}$) has no such element.

**Corollary 1.88 (Chinese Remainder Theorem).** *If $m, n$ are relatively prime, then there is a solution to the system*

$$x \equiv b \bmod m$$
$$x \equiv c \bmod n.$$

**Proof.** In the proof of Theorem 1.87, we showed that the map $f \colon \mathbb{Z} \to \mathbb{I}_m \times \mathbb{I}_n$, given by $a \mapsto ([a]_m, [a]_n)$, is surjective. But $([b]_m, [c]_n) = ([a]_m, [a]_n)$ says that $[a]_m = [b]_m$ and $[a]_n = [c]_n$; that is, $a \equiv b \bmod m$ and $a \equiv c \bmod n$. •

In light of Proposition 1.38, we may say that an element $a \in G$ has order $n$ if $\langle a \rangle \cong \mathbb{I}_n$. Theorem 1.87 can now be interpreted as saying that if $a$ and $b$ are commuting elements having relatively prime orders $m$ and $n$, then $ab$ has order $mn$. Let us give a direct proof of this result.

**Proposition 1.89.** *Let $G$ be a group, and let $a, b \in G$ be commuting elements of orders $m$ and $n$, respectively. If $(m, n) = 1$, then $ab$ has order $mn$.*

**Proof.** Since $a$ and $b$ commute, we have $(ab)^r = a^r b^r$ for all $r$, so that $(ab)^{mn} = a^{mn} b^{mn} = 1$. It suffices to prove that if $(ab)^k = 1$, then $mn \mid k$. If $1 = (ab)^k = a^k b^k$, then $a^k = b^{-k}$. Since $a$ has order $m$, we have $1 = a^{mk} = b^{-mk}$. Since $b$ has order $n$, Proposition 1.26 gives $n \mid mk$. As $(m, n) = 1$, however, we have $n \mid k$; a similar argument gives $m \mid k$. Finally, since $(m, n) = 1$, we have $mn \mid k$. Therefore, $mn \leq k$, and $mn$ is the order of $ab$. •

**Corollary 1.90.** *If $(m, n) = 1$, then $\phi(mn) = \phi(m)\phi(n)$, where $\phi$ is the Euler $\phi$-function.*

**Proof.** [29] Theorem 1.87 shows that the function $f \colon \mathbb{I}_{mn} \to \mathbb{I}_m \times \mathbb{I}_n$, given by $[a] \mapsto ([a]_m, [a]_n)$, is an isomorphism. This corollary will follow if we prove that $f(U(\mathbb{I}_{mn})) = U(\mathbb{I}_m) \times U(\mathbb{I}_n)$, for then

$$\phi(mn) = |U(\mathbb{I}_{mn})| = |f(U(\mathbb{I}_{mn}))|$$
$$= |U(\mathbb{I}_m) \times U(\mathbb{I}_n)| = |U(\mathbb{I}_m)| \cdot |U(\mathbb{I}_n)| = \phi(m)\phi(n).$$

If $[a] \in U(\mathbb{I}_{mn})$, then $[a][b] = [1]$ for some $[b] \in \mathbb{I}_{mn}$, and

$$f([ab]) = ([ab]_m, [ab]_n)$$
$$= ([a]_m[b]_m, [a]_n[b]_n) = ([a]_m, [a]_n)([b]_m, [b]_n) = ([1]_m, [1]_n);$$

that is, $[1]_m = [a]_m[b]_m$ and $[1]_n = [a]_n[b]_n$. Therefore, $f([a]) = ([a]_m, [a]_n) \in U(\mathbb{I}_m) \times U(\mathbb{I}_n)$, and $f(U(\mathbb{I}_{mn})) \subseteq U(\mathbb{I}_m) \times U(\mathbb{I}_n)$.

---

[29] See Exercise 2.41 on page 101 for a less cluttered proof.

For the reverse inclusion, if $f([c]) = ([c]_m, [c]_n) \in U(\mathbb{I}_m) \times U(\mathbb{I}_n)$, then we must show that $[c] \in U(\mathbb{I}_{mn})$. There is $[d]_m \in \mathbb{I}_m$ with $[c]_m[d]_m = [1]_m$, and there is $[e]_n \in \mathbb{I}_n$ with $[c]_n[e]_n = [1]_n$. Since $f$ is surjective, there is $b \in \mathbb{Z}$ with $([b]_m, [b]_n) = ([d]_m, [e]_n)$, so that $f([1]) = ([1]_m, [1]_n) = ([c]_m[b]_m, [c]_n[b]_n) = f([c][b])$. Since $f$ is an injection, $[1] = [c][b]$ and $[c] \in U(\mathbb{I}_{mn})$. •

**Corollary 1.91.**

(i) *If $p$ is prime, then $\phi(p^e) = p^e - p^{e-1} = p^e \left(1 - \frac{1}{p}\right)$.*

(ii) *If $n = p_1^{e_1} \cdots p_t^{e_t}$ is the prime factorization, where $p_1, \ldots, p_t$ are distinct primes, then*

$$\phi(n) = n\left(1 - \frac{1}{p_1}\right) \cdots \left(1 - \frac{1}{p_t}\right).$$

**Proof.** Part (i) holds because $(k, p^e) = 1$ if and only if $p \nmid k$, while part (ii) follows from Corollary 1.90. •

**Lemma 1.92.** *Let $G = \langle a \rangle$ be a cyclic group.*

(i) *Every subgroup $S$ of $G$ is cyclic.*

(ii) *If $|G| = n$, then $G$ has a unique subgroup of order $d$ for each divisor $d$ of $n$.*

**Proof.**

(i) We may assume that $S \neq \{1\}$. Each element $s \in S$, as every element of $G$, is a power of $a$. If $m$ is the smallest positive integer with $a^m \in S$, we claim that $S = \langle a^m \rangle$. Clearly, $\langle a^m \rangle \subseteq S$. For the reverse inclusion, let $s = a^k \in S$. By the Division Algorithm, $k = qm + r$, where $0 \leq r < m$. Hence, $s = a^k = a^{mq}a^r = a^r$. If $r > 0$, we contradict the minimality of $m$. Thus, $k = qm$ and $s = a^k = (a^m)^q \in \langle a^m \rangle$.

(ii) If $n = cd$, we show that $a^c$ has order $d$ (whence $\langle a^c \rangle$ is a subgroup of order $d$). Clearly $(a^c)^d = a^{cd} = a^n = 1$; we claim that $d$ is the smallest such power. If $(a^c)^m = 1$, where $m < d$, then $n \mid cm$, by Proposition 1.26; hence $cm = ns = dcs$ for some integer $s$, and $m = ds \geq d$, a contradiction.

To prove uniqueness, assume that $\langle x \rangle$ is a subgroup of order $d$ [every subgroup is cyclic, by part (i)]. Now $x = a^m$ and $1 = x^d = a^{md}$; hence $md = nk$ for some integer $k$. Therefore, $x = a^m = (a^{n/d})^k = (a^c)^k$, so that $\langle x \rangle \subseteq \langle a^c \rangle$. Since both subgroups have the same order $d$, it follows that $\langle x \rangle = \langle a^c \rangle$. •

The next theorem will be used to prove Theorem 2.46: the multiplicative group $\mathbb{I}_p^\times$ is cyclic if $p$ is prime. We will use Proposition 2.71(iii) in the next proof; it says that $n = \sum_{d \mid n} \phi(d)$ for every integer $n \geq 1$.

**Theorem 1.93.** *A group $G$ of order $n$ is cyclic if and only if, for each divisor $d$ of $n$, there is at most one cyclic subgroup of order $d$.*

**Proof.** If $G$ is cyclic, then the result follows from Lemma 1.92.

Conversely, define an equivalence relation on a group $G$ by $x \equiv y$ if $\langle x \rangle = \langle y \rangle$; that is, $x$ and $y$ are equivalent if they generate the same cyclic subgroup. Denote the equivalence class containing an element $x$ by $\mathrm{gen}(C)$, where $C = \langle x \rangle$; thus, $\mathrm{gen}(C)$ consists of all the generators of $C$. As usual, equivalence classes form a partition, and so $G$ is the disjoint union:

$$G = \bigcup_C \mathrm{gen}(C),$$

where $C$ ranges over all cyclic subgroups of $G$. In Theorem 1.39(ii), we proved that $|\mathrm{gen}(C)| = \phi(|C|)$, and so $|G| = \sum_C \phi(|C|)$.

By hypothesis, for any divisor $d$ of $n$, the group $G$ has at most one cyclic subgroup of order $d$. Therefore,

$$n = \sum_C |\mathrm{gen}(C)| = \sum_C \phi(|C|) \leq \sum_{d \mid n} \phi(d) = n,$$

the last equality being Proposition 2.71(iii). Hence, for every divisor $d$ of $n$, we must have $\phi(d)$ arising as $|\mathrm{gen}(C)|$ for some cyclic subgroup $C$ of $G$ of order $d$. In particular, $\phi(n)$ arises; there is a cyclic subgroup of order $n$, and so $G$ is cyclic.   •

Here is a variation of Theorem 1.93 (shown to me by D. Leep) which constrains the number of cyclic subgroups of prime order in a finite abelian group $G$. We remark that we must assume that $G$ is abelian, for the group $\mathbf{Q}$ of quaternions is a nonabelian group of order 8 having exactly one (cyclic) subgroup of order 2.

**Theorem 1.94.** *If $G$ is an abelian group of order $n$ having at most one cyclic subgroup of order $p$ for each prime divisor $p$ of $n$, then $G$ is cyclic.*

**Proof.** The proof is by induction on $n = |G|$, with the base step $n = 1$ obviously true. For the inductive step, note that the hypothesis is inherited by subgroups of $G$. We claim that there is some element $x$ in $G$ whose order is a prime divisor $p$ of $|G|$. Choose $y \in G$ with $y \neq 1$; its order $k$ is a divisor of $|G|$, by Lagrange's Theorem, and so $k = pm$ for some prime $p$. By Exercise 1.32 on page 27, the element $x = y^m$ has order $p$. Define $\theta : G \to G$ by $\theta : g \mapsto g^p$ ($\theta$ is a homomorphism because $G$ is abelian). Now $x \in \ker \theta$, so that $|\ker \theta| \geq p$. If $|\ker \theta| > p$, then there would be more than $p$ elements $g \in G$ satisfying $g^p = 1$, and this would force more than one subgroup of order $p$ in $G$. Therefore, $|\ker \theta| = p$. By the First Isomorphism Theorem, $G/\ker \theta \cong \mathrm{im}\,\theta \subseteq G$. Thus, $\mathrm{im}\,\theta$ is a subgroup of $G$ of order $n/p$ satisfying the inductive hypothesis, so there is an element $z \in \mathrm{im}\,\theta$ with $\mathrm{im}\,\theta = \langle z \rangle$. Moreover, since $z \in \mathrm{im}\,\theta$, there is $b \in G$ with $z = b^p$. There are now two cases. If $p \nmid n/p$, then $xz$ has order $p \cdot n/p = n$, by Proposition 1.89, and so $G = \langle xz \rangle$. If $p \mid n/p$, then Exercise 1.33 on page 27 shows that $b$ has order $n$, and $G = \langle b \rangle$.   •

### Exercises

\* **1.72.** Recall that $U(\mathbb{I}_m) = \{[r] \in \mathbb{I}_m : (r, m) = 1\}$ is a multiplicative group. Prove that $U(\mathbb{I}_9) \cong \mathbb{I}_6$ and $U(\mathbb{I}_{15}) \cong \mathbb{I}_4 \times \mathbb{I}_2$. (Theorem 4.24 says, for every finite abelian group $G$, that there exists an integer $m$ with $G$ isomorphic to a subgroup of $U(\mathbb{I}_m)$.)

**1.73.** (i) Let $H$ and $K$ be groups. Without using the First Isomorphism Theorem, prove that $H^* = \{(h, 1) : h \in H\}$ and $K^* = \{(1, k) : k \in K\}$ are normal subgroups of $H \times K$ with $H \cong H^*$ and $K \cong K^*$, and $f \colon H \to (H \times K)/K^*$, defined by $f(h) = (h, 1)K^*$, is an isomorphism.

   (ii) Use Proposition 1.85 to prove that $K^* \lhd (H \times K)$ and $(H \times K)/K^* \cong H$.
   **Hint.** Consider the function $f \colon H \times K \to H$ defined by $f \colon (h, k) \mapsto h$.

**1.74.** (i) Prove that every subgroup of $\mathbf{Q} \times \mathbb{I}_2$ is normal (see the discussion on page 44).

   (ii) Prove that there exists a nonnormal subgroup of $G = \mathbf{Q} \times \mathbb{I}_4$. Conclude that $G$ is not hamiltonian.

**1.75.** (i) Prove that $\mathrm{Aut}(\mathbf{V}) \cong S_3$ and that $\mathrm{Aut}(S_3) \cong S_3$. Conclude that nonisomorphic groups can have isomorphic automorphism groups.

   (ii) Prove that $\mathrm{Aut}(\mathbb{Z}) \cong \mathbb{I}_2$. Conclude that an infinite group can have a finite automorphism group.

**1.76.** (i) If $G$ is a group for which $\mathrm{Aut}(G) = \{1\}$, prove that $g^2 = 1$ for all $g \in G$.

   (ii) If $G$ is a group, prove that $\mathrm{Aut}(G) = \{1\}$ if and only if $|G| \leq 2$.
   **Hint.** By (i), $G$ can be viewed as a vector space over $\mathbb{F}_2$. You may use Corollary 5.50, which states that every $\mathrm{GL}(V) \neq \{1\}$ for every infinite-dimensional vector space $V$.

\* **1.77.** Prove that if $G$ is a group for which $G/Z(G)$ is cyclic, where $Z(G)$ denotes the center of $G$, then $G$ is abelian; that is, $G/Z(G) = \{1\}$.

**Hint.** If $G/Z(G)$ is cyclic, prove that a generator gives an element outside of $Z(G)$ which commutes with each element of $G$.

\* **1.78.** (i) Prove that $\mathbf{Q}/Z(\mathbf{Q}) \cong \mathbf{V}$, where $\mathbf{Q}$ is the group of quaternions and $\mathbf{V}$ is the four-group; conclude that the quotient of a group by its center can be abelian.

   (ii) Prove that $\mathbf{Q}$ has no subgroup isomorphic to $\mathbf{V}$. Conclude that the quotient $\mathbf{Q}/Z(\mathbf{Q})$ is not isomorphic to a subgroup of $\mathbf{Q}$.

**1.79.** Let $G$ be a finite group with $K \lhd G$. If $(|K|, [G : K]) = 1$, prove that $K$ is the unique subgroup of $G$ having order $|K|$.

**Hint.** If $H \subseteq G$ and $|H| = |K|$, what happens to elements of $H$ in $G/K$?

\* **1.80.** If $H$ and $K$ are subgroups of a group $G$, prove that $HK$ is a subgroup of $G$ if and only if $HK = KH$.

**Hint.** Use the fact that $H \subseteq HK$ and $K \subseteq HK$.

\* **1.81.** Let $G$ be a group and regard $G \times G$ as the direct product of $G$ with itself. If the multiplication $\mu \colon G \times G \to G$ is a group homomorphism, prove that $G$ must be abelian.

∗ **1.82.** Generalize Theorem 1.87 as follows. Let $G$ be a finite (additive) abelian group of order $mn$, where $(m, n) = 1$. Define

$$G_m = \{g \in G : \text{order } (g) \mid m\} \text{ and } G_n = \{h \in G : \text{order } (h) \mid n\}.$$

(i) Prove that $G_m$ and $G_n$ are subgroups with $G_m \cap G_n = \{0\}$.

(ii) Prove that $G = G_m + G_n = \{g + h : g \in G_m \text{ and } h \in G_n\}$.

(iii) Prove that $G \cong G_m \times G_n$.

∗ **1.83.** Let $G$ be a finite group, let $p$ be prime, and let $H$ be a normal subgroup of $G$. If both $|H|$ and $|G/H|$ are powers of $p$, prove that $|G|$ is a power of $p$.

**1.84.** If $H$ and $K$ are normal subgroups of a group $G$ with $HK = G$, prove that

$$G/(H \cap K) \cong (G/H) \times (G/K).$$

**Hint.** If $\varphi \colon G \to (G/H) \times (G/K)$ is defined by $x \mapsto (xH, xK)$, then $\ker \varphi = H \cap K$; moreover, we have $G = HK$, so that

$$\bigcup_a aH = HK = \bigcup_b bK.$$

**Definition.** If $H_1, \ldots, H_n$ are groups, then their ***direct product***

$$H_1 \times \cdots \times H_n$$

is the set of all $n$-tuples $(h_1, \ldots, h_n)$, where $h_i \in H_i$ for all $i$, with coordinatewise multiplication:

$$(h_1, \ldots, h_n)(h_1', \ldots, h_n') = (h_1 h_1', \ldots, h_n h_n').$$

∗ **1.85.** Let the prime factorization of an integer $m$ be $m = p_1^{e_1} \cdots p_n^{e_n}$.

(i) Generalize Theorem 1.87 by proving that

$$\mathbb{I}_m \cong \mathbb{I}_{p_1^{e_1}} \times \cdots \times \mathbb{I}_{p_n^{e_n}}.$$

(ii) Generalize Corollary 1.90 by proving that

$$U(\mathbb{I}_m) \cong U(\mathbb{I}_{p_1^{e_1}}) \times \cdots \times U(\mathbb{I}_{p_n^{e_n}}).$$

∗ **1.86.** Define $A, B \in \mathrm{GL}(2, \mathbb{Q})$ by $A = \left[\begin{smallmatrix} 0 & -1 \\ 1 & 0 \end{smallmatrix}\right]$ and $B = \left[\begin{smallmatrix} 0 & 1 \\ -1 & 1 \end{smallmatrix}\right]$. The quotient group $M = \langle A, B \rangle / N$, where $N = \langle \pm I \rangle$, is called the ***modular group***.

(i) Show that $a^2 = 1 = b^3$, where $a = AN$ and $b = BN$ in $M$, and prove that $ab$ has infinite order. (See Exercise 1.34 on page 27.)

(ii) Prove that $M \cong \mathrm{SL}(2, \mathbb{Z})/N$.

## Section 1.7. Group Actions

Groups of permutations were our first examples of abstract groups; the next result shows that abstract groups can be viewed as groups of permutations.

**Theorem 1.95 (Cayley).** *Every group $G$ is isomorphic to a subgroup of the symmetric group $S_G$. In particular, if $|G| = n$, then $G$ is isomorphic to a subgroup of $S_n$.*

**Proof.** For each $a \in G$, define "translation" $\tau_a \colon G \to G$ by $\tau_a(x) = ax$ for every $x \in G$ (if $a \neq 1$, then $\tau_a$ is not a homomorphism). For $a$, $b \in G$, $(\tau_a \tau_b)(x) = \tau_a(\tau_b(x)) = \tau_a(bx) = a(bx) = (ab)x$, by associativity, so that

$$\tau_a \tau_b = \tau_{ab}.$$

It follows that each $\tau_a$ is a bijection, for its inverse is $\tau_{a^{-1}}$:

$$\tau_a \tau_{a^{-1}} = \tau_{aa^{-1}} = \tau_1 = 1_G = \tau_{a^{-1}a},$$

and so $\tau_a \in S_G$.

Define $\varphi \colon G \to S_G$ by $\varphi(a) = \tau_a$. Rewriting,

$$\varphi(a)\varphi(b) = \tau_a \tau_b = \tau_{ab} = \varphi(ab),$$

so that $\varphi$ is a homomorphism. Finally, $\varphi$ is an injection. If $\varphi(a) = \varphi(b)$, then $\tau_a = \tau_b$, and hence $\tau_a(x) = \tau_b(x)$ for all $x \in G$; in particular, when $x = 1$, this gives $a = b$, as desired. The last statement follows from Exercise 1.46 on page 44, which says that if $X$ is a set with $|X| = n$, then $S_X \cong S_n$. $\bullet$

The reader may note, in the proof of Cayley's Theorem, that the permutation $\tau_a \colon x \mapsto ax$ is just the $a$th row of the multiplication table of $G$.

To tell the truth, Cayley's Theorem itself is only mildly interesting, but a generalization having the identical proof is more useful.

**Theorem 1.96 (Representation on Cosets).** *If $H$ is a subgroup of finite index $n$ in a group $G$, then there exists a homomorphism $\varphi \colon G \to S_n$ with $\ker \varphi \subseteq H$.*

**Proof.** We denote the family of all the left cosets of $H$ in $G$ by $G/H$, even though $H$ may not be a normal subgroup.

For each $a \in G$, define "translation" $\tau_a \colon G/H \to G/H$ by $\tau_a(xH) = axH$ for every $x \in G$. For $a$, $b \in G$,

$$(\tau_a \tau_b)(xH) = \tau_a(\tau_b(xH)) = \tau_a(bxH) = a(bxH) = (ab)xH,$$

by associativity, so that

$$\tau_a \tau_b = \tau_{ab}.$$

It follows that each $\tau_a$ is a bijection, for its inverse is $\tau_{a^{-1}}$:

$$\tau_a \tau_{a^{-1}} = \tau_{aa^{-1}} = \tau_1 = 1_{G/H} = \tau_{a^{-1}}\tau_a,$$

and so $\tau_a \in S_{G/H}$. Define $\varphi \colon G \to S_{G/H}$ by $\varphi(a) = \tau_a$. Rewriting,

$$\varphi(a)\varphi(b) = \tau_a \tau_b = \tau_{ab} = \varphi(ab),$$

so that $\varphi$ is a homomorphism. Finally, if $a \in \ker \varphi$, then $\varphi(a) = 1_{G/H}$, so that $\tau_a(xH) = xH$ for all $x \in G$; in particular, when $x = 1$, this gives $aH = H$, and $a \in H$, by Lemma 1.46(i). The result follows from Exercise 1.46 on page 44, for $|G/H| = n$, and so $S_{G/H} \cong S_n$. $\bullet$

When $H = \{1\}$, this is the Cayley Theorem, for then $\ker \varphi = \{1\}$ and $\varphi$ is an injection.

We are now going to classify all groups of order up to 7. By Example 1.59, every group of prime order $p$ is isomorphic to $\mathbb{I}_p$, and so, up to isomorphism, there is just one group of order $p$. Of the possible orders through 7, four are primes, namely, 2, 3, 5, and 7, and so we need look only at orders 4 and 6.

**Proposition 1.97.** *Every group $G$ of order 4 is isomorphic to either $\mathbb{I}_4$ or the four-group* **V**. *Moreover, $\mathbb{I}_4$ and* **V** *are not isomorphic.*

**Proof.** By Lagrange's Theorem, every element in $G$ has order 1, 2, or 4. If there is an element of order 4, then $G$ is cyclic. Otherwise, $x^2 = 1$ for all $x \in G$, so that Exercise 1.35 on page 27 shows that $G$ is abelian.

If distinct elements $x$ and $y$ in $G$ are chosen, neither being 1, then we quickly check that $xy \notin \{1, x, y\}$; hence, $G = \{1, x, y, xy\}$. It is easy to see that the bijection $f\colon G \to \mathbf{V}$, defined by $f(1) = 1$, $f(x) = (1\ 2)(3\ 4)$, $f(y) = (1\ 3)(2\ 4)$, and $f(xy) = (1\ 4)(2\ 3)$, is an isomorphism, for the product of any two non-identity elements is the third one. We have already seen, in Example 1.60, that $\mathbb{I}_4 \not\cong \mathbf{V}$.  •

Another proof of Proposition 1.97 uses Cayley's Theorem: $G$ is isomorphic to a subgroup of $S_4$, and it is not too difficult to show, using Table 1 on page 10, that every subgroup of $S_4$ of order 4 is either cyclic or isomorphic to the four-group.

**Proposition 1.98.** *If $G$ is a group of order 6, then $G$ is isomorphic to either $\mathbb{I}_6$ or $S_3$.*[30] *Moreover, $\mathbb{I}_6$ and $S_3$ are not isomorphic.*

**Proof.**[31] By Lagrange's Theorem, the only possible orders of nonidentity elements are 2, 3, and 6. Of course, $G \cong \mathbb{I}_6$ if $G$ has an element of order 6. Now Exercise 1.36 on page 27 shows that $G$ must contain an element of order 2, say, $t$. We distinguish two cases.

**Case 1.** $G$ is abelian.

If there is a second element of order 2, say, $a$, then it is easy to see, using $at = ta$, that $H = \{1, a, t, at\}$ is a subgroup of $G$. This contradicts Lagrange's Theorem, because 4 is not a divisor of 6. It follows that $G$ must contain an element $b$ of order 3. But $tb$ has order 6, by Proposition 1.89. Therefore, $G$ is cyclic if it is abelian.

**Case 2.** $G$ is not abelian.

If $G$ has no elements of order 3, then $x^2 = 1$ for all $x \in G$, and $G$ is abelian, by Exercise 1.35 on page 27. Therefore, $G$ contains an element $s$ of order 3 as well as the element $t$ of order 2.

Now $|\langle s \rangle| = 3$, so that $[G : \langle s \rangle] = |G|/|\langle s \rangle| = 6/3 = 2$, and so $\langle s \rangle$ is a normal subgroup of $G$, by Proposition 1.68(ii). Since $t = t^{-1}$, we have $tst \in \langle s \rangle$; hence,

---

[30] Cayley states this proposition in an article he wrote in 1854. However, in 1878, in the *American Journal of Mathematics*, he wrote, "The general problem is to find all groups of a given order $n$; ... if $n = 6$, there are three groups; a group

$$1, \alpha, \alpha^2, \alpha^3, \alpha^4, \alpha^5 \quad (\alpha^6 = 1),$$

and two more groups

$$1, \beta, \beta^2, \alpha, \alpha\beta, \alpha\beta^2 \qquad (\alpha^2 = 1, \beta^3 = 1),$$

viz., in the first of these $\alpha\beta = \beta\alpha$ while in the other of them, we have $\alpha\beta = \beta^2\alpha, \alpha\beta^2 = \beta\alpha$." Cayley's list is $\mathbb{I}_6$, $\mathbb{I}_2 \times \mathbb{I}_3$, and $S_3$; of course, $\mathbb{I}_2 \times \mathbb{I}_3 \cong \mathbb{I}_6$. Even Homer nods.

[31] We give another proof in Proposition 4.87.

$tst = s^i$ for $i = 0, 1$ or 2. Now $i \neq 0$, for $tst = s^0 = 1$ implies $s = 1$. If $i = 1$, then $s$ and $t$ commute, and this gives $st$ of order 6, as in Case 1 (which forces $G$ to be cyclic, hence abelian, contrary to our present hypothesis). Therefore, $tst = s^2 = s^{-1}$.

We construct an isomorphism $G \to S_3$. Let $H = \langle t \rangle$, and consider the homomorphism $\varphi : G \to S_{G/\langle t \rangle}$ given by

$$\varphi(g) : x\langle t \rangle \mapsto gx\langle t \rangle.$$

By Theorem 1.96, $\ker \varphi \subseteq \langle t \rangle$, so that either $\ker \varphi = \{1\}$ (and $\varphi$ is injective), or $\ker \varphi = \langle t \rangle$. Now $G/\langle t \rangle = \{\langle t \rangle, s\langle t \rangle, s^2\langle t \rangle\}$, and, in two-rowed notation,

$$\varphi(t) = \begin{pmatrix} \langle t \rangle & s\langle t \rangle & s^2\langle t \rangle \\ t\langle t \rangle & ts\langle t \rangle & ts^2\langle t \rangle \end{pmatrix}.$$

If $\varphi(t)$ is the identity permutation, then $ts\langle t \rangle = s\langle t \rangle$, so that $s^{-1}ts \in \langle t \rangle = \{1, t\}$, by Lemma 1.46. But now $s^{-1}ts = t$ (it cannot be 1); hence, $ts = st$, contradicting $t$ and $s$ not commuting. Therefore, $t \notin \ker \varphi$, and $\varphi \colon G \to S_{G/\langle t \rangle} \cong S_3$ is an injective homomorphism. Since both $G$ and $S_3$ have order 6, $\varphi$ must be a bijection, and so $G \cong S_3$.

It is clear that $\mathbb{I}_6$ and $S_3$ are not isomorphic, for $\mathbb{I}_6$ is abelian and $S_3$ is not. •

One consequence of this result is another proof that $\mathbb{I}_6 \cong \mathbb{I}_2 \times \mathbb{I}_3$ (Theorem 1.87).

Classifying groups of order 8 is more difficult, for we have not yet developed enough theory. Theorem 4.88 says there are only five nonisomorphic groups of order 8; three are abelian: $\mathbb{I}_8$, $\mathbb{I}_4 \times \mathbb{I}_2$, and $\mathbb{I}_2 \times \mathbb{I}_2 \times \mathbb{I}_2$; two are nonabelian: $D_8$ and $\mathbf{Q}$.

| Order of Group | Number of Groups |
|:---:|:---:|
| 2 | 1 |
| 4 | 2 |
| 8 | 5 |
| 16 | 14 |
| 32 | 51 |
| 64 | 267 |
| 128 | $2,328$ |
| 256 | $56,092$ |
| 512 | $10,494,213$ |
| 1024 | $49,487,365,422$ |

**Table 4.** Too many 2-groups.

We can continue this discussion for larger orders, but things soon get out of hand, as Table 4 shows (Besche–Eick–O'Brien, The groups of order at most 2000, *Electron. Res. Announc. AMS* 7 (2001), 1–4). Making a telephone directory of groups is not the way to study them.

Groups arose by abstracting the fundamental properties enjoyed by permutations. But there is an important feature of permutations that the axioms do not mention: permutations are functions. We shall see that there are interesting consequences when this feature is restored.

**Definition.** A group $G$ ***acts*** on a set $X$ if there is a function $G \times X \to X$, denoted by $(g, x) \mapsto gx$, such that

(i) $(gh)x = g(hx)$ for all $g$, $h \in G$ and $x \in X$,

(ii) $1x = x$ for all $x \in X$, where 1 is the identity in $G$.

If $G$ acts on $X$, we also call $X$ a ***$G$-set***.

If a group $G$ acts on a set $X$, then fixing the first variable, say $g$, gives a function $\alpha_g \colon X \to X$, namely, $\alpha_g \colon x \mapsto gx$. This function is a permutation of $X$, for its inverse is $\alpha_{g^{-1}}$:

$$\alpha_g \alpha_{g^{-1}} = \alpha_1 = 1_X = \alpha_{g^{-1}} \alpha_g.$$

It is easy to see that $\alpha \colon G \to S_X$, defined by $\alpha \colon g \mapsto \alpha_g$, is a homomorphism. Conversely, given any homomorphism $\varphi \colon G \to S_X$, define $gx = \varphi(g)(x)$. Thus, an action of a group $G$ on a set $X$ is merely another way of viewing a homomorphism $G \to S_X$.

Cayley's Theorem says that a group $G$ acts on itself by (left) translation, and its generalization, Theorem 1.96, shows that $G$ also acts on the family of (left) cosets of a subgroup $H$ by (left) translation.

**Example 1.99.** We show that $G$ acts on itself by conjugation. For each $g \in G$, define $\alpha_g \colon G \to G$ to be conjugation

$$\alpha_g(x) = gxg^{-1}.$$

To verify axiom (i), note that for each $x \in G$,

$$(\alpha_g \alpha_h)(x) = \alpha_g(\alpha_h(x)) = \alpha_g(hxh^{-1})$$
$$= g(hxh^{-1})g^{-1} = (gh)x(gh)^{-1} = \alpha_{gh}(x).$$

Therefore, $\alpha_g \alpha_h = \alpha_{gh}$. To prove axiom (ii), note that $\alpha_1(x) = 1x1^{-1} = x$ for each $x \in G$, and so $\alpha_1 = 1_G$.  ◀

The following definitions are fundamental.

**Definition.** If $G$ acts on $X$ and $x \in X$, then the ***orbit*** of $x$, denoted by $\mathcal{O}(x)$, is the subset

$$\mathcal{O}(x) = \{gx : g \in G\} \subseteq X;$$

the ***stabilizer*** of $x$, denoted by $G_x$, is the subgroup

$$G_x = \{g \in G : gx = x\} \subseteq G.$$

The ***orbit space***, denoted by $X/G$, is the set of all the orbits.

If $G$ acts on a set $X$, define a relation on $X$ by $x \equiv y$ in case there exists $g \in G$ with $y = gx$. It is easy to see that this is an equivalence relation whose equivalence classes are the orbits. The orbit space is the family of equivalence classes.

Let us find some orbits and stabilizers.

**Example 1.100.** Cayley's Theorem says that $G$ acts on itself by translations: $\tau_g \colon a \mapsto ga$. If $a \in G$, then the orbit $\mathcal{O}(a) = G$, for if $b \in G$, then $b = (ba^{-1})a = \tau_{ba^{-1}}(a)$. The stabilizer $G_a$ of $a \in G$ is $\{1\}$, for if $a = \tau_g(a) = ga$, then $g = 1$. We say that $G$ acts **transitively** on $X$ if there is only one orbit.

More generally, $G$ acts transitively on $G/H$ [the family of (left) cosets of a (not necessarily normal) subgroup $H$] by translations $\tau_g \colon aH \mapsto gaH$. The orbit $\mathcal{O}(aH) = G/H$, for if $bH \in G/H$, then $\tau_{ba^{-1}} \colon aH \mapsto bH$. The stabilizer $G_{aH}$ of the coset $aH$ is $aHa^{-1}$, for $gaH = aH$ if and only if $a^{-1}ga \in H$ if and only if $g \in aHa^{-1}$. ◄

**Example 1.101.** Let $X = \{1, 2, \ldots, n\}$, let $\alpha \in S_n$, and define the obvious action of the cyclic group $G = \langle \alpha \rangle$ on $X$ by $\alpha^k \cdot i = \alpha^k(i)$. If $i \in X$, then

$$\mathcal{O}(i) = \{\alpha^k(i) : 0 \le k < |G|\}.$$

Suppose the complete factorization of $\alpha$ is $\alpha = \beta_1 \cdots \beta_{t(\alpha)}$ and $i = i_1$ is moved by $\alpha$. If the cycle involving $i_1$ is $\beta_j = (i_1 \ i_2 \ \ldots \ i_r)$, then the proof of Theorem 1.6 shows that $i_{k+1} = \alpha^k(i_1)$ for all $k < r$. Therefore,

$$\mathcal{O}(i) = \{i_1, i_2, \ldots, i_r\},$$

where $i = i_1$. It follows that $|\mathcal{O}(i)| = r$.

The stabilizer $G_i$ of a number $i$ is $G$ if $\alpha$ fixes $i$; however, if $\alpha$ moves $i$, then $G_i$ depends on the size of the orbit $\mathcal{O}(i)$. For example, if $\alpha = (1\,2\,3)(4\,5)(6)$, then $G_6 = G$, $G_1 = \langle \alpha^3 \rangle$, and $G_4 = \langle \alpha^2 \rangle$. ◄

**Example 1.102.** Let $X = \{v_0, \ v_1, \ v_2, \ v_3\}$ be the vertices of a square, and let $G$ be the dihedral group $D_8$ acting on $X$, as in Figure 1.9 (for clarity, the vertices in the figure are labeled 0, 1, 2, 3 instead of $v_0, \ v_1, \ v_2, \ v_3$).

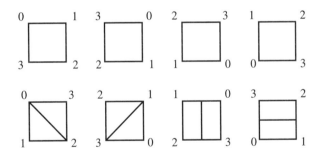

**Figure 1.9.** Dihedral group $D_8$.

Thus,

$$G = \{\text{rotations}\} \cup \{\text{reflections}\}$$
$$= \{(1), (0\,1\,2\,3), (0\,2)(1\,3), (0\,3\,2\,1)\} \cup \{(1\,3), (0\,2), (0\,1)(2\,3), (0\,3)(1\,2)\}.$$

For each vertex $v_i \in X$, there is some $g \in G$ with $gv_0 = v_i$; therefore, $\mathcal{O}(v_0) = X$ and $D_8$ acts transitively.

What is the stabilizer $G_{v_0}$ of $v_0$? Aside from the identity, only one $g \in D_8$ fixes $v_0$, namely, $g = (1\ 3)$; hence $G_{v_0}$ is a subgroup of order 2. (This example can be generalized to the dihedral group $D_{2n}$ acting on a regular $n$-gon.)  ◄

**Example 1.103.** When a group $G$ acts on itself by conjugation, then the orbit $\mathcal{O}(x)$ is

$$\{y \in G : y = axa^{-1} \text{ for some } a \in G\}.$$

Thus, $\mathcal{O}(x)$ is called the **conjugacy class** of $x$, and it is commonly denoted by $x^G$ (we have already mentioned conjugacy classes in Exercise 1.49 on page 44). For example, Theorem 1.9 shows that if $\alpha \in S_n$, then the conjugacy class of $\alpha$ consists of all the permutations in $S_n$ having the same cycle structure as $\alpha$. As a second example, an element $z$ lies in the center $Z(G)$ if and only if $z^G = \{z\}$; that is, no other elements in $G$ are conjugate to $z$.

If $x \in G$, then the stabilizer $G_x$ of $x$ is

$$C_G(x) = \{g \in G : gxg^{-1} = x\}.$$

This subgroup of $G$, consisting of all $g \in G$ that commute with $x$, is called the **centralizer** of $x$ in $G$.  ◄

**Example 1.104.** Every group $G$ acts on the set $X$ of all its subgroups, by conjugation: if $a \in G$, then $a$ acts by $H \mapsto aHa^{-1}$, where $H \subseteq G$.

If $H$ is a subgroup of a group $G$, then a **conjugate** of $H$ is a subgroup of $G$ of the form

$$aHa^{-1} = \{aha^{-1} : h \in H\},$$

where $a \in G$. Since conjugation $h \mapsto aha^{-1}$ is an injection $H \to G$ with image $aHa^{-1}$, it follows that conjugate subgroups of $G$ are isomorphic. For example, in $S_3$, all cyclic subgroups of order 2 are conjugate (for their generators are conjugate).

The orbit of a subgroup $H$ consists of all its conjugates; notice that $H$ is the only element in its orbit if and only if $H \lhd G$; that is, $aHa^{-1} = H$ for all $a \in G$. The stabilizer of $H$ is

$$N_G(H) = \{g \in G : gHg^{-1} = H\}.$$

This subgroup of $G$ is called the **normalizer** of $H$ in $G$. Of course, $H \lhd N_G(H)$; indeed, the normalizer is the largest subgroup of $G$ in which $H$ is normal.  ◄

We have already defined the centralizer of an element; we now define the centralizer of a subgroup.

**Definition.** If $H$ is a subgroup of a group $G$, then the **centralizer** of $H$ in $G$ is

$$C_G(H) = \{g \in G : gh = hg \text{ for all } h \in H\}.$$

It is easy to see that $C_G(H)$ is a subgroup of $G$, and $C_G(G) = Z(G)$. Note that $C_G(H) \subseteq N_G(H)$.

**Proposition 1.105** ($N/C$ **Lemma**).

(i) *If $H \subseteq G$, then $C_G(H) \lhd N_G(H)$ and there is an imbedding*

$$N_G(H)/C_G(H) \hookrightarrow \text{Aut}(H).$$

(ii) *$G/Z(G) \cong \text{Inn}(G)$, where $\text{Inn}(G)$ is the subgroup of $\text{Aut}(G)$ consisting of all the inner automorphisms.*

**Proof.**

(i) If $a \in G$, denote conjugation $g \mapsto aga^{-1}$ by $\gamma_a$. Define $\varphi\colon N_G(H) \to \text{Aut}(H)$ by $\varphi\colon a \mapsto \gamma_a | H$. Note that $\varphi$ is well-defined, for $\gamma_a | H \in \text{Aut}(H)$ because $a \in N_G(H)$. It is routine to check that $\varphi$ is a homomorphism. Now the following statements are equivalent: $a \in \ker \varphi$; $\gamma_a | H = 1_H$; $aha^{-1} = h$ for all $h \in H$; $a \in C_G(H)$. The First Isomorphism Theorem gives $C_G(H) \lhd N_G(H)$ and $N_G(H)/C_G(H) \cong \text{im } \varphi \subseteq \text{Aut}(H)$.

(ii) In the special case $H = G$, we have $N_G(H) = G$, $C_G(H) = Z(G)$, and $\text{im } \varphi = \text{Inn}(G)$.  •

**Remark.** We claim that $\text{Inn}(G) \lhd \text{Aut}(G)$. If $\varphi \in \text{Aut}(G)$ and $g \in G$, then

$$\varphi \gamma_a \varphi^{-1}\colon g \mapsto \varphi^{-1}g \mapsto a\varphi^{-1}ga^{-1} \mapsto \varphi(a)g\varphi(a^{-1}).$$

Thus, $\varphi \gamma_a \varphi^{-1} = \gamma_{\varphi(a)} \in \text{Inn}(G)$. Recall that an automorphism is called **outer** if it is not inner; the **outer automorphism group** is defined by $\text{Out}(G) = \text{Aut}(G)/\text{Inn}(G)$.  ◄

**Proposition 1.106.** *If $G$ acts on a set $X$, then $X$ is the disjoint union of the orbits. If $X$ is finite, then*

$$|X| = \sum_i |\mathcal{O}(x_i)|,$$

*where one $x_i$ is chosen from each orbit.*

**Proof.** As we have mentioned earlier, the relation on $X$, given by $x \equiv y$ if there exists $g \in G$ with $y = gx$, is an equivalence relation whose equivalence classes are the orbits. Therefore, the orbits partition $X$.

The count given in the second statement is correct: since the orbits are disjoint, no element in $X$ is counted twice.  •

Here is the connection between orbits and stabilizers.

**Theorem 1.107.** *If $G$ acts on a set $X$ and $x \in X$, then*

$$|\mathcal{O}(x)| = [G : G_x],$$

*the index of the stabilizer $G_x$ in $G$.*

**Proof.** Let $G/G_x$ denote the family of all the left cosets of $G_x$ in $G$. We will exhibit a bijection $\varphi\colon G/G_x \to \mathcal{O}(x)$, and this will give the result, since $|G/G_x| = [G : G_x]$. Define $\varphi\colon gG_x \mapsto gx$. Now $\varphi$ is well-defined: if $gG_x = hG_x$, then $h = gf$ for some $f \in G_x$; that is, $fx = x$; hence, $hx = gfx = gx$. Now $\varphi$ is an injection: if $gx = \varphi(gG_x) = \varphi(hG_x) = hx$, then $h^{-1}gx = x$; hence, $h^{-1}g \in G_x$, and $gG_x = hG_x$.

Lastly, $\varphi$ is a surjection: if $y \in \mathcal{O}(x)$, then $y = gx$ for some $g \in G$, and so $y = \varphi(gG_x)$.  •

In Example 1.102, $D_8$ acting on the four corners of a square, we saw that $|\mathcal{O}(v_0)| = 4$, $|G_{v_0}| = 2$, and $[G : G_{v_0}] = 8/2 = 4$. In Example 1.101, $G = \langle \alpha \rangle \subseteq S_n$ acting on $X = \{1, 2, \ldots, n\}$, we saw that if $\alpha = \beta_1 \cdots \beta_t$ is the complete factorization into disjoint cycles and $\ell$ occurs in the $r_j$-cycle $\beta_j$, then $r_j = |\mathcal{O}(\ell)|$. Theorem 1.107 says that $r_j$ is a divisor of the order $k$ of $\alpha$ (but Theorem 1.27 tells us more: $k$ is the lcm of the lengths of the cycles occurring in the factorization).

**Corollary 1.108.** *If a finite group $G$ acts on a set $X$, then the number of elements in any orbit is a divisor of $|G|$.*

**Proof.** This follows at once from Lagrange's Theorem.  •

Table 1 on page 10 displays the number of permutations in $S_4$ of each cycle structure; these numbers are 1, 6, 8, 6, 3. Note that each of these numbers is a divisor of $|S_4| = 24$. Table 2 on page 10 shows that the corresponding numbers for $S_5$ are 1, 10, 20, 30, 24, 20, and 15, and these are all divisors of $|S_5| = 120$. We now recognize these subsets as being conjugacy classes, and the next corollary explains why these numbers divide the group order.

**Corollary 1.109.** *If $x$ lies in a finite group $G$, then the number of conjugates of $x$ is the index of its centralizer:*

$$|x^G| = [G : C_G(x)],$$

*and hence it is a divisor of $|G|$.*

**Proof.** As in Example 1.103, the orbit of $x$ is its conjugacy class $x^G$, and the stabilizer $G_x$ is the centralizer $C_G(x)$.  •

**Proposition 1.110.** *If $H$ is a subgroup of a finite group $G$, then the number of conjugates of $H$ in $G$ is $[G : N_G(H)]$.*

**Proof.** As in Example 1.104, the orbit of $H$ is the family of all its conjugates, and the stabilizer is its normalizer $N_G(H)$.  •

There are some interesting applications of group actions to Combinatorics in the next section, but let us first apply group actions to Group Theory.

When we began classifying groups of order 6, it would have been helpful to be able to assert that any such group has an element of order 3 (we were able to use an earlier exercise to assert the existence of an element of order 2). We now prove that if $p$ is a prime divisor of $|G|$, where $G$ is a finite group, then $G$ contains an element of order $p$ (Proposition 1.84 proved the special case of this when $G$ is abelian).

**Theorem 1.111 (Cauchy).** *If $G$ is a finite group whose order is divisible by a prime $p$, then $G$ contains an element of order $p$.*

**Proof.** We prove the theorem by induction on $m \geq 1$, where $|G| = pm$. The base step $m = 1$ is true, for Lagrange's Theorem shows that every nonidentity element in a group of order $p$ has order $p$.

Let us now prove the inductive step. If $x \in G$, then the number of conjugates of $x$ is $|x^G| = [G : C_G(x)]$, where $C_G(x)$ is the centralizer of $x$ in $G$. If $x \notin Z(G)$, then $x^G$ has more than one element, and so $|C_G(x)| < |G|$. We are done if $p \mid |C_G(x)|$, for the inductive hypothesis gives an element of order $p$ in $C_G(x) \subseteq G$. Therefore, we may assume that $p \nmid |C_G(x)|$ for all noncentral $x \in G$. Better, since $p$ is prime and $|G| = [G : C_G(x)]|C_G(x)|$, Euclid's Lemma gives

$$p \mid [G : C_G(x)].$$

After recalling that $Z(G)$ consists of all those elements $x \in G$ with $|x^G| = 1$, we may use Proposition 1.106 to see that

$$|G| = |Z(G)| + \sum_i [G : C_G(x_i)],$$

where one $x_i$ is selected from each conjugacy class having more than one element. Since $|G|$ and all $[G : C_G(x_i)]$ are divisible by $p$, it follows that $|Z(G)|$ is divisible by $p$. But $Z(G)$ is abelian, and so Proposition 1.84 says that $Z(G)$, and hence $G$, contains an element of order $p$. •

**Definition.** The *class equation* of a finite group $G$ is

$$|G| = |Z(G)| + \sum_i [G : C_G(x_i)],$$

where one $x_i$ is selected from each conjugacy class having more than one element.

**Definition.** If $p$ is prime, then a group $G$ is called a *$p$-group* if every element has order a power of $p$.

**Proposition 1.112.** *A finite group $G$ is a $p$-group if and only if $|G| = p^n$ for some $n \geq 0$.*

**Proof.** Let $G$ be a finite $p$-group. If $|G| \neq p^n$, then there is some prime $q \neq p$ with $q \mid |G|$. By Cauchy's Theorem, $G$ contains an element of order $q$, a contradiction. The converse follows from Lagrange's Theorem. •

We have seen examples of groups whose center is trivial; for example, $Z(S_3) = \{1\}$. For finite $p$-groups, however, this is never true.

**Theorem 1.113.** *If $p$ is prime and $G \neq \{1\}$ is a finite $p$-group, then the center of $G$ is nontrivial: $Z(G) \neq \{1\}$.*

**Proof.** Consider the class equation

$$|G| = |Z(G)| + \sum_i [G : C_G(x_i)].$$

Each $C_G(x_i)$ is a proper subgroup of $G$, for $x_i \notin Z(G)$. Since $G$ is a $p$-group, $[G : C_G(x_i)]$ is a divisor of $|G|$, hence is itself a power of $p$. Thus, $p$ divides each of the terms in the class equation other than $|Z(G)|$, and so $p \mid |Z(G)|$ as well. Therefore, $Z(G) \neq \{1\}$.  •

McLain gave an example of an infinite $p$-group $G$ with $Z(G) = \{1\}$ (Robinson, *A Course in the Theory of Groups*, p. 362).

**Corollary 1.114.** *If $p$ is prime, then every group $G$ of order $p^2$ is abelian.*

**Proof.** If $G$ is not abelian, then its center $Z(G)$ is a proper subgroup, so that $|Z(G)| = 1$ or $p$, by Lagrange's Theorem. But Theorem 1.113 says that $Z(G) \neq \{1\}$, and so $|Z(G)| = p$. The center is always a normal subgroup, so that the quotient $G/Z(G)$ is defined; it has order $p$, and hence $G/Z(G)$ is cyclic. This contradicts Exercise 1.77 on page 59.  •

**Example 1.115.** Who would have guessed that Cauchy's Theorem (if $G$ is a group whose order is a multiple of a prime $p$, then $G$ has an element of order $p$) and Fermat's Theorem (if $p$ is prime, then $a^p \equiv a \bmod p$) can be proved simultaneously? The elementary yet ingenious proof of Cauchy's Theorem is due to McKay in 1959 (Montgomery–Ralston, *Selected Papers in Algebra*, p. 41); Mann showed me that McKay's argument also proves Fermat's Theorem.

If $G$ is a finite group and $p$ is prime, denote the cartesian product of $p$ copies of $G$ by $G^p$, and define

$$X = \{(a_0, a_1, \ldots, a_{p-1}) \in G^p : a_0 a_1 \cdots a_{p-1} = 1\}.$$

Note that $|X| = |G|^{p-1}$, for having chosen the last $p-1$ entries arbitrarily, the 0th entry must equal $(a_1 a_2 \cdots a_{p-1})^{-1}$. Introduce an action of $\mathbb{I}_p$ on $X$ by defining, for $0 \leq i \leq p-1$,

$$[i](a_0, a_1, \ldots, a_{p-1}) = (a_i, a_{i+1}, \ldots, a_{p-1}, a_0, a_1, \ldots, a_{i-1}).$$

The product of the entries in the new $p$-tuple is a conjugate of $a_0 a_1 \cdots a_{p-1}$:

$$a_i a_{i+1} \cdots a_{p-1} a_0 a_1 \cdots a_{i-1} = (a_0 a_1 \cdots a_{i-1})^{-1}(a_0 a_1 \cdots a_{p-1})(a_0 a_1 \cdots a_{i-1}).$$

This conjugate is 1 (for $g^{-1}1g = 1$), and so $[i](a_0, a_1, \ldots, a_{p-1}) \in X$. By Corollary 1.108, the size of every orbit of $X$ is a divisor of $|\mathbb{I}_p| = p$; since $p$ is prime, these sizes are either 1 or $p$. Now orbits with just one element consist of a $p$-tuple all of whose entries $a_i$ are equal, for all cyclic permutations of the $p$-tuple are the same. In other words, such an orbit corresponds to an element $a \in G$ with $a^p = 1$. Clearly, $(1, 1, \ldots, 1)$ is such an orbit; if it were the only such, then we would have

$$|G|^{p-1} = |X| = 1 + kp$$

for some $k \geq 0$; that is, $|G|^{p-1} \equiv 1 \bmod p$. If $p$ is a divisor of $|G|$, then we have a contradiction, for $|G|^{p-1} \equiv 0 \bmod p$. We have thus proved Cauchy's Theorem: if a prime $p$ is a divisor of $|G|$, then $G$ has an element of order $p$.

Choose a group $G$ of order $n$, say, $G = \mathbb{I}_n$, where $n$ is not a multiple of $p$. By Lagrange's Theorem, $G$ has no elements of order $p$, so that if $a^p = 1$, then $a = 1$. Therefore, the only orbit in $G^p$ of size 1 is $(1, 1, \ldots, 1)$, and so

$$n^{p-1} = |G|^{p-1} = |X| = 1 + kp;$$

that is, if $p$ is not a divisor of $n$, then $n^{p-1} \equiv 1 \bmod p$. Multiplying both sides by $n$, we have $n^p \equiv n \bmod p$, a congruence also holding when $p$ is a divisor of $n$; this is Fermat's Theorem. ◄

We have seen, in Proposition 1.70, that $A_4$ is a group of order 12 having no subgroup of order 6. Thus, the assertion that if $d$ is a divisor of $|G|$, then $G$ must have a subgroup of order $d$, is false. However, this assertion is true when $G$ is a $p$-group.

**Proposition 1.116.** *If $G$ is a group of order $p^\ell$, then $G$ has a normal subgroup of order $p^k$ for every $k \le \ell$.*

**Proof.** We prove the result by induction on $\ell \ge 0$. The base step is obviously true, and so we proceed to the inductive step. By Theorem 1.113, the center of $G$ is a nontrivial normal subgroup: $Z(G) \ne \{1\}$. Let $Z \subseteq Z(G)$ be a subgroup of order $p$; as any subgroup of $Z(G)$, the subgroup $Z$ is a normal subgroup of $G$. If $k \le \ell$, then $p^{k-1} \le p^{\ell-1} = |G/Z|$. By induction, $G/Z$ has a normal subgroup $H^*$ of order $p^{k-1}$. The Correspondence Theorem says there is a subgroup $H$ of $G$ containing $Z$ with $H^* = H/Z$; moreover, $H^* \lhd G/Z$ implies $H \lhd G$. But $|H/Z| = p^{k-1}$ implies $|H| = p^k$, as desired. •

Abelian groups (and the quaternions) have the property that every subgroup is normal. At the opposite pole are groups having no normal subgroups other than the two obvious ones: $\{1\}$ and $G$.

**Definition.** A group $G$ is called *simple* if $G \ne \{1\}$ and $G$ has no normal subgroups other than $\{1\}$ and $G$ itself.

**Proposition 1.117.** *An abelian group $G$ is simple if and only if it is finite and of prime order.*

**Proof.** If $G$ is finite of prime order $p$, then $G$ has no subgroups $H$ other than $\{1\}$ and $G$; otherwise Lagrange's Theorem would show that $|H|$ is a divisor of $p$. Therefore, $G$ is simple.

Conversely, assume that $G$ is simple. Since $G$ is abelian, every subgroup is normal, and so $G$ has no subgroups other than $\{1\}$ and $G$. Hence, if $x \in G$ and $x \ne 1$ (simple groups are nontrivial), then $\langle x \rangle = G$. If $x$ has infinite order, then all the powers of $x$ are distinct, and so $\langle x^2 \rangle \subsetneq \langle x \rangle$ is a forbidden subgroup of $\langle x \rangle$, a contradiction. Therefore, every $x \in G$ has finite order. If $x$ has (finite) order $m$ and $m$ is composite, say $m = k\ell$, then $\langle x^k \rangle$ is a proper nontrivial subgroup of $\langle x \rangle$, a contradiction. Therefore, $G = \langle x \rangle$ has prime order. •

**Corollary 1.118.** *A finite $p$-group $G$ is simple if and only if $|G| = p$.*

**Proof.** We claim that if $|G| > p$, then $G$ is not simple. If $G$ is abelian, then Proposition 1.117 shows that $G$ is not simple. Hence, we may assume that $G$ is not abelian; that is, its center, $Z(G)$, is a proper subgroup. By Theorem 1.113, we have $Z(G) \ne \{1\}$. But $Z(G) \lhd G$, and so $G$ is not simple. •

We are now going to show that $A_5$ is a nonabelian simple group (indeed, it is the smallest such; there is no nonabelian simple group of order less than 60). This was first proved by Galois, and it is the key to showing that quintic polynomials are not solvable by radicals.

Suppose that an element $x \in G$ has $k$ conjugates. If there is a subgroup $H \subseteq G$ with $x \in H \subseteq G$, how many conjugates does $x$ have in $H$? Since

$$x^H = \{hxh^{-1} : h \in H\} \subseteq \{gxg^{-1} : g \in G\} = x^G,$$

we have $|x^H| \leq |x^G|$. It is possible that there is strict inequality $|x^H| < |x^G|$. For example, if $G = S_3$, $x = (1\ 2)$, and $H = \langle x \rangle$, then $|x^H| = 1$ (because $H$ is abelian) and $|x^G| = 3$ (because all transpositions are conjugate in $S_3$). Let us now consider this question for $G = S_5$, $x = (1\ 2\ 3)$, and $H = A_5$.

**Lemma 1.119.** *All 3-cycles are conjugate in $A_5$.*

**Proof.** Let $G = S_5$, $\alpha = (1\ 2\ 3)$, and $H = A_5$. We know that $|\alpha^{S_5}| = 20$, for there are twenty 3-cycles in $S_5$, as we saw in Table 2 on page 10. Therefore, $20 = |S_5|/|C_{S_5}(\alpha)| = 120/|C_{S_5}(\alpha)|$, by Corollary 1.109, and so $|C_{S_5}(\alpha)| = 6$; that is, there are exactly six permutations in $S_5$ that commute with $\alpha$. Here they are:

$$(1),\ (1\ 2\ 3),\ (1\ 3\ 2),\ (4\ 5),\ (4\ 5)(1\ 2\ 3),\ (4\ 5)(1\ 3\ 2).$$

The last three of these are odd permutations, so that $|C_{A_5}(\alpha)| = 3$. We conclude that

$$|\alpha^{A_5}| = |A_5|/|C_{A_5}(\alpha)| = 60/3 = 20;$$

that is, all 3-cycles are conjugate to $\alpha = (1\ 2\ 3)$ in $A_5$.  •

This lemma can be generalized from $A_5$ to all $A_n$ for $n \geq 5$; see Exercise 1.99 on page 75.

**Lemma 1.120.** *If $n \geq 3$, every element in $A_n$ is a product of 3-cycles.*

**Proof.** If $\alpha \in A_n$, then $\alpha$ is a product of an even number of transpositions:

$$\alpha = \tau_1 \tau_2 \cdots \tau_{2q-1} \tau_{2q}.$$

Of course, we may assume that adjacent $\tau$'s are distinct. As the transpositions may be grouped in pairs $\tau_{2i-1}\tau_{2i}$, it suffices to consider products $\tau\tau'$, where $\tau$ and $\tau'$ are transpositions. If $\tau$ and $\tau'$ are not disjoint, then $\tau = (i\ j)$, $\tau' = (i\ k)$, and $\tau\tau' = (i\ k\ j)$; if $\tau$ and $\tau'$ are disjoint, then $\tau\tau' = (i\ j)(k\ \ell) = (i\ j)(j\ k)(j\ k)(k\ \ell) = (i\ j\ k)(j\ k\ \ell)$.  •

**Theorem 1.121.** *$A_5$ is a simple group.*

**Proof.** We shall show that if $H$ is a normal subgroup of $A_5$ and $H \neq \{(1)\}$, then $H = A_5$. Now if $H$ contains a 3-cycle, then normality forces $H$ to contain all its conjugates. By Lemma 1.119, $H$ contains every 3-cycle, and by Lemma 1.120, $H = A_5$. Therefore, it suffices to prove that $H$ contains a 3-cycle.

As $H \neq \{(1)\}$, it contains some $\sigma \neq (1)$. We may assume, after a harmless relabeling, that either $\sigma = (1\ 2\ 3)$, $\sigma = (1\ 2)(3\ 4)$, or $\sigma = (1\ 2\ 3\ 4\ 5)$. As we have just remarked, we are done if $\sigma$ is a 3-cycle.

If $\sigma = (1\ 2)(3\ 4)$, define $\tau = (1\ 2)(3\ 5)$. Now $H$ contains $(\tau\sigma\tau^{-1})\sigma^{-1}$, because it is a normal subgroup, and $\tau\sigma\tau^{-1}\sigma^{-1} = (3\ 5\ 4)$, as the reader should check. If $\sigma = (1\ 2\ 3\ 4\ 5)$, define $\rho = (1\ 3\ 2)$; now $H$ contains $\rho\sigma\rho^{-1}\sigma^{-1} = (1\ 3\ 4)$, as the reader should also check. We have shown that if $H \neq \{(1)\}$, then $H$ contains a 3-cycle and, hence, $H = A_5$. Therefore, $A_5$ is simple.   •

Without much more effort, we can prove that the alternating groups $A_n$ are simple for all $n \geq 5$. Observe that $A_4$ is not simple, for the four-group **V** is a normal subgroup of $A_4$.

**Lemma 1.122.** *$A_6$ is a simple group.*

**Proof.** Let $H \neq \{(1)\}$ be a normal subgroup of $A_6$; we must show that $H = A_6$. Assume that there is some $\alpha \in H$ with $\alpha \neq (1)$ that fixes some $i$, where $1 \leq i \leq 6$. Define

$$F = \{\sigma \in A_6 : \sigma(i) = i\}.$$

Note that $\alpha \in H \cap F$, so that $H \cap F \neq \{(1)\}$. The Second Isomorphism Theorem gives $H \cap F \lhd F$. But $F$ is simple, for $F \cong A_5$, and so the only normal subgroups in $F$ are $\{(1)\}$ and $F$. Since $H \cap F \neq \{(1)\}$, we have $H \cap F = F$; that is, $F \subseteq H$. It follows that $H$ contains a 3-cycle (for $F$ does), and so $H = A_6$, by Exercise 1.99 on page 75.

We may now assume that there is no $\alpha \in H$ with $\alpha \neq (1)$ that fixes some $i$ with $1 \leq i \leq 6$. If we consider the cycle structures of permutations in $A_6$, however, any such $\alpha$ must have cycle structure $(1\ 2)(3\ 4\ 5\ 6)$ or $(1\ 2\ 3)(4\ 5\ 6)$. In the first case, $\alpha^2 \in H$ is a nontrivial permutation that fixes 1 (and also 2), a contradiction. In the second case, $H$ contains $\alpha(\beta\alpha^{-1}\beta^{-1})$, where $\beta = (2\ 3\ 4)$, and it is easily checked that this is a nontrivial element in $H$ which fixes 1, another contradiction. Therefore, no such normal subgroup $H$ can exist, and so $A_6$ is a simple group.   •

**Theorem 1.123.** *$A_n$ is a simple group for all $n \geq 5$.*

**Proof.** If $H$ is a nontrivial normal subgroup of $A_n$, then $H \neq \{(1)\}$ and we must show that $H = A_n$. By Exercise 1.99 on page 75, it suffices to prove that $H$ contains a 3-cycle. If $\beta \in H$ is nontrivial, then there exists some $i$ that $\beta$ moves; say, $\beta(i) = j \neq i$. Choose a 3-cycle $\alpha$ that fixes $i$ and moves $j$. The permutations $\alpha$ and $\beta$ do not commute: $\beta\alpha(i) = \beta(i) = j$, while $\alpha\beta(i) = \alpha(j) \neq j$. It follows that $\gamma = (\alpha\beta\alpha^{-1})\beta^{-1}$ is a nontrivial element of $H$. But $\beta\alpha^{-1}\beta^{-1}$ is a 3-cycle, by Theorem 1.9, and so $\gamma = \alpha(\beta\alpha^{-1}\beta^{-1})$ is a product of two 3-cycles. Hence, $\gamma$ moves at most 6 symbols, say, $i_1, \ldots, i_6$ (if $\gamma$ moves fewer than 6 symbols, just adjoin others so we have a list of 6). Define

$$F = \{\sigma \in A_n : \sigma \text{ fixes all } i \neq i_1, \ldots, i_6\}.$$

Now $F \cong A_6$ and $\gamma \in H \cap F$. Hence, $H \cap F$ is a nontrivial normal subgroup of $F$. But $F$ is simple, being isomorphic to $A_6$, and so $H \cap F = F$; that is, $F \subseteq H$. Therefore, $H$ contains a 3-cycle, and so $H = A_n$; the proof is complete.   •

# Exercises

**1.87.** If $a$ and $b$ are elements in a group $G$, prove that $ab$ and $ba$ have the same order.

**Hint.** Use a conjugation.

**1.88.** Prove that if $G$ is a finite group of odd order, then no $x \in G$, other than $x = 1$, is conjugate to its inverse.

**Hint.** If $x$ is conjugate to $x^{-1}$, how many elements are in $x^G$?

**1.89.** Prove that no two of the following groups of order 8 are isomorphic:

$$\mathbb{I}_8; \quad \mathbb{I}_4 \times \mathbb{I}_2; \quad \mathbb{I}_2 \times \mathbb{I}_2 \times \mathbb{I}_2; \quad D_8; \quad \mathbf{Q}.$$

$*$ **1.90.** Show that $S_4$ has a subgroup isomorphic to $D_8$.

$*$ **1.91.** Prove that $S_4/\mathbf{V} \cong S_3$.

**Hint.** Use Proposition 1.98.

$*$ **1.92.**   (i) Prove that $A_4 \not\cong D_{12}$.

   **Hint.** Recall that $A_4$ has no element of order 6.

   (ii) Prove that $D_{12} \cong S_3 \times \mathbb{I}_2$.
   **Hint.** Each element $x \in D_{12}$ has a unique factorization of the form $x = b^i a^j$, where $b^6 = 1$ and $a^2 = 1$.

$*$ **1.93.**   (i) If $G$ is a group, then a normal subgroup $H \lhd G$ is called a ***maximal normal subgroup*** if there is no normal subgroup $K$ of $G$ with $H \subsetneq K \subsetneq G$. Prove that a normal subgroup $H$ is a maximal normal subgroup of $G$ if and only if $G/H$ is a simple group.

   (ii) Prove that every finite abelian group $G$ has a subgroup of prime index.
   **Hint.** Use Proposition 1.117.

   (iii) Prove that $A_6$ has no subgroup of prime index.

**1.94.**   (i) **(Landau)** Given a positive integer $n$ and a positive rational $q$, prove that there are only finitely many $n$-tuples $(i_1, \ldots, i_n)$ of positive integers with $q = \sum_{j=1}^{n} 1/i_j$.

   (ii) Prove, for every positive integer $n$, that there are only finitely many finite groups having exactly $n$ conjugacy classes.
   **Hint.** Use part (i) and the Class Equation.

**1.95.** Find $N_G(H)$ if $G = S_4$ and $H = \langle (1\ 2\ 3) \rangle$.

$*$ **1.96.** If $H$ is a proper subgroup of a finite group $G$, prove that $G$ is not the union of all the conjugates of $H$: that is, $G \neq \bigcup_{x \in G} xHx^{-1}$.

$*$ **1.97.**   (i) If $H$ is a subgroup of $G$ and $x \in H$, prove that

$$C_H(x) = H \cap C_G(x).$$

   (ii) If $H$ is a subgroup of index 2 in a finite group $G$ and $x \in H$, prove that either $|x^H| = |x^G|$ or $|x^H| = \frac{1}{2}|x^G|$, where $x^H$ is the conjugacy class of $x$ in $H$.
   **Hint.** Use the Second Isomorphism Theorem.

   (iii) Prove that there are two conjugacy classes of 5-cycles in $A_5$, each of which has 12 elements.

**Hint.** If $\alpha = (1\ 2\ 3\ 4\ 5)$, then $|C_{S_5}(\alpha)| = 5$ because $24 = 120/|C_{S_5}(\alpha)|$; hence $C_{S_5}(\alpha) = \langle \alpha \rangle$. What is $C_{A_5}(\alpha)$?

(iv) Prove that the conjugacy classes in $A_5$ have sizes 1, 12, 12, 15, and 20.

**1.98.** (i) Prove that every normal subgroup $H$ of a group $G$ is a union of conjugacy classes of $G$, one of which is $\{1\}$.

(ii) Use part (i) and Exercise 1.97 to give a second proof of the simplicity of $A_5$.

* **1.99.** (i) For all $n \geq 5$, prove that all 3-cycles are conjugate in $A_n$.

**Hint.** Show that $(1\ 2\ 3)$ and $(i\ j\ k)$ are conjugate by considering two cases: they are not disjoint (so they move at most 5 letters); they are disjoint.

(ii) Prove that if a normal subgroup $H$ of $A_n$ contains a 3-cycle, where $n \geq 5$, then $H = A_n$. (*Remark.* We have proved this in Lemma 1.120 when $n = 5$.)

**1.100.** Prove that the only normal subgroups of $S_4$ are $\{(1)\}$, $\mathbf{V}$, $A_4$, and $S_4$.

**Hint.** Use Theorem 1.9, checking the various cycle structures one at a time.

**1.101.** Prove that $A_5$ is a group of order 60 that has no subgroup of order 30.

* **1.102.** (i) Prove, for all $n \geq 5$, that the only normal subgroups of $S_n$ are $\{(1)\}$, $A_n$, and $S_n$.

**Hint.** If $H \lhd S_n$ is a proper subgroup and $H \neq A_n$, then $H \cap A_n = \{(1)\}$.

(ii) Prove that if $n \geq 3$, then $A_n$ is the only subgroup of $S_n$ of order $\frac{1}{2}n!$.
**Hint.** If $H$ is a second such subgroup, then $H \lhd S_n$ and $(H \cap A_n) \lhd A_n$.

(iii) Prove that $S_5$ is a group of order 120 having no subgroup of order 30.
**Hint.** Use the representation on the cosets of a supposed subgroup of order 30, as well as the simplicity of $A_5$.

(iv) Prove that $S_5$ contains no subgroup of order 40.

* **1.103.** (i) Let $\sigma, \tau \in S_5$, where $\sigma$ is a 5-cycle and $\tau$ is a transposition. Prove that $S_5 = \langle \sigma, \tau \rangle$.

(ii) Give an example showing that $S_n$, for some $n$, contains an $n$-cycle $\sigma$ and a transposition $\tau$ such that $\langle \sigma, \tau \rangle \neq S_n$.

* **1.104.** Let $G$ be a subgroup of $S_n$.

(i) If $G \cap A_n = \{1\}$, prove that $|G| \leq 2$.

(ii) If $G$ is a simple group with more than two elements, prove that $G \subseteq A_n$.

* **1.105.** (i) If $n \geq 5$, prove that $S_n$ has no subgroup of index $r$, where $2 < r < n$.

(ii) Prove that if $n \geq 5$, then $A_n$ has no subgroup of index $r$, where $2 \leq r < n$.

**1.106.** (i) Prove that if a simple group $G$ has a subgroup of index $n > 1$, then $G$ is isomorphic to a subgroup of $S_n$.

**Hint.** Kernels are normal subgroups.

(ii) Prove that an infinite simple group (such do exist) has no subgroups of finite index $n > 1$.
**Hint.** Use part (i).

∗ **1.107.** If $G$ is a group of order $n$, prove that $G$ is isomorphic to a subgroup of $\mathrm{GL}(n,\mathbb{Q})$.

**Hint.** If $\sigma \in S_n$, then the $n \times n$ **permutation matrix** $P_\sigma$ is the matrix obtained from the $n \times n$ identity matrix by permuting its columns via $\sigma$. Show that $\sigma \mapsto P_\sigma$ is an injective homomorphism $S_n \to \mathrm{GL}(n,\mathbb{Q})$.

∗ **1.108.** Let $G$ be a group with $|G| = mp$, where $p$ is prime and $1 < m < p$. Prove that $G$ is not simple.

**Hint.** Show that $G$ has a subgroup $H$ of order $p$, and use the representation of $G$ on the cosets of $H$.

**Remark.** We can now show that all but 11 of the numbers smaller than 60 are not orders of nonabelian simple groups (namely, 12, 18, 24, 30, 36, 40, 45, 48, 50, 54, 56). Theorem 1.113 eliminates all prime powers (for the center is always a normal subgroup), and this exercise eliminates all numbers of the form $mp$, where $p$ is prime and $m < p$.  ◀

∗ **1.109.**  (i) Let a group $G$ act on a set $X$, and suppose that $x, y \in X$ lie in the same orbit: $y = gx$ for some $g \in G$. Prove that $G_y = gG_xg^{-1}$.

  (ii) Let $G$ be a finite group acting on a set $X$; prove that if $x, y \in X$ lie in the same orbit, then $|G_x| = |G_y|$.

## Section 1.8. Counting

We now use groups to solve some difficult counting problems.

**Theorem 1.124 (Burnside's Lemma).**[32] *Let $G$ be a finite group acting on a finite set $X$. If $N$ is the number of orbits, then*

$$N = \frac{1}{|G|} \sum_{\tau \in G} \mathrm{Fix}(\tau),$$

*where* $\mathrm{Fix}(\tau)$ *is the number of* $x \in X$ *fixed by* $\tau$.

**Proof.** List the elements of $X$ as follows: choose $x_1 \in X$, and then list all the elements $x_1, x_2, \ldots, x_r$ in the orbit $\mathcal{O}(x_1)$; then choose $x_{r+1} \notin \mathcal{O}(x_1)$, and list the elements $x_{r+1}, x_{r+2}, \ldots$ in $\mathcal{O}(x_{r+1})$; continue this procedure until all the elements of $X$ are listed. Now list the elements $\tau_1, \tau_2, \ldots, \tau_n$ of $G$, and form Figure 1.10, where

$$f_{i,j} = \begin{cases} 1 & \text{if } \tau_i \text{ fixes } x_j \\ 0 & \text{if } \tau_i \text{ moves } x_j. \end{cases}$$

---

[32]Burnside himself attributed this lemma to Frobenius. To avoid the confusion that would be caused by changing a popular name, P. M. Neumann has suggested that it be called "not-Burnside's Lemma." Burnside was a fine mathematician, and there do exist theorems properly attributed to him. For example, Burnside proved Theorem 7.62, which implies that there are no simple groups of order $p^m q^n$, where $p$ and $q$ are primes.

| | $x_1$ | $x_2$ | $\cdots$ | $x_{r+1}$ | $x_{r+2}$ | $\cdots$ |
|---|---|---|---|---|---|---|
| $\tau_1$ | $f_{1,1}$ | $f_{1,2}$ | $\cdots$ | $f_{1,r+1}$ | $f_{1,r+2}$ | $\cdots$ |
| $\tau_2$ | $f_{2,1}$ | $f_{2,2}$ | $\cdots$ | $f_{2,r+1}$ | $f_{2,r+2}$ | $\cdots$ |
| $\tau_i$ | $f_{i,1}$ | $f_{i,2}$ | $\cdots$ | $f_{i,r+1}$ | $f_{i,r+2}$ | $\cdots$ |
| $\tau_n$ | $f_{n,1}$ | $f_{n,2}$ | $\cdots$ | $f_{n,r+1}$ | $f_{n,r+2}$ | $\cdots$ |

**Figure 1.10.** Burnside's Lemma.

Now $\mathrm{Fix}(\tau_i)$, the number of $x$ fixed by $\tau_i$, is the number of 1's in the $i$th row of the array; therefore, $\sum_{\tau \in G} \mathrm{Fix}(\tau)$ is the total number of 1's in the array. Let us now look at the columns. The number of 1's in the first column is the number of $\tau_i$ that fix $x_1$; by definition, these $\tau_i$ comprise $G_{x_1}$. Thus, the number of 1's in column 1 is $|G_{x_1}|$. Similarly, the number of 1's in column 2 is $|G_{x_2}|$. By Exercise 1.109 on page 76, $|G_{x_1}| = |G_{x_2}|$. By Theorem 1.107, the number of 1's in the $r$ columns labeled by the $x_i \in \mathcal{O}(x_1)$ is thus

$$r|G_{x_1}| = |\mathcal{O}(x_1)| \cdot |G_{x_1}| = (|G|/|G_{x_1}|)\, |G_{x_1}| = |G|.$$

The same is true for any other orbit: its columns collectively contain exactly $|G|$ 1's. Therefore, if there are $N$ orbits, there are $N|G|$ 1's in the array. We conclude that

$$\sum_{\tau \in G} \mathrm{Fix}(\tau) = N|G|. \quad \bullet$$

We are going to use Burnside's Lemma to solve problems of the following sort. How many striped flags are there having six stripes (of equal width) each of which can be colored red, white, or blue? Clearly, the two flags in Figure 1.11 are the same: the bottom flag is just the top one rotated about its center.

| $r$ | $w$ | $b$ | $r$ | $w$ | $b$ |
|---|---|---|---|---|---|

| $b$ | $w$ | $r$ | $b$ | $w$ | $r$ |
|---|---|---|---|---|---|

**Figure 1.11.** Striped flags.

Let $X$ be the set of all 6-tuples of colors; if $x \in X$, then

$$x = (c_1, c_2, c_3, c_4, c_5, c_6),$$

where each $c_i$ denotes either red, white, or blue. Let $\tau$ be the permutation that reverses all the indices:

$$\tau = \begin{pmatrix} 1 & 2 & 3 & 4 & 5 & 6 \\ 6 & 5 & 4 & 3 & 2 & 1 \end{pmatrix} = (1\,6)(2\,5)(3\,4)$$

(thus, $\tau$ "rotates" each 6-tuple $x$ of colored stripes). The cyclic group $G = \langle \tau \rangle$ acts on $X$; since $|G| = 2$, the orbit of any 6-tuple $x$ consists of either one or two elements: either $\tau$ fixes $x$ or it does not. Since a flag is unchanged by rotation, it is reasonable to identify a flag with an orbit of a 6-tuple. For example, the orbit consisting of the 6-tuples

$$(r, w, b, r, w, b) \quad \text{and} \quad (b, w, r, b, w, r)$$

describes the flag in Figure 1.11. The number of flags is thus the number $N$ of orbits; by Burnside's Lemma, $N = \frac{1}{2}[\text{Fix}((1)) + \text{Fix}(\tau)]$. The identity permutation $(1)$ fixes every $x \in X$, and so $\text{Fix}((1)) = 3^6$ (there are three colors). Now $\tau$ fixes a 6-tuple $x$ if it is a "palindrome," that is, if the colors in $x$ read the same forward as backward. For example,

$$x = (r, r, w, w, r, r)$$

is fixed by $\tau$. Conversely, if

$$x = (c_1, c_2, c_3, c_4, c_5, c_6)$$

is fixed by $\tau = (1\ 6)(2\ 5)(3\ 4)$, then $c_1 = c_6$, $c_2 = c_5$, and $c_3 = c_4$; that is, $x$ is a palindrome. It follows that $\text{Fix}(\tau) = 3^3$, for there are three choices for each of $c_1$, $c_2$, and $c_3$. The number of flags is thus

$$N = \frac{1}{2}(3^6 + 3^3) = 378.$$

Let us make the notion of coloring more precise.

**Definition.** If a group $G$ acts on $X = \{1, \ldots, n\}$, and $\mathcal{C}$ is a set of $q$ *colors*, then $G$ acts on the set $\mathcal{C}^n$ of all $n$-tuples of colors by

$$\tau(c_1, \ldots, c_n) = (c_{\tau 1}, \ldots, c_{\tau n}) \quad \text{for all } \tau \in G.$$

An orbit of $(c_1, \ldots, c_n) \in \mathcal{C}^n$ is called a *(q, G)-coloring* of $X$.

Color each square in a $4 \times 4$ grid red or black (adjacent squares may have the same color; indeed, one possibility is that all the squares have the same color).

| 1 | 2 | 3 | 4 |
|---|---|---|---|
| 5 | 6 | 7 | 8 |
| 9 | 10 | 11 | 12 |
| 13 | 14 | 15 | 16 |

| 13 | 9 | 5 | 1 |
|----|---|---|---|
| 14 | 10 | 6 | 2 |
| 15 | 11 | 7 | 3 |
| 16 | 12 | 8 | 4 |

**Figure 1.12.** Chessboard and a rotation.

If $X$ consists of the 16 squares in the grid, and $\mathcal{C}$ consists of the two colors red and black, then the cyclic group $G = \langle R \rangle$ of order 4 acts on $X$, where $R$ is clockwise rotation by $90°$; Figure 1.12 shows how $R$ acts: the right square is $R$'s action on the left square. In cycle notation,

$$R = (1,\ 4,\ 16,\ 13)(2,\ 8,\ 15,\ 9)(3,\ 12,\ 14,\ 5)(6,\ 7,\ 11,\ 10),$$

$$R^2 = (1,\ 16)(4,\ 13)(2,\ 15)(8,\ 9)(3,\ 14)(12,\ 5)(6,\ 11)(7,\ 10),$$

$$R^3 = (1,\ 13,\ 16,\ 4)(2,\ 9,\ 15,\ 8)(3,\ 5,\ 14,\ 12)(6,\ 10,\ 11,\ 7).$$

A red and black chessboard does not change when it is rotated; it is merely viewed from a different position. Thus, we may regard a chessboard as a 2-coloring of $X$; the orbit of a 16-tuple corresponds to the four ways of viewing the board.

By Burnside's Lemma, the number of chessboards is

$$\tfrac{1}{4}\Big[\mathrm{Fix}((1)) + \mathrm{Fix}(R) + \mathrm{Fix}(R^2) + \mathrm{Fix}(R^3)\Big].$$

Now $\mathrm{Fix}((1)) = 2^{16}$, for every 16-tuple is fixed by the identity. To compute $\mathrm{Fix}(R)$, note that squares 1, 4, 16, 13 must all have the same color in a 16-tuple fixed by $R$. Similarly, squares 2, 8, 15, 9 must have the same color, squares 3, 12, 14, 5 must have the same color, and squares 6, 7, 11, 10 must have the same color. We conclude that $\mathrm{Fix}(R) = 2^4$; note that the exponent 4 is the number of cycles in the complete factorization of $R$. A similar analysis shows that $\mathrm{Fix}(R^2) = 2^8$, for the complete factorization of $R^2$ has 8 cycles, and $\mathrm{Fix}(R^3) = 2^4$, because the cycle structure of $R^3$ is the same as that of $R$. Therefore, the number $N$ of chessboards is

$$N = \tfrac{1}{4}\Big[2^{16} + 2^4 + 2^8 + 2^4\Big] = 16{,}456.$$

We now show, as in the discussion of the $4 \times 4$ chessboard, that the cycle structure of a permutation $\tau$ allows one to calculate $\mathrm{Fix}(\tau)$.

**Lemma 1.125.** *Let $C$ be a set of $q$ colors, and let $G$ be a subgroup of $S_n$. If $\tau \in G$, then*

$$\mathrm{Fix}(\tau) = q^{t(\tau)},$$

*where $t(\tau)$ is the number of cycles in the complete factorization of $\tau$.*

**Proof.** Since $\tau(c_1, \ldots, c_n) = (c_{\tau 1}, \ldots, c_{\tau n}) = (c_1, \ldots, c_n)$, we see that $c_{\tau i} = c_i$ for all $i$, and so $\tau i$ has the same color as $i$. It follows, for all $k$, that $\tau^k i$ has the same color as $i$, that is, all points in the orbit of $i$ acted on by $\langle \tau \rangle$ have the same color. If the complete factorization of $\tau$ is $\tau = \beta_1 \cdots \beta_{t(\tau)}$, and $i$ occurs in $\beta_j$, then Example 1.101 shows that the orbit containing $i$ is the set of symbols occurring in $\beta_j$. Thus, for an $n$-tuple to be fixed by $\tau$, all the symbols involved in each of the $t(\tau)$ cycles must have the same color; as there are $q$ colors, there are thus $q^{t(\tau)}$ $n$-tuples fixed by $\tau$.   •

**Theorem 1.126.** *Let $G$ act on a finite set $X$. If $N$ is the number of $(q, G)$-colorings of $X$, then*

$$N = \frac{1}{|G|} \sum_{\tau \in G} q^{t(\tau)},$$

*where $t(\tau)$ is the number of cycles in the complete factorization of $\tau$.*

**Proof.** Rewrite Burnside's Lemma using Lemma 1.125.   •

There is a generalization of this technique, due to Pólya (Biggs, *Discrete Mathematics*, p. 403), giving formulas of the sort that count the number of red, white, blue, and green flags having 20 stripes exactly 7 of which are red and 5 of which are blue.

## Exercises

**1.110.** How many flags are there with $n$ stripes of equal width, each of which can be colored any one of $q$ given colors?

**Hint.** The parity of $n$ is relevant.

**1.111.** Let $X$ be the squares in an $n \times n$ grid, and let $\rho$ be a rotation by $90°$. Define a ***chessboard*** to be a $(q, G)$-coloring, where the cyclic group $G = \langle \rho \rangle$ of order 4 is acting on $X$. Show that the number of chessboards is

$$\frac{1}{4} \left( q^{n^2} + q^{\lfloor (n^2+1)/2 \rfloor} + 2q^{\lfloor (n^2+3)/4 \rfloor} \right),$$

where $\lfloor x \rfloor$ is the greatest integer in the number $x$.

**1.112.** Let $X$ be a disk divided into $n$ congruent circular sectors, and let $\rho$ be a rotation by $(360/n)°$. Define a ***roulette wheel*** to be a $(q, G)$-coloring, where the cyclic group $G = \langle \rho \rangle$ of order $n$ is acting on $X$. Prove that if $n = 6$, then there are $\frac{1}{6}(2q + 2q^2 + q^3 + q^6)$ roulette wheels having 6 sectors. (The formula for the number of roulette wheels with $n$ sectors is

$$\frac{1}{n} \sum_{d | n} \phi(n/d) q^d,$$

where $\phi$ is the Euler $\phi$-function.)

**1.113.** Let $X$ be the vertices of a regular $n$-gon, and let the dihedral group $G = D_{2n}$ act on $X$ [as the usual group of symmetries (Examples 1.30 and 1.31)]. Define a ***bracelet*** to be a $(q, G)$-coloring of a regular $n$-gon, and call each of its vertices a ***bead***. (Not only can we rotate a bracelet, we can also *flip* it: that is, turn it upside down by rotating it in space about a line joining two beads.)

(i) How many bracelets are there having 5 beads, each of which can be colored any one of $q$ available colors?

**Hint.** The group $G = D_{10}$ is acting. Use Example 1.31 to assign to each symmetry a permutation of the vertices, and then show that the number of bracelets is

$$\frac{1}{10}(q^5 + 4q + 5q^3).$$

(ii) How many bracelets are there having 6 beads, each of which can be colored any one of $q$ available colors?

**Hint.** The group $G = D_{12}$ is acting. Assign a permutation of the vertices to each symmetry, and then show that the number of bracelets is

$$\frac{1}{12}(q^6 + 2q^4 + 4q^3 + 3q^2 + 2q).$$

# Chapter 2

# Commutative Rings I

As Chapter 1, this chapter contains some material usually found in an earlier course. We begin by discussing commutative rings, after which we will give some of the first results about vector spaces (with scalars in any field) and linear transformations. The chapter ends with the construction of *splitting fields*, which leads to the existence and classification of all finite fields.

## Section 2.1. First Properties

We recall the definition of a ring.

**Definition.** A *ring*[1] $R$ is a set with two binary operations, addition and multiplication, such that

(i) $R$ is an abelian group under addition,

(ii) $a(bc) = (ab)c$ for every $a$, $b$, $c \in R$,

(iii) there is an element $1 \in R$ with $1a = a = a1$ for every $a \in R$,

(iv) **Distributivity**: $a(b + c) = ab + ac$ and $(b + c)a = ba + ca$ for every $a$, $b$, $c \in R$.

The element 1 in a ring $R$ has several names; it is called *one*, the *unit* of $R$, or the *identity* in $R$. We do not assume that $1 \neq 0$, but see Proposition 2.2(ii).

**Remark.** Some authors do not demand, as part of the definition, that rings have 1; they point to natural examples, such as the even integers or the integrable functions,

---

[1]This term was probably coined by Hilbert, in 1897, when he wrote *Zahlring*. One of the meanings of the word *ring*, in German as in English, is collection, as in the phrase "a ring of thieves." (It has also been suggested that Hilbert used this term because, for a ring of algebraic integers, an appropriate power of each element "cycles back" to being a linear combination of lower powers.)

where a function $f : [0, \infty) \to \mathbb{R}$ is **integrable** if

$$\int_0^{\infty} |f(x)| \, dx = \lim_{t \to \infty} \int_0^t |f(x)| \, dx < \infty.$$

It is not difficult to see that if $f$ and $g$ are integrable, then so are their pointwise sum $f + g$ and pointwise product $fg$. The only candidate for a unit is the constant function $E$ with $E(x) = 1$ for all $x \in [0, \infty)$ but, obviously, $E$ is not integrable. We do not recognize either of these systems as a ring (but see Exercise 2.2 on page 88).

The absence of a unit makes many constructions more complicated. For example, if $R$ is a "ring without unit" and $a \in R$, then defining $(a)$, the principal ideal generated by $a$, as $(a) = \{ra : r \in R\}$, leads to the possibility that $a \notin (a)$; thus, we must redefine $(a)$ to force $a$ inside. Polynomial rings become strange: if $R$ has no unit, then $x$ is not a polynomial (see our construction of polynomial rings on page 92). There are other (more important) reasons for wanting a unit (for example, the discussion of tensor products would become complicated), but these examples should suffice to show that not assuming a unit can lead to some awkwardness; therefore, we insist that rings do have units. ◀

Addition and multiplication in a ring $R$ are binary operations, so there are functions $\alpha : R \times R \to R$ and $\mu : R \times R \to R$ with

$$\alpha(r, r') = r + r' \in R \quad \text{and} \quad \mu(r, r') = rr' \in R$$

for all $r, r' \in R$. The law of substitution holds here, as it does for any operation: if $r = r'$ and $s = s'$, then $r + s = r' + s'$ and $rs = r's'$.

**Example 2.1.** If $R = \mathrm{Mat}_n(\mathbb{R})$ denotes the set of all $n \times n$ matrices with entries in $\mathbb{R}$, then $R$ is a ring with binary operations matrix addition and matrix multiplication. The unit in $R$ is the *identity matrix* $I = [\delta_{ij}]$, where $\delta_{ij}$ is the **Kronecker delta**: $\delta_{ij} = 0$ if $i \neq j$, and $\delta_{ii} = 1$ for all $i$. ◀

Here are some elementary results.

**Proposition 2.2.** *Let $R$ be a ring.*

(i) $0 \cdot a = 0 = a \cdot 0$ *for every $a \in R$.*

(ii) *If $1 = 0$, then $R$ consists of the single element $0$. In this case, $R$ is called the* **zero ring**.[2]

(iii) *If $-a$ is the additive inverse of $a$, then $(-1)(-a) = a = (-1)(-a)$. In particular, $(-1)(-1) = 1$.*

(iv) $(-1)a = -a = a(-1)$ *for every $a \in R$.*

(v) *If $n \in \mathbb{N}$ and $n1 = 0$, then $na = 0$ for all $a \in R$ [if $a \in R$ and $n \in \mathbb{N}$, then $na = a + a + \cdots + a$ ($n$ summands)].*[3]

**Proof.**

(i) $0 \cdot a = (0 + 0)a = (0 \cdot a) + (0 \cdot a)$. Now subtract $0 \cdot a$ from both sides.

---

[2] The zero ring is not a very interesting ring, but it does arise occasionally.

[3] Thus, $na$ is the additive version of the multiplicative notation $a^n$.

(ii) If $1 = 0$, then $a = 1 \cdot a = 0 \cdot a = 0$ for all $a \in R$.

(iii) $0 = 0(-a) = (-1 + 1)(-a) = (-1)(-a) + (-a)$. Now add $a$ to both sides.

(iv) Multiply both sides of $(-1)(-a) = a$ by $-1$, and use part (iii).

(v) $na = a + \cdots + a = (1 + \cdots + 1)a = (n1)a = 0 \cdot a = 0$. •

Informally, a *subring* $S$ of a ring $R$ is a ring contained in $R$ such that $S$ and $R$ have the same addition, multiplication, and unit.

**Definition.** A subset $S$ of a ring $R$ is a **subring** of $R$ if

(i) $1 \in S$,[4]

(ii) if $a, b \in S$, then $a - b \in S$,

(iii) if $a, b \in S$, then $ab \in S$.

We shall write $S \subsetneq R$ to denote $S$ being a **proper** subring; that is, $S \subseteq R$ is a subring and $S \neq R$.

**Proposition 2.3.** *A subring $S$ of a ring $R$ is itself a ring.*

**Proof.** Parts (i) and (ii) in the definition of subring say that $S$ is a subgroup of the additive group $R$ [part (i) shows that $S$ is nonempty], while part (iii) shows that multiplication is a binary operation when restricted to $S$. The other statements in the definition of ring are identities that hold for all elements in $R$ and, hence, hold in particular for the elements in $S$. For example, associativity $a(bc) = (ab)c$ holds for all $a, b, c \in R$, and so it holds for all $a, b, c \in S \subseteq R$. •

Of course, one advantage of the notion of subring is that fewer ring axioms need be checked to determine whether a subset of a ring is itself a ring.

**Example 2.4.** Let $n \geq 3$ be an integer; if $\zeta_n = e^{2\pi i/n}$ is a primitive $n$th root of unity, define

$$\mathbb{Z}[\zeta_n] = \{a_0 + a_1\zeta_n + a_2\zeta_n^2 + \cdots + a_{n-1}\zeta_n^{n-1} \in \mathbb{C} : a_i \in \mathbb{Z}\}.$$

(We assume that $n \geq 3$, for $\zeta_2 = -1$ and $\mathbb{Z}[\zeta_2] = \mathbb{Z}$.) When $n = 4$, then $\mathbb{Z}[\zeta_4] = \mathbb{Z}[i]$ is called the ring of **Gaussian integers**. It is easy to check that $\mathbb{Z}[\zeta_n]$ is a subring of $\mathbb{C}$ (to prove that $\mathbb{Z}[\zeta_n]$ is closed under multiplication, note that if $m \geq n$, then $m = qn + r$, where $0 \leq r < n$, and $\zeta_n^m = \zeta_n^r$). ◄

**Definition.** A ring $R$ is **commutative** if $ab = ba$ for all $a, b \in R$.

The sets $\mathbb{Z}$, $\mathbb{Q}$, $\mathbb{R}$, and $\mathbb{C}$ are commutative rings with the usual addition and multiplication (the ring axioms are verified in courses in the foundations of Mathematics). Also, $\mathbb{I}_m$, the integers mod $m$, is a commutative ring (FCAA, p. 225).

---

[4]Exercise 2.6(iii) on page 89 gives a natural example of a subset $S$ of a ring $R$ which is not a subring even though $S$ and $R$ have the same addition and the same multiplication; they have different units.

**Proposition 2.5 (Binomial Theorem).** *If $R$ is a commutative ring and $a, b \in R$, then*

$$(a + b)^n = \sum_{r=0}^{n} \binom{n}{r} a^r b^{n-r}.$$

**Proof.** The usual inductive proof is valid in this generality. We define $a^0 = 1$ for every element $a \in R$; in particular, $0^0 = 1$.  •

Example 2.1 can be generalized. If $k$ is a commutative ring, then $\mathrm{Mat}_n(k)$, the set of all $n \times n$ matrices with entries in $k$, is a ring.

**Corollary 2.6.** *If $N \in \mathrm{Mat}_n(\mathbb{I}_p)$, then $(I + N)^p = I + N^p$.*

**Proof.** The subring $R$ of $\mathrm{Mat}_n(\mathbb{I}_p)$ generated by $N$ is a commutative ring, and so the Binomial Theorem applies:

$$(I + N)^p = \sum_{r=0}^{p} \binom{p}{r} N^{p-r}.$$

Now $p \mid \binom{p}{r}$ whenever $0 < r < p$ (FCAA, p. 42), so that $\binom{p}{r} N^{p-r} = 0$ in $R$.  •

Unless we say otherwise,

**all rings in the rest of this chapter are commutative;**

we will return to noncommutative rings in Chapter 6.

**Definition.** A ***domain*** (often called an *integral domain*) is a commutative ring $R$ that satisfies two extra axioms:

  (i) $1 \neq 0$;

  (ii) ***Cancellation Law***: for all $a, b, c \in R$, if $ca = cb$ and $c \neq 0$, then $a = b$.

The familiar examples of commutative rings, $\mathbb{Z}$, $\mathbb{Q}$, $\mathbb{R}$, and $\mathbb{C}$, are domains; the zero ring is not a domain. The Gaussian integers $\mathbb{Z}[i]$ is a commutative ring, and Exercise 2.9 on page 90 shows that it is a domain.

**Proposition 2.7.** *A nonzero commutative ring $R$ is a domain if and only if the product of any two nonzero elements of $R$ is nonzero.*

**Proof.** $ab = ac$ if and only if $a(b - c) = 0$.  •

Elements $a, b \in R$ are called ***zero divisors*** if $ab = 0$ and $a \neq 0$, $b \neq 0$. Thus, domains have no zero divisors.

**Proposition 2.8.** *The commutative ring $\mathbb{I}_m$ is a domain if and only if $m$ is prime.*

**Proof.** If $m$ is not prime, then $m = ab$, where $1 < a, b < m$; hence, both $[a]$ and $[b]$ are not zero in $\mathbb{I}_m$, yet $[a][b] = [m] = [0]$. Conversely, let $m$ be prime. If $[a][b] = [ab] = [0]$, where $[a], [b] \neq [0]$, then $m \mid ab$. Now Euclid's Lemma gives $m \mid a$ or $m \mid b$; if, say, $m \mid a$, then $a = md$ and $[a] = [m][d] = [0]$, a contradiction.  •

**Example 2.9.**

(i) Let $\mathcal{F}(\mathbb{R})$ be the set of all functions $\mathbb{R} \to \mathbb{R}$ equipped with the operations of ***pointwise addition*** and ***pointwise multiplication***: given $f, g \in \mathcal{F}(\mathbb{R})$, define $f + g, fg \in \mathcal{F}(\mathbb{R})$ by

$$f + g \colon a \mapsto f(a) + g(a) \quad \text{and} \quad fg \colon a \mapsto f(a)g(a)$$

(notice that $fg$ is *not* their composite). Pointwise operations are the usual addition and multiplication of functions in Calculus.

We claim that $\mathcal{F}(\mathbb{R})$ with these operations is a commutative ring. Verification of the axioms is left to the reader with the following hint: the zero element in $\mathcal{F}(\mathbb{R})$ is the constant function $z$ with value $0$ [that is, $z(a) = 0$ for all $a \in \mathbb{R}$] and the unit is the constant function $\varepsilon$ with $\varepsilon(a) = 1$ for all $a \in \mathbb{R}$. We now show that $\mathcal{F}(\mathbb{R})$ is not a domain. Define $f$ and $g$ as drawn

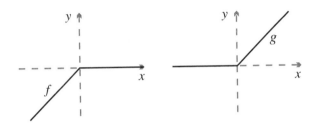

**Figure 2.1.** Zero divisors.

in Figure 2.1:

$$f(a) = \begin{cases} a & \text{if } a \leq 0 \\ 0 & \text{if } a \geq 0; \end{cases} \qquad g(a) = \begin{cases} 0 & \text{if } a \leq 0 \\ a & \text{if } a \geq 0. \end{cases}$$

Clearly, neither $f$ nor $g$ is zero (i.e., $f \neq z$ and $g \neq z$). On the other hand, for each $a \in \mathbb{R}$, $fg \colon a \mapsto f(a)g(a) = 0$, because at least one of the factors $f(a)$ or $g(a)$ is the number zero. Therefore, $fg = z$, and $\mathcal{F}(\mathbb{R})$ is not a domain.

(ii) All $C^\infty$-functions $f \colon \mathbb{R} \to \mathbb{R}$ (i.e., $f$ has an $n$th derivative $f^{(n)}$ for all $n \geq 0$) form a subring of $\mathcal{F}(\mathbb{R})$. The identity $\varepsilon$ is a constant function, hence is $C^\infty$, while the sum and product of $C^\infty$-functions are also $C^\infty$. This is proved with the *Leibniz formula*:[5]

$$(fg)^{(n)}(x) = \sum_{k=0}^{n} \binom{n}{k} f^{(k)}(x) g^{(n-k)}(x).$$

Hence, the $C^\infty$-functions form a commutative ring. ◄

As we saw in Propositions 2.2 and 2.5, some properties of ordinary arithmetic, that is, properties of the commutative ring $\mathbb{Z}$, hold in more generality. We now generalize some familiar definitions from $\mathbb{Z}$ to arbitrary commutative rings.

---

[5]It is easy to prove the Leibniz formula by induction on $n$, but it is not a special case of the Binomial Theorem.

**Definition.** Let $a$ and $b$ be elements of a commutative ring $R$. Then $a$ **divides** $b$ **in** $R$ (or $a$ is a **divisor** of $b$ or $b$ is a **multiple** of $a$), denoted by $a \mid b$, if there exists an element $c \in R$ with $b = ca$.

As an extreme example, if $0 \mid a$, then $a = 0 \cdot b$ for some $b \in R$. Since $0 \cdot b = 0$, however, we must have $a = 0$. Thus, $0 \mid a$ if and only if $a = 0$.

Notice that whether $a$ divides $b$ depends not only on the elements $a$ and $b$ but also on the ambient ring $R$. For example, 3 does divide 2 in $\mathbb{Q}$, for $2 = 3 \times \frac{2}{3}$ and $\frac{2}{3} \in \mathbb{Q}$; on the other hand, 3 does not divide 2 in $\mathbb{Z}$, because there is no *integer* $c$ with $3c = 2$.

**Definition.** An element $u$ in a commutative ring $R$ is called a **unit** if $u \mid 1$ in $R$, that is, if there exists $v \in R$ with $uv = 1$; the element $v$ is called the **inverse** of $u$ and $v$ is often denoted by $u^{-1}$.

Units are of interest because we can always divide by them: if $a \in R$ and $u$ is a unit in $R$ (so there is $v \in R$ with $uv = 1$), then

$$a = u(va)$$

is a factorization of $a$ in $R$, for $va \in R$; thus, it is reasonable to define the quotient $a/u$ as $va = u^{-1}a$. Whether an element $u \in R$ is a unit depends on the ambient ring $R$ (for being a unit means that $u \mid 1$ in $R$, and divisibility depends on $R$). For example, the number 2 is a unit in $\mathbb{Q}$, for $\frac{1}{2}$ lies in $\mathbb{Q}$ and $2 \times \frac{1}{2} = 1$, but 2 is not a unit in $\mathbb{Z}$, because there is no *integer* $v$ with $2v = 1$. In fact, the only units in $\mathbb{Z}$ are 1 and $-1$.

**Proposition 2.10.** *Let $R$ be a domain, and let $a, b \in R$ be nonzero. Then $a \mid b$ and $b \mid a$ if and only if $b = ua$ for some unit $u \in R$.*

**Proof.** If $b = ua$ and $a = vb$, then $b = ua = uvb$. Canceling, $1 = uv$.   ∎

There exist examples of commutative rings in which Proposition 2.10 is false, and so the hypothesis that $R$ be a domain is needed.

What are the units in $\mathbb{I}_m$?

**Proposition 2.11.** *If $a$ is an integer, then $[a]$ is a unit in $\mathbb{I}_m$ if and only if $a$ and $m$ are relatively prime. In fact, if $sa + tm = 1$, then $[a]^{-1} = [s]$.*

**Proof.** This follows from Proposition 1.52.   ∎

**Corollary 2.12.** *If $p$ is prime, then every nonzero $[a]$ in $\mathbb{I}_p$ is a unit.*

**Proof.** If $1 \le a < p$, then $(a, p) = 1$.   ∎

**Definition.** If $R$ is a nonzero commutative ring, then the **group of units** of $R$ is

$$U(R) = \{\text{all units in } R\}.$$

It is easy to check that $U(R)$ is a multiplicative group [this notation is consistent with that introduced for $U(\mathbb{I}_m)$ in Proposition 1.52]. It follows that a unit $u$ in $R$ has exactly one inverse in $R$, for each element in a group has a unique inverse.

There is an obvious difference between $\mathbb{Q}$ and $\mathbb{Z}$: every nonzero element of $\mathbb{Q}$ is a unit.

**Definition.** A ***field*** [6] $F$ is a commutative ring in which $1 \neq 0$ and every nonzero element $a$ is a unit; that is, there is $a^{-1} \in F$ with $a^{-1}a = 1$.

The first examples of fields are $\mathbb{Q}$, $\mathbb{R}$, and $\mathbb{C}$.

The definition of *field* can be restated in terms of the group of units; a commutative ring $R$ is a field if and only if $U(R) = R^{\times}$, the nonzero elements of $R$. To say this another way, $R$ is a field if and only if $R^{\times}$ is a multiplicative group [note that $U(R^{\times}) \neq \varnothing$ because we are assuming that $1 \neq 0$].

**Proposition 2.13.** *Every field $F$ is a domain.*

**Proof.** If $ab = ac$ and $a \neq 0$, then $b = a^{-1}(ab) = a^{-1}(ac) = c$.   •

The converse of this proposition is false, for $\mathbb{Z}$ is a domain that is not a field.

**Proposition 2.14.** *The commutative ring $\mathbb{I}_m$ is a field if and only if $m$ is prime.*

**Proof.** Corollary 2.12.   •

When $p$ is prime, we usually denote the field $\mathbb{I}_p$ by

$$\mathbb{F}_p.$$

In Exercise 2.8 on page 89, we will construct a field $\mathbb{F}_4$ with four elements. Given a prime $p$ and $n \geq 1$, we shall see, in Theorem 2.153 and Corollary 2.158, that there exist finite fields, unique up to isomorphism, having exactly $q = p^n$ elements; we will denote such fields by $\mathbb{F}_q$.

Every subring of a domain is itself a domain. Since fields are domains, it follows that every subring of a field is a domain. The converse is also true, and it is much more interesting: every domain is a subring of a field.

Given four elements $a$, $b$, $c$, and $d$ in a field $F$ with $b \neq 0$ and $d \neq 0$, assume that $ab^{-1} = cd^{-1}$. Multiply both sides by $bd$ to obtain $ad = bc$. In other words, were $ab^{-1}$ written as $a/b$, then we have just shown that $a/b = c/d$ implies $ad = bc$; that is, "cross multiplication" is valid. Conversely, if $ad = bc$ and both $b$ and $d$ are nonzero, then multiplication by $b^{-1}d^{-1}$ gives $ab^{-1} = cd^{-1}$, that is, $a/b = c/d$.

The proof of the next theorem is a straightforward generalization of the usual construction of the field of rational numbers $\mathbb{Q}$ from the domain of integers $\mathbb{Z}$.

**Theorem 2.15.** *If $R$ is a domain, then there is a field $F$ containing $R$ as a subring. Moreover, $F$ can be chosen so that, for each $f \in F$, there are $a$, $b \in R$ with $b \neq 0$ and $f = ab^{-1}$.*

---

[6]The derivation of the mathematical usage of the English term *field* (first used by Moore in 1893 in his article classifying the finite fields) as well as the German term *Körper* and the French term *corps* is probably similar to the derivation of the words *group* and *ring*: each word denotes a "realm" or a "collection of things." The word *domain* abbreviates the usual English translation *integral domain* of the German word *Integretätsbereich*, a collection of integers.

**Proof.** Define a relation $\equiv$ on $R \times R^\times$, where $R^\times$ is the set of all nonzero elements in $R$, by $(a, b) \equiv (c, d)$ if $ad = bc$. We claim that $\equiv$ is an equivalence relation. Verifications of reflexivity and symmetry are straightforward; here is the proof of transitivity. If $(a, b) \equiv (c, d)$ and $(c, d) \equiv (e, f)$, then $ad = bc$ and $cf = de$. But $ad = bc$ gives $adf = b(cf) = bde$. Canceling $d$, which is nonzero, gives $af = be$; that is, $(a, b) \equiv (e, f)$.

Denote the equivalence class of $(a, b)$ by $[a, b]$, define $F$ as the set of all equivalence classes, and equip $F$ with the following addition and multiplication (if we pretend that $[a, b]$ is the fraction $a/b$, then these are just the familiar formulas):

$$[a, b] + [c, d] = [ad + bc, bd] \qquad \text{and} \qquad [a, b][c, d] = [ac, bd].$$

First, since $b \neq 0$ and $d \neq 0$, we have $bd \neq 0$, because $R$ is a domain, and so the formulas make sense. Let us show that addition is well-defined. If $[a, b] = [a', b']$ (that is, $ab' = a'b$) and $[c, d] = [c', d']$ (that is, $cd' = c'd$), then we must show that $[ad + bc, bd] = [a'd' + b'c', b'd']$. But this is true:

$$(ad + bc)b'd' = ab'dd' + bb'cd' = a'bdd' + bb'c'd = (a'd' + b'c')bd.$$

A similar argument shows that multiplication is well-defined.

The verification that $F$ is a commutative ring is now routine: the zero element is $[0, 1]$, the unit is $[1, 1]$, and the additive inverse of $[a, b]$ is $[-a, b]$. It is easy to see that the family $R' = \{[a, 1] : a \in R\}$ is a subring of $F$, and we identify $a \in R$ with $[a, 1] \in R'$. To see that $F$ is a field, observe that if $[a, b] \neq [0, 1]$, then $a \neq 0$, and the inverse of $[a, b]$ is $[b, a]$.

Finally, if $b \neq 0$, then $[1, b] = [b, 1]^{-1}$, and so $[a, b] = [a, 1][b, 1]^{-1}$.  •

**Definition.** The field $F$ constructed from $R$ in Theorem 2.15 is called the *fraction field* of $R$; we denote it by $\mathrm{Frac}(R)$, and we denote $[a, b] \in \mathrm{Frac}(R)$ by $a/b$; in particular, the elements $[a, 1]$ of $R'$ are denoted by $a/1$ or, more simply, by $a$.

The fraction field of $\mathbb{Z}$ is $\mathbb{Q}$; that is, $\mathrm{Frac}(\mathbb{Z}) = \mathbb{Q}$.

**Definition.** A *subfield* of a field $K$ is a subring $k$ of $K$ that is also a field.

It is easy to see that a subset $k$ of a field $K$ is a subfield if and only if $k$ is a subring that is closed under inverses; that is, if $a \in k$ and $a \neq 0$, then $a^{-1} \in k$. It is also routine to see that any intersection of subfields of $K$ is itself a subfield of $K$ (note that the intersection is not equal to $\{0\}$ because $1$ lies in every subfield and all subfields have the same unit).

## Exercises

\* **2.1.** Prove that a ring $R$ has a unique $1$.

\* **2.2.** A *ring without unit* is an additive abelian group $R$ equipped with a (not necessarily commutative) associative multiplication that satisfies both distributive laws.

   (i) Prove that every additive abelian group $G$ is a ring without unit if we define $ab = 0$ for all $a, b \in G$.

(ii) Let $R$ be a ring without unit. As both $\mathbb{Z}$ and $R$ are additive abelian groups, so is their direct product $R^* = \mathbb{Z} \times R$. Define a multiplication on $R^*$ by

$$(m, r)(n, s) = (mn, ms + nr + rs),$$

where $ms = 0$ if $m = 0$, $ms$ is the sum of $s \in R$ with itself $m$ times if $m > 0$, and $ms$ is the sum of $-s$ with itself $|m|$ times if $m < 0$. Prove that $R^*$ is a ring [its unit is $(1, 0)$]. We say that $R^*$ arises from $R$ by **adjoining a unit**.

**2.3.** Let $R$ be a (not necessarily commutative) ring.

(i) If $(S_i)_{i \in I}$ is a family of subrings of $R$, prove that $\bigcap_{i \in I} S_i$ is also a subring of $R$.

(ii) If $X \subseteq R$ is a subset of $R$, define the **subring generated by** $X$, denoted by $\langle X \rangle$, to be the intersection of all the subrings of $R$ that contain $X$. Prove that $\langle X \rangle$ is the *smallest* subring containing $X$ in the following sense: if $S$ is a subring of $R$ and $X \subseteq S$, then $\langle X \rangle \subseteq S$.

**2.4.** (i) Prove that subtraction in $\mathbb{Z}$ is not an associative operation.

(ii) Give an example of a commutative ring $R$ in which subtraction is associative.

**2.5.** (i) If $R$ is a domain and $a \in R$ satisfies $a^2 = a$, prove that either $a = 0$ or $a = 1$.

(ii) Show that the commutative ring $\mathcal{F}(\mathbb{R})$ in Example 2.9 contains infinitely many elements $f$ with $f^2 = f$.

(iii) Find all the units in $\mathcal{F}(\mathbb{R})$.

$*$ **2.6.** (i) If $X$ is a set, prove that the Boolean group $\mathcal{B}(X)$ in Example 1.18 [whose elements are the subsets of $X$ and whose addition is $U + V = (U - V) \cup (V - U)$] is a commutative ring if we define multiplication to be intersection: $UV = U \cap V$. We call $\mathcal{B}(X)$ a **Boolean ring**.

**Hint.** You may use some standard facts of Set Theory: the distributive law: $U \cap (V \cup W) = (U \cap V) \cup (U \cap W)$; if $V'$ denotes the complement of $V$, then $U - V = U \cap V'$; the **De Morgan Law**: $(U \cap V)' = U' \cup V'$.

(ii) Prove that $\mathcal{B}(X)$ contains exactly one unit.

(iii) If $Y$ is a proper subset of $X$ (that is, $Y \subsetneq X$), show that the unit in $\mathcal{B}(Y)$ is distinct from the unit in $\mathcal{B}(X)$. Conclude that $\mathcal{B}(Y)$ is *not* a subring of $\mathcal{B}(X)$ even though, for all $A, B \subseteq Y$, addition and multiplication $A + B$ and $AB$ have the same meaning in $\mathcal{B}(Y)$ and $\mathcal{B}(X)$.

$*$ **2.7.** Generalize the construction of $\mathcal{F}(\mathbb{R})$: if $R$ is a nonzero commutative ring, let $\mathcal{F}(R)$ be the set of all functions from $R$ to $R$ with pointwise addition $f + g : r \mapsto f(r) + g(r)$ and pointwise multiplication $fg : r \mapsto f(r)g(r)$ for $r \in R$.

(i) Show that $\mathcal{F}(R)$ is a commutative ring.

(ii) Show that $\mathcal{F}(R)$ is not a domain.

(iii) Show that $\mathcal{F}(\mathbb{F}_2)$ has exactly four elements, and that $f + f = 0$ for every $f \in \mathcal{F}(\mathbb{F}_2)$.

$*$ **2.8. (Dean)** Define $\mathbb{F}_4$ to be all $2 \times 2$ matrices of the form

$$\begin{bmatrix} a & b \\ b & a + b \end{bmatrix},$$

where $a, b \in \mathbb{F}_2$.

(i) Prove that $\mathbb{F}_4$ is a commutative ring under the usual matrix operations of addition and multiplication.

(ii) Prove that $\mathbb{F}_4$ is a field with exactly four elements.

∗ **2.9.** Prove that the commutative ring $\mathbb{C}$ is a field, and conclude that the ring of Gaussian integers is a domain.

**2.10.** Prove that the only subring of $\mathbb{Z}$ is $\mathbb{Z}$ itself.

**2.11.**   (i) Prove that $R = \{a + b\sqrt{2} : a, b \in \mathbb{Z}\}$ is a domain.

(ii) Prove that $R = \{\frac{1}{2}(a + b\sqrt{2}) : a, b \in \mathbb{Z}\}$ is not a domain.

(iii) Prove that $R = \{a + b\alpha : a, b \in \mathbb{Z}\}$ is a domain, where $\alpha = \frac{1}{2}(1 + \sqrt{-19})$.
**Hint.** Use the fact that $\alpha$ is a root of $x^2 - x + 5$.

**2.12.** Show that $F = \{a + b\sqrt{2} : a, b \in \mathbb{Q}\}$ is a field.

**2.13.**   (i) Show that $F = \{a + bi : a, b \in \mathbb{Q}\}$ is a field.

(ii) Show that $F$ is the fraction field of the Gaussian integers.

**2.14.** Let $R$ be a (not necessarily commutative) ring.

(i) Define the ***circle operation*** $R \times R \to R$ by

$$a \circ b = a + b - ab.$$

Prove that the circle operation is associative and that $0 \circ a = a$ for all $a \in R$.

(ii) Prove that $R$ is a field if and only if $\{r \in R : r \neq 1\}$ is an abelian group under the circle operation.
**Hint.** If $a \neq 0$, then $a + 1 \neq 1$.

**2.15.** Find the multiplicative inverses of the nonzero elements of $\mathbb{I}_{11}$.

**2.16.** Prove that $\mathbb{Q}$ has no proper subfields.

**2.17.** Prove that every domain $R$ with a finite number of elements must be a field. (Using Proposition 2.8, this gives a new proof of sufficiency in Proposition 2.14.)

**Hint.** If $R^\times$ denotes the set of nonzero elements of $R$ and $r \in R^\times$, apply the Pigeonhole Principle after proving that multiplication by $r$ is an injection $R^\times \to R^\times$.

∗ **2.18.**   (i) For any field $k$, prove that the ***stochastic group*** $\Sigma(2, k)$, the set of all nonsingular $2 \times 2$ matrices with entries in $k$ whose column sums are 1, is a group under matrix multiplication.

(ii) Define the ***affine group*** $\mathrm{Aff}(1, k)$ to be the set of all $f : k \to k$ of the form $f(x) = ax + b$, where $a, b \in k$ and $a \neq 0$. Prove that $\Sigma(2, k) \cong \mathrm{Aff}(1, k)$ (see Exercise 1.53 on page 45).

(iii) If $k$ is a finite field with $q$ elements, prove that $|\Sigma(2, k)| = q(q - 1)$.

(iv) Prove that $\Sigma(2, \mathbb{F}_3) \cong S_3$.

∗ **2.19.** If $k$ is a field and $G$ is a group of order $n$, prove that $G$ is isomorphic to a subgroup of $\mathrm{GL}(n, k)$.

**Hint.** See Exercise 1.107 on page 76.

## Section 2.2. Polynomials

Even though the reader is familiar with polynomials, we now introduce them carefully. The key observation is that one should pay attention to where the coefficients of polynomials live.

**Definition.** If $R$ is a commutative ring, then a ***formal power series*** over $R$ is a sequence

$$\sigma = (s_0, s_1, s_2, \ldots, s_i, \ldots);$$

the entries $s_i \in R$, for all $i \geq 0$, are called the ***coefficients*** of $\sigma$.

To determine when two formal power series are equal, let us recognize that a sequence $\sigma$ is really a function $\sigma \colon \mathbb{N} \to R$, where $\mathbb{N}$ is the set of natural numbers, with $\sigma(i) = s_i$ for all $i \geq 0$. Thus, if $\tau = (t_0, t_1, t_2, \ldots, t_i, \ldots)$ is a formal power series over $R$, then $\sigma = \tau$ if and only if $\sigma(i) = \tau(i)$ for all $i \geq 0$; that is, $\sigma = \tau$ if and only if $s_i = t_i$ for all $i \geq 0$.

**Definition.** A ***polynomial*** over a commutative ring $R$ is a formal power series $\sigma = (s_0, s_1, \ldots, s_i, \ldots)$ over $R$ for which there exists some integer $n \geq 0$ with $s_i = 0$ for all $i > n$; that is,

$$\sigma = (s_0, s_1, \ldots, s_n, 0, 0, \ldots).$$

A polynomial has only finitely many nonzero coefficients. The ***zero polynomial***, denoted by $\sigma = 0$, is the sequence $\sigma = (0, 0, 0, \ldots)$.

**Definition.** If $\sigma = (s_0, s_1, \ldots, s_n, 0, 0, \ldots)$ is a nonzero polynomial, then there is $n \geq 0$ with $s_n \neq 0$ and $s_i = 0$ for all $i > n$. We call $s_n$ the ***leading coefficient*** of $\sigma$, we call $n$ the ***degree*** of $\sigma$, and we denote the degree by $n = \deg(\sigma)$.

The zero polynomial $0$ does not have a degree because it has no nonzero coefficients.[7]

**Notation.** If $R$ is a commutative ring, then $R[[x]]$ denotes the set of all formal power series over $R$, and $R[x] \subseteq R[[x]]$ denotes the set of all polynomials over $R$.

**Proposition 2.16.** *If $R$ is a commutative ring, then $R[[x]]$ is a commutative ring that contains $R[x]$ and $R$ as subrings.*

**Proof.** Let $\sigma = (s_0, s_1, \ldots)$ and $\tau = (t_0, t_1, \ldots)$ be formal power series over $R$. Define addition and multiplication by

$$\sigma + \tau = (s_0 + t_0, s_1 + t_1, \ldots, s_n + t_n, \ldots)$$

and

$$\sigma\tau = (c_0, c_1, c_2, \ldots),$$

where $c_k = \sum_{i+j=k} s_i t_j = \sum_{i=0}^{k} s_i t_{k-i}$. Verification of the axioms in the definition of commutative ring is routine, as is checking that $R[x]$ is a subring of $R[[x]]$. We identify $R$ with the subring $R' = \{(r, 0, 0, \ldots) : r \in R\} \subseteq R[x]$. $\bullet$

---

[7]Some authors define $\deg(0) = -\infty$, where $-\infty < n$ for every integer $n$ (this is sometimes convenient). We choose not to assign a degree to the zero polynomial $0$ because it often must be treated differently.

**Lemma 2.17.** *Let $R$ be a commutative ring and let $\sigma, \tau \in R[x]$ be nonzero polynomials.*

(i) *Either $\sigma\tau = 0$ or $\deg(\sigma\tau) \leq \deg(\sigma) + \deg(\tau)$.*

(ii) *If $R$ is a domain, then $\sigma\tau \neq 0$ and*

$$\deg(\sigma\tau) = \deg(\sigma) + \deg(\tau).$$

(iii) *If $\sigma, \tau \neq 0$ and $\tau \mid \sigma$ in $R[x]$, then $\deg(\tau) \leq \deg(\sigma)$.*

(iv) *If $R$ is a domain, then $R[x]$ is a domain.*

**Proof.** Let $\sigma = (s_0, s_1, \dots)$ and $\tau = (t_0, t_1, \dots)$ have degrees $m$ and $n$, respectively.

(i) If $k > m + n$, then each term in $\sum_i s_i t_{k-i}$ is 0 (for either $s_i = 0$ or $t_{k-i} = 0$).

(ii) Each term in $\sum_i s_i t_{m+n-i}$ is 0, with the possible exception of $s_m t_n$. Since $R$ is a domain, $s_m \neq 0$ and $t_n \neq 0$ imply $s_m t_n \neq 0$.

(iii) Immediate from part (ii).

(iv) This follows from part (ii), because the product of two nonzero polynomials is now nonzero. •

Here is the link between this discussion and the usual notation.

**Definition.** The ***indeterminate*** $x \in R[x]$ is

$$x = (0, 1, 0, 0, \dots).$$

One reason for our insisting that rings have units is that it enables us to define indeterminates.

**Lemma 2.18.**

(i) *If $\sigma = (s_0, s_1, \dots)$, then*

$$x\sigma = (0, s_0, s_1, \dots);$$

*that is, multiplying by $x$ shifts each coefficient one step to the right.*

(ii) *If $n \geq 0$, then $x^n$ is the polynomial having 0 everywhere except for 1 in the $n$th coordinate.*

(iii) *If $r \in R$, then*

$$(r, 0, 0, \dots)(s_0, s_1, \dots, s_j, \dots) = (rs_0, rs_1, \dots, rs_j, \dots).$$

**Proof.** Each is a routine computation using the definition of polynomial multiplication. •

If we identify $(r, 0, 0, \dots)$ with $r$, then Lemma 2.18(iii) reads

$$r(s_0, s_1, \dots, s_i, \dots) = (rs_0, rs_1, \dots, rs_i, \dots).$$

We can now recapture the usual notation.

**Proposition 2.19.** *If $\sigma = (s_0, s_1, \dots, s_n, 0, 0, \dots) \in R[x]$ has degree $n$, then*

$$\sigma = s_0 + s_1 x + s_2 x^2 + \cdots + s_n x^n.$$

**Proof.**

$$\sigma = (s_0, s_1, \ldots, s_n, 0, 0, \ldots)$$
$$= (s_0, 0, 0, \ldots) + (0, s_1, 0, \ldots) + \cdots + (0, 0, \ldots, s_n, 0, \ldots)$$
$$= s_0(1, 0, 0, \ldots) + s_1(0, 1, 0, \ldots) + \cdots + s_n(0, 0, \ldots, 1, 0, \ldots)$$
$$= s_0 + s_1 x + s_2 x^2 + \cdots + s_n x^n. \quad \bullet$$

We shall use this familiar (and standard) notation from now on. As is customary, we shall write

$$f(x) = s_0 + s_1 x + s_2 x^2 + \cdots + s_n x^n$$

instead of $\sigma = (s_0, s_1, \ldots, s_n, 0, 0, \ldots)$. We will denote formal power series by $s_0 + s_1 x + s_2 x^2 + \cdots$ or by $\sum_{n=0}^{\infty} s_n x^n$.

Here is some standard vocabulary associated with polynomials. If $f(x) = s_0 + s_1 x + s_2 x^2 + \cdots + s_n x^n$, where $s_n \neq 0$, then $s_0$ is called its **constant term** and, as we have already said, $s_n$ is called its *leading coefficient*. If its leading coefficient $s_n = 1$, then $f(x)$ is called **monic**. Every polynomial other than the zero polynomial 0 (having all coefficients 0) has a degree. A **constant polynomial** is either the zero polynomial or a polynomial of degree 0. Polynomials of degree 1, namely, $a + bx$ with $b \neq 0$, are called **linear**, polynomials of degree 2 are **quadratic**,[8] degree 3's are **cubic**, then **quartics**, **quintics**, and so on.

**Corollary 2.20.** *Formal power series (hence polynomials)* $s_0 + s_1 x + s_2 x^2 + \cdots$ *and* $t_0 + t_1 x + t_2 x^2 + \cdots$ *in* $R[[x]]$ *are equal if and only if* $s_i = t_i$ *for all* $i$.

**Proof.** This is merely a restatement of the definition of equality of sequences, rephrased in the usual notation for formal power series. $\quad \bullet$

We can now describe the usual role of $x$ in $f(x)$ as a variable. If $R$ is a commutative ring, each polynomial $f(x) = s_0 + s_1 x + s_2 x^2 + \cdots + s_n x^n \in R[x]$ defines a **polynomial function** $f \colon R \to R$ by evaluation: if $a \in R$, define $f(a) = s_0 + s_1 a + s_2 a^2 + \cdots + s_n a^n \in R$. The reader should realize that polynomials and polynomial functions are distinct objects. For example, if $R$ is a finite ring (e.g., $R = \mathbb{I}_m$), then there are only finitely many functions from $R$ to itself, and so there are only finitely many polynomial functions. On the other hand, there are infinitely many polynomials: for example, all the powers $1, x, x^2, \ldots, x^n, \ldots$ are distinct, by Corollary 2.20.

**Definition.** Let $k$ be a field. The fraction field $\mathrm{Frac}(k[x])$ of $k[x]$, denoted by $k(x)$, is called the **field of rational functions** over $k$.

**Proposition 2.21.** *If* $k$ *is a field, then the elements of* $k(x)$ *have the form* $f(x)/g(x)$, *where* $f(x), g(x) \in k[x]$ *and* $g(x) \neq 0$.

**Proof.** Theorem 2.15. $\quad \bullet$

---

[8]Quadratic polynomials are so called because the particular quadratic $x^2$ gives the area of a square (*quadratic* comes from the Latin word meaning "four," which is to remind us of the four-sided figure); similarly, cubic polynomials are so called because $x^3$ gives the volume of a cube. Linear polynomials are so called because the graph of a linear polynomial in $\mathbb{R}[x]$ is a line.

**Proposition 2.22.** *If $p$ is prime, then the field of rational functions $\mathbb{F}_p(x)$ is an infinite field containing $\mathbb{F}_p$ as a subfield.*

**Proof.** By Lemma 2.17(iv), $\mathbb{F}_p[x]$ is an infinite domain, because the powers $x^n$, for $n \in \mathbb{N}$, are distinct. Thus, its fraction field, $\mathbb{F}_p(x)$, is an infinite field containing $\mathbb{F}_p[x]$ as a subring. But $\mathbb{F}_p[x]$ contains $\mathbb{F}_p$ as a subring, by Proposition 2.16.    •

In spite of the difference between polynomials and polynomial functions (we shall see, in Corollary 2.43, that these objects coincide when the coefficient ring $R$ is an infinite field), $R[x]$ is often called the ring of all *polynomials over $R$ in one variable*. If we write $A = R[x]$, then the polynomial ring $A[y]$ is called the ring of all *polynomials over $R$ in two variables $x$ and $y$*, and it is denoted by $R[x, y]$. For example, the quadratic polynomial $ax^2 + bxy + cy^2 + dx + ey + f$ can be written $cy^2 + (bx + e)y + (ax^2 + dx + f)$, a polynomial in $y$ with coefficients in $R[x]$. By induction, we can form the commutative ring $R[x_1, x_2, \ldots, x_n]$ of all **polynomials in $n$ variables** over $R$,

$$R[x_1, x_2, \ldots, x_{n+1}] = \big( R[x_1, x_2, \ldots, x_n] \big)[x_{n+1}].$$

Lemma 2.17(iv) can now be generalized, by induction on $n$, to say that if $R$ is a domain, then so is $R[x_1, x_2, \ldots, x_n]$. Moreover, when $k$ is a field, we can describe $\mathrm{Frac}(k[x_1, x_2, \ldots, x_n])$ as all **rational functions in $n$ variables**

$$k(x_1, x_2, \ldots, x_n);$$

its elements have the form $f(x_1, x_2, \ldots, x_n)/g(x_1, x_2, \ldots, x_n)$, where $f$ and $g$ lie in $k[x_1, x_2, \ldots, x_n]$.

## Exercises

**2.20.** Prove that if $R$ is a commutative ring, then $R[x]$ is never a field.

**Hint.** If $x^{-1}$ exists, what is its degree?

* **2.21.**   (i) Let $R$ be a domain. Prove that if a polynomial in $R[x]$ is a unit, then it is a nonzero constant (the converse is true if $R$ is a field).

      **Hint.** Compute degrees.

  (ii) Show that $(2x + 1)^2 = 1$ in $\mathbb{I}_4[x]$. Conclude that $2x + 1$ is a unit in $\mathbb{I}_4[x]$, and that the hypothesis in part (i) that $R$ be a domain is necessary.

* **2.22.** Show that the polynomial function defined by the polynomial $x^p - x \in \mathbb{F}_p[x]$ is identically zero.

**Hint.** Use Fermat's Theorem.

* **2.23.** If $R$ is a commutative ring and $f(x) = \sum_{i=0}^n s_i x^i \in R[x]$ has degree $n \geq 1$, define its **derivative** $f'(x) \in R[x]$ by

$$f'(x) = s_1 + 2s_2 x + 3s_3 x^2 + \cdots + n s_n x^{n-1};$$

if $f(x)$ is a constant polynomial, define its derivative to be the zero polynomial.

Prove that the usual rules of Calculus hold:

$$(f + g)' = f' + g';$$
$$(rf)' = r(f') \quad \text{if } r \in R;$$
$$(fg)' = fg' + f'g;$$
$$(f^n)' = nf^{n-1}f' \quad \text{for all } n \geq 1.$$

\* **2.24.** Let $R$ be a commutative ring and let $f(x) \in R[x]$.

   (i) Prove that if $(x - a)^2 \mid f(x)$, then $(x - a) \mid f'(x)$ in $R[x]$.

   (ii) Prove that if $(x - a) \mid f(x)$ and $(x - a) \mid f'(x)$, then $(x - a)^2 \mid f(x)$.

**2.25.** (i) Prove that the derivative $D\colon R[x] \to R[x]$, given by $D\colon f \mapsto f'$, is a homomorphism of additive abelian groups.

   (ii) If $f(x) = ax^{2p} + bx^p + c \in \mathbb{F}_p[x]$, prove that $f'(x) = 0$.

   (iii) Prove that a polynomial $f(x) \in \mathbb{F}_p[x]$ has $f'(x) = 0$ if and only if there is a polynomial $g(x) = \sum a_n x^n$ with $f(x) = g(x^p)$; that is, $f(x) = \sum a_n x^{np} \in \mathbb{F}_p[x^p]$.

   (iv) What is $\ker D$, where $D\colon \mathbb{F}_p[x] \to \mathbb{F}_p[x]$? What is $\ker D$, where $D\colon \mathbb{Q}[x] \to \mathbb{Q}[x]$?

**2.26.** If $f(x) = a_0 + a_1 x + \cdots + a_n x^n \in \mathbb{Q}[x]$, define

$$\int f = a_0 x + \frac{1}{2} a_1 x^2 + \cdots + \frac{1}{n+1} a_n x^{n+1} \in \mathbb{Q}[x].$$

   (i) Prove that $\int\colon \mathbb{Q}[x] \to \mathbb{Q}[x]$ is a homomorphism of additive abelian groups.

   (ii) Prove that $D\int = 1_{\mathbb{Q}[x]}$ but that $\int D \neq 1_{\mathbb{Q}[x]}$.

\* **2.27.** Prove that if $R$ is a domain, then $R[[x]]$ is a domain.

**Hint.** If $\sigma = (s_0, s_1, \dots) \in R[[x]]$ is nonzero, define the **order** of $\sigma$, denoted by $\mathrm{ord}(\sigma)$, to be the smallest $n \geq 0$ for which $s_n \neq 0$. If $R$ is a domain and $\sigma, \tau \in R[[x]]$ are nonzero, prove that $\sigma\tau \neq 0$ and $\mathrm{ord}(\sigma\tau) = \mathrm{ord}(\sigma) + \mathrm{ord}(\tau)$.

\* **2.28.** (i) If $R$ is a domain and $\sigma = \sum_{n=0}^{\infty} x^n \in R[[x]]$, prove that $\sigma = 1/(1 - x)$ in $R[[x]]$; that is, $(1 - x)\sigma = 1$.

   (ii) Let $k$ be a field. Prove that a formal power series $\sigma \in k[[x]]$ is a unit if and only if its constant term is nonzero; that is, $\mathrm{ord}(\sigma) = 0$.

   (iii) Prove that if $\sigma \in k[[x]]$ and $\mathrm{ord}(\sigma) = n$, then $\sigma = x^n u$, where $u$ is a unit in $k[[x]]$.

**2.29.** (i) Let $R$ be a commutative ring. Call a sequence $(f_n(x))_{n \geq 0} = (\sum_i a_{ni} x^i)_{n \geq 0}$ of formal power series in $R[[x]]$ **summable** if, for each $i$, there are only finitely many nonzero $a_{ni}$. If $g(x) = \sum_i b_i x^i \in R[[x]]$ and $(f_n(x))_{n \geq 0}$ is summable, prove that $\sum_i (\sum_n a_{ni}) x^i$ is a formal power series in $R[[x]]$.

   (ii) If $h(x) = \sum_i c_i x^i \in R[[x]]$ and $c_0 = 0$, prove that $(h^n(x))_{n \geq 0}$ is summable. Conclude that the **composite**

$$(g \circ h)(x) = b_0 + b_1 h + b_2 h^2 + \cdots$$

   is a power series.

   (iii) Define $\log(1 + z) = \sum_{i \geq 1} (-1)^i z^i / i \in \mathbb{C}[[x]]$ and $\exp(z) = \sum_n z^n / n!$. Prove that the composite $\exp \circ \log = 1$.

## Section 2.3. Homomorphisms

Just as there are homomorphisms of groups, there are homomorphisms of rings.

**Definition.** If $A$ and $R$ are (not necessarily commutative) rings, a (**ring**) **homomorphism** is a function $f\colon A \to R$ such that

(i) $f(1) = 1$,

(ii) $f(a + a') = f(a) + f(a')$ for all $a$, $a' \in A$,

(iii) $f(aa') = f(a)f(a')$ for all $a$, $a' \in A$.

A ring homomorphism that is also a bijection is called an **isomorphism**. Rings $A$ and $R$ are called **isomorphic**, denoted by $A \cong R$, if there is an isomorphism $f\colon A \to R$.

We continue to focus on commutative rings.

**Example 2.23.**

(i) Let $R$ be a domain and let $F = \mathrm{Frac}(R)$ denote its fraction field. In Theorem 2.15 we said that $R$ is a subring of $F$, but that is not the truth; $R$ is not even a subset of $F$. We did find a subring $R'$ of $F$, however, that has a very strong resemblance to $R$, namely, $R' = \{[a, 1] : a \in R\} \subseteq F$. The function $f\colon R \to R'$, given by $f(a) = [a, 1] = a/1$, is an isomorphism.

(ii) In the proof of Proposition 2.16, we "identified" an element $r$ in a commutative ring $R$ with the constant polynomial $(r, 0, 0, \dots)$. We said that $R$ is a subring of $R[x]$, but that is not the truth. The subset $R' = \{(r, 0, 0, \dots) : r \in R\}$ is a subring of $R[x]$, and the function $f\colon R \to R'$, defined by $f(r) = (r, 0, 0, \dots)$, is an isomorphism.

(iii) If $S$ is a subring of a commutative ring $R$, then the inclusion $i\colon S \to R$ is a ring homomorphism because we have insisted that the identity 1 of $R$ lies in $S$. [See Exercise 2.6(iii) on page 89.]  ◄

**Example 2.24.**

(i) Complex conjugation $z = a + ib \mapsto \bar{z} = a - ib$ is a homomorphism $\mathbb{C} \to \mathbb{C}$, because $\bar{1} = 1, \overline{z + w} = \bar{z} + \bar{w}$, and $\overline{zw} = \bar{z}\,\bar{w}$; it is a bijection because $\bar{\bar{z}} = z$ (so that it is its own inverse), and so it is an isomorphism.

(ii) Here is an example of a homomorphism of rings that is not an isomorphism. Choose $m \geq 2$ and define $f\colon \mathbb{Z} \to \mathbb{I}_m$ by $f(n) = [n]$. Notice that $f$ is surjective (but not injective). More generally, if $R$ is a commutative ring with its unit denoted by $\varepsilon$, then the function $\chi\colon \mathbb{Z} \to R$, defined by $\chi(n) = n\varepsilon$, is a ring homomorphism.  ◄

The next theorem is of fundamental importance, and so we give full details in its proof. In language to be introduced later, it says that the polynomial ring $R[x_1, \dots, x_n]$ is the *free commutative R-algebra* generated by the indeterminates.

**Theorem 2.25.** *Let $R$ and $S$ be commutative rings, and let $\varphi\colon R \to S$ be a ring homomorphism. If $s_1,\ldots,s_n \in S$, then there exists a unique ring homomorphism*

$$\Phi\colon R[x_1,\ldots,x_n] \to S$$

*with $\Phi(x_i) = s_i$ for all $i$ and $\Phi(r) = \varphi(r)$ for all $r \in R$.*

**Proof.** The proof is by induction on $n \geq 1$. If $n = 1$, denote $x_1$ by $x$ and $s_1$ by $s$. Define $\Phi\colon R[x] \to S$ as follows: if $f(x) = \sum_i r_i x^i$, then

$$\Phi\colon r_0 + r_1 x + \cdots + r_n x^n \mapsto \varphi(r_0) + \varphi(r_1)s + \cdots + \varphi(r_n)s^n = \Phi(f)$$

($\Phi$ is well-defined because of Corollary 2.20, uniqueness of coefficients.) This formula shows that $\Phi(x) = s$ and $\Phi(r) = \varphi(r)$ for all $r \in R$.

Let us prove that $\Phi$ is a homomorphism. First, $\Phi(1) = \varphi(1) = 1$, because $\varphi$ is a homomorphism. Second, if $g(x) = a_0 + a_1 x + \cdots + a_m x^m$, then

$$\Phi(f + g) = \Phi\Big(\sum_i (r_i + a_i)x^i\Big) = \sum_i \varphi(r_i + a_i)s^i$$

$$= \sum_i (\varphi(r_i) + \varphi(a_i))s^i = \sum_i \varphi(r_i)s^i + \sum_i \varphi(a_i)s^i$$

$$= \Phi(f) + \Phi(g).$$

Third, let $f(x)g(x) = \sum_k c_k x^k$, where $c_k = \sum_{i+j=k} r_i a_j$. Then

$$\Phi(fg) = \Phi\Big(\sum_k c_k x^k\Big) = \sum_k \varphi(c_k)s^k$$

$$= \sum_k \varphi\Big(\sum_{i+j=k} r_i a_j\Big)s^k = \sum_k \Big(\sum_{i+j=k} \varphi(r_i)\varphi(a_j)\Big)s^k.$$

On the other hand,

$$\Phi(f)\Phi(g) = \Big(\sum_i \varphi(r_i)s^i\Big)\Big(\sum_j \varphi(a_j)s^j\Big) = \sum_k \Big(\sum_{i+j=k} \varphi(r_i)\varphi(a_j)\Big)s^k.$$

Uniqueness of $\Phi$ is obvious: if $\theta\colon R[x] \to S$ is a homomorphism with $\theta(x) = s$ and $\theta(r) = \varphi(r)$ for all $r \in R$, then $\theta(r_0 + r_1 x + \cdots + r_d x^d) = \varphi(r_0) + \varphi(r_1)s + \cdots + \varphi(r_d)s^d$. We have completed the proof of the base step. For the inductive step, define $A = R[x_1,\ldots,x_n]$; the inductive hypothesis gives a homomorphism $\psi\colon A \to S$ with $\psi(x_i) = s_i$ for all $i \leq n$ and $\psi(r) = \varphi(r)$ for all $r \in R$. The base step gives a homomorphism $\Psi\colon A[x_{n+1}] \to S$ with $\Psi(x_{n+1}) = s_{n+1}$ and $\Psi(a) = \psi(a)$ for all $a \in A$. The result follows because $R[x_1,\ldots,x_{n+1}] = A[x_{n+1}]$, $\Psi(x_i) = \psi(x_i) = s_i$ for all $i \leq n$, $\Psi(x_{n+1}) = \psi(x_{n+1}) = s_{n+1}$, and $\Psi(r) = \psi(r) = \varphi(r)$ for all $r \in R$. $\bullet$

**Definition.** If $R$ is a commutative ring and $a \in R$, then **evaluation at** $a$ is the function $e_a\colon R[x] \to R$, defined by $e_a(f(x)) = f(a)$; that is, $e_a(\sum_i r_i x^i) = \sum_i r_i a^i$.

**Corollary 2.26.** *If $R$ is a commutative ring, then evaluation $e_a\colon R[x] \to R$ is a ring homomorphism for every $a \in R$.*

**Proof.** If we set $R = S$, $\varphi = 1_R$, and $\Phi(x) = a$ in Theorem 2.25, then $\Phi = e_a$. $\bullet$

For example, if $R$ is a commutative ring and $a \in R$, then $f(x) = q(x)g(x) + r(x)$ in $R[x]$ implies $f(a) = q(a)g(a) + r(a)$ in $R$.

**Corollary 2.27.** *If $R$ and $S$ are commutative rings and $\varphi \colon R \to S$ is a ring homomorphism, then there is a ring homomorphism $\varphi^* \colon R[x] \to S[x]$ given by*

$$\varphi^* \colon r_0 + r_1 x + r_2 x^2 + \cdots \mapsto \varphi(r_0) + \varphi(r_1)x + \varphi(r_2)x^2 + \cdots .$$

*Moreover, $\varphi^*$ is an isomorphism if $\varphi$ is.*

**Proof.** That $\varphi^*$ is a ring homomorphism is a special case of Theorem 2.25. If $\varphi$ is an isomorphism, then $(\varphi^{-1})^*$ is the inverse of $\varphi^*$.   •

For example, $r \colon \mathbb{Z} \to \mathbb{F}_p$, reduction mod $p$, gives the ring homomorphism $r^* \colon \mathbb{Z}[x] \to \mathbb{F}_p[x]$ which reduces all coefficients mod $p$.

Certain properties of a ring homomorphism $f \colon A \to R$ follow from its being a homomorphism between the additive groups $A$ and $R$. For example, $f(0) = 0$, $f(-a) = -f(a)$, and $f(na) = nf(a)$ for all $n \in \mathbb{Z}$.

**Proposition 2.28.** *Let $f \colon A \to R$ be a ring homomorphism.*

(i) $f(a^n) = f(a)^n$ *for all $n \geq 0$ for all $a \in A$.*

(ii) *If $a \in R$ is a unit, then $f(a)$ is a unit and $f(a^{-1}) = f(a)^{-1}$, and so $f(U(A)) \subseteq U(R)$, where $U(A)$ is the group of units of $A$. Moreover, if $f$ is an isomorphism, then $U(A) \cong U(R)$.*

**Proof.**

(i) Induction on $n \geq 0$.

(ii) If $ab = 1$, then $1 = f(ab) = f(a)f(b)$.   •

**Definition.** If $f \colon A \to R$ is a ring homomorphism, then its **kernel** is

$$\ker f = \{a \in A \text{ with } f(a) = 0\}$$

and its **image** is

$$\operatorname{im} f = \{r \in R : r = f(a) \text{ for some } a \in R\}.$$

Notice that if we forget their multiplications, then the rings $A$ and $R$ are additive abelian groups and these definitions coincide with the group-theoretic ones.

Let $k$ be a commutative ring, let $a \in k$, and let $e_a \colon k[x] \to k$ be the evaluation homomorphism $f(x) \mapsto f(a)$. Now $e_a$ is always surjective, for if $b \in k$, then $b = e_a(f)$, where $f(x) = x - a + b$. By definition, $\ker e_a$ consists of all those polynomials $g(x)$ for which $g(a) = 0$.

The kernel of a group homomorphism is not merely a subgroup; it is a *normal* subgroup; that is, it is also closed under conjugation by any element in the ambient group. Similarly, if $R$ is not the zero ring, the kernel of a ring homomorphism $f \colon A \to R$ is almost a subring [$\ker f$ is not a subring because it never contains 1: $f(1) = 1 \neq 0$]; we shall see that $\ker f$ is closed under multiplication by any element in the ambient ring.

**Definition.** An *ideal* in a commutative ring $R$ is a subset $I$ of $R$ such that

(i) $0 \in I$,

(ii) if $a, b \in I$, then $a + b \in I$, [9]

(iii) if $a \in I$ and $r \in R$, then $ra \in I$.

The ring $R$ itself and $(0)$, the subset consisting of $0$ alone, are always ideals in a commutative ring $R$. An ideal $I \neq R$ is called a **proper ideal**.

**Proposition 2.29.** *If $f \colon A \to R$ is a ring homomorphism, then $\ker f$ is an ideal in $A$ and $\operatorname{im} f$ is a subring of $R$. Moreover, if $A$ and $R$ are not zero rings, then $\ker f$ is a proper ideal.*

**Proof.** $\ker f$ is an additive subgroup of $A$; moreover, if $u \in \ker f$ and $a \in A$, then $f(au) = f(a)f(u) = f(a) \cdot 0 = 0$. Hence, $\ker f$ is an ideal. If $R$ is not the zero ring, then $1 \neq 0$; hence, $\ker f$ is a proper ideal in $A$ (the identity $1 \notin \ker f$ because $f(1) = 1 \neq 0$). It is routine to check that $\operatorname{im} f$ is a subring of $R$. •

**Proposition 2.30.** *A ring homomorphism $f \colon A \to R$ is an injection if and only if $\ker f = (0)$.*

**Proof.** This follows from the corresponding result for group homomorphisms, for $f$ is a homomorphism from the additive group of $A$ to the additive group of $R$. •

**Example 2.31.**

(i) If an ideal $I$ in a commutative ring $R$ contains $1$, then $I = R$, for now $I$ contains $r = r1$ for every $r \in R$. Indeed, if $I$ contains a unit $u$, then $I = R$, for then $I$ contains $u^{-1}u = 1$.

(ii) It follows from (i) that if $R$ is a field, then the only ideals $I$ in $R$ are $(0)$ and $R$ itself: if $I \neq (0)$, it contains some nonzero element, and every nonzero element in a field is a unit.

Conversely, assume that $R$ is a nonzero commutative ring whose only ideals are $R$ itself and $(0)$. If $a \in R$ and $a \neq 0$, then $(a) = \{ra : r \in R\}$ is a nonzero ideal, and so $(a) = R$; hence, $1 \in R = (a)$. Thus, there is $r \in R$ with $1 = ra$; that is, $a$ has an inverse in $R$, and so $R$ is a field. ◀

**Corollary 2.32.** *If $k$ is a field and $f \colon k \to R$ is a ring homomorphism, where $R$ is not the zero ring, then $f$ is an injection.*

**Proof.** The only proper ideal in $k$ is $(0)$, by Example 2.31(ii), so $\ker f = (0)$. •

**Definition.** If $b_1, b_2, \ldots, b_n$ lie in $R$, then the set of all linear combinations

$$I = \{r_1 b_1 + r_2 b_2 + \cdots + r_n b_n : r_i \in R \text{ for all } i\}$$

is an ideal in $R$. We write $I = (b_1, b_2, \ldots, b_n)$ in this case, and we call $I$ the **ideal generated by** $b_1, b_2, \ldots, b_n$. In particular, if $n = 1$, then

$$I = (b) = \{rb : r \in R\}$$

---

[9] In contrast to the definition of subring, it suffices to assume that $a + b \in I$ instead of $a - b \in I$. If $I$ is an ideal and $b \in I$, then $(-1)b \in I$, and so $a - b = a + (-1)b \in I$.

is an ideal in $R$. The ideal $(b)$, also denoted by $Rb$ and consisting of all the multiples of $b$, is called the **principal ideal** generated by $b$.

Both $R$ and $(0)$ are principal ideals [note that $R = (1)$]. In $\mathbb{Z}$, the even integers form the principal ideal $(2)$.

**Definition.** Elements $a$ and $b$ in a commutative ring $R$ are **associates** if there exists a unit $u \in R$ with $b = ua$.

For example, in $\mathbb{Z}$, the only units are $\pm 1$, and so the associates of an integer $m$ are $\pm m$. If $k$ is a field, the only units in $k[x]$ are the nonzero constants, and so the associates of a polynomial $f(x) \in k[x]$ are the polynomials $uf(x)$, where $u \in k$ and $u \neq 0$. The only units in $\mathbb{Z}[x]$ are $\pm 1$, and the only associates of a polynomial $f(x) \in \mathbb{Z}[x]$ are $\pm f(x)$.

In any commutative ring $R$, associates $a$ and $b$ generate the same principal ideal; the converse may be false if $R$ is not a domain.

**Proposition 2.33.** *Let $R$ be a domain and let $a, b \in R$.*

(i) *$a \mid b$ and $b \mid a$ if and only if $a$ and $b$ are associates.*

(ii) *The principal ideals $(a)$ and $(b)$ are equal if and only if $a$ and $b$ are associates.*

**Proof.**

(i) If $a \mid b$ and $b \mid a$, there are $r, s \in R$ with $b = ra$ and $a = sb$, and so $b = ra = rsb$. If $b = 0$, then $a = 0$ (because $b \mid a$); if $b \neq 0$, then we may cancel it ($R$ is a domain) to obtain $1 = rs$. Hence, $r$ and $s$ are units, and $a$ and $b$ are associates. The converse is obvious.

(ii) If $(a) = (b)$, then $a \in (b)$; hence, $a = rb$ for some $r \in R$, and so $b \mid a$. Similarly, $b \in (a)$ implies $a \mid b$, and so (i) shows that $a$ and $b$ are associates.

Conversely, if $a = ub$, where $u$ is a unit, then $a \in (b)$ and $(a) \subseteq (b)$. Similarly, $b = u^{-1}a$ implies $(b) \subseteq (a)$, and so $(a) = (b)$.  •

## Exercises

**2.30.** (i) Let $\varphi \colon A \to R$ be a ring isomorphism, and let $\psi \colon R \to A$ be its inverse function. Show that $\psi$ is a ring isomorphism.

(ii) Show that the composite of two ring homomorphisms (isomorphisms) is again a ring homomorphism (isomorphism).

(iii) Show that $A \cong R$ defines an equivalence relation on any set of commutative rings.

**2.31.** If $R$ is a commutative ring, prove that $R[x, y] \cong R[y, x]$. In fact, prove that there is an isomorphism $\Phi$ with $\Phi(x) = y$, $\Phi(y) = x$, and $\Phi(r) = r$ for all $r \in R$.

**2.32.** If $(I_j)_{j \in J}$ is a family of ideals in a commutative ring $R$, prove that $\bigcap_{j \in J} I_j$ is an ideal in $R$.

\* **2.33.** If $R$ is a commutative ring and $c \in R$, prove that the function $\varphi \colon R[x] \to R[x]$, defined by $f(x) \mapsto f(x+c)$, is an isomorphism. In more detail, $\varphi(\sum_i s_i x^i) = \sum_i s_i (x+c)^i$.

**2.34.** (i) Prove that any two fields having exactly four elements are isomorphic.

$\hspace{1cm}$ **Hint.** First prove that $1 + 1 = 0$, and then show that the nonzero elements form a cyclic group of order 3 under multiplication.

$\hspace{0.5cm}$ (ii) Prove that the commutative rings $\mathbb{I}_4$ and $\mathbb{F}_4$ (the field with four elements in Exercise 2.8 on page 89) are not isomorphic.

**∗ 2.35.** (i) Let $k$ be a field that contains $\mathbb{F}_p$ as a subfield [e.g., $k = \mathbb{F}_p(x)$]. For every positive integer $n$, show that the function $\varphi_n\colon k \to k$, given by $\varphi(a) = a^{p^n}$, is a ring homomorphism.

$\hspace{0.5cm}$ (ii) Prove that every element $a \in \mathbb{F}_p$ has a $p$th root (i.e., there is $b \in \mathbb{F}_p$ with $a = b^p$).

**2.36.** If $R$ is a field, show that $R \cong \mathrm{Frac}(R)$. More precisely, show that the ring homomorphism $f\colon R \to \mathrm{Frac}(R)$, given by $r \mapsto [r, 1]$, is an isomorphism.

**∗ 2.37.** (i) If $A$ and $R$ are domains and $\varphi\colon A \to R$ is a ring isomorphism, prove that

$$[a, b] \mapsto [\varphi(a), \varphi(b)]$$

is a ring isomorphism $\mathrm{Frac}(A) \to \mathrm{Frac}(R)$.

$\hspace{0.5cm}$ (ii) Prove that if a field $k$ contains an isomorphic copy of $\mathbb{Z}$ as a subring, then $k$ must contain an isomorphic copy of $\mathbb{Q}$.

$\hspace{0.5cm}$ (iii) Let $R$ be a domain and let $\varphi\colon R \to k$ be an injective ring homomorphism, where $k$ is a field. Prove that there exists a unique ring homomorphism $\Phi\colon \mathrm{Frac}(R) \to k$ extending $\varphi$; that is, $\Phi|R = \varphi$.

**∗ 2.38.** If $R$ is a domain with $F = \mathrm{Frac}(R)$, prove that $\mathrm{Frac}(R[x]) \cong F(x)$.

**∗ 2.39.** (i) If $R$ and $S$ are commutative rings, show that their ***direct product*** $R \times S$ is also a commutative ring, where addition and multiplication in $R \times S$ are defined coordinatewise:

$$(r, s) + (r', s') = (r + r', s + s') \quad \text{and} \quad (r, s)(r', s') = (rr', ss').$$

$\hspace{0.5cm}$ (ii) Show that if $m$ and $n$ are relatively prime, then $\mathbb{I}_{mn} \cong \mathbb{I}_m \times \mathbb{I}_n$ as rings.
$\hspace{1cm}$ **Hint.** See Theorem 1.87.

$\hspace{0.5cm}$ (iii) If neither $R$ nor $S$ is the zero ring, show that $R \times S$ is not a domain.

$\hspace{0.5cm}$ (iv) Show that $R \times (0)$ is an ideal in $R \times S$.

$\hspace{0.5cm}$ (v) Show that $R \times (0)$ is a ring isomorphic to $R$, but it is not a subring of $R \times S$.

**∗ 2.40.** (i) Give an example of a commutative ring $R$ with nonzero ideals $I$ and $J$ such that $I \cap J = (0)$.

$\hspace{0.5cm}$ (ii) If $I$ and $J$ are nonzero ideals in a domain $R$, prove that $I \cap J \neq (0)$.

**∗ 2.41.** (i) If $R$ and $S$ are nonzero commutative rings, prove that

$$U(R \times S) = U(R) \times U(S),$$

$\hspace{1cm}$ where $U(R)$ is the group of units of $R$.

$\hspace{1cm}$ **Hint.** Show that $(r, s)$ is a unit in $R \times S$ if and only if $r$ is a unit in $R$ and $s$ is a unit in $S$.

$\hspace{0.5cm}$ (ii) Redo Exercise 1.72 on page 59 using part (i).

$\hspace{0.5cm}$ (iii) Use part (i) to give another proof of Corollary 1.90.

**2.42.** Let $F$ be the set of all $2 \times 2$ real matrices of the form

$$A = \begin{bmatrix} a & b \\ -b & a \end{bmatrix}.$$

(i)  Prove that $F$ is a field (with operations matrix addition and matrix multiplication).

(ii)  Prove that $\varphi \colon F \to \mathbb{C}$, defined by $\varphi(A) = a + ib$, is an isomorphism.

## Section 2.4. From Arithmetic to Polynomials

We are now going to see that, when $k$ is a field, virtually all the familiar theorems in $\mathbb{Z}$ have polynomial analogs in $k[x]$; moreover, the familiar proofs can be translated into proofs here. From now on, we will usually shorten "ring homomorphism" to "homomorphism" if it will not cause confusion.

The Division Algorithm for polynomials with coefficients in a field says that long division is possible.

**Theorem 2.34 (Division Algorithm).** *If $k$ is a field and $f(x), g(x) \in k[x]$ with $f(x) \neq 0$, then there are unique polynomials $q(x), r(x) \in k[x]$ with*

$$g(x) = q(x)f(x) + r(x),$$

*where either $r(x) = 0$ or $\deg(r) < \deg(f)$.*

**Proof.** We first prove the existence of such $q$ and $r$. If $f \mid g$, then $g = qf$ for some $q$; define the remainder $r = 0$, and we are done. If $f \nmid g$, then consider all (necessarily nonzero) polynomials of the form $g - qf$ as $q$ varies over $k[x]$. There is a polynomial $r = g - qf$ having least degree among all such polynomials. Since $g = qf + r$, it suffices to show that $\deg(r) < \deg(f)$. Write $f(x) = s_n x^n + \cdots + s_1 x + s_0$ and $r(x) = t_m x^m + \cdots + t_1 x + t_0$. Now $s_n \neq 0$ implies that $s_n$ is a unit, because $k$ is a field, and so $s_n^{-1}$ exists in $k$. If $\deg(r) \geq \deg(f)$, define

$$h(x) = r(x) - cx^{m-n} f(x),$$

where $c = t_m s_n^{-1}$. Note that $h = 0$ or $\deg(h) < \deg(r)$. If $h = 0$, then $r = cx^{m-n} f$ and

$$g = qf + r = qf + cx^{m-n} f = (q + cx^{m-n})f,$$

contradicting $f \nmid g$. If $h \neq 0$, then $\deg(h) < \deg(r)$ and

$$g - qf = r = h + cx^{m-n} f.$$

Thus, $g - (q + cx^{m-n})f = h$, contradicting $r$ being a polynomial of least degree having this form. Therefore, $\deg(r) < \deg(f)$.

To prove uniqueness of $q(x)$ and $r(x)$, assume that $g = q'f + r'$, where $\deg(r') < \deg(f)$. Then

$$(q - q')f = r' - r.$$

If $r' \neq r$, then each side has a degree. But $\deg((q - q')f) = \deg(q - q') + \deg(f) \geq \deg(f)$, while $\deg(r' - r) \leq \max\{\deg(r'), \deg(r)\} < \deg(f)$, a contradiction. Hence, $r' = r$ and $(q - q')f = 0$. As $k[x]$ is a domain and $f \neq 0$, it follows that $q - q' = 0$ and $q = q'$.  $\bullet$

**Definition.** If $f(x)$ and $g(x)$ are polynomials in $k[x]$, where $k$ is a field, then the polynomials $q(x)$ and $r(x)$ occurring in the Division Algorithm are called the *quotient* and the *remainder* after dividing $g(x)$ by $f(x)$.

The hypothesis that $k$ is a field is much too strong; the existence of quotient and remainder holds in $R[x]$ for any commutative ring $R$ as long as the leading coefficient of $f(x)$ is a unit in $R$; in particular, this is so whenever $f(x)$ is a monic polynomial. However, uniqueness of quotient and remainder may not hold if $R$ is not a domain (see Exercise 2.51(ii) on page 114).

**Corollary 2.35.** *Let $R$ be a commutative ring, and let $f(x) \in R[x]$ be a monic polynomial. If $g(x) \in R[x]$, then there exist $q(x), r(x) \in R[x]$ with*

$$g(x) = q(x)f(x) + r(x),$$

*where either $r(x) = 0$ or $\deg(r) < \deg(f)$.*

**Proof.** The proof of the Division Algorithm can be repeated here once we observe that $c = t_m s_n^{-1} = t_m \in R$, for $s_n = 1$ because $f(x)$ is monic.   •

The importance of the Division Algorithm arises from viewing the remainder as the obstruction to whether $f(x) \mid g(x)$; that is, whether $g \in (f)$. To see if $f \mid g$, first write $g = qf + r$ and then try to show that $r = 0$.

The ideals in $k[x]$ are quite simple when $k$ is a field.

**Theorem 2.36.** *If $k$ is a field, then every ideal $I$ in $k[x]$ is a principal ideal; that is, there is $d \in I$ with $I = (d)$. Moreover, if $I \neq (0)$, then $d$ can be chosen to be a monic polynomial.*

**Proof.** If $I = (0)$, then $I$ is a principal ideal with generator $0$. Otherwise, let $d$ be a polynomial in $I$ of least degree. We may assume that $d$ is monic (if $a_n$ is the leading coefficient of $d$, then $a_n \neq 0$, and $a_n^{-1} \in k$ because $k$ is a field; hence, $a_n^{-1}d$ is a monic polynomial in $I$ of the same degree as $d$).

Clearly, $(d) \subseteq I$. For the reverse inclusion, let $f \in I$. By the Division Algorithm, $f = qd + r$, where either $r = 0$ or $\deg(r) < \deg(d)$. But $r = f - qd \in I$; if $r \neq 0$, then we contradict $d$ being a polynomial in $I$ of minimal degree. Hence, $r = 0$, $f \in (d)$, and $I = (d)$.   •

It is not true that ideals in arbitrary commutative rings are always principal.

**Example 2.37.** Let $R = \mathbb{Z}[x]$, the commutative ring of all polynomials over $\mathbb{Z}$. It is easy to see that the set $I$ of all polynomials with even constant term is an ideal in $\mathbb{Z}[x]$. We show that $I$ is *not* a principal ideal.

Suppose there is $d(x) \in \mathbb{Z}[x]$ with $I = (d(x))$. The constant $2 \in I$, so that there is $f(x) \in \mathbb{Z}[x]$ with $2 = d(x)f(x)$. Since the degree of a product is the sum of the degrees of the factors, $0 = \deg(2) = \deg(d) + \deg(f)$. Since degrees are nonnegative, it follows that $\deg(d) = 0$ [i.e., $d(x)$ is a nonzero constant]. As constants here are integers, the candidates for $d(x)$ are $\pm 1$ and $\pm 2$. Suppose $d(x) = \pm 2$; since $x \in I$, there is $g(x) \in \mathbb{Z}[x]$ with $x = d(x)g(x) = \pm 2g(x)$. But every coefficient on the right side is even, while the coefficient of $x$ on the left side is 1. This contradiction gives

$d(x) = \pm 1$. By Example 2.31(ii), $I = \mathbb{Z}[x]$, another contradiction. Therefore, no such $d(x)$ exists; that is, $I$ is not a principal ideal.  ◄

We now turn our attention to roots of polynomials.

**Definition.** If $f(x) \in k[x]$, where $k$ is a field, then a **root** of $f(x)$ **in** $k$ is an element $a \in k$ with $f(a) = 0$.

**Remark.** The polynomial $f(x) = x^2 - 2$ has its coefficients in $\mathbb{Q}$, but we usually say that $\sqrt{2}$ is a root of $f(x)$ even though $\sqrt{2}$ is irrational; that is, $\sqrt{2} \notin \mathbb{Q}$. We shall see later, in Theorem 2.148, that for every polynomial $f(x) \in k[x]$, where $k$ is any field, there is a larger field $E$ that contains $k$ as a subfield and that contains all the roots of $f(x)$. For example, $x^2 - 2 \in \mathbb{F}_3[x]$ has no root in $\mathbb{F}_3$, but we shall see that a version of $\sqrt{2}$ does exist in some (finite) field containing $\mathbb{F}_3$.  ◄

**Lemma 2.38.** *Let $f(x) \in k[x]$, where $k$ is a field, and let $u \in k$. Then there is $q(x) \in k[x]$ with*

$$f(x) = q(x)(x - u) + f(u).$$

**Proof.** The Division Algorithm gives

$$f(x) = q(x)(x - u) + r;$$

the remainder $r$ is a constant because $x - u$ has degree 1. By Corollary 2.26, evaluation at $u$ is a ring homomorphism; hence, $f(u) = q(u)(u - u) + r$, and so $f(u) = r$.  •

There is a connection between roots and factoring.

**Proposition 2.39.** *If $f(x) \in k[x]$, where $k$ is a field, then $a$ is a root of $f(x)$ in $k$ if and only if $x - a$ divides $f(x)$ in $k[x]$.*

**Proof.** If $a$ is a root of $f(x)$ in $k$, then $f(a) = 0$ and Lemma 2.38 gives $f(x) = q(x)(x - a)$. Conversely, if $f(x) = g(x)(x - a)$, then evaluating at $a$ gives $f(a) = g(a)(a - a) = 0$.  •

**Theorem 2.40.** *Let $k$ be a field and let $f(x) \in k[x]$. If $f(x)$ has degree $n$, then $f(x)$ has at most $n$ roots in $k$.*

**Proof.** We prove the statement by induction on $n \geq 0$. If $n = 0$, then $f(x)$ is a nonzero constant, and so the number of its roots in $k$ is zero. Now let $n > 0$. If $f(x)$ has no roots in $k$, we are done, for $0 \leq n$. Otherwise, we may assume that $f(x)$ has a root $a \in k$. By Proposition 2.39,

$$f(x) = q(x)(x - a);$$

moreover, $q(x) \in k[x]$ has degree $n - 1$. If there is a root $b \in k$ with $b \neq a$, then applying the evaluation homomorphism $e_b$ gives

$$0 = f(b) = q(b)(b - a).$$

Since $b - a \neq 0$, we have $q(b) = 0$ (for $k$ is a field, hence a domain), so that $b$ is a root of $q(x)$. Now $\deg(q) = n - 1$, so that the inductive hypothesis says that $q(x)$ has at most $n - 1$ roots in $k$. Therefore, $f(x)$ has at most $n$ roots in $k$.  •

**Example 2.41.** Theorem 2.40 is not true for polynomials with coefficients in an arbitrary commutative ring $R$. For example, if $R = \mathbb{I}_8$, then the quadratic polynomial $x^2 - 1 \in \mathbb{I}_8[x]$ has four roots in $R$, namely, $[1], [3], [5]$, and $[7]$. On the other hand, Exercise 2.50 on page 114 says that Theorem 2.40 remains true if we assume that the coefficient ring $R$ is a domain. ◄

**Corollary 2.42.** *Every $n$th root of unity in $\mathbb{C}$ is equal to*

$$e^{2\pi i k/n} = \cos\left(\tfrac{2\pi k}{n}\right) + i\sin\left(\tfrac{2\pi k}{n}\right),$$

*where $k = 0, 1, 2, \ldots, n - 1$.*

**Proof.** Each of the $n$ different complex numbers $e^{2\pi i k/n}$ is an $n$th root of unity; that is, each is a root of $x^n - 1$. By Theorem 2.40, there can be no other complex roots. •

Recall that every polynomial $f(x) \in k[x]$ determines the polynomial function $k \to k$ that sends $a$ into $f(a)$ for all $a \in k$. In Exercise 2.22 on page 94, however, we saw that the nonzero polynomial $x^p - x \in \mathbb{F}_p[x]$ determines the constant function zero. This pathology vanishes when the field $k$ is infinite.

**Corollary 2.43.** *Let $k$ be an infinite field and let $f(x)$ and $g(x)$ be polynomials in $k[x]$. If $f(x)$ and $g(x)$ determine the same polynomial function [that is, $f(a) = g(a)$ for all $a \in k$], then $f(x) = g(x)$.*

**Proof.** If $f(x) \neq g(x)$, then the polynomial $h(x) = f(x) - g(x)$ is nonzero, so that it has some degree, say, $n$. Now every element of $k$ is a root of $h(x)$; since $k$ is infinite, $h(x)$ has more than $n$ roots, and this contradicts the theorem. •

This proof yields a more general result.

**Corollary 2.44.** *Let $k$ be a (possibly finite) field, let $f(x), g(x) \in k[x]$, and let $\deg(f) \leq \deg(g) = n$. If $f(a) = g(a)$ for $n + 1$ elements $a \in k$, then $f(x) = g(x)$.*

**Proof.** If $f \neq g$, then $\deg(f - g)$ is defined, $\deg(f - g) \leq n$, and $f - g$ has too many roots. •

We now generalize Corollary 2.43 to polynomials in several variables. Denote the $n$-tuple $(x_1, \ldots, x_n)$ by $X$.

**Proposition 2.45.** *Let $f(X), g(X) \in k[X] = k[x_1, \ldots, x_n]$, where $k$ is an infinite field.*

(i) *If $f(X)$ is nonzero, then there are $a_1, \ldots, a_n \in k$ with $f(a_1, \ldots, a_n) \neq 0$.*

(ii) *If $f(a_1, \ldots, a_n) = g(a_1, \ldots, a_n)$ for all $(a_1, \ldots, a_n) \in k^n$, then $f(X) = g(X)$.*

**Proof.**

(i) The proof is by induction on $n \geq 1$. If $n = 1$, then the result is Corollary 2.43, for if $f(a) = 0$ for all $a \in k$, then $f = 0$. For the inductive step, assume that

$$f(x_1, \ldots, x_{n+1}) = B_0 + B_1 x_{n+1} + B_2 x_{n+1}^2 + \cdots + B_r x_{n+1}^r,$$

where $B_i \in k[x_1, \ldots, x_n]$ and $B_r = B_r(x_1, \ldots, x_n) \neq 0$. By induction, there are $a_1, \ldots, a_n \in k$ with $B_r(a_1, \ldots, a_n) \neq 0$. Therefore, $f(a_1, \ldots, a_n, x_{n+1}) = B_0(a_1, \ldots, a_n) + B_1(a_1, \ldots, a_n) x_{n+1} + \cdots + B_r(a_1, \ldots, a_n) x_{n+1}^r \neq 0$ in $k[x_{n+1}]$. By the base step, there is $a \in k$ with $f(a_1, \ldots, a_n, a) \neq 0$.

(ii) The proof is by induction on $n \geq 1$; the base step is Corollary 2.43. For the inductive step, write

$$f(X, y) = \sum_i p_i(X) y^i \quad \text{and} \quad g(X, y) = \sum_i q_i(X) y^i,$$

where $X$ denotes $(x_1, \ldots, x_n)$. Suppose that $f(a, \beta) = g(a, \beta)$ for every $a \in k^n$ and every $\beta \in k$. For fixed $a \in k^n$, define $F_a(y) = \sum_i p_i(a) y^i$ and $G_a(y) = \sum_i q_i(a) y^i$. Since both $F_a(y)$ and $G_a(y)$ are in $k[y]$, the base step gives $p_i(a) = q_i(a)$ for all $a \in k^n$. By the inductive hypothesis, $p_i(X) = q_i(X)$ for all $i$, and hence

$$f(X, y) = \sum_i p_i(X) y^i = \sum_i q_i(X) y^i = g(X, y). \quad \bullet$$

Here is another nice application of Theorem 2.40.

**Theorem 2.46.** *Let $k$ be a field. If $G$ is a finite subgroup of the multiplicative group $k^\times$, then $G$ is cyclic. In particular, if $k$ itself is finite (e.g., $k = \mathbb{F}_p$), then $k^\times$ is cyclic.*

**Proof.** Let $d$ be a divisor of $|G|$. If there are two subgroups of $G$ of order $d$, say, $S$ and $T$, then $|S \cup T| > d$. But each $a \in S \cup T$ satisfies $a^d = 1$, by Lagrange's Theorem, and hence it is a root of $x^d - 1$. This contradicts Theorem 2.40, for this polynomial now has too many roots in $k$. Thus, $G$ is cyclic, by Theorem 1.93. $\quad \bullet$

**Definition.** If $k$ is a finite field, a generator of the cyclic group $k^\times$ is called a *primitive element* of $k$.

Although the multiplicative groups $\mathbb{F}_p^\times$ are cyclic, no explicit formula giving a primitive element of $\mathbb{F}_p^\times$ for all $p$, say, $[a(p)]$, is known.

**Corollary 2.47.** *If $p$ is prime, then the group of units $U(\mathbb{I}_p)$ is cyclic.*

**Proof.** We have been writing $\mathbb{F}_p$ instead of $\mathbb{I}_p$, and so this follows at once from Theorem 2.46. $\quad \bullet$

The definition of a greatest common divisor of polynomials is essentially the same as the corresponding definition for integers.

**Definition.** If $f(x)$ and $g(x)$ are polynomials in $k[x]$, where $k$ is a field, then a **common divisor** is a polynomial $c(x) \in k[x]$ with $c(x) \mid f(x)$ and $c(x) \mid g(x)$. If $f(x)$ and $g(x)$ in $k[x]$ are not both 0, define their **greatest common divisor**, abbreviated gcd, to be the monic common divisor having largest degree. If $f(x) = 0 = g(x)$, define their gcd $= 0$. The gcd of $f(x)$ and $g(x)$ is often denoted by $(f, g)$.

We will prove the uniqueness of the gcd in Corollary 2.49.

**Theorem 2.48.** *If $k$ is a field and $f(x)$, $g(x) \in k[x]$, then their* gcd $d(x)$ *is a linear combination of $f(x)$ and $g(x)$; that is, there are $s(x), t(x) \in k[x]$ with*

$$d(x) = s(x)f(x) + t(x)g(x).$$

**Proof.** The set $(f, g)$ of all linear combinations of $f$ and $g$ is an ideal in $k[x]$. The theorem is true if both $f$ and $g$ are 0, and so we may assume that there is a monic polynomial $d(x)$ with $(f, g) = (d)$, by Theorem 2.36. Of course, $d$ lying in $(f, g)$ must be a linear combination: $d = sf + tg$. We claim that $d$ is a gcd. Now $d$ is a common divisor, for $f, g \in (f, g) = (d)$. If $h$ is a common divisor of $f$ and $g$, then $f = f_1 h$ and $g = g_1 h$. Hence, $d = sf + tg = (sf_1 + tg_1)h$ and $h \mid d$. Therefore, $\deg(h) \leq \deg(d)$, and so $d$ is a monic common divisor of largest degree.   $\bullet$

The end of the last proof gives a characterization of gcd's in $k[x]$.

**Corollary 2.49.** *Let $k$ be a field and let $f(x)$, $g(x) \in k[x]$.*

(i) *A monic common divisor $d(x)$ is the gcd if and only if $d(x)$ is divisible by every common divisor; that is, if $h(x)$ is a common divisor, then $h(x) \mid d(x)$.*

(ii) *$f$ and $g$ have a unique gcd.*

**Proof.**

(i) The end of the proof of Theorem 2.48 shows that if $h(x)$ is a common divisor, then $h(x) \mid d(x)$. Conversely, if $h(x) \mid d(x)$, then $\deg(h) \leq \deg(d)$, and so $d$ is a common divisor of largest degree.

(ii) If $d$ and $d'$ are gcd's of $f$ and $g$, then $d \mid d'$ and $d' \mid d$, by part (i). Since both $d$ and $d'$ are monic, we must have $d = d'$, by Proposition 2.10.   $\bullet$

If $u$ is a unit, then every polynomial $f(x)$ is divisible by $u$ and by $uf(x)$. The analog of a prime number is a polynomial having only divisors of these trivial sorts.

**Definition.** An element $p$ in a domain $R$ is **irreducible** if $p$ is neither 0 nor a unit and, in every factorization $p = uv$ in $R$, either $u$ or $v$ is a unit.

For example, a prime $p \in \mathbb{Z}$ is an irreducible element, as is $-p$. We now describe irreducible polynomials $p(x) \in k[x]$, when $k$ is a field.

**Proposition 2.50.** *If $k$ is a field, then a polynomial $p(x) \in k[x]$ is irreducible if and only if $\deg(p) = n \geq 1$ and there is no factorization in $k[x]$ of the form $p(x) = g(x)h(x)$ in which both factors have degree smaller than $n$.*

**Proof.** We show first that $h(x) \in k[x]$ is a unit if and only if $\deg(h) = 0$. If $h(x)u(x) = 1$, then $\deg(h) + \deg(u) = \deg(1) = 0$; since degrees are nonnegative, we have $\deg(h) = 0$. Conversely, if $\deg(h) = 0$, then $h(x)$ is a nonzero constant; that is, $h \in k$; since $k$ is a field, $h$ has an inverse.

If $p(x)$ is irreducible, then its only factorizations are of the form $p(x) = g(x)h(x)$, where $g(x)$ or $h(x)$ is a unit; that is, where either $\deg(g) = 0$ or $\deg(h) = 0$. Hence, $p(x)$ has no factorization in which both factors have smaller degree.

Conversely, if $p(x)$ is not irreducible, it has a factorization $p(x) = g(x)h(x)$ in which neither $g$ nor $h$ is a unit; that is, since $k$ is a field, neither $g$ nor $h$ has degree 0. Therefore, $p(x)$ is a product of polynomials of smaller degree. •

If $k$ is not a field, however, then this characterization of irreducible polynomials no longer holds. For example, $2x + 2 = 2(x + 1)$ is not irreducible in $\mathbb{Z}[x]$, even though, in any factorization, one factor has degree 0 and the other degree 1, for 2 is not a unit in $\mathbb{Z}[x]$. When $k$ is a field, the units are the nonzero constants, but this is no longer true for more general rings of coefficients (for example, Exercise 2.21(ii) on page 94 says that $[2]x + [1]$ is a unit in $\mathbb{I}_4[x]$).

As the definition of divisibility depends on the ambient ring, so irreducibility of a polynomial $p(x) \in k[x]$ also depends on the field $k$. For example, $p(x) = x^2 + 1$ is irreducible in $\mathbb{R}[x]$, but it factors as $(x + i)(x - i)$ in $\mathbb{C}[x]$. On the other hand, a linear polynomial $f(x) \in k[x]$ is always irreducible [if $f = gh$, then $1 = \deg(f) = \deg(g) + \deg(h)$, and so one of $g$ or $h$ must have degree 0 while the other has degree $1 = \deg(f)$].

**Corollary 2.51.** *Let $k$ be a field and let $f(x) \in k[x]$ be a quadratic or cubic polynomial. Then $f(x)$ is irreducible in $k[x]$ if and only if $f(x)$ has no root in $k$.*

**Proof.** An irreducible polynomial of degree $> 1$ has no roots in $k$, by Proposition 2.39. Conversely, if $f(x)$ is not irreducible, then $f(x) = g(x)h(x)$, where neither $g$ nor $h$ is constant; thus, neither $g$ nor $h$ has degree 0. Since $\deg(f) = \deg(g) + \deg(h)$, at least one of the factors has degree 1 and, hence, has a root. •

It is easy to see that Corollary 2.51 can be false if $\deg(f) \geq 4$. For example, $f(x) = x^4 + 2x^2 + 1 = (x^2 + 1)^2$ factors in $\mathbb{R}[x]$, yet it has no roots in $\mathbb{R}$.

Let us now consider polynomials $f(x) \in \mathbb{Q}[x]$. If the coefficients of $f(x)$ happen to be integers, there is a useful lemma of Gauss comparing its factorizations in $\mathbb{Z}[x]$ and in $\mathbb{Q}[x]$.

**Theorem 2.52 (Gauss's Lemma).** *Let $f(x) \in \mathbb{Z}[x]$. If $f(x) = G(x)H(x)$ in $\mathbb{Q}[x]$, where $\deg(G), \deg(H) < \deg(f)$, then $f(x) = g(x)h(x)$ in $\mathbb{Z}[x]$, where $\deg(g) = \deg(G)$ and $\deg(h) = \deg(H)$.*

**Proof.** Clearing denominators, there are positive integers $n', n''$ such that $g(x) = n'G(x)$ and $h(x) = n''H(x)$. Setting $n = n'n''$, we have

$$nf(x) = n'G(x)n''H(x) = g(x)h(x) \text{ in } \mathbb{Z}[x].$$

If $p$ is a prime divisor of $n$, consider the map $\mathbb{Z}[x] \to \mathbb{F}_p[x]$ which reduces all coefficients mod $p$. The equation becomes

$$0 = \overline{g}(x)\overline{h}(x).$$

But $\mathbb{F}_p[x]$ is a domain, because $\mathbb{F}_p$ is a field, and so at least one of these factors, say, $\overline{g}(x)$, is 0; that is, all the coefficients of $g(x)$ are multiples of $p$. Therefore, we may write $g(x) = pg'(x)$, where all the coefficients of $g'(x)$ lie in $\mathbb{Z}$. If $n = pm$, then

$$pmf(x) = pg'(x)h(x) \text{ in } \mathbb{Z}[x].$$

Cancel $p$, and continue canceling primes until we reach a factorization $f(x) = g^*(x)h^*(x)$ in $\mathbb{Z}[x]$ [note that $\deg(g^*) = \deg(g)$ and $\deg(h^*) = \deg(h)$]. •

The contrapositive version of Gauss's Lemma is more convenient to use. If $f(x) \in \mathbb{Z}[x]$ has no factorization in $\mathbb{Z}[x]$ as a product of two polynomials, each having degree smaller than $\deg(f)$, then $f(x)$ is irreducible in $\mathbb{Q}[x]$.

It is easy to see that if $p(x)$ and $q(x)$ are irreducible polynomials, then $p(x) \mid q(x)$ if and only if there is a unit $u$ with $q(x) = up(x)$. If, in addition, both $p(x)$ and $q(x)$ are monic, then $p(x) \mid q(x)$ implies $p(x) = q(x)$.

**Lemma 2.53.** *Let $k$ be a field, let $p(x), f(x) \in k[x]$, and let $d(x) = (p, f)$ be their gcd. If $p(x)$ is a monic irreducible polynomial, then*

$$d(x) = \begin{cases} 1 & \text{if } p(x) \nmid f(x) \\ p(x) & \text{if } p(x) \mid f(x). \end{cases}$$

**Proof.** Since $d(x) \mid p(x)$, we have $d(x) = 1$ or $d(x) = p(x)$. •

**Theorem 2.54 (Euclid's Lemma).** *Let $k$ be a field and let $f(x), g(x) \in k[x]$. If $p(x)$ is an irreducible polynomial in $k[x]$, and $p(x) \mid f(x)g(x)$, then either*

$$p(x) \mid f(x) \quad \text{or} \quad p(x) \mid g(x).$$

*More generally, if $p(x) \mid f_1(x) \cdots f_n(x)$, then $p(x) \mid f_i(x)$ for some $i$.*

**Proof.** Assume that $p \mid fg$ but that $p \nmid f$. Since $p$ is irreducible, $(p, f) = 1$, and so $1 = sp + tf$ for some polynomials $s$ and $t$. Therefore,

$$g = spg + tfg.$$

But $p \mid fg$, by hypothesis, and so $p \mid g$. •

**Definition.** Two polynomials $f(x), g(x) \in k[x]$, where $k$ is a field, are called *relatively prime* if their gcd is 1.

**Corollary 2.55.** *Let $f(x), g(x), h(x) \in k[x]$, where $k$ is a field, and let $h(x)$ and $f(x)$ be relatively prime. If $h(x) \mid f(x)g(x)$, then $h(x) \mid g(x)$.*

**Proof.** The proof of Theorem 2.54 works here: since $(h, f) = 1$, we have $1 = sh + tf$, and so $g = shg + tfg$. But $fg = hh_1$ for some $h_1$, and so $g = h(sg + th_1)$. •

**Definition.** If $k$ is a field, then a rational function $f(x)/g(x) \in k(x)$ is in *lowest terms* if $f(x)$ and $g(x)$ are relatively prime.

**Proposition 2.56.** *If $k$ is a field, every nonzero $f(x)/g(x) \in k(x)$ can be put in lowest terms.*

**Proof.** If $f = df'$ and $g = dg'$, where $d = (f, g)$, then $f'$ and $g'$ are relatively prime, and so $f'/g'$ is in lowest terms. $\quad \bullet$

The next result allows us to compute gcd's.

**Theorem 2.57 (Euclidean Algorithm).** *If $k$ is a field and $f(x), g(x) \in k[x]$, then there are algorithms for computing $\gcd(f, g)$, as well as for finding a pair of polynomials $s(x)$ and $t(x)$ with*

$$(f, g) = s(x)f(x) + t(x)g(x).$$

**Proof.** The proof is essentially a repetition of the proof of the Euclidean Algorithm in $\mathbb{Z}$; just iterate the Division Algorithm:

$$g = q_1 f + r_1$$
$$f = q_2 r_1 + r_2$$
$$r_1 = q_3 r_2 + r_3$$

$$\vdots$$

$$r_{n-3} = q_{n-1} r_{n-2} + r_{n-1}$$
$$r_{n-2} = q_n r_{n-1} + r_n$$
$$r_{n-1} = q_{n+1} r_n.$$

Since the degrees of the remainders are strictly decreasing, this procedure must stop after a finite number of steps. The claim is that $d = r_n$ is the gcd, once it is made monic. We see that $d$ is a common divisor of $f$ and $g$ by back substitution: work from the bottom up. To see that $d$ is the gcd, work from the top down to show that if $c$ is any common divisor of $f$ and $g$, then $c \mid r_i$ for every $i$. Finally, to find $s$ and $t$ with $d = sf + tg$, again work from the bottom up:

$$r_n = r_{n-2} - q_n r_{n-1}$$
$$= r_{n-2} - q_n(r_{n-3} - q_{n-1} r_{n-2})$$
$$= (1 + q_{n-1}) r_{n-2} - q_n r_{n-3}$$

$$\vdots$$

$$= sf + tg \quad \bullet$$

For a discussion of *antanairesis*, which shows how the Euclidean Algorithm in $\mathbb{Z}$ arose, see FCAA, pp. 47–48.

Here is an unexpected bonus from the Euclidean Algorithm.

**Corollary 2.58.** *Let $k$ be a subfield of a field $K$, so that $k[x]$ is a subring of $K[x]$. If $f(x), g(x) \in k[x]$, then their gcd in $k[x]$ is equal to their gcd in $K[x]$.*

**Proof.** The Division Algorithm in $K[x]$ gives

$$g(x) = Q(x)f(x) + R(x),$$

where $Q(x), R(x) \in K[x]$; since $f(x), g(x) \in k[x]$, the Division Algorithm in $k[x]$ gives

$$g(x) = q(x)f(x) + r(x),$$

where $q(x), r(x) \in k[x]$. But the equation $g(x) = q(x)f(x) + r(x)$ also holds in $K[x]$ because $k[x] \subseteq K[x]$, so that the uniqueness of quotient and remainder in the Division Algorithm in $K[x]$ gives $Q(x) = q(x) \in k[x]$ and $R(x) = r(x) \in k[x]$. Therefore, the list of equations occurring in the Euclidean Algorithm in $K[x]$ is exactly the same list occurring in the Euclidean Algorithm in the smaller ring $k[x]$, and so the same gcd is obtained in both polynomial rings. •

In spite of the fact that there are more divisors with complex coefficients, the gcd of $x^3 - 2x^2 + x - 2$ and $x^4 - 1$ computed in $\mathbb{R}[x]$ is equal to their gcd computed in $\mathbb{C}[x]$.

**Corollary 2.59.** *If $f(x), g(x) \in \mathbb{R}[x]$ have no common root in $\mathbb{C}$, then $f(x), g(x)$ are relatively prime in $\mathbb{R}[x]$.*

**Proof.** Assume that $d = (f, g) \neq 1$, where $d(x) \in \mathbb{R}[x]$. By the Fundamental Theorem of Algebra, $d(x)$ has a complex root $\alpha$. By Corollary 2.58, $d(x)$ is the gcd $(f, g)$ in $\mathbb{C}[x]$. Since $(x - \alpha) \mid d(x)$ in $\mathbb{C}[x]$, we have $(x - \alpha) \mid f(x)$ and $(x - \alpha) \mid g(x)$; that is, $\alpha$ is a common root of $f$ and $g$. •

We remark that Corollary 2.59 is true more generally: we may replace $\mathbb{R}$ by any field $k$ using Kronecker's Theorem, Theorem 2.148: for every field $k$ and every $f(x) \in k[x]$, there exists a field $K$ containing $k$ and all the roots of $f(x)$; that is, there are $a, \alpha_i \in K$ with $f(x) = a \prod_i (x - \alpha_i)$ in $K[x]$.

The next result, an analog for polynomials of the Fundamental Theorem of Arithmetic, shows that irreducible polynomials are "building blocks" of arbitrary polynomials in the same sense that primes are building blocks of arbitrary integers. To avoid long sentences, we continue to allow "products" having only one factor.

**Theorem 2.60 (Unique Factorization).** *If $k$ is a field, then every polynomial $f(x) \in k[x]$ of degree $\geq 1$ is a product of a nonzero constant and monic irreducibles. Moreover, if $f(x)$ has two such factorizations,*

$$f(x) = ap_1(x) \cdots p_m(x) \quad and \quad f(x) = bq_1(x) \cdots q_n(x),$$

*that is, $a$ and $b$ are nonzero constants and the $p$'s and $q$'s are monic irreducibles, then $a = b$, $m = n$, and the $q$'s may be reindexed so that $q_i = p_i$ for all $i$.*

**Proof.** We prove the existence of a factorization for a polynomial $f(x)$ by induction on $\deg(f) \geq 1$. If $\deg(f) = 1$, then $f(x) = ax + c$, where $a \neq 0$, and $f(x) = a(x + a^{-1}c)$. As any linear polynomial, $x + a^{-1}c$ is irreducible, and so it is a product of irreducibles (in our present usage of "product"). Assume now that $\deg(f) \geq 1$. If the leading coefficient of $f(x)$ is $a$, write $f(x) = a(a^{-1}f(x))$. If $f(x)$ is irreducible, we are done, for $a^{-1}f(x)$ is monic. If $f(x)$ is not irreducible, then $f(x) = g(x)h(x)$, where $\deg(g) < \deg(f)$ and $\deg(h) < \deg(f)$. By the inductive hypothesis, there are factorizations $g(x) = bp_1(x) \cdots p_m(x)$ and $h(x) = cq_1(x) \cdots q_n(x)$, where $b, c \in k$ and the $p$'s and $q$'s are monic irreducibles. It follows that

$$f(x) = (bc)p_1(x) \cdots p_m(x)q_1(x) \cdots q_n(x).$$

To prove uniqueness, suppose that there is an equation

$$ap_1(x) \cdots p_m(x) = bq_1(x) \cdots q_n(x)$$

in which $a$ and $b$ are nonzero constants and the $p$'s and $q$'s are monic irreducibles. We prove, by induction on $M = \max\{m, n\} \geq 1$, that $a = b$, $m = n$, and the $q$'s may be reindexed so that $q_i = p_i$ for all $i$. For the base step $M = 1$, we have $ap_1(x) = bq_1(x)$. Now $a$ is the leading coefficient because $p_1(x)$ is monic, while $b$ is the leading coefficient because $q_1(x)$ is monic. Therefore, $a = b$, and canceling gives $p_1(x) = q_1(x)$. For the inductive step, the given equation shows that $p_m(x) \mid q_1(x) \cdots q_n(x)$. By Euclid's Lemma for polynomials, there is some $i$ with $p_m(x) \mid q_i(x)$. But $q_i(x)$, being monic irreducible, has no monic divisors other than 1 and itself, so that $q_i(x) = p_m(x)$. Reindexing, we may assume that $q_n(x) = p_m(x)$. Canceling this factor, we have $ap_1(x) \cdots p_{m-1}(x) = bq_1(x) \cdots q_{n-1}(x)$. By the inductive hypothesis, $a = b$, $m - 1 = n - 1$ (hence $m = n$) and, after reindexing, $q_i = p_i$ for all $i$.   •

We now collect like factors.

**Definition.** Let $f(x) \in k[x]$, where $k$ is a field. A **_prime factorization_** of $f(x)$ is

$$f(x) = ap_1(x)^{e_1} \cdots p_m(x)^{e_m},$$

where $a$ is a nonzero constant, the $p_i(x)$ are distinct monic irreducible polynomials, and $e_i \geq 0$ for all $i$.

Theorem 2.60 shows that if $\deg(f) \geq 1$, then $f$ has prime factorizations; moreover, if all the exponents $e_i > 0$, then the factors in this prime factorization are unique. The statement of Proposition 2.61 below illustrates the convenience of allowing some $e_i = 0$.

Let $k$ be a field, and assume that there are $a, r_1, \ldots, r_n \in k$ with

$$f(x) = a \prod_{i=1}^{n} (x - r_i).$$

If $r_1, \ldots, r_s$, where $s \leq n$, are the distinct roots of $f(x)$, then a prime factorization of $f(x)$ is

$$f(x) = a(x - r_1)^{e_1}(x - r_2)^{e_2} \cdots (x - r_s)^{e_s}.$$

We call $e_j$ the **_multiplicity_** of the root $r_j$. As linear polynomials are always irreducible, unique factorization shows that multiplicities of roots are well-defined.

Let $f(x), g(x) \in k[x]$, where $k$ is a field. As with integers, using zero exponents allows us to assume that the same irreducible factors occur in both prime factorizations:

$$f = p_1^{a_1} \cdots p_m^{a_m} \quad \text{and} \quad g = p_1^{b_1} \cdots p_m^{b_m}.$$

**Definition.** If $f$ and $g$ are elements in a commutative ring $R$, then a ***common multiple*** is an element $m \in R$ with $f \mid m$ and $g \mid m$. If $f$ and $g$ in $R$ are not both 0, define their ***least common multiple***, abbreviated $\operatorname{lcm}(f, g)$, to be a common multiple $c$ of them with $c \mid m$ for every common multiple $m$. If $f = 0 = g$, define their $\operatorname{lcm} = 0$. The lcm of $f$ and $g$ is often denoted by $[f, g]$.

Recall that we have denoted the gcd of polynomials $f$ and $g$ by $(f, g)$.

**Proposition 2.61.** *Let $f(x), g(x) \in k[x]$, where $k$ is a field, have prime factorizations $f(x) = p_1^{a_1} \cdots p_n^{a_n}$ and $g(x) = p_1^{b_1} \cdots p_n^{b_n}$ in $k[x]$.*

(i) *$f \mid g$ if and only if $a_i \le b_i$ for all $i$.*

(ii) *If $m_i = \min\{a_i, b_i\}$ and $M_i = \max\{a_i, b_i\}$, then*

$$(f, g) = p_1^{m_1} \cdots p_n^{m_n} \quad and \quad [f, g] = p_1^{M_1} \cdots p_n^{M_n}.$$

**Proof.**

(i) If $f \mid g$, then $g = fh$, where $h = p_1^{c_1} \cdots p_n^{c_n}$ and $c_i \ge 0$ for all $i$. Hence,

$$g(x) = p_1^{b_1} \cdots p_n^{b_n} = \left(p_1^{a_1} \cdots p_m^{a_m}\right)\left(p_1^{c_1} \cdots p_n^{c_n}\right) = p_1^{a_1 + c_1} \cdots p_n^{a_n + c_n}.$$

By uniqueness, $a_i + c_i = b_i$; hence, $a_i \le a_i + c_i = b_i$. Conversely, if $a_i \le b_i$, then there is $c_i \ge 0$ with $b_i = a_i + c_i$. It follows that $h = p_1^{c_1} \cdots p_n^{c_n} \in k[x]$ and $g = fh$.

(ii) Let $d(x) = p_1^{m_1} \cdots p_n^{m_n}$. Now $d$ is a common divisor, for $m_i \le a_i, b_i$. If $D(x) = p_1^{e_1} \cdots p_n^{e_n}$ is any other common divisor, then $0 \le e_i \le \min\{a_i, b_i\} = m_i$, and so $D \mid d$. Therefore, $\deg(D) \le \deg(d)$, and $d(x)$ is the gcd (for it is monic). The argument for lcm is similar.  •

**Corollary 2.62.** *If $k$ is a field and $f(x), g(x) \in k[x]$ are monic polynomials, then*

$$[f, g](f, g) = fg.$$

**Proof.** The result follows from Proposition 2.61, for $m_i + M_i = a_i + b_i$.  •

Since the Euclidean Algorithm computes the gcd in $k[x]$ when $k$ is a field, Corollary 2.62 computes the lcm.

---

## Exercises

**2.43.** (i) Let $f(x), g(x) \in \mathbb{Q}[x]$ with $f(x)$ monic. Write a pseudocode implementing the Division Algorithm with input $f(x), g(x)$ and output $q(x), r(x)$, the quotient and remainder.

(ii) Find the quotient and remainder by dividing $x^3 + 2x^2 - 8x + 6$ by $x - 1$ as you would in high school. Conclude that the Division Algorithm is not your old friend.

* **2.44.** If $R$ is a commutative ring, define a relation $\equiv$ on $R$ by $a \equiv b$ if there is a unit $u \in R$ with $b = ua$. Prove that if $a \equiv b$, then $(a) = (b)$, where $(a) = \{ra : r \in R\}$. Conversely, prove that if $R$ is a domain, then $(a) = (b)$ implies $a \equiv b$ (Proposition 2.10).

**2.45.** Let $R$ be a commutative ring and let $\mathcal{F}(R)$ be the commutative ring of all functions $f: R \to R$ with pointwise operations.

(i) Show that $R$ is isomorphic to the subring of $\mathcal{F}(R)$ consisting of all the constant functions.

(ii) If $f(x) \in R[x]$, let $f^\flat: R \to R$ be the polynomial function associated to $f(x)$; that is, $f^\flat: r \mapsto f(r)$. Show that the function $\varphi: R[x] \to \mathcal{F}(R)$, defined by $\varphi(f(x)) = f^\flat$, is a ring homomorphism.

(iii) Prove that $\varphi: R[x] \to \mathcal{F}(R)$ is injective if $R$ is an infinite field.

**2.46.** A student claims that $x - 1$ is not irreducible because $x - 1 = (\sqrt{x} + 1)(\sqrt{x} - 1)$ is a factorization. Explain the error of his ways.

**2.47.** Let $f(x) = x^2 + x + 1 \in \mathbb{F}_2[x]$. Prove that $f(x)$ is irreducible and that $f(x)$ has a root $\alpha \in \mathbb{F}_4$. Use the construction of $\mathbb{F}_4$ in Exercise 2.8 on page 89 to display $\alpha$ explicitly.

**2.48.** Find the gcd of $x^2 - x - 2$ and $x^3 - 7x + 6$ in $\mathbb{F}_5[x]$, and express it as a linear combination of them.

**Hint.** The answer is $x - 2$.

**2.49.** Prove the converse of Euclid's Lemma. Let $k$ be a field and let $f(x) \in k[x]$ be a polynomial of degree $\geq 1$; if, whenever $f(x)$ divides a product of two polynomials, it necessarily divides one of the factors, then $f(x)$ is irreducible.

\* **2.50.** Let $R$ be a domain. If $f(x) \in R[x]$ has degree $n$, prove that $f(x)$ has at most $n$ roots in $R$.

**Hint.** Use $\mathrm{Frac}(R)$.

\* **2.51.** (i) Let $f(x), g(x) \in R[x]$, where $R$ is a domain. If the leading coefficient of $f(x)$ is a unit in $R$, then the Division Algorithm gives a quotient $q(x)$ and a remainder $r(x)$ after dividing $g(x)$ by $f(x)$. Prove that $q(x)$ and $r(x)$ are uniquely determined by $g(x)$ and $f(x)$.

(ii) Give an example of a commutative ring $R$ and $f(x), g(x) \in R[x]$ with $f$ monic such that the remainder after dividing $g$ by $f$ is not unique; that is, there are $q, q', r, r' \in R[x]$ with $qf + r = g = q'f + r'$, $\deg(r) < \deg(f)$, $\deg(r') < \deg(f)$, and $r \neq r'$.

\* **2.52.** Let $k$ be a field, and let $f(x), g(x) \in k[x]$ be relatively prime. If $h(x) \in k[x]$, prove that $f \mid h$ and $g \mid h$ imply $fg \mid h$.

**2.53.** If $k$ is a field in which $1 + 1 \neq 0$, prove that $\sqrt{1 - x^2}$ is not a rational function over $k$.

**Hint.** Mimic the classical proof that $\sqrt{2}$ is irrational.

\* **2.54.** (i) Let $f(x) = (x - a_1) \cdots (x - a_n) \in k[x]$, where $k$ is a field. Show that $f(x)$ has **no repeated roots** (i.e., all the $a_i$ are distinct elements of $k$) if and only if $\gcd(f, f') = 1$, where $f'(x)$ is the derivative of $f$.

**Hint.** Use Exercise 2.24 on page 95.

(ii) Prove that if $p(x) \in \mathbb{Q}[x]$ is an irreducible polynomial, then $p(x)$ has no repeated roots in $\mathbb{C}$.

**Hint.** Corollary 2.58.

(iii) Let $k = \mathbb{F}_2(x)$. Prove that $f(t) = t^2 - x \in k[t]$ is an irreducible polynomial. (There is a field $K$ containing $k$ and $\alpha = \sqrt{x}$, and $f(t) = (t - \alpha)^2$ in $K[t]$.)

**2.55.** If $p$ is prime, prove that there are exactly $\frac{1}{3}(p^3 - p)$ monic irreducible cubic polynomials in $\mathbb{F}_p[x]$. (A formula for the number of monic irreducible polynomials of degree $n$ in $\mathbb{F}_p[x]$ is given on page 167.)

## Section 2.5.  Irreducibility

Although there are some techniques to help decide whether an integer is prime, the general problem is open and is very difficult. Similarly, it is very difficult to determine whether a polynomial is irreducible, but there are some useful techniques that frequently work.

Let $k$ be a field. Proposition 2.39 shows that if $f(x) \in k[x]$ and $r$ is a root of $f(x)$ in $k$, then $f(x)$ is not irreducible; there is a factorization $f(x) = (x - r)g(x)$ in $k[x]$. We saw, in Corollary 2.51, that this decides the matter for quadratic and cubic polynomials in $k[x]$: such polynomials are irreducible in $k[x]$ if and only if they have no roots in $k$. This is no longer true for polynomials of degree $\geq 4$, as $f(x) = (x^2 + 1)(x^2 + 1)$ in $\mathbb{R}[x]$ shows. The next theorem tests for rational roots.

**Theorem 2.63.** *If $f(x) = a_0 + a_1 x + \cdots + a_n x^n \in \mathbb{Z}[x] \subseteq \mathbb{Q}[x]$, then every rational root of $f(x)$ has the form $b/c$, where $b \mid a_0$ and $c \mid a_n$.*

*In particular, if $f(x) \in \mathbb{Z}[x]$ is monic, then every rational root of $f(x)$ is an integer.*

**Proof.** We may assume that a root $b/c$ is in lowest terms; that is, $(b, c) = 1$. Evaluating gives $0 = f(b/c) = a_0 + a_1 b/c + \cdots + a_n b^n/c^n$, and multiplying through by $c^n$ gives
$$0 = a_0 c^n + a_1 b c^{n-1} + \cdots + a_n b^n.$$
Hence, $a_0 c^n = b(-a_1 c^{n-1} - \cdots - a_n b^{n-1})$, so that $b \mid a_0 c^n$. Since $b$ and $c$ are relatively prime, it follows that $b$ and $c^n$ are relatively prime, and so Euclid's Lemma in $\mathbb{Z}$ gives $b \mid a_0$. Similarly, $a_n b^n = c(-a_{n-1} b^{n-1} - \cdots - a_0 c^{n-1})$, $c \mid a_n b^n$, and $c \mid a_n$. •

It follows from the second statement that if an integer $a$ is not the $n$th power of an integer, then $x^n - a$ has no rational roots; that is, $\sqrt[n]{a}$ is irrational. For example, $\sqrt{2}$ is irrational.

The next criterion for irreducibility uses the integers mod $p$.

**Theorem 2.64.** *Let $f(x) = a_0 + a_1 x + a_2 x^2 + \cdots + x^n \in \mathbb{Z}[x]$ be monic, and let $p$ be prime. If $\overline{f}(x) = [a_0] + [a_1]x + [a_2]x^2 + \cdots + x^n$ is irreducible in $\mathbb{F}_p[x]$, then $f(x)$ is irreducible in $\mathbb{Q}[x]$.*

**Proof.** Reducing coefficients mod $p$ is a special case of Corollary 2.27, for the natural map $\varphi: \mathbb{Z} \to \mathbb{F}_p$ gives a ring homomorphism $\varphi^*: \mathbb{Z}[x] \to \mathbb{F}_p[x]$, namely, $\varphi^*: f(x) \mapsto \overline{f}(x)$. Suppose that $f(x)$ factors in $\mathbb{Z}[x]$; say, $f(x) = g(x)h(x)$, where $\deg(g) < \deg(f)$ and $\deg(h) < \deg(f)$. Now $\overline{f}(x) = \overline{g}(x)\overline{h}(x)$ (for $\varphi^*$ is a ring homomorphism), so that $\deg(\overline{f}) = \deg(\overline{g}) + \deg(\overline{h})$. Now $\overline{f}(x)$ is monic, because

$f(x)$ is, and so $\deg(\overline{f}) = \deg(f)$.[10] Thus, both $\overline{g}(x)$ and $\overline{h}(x)$ have degrees less than $\deg(\overline{f})$, contradicting the irreducibility of $\overline{f}(x)$ in $\mathbb{F}_p[x]$. Therefore, $f(x)$ is not a product of polynomials in $\mathbb{Z}[x]$ of smaller degree, and so Gauss's Lemma says that $f(x)$ is irreducible in $\mathbb{Q}[x]$.    •

Theorem 2.64 says that if one can find a prime $p$ with $\overline{f}(x)$ irreducible in $\mathbb{F}_p[x]$, then $f(x)$ is irreducible in $\mathbb{Q}[x]$. Until now, the finite fields $\mathbb{F}_p$ have been oddities; $\mathbb{F}_p$ has appeared only as a curious artificial construct. Now the finiteness of $\mathbb{F}_p$ is a genuine advantage, for there are only a finite number of polynomials in $\mathbb{F}_p[x]$ of any given degree. In principle, then, one can test whether a polynomial of degree $n$ in $\mathbb{F}_p[x]$ is irreducible by just looking at *all* the possible factorizations of it.

The converse of Theorem 2.64 is false: $x^2 - 2$, irreducible in $\mathbb{Q}[x]$, factors mod 2. More spectacularly, Proposition 2.92 shows that $x^4 + 1$ is an irreducible polynomial in $\mathbb{Q}[x]$ which factors in $\mathbb{F}_p[x]$ for every prime $p$.

**Example 2.65.** We determine the irreducible polynomials in $\mathbb{F}_2[x]$ of small degree.

As always, the linear polynomials $x$ and $x + 1$ are irreducible.

There are four quadratics: $x^2$, $x^2 + x$, $x^2 + 1$, $x^2 + x + 1$ (more generally, there are $p^n$ monic polynomials of degree $n$ in $\mathbb{F}_p[x]$, for there are $p$ choices for each of the $n$ coefficients $a_0, \ldots, a_{n-1}$). Since each of the first three has a root in $\mathbb{F}_2$, there is only one irreducible quadratic, namely, $x^2 + x + 1$.

There are eight cubics, of which four are reducible because their constant term is 0. The remaining polynomials are

$$x^3 + 1, \qquad x^3 + x + 1, \qquad x^3 + x^2 + 1, \qquad x^3 + x^2 + x + 1.$$

Since 1 is a root of the first and fourth, the middle two are the only irreducible cubics.

There are 16 quartics, of which eight are reducible because their constant term is 0. Of the eight with nonzero constant term, those having an even number of nonzero coefficients have 1 as a root. There are now only four surviving polynomials $f(x)$, and each of them has no roots in $\mathbb{F}_2$; i.e., they have no linear factors. If $f(x) = g(x)h(x)$, then both $g(x)$ and $h(x)$ must be irreducible quadratics. But there is only one irreducible quadratic, namely, $x^2 + x + 1$, and so $(x^2 + x + 1)^2 = x^4 + x^2 + 1$ factors while the other three quartics are irreducible.

### Irreducible Polynomials of Low Degree over $\mathbb{F}_2$

degree 2:   $x^2 + x + 1$.
degree 3:   $x^3 + x + 1$;   $x^3 + x^2 + 1$.
degree 4:   $x^4 + x^3 + 1$;   $x^4 + x + 1$;   $x^4 + x^3 + x^2 + x + 1$.    ◄

**Example 2.66.** Here is a list of the monic irreducible quadratics and cubics in $\mathbb{F}_3[x]$. The reader can verify that the list is correct by first enumerating all such polynomials; there are 6 monic quadratics having nonzero constant term, and there are 18 monic cubics having nonzero constant term. It must then be checked which of these have 1 or $-1$ as a root (it is more convenient to write $-1$ instead of 2).

---

[10]The hypothesis that $f(x)$ be monic can be relaxed; we may assume instead that $p$ does not divide its leading coefficient.

### Monic Irreducible Quadratics and Cubics over $\mathbb{F}_3$

degree 2:     $x^2 + 1$;                    $x^2 + x - 1$;                    $x^2 - x - 1$.

degree 3:     $x^3 - x + 1$;               $x^3 + x^2 - x + 1$;             $x^3 - x^2 + 1$;
              $x^3 - x^2 + x + 1$;         $x^3 - x - 1$;                   $x^3 + x^2 - 1$;
              $x^3 + x^2 + x - 1$;         $x^3 - x^2 - x - 1$.    ◄

**Example 2.67.**

(i) We show that $f(x) = x^4 - 5x^3 + 2x + 3$ is an irreducible polynomial in $\mathbb{Q}[x]$. By Corollary 2.63, the only candidates for rational roots of $f(x)$ are $\pm 1$ and $\pm 3$, and the reader may check that none of these is a root. Since $f(x)$ is a quartic, one cannot yet conclude that $f(x)$ is irreducible, for it might be a product of (irreducible) quadratics.

Let us try the criterion of Theorem 2.64. Since $\overline{f}(x) = x^4 + x^3 + 1$ in $\mathbb{F}_2[x]$ is irreducible, by Example 2.65, it follows that $f(x)$ is irreducible in $\mathbb{Q}[x]$. [It was not necessary to check that $f(x)$ has no rational roots; irreducibility of $\overline{f}(x)$ is enough to conclude irreducibility of $f(x)$.]

(ii) Let $\Phi_5(x) = x^4 + x^3 + x^2 + x + 1 \in \mathbb{Q}[x]$. In Example 2.65, we saw that $\overline{\Phi}_5(x) = x^4 + x^3 + x^2 + x + 1$ is irreducible in $\mathbb{F}_2[x]$, and so $\Phi_5(x)$ is irreducible in $\mathbb{Q}[x]$.    ◄

**Definition.** A complex number $\alpha$ is called an ***algebraic integer*** if $\alpha$ is a root of a monic $f(x) \in \mathbb{Z}[x]$.

We note that it is crucial, in the definition of algebraic integer, that $f(x) \in \mathbb{Z}[x]$ be monic. Every algebraic number $z$, that is, every complex number $z$ that is a root of some polynomial $g(x) \in \mathbb{Q}[x]$, is necessarily a root of some polynomial $h(x) \in \mathbb{Z}[x]$; just clear the denominators of the coefficients of $g(x)$.

Of course, every ordinary integer is an algebraic integer. To contrast ordinary integers with more general algebraic integers, elements of $\mathbb{Z}$ may be called ***rational integers***. Theorem 2.63 can be rephrased: an algebraic integer is either an integer or it is irrational. The next two group-theoretic results will be used to characterize algebraic integers.

**Proposition 2.68.** *Let $S$ be a normal subgroup of a group $A$. If $S = \langle s_1, \ldots, s_m \rangle$ and $A/S = \langle u_1, \ldots, u_n \rangle$, then $A$ can be generated by $m + n$ elements.*

**Proof.** Let $\pi : A \to A/S$ be the natural map, and let $\pi(x_i) = u_i$ for all $i$. Now if $a \in A$, then $\pi(a)$ is a word in the $u$: there are $e_j = \pm 1$ with

$$\pi(a) = u_{i_1}^{e_1} \cdots u_{i_k}^{e_k}.$$

Hence, $a^{-1} x_{i_1}^{e_1} \cdots x_{i_k}^{e_k} \in \ker \pi = S$; that is, $a \in \langle s_1, \ldots, s_m, x_1, \ldots, x_n \rangle$.    •

**Proposition 2.69.** *If an abelian group $A$ can be generated by $n$ elements, then every subgroup $S \subseteq A$ can be generated by $n$ or fewer elements.*

**Remark.** This proposition is false for nonabelian groups. Corollary 4.95 shows that a subgroup of a finitely generated group need not be finitely generated.    ◄

**Proof.** We prove the statement by induction on $n \geq 1$. If $A$ is cyclic, then Lemma 1.92(i) says that every subgroup of $A$ is cyclic.

For the inductive step, let $A = \langle a_1, \ldots, a_{n+1} \rangle$, and define $A' = \langle a_1, \ldots, a_n \rangle$. Since $A$ is abelian, $A' \lhd A$, and the Second Isomorphism Theorem gives

$$S/(S \cap A') \cong (S + A')/A' \subseteq A/A'.$$

But $A/A' = \langle a_{n+1} + A' \rangle$ is cyclic, so that $S/(S \cap A')$ is also cyclic, by the base step. In view of the inductive hypothesis, $S \cap A' \subseteq A'$ can be generated by $n$ or fewer elements. Proposition 2.68 gives the result. $\bullet$

The next proposition shows that the sum and product of algebraic integers are themselves algebraic integers. If $\alpha$ and $\beta$ are algebraic integers, it is not too difficult to show there are monic polynomials in $\mathbb{Q}[x]$ having $\alpha + \beta$ and $\alpha\beta$ as roots, but it is harder to show there are such polynomials in $\mathbb{Z}[x]$ (see Exercise 5.21 on page 311).

**Proposition 2.70.** *Let $\alpha \in \mathbb{C}$ and define $\mathbb{Z}[\alpha] = \{g(\alpha) : g(x) \in \mathbb{Z}[x]\}$.*

(i) $\mathbb{Z}[\alpha]$ *is a subring of* $\mathbb{C}$.

(ii) $\alpha$ *is an algebraic integer if and only if $\mathbb{Z}[\alpha]$ is a finitely generated additive abelian group.*

(iii) *The set $\mathbb{A}$ of all the algebraic integers is a subring of $\mathbb{C}$, and $\mathbb{A} \cap \mathbb{Q} = \mathbb{Z}$.*

**Proof.**

(i) If $g(x) = 1$ is the constant polynomial, then $g(x) \in \mathbb{Z}[x]$; hence, $1 = g(\alpha)$ and so $1 \in \mathbb{Z}[\alpha]$. Suppose that $f(\alpha), g(\alpha) \in \mathbb{Z}[\alpha]$, where $f(x), g(x) \in \mathbb{Z}[x]$. Now $f(x) + g(x)$ and $f(x)g(x)$ lie in $\mathbb{Z}[x]$, so that $f(\alpha) + g(\alpha), f(\alpha)g(\alpha) \in \mathbb{Z}[\alpha]$. Therefore, $\mathbb{Z}[\alpha]$ is a subring of $\mathbb{C}$.

(ii) If $\alpha$ is an algebraic integer, there is a monic polynomial $f(x) \in \mathbb{Z}[x]$ having $\alpha$ as a root. We claim that if $\deg(f) = n$, then $\mathbb{Z}[\alpha] = G$, where $G$ is the set of all linear combinations $m_0 + m_1\alpha + \cdots + m_{n-1}\alpha^{n-1}$ with $m_i \in \mathbb{Z}$. Clearly, $G \subseteq \mathbb{Z}[\alpha]$. For the reverse inclusion, each element $u \in \mathbb{Z}[\alpha]$ has the form $u = g(\alpha)$, where $g(x) \in \mathbb{Z}[x]$. Since $f(x)$ is monic, Corollary 2.35 of the Division Algorithm gives $q(x), r(x) \in \mathbb{Z}[x]$ with $g(x) = q(x)f(x) + r(x)$, where either $r(x) = 0$ or $\deg(r) < \deg(f) = n$. Therefore,

$$u = g(\alpha) = q(\alpha)f(\alpha) + r(\alpha) = r(\alpha) \in G.$$

Thus, the additive group of $\mathbb{Z}[\alpha]$ is finitely generated.

Conversely, if the additive group of the commutative ring $\mathbb{Z}[\alpha]$ is finitely generated, that is, $\mathbb{Z}[\alpha] = \langle g_1, \ldots, g_m \rangle$ as an abelian group, then each $g_j$ is a $\mathbb{Z}$-linear combination of powers of $\alpha$. Let $m$ be the largest power of $\alpha$ occurring in any of these $g$s. Since $\mathbb{Z}[\alpha]$ is a commutative ring, $\alpha^{m+1} \in \mathbb{Z}[\alpha]$; hence, $\alpha^{m+1}$ can be expressed as a $\mathbb{Z}$-linear combination of smaller powers of $\alpha$; say, $\alpha^{m+1} = \sum_{i=0}^{m} b_i\alpha^i$, where $b_i \in \mathbb{Z}$. Therefore, $\alpha$ is a root of $f(x) = x^{m+1} - \sum_{i=0}^{m} b_i x^i$, which is a monic polynomial in $\mathbb{Z}[x]$, and so $\alpha$ is an algebraic integer.

(iii) Suppose $\alpha$ and $\beta$ are algebraic integers; let $\alpha$ be a root of a monic $f(x) \in \mathbb{Z}[x]$ of degree $n$, and let $\beta$ be a root of a monic $g(x) \in \mathbb{Z}[x]$ of degree $m$. Now

$\mathbb{Z}[\alpha\beta]$ is an additive subgroup of $G = \langle \alpha^i \beta^j : 0 \leq i < n, 0 \leq j < m \rangle$. Since $G$ is finitely generated, so is its subgroup $\mathbb{Z}[\alpha\beta]$, by Proposition 2.69, and so $\alpha\beta$ is an algebraic integer. Similarly, $\mathbb{Z}[\alpha + \beta]$ is an additive subgroup of $\langle \alpha^i \beta^j : i + j \leq n + m - 1 \rangle$, and so $\alpha + \beta$ is also an algebraic integer.

The last statement is Theorem 2.63: if an algebraic integer is not an integer, it is irrational.  •

This last proposition gives a technique for proving that an integer $a$ is a divisor of an integer $b$. If we can prove that $b/a$ is an algebraic integer, then it must be an integer, for it is obviously rational. This will actually be used in Chapter 7 to prove that the degrees of the irreducible characters of a finite group $G$ are divisors of $|G|$.

We now consider some special algebraic integers.

**Definition.** If $n \geq 1$ is a positive integer, then an $n$th **root of unity** in a field $k$ is an element $\zeta \in k$ with $\zeta^n = 1$.

Corollary 2.42 shows that the numbers $e^{2\pi i k/n} = \cos(2\pi k/n) + i \sin(2\pi k/n)$ for some $k$ with $0 \leq k \leq n - 1$ are *all* the complex $n$th roots of unity. Just as there are two square roots of a number $a$, namely, $\sqrt{a}$ and $-\sqrt{a}$, there are $n$ different $n$th roots of $a$, namely, $e^{2\pi i k/n} \sqrt[n]{a}$ for $k = 0, 1, \ldots, n - 1$.

Every $n$th root of unity is, of course, a root of the polynomial $x^n - 1$. Therefore,

$$x^n - 1 = \prod_{\zeta^n = 1} (x - \zeta).$$

If $\zeta$ is an $n$th root of unity and $n$ is the smallest positive integer for which $\zeta^n = 1$, we say that $\zeta$ is a **primitive** $n$th **root of unity**. For example, $i$ is an 8th root of unity (for $i^8 = 1$), but not a primitive 8th root of unity; $i$ is a primitive 4th root of unity. The $n$th roots of unity form a multiplicative group, and the primitive $n$th roots of unity are the generators of this group, by Theorem 1.39. It follows from Proposition 1.26 that if $\zeta$ is a primitive $d$th root of unity and $\zeta^n = 1$, then $d \mid n$.

**Definition.** If $d$ is a positive integer, then the $d$th **cyclotomic polynomial**[11] is defined by

$$\Phi_d(x) = \prod (x - \zeta),$$

where $\zeta$ ranges over all the *primitive* $d$th roots of unity.

**Proposition 2.71.** *Let $n$ be a positive integer and regard $x^n - 1 \in \mathbb{Z}[x]$. Then*

(i) *we have*

$$x^n - 1 = \prod_{d \mid n} \Phi_d(x),$$

*where $d$ ranges over all the positive divisors $d$ of $n$ [in particular, $\Phi_1(x)$ and $\Phi_n(x)$ occur].*

---

[11]Since $|zw| = |z|\,|w|$ for any complex numbers $z$ and $w$, it follows that if $\zeta$ is an $n$th root of unity, then $1 = |\zeta^n| = |\zeta|^n$, so that $|\zeta| = 1$ and $\zeta$ lies on the unit circle. The roots of $x^n - 1$ are the $n$th roots of unity which divide the unit circle into $n$ equal arcs. This explains the term *cyclotomic*, for its Greek origin means "circle splitting."

(ii) $\Phi_n(x)$ *is a monic polynomial in* $\mathbb{Z}[x]$ *and* $\deg(\Phi_n) = \phi(n)$, *the Euler* $\phi$-*function.*

(iii) *For every integer* $n \geq 1$, *we have*

$$n = \sum_{d|n} \phi(d).$$

**Proof.**

(i) For each divisor $d$ of $n$, collect all terms in the equation $x^n - 1 = \prod(x - \zeta)$ with $\zeta$ a primitive $d$th root of unity.

(ii) We prove that $\Phi_n(x) \in \mathbb{Z}[x]$ by induction on $n \geq 1$. The base step is true, for $\Phi_1(x) = x - 1 \in \mathbb{Z}[x]$. For the inductive step, let $f(x) = \prod_{d|n, d<n} \Phi_d(x)$, so that

$$x^n - 1 = f(x)\Phi_n(x).$$

By induction, each $\Phi_d(x)$ is a monic polynomial in $\mathbb{Z}[x]$, and so $f(x)$ is a monic polynomial in $\mathbb{Z}[x]$. Since $f(x)$ is monic, Corollary 2.35 says that the quotient $(x^n - 1)/f(x)$ is a monic polynomial in $\mathbb{Z}[x]$. Exercise 2.51 on page 114 says that quotients are unique; hence, $(x^n - 1)/f(x) = \Phi_n(x)$, and so $\Phi_n(x) \in \mathbb{Z}[x]$.

(iii) Immediate from part (ii).  ●

It follows from Proposition 2.71(i) that if $p$ is prime, then $x^p - 1 = \Phi_1(x)\Phi_p(x)$. Since $\Phi_1(x) = x - 1$, we have

$$\Phi_p(x) = x^{p-1} + x^{p-2} + \cdots + x + 1.$$

The next corollary will be used in Chapter 7 to prove a theorem of Wedderburn.

**Corollary 2.72.** *If* $q$ *is a positive integer and* $d$ *is a divisor of an integer* $n$ *with* $d < n$, *then* $\Phi_n(q)$ *is a divisor of both* $q^n - 1$ *and* $(q^n - 1)/(q^d - 1)$.

**Proof.** We have just seen that $x^n - 1 = \Phi_n(x)f(x)$, where $f(x)$ is a monic polynomial with integer coefficients. Setting $x = q$ gives an equation in integers: $q^n - 1 = \Phi_n(q)f(q) \in \mathbb{Z}$; that is, $\Phi_n(q)$ is a divisor of $q^n - 1$.

If $d$ is a divisor of $n$ and $d < n$, consider the equation $x^d - 1 = \prod(x - \zeta)$, where $\zeta$ ranges over the $d$th roots of unity. Notice that each such $\zeta$ is an $n$th root of unity, because $d$ is a divisor of $n$. Since $d < n$, collecting terms in the equation $x^n - 1 = \prod(x - \zeta)$ gives

$$x^n - 1 = \Phi_n(x)(x^d - 1)g(x),$$

where $g(x)$ is the product of all the cyclotomic polynomials $\Phi_\delta(x)$ for all divisors $\delta$ of $n$ with $\delta < n$ and with $\delta$ not a divisor of $d$. It follows from Proposition 2.71 that $g(x)$ is a monic polynomial with integer coefficients. Therefore, $g(q) \in \mathbb{Z}$ and

$$\frac{q^n - 1}{q^d - 1} = \Phi_n(q)g(q) \in \mathbb{Z}.  ●$$

If we regard complex numbers as points in the plane, then we may define the **dot product** of $z = a + ib$ and $w = c + id$ to be

$$z \cdot w = ac + bd.$$

The next result will be used in Chapter 7 to investigate character tables.

**Proposition 2.73.** *If $\varepsilon_1, \ldots, \varepsilon_n$ are complex roots of unity, where $n \geq 2$, then*

$$\left| \sum_{j=1}^{n} \varepsilon_j \right| \leq \sum_{j=1}^{n} |\varepsilon_j| = n.$$

*Moreover, there is equality if and only if all the $\varepsilon_j$ are equal.*

**Proof.** If $u, v$ are nonzero complex numbers, the Triangle Inequality says that $|u + v| \leq |u| + |v|$, with equality if and only if $u/v$ is a positive real. The *Extended Triangle Inequality* says, for nonzero complex numbers $u_1, \ldots, u_n$, that $|u_1 + \cdots + u_n| \leq |u_1| + \cdots + |u_n|$, with equality if and only if there is $z$ and positive real numbers $r_j$ with $u_j = r_j z$ for all $j$. Thus, if there is equality and $j \neq k$, then $u_j/u_k = r_j z / r_k z = r_j/r_k$; that is, $u_j = (r_j/r_k)u_k$. When the $u_j = \varepsilon_j$ are roots of unity, then $|\varepsilon_j| = 1 = |\varepsilon_k|$, $r_j/r_k = 1$, and $r_j = r_k$; that is, $\varepsilon_j = \varepsilon_k$ and all $\varepsilon_j$ are equal. •

As any linear polynomial over a field, the cyclotomic polynomial $\Phi_2(x) = x + 1$ is irreducible in $\mathbb{Q}[x]$; $\Phi_3(x) = x^2 + x + 1$ is irreducible in $\mathbb{Q}[x]$ because it has no rational roots; we saw, in Example 2.67, that $\Phi_5(x)$ is irreducible in $\mathbb{Q}[x]$. Let us introduce another irreducibility criterion in order to prove that $\Phi_p(x)$ is irreducible in $\mathbb{Q}[x]$ for all primes $p$.

**Lemma 2.74.** *Let $g(x) \in \mathbb{Z}[x]$. If there is $c \in \mathbb{Z}$ with $g(x + c)$ irreducible in $\mathbb{Z}[x]$, then $g(x)$ is irreducible in $\mathbb{Q}[x]$.*

**Proof.** By Theorem 2.25, the function $\varphi : \mathbb{Z}[x] \to \mathbb{Z}[x]$, given by

$$f(x) \mapsto f(x + c),$$

is an isomorphism [its inverse is $f(x) \mapsto f(x-c)$]. If $g(x) = s(x)t(x)$, then $g(x+c) = \varphi(g(x))$, and $\varphi(g) = \varphi(st) = \varphi(s)\varphi(t)$ is a forbidden factorization of $g(x + c)$. Therefore, Gauss's Lemma, Theorem 2.52, says that $g(x)$ is irreducible in $\mathbb{Q}[x]$. •

**Theorem 2.75 (Eisenstein Criterion).** *Let $f(x) = a_0 + a_1 x + \cdots + a_n x^n \in \mathbb{Z}[x]$. If there is a prime $p$ dividing $a_i$ for all $i < n$ but with $p \nmid a_n$ and $p^2 \nmid a_0$, then $f(x)$ is irreducible in $\mathbb{Q}[x]$.*

**Proof.** Assume, on the contrary, that

$$f(x) = (b_0 + b_1 x + \cdots + b_m x^m)(c_0 + c_1 x + \cdots + c_k x^k),$$

where $m < n$ and $k < n$; by Gauss's Lemma, we may assume that both factors lie in $\mathbb{Z}[x]$. Now $p \mid a_0 = b_0 c_0$, so that Euclid's Lemma in $\mathbb{Z}$ gives $p \mid b_0$ or $p \mid c_0$; since $p^2 \nmid a_0$, only one of them is divisible by $p$, say, $p \mid c_0$ but $p \nmid b_0$. By hypothesis, the leading coefficient $a_n = b_m c_k$ is not divisible by $p$, so that $p$ does not divide $c_k$ (or $b_m$). Let $c_r$ be the first coefficient not divisible by $p$ (so that $p$ does divide

$c_0, \ldots, c_{r-1}$). If $r < n$, then $p \mid a_r$, and so $b_0 c_r = a_r - (b_1 c_{r-1} + \cdots + b_r c_0)$ is also divisible by $p$. This contradicts Euclid's Lemma, for $p \mid b_0 c_r$, but $p$ divides neither factor. It follows that $r = n$; hence $n \geq k \geq r = n$, and so $k = n$, contradicting $k < n$. Therefore, $f(x)$ is irreducible in $\mathbb{Q}[x]$.   •

R. Singer found the following elegant proof of Eisenstein's Criterion (Montgomery–Ralston, *Selected Papers in Algebra*, p. 78).

**Proof.** Let $\varphi^* \colon \mathbb{Z}[x] \to \mathbb{F}_p[x]$ be the ring homomorphism that reduces coefficients mod $p$, and let $\overline{f}(x)$ denote $\varphi^*(f(x))$. If $f(x)$ is not irreducible in $\mathbb{Q}[x]$, then Gauss's Theorem gives polynomials $g(x), h(x) \in \mathbb{Z}[x]$ with $f(x) = g(x)h(x)$, where $g(x) = b_0 + b_1 x + \cdots + b_m x^m$, $h(x) = c_0 + c_1 x + \cdots + c_k x^k$, and $m, k > 0$. There is thus an equation $\overline{f}(x) = \overline{g}(x)\overline{h}(x)$ in $\mathbb{F}_p[x]$.

Since $p \nmid a_n$, we have $\overline{f}(x) \neq 0$; in fact, $\overline{f}(x) = u x^n$ for some unit $u \in \mathbb{F}_p$, because all its coefficients aside from its leading coefficient are 0. By unique factorization in $\mathbb{F}_p[x]$, we must have $\overline{g}(x) = v x^m$ and $\overline{h}(x) = w x^k$ (for units $v, w$ in $\mathbb{F}_p$), so that each of $\overline{g}(x)$ and $\overline{h}(x)$ has constant term 0. Thus, $[b_0] = 0 = [c_0]$ in $\mathbb{F}_p$; equivalently, $p \mid b_0$ and $p \mid c_0$. But $a_0 = b_0 c_0$, and so $p^2 \mid a_0$, a contradiction. Therefore, $f(x)$ is irreducible in $\mathbb{Q}[x]$.   •

**Theorem 2.76 (Gauss).** *For every prime $p$, the $p$th cyclotomic polynomial $\Phi_p(x)$ is irreducible in $\mathbb{Q}[x]$.*

**Proof.** Since $\Phi_p(x) = (x^p - 1)/(x - 1)$, we have

$$\Phi_p(x+1) = [(x+1)^p - 1]/x = x^{p-1} + \binom{p}{1} x^{p-2} + \binom{p}{2} x^{p-3} + \cdots + p.$$

Since $p$ is prime, we have $p \mid \binom{p}{i}$ for all $i$ with $0 < i < p$ (FCAA, p. 42); hence, Eisenstein's Criterion applies, and $\Phi_p(x+1)$ is irreducible in $\mathbb{Q}[x]$. By Lemma 2.74, $\Phi_p(x)$ is irreducible in $\mathbb{Q}[x]$.   •

**Remark.**

(i) The cyclotomic polynomial $\Phi_d(x)$ is irreducible in $\mathbb{Q}[x]$ for every (not necessarily prime) $d \geq 1$ (Tignol, *Galois' Theory of Algebraic Equations*, p. 198).

(ii) We do not say that $x^{n-1} + x^{n-2} + \cdots + x + 1$ is irreducible when $n$ is not prime. For example, when $n = 4$, $x^3 + x^2 + x + 1 = (x + 1)(x^2 + 1)$.

(iii) Gauss needed Theorem 2.76 in order to prove that every regular 17-gon can be constructed with straightedge and compass. In fact, he proved that if $p$ is a prime of the form $p = 2^{2^m} + 1$, where $m \geq 0$, then every regular $p$-gon can be constructed using straightedge and compass (such primes $p$ are called **Fermat primes**; the only known such are 3, 5, 17, 257, and 65537). See Tignol, *Galois' Theory of Algebraic Equations*, pp. 200–206.   ◄

---

## Exercises

\* **2.56.** Let $\zeta = e^{2\pi i/n}$ be a primitive $n$th root of unity.

(i) Prove that $x^n - 1 = (x-1)(x-\zeta)(x-\zeta^2)\cdots(x-\zeta^{n-1})$ and, if $n$ is odd, that $x^n + 1 = (x+1)(x+\zeta)(x+\zeta^2)\cdots(x+\zeta^{n-1})$.

(ii) For numbers $a$ and $b$, prove that $a^n - b^n = (a-b)(a-\zeta b)(a-\zeta^2 b)\cdots(a-\zeta^{n-1}b)$ and, if $n$ is odd, that $a^n + b^n = (a+b)(a+\zeta b)(a+\zeta^2 b)\cdots(a+\zeta^{n-1}b)$.
**Hint.** Set $x = a/b$ if $b \neq 0$.

\* **2.57.** Determine whether the following polynomials are irreducible in $\mathbb{Q}[x]$.

(i) $f(x) = 3x^2 - 7x - 5$.

(ii) $f(x) = 2x^3 - x - 6$.

(iii) $f(x) = 8x^3 - 6x - 1$.

(iv) $f(x) = x^3 + 6x^2 + 5x + 25$.

(v) $f(x) = x^4 + 8x + 12$.
**Hint.** In $\mathbb{F}_5[x]$, $f(x) = (x+1)g(x)$, where $g(x)$ is irreducible.

(vi) $f(x) = x^5 - 4x + 2$.

(vii) $f(x) = x^4 + x^2 + x + 1$.
**Hint.** Show that $f(x)$ has no roots in $\mathbb{F}_3$ and that a factorization of $f(x)$ as a product of quadratics would force impossible restrictions on the coefficients.

(viii) $f(x) = x^4 - 10x^2 + 1$.
**Hint.** Show that $f(x)$ has no rational roots and that a factorization of $f(x)$ as a product of quadratics would force impossible restrictions on the coefficients.

**2.58.** Is $x^5 + x + 1$ irreducible in $\mathbb{F}_2[x]$?
**Hint.** Use Example 2.65.

**2.59.** Let $f(x) = (x^p - 1)/(x-1)$, where $p$ is prime. Using the identity

$$f(x+1) = x^{p-1} + pq(x),$$

where $q(x) \in \mathbb{Z}[x]$ has constant term 1, prove that $\Phi_p(x^{p^n}) = x^{p^n(p-1)} + \cdots + x^{p^n} + 1$ is irreducible in $\mathbb{Q}[x]$ for all $n \geq 0$.

**2.60.** Use the Eisenstein Criterion to prove that if $a$ is a squarefree integer, then $x^n - a$ is irreducible in $\mathbb{Q}[x]$ for every $n \geq 1$. Conclude that there are irreducible polynomials in $\mathbb{Q}[x]$ of every degree $n \geq 1$.

**2.61.** Let $k$ be a field, and let $f(x) = a_0 + a_1 x + \cdots + a_n x^n \in k[x]$ have degree $n$ and nonzero constant term $a_0$. Prove that if $f(x)$ is irreducible, then so is $a_n + a_{n-1}x + \cdots + a_0 x^n$.

---

## Section 2.6. Euclidean Rings and Principal Ideal Domains

Let us consider the parallel discussions of divisibility in $\mathbb{Z}$ and in $k[x]$, where $k$ is a field. A glance at proofs of the existence of gcd's, Euclid's Lemma, and unique factorization suggests that the Division Algorithm is the key property of these rings yielding these results. We begin with a generalization of gcd that makes sense in any domain.

**Definition.** Let $R$ be a domain. If $a, b \in R$, then a **greatest common divisor** (gcd) of $a, b$ is a common divisor $d \in R$ which is divisible by every common divisor; that is, if $c \mid a$ and $c \mid b$, then $c \mid d$.

By Corollary 2.49, greatest common divisors in $k[x]$, where $k$ is a field, are still gcd's under this new definition. However, gcd's (when they exist) need not be unique; for example, it is easy to see that if $c$ is a gcd of $f$ and $g$, then so is $uc$ for any unit $u \in R$. In the special case $R = \mathbb{Z}$, we forced uniqueness by requiring the gcd to be positive; in the case $R = k[x]$, where $k$ is a field, we forced uniqueness by further requiring the gcd to be monic. Similarly, least common multiples (when they exist) need not be unique; if $c$ is an lcm of $f$ and $g$, then so is $uc$ for any unit $u \in R$.

For an example of a domain in which a pair of elements does not have a gcd, see Exercise 2.67 on page 133.

**Example 2.77.** Let $R$ be a domain. If $p, a \in R$ with $p$ irreducible, we claim that a gcd $d$ of $p$ and $a$ exists. If $p \mid a$, then $p$ is a gcd; if $p \nmid a$, then 1 is a gcd.  ◄

**Example 2.78.** Even if a gcd of a pair of elements $a, b$ in a domain $R$ exists, it need not be an $R$-linear combination of $a$ and $b$. For example, let $R = k[x, y]$, where $k$ is a field. It is easy to see that 1 is a gcd of $x$ and $y$; if there exist $s = s(x, y), t = t(x, y) \in k[x, y]$ with $1 = xs + yt$, then the ideal $(x, y)$ generated by $x$ and $y$ would not be proper. However, Theorem 2.25 gives a ring homomorphism $\varphi \colon k[x, y] \to k$ with $\varphi(x) = 0 = \varphi(y)$, so that $(x, y) \subseteq \ker \varphi$. But $\ker \varphi$ is a proper ideal, by Proposition 2.29, a contradiction.  ◄

Informally, a *Euclidean ring* is a domain having a division algorithm.

**Definition.** A **Euclidean ring** is a domain $R$ that is equipped with a function

$$\partial : R - \{0\} \to \mathbb{N},$$

called a **degree function**, such that

(i) $\partial(f) \leq \partial(fg)$ for all $f, g \in R$ with $f, g \neq 0$;

(ii) **Division Algorithm**: for all $f, g \in R$ with $f \neq 0$, there exist $q, r \in R$ with

$$g = qf + r,$$

where either $r = 0$ or $\partial(r) < \partial(f)$.

**Example 2.79.**

(i) Let $R$ have a degree function $\partial$ that is identically 0. If $f \in R$ and $f \neq 0$, condition (ii) gives an equation $1 = qf + r$ with $\partial(r) < \partial(f)$. This forces $r = 0$, for $\partial(r) < \partial(f) = 0$ is not possible. Therefore, $R$ is a field.

(ii) The set of integers $\mathbb{Z}$ is a Euclidean ring with degree function $\partial(m) = |m|$. In $\mathbb{Z}$, we have

$$\partial(mn) = |mn| = |m||n| = \partial(m)\partial(n).$$

(iii) When $k$ is a field, the domain $k[x]$ is a Euclidean ring with degree function the usual degree of a nonzero polynomial. In $k[x]$, we have

$$\partial(fg) = \deg(fg) = \deg(f) + \deg(g) = \partial(f) + \partial(g). \quad \blacktriangleleft$$

Since $\partial(mn) = \partial(m)\partial(n)$ in $\mathbb{Z}$ and $\partial(fg) = \partial(f) + \partial(g)$ in $k[x]$, the behavior of the degree of a product is not determined by the axioms in the definition of a degree function. If a degree function $\partial$ is multiplicative, that is, if $\partial(fg) = \partial(f)\partial(g)$, then $\partial$ is called a **norm**.

**Theorem 2.80.** *Let $R$ be a Euclidean ring.*

(i) *Every ideal $I$ in $R$ is a principal ideal.*

(ii) *Every $a, b \in R$ has a gcd, say $d$, that is a linear combination of $a$ and $b$:*

$$d = sa + tb,$$

*where $s, t \in R$.*

(iii) **Euclid's Lemma**: *If an irreducible element $p \in R$ divides a product $ab$, then either $p \mid a$ or $p \mid b$.*

(iv) **Unique Factorization**: *If $a \in R$ and $a = p_1 \cdots p_m$, where the $p_i$ are irreducible elements, then this factorization is unique in the following sense: if $a = q_1 \cdots q_k$, where the $q_j$ are irreducible elements, then $k = m$ and the $q$'s can be reindexed so that $p_i$ and $q_i$ are associates for all $i$.*

**Proof.**

(i) If $I = (0)$, then $I$ is the principal ideal generated by 0; therefore, we may assume that $I \neq (0)$. By the Least Integer Axiom, the set of all degrees of nonzero elements in $I$ has a smallest element, say, $n$; choose $d \in I$ with $\partial(d) = n$. Clearly, $(d) \subseteq I$, and so it suffices to prove the reverse inclusion. If $a \in I$, then there are $q, r \in R$ with $a = qd + r$, where either $r = 0$ or $\partial(r) < \partial(d)$. But $r = a - qd \in I$, and so $d$ having least degree implies that $r = 0$. Hence, $a = qd \in (d)$, and $I = (d)$.

(ii) This proof is essentially the same as that of Theorem 2.48. We may assume that at least one of $a$ and $b$ is not zero (otherwise, the gcd is 0 and the result is obvious). Consider the ideal $I$ of all the linear combinations:

$$I = \{sa + tb : s, t \text{ in } R\}.$$

Now $I$ is an ideal containing $a$ and $b$. By part (i), there is $d \in I$ with $I = (d)$. Since $a, b \in (d)$, we see that $d$ is a common divisor. Finally, if $c$ is a common divisor, then $a = ca'$ and $b = cb'$; hence, $c \mid d$, because $d = sa + tb = sca' + tcb' = c(sa' + tb')$. Thus, $d$ is a gcd of $a$ and $b$.

(iii) If $p \mid a$, we are done. If $p \nmid a$, then Example 2.77 says that 1 is a gcd of $p$ and $a$. Part (ii) gives $s, t \in R$ with $1 = sp + ta$, and multiplying by $b$,

$$b = spb + tab.$$

Since $p \mid ab$, it follows that $p \mid b$, as desired.

(iv) This proof is essentially that of Theorem 2.60. We prove, by induction on $M = \max\{m, k\}$, that if $p_1 \cdots p_m = pa = q_1 \cdots q_k$, where the $p$'s and $q$'s are irreducible, then $m = k$ and, after reindexing, $p_i$ and $q_i$ are associates for all $i$. If $M = 1$, then $p_1 = a = q_1$. For the inductive step, the given equation shows that $p_m \mid q_1 \cdots q_k$. By part (iii), Euclid's Lemma, there is some $i$ with $p_m \mid q_i$. But $q_i$ is irreducible, so there is a unit $u$ with $q_i = up_m$; that is, $q_i$ and $p_m$ are associates. Reindexing, we may assume that $q_k = up_m$; canceling, we have $p_1 \cdots p_{m-1} = q_1 \cdots (q_{k-1}u)$. Since $q_{k-1}u$ is irreducible, the inductive hypothesis gives $m - 1 = k - 1$ (hence, $m = k$) and, after reindexing, $p_i$ and $q_i$ are associates for all $i$. •

**Example 2.81.** The Gaussian integers $\mathbb{Z}[i]$ form a Euclidean ring whose degree function

$$\partial(a + bi) = a^2 + b^2$$

is a norm. To see that $\partial$ is multiplicative, note first that if $\alpha = a + bi$, then

$$\partial(\alpha) = \alpha\overline{\alpha},$$

where $\overline{\alpha} = a - bi$ is the complex conjugate of $\alpha$. It follows that $\partial(\alpha\beta) = \partial(\alpha)\partial(\beta)$ for all $\alpha, \beta \in \mathbb{Z}[i]$, because

$$\partial(\alpha\beta) = \alpha\beta\overline{\alpha\beta} = \alpha\beta\overline{\alpha}\,\overline{\beta} = \alpha\overline{\alpha}\beta\overline{\beta} = \partial(\alpha)\partial(\beta);$$

indeed, this is even true for all $\alpha, \beta \in \mathbb{Q}[i] = \{x + yi : x, y \in \mathbb{Q}\}$.

We now show that $\partial$ satisfies the first property of a degree function. If $\beta = c + id \in \mathbb{Z}[i]$ and $\beta \neq 0$, then

$$1 \leq \partial(\beta),$$

for $\partial(\beta) = c^2 + d^2$ is a positive integer; it follows that if $\alpha, \beta \in \mathbb{Z}[i]$ and $\beta \neq 0$, then

$$\partial(\alpha) \leq \partial(\alpha)\partial(\beta) = \partial(\alpha\beta).$$

Let us show that $\partial$ also satisfies the Division Algorithm. Given $\alpha, \beta \in \mathbb{Z}[i]$ with $\beta \neq 0$, regard $\alpha/\beta$ as an element of $\mathbb{C}$. Rationalizing the denominator gives $\alpha/\beta = \alpha\overline{\beta}/\beta\overline{\beta} = \alpha\overline{\beta}/\partial(\beta)$, so that

$$\alpha/\beta = x + yi,$$

where $x, y \in \mathbb{Q}$. Write $x = a + u$ and $y = b + v$, where $a, b \in \mathbb{Z}$ are integers closest to $x$ and $y$, respectively; thus, $|u|, |v| \leq \frac{1}{2}$. (If $x$ or $y$ has the form $m + \frac{1}{2}$, where $m$ is an integer, then there is a choice of nearest integer: $x = m + \frac{1}{2}$ or $x = (m+1) - \frac{1}{2}$; a similar choice arises if $x$ or $y$ has the form $m - \frac{1}{2}$.) It follows that

$$\alpha = \beta(a + bi) + \beta(u + vi).$$

Notice that $\beta(u + vi) \in \mathbb{Z}[i]$, for it is equal to $\alpha - \beta(a + bi)$. Finally, we have

$$\partial\big(\beta(u + vi)\big) = \partial(\beta)\partial(u + vi),$$

and so $\partial$ will be a degree function if $\partial(u+vi) < 1$. And this is so, for the inequalities $|u| \leq \frac{1}{2}$ and $|v| \leq \frac{1}{2}$ give $u^2 \leq \frac{1}{4}$ and $v^2 \leq \frac{1}{4}$, and hence $\partial(u + vi) = u^2 + v^2 \leq \frac{1}{4} + \frac{1}{4} = \frac{1}{2} < 1$. Therefore, $\partial(\beta(u + vi)) < \partial(\beta)$, and so $\mathbb{Z}[i]$ is a Euclidean ring whose degree function is a norm. ◄

We now show that quotients and remainders in $\mathbb{Z}[i]$ may not be unique. For example, let $\alpha = 3 + 5i$ and $\beta = 2$. Then $\alpha/\beta = \frac{3}{2} + \frac{5}{2}i$; the possible choices are

$$a = 1 \text{ and } u = \tfrac{1}{2} \quad \text{or} \quad a = 2 \text{ and } u = -\tfrac{1}{2};$$
$$b = 2 \text{ and } v = \tfrac{1}{2} \quad \text{or} \quad b = 3 \text{ and } v = -\tfrac{1}{2}.$$

Hence, there are four quotients and remainders after dividing $3 + 5i$ by $2$ in $\mathbb{Z}[i]$, for each of the remainders (e.g., $1 + i$) has degree $2 < 4 = \partial(2)$:

$$3 + 5i = 2(1 + 2i) + (1 + i);$$
$$= 2(1 + 3i) + (1 - i);$$
$$= 2(2 + 2i) + (-1 + i);$$
$$= 2(2 + 3i) + (-1 - i).$$

For a long time, it was believed that the reason for the parallel behavior of the rings $\mathbb{Z}$ and $k[x]$, for $k$ a field, was that they are both Euclidean rings. Nowadays, however, we regard the fact that every ideal in them is a principal ideal as more significant.

**Definition.** A *principal ideal domain* is a domain $R$ in which every ideal is a principal ideal. This term is usually abbreviated to PID.

**Example 2.82.**

(i) Every field is a PID [Example 2.31(ii)].

(ii) Theorem 2.80(i) shows that every Euclidean ring is a PID.

(iii) If $k$ is a field, then the ring of formal power series, $k[[x]]$, is a PID (Exercise 2.63 on page 132). ◄

**Theorem 2.83.** *The ring $\mathbb{Z}[i]$ of Gaussian integers is a principal ideal domain.*

**Proof.** Example 2.81 says that $\mathbb{Z}[i]$ is a Euclidean ring, and Theorem 2.80(i) says that it is a PID. •

The hypothesis of Theorem 2.80 can be weakened.

**Theorem 2.84.** *Let $R$ be a PID.*

(i) *Every $a, b \in R$ has a gcd, say $d$, that is a linear combination of $a$ and $b$:*

$$d = sa + tb,$$

*where $s, t \in R$.*

(ii) **Euclid's Lemma:** *If an irreducible element $p \in R$ divides a product $ab$, then either $p \mid a$ or $p \mid b$.*

(iii) **Unique Factorization:** *If $a \in R$ and $a = p_1 \cdots p_m$, where the $p_i$ are irreducible elements, then this factorization[12] is unique in the following sense: if $a = q_1 \cdots q_k$, where the $q_j$ are irreducible elements, then $k = m$ and the $q$'s can be reindexed so that $p_i$ and $q_i$ are associates for all $i$.*

---

[12]Prime factorizations in PIDs always exist, but we do not need this fact now. It is more convenient for us to prove it later (see Lemma 5.17).

**Proof.** The proof of Theorem 2.80 is valid here.   •

The converse of Example 2.82(ii) is false: there are PIDs that are not Euclidean rings, as we will see in the next example.

**Example 2.85.** If $\alpha = \frac{1}{2}(1 + \sqrt{-19})$, then it is shown in Algebraic Number Theory that the ring

$$\mathbb{Z}(\alpha) = \{a + b\alpha : a, b \in \mathbb{Z}\}$$

is a PID [$\mathbb{Z}(\alpha)$ is the ring of algebraic integers in the quadratic number field $\mathbb{Q}(\sqrt{-19})$]; see Dummit–Foote, *Abstract Algebra*, p. 283. In 1949, Motzkin proved that $\mathbb{Z}(\alpha)$ is not Euclidean by showing that it does not have a certain property enjoyed by all Euclidean rings.

**Definition.** An element $u$ in a domain $R$ is a ***universal side divisor*** if $u$ is not a unit and, for every $x \in R$, either $u \mid x$ or there is a unit $z \in R$ with $u \mid (x + z)$.

**Proposition 2.86.** *If $R$ is a Euclidean ring but not a field, then $R$ has a universal side divisor.*

**Proof.** Let $\partial$ be the degree function on $R$, and define

$$S = \{\partial(v) : v \neq 0 \text{ and } v \text{ is not a unit}\}.$$

Since $R$ is not a field, Example 2.79(i) shows that $S$ is a nonempty subset of the natural numbers and, hence, $S$ has a smallest element, say, $\partial(u)$. We claim that $u$ is a universal side divisor. If $x \in R$, there are elements $q$ and $r$ with $x = qu + r$, where either $r = 0$ or $\partial(r) < \partial(u)$. If $r = 0$, then $u \mid x$; if $r \neq 0$, then $r$ must be a unit, otherwise its existence contradicts $\partial(u)$ being the smallest number in $S$. We have shown that $u$ is a universal side divisor.   •

Motzkin showed that $\mathbb{Z}(\alpha) = \{a + b\alpha : a, b \in \mathbb{Z}\}$ has no universal side divisors, proving that this PID is not a Euclidean ring [see K. S. Williams, "Note on Non-Euclidean Principal Ideal Domains," *Math. Mag.* 48 (1975), pp. 176–177].   ◄

What are the units in the Gaussian integers?

**Proposition 2.87.** *Let $R$ be a Euclidean ring, not a field, whose degree function $\partial$ is a norm.*

  (i) *An element $\alpha \in R$ is a unit if and only if $\partial(\alpha) = 1$.*

  (ii) *If $\alpha \in R$ and $\partial(\alpha) = p$, where $p$ is a prime number, then $\alpha$ is irreducible.*

  (iii) *The only units in the ring $\mathbb{Z}[i]$ of Gaussian integers are $\pm 1$ and $\pm i$.*

**Proof.**

  (i) Since $1^2 = 1$, we have $\partial(1)^2 = \partial(1)$, so that $\partial(1) = 0$ or $\partial(1) = 1$. If $\partial(1) = 0$, then $\partial(a) = \partial(1a) = \partial(1)\partial(a) = 0$ for all $a \in R$; by Example 2.79(i), $R$ is a field, contrary to our hypothesis. We conclude that $\partial(1) = 1$.

     If $\alpha \in R$ is a unit, then there is $\beta \in R$ with $\alpha\beta = 1$. Therefore, $\partial(\alpha)\partial(\beta) = 1$. Since the values of $\partial$ are nonnegative integers, $\partial(\alpha) = 1$.

For the converse, we begin by showing that there is no nonzero element $\beta \in R$ with $\partial(\beta) = 0$. If such an element existed, the Division Algorithm would give $1 = q\beta + r$, where $q, r \in R$ and either $r = 0$ or $\partial(r) < \partial(\beta) = 0$. The inequality cannot occur, and so $r = 0$; that is, $\beta$ is a unit. But if $\beta$ is a unit, then $\partial(\beta) = 1$, as we have just proved, and this contradicts $\partial(\beta) = 0$.

Assume now that $\partial(\alpha) = 1$. The Division Algorithm gives $q, r \in R$ with

$$\alpha = q\alpha^2 + r,$$

where $r = 0$ or $\partial(r) < \partial(\alpha^2)$. As $\partial(\alpha^2) = \partial(\alpha)^2 = 1$, either $r = 0$ or $\partial(r) = 0$. But we have just seen that $\partial(r) = 0$ cannot occur, so that $r = 0$ and $\alpha = q\alpha^2$. It follows that $1 = q\alpha$, and so $\alpha$ is a unit.

(ii) If, on the contrary, $\alpha = \beta\gamma$, where neither $\beta$ nor $\gamma$ is a unit, then $p = \partial(\alpha) = \partial(\beta)\partial(\gamma)$. As $p$ is prime, either $\partial(\beta) = 1$ or $\partial(\gamma) = 1$. By part (i), either $\beta$ or $\gamma$ is a unit; that is, $\alpha$ is irreducible.

(iii) If $\alpha = a + bi \in \mathbb{Z}[i]$ is a unit, then $1 = \partial(\alpha) = a^2 + b^2$. This can happen if and only if $a^2 = 1$ and $b^2 = 0$ or $a^2 = 0$ and $b^2 = 1$; that is, $\alpha = \pm 1$ or $\alpha = \pm i$. •

We are now going to use the Gaussian integers to prove a number-theoretic result: the Two Squares Theorem. If $n$ is an odd number, then either $n \equiv 1 \bmod 4$ or $n \equiv 3 \bmod 4$; consequently, the odd prime numbers are divided into two classes. For example, 5, 13, 17 are congruent to 1 mod 4, while 3, 7, 11 are congruent to 3 mod 4.

**Lemma 2.88.** *Let $p$ be an odd prime.*

(i) *If $p \equiv 1 \bmod 4$, then the congruence $m^2 \equiv -1 \bmod p$ is solvable.*

(ii) *If $p \equiv 3 \bmod 4$, then one of the congruences $x^2 \equiv 2 \bmod p$ or $x^2 \equiv -2 \bmod p$ is solvable.*

**Proof.**

(i) If $G = (\mathbb{F}_p)^\times$ is the multiplicative group of nonzero elements in $\mathbb{F}_p$, then Theorem 2.46 says that $G$ is a cyclic group of order $p - 1 \equiv 0 \bmod 4$; that is, 4 is a divisor of $|G|$. By Lemma 1.92(ii), $G$ contains a cyclic subgroup $S$ of order 4. Thus, $S = \langle [m] \rangle$, where $[m]$ is the congruence class of $m$ mod $p$. Since $[m]$ has order 4, we have $[m^4] = [1]$. Moreover, $[m^2] \neq [1]$ (lest $[m]$ have order $\leq 2 < 4$), and so $[m^2] = [-1]$, for $[-1]$ is the unique element in $S$ of order 2. Therefore, $m^2 \equiv -1 \bmod p$.

(ii) If $G = (\mathbb{F}_p)^\times$ is the multiplicative group of nonzero elements in $\mathbb{F}_p$, then $|G| = p - 1 = 2m$, where $m$ is odd (if $m$ were even, say, $m = 2m'$, then $p - 1 = 4m'$, contradicting $p \equiv 3 \bmod 4$).[13] The subset $H$ of all elements of odd order is a subgroup of order $m$ (because $G$ is abelian), and it has index 2 in $G$, for $[G : H] = |G|/|H| = 2m/m$. Now $[-1] \notin H$, so that the cosets of $H$ are $H$ and $[-1]H$. Writing elements of $G$ without brackets,

$$G = H \cup -H.$$

---

[13]$G$ is cyclic, by Theorem 2.46, but we can complete this proof without using this fact.

It suffices to prove that one of $2, -2$ lies in $H$, for Exercise 1.62 on page 46 says that every element in a finite abelian group of odd order has a square root. Otherwise, both $2, -2$ lie in $-H$; that is, $2 = -a$ and $-2 = -b$, where $a, b \in H$. Now $a^2 = (-a)^2 = 4 = (-b)^2 = b^2$, while Lagrange's Theorem gives $a^m = 1 = b^m$. But $(2, m) = 1$, because $m$ is odd, and so there are integers $s, t$ with $1 = 2s + mt$. Thus,

$$a = a^{2s} a^{mt} = b^{2s} b^{mt} = b,$$

giving the contradiction $2 = -2$.  •

**Theorem 2.89 (Fermat's Two-Squares Theorem).**[14] *An odd prime $p$ is a sum of two squares,*

$$p = a^2 + b^2,$$

*where $a$ and $b$ are integers, if and only if $p \equiv 1 \bmod 4$.*

**Proof.** Assume that $p = a^2 + b^2$. Since $p$ is odd, $a$ and $b$ have different parity; say, $a$ is even and $b$ is odd. Hence, $a = 2m$ and $b = 2n + 1$, and

$$p = a^2 + b^2 = 4m^2 + 4n^2 + 4n + 1 \equiv 1 \bmod 4.$$

Conversely, assume that $p \equiv 1 \bmod 4$. By Lemma 2.88, there is an integer $m$ such that

$$p \mid (m^2 + 1).$$

In $\mathbb{Z}[i]$, there is a factorization $m^2 + 1 = (m + i)(m - i)$, and so

$$p \mid (m + i)(m - i) \text{ in } \mathbb{Z}[i].$$

If $p \mid (m \pm i)$ in $\mathbb{Z}[i]$, then there are integers $u$ and $v$ with $m \pm i = p(u + iv)$. Comparing the imaginary parts gives $pv = 1$, a contradiction. We conclude that $p$ does not satisfy the analog of Euclid's Lemma in Theorem 2.80(ii) (recall that $\mathbb{Z}[i]$ is a Euclidean ring); it follows that $p$ is not an irreducible element in $\mathbb{Z}[i]$. Hence, there is a factorization

$$p = \alpha\beta \text{ in } \mathbb{Z}[i]$$

in which neither $\alpha = a + ib$ nor $\beta = c + id$ is a unit. Therefore, taking norms gives an equation in $\mathbb{Z}$:

$$p^2 = \partial(p) = \partial(\alpha\beta) = \partial(\alpha)\partial(\beta) = (a^2 + b^2)(c^2 + d^2).$$

By Proposition 2.87, the only units in $\mathbb{Z}[i]$ are $\pm 1$ and $\pm i$, so that any nonzero Gaussian integer that is not a unit has norm $> 1$; therefore, $a^2 + b^2 \neq 1$ and $c^2 + d^2 \neq 1$. Euclid's Lemma now gives $p \mid (a^2 + b^2)$ or $p \mid (c^2 + d^2)$; the Fundamental Theorem of Arithmetic gives $p = a^2 + b^2$ (or $p = c^2 + d^2$), as desired.  •

We are going to determine all the irreducible elements in $\mathbb{Z}[i]$, but we first prove a lemma.

**Lemma 2.90.** *If $\alpha \in \mathbb{Z}[i]$ is irreducible, then there is a unique prime number $p$ with $\alpha \mid p$ in $\mathbb{Z}[i]$.*

---

[14]Fermat was the first to state this theorem, but the first published proof is due to Euler. Gauss proved that there is only one pair of natural numbers $a$ and $b$ with $p = a^2 + b^2$.

**Proof.** Note that if $\alpha \in \mathbb{Z}[i]$, then $\overline{\alpha} \in \mathbb{Z}[i]$; since $\partial(\alpha) = \alpha\overline{\alpha}$, we have $\alpha \mid \partial(\alpha)$. Now $\partial(\alpha) = p_1 \cdots p_n$, where the $p_i$ are prime numbers. As $\mathbb{Z}[i]$ is a Euclidean ring, Euclid's Lemma gives $\alpha \mid p_i$ for some $i$ (for $\alpha$ is irreducible). If $\alpha \mid q$ for some prime $q \neq p_i$, then $\alpha \mid (q, p_i) = 1$, forcing $\alpha$ to be a unit. This contradiction shows that $p_i$ is the unique prime number divisible by $\alpha$. $\bullet$

**Proposition 2.91.** *Let $\alpha = a + bi \in \mathbb{Z}[i]$ be neither 0 nor a unit. Then $\alpha$ is irreducible if and only if it satisfies one of the following three conditions:*

(i) *$\alpha$ is an associate of $1 + i$ : that is, $\alpha = 1 + i$, $1 - i$, $-1 + i$, or $-1 - i$;*

(ii) *$\alpha = up$, where $u$ is a unit in $\mathbb{Z}[i]$ and $p$ is a prime in $\mathbb{Z}$ of the form $p = 4m + 3$;*

(iii) *$\partial(\alpha) = a^2 + b^2$ is a prime in $\mathbb{Z}$ of the form $4m + 1$.*

**Proof.** By Lemma 2.90, there is a unique prime number $p$ divisible by $\alpha$ in $\mathbb{Z}[i]$; there is $\beta \in \mathbb{Z}[i]$ with $p = \alpha\beta$. Since $\partial$ is a norm, there is an equation in $\mathbb{Z}$

(1) $$\partial(\alpha)\partial(\beta) = \partial(p) = p^2,$$

so that $\partial(\alpha) = p$ or $\partial(\alpha) = p^2$; that is,

$$a^2 + b^2 = p \quad \text{or} \quad a^2 + b^2 = p^2.$$

Looking at $p \bmod 4$, we see that there are three possibilities (for $p \equiv 0 \bmod 4$ cannot occur).

(i) $p \equiv 2 \bmod 4$.

In this case, $p = 2$, and so $a^2 + b^2 = 2$ or $a^2 + b^2 = 4$. The latter case cannot occur (because $a$ and $b$ are integers), and the first case gives $\alpha = 1 + i$ (up to multiplication by units). The result follows from Proposition 2.87(iii).

(ii) $p \equiv 3 \bmod 4$.

In this case, $a^2 + b^2 = p$ cannot occur (the easy direction of Theorem 2.89 gives $p \equiv 1 \bmod 4$), so that $\partial(\alpha) = a^2 + b^2 = p^2$. Since $p \in \mathbb{Z}$, we have $\partial(p) = p^2$, so that Equation (1) gives $p^2\partial(\beta) = p^2$. Thus, $\partial(\beta) = 1$, $\beta$ is a unit [by Proposition 2.87(i)], and $p$ is irreducible in $\mathbb{Z}[i]$. For example, 3 is an irreducible element of this first type.

(iii) $p \equiv 1 \bmod 4$.

If $\partial(\alpha)$ is a prime $p$ (with $p \equiv 1 \bmod 4$), then $\alpha$ is irreducible, by Proposition 2.87(ii). Conversely, suppose $\alpha$ is irreducible. As $\partial(\alpha) = p$ or $\partial(\alpha) = p^2$, it suffices to eliminate the latter possibility. Since $\alpha \mid p$, we have $p = \alpha\beta$ for some $\beta \in \mathbb{Z}[i]$; hence, as in case (i), $\partial(\alpha) = p^2$ implies that $\beta$ is a unit. Now $\alpha\overline{\alpha} = p^2 = (\alpha\beta)^2$, so that $\overline{\alpha} = \alpha\beta^2$. But $\beta^2 = \pm 1$ [the four units in $\mathbb{Z}[i]$ have square $\pm 1$, by Proposition 2.87(iii)], contradicting $\overline{\alpha} \neq \pm\alpha$. Therefore, $\partial(\alpha) = p$. For example, $2 + i$ is an irreducible element of this third type.

The proof of the converse: an element $\alpha \in \mathbb{Z}[i]$ is irreducible if it satisfies one of the three stated conditions, is left to the reader. $\bullet$

We can now complete an earlier discussion.

**Proposition 2.92.** *The polynomial $f(x) = x^4 + 1$ is irreducible in $\mathbb{Q}[x]$, yet it factors in $\mathbb{F}_p[x]$ for every prime $p$.*

**Remark.** We shall give a less computational proof in Proposition 3.10.  ◄

**Proof.** To see that $f(x)$ is irreducible, it suffices to show that $f(x-1)$ is irreducible, by Lemma 2.74. But $f(x-1) = x^4 - 4x^3 + 6x^2 - 4x + 2$ is irreducible, by Eisenstein's Criterion.

Now $x^4 + 1$ factors in $\mathbb{F}_2[x]$ because 1 is a root. We claim that if $p$ is an odd prime, then there is a factorization

$$x^4 + 1 = (x^2 + ax + b)(x^2 - ax + c) \text{ in } \mathbb{F}_p[x].$$

Equate like coefficients; such a factorization exists if and only if there are $a, b, c \in \mathbb{Z}$ that satisfy the system of congruences:

$$c + b - a^2 \equiv 0 \bmod p$$
$$a(c - b) \equiv 0 \bmod p$$
$$bc \equiv 1 \bmod p.$$

We consider two cases. If $p \equiv 1 \bmod 4$, then choosing $a = 0$ and $-b = c$ satisfies the top two congruences, while the third congruence reads $-b^2 \equiv 1 \bmod p$. As $p \equiv 1 \bmod 4$, Lemma 2.88(i) says that $-1$ is a square mod $p$; that is, $b$ can be chosen to satisfy the system. If $p \equiv 3 \bmod 4$, then choosing either $b = 1 = c$ or $b = -1 = c$ satisfies the bottom two congruences, while the first congruence reads $2 - a^2 \equiv 0 \bmod p$ or $-2 - a^2 \equiv 0 \bmod p$. As $p \equiv 3 \bmod 4$, Lemma 2.88(ii) says that $a^2 \equiv 2 \bmod p$ or $a^2 \equiv -2 \bmod p$ is solvable, and so $a$ can be chosen to satisfy the system. •

## Exercises

**2.62.** Let $R$ be a PID; if $a, b \in R$, prove that their lcm exists.

* **2.63.** (i) Prove that every nonzero ideal in $k[[x]]$ is equal to $(x^n)$ for some $n \geq 0$.

(ii) If $k$ is a field, prove that the ring of formal power series $k[[x]]$ is a PID.
   **Hint.** Use Exercise 2.28 on page 95.

* **2.64.** If $k$ is a field, prove that the ideal $(x, y)$ in $k[x, y]$ is not a principal ideal.

**2.65.** For every $m \geq 1$, prove that every ideal in $\mathbb{I}_m$ is a principal ideal. (If $m$ is composite, then $\mathbb{I}_m$ is not a PID because it is not a domain.)

**Definition.** Let $k$ be a field. A ***common divisor*** of $a_1(x)$, $a_2(x)$, ..., $a_n(x)$ in $k[x]$ is a polynomial $c(x) \in k[x]$ with $c(x) \mid a_i(x)$ for all $i$; the ***greatest common divisor*** is the monic common divisor of largest degree. We write $c(x) = (a_1, a_2, \ldots, a_n)$. A ***least common multiple*** of several elements is defined similarly.

**2.66.** Let $k$ be a field, and let polynomials $a_1(x)$, $a_2(x)$, ..., $a_n(x)$ in $k[x]$ be given.

(i) Show that the greatest common divisor $d(x)$ of these polynomials has the form $\sum t_i(x)a_i(x)$, where $t_i(x) \in k[x]$ for $1 \leq i \leq n$.

(ii) Prove that $c(x) \mid d(x)$ for every monic common divisor $c(x)$ of the $a_i(x)$.

\* **2.67.** Prove that there are domains $R$ containing a pair of elements having no gcd (according to the definition of gcd on page 124).

**Hint.** Let $k$ be a field and let $R$ be the subring of $k[x]$ consisting of all polynomials having no linear term; that is, $f(x) \in R$ if and only if

$$f(x) = s_0 + s_2 x^2 + s_3 x^3 + \cdots .$$

Show that $x^5$ and $x^6$ have no gcd in $R$.

**2.68.** Prove that $R = \mathbb{Z}[\sqrt{2}] = \{a + b\sqrt{2} : a, b \in \mathbb{Z}\}$ is a Euclidean ring with $\partial(a + b\sqrt{2}) = |a^2 - 2b^2|$.

**2.69.** Let $\partial$ be the degree function of a Euclidean ring $R$. If $m, n \in \mathbb{N}$ and $m \geq 1$, prove that $\partial'$ is also a degree function on $R$, where

$$\partial'(x) = m\partial(x) + n$$

for all $x \in R$. Conclude that a Euclidean ring may have no elements of degree 0 or degree 1.

**2.70.** Let $R$ be a Euclidean ring with degree function $\partial$.

   (i) Prove that $\partial(1) \leq \partial(a)$ for all nonzero $a \in R$.

   (ii) Prove that a nonzero $u \in R$ is a unit if and only if $\partial(u) = \partial(1)$.

**2.71.** Let $R$ be a Euclidean ring, and assume that $b \in R$ is neither zero nor a unit. Prove, for every $i \geq 0$, that $\partial(b^i) < \partial(b^{i+1})$.

**Hint.** There are $q, r \in R$ with $b^i = qb^{i+1} + r$.

## Section 2.7. Vector Spaces

Linear Algebra is the study of vector spaces and their homomorphisms (linear transformations) with applications to systems of linear equations. Aside from its intrinsic value, it is a necessary tool in further investigation of rings. Most readers have probably had some course involving matrices, perhaps only with real or complex entries. Here, we do not emphasize computational aspects of the subject, such as Gaussian elimination, finding inverses, determinants, and eigenvalues. Instead, we discuss more theoretical properties of vector spaces with scalars in any field.

Dimension is a rather subtle idea. We think of a curve in the plane, that is, the image of a continuous function $f \colon \mathbb{R} \to \mathbb{R}^2$, as a one-dimensional subset of a two-dimensional ambient space. Imagine the confusion at the end of the nineteenth century when a "space-filling curve" was discovered: there exists a continuous function $f \colon \mathbb{R} \to \mathbb{R}^2$ with image the whole plane! We are going to describe a way of defining dimension that works for analogs of Euclidean space (there are topological ways of defining dimension of more general spaces).

**Definition.** If $k$ is a field, then a ***vector space over*** $k$ is an additive abelian group $V$ equipped with a function $k \times V \to V$, denoted by $(a, v) \mapsto av$ and called ***scalar multiplication***, such that, for all $a, b, 1 \in k$ and all $u, v \in V$,

   (i) $a(u + v) = au + av$,

   (ii) $(a + b)v = av + bv$,

(iii) $(ab)v = a(bv)$,

(iv) $1v = v$.

The elements of $V$ are called ***vectors*** and the elements of $k$ are called ***scalars***.[15]

**Example 2.93.**

(i) Euclidean space $V = \mathbb{R}^n$ is a vector space over $\mathbb{R}$. Vectors are $n$-tuples $(a_1, \ldots, a_n)$, where $a_i \in \mathbb{R}$ for all $i$. Picture a vector $v$ as an arrow from the origin to the point having coordinates $(a_1, \ldots, a_n)$. Addition is given by

$$(a_1, \ldots, a_n) + (b_1, \ldots, b_n) = (a_1 + b_1, \ldots, a_n + b_n);$$

geometrically, the sum of two vectors is described by the *parallelogram law*. Scalar multiplication is given by

$$av = a(a_1, \ldots, a_n) = (aa_1, \ldots, aa_n).$$

Scalar multiplication $v \mapsto av$ "stretches" $v$ by a factor $|a|$, reversing its direction when $a$ is negative (we put quotes around *stretches* because $av$ is shorter than $v$ when $|a| < 1$).

(ii) We generalize part (i). If $k$ is any field, define $V = k^n$, the set of all $n$-tuples $v = (a_1, \ldots, a_n)$, where $a_i \in k$ for all $i$. Addition is given by

$$(a_1, \ldots, a_n) + (b_1, \ldots, b_n) = (a_1 + b_1, \ldots, a_n + b_n),$$

and scalar multiplication is given by

$$av = a(a_1, \ldots, a_n) = (aa_1, \ldots, aa_n).$$

(iii) If $R$ is a commutative ring and a field $k$ is a subring, then $R$ is a vector space over $k$. Regard the elements of $R$ as vectors and the elements of $k$ as scalars; define scalar multiplication $av$, where $a \in k$ and $v \in R$, to be the given product of two elements in $R$. Notice that the axioms in the definition of vector space are just particular cases of some of the axioms of a ring.

For example, if $k$ is a field, then the polynomial ring $R = k[x]$ is a vector space over $k$. Vectors are polynomials $f(x)$, scalars are elements $a \in k$, and scalar multiplication gives the polynomial $af(x)$; that is, if

$$f(x) = b_n x^n + \cdots + b_1 x + b_0,$$

then

$$af(x) = ab_n x^n + \cdots + ab_1 x + ab_0.$$

Here is another example: if $E$ is a field and $k$ is a subfield, then $E$ is a vector space over $k$. ◀

Informally, a *subspace* of a vector space $V$ is a subset of $V$ that is a vector space under the addition and scalar multiplication in $V$.

---

[15]The word *vector* comes from the Latin word meaning "to carry"; vectors in Euclidean space carry the data of length and direction. The word *scalar* comes from regarding $v \mapsto av$ as a change of scale. The terms *scale* and *scalar* come from the Latin word meaning "ladder," for the rungs of a ladder are evenly spaced.

**Definition.** If $V$ is a vector space over a field $k$, then a **subspace** of $V$ is a subset $U$ of $V$ such that

(i) $0 \in U$,

(ii) $u, u' \in U$ imply $u + u' \in U$,

(iii) $u \in U$ and $a \in k$ imply $au \in U$.

It is easy to see that every subspace is itself a vector space.

**Example 2.94.**

(i) The extreme cases $U = V$ and $U = \{0\}$ (where $\{0\}$ denotes the subset consisting of the zero vector alone) are always subspaces of a vector space $V$. A subspace $U \subseteq V$ with $U \neq V$ is called a **proper subspace** of $V$; we may denote $U$ being a proper subspace by $U \subsetneq V$.

(ii) If $v \in \mathbb{R}^n$ is a nonzero vector, then the line $\ell$ through the origin,

$$\ell = \{av : a \in \mathbb{R}\},$$

is a subspace of $\mathbb{R}^n$. Similarly, if $v_1, v_2 \in \mathbb{R}^n$ are noncollinear, then a plane $\pi$ through the origin,

$$\pi = \{av_1 + bv_2 : a, b \in \mathbb{R}\},$$

is a subspace of $\mathbb{R}^n$.

(iii) If $m \leq n$ and $\mathbb{R}^m$ is regarded as the set of all those vectors in $\mathbb{R}^n$ whose last $n - m$ coordinates are 0, then $\mathbb{R}^m$ is a subspace of $\mathbb{R}^n$. For example, we may regard the plane $\mathbb{R}^2$ as all points $(x, y, 0)$ in $\mathbb{R}^3$.

(iv) If $k$ is a field, then a **homogeneous linear system over** $k$ of $m$ equations in $n$ unknowns is a set of equations

$$a_{11}x_1 + \cdots + a_{1n}x_n = 0$$
$$a_{21}x_1 + \cdots + a_{2n}x_n = 0$$
$$\vdots \qquad \qquad \vdots$$
$$a_{m1}x_1 + \cdots + a_{mn}x_n = 0,$$

where $a_{ji} \in k$. A **solution** of this system is a vector $c^\top = (c_1, \ldots, c_n)^\top \in k^n$ (we regard vectors in $k^n$ as $n \times 1$ columns), where $\sum_i a_{ji}c_i = 0$ for all $j$; a solution $c^\top$ is **nontrivial** if some $c_i \neq 0$. The set of all solutions forms a subspace of $k^n$, called the **solution space** (or *nullspace*) of the system. The $m \times n$ matrix $A = [a_{ij}]$ is called the **coefficient matrix** of the system, and the system can be written compactly as $Ax^\top = 0$.

In particular, we can solve systems of linear equations over $\mathbb{F}_p$, where $p$ is prime. This says that we can treat a system of congruences mod $p$ just as one treats an ordinary system of equations. For example, the system of congruences

$$3x - 2y + z \equiv 1 \bmod 7$$
$$x + y - 2z \equiv 0 \bmod 7$$
$$-x + 2y + z \equiv 4 \bmod 7$$

can be regarded as a system of equations over the field $\mathbb{F}_7$. This system can be solved just as in high school, for inverses mod 7 are now known: $[2][4] = [1]$; $[3][5] = [1]$; $[6][6] = [1]$. The solution is

$$(x, y, z) = ([5], [4], [1]). \quad \blacktriangleleft$$

**Definition.** A *list* in a vector space $V$ is an ordered set $X = v_1, \ldots, v_n$ of vectors in $V$.

More precisely, a list $X$ is a function $\varphi \colon \{1, 2, \ldots, n\} \to V$, for some $n \geq 1$, with $\varphi(i) = v_i$ for all $i$, and we denote this list by $X = \varphi(1), \ldots, \varphi(n)$. Thus, $X$ is ordered in the sense that there is a first vector $v_1$, a second vector $v_2$, and so forth.[16] A vector may appear several times on a list; that is, $\varphi$ need not be injective.

**Definition.** Let $V$ be a vector space over a field $k$. A *k-linear combination* of a list $X = v_1, \ldots, v_n$ in $V$ is a vector $v$ of the form

$$v = a_1 v_1 + \cdots + a_n v_n,$$

where $a_i \in k$ for all $i$.

**Definition.** If $X = v_1, \ldots, v_m$ is a list in a vector space $V$, then the *subspace spanned by X*,

$$\langle v_1, \ldots, v_m \rangle,$$

is the set of all the $k$-linear combinations of $v_1, \ldots, v_m$. We also say that $v_1, \ldots, v_m$ *spans* $\langle v_1, \ldots, v_m \rangle$. (We will consider infinite spanning sets in Chapter 5.)

**Lemma 2.95.** *Let $V$ be a vector space over a field $k$.*

(i) *Every intersection of subspaces of $V$ is itself a subspace.*

(ii) *If $X = v_1, \ldots, v_m$ is a list in $V$, then the intersection of all the subspaces of $V$ containing $\{v_1, \ldots, v_m\}$ is $\langle v_1, \ldots, v_m \rangle$, the subspace spanned by $v_1, \ldots, v_m$, and so $\langle v_1, \ldots, v_m \rangle$ is the **smallest** subspace of $V$ containing $\{v_1, \ldots, v_m\}$.*

**Proof.** Part (i) is routine. For (ii), let $\mathcal{S}$ denote the family of all the subspaces of $V$ containing $\{v_1, \ldots, v_m\}$; $V$ is a subspace in $\mathcal{S}$. We claim that

$$\bigcap_{S \in \mathcal{S}} S = \langle v_1, \ldots, v_m \rangle.$$

The inclusion $\subseteq$ is clear, because $\langle v_1, \ldots, v_m \rangle \in \mathcal{S}$. For the reverse inclusion, note that if $S \in \mathcal{S}$, then $S$ contains $v_1, \ldots, v_m$, and so it contains the set of all linear combinations of $v_1, \ldots, v_m$, namely, $\langle v_1, \ldots, v_m \rangle$. $\quad \bullet$

---

[16]For the purists, a similar notational trick defines an *n-tuple*; it is a function we choose to write using parentheses and commas: $(a_1, \ldots, a_n)$. Thus, a list is an $n$-tuple.

It follows from the second part of the lemma that the subspace spanned by a list $X = v_1, \ldots, v_m$ does not depend on the ordering of the vectors, but only on the set of vectors themselves. Were all terminology in algebra consistent, we would call $\langle v_1, \ldots, v_m \rangle$ the subspace *generated by X*. The reason for the different terms is that the theories of groups, rings, and vector spaces developed independently of each other.

**Example 2.96.**

(i) If $X = \varnothing$, then $\langle X \rangle = \bigcap_{S \in \mathcal{S}} S$, where $\mathcal{S}$ is the family of all the subspaces of $V$, for every subspace contains $\varnothing$. Thus, $\langle \varnothing \rangle = \{0\}$.

(ii) Let $V = \mathbb{R}^2$, let $e_1 = (1, 0)$, and let $e_2 = (0, 1)$. Now $V = \langle e_1, e_2 \rangle$, for if $v = (a, b) \in V$, then

$$\begin{aligned} v &= (a, 0) + (0, b) \\ &= a(1, 0) + b(0, 1) \\ &= ae_1 + be_2 \in \langle e_1, e_2 \rangle. \end{aligned}$$

(iii) If $k$ is a field and $V = k^n$, define $e_i$ as the $n$-tuple having 1 in the $i$th coordinate and 0's elsewhere. The reader may adapt the argument in (i) to show that $e_1, \ldots, e_n$ spans $k^n$.

(iv) A vector space $V$ need not be spanned by a finite list. For example, let $V = k[x]$, and suppose that $X = f_1(x), \ldots, f_m(x)$ is a finite list in $V$. If $d$ is the largest degree of any of the $f_i(x)$, then every (nonzero) $k$-linear combination of $f_1(x), \ldots, f_m(x)$ has degree at most $d$. Thus, $x^{d+1}$ is not a $k$-linear combination of vectors in $X$, and so $X$ does not span $k[x]$. ◀

The following definition makes sense even though the term *dimension* has not yet been defined.

**Definition.** A vector space $V$ is called *finite-dimensional* if it is spanned by a finite list; otherwise, $V$ is called *infinite-dimensional*.

Example 2.96(iii) shows that $k^n$ is finite-dimensional, while Example 2.96(iv) shows that $k[x]$ is infinite-dimensional. By Example 2.93(iii), $\mathbb{R}$ and $\mathbb{C}$ are vector spaces over $\mathbb{Q}$; both of them are infinite-dimensional (by Exercise 2.78 on page 145, finite-dimensional vector spaces over $\mathbb{Q}$ are countable).

**Proposition 2.97.** *If $V$ is a vector space, then the following conditions on a list $X = v_1, \ldots, v_m$ spanning $V$ are equivalent.*

(i) *$X$ is not a shortest spanning list.*

(ii) *Some $v_i$ is in the subspace spanned by the others; that is,*

$$v_i \in \langle v_1, \ldots, \widehat{v_i}, \ldots, v_m \rangle$$

*(if $v_1, \ldots, v_m$ is a list, then $v_1, \ldots, \widehat{v_i} \ldots, v_m$ is the shorter list with $v_i$ deleted).*

(iii) *There are scalars $a_1, \ldots, a_m$, not all zero, with*

$$\sum_{\ell=1}^{m} a_\ell v_\ell = 0.$$

**Proof.** (i) $\Rightarrow$ (ii). If $X$ is not a shortest spanning list, then one of the vectors in $X$, say $v_i$, can be thrown out, and the shorter list still spans. Thus, $v_i$ is a linear combination of the others.

(ii) $\Rightarrow$ (iii). If $v_i = \sum_{j \neq i} c_j v_j$, then define $a_i = -1 \neq 0$ and $a_j = c_j$ for all $j \neq i$.

(iii) $\Rightarrow$ (i). The given equation implies that one of the vectors, say, $v_i$, is a linear combination of the others. Deleting $v_i$ gives a shorter list, which still spans: if $v \in V$ is a linear combination of all the $v_j$ (including $v_i$), just substitute the expression for $v_i$ as a linear combination of the other $v_j$ and collect terms.   $\bullet$

**Definition.** A list $X = v_1, \ldots, v_m$ in a vector space $V$ is **linearly dependent** if there are scalars $a_1, \ldots, a_m$, not all zero, with $\sum_{\ell=1}^{m} a_\ell v_\ell = 0$; otherwise, $X$ is called **linearly independent**.

The empty set $\varnothing$ is defined to be linearly independent (we may interpret $\varnothing$ as a list of length 0).

Note that linear independence of a list $X = v_1, \ldots, v_m$ does not depend on the ordering of the vectors, but only on the set of vectors themselves.

**Example 2.98.**

(i) Any list $X = v_1, \ldots, v_m$ containing the zero vector is linearly dependent.

(ii) A list $v_1$ of length 1 is linearly dependent if and only if $v_1 = 0$; hence, a list $v_1$ of length 1 is linearly independent if and only if $v_1 \neq 0$.

(iii) A list $v_1, v_2$ is linearly dependent if and only if one of the vectors is a scalar multiple of the other.

(iv) If there is a repetition on the list $v_1, \ldots, v_m$ (that is, if $v_i = v_j$ for some $i \neq j$), then $v_1, \ldots, v_m$ is linearly dependent: define $c_i = 1$, $c_j = -1$, and all other $c = 0$. Therefore, if $v_1, \ldots, v_m$ is linearly independent, then all the vectors $v_i$ are distinct.   $\blacktriangleleft$

The contrapositive of Proposition 2.97 is worth stating.

**Corollary 2.99.** *If $X = v_1, \ldots, v_m$ is a list spanning a vector space $V$, then $X$ is a shortest spanning list if and only if $X$ is linearly independent.*

Linear independence has been defined indirectly, as not being linearly dependent. Because of the importance of linear independence, let us define it directly. A list $X = v_1, \ldots, v_m$ is **linearly independent** if, whenever a $k$-linear combination $\sum_{\ell=1}^{m} a_\ell v_\ell = 0$, then every $a_i = 0$. It follows that every sublist of a linearly independent list is itself linearly independent (this is one reason for decreeing that $\varnothing$ be linearly independent).

We have arrived at the notion we have been seeking.

**Definition.** A ***basis*** of a vector space $V$ is a linearly independent list that spans $V$.

Thus, bases are shortest spanning lists. Of course, all the vectors in a linearly independent list $v_1, \ldots, v_n$ are distinct, by Example 2.98(iv). Note that a list $X = v_1, \ldots, v_m$ being a basis does not depend on the ordering of the vectors, but only on the set of vectors themselves, for neither spanning nor linear independence depends on the ordering.

**Example 2.100.** In Example 2.96(iii), we saw that $X = e_1, \ldots, e_n$ spans $k^n$, where $e_i$ is the $n$-tuple having 1 in the $i$th coordinate and 0's elsewhere. It is easy to see that $X$ is linearly independent: $\sum_{i=1}^n a_i e_i = (a_1, \ldots, a_n)$, and $(a_1, \ldots, a_n) = (0, \ldots, 0)$ if and only if all $a_i = 0$. Hence, the list $e_1, \ldots, e_n$ is a basis; it is called the ***standard basis*** of $k^n$. ◄

**Proposition 2.101.** *Let $X = v_1, \ldots, v_n$ be a list in a vector space $V$ over a field $k$. Then $X$ is a basis if and only if each vector in $V$ has a unique expression as a $k$-linear combination of vectors in $X$.*

**Proof.** If a vector $v = \sum a_i v_i = \sum b_i v_i$, then $\sum (a_i - b_i) v_i = 0$, and so independence gives $a_i = b_i$ for all $i$; that is, the expression is unique.

Conversely, existence of an expression shows that the list of $v_i$ spans. Moreover, if $0 = \sum c_i v_i$ with not all $c_i = 0$, then the vector 0 does not have a unique expression as a linear combination of the $v_i$. ●

**Definition.** If $X = v_1, \ldots, v_n$ is a basis of a vector space $V$ and $v \in V$, then there are unique scalars $a_1, \ldots, a_n$ with $v = \sum_{i=1}^n a_i v_i$. The $n$-tuple $(a_1, \ldots, a_n)$ is called the ***coordinate list*** of a vector $v \in V$ relative to the basis $X$.

Observe that if $v_1, \ldots, v_n$ is the standard basis of $V = k^n$, then this coordinate list coincides with the usual coordinate list.

Coordinates are the reason we have defined bases as lists and not as subsets. If $v_1, \ldots, v_n$ is a basis of a vector space $V$ over a field $k$, then each vector $v \in V$ has a unique expression

$$v = a_1 v_1 + a_2 v_2 + \cdots + a_n v_n,$$

where $a_i \in k$ for all $i$. Since there is a first vector $v_1$, a second vector $v_2$, and so forth, the coefficients in this $k$-linear combination determine a unique $n$-tuple $(a_1, a_2, \ldots, a_n)$. Were a basis merely a subset of $V$ and not a list (i.e., an ordered subset), then there would be $n!$ coordinate lists for every vector.

We are going to define the *dimension* of a vector space $V$ to be the number of vectors in a basis. Two questions arise at once.

(i) Does every vector space have a basis?

(ii) Do all bases of a vector space have the same number of elements?

The first question is easy to answer; the second needs some thought.

**Theorem 2.102.** *Every finite-dimensional*[17] *vector space $V$ has a basis.*

**Proof.** A finite spanning list $X$ exists, since $V$ is finite-dimensional. If it is linearly independent, it is a basis; if not, $X$ can be shortened to a spanning sublist $X'$, by Proposition 2.97. If $X'$ is linearly independent, it is a basis; if not, $X'$ can be shortened to a spanning sublist $X''$. Eventually, we arrive at a shortest spanning sublist, which is independent, by Corollary 2.99, and hence it is a basis. •

We can now prove Invariance of Dimension, one of the most important results about vector spaces.

**Lemma 2.103.** *Let $u_1, \ldots, u_n$ and $v_1, \ldots, v_m$ be lists in a vector space $V$, and let $v_1, \ldots, v_m \in \langle u_1, \ldots, u_n \rangle$. If $m > n$, then $v_1, \ldots, v_m$ is linearly dependent.*

**Proof.** The proof is by induction on $n \geq 1$.

If $n = 1$, then there are at least two vectors $v_1, v_2$ and $v_1 = a_1 u_1$ and $v_2 = a_2 u_1$. If $u_1 = 0$, then $v_1 = 0$ and the list of $v$'s is linearly dependent. Suppose $u_1 \neq 0$. We may assume that $v_1 \neq 0$, or we are done; hence, $a_1 \neq 0$. Therefore, $v_1, v_2$ is linearly dependent, for $v_2 - a_2 a_1^{-1} v_1 = 0$, and hence the larger list $v_1, \ldots, v_m$ is linearly dependent.

Let us prove the inductive step. There are equations, for $i = 1, \ldots, m$,

$$v_i = a_{i1} u_1 + \cdots + a_{in} u_n.$$

We may assume that some $a_{i1} \neq 0$; otherwise $v_1, \ldots, v_m \in \langle u_2, \ldots, u_n \rangle$, and the inductive hypothesis applies. Changing notation if necessary (that is, by re-ordering the $v$'s), we may assume that $a_{11} \neq 0$. For each $i \geq 2$, define

$$v_i' = v_i - a_{i1} a_{11}^{-1} v_1 \in \langle u_2, \ldots, u_n \rangle$$

[if we write $v_i'$ as a linear combination of the $u$'s, then $a_{i1} - (a_{i1} a_{11}^{-1}) a_{11} = 0$ is the coefficient of $u_1$]. Since $m - 1 > n - 1$, the inductive hypothesis gives scalars $b_2, \ldots, b_m$, not all 0, with

$$b_2 v_2' + \cdots + b_m v_m' = 0.$$

Rewrite this equation using the definition of $v_i'$:

$$\left( -\sum_{i \geq 2} b_i a_{i1} a_{11}^{-1} \right) v_1 + b_2 v_2 + \cdots + b_m v_m = 0.$$

Not all the coefficients are 0, and so $v_1, \ldots, v_m$ is linearly dependent. •

The following familiar fact illustrates the intimate relation between Linear Algebra and systems of linear equations.

**Corollary 2.104.** *A homogeneous system of linear equations over a field $k$ with more unknowns than equations has a nontrivial solution.*

---

[17]The definitions of spanning and linear independence can be extended to infinite-dimensional vector spaces, and Theorem 5.47 says that bases always exist. It turns out that a basis of $k[x]$ is $1, x, x^2, \ldots, x^n, \ldots$.

**Proof.** An $n$-tuple $(b_1, \ldots, b_n)^\top \in k^n$ is a solution of a system

$$a_{11}x_1 + \cdots + a_{1n}x_n = 0$$

$$\vdots \qquad \vdots \qquad \vdots$$

$$a_{m1}x_1 + \cdots + a_{mn}x_n = 0$$

if $a_{i1}b_1 + \cdots + a_{in}b_n = 0$ for all $i$. Rewrite this as $b_1a_{i1} + \cdots + b_na_{in} = 0$, and sum over all $i$: if $\gamma_1, \ldots, \gamma_n \in k^m$ are the columns of the coefficient matrix $[a_{ij}]$, then

$$b_1\gamma_1 + \cdots + b_n\gamma_n = 0.$$

Now $k^m$ can be spanned by $m$ vectors (the standard basis, for example). Since $n > m$, by hypothesis, Lemma 2.103 shows that the list $\gamma_1, \ldots, \gamma_n$ is linearly dependent; there are scalars $c_1, \ldots, c_n$, not all zero, with $c_1\gamma_1 + \cdots + c_n\gamma_n = 0$. Therefore, $c^\top = (c_1, \ldots, c_n)^\top$ is a nontrivial solution of the system.  •

**Theorem 2.105 (Invariance of Dimension).** *If $X = x_1, \ldots, x_n$ and $Y = y_1, \ldots, y_m$ are bases of a vector space $V$, then $m = n$.*

**Proof.** Suppose that $m \neq n$. If $n < m$, then $y_1, \ldots, y_m \in \langle x_1, \ldots, x_n \rangle$, because $X$ spans $V$, and Lemma 2.103 gives $Y$ linearly dependent, a contradiction. A similar contradiction arises if $m < n$, and so $m = n$.  •

It is now permissible to make the following definition.

**Definition.** The ***dimension*** of a finite-dimensional vector space $V$ over a field $k$, denoted by $\dim_k(V)$ or $\dim(V)$, is the number of elements in a basis of $V$.

**Example 2.106.**

(i) Example 2.100 shows that $k^n$ has dimension $n$, which agrees with our intuition when $k = \mathbb{R}$. Thus, the plane $\mathbb{R} \times \mathbb{R}$ is two-dimensional!

(ii) If $V = \{0\}$, then $\dim(V) = 0$, for there are no elements in its basis $\varnothing$. (This is another good reason for defining $\varnothing$ to be linearly independent.)

(iii) Let $X = \{x_1, \ldots, x_n\}$ be a finite set. Define

$$k^X = \{\text{functions } f \colon X \to k\}.$$

Now $k^X$ is a vector space if we define addition $k^X \times k^X \to k^X$ by

$$(f, g) \mapsto f + g \colon x \mapsto f(x) + g(x)$$

and scalar multiplication $k \times k^X \to k^X$ by

$$(a, f) \mapsto af \colon x \mapsto af(x).$$

It is easy to check that the set of $n$ functions of the form $f_x$, where $x \in X$, defined by

$$f_x(y) = \begin{cases} 1 & \text{if } y = x; \\ 0 & \text{if } y \neq x, \end{cases}$$

form a basis, and so $\dim(k^X) = n = |X|$.

This is not a new example: since an $n$-tuple $(a_1, \ldots, a_n)$ is really a function $f \colon \{1, \ldots, n\} \to k$ with $f(i) = a_i$ for all $i$, the functions $f_x$ comprise the standard basis. ◄

Here is a second proof of Invariance of Dimension; it will be used in Chapter 5 to adapt the notion of dimension to the notion of *transcendence degree*. We begin with a modification of the proof of Proposition 2.97.

**Lemma 2.107.** *If $X = v_1, \ldots, v_n$ is a linearly dependent list of vectors in a vector space $V$, then there exists $v_r$ with $r \geq 1$ with $v_r \in \langle v_1, v_2, \ldots, v_{r-1} \rangle$ [when $r = 1$, we interpret $\langle v_1, \ldots, v_{r-1} \rangle$ to mean $\{0\}$].*

**Remark.** Let us compare Proposition 2.97 with this one. The earlier result says that if $v_1, v_2, v_3$ is linearly dependent, then either $v_1 \in \langle v_2, v_3 \rangle, v_2 \in \langle v_1, v_3 \rangle$, or $v_3 \in \langle v_1, v_2 \rangle$. This lemma says that either $v_1 \in \{0\}$, $v_2 \in \langle v_1 \rangle$, or $v_3 \in \langle v_1, v_2 \rangle$. ◄

**Proof.** Let $r$ be the largest integer for which $v_1, \ldots, v_{r-1}$ is linearly independent. If $v_1 = 0$, then $v_1 \in \{0\}$, and we are done. If $v_1 \neq 0$, then $r \geq 2$; since $v_1, v_2, \ldots, v_n$ is linearly dependent, we have $r - 1 < n$. As $r - 1$ is largest, the list $v_1, v_2, \ldots, v_r$ is linearly dependent. There are thus scalars $a_1, \ldots, a_r$, not all zero, with $a_1 v_1 + \cdots + a_r v_r = 0$. In this expression, we must have $a_r \neq 0$, for otherwise $v_1, \ldots, v_{r-1}$ would be linearly dependent. Therefore,

$$v_r = \sum_{i=1}^{r-1} (-a_r^{-1}) a_i v_i \in \langle v_1, \ldots, v_{r-1} \rangle. \quad \bullet$$

**Lemma 2.108 (Exchange Lemma).** *If $X = x_1, \ldots, x_m$ is a basis of a vector space $V$ and $y_1, \ldots, y_n$ is a linearly independent list in $V$, then $n \leq m$.*

**Proof.** We begin by showing that one of the $x$'s in $X$ can be replaced by $y_n$ so that the new list still spans $V$. Now $y_n \in \langle X \rangle$, since $X$ spans $V$, so that the list

$$y_n, x_1, \ldots, x_m$$

is linearly dependent, by Proposition 2.97. Since the list $y_1, \ldots, y_n$ is linearly independent, $y_n \notin \langle 0 \rangle$. By Lemma 2.107, there is some $i$ with $x_i = a y_n + \sum_{j<i} a_j x_j$. Throwing out $x_i$ and replacing it by $y_n$ gives a spanning list

$$X' = y_n, x_1, \ldots, \widehat{x_i}, \ldots, x_m$$

[if $v = \sum_{j=1}^{m} b_j x_j$, then (as in the proof of Proposition 2.97) replace $x_i$ by its expression as a $k$-linear combination of the other $x$'s and $y_n$, and then collect terms].

Now repeat this argument for the spanning list $y_{n-1}, y_n, x_1, \ldots, \widehat{x_i}, \ldots, x_m$. The options offered by Lemma 2.107 for this linearly dependent list are $y_n \in \langle y_{n-1} \rangle$, $x_1 \in \langle y_{n-1}, y_n \rangle$, $x_2 \in \langle y_{n-1}, y_n, x_1 \rangle$, and so forth. Since $Y$ is linearly independent, so is its sublist $y_{n-1}, y_n$, and the first option $y_n \in \langle y_{n-1} \rangle$ is not feasible. It follows that the disposable vector (provided by Lemma 2.107) must be one of the remaining $x$'s, say $x_\ell$. After throwing out $x_\ell$, we have a new spanning list $X''$. Repeat this construction of spanning lists; each time a new $y$ is adjoined as the first vector, an $x$ is thrown out, for the option $y_i \in \langle y_{i+1}, \ldots, y_n \rangle$ is not feasible. If $n > m$, that is, if there are more $y$'s than $x$'s, then this procedure ends with a spanning list consisting of $m$ $y$'s (one for each of the $m$ $x$'s thrown out) and no $x$'s. Thus a proper sublist $y_1, \ldots, y_m$ of $Y$ spans $V$, contradicting the linear independence of $Y$. Therefore, $n \leq m$.  •

**Theorem 2.109 (Invariance of Dimension again).** *If $X = x_1, \ldots, x_m$ and $Y = y_1, \ldots, y_n$ are bases of a vector space $V$, then $m = n$.*

**Proof.** By Lemma 2.108, viewing $X$ as a basis with $m$ elements and $Y$ as a linearly independent list with $n$ elements gives the inequality $n \leq m$; viewing $Y$ a basis and $X$ as a linearly independent list gives the reverse inequality $m \leq n$. Therefore, $m = n$, as desired.  •

We have constructed bases as shortest spanning lists; we are now going to construct them as longest linearly independent lists.

**Definition.** A *maximal* (or *longest*) linearly independent list $u_1, \ldots, u_m$ in a vector space $V$ is a linearly independent list for which there is no vector $v \in V$ with $u_1, \ldots, u_m, v$ linearly independent.

**Lemma 2.110.** *Let $X = u_1, \ldots, u_m$ be a linearly independent list in a vector space $V$. If $X$ does not span $V$, then there exists $v \in V$ such that the list $X' = u_1, \ldots, u_m, v$ is linearly independent.*

**Proof.** Since $X$ does not span $V$, there exists $v \in V$ with $v \notin \langle u_1, \ldots, u_m \rangle$. By Proposition 2.97(ii), the longer list $X'$ is linearly independent.  •

**Proposition 2.111.** *Let $V$ be a finite-dimensional vector space; say, $\dim(V) = n$.*

   (i) *There exist maximal linearly independent lists in $V$.*

   (ii) *Every maximal linearly independent list $X$ is a basis of $V$.*

**Proof.**

   (i) If a linearly independent list $X = x_1, \ldots, x_r$ is not a basis, then it does not span: there is $w \in V$ with $w \notin \langle x_1, \ldots, x_r \rangle$. By Lemma 2.110, the longer list $X' = x_1, \ldots, x_r, w$ is linearly independent. If $X'$ is a basis, we are done; otherwise, repeat and construct a longer list. If this process does not stop, then there is a linearly independent list having $n+1$ elements. Comparing this list with a basis of $V$, we contradict the inequality in the Exchange Lemma.

   (ii) If a maximal linearly independent list $X$ is not a basis, then Lemma 2.110 constructs a larger linearly independent list, contradicting the maximality of $X$.  •

**Corollary 2.112.** *Let $V$ be a vector space with $\dim(V) = n$.*

   (i) *Any list of $n$ vectors that spans $V$ must be linearly independent.*

   (ii) *Any linearly independent list of $n$ vectors must span $V$.*

**Proof.**

 (i) Were a list linearly dependent, it could be shortened to give a basis; this basis is too small.

 (ii) If a list does not span, it could be lengthened to give a basis; this basis is too large. •

**Proposition 2.113.** *Let $V$ be a finite-dimensional vector space. If $Z = u_1, \ldots, u_m$ is a linearly independent list in $V$, then $Z$ can be extended to a basis: there are vectors $v_{m+1}, \ldots, v_n$ such that $u_1, \ldots, u_m, v_{m+1}, \ldots, v_n$ is a basis of $V$.*

**Proof.** Iterated use of Lemma 2.110 [as in the proof of Proposition 2.111(i)] shows that $Z$ can be extended to a maximal linearly independent set $X$ in $V$. But Proposition 2.111(ii) says that $X$ is a basis. •

**Corollary 2.114.** *If $\dim(V) = n$, then any list of $n+1$ or more vectors is linearly dependent.*

**Proof.** Otherwise, such a list could be extended to a basis having too many elements. •

**Corollary 2.115.** *Let $U$ be a subspace of a vector space $V$, where $\dim(V) = n$.*

 (i) *$U$ is finite-dimensional and $\dim(U) \leq \dim(V)$.*

 (ii) *If $\dim(U) = \dim(V)$, then $U = V$.*

**Proof.**

 (i) Any linearly independent list in $U$ is also a linearly independent list in $V$. Hence, there exists a maximal linearly independent list $X = u_1, \ldots, u_m$ in $U$. By Proposition 2.111, $X$ is a basis of $U$; hence, $U$ is finite-dimensional and $\dim(U) = m \leq n$.

 (ii) If $\dim(U) = \dim(V)$, then a basis of $U$ is already a basis of $V$ (otherwise it could be extended to a basis of $V$ that would be too large). •

## Exercises

**2.72.** If the only subspaces of a vector space $V$ are $\{0\}$ and $V$ itself, prove that $\dim(V) \leq 1$.

**2.73.** Prove, in the presence of all the other axioms in the definition of vector space, that the commutative law for vector addition is redundant; that is, if $V$ satisfies all the other axioms, then $u + v = v + u$ for all $u, v \in V$.

**Hint.** If $u, v \in V$, evaluate $-[(-v) + (-u)]$ in two ways.

**2.74.** If $V$ is a vector space over $\mathbb{F}_2$ and $v_1 \neq v_2$ are nonzero vectors in $V$, prove that $v_1, v_2$ is linearly independent. Is this true for vector spaces over any other field?

**2.75.** Prove that the columns of an $m \times n$ matrix $A$ over a field $k$ are linearly dependent in $k^m$ if and only if the homogeneous system $Ax = 0$ has a nontrivial solution.

**2.76.** If $U$ is a subspace of a vector space $V$ over a field $k$, define a scalar multiplication on the (additive) quotient *group* $V/U$ by

$$\alpha(v + U) = \alpha v + U,$$

where $\alpha \in k$ and $v \in V$. Prove that this is a well-defined function that makes $V/U$ into a vector space over $k$ ($V/U$ is called a **quotient space**).

**2.77.** If $V$ is a finite-dimensional vector space and $U$ is a subspace, prove that

$$\dim(U) + \dim(V/U) = \dim(V).$$

**Hint.** Prove that if $v_1 + U, \ldots, v_r + U$ is a basis of $V/U$, then the list $v_1, \ldots, v_r$ is linearly independent.

\* **2.78.** Prove that every finite-dimensional vector space over a countable field is countable.

**Definition.** If $U$ and $W$ are subspaces of a vector space $V$, define

$$U + W = \{u + w : u \in U \text{ and } w \in W\}.$$

\* **2.79.** (i) Prove that $U + W$ is a subspace of $V$.

(ii) If $U$ and $U'$ are subspaces of a finite-dimensional vector space $V$, prove that

$$\dim(U) + \dim(U') = \dim(U \cap U') + \dim(U + U').$$

**Hint.** Take a basis of $U \cap U'$ and extend it to bases of $U$ and of $U'$.

**Definition.** Let $V$ be a vector space having subspaces $U$ and $W$. Then $V$ is the **direct sum**, $V = U \oplus W$, if $U \cap W = \{0\}$ and $V = U + W$.

\* **2.80.** If $U$ and $W$ are finite-dimensional vector spaces over a field $k$, prove that

$$\dim(U \oplus W) = \dim(U) + \dim(W).$$

**2.81.** Let $U$ be a subspace of a finite-dimensional vector space $V$. Prove that there exists a subspace $W$ of $V$ with $V = U \oplus W$.

**Hint.** Extend a basis $X$ of $U$ to a basis $X'$ of $V$, and define $W = \langle X' - X \rangle$.

## Section 2.8. Linear Transformations and Matrices

Homomorphisms between vector spaces are called *linear transformations*.

**Definition.** If $V$ and $W$ are vector spaces over a field $k$, then a **linear transformation** is a function $T \colon V \to W$ such that, for all vectors $u, v \in V$ and all scalars $a \in k$,

(i) $T(u + v) = T(u) + T(v)$,

(ii) $T(av) = aT(v)$.

We say that a linear transformation $T \colon V \to W$ is an **isomorphism** (or is **nonsingular**) if it is a bijection. Two vector spaces $V$ and $W$ over $k$ are **isomorphic**, denoted by $V \cong W$, if there exists an isomorphism $T \colon V \to W$.

If we forget the scalar multiplication, then a vector space is an (additive) abelian group and a linear transformation $T$ is a group homomorphism; thus, $T(0) = 0$. It is easy to see that $T$ preserves all $k$-linear combinations:

$$T(a_1 v_1 + \cdots + a_m v_m) = a_1 T(v_1) + \cdots + a_m T(v_m).$$

**Example 2.116.**

(i) The identity function $1_V \colon V \to V$ on any vector space $V$ is a nonsingular linear transformation.

(ii) If $\theta$ is an angle, then rotation about the origin by $\theta$ is a linear transformation $R_\theta \colon \mathbb{R}^2 \to \mathbb{R}^2$. The function $R_\theta$ preserves addition because it takes parallelograms to parallelograms, and it preserves scalar multiplication because it preserves the lengths of arrows [see Example 2.93(i)]. Every rotation is nonsingular: the inverse of $R_\theta$ is $R_{-\theta}$.

(iii) If $V$ and $W$ are vector spaces over a field $k$, write $\operatorname{Hom}_k(V, W)$ for the set of all linear transformations $V \to W$. Define *addition* $S + T$ by $v \mapsto S(v) + T(v)$ for all $v \in V$, and define *scalar multiplication* $aT \colon V \to W$, where $a \in k$, by $v \mapsto a[T(v)]$ for all $v \in V$. Both $S + T$ and $aT$ are linear transformations, and $\operatorname{Hom}_k(V, W)$ is a vector space over $k$.

(iv) Regard elements of $k^n$ as $n \times 1$ column vectors. If $A$ is an $m \times n$ matrix with entries in $k$, then $T \colon k^n \to k^m$, given by $v \mapsto Av$ (where $Av$ is the $m \times 1$ column vector given by matrix multiplication), is a linear transformation.  ◀

**Definition.** If $V$ is a vector space over a field $k$, then the **general linear group**, denoted by $\operatorname{GL}(V)$, is the set of all nonsingular linear transformations $V \to V$.

The composite $ST$ of linear transformations $S$ and $T$ is again a linear transformation, and $ST$ is an isomorphism if both $S$ and $T$ are; moreover, the inverse of an isomorphism is again a linear transformation. It follows that $\operatorname{GL}(V)$ is a group with composition as operation, for composition of functions is always associative.

Just as a group homomorphism or a ring homomorphism, each linear transformation has a kernel and an image.

**Definition.** If $T \colon V \to W$ is a linear transformation, then the **kernel** (or **null space**) of $T$ is

$$\ker T = \{v \in V : T(v) = 0\},$$

and the **image** (or **range**) of $T$ is

$$\operatorname{im} T = \{w \in W : w = T(v) \text{ for some } v \in V\}.$$

As in Example 2.116(iv), an $m \times n$ matrix $A$ with entries in a field $k$ determines a linear transformation $k^n \to k^m$, namely, $y \mapsto Ay$, where $y$ is an $n \times 1$ column vector. The kernel of this linear transformation is usually called the *solution space* of $A$ [see Example 2.94(iv)].

The proof of the next proposition is straightforward.

**Proposition 2.117.** *Let $T \colon V \to W$ be a linear transformation.*

(i) $\ker T$ *is a subspace of $V$ and* $\operatorname{im} T$ *is a subspace of $W$.*

(ii) $T$ *is injective if and only if* $\ker T = \{0\}$.

We can now interpret the fact that a homogeneous system over a field $k$ with $m$ equations in $n$ unknowns has a nontrivial solution if $m < n$. If $A$ is the $m \times n$ coefficient matrix of the system, then $T\colon x \mapsto Ax$ is a linear transformation $k^n \to k^m$. If there is only the trivial solution, then $\ker T = \{0\}$, so that $k^n$ is isomorphic to a subspace of $k^m$, contradicting Corollary 2.115(i): if $U \subseteq V$, then $\dim(U) \leq \dim(V)$.

**Lemma 2.118.** *Let* $T\colon V \to W$ *be a linear transformation.*

(i) *If* $T$ *is an isomorphism, then for every basis* $X = v_1, v_2, \ldots, v_n$ *of* $V$, *the list* $T(X) = T(v_1), T(v_2), \ldots, T(v_n)$ *is a basis of* $W$.

(ii) *Conversely, if there exists some basis* $X = v_1, v_2, \ldots, v_n$ *of* $V$ *for which* $T(X) = T(v_1), T(v_2), \ldots, T(v_n)$ *is a basis of* $W$, *then* $T$ *is an isomorphism.*

**Proof.**

(i) Let $T$ be an isomorphism. If $\sum c_i T(v_i) = 0$, then $T(\sum c_i v_i) = 0$, and so $\sum c_i v_i \in \ker T = \langle 0 \rangle$. Hence each $c_i = 0$, because $X$ is linearly independent, and so $T(X)$ is linearly independent. If $w \in W$, then the surjectivity of $T$ provides $v \in V$ with $w = T(v)$. But $v = \sum a_i v_i$, and so $w = T(v) = T(\sum a_i v_i) = \sum a_i T(v_i)$. Therefore, $T(X)$ spans $W$, and so it is a basis of $W$.

(ii) Let $w \in W$. Since $T(v_1), \ldots, T(v_n)$ is a basis of $W$, we have $w = \sum c_i T(v_i) = T(\sum c_i v_i)$, and so $T$ is surjective. If $\sum c_i v_i \in \ker T$, then $\sum c_i T(v_i) = 0$, and so linear independence gives all $c_i = 0$; hence, $\sum c_i v_i = 0$ and $\ker T = \langle 0 \rangle$. Therefore, $T$ is an isomorphism.   •

Recall the Pigeonhole Principle, Exercise 1.8 on page 14: a function $f\colon X \to X$ on a finite set $X$ is an injection if and only if it is a surjection. Here is the Linear Algebra version.

**Proposition 2.119 (Pigeonhole Principle).** *Let* $V$ *be a finite-dimensional vector space with* $\dim(V) = n$, *and let* $T\colon V \to V$ *be a linear transformation. The following statements are equivalent:*

(i) $T$ *is nonsingular;*

(ii) $T$ *is surjective;*

(iii) $T$ *is injective.*

**Proof.**

(i) $\Rightarrow$ (ii) This implication is obvious.

(ii) $\Rightarrow$ (iii) Let $v_1, \ldots, v_n$ be a basis of $V$. Since $T$ is surjective, there are vectors $u_1, \ldots, u_n$ with $Tu_i = v_i$ for all $i$. We claim that $u_1, \ldots, u_n$ is linearly independent. If there are scalars $c_1, \ldots, c_n$, not all zero, with $\sum c_i u_i = 0$, then we obtain a dependency relation $0 = \sum c_i T(u_i) = \sum c_i v_i$, a contradiction. By Corollary 2.112(ii), $u_1, \ldots, u_n$ is a basis of $V$. To show that $T$ is injective, it suffices to show that $\ker T = \langle 0 \rangle$. Suppose that $T(v) = 0$. Now $v = \sum c_i u_i$, and so $0 = T \sum c_i u_i = \sum c_i v_i$; hence, linear independence of $v_1, \ldots, v_n$ gives all $c_i = 0$, and so $v = 0$. Therefore, $T$ is injective.

(iii) $\Rightarrow$ (i) Let $v_1, \ldots, v_n$ be a basis of $V$. If $c_1, \ldots, c_n$ are scalars, not all 0, then $\sum c_i v_i \neq 0$, for a basis is linearly independent. Since $T$ is injective, it follows that $\sum c_i T v_i \neq 0$, and so $T v_1, \ldots, T v_n$ is linearly independent. Therefore, Corollary 2.112(ii) shows that $T$ is nonsingular.  $\bullet$

We now show how to construct linear transformations $T \colon V \to W$, where $V$ and $W$ are vector spaces over a field $k$. The next theorem says that there is a linear transformation that can do anything to a basis; moreover, such a linear transformation is unique.

**Theorem 2.120.** *Let $V$ and $W$ be vector spaces over a field $k$.*

(i) *If $v_1, \ldots, v_n$ is a basis of $V$ and $u_1, \ldots, u_n$ is a list in $W$, then there exists a unique linear transformation $T \colon V \to W$ with $T(v_i) = u_i$ for all $i$.*

(ii) *If linear transformations $S, T \colon V \to W$ agree on a basis, then $S = T$.*

**Proof.** By Theorem 2.101, each $v \in V$ has a unique expression of the form $v = \sum_i a_i v_i$, and so $T \colon V \to W$, given by $T(v) = \sum a_i u_i$, is a (well-defined) function. It is now a routine verification to check that $T$ is a linear transformation.

To prove uniqueness of $T$, assume that $S \colon V \to W$ is a linear transformation with $S(v_i) = u_i = T(v_i)$ for all $i$. If $v \in V$, then $v = \sum a_i v_i$ and

$$S(v) = S\left(\sum a_i v_i\right) = \sum S(a_i v_i) = \sum a_i S(v_i) = \sum a_i T(v_i) = T(v).$$

Since $v$ is arbitrary, $S = T$.  $\bullet$

The statement of Theorem 2.120 can be pictured. The list $u_1, \ldots, u_n$ in $W$ gives the function $f \colon X = \{v_1, \ldots, v_n\} \to W$ defined by $f(v_i) = u_i$ for all $i$; the vertical arrow $X \to V$ is the inclusion:

$$
\begin{array}{ccc}
V & & \\
\uparrow & \searrow{\scriptstyle T} & \\
X & \xrightarrow[f]{} & W
\end{array}
$$

**Theorem 2.121.** *If $V$ is an $n$-dimensional vector space over a field $k$, then $V$ is isomorphic to $k^n$.*

**Proof.** Choose a basis $v_1, \ldots, v_n$ of $V$. If $e_1, \ldots, e_n$ is the standard basis of $k^n$, then Theorem 2.120(i) says that there is a linear transformation $T \colon V \to k^n$ with $T(v_i) = e_i$ for all $i$; by Lemma 2.118, $T$ is an isomorphism.  $\bullet$

Theorem 2.121 does more than say that every finite-dimensional vector space is essentially the familiar vector space of all $n$-tuples. It says that a choice of basis in $V$ is tantamount to choosing coordinate lists for every vector in $V$. The freedom to change coordinates is important because the usual coordinates may not be the most convenient ones for a given problem, as the reader has seen (in a Calculus course) when rotating axes to simplify the equation of a conic section.

**Corollary 2.122.** *Two finite-dimensional vector spaces $V$ and $W$ over a field $k$ are isomorphic if and only if $\dim(V) = \dim(W)$.*

**Proof.** Assume that there is an isomorphism $T: V \to W$. If $X = v_1, \ldots, v_n$ is a basis of $V$, then Lemma 2.118 says that $T(v_1), \ldots, T(v_n)$ is a basis of $W$. Therefore, $\dim(W) = n = \dim(V)$.

If $n = \dim(V) = \dim(W)$, there are isomorphisms $T: V \to k^n$ and $S: W \to k^n$, by Theorem 2.121, and the composite $S^{-1}T: V \to W$ is an isomorphism. •

Linear transformations defined on $k^n$ are easy to describe.

**Theorem 2.123.** *If $T: k^n \to k^m$ is a linear transformation, then there exists a unique $m \times n$ matrix $A$ such that*

$$T(y) = Ay$$

*for all $y \in k^n$ (here, $y$ is an $n \times 1$ column matrix and $Ay$ is matrix multiplication).*

**Proof.** If $e_1, \ldots, e_n$ is the standard basis of $k^n$ and $e'_1, \ldots, e'_m$ is the standard basis of $k^m$, define $A = [a_{ij}]$ to be the matrix whose $j$th column is the coordinate list of $T(e_j)$. If $S: k^n \to k^m$ is defined by $S(y) = Ay$, then $S = T$ because both agree on a basis: $T(e_j) = \sum_i a_{ij}e_i = Ae_j$. Uniqueness of $A$ follows from Theorem 2.120(ii): if $T(y) = By$ for all $y$, then $Be_j = T(e_j) = Ae_j$ for all $j$; that is, the columns of $A$ and $B$ are the same. •

Theorem 2.123 establishes the connection between linear transformations and matrices, and the definition of matrix multiplication arises from applying this construction to the composite of two linear transformations.

**Definition.** Let $X = v_1, \ldots, v_n$ be a basis of $V$ and let $Y = w_1, \ldots, w_m$ be a basis of $W$. If $T: V \to W$ is a linear transformation, then the ***matrix of $T$*** is the $m \times n$ matrix $A = [a_{ij}]$ whose $j$th column $a_{1j}, a_{2j}, \ldots, a_{mj}$ is the coordinate list of $T(v_j)$ determined by the $w$'s: $T(v_j) = \sum_{i=1}^{m} a_{ij}w_i$.

Since the matrix $A$ depends on the choice of bases $X$ and $Y$, we will write

$$A = {}_Y[T]_X$$

when it is necessary to display them.

**Remark.** Let $T: V \to W$ be a linear transformation, and let $X = v_1, \ldots, v_n$ and $Y = w_1, \ldots, w_m$ be bases of $V$ and $W$, respectively. The matrix for $T$ is set up from the explicit equation

$$T(v_j) = a_{1j}w_1 + a_{2j}w_2 + \cdots + a_{mj}w_m;$$

that is, $T(v_j) = \sum_i a_{ji}w_i$. The indices are reversed; why not write $T(v_j) = \sum_i a_{ji}w_j$? To ask this question another way: why did we choose the coordinates of $T(v_j)$ to be the $j$th column of the associated matrix and not the $j$th row?

Consider the important special case $T: k^n \to k^m$ in Example 2.116(iv) given by $T(y) = Ay$, where $A$ is an $m \times n$ matrix and $y$ is an $n \times 1$ column vector. If $e_1, \ldots, e_n$ and $e'_1, \ldots, e'_m$ are the standard bases of $k^n$ and $k^m$, respectively, then the definition of matrix multiplication says that $T(e_j) = Ae_j$ is the $j$th column of $A$. But

$$Ae_j = a_{1j}e_1' + a_{2j}e_2' + \cdots + a_{mj}e_m';$$

that is, the coordinates of $T(e_j) = Ae_j$ with respect to the basis $e_1', \ldots, e_m'$ are $(a_{1j}, \ldots, a_{mj})$. Therefore, the matrix associated to $T$ is the original matrix $A$.  ◄

In case $V = W$, we often let the bases $X = v_1, \ldots, v_n$ and $Y = w_1, \ldots, w_m$ coincide. If $1_V : V \to V$, given by $v \mapsto v$, is the identity linear transformation, then $_X[1_V]_X$ is the $n \times n$ **identity matrix** $I_n$ (usually, the subscript $n$ is omitted), defined by

$$I = [\delta_{ij}],$$

where $\delta_{ij}$ is the Kronecker delta. Thus, $I$ has 1's on the diagonal and 0's elsewhere. On the other hand, if $X$ and $Y$ are different bases, then $_Y[1_V]_X$ is not the identity matrix. The matrix $_Y[1_V]_X$ is called the **transition matrix** from $X$ to $Y$; its columns are the coordinate lists of the $v$'s with respect to the $w$'s.

In Theorem 2.126, we shall prove that matrix multiplication arises from composition of linear transformations. If $T : V \to W$ has matrix $A$ and $S : W \to U$ has matrix $B$, then the linear transformation $ST : V \to U$ has matrix $BA$.

**Example 2.124.**

(i) Let $X = \varepsilon_1, \varepsilon_2 = (1,0), (0,1)$ be the standard basis of $\mathbb{R}^2$. If $T : \mathbb{R}^2 \to \mathbb{R}^2$ is rotation by $90°$, then $T : \varepsilon_1 \mapsto \varepsilon_2$ and $\varepsilon_2 \mapsto -\varepsilon_1$. Hence, the matrix of $T$ relative to $X$ is

$$_X[T]_X = \begin{bmatrix} 0 & -1 \\ 1 & 0 \end{bmatrix}.$$

If we re-order $X$ to obtain the new basis $Y = \eta_1, \eta_2$, where $\eta_1 = \varepsilon_2$ and $\eta_2 = \varepsilon_1$, then $T(\eta_1) = T(\varepsilon_2) = -\varepsilon_1 = -\eta_2$ and $T(\eta_2) = T(\varepsilon_1) = \varepsilon_2 = \eta_1$. The matrix of $T$ relative to $Y$ is

$$_Y[T]_Y = \begin{bmatrix} 0 & 1 \\ -1 & 0 \end{bmatrix}.$$

(ii) Let $k$ be a field, let $T : V \to V$ be a linear transformation on a two-dimensional vector space, and assume that there is some vector $v \in V$ with $T(v)$ not a scalar multiple of $v$. The assumption on $v$ says that the list $X = v, T(v)$ is linearly independent, by Example 2.98(iii), and hence it is a basis of $V$ [because $\dim(V) = 2$]. Write $v_1 = v$ and $v_2 = Tv$.

We compute $_X[T]_X$:

$$T(v_1) = v_2 \quad \text{and} \quad T(v_2) = av_1 + bv_2$$

for some $a, b \in k$. We conclude that

$$_X[T]_X = \begin{bmatrix} 0 & a \\ 1 & b \end{bmatrix}. \quad ◄$$

The following proposition is a paraphrase of Theorem 2.120(i).

**Proposition 2.125.** *Let $V$ and $W$ be vector spaces over a field $k$, and let $X = v_1, \ldots, v_n$ and $Y = w_1, \ldots, w_m$ be bases of $V$ and $W$, respectively. If $\operatorname{Hom}_k(V, W)$ denotes the set of all linear transformations $T \colon V \to W$, and $\operatorname{Mat}_{m \times n}(k)$ denotes the set of all $m \times n$ matrices with entries in $k$, then the function $T \mapsto {}_Y[T]_X$ is a bijection $F \colon \operatorname{Hom}_k(V, W) \to \operatorname{Mat}_{m \times n}(k)$.*

**Remark.** See Corollary 2.127 below. ◄

**Proof.** Given a matrix $A$, its columns define vectors in $W$; in more detail, if the $j$th column of $A$ is $(a_{1j}, \ldots, a_{mj})$, define $z_j = \sum_{i=1}^{m} a_{ij} w_i$. By Theorem 2.120(i), there exists a linear transformation $T \colon V \to W$ with $T(v_j) = z_j$ and ${}_Y[T]_X = A$. Therefore, $F$ is surjective.

To see that $F$ is injective, suppose that ${}_Y[T]_X = A = {}_Y[S]_X$. Since the columns of $A$ determine $T(v_j)$ and $S(v_j)$ for all $j$, Theorem 2.120(ii) gives $S = T$. •

The next theorem shows where the definition of matrix multiplication comes from: the product of two matrices is the matrix of a composite.

**Theorem 2.126.** *Let $T \colon V \to W$ and $S \colon W \to U$ be linear transformations. Choose bases $X = x_1, \ldots, x_n$ of $V$, $Y = y_1, \ldots, y_m$ of $W$, and $Z = z_1, \ldots, z_\ell$ of $U$. Then*

$$ {}_Z[S \circ T]_X = \big({}_Z[S]_Y\big)\big({}_Y[T]_X\big), $$

*where the product on the right is matrix multiplication.*

**Proof.** Let ${}_Y[T]_X = [a_{ij}]$, so that $T(x_j) = \sum_p a_{pj} y_p$, and let ${}_Z[S]_Y = [b_{qp}]$, so that $S(y_p) = \sum_q b_{qp} z_q$. Then

$$ ST(x_j) = S(T(x_j)) = S\Big(\sum_p a_{pj} y_p\Big) $$

$$ = \sum_p a_{pj} S(y_p) = \sum_p \sum_q a_{pj} b_{qp} z_q = \sum_q c_{qj} z_q, $$

where $c_{qj} = \sum_p b_{qp} a_{pj}$. Therefore,

$$ {}_Z[ST]_X = [c_{qj}] = \big({}_Z[S]_Y\big)\big({}_Y[T]_X\big). \quad • $$

**Corollary 2.127.** *If $X$ is a basis of an $n$-dimensional vector space $V$ over a field $k$, then $F \colon \operatorname{Hom}_k(V, V) \to \operatorname{Mat}_n(k)$, given by $T \mapsto {}_X[T]_X$, is an isomorphism of rings.*

**Proof.** The function $F$ is a bijection, by Proposition 2.125. It is easy to see that $F(1_V) = I$ and $F(T + S) = F(T) + F(S)$, while $F(TS) = F(T)F(S)$ follows from Theorem 2.126. Therefore, $F$ is an isomorphism of rings. •

**Corollary 2.128.** *Matrix multiplication is associative.*

**Proof.** Let $A$ be an $m \times n$ matrix, let $B$ be an $n \times p$ matrix, and let $C$ be a $p \times q$ matrix. By Theorem 2.120(i), there are linear transformations

$$ k^q \xrightarrow{T} k^p \xrightarrow{S} k^n \xrightarrow{R} k^m $$

with $C = [T]$, $B = [S]$, and $A = [R]$.

Then
$$[R \circ (S \circ T)] = [R][S \circ T] = [R]([S][T]) = A(BC).$$
On the other hand,
$$[(R \circ S) \circ T] = [R \circ S][T] = ([R][S])[T] = (AB)C.$$
Since composition of functions is associative, $R \circ (S \circ T) = (R \circ S) \circ T$, and so
$$A(BC) = [R \circ (S \circ T)] = [(R \circ S) \circ T] = (AB)C. \quad \bullet$$

The connection with composition of linear transformations is the real reason why matrix multiplication is associative. Recall that an $n \times n$ matrix $P$ is called **nonsingular** if there is an $n \times n$ matrix $Q$ with $PQ = I = QP$. If such a matrix $Q$ exists, it is unique, and it is denoted by $P^{-1}$.

**Corollary 2.129.** *Let $T \colon V \to W$ be a linear transformation of vector spaces $V$ and $W$ over a field $k$, and let $X$ and $Y$ be bases of $V$ and $W$, respectively. If $T$ is an isomorphism, then the matrix of $T^{-1}$ is the inverse of the matrix of $T$ :*
$$_X[T^{-1}]_Y = (_Y[T]_X)^{-1}.$$

**Proof.** We have $I = {}_Y[1_W]_Y = (_Y[T]_X)(_X[T^{-1}]_Y)$, and so Theorem 2.126 gives $I = {}_X[1_V]_X = (_X[T^{-1}]_Y)(_Y[T]_X). \quad \bullet$

The next corollary determines all the matrices arising from the same linear transformation as we vary bases.

**Corollary 2.130.** *Let $T \colon V \to V$ be a linear transformation on a vector space $V$ over a field $k$. If $X$ and $Y$ are bases of $V$, then there is a nonsingular matrix $P$ (namely, the transition matrix $P = {}_Y[1_V]_X$) with entries in $k$ so that*
$$_Y[T]_Y = P(_X[T]_X)P^{-1}.$$
*Conversely, if $B = PAP^{-1}$, where $B, A$, and $P$ are $n \times n$ matrices with $P$ nonsingular, then there is a linear transformation $T \colon k^n \to k^n$ and bases $X$ and $Y$ of $k^n$ such that $B = {}_Y[T]_Y$ and $A = {}_X[T]_X$.*

**Proof.** The first statement follows from Theorem 2.126 and associativity:
$$_Y[T]_Y = {}_Y[1_V T 1_V]_Y = (_Y[1_V]_X)(_X[T]_X)(_X[1_V]_Y).$$
Set $P = {}_Y[1_V]_X$ and note that Corollary 2.129 gives $P^{-1} = {}_X[1_V]_Y$.

For the converse, let $E = e_1, \ldots, e_n$ be the standard basis of $k^n$, and define $T \colon k^n \to k^n$ by $T(e_j) = Ae_j$ (remember that vectors in $k^n$ are column vectors, so that $Ae_j$ is matrix multiplication; indeed, $Ae_j$ is the $j$th column of $A$). It follows that $A = {}_E[T]_E$. Now define a basis $Y = y_1, \ldots, y_n$ by $y_j = P^{-1}e_j$; that is, the vectors in $Y$ are the columns of $P^{-1}$. Note that $Y$ is a basis because $P^{-1}$ is nonsingular. It suffices to prove that $B = {}_Y[T]_Y$; that is, $T(y_j) = \sum_i b_{ij} y_i$, where $B = [b_{ij}]$:
$$T(y_j) = Ay_j = AP^{-1}e_j = P^{-1}Be_j$$
$$= P^{-1} \sum_i b_{ij} e_i = \sum_i b_{ij} P^{-1} e_i = \sum_i b_{ij} y_i. \quad \bullet$$

**Definition.** Two $n \times n$ matrices $B$ and $A$ with entries in a field $k$ are **similar** if there is a nonsingular matrix $P$ with entries in $k$ such that $B = PAP^{-1}$.

Corollary 2.130 says that two matrices arise from the same linear transformation on a vector space $V$ (from different choices of bases) if and only if they are similar. In Chapter 8, we will see how to determine whether two given matrices are similar.

The next corollary shows that "one-sided inverses" are enough.

**Corollary 2.131.** *If $A$ and $B$ are $n \times n$ matrices with $AB = I$, then $BA = I$. Therefore, $A$ is nonsingular with inverse $B$.*

**Proof.** There are linear transformations $T, S \colon k^n \to k^n$ with $[T] = A$ and $[S] = B$, and $AB = I$ gives

$$[TS] = [T][S] = [1_{k^n}].$$

Since $T \mapsto [T]$ is a bijection, by Proposition 2.125, it follows that $TS = 1_{k^n}$. By Set Theory, $T$ is a surjection and $S$ is an injection. But the Pigeonhole Principle, Proposition 2.119, says that both $T$ and $S$ are nonsingular, so that $S = T^{-1}$ and $TS = 1_{k^n} = ST$. Therefore, $I = [ST] = [S][T] = BA$, as desired.  •

**Definition.** The set of all nonsingular $n \times n$ matrices with entries in $k$ is denoted by $\mathrm{GL}(n, k)$.

Now that we have proven associativity, it is easy to prove that $\mathrm{GL}(n, k)$ is a group under matrix multiplication.

A choice of basis gives an isomorphism between the general linear group and the group of nonsingular matrices.

**Proposition 2.132.** *If $V$ is an $n$-dimensional vector space over a field $k$ and $X$ is a basis of $V$, then $f \colon \mathrm{GL}(V) \to \mathrm{GL}(n, k)$, given by $f(T) = {}_X[T]_X$, is a group isomorphism.*

**Proof.** By Corollary 2.127, the function $F \colon T \mapsto {}_X[T]_X$ is a ring isomorphism $\mathrm{Hom}_k(V, V) \to \mathrm{Mat}_n(k)$, and so Proposition 2.28(ii) says that the restriction of $F$ gives an isomorphism $U(\mathrm{Hom}_k(V, V)) \to U(\mathrm{Mat}_n(k))$ between the groups of units of these rings. Now $T \colon V \to V$ is a unit if and only if it is nonsingular, while Corollary 2.129 shows that $F(T) = f(T)$ is a nonsingular matrix.  •

The center of the general linear group is easily identified; we now generalize Exercise 1.64 on page 46.

**Definition.** A linear transformation $T \colon V \to V$ is a **scalar transformation** if there is $c \in k$ with $T(v) = cv$ for all $v \in V$; that is, $T = c1_V$. An $n \times n$ matrix $A$ is a **scalar matrix** if $A = cI$, where $c \in k$ and $I$ is the identity matrix.

A scalar transformation $T = c1_V$ is nonsingular if and only if $c \neq 0$ (its inverse is $c^{-1}1_V$).

**Corollary 2.133.**

(i) *The center of the group* $\mathrm{GL}(V)$ *consists of all the nonsingular scalar transformations.*

(ii) *The center of the group* $\mathrm{GL}(n,k)$ *consists of all the nonsingular scalar matrices.*

**Proof.**

(i) If $T \in \mathrm{GL}(V)$ is not scalar, then Example 2.124(ii) shows that there exists $v \in V$ with $v, T(v)$ linearly independent. By Proposition 2.111, there is a basis $v, T(v), u_3, \ldots, u_n$ of $V$. It is easy to see that $v, v + T(v), u_3, \ldots, u_n$ is also a basis of $V$, and so there is a nonsingular linear transformation $S$ with $S(v) = v$, $S(T(v)) = v + T(v)$, and $S(u_i) = u_i$ for all $i$. Now $S$ and $T$ do not commute, for $ST(v) = v + T(v)$ while $TS(v) = T(v)$. Therefore, $T$ is not in the center of $\mathrm{GL}(V)$.

(ii) If $f \colon G \to H$ is any group isomorphism between groups $G$ and $H$, then $f(Z(G)) = Z(H)$. In particular, if $T = c1_V$ is a nonsingular scalar transformation, then $[T]$ is in the center of $\mathrm{GL}(n,k)$. But $[T] = cI$ is a scalar matrix: if $X = v_1, \ldots, v_n$ is a basis of $V$, then $T(v_i) = cv_i$ for all $i$. $\quad\bullet$

---

# Exercises

**2.82.** If $U$ and $W$ are vector spaces over a field $k$, define their (external) ***direct sum***

$$U \oplus W = \{(u, w) : u \in U \text{ and } w \in W\}$$

with addition $(u, w) + (u', w') = (u + u', w + w')$ and scalar multiplication $\alpha(u, w) = (\alpha u, \alpha w)$ for all $\alpha \in k$. (Compare this definition with that on page 145.)

Let $V$ be a vector space with subspaces $U$ and $W$ such that $U \cap W = \{0\}$ and $U + W = \{u + w : u \in U \text{ and } w \in W\} = V$. Prove that $V \cong U \oplus W$.

\* **2.83.** Recall Example 2.116(iii): if $V$ and $W$ are vector spaces over a field $k$, then $\mathrm{Hom}_k(V, W)$ is a vector space over $k$,

(i) If $V$ and $W$ are finite-dimensional, prove that

$$\dim(\mathrm{Hom}_k(V, W)) = \dim(V)\dim(W).$$

(ii) The ***dual space*** $V^*$ of a vector space $V$ over $k$ is defined by

$$V^* = \mathrm{Hom}_k(V, k).$$

If $\dim(V) = n$, prove that $\dim(V^*) = n$, and hence that $V^* \cong V$.

(iii) If $X = v_1, \ldots, v_n$ is a basis of $V$, define $\delta_1, \ldots, \delta_n \in V^*$ by

$$\delta_i(v_j) = \begin{cases} 0 & \text{if } j \neq i \\ 1 & \text{if } j = i. \end{cases}$$

Prove that $\delta_1, \ldots, \delta_n$ is a basis of $V^*$ (it is called the ***dual basis*** arising from $v_1, \ldots, v_n$).

**2.84.** If $A = \left[\begin{smallmatrix} a & b \\ c & d \end{smallmatrix}\right]$, define $\det(A) = ad - bc$. If $V$ is a vector space with basis $X = v_1, v_2$, define $T\colon V \to V$ by $T(v_1) = av_1 + bv_2$ and $T(v_2) = cv_1 + dv_2$. Prove that $T$ is nonsingular if and only if $\det({}_X[T]_X) \neq 0$.

**Hint.** You may assume the following (easily proved) fact of Linear Algebra: given a system of linear equations with coefficients in a field,

$$ax + by = p$$
$$cx + dy = q,$$

there exists a unique solution if and only if $ad - bc \neq 0$.

**2.85.** Let $U$ be a subspace of a vector space $V$.

(i) Prove that the **natural map** $\pi\colon V \to V/U$, given by $v \mapsto v + U$, is a linear transformation with kernel $U$. (Quotient spaces were defined in Exercise 2.76 on page 145.)

(ii) **(First Isomorphism Theorem for Vector Spaces)** Prove that if $T\colon V \to W$ is a linear transformation, then $\ker T$ is a subspace of $V$ and $\varphi\colon V/\ker T \to \operatorname{im} T$, given by $\varphi\colon v + \ker T \mapsto T(v)$, is an isomorphism.

\* **2.86.** Let $V$ be a finite-dimensional vector space over a field $k$, and let $\mathcal{B}$ denote the family of all the bases of $V$. Prove that $\mathcal{B}$ is a transitive $\mathrm{GL}(V)$-set.

**Hint.** Use Theorem 2.120(i).

**2.87.** (i) If $V$ is an $n$-dimensional vector space over a field $k$, prove that $\mathrm{GL}(n, k)$ acts on $\operatorname{Mat}_n(k)$ by conjugation: if $P \in \mathrm{GL}(n, k)$ and $A \in \operatorname{Mat}_n(k)$, then the action takes $A$ to $PAP^{-1}$.

(ii) Prove that there is a bijection from the orbit space $\operatorname{Mat}_n(k)/\mathrm{GL}(n, k)$ (the family of all orbits) to $\operatorname{Hom}_k(V, V)$.

\* **2.88.** An $n \times n$ matrix $N$ with entries in a field $k$ is **strictly upper triangular** if all entries of $N$ above and on its diagonal are 0.

(i) Prove that the sum and product of strictly upper triangular matrices is again strictly upper triangular.

(ii) Prove that if $N$ is strictly upper triangular, then $N^n = 0$.
   **Hint.** Let $e_1, \ldots, e_n$ be the standard basis of $k^n$ (regarded as column vectors), and define $T\colon k^n \to k^n$ by $T(e_i) = Ne_i$. Show that $T^i(e_j) = 0$ for all $j \leq i$ and $T(e_{i+1}) \in \langle e_1, \ldots, e_i \rangle$, and conclude that $T^n(e_i) = 0$ for all $i$.

**2.89.** Define the **rank** of a linear transformation $T\colon V \to W$ between vector spaces over a field $k$ by

$$\operatorname{rank}(T) = \dim_k(\operatorname{im} T).$$

(i) Regard the columns of an $m \times n$ matrix $A$ as $m$-tuples, and define the **column space** of $A$ to be the subspace of $k^m$ spanned by the columns; define the **rank** of $A$, denoted by $\operatorname{rank}(A)$, to be the dimension of the column space. If $T\colon k^n \to k^m$ is the linear transformation defined by $T(X) = AX$, where $X$ is an $n \times 1$ vector, prove that

$$\operatorname{rank}(A) = \operatorname{rank}(T).$$

(ii) If $A$ is an $m \times n$ matrix and $B$ is a $p \times m$ matrix, prove that

$$\operatorname{rank}(BA) \leq \operatorname{rank}(A).$$

(iii) Prove that similar $n \times n$ matrices have the same rank.

---

## Section 2.9. Quotient Rings and Finite Fields

Let us return to commutative rings. The Fundamental Theorem of Algebra states that every nonconstant polynomial in $\mathbb{C}[x]$ is a product of linear polynomials in $\mathbb{C}[x]$; that is, $\mathbb{C}$ contains all the roots of every polynomial in $\mathbb{C}[x]$. We are going to prove a *local* analog of the Fundamental Theorem of Algebra for polynomials over an arbitrary field $k$: given a polynomial $f(x) \in k[x]$, there is some field $K$ containing $k$ that also contains all the roots of $f(x)$ (we call this a *local analog*, for even though the larger field $K$ contains all the roots of the polynomial $f(x)$, it may not contain roots of other polynomials in $k[x]$). The main idea behind the construction of $K$ involves quotient rings, a construction akin to quotient groups.

Let $I$ be an ideal in a commutative ring $R$. If we forget the multiplication, then $I$ is a subgroup of the additive group $R$; since $R$ is an abelian group, the subgroup $I$ is necessarily normal, and so the quotient group $R/I$ is defined. Recall Lemma 1.46(i), which we now write in additive notation: $a + I = b + I$ in $R/I$ if and only if $a - b \in I$.

**Theorem 2.134.** *If $I$ is an ideal in a commutative ring $R$, then the additive abelian group $R/I$ can be made into a commutative ring in such a way that the **natural map** $\pi\colon R \to R/I$, given by $a \mapsto a + I$, is a surjective ring homomorphism.*

**Proof.** Define multiplication on the additive abelian group $R/I$ by

$$(a + I)(b + I) = ab + I.$$

To see that this is a well-defined function $R/I \times R/I \to R/I$, assume that $a + I = a' + I$ and $b + I = b' + I$; that is, $a - a' \in I$ and $b - b' \in I$. We must show that $(a' + I)(b' + I) = a'b' + I = ab + I$; that is, $ab - a'b' \in I$. This is true:

$$ab - a'b' = ab - a'b + a'b - a'b' = (a - a')b + a'(b - b') \in I.$$

To verify that $R/I$ is a commutative ring, it suffices to show that $1 + I$ is the unit, multiplication is associative and commutative, and the distributive law holds. Proofs of these properties are routine, for they are inherited from the corresponding property in $R$. For example, multiplication in $R/I$ is commutative because

$$(a + I)(b + I) = ab + I = ba + I = (b + I)(a + I).$$

Finally, $\pi\colon R \to R/I$ is a ring homomorphism. We know that $\pi$ preserves addition and that $\pi(1) = 1 + I$. To see that $\pi(a)\pi(b) = \pi(ab)$, rewrite the equation $(a + I)(b + I) = ab + I$ using the definition of $\pi$, namely, $a + I = \pi(a)$. Finally, $\pi$ is surjective because $a + I = \pi(a)$.  •

**Definition.** The commutative ring $R/I$ constructed in Theorem 2.134 is called the **quotient ring** of $R$ modulo $I$ (usually pronounced $R$ mod $I$).

We saw, in Example 1.74, that the additive abelian group $\mathbb{Z}/(m)$ is identical to $\mathbb{I}_m$. They have the same elements: the coset $a + (m)$ and the congruence class $[a]$ are the same subset of $\mathbb{Z}$; they have the same addition:

$$a + (m) + b + (m) = a + b + (m) = [a + b] = [a] + [b].$$

We can now see that the quotient *ring* $\mathbb{Z}/(m)$ coincides with the commutative *ring* $\mathbb{I}_m$, for the two multiplications coincide as well:

$$\big(a + (m)\big)\big(b + (m)\big) = ab + (m) = [ab] = [a][b].$$

Here is the converse of Proposition 2.29.

**Corollary 2.135.** *If $I$ is an ideal in a commutative ring $R$, then there are a commutative ring $A$ and a ring homomorphism $\pi\colon R \to A$ with $I = \ker \pi$.*

**Proof.** The natural map $\pi\colon R \to R/I$ is a ring homomorphism and $I = \ker \pi$.   •

**Theorem 2.136 (First Isomorphism Theorem for Rings).** *If $f\colon R \to A$ is a ring homomorphism, then $\ker f$ is an ideal in $R$, $\operatorname{im} f$ is a subring of $A$, and*

$$R/\ker f \cong \operatorname{im} f.$$

**Proof.** Let $I = \ker f$. We have already seen, in Proposition 2.29, that $I$ is an ideal in $R$ and that $\operatorname{im} f$ is a subring of $A$.

If we forget the multiplication in the rings, then the proof of the First Isomorphism Theorem for Groups, Theorem 1.76, shows that $\varphi\colon R/I \to \operatorname{im} f$, defined by $\varphi(r+I) = f(r)$, is an isomorphism of additive groups. Since $\varphi(1+I) = f(1) = 1$, it suffices to prove that $\varphi$ preserves multiplication. But $\varphi\big((r+I)(s+I)\big) = \varphi(rs+I) = f(rs) = f(r)f(s) = \varphi(r + I)\varphi(s + I)$. Therefore, $\varphi$ is a ring isomorphism.   •

For rings as for groups, the First Isomorphism Theorem creates an isomorphism from a homomorphism once we know its kernel and image. This says that there is no significant difference between a quotient ring and the image of a homomorphism. There are analogs for commutative rings of the Second and Third Isomorphism Theorems for Groups, but they are less useful for rings than their group analogs are.

**Definition.** If $k$ is a field, the intersection of all the subfields of $k$ is called the *prime field* of $k$.

Recall that if $X$ is a subset of a field, then $\langle X \rangle$, the *subfield generated by $X$*, is the intersection of all the subfields containing $X$; it is the smallest such subfield in the sense that any subfield $F$ containing $X$ must contain $\langle X \rangle$. Thus, the prime field is the subfield generated by 1. The prime field of $\mathbb{C}$ and of $\mathbb{R}$ is $\mathbb{Q}$, for every subfield of $\mathbb{C}$ contains $\mathbb{Q}$; the prime field of $\mathbb{F}_p(x)$, or of any finite field, is $\mathbb{F}_p$.

**Proposition 2.137.** *If $k$ is a field, then its prime field is isomorphic to $\mathbb{Q}$ or to $\mathbb{F}_p$ for some prime $p$.*

**Proof.** Consider the ring homomorphism $\chi\colon \mathbb{Z} \to k$, defined by $\chi(n) = n\varepsilon$, where $\varepsilon$ denotes the unit in $k$. Since every ideal in $\mathbb{Z}$ is principal, there is an integer $m$ with $\ker \chi = (m)$. If $m = 0$, then $\chi$ is an injection, and so there is an isomorphic copy of

$\mathbb{Z}$ that is a subring of $k$. By Exercise 2.37(ii), there is a field $Q \cong \mathrm{Frac}(\mathbb{Z}) = \mathbb{Q}$ with $\mathrm{im}\,\chi \subseteq Q \subseteq k$. Now $Q$ is the prime field of $k$, for it is the subfield generated by $\varepsilon$. If $m \neq 0$, the First Isomorphism Theorem gives $\mathbb{I}_m = \mathbb{Z}/(m) \cong \mathrm{im}\,\chi \subseteq k$. Since $k$ is a field, $\mathrm{im}\,\chi$ is a domain, and so Proposition 2.8 gives $m$ prime. If we now write $p$ instead of $m$, then $\mathrm{im}\,\chi \cong \mathbb{F}_p$ is the prime field of $k$, for it is the subfield generated by $\varepsilon$.   •

This last result is the first step in classifying different types of fields.

**Definition.** A field $k$ has *characteristic* 0 if its prime field is isomorphic to $\mathbb{Q}$; it has *characteristic* $p$ if its prime field is isomorphic to $\mathbb{F}_p$ for some prime $p$.

The fields $\mathbb{Q}$, $\mathbb{R}$, $\mathbb{C}$ have characteristic 0, as does any subfield of them; every finite field has characteristic $p$ for some prime $p$, as does $\mathbb{F}_p(x)$, the ring of all rational functions over $\mathbb{F}_p$.

**Proposition 2.138.** *Let $k$ be a field of characteristic $p > 0$. If $ma = 0$, where $m \in \mathbb{Z}$ and $a \in k$, then either $a = 0$ or $p \mid m$. Hence, $pa = 0$ for all $a \in k$.*

**Proof.** Since $k$ has characteristic $p$, we have $p \cdot 1 = 0$, where 1 is the unit in $k$. Hence, if $a = 0$ or $p \mid m$, then Proposition 2.2(v) gives $ma = 0$. Conversely, assume that $ma = 0$. If $a \neq 0$, then $0 = m \cdot 1 = maa^{-1}$. This is an equation in the prime field of $k$, which is isomorphic to $\mathbb{F}_p$. Thus, $m \equiv 0 \bmod p$, and $p \mid m$.   •

**Proposition 2.139.** *If $k$ is a finite field, then $|k| = p^n$ for some prime $p$ and some $n \geq 1$.*

**Proof.** The prime field of $k$ is isomorphic to $\mathbb{F}_p$ for some prime $p$. Now $k$ is a finite-dimensional vector space over $\mathbb{F}_p$ and, if $\dim_{\mathbb{F}_p}(k) = n$, then $|k| = p^n$.   •

**Remark.** Here is a proof of the last proposition using Group Theory. Assume that $k$ is a finite field whose order $|k|$ is divisible by distinct primes $p$ and $q$. By Proposition 1.84, Cauchy's Theorem for finite abelian groups, the additive group $k$ contains elements $a$ and $b$ of order $p$ and $q$, respectively. If $\varepsilon$ denotes the unit in $k$, then the elements $p\varepsilon$ (the sum of $\varepsilon$ with itself $p$ times) and $q\varepsilon$ satisfy $(p\varepsilon)a = pa = 0$ and $(q\varepsilon)b = qb = 0$. Since $k$ is a field, it is a domain, and so

$$p\varepsilon = 0 = q\varepsilon.$$

But $(p, q) = 1$, so there are integers $s$ and $t$ with $sp + tq = 1$. Hence, $\varepsilon = s(p\varepsilon) + t(q\varepsilon) = 0$, and this is a contradiction. Therefore, $|k|$ has only one prime divisor, say, $p$, and so $|k|$ is a power of $p$.   ◀

**Proposition 2.140.** *If $k$ is a field and $I = (f(x))$, where $f(x)$ is a nonzero polynomial in $k[x]$, then the following are equivalent:*

   (i) *$f(x)$ is irreducible;*

   (ii) *$k[x]/I$ is a field;*

   (iii) *$k[x]/I$ is a domain.*

**Proof.**

(i) $\Rightarrow$ (ii) Assume that $f(x)$ is irreducible. Since $I = (f)$ is a proper ideal, the unit in $k[x]/I$, namely, $1 + I$, is not zero. If $g(x) + I \in k[x]/I$ is nonzero, then $g(x) \notin I$: that is, $g(x)$ is not a multiple of $f(x)$ or, to say it another way, $f \nmid g$. By Lemma 2.53, $f$ and $g$ are relatively prime, and there are polynomials $s$ and $t$ with $sg + tf = 1$. Thus, $sg - 1 \in I$, so that $1 + I = sg + I = (s+I)(g+I)$. Therefore, every nonzero element of $k[x]/I$ has an inverse, and $k[x]/I$ is a field.

(ii) $\Rightarrow$ (iii) Every field is a domain.

(iii) $\Rightarrow$ (i) Assume that $k[x]/I$ is a domain. If $f(x)$ is not irreducible, then $f(x) = g(x)h(x)$ in $k[x]$, where $\deg(g) < \deg(f)$ and $\deg(h) < \deg(f)$. Recall that the zero in $k[x]/I$ is $0 + I = I$. Thus, if $g + I = I$, then $g \in I = (f)$ and $f \mid g$, contradicting $\deg(g) < \deg(f)$. Similarly, $h + I \neq I$. However, the product $(g + I)(h + I) = f + I = I$ is zero in the quotient ring, which contradicts $k[x]/I$ being a domain. Therefore, $f(x)$ is irreducible. $\bullet$

The structure of general quotient rings $R/I$ can be complicated, but for special choices of $R$ and $I$, the commutative ring $R/I$ can be easily described. For example, when $p(x)$ is an irreducible polynomial, the following proposition gives a complete description of the field $k[x]/(p(x))$.

**Proposition 2.141.** *Let $k$ be a field, let $p(x)$ be a monic irreducible polynomial in $k[x]$ of degree $d$, let $K = k[x]/I$, where $I = (p(x))$, and let $\beta = x + I \in K$.*

(i) *$K$ is a field and $k' = \{a + I : a \in k\}$ is a subfield of $K$ isomorphic to $k$. Therefore, if $k'$ is identified with $k$, then $k$ is a subfield of $K$.*

(ii) *$\beta$ is a root of $p(x)$ in $K$.*

(iii) *If $g(x) \in k[x]$ and $\beta$ is a root of $g(x)$, then $p(x) \mid g(x)$ in $k[x]$.*

(iv) *$p(x)$ is the unique monic irreducible polynomial in $k[x]$ having $\beta$ as a root.*

(v) *The list $1, \beta, \beta^2, \ldots, \beta^{d-1}$ is a basis of $K$ as a vector space over $k$, and so $\dim_k(K) = d$.*

**Proof.**

(i) The quotient ring $K = k[x]/I$ is a field, by Proposition 2.140 [since $p(x)$ is irreducible], and Corollary 2.32 says that the restriction of the natural map $a \mapsto a + I$ is an isomorphism $k \to k'$.

(ii) Let $p(x) = a_0 + a_1 x + \cdots + a_{d-1} x^{d-1} + x^d$, where $a_i \in k$ for all $i$. In $K = k[x]/I$, we have

$$
\begin{aligned}
p(\beta) &= (a_0 + I) + (a_1 + I)\beta + \cdots + (1 + I)\beta^d \\
&= (a_0 + I) + (a_1 + I)(x + I) + \cdots + (1 + I)(x + I)^d \\
&= (a_0 + I) + (a_1 x + I) + \cdots + (1 x^d + I) \\
&= a_0 + a_1 x + \cdots + x^d + I \\
&= p(x) + I = I,
\end{aligned}
$$

because $I = (p(x))$. But $I = 0 + I$ is the zero element of $K = k[x]/I$, and so $\beta$ is a root of $p(x)$.

(iii) If $p(x) \nmid g(x)$ in $k[x]$, then their gcd is 1, because $p(x)$ is irreducible. Therefore, there are $s(x), t(x) \in k[x]$ with $1 = sp + tg$. Since $k[x] \subseteq K[x]$, we may regard this as an equation in $K[x]$. Evaluating at $\beta$ gives the contradiction $1 = 0$.

(iv) Let $h(x) \in k[x]$ be a monic irreducible polynomial having $\beta$ as a root. By part (iii), we have $p(x) \mid h(x)$. Since $h(x)$ is irreducible, we have $h(x) = cp(x)$ for some constant $c$; since $h(x)$ and $p(x)$ are monic, we have $c = 1$ and $h(x) = p(x)$.

(v) Every element of $K$ has the form $f(x) + I$, where $f(x) \in k[x]$. By the Division Algorithm, there are polynomials $q(x), r(x) \in k[x]$ with $f = qp + r$ and either $r(x) = 0$ or $\deg(r) < d = \deg(p)$. Since $f - r = qp \in I$, it follows that $f(x) + I = r(x) + I$. If $r(x) = b_0 + b_1 x + \cdots + b_{d-1} x^{d-1}$, where $b_i \in k$ for all $i$, then we see, as in the proof of part (ii), that $r(x) + I = b_0 + b_1 \beta + \cdots + b_{d-1} \beta^{d-1}$. Therefore, $1, \beta, \beta^2, \ldots, \beta^{d-1}$ spans $K$.

By Proposition 2.101, it suffices to prove uniqueness. Suppose that
$$b_0 + b_1 \beta + \cdots + b_{d-1} \beta^{n-1} = c_0 + c_1 \beta + \cdots + c_{d-1} \beta^{d-1}.$$
Define $g(x) \in k[x]$ by $g(x) = \sum_{i=0}^{d-1} (b_i - c_i) x^i$; if $g(x) = 0$, we are done. If $g(x) \neq 0$, then $\deg(g)$ is defined, and $\deg(g) < d = \deg(p)$. On the other hand, $\beta$ is a root of $g(x)$, and so part (iii) gives $p(x) \mid g(x)$; hence, $\deg(p) \leq \deg(g)$, and this is a contradiction. It follows that $1, \beta, \beta^2, \ldots, \beta^{d-1}$ is a basis of $K$ as a vector space over $k$, and this gives $\dim_k(K) = d$. •

**Definition.** If $K$ is a field containing $k$ as a subfield, then $K$ is called an **extension field** of $k$, and we write "$K/k$ is an extension field."[18] An extension field $K/k$ is a **finite extension** if $K$ is a finite-dimensional vector space over $k$. The dimension of $K$, denoted by
$$[K : k],$$
is called the **degree** of $K/k$.

Proposition 2.141(v) shows why $[K : k]$ is called the degree of $K/k$.

**Example 2.142.** The polynomial $x^2 + 1 \in \mathbb{R}[x]$ is irreducible, and so $K = \mathbb{R}[x]/(x^2 + 1)$ is an extension field $K/\mathbb{R}$ of degree 2. If $\beta$ is a root of $x^2 + 1$, then $\beta^2 = -1$; moreover, every element of $K$ has a unique expression of the form $a + b\beta$, where $a, b \in \mathbb{R}$. Clearly, this is another construction of $\mathbb{C}$ (which we have been viewing as the points in the plane equipped with a certain addition and multiplication).

Here is a natural way to construct an isomorphism $K \to \mathbb{C}$. Consider the map $\varphi \colon \mathbb{R}[x] \to \mathbb{C}$ given by Theorem 2.25: $\varphi \colon f(x) \mapsto f(i)$. First, $\varphi$ is surjective, for $a + ib = \varphi(a + bx) \in \operatorname{im} \varphi$. Second, $\ker \varphi = \{f(x) \in \mathbb{R}[x] : f(i) = 0\}$, the set of all polynomials in $\mathbb{R}[x]$ having $i$ as a root. We know that $x^2 + 1 \in \ker \varphi$, so that $(x^2 + 1) \subseteq \ker \varphi$. For the reverse inclusion, if $g(x) \in \ker \varphi$, then $i$ is a root of $g(x)$

---

[18]This notation should not be confused with the notation for a quotient ring, for a field $K$ has no interesting ideals; in particular, if $k \subsetneq K$, then $k$ is not an ideal in $K$.

and $g(x) \in (x^2 + 1)$, by Proposition 2.141(iii). Therefore, $\ker \varphi = (x^2 + 1)$, and the First Isomorphism Theorem gives $\mathbb{R}[x]/(x^2 + 1) \cong \mathbb{C}$.

Viewing $\mathbb{C}$ as a quotient ring allows us to view its multiplication in a new light: first treat $i$ as a variable and then impose the condition $i^2 = -1$; that is, first multiply in $\mathbb{R}[x]$ and then reduce mod $(x^2 + 1)$. Thus, to compute $(a + bi)(c + di)$, first write $ac + (ad + bc)i + bdi^2$, and then observe that $i^2 = -1$. More generally, if $\beta$ is a root of an irreducible $p(x) \in k[x]$, then the easiest way to multiply

$$(b_0 + b_1\beta + \cdots + b_{n-1}\beta^{n-1})(c_0 + c_1\beta + \cdots + c_{n-1}\beta^{n-1})$$

in the quotient ring $k[x]/(p(x))$ is to regard the factors as polynomials in an indeterminate $\beta$, multiply them, and then impose the condition that $p(\beta) = 0$. ◄

The first step in classifying fields involves their characteristic. Here is the second step.

**Definition.** Let $K/k$ be an extension field. An element $\alpha \in K$ is **algebraic** over $k$ if there is some nonzero polynomial $f(x) \in k[x]$ having $\alpha$ as a root; otherwise, $\alpha$ is **transcendental** over $k$. An extension $K/k$ is **algebraic** if every $\alpha \in K$ is algebraic over $k$.

When a real number is called transcendental, it usually means that it is transcendental over $\mathbb{Q}$. For example, $\pi$ and $e$ are transcendental numbers.

**Proposition 2.143.** *If $K/k$ is a finite extension field, then $K/k$ is an algebraic extension.*

**Proof.** By definition, $K/k$ finite means that $[K : k] = n < \infty$; that is, $K$ has dimension $n$ as a vector space over $k$. By Corollary 2.114, the list of $n + 1$ vectors $1, \alpha, \alpha^2, \ldots, \alpha^n$ is dependent: there are $c_0, c_1, \ldots, c_n \in k$, not all 0, with $\sum c_i \alpha^i = 0$. Thus, the polynomial $f(x) = \sum c_i x^i$ is not the zero polynomial, and $\alpha$ is a root of $f(x)$. Therefore, $\alpha$ is algebraic over $k$. •

The converse of this last proposition is not true. We shall see that the set $\mathbb{A}$ of all complex numbers algebraic over $\mathbb{Q}$ is an algebraic extension of $\mathbb{Q}$ that is not a finite extension.

**Definition.** If $K/k$ is an extension and $\alpha \in K$, then $k(\alpha)$ is the intersection of all those subfields of $K$ containing $k$ and $\alpha$; we call $k(\alpha)$ the subfield of $K$ obtained by **adjoining** $\alpha$ to $k$ (instead of the subfield generated by $k$ and $\alpha$).

More generally, if $A$ is a (possibly infinite) subset of $K$, define $k(A)$ to be the intersection of all the subfields of $K$ containing $k \cup A$; we call $k(A)$ the subfield of $K$ obtained by **adjoining** $A$ to $k$. In particular, if $A = \{z_1, \ldots, z_n\}$ is a finite subset, then we may denote $k(A)$ by $k(z_1, \ldots, z_n)$.

It is clear that $k(A)$ is the smallest subfield of $K$ containing $k$ and $A$; that is, if $B$ is any subfield of $K$ containing $k$ and $A$, then $k(A) \subseteq B$.

We now show that the field $k[x]/(p(x))$, where $p(x) \in k[x]$ is irreducible, is intimately related to adjunction.

**Theorem 2.144.**

(i) *If $K/k$ is an extension and $\alpha \in K$ is algebraic over $k$, then there is a unique monic irreducible polynomial $p(x) \in k[x]$ having $\alpha$ as a root. Moreover, if $I = (p(x))$, then $k[x]/I \cong k(\alpha)$; indeed, there exists an isomorphism*

$$\varphi : k[x]/I \to k(\alpha)$$

*with $\varphi(x + I) = \alpha$ and $\varphi(c + I) = c$ for all $c \in k$.*

(ii) *If $\alpha' \in K$ is another root of $p(x)$, then there is an isomorphism*

$$\theta : k(\alpha) \to k(\alpha')$$

*with $\theta(\alpha) = \alpha'$ and $\theta(c) = c$ for all $c \in k$.*

**Proof.**

(i) Consider the map $\varphi \colon k[x] \to K$ given by Theorem 2.25: $\varphi \colon f(x) \mapsto f(\alpha)$. Now $\operatorname{im} \varphi$ is the subring of $K$ consisting of all polynomials in $\alpha$, that is, all elements of the form $f(\alpha)$ with $f(x) \in k[x]$, while $\ker \varphi$ is the ideal in $k[x]$ consisting of all those $f(x) \in k[x]$ having $\alpha$ as a root. Since every ideal in $k[x]$ is a principal ideal, we have $\ker \varphi = (p(x))$ for some monic polynomial $p(x) \in k[x]$. But $k[x]/(p(x)) \cong \operatorname{im} \varphi$, which is a domain, and so $p(x)$ is irreducible, by Proposition 2.140. This same proposition says that $k[x]/(p(x))$ is a field, and so the First Isomorphism Theorem gives $k[x]/(p(x)) \cong \operatorname{im} \varphi$; that is, $\operatorname{im} \varphi$ is a subfield of $K$ containing $k$ and $\alpha$. Since every subfield of $K$ that contains $k$ and $\alpha$ must contain $\operatorname{im} \varphi$, we have $\operatorname{im} \varphi = k(\alpha)$. We have proved everything in the statement except the uniqueness of $p(x)$; but this follows from Proposition 2.141(iv).

(ii) As in part (i), there are isomorphisms $\varphi \colon k[x]/I \to k(\alpha)$ and $\psi \colon k[x]/I \to k(\alpha')$ with $\varphi(c+I) = c$ and $\psi(c) = c+I$ for all $c \in k$; moreover, $\varphi \colon x+I \mapsto \alpha$ and $\psi \colon x+I \mapsto \alpha'$. The composite $\theta = \psi\varphi^{-1}$ is the desired isomorphism. •

**Definition.** If $K/k$ is an extension field and $\alpha \in K$ is algebraic over $k$, then the unique monic irreducible polynomial $p(x) \in k[x]$ having $\alpha$ as a root is called the ***minimal polynomial*** of $\alpha$ over $k$, and it is denoted by

$$\operatorname{irr}(\alpha, k) = p(x).$$

The minimal polynomial $\operatorname{irr}(\alpha, k)$ does depend on $k$. For example, $\operatorname{irr}(i, \mathbb{R}) = x^2 + 1$, while $\operatorname{irr}(i, \mathbb{C}) = x - i$.

Suppose now that $\alpha$ is an algebraic integer; that is, there is a monic polynomial $f(x) \in \mathbb{Z}[x]$ having $\alpha$ as a root. Since $\mathbb{Z}[x] \subseteq \mathbb{Q}[x]$, every algebraic integer $\alpha$ has a unique minimal polynomial $m(x) = \operatorname{irr}(\alpha, \mathbb{Q}) \in \mathbb{Q}[x]$, and $m(x)$ is irreducible in $\mathbb{Q}[x]$.

**Corollary 2.145.** *If $\alpha$ is an algebraic integer, then $\operatorname{irr}(\alpha, \mathbb{Q})$ lies in $\mathbb{Z}[x]$.*

**Proof.** Let $p(x) \in \mathbb{Z}[x]$ be the monic polynomial of least degree having $\alpha$ as a root. If $p(x) = G(x)H(x)$ in $\mathbb{Q}[x]$, where $\deg(G) < \deg(p)$ and $\deg(H) < \deg(p)$, then $\alpha$ is a root of either $G(x)$ or $H(x)$. By Gauss's Theorem 2.52, there is a factorization $p(x) = g(x)h(x)$ in $\mathbb{Z}[x]$ with $\deg(g) = \deg(G)$ and $\deg(h) = \deg(H)$; in fact, there are rationals $c$ and $d$ with $g(x) = cG(x)$ and $h(x) = dH(x)$. If $a$ is the leading coefficient of $g(x)$ and $b$ is the leading coefficient of $h(x)$, then $ab = 1$, for $p(x)$ is monic. Therefore, we may assume that $a = 1 = b$, for $a, b \in \mathbb{Z}$ (the only other option is $a = -1 = b$); that is, we may assume that both $g(x)$ and $h(x)$ are monic. Since $\alpha$ is a root of $g(x)$ or $h(x)$, we have contradicted $p(x)$ being a monic polynomial in $\mathbb{Z}[x]$ of least degree having $\alpha$ as a root. It follows that $p(x) = \operatorname{irr}(\alpha, \mathbb{Q})$, for the latter is the unique monic irreducible polynomial in $\mathbb{Q}[x]$ having $\alpha$ as a root. •

**Remark.** We define the (algebraic) **conjugates** of $\alpha$ to be the roots of $\operatorname{irr}(\alpha, \mathbb{Q})$, and we define the **norm** of $\alpha$ to be the absolute value of the product of the conjugates of $\alpha$. Of course, the norm of $\alpha$ is just the absolute value of the constant term of $\operatorname{irr}(\alpha, \mathbb{Q})$, and so it is an (ordinary) integer. Norms are very useful in Algebraic Number Theory, as we have seen in the proof of Theorem 2.89, Fermat's Two-Squares Theorem. ◀

The following formula is quite useful, especially when proving a theorem by induction on degrees.

**Theorem 2.146.** *Let $k \subseteq E \subseteq K$ be fields, with $E$ a finite extension of $k$ and $K$ a finite extension of $E$. Then $K$ is a finite extension of $k$, and*

$$[K : k] = [K : E][E : k].$$

**Proof.** If $A = a_1, \ldots, a_n$ is a basis of $E$ over $k$ and $B = b_1, \ldots, b_m$ is a basis of $K$ over $E$, then it suffices to prove that a list $X$ of all $a_i b_j$ is a basis of $K$ over $k$.

To see that $X$ spans $K$, take $u \in K$. Since $B$ is a basis of $K$ over $E$, there are scalars $\lambda_j \in E$ with $u = \sum_j \lambda_j b_j$. Since $A$ is a basis of $E$ over $k$, there are scalars $\mu_{ji} \in k$ with $\lambda_j = \sum_i \mu_{ji} a_i$. Therefore, $u = \sum_{ij} \mu_{ji} a_i b_j$, and $X$ spans $K$ over $k$.

To prove that $X$ is linearly independent over $k$, assume that there are scalars $\mu_{ji} \in k$ with $\sum_{ij} \mu_{ji} a_i b_j = 0$. If we define $\lambda_j = \sum_i \mu_{ji} a_i$, then $\lambda_j \in E$ and $\sum_j \lambda_j b_j = 0$. Since $B$ is linearly independent over $E$, it follows that

$$0 = \lambda_j = \sum_i \mu_{ji} a_i$$

for all $j$. Since $A$ is linearly independent over $k$, it follows that $\mu_{ji} = 0$ for all $j$ and $i$, as desired. •

There are several classical problems in Euclidean Geometry: trisecting an angle; duplicating the cube (given a cube with side length 1, construct a cube whose volume is 2); squaring the circle (given a circle of radius 1, construct a square whose area is equal to the area of the circle). In short, the problems ask whether geometric constructions can be made using only a straightedge (ruler) and compass according to certain rules. Theorem 2.146 has a beautiful application in proving the unsolvability of these classical problems. See a sketch of the proof in Kaplansky, *Field and Rings*, pp. 8–9, or the more detailed account in FCAA, pp. 332–344.

**Example 2.147.** Let $f(x) = x^4 - 10x^2 + 1 \in \mathbb{Q}[x]$. If $\beta$ is a root of $f(x)$, then the quadratic formula gives $\beta^2 = 5 \pm 2\sqrt{6}$. But the identity $a + 2\sqrt{ab} + b = \left(\sqrt{a} + \sqrt{b}\right)^2$

gives $\beta = \pm(\sqrt{2} + \sqrt{3})$. Similarly, $5 - 2\sqrt{6} = (\sqrt{2} - \sqrt{3})^2$, so that the roots of $f(x)$ are

$$\sqrt{2} + \sqrt{3}, \quad -\sqrt{2} - \sqrt{3}, \quad \sqrt{2} - \sqrt{3}, \quad -\sqrt{2} + \sqrt{3}.$$

By Theorem 2.63, the only possible rational roots of $f(x)$ are $\pm 1$, and so we have just proved that these roots are irrational.

We claim that $f(x)$ is irreducible in $\mathbb{Q}[x]$. If $g(x)$ is a quadratic factor of $f(x)$ in $\mathbb{Q}[x]$, then

$$g(x) = (x - a\sqrt{2} - b\sqrt{3})(x - c\sqrt{2} - d\sqrt{3}),$$

where $a, b, c, d \in \{1, -1\}$. Multiplying,

$$g(x) = x^2 - \big((a + c)\sqrt{2} + (b + d)\sqrt{3}\big)x + 2ac + 3bd + (ad + bc)\sqrt{6}.$$

We check easily that $(a+c)\sqrt{2}+(b+d)\sqrt{3}$ is rational if and only if $a+c = 0 = b+d$; but these equations force $ad+bc \neq 0$, and so the constant term of $g(x)$ is not rational. Therefore, $g(x) \notin \mathbb{Q}[x]$, and so $f(x)$ is irreducible in $\mathbb{Q}[x]$. If $\beta = \sqrt{2} + \sqrt{3}$, then $f(x) = \mathrm{irr}(\beta, \mathbb{Q})$.

Consider the field $E = \mathbb{Q}(\beta) = \mathbb{Q}(\sqrt{2} + \sqrt{3})$. There is a tower of fields $\mathbb{Q} \subseteq E \subseteq F$, where $F = \mathbb{Q}(\sqrt{2}, \sqrt{3})$, and so

$$[F : \mathbb{Q}] = [F : E][E : \mathbb{Q}],$$

by Theorem 2.146. Since $E = \mathbb{Q}(\beta)$ and $\beta$ is a root of an irreducible polynomial of degree 4, namely, $f(x)$, we have $[E : \mathbb{Q}] = 4$. On the other hand,

$$[F : \mathbb{Q}] = [F : \mathbb{Q}(\sqrt{2})][\mathbb{Q}(\sqrt{2}) : \mathbb{Q}].$$

Now $[\mathbb{Q}(\sqrt{2}) : \mathbb{Q}] = 2$, because $\sqrt{2}$ is a root of the irreducible quadratic $x^2 - 2$ in $\mathbb{Q}[x]$. We claim that $[F : \mathbb{Q}(\sqrt{2})] \leq 2$. The field $F$ arises by adjoining $\sqrt{3}$ to $\mathbb{Q}(\sqrt{2})$; either $\sqrt{3} \in \mathbb{Q}(\sqrt{2})$, in which case the degree is 1, or $x^2 - 3$ is irreducible in $\mathbb{Q}(\sqrt{2})[x]$, in which case the degree is 2 (in fact, the degree is 2). It follows that $[F : \mathbb{Q}] \leq 4$, and so the equation $[F : \mathbb{Q}] = [F : E][E : \mathbb{Q}]$ gives $[F : E] = 1$; that is, $F = E$.

Let us note that $F$ arises from $\mathbb{Q}$ by adjoining all the roots of $f(x)$, and it also arises from $\mathbb{Q}$ by adjoining all the roots of $g(x) = (x^2 - 2)(x^2 - 3)$.  ◀

We now prove two important results: the first, due to Kronecker, says that if $f(x) \in k[x]$, where $k$ is any field, then there is some larger field $E$ that contains $k$ and all the roots of $f(x)$; the second, due to Galois, constructs finite fields other than $\mathbb{F}_p$.

**Theorem 2.148 (Kronecker).** *If $k$ is a field and $f(x) \in k[x]$, there exists a field $K$ containing $k$ as a subfield with $f(x)$ a product of linear polynomials in $K[x]$.*

**Proof.** The proof is by induction on $\deg(f)$. If $\deg(f) = 1$, then $f(x)$ is linear and we can choose $K = k$. If $\deg(f) > 1$, write $f(x) = p(x)g(x)$, where $p(x)$ is irreducible. Now Proposition 2.141(i) provides a field $F$ containing $k$ and a root $z$ of $p(x)$. Hence, in $F[x]$, we have $p(x) = (x - z)h(x)$ and $f(x) = (x - z)h(x)g(x)$. By induction, there is a field $K$ containing $F$ (and hence $k$) so that $h(x)g(x)$, and hence $f(x)$, is a product of linear factors in $K[x]$.  •

For the familiar fields $\mathbb{Q}$, $\mathbb{R}$, and $\mathbb{C}$, Kronecker's Theorem offers nothing new. The Fundamental Theorem of Algebra, first proved by Gauss in 1799 (completing earlier attempts of Euler and of Lagrange), says that every nonconstant $f(x) \in \mathbb{C}[x]$ has a root in $\mathbb{C}$; it follows, by induction on the degree of $f(x)$, that all the roots of $f(x)$ lie in $\mathbb{C}$; that is, $f(x) = a(x - r_1) \cdots (x - r_n)$, where $a \in \mathbb{C}$ and $r_j \in \mathbb{C}$ for all $j$. On the other hand, if $k = \mathbb{F}_p$ or $k = \mathbb{C}(x) = \operatorname{Frac}(\mathbb{C}[x])$, the Fundamental Theorem does not apply; but Kronecker's Theorem does apply to tell us, for any given $f(x)$, that there is always some larger field $E$ containing all the roots of $f(x)$. For example, there is some field containing $\mathbb{C}(x)$ and $\sqrt{x}$. We will prove a general version of the Fundamental Theorem in Chapter 5; Theorem 5.57 says that every field $k$ is a subfield of an ***algebraically closed*** field $K$, that is, $K$ is a field containing $k$ such that every $f(x) \in K[x]$ is a product of linear polynomials in $K[x]$. In contrast, Kronecker's Theorem gives roots of only one polynomial at a time.

When we defined, on page 161, the field $k(A)$ obtained from a field $k$ by adjoining a set $A$, we assumed that $A \subseteq K$ for some extension field $K/k$. In light of Kronecker's Theorem, we may now speak of the field $k(A)$ obtained by adjoining all the roots $A = \{z_1, \ldots, z_n\}$ of some $f(x) \in k[x]$ without having to assume, a priori, that there is some field extension $K/k$ containing $A$.

**Definition.** If $K/k$ is an extension field and $f(x) \in k[x]$ is nonconstant, then $f(x)$ ***splits over $K$*** if $f(x) = a(x - z_1) \cdots (x - z_n)$, where $z_1, \ldots, z_n$ are in $K$ and $a \in k$. An extension field $E/k$ is called a ***splitting field*** of $f(x)$ ***over*** $k$ if $f(x)$ splits over $E$, but $f(x)$ does not split over any proper subfield of $E$.

Consider $f(x) = x^2 + 1 \in \mathbb{Q}[x]$. The roots of $f(x)$ are $\pm i$, and so $f(x)$ splits over $\mathbb{C}$; that is, $f(x) = (x - i)(x + i)$ is a product of linear polynomials in $\mathbb{C}[x]$. However, $\mathbb{C}$ is not a splitting field of $f(x)$ over $\mathbb{Q}$; there are proper subfields of $\mathbb{C}$ containing $\mathbb{Q}$ and all the roots of $f(x)$. For example, $\mathbb{Q}(i)$ is such a subfield; in fact, it is the splitting field of $f(x)$ over $\mathbb{Q}$. Note that a splitting field of a polynomial $g(x) \in k[x]$ depends on $k$ as well as on $g(x)$. The splitting field of $x^2 + 1$ over $\mathbb{Q}$ is $\mathbb{Q}(i)$, while the splitting field of $x^2 + 1$ over $\mathbb{R}$ is $\mathbb{R}(i) = \mathbb{C}$.

In Example 2.147, we proved that $E = \mathbb{Q}(\sqrt{2} + \sqrt{3})$ is a splitting field of $f(x) = x^4 - 10x^2 + 1$, as well as a splitting field of $g(x) = (x^2 - 2)(x^2 - 3)$.

The existence of splitting fields is an easy consequence of Kronecker's Theorem.

**Corollary 2.149.** *If $k$ is a field and $f(x) \in k[x]$, then a splitting field of $f(x)$ over $k$ exists.*

**Proof.** By Kronecker's Theorem, there is an extension field $K/k$ such that $f(x)$ splits in $K[x]$; say, $f(x) = a(x - \alpha_1) \cdots (x - \alpha_n)$. The subfield $E = k(\alpha_1, \ldots, \alpha_n)$ of $K$ is a splitting field of $f(x)$ over $k$ (a proper subfield of $E$ omits some $\alpha_i$). $\quad \bullet$

A splitting field of $f(x) \in k[x]$ is a *smallest* field extension $E/k$ containing all the roots of $f(x)$. We say "a" splitting field instead of "the" splitting field because it is not obvious that any two splitting fields of $f(x)$ over $k$ are isomorphic. Analysis of this technical point will not only prove uniqueness of splitting fields

(Theorem 2.157), it will enable us to prove Corollary 2.158: any two finite fields with the same number of elements are isomorphic.

**Example 2.150.** Let $k$ be a field and let $E = k(y_1, \ldots, y_n)$ be the rational function field in $n$ variables $y_1, \ldots, y_n$ over $k$; that is, $E = \operatorname{Frac}(k[y_1, \ldots, y_n])$, the fraction field of the ring of polynomials in $n$ variables. The **general polynomial of degree** $n$ over $k$ is defined to be

$$f(x) = \prod_i (x - y_i) \in E[x].$$

The coefficients of $f(x) = (x - y_1)(x - y_2) \cdots (x - y_n)$, which we denote by $a_i$, can be given explicitly [see Equations (1) on page 173] in terms of the $y$'s. Notice that $E$ is a splitting field of $f(x)$ over the field $K = k(a_0, \ldots, a_{n-1})$, for it arises from $K$ by adjoining all the roots of $f(x)$, namely, all the $y$'s.  ◄

**Example 2.151.** Let $f(x) = x^n - 1 \in k[x]$ for some field $k$, and let $E/k$ be a splitting field. In Theorem 2.46, we saw that the group $\Gamma_n$ of all $n$th roots of unity in $E$ is a cyclic group: $\Gamma_n = \langle \omega \rangle$, where $\omega$ is a primitive $n$th root of unity. It follows that $k(\omega) = E$ is a splitting field of $f(x)$, for every $n$th root of unity is a power of $\omega$.  ◄

Here is another application of Kronecker's Theorem.

**Proposition 2.152.** *Let $p$ be prime, and let $k$ be a field. If $f(x) = x^p - c \in k[x]$ and $\alpha$ is a $p$th root of $c$ (in some splitting field), then either $f(x)$ is irreducible in $k[x]$ or $c$ has a $p$th root in $k$. In either case, if $k$ contains the $p$th roots of unity, then $k(\alpha)$ is a splitting field of $f(x)$.*

**Proof.** By Kronecker's Theorem, there exists an extension field $K/k$ that contains all the roots of $f(x)$; that is, $K$ contains all the $p$th roots of $c$. If $\alpha^p = c$, then every such root has the form $\omega\alpha$, where $\omega$ is a $p$th root of unity; that is, $\omega$ is a root of $x^p - 1$.

If $f(x)$ is not irreducible in $k[x]$, then there is a factorization $f(x) = g(x)h(x)$ in $k[x]$, where $g(x)$ is a nonconstant polynomial with $d = \deg(g) < \deg(f) = p$. Now the constant term $b$ of $g(x)$ is, to sign, the product of some of the roots of $f(x)$:

$$\pm b = \alpha^d \omega,$$

where $\omega$, which is a product of $d$ $p$th roots of unity, is itself a $p$th root of unity. It follows that

$$(\pm b)^p = (\alpha^d \omega)^p = \alpha^{dp} = c^d.$$

But $p$ being prime and $d < p$ forces $(d, p) = 1$; hence, there are integers $s$ and $t$ with $1 = sd + tp$. Hence,

$$c = c^{sd+tp} = c^{sd}c^{tp} = (\pm b)^{ps}c^{tp} = [(\pm b)^s c^t]^p.$$

Therefore, $c$ has a $p$th root in $k$.

We now assume that $k$ contains $\Omega$, all the $p$th roots of unity. If $\alpha \in K$ is a $p$th root of $c$, then $f(x) = \prod_{\omega \in \Omega}(x - \omega\alpha)$ shows that $f(x)$ splits over $K$ and that $k(\alpha)$ is a splitting field of $f(x)$ over $k$.  •

We are now going to construct the finite fields. My guess is that Galois knew that $\mathbb{C}$ can be constructed by adjoining a root of a polynomial, namely, $x^2+1$, to $\mathbb{R}$, and so it was natural for him to adjoin a root of a polynomial to $\mathbb{F}_p$. Note, however, that Kronecker's Theorem was not proved until a half century after Galois's death.

**Theorem 2.153 (Galois).** *If $p$ is prime and $n$ is a positive integer, then there exists a field having exactly $p^n$ elements.*

**Proof.** Write $q = p^n$, and consider the polynomial

$$g(x) = x^q - x \in \mathbb{F}_p[x].$$

By Kronecker's Theorem, there is a field extension $K/\mathbb{F}_p$ with $g(x)$ a product of linear factors in $K[x]$. Define

$$E = \{\alpha \in K : g(\alpha) = 0\};$$

that is, $E$ is the set of all the roots of $g(x)$. Since the derivative $g'(x) = qx^{q-1} - 1 = p^n x^{q-1} - 1 = -1$, we have $\gcd(g, g') = 1$. By Exercise 2.54 on page 114, all the roots of $g(x)$ are distinct; that is, $E$ has exactly $q = p^n$ elements.

The theorem will follow if $E$ is a subfield of $K$. Of course, $1 \in E$. If $a, b \in E$, then $a^q = a$ and $b^q = b$. Therefore, $(ab)^q = a^q b^q = ab$, and $ab \in E$. By Exercise 2.35 on page 101, $(a-b)^q = a^q - b^q = a - b$, so that $a - b \in E$. Finally, if $a \neq 0$, then the cancellation law applied to $a^q = a$ gives $a^{q-1} = 1$, and so the inverse of $a$ is $a^{q-2}$ (which lies in $E$ because $E$ is closed under multiplication). $\bullet$

Recall Theorem 2.46: the multiplicative group of a finite field $k$ is a cyclic group. A generator $\alpha$ of this group is called a ***primitive element***; that is, every nonzero element of $k$ is a power of $\alpha$.

**Corollary 2.154.** *For every prime $p$ and every integer $n \geq 1$, there exists an irreducible polynomial $g(x) \in \mathbb{F}_p[x]$ of degree $n$. In fact, if $\alpha$ is a primitive element of $\mathbb{F}_{p^n}$, then its minimal polynomial $g(x) = \mathrm{irr}(\alpha, \mathbb{F}_p)$ has degree $n$.*

**Remark.** An easy modification of the proof replaces $\mathbb{F}_p$ by any finite field. ◀

**Proof.** Let $E/\mathbb{F}_p$ be an extension field with $p^n$ elements, and let $\alpha \in E$ be a primitive element. Clearly, $\mathbb{F}_p(\alpha) = E$, for it contains every power of $\alpha$, hence every nonzero element of $E$. By Theorem 2.144(i), $g(x) = \mathrm{irr}(\alpha, \mathbb{F}_p) \in \mathbb{F}_p[x]$ is an irreducible polynomial having $\alpha$ as a root. If $\deg(g) = d$, then Proposition 2.141(v) gives $[\mathbb{F}_p[x]/(g(x)) : \mathbb{F}_p] = d$; but $\mathbb{F}_p[x]/(g(x)) \cong \mathbb{F}_p(\alpha) = E$, by Theorem 2.144(i), so that $[E : \mathbb{F}_p] = n$. Therefore, $n = d$, and so $g(x)$ is an irreducible polynomial of degree $n$. $\bullet$

This corollary can also be proved by counting. If $m = p_1^{e_1} \cdots p_n^{e_n}$, define the ***Möbius function*** by

$$\mu(m) = \begin{cases} 1 & \text{if } m = 1; \\ 0 & \text{if any } e_i > 1; \\ (-1)^n & \text{if } 1 = e_1 = e_2 = \cdots = e_n. \end{cases}$$

If $N_n$ is the number of irreducible polynomials in $\mathbb{F}_p[x]$ of degree $n$, then

$$N_n = \frac{1}{n} \sum_{d|n} \mu(d) p^{n/d}.$$

An elementary proof can be found in G. J. Simmons, The Number of Irreducible Polynomials of Degree $n$ over $\mathrm{GF}(p)$, *Amer. Math. Monthly* 77 (1970), pp. 743–745.

**Example 2.155.**

(i) In Exercise 2.8 on page 89, we constructed a field with four elements:

$$\mathbb{F}_4 = \left\{ \left[\begin{smallmatrix} a & b \\ b & a+b \end{smallmatrix}\right] : a, b \in \mathbb{F}_2 \right\}.$$

On the other hand, we may construct a field of order 4 as the quotient $F = \mathbb{F}_2[x]/(q(x))$, where $q(x) \in \mathbb{F}_2[x]$ is the irreducible polynomial $x^2 + x + 1$. By Proposition 2.141(v), $F$ is a field consisting of all $a + b\beta$, where $\beta = x + (q(x))$ is a root of $q(x)$ and $a, b \in \mathbb{F}_2$. Since $\beta^2 + \beta + 1 = 0$, we have $\beta^2 = -\beta - 1 = \beta + 1$; moreover, $\beta^3 = \beta\beta^2 = \beta(\beta + 1) = \beta^2 + \beta = 1$. It is now easy to see that there is a ring isomorphism $\varphi : \mathbb{F}_4 \to F$ with $\varphi\left(\left[\begin{smallmatrix} a & b \\ b & a+b \end{smallmatrix}\right]\right) = a + b\beta$.

(ii) According to the table in Example 2.66 on page 116, there are three monic irreducible quadratics in $\mathbb{F}_3[x]$, namely,

$$p(x) = x^2 + 1, \quad q(x) = x^2 + x - 1, \quad \text{and} \quad r(x) = x^2 - x - 1;$$

each gives rise to a field with $9 = 3^2$ elements. Let us look at the first two in more detail. Proposition 2.141(v) says that $E = \mathbb{F}_3[x]/(p(x))$ is given by

$$E = \{a + b\alpha : \text{ where } \alpha^2 + 1 = 0\}.$$

Similarly, if $F = \mathbb{F}_3[x]/(q(x))$, then

$$F = \{a + b\beta : \text{ where } \beta^2 + \beta - 1 = 0\}.$$

These two fields are isomorphic. The map $\varphi \colon E \to F$ (found by trial and error), defined by $\varphi(a + b\alpha) = a + b(1 - \beta)$, is an isomorphism.

Now $\mathbb{F}_3[x]/(x^2 - x - 1)$ is also a field with nine elements, and we shall soon see that it is isomorphic to both of the two fields $E$ and $F$ just given (Corollary 2.158).

(iii) In Example 2.66, we exhibited eight monic irreducible cubics $p(x) \in \mathbb{F}_3[x]$; each of them gives rise to a field $\mathbb{F}_3[x]/(p(x))$ having $27 = 3^3$ elements.  ◄

We are now going to solve the isomorphism problem for finite fields.

**Lemma 2.156.** *Let $\varphi \colon k \to k'$ be an isomorphism of fields, and let $\varphi^* \colon k[x] \to k'[x]$ be the ring isomorphism of Corollary 2.27: $\varphi^* \colon g(x) = a_0 + a_1 x + \cdots + a_n x^n \mapsto g^*(x) = \varphi(a_0) + \varphi(a_1)x + \cdots + \varphi(a_n)x^n$. Let $f(x) \in k[x]$ and $f^*(x) = \varphi^*(f) \in k'[x]$. If $E$ is a splitting field of $f(x)$ over $k$ and $E'$ is a splitting field of $f^*(x)$ over $k'$, then there is an isomorphism $\Phi \colon E \to E'$ extending $\varphi$:*

$$
\begin{array}{ccc}
E & \overset{\Phi}{-\!\!\!-\!\!\!\rightarrow} & E' \\
\big| & & \big| \\
k & \underset{\varphi}{\longrightarrow} & k'
\end{array}
$$

**Proof.** The proof is by induction on $d = [E : k]$. If $d = 1$, then $f(x)$ is a product of linear polynomials in $k[x]$, and it follows easily that $f^*(x)$ is also a product of linear polynomials in $k'[x]$. Therefore, $E' = k'$, and we may set $\Phi = \varphi$.

For the inductive step, choose a root $z$ of $f(x)$ in $E$ that is not in $k$, and let $p(x) = \mathrm{irr}(z, k)$ be the minimal polynomial of $z$ over $k$. Now $\deg(p) > 1$, because $z \notin k$; moreover, $[k(z) : k] = \deg(p)$, by Proposition 2.141(v). Let $z'$ be a root of $p^*(x)$ in $E'$, and let $p^*(x) = \mathrm{irr}(z', k')$ be the corresponding monic irreducible polynomial in $k'[x]$.

By a straightforward generalization[19] of Proposition 2.144(ii), there is an isomorphism $\widetilde{\varphi}: k(z) \to k'(z')$ extending $\varphi$ with $\widetilde{\varphi}: z \mapsto z'$. We may regard $f(x)$ as a polynomial with coefficients in $k(z)$, for $k \subseteq k(z)$ implies $k[x] \subseteq k(z)[x]$. We claim that $E$ is a splitting field of $f(x)$ over $k(z)$; that is,

$$E = k(z)(z_1, \ldots, z_n),$$

where $z_1, \ldots, z_n$ are the roots of $f(x)/(x - z)$; after all,

$$E = k(z, z_1, \ldots, z_n) = k(z)(z_1, \ldots, z_n).$$

Similarly, $E'$ is a splitting field of $f^*(x)$ over $k'(z')$. But $[E : k(z)] < [E : k]$, by Theorem 2.146, so that the inductive hypothesis gives an isomorphism $\Phi: E \to E'$ that extends $\widetilde{\varphi}$ and, hence, $\varphi$.  •

**Theorem 2.157.** *If $k$ is a field and $f(x) \in k[x]$, then any two splitting fields of $f(x)$ over $k$ are isomorphic via an isomorphism that fixes $k$ pointwise.*

**Proof.** Let $E$ and $E'$ be splitting fields of $f(x)$ over $k$. If $\varphi$ is the identity, then Lemma 2.156 applies at once.  •

It is remarkable that the next theorem was not proved until the 1890s, 60 years after Galois discovered finite fields.

**Corollary 2.158 (Moore).** *Any two finite fields having exactly $p^n$ elements are isomorphic.*

**Proof.** If $E$ is a field with $q = p^n$ elements, then Lagrange's Theorem applied to the multiplicative group $E^\times$ shows that $a^{q-1} = 1$ for every $a \in E^\times$. It follows that every element of $E$ is a root of $f(x) = x^q - x \in \mathbb{F}_p[x]$, and so $E$ is a splitting field of $f(x)$ over $\mathbb{F}_p$.  •

Finite fields are often called ***Galois fields*** in honor of their discoverer. In light of Corollary 2.158, we may speak of *the* field with $q$ elements, where $q = p^n$ is a power of a prime $p$, and we denote it by

$$\mathbb{F}_q.$$

---

[19] Proving this generalization earlier would have involved introducing all the notation in the present hypothesis, and so it would have made a simple result appear complicated. The isomorphism $\varphi: k \to k'$ induces an isomorphism $\varphi^*: k[x] \to k'[x]$, which takes $p(x)$ to some polynomial $p^*(x)$, and $\varphi^*$ induces an isomorphism $k[x]/(p(x)) \to k'[x]/(p^*(x))$.

# Exercises

**2.90.** Prove that if $I = (0)$, then $R/I \cong R$.

**2.91.** Prove the **Third Isomorphism Theorem for Rings**: if $R$ is a commutative ring having ideals $I \subseteq J$, then $J/I$ is an ideal in $R/I$ and there is a ring isomorphism $(R/I)/(J/I) \cong R/J$.

**2.92.** For every commutative ring $R$, prove that $R[x]/(x) \cong R$.

**2.93.** Prove that $\mathbb{F}_3[x]/(x^3 - x^2 + 1) \cong \mathbb{F}_3[x]/(x^3 - x^2 + x + 1)$ without using Corollary 2.158.

**2.94.** Let $h(x), p(x) \in k[x]$ be monic polynomials, where $k$ is a field. If $p(x)$ is irreducible and every root of $h(x)$ (in an appropriate splitting field) is also a root of $p(x)$, prove that $h(x) = p(x)^m$ for some integer $m \geq 1$.

**Hint.** Use induction on $\deg(h)$.

**2.95.** **(Chinese Remainder Theorem)** (i) Prove that if $k$ is a field and $f(x)$, $f'(x) \in k[x]$ are relatively prime, then given $b(x), b'(x) \in k[x]$, there exists $c(x) \in k[x]$ with

$$c - b \in (f) \quad \text{and} \quad c - b' \in (f');$$

moreover, if $d(x)$ is another common solution, then $c - d \in (ff')$.

(ii) Prove that if $k$ is a field and $f(x)$, $g(x) \in k[x]$ are relatively prime, then

$$k[x]/(f(x)g(x)) \cong k[x]/(f(x)) \times k[x]/(g(x)).$$

**Hint.** See the proof of Theorem 1.87.

**2.96.** (i) Prove that a field $K$ cannot have subfields $k'$ and $k''$ with $k' \cong \mathbb{Q}$ and $k'' \cong \mathbb{F}_p$ for some prime $p$.

(ii) Prove that a field $K$ cannot have subfields $k'$ and $k''$ with $k' \cong \mathbb{F}_p$ and $k'' \cong \mathbb{F}_q$, where $p \neq q$ are primes.

**2.97.** Prove that the stochastic group $\Sigma(2, \mathbb{F}_4) \cong A_4$ (see Exercise 2.18 on page 90).

**2.98.** Let $f(x) = s_0 + s_1 x + \cdots + s_{n-1} x^{n-1} + x^n \in k[x]$, where $k$ is a field, and suppose that $f(x) = (x - \alpha_1)(x - \alpha_2) \cdots (x - \alpha_n)$. Prove that $s_{n-1} = -(\alpha_1 + \alpha_2 + \cdots + \alpha_n)$ and that $s_0 = (-1)^n \alpha_1 \alpha_2 \cdots \alpha_n$. Conclude that the sum and product of all the roots of $f(x)$ lie in $k$.

**2.99.** Write addition and multiplication tables for the field $\mathbb{F}_8$ with eight elements using an irreducible cubic over $\mathbb{F}_2$.

**2.100.** Let $k \subseteq K \subseteq E$ be fields. Prove that if $E$ is a finite extension of $k$, then $E$ is a finite extension of $K$ and $K$ is a finite extension of $k$.

**2.101.** Let $k \subseteq F \subseteq K$ be a tower of fields, and let $z \in K$. Prove that if $k(z)/k$ is finite, then $[F(z) : F] \leq [k(z) : k]$. In particular, $[F(z) : F]$ is finite.

**Hint.** Use Proposition 2.141 to obtain an irreducible polynomial $p(x) \in k[x]$; the polynomial $p(x)$ may factor in $K[x]$.

**2.102.** (i) Is $\mathbb{F}_4$ a subfield of $\mathbb{F}_8$?

(ii) For any prime $p$, prove that if $\mathbb{F}_{p^n}$ is a subfield of $\mathbb{F}_{p^m}$, then $n \mid m$ (the converse is also true, as we shall see later).

**Hint.** View $\mathbb{F}_{p^m}$ as a vector space over $\mathbb{F}_{p^n}$.

**2.103.** Let $K/k$ be an extension field. If $A \subseteq K$ and $u \in k(A)$, prove that there are $a_1, \ldots, a_n \in A$ with $u \in k(a_1, \ldots, a_n)$.

**2.104.** Let $E/k$ be a field extension. If $v \in E$ is algebraic over $k$, prove that $v^{-1}$ is algebraic over $k$.

# Chapter 3

# Galois Theory

## Section 3.1. Insolvability of the Quintic

This chapter will discuss what is nowadays called *Galois Theory* (it was originally called *Theory of Equations*), the interrelation between extension fields and certain groups associated to them, called *Galois groups*. Informally, we say that a polynomial is *solvable by radicals* if there is a generalization of the quadratic formula that gives its roots. Galois Theory will enable us to prove the theorem of Abel–Ruffini (there are polynomials of degree 5 that are not solvable by radicals) as well as Galois's Theorem describing all those polynomials (over a field of characteristic 0) which are solvable by radicals. Another corollary of this theory is a proof of the Fundamental Theorem of Algebra.

By Kronecker's Theorem, Theorem 2.148, for each monic $f(x) \in k[x]$, where $k$ is a field, there is an extension field $K/k$ and (not necessarily distinct) roots $z_1, \ldots, z_n \in K$ with

$$f(x) = x^n + a_{n-1}x^{n-1} + \cdots + a_1 x + a_0 = (x - z_1) \cdots (x - z_n).$$

By induction on $n \geq 1$, we can easily generalize Exercise 2.98 on page 170:

$$(1) \quad \begin{cases} a_{n-1} = -\sum_i z_i \\ a_{n-2} = \sum_{i<j} z_i z_j \\ a_{n-3} = -\sum_{i<j<k} z_i z_j z_k \\ \quad\vdots \\ a_0 = (-1)^n z_1 z_2 \cdots z_n. \end{cases}$$

**Definition.** The ***elementary symmetric functions*** of $n$ variables are the polynomials, for $j = 1, \ldots, n$,

$$e_j(x_1, \ldots, x_n) = \sum_{i_1 < \cdots < i_j} x_{i_1} \cdots x_{i_j}.$$

Equations (1) show that if $z_1, \ldots, z_n$ are the roots of $x^n + a_{n-1}x^{n-1} + \cdots + a_0$, then

$$e_j(z_1, \ldots, z_n) = (-1)^j a_{n-j}.$$

In particular, $-a_{n-1}$ is the sum of the roots and $(-1)^n a_0$ is the product of the roots.

Given the coefficients $a_0, \ldots, a_{n-1}$ of $f(x)$, can we find its roots? That is, can we solve the system (1) of $n$ equations in $n$ unknowns? If $n = 2$, the answer is yes: the quadratic formula works. If $n = 3$ or 4, the answer is still yes, for the cubic and quartic formulas work. But if $n \geq 5$, we shall see that no *analogous* solution exists. We do not say that no solution of system (1) exists if $n \geq 5$. It is quite possible that there are ways of finding the roots of a quintic polynomial if we do not limit ourselves to formulas involving only field operations and extraction of roots. Indeed, we can find the roots by *Newton's method*: if $r$ is a real root of a polynomial $f(x)$ and $x_0$ is a "good" approximation to $r$, then $r = \lim_{n \to \infty} x_n$, where $x_n$ is defined recursively by $x_{n+1} = x_n - f(x_n)/f'(x_n)$ for all $n \geq 0$. There is a method of Hermite finding roots of quintics using elliptic modular functions, and there are methods for finding the roots of many polynomials of higher degree using hypergeometric functions (see King, *Beyond the Quartic Equation*).

If $n \geq 5$, Abel proved (Theorem 3.28) that there are polynomials of degree $n$ that are not solvable by radicals (we will define this notion more carefully later). The key observation is that symmetry is present. Recall from Chapter 1 that if $\Omega$ is a polygon in the plane $\mathbb{R}^2$, then its *symmetry group* $\Sigma(\Omega)$ consists of all those isometries $\varphi \colon \mathbb{R}^2 \to \mathbb{R}^2$ of the plane for which $\varphi(\Omega) = \Omega$. Moreover, isometries $\varphi \in \Sigma(\Omega)$ are completely determined by their values on the vertices of $\Omega$; indeed, if $\Omega$ has $n$ vertices, then $\Sigma(\Omega)$ is isomorphic to a subgroup of $S_n$. We will set up an analogy with symmetry groups in which polynomials play the role of polygons, a splitting field of a polynomial plays the role of the plane $\mathbb{R}^2$, and an *automorphism fixing $k$* plays the role of an isometry. We will see, on page 181, that regular polygons correspond to irreducible polynomials.

**Definition.** Let $E/k$ be an extension field. An ***automorphism*** of $E$ is an isomorphism $\sigma \colon E \to E$; an automorphism $\sigma$ of $E$ ***fixes*** $k$ if $\sigma(a) = a$ for every $a \in k$.

Note that an extension field $E/k$ is a vector space over $k$ and, if $\sigma \colon E \to E$ fixes $k$, then $\sigma$ is a $k$-linear transformation $[\sigma(ae) = \sigma(a)\sigma(e) = a\sigma(e)$ for all $a \in k$ and $e \in E]$. For example, a splitting field of $f(x) = x^2 + 1$ over $\mathbb{Q}$ is $E = \mathbb{Q}(i)$, and complex conjugation $\sigma \colon a \mapsto \bar{a}$ is an example of an automorphism of $E$ fixing $\mathbb{Q}$.

**Proposition 3.1.** *Let $k$ be a field, let*

$$f(x) = x^n + a_{n-1}x^{n-1} + \cdots + a_1 x + a_0 \in k[x],$$

and let $E = k(z_1, \ldots, z_n)$ be a splitting field of $f(x)$ over $k$. If $\sigma: E \to E$ is an automorphism fixing $k$, then $\sigma$ permutes the set of roots $\{z_1, \ldots, z_n\}$ of $f(x)$.

**Proof.** If $z$ is a root of $f(x)$, then

$$0 = f(z) = z^n + a_{n-1}z^{n-1} + \cdots + a_1 z + a_0.$$

Applying $\sigma$ to this equation gives

$$\begin{aligned}
0 &= \sigma(z)^n + \sigma(a_{n-1})\sigma(z)^{n-1} + \cdots + \sigma(a_1)\sigma(z) + \sigma(a_0) \\
&= \sigma(z)^n + a_{n-1}\sigma(z)^{n-1} + \cdots + a_1\sigma(z) + a_0 \\
&= f(\sigma(z)),
\end{aligned}$$

because $\sigma$ fixes $k$. Therefore, $\sigma(z)$ is a root of $f(x)$; thus, if $\Omega$ is the set of all the roots, then $\sigma|\Omega: \Omega \to \Omega$, where $\sigma|\Omega$ is the restriction. But $\sigma|\Omega$ is injective (because $\sigma$ is), so that $\sigma|\Omega$ is a permutation of $\Omega$, by the Pigeonhole Principle. •

Here is the analog of the symmetry group $\Sigma(\Omega)$ of a polygon $\Omega$.

**Definition.** The **Galois group** of an extension field $E/k$, denoted by $\mathrm{Gal}(E/k)$, is the set of all those automorphisms of $E$ that fix $k$. If $f(x) \in k[x]$ and $E = k(z_1, \ldots, z_n)$ is a splitting field of $f(x)$ over $k$, then the **Galois group** of $f(x)$ over $k$ is defined to be $\mathrm{Gal}(E/k)$.

It is easy to check that $\mathrm{Gal}(E/k)$ is a group with operation composition of functions. This definition is due to E. Artin, in keeping with his and Noether's emphasis on "abstract" Algebra. Galois's original version (a group isomorphic to this one) was phrased, not in terms of automorphisms, but in terms of certain permutations of the roots of a polynomial (Tignol, *Galois' Theory of Algebraic Equations*, pp. 295–302). Note that $\mathrm{Gal}(E/k)$ is independent of the choice of splitting field $E$, by Theorem 2.157.

The following lemma will be used several times.

**Lemma 3.2.** *Let* $\sigma \in \mathrm{Gal}(E/k)$, *where* $E = k(z_1, \ldots, z_n)$. *If* $\sigma(z_i) = z_i$ *for all* $i$, *then* $\sigma$ *is the identity* $1_E$.

**Proof.** We prove this lemma by induction on $n \geq 1$. If $n = 1$, then each $u \in E$ has the form $u = f(z_1)/g(z_1)$, where $f(x)$, $g(x) \in k[x]$ and $g(z_1) \neq 0$. But $\sigma$ fixes $z_1$ as well as the coefficients of $f(x)$ and of $g(x)$, so that $\sigma$ fixes all $u \in E$. For the inductive step, write $K = k(z_1, \ldots, z_{n-1})$, and note that $E = K(z_n)$ [for $K(z_n)$ is the smallest subfield containing $k$ and $z_1, \ldots, z_{n-1}, z_n$]. Having noted this, the inductive step is just a repetition of the base step with $k$ replaced by $K$. •

**Theorem 3.3.** *If* $f(x) \in k[x]$ *has degree* $n$, *then its Galois group* $\mathrm{Gal}(E/k)$ *is isomorphic to a subgroup of* $S_n$.

**Proof.** Let $X = \{z_1, \ldots, z_n\}$. If $\sigma \in \mathrm{Gal}(E/k)$, then Proposition 3.1 shows that its restriction $\sigma|X$ is a permutation of $X$; that is, $\sigma|X \in S_X$. Define $\varphi: \mathrm{Gal}(E/k) \to S_X$ by $\varphi: \sigma \mapsto \sigma|X$. To see that $\varphi$ is a homomorphism, note that both $\varphi(\sigma\tau)$ and $\varphi(\sigma)\varphi(\tau)$ are functions $X \to X$ that agree on each $z_i \in X$: $\varphi(\sigma\tau): z_i \mapsto (\sigma\tau)(z_i)$, while $\varphi(\sigma)\varphi(\tau): z_i \mapsto \sigma(\tau(z_i))$, and these are the same.

The image of $\varphi$ is a subgroup of $S_X \cong S_n$. The kernel of $\varphi$ is the set of all $\sigma \in \mathrm{Gal}(E/k)$ with $\sigma|X = 1_X$; that is, $\sigma$ fixes each of the roots $z_i$. As $\sigma$ also fixes $k$, by the definition of Galois group, Lemma 3.2 gives $\ker \varphi = \{1\}$. Therefore, $\varphi$ is injective. •

If $f(x) = x^2 + 1 \in \mathbb{Q}[x]$, then complex conjugation $\sigma$ is an automorphism of its splitting field $\mathbb{Q}(i)$ (for $\sigma$ interchanges the roots $i$ and $-i$); since $\sigma$ fixes $\mathbb{Q}$, we have $\sigma \in G = \mathrm{Gal}(\mathbb{Q}(i)/\mathbb{Q})$. Now $G$ is a subgroup of the symmetric group $S_2$, which has order 2; it follows that $G = \langle \sigma \rangle \cong \mathbb{I}_2$. The reader should regard the elements of any Galois group $\mathrm{Gal}(E/k)$ as generalizations of complex conjugation.

In order to compute the order of the Galois group, we must first discuss *separability*.

**Lemma 3.4.** *If $k$ is a field of characteristic 0, then every irreducible polynomial $p(x) \in k[x]$ has no repeated roots.*

**Proof.** Let $f(x) \in k[x]$ be a (not necessarily irreducible) polynomial. In Exercise 2.54 on page 114, we saw that $f(x)$ has no repeated roots if and only if the gcd $(f, f') = 1$, where $f'(x)$ is the derivative of $f(x)$.

Now consider $p(x) \in k[x]$; we may assume that $p(x)$ is monic of degree $d \geq 1$. The coefficient $dx^{d-1}$ of $p'(x)$ is nonzero, because $k$ has characteristic 0, and so $p'(x) \neq 0$. Since $p(x)$ is irreducible, its only divisors are constants and associates; as $p'(x)$ has smaller degree, it is not an associate of $p(x)$, and so gcd $(p, p') = 1$. •

**Definition.** An *irreducible* polynomial $p(x)$ is **separable** if it has no repeated roots. An arbitrary polynomial $f(x)$ is **separable** if each of its irreducible factors has no repeated roots; otherwise, it is **inseparable**.

Recall Theorem 2.144(i): if $E/k$ is an extension field and $\alpha \in E$ is algebraic over $k$, then there is a unique monic irreducible polynomial $\mathrm{irr}(\alpha, k) \in k[x]$, called its *minimal polynomial*, having $\alpha$ as a root.

**Definition.** Let $E/k$ be an algebraic extension. An element $\alpha \in E$ is **separable** if either $\alpha$ is transcendental over $k$ or $\alpha$ is algebraic over $k$ and its minimal polynomial $\mathrm{irr}(\alpha, k)$ is separable; that is, $\mathrm{irr}(\alpha, k)$ has no repeated roots. A extension field $E/k$ is **separable** if each of its elements is separable; $E/k$ is **inseparable** if it is not separable.

Lemma 3.4 shows that every extension field $E/k$ is separable if $k$ has characteristic 0. If $E$ is a finite field with $p^n$ elements, then Lagrange's Theorem (for the multiplicative group $E^\times$) shows that every element of $E$ is a root of $g(x) = x^{p^n} - x$. We saw, in the proof of Theorem 2.153 (the existence of finite fields with $p^n$ elements), that $g(x)$ has no repeated roots. It follows that if $k \subseteq E$, then $E/k$ is separable, for if $\alpha \in E$, then $\mathrm{irr}(\alpha, k)$ is a divisor of $g(x)$.

**Example 3.5.** Here is an example of an inseparable extension. Let $k = \mathbb{F}_p(t) = \mathrm{Frac}(\mathbb{F}_p[t])$, and let $E = k(\alpha)$, where $\alpha$ is a root of $f(x) = x^p - t$; that is, $\alpha^p = t$. In $E[x]$, we have

$$f(x) = x^p - t = x^p - \alpha^p = (x - \alpha)^p.$$

If we show that $\alpha \notin k$, then $f(x)$ is irreducible (by Proposition 2.152), hence $f(x) = \operatorname{irr}(\alpha, k)$ is an inseparable polynomial, and so $E/k$ is inseparable. If, on the contrary, $\alpha \in k$, then there are $g(t), h(t) \in \mathbb{F}_p[t]$ with $\alpha = g(t)/h(t)$. Hence, $g = \alpha h$ and $g^p = \alpha^p h^p = t h^p$, so that

$$\deg(g^p) = \deg(t h^p) = 1 + \deg(h^p).$$

But $p \mid \deg(g^p)$ and $p \mid \deg(h^p)$, and this gives a contradiction. ◀

We will study separability and inseparability more thoroughly in Chapter 5.

**Example 3.6.** Let $n$ be a positive integer. Theorem 2.46 says that every finite subgroup of the multiplicative group of a field $E$ is cyclic; hence, the group $\Gamma_n(E)$ of all the $n$th roots of unity in $E$ is cyclic; any generator of this group, say, $\omega$, is called a *primitive $n$th root of unity*. Let $f(x) = x^n - 1 \in k[x]$, where $k$ is a field. What is the order of $\Gamma_n(\mathrm{E})$ if $E/k$ is a splitting field of $f(x)$? If the characteristic of $k$ is 0, we know that $f(x)$ has $n$ distinct roots [by Exercise 2.54 on page 114, for $(f, f') = 1$]. Thus, $|\Gamma_n(E)| = n$ and a primitive $n$th root of unity $\omega$ has order $n$. Since every extension field of characteristic 0 is separable, $\omega$ is a separable element.

Suppose the characteristic of $k$ is a prime $p$. Write $n = p^e m$, where $(m, p) = 1$. If $g(x) = x^m - 1$, then $m x^{m-1} \neq 0$ [because $(m, p) = 1$] and $(g, g') = 1$; hence, $g(x)$ has no repeated roots, and $E$ contains $m$ distinct $m$th roots of unity. We claim that $|\Gamma_n(E)| = m$; that is, there are no other $n$th roots of unity in $E$. If $\beta$ is an $n$th root of unity, then $1 = \beta^n = (\beta^m)^{p^e}$; that is, $\beta^m$ is a root of $x^{p^e} - 1$. But $x^{p^e} - 1 = (x - 1)^{p^e}$, because $k$ has characteristic $p$, so that $\beta^m = 1$. If $\omega$ is a primitive $n$th root of unity, $\operatorname{irr}(\omega, k) \mid x^m - 1$. Hence, all its roots are distinct, and so $\omega$ is a separable element in this case as well. ◀

Separability of $E/k$ allows us to find the order of $\operatorname{Gal}(E/k)$.

**Theorem 3.7.**

(i) *Let $\varphi\colon k \to k'$ be an isomorphism of fields, and let $\varphi^*\colon k[x] \to k'[x]$ be the ring isomorphism of* Corollary 2.27: $\varphi^*\colon g(x) = a_0 + a_1 x + \cdots + a_n x^n \mapsto g^*(x) = \varphi(a_0) + \varphi(a_1)x + \cdots + \varphi(a_n)x^n$; *let $f(x) \in k[x]$ and $f^*(x) = \varphi^*(f) \in k'[x]$. If $E/k$ is a splitting field of $f(x)$ and $E^*/k'$ is a splitting field of $f^*(x)$, then there are exactly $[E : k]$ isomorphisms $\Phi\colon E \to E^*$ that extend $\varphi$:*

$$
\begin{array}{ccc}
E & \overset{\Phi}{\dashrightarrow} & E^* \\
| & & | \\
k & \underset{\varphi}{\longrightarrow} & k'
\end{array}
$$

(ii) *If $E/k$ is a splitting field of a separable $f(x) \in k[x]$, then*

$$|\operatorname{Gal}(E/k)| = [E : k].$$

**Proof.**

(i) The proof, by induction on $[E : k]$, modifies that of Lemma 2.156. The base step $[E : k] = 1$ gives $E = k$, and there is only one extension $\Phi$ of $\varphi$, namely,

$\varphi$ itself. If $[E : k] > 1$, let $f(x) = p(x)g(x)$, where $p(x)$ is an irreducible factor of largest degree, say, $d$. We may assume that $d > 1$; otherwise $f(x)$ splits over $k$ and $[E : k] = 1$. Choose a root $\alpha$ of $p(x)$ [note that $\alpha \in E$ because $E$ is a splitting field of $f(x) = p(x)g(x)$]. If $\widetilde{\varphi} : k(\alpha) \to E^*$ is any extension of $\varphi$, then $\varphi(\alpha)$ is a root $\alpha^*$ of $p^*(x)$, by Proposition 3.1; since $f^*(x)$ is separable, $p^*(x)$ has exactly $d$ roots $\alpha^* \in E^*$; by Lemma 3.2 and Theorem 2.144(ii), there are exactly $d$ isomorphisms $\widehat{\varphi} : k(\alpha) \to k'(\alpha^*)$ extending $\varphi$, one for each $\alpha^*$. Now $E$ is also a splitting field of $f(x)$ over $k(\alpha)$, because adjoining all the roots of $f(x)$ to $k(\alpha)$ still produces $E$; similarly, $E^*$ is a splitting field of $f^*(x)$ over $k'(\alpha^*)$. Now $[E : k(\alpha)] < [E : k]$, because $[E : k(\alpha)] = [E : k]/d$, so that induction shows that each of the $d$ isomorphisms $\widehat{\varphi}$ has exactly $[E : k]/d$ extensions $\Phi : E \to E^*$. Thus, we have constructed $[E : k]$ isomorphisms extending $\varphi$. But there are no others, because every $\tau$ extending $\varphi$ has $\tau|k(\alpha) = \widehat{\varphi}$ for some $\widehat{\varphi} : k(\alpha) \to k'(\alpha^*)$.

(ii) In part (i), take $k = k'$, $E = E^*$, and $\varphi = 1_k$.    $\bullet$

**Example 3.8.** The separability hypothesis in Theorem 3.7(ii) is necessary. In Example 3.5, we saw that if $k = \mathbb{F}_p(t)$ and $\alpha$ is a root of $x^p - t$, then $E = k(\alpha)$ is an inseparable extension. Moreover, $x^p - t = (x - \alpha)^p$, so that $\alpha$ is the only root of this polynomial. Hence, if $\sigma \in \mathrm{Gal}(E/k)$, then Proposition 3.1 shows that $\sigma(\alpha) = \alpha$. Therefore, $\mathrm{Gal}(E/k) = \{1\}$, by Lemma 3.2, and so $|\mathrm{Gal}(E/k)| = 1 < p = [E : k]$ in this case.    ◀

**Corollary 3.9.** *Let $E/k$ be a splitting field of a separable polynomial $f(x) \in k[x]$ of degree $n$. If $f(x)$ is irreducible, then $n \mid |\mathrm{Gal}(E/k)|$.*

**Proof.** By Theorem 3.7(ii), $|\mathrm{Gal}(E/k)| = [E : k]$. Let $\alpha \in E$ be a root of $f(x)$. Since $f(x)$ is irreducible, $[k(\alpha) : k] = n$, by Proposition 2.141(v), and

$$[E : k] = [E : k(\alpha)][k(\alpha) : k] = n[E : k(\alpha)].    \bullet$$

We will give another proof of Corollary 3.9 on page 181. In Proposition 3.41, we shall see that a splitting field of a separable polynomial is a separable extension.

Here is another proof of Proposition 2.92.

**Proposition 3.10.** *The polynomial $f(x) = x^4 + 1$ is irreducible in $\mathbb{Q}[x]$, yet it factors in $\mathbb{F}_p[x]$ for every prime $p$.*

**Proof.** We proved that $x^4 + 1$ is irreducible in $\mathbb{Q}[x]$ in Proposition 2.92; we now show, for all primes $p$, that $x^4 + 1$ factors in $\mathbb{F}_p[x]$. If $p = 2$, then $x^4 + 1 = (x + 1)^4$, and so we may assume that $p$ is an odd prime. It is easy to check that every square $n^2$ is congruent to 0, 1, or 4 mod 8; since $p$ is odd, we must have $p^2 \equiv 1$ mod 8. Therefore, $|(\mathbb{F}_{p^2})^\times| = p^2 - 1$ is divisible by 8. But $(\mathbb{F}_{p^2})^\times$ is a cyclic group, by Theorem 2.46, and so it has a (cyclic) subgroup of order 8, by Lemma 1.92. It follows that $\mathbb{F}_{p^2}$ contains all the 8th roots of unity; in particular, $\mathbb{F}_{p^2}$ contains all the roots of $x^4 + 1$. Hence, the splitting field $E_p$ of $x^4 + 1$ over $\mathbb{F}_p$ is $\mathbb{F}_{p^2}$, and so $[E_p : \mathbb{F}_p] = 2$. But if $x^4 + 1$ were irreducible in $\mathbb{F}_p[x]$, then $4 \mid [E_p : \mathbb{F}_p]$, by Corollary 3.9. Therefore, $x^4 + 1$ factors in $\mathbb{F}_p[x]$ for every prime $p$.    $\bullet$

Here are some computations of Galois groups of specific polynomials in $\mathbb{Q}[x]$.

**Example 3.11.**

(i) Let $f(x) = x^3 - 1 \in \mathbb{Q}[x]$. Now $f(x) = (x-1)(x^2 + x + 1)$, where $x^2 + x + 1$ is irreducible (the quadratic formula shows that its roots $\omega$ and $\overline{\omega}$ do not lie in $\mathbb{Q}$). The splitting field of $f(x)$ is $\mathbb{Q}(\omega)$, for $\omega^2 = \overline{\omega}$, and so $[\mathbb{Q}(\omega) : \mathbb{Q}] = 2$. Therefore, $|\operatorname{Gal}(\mathbb{Q}(\omega)/\mathbb{Q})| = 2$, by Theorem 3.7(ii), and it is cyclic of order 2. Its nontrivial element is complex conjugation.

(ii) Let $f(x) = x^2 - 2 \in \mathbb{Q}[x]$. Now $f(x)$ is irreducible with roots $\pm\sqrt{2}$, so that $E = \mathbb{Q}(\sqrt{2})$ is a splitting field. By Theorem 3.7(ii), $|\operatorname{Gal}(E/\mathbb{Q})| = 2$. Now every element of $E$ has a unique expression of the form $a + b\sqrt{2}$, where $a, b \in \mathbb{Q}$ [Proposition 2.141(v)], and it is easily seen that $\sigma \colon E \to E$, defined by $\sigma \colon a + b\sqrt{2} \mapsto a - b\sqrt{2}$, is an automorphism of $E$ fixing $\mathbb{Q}$. Therefore, $\operatorname{Gal}(E/\mathbb{Q}) = \langle \sigma \rangle$, where $\sigma$ interchanges $\sqrt{2}$ and $-\sqrt{2}$.

(iii) Let $g(x) = x^3 - 2 \in \mathbb{Q}[x]$. The roots of $g(x)$ are $\beta$, $\omega\beta$, and $\omega^2\beta$, where $\beta = \sqrt[3]{2}$, the real cube root of 2, and $\omega$ is a primitive cube root of unity. It is easy to see that the splitting field of $g(x)$ is $E = \mathbb{Q}(\beta, \omega)$. Note that

$$[E : \mathbb{Q}] = [E : \mathbb{Q}(\beta)][\mathbb{Q}(\beta) : \mathbb{Q}] = 3[E : \mathbb{Q}(\beta)],$$

for $g(x)$ is irreducible over $\mathbb{Q}$ (it is a cubic having no rational roots). Now $E \neq \mathbb{Q}(\beta)$, for every element in $\mathbb{Q}(\beta)$ is real, while the complex number $\omega$ is not real. Therefore, $[E : \mathbb{Q}] = |\operatorname{Gal}(E/\mathbb{Q})| > 3$. On the other hand, we know that $\operatorname{Gal}(E/\mathbb{Q})$ is isomorphic to a subgroup of $S_3$, and so we must have $\operatorname{Gal}(E/\mathbb{Q}) \cong S_3$.

(iv) We examined $f(x) = x^4 - 10x^2 + 1 \in \mathbb{Q}[x]$ in Example 2.147, when we saw that $f(x)$ is irreducible; in fact, $f(x) = \operatorname{irr}(\beta, \mathbb{Q})$, where $\beta = \sqrt{2} + \sqrt{3}$. If $E = \mathbb{Q}(\beta)$, then $[E : \mathbb{Q}] = 4$; moreover, $E$ is a splitting field of $f(x)$, where the other roots of $f(x)$ are $-\sqrt{2} - \sqrt{3}$, $-\sqrt{2} + \sqrt{3}$, and $\sqrt{2} - \sqrt{3}$. It follows from Theorem 3.7(ii) that if $G = \operatorname{Gal}(E/\mathbb{Q})$, then $|G| = 4$; hence, either $G \cong \mathbb{I}_4$ or $G \cong \mathbf{V}$.

We also saw, in Example 2.147, that $E$ contains $\sqrt{2}$ and $\sqrt{3}$. If $\sigma$ is an automorphism of $E$ fixing $\mathbb{Q}$, then $\sigma(\sqrt{2}) = u\sqrt{2}$, where $u = \pm 1$, because $\sigma(\sqrt{2})^2 = 2$. Therefore, $\sigma^2(\sqrt{2}) = \sigma(u\sqrt{2}) = u\sigma(\sqrt{2}) = u^2\sqrt{2} = \sqrt{2}$; similarly, $\sigma^2(\sqrt{3}) = \sqrt{3}$. If $\alpha$ is a root of $f(x)$, then $\alpha = u\sqrt{2} + v\sqrt{3}$, where $u, v = \pm 1$. Hence,

$$\sigma^2(\alpha) = u\sigma^2(\sqrt{2}) + v\sigma^2(\sqrt{3}) = u\sqrt{2} + v\sqrt{3} = \alpha.$$

Lemma 3.2 gives $\sigma^2 = 1_E$ for all $\sigma \in \operatorname{Gal}(E/\mathbb{Q})$, and so $\operatorname{Gal}(E/\mathbb{Q}) \cong \mathbf{V}$.

Here is another way to compute $G = \operatorname{Gal}(E/\mathbb{Q})$. We saw in Example 2.147 that $E = \mathbb{Q}(\sqrt{2} + \sqrt{3}) = \mathbb{Q}(\sqrt{2}, \sqrt{3})$ is also a splitting field of $g(x) = (x^2 - 2)(x^2 - 3)$ over $\mathbb{Q}$. By Proposition 2.144(ii), there is an automorphism $\varphi \colon \mathbb{Q}(\sqrt{2}) \to \mathbb{Q}(\sqrt{2})$ taking $\sqrt{2} \mapsto -\sqrt{2}$. But $\sqrt{3} \notin \mathbb{Q}(\sqrt{2})$, as we noted in Example 2.147, so that $x^2 - 3$ is irreducible over $\mathbb{Q}(\sqrt{2})$. Lemma 2.156 shows that $\varphi$ extends to an automorphism $\Phi \colon E \to E$; of course, $\Phi \in \operatorname{Gal}(E/\mathbb{Q})$. There are two possibilities: $\Phi(\sqrt{3}) = \pm\sqrt{3}$. Indeed, it is now easy to see that the elements of $\mathrm{Gal}(E/\mathbb{Q})$ correspond to the four-group, consisting of the identity and the permutations (in cycle notation)

$$(\sqrt{2}, \, -\sqrt{2})(\sqrt{3}, \, \sqrt{3}), \quad (\sqrt{2}, \, -\sqrt{2})(\sqrt{3}, \, -\sqrt{3}), \quad (\sqrt{2}, \, \sqrt{2})(\sqrt{3}, \, -\sqrt{3}). \quad \blacktriangleleft$$

Here are two more general computations of Galois groups.

**Proposition 3.12.** *If $m$ is a positive integer, $k$ is a field, and $E$ is a splitting field of $x^m - 1$ over $k$, then $\mathrm{Gal}(E/k)$ is abelian; in fact, $\mathrm{Gal}(E/k)$ is isomorphic to a subgroup of the (multiplicative) group of units $U(\mathbb{I}_m) = \{[i] \in \mathbb{I}_m : (i, m) = 1\}$.*

**Proof.** By Example 2.151, $E = k(\omega)$, where $\omega$ is a primitive $m$th root of unity, and so $E = k(\omega)$. The group $\Gamma_m$ of all roots of $x^m - 1$ in $E$ is cyclic (with generator $\omega$) and, if $\sigma \in \mathrm{Gal}(E/k)$, then its restriction to $\Gamma_m$ is an automorphism of $\Gamma_m$. Hence, $\sigma(\omega) = \omega^i$ must also be a generator of $\Gamma_m$; that is, $(i, m) = 1$, by Theorem 1.39(ii). It is easy to see that $i$ is uniquely determined mod $m$, so that the function $\theta \colon \mathrm{Gal}(k(\omega)/k) \to U(\mathbb{I}_m)$, given by $\theta(\sigma) = [i]$ if $\sigma(\omega) = \omega^i$, is well-defined. Now $\theta$ is a homomorphism, for if $\tau(\omega) = \omega^j$, then

$$\tau\sigma(\omega) = \tau(\omega^i) = (\omega^i)^j = \omega^{ij}.$$

Therefore, Lemma 3.2 shows that $\theta$ is injective.  •

**Remark.** We cannot conclude more from the last proposition; Theorem 4.24 says, given any finite abelian group $G$, that there is an integer $m$ with $G$ isomorphic to a subgroup of $U(\mathbb{I}_m)$. However, if $p$ is prime, then $\mathrm{Gal}(E/k) \cong U(\mathbb{I}_p)$.  $\blacktriangleleft$

**Theorem 3.13.** *If $p$ is prime, then*

$$\mathrm{Gal}(\mathbb{F}_{p^n}/\mathbb{F}_p) \cong \mathbb{I}_n,$$

*and a generator is the **Frobenius automorphism** $\mathrm{Fr} \colon u \mapsto u^p$.*

**Proof.** Let $q = p^n$, and let $G = \mathrm{Gal}(\mathbb{F}_q/\mathbb{F}_p)$. Since $\mathbb{F}_q$ has characteristic $p$, we have $(a + b)^p = a^p + b^p$, and so the Frobenius $\mathrm{Fr}$ is a homomorphism of fields. As any homomorphism of fields, $\mathrm{Fr}$ is injective; as $\mathbb{F}_q$ is finite, $\mathrm{Fr}$ must be an automorphism, by the Pigeonhole Principle; that is, $\mathrm{Fr} \in G$ ($\mathrm{Fr}$ fixes $\mathbb{F}_p$, by Fermat's Theorem).

If $\pi \in \mathbb{F}_q$ is a primitive element, then $d(x) = \mathrm{irr}(\pi, \mathbb{F}_p)$ has degree $n$, by Corollary 2.154, and so $|G| = n$, by Theorem 3.7(ii). It suffices to prove that the order $j$ of $\mathrm{Fr}$ is not less than $n$. But if $\mathrm{Fr}^j = 1_{\mathbb{F}_q}$ for $j < n$, then $u^{p^j} = u$ for all of the $q = p^n$ elements $u \in \mathbb{F}_q$, giving too many roots of the polynomial $x^{p^j} - x$.  •

The following nice corollary of Lemma 2.156 says, in our analogy between Galois Theory and symmetry of polygons, that irreducible polynomials correspond to regular polygons.

**Proposition 3.14.** *Let $k$ be a field, let $f(x) \in k[x]$, and let $E/k$ be a splitting field of $f(x)$. If $f(x)$ has no repeated roots, then $f(x)$ is irreducible if and only if $\mathrm{Gal}(E/k)$ acts transitively on the roots of $f(x)$.*

**Proof.** Assume that $f(x)$ is irreducible, and let $\alpha, \beta \in E$ be roots of $f(x)$. By Theorem 2.144(i), there is an isomorphism $\varphi : k(\alpha) \to k(\beta)$ with $\varphi(\alpha) = \beta$ and which fixes $k$. Lemma 2.156 shows that $\varphi$ extends to an automorphism $\Phi$ of $E$ that fixes $k$; that is, $\Phi \in \mathrm{Gal}(E/k)$. Now $\Phi(\alpha) = \varphi(\alpha) = \beta$, and so $\mathrm{Gal}(E/k)$ acts transitively on the roots.

Conversely, assume that $\mathrm{Gal}(E/k)$ acts transitively on the roots of $f(x)$. Let $f(x) = p_1(x) \cdots p_t(x)$ be a factorization into irreducibles in $k[x]$, where $t \geq 2$. Choose a root $\alpha \in E$ of $p_1(x)$ and a root $\beta \in E$ of $p_2(x)$; note that $\beta$ is not a root of $p_1(x)$, because $f(x)$ has no repeated roots. By hypothesis, there is $\sigma \in \mathrm{Gal}(E/k)$ with $\sigma(\alpha) = \beta$. Now $\sigma$ permutes the roots of $p_1(x)$, by Proposition 3.1, contradicting $\beta$ not being a root of $p_1(x)$. Hence, $t = 1$ and $f(x)$ is irreducible. $\bullet$

We can now give another proof of Corollary 3.9. Theorem 1.107 says that if $X$ is a $G$-set, then $|G| = |\mathcal{O}(x)||G_x|$, where $\mathcal{O}(x)$ is the orbit of $x \in X$ and $G_x$ is the stabilizer of $x$. In particular, if $X$ is a transitive $G$-set, then $|X|$ is a divisor of $|G|$. Let $f(x) \in k[x]$ be a separable irreducible polynomial of degree $n$, and let $E/k$ be its splitting field. If $X$ is the set of roots of $f(x)$, then $X$ is a transitive $\mathrm{Gal}(E/k)$-set, by Proposition 3.14, and so $n = \deg(f) = |X|$ is a divisor of $|\mathrm{Gal}(E/k)|$.

The analogy is complete.

**3.1.1. Classical Formulas and Solvability by Radicals.** Here is our basic strategy. First, we will translate the classical formulas (giving the roots of polynomials of degree at most 4) in terms of subfields of a splitting field $E$ over $k$. Second, this translation into the language of fields will further be translated into the language of groups: if there is a formula for the roots of $f(x)$, then $\mathrm{Gal}(E/k)$ must be a *solvable* group (which we will soon define). Finally, polynomials of degree at least 5 can have Galois groups that are not solvable. The conclusion is that there are polynomials of degree 5 having no formula analogous to the classical formulas which gives their roots. Without further ado, here is the translation of the existence of a formula for the roots of a polynomial in terms of subfields of a splitting field.

**Definition.** A *pure extension* of *type* $m$ is an extension field $k(u)/k$, where $u^m \in k$ for some $m \geq 1$. An extension field $K/k$ is a *radical extension* if there is a tower of intermediate fields

$$k = K_0 \subseteq K_1 \subseteq \cdots \subseteq K_t = K$$

in which each $K_{i+1}/K_i$ is a pure extension.

Of course, every pure extension is a radical extension. If $u^m = a \in k$, then $k(u)$ arises from $k$ by adjoining an $m$th root of $a$. If $k \subseteq \mathbb{C}$, there are $m$ different $m$th roots of $a$, namely, $u, \omega u, \omega^2 u, \ldots, \omega^{m-1} u$, where $\omega = e^{2\pi i/m}$ is a primitive

$m$th root of unity. More generally, if $k$ contains the $m$th roots of unity, then a pure extension $k(u)$ of type $m$ (that is, $u^m = a \in k$) is a splitting field of $x^m - a$. Not every subfield $k$ of $\mathbb{C}$ contains all the roots of unity; for example, 1 and $-1$ are the only roots of unity in $\mathbb{Q}$. Since we seek formulas involving extraction of roots, it will eventually be convenient to assume that $k$ contains appropriate roots of unity.

When we say that there is a *formula* for the roots of a polynomial $f(x)$ analogous to the quadratic formula, we mean that there is some expression giving the roots of $f(x)$ in terms of its coefficients: this expression may involve field operations, constants, and extraction of roots, but it should not involve other operations such as cosine, definite integral, or limit, for example. We maintain that the intuitive idea of formula just described is captured by the following definition.

**Definition.** Let $f(x) \in k[x]$ have a splitting field $E$. We say that $f(x)$ is *solvable by radicals* if there is a radical extension

$$k = K_0 \subseteq K_1 \subseteq \cdots \subseteq K_t$$

with $E \subseteq K_t$.

By Exercise 3.1 on page 191, solvability by radicals does not depend on the choice of splitting field.

**Example 3.15.**

   (i) For every field $k$ and every $n \geq 1$, we show that $f(x) = x^n - 1 \in k[x]$ is solvable by radicals. By Example 2.151, a splitting field of $x^n - 1$ is $E = k(\omega)$, where $\omega$ is a primitive $n$th root of unity (if $p \mid n$, then a smaller power of $\omega$ equals 1). Thus, $E/k$ is a pure extension and, hence, a radical extension.

  (ii) Let $p$ be a prime and let $k$ contain all $p$th roots of unity (if $k$ has characteristic $p$, this is automatically true). If $k(u)/k$ is a pure extension of type $p$, then we claim that $k(u)$ is a splitting field of $f(x) = x^p - u^p$. If $k$ has characteristic $p$, then $x^p - u^p = (x - u)^p$, and $f(x)$ splits over $k(u)$; otherwise, $k$ contains a primitive $p$th root of unity, $\omega$, and $f(x) = \prod_i (x - \omega^i u)$. Note that $f(x)$ is separable if characteristic $k \neq p$.  ◄

Let us further illustrate this definition by considering the classical formulas for polynomials of small degree.

### Quadratics

If $f(x) = x^2 + bx + c$, then the quadratic formula gives its roots as

$$\tfrac{1}{2}\left(-b \pm \sqrt{b^2 - 4c}\right).$$

Let $k = \mathbb{Q}(b, c)$. Define $K_1 = k(u)$, where $u = \sqrt{b^2 - 4c}$. Then $K_1$ is a radical extension of $k$ (even a pure extension), for $u^2 \in k$. Moreover, the quadratic formula implies that $K_1$ is the splitting field of $f(x)$, and so $f(x)$ is solvable by radicals.

## Cubics

Let $f(X) = X^3 + bX^2 + cX + d$, and let $k = \mathbb{Q}(b, c, d)$. Recall that the change of variable $X = x - \frac{1}{3}b$ yields a new polynomial $\widetilde{f}(x) = x^3 + qx + r \in k[x]$ having the same splitting field $E$ [for if $u$ is a root of $\widetilde{f}(x)$, then $u - \frac{1}{3}b$ is a root of $f(x)$]; it follows that $\widetilde{f}(x)$ is solvable by radicals if and only if $f(x)$ is. The cubic formula gives the roots of $\widetilde{f}(x)$ as

$$g + h, \quad \omega g + \omega^2 h, \quad \text{and} \quad \omega^2 g + \omega h,$$

where $g^3 = \frac{1}{2}\left(-r + \sqrt{R}\right)$, $h = -q/3g$, $R = r^2 + \frac{4}{27}q^3$, and $\omega$ is a primitive cube root of unity. Because of the constraint $gh = -\frac{1}{3}q$, each of these has a "mate," namely, $h = -q/(3g)$, $-q/(3\omega g) = \omega^2 h$, and $-q/(3\omega^2 g) = \omega h$.

Let us show that $\widetilde{f}(x)$ is solvable by radicals. Define $K_1 = k(\sqrt{R})$, where $R = r^2 + \frac{4}{27}q^3$, and define $K_2 = K_1(\alpha)$, where $\alpha^3 = \frac{1}{2}(-r + \sqrt{R})$. The cubic formula shows that $K_2$ contains the root $\alpha + \beta$ of $\widetilde{f}(x)$, where $\beta = -q/3\alpha$. Finally, define $K_3 = K_2(\omega)$, where $\omega^3 = 1$. The other roots of $\widetilde{f}(x)$ are $\omega\alpha + \omega^2\beta$ and $\omega^2\alpha + \omega\beta$, both of which lie in $K_3$, and so $E \subseteq K_3$.

A splitting field $E$ need not equal $K_3$. If $f(x) \in \mathbb{Q}[x]$ is an irreducible cubic all of whose roots are real, then $E \subseteq \mathbb{R}$. As any cubic, $f(x)$ is solvable by radicals, and so there is a radical extension $K_t/\mathbb{Q}$ with $E \subseteq K_t$. The so-called *Casus Irreducibilis* (Theorem 3.67) says that any radical extension $K_t/\mathbb{Q}$ containing $E$ is not contained in $\mathbb{R}$. Therefore, $E \neq K_t$. In down-to-earth language, any formula for the roots of an irreducible cubic in $\mathbb{Q}[x]$ having all roots real requires the presence of complex numbers!

## Quartics

Let $f(X) = X^4 + bX^3 + cX^2 + dX + e$, and let $k = \mathbb{Q}(b, c, d, e)$. The change of variable $X = x - \frac{1}{4}b$ yields a new polynomial $\widetilde{f}(x) = x^4 + qx^2 + rx + s \in k[x]$; moreover, the splitting field $E$ of $f(x)$ is equal to the splitting field of $\widetilde{f}(x)$, for if $u$ is a root of $\widetilde{f}(x)$, then $u - \frac{1}{4}b$ is a root of $f(x)$. Factor $\widetilde{f}(x)$ in $\mathbb{C}[x]$:

$$\widetilde{f}(x) = x^4 + qx^2 + rx + s = (x^2 + jx + \ell)(x^2 - jx + m),$$

and determine $j$, $\ell$, and $m$. Now $j^2$ is a root of the *resolvent cubic* defined on page 3:

$$(j^2)^3 + 2q(j^2)^2 + (q^2 - 4s)j^2 - r^2.$$

The cubic formula gives $j^2$, from which we can determine $m$ and $\ell$, and hence the roots of the quartic.

Define pure extensions

$$k = K_0 \subseteq K_1 \subseteq K_2 \subseteq K_3,$$

as in the cubic case, so that $j^2 \in K_3$. Define $K_4 = K_3(j)$ (so that $\ell, m \in K_4$). Finally, define $K_5 = K_4\left(\sqrt{j^2 - 4\ell}\right)$ and $K_6 = K_5\left(\sqrt{j^2 - 4m}\right)$ [giving roots of the quadratic factors $x^2 + jx + \ell$ and $x^2 - jx + m$ of $\widetilde{f}(x)$]. The quartic formula gives $E \subseteq K_6$.

We have just seen that quadratics, cubics, and quartics in $\mathbb{Q}[x]$ are solvable by radicals. Conversely, let $f(x) \in k[x]$ have splitting field $E/k$. If $f(x)$ is solvable by radicals, we claim that there is a formula which expresses its roots in terms of its coefficients. Suppose that

$$k = K_0 \subseteq K_1 \subseteq \cdots \subseteq K_t$$

is a tower of pure extensions with $E \subseteq K_t$. Let $z$ be a root of $f(x)$. Now $z \in K_t = K_{t-1}(u)$, where $u$ is an $m$th root of some element $\alpha \in K_{t-1}$; hence, $z$ can be expressed in terms of $u$ and $K_{t-1}$; that is, $z$ can be expressed in terms of $\sqrt[m]{\alpha}$ and $K_{t-1}$. But $K_{t-1} = K_{t-2}(v)$, where some power of $v$ lies in $K_{t-2}$. Hence, $z$ can be expressed in terms of $u$, $v$, and $K_{t-2}$. Ultimately, $z$ is expressed by a formula analogous to the classical formulas.

**3.1.2. Translation into Group Theory.** The second stage of the strategy involves investigating the effect of $f(x)$ being solvable by radicals on its Galois group.

Suppose that $k(u)/k$ is a pure extension of type 6; that is, $u^6 \in k$. Now $k(u^3)/k$ is a pure extension of type 2, for $(u^3)^2 = u^6 \in k$, and $k(u)/k(u^3)$ is obviously a pure extension of type 3. Thus, $k(u)/k$ can be replaced by a tower of pure extensions $k \subseteq k(u^3) \subseteq k(u)$ of types 2 and 3. More generally, we may assume, given a tower of pure extensions, that each field is of prime type over its predecessor: if $k \subseteq k(u)$ is of type $m$, then factor $m = p_1 \cdots p_q$, where the $p$'s are (not necessarily distinct) primes, and replace $k \subseteq k(u)$ by

$$k \subseteq k(u^{m/p_1}) \subseteq k(u^{m/p_1 p_2}) \subseteq \cdots \subseteq k(u).$$

**Definition.** An extension field $E/k$ is called **normal** if it is the splitting field of a polynomial in $k[x]$.

**Example 3.16.** If $E/\mathbb{Q}$ is the splitting field of $x^3 - 2$, then $E$ contains $\alpha, \omega\alpha$, and $\omega^2\alpha$, where $\alpha = \sqrt[3]{2}$ and $\omega = e^{2\pi i/3}$. The extension $\mathbb{Q}(\omega)/\mathbb{Q}$ is normal (it is the splitting field of $x^3 - 1$), but the extensions $\mathbb{Q}(\alpha)/\mathbb{Q}$, $\mathbb{Q}(\omega\alpha)/\mathbb{Q}$, and $\mathbb{Q}(\omega^2\alpha)/\mathbb{Q}$ are not normal. Notice that the subfields $\mathbb{Q}(\alpha)$, $\mathbb{Q}(\omega\alpha)$, and $\mathbb{Q}(\omega^2\alpha)$ are isomorphic; in fact, the automorphism $\sigma \in \text{Gal}(E/\mathbb{Q})$ with $\sigma(\alpha) = \omega\alpha$ is an isomorphism $\mathbb{Q}(\alpha) \to \mathbb{Q}(\omega\alpha)$. ◄

Here is a key result allowing us to translate solvability by radicals into the language of Galois groups (it also shows why *normal extension fields* are so called).

**Theorem 3.17.** *Let $k \subseteq B \subseteq E$ be a tower of fields. If $B/k$ and $E/k$ are normal extensions, then $\sigma(B) = B$ for all $\sigma \in \text{Gal}(E/k)$, $\text{Gal}(E/B) \lhd \text{Gal}(E/k)$, and*

$$\text{Gal}(E/k)/\text{Gal}(E/B) \cong \text{Gal}(B/k).$$

**Proof.** Since $B/k$ is a normal extension, it is a splitting field of some $f(x)$ in $k[x]$; that is, $B = k(z_1, \ldots, z_t) \subseteq E$, where $z_1, \ldots, z_t$ are the roots of $f(x)$. If $\sigma \in \text{Gal}(E/k)$, the restriction of $\sigma$ to $B$ is an automorphism of $B$, and it thus permutes $z_1, \ldots, z_t$, by Proposition 3.1(i) (for $\sigma$ fixes $k$); hence, $\sigma(B) = B$. Define $\rho \colon \text{Gal}(E/k) \to \text{Gal}(B/k)$ by $\sigma \mapsto \sigma|B$. It is easy to see, as in the proof of Theorem 3.3, that $\rho$ is a homomorphism and $\ker \rho = \text{Gal}(E/B)$; thus, $\text{Gal}(E/B) \lhd \text{Gal}(E/k)$. But $\rho$ is surjective: if $\tau \in \text{Gal}(B/k)$, then Lemma 2.156 applies to show that there is $\sigma \in \text{Gal}(E/k)$ extending $\tau$ [i.e., $\rho(\sigma) = \sigma|B = \tau$]. The First Isomorphism Theorem completes the proof. •

The next technical result will be needed when we apply Theorem 3.17.

**Lemma 3.18.**

(i) *If $B = k(u_1, \ldots, u_t)/k$ is a finite extension field, then there is a normal extension $E/k$ containing $B$; that is, $E$ is a splitting field of some $f(x) \in k[x]$. If each $u_i$ is separable over $k$, then $f(x)$ is a separable polynomial and, if $G = \text{Gal}(E/k)$, then*

$$E = k(\sigma(u_1), \ldots, \sigma(u_t) : \sigma \in G).$$

(ii) *If $B/k$ is a radical extension, then the normal extension $E/k$ is a radical extension.*

**Proof.**

(i) By Theorem 2.144(i), there are irreducible polynomials $p_i(x) = \text{irr}(\alpha_i, k) \in k[x]$, for $i = 1, \ldots, t$, with $p_i(u_i) = 0$. Define $E$ to be a splitting field of $f(x) = p_1(x) \cdots p_t(x)$ over $k$. Since $u_i \in E$ for all $i$, we have $B = k(u_1, \ldots, u_t) \subseteq E$. If each $u_i$ is separable over $k$, then each $p_i(x)$ is a separable polynomial, and hence $f(x)$ is a separable polynomial.

For each pair of roots $u$ and $u'$ of any $p_i(x)$, Theorem 2.144(ii) gives an isomorphism $\gamma \colon k(u) \to k(u')$ which fixes $k$ and which takes $u \mapsto u'$. By Lemma 2.156, each such $\gamma$ extends to an automorphism $\sigma \in G = \text{Gal}(E/k)$. Thus, $f(x)$ splits over $k(\sigma(u_1), \ldots, \sigma(u_t) : \sigma \in G)$. But $E/k$ is a splitting field of $f(x)$ over $k$ and $k(\sigma(u_1), \ldots, \sigma(u_t) : \sigma \in G) \subseteq E$; hence,

$$E = k(\sigma(u_1), \ldots, \sigma(u_t) : \sigma \in G),$$

because a splitting field is the smallest field over which $f(x)$ splits.

(ii) Assume now that $B/k$ is a radical extension; say, $B = k(v_1, \ldots, v_s)$, where

$$k \subseteq k(v_1) \subseteq k(v_1, v_2) \subseteq \cdots \subseteq k(v_1, \ldots, v_s) = B$$

and each $k(v_1, \ldots, v_{i+1})/k(v_1, \ldots, v_i)$ is a pure extension; of course, $\sigma(B) = k(\sigma(v_1), \ldots, \sigma(v_s))$ is a radical extension of $k$ for every $\sigma \in G$. We now show that $E = k(\sigma(v_1), \ldots, \sigma(v_s) : \sigma \in G)$ is a radical extension of $k$. Define

$$B_1 = k(\sigma(v_1) : \sigma \in G).$$

Now if $G = \{1, \sigma, \tau, \ldots\}$, then the tower

$$k \subseteq k(v_1) \subseteq k(v_1, \sigma(v_1)) \subseteq k(v_1, \sigma(v_1), \tau(v_1)) \subseteq \cdots \subseteq B_1$$

displays $B_1$ as a radical extension of $k$. For example, $v_1^m$ lies in $k$, and so $\tau(v_1)^m = \tau(v_1^m)$ lies in $\tau(k) = k$; since $k \subseteq k(v_1, \sigma(v_1))$, we have $\tau(v_1)^m \in k(v_1, \sigma(v_1))$. Having defined $B_1$, define $B_{i+1}$ inductively:

$$B_{i+1} = B_i(\sigma(v_{i+1}) : \sigma \in G).$$

Assume, by induction, that $B_i/k$ is a radical extension and that $\sigma(B_i) \subseteq B_i$ for all $\sigma \in G$. Now $B_{i+1}/B_i$ is a radical extension: for example, $v_{i+1}^n \in B_i$, so that $\sigma(v_{i+1})^n \in \sigma(B_i) \subseteq B_i$. Thus, every $B_i$ is a radical extension of $k$ and, therefore, $E = B_s$ is a radical extension of $k$.   •

We can now give the heart of the translation we have been seeking: a radical extension $E/k$ gives rise to a sequence of subgroups of $\mathrm{Gal}(E/k)$.

**Lemma 3.19.** *Let*

$$k = K_0 \subseteq K_1 \subseteq K_2 \subseteq \cdots \subseteq K_t$$

*be a tower with each $K_i/K_{i-1}$ a pure extension of prime type $p_i$. If $K_t/k$ is a normal extension and $k$ contains all the $p_i$th roots of unity, for $i = 1, \ldots, t$, then there is a sequence of subgroups*

$$\mathrm{Gal}(K_t/k) = G_0 \supseteq G_1 \supseteq G_2 \supseteq \cdots \supseteq G_t = \{1\},$$

*with each $G_{i+1} \lhd G_i$ and $G_i/G_{i+1}$ cyclic of prime order $p_{i+1}$ or $\{1\}$.*

**Proof.** For each $i$, define $G_i = \mathrm{Gal}(K_t/K_i)$. It is clear that

$$\mathrm{Gal}(K_t/k) = G_0 \supseteq G_1 \supseteq G_2 \supseteq \cdots \supseteq G_t = \{1\}$$

is a sequence of subgroups. Now $K_1 = k(u)$, where $u^{p_1} \in k$; since $k$ contains all the $p_1$th roots of unity, Example 3.15(ii) says that $K_1/k$ is a splitting field of the polynomial $f(x) = x^{p_1} - u^{p_1}$. Theorem 3.17 now applies: $G_1 = \mathrm{Gal}(K_t/K_1)$ is a normal subgroup of $G_0 = \mathrm{Gal}(K_t/k)$ and $G_0/G_1 \cong \mathrm{Gal}(K_1/k)$. Now Example 3.15(ii) also says that if characteristic $k \neq p_1$, then $f(x)$ is separable. By Theorem 3.7(ii), $G_0/G_1 \cong \mathbb{I}_{p_1}$. If characteristic $k = p_1$, then Example 3.8 shows that $G_0/G_1 \cong \mathrm{Gal}(K_1/k) = \{1\}$. This argument can be repeated for each $i$.   •

We have been led to the following definition.

**Definition.** A **normal series**[1] of a group $G$ is a sequence of subgroups

$$G = G_0 \supseteq G_1 \supseteq G_2 \supseteq \cdots \supseteq G_t = \{1\}$$

with each $G_{i+1}$ a normal subgroup of $G_i$; the **factor groups** of this series are the quotient groups

$$G_0/G_1, \; G_1/G_2, \; \ldots, \; G_{t-1}/G_t.$$

The **length** of this series is the number of nontrivial factor groups.

A finite group $G$ is called **solvable** if it has a normal series each of whose factor groups has prime order (a possibly infinite group $G$ is called *solvable* if it has a normal series with abelian factor groups. Exercise 4.39 on page 261 shows that the two definitions coincide when $G$ is finite).

In this language, Lemma 3.19 says that $\mathrm{Gal}(K_t/k)$ is a solvable group if $K_t$ is a radical extension of $k$ and $k$ contains appropriate roots of unity.

---

[1]This terminology is not quite standard. We know that normality is not transitive; that is, if $H \subseteq K$ are subgroups of a group $G$, then $H \lhd K$ and $K \lhd G$ do not force $H \lhd G$. A subgroup $H \subseteq G$ is called a **subnormal subgroup** if there is a chain $G = G_0 \supseteq G_1 \supseteq \cdots \supseteq G_t = H$ with $G_i \lhd G_{i-1}$ for all $i \geq 1$. Normal series as defined in the text are called **subnormal series** by some authors; they reserve the name *normal series* for those series in which each $G_i$ is a normal subgroup of the big group $G$.

**Example 3.20.**

(i) By Exercise 1.93(ii) on page 74, every finite abelian group $G$ has a (necessarily normal) subgroup of prime index. It follows, by induction on $|G|$, that every finite abelian group is solvable.

(ii) Let us see that $S_4$ is a solvable group. Consider the chain of subgroups

$$S_4 \supseteq A_4 \supseteq \mathbf{V} \supseteq W \supseteq \{1\},$$

where $\mathbf{V}$ is the four-group and $W$ is any subgroup of $\mathbf{V}$ of order 2. Note, since $\mathbf{V}$ is abelian, that $W$ is a normal subgroup of $\mathbf{V}$. Now $|S_4/A_4| = |S_4|/|A_4| = 24/12 = 2$, $|A_4/\mathbf{V}| = |A_4|/|\mathbf{V}| = 12/4 = 3$, $|\mathbf{V}/W| = |\mathbf{V}|/|W| = 4/2 = 2$, and $|W/\{1\}| = |W| = 2$. Since each factor group has prime order, $S_4$ is solvable. In Example 3.24, we shall see that $S_5$ is not a solvable group.

(iii) A nonabelian simple group $G$, for example, $G = A_5$, is not solvable, for its only proper normal subgroup is $\{1\}$, and $G/\{1\} \cong G$ is not cyclic of prime order. ◄

The awkward hypothesis about roots of unity in the next lemma will soon be removed.

**Lemma 3.21.** *Let $k$ be a field, let $f(x) \in k[x]$ be solvable by radicals, and let $k = K_0 \subseteq K_1 \subseteq \cdots \subseteq K_t$ be a tower with $K_i/K_{i-1}$ a pure extension of prime type $p_i$ for all $i$. If $K_t$ contains a splitting field $E$ of $f(x)$ and $k$ contains all the $p_i$th roots of unity, then the Galois group $\mathrm{Gal}(E/k)$ is a quotient of a solvable group.*

**Proof.** By Lemma 3.18, we may assume that $K_t$ is a normal extension of $k$. The hypothesis on $k$ allows us to apply Lemma 3.19 to see that $\mathrm{Gal}(K_t/k)$ is a solvable group. Since $E$ and $K_t$ are splitting fields over $k$, Theorem 3.17 shows that $\mathrm{Gal}(K_t/E) \triangleleft \mathrm{Gal}(K_t/k)$ and $\mathrm{Gal}(K_t/k)/\mathrm{Gal}(K_t/E) \cong \mathrm{Gal}(E/k)$, as desired. •

**Proposition 3.22.** *Every quotient of a solvable group $G$ is itself a solvable group.*

**Proof.** Let $G = G_0 \supseteq G_1 \supseteq G_2 \supseteq \cdots \supseteq G_t = \{1\}$ be a sequence of subgroups as in the definition of solvable group. If $N \triangleleft G$, we must show that $G/N$ is solvable. Now $G_i N$ is a subgroup of $G$ for all $i$, and so there is a sequence of subgroups

$$G = G_0 N \supseteq G_1 N \supseteq \cdots \supseteq G_t N = N \supseteq \{1\}.$$

To see that this is a normal series, we claim, with obvious notation, that

$$(g_i n)G_{i+1}N(g_i n)^{-1} \subseteq g_i G_{i+1} N g_i^{-1} = g_i G_{i+1} g_i^{-1} N \subseteq G_{i+1}N.$$

The first inclusion holds because $n(G_{i+1}N)n^{-1} \subseteq NG_{i+1}N \subseteq (G_{i+1}N)(G_{i+1}N) = G_{i+1}N$ (for $G_{i+1}N$ is a subgroup). The equality holds because $Ng_i^{-1} = g_i^{-1}N$ (for $N \triangleleft G$, and so its right cosets coincide with its left cosets). The last inclusion holds because $G_{i+1} \triangleleft G_i$.

The Second Isomorphism Theorem gives

$$\frac{G_i}{G_i \cap (G_{i+1}N)} \cong \frac{G_i(G_{i+1}N)}{G_{i+1}N} = \frac{G_i N}{G_{i+1}N},$$

the last equation holding because $G_i G_{i+1} = G_i$. Since $G_{i+1} \lhd G_i \cap G_{i+1} N$, the Third Isomorphism Theorem gives a surjection $G_i/G_{i+1} \to G_i/[G_i \cap G_{i+1} N]$, and so the composite is a surjection $G_i/G_{i+1} \to G_i N/G_{i+1} N$. As $G_i/G_{i+1}$ is cyclic of prime order, its image is either cyclic of prime order or trivial. Therefore, $G/N$ is a solvable group. $\quad \bullet$

**Proposition 3.23.** *Every subgroup $H$ of a solvable group $G$ is solvable.*

**Proof.** Since $G$ is solvable, there is a sequence of subgroups

$$G = G_0 \supseteq G_1 \supseteq \cdots \supseteq G_t = \{1\}$$

with $G_i$ normal in $G_{i-1}$ and $G_{i-1}/G_i$ cyclic for all $i$. Consider the sequence of subgroups

$$H = H \cap G_0 \supseteq H \cap G_1 \supseteq \cdots \supseteq H \cap G_t = \{1\}.$$

This is a normal series: if $h_{i+1} \in H \cap G_{i+1}$ and $g_i \in H \cap G_i$, then $g_i h_{i+1} g_i^{-1} \in H$, for $g_i, h_{i+1} \in H$; also, $g_i h_{i+1} g_i^{-1} \in G_{i+1}$ because $G_{i+1}$ is normal in $G_i$. Therefore, $g_i h_{i+1} g_i^{-1} \in H \cap G_{i+1}$, and so $H \cap G_{i+1} \lhd H \cap G_i$. Finally, the Second Isomorphism Theorem gives

$$(H \cap G_i)/(H \cap G_{i+1}) = (H \cap G_i)/[(H \cap G_i) \cap G_{i+1}]$$
$$\cong G_{i+1}(H \cap G_i)/G_{i+1}.$$

But the last quotient group is a subgroup of $G_i/G_{i+1}$. Since the only subgroups of a cyclic group $C$ of prime order are $C$ and $\{1\}$, it follows that the nontrivial factor groups $(H \cap G_i)/(H \cap G_{i+1})$ are cyclic of prime order. Therefore, $H$ is a solvable group. $\quad \bullet$

**Example 3.24.** In Example 3.20(ii), we showed that $S_4$ is a solvable group. On the other hand, if $n \geq 5$, then the symmetric group $S_n$ is not solvable. Otherwise, each of its subgroups would also be solvable. But $A_5 \subseteq S_5 \subseteq S_n$, and the simple group $A_5$ is not solvable, by Example 3.20(iii). $\quad \blacktriangleleft$

**Proposition 3.25.** *If $H \lhd G$ and both $H$ and $G/H$ are solvable groups, then $G$ is solvable.*

**Proof.** Since $G/H$ is solvable, there is a normal series

$$G/H \supseteq K_1^* \supseteq K_2^* \supseteq \cdots \supseteq K_m^* = \{1\}$$

having factor groups of prime order. By the Correspondence Theorem for Groups, there are subgroups $K_i$ of $G$,

$$G \supseteq K_1 \supseteq K_2 \supseteq \cdots \supseteq K_m = H,$$

with $K_i/H = K_i^*$ and $K_{i+1} \lhd K_i$ for all $i$. By the Third Isomorphism Theorem,

$$K_i^*/K_{i+1}^* \cong K_i/K_{i+1}$$

for all $i$, and so $K_i/K_{i+1}$ is cyclic of prime order for all $i$.

Since $H$ is solvable, there is a normal series

$$H = H_0 \supseteq H_1 \supseteq \cdots \supseteq H_q = \{1\}$$

having factor groups of prime order. Splice these two series together,

$$G \supseteq K_1 \supseteq \cdots \supseteq K_m = H_0 \supseteq H_1 \supseteq \cdots \supseteq H_q = \{1\},$$

to obtain a normal series of $G$ having factor groups of prime order (note that $H \lhd G$ implies $H_0 = H \lhd K_m$).  •

**Corollary 3.26.** *If $H$ and $K$ are solvable groups, then $H \times K$ is solvable.*

**Proof.** The result follows from Proposition 3.25 because $(H \times K)/H \cong K$.  •

We return to fields, for we can now give the main criterion that a polynomial be solvable by radicals.

**Theorem 3.27 (Galois).** *Let $f(x) \in k[x]$, where $k$ is a field, and let $E$ be a splitting field of $f(x)$ over $k$. If $f(x)$ is solvable by radicals, then its Galois group $\mathrm{Gal}(E/k)$ is a solvable group.*

**Remark.** The converse of this theorem is false if $k$ has characteristic $p > 0$ (Theorem 3.60), but it is true when $k$ has characteristic 0 (Corollary 3.57).  ◄

**Proof.** Let $p_1, \ldots, p_t$ be the types of the pure extensions occurring in the radical extension arising from $f(x)$ being solvable by radicals. Define $m$ to be the product of all these $p_i$, define $E^*$ to be a splitting field of $x^m - 1$ over $E$, and define $k^* = k(\Omega)$, where $\Omega$ is the set of all $m$th roots of unity in $E^*$. Now $E^*/k^*$ is a normal extension, for it is a splitting field of $f(x)$ over $k^*$, and so $\mathrm{Gal}(E^*/k^*)$ is solvable, by Lemma 3.21. Consider the tower $k \subseteq k^* \subseteq E^*$:

since $k^*/k$ is normal, Theorem 3.17 gives $\mathrm{Gal}(E^*/k^*) \lhd \mathrm{Gal}(E^*/k)$ and

$$\mathrm{Gal}(E^*/k)/\mathrm{Gal}(E^*/k^*) \cong \mathrm{Gal}(k^*/k).$$

Now $\mathrm{Gal}(E^*/k^*)$ is solvable, while $\mathrm{Gal}(k^*/k)$ is abelian, hence solvable, by Proposition 3.12; therefore, $\mathrm{Gal}(E^*/k)$ is solvable, by Proposition 3.25. Finally, we may use Theorem 3.17 once again, for the tower $k \subseteq E \subseteq E^*$ satisfies the hypothesis that both $E$ and $E^*$ are normal [$E^*$ is a splitting field of $(x^m - 1)f(x)$]. It follows that $\mathrm{Gal}(E^*/k)/\mathrm{Gal}(E^*/E) \cong \mathrm{Gal}(E/k)$, and so $\mathrm{Gal}(E/k)$, being a quotient of a solvable group, is solvable.  •

Recall that if $k$ is a field and $E = k(y_1, \ldots, y_n) = \mathrm{Frac}(k[y_1, \ldots, y_n])$ is the field of rational functions, then the *general polynomial of degree $n$ over $k$* is

$$(x - y_1)(x - y_2) \cdots (x - y_n).$$

Galois's Theorem is strong enough to prove that there is no generalization of the quadratic formula for the general quintic polynomial.

**Theorem 3.28 (Abel–Ruffini).** *If $n \geq 5$, the general polynomial*

$$f(x) = (x - y_1)(x - y_2) \cdots (x - y_n)$$

*over a field $k$ is not solvable by radicals.*

**Proof.** In Example 2.150, we saw that if $E = k(y_1, \ldots, y_n)$ is the field of all rational functions in $n$ variables with coefficients in a field $k$, and if $F = k(a_0, \ldots, a_{n-1})$, where the $a_i$ are the coefficients of $f(x)$, then $E$ is the splitting field of $f(x)$ over $F$.

We claim that $\mathrm{Gal}(E/F) \cong S_n$. Recall Exercise 2.37 on page 101: if $A$ and $R$ are domains and $\varphi \colon A \to R$ is an isomorphism, then $a/b \mapsto \varphi(a)/\varphi(b)$ is an isomorphism $\mathrm{Frac}(A) \to \mathrm{Frac}(R)$. Now if $\sigma \in S_n$, then Theorem 2.25 gives an automorphism $\widetilde{\sigma}$ of $k[y_1, \ldots, y_n]$, defined by $\widetilde{\sigma} \colon f(y_1, \ldots, y_n) \mapsto f(y_{\sigma 1}, \ldots, y_{\sigma n})$; that is, $\widetilde{\sigma}$ just permutes the variables. Thus, $\widetilde{\sigma}$ extends to an automorphism $\sigma^*$ of $E = \mathrm{Frac}(k[y_1, \ldots, y_n])$, and Equations (1) on page 173 show that $\sigma^*$ fixes $F$; hence, $\sigma^* \in \mathrm{Gal}(E/F)$. Using Lemma 3.2, it is easy to see that $\sigma \mapsto \sigma^*$ is an injection $S_n \to \mathrm{Gal}(E/F)$, so that $|S_n| \leq |\mathrm{Gal}(E/F)|$. On the other hand, Theorem 3.3 shows that $\mathrm{Gal}(E/F)$ can be imbedded in $S_n$, giving the reverse inequality $|\mathrm{Gal}(E/F)| \leq |S_n|$. Therefore, $\mathrm{Gal}(E/F) \cong S_n$. But $S_n$ is not a solvable group if $n \geq 5$, by Example 3.24, and so Theorem 3.27 shows that $f(x)$ is not solvable by radicals. $\quad \bullet$

Some quintics in $\mathbb{Q}[x]$ are solvable by radicals; for example, Example 3.15 says that $x^5 - 1$ is solvable by radicals. Here is an explicit example of a quintic polynomial in $\mathbb{Q}[x]$ that is not solvable by radicals.

**Corollary 3.29.** $f(x) = x^5 - 4x + 2 \in \mathbb{Q}[x]$ *is not solvable by radicals.*

**Proof.** By Eisenstein's criterion (Theorem 2.75), $f(x)$ is irreducible over $\mathbb{Q}$. We now use some Calculus. There are exactly two real roots of the derivative $f'(x) = 5x^4 - 4$, namely, $\pm\sqrt[4]{4/5} \sim \pm .946$, and so $f(x)$ has two critical points. Now $f(\sqrt[4]{4/5}) < 0$ and $f(-\sqrt[4]{4/5}) > 0$, so that $f(x)$ has one relative maximum and one relative minimum. It follows easily that $f(x)$ has exactly three real roots.

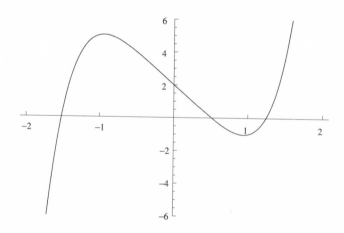

**Figure 3.1.** $f(x) = x^5 - 4x + 2$.

Let $E/\mathbb{Q}$ be the splitting field of $f(x)$ contained in $\mathbb{C}$. The restriction of complex conjugation to $E$, call it $\tau$, interchanges the two complex roots while it fixes the three real roots. Thus, if $X$ is the set of five roots of $f(x)$, then $\tau$ is a transposition in $S_X$. The Galois group $\text{Gal}(E/\mathbb{Q})$ of $f(x)$ is isomorphic to a subgroup $G \subseteq S_X$. Corollary 3.9 gives $|G| = [E : \mathbb{Q}]$ divisible by 5, so that $G$ contains an element $\sigma$ of order 5, by Cauchy's Theorem (Theorem 1.111). Now $\sigma$ must be a 5-cycle, for the only elements of order 5 in $S_X \cong S_5$ are 5-cycles. But Exercise 1.103 on page 75 says that $S_5$ is generated by any transposition and any 5-cycle. Since $G \supseteq \langle \sigma, \tau \rangle$, we have $G = S_X$. By Example 3.24, $\text{Gal}(E/\mathbb{Q}) \cong S_5$ is not a solvable group, and Theorem 3.27 says that $f(x)$ is not solvable by radicals. $\bullet$

## Exercises

∗ **3.1.** Prove that solvability by radicals does not depend on the choice of splitting field: if $E/k$ and $E'/k$ are splitting fields of $f(x) \in k[x]$ and there is a radical extension $K_t/k$ with $E \subseteq K_t$, prove that there is a radical extension $K'_r/k$ with $E' \subseteq K'_r$.

∗ **3.2.** Let $f(x) \in E[x]$ be monic, where $E$ is a field, and let $\sigma \colon E \to E$ be an automorphism. If $f(x)$ splits and $\sigma$ fixes every root of $f(x)$, prove that $\sigma$ fixes every coefficient of $f(x)$.

∗ **3.3.** **(Accessory Irrationalities)** Let $E/k$ be a splitting field of $f(x) \in k[x]$ with Galois group $G = \text{Gal}(E/k)$. Prove that if $k^*/k$ is an extension field and $E^*$ is a splitting field of $f(x)$ over $k^*$, then $\sigma \mapsto \sigma|E$ is an injective homomorphism $\text{Gal}(E^*/k^*) \to \text{Gal}(E/k)$.

**Hint.** If $\sigma \in \text{Gal}(E^*/k^*)$, then $\sigma$ permutes the roots of $f(x)$, so that $\sigma|E \in \text{Gal}(E/k)$.

**3.4.** (i) Let $K/k$ be an extension field, and let $f(x) \in k[x]$ be a separable polynomial. Prove that $f(x)$ is a separable polynomial when viewed as a polynomial in $K[x]$.

(ii) Let $k$ be a field, and let $f(x), g(x) \in k[x]$. Prove that if both $f(x)$ and $g(x)$ are separable polynomials, then their product $f(x)g(x)$ is also a separable polynomial.

**3.5.** Let $k$ be a field and let $f(x) \in k[x]$ be a separable polynomial. If $E/k$ is a splitting field of $f(x)$, prove that every root of $f(x)$ in $E$ is a separable element over $k$.

**3.6.** (i) Let $K/k$ be an extension field that is a splitting field of a polynomial $f(x) \in k[x]$. If $p(x) \in k[x]$ is a monic irreducible polynomial with no repeated roots and

$$p(x) = g_1(x) \cdots g_r(x) \text{ in } K[x],$$

where the $g_i(x)$ are monic irreducible polynomials in $K[x]$, prove that all the $g_i(x)$ have the same degree. Conclude that $\deg(p) = r \deg(g_i)$.

**Hint.** In some splitting field $E/K$ of $p(x)f(x)$, let $\alpha$ be a root of $g_i(x)$ and $\beta$ be a root of $g_j(x)$, where $i \neq j$. There is an isomorphism $\varphi \colon k(\alpha) \to k(\beta)$ with $\varphi(\alpha) = \beta$, which fixes $k$ and which admits an extension to $\Phi \colon E \to E$. Show that $\Phi|K$ induces an automorphism of $K[x]$ taking $g_i(x)$ to $g_j(x)$.

(ii) Let $E/k$ be a finite extension field. Prove that $E/k$ is a normal extension if and only if every irreducible $p(x) \in k[x]$ having a root in $E$ splits in $E[x]$. (Compare with Theorem 3.36 which uses a separability hypothesis.)

**Hint.** Use part (i).

**3.7.** (i) Give an example of a group $G$ having a subnormal subgroup that is not a normal subgroup.

(ii) Give an example of a group $G$ having a subgroup that is not a subnormal subgroup.

\* **3.8.** Prove that a finite solvable group $G \neq \{1\}$ has a normal subgroup of index $p$ for some prime $p$. (Compare Exercise 1.93 on page 74.)

**3.9.** Prove that the following statements are equivalent for $f(x) = ax^2 + bx + c \in \mathbb{Q}[x]$.

   (i) $f(x)$ is irreducible in $\mathbb{Q}[x]$.

   (ii) $\sqrt{b^2 - 4ac}$ is not rational.

   (iii) $\mathrm{Gal}(\mathbb{Q}(\sqrt{b^2 - 4ac})/\mathbb{Q})$ has order 2.

\* **3.10.** Let $k$ be a field, let $f(x) \in k[x]$ be a polynomial of degree $p$, where $p$ is prime, and let $E/k$ be a splitting field of $f(x)$. Prove that if $\mathrm{Gal}(E/k) \cong \mathbb{I}_p$, then $f(x)$ is irreducible.
**Hint.** Show that $f(x)$ has no repeated roots.

\* **3.11.** Generalize Theorem 3.13: prove that if $E$ is a finite field and $k \subseteq E$ is a subfield, then $\mathrm{Gal}(E/k)$ is cyclic.

## Section 3.2. Fundamental Theorem of Galois Theory

Galois Theory analyzes the relation between an algebraic extension $E$ of a field $k$ and the subgroups of its Galois group $\mathrm{Gal}(E/k)$. This analysis will enable us to prove the converse of Galois's Theorem: if $k$ is a field of characteristic 0 and $f(x) \in k[x]$ has a solvable Galois group, then $f(x)$ is solvable by radicals; it will also yield a proof of the Fundamental Theorem of Algebra.

Let $E$ be a field and let $\mathrm{Aut}(E)$ be the group of all (field) automorphisms of $E$. If $k$ is any subfield of $E$, then the Galois group $\mathrm{Gal}(E/k)$ is a subgroup of $\mathrm{Aut}(E)$, and so it acts on $E$. We have already seen several theorems about Galois groups whose hypothesis involves a normal extension $E/k$. It turns out that the way to understand normal extensions $E/k$ is to examine them in the context of this action of $\mathrm{Gal}(E/k)$ on $E$ and separability.

What elements of $E$ are stabilized by every $\sigma$ in some subset $H$ of $\mathrm{Aut}(E)$?

**Definition.** If $E$ is a field and $H$ is a subset[2] of $\mathrm{Aut}(E)$, then the **fixed field** of $H$ is defined by

$$E^H = \{a \in E : \sigma(a) = a \text{ for all } \sigma \in H\}.$$

It is easy to see that if $\sigma \in \mathrm{Aut}(E)$, then $E^\sigma = \{a \in E : \sigma(a) = a\}$ is a subfield of $E$; it follows that $E^H$ is a subfield of $E$, for

$$E^H = \bigcap_{\sigma \in H} E^\sigma.$$

**Example 3.30.** If $k$ is a subfield of $E$ and $G = \mathrm{Gal}(E/k)$, then $k \subseteq E^G$, but this inclusion can be strict. For example, let $E = \mathbb{Q}(\sqrt[3]{2}) \subseteq \mathbb{R}$. If $\sigma \in G = \mathrm{Gal}(E/\mathbb{Q})$, then $\sigma$ must fix $\mathbb{Q}$, and so it permutes the roots of $f(x) = x^3 - 2$. But the other

---

[2]The most important instance of a fixed field $E^H$ arises when $H$ is a subgroup of $\mathrm{Aut}(E)$, but we will meet cases in which it is merely a subset; for example, $H = \{\sigma\}$.

two roots of $f(x)$ are not real, so that $\sigma(\sqrt[3]{2}) = \sqrt[3]{2}$. Lemma 3.2 gives $\sigma = 1_G$; that is, $E^G = E$. Note that $E$ is not a splitting field of $f(x)$. ◄

The proof of the following proposition is almost obvious.

**Proposition 3.31.** *If $E$ is a field, then the function from subsets of $\mathrm{Aut}(E)$ to subfields of $E$, given by $H \mapsto E^H$, is* **order-reversing***: if $H \subseteq L \subseteq \mathrm{Aut}(E)$, then $E^L \subseteq E^H$.*

**Proof.** If $a \in E^L$, then $\sigma(a) = a$ for all $\sigma \in L$. Since $H \subseteq L$, it follows, in particular, that $\sigma(a) = a$ for all $\sigma \in H$. Hence, $E^L \subseteq E^H$. •

Our immediate goal is to determine the degree $[E : E^G]$, where $G \subseteq \mathrm{Aut}(E)$. To this end, we introduce the notion of characters.

**Definition.** A **character**[3] of a group $G$ in a field $E$ is a (group) homomorphism $\sigma \colon G \to E^\times$, where $E^\times$ denotes the multiplicative group of nonzero elements of the field $E$.

If $\sigma \in \mathrm{Aut}(E)$, then its restriction $\sigma|E^\times \colon E^\times \to E^\times$ is a character in $E$. In particular, if $k$ is a subfield of $E$, then every $\sigma \in \mathrm{Gal}(E/k)$ gives a character in $E$.

**Definition.** Let $E$ be a field and let $G \subseteq \mathrm{Aut}(E)$. A list $\sigma_1, \ldots, \sigma_n$ of characters of $G$ in $E$ is **independent** if, whenever $\sum_i c_i \sigma_i(x) = 0$, for $c_1, \ldots, c_n \in E$ and all $x \in G$, then all the $c_i = 0$.

In Example 2.106(iii), we saw that the set $V$ of all the functions from a set $X$ to a field $E$ is a vector space over $E$: addition of functions is defined by

$$\sigma + \tau \colon x \mapsto \sigma(x) + \tau(x),$$

and scalar multiplication is defined, for $c \in E$, by

$$c\sigma \colon x \mapsto c\sigma(x).$$

Independence of characters, as just defined, is linear independence in the vector space $V$ when $X$ is the group $G$.

**Proposition 3.32 (Dedekind).** *Every list $\sigma_1, \ldots, \sigma_n$ of distinct characters of a group $G$ in a field $E$ is independent.*

**Proof.** The proof is by induction on $n \geq 1$. The base step $n = 1$ is true, for if $c\sigma(x) = 0$ for all $x \in G$, then either $c = 0$ or $\sigma(x) = 0$; but $\sigma(x) \neq 0$, because $\mathrm{im}\,\sigma \subseteq E^\times$.

Assume that $n > 1$; if the characters are not independent, there are $c_i \in E$, not all zero, with

(1) $$c_1\sigma_1(x) + \cdots + c_{n-1}\sigma_{n-1}(x) + c_n\sigma_n(x) = 0$$

---

[3]This definition gives a special case of *character* in Representation Theory: if $\sigma \colon G \to \mathrm{GL}(n, E)$ is a homomorphism, then its **character** $\chi_\sigma \colon G \to E$ is defined, for $x \in G$, by

$$\chi_\sigma(x) = \mathrm{tr}(\sigma(x)),$$

where the trace, $\mathrm{tr}(A)$, of an $n \times n$ matrix $A$ is the sum of its diagonal entries. If $n = 1$, then $\mathrm{GL}(1, E) = E^\times$ and $\chi_\sigma(x) = \sigma(x)$ is called a **linear character**.

for all $x \in G$. We may assume that all $c_i \neq 0$, for if some $c_i = 0$, then the inductive hypothesis can be invoked to reach a contradiction. Multiplying by $c_n^{-1}$ if necessary, we may assume that $c_n = 1$. Since $\sigma_n \neq \sigma_1$, there exists $y \in G$ with $\sigma_1(y) \neq \sigma_n(y)$. In Equation (1), replace $x$ by $yx$ to obtain

$$c_1 \sigma_1(y) \sigma_1(x) + \cdots + c_{n-1} \sigma_{n-1}(y) \sigma_{n-1}(x) + \sigma_n(y) \sigma_n(x) = 0,$$

for $\sigma_i(yx) = \sigma_i(y)\sigma_i(x)$. Now multiply this equation by $\sigma_n(y)^{-1}$ to obtain the equation

$$c_1 \sigma_n(y)^{-1} \sigma_1(y) \sigma_1(x) + \cdots + c_{n-1} \sigma_n(y)^{-1} \sigma_{n-1}(y) \sigma_{n-1}(x) + \sigma_n(x) = 0.$$

Subtract this last equation from Equation (1) to obtain a sum of $n - 1$ terms:

$$c_1 \big[1 - \sigma_n(y)^{-1} \sigma_1(y)\big] \sigma_1(x) + c_2 \big[1 - \sigma_n(y)^{-1} \sigma_2(y)\big] \sigma_2(x) + \cdots = 0.$$

By induction, each of the coefficients $c_i[1 - \sigma_n(y)^{-1}\sigma_i(y)] = 0$. Now $c_i \neq 0$, and so $\sigma_n(y)^{-1}\sigma_i(y) = 1$ for all $i < n$. In particular, $\sigma_n(y) = \sigma_1(y)$, contradicting the definition of $y$. •

**Lemma 3.33.** *If $G = \{\sigma_1, \ldots, \sigma_n\}$ is a set of $n$ distinct automorphisms of a field $E$, then*

$$[E : E^G] \geq n.$$

**Proof.** Suppose, on the contrary, that $[E : E^G] = r < n$, and let $\alpha_1, \ldots, \alpha_r$ be a basis of $E/E^G$. Consider the homogeneous linear system over $E$ of $r$ equations in $n$ unknowns:

$$\sigma_1(\alpha_1)x_1 + \cdots + \sigma_n(\alpha_1)x_n = 0$$
$$\sigma_1(\alpha_2)x_1 + \cdots + \sigma_n(\alpha_2)x_n = 0$$
$$\vdots \qquad \vdots \qquad \qquad \vdots$$
$$\sigma_1(\alpha_r)x_1 + \cdots + \sigma_n(\alpha_r)x_n = 0.$$

Since $r < n$, there are more unknowns than equations, and Corollary 2.104 gives a nontrivial solution $(c_1, \ldots, c_n)$ in $E^n$.

We are now going to show that $\sigma_1(\beta)c_1 + \cdots + \sigma_n(\beta)c_n = 0$ for every $\beta \in E^\times$, which will contradict the independence of the characters $\sigma_1|E^\times, \ldots, \sigma_n|E^\times$. Since $\alpha_1, \ldots, \alpha_r$ is a basis of $E$ over $E^G$, each $\beta \in E$ can be written

$$\beta = \sum b_i \alpha_i,$$

where $b_i \in E^G$. Multiply the $i$th row of the system by $\sigma_1(b_i)$ to obtain the system with $i$th row

$$\sigma_1(b_i)\sigma_1(\alpha_i)c_1 + \cdots + \sigma_1(b_i)\sigma_n(\alpha_i)c_n = 0.$$

But $\sigma_1(b_i) = b_i = \sigma_j(b_i)$ for all $i, j$, because $b_i \in E^G$. Thus, the system has $i$th row

$$\sigma_1(b_i\alpha_i)c_1 + \cdots + \sigma_n(b_i\alpha_i)c_n = 0.$$

Adding all the rows gives

$$\sigma_1(\beta)c_1 + \cdots + \sigma_n(\beta)c_n = 0,$$

contradicting the independence of the characters. •

**Proposition 3.34.** *If $G = \{\sigma_1, \ldots, \sigma_n\}$ is a subgroup of $\mathrm{Aut}(E)$, then*

$$[E : E^G] = |G|.$$

**Proof.** In light of Lemma 3.33, it suffices to prove that $[E : E^G] \leq |G|$. If, on the contrary, $[E : E^G] > n$, let $\omega_1, \ldots, \omega_{n+1}$ be a linearly independent list of vectors in $E$ over $E^G$. Consider the system of $n$ equations in $n + 1$ unknowns:

$$\sigma_1(\omega_1)x_1 + \cdots + \sigma_1(\omega_{n+1})x_{n+1} = 0$$

$$\vdots \qquad\qquad \vdots$$

$$\sigma_n(\omega_1)x_1 + \cdots + \sigma_n(\omega_{n+1})x_{n+1} = 0.$$

Corollary 2.104 gives nontrivial solutions over $E$, which we proceed to normalize. Choose a nontrivial solution $(\beta_1, \ldots, \beta_r, 0, \ldots, 0)$ having the smallest number $r$ of nonzero components (by reindexing the $\omega_i$, we may assume that all nonzero components come first). Note that $r \neq 1$, lest $\sigma_1(\omega_1)\beta_1 = 0$ imply $\beta_1 = 0$, contradicting $(\beta_1, 0, \ldots, 0)$ being nontrivial. Multiplying by its inverse if necessary, we may assume that $\beta_r = 1$. Not all $\beta_i \in E^G$, lest the row corresponding to $\sigma = 1_E$ violate the linear independence of $\omega_1, \ldots, \omega_{n+1}$. Our last assumption is that $\beta_1$ does not lie in $E^G$ (this, too, can be accomplished by reindexing the $\omega_i$); thus, there is some $\sigma_k$ with $\sigma_k(\beta_1) \neq \beta_1$. Since $\beta_r = 1$, the original system has $j$th row

$$(2) \qquad\qquad \sigma_j(\omega_1)\beta_1 + \cdots + \sigma_j(\omega_{r-1})\beta_{r-1} + \sigma_j(\omega_r) = 0.$$

Apply $\sigma_k$ to this system to obtain

$$\sigma_k\sigma_j(\omega_1)\sigma_k(\beta_1) + \cdots + \sigma_k\sigma_j(\omega_{r-1})\sigma_k(\beta_{r-1}) + \sigma_k\sigma_j(\omega_r) = 0.$$

Since $G$ is a group, $\sigma_k\sigma_1, \ldots, \sigma_k\sigma_n$ is just a permutation of $\sigma_1, \ldots, \sigma_n$. Setting $\sigma_k\sigma_j = \sigma_i$, the system has $i$th row

$$\sigma_i(\omega_1)\sigma_k(\beta_1) + \cdots + \sigma_i(\omega_{r-1})\sigma_k(\beta_{r-1}) + \sigma_i(\omega_r) = 0.$$

Subtract this from the $i$th row of Equation (2) to obtain a new system with $i$th row

$$\sigma_i(\omega_1)\big[\beta_1 - \sigma_k(\beta_1)\big] + \cdots + \sigma_i(\omega_{r-1})\big[\beta_{r-1} - \sigma_k(\beta_{r-1})\big] = 0.$$

Since $\beta_1 - \sigma_k(\beta_1) \neq 0$, we have found a nontrivial solution of the original system having fewer than $r$ nonzero components, a contradiction. •

These ideas give a result needed in the proof of the Fundamental Theorem of Galois Theory.

**Theorem 3.35.** *If $G$ and $H$ are finite subgroups of $\mathrm{Aut}(E)$ with $E^G = E^H$, then $G = H$.*

**Proof.** We first show that $\sigma \in \mathrm{Aut}(E)$ fixes $E^G$ if and only if $\sigma \in G$. Clearly, $\sigma$ fixes $E^G$ if $\sigma \in G$. Suppose, conversely, that $\sigma$ fixes $E^G$ but $\sigma \notin G$. If $|G| = n$, then

$$n = |G| = [E : E^G],$$

by Proposition 3.34. Since $\sigma$ fixes $E^G$, we have $E^G \subseteq E^{G \cup \{\sigma\}}$. But the reverse inequality always holds, by Proposition 3.31, so that $E^G = E^{G \cup \{\sigma\}}$. Hence,

$$n = [E : E^G] = [E : E^{G \cup \{\sigma\}}] \geq |G \cup \{\sigma\}| = n + 1,$$

by Lemma 3.33, a contradiction.

If $\sigma \in H$, then $\sigma$ fixes $E^H = E^G$, and hence $\sigma \in G$; that is, $H \subseteq G$; the reverse inclusion is proved the same way, and so $H = G$.  •

Here is the characterization we have been seeking. Recall that a normal extension is a splitting field of some polynomial; we now characterize splitting fields of separable polynomials.

**Theorem 3.36.** *If $E/k$ is a finite extension with Galois group $G = \mathrm{Gal}(E/k)$, then the following statements are equivalent.*

(i) *$E$ is a splitting field of some separable polynomial $f(x) \in k[x]$.*

(ii) *$k = E^G$.*

(iii) *If an irreducible $p(x) \in k[x]$ has a root in $E$, then it is separable and splits in $E[x]$.*

**Proof.**

(i) $\Rightarrow$ (ii) By Theorem 3.7(ii), $|G| = [E : k]$. But Proposition 3.34 gives $|G| = [E : E^G]$; hence,

$$[E : k] = [E : E^G].$$

Since $k \subseteq E^G$, we have $[E : k] = [E : E^G][E^G : k]$, so that $[E^G : k] = 1$ and $k = E^G$.

(ii) $\Rightarrow$ (iii) Let $p(x) \in k[x]$ be an irreducible polynomial having a root $\alpha$ in $E$, and let the distinct elements of the set $\{\sigma(\alpha) : \sigma \in G\}$ be $\alpha_1, \ldots, \alpha_n$. Define $g(x) \in E[x]$ by

$$g(x) = \prod (x - \alpha_i).$$

Now each $\sigma \in G$ permutes the $\alpha_i$, so that each $\sigma$ fixes each of the coefficients of $g(x)$ (for they are elementary symmetric functions of the roots); that is, the coefficients of $g(x)$ lie in $E^G = k$. Hence, $g(x)$ is a polynomial in $k[x]$ having no repeated roots. Now $p(x)$ and $g(x)$ have a common root in $E$, and so their gcd in $E[x]$ is not 1, by Corollary 2.59. Since $p(x)$ is irreducible, it must divide $g(x)$. Therefore, $p(x)$ has no repeated roots; that is, $p(x)$ is separable. Finally, $g(x) = p(x)$, for they are monic polynomials of the same degree having the same roots. Hence, $p(x)$ splits in $E[x]$.

(iii) $\Rightarrow$ (i) Choose $\alpha_1 \in E$ with $\alpha_1 \notin k$. Since $E/k$ is a finite extension, $\alpha_1$ must be algebraic over $k$; let $p_1(x) = \mathrm{irr}(\alpha_1, k) \in k[x]$ be its minimal polynomial. By hypothesis, $p_1(x)$ is a separable polynomial that splits over $E$; let $K_1 \subseteq E$ be its splitting field. If $K_1 = E$, we are done. Otherwise, choose $\alpha_2 \in E$ with $\alpha_2 \notin K_1$. By hypothesis, there is a separable irreducible $p_2(x) \in k[x]$ having $\alpha_2$ as a root. Let $K_2 \subseteq E$ be the splitting field of $p_1(x)p_2(x)$, a separable polynomial. If $K_2 = E$, we are done; otherwise, repeat this construction. This process must end with $K_m = E$ for some $m$ because $E/k$ is finite. Thus, $E$ is a splitting field of the separable polynomial $p_1(x) \cdots p_m(x)$.  •

**Definition.** A finite extension field $E/k$ is a ***Galois extension***[4] if it satisfies any of the equivalent conditions in Theorem 3.36.

**Example 3.37.** If $B/k$ is a finite separable extension and $E/B$ is the radical extension of $B$ constructed in Lemma 3.18, then Theorem 3.36(i) shows that $E/k$ is a Galois extension. ◄

**Corollary 3.38.** *If $E/k$ is a finite Galois extension and $B$ is an **intermediate field** (that is, a subfield $B$ with $k \subseteq B \subseteq E$), then $E/B$ is a Galois extension.*

**Proof.** We know that $E$ is a splitting field of some separable polynomial $f(x) \in k[x]$; that is, $E = k(\alpha_1, \ldots, \alpha_n)$, where $\alpha_1, \ldots, \alpha_n$ are the roots of $f(x)$. Since $k \subseteq B \subseteq E$, we have $E = B(\alpha_1, \ldots, \alpha_n)$, and $f(x) \in B[x]$. •

We do not say that if $E/k$ is a finite Galois extension and $B/k$ is an intermediate field, then $B/k$ is a Galois extension, for this may not be true. In Example 3.11(iii), we saw that $E = \mathbb{Q}(\sqrt[3]{2}, \omega)$ is a splitting field of $x^3 - 2$ over $\mathbb{Q}$, where $\omega$ is a primitive cube root of unity, and so it is a Galois extension. However, the intermediate field $B = \mathbb{Q}(\sqrt[3]{2})$ is not a Galois extension, for $x^3 - 2$ is an irreducible polynomial having a root in $B$, yet it does not split in $B[x]$.

The next proposition determines when an intermediate field $B$ is a Galois extension.

**Definition.** Let $E/k$ be a Galois extension and let $B$ be an intermediate field. A ***conjugate*** of $B$ is an intermediate field of the form

$$\sigma(B) = \{\sigma(b) : b \in B\}$$

for some $\sigma \in \mathrm{Gal}(E/k)$.

**Proposition 3.39.** *If $E/k$ is a finite Galois extension, then an intermediate field $B$ is a Galois extension of $k$ if and only if $B$ has no conjugates other than $B$ itself.*

**Proof.** Assume that $\sigma(B) = B$ for all $\sigma \in G$, where $G = \mathrm{Gal}(E/k)$. Let $p(x) \in k[x]$ be an irreducible polynomial having a root $\beta$ in $B$. Since $B \subseteq E$ and $E/k$ is Galois, $p(x)$ is a separable polynomial and it splits in $E[x]$. If $\beta' \in E$ is another root of $p(x)$, there exists an isomorphism $\sigma \in G$ with $\sigma(\beta) = \beta'$ (for $G$ acts transitively on the roots of an irreducible polynomial, by Proposition 3.14). Therefore, $\beta' = \sigma(\beta) \in \sigma(B) = B$, so that $p(x)$ splits in $B[x]$. Therefore, $B/k$ is a Galois extension.

The converse follows from Theorem 3.17: since $B/k$ is a splitting field of some (separable) polynomial $f(x)$ over $k$, it is a normal extension. •

In Example 2.150, we considered $E = k(y_1, \ldots, y_n)$, the rational function field in $n$ variables with coefficients in a field $k$, and its subfield $K = k(a_0, \ldots, a_{n-1})$, where

$$f(x) = (x - y_1)(x - y_2) \cdots (x - y_n) = a_0 + a_1 x + \cdots + a_{n-1} x^{n-1} + x^n$$

is the general polynomial of degree $n$ over $k$. We saw that $E$ is a splitting field of $f(x)$ over $K$, for it arises from $K$ by adjoining to it all the roots of $f(x)$, namely,

---

[4] Infinite extension fields may be Galois extensions; we shall define them in Chapter 6.

$Y = \{y_1, \ldots, y_n\}$. Since every permutation of $Y$ extends to an automorphism of $E$, by Theorem 2.25, we may regard $S_n$ as a subgroup of $\operatorname{Aut}(E)$. The elements of $K$ are called the *symmetric functions* in $n$ variables over $k$.

**Definition.** A rational function $g(y_1, \ldots, y_n)/h(y_1, \ldots, y_n) \in k(y_1, \ldots, y_n)$ is a ***symmetric function*** if it is unchanged by permuting its variables: for every $\sigma \in S_n$, we have $g(y_{\sigma 1}, \ldots, y_{\sigma n})/h(y_{\sigma 1}, \ldots, y_{\sigma n}) = g(y_1, \ldots, y_n)/h(y_1, \ldots, y_n)$.

The *elementary symmetric functions* are the *polynomials*, for $j = 1, \ldots, n$,

$$e_j(y_1, \ldots, y_n) = \sum_{i_1 < \cdots < i_j} y_{i_1} \cdots y_{i_j}.$$

We have seen that if $a_j$ is the $j$th coefficient of the general polynomial of degree $n$, then $a_j = (-1)^j e_{n-j}(y_1, \ldots, y_n)$. We now prove that $K = k(e_1, \ldots, e_n) = E^{S_n}$.

**Theorem 3.40 (Fundamental Theorem of Symmetric Functions).** *If $k$ is a field, every symmetric function in $k(y_1, \ldots, y_n)$ is a rational function in the elementary symmetric functions $e_1, \ldots, e_n$.*

**Proof.** Let $K = k(e_1, \ldots, e_n) \subseteq E = k(y_1, \ldots, y_n)$. As we saw in Example 2.150, $E$ is the splitting field of the general polynomial $f(x)$ of degree $n$:

$$f(x) = \prod_{i=1}^{n}(x - y_i).$$

As $f(x)$ is a separable polynomial, $E/K$ is a Galois extension. We saw, in the proof of the Abel–Ruffini Theorem, that $\operatorname{Gal}(E/K) \cong S_n$. Therefore, $E^{S_n} = K$, by Theorem 3.36. But $g(y_1, \ldots, y_n)/h(y_1, \ldots, y_n) \in E^{S_n}$ if and only if it is unchanged by permuting its variables; that is, it is a symmetric function.   •

There is a useful variation of Theorem 3.40. The ***Fundamental Theorem of Symmetric Polynomials*** says that every symmetric *polynomial* $f \in k[x_1, \ldots, x_n]$ lies in $k[e_1, \ldots, e_n]$; that is, $f$ is a polynomial (not merely a rational function) in the elementary symmetric functions. There is a proof of this in van der Waerden, *Modern Algebra* I, pp. 78–81, but we find it more natural to prove it, in Exercise 5.79 on page 379, using the Division Algorithm for polynomials in several variables.

**Definition.** If $A$ and $B$ are subfields of a field $E$, then their ***compositum***, denoted by $A \vee B$, is the intersection of all the subfields of $E$ containing $A \cup B$.

It is easy to see that $A \vee B$ is the smallest subfield of $E$ containing both $A$ and $B$. For example, if $E/k$ is an extension field with intermediate fields $A = k(\alpha_1, \ldots, \alpha_n)$ and $B = k(\beta_1, \ldots, \beta_m)$, then their compositum is

$$k(\alpha_1, \ldots, \alpha_n) \vee k(\beta_1, \ldots, \beta_m) = k(\alpha_1, \ldots, \alpha_n, \beta_1, \ldots, \beta_m).$$

**Proposition 3.41.**

(i) *Every finite Galois extension $E/k$ is separable.*

(ii) *If $E/k$ is a (not necessarily finite) algebraic extension and $S \subseteq E$ is a (possibly infinite)[5] set of separable elements, then $k(S)/k$ is separable.*

(iii) *Let $E/k$ be a (not necessarily finite) algebraic extension, where $k$ is a field, and let $A$ and $B$ be intermediate fields. If both $A/k$ and $B/k$ are separable, then their compositum $A \vee B$ is also a separable extension of $k$.*

**Proof.**

(i) If $\beta \in E$, then $p(x) = \text{irr}(\beta, k) \in k[x]$ is an irreducible polynomial in $k[x]$ having a root in $E$. By Theorem 3.36(iii), $p(x)$ is a separable polynomial (which splits in $E[x]$). Therefore, $\beta$ is separable over $k$, and $E/k$ is separable.

(ii) Let us first consider the case when $S$ is finite; that is, $B = k(\alpha_1, \ldots, \alpha_t)$ is a finite extension, where each $\alpha_i$ is separable over $k$. By Lemma 3.18(i), there is an extension field $E/B$ that is a splitting field of some separable polynomial $f(x) \in k[x]$; hence, $E/k$ is a Galois extension, by Theorem 3.36(i). By part (i), $E/k$ is separable; that is, for all $\alpha \in E$, the polynomial $\text{irr}(\alpha, k)$ has no repeated roots. In particular, $\text{irr}(\alpha, k)$ has no repeated roots for all $\alpha \in B$, and so $B/k$ is separable.

We now consider the general case. If $\alpha \in k(S)$, then Exercise 2.103 on page 171 says that there are finitely many elements $\alpha_1, \ldots, \alpha_n \in S$ with $\alpha \in B = k(\alpha_1, \ldots, \alpha_n)$. As we have just seen, $B/k$ is separable, and so $\alpha$ is separable over $k$. As $\alpha$ is an arbitrary element of $k(S)$, it follows that $k(S)/k$ is separable.

(iii) Apply part (ii) to the subset $S = A \cup B$, for $A \vee B = k(A \cup B)$. •

We are now going to show, when $E/k$ is a finite Galois extension, that the intermediate fields are classified by the subgroups of $\text{Gal}(E/k)$.

We begin with some general definitions.

**Definition.** A set $X$ is a ***partially ordered set*** if it has a binary relation $x \preceq y$ defined on it that satisfies, for all $x, y, z \in X$,

(i) *Reflexivity*: $x \preceq x$;

(ii) *Antisymmetry*: if $x \preceq y$, and $y \preceq x$, then $x = y$;

(iii) *Transitivity*: if $x \preceq y$ and $y \preceq z$, then $x \preceq z$.

An element $c$ in a partially ordered set $X$ is an ***upper bound*** of a pair $a, b \in X$ if $a \preceq c$ and $b \preceq c$; an element $d \in X$ is a ***least upper bound*** of $a, b$ if $d$ is an upper bound and $d \preceq c$ for every upper bound $c$ of $a$ and $b$. ***Lower bounds*** and ***greatest lower bounds*** are defined similarly, everywhere reversing the inequalities.

We shall return to partially ordered sets in Chapter 5 (when we discuss Zorn's Lemma) and Chapter 6 (when we discuss inverse and direct limits). Here, we are more interested in special partially ordered sets called *lattices*.

---

[5]Remember that transcendental elements are always separable, by definition. When $E/k$ is not algebraic, this statement is true if finitely many transcendental elements are adjoined, but it may be false if infinitely many transcendental elements are adjoined.

**Definition.** A *lattice* is a partially ordered set $\mathcal{L}$ in which every pair of elements $a, b \in \mathcal{L}$ has a greatest lower bound $a \wedge b$ and a least upper bound $a \vee b$.

**Example 3.42.**

(i) If $U$ is a set, define $\mathcal{L}$ to be the family of all the subsets of $U$, and define a partial order $A \preceq B$ by $A \subseteq B$. Then $\mathcal{L}$ is a lattice, where $A \wedge B = A \cap B$ and $A \vee B = A \cup B$.

(ii) If $G$ is a group, define $\mathcal{L} = \mathrm{Sub}(G)$ to be the family of all the subgroups of $G$, and define $A \preceq B$ to mean $A \subseteq B$; that is, $A$ is a subgroup of $B$. Then $\mathcal{L}$ is a lattice, where $A \wedge B = A \cap B$ and $A \vee B$ is the subgroup generated by $A \cup B$.

(iii) If $E/k$ is an extension field, define $\mathcal{L} = \mathrm{Int}(E/k)$ to be the family of all the intermediate fields, and define $K \preceq B$ to mean $K \subseteq B$; that is, $K$ is a subfield of $B$. Then $\mathcal{L}$ is a lattice, where $A \wedge B = A \cap B$ and $A \vee B$ is the compositum of $A$ and $B$.

(iv) If $n$ is a positive integer, define $\mathrm{Div}(n)$ to be the set of all the positive divisors of $n$. Then $\mathrm{Div}(n)$ is a partially ordered set if one defines $d \preceq d'$ to mean $d \mid d'$. Here, $d \wedge d' = \gcd(d, d')$ and $d \vee d' = \mathrm{lcm}(d, d')$. ◄

**Definition.** Let $\mathcal{L}$ and $\mathcal{L}'$ be partially ordered sets. A function $f : \mathcal{L} \to \mathcal{L}'$ is called *order-reversing* if $a \preceq b$ in $\mathcal{L}$ implies $f(b) \preceq f(a)$ in $\mathcal{L}'$.

**Example 3.43.** There exist lattices $\mathcal{L}$ and $\mathcal{L}'$ and an order-reversing bijection $\varphi : \mathcal{L} \to \mathcal{L}'$ whose inverse $\varphi^{-1} : \mathcal{L}' \to \mathcal{L}$ is not order-reversing. For example, consider the lattices

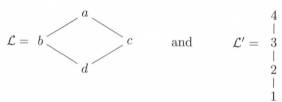

The bijection $\varphi : \mathcal{L} \to \mathcal{L}'$, defined by

$$\varphi(a) = 1, \quad \varphi(b) = 2, \quad \varphi(c) = 3, \quad \varphi(d) = 4,$$

is an order-reversing bijection, but its inverse $\varphi^{-1} : \mathcal{L}' \to \mathcal{L}$ is not order-reversing, because $2 \preceq 3$ but $c = \varphi^{-1}(3) \npreceq \varphi^{-1}(2) = b$. ◄

The De Morgan laws say that if $A$ and $B$ are subsets of a set $X$, then

$$(A \cap B)' = A' \cup B' \quad \text{and} \quad (A \cup B)' = A' \cap B',$$

where $A'$ denotes the complement of $A$. These identities are generalized in the next lemma.

**Lemma 3.44.** *Let $\mathcal{L}$ and $\mathcal{L}'$ be lattices, and let $\varphi : \mathcal{L} \to \mathcal{L}'$ be a bijection such that both $\varphi$ and $\varphi^{-1}$ are order-reversing. Then*

$$\varphi(a \wedge b) = \varphi(a) \vee \varphi(b) \quad \text{and} \quad \varphi(a \vee b) = \varphi(a) \wedge \varphi(b).$$

**Proof.** Since $a, b \preceq a \vee b$, we have $\varphi(a \vee b) \preceq \varphi(a), \varphi(b)$; that is, $\varphi(a \vee b)$ is a lower bound of $\varphi(a), \varphi(b)$. It follows that $\varphi(a \vee b) \preceq \varphi(a) \wedge \varphi(b)$.

For the reverse inequality, surjectivity of $\varphi$ gives $c \in \mathcal{L}$ with $\varphi(a) \wedge \varphi(b) = \varphi(c)$. Now $\varphi(c) = \varphi(a) \wedge \varphi(b) \preceq \varphi(a), \varphi(b)$. Applying $\varphi^{-1}$, which is also order-reversing, we have $a, b \preceq c$. Hence, $c$ is an upper bound of $a, b$, so that $a \vee b \preceq c$. Therefore, $\varphi(a \vee b) \succeq \varphi(c) = \varphi(a) \wedge \varphi(b)$. A similar argument proves the other half of the statement. •

Recall Example 3.42: if $G$ is a group, then $\mathrm{Sub}(G)$ is the lattice of all its subgroups and, if $E/k$ is an extension field, then $\mathrm{Int}(E/k)$ is the lattice of all the intermediate fields.

**Theorem 3.45 (Fundamental Theorem of Galois Theory).** *Let $E/k$ be a finite[6] Galois extension with Galois group $G = \mathrm{Gal}(E/k)$.*

(i) *The function $\gamma \colon \mathrm{Sub}(\mathrm{Gal}(E/k)) \to \mathrm{Int}(E/k)$, defined by*

$$\gamma \colon H \mapsto E^H,$$

*is an order-reversing bijection whose inverse, $\delta \colon \mathrm{Int}(E/k) \to \mathrm{Sub}(\mathrm{Gal}(E/k))$, is the order-reversing bijection*

$$\delta \colon B \mapsto \mathrm{Gal}(E/B).$$

(ii) *For every $B \in \mathrm{Int}(E/k)$ and $H \in \mathrm{Sub}(\mathrm{Gal}(E/k))$,*

$$E^{\mathrm{Gal}(E/B)} = B \quad and \quad \mathrm{Gal}(E/E^H) = H.$$

(iii) *For every $H, K \in \mathrm{Sub}(\mathrm{Gal}(E/k))$ and $A, B \in \mathrm{Int}(E/k)$,*

$$E^{H \vee K} = E^H \cap E^K;$$
$$E^{H \cap K} = E^H \vee E^K;$$
$$\mathrm{Gal}(E/(A \vee B)) = \mathrm{Gal}(E/A) \cap \mathrm{Gal}(E/B);$$
$$\mathrm{Gal}(E/(A \cap B)) = \mathrm{Gal}(E/A) \vee \mathrm{Gal}(E/B).$$

(iv) *For every $B \in \mathrm{Int}(E/k)$ and $H \in \mathrm{Sub}(\mathrm{Gal}(E/k))$,*

$$[B : k] = [G : \mathrm{Gal}(E/B)] \quad and \quad [G : H] = [E^H : k].$$

(v) *If $B \in \mathrm{Int}(E/k)$, then $B/k$ is a Galois extension if and only if $\mathrm{Gal}(E/B)$ is a normal subgroup of $G$.*

**Proof.**

(i) Proposition 3.31 proves that $\gamma$ is order-reversing, and it is also easy to prove that $\delta$ is order-reversing. Now injectivity of $\gamma$ is proved in Theorem 3.35, so that it suffices to prove that $\gamma\delta \colon \mathrm{Int}(E/k) \to \mathrm{Int}(E/k)$ is the identity;[7] it will follow that $\gamma$ is a bijection with inverse $\delta$. If $B$ is an intermediate field, then $\delta\gamma \colon B \mapsto E^{\mathrm{Gal}(E/B)}$. But $E/E^B$ is a Galois extension, by Corollary 3.38, and so $E^{\mathrm{Gal}(E/B)} = B$, by Theorem 3.36.

(ii) This is just the statement that $\gamma\delta$ and $\delta\gamma$ are identity functions.

---

[6]There is a generalization to infinite Galois extensions; see Theorem 6.175.

[7]If $f \colon X \to Y$ and $g \colon Y \to X$, then $gf = 1_X$ implies that $g$ is surjective and $f$ is injective.

(iii) These statements follow from Lemma 3.44.

(iv) By Theorem 3.7(ii) and the fact that $E/B$ is a Galois extension,

$$[B : k] = [E : k]/[E : B] = |G|/|\operatorname{Gal}(E/B)| = [G : \operatorname{Gal}(E/B)].$$

Thus, the degree of $B/k$ is the index of its Galois group in $G$. The second equation follows from this one; take $B = E^H$, noting that (ii) gives $\operatorname{Gal}(E/E^H) = H$:

$$[E^H : k] = [G : \operatorname{Gal}(E/E^H)] = [G : H].$$

(v) It follows from Theorem 3.17 that $\operatorname{Gal}(E/B) \lhd G$ when $B/k$ is a Galois extension (both $B/k$ and $E/k$ are normal extensions). For the converse, let $H = \operatorname{Gal}(E/B)$, and assume that $H \lhd G$. Now $E^H = E^{\operatorname{Gal}(E/B)} = B$, by (ii), and so it suffices to prove that $\sigma(E^H) = E^H$ for every $\sigma \in G$, by Proposition 3.39. Suppose now that $a \in E^H$; that is, $\eta(a) = a$ for all $\eta \in H$. If $\sigma \in G$, then we must show that $\eta(\sigma(a)) = \sigma(a)$ for all $\eta \in H$. Now $H \lhd G$ says that if $\eta \in H$ and $\sigma \in G$, then there is $\eta' \in H$ with $\eta\sigma = \sigma\eta'$ (of course, $\eta' = \sigma^{-1}\eta\sigma$). But

$$\eta\sigma(a) = \sigma\eta'(a) = \sigma(a),$$

because $\eta'(a) = a$, as desired. Therefore, $B/k = E^H/k$ is Galois.   •

**Example 3.46.** We use our discussion of $f(x) = x^3 - 2 \in \mathbb{Q}[x]$ in Example 3.16 to illustrate the Fundamental Theorem. The roots of $f(x)$ are $\alpha_1 = \beta$, $\alpha_2 = \omega\beta$, and $\alpha_3 = \omega^2\beta$, where $\beta = \sqrt[3]{2}$ and $\omega$ is a primitive cube root of unity. By Example 3.11(iii), the splitting field is $E = \mathbb{Q}(\beta, \omega)$ and $\operatorname{Gal}(E/\mathbb{Q}) \cong S_3$.

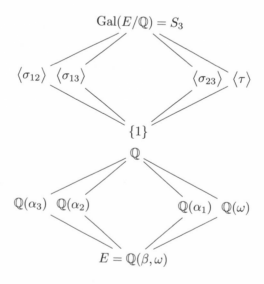

**Figure 3.2.** Sub($\operatorname{Gal}(E/\mathbb{Q})$) and Int($E/\mathbb{Q}$).

The top of Figure 3.2 shows the lattice of subgroups of $\operatorname{Gal}(E/\mathbb{Q})$: $\sigma_{ij}$ denotes the automorphism that interchanges $\alpha_i$, $\alpha_j$, where $i, j \in \{1, 2, 3\}$, and fixes the other root; $\tau$ denotes the automorphism sending $\alpha_1 \mapsto \alpha_2$, $\alpha_2 \mapsto \alpha_3$, and $\alpha_3 \mapsto \alpha_1$.

The bottom of Figure 3.2 shows the lattice of intermediate fields (without the Fundamental Theorem, it would not be obvious that these are the only such).

We compute fixed fields. If $\sigma = \sigma_{12}$, what is $E^{\langle \sigma \rangle}$? Now

$$\sigma(\alpha_1) = \sigma(\beta) = \omega\beta \quad \text{and} \quad \sigma(\alpha_2) = \sigma(\omega\beta) = \beta.$$

Hence,

$$\sigma(\alpha_2/\alpha_1) = \sigma(\omega\beta/\beta) = \sigma(\omega).$$

On the other hand,

$$\sigma(\alpha_2/\alpha_1) = \sigma(\alpha_2)/\sigma(\alpha_1) = \beta/\omega\beta = \omega^2.$$

Therefore, $\sigma(\omega) = \omega^2$, so that $\omega \notin E^{\langle \sigma \rangle}$. We conclude that $E^{\langle \sigma \rangle} = \mathbb{Q}(\alpha_3)$.

What is $E^{\langle \tau \rangle}$? We note that it contains no root $\alpha_i$. On the other hand,

$$\sigma(\omega) = \sigma(\alpha_2/\alpha_1) = \sigma(\alpha_2)/\sigma(\alpha_1) = \omega^2\beta/\omega\beta = \omega,$$

so that $\omega \in E^{\langle \tau \rangle}$. Thus, $E^{\langle \tau \rangle} = \mathbb{Q}(\omega)$. Note, as the Fundamental Theorem predicts, that $\mathbb{Q}(\omega)/\mathbb{Q}$ is a normal extension, for it corresponds to the normal subgroup $\langle \tau \rangle$ of $\mathrm{Gal}(E/\mathbb{Q})$; that is, $A_3 \lhd S_3$ [of course, $\mathbb{Q}(\omega)/\mathbb{Q}$ is the splitting field of $x^3 - 1$]. ◄

Here are some corollaries.

**Theorem 3.47.** *If $E/k$ is a finite Galois extension whose Galois group is abelian, then every intermediate field is a Galois extension.*

**Proof.** Every subgroup of an abelian group is a normal subgroup. •

**Corollary 3.48.** *A finite Galois extension $E/k$ has only finitely many intermediate fields.*

**Proof.** The finite group $\mathrm{Gal}(E/k)$ has only finitely many subgroups. •

**Definition.** An extension field $E/k$ is a ***simple extension*** if there is $u \in E$ with $E = k(u)$.

The following theorem characterizes simple extensions.

**Theorem 3.49 (Steinitz).** *A finite extension $E/k$ is simple if and only if it has only finitely many intermediate fields.*

**Proof.** Assume that $E/k$ is a simple extension, so that $E = k(u)$; let $p(x) = \mathrm{irr}(u, k) \in k[x]$ be its minimal polynomial. If $B$ is any intermediate field, let

$$q(x) = \mathrm{irr}(u, B) = b_0 + b_1 x + \cdots + b_{n-1}x^{n-1} + x^n \in B[x]$$

be the minimal polynomial of $u$ over $B$, and define

$$B' = k(b_0, \ldots, b_{n-1}) \subseteq B.$$

Note that $q(x)$ is an irreducible polynomial over the smaller field $B'$. Now

$$E = k(u) \subseteq B'(u) \subseteq B(u) \subseteq E,$$

so that $B'(u) = E = B(u)$. Hence, $[E : B] = [B(u) : B]$ and $[E : B'] = [B'(u) : B']$. But each of these is equal to $\deg(q)$, by Proposition 2.141(v), so that $[E : B] = \deg(q) = [E : B']$. Since $B' \subseteq B$, it follows that $[B : B'] = 1$; that is,

$$B = B' = k(b_0, \ldots, b_{n-1}).$$

We have characterized $B$ in terms of the coefficients of $q(x)$, a monic divisor of $p(x) = \mathrm{irr}(u, k)$ in $E[x]$. But $p(x)$ has only finitely many monic divisors, and hence there are only finitely many intermediate fields.

Conversely, assume that $E/k$ has only finitely many intermediate fields. If $k$ is a finite field, then we know that $E/k$ is a simple extension (take $u$ to be a primitive element); therefore, we may assume that $k$ is infinite. Since $E/k$ is a finite extension, there are elements $u_1, \ldots, u_n$ with $E = k(u_1, \ldots, u_n)$. By induction on $n \geq 1$, it suffices to prove that $E = k(u, v)$ is a simple extension. Now there are infinitely many elements $c \in E$ of the form $c = u + tv$, where $t \in k$, for $k$ is now infinite. Since there are only finitely many intermediate fields, there are, in particular, only finitely many fields of the form $k(c)$. By the Pigeonhole Principle, there exist distinct $t, t' \in k$ with $k(c) = k(c')$, where $c' = u + t'v$. Clearly, $k(c) \subseteq k(u, v)$. For the reverse inclusion, the field $k(c) = k(c')$ contains $c - c' = (t - t')v$, so that $v \in k(c)$ (because $t - t' \neq 0$). Hence, $u = c - tv \in k(c)$, and so $k(c) = k(u, v)$.   •

An immediate consequence is that every Galois extension is simple; in fact, even more is true.

**Theorem 3.50 (Theorem of the Primitive Element).** *If $B/k$ is a finite separable extension, then there is $u \in B$ with $B = k(u)$. In particular, if $k$ has characteristic 0, then every finite extension $B/k$ is a simple extension.*

**Proof.** By Example 3.37, the radical extension $E/k$ constructed in Lemma 3.18 is a Galois extension having $B$ as an intermediate field, so that Corollary 3.48 says that the extension $E/k$ has only finitely many intermediate fields. It follows at once that the extension field $B/k$ has only finitely many intermediate fields, and so Steinitz's Theorem says that $B/k$ has a primitive element.   •

The Theorem of the Primitive Element was known to Lagrange, and Galois used a modification of it to construct the original version of the Galois group.

We now turn to finite fields.

**Theorem 3.51.** *The finite field $\mathbb{F}_q$, where $q = p^n$, has exactly one subfield of order $p^d$ for every divisor $d$ of $n$, and no others.*

**Proof.** First, $\mathbb{F}_q/\mathbb{F}_p$ is a Galois extension, for it is a splitting field of the separable polynomial $x^q - x$ (all the roots of $x^q - x$ are distinct). Now $G = \mathrm{Gal}(\mathbb{F}_q/\mathbb{F}_p)$ is cyclic of order $n$, by Theorem 3.13. Since a cyclic group of order $n$ has exactly one subgroup of order $d$ for every divisor $d$ of $n$, by Lemma 1.92, it follows that $G$ has exactly one subgroup $H$ of index $n/d$. Therefore, there is only one intermediate field, namely, $E^H$, with $[E^H : \mathbb{F}_p] = [G : H] = n/d$, and $E^H = \mathbb{F}_{p^{n/d}}$.   •

The Fundamental Theorem of Algebra was first proved by Gauss in 1799. Here is an algebraic proof which uses the Fundamental Theorem of Galois Theory (as well as a Sylow Theorem we will prove in Chapter 4).

We assume only that $\mathbb{R}$ satisfies a weak form of the Intermediate Value Theorem: if $f(x) \in \mathbb{R}[x]$ and there exist $a, b \in \mathbb{R}$ such that $f(a) > 0$ and $f(b) < 0$, then $f(x)$ has a real root.

(i) *Every positive real number $r$ has a real square root.*

If $f(x) = x^2 - r$, then $f(1 + r) = (1 + r)^2 - r = 1 + r + r^2 > 0$, and $f(0) = -r < 0$.

(ii) *Every quadratic $g(x) \in \mathbb{C}[x]$ has a complex root.*

First, every complex number $z$ has a complex square root: when $z$ is written in polar form $z = re^{i\theta}$, where $r \geq 0$, then $\sqrt{z} = \sqrt{r}e^{i\theta/2}$. The quadratic formula gives the (complex) roots of $g(x)$.

(iii) *The field $\mathbb{C}$ has no extension fields of degree 2.*

Such an extension field would contain an element whose minimal polynomial is an irreducible quadratic in $\mathbb{C}[x]$; but item (ii) shows that no such polynomial exists.

(iv) *Every $f(x) \in \mathbb{R}[x]$ having odd degree has a real root.*

Let $f(x) = a_0 + a_1 x + \cdots + a_{n-1}x^{n-1} + x^n \in \mathbb{R}[x]$. Define $t = 1 + \sum |a_i|$. Now $|a_i| \leq t - 1$ for all $i$ and, if $h(x) = f(x) - x^n$, then $|h(t)| < t^n$:

$$|h(t)| = \left| a_0 + a_1 t + \cdots + a_{n-1}t^{n-1} \right|$$
$$\leq (t - 1)\left(1 + t + \cdots + t^{n-1}\right) = t^n - 1 < t^n.$$

Therefore, $-t^n < -|h(t)| \leq h(t)$ and $0 = -t^n + t^n < h(t) + t^n = f(t)$. A similar argument shows that $|h(-t)| < t^n$, so that

$$f(-t) = h(-t) + (-t)^n < t^n + (-t)^n.$$

When $n$ is odd, $(-t)^n = -t^n$, and so $f(-t) < t^n - t^n = 0$. Therefore, the Intermediate Value Theorem provides a real number $r$ with $f(r) = 0$; that is, $f(x)$ has a real root.

(v) *There is no extension field $E/\mathbb{R}$ of odd degree $> 1$.*

If $u \in E$, then its minimal polynomial $\mathrm{irr}(u, \mathbb{R})$ must have even degree, by item (iv), so that $[\mathbb{R}(u) : \mathbb{R}]$ is even. Hence $[E : \mathbb{R}] = [E : \mathbb{R}(u)][\mathbb{R}(u) : \mathbb{R}]$ is even.

**Theorem 3.52 (Fundamental Theorem of Algebra).** *Every nonconstant $f(x)$ in $\mathbb{C}[x]$ has a complex root.*

**Proof.** If $g(x) = \sum a_i x^i \in \mathbb{C}[x]$, define $\overline{g}(x) = \sum \overline{a}_i x^i$, where $\overline{a}_i$ is the complex conjugate of $a_i$. Now $g(x)\overline{g}(x) = \sum c_k x^k$, where $c_k = \sum_{i+j=k} a_i \overline{a}_j$; hence, $\overline{c}_k = c_k$ and $g(x)\overline{g}(x) \in \mathbb{R}[x]$. We claim that if $g(x)\overline{g}(x)$ has a (complex) root, say $z$, then $g(x)$ has a root. Since $g(z)\overline{g}(z) = 0$, either $g(z) = 0$ and $z$ is a root of $g(x)$, or $\overline{g}(z) = 0$. In the latter case, $z$ is a root of $\overline{g}(x)$, and so $\overline{z}$ is a root of $g(x)$. In either event, $g(x)$ has a root.

It suffices to prove that every nonconstant monic polynomial $f(x)$ with real coefficients has a complex root. Let $E/\mathbb{R}$ be a splitting field of $(x^2 + 1)f(x)$; of course, $\mathbb{C}$ is an intermediate field. Since $\mathbb{R}$ has characteristic 0, $E/\mathbb{R}$ is a Galois extension; let $G = \text{Gal}(E/\mathbb{R})$ be its Galois group. Now $|G| = 2^m \ell$, where $m \geq 0$ and $\ell$ is odd. By the Sylow Theorem (Theorem 4.39), $G$ has a subgroup $H$ of order $2^m$; let $B = E^H$ be the corresponding intermediate field. By the Fundamental Theorem of Galois Theory, the degree $[B : \mathbb{R}]$ is equal to the index $[G : H] = \ell$. But we have seen, in item (v), that $\mathbb{R}$ has no extension field of odd degree greater than 1; hence $\ell = 1$ and $G$ is a 2-group. Now $E/\mathbb{C}$ is also a Galois extension, and $\text{Gal}(E/\mathbb{C}) \subseteq G$ is also a 2-group. If this group is nontrivial, then it has a subgroup $K$ of index 2. By the Fundamental Theorem once again, the intermediate field $E^K$ is an extension field of $\mathbb{C}$ of degree 2, contradicting item (iii). We conclude that $[E : \mathbb{C}] = 1$; that is, $E = \mathbb{C}$. But $E$ is a splitting field of $f(x)$ over $\mathbb{C}$, and so $f(x)$ has a complex root.   •

We now prove the converse of Galois's Theorem (which holds only in characteristic 0): if the Galois group of a polynomial $f(x)$ is solvable, then $f(x)$ is solvable by radicals. In order to prove that certain extension fields are pure extensions, we will use the *norm*.

**Definition.** If $E/k$ is a Galois extension and $u \in E^\times$, the nonzero elements of $E$, define the **norm** of $u$ by

$$N(u) = \prod_{\sigma \in \text{Gal}(E/k)} \sigma(u).$$

If $E = \mathbb{Q}(i)$, then $\text{Gal}(E/\mathbb{Q}) = \langle \tau \rangle$, where $\tau \colon z \mapsto \overline{z}$ is complex conjugation. Here, $N(u) = z\overline{z}$, which is the norm that was used in the proof of Fermat's Two–Squares Theorem, Theorem 2.89. More generally, this norm coincides with the norm of an algebraic integer defined on page 163.

Here are some preliminary properties of the norm, whose simple proofs are left to the reader.

  (i) If $u \in E^\times$, then $N(u) \in k^\times$ (because $N(u) \in E^G = k$).
  (ii) $N(uv) = N(u)N(v)$, so that $N \colon E^\times \to k^\times$ is a homomorphism.
  (iii) If $a \in k^\times \subseteq E^\times$, then $N(a) = a^n$, where $n = [E : k]$.
  (iv) If $\sigma \in G$ and $u \in E^\times$, then $N(\sigma(u)) = N(u)$.

Given a homomorphism, we always ask about its kernel and image. The image of the norm is not easy to compute; the next result (which was the ninetieth theorem in Hilbert's 1897 exposition of Algebraic Number Theory) computes the kernel of the norm in a special case.

**Theorem 3.53 (Hilbert's Theorem 90).** *Let $E/k$ be a Galois extension whose Galois group $G = \text{Gal}(E/k)$ is cyclic of order $n$, say, with generator $\sigma$. If $u \in E^\times$, then $N(u) = 1$ if and only if there exists $v \in E^\times$ with $u = v\sigma(v)^{-1}$.*

**Proof.** If $u = v\sigma(v)^{-1}$, then

$$N(u) = N(v\sigma(v)^{-1}) = N(v)N(\sigma(v)^{-1}) = N(v)N(\sigma(v))^{-1} = N(v)N(v)^{-1} = 1.$$

Conversely, let $N(u) = 1$. Define "partial norms" in $E^\times$:

$$\delta_0 = u,$$
$$\delta_1 = u\sigma(u),$$
$$\delta_2 = u\sigma(u)\sigma^2(u),$$
$$\vdots$$
$$\delta_{n-1} = u\sigma(u) \cdots \sigma^{n-1}(u).$$

Note that $\delta_{n-1} = N(u) = 1$. It is easy to see that

(3) $$u\sigma(\delta_i) = \delta_{i+1} \text{ for all } 0 \le i \le n - 2.$$

By independence of the characters $1, \sigma, \sigma^2, \ldots, \sigma^{n-1}$, there exists $y \in E$ with

$$\delta_0 y + \delta_1 \sigma(y) + \cdots + \delta_{n-2}\sigma^{n-2}(y) + \sigma^{n-1}(y) \neq 0;$$

call this sum $z$. Using Equation (3), we easily check that

$$\sigma(z) = \sigma(\delta_0)\sigma(y) + \sigma(\delta_1)\sigma^2(y) + \cdots + \sigma(\delta_{n-2})\sigma^{n-1}(y) + \sigma^n(y)$$
$$= u^{-1}\delta_1\sigma(y) + u^{-1}\delta_2\sigma^2(y) + \cdots + u^{-1}\delta_{n-1}\sigma^{n-1}(y) + y$$
$$= u^{-1}\Big(\delta_1\sigma(y) + \delta_2\sigma^2(y) + \cdots + \delta_{n-1}\sigma^{n-1}(y)\Big) + u^{-1}\delta_0 y$$
$$= u^{-1}z. \quad \bullet$$

**Corollary 3.54.** *Let $E/k$ be a Galois extension of prime degree $p$. If $k$ contains a primitive $p$th root of unity $\omega$, then $E = k(z)$, where $z^p \in k$, and so $E/k$ is a pure extension of type $p$.*

**Proof.** The Galois group $G = \mathrm{Gal}(E/k)$ has order $p$, hence is cyclic; let $\sigma$ be a generator. Observe that $N(\omega) = \omega^p = 1$, because $\omega \in k$. By Hilbert's Theorem 90, we have $\omega = z\sigma(z)^{-1}$ for some $z \in E$. Hence $\sigma(z) = \omega^{-1}z$. Thus, $\sigma(z^p) = (\omega^{-1}z)^p = z^p$, and so $z^p \in E^G$, because $\sigma$ generates $G$; since $E/k$ is Galois, however, we have $E^G = k$, so that $z^p \in k$. Note that $z \notin k$, lest $\omega = 1$, so that $k(z) \neq k$ is an intermediate field. Therefore $E = k(z)$, because $[E : k] = p$ is prime, and hence $E$ has no proper intermediate fields. $\quad \bullet$

We confess that we have presented Hilbert's Theorem 90 not only because of its corollary, which will be used to prove Galois's Theorem, but also because it is a well-known result that is an early instance of Homological Algebra. Here is an elegant proof of Corollary 3.54 which does not use Hilbert's Theorem 90 (we warn the reader that it does use eigenvalues, a topic we have not yet introduced).

**Proposition 3.55 (= Corollary 3.54).** *Let $E/k$ be a Galois extension of prime degree $p$. If $k$ contains a primitive $p$th root of unity $\omega$, then $E = k(z)$, where $z^p \in k$, and so $E/k$ is a pure extension of type $p$.*

**Proof. (Houston)** Since $E/k$ is a Galois extension of degree $p$, its Galois group $G = \mathrm{Gal}(E/k)$ has order $p$, and hence it is cyclic: $G = \langle \sigma \rangle$. We view $\sigma \colon E \to E$ as a linear transformation. Now $\sigma$ satisfies the polynomial $x^p - 1$, because $\sigma^p = 1_E$, by Lagrange's Theorem. But $\sigma$ satisfies no polynomial of smaller degree, lest we contradict independence of the characters $1, \sigma, \sigma^2, \ldots, \sigma^{p-1}$. Therefore, $x^p - 1$ is the minimal polynomial of $\sigma$, and so every $p$th root of unity $\omega$ is an eigenvalue of $\sigma$. Since $\omega^{-1} \in k$, by hypothesis, there is some eigenvector $z \in E$ of $\sigma$ with $\sigma(z) = \omega^{-1}z$ (note that $z \notin k$ because it is not fixed by $\sigma$). Hence, $\sigma(z^p) = (\sigma(z))^p = (\omega^{-1})^p z^p = z^p$, from which it follows that $z^p \in E^G = k$. Now $p = [E : k] = [E : k(z)][k(z) : k]$; since $p$ is prime and $[k(z) : k] \neq 1$, we have $[E : k(z)] = 1$; that is, $E = k(z)$, and so $E/k$ is a pure extension. $\bullet$

**Theorem 3.56 (Galois).** *Let $k$ be a field of characteristic 0, let $E/k$ be a Galois extension, and let $G = \mathrm{Gal}(E/k)$ be a solvable group. Then $E$ can be imbedded in a radical extension of $k$.*

**Proof.** Since $G$ is solvable, Exercise 3.8 on page 192 says that it has a normal subgroup $H$ of prime index, say, $p$. Let $\omega$ be a primitive $p$th root of unity, which exists in some extension field because $k$ has characteristic 0.

Case (i): $\omega \in k$. We prove the statement by induction on $[E : k]$. The base step is obviously true, for $k = E$ is a radical extension of itself. For the inductive step, consider the intermediate field $E^H$. Now $E/E^H$ is a Galois extension, by Corollary 3.38, and $\mathrm{Gal}(E/E^H)$ is solvable, being a subgroup of the solvable group $G$. Since $[E : E^H] < [E : k]$, the inductive hypothesis gives a radical tower $E^H \subseteq R_1 \subseteq \cdots \subseteq R_t$, where $E \subseteq R_t$. Now $E^H/k$ is a Galois extension, for $H \triangleleft G$, and its index $[G : H] = p = [E^H : k]$, by the Fundamental Theorem. Corollary 3.54 now applies to give $E^H = k(z)$, where $z^p \in k$; that is, $E^H/k$ is a pure extension. Hence, the radical tower above can be lengthened by adding the prefix $k \subseteq E^H$, thus displaying $R_t/k$ as a radical extension.

Case (ii): General case. Let $k^* = k(\omega)$, and define $E^* = E(\omega)$. We claim that $E^*/k$ is a Galois extension. Since $E/k$ is a Galois extension, it is the splitting field of some separable $f(x) \in k[x]$, and so $E^*$ is a splitting field over $k$ of $f(x)(x^p - 1)$. But $x^p - 1$ is separable, because $k$ has characteristic 0, and so $E^*/k$ is a Galois extension. Therefore, $E^*/k^*$ is also a Galois extension, by Corollary 3.38. Let $G^* = \mathrm{Gal}(E^*/k^*)$. By Exercise 3.3 on page 191 (Accessory Irrationalities), there is an injection $\psi \colon G^* \to G = \mathrm{Gal}(E/k)$, so that $G^*$ is solvable, being isomorphic to a subgroup of a solvable group. Since $\omega \in k^*$, the first case says that there is a radical tower $k^* \subseteq R_1^* \subseteq \cdots \subseteq R_m^*$ with $E \subseteq E^* \subseteq R_m^*$. But $k^* = k(\omega)$ is a pure extension, so that this last radical tower can be lengthened by adding the prefix $k \subseteq k^*$, thus displaying $R_m^*/k$ as a radical extension. $\bullet$

**Corollary 3.57 (Galois).** *If $k$ is a field of characteristic 0 and $f(x) \in k[x]$, then $f(x)$ is solvable by radicals if and only if the Galois group of $f(x)$ is a solvable group.*

**Remark.** A counterexample in characteristic $p$ is given in Theorem 3.60. $\blacktriangleleft$

**Proof.** Let $E/k$ be a splitting field of $f(x)$ and let $G = \text{Gal}(E/k)$. Since $G$ is solvable, Theorem 3.56 says that there is a radical extension $R/k$ with $E \subseteq R$; that is, $f(x)$ is solvable by radicals. The converse is Theorem 3.27. •

We now have another proof of the existence of the classical formulas.

**Corollary 3.58.** *If $k$ has characteristic $0$, then every $f(x) \in k[x]$ with $\deg(f) \leq 4$ is solvable by radicals.*

**Proof.** If $G$ is the Galois group of $f(x)$, then $G$ is isomorphic to a subgroup of $S_4$. But $S_4$ is a solvable group, and so every subgroup of $S_4$ is also solvable. By Corollary 3.57, $f(x)$ is solvable by radicals. •

Suppose we know the Galois group $G$ of a polynomial $f(x) \in \mathbb{Q}[x]$ and that $G$ is solvable. Can we use this information to find the roots of $f(x)$? The answer is affirmative; we suggest the reader look at the book by Gaal, *Classical Galois Theory with Examples*, to see how this is done.

In 1827, Abel proved that if the Galois group of a polynomial $f(x)$ is commutative, then $f(x)$ is solvable by radicals (of course, Galois groups had not yet been defined). This result was superseded by Galois's Theorem, proved in 1830 (for abelian groups are solvable), but it is the reason why abelian groups are so called.

A deep theorem of Feit and Thompson (1963) says that every group of odd order is solvable. It follows that if $k$ is a field of characteristic $0$ and $f(x) \in k[x]$ is a polynomial whose Galois group has odd order or, equivalently, whose splitting field has odd degree over $k$, then $f(x)$ is solvable by radicals.

The next proposition gives an example showing that the converse of Galois's Theorem is false in prime characteristic.

**Lemma 3.59.** *The polynomial $f(x) = x^p - x - t \in \mathbb{F}_p[t]$ has no roots in $\mathbb{F}_p(t)$, the field of rational functions over $\mathbb{F}_p$.*

**Proof.** If there is a root $\alpha$ of $f(x)$ lying in $\mathbb{F}_p(t)$, then there are $g(t), h(t) \in \mathbb{F}_p[t]$ with $\alpha = g(t)/h(t)$; we may assume that $(g, h) = 1$. Since $\alpha$ is a root of $f(x)$, we have $(g/h)^p - (g/h) = t$; clearing denominators, there is an equation

$$g^p - h^{p-1}g = th^p$$

in $\mathbb{F}_p[t]$. Hence, $g \mid th^p$. Since $(g, h) = 1$, we have $g \mid t$, so that $g(t) = at$ or $g(t)$ is a constant, say, $g(t) = b$, where $a, b \in \mathbb{F}_p$. Transposing $h^{p-1}g$ in the displayed equation shows that $h \mid g^p$; but $(g, h) = 1$ forces $h$ to be a constant. We conclude that if $\alpha = g/h$, then $\alpha = at$ or $\alpha = b$. In the first case,

$$0 = \alpha^p - \alpha - t = (at)^p - (at) - t = a^p t^p - at - t$$
$$= at^p - at - t \quad \text{(by Fermat's Theorem in } \mathbb{F}_p)$$
$$= t(at^{p-1} - a - 1).$$

Hence, $at^{p-1} - a - 1 = 0$. But $a \neq 0$, and this contradicts $t$ being transcendental over $\mathbb{F}_p$. In the second case, $\alpha = b \in \mathbb{F}_p$. But $b$ is not a root of $f(x)$, for $f(b) = b^p - b - t = -t$, by Fermat's Theorem. Thus, no root $\alpha$ of $f(x)$ can lie in $\mathbb{F}_p(t)$. •

**Theorem 3.60.** Let $k = \mathbb{F}_p(t)$, where $p$ is prime. The Galois group of $f(x) = x^p - x - t$ over $k$ is cyclic of order $p$, but $f(x)$ is not solvable by radicals over $k$.

**Proof.** Let $\alpha$ be a root of $f(x)$. It is easy to see that the roots of $f(x)$ are $\alpha + i$, where $0 \leq i < p$, for Fermat's Theorem gives $i^p = i$ in $\mathbb{F}_p$, and so

$$f(\alpha + i) = (\alpha + i)^p - (\alpha + i) - t = \alpha^p + i^p - \alpha - i - t = \alpha^p - \alpha - t = 0.$$

It follows that $f(x)$ is a separable polynomial and that $k(\alpha)$ is a splitting field of $f(x)$ over $k$. We claim that $f(x)$ is irreducible in $k[x]$. Suppose that $f(x) = g(x)h(x)$, where

$$g(x) = x^d + c_{d-1}x^{d-1} + \cdots + c_0 \in k[x]$$

and $0 < d < \deg(f) = p$; then $g(x)$ is a product of $d$ factors of the form $\alpha + i$. Now $-c_{d-1} \in k$ is the sum of the roots: $-c_{d-1} = d\alpha + j$, where $j \in \mathbb{F}_p$, and so $d\alpha \in k$. Since $0 < d < p$, however, $d \neq 0$ in $k$, and this forces $\alpha \in k$, contradicting Lemma 3.59. Therefore, $f(x)$ is an irreducible polynomial in $k[x]$. Since $\deg(f) = p$, we have $[k(\alpha) : k] = p$ and, since $f(x)$ is separable, $|\operatorname{Gal}(k(\alpha)/k)| = [k(\alpha) : k] = p$. Therefore, $\operatorname{Gal}(k(\alpha)/k) \cong \mathbb{I}_p$.

It will be convenient to have certain roots of unity available. Let $\Omega$ be the set of all $q$th roots of unity, where $q < p$ is a prime divisor of $p!$. We claim that $\alpha \notin k(\Omega)$. On the one hand, if $n = \prod_{q<p} q$, then $\Omega$ is contained in the splitting field of $x^n - 1$, and so $[k(\Omega) : k] \mid n!$, by Theorem 3.3. It follows that $p \nmid [k(\Omega) : k]$. On the other hand, if $\alpha \in k(\Omega)$, then $k(\alpha) \subseteq k(\Omega)$ and $[k(\Omega) : k] = [k(\Omega) : k(\alpha)][k(\alpha) : k] = p[k(\Omega) : k(\alpha)]$. Hence, $p \mid [k(\Omega) : k]$, and this is a contradiction.

If $f(x)$ were solvable by radicals over $k(\Omega)$, there would be a radical extension

$$k(\Omega) = B_0 \subseteq B_1 \subseteq \cdots \subseteq B_r$$

with $k(\Omega, \alpha) \subseteq B_r$. We may assume, for each $i \geq 1$, that $B_i/B_{i-1}$ is of prime type; that is, $B_i = B_{i-1}(u_i)$, where $u_i^{q_i} \in B_{i-1}$ and $q_i$ is prime. There is some $j \geq 1$ with $\alpha \in B_j$ but $\alpha \notin B_{j-1}$. Simplifying notation, we set $u_j = u$, $q_j = q$, $B_{j-1} = B$, and $B_j = B'$. Thus, $B' = B(u)$, $u^q = b \in B$, $\alpha \in B'$, and $\alpha, u \notin B$. We claim that $f(x) = x^p - x - t$, which we know to be irreducible in $k[x]$, is also irreducible in $B[x]$. By Accessory Irrationalities [Exercise 3.3 on page 191], restriction gives an injection $\operatorname{Gal}(B(\alpha)/B) \to \operatorname{Gal}(k(\alpha)/k)) \cong \mathbb{I}_p$. If $\operatorname{Gal}(B(\alpha)/B) = \{1\}$, then $B(\alpha) = B$ and $\alpha \in B$, a contradiction. Therefore, $\operatorname{Gal}(B(\alpha)/B) \cong \mathbb{I}_p$, and $f(x)$ is irreducible in $B[x]$, by Exercise 3.10 on page 192.

Since $u \notin B'$ and $B$ contains all the $q$th roots of unity, Proposition 2.152 shows that $x^q - b$ is irreducible in $B[x]$, for it does not split in $B[x]$. Now $B' = B(u)$ is a splitting field of $x^q - b$, and so $[B' : B] = q$. We have $B \subsetneq B(\alpha) \subseteq B'$, and

$$q = [B' : B] = [B' : B(\alpha)][B(\alpha) : B].$$

Since $q$ is prime, $[B' : B(\alpha)] = 1$; that is, $B' = B(\alpha)$, and so $q = [B' : B]$. As $\alpha$ is a root of the irreducible polynomial $f(x) = x^p - x - t \in B[x]$, we have $[B(\alpha) : B] = p$; therefore, $q = p$. Now $B(u) = B' = B(\alpha)$ is a separable extension, by Proposition 3.41, for $\alpha$ is a separable element. It follows that $u \in B'$ is also a separable element, contradicting $\operatorname{irr}(u, B) = x^q - b = x^p - b = (x - u)^p$ having repeated roots.

We have shown that $f(x)$ is not solvable by radicals over $k(\Omega)$. It follows that $f(x)$ is not solvable by radicals over $k$, for if there were a radical extension $k = R_0 \subseteq R_1 \subseteq \cdots \subseteq R_t$ with $k(\alpha) \subseteq R_t$, then $k(\Omega) = R_0(\Omega) \subseteq R_1(\Omega) \subseteq \cdots \subseteq R_t(\Omega)$ would show that $f(x)$ is solvable by radicals over $k(\Omega)$, a contradiction. •

## Exercises

* **3.12.** Let $k$ be a field, let $f(x) \in k[x]$ be a separable polynomial, and let $E/k$ be a splitting field of $f(x)$. Assume further that there is a factorization $f(x) = g(x)h(x)$ in $k[x]$, and that $B/k$ and $C/k$ are intermediate fields that are splitting fields of $g(x)$ and $h(x)$, respectively.

(i) Prove that $\mathrm{Gal}(E/B), \mathrm{Gal}(E/C)$ are normal subgroups of $\mathrm{Gal}(E/k)$.

(ii) Prove that $\mathrm{Gal}(E/B) \cap \mathrm{Gal}(E/C) = \{1\}$.

(iii) If $B \cap C = k$, prove that $\mathrm{Gal}(E/B)\,\mathrm{Gal}(E/C) = \mathrm{Gal}(E/k)$. Use Proposition 1.86 and Theorem 3.17 to show, in this case, that

$$\mathrm{Gal}(E/k) \cong \mathrm{Gal}(B/k) \times \mathrm{Gal}(C/k).$$

[Note that $\mathrm{Gal}(B/k)$ is not a subgroup of $\mathrm{Gal}(E/k)$.]

(iv) Use (iii) to give another proof that $\mathrm{Gal}(E/\mathbb{Q}) \cong \mathbf{V}$, where $E = \mathbb{Q}(\sqrt{2} + \sqrt{3})$ [see Example 2.147 on page 163].

(v) Let $f(x) = (x^3 - 2)(x^3 - 3) \in \mathbb{Q}[x]$. If $B/\mathbb{Q}$ and $C/\mathbb{Q}$ are the splitting fields of $x^3 - 2$ and $x^3 - 3$ inside $\mathbb{C}$, prove that $\mathrm{Gal}(E/\mathbb{Q}) \not\cong \mathrm{Gal}(B/\mathbb{Q}) \times \mathrm{Gal}(C/\mathbb{Q})$, where $E$ is the splitting field of $f(x)$ contained in $\mathbb{C}$.

**3.13.** Let $k$ be a field of characteristic 0, and let $f(x) \in k[x]$ be a polynomial of degree 5 with splitting field $E/k$. Prove that $f(x)$ is solvable by radicals if and only if $[E : k] < 60$.

**3.14.** Let $E/k$ be a Galois extension with $\mathrm{Gal}(E/k)$ cyclic of order $n$. If $\varphi \colon \mathrm{Int}(E/k) \to \mathrm{Div}(n)$ is defined by $\varphi(L) = [L : k]$, prove that $\varphi$ is an order-preserving lattice isomorphism [see Example 3.42(iv)].

**3.15.** Use Theorem 3.51 to prove that $\mathbb{F}_{p^m}$ is a subfield of $\mathbb{F}_{p^n}$ if and only if $m \mid n$.

**3.16.** Find all finite fields $k$ whose subfields form a *chain*; that is, if $k'$ and $k''$ are subfields of $k$, then either $k' \subseteq k''$ or $k'' \subseteq k'$.

**3.17.** (i) Let $k$ be an infinite field, let $f(x) \in k[x]$ be a separable polynomial, and let $E = k(\alpha_1, \ldots, \alpha_n)$, where $\alpha_1, \ldots, \alpha_n$ are the roots of $f(x)$. Prove that there are $c_i \in k$ so that $E = k(\beta)$, where $\beta = c_1\alpha_1 + \cdots + c_n\alpha_n$.

**Hint.** Use the proof of Steinitz's Theorem.

(ii) **(Janusz)** Let $k$ be a finite field and let $k(\alpha, \beta)/k$ be finite. If $k(\alpha) \cap k(\beta) = k$, prove that $E = k(\alpha + \beta)$. (This result is false in general. For example, N. Boston used the computer algebra system MAGMA to show that there are primitive elements $\alpha$ of $\mathbb{F}_{2^6}$ and $\beta$ of $\mathbb{F}_{2^{10}}$ such that $\mathbb{F}_2(\alpha, \beta) = \mathbb{F}_{2^{30}}$ while $\mathbb{F}_2(\alpha + \beta) = \mathbb{F}_{2^{15}}$.)

**Hint.** Use Proposition 2.61(ii).

**3.18.** Let $E/k$ be a finite Galois extension with Galois group $G = \mathrm{Gal}(E/k)$. Define the *trace* $T: E \to E$ by

$$T(u) = \sum_{\sigma \in G} \sigma(u).$$

(i) Prove that $\mathrm{im}\, T \subseteq k$ and that $T(u + v) = T(u) + T(v)$ for all $u, v \in E$.

(ii) Use independence of characters to prove that $T$ is not identically zero.

**3.19.** Let $E/k$ be a Galois extension with $[E : k] = n$ and with cyclic Galois group $G = \mathrm{Gal}(E/k)$, say, $G = \langle \sigma \rangle$. Define $\tau = \sigma - 1_E$, and prove that $\mathrm{im}\, \tau = \ker T$, where $T: E \to E$ is the trace. Conclude, in this case, that the **Trace Theorem** is true:

$$\ker T = \{a \in E : a = \sigma(u) - u \text{ for some } u \in E\}.$$

**Hint.** Show that $\ker \tau = k$, so that $\dim(\mathrm{im}\, \tau) = n - 1 = \dim(\ker T)$.

**3.20.** Let $k$ be a field of characteristic $p > 0$, and let $E/k$ be a Galois extension having a cyclic Galois group $G = \langle \sigma \rangle$ of order $p$. Using the Trace Theorem, prove that there is an element $u \in E$ with $\sigma(u) - u = 1$. Prove that $E = k(u)$ and that there is $c \in k$ with $\mathrm{irr}(u, k) = x^p - x - c$. (This is an additive version of Hilbert's Theorem 90.)

## Section 3.3. Calculations of Galois Groups

The *discriminant* of a polynomial is useful for computing its Galois group.

**Definition.** If $f(x) = \prod_i (x - \alpha_i) \in k[x]$, where $k$ is a field, define

$$\Delta = \Delta(f) = \prod_{i<j} (\alpha_i - \alpha_j),$$

and define the **discriminant** to be

$$D = D(f) = \Delta^2 = \prod_{i<j} (\alpha_i - \alpha_j)^2.$$

The product $\Delta = \prod_{i<j}(\alpha_i - \alpha_j)$ has one factor $\alpha_i - \alpha_j$ for each distinct pair of indices $(i, j)$ (the inequality $i < j$ prevents a pair of indices from occurring twice). It is clear that $f(x)$ has repeated roots if and only if its discriminant $D(f) = 0$. If $E/k$ is a splitting field of $f(x)$, then each $\sigma \in \mathrm{Gal}(E/k)$ permutes the roots, and so $\sigma$ permutes all the distinct pairs. However, it may happen that $i < j$ while the subscripts involved in $\sigma(\alpha_i) - \sigma(\alpha_j)$ are in reverse order. For example, suppose the roots of a cubic are $\alpha_1, \alpha_2$, and $\alpha_3$. If there is $\sigma \in G$ with $\sigma(\alpha_1) = \alpha_2$, $\sigma(\alpha_2) = \alpha_1$, and $\sigma(\alpha_3) = \alpha_3$ [that is, $\sigma$ is a transposition], then

$$\sigma(\Delta) = \big(\sigma(\alpha_1) - \sigma(\alpha_2)\big)\big(\sigma(\alpha_1) - \sigma(\alpha_3)\big)\big(\sigma(\alpha_2) - \sigma(\alpha_3)\big)$$

$$= (\alpha_2 - \alpha_1)(\alpha_2 - \alpha_3)(\alpha_1 - \alpha_3) = -(\alpha_1 - \alpha_2)(\alpha_2 - \alpha_3)(\alpha_1 - \alpha_3) = -\Delta.$$

Each term $\alpha_i - \alpha_j$ occurs in $\sigma(\Delta)$, but with a possible sign change. We conclude, for all $\sigma \in \mathrm{Gal}(E/k)$, that $\sigma(\Delta) = \pm\Delta$. It is natural to consider $\Delta^2$ rather than $\Delta$, for $\Delta$ depends not only on the roots of $f(x)$, but also on the order in which they are listed, whereas $D = \Delta^2$ does not depend on the listing of the roots. For a connection between discriminants and the alternating group $A_n$, see Proposition 3.63. In fact, $\sigma(\Delta) = \mathrm{sgn}(\sigma)\Delta$.

**Proposition 3.61.** *If $f(x) \in k[x]$ is a separable polynomial, then its discriminant $D(f)$ lies in $k$.*

**Proof.** Let $E/k$ be a splitting field of $f(x)$; since $f(x)$ is separable, Theorem 3.36 applies to show that $E/k$ is a Galois extension. Each $\sigma \in \mathrm{Gal}(E/k)$ permutes the roots $u_1, \ldots, u_n$ of $f(x)$, and $\sigma(\Delta) = \pm\Delta$, as we have just seen. Therefore,

$$\sigma(D) = \sigma(\Delta^2) = \sigma(\Delta)^2 = (\pm\Delta)^2 = D,$$

so that $D \in E^G$. But $E/k$ is a Galois extension, so that $E^G = k$ and $D \in k$. $\bullet$

If $f(x) = x^2 + bx + c \in k[x]$, where $k$ is a field of characteristic $\neq 2$, then the quadratic formula gives the roots of $f(x)$:

$$\alpha = \tfrac{1}{2}\left(-b + \sqrt{b^2 - 4c}\right) \quad \text{and} \quad \beta = \tfrac{1}{2}\left(-b - \sqrt{b^2 - 4c}\right).$$

It follows that

$$D = \Delta^2 = (\alpha - \beta)^2 = b^2 - 4c.$$

If $f(x)$ is a cubic with roots $\alpha$, $\beta$, $\gamma$, then

$$D = \Delta^2 = (\alpha - \beta)^2 (\alpha - \gamma)^2 (\beta - \gamma)^2;$$

it is not obvious how to compute the discriminant $D$ from the coefficients of $f(x)$ (see Theorem 3.62(ii) below).

**Definition.** A polynomial $f(x) = x^n + c_{n-1}x^{n-1} + \cdots + c_0 \in k[x]$ is **reduced** if $c_{n-1} = 0$. If characteristic$(k) \nmid n$, then the **associated reduced polynomial** of $f(x)$ is

$$\widetilde{f}(x) = f(x - \tfrac{1}{n}c_{n-1}).$$

If $f(x) = x^n + c_{n-1}x^{n-1} + \cdots + c_0 \in k[x]$ and $\beta \in k$ is a root of $\widetilde{f}(x)$, then

$$0 = \widetilde{f}(\beta) = f(\beta - \tfrac{1}{n}c_{n-1}).$$

Hence, $\beta$ is a root of $\widetilde{f}(x)$ if and only if $\beta - \tfrac{1}{n}c_{n-1}$ is a root of $f(x)$.

**Theorem 3.62.** *Let $k$ be a field of characteristic $0$.*

(i) *A polynomial $f(x) \in k[x]$ and its associated reduced polynomial $\widetilde{f}(x)$ have the same discriminant: $D(f) = D(\widetilde{f})$.*

(ii) *The discriminant of a reduced cubic $\widetilde{f}(x) = x^3 + qx + r$ is*

$$D = D(\widetilde{f}) = -4q^3 - 27r^2.$$

**Proof.**

(i) If the roots of $f(x) = \sum c_i x^i$ are $\alpha_1, \ldots, \alpha_n$, then the roots of $\widetilde{f}(x)$ are $\beta_1, \ldots, \beta_n$, where $\beta_i = \alpha_i + \tfrac{1}{n}c_{n-1}$. Therefore, $\beta_i - \beta_j = \alpha_i - \alpha_j$ for all $i, j$,

$$\Delta(f) = \prod_{i<j}(\alpha_i - \alpha_j) = \prod_{i<j}(\beta_i - \beta_j) = \Delta(\widetilde{f}),$$

and so the discriminants, which are the squares of these, are equal.

(ii) The cubic formula gives the roots of $\widetilde{f}(x)$ as

$$\alpha = g + h, \quad \beta = \omega g + \omega^2 h, \quad \text{and} \quad \gamma = \omega^2 g + \omega h,$$

where $g = \left[\frac{1}{2}\left(-r + \sqrt{R}\right)\right]^{1/3}$, $h = -q/3g$, $R = r^2 + \frac{4}{27}q^3$, and $\omega$ is a cube root of unity. Because $\omega^3 = 1$, we have

$$\begin{aligned}
\alpha - \beta &= (g + h) - (\omega g + \omega^2 h) \\
&= (g - \omega^2 h) - (\omega g - h) \\
&= (g - \omega^2 h) - (g - \omega^2 h)\omega \\
&= (g - \omega^2 h)(1 - \omega).
\end{aligned}$$

Similar calculations give

$$\alpha - \gamma = (g + h) - (\omega^2 g + \omega h) = (g - \omega h)(1 - \omega^2)$$

and

$$\beta - \gamma = (\omega g + \omega^2 h) - (\omega^2 g + \omega h) = (g - h)\omega(1 - \omega).$$

It follows that

$$\Delta = (g - h)(g - \omega h)(g - \omega^2 h)\omega(1 - \omega^2)(1 - \omega)^2.$$

By Exercise 3.21 on page 221, we have $\omega(1 - \omega^2)(1 - \omega)^2 = 3i\sqrt{3}$; moreover, the identity

$$x^3 - 1 = (x - 1)(x - \omega)(x - \omega^2),$$

with $x = g/h$, gives

$$(g - h)(g - \omega h)(g - \omega^2 h) = g^3 - h^3 = \sqrt{R}$$

(we saw on page 2 that $g^3 - h^3 = \sqrt{R}$). Therefore, $\Delta = 3i\sqrt{3}\sqrt{R}$, and

$$D = \Delta^2 = -27R = -27r^2 - 4q^3. \quad \bullet$$

**Remark.** Let $k$ be a field, and let $f(x) = a_m x^m + a_{m-1}x^{m-1} + \cdots + a_1 x + a_0$ and $g(x) = b_n x^n + b_{n-1}x^{n-1} + \cdots + b_1 x + b_0 \in k[x]$ have degrees $m$ and $n$, respectively. Their **resultant** is defined as

$$\text{Res}(f, g) = \det(M),$$

where $M = M(f, g)$ is the $(m + n) \times (m + n)$ matrix

$$M = \begin{bmatrix}
a_m & a_{m-1} & \cdots & a_1 & a_0 & & & \\
 & a_m & a_{m-1} & \cdots & a_1 & a_0 & & \\
 & & a_m & a_{m-1} & \cdots & a_1 & a_0 & \\
 & & & \cdots & & & & \\
b_n & b_{n-1} & \cdots & b_1 & b_0 & & & \\
 & b_n & b_{n-1} & \cdots & b_1 & b_0 & & \\
 & & b_n & b_{n-1} & \cdots & b_1 & b_0 & \\
 & & & \cdots & & & &
\end{bmatrix};$$

there are $n$ rows for the coefficients $a_i$ of $f(x)$ and $m$ rows for the coefficients $b_j$ of $g(x)$; all the entries other than those shown are assumed to be 0. It can be proved that $\text{Res}(f, g) = 0$ if and only if $f$ and $g$ have a nonconstant common divisor

(Jacobson, *Basic Algebra* I, p. 309). We mention the resultant here because the discriminant can be computed in terms of it:

$$D(f) = (-1)^{n(n-1)/2} \mathrm{Res}(f, f'),$$

where $f'(x)$ is the derivative of $f(x)$ [see van der Waerden, *Modern Algebra* I, pp. 83–88, or Dummit–Foote, *Abstract Algebra*, pp. 600–602]. ◀

Here is a way to use the discriminant in computing Galois groups.

**Proposition 3.63.** *Let $k$ be a field of characteristic $\neq 2$, let $f(x) \in k[x]$ be a polynomial of degree $n$ with no repeated roots, and let $D = \Delta^2$ be its discriminant. Let $E/k$ be a splitting field of $f(x)$, and let $G = \mathrm{Gal}(E/k)$ be regarded as a subgroup of $S_n$ (as in Theorem 3.3).*

(i) *If $H = A_n \cap G$, then $E^H = k(\Delta)$.*

(ii) *$G$ is a subgroup of $A_n$ if and only if $\Delta = \sqrt{D} \in k$.*

**Proof.**

(i) The Second Isomorphism Theorem gives $H = (G \cap A_n) \triangleleft G$ and

$$[G : H] = [G : A_n \cap G] = [A_n G : A_n] \leq [S_n : A_n] = 2.$$

By the Fundamental Theorem of Galois Theory (which applies because $f(x)$ has no repeated roots, hence is separable), $[E^H : k] = [G : H]$, so that $[E^H : k] = [G : H] \leq 2$. By Exercise 3.25 on page 221, we have $k(\Delta) \subseteq E^{A_n}$, and so $k(\Delta) \subseteq E^H$. Therefore,

$$[E^H : k] = [E^H : k(\Delta)][k(\Delta) : k] \leq 2.$$

There are two cases. If $[E^H : k] = 1$, then each factor in the displayed equation is 1; in particular, $[E^H : k(\Delta)] = 1$ and $E^H = k(\Delta)$. If $[E^H : k] = 2$, then $[G : H] = 2$ and there exists $\sigma \in G$, $\sigma \notin A_n$, so that $\sigma(\Delta) = -\Delta$. Now $\Delta \neq 0$, because $f(x)$ has no repeated roots, and $-\Delta \neq \Delta$, because $k$ does not have characteristic 2. Hence, $\Delta \notin E^G = k$ and $[k(\Delta) : k] > 1$. It follows from the displayed inequality that $[E^H : k(\Delta)] = 1$ and $E^H = k(\Delta)$.

(ii) The following are equivalent: $G \subseteq A_n$; $H = G \cap A_n = G$; $E^H = E^G = k$. Since $E^H = k(\Delta)$, by part (i), $E^H = k$ is equivalent to $k(\Delta) = k$; that is, $\Delta = \sqrt{D} \in k$. •

We can now show how to compute Galois groups of polynomials over $\mathbb{Q}$ of low degree.

If $f(x) \in \mathbb{Q}[x]$ is quadratic, then its Galois group has order either 1 or 2 (because the symmetric group $S_2$ has order 2). The Galois group has order 1 if $f(x)$ splits; it has order 2 if $f(x)$ does not split; that is, if $f(x)$ is irreducible.

If $f(x) \in \mathbb{Q}[x]$ is a cubic having a rational root, then its Galois group $G$ is the same as that of its quadratic factor. Otherwise $f(x)$ is irreducible; since $|G|$ is now a multiple of 3, by Corollary 3.9, and $G \subseteq S_3$, it follows that either $G \cong A_3 \cong \mathbb{I}_3$ or $G \cong S_3$.

**Proposition 3.64.** *Let* $f(x) \in \mathbb{Q}[x]$ *be an irreducible cubic with Galois group* $G$ *and discriminant* $D$.

(i) $f(x)$ *has exactly one real root if and only if* $D < 0$, *in which case* $G \cong S_3$.

(ii) $f(x)$ *has three real roots if and only if* $D > 0$. *In this case, either* $\sqrt{D} \in \mathbb{Q}$ *and* $G \cong \mathbb{I}_3$ *or* $\sqrt{D} \notin \mathbb{Q}$ *and* $G \cong S_3$.

**Proof.** Note first that $D \neq 0$, for irreducible polynomials over $\mathbb{Q}$ have no repeated roots because $\mathbb{Q}$ has characteristic 0. Let $E/\mathbb{Q}$ be the splitting field of $f(x)$.

(i) Suppose that $f(x)$ has one real root $\alpha$ and two complex roots: $\beta = u + iv$ and $\overline{\beta} = u - iv$, where $u, v \in \mathbb{R}$. Since $\beta - \overline{\beta} = 2iv$ and $\alpha = \overline{\alpha}$, we have

$$\Delta = (\alpha - \beta)(\alpha - \overline{\beta})(\beta - \overline{\beta}) = (\alpha - \beta)(\overline{\alpha - \beta})(\beta - \overline{\beta}) = 2iv|\alpha - \beta|^2,$$

and so $D = \Delta^2 = -4v^2|\alpha - \beta|^4 < 0$. Now $E \neq \mathbb{Q}(\alpha)$, because $\beta \in E$ is not real, so that $[E : \mathbb{Q}] = 6$ and $G \cong S_3$.

(ii) If $f(x)$ has three real roots, then $\Delta$ is real, $D = \Delta^2 > 0$, and $\sqrt{D}$ is real. By Proposition 3.63(ii), $G \cong A_3 \cong \mathbb{I}_3$ if and only if $\sqrt{D}$ is rational, and $G \cong S_3$ if $\sqrt{D}$ is irrational. ●

**Example 3.65.** The polynomial $f(x) = x^3 - 2 \in \mathbb{Q}[x]$ is irreducible, by Eisenstein's Criterion. Its discriminant is $D = -108$, and so its Galois group is $S_3$, by part (i) of the proposition.

The polynomial $x^3 - 4x + 2 \in \mathbb{Q}[x]$ is irreducible, by Eisenstein's Criterion; its discriminant is $D = 148$, and so it has three real roots. Since $\sqrt{148} = 2\sqrt{37}$ is irrational, the Galois group is $S_3$.

The polynomial $f(x) = x^3 - 48x + 64 \in \mathbb{Q}[x]$ is irreducible, by Theorem 2.63 (it has no rational roots); the discriminant is $D = 2^{12}3^4$, and so $f(x)$ has three real roots. Since $\sqrt{D} = 2^6 3^2$ is rational, the Galois group is $A_3 \cong \mathbb{I}_3$. ◀

**Corollary 3.66.** *Let* $f(x) = x^3 + qx + r \in \mathbb{C}[x]$ *have discriminant* $D$ *and roots* $u, v$ *and* $w$. *If* $F = \mathbb{Q}(q, r)$, *then* $F(u, \sqrt{D})$ *is a splitting field of* $f(x)$ *over* $F$.

**Proof.** Let $E = F(u, v, w)$ be a splitting field of $f(x)$, and let $K = F(u, \sqrt{D})$. Now $K \subseteq E$, for the definition of discriminant gives $\sqrt{D} = \pm(u - v)(u - w)(v - w) \in E$. For the reverse inclusion, it suffices to prove that $v \in K$ and $w \in K$. Since $u \in K$ is a root of $f(x)$, there is a factorization

$$f(x) = (x - u)g(x) \text{ in } K[x].$$

Now the roots of the quadratic $g(x)$ are $v$ and $w$, so that

$$g(x) = (x - v)(x - w) = x^2 - (v + w)x + vw.$$

Since $g(x)$ has its coefficients in $K$ and $u \in K$, we have

$$g(u) = (u - v)(u - w) \in K.$$

Therefore,

$$v - w = (u - v)(u - w)(v - w)/(u - v)(u - w)$$

$$= \pm \sqrt{D}/(u - v)(u - w) \in K.$$

On the other hand, $v + w \in K$, because it is a coefficient of $g(x)$ and $g(x) \in K[x]$. But we have just seen that $v - w \in K$; hence, $v, w \in K$ and $E = F(u, v, w) \subseteq K = F(u, \sqrt{D})$. Therefore, $F(u, v, w) = F(u, \sqrt{D})$. •

In Example 1.2 on page 2, we observed that the cubic formula giving the roots of $f(x) = x^3 + qx + r$ involves $\sqrt{R}$, where $R = r^2 + 4q^3/27$. Thus, when $R$ is negative, every root of $f(x)$ involves complex numbers. Since every cubic $f(x)$ has at least one real root, this phenomenon disturbed mathematicians of the sixteenth century, and they spent much time trying to rewrite specific formulas to eliminate complex numbers. The next theorem shows why such attempts were doomed to fail. On the other hand, these attempts ultimately led to a greater understanding of numbers in general and of complex numbers in particular.

**Theorem 3.67 (Casus Irreducibilis).** *If* $f(x) = x^3 + qx + r \in \mathbb{Q}[x]$ *is an irreducible cubic having three real roots* $u, v$, *and* $w$, *then any radical extension* $K_t/\mathbb{Q}$ *containing the splitting field of* $f(x)$ *is not real; that is,* $K_t \not\subseteq \mathbb{R}$.

**Proof.** Let $F = \mathbb{Q}(q, r)$, let $E = F(u, v, w)$ be a splitting field of $f(x)$, and let

$$F = K_0 \subseteq K_1 \subseteq \cdots \subseteq K_t$$

be a radical tower with $E \subseteq K_t$.

Since all the roots $u, v$ and $w$ are real,

$$D = [(u - v)(u - w)(v - w)]^2 \geq 0,$$

and so $\sqrt{D}$ is real. There is no loss in generality in assuming that $\sqrt{D}$ has been adjoined first:

$$K_1 = F(\sqrt{D}).$$

We claim that $f(x)$ remains irreducible in $K_1[x]$. If not, then $K_1$ contains a root of $f(x)$, say, $u$. Now $w \in K_1(v)$, because $x - w = f(x)/(x - u)(x - v) \in K_1(v)[x]$, and hence $E \subseteq K_1(v)$. The reverse inclusion holds, for $E$ contains $v$ and $\sqrt{D} = (u - v)(u - w)(v - w)$; thus, $E = K_1(v)$. Now $[E : K_1] \leq 2$ and $[K_1 : F] \leq 2$, so that $[E : F] = [E : K_1][K_1 : F]$ is a divisor of 4. By Corollary 3.9, the irreducibility of $f(x)$ over $F$ gives $3 \mid [E : F]$. This contradiction shows that $f(x)$ is irreducible in $K_1[x]$.

We may assume that each pure extension $K_{i+1}/K_i$ in the radical tower is of prime type. As $f(x)$ is irreducible in $K_1[x]$ and splits in $K_t[x]$ (because $E \subseteq K_t$), there is a first pure extension $K_{j+1}/K_j$ with $f(x)$ irreducible in $K_j[x]$ and factoring in $K_{j+1}[x]$. By hypothesis, $K_{j+1} = K_j(\alpha)$, where $\alpha$ is a root of $x^p - c$ for some prime $p$ and some $c \in K_j$. By Proposition 2.152, either $x^p - c$ is irreducible over $K_j$ or $c$ is a $p$th power in $K_j$. In the latter case, we have $K_{j+1} = K_j$, contradicting

$f(x)$ being irreducible over $K_j$ but not over $K_{j+1}$. Therefore, $x^p - c$ is irreducible over $K_j$, so that

$$[K_{j+1} : K_j] = p.$$

Since $f(x)$ factors over $K_{j+1}$, there is a root of $f(x)$ lying in it, say,

$$u \in K_{j+1};$$

hence, $K_j \subseteq K_j(u) \subseteq K_{j+1}$. But $f(x)$ is an irreducible cubic over $K_j$, so that $3 \mid [K_{j+1} : K_j] = p$, by Corollary 3.9. It follows that $p = 3$ and

$$K_{j+1} = K_j(u).$$

Now $K_{j+1}$ contains $u$ and $\sqrt{D}$, so that $K_j \subseteq E = F(u, \sqrt{D}) \subseteq K_{j+1}$, by Corollary 3.66. Since $[K_{j+1} : K_j]$ has no proper intermediate subfields (Corollary 3.9 again), we have $K_{j+1} = E$. Thus, $K_{j+1}$ is a splitting field of $f(x)$ over $K_j$, and hence $K_{j+1}$ is a Galois extension of $K_j$. The polynomial $x^3 - c$ (remember that $p = 3$) has a root, namely $\alpha$, in $K_{j+1}$, so that Theorem 3.36 says that $K_{j+1}$ contains the other roots $\omega\alpha$ and $\omega^2\alpha$ as well, where $\omega$ is a primitive cube root of unity. But this gives $\omega = (\omega\alpha)/\alpha \in K_{j+1}$, which is a contradiction because $\omega$ is not real while $K_{j+1} \subseteq K_t \subseteq \mathbb{R}$.   •

Before examining quartics, we cite Exercise 4.26 on page 251. If $d$ is a divisor of $|S_4| = 24$, then $S_4$ has a subgroup of order $d$; moreover, $\mathbf{V}$ and $\mathbb{I}_4$ are nonisomorphic subgroups of order 4, but any two subgroups of order $d \neq 4$ are isomorphic. We conclude that the Galois group $G$ of a quartic is determined, up to isomorphism, by its order unless $|G| = 4$.

Consider a (reduced) quartic $f(x) = x^4 + qx^2 + rx + s \in \mathbb{Q}[x]$; let $E/\mathbb{Q}$ be its splitting field and let $G = \text{Gal}(E/\mathbb{Q})$ be its Galois group [by Exercise 3.22(ii) on page 221, a polynomial and its associated reduced polynomial have the same Galois group]. If $f(x)$ has a rational root $\alpha$, then $f(x) = (x - \alpha)c(x)$, and its Galois group is the same as that of the cubic factor $c(x)$; but Galois groups of cubics have already been discussed. Suppose that $f(x) = h(x)\ell(x)$ is the product of two irreducible quadratics; let $\alpha$ be a root of $h(x)$ and let $\beta$ be a root of $\ell(x)$. If $\mathbb{Q}(\alpha) \cap \mathbb{Q}(\beta) = \mathbb{Q}$, then Exercise 3.12(iii) on page 211 shows that $G \cong \mathbf{V}$, the four group; otherwise, $\alpha \in \mathbb{Q}(\beta)$, so that $\mathbb{Q}(\beta) = \mathbb{Q}(\alpha, \beta) = E$, and $G$ has order 2.

We are left with the case of $f(x)$ irreducible. The basic idea now is to compare $G$ with the four group $\mathbf{V}$, namely, the normal subgroup of $S_4$,

$$\mathbf{V} = \big\{(1), (1\ 2)(3\ 4), (1\ 3)(2\ 4), (1\ 4)(2\ 3)\big\},$$

so that we can identify the fixed field of $\mathbf{V} \cap G$. If the four roots of $f(x)$ are $\alpha_1$, $\alpha_2$, $\alpha_3$, $\alpha_4$ [Proposition 3.69(ii) shows that these are distinct], consider the numbers:

(1)
$$\begin{cases} u = (\alpha_1 + \alpha_2)(\alpha_3 + \alpha_4), \\ v = (\alpha_1 + \alpha_3)(\alpha_2 + \alpha_4), \\ w = (\alpha_1 + \alpha_4)(\alpha_2 + \alpha_3) \end{cases}$$

It is clear that if $\sigma \in \mathbf{V} \cap G$, then $\sigma$ fixes $u$, $v$, and $w$. Conversely, if $\sigma \in S_4$ fixes $u = (\alpha_1 + \alpha_2)(\alpha_3 + \alpha_4)$, then

$$\sigma \in \mathbf{V} \cup \big\{(1\ 2), (3\ 4), (1\ 3\ 2\ 4), (1\ 4\ 2\ 3)\big\}.$$

However, none of the last four permutations fixes both $v$ and $w$, and so $\sigma \in G$ fixes each of $u$, $v$, $w$ if and only if $\sigma \in \mathbf{V} \cap G$. Therefore,

$$E^{\mathbf{V} \cap G} = \mathbb{Q}(u, v, w).$$

**Definition.** The *resolvent cubic* of $f(x) = x^4 + qx^2 + rx + s$ is

$$g(x) = (x - u)(x - v)(x - w),$$

where $u, v, w$ are the numbers defined in Equations (1).

**Proposition 3.68.** *The resolvent cubic of $f(x) = x^4 + qx^2 + rx + s$ is*

$$g(x) = x^3 - 2qx^2 + (q^2 - 4s)x + r^2.$$

**Proof.** If $f(x) = (x^2 + jx + \ell)(x^2 - jx + m)$, then we saw, in our discussion of the quartic formula on page 3, that $j^2$ is a root of

$$h(x) = x^3 + 2qx^2 + (q^2 - 4s)x - r^2,$$

a polynomial differing from the claimed expression for $g(x)$ only in the sign of its quadratic and constant terms. Thus, a number $\beta$ is a root of $h(x)$ if and only if $-\beta$ is a root of $g(x)$.

Let the four roots $\alpha_1, \alpha_2, \alpha_3, \alpha_4$ of $f(x)$ be indexed so that $\alpha_1, \alpha_2$ are the roots of $x^2 + jx + \ell$ and $\alpha_3, \alpha_4$ are the roots of $x^2 - jx + m$. Then $j = -(\alpha_1 + \alpha_2)$ and $-j = -(\alpha_3 + \alpha_4)$; therefore,

$$u = (\alpha_1 + \alpha_2)(\alpha_3 + \alpha_4) = -j^2$$

and $-u$ is a root of $h(x)$ since $h(j^2) = 0$.

Now factor $f(x)$ into two quadratics, say,

$$f(x) = (x^2 + \widetilde{j}x + \widetilde{\ell})(x^2 - \widetilde{j}x + \widetilde{m}),$$

where $\alpha_1, \alpha_3$ are the roots of the first factor and $\alpha_2, \alpha_4$ are the roots of the second. The same argument as before now shows that

$$v = (\alpha_1 + \alpha_3)(\alpha_2 + \alpha_4) = -\widetilde{j}^{\,2};$$

hence $-v$ is a root of $h(x)$. Similarly, $-w = -(\alpha_1 + \alpha_4)(\alpha_2 + \alpha_3)$ is a root of $h(x)$. Therefore,

$$h(x) = (x + u)(x + v)(x + w),$$

and so

$$g(x) = (x - u)(x - v)(x - w)$$

is obtained from $h(x)$ by changing the sign of the quadratic and constant terms. $\bullet$

**Proposition 3.69.**

(i) *The discriminant $D(f)$ of a quartic polynomial $f(x) \in \mathbb{Q}[x]$ is equal to the discriminant $D(g)$ of its resolvent cubic $g(x)$.*

(ii) *If $f(x)$ is irreducible, then $g(x)$ has no repeated roots.*

**Proof.**

(i) One checks easily that
$$u - v = \alpha_1\alpha_3 + \alpha_2\alpha_4 - \alpha_1\alpha_2 - \alpha_3\alpha_4 = -(\alpha_1 - \alpha_4)(\alpha_2 - \alpha_3).$$
Similarly,
$$u - w = -(\alpha_1 - \alpha_3)(\alpha_2 - \alpha_4) \quad \text{and} \quad v - w = (\alpha_1 - \alpha_2)(\alpha_3 - \alpha_4).$$
We conclude that
$$D(g) = [(u - v)(u - w)(v - w)]^2 = \left[-\prod_{i<j}(\alpha_i - \alpha_j)\right]^2 = D(f).$$

(ii) If $f(x)$ is irreducible, then it has no repeated roots (it is separable because $\mathbb{Q}$ has characteristic 0), and so $D(f) \neq 0$. But $D(g) = D(f) \neq 0$, and so $g(x)$ has no repeated roots.  •

In the notation of Equations (1) on page 219, if $f(x)$ is an irreducible quartic, then $u, v, w$ are distinct, and so our discussion there gives $E^{\mathbf{V} \cap G} = \mathbb{Q}(u, v, w)$, where $G = \mathrm{Gal}(E/\mathbb{Q})$ is the Galois group of $f(x)$. We can almost compute $G$; there is one ambiguous case. The resolvent cubic contains much information about the Galois group of the irreducible quartic from which it comes.

**Proposition 3.70.** *Let $f(x) \in \mathbb{Q}[x]$ be an irreducible quartic. Let $G$ be its Galois group, $D$ its discriminant, $g(x)$ its resolvent cubic, and $m$ the order of the Galois group of $g(x)$.*

(i) *If $m = 6$, then $G \cong S_4$. In this case, $g(x)$ is irreducible and $\sqrt{D}$ is irrational.*

(ii) *If $m = 3$, then $G \cong A_4$. In this case, $g(x)$ is irreducible and $\sqrt{D}$ is rational.*

(iii) *If $m = 1$, then $G \cong \mathbf{V}$. In this case, $g(x)$ splits in $\mathbb{Q}[x]$.*

(iv) *If $m = 2$, then $G \cong D_8$ or $G \cong \mathbb{I}_4$. In this case, $g(x)$ has an irreducible quadratic factor.*

**Proof.** We have seen that $E^{\mathbf{V} \cap G} = \mathbb{Q}(u, v, w)$. By the Fundamental Theorem of Galois Theory,
$$[G : \mathbf{V} \cap G] = [E^{\mathbf{V} \cap G} : \mathbb{Q}] = [\mathbb{Q}(u, v, w) : \mathbb{Q}] = |\mathrm{Gal}(\mathbb{Q}(u, v, w)/\mathbb{Q})| = m.$$
Since $f(x)$ is irreducible, $|G|$ is divisible by 4, by Corollary 3.9, and the group-theoretic statements follow from Exercise 3.28 on page 222. Finally, in the first two cases, $|G|$ is divisible by 12, and Proposition 3.63(ii) shows whether $G \cong S_4$ or $G \cong A_4$. The conditions on $g(x)$ in the last two cases are easy to see.  •

**Example 3.71.**

(i) Let $f(x) = x^4 - 4x + 2 \in \mathbb{Q}[x]$; $f(x)$ is irreducible, by Eisenstein's criterion. [Alternatively, we can see that $f(x)$ has no rational roots, using Theorem 2.63, and then show that $f(x)$ has no irreducible quadratic factors by examining conditions imposed on its coefficients.] By Proposition 3.68, the resolvent cubic is
$$g(x) = x^3 - 8x + 16.$$

Now $g(x)$ is irreducible (for $g(x) = x^3 + 2x + 1$ in $\mathbb{F}_5[x]$, and the latter polynomial is irreducible because it has no roots in $\mathbb{F}_5$). The discriminant of $g(x)$ is $-4864$, so that Theorem 3.64(i) shows that the Galois group of $g(x)$ is $S_3$, hence has order 6. Theorem 3.70(i) now shows that $G \cong S_4$.

(ii) Let $f(x) = x^4 - 10x^2 + 1 \in \mathbb{Q}[x]$; $f(x)$ is irreducible, by Example 2.147. By Proposition 3.68, the resolvent cubic is

$$x^3 + 20x^2 + 96x = x(x+8)(x+12).$$

In this case, $\mathbb{Q}(u, v, w) = \mathbb{Q}$ and $m = 1$. Therefore, $G \cong \mathbf{V}$. [This should not be a surprise once we recall Example 2.147, for $f(x)$ is the irreducible polynomial of $\alpha = \sqrt{2} + \sqrt{3}$, where $\mathbb{Q}(\alpha) = \mathbb{Q}(\sqrt{2}, \sqrt{3})$.] ◀

An interesting open question is the ***inverse Galois problem***: which finite abstract groups $G$ are isomorphic to $\mathrm{Gal}(E/\mathbb{Q})$, where $E/\mathbb{Q}$ is a Galois extension? Hilbert proved that the symmetric groups $S_n$ are such Galois groups, and Shafarevich proved that every solvable group is a Galois group (Neukirch–Schmidt–Wingberg, *Cohomology of Number Fields*, Chapter IX, §6). After the classification of the finite simple groups in the 1980s, it was shown that most simple groups are Galois groups. For more information, the reader is referred to Malle–Matzat, *Inverse Galois Theory* and Serre, *Topics in Galois Theory*.

## Exercises

**∗ 3.21.** Prove that $\omega(1 - \omega^2)(1 - \omega)^2 = 3i\sqrt{3}$, where $\omega = e^{2\pi i/3}$.

**∗ 3.22.** (i) Prove that if $a \neq 0$, then $f(x)$ and $af(x)$ have the same discriminant and the same Galois group. Conclude that it is no loss in generality to restrict our attention to monic polynomials when computing Galois groups.

(ii) Let $k$ be a field of characteristic 0. Prove that a polynomial $f(x) \in k[x]$ and its associated reduced polynomial $\widetilde{f}(x)$ have the same Galois group.

**3.23.** (i) Let $k$ be a field of characteristic 0. If $f(x) = x^3 + ax^2 + bx + c \in k[x]$, then its associated reduced polynomial is $x^3 + qx + r$, where

$$q = b - \tfrac{1}{3}a^2 \quad \text{and} \quad r = \tfrac{2}{27}a^3 - \tfrac{1}{3}ab + c.$$

(ii) Show that the discriminant of $f(x)$ is

$$D = a^2b^2 - 4b^3 - 4a^3c - 27c^2 + 18abc.$$

**3.24.** Find the Galois group of the cubic polynomial arising from the castle problem in Exercise 1.2 on page 4.

**∗ 3.25.** If $\sigma \in S_n$ and $f(x_1, \ldots, x_n) \in k[x_1, \ldots, x_n]$, where $k$ is a field, define

$$(\sigma f)(x_1, \ldots, x_n) = f(x_{\sigma 1}, \ldots, x_{\sigma n}).$$

(i) Prove that $(\sigma, f(x_1, \ldots, x_n)) \mapsto \sigma f$ is an action of $S_n$ on $k[x_1, \ldots, x_n]$.

(ii) Let $\Delta = \Delta(x_1, \ldots, x_n) = \prod_{i<j}(x_i - x_j)$ (on page 212, we saw that $\sigma \Delta = \pm \Delta$ for all $\sigma \in S_n$). If $\sigma \in S_n$, prove that $\sigma \in A_n$ if and only if $\sigma \Delta = \Delta$.

**Hint.** Define $\varphi\colon S_n \to G$, where $G$ is the multiplicative group $\{1, -1\}$, by
$$\varphi(\sigma) = \begin{cases} 1 & \text{if } \sigma\Delta = \Delta; \\ -1 & \text{if } \sigma\Delta = -\Delta. \end{cases}$$
Prove that $\varphi$ is a homomorphism, and that $\ker \varphi = A_n$.

**3.26.** Prove that if $f(x) \in \mathbb{Q}[x]$ is an irreducible quartic whose discriminant has a rational square root, then the Galois group of $f(x)$ has order 4 or 12.

**3.27.** Let $f(x) = x^4 + rx + s \in \mathbb{Q}[x]$ have Galois group $G$.

  (i) Prove that the discriminant of $f(x)$ is $-27r^4 + 256s^3$.

  (ii) Prove that if $s < 0$, then $G$ is not isomorphic to a subgroup of $A_4$.

  (iii) Prove that $f(x) = x^4 + x + 1$ is irreducible and that $G \cong S_4$.

$*$ **3.28.** Let $G$ be a subgroup of $S_4$ with $|G|$ a multiple of 4; define $m = |G/(G \cap \mathbf{V})|$.

  (i) Prove that $m$ is a divisor of 6.

  (ii) If $m = 6$, then $G = S_4$; if $m = 3$, then $G = A_4$; if $m = 1$, then $G = \mathbf{V}$; if $m = 2$, then $G \cong D_8$, $G \cong \mathbb{I}_4$, or $G \cong \mathbf{V}$.

  (iii) Let $G$ be a subgroup of $S_4$, and let $G$ act transitively on $X = \{1, 2, 3, 4\}$. If $|G/(\mathbf{V} \cap G)| = 2$, prove that $G \cong D_8$ or $G \cong \mathbb{I}_4$. [If we merely assume that $G$ acts transitively on $X$, then $|G|$ is a multiple of 4 (Corollary 3.9). The added hypothesis $|G/(\mathbf{V} \cap G)| = 2$ removes the possibility $G \cong \mathbf{V}$ when $m = 2$ in Exercise 3.28.]

**3.29.** Compute the Galois group over $\mathbb{Q}$ of $x^4 + x^2 - 6$.

**3.30.** Compute the Galois group over $\mathbb{Q}$ of $f(x) = x^4 + x^2 + x + 1$.

**Hint.** Use Example 2.66 to prove irreducibility of $f(x)$, and prove irreducibility of the resolvent cubic by reducing mod 2.

**3.31.** Compute the Galois group over $\mathbb{Q}$ of $f(x) = 4x^4 + 12x + 9$.

**Hint.** Prove that $f(x)$ is irreducible in two steps: first show that it has no rational roots, and then use Descartes's method (on page 183) to show that $f(x)$ is not the product of two quadratics over $\mathbb{Q}$.

# Chapter 4

# Groups II

We now investigate the structure of groups. Finite abelian groups turn out to be rather uncomplicated: they are direct sums of cyclic groups. Returning to nonabelian groups, the Sylow Theorems show, for any prime $p$, that finite groups $G$ have subgroups of order $p^e$, where $p^e$ is the largest power of $p$ dividing $|G|$, and any two such subgroups are isomorphic. The ideas of normal series and solvability that arose in Galois Theory lead to the Jordan–Hölder Theorem, which shows that simple groups are, in a certain sense, building blocks of finite groups. Consequently, we display more examples of simple groups to accompany the cyclic groups of prime order and the alternating groups $A_n$, for $n \geq 5$, which we have already proved to be simple. Free groups and presentations are then introduced, for they are useful in constructing and describing arbitrary groups. The chapter ends with a proof that every subgroup of a free group is itself a free group.

## Section 4.1. Finite Abelian Groups

We continue our study of groups by classifying all finite abelian groups; as is customary, we use additive notation for the binary operation in these groups.

**4.1.1. Direct Sums.** Groups in this subsection are arbitrary, possibly infinite, abelian groups.

There are two ways to describe the *direct sum* of abelian groups $S_1, \ldots, S_n$. The easiest version is sometimes called their **external direct sum**, which we temporarily denote by $S_1 \times \cdots \times S_n$; its elements are all $n$-tuples $(s_1, \ldots, s_n)$, where $s_i \in S_i$ for $i = 1, \ldots, n$, and its binary operation is coordinatewise addition:

$$(s_1, \ldots, s_n) + (s_1', \ldots, s_n') = (s_1 + s_1', \ldots, s_n + s_n').$$

However, the most useful version, isomorphic to $S_1 \times \cdots \times S_n$, is sometimes called their **internal direct sum**; it is the additive version of the statement of Proposition 1.86 involving subgroups $S_i$ of a given group $G$. Recall that the subgroup of a group $G$ generated by subgroups $S$ and $T$ is denoted by $S \vee T$. If $G$ is an *abelian* group written additively, it is easy to see that

$$S \vee T = S + T = \{s + t : s \in S \text{ and } t \in T\}.$$

**Definition.** If $S$ and $T$ are subgroups of an abelian group $G$ with $S + T = G$ and $S \cap T = \{0\}$, then $G$ is the (*internal*) **direct sum**, denoted by

$$G = S \oplus T.$$

Here are several characterizations of internal direct sum.

**Proposition 4.1.** *The following statements are equivalent for an abelian group $G$ and subgroups $S$ and $T$ of $G$. Let $i \colon S \to G$ and $j \colon T \to G$ be inclusions.*

(i) $G = S \oplus T$.

(ii) *For each $g \in G$, there are unique $s \in S$ and $t \in T$ with $g = s + t$.*

(iii) *There are homomorphisms $p \colon G \to S$ and $q \colon G \to T$ (called **projections**) such that*

$$pi = 1_S, \quad qj = 1_T, \quad pj = 0, \quad qi = 0, \quad \text{and} \quad ip + jq = 1_G.$$

**Remark.** The equations $pi = 1_S$ and $qj = 1_T$ imply that the functions $p$ and $q$ are surjective. ◀

**Proof.**

(i) $\Rightarrow$ (ii) By hypothesis, $G = S + T$, so that each $g \in G$ has an expression of the form $g = s + t$ with $s \in S$ and $t \in T$. To see that this expression is unique, suppose also that $g = s' + t'$, where $s' \in S$ and $t' \in T$. Then $s + t = s' + t'$ gives $s - s' = t' - t \in S \cap T = \{0\}$. Therefore, $s = s'$ and $t = t'$, as desired.

(ii) $\Rightarrow$ (iii) If $g \in G$, then $g = s + t$, where $s \in S$ and $t \in T$ are unique. The functions $p$ and $q$, given by $p(g) = s$ and $q(g) = t$, are well-defined because of the uniqueness hypothesis, and it is easily checked that $p$ and $q$ are homomorphisms. It is routine to check the equations in the statement.

(iii) $\Rightarrow$ (i) If $g \in G$, then $g = ipg + jqg \in \operatorname{im} i + \operatorname{im} j = S + T$. If $g \in S$, then $g = is$ and $qg = qis = 0$; if $g \in T$, then $g = jt$ and $pg = pjt = 0$. Hence, if $g \in S \cap T$, then $g = ipg + jqg = 0$ and $S \cap T = \{0\}$. Since $S + T = G$, we have $G = S \oplus T$. •

In light of the next corollary, we will omit the adjectives *external* and *internal* when speaking of direct sums of two groups, but our viewpoint is almost always internal.

**Corollary 4.2.**

(i) *If an abelian group $G$ is an internal direct sum, $G = S \oplus T$, then*

$$S \times T \cong S \oplus T$$

*via $(s, t) \mapsto s + t$.*

(ii) *Conversely, every external direct sum is an internal direct sum: given abelian groups $S$ and $T$, then*
$$S \times T = S' \oplus T',$$
*where $S' = \{(s,0) : s \in S\} \cong S$ and $T' = \{(0,t) : t \in T\} \cong T$.*

**Proof.**

(i) Define $f \colon S \times T \to S \oplus T$ by $f \colon (s,t) \mapsto s + t$. Now $f$ is a homomorphism: $f \colon (s,t) + (s',t') = (s + s', t + t') \mapsto s + s' + t + t'$; on the other hand, $f(s,t) + f(s',t') = s + t + s' + t'$. These are equal because $S \oplus T$ abelian gives $t + s' = s' + t$. Finally, $f$ is an isomorphism, for its inverse $s + t \mapsto (s,t)$ is well-defined because of uniqueness of expression.

(ii) The subgroup $S' \subseteq S \times T$ is isomorphic to $S$ via $(s,0) \mapsto s$; similarly, $T' \cong T$ via $(0,t) \mapsto t$ (these maps are analogs of the inclusions $i$ and $j$ in Proposition 4.1). Now $S' + T' = S \times T$, for $(s,t) = (s,0) + (0,t) \in S' + T'$. Clearly, $S' \cap T' = \{(0,0)\}$, and so $S \times T = S' \oplus T'$. •

We define the external direct sum $S_1 \times \cdots \times S_n$ of groups $S_1, \ldots, S_n$ as their cartesian product with coordinatewise addition; we define the internal direct sum of several subgroups inductively.

**Definition.** If $S_1, \ldots, S_n$ are subgroups of an abelian group $G$, define their *internal direct sum* by induction on $n \geq 2$:
$$S_1 \oplus \cdots \oplus S_{n+1} = (S_1 \oplus \cdots \oplus S_n) \oplus S_{n+1}.$$
The direct sum $S_1 \oplus \cdots \oplus S_n$ is also denoted by $\bigoplus_{i=1}^n S_i$.

The definition of direct sum uses a list $S_1, \ldots, S_n$ of subgroups, but Proposition 4.4(iii) below shows that the direct sum does not depend on an ordering of the subgroups. We now generalize Proposition 4.1 and its corollary.

**Lemma 4.3.** *Let $G = S_1 + \cdots + S_n$ be an abelian group, where $S_1, \ldots, S_n$ are subgroups; that is, for each $g \in G$, there exist $s_i \in S_i$ with $g = s_1 + \cdots + s_n$. Then all such expressions are unique if and only if*
$$S_i \cap (S_1 + \cdots + \widehat{S}_i + \cdots + S_n) = \{0\}$$
*for each $i$, where $\widehat{S}_i$ means that the term $S_i$ is omitted from the sum.*

**Proof.** Suppose that $x \in S_i \cap (S_1 + \cdots + \widehat{S}_i + \cdots + S_n)$. Then $x = s_i \in S_i$ and $s_i = \sum_{j \neq i} s_j$, where $s_j \in S_j$. The element $0$ has two expressions: $0 = -s_i + \sum_{j \neq i} s_j$ and $0 = 0 + 0 + \cdots + 0$. By uniqueness, $-s_i = 0$ and all $s_j = 0$; hence, $x = s_i = 0$.

Conversely, assume that $s_1 + \cdots + s_n = s'_1 + \cdots + s'_n$, where $s_i, s'_i \in S_i$ for all $i$. Then $s_i - s'_i \in S_i \cap (S_1 + \cdots + \widehat{S}_i + \cdots + S_n) = \{0\}$; that is, $s_i = s'_i$ for all $i$. •

**Proposition 4.4.** *Let $G = S_1 + \cdots + S_n$, where the $S_i$ are subgroups, and let $j_i \colon S_i \to G$ be inclusions. The following conditions are equivalent.*

(i) $G = S_1 \oplus \cdots \oplus S_n$.

(ii) *Every $g \in G$ has a unique expression of the form $g = s_1 + \cdots + s_n$, where $s_i \in S_i$ for all $i$.*

(iii) *For each $i$, $S_i \cap (S_1 + \cdots + \widehat{S}_i + \cdots + S_n) = \{0\}$.*

(iv) *There are homomorphisms $p_i \colon G \to S_i$ for all $i$ (called **projections**) such that*

$$p_i j_i = 1_{S_i}, \quad p_k j_i = 0 \text{ for } k \neq i, \text{ and } \sum_i j_i p_i = 1_G.$$

**Proof.**

(i) $\Rightarrow$ (ii) The proof is by induction on $n \geq 2$. The base step is Proposition 4.1. For the inductive step, define $T = S_1 + \cdots + S_n$, so that $G = T \oplus S_{n+1}$ (this is the inductive definition). By the case $n = 2$, each $g \in G$ has a unique expression of the form $g = t + s_{n+1}$, where $t \in T$ and $s_{n+1} \in S_{n+1}$. But the inductive hypothesis says that $t$ has a unique expression of the form $t = s_1 + \cdots + s_n$, where $s_i \in S_i$ for all $i \leq n$, as desired.

(ii) $\Rightarrow$ (iii) This follows from Lemma 4.3.

(iii) $\Rightarrow$ (iv) By Lemma 4.3, we may assume item (ii). Thus, uniqueness of expression says, for each $i$, that the functions $p_i \colon G \to S_i$, given by $p_i \colon g = s_1 + \cdots + s_n \mapsto s_i$, are well-defined. Verification of the displayed equations is routine.

(iv) $\Rightarrow$ (i) The proof is by induction on $n \geq 2$; the base step is Proposition 4.1. For the inductive step, define $T = S_1 + \cdots + S_n$, define $j \colon T \to G$ to be the inclusion, and define $p \colon G \to T$ by $s_1 + \cdots + s_n + s_{n+1} \mapsto s_1 + \cdots + s_n$. By the base step, $G = T \oplus S_{n+1}$ and, by the inductive hypothesis, $T = S_1 \oplus \cdots \oplus S_n$. Therefore, $G = [S_1 \oplus \cdots \oplus S_n] \oplus S_{n+1}$. •

The next corollary allows us to omit the adjectives *external* and *internal*.

**Corollary 4.5.**

(i) *If an abelian group $G$ is an internal direct sum, $G = S_1 \oplus \cdots \oplus S_n$, then*

$$S_1 \times \cdots \times S_n \cong S_1 \oplus \cdots \oplus S_n$$

*via $(s_1, \ldots, s_n) \mapsto s_1 + \cdots + s_n$.*

(ii) *Conversely, the external direct sum is an internal direct sum,*

$$S_1 \times \cdots \times S_n = S_1' \oplus \cdots \oplus S_n',$$

*where $S_i' = \{(0, \ldots, 0, s_i, 0, \ldots 0) : s_i \in S_i\} \cong S_i$ for all $i$.*

**Proof.**

(i) Define $f \colon S_1 \times \cdots \times S_n \to G$ by $f \colon (s_1, \ldots, s_n) \mapsto s_1 + \cdots + s_n$. As in the proof of Corollary 4.2, $f$ is a homomorphism. Moreover, $f$ is an isomorphism, for its inverse $s_1 + \cdots + s_n \mapsto (s_1, \ldots, s_n)$ is well-defined because of uniqueness of expression.

(ii) Now $S'_i \subseteq S_1 \times \cdots \times S_n$ is isomorphic to $S_i$ via $(0, \ldots, 0, s_i, 0, \ldots 0) \mapsto s_i$. Moreover, $S_1 \times \cdots \times S_n = S'_1 + \cdots + S'_n$, for $(s_1, \ldots, s_n) = (s_1, 0, \ldots, 0) + \cdots + (0, \ldots, 0, s_n)$, while $S'_i \cap (S'_1 + \cdots + \widehat{S_i} + \cdots + S'_n) = \{(0, \ldots, 0)\}$. •

The generators of a direct sum of cyclic groups enjoy a type of independence.

**Corollary 4.6.** *Let* $G = \langle y_1, \ldots, y_n \rangle$. *Then* $\sum_i m_i y_i = 0$ *in* $G$ *implies* $m_i y_i = 0$ *for all* $i$ *if and only if*

$$G = \langle y_1 \rangle \oplus \cdots \oplus \langle y_n \rangle.$$

**Proof.** We use Proposition 4.4(iii) to show that $G$ is a direct sum. If

$$g \in \langle y_i \rangle \cap \langle y_1, \ldots, \widehat{y_i}, \ldots, y_n \rangle,$$

there are $m_i, m_j \in \mathbb{Z}$ with $m_i y_i = g = \sum_{j \neq i} m_j y_j$, and so $-m_i y_i + \sum_{j \neq i} m_j y_j = 0$. By hypothesis, each summand is 0; in particular, $g = m_i y_i = 0$, as desired.

Conversely, suppose that $G = \langle y_1 \rangle \oplus \cdots \oplus \langle y_n \rangle$. If $\sum_i m_i y_i = 0$, then uniqueness of expression, Proposition 4.4(ii), gives $m_i y_i = 0$ for each $i$. •

**Example 4.7.** Linear independence in a vector space is intimately related to direct sums of subspaces. View an $n$-dimensional vector space $V$ over a field $k$ merely as an additive abelian group by forgetting its scalar multiplication. If $X = v_1, \ldots, v_n$ is a linearly independent list in $V$, we claim that

$$V = \langle v_1 \rangle \oplus \cdots \oplus \langle v_n \rangle,$$

where $\langle v_i \rangle = \{r v_i : r \in k\}$ is the one-dimensional subspace spanned by $v_i$. Each $v \in V$ has a unique expression of the form $v = a_1 v_1 + \cdots + a_n v_n$, where $a_i v_i \in \langle v_i \rangle$. Thus, $V$ is a direct sum, by Proposition 4.4(ii).

Conversely, if $X = v_1, \ldots, v_n$ is a list in a vector space $V$ over a field $k$ and the subspace it generates is a direct sum of one-dimensional subspaces, $\langle v_1 \rangle \oplus \cdots \oplus \langle v_n \rangle$, then $X$ is linearly independent. By uniqueness of expression, $\sum_i a_i v_i = 0$ in $V$ implies $a_i v_i = 0$ for each $i$, where $a_i \in k$. But $a_i v_i = 0$ holds in a vector space, where $a_i \in k$ and $v \in V$, if and only if $a_i = 0$ or $v_i = 0$. Therefore, $X = v_1, \ldots, v_n$ is a linearly independent list. ◀

Given subgroups $S_1, \ldots, S_n$ of an abelian group $G$, then Proposition 4.4(iii) says that $\langle S_1, \ldots, S_n \rangle = S_1 \oplus \cdots \oplus S_n$ if $S_i \cap (S_1 + \cdots + \widehat{S_i} + \cdots + S_n) = \{0\}$ for all $i$. A common mistake is to say that it suffices to assume that $S_i \cap S_j = \{0\}$ for all $i \neq j$; the following example shows that this is not enough.

**Example 4.8.** As in Example 4.7, view a two-dimensional vector space $V$ over a field $k$ as an additive abelian group, and let $x, y$ be a basis. It is easy to see that the intersection of any two of the one-dimensional subspaces $\langle x \rangle$, $\langle y \rangle$, and $\langle x+y \rangle$ is $\{0\}$. On the other hand, $V \neq \langle x \rangle \oplus \langle y \rangle \oplus \langle x+y \rangle$, because $[\langle x \rangle \oplus \langle y \rangle] \cap \langle x+y \rangle \neq \{0\}$. ◀

Now that we have examined finite direct sums, we can generalize Proposition 1.85 from two summands to a finite number of summands. Although we state the next result for abelian groups, it should be clear that the proof works for non-abelian groups as well if we assume that the subgroups $H_i$ are normal subgroups (Exercise 4.1 on page 241).

**Proposition 4.9.** *If $G_1, \ldots, G_n$ are (abelian) groups and $H_i \subseteq G_i$ are subgroups, then*

$$(G_1 \times \cdots \times G_n)/(H_1 \times \cdots \times H_n) \cong (G_1/H_1) \times \cdots \times (G_n/H_n).$$

**Proof.** The function $f : G_1 \times \cdots \times G_n \to (G_1/H_1) \times \cdots \times (G_n/H_n)$, given by $f \colon (g_1, \ldots, g_n) \mapsto (g_1 + H_1, \ldots, g_n + H_n)$, is a surjective homomorphism with $\ker f = H_1 \times \cdots \times H_n$. The First Isomorphism Theorem gives the result.   •

If $G$ is an abelian group and $m$ is an integer, let us write

$$mG = \{ma : a \in G\}.$$

It is easy to see that $mG$ is a subgroup of $G$ (Exercise 4.2 on page 241 says that $mG$ may not be a subgroup if $G$ is nonabelian).

**Proposition 4.10.** *If $G$ is an abelian group and $p$ is prime, then $G/pG$ is a vector space over $\mathbb{F}_p$.*

**Proof.** If $[r] \in \mathbb{F}_p = \mathbb{I}_p$ and $a \in G$, define scalar multiplication

$$[r](a + pG) = ra + pG.$$

This formula is well-defined: if $r' \equiv r \bmod p$, then $r' = r + pm$ for some integer $m$, and so

$$r'a + pG = ra + pma + pG = ra + pG,$$

because $pma \in pG$; hence, $[r'](a + pG) = [r](a + pG)$. It is now routine to check that the axioms for a vector space do hold.   •

Direct sums of copies of $\mathbb{Z}$ arise often enough to have their own name.

**Definition.** If $\langle x_1 \rangle, \ldots, \langle x_n \rangle$ are infinite cyclic groups, then their direct sum

$$F = \langle x_1 \rangle \oplus \cdots \oplus \langle x_n \rangle$$

is called a ***free abelian group*** with ***basis*** the list[1] $X = x_1, \ldots, x_n$. More generally, any group isomorphic to $F$ is called a free abelian group.[2]

For example, $\mathbb{Z}^m = \mathbb{Z} \times \cdots \times \mathbb{Z}$, the group of all $m$-tuples $(n_1, \ldots, n_m)$ of integers, is a free abelian group. One basis of $\mathbb{Z}^m$ is the ***standard basis*** $e_1, \ldots, e_m$, where $e_i$ is the $m$-tuple having 1 in the $i$th place and 0's elsewhere.

**Proposition 4.11.** $\mathbb{Z}^m \cong \mathbb{Z}^n$ *if and only if $m = n$.*

---

[1]We defined bases of vector spaces as lists, not as subsets, for it is important to distinguish ordered subsets (lists) from subsets. This same fussiness persists in the discussion of free abelian groups.

[2]We will consider free abelian groups with infinite bases on page 450.

**Proof.** Only necessity needs proof. Note first that if an abelian group $G$ is a direct sum, $G = G_1 \oplus \cdots \oplus G_n$, then $2G = 2G_1 \oplus \cdots \oplus 2G_n$. It follows from Proposition 4.9 that

$$G/2G \cong (G_1/2G_1) \oplus \cdots \oplus (G_n/2G_n).$$

In particular, if $G = \mathbb{Z}^n$, then $|G/2G| = 2^n$. Finally, if $\mathbb{Z}^n \cong \mathbb{Z}^m$, then $\mathbb{Z}^n/2\mathbb{Z}^n \cong \mathbb{Z}^m/2\mathbb{Z}^m$ and $2^n = 2^m$. We conclude that $n = m$. •

**Corollary 4.12.** *If $F$ is a free abelian group, then any two (finite) bases of $F$ have the same number of elements.*

**Proof.** If $x_1, \ldots, x_n$ is a basis of $F$, then $F \cong \mathbb{Z}^n$, and if $y_1, \ldots, y_m$ is another basis of $F$, then $F \cong \mathbb{Z}^m$. By the proposition, $m = n$. •

**Definition.** If $F$ is a free abelian group with basis $x_1, \ldots, x_n$, then $n$ is called the *rank* of $F$, and we write

$$\mathrm{rank}(F) = n.$$

Corollary 4.12 says that $\mathrm{rank}(F)$ is well-defined; that is, it does not depend on the choice of basis. In this language, Proposition 4.11 says that two free abelian groups are isomorphic if and only if they have the same rank. Thus, the rank of a free abelian group plays the same role as the dimension of a vector space.

Recall Theorem 2.120: *Let $v_1, \ldots, v_n$ be a basis of a vector space $V$. If $W$ is a vector space and $u_1, \ldots, u_n$ is a list in $W$, then there exists a unique linear transformation $T \colon V \to W$ with $T(v_i) = u_i$ for all $i$.* The next theorem shows that bases of free abelian groups play the same role as bases of vector spaces (and look at Theorem 2.25 as well).

**Theorem 4.13 (Freeness Property).** *Let $F$ be a free abelian group with basis $X = x_1, \ldots, x_n$. If $G$ is any abelian group and $\gamma \colon X \to G$ is any function, then there exists a unique homomorphism $h \colon F \to G$ with $h(x_i) = \gamma(x_i)$ for all $x_i$.*

**Proof.** Every element $a \in F$ has a unique expression of the form $a = \sum_{i=1}^{n} m_i x_i$, where $m_i \in \mathbb{Z}$. This uniqueness implies that $h \colon F \to G$, given by

$$h(a) = \sum_{i=1}^{n} m_i \gamma(x_i),$$

is well-defined. It is easy to see that $h$ is a homomorphism. If $h' \colon F \to G$ is a homomorphism with $h'(x_i) = h(x_i)$ for all $i$, then $h' = h$, for two homomorphisms that agree on a set of generators must be equal. •

The freeness property characterizes free abelian groups.

**Proposition 4.14.** *Let $X = x_1, \ldots, x_n$ be a list in an abelian group $A$, and let $A$ have the freeness property: for every abelian group $G$ and every function $\gamma\colon X \to G$, there exists a unique homomorphism $g\colon A \to G$ with $g(x_i) = \gamma(x_i)$ for all $x_i$. Then $A$ is a free abelian group of rank $n$ with basis $X$.*

**Proof.** Let $Y = e_1, \ldots, e_n$ be a basis of $\mathbb{Z}^n$, and let $k\colon Y \to \mathbb{Z}^n$ be the inclusion. Consider the diagram

where $j\colon X \to A$ is the inclusion; define $q(x_i) = e_i$ and $p(e_i) = x_i$ for all $i$. By the freeness property, there is a map $g\colon A \to \mathbb{Z}^n$ with $gj = kq$ (for $kq\colon X \to \mathbb{Z}^n$). Since $\mathbb{Z}^n$ is a free abelian group with basis $Y$, there is a map $h\colon \mathbb{Z}^n \to A$ with $hk = jp$, by Theorem 4.13. To see that $g\colon A \to \mathbb{Z}^n$ is an isomorphism, consider the diagram

$$
\begin{array}{c}
A \\
\uparrow j \quad \searrow^{hg} \\
X \xrightarrow{\quad j \quad} A
\end{array}
$$

Now $hgj = hkq = jpq = j$. By hypothesis, $hg$ is the unique such homomorphism. But $1_A$ is another such, and so $hg = 1_A$. A similar diagram shows that the other composite $gh = 1_{\mathbb{Z}^n}$, and so $g$ and $h$ are isomorphisms. Finally, that $\mathbb{Z}^n$ is free with basis $Y$ implies that $A$ is free with basis $X = h(Y)$.  •

**4.1.2. Basis Theorem.** It is convenient to analyze finite abelian groups *locally*; that is, one prime at a time. Recall that a *p-group* is a group $G$ each of whose elements has order a power of $p$; if $G$ is finite, this is equivalent to $|G| = p^k$ for some $k \geq 0$. When working wholly in the context of abelian groups, $p$-groups are usually called *p-primary groups*.

**Definition.** Let $p$ be a prime. An abelian group $G$ is *p-**primary*** if, for each $a \in G$, there is $n \geq 1$ with $p^n a = 0$. If we do not want to specify the prime $p$, we merely say that $G$ is ***primary*** (instead of $p$-primary).

If $G$ is any abelian group, then its *p-**primary component*** is

$$G_p = \{a \in G : p^n a = 0 \text{ for some } n \geq 1\}.$$

It is easy to see, for every prime $p$, that $G_p$ is a subgroup of $G$ (this is not the case if $G$ is not abelian; for example, $G_2$ is not a subgroup when $G = S_3$).

**Theorem 4.15 (Primary Decomposition).**

(i) *Every finite abelian group $G$ is the direct sum of its p-primary components:*

$$G = G_{p_1} \oplus \cdots \oplus G_{p_n}.$$

(ii) *Two finite abelian groups $G$ and $G'$ are isomorphic if and only if $G_p \cong G'_p$ for every prime $p$.*

**Proof.**

(i) Let $x \in G$ be nonzero, and let its order be $d$. By the Fundamental Theorem of Arithmetic, there are distinct primes $p_1, \ldots, p_n$ and positive exponents $e_1, \ldots, e_n$ with

$$d = p_1^{e_1} \cdots p_n^{e_n}.$$

Define $r_i = d/p_i^{e_i}$, so that $p_i^{e_i} r_i = d$. It follows that $r_i x \in G_{p_i}$ for each $i$ (because $dx = 0$). But the gcd of $r_1, \ldots, r_n$ is 1 (the only possible prime divisors of $d$ are $p_1, \ldots, p_n$, and no $p_i$ is a common divisor because $p_i \nmid r_i$); hence, there are integers $s_1, \ldots, s_n$ with $1 = \sum_i s_i r_i$. Therefore,

$$x = \sum_i s_i r_i x \in G_{p_1} + \cdots + G_{p_n}.$$

Write $H_i = G_{p_1} + \cdots + \widehat{G}_{p_i} + \cdots + G_{p_n}$. By Proposition 4.4(iii), it suffices to prove, for all $i$, that

$$G_{p_i} \cap H_i = \{0\}.$$

If $x \in G_{p_i} \cap H_i$, then $p_i^\ell x = 0$ for some $\ell \geq 0$ (since $x \in G_{p_i}$) and $ux = 0$ for some $u = \prod_{j \neq i} p_j^{g_j}$ (since $x \in H_i$, we have $x = \sum_{j \neq i} y_j$ and $p_j^{g_j} y_j = 0$). But $p_i^\ell$ and $u$ are relatively prime, so there exist integers $s$ and $t$ with $1 = sp_i^\ell + tu$. Therefore,

$$x = (sp_i^\ell + tu)x = sp_i^\ell x + tux = 0.$$

(ii) If $f\colon G \to G'$ is a homomorphism, then $f(G_p) \subseteq G'_p$ for every prime $p$, for if $p^\ell a = 0$, then $0 = f(p^\ell a) = p^\ell f(a)$. If $f$ is an isomorphism, then $f^{-1}\colon G' \to G$ is also an isomorphism [so that $f^{-1}(G'_p) \subseteq G_p$ for all $p$]. It follows that each restriction $f|G_p\colon G_p \to G'_p$ is an isomorphism, with inverse $f^{-1}|G'_p$.

Conversely, if there are isomorphisms $f_p\colon G_p \to G'_p$ for all $p$, then there is an isomorphism $\varphi\colon \bigoplus_p G_p \to \bigoplus_p G'_p$, given by $\sum_p a_p \mapsto \sum_p f_p(a_p)$. •

The next type of subgroup will play an important role.

**Definition.** Let $p$ be prime and let $G$ be a $p$-primary abelian group. A subgroup $S \subseteq G$ is a **pure subgroup**[3] if, for all $n \geq 0$,

$$S \cap p^n G = p^n S.\,^4$$

The inclusion $S \cap p^n G \supseteq p^n S$ is true for every subgroup $S \subseteq G$, and so it is only the reverse inclusion $S \cap p^n G \subseteq p^n S$ that is significant. It says that if $s \in S$ satisfies an equation $s = p^n a$ for some $a \in G$, then there exists $s' \in S$ with $s = p^n s'$.

---

[3]Recall that *pure extensions* $k(u)/k$ arose in our discussion of solvability by radicals on page 181; in such an extension, the adjoined element $u$ satisfies the equation $u^n = a$ for some $a \in k$. Pure subgroups are defined in terms of similar equations (written additively), and they are probably so called because of this.

[4]If $G$ is not a primary group, then a pure subgroup $S \subseteq G$ is defined to be a subgroup that satisfies $S \cap mG = mS$ for all $m \in \mathbb{Z}$ (see Exercises 4.4 and 4.6 on page 241).

**Example 4.16.**

(i) Every direct summand $S$ of $G$ is a pure subgroup. Let $G = S \oplus T$ and $s \in S$. If $s = p^n(u + v)$ for $u \in S$ and $v \in T$, then $p^n v = s - p^n v \in S \cap T = \{0\}$, and $s = p^n u$. (The converse, every pure subgroup $S$ of a group $G$ is a direct summand, is true when $G$ is finite [see Exercise 4.5 on page 241], but it may be false when $G$ is infinite [see Exercise 4.6 on page 241].)

(ii) If $G = \langle a \rangle$ is a cyclic group of order $p^2$, where $p$ is prime, then $S = \langle pa \rangle$ is not a pure subgroup of $G$, for $s = pa \in S$, but there is no element $s' \in S$ with $s = ps'$ (because $s' = mpa$, for $m \in \mathbb{Z}$, and so $ps' = mp^2 a = 0$).  ◄

**Lemma 4.17.** *If $p$ is prime and $G$ is a finite $p$-primary abelian group, then $G$ has a nonzero pure cyclic subgroup. Indeed, if $y$ is an element of largest order in $G$, then $\langle y \rangle$ is a pure cyclic subgroup.*

**Proof.** Since $G$ is finite, there exists $y \in G$ of largest order, say, $p^\ell$. We claim that $S = \langle y \rangle$ is a pure subgroup of $G$.

If $s \in S$, then $s = mp^t y$, where $t \geq 0$ and $p \nmid m$. Suppose that

$$s = p^n a$$

for some $a \in G$; an element $s' \in S$ with $s = mp^t y = p^n s'$ must be found. We may assume that $n < \ell$: otherwise, $s = p^n a = 0$ (since $y$ has largest order $p^\ell$, we have $p^\ell g = 0$ for all $g \in G$), and we may choose $s' = 0$.

We claim that $t \geq n$. If $t < n$, then

$$p^\ell a = p^{\ell-n} p^n a = p^{\ell-n} s = p^{\ell-n} mp^t y = mp^{\ell-n+t} y.$$

But $p \nmid m$ and $\ell - n + t < \ell$, because $-n + t < 0$, and so $p^\ell a \neq 0$, contradicting $y$ having largest order. Thus, $t \geq n$, and we can define $s' = mp^{t-n} y$. Now $s' \in S$ and

$$p^n s' = p^n mp^{t-n} y = mp^t y = s,$$

so that $S$ is a pure subgroup.  •

**Definition.** If $p$ is prime and $G$ is a finite $p$-primary abelian group, then

$$d(G) = \dim(G/pG).$$

Observe that $d$ is additive over direct sums,

$$d(G \oplus H) = d(G) + d(H),$$

for Proposition 1.85 gives

$$(G \oplus H)/p(G \oplus H) = (G \oplus H)/(pG \oplus pH) \cong (G/pG) \oplus (H/pH).$$

The dimension of the left side is $d(G \oplus H)$ and the dimension of the right-hand side is $d(G) + d(H)$, for the union of a basis of $G/pG$ and a basis of $H/pH$ is a basis of $(G/pG) \oplus (H/pH)$.

There are nonzero $p$-primary abelian groups $H$ with $d(H) = 0$: for example, if $H$ is the **Prüfer group** $\mathbb{Z}(p^\infty)$, defined on page 647, then $H = pH$; that is, $d(H) = 0$. Thus, if $G$ is any $p$-primary abelian group, then $d(G \oplus H) = d(G)$.

Exercise 4.3 on page 241 shows that if $G$ is a *finite* abelian group, then $d(G) = 0$ if and only if $G = \{0\}$.

Finite $p$-primary abelian groups $G$ with $d(G) = 1$ are easily characterized.

**Lemma 4.18.** *If $G$ is a finite $p$-primary abelian group, then $d(G) = 1$ if and only if $G$ is a nonzero cyclic group.*

**Proof.** If $G$ is a nonzero cyclic group, then so is any nonzero quotient of $G$; in particular, $G/pG$ is cyclic. Now $G/pG \neq \{0\}$, by Exercise 4.3 on page 241, and so $\dim(G/pG) = 1$.

Conversely, if $d(G) = 1$, then $G/pG \cong \mathbb{I}_p$; hence $G/pG$ is cyclic, say, $G/pG = \langle z + pG \rangle$. Of course, $G \neq \{0\}$, and we are done if $G = \langle z \rangle$. Assume, on the contrary, that $\langle z \rangle$ is a proper subgroup of $G$. The Correspondence Theorem says that $pG$ is a maximal subgroup of $G$ (for $\mathbb{I}_p$ is a simple group). We claim that $pG$ is the only maximal subgroup of $G$. If $L \subseteq G$ is any maximal subgroup, then $G/L \cong \mathbb{I}_p$, for $G/L$ is a simple abelian group and, hence, has order $p$ (abelian simple groups are precisely the cyclic groups of prime order, by Proposition 1.117). It follows that if $a \in G$, then $p(a + L) = 0$ in $G/L$, and so $pa \in L$; that is, $pG \subseteq L$. But here $pG$ is a maximal subgroup, so that $pG = L$. As every proper subgroup is contained in a maximal subgroup, every proper subgroup of $G$ is contained in $pG$. In particular, $\langle z \rangle \subseteq pG$, so that the generator $z + pG$ of $G/pG$ is zero, a contradiction. Therefore, $G = \langle z \rangle$ is a nonzero cyclic group.  •

We need one more lemma before proving the main theorem.

**Lemma 4.19.** *Let $G$ be a finite $p$-primary abelian group.*

  (i) *If $S \subseteq G$, then $d(G/S) \leq d(G)$.*

  (ii) *If $S$ is a pure subgroup of $G$, then $d(G) = d(S) + d(G/S)$.*

**Proof.**

  (i) By the Correspondence Theorem, $p(G/S) = (pG + S)/S$, so that
$$(G/S)/p(G/S) = (G/S)/[(pG + S)/S] \cong G/(pG + S),$$
by the Third Isomorphism Theorem. Since $pG \subseteq pG + S$, there is a surjective homomorphism (of vector spaces over $\mathbb{F}_p$),
$$G/pG \to G/(pG + S),$$
namely, $g + pG \mapsto g + (pG + S)$. Hence,
$$\dim(G/pG) \geq \dim(G/(pG + S)) = d(G/S).$$

  (ii) We now analyze $(pG + S)/pG$, the kernel of $G/pG \to G/(pG + S)$. By the Second Isomorphism Theorem,
$$(pG + S)/pG \cong S/(S \cap pG).$$
Since $S$ is a pure subgroup, $S \cap pG = pS$; therefore,
$$(pG + S)/pG \cong S/pS,$$

and so $\dim[(pG + S)/pG] = d(S)$. But if $W$ is a subspace of a finite-dimensional vector space $V$, then $\dim(V) = \dim(W) + \dim(V/W)$, by Exercise 2.77 on page 145. Hence, for $V = G/pG$ and $W = (pG + S)/pG$, we have $d(G) = d(S) + d(G/S)$. •

**Theorem 4.20 (Basis Theorem).** *Every finite abelian group $G$ is a direct sum of primary cyclic groups.*

**Proof.** By the Primary Decomposition, Theorem 4.15, we may assume that $G$ is $p$-primary for some prime $p$. We prove that $G$ is a direct sum of cyclic groups by induction on $d(G) \geq 1$. The base step is Lemma 4.18, which shows that $G$ must be cyclic in this case.

For the inductive step, Lemma 4.17 says that there exists a nonzero pure cyclic subgroup $S \subseteq G$, and Lemma 4.19 says that

$$d(G/S) = d(G) - d(S) = d(G) - 1 < d(G).$$

By induction, $G/S$ is a direct sum of cyclic groups, say,

$$G/S = \bigoplus_{i=1}^{q} \langle \overline{x}_i \rangle,$$

where $\overline{x}_i = x_i + S$.

Let $g \in G$ and let $\overline{g} = g + S$ in $G/S$ have order $p^\ell$. We claim that there is a *lifting* $z \in G$ (that is, $z + S = \overline{g} = g + S$) such that

$$\text{order } z = \text{ order } \overline{g}.$$

Now $g$ has order $p^n$, where $n \geq \ell$. But $p^\ell(g + S) = p^\ell \overline{g} = 0$ in $G/S$, so there is some $s \in S$ with $p^\ell g = s$. By purity, there is $s' \in S$ with $p^\ell g = p^\ell s'$. If we define $z = g - s'$, then $p^\ell z = 0$ and $z + S = g + S = \overline{g}$. If $z$ has order $p^m$, then $m \geq \ell$ because $z \mapsto \overline{g}$; since $p^\ell z = 0$, the order of $z$ is equal to $p^\ell$.

For each $i$, choose a lifting $z_i \in G$ with order $z_i = $ order $\overline{x}_i$, and define $T$ by

$$T = \langle z_1, \ldots, z_q \rangle.$$

Now $S + T = G$, because $G$ is generated by $S$ and the $z_i$. To see that $G = S \oplus T$, it suffices to prove that $S \cap T = \{0\}$. If $y \in S \cap T$, then $y = \sum_i m_i z_i$, where $m_i \in \mathbb{Z}$. Now $y \in S$, and so $\sum_i m_i \overline{x}_i = 0$ in $G/S$. Since $G/S$ is the direct sum $\langle \overline{x}_1 \rangle \oplus \cdots \oplus \langle \overline{x}_n \rangle$, Corollary 4.6 says that each $m_i \overline{x}_i = 0$. Therefore, $m_i z_i = 0$ for all $i$, and hence $y = 0$.

Finally, $G = S \oplus T$ implies $d(G) = d(S) + d(T) = 1 + d(T)$, so that $d(T) < d(G)$. By induction, $T$ is a direct sum of cyclic groups, and this completes the proof. •

Here is a shorter proof of the Basis Theorem. One reason we have given the longer proof above is that it fits well with the upcoming proof of the Fundamental Theorem.

**Lemma 4.21.** *A finite $p$-primary abelian group $G$ is cyclic if and only if it has a unique subgroup of order $p$.*

**Proof.** Recall Theorem 1.93: if $G$ is an abelian group of order $n$ having at most one cyclic subgroup of order $p$ for every prime divisor $p$ of $n$, then $G$ is cyclic. The lemma follows at once when $n$ is a power of $p$. The converse is Lemma 1.92. •

**Remark.** We cannot remove the hypothesis that $G$ be abelian, for the group **Q** of quaternions is a 2-group having a unique subgroup of order 2. However, if $G$ is a (possibly nonabelian) finite $p$-group having a unique subgroup of order $p$, then $G$ is either cyclic or generalized quaternion (the latter groups are defined on page 271). A proof of this last result can be found in Rotman, *An Introduction to the Theory of Groups*, pp. 121–122.

The finiteness hypothesis cannot be removed, for the Prüfer group $\mathbb{Z}(p^\infty)$ is an infinite abelian $p$-primary group having a unique subgroup of order $p$ [see Proposition 9.25(iii)]. ◄

**Lemma 4.22.** *Let $G$ be a finite $p$-primary abelian group. If $a$ is an element of largest order in $G$, then $A = \langle a \rangle$ is a direct summand of $G$.*

**Remark.** There is a shorter proof of this, using the Basis Theorem and the fact (proved in Lemma 4.17) that $A$ is a pure subgroup. However, we want this alternative proof of the Basis Theorem to be self-contained. ◄

**Proof.** The proof is by induction on $|G| \geq 1$; the base step is trivially true. We may assume that $G$ is not cyclic, for any group is a direct summand of itself (with complementary summand $\{0\}$). Now $A = \langle a \rangle$ has a unique subgroup of order $p$; call it $C$. By Lemma 4.21, $G$ contains another subgroup of order $p$, say $C'$. Of course, $A \cap C' = \{0\}$. By the Second Isomorphism Theorem, $(A + C')/C' \cong A/(A \cap C') \cong A$ is a cyclic subgroup of $G/C'$. But no homomorphic image of $G$ can have a cyclic subgroup of order greater than $|A|$ (for no element of an image can have order larger than the order of $a$). Therefore, $(A + C')/C'$ is a cyclic subgroup of $G/C'$ of largest order and, by the inductive hypothesis, it is a direct summand; the Correspondence Theorem gives a subgroup $B/C'$, with $C' \subseteq B \subseteq G$, such that

$$G/C' = ((A + C')/C') \oplus (B/C').$$

We claim that $G = A \oplus B$. Clearly, $G = A + C' + B = A + B$ (for $C' \subseteq B$), while $A \cap B \subseteq A \cap ((A + C') \cap B) \subseteq A \cap C' = \{0\}$. •

**Theorem 4.23 (Basis Theorem Again).** *Every finite abelian group $G$ is a direct sum of primary cyclic groups.*

**Proof.** The proof is by induction on $|G| \geq 1$, and the base step is obviously true. To prove the inductive step, let $p$ be a prime divisor of $|G|$. Now $G = G_p \oplus H$, where $p \nmid |H|$ (either we can invoke the Primary Decomposition or reprove this special case of it). By induction, $H$ is a direct sum of primary cyclic groups. If $G_p$ is cyclic, we are done. Otherwise, Lemma 4.22 applies to write $G_p = A \oplus B$, where $A$ is primary cyclic. By the inductive hypothesis, $B$ is a direct sum of primary cyclic groups, and the theorem is proved. •

The shortest proof of the Basis Theorem that I know is due to Navarro, On the fundamental theorem of finite abelian groups, *Amer. Math. Monthly* 110 (2003), pp. 153–154. Another short proof is due to Rado, A proof of the basis theorem for finitely generated Abelian groups, *J. London Math. Soc.* 26 (1951), pp. 75–76, erratum, 160.

Here is a nice application of the Basis Theorem. The proof uses Dirichlet's Theorem on primes in arithmetic progressions: if $(a, d) = 1$, then there are infinitely many primes of the form $a + nd$ (Borevich–Shafarevich, *Number Theory*, p. 339).

**Theorem 4.24.** *If $G$ is a finite abelian group, then there exists an integer $m$ such that $G$ is isomorphic to a subgroup of $U(\mathbb{I}_m) = \{[k] : (k, m) = 1\}$.*

**Proof.** Consider the special case when $G$ is a cyclic group of order $d$. By Dirichlet's Theorem, there is a prime $p$ of the form $1 + nd$, and so $d \mid (p - 1)$. Now the group of units $U(\mathbb{I}_p)$ is a cyclic group of order $p - 1$, by Corollary 2.47, and so it contains a cyclic subgroup of order $d$, by Lemma 1.92. Thus, $G$ is isomorphic to a subgroup of $U(\mathbb{I}_p)$ in this case.

By the Basis Theorem, $G \cong \bigoplus_{i=1}^{n} C_i$, where $C_i$ is a cyclic group of order $d_i$, say. By Dirichlet's Theorem, for each $i \leq n$, there exists a prime $p_i$ with $p_i \equiv 1 \bmod d_i$. Moreover, since there are infinitely many such primes for each $i$, we may assume that the primes $p_1, \ldots, p_n$ are distinct. By Theorem 1.87 (essentially, the Chinese Remainder Theorem), $\mathbb{I}_m \cong \mathbb{I}_{p_1} \oplus \cdots \oplus \mathbb{I}_{p_n}$, where $m = p_1 \cdots p_n$, and so

$$U(\mathbb{I}_m) \cong U(\mathbb{I}_{p_1}) \oplus \cdots \oplus U(\mathbb{I}_{p_n}).$$

Since $C_i$ is isomorphic to a subgroup of $U(\mathbb{I}_{p_i})$ for all $i$, we have $G \cong \bigoplus_i C_i$ isomorphic to a subgroup of $\bigoplus_i U(\mathbb{I}_{p_i}) \cong U(\mathbb{I}_m)$.  •

**4.1.3. Fundamental Theorem.** When are two finite abelian groups $G$ and $G'$ isomorphic? By the Basis Theorem, these groups are direct sums of cyclic groups, and so our first guess is that $G \cong G'$ if they have the same number of cyclic summands of each type. But this hope is dashed by Theorem 1.87, which says that if $m$ and $n$ are relatively prime, then $\mathbb{I}_{mn} \cong \mathbb{I}_m \times \mathbb{I}_n$; for example, $\mathbb{I}_6 \cong \mathbb{I}_2 \times \mathbb{I}_3$. Thus, we retreat and try to count *primary* cyclic summands. But how can we do this? Why should two decompositions of a finite $p$-primary group have the same number of summands of order $p^2$ or $p^{17}$? We are asking whether there is a unique factorization theorem here, analogous to the Fundamental Theorem of Arithmetic.

Before stating the next lemma, recall that we have defined

$$d(G) = \dim(G/pG).$$

In particular, $d(pG) = \dim(pG/p^2 G)$ and, more generally,

$$d(p^n G) = \dim(p^n G / p^{n+1} G).$$

**Lemma 4.25.** *Let $G$ be a finite $p$-primary abelian group, where $p$ is prime, and let $G = \bigoplus_j C_j$, where each $C_j$ is cyclic. If $b_n \geq 0$ is the number of summands $C_j$ having order $p^n$, then there is an integer $t$ with*

$$d(p^n G) = b_{n+1} + b_{n+2} + \cdots + b_t.$$

**Proof.** Let $B_n$ be the direct sum of all $C_j$, if any, of order $p^n$. Since $G$ is finite, there is some $t$ with

$$G = B_1 \oplus B_2 \oplus \cdots \oplus B_t.$$

Now

$$p^n G = p^n B_{n+1} \oplus \cdots \oplus p^n B_t,$$

because $p^n B_j = \{0\}$ for all $j \leq n$. Similarly,

$$p^{n+1} G = p^{n+1} B_{n+2} \oplus \cdots \oplus p^{n+1} B_t.$$

By Proposition 4.9, $p^n G / p^{n+1} G$ is isomorphic to

$$\left[ p^n B_{n+1} / p^{n+1} B_{n+1} \right] \oplus \left[ p^n B_{n+2} / p^{n+1} B_{n+2} \right] \oplus \cdots \oplus \left[ p^n B_t / p^{n+1} B_t \right].$$

By Exercise 4.10 on page 242, $d(p^n B_m / p^{n+1} B_m) = d(p^n B_m) = b_m$ for all $n < m$; since $d$ is additive over direct sums, we have $d(p^n G) = b_{n+1} + b_{n+2} + \cdots + b_t$. •

The numbers $b_n$ can now be described in terms of $G$.

**Definition.** Let $G$ be a finite $p$-primary abelian group, where $p$ is prime. For $n \geq 0$, define[5]

$$U_p(n, G) = d(p^n G) - d(p^{n+1} G).$$

Lemma 4.25 shows that $d(p^n G) = b_{n+1} + \cdots + b_t$ and $d(p^{n+1} G) = b_{n+2} + \cdots + b_t$, so that $U_p(n, G) = b_{n+1}$.

**Theorem 4.26.** *If $p$ is prime, any two decompositions of a finite $p$-primary abelian group $G$ into direct sums of cyclic groups have the same number of cyclic summands of each type. More precisely, for each $n \geq 0$, the number of cyclic summands having order $p^{n+1}$ is $U_p(n, G)$.*

**Proof.** By the Basis Theorem, there exist cyclic subgroups $C_i$ with $G = \bigoplus_i C_i$. Lemma 4.25 shows, for each $n \geq 0$, that the number of $C_i$ having order $p^{n+1}$ is $U_p(n, G)$, a number that is defined without any mention of the given decomposition of $G$ into a direct sum of cyclics. Thus, if $G = \bigoplus_j D_j$ is another decomposition of $G$, where each $D_j$ is cyclic, then the number of $D_j$ having order $p^{n+1}$ is also $U_p(n, G)$, as desired. •

**Corollary 4.27.** *If $G$ and $G'$ are finite $p$-primary abelian groups, then $G \cong G'$ if and only if $U_p(n, G) = U_p(n, G')$ for all $n \geq 0$.*

**Proof.** If $\varphi : G \to G'$ is an isomorphism, then $\varphi(p^n G) = p^n G'$ for all $n \geq 0$, and so $\varphi$ induces isomorphisms of the $\mathbb{F}_p$-vector spaces $p^n G / p^{n+1} G \cong p^n G' / p^{n+1} G'$, for all $n \geq 0$, by $p^n g + p^{n+1} G \mapsto p^n \varphi(g) + p^{n+1} G'$. Thus, their dimensions are the same; that is, $U_p(n, G) = U_p(n, G')$.

Conversely, assume that $U_p(n, G) = U_p(n, G')$ for all $n \geq 0$. If $G = \bigoplus_i C_i$ and $G' = \bigoplus_j C_j'$, where the $C_i$ and $C_j'$ are cyclic, then Lemma 4.25 shows that the number of summands of each type is the same, and so it is a simple matter to construct an isomorphism $G \to G'$. •

---

[5]A theorem of Ulm classifies all *countable* $p$-primary abelian groups, using *Ulm invariants* which generalize $U_n(n, G)$. Our proof of the Fundamental Theorem is an adaptation of the proof of Ulm's Theorem given in Kaplansky, *Infinite Abelian Groups*, p. 27.

**Definition.** If $G$ is a $p$-primary abelian group, then its *elementary divisors* are the numbers in the sequence

$$U_p(0, G) \ p\text{'s}, \ U_p(1, G) \ p^2\text{'s}, \ \ldots, \ U_p(t-1, G) \ p^t\text{'s},$$

where $p^t$ is the largest order of a cyclic summand of $G$.

If $G$ is a finite abelian group, then its *elementary divisors* are the elementary divisors of all its primary components.

**Theorem 4.28 (Fundamental Theorem of Finite Abelian Groups).** *Two finite abelian groups $G$ and $G'$ are isomorphic if and only if they have the same elementary divisors; that is, any two decompositions of $G$ and $G'$ into direct sums of primary cyclic groups have the same number of such summands of each order.*

**Proof.** By the Primary Decomposition, Theorem 4.15(ii), $G \cong G'$ if and only if, for each prime $p$, their primary components are isomorphic. The result now follows from Corollary 4.27. •

**Example 4.29.** How many abelian groups are there of order 72? Now $72 = 2^3 3^2$, so that any abelian group of order 72 is the direct sum of groups of order 8 and order 9. There are three groups of order 8, described by the elementary divisors

$$(2, 2, 2), \quad (2, 4), \quad \text{and} \quad (8);$$

there are two groups of order 9, described by the elementary divisors

$$(3, 3) \quad \text{and} \quad (9).$$

Therefore, up to isomorphism, there are six abelian groups of order 72. ◄

Here is a second type of decomposition of a finite abelian group into a direct sum of cyclics that does not mention primary groups.

**Proposition 4.30.** *Every finite abelian group $G$ is a direct sum of cyclic groups,*

$$G = J(c_1) \oplus J(c_2) \oplus \cdots \oplus J(c_r),$$

*where $r \geq 1$, $J(c_i)$ is a cyclic group of order $c_i$, and*

$$c_1 \mid c_2 \mid \cdots \mid c_r.$$

**Proof.** Let $p_1, \ldots, p_n$ be the prime divisors of $|G|$. By the Basis Theorem, we have, for each $p_i$,

$$G_{p_i} = J(p_i^{e_{i1}}) \oplus J(p_i^{e_{i2}}) \oplus \cdots \oplus J(p_i^{e_{ir}}).$$

We may assume that $0 \leq e_{i1} \leq e_{i2} \leq \cdots \leq e_{ir}$; moreover, we may allow "dummy" exponents $e_{ij} = 0$ so that the same last index $r$ can be used for all $i$. Define

$$c_j = p_1^{e_{1j}} p_2^{e_{2j}} \cdots p_n^{e_{nj}}.$$

It is plain that $c_1 \mid c_2 \mid \cdots \mid c_r$. Finally, Theorem 1.87 shows, for every $j$, that

$$J(p_1^{e_{1j}}) \oplus J(p_2^{e_{2j}}) \oplus \cdots \oplus J(p_n^{e_{nj}}) \cong J(c_j). \quad •$$

**Definition.** If $G$ is a finite abelian group,[6] then its ***exponent*** is the smallest positive integer $m$ for which $mG = \{0\}$.

**Corollary 4.31.** *If $G$ is a finite abelian group and $G = J(c_1) \oplus J(c_2) \oplus \cdots \oplus J(c_r)$, $J(c_i)$ is a cyclic group of order $c_i$, and $c_1 \mid c_2 \mid \cdots \mid c_r$, then $c_r$ is the exponent of $G$.*

**Proof.** Since $c_i \mid c_r$ for all $i$, we have $c_r J(c_i) = 0$ for all $i$, and so $c_r G = \{0\}$. On the other hand, there is no number $e$ with $1 \leq e < c_r$ with $eJ(c_r) = \{0\}$, and so $c_r$ is the smallest positive integer annihilating $G$. •

**Corollary 4.32.** *Every noncyclic finite abelian group $G$ has a subgroup isomorphic to $\mathbb{I}_c \oplus \mathbb{I}_c$ for some $c > 1$.*

**Proof.** By Proposition 4.30, $G \cong \mathbb{I}_{c_1} \oplus \mathbb{I}_{c_2} \oplus \cdots \oplus \mathbb{I}_{c_r}$, where $r \geq 2$, because $G$ is not cyclic. Since $c_1 \mid c_2$, the cyclic group $\mathbb{I}_{c_2}$ contains a subgroup isomorphic to $\mathbb{I}_{c_1}$, and so $G$ has a subgroup isomorphic to $\mathbb{I}_{c_1} \oplus \mathbb{I}_{c_1}$. •

Let us return to the structure of finite abelian groups.

**Definition.** If $G$ is a finite abelian group and

$$G = J(c_1) \oplus J(c_2) \oplus \cdots \oplus J(c_r),$$

where $r \geq 1$, $J(c_j)$ is a cyclic group of order $c_j > 1$, and $c_1 \mid c_2 \mid \cdots \mid c_r$, then $c_1, c_2, \ldots, c_r$ are called the ***invariant factors*** of $G$.

**Corollary 4.33.** *If $G$ is a finite abelian group with invariant factors $c_1, \ldots, c_r$ and elementary divisors $\{p_i^{e_{ij}}\}$, then $|G| = \prod_{j=1}^{r} c_j = \prod_{i,j} p_i^{e_{ij}}$, and its exponent is $c_r$.*

**Proof.** We have

$$G \cong \mathbb{Z}/(c_1) \oplus \cdots \oplus \mathbb{Z}/(c_r) \cong \mathbb{I}_{c_1} \oplus \cdots \oplus \mathbb{I}_{c_r}.$$

It follows that $|G| = \prod_{j=1}^{r} c_j = \prod_{i,j} p_i^{e_{ij}}$ (for $c_j = p_1^{e_{1j}} p_2^{e_{2j}} \cdots p_n^{e_{nj}}$). That $c_r$ is the exponent was proved in Corollary 4.31. •

**Example 4.34.** The proof of Theorem 4.30 shows how to construct the invariant factors of a finite abelian group $G$ from its elementary divisors. We illustrate this for the group $G = \mathbb{I}_2 \oplus \mathbb{I}_2 \oplus \mathbb{I}_2 \oplus \mathbb{I}_3 \oplus \mathbb{I}_3$; its elementary divisors for the primes 2 and 3 are, respectively, $(2, 2, 2)$ and $(3, 3)$. Make both sequences the same length by adding an extra 1 (arising from a dummy exponent 0); thus, we replace $(3, 3)$ by $(1, 3, 3)$. The invariant factors of $G$ are $(2 \times 1, 2 \times 3, 2 \times 3) = (2, 6, 6)$. If $G = \mathbb{I}_2 \oplus \mathbb{I}_2 \oplus \mathbb{I}_2 \oplus \mathbb{I}_9$, its elementary divisors are $(2, 2, 2)$ and 9 [that is, $(2, 2, 2)$ and $(1, 1, 9)$], and its invariant factors are $(2, 2, 18)$. In Example 4.29, we displayed the

---

[6]This definition applies to nonabelian groups $G$ as well; it is the smallest positive integer $m$ with $x^m = 1$ for all $x \in G$.

elementary divisors of abelian groups of order 72; here are their invariant factors:

$$\textit{elementary divisors} \leftrightarrow \textit{invariant factors}$$

$$(2,2,2,3,3) = (2,2,2,1,3,3) \leftrightarrow 2 \mid 6 \mid 6$$
$$(2,4,3,3) \leftrightarrow 6 \mid 12$$
$$(8,3,3) = (1,8,3,3) \leftrightarrow 3 \mid 24$$
$$(2,2,2,9) = (2,2,2,1,1,9) \leftrightarrow 2 \mid 2 \mid 18$$
$$(2,4,9) = (2,4,1,9) \leftrightarrow 2 \mid 36$$
$$(8,9) \leftrightarrow 72 \quad \blacktriangleleft$$

**Theorem 4.35 (Invariant Factors).** *Two finite abelian groups are isomorphic if and only they have the same invariant factors.*

**Proof.** Given the elementary divisors of $G$, we can construct invariant factors, as in the proof of Proposition 4.30:

$$c_j = p_1^{e_{1j}} p_2^{e_{2j}} \cdots p_n^{e_{nj}},$$

where those factors $p_i^{e_{i1}}, p_i^{e_{i2}}, \ldots$ not equal to $p_i^0 = 1$ are the elementary divisors of the $p_i$-primary component of $G$. Thus, the invariant factors depend only on $G$ because they are defined in terms of the elementary divisors.

To prove isomorphism, it suffices, by the Fundamental Theorem, to show that the elementary divisors can be computed from the invariant factors. Since $c_j = p_1^{e_{1j}} p_2^{e_{2j}} \cdots p_n^{e_{nj}}$, the Fundamental Theorem of Arithmetic shows that $c_j$ determines all the prime powers $p_i^{e_{ij}}$ that are distinct from 1; that is, the invariant factors $c_j$ determine the elementary divisors. •

In Example 4.34, we started with elementary divisors and computed invariant factors of groups of order 72. Let us now start with invariant factors and compute elementary divisors:

$$\textit{invariant factors} \leftrightarrow \textit{elementary divisors}$$

$$2 \mid 6 \mid 6 = 2 \mid 2 \cdot 3 \mid 2 \cdot 3 \leftrightarrow (2,2,2,3,3)$$
$$6 \mid 12 = 2 \cdot 3 \mid 2^2 \cdot 3 \leftrightarrow (2,4,3,3)$$
$$3 \mid 24 = 3 \mid 2^3 \cdot 3 \leftrightarrow (8,3,3)$$
$$2 \mid 2 \mid 18 = 2 \mid 2 \mid 2 \cdot 3^2 \leftrightarrow (2,2,2,9)$$
$$2 \mid 36 = 2 \mid 2^2 \cdot 3^2 \leftrightarrow (2,4,9)$$
$$72 = 2^3 \cdot 3^2 \leftrightarrow (8,9).$$

The results of this section will be generalized in Chapter 8, from finite abelian groups to *finitely generated* abelian groups (an abelian group $G$ is finitely generated if it is isomorphic to a quotient of $\mathbb{Z}^m$ for some $m \geq 0$). The Basis Theorem generalizes: every finitely generated abelian group $G$ is a direct sum of cyclic groups, $G = T \oplus F$, where $T$ is finite and $F$ is free abelian. The Fundamental Theorem also generalizes: given two decompositions of $G$ into direct sums of primary cyclic and infinite cyclic groups, the number of summands of each kind is the same in both decompositions. The Basis Theorem is no longer true for abelian groups that are not finitely generated; for example, the additive group $\mathbb{Q}$ of rational numbers is not a direct sum of cyclic groups.

## Exercises

* **4.1.** (i) Let $G$ be an arbitrary, possibly nonabelian, group, and let $S$ and $T$ be normal subgroups of $G$. Prove that if $S \cap T = \{1\}$, then $st = ts$ for all $s \in S$ and $t \in T$.

   **Hint.** Show that $sts^{-1}t^{-1} \in S \cap T$.

   (ii) Prove that Proposition 4.4 holds for nonabelian groups $G$ if we assume that all the subgroups $S_i$ are normal subgroups.

* **4.2.** Give an example of a nonabelian group $G$ for which $G^m = \{a^m : a \in G\}$ is not a subgroup.

* **4.3.** Let $G$ be a $p$-primary abelian group. If $G = pG$, prove that either $G = \{0\}$ or $G$ is infinite.

* **4.4.** Let $G$ be an abelian group, not necessarily primary. Define a subgroup $S \subseteq G$ to be a ***pure subgroup*** if, for all $m \in \mathbb{Z}$,

$$S \cap mG = mS.$$

Prove that if $G$ is a $p$-primary abelian group, then a subgroup $S \subseteq G$ is pure as just defined if and only if $S \cap p^n G = p^n S$ for all $n \geq 0$ (the definition on page 231).

* **4.5.** Prove that a subgroup of a finite abelian group is a direct summand if and only if it is a pure subgroup.

   **Hint.** Modify the proof of the Basis Theorem, Theorem 4.20.

* **4.6.** Let $G$ be a possibly infinite abelian group.

   (i) Prove that every direct summand $S$ of $G$ is a pure subgroup.

   Define the ***torsion[7] subgroup[8]*** $tG$ of $G$ as

   $$tG = \{a \in G : a \text{ has finite order}\}.$$

   An abelian group $G$ is ***torsion*** if $G = tG$; that is, every element in $G$ has finite order.

   (ii) Prove that $tG$ is a pure subgroup of $G$. (There exist abelian groups $G$ whose torsion subgroup $tG$ is not a direct summand [see Exercise 8.3(iii) on page 651]; hence, a pure subgroup need not be a direct summand.)

   (iii) Prove that $G/tG$ is an abelian group in which every nonzero element has infinite order.

* **4.7.** Let $p$ be prime and let $q$ be relatively prime to $p$.

   (i) Prove that if $G$ is a $p$-primary group and $g \in G$, then there exists $x \in G$ with $qx = g$.

---

[7]This terminology comes from Algebraic Topology. To each space $X$, a sequence of abelian groups is assigned, called *homology groups*, and if $X$ is "twisted," then there are elements of finite order in some of these groups.

[8]If $G$ is not abelian, then $tG$ need not be a subgroup; see Exercise 1.34 on page 27.

(ii) Prove (i) without assuming that $G$ is abelian. Let $G$ be a group in which every element has order a power of $p$. If $g \in G$, then there exists $x \in G$ with $x^q = g$.

* **4.8.** Let $G = \langle a \rangle$ be a cyclic group of finite order $m$. Prove that $G/nG$ is a cyclic group of order $d$, where $d = (m, n)$.

* **4.9.** For an abelian group $G$ and a positive integer $n$, define
$$G[n] = \{g \in G : g^n = 1\}.$$
If $G = \langle a \rangle$ has order $m$, prove that $G[n] = \langle a^{m/d} \rangle$, where $d = (m, n)$, and conclude that $G[n] \cong \mathbb{I}_d$.

* **4.10.** Prove that if $B = B_m = \langle x_1 \rangle \oplus \cdots \oplus \langle x_{b_m} \rangle$ is a direct sum of $b_m$ cyclic groups of order $p^m$, then for $n < m$, the cosets $p^n x_i + p^{n+1} B$ for $1 \le i \le b_m$ are a basis for $p^n B/p^{n+1} B$. Conclude that $d(p^n B_m) = b_m$ when $n < m$. [Recall that if $G$ is a finite abelian group, then $G/pG$ is a vector space over $\mathbb{F}_p$ and $d(G) = \dim(G/pG)$.]

**4.11.** The proof of Theorem 4.20 contains the following result: if $S$ is a pure subgroup of a $p$-primary abelian group $G$, then every $g + S \in G/S$ has a lifting $g \in G$ with $g$ and $g + S$ having the same order. Prove the converse: if $S$ is a subgroup of $G$ such that every element of $G/S$ has a lifting of the same order, then $S$ is a pure subgroup.

**4.12.** If $G$ is a finite abelian group (not necessarily primary) and $x \in G$ has maximal order (that is, no element in $G$ has larger order), prove that $\langle x \rangle$ is a direct summand of $G$.

* **4.13.**   (i) If $G$ and $H$ are finite abelian groups, prove, for all primes $p$ and all $n \ge 0$, that
$$U_p(n, G \oplus H) = U_p(n, G) + U_p(n, H).$$

  (ii) If $A$, $B$, and $C$ are finite abelian groups, prove that $A \oplus B \cong A \oplus C$ implies $B \cong C$.

  (iii) If $A$ and $B$ are finite abelian groups, prove that $A \oplus A \cong B \oplus B$ implies $A \cong B$.

**4.14.** If $n$ is a positive integer, then a **partition of** $n$ is a sequence of positive integers $i_1 \le i_2 \le \cdots \le i_r$ with $i_1 + i_2 + \cdots + i_r = n$. If $p$ is prime, prove that the number of nonisomorphic abelian groups of order $p^n$ is equal to the number of partitions of $n$.

**4.15.** Prove that there are, up to isomorphism, exactly 14 abelian groups of order 288.

**4.16.** Prove the uniqueness assertion in the Fundamental Theorem of Arithmetic by applying the Fundamental Theorem of Finite Abelian Groups to $G = \mathbb{I}_n$.

**4.17.**   (i) If $G$ is a finite abelian group, define
$$\nu_k(G) = \text{ the number of elements in } G \text{ of order } k.$$

  Prove that two finite abelian groups $G$ and $G'$ are isomorphic if and only if $\nu_k(G) = \nu_k(G')$ for all integers $k$.

  **Hint.** If $B$ is a direct sum of $k$ copies of a cyclic group of order $p^n$, then how many elements of order $p^n$ are in $B$?

  (ii) Give an example of two nonisomorphic not necessarily abelian finite groups $G$ and $G'$ for which $\nu_k(G) = \nu_k(G')$ for all integers $k$.
  **Hint.** Take $G$ of order $p^3$ (Proposition 4.48).

**4.18.** Prove that the additive group $\mathbb{Q}$ is not a direct sum: $\mathbb{Q} \ne A \oplus B$, where $A$ and $B$ are nonzero subgroups.

**4.19.** Let $G = B_1 \oplus B_2 \oplus \cdots \oplus B_t$, where the $B_i$ are subgroups of $G$.

  (i) Prove that $G[p] = B_1[p] \oplus B_2[p] \oplus \cdots \oplus B_t[p]$.

(ii) Prove, for all $n \geq 0$, that

$$p^n G \cap G[p] = (p^n G \cap B_1[p]) \oplus (p^n G \cap B_2[p]) \oplus \cdots \oplus (p^n G \cap B_t[p])$$
$$= (p^n B_1 \cap B_1[p]) \oplus (p^n B_2 \cap B_2[p]) \oplus \cdots \oplus (p^n B_t \cap B_t[p]).$$

(iii) If $G$ is a finite $p$-primary abelian group, prove, for all $n \geq 0$, that

$$U_p(n, G) = \dim \left( \frac{p^n G \cap G[p]}{p^{n+1} G \cap G[p]} \right).$$

## Section 4.2. Sylow Theorems

We return to nonabelian groups, and so we revert to the multiplicative notation. The Sylow theorems introduce Sylow subgroups, which are analogs, for finite nonabelian groups, of the primary components of finite abelian groups.

Recall that a group $G$ is called *simple* if $G \neq \{1\}$ and it has no normal subgroups other than $\{1\}$ and $G$ itself. We saw, in Proposition 1.117, that the abelian simple groups are precisely the cyclic groups $\mathbb{I}_p$ of prime order $p$, and we saw, in Theorem 1.123, that $A_n$ is a nonabelian simple group for all $n \geq 5$. In fact, $A_5$ is the nonabelian simple group of smallest order. How can we prove that a nonabelian group $G$ of order less than $60 = |A_5|$ is not simple? Exercise 1.108 on page 76 states that if $G$ is a group of order $mp$, where $p$ is prime and $1 < m < p$, then $G$ is not simple. This exercise shows that many of the numbers less than 60 are not orders of simple groups. After throwing out all prime powers (finite nonabelian $p$-groups are never simple, by Corollary 1.118), the only remaining possibilities are

$$12, 18, 24, 30, 36, 40, 45, 48, 50, 54, 56.$$

The solution to Exercise 1.108 uses Cauchy's Theorem, which says that $G$ has a subgroup of order $p$. We shall see that if $G$ has a subgroup of prime power order $p^e$ instead of $p$, then the exercise can be generalized and the list of candidates can be shortened. Proposition 4.44 uses this result to show that $A_5$ is, indeed, the smallest nonabelian simple group.

The first book on Group Theory, Jordan's *Traité des Substitutions et des Équations Algébriques*, was published in 1870; more than half of it is devoted to Galois Theory, then called the Theory of Equations. At about the same time, but too late for publication in Jordan's book, three fundamental theorems were discovered. In 1868, Schering proved the Basis Theorem: every finite abelian group is a direct product of primary cyclic groups. In 1870, Kronecker, unaware of Schering's proof, also proved this result. In 1878, Frobenius and Stickelberger proved the Fundamental Theorem of Finite Abelian Groups. In 1872, Sylow showed, for every finite group $G$ and every prime $p$, that if $p^e$ is the largest power of $p$ dividing $|G|$, then $G$ has a subgroup of order $p^e$ (nowadays called a *Sylow p-subgroup*). We will use such subgroups to generalize Exercise 1.108.

Our strategy for proving the Sylow Theorems works best if we adopt the following definition.

**Definition.** Let $p$ be prime. A ***Sylow p-subgroup*** of a finite group $G$ is a maximal $p$-subgroup $P$.

Maximality means that if $Q$ is a $p$-subgroup of $G$ and $P \subseteq Q$, then $P = Q$.

It follows from Lagrange's Theorem that if $p^e$ is the largest power of $p$ dividing the order of a group $G$, then a subgroup of $G$ of order $p^e$, should it exist, is a maximal $p$-subgroup. It is not clear, however, that $G$ has any subgroups of order $p^e$, but it is clear that maximal $p$-subgroups always exist. We shall prove, in Theorem 4.39, that Sylow $p$-subgroups do have order $p^e$.

Let us show that if $S$ is any $p$-subgroup of $G$ (perhaps $S = \{1\}$), then there exists a Sylow $p$-subgroup $P$ containing $S$. If there is no $p$-subgroup strictly containing $S$, then $S$ itself is a Sylow $p$-subgroup. Otherwise, there is a $p$-subgroup $P_1$ with $S \subsetneq P_1$. If $P_1$ is maximal, it is Sylow, and we are done. Otherwise, there is some $p$-subgroup $P_2$ with $P_1 \subsetneq P_2$. This procedure of producing larger and larger $p$-subgroups $P_i$ must end after a finite number of steps because $|P_i| \le |G|$ for all $i$; the largest $P_i$ must, therefore, be a Sylow $p$-subgroup.

Recall that a *conjugate* of a subgroup $H \subseteq G$ is a subgroup of $G$ of the form

$$aHa^{-1} = \{aha^{-1} : h \in H\},$$

where $a \in G$. The *normalizer* of $H$ in $G$ is the subgroup

$$N_G(H) = \{a \in G : aHa^{-1} = H\},$$

and Proposition 1.110 states that if $H$ is a subgroup of a finite group $G$, then the number of conjugates of $H$ in $G$ is $[G : N_G(H)]$.

It is obvious that $H \lhd N_G(H)$, and so the quotient group $N_G(H)/H$ is defined.

**Lemma 4.36.** *Let $P$ be a Sylow $p$-subgroup of a finite group $G$.*

(i) *Every conjugate of $P$ is also a Sylow $p$-subgroup of $G$.*

(ii) $|N_G(P)/P|$ *is prime to $p$.*

(iii) *If $a \in G$ has order some power of $p$ and $aPa^{-1} = P$, then $a \in P$.*

**Proof.**

(i) If $a \in G$ and $aPa^{-1}$ is not a Sylow $p$-subgroup of $G$, then there is a $p$-subgroup $Q$ with $aPa^{-1} \subsetneq Q$. But $P \subsetneq a^{-1}Qa$, contradicting the maximality of $P$.

(ii) If $p$ divides $|N_G(P)/P|$, then Cauchy's Theorem shows that $N_G(P)/P$ contains an element $aP$ of order $p$, and hence $N_G(P)/P$ contains a subgroup $S^* = \langle aP \rangle$ of order $p$. By the Correspondence Theorem, there is a subgroup $S$ with $P \subseteq S \subseteq N_G(P)$ such that $S/P \cong S^*$. But $S$ is a $p$-subgroup of $N_G(P) \subseteq G$ (by Exercise 1.83 on page 60) strictly larger than $P$, contradicting the maximality of $P$. We conclude that $p$ does not divide $|N_G(P)/P|$.

(iii) By the definition of normalizer, the element $a$ lies in $N_G(P)$. If $a \notin P$, then the coset $aP$ is a nontrivial element of $N_G(P)/P$ having order some power of $p$; in light of part (ii), this contradicts Lagrange's Theorem.  •

Since every conjugate of a Sylow $p$-subgroup is a Sylow $p$-subgroup, it is natural to let $G$ act on the Sylow $p$-subgroups by conjugation.

**Theorem 4.37 (Sylow).** *Let $G$ be a finite group of order $p_1^{e_1} \cdots p_t^{e_t}$, and let $P$ be a Sylow $p$-subgroup of $G$ for some prime $p = p_j$.*

(i) *Every Sylow $p$-subgroup is conjugate to $P$.*[9]

(ii) *If there are $r_j$ Sylow $p_j$-subgroups, then $r_j$ is a divisor of $|G|/p_j^{e_j}$ and*

$$r_j \equiv 1 \bmod p_j.$$

**Proof.** Let $X = \{P_1, \ldots, P_{r_j}\}$ be the set of all the conjugates of $P$, where we have denoted $P$ by $P_1$. If $Q$ is any Sylow $p$-subgroup of $G$, then $Q$ acts on $X$ by conjugation: if $a \in Q$, then it sends

$$P_i = g_i P g_i^{-1} \mapsto a\big(g_i P g_i^{-1}\big)a^{-1} = (ag_i)P(ag_i)^{-1} \in X.$$

By Corollary 1.108, the number of elements in any orbit is a divisor of $|Q|$; that is, every orbit has size some power of $p$ (because $Q$ is a $p$-group). If there is an orbit of size 1, then there is some $P_i$ with $aP_ia^{-1} = P_i$ for all $a \in Q$. By Lemma 4.36, we have $a \in P_i$ for all $a \in Q$; that is, $Q \subseteq P_i$. But $Q$, being a Sylow $p$-subgroup, is a maximal $p$-subgroup of $G$, and so $Q = P_i$. In particular, if $Q = P_1$, then there is only one orbit of size 1, namely, $\{P_1\}$, and all the other orbits have sizes that are honest powers of $p$. We conclude that $|X| = r_j \equiv 1 \bmod p_j$.

Suppose now that there is some Sylow $p$-subgroup $Q$ that is not a conjugate of $P$; thus, $Q \neq P_i$ for any $i$. Again, we let $Q$ act on $X$, and again, we ask if there is an orbit of size 1, say, $\{P_k\}$. As in the previous paragraph, this implies $Q = P_k$, contrary to our present assumption that $Q \notin X$. Hence, there are no orbits of size 1, which says that each orbit has size an honest power of $p$. It follows that $|X| = r_j$ is a multiple of $p$; that is, $r_j \equiv 0 \bmod p_j$, which contradicts the congruence $r_j \equiv 1 \bmod p_j$. Therefore, no such $Q$ can exist, and so all Sylow $p$-subgroups are conjugate to $P$.

Finally, since all Sylow $p$-subgroups are conjugate, we have $r_j = [G : N_G(P)]$, and so $r_j$ is a divisor of $|G|$. But $r_j \equiv 1 \bmod p_j$ implies $(r_j, p_j^{e_j}) = 1$, so that Euclid's Lemma gives $r_j$ a divisor of $|G|/p_j^{e_j}$.  •

**Corollary 4.38.** *A finite group $G$ has a unique Sylow $p$-subgroup $P$ for some prime $p$ if and only if $P \triangleleft G$.*

**Proof.** Assume that $P$, a Sylow $p$-subgroup of $G$, is unique. For each $a \in G$, the conjugate $aPa^{-1}$ is also a Sylow $p$-subgroup; by uniqueness, $aPa^{-1} = P$ for all $a \in G$, and so $P \triangleleft G$.

Conversely, assume that $P \triangleleft G$. If $Q$ is any Sylow $p$-subgroup, then $Q = aPa^{-1}$ for some $a \in G$; but $aPa^{-1} = P$, by normality, and so $Q = P$.  •

The next result shows that the order of a Sylow $p$-subgroup of a group $G$ is the largest power of $p$ dividing $|G|$.

---

[9]It follows that all Sylow $p$-subgroups are isomorphic.

**Theorem 4.39 (Sylow).** *If $G$ is a finite group of order $p^e m$, where $p$ is prime and $p \nmid m$, then every Sylow $p$-subgroup $P$ of $G$ has order $p^e$.*

**Proof.** We first show that $p \nmid [G : P]$. Now $[G : P] = [G : N_G(P)][N_G(P) : P]$. The first factor, $[G : N_G(P)] = r$, is the number of conjugates of $P$ in $G$, and so $p$ does not divide $[G : N_G(P)]$ because $r \equiv 1 \bmod p$. The second factor, $[N_G(P) : P] = |N_G(P)/P|$, is also not divisible by $p$, by Lemma 4.36. Therefore, $p$ does not divide $[G : P]$, by Euclid's Lemma.

Now $|P| = p^k$ for some $k \leq e$, and so $[G : P] = |G|/|P| = p^e m/p^k = p^{e-k}m$. Since $p$ does not divide $[G : P]$, we must have $k = e$; that is, $|P| = p^e$.    •

**Example 4.40.**

(i) Let $G = S_4$. Since $|S_4| = 24 = 2^3 3$, a Sylow 2-subgroup $P$ of $S_4$ has order 8. We have seen, in Exercise 1.90 on page 74, that $S_4$ contains a copy of the dihedral group $D_8$ (the symmetries of a square). The Sylow Theorem says that all subgroups of $S_4$ of order 8 are conjugate, hence isomorphic; thus, $P \cong D_8$. Moreover, the number $r$ of Sylow 2-subgroups is a divisor of 24 congruent to 1 mod 2; that is, $r$ is an odd divisor of 24. Since $r \neq 1$ (Exercise 4.20 on page 250), there are exactly three Sylow 2-subgroups.

(ii) If $G$ is a finite abelian group, then a Sylow $p$-subgroup is just its $p$-primary component (since $G$ is abelian, every subgroup is normal, and so there is a unique Sylow $p$-subgroup for every prime $p$).    ◄

Here is a second proof of the last Sylow Theorem.

**Theorem 4.41.** *If $G$ is a finite group of order $p^e m$, where $p$ is a prime and $p \nmid m$, then $G$ has a subgroup of order $p^e$.*

**Proof. (Wielandt)** If $X$ is the family of all those *subsets* of $G$ having exactly $p^e$ elements, then $|X| = \binom{p^e m}{p^e}$, and $p \nmid |X|$.[10] Now $G$ acts on $X$: define $gB$, for $g \in G$ and $B \in X$, by

$$gB = \{gb : b \in B\}.$$

If $p$ divides $|\mathcal{O}(B)|$ for every $B \in X$, where $\mathcal{O}(B)$ is the orbit of $B$, then $p$ is a divisor of $|X|$, for $X$ is the disjoint union of orbits (Proposition 1.106). As $p \nmid |X|$, there exists a subset $B$ with $|B| = p^e$ and with $|\mathcal{O}(B)|$ not divisible by $p$. If $G_B$ is the stabilizer of this subset, Theorem 1.107 gives $[G : G_B] = |\mathcal{O}(B)|$, and so $|G| = |G_B| \cdot |\mathcal{O}(B)|$. Since $p^e \mid |G|$ and $(p^e, |\mathcal{O}(B)|) = 1$, Euclid's Lemma gives $p^e \mid |G_B|$. Therefore, $p^e \leq |G_B|$.

For the reverse inequality, choose an element $b \in B$. Now $G_B b$ is a right coset of $G_B$, and so $|G_B| = |G_B b|$. But $G_B b \subseteq B$, because $G_B b = \{gb : g \in G_B\}$ and $gb \in gB \subseteq B$ (for $g \in G_B$). Therefore, $|G_B| \leq |B| = p^e$. We conclude that $G_B$ is a subgroup of $G$ of order $p^e$.    •

---

[10]If $n = p^e m$, where $p$ is a prime not dividing $e$, then $p \nmid \binom{n}{p^e}$; otherwise, cross multiply and use Euclid's Lemma.

**Proposition 4.42.** *A finite group $G$ all of whose Sylow subgroups are normal is the direct product of its Sylow subgroups.*[11]

**Proof.** Let $|G| = p_1^{e_1} \cdots p_t^{e_t}$ and let $G_{p_i}$ be the Sylow $p_i$-subgroup of $G$. We use Exercise 4.1 on page 241, the generalization of Proposition 4.4 to nonabelian groups. The subgroup $S$ generated by all the Sylow subgroups is $G$, for $p_i^{e_i} \mid |S|$ for all $i$, hence, $|S| = |G|$. Finally, if $x \in G_{p_i} \cap \langle \bigcup_{j \neq i} G_{p_j} \rangle$, then $x = s_i \in G_{p_i}$ and $x = \prod_{j \neq i} s_j$, where $s_j \in G_{p_j}$. Now $x^{p_i^n} = 1$ for some $n \geq 1$. On the other hand, there is some power of $p_j$, say $q_j$, with $s_j^{q_j} = 1$ for all $j$. Since the $s_j$ commute with each other, by Exercise 4.1(i), we have $1 = x^q = (\prod_{j \neq i} s_j)^q$, where $q = \prod_{j \neq i} q_j$. Since $(p_i^n, q) = 1$, there are integers $u$ and $v$ with $1 = up_i^n + vq$, and so $x = x^1 = x^{up_i^n} x^{vq} = 1$. Thus, $G$ is the direct product of its Sylow subgroups. $\bullet$

We can now generalize Exercise 1.108 on page 76 and its solution.

**Lemma 4.43.** *There is no nonabelian simple group $G$ of order $|G| = p^e m$, where $p$ is prime, $p \nmid m$, and $p^e \nmid (m-1)!$.*

**Proof.** Suppose that such a simple group $G$ exists. By Sylow's Theorem, $G$ contains a subgroup $P$ of order $p^e$, hence of index $m$. We may assume that $m > 1$, for nonabelian $p$-groups are never simple, by Corollary 1.118. By Theorem 1.96, there exists a homomorphism $\varphi \colon G \to S_m$ with $\ker \varphi \subseteq P$. Since $G$ is simple, however, it has no proper normal subgroups; hence $\ker \varphi = \{1\}$ and $\varphi$ is an injection; that is, $G \cong \varphi(G) \subseteq S_m$. By Lagrange's Theorem, $p^e m \mid m!$, and so $p^e \mid (m-1)!$, contrary to the hypothesis. $\bullet$

**Proposition 4.44.** *There are no nonabelian simple groups of order less than 60.*

**Proof.** The reader may now check that the only integers $n$ between 2 and 59, neither a prime power nor having a factorization of the form $n = p^e m$ as in the statement of Lemma 4.43, are $n = 30, 40$, and 56, and so these three numbers are the only candidates for orders of nonabelian simple groups of order $< 60$.

Suppose there is a simple group $G$ of order 30. Let $P$ be a Sylow 5-subgroup of $G$, so that $|P| = 5$. The number $r_5$ of conjugates of $P$ is a divisor of 30 and $r_5 \equiv 1 \bmod 5$. Now $r_5 \neq 1$ lest $P \triangleleft G$, so that $r_5 = 6$. By Lagrange's Theorem, the intersection of any two of these is trivial (intersections of Sylow subgroups can be more complicated; see Exercise 4.21 on page 250). There are four nonidentity elements in each of these subgroups, and so there are $6 \times 4 = 24$ nonidentity elements in their union. Similarly, the number $r_3$ of Sylow 3-subgroups of $G$ is 10 (for $r_3 \neq 1$, $r_3$ is a divisor of 30, and $r_3 \equiv 1 \bmod 3$). There are two nonidentity elements in each of these subgroups, and so the union of these subgroups has 20 nonidentity elements. We have exceeded 30, the number of elements in $G$, and so $G$ cannot be simple.

Let $G$ be a group of order 40, and let $P$ be a Sylow 5-subgroup of $G$. If $r$ is the number of conjugates of $P$, then $r \mid 40$ and $r \equiv 1 \bmod 5$. These conditions force $r = 1$, so that $P \triangleleft G$; therefore, no simple group of order 40 can exist.

---

[11]Such groups $G$ are called ***nilpotent***.

Finally, suppose there is a simple group $G$ of order 56. If $P$ is a Sylow 7-subgroup of $G$, then $P$ must have $r_7 = 8$ conjugates (for $r_7 \mid 56$ and $r_7 \equiv 1 \mod 7$). Since these groups are cyclic of prime order, the intersection of any pair of them is $\{1\}$, and so there are 48 nonidentity elements in their union. Thus, adding the identity, we have accounted for 49 elements of $G$. Now a Sylow 2-subgroup $Q$ has order 8, and so it contributes seven more nonidentity elements, giving 56 elements. But there is a second Sylow 2-subgroup, lest $Q \lhd G$, and we have exceeded our quota. Therefore, there is no simple group of order 56.  •

The order of the next nonabelian simple group is 168.

The "converse" of Lagrange's Theorem is false: if $G$ is a finite group of order $n$ and $d \mid n$, then $G$ may not have a subgroup of order $d$. For example, we proved, in Proposition 1.70, that the alternating group $A_4$ is a group of order 12 having no subgroup of order 6.

**Proposition 4.45.** *Let $G$ be a finite group. If $p$ is prime and $p^k$ divides $|G|$, then $G$ has a subgroup of order $p^k$.*

**Proof.** If $|G| = p^e m$, where $p \nmid m$, then a Sylow $p$-subgroup $P$ of $G$ has order $p^e$. Hence, if $p^k$ divides $|G|$, then $p^k$ divides $|P|$. By Proposition 1.116, $P$ has a subgroup of order $p^k$; a fortiori, $G$ has a subgroup of order $p^k$.  •

What examples of $p$-groups have we seen? Of course, cyclic groups of order $p^n$ are $p$-groups, as is any direct product of copies of these. By the Fundamental Theorem, this describes all (finite) abelian $p$-groups. The only nonabelian examples we have seen so far are the dihedral groups $D_{2n}$ (which are 2-groups when $n$ is a power of 2) and the quaternions $\mathbf{Q}$ of order 8 (of course, for every 2-group $A$, the direct products $D_8 \times A$ and $\mathbf{Q} \times A$ are also nonabelian 2-groups). Here are some new examples.

**Definition.** A ***unitriangular*** matrix over a field $k$ is an upper triangular matrix each of whose diagonal terms is 1. Define $\mathrm{UT}(n, k)$ to be the set of all $n \times n$ unitriangular matrices over $k$.

For example, $\begin{bmatrix} a_{11} & a_{12} & a_{13} \\ 0 & a_{22} & a_{23} \\ 0 & 0 & a_{33} \end{bmatrix}$ is upper triangular, and $\begin{bmatrix} 1 & a_{12} & a_{13} \\ 0 & 1 & a_{23} \\ 0 & 0 & 1 \end{bmatrix}$ is unitriangular.

**Remark.** We can generalize this definition by allowing $k$ to be any commutative ring. For example, the group $\mathrm{UT}(n, \mathbb{Z})$ is an interesting group (it is a finitely generated torsion-free nilpotent group).  ◄

**Proposition 4.46.** *If $k$ is a field, then $\mathrm{UT}(n, k)$ is a subgroup of $\mathrm{GL}(n, k)$.*

**Proof.** Of course, the identity matrix $I$ is unitriangular, so that $I \in \mathrm{UT}(n, k)$. If $A \in \mathrm{UT}(n, k)$, then $A = I + N$, where $N$ is *strictly* upper triangular; that is, $N$ is an upper triangular matrix having only 0's on its diagonal. Note that $N^n = 0$, by Exercise 2.88(ii) on page 155.

Exercise 2.88(i) says that if $N, M$ are strictly upper triangular, then so are $N + M$ and $NM$. Hence, $(I + N)(I + M) = I + (N + M + NM)$ is unitriangular, and $\mathrm{UT}(n, k)$ is closed. Now unitriangular matrices are nonsingular because $\det(I + N) = 1$, but a proof of this without determinants will also show that $A^{-1} = (I + N)^{-1}$ is unitriangular. Recalling the power series expansion $1/(1 + x) = 1 - x + x^2 - x^3 + \cdots$, we define $B = I - N + N^2 - N^3 + \cdots$ (this series stops after $n - 1$ terms because $N^n = 0$). The reader may now check that $BA = B(I + N) = I$, so that $B = A^{-1}$. Moreover, $N$ strictly upper triangular implies that $-N + N^2 - N^3 + \cdots \pm N^{n-1}$ is also strictly upper triangular, and so $A^{-1} = B$ is unitriangular. Therefore, $\mathrm{UT}(n, k)$ is a subgroup of $\mathrm{GL}(n, k)$. •

**Proposition 4.47.** *Let $q = p^e$, where $p$ is prime. For each $n \geq 2$, $\mathrm{UT}(n, \mathbb{F}_q)$ is a p-group of order $q^{\binom{n}{2}} = q^{n(n-1)/2}$.*

**Proof.** The number of entries in an $n \times n$ unitriangular matrix lying strictly above the diagonal is $\binom{n}{2} = \frac{1}{2}n(n-1)$ (throw away $n$ diagonal entries from the total of $n^2$ entries; half of the remaining $n^2 - n$ entries are above the diagonal). Since each of these entries can be any element of $\mathbb{F}_q$, there are exactly $q^{\binom{n}{2}}$ $n \times n$ unitriangular matrices over $\mathbb{F}_q$, and so this is the order of $\mathrm{UT}(n, \mathbb{F}_q)$. •

Recall Exercise 1.35 on page 27: if $G$ is a group and $x^2 = 1$ for all $x \in G$, then $G$ is abelian. We now ask whether a group $G$ satisfying $x^p = 1$ for all $x \in G$, where $p$ is an odd prime, must also be abelian.

**Proposition 4.48.** *If $p$ is an odd prime, then there exists a nonabelian group $G$ of order $p^3$ with $x^p = 1$ for all $x \in G$.*

**Proof.** If $G = \mathrm{UT}(3, \mathbb{F}_p)$, then $|G| = p^3$. Now $G$ is not abelian; for example, the matrices $\begin{bmatrix} 1 & 1 & 0 \\ 0 & 1 & 1 \\ 0 & 0 & 1 \end{bmatrix}$ and $\begin{bmatrix} 1 & 0 & 1 \\ 0 & 1 & 1 \\ 0 & 0 & 1 \end{bmatrix}$ do not commute. If $A \in G$, then $A = I + N$, where $N$ is strictly upper triangular; since $p$ is an odd prime, $p \geq 3$, and $N^p = 0$. Finally, Corollary 2.6 says that

$$A^p = (I + N)^p = I^p + N^p = I. \quad •$$

**Theorem 4.49.** *Let $\mathbb{F}_q$ denote the finite field with $q$ elements. Then*

$$|\mathrm{GL}(n, \mathbb{F}_q)| = (q^n - 1)(q^n - q)(q^n - q^2) \cdots (q^n - q^{n-1}).$$

**Proof.** Let $V$ be an $n$-dimensional vector space over $\mathbb{F}_q$. We show first that there is a bijection $\Phi \colon \mathrm{GL}(n, \mathbb{F}_q) \to \mathcal{B}$, where $\mathcal{B}$ is the set of all bases of $V$.[12] Choose, once for all, a basis $e_1, \ldots, e_n$ of $V$. If $T \in \mathrm{GL}(n, \mathbb{F}_q)$, define

$$\Phi(T) = Te_1, \ldots, Te_n.$$

By Lemma 2.118, $\Phi(T) \in \mathcal{B}$ because $T$, being nonsingular, carries a basis into a basis. But $\Phi$ is a bijection, for given a basis $v_1, \ldots, v_n$, there is a unique linear transformation $S$, necessarily nonsingular (by Lemma 2.118), with $Se_i = v_i$ for all $i$ (by Theorem 2.120).

---

[12]Recall that a basis of a finite-dimensional vector space is an *ordered set* of vectors, not merely a *set* of vectors. This distinction is critical in this proof.

Our problem now is to count the number of bases $v_1, \ldots, v_n$ of $V$. There are $q^n$ vectors in $V$, and so there are $q^n - 1$ candidates for $v_1$ (the zero vector is not a candidate). Having chosen $v_1$, we see that the candidates for $v_2$ are those vectors not in $\langle v_1 \rangle$, the subspace spanned by $v_1$; there are thus $q^n - q$ candidates for $v_2$. More generally, having chosen a linearly independent list $v_1, \ldots, v_i$, we see that $v_{i+1}$ can be any vector not in $\langle v_1, \ldots, v_i \rangle$. Thus, there are $q^n - q^i$ candidates for $v_{i+1}$. The result follows by induction on $n$. $\quad\bullet$

**Theorem 4.50.** *If $p$ is prime and $q = p^m$, then the unitriangular group $\mathrm{UT}(n, \mathbb{F}_q)$ is a Sylow $p$-subgroup of $\mathrm{GL}(n, \mathbb{F}_q)$.*

**Proof.** Since $q^n - q^i = q^i(q^{n-i} - 1)$, the highest power of $p$ dividing $|\mathrm{GL}(n, \mathbb{F}_q)|$ is

$$qq^2q^3 \cdots q^{n-1} = q^{\binom{n}{2}}.$$

But $|\mathrm{UT}(n, \mathbb{F}_q)| = q^{\binom{n}{2}}$, and so $\mathrm{UT}(n, \mathbb{F}_q)$ must be a Sylow $p$-subgroup. $\quad\bullet$

**Corollary 4.51.** *If $p$ is prime and $G$ is a finite $p$-group, then $G$ is isomorphic to a subgroup of the unitriangular group $\mathrm{UT}(|G|, \mathbb{F}_p)$.*

**Proof.** A modest generalization of Exercise 1.107 on page 76 shows, for any field $k$, that every group of order $m$ can be imbedded in $\mathrm{GL}(m, k)$. In particular, $G$ can be imbedded in $\mathrm{GL}(m, \mathbb{F}_p)$. Now $G$ is a $p$-group, and so it is contained in a Sylow $p$-subgroup $P$ of $\mathrm{GL}(m, \mathbb{F}_p)$, for every $p$-subgroup lies in some Sylow $p$-subgroup. Since all Sylow $p$-subgroups are conjugate, there is $a \in \mathrm{GL}(m, \mathbb{F}_p)$ with $P = a\left(\mathrm{UT}(m, \mathbb{F}_p)\right) a^{-1}$. Therefore,

$$G \cong a^{-1}Ga \subseteq a^{-1}Pa \subseteq \mathrm{UT}(m, \mathbb{F}_p). \quad \bullet$$

A natural question is to find the Sylow subgroups of symmetric groups. This can be done, and the answer is in terms of a construction called *wreath product* (Rotman, *An Introduction to the Theory of Groups*, p. 176).

## Exercises

$*$ **4.20.** Show that $S_4$ has more than one Sylow 2-subgroup.

$*$ **4.21.** Give an example of a finite group $G$ having Sylow $p$-subgroups (for some prime $p$) $P, Q$, and $R$ such that $P \cap Q = \{1\}$ and $P \cap R \neq \{1\}$.
**Hint.** Consider $S_3 \times S_3$.

$*$ **4.22.** A subgroup $H$ of a group $G$ is called **characteristic** if $\varphi(H) \subseteq H$ for every isomorphism $\varphi\colon G \to G$. A subgroup $S$ of a group $G$ is called **fully invariant** if $\varphi(S) \subseteq S$ for every homomorphism $\varphi\colon G \to G$.

(i) Prove that every fully invariant subgroup is a characteristic subgroup, and that every characteristic subgroup is a normal subgroup.

(ii) Prove that the commutator subgroup, $G'$, is a normal subgroup of $G$ by showing that it is a fully invariant subgroup.

(iii) Give an example of a group $G$ having a normal subgroup $H$ that is not a characteristic subgroup.

(iv) Prove that $Z(G)$, the center of a group $G$, is a characteristic subgroup (and so $Z(G) \triangleleft G$), but that it need not be a fully invariant subgroup.
**Hint.** Let $G = S_3 \times \mathbb{I}_2$.

(v) For any group $G$, prove that if $H \triangleleft G$, then $Z(H) \triangleleft G$.

**4.23.** If $G$ is an abelian group, prove, for all positive integers $m$, that $mG$ and $G[m]$ are fully invariant subgroups.

$*$ **4.24. (Frattini Argument).** Let $K$ be a normal subgroup of a finite group $G$. If $P$ is a Sylow $p$-subgroup of $K$ for some prime $p$, prove that

$$G = K N_G(P),$$

where $K N_G(P) = \{ab : a \in K \text{ and } b \in N_G(P)\}$.
**Hint.** If $g \in G$, then $gPg^{-1}$ is a Sylow $p$-subgroup of $K$, and so it is conjugate to $P$ in $K$.

**4.25.** Prove that $\mathrm{UT}(3, \mathbb{F}_2) \cong D_8$; conclude that $D_8$ is a Sylow 2-subgroup of $\mathrm{GL}(3, \mathbb{F}_2)$.
**Hint.** You may use the fact that the only nonabelian groups of order 8 are $D_8$ and $\mathbf{Q}$.

$*$ **4.26.** (i) Prove that if $d$ is a positive divisor of 24, then $S_4$ has a subgroup of order $d$.

(ii) If $d \neq 4$, prove that any two subgroups of $S_4$ having order $d$ are isomorphic.

**4.27.** Prove that a Sylow 2-subgroup of $A_5$ has exactly five conjugates.

**4.28.** (i) Find a Sylow 3-subgroup of $S_6$.
**Hint.** $\{1, 2, 3, 4, 5, 6\} = \{1, 2, 3\} \cup \{4, 5, 6\}$.

(ii) Show that a Sylow 2-subgroup of $S_6$ is isomorphic to $D_8 \times \mathbb{I}_2$.
**Hint.** $\{1, 2, 3, 4, 5, 6\} = \{1, 2, 3, 4\} \cup \{5, 6\}$.

$*$ **4.29.** (i) Prove that a Sylow 2-subgroup of $A_6$ is isomorphic to $D_8$.

(ii) Prove that a Sylow 3-subgroup of $A_6$ is isomorphic to $\mathbb{I}_3 \times \mathbb{I}_3$.

(iii) Prove that the normalizer of a Sylow 5-subgroup of $A_6$ is isomorphic to $D_{10}$.

**4.30.** Let $Q$ be a normal $p$-subgroup of a finite group $G$. Prove that $Q \subseteq P$ for every Sylow $p$-subgroup $P$ of $G$.
**Hint.** Use the fact that any other Sylow $p$-subgroup of $G$ is conjugate to $P$.

**4.31.** (i) Let $G$ be a finite group and let $P$ be a Sylow $p$-subgroup of $G$. If $H \triangleleft G$, prove that $HP/H$ is a Sylow $p$-subgroup of $G/H$ and $H \cap P$ is a Sylow $p$-subgroup of $H$.
**Hint.** Show that $[G/H : HP/H]$ and $[H : H \cap P]$ are prime to $p$.

(ii) Let $P$ be a Sylow $p$-subgroup of a finite group $G$. Give an example of a subgroup $H$ of $G$ with $H \cap P$ not a Sylow $p$-subgroup of $H$.
**Hint.** Choose a subgroup $H$ of $S_4$ with $H \cong S_3$, and find a Sylow 3-subgroup $P$ of $S_4$ with $H \cap P = \{1\}$.

$*$ **4.32.** Let $G$ be a group of order 90.

(i) If a Sylow 5-subgroup $P$ of $G$ is not normal, prove that it has six conjugates.
**Hint.** If $P$ has 18 conjugates, there are 72 elements in $G$ of order 5. Show that $G$ has more than 18 other elements.

(ii) Prove that $G$ is not simple.

> **Hint.** Use Exercises 1.104 and 1.105(ii) on page 75.

**4.33.** Prove that there is no simple group of order 96, 120, 150, 300, 312, or 1000.

## Section 4.3. Solvable Groups

Galois introduced groups to investigate polynomials in $k[x]$, where $k$ is a field of characteristic 0, and he proved that such a polynomial is solvable by radicals if and only if its Galois group is a solvable group. Solvable groups are an interesting family of groups in their own right, and we now examine them a bit more.

Recall that a *normal series* of a group $G$ is a finite sequence of subgroups, $G = G_0, G_1, \ldots, G_n$, with

$$G = G_0 \supseteq G_1 \supseteq \cdots \supseteq G_n = \{1\}$$

and $G_{i+1} \lhd G_i$ for all $i$. The *factor groups* of the series are the groups

$$G_0/G_1, G_1/G_2, \ldots, G_{n-1}/G_n,$$

the *length* of the series is the number of strict inclusions (equivalently, the length is the number of nontrivial factor groups), and $G$ is *solvable* if it has a normal series whose factor groups are cyclic of prime order.

We begin with a technical result that generalizes the Second Isomorphism Theorem; it is useful when comparing different normal series of a group.

**Lemma 4.52 (Zassenhaus Lemma).** *Given four subgroups $A \lhd A^*$ and $B \lhd B^*$ of a group $G$, then $A(A^* \cap B) \lhd A(A^* \cap B^*)$, $B(B^* \cap A) \lhd B(B^* \cap A^*)$, and there is an isomorphism*

$$\frac{A(A^* \cap B^*)}{A(A^* \cap B)} \cong \frac{B(B^* \cap A^*)}{B(B^* \cap A)}.$$

**Remark.** The isomorphism is symmetric in the sense that the right side is obtained from the left by interchanging the symbols $A$ and $B$.

The Zassenhaus Lemma is sometimes called the *Butterfly Lemma* because of the following picture. I confess that I have never liked this picture; it doesn't remind me of a butterfly, and it doesn't help me understand or remember the proof.

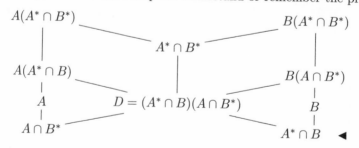

**Proof.** We claim that $(A \cap B^*) \lhd (A^* \cap B^*)$: that is, if $c \in A \cap B^*$ and $x \in A^* \cap B^*$, then $xcx^{-1} \in A \cap B^*$. Now $xcx^{-1} \in A$ because $c \in A$, $x \in A^*$, and $A \lhd A^*$; but also $xcx^{-1} \in B^*$, because $c, x \in B^*$. Hence, $(A \cap B^*) \lhd (A^* \cap B^*)$; similarly,

$(A^* \cap B) \lhd (A^* \cap B^*)$. Therefore, the subset $D$, defined by $D = (A \cap B^*)(A^* \cap B)$, is a normal subgroup of $A^* \cap B^*$, because it is generated by two normal subgroups.

Using the symmetry in the remark, it suffices to show that there is an isomorphism

$$\frac{A(A^* \cap B^*)}{A(A^* \cap B)} \to \frac{A^* \cap B^*}{D}.$$

Define $\varphi : A(A^* \cap B^*) \to (A^* \cap B^*)/D$ by $\varphi : ax \mapsto xD$, where $a \in A$ and $x \in A^* \cap B^*$. Now $\varphi$ is well-defined: if $ax = a'x'$, where $a' \in A$ and $x' \in A^* \cap B^*$, then $(a')^{-1}a = x'x^{-1} \in A \cap (A^* \cap B^*) = A \cap B^* \subseteq D$. Also, $\varphi$ is a homomorphism: $axa'x' = a''xx'$, where $a'' = a(xa'x^{-1}) \in A$ (because $A \lhd A^*$), and so $\varphi(axa'x') = \varphi(a''xx') = xx'D = \varphi(ax)\varphi(a'x')$. It is routine to check that $\varphi$ is surjective and that $\ker \varphi = A(A^* \cap B)$. The First Isomorphism Theorem completes the proof. •

The Zassenhaus Lemma implies the Second Isomorphism Theorem: if $S$ and $T$ are subgroups of a group $G$ with $T \lhd G$, then $TS/T \cong S/(S \cap T)$; set $A^* = G$, $A = T$, $B^* = S$, and $B = S \cap T$.

**Definition.** A *composition series* is a normal series all of whose nontrivial factor groups are simple. The list of nontrivial factor groups of a composition series is called the list of *composition factors* of $G$. The *length* of a composition series is the number of nontrivial factor groups.

A group need not have a composition series; for example, the abelian group $\mathbb{Z}$ has no composition series.

**Proposition 4.53.** *Every finite group $G$ has a composition series.*

**Proof.** Let $G$ be a least criminal; that is, assume that $G$ is a finite group of smallest order that does not have a composition series. Now $G$ is not simple, otherwise $G \supsetneq \{1\}$ is a composition series. Hence, $G$ has a proper normal subgroup $H$. Since $G$ is finite, we may assume that $H$ is a maximal normal subgroup, so that $G/H$ is a simple group. But $|H| < |G|$, so that $H$ has a composition series: say, $H = H_0 \supsetneq H_1 \supsetneq \cdots \supsetneq \{1\}$. Hence, $G \supsetneq H_0 \supsetneq H_1 \supsetneq \cdots \supsetneq \{1\}$ is a composition series for $G$, a contradiction. •

A group $G$ is solvable if it has a normal series with factor groups cyclic of prime order. As cyclic groups of prime order are simple groups, a normal series as in the definition of solvable group is a composition series, and so the composition factors of $G$ are cyclic groups of prime order.

Here are two composition series of $G = \langle a \rangle$, a cyclic group of order 30 (note that normality of subgroups is automatic because $G$ is abelian). The first is

$$G = \langle a \rangle \supseteq \langle a^2 \rangle \supseteq \langle a^{10} \rangle \supseteq \{1\};$$

the factor groups of this series are $\langle a \rangle / \langle a^2 \rangle \cong \mathbb{I}_2$, $\langle a^2 \rangle / \langle a^{10} \rangle \cong \mathbb{I}_5$, and $\langle a^{10} \rangle / \{1\} \cong \langle a^{10} \rangle \cong \mathbb{I}_3$ (see Example 1.83 on page 54). Another normal series is

$$G = \langle a \rangle \supseteq \langle a^5 \rangle \supseteq \langle a^{15} \rangle \supseteq \{1\};$$

the factor groups of this series are $\langle a \rangle / \langle a^5 \rangle \cong \mathbb{I}_5$, $\langle a^5 \rangle / \langle a^{15} \rangle \cong \mathbb{I}_3$, and $\langle a^{15} \rangle / \{1\} \cong \langle a^{15} \rangle \cong \mathbb{I}_2$. Notice that the same factor groups arise, although the order in which they arise is different. We will see that this phenomenon always occurs: different composition series of the same group have the same factor groups. This is the *Jordan–Hölder Theorem*, and the next definition makes its statement more precise.

**Definition.** Two normal series of a group $G$ are **equivalent** if there is a bijection between the lists of nontrivial factor groups of each so that corresponding factor groups are isomorphic.

The Jordan–Hölder Theorem says that any two composition series of a group are equivalent. It is more efficient to prove a more general theorem, due to Schreier.

**Definition.** A **refinement** of a normal series of a group $G$ is a normal series $G = N_0, \ldots, N_k = \{1\}$ having the original series as a subsequence.

In other words, a refinement of a normal series is a normal series obtained from the original one by inserting more subgroups.

Notice that a composition series admits only insignificant refinements; one can merely repeat terms (if $G_i / G_{i+1}$ is simple, then it has no proper nontrivial normal subgroups and, hence, there is no intermediate subgroup $L$ with $G_i \supsetneq L \supsetneq G_{i+1}$ and $L \lhd G_i$). Therefore, any refinement of a composition series is equivalent to the original composition series.

**Theorem 4.54 (Schreier Refinement Theorem).** *Any two normal series*

$$G = G_0 \supseteq G_1 \supseteq \cdots \supseteq G_n = \{1\}$$

*and*

$$G = N_0 \supseteq N_1 \supseteq \cdots \supseteq N_k = \{1\}$$

*of a group $G$ have equivalent refinements.*

**Proof.** We insert a copy of the second series between each pair of adjacent terms in the first series. In more detail, for each $i \geq 0$, define

$$G_{ij} = G_{i+1}(G_i \cap N_j)$$

(this is a subgroup, by Proposition 1.72(i), because $G_{i+1} \lhd G_i$). Since $N_0 = G$, we have

$$G_{i0} = G_{i+1}(G_i \cap N_0) = G_{i+1}G_i = G_i,$$

and since $N_k = \{1\}$, we have

$$G_{ik} = G_{i+1}(G_i \cap N_k) = G_{i+1}.$$

Therefore, the series of $G_i$ is a subsequence of the series of $G_{ij}$:

$$\cdots \supseteq G_i = G_{i0} \supseteq G_{i1} \supseteq G_{i2} \supseteq \cdots \supseteq G_{ik} = G_{i+1} \supseteq \cdots .$$

Similarly, the second series of $N_j$ is a subsequence of the series

$$N_{ji} = N_{j+1}(N_j \cap G_i).$$

Both doubly indexed sequences have $nk$ terms. For each $i, j$, the Zassenhaus Lemma, for the four subgroups $G_{i+1} \lhd G_i$ and $N_{j+1} \lhd N_j$, says both subsequences are normal series, hence are refinements, and there is an isomorphism

$$\frac{G_{i+1}(G_i \cap N_j)}{G_{i+1}(G_i \cap N_{j+1})} \cong \frac{N_{j+1}(N_j \cap G_i)}{N_{j+1}(N_j \cap G_{i+1})};$$

that is,

$$G_{i,j}/G_{i,j+1} \cong N_{j,i}/N_{j,i+1}.$$

The association $G_{i,j}/G_{i,j+1} \mapsto N_{j,i}/N_{j,i+1}$ is a bijection showing that the two refinements are equivalent. •

**Theorem 4.55 (Jordan–Hölder Theorem[13]).** *Any two composition series of a group $G$ are equivalent. In particular, the length of a composition series, if one exists, is an invariant of $G$.*

**Proof.** As we remarked earlier, any refinement of a composition series is equivalent to the original composition series. It now follows from Schreier's Theorem that any two composition series are equivalent. •

Here is a new proof of the Fundamental Theorem of Arithmetic.

**Corollary 4.56.** *Every integer $n \geq 2$ has a factorization into primes, and the prime factors and their multiplicities are uniquely determined by $n$.*

**Proof.** Since the group $\mathbb{I}_n$ is finite, it has a composition series; let $S_1, \ldots, S_t$ be the factor groups. Now an abelian group is simple if and only if it is of prime order, by Proposition 1.117; since $n = |\mathbb{I}_n|$ is the product of the orders of the factor groups (Exercise 4.37 on page 261), we have proved that $n$ is a product of primes. Moreover, the Jordan–Hölder Theorem gives the uniqueness of the (prime) orders of the factor groups and their multiplicities. •

**Example 4.57.**

(i) Nonisomorphic groups can have the same composition factors. For example, both $\mathbb{I}_4$ and $\mathbf{V}$ have composition series whose factor groups are $\mathbb{I}_2, \mathbb{I}_2$.

(ii) Let $G = \mathrm{GL}(2, \mathbb{F}_4)$ be the general linear group of all $2 \times 2$ nonsingular matrices with entries in the field $\mathbb{F}_4$ with four elements. Now $\det \colon G \to (\mathbb{F}_4)^\times$, where $(\mathbb{F}_4)^\times \cong \mathbb{I}_3$ is the multiplicative group of nonzero elements of $\mathbb{F}_4$. Since $\ker \det = \mathrm{SL}(2, \mathbb{F}_4)$, the special linear group consisting of those matrices of determinant 1, there is a normal series

$$G = \mathrm{GL}(2, \mathbb{F}_4) \supseteq \mathrm{SL}(2, \mathbb{F}_4) \supseteq \{1\}.$$

The factor groups of this normal series are $\mathbb{I}_3$ and $\mathrm{SL}(2, \mathbb{F}_4)$. It is true that $\mathrm{SL}(2, \mathbb{F}_4)$ is a nonabelian simple group [in fact, Corollary 4.71 says that $\mathrm{SL}(2, \mathbb{F}_4) \cong A_5$], and so this series is a composition series. We cannot yet conclude that $G$ is not solvable, for the definition of solvability requires that there

---

[13]In 1868, Jordan proved that the orders of the factor groups of a composition series depend only on $G$ and not on the composition series; in 1889, Hölder proved that the factor groups themselves, up to isomorphism, do not depend on the composition series.

be some composition series, not necessarily this one, having factor groups of prime order. However, the Jordan–Hölder Theorem says that if one composition series of $G$ has all its factor groups of prime order, then so does every other composition series. We may now conclude that $\mathrm{GL}(2, \mathbb{F}_4)$ is not a solvable group. ◀

Let us discuss the significance of the Jordan–Hölder Theorem.

**Definition.** If $G$ is a group and $K \lhd G$, then $G$ is an ***extension*** of $K$ by $G/K$.

For example, Exercise 1.83 on page 60 says that an extension of one $p$-group by another $p$-group is itself a $p$-group, and Proposition 3.25 says that any extension of one solvable group by another is itself a solvable group. However, it is not true that an extension of one abelian group by another is an abelian group.

The study of extensions involves the inverse question: how much of $G$ can be recovered from a normal subgroup $K$ and the quotient $Q = G/K$? For example, we do know that if $K$ and $Q$ are finite, then $|G| = |K||Q|$.

**Example 4.58.**

(i) The direct product $K \times Q$ is an extension of $K$ by $Q$ (and $K \times Q$ is an extension of $Q$ by $K$).

(ii) Both $S_3$ and $\mathbb{I}_6$ are extensions of $\mathbb{I}_3$ by $\mathbb{I}_2$. On the other hand, $\mathbb{I}_6$ is an extension of $\mathbb{I}_2$ by $\mathbb{I}_3$, but $S_3$ is not, for $S_3$ contains no *normal* subgroup of order 2. ◀

We have just seen, for any given pair of groups $K$ and $Q$, that an extension of $K$ by $Q$ always exists (the direct product), but there may be nonisomorphic such extensions. Hence, an extension of $K$ by $Q$ may be viewed as a "product" of $K$ and $Q$, but this product is not single-valued. The ***extension problem*** is to classify all possible extensions of a given pair of groups $K$ and $Q$.

Suppose that a group $G$ has a normal series

$$G = K_0 \supseteq K_1 \supseteq \cdots \supseteq K_{n-1} \supseteq K_n = \{1\}$$

with factor groups $Q_1, \ldots, Q_n$, where

$$Q_i = K_{i-1}/K_i$$

for all $i \geq 1$. Now $K_n = \{1\}$, so that $K_{n-1} = Q_n$, but something more interesting occurs next: $K_{n-2}/K_{n-1} = Q_{n-1}$, so that $K_{n-2}$ is an extension of $K_{n-1}$ by $Q_{n-1}$. If we could solve the extension problem, then we could recapture $K_{n-2}$ from $K_{n-1}$ and $Q_{n-1}$—that is, from $Q_n$ and $Q_{n-1}$. Next, observe that $K_{n-3}/K_{n-2} = Q_{n-2}$, so that $K_{n-3}$ is an extension of $K_{n-2}$ by $Q_{n-2}$. If we could solve the extension problem, then we could recapture $K_{n-3}$ from $K_{n-2}$ and $Q_{n-2}$; that is, we could recapture $K_{n-3}$ from $Q_n, Q_{n-1}$, and $Q_{n-2}$. Climbing up the composition series in this way, we could recapture $G = K_0$ from $Q_n, Q_{n-1}, \ldots, Q_1$; that is, $G$ is a "product" of the factor groups. If the normal series is a composition series, then the Jordan–Hölder Theorem is a unique factorization theorem: the factors in this product, namely, the composition factors of $G$, are uniquely determined by $G$. Therefore, we could survey all finite groups if we knew the finite simple groups and we could solve the extension problem. All the finite simple groups are known; the proof of the **Classification Theorem of Finite Simple Groups** was completed in the first decade of the 21st century; this theorem, one of the deepest theorems in Mathematics, gives a complete list of all the finite simple groups, along with interesting properties of them. In a sense, the extension problem has also been solved. In Chapter 9, we will give Schreier's solution describing all possible multiplication tables for extensions; this study leads to *cohomology of groups* and the *Schur–Zassenhaus Lemma*. On the other hand, the extension problem is unsolved in that no one knows a way, given $K$ and $Q$, to compute the exact number of nonisomorphic extensions of $K$ by $Q$.

We now pass from general groups (whose composition factors are arbitrary simple groups) to solvable groups (whose composition factors are cyclic groups of prime order; cyclic groups of prime order are simple in every sense of the word). Even though solvable groups arose in determining those polynomials that are solvable by radicals, there are purely group-theoretic theorems about solvable groups making no direct reference to Galois Theory and polynomials. For example, a theorem of Burnside says that if $|G| = p^m q^n$, where $p$ and $q$ are prime, then $G$ is solvable. The remarkable *Feit–Thompson Theorem* states that every group of odd order is solvable (Exercise 4.41 on page 262 says that this is equivalent to every nonabelian finite simple group having even order).

Solvability of a group is preserved by standard group-theoretic constructions. For example, every quotient of a solvable group is itself a solvable group (Proposition 3.22), and every subgroup of a solvable group is itself solvable (Proposition 3.23). An extension of one solvable group by another is itself solvable (Proposition 3.25): if $K \lhd G$ and both $K$ and $G/K$ are solvable, then $G$ is solvable, and a direct product of solvable groups is itself solvable (Corollary 3.26).

**Proposition 4.59.** *Every finite p-group $G$ is solvable.*

**Proof.** If $G$ is abelian, then $G$ is solvable. Otherwise, its center, $Z(G)$, is a proper nontrivial normal abelian subgroup, by Theorem 1.113. Now $Z(G)$ is solvable because it is abelian, and $G/Z(G)$ is solvable, by induction on $|G|$, and so $G$ is solvable, by Proposition 3.25. •

It follows, of course, that a direct product of finite $p$-groups, for various primes $p$, is solvable. Thus, finite abelian groups are solvable.

**Definition.** If $G$ is a group and $x, y \in G$, then their **commutator** $[x, y]$ is the element
$$[x, y] = xyx^{-1}y^{-1}.$$
If $X$ and $Y$ are subgroups of a group $G$, then $[X, Y]$ is defined by
$$[X, Y] = \langle [x, y] : x \in X \text{ and } y \in Y \rangle.$$
In particular, the **commutator subgroup** $G'$ of a group $G$ is
$$G' = [G, G],$$

the subgroup generated by all the commutators.[14]

It is clear that two elements $x$ and $y$ in a group $G$ commute if and only if their commutator $[x, y]$ is 1. The next proposition generalizes this observation.

**Proposition 4.60.** *Let $G$ be a group.*

(i) *The commutator subgroup $G'$ is a normal subgroup of $G$, and $G/G'$ is abelian.*

(ii) *If $H \lhd G$ and $G/H$ is abelian, then $G' \subseteq H$.*

**Proof.**

(i) The inverse of a commutator $xyx^{-1}y^{-1}$ is itself a commutator: $[x, y]^{-1} = yxy^{-1}x^{-1} = [y, x]$. Therefore, each element of $G'$ is a product of commutators. But any conjugate of a commutator (and hence, a product of commutators) is another such:

$$a[x, y]a^{-1} = a(xyx^{-1}y^{-1})a^{-1}$$
$$= axa^{-1}aya^{-1}ax^{-1}a^{-1}ay^{-1}a^{-1}$$
$$= [axa^{-1}, aya^{-1}].$$

Therefore, $G' \lhd G$. [Alternatively, $G' \lhd G$ because it is fully invariant: if $\varphi \colon G \to G$ is a homomorphism, then $\varphi([x, y]) = [\varphi(x), \varphi(y)] \in G'$.]

If $aG', bG' \in G/G'$, then

$$aG'bG'(aG')^{-1}(bG')^{-1} = aba^{-1}b^{-1}G' = [a, b]G' = G',$$

and so $G/G'$ is abelian.

(ii) Suppose that $H \lhd G$ and $G/H$ is abelian. If $a, b \in G$, then $aHbH = bHaH$; that is, $abH = baH$, and so $b^{-1}a^{-1}ba \in H$. As every commutator has the form $b^{-1}a^{-1}ba$, we have $G' \subseteq H$.   •

**Example 4.61.**

(i) A group $G$ is abelian if and only if $G' = \{1\}$.

(ii) If $G$ is a simple group, then $G' = \{1\}$ or $G' = G$, for $G'$ is a normal subgroup. The first case occurs when $G$ has prime order; the second case occurs otherwise. In particular, $(A_n)' = A_n$ for all $n \geq 5$. (A group $G$ for which $G' = G$ is called ***perfect***; thus, every nonabelian simple group is perfect.)

(iii) What is $(S_n)'$? Since $S_n/A_n \cong \mathbb{I}_2$ is abelian, Proposition 4.60 shows that $(S_n)' \subseteq A_n$. For the reverse inclusion, first note that $(S_n)' \cap A_n \lhd A_n$; hence, if $n \geq 5$, simplicity of $A_n$ implies that this intersection is trivial or $A_n$. But $(S_n)' \cap A_n \neq \{(1)\}$, so $(S_n)' \cap A_n = A_n$ and $A_n \subseteq (S_n)'$. Therefore, $(S_n)' = A_n$ for all $n \geq 5$. Exercise 4.36 on page 261 shows that the equality $(S_n)' = A_n$ also holds for $n = 2, 3, 4$.   ◄

Let us iterate the formation of the commutator subgroup.

---

[14]The subset consisting of all the commutators need not be closed under products, and so the *set* of all commutators may not be a subgroup. The smallest group in which a product of two commutators is not a commutator has order 96. Also, see Carmichael's exercise on page 270.

**Definition.** The *derived series* of $G$ is
$$G = G^{(0)} \supseteq G^{(1)} \supseteq G^{(2)} \supseteq \cdots \supseteq G^{(i)} \supseteq G^{(i+1)} \supseteq \cdots,$$
where $G^{(0)} = G$, $G^{(1)} = G'$ and, more generally, $G^{(i+1)} = (G^{(i)})' = [G^{(i)}, G^{(i)}]$ for all $i \geq 0$.

It is easy to prove, by induction on $i \geq 0$, that $G^{(i)}$ is fully invariant, which implies that $G^{(i)} \lhd G$. It follows from Exercise 1.60 on page 46 that $G^{(i)} \lhd G^{(i-1)}$, and so the derived series is a normal series. We now prove that $G$ is solvable if and only if its derived series reaches $\{1\}$.

**Proposition 4.62.**

(i) *A finite group $G$ is solvable if and only if it has a normal series with abelian factor groups.*

(ii) *A finite group $G$ is solvable if and only if there is some n with $G^{(n)} = \{1\}$.*

**Proof.**

(i) If $G$ is solvable, then it has a normal series whose factor groups $G_i/G_{i+1}$ are all cyclic of prime order, hence are abelian.

Conversely, if $G$ has a normal series with abelian factor groups, then the factor groups of any refinement are also abelian. In particular, the factor groups of a composition series of $G$, which exists because $G$ is finite, are abelian simple groups; hence, they are cyclic of prime order, and so $G$ is solvable.

(ii) Assume that $G$ is solvable, so there is a normal series
$$G = G_0 \supseteq G_1 \supseteq \cdots \supseteq G_n = \{1\}$$
whose factor groups $G_i/G_{i+1}$ are abelian. We show, by induction on $i \geq 0$, that $G^{(i)} \subseteq G_i$. Since $G^{(0)} = G = G_0$, the base step is obviously true. For the inductive step, since $G_i/G_{i+1}$ is abelian, Proposition 4.60 gives $(G_i)' \subseteq G_{i+1}$. On the other hand, the inductive hypothesis gives $G^{(i)} \subseteq G_i$, which implies that
$$G^{(i+1)} = (G^{(i)})' \subseteq (G_i)' \subseteq G_{i+1}.$$
In particular, $G^{(n)} \subseteq G_n = \{1\}$, which is what we wished to show.

Conversely, if $G^{(n)} = \{1\}$, then the derived series is a normal series (a normal series must end with $\{1\}$) with abelian factor groups, and so part (i) gives $G$ solvable. •

For example, the derived series of $G = S_4$ is easily seen to be
$$S_4 \supsetneq A_4 \supsetneq \mathbf{V} \supsetneq \{(1)\}.$$

Our earlier definition of solvability applies only to finite groups, whereas the characterization in Proposition 4.62 makes sense for all groups, possibly infinite. Nowadays, most authors call a group solvable if its derived series reaches $\{1\}$ after a finite number of steps. With this new definition, every abelian group is solvable, whereas it is easy to see that abelian groups are solvable in the sense of the original definition if and only if they are finite. In Exercise 4.39 on page 261, the reader will be asked to prove, using the criterion in Proposition 4.62, that subgroups, quotient groups, and extensions of solvable groups are also solvable (in the new, generalized, sense).

We state two interesting theorems of P. Hall. The first generalizes the Sylow Theorems.

**Theorem.** *If $G$ is a finite solvable group of order $ab$, where $(a, b) = 1$, then $G$ has a subgroup of order $a$, and any two such subgroups are conjugate.*

**Proof.** Rotman, *An Introduction to the Theory of Groups*, p. 108.   •

**Definition.** If $G$ is a group of order $p^k m$, where $p$ is prime and $p \nmid m$, then a *p-complement* is a subgroup of order $m$.

In general, $p$-complements do not exist. By Exercise 9.11 on page 765, every group of order 15 is cyclic. Hence, $A_5$, a group of order $60 = 2^2 \cdot 15$, has no 2-complement: if it had a subgroup of order 15, it would have an element of order 15.

**Theorem.** *A finite group $G$ has a p-complement for every prime divisor $p$ of $|G|$ if and only if $G$ is solvable.*

**Proof.** Ibid., p. 110.   •

There are other interesting classes of groups defined in terms of normal series. One of the most interesting such classes consists of *nilpotent* groups.

**Definition.** The *descending central series* of a group $G$ is

$$G = \gamma_1(G) \supseteq \gamma_2(G) \supseteq \cdots ,$$

where $\gamma_{i+1}(G) = [\gamma_i(G), G]$. A group $G$ is called *nilpotent* if the lower central series reaches $\{1\}$; that is, if $\gamma_n(G) = \{1\}$ for some $n$.

Note that $\gamma_2(G) = G'$, but the derived series and the lower central series may differ afterward; for example, $\gamma_3(G) = [G', G] \supseteq G^{(2)}$, with strict inequality possible.

Finite nilpotent groups are characterized by Proposition 4.42: they are the groups that are direct products of their Sylow subgroups, and so one regards finite nilpotent groups as generalized $p$-groups. Some examples of nilpotent groups are $\mathrm{UT}(n, \mathbb{F}_q)$, $\mathrm{UT}(n, \mathbb{Z})$ (unitriangular groups over $\mathbb{Z}$), and the Frattini subgroup $\Phi(G)$ (defined in Exercise 4.47 on page 262) of a finite group $G$. We can prove results, for infinite nilpotent groups as well as for finite ones, such as those in Exercise 4.48 on page 263: every subgroup and every quotient of a finite nilpotent group $G$ is again nilpotent; if $G/Z(G)$ is nilpotent, then so is $G$; every normal subgroup $H$ intersects $Z(G)$ nontrivially. However, an extension of one nilpotent group by another need not be nilpotent. For example, the symmetric group $S_3$ is not nilpotent, but it is an extension of $\mathbb{I}_3$ by $\mathbb{I}_2$.

There is a uniqueness theorem for direct products.

**Definition.** A group $G$ is *indecomposable* if $G \neq \{1\}$ and $G = H \times K$ implies that $H = \{1\}$ or $K = \{1\}$.

For example, simple groups are indecomposable, as are cyclic groups of prime power order. The *Krull–Schmidt Theorem* says that if $G$ is finite (more generally, if $G$ has both the ascending and descending chain conditions on its normal subgroups; that is, every ascending chain of normal subgroups [and every descending chain of normal subgroups] is eventually constant), then it has a unique factorization as a direct product of indecomposables.

**Theorem (Krull–Schmidt).** *Let $G$ be a finite group (more generally, let $G$ have both chain conditions on its normal subgroups). If*

$$G = H_1 \times \cdots \times H_s = K_1 \times \cdots \times K_t,$$

*where the $H_i, K_j$ are indecomposable, then $s = t$ and, after reindexing, $H_i \cong K_i$ for all $i$. Moreover, there is a **replacement property**: given any $r$ between 1 and $s$, the reindexing may be chosen so that*

$$G = H_1 \times \cdots \times H_r \times K_{r+1} \times \cdots \times K_s.$$

**Proof.** Rotman, *An Introduction to the Theory of Groups*, p. 149. •

The Krull–Schmidt Theorem gives another proof of the Fundamental Theorem of Finite Abelian Groups if we assume the Basis Theorem.

## Exercises

**4.34.** Let $p$ be prime and let $G$ be a nonabelian group of order $p^3$. Prove that $Z(G) = G'$.

**Hint.** Show first that both subgroups have order $p$.

**4.35.** Prove that if $H$ is a subgroup of a group $G$ and $G' \subseteq H$, then $H \triangleleft G$.

**Hint.** Use the Correspondence Theorem.

* **4.36.** (i) Prove that $(S_n)' = A_n$ for $n = 2, 3, 4$ [see Example 4.61(ii) for $n \geq 5$].

   (ii) Prove that $(\mathrm{GL}(n, k))' \subseteq \mathrm{SL}(n, k)$. (The reverse inclusion is also true; see Exercise 4.57 on page 270 for the case $n = 2$.)

* **4.37.** If $G$ is a finite group and

$$G = G_0 \supseteq G_1 \supseteq \cdots \supseteq G_n = \{1\}$$

is a normal series, prove that the order of $G$ is the product of the orders of the factor groups:

$$|G| = \prod_{i=0}^{n-1} |G_i/G_{i+1}|.$$

**4.38.** Prove that any two finite solvable groups of the same order have the same composition factors.

* **4.39.** Let $G$ be an arbitrary, possibly infinite, group.

   (i) Prove that if $H \subseteq G$, then $H^{(i)} \subseteq G^{(i)}$ for all $i$. Conclude, using Proposition 4.62, that every subgroup of a solvable group is solvable.

(ii) Prove that if $f: G \to K$ is a surjective homomorphism, then

$$f(G^{(i)}) = K^{(i)}$$

for all $i$. Conclude, using Proposition 4.62, that every quotient of a solvable group is also solvable.

(iii) For every group $G$, prove, by double induction, that

$$G^{(m+n)} = (G^{(m)})^{(n)}.$$

(iv) Prove, using Proposition 4.62, that if $H \lhd G$ and both $H$ and $G/H$ are solvable, then $G$ is solvable.

**4.40.** Let $p$ and $q$ be primes.

(i) Prove that every group of order $pq$ is solvable.
**Hint.** If $p = q$, then $G$ is abelian. If $p < q$, then a divisor $r$ of $pq$ for which $r \equiv 1 \bmod q$ must equal 1.

(ii) Prove that every group $G$ of order $p^2q$ is solvable.
**Hint.** If $G$ is not simple, use Proposition 3.25. If $p > q$, then $r \equiv 1 \bmod p$ forces $r = 1$. If $p < q$, then $r = p^2$ and there are more than $p^2q$ elements in $G$.

* **4.41.** Show that the Feit–Thompson Theorem, "Every finite group of odd order is solvable," is equivalent to "Every nonabelian finite simple group has even order."
**Hint.** For sufficiency, choose a "least criminal": a nonsolvable group $G$ of smallest odd order. By hypothesis, $G$ is not simple, and so it has a proper nontrivial normal subgroup.

**4.42.** (i) Prove that the infinite cyclic group $\mathbb{Z}$ does not have a composition series.

(ii) Prove that an *abelian* group $G$ has a composition series if and only if $G$ is finite.

**4.43.** Prove that if $G$ is a finite group and $H \lhd G$, then there is a composition series of $G$ one of whose terms is $H$.
**Hint.** Use Schreier's Theorem.

**4.44.** (i) Prove that if $S$ and $T$ are solvable subgroups of a group $G$ and $S \lhd G$, then $ST$ is a solvable subgroup of $G$.
**Hint.** Use the Second Isomorphism Theorem.

(ii) If $G$ is a finite group, define $\mathcal{S}(G)$ to be the subgroup of $G$ generated by all normal solvable subgroups of $G$. Prove that $\mathcal{S}(G)$ is the unique maximal normal solvable subgroup of $G$ and that $G/\mathcal{S}(G)$ has no nontrivial normal solvable subgroups.

**4.45.** (i) Prove that the dihedral groups $D_{2n}$ are solvable.

(ii) Give a composition series for $D_{2n}$.

**4.46.** (**Rosset**) Let $G$ be a group containing elements $x$ and $y$ such that the orders of $x$, $y$, and $xy$ are pairwise relatively prime; prove that $G$ is not solvable.

* **4.47.** (i) If $G$ is a finite group, then its **Frattini subgroup**, denoted by $\Phi(G)$, is defined to be the intersection of all the maximal subgroups of $G$. Prove that $\Phi(G)$ is a characteristic subgroup, and hence it is a normal subgroup of $G$.

(ii) Prove that if $p$ is prime and $G$ is a finite abelian $p$-group, then $\Phi(G) = pG$. [The **Burnside Basis Theorem** says that if $G$ is any finite $p$-group, then $G/\Phi(G)$ is a vector space over $\mathbb{F}_p$, and its dimension is the minimum number of generators of $G$ (Rotman, *An Introduction to the Theory of Groups*, p. 124)].

* **4.48.** (i) If $G$ is a nilpotent group, prove that its center $Z(G) \neq \{1\}$.

   (ii) If $G$ is a group with $G/Z(G)$ nilpotent, prove that $G$ is nilpotent.

   (iii) If $G$ is a nilpotent group, prove that every subgroup and every quotient group of $G$ is also nilpotent.

   (iv) If $G$ is a finite $p$-group and $H \triangleleft G$, prove that $H \cap Z(G) \neq \{1\}$. (The generalization of this result to finite nilpotent groups is true.)

**4.49.** Let $\mathfrak{A}$ denote the class of all abelian groups, $\mathfrak{N}$ the class of all nilpotent groups, and $\mathfrak{S}$ the class of all solvable groups.

   (i) Prove that $\mathfrak{A} \subseteq \mathfrak{N} \subseteq \mathfrak{S}$.

   (ii) Show that each of the inclusions in part (i) is strict; that is, there is a nilpotent group that is not abelian, and there is a solvable group that is not nilpotent.

**4.50.** If $G$ is a group and $g, x \in G$, write $g^x = xgx^{-1}$.

   (i) Prove, for all $x, y, z \in G$, that $[x, yz] = [x, y][x, z]^y$ and $[xy, z] = [y, z]^x [x, z]$.

   (ii) **(Jacobi Identity)** If $x, y, z \in G$ are elements in a group $G$, define
   $$[x, y, z] = [x, [y, z]].$$

   Prove that
   $$[x, y^{-1}, z]^y [y, z^{-1}, x]^z [z, x^{-1}, y]^x = 1.$$

**4.51.** If $H, K, L$ are subgroups of a group $G$, define
$$[H, K, L] = \langle \{[h, k, \ell] : h \in H, k \in K, \ell \in L\} \rangle.$$

   (i) Prove that if $[H, K, L] = \{1\} = [K, L, H]$, then $[L, H, K] = \{1\}$.

   (ii) **(Three Subgroups Lemma)** If $N \triangleleft G$ and $[H, K, L][K, L, H] \subseteq N$, prove that
   $$[L, H, K] \subseteq N.$$

   (iii) Prove that if $G$ is a group with $G = G'$, then $G/Z(G)$ is centerless.
   **Hint.** If $\pi \colon G \to G/Z(G)$ is the natural map, define $\zeta^2(G) = \pi^{-1}(Z(G/Z(G)))$. Use the Three Subgroups Lemma with $L = \zeta^2(G)$ and $H = K = G$.

   (iv) Prove, for all $i, j$, that $[\gamma_i(G), \gamma_j(G)] \subseteq \gamma_{i+j}(G)$.

## Section 4.4. Projective Unimodular Groups

The Jordan–Hölder Theorem shows that simple groups can be viewed as building blocks of finite groups. As a practical matter, one can often reduce a problem about finite groups to a problem about finite simple groups (see our proof of Proposition 7.64 for such an example). This empirical fact says that a knowledge of simple groups is very useful. Now the only simple groups we have seen so far are cyclic groups of prime order and the alternating groups $A_n$ for $n \geq 5$. We are going to show that certain finite groups of matrices are simple, and we begin by considering some matrices that play the same role for $2 \times 2$ linear groups as the 3-cycles play for the alternating groups.

**Definition.** A ***transvection***[15] over a field $k$ is a matrix of the form

$$B_{12}(r) = \begin{bmatrix} 1 & r \\ 0 & 1 \end{bmatrix} \quad \text{or} \quad B_{21}(r) = \begin{bmatrix} 1 & 0 \\ r & 1 \end{bmatrix},$$

where $r \in k$ and $r \neq 0$.

Let $A$ be a $2 \times 2$ matrix. It is easy to see that $B_{12}(r)A$ is the matrix obtained from $A$ by replacing Row(1) by Row(1) $+ r$Row(2), and that $B_{21}(r)A$ is the matrix obtained from $A$ by replacing Row(2) by Row(2) $+ r$Row(1).

**Lemma 4.63.** *If $k$ is a field and $A \in \mathrm{GL}(2, k)$, then*

$$A = UD,$$

*where $U$ is a product of transvections, $D = \mathrm{diag}\{1, d\} = \begin{bmatrix} 1 & 0 \\ 0 & d \end{bmatrix}$, and $d = \det(A)$.*

**Proof.** Let $A = \begin{bmatrix} p & q \\ r & s \end{bmatrix}$. If $r \neq 0$, then $B_{12}(r^{-1}(1-p))A = \begin{bmatrix} 1 & x \\ r & s \end{bmatrix}$, where $x = q + r^{-1}(1-p)s$. Now $B_{21}(-r)\begin{bmatrix} 1 & x \\ r & s \end{bmatrix} = \begin{bmatrix} 1 & x \\ 0 & d \end{bmatrix}$, where $d = \det(A)$ [because transvections have determinant 1]. Finally, $B_{12}(-d^{-1})\begin{bmatrix} 1 & x \\ 0 & d \end{bmatrix} = \begin{bmatrix} 1 & 0 \\ 0 & d \end{bmatrix} = D$. Thus, $WA = D$, where $W$ is a product of transvections. Since the inverse of a transvection is also a transvection, we have $A = W^{-1}D$, which is the factorization we seek.

If $r = 0$, then $p \neq 0$ (because $A$ is nonsingular), and $B_{21}(1)A$ replaces Row(2) of $A$ by Row(2) + Row(1) $= [p \ q + s]$, which returns us to the first case. $\bullet$

Recall that $\mathrm{SL}(2, k)$ is the subgroup of $\mathrm{GL}(2, k)$ consisting of all matrices of determinant 1. If $k$ is a finite field, then $k \cong \mathbb{F}_q$, where $q = p^n$ and $p$ is prime; we may denote $\mathrm{GL}(2, \mathbb{F}_q)$ by $\mathrm{GL}(2, q)$ and, similarly, we may denote $\mathrm{SL}(2, \mathbb{F}_q)$ by $\mathrm{SL}(2, q)$.

**Proposition 4.64.**

  (i) *If $k$ is a field, then $\mathrm{SL}(2, k)$ is generated by transvections.*

  (ii) *If $k$ is a field, then $\mathrm{GL}(2, k)/\mathrm{SL}(2, k) \cong k^\times$, where $k^\times$ is the multiplicative group of nonzero elements of $k$.*

  (iii) *If $k = \mathbb{F}_q$, then*

$$|\mathrm{SL}(2, \mathbb{F}_q)| = (q+1)q(q-1).$$

**Proof.**

  (i) If $A \in \mathrm{SL}(2, k)$, then Lemma 4.63 gives a factorization $A = UD$, where $U$ is a product of transvections and $D = \mathrm{diag}\{1, d\}$, where $d = \det(A)$. Since $A \in \mathrm{SL}(2, k)$, we have $d = \det(A) = 1$, and so $A = U$.

---

[15] Most group theorists define a $2 \times 2$ transvection as a matrix that is *similar* to $B_{12}(r)$ or $B_{21}(r)$ [that is, a conjugate of $B_{12}(r)$ or $B_{21}(r)$ in $\mathrm{GL}(2, k)$]. The word *transvection* is a synonym for *transporting*, and its usage in this context is probably due to E. Artin, who gives the following definition in his book *Geometric Algebra*: "An element $\tau \in \mathrm{GL}(V)$, where $V$ is an $n$-dimensional vector space, is called a ***transvection*** if it keeps every vector of some hyperplane $H$ fixed and moves any vector $x \in V$ by some vector of $H$; that is, $\tau(x) - x \in H$." In our case, $B_{12}(r)$ fixes the "$x$-axis" and $B_{21}(r)$ fixes the "$y$-axis."

(ii) If $a \in k^{\times}$, then the matrix $\mathrm{diag}\{1,a\}$ has determinant $a$, hence is nonsingular, and so the map $\det\colon \mathrm{GL}(2,k) \to k^{\times}$ is surjective. The definition of $\mathrm{SL}(2,k)$ shows that it is the kernel of $\det$, and so the First Isomorphism Theorem gives the result.

(iii) If $H$ is a normal subgroup of a finite group $G$, then Lagrange's Theorem gives $|H| = |G|/|G/H|$. In particular,

$$|\mathrm{SL}(2,\mathbb{F}_q)| = |\mathrm{GL}(2,\mathbb{F}_q)|/|\mathbb{F}_q^{\times}|.$$

But $|\mathrm{GL}(2,\mathbb{F}_q)| = (q^2-1)(q^2-q)$, by Theorem 4.49, and $|\mathbb{F}_q^{\times}| = q-1$. Hence, $|\mathrm{SL}(2,\mathbb{F}_q)| = (q+1)q(q-1)$. •

We now compute the center of these matrix groups. If $V$ is a two-dimensional vector space over $k$, then we proved, in Proposition 2.132, that $\mathrm{GL}(2,k) \cong \mathrm{GL}(V)$, the group of all nonsingular linear transformations on $V$. Moreover, Corollary 2.133 identifies the center with the scalar transformations.

**Proposition 4.65.** *The center of* $\mathrm{SL}(2,k)$, *denoted by* $\mathrm{SZ}(2,k)$, *consists of all scalar matrices* $aI = \begin{bmatrix} a & 0 \\ 0 & a \end{bmatrix}$ *with* $a^2 = 1$.

**Remark.** Here we see that $\mathrm{SZ} = \mathrm{SL} \cap Z(\mathrm{GL})$, but it is not true in general that $H \subseteq G$ implies $Z(H) = H \cap Z(G)$ (even when $H$ is normal). We always have $H \cap Z(G) \subseteq Z(H)$, but this inclusion may be strict. For example, if $G = S_3$ and $H = A_3 \cong \mathbb{I}_3$, then $Z(A_3) = A_3$ while $A_3 \cap Z(S_3) = \{1\}$. ◄

**Proof.** It is more convenient here to use linear transformations than matrices. Assume that $T \in \mathrm{SL}(2,k)$ is not a scalar transformation. Thus, there is a nonzero vector $v \in V$ with $Tv$ not a scalar multiple of $v$. It follows that the list $v, Tv$ is linearly independent and, since $\dim(V) = 2$, that it is a basis of $V$. Define $S\colon V \to V$ by $S(v) = v$ and $S(Tv) = v + Tv$. Notice, relative to the basis $v, Tv$, that $S$ has matrix $B_{12}(1)$, so that $\det(S) = 1$. Now $T$ and $S$ do not commute, for $TS(v) = Tv$ while $ST(v) = v + Tv$. It follows that the center must consist of scalar transformations. In matrix terms, the center consists of scalar matrices $A = \mathrm{diag}\{a,a\}$ with $a^2 = \det(A) = 1$. •

**Definition.** The *projective unimodular*[16] *group* is the quotient group

$$\mathrm{PSL}(2,k) = \mathrm{SL}(2,k)/\mathrm{SZ}(2,k).$$

For any field $k$, if $c \in k$ satisfies $c^2 = 1$, then $c = \pm 1$. In particular, if $k = \mathbb{F}_q$, where $q$ is a power of 2, then $\mathbb{F}_q$ has characteristic 2 and $c^2 = 1$ implies $c = 1$. Hence, $\mathrm{SZ}(2,q) = \{I\}$ and so $\mathrm{PSL}(2,2^n) = \mathrm{SL}(2,2^n)$.

**Proposition 4.66.**

$$|\mathrm{PSL}(2,q)| = \begin{cases} \frac{1}{2}(q+1)q(q-1) & \text{if } q = p^n \text{ and } p \text{ is an odd prime;} \\ (q+1)q(q-1) & \text{if } q = 2^n. \end{cases}$$

---

[16]A matrix is called *unimodular* if it has determinant 1. The adjective *projective* arises because this group turns out to consist of automorphisms of a projective plane.

**Proof.** Proposition 4.64(iii) gives $|\mathrm{PSL}(2,q)| = (q+1)q(q-1)/|\mathrm{SZ}(2,q)|$, and Proposition 4.65 gives

$$|\mathrm{SZ}(2,q)| = |\{a \in \mathbb{F}_q : a^2 = 1\}|.$$

Now $\mathbb{F}_q^\times$ is a cyclic group of order $q-1$, by Theorem 2.46. If $q$ is odd, then $q-1$ is even, and the cyclic group $\mathbb{F}_q^\times$ has a unique subgroup of order 2; if $q$ is a power of 2, then we noted, just before the statement of this proposition, that $\mathrm{SZ}(2,q) = \{I\}$. Therefore, $|SZ(2,q)| = 2$ if $q$ is a power of an odd prime, and $|SZ(2,q)| = 1$ if $q$ is a power of 2. •

We are now going to prove that the groups $\mathrm{PSL}(2,q)$ are simple for all prime powers $q \geq 4$. As we said earlier, the transvections will play the role of the 3-cycles (Exercise 1.99 on page 75).

**Lemma 4.67.** *If $H$ is a normal subgroup of $\mathrm{SL}(2,q)$ containing a transvection $B_{12}(r)$ or $B_{21}(r)$, then $H = \mathrm{SL}(2,q)$.*

**Proof.** Note first that if $U = \begin{bmatrix} 0 & -1 \\ 1 & 0 \end{bmatrix}$, then $\det(U) = 1$ and $U \in \mathrm{SL}(2,q)$; since $H$ is a normal subgroup, $U B_{12}(r) U^{-1}$ also lies in $H$. But $U B_{12}(r) U^{-1} = B_{21}(-r)$, from which it follows that $H$ contains a transvection of the form $B_{12}(r)$ if and only if it contains a transvection of the form $B_{21}(-r)$. Since SL is generated by the transvections, it suffices to show that every transvection $B_{12}(r)$ lies in $H$.

The following conjugate of $B_{12}(r)$ lies in $H$ (because $H$ is normal):

$$\begin{bmatrix} \alpha & \beta \\ 0 & \alpha^{-1} \end{bmatrix} \begin{bmatrix} 1 & r \\ 0 & 1 \end{bmatrix} \begin{bmatrix} \alpha^{-1} & -\beta \\ 0 & \alpha \end{bmatrix} = \begin{bmatrix} 1 & r\alpha^2 \\ 0 & 1 \end{bmatrix} = B_{12}(r\alpha^2).$$

Define

$$G = \{0\} \cup \{u \in \mathbb{F}_q : B_{12}(u) \in H\}.$$

We have just shown that $r\alpha^2 \in G$ for all $\alpha \in \mathbb{F}_q$. It is easy to check that $G$ is a subgroup of the additive group of $\mathbb{F}_q$ and, hence, it contains all the elements of the form $u = r(\alpha^2 - \beta^2)$, where $\alpha, \beta \in k$. We claim that $G = \mathbb{F}_q$, which will complete the proof.

If $q$ is odd, then each $w \in \mathbb{F}_q$ is a difference of squares:

$$w = [\tfrac{1}{2}(w+1)]^2 - [\tfrac{1}{2}(w-1)]^2.$$

Hence, if $u \in \mathbb{F}_q$, there are $\alpha, \beta \in \mathbb{F}_q$ with $r^{-1}u = \alpha^2 - \beta^2$, and so $u = r(\alpha^2 - \beta^2) \in G$; therefore, $G = \mathbb{F}_q$. If $q = 2^m$, then the function $u \mapsto u^2$ is an injection $\mathbb{F}_q \to \mathbb{F}_q$ (for if $u^2 = v^2$, then $0 = u^2 - v^2 = (u-v)^2$, and $u = v$). The Pigeonhole Principle says that this function is surjective, and so every element $u$ has a square root in $\mathbb{F}_q$. In particular, there is $\alpha \in \mathbb{F}_q$ with $r^{-1}u = \alpha^2$, and $u = r\alpha^2 \in G$. •

We need a short technical lemma before giving the main result.

**Lemma 4.68.** *Let $H$ be a normal subgroup of $\mathrm{SL}(2,q)$. If $A \in H$ is similar to $R = \begin{bmatrix} \alpha & \beta \\ \gamma & \delta \end{bmatrix}$, where $R \in \mathrm{GL}(2,q)$, then there is $u \in \mathbb{F}_q$ so that $H$ contains*

$$\begin{bmatrix} \alpha & u^{-1}\beta \\ u\gamma & \delta \end{bmatrix}.$$

**Proof.** By hypothesis, there is a matrix $P \in \mathrm{GL}(2, q)$ with $R = PAP^{-1}$. There is a matrix $U \in \mathrm{SL}$ and a diagonal matrix $D = \mathrm{diag}\{1, u\}$ with $P^{-1} = UD$, by Lemma 4.63. Therefore, $A = UDRD^{-1}U^{-1}$; since $H \lhd \mathrm{SL}$, we have $DRD^{-1} = U^{-1}AU \in H$. But

$$
DRD^{-1} = \begin{bmatrix} 1 & 0 \\ 0 & u \end{bmatrix} \begin{bmatrix} \alpha & \beta \\ \gamma & \delta \end{bmatrix} \begin{bmatrix} 1 & 0 \\ 0 & u^{-1} \end{bmatrix} = \begin{bmatrix} \alpha & u^{-1}\beta \\ u\gamma & \delta \end{bmatrix}. \quad \bullet
$$

The next theorem was proved by Jordan in 1870 for $q$ prime. In 1893, after Cole had discovered a simple group of order 504, Moore recognized Cole's group as $\mathrm{PSL}(2, 8)$, and he then proved the simplicity of $\mathrm{PSL}(2, q)$ for all prime powers $q \geq 4$. Jordan proved, for all $m \geq 3$, that $\mathrm{PSL}(m, p)$ is simple for all primes $p$. In 1897, Dickson proved that $\mathrm{PSL}(m, q)$ is simple for all prime powers $q$.

We are going to use Corollary 2.130: two $n \times n$ matrices $A$ and $B$ over a field $k$ are similar (that is, there exists a nonsingular matrix $P$ with $B = PAP^{-1}$) if and only if they both arise from a single linear transformation $\varphi \colon k^n \to k^n$ relative to two choices of bases of $k^n$. Of course, two nonsingular $n \times n$ matrices $A$ and $B$ over a field $k$ are similar if and only if they are conjugate elements in the group $\mathrm{GL}(n, k)$.

**Theorem 4.69 (Jordan–Moore).** *The groups* $\mathrm{PSL}(2, q)$ *are simple for all prime powers* $q \geq 4$.

**Remark.** It is true that $\mathrm{PSL}(2, k)$ is a simple group for every infinite field $k$. By Proposition 4.66, $|\mathrm{PSL}(2, 2)| = 6$ and $|\mathrm{PSL}(2, 3)| = 12$, so that neither of these groups is simple. ◄

**Proof.** It suffices to prove that a normal subgroup $H$ of $\mathrm{SL}(2, q)$ that contains a matrix not in the center $\mathrm{SZ}(2, q)$ must be all of $\mathrm{SL}(2, q)$.

Suppose, first, that $H$ contains a matrix $A = \begin{bmatrix} \alpha & 0 \\ \beta & \alpha^{-1} \end{bmatrix}$, where $\alpha \neq \pm 1$; that is, $\alpha^2 \neq 1$. If $B = B_{21}(1)$, then $H$ contains the commutator $BAB^{-1}A^{-1} = B_{21}(1 - \alpha^{-2})$, which is a transvection because $1 - \alpha^{-2} \neq 0$. Therefore, $H = \mathrm{SL}(2, q)$, by Lemma 4.67.

To complete the proof, we need only show that $H$ contains a matrix whose top row is $[\alpha \ 0]$, where $\alpha \neq \pm 1$. By hypothesis, there is some matrix $M \in H$ that is not a scalar matrix. Let $\varphi \colon k^2 \to k^2$ be the linear transformation given by $\varphi(v) = Mv$, where $v$ is a $2 \times 1$ column vector. If $\varphi(v) = c_v v$ for all $v$, where $c_v \in k$, then the matrix $[\varphi]$ relative to any basis of $k^2$ is a diagonal matrix. In this case, $M$ is similar to a diagonal matrix $D = \mathrm{diag}\{\alpha, \beta\}$, and Lemma 4.68 says that $D \in H$. Since $M \notin \mathrm{SZ}(2, q)$, we must have $\alpha \neq \beta$. But $\alpha\beta = \det(M) = 1$, and so $\alpha \neq \pm 1$. Therefore, $D$ is a matrix in $H$ of the desired form.

In the remaining case, there is a vector $v$ with $\varphi(v)$ not a scalar multiple of $v$, and we saw in Example 2.124(ii) that $M$ is similar to a matrix of the form $\begin{bmatrix} 0 & a \\ 1 & b \end{bmatrix}$; we must have $a = -1$ because $M$ has determinant 1. Lemma 4.68 now says that there is some $u \in k$ with

$$
D = \begin{bmatrix} 0 & -u^{-1} \\ u & b \end{bmatrix} \in H.
$$

If $T = \text{diag}\{\alpha, \alpha^{-1}\}$ (where $\alpha$ will be chosen in a moment), then the commutator

$$V = (TDT^{-1})D^{-1} = \begin{bmatrix} \alpha^2 & 0 \\ ub(\alpha^{-2} - 1) & \alpha^{-2} \end{bmatrix} \in H.$$

We are done if $\alpha^2 \neq \pm 1$; that is, if there is some nonzero $\alpha \in k$ with $\alpha^4 \neq 1$. If $q > 5$, then such an element $\alpha$ exists, for the polynomial $x^4 - 1$ has at most four roots in a field. If $q = 4$, then every $\alpha \in \mathbb{F}_4$ is a root of the equation $x^4 - x$; that is, $\alpha^4 = \alpha$. Hence, if $\alpha \neq 1$, then $\alpha^4 \neq 1$.

Only the case $q = 5$ remains. The entry $b$ in $D$ shows up in the lower left corner $v = ub(\alpha^{-2} - 1)$ of the commutator $V$. There are two subcases depending on whether $b \neq 0$ or $b = 0$. In the first subcase, choose $\alpha = 2$ so that $\alpha^{-2} = 4 = \alpha^2$ and $v = (4 - 1)ub = 3ub \neq 0$. Now $H$ contains $V^2 = B_{21}(-2v)$, which is a transvection because $-2v = -6ub = 4ub \neq 0$. Finally, if $b = 0$, then $D$ has the form

$$D = \begin{bmatrix} 0 & -u^{-1} \\ u & 0 \end{bmatrix}.$$

Conjugating $D$ by $B_{12}(y)$ for $y \in \mathbb{F}_5$ gives a matrix $B_{12}(y)DB_{12}(-y) \in H$ whose top row is

$$\begin{bmatrix} uy & -uy^2 - u^{-1} \end{bmatrix}.$$

If we choose $y = 2u^{-1}$, then the top row is $[2\ 0]$, and the proof is complete. •

Here are the first few orders of these simple groups:

$$|\text{PSL}(2,4)| = 60;$$
$$|\text{PSL}(2,5)| = 60;$$
$$|\text{PSL}(2,7)| = 168;$$
$$|\text{PSL}(2,8)| = 504;$$
$$|\text{PSL}(2,9)| = 360;$$
$$|\text{PSL}(2,11)| = 660.$$

These numbers are the only orders of nonabelian simple groups under 1000. Some of these, namely, 60 and 360, coincide with orders of alternating groups. The next proposition shows that $\text{PSL}(2,4) \cong A_5 \cong \text{PSL}(2,5)$, and Exercise 4.54 on page 270 shows that $\text{PSL}(2,9) \cong A_6$. [There do exist nonisomorphic simple groups of the same order: $A_8$ and $\text{PSL}(3,4)$ are such (Rotman, *An Introduction to the Theory of Groups*, p. 233); they have order $\frac{1}{2}8! = 20,160$.]

**Proposition 4.70.** *If $G$ is a simple group of order 60, then $G \cong A_5$.*

**Proof.** It suffices to show that $G$ has a subgroup $H$ of index 5, for then Theorem 1.96, the representation on the cosets of $H$, gives a homomorphism $\varphi : G \to S_5$ with $\ker \varphi \subseteq H$. As $G$ is simple, the proper normal subgroup $\ker \varphi$ is equal to $\{1\}$, and so $G$ is isomorphic to a subgroup of $S_5$ of order 60. By Exercise 1.102(ii) on page 75, $A_5$ is the only subgroup of $S_5$ of order 60, and so $G \cong A_5$.

Suppose that $P$ and $Q$ are Sylow 2-subgroups of $G$ with $P \cap Q \neq \{1\}$; choose $x \in P \cap Q$ with $x \neq 1$. Now $P$ has order 4, hence is abelian, and so $4 \mid |C_G(x)|$, by

Lagrange's Theorem. Indeed, since both $P$ and $Q$ are abelian, the subset $P \cup Q$ is contained in $C_G(x)$, so that $|C_G(x)| \geq |P \cup Q| > 4$. Therefore, $|C_G(x)|$ is a proper multiple of 4 which is also a divisor of 60: either $|C_G(x)| = 12$, $|C_G(x)| = 20$, or $|C_G(x)| = 60$. The second case cannot occur lest $C_G(x)$ have index 3, and representing $G$ on its cosets would show that $G$ is isomorphic to a subgroup of $S_3$; the third case cannot occur lest $x \in Z(G) = \{1\}$. Therefore, $C_G(x)$ is a subgroup of $G$ of index 5, and we are done in this case. We may now assume that every two Sylow 2-subgroups of $G$ intersect in $\{1\}$.

A Sylow 2-subgroup $P$ of $G$ has $r = [G : N_G(P)]$ conjugates, where $r = 3, 5$, or 15. Now $r \neq 3$ ($G$ has no subgroup of index 3). We show that $r = 15$ is not possible by counting elements. Each Sylow 2-subgroup contains three nonidentity elements. Since any two Sylow 2-subgroups intersect trivially (as we have seen above), their union contains $15 \times 3 = 45$ nonidentity elements. Now a Sylow 5-subgroup of $G$ must have six conjugates (the number $r_5$ of them is a divisor of 60 satisfying $r_5 \equiv 1 \mod 5$). But Sylow 5-subgroups are cyclic of order 5, so that the intersection of any pair of them is $\{1\}$, and so the union of them contains $6 \times 4 = 24$ nonidentity elements. We have exceeded the number of elements in $G$, and so this case cannot occur. •

**Corollary 4.71.** $\text{PSL}(2, 4) \cong A_5 \cong \text{PSL}(2, 5)$.

**Proof.** All three groups are simple and have order 60. •

Theorem 4.69 generalizes to $m \times m$ matrices.

**Definition.** If $m \geq 2$ and $k$ is a field, then the ***projective unimodular group*** is the quotient group

$$\text{PSL}(m, k) = \text{SL}(m, k)/\text{SZ}(m, k),$$

where $\text{SZ}(m, k)$ is the normal subgroup of all diagonal unimodular matrices.

**Theorem (Jordan–Dickson).**

(i) *If $d = (m, q - 1)$, then $|\text{PSL}(m, q)| = \frac{1}{d}(q^m - 1)(q^m - q) \cdots (q^m - q^{m-1})$.*

(ii) *For every $m \geq 3$ and every field $k$, the group $\text{PSL}(m, k)$ is simple.*

**Proof.** Ibid, p. 223 and p. 232. •

In addition to the cyclic groups of prime order, the alternating groups, and the projective unimodular groups, there are several other infinite families of finite simple groups, collectively called the simple groups of ***Lie type***. The *Classification Theorem* says that every finite simple group either lies in one of these families or it is one of 26 ***sporadic*** simple groups, the largest of which is the "Monster" of order approximately $8.08 \times 10^{53}$. We refer the interested reader to the books by E. Artin, Carter (as well as the chapter by Carter in the book edited by Kostrikin–Shafarevich), Conway et al, Dieudonné, and Gorenstein–Lyons–Solomon.

# Exercises

**4.52.** Give a composition series for $GL(2, 5)$ and list its factor groups.

**4.53.** (i) Prove that $PSL(2, 2) \cong S_3$.

(ii) Prove that $PSL(2, 3) \cong A_4$.

∗ **4.54.** Prove that any simple group $G$ of order 360 is isomorphic to $A_6$. Conclude that $PSL(2, 9) \cong A_6$.

**Hint.** Let $N(P)$ be the normalizer of a Sylow 5-subgroup $P$. Prove that there is an element $\alpha$ of order 3 such that $A = \langle N(P), \alpha \rangle$ has order 120. Since $N(P) \cong D_{10}$, by Exercise 4.29(iii) on page 251, we have $|A| = 30, 60,$ or $360$.

**4.55.** (i) Prove that $SL(2, 5)$ is not solvable.

(ii) Show that a Sylow 2-subgroup of $SL(2, 5)$ is isomorphic to $\mathbf{Q}$, the quaternion group of order 8.

(iii) Prove that the Sylow $p$-subgroups of $SL(2, 5)$ are cyclic if $p$ is an odd prime. Conclude, for every prime divisor $p$ of $|SL(2, 5)|$, that all the Sylow $p$-subgroups of $SL(2, 5)$ have a unique subgroup of order $p$.

**4.56.** Prove that $GL(2, 7)$ is not solvable.

∗ **4.57.** (i) Prove that $SL(2, q)$ is the commutator subgroup of $GL(2, q)$ for all prime powers $q \geq 4$.

(ii) What is the commutator subgroup of $GL(2, q)$ when $q = 2$ and when $q = 3$?

**4.58.** Let $\pi$ be a primitive element of $\mathbb{F}_8$.

(i) What is the order of $\left[\begin{smallmatrix} \pi & 0 \\ 1 & \pi \end{smallmatrix}\right]$ considered as an element of $GL(2, 8)$?

(ii) What is the order of $\left[\begin{smallmatrix} \pi & 0 & 0 \\ 1 & \pi & 0 \\ 0 & 1 & \pi \end{smallmatrix}\right]$ considered as an element of $GL(3, 8)$?

**Hint.** Show that if $N = \left[\begin{smallmatrix} 0 & 0 & 0 \\ 1 & 0 & 0 \\ 0 & 1 & 0 \end{smallmatrix}\right]$, then $N^2 = \left[\begin{smallmatrix} 0 & 0 & 0 \\ 0 & 0 & 0 \\ 1 & 0 & 0 \end{smallmatrix}\right]$ and $N^3 = 0$, and use the Binomial Theorem to show that if $A = \left[\begin{smallmatrix} \pi & 0 & 0 \\ 1 & \pi & 0 \\ 0 & 1 & \pi \end{smallmatrix}\right]$, then $A^m = \pi^m I + m\pi^{m-1} N + \binom{m}{2}\pi^{m-2} N^2$.

## Section 4.5. Free Groups and Presentations

How can we describe a group? By Cayley's Theorem, a finite group $G$ is isomorphic to a subgroup of the symmetric group $S_n$, where $n = |G|$, and so $G$ can always be defined as a subgroup of $S_n$ generated by certain permutations. An example of this kind of construction occurs as an exercise in Carmichael, *An Introduction to the Theory of Groups*, p. 39:

Let $G$ be the subgroup of $S_{16}$ generated by the following permutations:

$$(a\ c)(b\ d); \quad (e\ g)(f\ h);$$
$$(i\ k)(j\ \ell); \quad (m\ o)(n\ p);$$
$$(a\ c)(e\ g)(i\ k); \quad (a\ b)(c\ d)(m\ o);$$
$$(e\ f)(g\ h)(m\ n)(o\ p); \quad (i\ j)(k\ \ell).$$

Prove that $|G| = 256$, $|G'| = 16$, $\alpha = (i\ k)(j\ \ell)(m\ o)(n\ p) \in G'$, but $\alpha$ is not a commutator.[17]

A second way of describing a group $G$ is by replacing $S_n$ with $\mathrm{GL}(n, k)$ for some $n \geq 2$ and some field $k$. Exercise 1.107 on page 76 says that every group of order $n$ can be imbedded in $\mathrm{GL}(n, \mathbb{Q})$. It is easy to see that $\mathbb{Q}$ can be replaced by $k$; in fact, the set of all $n \times n$ permutation matrices (whose entries are 0's and 1's) is a subgroup of $\mathrm{GL}(n, k)$ isomorphic to $S_n$ [a given group $G$ of order $n$ can often be imbedded in $\mathrm{GL}(m, k)$ for $m < n$ if we use entries in $k$ other than 0 and 1; for example, on page 43, we defined the quaternions $\mathbf{Q}$, a group of order 8, as a subgroup of $\mathrm{GL}(2, \mathbb{C})$]. For relatively small groups, descriptions in terms of permutations or matrices are useful, but when $n$ is large, such descriptions are impractical.

We can also describe groups as being generated by elements subject to certain relations. For example, the dihedral group $D_{2n}$ can be characterized as a group of order $2n$ that can be generated by two elements $a$ and $b$ such that $a^n = 1 = b^2$ and $bab = a^{-1}$. Consider the following definition.

**Definition.** A group $G$ is of **type** $\mathbb{T}(x, y \mid x^{2^{n-1}}, yxy^{-1}x, y^{-2}x^{2^{n-2}})$ if $n \geq 2$ and $G$ is generated by two elements $a$ and $b$ such that

$$a^{2^{n-1}} = 1, \ bab^{-1} = a^{-1}, \ \text{and} \ b^2 = a^{2^{n-2}}.$$

For $n = 3$, the quaternion group $\mathbf{Q}$ is of type $\mathbb{T}(x, y \mid x^4, yxy^{-1}x, y^{-2}x^2)$, by Exercise 1.67 on page 46, but so are the cyclic groups $\langle a \rangle$ of orders 4 or 2. Indeed, the trivial group $\{1\}$ is also a group of this type. Consider the next definition.

**Definition.** The **generalized quaternions** $\mathbf{Q}_n$ is a group of order $2^n$ and of type $\mathbb{T}(x, y \mid x^{2^{n-1}}, yxy^{-1}x, y^{-2}x^{2^{n-2}})$.

Does a generalized quaternion group of order $2^n$ exist? If $n = 3$, the answer is affirmative; the group $\mathbf{Q}$ of quaternions is equal to $\mathbf{Q}_3$. Is there a generalized quaternion group of order 16? More generally, can we define other types of groups, and will there always exist groups of these types? In the 1880s, von Dyck invented *free groups*, which are the key to dealing with these questions. Here is a modern definition of free group, mimicking the freeness property of free abelian groups (Theorem 4.13), and of polynomial rings (Theorem 2.25) which, in turn, are modeled on Theorem 2.120, the fundamental result of Linear Algebra enabling us to describe linear transformations by matrices.

---

[17]Carmichael posed this exercise in the 1930s, before the era of high-speed computers, and he expected his readers to solve it by hand.

**Definition.** Let $X$ be a subset[18] of a group $F$. Then $F$ is a ***free group*** with ***basis*** $X$ if, for every group $G$ and every function $f: X \to G$, there exists a unique homomorphism $\varphi: F \to G$ with $\varphi(x) = f(x)$ for all $x \in X$:

$$
\begin{array}{ccc}
F & & \\
\uparrow & \searrow^{\varphi} & \\
X & \xrightarrow{f} & G
\end{array}
$$

If $F$ is free with basis $X$, we may denote $F$ by $F(X)$; in particular, if $X = \{x_1, \ldots, x_n\}$, we may denote $F$ by $F(x_1, \ldots, x_n)$.

Thus, if $F$ is free with basis $X$ and $G$ is any group, specifying values $\varphi(x) \in G$ for every $x \in X$ defines a function $\varphi: X \to G$ which extends to a unique homomorphism $\widetilde{\varphi}: F \to G$.

Suppose a free group $F(x, y)$ with basis $\{x, y\}$ exists; let $N$ be the normal subgroup of $F$ generated by $\{x^{2^{n-1}}, yxy^{-1}x, y^{-2}x^{2^{n-2}}\}$, and define $\mathbf{Q}_n = F/N$. It is clear that $F/N$ is a group of type $\mathbb{T}(x, y \mid x^{2^{n-1}}, yxy^{-1}x, y^{-2}x^{2^{n-2}})$: it is generated by two elements $a = xN$ and $b = yN$ that satisfy the equations in the definition. It is not clear whether $F/N$ is a generalized quaternion group, for we do not know if it has order $2^n$ (it is not even obvious that $F/N$ is finite!). The answer is affirmative, and it is proved in Proposition 4.85.

The first question, then, is whether free groups exist. The idea of the construction is simple and natural; we begin with a discussion of semigroups and monoids.

**Definition.** Let $X$ be a set, called an ***alphabet***. If $n$ is a positive integer, a ***word*** $w$ on $X$ of ***length*** $n \geq 1$ is a function $w: \{1, 2, \ldots, n\} \to X$; denote its length by $|w| = n$. Let 1 be a symbol not in $X$; the symbol 1, called the ***empty word***, is a ***word*** of ***length*** 0.

Thus, a word $w$ on $X$ of length $n \geq 1$ is an $n$-tuple lying in $X^n$. Since $1 \notin X$, the empty word 1 does not occur in the spelling of a nonempty word. In practice, we shall write a nonempty word $w$ as follows: if $w(i) = a_i \in X$, then

$$
w = a_1 \cdots a_n.
$$

The definition of equality of functions reads here as follows. If $u = a_1 \cdots a_n$ and $v = a'_1 \cdots a'_m$ are nonempty words on $X$, then $u = v$ if and only if $n = m$ and $a_i = a'_i$ for all $i$; thus, every word on $X$ has a unique spelling.

Recall that a *semigroup* is a set having an associative binary operation, and a *monoid* is a semigroup having a (two-sided) identity element 1.

**Definition.** If $S$ and $S'$ are semigroups, then a ***homomorphism*** is a function $f: S \to S'$ such that $f(xy) = f(x)f(y)$. If $S$ and $S'$ are monoids, then a ***homomorphism*** $f: S \to S'$ is a semigroup homomorphism with $f(1) = 1$.

---

[18]The subset $X$ may be infinite. When $X$ is finite, it is convenient, as in our earlier discussion of bases of vector spaces and bases of free abelian groups, to define bases as *lists* (ordered sets) in $F$ rather than as mere subsets.

Of course, every group is a monoid, and a homomorphism between groups is a homomorphism of them *qua* monoids. Every ring $R$ is a monoid under multiplication, which is noncommutative if $R$ is. Recall Corollary 1.25: every semigroup enjoys generalized associativity.

**Example 4.72.** If $X$ is a set, let $X^*$ denote the set of all words on the alphabet $X$, including the empty word 1 (if $X = \varnothing$, then $X^*$ consists of only the empty word). Define a binary operation on $X^*$, called ***juxtaposition***. Define $1u = u = u1$ for every word $u$. If $u = a_1 \cdots a_n$ and $v = a_1' \cdots a_m'$ are nonempty words on $X$, define

$$uv = a_1 \cdots a_n a_1' \cdots a_m'.$$

Note that $|uv| = |u| + |v|$. Juxtaposition is associative: if $w = a_1'' \cdots a_k''$, then both $(uv)w$ and $u(vw)$ are $(n+m+k)$-tuples whose $j$th coordinates are equal for all $j$. Thus, $X^*$ is a noncommutative monoid whose identity is the empty word. ◄

Given a set $X$, let $X^{-1}$ be a set which is disjoint from $X$ and which is equipped with a bijection $X \to X^{-1}$, denoted by $x \mapsto x^{-1}$. We assume that the empty word lies in neither $X$ nor $X^{-1}$. Now define a new alphabet

$$X \cup X^{-1}.$$

The nonempty words $w \in (X \cup X^{-1})^*$ have a unique expression

$$w = x_1^{e_1} \cdots x_n^{e_n},$$

where $x_i \in X$ and $e_i = \pm 1$ (we may denote $x \in X$ by $x^1$). Since $1 \notin X \cup X^{-1}$, the empty word 1 does not occur in the spelling of a nonempty word. We write

$$X^{**} = (X \cup X^{-1})^*.$$

**Definition.** The ***inverse*** of a nonempty word $w = x_1^{e_1} \cdots x_n^{e_n}$ is

$$w^{-1} = (x_1^{e_1} \cdots x_n^{e_n})^{-1} = x_n^{-e_n} \cdots x_1^{-e_1}.$$

The empty word 1 is defined to be its own inverse.

It follows that $(w^{-1})^{-1} = w$ for every word $w$. Notice that the product $xx^{-1}$ in $X^{**}$ is a word of length 2; in particular, it is not the empty word 1, which has length 0.

The monoid $X^{**}$ is our first approximation to a construction of a free group with basis $X$. That $xx^{-1} \neq 1$ shows that $X^{**}$ is not a group.

**Definition.** A ***subword*** of a nonempty word $w = x_n^{e_n} \cdots x_1^{e_1} \in X^{**}$ is either the empty word or a word of the form $u = x_r^{e_r} \cdots x_s^{e_s}$, where $1 \leq r \leq s \leq n$.

Thus, if $x_r^{e_r} \cdots x_s^{e_s}$ is a subword of $w$, then we may write $w = A x_r^{e_r} \cdots x_s^{e_s} B$, where $A$ and $B$ are subwords of $w$. If $r = 1$, then $A = 1$ and $x_r^{e_r} \cdots x_s^{e_s}$ is an *initial segment* of $w$; if $s = n$, then $B = 1$ and $x_r^{e_r} \cdots x_s^{e_s}$ is a *terminal segment* of $w$.

The most important words are *reduced* words.

**Definition.** A word $w \in X^{**}$ is ***reduced*** if $w = 1$ or if $w$ has no subwords of the form $xx^{-1}$ or $x^{-1}x$ for some $x \in X$.

Note that every subword of a reduced word is itself reduced.

**Definition.** If $w \in X^{**}$ is not reduced, then it contains a subword of the form $x^e x^{-e}$, where $x \in X$ and $e = \pm 1$. We may write $w = A x^e x^{-e} B$. If $w_1 = AB$, then we say that $w \to w_1$ is an *elementary cancellation*. If a word $w$ is not reduced, then a *reduction* of $w$ is a finite sequence of elementary cancellations

$$w \to w_1 \to \cdots \to w_r$$

with $w_r$ reduced.

If $w$ is any word, then there is a reduction $ww^{-1} \to w_1 \to \cdots \to 1$.

**Lemma 4.73.** *If $w \in X^{**}$, then either $w$ is reduced or there is a reduction*

$$w \to w_1 \to \cdots \to w_r.$$

**Proof.** We use induction on $|w| \geq 0$. If $w$ is not reduced and $w \to w_1$ is an elementary cancellation, then $|w_1| = |w| - 2$. By induction, either $w_1$ is reduced, or there is a reduction $w_1 \to \cdots \to w_r$, and so $w \to w_1 \to \cdots \to w_r$ is a reduction of $w$.  •

A word $w$ that is not reduced can have many reductions.

It is natural to try to define a free group with basis $X$ as the set $\mathcal{R}$ of all *reduced* words in $X^{**}$, with juxtaposition as the binary operation, but this is not good enough; $\mathcal{R}$ is not closed, for $u$ and $v$ reduced does not imply that $uv$ is reduced. The obvious way to fix this is to change the operation from juxtaposition to juxtaposition followed by a reduction. The following lemma shows that this new binary operation is well-defined.

**Lemma 4.74.** *Let $X$ be a set and let $w \in X^{**}$. If*

$$w \to w_1 \to w_2 \to \cdots \to w_r \quad and \quad w \to w_1' \to w_2' \to \cdots \to w_q'$$

*are reductions, then $w_r = w_q'$.*

**Proof.** The proof is by induction on $|w| \geq 0$; the base step is obviously true.

We claim that either $w_1 = w_1'$ or there is $z \in X^{**}$ and elementary cancellations $w_1 \to z$ and $w_1' \to z$. We distinguish two cases. Suppose $w = Ass^{-1}Btt^{-1}C$, where $s, t \in X \cup X^{-1}$, $w_1 = ABtt^{-1}C$, and $w_1' = Ass^{-1}BC$. In this case, set $z = ABC$. The other case is $w = Ass^{-1}sB$; the elementary cancellation $w \to w_1$ deletes $ss^{-1}$; the elementary cancellation $w \to w_1'$ deletes $s^{-1}s$. In this case, $w_1 = AsB = w_1'$. Thus, the claim is true.

For the inductive step, choose a reduction $z \to \cdots \to w_d''$. The inductive hypothesis applies to the reductions $w_1 \to w_2 \to \cdots \to w_r$ and $w_1 \to z \to \cdots \to w_d''$, because $|w_1| < |w|$; hence, $w_r = w_d''$. Similarly, the inductive hypothesis applies to $w_1' \to z \to \cdots \to w_q'$ and $w_1' \to w_2' \to \cdots \to w_d''$, because $|w_1'| < |w|$. Hence, $w_d = w_q'$, and so $w_r = w_q'$.  •

In light of Lemma 4.74, all reductions of a word $w \in X^{**}$ end with the same reduced word, say, $w_r$. We denote this reduced word by

$$\mathrm{red}(w) = w_r.$$

**Corollary 4.75.** *If $F(X)$ is the set of all reduced words on $X$, then*

$$u * v = \mathrm{red}(uv)$$

*is a well-defined binary operation on $F(X)$*

**Proof.** Immediate from Lemma 4.74.  •

If $u, v$ are reduced words for which $uv$ is also reduced, then $u * v = uv$. In particular, if $u = x_1^{e_1} \cdots x_n^{e_n}$ is reduced, then $u = x_1^{e_1} * u'$, where $u' = x_2^{e_2} \cdots x_n^{e_n}$.

**Theorem 4.76.** *If $X$ is a set, then the set $F(X)$ of all reduced words on $X$ with binary operation $u * v = \mathrm{red}(uv)$ is a free group with basis $X$.*

**Proof.** It is easy to see that the empty word 1 is the identity element and that the inverses defined on page 273 satisfy the group axiom for inverses. Only associativity need be checked. Given reduced words $u, v, y$, define $w = uvy$ in $X^{**}$ [we do not need parentheses, for $X^{**}$ is a monoid, and its multiplication is associative]. But Lemma 4.74 says that the reductions $w = (uv)y \to \cdots \to (u * v)y \to \cdots \to w_r$ and $w = u(vy) \to \cdots \to u(v * y) \to \cdots \to w'_q$ have the same reduced ending: $w_r = w'_q$; that is, $(u * v) * y = u * (v * y)$. Therefore, $F(X)$ is a group.

Let $G$ be a group and let $f \colon X \to G$ be a function. If $u \in F(X)$, then $u$ is reduced, and it has a unique expression $u = x_1^{e_1} \cdots x_n^{e_n}$. Define a function $\varphi \colon F(X) \to G$ by $\varphi(1) = 1$ and

$$\varphi(u) = \varphi(x_1^{e_1} \cdots x_n^{e_n}) = f(x_1)^{e_1} \cdots f(x_n)^{e_n}.$$

It suffices to prove that $\varphi$ is a homomorphism, for uniqueness of $\varphi$ will follow from $X$ generating $F(X)$. If $u, v \in F(X)$, we prove that $\varphi(u*v) = \varphi(u)\varphi(v)$ by induction on $|u| + |v| \geq 0$. The base step $|u| + |v| = 0$ is true, for $\varphi(1 * 1) = \varphi(1) = 1 = \varphi(1)\varphi(1)$. In fact, $\varphi(1 * v) = \varphi(1)\varphi(v)$, so that we may assume that $|u| \geq 1$ in proving the inductive step. Write the reduced word $u = x^e u'$, where $u' = x_2^{e_2} \cdots x_n^{e_n}$, and note, since $u$ is reduced, that

$$\varphi(u) = f(e)^e f(x_2)^{e_2} \cdots f(x_n)^{e_n} = \varphi(x^e)\varphi(u').$$

Now $u * v = x^e * (u' * v)$; write the reduced word $u' * v = z_1^{c_1} \cdots z_t^{c_t}$, where $z_1, \ldots, z_t \in X$. There are two cases. If $x^e \neq z_1^{-c_1}$, then $x^e z_1^{c_1} \cdots z_t^{c_t}$ is reduced, and the formula defining $\varphi$ gives

$$\begin{aligned}
\varphi(u * v) &= \varphi(x^e * u' * v) = \varphi(x^e z_1^{c_1} \cdots z_t^{c_t}) \\
&= f(x)^e f(z_1)^{c_1} \cdots f(z_t)^{c_t} = \varphi(x^e)\varphi(u' * v) \\
&= \varphi(x^e)\varphi(u')\varphi(v) = \varphi(u)\varphi(v)
\end{aligned}$$

(the inductive hypothesis gives the penultimate equality).

If $x^e = z_1^{-c_1}$, then $u * v = x^e * u' * v = z_2^{c_2} \cdots z_t^{c_t}$. Hence,

$$\begin{aligned}
\varphi(u * v) &= \varphi(x^e * u' * v) = \varphi(z_2^{c_2} \cdots z_t^{c_t}) \\
&= f(z_2)^{c_2} \cdots f(z_t)^{c_t} = [f(x)^e f(x_1)^{e_1}]f(z_2)^{c_2} \cdots f(z_t)^{c_t} \\
&= \varphi(x^e)\varphi(u' * v) = \varphi(x^e)\varphi(u')\varphi(v) = \varphi(u)\varphi(v)
\end{aligned}$$

because $f(x)^e f(x_1)^{e_1} = 1$. Therefore, $\varphi$ is a homomorphism, and $F(X)$ is a free group with basis $X$.  •

**Remark.** From now on, multiplication $u * v = \mathrm{red}(uv)$ will be denoted by $uv$.  ◀

Here are descriptions of some other proofs of the existence of free groups.

(i) In the first edition of this book, a free group on $X$ was constructed from $X^{**}$ in a different way. It is important to assign a unique reduced word to every (not necessarily reduced) word $w$ on $X \cup X^{-1}$. Define two elementary operations on $X^{**}$: the first is elementary cancellation, as above; the second is *elementary insertion*, which changes $w = AB$ into $w' = Ax^e x^{-e} B$ for some $x \in X$ and $e = \pm 1$. Define $w \sim w'$ if there is a finite sequence of elementary operations (cancellations and insertions) from $w$ to $w'$. This is an equivalence relation (insertions are introduced to have symmetry). One proves that each equivalence class $[w]$ contains a unique reduced word (Rotman, *An Introduction to the Theory of Groups*, p. 233). The existence of a free group is proved as follows: elements are equivalence classes $[u]$ of words, and the product $[u][v]$ is the class of the unique reduced word in $[uv]$.

(ii) There is a shorter proof of the existence of the free group with basis a given set $X$, due to Barr (it is the Adjoint Functor Theorem of Category Theory specialized to the category of groups; see Montgomery–Ralston, *Selected Papers in Algebra*, pp. 2–5). We have not given this proof here because it does not describe the elements of $F(X)$ as words in $X$, and this description is very important in studying and using free groups.

(iii) Another construction is called the *van der Waerden trick*. Let $R$ be the set of all reduced words on $X$ [of course, this is the underlying set of $F(X)$]. For each $x \in X$, consider the functions $|x|: R \to R$ and $|x^{-1}|: R \to R$, defined as follows. If $\epsilon = \pm 1$, then

$$|x^{\epsilon}|(x_1^{e_1} \cdots x_n^{e_n}) = x^{\epsilon} x_1^{e_1} \cdots x_n^{e_n} \quad \text{if } x^{\epsilon} \neq x_1^{e_1};$$
$$|x^{-\epsilon}|(x_1^{e_1} \cdots x_n^{e_n}) = x_2^{e_2} \cdots x_n^{e_n} \quad \text{if } x^{\epsilon} = x_1^{e_1}.$$

It turns out that $|x|$ is a permutation of $R$ (its inverse is $|x^{-1}|$), and the subgroup $F$ of the symmetric group $S_R$ generated by $|X| = \{|x| : x \in X\}$ is a free group with basis $|X|$.

(iv) There are topological proofs. A *pointed space* is an ordered pair $(X, x_0)$, where $X$ is a topological space and $x_0 \in X$; we call $x_0$ the *basepoint*. A *pointed map* $f: (X, x_0) \to (Y, y_0)$ is a continuous map $f: X \to Y$ with $f(x_0) = y_0$. The *fundamental group* $\pi_1(X, x_0)$ is the set of all (pointed) homotopy classes of pointed maps $(S^1, 1) \to (X, x_0)$, where $S^1$ is the unit circle $\{e^{2\pi i x} : x \in \mathbb{R}\}$ with basepoint $1 = e^0$. Given an indexed family of circles, $(S_i^1, b_0)_{i \in I}$, any two intersecting only in their common basepoint $b_0$, then their union (suitably topologized) is called a *bouquet of $I$ circles*. For example, a figure 8 is a bouquet of two circles. Then the fundamental group $\pi_1(B, b_0)$ is a free group with basis a set of cardinality $|I|$ (Rotman, *An Introduction to the Theory of Groups*, p. 376).

(v) If $X$ is a *graph* (a one-dimensional space constructed of edges and vertices), then $\pi_1(X)$ is also a free group (Serre, *Trees*, p. 23, where it is shown that every connected graph has the homotopy type of a bouquet of circles).

The free group $F$ with basis $X$ that we have just constructed is generated by $X$. Are any two free groups with basis $X$ isomorphic?

**Proposition 4.77.**

(i) *Let $X_1$ be a basis of a free group $F_1$ and let $X_2$ be a basis of a free group $F_2$. If there is a bijection $f\colon X_1 \to X_2$, then $F_1 \cong F_2$; indeed, there is an isomorphism $\varphi\colon F_1 \to F_2$ extending $f$.*

(ii) *If $F(X)$ is a free group with basis $X$, then $F(X)$ is generated by $X$.*

**Proof.**

(i) The following diagram, in which the vertical arrows are inclusions, will help the reader follow the proof:

$$
\begin{array}{ccc}
F_1 & \underset{\varphi_2}{\overset{\varphi_1}{\rightleftarrows}} & F_2 \\
\big\uparrow & & \big\uparrow \\
X_1 & \underset{f^{-1}}{\overset{f}{\rightleftarrows}} & X_2
\end{array}
$$

We may regard $f$ as having target $F_2$, because $X_2 \subseteq F_2$; since $F_1$ is a free group with basis $X_1$, there is a homomorphism $\varphi_1\colon F_1 \to F_2$ extending $f$. Similarly, there exists a homomorphism $\varphi_2\colon F_2 \to F_1$ extending $f^{-1}$. It follows that the composite $\varphi_2\varphi_1\colon F_1 \to F_1$ is a homomorphism extending $1_X$. But the identity $1_{F_1}$ also extends $1_X$, so that uniqueness of the extension gives $\varphi_2\varphi_1 = 1_{F_1}$. In the same way, we see that the other composite $\varphi_1\varphi_2 = 1_{F_2}$, and so $\varphi_1$ is an isomorphism.

(ii) If $F(X)$ is the free group with basis $X$ constructed in Theorem 4.76, then $X$ generates $F(X)$. By part (i), there is an isomorphism $\varphi\colon F(X) \to F$ with $\varphi(X) = X$ (take $f\colon X \to X$ to be the identity $1_X$). Since $X$ generates $F(X)$, $\varphi(X) = X$ generates $\operatorname{im} \varphi = F$; that is, $X$ generates $F$.  •

There is a notion of rank for free groups, but we must first check that all bases in a free group have the same number of elements (which might be an infinite cardinal).

**Lemma 4.78.** *If $F$ is a free group with basis $X$, then $F/F'$ is a free abelian group with basis $X' = \{xF' : x \in X\}$, where $F'$ is the commutator subgroup of $F$.*

**Proof.** We begin by noting that $X'$ generates $F/F'$; this follows from Proposition 4.77(ii), which says that $X$ generates $F$. We prove that $F/F'$ is a free abelian group with basis $X'$ by using the criterion in Proposition 4.14. Consider the following diagram:

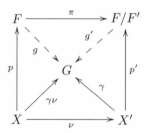

Here, $G$ is an arbitrary abelian group, $p$ and $p'$ are inclusions, $\pi$ is the natural map, $\nu\colon x \mapsto xF'$, and $\gamma\colon X' \to G$ is a function. Let $g\colon F \to G$ be the unique homomorphism with $gp = \gamma\nu$ given by the definition of free group (for $\gamma\nu\colon X \to G$ is a function), and define $g'\colon F/F' \to G$ by $wF' \mapsto g(w)$ ($g'$ is well-defined because $G$ abelian forces $F' \subseteq \ker g$). Now $g'p' = \gamma$, for

$$g'p'\nu = g'\pi p = gp = \gamma\nu;$$

since $\nu$ is a surjection, it follows that $g'p' = \gamma$. Finally, $g'$ is the unique such map, for if $g''$ satisfies $g''p' = \gamma$, then $g'$ and $g''$ agree on the generating set $X'$, hence they are equal.  •

**Proposition 4.79.** *Let $F$ be the free group with basis $X$. If $|X| = n$, then every basis of $F$ has $n$ elements.*

**Proof.** By Lemma 4.78, $F/F'$ is a free abelian group of rank $n$. On the other hand, if $Y$ is a basis of $F$ and $|Y| = m$, then $F/F'$ is a free abelian group of rank $m$. By Proposition 4.11, we have $m = n$.  •

The reader may show, using Theorem 5.52, that Proposition 4.79 is true even if a basis is infinite: any two bases of a free group have the same cardinal. The following definition now makes sense.

**Definition.** The *rank* of a free group $F$, denoted by $\mathrm{rank}(F)$, is the number of elements in a basis of $F$.

Proposition 4.77(i) can now be rephrased: two free groups are isomorphic if and only if they have the same rank. A free group $F$ of finite rank $n$ is often denoted by $F_n$; if $X = \{x_1, \dots, x_n\}$ is a basis of $F$, we may also write $F_n(X)$ or $F(x_1, \dots, x_n)$.

We are now going to prove that there is a rich supply of homomorphisms from free groups to finite groups. Before stating the result, let us give an example of a construction in its proof. Let $F = F(x, y)$ be the free group with basis $x, y$ and let

$$g = x^{-1}yyx^{-1}x^{-1}y^{-1}xy.$$

Label the 8 positions from right to left: $x$ or $x^{-1}$ occurs in positions 2, 4, 5, 8 and $y$ or $y^{-1}$ occurs in positions 1, 3, 6, 7. We are going to construct two permutations

$\alpha_x, \alpha_y \in S_9$ in two stages. Here is the first stage for $\alpha_x$.

$$\alpha_x = \begin{pmatrix} 1 & 2 & 3 & 4 & 5 & 6 & 7 & 8 & 9 \\ & 3 & & 4 & 5 & & & 8 & \end{pmatrix}.$$

These values arise as follows: the factor $x$ occurs in position 2, and $\alpha_x \colon 2 \mapsto 3$. On the other hand, $x^{-1}$ occurs in positions 4, 5, and 8, and $\alpha_x \colon 5 \mapsto 4$, $6 \mapsto 5$, and $9 \mapsto 8$; that is, $\alpha_x^{-1} \colon 4 \mapsto 5$, $5 \mapsto 6$, and $8 \mapsto 9$. The second stage uses Exercise 1.9 on page 15: we can assign values to the other integers between 1 and 9 to get a permutation in $S_9$. Here is the analogous construction for $\alpha_y$; its first stage is

$$\alpha_y = \begin{pmatrix} 1 & 2 & 3 & 4 & 5 & 6 & 7 & 8 & 9 \\ 2 & & 3 & & & 7 & 8 & & \end{pmatrix}.$$

As with $\alpha_x$, Exercise 1.9 applies to give a permutation $\alpha_y \in S_9$. The composite

$$\gamma = \alpha_x^{-1}\alpha_y\alpha_y\alpha_x^{-1}\alpha_x^{-1}\alpha_y^{-1}\alpha_x\alpha_y$$

sends $1 \mapsto 2 \mapsto 3 \mapsto 4 \mapsto 5 \mapsto 6 \mapsto 7 \mapsto 8 \mapsto 9$, and so $\gamma \neq (1)$.

**Proposition 4.80.** *If $F$ is a free group, $g \in F$, and $g \neq 1$, then there is a finite group $S$ and a homomorphism $\varphi \colon F \to S$ with $\varphi(g) \neq 1$.*

**Proof.** Let $X$ be a basis of $F$, and write $g$ as a reduced word:

$$g = x_{i_n}^{e_n} \cdots x_{i_2}^{e_2} x_{i_1}^{e_1},$$

where $x_{i_k} \in X$ and $e_k = \pm 1$ for $k = 1, \ldots, n$. There are $m \leq n$ distinct basis elements occurring in this word, say, $x_1, \ldots, x_j, \ldots, x_m$ (thus, each $j$ occurs at least once as some $i_k$). For each such basis element $x_j$, we are going to construct a permutation $\alpha_j \in S_{n+1}$. Consider the set of all positions $k$ where $x_j$ or $x_j^{-1}$ occurs. If $x_j$ occurs, define $\alpha_j(k) = k + 1$; if $x_i^{-1}$ occurs, define $\alpha_j(k+1) = k$; that is, $\alpha_j^{-1}(k) = k + 1$. We have defined $\alpha_j$ on a subset of $\{1, 2, \ldots, n+1\}$. [There would be a problem if $g$ were not reduced. Suppose that $x_{i_{k+1}} = x_j = x_{i_k}$. If $e_{k+1} = 1$ and $e_k = -1$, then $\alpha_j$ is not well-defined, for $\alpha_j(k+1) = k$ and $\alpha_j(k+1) = k + 2$; if $e_{k+1} = -1$ and $e_k = 1$, then $\alpha_j$ is not injective, for $\alpha_j(k) = k + 1 = \alpha_j(k+2)$.] Exercise 1.9 on page 15 completes each $\alpha_j$ to a permutation in $S_{n+1}$.

Since $F$ is free with basis $X$, we can define a homomorphism $\varphi \colon F \to S_{n+1}$ by specifying its values on the basis $X$. If $x \in X$ is an $x_j$ occurring in the spelling of $g$, define $\varphi(x_j) = \alpha_j$; if $x$ is not involved in $g$, define $\varphi(x) = 1$. We have

$$\varphi(g) = \alpha_{i_n}^{e_n} \cdots \alpha_{i_2}^{e_2} \alpha_{i_1}^{e_1}$$

(remember that each $i_k$ indexes one of the basis elements $x_j$). But $\varphi(g) \neq (1)$, for it sends $1 \mapsto n + 1$. •

**Definition.** A group $G$ is **residually finite** if the intersection of all its normal subgroups of finite index is trivial.

Every finite group is residually finite; however, the group $\mathbb{Q}$ of rationals is not.

**Corollary 4.81.** *Every free group is residually finite.*

**Proof.** It suffices to prove that if $g \neq 1$, then there is some normal subgroup $K$ of finite index with $g \notin K$. By Proposition 4.80, there is a finite group $S$ and a homomorphism $\varphi \colon F \to S$ with $\varphi(g) \neq 1$. Hence, $g \notin \ker \varphi$. But $S$ finite forces $K = \ker \varphi$ to have finite index.  •

Let us return to describing groups.

**Proposition 4.82.** *Every group $G$ is a quotient of a free group.*

**Proof.** Let $X$ be a set for which there exists a bijection $f \colon X \to G$ (for example, we could take $X$ to be the underlying set of $G$ and $f = 1_G$), and let $F$ be the free group with basis $X$. There exists a homomorphism $\varphi \colon F \to G$ extending $f$, and $\varphi$ is surjective because $f$ is. Therefore, $G \cong F/\ker \varphi$.  •

**Definition.** Let $X$ be a set, $F = F(X)$ the free group with basis $X$, and $R \subseteq F$ a set of words on $X$. A group $G$ has a **presentation**,
$$G = (X \mid R),$$
if $G \cong F/N$, where $N$ is the normal subgroup of $F$ generated by $R$; that is, $N$ is the subgroup of $F$ generated by all conjugates of elements of $R$.

We call the set $X$ **generators** and the set $R$ **relations**. (The term *generators* is now being used in a generalized sense, for $X$ is not a subset of $G$; the subset $\{xN \colon x \in X\}$ does generate $F/N$ in the usual sense.)

Proposition 4.82 says that every group has a presentation.

**Example 4.83.**

(i) A group has many presentations. For example, $G = \mathbb{I}_6$ has presentations
$$(x \mid x^6) \quad \text{and} \quad (a, b \mid a^3, b^2, aba^{-1}b^{-1}).$$
Notice what this means: there are isomorphisms $\mathbb{I}_6 \cong F(x)/\langle x^6 \rangle$ and $\mathbb{I}_6 \cong F(a, b)/N$, where $N$ is the normal subgroup generated by $a^3, b^2, aba^{-1}b^{-1}$. The relation $aba^{-1}b^{-1}$ says that $a$ and $b$ commute. If we replace this commutator by $abab$, then we have a presentation of $S_3$, for now we have $bab = a^{-1}$. If we delete this relation, we obtain a presentation of the infinite *modular group* $M$ defined in Exercise 1.86 on page 60.

(ii) The free group with basis $X$ has a presentation
$$(X \mid \varnothing).$$
A free group is so called precisely because it has a presentation with no relations.  ◄

A word on notation. Often, we write the relations in a presentation as equations. Thus, the relations
$$a^3, \quad b^2, \quad aba^{-1}b^{-1}$$
in the second presentation of $\mathbb{I}_6$ may also be written as
$$a^3 = 1, \quad b^2 = 1, \quad ab = ba.$$

**Definition.** A group $G$ is **_finitely generated_** if it has a presentation $(X \mid R)$ with $X$ finite. A group $G$ is called **_finitely presented_** if it has a presentation $(X \mid R)$ in which both $X$ and $R$ are finite.

It is easy to see that a group $G$ is finitely generated if and only if there exists a finite subset $A \subseteq G$ with $G = \langle A \rangle$. Of course, every finitely generated free group is finitely presented. There do exist finitely generated groups that are not finitely presented (Rotman, *An Introduction to the Theory of Groups*, p. 417).

**Remark.** There are interesting connections between Group Theory and Algebraic Topology. A finite **_simplicial complex_** is a topological space that can be *triangulated* in the sense that it is the union of finitely many vertices, edges, triangles, tetrahedra, and so forth. We can prove that a group $G$ is finitely presented if and only if there is a finite simplicial complex $X$ with $G \cong \pi_1(X)$ (Ibid., p. 400).

Quite often, a group arises from a presentation. For example, if $X$ is a simplicial complex containing subcomplexes $Y_1$ and $Y_2$ such that $Y_1 \cup Y_2 = X$ and $Y_1 \cap Y_2$ is connected, then **_van Kampen's Theorem_** says that a presentation of $\pi_1(X)$ can be given if we know presentations of $\pi_1(Y_1)$ and $\pi_1(Y_2)$ (Ibid., p. 396).  ◄

A fundamental problem is how to determine whether two presentations give isomorphic groups. It can be proved that no algorithm can exist that solves this problem. Indeed, it is an undecidable problem whether a presentation defines the (trivial) group of order 1 (Ibid., p. 469).

We can now generalize the notion of the *type* of a group on page 271.

**Definition.** Let $F$ be the free group with basis $X$, and let $R \subseteq F$. A group $G$ is of **_type_** $\mathbb{T}(X \mid R)$ if there is a surjective homomorphism $\varphi \colon F \to G$ with $\varphi(r) = 1$ for all $r \in R$.

A group $G$ with presentation $G = (X \mid R)$ obviously has type $\mathbb{T}(X \mid R)$, but the converse is false. For example, the trivial group $\{1\}$ has type $\mathbb{T}(X \mid R)$ for every ordered pair $(X \mid R)$. Here is the connection between presentations and types.

**Theorem 4.84 (von Dyck's Theorem).**

(i) *If groups $G$ and $H$ have presentations*

$$G = (X \mid R) \quad \text{and} \quad H = (X \mid R \cup S),$$

*then $H$ is a quotient of $G$. In particular, if $H$ is a group of type $\mathbb{T}(X \mid R)$, then $H$ is a quotient of $G$.*

(ii) *Let $G = (X \mid R)$ and $H$ be a group of type $\mathbb{T}(X \mid R)$. If $G$ is finite and $|G| = |H|$, then $G \cong H$.*

**Proof.**

(i) Let $F$ be the free group with basis $X$. If $N$ is the normal subgroup of $F$ generated by $R$ and $K$ is the normal subgroup generated by $R \cup S$, then $N \subseteq K$. Recall the proof of the Third Isomorphism Theorem: the function $\psi \colon F/N \to F/K$, given by $\psi \colon fN \mapsto fK$, is a surjective homomorphism [with

$\ker \psi = K/N$, so that $(F/N)/(K/N) \cong F/K$]; that is, $H = F/K$ is a quotient of $G = F/N$. In particular, to say that $H$ has type $\mathbb{T}(X \mid R)$ is to say that it satisfies all the relations holding in $G$.

(ii) Since $G$ is finite, the Pigeonhole Principle says that the surjective homomorphism $\psi \colon G \to H$ in part (i) is an isomorphism. $\quad \bullet$

Note that if $G = (X \mid R)$ is a finite group, then von Dyck's Theorem implies that $|G| \geq |H|$ for every group $H$ of type $\mathbb{T}(X \mid R)$.

**Proposition 4.85.** *For every $n \geq 3$, the generalized quaternion group $\mathbf{Q}_n$ exists: the group with presentation*

$$\mathbf{Q}_n = \left( a, b \mid a^{2^{n-1}} = 1, bab^{-1} = a^{-1}, b^2 = a^{2^{n-2}} \right)$$

*has order $2^n$.*

**Proof.** The cyclic subgroup $\langle a \rangle$ in $\mathbf{Q}_n$ has order at most $2^{n-1}$, because $a^{2^{n-1}} = 1$. The relation $bab^{-1} = a^{-1}$ implies that $\langle a \rangle \triangleleft \mathbf{Q}_n = \langle a, b \rangle$, so that $\mathbf{Q}_n/\langle a \rangle$ is generated by the image of $b$. Finally, the relation $b^2 = a^{2^{n-2}}$ shows that $|\mathbf{Q}_n/\langle a \rangle| \leq 2$. Hence,

$$|\mathbf{Q}_n| \leq |\langle a \rangle||\mathbf{Q}_n/\langle a \rangle| \leq 2^{n-1} \cdot 2 = 2^n.$$

We prove the reverse inequality by constructing a concrete group $H_n$ of type $\mathbb{T}(x, y \mid x^{2^{n-1}}, yxy^{-1}x, y^{-2}x^{2^{n-2}})$. Consider the complex matrices $A = \begin{bmatrix} \omega & 0 \\ 0 & \omega^{-1} \end{bmatrix}$ and $B = \begin{bmatrix} 0 & 1 \\ -1 & 0 \end{bmatrix}$, where $\omega$ is a primitive $2^{n-1}$th root of unity, and let $H_n = \langle A, B \rangle \subseteq \mathrm{GL}(2, \mathbb{C})$. We claim that $A$ and $B$ satisfy the necessary relations. For all $i \geq 1$,

$$A^{2^i} = \begin{bmatrix} \omega^{2^i} & 0 \\ 0 & \omega^{-2^i} \end{bmatrix},$$

so that $A^{2^{n-1}} = I$; indeed, $A$ has order $2^{n-1}$. Moreover, $B^2 = \begin{bmatrix} -1 & 0 \\ 0 & -1 \end{bmatrix} = A^{2^{n-2}}$ and $BAB^{-1} = \begin{bmatrix} \omega^{-1} & 0 \\ 0 & \omega \end{bmatrix} = A^{-1}$. Notice that $A$ and $B$ do not commute; hence, $B \notin \langle A \rangle$, and so the cosets $\langle A \rangle$ and $B\langle A \rangle$ are distinct. Since $A$ has order $2^{n-1}$, it follows that

$$|H_n| \geq |\langle A \rangle \cup B\langle A \rangle| = 2^{n-1} + 2^{n-1} = 2^n.$$

By von Dyck's Theorem, $2^n \leq |H_n| \leq |\mathbf{Q}_n| \leq 2^n$. Therefore, $|\mathbf{Q}_n| = 2^n$, and Theorem 4.84(ii) gives $\mathbf{Q}_n \cong H_n$. $\quad \bullet$

In Exercise 1.66 on page 46, we gave a concrete construction of the dihedral group $D_{2n}$, and we can use that group—as in the last proof—to give a presentation of it.

**Proposition 4.86.** *The dihedral group $D_{2n}$ has a presentation*

$$D_{2n} = (a, b \mid a^n = 1, b^2 = 1, bab = a^{-1}).$$

**Proof.** Let $D_{2n}$ denote the group defined by the presentation, and let $C_{2n}$ be the group of order $2n$ constructed in Exercise 1.66 on page 46. By von Dyck's Theorem, $|D_{2n}| \geq |C_{2n}| = 2n$. We prove the reverse inequality. The cyclic subgroup $\langle a \rangle$ in $D_{2n}$ has order at most $n$, because $a^n = 1$. The relation $bab^{-1} = a^{-1}$ implies that

$\langle a \rangle \lhd D_{2n} = \langle a, b \rangle$, so that $D_{2n}/\langle a \rangle$ is generated by the image of $b$. Finally, the relation $b^2 = 1$ shows that $|D_{2n}/\langle a \rangle| \leq 2$. Hence, $|D_{2n}| \leq |\langle a \rangle||D_{2n}/\langle a \rangle| \leq 2n$, and $|D_{2n}| = 2n$. Therefore, Theorem 4.84(ii) gives $D_{2n} \cong C_{2n}$. •

In Chapter 1, we classified the groups of order 7 or less. Since groups of prime order are cyclic, it was only a question of classifying the groups of orders 4 and 6. The proof we gave, in Proposition 1.98, that every nonabelian group of order 6 is isomorphic to $S_3$ was rather complicated, analyzing the representation of a group on the cosets of a cyclic subgroup. Here is a proof in the present spirit.

**Proposition 4.87.** *If $G$ is a nonabelian group of order 6, then $G \cong S_3$.*

**Proof.** As in the proof of Proposition 1.98, $G$ must contain elements $a$ and $b$ of orders 3 and 2, respectively. Now $\langle a \rangle \lhd G$, because it has index 2, and so either $bab^{-1} = a$ or $bab^{-1} = a^{-1}$. The first possibility cannot occur, because $G$ is not abelian. Therefore, $G$ has type $\mathbb{T}(a, b \mid a^3, b^2, bab = a^{-1})$, and so Theorem 4.84(ii) gives $D_6 \cong G$ (of course, $D_6 \cong S_3$). •

We can now classify the groups of order 8.

**Theorem 4.88.** *Every group $G$ of order 8 is isomorphic to*

$$D_8, \quad \mathbf{Q}, \quad \mathbb{I}_8, \quad \mathbb{I}_4 \oplus \mathbb{I}_2, \quad or \quad \mathbb{I}_2 \oplus \mathbb{I}_2 \oplus \mathbb{I}_2.$$

*Moreover, no two of the displayed groups are isomorphic.*

**Proof.** If $G$ is abelian, then the Basis Theorem shows that $G$ is a direct sum of cyclic groups, and the Fundamental Theorem shows that the only such groups are those listed. Therefore, we may assume that $G$ is not abelian.

Now $G$ cannot have an element of order 8, lest it be cyclic, hence abelian; moreover, not every nonidentity element can have order 2, lest $G$ be abelian, by Exercise 1.35 on page 27. We conclude that $G$ must have an element $a$ of order 4; hence, $\langle a \rangle$ has index 2, and so $\langle a \rangle \lhd G$. Choose $b \in G$ with $b \notin \langle a \rangle$; note that $G = \langle a, b \rangle$ because $\langle a \rangle$, having index 2, must be a maximal subgroup. Now $b^2 \in \langle a \rangle$, because $G/\langle a \rangle$ is a group of order 2, and so $b^2 = a^i$, where $0 \leq i \leq 3$. We cannot have $b^2 = a$ or $b^2 = a^3 = a^{-1}$ lest $b$ have order 8. Therefore, either

$$b^2 = a^2 \quad or \quad b^2 = 1.$$

Furthermore, $bab^{-1} \in \langle a \rangle$, by normality, and so $bab^{-1} = a$ or $bab^{-1} = a^{-1}$ (for $bab^{-1}$ has the same order as $a$). But $bab^{-1} = a$ says that $a$ and $b$ commute, which implies that $G$ is abelian. We conclude that $bab^{-1} = a^{-1}$. Therefore, there are only two possibilities:

$$a^4 = 1, \ b^2 = a^2, \ bab^{-1} = a^{-1} \quad or \quad a^4 = 1, \ b^2 = 1, \ bab^{-1} = a^{-1}.$$

The first equations give relations of a presentation for $\mathbf{Q}$, by Proposition 4.85, while the second equations give relations of a presentation of $D_8$, by Proposition 4.86. Now von Dyck's Theorem gives a surjective homomorphism $\mathbf{Q} \to G$ or $D_8 \to G$, as $|G| = 8$. Theorem 4.84(ii) says that these homomorphisms must be isomorphisms.

Finally, Exercise 1.69 on page 47 shows that $\mathbf{Q}$ and $D_8$ are not isomorphic. •

The reader may continue this classification of the groups $G$ of small order $|G| \leq 15$; the results are displayed in Table 1.

By Corollary 1.114, every group of order $p^2$, where $p$ is prime, is abelian, and so every group of order 9 is abelian. By the Fundamental Theorem of Finite Abelian Groups, there are only two such groups up to isomorphism: $\mathbb{I}_9$ and $\mathbb{I}_3 \times \mathbb{I}_3$. If $p$ is prime, then every group of order $2p$ is either cyclic or dihedral (Exercise 4.64 on page 285). Thus, there are only two groups of order 10 and only two groups of order 14. There are five groups of order 12 (Rotman, *An Introduction to the Theory of Groups*, p. 84). Two of these are abelian: $\mathbb{I}_{12} \cong \mathbb{I}_4 \times \mathbb{I}_3$ and $\mathbb{I}_3 \times \mathbf{V}$; the nonabelian groups of order 12 are $D_{12} \cong S_3 \times \mathbb{I}_2$, $A_4$, and a group $T$ having the presentation

$$T = \left( a, b \mid a^6 = 1, b^2 = a^3 = (ab)^2 \right)$$

[Exercise 4.65 on page 285 realizes $T$ as a group of matrices]. The group $T$, sometimes called a ***dicyclic group*** of type $(2, 2, 3)$, is an example of a *semidirect product*; it is discussed in Example 9.13. A group of order $pq$, where $p < q$ are primes and $q \not\equiv 1 \bmod p$, must be cyclic, and so there is only one group of order 15 [Ibid., p. 83]. There are fourteen nonisomorphic groups of order 16, so this is a good place to stop.

| Order | Groups |
|-------|--------|
| 4     | $\mathbb{I}_4$, $\mathbf{V}$ |
| 6     | $\mathbb{I}_6$, $S_3$ |
| 8     | $\mathbb{I}_8$, $\mathbb{I}_4 \times \mathbb{I}_2$, $\mathbb{I}_2 \times \mathbb{I}_2 \times \mathbb{I}_2$, $D_8$, $\mathbf{Q}$ |
| 9     | $\mathbb{I}_9$, $\mathbb{I}_3 \times \mathbb{I}_3$ |
| 10    | $\mathbb{I}_{10}$, $D_{10}$ |
| 12    | $\mathbb{I}_{12}$, $\mathbb{I}_3 \times \mathbf{V}$, $D_{12}$, $A_4$, $T$ |
| 14    | $\mathbb{I}_{14}$, $D_{14}$ |
| 15    | $\mathbb{I}_{15}$ |

**Table 1.** Groups of small order

## Exercises

\* **4.59.** Let $F$ be a free group with basis $X$ and let $A \subseteq X$. Prove that if $N$ is the normal subgroup of $F$ generated by $A$, then $F/N$ is a free group.

\* **4.60.** Let $F$ be a free group.

   (i) Prove that $F$ has no elements of finite order (other than 1).

   (ii) Prove that a free group $F$ is abelian if and only if $\operatorname{rank}(F) \leq 1$.
       **Hint.** Map a free group of rank $\geq 2$ onto a nonabelian group.

   (iii) Prove that if $\operatorname{rank}(F) \geq 2$, then $Z(F) = \{1\}$, where $Z(F)$ is the center of $F$.

**4.61.** Prove that a free group is solvable if and only if it is infinite cyclic (use the definition of solvable on page 259).

**4.62.** (i) If $G$ is a finitely generated group and $n$ is a positive integer, prove that $G$ has only finitely many subgroups of index $n$.

**Hint.** Consider homomorphisms $G \to S_n$.

(ii) If $H$ and $K$ are subgroups of finite index in a group $G$, prove that $H \cap K$ also has finite index in $G$.

**4.63.** (i) Prove that each of the generalized quaternion groups $\mathbf{Q}_n$ has a unique subgroup of order 2, namely, $\langle b^2 \rangle$, and this subgroup is the center $Z(\mathbf{Q}_n)$.

(ii) Prove that $\mathbf{Q}_n/Z(\mathbf{Q}_n) \cong D_{2^{n-1}}$.

$*$ **4.64.** If $p$ is prime, prove that every group $G$ of order $2p$ is either cyclic or isomorphic to $D_{2p}$.

**Hint.** By Cauchy's Theorem, $G$ must contain an element $a$ of order $p$, and $\langle a \rangle \lhd G$ because it has index 2.

$*$ **4.65.** Let $G$ be the subgroup of $\mathrm{GL}(2,\mathbb{C})$ generated by $\left[\begin{smallmatrix} \omega & 0 \\ 0 & \omega^2 \end{smallmatrix}\right]$ and $\left[\begin{smallmatrix} 0 & i \\ i & 0 \end{smallmatrix}\right]$, where $\omega = e^{2\pi i/3}$ is a primitive cube root of unity.

(i) Prove that $G$ is a group of order 12 that is not isomorphic to $A_4$ or to $D_{12}$.

(ii) Prove that $G$ is isomorphic to the group $T$ on page 284.

**4.66.** Prove that every finite group is finitely presented.

**4.67.** Compute the order of the group $G$ with the presentation

$$G = \left( a, b, c, d \mid bab^{-1} = a^2, bdb^{-1} = d^2, c^{-1}ac = b^2, dcd^{-1} = c^2, bd = db \right).$$

**4.68.** (i) If $X$ is a set, prove that $X^*$ is the set of all *positive* words $w$ on $X$; that is, $X^*$ is the subset of $X^{**}$ consisting of the empty word 1 and all $x_1^{e_1} \cdots x_n^{e_n}$ with all $e_i = 1$.

(ii) Define a ***free monoid***, and prove that $X^*$ is the free monoid with basis $X$.

## Section 4.6. Nielsen–Schreier Theorem

We are now going to prove one of the most fundamental results about free groups: every subgroup is also free. Nielsen proved, in 1921, that finitely generated subgroups of free groups are free.[19] Even if a free group $F$ has finite rank, Nielsen's proof does not show that every subgroup of $F$ is free, for subgroups of a finitely generated free group need not be finitely generated (Corollary 4.95). The finiteness hypothesis was removed by Schreier in 1926, and the subgroup theorem is called the Nielsen–Schreier Theorem.

A second type of proof was found by Baer and Levi in 1933. It uses a correspondence, analogous to that between Galois groups and intermediate fields, between *covering spaces* $\widetilde{X}$ of a topological space $X$ and subgroups of its fundamental group

---

[19]If $S$ is a finitely generated subgroup of a free group, then Nielsen's proof shows that $S$ is free by giving an algorithm, analogous to Gaussian elimination in Linear Algebra, that replaces a finite generating set with a basis of $S$ (Lyndon–Schupp, *Combinatorial Group Theory*, pp. 4–13). This theoretical algorithm has evolved into the *Schreier–Sims algorithm*, an efficient way to compute the order of a subgroup $H \subseteq S_n$ when a generating set of $H$ is given.

$\pi_1(X)$. It turns out that $\pi_1(\widetilde{X})$ is isomorphic to a subgroup of $\pi_1(X)$. Conversely, given any subgroup $S \subseteq \pi_1(X)$, there exists a covering space $\widetilde{X}_S$ of $X$ for which $\pi_1(\widetilde{X}_S)$ is isomorphic to $S$:

If a space $Y$ is a *graph*, then every covering space $\widetilde{Y}$ of $Y$ is also a graph; moreover, the fundamental group $\pi_1(Y)$ of a graph $Y$ is a free group. Once these facts are established, the proof proceeds as follows. Given a free group $F$, there is a graph $Y$ (a *bouquet of circles*) with $F \cong \pi_1(Y)$; given a subgroup $S \subseteq F$, we know that $S \cong \pi_1(\widetilde{Y}_S)$. But $\widetilde{Y}_S$ is also a graph, so that $\pi_1(\widetilde{Y}_S)$ is free and, hence, $S$ is free. There are versions of this proof that avoid topology (Rotman, *An Introduction to the Theory of Groups*, pp. 377–384). There are interesting variations of this idea. One such involves trees (which arise as *universal covering spaces* of connected graphs); Serre characterizes free groups by their action on trees (Serre, *Trees*, p. 27), which he then uses to prove the Nielsen–Schreier Theorem (Ibid, p. 29). A second variation is due to Higgins, *Notes on Categories and Groupoids*, pp. 117–118.

We prove the Nielsen–Schreier Theorem following Weir, The Reidemeister–Schreier and Kuroš Subgroup Theorems, *Mathematika* 3 (1956), 47–55, because it requires less preparation than the others. The idea arises from a proof of the Reidemeister–Schreier Theorem [which gives presentations of subgroups of a group $G$ in terms of a presentation of $G$ (Lyndon–Schupp, *Combinatorial Group Theory*, pp. 102–104)].

**Definition.** Let $S$ be a subgroup of a group $G$. A (right) **transversal** $\ell$ of $S$ in $G$ is a subset of $G$ consisting of exactly one element $\ell(Sb) \in Sb$ from every right coset $Sb$, and with $\ell(S) = 1$.

Here is an outline of the proof. We will show that a subgroup $S$ of a free group $F$ is itself free by exhibiting a basis of it. Since all we know about $S$ is that it lies in $F$, it is reasonable to consider the cosets of $S$ in $F$ and, hence, transversals of $S$ in $F$. Consideration of any fixed transversal allows us to exhibit a generating set $A$ of $S$ (Corollary 4.90). After examining relations among these generators, we give a presentation of $S$ (Lemma 4.91). This presentation is greatly simplified by using a special kind of transversal, and it is then shown that a certain subset of $A$ is a basis of $S$.

Let $F$ be a free group with basis $X$, let $S$ be a subgroup of $F$, and let $\ell$ be a transversal of $S$ in $F$. If $\ell(Sb) = u$ and $x \in X$, then $ux \in Sbx$; hence, if $\ell(Sbx) = v$, then $uxv^{-1} \in S$; that is, both $\ell(Sb)x$ and $\ell(Sbx)$ lie in the coset $Sbx$. Define

$$[Sb, x] = \ell(Sb)x\ell(Sbx)^{-1} \in S.$$

Define $Y$ to be the free group on new symbols $y_{Sb,x}$, where $x \in X$ and $Sb$ varies over all cosets of $S$ in $F$, and define $\varphi : Y \to S$ to be the homomorphism given by

$$\varphi: y_{Sb,x} \mapsto [Sb, x].$$

For each coset $Sb$, we define a **coset function** $F \to Y$, denoted by $u \mapsto u^{Sb}$. These coset functions are defined simultaneously, by induction on $|u| \geq 0$, where $u$ is a reduced word on $X$. For all $x \in X$ and all cosets $Sb$, define

$$1^{Sb} = 1, \quad x^{Sb} = y_{Sb,x}, \quad \text{and} \quad (x^{-1})^{Sb} = \left(x^{Sbx^{-1}}\right)^{-1}.$$

If $u = x^{\varepsilon} v$ is a reduced word of length $n + 1$, where $\varepsilon = \pm 1$ and $|v| = n \geq 1$, define

$$u^{Sb} = (x^{\varepsilon})^{Sb} v^{Sbx^{\varepsilon}}.$$

**Lemma 4.89.**

(i) *For all $u, v \in F$, the coset functions satisfy $(uv)^{Sb} = u^{Sb} v^{Sbu}$.*

(ii) *For all $u \in F$, $(u^{-1})^{Sb} = (u^{Sbu^{-1}})^{-1}$.*

(iii) *For all $u \in F$,*

$$\varphi(u^{Sb}) = \ell(Sb) u \ell(Sbu)^{-1}.$$

(iv) *The function $\theta \colon S \to Y$, given by $\theta \colon u \mapsto u^S$, is a homomorphism, and*

$$\varphi\theta = 1_S.$$

**Proof.**

(i) The proof is by induction on $|u|$, where $u$ is reduced. If $|u| = 0$, then $u = 1$ and $(uv)^{Sb} = v^{Sb}$; on the other hand, $1^{Sb} v^{Sb1} = v^{Sb}$.

For the inductive step, write $u = x^{\varepsilon} w$. Then

$$\begin{aligned}
(uv)^{Sb} &= (x^{\varepsilon})^{Sb} (wv)^{Sbx^{\varepsilon}} &&\text{(definition of coset functions)} \\
&= (x^{\varepsilon})^{Sb} w^{Sbx^{\varepsilon}} v^{Sbx^{\varepsilon}w} &&\text{(inductive hypothesis)} \\
&= (x^{\varepsilon})^{Sb} w^{Sbx^{\varepsilon}} v^{Sbu} = (x^{\varepsilon} w)^{Sb} v^{Sbu} = u^{Sb} v^{Sbu}.
\end{aligned}$$

(ii) The result follows from $1 = 1^{Sb} = (u^{-1}u)^{Sb} = (u^{-1})^{Sb} u^{Sbu^{-1}}$.

(iii) Recall that $\varphi \colon Y \to S$ is the homomorphism with $\varphi \colon y_{Sb,x} \mapsto [Sb, x] = \ell(Sb) x \ell(Sbx)^{-1}$; the formula in the statement says that $\varphi$ replaces $x^{\varepsilon}$ in $\ell(Sb) x \ell(Sbx)^{-1}$ by $u$. The proof is by induction on $|u| \geq 0$. For the base step, $\varphi(1^{Sb}) = \varphi(1) = 1$, while $\ell(S) 1 \ell(S1)^{-1} = 1$. For the inductive step, write $u = x^{\varepsilon} v$, where $u$ is reduced. Then

$$\begin{aligned}
\varphi(u^{Sb}) &= \varphi((x^{\varepsilon}v)^{Sb}) = \varphi((x^{\varepsilon})^{Sb} v^{Sbx^{\varepsilon}}) \\
&= \varphi((x^{\varepsilon})^{Sb}) \varphi(v^{Sbx^{\varepsilon}}) = \varphi((x^{\varepsilon})^{Sb}) \ell(Sbx^{\varepsilon}) v \ell(Sbx^{\varepsilon}v)^{-1},
\end{aligned}$$

the last equation following from the inductive hypothesis. There are now two cases, depending on the sign $\varepsilon$. If $\varepsilon = +1$, then

$$\begin{aligned}
\varphi(u^{Sb}) &= \ell(Sb) x \ell(Sbx)^{-1} \ell(Sbx) v \ell(Sbxv)^{-1} \\
&= \ell(Sb) x v \ell(Sbxv)^{-1} = \ell(Sb) u \ell(Sbu)^{-1}.
\end{aligned}$$

If $\varepsilon = -1$, then

$$\varphi(u^{Sb}) = \varphi((y_{Sbx^{-1},x})^{-1})\ell(Sbx^{-1})v\ell(Sbx^{-1}v)^{-1}$$
$$= \left(\ell(Sbx^{-1})x\ell(Sbx^{-1}x)^{-1}\right)^{-1}\ell(Sbx^{-1})v\ell(Sbx^{-1}v)^{-1}$$
$$= \ell(Sb)x^{-1}\ell(Sbx^{-1})^{-1}\ell(Sbx^{-1})v\ell(Sbx^{-1}v)^{-1}$$
$$= \ell(Sb)x^{-1}v\ell(Sbx^{-1}v)^{-1} = \ell(Sb)u\ell(Sbu)^{-1}.$$

(iv) For $u \in S$, define $\theta\colon S \to Y$ by $\theta : u \mapsto u^S$ (of course, $\theta$ is the restriction to $S$ of the coset function $u \mapsto u^{Sb}$ when $b = 1$). Now, if $u, v \in S$, then

$$\theta(uv) = (uv)^S = u^S v^{Su} = u^S v^S = \theta(u)\theta(v),$$

because $Su = S$ when $u \in S$. Therefore, $\theta$ is a homomorphism. Moreover, if $u \in S$, then (iii) gives $\varphi\theta(u) = \varphi(u^S) = \ell(S1)u\ell(S1u)^{-1} = u$. $\bullet$

**Corollary 4.90.** *If $S$ is a subgroup of a free group $F$ and $\ell$ is a transversal of $S$ in $F$, then the set of all $[Sb, x] = \ell(Sb)x\ell(Sbx)^{-1}$ generates $S$.*

**Proof.** Since the composite $\varphi\theta = 1_S$, the function $\varphi\colon Y \to S$ is surjective; hence, the images $[Sb, x]$ of the generators $y_{Sb,x}$ of $Y$ generate $\operatorname{im}\varphi = S$. $\bullet$

**Lemma 4.91.** *If $F$ is free with basis $X$ and $\ell$ is a transversal of a subgroup $S$ in $F$, then a presentation of $S$ is*

$$S = (y_{Sb,x} \mid \ell(Sb)^S \text{ for all } x \in X \text{ and all cosets } Sb).$$

**Proof.** Let $N$ be the normal subgroup of $Y$ generated by all $\ell(Sb)^S$, and let $K = \ker\varphi$. By Lemma 4.89(iv), $\theta\colon S \to Y$ is a homomorphism with $\varphi\theta = 1_S$ (where $\varphi\colon y_{Sb,x} \mapsto [Sb, x]$ and $\theta\colon u \mapsto u^S$). It follows from Exercise 4.73(ii) on page 293 that $K$ is the normal subgroup of $Y$ generated by $\{y^{-1}\rho(y) : y \in Y\}$, where $\rho = \theta\varphi$. By Lemma 4.89(i),

$$y_{Sb,x}^{-1}\rho(y_{Sb,x}) = y_{Sb,x}^{-1}\left(\ell(Sb)x\ell(Sbx)^{-1}\right)^S$$
$$= y_{Sb,x}^{-1}\ell(Sb)^S x^{Sb}\left(\ell(Sbx)^{-1}\right)^{Sbx}$$
$$= \left(y_{Sb,x}^{-1}\ell(Sb)^S y_{Sb,x}\right)\left(\ell(Sbx)^{-1}\right)^{Sbx},$$

for $x^{Sb} = y_{Sb,x}$ is part of the definition of the coset function $u \mapsto u^{Sb}$. Therefore,

$$(1) \qquad\qquad y_{Sb,x}^{-1}\rho(y_{Sb,x}) = \left(y_{Sb,x}^{-1}\ell(Sb)^S y_{Sb,x}\right)\left(\ell(Sbx)^S\right)^{-1},$$

because Lemma 4.89(ii) gives $(\ell(Sbx)^{-1})^{Sbx} = (\ell(Sbx)^S)^{-1}$. It follows from Equation (1) that $y_{Sb,x}^{-1}\rho(y_{Sb,x}) \in N$, and so $K \subseteq N$. For the reverse inclusion, Equation (1) says that $\ell(Sb)^S \in K$ if and only if $\ell(Sbx)^S \in K$. Therefore, the desired inclusion can be proved by induction on $|\ell(Sb)|$, and so $K = N$. Therefore, $S \cong Y/\ker\varphi = Y/K$. $\bullet$

We now choose a special transversal.

**Definition.** Let $F$ be a free group with basis $X$ and let $S$ be a subgroup of $F$. A *Schreier transversal* is a transversal $\ell$ with the property that if $\ell(Sb) = x_1^{\varepsilon_1} x_2^{\varepsilon_2} \cdots x_n^{\varepsilon_n}$ is a reduced word, then every initial segment $x_1^{\varepsilon_1} x_2^{\varepsilon_2} \cdots x_k^{\varepsilon_k}$, for $1 \leq k \leq n$, is also in the transversal.

**Lemma 4.92.** *A Schreier transversal exists for every subgroup $S$ of $F$.*

**Proof.** Define the *length* $|Sb|$ of a coset $Sb$ to be the minimum length of the elements $sb \in Sb$. We prove, by induction on the length $|Sb|$, that there is a representative $\ell(Sb) \in Sb$ such that all its initial segments are representatives of cosets of shorter length. Begin by defining $\ell(S) = 1$. For the inductive step, let $|Sz| = n + 1$ and let $ux^\varepsilon \in Sz$, where $\varepsilon = \pm 1$ and $|ux^\varepsilon| = n + 1$. Now $|Su| = n$, for if its length were $m < n$, it would have a representative $v$ of length $m$, and then $vx^\varepsilon$ would be a representative of $Sz$ of length $< n + 1$. By induction, there exists $b = \ell(Su)$ with every initial segment also a representative; define $\ell(Sz) = bx^\varepsilon$. •

Here is the result we have been seeking.

**Theorem 4.93 (Nielsen–Schreier).** *Every subgroup $S$ of a free group $F$ is free. In fact, if $X$ is a basis of $F$ and $\ell$ is a Schreier transversal of $S$ in $F$, then a basis for $S$ consists of all $[Sb, x] = \ell(Sb)x\ell(Sbx)^{-1}$ that are not $1$.*

**Proof.** Recall that $Y$ is the free group with basis all symbols $y_{Sb,x}$ and $\varphi \colon Y \to S$ is given by $y_{Sb,x} \mapsto [Sb, x]$. We have seen, in the proof of Lemma 4.91, that $S \cong Y/\ker\varphi$ and, also, $\ker\varphi$ is the normal subgroup of $Y$ generated by all $\ell(Sb)^S$. By Exercise 4.59 on page 284, it suffices to show that $\ker\varphi$ is equal to the normal subgroup $T$ of $Y$ generated by all *special* $y_{Sb,x}$; that is, by those $y_{Sb,x}$ for which $[Sb, x] = 1$. Clearly, $T \subseteq \ker\varphi$, and so it suffices to prove the reverse inclusion. We prove, by induction on the length $|\ell(Sv)|$, that $\ell(Sv)^S$ is a word on the special $y_{Sb,x}$. If $|\ell(Sv)| = 0$, then $\ell(Sv) = \ell(S) = 1$, which is a word on the special $y_{Sb,x}$. If $|\ell(Sv)| > 0$, then $\ell(Sv) = ux^\varepsilon$, where $\varepsilon = \pm 1$ and $|u| < |\ell(Sv)|$. Since $\ell$ is a Schreier transversal, $u$ is also a representative: $u = \ell(Su)$. By Lemma 4.89(i),

$$\ell(Sv)^S = u^S(x^\varepsilon)^{Su}.$$

By induction, $u^S$ is a word on the special $y_{Sb,x}$, and hence $u^S \in T$.

It remains to prove that $(x^\varepsilon)^{Su}$ is a word on the special $y_{Sb,x}$. If $\varepsilon = +1$, then $(x^\varepsilon)^{Su} = x^{Su} = y_{Su,x}$. But $\ell(Sux) = ux$, because $v = ux$ and $\ell$ is a Schreier transversal, so that

$$\varphi(y_{Su,x}) = t_{Su,x} = \ell(Su)x\ell(Sux)^{-1} = ux(ux)^{-1} = 1.$$

Therefore, $y_{Su,x}$ is special and $x^{Su}$ lies in $T$. If $\varepsilon = -1$, then the definition of coset functions gives

$$(x^{-1})^{Su} = (x^{Sux^{-1}})^{-1} = (y_{Sux^{-1},x})^{-1}.$$

Hence,

$$\varphi((x^{-1})^{Su}) = (t_{Sux^{-1},x})^{-1} = [\ell(Sux^{-1})x\ell(Sux^{-1}x)]^{-1} = [\ell(Sux^{-1})x\ell(Su)]^{-1}.$$

Since $\ell$ is a Schreier transversal, we have $\ell(Su) = u$ and $\ell(Sux^{-1}) = \ell(Sv) = v = ux^{-1}$. Hence,

$$\varphi((x^{-1})^{Su}) = [(ux^{-1})xu^{-1}]^{-1} = 1.$$

Therefore, $y_{Sux^{-1},x}$ is special, $(x^{-1})^{Su} \in T$, and the proof is complete.   •

Here is a nice application of the Nielsen–Schreier Theorem.

**Corollary 4.94.** *Let $F$ be a free group, and let $u, v \in F$. Then $u$ and $v$ commute if and only if there is $z \in F$ with $u, v \in \langle z \rangle$.*

**Proof.** Sufficiency is obvious; if both $u, v \in \langle z \rangle$, then they lie in an abelian subgroup, and hence they commute.

Conversely, the Nielsen–Schreier Theorem says that the subgroup $\langle u, v \rangle$ is free. On the other hand, the condition that $u$ and $v$ commute says that $\langle u, v \rangle$ is abelian. But an abelian free group is cyclic, by Exercise 4.60 on page 284; therefore, $\langle u, v \rangle \cong \langle z \rangle$ for some $z \in G$.   •

The next result shows, in contrast to abelian groups, that a subgroup of a finitely generated group need not be finitely generated.

**Corollary 4.95.** *If $F$ is a free group of rank $2$, then its commutator subgroup $F'$ is a free group of infinite rank.*

**Proof.** Let $\{x, y\}$ be a basis of $F$. Since $F/F'$ is free abelian with basis $\{xF', yF'\}$, by Lemma 4.78, every coset $F'b$ has a unique representative of the form $x^m y^n$, where $m, n \in \mathbb{Z}$; it follows that the transversal choosing $\ell(F'b) = x^m y^n$ is a Schreier transversal, for every subword of $x^m y^n$ is a word of the same form. If $n > 0$, then $\ell(F'y^n) = y^n$, but $\ell(F'y^n x) = xy^n \neq y^n x$. Therefore, there are infinitely many elements $[Sy^n, x] = \ell(F'y^n)x\ell(F'y^n x)^{-1} \neq 1$, and so the result follows from the Nielsen–Schreier Theorem.   •

An arbitrary subgroup of a finitely generated free group need not be finitely generated, but a subgroup of finite index must be finitely generated.

**Corollary 4.96.** *If $F$ is a free group of finite rank $n$, then every subgroup $S$ of $F$ having finite index $j$ is also finitely generated. In fact, $\mathrm{rank}(S) = jn - j + 1$.*

**Remark.** In geometric proofs of the Nielsen–Schreier Theorem, one sees that $\mathrm{rank}(S) = 1 - \chi(\widetilde{X}_S)$, where $\chi$ is the Euler–Poincaré characteristic.   ◄

**Proof.** Let $X = \{x_1, \ldots, x_n\}$ be a basis of $F$ and let $\ell$ be a Schreier transversal. By Theorem 4.93, a basis of $S$ consists of all those elements $[Sb, x]$ not equal to 1. There are $j$ choices for $Sb$ and $n$ choices for $x$, and so there are at most $jn$ elements in a basis of $S$. Therefore, $\mathrm{rank}(S) \leq jn$, and so $S$ is finitely generated.

Call an ordered pair $(Sb, x)$ **trivial** if $[Sb, x] = 1$; that is, if $\ell(Sb)x = \ell(Sbx)$. We will show that there is a bijection $\psi$ between the family of cosets $\{Sb : b \notin S\}$ and the trivial ordered pairs, so that there are $j - 1$ trivial ordered pairs. It will then follow that $\mathrm{rank}(S) = jn - (j - 1) = jn - j + 1$.

Let $\ell(Sb) = b$; since $Sb \neq S$, we have $b = ux^\varepsilon$, where $\varepsilon = \pm 1$ and $u \in \ell$ ($u = 1$ is possible); that is, $\ell(Su) = u$. Define $\psi(Sb)$ as follows, where $b = ux^\varepsilon$:

$$\psi(Sb) = \psi(Sux^\varepsilon) = \begin{cases} (Su, x) & \text{if } \varepsilon = +1; \\ (Sux^{-1}, x) & \text{if } \varepsilon = -1. \end{cases}$$

Note that $\psi(Sux^\varepsilon)$ is a trivial ordered pair: if $\varepsilon = +1$, then $(Su, x)$ is trivial, for $\ell(Su)x\ell(Sux)^{-1} = ux(ux)^{-1} = 1$; if $\varepsilon = -1$, then $(Sux^{-1}, x)$ is trivial, for $(ux^{-1})x = u$ and $\ell(Sux^{-1})x\ell(Sux^{-1}x)^{-1} = (ux^{-1})xu^{-1} = 1$.

To see that $\psi$ is injective, suppose that $\psi(Sb) = \psi(Sc)$, where $b = ux^\varepsilon$ and $c = vy^\eta$; we assume that $x, y$ lie in the given basis of $F$ and that $\varepsilon = \pm 1$ and $\eta = \pm 1$. There are four possibilities, depending on the signs of $\varepsilon$ and $\eta$. If $\varepsilon = +1 = \eta$, then $(Su, x) = (Sv, y)$; hence, $x = y$ and $Su = Sv$; that is, $u = v$, for $\ell$ is a Schreier transversal and $u, v \in \ell$. Thus, $b = ux = vy = c$ and $Sb = Sc$. Similar calculations show that $Sb = Sc$ when $\varepsilon = -1 = \eta$ and when $\varepsilon, \eta$ have opposite sign.

To see that $\psi$ is surjective, take a trivial ordered pair $(Sw, x)$; that is, $\ell(Sw)x = wx = \ell(Swx)$. Now $w = ux^\varepsilon$, where $u \in \ell$ and $\varepsilon = \pm 1$. If $\varepsilon = +1$, then $w$ does not end with $x^{-1}$, and $\psi(Swx) = (Sw, x)$. If $\varepsilon = -1$, then $w$ does end with $x^{-1}$, and so $\psi(Su) = (Sux^{-1}, x) = (Sw, x)$. $\bullet$

**Corollary 4.97.** *There exist nonisomorphic finitely presented groups $G$ and $H$ each of which is isomorphic to a subgroup of the other.*

**Proof.** If $G$ is a free group of rank 2 and $H$ is a free group of rank 3, then $G \not\cong H$. Both $G$ and $H$ are finitely generated free groups and, hence, are finitely presented. Clearly, $G$ is isomorphic to a subgroup of $H$. On the other hand, the commutator subgroup $G'$ is free of infinite rank, and so $G'$, hence $G$, contains a free subgroup of rank 3; that is, $H$ is isomorphic to a subgroup of $G$. $\bullet$

We are at the beginning of a rich subject, called **Combinatorial Group Theory**, which investigates what properties of groups follow from constraints on their presentations; for example, can a finite group have a presentation with the same number of generators as relations? One of the most remarkable results is the unsolvability of the word problem. A group $G$ has a **solvable word problem** if it has a presentation $G = (X \mid R)$ for which there exists an algorithm to determine whether an arbitrary word $w$ on $X$ is equal to the identity element in $G$ (if $X$ and $R$ are finite, it can be proved that this property is independent of the choice of presentation). In the late 1950s, P. S. Novikov and Boone, independently, proved that there exists a finitely presented group $G$ that does not have a solvable word problem (Rotman, *An Introduction to the Theory of Groups*, p. 431). Other problems involve finding presentations for known groups, as we have done for $\mathbf{Q}_n$ and $D_{2n}$; an excellent reference for such questions is Coxeter–Moser, *Generators and Relations for Discrete Groups*; see also Johnson, *Topics in the Theory of Group Presentations*.

Another problem is whether a group defined by a presentation is finite or infinite. For example, **Burnside's problem** asks whether a finitely generated group $G$ of **finite exponent** $m$, that is, $x^m = 1$ for all $x \in G$, must be finite [Burnside had proved that if such a group $G$ happens to be a subgroup of $\mathrm{GL}(n, \mathbb{C})$ for some $n$, then $G$ is finite (Robinson, *A Course in the Theory of Groups*, p. 221)]. The answer in general, however, is negative; such a group can be infinite. This was first proved for $m$ odd and large, in 1968, by Novikov and Adyan, in a long and complicated paper. Using a geometric technique involving *van Kampen diagrams*, Ol'shanskii gave a much shorter and simpler proof in 1982 (Ol'shanskii, *Geometry of Defining Relations in Groups*). Finally, Ivanov [The Free Burnside Groups of Sufficiently Large Exponents, *Internat. J. Algebra Comput.* 4 (1994), (ii) + 308 pp.] completed the solution by showing that the presented group can be infinite when $m$ is even and large. It is an open question whether a *finitely presented* group of finite exponent must be finite.

The interaction between presentations and algorithms is both theoretical and practical. A theorem of Higman (Rotman, *An Introduction to the Theory of Groups*, p. 451) states that a finitely generated group $G$ can be imbedded as a subgroup of a finitely presented group $H$ if and only if $G$ is *recursively presented*: there is a presentation of $G$ whose relations can be given by an algorithm. On the practical side, many efficient algorithms solving group-theoretic problems have been implemented (Sims, *Computation with Finitely Presented Groups*). The first such algorithm was **coset enumeration** (Lyndon–Schupp, *Combinatorial Group Theory*, pp. 163–167), which computes the order of a group $G$, defined by a presentation, provided that $|G|$ is finite [unfortunately, there can be no algorithm to determine, in advance, whether $G$ is finite (Rotman, *An Introduction to the Theory of Groups*, p. 469)].

Combinatorial Group Theory has developed into **Geometric Group Theory**, which owes much to the study of *hyperbolic groups* introduced by Gromov (*Hyperbolic Groups, Essays in Group Theory*, MSRI Publications 8, Springer, 1987). A **Cayley graph** of a finitely generated group $G$ is a directed graph depending on a given finite generating set $S$: its vertices are the elements of $G$, and there is an edge from $g$ to $h$ if $h = gs$ for some $s \in S$. For example, the Cayley graph of a free group relative to a basis is a tree. As we mentioned earlier, there is a proof of the Nielsen–Schreier Theorem using this viewpoint (Serre, *Trees*, p. 29). A **hyperbolic group** is a finitely presented group having a Cayley graph which, when viewed as a metric space, resembles hyperbolic space. Examples of hyperbolic groups are finite groups, free groups, groups having an infinite cyclic group of finite index, fundamental groups of negatively curved surfaces. The free abelian group $\mathbb{Z}^2$ is an example of a group that is not hyperbolic. It is known that a subgroup of a hyperbolic group need not be hyperbolic; for example, since hyperbolic groups are finitely generated (even finitely presented), the commutator subgroup of a free group $F_2$ is not hyperbolic. In 1999, Brady gave an example of a hyperbolic group containing a finitely presented subgroup that is not hyperbolic. The proof involves geometric techniques such as Morse Theory, analogous to those arising in Differential Geometry. In 1986, Culler and Vogtmann introduced *outer space*, the boundary of a compactification of a Cayley graph (analogous to Teichmuller spaces in Complex Variables). They obtained information about automorphism groups of free groups by proving that $\mathrm{Out}(F_n) = \mathrm{Aut}(F_n)/\mathrm{Inn}(F_n)$ acts nicely on outer space; for example, $\mathrm{Out}(F_n)$ has a subgroup of finite index whose *cohomological dimension* is $2n-3$. The reader should also browse in Collins–Grigorchuk–Kurchanov–Zieschang,

*Combinatorial Group Theory and Applications to Geometry* and Stillwell, *Classical Topology and Combinatorial Group Theory.*

# Exercises

**4.69.** Let $G$ be a finitely generated group, and let $H \subseteq G$ have finite index. Prove that $H$ is finitely generated.

**4.70.** Prove that if $F$ is free of finite rank $n \geq 2$, then its commutator subgroup $F'$ is free of infinite rank.

**4.71.** Let $G$ be a finite group that is not cyclic. If $G \cong F/S$, where $F$ is a free group of finite rank, prove that $\text{rank}(S) > \text{rank}(F)$.

**4.72.** Prove that if $G$ is a finite group generated by two elements $a, b$ having order 2, then $G \cong D_{2n}$ for some $n \geq 2$.

**Remark.** There is an infinite group, $D_\infty$, called the **infinite dihedral group**, which is generated by two elements of order 2; it has the presentation

$$D_\infty = (a, b \mid a^2, b^2).$$

The reader should compare $D_\infty$ with the modular group $M$, defined in Exercise 1.86 on page 60, which is an infinite group having the presentation

$$M = (a, b \mid a^3, b^2). \quad \blacktriangleleft$$

∗ **4.73.** Let $Y$ and $S$ be groups, and let $\varphi \colon Y \to S$ and $\theta \colon S \to Y$ be homomorphisms with $\varphi\theta = 1_S$.

   (i) If $\rho \colon Y \to Y$ is defined by $\rho = \theta\varphi$, prove that $\rho\rho = \rho$ and $\rho(a) = a$ for every $a \in \text{im}\,\theta$. (The homomorphism $\rho$ is called a **retraction**.)

   (ii) If $K$ is the normal subgroup of $Y$ generated by all $y^{-1}\rho(y)$ for $y \in Y$, prove that $K = \ker\varphi$.
   **Hint.** Note that $\ker\varphi = \ker\rho$ because $\theta$ is an injection. Use the equation $y = \rho(y)(\rho(y)^{-1})y$ for all $y \in Y$.

# Chapter 5

# Commutative Rings II

Our main interest in this chapter is the study of polynomials in several variables. As usual, it is simpler to begin by looking at a more general setting—in this case, commutative rings—before getting involved with polynomial rings. It turns out that the nature of the ideals in a commutative ring is important: for example, we have already seen that gcd's exist in PIDs, while this may not be true in other commutative rings. We will see that polynomial rings are UFDs; that is, their elements have unique factorization into irreducibles, and gcd's exist in all UFDs. Three special types of ideals—prime ideals, maximal ideals, and finitely generated ideals—are the most interesting. A commutative ring is called *noetherian* if every ideal is finitely generated, and Hilbert's Basis Theorem shows that $k[x_1, \ldots, x_n]$ is noetherian when $k$ is a field. Next, we introduce Zorn's Lemma, and we then collect several interesting applications of it, such as the existence and uniqueness of algebraic closures of fields and the existence of transcendence bases. The next step introduces a geometric viewpoint in which ideals correspond to certain affine subsets called *varieties*; this discussion involves the Nullstellensatz and primary decompositions. Finally, the last section introduces the idea of *Gröbner bases*, which extends the Division Algorithm from $k[x]$ to $k[x_1, \ldots, x_n]$ and which yields a practical algorithm for deciding many problems that can be encoded in terms of polynomials in several variables.

## Section 5.1. Prime Ideals and Maximal Ideals

A great deal of Number Theory involves divisibility: given two integers $a, b$, when does $a \mid b$; that is, when is $a$ a divisor of $b$? This question translates into a question about principal ideals, for $a \mid b$ if and only if $(b) \subseteq (a)$. We now introduce two especially interesting types of ideals: *prime ideals*, which are related to Euclid's Lemma, and *maximal ideals*.

Let us begin with the analog of the Correspondence Theorem for Groups.

**Proposition 5.1 (Correspondence Theorem for Rings).** *If $I$ is a proper ideal in a commutative ring $R$, then there is an inclusion-preserving bijection $\varphi$ from the set of all ideals $J$ in $R$ containing $I$ to the set of all ideals in $R/I$, given by*

$$\varphi \colon J \mapsto J/I = \{a + I : a \in J\}.$$

**Proof.** If we forget its multiplication, the commutative ring $R$ is merely an additive abelian group and its ideal $I$ is a (normal) subgroup. The Correspondence Theorem for Groups, Theorem 1.82, now applies to the natural map $\pi \colon R \to R/I$, and it gives an inclusion-preserving bijection

$$\Phi : \{\text{all subgroups of } R \text{ containing } I\} \to \{\text{all subgroups of } R/I\},$$

where $\Phi(J) = \pi(J) = J/I$.

If $J$ is an ideal, then $\Phi(J)$ is also an ideal, for if $r \in R$ and $a \in J$, then $ra \in J$, and

$$(r + I)(a + I) = ra + I \in J/I.$$

Let $\varphi$ be the restriction of $\Phi$ to the set of intermediate ideals; $\varphi$ is an injection because $\Phi$ is an injection. To see that $\varphi$ is surjective, let $J^*$ be an ideal in $R/I$. Now $\pi^{-1}(J^*)$ is an intermediate ideal in $R$, for it contains $I = \pi^{-1}((0))$, and $\varphi(\pi^{-1}(J^*)) = \pi(\pi^{-1}(J^*)) = J^*.$[1]  $\bullet$

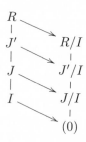

**Figure 5.1.** Correspondence Theorem.

Usually, the Correspondence Theorem for Rings is invoked, tacitly, by saying that every ideal in the quotient ring $R/I$ has the form $J/I$ for some unique ideal $J$ with $I \subseteq J \subseteq R$.

**Example 5.2.** Let $I = (m)$ be a nonzero ideal in $\mathbb{Z}$. If $J$ is an ideal in $\mathbb{Z}$ containing $I$, then $J = (a)$ for some $a \in \mathbb{Z}$ (because $\mathbb{Z}$ is a PID). Since $(m) \subseteq (a)$ if and only if $a \mid m$, the Correspondence Theorem for Rings shows that every ideal in the ring $\mathbb{Z}/I = \mathbb{I}_m$ has the form $J/I = ([a])$ for some divisor $a$ of $m$.  $\blacktriangleleft$

**Definition.** An ideal $I$ in a commutative ring $R$ is called a **prime ideal** if it is a proper ideal, that is, $I \neq R$, and $ab \in I$ implies that $a \in I$ or $b \in I$.

**Example 5.3.**

(i) The ideal $(0)$ is a prime ideal in a ring $R$ if and only if $R$ is a domain.

---

[1]If $X$ and $Y$ are sets, $f \colon X \to Y$ is a function, and $S$ is a subset of $Y$, then $ff^{-1}(S) \subseteq S$; if $f$ is surjective, then $ff^{-1}(S) = S$.

(ii) We claim that the prime ideals in $\mathbb{Z}$ are precisely the ideals $(p)$, where either $p = 0$ or $p$ is prime. Since $m$ and $-m$ generate the same principal ideal, we may restrict our attention to nonnegative generators. If $p = 0$, then the result follows from item (i), for $\mathbb{Z}$ is a domain. If $p > 0$, we show first that $(p)$ is a proper ideal; otherwise, $1 \in (p)$, and there would be an integer $a$ with $ap = 1$, a contradiction.

Let $p \in \mathbb{Z}$ be prime. If $p \mid ab$, where $a, b \in \mathbb{Z}$, then $ab \in (p)$. Euclid's Lemma says that $p \mid a$ or $p \mid b$; that is, $a \in (p)$ or $b \in (p)$. Thus, $(p)$ is a prime ideal in $\mathbb{Z}$. Conversely, if $m > 1$ is not prime, then it has a factorization $m = ab$ with $0 < a < m$ and $0 < b < m$; thus, neither $a$ nor $b$ is a multiple of $m$, and so neither lies in $(m)$. But $ab = m \in (m)$, and so $(m)$ is not a prime ideal. ◄

**Proposition 5.4.** *An ideal $I$ in $R$ is a prime ideal if and only if $R/I$ is a domain.*

**Proof.** If $I$ is a prime ideal, then $I$ is a proper ideal; hence, $R/I$ is not the zero ring, and so $1 + I \neq 0 + I$. If $(a + I)(b + I) = 0 + I$, then $ab \in I$. Hence, $a \in I$ or $b \in I$; that is, $a + I = 0 + I$ or $b + I = 0 + I$, which says that $R/I$ is a domain.

Conversely, if $R/I$ is a domain, then $R/I$ is not the zero ring, so that $I$ is a proper ideal. Moreover, $(a + I)(b + I) = 0 + I$ in $R/I$ implies that $a + I = 0 + I$ or $b + I = 0 + I$; that is, $a \in I$ or $b \in I$. Hence, $I$ is a prime ideal. •

**Definition.** If $I$ and $J$ are ideals in a commutative ring $R$, then

$$IJ = \left\{ \text{all finite sums } \sum_{\ell} a_\ell b_\ell : a_\ell \in I \text{ and } b_\ell \in J \right\}.$$

It is easy to see that $IJ$ is an ideal in $R$, and that $IJ \subseteq I \cap J$.

**Proposition 5.5.** *Let $P$ be a prime ideal in a commutative ring $R$. If $I$ and $J$ are ideals with $IJ \subseteq P$, then $I \subseteq P$ or $J \subseteq P$.*

**Proof.** If, on the contrary, $I \not\subseteq P$ and $J \not\subseteq P$, then there are $a \in I$ and $b \in J$ with $a, b \notin P$. But $ab \in IJ \subseteq P$, contradicting $P$ being prime. •

The characterization of prime ideals in $\mathbb{Z}$ in Example 5.3(ii) generalizes to PIDs.

**Proposition 5.6.**

(i) *An ideal $I$ in a PID $R$ is a prime ideal if and only if $I = (0)$ or $I = (p)$ for $p$ irreducible.*

(ii) *An ideal $I = (p(x))$ in $k[x]$, where $k$ is a field, is a prime ideal if and only if $I = (0)$ or $I = (p(x))$ for $p(x)$ irreducible.*

**Proof.**

(i) Since $R$ is a domain, the zero ideal $(0)$ is a prime ideal. Suppose that $p$ is irreducible. First, $(p)$ is a proper ideal: if $(p) = R$, then $1 \in (p)$ and $p$ is a unit, a contradiction. Second, if $ab \in (p)$, then $p \mid ab$, and so Euclid's Lemma in $R$ [Theorem 2.84(ii)] gives $p \mid a$ or $p \mid b$. Thus, $a \in (p)$ or $b \in (p)$. Therefore, $(p)$ is a prime ideal.

Conversely, assume that $(p)$ is a nonzero prime ideal. If $p$ is not irreducible, then $p = ab$, where neither $a$ nor $b$ is a unit. Since $ab \in (p)$, we must have $a$ or $b$ in $(p)$. But if $a \in (p)$, then $a = pa'$ for some $a' \in R$. Hence, $p = ab = pa'b$ and $1 = a'b$, contradicting $b$ not being a unit.

(ii) If $k$ is a field, then $k[x]$ is a PID.  •

The next proposition says that prime ideals in a quotient ring $R/I$ correspond to prime ideals in $R/I$ containing $I$.

**Proposition 5.7.** *Let $R$ and $A$ be commutative rings, and let $f\colon R \to A$ be a ring homomorphism. If $Q$ is a prime ideal in $A$, then $f^{-1}(Q)$ is a prime ideal in $R$.*

**Proof.** Let $P = f^{-1}(Q)$. If $1 \in P$, then $1 = f(1) \in f(P) = ff^{-1}(Q) \subseteq Q$; as prime ideals are proper ideals, this is a contradiction. Hence, $1 \notin P$, and $P$ is a proper ideal. If $ab \in P$, then $f(a)f(b) = f(ab) \in f(P) \subseteq Q$. Since $Q$ is prime, either $f(a)$ or $f(b)$ is in $Q$, and so either $a$ or $b$ lies in $f^{-1}(Q) = P$.  •

Inverse image is better behaved than forward image: the image of a prime ideal under a ring homomorphism may *not* be a prime ideal. For example, $(0)$ is a prime ideal in $\mathbb{Z}$, because $\mathbb{Z}$ is a domain, but its image $(0)$ under the natural map $\pi\colon \mathbb{Z} \to \mathbb{I}_4$ is not a prime ideal, because $\mathbb{I}_4$ is not a domain.

Let $I$ be an ideal in a commutative ring $R$; we may write $I \subsetneq R$ if $I$ is a proper ideal. More generally, for ideals $I$ and $J$, we may write $I \subsetneq J$ if $I \subseteq J$ and $I \neq J$.

Here is a second interesting type of ideal.

**Definition.** An ideal $I$ in a commutative ring $R$ is a ***maximal ideal*** if $I$ is a proper ideal and there is no proper ideal $J$ with $I \subsetneq J \subsetneq R$.

Thus, if $I$ is a maximal ideal in a commutative ring $R$ and $J$ is a proper ideal with $I \subseteq J$, then $I = J$. Does every commutative ring $R$ contain a maximal ideal? The (positive) answer to this question involves *Zorn's Lemma*, which we will discuss in Section 5.4.

**Example 5.8.** We claim that the ideal $(0)$ is a maximal ideal in a nonzero commutative ring $R$ if and only if $R$ is a field. Suppose $(0)$ is a maximal ideal. If $a \neq 0$, then $(0) \subsetneq (a) \subseteq R$; by maximality, $(a) = R$. Hence, $1 \in (a)$, $a$ is a unit, and so $R$ is a field. Conversely, let $R$ be a field and let $I$ be a nonzero ideal. If $a \in I$ is nonzero, then $a$ is a unit; hence, $1 \in I$ and $I = R$.  ◀

**Proposition 5.9.** *A proper ideal $I$ in a nonzero commutative ring $R$ is a maximal ideal if and only if $R/I$ is a field.*

**Proof.** The Correspondence Theorem for Rings shows that $I$ is a maximal ideal if and only if $R/I$ has no ideals other than $(0)$ and $R/I$ itself; Example 5.8 shows that this property holds if and only if $R/I$ is a field.  •

The last proof really shows that maximal ideals $J/I$ in $R/I$ correspond to maximal ideals $J$ in $R$ containing $I$.

**Corollary 5.10.** *Every maximal ideal $I$ in a commutative ring $R$ is a prime ideal.*

**Proof.** If $I$ is a maximal ideal, then $R/I$ is a field. Since every field is a domain, $R/I$ is a domain, and so $I$ is a prime ideal. •

**Example 5.11.** The converse of Corollary 5.10 is false. For example, consider the principal ideal $(x)$ in $\mathbb{Z}[x]$. By Exercise 2.92 on page 170, we have

$$\mathbb{Z}[x]/(x) \cong \mathbb{Z}.$$

Since $\mathbb{Z}$ is a domain, $(x)$ is a prime ideal in $\mathbb{Z}[x]$, but $(x)$ is not a maximal ideal because $\mathbb{Z}$ is not a field. It is not difficult to exhibit a proper ideal $J$ strictly containing $(x)$. In fact, $I = (x) \subsetneq J = \{f(x) \in \mathbb{Z}[x] : f(x) \text{ has even constant term}\}$. Note that $J$ is a maximal ideal, for $\mathbb{Z}[x]/J \cong \mathbb{F}_2$ is a field. ◀

The converse of Corollary 5.10 is true when $R$ is a PID.

**Theorem 5.12.** *If $R$ is a PID, then every nonzero prime ideal $I$ is a maximal ideal.*

**Proof.** Since $R$ is a PID, Proposition 5.6 says that $I = (p)$ for some irreducible element $p$. If $(p) \subseteq J = (a)$, then $a \mid p$. Hence, either $a$ and $p$ are associates, in which case $(a) = (p)$, or $a$ is a unit, in which case $J = (a) = R$. Therefore, $I = (p)$ is a maximal ideal. •

It now follows from Example 5.11 that $\mathbb{Z}[x]$ is not a PID. Here is a second proof of Proposition 2.140.

**Corollary 5.13.** *If $k$ is a field and $p(x) \in k[x]$ is irreducible, then the quotient ring $k[x]/(p(x))$ is a field.*

**Proof.** Since $p(x)$ is irreducible, the principal ideal $I = (p(x))$ is a nonzero prime ideal; since $k[x]$ is a PID, $I$ is a maximal ideal, and so $k[x]/I$ is a field. •

If $k$ is a field and $(a_1, \ldots, a_n) \in k^n$, then $(x_1 - a_1, \ldots, x_n - a_n)$ is a maximal ideal in $k[x_1, \ldots, x_n]$, by Exercise 5.4 on page 301. Conversely, Theorem 5.100 says that every maximal ideal in $\mathbb{C}[x_1, \ldots, x_n]$ has the form $(x_1 - a_1, \ldots, x_n - a_n)$; that is, there is a bijection between $\mathbb{C}^n$ and the set of all maximal ideals in $\mathbb{C}[x_1, \ldots, x_n]$. In Theorem 10.77, we shall see that this remains true if $\mathbb{C}$ is replaced by any algebraically closed field. On the other hand, the prime ideals in the polynomial ring $k[x_1, \ldots, x_n]$, where $k$ is an arbitrary field, can be quite complicated.

**Example 5.14.** Let $k$ be a field, and let $a = (a_1, \ldots, a_n) \in k^n$. The *evaluation map* $e_a : k[x_1, \ldots, x_n] \to k$, defined by

$$e_a : f(x_1, \ldots, x_n) \mapsto f(a) = f(a_1, \ldots, a_n),$$

is surjective (if $b \in k$, take $f$ to be the constant polynomial $b$), so that $\ker e_a$ is a maximal ideal. Now $I = (x_1 - a_1, \ldots, x_n - a_n) \subseteq \ker e_a$. Since $I$ is a maximal ideal, by Exercise 5.4, it follows that $\ker e_a = I = (x_1 - a_1, \ldots, x_n - a_n)$ is a maximal ideal. ◀

The proof of the next result is taken from Kaplansky, *Commutative Rings*.

**Proposition 5.15.** *Let $B$ be a subset of a commutative ring $R$ that is closed under addition and multiplication ($B$ may not be an ideal).*

(i) *Let $J_1, \ldots, J_n$ be ideals in $R$, at least $n - 2$ of which are prime. If $B \subseteq J_1 \cup \cdots \cup J_n$, then $B$ is contained in some $J_i$.*

(ii) *Let $I$ be an ideal in $R$ with $I \subsetneq B$. If there are prime ideals $P_1, \ldots, P_n$ such that the set-theoretic complement $B - I \subseteq P_1 \cup \cdots \cup P_n$, then $B \subseteq P_i$ for some $i$.*

**Proof.**

(i) The proof is by induction on $n \geq 2$. For the base step $n = 2$, neither of the ideals $J_1$ or $J_2$ need be prime. If $B \not\subseteq J_2$, then there is $b_1 \in B$ with $b_1 \notin J_2$; since $B \subseteq J_1 \cup J_2$, we must have $b_1 \in J_1$. Similarly, if $B \not\subseteq J_1$, there is $b_2 \in B$ with $b_2 \notin J_1$ and $b_2 \in J_2$. However, if $y = b_1 + b_2$, then $y \notin J_1$: otherwise, $b_2 = y - b_1 \in J_1$ (because both $y$ and $b_1$ are in $J_1$), a contradiction. Similarly, $y \notin J_2$, contradicting $B \subseteq J_1 \cup J_2$.

For the inductive step, assume that $B \subseteq J_1 \cup \cdots \cup J_{n+1}$, where at least $n - 1 = (n + 1) - 2$ of the $J_i$ are prime ideals. Let

$$D_i = J_1 \cup \cdots \cup \widehat{J_i} \cup \cdots \cup J_{n+1}.$$

Since $D_i$ is a union of $n$ ideals at least $(n - 1) - 1 = n - 2$ of which are prime, the inductive hypothesis allows us to assume that $B \not\subseteq D_i$ for all $i$. Hence, for all $i$, there exists $b_i \in B$ with $b_i \notin D_i$; since $B \subseteq D_i \cup J_i$, we must have $b_i \in J_i$. Now $n \geq 3$, so that at least one of the $J_i$ is a prime ideal; for notation, assume that $J_1$ is prime. Consider the element

$$y = b_1 + b_2 b_3 \cdots b_{n+1}.$$

Since all $b_i \in B$ and $B$ is closed under addition and multiplication, $y \in B$. Now $y \notin J_1$; otherwise, $b_2 b_3 \cdots b_{n+1} = y - b_1 \in J_1$. Since $J_1$ is prime, some $b_i \in J_1$. This is a contradiction, for $b_i \notin D_i \supseteq J_1$. If $i > 1$ and $y \in J_i$, then $b_2 b_3 \cdots b_{n+1} \in J_i$, because $J_i$ is an ideal, and so $b_1 = y - b_2 b_3 \cdots b_{n+1} \in J_i$. This cannot be, for $b_1 \notin D_1 \supseteq J_i$. Therefore, $y \notin J_i$ for any $i$, contradicting $B \subseteq J_1 \cup \cdots \cup J_{n+1}$.

(ii) The hypothesis gives $B \subseteq I \cup P_1 \cup \cdots \cup P_n$, so that part (i) gives $B \subseteq I$ or $B \subseteq P_i$. Since $I$ is a proper subset of $B$, the first possibility cannot occur.  •

## Exercises

**5.1.**   (i) Find all the maximal ideals in $\mathbb{Z}$.

(ii) Find all the maximal ideals in $\mathbb{R}[x]$; that is, describe those $g(x) \in \mathbb{R}[x]$ for which $(g)$ is a maximal ideal.

(iii) Find all the maximal ideals in $\mathbb{C}[x]$.

\* **5.2.** Let $I$ be an ideal in a commutative ring $R$. If $J^*$ and $L^*$ are ideals in $R/I$, prove that there exist ideals $J$ and $L$ in $R$ containing $I$ such that $J/I = J^*$, $L/I = L^*$, and $(J \cap L)/I = J^* \cap L^*$. Conclude that if $J^* \cap L^* = (0)$, then $J \cap L = I$.

**Hint.** Use the Correspondence Theorem for Rings.

\* **5.3.** (i) Give an example of a commutative ring containing two prime ideals $P$ and $Q$ for which $P \cap Q$ is not a prime ideal.

    (ii) If $P_1 \supseteq P_2 \supseteq \cdots \supseteq P_n \supseteq P_{n+1} \supseteq \cdots$ is a decreasing sequence of prime ideals in a commutative ring $R$, prove that.$\bigcap_{n \geq 1} P_n$ is a prime ideal.

\* **5.4.** (i) If $k$ is a field and $a_1, \ldots, a_n \in k$, prove that $(x_1 - a_1, \ldots, x_n - a_n)$ is a maximal ideal in $k[x_1, \ldots, x_n]$.

    (ii) Prove that if $x_i - b \in (x_1 - a_1, \ldots, x_n - a_n)$ for some $i$, where $b \in k$, then $b = a_i$.

    (iii) Prove that $\mu \colon k^n \to \{\text{maximal ideals in } k[x_1, \ldots, x_n]\}$, given by

$$\mu \colon (a_1, \ldots, a_n) \mapsto (x_1 - a_1, \ldots, x_n - a_n),$$

is an injection, and give an example of a field $k$ for which $\mu$ is not a surjection.

\* **5.5.** Prove that if $P$ is a prime ideal in a commutative ring $R$ and $r^n \in P$ for some $r \in R$ and $n \geq 1$, then $r \in P$.

**5.6.** Prove that the ideal $(x^2 - 2, y^2 + 1, z)$ in $\mathbb{Q}[x, y, z]$ is a proper ideal.

\* **5.7.** (i) Call a subset $S$ of a commutative ring $R$ **multiplicative** if $0 \notin S$, $1 \in S$, and, if $s, s' \in S$, then $ss' \in S$; that is, $S$ is a monoid under multiplication with $0 \notin S$. Prove that an ideal $J$ which is maximal with the property that $J \cap S = \varnothing$ is a prime ideal. (The existence of such an ideal $J$ is proved, using Zorn's Lemma, in Exercise 5.39 on page 321.)

    (ii) Let $S$ be a multiplicative subset of a commutative ring $R$, and suppose that there is an ideal $I$ with $I \cap S = \varnothing$. If $P$ is an ideal maximal such that $I \subseteq P$ and $P \cap S = \varnothing$, prove that $P$ is a prime ideal.

\* **5.8.** (i) If $I$ and $J$ are ideals in a commutative ring $R$, prove that $IJ \subseteq I \cap J$. Recall that $IJ = \{\text{all finite sums } \sum_\ell a_\ell b_\ell \colon a_\ell \in I \text{ and } b_\ell \in J\}$.

    (ii) If $I = (2) = J$ is the ideal of even integers in $\mathbb{Z}$, prove that $I^2 = IJ \subsetneq I \cap J = I$.

    (iii) Let $P$ be a prime ideal and let $Q_1, \ldots Q_r$ be ideals. Prove that if $Q_1 \cap \cdots \cap Q_r \subseteq P$, then $Q_i \subseteq P$ for some $i$.

\* **5.9.** Let $I$ and $J$ be ideals in a commutative ring $R$.

    (i) Prove that the map $R/(I \cap J) \to R/I \times R/J$, given by $\varphi \colon r \mapsto (r + I, r + J)$, is an injection.

    (ii) Call $I$ and $J$ **coprime** if $I + J = R$. Prove that if $I$ and $J$ are coprime, then the ring homomorphism $\varphi \colon R/(I \cap J) \to R/I \times R/J$ in part (i) is a surjection.
    **Hint.** If $I$ and $J$ are coprime, there are $a \in I$ and $b \in J$ with $1 = a + b$. If $r, r' \in R$, prove that$(d + I, d + J) = (r + I, r' + J) \in R/I \times R/J$, where $d = r'a + rb$.

    (iii) Generalize the **Chinese Remainder Theorem** as follows. Let $R$ be a commutative ring and let $I_1, \ldots, I_n$ be pairwise coprime ideals; that is, $I_i$ and $I_j$ are coprime for all $i \neq j$. Prove that if $a_1, \ldots, a_n \in R$, then there exists $r \in R$ with $r + I_i = a_i + I_i$ for all $i$.

(iv) If $I$ and $J$ are coprime ideals in a commutative ring $R$, prove that
$$I \cap J = IJ.$$

**5.10.** Let $R$ be a commutative ring; given an ideal $I$ and a subset $S \subseteq R$, the **colon ideal** $(I : S)$ (or **ideal quotient**) is defined by
$$(I : S) = \{r \in R : rs \in I \text{ for all } s \in S\}.$$

(i) Prove that $(I : S)$ is an ideal.

(ii) If $J = (S)$ is the ideal generated by $S$, prove that $(I : S) = (I : J)$.

(iii) Let $R$ be a domain and let $a, b \in R$, where $b \neq 0$. If $I = (ab)$ and $J = (b)$, prove that $(I : J) = (a)$.

∗ **5.11.**   (i) If $I, J$ are ideals in a commutative ring $R$, prove that $I \subseteq (I : J)$ and that $J(I : J) \subseteq I$.

(ii) Prove that if $I = Q_1 \cap \cdots \cap Q_r$, then
$$(I : J) = (Q_1 : J) \cap \cdots \cap (Q_r : J).$$

(iii) If $I$ is an ideal in a commutative ring $R$ and $J = J_1 + \cdots + J_n$ is a sum of ideals, prove that
$$(I : J) = (I : J_1) \cap \cdots \cap (I : J_n).$$

**5.12.** A **Boolean ring** is a ring $R$ in which $a^2 = a$ for all $a \in R$.

(i) Prove that every Boolean ring is commutative.

(ii) Prove that every prime ideal in a Boolean ring is a maximal ideal.
**Hint.** When is a Boolean ring a domain?

**5.13.** A commutative ring $R$ is a **local ring** if it has a unique maximal ideal.

(i) If $p$ is a prime, prove that the ring of $p$-**adic fractions**,
$$\mathbb{Z}_{(p)} = \{a/b \in \mathbb{Q} : p \nmid b\},$$
is a local ring.

(ii) If $k$ is a field, prove that the ring $k[[x]]$ of all power series is a local ring.

(iii) If $R$ is a local ring with unique maximal ideal $\mathfrak{m}$, prove that $a \in R$ is a unit if and only if $a \notin \mathfrak{m}$.
**Hint.** You may assume Exercise 5.40 on page 321: every nonunit in a commutative ring lies in some maximal ideal.

## Section 5.2. Unique Factorization Domains

We have proved unique factorization theorems in $\mathbb{Z}$ and in $k[x]$, where $k$ is a field. We are now going to prove a common generalization: every PID has a unique factorization theorem. We will then prove a theorem of Gauss: if $R$ has a unique factorization theorem, then so does $R[x]$. A corollary is that there is a unique factorization theorem in the ring $k[x_1, \ldots, x_n]$ of all polynomials in several variables over a field $k$, and an immediate consequence is that any two polynomials in several variables have a gcd.

We begin by generalizing some earlier definitions. Recall that an element $p$ in a domain $R$ is *irreducible* if it is neither 0 nor a unit and its only factors are units or associates of $p$. For example, the irreducibles in $\mathbb{Z}$ are the numbers $\pm p$, where $p$ is a prime, and the irreducibles in $k[x]$, where $k$ is a field, are the irreducible polynomials $p(x)$; that is, $\deg(p) \geq 1$ and $p(x)$ has no factorization $p(x) = f(x)g(x)$ where $\deg(f) < \deg(p)$ and $\deg(g) < \deg(p)$. This characterization of irreducible polynomials does not persist in rings $R[x]$ when $R$ is not a field. For example, in $\mathbb{Z}[x]$, the polynomial $f(x) = 2x + 2$ cannot be factored into two polynomials, each having degree smaller than $\deg(f) = 1$, yet $f(x)$ is not irreducible: in the factorization $2x + 2 = 2(x + 1)$, neither 2 nor $x + 1$ is a unit.

Here is the generalization we are seeking.

**Definition.** A domain $R$ is a **UFD** (*unique factorization domain* or *factorial ring*) if

  (i) every $r \in R$, neither 0 nor a unit, is a product of irreducibles;

  (ii) if $p_1 \cdots p_m = q_1 \cdots q_n$, where all $p_i$ and $q_j$ are irreducible, then $m = n$ and there is a permutation $\sigma \in S_n$ with $p_i$ and $q_{\sigma(i)}$ associates for all $i$.

When we proved that $\mathbb{Z}$ and $k[x]$, for $k$ a field, have unique factorization into irreducibles, we did not mention associates because, in each case, irreducible elements were always replaced by favorite choices of associates: in $\mathbb{Z}$, *positive* irreducibles (i.e., primes) were chosen; in $k[x]$, *monic* irreducible polynomials were chosen. The reader should see, for example, that the statement: "$\mathbb{Z}$ is a UFD" is just a restatement of the Fundamental Theorem of Arithmetic.

**Proposition 5.16.** *Let $R$ be a domain in which every $r \in R$, neither 0 nor a unit, is a product of irreducibles. Then $R$ is a UFD if and only if $(p)$ is a prime ideal in $R$ for every irreducible element $p \in R$.*[2]

**Proof.** Assume that $R$ is a UFD. If $a, b \in R$ and $ab \in (p)$, then there is $r \in R$ with

$$ab = rp.$$

Factor each of $a, b$, and $r$ into irreducibles; by unique factorization, the left side of the equation must involve an associate of $p$. This associate arose as a factor of $a$ or $b$, and hence $a \in (p)$ or $b \in (p)$. Therefore, $(p)$ is a prime ideal.

The proof of the converse is merely an adaptation of the proof of the Fundamental Theorem of Arithmetic. Assume that

$$p_1 \cdots p_m = q_1 \cdots q_n,$$

where $p_i$ and $q_j$ are irreducible elements. We prove, by induction on $\max\{m, n\} \geq 1$, that $n = m$ and the $q$'s can be reindexed so that $q_i$ and $p_i$ are associates for all $i$. If $\max\{m, n\} = 1$, then $p_1 = q_1$, and the base step is obviously true. For the inductive step, the given equation shows that $p_1 \mid q_1 \cdots q_n$. By hypothesis, $(p_1)$ is a prime ideal (this is the analog of Euclid's Lemma), and so there is some $q_j$ with $p_1 \mid q_j$. But $q_j$, being irreducible, has no divisors other than units and associates, so that

---

[2]An element $p$ for which $(p)$ is a nonzero prime ideal is often called a *prime element*. Such elements have the property that $p \mid ab$ implies $p \mid a$ or $p \mid b$.

$q_j$ and $p_1$ are associates: $q_j = up_1$ for some unit $u$. Canceling $p_1$ from both sides, we have $p_2 \cdots p_m = uq_1 \cdots \widehat{q_j} \cdots q_n$. By the inductive hypothesis, $m - 1 = n - 1$ (so that $m = n$) and, after possible reindexing, $q_i$ and $p_i$ are associates for all $i$.  •

The proofs we have given that $\mathbb{Z}$ and $k[x]$, where $k$ is a field, are UFDs involve the Division Algorithm; as a consequence, it is not difficult to generalize them to prove that every Euclidean ring is a UFD. Since there are PIDs that are not Euclidean (Example 2.85), we now show that every PID is, in fact, a UFD. The proof uses a new idea: chains of ideals.

**Lemma 5.17.**

(i) *If $R$ is a commutative ring and*
$$I_1 \subseteq I_2 \subseteq \cdots \subseteq I_n \subseteq I_{n+1} \subseteq \cdots$$
*is an ascending chain of ideals in $R$, then $J = \bigcup_{n \geq 1} I_n$ is an ideal in $R$.*

(ii) *If $R$ is a PID, then it has no infinite strictly ascending chain of ideals*
$$I_1 \subsetneq I_2 \subsetneq \cdots \subsetneq I_n \subsetneq I_{n+1} \subsetneq \cdots.$$

(iii) *If $R$ is a PID and $r \in R$ is neither $0$ nor a unit, then $r$ is a product of irreducibles.*

**Proof.**

(i) We claim that $J$ is an ideal. If $a \in J$, then $a \in I_n$ for some $n$; if $r \in R$, then $ra \in I_n$, because $I_n$ is an ideal; hence, $ra \in J$. If $a, b \in J$, then there are ideals $I_n$ and $I_m$ with $a \in I_n$ and $b \in I_m$; since the chain is ascending, we may assume that $I_n \subseteq I_m$, and so $a, b \in I_m$. As $I_m$ is an ideal, $a + b \in I_m$ and, hence, $a + b \in J$. Therefore, $J$ is an ideal.

(ii) If, on the contrary, an infinite strictly ascending chain exists, then define $J = \bigcup_{n \geq 1} I_n$. By (i), $J$ is an ideal; since $R$ is a PID, we have $J = (d)$ for some $d \in J$. Now $d$ got into $J$ by being in $I_n$ for some $n$. Hence
$$J = (d) \subseteq I_n \subsetneq I_{n+1} \subseteq J,$$
and this is a contradiction.

(iii) A divisor $r$ of an element $a \in R$ is called a *proper divisor* of $a$ if $r$ is neither a unit nor an associate of $a$. If $r$ is a divisor of $a$, then $(a) \subseteq (r)$; if $r$ is a proper divisor, then $(a) \subsetneq (r)$, for if the inequality is not strict, then $(a) = (r)$, and this forces $a$ and $r$ to be associates, by Proposition 2.33.

Call a nonzero nonunit $a \in R$ *good* if it is a product of irreducibles; call it *bad* otherwise. We must show that there are no bad elements. If $a$ is bad, it is not irreducible, and so $a = rs$, where both $r$ and $s$ are proper divisors. But the product of good elements is good, and so at least one of the factors, say $r$, is bad. The first paragraph shows that $(a) \subsetneq (r)$. It follows, by induction, that there exists a sequence $a_1 = a, a_2 = r, a_3, \ldots, a_n, \ldots$ of bad elements with each $a_{n+1}$ a proper divisor of $a_n$, and this sequence yields a strictly ascending chain
$$(a_1) \subsetneq (a_2) \subsetneq \cdots \subsetneq (a_n) \subsetneq (a_{n+1}) \subsetneq \cdots,$$
contradicting part (i) of this lemma.  •

**Theorem 5.18.** *If $R$ is a* PID, *then $R$ is a* UFD. *In particular, every Euclidean ring is a* UFD.

**Proof.** In view of the last two results, it suffices to prove that $(p)$ is a prime ideal whenever $p$ is irreducible; this was shown in Proposition 5.6. •

Recall, given a finite number of elements $a_1, \ldots, a_n$ in a domain $R$, that a *common divisor* is an element $c \in R$ with $c \mid a_i$ for all $i$; a *greatest common divisor* or *gcd* is a common divisor $d$ with $c \mid d$ for every common divisor $c$. Even in the familiar examples of $\mathbb{Z}$ and $k[x]$, gcd's are not unique unless an extra condition is imposed. For example, in $k[x]$, where $k$ is a field, we imposed the condition that nonzero gcd's are monic polynomials. In a general PID, elements may not have favorite associates. However, there is some uniqueness. If $R$ is a domain, then it is easy to see that if $d$ and $d'$ are gcd's of elements $a_1, \ldots, a_n$, then $d \mid d'$ and $d' \mid d$. It follows from Proposition 2.33 that $d$ and $d'$ are associates and, hence, that $(d) = (d')$. Thus, gcd's are not unique, but they all generate the same principal ideal.

The idea in Proposition 2.61 carries over to show that gcd's do exist in UFDs.

**Proposition 5.19.** *If $R$ is a* UFD, *then a gcd $(a_1, \ldots, a_n)$ of any finite set of elements $a_1, \ldots, a_n$ in $R$ exists.*

**Proof.** We prove first that a gcd of two elements $a$ and $b$ exists. There are distinct irreducibles $p_1, \ldots, p_t$ with

$$a = p_1^{e_1} p_2^{e_2} \cdots p_t^{e_t} \quad \text{and} \quad b = p_1^{f_1} p_2^{f_2} \cdots p_t^{f_t},$$

where $e_i \geq 0$ and $f_i \geq 0$ for all $i$. It is easy to see that if $c \mid a$, then the factorization of $c$ into irreducibles is $c = w p_1^{g_1} p_2^{g_2} \cdots p_t^{g_t}$, where $0 \leq g_i \leq e_i$ for all $i$ and $w$ is a unit. Thus, $c$ is a common divisor of $a$ and $b$ if and only if $g_i \leq m_i$ for all $i$, where

$$m_i = \min\{e_i, f_i\}.$$

It is now clear that $p_1^{m_1} p_2^{m_2} \cdots p_t^{m_t}$ is a gcd of $a$ and $b$.

More generally, if $a_i = u_i p_1^{e_{i1}} p_2^{e_{i2}} \cdots p_t^{e_{it}}$, where $e_{ij} \geq 0$ and $i = 1, \ldots, n$ and $u_i$ are units, then

$$d = p_1^{\mu_1} p_2^{\mu_2} \cdots p_t^{\mu_t}$$

is a gcd of $a_1, \ldots, a_n$, where $\mu_j = \min\{e_{1j}, e_{2j}, \ldots, e_{nj}\}$. •

We caution the reader that we have *not* proved that a gcd of elements $a_1, \ldots, a_n$ is a linear combination of them; indeed, this may not be true (see Exercise 5.18 on page 311).

Recall that if $a_1, \ldots, a_n$ are elements in a commutative ring $R$, not all zero, then their lcm, *least common multiple*, is a common multiple $c$ with $c \mid m$ for every common multiple $m$. Least common multiples exist in UFDs.

**Proposition 5.20.** *Let $R$ be a* UFD, *and let $a_1, \ldots, a_n$ in $R$. An* lcm $[a_1, \ldots, a_n]$ *of $a_1, \ldots, a_n$ exists, and*

$$ab = (a_1, \ldots, a_n)[a_1, \ldots, a_n].$$

**Proof.** We may assume that all $a_i \neq 0$. If $a, b \in R$, there are distinct irreducibles $p_1, \ldots, p_t$ with
$$a = p_1^{e_1} p_2^{e_2} \cdots p_t^{e_t} \quad \text{and} \quad b = p_1^{f_1} p_2^{f_2} \cdots p_t^{f_t},$$
where $e_i \geq 0$ and $f_i \geq 0$ for all $i$. The reader may adapt the proof of Proposition 2.61 to prove that if $M_i = \max\{e_i, f_i\}$, then $p_1^{M_1} p_2^{M_2} \cdots p_t^{M_t}$ is an lcm of $a$ and $b$. The first statement follows by induction on $n$. The second statement is an adaptation of the proof of Corollary 2.62. •

**Example 5.21.** Let $k$ be a field and let $R$ be the subring of $k[x]$ consisting of all polynomials $f(x) \in k[x]$ having no linear term; that is, $f(x) = a_0 + a_2 x^2 + \cdots + a_n x^n$. In Exercise 2.67 on page 133, we showed that $x^5$ and $x^6$ have no gcd in $R$. It now follows from Proposition 5.19 that $R$ is not a UFD. ◄

**Definition.** Elements $a_1, \ldots, a_n$ in a UFD $R$ are called ***relatively prime*** if their gcd is a unit; that is, if every common divisor of $a_1, \ldots, a_n$ is a unit.

We are now going to prove that if $R$ is a UFD, then so is $R[x]$. Recall Exercise 2.21 on page 94: if $R$ is a domain, then the units in $R[x]$ are the units in $R$.

**Definition.** A polynomial $f(x) = a_n x^n + \cdots + a_1 x + a_0 \in R[x]$, where $R$ is a UFD, is called ***primitive*** if its coefficients are relatively prime; that is, the only common divisors of $a_n, \ldots, a_1, a_0$ are units.

Of course, every monic polynomial is primitive. Observe that if $f(x)$ is not primitive, then there exists an irreducible $q \in R$ that divides each of its coefficients: if the gcd is a nonunit $d$, then take for $q$ any irreducible factor of $d$.

**Example 5.22.** We claim that if $R$ is a UFD, then every irreducible $p(x) \in R[x]$ of positive degree is primitive. Otherwise, there is an irreducible $q \in R$ with $p(x) = qg(x)$; note that $\deg(q) = 0$ because $q \in R$. Since $p(x)$ is irreducible, its only factors are units and associates; since $q$ is not a unit, it must be an associate of $p(x)$. But every unit in $R[x]$ has degree 0 (i.e., is a constant), for $uv = 1$ implies $\deg(u) + \deg(v) = \deg(1) = 0$; hence, associates in $R[x]$ have the same degree. Therefore, $q$ is not an associate of $p(x)$, for the latter has positive degree, and so $p(x)$ is primitive. ◄

We begin with a technical lemma.

**Lemma 5.23 (Gauss).** *If $R$ is a UFD and $f(x), g(x) \in R[x]$ are both primitive, then their product $f(x)g(x)$ is also primitive.*

**Proof.** If $f(x)g(x)$ is not primitive, there is an irreducible $p \in R$ which divides all its coefficients. Let $P = (p)$ and let $\pi: R \to R/P$ be the natural map $a \mapsto a + P$. Proposition 2.27 shows that the function $\widetilde{\pi}: R[x] \to (R/P)[x]$, which replaces each coefficient $c$ of a polynomial by $\pi(c)$, is a ring homomorphism. Now $\widetilde{\pi}(fg) = 0$ in $(R/(p))[x]$. Since $P$ is a prime ideal, both $R/P$ and $(R/P)[x]$ are domains. But neither $\widetilde{\pi}(f)$ nor $\widetilde{\pi}(g)$ is 0 in $(R/P)[x]$, because $f$ and $g$ are primitive, and this contradicts $(R/P)[x]$ being a domain. •

**Lemma 5.24.** *Let $R$ be a UFD, let $Q = \operatorname{Frac}(R)$, and let $f(x) \in Q[x]$ be nonzero.*

(i) *There is a factorization*

$$f(x) = c(f)f^*(x),$$

*where* $c(f) \in Q$ *and* $f^*(x) \in R[x]$ *is primitive. This factorization is unique in the sense that if* $f(x) = qg^*(x)$, *where* $q \in Q$ *and* $g^*(x) \in R[x]$ *is primitive, then there is a unit* $w \in R$ *with* $q = wc(f)$ *and* $f^*(x) = wg^*(x)$.

(ii) *If* $f(x), g(x) \in R[x]$, *then* $c(fg)$ *and* $c(f)c(g)$ *are associates in* $R$ *and* $(fg)^*$ *and* $f^*g^*$ *are associates in* $R[x]$.

(iii) *Let* $f(x) \in Q[x]$ *have a factorization* $f(x) = qg^*(x)$, *where* $q \in Q$ *and* $g^*(x) \in R[x]$ *is primitive. Then* $f(x) \in R[x]$ *if and only if* $q \in R$.

(iv) *Let* $g^*(x), f(x) \in R[x]$. *If* $g^*(x)$ *is primitive and* $g^*(x) \mid bf(x)$, *where* $b \in R$ *and* $b \neq 0$, *then* $g^*(x) \mid f(x)$.

**Proof.**

(i) Clearing denominators, there is $b \in R$ with $bf(x) \in R[x]$. If $d$ is the gcd of the coefficients of $bf(x)$, then $(b/d)f(x) \in R[x]$ is a primitive polynomial. If we define $c(f) = d/b$ and $f^*(x) = c(f)f(x)$, then $f^*(x)$ is primitive and $f(x) = c(f)f^*(x)$.

To prove uniqueness, suppose that $c(f)f^*(x) = f(x) = qg^*(x)$, where $c(f), q \in Q$ and $f^*(x), g^*(x) \in R[x]$ are primitive. Exercise 5.14 on page 311 allows us to write $q/c(f)$ in lowest terms: $q/c(f) = u/v$, where $u$ and $v$ are relatively prime elements of $R$. The equation $vf^*(x) = ug^*(x)$ holds in $R[x]$; equating like coefficients, we see that $v$ is a common divisor of all the coefficients of $ug^*(x)$. Since $u$ and $v$ are relatively prime, Exercise 5.15(i) on page 311 says that $v$ is a common divisor of all the coefficients of $g^*(x)$. But $g^*(x)$ is primitive, and so $v$ is a unit. A similar argument shows that $u$ is a unit. Therefore, $q/c(f) = u/v$ is a unit in $R$, call it $w$; we have $q = wc(f)$ and $f^*(x) = wg^*(x)$.

(ii) There are two factorizations of $f(x)g(x)$ in $R[x]$:

$$f(x)g(x) = c(fg)(fg)^*$$
$$f(x)g(x) = c(f)f^*(x)c(g)g^*(x) = c(f)c(g)f^*(x)g^*(x).$$

Since the product of primitive polynomials is primitive, each of these is a factorization as in part (i); the uniqueness assertion there says that $c(fg)$ is an associate of $c(f)c(g)$ and $(fg)^*$ is an associate of $f^*g^*$.

(iii) If $q \in R$, then it is obvious that $f(x) = qg^*(x) \in R[x]$. Conversely, if $f(x) \in R[x]$, then there is no need to clear denominators, and so $c(f) = d \in R$, where $d$ is the gcd of the coefficients of $f(x)$. Thus, $f(x) = df^*(x)$. By uniqueness, there is a unit $w \in R$ with $q = wd \in R$.

(iv) Since $bf = hg^*$, we have $bc(f)f^* = c(h)h^*g^* = c(h)(hg)^*$. By uniqueness, $f^*$, $(hg)^*$, and $h^*g^*$ are associates, and so $g^* \mid f^*$. But $f = c(f)f^*$, and so $g^* \mid f$. •

**Definition.** Let $R$ be a UFD with $Q = \text{Frac}(R)$. If $f(x) \in Q[x]$, there is a factorization $f(x) = c(f)f^*(x)$, where $c(f) \in Q$ and $f^*(x) \in R[x]$ is primitive. We call $c(f)$ the **content** of $f(x)$ and $f^*(x)$ the **associated primitive polynomial.**

In light of Lemma 5.24(i), both $c(f)$ and $f^*(x)$ are essentially unique.

**Theorem 5.25 (Gauss).** *If $R$ is a UFD, then $R[x]$ is also a UFD.*

**Proof.** We show, by induction on $\deg(f)$, that every $f(x) \in R[x]$, neither zero nor a unit, is a product of irreducibles. The base step $\deg(f) = 0$ is true, because $f(x)$ is a constant, hence lies in $R$, and hence is a product of irreducibles (for $R$ is a UFD). For the inductive step $\deg(f) > 0$, we have $f(x) = c(f)f^*(x)$, where $c(f) \in R$ and $f^*(x)$ is primitive. Now $c(f)$ is either a unit or a product of irreducibles, by the base step. If $f^*(x)$ is irreducible, we are done. Otherwise, $f^*(x) = g(x)h(x)$, where neither $g$ nor $h$ is a unit. Since $f^*(x)$ is primitive, however, neither $g$ nor $h$ is a constant; therefore, each of these has degree less than $\deg(f^*) = \deg(f)$, and so each is a product of irreducibles, by the inductive hypothesis.

Proposition 5.16 now applies: it suffices to show that if $p(x) \in R[x]$ is irreducible, then $(p(x))$ is a prime ideal in $R[x]$; that is, if $p \mid fg$, then $p \mid f$ or $p \mid g$. Let us assume that $p(x) \nmid f(x)$.

(i) Suppose that $\deg(p) = 0$. Now $f(x) = c(f)f^*(x)$ and $g(x) = c(g)g^*(x)$, where $f^*(x), g^*(x)$ are primitive and $c(f), c(g) \in R$, by Lemma 5.24(iii). Since $p \mid fg$, we have

$$p \mid c(f)c(g)f^*(x)g^*(x).$$

Write $f^*(x)g^*(x) = \sum_i a_i x^i$, where $a_i \in R$, so that $p \mid c(f)c(g)a_i$ in $R$ for all $i$. Now $f^*(x)g^*(x)$ is primitive, so there is some $i$ with $p \nmid a_i$ in $R$. Since $R$ is a UFD, Proposition 5.16 says that $p$ generates a prime ideal in $R$; that is, if $s, t \in R$ and $p \mid st$ in $R$, then $p \mid s$ or $p \mid t$. In particular, $p \mid c(f)c(g)$ in $R$; in fact, $p \mid c(f)$ or $p \mid c(g)$. If $p \mid c(f)$, then $p$ divides $c(f)f^*(x) = f(x)$, a contradiction. Therefore, $p \mid c(g)$ and, hence, $p \mid g(x)$; we have shown that $p$ generates a prime ideal in $R[x]$.

(ii) Suppose that $\deg(p) > 0$. Let

$$(p, f) = \{s(x)p(x) + t(x)f(x) \colon s(x), t(x) \in R[x]\};$$

of course, $(p, f)$ is an ideal in $R[x]$ containing $p(x)$ and $f(x)$. Choose $m(x) \in (p, f)$ of minimal degree. If $Q = \text{Frac}(R)$ is the fraction field of $R$, then the Division Algorithm in $Q[x]$ gives polynomials $q'(x), r'(x) \in Q[x]$ with

$$f(x) = m(x)q'(x) + r'(x),$$

where either $r'(x) = 0$ or $\deg(r') < \deg(m)$. Clearing denominators, there is a constant $b \in R$ and polynomials $q(x), r(x) \in R[x]$ with

$$bf(x) = q(x)m(x) + r(x),$$

where $r(x) = 0$ or $\deg(r) < \deg(m)$. Since $m \in (p, f)$, there are polynomials $s(x), t(x) \in R[x]$ with $m = sp + tf$; hence $r = bf - qm \in (p, f)$. Since $m$ has minimal degree in $(p, f)$, we must have $r = 0$; that is, $bf(x) = m(x)q(x)$, and so $bf(x) = c(m)m^*(x)q(x)$. But $m^*(x)$ is primitive, and $m^*(x) \mid bf(x)$, so that $m^*(x) \mid f(x)$, by Lemma 5.24(iv). A similar argument, replacing $f(x)$ by $p(x)$ (that is, beginning with an equation $b''p(x) = q''(x)m(x)+r''(x)$ for some constant $b''$), gives $m^*(x) \mid p(x)$. Since $p(x)$ is irreducible, its only factors are units and associates. If $m^*(x)$ were an associate of $p(x)$, then $p(x) \mid f(x)$ (because $p \mid m^*$ and $m^* \mid f$), contrary to our assumption that $p \nmid f$. Hence, $m^*(x)$ must be a unit; that is, $m(x) = c(m) \in R$, and so $(p, f)$ contains the nonzero constant $c(m)$. Now $c(m) = sp+tf$, and so $c(m)g = spg+tfg$. Since $p(x) \mid f(x)g(x)$, we have $p(x) \mid c(m)g(x)$. But $p(x)$ is primitive, because it is irreducible, by Example 5.22, and so Lemma 5.24(iv) gives $p(x) \mid g(x)$. •

**Corollary 5.26.** *If $k$ is a field, then $k[x_1, \ldots, x_n]$ is a UFD.*

**Proof.** The proof is by induction on $n \geq 1$. We proved, in Theorem 2.60, that the polynomial ring $k[x_1]$ in one variable is a UFD. For the inductive step, recall that $k[x_1, \ldots, x_n, x_{n+1}] = R[x_{n+1}]$, where $R = k[x_1, \ldots, x_n]$. By induction, $R$ is a UFD and, by Theorem 5.25, so is $R[x_{n+1}]$. •

**Corollary 5.27.** *If $k$ is a field, then $p = p(x_1, \ldots, x_n) \in k[x_1, \ldots, x_n]$ is irreducible if and only if $p$ generates a prime ideal in $k[x_1, \ldots, x_n]$.*

**Proof.** Proposition 5.16 applies because $k[x_1, \ldots, x_n]$ is a UFD. •

Proposition 5.19 shows that if $k$ is a field, then gcd's exist in $k[x_1, \ldots, x_n]$.

**Corollary 5.28 (Gauss's Lemma).** *Let $R$ be a UFD, let $Q = \mathrm{Frac}(R)$, and let $f(x) \in R[x]$. If $f(x) = G(x)H(x)$ in $Q[x]$, then there is a factorization*

$$f(x) = g(x)h(x) \text{ in } R[x],$$

*where $\deg(g) = \deg(G)$ and $\deg(h) = \deg(H)$; in fact, $G(x)$ is a constant multiple of $g(x)$ and $H(x)$ is a constant multiple of $h(x)$. Therefore, if $f(x)$ does not factor into polynomials of smaller degree in $R[x]$, then $f(x)$ is irreducible in $Q[x]$.*

**Proof.** By Lemma 5.24(i), the factorization $f(x) = G(x)H(x)$ in $Q[x]$ gives $q, q' \in Q$ with

$$f(x) = qG^*(x)q'H^*(x) \text{ in } Q[x],$$

where $G^*(x), H^*(x) \in R[x]$ are primitive. But $G^*(x)H^*(x)$ is primitive, by Gauss's Lemma 5.23. Since $f(x) \in R[x]$, Lemma 5.24(iii) applies to say that the equation $f(x) = qq'[G^*(x)H^*(x)]$ forces $qq' \in R$. Therefore, $qq'G^*(x) \in R[x]$, and a factorization of $f(x)$ in $R[x]$ is $f(x) = [qq'G^*(x)]H^*(x)$. •

The special case $R = \mathbb{Z}$ and $Q = \mathbb{Q}$ was proved in Theorem 2.52.

Here is a second proof of Gauss's Lemma, in the style of the proof of Lemma 5.23 showing that the product of primitive polynomials is primitive.

**Proof.** Clearing denominators, we may assume there is $r \in R$ with

$$rf(x) = g(x)h(x) \text{ in } R[x]$$

[in more detail, there are $r', r'' \in R$ with $g(x) = r'G(x)$ and $h(x) = r''H(x)$; set $r = r'r''$]. If $p$ is an irreducible divisor of $r$ and $P = (p)$, consider the map $R[x] \to (R/P)[x]$ which reduces all coefficients mod $P$. The equation becomes

$$0 = \bar{g}(x)\bar{h}(x).$$

But $(R/P)[x]$ is a domain because $R/P$ is (Proposition 5.16), and so at least one of these factors, say, $\bar{g}(x)$, is 0; that is, all the coefficients of $g(x)$ are multiples of $p$. Therefore, we may write $g(x) = pg'(x)$, where all the coefficients of $g'(x)$ lie in $R$. If $r = ps$, then

$$psf(x) = pg'(x)h(x) \text{ in } R[x].$$

Cancel $p$, and continue canceling irreducibles until we reach a factorization $f(x) = g^*(x)h^*(x)$ in $R[x]$ [note that $\deg(g^*) = \deg(g)$ and $\deg(h^*) = \deg(h)$].  •

**Example 5.29.** We claim that $f(x,y) = x^2 + y^2 - 1 \in k[x,y]$ is irreducible, where $k$ is a field. Write $Q = k(y) = \mathrm{Frac}(k[y])$, and view $f(x,y) \in Q[x]$. Now the quadratic $g(x) = x^2 + (y^2 - 1)$ is irreducible in $Q[x]$ if and only if it has no roots in $Q = k(y)$, and this is so, by Exercise 2.53 on page 114. Hence, Proposition 5.16 shows that $(x^2 + y^2 - 1)$ is a prime ideal, for it is generated by an irreducible polynomial in $Q[x] = k[x,y]$.  ◀

Irreducibility of a polynomial in several variables is more difficult to determine than irreducibility of a polynomial of one variable, but here is one criterion.

**Proposition 5.30.** *Let $k$ be a field, and view $f(x_1, \ldots, x_n) \in k[x_1, \ldots, x_n]$ as a polynomial in $R[x_n]$, where $R = k[x_1, \ldots, x_{n-1}]$:*

$$f(x_n) = a_0(x_1, \ldots, x_{n-1}) + a_1(x_1, \ldots, x_{n-1})x_n + \cdots + a_m(x_1, \ldots, x_{n-1})x_n^m.$$

*If $f(x_n)$ is primitive and cannot be factored into two polynomials of lower degree in $R[x_n]$, then $f(x_1, \ldots, x_n)$ is irreducible in $k[x_1, \ldots, x_n]$.*

**Proof.** Suppose that $f(x_n) = g(x_n)h(x_n)$ in $R[x_n]$; by hypothesis, the degrees of $g$ and $h$ in $x_n$ cannot both be less than $\deg(f)$; say, $\deg(g) = 0$. It follows, because $f$ is primitive, that $g$ is a unit in $k[x_1, \ldots, x_{n-1}]$. Therefore, $f(x_1, \ldots, x_n)$ is irreducible in $R[x_n] = k[x_1, \ldots, x_n]$.  •

Of course, the proposition applies to any variable $x_i$, not just to $x_n$.

**Corollary 5.31.** *If $k$ is a field and $g(x_1, \ldots, x_n), h(x_1, \ldots, x_n) \in k[x_1, \ldots, x_n]$ are relatively prime, then $f(x_1, \ldots, x_n, y) = yg(x_1, \ldots, x_n) + h(x_1, \ldots, x_n)$ is irreducible in $k[x_1, \ldots, x_n, y]$.*

**Proof.** Let $R = k[x_1, \cdots, x_n]$. Note that $f$ is primitive in $R[y]$, because $(g, h) = 1$ forces any divisor of its coefficients $g, h$ to be a unit. Since $f$ is linear in $y$, it is not the product of two polynomials in $R[y]$ of smaller degree, and hence Proposition 5.30 shows that $f$ is irreducible in $R[y] = k[x_1, \ldots, x_n, y]$.  •

For example, $xy^2 + z$ is an irreducible polynomial in $k[x, y, z]$ because it is a primitive polynomial that is linear in $x$.

**Example 5.32.** The polynomials $x$ and $y^2 + z^2 - 1$ are relatively prime in $\mathbb{R}[x, y, z]$, so that $f(x, y, z) = x^2 + y^2 + z^2 - 1$ is irreducible, by Corollary 5.31. Since $\mathbb{R}[x, y, z]$ is a UFD, Corollary 5.27 gives $(f)$ a prime ideal, hence $\mathbb{R}[x, y, z]/(x^2 + y^2 + z^2 - 1)$ is a domain. ◀

# Exercises

\* **5.14.** Let $R$ be a UFD and let $Q = \operatorname{Frac}(R)$ be its fraction field. Prove that each nonzero $a/b \in Q$ has an expression in lowest terms; that is, $a$ and $b$ are relatively prime.

\* **5.15.** Let $R$ be a UFD.

(i) If $a, b, c \in R$ and $a$ and $b$ are relatively prime, prove that $a \mid bc$ implies $a \mid c$.

(ii) If $a, c_1, \ldots, c_n \in R$ and $c_i \mid a$ for all $i$, prove that $c \mid a$, where $c = \operatorname{lcm}\{c_1, \ldots, c_n\}$.

**5.16.** If $R$ is a domain, prove that the only units in $R[x_1, \ldots, x_n]$ are units in $R$. On the other hand, prove that $2x + 1$ is a unit in $\mathbb{I}_4[x]$.

**5.17.** Prove that a UFD $R$ is a PID if and only if every nonzero prime ideal is a maximal ideal.

\* **5.18.** (i) Prove that $x$ and $y$ are relatively prime in $k[x, y]$, where $k$ is a field.

(ii) Prove that $1$ is not a linear combination of $x$ and $y$ in $k[x, y]$.

**5.19.** (i) Prove that $\mathbb{Z}[x_1, \ldots, x_n]$ is a UFD for all $n \geq 1$.

(ii) If $R$ is a field, prove that the ring of polynomials in infinitely many variables, $R = k[x_1, x_2, \ldots, x_n, \ldots]$, is also a UFD.
**Hint.** We have not given a formal definition of $R$ (it will be given in Exercise 6.107 on page 513 and also on page 713), but, for the purposes of this exercise, regard $R$ as the union of the ascending chain of subrings $k[x_1] \subsetneq k[x_1, x_2] \subsetneq \cdots \subsetneq k[x_1, x_2, \ldots, x_n] \subsetneq \cdots$.

**5.20.** Let $k$ be a field and let $f(x_1, \ldots, x_n) \in k[x_1, \ldots, x_n]$ be a primitive polynomial in $R[x_n]$, where $R = k[x_1, \ldots, x_{n-1}]$. If $f$ is either quadratic or cubic in $x_n$, prove that $f$ is irreducible in $k[x_1, \ldots, x_n]$ if and only if $f$ has no roots in $k(x_1, \ldots, x_{n-1})$.

\* **5.21.** Let $\alpha \in \mathbb{C}$ be a root of $f(x) \in \mathbb{Z}[x]$. If $f(x)$ is monic, prove that the minimal polynomial $p(x) = \operatorname{irr}(\alpha, \mathbb{Q})$ lies in $\mathbb{Z}[x]$.
**Hint.** Use Lemma 5.24.

**5.22.** Let $R$ be a UFD with $Q = \operatorname{Frac}(R)$. If $f(x) \in R[x]$, prove that $f(x)$ is irreducible in $R[x]$ if and only if $f(x)$ is primitive and $f(x)$ is irreducible in $Q[x]$.

**5.23.** Prove that $f(x, y) = xy^3 + x^2 y^2 - x^5 y + x^2 + 1$ is an irreducible polynomial in $\mathbb{R}[x, y]$.

\* **5.24.** Let $D = \det\left(\begin{bmatrix} x & y \\ z & w \end{bmatrix}\right)$, so that $D$ lies in the polynomial ring $\mathbb{Z}[x, y, z, w]$.

(i) Prove that $(D)$ is a prime ideal in $\mathbb{Z}[x, y, z, w]$.
**Hint.** Prove first that $D$ is an irreducible element.

(ii) Prove that $\mathbb{Z}[x, y, z, w]/(D)$ is not a UFD. [This is another example of a domain that is not a UFD. In Example 5.21, we saw that if $k$ is a field, then the subring $R \subseteq k[x]$ consisting of all polynomials having no linear term is not a UFD.]

---

## Section 5.3. Noetherian Rings

*Hilbert's Basis Theorem* states one of the most important properties of $k[x_1, \ldots, x_n]$ for $k$ a field: every ideal can be generated by a finite number of elements. This finiteness property is intimately related to chains of ideals; it arose in the proof of Lemma 5.17 when we proved that PIDs are UFDs (I apologize for so many acronyms, but here comes another one!).

**Definition.** A commutative ring $R$ satisfies **ACC** (*ascending chain condition*) if every ascending chain of ideals

$$I_1 \subseteq I_2 \subseteq \cdots \subseteq I_n \subseteq \cdots$$

*stops*; that is, the sequence is constant from some point on: there is an integer $N$ with $I_N = I_{N+1} = I_{N+2} = \cdots$.

Lemma 5.17(ii) shows that every PID satisfies ACC.

Here is an important type of ideal.

**Definition.** If $U$ is a subset of a commutative ring $R$, then $(U)$, the **ideal generated by** $U$, is the set of all finite linear combinations

$$(U) = \left\{ \sum_{\text{finite}} r_i u_i : r_i \in R \text{ and } u_i \in U \right\}.$$

We say that an ideal $I$ is **finitely generated** if $I = (U)$, where $U = \{u_1, \ldots, u_n\}$ is finite. We abbreviate $I = (\{u_1, \ldots, u_n\})$ to

$$I = (u_1, \ldots, u_n),$$

and we say that the ideal $I$ is **generated by** $u_1, \ldots, u_n$.

A set of generators $u_1, \ldots, u_n$ of an ideal $I$ is sometimes called a **basis** of $I$ (even though this is a weaker notion than that of a basis of a vector space—we do not assume that the coefficients $r_i$ in the expression $c = \sum r_i u_i$ are uniquely determined by $c$).

Of course, every ideal $I$ in a PID is finitely generated, for it can be generated by one element.

**Proposition 5.33.** *The following conditions are equivalent for a commutative ring $R$.*

(i) *$R$ satisfies* ACC.

(ii) *$R$ satisfies the **maximum condition**: every nonempty family $\mathcal{F}$ of ideals in $R$ has a maximal element; that is, there is some $M \in \mathcal{F}$ for which there is no $I \in \mathcal{F}$ with $M \subsetneq I$.*

(iii) *Every ideal in $R$ is finitely generated.*

**Proof.** (i) $\Rightarrow$ (ii) Let $\mathcal{F}$ be a nonempty family of ideals in $R$, and assume that $\mathcal{F}$ has no maximal element. Choose $I_1 \in \mathcal{F}$. Since $I_1$ is not a maximal element, there is $I_2 \in \mathcal{F}$ with $I_1 \subsetneq I_2$. Now $I_2$ is not a maximal element in $\mathcal{F}$, and so there is $I_3 \in \mathcal{F}$ with $I_2 \subsetneq I_3$. Continuing in this way constructs an ascending chain of ideals in $R$ that does not stop, contradicting ACC.

(ii) $\Rightarrow$ (iii) Let $I$ be an ideal in $R$, and define $\mathcal{F}$ to be the family of all the finitely generated ideals contained in $I$; of course, $\mathcal{F} \neq \varnothing$, for $(0) \in \mathcal{F}$. By hypothesis, there exists a maximal element $M \in \mathcal{F}$. Now $M \subseteq I$ because $M \in \mathcal{F}$. If $M \subsetneq I$, then there is $a \in I$ with $a \notin M$. The ideal

$$J = \{m + ra : m \in M \text{ and } r \in R\} \subseteq I$$

is finitely generated, and so $J \in \mathcal{F}$; but $M \subsetneq J$, contradicting the maximality of $M$. Therefore, $M = I$, and $I$ is finitely generated.

(iii) $\Rightarrow$ (i) Assume that every ideal in $R$ is finitely generated, and let

$$I_1 \subseteq I_2 \subseteq \cdots \subseteq I_n \subseteq \cdots$$

be an ascending chain of ideals in $R$. By Lemma 5.17(i), the ascending union $J = \bigcup_{n \geq 1} I_n$ is an ideal. By hypothesis, there are elements $a_i \in J$ with $J = (a_1, \ldots, a_q)$. Now $a_i$ got into $J$ by being in $I_{n_i}$ for some $n_i$. If $N$ is the largest $n_i$, then $I_{n_i} \subseteq I_N$ for all $i$; hence, $a_i \in I_N$ for all $i$, and so

$$J = (a_1, \ldots, a_q) \subseteq I_N \subseteq J.$$

It follows that if $n \geq N$, then $J = I_N \subseteq I_n \subseteq J$, so that $I_n = J$; therefore, the chain stops, and $R$ has ACC. $\bullet$

We now give a name to a commutative ring that satisfies any of the three equivalent conditions in the proposition.

**Definition.** A commutative ring $R$ is called ***noetherian***[3] if every ideal in $R$ is finitely generated.

We shall soon see that $k[x_1, \ldots, x_n]$ is noetherian whenever $k$ is a field. On the other hand, here is an example of a commutative ring that is not noetherian.

**Example 5.34.** Let $R = \mathcal{F}(\mathbb{R})$ be the ring of all real-valued functions on the reals under pointwise operations (see Example 2.9). For every positive integer $n$,

$$I_n = \{f : \mathbb{R} \to \mathbb{R} : f(x) = 0 \text{ for all } x \geq n\}$$

is an ideal and $I_n \subsetneq I_{n+1}$ for all $n$. Therefore, $R$ does not satisfy ACC, and so $R$ is not noetherian. Exercise 5.28 on page 316 asks you to prove that the family $\{I_n : n \geq 1\}$ does not have a maximal element, and that $I = \bigcup_n I_n$ is not finitely generated. $\blacktriangleleft$

Here is an application of the maximum condition.

---

[3]This name honors Emmy Noether (1882–1935), who introduced chain conditions in 1921.

**Corollary 5.35.** *If $I$ is a proper ideal in a noetherian ring $R$, then there exists a maximal ideal $M$ in $R$ containing $I$. In particular, every noetherian ring has maximal ideals.*[4]

**Proof.** Let $\mathcal{F}$ be the family of all those proper ideals in $R$ which contain $I$; note that $\mathcal{F} \neq \varnothing$ because $I \in \mathcal{F}$. Since $R$ is noetherian, the maximum condition gives a maximal element $M$ in $\mathcal{F}$. We must still show that $M$ is a maximal ideal in $R$ (that is, that $M$ is a maximal element in the larger family $\mathcal{F}'$ consisting of all the proper ideals in $R$). This is clear: if there is a proper ideal $J$ with $M \subseteq J$, then $I \subseteq J$, and $J \in \mathcal{F}$. Hence, maximality of $M$ gives $M = J$, and so $M$ is a maximal ideal in $R$.   •

Here is one way to construct a new noetherian ring from an old one.

**Corollary 5.36.** *If $R$ is a noetherian ring and $I$ is an ideal in $R$, then $R/I$ is also noetherian.*

**Proof.** If $A$ is an ideal in $R/I$, then the Correspondence Theorem for Rings provides an ideal $J$ in $R$ with $J/I = A$. Since $R$ is noetherian, the ideal $J$ is finitely generated, say, $J = (b_1, \ldots, b_n)$, and so $A = J/I$ is also finitely generated (by the cosets $b_1 + I, \ldots, b_n + I$). Therefore, $R/I$ is noetherian.   •

The following anecdote is well known. Around 1890, Hilbert proved the famous Hilbert Basis Theorem, showing that every ideal in $\mathbb{C}[x_1, \ldots, x_n]$ is finitely generated. As we will see, the proof is nonconstructive in the sense that it does not give an explicit set of generators of an ideal. It is reported that when P. Gordan, one of the leading algebraists of the time, first saw Hilbert's proof, he said, "This is not Mathematics, but theology!" On the other hand, Gordan said, in 1899 when he published a simplified proof of Hilbert's Theorem, "I have convinced myself that theology also has its advantages."

**Lemma 5.37.** *A commutative ring $R$ is noetherian if and only if, for every sequence $a_1, \ldots, a_n, \ldots$ of elements in $R$, there exist $m \geq 1$ and $r_1, \ldots, r_m \in R$ with $a_{m+1} = r_1 a_1 + \cdots + r_m a_m$.*

**Proof.** Assume that $R$ is noetherian and that $a_1, \ldots, a_n, \ldots$ is a sequence of elements in $R$. If $I_n$ is the ideal generated by $a_1, \ldots, a_n$, then there is an ascending chain of ideals, $I_1 \subseteq I_2 \subseteq \cdots$. By ACC, there exists $m \geq 1$ with $I_m = I_{m+1}$. Therefore, $a_{m+1} \in I_{m+1} = I_m$, and so there are $r_i \in R$ with $a_{m+1} = r_1 a_1 + \cdots + r_m a_m$.

Conversely, suppose that $R$ satisfies the condition on sequences of elements. If $R$ is not noetherian, then there is an ascending chain of ideals $I_1 \subseteq I_2 \subseteq \cdots$ that does not stop. Deleting any repetitions if necessary, we may assume that $I_n \subsetneq I_{n+1}$ for all $n$. For each $n$, choose $a_{n+1} \in I_{n+1}$ with $a_{n+1} \notin I_n$. By hypothesis, there exist $m$ and $r_i \in R$ for $i \leq m$ with $a_{m+1} = \sum_{i \leq m} r_i a_i \in I_m$. This contradiction implies that $R$ is noetherian.   •

---

[4]This corollary is true without assuming that $R$ is noetherian, but the proof of the general result needs Zorn's Lemma; see Theorem 5.43.

**Theorem 5.38 (Hilbert Basis Theorem).** *If $R$ is a commutative noetherian ring, then $R[x]$ is also noetherian.*

**Proof. (Sarges)** Assume that $I$ is an ideal in $R[x]$ that is not finitely generated; of course, $I \neq (0)$. Define $f_0(x)$ to be a polynomial in $I$ of minimal degree and define, inductively, $f_{n+1}(x)$ to be a polynomial of minimal degree in $I - (f_0, \ldots, f_n)$. Note that $f_n(x)$ exists for all $n \geq 0$: if $I - (f_0, \ldots, f_n)$ were empty, then $I$ would be finitely generated. It is clear that

$$\deg(f_0) \leq \deg(f_1) \leq \deg(f_2) \leq \cdots .$$

Let $a_n$ denote the leading coefficient of $f_n(x)$. Lemma 5.37 gives an integer $m$ with $a_{m+1} \in (a_0, \ldots, a_m)$: there are $r_i \in R$ with $a_{m+1} = r_0 a_0 + \cdots + r_m a_m$. Define

$$f^*(x) = f_{m+1}(x) - \sum_{i=0}^{m} x^{d_{m+1} - d_i} r_i f_i(x),$$

where $d_i = \deg(f_i)$. Now $f^* \in I - (f_0, \ldots, f_m)$ for, otherwise, $f_{m+1} \in (f_0, \ldots, f_m)$. We claim that $\deg(f^*) < \deg(f_{m+1})$. If $f_i(x) = a_i x^{d_i} +$ lower terms, then

$$f^*(x) = f_{m+1}(x) - \sum_{i=0}^{m} x^{d_{m+1} - d_i} r_i f_i(x)$$

$$= (a_{m+1} x^{d_{m+1}} + \text{lower terms}) - \sum_{i=0}^{m} x^{d_{m+1} - d_i} r_i (a_i x^{d_i} + \text{lower terms}).$$

The leading term being subtracted is thus $\sum_{i=0}^{m} r_i a_i x^{d_{m+1}} = a_{m+1} x^{d_{m+1}}$. We have contradicted $f_{m+1}(x)$ having minimal degree among polynomials in $I$ not in $(f_0, \ldots, f_m)$. •

**Corollary 5.39.**

  (i) *If $k$ is a field, then $k[x_1, \ldots, x_n]$ is noetherian.*

 (ii) *The ring $\mathbb{Z}[x_1, \ldots, x_n]$ is noetherian.*

(iii) *For any ideal $I$ in $k[x_1, \ldots, x_n]$, where $k = \mathbb{Z}$ or $k$ is a field, the quotient ring $k[x_1, \ldots, x_n]/I$ is noetherian.*

**Proof.** The proofs of the first two items are by induction on $n \geq 1$, using the theorem, while the proof of item (iii) follows from Corollary 5.36. •

## Exercises

**5.25.** (i) Give an example of a noetherian ring $R$ containing a subring that is not noetherian.

(ii) Give an example of a commutative ring $R$ containing proper ideals $I \subsetneq J \subsetneq R$ with $J$ finitely generated but with $I$ not finitely generated.

**5.26.** Let $R$ be a noetherian domain such that every $a, b \in R$ has a gcd that is an $R$-linear combination of $a$ and $b$. Prove that $R$ is a PID. (The noetherian hypothesis is necessary, for there exist non-noetherian domains, called *Bézout rings*, in which every finitely generated ideal is principal.)

**Hint.** Use induction on the number of generators of an ideal.

**5.27.** Give a proof not using Proposition 5.33 that every nonempty family $\mathcal{F}$ of ideals in a PID $R$ has a maximal element.

* **5.28.** Example 5.34 shows that $R = \mathcal{F}(\mathbb{R})$, the ring of all functions on $\mathbb{R}$ under pointwise operations, does not satisfy ACC.

    (i) Show that the family of ideals $\{I_n : n \geq 1\}$ in that example does not have a maximal element.

    (ii) Prove that $I = \bigcup_{n \geq 1} I_n$ is an ideal that is not finitely generated.

**5.29.** If $R$ is a commutative ring, define the ring of formal power series in several variables inductively:

$$R[[x_1, \ldots, x_{n+1}]] = A[[x_{n+1}]],$$

where $A = R[[x_1, \ldots, x_n]]$. Prove that if $R$ is a noetherian ring, then $R[[x_1, \ldots, x_n]]$ is also a noetherian ring.

**Hint.** If $n = 1$, use Exercise 2.63 on page 132; when $n \geq 1$, use the proof of the Hilbert Basis Theorem, but replace the degree of a polynomial by the *order* of a formal power series (the order of a nonzero formal power series $\sum c_i x^i$ is defined to be $n$, where $n$ is the smallest $i$ with $c_i \neq 0$; see Exercise 2.27 on page 95).

**5.30.** Let

$$S^2 = \{(x, y, z) \in \mathbb{R}^3 : x^2 + y^2 + z^2 = 1\}$$

be the 2-sphere in $\mathbb{R}^3$. Prove that

$$I = \{f(x, y, z) \in \mathbb{R}[x, y, z] : f(a, b, c) = 0 \text{ for all } (a, b, c) \in S^2\}$$

is a finitely generated ideal in $\mathbb{R}[x, y, z]$.

**5.31.** If $R$ and $S$ are noetherian, prove that their direct product $R \times S$ is also noetherian.

**5.32.** Call a commutative ring $R$ that is also a vector space over a field $k$ a ***commutative k-algebra*** if

$$(\alpha u)v = \alpha(uv) = u(\alpha v)$$

for all $\alpha \in k$ and $u, v \in R$.

    (i) Prove that if a commutative $k$-algebra $R$ is finite-dimensional over $k$, then $R$ is noetherian.

    (ii) Prove that if the additive group of a commutative ring $R$ is free abelian of finite rank, then $R$ is noetherian.

## Section 5.4. Zorn's Lemma and Applications

Dealing with infinite sets often requires appropriate tools of Set Theory. In this section, we present Zorn's Lemma, the most useful such tool, as well as some applications of it to vector spaces, commutative rings, and fields.

### 5.4.1. Zorn's Lemma.

**Definition.** If $A$ is a set, let $(2^A)^\#$ denote the family of all its nonempty subsets. The ***Axiom of Choice*** states that if $A$ is a nonempty set, then there exists a function $\beta : (2^A)^\# \to A$ with $\beta(S) \in S$ for every nonempty subset $S$ of $A$. Such a function $\beta$ is called a ***choice function***.

Informally, the Axiom of Choice is a harmless looking statement; it says that we can simultaneously choose one element from each nonempty subset of a set. The Axiom of Choice is easy to accept, and it is one of the standard axioms of Set Theory; in fact, it is equivalent to the statement that the cartesian product of nonempty sets is itself nonempty.[5] But the Axiom of Choice is not convenient to use as it stands. There are various equivalent forms of it that are more useful, and we now discuss the most popular of them, *Zorn's Lemma* and the *Well-Ordering Principle*.

Recall that a set $X$ is *partially ordered* if there is a relation $x \preceq y$ defined on $X$ that is reflexive, antisymmetric, and transitive. We will state Zorn's Lemma after giving several definitions.

**Definition.** An element $m$ in a partially ordered set $X$ is a ***maximal element*** if there is no $x \in X$ for which $m \prec x$; that is,

$$\text{if } m \preceq x, \text{ then } m = x.$$

### Example 5.40.

(i) A partially ordered set may have no maximal elements. For example, $\mathbb{R}$, with its usual ordering, has no maximal elements.

(ii) A partially ordered set may have many maximal elements. For example, if $A$ is a nonempty set and $X = 2^A$ is the family of all the proper subsets of $A$ partially ordered by inclusion, then a subset $S \subseteq A$ is a maximal element of $X$ if and only if $S = A - \{a\}$ for some $a \in A$; that is, $S$ is the complement of a point.

(iii) If $X$ is the family of all the proper ideals in a commutative ring $R$, partially ordered by inclusion, then a maximal element in $X$ is a maximal ideal.  ◄

Zorn's Lemma gives a condition that guarantees the existence of maximal elements.

**Definition.** A partially ordered set $X$ is a ***chain*** (or is ***simply ordered*** or is ***totally ordered***) if, for all $x, y \in X$, either $x \preceq y$ or $y \preceq x$.

The set of real numbers $\mathbb{R}$ with its usual ordering is a chain.

Recall that an *upper bound* of a nonempty subset $Y$ of a partially ordered set $X$ is an element $x_0 \in X$, not necessarily in $Y$, with $y \preceq x_0$ for every $y \in Y$.

---

[5]A set $X$ is ***nonempty*** if there exists an element $x \in X$. We can prove, by induction on $n$, that a cartesian product of $n$ nonempty sets is nonempty. Thus, the Axiom of Choice is only interesting for infinite sets $A$.

**Zorn's Lemma.** *If $X$ is a nonempty partially ordered set in which every chain has an upper bound in $X$, then $X$ has a maximal element.*

We now introduce some definitions to enable us to state the Well-Ordering Principle.

**Definition.** A partially ordered set $X$ is **well-ordered** if every nonempty subset $S$ of $X$ contains a **smallest element**; that is, there is $s_0 \in S$ with

$$s_0 \preceq s \text{ for all } s \in S.$$

The set of natural numbers $\mathbb{N}$ is well-ordered (this is precisely what the Least Integer Axiom in Chapter 1 states), but the set $\mathbb{Z}$ of all integers is not well-ordered, because the negative integers form a nonempty subset having no smallest element. Note that every well-ordered set $X$ is a chain: if $x, y \in X$, then the nonempty subset $\{x, y\}$ has a least element, say, $x$, and so $x \preceq y$.

**Well-Ordering Principle.** *Every set $X$ has some well-ordering of its elements.*

If $X$ happens to be a partially ordered set, then a well-ordering, whose existence is asserted by the Well-Ordering Principle, may have nothing to do with the original partial ordering. For example, $\mathbb{Z}$ can be well-ordered:

$$0 \preceq 1 \preceq -1 \preceq 2 \preceq -2 \preceq \cdots.$$

**Theorem 5.41.** *The following statements are equivalent.*

(i) *Zorn's Lemma.*

(ii) *The Well-Ordering Principle.*

(iii) *The Axiom of Choice.*

**Proof.** Kaplansky, *Set Theory and Metric Spaces*, Chapter 3.   •

Henceforth, we shall assume, unashamedly, that all these statements are true, and we will use any of them whenever convenient.

The next proposition is frequently used in verifying that the hypothesis of Zorn's Lemma does hold.

**Proposition 5.42.** *If $C$ is a chain and $S = \{c_1, \ldots, c_n\}$ is a finite subset of $C$, then there exists some $c_i$ with $c_j \preceq c_i$ for all $c_j \in S$.*

**Proof.** The proof is by induction on $n \geq 1$. The base step is trivially true. Let $S = \{c_1, \ldots, c_{n+1}\}$. The inductive hypothesis provides $c_i$, for $1 \leq i \leq n$, with $c_j \preceq c_i$ for all $c_j \in S - \{c_{n+1}\}$. Since $C$ is a chain, either $c_i \preceq c_{n+1}$ or $c_{n+1} \preceq c_i$. Either case provides a largest element of $S$.   •

Let us illustrate how Zorn's Lemma is used. We have already proved the next result for noetherian rings using the maximal condition holding there.

**Theorem 5.43.** *If $R$ is a nonzero commutative ring, then $R$ has a maximal ideal. Indeed, every proper ideal $I$ in $R$ is contained in a maximal ideal.*

**Proof.** The second statement implies the first, for if $R$ is a nonzero ring, then the ideal $(0)$ is a proper ideal, and so there exists a maximal ideal in $R$ containing it.

Let $X$ be the family of all the proper ideals containing $I$ partially ordered by inclusion (note that $X \neq \varnothing$ because $I \in X$). A maximal element of $X$, if one exists, is a maximal ideal in $R$, for there is no proper ideal strictly containing it.

Let $\mathcal{C}$ be a chain in $X$; thus, given $I, J \in \mathcal{C}$, either $I \subseteq J$ or $J \subseteq I$. We claim that $I^* = \bigcup_{I \in \mathcal{C}} I$ is an upper bound of $\mathcal{C}$. Clearly, $I \subseteq I^*$ for all $I \in \mathcal{C}$, so that it remains to prove that $I^*$ is a proper ideal. Lemma 5.17(i) shows that $I^*$ is an ideal; let us show that $I^*$ is a proper ideal. If $I^* = R$, then $1 \in I^*$; now 1 got into $I^*$ because $1 \in I$ for some $I \in \mathcal{C}$, and this contradicts $I$ being a proper ideal.

We have verified that every chain in $X$ has an upper bound. Hence, Zorn's Lemma provides a maximal element in $X$, as desired.  •

**Remark.**

(i) Commutativity of multiplication is not used in the proof of Theorem 5.43. Thus, using terminology not yet introduced, every left (or right) ideal in a ring is contained in a maximal left (or right) ideal.

(ii) Theorem 5.43 would be false if the definition of ring $R$ did not insist on $R$ containing 1. An example of such a "ring without unit" is any additive abelian group $G$ with multiplication defined by $ab = 0$ for all $a, b \in G$. The usual definition of *ideal* makes sense, and it is easy to see that a subset $S \subseteq G$ is an ideal if and only if it is a subgroup. Thus, a maximal ideal $S$ is just a maximal subgroup; that is, $G/S$ has no proper subgroups, which says that $G/S$ is a simple abelian group. But an abelian group is simple if and only if it is a finite group of prime order, so that $S$ is a maximal ideal in $G$ if and only if $|G/S| = p$ for some prime $p$. Now the additive abelian group $G = \mathbb{Q}$ has no nonzero finite quotients (for all $x \in G/S$, there is $y \in G/S$ with $x = py$), and so it has no maximal subgroups. Therefore, the "ring without unit" $G$ has no maximal ideals.  ◄

We emphasize the necessity of checking, when applying Zorn's Lemma to a partially ordered set $X$, that $X$ be nonempty; after all, the conclusion of Zorn's Lemma is that there exists a certain kind of element in $X$. For example, a careless person might claim that Zorn's Lemma can be used to prove that there is a maximal uncountable subset of $\mathbb{Z}$. Define $X$ to be the set of all the uncountable subsets of $\mathbb{Z}$, and partially order $X$ by inclusion. If $C$ is a chain in $X$, then it is clear that the uncountable subset $S^* = \bigcup_{S \subseteq C} S$ is an upper bound of $C$, for $S \subseteq S^*$ for every $S \in C$. Therefore, Zorn's Lemma provides a maximal element in $X$, which must be a maximal uncountable subset of $\mathbb{Z}$. The flaw, of course, is that $X = \varnothing$ (for every subset of a countable set is itself countable).

The next application characterizes noetherian rings in terms of their prime ideals.

**Lemma 5.44.** *Let $R$ be a commutative ring and let $\mathcal{F}$ be the family of all those ideals in $R$ that are not finitely generated. If $\mathcal{F} \neq \varnothing$, then $\mathcal{F}$ has a maximal element.*

**Proof.** Partially order $\mathcal{F}$ by inclusion. It suffices, by Zorn's Lemma, to prove that if $\mathcal{C}$ is a chain in $\mathcal{F}$, then $I^* = \bigcup_{I \in \mathcal{C}} I$ is not finitely generated. If, on the contrary, $I^* = (a_1, \ldots, a_n)$, then $a_j \in I_j$ for some $I_j \in \mathcal{C}$. But $\mathcal{C}$ is a chain, and so one of the ideals $I_1, \ldots, I_n$, call it $I_0$, contains the others, by Proposition 5.42. It follows that $I^* = (a_1, \ldots, a_n) \subseteq I_0$. The reverse inclusion is clear, for $I \subseteq I^*$ for all $I \in \mathcal{C}$. Therefore, $I_0 = I^*$ is finitely generated, contradicting $I_0 \in \mathcal{F}$.  •

**Theorem 5.45 (I. S. Cohen).** *A commutative ring $R$ is noetherian if and only if every prime ideal in $R$ is finitely generated.*

**Proof.** Only sufficiency needs proof. Assume that every prime ideal is finitely generated, and let $\mathcal{F}$ be the family of all those ideals in $R$ that are not finitely generated. If $\mathcal{F} \neq \varnothing$, then the lemma provides an ideal $I$ that is not finitely generated and is maximal such. We will show that $I$ is a prime ideal. With the hypothesis that every prime ideal is finitely generated, this contradiction will show that $\mathcal{F} = \varnothing$ and, hence, that $R$ is noetherian.

Suppose that $ab \in I$ but $a \notin I$ and $b \notin I$. Since $a \notin I$, the ideal $I + Ra$ is strictly larger than $I$, and so $I + Ra$ is finitely generated; indeed, we may assume that

$$I + Ra = (i_1 + r_1 a, \ldots, i_n + r_n a),$$

where $i_k \in I$ and $r_k \in R$ for all $k$. Consider $J = (I : a) = \{x \in R : xa \in I\}$. Now $I + Rb \subseteq J$; since $b \notin I$, we have $I \subsetneq J$, and so $J$ is finitely generated. We claim that $I = (i_1, \ldots, i_n, Ja)$. Clearly, $(i_1, \ldots, i_n, Ja) \subseteq I$, for every $i_k \in I$ and $Ja \subseteq I$. For the reverse inclusion, if $z \in I \subseteq I + Ra$, there are $u_k \in R$ with $z = \sum_k u_k (i_k + r_k a)$. Then $(\sum_k u_k r_k)a = z - \sum_k u_k i_k \in I$, so that $\sum_k u_k r_k \in J$. Hence, $z = \sum_k u_k i_k + (\sum_k u_k r_k)a \in (i_1, \ldots, i_n, Ja)$. It follows that $I = (i_1, \ldots, i_n, Ja)$ is finitely generated, a contradiction, and so $I$ is a prime ideal.  •

A theorem of Krull, Corollary 10.175, says that noetherian rings have DCC (*descending chain condition*) on prime ideals: every descending series of ideals

$$I_1 \supseteq I_2 \supseteq \cdots \supseteq I_n \supseteq \cdots$$

is constant from some point on.

# Exercises

**5.33.** Prove that the Axiom of Choice is true if and only if, for every nonempty index set $I$, every cartesian product $\prod_{i \in I} X_i$ of nonempty sets $X_i$ is nonempty.

**5.34.** Prove that the Well-Ordering Principle implies the Axiom of Choice.

∗ **5.35.**   Recall that if $S$ is a subset of a partially ordered set $X$, then the **least upper bound** of $S$ (should it exist) is an upper bound $m$ of $S$ such that $m \preceq u$ for every upper bound $u$ of $S$. If $X$ is the following partially ordered set:

(in which $d \preceq a$ is indicated by a line joining $a$ and $d$ with $a$ higher than $d$), prove that the subset $S = \{c, d\}$ has an upper bound but no least upper bound.

**5.36.** Let $G$ be an abelian group, and let $S \subseteq G$ be a subgroup.

    (i) Prove that there exists a subgroup $H$ of $G$ maximal with the property that $H \cap S = \{0\}$. Is this true if $G$ is not abelian?

    (ii) If $H$ is maximal with $H \cap S = \{0\}$, prove that $G/(H + S)$ is torsion.

$*$ **5.37.** Call a subset $C$ of a partially ordered set $X$ *cofinal* if, for each $x \in X$, there exists $c \in C$ with $x \preceq c$.

    (i) Prove that $\mathbb{Q}$ and $\mathbb{Z}$ are cofinal subsets of $\mathbb{R}$.

    (ii) Prove that every chain $X$ contains a well-ordered cofinal subset.
       **Hint.** Use Zorn's Lemma on the family of all the well-ordered subsets of $X$.

    (iii) Prove that every well-ordered subset in $X$ has an upper bound if and only if every chain in $X$ has an upper bound.

**5.38.** Prove that every commutative ring $R$ has a *minimal prime ideal*, that is, a prime ideal $I$ for which there is no prime ideal $P$ with $P \subsetneq I$.

**Hint.** Partially order the set of all prime ideals by *reverse inclusion*: $P \preceq Q$ means $P \supseteq Q$. See Exercise 5.3 on page 301.

$*$ **5.39.** Recall that a subset $S$ of a commutative ring $R$ is *multiplicative* if $0 \notin S$, $1 \in S$, and $s, s' \in S$ implies $ss' \in S$. Complete Exercise 5.7 on page 301 by proving that if $S$ is a multiplicative set with $S \cap I = \varnothing$, where $I$ is an ideal in $R$, then there exists an ideal $J$ maximal such that $I \subseteq J$ and $J \cap S = \varnothing$.

$*$ **5.40.** Prove that every nonunit in a commutative ring lies in some maximal ideal. [This result was used to solve Exercise 5.13(iii) on page 302.]

**5.4.2. Vector Spaces.** Here are some applications of Zorn's Lemma to Linear Algebra. We begin by generalizing the usual definition of a basis of a vector space so that it applies to all, not necessarily finite-dimensional, vector spaces.

**Definition.** Let $V$ be a vector space over a field $k$, and let $Y \subseteq V$ be a (possibly infinite) subset.[6]

    (i) $Y$ is *linearly independent* if every finite subset of $Y$ is linearly independent.

    (ii) $Y$ *spans* $V$ if each $v \in V$ is a linear combination of finitely[7] many elements of $Y$. We write $V = \langle Y \rangle$ if $V$ is spanned by $Y$.

---

    [6]When dealing with infinite bases, it is more convenient to work with subsets instead of with lists, that is, ordered subsets. We noted in Chapter 2 that whether a finite list $x_1, \ldots, x_n$ of vectors is a basis depends only on the subset $\{x_1, \ldots, x_n\}$ and not upon its ordering.

    [7]Only finite sums of elements in $V$ are allowed. Without limits, convergence of infinite series does not make sense, and so a sum with infinitely many nonzero terms is not defined.

(iii) A **basis** of a vector space $V$ is a linearly independent subset that spans $V$.

We say that **almost all** elements of a set $Y$ have a certain property if there are at most finitely many $y \in Y$ which do not enjoy this property. For example, let $Y = \{y_i : i \in I\}$ be a subset of a vector space. To say that $\sum a_i y_i = 0$ for almost all $a_i = 0$ means that only finitely many $a_i$ can be nonzero. Thus, $Y$ is linearly independent if, whenever $\sum a_i y_i = 0$, where almost all $a_i = 0$, then all $a_i = 0$.

**Example 5.46.** Let $k$ be a field, and regard $V = k[x]$ as a vector space over $k$. We claim that
$$Y = \{1, x, x^2, \ldots, x^n, \ldots\}$$
is a basis of $V$. Now $Y$ spans $V$, for every polynomial of degree $d \geq 0$ is a $k$-linear combination of $1, x, x^2, \ldots, x^d$. Also, $Y$ is linearly independent, because there are no scalars $a_0, a_1, \ldots, a_n$, not all 0, with $\sum_{i=0}^{n} a_i x^i = 0$ (the polynomial $f(x) = \sum_{i=0}^{n} a_i x^i$ is the zero polynomial if and only if all its coefficients are 0). Therefore, $Y$ is a basis of $V$.  ◄

**Theorem 5.47.** *Every vector space $V$ over a field $k$ has a basis. Indeed, every linearly independent subset $B$ of $V$ is contained in a basis of $V$; that is, there is a subset $B'$ so that $B \cup B'$ is a basis of $V$.*

**Proof.** Note that the first statement follows from the second, for $B = \varnothing$ is a linearly independent subset contained in any basis.

Let $X$ be the family of all the linearly independent subsets of $V$ containing $B$. The family $X$ is nonempty, for $B \in X$. Partially order $X$ by inclusion. We use Zorn's Lemma to prove the existence of a maximal element in $X$. Let $\mathcal{B} = (B_j)_{j \in J}$ be a chain of $X$. Thus, each $B_j$ is a linearly independent subset containing $B$ and, for all $i, j \in J$, either $B_j \subseteq B_i$ or $B_i \subseteq B_j$. Proposition 5.42 says that if $B_{j_1}, \ldots, B_{j_n}$ is any *finite* family of $B_j$'s, then one contains all of the others.

Let $B^* = \bigcup_{j \in J} B_j$. Clearly, $B^*$ contains $B$ and $B_j \subseteq B^*$ for all $j \in J$. Thus, $B^*$ is an upper bound of $\mathcal{B}$ if it belongs to $X$, that is, if $B^*$ is a linearly independent subset of $V$. If $B^*$ is not linearly independent, then it has a finite subset $y_{i_1}, \ldots, y_{i_m}$ that is linearly dependent. How did $y_{i_k}$ get into $B^*$? Answer: $y_{i_k} \in B_{j_k}$ for some index $j_k$. Since there are only finitely many $y_{i_k}$, Proposition 5.42 applies again: there exists $B_{j_0}$ containing all the $B_{i_k}$; that is, $y_{i_1}, \ldots, y_{i_m} \in B_{j_0}$. But $B_{j_0}$ is linearly independent, by hypothesis, and this is a contradiction. Therefore, $B^*$ is an upper bound of the chain $\mathcal{B}$. Thus, every chain in $X$ has an upper bound and, hence, Zorn's Lemma applies to say that there exists a maximal element in $X$.

Let $M$ be a maximal element in $X$. Since $M$ is linearly independent, it suffices to show that it spans $V$ (for then $M$ is a basis of $V$ containing $B$). If $M$ does not span $V$, then there is $v_0 \in V$ with $v_0 \notin \langle M \rangle$, the subspace spanned by $M$. By Lemma 2.110, the subset $M^* = M \cup \{v_0\}$ is linearly independent, contradicting the maximality of $M$. Therefore, $M$ spans $V$, and so it is a basis of $V$. The last statement follows if we define $B' = M - B$.  •

Recall that a subspace $W$ of a vector space $V$ is a *direct summand* if there is a subspace $W'$ of $V$ with $\{0\} = W \cap W'$ and $V = W + W'$ (i.e., each $v \in V$ can be written as $v = w + w'$, where $w \in W$ and $w' \in W'$). We say that $V$ is the *direct sum* of $W$ and $W'$, and we write $V = W \oplus W'$.

**Corollary 5.48.** *Every subspace $W$ of a vector space $V$ is a direct summand.*

**Proof.** Let $B$ be a basis of $W$. By the theorem, there is a subset $B'$ with $B \cup B'$ a basis of $V$. It is straightforward to check that $V = W \oplus \langle B' \rangle$, where $\langle B' \rangle$ denotes the subspace spanned by $B'$. •

The proof of Theorem 5.47 is typical of proofs using Zorn's Lemma. After obtaining a maximal element, the argument is completed indirectly: if the desired result were false, then a maximal element could be enlarged.

We can now generalize Theorem 2.120 to infinite-dimensional vector spaces.

**Theorem 5.49.** *Let $V$ and $W$ be vector spaces over a field $k$. If $X$ is a basis of $V$ and $f \colon X \to W$ is a function, then there exists a unique linear transformation $T \colon V \to W$ with $T(x) = f(x)$ for all $x \in X$.*

**Proof.** As in the proof of Proposition 2.101, each $v \in V$ has a unique expression of the form $v = \sum_i a_i x_i$, where $x_1, \dots, x_n \in X$ and $a_i \in k$, and so $T \colon V \to W$, given by $T(v) = \sum a_i f(x_i)$, is a (well-defined) function. It is routine to check that $T$ is a linear transformation and that it is the unique such extending $f$. •

**Corollary 5.50.** *If $V$ is an infinite-dimensional vector space over a field $k$, then $\mathrm{GL}(V) \neq \{1\}$.*

**Proof.** Let $X$ be a basis of $V$, and choose distinct elements $y, z \in X$. By Theorem 5.49, there exists a linear transformation $T \colon V \to V$ with $T(y) = z$, $T(z) = y$, and $T(x) = x$ for all $x \in X - \{y, z\}$. Now $T$ is nonsingular, because $T^2 = 1_V$. •

**Example 5.51.**

(i) The field of real numbers $\mathbb{R}$ is a vector space over $\mathbb{Q}$; a basis is usually called a **Hamel basis**, and it is useful in constructing analytic counterexamples. For example, we may use a Hamel basis to prove the existence of an everywhere discontinuous additive function $f \colon \mathbb{R} \to \mathbb{R}$:

$$f(x + y) = f(x) + f(y).$$

Here is a sketch of a proof, using infinite cardinal numbers, that such discontinuous functions $f$ exist. By Theorem 5.49, if $B$ is a (possibly infinite) basis of a vector space $V$, then any function $f \colon B \to V$ extends to a linear transformation $F \colon V \to V$; namely, $F(\sum r_i b_i) = \sum r_i f(b_i)$. A Hamel basis has cardinal $c = |\mathbb{R}|$, and so there are $c^c = 2^c > c$ functions $f \colon \mathbb{R} \to \mathbb{R}$ satisfying $f(x + y) = f(x) + f(y)$, for every linear transformation is additive. On the other hand, every continuous function $\mathbb{R} \to \mathbb{R}$ is determined by its values on $\mathbb{Q}$, which is countable. It follows that there are only $\aleph_0^{\aleph_0} = c$ continuous functions $\mathbb{R} \to \mathbb{R}$. Therefore, there exists an additive function $f \colon \mathbb{R} \to \mathbb{R}$ and a real number $u$ with $f$ discontinuous at $u$: there is some $\epsilon > 0$ such that, for every $\delta > 0$, there is $v \in \mathbb{R}$ with $|v - u| < \delta$ and $|f(v) - f(u)| \geq \epsilon$. We now show that $f$ is discontinuous at every $w \in \mathbb{R}$. The

identity $v - u = (v + w - u) - w$ gives $|(v + w - u) - w| < \delta$, and the identity
$f(v + w - u) - f(w) = f(v) - f(u)$ gives $|f(v + w - u) - f(w)| \geq \epsilon$.

(ii) A Hamel basis $H$ can be used to construct a nonmeasurable subset of $\mathbb{R}$ (in
the sense of Lebesgue): if $H'$ is obtained from $H$ by removing one element,
then the subspace over $\mathbb{Q}$ spanned by $H'$ is nonmeasurable (Kharazishvili,
*Nonmeasurable Sets and Functions*, North Holland Mathematics Studies 195,
Elsevier, Amsterdam, 2004, p. 35).

(iii) A Hamel basis $H$ of $\mathbb{R}$ (viewed as a vector space over $\mathbb{Q}$) can be used to give
a positive definite inner product on $\mathbb{R}$ all of whose values are rational.

> **Definition.** An ***inner product*** on a vector space $V$ over a field $k$ is a
> function $V \times V \to k$, whose values are denoted by $(v, w)$, such that
> (a) $(v + v', w) = (v, w) + (v', w)$ for all $v, v', w \in V$;
> (b) $(\alpha v, w) = \alpha(v, w)$ for all $v, w \in V$ and $\alpha \in k$;
> (c) $(v, w) = (w, v)$ for all $v, w \in V$.
> An inner product is ***positive definite*** if $(v, v) \geq 0$ for all $v \in V$ and
> $(v, v) \neq 0$ whenever $v \neq 0$.

> Using zero coefficients if necessary, for each $v, w \in \mathbb{R}$, there are $h_i \in H$
> and rationals $a_i$ and $b_i$ with $v = \sum a_i h_i$ and $w = \sum b_i h_i$ (the nonzero $a_i$ and
> nonzero $b_i$ are uniquely determined by $v$ and $w$, respectively). Define
> $$(v, w) = \sum a_i b_i;$$
> note that the sum has only finitely many nonzero terms. It is routine to check
> that we have defined a positive definite inner product. (Fixing a value of the
> first coordinate, say, $(5, \square)\colon \mathbb{R} \to \mathbb{Q}$, gives another example of an additive
> function on $\mathbb{R}$ that is not continuous.)  ◀

There is a notion of dimension for infinite-dimensional vector spaces; of course,
dimension will now be an infinite cardinal number. In the following proof, we
shall cite and use several facts about cardinals. Recall that we denote the cardinal
number of a set $X$ by $|X|$.

**Theorem 5.52.** *Let $k$ be a field and let $V$ be a vector space over $k$.*

(i) *Any two bases of $V$ have the same number of elements (that is, they have the
same cardinal number); this cardinal, called the **dimension** of $V$, is denoted
by $\dim(V)$.*

(ii) *Vector spaces $V$ and $V'$ over $k$ are isomorphic if and only if $\dim(V) =
\dim(V')$.*

**Proof.**

(i) Let $B$ and $B'$ be bases of $V$. If $B$ is finite, then $V$ is finite-dimensional, and
hence $B'$ is also finite (Corollary 2.115); moreover, Invariance of Dimension,
Theorem 2.109, says that $|B| = |B'|$. Therefore, we may assume that both $B$
and $B'$ are infinite.

Each $v \in V$ has a unique expression of the form $v = \sum_{b \in B} \alpha_b b$, where
$\alpha_b \in k$ and almost all $\alpha_b = 0$. Define the ***support*** of $v$ (with respect to $B$)

by $\operatorname{supp}_B(v) = \{b \in B : \alpha_b \neq 0\}$; thus, $\operatorname{supp}_B(v)$ is a finite subset of $B$ for every $v \in V$. Define $f \colon B' \to \operatorname{Fin}(B)$, the family of all finite subsets of $B$, by $f(b') = \operatorname{supp}_{B'}(b')$. Note that if $\operatorname{supp}_{B'}(b') = \{b_1, \ldots, b_n\}$, then $b' \in \langle b_1, \ldots, b_n \rangle = \langle \operatorname{supp}_{B'}(b') \rangle$, the subspace spanned by $\operatorname{supp}_{B'}(b')$. Since $\langle \operatorname{supp}_{B'}(b') \rangle$ has dimension $n$, it contains at most $n$ elements of $B'$, because $B'$ is independent (Corollary 2.114). Therefore, $f^{-1}(T)$ is finite for every $T \in B$ [of course, $f^{-1}(T) = \varnothing$ is possible]. Now $|B'| \leq |\operatorname{Fin}(B)| = |B|$.[8] Interchanging the roles of $B$ and $B'$ gives the reverse inequality $|B| \leq |B'|$, and so $|B| = |B'|$.[9]

(ii) Adapt the proof of the finite-dimensional version, Corollary 2.122. ●

# Exercises

**5.41.** (i) If $S$ is a subspace of a vector space $V$, prove that there exists a subspace $W$ of $V$ maximal with the property that $W \cap S = \{0\}$.

(ii) Prove that $V = W \oplus S$.

**5.42.** Regard $\mathbb{R}$ as a vector space over $\mathbb{Q}$. If $P$ is the set of primes in $\mathbb{Z}$, prove that $\{\sqrt{p} : p \in P\}$ is linearly independent.

**5.43.** If $k$ is a countable field and $V$ is a vector space over $k$ of countable dimension, prove that $V$ is countable. Conclude that $\dim_{\mathbb{Q}}(\mathbb{R})$ is uncountable.

**5.4.3. Algebraic Closure.** Our next application involves algebraic closures of fields. Recall that a field extension $K/k$ is *algebraic* if every $a \in K$ is a root of some nonzero polynomial $f(x) \in k[x]$; that is, $K/k$ is an algebraic extension if every element $a \in K$ is algebraic over $k$.

We have already discussed algebraic extensions in Proposition 2.141, and the following proposition adds a bit more.

**Proposition 5.53.** *Let $K/k$ be an extension.*

(i) *If $z \in K$, then $z$ is algebraic over $k$ if and only if $k(z)/k$ is finite.*

(ii) *If $z_1, z_2, \ldots, z_n \in K$ are algebraic over $k$, then $k(z_1, z_2, \ldots, z_n)/k$ is finite.*

(iii) *If $y, z \in K$ are algebraic over $k$, then $y + z$, $yz$, and $y^{-1}$ (if $y \neq 0$) are also algebraic over $k$.*

(iv) *Define*
$$(K/k)_{\mathrm{alg}} = \{z \in K : z \text{ is algebraic over } k\}.$$
*Then $(K/k)_{\mathrm{alg}}$ is a subfield of $K$.*

---

[8]We use two elementary facts: (i) if $X$ is infinite and $f \colon X \to Y$ is a function which is finite-to-one (that is, $f^{-1}(y)$ is finite for all $y \in Y$), then $|X| \leq |Y|$; (ii) if $Y$ is infinite, then $|\operatorname{Fin}(Y)| = |Y|$.

[9]If $X$ and $Y$ are sets with $|X| \leq |Y|$ and $|Y| \leq |X|$, then $|X| = |Y|$. This is usually called the *Schroeder–Mac Lane Theorem*; see Birkhoff–Mac Lane, *A Survey of Modern Algebra*, p. 387.

**Proof.**

(i) If $k(z)/k$ is finite, then Proposition 2.141(i) shows that $z$ is algebraic over $k$. Conversely, if $z$ is algebraic over $k$, then Proposition 2.141(v) shows that $k(z)/k$ is finite.

(ii) We prove this by induction on $n \geq 1$; the base step is part (i). For the inductive step, there is a tower of fields

$$k \subseteq k(z_1) \subseteq k(z_1, z_2) \subseteq \cdots \subseteq k(z_1, \ldots, z_n) \subseteq k(z_1, \ldots, z_{n+1}).$$

Now $[k(z_{n+1}) : k]$ is finite (by Theorem 2.144); say, $[k(z_{n+1}) : k] = d$, where $d$ is the degree of the monic irreducible polynomial in $k[x]$ having $z_{n+1}$ as a root. Since $z_{n+1}$ satisfies a polynomial of degree $d$ over $k$, it satisfies a polynomial of degree $d' \leq d$ over the larger field $F = k(z_1, \ldots, z_n)$:

$$d' = [k(z_1, \ldots, z_{n+1}) : k(z_1, \ldots, z_n)] = [F(z_{n+1}) : F] \leq [k(z_{n+1}) : k] = d.$$

Therefore,

$$[k(z_1, \ldots, z_{n+1}) : k] = [F(z_{n+1}) : k] = [F(z_{n+1}) : F][F : k] \leq d[F : k] < \infty,$$

because $[F : k] = [k(z_1, \ldots, z_n) : k]$ is finite, by the inductive hypothesis.

(iii) Now $k(y, z)/k$ is finite, by part (ii). Therefore, $k(y+z) \subseteq k(y, z)$ and $k(yz) \subseteq k(y, z)$ are also finite, for any subspace of a finite-dimensional vector space is itself finite-dimensional [Corollary 2.115]. By part (i), $y + z$, $yz$, and $y^{-1}$ are algebraic over $k$.

(iv) This follows at once from part (iii). $\bullet$

**Definition.** Given the extension $\mathbb{C}/\mathbb{Q}$, define the **algebraic numbers** by

$$\mathbb{A} = (\mathbb{C}/\mathbb{Q})_{\mathrm{alg}}.$$

Thus, $\mathbb{A}$ consists of all those complex numbers that are roots of nonzero polynomials in $\mathbb{Q}[x]$, and the proposition shows that $\mathbb{A}$ is a subfield of $\mathbb{C}$ that is algebraic over $\mathbb{Q}$.

**Example 5.54.** We claim that $\mathbb{A}/\mathbb{Q}$ is an algebraic extension that is not finite. Suppose, on the contrary, that $[\mathbb{A} : \mathbb{Q}] = n$ for some integer $n$. There exist irreducible polynomials in $\mathbb{Q}[x]$ of degree $n + 1$; for example, $p(x) = x^{n+1} - 2$. If $\alpha$ is a root of $p(x)$, then $\alpha \in \mathbb{A}$, and so $\mathbb{Q}(\alpha) \subseteq \mathbb{A}$. Thus,

$$n = [\mathbb{A} : \mathbb{Q}] = [\mathbb{A} : \mathbb{Q}(\alpha)][\mathbb{Q}(\alpha) : \mathbb{Q}] \geq n + 1,$$

a contradiction. $\blacktriangleleft$

**Lemma 5.55.**

(i) If $k \subseteq K \subseteq E$ is a tower of fields with $E/K$ and $K/k$ algebraic, then $E/k$ is also algebraic.

(ii) Let

$$K_0 \subseteq K_1 \subseteq \cdots \subseteq K_n \subseteq K_{n+1} \subseteq \cdots$$

be an ascending tower of fields. If $K_{n+1}/K_n$ is algebraic for all $n \geq 0$, then $K^* = \bigcup_{n \geq 0} K_n$ is a field algebraic over $K_0$.

(iii) *Let $K = k(A)$; that is, $K$ is obtained from $k$ by adjoining the elements in a (possibly infinite) set $A$. If each element $a \in A$ is algebraic over $k$, then $K/k$ is an algebraic extension.*

**Proof.**

(i) Let $e \in E$; since $E/K$ is algebraic, there is some $f(x) = \sum_{i=0}^{n} a_i x^i \in K[x]$ having $e$ as a root. If $F = k(a_0, \dots, a_n)$, then $e$ is algebraic over $F$, and so $k(a_0, \dots, a_n, e) = F(e)$ is a finite extension of $F$; that is, $[F(e) : F]$ is finite. Since $K/k$ is an algebraic extension, each $a_i$ is algebraic over $k$, and Proposition 5.53(ii) shows that the intermediate field $F$ is finite-dimensional over $k$; that is, $[F : k]$ is finite,

$$[k(a_0, \dots, a_n, e) : k] = [F(e) : k] = [F(e) : F][F : k] < \infty,$$

and so $e$ is algebraic over $k$, by Proposition 5.53(i). Hence $E/k$ is algebraic.

(ii) If $y, z \in K^*$, then they are there because $y \in K_m$ and $z \in K_n$; we may assume that $m \le n$, so that both $y, z \in K_n \subseteq K^*$. Since $K_n$ is a field, it contains $y + z$, $yz$, and $y^{-1}$ if $y \ne 0$. Therefore, $K^*$ is a field.

If $z \in K^*$, then $z$ must lie in $K_n$ for some $n$. But $K_n/K_0$ is algebraic, by an obvious inductive generalization of part (i), and so $z$ is algebraic over $K_0$. Since every element of $K^*$ is algebraic over $K_0$, the extension $K^*/K_0$ is algebraic.

(iii) Let $z \in k(A)$; by Exercise 2.103 on page 171, there is an expression for $z$ involving $k$ and finitely many elements of $A$; say, $a_1, \dots, a_m$. Hence, $z \in k(a_1, \dots, a_m)$. By Proposition 5.53(ii), $k(z)/k$ is finite and hence $z$ is algebraic over $k$. •

**Definition.** A field $K$ is ***algebraically closed*** if every nonconstant $f(x) \in K[x]$ has a root in $K$. An ***algebraic closure*** of a field $k$ is an algebraic extension $\bar{k}$ of $k$ that is algebraically closed.

The algebraic closure of $\mathbb{Q}$ turns out to be the algebraic numbers: $\overline{\mathbb{Q}} = \mathbb{A}$.

The Fundamental Theorem of Algebra says that $\mathbb{C}$ is algebraically closed; moreover, $\mathbb{C}$ is an algebraic closure of $\mathbb{R}$. We have already proved this in Theorem 3.52, but the simplest proof of the Fundamental Theorem is probably that using Liouville's Theorem in Complex Variables: every bounded entire function is constant. If $f(x) \in \mathbb{C}[x]$ had no roots, then $1/f(x)$ would be a bounded entire function that is not constant.

There are two main results here. First, every field has an algebraic closure; second, any two algebraic closures of a field are isomorphic. Our proof of existence will make use of "big" polynomial rings: we assume that if $k$ is a field and $T$ is an infinite set, then there is a polynomial ring $k[T]$ having one indeterminate for each $t \in T$. We have already constructed $k[T]$ when $T$ is finite, and the infinite case is essentially a union of $k[U]$, where $U$ ranges over all the finite subsets of $T$. Constructions of $k[T]$ for infinite $T$ will be given in Exercise 6.107 on page 513 and in Proposition 8.109.

**Lemma 5.56.** *Let $k$ be a field, and let $k[T]$ be the polynomial ring in a set $T$ of indeterminates. If $t_1, \ldots, t_n \in T$ are distinct, where $n \geq 2$, and $f_i(t_i) \in k[t_i] \subseteq k[T]$ are nonconstant polynomials, then the ideal $I = (f_1(t_1), \ldots, f_n(t_n))$ in $k[T]$ is a proper ideal.*

**Remark.** If $n = 2$, then $f_1(t_1)$ and $f_2(t_2)$ are relatively prime, and this lemma says that 1 is not a linear combination of them. In contrast, $k[t_1]$ is a PID, and relatively prime polynomials of a single variable do generate $k[t_1]$. ◄

**Proof.** If $I$ is not a proper ideal in $k[T]$, then there exist $h_i(T) \in k[T]$ with

$$1 = h_1(T)f_1(t_1) + \cdots + h_n(T)f_n(t_n).$$

Consider the field extension $k(\alpha_1, \ldots, \alpha_n)$, where $\alpha_i$ is a root of $f_i(t_i)$ for $i = 1, \ldots, n$ (the $f_i$ are not constant). Denote the variables involved in the $h_i(T)$ other than $t_1, \ldots, t_n$, if any, by $t_{n+1}, \ldots, t_m$. Evaluating when $t_i = \alpha_i$ if $i \leq n$ and $t_i = 0$ if $i \geq n+1$ (by Corollary 2.26, evaluation is a ring homomorphism $k[T] \to k(\alpha_1, \ldots, \alpha_n)$), the right side is 0, and we have the contradiction $1 = 0$. •

**Theorem 5.57.** *Given a field $k$, there exists an algebraic closure $\overline{k}$ of $k$.*

**Proof.** Let $T$ be a set in bijective correspondence with the family of nonconstant polynomials in $k[x]$. Let $R = k[T]$ be the big polynomial ring, and let $I$ be the ideal in $R$ generated by all elements of the form $f(t_f)$, where $t_f \in T$; that is, if

$$f(x) = x^n + a_{n-1}x^{n-1} + \cdots + a_0,$$

where $a_i \in k$, then

$$f(t_f) = (t_f)^n + a_{n-1}(t_f)^{n-1} + \cdots + a_0.$$

We claim that the ideal $I$ is proper; if not, $1 \in I$, and there are distinct $t_1, \ldots, t_n \in T$ and polynomials $h_1(T), \ldots, h_n(T) \in k[T]$ with $1 = h_1(T)f_1(t_1) + \cdots + h_n(T)f_n(t_n)$, contradicting Lemma 5.56. Therefore, there is a maximal ideal $M$ in $R$ containing $I$, by Theorem 5.43. Define $K = R/M$. The proof is now completed in a series of steps.

(i)  *$K/k$ is a field extension.*

We know that $K = R/M$ is a field because $M$ is a maximal ideal. Let $i: k \to k[T]$ be the ring map taking $a \in k$ to the constant polynomial $a$, and let $\theta$ be the composite $k \overset{i}{\to} k[T] = R \overset{\text{nat}}{\to} R/M = K$. Now $\theta$ is injective, by Corollary 2.32, because $k$ is a field. We identify $k$ with $\operatorname{im} \theta \subseteq K$.

(ii)  *Every nonconstant $f(x) \in k[x]$ splits in $K[x]$.*

By definition, there is $t_f \in T$ with $f(t_f) \in I \subseteq M$, and the coset $t_f + M \in R/M = K$ is a root of $f(x)$. It now follows by induction on degree that $f(x)$ splits over $K$.

(iii)  *The extension $K/k$ is algebraic.*

By Lemma 5.55(iii), it suffices to show that each $t_f + M$ is algebraic over $k$ [for $K = k(\text{all } t_f + M)$]; but this is obvious, for $t_f$ is a root of $f(x) \in k[x]$.

We complete the proof as follows. Let $k_1 = K$ and construct $k_{n+1}$ from $k_n$ in the same way $K$ is constructed from $k$. There is a tower of fields $k = k_0 \subseteq k_1 \subseteq \cdots \subseteq k_n \subseteq k_{n+1} \subseteq \cdots$ with each extension $k_{n+1}/k_n$ algebraic and with every nonconstant polynomial in $k_n[x]$ having a root in $k_{n+1}$. By Lemma 5.55(ii), $E = \bigcup_n k_n$ is an algebraic extension of $k$. We claim that $E$ is algebraically closed. If $g(x) = \sum_{i=0}^{m} e_i x^i \in E[x]$ is a nonconstant polynomial, then it has only finitely many coefficients $e_0, \ldots, e_m$, and so there is some $k_q$ that contains them all. It follows that $g(x) \in k_q[x]$ and so $g(x)$ has a root in $k_{q+1} \subseteq E$, as desired. Therefore, $E$ is an algebraic closure of $k$. •

**Remark.** It turns out that $K = k_1$ is algebraically closed, but a proof is tricky. See I. M. Isaacs, Roots of Polynomials in Algebraic Extensions of Fields, *American Mathematical Monthly* 87 (1980), 543–544. ◄

**Corollary 5.58.** *If $k$ is a countable field, then it has a countable algebraic closure. In particular, the algebraic closures of the prime fields $\mathbb{Q}$ and $\mathbb{F}_p$ are countable.*

**Proof.** If $k$ is countable, then the set $T$ of all nonconstant polynomials is countable, say, $T = \{t_1, t_2, \ldots\}$, because $k[x]$ is countable. Hence, $k[T] = \bigcup_{\ell \geq 1} k[t_1, \ldots, t_\ell]$ is countable, as is its quotient $k_1$ (our notation is that in the proof of Theorem 5.57; thus, $\bigcup_{n \geq 1} k_n$ is an algebraic closure of $k$). It follows, by induction on $n \geq 1$, that every $k_n$ is countable. Finally, a countable union of countable sets is itself countable, so that an algebraic closure of $k$ is countable. •

We are now going to prove uniqueness of an algebraic closure.

**Definition.** If $F/k$ and $K/k$ are field extensions, then a **$k$-map** is a ring homomorphism $\varphi : F \to K$ that fixes $k$ pointwise.

Recall Proposition 3.1: if $K/k$ is a field extension, $\varphi : K \to K$ is a $k$-map, and $f(x) \in k[x]$, then $\varphi$ permutes all the roots of $f(x)$ that lie in $K$.

**Lemma 5.59.** *If $K/k$ is an algebraic extension, then every $k$-map $\varphi : K \to K$ is an automorphism of $K$.*

**Proof.** By Corollary 2.32, the $k$-map $\varphi$ is injective. To see that $\varphi$ is surjective, let $a \in K$. Since $K/k$ is algebraic, there is an irreducible polynomial $p(x) \in k[x]$ having $a$ as a root. As we have just remarked, the $k$-map $\varphi$ permutes the set $A$ of all those roots of $p(x)$ that lie in $K$. Therefore, $a \in \varphi(A) \subseteq \operatorname{im} \varphi$. •

The next lemma will use Zorn's Lemma by partially ordering a family of functions. Since a function is essentially a set (its graph), it is reasonable to take a union of functions in order to obtain an upper bound; we give details below.

**Lemma 5.60.** *Let $k$ be a field and let $\overline{k}/k$ be an algebraic closure. If $F/k$ is an algebraic extension, then there is an injective $k$-map $\psi : F \to \overline{k}$.*

**Proof.** If $E$ is an intermediate field, $k \subseteq E \subseteq F$, let us call an ordered pair $(E, f)$ an *approximation* if $f : E \to \overline{k}$ is a $k$-map. In the following diagram, all arrows other than $f$ are inclusions:

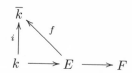

Define $X = \{$approximations $(E, f) : k \subseteq E \subseteq F\}$. Note that $X \neq \varnothing$ because $(k, i) \in X$. Partially order $X$ by

$$(E, f) \preceq (E', f') \text{ if } E \subseteq E' \text{ and } f'|E = f.$$

That the restriction $f'|E$ is $f$ means that $f'$ *extends* $f$; that is, the two functions agree whenever possible: $f'(u) = f(u)$ for all $u \in E$.

It is easy to see that an upper bound of a chain

$$\mathcal{S} = \{(E_j, f_j) : j \in J\}$$

is given by $(\bigcup E_j, \bigcup f_j)$. That $\bigcup E_j$ is an intermediate field is, by now, a routine argument. We can take the union of the graphs of the $f_j$, but here is a more down-to-earth description of $\Phi = \bigcup f_j$: if $u \in \bigcup E_j$, then $u \in E_{j_0}$ for some $j_0$, and $\Phi \colon u \mapsto f_{j_0}(u)$. Note that $\Phi$ is well-defined: if $u \in E_{j_1}$, we may assume, for notation, that $E_{j_0} \subseteq E_{j_1}$, and then $f_{j_1}(u) = f_{j_0}(u)$ because $f_{j_1}$ extends $f_{j_0}$. Observe that $\Phi$ is a $k$-map because all the $f_j$ are.

By Zorn's Lemma, there exists a maximal element $(E_0, f_0)$ in $X$. We claim that $E_0 = F$, and this will complete the proof (take $\psi = f_0$). If $E_0 \subsetneq F$, then there is $a \in F$ with $a \notin E_0$. Since $F/k$ is algebraic, we have $F/E_0$ algebraic, and there is an irreducible $p(x) \in E_0[x]$ having $a$ as a root. Since $\overline{k}/k$ is algebraic and $\overline{k}$ is algebraically closed, we have a factorization in $\overline{k}[x]$:

$$f_0^*(p(x)) = \prod_{i=1}^{n}(x - b_i),$$

where $f_0^* \colon E_0[x] \to \overline{k}[x]$ is the map $f_0^* \colon e_0 + \cdots + e_n x^n \mapsto f_0(e_0) + \cdots + f_0(e_n)x^n$. If all the $b_i$ lie in $f_0(E_0) \subseteq \overline{k}$, then $f_0^{-1}(b_i) \in E_0 \subseteq F$ for all $i$, and there is a factorization of $p(x)$ in $F[x]$, namely, $p(x) = \prod_{i=1}^{n}[x - f_0^{-1}(b_i)]$. But $a \notin E_0$ implies $a \neq f_0^{-1}(b_i)$ for any $i$. Thus, $x - a$ is another factor of $p(x)$ in $F[x]$, contrary to unique factorization. We conclude that there is some $b_i \notin \operatorname{im} f_0$. By Theorem 2.144(i), we may define $f_1 \colon E_0(a) \to \overline{k}$ by

$$c_0 + c_1 a + c_2 a^2 + \cdots \mapsto f_0(c_0) + f_0(c_1)b_i + f_0(c_2)b_i^2 + \cdots .$$

A straightforward check shows that $f_1$ is a (well-defined) $k$-map extending $f_0$. Hence, $(E_0, f_0) \prec (E_0(a), f_1)$, contradicting the maximality of $(E_0, f_0)$. This completes the proof.  •

**Theorem 5.61.** *Any two algebraic closures of a field $k$ are isomorphic via a $k$-map.*

**Proof.** Let $K$ and $L$ be two algebraic closures of a field $k$. By Lemma 5.60, there are $k$-maps $\psi \colon K \to L$ and $\theta \colon L \to K$. By Lemma 5.59, both composites $\theta\psi \colon K \to K$ and $\psi\theta \colon L \to L$ are automorphisms. It follows that $\psi$ (and $\theta$) is a $k$-isomorphism.  •

It is now permissible to speak of *the* algebraic closure of a field.

**5.4.4. Lüroth's Theorem.** We now investigate the structure of arbitrary fields, beginning with *simple transcendental extensions* $k(x)$, where $k$ is a field and $x$ is transcendental over $k$; that is, we examine the function field $k(x)$. Zorn's Lemma is not needed in the present subsection.

**Definition.** If $\varphi \in k(x)$, then there are polynomials $g(x), h(x) \in k[x]$ with $(g, h) = 1$ and $\varphi = g(x)/h(x)$. Define the *degree* of $\varphi$ by

$$\text{degree}(\varphi) = \max\{\deg(g), \deg(h)\}.$$

A rational function $\varphi \in k(x)$ is called a *linear fractional transformation* if

$$\varphi = \frac{ax + b}{cx + d},$$

where $a, b, c, d \in k$ and $ad - bc \neq 0$. Let

$$\text{LF}(k)$$

denote the group of all linear fractional transformations in $k(x)$ with binary operation composition: if $\varphi \colon x \mapsto (ax + b)/(cx + d)$ and $\psi \colon x \mapsto (rx + s)/(tx + u)$, then

$$\psi\varphi \colon x \mapsto \frac{r\varphi(x) + s}{t\varphi(x) + u} = \frac{(ra + sc)x + (rb + sd)}{(ta + ud)x + (tb + ud)}.$$

Now $\varphi \in k(x)$ has degree 0 if and only if $\varphi$ is a constant (that is, $\varphi \in k$), while Exercise 5.44 on page 341 says that $\varphi \in k(x)$ has degree 1 if and only if $\varphi$ is a linear fractional transformation.

**Proposition 5.62.** *If $\varphi \in k(x)$ is nonconstant, then $\varphi$ is transcendental over $k$ and $k(x)$ is a finite extension of $k(\varphi)$ with*

$$[k(x) : k(\varphi)] = \text{degree}(\varphi).$$

*Moreover, if $\varphi = g(x)/h(x)$ and $(g, h) = 1$, then*

$$\text{irr}(x, k(\varphi)) = g(y) - \varphi h(y),$$

*where $\varphi h(y)$ denotes the product of $\varphi$ and $h(y)$ in $k(\varphi)[y]$.*

**Proof.** Let $g(x) = \sum a_i x^i$ and $h(x) = \sum b_i x^i \in k[x]$. Define

$$\theta(y) = g(y) - \varphi h(y).$$

Now $\theta(y)$ is a polynomial in $k(\varphi)[y]$: $\theta(y) = \sum a_i y^i - \varphi \sum b_i y^i = \sum (a_i - \varphi b_i) y^i$. If $\theta(y)$ were the zero polynomial, then all its coefficients would be 0. But if $b_i$ is a nonzero coefficient of $h(y)$, then $a_i - \varphi b_i = 0$ gives $\varphi = a_i/b_i$, contradicting $\varphi$ not being a constant; that is, $\varphi \notin k$. We compute the degree of $\theta(y)$:

$$\deg(\theta) = \deg(g(y) - \varphi h(y)) = \max\{\deg(g), \deg(h)\} = \text{degree}(\varphi).$$

Now $x$ is a root of $\theta(y)$, because $\varphi = g/h$, so that $x$ is algebraic over $k(\varphi)$. Were $\varphi$ algebraic over $k$, then $k(\varphi)/k$ would be finite, giving $[k(x) : k] = [k(x) : k(\varphi)][k(\varphi) : k]$ finite, a contradiction. Therefore, $\varphi$ is transcendental over $k$.

We claim that $\theta(y)$ is an irreducible polynomial in $k(\varphi)[y]$. If not, then $\theta(y)$ factors in $k[\varphi][y]$, by Gauss's Corollary 5.28. But $\theta(y) = g(y) - \varphi h(y)$ is linear in $\varphi$, and so Corollary 5.31 shows that $\theta(y)$ is irreducible. Finally, since $\deg(\theta) =$ degree$(\varphi)$, we have $[k(x) : k(\varphi)] = \text{degree}(\varphi)$.   •

**Corollary 5.63.** *Let $\varphi \in k(x)$, where $k(x)$ is the field of rational functions over a field $k$. Then $k(\varphi) = k(x)$ if and only if $\varphi$ is a linear fractional transformation.*

**Proof.** By Proposition 5.62, $k(\varphi) = k(x)$ if and only if degree$(\varphi) = 1$; that is, $\varphi$ is a linear fractional transformation.   •

Define a map $\zeta \colon \text{GL}(2, k) \to \text{LF}(k)$ by $\begin{bmatrix} a & b \\ c & d \end{bmatrix} \mapsto (ax + b)/(cx + d)$. It is easily checked that $\zeta$ is a homomorphism of groups. In Exercise 5.45 on page 341, the reader will prove that $\ker \zeta = Z(2, k)$, the center of $\text{GL}(2, k)$ consisting of all nonzero $2 \times 2$ scalar matrices. Hence, if

$$\text{PGL}(2, k) = \text{GL}(2, k)/Z(2, k),$$

then $\text{LF}(k) \cong \text{PGL}(2, k)$.

**Corollary 5.64.** *If $k(x)$ is the field of rational functions over a field $k$, then*

$$\text{Gal}(k(x)/k) \cong \text{LF}(k) \cong \text{PGL}(2, k).$$

**Proof.** Let $\sigma \colon k(x) \to k(x)$ be an automorphism of $k(x)$ fixing $k$. Since $k(\sigma(x)) = k(x)$, Corollary 5.63 says that $\sigma(x)$ is a linear fractional transformation. Define $\gamma \colon \text{Gal}(k(x)/k) \to \text{LF}(k)$ by $\gamma \colon \sigma \mapsto \sigma(x)$. Now $\gamma$ is a homomorphism: $\gamma(\sigma\tau) = \gamma(\sigma)\gamma(\tau)$, because $(\sigma\tau)(x) = \sigma(x)\tau(x)$ (remember that the binary operation in $\text{LF}(k)$ is composition). Finally, $\gamma$ is an isomorphism: $\gamma^{-1}$ is the function assigning, to any linear fractional transformation $\varphi = (ax + b)/(cx + d)$, the automorphism of $k(x)$ that sends $x$ to $\varphi$.   •

**Theorem 5.65 (Lüroth's Theorem).** *If $k(x)$ is a simple transcendental extension, then every intermediate field $B$ with $k \subsetneq B \subseteq k(x)$ is also a simple transcendental extension of $k$: there is $\varphi \in B$ with $B = k(\varphi)$.*

**Proof.** We claim that $x$ is algebraic over $B$. If $\psi \in B$ is not constant, then Proposition 5.62 gives $[k(x) : k(\psi)] = \text{degree}(\psi) < \infty$. Since $[k(x) : B][B : k(\psi)] = [k(x) : k(\psi)]$, we have $[k(x) : B] < \infty$, and $x$ is algebraic over $B$. Let

$$\text{irr}(x, B) = y^n + \varphi_{n-1}y^{n-1} + \cdots + \varphi_0 \in B[y].$$

The proof of Proposition 5.62 shows that if $B = k(\varphi)$, where $\varphi \in k(x)$, then $\varphi$ is a coefficient of $\text{irr}(x, k(\varphi))$. The proof of Lüroth's Theorem is a converse, showing that there is some coefficient $\varphi = \varphi_\ell$ of $\text{irr}(x, B)$ with degree$(\varphi) = 1$; it will then follow that $B = k(\varphi)$. We analyze $\text{irr}(x, B)$.

Each coefficient $\varphi_\ell \in B \subseteq k(x)$ is a rational function, say, $\varphi_\ell = g_\ell(x)/h_\ell(x)$, where $g_\ell(x), h_\ell(x) \in k[x]$; we may assume, for all $\ell$, that $g_\ell/h_\ell$ is in lowest terms; that is, $(g_\ell, h_\ell) = 1$. Let

$$f(x) = \text{lcm}\{h_0, \ldots, h_{n-1}\} \in k[x].$$

Thus, there are $u_\ell(x) \in k[x]$, for all $\ell$, with $f = u_\ell h_\ell$. Now $\gcd\{u_0, \ldots, u_{n-1}\} = 1$: if $d = \gcd$ is not constant, then $d \mid f$, and $f/d$ would be a common multiple of $h_0, \ldots, h_{n-1}$ with $\deg(f/d) < \deg(f)$. Consider the polynomial

$$i(x, y) = f(x)\operatorname{irr}(x, B) = f(x)y^n + \varphi_{n-1}y^{n-1} + \cdots + \varphi_0.$$

Note that the coefficients lie in $k[x]$; in particular, $f\varphi_\ell = fg_\ell/h_\ell = (u_\ell h_\ell)g_\ell/h_\ell = u_\ell g_\ell$. Thus,

$$i(x, y) = f(x)y^n + u_{n-1}g_{n-1}y^{n-1} + \cdots + u_0 g_0.$$

We show that $i(x, y)$ is a primitive polynomial. Since $f(x) = f(x)^{-1}\big(f(x)\operatorname{irr}(x, B)\big)$, it will then follow from Lemma 5.24(i) that $f(x)^{-1}$ is the content of $\operatorname{irr}(x, B)$ and $\operatorname{irr}(x, B)^* = i(x, y) = f(x)\operatorname{irr}(x, B)$. Suppose there is an irreducible $p(x) \in k[x]$ dividing $f(x)$ and each $u_\ell g_\ell$. Recall that $f = u_\ell h_\ell$ for every $\ell$. If $p \nmid u_j$ for some $j$, then $p \mid h_j$, by Euclid's Lemma in $k[x]$. It follows that $p \nmid g_j$, because $(g_j, h_j) = 1$. Since, $p \mid u_j g_j$, Euclid's Lemma gives $p \mid u_j$, a contradiction. We conclude that $p \mid u_\ell$ for all $\ell$, which contradicts $\gcd\{u_0, \ldots, u_{n-1}\} = 1$. Therefore, $i(x, y)$ is primitive.

Denote the highest exponent of $y$ or $x$ occurring in a polynomial $a(x, y)$ by

$$\deg_y(a) \quad \text{or} \quad \deg_x(a).$$

Since $i(x, y) = f(x)y^n + \sum_{\ell=0}^{n-1} f(x)\varphi_\ell y^\ell$, we have $n = \deg_y(i)$ and $m = \deg_x(i) = \max_\ell\{\deg(f), \deg(f\varphi_\ell)\}$. Now $h_\ell(x) \mid f(x)$ for all $\ell$, so that $\deg(h_\ell) \le \deg(f) \le m$ [for $f(x)$ is a coefficient of $i(x, y)$]. But $f\varphi_\ell = u_\ell g_\ell \in k[x]$, so that $\deg(g_\ell) \le \deg(u_\ell g_\ell) = \deg(f\varphi_\ell) \le m$. We conclude that $\deg(g_\ell) \le m$ and $\deg(h_\ell) \le m$.

Some coefficient $\varphi_j = g_j/h_j$ of $\operatorname{irr}(x, B)$ is not constant, lest $x$ be algebraic over $k$. Omit the subscript $j$, and write $\varphi = \varphi_j$, $g(x) = g_j(x)$, and $h(x) = h_j(x)$:

$$\varphi = g(x)/h(x) \in B.$$

If $g(x) = \sum_i b_i x^i$, write $g(y) = \sum_i b_i y^i$. Now $x$ is a root of $g(y) - \varphi h(y) = g(y) - g(x)h(x)^{-1}h(y) \in B[y]$; hence, $\operatorname{irr}(x, B)$ divides $g(y) - \varphi h(y)$ in $B[y] \subseteq k(x)[y]$. Therefore, there is $q(x, y) \in k(x)[y]$ with

$$(1) \qquad\qquad \operatorname{irr}(x, B)q(x, y) = g(y) - \varphi h(y).$$

Since $g(y) - \varphi h(y) = h(x)^{-1}\big(h(x)g(y) - g(x)h(y)\big)$, the content $c\big(g(y) - \varphi h(y)\big)$ is $h(x)^{-1}$ and the associated primitive polynomial $\big(g(y) - \varphi h(y)\big)^*$ is

$$\Phi(x, y) = h(x)g(y) - g(x)h(y).$$

Notice that $\Phi(x, y) \in k[x][y]$ and that $\Phi(y, x) = -\Phi(x, y)$.

Rewrite Equation (1) by replacing the right hand side and each factor on the left hand side by factorizations into (content)×(primitive):

$$f(x)^{-1}i(x, y)c(q)q(x, y)^* h(x) = \Phi(x, y)$$

[remember that $\operatorname{irr}(x, B) = f(x)^{-1}i(x, y)$ and $g(y) - \varphi h(y) = h(x)^{-1}\Phi(x, y)$]. The product $i(x, y)q(x, y)^*$ is primitive, by Gauss's Lemma 5.23. But $\Phi(x, y) \in k[x][y]$, so that Lemma 5.24(iii) gives $f(x)^{-1}c(q)h(x) \in k[x]$. We now define $q^{**}(x, y) = f(x)^{-1}c(q)h(x)q(x, y)$, so that $q^{**}(x, y) \in k[x, y]$ and

$$(2) \qquad\qquad i(x, y)q^{**}(x, y) = \Phi(x, y) \quad \text{in } k[x, y].$$

Let us compute degrees in Equation (2): the degree in $x$ of the left hand side is

(3)        $\deg_x(iq^{**}) = \deg_x(i) + \deg_x(q^{**}) = m + \deg_x(q^{**})$,

while the degree in $x$ of the right hand side is

(4)        $\deg_x(\Phi) = \max\{\deg(g), \deg(h)\} \le m$,

as we saw above. We conclude that $m + \deg_x(q^{**}) \le m$ and $\deg_x(q^{**}) = 0$; that is, $q^{**}(x, y)$ is a function of $y$ alone. But $\Phi(x, y)$ is a primitive polynomial in $x$, so that the symmetry $\Phi(y, x) = -\Phi(x, y)$ shows that it is also a primitive polynomial in $y$. Hence, $q^{**}$ is a constant, and so $i(x, y)$ and $\Phi(x, y)$ are associates in $k[x, y]$. Thus, Equation (3) gives $\deg_x(\Phi) = \deg_x(i) = m$, and Equation (4) now gives

$$m = \deg_x(\Phi) = \max\{\deg(g), \deg(h)\}.$$

Symmetry of $\Phi$ also gives $\deg_y(\Phi) = \deg_x(\Phi)$, and so

$$n = \deg_y(\Phi) = \deg_x(\Phi) = m = \max\{\deg(g), \deg(h)\}.$$

By definition, $\mathrm{degree}(\varphi) = \max\{\deg(g), \deg(h)\} = m$; hence, Proposition 5.62 gives $[k(x) : k(\varphi)] = m$. Finally, since $\varphi \in B$, we have $[k(x) : k(\varphi)] = [k(x) : B][B : k(\varphi)]$. As $[k(x) : B] = n = m$, this forces $[B : k(\varphi)] = 1$; that is, $B = k(\varphi)$.  •

Let us quote van der Waerden, *Modern Algebra* I, p. 199.

The significance of Lüroth's theorem in geometry is as follows:
A plane (irreducible) algebraic curve $F(\xi, \eta) = 0$ is called *rational* if its points, except a finite number of them, can be represented in terms of rational parametric equations:

$$\xi = f(t),$$
$$\eta = g(t).$$

It may happen that every point of the curve (perhaps with a finite number of exceptions) belongs to several values of $t$. (Example: If we put

$$\xi = t^2,$$
$$\eta = t^2 + 1,$$

the same point belongs to $t$ and $-t$.) But by means of Lüroth's theorem this can always be avoided by a suitable choice of the parameter. For let $\Delta$ be a field containing the coefficients of the functions $f, g$, and let $t$, for the present, be an indeterminate. $\Sigma = \Delta(f, g)$ is a subfield of $\Delta(t)$. If $t'$ is a primitive element of $\Sigma$, we have, for example,

$$f(t) = f_1(t') \quad \text{(rational)}$$
$$g(t) = g_1(t') \quad \text{(rational)}$$
$$t' = \varphi(f, g) = \varphi(\xi, \eta),$$

and we can verify easily that the new parametrization

$$\xi = f_1(t'),$$
$$\eta = g_1(t')$$

represents the same curve, while the denominator of the function $\varphi(x, y)$ vanishes only at a finite number of points of the curve so that to all points of the curve (apart from a finite number of them) there belongs only *one* $t'$-value.

The generalization of Lüroth's Theorem to several variables is best posed geometrically: can the term *curve* in van der Waerden's account be replaced by *surface* or *higher-dimensional variety*? There is a theorem of Castelnuovo that gives a positive answer for certain surfaces over an algebraically closed field, but there are negative examples (over other fields) in all dimensions $\geq 2$.

**5.4.5. Transcendence.** We now consider more general field extensions.

**Definition.** Let $E/k$ be a field extension. A subset $U$ of $E$ is **algebraically dependent** over $k$ if there exists a finite subset $\{u_1, \ldots, u_n\} \subseteq U$ and a nonzero polynomial $f(x_1, \ldots, x_n) \in k[x_1, \ldots, x_n]$ with $f(u_1, \ldots, u_n) = 0$. A subset $B$ of $E$ is **algebraically independent** if it is not algebraically dependent. A field extension $E/k$ is **purely transcendental** if either $E = k$ or $E$ contains an algebraically independent subset $B$ and $E = k(B)$.

Let $E/k$ be a field extension, let $u_1, \ldots, u_n \in E$, and let $\varphi \colon k[x_1, \ldots, x_n] \to E$ be the evaluation map; that is, $\varphi$ is the homomorphism sending $f(x_1, \ldots, x_n)$ to $f(u_1, \ldots, u_n)$ for all $f(x_1, \ldots, x_n) \in k[x_1, \ldots, x_n]$. Now $\{u_1, \ldots, u_n\}$ is algebraically dependent if and only if $\ker \varphi \neq (0)$. If $\{u_1, \ldots, u_n\}$ is algebraically independent, then $\varphi$ extends to an isomorphism $\widetilde{\varphi} \colon k(x_1, \ldots, x_n) \to k(u_1, \ldots, u_n) \subseteq E$, where $k(x_1, \ldots, x_n)$ is the field of rational functions $\mathrm{Frac}(k[x_1, \ldots, x_n])$:

$$
\begin{array}{ccc}
k(x_1, \ldots, x_n) & \xrightarrow{\ \widetilde{\varphi}\ } & \mathrm{Frac}(E) = E \\[4pt]
\big\uparrow & & \big\uparrow \\[4pt]
k[x_1, \ldots, x_n] & \xrightarrow[\ \varphi\ ]{} & E
\end{array}
$$

In particular, if $\{u_1, \ldots, u_n\}$ is algebraically independent and $E = k(u_1, \ldots, u_n)$, then $\widetilde{\varphi}$ is an isomorphism $k(x_1, \ldots, x_n) \to k(u_1, \ldots, u_n)$ with $x_i \mapsto u_i$ for all $i$. Therefore, a purely transcendental extension $k(u_1, \ldots, u_n)/k$ is isomorphic to the **function field in $n$ variables**.

Since algebraically dependent subsets are necessarily nonempty, it follows that the empty subset $\varnothing$ is algebraically independent. A singleton $\{e\} \subseteq E$ is algebraically dependent if $e$ is algebraic over $k$; that is, $e$ is a root of a nonconstant polynomial over $k$, and it is algebraically independent if $e$ is transcendental over $k$, in which case $k(e) \cong k(x)$.

Proposition 2.97 says that if $V$ is a vector space and $X = v_1, \ldots, v_m$ is a list in $V$, then $X$ is linearly dependent if and only if some $v_i$ is in the subspace spanned by the others. Here is an analog of this for algebraic dependence.

**Proposition 5.66.** *Let $E/k$ be a field extension. Then $U \subseteq E$ is algebraically dependent over $k$ if and only if there is $v \in U$ with $v$ algebraic over $k(U - \{v\})$.*

**Proof.** If $U$ is algebraically dependent over $k$, then there is a finite algebraically dependent subset $\{u_1, \ldots, u_n\} \subseteq U$; thus, we may assume that $U$ is finite. We prove, by induction on $n \geq 1$, that some $u_i$ is algebraic over $k(U - \{u_i\})$. If $n = 1$, then there is some nonzero $f(x) \in k[x]$ with $f(u_1) = 0$; that is, $u_1$ is algebraic over $k$. But $U - \{u_1\} = \varnothing$, and so $u_1$ is algebraic over $k(U - \{u_1\}) = k(\varnothing) = k$. For the inductive step, let $U = \{u_1, \ldots, u_{n+1}\}$ be algebraically dependent. We may assume that $\{u_1, \ldots, u_n\}$ is algebraically independent; otherwise, the inductive hypothesis gives some $u_j$, for $1 \leq j \leq n$, which is algebraic over $k(u_1, \ldots, \widehat{u}_j, \ldots, u_n)$ and, hence, algebraic over $k(U - \{u_j\})$. Since $U$ is algebraically dependent, there is a nonzero $f(X, y) \in k[x_1, \ldots, x_n, y]$ with $f(u_1, \ldots, u_n, u_{n+1}) = 0$, where $X = (x_1, \ldots, x_n)$ and $y$ is a new variable. We may write $f(X, y) = \sum_i g_i(X)y^i$, where $g_i(X) \in k[X]$ (because $k[X, y] = k[X][y]$). Since $f(X, y) \neq 0$, some $g_i(X) \neq 0$, and it follows from the algebraic independence of $\{u_1, \ldots, u_n\}$ that $g_i(u_1, \ldots, u_n) \neq 0$. Therefore, $h(y) = \sum_i g_i(u_1, \ldots, u_n)y^i \in k(U)[y]$ is not the zero polynomial. But $0 = f(u_1, \ldots, u_n, u_{n+1}) = h(u_{n+1})$, so that $u_{n+1}$ is algebraic over $k(u_1, \ldots, u_n)$.

For the converse, assume that $v$ is algebraic over $k(U - \{v\})$. We may assume that $U - \{v\}$ is finite, say, $U - \{v\} = \{u_1, \ldots, u_n\}$, where $n \geq 0$ (if $n = 0$, we mean that $U - \{v\} = \varnothing$). We prove, by induction on $n \geq 0$, that $U$ is algebraically dependent. If $n = 0$, then $v$ is algebraic over $k$, and so $\{v\}$ is algebraically dependent. For the inductive step, let $U - \{u_{n+1}\} = \{u_1, \ldots, u_n\}$. We may assume that $U - \{u_{n+1}\} = \{u_1, \ldots, u_n\}$ is algebraically independent, for otherwise $U - \{u_{n+1}\}$, and hence its superset $U$, is algebraically dependent. By hypothesis, there is a nonzero polynomial $f(y) = \sum_i c_i y^i \in k(u_1, \ldots, u_n)[y]$ with $f(u_{n+1}) = 0$. As $f(y) \neq 0$, we may assume that at least one of its coefficients is nonzero. For all $i$, the coefficient $c_i \in k(u_1, \ldots, u_n)$, so there are rational functions $c_i(x_1, \ldots, x_n)$ with $c_i(u_1, \ldots, u_n) = c_i$ [because $k(u_1, \ldots, u_n) \cong k(x_1, \ldots, x_n)$, the function field in $n$ variables]. Since $f(u_{n+1}) = 0$, we may clear denominators and assume that each $c_i(x_1, \ldots, x_n)$ is a polynomial in $k[x_1, \ldots, x_n]$. Moreover, that some $c_i(u_1, \ldots, u_n) \neq 0$ implies $c_i(x_1, \ldots, x_n) \neq 0$. Hence,

$$c(x_1, \ldots, x_n, y) = \sum_i c_i(x_1, \ldots, x_n)y^i$$

is nonzero and, therefore, $\{u_1, \ldots, u_{n+1}\}$ is algebraically dependent.  •

There is a strong parallel between linear dependence in a vector space and algebraic dependence in a field. The analog of a basis in a vector space is a *transcendence basis* in a field; the analog of dimension is *transcendence degree*. In fact, both discussions are special cases of theorems about *dependency relations*. We present the general discussion here. The reader should be aware that these two special cases are the most interesting ones, but we will apply the general theorems to vector spaces over division rings in Exercise 7.2 on page 532.

Recall that a *relation* $R$ from a set $Y$ to a set $Z$ is a subset $R \subseteq Y \times Z$: we write $yRz$ instead of $(y, z) \in R$. In particular, if $\Omega$ is a set, $2^\Omega$ is the family of all its subsets, and $\preceq$ is a relation from $\Omega$ to $2^\Omega$, then $\preceq$ is a relation between elements $u$ of $\Omega$ and subsets $S \subseteq \Omega$. We write $u \preceq S$ instead of $(u, S) \in \preceq$, and we call $\preceq$ a *relation on $\Omega$*.

**Definition.** A **dependency relation on a set** $\Omega$ is a relation $\preceq$ from $\Omega$ to $2^\Omega$, pronounced "is dependent on," satisfying the following **Dependency Axioms**:

   (i) if $S \subseteq \Omega$ and $u \in S$, then $u \preceq S$;

   (ii) if $u \preceq S$, then there exists a finite subset $S' \subseteq S$ with $u \preceq S'$;

   (iii) (**Transitivity**) if $u \preceq S$ and, for some $T \subseteq \Omega$, we have $s \preceq T$ for every $s \in S$, then $u \preceq T$;

   (iv) (**Exchange Axiom**) if $u \preceq S$ and $u \npreceq S - \{v\}$,[10] then $v \preceq (S - \{v\}) \cup \{u\}$.

The Transitivity Axiom says that if $x$ is dependent on $S$ and each element of $S$ is dependent on $T$, then $x$ is dependent on $T$. The Exchange Axiom says that if $u$ is dependent on $S = \{v, s_1, \ldots, s_n\}$ but not on $\{s_1, \ldots, s_n\}$, then $v$ can be exchanged with $u$ when $S$ is exchanged with $\{u, s_1, \ldots, s_n\}$.

**Example 5.67.** If $\Omega$ is a vector space, define $u \preceq S$ to mean $u \in \langle S \rangle$, the subspace spanned by $S$. We claim that $\preceq$ is a dependency relation. The first three Dependency Axioms are easily checked. We verify the Exchange Axiom. Suppose that $u \preceq S$ and $u \npreceq S - \{v\}$; that is, $u \in \langle S \rangle$, but $u \notin \langle S - \{v\} \rangle$. Thus, there are scalars $a, a_i$ with $u = av + \sum_i a_i s_i$, where $s_i \in S - \{v\}$. Now $a \neq 0$, lest $u \in \langle S - \{v\} \rangle$. Hence, $v = a^{-1}(u - \sum_i a_i s_i) \in \langle S - \{v\}, u \rangle$; that is, $v \preceq (S - \{v\}) \cup \{u\}$. ◀

**Lemma 5.68.** *If $E/k$ is a field extension, then $u \preceq S$, defined by $u$ being algebraic over $k(S)$, is a dependency relation on $E$.*

**Proof.** It is easy to check the first two Dependency Axioms, and we now verify the Transitivity Axiom. If $u \preceq S$, then $u$ is algebraic over $k(S)$; that is, $u \in (E/k(S))_{\text{alg}} = \{e \in E : e \text{ is algebraic over } k(S)\}$. Suppose there is some $T \subseteq E$ with $s \preceq T$ for every $s \in S$; that is, $S \subseteq (E/k(T))_{\text{alg}}$. It follows from Lemma 5.55(iii) that $(E/k(S))_{\text{alg}} \subseteq (E/k(T))_{\text{alg}}$; that is, $u \in (E/k(T))_{\text{alg}}$, and so $u \preceq T$.

The Exchange Axiom assumes that $u \preceq S$ [that is, $u$ is algebraic over $k(S)$] and $u$ is transcendental over $k(S - \{v\})$ [that is, $u \npreceq S - \{v\}$]. Note that $v \in S$ [lest $S - \{v\} = S$] and $u \notin S$ [lest $u$ be algebraic over $k(S - \{v\})$]. Let us apply Proposition 5.66 to the subsets $U' = \{u, v\}$ and $S' = S - \{v\}$ of $E$ and the subfield $k' = k(S')$. With this notation, $k'(U' - \{u\}) = k'(v) = k(S', v) = k(S)$, so that $u$ algebraic over $k(S)$ can be restated as $u$ algebraic over $k'(U' - \{u\})$. Thus, Proposition 5.66 says that $U' = \{u, v\}$ is algebraically dependent over $k' = k(S')$: there is a nonzero polynomial $f(x, y) \in k(S')[x, y]$ with $f(u, v) = 0$. In more

---

[10]Of course, $v \in S$; otherwise, $S - \{v\} = S$. Note also that $u \neq v$.

detail, $f(x, y) = g_0(x) + g_1(x)y + \cdots + g_n(x)y^n$, where $g_i(x) \in k(S')[x]$; that is, the coefficients of all $g_i(x)$ do not involve $u, v$. Define $h(y) = f(u, y) = \sum_i g_i(u)y^i \in k(S', u)[y]$. Now $h(y)$ is not the zero polynomial: some $g_i(u) \neq 0$ because $u$ is transcendental over $k(S - \{v\}) = k(S')$. But $h(v) = f(u, v) = 0$. Therefore, $v$ is algebraic over $k(S - \{v\}, u)$; that is, $v \preceq S - \{v\} \cup \{u\}$.   •

We can rephrase the notion of linear dependence in a vector space using the dependency relation in Example 5.67. A subset $S$ in a vector space is linearly dependent if there is $s \in S$ with $s \in \langle S - \{s\} \rangle$; that is, $s \preceq S - \{s\}$.

**Definition.** Let $\preceq$ be a dependency relation on a set $\Omega$. Call a subset $S \subseteq \Omega$ **dependent** if there exists $s \in S$ with $s \preceq S - \{s\}$; call $S$ **independent** if it is not dependent.

Note that $\varnothing$ is independent, for dependent subsets have elements. If $S \neq \varnothing$, then $S$ is independent if and only if $s \npreceq S - \{s\}$ for all $s \in S$. It follows that every subset of an independent set is itself independent.

In light of Lemma 5.68, an algebraically dependent subset of a field (as defined on page 335 and characterized by Proposition 5.66) is dependent in the sense just defined: it arises from the dependency relation $u \preceq S$ defined by $u$ being algebraic over $k(S)$.

**Definition.** Let $\preceq$ be a dependency relation on a set $\Omega$. We say that a subset $S$ **generates** $\Omega$ if $x \preceq S$ for all $x \in \Omega$. A **basis** of $\Omega$ is an independent subset that generates $\Omega$.

**Lemma 5.69.**

(i) *Let $\preceq$ be a dependency relation on a set $\Omega$. If $T \subseteq \Omega$ is independent and $z \npreceq T$ for some $z \in \Omega$, then $T \cup \{z\} \supsetneq T$ is a strictly larger independent subset.*

(ii) *Let $E/k$ be a field extension. If $T \subseteq E$ is algebraically independent over $k$ and $z \in E$ is transcendental over $k(T)$, then $T \cup \{z\}$ is algebraically independent.*

**Proof.**

(i) Since $z \npreceq T$, Dependency Axiom (i) gives $z \notin T$, and so $T \subsetneq T \cup \{z\}$; it follows that $(T \cup \{z\}) - \{z\} = T$. If $T \cup \{z\}$ is dependent, then there exists $t \in T \cup \{z\}$ with $t \preceq (T \cup \{z\}) - \{t\}$. If $t = z$, then $z \preceq T \cup \{z\} - \{z\} = T$, contradicting $z \npreceq T$. Therefore, $t \in T$. Since $T$ is independent, $t \npreceq T - \{t\}$. If we set $S = T \cup \{z\} - \{t\}$, $t = x$, and $y = z$ in the Exchange Axiom, we conclude that $z \preceq (T \cup \{z\} - \{t\}) - \{z\} \cup \{t\} = T$, contradicting the hypothesis $z \npreceq T$. Therefore, $T \cup \{z\}$ is independent.

(ii) By Proposition 5.66, this is a special case of (i).   •

**Definition.** If $E/k$ is a field extension, then a **transcendence basis** is a maximal algebraically independent subset of $E$ over $k$.

**Theorem 5.70.**

(i) If $\preceq$ is a dependency relation on a set $\Omega$, then $\Omega$ has a basis. In fact, every independent subset $B$ of $\Omega$ is part of a basis.

(ii) If $E/k$ is a field extension, then $E$ has a transcendence basis. In fact, every algebraically independent subset is part of a transcendence basis.

**Proof.**

(i) Since the empty set $\varnothing$ is independent, the second statement implies the first.

We use Zorn's Lemma to prove the existence of maximal independent subsets of $\Omega$ containing $B$. Let $X$ be the family of all independent subsets of $\Omega$ containing $B$, partially ordered by inclusion. Note that $X$ is nonempty, for $B \in X$. Suppose that $\mathcal{B} = (B_j)_{j \in J}$ is a chain in $X$. It is clear that $B^* = \bigcup_{j \in J} B_j$ is an upper bound of $\mathcal{B}$ if it lies in $X$, that is, if $B^*$ is independent. If, on the contrary, $B^*$ is dependent, then there is $y \in B^*$ with $y \preceq B^* - \{y\}$. By Dependency Axiom (ii), there is a finite subset $\{x_1, \ldots, x_n\} \subseteq B^* - \{y\}$ with $y \preceq \{x_1, \ldots, x_n\} - \{y\}$. Now there is $B_{j_0} \in \mathcal{B}$ with $y \in B_{j_0}$, and, for each $i$ with $1 \le i \le n$, there is $B_{j_i} \in \mathcal{B}$ with $x_i \in B_{j_i}$. Since $\mathcal{B}$ is a chain, one of these, call it $B'$, contains all the others, and the dependent set $\{y, x_1, \ldots, x_n\}$ is contained in $B'$. But since $B'$ is independent, so are its subsets, and this is a contradiction. Zorn's Lemma now provides a maximal element $M$ of $X$; that is, $M$ is a maximal independent subset of $\Omega$ containing $B$. If $M$ is not a basis, then there exists $x \in \Omega$ with $x \not\preceq M$. By Lemma 5.69, $M \cup \{x\}$ is an independent set strictly larger than $M$, contradicting the maximality of $M$.

(ii) This is a special case of (i).   •

**Theorem 5.71.** If $B$ is a transcendence basis, then $k(B)/k$ is purely transcendental and $E/k(B)$ is algebraic.

**Proof.** By Theorem 5.70, it suffices to show that if $B$ is a transcendence basis, then $E/k(B)$ is algebraic. If not, then there exists $u \in E$ with $u$ transcendental over $k(B)$. By Lemma 5.69, $B \cup \{u\}$ is algebraically independent, and this contradicts the maximality of $B$.   •

We now generalize the proof of Lemma 2.108, the Exchange Lemma, and its application to Invariance of Dimension, Theorem 2.109.

**Theorem 5.72.**

(i) If $\Omega$ is a set with a dependency relation $\preceq$, then any two bases $B$ and $C$ have the same cardinality.

(ii) If $B$ and $C$ are transcendence bases of a field extension $E/k$, then $|B| = |C|$.

**Proof.**

(i) If $B = \varnothing$, we claim that $C = \varnothing$. Otherwise, there exists $y \in C$ and, since $C$ is independent, $y \not\preceq C - \{y\}$. But $y \preceq B = \varnothing$ and $\varnothing \subseteq C - \{y\}$, so that Dependency Axiom (iii) gives $y \preceq C - \{y\}$, a contradiction. Therefore, we may assume that both $B$ and $C$ are nonempty.

Now assume that $B$ is finite; say, $B = \{x_1, \ldots, x_n\}$. We prove, by induction on $k \geq 0$, that there exists $\{y_1, \ldots, y_{k-1}\} \subseteq C$ with

$$B_k = \{y_1, \ldots, y_{k-1}, x_k, \ldots, x_n\}$$

a basis; that is, the elements $x_1 \ldots, x_{k-1}$ in $B$ can be exchanged with elements $y_1, \ldots, y_{k-1} \in C$ so that $B_k$ is a basis. We define $B_0 = B$, and we interpret the base step to mean that if none of the elements of $B$ are exchanged, then $B = B_0$ is a basis; this is obviously true. For the inductive step, assume that $B_k = \{y_1, \ldots, y_{k-1}, x_k, \ldots, x_n\}$ is a basis. We claim that there is $y \in C$ with $y \npreceq B_k - \{x_k\}$. Otherwise, $y \preceq B_k - \{x_k\}$ for all $y \in C$. But $x_k \preceq C$, because $C$ is a basis, and so Dependency Axiom (iii) gives $x_k \preceq B_k - \{x_k\}$, contradicting the independence of $B_k$. Hence, we may choose $y_k \in C$ with $y_k \npreceq B_k - \{x_k\}$. By Lemma 5.69, the set $B_{k+1}$, defined by

$$B_{k+1} = (B_k - \{x_k\}) \cup \{y_k\} = \{y_1, \ldots, y_k, x_{k+1}, \ldots, x_n\},$$

is independent. To see that $B_{k+1}$ is a basis, it suffices to show that it generates $\Omega$. Now $y_k \preceq B_k$ (because $B_k$ is a basis), and $y_k \npreceq B_k - \{x_k\}$; the Exchange Axiom gives $x_k \preceq (B_k - \{x_k\}) \cup \{y_k\} = B_{k+1}$. By Dependency Axiom (i), all the other elements of $B_k$ are dependent on $B_{k+1}$. Now each element of $\Omega$ is dependent on $B_k$, and each element of $B_k$ is dependent on $B_{k+1}$. By Dependency Axiom (iii), $B_{k+1}$ generates $\Omega$.

If $|C| > n = |B|$, that is, if there are more $y$'s than $x$'s, then $B_n \subsetneq C$. Thus a proper subset of $C$ generates $\Omega$, contradicting the independence of $C$. Therefore, $|C| \leq |B|$. It follows that $C$ is finite, and so the preceding argument can be repeated, interchanging the roles of $B$ and $C$. Hence, $|B| \leq |C|$, and we conclude that $|B| = |C|$ if $\Omega$ has a finite basis.

When $B$ is infinite, the reader may complete the proof by adapting the proof of Theorem 5.52. In particular, replace $\mathrm{supp}(u)$ in that proof by the smallest finite subset satisfying Dependency Axiom (ii).

(ii) This is a special case of (i).   •

Theorem 5.72 shows that the following analog of dimension is well-defined.

**Definition.** The ***transcendence degree*** of $E/k$ is defined by

$$\mathrm{trdeg}(E/k) = |B|,$$

where $B$ is a transcendence basis of $E/k$.

**Example 5.73.**

(i) If $E/k$ is a field extension, then $\mathrm{trdeg}(E/k) = 0$ if and only if $E/k$ is algebraic.

(ii) If $E = k(x_1, \ldots, x_n)$ is the function field in $n$ variables over a field $k$, then $\mathrm{trdeg}(E/k) = n$, because $\{x_1, \ldots, x_n\}$ is a transcendence basis of $E$.   ◄

Here is a small application of transcendence degree.

**Proposition 5.74.** *There are nonisomorphic fields each of which is isomorphic to a subfield of the other.*

**Proof.** Clearly, $\mathbb{C}$ is isomorphic to a subfield of $\mathbb{C}(x)$. However, we claim that $\mathbb{C}(x)$ is isomorphic to a subfield of $\mathbb{C}$. Let $B$ be a transcendence basis of $\mathbb{C}$ over $\mathbb{Q}$, and discard one of its elements, say, $b$. The algebraic closure $F$ of $\mathbb{Q}(B - \{b\})$ is a proper subfield of $\mathbb{C}$, for $b \notin F$; in fact, $b$ is transcendental over $F$, by Proposition 5.66. Hence, $F \cong \mathbb{C}$, by Exercise 5.51 on page 341, and so $F(b) \cong \mathbb{C}(x)$. Therefore, each of $\mathbb{C}$ and $\mathbb{C}(x)$ is isomorphic to a subfield of the other. On the other hand $\mathbb{C}(x) \ncong \mathbb{C}$, because $\mathbb{C}(x)$ is not algebraically closed.   •

## Exercises

∗ **5.44.** Prove that $\varphi \in k(x)$ has degree 1 if and only if $\varphi$ is a linear fractional transformation.

∗ **5.45.** For any field $k$, define a map $\zeta \colon \mathrm{GL}(2, k) \to \mathrm{LF}(k)$ by

$$\zeta \colon \left[\begin{smallmatrix} a & b \\ c & d \end{smallmatrix}\right] \mapsto (ax + b)/(cx + d).$$

(i) Prove that $\zeta$ is a surjective group homomorphism.

(ii) Prove that $\ker \zeta = Z(2, k)$, the subgroup of $\mathrm{GL}(2, k)$ consisting of all nonzero scalar matrices $[Z(2, k)$ is the center of $\mathrm{GL}(2, k)]$.

**5.46.** Prove that the set $\mathbb{A}$ of all algebraic numbers is an algebraic closure of $\mathbb{Q}$.

**5.47.** Consider the tower $\mathbb{Q} \subseteq \mathbb{Q}(x) \subseteq \mathbb{Q}(x, x + \sqrt{2}) = E$. Prove that $\{x, x + \sqrt{2}\}$ is algebraically independent over $\mathbb{Q}$ and $\mathrm{trdeg}(E/\mathbb{Q}) = 2$.

**5.48.** Prove that there is no intermediate field $K$ with $\mathbb{Q} \subseteq K \subsetneq \mathbb{C}$ with $\mathbb{C}/K$ purely transcendental. Conclude that a field extension $E/k$ may not have an intermediate field $K$ with $K/k$ algebraic and $E/K$ purely transcendental.

**5.49.** If $E = k(X)$ is an extension of a field $k$ and every pair $u, v \in X$ is algebraically dependent, prove that $\mathrm{trdeg}(E/k) \leq 1$. Conclude that if

$$k \subseteq k_1 \subseteq k_2 \subseteq \cdots$$

is a tower of fields with $\mathrm{trdeg}(k_n/k) = 1$ for all $n \geq 1$, then $\mathrm{trdeg}(k^*/k) = 1$, where $k^* = \bigcup_{n \geq 1} k_n$.

**5.50.** Prove that if $k$ is the prime field of a field $E$ and $\mathrm{trdeg}(E/k) \leq \aleph_0$, then $E$ is countable.

∗ **5.51.** (i) Prove that two algebraically closed fields of the same characteristic are isomorphic if and only if they have the same transcendence degree over their prime fields.

**Hint.** Use Lemma 5.60.

(ii) Prove that $\mathrm{trdeg}(\mathbb{C}/\mathbb{Q}) = c$, where $c = |\mathbb{R}|$.

(iii) Prove that a field $F$ is isomorphic to $\mathbb{C}$ if and only if $F$ has characteristic 0, it is algebraically closed, and $\mathrm{trdeg}(F/\mathbb{Q}) = c$.

∗ **5.52.** (i) If $k \subseteq F \subseteq E$ is a tower of fields, prove that

$$\mathrm{trdeg}(E/k) = \mathrm{trdeg}(E/F) + \mathrm{trdeg}(F/k).$$

**Hint.** Prove that if $X$ is a transcendence basis of $F/k$ and $Y$ is a transcendence basis of $E/F$, then $X \cup Y$ is a transcendence basis for $E/k$.

(ii) Let $E/k$ be a field extension, and let $K$ and $L$ be intermediate fields. Prove that

$$\operatorname{trdeg}(K \vee L) + \operatorname{trdeg}(K \cap L) = \operatorname{trdeg}(K) + \operatorname{trdeg}((L),$$

where $K \vee L$ is the compositum.

**Hint.** Extend a transcendence basis of $K \cap L$ to a transcendence basis of $K$ and to a transcendence basis of $L$.

**5.4.6. Separability.** We continue our investigation into the structure of fields by considering separability in more detail. Recall that an element $\alpha \in E$ is *separable over $k$*, where $E/k$ is a field extension, if either $\alpha$ is transcendental over $k$ or $\operatorname{irr}(\alpha, k)$ is a separable polynomial;[11] that is, $\operatorname{irr}(\alpha, k)$ has no repeated roots. An extension $E/k$ is *separable* if every $\alpha \in E$ is separable over $k$; otherwise, it is *inseparable*. We will see that if $k$ has characteristic 0, then every extension $E/k$ is separable; thus, this subsection is most interesting for fields of characteristic $p$.

**Proposition 5.75.** *Let $f(x) \in k[x]$ be a nonconstant polynomial, where $k$ is a field, let $E/k$ be a splitting field of $f(x)$, and let $f'(x)$ be the derivative of $f(x)$.*

(i) *$f(x)$ has repeated roots if and only if $(f, f') \neq 1$.*

(ii) *If $k$ has characteristic $p > 0$, then $f'(x) = 0$ if and only if $f(x) \in k[x^p]$.*

(iii) *If $f'(x) = 0$, then (ii) implies (i). Conversely, if $k$ has characteristic $p > 0$ and $f(x) \in k[x]$ is irreducible, then (i) implies (ii).*

   *In particular, if $k$ has characteristic $p$ and $f(x) \in k[x]$ is irreducible, then all four conditions in (i) and (ii) are equivalent.*

**Proof.**

(i) If $f(x)$ has repeated roots, then there is $\alpha \in E$ with $f(x) = (x - \alpha)^2 g(x)$ in $E[x]$. Hence, $f'(x) = 2(x - \alpha)g(x) + (x - \alpha)^2 g'(x)$; thus, $x - \alpha$ is a common divisor of $f(x)$ and $f'(x)$, and so $(f, f') \neq 1$.

   Conversely, it suffices to work in $E/k$, for Corollary 2.58 says that the gcd $(f, f')$ does not depend on whether it is computed in $k[x]$ or in $E[x]$. If $x - \alpha$ is a divisor of $(f, f')$, then $f(x) = (x - \alpha)u(x)$ and $f'(x) = (x - \alpha)v(x)$. The product rule gives $f' = u + (x - \alpha)u'$, so that $u = (x - \alpha)(v - u')$. Therefore,

$$f(x) = (x - \alpha)u(x) = (x - \alpha)^2(v(x) - u'(x)),$$

and so $f(x)$ has a repeated root.

(ii) If $f(x) = \sum_i a_i x^i$, then $f'(x) = \sum_i i a_i x^{i-1}$. Suppose that $f'(x) = 0$. If $f(x) = 0$, then $f(x) \in k[x^p]$; if $f(x) \neq 0$, then some $a_i \neq 0$. Since $i a_i = 0$, Proposition 2.138 says that $p \mid i$. Thus, the only nonzero terms $a_i x^i$ in $f(x)$ have $i$ a multiple of $p$; that is, $f(x) \in k[x^p]$.

   Conversely, if $f(x) \in k[x^p]$, then $f(x) = \sum_j a_{pj} x^{pj}$. Therefore, $f'(x) = \sum_j pj a_{pj} x^{pj-1} = 0$.

---

[11] Recall that a polynomial is *separable* if each of its irreducible factors has no repeated roots.

(iii) If $f'(x) = 0$, then $(f, f') = (f, 0) = f$; since $f(x)$ is not constant, part (i) says that $(f, f') \neq 1$. Thus, (ii) implies (i).

Suppose that $f(x)$ is irreducible. Assume $(f, f') \neq 1$, as in (i). If $f'(x) \neq 0$, then $\deg(f') < \deg(f)$, and so $(f, f')$ is a monic divisor of $f$ having degree smaller than $\deg(f)$. Since $f$ is irreducible, we have $(f, f') = 1$, a contradiction. Therefore, $f'(x) = 0$, which is the condition in (ii).   •

**Corollary 5.76.** *If $k$ is a field of characteristic $p > 0$ and $f(x) \in k[x]$, then there exists $e \geq 0$ and a polynomial $g(x) \in k[x]$ with $g(x) \notin k[x^p]$,*

$$f(x) = g(x^{p^e}),$$

*and $\deg(f) = p^e \deg(g)$. Moreover, if $f(x)$ is irreducible, then $g(x)$ is irreducible and separable.*

**Proof.** If $f(x) \notin k[x^p]$, define $g(x) = f(x)$; if $f(x) \in k[x^p]$, there is $f_1(x) \in [x]$ with $f(x) = f_1(x^p)$. Note that $\deg(f) = p \deg(f_1)$. If $f_1(x) \notin k[x^p]$, define $g(x) = f_1(x)$; otherwise, there is $f_2(x) \in k[x]$ with $f_1(x) = f_2(x^p)$; that is,

$$f(x) = f_1(x^p) = f_2(x^{p^2}).$$

Since $\deg(f) > \deg(f_1) > \cdots$, iteration of this procedure must end after a finite number $e$ of steps. Thus, $f(x) = g(x^{p^e})$, where $g(x)$, defined by $g(x) = f_e(x)$, does not lie in $k[x^p]$. If, now, $f(x)$ is irreducible, then $f_1(x)$ is irreducible, for a factorization of $f_1(x)$ would give a factorization of $f(x)$. It follows that $f_1(x), f_2(x), \ldots$ are all irreducible; in particular, $f_e(x)$ is irreducible. Finally, $g(x)$ is separable, by Proposition 5.75(iii), for it has no repeated roots.   •

**Definition.** Let $k$ be a field of characteristic $p > 0$, and let $f(x) \in k[x]$. If $f(x) = g(x^{p^e})$, where $g(x) \in k[x]$ but $g(x) \notin k[x^p]$, then

$$\deg(f) = p^e \deg(g).$$

We call $p^e$ the **degree of inseparability** of $f(x)$, and we call $\deg(g)$ the **reduced degree** of $f(x)$.

**Example 5.77.** Let $f(x) = x^{p^3} + x^p + t \in \mathbb{F}_p(t)[x]$. If $g(x) = x^{p^2} + x + t$, then $f(x) = g(x^p)$. Now $g(x)$ is separable, for $g'(x) = 1 \neq 0$. Therefore, $f(x)$ has degree of inseparability $p$ and reduced degree $p^2$.   ◄

If $k$ is a field of prime characteristic $p > 0$, then the Frobenius map $\mathrm{Fr} \colon k \to k$, defined by $\mathrm{Fr} \colon \alpha \mapsto \alpha^p$, is a homomorphism [because $(\alpha + \beta)^p = \alpha^p + \beta^p$]. As any homomorphism of fields, $\mathrm{Fr}$ is an injection. Denote $\mathrm{im}\,\mathrm{Fr}$ by $k^p$, so that $k^p$ is the subfield of $k$ consisting of all the $p$th powers of elements in $k$:

$$k^p = \mathrm{im}\,\mathrm{Fr} = \{a^p : a \in k\}.$$

To say that $\mathrm{Fr}$ is surjective, that is, $k = k^p$, is to say that every element in $k$ has a $p$th root in $k$.

**Definition.** A field $k$ is called **perfect** if either $k$ has characteristic $0$ or if $k$ has characteristic $p > 0$ and $k = k^p$.

The Pigeonhole Principle, Exercise 1.8 on page 14, shows that every finite field $F$ is perfect, for Fr: $F \to F$ injective implies Fr surjective. Thus, every prime field is perfect.

Existence of $p$th roots in $k$ is closely related to separability.

**Proposition 5.78.**

(i) *A field $k$ is perfect if and only if every polynomial in $k[x]$ is separable.*

(ii) *Every algebraic extension $E/k$ of a perfect field $k$ is a separable extension.*

(iii) *The algebraic closure $\overline{\mathbb{F}}_p$ of $\mathbb{F}_p$ is perfect, and $\overline{\mathbb{F}}_p/\mathbb{F}_p$ is a separable extension.*

**Proof.**

(i) If $k$ has characteristic 0, then Lemma 3.4 shows that every polynomial in $k[x]$ is separable. Assume now that $k$ has characteristic $p > 0$ and that $f(x) \in k[x]$ is inseparable. By Proposition 5.75, $f(x) \in k[x^p]$, so that $f(x) = \sum_i a_i x^{pi}$. If every element in $k$ has a $p$th root, then $a_i = b_i^p$ for $b_i \in k$. Hence,

$$f(x) = \sum_i b_i^p x^{pi} = \left( \sum_i b_i x^i \right)^p,$$

and so $f(x)$ is not irreducible. In other words, if $k = k^p$, then every irreducible polynomial in $k[x]$ is separable and, hence, every polynomial is separable.

Conversely, assume that every polynomial in $k[x]$ is separable. If $k$ has characteristic 0, there is nothing to prove. If $k$ has characteristic $p > 0$ and $a \in k$, then $x^p - a$ has repeated roots; since our hypothesis says that irreducible polynomials are separable, $x^p - a$ must factor. Proposition 2.152 now says that $a$ has a $p$th root in $k$; that is, $a \in k^p$. Therefore, $k = k^p$, and so $k$ is perfect.

(ii) If $E/k$ is an algebraic extension, then every $\alpha \in E$ has a minimum polynomial $\mathrm{irr}(\alpha, k)$; since $\mathrm{irr}(\alpha, k)$ is a separable polynomial, by part (i), $\alpha$ is separable over $k$, and so $E/k$ is a separable extension.

(iii) We know that $\mathbb{F}_p$ is perfect, and the result follows from part (ii).  •

We will soon need the following variant of Proposition 2.152.

**Lemma 5.79.** *Let $p$ be a prime, let $e \geq 0$, and let $k$ be a field of characteristic $p > 0$. If $c \in k$ and $c \notin k^p$, then $f(x) = x^{p^e} - c$ is irreducible in $k[x]$.*

**Proof.** The proof is by induction on $e \geq 0$, the base step being true because every linear polynomial is irreducible. For the inductive step, suppose the statement is false. Let $g(x) \in k[x]$ be irreducible, and let $g(x)^m$, for $m \geq 1$, be the highest power of $g(x)$ dividing $f(x)$:

$$x^{p^e} - c = g(x)^m h(x),$$

where $(g(x), h(x)) = 1$. Take the derivative: $0 = mg(x)^{m-1}g'(x)h(x) + g(x)^m h'(x)$; now divide by $g(x)^{m-1}$:

$$0 = mg'(x)h(x) + g(x)h'(x).$$

Thus, $h(x) \mid h'(x)$, because $(g, h) = 1$. If $h'(x) \neq 0$, then $\deg(h')$ is defined and $\deg(h') < \deg(h)$, a contradiction; hence, $h'(x) = 0$ and $mg'(x)h(x) = 0$. Since $h(x) \neq 0$, we have

(5)
$$mg'(x) = 0,$$

which says that $(g^m(x))' = 0$. Proposition 5.75 now gives

$$g^m(x) = g_1(x^p) \text{ and } h(x) = h_1(x^p), \quad \text{where } g_1(x), h_1(x) \in k[x].$$

Therefore,

$$x^{p^e} - c = g(x)^m h(x) = g_1(x^p)h_1(x^p),$$

and so, replacing $x^p$ by $x$, we have

$$x^{p^{e-1}} - c = g_1(x)h_1(x).$$

Since $x^{p^{e-1}} - c$ is irreducible, by the inductive hypothesis, one of $g_1$, $h_1$ must be constant. But if $g_1(x)$ is constant, then $g_1(x^p)$ is constant and $g^m(x)$ is constant, a contradiction. Therefore, $h_1(x)$ is constant; absorb it into $g_1(x)$. Hence, $x^{p^{e-1}} - c = g_1(x)$, and so

$$x^{p^e} - c = g_1(x^p) = g(x)^m.$$

If $p \mid m$, then $x^{p^e} - c = (g(x)^{m/p})^p$, and so all the coefficients lie in $k^p$, contradicting $c \notin k^p$; therefore, $p \nmid m$. Equation (5) now gives $g'(x) = 0$, so that $g(x) \in k[x^p]$ [by Proposition 5.75]; say, $g(x) = g_2(x^p)$. This forces $m = 1$, because $x^{p^e} - c = g(x)^m$ gives $x^{p^{e-1}} - c = g_2(x)^m$, which is a forbidden factorization of the irreducible $x^{p^{e-1}} - c$. •

If $E/k$ is a field extension, where $k$ has characteristic $p$, then $k^p \subseteq E^p$, but we do not know whether $k \subseteq E^p$; that is, $E^p$ may not be an intermediate field of $E/k$ (for example, take $E = k$). Denote the subfield of $E$ obtained by adjoining $E^p$ to $k$ by $k(E^p)$.

**Proposition 5.80.**

(i) Let $k \subseteq B \subseteq E$ be a tower of fields with $E/k$ algebraic. If $E/k$ is separable, then $E/B$ is separable.

(ii) Let $E/k$ be an algebraic field extension, where $k$ has characteristic $p > 0$. If $E/k$ is a separable extension, then $E = k(E^p)$. Conversely, if $E/k$ is finite and $E = k(E^p)$, then $E/k$ is separable.

**Proof.**

(i) If $\alpha \in E$, then $\alpha$ is algebraic over $B$, and $\text{irr}(\alpha, B) \mid \text{irr}(\alpha, k)$ in $B[x]$, for their gcd is not 1 and $\text{irr}(\alpha, B)$ is irreducible. Since $\text{irr}(\alpha, k)$ has no repeated roots, $\text{irr}(\alpha, B)$ has no repeated roots, and hence $\text{irr}(\alpha, B)$ is a separable polynomial. Therefore, $E/B$ is a separable extension.

(ii) Let $E/k$ be a separable extension. Now $k(E^p) \subseteq E$, so that $E/k(E^p)$ is a separable extension, by part (i). If $\beta \in E$, then $\beta^p \in E^p \subseteq k(E^p)$; say, $\beta^p = \alpha$. If $\beta \notin k(E^p)$, then $\text{irr}(\beta, k(E^p))$ divides $x^p - \alpha$ in $(k(E^p))[x]$. Hence, $\text{irr}(\beta, k(E^p))$ is not separable, for it has repeated roots $[x^p - \alpha = (x - \beta)^p]$, contradicting separability of $E/k(E^p)$. Therefore, $\beta \in k(E^p)$ and $E = k(E^p)$.

Conversely, suppose that $E = k(E^p)$. We begin by showing that if $\beta_1, \ldots, \beta_s$ is a linearly independent list in $E$ (where $E$ is now viewed only as a vector space over $k$), then $\beta_1^p, \ldots, \beta_s^p$ is also linearly independent over $k$. Extend $\beta_1, \ldots, \beta_s$ to a basis $\beta_1, \ldots, \beta_n$ of $E$, where $n = [E : k]$. Now $\beta_1^p, \ldots, \beta_n^p$ spans $E^p$ over $k^p$, for if $\eta \in E$, then $\eta = \sum_i a_i \beta_i$, where $a_i \in k$, and hence $\eta^p = \sum_i a_i^p \beta_i^p$. Now take any element $\gamma \in E$. Since $E = k(E^p)$, we have $\gamma = \sum_j c_j \eta_j$, where $c_j \in k$ and $\eta_j \in E^p$. But $\eta_j = \sum_i a_{ji}^p \beta_i^p$ for $a_{ji} \in k$, as we have just seen, so that $\gamma = \sum_i (\sum_j c_j a_{ji}^p) \beta_i^p$; that is, $\beta_1^p, \ldots, \beta_n^p$ spans $E$ over $k$. Since $\dim_k(E) = n$, this list is a basis, and hence its sublist $\beta_1^p, \ldots, \beta_s^p$ must be linearly independent over $k$.

Each $\alpha \in E$ is algebraic over $k$, for $E/k$ is finite. If $\mathrm{irr}(\alpha, k)$ has degree $m$, then $1, \alpha, \alpha^2, \ldots, \alpha^m$ is linearly dependent over $k$, while $1, \alpha, \alpha^2, \ldots, \alpha^{m-1}$ is linearly independent. If $\alpha$ is inseparable, then $\mathrm{irr}(\alpha, k) = f_e(x^{p^e})$ and $m = p^e r$, where $r$ is the reduced degree of $\mathrm{irr}(\alpha, k)$. Since $r = m/p^e < m$, we have $1, \alpha, \alpha^2, \ldots, \alpha^r$ linearly independent over $k$. But $\alpha^{p^e}$ is a root of $f_e(x)$, so there is a nontrivial dependency relation on $1, \alpha^{p^e}, \alpha^{2p^e}, \ldots, \alpha^{rp^e}$ (because $rp^e = m$). We have seen, in the preceding paragraph, that linear independence of $1, \alpha, \alpha^2, \ldots, \alpha^r$ implies linear independence of $1, \alpha^{p^e}, \alpha^{2p^e}, \ldots, \alpha^{rp^e}$. This contradiction shows that $\alpha$ must be separable over $k$. •

**Corollary 5.81.** *Let $E/k$ be a finite separable extension, where $k$ has characteristic $p$. If a list $\beta_1, \ldots, \beta_r$ in $E$ is linearly independent over $k$, then for all $e \geq 1$, the list $\beta_1^{p^e}, \ldots, \beta_r^{p^e}$ is also linearly independent over $k$.*

**Proof.** The proof is by induction on $e \geq 1$, with the hypothesis of separability used in the form $E = k(E^p)$, as in the proof of Proposition 5.80(ii). •

**Corollary 5.82.** *If $k \subseteq B \subseteq E$ is a tower of algebraic extensions, then $B/k$ and $E/B$ are separable extensions if and only if $E/k$ is a separable extension.*

**Proof.** Since $B/k$ and $E/B$ are separable, Proposition 5.80(ii) gives $B = k(B^p)$ and $E = B(E^p)$. Therefore,

$$E = B(E^p) = k(B^p)(E^p) = k(B^p \cup E^p) = k(E^p) \subseteq E,$$

because $B^p \subseteq E^p$. Therefore, $E/k$ is separable, by Proposition 5.80(ii).

Conversely, if every element of $E$ is separable over $k$, we have, in particular, that each element of $B$ is separable over $k$; hence, $B/k$ is a separable extension. Finally, Proposition 5.80(i) shows that $E/B$ is a separable extension. •

**Proposition 5.83.** *If $E/k$ is an algebraic extension, define*

$$E_s = \{\alpha \in E : \alpha \text{ is separable over } k\};$$

*then $E_s$ is an intermediate field that is the unique maximal separable extension of $k$ contained in $E$.*

**Proof.** This follows from Proposition 3.41(ii), for if $\alpha, \beta$ are separable over $k$, then $k(\alpha, \beta)/k$ is separable, and hence $\alpha + \beta$, $\alpha\beta$, and $\alpha^{-1}$ are all separable over $k$. •

Not surprisingly, if $E/k$ is an algebraic extension, then the extension $E/E_s$ has a special property. Of course, $E_s$ is of interest only when $k$ has characteristic $p > 0$ (otherwise, $E_s = E$).

The next type of extension is "complementary" to separable extensions.

**Definition.** Let $E/k$ be a field extension, where $k$ has characteristic $p > 0$. Then $E/k$ is a *purely inseparable extension* if $E/k$ is algebraic and, for every $\alpha \in E$, there is $e \geq 0$ with $\alpha^{p^e} \in k$.

If $E/k$ is a purely inseparable extension and $B$ is an intermediate field, then it is clear that $E/B$ is purely inseparable.

**Proposition 5.84.** *If $E/k$ is an algebraic field extension, where $k$ has characteristic $p > 0$, then $E/E_s$ is a purely inseparable extension; moreover, if $\alpha \in E$, then* $\mathrm{irr}(\alpha, E_s) = x^{p^m} - c$ *for some $c \in k$ and $m \geq 0$.*

**Proof.** If $\alpha \in E$, write $\mathrm{irr}(\alpha, k) = f_e(x^{p^e})$, where $e \geq 0$ and $f_e(x) \in k[x]$ is a separable polynomial. It follows that $\alpha^{p^e}$ is separable over $k$ and $\alpha^{p^e} \in E_s$. If $\alpha \notin E_s$, choose $m$ minimal with $\alpha^{p^m} \in E_s$. If $c = \alpha^{p^m}$, then $\alpha$ is a root of $x^{p^m} - c$, which is irreducible, by Lemma 5.79; hence, $\mathrm{irr}(\alpha, E_s) = x^{p^m} - c$. $\bullet$

**Definition.** If $E/k$ is a finite extension, then $[E : k]_s = [E_s : k]$ is called the *separability degree*, and $[E : k]_i = [E : E_s]$ is called the *inseparability degree*.

Note that $E/k$ is separable if and only if $[E : k]_i = 1$. It is clear that

$$[E : k] = [E : k]_s[E : k]_i.$$

**Proposition 5.85.** *Let $E/k$ be a finite extension, where $k$ is a field of characteristic $p > 0$. If $E/k$ is purely inseparable, then $[E : k] = p^e$ for some $e \geq 0$. Hence, for some $e \geq 0$,*

$$[E : k]_i = [E : E_s] = p^e.$$

**Proof.** If $\alpha \in E$, then $\alpha$ is purely inseparable over $k$; if $\alpha$ is not constant, then $\mathrm{irr}(\alpha, E_s) = x^{p^m} - c$ for some $c \in k$, where $m \geq 1$. Therefore,

$$[E : k] = [E : k(\alpha)][k(\alpha) : k] = [E : k(\alpha)]p^m.$$

Now $[E : k(\alpha)] < [E : k]$; since $E/k(\alpha)$ is purely inseparable, the proof can be completed by induction. The second statement follows from Proposition 5.84, for $E$ is purely inseparable over $E_s$. $\bullet$

**Proposition 5.86.** *If $k \subseteq B \subseteq E$ is a tower of finite extensions, where $k$ is a field of characteristic $p > 0$, then*

$$[E : k]_s = [E : B]_s[B : k]_s \quad and \quad [E : k]_i = [E : B]_i[B : k]_i.$$

**Proof.** In light of the equation $[E : k] = [E : k]_s[E : k]_i$, it suffices to prove $[E : k]_s = [E : B]_s[B : k]_s$.

The notation $B_s$ is unambiguous, but here the notation $E_s$ is ambiguous. We write $E_s$ to denote the intermediate field consisting of all those elements of $E$ that are separable over $k$, and we write

$$E_B = \{\alpha \in E : \alpha \text{ is separable over } B\}.$$

We have $k \subseteq B_s \subseteq E_s \subseteq E_B \subseteq E$; let us see that $E_s \subseteq E_B$. If $\alpha \in E$ is separable over $k$, then $\mathrm{irr}(\alpha, k)$ has no repeated roots. Hence, $\alpha$ is separable over $B$, because $\mathrm{irr}(\alpha, B) \mid \mathrm{irr}(\alpha, k)$ in $B[x]$, and so $\alpha \in E_B$. With this notation,

$$[E : k]_s = [E_s : k], \quad [E : B]_s = [E_B : B], \quad \text{and} \quad [B : k]_s = [B_s : k].$$

Now $[E : k]_s = [E_s : k] = [E_s : B_s][B_s : k] = [E_s : B_s][B : k]_s$. Thus, it suffices to prove that $[E_s : B_s] = [E_B : B]$, for $[E_B : B] = [E : B]_s$.

We show that $[E_s : B_s] \leq [E_B : B]$ by proving that a list $\beta_1, \dots, \beta_r$ in $E_s \subseteq E_B$ linearly independent over $B_s$ is also linearly independent over $B$. Suppose that $\sum b_i \beta_i = 0$, where $b_i \in B$ are not all 0. For all $e \geq 0$, we have $0 = (\sum b_i \beta_i)^{p^e} = \sum b_i^{p^e} \beta_i^{p^e}$. But there is $e \geq 0$ with $b_i^{p^e} \in B_s$ for all $i$, because $B/B_s$ is purely inseparable, and so the list $\beta_1^{p^e}, \dots, \beta_r^{p^e}$ is linearly dependent over $B_s$, contradicting Corollary 5.81 (for $E_s/B_s$ is a separable extension). For the reverse inequality $[E_s : B_s] \geq [E_B : B]$, take a list $\gamma_1, \dots, \gamma_t$ in $E_B$ that is linearly independent over $B$. Since $E_B/E_s$ is purely inseparable (it is an intermediate field of $E/E_s$), there is $e \geq 0$ with $\gamma_i^{p^e} \in E_s$ for all $i$. But $E_s/B$ is a separable extension, so that Corollary 5.81 gives $\gamma_1^{p^e}, \dots, \gamma_t^{p^e}$ linearly independent over $B$; a fortiori, $\gamma_1^{p^e}, \dots, \gamma_t^{p^e}$ is linearly independent over $B_s$. Therefore, $[E_s : B_s] = [E_B : B]$.  •

## Exercises

∗ **5.53.** Prove that if $E/k$ is an algebraic extension and $\beta \in E$ is both separable and purely inseparable, then $\beta \in k$.

**5.54.** Let $k$ be a field of characteristic $p > 0$, and let $f(x) = x^{2p} - x^p + t \in k(t)[x]$.

  (i) Prove that $f(x)$ is an irreducible polynomial in $k(t)[x]$.

  (ii) Prove that $f(x)$ is inseparable.

  (iii) Prove that there exists an algebraic extension $E/k(t)$ for which there is no intermediate field $E_i$ with $E_i/k$ purely inseparable and $E/E_i$ separable. (Compare with Corollary 5.84 and Proposition 3.41.)

## Section 5.5. Varieties

Analytic Geometry gives pictures of equations. For example, we picture a function $f \colon \mathbb{R} \to \mathbb{R}$ as its graph, which consists of all the ordered pairs $(a, f(a))$ in the plane; that is, $f$ is the set of all the solutions $(a, b) \in \mathbb{R}^2$ of

$$g(x, y) = y - f(x) = 0.$$

We can also picture equations that are not graphs of functions. For example, the set of all the zeros of the polynomial

$$h(x, y) = x^2 + y^2 - 1$$

is the unit circle. Simultaneous solutions in $\mathbb{R}^2$ of several polynomials of two variables can also be pictured; indeed, simultaneous solutions of several polynomials of $n$ variables can be pictured in $\mathbb{R}^n$. There is a very strong connection between the rings $k[x_1, \ldots, x_n]$ and the geometry of subsets of $k^n$ going far beyond this. The key idea is Exercise 5.4 on page 301, which says that if $(a_1, \ldots, a_n) \in k^n$, then $(x_1 - a_1, \ldots, x_n - a_n)$ is a maximal ideal in $k[x_1, \ldots, x_n]$ (we shall see, in Theorem 5.100, that the converse is true when $k$ is algebraically closed). Pictures in $k^n$ are called *varieties*, and they can be described in terms of ideals. The interplay between $k[x_1, \ldots, x_n]$ and varieties has evolved into what is nowadays called *Algebraic Geometry*, and this section may be regarded as an introduction to this subject.

**5.5.1. Varieties and Ideals.** Let $k$ be a field and let $k^n$ denote the set of all $n$-tuples:

$$k^n = \big\{ a = (a_1, \ldots, a_n) \colon a_i \in k \text{ for all } i \big\}.$$

We use the abbreviation

$$X = (x_1, \ldots, x_n),$$

so that the polynomial ring $k[x_1, \ldots, x_n]$ in several variables may be denoted by $k[X]$ and a polynomial $f(x_1, \ldots, x_n)$ in $k[X]$ may be abbreviated by $f(X)$.

Polynomials $f(X) \in k[X]$ determine polynomial functions $k^n \to k$.

**Definition.** If $f(X) \in k[X]$, its ***polynomial function*** $f^\flat \colon k^n \to k$ is defined by evaluation:

$$f^\flat \colon (a_1, \ldots, a_n) \mapsto f(a_1, \ldots, a_n).$$

In Proposition 2.45(ii), we proved that if $k$ is an infinite field and $f^\flat = g^\flat$, then $f(X) = g(X)$. Recall that algebraically closed fields are infinite.

***For the remainder of this section, we assume that all fields are infinite.***

Consequently, we drop the $f^\flat$ notation and identify polynomials with their polynomial functions.

**Definition.** If $f(X) \in k[X] = k[x_1, \ldots, x_n]$ and $f(a) = 0$, where $a \in k^n$, then $a$ is called a ***zero*** of $f(X)$. [If $f(x)$ is a polynomial in one variable, then a zero of $f(x)$ is usually called a *root*[12] of $f(x)$.]

**Proposition 5.87.** *If $k$ is an algebraically closed field and $f(X) \in k[X]$ is not a constant, then $f(X)$ has a zero.*

**Proof.** We prove the result by induction on $n \geq 1$, where $X = (x_1, \ldots, x_n)$. The base step follows at once from our assuming that $k^1 = k$ is algebraically closed. As in the proof of Proposition 2.45(ii), write

$$f(X, y) = \sum_i g_i(X) y^i.$$

---

[12]The etymology of *root* is discussed in FCAA, pp. 33–34.

For each $a \in k^n$, define $f_a(y) = \sum_i g_i(a)y^i$. If $f(X, y)$ has no zeros, then each $f_a(y) \in k[y]$ has no zeros, and the base step says that $f_a(y)$ is a nonzero constant for all $a \in k^n$. Thus, $g_i(a) = 0$ for all $i > 0$ and all $a \in k^n$. By Proposition 2.45(ii), which applies because algebraically closed fields are infinite, $g_i(X) = 0$ for all $i > 0$, and so $f(X, y) = g_0(X)y^0 = g_0(X)$. By the inductive hypothesis, $g_0(X)$ is a nonzero constant, and the proof is complete.   •

Here are some general definitions describing solution sets of polynomials.

**Definition.** If $F$ is a subset of $k[X] = k[x_1, \ldots, x_n]$, then the ***variety*** [13],[14] defined by $F$ is

$$\mathrm{Var}(F) = \{a \in k^n : f(a) = 0 \text{ for every } f(X) \in F\};$$

thus, $\mathrm{Var}(F)$ consists of all those $a \in k^n$ which are zeros of every $f(X) \in F$.

**Example 5.88.**

(i) Assume that $k$ is algebraically closed; Proposition 5.87 now says that if $f(X) \in k[X]$ is not constant, then $\mathrm{Var}(f) \neq \varnothing$.

(ii) Here are some varieties defined by two equations:

$$\mathrm{Var}(x, y) = \{(a, b) \in k^2 : x = 0 \text{ and } y = 0\} = \{(0, 0)\}$$

and

$$\mathrm{Var}(xy) = x\text{-axis} \cup y\text{-axis}.$$

(iii) Here is an example in higher-dimensional space. Let $A$ be an $m \times n$ matrix with entries in $k$. A system of $m$ equations in $n$ unknowns,

$$AX = B,$$

where $B$ is an $n \times 1$ column matrix, defines a variety, $\mathrm{Var}(AX = B)$, which is a subset of $k^n$. Of course, $AX = B$ is really a shorthand for a set of $m$ linear equations in $n$ variables, and $\mathrm{Var}(AX = B)$ is usually called the ***solution set*** of the system $AX = B$. When this system is homogeneous, that is, when $B = 0$, then $\mathrm{Var}(AX = 0)$ is a subspace of $k^n$, called the ***solution space*** of the system.   ◄

The next result shows, as far as varieties are concerned, that we may just as well assume that the subsets $F$ of $k[X]$ are ideals of $k[X]$.

**Proposition 5.89.** *Let $k$ be a field, and let $F$ and $G$ be subsets of $k[X]$.*

(i) *If $F \subseteq G \subseteq k[X]$, then $\mathrm{Var}(G) \subseteq \mathrm{Var}(F)$.*

(ii) *If $F \subseteq k[X]$ and $I = (F)$ is the ideal generated by $F$, then*

$$\mathrm{Var}(F) = \mathrm{Var}(I).$$

---

[13]There is some disagreement about the usage of this term. Some call this an *affine variety*, in contrast to the analogous *projective variety*. Many insist that varieties should be *irreducible*, which we will define later in this section.

[14]The term *variety* arose in 1869 as E. Beltrami's translation of the German term *Mannigfaltigkeit* used by Riemann; nowadays, this term is usually translated as *manifold*. (The modern notion of variety is *scheme*, and a scheme is an algebraic analogue of the modern notion of manifold!)

**Proof.**

(i) If $a \in \mathrm{Var}(G)$, then $g(a) = 0$ for all $g(X) \in G$; since $F \subseteq G$, it follows, in particular, that $f(a) = 0$ for all $f(X) \in F$.

(ii) Since $F \subseteq (F) = I$, we have $\mathrm{Var}(I) \subseteq \mathrm{Var}(F)$, by part (i). For the reverse inclusion, let $a \in \mathrm{Var}(F)$, so that $f(a) = 0$ for every $f(X) \in F$. If $g(X) \in I$, then $g(X) = \sum_i r_i(X) f_i(X)$, where $r_i(X) \in k[X]$ and $f_i(X) \in F$; hence, $g(a) = \sum_i r_i(a) f_i(a) = 0$ and $a \in \mathrm{Var}(I)$. $\bullet$

It follows that not every subset of $k^n$ is a variety. For example, if $n = 1$, then $k[x]$ is a PID. Hence, if $F$ is a subset of $k[x]$, then $(F) = (g(x))$ for some $g(x) \in k[x]$, and so

$$\mathrm{Var}(F) = \mathrm{Var}((F)) = \mathrm{Var}((g)) = \mathrm{Var}(g).$$

But if $g(x) \neq 0$, then it has only a finite number of roots, and so $\mathrm{Var}(F)$ is finite. Thus, for infinite fields $k$, most subsets of $k^1 = k$ are not varieties.

In spite of our wanting to draw pictures in the plane, there is a major defect with $k = \mathbb{R}$: some polynomials have no zeros. For example, $f(x) = x^2 + 1$ has no real roots, and so $\mathrm{Var}(x^2 + 1) = \varnothing$. More generally, $g(x_1, \ldots, x_n) = x_1^2 + \cdots + x_n^2 + 1$ has no zeros in $\mathbb{R}^n$, and so $\mathrm{Var}(g) = \varnothing$. It is natural to want the simplest varieties, those defined by a single nonconstant polynomial, to be nonempty. For polynomials in one variable over a field $k$, this amounts to saying that $k$ is algebraically closed. In light of Proposition 5.87, we know that $\mathrm{Var}(f) \neq \varnothing$ for every nonconstant $f(X)$ in several variables over an algebraically closed field. Of course, varieties are of interest for all fields $k$, but it makes more sense to consider the simplest case before trying to understand more complicated problems. On the other hand, many of the first results are valid for any field $k$. Thus, even though we may state weaker hypotheses, the reader may always assume (the most important case here) that $k$ is algebraically closed.

Here are some elementary properties of Var.

**Proposition 5.90.** *Let $k$ be a field.*

(i) $\mathrm{Var}(1) = \varnothing$ *and* $\mathrm{Var}(0) = k^n$, *where $0$ is the zero polynomial.*

(ii) *If $I$ and $J$ are ideals in $k[X]$, then*

$$\mathrm{Var}(IJ) = \mathrm{Var}(I \cap J) = \mathrm{Var}(I) \cup \mathrm{Var}(J),$$

*where $IJ = \left\{ \sum_i f_i(X) g_i(X) \colon f_i(X) \in I \text{ and } g_i(X) \in J \right\}$.*

(iii) *If $\{I_\ell \colon \ell \in L\}$ is a family of ideals in $k[X]$, then $\mathrm{Var}\left( \sum_\ell I_\ell \right) = \bigcap_\ell \mathrm{Var}(I_\ell)$, where $\sum_\ell I_\ell$ is the set of all finite sums of the form $\sum_\ell r_\ell$ with $r_\ell \in I_\ell$.*

**Proof.**

(i) That $\mathrm{Var}(1) = \varnothing$ is clear, for the constant polynomial 1 has no zeros. That $\mathrm{Var}(0) = k^n$ is clear, for every point $a$ is a zero of the zero polynomial.

(ii) Since $IJ \subseteq I \cap J$, it follows that $\mathrm{Var}(IJ) \supseteq \mathrm{Var}(I \cap J)$; since $IJ \subseteq I$, it follows that $\mathrm{Var}(IJ) \supseteq \mathrm{Var}(I)$. Similarly, $\mathrm{Var}(IJ) \supseteq \mathrm{Var}(J)$. Hence,

$$\mathrm{Var}(IJ) \supseteq \mathrm{Var}(I \cap J) \supseteq \mathrm{Var}(I) \cup \mathrm{Var}(J).$$

To complete the proof, it suffices to show that $\text{Var}(I) \cup \text{Var}(J) \supseteq \text{Var}(IJ)$. If $a \notin \text{Var}(I) \cup \text{Var}(J)$, then there exist $f(X) \in I$ and $g(X) \in J$ with $f(a) \neq 0$ and $g(a) \neq 0$. But $f(X)g(X) \in IJ$ and $(fg)(a) = f(a)g(a) \neq 0$, because fields are domains. Therefore, $a \notin \text{Var}(IJ)$, as desired.

(iii) For each $\ell$, the inclusion $I_\ell \subseteq \sum_\ell I_\ell$ gives $\text{Var}\left(\sum_\ell I_\ell\right) \subseteq \text{Var}(I_\ell)$, and so

$$\text{Var}\left(\sum_\ell I_\ell\right) \subseteq \bigcap_\ell \text{Var}(I_\ell).$$

For the reverse inclusion, if $g(X) \in \sum_\ell I_\ell$, then there are finitely many $\ell$ with $g(X) = \sum_\ell f_\ell$, where $f_\ell(X) \in I_\ell$. Therefore, if $a \in \bigcap_\ell \text{Var}(I_\ell)$, then $f_\ell(a) = 0$ for all $\ell$, and so $g(a) = 0$; that is, $a \in \text{Var}\left(\sum_\ell I_\ell\right)$. •

**Corollary 5.91.** *If $k$ is a field, then $k^n$ is a topological space whose closed sets are the varieties.*

**Proof.** The different parts of Proposition 5.90 verify the axioms for closed sets that define a topology. •

**Definition.** The **Zariski topology** on $k^n$ is the topology whose closed sets are the varieties.

The usual way of regarding $\mathbb{R} = \mathbb{R}^1$ as a topological space has many closed sets; for example, every closed interval is a closed set. In contrast, the only Zariski closed sets in $\mathbb{R}$, aside from $\mathbb{R}$ itself, are finite sets.

**Definition.** A **hypersurface** in $k^n$ is a subset of the form $\text{Var}(f)$ for some non-constant $f(X) \in k[X]$.

**Corollary 5.92.** *Every variety $\text{Var}(I)$ in $k^n$ is the intersection of finitely many hypersurfaces.*

**Proof.** By the Hilbert Basis Theorem, there are $f_1, \dots, f_t \in k[X]$ with $I = (f_1, \dots, f_t) = \sum_i (f_i)$. By Proposition 5.90(iii), we have $\text{Var}(I) = \bigcap_i \text{Var}(f_i)$. •

Given an ideal $I$ in $k[X]$, we have just defined its variety $\text{Var}(I) \subseteq k^n$. We now reverse direction: given a subset $A \subseteq k^n$, we assign an ideal in $k[X]$ to it; in particular, we assign an ideal to every variety.

**Definition.** If $A \subseteq k^n$, define its **coordinate ring** $k[A]$ to be the commutative ring

$$k[A] = \{f|A : f(X) \in k[X]\}$$

under pointwise operations.[15] If $A \subseteq k^n$, define

$$\text{Id}(A) = \{f(X) \in k[X] = k[x_1, \dots, x_n] : f(a) = 0 \text{ for every } a \in A\}.$$

The Hilbert Basis Theorem tells us that $\text{Id}(A)$ is always a finitely generated ideal.

---

[15] Recall that we are assuming that all fields $k$ are infinite, and so we can identify polynomials $f(X) \in k[X]$ with polynomial functions $f \colon k^n \to k$.

**Proposition 5.93.** *If $A \subseteq k^n$, then there is an isomorphism*

$$k[X]/\operatorname{Id}(A) \cong k[A],$$

*where $k[A]$ is the coordinate ring of $A$.*

**Proof.** The restriction map res: $k[X] \to k[A]$ is a surjection with kernel $\operatorname{Id}(A)$, and so the result follows from the First Isomorphism Theorem. [Thus, if two polynomials $f, g$ agree on $A$, then $f|A = g|A$ and $f + \operatorname{Id}(A) = g + \operatorname{Id}(A)$.] •

Although the definition of $\operatorname{Var}(F)$ makes sense for any subset $F$ of $k[X]$, it is most interesting when $F$ is an ideal. Similarly, although the definition of $\operatorname{Id}(A)$ makes sense for any subset $A$ of $k^n$, it is most interesting when $A$ is a variety. After all, varieties are comprised of solutions of (polynomial) equations, which is what we care about.

**Proposition 5.94.** *Let $k$ be an infinite field.*

(i) $\operatorname{Id}(\varnothing) = k[X]$ *and* $\operatorname{Id}(k^n) = (0)$.

(ii) *If $A \subseteq B$ are subsets of $k^n$, then* $\operatorname{Id}(B) \subseteq \operatorname{Id}(A)$.

(iii) *If $\{A_\ell : \ell \in L\}$ is a family of subsets of $k^n$, then* $\operatorname{Id}\left(\bigcup_\ell A_\ell\right) = \bigcap_\ell \operatorname{Id}(A_\ell)$.

**Proof.**

(i) By definition, $f(X) \in \operatorname{Id}(A)$ for some subset $A \subseteq k^n$ if and only if $f(a) = 0$ for all $a \in A$; hence, if $f(X) \notin \operatorname{Id}(A)$, then there exists $a \in A$ with $f(a) \neq 0$. In particular, if $A = \varnothing$, every $f(X) \in k[X]$ must lie in $\operatorname{Id}(\varnothing)$, for there are no elements $a \in \varnothing$. Therefore, $\operatorname{Id}(\varnothing) = k[X]$.

If $f(X) \in \operatorname{Id}(k^n)$, then $f^\flat = 0^\flat$, and so $f(X) = 0$, by Proposition 2.45(ii).

(ii) If $f(X) \in \operatorname{Id}(B)$, then $f(b) = 0$ for all $b \in B$; in particular, $f(a) = 0$ for all $a \in A$, because $A \subseteq B$, and so $f(X) \in \operatorname{Id}(A)$.

(iii) Since $A_\ell \subseteq \bigcup_\ell A_\ell$, we have $\operatorname{Id}(A_\ell) \supseteq \operatorname{Id}\left(\bigcup_\ell A_\ell\right)$ for all $\ell$; hence, $\bigcap_\ell \operatorname{Id}(A_\ell) \supseteq \operatorname{Id}\left(\bigcup_\ell A_\ell\right)$. For the reverse inclusion, suppose that $f(X) \in \bigcap_\ell \operatorname{Id}(A_\ell)$; that is, $f(a_\ell) = 0$ for all $\ell$ and all $a_\ell \in A_\ell$. If $b \in \bigcup_\ell A_\ell$, then $b \in A_\ell$ for some $\ell$, and hence $f(b) = 0$; therefore, $f(X) \in \operatorname{Id}\left(\bigcup_\ell A_\ell\right)$. •

We would like to have a formula for $\operatorname{Id}(A \cap B)$. Certainly, it is not true that $\operatorname{Id}(A \cap B) = \operatorname{Id}(A) \cup \operatorname{Id}(B)$, for the union of two ideals is almost never an ideal.

The next idea arises in characterizing those ideals of the form $\operatorname{Id}(V)$ when $V$ is a variety.

**Definition.** If $I$ is an ideal in a commutative ring $R$, then its **radical** is

$$\operatorname{radical}(I) = \sqrt{I} = \{r \in R : r^m \in I \text{ for some integer } m \geq 1\}.$$

An ideal $I$ is called a **radical ideal** [16] if $\sqrt{I} = I$.

---

[16]This term is appropriate, for if $r^m \in I$, then its $m$th root $r$ also lies in $I$.

Exercise 5.56 on page 367 asks you to prove that $\sqrt{I}$ is an ideal. It is easy to see that $I \subseteq \sqrt{I}$, and so an ideal $I$ is a radical ideal if and only if $\sqrt{I} \subseteq I$. For example, every prime ideal $P$ is a radical ideal, for if $f^n \in P$, then $f \in P$. It is easy to give an example of an ideal that is not radical: $I = (x^2)$ is not a radical ideal because $x^2 \in I$ and $x \notin I$.

**Definition.** An element $a$ in a ring $R$ is called **nilpotent** if $a \neq 0$ and there is some $n \geq 1$ with $a^n = 0$.

Note that $I$ is a radical ideal in a commutative ring $R$ if and only if $R/I$ has no nilpotent elements. A commutative ring having no nilpotent elements is called **reduced**.

**Proposition 5.95.** *If an ideal $I = \mathrm{Id}(A)$ for some $A \subseteq k^n$, then it is a radical ideal. Hence, the coordinate ring $k[A]$ has no nilpotent elements.*

**Proof.** Since $I \subseteq \sqrt{I}$ is always true, it suffices to check the reverse inclusion. By hypothesis, $I = \mathrm{Id}(A)$ for some $A \subseteq k^n$; hence, if $f \in \sqrt{I}$, then $f^m \in I = \mathrm{Id}(A)$; that is, $f(a)^m = 0$ for all $a \in A$. But the values of $f(a)^m$ lie in the field $k$, so that $f(a)^m = 0$ implies $f(a) = 0$; that is, $f \in \mathrm{Id}(A) = I$.   •

**Proposition 5.96.**

 (i) *If $I$ and $J$ are ideals, then $\sqrt{I \cap J} = \sqrt{I} \cap \sqrt{J}$.*

 (ii) *If $I$ and $J$ are radical ideals, then $I \cap J$ is a radical ideal.*

**Proof.**

 (i) If $f \in \sqrt{I \cap J}$, then $f^m \in I \cap J$ for some $m \geq 1$. Hence, $f^m \in I$ and $f^m \in J$, and so $f \in \sqrt{I}$ and $f \in \sqrt{J}$; that is, $f \in \sqrt{I} \cap \sqrt{J}$.

 For the reverse inclusion, assume that $f \in \sqrt{I} \cap \sqrt{J}$, so that $f^m \in I$ and $f^q \in J$. We may assume that $m \geq q$, and so $f^m \in I \cap J$; that is, $f \in \sqrt{I \cap J}$.

 (ii) If $I$ and $J$ are radical ideals, then $I = \sqrt{I}$ and $J = \sqrt{J}$; by part (i),

$$I \cap J \subseteq \sqrt{I \cap J} = \sqrt{I} \cap \sqrt{J} = I \cap J.   •$$

**5.5.2. Nullstellensatz.** We are now going to prove Hilbert's *Nullstellensatz*[17] for $\mathbb{C}[X]$. The reader will see that the proof we will give generalizes to any uncountable algebraically closed field. The theorem is actually true for all algebraically closed fields (we shall prove it in Chapter 10), and so the proof in this section does not, alas, cover the algebraic closures of the prime fields, for example, which are countable.

**Lemma 5.97.** *Let $k$ be a field and let $\varphi \colon k[X] \to k$ be a surjective ring homomorphism which fixes $k$ pointwise. If $J = \ker \varphi$, then $\mathrm{Var}(J) \neq \varnothing$.*

**Proof.** Let $\varphi(x_i) = a_i \in k$ and let $a = (a_1, \ldots, a_n) \in k^n$. If

$$f(X) = \sum_{\alpha_1, \ldots, \alpha_n} c_{\alpha_1, \ldots, \alpha_n} x_1^{\alpha_1} \cdots x_n^{\alpha_n} \in k[X],$$

---

[17]The German word *Nullstelle* means *root* or *zero*, and so *Nullstellensatz* means the *theorem of zeros*.

then

$$\varphi(f(X)) = \sum_{\alpha_1,\ldots,\alpha_n} c_{\alpha_1,\ldots,\alpha_n} \varphi(x_1)^{\alpha_1} \cdots \varphi(x_n)^{\alpha_n}$$

$$= \sum_{\alpha_1,\ldots,\alpha_n} c_{\alpha_1,\ldots,\alpha_n} a_1^{\alpha_1} \cdots a_n^{\alpha_n} = f(a_1,\ldots,a_n) = f(a).$$

Hence, if $f(X) \in J = \ker \varphi$, then $f(a) = 0$, and so $a \in \mathrm{Var}(J)$. •

The next proof will use a bit of cardinality.

**Theorem 5.98 (Weak Nullstellensatz over $\mathbb{C}$).** *If $f_1(X),\ldots,f_t(X) \in \mathbb{C}[X]$, then $I = (f_1,\ldots,f_t)$ is a proper ideal in $\mathbb{C}[X]$ if and only if $\mathrm{Var}(I) \neq \varnothing$.*

**Proof.** If $\mathrm{Var}(I) \neq \varnothing$, then $I$ is a proper ideal, because $\mathrm{Var}(\mathbb{C}[X]) = \varnothing$.

For the converse, suppose that $I$ is a proper ideal. By Corollary 5.35, there is a maximal ideal $M$ containing $I$, and so $K = \mathbb{C}[X]/M$ is a field. It is plain that the natural map $\mathbb{C}[X] \to \mathbb{C}[X]/M = K$ carries $\mathbb{C}$ to itself, so that $K/\mathbb{C}$ is an extension field; it follows that $K$ is a vector space over $\mathbb{C}$. Now $\mathbb{C}[X]$ has countable dimension, as a $\mathbb{C}$-space, for a basis consists of all the monic monomials $1, x, x^2, x^3, \ldots$. Therefore, $\dim_{\mathbb{C}}(K)$ is countable (possibly finite), for it is a quotient of $\mathbb{C}[X]$.

Suppose that $K$ is a proper extension of $\mathbb{C}$; that is, there is some $t \in K$ with $t \notin \mathbb{C}$. Since $\mathbb{C}$ is algebraically closed, $t$ cannot be algebraic over $\mathbb{C}$, and so it is transcendental. Consider the subset $B$ of $K$,

$$B = \{1/(t - c) \colon c \in \mathbb{C}\}$$

(note that $t - c \neq 0$ because $t \notin \mathbb{C}$). The set $B$ is uncountable, for it is indexed by the uncountable set $\mathbb{C}$. We claim that $B$ is linearly independent over $\mathbb{C}$; if so, then the fact that $\dim_{\mathbb{C}}(K)$ is countable is contradicted, and we will conclude that $K = \mathbb{C}$. If $B$ is linearly dependent, there are nonzero $a_1,\ldots,a_r \in \mathbb{C}$ and distinct $c_1,\ldots,c_r \in \mathbb{C}$ with $\sum_{i=1}^r a_i/(t - c_i) = 0$. Clearing denominators, we have a polynomial $h(t) \in \mathbb{C}[t]$:

$$h(t) = \sum_i a_i(t - c_1) \cdots \widehat{(t - c_i)} \cdots (t - c_r) = 0.$$

Now $h(c_1) = a_1(c_1 - c_2) \cdots (c_1 - c_r) \neq 0$, so that $h(t)$ is not the zero polynomial. But this contradicts $t$ being transcendental; therefore, $K = \mathbb{C}$. Lemma 5.97 now applies to show that $\mathrm{Var}(M) \neq \varnothing$. But $\mathrm{Var}(M) \subseteq \mathrm{Var}(I)$, and this completes the proof. •

Consider the special case of this theorem for $I = (f(x)) \subseteq \mathbb{C}[x]$, where $f(x)$ is not a constant. To say that $\mathrm{Var}(f) \subseteq \mathbb{C}$ is nonempty is to say that $f(x)$ has a complex root. Thus, the Weak Nullstellensatz is a generalization to several variables of the Fundamental Theorem of Algebra.

The following proof of Hilbert's Nullstellensatz uses the "Rabinowitch trick" of imbedding a polynomial ring in $n$ variables into a polynomial ring in $n+1$ variables.

**Theorem 5.99 (Nullstellensatz).** *If $I$ is an ideal in $\mathbb{C}[X]$, then $\mathrm{Id}(\mathrm{Var}(I)) = \sqrt{I}$. Thus, $f$ vanishes on $\mathrm{Var}(I)$ if and only if $f^m \in I$ for some $m \geq 1$.*

**Proof.** The inclusion $\mathrm{Id}(\mathrm{Var}(I)) \supseteq \sqrt{I}$ is obviously true, for if $f^m(a) = 0$ for some $m \geq 1$ and all $a \in \mathrm{Var}(I)$, then $f(a) = 0$ for all $a$, because $f(a) \in \mathbb{C}$.

For the converse, assume that $h \in \mathrm{Id}(\mathrm{Var}(I))$, where $I = (f_1, \ldots, f_t)$; that is, if $f_i(a) = 0$ for all $i$, where $a \in \mathbb{C}^n$, then $h(a) = 0$. We must show that some power of $h$ lies in $I$. Of course, we may assume that $h$ is not the zero polynomial. Let us regard

$$\mathbb{C}[x_1, \ldots, x_n] \subseteq \mathbb{C}[x_1, \ldots, x_n, y];$$

thus, every $f_i(x_1, \ldots, x_n)$ is regarded as a polynomial in $n + 1$ variables that does not depend on the last variable $y$. We claim that the polynomials

$$f_1, \ \ldots, \ f_t, \ 1 - yh$$

in $\mathbb{C}[x_1, \ldots, x_n, y]$ have no common zeros. If $(a_1, \ldots, a_n, b) \in \mathbb{C}^{n+1}$ is a common zero, then $a = (a_1, \ldots, a_n) \in \mathbb{C}^n$ is a common zero of $f_1, \ldots, f_t$, and so $h(a) = 0$. But now $1 - bh(a) = 1 \neq 0$. The weak Nullstellensatz now applies to show that the ideal $(f_1, \ldots, f_t, 1 - yh)$ in $\mathbb{C}[x_1, \ldots, x_n, y]$ is not a proper ideal. Therefore, there are $g_1, \ldots, g_{t+1} \in \mathbb{C}[x_1, \ldots, x_n, y]$ with

$$1 = f_1 g_1 + \cdots + f_t g_t + (1 - yh)g_{t+1}.$$

Let $d_i$ be the degree in $y$ of $g_i(x_1, \ldots, x_n, y)$. Make the substitution $y = 1/h$, so that the last term involving $g_{t+1}$ vanishes. Rewriting, $g_i(X, y) = \sum_{j=0}^{d_i} u_j(X)y^j$, and so $g_i(X, h^{-1}) = \sum_{j=0}^{d_i} u_j(X)h^{-j}$. It follows that

$$h^{d_i} g_i(X, h^{-1}) \in \mathbb{C}[X].$$

Therefore, if $m = \max\{d_1, \ldots, d_t\}$, then

$$h^m = (h^m g_1)f_1 + \cdots + (h^m g_t)f_t \in I. \quad \bullet$$

**Theorem 5.100.** *Every maximal ideal $M$ in $\mathbb{C}[x_1, \ldots, x_n]$ has the form*

$$M = (x_1 - a_1, \ldots, x_n - a_n) = \mathrm{Id}(a)$$

*for some $a = (a_1, \ldots, a_n) \in \mathbb{C}^n$.*

**Proof.** By Theorem 5.99, $\mathrm{Id}(\mathrm{Var}(M)) = \sqrt{M} = M$, because $M$ is a maximal, hence prime, ideal. Since $M$ is a proper ideal, we have $\mathrm{Var}(M) \neq \varnothing$, by Theorem 5.98; that is, there is $a = (a_1, \ldots, a_n) \in \mathbb{C}^n$ with $f(a) = 0$ for all $f \in M$. Hence, $a \in \mathrm{Var}(M)$, and Proposition 5.94(ii) gives $M = \mathrm{Id}(\mathrm{Var}(M)) \subseteq \mathrm{Id}(a)$. Since $\mathrm{Id}(a)$ does not contain any nonzero constant, it is a proper ideal, and so maximality of $M$ gives $M = \mathrm{Id}(a) = \{f(X) \in \mathbb{C}[X] : f(a) = 0\}$. If $f_i(X) = x_i - a_i$, then $f_i(a) = 0$, so that $(f_1, \ldots, f_n) = (x_1 - a_1, \ldots, x_n - a_n) \subseteq \mathrm{Id}(a)$. But $(x_1 - a_1, \ldots, x_n - a_n)$ is a maximal ideal, by Exercise 5.4 on page 301, so that $(x_1 - a_1, \ldots, x_n - a_n) = \mathrm{Id}(a) = M$. $\quad \bullet$

Thus, $a \mapsto \mathrm{Id}(a)$ is a bijection from $\mathbb{C}^n$ to the maximal ideals in $\mathbb{C}[x_1, \ldots, x_n]$.

We continue the study of the operators Var and Id.

**Proposition 5.101.** *Let $k$ be any field.*

(i) *For every subset $F \subseteq k^n$,*

$$\mathrm{Var}(\mathrm{Id}(F)) \supseteq F.$$

(ii) *For every ideal $I \subseteq k[X]$,*

$$\mathrm{Id}(\mathrm{Var}(I)) \supseteq I.$$

(iii) *If $V$ is a variety of $k^n$, then $\mathrm{Var}(\mathrm{Id}(V)) = V$.*

(iv) *If $F \subseteq k^n$, then $\mathrm{Var}(\mathrm{Id}(F)) = \overline{F}$, the **Zariski closure** of $F$, that is, the intersection of all those varieties containing $F$.*

(v) *If $V \subseteq V^* \subseteq k^n$ are varieties, then*

$$V^* = V \cup \overline{V^* - V},$$

*the Zariski closure of $V^* - V$.*

**Proof.**

(i) This result is almost a tautology. If $a \in F$, then $g(a) = 0$ for all $g(X) \in \mathrm{Id}(F)$. But every $g(X) \in \mathrm{Id}(F)$ annihilates $F$, by definition of $\mathrm{Id}(F)$, and so $a \in \mathrm{Var}(\mathrm{Id}(F))$. Therefore, $\mathrm{Var}(\mathrm{Id}(F)) \supseteq F$.

(ii) Again, we merely look at the definitions. If $f(X) \in I$, then $f(a) = 0$ for all $a \in \mathrm{Var}(I)$; hence, $f(X)$ is surely one of the polynomials annihilating $\mathrm{Var}(I)$.

(iii) If $V$ is a variety, then $V = \mathrm{Var}(J)$ for some ideal $J$ in $k[X]$. Now

$$\mathrm{Var}(\mathrm{Id}(\mathrm{Var}(J))) \supseteq \mathrm{Var}(J),$$

by part (i). Also, part (ii) gives $\mathrm{Id}(\mathrm{Var}(J)) \supseteq J$, and applying Proposition 5.89(i) gives the reverse inclusion

$$\mathrm{Var}(\mathrm{Id}(\mathrm{Var}(J))) \subseteq \mathrm{Var}(J).$$

Therefore, $\mathrm{Var}(\mathrm{Id}(\mathrm{Var}(J))) = \mathrm{Var}(J)$; that is, $\mathrm{Var}(\mathrm{Id}(V)) = V$.

(iv) By Proposition 5.90(iii), $\overline{F} = \bigcap_{V \supseteq F} V$ is a variety containing $F$. Since $\mathrm{Var}(\mathrm{Id}(F))$ is a variety containing $F$, it is one of varieties $V$ being intersected to form $\overline{F}$, and so $\overline{F} \subseteq \mathrm{Var}(\mathrm{Id}(F))$. For the reverse inclusion, it suffices to prove that if $V$ is any variety containing $F$, then $V \supseteq \mathrm{Var}(\mathrm{Id}(F))$. If $V \supseteq F$, then $\mathrm{Id}(V) \subseteq \mathrm{Id}(F)$, and $V = \mathrm{Var}(\mathrm{Id}(V)) \supseteq \mathrm{Var}(\mathrm{Id}(F))$.

(v) Since $V^* - V \subseteq V^*$, we have $\overline{V^* - V} \subseteq \overline{V^*} = V^*$. By hypothesis, $V \subseteq V^*$, and so $V \cup \overline{V^* - V} \subseteq V^*$. For the reverse inclusion, there is an equation of subsets, $V^* = V \cup (V^* - V)$. Taking closures,

$$V^* = \overline{V^*} = \overline{V} \cup \overline{V^* - V} = V \cup \overline{V^* - V},$$

because $V = \overline{V}$. $\bullet$

**Corollary 5.102.**

(i) *If $V_1$ and $V_2$ are varieties over any field $k$ and $\mathrm{Id}(V_1) = \mathrm{Id}(V_2)$, then $V_1 = V_2$.*

(ii) *If $I_1$ and $I_2$ are radical ideals in $\mathbb{C}[x]$ and $\mathrm{Var}(I_1) = \mathrm{Var}(I_2)$, then $I_1 = I_2$.*

**Proof.**

(i) If $\mathrm{Id}(V_1) = \mathrm{Id}(V_2)$, then $\mathrm{Var}(\mathrm{Id}(V_1)) = \mathrm{Var}(\mathrm{Id}(V_2))$; it now follows from Proposition 5.101(iii) that $V_1 = V_2$.

(ii) If $\mathrm{Var}(I_1) = \mathrm{Var}(I_2)$, then $\mathrm{Id}(\mathrm{Var}(I_1)) = \mathrm{Id}(\mathrm{Var}(I_2))$. By the Nullstellensatz, $\sqrt{I_1} = \sqrt{I_2}$; since $I_1$ and $I_2$ are radical ideals, we have $I_1 = I_2$.  •

We can now give a geometric interpretation of colon ideals.

**Proposition 5.103.** *Let $I$ be a radical ideal in $\mathbb{C}[X]$. Then, for every ideal $J$,*

$$\mathrm{Var}((I : J)) = \overline{\mathrm{Var}(I) - \mathrm{Var}(J)}.$$

**Proof.** We first show that $\mathrm{Var}((I : J)) \supseteq \overline{\mathrm{Var}(I) - \mathrm{Var}(J)}$. If $f \in (I : J)$, then $fg \in I$ for all $g \in J$. Hence, if $x \in \mathrm{Var}(I)$, then $f(x)g(x) = 0$ for all $g \in J$. However, if $x \notin \mathrm{Var}(J)$, then there is $g \in J$ with $g(x) \neq 0$. Since $\mathbb{C}[X]$ is a domain, we have $f(x) = 0$ for all $x \in \mathrm{Var}(I) - \mathrm{Var}(J)$; that is, $f \in \mathrm{Id}(\mathrm{Var}(I) - \mathrm{Var}(J))$. Thus, $(I : J) \subseteq \mathrm{Id}(\mathrm{Var}(I) - \mathrm{Var}(J))$, and so

$$\mathrm{Var}((I : J)) \supseteq \mathrm{Var}(\mathrm{Id}(\mathrm{Var}(I) - \mathrm{Var}(J))) = \overline{\mathrm{Var}(I) - \mathrm{Var}(J)},$$

by Proposition 5.101(iv).

For the reverse inclusion, take $x \in \mathrm{Var}((I : J))$. Thus, if $f \in (I : J)$, then $f(x) = 0$; that is,

$$\text{if } fg \in I \text{ for all } g \in J, \text{ then } f(x) = 0.$$

Suppose now that $h \in \mathrm{Id}(\mathrm{Var}(I) - \mathrm{Var}(J))$. If $g \in J$, then $hg$ vanishes on $\mathrm{Var}(J)$ (because $g$ does); on the other hand, $hg$ vanishes on $\mathrm{Var}(I) - \mathrm{Var}(J)$ (because $h$ does). It follows that $hg$ vanishes on $\mathrm{Var}(J) \cup (\mathrm{Var}(I) - \mathrm{Var}(J)) = \mathrm{Var}(I)$; hence, $hg \in \sqrt{I} = I$ for all $g \in J$, because $I$ is a radical ideal, and so $h \in (I : J)$. Therefore, $h(x) = 0$ for all $h \in (I : J)$, which gives $x \in \mathrm{Var}(\mathrm{Id}(\mathrm{Var}(I) - \mathrm{Var}(J))) = \overline{\mathrm{Var}(I) - \mathrm{Var}(J)}$, as desired.  •

**5.5.3. Irreducible Varieties.** Can a variety be decomposed into simpler subvarieties?

**Definition.** A variety $V$ over a field $k$ is ***irreducible*** if it is not a union of two proper subvarieties; that is, $V \neq W' \cup W''$, where both $W'$ and $W''$ are nonempty varieties.

**Proposition 5.104.** *Let $k$ be any field. Every variety $V$ in $k^n$ is a union of finitely many irreducible subvarieties:*

$$V = V_1 \cup V_2 \cup \cdots \cup V_m.$$

**Proof.** Call a variety $W \in k^n$ *good* if it is irreducible or a union of finitely many irreducible subvarieties; otherwise, call $W$ *bad*. We must show that there are no bad varieties. If $W$ is bad, it is not irreducible, and so $W = W' \cup W''$, where both $W'$ and $W''$ are proper subvarieties. But a union of good varieties is good, and so at least one of $W'$ and $W''$ is bad; say, $W'$ is bad, and rename it $W' = W_1$. Repeat this construction for $W_1$ to get a bad subvariety $W_2$. It follows by induction that there exists a strictly descending sequence

$$W \supsetneqq W_1 \supsetneqq \cdots \supsetneqq W_n \supsetneqq \cdots$$

of bad subvarieties. Since the operator Id reverses inclusions, there is a strictly increasing chain of ideals [the inclusions are strict because of Corollary 5.102(i)]

$$\mathrm{Id}(W) \subsetneqq \mathrm{Id}(W_1) \subsetneqq \cdots \subsetneqq \mathrm{Id}(W_n) \subsetneqq \cdots,$$

contradicting the Hilbert Basis Theorem. Therefore, every variety is good. •

Irreducible varieties over $\mathbb{C}$ have a nice characterization.

**Proposition 5.105.** *A variety $V$ in $\mathbb{C}^n$ is irreducible if and only if $\mathrm{Id}(V)$ is a prime ideal in $\mathbb{C}[X]$. Hence, the coordinate ring $\mathbb{C}[V]$ of an irreducible variety $V$ is a domain.*

**Proof.** Assume that $V$ is an irreducible variety. It suffices to show that if $f_1(X)$, $f_2(X) \notin \mathrm{Id}(V)$, then $f_1(X)f_2(X) \notin \mathrm{Id}(V)$. Define, for $i = 1, 2$,

$$W_i = V \cap \mathrm{Var}(f_i(X)).$$

Note that each $W_i$ is a subvariety of $V$, for it is the intersection of two varieties; moreover, since $f_i(X) \notin \mathrm{Id}(V)$, there is some $a_i \in V$ with $f_i(a_i) \neq 0$, and so $W_i$ is a proper subvariety of $V$. Since $V$ is irreducible, we cannot have $V = W_1 \cup W_2$. Thus, there is some $b \in V$ that is not in $W_1 \cup W_2$; that is, $f_1(b) \neq 0 \neq f_2(b)$. Therefore, $f_1(b)f_2(b) \neq 0$, hence $f_1(X)f_2(X) \notin \mathrm{Id}(V)$, and so $\mathrm{Id}(V)$ is a prime ideal.

Conversely, assume that $\mathrm{Id}(V)$ is a prime ideal. Suppose that $V = V_1 \cup V_2$, where $V_1$ and $V_2$ are subvarieties. If $V_2 \subsetneqq V$, then we must show that $V = V_1$. Now

$$\mathrm{Id}(V) = \mathrm{Id}(V_1) \cap \mathrm{Id}(V_2) \supseteq \mathrm{Id}(V_1)\,\mathrm{Id}(V_2);$$

the equality is given by Proposition 5.94, and the inequality is given by Exercise 5.8 on page 301. Since $\mathrm{Id}(V)$ is a prime ideal, Proposition 5.5 says that $\mathrm{Id}(V_1) \subseteq \mathrm{Id}(V)$ or $\mathrm{Id}(V_2) \subseteq \mathrm{Id}(V)$. But $V_2 \subsetneqq V$ implies $\mathrm{Id}(V_2) \supsetneqq \mathrm{Id}(V)$, and we conclude that $\mathrm{Id}(V_1) \subseteq \mathrm{Id}(V)$. Now the reverse inequality $\mathrm{Id}(V_1) \supseteq \mathrm{Id}(V)$ holds as well, because $V_1 \subseteq V$, and so $\mathrm{Id}(V_1) = \mathrm{Id}(V)$. Therefore, $V_1 = V$, by Corollary 5.102, and so $V$ is irreducible. •

**Remark.** The Nullstellensatz shows that $V \mapsto \mathrm{Id}(V)$ is a bijection from varieties in $\mathbb{C}^n$ to radical ideals in $\mathbb{C}[X]$. Proposition 5.105 shows that the restriction to irreducible varieties is a bijection to $\mathrm{Spec}(\mathbb{C}[X])$. In light of this, one transfers the Zariski topology on $\mathbb{C}^n$ to $\mathrm{Spec}(\mathbb{C}[X])$: if $I$ is an ideal, define

$$V(I) = \{\mathfrak{p} \in \mathrm{Spec}(k[X]) : \mathfrak{p} \supseteq I\}.$$

This construction is extended to arbitrary commutative rings $R$ [recall that $\mathrm{Spec}(R)$ is the family of all its prime ideals]. The ***Zariski topology*** on $\mathrm{Spec}(R)$ is defined by decreeing that the closed subsets are subsets of the form $V(I)$: the set of all prime ideals containing an ideal $I$. Exercise 5.61 on page 368 asks you to give a direct proof that these closed sets are, in fact, a topology on $\mathrm{Spec}(R)$. ◄

We now consider whether the irreducible subvarieties in the decomposition of a variety over an arbitrary field $k$ into a union of irreducible varieties are uniquely determined. There is one obvious way to arrange nonuniqueness. If $P \subsetneqq Q$ in $k[X]$ are two prime ideals [for example, if $P = (x)$ and $Q = (x, y)$ in $k[x, y]$], then $\operatorname{Var}(Q) \subsetneqq \operatorname{Var}(P)$; if $\operatorname{Var}(P)$ is a subvariety of a variety $V$, say, $V = \operatorname{Var}(P) \cup V_2 \cup \cdots \cup V_m$, then $\operatorname{Var}(Q)$ can be one of the $V_i$ or it can be left out.

**Definition.** A decomposition $V = V_1 \cup \cdots \cup V_m$ is an ***irredundant union*** if no $V_i$ can be omitted; that is, for all $i$,

$$V \neq V_1 \cup \cdots \cup \widehat{V_i} \cup \cdots \cup V_m.$$

**Proposition 5.106.** *Every variety $V$ over an arbitrary field $k$ is an irredundant union of irreducible subvarieties*

$$V = V_1 \cup \cdots \cup V_m;$$

*moreover, the irreducible subvarieties $V_i$ are uniquely determined by $V$.*

**Proof.** By Proposition 5.104, $V$ is a union of finitely many irreducible subvarieties; say, $V = V_1 \cup \cdots \cup V_m$. If $m$ is chosen minimal, then this union must be irredundant.

We now prove uniqueness. Suppose that $V = W_1 \cup \cdots \cup W_s$ is an irredundant union of irreducible subvarieties. Let $X = \{V_1, \ldots, V_m\}$ and let $Y = \{W_1, \ldots, W_s\}$; we shall show that $X = Y$. If $V_i \in X$, we have

$$V_i = V_i \cap V = \bigcup_j (V_i \cap W_j).$$

Now $V_i \cap W_j \neq \varnothing$ for some $j$; since $V_i$ is irreducible, there is only one such $W_j$. Therefore, $V_i = V_i \cap W_j$, and so $V_i \subseteq W_j$. The same argument applied to $W_j$ shows that there is exactly one $V_\ell$ with $W_j \subseteq V_\ell$. Hence,

$$V_i \subseteq W_j \subseteq V_\ell.$$

Since the union $V_1 \cup \cdots \cup V_m$ is irredundant, we must have $V_i = V_\ell$, and so $V_i = W_j = V_\ell$; that is, $V_i \in Y$ and $X \subseteq Y$. The reverse inclusion is proved in the same way.  •

**Definition.** An intersection $I = J_1 \cap \cdots \cap J_m$ is ***irredundant*** if no $J_i$ can be omitted; that is, for all $i$,

$$I \neq J_1 \cap \cdots \cap \widehat{J_i} \cap \cdots \cap J_m.$$

**Corollary 5.107.** *Every radical ideal $J$ in $\mathbb{C}[X]$ is an irredundant intersection of prime ideals:*

$$J = P_1 \cap \cdots \cap P_m.$$

*Moreover, the prime ideals $P_i$ are uniquely determined by $J$.*

**Remark.** This corollary is generalized in Exercise 5.66 on page 369: an ideal in an arbitrary commutative noetherian ring is a radical ideal if and only if it is an intersection of finitely many prime ideals.  ◄

**Proof.** Since $J$ is a radical ideal, there is a variety $V$ with $J = \text{Id}(V)$. Now $V$ is an irredundant union of irreducible subvarieties,

$$V = V_1 \cup \cdots \cup V_m,$$

so that

$$J = \text{Id}(V) = \text{Id}(V_1) \cap \cdots \cap \text{Id}(V_m).$$

By Proposition 5.105, $V_i$ irreducible implies $\text{Id}(V_i)$ is prime, and so $J$ is an intersection of prime ideals. This is an irredundant intersection, for if there is $\ell$ with $J = \text{Id}(V) = \bigcap_{j \neq \ell} \text{Id}(V_j)$, then

$$V = \text{Var}(\text{Id}(V)) = \bigcup_{j \neq \ell} \text{Var}(\text{Id}(V_j)) = \bigcup_{j \neq \ell} V_j,$$

contradicting the given irredundancy of the union.

Uniqueness is proved similarly. If $J = \text{Id}(W_1) \cap \cdots \cap \text{Id}(W_s)$, where each $\text{Id}(W_i)$ is a prime ideal (hence is a radical ideal), then each $W_i$ is an irreducible variety. Applying Var expresses $V = \text{Var}(\text{Id}(V)) = \text{Var}(J)$ as an irredundant union of irreducible subvarieties, and the uniqueness of this decomposition gives the uniqueness of the prime ideals in the intersection.  •

**5.5.4. Primary Decomposition.** Given an ideal $I$ in $\mathbb{C}[X]$, how can we find the irreducible components $C_i$ of $\text{Var}(I)$? To ask the question another way, what are the prime ideals $P_i$ with $C_i = \text{Var}(P_i)$? The first guess is that $I = P_1 \cap \cdots \cap P_r$, but this is easily seen to be incorrect: an ideal need not be an intersection of prime ideals. For example, in $\mathbb{C}[x]$, the ideal $((x-1)^2)$ is not an intersection of prime ideals. In light of the Nullstellensatz, we can replace the prime ideals $P_i$ by ideals $Q_i$ with $\sqrt{Q_i} = P_i$, for $\text{Var}(P_i) = \text{Var}(Q_i)$. We are led to the notion of *primary ideal*, defined soon, and the *Primary Decomposition Theorem*, which states that every ideal in a commutative noetherian ring, not merely in $\mathbb{C}[X]$, is an intersection of primary ideals.

**Definition.** An ideal $Q$ in a commutative ring $R$ is **primary** if it is a proper ideal such that $ab \in Q$ (where $a, b \in R$) and $b \notin Q$ implies $a^n \in Q$ for some $n \geq 1$.

It is clear that every prime ideal is primary. Moreover, in $\mathbb{Z}$, the ideal $(p^e)$, where $p$ is prime and $e \geq 2$, is a primary ideal that is not a prime ideal. Example 5.112 below shows that this example is misleading: there are primary ideals that are not powers of prime ideals; there are powers of prime ideals that are not primary ideals.

**Proposition 5.108.** *If $Q$ is a primary ideal in a commutative ring, then its radical $P = \sqrt{Q}$ is a prime ideal. Moreover, if $Q$ is primary, then $ab \in Q$ and $a \notin Q$ implies $b \in P$.*

**Proof.** Assume that $ab \in \sqrt{Q}$, so that $(ab)^m = a^m b^m \in Q$ for some $m \geq 1$. If $a \notin \sqrt{Q}$, then $a^m \notin Q$. Since $Q$ is primary, it follows that some power of $b^m$, say, $b^{mn} \in Q$; that is, $b \in \sqrt{Q}$. We have proved that $\sqrt{Q}$ is prime, as well as the second statement.  •

If $Q$ is primary and $P = \sqrt{Q}$, then we often call $Q$ a *P-**primary ideal***, and we say that $Q$ and $P$ ***belong*** to each other.

We now prove that the properties in Proposition 5.108 characterize primary ideals.

**Proposition 5.109.** *Let $J$ and $T$ be ideals in a commutative ring. If*

(i) $J \subseteq T$,

(ii) $t \in T$ *implies there is some $m \geq 1$ with $t^m \in J$,*

(iii) *if $ab \in J$ and $a \notin J$, then $b \in T$,*

*then $J$ is a primary ideal with radical $T$.*

**Proof.** First, $J$ is a primary ideal, for if $ab \in J$ and $a \notin J$, then item (iii) gives $b \in T$, and item (ii) gives $b^m \in J$. It remains to prove that $T = \sqrt{J}$. Now item (ii) gives $T \subseteq \sqrt{J}$. For the reverse inclusion, if $r \in \sqrt{J}$, then $r^m \in J$; choose $m$ minimal. If $m = 1$, then item (i) gives $r \in J \subseteq T$, as desired. If $m > 1$, then $rr^{m-1} \in J$; since $r^{m-1} \notin J$, item (iii) gives $r \in T$. Therefore, $T = \sqrt{J}$.  •

Let $R$ be a commutative ring, and let $M$ be an ideal. Each $a \in R$ defines an $R$-map $a_M \colon M \to M$ by $a_M \colon m \mapsto am$.

**Lemma 5.110.** *Let $Q$ be an ideal in a commutative ring $R$. Then $Q$ is a primary ideal if and only if, for each $a \in R$, the map $a_{R/Q} \colon R/Q \to R/Q$, given by $r + Q \mapsto ar + Q$, is either an injection or is nilpotent $[(a_{R/Q})^n = 0$ for some $n \geq 1]$.*

**Proof.** Assume that $Q$ is primary. If $a \in R$ and $a_{R/Q}$ is not an injection, then there is $b \in R$ with $b \notin Q$ and $a_{R/Q}(b+Q) = ab+Q = Q$; that is, $ab \in Q$. We must prove that $a_{R/Q}$ is nilpotent. Since $Q$ is primary, there is $n \geq 1$ with $a^n \in Q$; hence, $a^n r \in Q$ for all $r \in R$, because $Q$ is an ideal. Thus, $(a_{R/Q})^n(r+Q) = a^n r + Q = Q$ for all $r \in R$, and $(a_{R/Q})^n = 0$; that is, $a_{R/Q}$ is nilpotent.

Conversely, assume that every $a_{R/Q}$ is either injective or nilpotent. Suppose that $ab \in Q$ and $a \notin Q$. Then $b_{R/Q}$ is not injective, for $a + Q \in \ker b_{R/Q}$. By hypothesis, $(b_{R/Q})^n = 0$ for some $n \geq 1$; that is, $b^n r \in Q$ for all $r \in R$. Setting $r = 1$ gives $b^n \in Q$, and so $Q$ is primary.  •

The next result gives a way of constructing primary ideals.

**Proposition 5.111.** *If $P$ is a maximal ideal in a commutative ring $R$ and $Q$ is an ideal with $P^e \subseteq Q \subseteq P$ for some $e \geq 0$, then $Q$ is a $P$-primary ideal. In particular, every power of a maximal ideal is primary.*

**Proof.** We show, for each $a \in R$, that $a_{R/Q}$ is either nilpotent or injective. Suppose first that $a \in P$. In this case, $a^e \in P^e \subseteq Q$; hence, $a^e b \in Q$ for all $b \in R$, and so $(a_{R/Q})^e = 0$; that is, $a_{R/Q}$ is nilpotent. Now assume that $a \notin P$; we are going to show that $a + Q$ is a unit in $R/Q$, which implies that $a_{R/Q}$ is injective. Since $P$ is a maximal ideal, the ring $R/P$ is a field; since $a \notin P$, the element $a + P$ is a unit in $R/P$: there are $a' \in R$ and $z \in P$ with $aa' = 1 - z$. Now $z + Q$ is a nilpotent element of $R/Q$, for $z^e \in P^e \subseteq Q$. Thus, $1 - z + Q$ is a unit in $R/Q$

(its inverse is $1 + z + \cdots + z^{e-1}$). It follows that $a + Q$ is a unit in $R/Q$, for $aa' + Q = 1 - z + Q$. The result now follows from Lemma 5.110. Finally, $Q$ belongs to $P$, for $P = \sqrt{P^e} \subseteq \sqrt{Q} \subseteq \sqrt{P} = P$. •

**Example 5.112.**

(i) We now show that a power of a prime ideal need not be primary. Suppose that $R$ is a commutative ring containing elements $a, b, c$ such that $ab = c^2$, $P = (a, c)$ is a prime ideal, $a \notin P^2$, and $b \notin P$. Now $ab = c^2 \in P^2$; were $P^2$ primary, then $a \notin P^2$ would imply that $b \in \sqrt{P^2} = P$, and this is not so. We construct such a ring $R$ as follows. Let $k$ be a field, and define $R = k[x, y, z]/(xy - z^2)$ (note that $R$ is noetherian). Define $a, b, c \in R$ to be the cosets of $x, y, z$, respectively. Now $P = (a, c)$ is a prime ideal, for the Third Isomorphism Theorem for Rings, Exercise 2.91 on page 170, gives

$$R/(a, c) = \frac{k[x, y, z]/(xy - z^2)}{(x, z)/(xy - z^2)} \cong \frac{k[x, y, z]}{(x, z)} \cong k[y],$$

which is a domain. The equation $ab = c^2$ obviously holds in $R$. Were $a \in P^2$, then lifting this relation to $k[x, y, z]$ would yield an equation

$$x = f(x, y, z)x^2 + g(x, y, z)xz + h(x, y, z)z^2 + \ell(x, y, z)(xy - z^2).$$

Setting $y = 0 = z$ (i.e., using the evaluation homomorphism $k[x, y, z] \to k[x]$) gives the equation $x = f(x, 0, 0)x^2$ in $k[x]$, a contradiction. A similar argument shows that $b \notin P$.

(ii) We use Proposition 5.111 to show that there are primary ideals $Q$ that are not powers of prime ideals. Let $R = k[x, y]$, where $k$ is a field. The ideal $P = (x, y)$ is maximal, hence prime (for $R/P \cong k$); moreover,

$$P^2 \subsetneq (x^2, y) \subsetneq (x, y) = P$$

[the strict inequalities follow from $x \notin (x^2, y)$ and $y \notin P^2$]. Thus, $Q = (x^2, y)$ is not a power of $P$; indeed, we show that $Q \neq L^e$, where $L$ is a prime ideal. If $Q = L^e$, then $P^2 \subseteq L^e \subseteq P$, hence $\sqrt{P^2} \subseteq \sqrt{L^e} \subseteq \sqrt{P}$, and so $P \subseteq L \subseteq P$, a contradiction. ◄

We now generalize Corollary 5.107 by proving that every ideal in a noetherian ring, in particular, in $k[X]$ for $k$ a field, is an intersection of primary ideals. This result, along with uniqueness properties, was first proved by E. Lasker;[18] his proof was later simplified by E. Noether. Note that we will be working in arbitrary noetherian rings, not merely in $k[X]$.

**Definition.** A *primary decomposition* of an ideal $I$ in a commutative ring $R$ is a finite family of primary ideals $Q_1, \ldots, Q_r$ with

$$I = Q_1 \cap Q_2 \cap \cdots \cap Q_r.$$

**Theorem 5.113 (Lasker–Noether I).** *If $R$ is a commutative noetherian ring, then every proper ideal $I$ in $R$ has a primary decomposition.*

---

[18] Emanuel Lasker was also the World Chess Champion in 1894–1910.

**Proof.** Let $\mathcal{F}$ be the family of all those proper ideals in $R$ that do not have a primary decomposition; we must show that $\mathcal{F}$ is empty. Since $R$ is noetherian, if $\mathcal{F} \neq \varnothing$, then it has a maximal element, say, $J$. Of course, $J$ is not primary, and so there exists $a \in R$ with $a_{R/J} \colon R/J \to R/J$ neither injective nor nilpotent. The ascending chain of ideals of $R/J$,

$$\ker a_{R/J} \subseteq \ker (a_{R/J})^2 \subseteq \ker (a_{R/J})^3 \subseteq \cdots,$$

must stop (because $R/J$, being a quotient of the noetherian ring $R$, is itself noetherian); there is $m \geq 1$ with $\ker(a_{R/J}^{\ell}) = \ker(a_{R/J}^m)$ for all $\ell \geq m$. Denote $(a_{R/J})^m$ by $\varphi$, so that $\ker(\varphi^2) = \ker \varphi$. Note that $\ker \varphi \neq (0)$, because $(0) \subsetneq \ker a_{R/J} \subseteq \ker(a_{R/J})^m = \ker \varphi$, and that $\operatorname{im} \varphi = \operatorname{im}(a_{R/J})^m \neq (0)$, because $a_{R/J}$ is not nilpotent. We claim that

$$\ker \varphi \cap \operatorname{im} \varphi = (0).$$

If $x \in \ker \varphi \cap \operatorname{im} \varphi$, then $\varphi(x) = 0$ and $x = \varphi(y)$ for some $y \in R/J$. But $\varphi(x) = \varphi(\varphi(y)) = \varphi^2(y)$, so that $y \in \ker(\varphi^2) = \ker \varphi$ and $x = \varphi(y) = 0$.

If $\pi \colon R \to R/J$ is the natural map, then $A = \pi^{-1}(\ker \varphi)$ and $A' = \pi^{-1}(\operatorname{im} \varphi)$ are ideals of $R$ with $A \cap A' = J$. It is obvious that $A$ is a proper ideal; we claim that $A'$ is also proper. Otherwise, $A' = R$, so that $A \cap A' = A$; but $A \cap A' = J$, as we saw above, and $A \neq J$, a contradiction. Since $A$ and $A'$ are strictly larger than $J$, neither of them lies in $\mathcal{F}$: there are primary decompositions $A = Q_1 \cap \cdots \cap Q_m$ and $A' = Q_1' \cap \cdots \cap Q_n'$. Therefore,

$$J = A \cap A' = Q_1 \cap \cdots \cap Q_m \cap Q_1' \cap \cdots \cap Q_n',$$

contradicting $J$ not having a primary decomposition (for $J \in \mathcal{F}$).  •

**Definition.** A primary decomposition $I = Q_1 \cap \cdots \cap Q_r$ is ***irredundant*** if no $Q_i$ can be omitted; for all $i$,

$$I \neq Q_1 \cap \cdots \cap \widehat{Q}_i \cap \cdots \cap Q_r.$$

The prime ideals $P_1 = \sqrt{Q_1}, \ldots, P_r = \sqrt{Q_r}$ are called the ***associated prime ideals*** of the irredundant primary decomposition.

It is clear that any primary decomposition can be made irredundant by throwing away, one at a time, any primary ideals that contain the intersection of the others.

**Theorem 5.114 (Lasker–Noether II).** *If $I$ is an ideal in a noetherian ring $R$, then any two irredundant primary decompositions of $I$ have the same set of associated prime ideals. Hence, the associated prime ideals are uniquely determined by $I$.*

**Proof.** Let $I = Q_1 \cap \cdots \cap Q_r$ be an irredundant primary decomposition, and let $P_i = \sqrt{Q_i}$. We are going to prove that a prime ideal $P$ in $R$ is equal to some $P_i$ if and only if there is $c \notin I$ with $(I : c)$ a $P$-primary ideal. This will suffice, for the colon ideal $(I : c)$ is defined solely in terms of $I$ and not in terms of any primary decomposition.

Given $P_i$, there exists $c \in \bigcap_{j \neq i} Q_j$ with $c \notin Q_i$, because of irredundancy; we show that $(I : c_i)$ is $P_i$-primary. Recall Proposition 5.109: the following three conditions: (i) $(I : c) \subseteq P_i$; (ii) $b \in P_i$ implies there is some $m \geq 1$ with $b^m \in (I : c)$; (iii) if $ab \in (I : c)$ and $a \notin (I : c)$, imply that $b \in P_i$ and $(I : c)$ is $P_i$-primary.

To see (i), take $u \in (I : c)$; then $uc \in I \subseteq P_i$. As $c \notin Q_i$, we have $u \in P_i$, by Proposition 5.108. To prove (ii), we first show that $Q_i \subseteq (I : c)$. If $a \in Q_i$, then $ca \in Q_i$, since $Q_i$ is an ideal. If $j \neq i$, then $c \in Q_j$, and so $ca \in Q_j$. Therefore, $ca \in Q_1 \cap \cdots \cap Q_r = I$, and so $a \in (I : c)$. If, now, $b \in P_i$, then $b^m \in Q_i \subseteq (I : c)$. Finally, we establish (iii) by proving its contrapositive: if $xy \in (I : c)$ and $x \notin P_i$, then $y \in (I : c)$. Thus, assume that $xyc \in I$; since $I \subseteq Q_i$ and $x \notin P_i = \sqrt{Q_i}$, we have $yc \in Q_i$. But $yc \in Q_j$ for all $j \neq i$, for $c \in Q_j$. Therefore, $yc \in Q_1 \cap \cdots \cap Q_r = I$, and so $y \in (I : c)$. We conclude that $(I : c)$ is $P_i$-primary.

Conversely, assume that there is an element $c \notin I$ and a prime ideal $P$ such that $(I : c)$ is $P$-primary. We must show that $P = P_i$ for some $i$. Exercise 5.11(ii) on page 302 gives $(I : c) = (Q_1 : c) \cap \cdots \cap (Q_r : c)$. Therefore, by Proposition 5.96,

$$P = \sqrt{(I : c)} = \sqrt{(Q_1 : c)} \cap \cdots \cap \sqrt{(Q_r : c)}.$$

If $c \in Q_i$, then $(Q_i : c) = R$; if $c \notin Q_i$, then, as we saw in the first part of this proof, $(Q_i : c)$ is $P_i$-primary. Thus, there is $s \leq r$ with

$$P = \sqrt{(Q_{i_1} : c)} \cap \cdots \cap \sqrt{(Q_{i_s} : c)} = P_{i_1} \cap \cdots \cap P_{i_s}.$$

Of course, $P \subseteq P_{i_j}$ for all $j$. On the other hand, Exercise 5.8(iii) on page 301 gives $P_{i_j} \subseteq P$ for some $j$, and so $P = P_{i_j}$, as desired. •

**Example 5.115.**

(i) Let $R = \mathbb{Z}$, let $(n)$ be a nonzero proper ideal, and let $n = p_1^{e_1} \cdots p_t^{e_t}$ be the prime factorization. Then

$$(n) = (p_1^{e_1}) \cap \cdots \cap (p_t^{e_t})$$

is an irredundant primary decomposition.

(ii) Let $R = k[x, y]$, where $k$ is a field. Define $Q_1 = (x)$ and $Q_2 = (x, y)^2$. Note that $Q_1$ is prime, and hence $Q_1$ is $P_1$-primary for $P_1 = Q_1$. Also, $P_2 = (x, y)$ is a maximal ideal, and so $Q_2 = P_2^2$ is $P_2$-primary, by Proposition 5.111. Define $I = Q_1 \cap Q_2$. This primary decomposition of $I$ is irredundant. The associated primes of $I$ are thus $\{P_1, P_2\}$. ◄

There is a second uniqueness result that describes a *normalized* primary decomposition, but we precede it by a lemma.

**Lemma 5.116.** *If $P$ is a prime ideal and $Q_1, \ldots, Q_n$ are $P$-primary ideals, then $Q_1 \cap \cdots \cap Q_n$ is also a $P$-primary ideal.*

**Proof.** We verify that the three items in the hypothesis of Proposition 5.109 hold for $I = Q_1 \cap \cdots \cap Q_n$. Clearly, $I \subseteq P$. Second, if $b \in P$, then $b^{m_i} \in Q_i$ for all $i$, because $Q_i$ is $P$-primary. Hence, $b^m \in I$, where $m = \max\{m_1, \ldots, m_n\}$. Finally, assume that $ab \in I$. If $a \notin I$, then $a \notin Q_i$ for some $i$. As $Q_i$ is $P$-primary, $ab \in I \subseteq Q_i$ and $a \notin Q_i$ imply $b \in P$. Therefore, $I$ is $P$-primary. •

**Definition.** A primary decomposition $I = Q_1 \cap \cdots \cap Q_r$ is **normal** if it is irredundant and all the prime ideals $P_i = \sqrt{Q_i}$ are distinct.

**Corollary 5.117.** *If $R$ is a noetherian ring, then every proper ideal in $R$ has a normal primary decomposition.*

**Proof.** By Theorem 5.113, every proper ideal $I$ has a primary decomposition, say,

$$I = Q_1 \cap \cdots \cap Q_r,$$

where $Q_i$ is $P_i$-primary. If $P_r = P_i$ for some $i < r$, then $Q_i$ and $Q_r$ can be replaced by $Q' = Q_i \cap Q_r$, which is primary, by Lemma 5.116. Iterating, we eventually arrive at a primary decomposition with all prime ideals distinct. If this decomposition is not irredundant, remove primary ideals from it, one at a time, to obtain a normal primary decomposition. $\quad\bullet$

**Definition.** If $I = Q_1 \cap \cdots \cap Q_r$ is a normal primary decomposition, then the minimal prime ideals $P_i = \sqrt{Q_i}$ are called **isolated** prime ideals; the other prime ideals, if any, are called **embedded**.

Consider the ideal $I = (x) \cap (x, y)^2$ in $k[x, y]$, where $k$ is a field. We gave an irredundant primary decomposition of $I$ in Example 5.115. The associated primes are $(x)$ and $(x, y)$, so that $(x)$ is an isolated prime and $(x, y)$ is an embedded prime.

**Definition.** A prime ideal $P$ is **minimal** over an ideal $I$ if $I \subseteq P$ and there is no prime ideal $P'$ with $I \subseteq P' \subsetneq P$.

**Corollary 5.118.** *Let $I$ be an ideal in a noetherian ring $R$.*

  (i) *Any two normal primary decompositions of $I$ have the same set of isolated prime ideals, and so the isolated prime ideals are uniquely determined by $I$.*

 (ii) *$I$ has only finitely many minimal prime ideals.*

(iii) *A noetherian ring has only finitely many minimal prime ideals.*

**Proof.**

  (i) Let $I = Q_1 \cap \cdots \cap Q_n$ be a normal primary decomposition. If $P$ is any prime ideal containing $I$, then

$$P \supseteq I = Q_1 \cap \cdots \cap Q_n \supseteq Q_1 \cdots Q_n.$$

   Now $P \supseteq Q_i$ for some $i$, by Proposition 5.5, and so $P \supseteq \sqrt{Q_i} = P_i$. In other words, any prime ideal containing $I$ must contain an isolated associated prime ideal. Hence, the isolated primes are the minimal elements in the set of associated primes of $I$; by Theorem 5.114, they are uniquely determined by $I$.

 (ii) As in part (i), any prime ideal $P$ containing $I$ must contain an isolated prime of $I$. Hence, if $P$ is minimal over $I$, then $P$ must equal an isolated prime ideal of $I$. The result follows, for $I$ has only finitely many isolated prime ideals.

(iii) This follows from part (ii) taking $I = (0)$. $\quad\bullet$

Here are some natural problems arising as these ideas are investigated further. First, what is the dimension of a variety over a field $k$? There are several candidates, and the prime ideals containing it are the key. If $V$ is a variety, then its dimension is the length of a longest chain of prime ideals in its coordinate ring $k[V]$ (which, by the Correspondence Theorem for Rings on page 295, is the length of a longest chain of prime ideals above $\mathrm{Id}(V)$ in $k[X]$).

It turns out to be more convenient to work in a larger *projective space* arising from $k^n$ by adjoining a "hyperplane at infinity." For example, a projective plane arises from the usual plane by adjoining a line at infinity (it is the "horizon" where all parallel lines meet). To distinguish it from projective space, $k^n$ is called *affine space*, for it consists of the "finite points"—that is, not the points at infinity. If we study varieties in projective space, now defined as zeros of a set of *homogeneous* polynomials, then it is often true that many separate affine cases become part of one simpler projective property. For example, if $C \subseteq k^n$, define $\deg(C)$ to be the largest number of points in $C \cap \ell$ as $\ell$ varies over all lines. If $C = \mathrm{Var}(f)$ is a curve arising from a polynomial $f$ of degree $d$, we would like to have $\deg(C) = d$, but there are several problems. First, we must demand that the coefficient field be algebraically closed, lest $\mathrm{Var}(f) = \varnothing$ give $\deg(C) = 0$. Second, there may be multiple roots, and so some intersections may have to be counted with a certain *multiplicity*. *Bézout's Theorem* states that if $C$ and $C'$ are two curves in $\mathbb{C}^2$, then $|C \cap C'| = \deg(C)\deg(C')$ (where points are counted with multiplicities). Defining multiplicities for intersections of higher-dimensional varieties is very subtle.

Finally, there is a deep analogy between differentiable manifolds and varieties. An $n$-*manifold* is a Hausdorff space $M$ each of whose points has an open neighborhood homeomorphic to $\mathbb{R}^n$; that is, it is a union of open replicas of Euclidean space glued together in a coherent way; $M$ is *differentiable* if it has a tangent space at each of its points. For example, a torus $T$ (i.e., a doughnut) is a differentiable manifold. A variety $V$ can be identified with its coordinate ring $k[V]$, and neighborhoods of its points can be described "locally", using what is called a *sheaf* of local rings over $\mathrm{Spec}(k[V])$. If we "glue" sheaves together along open subsets, we obtain a *scheme*, and schemes are the modern way to treat varieties.

## Exercises

**5.55.** Prove that if an element $a$ in a commutative ring $R$ is nilpotent, then $1 + a$ is a unit.

**Hint.** Consider the formal power series for $1/(1+a)$.

∗ **5.56.** Prove that the radical $\sqrt{I}$ of an ideal $I$ in a commutative ring $R$ is an ideal.

**Hint.** If $f^r \in I$ and $g^s \in I$, prove that $(f+g)^{r+s} \in I$.

**5.57.** If $R$ is a commutative ring, then its *nilradical* $\mathrm{nil}(R)$ is defined to be the intersection of all the prime ideals in $R$. Prove that $\mathrm{nil}(R)$ is the set of all the nilpotent elements in $R$:

$$\mathrm{nil}(R) = \{r \in R : r^m = 0 \text{ for some } m \geq 1\}.$$

**Hint.** If $r \in R$ is not nilpotent, use Exercise 5.7 on page 301 to show that there is some prime ideal not containing $r$.

**5.58.** (i) Show that $x^2 + y^2$ is irreducible in $\mathbb{R}[x, y]$, and conclude that $(x^2 + y^2)$ is a prime, hence radical, ideal in $\mathbb{R}[x, y]$.

(ii) Prove that $\mathrm{Var}(x^2 + y^2) = \{(0, 0)\}$.

(iii) Prove that $\mathrm{Id}(\mathrm{Var}(x^2 + y^2)) \supsetneq (x^2 + y^2)$, and conclude that the radical ideal $(x^2 + y^2)$ in $\mathbb{R}[x, y]$ is not of the form $\mathrm{Id}(V)$ for some variety $V$. Conclude that the Nullstellensatz may fail in $k[X]$ if $k$ is not algebraically closed.

(iv) Prove that $(x^2 + y^2) = (x + iy) \cap (x - iy)$ in $\mathbb{C}[x, y]$.

(v) Prove that $\mathrm{Id}(\mathrm{Var}(x^2 + y^2)) = (x^2 + y^2)$ in $\mathbb{C}[x, y]$.

**5.59.** Let $f_1(X), \ldots, f_t(X) \in \mathbb{C}[X]$. Prove that $\mathrm{Var}(f_1, \ldots, f_t) = \varnothing$ if and only if there are $h_1, \ldots, h_t \in \mathbb{C}[X]$ such that

$$1 = \sum_{i=1}^{t} h_i(X) f_i(X).$$

* **5.60.** Let $I = \big(f_1(X), \ldots, f_t(X)\big) \subseteq \mathbb{C}[X]$. For every $g(X) \in \mathbb{C}[X]$, prove that $g \in \sqrt{I} \subseteq \mathbb{C}[X]$ if and only if $(f_1, \ldots, f_t, 1 - yg)$ is not a proper ideal in $\mathbb{C}[X, y]$.

**Hint.** Use the Rabinowitch trick.

* **5.61.** Let $R$ be a commutative ring, and let $\mathrm{Spec}(R)$ denote the set of all the prime ideals in $R$. If $I$ is an ideal in $R$, define

$$\overline{I} = \{\text{all the prime ideals in } R \text{ containing } I\}.$$

Give direct proofs of the following statements.

(i) $\overline{(0)} = \mathrm{Spec}(R)$.

(ii) $\overline{R} = \varnothing$.

(iii) $\overline{\sum_{\ell} I_{\ell}} = \bigcap_{\ell} \overline{I_{\ell}}$.

(iv) $\overline{I \cap J} = \overline{IJ} = \overline{I} \cup \overline{J}$.

Conclude that $\mathrm{Spec}(R)$ is a topological space whose closed subsets are the Zariski closed sets as defined in the remark on page 359.

**5.62.** Prove that an ideal $P$ in $\mathrm{Spec}(R)$ is closed (that is, the one-point set $\{P\}$ is a Zariski closed set) if and only if $P$ is a maximal ideal. Is $\mathrm{Spec}(\mathbb{Z})$ Hausdorff?

**5.63.** Let $f \colon R \to A$ be a ring homomorphism, and define $f^* \colon \mathrm{Spec}(A) \to \mathrm{Spec}(R)$ by $f^*(Q) = f^{-1}(Q)$, where $Q$ is any prime ideal in $A$. Prove that $f^*$ is a continuous function. [Recall that $f^{-1}(Q)$ is a prime ideal, by Proposition 5.7.]

**5.64.** Prove that the function $\varphi \colon k^n \to \mathrm{Spec}(k[x_1, \ldots, x_n])$, given by

$$\varphi \colon (a_1, \ldots, a_n) \mapsto (x_1 - a_1, \ldots, x_n - a_n),$$

is a continuous injection [where $k = \mathbb{C}$ or $k$ is an (uncountable) algebraically closed field and both $k^n$ and $\mathrm{Spec}(k[x_1, \ldots, x_n])$ are equipped with the Zariski topology].

**5.65.** Prove that any descending chain

$$F_1 \supseteq F_2 \supseteq \cdots \supseteq F_m \supseteq F_{m+1} \supseteq \cdots$$

of Zariski closed sets in $k^n$ (where $k$ is a field) stops; there is some $t$ with $F_t = F_{t+1} = \cdots$.

**∗ 5.66.** If $R$ is a commutative noetherian ring, prove that an ideal $I$ in $R$ is a radical ideal if and only if $I = P_1 \cap \cdots \cap P_r$, where the $P_i$ are prime ideals.

**5.67.** Give an example of a commutative ring $R$ containing an ideal $I$ that is not primary and whose radical $\sqrt{I}$ is prime.

**Hint.** Take $R = k[x, y]$, where $k$ is a field, and $I = (x^2, xy)$.

**5.68.** Let $R = k[x, y]$, where $k$ is a field, and let $I = (x^2, y)$. For each $a \in k$, prove that $I = (x) \cap (y + ax, x^2)$ is an irredundant primary decomposition. Conclude that the primary ideals in an irredundant primary decomposition of an ideal need not be unique.

## Section 5.6. Algorithms in $k[x_1, \ldots, x_n]$

Computations and algorithms are useful, if for no other reason than to serve as data from which we might conjecture theorems. But algorithms can do more than provide data in particular cases. For example, the Euclidean Algorithm is used in an essential way in proving that if $K/k$ is a field extension and $f(x), g(x) \in k[x]$, then their gcd in $K[x]$ is equal to their gcd in $k[x]$.

Given two polynomials $f(x), g(x) \in k[x]$ with $g(x) \neq 0$, where $k$ is a field, when is $g(x)$ a divisor of $f(x)$? The Division Algorithm gives unique polynomials $q(x), r(x) \in k[x]$ with

$$f(x) = q(x)g(x) + r(x),$$

where $r = 0$ or $\deg(r) < \deg(g)$, and $g \mid f$ if and only if the remainder $r = 0$. Let us look at this formula from a different point of view. To say that $g \mid f$ is to say that $f \in (g)$, the principal ideal generated by $g(x)$. Thus, the remainder $r$ is the obstruction to $f$ lying in this ideal; that is, $f \in (g)$ if and only if $r = 0$. Now consider the *membership problem*. Given polynomials

$$f(x), g_1(x), \ldots, g_m(x) \in k[x],$$

where $k$ is a field, when is $f \in I = (g_1, \ldots, g_m)$? The Euclidean Algorithm finds $d = \gcd\{g_1, \ldots, g_m\}$,[19] and $I = (d)$. Thus, the two classical algorithms combine to give an algorithm determining whether $f \in I = (g_1, \ldots, g_m) = (d)$.

We now ask whether there is an algorithm in $k[x_1, \ldots, x_n] = k[X]$ to determine, given $f(X), g_1(X), \ldots, g_m(X) \in k[X]$, whether $f \in (g_1, \ldots, g_m)$. A generalized Division Algorithm in $k[X]$ should be an algorithm yielding

$$r(X), a_1(X), \ldots, a_m(X) \in k[X],$$

with $r(X)$ unique, such that

$$f = a_1 g_1 + \cdots + a_m g_m + r$$

and $f \in (g_1, \ldots, g_m)$ if and only if $r = 0$. Since $(g_1, \ldots, g_m)$ consists of all the linear combinations of the $g$'s, such an algorithm would say that the remainder $r$ is the obstruction to $f$ lying in $(g_1, \ldots, g_m)$.

We are going to show that both the Division Algorithm and the Euclidean Algorithm can be extended to polynomials in several variables. Even though these

---

[19]Use induction on $m \geq 2$ to find $d' = \gcd\{g_1, \ldots, g_{m-1}\}$; then $d = \gcd\{d', g_m\}$.

results are elementary, they were discovered only recently, in 1965, by B. Buchberger. Algebra has always dealt with algorithms, but the power and beauty of the axiomatic method has dominated the subject ever since Cayley and Dedekind in the second half of the nineteenth century. After the invention of the transistor in 1948, high-speed calculation became a reality, and old complicated algorithms, as well as new ones, could be implemented; a higher order of computing had entered Algebra. Most likely, the development of Computer Science is a major reason why generalizations of the classical algorithms, from polynomials in one variable to polynomials in several variables, are only now being discovered. This is a dramatic illustration of the impact of external ideas on Mathematics.

**5.6.1. Monomial Orders.** The most important feature of the Division Algorithm in $k[x]$, where $k$ is a field, is that the remainder $r(x)$ has small degree. Without the inequality $\deg(r) < \deg(g)$, the result would be virtually useless; after all, given any $Q(x) \in k[x]$, there is an equation

$$f(x) = Q(x)g(x) + [f(x) - Q(x)g(x)].$$

When dividing $f(x)$ by $g(x)$ in $k[x]$, one usually arranges the monomials in $f(x)$ in descending order, according to degree:

$$f(x) = c_n x^n + c_{n-1} x^{n-1} + \cdots + c_2 x^2 + c_1 x + c_0.$$

Consider a polynomial in several variables:

$$f(X) = f(x_1, \ldots, x_n) = \sum c_{(\alpha_1, \ldots, \alpha_n)} x_1^{\alpha_1} \cdots x_n^{\alpha_n},$$

where $c_{(\alpha_1, \ldots, \alpha_n)} \in k$ and $\alpha_i \geq 0$ for all $i$. We will abbreviate $(\alpha_1, \ldots, \alpha_n)$ to $\alpha$ and $x_1^{\alpha_1} \cdots x_n^{\alpha_n}$ to $X^\alpha$, so that $f(X)$ can be written more compactly as

$$f(X) = \sum_\alpha c_\alpha X^\alpha.$$

Our aim is to arrange the monomials involved in $f(X)$ in a reasonable way.

**Definition.** The *degree* of a nonzero monomial $c x_1^{\alpha_1} \cdots x_n^{\alpha_n} = cX^\alpha \in k[X] = k[x_1, \ldots, x_n]$ is the $n$-tuple $\alpha = (\alpha_1, \ldots, \alpha_n) \in \mathbb{N}^n$. We write

$$\mathrm{DEG}(cX^\alpha) = \alpha.$$

The *weight* $|\alpha|$ of $cX^\alpha$ is the sum $|\alpha| = \alpha_1 + \cdots + \alpha_n \in \mathbb{N}$.

The set $\mathbb{N}^n$, consisting of all the $n$-tuples $\alpha = (\alpha_1, \ldots, \alpha_n)$ of natural numbers, is a commutative monoid, where addition is coordinatewise:

$$(\alpha_1, \ldots, \alpha_n) + (\beta_1, \ldots, \beta_n) = (\alpha_1 + \beta_1, \ldots, \alpha_n + \beta_n).$$

**Proposition 5.119.** *Let $\Omega$ be a well-ordered set.*

(i) *$\Omega$ is a chain; that is, if $x, y \in \Omega$, then either $x \preceq y$ or $y \preceq x$.*

(ii) *Every strictly decreasing sequence in $\Omega$ is finite.*

**Proof.**

(i) The subset $\{x, y\}$ has a smallest element, which must be either $x$ or $y$. In the first case, $x \preceq y$; in the second case, $y \preceq x$.

(ii) Assume that there is an infinite strictly decreasing sequence, say,

$$x_1 \succ x_2 \succ x_3 \succ \cdots .$$

Since $\Omega$ is well-ordered, the subset consisting of all the $x_i$ has a smallest element, say, $x_n$. But $x_{n+1} \prec x_n$, a contradiction. •

The second property of well-ordered sets will be used in showing that an algorithm eventually stops. Given $f(x), g(x) \in k[x]$, the Division Algorithm yielding $q, r \in k[x]$ with $f = qg + r$ and either $r = 0$ or $\deg(r) < \deg(g)$ proceeds by lowering the degree of $f$ at each step; the Euclidean Algorithm proceeds by lowering the degree of certain remainders. If the algorithm yielding the gcd does not stop at a given step, then the natural number associated to the next step—the degree of an associated polynomial—is strictly smaller. Since the set $\mathbb{N}$ of natural numbers, equipped with the usual inequality $\leq$, is well-ordered, any strictly decreasing sequence of natural numbers must be finite; that is, the algorithm stops after a finite number of steps.

We are interested in orderings of degrees that are compatible with addition in the monoid $\mathbb{N}^n$.

**Definition.** A **monomial order** is a well-ordering of $\mathbb{N}^n$ such that

$$\alpha \preceq \beta \quad \text{implies} \quad \alpha + \gamma \preceq \beta + \gamma$$

for all $\alpha, \beta, \gamma \in \mathbb{N}^n$.

A monomial order on $\mathbb{N}^n$ gives a well-ordering of monomials in $k[x_1, \ldots, x_n]$: define

$$X^\alpha \preceq X^\beta$$

if $\alpha \preceq \beta$. Thus, monomials are ordered according to their degrees: $X^\alpha \preceq X^\beta$ if $\mathrm{DEG}(X^\alpha) \preceq \mathrm{DEG}(X^\beta)$. We now extend this definition of degree from monomials to polynomials.

**Definition.** If $\mathbb{N}^n$ is equipped with a monomial order, then every $f(X) \in k[X] = k[x_1, \ldots, x_n]$ can be written with its largest monomial first, followed by its other, smaller, monomials in descending order: $f(X) = c_\alpha X^\alpha +$ lower monomials. Define its **leading monomial**[20] to be

$$\mathrm{LM}(f) = c_\alpha X^\alpha$$

and its **degree** to be

$$\mathrm{DEG}(f) = \alpha = \mathrm{DEG}(c_\alpha X^\alpha) = \mathrm{DEG}(\mathrm{LM}(f)).$$

Call $f(X)$ **monic** if $\mathrm{LM}(f) = X^\alpha$; that is, if $c_\alpha = 1$.

There are many examples of monomial orders, but we shall give only the two most popular ones. Here is the first example.

**Definition.** The **lexicographic order** on $\mathbb{N}^n$ is defined by $\alpha \preceq_{\mathrm{lex}} \beta$ if either $\alpha = \beta$ or the first nonzero coordinate in $\beta - \alpha$ is positive.[21]

---

[20]The leading monomial if often called the **leading term**; it is then denoted by LT.

[21]The difference $\beta - \alpha$ may not lie in $\mathbb{N}^n$, but it does lie in $\mathbb{Z}^n$.

In other words, if $\alpha \prec_{\mathrm{lex}} \beta$, their first $i-1$ coordinates agree for some $i \geq 1$ [that is, $\alpha_1 = \beta_1, \ldots, \alpha_{i-1} = \beta_{i-1}$] and there is strict inequality $\alpha_i < \beta_i$.

The term *lexicographic* refers to the standard ordering in a dictionary. For example, the following 8-letter German words are increasing in lexicographic order (the letters are ordered $a < b < c < \cdots < z$):

ausgehen

ausladen

auslagen

auslegen

bedeuten

**Proposition 5.120.** *The lexicographic order on $\mathbb{N}^n$ is a monomial order.*

**Proof.** First, we show that the lexicographic order is a partial order. The relation $\preceq_{\mathrm{lex}}$ is reflexive, for its definition shows that $\alpha \preceq_{\mathrm{lex}} \alpha$. To prove antisymmetry, assume that $\alpha \preceq_{\mathrm{lex}} \beta$ and $\beta \preceq_{\mathrm{lex}} \alpha$. If $\alpha \neq \beta$, there is a first coordinate, say the $i$th, where they disagree. For notation, we may assume that $\alpha_i < \beta_i$. But this contradicts $\beta \preceq_{\mathrm{lex}} \alpha$. To prove transitivity, suppose that $\alpha \prec_{\mathrm{lex}} \beta$ and $\beta \prec_{\mathrm{lex}} \gamma$ (it suffices to consider strict inequality). Now $\alpha_1 = \beta_1, \ldots, \alpha_{i-1} = \beta_{i-1}$ and $\alpha_i < \beta_i$. Let $\gamma_p$ be the first coordinate with $\beta_p < \gamma_p$. If $p < i$, then

$$\gamma_1 = \beta_1 = \alpha_1, \ldots, \gamma_{p-1} = \beta_{p-1} = \alpha_{p-1}, \; \alpha_p = \beta_p < \gamma_p;$$

if $p \geq i$, then

$$\gamma_1 = \beta_1 = \alpha_1, \ldots, \gamma_{i-1} = \beta_{i-1} = \alpha_{i-1}, \; \alpha_i < \beta_i = \gamma_i.$$

In either case, the first nonzero coordinate of $\gamma - \alpha$ is positive; that is, $\alpha \prec_{\mathrm{lex}} \gamma$.

Next, we show that the lexicographic order is a well-order. If $S$ is a nonempty subset of $\mathbb{N}^n$, define

$$C_1 = \{\text{all first coordinates of } n\text{-tuples in } S\},$$

and define $\delta_1$ to be the smallest number in $C_1$ (note that $C_1$ is a nonempty subset of the well-ordered set $\mathbb{N}$). Define

$$C_2 = \{\text{all second coordinates of } n\text{-tuples } (\delta_1, \alpha_2, \ldots, \alpha_n) \in S\}.$$

Since $C_2 \neq \varnothing$, it contains a smallest number, $\delta_2$. Similarly, for all $i < n$, define $C_{i+1}$ as all the $(i+1)$th coordinates of those $n$-tuples in $S$ whose first $i$ coordinates are $(\delta_1, \delta_2, \ldots, \delta_i)$, and define $\delta_{i+1}$ to be the smallest number in $C_{i+1}$. By construction, the $n$-tuple $\delta = (\delta_1, \delta_2, \ldots, \delta_n)$ lies in $S$; moreover, if $\alpha = (\alpha_1, \alpha_2, \ldots, \alpha_n) \in S$, then

$$\alpha - \delta = (\alpha_1 - \delta_1, \alpha_2 - \delta_2, \ldots, \alpha_n - \delta_n)$$

has all its coordinates nonnegative. Hence, if $\alpha \neq \delta$, then its first nonzero coordinate is positive, and so $\delta \prec_{\mathrm{lex}} \alpha$. Therefore, the lexicographic order is a well-order.

Assume that $\alpha \preceq_{\mathrm{lex}} \beta$; we claim that

$$\alpha + \gamma \preceq_{\mathrm{lex}} \beta + \gamma$$

for all $\gamma \in \mathbb{N}$. If $\alpha = \beta$, then $\alpha + \gamma = \beta + \gamma$. If $\alpha \prec_{\text{lex}} \beta$, then the first nonzero coordinate of $\beta - \alpha$ is positive. But

$$(\beta + \gamma) - (\alpha + \gamma) = \beta - \alpha,$$

and so $\alpha + \gamma \prec_{\text{lex}} \beta + \gamma$. Therefore, $\preceq_{\text{lex}}$ is a monomial order. $\bullet$

**Remark.** If $\Omega$ is any well-ordered set with order $\preceq$, then the lexicographic order on $\Omega^n$ can be defined by $a = (a_1, \ldots, a_n) \preceq_{\text{lex}} b = (b_1, \ldots, b_n)$ if either $a = b$ or they first disagree in the $i$th coordinate and $a_i \prec b_i$. It is straightforward to generalize Proposition 5.120 by replacing $\mathbb{N}^n$ with $\Omega^n$. ◄

If $\preceq$ is a monomial order on $\mathbb{N}^n$, then monomials in $k[X]$ are well-ordered by $X^\alpha \preceq X^\beta$ if $\alpha \preceq \beta$. In particular, $x_1 \succ x_2 \succ x_3 \succ \cdots$ in the lexicographic order, for

$$(1, 0, \ldots, 0) \succ (0, 1, 0, \ldots, 0) \succ \cdots \succ (0, 0, \ldots, 1).$$

Permutations of the variables $x_{\sigma(1)}, \ldots, x_{\sigma(n)}$ can arise from different lexicographic orders on $\mathbb{N}^n$.

When we were constructing the free group with basis a set $X$, we defined a monoid $\mathcal{W}(X)$: its elements are the empty word together with all the words $x_1^{e_1} \cdots x_p^{e_p}$ on $X$, where $p \geq 1$ and $e_i = \pm 1$ for all $i$; its binary operation is juxtaposition. Of more interest here is the submonoid $\mathcal{W}^+(X) \subseteq \mathcal{W}(X)$ consisting of all the *positive* words on $X$:

$$\mathcal{W}^+(X) = \{x_1 \cdots x_p \in \mathcal{W}(X) : x_i \in X \text{ and } p \geq 0\}.$$

In contrast to $\mathbb{N}^n$, in which all words have length $n$, the monoid $\mathcal{W}^+(X)$ has words of different lengths.

**Corollary 5.121.** *If $\Omega$ is a well-ordered set, then the monoid $\mathcal{W}^+(\Omega)$ is well-ordered in the lexicographic order (which we also denote by $\preceq_{\text{lex}}$).*

**Proof.** We will only give a careful definition of the lexicographic order here; the proof that it is a well-order is left to the reader. First, define $1 \preceq_{\text{lex}} w$ for all $w \in \mathcal{W}^+(\Omega)$. Next, given words $u = x_1 \cdots x_p$ and $v = y_1 \cdots y_q$ in $\mathcal{W}^+(\Omega)$, make them the same length by adjoining 1's at the end of the shorter word, and rename them $u'$ and $v'$ in $\mathcal{W}^+(\Omega)$. If $m \geq \max\{p, q\}$, we may regard $u', v', \in \Omega^m$, and we define $u \preceq_{\text{lex}} v$ if $u' \preceq_{\text{lex}} v'$ in $\Omega^m$. (This is the word order commonly used in dictionaries, where a blank precedes any letter: for example, *muse* precedes *museum*.) $\bullet$

**Definition.** Given a monomial order on $\mathbb{N}^n$, each polynomial $f(X) = \sum_\alpha c_\alpha X^\alpha \in k[X] = k[x_1, \ldots, x_n]$ can be written with the degrees of its monomials in descending order: $\alpha_1 \succ \alpha_2 \succ \cdots \succ \alpha_p$. Define

$$\text{word}(f) = \alpha_1 \cdots \alpha_p \in \mathcal{W}^+(\mathbb{N}^n).$$

In light of Corollary 5.121, it makes sense to write

$$\text{word}(f) \preceq_{\text{lex}} \text{word}(g).$$

The next lemma considers the change in word($f$) after replacing a monomial $c_\beta X^\beta$ in $f(X)$, not necessarily the leading monomial, by a polynomial $h$ with $\text{DEG}(h) \prec \beta$.

**Lemma 5.122.** *Given a monomial order on $\mathbb{N}^n$, let $f(X), h(X) \in k[X]$, let $c_\beta X^\beta$ be a nonzero monomial in $f(X)$, and let $\text{DEG}(h) \prec \beta$.*

(i)  *word$\big(f(X) - c_\beta X^\beta + h(X)\big) \prec_{\text{lex}}$ word$(f)$ in $\mathcal{W}^+(X)$.*

(ii) *Any sequence of steps of the form*

$$f(X) \to f(X) - c_\beta X^\beta + h(X),$$

*where $c_\beta X^\beta$ is a nonzero monomial in $f(X)$ and $\text{DEG}(h) \prec \beta$, must be finite.*

**Proof.**

(i) The result is clearly true if $c_\beta X^\beta = \text{LM}(f)$, and so we may assume that $\beta \prec \text{DEG}(f)$. Write $f(X) = f'(X) + c_\beta X^\beta + f''(X)$, where $f'(X)$ is the sum of all monomials in $f(X)$ with $\text{DEG} \succ \beta$ and $f''(X)$ is the sum of all monomials in $f(X)$ with $\text{DEG} \prec \beta$. The sum of the monomials in $f(X) - c_\beta X^\beta + h(X)$ having $\text{DEG} \succ \beta$ is $f'(X)$, and the sum of the lower monomials is $f''(X) + h(X)$. Now $\text{DEG}(f'' + h) = \gamma \prec \beta$, by Exercise 5.73 on page 376. Therefore, the leading monomials of $f(X)$ and $f(X) - c_\beta X^\beta + h(X)$ agree, while the next monomial in $f(X) - c_\beta X^\beta + h(X)$ has $\text{DEG } \gamma \prec \beta$. The definition of the lexicographic order on $\mathcal{W}^+(\mathbb{N}^n)$ now gives $f(X) \succ_{\text{lex}} f(X) - c_\beta X^\beta + h(X)$, for the first disagreement occurs in the $\beta$th position: word$(f) = \alpha_1 \cdots \alpha_i \beta \cdots$ and word$\big(f(X) - c_\beta X^\beta + g(X)\big) = \alpha_1 \cdots \alpha_i \gamma \cdots$, where $\beta \succ \gamma$.

(ii) By part (i), word$(f) \succ_{\text{lex}}$ word$\big(f(X) - c_\beta X^\beta + g(X)\big)$. Since $\mathcal{W}^+(\mathbb{N}^n)$ is well-ordered, it follows from Proposition 5.119 that any sequence of steps of the form $f(X) \to f(X) - c_\beta X^\beta + g(X)$ must be finite.  •

The classical Division Algorithm is a sequence of steps in which the leading monomial of a polynomial is replaced by a polynomial of smaller degree. The Division Algorithm for polynomials in several variables is also a sequence of steps, but a step may involve replacing a monomial, not necessarily the leading monomial, by a polynomial of smaller degree. This is the reason we have introduced $\mathcal{W}^+(\mathbb{N}^n)$, for an induction on $\text{DEG}$ is not strong enough to prove that a sequence of such replacements must stop.

Here is a second monomial order. Recall that if $\alpha = (\alpha_1, \ldots, \alpha_n) \in \mathbb{N}^n$, then its *weight* is $|\alpha| = \alpha_1 + \cdots + \alpha_n$.

**Definition.** The ***degree-lexicographic order*** on $\mathbb{N}^n$ is defined by $\alpha \preceq_{\text{dlex}} \beta$ if either $\alpha = \beta$, or $|\alpha| < |\beta|$, or $|\alpha| = |\beta|$ and the first nonzero coordinate in $\beta - \alpha$ is positive.

It would be more natural for us to call this the *weight-lexicographic order*. In other words, given $(\alpha_1, \ldots, \alpha_n) = \alpha \neq \beta = (\beta_1, \ldots, \beta_n)$, first check weights: if $|\alpha| < |\beta|$, then $\alpha \preceq_{\text{dlex}} \beta$; if there is a tie, that is, if $\alpha$ and $\beta$ have the same weight, then order them lexicographically. For example, $(1, 2, 3, 0) \prec_{\text{dlex}} (0, 2, 5, 0)$ and $(1, 2, 3, 4) \prec_{\text{dlex}} (1, 2, 5, 2)$.

**Proposition 5.123.** *The degree-lexicographic order* $\preceq_{\mathrm{dlex}}$ *is a monomial order on* $\mathbb{N}^n$.

**Proof.** It is routine to show that $\preceq_{\mathrm{dlex}}$ is a partial order on $\mathbb{N}^n$. To see that it is a well-order, let $S$ be a nonempty subset of $\mathbb{N}^n$. The weights of elements in $S$ form a nonempty subset of $\mathbb{N}$, and so there is a smallest such weight, say, $t$. The nonempty subset of all $\alpha \in S$ having weight $t$ has a smallest element, because the degree-lexicographic order $\preceq_{\mathrm{dlex}}$ coincides with the lexicographic order $\preceq_{\mathrm{lex}}$ on this subset. Hence, there is a smallest element in $S$ in the degree-lexicographic order.

Assume that $\alpha \preceq_{\mathrm{dlex}} \beta$ and $\gamma \in \mathbb{N}^n$. Now $|\alpha + \gamma| = |\alpha| + |\gamma|$, so that $|\alpha| = |\beta|$ implies $|\alpha + \gamma| = |\beta + \gamma|$ and $|\alpha| < |\beta|$ implies $|\alpha + \gamma| < |\beta + \gamma|$; in the latter case, Proposition 5.120 shows that $\alpha + \gamma \preceq_{\mathrm{dlex}} \beta + \gamma$. •

The next proposition shows, with respect to any monomial order, that polynomials in several variables behave like polynomials in a single variable.

**Proposition 5.124.** *Let* $\preceq$ *be a monomial order on* $\mathbb{N}^n$, *and let* $f(X), g(X), h(X) \in k[X] = k[x_1, \ldots, x_n]$, *where* $k$ *is a field.*

(i) *If* $\mathrm{DEG}(f) = \mathrm{DEG}(g)$, *then* $\mathrm{LM}(g) \mid \mathrm{LM}(f)$.

(ii) $\mathrm{LM}(hg) = \mathrm{LM}(h)\mathrm{LM}(g)$.

(iii) *If* $\mathrm{DEG}(f) = \mathrm{DEG}(hg)$, *then* $\mathrm{LM}(g) \mid \mathrm{LM}(f)$.

**Proof.**

(i) If $\mathrm{DEG}(f) = \alpha = \mathrm{DEG}(g)$, then $\mathrm{LM}(f) = cX^\alpha$ and $\mathrm{LM}(g) = dX^\alpha$. Since $k$ is a field, $\mathrm{LM}(g) \mid \mathrm{LM}(f)$ [and also $\mathrm{LM}(f) \mid \mathrm{LM}(g)$].

(ii) Let $\mathrm{DEG}(g) = \gamma$, so that $g(X) = bX^\gamma + \text{lower monomials}$, and let $\mathrm{DEG}(h) = \beta$, so that $h(X) = cX^\beta + \text{lower monomials}$; thus, $\mathrm{LM}(g) = bX^\beta$ and $\mathrm{LM}(h) = cX^\gamma$. Clearly, $cbX^{\gamma+\beta}$ is a nonzero monomial in $h(X)g(X)$. To see that it is the leading monomial, let $c_\mu X^\mu$ be a monomial in $h(X)$ with $\mu \prec \gamma$, and let $b_\nu X^\nu$ be a monomial in $g(X)$ with $\nu \prec \beta$. Now $\mathrm{DEG}(c_\mu X^\mu b_\nu X^\nu) = \mu + \nu$; since $\preceq$ is a monomial order, we have $\mu + \nu \prec \gamma + \nu \prec \gamma + \beta$. Thus, $cbX^{\gamma+\beta}$ is the monomial in $h(X)g(X)$ with largest degree.

(iii) Since $\mathrm{DEG}(f) = \mathrm{DEG}(hg)$, part (i) gives $\mathrm{LM}(hg) \mid \mathrm{LM}(f)$ and part (ii) gives $\mathrm{LM}(h)\mathrm{LM}(g) = \mathrm{LM}(hg)$; hence, $\mathrm{LM}(g) \mid \mathrm{LM}(f)$. •

# Exercises

**5.69.** Give an example of a well-ordered set $X$ containing an element $u$ having infinitely many predecessors.

**5.70.** Every subset $X \subseteq \mathbb{R}$ is a chain. Prove that $X$ is countable if it is well-ordered.

**Hint.** There is a rational number between any two real numbers.

**5.71.** (i) Write the first 10 monic monomials in $k[x, y]$ in lexicographic order and in degree-lexicographic order.

(ii) Write all the monic monomials in $k[x, y, z]$ of weight at most 2 in lexicographic order and in degree-lexicographic order.

*  **5.72.** (i) Let $(X, \preceq)$ and $(Y, \preceq')$ be well-ordered sets, where $X$ and $Y$ are disjoint. Define a binary relation $\leq$ on $X \cup Y$ by

$$x_1 \leq x_2 \quad \text{if } x_1, x_2 \in X \text{ and } x_1 \preceq x_2,$$

$$y_1 \leq y_2 \quad \text{if } y_1, y_2 \in Y \text{ and } y_1 \preceq' y_2,$$

$$x \leq y \quad \text{if } x \in X \text{ and } y \in Y.$$

Prove that $(X \cup Y, \leq)$ is a well-ordered set.

(ii) If $r \leq n$, we may regard $\mathbb{N}^r$ as the subset of $\mathbb{N}^n$ consisting of all $n$-tuples of the form $(n_1, \ldots, n_r, 0, \ldots, 0)$, where $n_i \in \mathbb{N}$ for all $i \leq r$. Prove that there exists a monomial order on $\mathbb{N}^n$ in which $a \prec b$ whenever $\alpha \in \mathbb{N}^r$ and $\beta \in \mathbb{N}^n - \mathbb{N}^r$.
**Hint.** Consider the lex order on $k[x_1, \ldots, x_n]$ in which $x_1 \prec x_2 \prec \cdots \prec x_n$.

*  **5.73.** Let $\preceq$ be a monomial order on $\mathbb{N}^n$, and let $f(X), g(X) \in k[X] = k[x_1, \ldots, x_n]$ be nonzero polynomials. Prove that if $f + g \neq 0$, then

$$\mathrm{DEG}(f + g) \preceq \max\{\mathrm{DEG}(f), \mathrm{DEG}(g)\},$$

and that strict inequality can occur only if $\mathrm{DEG}(f) = \mathrm{DEG}(g)$.

---

**5.6.2. Division Algorithm.** We are now going to use monomial orders to give a Division Algorithm for polynomials in several variables.

**Definition.** Let $\preceq$ be a monomial order on $\mathbb{N}^n$ and let $f(X), g(X) \in k[X] = k[x_1, \ldots, x_n]$. If there is a nonzero monomial $c_\beta X^\beta$ in $f(X)$ with $\mathrm{LM}(g) \mid c_\beta X^\beta$, then *reduction*

$$f(X) \xrightarrow{g} f'(X) = f(X) - \frac{c_\beta X^\beta}{\mathrm{LM}(g)} g(X)$$

is the replacement of $f(X)$ by $f'(X)$.

Reduction uses $g$ to eliminate a monomial of degree $\beta$ from $f$. Now $g(X) = bX^\gamma + \text{lower terms}$. If $\mathrm{LM}(g) = bX^\gamma$, then $\mathrm{LM}(g) \mid c_\beta X^\beta$ implies $\gamma \preceq \beta$. Hence,

$$(1) \qquad \frac{c_\beta X^\beta}{\mathrm{LM}(g)} g(X) = \frac{c_\beta X^{\beta - \gamma}}{b} (bX^\gamma + \text{lower terms}) = c_\beta X^\beta - h(X),$$

where $\mathrm{DEG}(h) \prec \beta$. Thus,

$$f'(X) = f(X) - \frac{c_\beta X^\beta}{\mathrm{LM}(g)} g(X) = f(X) - c_\beta X^\beta + h(X).$$

When $\beta = \mathrm{DEG}(f)$, it replaces the leading monomial $\mathrm{LM}(f)$; when $\beta \prec \mathrm{DEG}(f)$, reduction is a replacement as in Lemma 5.122.

**Proposition 5.125.** *Let $\preceq$ be a monomial order on $\mathbb{N}^n$, let $f(X), g(X) \in k[X] = k[x_1, \ldots, x_n]$, and let $c_\beta X^\beta$ be a nonzero monomial in $f(X)$ with $\mathrm{LM}(g) \mid c_\beta X^\beta$; define $f'(X) = f(X) - \frac{c_\beta X^\beta}{\mathrm{LM}(g)} g(X)$.*

(i) *If $\beta = \mathrm{DEG}(f)$, then either $f'(X) = 0$ or $\mathrm{DEG}(f') \prec \mathrm{DEG}(f)$.*

(ii) *If $\beta \prec \mathrm{DEG}(f)$, then $\mathrm{DEG}(f') = \mathrm{DEG}(f)$.*

*In either case,*

$$\mathrm{DEG}\Big(\frac{c_\beta X^\beta}{\mathrm{LM}(g)} g(X)\Big) \preceq \mathrm{DEG}(f).$$

**Proof.** We have seen, in Equation (1), that reduction replaces a monomial of degree $\beta$ either with 0 or with a polynomial $h(X)$ having $\mathrm{DEG}(h) \prec \beta$. In case (i), $\beta = \mathrm{DEG}(f)$, then $\mathrm{DEG}(f') \prec \mathrm{DEG}(f)$; in case (ii), $\beta \prec \mathrm{DEG}(f)$, we have $\mathrm{DEG}(f') = \mathrm{DEG}(f)$. It is now easy to see that the last stated inequality holds. •

**Definition.** Let $\{g_1, \ldots, g_m\}$ be a set of polynomials in $k[X]$. A polynomial $r(X)$ is **reduced mod** $\{g_1, \ldots, g_m\}$ if either $r(X) = 0$ or no $\mathrm{LM}(g_i)$ divides any nonzero monomial in $r(X)$.

Here is the Division Algorithm for polynomials in several variables. Because the algorithm requires the "divisor polynomials" $\{g_1, \ldots, g_m\}$ to be used in a specific order (after all, an algorithm must give explicit directions), we will be using an $m$-tuple of polynomials instead of a subset of polynomials. We use the notation $[g_1, \ldots, g_m]$ for the $m$-tuple whose $i$th entry is $g_i$, because the usual notation $(g_1, \ldots, g_m)$ would be confused with the notation for the ideal $(g_1, \ldots, g_m)$ generated by the $g_i$.

**Theorem 5.126 (Division Algorithm in** $k[x_1, \ldots, x_n]$**).** *Let $\preceq$ be a monomial order on $\mathbb{N}^n$, and let $k[X] = k[x_1, \ldots, x_n]$. If $f(X) \in k[X]$ and $G = [g_1(X), \ldots, g_m(X)]$ is an $m$-tuple of polynomials in $k[X]$, then there is an algorithm giving polynomials $r(X), a_1(X), \ldots, a_m(X) \in k[X]$ with*

$$f = a_1 g_1 + \cdots + a_m g_m + r,$$

*where $r$ is reduced* $\mod \{g_1, \ldots, g_m\}$*, and $a_i g_i = 0$ or $\mathrm{DEG}(a_i g_i) \preceq \mathrm{DEG}(f)$ for all $i$.*

**Proof.** Once a monomial order is chosen, so that leading monomials and degrees are defined, the algorithm is a straightforward generalization of the Division Algorithm in one variable. First, apply reductions of the form $h \xrightarrow{g_1} h'$ as many times as possible, then apply reductions of the form $h \xrightarrow{g_2} h'$, then $h \xrightarrow{g_1} h'$ again, etc. Here is a pseudocode describing the algorithm more precisely:

Input: $f(X) = \sum_\beta c_\beta X^\beta$, $[g_1, \ldots, g_m]$
Output: $r, a_1, \ldots, a_m$
$r := f$; $a_i := 0$
WHILE $f$ is not reduced mod $\{g_1, \ldots, g_m\}$ DO
    select smallest $i$ with $\mathrm{LM}(g_i) \mid c_\beta X^\beta$ for some $\beta$ maximal such that
    $f - [c_\beta X^\beta / \mathrm{LM}(g_i)] g_i := f$
    $a_i + [c_\beta X^\beta / \mathrm{LM}(g_i)] := a_i$
END WHILE

At each step $h_j \xrightarrow{g_i} h_{j+1}$ of the algorithm,

$$\mathrm{word}(h_j) \succ_{\mathrm{lex}} \mathrm{word}(h_{j+1}) \text{ in } \mathcal{W}^+(\mathbb{N}^n),$$

by Lemma 5.122, and so the algorithm does stop, because $\preceq_{\mathrm{lex}}$ is a well-order on $\mathcal{W}^+(\mathbb{N}^n)$. Obviously, the output $r(X)$ is reduced mod $\{g_1, \ldots, g_m\}$, for if $r(X)$ has a monomial divisible by some $\mathrm{LM}(g_i)$, then one further reduction is possible.

Finally, each monomial in $a_i(X)$ has the form $c_\beta X^\beta / \mathrm{LM}(g_i)$ for some intermediate output $h(X)$ (as one sees in the pseudocode). It now follows from Proposition 5.125 that either $a_i g_i = 0$ or $\mathrm{DEG}(a_i g_i) \preceq \mathrm{DEG}(f)$.    •

**Definition.** Given a monomial order on $\mathbb{N}^n$, a polynomial $f(X) \in k[X]$, and an $m$-tuple $G = [g_1, \ldots, g_m]$, we call the output $r(X)$ of the Division Algorithm the *remainder of f mod G*.

The remainder $r$ of $f$ mod $G$ is reduced mod $\{g_1, \ldots, g_m\}$, and $f - r \in I = (g_1, \ldots, g_m)$. The Division Algorithm requires that $G$ be an $m$-tuple, because of the command

$$\text{select smallest } i \text{ with } \mathrm{LM}(g_i) \mid c_\beta X^\beta \text{ for some } \beta$$

specifying the order of reductions. The next example shows that the remainder may depend not only on the set of polynomials $\{g_1, \ldots, g_m\}$ but also on the ordering of the coordinates in the $m$-tuple $G = [g_1, \ldots, g_m]$. That is, if $\sigma \in S_m$ is a permutation and $G_\sigma = [g_{\sigma(1)}, \ldots, g_{\sigma(m)}]$, then the remainder $r_\sigma$ of $f$ mod $G_\sigma$ may not be the same as the remainder $r$ of $f$ mod $G$. Even worse, it is possible that $r \neq 0$ and $r_\sigma = 0$, so that the remainder mod $G$ is not the obstruction to $f$ being in the ideal $(g_1, \ldots, g_m)$. We illustrate this phenomenon in the next example, and we will deal with it in the next section.

**Example 5.127.** Let $f(x, y, z) = x^2 y^2 + xy$, and let $G = [g_1, g_2, g_3]$, where

$$g_1 = y^2 + z^2$$
$$g_2 = x^2 y + yz$$
$$g_3 = z^3 + xy.$$

We use the degree-lexicographic order on $\mathbb{N}^3$. Now $y^2 = \mathrm{LM}(g_1) \mid \mathrm{LM}(f) = x^2 y^2$, and so $f \xrightarrow{g_1} h$, where $h = f - \frac{x^2 y^2}{y^2}(y^2 + z^2) = -x^2 z^2 + xy$. The polynomial $-x^2 z^2 + xy$ is reduced mod $G$, because neither $-x^2 z^2$ nor $xy$ is divisible by any of the leading monomials $\mathrm{LM}(g_1) = y^2$, $\mathrm{LM}(g_2) = x^2 y$, or $\mathrm{LM}(g_3) = z^3$.

On the other hand, let us apply the Division Algorithm using the 3-tuple $G' = [g_2, g_1, g_3]$. The first reduction gives $f \xrightarrow{g_2} h'$, where

$$h' = f - \frac{x^2 y^2}{x^2 y}(x^2 y + yz) = -y^2 z + xy.$$

Now $h'$ is not reduced, and reducing mod $g_1$ gives

$$h' - \frac{-y^2 z}{y^2}(y^2 + z^2) = z^3 + xy.$$

But $z^3 + xy = g_3$, and so $z^3 + xy \xrightarrow{g_3} 0$.

Thus, the remainder depends on the ordering of the divisor polynomials $g_i$ in the $m$-tuple.

For a simpler example of different remainders (but with neither remainder 0), see Exercise 5.74.    ◄

# Exercises

* **5.74.** Let $G = [x - y, x - z]$ and $G' = [x - z, x - y]$. Show that the remainder of $x$ mod $G$ (degree-lexicographic order) is distinct from the remainder of $x$ mod $G'$.

**5.75.** Use the degree-lexicographic order in this exercise.

(i) Find the remainder of $x^7 y^2 + x^3 y^2 - y + 1 \bmod [xy^2 - x, x - y^3]$.

(ii) Find the remainder of $x^7 y^2 + x^3 y^2 - y + 1 \bmod [x - y^3, xy^2 - x]$.

**5.76.** Use the degree-lexicographic order in this exercise.

(i) Find the remainder of $x^2 y + xy^2 + y^2 \bmod [y^2 - 1, xy - 1]$.

(ii) Find the remainder of $x^2 y + xy^2 + y^2 \bmod [xy - 1, y^2 - 1]$.

* **5.77.** Let $X^\alpha$ be a monomial, and let $f(X), g(X) \in k[X]$ be polynomials none of whose monomials is divisible by $X^\alpha$. Prove that none of the monomials in $f(X) - g(X)$ is divisible by $X^\alpha$.

**5.78.** Let $f(X) = \sum_\alpha c_\alpha X^\alpha \in k[X]$, where $k$ is a field and $X = (x_1, \ldots, x_n)$, be symmetric; that is, for all permutations $\sigma \in S_n$,

$$f(x_{\sigma 1}, \ldots, x_{\sigma n}) = f(x_1, \ldots, x_n).$$

If a monomial $c_\alpha x_1^{\alpha_1} \cdots x_n^{\alpha_n}$ in $f(X)$ occurs with nonzero coefficient $c_\alpha$, prove that every monomial $x_{\sigma 1}^{\alpha_1} \cdots x_{\sigma n}^{\alpha_n}$, where $\sigma \in S_n$, also occurs in $f(X)$ with nonzero coefficient.

* **5.79.** Let $\mathbb{N}^n$ be equipped with the degree-lexicographic order, let $X = (x_1, \ldots, x_n)$, and let $k(X) = k[x_1, \ldots, x_n]$, where $k$ is a field.

(i) If $f(X) = \sum_\alpha c_\alpha X^\alpha \in k[X]$ is symmetric and $\mathrm{DEG}(f) = \beta = (\beta_1, \ldots, \beta_n)$, prove that $\beta_1 \geq \beta_2 \geq \cdots \geq \beta_n$.

(ii) If $e_1, \ldots, e_n$ are the elementary symmetric polynomials, prove that

$$\mathrm{DEG}(e_i) = (1, \ldots, 1, 0, \ldots, 0),$$

where there are $i$ 1's.

(iii) Let $(\gamma_1, \ldots, \gamma_n) = (\beta_1 - \beta_2, \beta_2 - \beta_3, \ldots, \beta_{n-1} - \beta_n, \beta_n)$. Prove that if $g(x_1, \ldots, x_n) = x_1^{\gamma_1} \cdots x_n^{\gamma_n}$, then $g(e_1, \ldots, e_n)$ is symmetric and $\mathrm{DEG}(g) = \beta$.

(iv) **(Fundamental Theorem of Symmetric Polynomials)** Prove that if $k$ is a field, then every symmetric polynomial $f(X) \in k[X]$ is a *polynomial* in the elementary symmetric functions $e_1, \ldots, e_n$. (Compare with Theorem 3.40.)
**Hint.** Prove that $h(X) = f(X) - c_\beta g(e_1, \ldots, e_n)$ is symmetric and $\mathrm{DEG}(h) < \beta$.

## Section 5.7. Gröbner Bases

We will assume in this section that $\mathbb{N}^n$ is equipped with some monomial order (the reader may use the degree-lexicographic order), so that degrees are defined and the Division Algorithm makes sense.

We have seen that the remainder of $f \bmod [g_1, \ldots, g_m]$ obtained from the Division Algorithm can depend on the order in which the $g_i$ are listed. Informally, a *Gröbner basis* $\{g_1, \ldots, g_m\}$ of the ideal $I = (g_1, \ldots, g_m)$ is a generating set such that, for any of the $m$-tuples $G$ formed from the $g_i$, the remainder of $f$ mod $G$ is always the obstruction to whether $f$ lies in $I$. We define Gröbner bases using a property that is more easily checked, and we then show, in Proposition 5.128, that they are characterized by the more interesting obstruction property just mentioned.

**Definition.** A *set* of polynomials $\{g_1, \ldots, g_m\}$ is a **Gröbner basis**[22] of the ideal $I = (g_1, \ldots, g_m)$ if, for each nonzero $f \in I$, there is some $g_i$ with $\mathrm{LM}(g_i) \mid \mathrm{LM}(f)$.

Note that a Gröbner basis is a *set* of polynomials, not an $m$-tuple of polynomials. Example 5.127 shows that

$$\{y^2 + z^2, x^2 y + yz, z^3 + xy\}$$

is not a Gröbner basis of the ideal $I = (y^2 + z^2, x^2 y + yz, z^3 + xy)$.

**Proposition 5.128.** *A set $\{g_1, \ldots, g_m\}$ of polynomials is a Gröbner basis of $I = (g_1, \ldots, g_m)$ if and only if, for each $m$-tuple $G_\sigma = [g_{\sigma(1)}, \ldots, g_{\sigma(m)}]$, where $\sigma \in S_m$, every $f \in I$ has remainder $0$ mod $G_\sigma$.*

**Proof.** Assume there is some permutation $\sigma \in S_m$ and some $f \in I$ whose remainder mod $G_\sigma$ is not 0. Among all such polynomials, choose $f$ of minimal degree. Since $\{g_1, \ldots, g_m\}$ is a Gröbner basis, $\mathrm{LM}(g_i) \mid \mathrm{LM}(f)$ for some $i$; select the smallest $\sigma(i)$ for which there is a reduction $f \xrightarrow{g_{\sigma(i)}} h$, and note that $h \in I$. Since $\mathrm{DEG}(h) \prec \mathrm{DEG}(f)$, by Proposition 5.125, the Division Algorithm gives a sequence of reductions $h = h_0 \to h_1 \to h_2 \to \cdots \to h_p = 0$. But the Division Algorithm for $f$ adjoins $f \to h$ at the front, showing that 0 is the remainder of $f$ mod $G_\sigma$, a contradiction.

Conversely, let $\{g_1, \ldots, g_m\}$ be a Gröbner basis of $I = (g_1, \ldots, g_m)$. If there is a nonzero $f \in I$ with $\mathrm{LM}(g_i) \nmid \mathrm{LM}(f)$ for every $i$, then in any reduction $f \xrightarrow{g_i} h$, we have $\mathrm{LM}(h) = \mathrm{LM}(f)$. Hence, if $G = [g_1, \ldots, g_m]$, the Division Algorithm mod $G$ gives reductions $f \to h_1 \to h_2 \to \cdots \to h_p = r$ in which $\mathrm{LM}(r) = \mathrm{LM}(f)$. Therefore, $r \neq 0$; that is, the remainder of $f$ mod $G$ is not zero, and this is a contradiction.  •

**Corollary 5.129.** *Let $I = (g_1, \ldots, g_m)$ be an ideal, let $\{g_1, \ldots, g_m\}$ be a Gröbner basis of $I$, and let $G = [g_1, \ldots, g_m]$ be any $m$-tuple formed from the $g_i$. If $f(X) \in k[X]$, then there is a unique $r(X) \in k[X]$, which is reduced mod $G$, such that $f - r \in I$; in fact, $r$ is the remainder of $f$ mod $G$.*

**Proof.** The Division Algorithm gives polynomials $a_1, \ldots, a_m$ and a polynomial $r$ reduced mod $G$ with $f = a_1 g_1 + \cdots + a_m g_m + r$; clearly, $f - r = a_1 g_1 + \cdots + a_m g_m \in I$.

To prove uniqueness, suppose that $r$ and $r'$ are reduced mod $G$ and that $f - r$ and $f - r'$ lie in $I$, so that $(f - r') - (f - r) \overset{\cdot}{=} r - r' \in I$. Since $r$ and $r'$ are reduced mod $G$, none of their monomials is divisible by any $\mathrm{LM}(g_i)$. If $r - r' \neq 0$, then Exercise 5.77 on page 379 says that no monomial in $r - r'$ is divisible by any

---

[22]It was B. Buchberger who, in his dissertation, defined Gröbner bases and proved their main properties. He named these bases to honor his thesis advisor, W. Gröbner.

$\mathrm{LM}(g_i)$; in particular, $\mathrm{LM}(r-r')$ is not divisible by any $\mathrm{LM}(g_i)$, and this contradicts Proposition 5.128. Therefore, $r = r'$. •

The next corollary shows that Gröbner bases resolve the problem of different remainders in the Division Algorithm arising from different permutations of $g_1, \ldots, g_m$.

**Corollary 5.130.** *Let $I = (g_1, \ldots, g_m)$ be an ideal, let $\{g_1, \ldots, g_m\}$ be a Gröbner basis of $I$, and let $G$ be the $m$-tuple $G = [g_1, \ldots, g_m]$.*

(i) *If $f(X) \in k[X]$ and $G_\sigma = [g_{\sigma(1)}, \ldots, g_{\sigma(m)}]$, where $\sigma \in S_m$ is a permutation, then the remainder of $f$ mod $G$ is equal to the remainder of $f$ mod $G_\sigma$.*

(ii) *A polynomial $f \in I$ if and only if $f$ has remainder $0$ mod $G$.*

**Proof.**

(i) If $r$ is the remainder of $f$ mod $G$, then Corollary 5.129 says that $r$ is the unique polynomial, reduced mod $G$, with $f - r \in I$; similarly, the remainder $r_\sigma$ of $f$ mod $G_\sigma$ is the unique polynomial, reduced mod $G_\sigma$, with $f - r_\sigma \in I$. The uniqueness assertion in Corollary 5.129 gives $r = r_\sigma$.

(ii) Proposition 5.128 shows that if $f \in I$, then its remainder is $0$. For the converse, if $r$ is the remainder of $f$ mod $G$, then $f = q + r$, where $q \in I$. Hence, if $r = 0$, then $f \in I$. •

**5.7.1. Buchberger's Algorithm.** There are several obvious questions. Do Gröbner bases exist and, if they do, are they unique? Given an ideal $I$ in $k[X]$, is there an algorithm to find a Gröbner basis of $I$?

The notion of *S-polynomial* will allow us to recognize a Gröbner basis, but we first introduce some notation.

**Definition.** If $\alpha = (\alpha_1, \ldots, \alpha_n)$ and $\beta = (\beta_1, \ldots, \beta_n)$ are in $\mathbb{N}^n$, define

$$\alpha \vee \beta = \mu,$$

where $\mu = (\mu_1, \ldots, \mu_n)$ is given by $\mu_i = \max\{\alpha_i, \beta_i\}$.

Note that $X^{\alpha \vee \beta}$ is the least common multiple of the monomials $X^\alpha$ and $X^\beta$.

**Definition.** Let $f(X), g(X) \in k[X]$. If $\mathrm{LM}(f) = a_\alpha X^\alpha$ and $\mathrm{LM}(g) = b_\beta X^\beta$, define

$$L(f, g) = X^{\alpha \vee \beta}.$$

The *S-polynomial* $S(f, g)$ is defined by

$$S(f, g) = \frac{L(f, g)}{\mathrm{LM}(f)} f - \frac{L(f, g)}{\mathrm{LM}(g)} g.$$

Note that $S(f, g) = -S(g, f)$.

Let $f(X) = aX^\alpha + f'(X)$ and $g(X) = bX^\beta + g'(X)$, where $\text{DEG}(f') \prec \alpha$ and $\text{DEG}(g') \prec \beta$. If $\beta \preceq \alpha$, then

$$
\begin{aligned}
S(f, g) &= \frac{L(f, g)}{\text{LM}(f)} f - \frac{L(f, g)}{\text{LM}(g)} g \\
&= a_\alpha^{-1} X^{(\alpha \vee \beta) - \alpha} f - b_\beta^{-1} X^{(\alpha \vee \beta) - \beta} g \\
&= [X^\alpha + a_\alpha^{-1} X^{(\alpha \vee \beta) - \alpha} f'] - [X^\alpha + b_\beta^{-1} X^{(\alpha \vee \beta) - \beta} g'] \\
&= a_\alpha^{-1} X^{(\alpha \vee \beta) - \alpha} f' - b_\beta^{-1} X^{(\alpha \vee \beta) - \beta} g' \\
&= \frac{L(f, g)}{\text{LM}(f)} f' - \frac{L(f, g)}{\text{LM}(g)} g'.
\end{aligned}
$$

Thus, either $S(f, g) = 0$ or $\text{DEG}(S(f, g)) \prec \max\{\text{DEG}(f), \text{DEG}(g)\}$.

**Example 5.131.** We show that if $f = X^\alpha$ and $g = X^\beta$ are monomials, then $S(f, g) = 0$. Since $f$ and $g$ are monomials, we have $\text{LM}(f) = f$ and $\text{LM}(g) = g$. Hence,

$$
S(f, g) = \frac{L(f, g)}{\text{LM}(f)} f - \frac{L(f, g)}{\text{LM}(g)} g = \frac{X^{\alpha \vee \beta}}{f} f - \frac{X^{\alpha \vee \beta}}{g} g = 0. \quad \blacktriangleleft
$$

The following technical lemma indicates why $S$-polynomials are relevant. It gives a condition when a polynomial can be rewritten as a linear combination of $S$-polynomials with monomial coefficients.

**Lemma 5.132.** Let $g_1(X), \ldots, g_\ell(X) \in k[X] = k[x_1, \ldots, x_n]$. Given monomials $c_j X^{\alpha(j)}$, where $\alpha(j) \in \mathbb{N}^n$, let $h(X) = \sum_{j=1}^\ell c_j X^{\alpha(j)} g_j(X)$.

Let $\delta \in \mathbb{N}^n$. If $\text{DEG}(h) \prec \delta$ and $\text{DEG}(c_j X^{\alpha(j)} g_j(X)) = \delta$ for all $j < \ell$, then there are $d_j \in k$ with

$$
h(X) = \sum_j d_j X^{\delta - \mu(j)} S(g_j, g_{j+1}),
$$

where $\mu(j) = \text{DEG}(g_j) \vee \text{DEG}(g_{j+1})$, and for all $j < \ell$,

$$
\text{DEG}\big(X^{\delta - \mu(j)} S(g_j, g_{j+1})\big) \prec \delta.
$$

**Proof.** Let $\text{LM}(g_j) = b_j X^{\beta(j)}$, so that $\text{LM}(c_j X^{\alpha(j)} g_j(X)) = c_j b_j X^\delta$. The coefficient of $X^\delta$ in $h(X)$ is thus $\sum_j c_j b_j$. Since $\text{DEG}(h) \prec \delta$, we must have $\sum_j c_j b_j = 0$. Define monic polynomials

$$
u_j(X) = b_j^{-1} X^{\alpha(j)} g_j(X).
$$

There is a telescoping sum

$$
\begin{aligned}
h(X) = \sum_{j=1}^\ell c_j X^{\alpha(j)} g_j(X) &= \sum_{j=1}^\ell c_j b_j u_j \\
&= c_1 b_1 (u_1 - u_2) + (c_1 b_1 + c_2 b_2)(u_2 - u_3) + \cdots \\
&\quad + (c_1 b_1 + \cdots + c_{\ell-1} b_{\ell-1})(u_{\ell-1} - u_\ell) \\
&\quad + (c_1 b_1 + \cdots + c_\ell b_\ell) u_\ell.
\end{aligned}
$$

Now the last monomial $(c_1 b_1 + \cdots + c_\ell b_\ell) u_\ell = 0$ because $\sum_j c_j b_j = 0$. We have $\alpha(j) + \beta(j) = \delta$, since $\mathrm{DEG}(c_j X^{\alpha(j)} g_j(X)) = \delta$, so that $X^{\beta(j)} \mid X^\delta$ for all $j$. Hence, for all $j < \ell$, we have $\mathrm{lcm}\{X^{\beta(j)}, X^{\beta(j+1)}\} = X^{\beta(j) \vee \beta(j+1)} \mid X^\delta$; that is, if we write $\mu(j) = \beta(j) \vee \beta(j+1)$, then $\delta - \mu(j) \in \mathbb{N}^n$. But

$$X^{\delta - \mu(j)} S(g_j, g_{j+1}) = X^{\delta - \mu(j)} \left( \frac{X^{\mu(j)}}{\mathrm{LM}(g_j)} g_j(X) - \frac{X^{\mu(j)}}{\mathrm{LM}(g_{j+1})} g_{j+1}(X) \right)$$
$$= \frac{X^\delta}{\mathrm{LM}(g_j)} g_j(X) - \frac{X^\delta}{\mathrm{LM}(g_{j+1})} g_{j+1}(X)$$
$$= b_j^{-1} X^{\alpha(j)} g_j - b_{j+1}^{-1} X^{\alpha(j+1)} g_{j+1}$$
$$= u_j - u_{j+1}.$$

Substituting this equation into the telescoping sum gives a sum of the desired form, where $d_j = c_1 b_1 + \cdots + c_j b_j$:

$$h(X) = c_1 b_1 X^{\delta - \mu(1)} S(g_1, g_2) + (c_1 b_1 + c_2 b_2) X^{\delta - \mu(2)} S(g_2, g_3) + \cdots$$
$$+ (c_1 b_1 + \cdots + c_{\ell-1} b_{\ell-1}) X^{\delta - \mu(\ell-1)} S(g_{\ell-1}, g_\ell).$$

Finally, since both $u_j$ and $u_{j+1}$ are monic with leading monomial of $\mathrm{DEG}\ \delta$, we have $\mathrm{DEG}(u_j - u_{j+1}) \prec \delta$. But we have shown that $u_j - u_{j+1} = X^{\delta - \mu(j)} S(g_j, g_{j+1})$, and so $\mathrm{DEG}(X^{\delta - \mu(j)} S(g_j, g_{j+1})) \prec \delta$, as desired.   •

Let $I = (g_1, \ldots, g_m)$. By Proposition 5.128, $\{g_1, \ldots, g_m\}$ is a Gröbner basis of the ideal $I$ if every $f \in I$ has remainder 0 mod $G$ (where $G$ is any $m$-tuple formed by ordering the $g_i$). The importance of the next theorem lies in its showing that it is necessary to compute the remainders of only finitely many polynomials, namely, the $S$-polynomials $S(g_p, g_q)$, to determine whether $\{g_1, \ldots, g_m\}$ is a Gröbner basis.

**Theorem 5.133 (Buchberger).** *A set $\{g_1, \ldots, g_m\}$ is a Gröbner basis of $I = (g_1, \ldots, g_m)$ if and only if $S(g_p, g_q)$ has remainder 0 mod $G$ for all $p, q$, where $G = [g_1, \ldots, g_m]$.*

**Proof.** Clearly, $S(g_p, g_q)$, being a linear combination of $g_p$ and $g_q$, lies in $I$. Hence, if $G = \{g_1, \ldots, g_m\}$ is a Gröbner basis, then $S(g_p, g_q)$ has remainder 0 mod $G$, by Proposition 5.128.

Conversely, assume that $S(g_p, g_q)$ has remainder 0 mod $G$ for all $p, q$; we must show that every $f \in I$ has remainder 0 mod $G$. By Proposition 5.128, it suffices to show that if $f \in I$, then $\mathrm{LM}(g_i) \mid \mathrm{LM}(f)$ for some $i$. Suppose there is $f \in I$ for which this is false. Since $f \in I = (g_1, \ldots, g_m)$, we may write $f = \sum_i h_i g_i$, and so

$$\mathrm{DEG}(f) \preceq \max_i \{\mathrm{DEG}(h_i g_i)\}.$$

If $\mathrm{DEG}(f) = \mathrm{DEG}(h_i g_i)$ for some $i$, then Proposition 5.124 gives $\mathrm{LM}(g_i) \mid \mathrm{LM}(f)$, a contradiction. Hence, we may assume strict inequality: $\mathrm{DEG}(f) \prec \max_i \{\mathrm{DEG}(h_i g_i)\}$.

The polynomial $f$ may be written as a linear combination of the $g_i$ in many ways. Of all the expressions of the form $f = \sum_i h_i g_i$, choose one in which $\delta = \max_i \{\mathrm{DEG}(h_i g_i)\}$ is minimal (which is possible because $\preceq$ is a well-order). We are done if $\mathrm{DEG}(f) = \delta$, as we have seen above; therefore, we may assume that there is strict inequality: $\mathrm{DEG}(f) \prec \delta$. Write

$$(1) \qquad f = \sum_{j,\, \mathrm{DEG}(h_j g_j)=\delta} h_j g_j + \sum_{\ell,\, \mathrm{DEG}(h_\ell g_\ell) \prec \delta} h_\ell g_\ell.$$

If $\mathrm{DEG}(\sum_j h_j g_j) = \delta$, then $\mathrm{DEG}(f) = \delta$, a contradiction; hence, $\mathrm{DEG}(\sum_j h_j g_j) \prec \delta$. But the coefficient of $X^\delta$ in this sum is obtained from its leading monomials, so that

$$\mathrm{DEG}\Big(\sum_j \mathrm{LM}(h_j) g_j\Big) \prec \delta.$$

Now $\sum_j \mathrm{LM}(h_j) g_j$ is a polynomial satisfying the hypotheses of Lemma 5.132, and so there are constants $d_j$ and degrees $\mu(j)$ so that

$$(2) \qquad \sum_j \mathrm{LM}(h_j) g_j = \sum_j d_j X^{\delta - \mu(j)} S(g_j, g_{j+1}),$$

where $\mathrm{DEG}\big(X^{\delta-\mu(j)} S(g_j, g_{j+1})\big) \prec \delta.$[23]

Since each $S(g_j, g_{j+1})$ has remainder 0 mod $G$, the Division Algorithm gives $a_{ji}(X) \in k[X]$ with

$$S(g_j, g_{j+1}) = \sum_i a_{ji} g_i,$$

where $\mathrm{DEG}(a_{ji} g_i) \preceq \mathrm{DEG}(S(g_j, g_{j+1}))$ for all $j, i$. It follows that

$$X^{\delta - \mu(j)} S(g_j, g_{j+1}) = \sum_i X^{\delta - \mu(j)} a_{ji} g_i.$$

Therefore, Lemma 5.132 gives

$$(3) \qquad \mathrm{DEG}(X^{\delta-\mu(j)} a_{ji}) \preceq \mathrm{DEG}(X^{\delta-\mu(j)} S(g_j, g_{j+1})) \prec \delta.$$

Substituting into Equation (2), we have

$$\sum_j \mathrm{LM}(h_j) g_j = \sum_j d_j X^{\delta-\mu(j)} S(g_j, g_{j+1})$$
$$= \sum_j d_j \Big(\sum_i X^{\delta-\mu(j)} a_{ji} g_i\Big)$$
$$= \sum_i \Big(\sum_j d_j X^{\delta-\mu(j)} a_{ji}\Big) g_i.$$

If we denote $\sum_j d_j X^{\delta-\mu(j)} a_{ji}$ by $h'_i$, then

$$(4) \qquad \sum_j \mathrm{LM}(h_j) g_j = \sum_i h'_i g_i,$$

where, by Equation (3), $\mathrm{DEG}(h'_i g_i) \prec \delta$ for all $i$.

---

[23] The reader may wonder why we consider all $S$-polynomials $S(g_p, g_q)$ instead of only those of the form $S(g_i, g_{i+1})$. The answer is that the remainder condition is applied only to those $h_j g_j$ for which $\mathrm{DEG}(h_j g_j) = \delta$, and so the indices viewed as $i$'s need not be consecutive.

Finally, we substitute the expression in Equation (4) into Equation (1):

$$f = \sum_{\substack{j \\ \text{DEG}(h_j g_j)=\delta}} h_j g_j \quad + \quad \sum_{\substack{\ell \\ \text{DEG}(h_\ell g_\ell) \prec \delta}} h_\ell g_\ell$$

$$= \sum_{\substack{j \\ \text{DEG}(h_j g_j)=\delta}} \text{LM}(h_j) g_j \quad + \quad \sum_{\substack{j \\ \text{DEG}(h_j g_j)=\delta}} [h_j - \text{LM}(h_j)] g_j \quad + \quad \sum_{\substack{\ell \\ \text{DEG}(h_\ell g_\ell) \prec \delta}} h_\ell g_\ell$$

$$= \sum_{i} h_i' g_i \quad + \quad \sum_{\substack{j \\ \text{DEG}(h_j g_j)=\delta}} [h_j - \text{LM}(h_j)] g_j \quad + \quad \sum_{\substack{\ell \\ \text{DEG}(h_\ell g_\ell) \prec \delta}} h_\ell g_\ell.$$

We have rewritten $f$ as a linear combination of the $g_i$ in which each monomial has degree DEG strictly smaller than $\delta$, contradicting the minimality of $\delta$. This completes the proof. •

**Definition.** A ***monomial ideal*** in $k[X] = k[x_1, \ldots, x_n]$ is an ideal $I$ that is generated by monomials; that is, $I = (X^{\alpha(1)}, \ldots, X^{\alpha(q)})$, where $\alpha(j) \in \mathbb{N}^n$ for $j = 1, \ldots, q$.

**Lemma 5.134.** *Let $I = (X^{\alpha(1)}, \ldots, X^{\alpha(q)})$ be a monomial ideal.*

(i) *Let $f(X) = \sum_\beta c_\beta X^\beta$. Then $f(X) \in I$ if and only if, for each nonzero $c_\beta X^\beta$, there is $j$ with $X^{\alpha(j)} \mid c_\beta X^\beta$.*

(ii) *If $G = [g_1, \ldots, g_m]$ and $r$ is reduced mod $G$, then $r$ does not lie in the monomial ideal $(\text{LM}(g_1), \ldots, \text{LM}(g_m))$.*

**Proof.**

(i) If each monomial in $f$ is divisible by some $X^{\alpha(i)}$, then just collect terms (for each $i$) to see that $f \in I$. Conversely, if $f \in I$, then $f = \sum_i a_i(X) X^{\alpha(i)}$, where $a_i(X) \in k[X]$. Expand this expression to see that every monomial in $f$ is divisible by some $X^{\alpha(i)}$.

(ii) The definition of being reduced mod $G$ says that no monomial in $r(X)$ is divisible by any $\text{LM}(g_i)$. Hence, $r \notin (\text{LM}(g_1), \ldots, \text{LM}(g_m))$, by part (i). •

**Corollary 5.135.** *If $I = (f_1, \ldots, f_s)$ is a monomial ideal in $k[X]$, that is, each $f_i$ is a monomial, then $\{f_1, \ldots, f_s\}$ is a Gröbner basis of $I$.*

**Proof.** By Example 5.131, the $S$-polynomial of any pair of monomials is 0. •

Here is the main result.

**Theorem 5.136 (Buchberger's Algorithm).** *Every ideal $I = (f_1, \ldots, f_s)$ in $k[X]$ has a Gröbner basis[24] which can be computed by an algorithm.*

**Proof.** Here is a pseudocode for an algorithm.

Input: $B = \{f_1, \ldots, f_s\} \quad G = [f_1, \ldots, f_s]$

---

[24]A nonconstructive proof of the existence of a Gröbner basis can be given using the proof of the Hilbert Basis Theorem; for example, see Section 2.5 of the book by Cox, Little, and O'Shea (they give a constructive proof in Section 2.7).

Output: a Gröbner basis $B = \{g_1, \ldots, g_m\}$
containing $\{f_1, \ldots, f_s\}$
$B := \{f_1, \ldots, f_s\}; \quad G := [f_1, \ldots, f_s]$
REPEAT
  $B' := B; \quad G' := G$
  FOR each pair $g, g'$ with $g \neq g' \in B'$ DO
    $r :=$ remainder of $S(g, g') \bmod G'$
    IF $r \neq 0$ THEN
      $B := B \cup \{r\}; \quad G' := [g_1, \ldots, g_m, r]$
    END IF
  END FOR
UNTIL $B = B'$

Now each loop of the algorithm enlarges a subset $B \subseteq I = (g_1, \ldots, g_m)$ by adjoining the remainder mod $G$ of one of its $S$-polynomials $S(g, g')$. As $g, g' \in I$, the remainder $r$ of $S(g, g')$ lies in $I$, and so the larger set $B \cup \{r\}$ is contained in $I$.

The only obstruction to the algorithm stopping at some $B'$ is if some $S(g, g')$ does not have remainder 0 mod $G'$. Thus, if the algorithm stops, then Theorem 5.133 shows that $B'$ is a Gröbner basis.

To see that the algorithm does stop, suppose a loop starts with $B'$ and ends with $B$. Since $B' \subseteq B$, we have an inclusion of monomial ideals

$$(\mathrm{LM}(g') \colon g' \in B') \subseteq (\mathrm{LM}(g) \colon g \in B).$$

We claim that if $B' \subsetneq B$, then there is also a strict inclusion of ideals. Suppose that $r$ is a nonzero remainder of some $S$-polynomial mod $B'$, and that $B = B' \cup \{r\}$. By definition, the remainder $r$ is reduced mod $G'$, and so no monomial in $r$ is divisible by $\mathrm{LM}(g')$ for any $g' \in B'$; in particular, $\mathrm{LM}(r)$ is not divisible by any $\mathrm{LM}(g')$. Hence, $\mathrm{LM}(r) \notin (\mathrm{LM}(g') \colon g' \in B')$, by Lemma 5.134. On the other hand, we do have $\mathrm{LM}(r) \in (\mathrm{LM}(g) \colon g \in B)$. Therefore, if the algorithm does not stop, there is an infinite strictly ascending chain of ideals in $k[X]$, which contradicts the Hilbert Basis Theorem, for $k[X]$ has ACC.   •

**Example 5.137.** The reader may show that $B' = \{y^2 + z^2, x^2y + yz, z^3 + xy\}$ is not a Gröbner basis because $S(y^2 + z^2, x^2y + yz) = x^2z^2 - y^2z$ does not have remainder 0 mod $G'$. However, adjoining $x^2z^2 - y^2z$ does give a Gröbner basis $B$ because all $S$-polynomials in $B$ have remainder 0 mod $B'$.   ◄

Theoretically, Buchberger's algorithm computes a Gröbner basis, but the question arises how practical it is. In very many cases, it does compute in a reasonable amount of time; on the other hand, there are examples in which it takes a very long time to produce its output. The efficiency of Buchberger's Algorithm is discussed in Cox–Little–O'Shea, *Ideals, Varieties, and Algorithms*, Section 2.9.

**Corollary 5.138.**

(i) *If $I = (f_1, \ldots, f_t)$ is an ideal in $k[X]$, then there is an algorithm to determine whether a polynomial $h(X) \in k[X]$ lies in $I$.*

(ii) *If $I = (f_1, \ldots, f_t)$ and $I' = (f'_1, \ldots, f'_s)$ are ideals in $k[X]$, then there is an algorithm to determine whether $I = I'$.*

**Proof.**

(i) Use Buchberger's algorithm to find a Gröbner basis $B$ of $I$, and then use the Division Algorithm to compute the remainder of $h$ mod $G$ (where $G$ is any $m$-tuple arising from ordering the polynomials in $B$). By Corollary 5.130(ii), $h \in I$ if and only if $r = 0$.

(ii) Use Buchberger's algorithm to find Gröbner bases $\{g_1, \ldots, g_m\}$, $\{g'_1, \ldots, g'_p\}$ of $I$, $I'$, respectively. By part (i), there is an algorithm to determine whether each $g'_j \in I$, and $I' \subseteq I$ if each $g'_j \in I$. Similarly, there is an algorithm to determine the reverse inclusion, and so there is an algorithm to determine whether $I = I'$. •

One must be careful here. Corollary 5.138 does not begin by saying "If $I$ is an ideal in $k[X]$"; instead, it specifies a basis: $I = (f_1, \ldots, f_t)$. The reason, of course, is that Buchberger's Algorithm requires a basis as input. For example, the algorithm cannot be used directly to check whether a polynomial $f(X)$ lies in the radical $\sqrt{I}$, for we do not have a basis of $\sqrt{I}$. The book of Becker–Weispfenning, *Gröbner Bases*, p. 393, gives an algorithm computing a basis of $\sqrt{I}$ when the field $k$ of coefficients satisfies certain conditions.

No algorithm is known that computes the associated primes of an ideal, although there are algorithms to do some special cases of this general problem. If an ideal $I$ has a primary decomposition $I = Q_1 \cap \cdots \cap Q_r$, then the associated prime $P_i$ has the form $\sqrt{(I : c_i)}$ for any $c_i \in \bigcap_{j \neq i} Q_j$ and $c_i \notin Q_i$. Now there is an algorithm computing a basis of colon ideals (Becker–Weispfenning, *Gröbner Bases*, p. 266); thus, we could compute $P_i$ if there were an algorithm finding elements $c_i$. A survey of applications of Gröbner bases to various parts of Mathematics can be found in Buchberger–Winkler, *Gröbner Bases and Applications*.

A Gröbner basis $B = \{g_1, \ldots, g_m\}$ can be too large. For example, it follows from Proposition 5.128 that if $f \in I$, then $B \cup \{f\}$ is also a Gröbner basis of $I$; thus, we seek Gröbner bases that are, in some sense, minimal.

**Definition.** A basis $\{g_1, \ldots, g_m\}$ of an ideal $I$ is ***reduced*** if

(i) each $g_i$ is monic;

(ii) each $g_i$ is reduced mod $\{g_1, \ldots, \widehat{g_i}, \ldots, g_m\}$.

Exercise 5.84 on page 390 gives an algorithm for computing a reduced basis for every ideal $(f_1, \ldots, f_t)$. When combined with the algorithm in Exercise 5.85 on page 390, it shrinks a Gröbner basis to a *reduced* Gröbner basis. It can be proved (Becker–Weispfenning, *Gröbner Bases*, p. 209) that a reduced Gröbner basis of an ideal is unique.

In the special case when each $f_i(X)$ is linear, that is,

$$f_i(X) = a_{i1}x_1 + \cdots + a_{in}x_n,$$

the common zeros $\text{Var}(f_1, \ldots, f_t)$ are the solutions of a homogeneous system of $t$ equations in $n$ unknowns. If $A = [a_{ij}]$ is the $t \times n$ matrix of coefficients, then it can be shown that the reduced Gröbner basis corresponds to the row reduced echelon form for the matrix $A$ (Ibid., Section 10.5).

Another special case occurs when $f_1, \ldots, f_t$ are polynomials in one variable. The reduced Gröbner basis obtained from $\{f_1, \ldots, f_t\}$ turns out to be their gcd, and so the Euclidean Algorithm has been generalized to polynomials in several variables (Ibid., p. 217, last paragraph).

We end this chapter by showing how to find a basis of an intersection of ideals. Given a system of polynomial equations in several variables, one way to find solutions is to eliminate variables (van der Waerden, *Modern Algebra* II, Chapter XI). Given an ideal $I \subseteq k[X]$, we are led to an ideal in a subset of the indeterminates, which is essentially the intersection of $\text{Var}(I)$ with a lower-dimensional plane.

**Definition.** Let $k$ be a field and let $I \subseteq k[X, Y]$ be an ideal, where $k[X, Y]$ is the polynomial ring in disjoint sets of variables $X \cup Y$. The ***elimination ideal*** $I_X$ is defined by $I_X = I \cap k[X]$.

For example, if $I = (x^2, xy)$, then a Gröbner basis is $\{x^2, xy\}$ (by Corollary 5.135, because its generators are monomials), and $I_x = (x^2) \subseteq k[x]$, while $I_y = (0)$.

**Proposition 5.139.** *Let $k$ be a field and let $k[X] = k[x_1, \ldots, x_n]$ have a monomial order for which $x_1 \succ x_2 \succ \cdots \succ x_n$ (for example, the lexicographic order) and, for a fixed $p > 1$, let $Y = x_p, \ldots, x_n$. If $I \subseteq k[X]$ has a Gröbner basis $G = \{g_1, \ldots, g_m\}$, then $G \cap I_Y$ is a Gröbner basis for the elimination ideal $I_Y = I \cap k[x_p, \ldots, x_n]$.*

**Proof.** Recall that $\{g_1, \ldots, g_m\}$ being a Gröbner basis of $I = (g_1, \ldots, g_m)$ means that for each nonzero $f \in I$, there is $g_i$ with $\text{LM}(g_i) \mid \text{LM}(f)$. Let $f(x_p, \ldots, x_n) \in I_Y$ be nonzero. Since $I_Y \subseteq I$, there is some $g_i(X)$ with $\text{LM}(g_i) \mid \text{LM}(f)$; hence, $\text{LM}(g_i)$ involves only the "later" variables $x_p, \ldots, x_n$. Let $\text{DEG}(\text{LM}(g_i)) = \beta$. If $g_i$ has a monomial $c_\alpha X^\alpha$ involving "early" variables $x_i$ with $i < p$, then $\alpha \succ \beta$, because $x_1 \succ \cdots \succ x_p \succ \cdots \succ x_n$. This is a contradiction, for $\beta$, the degree of the leading monomial of $g_i$, is greater than the degree of any other monomial in $g_i$. It follows that $g_i \in k[x_p, \ldots, x_n]$. Exercise 5.83 on page 389 shows that $G \cap k[x_p, \ldots, x_n]$ is a Gröbner basis for $I_Y = I \cap k[x_p, \ldots, x_n]$. $\bullet$

We can now give Gröbner bases of intersections of ideals.

**Proposition 5.140.** *Let $k$ be a field, and let $I_1, \ldots, I_t$ be ideals in $k[X]$, where $X = x_1, \ldots, x_n$.*

(i) *Consider the polynomial ring $k[X, y_1, \ldots, y_t]$ in $n + t$ indeterminates. If $J$ is the ideal in $k[X, y_1, \ldots, y_t]$ generated by $1 - (y_1 + \cdots + y_t)$ and $y_j I_j$, for all $j$, then $\bigcap_{j=1}^t I_j = J_X$.*

(ii) *Given Gröbner bases of $I_1, \ldots, I_t$, a Gröbner basis of $\bigcap_{j=1}^t I_j$ can be computed.*

**Proof.**

(i) If $f = f(X) \in J_X = J \cap k[X]$, then $f \in J$, and so there is an equation

$$f(X) = g(X, Y)\Big(1 - \sum y_j\Big) + \sum_j h_j(X, y_1, \ldots, y_t)y_j q_j(X),$$

where $g, h_j \in k[X, Y]$ and $q_j \in I_j$. If $y_j = 1$ and $y_\ell = 0$ for $\ell \neq j$, then $f = h_j(X, 0, \ldots, 1, \ldots, 0)q_j(X)$. Note that $h_j(X, 0, \ldots, 1, \ldots, 0) \in k[X]$, and so $f \in I_j$. As $j$ was arbitrary, we have $f \in \bigcap I_j$, and so $J_X \subseteq \bigcap I_j$. For the reverse inclusion, $f \in \bigcap I_j$ implies $f \in J_X$, for $f = f\big(1 - \sum y_j\big) + \sum_j y_j f$.

(ii) This follows from part (i) and Proposition 5.139 if we use a monomial order in which all the variables in $X$ precede the variables in $Y$. $\bullet$

**Example 5.141.** Consider the ideal $I = (x) \cap (x^2, xy, y^2) \subseteq k[x, y]$, where $k$ is a field. Even though it is not difficult to find a basis of $I$ by hand, we shall use Gröbner bases to illustrate Proposition 5.140. Let $u$ and $v$ be new variables, and define $J = (1 - u - v, ux, vx^2, vxy, vy^2) \subseteq k[x, y, u, v]$. The first step is to find a Gröbner basis of $J$; we use the lexicographic monomial order with $x \prec y \prec u \prec v$. Since the $S$-polynomial of two monomials is 0 (Example 5.131), Buchberger's algorithm quickly gives a Gröbner basis[25] $G$ of $J$:

$$G = \{v + u - 1, x^2, yx, ux, uy^2 - y^2\}.$$

It follows from Proposition 5.139 that a Gröbner basis of $I$ is $G \cap k[x, y]$: all those elements of $G$ that do not involve the variables $u$ and $v$. Thus,

$$I = (x) \cap (x^2, xy, y^2) = (x^2, xy). \quad \blacktriangleleft$$

---

## Exercises

Use the degree-lexicographic monomial order in the following exercises.

**5.80.** Let $I = (y - x^2, z - x^3)$.

(i) Order $x \prec y \prec z$, and let $\preceq_{\text{lex}}$ be the corresponding monomial order on $\mathbb{N}^3$. Prove that $[y - x^2, z - x^3]$ is not a Gröbner basis of $I$.

(ii) Order $y \prec z \prec x$, and let $\preceq_{\text{lex}}$ be the corresponding monomial order on $\mathbb{N}^3$. Prove that $[y - x^2, z - x^3]$ is a Gröbner basis of $I$.

**5.81.** Find a Gröbner basis of $I = (x^2 - 1, xy^2 - x)$ and of $J = (x^2 + y, x^4 + 2x^2y + y^2 + 3)$.

**5.82.** (i) Find a Gröbner basis of $I = (xz, xy - z, yz - x)$. Does $x^3 + x + 1$ lie in $I$?

(ii) Find a Gröbner basis of $I = (x^2 - y, y^2 - x, x^2y^2 - xy)$. Does $x^4 + x + 1$ lie in $I$?

$*$ **5.83.** Let $I$ be an ideal in $k[X]$, where $k$ is a field and $k[X]$ has a monomial order. Prove that if a set of polynomials $\{g_1, \ldots, g_m\} \subseteq I$ has the property that, for each nonzero $f \in I$, there is some $g_i$ with $\text{LM}(g_i) \mid \text{LM}(f)$, then $I = (g_1, \ldots, g_m)$. Conclude, in the definition of Gröbner basis, that one need not assume that $I$ is generated by $g_1, \ldots, g_m$.

---

[25]This is actually the reduced Gröbner basis given by Exercise 5.85 on page 390.

* **5.84.** Show that the following pseudocode gives a reduced basis $Q$ of an ideal $I = (f_1, \ldots, f_t)$:

> Input: $P = [f_1, \ldots, f_t]$
> Output: $Q = [q_1, \ldots, q_s]$
> $Q := P$
> WHILE there is $q \in Q$ which is
>         not reduced mod $Q - \{q\}$ DO
>    select $q \in Q$ which is not reduced mod $Q - \{q\}$
>    $Q := Q - \{q\}$
>    $h :=$ the remainder of $q$ mod $Q$
>    IF $h \neq 0$ THEN
>      $Q := Q \cup \{h\}$
>    END IF
> END WHILE
> make all $q \in Q$ monic

**5.85.** Show that the following pseudocode replaces a Gröbner basis $G$ with a reduced Gröbner basis $H$:

> Input: $G = \{g_1, \ldots, g_m\}$
> Output: $H$
> $H := \varnothing; \quad F := G$
> WHILE $F \neq \varnothing$ DO
>    select $f'$ from $F$
>    $F := F - \{f'\}$
>    IF $\mathrm{LM}(f) \nmid \mathrm{LM}(f')$ for all $f \in F$ AND
>        $\mathrm{LM}(h) \nmid \mathrm{LM}(f')$ for all $h \in H$ THEN
>      $H := H \cup \{f'\}$
>    END IF
> END WHILE
> apply the algorithm in Exercise 5.84 to $H$

*Chapter 6*

# Rings

We now introduce *R-modules*, where $R$ is a (not necessarily commutative) ring; formally, they generalize vector spaces in the sense that scalars are allowed to be in a ring $R$ instead of in a field. In Chapter 8, we shall see, when $R$ is a PID, that the classification of finitely generated $R$-modules simultaneously gives a classification of all finitely generated abelian groups as well as the classification of square matrices over a field by canonical forms. In Chapter 7, modules over noncommutative rings will be used, in an essential way, to prove a theorem of Burnside: every finite group of order $p^m q^n$, where $p$ and $q$ are prime, is solvable. In the last two chapters, modules will be used to obtain theorems in Group Theory, Number Theory, and Commutative Algebra.

### Section 6.1. Modules

We have concentrated on commutative rings in earlier chapters; we now investigate noncommutative rings. Here are some examples of rings that are not commutative.

**Example 6.1.**

(i) If $k$ is any nonzero commutative ring, then $\text{Mat}_n(k)$, all $n \times n$ matrices with entries in $k$, is a ring under matrix multiplication and matrix addition; it is commutative if and only if $n = 1$.

Let $k$ be any, not necessarily commutative, ring. If $A = [a_{ip}]$ is an $n \times \ell$ matrix and $B = [b_{pj}]$ is an $\ell \times m$ matrix, then their product $AB$ is defined to be the $n \times m$ matrix whose $ij$ entry has the usual formula: $(AB)_{ij} = \sum_p a_{ip} b_{pj}$; just make sure that entries in $A$ always appear on the left and that entries of $B$ always appear on the right. Thus, $\text{Mat}_n(k)$ is a ring, even if $k$ is not commutative.

(ii) If $G$ is a group (whose binary operation is written multiplicatively), we define the ***group ring*** $\mathbb{Z}G$ as follows. Its additive group is the free abelian group having a basis labeled by the elements of $G$; thus, each element has a unique

391

expression of the form $\sum_{g \in G} a_g g$, where $a_g \in \mathbb{Z}$ for all $g \in G$ and *almost all* $a_g = 0$; that is, only finitely many $a_g$ can be nonzero. If $g$ and $h$ are basis elements (i.e., if $g, h \in G$), define their product in $\mathbb{Z}G$ to be their product $gh$ in $G$, while $ag = ga$ whenever $a \in \mathbb{Z}$ and $g \in G$. The product of any two elements of $\mathbb{Z}G$ is defined by extending by linearity:

$$\left( \sum_{g \in G} a_g g \right) \left( \sum_{h \in G} b_h h \right) = \sum_{z \in G} \left( \sum_{gh=z} a_g b_h \right) z.$$

The group ring $\mathbb{Z}G$ is commutative if and only if the group $G$ is abelian.

If $k$ is a field, then there is a similar construction of the **group algebra** $kG$; the additive group of $kG$ is now a vector space over $k$ having a basis labeled by the elements of $G$. Example 6.6 generalizes both of these examples: if $k$ is a commutative ring and $G$ is a group, it constructs the group algebra $kG$.

(iii) An **endomorphism** of an abelian group $A$ is a homomorphism $f \colon A \to A$. The **endomorphism ring** of $A$, denoted by $\operatorname{End}(A)$, is the set of all endomorphisms under pointwise addition

$$f + g \colon a \mapsto f(a) + g(a),$$

and composition as multiplication. It is easy to check that $\operatorname{End}(A)$ is always a ring, and simple examples show that it may not be commutative. For example, there are endomorphisms of $\mathbb{Z} \oplus \mathbb{Z}$ which do not commute. [In fact, $\operatorname{End}(\mathbb{Z} \oplus \mathbb{Z}) \cong \operatorname{Mat}_2(\mathbb{Z})$.]

(iv) We can define the polynomial ring $k[x]$ when $k$ is any, not necessarily commutative, ring, if we insist that the indeterminate $x$ commutes with constants.

(v) Let $k$ be a ring, and let $\sigma \colon k \to k$ be a ring endomorphism. Define a new multiplication on polynomials $k[x] = \{ \sum_i a_i x^i : a_i \in k \}$ satisfying

$$xa = \sigma(a)x.$$

Thus, multiplication of two polynomials is now given by

$$\left( \sum_i a_i x^i \right) \left( \sum_j b_j x^j \right) = \sum_r c_r x^r,$$

where $c_r = \sum_{i+j=r} a_i \sigma^i(b_j)$. It is a routine exercise to show that $k[x]$, equipped with this new multiplication, is a not necessarily commutative ring. We denote this ring by $k[x; \sigma]$, and we call it a ring of **skew polynomials**.

(vi) If $R_1, \ldots, R_t$ are rings, then their **direct product**,

$$R = R_1 \times \cdots \times R_t,$$

is the cartesian product with coordinatewise addition and multiplication:

$$(r_i) + (r_i') = (r_i + r_i') \quad \text{and} \quad (r_i)(r_i') = (r_i r_i');$$

we have abbreviated $(r_1, \ldots, r_t)$ to $(r_i)$. It is easy to see that $R = R_1 \times \cdots \times R_t$ is a ring. Let us identify $r_i \in R_i$ with the "vector" whose $i$th coordinate is $r_i$ and whose other coordinates are 0. If $i \neq j$, then $r_i r_j = 0$.

(vii) A *division ring* $D$ (or *skew field*) is a "possibly noncommutative field"; that is, $D$ is a ring in which $1 \neq 0$ and every nonzero element $a \in D$ has a multiplicative inverse: there exists $a' \in D$ with $aa' = 1 = a'a$. Equivalently, a ring $D$ is a division ring if the set $D^\times$ of its nonzero elements forms a group under multiplication. Of course, fields are division rings; here is a noncommutative example.

Let $\mathbb{H}$ be a four-dimensional vector space over $\mathbb{R}$, and label a basis $1, i, j, k$. Thus, a typical element $h$ in $\mathbb{H}$ is

$$h = a + bi + cj + dk,$$

where $a, b, c, d \in \mathbb{R}$. Define multiplication of basis elements as follows:

$$i^2 = j^2 = k^2 = -1;$$

$$ij = k = -ji; \quad jk = i = -kj; \quad ki = j = -ik;$$

we insist that every $a \in \mathbb{R}$ commutes with $1, i, j, k$ and $1h = h = h1$ for all $h \in \mathbb{H}$. Finally, if multiplication of arbitrary elements is defined by extending by linearity, then it is straightforward to check that $\mathbb{H}$ is a ring; it is called the (real) *quaternions*.[1] To see that $\mathbb{H}$ is a division ring, it suffices to find inverses of nonzero elements. Define the *conjugate* of $u = a + bi + cj + dk \in \mathbb{H}$ by

$$\overline{u} = a - bi - cj - dk;$$

we see easily that

$$u\overline{u} = a^2 + b^2 + c^2 + d^2.$$

Hence, $u\overline{u} \neq 0$ when $u \neq 0$, and so

$$u^{-1} = \overline{u}/u\overline{u} = \overline{u}/(a^2 + b^2 + c^2 + d^2).$$

It is not difficult to prove that conjugation is an additive isomorphism satisfying

$$\overline{uw} = \overline{w}\,\overline{u}.$$

Just as the Gaussian integers were used to prove Fermat's Two-Squares Theorem (Theorem 2.89)—an odd prime $p$ is a sum of two squares if and only if $p \equiv 1 \bmod 4$—so, too, can the quaternions be used to prove Lagrange's Theorem that every positive integer is the sum of four squares (Samuel, *Algebraic Theory of Numbers*, pp. 82–85).

The only property of the field $\mathbb{R}$ we have used in constructing $\mathbb{H}$ is that a sum of nonzero squares is nonzero; any subfield of $\mathbb{R}$ has this property, but $\mathbb{C}$ does not. For example, there is a division ring of rational quaternions. We shall construct other examples of division rings in Chapter 9 when we discuss *crossed product algebras*. ◄

---

[1] The quaternions were discovered in 1843 by W. R. Hamilton when he was seeking a generalization of the complex numbers to model some physical phenomena. He had hoped to construct a three-dimensional algebra for this purpose, but he succeeded only when he saw that dimension 3 should be replaced by dimension 4. This is why Hamilton called $\mathbb{H}$ the *quaternions*, and this division ring is denoted by $\mathbb{H}$ to honor Hamilton. The reader may check that the subset $\{\pm 1, \pm i, \pm j, \pm k\}$ is a multiplicative group isomorphic to the group $\mathbf{Q}$ of quaternions (see Exercise 6.14 on page 416).

A *subring* $S$ of a ring $R$ is a ring contained in $R$ so that $1 \in S$ and if $s, s' \in S$, then their sum $s + s'$ and product $ss'$ have the same meaning in $S$ as in $R$. Recall the formal definition.

**Definition.** A ***subring*** $S$ of a ring $R$ is a subset of $R$ such that

(i) $1 \in S$;

(ii) if $a, b \in S$, then $a - b \in S$;

(iii) if $a, b \in S$, then $ab \in S$.

**Example 6.2.**

(i) The **center** of a ring $R$, denoted by $Z(R)$, is the set of all those elements $z \in R$ commuting with everything:

$$Z(R) = \{z \in R : zr = rz \text{ for all } r \in R\}.$$

It is easy to see that $Z(R)$ is a subring of $R$. If $k$ is a commutative ring, then $k \subseteq Z(kG)$. Exercise 6.9 on page 415 asks you to prove that the center of a matrix ring, $Z(\mathrm{Mat}_n(R))$, is the set of all **scalar matrices** $aI$, where $a \in Z(R)$ and $I$ is the identity matrix. Exercise 6.10 on page 415 says that $Z(\mathbb{H}) = \{a1 : a \in \mathbb{R}\}$.

(ii) If $D$ is a division ring, then its center, $Z(D)$, is a field. Moreover, if $D^\times$ is the multiplicative group of the nonzero elements of $D$, then $Z(D^\times) = Z(D)^\times$; that is, the center of the multiplicative group $D^\times$ consists of the nonzero elements of $Z(D)$. ◄

Here are two nonexamples.

**Example 6.3.**

(i) Define $S = \{a + ib : a, b \in \mathbb{Z}\} \subseteq \mathbb{C}$. Define addition in $S$ to coincide with addition in $\mathbb{C}$, but define multiplication in $S$ by

$$(a + bi)(c + di) = ac + (ad + bc)i$$

(thus, $i^2 = 0$ in $S$, whereas $i^2 \neq 0$ in $\mathbb{C}$). It is easy to check that $S$ is a ring, but it is not a subring of $\mathbb{C}$.

(ii) If $R = \mathbb{Z} \times \mathbb{Z}$, the direct product with coordinatewise operations, then its unit is $(1, 1)$. Let

$$S = \{(n, 0) \in \mathbb{Z} \times \mathbb{Z} : n \in \mathbb{Z}\}.$$

It is easily checked that $S$ is closed under addition and multiplication; indeed, $S$ is a ring, for $(1, 0)$ is the unit in $S$. However, $S$ is not a subring of $R$ because $S$ does not contain the unit of $R$. ◄

An immediate complication arising from noncommutativity is that the notion of ideal splinters into three notions. There are now left ideals, right ideals, and two-sided ideals.

**Definition.** Let $R$ be a ring, and let $I$ be an additive subgroup of $R$. Then $I$ is a **left ideal** if $a \in I$ and $r \in R$ implies $ra \in I$, while $I$ is a **right ideal** if $ar \in I$. We say that $I$ is a **two-sided ideal** if it is both a left ideal and a right ideal.

**Example 6.4.** In $\text{Mat}_2(\mathbb{R})$, the equation $\begin{bmatrix} a & b \\ c & d \end{bmatrix} \begin{bmatrix} r & 0 \\ s & 0 \end{bmatrix} = \begin{bmatrix} * & 0 \\ * & 0 \end{bmatrix}$ shows that the "first columns" (that is, the matrices that are 0 off the first column), form a left ideal (the "second columns" also form a left ideal); neither of these left ideals is a right ideal. The equation $\begin{bmatrix} r & s \\ 0 & 0 \end{bmatrix} \begin{bmatrix} a & b \\ c & d \end{bmatrix} = \begin{bmatrix} * & * \\ 0 & 0 \end{bmatrix}$ shows that the "first rows" (that is, the matrices that are 0 off the first row) form a right ideal (the "second rows" also form a right ideal); neither of these right ideals is a left ideal. The only two-sided ideals are $\{0\}$ and $\text{Mat}_2(\mathbb{R})$ itself. This example generalizes, in the obvious way, to give examples of one-sided ideals in $\text{Mat}_n(k)$ for all $n \geq 2$ and every ring $k$. ◀

**Example 6.5.** In a direct product of rings, $R = R_1 \times \cdots \times R_t$, each $R_j$ is identified with
$$R_j = \big\{ (0, \ldots, 0, r_j, 0, \ldots, 0) : r_j \in R_j \big\},$$
where $r_j$ occurs in the $j$th coordinate. It is easy to see that each such $R_j$ is a two-sided ideal in $R$ (for if $j \neq i$, then $r_j r_i = 0$ and $r_i r_j = 0$). Moreover, any left or right ideal in $R_j$ is also a left or right ideal in $R$. ◀

Two-sided ideals arise from homomorphisms; we recall the definition.

**Definition.** If $R$ and $S$ are rings, then a **ring homomorphism** (or *ring map*) is a function $\varphi \colon R \to S$ such that, for all $r, r' \in R$,

(i) $\varphi(r + r') = \varphi(r) + \varphi(r')$;

(ii) $\varphi(rr') = \varphi(r)\varphi(r')$;

(iii) $\varphi(1) = 1$.

A **ring isomorphism** is a ring homomorphism that is also a bijection.

**Example 6.6.** If $G$ is a group and $k$ is a commutative ring, let
$$\mathcal{F}(G, k) = \{\varphi \colon G \to k : \varphi(g) = 0 \text{ for almost all } g \in G\}.$$

Equip $\mathcal{F}(G, k)$ with pointwise addition, and define a binary operation, called **convolution**, by
$$\varphi\psi(g) = \sum_{x \in G} \varphi(x)\psi(x^{-1}g).$$

In Exercise 6.21 on page 417, the reader will check that $\mathcal{F}(G, k)$ is a ring. If $u \in G$, define $\varphi_u(g) = 0$ for $g \neq u$ while $\varphi_u(u) = 1$. If $k = \mathbb{Z}$ or $k$ is a field, we defined the group ring $kG$ in Example 6.1(ii). Exercise 6.21 also says that $kG \to \mathcal{F}(G, k)$, defined by $u \mapsto \varphi_u$, is a ring isomorphism. ◀

If $\varphi \colon R \to S$ is a ring homomorphism, then the **kernel** is defined as usual:
$$\ker \varphi = \{r \in R : \varphi(r) = 0\}.$$

The **image** is also defined as usual:
$$\text{im } \varphi = \{s \in S : s = \varphi(r) \text{ for some } r \in R\}.$$

The kernel of a ring map $\varphi$ is always a two-sided ideal, for if $\varphi(a) = 0$ and $r \in R$, then

$$\varphi(ra) = \varphi(r)\varphi(a) = 0 = \varphi(a)\varphi(r) = \varphi(ar);$$

hence, $a \in \ker\varphi$ implies that both $ra$ and $ar$ lie in $\ker\varphi$. On the other hand, $\operatorname{im}\varphi$ is only a subring of $S$.

We can form the **quotient ring** $R/I$ when $I$ is a two-sided ideal, because the multiplication on the quotient abelian group $R/I$, given by $(r+I)(s+I) = rs+I$, is well-defined: if $r+I = r'+I$ and $s+I = s'+I$, then $rs+I = r's'+I$. That is, if $r - r' \in I$ and $s - s' \in I$, then $rs - r's' \in I$. To see this, note that

$$rs - r's' = rs - rs' + rs' - r's' = r(s - s') + (r - r')s \in I,$$

for both $s-s'$ and $r-r'$ lie in $I$ and each term on the right side also lies in $I$, because $I$ is a two-sided ideal. It is easy to see that the **natural map** $\pi\colon R \to R/I$, defined (as usual) by $r \mapsto r+I$, is a ring map. It is routine to check that the Isomorphism Theorems and the Correspondence Theorem hold for all rings.

**Example 6.7.** Here is an example in which $R/I$ is not a ring when $I$ is not a two-sided ideal. Let $R = \operatorname{Mat}_2(\mathbb{R})$ and let $I$ be the left ideal of first columns (see Example 6.4). Set $A = \left[\begin{smallmatrix} 0 & 1 \\ 2 & 1 \end{smallmatrix}\right]$, $A' = \left[\begin{smallmatrix} 0 & 1 \\ 0 & 1 \end{smallmatrix}\right]$, $B = \left[\begin{smallmatrix} 1 & 1 \\ 1 & 0 \end{smallmatrix}\right]$, and $B' = \left[\begin{smallmatrix} 0 & 1 \\ 0 & 0 \end{smallmatrix}\right]$. Note that $A - A' \in I$ and $B - B' \in I$. However, $AB = \left[\begin{smallmatrix} 1 & 0 \\ 3 & 2 \end{smallmatrix}\right]$ and $A'B' = \left[\begin{smallmatrix} 0 & 0 \\ 0 & 1 \end{smallmatrix}\right]$, so that $AB - A'B' \notin I$. ◄

A *left $R$-module* is just a "vector space over a ring $R$"; that is, in the definition of vector space, allow the scalars to be in $R$ instead of in a field.

**Definition.** Let $R$ be a ring. A **left $R$-module** is an (additive) abelian group $M$ equipped with a **scalar multiplication** $R \times M \to M$, denoted by

$$(r, m) \mapsto rm,$$

such that the following axioms hold for all $m, m' \in M$ and all $r, r', 1 \in R$:

   (i) $r(m + m') = rm + rm'$.

   (ii) $(r + r')m = rm + r'm$.

   (iii) $(rr')m = r(r'm)$.

   (iv) $1m = m$.

We often denote a left $R$-module $M$ by $_RM$.

An abelian group $M$ is a **right $R$-module** if there is a **scalar multiplication** $M \times R \to M$, denoted by

$$(m, r) \mapsto mr,$$

such that the following axioms hold for all $m, m' \in M$ and $r, r', 1 \in R$:

   (i) $(m + m')r = mr + m'r$.

   (ii) $m(r + r') = mr + mr'$.

   (iii) $m(rr') = (mr)r'$.

   (iv) $m1 = m$.

We often denote a right $R$-module $M$ by $M_R$.

Of course, there is nothing to prevent us from denoting the scalar multiplication in a right $R$-module by $(m, r) \mapsto rm$. If we do so, then we see that only axiom (iii) differs from the axioms for a left $R$-module; the right version now reads

$$(rr')m = r'(rm).$$

If $R$ is commutative, however, this distinction vanishes, for $(rr')m = (r'r)m = r'(rm)$. Thus, when $R$ is commutative, we will omit the adjective left or right; we will merely say that an abelian group $M$ equipped with scalars in $R$ is an $R$-module.

Here are some examples of $R$-modules with $R$ commutative.

**Example 6.8.**

(i) Every vector space over a field $k$ is a $k$-module.

(ii) The Laws of Exponents, Proposition 1.23, say that every abelian group is a $\mathbb{Z}$-module.

(iii) Every commutative ring $R$ is a module over itself if we define scalar multiplication $R \times R \to R$ to be the given multiplication of elements of $R$. More generally, every ideal $I$ in $R$ is an $R$-module, for if $i \in I$ and $r \in R$, then $ri \in I$.

(iv) Let $T: V \to V$ be a linear transformation on a finite-dimensional vector space $V$ over a field $k$. The vector space $V$ can be made into a $k[x]$-module by defining scalar multiplication $k[x] \times V \to V$ as follows: if $f(x) = \sum_{i=0}^{m} c_i x^i$ lies in $k[x]$, then

$$f(x)v = \left(\sum_{i=0}^{m} c_i x^i\right)v = \sum_{i=0}^{m} c_i T^i(v),$$

where $T^0$ is the identity map $1_V$, $T^1 = T$, and $T^i$ is the composite of $T$ with itself $i$ times if $i \geq 2$. We denote $V$ viewed as a $k[x]$-module by $V^T$.

Here is a special case of this construction. Let $A$ be an $n \times n$ matrix with entries in $k$, and let $T: k^n \to k^n$ be the linear transformation $T(w) = Aw$, where $w$ is an $n \times 1$ column vector and $Aw$ is matrix multiplication. Now the vector space $k^n$ becomes a $k[x]$-module by defining scalar multiplication $k[x] \times k^n \to k^n$ as follows: if $f(x) = \sum_{i=0}^{m} c_i x^i \in k[x]$, then

$$f(x)w = \left(\sum_{i=0}^{m} c_i x^i\right)w = \sum_{i=0}^{m} c_i A^i w,$$

where $A^0 = I$ is the identity matrix, $A^1 = A$, and $A^i$ is the $i$th power of $A$ if $i \geq 2$. We now show that $(k^n)^T = (k^n)^A$. Both modules are comprised of the same elements (namely, all $n$-tuples), and the scalar multiplications coincide: in $(k^n)^T$, we have $xw = T(w)$; in $(k^n)^A$, we have $xw = Aw$; these are the same because $T(w) = Aw$.

(v) The construction in part (iv) can be generalized. Let $k$ be a commutative ring, $M$ a $k$-module, and $\varphi: M \to M$ a $k$-map. Then $M$ becomes a $k[x]$-module, denoted by $M^\varphi$, if we define

$$\left(\sum_{i=0}^{m} c_i x^i\right) m = \sum_{i=0}^{m} c_i \varphi^i(m),$$

where $f(x) = \sum_{i=0}^{m} c_i x^i \in k[x]$ and $m \in M$.  ◀

Here are some examples of modules over noncommutative rings.

**Example 6.9.**

(i) Left ideals in a ring $R$ are left $R$-modules, while right ideals in $R$ are right $R$-modules. Thus, we see that left $R$-modules and right $R$-modules are distinct entities (Example 6.4).

(ii) If $S$ is a subring of a ring $R$, then $R$ is a left and a right $S$-module, where scalar multiplication is just the given multiplication of elements of $R$. For example, if $k$ is a (not necessarily commutative) ring, then $k[X]$ is a left $k$-module; if $k$ is a field, then $k[X]$ is a vector space over $k$.

(iii) If $A$ is an abelian group, then $A$ is a left $\operatorname{End}(A)$-module, where scalar multiplication $\operatorname{End}(A) \times A \to A$ is defined by $(f, a) \mapsto f(a)$. Let us check the associativity axiom (iii) in the definition of module. We use extra-fussy notation here: write $f \circ g$ to denote the composite [which is the product of $f$ and $g$ in $\operatorname{End}(A)$], and write $f * a$ to denote the action of $f$ on $a$ [so that $f * a = f(a)$]. Now $(fg) * a = (f \circ g) * a = (f \circ g)(a) = f(g(a))$, while $f * (g * a) = f * (g(a)) = f(g(a))$. Thus, $(fg) * a = f * (g * a)$.

(iv) Let $E/k$ be a field extension with Galois group $G = \operatorname{Gal}(E/k)$. Then $E$ is a left $kG$-module: if $e \in E$, then

$$\left(\sum_{\sigma \in G} a_\sigma \sigma\right)(e) = \sum_{\sigma \in G} a_\sigma \sigma(e).$$

(v) Let $G$ be a group, let $k$ be a commutative ring, and let $A$ be a left $kG$-module. Define a new action of $G$ on $A$, denoted by $g * a$, by

$$g * a = g^{-1} a,$$

where $a \in A$ and $g \in G$. For an arbitrary element of $kG$, define

$$\left(\sum_{g \in G} m_g g\right) * a = \sum_{g \in G} m_g g^{-1} a.$$

It is easy to see that $A$ is a right $kG$-module under this new action; that is, if $u \in kG$ and $a \in A$, the function $A \times kG \to A$, given by $(a, u) \mapsto u * a$, satisfies the axioms in the definition of right module. Of course, we usually write $au$ instead of $u * a$. Thus, a $kG$-module can be viewed as either a left or a right $kG$-module.  ◀

Here is the appropriate notion of homomorphism.

**Definition.** If $R$ is a ring and $M$ and $N$ are both left $R$-modules [or both right $R$-modules], then a function $f \colon M \to N$ is an $R$-**homomorphism** (or $R$-**map**) if

(i) $f(m + m') = f(m) + f(m')$;

(ii) $f(rm) = rf(m)$   [or $f(mr) = f(m)r$]

for all $m, m' \in M$ and all $r \in R$.

If an $R$-homomorphism is a bijection, then it is called an **$R$-isomorphism**; we call $R$-modules $M$ and $N$ **isomorphic**, denoted by $M \cong N$, if there is some $R$-isomorphism $f\colon M \to N$.

Note that the composite of $R$-homomorphisms is an $R$-homomorphism and, if $f$ is an $R$-isomorphism, then its inverse function $f^{-1}$ is also an $R$-isomorphism.

**Example 6.10.**

(i) If $R$ is a field, then $R$-modules are vector spaces and $R$-maps are linear transformations. Isomorphisms here are nonsingular linear transformations.

(ii) By Example 6.8(ii), $\mathbb{Z}$-modules are just abelian groups, and Lemma 1.58 shows that every homomorphism of (abelian) groups is a $\mathbb{Z}$-map.

(iii) If $M$ is a left $R$-module and $r \in Z(R)$, then **multiplication by $r$** (or *homothety by $r$*) is the function $\mu_r\colon M \to M$ given by $m \mapsto rm$.

The functions $\mu_r$ are $R$-maps because $r$ lies in the center $Z(R)$: if $a \in R$ and $m \in M$, then $\mu_r(am) = ram$ while $a\mu_r(m) = arm$.

(iv) Let $T\colon V \to V$ be a linear transformation on a vector space $V$ over a field $k$, let $v_1, \ldots, v_n$ be a basis of $V$, and let $A$ be the matrix of $T$ relative to this basis. We now show that the two $k[x]$-modules $V^T$ and $(k^n)^A$ are isomorphic.

Define $\varphi\colon V \to k^n$ by $\varphi(v_i) = e_i$, where $e_1, \ldots, e_n$ is the standard basis of $k^n$; the linear transformation $\varphi$ is an isomorphism of vector spaces. To see that $\varphi$ is a $k[x]$-map, it suffices to prove that $\varphi(f(x)v) = f(x)\varphi(v)$ for all $f(x) \in k[x]$ and all $v \in V$. Now

$$\varphi(xv_i) = \varphi(T(v_i)) = \varphi\left(\sum a_{ji}v_j\right) = \sum a_{ji}\varphi(v_j) = \sum a_{ji}e_j,$$

which is the $i$th column of $A$. On the other hand,

$$x\varphi(v_i) = A\varphi(v_i) = Ae_i,$$

which is also the $i$th column of $A$. It follows that $\varphi(xv) = x\varphi(v)$ for all $v \in V$, and we can easily prove, by induction on $\deg(f)$, that $\varphi(f(x)v) = f(x)\varphi(v)$ for all $f(x) \in k[x]$ and all $v \in V$. ◄

The next proposition generalizes the last example.

**Proposition 6.11.** *Let $V$ be a vector space over a field $k$, and let $T, S\colon V \to V$ be linear transformations. Then the $k[x]$-modules $V^T$ and $V^S$ in Example 6.8(iv) are $k[x]$-isomorphic if and only if there is a vector space isomorphism $\varphi\colon V \to V$ with*

$$S = \varphi T \varphi^{-1}.$$

**Proof.** If $\varphi\colon V^T \to V^S$ is a $k[x]$-isomorphism, then $\varphi\colon V \to V$ is an isomorphism of vector spaces with

$$\varphi(f(x)v) = f(x)\varphi(v)$$

for all $v \in V$ and all $f(x) \in k[x]$. In particular, if $f(x) = x$, then

$$\varphi(xv) = x\varphi(v).$$

But the definition of scalar multiplication in $V^T$ is $xv = T(v)$, while the definition of scalar multiplication in $V^S$ is $xv = S(v)$. Hence, for all $v \in V$, we have

$$\varphi(T(v)) = S(\varphi(v)).$$

Therefore,

$$\varphi T = S\varphi.$$

As $\varphi$ is an isomorphism, we have the desired equation $S = \varphi T \varphi^{-1}$.

Conversely, we may assume that $\varphi(f(x)v) = f(x)\varphi(v)$ in the special cases $\deg(f) \leq 1$:

$$\varphi(xv) = \varphi T(v) = S\varphi(v) = x\varphi(v).$$

Next, an easy induction shows that $\varphi(x^n v) = x^n \varphi(v)$, and a second easy induction, on $\deg(f)$, shows that $\varphi(f(x)v) = f(x)\varphi(v)$ for all $f(x) \in k[x]$.   •

It is worthwhile to make a special case of the proposition explicit. The next corollary shows how comfortably similarity of matrices fits into the language of modules (and we will see, in Chapter 8, how this contributes to finding canonical forms for matrices).

**Corollary 6.12.** *Let $k$ be a field, and let $A$ and $B$ be $n \times n$ matrices with entries in $k$. Then the $k[x]$-modules $(k^n)^A$ and $(k^n)^B$ in Example 6.8(iv) are $k[x]$-isomorphic if and only if $A$ and $B$ are similar; that is, there is a nonsingular matrix $P$ with*

$$B = PAP^{-1}.$$

**Proof.** Define $T: k^n \to k^n$ by $T(y) = Ay$, where $y \in k^n$ is a column; by Example 6.8(iv), the $k[x]$-module $(k^n)^T = (k^n)^A$. Similarly, define $S: k^n \to k^n$ by $S(y) = By$, and denote the corresponding $k[x]$-module by $(k^n)^B$. The proposition now gives an isomorphism $\varphi: V^T \to V^S$ with

$$\varphi(Ay) = B\varphi(y).$$

By Proposition 2.123, there is an $n \times n$ matrix $P$ with $\varphi(y) = Py$ for all $y \in k^n$ (which is nonsingular because $\varphi$ is an isomorphism). Therefore,

$$PAy = BPy$$

for all $y \in k^n$, and so

$$PA = BP;$$

hence, $B = PAP^{-1}$.

Conversely, the nonsingular matrix $P$ gives an isomorphism $\varphi: k^n \to k^n$ by $\varphi(y) = Py$ for all $y \in k^n$. The proposition now shows that $\varphi: (k^n)^A \to (k^n)^B$ is a $k[x]$-module isomorphism.   •

Homomorphisms can be added.

**Definition.** If $M$ and $N$ are left $R$-modules, then
$$\mathrm{Hom}_R(M, N) = \{\text{all } R\text{-homomorphisms } M \to N\}.$$
If $f, g \in \mathrm{Hom}_R(M, N)$, define $f + g \colon M \to N$ by
$$f + g \colon m \mapsto f(m) + g(m).$$

**Proposition 6.13.**

   (i) *If $M, N$ are left $R$-modules over a ring $R$, then $\mathrm{Hom}_R(M, N)$ is an abelian group, where addition has just been defined. Moreover, there are distributive laws: if $p \colon M' \to M$ and $q \colon N \to N'$, then*
$$(f + g)p = fp + gp \quad \text{and} \quad q(f + g) = qf + qg$$
*for all $f, g \in \mathrm{Hom}_R(M, N)$.*

   (ii) *If $R$ is commutative, then $\mathrm{Hom}_R(M, N)$ is an $R$-module.*

**Proof.** The proof of (i) is straightforward, but (ii) needs some details. Let $R$ be a commutative ring, and let $f \colon M \to N$ be an $R$-map. If $r \in R$, define a function $rf \colon M \to N$ by
$$rf \colon m \mapsto f(rm).$$
We claim that $rf$ is an $R$-map: if $a \in R$, then $(rf)(am) = a[(rf)(m)]$. Now $(rf)(am) = f(ram) = raf(m)$, while $a[(rf)(m)] = af(rm) = arf(m)$; these are equal because $ra = ar$. The reader may now check that $R \times \mathrm{Hom}_R(M, N) \to \mathrm{Hom}_R(M, N)$, defined by $(r, f) \mapsto rf$, is a scalar multiplication making $\mathrm{Hom}_R(M, N)$ into an $R$-module. $\bullet$

   We are now going to show that ring elements can be regarded as operators (that is, as endomorphisms) on an abelian group.

**Definition.** A ***representation*** of a ring $R$ is a ring homomorphism
$$\sigma \colon R \to \mathrm{End}(M),$$
where $M$ is an abelian group.

   Representations of rings can be translated into the language of modules.

**Proposition 6.14.** *Every representation $\sigma \colon R \to \mathrm{End}(M)$, where $M$ is an abelian group, equips $M$ with the structure of a left $R$-module. Conversely, every left $R$-module $M$ determines a representation $\sigma \colon R \to \mathrm{End}(M)$.*

**Proof.** Given a homomorphism $\sigma \colon R \to \mathrm{End}(M)$, denote $\sigma(r) \colon M \to M$ by $\sigma_r$, and define scalar multiplication $R \times M \to M$ by
$$rm = \sigma_r(m),$$
where $m \in M$. A routine calculation shows that $M$, equipped with this scalar multiplication, is a left $R$-module.

   Conversely, assume that $M$ is a left $R$-module. If $r \in R$, then $m \mapsto rm$ defines an endomorphism $T_r \colon M \to M$. It is easily checked that the function $\sigma \colon R \to \mathrm{End}(M)$, given by $\sigma \colon r \mapsto T_r$, is a representation. $\bullet$

**Definition.** A left $R$-module is called **_faithful_** if, for all $r \in R$, whenever $rm = 0$ for all $m \in M$, we have $r = 0$.

Of course, $M$ being faithful merely says that the representation $\sigma \colon R \to \mathrm{End}(M)$ (given in Proposition 6.14) is an injection. When $R = \mathbb{Z}$, an abelian group $A$ is faithful if there is no positive integer $m$ with $mA = \{0\}$; that is, $A$ does not have an exponent.

Instead of stating definitions and results for left $R$-modules and then saying that similar statements hold for right $R$-modules, let us now show that it suffices to consider left modules only.

**Definition.** Let $R$ be a ring with multiplication $\mu \colon R \times R \to R$. Define the **_opposite ring_** to be the ring $R^{\mathrm{op}}$ whose additive group is the same as the additive group of $R$, but whose multiplication $\mu^{\mathrm{op}} \colon R \times R \to R$ is defined by $\mu^{\mathrm{op}}(r, s) = \mu(s, r) = sr$.

Thus, we have merely reversed the order of multiplication. It is straightforward to check that $R^{\mathrm{op}}$ is a ring; it is obvious that $(R^{\mathrm{op}})^{\mathrm{op}} = R$; moreover, $R = R^{\mathrm{op}}$ if and only if $R$ is commutative.

**Proposition 6.15.** *Every right $R$-module $M$ is a left $R^{\mathrm{op}}$-module, and every left $R$-module is a right $R^{\mathrm{op}}$-module.*

**Proof.** We will be ultra-fussy in this proof. To say that $M$ is a right $R$-module is to say that there is a function $\sigma \colon M \times R \to M$, denoted by $\sigma(m, r) = mr$. If $\mu \colon R \times R \to R$ is the given multiplication in $R$, then axiom (iii) in the definition of right $R$-module says that

$$\sigma(m, \mu(r, r')) = \sigma(\sigma(m, r), r').$$

To obtain a left $R$-module, define $\sigma' \colon R \times M \to M$ by $\sigma'(r, m) = \sigma(m, r)$. To see that $M$ is a left $R^{\mathrm{op}}$-module, it is only a question of checking axiom (iii), which reads, in the fussy notation,

$$\sigma'(\mu^{\mathrm{op}}(r, r'), m) = \sigma'(r, \sigma'(r', m)).$$

But

$$\sigma'(\mu^{\mathrm{op}}(r, r'), m) = \sigma(m, \mu^{\mathrm{op}}(r, r')) = \sigma(m, \mu(r', r)) = m(r'r),$$

while the right side is

$$\sigma'(r, \sigma'(r', m)) = \sigma(\sigma'(r', m), r) = \sigma(\sigma(m, r'), r) = (mr')r.$$

Thus, the two sides are equal because $M$ is a right $R$-module.

The second half of the proposition now follows because a right $R^{\mathrm{op}}$-module $M$ is a left $(R^{\mathrm{op}})^{\mathrm{op}}$-module; that is, $M$ is a left $R$-module, for $(R^{\mathrm{op}})^{\mathrm{op}} = R$.  •

By Proposition 6.15, any theorem about left $R$-modules, as $R$ varies over all rings, is, in particular, a theorem about left $R^{\mathrm{op}}$-modules; that is, it is also a theorem about right $R$-modules. From now on, we will state theorems for left $R$-modules, tacitly realizing that they hold for right $R$-modules as well.

Let us now see that opposite rings are more than an expository device; they do occur in nature. An ***anti-isomorphism*** $\varphi\colon R \to A$, where $R$ and $A$ are rings, is an additive bijection such that

$$\varphi(rs) = \varphi(s)\varphi(r).$$

It is easy to see that $R$ and $A$ are anti-isomorphic if and only if $R \cong A^{\mathrm{op}}$.

In Example 6.1(iii), we defined $\mathrm{End}(A)$, where $A$ is an abelian group, as the set of all homomorphisms $A \to A$; it is a ring under pointwise addition and composition as multiplication. We generalize this construction. If $M$ is a left $R$-module, then an $R$-map $f\colon M \to M$ is called an $R$-***endomorphism*** of $M$. The ***endomorphism ring***, denoted by $\mathrm{End}_R(M)$, is the set of all $R$-endomorphisms of $M$. As a set, $\mathrm{End}_R(M) = \mathrm{Hom}_R(M, M)$, which is an additive abelian group. Now define multiplication to be composition: if $f, g\colon M \to M$, then $fg\colon m \mapsto f(g(m))$.

If $M$ is regarded as an abelian group, then we may write $\mathrm{End}_{\mathbb{Z}}(M)$ for the endomorphism ring $\mathrm{End}(M)$ (with no subscript) defined in Example 6.1(iii). Note that $\mathrm{End}_R(M)$ is a subring of $\mathrm{End}_{\mathbb{Z}}(M)$. It is shown, in Example 6.9(iii), that an abelian group $M$ is always a left $\mathrm{End}(M)$-module. It follows that if $R$ is a ring and $M$ is a left $R$-module, then $M$ is a left $\mathrm{End}_R(M)$-module.

**Proposition 6.16.** *If a ring $R$ is regarded as a left module over itself, then there is an isomorphism of rings*

$$\mathrm{End}_R(R) \cong R^{\mathrm{op}}.$$

**Proof.** Define $\varphi\colon \mathrm{End}_R(R) \to R$ by $\varphi(f) = f(1)$; it is routine to check that $\varphi$ is an isomorphism of additive abelian groups. Now $\varphi(f)\varphi(g) = f(1)g(1)$. On the other hand, $\varphi(fg) = (f \circ g)(1) = f(g(1))$. But if we write $r = g(1)$, then $f(g(1)) = f(r) = f(r \cdot 1) = rf(1)$, because $f$ is an $R$-map, and so $f(g(1)) = rf(1) = g(1)f(1)$. Therefore,

$$\varphi(fg) = \varphi(g)\varphi(f).$$

We have shown that $\varphi\colon \mathrm{End}_R(R) \to R$ is an additive bijection that reverses multiplication. Composing $\varphi$ with an anti-isomorphism $R \to R^{\mathrm{op}}$ gives a ring isomorphism $\mathrm{End}_R(R) \to R^{\mathrm{op}}$. $\bullet$

If $k$ is a commutative ring, then transposition, $A \mapsto A^{\top}$, is an anti-isomorphism $\mathrm{Mat}_n(k) \to \mathrm{Mat}_n(k)$, because $(AB)^{\top} = B^{\top}A^{\top}$; therefore, $\mathrm{Mat}_n(k) \cong [\mathrm{Mat}_n(k)]^{\mathrm{op}}$. However, when $k$ is not commutative, the formula $(AB)^{\top} = B^{\top}A^{\top}$ no longer holds. For example,

$$\left( \begin{bmatrix} a & b \\ c & d \end{bmatrix} \begin{bmatrix} p & q \\ r & s \end{bmatrix} \right)^{\top} = \begin{bmatrix} ap + br & aq + bs \\ cp + dr & cq + ds \end{bmatrix}^{\top},$$

while

$$\begin{bmatrix} p & q \\ r & s \end{bmatrix}^{\top} \begin{bmatrix} a & b \\ c & d \end{bmatrix}^{\top} = \begin{bmatrix} p & r \\ q & s \end{bmatrix} \begin{bmatrix} a & c \\ b & d \end{bmatrix}$$

has $pa + rb \neq ap + br$ as its $1, 1$ entry.

**Proposition 6.17.** *If $R$ is any ring, then*

$$[\mathrm{Mat}_n(R)]^{\mathrm{op}} \cong \mathrm{Mat}_n(R^{\mathrm{op}}).$$

**Proof.** We claim that transposition $A \mapsto A^\top$ is an isomorphism of rings,

$$[\mathrm{Mat}_n(R)]^{\mathrm{op}} \to \mathrm{Mat}_n(R^{\mathrm{op}}).$$

First, it follows from $(A^\top)^\top = A$ that $A \mapsto A^\top$ is a bijection. Let us set notation. If $M = [m_{ij}]$ is a matrix, its $ij$ entry $m_{ij}$ may also be denoted by $(M)_{ij}$. Denote the multiplication in $R^{\mathrm{op}}$ by $a * b$, where $a * b = ba$, and denote the multiplication in $[\mathrm{Mat}_n(R)]^{\mathrm{op}}$ by $A * B$, where $(A * B)_{ij} = (BA)_{ij} = \sum_k b_{ik}a_{kj} \in R$. We must show that $(A * B)^\top = A^\top B^\top$ in $\mathrm{Mat}_n(R^{\mathrm{op}})$. In $[\mathrm{Mat}_n(R)]^{\mathrm{op}}$, we have

$$(A * B)^\top_{ij} = (BA)^\top_{ij} = (BA)_{ji} = \sum_k b_{jk}a_{ki}.$$

In $\mathrm{Mat}_n(R^{\mathrm{op}})$, we have

$$(A^\top B^\top)_{ij} = \sum_k (A^\top)_{ik} * (B^\top)_{kj} = \sum_k (A)_{ki} * (B)_{jk} = \sum_k a_{ki} * b_{jk} = \sum_k b_{jk}a_{ki}.$$

Therefore, $(A * B)^\top = A^\top B^\top$ in $\mathrm{Mat}_n(R^{\mathrm{op}})$, as desired. •

We now show that constructions made for abelian groups and for vector spaces can also be made for modules. Informally, a *submodule* $S$ is an $R$-module contained in a larger $R$-module $M$ such that if $s, s' \in S$ and $r \in R$, then $s + s'$ and $rs$ have the same meaning in $S$ as in $M$.

**Definition.** If $M$ is a left $R$-module, then a **submodule** $N$ of $M$, denoted by $N \subseteq M$, is an additive subgroup $N$ of $M$ closed under scalar multiplication: $rn \in N$ whenever $n \in N$ and $r \in R$.

**Example 6.18.**

(i) Both $\{0\}$ and $M$ are submodules of a left $R$-module $M$. A **proper submodule** of $M$ is a submodule $N \subseteq M$ with $N \neq M$. In this case, we may write $N \subsetneq M$.

(ii) If a ring $R$ is viewed as a left module over itself, then a submodule of $R$ is a left ideal; $I$ is a proper submodule when it is a proper ideal.

(iii) A submodule of a $\mathbb{Z}$-module (i.e., of an abelian group) is a subgroup, and a submodule of a vector space is a subspace.

(iv) A submodule $W$ of $V^T$, where $T \colon V \to V$ is a linear transformation, is a subspace $W$ of $V$ with $T(W) \subseteq W$ (it is clear that a submodule has this property; the converse is left as an exercise for the reader). Such a subspace is called an **invariant subspace**.

(v) If $M$ is a left $R$-module over a ring $R$ and $r \in Z(R)$, then

$$rM = \{rm : m \in M\}$$

is a submodule of $M$. If $r$ is an element of $R$ not in the center of $R$, let $J = Rr = \{sr : s \in R\}$ ($J$ is the left ideal generated by $r$). Now

$$JM = \{am : a \in J \text{ and } m \in M\}$$

is a submodule. We illustrate these constructions. Let $R = \mathrm{Mat}_2(k)$, where $k$ is a field, let $r = \left[\begin{smallmatrix} 1 & 0 \\ 0 & 0 \end{smallmatrix}\right]$ ($r \notin Z(R)$), and let $M = {}_R R$ (that is, $R$ viewed as a left $R$-module). Now $rM = \left\{\left[\begin{smallmatrix} * & * \\ 0 & 0 \end{smallmatrix}\right]\right\}$, which is a right ideal but not a left ideal; hence, $rM$ is not a submodule of $M$. On the other hand, if $J = Rr$, then $JM = \{[\begin{smallmatrix} * & 0 \\ * & 0 \end{smallmatrix}]\} = J$ is a submodule of $M$.

More generally, if $J$ is any left ideal in $R$ and $M$ is a left $R$-module, then

$$JM = \Big\{\sum_i j_i m_i : j_i \in J \text{ and } m_i \in M\Big\}$$

is a submodule of $M$.

(vi) If $(S_i)_{i \in I}$ is a family of submodules of a left $R$-module $M$, then $\bigcap_{i \in I} S_i$ is a submodule of $M$.

(vii) If $X$ is a subset of a left $R$-module $M$, then

$$\langle X \rangle = \Big\{\sum_{\text{finite}} r_i x_i : r_i \in R \text{ and } x_i \in X\Big\},$$

the set of all $R$-***linear combinations*** of elements in $X$, is called the ***submodule generated by*** $X$ (see Exercise 6.3 on page 414 for a characterization of $\langle X \rangle$). A left $R$-module $M$ is ***finitely generated*** if $M$ is generated by a finite set; that is, there is a finite subset $X = \{x_1, \dots, x_n\} \subseteq M$ with $M = \langle X \rangle$. If $X = \{x\}$ is a single element, then $\langle x \rangle = Rx$ is called the ***cyclic submodule generated by*** $x$. For example, a vector space is finitely generated if and only if it is finite-dimensional.

(viii) If $S$ and $T$ are submodules of a left $R$-module $M$, then

$$S + T = \{s + t : s \in S \text{ and } t \in T\}$$

is a submodule of $M$ which contains $S$ and $T$. Indeed, it is the submodule generated by $S \cup T$.

(ix) Recall that a field extension $E/k$ with Galois group $G = \mathrm{Gal}(E/k)$ is a left $kG$-module [Example 6.9(iv)]. We say that $E/k$ has a ***normal basis*** if $E$ is a cyclic left $kG$-module. Theorem 10.100 says that every Galois extension $E/k$ has a normal basis.  ◀

We continue extending definitions from abelian groups and vector spaces to modules.

**Definition.** If $f \colon M \to N$ is an $R$-map between left $R$-modules, then its ***kernel*** is

$$\ker f = \{m \in M : f(m) = 0\}$$

and its ***image*** is

$$\operatorname{im} f = \{n \in N : \text{there exists } m \in M \text{ with } n = f(m)\}.$$

It is routine to check that $\ker f$ is a submodule of $M$ and that $\operatorname{im} f$ is a submodule of $N$. Suppose that $M = \langle X \rangle$; that is, $M$ is generated by a subset $X$. Suppose further that $N$ is a module and that $f, g \colon M \to N$ are $R$-homomorphisms. If $f$ and $g$ agree on $X$ [that is, if $f(x) = g(x)$ for all $x \in X$], then $f = g$. The reason is that $f - g \colon M \to N$, defined by $f - g \colon m \mapsto f(m) - g(m)$, is an $R$-homomorphism with $X \subseteq \ker(f - g)$. Therefore, $M = \langle X \rangle \subseteq \ker(f - g)$, and so $f - g$ is identically zero; that is, $f = g$.

**Definition.** If $N$ is a submodule of a left $R$-module $M$, then the **quotient module** is the quotient group $M/N$ (remember that $M$ is an abelian group and $N$ is a subgroup) equipped with the scalar multiplication

$$r(m + N) = rm + N.$$

The **natural map** $\pi\colon M \to M/N$, given by $m \mapsto m + N$, is easily seen to be an $R$-map.

Scalar multiplication in the definition of quotient module is well-defined: if $m + N = m' + N$, then $m - m' \in N$, hence $r(m - m') \in N$ (because $N$ is a submodule), and so $rm - rm' \in N$ and $rm + N = rm' + N$.

**Definition.** If $f\colon M \to N$ is a map, its **cokernel** is

$$\operatorname{coker} f = N/\operatorname{im} f.$$

A map $f\colon M \to N$ is surjective if and only if $\operatorname{coker} f = \{0\}$.

**Theorem 6.19 (First Isomorphism Theorem).** *If $f\colon M \to N$ is an $R$-map of left $R$-modules, then there is an $R$-isomorphism*

$$\varphi\colon M/\ker f \to \operatorname{im} f$$

*given by*

$$\varphi\colon m + \ker f \mapsto f(m).$$

**Proof.** If we view $M$ and $N$ only as abelian groups, then the First Isomorphism Theorem for Groups says that $\varphi\colon M/\ker f \to \operatorname{im} f$ is an isomorphism of abelian

$$
\begin{array}{ccc}
M & \xrightarrow{\ f\ } & N \\
{\scriptstyle \pi}\big\downarrow & & \big\uparrow{\scriptstyle \text{inc}} \\
M/\ker f & \xrightarrow[\varphi]{} & \operatorname{im} f
\end{array}
$$

groups. But $\varphi$ is an $R$-map: $\varphi(r(m + \ker f)) = \varphi(rm + \ker f) = f(rm)$; since $f$ is an $R$-map, however, $f(rm) = rf(m) = r\varphi(m + \ker f)$, as desired.   •

The Second and Third Isomorphism Theorems are corollaries of the first one.

**Theorem 6.20 (Second Isomorphism Theorem).** *If $S$ and $T$ are submodules of a left $R$-module $M$, then there is an $R$-isomorphism*

$$S/(S \cap T) \to (S + T)/T.$$

**Proof.** Let $\pi\colon M \to M/T$ be the natural map, so that $\ker \pi = T$; define $h = \pi|S$, so that $h\colon S \to M/T$. Now $\ker h = S \cap T$ and $\operatorname{im} h = (S+T)/T$ [for $(S+T)/T$ consists of all those cosets in $M/T$ having a representative in $S$]. The First Isomorphism Theorem now applies.   •

**Theorem 6.21 (Third Isomorphism Theorem).** *If $T \subseteq S \subseteq M$ is a tower of submodules, then $S/T$ is a submodule of $M/T$ and there is an $R$-isomorphism*

$$(M/T)/(S/T) \to M/S.$$

**Proof.** Define the map $g \colon M/T \to M/S$ to be ***enlargement of coset***; that is,

$$g \colon m + T \mapsto m + S.$$

Now $g$ is well-defined: if $m + T = m' + T$, then $m - m' \in T \subseteq S$ and $m + S = m' + S$. Moreover, $\ker g = S/T$ and $\operatorname{im} g = M/S$. Again, the First Isomorphism Theorem completes the proof. •

If $f \colon M \to N$ is a map of modules and $S \subseteq N$, then the reader may check that

$$f^{-1}(S) = \{m \in M \colon f(m) \in S\}$$

is a submodule of $M$ containing $\ker f$.

**Theorem 6.22 (Correspondence Theorem).** *If $T$ is a submodule of a left $R$-module $M$, then*

$$\varphi \colon \{intermediate\ submodules\ T \subseteq S \subseteq M\} \to \{submodules\ of\ M/T\},$$

*given by $\varphi \colon S \mapsto S/T$, is a bijection. Moreover, $S \subseteq S'$ in $M$ if and only if $S/T \subseteq S'/T$ in $M/T$.*

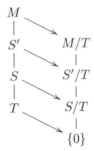

**Proof.** Since every module is an additive abelian group, every submodule is a subgroup, and so the Correspondence Theorem for Groups, Theorem 1.82, shows that $\varphi$ is an injection that preserves inclusions: $S \subseteq S'$ in $M$ if and only if $S/T \subseteq S'/T$ in $M/T$. The remainder of this proof is an adaptation of the proof of Proposition 5.1; we need check only that additive homomorphisms here are $R$-maps, and this is straightforward. •

**Proposition 6.23.** *A left $R$-module $M$ is cyclic if and only if $M \cong R/I$ for some left ideal $I$.*

**Proof.** If $M$ is cyclic, then $M = \langle m \rangle$ for some $m \in M$. Define $f \colon R \to M$ by $f(r) = rm$. Now $f$ is surjective, since $M$ is cyclic, and its kernel is some left ideal $I$. The First Isomorphism Theorem gives $R/I \cong M$.

Conversely, $R/I$ is cyclic with generator $1 + I$, and any module isomorphic to a cyclic module is itself cyclic. •

**Definition.** A left $R$-module $M$ is ***simple*** (or ***irreducible***) if $M \neq \{0\}$ and $M$ has no proper nonzero submodules; that is, the only submodules of $M$ are $\{0\}$ and $M$.

**Example 6.24.** By Proposition 1.117, an abelian group $G$ is simple if and only if $G \cong \mathbb{I}_p$ for some prime $p$. ◄

**Corollary 6.25.** *A left $R$-module $M$ is simple if and only if $M \cong R/I$, where $I$ is a maximal left ideal.*

**Proof.** This follows from the Correspondence Theorem. •

Thus, the existence of maximal left ideals guarantees the existence of simple left $R$-modules.

The notion of direct sum, already discussed for vector spaces and for abelian groups, extends to modules. Recall that an abelian group $G$ is an *internal* direct sum of subgroups $S$ and $T$ if $S + T = G$ and $S \cap T = \{0\}$, while an *external* direct sum is the abelian group whose underlying set is the cartesian product $S \times T$ and whose binary operation is pointwise addition; both versions give isomorphic abelian groups. The internal-external viewpoints persist for modules.

**Definition.** If $S$ and $T$ are left $R$-modules over a ring $R$, then their **external direct sum**, denoted[2] by $S \times T$, is the cartesian product $S \times T$ with coordinatewise operations:

$$(s, t) + (s', t') = (s + s', t + t');$$
$$r(s, t) = (rs, rt),$$

where $s, s' \in S$, $t, t' \in T$, and $r \in R$.

There are injective $R$-maps $\lambda_S \colon S \to S \times T$ and $\lambda_T \colon T \to S \times T$ given, respectively, by $\lambda_S \colon s \mapsto (s, 0)$ and $\lambda_T \colon t \mapsto (0, t)$.

**Proposition 6.26.** *The following statements are equivalent for left $R$-modules $M$, $S$, and $T$.*

(i) $S \times T \cong M$.

(ii) *There exist injective $R$-maps $i \colon S \to M$ and $j \colon T \to M$ such that*

$$M = \operatorname{im} i + \operatorname{im} j \quad \text{and} \quad \operatorname{im} i \cap \operatorname{im} j = \{0\}.$$

(iii) *There exist $R$-maps $i \colon S \to M$ and $j \colon T \to M$ such that, for every $m \in M$, there are unique $s \in S$ and $t \in T$ with*

$$m = is + jt.$$

(iv) *There are $R$-maps $i \colon S \to M$, $j \colon T \to M$, $p \colon M \to S$, and $q \colon M \to T$ such that*

$$pi = 1_S, \quad qj = 1_T, \quad pj = 0, \quad qi = 0, \quad \text{and} \quad ip + jq = 1_M.$$

**Remark.** The maps $i$ and $j$ are called **injections**, and the maps $p$ and $q$ are called **projections**. The equations $pi = 1_S$ and $qj = 1_T$ show that the maps $i$ and $j$ must be injective (so that $\operatorname{im} i \cong S$ and $\operatorname{im} j \cong T$) and the maps $p$ and $q$ must be surjective. ◀

**Proof.** Adapt the proof of Proposition 4.1. •

---

[2]Our notation $S \times T$ for external direct sum is a temporary one, so that we may contrast external and internal versions of direct sum.

*Internal direct sum* is the most important instance of a module isomorphic to a direct sum.

**Definition.** Let $S$ and $T$ be submodules of a left $R$-module $M$. Then $M$ is their ***internal direct sum*** if $M \cong S \times T$ with $i\colon S \to M$ and $j\colon T \to M$ the inclusions. We denote an internal direct sum by

$$M = S \oplus T.$$

In this section only, we will use the notation $S \times T$ to denote the external direct sum (whose underlying set is the cartesian product of all ordered pairs) and the notation $M = S \oplus T$ to denote the internal direct sum ($S$ and $T$ submodules of $M$ as just defined). Soon, we shall write as the mathematical world writes: the notation $S \oplus T$ will be used for either version of direct sum.

Here is a restatement of Proposition 6.26 for internal direct sums.

**Corollary 6.27.** *The following conditions are equivalent for a left $R$-module $M$ with submodules $S$ and $T$.*

(i) $M = S \oplus T$.

(ii) $S + T = M$ *and* $S \cap T = \{0\}$.

(iii) *Each $m \in M$ has a unique expression of the form $m = s + t$ for $s \in S$ and $t \in T$.*

**Proof.** Immediate from Proposition 6.26 by taking $i$ and $j$ to be inclusions. $\quad \bullet$

**Definition.** A submodule $S$ of a left $R$-module $M$ is a ***direct summand*** of $M$ if there exists a submodule $T$ of $M$ with $M = S \oplus T$.

The next corollary will connect direct summands with a special type of homomorphism.

**Definition.** Let $S$ be a submodule of a left $R$-module $M$. Then $S$ is a ***retract*** of $M$ if there exists an $R$-homomorphism $\rho\colon M \to S$, called a ***retraction***, with $\rho(s) = s$ for all $s \in S$.

Retractions of nonabelian groups arose in Exercise 4.73 on page 293.

**Corollary 6.28.** *A submodule $S$ of a left $R$-module $M$ is a direct summand if and only if there exists a retraction $\rho\colon M \to S$.*

**Proof.** In this case, we let $i\colon S \to M$ be the inclusion. We show that $M = S \oplus T$, where $T = \ker \rho$. If $m \in M$, then $m = (m - \rho m) + \rho m$. Plainly, $\rho m \in \operatorname{im} \rho = S$. On the other hand, $\rho(m - \rho m) = \rho m - \rho \rho m = 0$, because $\rho m \in S$ and so $\rho(\rho m) = \rho m$. Therefore, $M = S + T$.

If $m \in S$, then $\rho m = m$; if $m \in T = \ker \rho$, then $\rho m = 0$. Hence, if $m \in S \cap T$, then $m = 0$. Therefore, $S \cap T = \{0\}$, and $M = S \oplus T$.

For the converse, if $M = S \oplus T$, then each $m \in M$ has a unique expression of the form $m = s + t$, where $s \in S$ and $t \in T$, and it is easy to check that $\rho\colon M \to S$, defined by $\rho\colon s + t \mapsto s$, is a retraction $M \to S$. $\quad \bullet$

**Corollary 6.29.** *If $M = S \oplus T$ and $S \subseteq A \subseteq M$, then $A = S \oplus (A \cap T)$.*

**Proof.** Let $\rho\colon M \to S$ be the retraction $s + t \mapsto s$. Since $S \subseteq A$, the restriction $\rho|A\colon A \to S$ is a retraction with $\ker(\rho|A) = A \cap T$.   •

The direct sum construction can be extended to finitely[3] many modules. There are also external and internal versions.

**Definition.** Let $S_1, \ldots, S_n$ be left $R$-modules. Define the ***external direct sum***

$$S_1 \times \cdots \times S_n$$

to be the left $R$-module whose underlying set is the cartesian product $S_1 \times \cdots \times S_n$ and whose operations are

$$(s_1, \ldots, s_n) + (s_1', \ldots, s_n') = (s_1 + s_1', \ldots, s_n + s_n'),$$
$$r(s_1, \ldots, s_n) = (rs_1, \ldots, rs_n).$$

Let $M$ be a left $R$-module, and let $S_1, \ldots, S_n$ be submodules of $M$. Define $M$ to be the ***internal direct sum***

$$M = S_1 \oplus \cdots \oplus S_n$$

if each $m \in M$ has a unique expression of the form $m = s_1 + \cdots + s_n$, where $s_i \in S_i$ for all $i = 1, \ldots, n$.

For example, if $V$ is an $n$-dimensional vector space over a field $k$ and $v_1, \ldots, v_n$ is a basis, then

$$V = \langle v_1 \rangle \oplus \cdots \oplus \langle v_n \rangle.$$

We let the reader prove that the internal and external versions, when the former is defined, are isomorphic.

If $S_1, \ldots, S_n$ are submodules of a module $M$, when is $\langle S_1, \ldots, S_n \rangle$, the submodule generated by the $S_i$, equal to their direct sum? A common mistake is to say that it is enough to assume that $S_i \cap S_j = \{0\}$ for all $i \neq j$, but Example 4.8 on page 227 shows that this is not enough.

**Proposition 6.30.** *Let $M = S_1 + \cdots + S_n$, where the $S_i$ are submodules; that is, each $m \in M$ has a (not necessarily unique) expression of the form*

$$m = s_1 + \cdots + s_n,$$

*where $s_i \in S_i$ for all $i$. Then $M = S_1 \oplus \cdots \oplus S_n$ if and only if, for each $i$,*

$$S_i \cap \left( S_1 + \cdots + \widehat{S_i} + \cdots + S_n \right) = \{0\},$$

*where $\widehat{S_i}$ means that the term $S_i$ is omitted from the sum.*

**Proof.** Adapt the proof of Proposition 4.4. See Exercise 6.110 on page 514 for the generalization of this proposition for infinitely many submodules.   •

Here is the module analog of Proposition 4.9.

---

[3]There are several ways to extend this notion to infinitely many modules; the most useful, *direct sum* and *direct product*, are defined on page 427.

**Proposition 6.31.** *If $M_1, M_2, \ldots, M_n$ are left $R$-modules and $S_i \subseteq M_i$ are submodules, then*

$$(M_1 \times \cdots \times M_n)/(S_1 \times \cdots \times S_n) \cong (M_1/S_1) \times \cdots \times (M_n/S_n).$$

**Proof.** The function $\varphi \colon M_1 \times \cdots \times M_n \to (M_1/S_1) \times \cdots \times (M_n/S_n)$, given by

$$\varphi \colon (m_1, \ldots, m_n) \mapsto (m_1 + S_1, \ldots, m_n + S_n),$$

is a surjective homomorphism with kernel $S_1 \times \cdots \times S_n$. $\quad \bullet$

The following terminology, coined by the algebraic topologist Hurewicz, comes from Advanced Calculus, where a differential form $\omega$ is called **closed** if $d\omega = 0$ and it is called **exact** if $\omega = dh$ for some function $h$ (see Proposition 8.123 on page 726). It is interesting to look at Hurewicz–Wallman, *Dimension Theory*, Chapter VIII, which was written just before this coinage. Many results there would have been much simpler to state and to digest had the term *exact* been available.

**Definition.** A sequence of $R$-maps and left $R$-modules

$$\cdots \to M_{n+1} \xrightarrow{f_{n+1}} M_n \xrightarrow{f_n} M_{n-1} \to \cdots$$

is called an **exact sequence** if $\operatorname{im} f_{n+1} = \ker f_n$ for all $n$.

Observe that there is no need to label an arrow $\{0\} \xrightarrow{f} A$ or $B \xrightarrow{g} \{0\}$ for, in either case, such maps are unique: either $f \colon 0 \mapsto 0$ or $g$ is the zero map $g(b) = 0$ for all $b \in B$.

Here are some simple consequences of a sequence of homomorphisms being exact.

**Proposition 6.32.**

(i) *A sequence $0 \to A \xrightarrow{f} B$ is exact if and only if $f$ is injective.*[4]

(ii) *A sequence $B \xrightarrow{g} C \to 0$ is exact if and only if $g$ is surjective.*

(iii) *A sequence $0 \to A \xrightarrow{h} B \to 0$ is exact if and only if $h$ is an isomorphism.*

**Proof.**

(i) The image of $0 \to A$ is $\{0\}$, so that exactness gives $\ker f = \{0\}$, and so $f$ is injective. Conversely, given $f \colon A \to B$, there is an exact sequence $\ker f \to A \xrightarrow{f} B$. If $f$ is injective, then $\ker f = \{0\}$.

(ii) The kernel of $C \to 0$ is $C$, so that exactness of $B \xrightarrow{g} C \to 0$ gives $\operatorname{im} g = C$, and so $g$ is surjective. Conversely, given $g \colon B \to C$, there is an exact sequence $B \xrightarrow{g} C \to C/\operatorname{im} g$ (Exercise 6.26). If $g$ is surjective, then $C = \operatorname{im} g$ and $\operatorname{coker} g = C/\operatorname{im} g = \{0\}$.

(iii) Part (i) shows that $h$ is injective if and only if $0 \to A \xrightarrow{h} B$ is exact; part (ii) shows that $h$ is surjective if and only if $A \xrightarrow{h} B \to 0$ is exact. Therefore, $h$ is an isomorphism if and only if the sequence $0 \to A \xrightarrow{h} B \to 0$ is exact. $\quad \bullet$

---

[4]In displays, we usually write 0 instead of $\{0\}$.

**Definition.** A *short exact sequence* is an exact sequence of the form

$$0 \to A \xrightarrow{f} B \xrightarrow{g} C \to 0.$$

We also call this short exact sequence an *extension* of $A$ by $C$.

An extension is a short exact sequence, but we often call its middle module $B$ an extension of $A$ by $C$ as well (so do most people).

The Second and Third Isomorphism Theorems can be restated in the language of exact sequences.

**Proposition 6.33.**

(i) *If* $0 \to A \xrightarrow{f} B \xrightarrow{g} C \to 0$ *is a short exact sequence, then*

$$A \cong \operatorname{im} f \quad and \quad B/\operatorname{im} f \cong C.$$

(ii) *If* $T \subseteq S \subseteq M$ *is a tower of submodules, then there is an exact sequence*

$$0 \to S/T \xrightarrow{f} M/T \xrightarrow{g} M/S \to 0.$$

**Proof.**

(i) Since $f$ is injective, it is an isomorphism $A \to \operatorname{im} f$. The First Isomorphism Theorem gives $B/\ker g \cong \operatorname{im} g$. By exactness, however, $\ker g = \operatorname{im} f$ and $\operatorname{im} g = C$; therefore, $B/\operatorname{im} f \cong C$.

(ii) Define $f \colon S/T \to M/T$ to be the inclusion, and define $g \colon M/T \to M/S$ to be "enlargement of coset" $g \colon m + T \mapsto m + S$. As in the proof of Theorem 6.21, $g$ is surjective, and $\ker g = S/T = \operatorname{im} f$. •

In the special case when $A$ is a submodule of $B$ and $f \colon A \to B$ is the inclusion, exactness of $0 \to A \xrightarrow{f} B \xrightarrow{g} C \to 0$ gives $B/A \cong C$.

**Definition.** A short exact sequence

$$0 \to A \xrightarrow{i} B \xrightarrow{p} C \to 0$$

is *split* if there exists a map $j \colon C \to B$ with $pj = 1_C$.

**Proposition 6.34.** *If an exact sequence*

$$0 \to A \xrightarrow{i} B \xrightarrow{p} C \to 0$$

*is split, then* $B \cong A \oplus C$.

**Remark.** Exercise 6.32 on page 418 characterizes split short exact sequences. ◀

**Proof.** We show that $B = \operatorname{im} i \oplus \operatorname{im} j$, where $j \colon C \to B$ satisfies $pj = 1_C$. If $b \in B$, then $pb \in C$ and $b - jpb \in \ker p$, for $p(b - jpb) = pb - pj(pb) = 0$ because $pj = 1_C$. By exactness, there is $a \in A$ with $ia = b - jpb$. It follows that $B = \operatorname{im} i + \operatorname{im} j$. It remains to prove that $\operatorname{im} i \cap \operatorname{im} j = \{0\}$. If $ia = x = jc$, then $px = pia = 0$, because $pi = 0$, whereas $px = pjc = c$, because $pj = 1_C$. Therefore, $x = jc = 0$, and so $B \cong A \oplus C$. •

**Example 6.35.** The converse of the last proposition is not true: there exist exact sequences $0 \to A \to B \to C \to 0$ with $B \cong A \oplus C$ which are not split. Let $A = \langle a \rangle$, $B = \langle b \rangle$, and $C = \langle c \rangle$ be cyclic groups of orders 2, 4, and 2, respectively. If $i \colon A \to B$ is defined by $i(a) = 2b$ and $p \colon B \to C$ is defined by $p(b) = c$, then $0 \to A \xrightarrow{i} B \xrightarrow{p} C \to 0$ is an exact sequence that is not split: $\operatorname{im} i = \langle 2b \rangle$ is not even a pure subgroup of $B$. By Exercise 6.25 on page 417, for any abelian group $M$, there is an exact sequence

$$0 \to A \xrightarrow{i'} B \oplus M \xrightarrow{p'} C \oplus M \to 0,$$

where $i'(a) = (2b, 0)$ and $p'(b, m) = (c, m)$, and this sequence does not split either. If we choose $M = \mathbb{I}_4[x] \oplus \mathbb{I}_2[x]$ (the direct summands are the polynomial rings over $\mathbb{I}_4$ and $\mathbb{I}_2$, respectively), then $A \oplus (C \oplus M) \cong B \oplus M$. (For readers who are familiar with infinite direct sums, which we introduce later in this chapter, $M$ is the direct sum of infinitely many copies of $\mathbb{I}_4 \oplus \mathbb{I}_2$.) ◀

We have already considered commutative noetherian rings. For example, the Hilbert Basis Theorem says that if $R$ is noetherian, then so is the polynomial ring $R[x]$. Here is a noncommutative generalization.

**Definition.** A ring $R$ is **left noetherian** if every left ideal in $R$ is finitely generated. Similarly, a ring $R$ is **right noetherian** if every right ideal is finitely generated.[5]

Exercise 6.16 on page 416 gives an example of a left noetherian ring that is not right noetherian.

**Proposition 6.36.** *A ring $R$ is left noetherian if and only if every submodule of a finitely generated left $R$-module $M$ is itself finitely generated.*

**Remark.** Proposition 2.69 says that if $A$ is an abelian group that can be generated by $n$ elements, then every subgroup can be generated by $n$ or fewer elements. It is easy to generalize this to finitely generated $R$-modules, where $R$ is a PID. However, this is not true more generally. For example, if $R$ is not a PID, there there is some ideal $I$ that is not principal. Thus, $R$ has one generator while its submodule $I$ cannot be generated by one element. ◀

**Proof.** Assume that every submodule of a finitely generated left $R$-module is finitely generated. In particular, every submodule of $R$, which is a cyclic left $R$-module and hence is finitely generated, is finitely generated. But submodules of $R$ are left ideals, and so every left ideal is finitely generated; that is, $R$ is left noetherian.

We prove the converse by induction on $n \geq 1$, where $M = \langle x_1, \ldots, x_n \rangle$. If $n = 1$, then $M$ is cyclic, and so Proposition 6.23 gives $M \cong R/I$ for some left ideal $I$. If $S$ is a submodule of $M$, then the Correspondence Theorem gives a left ideal $J$ with $I \subseteq J \subseteq R$ and $S \cong J/I$. But $R$ is left noetherian, so that $J$, and hence $S \cong J/I$, is finitely generated.

---

[5]There is an important characterization of left noetherian rings in Proposition 7.6 in terms of chain conditions.

If $n \geq 1$ and $M = \langle x_1, \ldots, x_n, x_{n+1} \rangle$, consider the exact sequence

$$0 \to M' \xrightarrow{i} M \xrightarrow{p} M'' \to 0,$$

where $M' = \langle x_1, \ldots, x_n \rangle$, $M'' = M/M'$, $i$ is the inclusion, and $p$ is the natural map. Note that $M''$ is cyclic, being generated by $x_{n+1} + M'$. If $S \subseteq M$ is a submodule, there is an exact sequence

$$0 \to S \cap M' \to S \to S/(S \cap M') \to 0.$$

Now $S \cap M' \subseteq M'$, and hence it is finitely generated, by the inductive hypothesis. Furthermore, $S/(S \cap M') \cong (S + M')/M' \subseteq M/M'$, so that $S/(S \cap M')$ is finitely generated, by the base step. Using Exercise 6.30 on page 418, we conclude that $S$ is finitely generated   •

We can augment Proposition 2.70.

**Proposition 6.37.**

(i) *If $M$ is a finitely generated abelian group that is a faithful left $R$-module for some ring $R$, then the additive group of $R$ is finitely generated.*

(ii) *Let $\mathbb{Z}[\alpha]$ be the subring of $\mathbb{C}$ generated by a complex number $\alpha$. If there is a faithful $\mathbb{Z}[\alpha]$-module $M$ that is finitely generated as an abelian group, then $\alpha$ is an algebraic integer.*

**Proof.**

(i) The ring $R$ is isomorphic to a subring of $\mathrm{End}_{\mathbb{Z}}(M)$, by Proposition 6.14. Since $M$ is finitely generated, Exercise 6.47 on page 449 shows that $\mathrm{End}_{\mathbb{Z}}(M) = \mathrm{Hom}_{\mathbb{Z}}(M, M)$ is finitely generated. Therefore, the additive group of $R$ is finitely generated, by Proposition 2.70.

(ii) It suffices to prove that the ring $\mathbb{Z}[\alpha]$ is finitely generated as an abelian group, by Proposition 2.70, and this follows from part (i).   •

## Exercises

**6.1.** Let $R$ be a ring. Call an (additive) abelian group $M$ an ***almost left $R$-module*** if there is a function $R \times M \to M$ satisfying all the axioms of a left $R$-module except axiom (iv): we do not assume that $1m = m$ for all $m \in M$. Prove that $M = M_1 \oplus M_0$, where $M_1 = \{m \in M : 1m = m\}$ and $M_0 = \{m \in M : rm = 0 \text{ for all } r \in R\}$ are subgroups of $M$ that are almost left $R$-modules; in fact, $M_1$ is a left $R$-module.

* **6.2.** Let $R$ be a nonzero ring, and let $a \in R$ not have a left inverse; that is, there is no $b \in R$ with $ba = 1$. Prove that there is a maximal left ideal in $R$ containing $a$.

* **6.3.** (i) If $X$ is a subset of a module $M$, prove that $\langle X \rangle$, the submodule of $M$ generated by $X$ [as defined in Example 6.18(vii)], is equal to $\bigcap S$, where the intersection ranges over all those submodules $S \subseteq M$ containing $X$.

(ii) Prove that $\langle X \rangle$ is the ***smallest*** submodule containing $X$: if $S$ is any submodule of $M$ with $X \subseteq S$, then $\langle X \rangle \subseteq S$.

**6.4.** Prove that if $f\colon M \to N$ is an $R$-map and $K$ is a submodule of $M$ with $K \subseteq \ker f$, then $f$ induces an $R$-map $\overline{f}\colon M/K \to N$ by $\overline{f}\colon m + K \mapsto f(m)$.

* **6.5.** (i) Let $I$ be a two-sided ideal in a ring $R$. Prove that an abelian group $M$ is a left $(R/I)$-module if and only if it is a left $R$-module that is annihilated by $I$.

(ii) Let $R$ be a commutative ring and let $J$ be an ideal in $R$. Recall that if $M$ is an $R$-module, then $JM = \{\sum_i j_i m_i \colon j_i \in J \text{ and } m_i \in M\}$ is a submodule of $M$. Prove that $M/JM$ is an $(R/J)$-module if we define scalar multiplication

$$(r + J)(m + JM) = rm + JM.$$

Conclude that if $JM = \{0\}$, then $M$ itself is an $(R/J)$-module; in particular, if $J$ is a maximal ideal in $R$ and $JM = \{0\}$, then $M$ is a vector space over $R/J$.

(iii) Let $R$ be a commutative ring, and let $F = \bigoplus_{i=1}^{n} \langle b_i \rangle$ be a *free* $R$-module, where $R \cong \langle b_i \rangle$ via $r \mapsto rb_i$; that is, $F$ is a direct sum of cyclic $R$-modules, each isomorphic to $R$. If $M$ is a maximal ideal in $R$, prove that the cosets $(b_i + MF)_{1 \le i \le n}$ form a basis of the vector space $F/MF$ over the field $R/M$. Conclude that any two decompositions of $F$ into direct sums of copies of $R$ have the same number of summands.

* **6.6.** Let $R$ be the set of all complex matrices of the form $\left[\begin{smallmatrix} a & b \\ -\overline{b} & \overline{a} \end{smallmatrix}\right]$, where $\overline{a}$ denotes the complex conjugate of $a$. Prove that $R$ is a subring of $\mathrm{Mat}_2(\mathbb{C})$ and that $R \cong \mathbb{H}$, where $\mathbb{H}$ is the division ring of quaternions.

* **6.7.** (i) If $k$ is a commutative ring and $G$ is a cyclic group of finite order $n$, prove that $kG \cong k[x]/(x^n - 1)$.

(ii) If $k$ is a domain,[6] define the ring of *__Laurent polynomials__* as the subring of $k(x)$ consisting of all rational functions of the form $f(x)/x^n$ for $n \in \mathbb{Z}$. If $G$ is infinite cyclic, prove that $kG$ is isomorphic to Laurent polynomials.

**6.8.** Let $R$ be a four-dimensional vector space over $\mathbb{C}$ with basis $1, i, j, k$. Define a multiplication on $R$ so that these basis elements satisfy the same identities satisfied in the quaternions $\mathbb{H}$ [see Example 6.1(vii)]. Prove that $R$ is not a division ring.

* **6.9.** (i) If $R$ is a ring, possibly noncommutative, prove that $\mathrm{Mat}_n(R)$ is a ring.

(ii) Prove that the center of a matrix ring $\mathrm{Mat}_n(R)$ is the set of all scalar matrices $aI$, where $a \in Z(R)$ and $I$ is the identity matrix.

* **6.10.** Prove that $Z(\mathbb{H}) = \{a1 : a \in \mathbb{R}\}$.

* **6.11.** Let $R = R_1 \times \cdots \times R_m$ be a direct product of rings.

(i) Prove that $R^{\mathrm{op}} = R_1^{\mathrm{op}} \times \cdots \times R_m^{\mathrm{op}}$.

(ii) Prove that $Z(R) = Z(R_1) \times \cdots \times Z(R_m)$.

(iii) If $k$ is a field and

$$R = \mathrm{Mat}_{n_1}(k) \times \cdots \times \mathrm{Mat}_{n_m}(k),$$

prove that $\dim_k(Z(R)) = m$.

---

[6]Laurent series over an arbitrary commutative ring $k$ can be defined using localization at the multiplicative subset $\{x^n : n \ge 0\}$.

∗ **6.12.** If $\Delta$ is a division ring, prove that $\Delta^{\mathrm{op}}$ is also a division ring.

∗ **6.13.** An element $a$ in a ring $R$ has a **left inverse** if there is $u \in R$ with $ua = 1$, and it has a **right inverse** if there is $w \in R$ with $aw = 1$.

   (i) Prove that if $a \in R$ has both a left inverse $u$ and a right inverse $w$, then $u = w$.

   (ii) Give an example of a ring $R$ in which an element $a$ has two distinct left inverses.
     **Hint.** Define $R = \mathrm{End}_k(V)$, where $V$ is a vector space over a field $k$ with basis $\{b_n : n \geq 1\}$, and define $a \in R$ by $a(b_n) = b_{n+1}$ for all $n \geq 1$.

   (iii) **(Kaplansky)** Let $R$ be a ring, and let $a, u, v \in R$ satisfy $ua = 1 = va$. If $u \neq v$, prove that $a$ has infinitely many left inverses. Conclude that each element in a finite ring has at most one left inverse.
     **Hint.** Are the elements $u + a^n(1 - au)$ distinct?

∗ **6.14.** Write the elements of the group $\mathbf{Q}$ of quaternions as

$$1, \ \overline{1}, \ i, \ \overline{i}, \ j, \ \overline{j}, \ k, \ \overline{k},$$

and define a linear transformation $\varphi : \mathbb{R}\mathbf{Q} \to \mathbb{H}$ by

$$\varphi(x) = x \quad \text{and} \quad \varphi(\overline{x}) = -x \quad \text{for } x = 1, i, j, k.$$

Prove that $\varphi$ is a surjective ring map, and conclude that there is an isomorphism of rings $\mathbb{R}\mathbf{Q}/\ker\varphi \cong \mathbb{H}$. (See Example 7.117 for a less computational proof.)

**6.15.** (i) For $k$ a field and $G$ a finite group, prove that $(kG)^{\mathrm{op}} \cong kG$.

   (ii) Prove that $\mathbb{H}^{\mathrm{op}} \cong \mathbb{H}$, where $\mathbb{H}$ is the division ring of real quaternions.

∗ **6.16. (Small)** (i) Prove that the ring of all matrices of the form $\left[\begin{smallmatrix} a & 0 \\ b & c \end{smallmatrix}\right]$, where $a \in \mathbb{Z}$ and $b, c \in \mathbb{Q}$, is left noetherian but not right noetherian.

   (ii) Give an example of a ring $R$ for which $R^{\mathrm{op}} \not\cong R$.

∗ **6.17. (Bass)** Recall that a family $(A_i)_{i \in I}$ of left $R$-modules is a *chain* if, for each $i, j \in I$, either $A_i \subseteq A_j$ or $A_j \subseteq A_i$. Prove that a left $R$-module $M$ is finitely generated if and only if the union of every chain of proper submodules of $M$ is a proper submodule.

∗ **6.18.** (i) Prove that the following conditions are equivalent for a left $R$-module $M$: (a) every submodule of $M$ is finitely generated; (b) $M$ has ACC on submodules; (c) $M$ has the maximum condition on submodules. (See Proposition 5.33.)

   (ii) If $R$ is a left noetherian ring and $M$ is a finitely generated left $R$-module, prove that $M$ satisfies the conditions in part (i).

**6.19.** (i) If $R$ is a ring, $r \in R$, and $k \subseteq Z(R)$ is a subring, prove that the subring generated by $r$ and $k$ is commutative.

   (ii) If $\Delta$ is a division ring, $r \in \Delta$, and $k \subseteq Z(\Delta)$ is a subring, prove that the subdivision ring generated by $r$ and $k$ is a (commutative) field.

**6.20.** If $R$ is a ring in which $x^2 = x$ for every $x \in R$, prove that $R$ is commutative. (A Boolean ring is an example of such a ring.)

**Remark.** There are vast generalizations of this result. Here are two such examples. If $R$ is a ring for which there exists an integer $n > 1$ such that $x^n - x \in Z(R)$ for all $x \in R$, then $R$ is commutative. If $R$ is a ring such that, for all $x, y \in R$, there exists $n = n(x, y)$ with $(xy - yx)^n = xy - yx$, then $R$ is commutative. (See Herstein, *Noncommutative Rings*, Chapter 3.) ◄

* **6.21.** In Example 6.6, we defined $\mathcal{F}(G, k)$, where $G$ is a group and $k$ is a commutative ring, as all the functions $\varphi \colon G \to k$ with $\varphi(x) = 0$ for almost all $x \in G$.

  (i) Prove that $\mathcal{F}(G, k)$ is a ring under pointwise addition and *convolution*:
  $$\varphi\psi(g) = \sum_{x \in G} \varphi(x)\psi(x^{-1}g).$$

  (ii) If $u \in G$, define $\varphi_u \in \mathcal{F}(G, k)$ by $\varphi_u(g) = 0$ for $g \neq u$ while $\varphi_u(u) = 1$. Prove that $\Phi \colon kG \to \mathcal{F}(G, k)$, given by $\Phi \colon u \mapsto \varphi_u$, is a ring isomorphism when $k = \mathbb{Z}$ or $k$ is a field.

**6.22.** Let $M$ be a nonzero $R$-module over a commutative ring $R$. If $m \in M$, define its **order ideal** by
$$\operatorname{ord}(m) = \{r \in R : rm = 0\}.$$
Prove that every maximal element in $\mathcal{X} = \{\operatorname{ord}(m) : m \in M \text{ and } m \neq 0\}$ is a prime ideal.

**6.23.** Let $A \xrightarrow{f} B \xrightarrow{g} C$ be a sequence of module maps. Prove that $gf = 0$ if and only if $\operatorname{im} f \subseteq \ker g$. Give an example of such a sequence that is not exact.

**6.24.** If $0 \to M \to 0$ is an exact sequence, prove that $M = \{0\}$.

* **6.25.** Let $0 \to A \to B \to C \to 0$ be a short exact sequence of modules. If $M$ is any module, prove that there are exact sequences
  $$0 \to A \oplus M \to B \oplus M \to C \to 0$$

and
  $$0 \to A \to B \oplus M \to C \oplus M \to 0.$$

* **6.26.** If $f \colon M \to N$ is a map, prove that there is an exact sequence
  $$0 \to \ker f \to M \xrightarrow{f} N \to \operatorname{coker} f \to 0.$$

**6.27.** If $A \xrightarrow{f} B \to C \xrightarrow{h} D$ is an exact sequence, prove that $f$ is surjective if and only if $h$ is injective.

**6.28.** If $A \xrightarrow{f} B \xrightarrow{g} C \xrightarrow{h} D \xrightarrow{k} E$ is exact, prove that there is an exact sequence
$$0 \to \operatorname{coker} f \xrightarrow{\alpha} C \xrightarrow{\beta} \ker k \to 0,$$
where $\alpha \colon b + \operatorname{im} f \mapsto gb$ and $\beta \colon c \mapsto hc$.

* **6.29.** (i) Let $\to A_{n+1} \xrightarrow{d_{n+1}} A_n \xrightarrow{d_n} A_{n-1} \to$ be an exact sequence, and let $\operatorname{im} d_{n+1} = K_n = \ker d_n$ for all $n$. Prove that
  $$0 \to K_n \xrightarrow{i_n} A_n \xrightarrow{d'_n} K_{n-1} \to 0$$

  is an exact sequence for all $n$, where $i_n$ is the inclusion and $d'_n$ is obtained from $d_n$ by changing its target. We say that the original sequence has been **factored** into these short exact sequences.

  (ii) Let
  $$\to A_1 \xrightarrow{f_1} A_0 \xrightarrow{f_0} K \to 0 \quad \text{and} \quad 0 \to K \xrightarrow{g_0} B_0 \xrightarrow{g_1} B_1 \to$$
  be exact sequences. Prove that
  $$\to A_1 \xrightarrow{f_1} A_0 \xrightarrow{g_0 f_0} B_0 \xrightarrow{g_1} B_1 \to$$
  is an exact sequence. We say that the original two sequences have been **spliced** to form the new exact sequence.

∗ **6.30.** Generalize Proposition 2.68 to modules. Let $0 \to A \overset{i}{\to} B \overset{p}{\to} C \to 0$ be a short exact sequence of modules.

    (i) Assume that $A = \langle X \rangle$ and $C = \langle Y \rangle$. For each $y \in Y$, choose $y' \in B$ with $p(y') = y$. Prove that
$$B = \langle i(X) \cup \{y' : y \in Y\} \rangle.$$

    (ii) Prove that if both $A$ and $C$ are finitely generated, then $B$ is finitely generated. More precisely, prove that if $A$ can be generated by $m$ elements and $C$ can be generated by $n$ elements, then $B$ can be generated by $m + n$ elements.

**6.31.** Prove that every short exact sequence of vector spaces is split.

∗ **6.32.** Prove that a short exact sequence
$$0 \to A \overset{i}{\to} B \overset{p}{\to} C \to 0$$
splits if and only if there exists $q \colon B \to A$ with $qi = 1_A$.

---

## Section 6.2. Categories

Eilenberg and Mac Lane invented categories and functors in the 1940s by distilling ideas that had arisen in Algebraic Topology, where topological spaces and continuous maps are studied by means of certain algebraic systems (homology groups, cohomology rings, homotopy groups) associated to them. Categorical notions have proven to be valuable in purely algebraic contexts as well; indeed, it is fair to say that the recent great strides in Algebraic Geometry and Arithmetic Geometry (for example, the positive solution of Fermat's Last Theorem) could not have occurred outside a categorical setting.

Imagine a Set Theory whose primitive terms, instead of *set* and *element*, are *set* and *function*. How could we define bijection, cartesian product, union, and intersection? Category Theory will force us to think in this way. Now categories are the context for discussing general properties of systems such as groups, rings, vector spaces, modules, sets, and topological spaces, in tandem with their respective transformations: homomorphisms, functions, and continuous maps. Here are two basic reasons for studying categories: the first is that they are needed to define functors and natural transformations; the other is that categories will force us to regard a module, for example, not in isolation, but in a context serving to relate it to all other modules (for example, we will define certain modules as solutions to *universal mapping problems*).

There are well-known set-theoretic "paradoxes" showing that contradictions arise if we are not careful about how the undefined terms *set* and *element* are used. For example, **Russell's paradox** gives a contradiction arising from regarding every collection as a set. Define a *Russell set* to be a set $S$ that is not a member of itself; that is, $S \notin S$, and define $R$ to be the collection of all Russell sets. Either $R$ is a Russell set or it is not a Russell set. On the one hand, if $R \in R$, then $R$ is not a Russell set—the criterion being $R \notin R$; but the only members of $R$ are Russell sets, so that $R$ is a Russell set, a contradiction. On the other hand, if $R \notin R$, then $R$ is a Russell set, and so it belongs to $R$, for $R$ contains every Russell set;

that is, $R \in R$ and so $R$ is not a Russell set, another contradiction.[7] We conclude that some conditions are needed to determine which collections are allowed to be sets; such conditions are given in the ***Zermelo–Fraenkel axioms*** for Set Theory, specifically, by the ***Axiom of Comprehension***. The collection $R$ is not a set, and this is one way to resolve the Russell paradox.

Let us give a bit more detail. The Zermelo–Fraenkel axioms have primitive terms *class* and $\in$ and rules for constructing classes, as well as for constructing certain special classes, called ***sets***. For example, finite classes and the natural numbers $\mathbb{N}$ are assumed to be sets. A class is called ***small*** if it has a cardinal number, and it is a theorem that a class is a set if and only if it is small. A class that is not a set is called a ***proper class***. For example, $\mathbb{N}$, $\mathbb{Z}$, $\mathbb{Q}$, $\mathbb{R}$, and $\mathbb{C}$ are sets, the collection of all sets is a proper class, and the collection $R$ of all Russell classes is not even a class. For a more complete discussion, see Mac Lane, *Categories for the Working Mathematician*, pp. 21–24, and Herrlich–Strecker, *Category Theory*, Chapter II and the Appendix. We quote Herrlich and Strecker, p. 331.

> There are two important points (in different approaches to Category Theory). ... First, there is no such thing as *the* category **Sets** of all sets. If one approaches Set Theory from a naive standpoint, inconsistencies will arise, and approaching it from any other standpoint requires an axiom scheme, so that the properties of **Sets** will depend upon the foundation chosen. ... The second point is that (there is) a foundation that allows us to perform all of the categorical-theoretical constructions that at the moment seem desirable. If at some later time different constructions that cannot be performed within this system are needed, then the foundation should be expanded to accommodate them, or perhaps should be replaced entirely. After all, the purpose of foundations is not to arbitrarily restrict inquiry, but to provide a framework wherein one can legitimately perform those constructions and operations that are mathematically interesting and useful, so long as they are not inconsistent within themselves.

We will be rather relaxed about Set Theory. As a practical matter, when an alleged class arises, there are three possibilities: it is a set; it is a proper class; it is not a class at all. In this book, we will not worry about the possibility that an alleged class is not a class.

**Definition.** A ***category*** $\mathcal{C}$ consists of three ingredients: a class obj$(\mathcal{C})$ of ***objects***, a *set* of ***morphisms*** $\operatorname{Hom}(A, B)$ for every ordered pair $(A, B)$ of objects, and ***composition*** $\operatorname{Hom}(A, B) \times \operatorname{Hom}(B, C) \to \operatorname{Hom}(A, C)$, denoted by

$$(f, g) \mapsto gf,$$

for every ordered triple $(A, B, C)$ of objects. [We often write $f \colon A \to B$ or $A \xrightarrow{f} B$ to denote $f \in \operatorname{Hom}(A, B)$.] These ingredients are subject to the following axioms.

---

[7]Compare this argument with the proof that $|2^X| > |X|$ for a set $X$. If, on the contrary, $|2^X| = |X|$, there is a bijection $\varphi \colon 2^X \to X$, and then each $x \in X$ has the form $\varphi(S)$ for a unique subset $S \subseteq X$. Considering whether $\varphi(S^*) \in S^*$, where $S^* = \{x = \varphi(S) : \varphi(S) \notin S\}$, gives a contradiction.

(i) The Hom sets are pairwise disjoint;[8] that is, each morphism $f \in \operatorname{Hom}(A, B)$ has a unique **domain** $A$ and a unique **target** $B$.

(ii) For each object $A$, there is an **identity morphism** $1_A \in \operatorname{Hom}(A, A)$ such that

$$f1_A = f \text{ and } 1_B f = f \text{ for all } f\colon A \to B.$$

(iii) Composition is associative: given morphisms

$$A \xrightarrow{f} B \xrightarrow{g} C \xrightarrow{h} D,$$

we have

$$h(gf) = (hg)f.$$

The important notion, in this circle of ideas, is not category but *functor*, which will be introduced in the next section. Categories are necessary because they are an essential ingredient in the definition of functor. A similar situation occurs in Linear Algebra: linear transformation is the important notion, but we must first consider vector spaces in order to define it.

The following examples explain certain fine points in the definition of category.

**Example 6.38.**

(i) $\mathcal{C} = \mathbf{Sets}$. The objects in this category are sets (not proper classes), morphisms are functions, and composition is the usual composition of functions.

A standard result of Set Theory is that $\operatorname{Hom}(A, B)$, the class of all functions from a set $A$ to a set $B$, is a set. That Hom sets are pairwise disjoint is just the reflection of the definition of equality of functions given in Chapter 1: in order that two functions be equal, they must, first, have the same domains and the same targets (and, of course, they must have the same graphs).

(ii) $\mathcal{C} = \mathbf{Groups}$. Here, objects are groups, morphisms are homomorphisms, and composition is the usual composition (homomorphisms are functions).

(iii) $\mathcal{C} = \mathbf{ComRings}$. Here, objects are commutative rings, morphisms are ring homomorphisms, and composition is the usual composition.

(iv) $\mathcal{C} = {}_R\mathbf{Mod}$. The objects in this category are left $R$-modules over a ring $R$, morphisms are $R$-homomorphisms, and composition is the usual composition. We denote the sets $\operatorname{Hom}(A, B)$ in ${}_R\mathbf{Mod}$ by

$$\operatorname{Hom}_R(A, B).$$

If $R = \mathbb{Z}$, then we often write

$$_{\mathbb{Z}}\mathbf{Mod} = \mathbf{Ab}$$

to remind ourselves that $\mathbb{Z}$-modules are just abelian groups.

---

[8]In the unlikely event that some particular candidate for a category does not have disjoint Hom sets, we can force pairwise disjointness by redefining $\operatorname{Hom}(A, B)$ as $Hom(A, B) = \{A\} \times \operatorname{Hom}(A, B) \times \{B\}$, so that each morphism $f \in \operatorname{Hom}(A, B)$ is relabeled as $(A, f, B)$. If $(A, B) \neq (A', B')$, then $Hom(A, B)$ and $Hom(A', B')$ are disjoint.

(v) $\mathcal{C} = \mathbf{Mod}_R$. The objects in this category are right $R$-modules over a ring $R$, morphisms are $R$-homomorphisms, and composition is the usual composition. The Hom sets in $\mathbf{Mod}_R$ are also denoted by

$$\mathrm{Hom}_R(A, B).$$

(vi) $\mathcal{C} = \mathbf{PO}(X)$. If $X$ is a partially ordered set, regard it as a category whose objects are the elements of $X$, whose Hom sets are either empty or have only one element:

$$\mathrm{Hom}(x, y) = \begin{cases} \varnothing & \text{if } x \npreceq y \\ \{\kappa_y^x\} & \text{if } x \preceq y \end{cases}$$

(the symbol $\kappa_y^x$ denotes the unique element in the Hom set when $x \preceq y$) and whose composition is given by

$$\kappa_z^y \kappa_y^x = \kappa_z^x.$$

Note that $1_x = \kappa_x^x$, by reflexivity, while composition makes sense because $\preceq$ is transitive.[9]

We insisted, in the definition of category, that $\mathrm{Hom}(A, B)$ be a set, but we left open the possibility that it be empty. The category $\mathbf{PO}(X)$ is an example in which this possibility occurs. [Not every Hom set in a category $\mathcal{C}$ can be empty, for $\mathrm{Hom}(A, A) \neq \varnothing$ for every object $A \in \mathcal{C}$ because it contains the identity morphism $1_A$.]

(vii) $\mathcal{C} = \mathcal{C}(G)$. If $G$ is a group, then the following description defines a category $\mathcal{C}(G)$: there is only one object, denoted by $*$, $\mathrm{Hom}(*, *) = G$, and composition

$$\mathrm{Hom}(*, *) \times \mathrm{Hom}(*, *) \to \mathrm{Hom}(*, *);$$

that is, $G \times G \to G$, is the given multiplication in $G$. We leave verification of the axioms to the reader.[10]

The category $\mathcal{C}(G)$ has an unusual property. Since $*$ is merely an object, not a set, there are no *functions* $* \to *$ defined on it; thus, morphisms here are not functions. Another curious property of this category is also a consequence of there being only one object: there are no proper "subobjects" here.

(viii) There are many interesting nonalgebraic examples of categories. For example, $\mathcal{C} = \mathbf{Top}$, the category with objects all topological spaces, morphisms all continuous functions, and usual composition. One step in verifying that $\mathbf{Top}$ is a category is showing that the composite of continuous functions is continuous. ◄

Here is how to translate *isomorphism* into categorical language.

**Definition.** A morphism $f \colon A \to B$ in a category $\mathcal{C}$ is an ***isomorphism*** if there exists a morphism $g \colon B \to A$ in $\mathcal{C}$ with

$$gf = 1_A \quad \text{and} \quad fg = 1_B.$$

---

[9]A nonempty set $X$ is called ***quasi-ordered*** if it has a relation $x \preceq y$ that is reflexive and transitive (if, in addition, this relation is antisymmetric, then $X$ is partially ordered). $\mathbf{PO}(X)$ is a category for every quasi-ordered set.

[10]That every element in $G$ have an inverse is not needed to prove that $\mathcal{C}(G)$ is a category, and $\mathcal{C}(G)$ is a category for every monoid $G$.

The morphism $g$ is called the **inverse** of $f$.

It is easy to see that an inverse of an isomorphism is unique.

Identity morphisms in a category are always isomorphisms. If $\mathcal{C} = \mathbf{PO}(X)$, where $X$ is a partially ordered set, then the only isomorphisms are identities; if $\mathcal{C} = \mathcal{C}(G)$, where $G$ is a group [see Example 6.38(vii)], then every morphism is an isomorphism. If $\mathcal{C} = \mathbf{Sets}$, then isomorphisms are bijections; if $\mathcal{C} = \mathbf{Groups}$, $\mathcal{C} = {}_R\mathbf{Mod}$, $\mathcal{C} = \mathbf{Mod}_R$, or $\mathcal{C} = \mathbf{ComRings}$, then isomorphisms are isomorphisms in the usual sense; if $\mathcal{C} = \mathbf{Top}$, then isomorphisms are homeomorphisms.

Let us give a name to a feature of the categories ${}_R\mathbf{Mod}$ and $\mathbf{Mod}_R$ (which we saw in Proposition 6.13) that is not shared by more general categories: homomorphisms can be added.

**Definition.** A category $\mathcal{C}$ is **pre-additive** if every $\operatorname{Hom}(A, B)$ is equipped with a binary operation making it an (additive) abelian group for which the distributive laws hold: for all $f, g \in \operatorname{Hom}(A, B)$,

(i) if $p\colon B \to B'$, then
$$p(f + g) = pf + pg \in \operatorname{Hom}(A, B');$$

(ii) if $q\colon A' \to A$, then
$$(f + g)q = fq + gq \in \operatorname{Hom}(A', B).$$

In Exercise 6.35 on page 434, it is shown that **Groups** does not have the structure of a pre-additive category.

A category is defined in terms of objects and morphisms; its objects need not be sets, and its morphisms need not be functions [$\mathcal{C}(G)$ in Example 6.38(vii) is such a category]. We now set ourselves the exercise of trying to describe various constructions in **Sets** or in ${}_R\mathbf{Mod}$ so that they make sense in arbitrary categories.

In Proposition 6.26(iii), we gave the following characterization of *direct sum* $M = A \oplus B$: there are homomorphisms $p\colon M \to A$, $q\colon M \to B$, $i\colon A \to M$, and $j\colon B \to M$ such that
$$pi = 1_A, \ qj = 1_B, \ pj = 0, \ qi = 0, \quad \text{and} \quad ip + jq = 1_M.$$

Even though this description of direct sum is phrased in terms of arrows, it is not general enough to make sense in every category; morphisms can be added because ${}_R\mathbf{Mod}$ is pre-additive, but they cannot be added in **Sets**, for example. In Corollary 6.28, we gave another description of direct sum in terms of arrows: there is a map $\rho\colon M \to S$ with $\rho s = s$; moreover, $\ker \rho = \operatorname{im} j$, $\operatorname{im} \rho = \operatorname{im} i$, and $\rho(s) = s$ for every $s \in \operatorname{im} \rho$. This description does not make sense in arbitrary categories because the image of a morphism may fail to be defined. For example, the morphisms in $\mathcal{C}(G)$ [see Example 6.38(vii)] are elements in $\operatorname{Hom}(*, *) = G$, not functions, and so the image of a morphism has no obvious meaning. Thus, we have to think a bit more in order to find the appropriate categorical description (see Proposition 6.39). On the other hand, we can define *direct summand* categorically using *retracts*: recall that an object $S$ is (isomorphic to) a retract of an object $M$ if there exist morphisms $i\colon S \to M$ and $\rho\colon M \to S$ with $\rho i = 1_S$.

One of the nice aspects of thinking in a categorical way is that it enables us to see analogies we might not have recognized before. For example, we shall soon see that direct sum in $_R\mathbf{Mod}$ is the same notion as disjoint union in **Sets**.

We begin with a very formal definition.

**Definition.** A *diagram* in a category $\mathcal{C}$ is a directed multigraph[11] whose vertices are objects in $\mathcal{C}$ and whose arrows are morphisms in $\mathcal{C}$.

For example, here are two diagrams in a category:

If we think of an arrow as a "one-way street," then a *path* in a diagram is a "walk" from one vertex to another taking care never to walk the wrong way. A path in a diagram may be regarded as a composite of morphisms.

**Definition.** A diagram *commutes* if, for each pair of vertices $A$ and $B$, any two paths from $A$ to $B$ are equal; that is, the composites are the same morphism.

For example, the triangular diagram above commutes if $gf = h$ and $kf = h$, and the square diagram above commutes if $gf = f'g'$. The term *commutes* in this context arises from this last example.

If $A$ and $B$ are subsets of a set $S$, then their intersection is defined:

$$A \cap B = \{s \in S : s \in A \text{ and } s \in B\}$$

[if two sets are not given as subsets, then their intersection may not be what one expects: for example, if $\mathbb{Q}$ is defined as all equivalence classes of ordered pairs $(m, n)$ of integers with $n \neq 0$, then $\mathbb{Z} \cap \mathbb{Q} = \varnothing$].

We can force two overlapping subsets $A$ and $B$ to be disjoint by "disjointifying" them. Consider the cartesian product $(A \cup B) \times \{1, 2\}$ and its subsets $A' = A \times \{1\}$ and $B' = B \times \{2\}$. It is plain that $A' \cap B' = \varnothing$, for a point in the intersection would have coordinates $(a, 1) = (b, 2)$; this cannot be, for their second coordinates are not equal. We call $A' \cup B'$ the *disjoint union* of $A$ and $B$. Let us take note of the functions $\alpha \colon A \to A'$ and $\beta \colon B \to B'$, given by $\alpha \colon a \mapsto (a, 1)$ and $\beta \colon b \mapsto (b, 2)$. We denote the disjoint union $A' \cup B'$ by $A \sqcup B$.

If there are functions $f \colon A \to X$ and $g \colon B \to X$, for some set $X$, then there is a unique function $\theta \colon A \sqcup B \to X$ given by $\theta\alpha = f$ and $\theta\beta = g$. The function $\theta$ is well-defined because $A$ and $B$ are disjoint.

Here is a way to describe this construction *categorically* (i.e., with diagrams).

**Definition.** If $A$ and $B$ are objects in a category $\mathcal{C}$, then their *coproduct*, denoted by $A \sqcup B$, is an object $C$ in obj$(\mathcal{C})$ together with *injections* $\alpha \colon A \to A \sqcup B$ and

---

[11]A *directed multigraph* consists of a set $V$, called *vertices* and, for each ordered pair $(u, v) \in V \times V$, a (possibly empty) set arr$(u, v)$, called *arrows* from $u$ to $v$.

$\beta\colon B \to A \sqcup B$, such that, for every object $X$ in $\mathcal{C}$ and every pair of morphisms $f\colon A \to X$ and $g\colon B \to X$, there exists a unique morphism $\theta\colon A \sqcup B \to X$ making the following diagram commute (i.e., $\theta\alpha = f$ and $\theta\beta = g$):

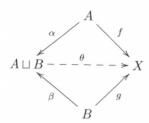

Here is a formal proof that the set $A \sqcup B = A' \cup B' \subseteq (A \cup B) \times \{1, 2\}$ just constructed is a coproduct in **Sets**. If $X$ is any set and $f\colon A \to X$ and $g\colon B \to X$ are any given functions, then there exists a function $\theta\colon A \sqcup B \to X$ that extends both $f$ and $g$: if $c \in A \sqcup B$, then either $c = (a, 1) \in A'$ or $c = (b, 2) \in B'$; define $\theta((a, 1)) = f(a)$ and define $\theta((b, 2)) = g(b)$, so that $\theta\alpha = f$ and $\theta\beta = g$. Let us show that $\theta$ is the unique function on $A \sqcup B$ extending both $f$ and $g$. If $\psi\colon A \sqcup B \to X$ satisfies $\psi\alpha = f$ and $\psi\beta = g$, then

$$\psi(\alpha(a)) = \psi((a, 1)) = f(a) = \theta((a, 1))$$

and, similarly,

$$\psi((b, 2)) = g(b).$$

Therefore, $\psi$ agrees with $\theta$ on $A' \cup B' = A \sqcup B$, and so $\psi = \theta$.

We do not assert that coproducts always exist; in fact, it is easy to construct examples of categories in which a pair of objects does not have a coproduct (Exercise 6.34 on page 433). The formal proof just given, however, shows that coproducts do exist in **Sets**, where they are disjoint unions. Coproducts exist in the category of groups, and they are called *free products*. Free groups turn out to be free products of infinite cyclic groups (analogous to free abelian groups being direct sums of infinite cyclic groups) (Rotman, *An Introduction to the Theory of Groups*, p. 388). A theorem of Kurosh states that every subgroup of a free product is itself a free product (Ibid., p. 392).

**Proposition 6.39.** *If $A$ and $B$ are $R$-modules, then a coproduct in $_R$**Mod** exists, and it is the (external) direct sum $C = A \oplus B$.*

**Proof.** The statement of the proposition is not complete, for a coproduct requires injection morphisms $\alpha$ and $\beta$. The underlying set of external direct sum $C$ is the cartesian product $A \times B$, so that we may define $\alpha\colon A \to C$ by $\alpha\colon a \mapsto (a, 0)$ and $\beta\colon B \to C$ by $\beta\colon b \mapsto (0, b)$.

Now let $X$ be a module, and let $f\colon A \to X$ and $g\colon B \to X$ be homomorphisms. Define $\theta\colon C \to X$ by $\theta\colon (a, b) \mapsto f(a) + g(b)$. First, the diagram commutes: if $a \in A$, then $\theta\alpha(a) = \theta((a, 0)) = f(a)$ and, similarly, if $b \in B$, then $\theta\beta(b) = \theta((0, b)) = g(b)$. Finally, $\theta$ is unique. If $\psi\colon C \to X$ makes the diagram commute, then $\psi((a, 0)) = f(a)$ for all $a \in A$ and $\psi((0, b)) = g(b)$ for all $b \in B$. Since $\psi$ is a homomorphism, we have

$$\psi((a,b)) = \psi((a,0) + (0,b)) = \psi((a,0)) + \psi((0,b)) = f(a) + g(b).$$

Therefore, $\psi = \theta$.   •

A similar proof shows that coproducts exist in $\mathbf{Mod}_R$.

We can give an explicit formula for the map $\theta$ in the proof of Proposition 6.39. If $f \colon A \to X$ and $g \colon B \to X$ are maps, then $\theta \colon A \oplus B \to X$ is given by

$$\theta \colon (a,b) \mapsto f(a) + g(b).$$

**Proposition 6.40.** *If $\mathcal{C}$ is a category and $A$ and $B$ are objects in $\mathcal{C}$, then any two coproducts of $A$ and $B$, should they exist, are isomorphic.*

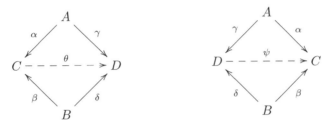

**Proof.** Suppose that $C$ and $D$ are coproducts of $A$ and $B$. In more detail, assume that $\alpha \colon A \to C$, $\beta \colon B \to C$, $\gamma \colon A \to D$, and $\delta \colon B \to D$ are injection morphisms. If, in the defining diagram for $C$, we take $X = D$, then there is a morphism $\theta \colon C \to D$ making the left diagram commute. Similarly, if, in the defining diagram for $D$, we take $X = C$, we obtain a morphism $\psi \colon D \to C$ making the right diagram commute.

Consider now the following diagram, which arises from the juxtaposition of the previous two diagrams:

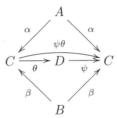

This diagram commutes because $\psi\theta\alpha = \psi\gamma = \alpha$ and $\psi\theta\beta = \psi\delta = \beta$. But plainly, the identity morphism $1_C \colon C \to C$ also makes this diagram commute. By the uniqueness of the dashed arrow in the defining diagram for coproduct, $\psi\theta = 1_C$. The same argument, mutatis mutandis, shows that $\theta\psi = 1_D$. We conclude that $\theta \colon C \to D$ is an isomorphism.   •

Informally, an object $S$ in a category $\mathcal{C}$ is called a ***solution*** to a ***universal mapping problem*** if $S$ is defined by a diagram showing, whenever we vary an object $X$ and various morphisms, that there exists a unique morphism making some subdiagrams commute. The "metatheorem" is that solutions, if they exist, are unique up to unique isomorphism. The proof just given is a prototype for proving the metatheorem[12] (if we wax categorical, then the statement of the metatheorem can be made precise, and we can then prove it; see Mac Lane, *Categories for the Working Mathematician*, Chapter III, for appropriate definitions, statement, and proof). Indeed, we have already done this in our proof of Lemma 4.78 when we proved that the rank of a nonabelian free group is well-defined. The strategy of such a proof involves two steps. First, if $C$ and $C'$ are solutions, get morphisms $\theta\colon C \to C'$ and $\psi\colon C' \to C$ by setting $X = C'$ in the diagram showing that $C$ is a solution, and by setting $X = C$ in the corresponding diagram showing that $C'$ is a solution. Second, set $X = C$ in the diagram for $C$ and show that both $\psi\theta$ and $1_C$ are "dashed" morphisms making the diagram commute; as such a dashed morphism is unique, conclude that $\psi\theta = 1_C$. Similarly, the other composite $\theta\psi = 1_{C'}$, and so $\theta$ is an isomorphism.

Here is a construction "dual" to coproduct.

**Definition.** If $A$ and $B$ are objects in a category $\mathcal{C}$, then their **product**, denoted by $A \sqcap B$, is an object $P \in \mathrm{obj}(\mathcal{C})$ and **projections** $p\colon P \to A$ and $q\colon P \to B$, such that, for every object $X \in \mathcal{C}$ and every pair of morphisms $f\colon X \to A$ and $g\colon X \to B$, there exists a unique morphism $\theta\colon X \to P$ making the following diagram commute.

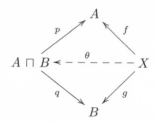

The cartesian product $P = A \times B$ of two sets $A$ and $B$ is the categorical product in **Sets**. Define $p\colon A \times B \to A$ by $p\colon (a, b) \mapsto a$ and define $q\colon A \times B \to B$ by $q\colon (a, b) \mapsto b$. If $X$ is a set and $f\colon X \to A$ and $g\colon X \to B$ are functions, then the reader may show that $\theta\colon X \to A \times B$, defined by $\theta\colon x \mapsto (f(x), g(x)) \in A \times B$, satisfies the necessary conditions.

**Proposition 6.41.** *If $A$ and $B$ are objects in a category $\mathcal{C}$, then any two products of $A$ and $B$, should they exist, are isomorphic.*

**Proof.** Adapt the proof of the prototype, Proposition 6.40  •

Reversing the arrows in the defining diagram for coproduct gives the defining diagram for product. A similar reversal of arrows can be seen in Exercise 6.69 on page 460: the diagram characterizing surjections in $_R\mathbf{Mod}$ is obtained by reversing all the arrows in the diagram characterizing injections. If $S$ is a solution to a universal mapping problem posed by a diagram $\mathcal{D}$, let $\mathcal{D}'$ be the diagram obtained from $\mathcal{D}$ by reversing all its arrows. If $S'$ is a solution to the universal mapping problem posed by $\mathcal{D}'$, then we call $S$ and $S'$ **duals**. There are examples of categories in which an object and its dual object both exist, and there are examples in which an object exists but its dual does not.

---

[12] Another prototype is given in Exercise 6.42 on page 435.

What is the product of two modules?

**Proposition 6.42.** *If $R$ is a ring and $A$ and $B$ are left $R$-modules, then their (categorical) product $A \sqcap B$ exists in $_R\mathbf{Mod}$; in fact,*

$$A \sqcap B \cong A \oplus B \cong A \sqcup B.$$

**Remark.** Thus, the product and coproduct of two objects, though distinct in **Sets**, coincide in $_R\mathbf{Mod}$.  ◄

**Proof.** In Proposition 6.26(iii), we characterized $M \cong A \sqcup B$ by the existence of projection and injection morphisms $A \underset{p}{\overset{i}{\rightleftarrows}} M \overset{q}{\underset{j}{\rightleftarrows}} B$ satisfying the equations

$$pi = 1_A, \quad qj = 1_B, \quad pj = 0, \quad qi = 0, \quad \text{and} \quad ip + jq = 1_M.$$

If $X$ is a module and $f \colon X \to A$ and $g \colon X \to B$ are homomorphisms, define $\theta \colon X \to A \sqcup B$ by $\theta(x) = if(x) + jg(x)$. The product diagram

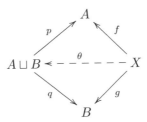

commutes because $p\theta(x) = pif(x) + pjg(x) = pif(x) = f(x)$ for all $x \in X$ (using the given equations) and, similarly, $q\theta(x) = g(x)$. To prove uniqueness of $\theta$, note that the equation $ip + jq = 1_{A \sqcup B}$ gives

$$\psi = ip\psi + jq\psi = if + jg = \theta. \quad \bullet$$

Exercise 6.36 on page 434 shows that products in **Groups** are direct products.

There are (at least) two ways to extend the notion of direct sum of modules from two summands to an indexed family of summands.

**Definition.** Let $R$ be a ring and let $(A_i)_{i \in I}$ be an indexed family of left $R$-modules. The ***direct product*** $\prod_{i \in I} A_i$ is the cartesian product [i.e., the set of all $I$-tuples[13] $(a_i)$ whose $i$th coordinate $a_i$ lies in $A_i$ for every $i$] with coordinatewise addition and scalar multiplication:

$$(a_i) + (b_i) = (a_i + b_i)$$
$$r(a_i) = (ra_i),$$

where $r \in R$ and $a_i, b_i \in A_i$ for all $i$.

The ***direct sum***, denoted by $\bigoplus_{i \in I} A_i$ (or by $\sum_{i \in I} A_i$), is the submodule of $\prod_{i \in I} A_i$ consisting of all $(a_i)$ having only finitely many nonzero coordinates.

---

[13] An $I$-*tuple* is a function $f \colon I \to \bigcup_i A_i$ with $f(i) \in A_i$ for all $i \in I$.

Given a family $(A_j)_{j \in I}$ of left $R$-modules, define ***injections*** $\alpha_i \colon A_i \to \bigoplus_j A_j$ by setting $\alpha_i(a_i)$ to be the $I$-tuple whose $i$th coordinate is $a_i$ and whose other coordinates are 0. Each $m \in \bigoplus_{i \in I} A_i$ has a unique expression of the form

$$m = \sum_{i \in I} \alpha_i(a_i),$$

where $a_i \in A_i$ and almost all $a_i = 0$; that is, only finitely many $a_i$ can be nonzero.

Note that if the index set $I$ is finite, then $\prod_{i \in I} A_i = \bigoplus_{i \in I} A_i$. On the other hand, when $I$ is infinite and infinitely many $A_i \neq 0$, then the direct sum is a proper submodule of the direct product (they are almost never isomorphic).

We now extend the definitions of coproduct and product to a family of objects.

**Definition.** Let $\mathcal{C}$ be a category, and let $(A_i)_{i \in I}$ be a family of objects in $\mathcal{C}$ indexed by a set $I$. A ***coproduct*** is an ordered pair $(C, \{\alpha_i \colon A_i \to C\})$, consisting of an object $C$ and a family $(\alpha_i \colon A_i \to C)_{i \in I}$ of ***injections***, that satisfies the following property. For every object $X$ equipped with morphisms $f_i \colon A_i \to X$, there exists a unique morphism $\theta \colon C \to X$ making the following diagram commute for each $i$:

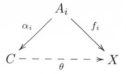

A coproduct, should it exist, is denoted by $\bigsqcup_{i \in I} A_i$; it is unique up to isomorphism.

We sketch the existence of the disjoint union of sets $(A_i)_{i \in I}$. First form the set $B = (\bigcup_{i \in I} A_i) \times I$, and then define

$$A_i' = \{(a_i, i) \in B : a_i \in A_i\}.$$

Then the ***disjoint union*** is $\bigsqcup_{i \in I} A_i = \bigcup_{i \in I} A_i'$ (of course, the disjoint union of two sets is a special case of this construction). The reader may show that $\bigsqcup_i A_i$ together with the functions $\alpha_i \colon A_i \to \bigsqcup_i A_i$, given by $\alpha_i \colon a_i \mapsto (a_i, i) \in \bigsqcup_i A_i$ (where $a_i \in A_i$), comprise the coproduct in **Sets**; that is, we have described a solution to the universal mapping problem.

**Proposition 6.43.** *If $(A_i)_{i \in I}$ is a family of left $R$-modules, then the direct sum $\bigoplus_{i \in I} A_i$ is their coproduct in $_R\mathbf{Mod}$.*

**Proof.** The statement of the proposition is not complete, for a coproduct requires injection morphisms $\alpha_i$. Denote $\bigoplus_{i \in I} A_i$ by $C$, and define $\alpha_i \colon A_i \to C$ as follows: if $a_i \in A_i$, then $\alpha_i(a_i) \in C$ is the $I$-tuple whose $i$th coordinate is $a_i$ and whose other coordinates are zero.

Now let $X$ be a module and, for each $i \in I$, let $f_i \colon A_i \to X$ be homomorphisms. Define $\theta \colon C \to X$ by $\theta \colon (a_i) \mapsto \sum_i f_i(a_i)$ (note that this makes sense, for only finitely many $a_i$ are nonzero). First, the diagram commutes: if $a_i \in A_i$, then $\theta\alpha_i(a_i) = f_i(a_i)$. Finally, $\theta$ is unique. If $\psi \colon C \to X$ makes the diagram commute, then $\psi((a_i)) = f_i(a_i)$. Since $\psi$ is a homomorphism, we have

$$\psi((a_i)) = \psi\left(\sum_i \alpha_i(a_i)\right) = \sum_i \psi\alpha_i(a_i) = \sum_i f_i(a_i).$$

Therefore, $\psi = \theta$.  •

Let us make the formula for $\theta$ explicit. If $f_i \colon A_i \to X$ are given homomorphisms, then $\theta \colon \bigoplus_{i \in I} A_i \to X$ is given by

$$\theta \colon (a_i) \mapsto \sum_{i \in I} f_i(a_i);$$

there are only finitely many nonzero terms in the sum $\sum_{i \in I} f_i(a_i)$ because almost all $a_i = 0$.

Here is the dual notion.

**Definition.** Let $\mathcal{C}$ be a category, and let $(A_i)_{i \in I}$ be a family of objects in $\mathcal{C}$ indexed by a set $I$. A **product** is an ordered pair $(C, \{p_i \colon C \to A_i\})$, consisting of an object $C$ and a family $(p_i \colon C \to A_i)_{i \in I}$ of **projections**, that satisfies the following condition. For every object $X$ equipped with morphisms $f_i \colon X \to A_i$, there exists a unique morphism $\theta \colon X \to C$ making the following diagram commute for each $i$:

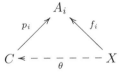

Should it exist, a product is denoted by $\prod_{i \in I} A_i$, and it is unique up to isomorphism.

We let the reader prove that cartesian product is the product in **Sets**.

**Proposition 6.44.** *If $(A_i)_{i \in I}$ is a family of left $R$-modules, then the direct product $C = \prod_{i \in I} A_i$ is their product in $_R\mathbf{Mod}$.*

**Proof.** The statement of the proposition is not complete, for a product requires projections. For each $j \in I$, define $p_j \colon C \to A_j$ by $p_j \colon (a_i) \mapsto a_j \in A_j$.

Now let $X$ be a module and, for each $i \in I$, let $f_i \colon X \to A_i$ be a homomorphism. Define $\theta \colon X \to C$ by $\theta \colon x \mapsto (f_i(x))$. First, the diagram commutes: if $x \in X$, then $p_i\theta(x) = f_i(x)$. Finally, $\theta$ is unique. If $\psi \colon X \to C$ makes the diagram commute, then $p_i\psi(x) = f_i(x)$ for all $i$; that is, for each $i$, the $i$th coordinate of $\psi(x)$ is $f_i(x)$, which is also the $i$th coordinate of $\theta(x)$. Therefore, $\psi(x) = \theta(x)$ for all $x \in X$, and so $\psi = \theta$.  •

An explicit formula for the map $\theta \colon X \to \prod_{i \in I} A_i$ is $\theta \colon x \mapsto (f_i(x))$.

The categorical viewpoint makes the proof of the next theorem straightforward.

**Theorem 6.45.** *Let $R$ be a ring.*

(i) *For every left $R$-module $A$ and every family $(B_i)_{i \in I}$ of left $R$-modules,*

$$\operatorname{Hom}_R\Big(A, \prod_{i \in I} B_i\Big) \cong \prod_{i \in I} \operatorname{Hom}_R(A, B_i),$$

*via the $\mathbb{Z}$-isomorphism[14] $\varphi \colon f \mapsto (p_i f)$ ($p_i$ are the projections of the product $\prod_{i \in I} B_i$).*

(ii) *For every left $R$-module $B$ and every family $(A_i)_{i \in I}$ of $R$-modules,*

$$\operatorname{Hom}_R\Big(\bigoplus_{i \in I} A_i, B\Big) \cong \prod_{i \in I} \operatorname{Hom}_R(A_i, B),$$

*via the $R$-isomorphism $f \mapsto (f\alpha_i)$ ($\alpha_i$ are the injections of the sum $\bigoplus_{i \in I} A_i$).*

(iii) *If $A, A', B$, and $B'$ are left $R$-modules. then there are $\mathbb{Z}$-isomorphisms*

$$\operatorname{Hom}_R(A, B \oplus B') \cong \operatorname{Hom}_R(A, B) \oplus \operatorname{Hom}_R(A, B')$$

*and*

$$\operatorname{Hom}_R(A \oplus A', B) \cong \operatorname{Hom}_R(A, B) \oplus \operatorname{Hom}_R(A', B).$$

**Proof.**

(i) It is easy to see that $\varphi$ is additive; let us see that $\varphi$ is surjective. If $(f_i) \in \prod_i \operatorname{Hom}_R(A, B_i)$, then $f_i \colon A \to B_i$ for every $i$.

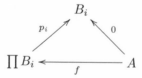

By Proposition 6.44, $\prod B_i$ is the product in $_R\mathbf{Mod}$, and so there is a unique $R$-map $\theta \colon A \to \prod B_i$ with $p_i \theta = f_i$ for all $i$. Thus, $(f_i) = \varphi(\theta)$ and $\varphi$ is surjective.

To see that $\varphi$ is injective, suppose that $f \in \ker \varphi$; that is, $0 = \varphi(f) = (p_i f)$. Thus, $p_i f = 0$ for every $i$. Hence, the following diagram containing $f$ commutes:

$$
\begin{array}{ccc}
 & B_i & \\
{}^{p_i}\nearrow & & \nwarrow{}^{0} \\
\prod B_i \longleftarrow{}_{\;f\;} & & A
\end{array}
$$

But the zero homomorphism also makes this diagram commute, and so the uniqueness of the arrow $A \to \prod B_i$ gives $f = 0$.

(ii) This proof, similar to that of part (i), is left to the reader.

(iii) When the index set is finite, direct sum and direct product of modules are equal. $\bullet$

---

[14]There are certain cases when the abelian group $\operatorname{Hom}_R(A, B)$ is a module; in these cases, the $\mathbb{Z}$-isomorphisms in parts (i), (ii), and (iii) are module isomorphisms (see Theorem 6.65).

There are examples showing that $\operatorname{Hom}_R(A, \bigoplus_i B_i) \not\cong \bigoplus_i \operatorname{Hom}_R(A, B_i)$ (see Exercise 6.39 on page 434) and $\operatorname{Hom}_R(\prod_i A_i, B) \not\cong \prod_i \operatorname{Hom}_R(A_i, B)$.

Let $\Pi = \prod_{n \geq 1} \langle e_n \rangle$, where each $\langle e_n \rangle$ is infinite cyclic. Call a torsion-free abelian group $S$ **slender** if, for every homomorphism $f \colon \Pi \to S$, we have $f(e_n) = 0$ for large $n$. Sąsiada proved that a countable torsion-free abelian group is slender if and only if it is reduced (Fuchs, *Infinite Abelian Groups* II, p. 159), and Fuchs proved that any direct sum of slender groups is slender (Ibid., p. 160). Here is a remarkable theorem of Łoś (Ibid., p 162). If $S$ is slender and $(A_i)_{i \in I}$ is a family of torsion-free abelian groups, where $I$ is not a measurable cardinal,[15] then there is an isomorphism

$$\varphi \colon \operatorname{Hom}(\prod_{i \in I} A_i, S) \to \bigoplus_{i \in I} \operatorname{Hom}(A_i, S).$$

In fact, if $f \colon \prod_{i \in I} A_i \to S$, then there is a finite subset $A_{i_1}, \ldots, A_{i_n}$ with $\varphi(f) = f|(A_{i_1} \oplus \cdots \oplus A_{i_n})$. In particular,

$$\operatorname{Hom}_{\mathbb{Z}}(\prod_{i \in \mathbb{N}} \mathbb{Z}_i, \mathbb{Z}) \cong \bigoplus_{i \in \mathbb{N}} \mathbb{Z}_i \quad \text{and} \quad \operatorname{Hom}_{\mathbb{Z}}(\bigoplus_{i \in \mathbb{N}} \mathbb{Z}_i, \mathbb{Z}) \cong \prod_{i \in \mathbb{N}} \mathbb{Z}_i.$$

We now present two dual constructions, *pullbacks* and *pushouts*, that are often useful. We shall see, in **Sets**, that intersections are pullbacks and unions are pushouts.

**Definition.** Given two morphisms $f \colon B \to A$ and $g \colon C \to A$ in a category $\mathcal{C}$, a **solution** is an ordered triple $(D, \alpha, \beta)$ making the left-hand diagram commute. A **pullback** (or *fibered product*) is a solution $(D, \alpha, \beta)$ that is "best" in the following sense: for every solution $(X, \alpha', \beta')$, there exists a unique morphism $\theta \colon X \to D$ making the right-hand diagram above commute.

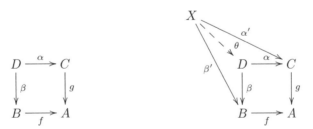

**Example 6.46.** We show that kernel is a pullback. More precisely, if $f \colon B \to A$ is a homomorphism in $_R\mathbf{Mod}$, then the pullback of the first diagram in Figure 6.1 is $(\ker f, 0, i)$, where $i \colon \ker f \to B$ is the inclusion. Let $i' \colon X \to B$ be a map with $fi' = 0$; then $fi'x = 0$ for all $x \in X$, and so $i'x \in \ker f$. If we define $\theta \colon X \to \ker f$ to be the map obtained from $i'$ by changing its target, then the diagram commutes: $i\theta = i'$. To prove uniqueness of the map $\theta$, suppose that $\theta' \colon X \to \ker f$ satisfies $i\theta' = i'$. Since $i$ is the inclusion, $\theta'x = i'x = \theta x$ for all $x \in X$, and so $\theta' = \theta$. Thus, $(\ker f, 0, i)$ is a pullback. Compare with Exercise 6.88 on page 486. ◀

---

[15]A cardinal number $d$ is **measurable** if $d$ is uncountable and every set of cardinal $d$ has a countably additive measure whose only values are 0 and 1. It is unknown whether measurable cardinals exist.

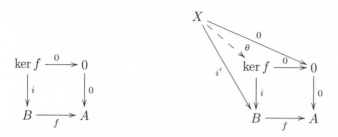

**Figure 6.1.** Kernel as pullback.

Pullbacks, when they exist, are unique up to isomorphism; the proof is in the same style as the proof of Proposition 6.40 that coproducts are unique.

**Proposition 6.47.** *The pullback of two maps $f\colon B \to A$ and $g\colon C \to A$ in $_R\mathbf{Mod}$ exists.*

**Proof.** Define
$$D = \{(b,c) \in B \sqcup C : f(b) = g(c)\},$$
define $\alpha\colon D \to C$ to be the restriction of the projection $(b,c) \mapsto c$, and define $\beta\colon D \to B$ to be the restriction of the projection $(b,c) \mapsto b$. It is easy to see that $(D, \alpha, \beta)$ is a solution.

If $(X, \alpha', \beta')$ is another solution, define a map $\theta\colon X \to D$ by $\theta\colon x \mapsto (\beta'(x), \alpha'(x))$. The values of $\theta$ do lie in $D$, for $f\beta'(x) = g\alpha'(x)$ because $X$ is a solution. We let the reader prove that the diagram commutes and that $\theta$ is unique. •

**Example 6.48.** That $B$ and $C$ are subsets of a set $A$ can be restated as saying that there are inclusion maps $i\colon B \to A$ and $j\colon C \to A$. The reader will enjoy proving that the pullback $D$ exists in **Sets**, and that $D = B \cap C$. ◄

Here is the dual construction.

**Definition.** Given two morphisms $f\colon A \to B$ and $g\colon A \to C$ in a category $\mathcal{C}$, a **solution** is an ordered triple $(D, \alpha, \beta)$ making the left-hand diagram commute. A **pushout** (or *fibered sum*) is a solution $(D, \alpha, \beta)$ that is "best" in the following sense: for every solution $(X, \alpha', \beta')$, there exists a unique morphism $\theta\colon D \to X$ making the right-hand diagram commute.

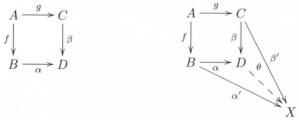

**Example 6.49.** We show that cokernel is a pushout. More precisely, if $f\colon A \to B$ is a homomorphism in $_R\mathbf{Mod}$, then the pushout of the first diagram in Figure 6.2

is (coker $f, \pi, 0$), where $\pi \colon B \to \operatorname{coker} f$ is the natural map. The verification that cokernel is a pushout is similar to that in Example 6.46, ◄

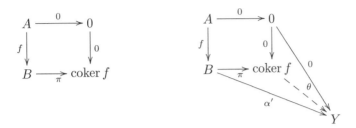

**Figure 6.2.** Cokernel as pushout.

Again, pushouts are unique up to isomorphism when they exist.

**Proposition 6.50.** *The pushout of two maps $f \colon A \to B$ and $g \colon A \to C$ in $_R\mathbf{Mod}$ exists.*

**Proof.** It is easy to see that

$$S = \{ \big( f(a), -g(a) \big) \in B \sqcup C : a \in A \}$$

is a submodule of $B \sqcup C$. Define $D = (B \sqcup C)/S$, define $\alpha \colon B \to D$ by $b \mapsto (b, 0) + S$, and define $\beta \colon C \to D$ by $c \mapsto (0, c) + S$. It is easy to see that $(D, \alpha, \beta)$ is a solution.

Given another solution $(X, \alpha', \beta')$, define the map $\theta \colon D \to X$ by $\theta \colon (b, c) + S \mapsto \alpha'(b) + \beta'(c)$. Again, we let the reader prove commutativity of the diagram and uniqueness of $\theta$.  •

Pushouts in **Groups** are quite interesting; for example, the pushout of two injective homomorphisms is called a *free product with amalgamation*.

**Example 6.51.** If $B$ and $C$ are subsets of a set $A$, then there are inclusion maps $i \colon B \cap C \to B$ and $j \colon B \cap C \to B$. The reader will enjoy proving that the pushout $D$ exists in **Sets**, and that $D$ is their union $B \cup C$.  ◄

## Exercises

**6.33.** (i) Prove, in every category $\mathcal{C}$, that each object $A \in \mathcal{C}$ has a unique identity morphism.

  (ii) If $f$ is an isomorphism in a category, prove that its inverse is unique.

∗ **6.34.** (i) Let $X$ be a partially ordered set, and let $a, b \in X$. Show, in $\mathbf{PO}(X)$ [defined in Example 6.38(vi)], that the coproduct $a \sqcup b$ is the least upper bound of $a$ and $b$, and that the product $a \sqcap b$ is the greatest lower bound.

  (ii) Let $Y$ be a set, let $2^Y$ denote the family of all its subsets, and regard $2^Y$ as a partially ordered set under inclusion. If $A$ and $B$ are subsets of $Y$, show, in $\mathbf{PO}(2^Y)$, that the coproduct $A \sqcup B = A \cup B$ and that the product $A \sqcap B = A \cap B$.

(iii) Give an example of a category in which there are two objects whose coproduct does not exist.

**Hint.** See Exercise 5.35 on page 320.

* **6.35.** (i) Prove that **Groups** is not a pre-additive category.

   **Hint.** If $G$ is not abelian and $f, g \colon G \to G$ are homomorphisms, show that the function $x \mapsto f(x)g(x)$ may not be a homomorphism.

   (ii) Prove that **ComRings** is not a pre-additive category.

* **6.36.** If $A$ and $B$ are (not necessarily abelian) groups, prove that $A \sqcap B = A \times B$ (direct product) in **Groups**.

**6.37.** If $G$ is a finite abelian group, prove that $\operatorname{Hom}_{\mathbb{Z}}(\mathbb{Q}, G) = 0$.

**6.38.** Generalize Proposition 6.31 for infinite index sets. Let $(M_i)_{i \in I}$ be a family of modules and, for each $i$, let $N_i$ be a submodule of $M_i$. Prove that

$$\left( \bigoplus_i M_i \right) \Big/ \left( \bigoplus_i N_i \right) \cong \bigoplus_i (M_i / N_i).$$

* **6.39.** (i) Prove, for every abelian group $A$, that $n \operatorname{Hom}(A, \mathbb{I}_n) = \{0\}$; that is, $nf = 0$ for every homomorphism $f \colon A \to \mathbb{I}_n$.

   (ii) Let $A = \bigoplus_{n \geq 2} \mathbb{I}_n$. Prove that $\operatorname{Hom}(A, \bigoplus_n \mathbb{I}_n) \ncong \bigoplus_n \operatorname{Hom}(A, \mathbb{I}_n)$.

   **Hint.** The right-hand side is a torsion group, but the element $1_A$ on the left-hand side has infinite order.

* **6.40.** Given a map $\sigma \colon \prod B_i \to \prod C_j$, find a map $\widetilde{\sigma}$ making the following diagram commute:

$$
\begin{array}{ccc}
\operatorname{Hom}(A, \prod B_i) & \xrightarrow{\ \sigma\ } & \operatorname{Hom}(A, \prod C_j) \\
{\scriptstyle \tau} \downarrow & & \downarrow {\scriptstyle \tau'} \\
\prod \operatorname{Hom}(A, B_i) & \dashrightarrow[\widetilde{\sigma}] & \prod \operatorname{Hom}(A, C_j)
\end{array}
$$

where $\tau$ and $\tau'$ are the isomorphisms of Theorem 6.45(i).

**Hint.** If $f \in \operatorname{Hom}(A, \prod B_i)$, define $\widetilde{\sigma} \colon (f_i) \mapsto (p_j \sigma f)$; that is, the $j$th coordinate of $\widetilde{\sigma}(f_i)$ is the $j$th coordinate of $\sigma(f) \in \prod C_j$.

* **6.41.** (i) Given a pushout diagram in $_R\mathbf{Mod}$

$$
\begin{array}{ccc}
A & \xrightarrow{\ g\ } & C \\
{\scriptstyle f} \downarrow & & \downarrow {\scriptstyle \beta} \\
B & \xrightarrow{\ \alpha\ } & D
\end{array}
$$

prove that $g$ injective implies $\alpha$ injective, and that $g$ surjective implies $\alpha$ surjective. Thus, parallel arrows have the same properties.

(ii) Given a pullback diagram in $_R\mathbf{Mod}$

$$
\begin{array}{ccc}
D & \xrightarrow{\ \alpha\ } & C \\
{\scriptstyle \beta} \downarrow & & \downarrow {\scriptstyle g} \\
B & \xrightarrow{\ f\ } & A
\end{array}
$$

prove that $f$ injective implies $\alpha$ injective, and that $f$ surjective implies $\alpha$ surjective. Thus, parallel arrows have the same properties.

**Definition.** An object $A$ in a category $\mathcal{C}$ is called an ***initial object*** if, for every object $C$ in $\mathcal{C}$, there exists a unique morphism $A \to C$.

An object $\Omega$ in a category $\mathcal{C}$ is called a ***terminal object*** if, for every object $C$ in $\mathcal{C}$, there exists a unique morphism $C \to \Omega$.

$*$ **6.42.** (i) Prove the uniqueness of initial and terminal objects, if they exist. Give an example of a category which contains no initial object. Give an example of a category that contains no terminal object.

(ii) If $\Omega$ is a terminal object in a category $\mathcal{C}$, prove, for any $G \in \text{obj}(\mathcal{C})$, that the projections $\lambda \colon G \sqcap \Omega \to G$ and $\rho \colon \Omega \sqcap G \to G$ are isomorphisms.

(iii) Let $A$ and $B$ be objects in a category $\mathcal{C}$. Define a new category $\mathcal{C}'$ whose objects are diagrams $A \xrightarrow{\alpha} C \xleftarrow{\beta} B$, where $C$ is an object in $\mathcal{C}$ and $\alpha$ and $\beta$ are morphisms in $\mathcal{C}$. Define a morphism in $\mathcal{C}'$ to be a morphism $\theta$ in $\mathcal{C}$ that makes the following diagram commute:

$$
\begin{array}{ccccc}
A & \xrightarrow{\alpha} & C & \xleftarrow{\beta} & B \\
{\scriptstyle 1_A}\downarrow & & \downarrow{\scriptstyle \theta} & & \downarrow{\scriptstyle 1_B} \\
A & \xrightarrow{\alpha'} & C' & \xleftarrow{\beta'} & B
\end{array}
$$

There is an obvious candidate for composition. Prove that $\mathcal{C}'$ is a category.

(iv) Prove that an initial object in $\mathcal{C}'$ is a coproduct in $\mathcal{C}$, and use this to give another proof of Proposition 6.40, the uniqueness of coproduct (should it exist).

(v) Give an analogous construction showing that product is a terminal object in a suitable category, and give another proof of Proposition 6.41.

$*$ **6.43.** A ***zero object*** in a category $\mathcal{C}$ is an object $Z$ that is both an initial object and a terminal object.

(i) Prove that $\{0\}$ is a zero object in $_R\mathbf{Mod}$.

(ii) Prove that $\varnothing$ is an initial object in **Sets**.

(iii) Prove that any one-point set is a terminal object in **Sets**.

(iv) Prove that a zero object does not exist in **Sets**.

**6.44.** (i) Assuming that coproducts exist, prove associativity:
$$ A \sqcup (B \sqcup C) \cong (A \sqcup B) \sqcup C. $$

(ii) Assuming that products exist, prove associativity:
$$ A \sqcap (B \sqcap C) \cong (A \sqcap B) \sqcap C. $$

**6.45.** Let $C_1, C_2, D_1, D_2$ be objects in a category $\mathcal{C}$.

(i) If there are morphisms $f_i \colon C_i \to D_i$, for $i = 1, 2$, and $C_1 \sqcap C_2$ and $D_1 \sqcap D_2$ exist, prove that there exists a unique morphism $f_1 \sqcap f_2$ making the following diagram commute:

$$C_1 \sqcap C_2 \xrightarrow{f_1 \sqcap f_2} D_1 \sqcap D_2$$

$$\downarrow{p_i} \qquad\qquad \downarrow{q_i}$$

$$C_i \xrightarrow{\quad f_i \quad} D_i$$

where $p_i$ and $q_i$ are projections.

(ii) If there are morphisms $g_i \colon X \to C_i$, where $X$ is an object in $\mathcal{C}$ and $i = 1, 2$, prove that there is a unique morphism $(g_1, g_2)$ making the following diagram commute:

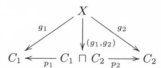

where the $p_i$ are projections.

**Hint.** First define an analog of the diagonal $\Delta_X \colon X \to X \times X$ in **Sets**, given by $x \mapsto (x, x)$, and then define $(g_1, g_2) = (g_1 \sqcap g_2)\Delta_X$.

**6.46.** Let $\mathcal{C}$ be a category having finite products and a terminal object $\Omega$. A ***group object*** in $\mathcal{C}$ is a quadruple $(G, \mu, \eta, \epsilon)$, where $G$ is an object in $\mathcal{C}$, $\mu \colon G \sqcap G \to G$, $\eta \colon G \to G$, and $\epsilon \colon \Omega \to G$ are morphisms, so that the following diagrams commute:

*Associativity*:

$$G \sqcap G \sqcap G \xrightarrow{1 \sqcap \mu} G \sqcap G$$

$$\downarrow{\mu \sqcap 1} \qquad\qquad \downarrow{\mu}$$

$$G \sqcap G \xrightarrow{\quad \mu \quad} G$$

*Identity*:

$$G \sqcap \Omega \xrightarrow{1 \sqcap \epsilon} G \sqcap G \xleftarrow{\epsilon \sqcap 1} \Omega \sqcap G$$

$$\searrow{\lambda} \qquad \downarrow{\mu} \qquad \swarrow{\rho}$$

$$G$$

where $\lambda$ and $\rho$ are the isomorphisms in Exercise 6.42 on page 435.

*Inverse*:

$$G \xrightarrow{(1,\eta)} G \sqcap G \xleftarrow{(\eta,1)} G$$

$$\downarrow{\omega} \qquad\qquad \downarrow{\mu} \qquad\qquad \downarrow{\omega}$$

$$\Omega \xrightarrow{\epsilon} G \xleftarrow{\epsilon} \Omega$$

where $\omega \colon G \to \Omega$ is the unique morphism to the terminal object.

  (i) Prove that a group object in **Sets** is a group.

 (ii) Prove that a group object in **Groups** is an abelian group.
      **Hint.** Use Exercise 1.81 on page 59.

(iii) Prove that a group object in **Top**$_2$, the category of all Hausdorff topological spaces, is a topological group.

**Remark.** We can now define ***cogroups***, the dual of groups.  ◀

## Section 6.3. Functors

Functors[16] are homomorphisms of categories.

**Definition.** If $\mathcal{C}$ and $\mathcal{D}$ are categories, then a **_functor_** $T\colon \mathcal{C} \to \mathcal{D}$ is a function such that

    (i) if $A \in \mathrm{obj}(\mathcal{C})$, then $T(A) \in \mathrm{obj}(\mathcal{D})$;

    (ii) if $f\colon A \to A'$ in $\mathcal{C}$, then $T(f)\colon T(A) \to T(A')$ in $\mathcal{D}$;

    (iii) if $A \xrightarrow{f} A' \xrightarrow{g} A''$ in $\mathcal{C}$, then $T(A) \xrightarrow{T(f)} T(A') \xrightarrow{T(g)} T(A'')$ in $\mathcal{D}$ and
$$T(gf) = T(g)T(f);$$

    (iv) for every $A \in \mathrm{obj}(\mathcal{C})$,
$$T(1_A) = 1_{T(A)}.$$

**Example 6.52.**

    (i) If $\mathcal{C}$ is a category, then the **_identity functor_** $1_{\mathcal{C}}\colon \mathcal{C} \to \mathcal{C}$ is defined by
$$1_{\mathcal{C}}(A) = A \text{ for all objects } A,$$

and
$$1_{\mathcal{C}}(f) = f \text{ for all morphisms } f.$$

    (ii) If $\mathcal{C}$ is a category and $A \in \mathrm{obj}(\mathcal{C})$, then the **Hom _functor_** $T_A\colon \mathcal{C} \to \mathbf{Sets}$ is defined by
$$T_A(B) = \mathrm{Hom}(A, B) \text{ for all } B \in \mathrm{obj}(\mathcal{C}),$$
and if $f\colon B \to B'$ in $\mathcal{C}$, then $T_A(f)\colon \mathrm{Hom}(A, B) \to \mathrm{Hom}(A, B')$ is given by
$$T_A(f)\colon h \mapsto fh.$$

We call $T_A(f)$ the **_induced map_**, and we denote it by
$$T_A(f) = f_*\colon h \mapsto fh.$$

Because of the importance of this example, we will verify the parts of the definition in detail. First, the very definition of category says that $\mathrm{Hom}(A, B)$ is a set. Note that the composite $fh$ makes sense:

$$A \underset{h}{\overset{fh}{\rightrightarrows}} B \xrightarrow{f} B'.$$

Suppose now that $g\colon B' \to B''$. Let us compare the functions
$$(gf)_* \text{ and } g_*f_*\colon \mathrm{Hom}(A, B) \to \mathrm{Hom}(A, B'').$$

If $h \in \mathrm{Hom}(A, B)$, i.e., if $h\colon A \to B$, then
$$(gf)_*\colon h \mapsto (gf)h;$$

on the other hand,
$$g_*f_*\colon h \mapsto fh \mapsto g(fh),$$

---

[16]The term *functor* was coined by the philosopher R. Carnap, and S. Mac Lane thought it was the appropriate term in this context.

as desired. Finally, if $f$ is the identity map $1_B \colon B \to B$, then

$$(1_B)_* \colon h \mapsto 1_B h = h$$

for all $h \in \mathrm{Hom}(A, B)$, so that $(1_B)_* = 1_{\mathrm{Hom}(A,B)}$.

Theorem 6.45(i) says that $T_A\left(\prod_i B_i\right) \cong \prod_i T_A(B_i)$; that is, $T_A$ preserves products. We usually denote $T_A$ by

$$\mathrm{Hom}(A, \square).$$

(iii) If $R$ is a commutative ring and $A$ is an $R$-module, then the Hom functor $T_A \colon {}_R\mathbf{Mod} \to \mathbf{Sets}$ has more structure. We saw, in Proposition 6.13(ii), that $\mathrm{Hom}_R(A, B)$ is an $R$-module. We now show that if $f \colon B \to B'$, then the induced map $f_* \colon \mathrm{Hom}_R(A, B) \to \mathrm{Hom}_R(A, B')$, given by $h \mapsto fh$, is an $R$-map. First, $f_*$ is additive: if $h, h' \in \mathrm{Hom}(A, B)$, then for all $a \in A$,

$$f_*(h + h') = f(h + h') \colon a \mapsto f(ha + h'a)$$
$$= fha + fh'a = (f_*(h) + f_*(h'))(a),$$

so that $f_*(h + h') = f_*(h) + f_*(h')$. Second, $f_*$ preserves scalars. Recall that if $r \in R$ and $h \in \mathrm{Hom}(A, B)$, then $rh \colon a \mapsto h(ra)$. Thus,

$$f_*(rh) \colon a \mapsto f(rh)(a) = fh(ra),$$

while

$$rf_*(h) = rfh \colon a \mapsto fh(ra).$$

Therefore, $f_*(rh) = (rf)_*(h)$.

We have shown, when $R$ is commutative, that $T_A \colon {}_R\mathbf{Mod} \to {}_R\mathbf{Mod}$; that is, $T_A$ is a functor with values in ${}_R\mathbf{Mod}$. In particular, if $R$ is a field, then the $\mathrm{Hom}_R$'s are vector spaces and the induced maps are linear transformations.

(iv) Let $\mathcal{C}$ be a category, and let $A \in \mathrm{obj}(\mathcal{C})$. Define $T \colon \mathcal{C} \to \mathcal{C}$ by $T(C) = A$ for every $C \in \mathrm{obj}(\mathcal{C})$, and $T(f) = 1_A$ for every morphism $f$ in $\mathcal{C}$. Then $T$ is a functor, called the **constant functor** at $A$.

(v) If $\mathcal{C} = \mathbf{Groups}$, define the **forgetful functor** $U \colon \mathbf{Groups} \to \mathbf{Sets}$ as follows: $U(G)$ is the "underlying" set of a group $G$ and $U(f)$ is a homomorphism $f$ regarded as a mere function. Strictly speaking, a group is an ordered pair $(G, \mu)$, where $G$ is its (underlying) set and $\mu \colon G \times G \to G$ is its operation, and $U((G, \mu)) = G$; the functor $U$ "forgets" the operation and remembers only the set.

There are many variants. For example, an $R$-module is an ordered triple $(M, \alpha, \sigma)$, where $M$ is a set, $\alpha \colon M \times M \to M$ is addition, and $\sigma \colon R \times M \to M$ is scalar multiplication. There are forgetful functors $U' \colon {}_R\mathbf{Mod} \to \mathbf{Ab}$ with $U'((M, \alpha, \sigma)) = (M, \alpha)$, and $U'' \colon {}_R\mathbf{Mod} \to \mathbf{Sets}$ with $U''((M, \alpha, \sigma)) = M$, for example.

(vi) The fundamental group $\pi_1$ is a functor from the category of spaces (with basepoint) to the category of groups. This example illustrates why, when defining functions, we have to be so fussy about targets; that is, if $f \colon X \to Y$, $Y \subseteq Z$, and $j \colon Y \to Z$ is the inclusion, then $f \neq jf$. The identity $f \colon S^1 \to S^1$ induces the identity homomorphism on $\pi_1(S^1) \cong \mathbb{Z}$, while $jf \colon S^1 \to \mathbb{R}^2$

induces $\pi_1(jf) = \pi_1(j)\pi_1(f) = 0$, because $\pi_1(\mathbb{R}^2) = \{0\}$. Similarly, we must also be fussy about domains of functions. ◄

The following result is useful, even though it is very easy to prove.

**Proposition 6.53.** *If $T: \mathcal{C} \to \mathcal{D}$ is a functor and $f: A \to B$ is an isomorphism in $\mathcal{C}$, then $T(f)$ is an isomorphism in $\mathcal{D}$.*

**Proof.** If $g$ is the inverse of $f$, apply $T$ to the equations

$$gf = 1_A \text{ and } fg = 1_B. \quad \bullet$$

This proposition illustrates, admittedly at a low level, the reason why it is useful to give categorical definitions: functors can recognize definitions phrased solely in terms of objects, morphisms, and diagrams. How could we prove this result in **Ab** if we only regard an isomorphism as a homomorphism that is an injection and a surjection?

A second type of functor reverses the direction of arrows.

**Definition.** If $\mathcal{C}$ and $\mathcal{D}$ are categories, then a ***contravariant functor*** $T: \mathcal{C} \to \mathcal{D}$ is a function such that

  (i) if $C \in \mathrm{obj}(\mathcal{C})$, then $T(C) \in \mathrm{obj}(\mathcal{D})$;

 (ii) if $f: C \to C'$ in $\mathcal{C}$, then $T(f): T(C') \to T(C)$ in $\mathcal{D}$;

(iii) if $C \xrightarrow{f} C' \xrightarrow{g} C''$ in $\mathcal{C}$, then $T(C'') \xrightarrow{T(g)} T(C') \xrightarrow{T(f)} T(C)$ in $\mathcal{D}$ and

$$T(gf) = T(f)T(g);$$

(iv) for every $A \in \mathrm{obj}(\mathcal{C})$,

$$T(1_A) = 1_{T(A)}.$$

To distinguish them from contravariant functors, the functors defined earlier are often called ***covariant functors***.

**Example 6.54.**

  (i) If $\mathcal{C}$ is a category and $B \in \mathrm{obj}(\mathcal{C})$, then the ***contravariant*** **Hom** ***functor*** $T^B: \mathcal{C} \to \textbf{Sets}$ is defined, for all $C \in \mathrm{obj}(\mathcal{C})$, by

$$T^B(C) = \mathrm{Hom}(C, B)$$

and, if $f: C \to C'$ in $\mathcal{C}$, then $T^B(f): \mathrm{Hom}(C', B) \to \mathrm{Hom}(C, B)$ is given by

$$T^B(f): h \mapsto hf.$$

We call $T^B(f)$ the ***induced map***, and we denote it by

$$T^B(f) = f^*: h \mapsto hf.$$

Because of the importance of this example, we verify the axioms, showing that $T^B$ is a (contravariant) functor. Note that the composite $hf$ makes sense:

$$C \underset{f}{\overset{hf}{\rightrightarrows}} C' \xrightarrow{h} B.$$

Given homomorphisms

$$C \xrightarrow{f} C' \xrightarrow{g} C'',$$

let us compare the functions

$$(gf)^* \text{ and } f^*g^* \colon \operatorname{Hom}(C'', B) \to \operatorname{Hom}(C, B).$$

If $h \in \operatorname{Hom}(C'', B)$ (i.e., if $h \colon C'' \to B$), then

$$(gf)^* \colon h \mapsto h(gf);$$

on the other hand,

$$f^*g^* \colon h \mapsto hg \mapsto (hg)f,$$

as desired. Finally, if $f$ is the identity map $1_C \colon C \to C$, then

$$(1_C)^* \colon h \mapsto h1_C = h$$

for all $h \in \operatorname{Hom}(C, B)$, so that $(1_C)^* = 1_{\operatorname{Hom}(C,B)}$.

Theorem 6.45(ii) says that the contravariant functor $T^B$ converts sums to products: $T^B(\bigoplus_i A_i) \cong \prod_i T^B(A_i)$. We usually denote $T^B$ by

$$\operatorname{Hom}(\square, B).$$

(ii) If $R$ is a commutative ring and $B$ is an $R$-module, then the contravariant Hom functor $T^B \colon {}_R\mathbf{Mod} \to \mathbf{Sets}$ has more structure. Proposition 6.13(ii) says that $\operatorname{Hom}_R(A, B)$ is an $R$-module. We show that if $f \colon C \to C'$ is an $R$-map, then the induced map $f^* \colon \operatorname{Hom}_R(C', B) \to \operatorname{Hom}_R(C, B)$, given by $h \mapsto hf$, is an $R$-map between $R$-modules. First, $f^*$ is additive: if $g, h \in \operatorname{Hom}(C', B)$, then for all $c' \in C'$,

$$f^*(g + h) = (g + h)f \colon c' \mapsto (g + h)f(c')$$
$$= gfc' + hfc' = (f^*(g) + f^*(h))(c'),$$

so that $f^*(g + h) = f^*(g) + f^*(h)$. Second, $f^*$ preserves scalars. Recall that if $r \in R$ and $h \in \operatorname{Hom}(A, B)$, then $rh \colon a \mapsto h(ra)$. Thus,

$$f^*(rh) \colon c' \mapsto (rh)f(c') = h(rf(c')),$$

while

$$rf^*(h) = r(hf) \colon c' \mapsto hf(rc').$$

These are the same, because $rf(c') = f(rc')$, and so $f^*(rh) = rf^*(h)$.

We have shown, when $R$ is commutative, that $T^A \colon {}_R\mathbf{Mod} \to {}_R\mathbf{Mod}$; that is, $T^A$ is a contravariant functor with values in ${}_R\mathbf{Mod}$. In particular, if $R$ is a field and $V, W$ are vector spaces over $R$, then $\operatorname{Hom}_R(V, W)$ is also a vector space and induced maps are linear transformations. A special case of this is the **dual space functor** $\operatorname{Hom}_R(\square, R)$. ◄

It is easy to see, as in Proposition 6.53, that contravariant functors preserve isomorphisms; that is, if $T \colon \mathcal{C} \to \mathcal{D}$ is a contravariant functor and $f \colon C \to C'$ is an isomorphism in $\mathcal{C}$, then $T(f) \colon T(C') \to T(C)$ is an isomorphism in $\mathcal{D}$.

The following construction plays the same role for categories and functors as opposite rings play for left and right modules.

**Definition.** If $\mathcal{C}$ is a category, its *opposite category* $\mathcal{C}^{\mathrm{op}}$ has objects $\mathrm{obj}(\mathcal{C}^{\mathrm{op}}) = \mathrm{obj}(\mathcal{C})$, morphisms $\mathrm{Hom}_{\mathcal{C}^{\mathrm{op}}}(A, B) = \mathrm{Hom}_{\mathcal{C}}(B, A)$ [we may write morphisms in $\mathcal{C}^{\mathrm{op}}$ as $f^{\mathrm{op}}$, where $f$ is a morphism in $\mathcal{C}$], and composition the reverse of that in $\mathcal{C}$; that is, $f^{\mathrm{op}}g^{\mathrm{op}} = (gf)^{\mathrm{op}}$ when $A \xrightarrow{f} B \xrightarrow{g} C$ in $\mathcal{C}$.

It is routine to check that $\mathcal{C}^{\mathrm{op}}$ is a category. We illustrate composition in $\mathcal{C}^{\mathrm{op}}$: a diagram $C \xrightarrow{g^{\mathrm{op}}} B \xrightarrow{f^{\mathrm{op}}} A$ in $\mathcal{C}^{\mathrm{op}}$ corresponds to $A \xrightarrow{f} B \xrightarrow{g} C$ in $\mathcal{C}$. Opposite categories are hard to visualize. If $\mathcal{C} = \mathbf{Sets}$, for example, the set $\mathrm{Hom}_{\mathbf{Sets}^{\mathrm{op}}}(X, \varnothing)$ for any set $X$ has exactly one element, namely, $i^{\mathrm{op}}$, where $i$ is the inclusion $\varnothing \to X$ in $\mathbf{Sets}$. But $i^{\mathrm{op}} \colon X \to \varnothing$ cannot be a function, for there are no functions from a nonempty set $X$ to $\varnothing$.

If $T \colon \mathcal{C} \to \mathcal{D}$ is a functor, define $T^{\mathrm{op}} \colon \mathcal{C}^{\mathrm{op}} \to \mathcal{D}^{\mathrm{op}}$ by $T^{\mathrm{op}}(C) = T(C)$ for all $C \in \mathrm{obj}(\mathcal{C})$ and $T^{\mathrm{op}}(f^{\mathrm{op}}) = T(f)^{\mathrm{op}}$ for all morphisms $f$ in $\mathcal{C}$. It is easy to show that $T^{\mathrm{op}}$ is a functor $\mathcal{C}^{\mathrm{op}} \to \mathcal{D}^{\mathrm{op}}$ having the same variance as $T$. For example, if $T$ is covariant, then

$$T^{\mathrm{op}}(f^{\mathrm{op}}g^{\mathrm{op}}) = T^{\mathrm{op}}([gf]^{\mathrm{op}}) = T(gf)^{\mathrm{op}}$$
$$= [TgTf]^{\mathrm{op}} = [Tf]^{\mathrm{op}}[Tg]^{\mathrm{op}} = T^{\mathrm{op}}(f^{\mathrm{op}})T^{\mathrm{op}}(g^{\mathrm{op}}).$$

**Definition.** If $\mathcal{C}$ and $\mathcal{D}$ are pre-additive categories, then a functor $T \colon \mathcal{C} \to \mathcal{D}$, of either variance, is called an *additive functor* if, for every pair of morphisms $f, g \colon A \to B$, we have

$$T(f + g) = T(f) + T(g).$$

Hom functors $_R\mathbf{Mod} \to \mathbf{Ab}$ of either variance are additive functors. Every covariant functor $T \colon \mathcal{C} \to \mathcal{D}$ gives rise to functions

$$T_{AB} \colon \mathrm{Hom}(A, B) \to \mathrm{Hom}(TA, TB),$$

for every $A$ and $B$, defined by $h \mapsto T(h)$. If $T$ is an additive functor between pre-additive categories, then each $T_{AB}$ is a homomorphism of abelian groups; the analogous statement for contravariant functors is also true.

Here is a modest generalization of Corollary 6.65(iii).

**Proposition 6.55.** *If* $T \colon {}_R\mathbf{Mod} \to \mathbf{Ab}$ *is an additive functor of either variance, then* $T$ *preserves finite direct sums:*

$$T(A_1 \oplus \cdots \oplus A_n) \cong T(A_1) \oplus \cdots \oplus T(A_n).$$

**Proof.** By induction, it suffices to prove that $T(A \oplus B) \cong T(A) \oplus T(B)$. Proposition 6.26(iii) characterizes $M = A \oplus B$ by maps $p \colon M \to A$, $q \colon M \to B$, $i \colon A \to M$, and $j \colon B \to M$ such that $pi = 1_A$, $qj = 1_B$, $pj = 0$, $qi = 0$, and $ip + jq = 1_M$. Since $T$ is an additive functor, Exercise 6.49 on page 449 gives $T(0) = 0$, and so $T$ preserves these equations. $\bullet$

We have just seen that additive functors $T \colon {}_R\mathbf{Mod} \to \mathbf{Ab}$ preserve the direct sum of two modules:

$$T(A \oplus C) = T(A) \oplus T(C).$$

If we regard such a direct sum as a split short exact sequence, then we may rephrase this by saying that if

$$0 \to A \xrightarrow{i} B \xrightarrow{p} C \to 0$$

is a split short exact sequence, then so is

$$0 \to T(A) \xrightarrow{T(i)} T(B) \xrightarrow{T(p)} T(C) \to 0.$$

This leads us to a more general question: if

$$0 \to A \xrightarrow{i} B \xrightarrow{p} C \to 0$$

is any, not necessarily split, short exact sequence, is

$$0 \to T(A) \xrightarrow{T(i)} T(B) \xrightarrow{T(p)} T(C) \to 0$$

also an exact sequence? Here is the answer for covariant Hom functors (there is no misprint in the statement of the theorem: "$\to 0$" should not appear at the end of the sequences, and we shall discuss this point after the proof).

**Theorem 6.56.** *If* $0 \to A \xrightarrow{i} B \xrightarrow{p} C$ *is an exact sequence of R-modules and X is an R-module, then there is an exact sequence*

$$0 \to \mathrm{Hom}_R(X, A) \xrightarrow{i_*} \mathrm{Hom}_R(X, B) \xrightarrow{p_*} \mathrm{Hom}_R(X, C).$$

**Proof.**

(i) $\ker i_* = \{0\}$.

If $f \in \ker i_*$, then $f \colon X \to A$ and $i_*(f) = 0$; that is,

$$if(x) = 0 \text{ for all } x \in X.$$

Since $i$ is injective, $f(x) = 0$ for all $x \in X$, and so $f = 0$.

(ii) $\operatorname{im} i_* \subseteq \ker p_*$.

If $g \in \operatorname{im} i_*$, then $g \colon X \to B$ and $g = i_*(f) = if$ for some $f \colon X \to A$. But $p_*(g) = pg = pif = 0$, because exactness of the original sequence, namely, $\operatorname{im} i = \ker p$, implies $pi = 0$.

(iii) $\ker p_* \subseteq \operatorname{im} i_*$.

If $g \in \ker p_*$, then $g \colon X \to B$ and $p_*(g) = pg = 0$. Hence, $pg(x) = 0$ for all $x \in X$, so that $g(x) \in \ker p = \operatorname{im} i$. Thus, $g(x) = i(a)$ for some $a \in A$; since $i$ is injective, this element $a$ is unique. Hence, the function $f \colon X \to A$, given by $f(x) = a$ if $g(x) = i(a)$, is well-defined. It is easy to check that $f \in \mathrm{Hom}_R(X, A)$; that is, $f$ is an $R$-homomorphism. Since

$$g(x + x') = g(x) + g(x') = i(a) + i(a') = i(a + a'),$$

we have

$$f(x + x') = a + a' = f(x) + f(x').$$

A similar argument shows that $f(rx) = rf(x)$ for all $r \in R$. But $i_*(f) = if$ and $if(x) = i(a) = g(x)$ for all $x \in X$; that is, $i_*(f) = g$, and so $g \in \operatorname{im} i_*$. •

**Example 6.57.** Even if the map $p\colon B \to C$ in the original exact sequence is assumed to be surjective, the functored sequence need not end with "$\to 0$;" that is, $p_*\colon \mathrm{Hom}_R(X, B) \to \mathrm{Hom}_R(X, C)$ may fail to be surjective.

The abelian group $\mathbb{Q}/\mathbb{Z}$ consists of cosets $q + \mathbb{Z}$ for $q \in \mathbb{Q}$, and it is easy to see that its element $\frac{1}{2} + \mathbb{Z}$ has order 2. It follows that $\mathrm{Hom}_{\mathbb{Z}}(\mathbb{I}_2, \mathbb{Q}/\mathbb{Z}) \neq \{0\}$, for it contains the nonzero homomorphism $[1] \mapsto \frac{1}{2} + \mathbb{Z}$.

Apply the functor $\mathrm{Hom}_{\mathbb{Z}}(\mathbb{I}_2, \square)$ to

$$0 \to \mathbb{Z} \xrightarrow{i} \mathbb{Q} \xrightarrow{p} \mathbb{Q}/\mathbb{Z} \to 0,$$

where $i$ is the inclusion and $p$ is the natural map. We have just seen that

$$\mathrm{Hom}_{\mathbb{Z}}(\mathbb{I}_2, \mathbb{Q}/\mathbb{Z}) \neq \{0\};$$

on the other hand, $\mathrm{Hom}_{\mathbb{Z}}(\mathbb{I}_2, \mathbb{Q}) = \{0\}$ because $\mathbb{Q}$ has no (nonzero) elements of finite order. Therefore, the induced map $p_*\colon \mathrm{Hom}_{\mathbb{Z}}(\mathbb{I}_2, \mathbb{Q}) \to \mathrm{Hom}_{\mathbb{Z}}(\mathbb{I}_2, \mathbb{Q}/\mathbb{Z})$ cannot be surjective. ◄

**Definition.** A covariant functor $T\colon {}_R\mathbf{Mod} \to \mathbf{Ab}$ is called *left exact* if exactness of

$$0 \to A \xrightarrow{i} B \xrightarrow{p} C$$

implies exactness of

$$0 \to T(A) \xrightarrow{T(i)} T(B) \xrightarrow{T(p)} T(C).$$

Thus, Theorem 6.56 shows that covariant Hom functors $\mathrm{Hom}_R(X, \square)$ are left exact functors. Investigation of the cokernel of $\mathrm{Hom}_R(X, \square)$ is done in Homological Algebra; it is related to a functor called $\mathrm{Ext}_R^1(X, \square)$.

There is an analogous result for contravariant Hom functors.

**Theorem 6.58.** *If $A \xrightarrow{i} B \xrightarrow{p} C \to 0$ is an exact sequence of $R$-modules and $Y$ is an $R$-module, then there is an exact sequence*

$$0 \to \mathrm{Hom}_R(C, Y) \xrightarrow{p^*} \mathrm{Hom}_R(B, Y) \xrightarrow{i^*} \mathrm{Hom}_R(A, Y).$$

**Proof.**

(i) $\ker p^* = \{0\}$.

　　If $h \in \ker p^*$, then $h\colon C \to Y$ and $0 = p^*(h) = hp$. Thus, $h(p(b)) = 0$ for all $b \in B$, so that $h(c) = 0$ for all $c \in \mathrm{im}\, p$. Since $p$ is surjective, $\mathrm{im}\, p = C$, and $h = 0$.

(ii) $\mathrm{im}\, p^* \subseteq \ker i^*$.

　　If $g \in \mathrm{Hom}_R(C, Y)$, then $i^* p^*(g) = (pi)^*(g) = 0$, because exactness of the original sequence, namely, $\mathrm{im}\, i = \ker p$, implies $pi = 0$.

(iii) $\ker i^* \subseteq \mathrm{im}\, p^*$.

　　If $g \in \ker i^*$, then $g\colon B \to Y$ and $i^*(g) = gi = 0$. If $c \in C$, then $c = p(b)$ for some $b \in B$, because $p$ is surjective. Define $f\colon C \to Y$ by $f(c) = g(b)$ if $c = p(b)$. Note that $f$ is well-defined: if $p(b) = p(b')$, then $b - b' \in \ker p = \mathrm{im}\, i$, so that $b - b' = i(a)$ for some $a \in A$. Hence,

$$g(b) - g(b') = g(b - b') = gi(a) = 0,$$

because $gi = 0$. The reader may check that $f$ is an $R$-map. Finally,

$$p^*(f) = fp = g,$$

for $c = p(b)$ implies $g(b) = f(c) = f(p(b))$. Therefore, $g \in \operatorname{im} p^*$.  •

**Example 6.59.** Even if the map $i: A \to B$ in the original exact sequence is assumed to be injective, the functored sequence need not end with "$\to 0$;" that is, $i^*: \operatorname{Hom}_R(B, Y) \to \operatorname{Hom}_R(A, Y)$ may fail to be surjective.

We claim that $\operatorname{Hom}_{\mathbb{Z}}(\mathbb{Q}, \mathbb{Z}) = \{0\}$. Suppose that $f: \mathbb{Q} \to \mathbb{Z}$ and $f(a/b) \neq 0$ for some $a/b \in \mathbb{Q}$. If $f(a/b) = m$, then, for all $n > 0$,

$$nf(a/nb) = f(na/nb) = f(a/b) = m.$$

Thus, $m$ is divisible by every positive integer $n$. Therefore, $m = 0$, lest we contradict the Fundamental Theorem of Arithmetic, and so $f = 0$.

If we apply the functor $\operatorname{Hom}_{\mathbb{Z}}(\square, \mathbb{Z})$ to the short exact sequence

$$0 \to \mathbb{Z} \xrightarrow{i} \mathbb{Q} \xrightarrow{p} \mathbb{Q}/\mathbb{Z} \to 0,$$

where $i$ is the inclusion and $p$ is the natural map, then the induced map

$$i^*: \operatorname{Hom}_{\mathbb{Z}}(\mathbb{Q}, \mathbb{Z}) \to \operatorname{Hom}_{\mathbb{Z}}(\mathbb{Z}, \mathbb{Z})$$

cannot be surjective, for $\operatorname{Hom}_{\mathbb{Z}}(\mathbb{Q}, \mathbb{Z}) = \{0\}$ while $\operatorname{Hom}_{\mathbb{Z}}(\mathbb{Z}, \mathbb{Z}) \neq \{0\}$, because it contains $1_{\mathbb{Z}}$.  ◄

**Definition.** A contravariant functor $T: {}_R\mathbf{Mod} \to \mathbf{Ab}$ is called **left exact** if exactness of

$$A \xrightarrow{i} B \xrightarrow{p} C \to 0$$

implies exactness of

$$0 \to T(C) \xrightarrow{T(p)} T(B) \xrightarrow{T(i)} T(A).$$

Thus, Theorem 6.58 shows that contravariant Hom functors $\operatorname{Hom}_R(\square, Y)$ are left exact functors.[17]

Here is a converse of Theorem 6.58; a dual statement holds for covariant Hom functors.

**Proposition 6.60.** *Let $i: B' \to B$ and $p: B \to B''$ be $R$-maps, where $R$ is a commutative ring. If*

$$0 \to \operatorname{Hom}_R(B'', M) \xrightarrow{p^*} \operatorname{Hom}_R(B, M) \xrightarrow{i^*} \operatorname{Hom}_R(B', M)$$

*is an exact sequence for every $R$-module $M$, then so is*

$$B' \xrightarrow{i} B \xrightarrow{p} B'' \to 0.$$

**Proof.**

(i) $p$ is surjective.

Let $M = B''/\operatorname{im} p$ and let $f: B'' \to B''/\operatorname{im} p$ be the natural map, so that $f \in \operatorname{Hom}(B'', M)$. Then $p^*(f) = fp = 0$, so that $f = 0$, because $p^*$ is injective. Therefore, $B''/\operatorname{im} p = 0$, and $p$ is surjective.

---

[17]These functors are called *left exact* because the functored sequences have $0 \to$ on the left.

(ii) $\operatorname{im} i \subseteq \ker p$.

Since $i^* p^* = 0$, we have $0 = (pi)^*$. Hence, if $M = B''$ and $g = 1_{B''}$, so that $g \in \operatorname{Hom}(B'', M)$, then $0 = (pi)^* g = gpi = pi$, and so $\operatorname{im} i \subseteq \ker p$.

(iii) $\ker p \subseteq \operatorname{im} i$.

Now choose $M = B / \operatorname{im} i$ and let $h \colon B \to M$ be the natural map, so that $h \in \operatorname{Hom}(B, M)$. Clearly, $i^* h = hi = 0$, so that exactness of the Hom sequence gives an element $h' \in \operatorname{Hom}_R(B'', M)$ with $p^*(h') = h'p = h$. We have $\operatorname{im} i \subseteq \ker p$, by part (ii); hence, if $\operatorname{im} i \neq \ker p$, there is an element $b \in B$ with $b \notin \operatorname{im} i$ and $b \in \ker p$. Thus, $hb \neq 0$ and $pb = 0$, which gives the contradiction $hb = h'pb = 0$.  •

**Definition.** A covariant functor $T \colon {}_R\mathbf{Mod} \to \mathbf{Ab}$ is an **exact functor** if exactness of

$$0 \to A \xrightarrow{i} B \xrightarrow{p} C \to 0$$

implies exactness of

$$0 \to T(A) \xrightarrow{T(i)} T(B) \xrightarrow{T(p)} T(C) \to 0.$$

An exact contravariant functor is defined similarly.

In the next section, we will see that Hom functors are exact functors for certain choices of modules.

If $A$ and $B$ are left $R$-modules, then $\operatorname{Hom}_R(A, B)$ is an abelian group. However, if $R$ is a commutative ring, then Proposition 6.13(ii) says that $\operatorname{Hom}_R(A, B)$ is an $R$-module. We now show, for any ring $R$, that $\operatorname{Hom}_R(A, B)$ is a module if $A$ or $B$ has extra structure.

**Definition.** Let $R$ and $S$ be rings and let $M$ be an abelian group. Then $M$ is an $(R, S)$-**bimodule**, denoted by

$$_R M_S,$$

if $M$ is a left $R$-module, a right $S$-module, and the two scalar multiplications are related by an associative law:

$$r(ms) = (rm)s$$

for all $r \in R$, $m \in M$, and $s \in S$.

If $M$ is an $(R, S)$-bimodule, it is permissible to write $rms$ with no parentheses, for the definition of bimodule says that the two possible associations agree.

**Example 6.61.**

(i) Every ring $R$ is an $(R, R)$-bimodule; the extra identity is just the associativity of multiplication in $R$.

(ii) Every two-sided ideal in a ring $R$ is an $(R, R)$-bimodule.

(iii) If $M$ is a left $R$-module (i.e., if $M = {}_R M$), then $M$ is an $(R, \mathbb{Z})$-bimodule; that is, $M = {}_R M_{\mathbb{Z}}$. Similarly, a right $R$-module $N$ is a bimodule ${}_{\mathbb{Z}} N_R$.

(iv) If $R$ is commutative, then every left (or right) $R$-module is an $(R, R)$-bimodule. In more detail, if $M = {}_R M$, define a new scalar multiplication $M \times R \to M$ by $(m, r) \mapsto rm$. To see that $M$ is a right $R$-module, we must show that $m(rr') = (mr)r'$, that is, $(rr')m = r'(rm)$, and this is so because $rr' = r'r$. Finally, $M$ is an $(R, R)$-bimodule because both $r(mr')$ and $(rm)r'$ are equal to $(rr')m$.

(v) In Example 6.9(v), we made any left $kG$-module $M$ into a right $kG$-module by defining $mg = g^{-1}m$ for every $m \in M$ and every $g$ in the group $G$. Even though $M$ is both a left and right $kG$-module, it is usually not a $(kG, kG)$-bimodule because the required associativity formula may not hold. For example, let $G$ be a nonabelian group, and let $g, h \in G$ be noncommuting elements. If $m \in M$, then $g(mh) = g(h^{-1}m) = (gh^{-1})m$; on the other hand, $(gm)h = h^{-1}(gm) = (h^{-1}g)m$. In particular, if $M = kG$ and $m = 1$, then $g(1h) = gh^{-1}$, while $(g1)h = h^{-1}g$. Therefore, $g(1h) \neq (g1)h$, and $kG$ is not a $(kG, kG)$-bimodule. ◄

We now show that $\operatorname{Hom}_R(A, B)$ is a module when one of the modules $A$ and $B$ is also a bimodule. The reader should bookmark this page, for the following technical result will be used often.

**Proposition 6.62.** *Let $R$ and $S$ be rings.*

(i) *Let ${}_R A_S$ be a bimodule and ${}_R B$ be a left $R$-module. Then $\operatorname{Hom}_R(A, B)$ is a left $S$-module if we define $sf \colon a \mapsto f(as)$, and $\operatorname{Hom}_R(A, \square) \colon {}_R\mathbf{Mod} \to {}_S\mathbf{Mod}$ is a functor.*

(ii) *Let ${}_R A_S$ be a bimodule and $B_S$ be a right $S$-module. Then $\operatorname{Hom}_S(A, B)$ is a right $R$-module if we define $fr \colon a \mapsto f(ra)$, and $\operatorname{Hom}_S(A, \square) \colon \mathbf{Mod}_S \to \mathbf{Mod}_R$ is a functor.*

(iii) *Let ${}_S B_R$ be a bimodule and $A_R$ be a right $R$-module. Then $\operatorname{Hom}_R(A, B)$ is a left $S$-module if we define $sf \colon a \mapsto s[f(a)]$, and $\operatorname{Hom}_R(\square, B) \colon \mathbf{Mod}_R \to {}_S\mathbf{Mod}$ is a functor.*

(iv) *Let ${}_S B_R$ be a bimodule and ${}_S A$ be a left $S$-module. Then $\operatorname{Hom}_S(A, B)$ is a right $R$-module if we define $fr \colon a \mapsto f(a)r$, and $\operatorname{Hom}_S(A, \square) \colon {}_S\mathbf{Mod} \to \mathbf{Mod}_R$ is a functor.*

**Proof.** We only prove (i); the proofs of the other parts are left to the reader. First, $as$ makes sense because $A$ is a right $S$-module, and so $f(as)$ is defined. To see that $\operatorname{Hom}_R(A, B)$ is a left $S$-module, we compare $(ss')f$ and $s(s'f)$, where $s, s' \in S$ and $f \colon A \to B$. Now $(ss')f \colon a \mapsto f(a(ss'))$, while $s(s'f) \colon a \mapsto (s'f)(as) = f((as)s')$. But $a(ss') = (as)s'$ because $A$ is a bimodule.

To see that the functor $\operatorname{Hom}_R(A, \square)$ takes values in ${}_S\mathbf{Mod}$, we must show that if $g \colon B \to B'$ is an $R$-map, then $g_* \colon \operatorname{Hom}_R(A, B) \to \operatorname{Hom}_R(A, B')$, given by $f \mapsto gf$, is an $S$-map; that is, $g_*(sf) = s(g_*f)$ for all $s \in S$ and $f \colon A \to B$. Now $g_*(sf) \colon a \mapsto g((sf)a) = g(f(as))$, and $s(g_*f) \colon a \mapsto (g_*f)(as) = gf(as) = g(f(as))$, as desired. •

For example, every ring $R$ is a $(\mathbb{Z}, R)$-bimodule. Hence, for any abelian group $D$, Proposition 6.62(i) shows that $\operatorname{Hom}_{\mathbb{Z}}(R, D)$ is a left $R$-module.

**Remark.** Let $f\colon A \to B$ be an $R$-map. Suppose we write $f(a)$ when $A$ is a right $R$-module and $(a)f$ when $A$ is a left $R$-module (that is, write the function symbol $f$ on the side opposite the scalar action). With this notation, each of the four parts of this exercise is an associative law. For example, in part (i) with both $A$ and $B$ left $R$-modules, writing $sf$ for $s \in S$, we have $a(sf) = (as)f$. Similarly, in part (ii), we define $fr$, for $r \in R$ so that $(fr)a = f(ra)$. ◄

**Corollary 6.63.** *If $R$ is commutative and $A, B$ are $R$-modules, then $\operatorname{Hom}_R(A, B)$ is an $R$-module if we define $rf\colon a \mapsto f(ra)$. In this case, $\operatorname{Hom}_R(A, \Box)\colon {}_R\mathbf{Mod} \to {}_R\mathbf{Mod}$ and $\operatorname{Hom}_R(\Box, B)\colon {}_R\mathbf{Mod} \to {}_R\mathbf{Mod}$ are functors.*

**Proof.** When $R$ is commutative, Example 6.61(iv) shows that $R$-modules are $(R, R)$-bimodules. Thus, Proposition 6.62 generalizes Proposition 6.13(ii). •

**Corollary 6.64.** *If $R$ is a ring and $M$ is a left $R$-module, then $\operatorname{Hom}_R(R, M)$ is a left $R$-module and*

$$\varphi_M\colon \operatorname{Hom}_R(R, M) \to M,$$

*given by $\varphi_M\colon f \mapsto f(1)$, is an $R$-isomorphism.*

**Proof.** Since $R$ is an $(R, R)$-bimodule, Proposition 6.62(i) says that $\operatorname{Hom}_R(R, M)$ is a left $R$-module if scalar multiplication $R \times \operatorname{Hom}_R(R, M) \to \operatorname{Hom}_R(R, M)$ is defined by $(r, f) \mapsto f_r$, where $f_r(a) = f(ar)$ for all $a \in R$.

It is easy to check that $\varphi_M$ is an additive function. To see that $\varphi_M$ is an $R$-homomorphism, note that

$$\varphi_M(rf) = (rf)(1) = f(1r) = f(r) = r[f(1)] = r\varphi_M(f),$$

because $f$ is an $R$-map. Consider the function $M \to \operatorname{Hom}_R(R, M)$ defined as follows: if $m \in M$, then $f_m\colon R \to M$ is given by $f_m(r) = rm$; it is easy to see that $f_m$ is an $R$-homomorphism, and that $m \mapsto f_m$ is the inverse of $\varphi_M$. •

In the presence of bimodules, the $\mathbb{Z}$-isomorphisms in Theorem 6.45 are module isomorphisms.

**Theorem 6.65.**

(i) *If ${}_R A_S$ is a bimodule and $(B_i)_{i \in I}$ is a family of left $R$-modules, then the $\mathbb{Z}$-isomorphism*

$$\varphi\colon \operatorname{Hom}_R\left(A, \prod_{i \in I} B_i\right) \cong \prod_{i \in I} \operatorname{Hom}_R(A, B_i),$$

*given by $\varphi\colon f \mapsto (p_i f)$ ($p_i$ are the projections of the product $\prod_{i \in I} B_i$), is an $S$-isomorphism.*

(ii) *If ${}_R A_S$ is a bimodule and $B, B'$ are left $R$-modules, then the $\mathbb{Z}$-isomorphism*

$$\operatorname{Hom}_R(A, B \oplus B') \cong \operatorname{Hom}_R(A, B) \oplus \operatorname{Hom}_R(A, B')$$

*is an $S$-isomorphism.*

(iii) *If $R$ is commutative, $A$ is an $R$-module, and $(B_i)_{i \in I}$ is a family of $R$-modules, then*

$$\varphi \colon \operatorname{Hom}_R\Big(A, \prod_{i \in I} B_i\Big) \cong \prod_{i \in I} \operatorname{Hom}_R(A, B_i)$$

*is an $R$-isomorphism.*

(iv) *If $R$ is commutative and $A, B, B'$ are $R$-modules, then the $\mathbb{Z}$-isomorphism*

$$\operatorname{Hom}_R(A, B \oplus B') \cong \operatorname{Hom}_R(A, B) \oplus \operatorname{Hom}_R(A, B')$$

*is an $R$-isomorphism.*

**Remark.** There is a similar result involving the isomorphism

$$\varphi \colon \operatorname{Hom}_R\Big(\bigoplus_{i \in I} A_i, B\Big) \cong \prod_{i \in I} \operatorname{Hom}_R(A_i, B). \quad \blacktriangleleft$$

**Proof.** To prove (i), we must show that $\varphi(sf) = s\varphi(f)$ for all $s \in S$ and $f \colon A \to \prod B_i$. Now $\varphi(sf) = (p_i(sf))$, the $I$-tuple whose $i$th coordinate is $p_i(sf)$. On the other hand, since $S$ acts coordinatewise on an $I$-tuple $(g_i)$ by $s(g_i) = (sg_i)$, we have $s\varphi(f) = (s(p_if))$. Thus, we must show that $p_i(sf) = s(p_if)$ for all $i$. Note that both of these are maps $A \to B_i$. If $a \in A$, then $p_i(sf) \colon a \mapsto p_i[(sf)(a)] = p_i(f(as))$, and $s(p_if) \colon a \mapsto (p_if)(as) = p_i(f(as))$, as desired.

Part (ii) is a special case of (i): when the index set if finite, direct sum and direct product of modules are equal. Parts (iii) and (iv) are special cases of (i) and (ii), for all $R$-modules are $(R, R)$-bimodules when $R$ is commutative. •

**Example 6.66.**

(i) A ***linear functional*** on a vector space $V$ over a field $k$ is a linear transformation $\varphi \colon V \to k$ [after all, $k$ is a (one-dimensional) vector space over itself]. For example, if

$$V = \{\text{continuous } f \colon [0, 1] \to \mathbb{R}\},$$

then integration, $f \mapsto \int_0^1 f(t)\, dt$, is a linear functional on $V$. Recall that if $V$ is a vector space over a field $k$, then its ***dual space*** is

$$V^* = \operatorname{Hom}_k(V, k).$$

By Corollary 6.63, $V^*$ is also a $k$-module; that is, $V^*$ is a vector space over $k$.

If $\dim(V) = n < \infty$, then Example 4.7 shows that $V = V_1 \oplus \cdots \oplus V_n$, where each $V_i$ is one-dimensional. By Theorem 6.65(iii), $V^* \cong \bigoplus_i \operatorname{Hom}_k(V_i, k)$ is a direct sum of $n$ one-dimensional spaces [for Corollary 6.64 gives $\operatorname{Hom}_k(k, k) \cong k$], and so $\dim(V^*) = \dim(V) = n$, by Exercise 6.48. Therefore, a finite-dimensional vector space and its dual space are isomorphic. It follows that the double dual, $V^{**}$, defined as $(V^*)^*$, is isomorphic to $V$ when $V$ is finite-dimensional.

(ii) There are variations of dual spaces. In Functional Analysis, one encounters topological real vector spaces $V$, so that it makes sense to speak of *continuous* linear functionals. The *topological dual* $V^*$ consists of all the continuous linear functionals, and it is important to know whether a space $V$ is *reflexive*; that is, whether the analog of the isomorphism $V \to V^{**}$ for finite-dimensional spaces is a homeomorphism for these spaces. For example, that *Hilbert space is reflexive* is one of its important properties.  ◄

## Exercises

∗ **6.47.** If $M$ is a finitely generated abelian group, prove that the additive group of the ring $\text{End}(M)$ is a finitely generated abelian group.

**Hint.** There is a finitely generated free abelian group $F$ mapping onto $M$; apply $\text{Hom}(\square, M)$ to $F \to M \to 0$ to obtain an injection $0 \to \text{Hom}(M, M) \to \text{Hom}(F, M)$. But $\text{Hom}(F, M)$ is a finite direct sum of copies of $M$.

∗ **6.48.** Let $v_1, \ldots, v_n$ be a basis of a vector space $V$ over a field $k$, so that every $v \in V$ has a unique expression $v = a_1 v_1 + \cdots + a_n v_n$, where $a_i \in k$ for $i = 1, \ldots, n$. Recall Exercise 2.83 on page 154. For each $i$, the function $v_i^*: V \to k$, defined by $v_i^*: v \mapsto a_i$, lies in the dual space $V^*$, and the list $v_1^*, \ldots, v_n^*$ is a basis of $V^*$ (called the **dual basis** of $v_1, \ldots, v_n$).

If $f: V \to V$ is a linear transformation, let $A$ be the matrix of $f$ with respect to a basis $v_1, \ldots, v_n$ of $V$; that is, the $i$th column of $A$ consists of the coordinate list of $f(v_i)$ with respect to the given basis. Prove that the matrix of the induced map $f^*: V^* \to V^*$ with respect to the dual basis is $A^\top$, the transpose of $A$.

∗ **6.49.** If $T: {}_R\mathbf{Mod} \to \mathbf{Ab}$ is an additive functor, of either variance, prove that $T(\{0\}) = \{0\}$ and that $T(0) = 0$, where $0$ denotes a zero morphism.

∗ **6.50.** Give an example of a covariant functor that does not preserve coproducts.

**6.51.** Let $\mathcal{A} \xrightarrow{S} \mathcal{B} \xrightarrow{T} \mathcal{C}$ be functors. Prove that the composite $\mathcal{A} \xrightarrow{TS} \mathcal{C}$ is a functor that is covariant if the variances of $S$ and $T$ are the same, and contravariant if the variances of $S$ and $T$ are different.

**6.52.** Define $F: \mathbf{ComRings} \to \mathbf{ComRings}$ on objects by $F(R) = R[x]$, and on ring homomorphisms $\varphi: R \to S$ by $F(\varphi): \sum_i a_i x^i \mapsto \sum_i \varphi(a_i) x^i$. Prove that $F$ is a functor.

**6.53.** Prove that there is a functor $\mathbf{Groups} \to \mathbf{Ab}$ taking each group $G$ to $G/G'$, where $G'$ is its commutator subgroup.

∗ **6.54.**  (i) If $X$ is a set and $k$ is a field, define the vector space $k^X$ to be the set of all functions $X \to k$ under pointwise operations. Prove that there is a functor $F: \mathbf{Sets} \to {}_k\mathbf{Mod}$ with $F(X) = k^X$.

  (ii) If $X$ is a set, define $F(X)$ to be the free group with basis $X$. Prove that there is a functor $F: \mathbf{Sets} \to \mathbf{Groups}$ with $F: X \mapsto F(X)$.

**6.55.** Let $R$ be a ring, and let $M, N$ be right $R$-modules. If $f \in \text{Hom}_R(M, N)$ and $r \in R$, define $rf: M \to N$ by $rf: m \mapsto f(mr)$.

  (i) Prove that if $r, s \in R$, then $(rs)f = r(sf)$ for all $f \in \text{Hom}_R(M, N)$.

  (ii) Show that $\text{Hom}_R(M, N)$ need not be a left $R$-module.

∗ **6.56.** (**Change of Rings**). Let $k, k^*$ be commutative rings, let $\varphi: k \to k^*$ be a ring homomorphism, and let $M$ be a left $k^*$-module.

  (i) Prove that $M^*$ is a $k$-module, denoted by ${}_\varphi M^*$ and called an **induced module**, if we define $rm^* = \varphi(r)m^*$ for all $r \in k$ and $m^* \in M^*$.

(ii) Prove that every $k^*$-map $f^*\colon M^* \to N^*$ is a $k$-map $_\varphi M^* \to _\varphi N^*$.

(iii) Use parts (i) and (ii) to prove that $\varphi$ induces an additive exact functor

$$_\varphi\square\colon {}_{k^*}\mathbf{Mod} \to {}_k\mathbf{Mod}$$

with $M^* \mapsto {}_\varphi M^*$. We call $_\varphi\square$ a *change of rings functor*.

## Section 6.4. Free and Projective Modules

The simplest modules are *free modules* and, as for abelian groups, every module is a quotient of a free module; that is, every module has a presentation by generators and relations. Projective modules are generalizations of free modules, and they, too, turn out to be useful.

**Definition.** A left $R$-module $F$ is called a *free* left $R$-module if $F$ is isomorphic to a direct sum of copies of $R$: that is, there is a (possibly infinite) index set $I$ with

$$F = \bigoplus_{i \in I} R_i,$$

where $R_i = \langle b_i \rangle \cong R$ for all $i$. We call $B = (b_i)_{i \in I}$ a *basis* of $F$.

A free $\mathbb{Z}$-module is a free abelian group (the free abelian groups defined in Chapter 4 are special cases of free $\mathbb{Z}$-modules, namely, those having finite bases). Every ring $R$, when considered as a left module over itself, is a free left $R$-module (with basis the one-point set $\{1\}$).

From our discussion of direct sums, we know that each $m \in F$ has a unique expression of the form

$$m = \sum_{i \in I} r_i b_i,$$

where $r_i \in R$ and almost all $r_i = 0$. A basis of a free module has a strong resemblance to a basis of a vector space. Indeed, it is easy to see that a vector space $V$ over a field $k$ is a free $k$-module and the two notions of basis coincide in this case.

Here is a generalization of Theorem 2.120 from finite-dimensional vector spaces to arbitrary free modules (in particular, to infinite-dimensional vector spaces).

**Proposition 6.67.** *Let $F$ be a free left $R$-module with basis $B$. For every left $R$-module $M$ and every function $\gamma\colon B \to M$, there exists a unique $R$-map $g\colon F \to M$ with $g(b) = \gamma(b)$ for all $b \in B$.*

$$
\begin{array}{ccc}
F & & \\
\uparrow & \searrow{\scriptstyle g} & \\
B & \xrightarrow{\ \gamma\ } & M
\end{array}
$$

**Proof.** Every element $v \in F$ has a unique expression of the form $v = \sum_{b \in B} r_b b$, where $r_b \in R$ and almost all $r_b = 0$. Define $g\colon F \to M$ by $g(v) = \sum_{b \in B} r_b \gamma(b)$. To prove uniqueness, suppose that $\theta\colon F \to M$ is an $R$-map with $\theta(b) = \gamma(b)$ for all $b \in B$. Thus, the maps $\theta$ and $g$ agree on a generating set $B$, and so $\theta = g$. $\quad\bullet$

**Proposition 6.68.** *If $R$ is a nonzero commutative ring, then any two bases of a free $R$-module $F$ have the same cardinality.*

**Proof.** Choose a maximal ideal $I$ in $R$ (which exists, by Theorem 5.43). If $B$ is a basis of the free $R$-module $F$, then Exercise 6.5 on page 415 shows that the set of cosets $(b + IF)_{b \in B}$ is a basis of the vector space $F/IF$ over the field $R/I$. If $Y$ is another basis of $F$, then the same argument gives $(y + IF)_{y \in Y}$, a basis of $F/IF$. But any two bases of a vector space have the same size (which is the dimension of the space), and so $|B| = |Y|$, by Theorem 5.52. •

**Definition.** If $F$ is a free $k$-module, where $k$ is a commutative ring, then the number of elements in a basis is called the **rank** of $F$.

Proposition 6.68 shows that the *rank* of free modules over commutative rings is well-defined. Of course, rank is the analog of dimension.

**Corollary 6.69.** *If $R$ is a nonzero commutative ring, then free $R$-modules $F$ and $F'$ are isomorphic if and only if* $\mathrm{rank}(F) = \mathrm{rank}(F')$.

**Proof.** Suppose that $\varphi \colon F \to F'$ is an isomorphism. If $B$ is a basis of $F$, then it is easy to see that $\varphi(B)$ is a basis of $F'$. But any two bases of the free module $F'$ have the same size, namely, $\mathrm{rank}(F')$, by Proposition 6.68. Hence, $\mathrm{rank}(F') = \mathrm{rank}(F)$.

Conversely, let $B$ be a basis of $F$, let $B'$ be a basis of $F'$, and let $\gamma \colon B \to B'$ be a bijection. Composing $\gamma$ with the inclusion $B' \to F'$, we may assume that $\gamma \colon B \to F'$. By Proposition 6.67, there is a unique $R$-map $\varphi \colon F \to F'$ extending $\gamma$. Similarly, we may regard $\gamma^{-1} \colon B' \to B$ as a function $B' \to F$, and there is a unique $\psi \colon F' \to F$ extending $\gamma^{-1}$. Finally, both $\psi\varphi$ and $1_F$ extend $1_B$, so that $\psi\varphi = 1_F$. Similarly, the other composite is $1_{F'}$, and so $\varphi \colon F \to F'$ is an isomorphism. (The astute reader will notice a strong resemblance of this proof to the uniqueness of a solution to a universal mapping problem.) •

**Definition.** We say that a ring $R$ has **IBN** (invariant basis number) if $R^m \cong R^n$ implies $m = n$ for all $m, n \in \mathbb{N}$.

Thus, every commutative ring has IBN. It can be shown that rank is well-defined for free left $R$-modules when $R$ is *left noetherian*; that is, if every left ideal in $R$ is finitely generated (Rotman, *An Introduction to Homological Algebra*, p. 113). However, there do exist noncommutative rings $R$ such that $R \cong R \oplus R$ as left $R$-modules [Ibid., p. 58; if $V$ is an infinite-dimensional vector space over a field $k$, then $R = \mathrm{End}_k(V)$ is such a ring], and so the notion of rank is not always defined. The reason the proof of Proposition 6.68(i) fails for noncommutative rings $R$ is that $R/I$ need not be a division ring if $I$ is a maximal two-sided ideal (Exercise 6.59 on page 459). The next proposition will enable us to use free modules to describe arbitrary modules.

**Proposition 6.70.** *Every left $R$-module $M$ is a quotient of a free left $R$-module $F$. Moreover, $M$ is finitely generated if and only if $F$ can be chosen to be finitely generated.*

**Proof.** Let $F$ be the direct sum of $|M|$ copies of $R$ (so $F$ is a free left $R$-module), and let $(x_m)_{m \in M}$ be a basis of $F$. By Proposition 6.67, there is an $R$-map $g \colon F \to M$ with $g(x_m) = m$ for all $m \in M$. Obviously, $g$ is a surjection, and so $F/\ker g \cong M$.

If $M$ is finitely generated, then $M = \langle m_1, \ldots, m_n \rangle$. If we choose $F$ to be the free left $R$-module with basis $\{x_1, \ldots, x_n\}$, then the map $g \colon F \to M$ with $g(x_i) = m_i$ is a surjection, for

$$\operatorname{im} g = \langle g(x_1), \ldots, g(x_n) \rangle = \langle m_1, \ldots, m_n \rangle = M.$$

The converse is obvious, for any image of a finitely generated module is itself finitely generated  •

The last proposition can be used to construct modules with prescribed properties. For example, let us consider $\mathbb{Z}$-modules (i.e., abelian groups). The group $\mathbb{Q}/\mathbb{Z}$ contains an element $a$ of order 2 satisfying the equations $a = 2^n a_n$ for all $n \geq 1$; take $a = \frac{1}{2} + \mathbb{Z}$ and $a_n = (\frac{1}{2})^{n+1} + \mathbb{Z}$. Of course, $\operatorname{Hom}_{\mathbb{Z}}(\mathbb{Q}, \mathbb{Q}/\mathbb{Z}) \neq \{0\}$ because it contains the natural map. Is there an abelian group $G$ with $\operatorname{Hom}_{\mathbb{Z}}(\mathbb{Q}, G) = \{0\}$ that contains an element $a$ of order 2 satisfying the equations $a = 2^n a_n$ for all $n \geq 1$? Let $F$ be the free abelian group with basis

$$\{a, b_1, b_2, \ldots, b_n, \ldots\}$$

and relations

$$\{2a, a - 2^n b_n, n \geq 1\};$$

that is, let $K$ be the subgroup of $F$ generated by $\{2a, a - 2^n b_n, n \geq 1\}$. Exercise 6.67 on page 459 asks the reader to verify that $G = F/K$ satisfies the desired properties. This construction is a special case of defining an $R$-module by *generators and relations* (we have done this for nonabelian groups in Chapter 4).

**Definition.** Let $X = (x_i)_{i \in I}$ be a basis of a free left $R$-module $F$, and let $Y = (\sum_i r_{ji} x_i)_{j \in J}$ be a subset of $F$. If $K$ is the submodule of $F$ generated by $Y$, then we say that the module $M = F/K$ has **generators** $X$ and **relations** $Y$. We also say that the ordered pair $(X \mid Y)$ is a **presentation** of $M$.

We will return to presentations at the end of this section, but let us now focus on the key property of bases, Lemma 6.67 (which holds for free modules as well as for vector spaces), in order to get a theorem about free modules that does not mention bases.

**Theorem 6.71.** *If $R$ is a ring and $F$ is a free left $R$-module, then for every surjection $p \colon A \to A''$ and each $h \colon F \to A''$, there exists a homomorphism $g$ making the following diagram commute:*

$$
\begin{array}{ccc}
 & F & \\
{\scriptstyle g}\swarrow & \downarrow {\scriptstyle h} & \\
A \xrightarrow{\ p\ } & A'' \longrightarrow & 0
\end{array}
$$

**Proof.** Let $B = (b_i)_{i \in I}$ be a basis of $F$. Since $p$ is surjective, there is $a_i \in A$ with $p(a_i) = h(b_i)$ for all $i$. There is an $R$-map $g \colon F \to A$ with $g(b_i) = a_i$ for all $i$, by

Proposition 6.67. Now $pg(b_i) = p(a_i) = h(b_i)$, so that $pg$ agrees with $h$ on the basis $B$; it follows that $pg = h$ on $\langle B \rangle = F$; that is, $pg = h$. •

**Definition.** We call a map $g \colon F \to A$ with $pg = h$ (in the diagram in Theorem 6.71) a *lifting* of $h$.

If $C$ is any, not necessarily free, module, then a lifting $g$ of $h$, should one exist, need not be unique. Since $pi = 0$, where $i \colon \ker p \to A$ is the inclusion, other liftings are $g + if$ for any $f \in \mathrm{Hom}_R(C, \ker p)$. Indeed, this is obvious from the exact sequence

$$0 \to \mathrm{Hom}(C, \ker p) \xrightarrow{\; i_* \;} \mathrm{Hom}(C, A) \xrightarrow{\; p_* \;} \mathrm{Hom}(C, A'').$$

Any two liftings of $h$ differ by a map in $\ker p_* = \mathrm{im}\, i_* \subseteq \mathrm{Hom}(C, A)$.

We now promote this (basis-free) property of free modules to a definition.

**Definition.** A left $R$-module $P$ is *projective* if, whenever $p$ is surjective and $h$ is any map, there exists a lifting $g$; that is, there exists a map $g$ making the following diagram commute:

$$
\begin{array}{ccc}
 & & P \\
 & {}^{g}\swarrow & \downarrow h \\
A & \xrightarrow{\; p \;} A'' & \longrightarrow 0
\end{array}
$$

**Remark.** The definition of projective module can be generalized to define a projective object in more general categories if we can translate *surjection* into the language of categories. For example, if we define surjections in **Groups** to be the usual surjections, then we can define projectives there. Exercise 6.57 on page 459 says that a group $G$ is projective in **Groups** if and only if it is a free group. ◀

We know that every free left $R$-module is projective; is the converse true? Is every projective $R$-module free? We shall see that the answer depends on the ring $R$. Note that if projective left $R$-modules happen to be free, then free modules are characterized without having to refer to a basis.

Let us now see that projective modules arise in a natural way. We know that the Hom functors are left exact; that is, for any module $P$, applying $\mathrm{Hom}_R(P, \square)$ to an exact sequence

$$0 \to A' \xrightarrow{\; i \;} A \xrightarrow{\; p \;} A''$$

gives an exact sequence

$$0 \to \mathrm{Hom}_R(P, A') \xrightarrow{\; i_* \;} \mathrm{Hom}_R(P, A) \xrightarrow{\; p_* \;} \mathrm{Hom}_R(P, A'').$$

**Proposition 6.72.** *A left $R$-module $P$ is projective if and only if $\mathrm{Hom}_R(P, \square)$ is an exact functor.*

**Remark.** Since $\mathrm{Hom}_R(P, \square)$ is a left exact functor, the thrust of the proposition is that $p_*$ is surjective whenever $p$ is surjective. ◀

**Proof.** If $P$ is projective, then given $h\colon P \to A''$, there exists a lifting $g\colon P \to A$ with $pg = h$. Thus, if $h \in \operatorname{Hom}_R(P, A'')$, then $h = pg = p_*(g) \in \operatorname{im} p_*$, and so $p_*$ is surjective. Hence, $\operatorname{Hom}(P, \square)$ is an exact functor.

For the converse, assume that $\operatorname{Hom}(P, \square)$ is an exact functor, so that $p_*$ is surjective: if $h \in \operatorname{Hom}_R(P, A'')$, there exists $g \in \operatorname{Hom}_R(P, A)$ with $h = p_*(g) = pg$. This says that given $p$ and $h$, there exists a lifting $g$ making the diagram commute; that is, $P$ is projective.  •

**Proposition 6.73.** *A left $R$-module $P$ is projective if and only if every short exact sequence $0 \to A \xrightarrow{i} B \xrightarrow{p} P \to 0$ is split.*

**Proof.** If $P$ is projective, then there exists $j\colon P \to B$ making the following diagram commute; that is, $pj = 1_P$.

$$
\begin{array}{ccc}
 & & P \\
 & {\scriptstyle j}\swarrow & \downarrow{\scriptstyle 1_P} \\
B & \xrightarrow{\ p\ } & P \xrightarrow{\quad} 0
\end{array}
$$

Corollary 6.28 now gives the result.

Conversely, assume that every short exact sequence ending with $P$ splits. Consider the left-hand diagram below with $p$ surjective. Now form the pullback. By

Exercise 6.41 on page 434, surjectivity of $p$ in the pullback diagram gives surjectivity of $\alpha$. By hypothesis, there is a map $j\colon P \to D$ with $\alpha j = 1_P$. Define $g\colon P \to B$ by $g = \beta j$. We check: $pg = p\beta j = f\alpha j = f1_P = f$. Therefore, $P$ is projective.  •

We restate one half of this proposition so that the word *exact* is not mentioned.

**Corollary 6.74.** *Let $A$ be a submodule of a module $B$. If $B/A$ is projective, then $A$ has a **complement**: there is a submodule $C$ of $B$ with $C \cong B/A$ and $B = A \oplus C$.*

**Proposition 6.75.**

   (i) *If $(P_i)_{i \in I}$ is a family of projective left $R$-modules, then their direct sum $\bigoplus_{i \in I} P_i$ is also projective.*

   (ii) *Every direct summand $S$ of a projective module $P$ is projective.*

**Proof.**

   (i) Consider the left-hand diagram below. If $\alpha_j\colon P_j \to \bigoplus P_i$ is an injection of the direct sum, then $h\alpha_j$ is a map $P_j \to C$, and so projectivity of $P_j$ gives a map $g_j\colon P_j \to B$ with $pg_j = h\alpha_j$. Since $\bigoplus P_i$ is a coproduct, there is a map

$\theta\colon \bigoplus P_i \to B$ with $\theta\alpha_j = g_j$ for all $j$. Hence, $p\theta\alpha_j = pg_j = h\alpha_j$ for all $j$, and so $p\theta = h$. Therefore, $\bigoplus P_i$ is projective.

(ii) Suppose that $S$ is a direct summand of a projective module $P$, so there are maps $q\colon P \to S$ and $i\colon S \to P$ with $qi = 1_S$. Now consider the diagram

$$
\begin{array}{ccc}
P & \underset{i}{\overset{q}{\rightleftarrows}} & S \\
{\scriptstyle h}\big\downarrow & \,\,\swarrow\,{\scriptstyle g} & \big\downarrow{\scriptstyle f} \\
B & \xrightarrow{\,\,p\,\,} & C \longrightarrow 0
\end{array}
$$

where $p$ is surjective. The composite $fq$ is a map $P \to C$; since $P$ is projective, there is a map $h\colon P \to B$ with $ph = fq$. Define $g\colon S \to B$ by $g = hi$. It remains to prove that $pg = f$. But $pg = phi = fqi = f1_S = f$.  •

**Theorem 6.76.** *A left $R$-module $P$ is projective if and only if it is a direct summand of a free left $R$-module.*

**Proof.** Sufficiency follows from Proposition 6.75, for free modules are projective, and every direct summand of a projective is itself projective.

Conversely, assume that $P$ is projective. By Proposition 6.70, every module is a quotient of a free module. Thus, there is a free module $F$ and a surjection $g\colon F \to P$, and so there is an exact sequence $0 \to \ker g \to F \xrightarrow{g} P \to 0$. Proposition 6.73 now shows that $P$ is a direct summand of $F$.  •

Theorem 6.76 gives another proof of Proposition 6.75. To prove (i), note that if $P_i$ is projective, then there are $Q_i$ with $P_i \oplus Q_i = F_i$, where $F_i$ is free. Thus,

$$\bigoplus_i (P_i \oplus Q_i) = \bigoplus_i P_i \oplus \bigoplus_i Q_i = \bigoplus_i F_i.$$

But, obviously, a direct sum of free modules is free. To prove (ii), note that if $P$ is projective, then there is a module $Q$ with $P \oplus Q = F$. If $S \oplus T = P$, then $S \oplus (T \oplus Q) = P \oplus Q = F$.

We can now give an example of a (commutative) ring $R$ and a projective $R$-module that is not free.

**Example 6.77.** The ring $R = \mathbb{I}_6$ is the direct sum of two ideals:

$$\mathbb{I}_6 = J \oplus I,$$

where $J = \{[0], [2], [4]\} \cong \mathbb{I}_3$ and $I = \{[0], [3]\} \cong \mathbb{I}_2$. Now $\mathbb{I}_6$ is a free module over itself, and so $J$ and $I$, being direct summands of a free module, are projective $\mathbb{I}_6$-modules. Neither $J$ nor $I$ can be free, however. After all, a (finitely generated) free $\mathbb{I}_6$-module $F$ is a direct sum of, say, $n$ copies of $\mathbb{I}_6$, and so $F$ has $6^n$ elements. Therefore, $J$ is too small to be free, for it has only three elements. ◄

Describing projective $R$-modules is a problem very much dependent on the ring $R$. In Chapter 8, for example, we will prove that if $R$ is a PID, then every submodule of a free module is itself free. It will then follow from Theorem 6.76 that every projective $R$-module is free in this case. A much harder result is that if $R = k[x_1, \dots, x_n]$ is the polynomial ring in $n$ variables over a field $k$, then every projective $R$-module is also free; this theorem, implicitly conjectured[18] by Serre, was proved, independently, by Quillen and by Suslin (see Rotman, *An Introduction to Homological Algebra*, pp. 203–211, for a proof). There is a proof of the Quillen–Suslin Theorem using Gröbner bases, due to Fitchas and Galligo, Nullstellensatz effectif et conjecture de Serre (théorème de Quillen–Suslin) pour le calcul formel, *Math. Nachr.* 149 (1990), 231–253.

There are domains having projective modules that are not free. For example, if $R$ is the ring of all the algebraic integers in an *algebraic number field $E$* (that is, $E/\mathbb{Q}$ is a field extension of finite degree), then every ideal in $R$ is a projective $R$-module. There are such rings $R$ that are not PIDs, and any ideal in $R$ that is not principal is a projective module that is not free (we will see this in Chapter 10 when we discuss *Dedekind rings*).

Here is another characterization of projective modules. Note that if $A$ is a free left $R$-module with basis $(a_i)_{i \in I}$, then each $x \in A$ has a unique expression $x = \sum_{i \in I} r_i a_i$, and so there are $R$-maps $\varphi_i \colon A \to R$ given by $\varphi_i \colon x \mapsto r_i$.

**Proposition 6.78.** *A left $R$-module $A$ is projective if and only if there exist elements $(a_i)_{i \in I}$ in $A$ and $R$-maps $(\varphi_i \colon A \to R)_{i \in I}$ such that*

(i) *for each $x \in A$, almost all $\varphi_i(x) = 0$;*

(ii) *for each $x \in A$, we have $x = \sum_{i \in I} (\varphi_i x) a_i$.*

*Moreover, $A$ is generated by $(a_i)_{i \in I}$ in this case.*

**Proof.** If $A$ is projective, there is a free left $R$-module $F$ and a surjective $R$-map $\psi \colon F \to A$. Since $A$ is projective, there is an $R$-map $\varphi \colon A \to F$ with $\psi \varphi = 1_A$, by Proposition 6.73. Let $(e_i)_{i \in I}$ be a basis of $F$, and define $a_i = \psi(e_i)$. Now if $x \in A$, then there is a unique expression $\varphi(x) = \sum_i r_i e_i$, where $r_i \in R$ and almost all $r_i = 0$. Define $\varphi_i \colon A \to R$ by $\varphi_i(x) = r_i$. Of course, given $x$, we have $\varphi_i(x) = 0$ for almost all $i$. Since $\psi$ is surjective, $A$ is generated by $\big(a_i = \psi(e_i)\big)_{i \in I}$. Finally,

$$x = \psi \varphi(x) = \psi\Big(\sum r_i e_i\Big) = \sum r_i \psi(e_i) = \sum (\varphi_i x) \psi(e_i) = \sum (\varphi_i x) a_i.$$

Conversely, given $(a_i)_{i \in I} \subseteq A$ and a family of $R$-maps $(\varphi_i \colon A \to R)_{i \in I}$ as in the statement, define $F$ to be the free left $R$-module with basis $(e_i)_{i \in I}$, and define an $R$-map $\psi \colon F \to A$ by $\psi \colon e_i \mapsto a_i$. It suffices to find an $R$-map $\varphi \colon A \to F$ with $\psi \varphi = 1_A$, for then $A$ is (isomorphic to) a retract (i.e., $A$ is a direct summand of

---

[18]On page 243 of "Faisceaux Algèbriques Cohérents," *Annals of Mathematics* 61 (1955), 197–278, Serre writes "... on ignore s'il existe des $A$-modules projectifs de type fini qui ne soient pas libres." Here, $A = k[x_1, \dots, x_n]$.

$F$), and hence $A$ is projective. Define $\varphi$ by $\varphi(x) = \sum_i(\varphi_i x)e_i$, for $x \in A$. The sum is finite, by condition (i), and so $\varphi$ is well-defined. By condition (ii),

$$\psi\varphi(x) = \psi \sum (\varphi_i x)e_i = \sum (\varphi_i x)\psi(e_i) = \sum (\varphi_i x)a_i = x;$$

that is, $\psi\varphi = 1_A$. •

**Definition.** If $A$ is a left $R$-module, then a subset $(a_i)_{i \in I}$ of $A$ and a family of $R$-maps $(\varphi_i : A \to R)_{i \in I}$ satisfying the conditions in Proposition 6.78 is called a **projective basis**.

An interesting application of projective bases is a proof of a result of Bkouche. Let $X$ be a locally compact Hausdorff space, let $C(X)$ be the ring of all continuous real-valued functions on $X$, and let $J$ be the ideal in $C(X)$ consisting of all such functions having compact support. Then $X$ is a paracompact space if and only if $J$ is a projective $C(X)$-module [Finney–Rotman, Paracompactness of locally compact Hausdorff spaces, *Michigan Math. J.* 17 (1970), 359–361].

Let us return to presentations of modules.

**Definition.** A left $R$-module $M$ is **finitely presented** if it has a presentation $(X \mid Y)$ in which both $X$ and $Y$ are finite.

If a left $R$-module $M$ is finitely presented, there is a short exact sequence

$$0 \to K \to F \to M \to 0,$$

where $F$ is free and both $K$ and $F$ are finitely generated. Equivalently, $M$ is finitely presented if there is an exact sequence

$$F' \to F \to M \to 0,$$

where both $F'$ and $F$ are finitely generated free modules (just map a finitely generated free module $F'$ onto $K$). Note that the second exact sequence does not begin with "$0 \to$."

**Proposition 6.79.** *If $R$ is a left noetherian ring, then every finitely generated left $R$-module $M$ is finitely presented.*

**Proof.** There is a surjection $\varphi : F \to M$, where $F$ is a finitely generated free left $R$-module. Since $R$ is left noetherian, Proposition 6.36 says that every submodule of $F$ is finitely generated. In particular, $\ker\varphi$ is finitely generated, and so $M$ is finitely presented. •

Every finitely presented left $R$-module is finitely generated, but we will soon see that the converse may be false. We begin by comparing two presentations of a module (we generalize a bit by replacing free modules with projectives).

**Proposition 6.80 (Schanuel's Lemma).** *Given exact sequences of left $R$-modules*

$$0 \to K \xrightarrow{i} P \xrightarrow{\pi} M \to 0$$

*and*

$$0 \to K' \xrightarrow{i'} P' \xrightarrow{\pi'} M \to 0,$$

*where $P$ and $P'$ are projective, there is an $R$-isomorphism*

$$K \oplus P' \cong K' \oplus P.$$

**Proof.** Consider the diagram with exact rows

$$
\begin{array}{ccccccccc}
0 & \longrightarrow & K & \xrightarrow{\ i\ } & P & \xrightarrow{\ \pi\ } & M & \longrightarrow & 0 \\
 & & \downarrow{\scriptstyle\alpha} & & \downarrow{\scriptstyle\beta} & & \downarrow{\scriptstyle 1_M} & & \\
0 & \longrightarrow & K' & \xrightarrow[\ i'\ ]{} & P' & \xrightarrow[\ \pi'\ ]{} & M & \longrightarrow & 0
\end{array}
$$

Since $P$ is projective, there is a map $\beta \colon P \to P'$ with $\pi'\beta = \pi$; that is, the right square in the diagram commutes. We now show that there is a map $\alpha \colon K \to K'$ making the other square commute. If $x \in K$, then $\pi'\beta ix = \pi i x = 0$, because $\pi i = 0$. Hence, $\beta i x \in \ker \pi' = \operatorname{im} i'$; thus, there is $x' \in K'$ with $i'x' = \beta i x$; moreover, $x'$ is unique because $i'$ is injective. Therefore, $\alpha \colon x \mapsto x'$ is a well-defined function $\alpha \colon K \to K'$ that makes the first square commute. The reader can show that $\alpha$ is an $R$-map. Consider the sequence

$$0 \to K \xrightarrow{\theta} P \oplus K' \xrightarrow{\psi} P' \to 0,$$

where $\theta \colon x \mapsto (ix, \alpha x)$ and $\psi \colon (u, x') \mapsto \beta u - i'x'$, for $x \in K$, $u \in P$, and $x' \in K'$. This sequence is exact; the straightforward calculation, using commutativity of the diagram and exactness of its rows, is left to the reader. But this sequence splits, because $P'$ is projective, so that $P \oplus K' \cong K \oplus P'$. •

**Corollary 6.81.** *If $M$ is a finitely presented left $R$-module and*

$$0 \to K \to F \to M \to 0$$

*is an exact sequence, where $F$ is a finitely generated free left $R$-module, then $K$ is finitely generated.*

**Proof.** Since $M$ is finitely presented, there is an exact sequence

$$0 \to K' \to F' \to M \to 0$$

with $F'$ free and with both $F'$ and $K'$ finitely generated. By Schanuel's Lemma, $K \oplus F' \cong K' \oplus F$. Now $K' \oplus F$ is finitely generated because both summands are, so that the left side is also finitely generated. But $K$, being a summand, is also a homomorphic image of $K \oplus F'$, and hence it is finitely generated. •

We can now give an example of a finitely generated module that is not finitely presented.

**Example 6.82.** Let $R$ be a commutative ring that is not noetherian; that is, $R$ contains an ideal $I$ that is not finitely generated (Example 5.34). We claim that the $R$-module $M = R/I$ is finitely generated but not finitely presented. Of course, $M$ is finitely generated; it is even cyclic. If $M$ were finitely presented, then there would be an exact sequence $0 \to K \to F \to M \to 0$ with $F$ free and both $K$ and $F$ finitely generated. Comparing this with the exact sequence $0 \to I \to R \to M \to 0$, as in Corollary 6.81, gives $I$ finitely generated, a contradiction. Therefore, $M$ is not finitely presented. ◀

## Exercises

∗ **6.57.** Prove that a group $G$ is projective in **Groups** if and only if $G$ is a free group. (Projectives in **Groups** are defined if we decree that *surjections* in this category are the usual surjections.)

∗ **6.58.** Let $R$ be a ring and let $S$ be a nonzero submodule of a free right $R$-module. Prove that if $a \in R$ is not a right zero-divisor (i.e., there is no nonzero $b \in R$ with $ba = 0$), then $Sa \neq \{0\}$.

∗ **6.59.** (i) If $k$ is a field, prove that the only two-sided ideals in $\mathrm{Mat}_2(k)$ are $(0)$ and the whole ring.

  (ii) Let $p$ be a prime and let $\varphi \colon \mathrm{Mat}_2(\mathbb{Z}) \to \mathrm{Mat}_2(\mathbb{F}_p)$ be the ring homomorphism which reduces entries mod $p$. Prove that $\ker \varphi$ is a maximal two-sided ideal in $\mathrm{Mat}_2(\mathbb{Z})$ and that $\mathrm{im}\, \varphi$ is not a division ring.

∗ **6.60.** (i) Prove that if a ring $R$ has IBN, then so does $R/I$ for every proper two-sided ideal $I$.

  (ii) If $F_\infty$ is the free abelian group with basis $(x_j)_{j \geq 0}$, prove that $\mathrm{End}(F_\infty)$ is isomorphic to the ring of all column-finite $\aleph_0 \times \aleph_0$ matrices with entries in $\mathbb{Z}$.

  (iii) Prove that $\mathrm{End}(F_\infty)$ does not have IBN.
  **Hint.** You may use the fact that $\mathrm{End}_k(V)$ does not have IBN, where $V$ is an infinite-dimensional vector space over a field $k$.

**6.61.** Let $M$ be a free $R$-module, where $R$ is a domain. Prove that if $rm = 0$, where $r \in R$ and $m \in M$, then either $r = 0$ or $m = 0$. (This is false if $R$ is not a domain.)

**6.62.** Use left exactness of Hom to prove that if $G$ is an abelian group, then $\mathrm{Hom}_{\mathbb{Z}}(\mathbb{I}_n, G) \cong G[n]$, where $G[n] = \{g \in G : ng = 0\}$.

∗ **6.63.** If $R$ is a domain but not a field and $Q = \mathrm{Frac}(R)$, prove that $\mathrm{Hom}_R(Q, R) = \{0\}$.

**6.64.** Prove that every left exact covariant functor $T \colon {}_R\mathbf{Mod} \to \mathbf{Ab}$ preserves pullbacks. Conclude that if $B$ and $C$ are submodules of a module $A$, then for every module $M$, we have
$$\mathrm{Hom}_R(M, B \cap C) = \mathrm{Hom}_R(M, B) \cap \mathrm{Hom}_R(M, C).$$

**6.65.** Given a set $X$, prove that there exists a free $R$-module $F$ with a basis $B$ for which there is a bijection $\varphi \colon B \to X$.

∗ **6.66.** (i) Prove that every vector space $V$ over a field $k$ is a free $k$-module.

  (ii) Prove that a subset $B$ of $V$ is a basis of $V$ considered as a vector space if and only if $B$ is a basis of $V$ considered as a free $k$-module.

∗ **6.67.** Define $G$ to be the abelian group having the presentation $(X \mid Y)$, where
$$X = \{a, b_1, b_2, \ldots, b_n, \ldots\} \quad \text{and} \quad Y = \{2a, a - 2^n b_n, n \geq 1\}.$$
Thus, $G = F/K$, where $F$ is the free abelian group with basis $X$ and $K = \langle Y \rangle$.

  (i) Prove that $a + K \in G$ is nonzero.

  (ii) Prove that $z = a + K$ satisfies equations $z = 2^n y_n$, where $y_n \in G$ and $n \geq 1$, and that $z$ is the unique such element of $G$.

  (iii) Prove that there is an exact sequence $0 \to \langle a \rangle \to G \to \bigoplus_{n \geq 1} \mathbb{I}_{2^n} \to 0$.

(iv) Prove that $\mathrm{Hom}_{\mathbb{Z}}(\mathbb{Q}, G) = \{0\}$ by applying $\mathrm{Hom}_{\mathbb{Z}}(\mathbb{Q}, \square)$ to the exact sequence in part (iii).

**6.68.** (i) If $R$ is a domain and $I$ and $J$ are nonzero ideals in $R$, prove that $I \cap J \neq (0)$.

(ii) Let $R$ be a domain and let $I$ be an ideal in $R$ that is a free $R$-module; prove that $I$ is a principal ideal.

∗ **6.69.** Let $\varphi \colon B \to C$ be an $R$-map of left $R$-modules.

(i) Prove that $\varphi$ is injective if and only if $\varphi$ can be canceled from the left; that is, for all modules $A$ and all maps $f, g \colon A \to B$, we have $\varphi f = \varphi g$ implies $f = g$:

$$A \underset{g}{\overset{f}{\rightrightarrows}} B \xrightarrow{\varphi} C.$$

(ii) Prove that $\varphi$ is surjective if and only if $\varphi$ can be canceled from the right; that is, for all $R$-modules $D$ and all $R$-maps $h, k \colon C \to D$, we have $h\varphi = k\varphi$ implies $h = k$:

$$B \xrightarrow{\varphi} C \underset{k}{\overset{h}{\rightrightarrows}} D.$$

∗ **6.70. (Eilenberg–Moore)** Let $G$ be a (possibly nonabelian) group.

(i) If $H$ is a proper subgroup of a group $G$, prove that there exists a group $L$ and distinct homomorphisms $f, g \colon G \to L$ with $f|H = g|H$.

**Hint.** Define $L = S_X$, where $X$ denotes the family of all the left cosets of $H$ in $G$ together with an additional element, denoted $\infty$. If $a \in G$, define $f(a) = f_a \in S_X$ by $f_a(\infty) = \infty$ and $f_a(bH) = abH$. Define $g \colon G \to S_X$ by $g = \gamma \circ f$, where $\gamma \in S_X$ is conjugation by the transposition $(H, \infty)$.

(ii) If $A$ and $G$ are groups, prove that a homomorphism $\varphi \colon A \to G$ is surjective if and only if $\varphi$ can be canceled from the right; that is, for all groups $L$ and all maps $f, g \colon G \to L$, we have $f\varphi = g\varphi$ implies $f = g$:

$$B \xrightarrow{\varphi} G \underset{g}{\overset{f}{\rightrightarrows}} L.$$

## Section 6.5. Injective Modules

There is another type of module that also turns out to be interesting. Injective modules are duals of projective modules in that both of these terms are characterized by diagrams, and the diagram for injectivity is obtained from the diagram for projectivity by reversing all arrows. The basic reason injective modules are interesting will not be seen until Chapter 9, when we discuss Homological Algebra. In the meantime, we will see here that injective $\mathbb{Z}$-modules are quite familiar.

**Definition.** A left $R$-module $E$ is **injective** if $\mathrm{Hom}_R(\square, E)$ is an exact contravariant functor.

We will give examples of injective modules after we establish some of their properties. Of course, $E = \{0\}$ is injective.

The next proposition is the dual of Proposition 6.72.

**Proposition 6.83.** *A left R-module $E$ is injective if and only if, given any map $f\colon A \to E$ and an injection $i\colon A \to B$, there exists $g\colon B \to E$ making the following diagram commute:*

$$
\begin{array}{ccc}
 & & E \\
 & \nearrow f \quad \nwarrow g & \\
0 \longrightarrow A & \xrightarrow{\ i\ } & B
\end{array}
$$

**Remark.** In words, homomorphisms from a submodule into $E$ can always be extended to homomorphisms from the big module into $E$.

Since the contravariant functor $\mathrm{Hom}_R(\square, E)$ is left exact for any module $E$, the thrust of the proposition is that $i^*$ is surjective whenever $i$ is an injection. ◄

**Proof.** If $E$ is an injective left $R$-module, then $\mathrm{Hom}_R(\square, E)$ is an exact functor, so that $i^*$ is surjective. Therefore, if $f \in \mathrm{Hom}_R(A, E)$, there exists $g \in \mathrm{Hom}_R(B, E)$ with $f = i^*(g) = gi$; that is, the diagram commutes.

For the converse, if $E$ satisfies the diagram condition, then given $f\colon A \to E$, there exists $g\colon B \to E$ with $gi = f$. Thus, if $f \in \mathrm{Hom}_R(A, E)$, then $f = gi = i^*(g) \in \mathrm{im}\, i^*$, and so $i^*$ is surjective. Hence, $\mathrm{Hom}(\square, E)$ is an exact functor, and so $E$ is injective. •

The next result is the dual of Proposition 6.73.

**Proposition 6.84.** *A left R-module $E$ is injective if and only if every short exact sequence $0 \to E \xrightarrow{i} B \xrightarrow{p} C \to 0$ splits.*

**Proof.** If $E$ is injective, then there exists $q\colon B \to E$ making the following diagram commute; that is, $qi = 1_E$:

$$
\begin{array}{ccc}
 & & E \\
 & \nearrow 1_E \quad \nwarrow q & \\
0 \longrightarrow E & \xrightarrow{\ i\ } & B
\end{array}
$$

Exercise 6.32 on page 418 now gives the result.

Conversely, assume every exact sequence beginning with $E$ splits. The pushout of the left-hand diagram below is the right-hand diagram. By Exercise 6.41 on

page 434, the map $\alpha$ is an injection, so that $0 \to E \to D \to \mathrm{coker}\,\alpha \to 0$ splits; that is, there is $q\colon D \to E$ with $q\alpha = 1_E$. If we define $g\colon B \to E$ by $g = q\beta$, then the original diagram commutes: $gi = q\beta i = q\alpha f = 1_E f = f$. Therefore, $E$ is injective. •

This proposition can be restated without mentioning the word *exact*.

**Corollary 6.85.** *If an injective left $R$-module $E$ is a submodule of a left $R$-module $M$, then $E$ is a direct summand of $M$: there is a submodule $S$ of $M$ with $M = E \oplus S$.*

**Proposition 6.86.** *Every direct summand of an injective module $E$ is injective.*

**Proof.** Suppose that $S$ is a direct summand of an injective module $E$, so there are maps $q: E \to S$ and $i: S \to E$ with $qi = 1_S$. Now consider the diagram

$$
\begin{array}{ccc}
S & \underset{q}{\overset{i}{\rightleftarrows}} & E \\
{\scriptstyle f}\big\uparrow & {\scriptstyle g} & \big\uparrow {\scriptstyle h} \\
0 \longrightarrow A & \underset{j}{\longrightarrow} & B
\end{array}
$$

where $j$ is injective. The composite $if$ is a map $A \to E$; since $E$ is injective, there is a map $h: B \to E$ with $hj = if$. Define $g: B \to S$ by $g = qh$. It remains to prove that $gj = f$. But $gj = qhj = qif = 1_S f = f$.  •

**Proposition 6.87.** *Let $(E_i)_{i \in I}$ be a family of left $R$-modules. Then $\prod_{i \in I} E_i$ is injective if and only if each $E_i$ is injective.*

**Proof.** Consider the diagram

$$
\begin{array}{c}
E \\
{\scriptstyle f}\big\uparrow \\
0 \longrightarrow A \underset{\kappa}{\longrightarrow} B
\end{array}
$$

where $E = \prod E_i$ and $\kappa: A \to B$ is an injection. Let $p_i: E \to E_i$ be the $i$th projection. Since $E_i$ is injective, there is $g_i: B \to E_i$ with $g_i \kappa = p_i f$. Now define $g: B \to E$ by $g: b \mapsto (g_i(b))$. The map $g$ extends $f$, for if $b = \kappa a$, then $g(\kappa a) = (g_i(\kappa a)) = (p_i f a) = f a$, because $x = (p_i x)$ is true for every $x$ in the product.

The converse follows from Proposition 6.86.  •

**Corollary 6.88.** *A finite direct sum of injective left $R$-modules is injective.*

**Proof.** The direct sum of finitely many modules is their direct product.  •

The following theorem is very useful.

**Theorem 6.89 (Baer Criterion).** *A left $R$-module $E$ is injective if and only if every $R$-map $f: I \to E$, where $I$ is a left ideal in $R$, can be extended to $R$.*

$$
\begin{array}{c}
E \\
{\scriptstyle f}\big\uparrow \quad {\scriptstyle g} \\
0 \longrightarrow I \underset{i}{\longrightarrow} R
\end{array}
$$

**Proof.** Since left ideals $I$ are submodules of $R$, the existence of extensions $g$ of $f$ is just a special case of the definition of injectivity of $E$.

Consider the diagram with exact row

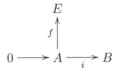

For notational convenience, let us assume that $i$ is the inclusion [this assumption amounts to permitting us to write $a$ instead of $i(a)$ whenever $a \in A$]. As in the proof of Lemma 5.60, we are going to use Zorn's Lemma on approximations to an extension of $f$. More precisely, let $X$ be the set of all ordered pairs $(A', g')$, where $A \subseteq A' \subseteq B$ and $g' \colon A' \to E$ extends $f$; that is, $g'|A = f$. Note that $X \neq \varnothing$ because $(A, f) \in X$. Partially order $X$ by defining

$$(A', g') \preceq (A'', g'')$$

to mean $A' \subseteq A''$ and $g''$ extends $g'$. The reader may supply the argument that Zorn's Lemma applies, and so there exists a maximal element $(A_0, g_0)$ in $X$. If $A_0 = B$, we are done, and so we may assume that there is some $b \in B$ with $b \notin A_0$.

Define
$$I = \{r \in R \colon rb \in A_0\}.$$
It is easy to see that $I$ is an ideal in $R$. Define $h \colon I \to E$ by
$$h(r) = g_0(rb).$$
By hypothesis, there is a map $h^* \colon R \to E$ extending $h$. Now define $A_1 = A_0 + \langle b \rangle$ and $g_1 \colon A_1 \to E$ by
$$g_1(a_0 + rb) = g_0(a_0) + rh^*(1),$$
where $a_0 \in A_0$ and $r \in R$.

Let us show that $g_1$ is well-defined. If $a_0 + rb = a_0' + r'b$, then $(r - r')b = a_0' - a_0 \in A_0$; it follows that $r - r' \in I$. Therefore, $g_0((r - r')b)$ and $h(r - r')$ are defined, and we have
$$g_0(a_0' - a_0) = g_0((r - r')b) = h(r - r') = h^*(r - r') = (r - r')h^*(1).$$
Thus, $g_0(a_0') - g_0(a_0) = rh^*(1) - r'h^*(1)$ and $g_0(a_0') + r'h^*(1) = g_0(a_0) + rh^*(1)$, as desired. Clearly, $g_1(a_0) = g_0(a_0)$ for all $a_0 \in A_0$, so that the map $g_1$ extends $g_0$. We conclude that $(A_0, g_0) \prec (A_1, g_1)$, contradicting the maximality of $(A_0, g_0)$. Therefore, $A_0 = B$, the map $g_0$ is a lifting of $f$, and $E$ is injective. $\bullet$

We have not yet presented any nonzero examples of injective modules, but we can now give some.

**Proposition 6.90.** *Let $R$ be a domain and let $Q = \mathrm{Frac}(R)$.*

   (i) *If $f \colon I \to Q$ is an $R$-map, where $I$ is an ideal in $R$, then there is $c \in Q$ with $f(a) = ca$ for all $a \in I$.*

   (ii) *$Q$ is an injective $R$-module.*

   (iii) *If $g \colon Q \to Q$ is an $R$-map, there is $c \in Q$ with $g(x) = cx$ for all $x \in Q$.*

**Proof.**

(i) If $a, b \in I$ are nonzero, then $f(ab)$ is defined (because $I$ is an ideal) and $af(b) = f(ab) = bf(a)$ (because $f$ is an $R$-map). Hence,

$$f(a)/a = f(b)/b.$$

If $c \in Q$ denotes their common value, then $f(a)/a = c$ and $f(a) = ca$ for all $a \in I$.

(ii) By the Baer Criterion, it suffices to extend an $R$-map $f \colon I \to Q$, where $I$ is an ideal in $R$, to all of $R$. By (i), there is $c \in Q$ with $f(a) = ca$ for all $a \in I$; define $g \colon R \to Q$ by

$$g(r) = cr$$

for all $r \in R$. It is obvious that $g$ is an $R$-map extending $f$, and so $Q$ is an injective $R$-module.

(iii) Let $g \colon Q \to Q$ be an $R$-map, and let $f = g|R \colon R \to Q$. By (i) with $I = R$, there is $c \in Q$ with $f(a) = g(a) = ca$ for all $a \in R$. Now if $x \in Q$, then $x = a/b$ for $a, b \in R$. Hence, $bx = a$ and $g(bx) = g(a)$. But $g(bx) = bg(x)$, because $g$ is an $R$-map. Therefore, $g(x) = ca/b = cx$.  •

**Definition.** Let $R$ be a domain. Then an $R$-module $D$ is **divisible** if, for each $d \in D$ and nonzero $r \in R$, there exists $d' \in D$ with $d = rd'$.

**Example 6.91.** Let $R$ be a domain.

(i) $\mathrm{Frac}(R)$ is a divisible $R$-module.

(ii) Every direct sum of divisible $R$-modules is divisible. Hence, every vector space over $\mathrm{Frac}(R)$ is a divisible $R$-module.

(iii) Every quotient of a divisible $R$-module is divisible.  ◀

**Lemma 6.92.** *If $R$ is a domain, then every injective $R$-module $E$ is divisible.*

**Proof.** Assume that $E$ is injective. Let $e \in E$ and let $r_0 \in R$ be nonzero; we must find $x \in E$ with $e = r_0 x$. Define $f \colon (r_0) \to E$ by $f(rr_0) = re$ (note that $f$ is well-defined: since $R$ is a domain, $rr_0 = r'r_0$ implies $r = r'$). Since $E$ is injective, there exists $h \colon R \to E$ extending $f$. In particular,

$$e = f(r_0) = h(r_0) = r_0 h(1),$$

so that $x = h(1)$ is the element in $E$ required by the definition of divisible.  •

We now prove the converse of Lemma 6.92 for PIDs.

**Corollary 6.93.** *If $R$ is a PID, then an $R$-module $E$ is injective if and only if it is divisible.*

**Proof.** Assume that $E$ is divisible. By the Baer criterion, Theorem 6.89, it suffices to extend maps $f \colon I \to E$ to all of $R$. Since $R$ is a PID, $I$ is principal; say, $I = (r_0)$ for some $r_0 \in I$. Since $E$ is divisible, there exists $e \in E$ with $r_0 e = f(r_0)$. Define $h \colon R \to E$ by $h(r) = re$. It is easy to see that $h$ is an $R$-map extending $f$, and so $E$ is injective.  •

**Remark.** Proposition 10.123 says that a domain is a *Dedekind ring* if and only if every divisible module is injective. Hence, if $R$ is a domain that is not Dedekind, then there exist injective $R$-modules having quotients that are not injective. ◄

**Example 6.94.** In light of Example 6.91, the following abelian groups are injective $\mathbb{Z}$-modules:

$$\mathbb{Q}, \quad \mathbb{R}, \quad \mathbb{C}, \quad \mathbb{Q}/\mathbb{Z}, \quad \mathbb{R}/\mathbb{Z}, \quad S^1,$$

where $S^1$ is the circle group; that is, the multiplicative group of all complex numbers $z$ with $|z| = 1$. ◄

Proposition 6.70 says, for any ring $R$, that every left $R$-module is a quotient of a projective left $R$-module (actually, it is a stronger result: every module is a quotient of a free left $R$-module).

**Corollary 6.95.** *Every abelian group $M$ can be imbedded as a subgroup of some injective abelian group.*

**Proof.** By Proposition 6.70, there is a free abelian group $F = \bigoplus_i \mathbb{Z}_i$ with $M = F/K$ for some $K \subseteq F$. Now

$$M = F/K = \Big(\bigoplus_i \mathbb{Z}_i\Big)/K \subseteq \Big(\bigoplus_i \mathbb{Q}_i\Big)/K,$$

where we have merely imbedded each copy $\mathbb{Z}_i$ of $\mathbb{Z}$ into a copy $\mathbb{Q}_i$ of $\mathbb{Q}$. But Example 6.91 gives divisibility of each $\mathbb{Q}_i$, of $\bigoplus_i \mathbb{Q}_i$, and of the quotient $(\bigoplus_i \mathbb{Q}_i)/K$. By Corollary 6.93, $(\bigoplus_i \mathbb{Q}_i)/K$ is injective. •

Writing a module as a quotient of a free module is the essence of describing it by generators and relations. We may think of Corollary 6.95 as dualizing this idea for abelian groups. The next theorem generalizes this corollary to left $R$-modules for any ring $R$, but its proof uses Proposition 6.138 (which follows from the Adjoint Isomorphism): if $R$ is a ring and $D$ is a divisible abelian group, then $\mathrm{Hom}_{\mathbb{Z}}(R, D)$ (which is a left $R$-module, as we noted on page 447) is an injective left $R$-module.

**Theorem 6.96.** *For every ring $R$, every left $R$-module $M$ can be imbedded as a submodule of some injective left $R$-module.*

**Proof.** If we regard $M$ as an abelian group, then Corollary 6.95 says that there is a divisible abelian group $D$ and an injective $\mathbb{Z}$-map $j : M \to D$. For $m \in M$, the function $f_m : r \mapsto j(rm)$ lies in $\mathrm{Hom}_{\mathbb{Z}}(R, D)$, and it is easy to see that $\varphi : m \mapsto f_m$ is an injective $R$-map $M \to \mathrm{Hom}_{\mathbb{Z}}(R, D)$ [recall that $\mathrm{Hom}_{\mathbb{Z}}(R, D)$ is a left $R$-module with scalar multiplication defined by $sf : R \to D$, where $sf : r \mapsto f(rs)$]. This completes the proof, for $\mathrm{Hom}_{\mathbb{Z}}(R, D)$ is an injective left $R$-module, by Proposition 6.138. •

This last theorem can be improved, for there is a smallest injective module containing any given module, called its ***injective envelope*** (Rotman, *An Introduction to Homological Algebra*, p. 127).

**Proposition 6.97.**

(i) *If $R$ is a left noetherian ring and $(E_i)_{i \in I}$ is a family of injective $R$-modules, then $\bigoplus_{i \in I} E_i$ is an injective $R$-module.*

(ii) **(Bass–Papp)** *If $R$ is a ring for which every direct sum of injective left $R$-modules is injective, then $R$ is left noetherian.*

**Proof.**

(i) By the Baer criterion, Theorem 6.89, it suffices to complete the diagram

where $J$ is an ideal in $R$. Since $R$ is noetherian, $J$ is finitely generated, say, $J = (a_1, \ldots, a_n)$. For $k = 1, \ldots, n$, $f(a_k) \in \bigoplus_{i \in I} E_i$ has only finitely many nonzero coordinates, occurring, say, at indices in $S(a_k) \subseteq I$. Thus, $S = \bigcup_{k=1}^{n} S(a_k)$ is a finite set, and so $\operatorname{im} f \subseteq \bigoplus_{i \in S} E_i$; by Corollary 6.88, this finite sum is injective. Hence, there is an $R$-map $g' : R \to \bigoplus_{i \in S} E_i$ extending $f$. Composing $g'$ with the inclusion of $\bigoplus_{i \in S} E_i$ into $\bigoplus_{i \in I} E_i$ completes the given diagram.

(ii) We show that if $R$ is not left noetherian, then there is a left ideal $I$ and an $R$-map to a sum of injectives that cannot be extended to $R$. Since $R$ is not left noetherian, there is a strictly ascending chain of left ideals $I_1 \subsetneq I_2 \subsetneq \cdots$; let $I = \bigcup I_n$. We note that $I/I_n \neq \{0\}$ for all $n$. By Theorem 6.96, we may imbed $I/I_n$ in an injective left $R$-module $E_n$; we claim that $E = \bigoplus_n E_n$ is not injective.

Let $\pi_n : I \to I/I_n$ be the natural map. For each $a \in I$, note that $\pi_n(a) = 0$ for large $n$ (because $a \in I_n$ for some $n$), and so the $R$-map $f : I \to \prod(I/I_n)$, defined by

$$f : a \mapsto (\pi_n(a)),$$

does have its image in $\bigoplus_n (I/I_n)$; that is, for each $a \in I$, almost all the coordinates of $f(a)$ are 0. Composing with the inclusion $\bigoplus(I/I_n) \to \bigoplus E_n = E$, we may regard $f$ as a map $I \to E$. If there is an $R$-map $g : R \to E$ extending $f$, then $g(1)$ is defined; say, $g(1) = (x_n)$. Choose an index $m$ and choose $a \in I$ with $a \notin I_m$; since $a \notin I_m$, we have $\pi_m(a) \neq 0$, and so $g(a) = f(a)$ has nonzero $m$th coordinate $\pi_m(a)$. But $g(a) = ag(1) = a(x_n) = (ax_n)$, so that $\pi_m(a) = ax_m$. It follows that $x_m \neq 0$ for all $m$, and this contradicts $g(1)$ lying in the direct sum $E = \bigoplus E_n$.  •

The next result gives a curious example of an injective module; we use it to give another proof of the Basis Theorem of Finite Abelian Groups.

**Proposition 6.98.** *Let $R$ be a PID, let $a \in R$ be neither zero nor a unit, and let $J = (a)$. Then $R/J$ is an injective $R/J$-module.*

**Proof.** By the Correspondence Theorem, every ideal in $R/J$ has the form $I/J$ for some ideal $I$ in $R$ containing $J$. Now $I = (b)$ for some $b \in I$, so that $I/J$ is cyclic with generator $x = b + J$. Since $(a) \subseteq (b)$, we have $a = rb$ for some $r \in R$. We are going to use the Baer criterion, Theorem 6.89, to prove that $R/J$ is injective.

Assume that $f : I/J \to R/J$ is an $R/J$-map, and write $f(b+J) = s+J$ for some $s \in R$. Since $r(b+J) = rb+J = a+J = 0$, we have $rf(b+J) = r(s+J) = rs+J = 0$, and so $rs \in J = (a)$. Hence, there is some $r' \in R$ with $rs = r'a = r'br$; canceling $r$ gives $s = r'b$. Thus,
$$f(b + J) = s + J = r'b + J.$$
Define $h : R/J \to R/J$ to be multiplication by $r'$; that is, $h : u + J \mapsto r'u + J$. The displayed equation gives $h(b + J) = f(b + J)$, so that $h$ does extend $f$. Therefore, $R/J$ is injective. $\bullet$

**Corollary 6.99 (Basis Theorem).** *Every finite abelian group $G$ is a direct sum of cyclic groups.*

**Proof.** By the Primary Decomposition, we may assume that $G$ is a $p$-primary group for some prime $p$. If $p^n$ is the largest order of elements in $G$, then $p^n g = 0$ for all $g \in G$, and so $G$ is an $\mathbb{I}_{p^n}$-module. If $x \in G$ has order $p^n$, then $S = \langle x \rangle \cong \mathbb{I}_{p^n}$. Hence, $S$ is injective, by Proposition 6.98. But injective submodules are always direct summands, and so $G = S \oplus T$ for some submodule $T$.[19] By induction on $|G|$, the complement $T$ is a direct sum of cyclic groups. $\bullet$

---

## Exercises

* **6.71.** Recall that an abelian group $A$ is *torsion-free* if it has no nonzero elements of finite order. Prove that the following conditions are equivalent for an abelian group $A$.

  (i) $A$ is torsion-free and divisible;

  (ii) $A$ a vector space over $\mathbb{Q}$;

  (iii) for every positive integer $n$, the multiplication map $\mu_n : A \to A$, given by $a \mapsto na$, is an isomorphism.

* **6.72.** (i) Prove that a left $R$-module $E$ is injective if and only if, for every left ideal $I$ in $R$, every short exact sequence $0 \to E \to B \to I \to 0$ of left $R$-modules splits.

  (ii) If $R$ is a domain, prove that torsion-free divisible $R$-modules are injective.

**6.73.** Prove the dual of Schanuel's Lemma. Given exact sequences
$$0 \to M \xrightarrow{i} E \xrightarrow{p} Q \to 0 \text{ and } 0 \to M \xrightarrow{i'} E' \xrightarrow{p'} Q' \to 0,$$
where $E$ and $E'$ are injective, then there is an isomorphism $Q \oplus E' \cong Q' \oplus E$.

**6.74.** (i) Prove that every vector space over a field $k$ is an injective $k$-module.

  (ii) Prove that if $0 \to U \to V \to W \to 0$ is an exact sequence of vector spaces, then the corresponding sequence of dual spaces $0 \to W^* \to V^* \to U^* \to 0$ is also exact.

**6.75.** (i) Prove that if a domain $R$ is an injective $R$-module, then $R$ is a field.

---

[19]Lemma 4.22 gives another proof of this fact.

(ii) Let $R$ be a domain that is not a field, and let $M$ be an $R$-module that is both injective and projective. Prove that $M = \{0\}$.

(iii) Prove that $\mathbb{I}_6$ is simultaneously an injective and a projective module over itself.

\* **6.76.** Prove that every torsion-free abelian group $A$ can be imbedded as a subgroup of a vector space over $\mathbb{Q}$.

**Hint.** Imbed $A$ in a divisible abelian group $D$, and show that $A \cap tD = \{0\}$, where $tD = \{d \in D : d \text{ has finite order}\}$.

\* **6.77.** Let $A$ and $B$ be abelian groups and let $\mu \colon A \to A$ be the multiplication map $a \mapsto na$.

(i) Prove that the induced maps

$$\mu_* \colon \operatorname{Hom}_{\mathbb{Z}}(A, B) \to \operatorname{Hom}_{\mathbb{Z}}(A, B) \quad \text{and} \quad \mu^* \colon \operatorname{Hom}_{\mathbb{Z}}(B, A) \to \operatorname{Hom}_{\mathbb{Z}}(B, A)$$

are also multiplication by $n$.

(ii) Prove that $\operatorname{Hom}_{\mathbb{Z}}(\mathbb{Q}, A)$ and $\operatorname{Hom}_{\mathbb{Z}}(A, \mathbb{Q})$ are vector spaces over $\mathbb{Q}$.

**6.78.** Give an example of two injective submodules of a module whose intersection is not injective.

**Hint.** Define abelian groups $A \cong \mathbb{Z}(p^\infty) \cong A'$:

$$A = (a_n, n \geq 0 | pa_0 = 0, pa_{n+1} = a_n) \text{ and } A' = (a'_n, n \geq 0 | pa'_0 = 0, pa'_{n+1} = a'_n).$$

In $A \oplus A'$, define $E = A \oplus \{0\}$ and $E' = \langle \{(a_{n+1}, a'_n) : n \geq 0\} \rangle$.

\* **6.79.** (**Pontrjagin Duality**) If $G$ is an abelian group, its ***Pontrjagin dual*** is the group

$$G^* = \operatorname{Hom}_{\mathbb{Z}}(G, \mathbb{Q}/\mathbb{Z}).$$

(Pontrjagin duality extends to locally compact abelian topological groups $G$, and the dual $G^*$ consists of all continuous homomorphisms $G \to \mathbb{R}/\mathbb{Z}$.)

(i) Prove that if $G$ is an abelian group and $a \in G$ is nonzero, then there is a homomorphism $f \colon G \to \mathbb{Q}/\mathbb{Z}$ with $f(a) \neq 0$.

(ii) Prove that $\mathbb{Q}/\mathbb{Z}$ is an injective abelian group.

(iii) Prove that if $0 \to A \to G \to B \to 0$ is an exact sequence of abelian groups, then so is $0 \to B^* \to G^* \to A^* \to 0$.

(iv) If $G \cong \mathbb{I}_n$, prove that $G^* \cong G$.

(v) If $G$ is a finite abelian group, prove that $G^* \cong G$.

(vi) Prove that if $G$ is a finite abelian group and $G/H$ is a quotient group of $G$, then $G/H$ is isomorphic to a subgroup of $G$. [The analogous statement for nonabelian groups is false: if $\mathbf{Q}$ is the group of quaternions, then $\mathbf{Q}/Z(\mathbf{Q}) \cong \mathbf{V}$, where $\mathbf{V}$ is the four-group; but $\mathbf{Q}$ has only one element of order 2 while $\mathbf{V}$ has three elements of order 2. This exercise is also false for infinite abelian groups: since $\mathbb{Z}$ has no element of order 2, it has no subgroup isomorphic to $\mathbb{Z}/2\mathbb{Z} \cong \mathbb{I}_2$.]

## Section 6.6. Tensor Products

One of the most compelling reasons to study *tensor products* comes from Algebraic Topology. We assign to every topological space $X$ a sequence of *homology groups*, $H_n(X)$ for $n \geq 0$, that are of basic importance. The *Künneth formula* computes the homology groups of the cartesian product $X \times Y$ of two topological spaces in terms of the tensor product of the homology groups of the factors $X$ and $Y$. Tensor products are also useful in many areas of Algebra. For example, they are involved in bilinear forms, the Adjoint Isomorphism, free algebras, exterior algebra, and determinants. They are especially interesting in Representation Theory (as we shall see in the next chapter), where they are used to construct *induced representations* (which extend representations of subgroups to representations of the whole group).

Consider the following general problem: if $A$ is a subring of a ring $R$, can we construct an $R$-module from an $A$-module $M$? Here is a naive approach. If $M$ is generated as an $A$-module by a set $X$, each $m \in M$ has an expression of the form $m = \sum_i a_i x_i$, where $a_i \in A$ and $x_i \in X$. Perhaps we can construct an $R$-module containing $M$ by taking all expressions of the form $\sum_i r_i x_i$, where $r_i \in R$. This simple idea is doomed to failure. For example, a cyclic group $G = \langle g \rangle$ of finite order $n$ is a $\mathbb{Z}$-module; can we make it into a $\mathbb{Q}$-module? A $\mathbb{Q}$-module $V$ is a vector space over $\mathbb{Q}$, and it is easy to see, when $v \in V$ and $q \in \mathbb{Q}$, that $qv = 0$ if and only if $q = 0$ or $v = 0$. If we could create a rational vector space $V$ containing $G$ in the naive way just described, then $ng = 0$ would imply $g = 0$ in $V$! Our idea of adjoining scalars to obtain a module over a larger ring still has merit but, plainly, we cannot be so cavalier about its construction. The proper way to deal with such matters is to use tensor products.

**Definition.** Let $R$ be a ring, let $A_R$ be a right $R$-module, let $_R B$ be a left $R$-module, and let $G$ be an (additive) abelian group. A function $f \colon A \times B \to G$ is called $R$-**biadditive** if, for all $a, a' \in A$, $b, b' \in B$, and $r \in R$, we have

$$f(a + a', b) = f(a, b) + f(a', b);$$
$$f(a, b + b') = f(a, b) + f(a, b');$$
$$f(ar, b) = f(a, rb).$$

Let $R$ be *commutative* and let $A$, $B$, and $M$ be $R$-modules. Then a biadditive function $f \colon A \times B \to M$ is called $R$-**bilinear** if

$$f(ar, b) = f(a, rb) = rf(a, b).$$

**Example 6.100.**

(i) If $R$ is a ring, then its multiplication $\mu \colon R \times R \to R$ is $R$-biadditive; the first two axioms are the right and left distributive laws, while the third axiom is associativity:

$$\mu(ar, b) = (ar)b = a(rb) = \mu(a, rb).$$

If $R$ is a commutative ring, then $\mu$ is $R$-bilinear, for $(ar)b = a(rb) = r(ab)$.

(ii) If $_R M$ is a left $R$-module, then its scalar multiplication $\sigma \colon R \times M \to M$ is $R$-biadditive; if $R$ is a commutative ring, then $\sigma$ is $R$-bilinear.

(iii) If $M_R$ is a right $R$-module and $_RN_R$ is an $(R, R)$-bimodule, then Proposition 6.62(iii) shows that $\operatorname{Hom}_R(M, N)$ is a left $R$-module: if $f \in \operatorname{Hom}_R(M, N)$ and $r \in R$, define $rf \colon M \to N$ by

$$rf \colon m \mapsto r[f(m)].$$

We can now see that ***evaluation*** $e \colon M \times \operatorname{Hom}_R(M, N) \to N$, given by $(m, f) \mapsto f(m)$, is $R$-biadditive.

The dual space $V^*$ of a vector space $V$ over a field $k$ gives a special case of this construction: evaluation $V \times V^* \to k$ is $k$-bilinear.

(iv) If $G^* = \operatorname{Hom}_{\mathbb{Z}}(G, \mathbb{Q}/\mathbb{Z})$ is the Pontrjagin dual of an abelian group $G$, then evaluation $G \times G^* \to \mathbb{Q}/\mathbb{Z}$ is $\mathbb{Z}$-bilinear. ◀

Tensor products convert biadditive functions into linear ones.

**Definition.** Given a ring $R$ and modules $A_R$ and $_RB$, then their ***tensor product*** is an abelian group $A \otimes_R B$ and an $R$-biadditive function

$$h \colon A \times B \to A \otimes_R B$$

such that, for every abelian group $G$ and every $R$-biadditive $f \colon A \times B \to G$, there exists a unique $\mathbb{Z}$-homomorphism $\widetilde{f} \colon A \otimes_R B \to G$ making the following diagram commute:

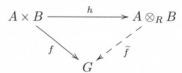

If a tensor product of $A$ and $B$ exists, then it is unique up to isomorphism, for it has been defined as a solution to a universal mapping problem (see the proof of Proposition 6.40 on page 425).

Quite often, $A \otimes_R B$ is denoted by $A \otimes B$ when $R = \mathbb{Z}$.

**Proposition 6.101.** *If $R$ is a ring and $A_R$ and $_RB$ are modules, then their tensor product exists.*

**Proof.** Let $F$ be the free abelian group with basis $A \times B$; that is, $F$ is free on all ordered pairs $(a, b)$, where $a \in A$ and $b \in B$. Define $S$ to be the subgroup of $F$ generated by all elements of the following types:

$$(a, b + b') - (a, b) - (a, b');$$
$$(a + a', b) - (a, b) - (a', b);$$
$$(ar, b) - (a, rb).$$

Define $A \otimes_R B = F/S$, denote the coset $(a, b) + S$ by $a \otimes b$, and define

$$h \colon A \times B \to A \otimes_R B \quad \text{by} \quad h \colon (a, b) \mapsto a \otimes b$$

(thus, $h$ is the restriction of the natural map $F \to F/S$). We have the following identities in $A \otimes_R B$:

$$a \otimes (b + b') = a \otimes b + a \otimes b';$$
$$(a + a') \otimes b = a \otimes b + a' \otimes b;$$
$$ar \otimes b = a \otimes rb.$$

It is now obvious that $h$ is $R$-biadditive.

Consider the following diagram, where $G$ is an abelian group and $f$ is $R$-biadditive:

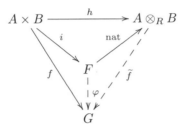

where $i \colon A \times B \to F$ is the inclusion. Since $F$ is free abelian with basis $A \times B$, there exists a homomorphism $\varphi \colon F \to G$ with $\varphi(a, b) = f(a, b)$ for all $(a, b)$; now $S \subseteq \ker \varphi$ because $f$ is $R$-biadditive, and so $\varphi$ induces a map $\widetilde{f} \colon A \otimes_R B \to G$ (because $A \otimes_R B = F/S$) by

$$\widetilde{f}(a \otimes b) = \widetilde{f}((a, b) + S) = \varphi(a, b) = f(a, b).$$

This equation may be rewritten as $\widetilde{f}h = f$; that is, the diagram commutes. Finally, $\widetilde{f}$ is unique because $A \otimes_R B$ is generated by the set of all $a \otimes b$'s.  •

Since $A \otimes_R B$ is generated by the elements of the form $a \otimes b$, every $u \in A \otimes_R B$ has the form

$$u = \sum_i a_i \otimes b_i.$$

This expression for $u$ is not unique; for example, there are expressions

$$0 = a \otimes (b + b') - a \otimes b - a \otimes b'$$
$$= (a + a') \otimes b - a \otimes b - a' \otimes b$$
$$= ar \otimes b - a \otimes rb.$$

Therefore, given some abelian group $G$, we must be suspicious of a *definition* of a map $g \colon A \otimes_R B \to G$ that is given by specifying $g$ on the generators $a \otimes b$; such a "function" $g$ may not be well-defined because elements have many expressions in terms of these generators. In essence, $g$ is only defined on $F$ (the free abelian group with basis $A \times B$), and we must still show that $g(S) = \{0\}$, because $A \otimes_R B = F/S$. The simplest (and safest!) procedure is to define an $R$-biadditive function on $A \times B$, and it will yield a (well-defined) homomorphism with domain $A \otimes_R B$. We illustrate this procedure in the next proofs.

**Proposition 6.102.** *Let* $f\colon A_R \to A'_R$ *and* $g\colon {}_R B \to {}_R B'$ *be maps of right R-modules and left R-modules, respectively. Then there is a unique $\mathbb{Z}$-homomorphism, denoted by* $f \otimes g\colon A \otimes_R B \to A' \otimes_R B'$, *with*

$$f \otimes g\colon a \otimes b \mapsto f(a) \otimes g(b).$$

**Proof.** The function $\varphi\colon A \times B \to A' \otimes_R B'$, given by $(a,b) \mapsto f(a) \otimes g(b)$, is easily seen to be an $R$-biadditive function. For example,

$$\varphi\colon (ar, b) \mapsto f(ar) \otimes g(b) = f(a)r \otimes g(b)$$

and

$$\varphi\colon (a, rb) \mapsto f(a) \otimes g(rb) = f(a) \otimes rg(b);$$

these are equal because of the identity $a'r \otimes b' = a' \otimes rb'$ in $A' \otimes_R B'$. The biadditive function $\varphi$ yields a unique homomorphism $A \otimes_R B \to A' \otimes_R B'$ taking

$$a \otimes b \mapsto f(a) \otimes g(b). \quad \bullet$$

**Corollary 6.103.** *Given maps of right R-modules,* $A \xrightarrow{f} A' \xrightarrow{f'} A''$, *and maps of left R-modules,* $B \xrightarrow{g} B' \xrightarrow{g'} B''$, *we have*

$$(f' \otimes g')(f \otimes g) = f'f \otimes g'g.$$

**Proof.** Both maps take $a \otimes b \mapsto f'f(a) \otimes g'g(b)$, and so the uniqueness of such a homomorphism gives the desired equation. $\quad \bullet$

**Theorem 6.104.** *Given* $A_R$, *there is an additive functor* $F_A\colon {}_R\mathbf{Mod} \to \mathbf{Ab}$, *defined by*

$$F_A(B) = A \otimes_R B \quad and \quad F_A(g) = 1_A \otimes g,$$

*where* $g\colon B \to B'$ *is a map of left R-modules.*

**Proof.** First, note that $F_A$ preserves identities: $F_A(1_B) = 1_A \otimes 1_B$ is the identity $1_{A \otimes B}$, because it fixes every generator $a \otimes b$. Second, $F_A$ preserves composition:

$$F_A(g'g) = 1_A \otimes g'g = (1_A \otimes g')(1_A \otimes g) = F_A(g')F_A(g),$$

by Corollary 6.103. Therefore, $F_A$ is a functor.

To see that $F_A$ is additive, we must show that $F_A(g + h) = F_A(g) + F_A(h)$, where $g, h\colon B \to B'$; that is, $1_A \otimes (g + h) = 1_A \otimes g + 1_A \otimes h$. This is also easy, for both these maps send $a \otimes b \mapsto a \otimes g(b) + a \otimes h(b)$. $\quad \bullet$

We denote the functor $F_A\colon {}_R\mathbf{Mod} \to \mathbf{Ab}$ by

$$A \otimes_R \square\,.$$

Of course, there is a similar result if we fix a left $R$-module $B$: there is an additive functor $\square \otimes_R B\colon \mathbf{Mod}_R \to \mathbf{Ab}$.

**Corollary 6.105.** *If* $f\colon M \to M'$ *and* $g\colon N \to N'$ *are, respectively, isomorphisms of right and left R-modules, then* $f \otimes g\colon M \otimes_R N \to M' \otimes_R N'$ *is an isomorphism of abelian groups.*

**Proof.** Now $f \otimes 1_{N'}$ is the value of the functor $F_{N'}$ on the isomorphism $f$, and hence $f \otimes 1_{N'}$ is an isomorphism; similarly, $1_M \otimes g$ is an isomorphism. By Corollary 6.103, we have $f \otimes g = (f \otimes 1_{N'})(1_M \otimes g)$. Therefore, $f \otimes g$ is an isomorphism, being the composite of isomorphisms. •

In general, the tensor product of two modules is only an abelian group; is it ever a module? In Proposition 6.62, we saw that $\operatorname{Hom}_R(M, N)$ has a module structure when one of the variables is a bimodule. Here is the analogous result for tensor product.

**Proposition 6.106.**

(i) *Given a bimodule $_S A_R$ and a left module $_R B$, the tensor product $A \otimes_R B$ is a left $S$-module, where $s(a \otimes b) = (sa) \otimes b$.*

(ii) *Given $A_R$ and $_R B_S$, the tensor product $A \otimes_R B$ is a right $S$-module, where $(a \otimes b)s = a \otimes (bs)$.*

**Proof.** For fixed $s \in S$, the multiplication $\mu_s \colon A \to A$, defined by $a \mapsto sa$, is an $R$-map, for $A$ being a bimodule gives

$$\mu_s(ar) = s(ar) = (sa)r = \mu_s(a)r.$$

If $F = \square \otimes_R B \colon \mathbf{Mod}_R \to \mathbf{Ab}$, then $F(\mu_s) \colon A \otimes_R B \to A \otimes_R B$ is a (well-defined) $\mathbb{Z}$-homomorphism. Thus, $F(\mu_s) = \mu_s \otimes 1_B \colon a \otimes b \mapsto (sa) \otimes b$, and so the formula in the statement of the lemma makes sense. It is now straightforward to check that the module axioms do hold for $A \otimes_R B$. •

For example, if $V$ and $W$ are vector spaces over a field $k$, then their tensor product $V \otimes_k W$ is also a vector space over $k$.

**Corollary 6.107.**

(i) *Given a bimodule $_S A_R$, then the functor $F_A = A \otimes_R \square \colon {}_R\mathbf{Mod} \to \mathbf{Ab}$ actually takes values in $_S\mathbf{Mod}$.*

(ii) *If $R$ is a commutative ring, then $A \otimes_R B$ is an $R$-module, where*

$$r(a \otimes b) = (ra) \otimes b = a \otimes rb$$

*for all $r \in R$, $a \in A$, and $b \in B$.*

(iii) *If $R$ is a commutative ring, $r \in R$, and $\mu_r \colon B \to B$ is multiplication by $r$, then $1_A \otimes \mu_r \colon A \otimes_R B \to A \otimes_R B$ is also multiplication by $r$.*

**Proof.**

(i) We know, by Proposition 6.106, that $A \otimes_R B$ is a left $S$-module, where $s(a \otimes b) = (sa) \otimes b$, and so it suffices to show that if $g \colon B \to B'$ is a map of left $R$-modules, then $F_A(g) = 1_A \otimes g$ is an $S$-map. But

$$
\begin{aligned}
(1_A \otimes g)[s(a \otimes b)] &= (1_A \otimes g)[(sa) \otimes b] \\
&= (sa) \otimes gb \\
&= s(a \otimes gb) \qquad \text{by Proposition 6.106} \\
&= s(1_A \otimes g)(a \otimes b).
\end{aligned}
$$

(ii) Since $R$ is commutative, we may regard $A$ as an $(R, R)$-bimodule by defining $ar = ra$. Proposition 6.106 now gives

$$
r(a \otimes b) = (ra) \otimes b = (ar) \otimes b = a \otimes rb.
$$

(iii) This statement merely sees the last equation $a \otimes rb = r(a \otimes b)$ from a different viewpoint:

$$
(1_A \otimes \mu_r)(a \otimes b) = a \otimes rb = r(a \otimes b). \quad \bullet
$$

Recall Corollary 6.64: if $M$ is a left $R$-module, then $\mathrm{Hom}_R(R, M)$ is also a left $R$-module, and there is an $R$-isomorphism $\varphi_M \colon \mathrm{Hom}_R(R, M) \to M$. Here is the analogous result for tensor product.

**Proposition 6.108.** *For every left $R$-module $M$, there is an $R$-isomorphism*

$$
\theta_M \colon R \otimes_R M \to M
$$

*given by* $\theta_M \colon r \otimes m \mapsto rm$.

**Proof.** The function $R \times M \to M$, given by $(r, m) \mapsto rm$, is $R$-biadditive, and so there is an $R$-homomorphism $\theta \colon R \otimes_R M \to M$ with $r \otimes m \mapsto rm$ [we are using the fact that $R$ is an $(R, R)$-bimodule]. To see that $\theta$ is an $R$-isomorphism, it suffices to find a $\mathbb{Z}$-homomorphism $f \colon M \to R \otimes_R M$ with $\theta f$ and $f\theta$ identity maps (for it is now only a question of whether the *function* $\theta$ is a bijection). Such a $\mathbb{Z}$-map is given by $f \colon m \mapsto 1 \otimes m$. $\quad \bullet$

After a while, we see that proving properties of tensor products is just a matter of showing that the obvious maps are, indeed, well-defined functions.

We have made some progress in our original problem: given a left $k$-module $M$, where $k$ is a subring of a ring $K$, we can create a left $K$-module from $M$ by *extending scalars*; that is, Proposition 6.106 shows that $K \otimes_k M$ is a left $K$-module, for $K$ is a $(K, k)$-bimodule. The following special case of extending scalars is important in Representation Theory. If $H$ is a subgroup of a group $G$ and $V$ is a left $kH$-module, then the *induced module* $V^G = kG \otimes_{kH} V$ is a left $kG$-module, by Proposition 6.106. Note that $kG$ is a right $kH$-module (it is even a right $kG$-module), and so the tensor product $kG \otimes_{kH} V$ makes sense.

We have defined $R$-biadditive functions for arbitrary, possibly noncommutative, rings $R$, whereas we have defined $R$-bilinear functions only for commutative rings. Tensor product was defined as the solution of a certain universal mapping problem involving $R$-biadditive functions; we now consider the analogous problem for $R$-bilinear functions when $R$ is commutative.

Here is a provisional definition, soon to be seen unnecessary.

**Definition.** If $k$ is a commutative ring, then a $k$-***bilinear product*** is a $k$-module $X$ and a $k$-bilinear function $h\colon A \times B \to X$ such that, for every $k$-module $M$ and every $k$-bilinear function $g\colon A \times B \to M$, there exists a unique $k$-homomorphism $\widetilde{g}\colon X \to M$ making the following diagram commute:

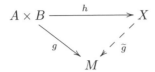

The next result shows that $k$-bilinear products exist, but that they are nothing new.

**Proposition 6.109.** *If $k$ is a commutative ring and $A$ and $B$ are $k$-modules, then the $k$-module $A \otimes_k B$ is a $k$-bilinear product.*

**Proof.** We show that $X = A \otimes_k B$ provides the solution if we define $h(a,b) = a \otimes b$; note that $h$ is also $k$-bilinear, thanks to Corollary 6.107. Since $g$ is $k$-bilinear, it is $k$-biadditive, and so there does exist a $\mathbb{Z}$-homomorphism $\widetilde{g}\colon A \otimes_k B \to M$ with $\widetilde{g}(a \otimes b) = g(a, b)$ for all $(a, b) \in A \times B$. We need only show that $\widetilde{g}$ is a $k$-map. If $u \in k$,

$$\begin{aligned}
\widetilde{g}(u(a \otimes b)) &= \widetilde{g}((ua) \otimes b) \\
&= g(ua, b) \\
&= ug(a, b) \qquad \text{for } g \text{ is } k\text{-bilinear} \\
&= u\widetilde{g}(a \otimes b). \quad \bullet
\end{aligned}$$

As a consequence of the proposition, the term *bilinear product* is unnecessary, and we shall call it the *tensor product* instead.

The next theorem says that tensor product preserves arbitrary direct sums.

**Theorem 6.110.** *Given a right module $A_R$ and left $R$-modules $\{_R B_i : i \in I\}$, there is a $\mathbb{Z}$-isomorphism*

$$\varphi\colon A \otimes_R \left( \bigoplus_{i \in I} B_i \right) \to \bigoplus_{i \in I}(A \otimes_R B_i)$$

*with $\varphi\colon a \otimes (b_i) \mapsto (a \otimes b_i)$. Moreover, if $R$ is commutative, then $\varphi$ is an $R$-isomorphism.*

**Proof.** Since the function $f\colon A \times \left( \bigoplus_i B_i \right) \to \bigoplus_i(A \otimes_R B_i)$, given by $f\colon (a, (b_i)) \mapsto (a \otimes b_i)$, is $R$-biadditive, there exists a $\mathbb{Z}$-homomorphism

$$\varphi\colon A \otimes_R \left( \bigoplus_i B_i \right) \to \bigoplus_i(A \otimes_R B_i)$$

with $\varphi\colon a \otimes (b_i) \mapsto (a \otimes b_i)$. If $R$ is commutative, then $A \otimes_R \left( \bigoplus_{i \in I} B_i \right)$ and $\bigoplus_{i \in I}(A \otimes_R B_i)$ are $R$-modules and $\varphi$ is an $R$-map (for $\varphi$ is the function given by the universal mapping problem in Proposition 6.109).

To see that $\varphi$ is an isomorphism, we give its inverse. Denote the injection $B_j \to \bigoplus_i B_i$ by $\lambda_j$ [where $\lambda_j(b_j) \in \bigoplus_i B_i$ has $j$th coordinate $b_j$ and all other coordinates 0], so that $1_A \otimes \lambda_j \colon A \otimes_R B_j \to A \otimes_R (\bigoplus_i B_i)$. That direct sum is the coproduct in $_R\mathbf{Mod}$ gives a homomorphism $\theta \colon \bigoplus_i (A \otimes_R B_i) \to A \otimes_R (\bigoplus_i B_i)$ with $\theta \colon (a \otimes b_i) \mapsto a \otimes \sum_i \lambda_i(b_i)$. It is now routine to check that $\theta$ is the inverse of $\varphi$, so that $\varphi$ is an isomorphism. $\quad\bullet$

**Example 6.111.** Let $k$ be a field and let $V$ and $W$ be $k$-modules; that is, $V$ and $W$ are vector spaces over $k$. Now $W$ is a free $k$-module; say, $W = \bigoplus_{i \in I} \langle w_i \rangle$, where $(w_i)_{i \in I}$ is a basis of $W$. Therefore, $V \otimes_k W \cong \bigoplus_{i \in I} V \otimes_k \langle w_i \rangle$. Similarly, $V = \bigoplus_{j \in J} \langle v_j \rangle$, where $(v_j)_{j \in J}$ is a basis of $V$ and $V \otimes_k \langle w_i \rangle \cong \bigoplus_{j \in J} \langle v_j \rangle \otimes_k \langle wi \rangle$ for each $i$. But the one-dimensional vector spaces $\langle v_j \rangle$ and $\langle w_i \rangle$ are isomorphic to $k$, and Proposition 6.108 gives $\langle v_j \rangle \otimes_k \langle w_i \rangle \cong \langle v_j \otimes w_i \rangle$. Hence, $V \otimes_k W$ is a vector space over $k$ having $(v_j \otimes w_i)_{(j,i) \in J \times I}$ as a basis. In case both $V$ and $W$ are finite-dimensional, we have

$$\dim(V \otimes_k W) = \dim(V) \dim(W). \quad \blacktriangleleft$$

**Example 6.112.** We now show that there may exist elements in a tensor product $V \otimes_k V$ that cannot be written in the form $u \otimes w$ for $u, w \in V$.

Let $v_1, v_2$ be a basis of a two-dimensional vector space $V$ over a field $k$. As in Example 6.111, a basis for $V \otimes_k V$ is

$$v_1 \otimes v_1, \; v_1 \otimes v_2, \; v_2 \otimes v_1, \; v_2 \otimes v_2.$$

We claim that there do not exist $u, w \in V$ with $v_1 \otimes v_2 + v_2 \otimes v_1 = u \otimes w$. Otherwise, write $u$ and $w$ in terms of $v_1$ and $v_2$:

$$
\begin{aligned}
v_1 \otimes v_2 + v_2 \otimes v_1 &= u \otimes w \\
&= (av_1 + bv_2) \otimes (cv_1 + dv_2) \\
&= acv_1 \otimes v_1 + adv_1 \otimes v_2 + bcv_2 \otimes v_1 + bdv_2 \otimes v_2.
\end{aligned}
$$

By linear independence of the basis,

$$ac = 0 = bd \quad \text{and} \quad ad = 1 = bc.$$

The first equation gives $a = 0$ or $c = 0$, and either possibility, when substituted into the second equation, gives $0 = 1$. $\quad \blacktriangleleft$

As a consequence of Theorem 6.110, if

$$0 \to B' \xrightarrow{i} B \xrightarrow{p} B'' \to 0$$

is a split short exact sequence of left $R$-modules, then, for every right $R$-module $A$,

$$0 \to A \otimes_R B' \xrightarrow{1_A \otimes i} A \otimes_R B \xrightarrow{1_A \otimes p} A \otimes_R B'' \to 0$$

is also a split short exact sequence. What if the exact sequence is not split?

**Theorem 6.113 (Right Exactness).** *Let $A$ be a right $R$-module, and let*

$$B' \xrightarrow{i} B \xrightarrow{p} B'' \to 0$$

*be an exact sequence of left R-modules. Then*

$$A \otimes_R B' \xrightarrow{1_A \otimes i} A \otimes_R B \xrightarrow{1_A \otimes p} A \otimes_R B'' \to 0$$

*is an exact sequence of abelian groups.*

**Remark.**

(i) The absence of $0 \to$ at the beginning of the sequence will be discussed after this proof.

(ii) We will give a nicer proof of this theorem, in Proposition 6.136, once we prove the Adjoint Isomorphism.  ◄

**Proof.** There are three things to check.

(i) $\operatorname{im}(1 \otimes i) \subseteq \ker(1 \otimes p)$.

It suffices to prove that the composite is 0; but

$$(1 \otimes p)(1 \otimes i) = 1 \otimes pi = 1 \otimes 0 = 0.$$

(ii) $\ker(1 \otimes p) \subseteq \operatorname{im}(1 \otimes i)$.

Let $E = \operatorname{im}(1 \otimes i)$. By part (i), $E \subseteq \ker(1 \otimes p)$, and so $1 \otimes p$ induces a map $\widetilde{p} \colon (A \otimes B)/E \to A \otimes B''$ with

$$\widetilde{p} \colon a \otimes b + E \mapsto a \otimes pb,$$

where $a \in A$ and $b \in B$. Now if $\pi \colon A \otimes B \to (A \otimes B)/E$ is the natural map, then

$$\widetilde{p}\pi = 1 \otimes p,$$

for both send $a \otimes b \mapsto a \otimes pb$.

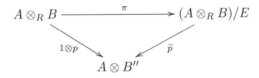

Suppose we show that $\widetilde{p}$ is an isomorphism. Then

$$\ker(1 \otimes p) = \ker \widetilde{p}\pi = \ker \pi = E = \operatorname{im}(1 \otimes i),$$

and we are done. To see that $\widetilde{p}$ is, indeed, an isomorphism, we construct its inverse $A \otimes B'' \to (A \otimes B)/E$. Define

$$f \colon A \times B'' \to (A \otimes B)/E$$

as follows. If $b'' \in B''$, there is $b \in B$ with $pb = b''$, because $p$ is surjective; let

$$f \colon (a, b'') \mapsto a \otimes b.$$

Now $f$ is well-defined: if $pb_1 = b''$, then $p(b - b_1) = 0$ and $b - b_1 \in \ker p = \operatorname{im} i$. Thus, there is $b' \in B'$ with $ib' = b - b_1$; hence $a \otimes (b - b_1) = a \otimes ib' \in \operatorname{im}(1 \otimes i) = E$. Clearly, $f$ is $R$-biadditive, and so the definition of tensor product gives a homomorphism $\widetilde{f} \colon A \otimes B'' \to (A \otimes B)/E$ with $\widetilde{f}(a \otimes b'') = a \otimes b + E$. The reader may check that $\widetilde{f}$ is the inverse of $\widetilde{p}$, as desired.

(iii) $1 \otimes p$ is surjective.

If $\sum a_i \otimes b_i'' \in A \otimes B''$, then there exist $b_i \in B$ with $pb_i = b_i''$ for all $i$, for $p$ is surjective. But

$$1 \otimes p : \sum a_i \otimes b_i \mapsto \sum a_i \otimes pb_i = \sum a_i \otimes b_i''. \quad \bullet$$

A similar statement holds for the functor $\square \otimes_R B$. If $B$ is a left $R$-module and

$$A' \xrightarrow{i} A \xrightarrow{p} A'' \to 0$$

is a short exact sequence of right $R$-modules, then the following sequence is exact:

$$A' \otimes_R B \xrightarrow{i \otimes 1_B} A \otimes_R B \xrightarrow{p \otimes 1_B} A'' \otimes_R B \to 0.$$

**Definition.** A (covariant) functor $T : {}_R\mathbf{Mod} \to \mathbf{Ab}$ is called *right exact* if exactness of a sequence of left $R$-modules

$$B' \xrightarrow{i} B \xrightarrow{p} B'' \to 0$$

implies exactness of the sequence

$$T(B') \xrightarrow{T(i)} T(B) \xrightarrow{T(p)} T(B'') \to 0.$$

There is a similar definition for covariant functors $\mathbf{Mod}_R \to \mathbf{Ab}$.

In this terminology, the functors $A \otimes_R \square$ and $\square \otimes_R B$ are right exact functors.

The next example illustrates the absence of "$0 \to$" in Theorem 6.113.

**Example 6.114.** Consider the exact sequence of abelian groups

$$0 \to \mathbb{Z} \xrightarrow{i} \mathbb{Q} \to \mathbb{Q}/\mathbb{Z} \to 0,$$

where $i$ is the inclusion. By right exactness, there is an exact sequence

$$\mathbb{I}_2 \otimes \mathbb{Z} \xrightarrow{1 \otimes i} \mathbb{I}_2 \otimes \mathbb{Q} \to \mathbb{I}_2 \otimes (\mathbb{Q}/\mathbb{Z}) \to 0$$

(we have abbreviated $\otimes_\mathbb{Z}$ to $\otimes$). Now $\mathbb{I}_2 \otimes \mathbb{Z} \cong \mathbb{I}_2$, by Proposition 6.108. On the other hand, if $a \otimes q$ is a generator of $\mathbb{I}_2 \otimes \mathbb{Q}$, then

$$a \otimes q = a \otimes (2q/2) = 2a \otimes (q/2) = 0 \otimes (q/2) = 0.$$

Therefore, $\mathbb{I}_2 \otimes \mathbb{Q} = \{0\}$, and so $1 \otimes i$ cannot be an injection. ◄

We have seen that if $B'$ is a submodule of a left $R$-module $B$, then $A \otimes_R B'$ may not be a submodule of $A \otimes_R B$. Clearly, this is related to our initial problem of imbedding an abelian group $G$ in a vector space over $\mathbb{Q}$. In Chapter 9, we shall consider $\ker(A \otimes_R B' \xrightarrow{1_A \otimes i} A \otimes_R B)$, where $i : B' \to B$ is inclusion, using the functor $\mathrm{Tor}_1^R(A, \square)$ of Homological Algebra.

The next proposition helps compute tensor products.

**Proposition 6.115.** *For every abelian group $B$, we have $\mathbb{I}_n \otimes_\mathbb{Z} B \cong B/nB$.*

**Proof.** If $A$ is a finite cyclic group of order $n$, there is an exact sequence

$$0 \to \mathbb{Z} \xrightarrow{\mu_n} \mathbb{Z} \xrightarrow{p} A \to 0,$$

where $\mu_n$ is multiplication by $n$. Tensoring by an abelian group $B$ gives exactness of

$$\mathbb{Z} \otimes_\mathbb{Z} B \xrightarrow{\mu_n \otimes 1_B} \mathbb{Z} \otimes_\mathbb{Z} B \xrightarrow{p \otimes 1_B} A \otimes_\mathbb{Z} B \to 0.$$

Consider the diagram

$$
\begin{array}{ccccccc}
\mathbb{Z} \otimes_\mathbb{Z} B & \xrightarrow{\mu_n \otimes 1_B} & \mathbb{Z} \otimes_\mathbb{Z} B & \xrightarrow{p \otimes 1_B} & A \otimes_\mathbb{Z} B & \longrightarrow & 0 \\
\downarrow{\scriptstyle \theta} & & \downarrow{\scriptstyle \theta} & & & & \\
B & \xrightarrow[\mu_n]{} & B & \xrightarrow[\pi]{} & B/nB & \longrightarrow & 0
\end{array}
$$

where $\theta \colon \mathbb{Z} \otimes_\mathbb{Z} B \to B$ is the isomorphism of Proposition 6.108, namely, $\theta \colon m \otimes b \mapsto mb$, where $m \in \mathbb{Z}$ and $b \in B$. This diagram commutes, for both composites take $m \otimes b \mapsto nmb$. Proposition 6.116, the next, very general, result, will yield $A \otimes_\mathbb{Z} B \cong B/nB$ when applied to this diagram.   •

**Proposition 6.116.** *Given a commutative diagram with exact rows in which $f$ is a surjection and $g$ is an isomorphism,*

$$
\begin{array}{ccccccc}
A' & \xrightarrow{i} & A & \xrightarrow{p} & A'' & \longrightarrow & 0 \\
\downarrow{\scriptstyle f} & & \downarrow{\scriptstyle g} & & \vdots\,{\scriptstyle h} & & \\
B' & \xrightarrow[j]{} & B & \xrightarrow[q]{} & B'' & \longrightarrow & 0
\end{array}
$$

*there exists a unique isomorphism $h \colon A'' \to B''$ making the augmented diagram commute.*

**Proof.** If $a'' \in A''$, then there is $a \in A$ with $p(a) = a''$ because $p$ is surjective. Define $h(a'') = qg(a)$. Of course, we must show that $h$ is well-defined; that is, if $u \in A$ satisfies $p(u) = a''$, then $qg(u) = qg(a)$. Since $p(a) = p(u)$, we have $p(a - u) = 0$, so that $a - u \in \ker p = \operatorname{im} i$, by exactness. Hence, $a - u = i(a')$, for some $a' \in A'$. Thus, $qg(a - u) = qgi(a') = qjf(a') = 0$, because $qj = 0$. Therefore, $h$ is well-defined.

To prove uniqueness of $h$, suppose that $h' \colon A'' \to B''$ satisfies $h'p = qg$. If $a'' \in A''$, choose $a \in A$ with $pa = a''$; then $h'a'' = h'pa = qga = ha''$.

To see that $h$ is an injection, suppose that $h(a'') = 0$. Now $0 = ha'' = qga$, where $pa = a''$; hence, $ga \in \ker q = \operatorname{im} j$, and so $ga = jb'$ for some $b' \in B'$. Since $f$ is surjective, there is $a' \in A'$ with $fa' = b'$. Commutativity of the first square gives $gia' = jfa' = jb' = ga$. Since $g$ is an injective, we have $ia' = a$. Therefore, $0 = pia' = pa = a''$ and $h$ is injective.

To see that $h$ is a surjection, let $b'' \in B''$. Since $q$ is surjective, there is $b \in B$ with $qb = b''$; since $g$ is surjective, there is $a \in A$ with $qa = b$. Commutativity of the second square gives $h(pa) = qga = qb = b''$.   •

The proof of the last proposition is an example of ***diagram chasing***. Such proofs appear long, but they are, in truth, quite routine. We select an element and, at each step, there is essentially only two things to do: either apply a map to it or take an inverse image of it. The proof of the dual proposition is another example of this sort of thing.

**Proposition 6.117.** *Given a commutative diagram with exact rows in which $g$ is an isomorphism and $h$ is an injection,*

$$
\begin{array}{ccccccccc}
0 & \longrightarrow & A' & \overset{i}{\longrightarrow} & A & \overset{p}{\longrightarrow} & A'' & & \\
 & & \big\downarrow{\scriptstyle f} & & \big\downarrow{\scriptstyle g} & & \big\downarrow{\scriptstyle h} & & \\
0 & \longrightarrow & B' & \underset{j}{\longrightarrow} & B & \underset{q}{\longrightarrow} & B'' & &
\end{array}
$$

*there exists a unique isomorphism $f\colon A' \to B'$ making the augmented diagram commute.*

**Proof.** A diagram chase.   •

A tensor product of two nonzero modules can be zero. The following proposition generalizes the computation in Example 6.114.

**Proposition 6.118.** *If $D$ is a divisible abelian group and $T$ is an abelian group with every element of finite order, then $D \otimes_{\mathbb{Z}} T = \{0\}$.*

**Proof.** It suffices to show that each generator $d \otimes t$, where $d \in D$ and $t \in T$, is $0$ in $D \otimes_{\mathbb{Z}} T$. As $t$ has finite order, there is a nonzero integer $n$ with $nt = 0$. Since $D$ is divisible, there exists $d' \in D$ with $d = nd'$. Hence,

$$d \otimes t = nd' \otimes t = d' \otimes nt = d' \otimes 0 = 0. \quad •$$

We now understand why we cannot make a finite cyclic group $G$ into a $\mathbb{Q}$-module. Even though $0 \to \mathbb{Z} \to \mathbb{Q}$ is exact, the sequence $0 \to \mathbb{Z} \otimes_{\mathbb{Z}} G \to \mathbb{Q} \otimes_{\mathbb{Z}} G$ is not exact; since $\mathbb{Z} \otimes_{\mathbb{Z}} G = G$ and $\mathbb{Q} \otimes_{\mathbb{Z}} G = \{0\}$, the group $G$ cannot be imbedded into $\mathbb{Q} \otimes_{\mathbb{Z}} G$.

**Corollary 6.119.** *If $D$ is a nonzero divisible abelian group with every element of finite order (e.g., $D = \mathbb{Q}/\mathbb{Z}$), then there is no multiplication $D \times D \to D$ making $D$ a ring.*

**Proof.** Assume, on the contrary, that there is a multiplication $\mu\colon D \times D \to D$ making $D$ a ring. If $1$ is the identity, we have $1 \neq 0$, lest $D$ be the zero ring. Since multiplication in a ring is $\mathbb{Z}$-bilinear, there is a homomorphism $\widetilde{\mu}\colon D \otimes_{\mathbb{Z}} D \to D$ with $\widetilde{\mu}(d \otimes d') = \mu(d, d')$ for all $d, d' \in D$. In particular, if $d \neq 0$, then $\widetilde{\mu}(d \otimes 1) = \mu(d, 1) = d \neq 0$. But $D \otimes_{\mathbb{Z}} D = \{0\}$, by Proposition 6.118, so that $\widetilde{\mu}(d \otimes 1) = 0$. This contradiction shows that no multiplication $\mu$ on $D$ exists.   •

In contrast to the Hom functors, the tensor functors obey certain commutativity and associativity laws.

**Proposition 6.120 (Commutativity).** *If $k$ is a commutative ring and $M$ and $N$ are $k$-modules, then there is a $k$-isomorphism*

$$\tau\colon M \otimes_k N \to N \otimes_k M$$

*with $\tau\colon m \otimes n \mapsto n \otimes m$.*

**Proof.** First, Corollary 6.107 shows that both $M \otimes_k N$ and $N \otimes_k M$ are $k$-modules. Consider the diagram

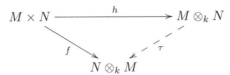

where $f(m,n) = n \otimes m$. It is easy to see that $f$ is $k$-bilinear, and so there is a unique $k$-map $\tau\colon M \otimes_k N \to N \otimes_k M$ with $\tau\colon m \otimes n \mapsto n \otimes m$. Similarly, there is a $k$-map $\tau'\colon N \otimes_k M \to M \otimes_k N$ with $\tau'\colon n \otimes m \mapsto m \otimes n$. Clearly, $\tau'$ is the inverse of $\tau$; that is, $\tau$ is a $k$-isomorphism. $\quad\bullet$

**Proposition 6.121 (Associativity).** *Given $A_R, {}_RB_S$, and ${}_SC$, there is an isomorphism*

$$\theta\colon A \otimes_R (B \otimes_S C) \cong (A \otimes_R B) \otimes_S C$$

*given by*

$$a \otimes (b \otimes c) \mapsto (a \otimes b) \otimes c.$$

**Proof.** Define a **triadditive** function $f\colon A \times B \times C \to G$, where $G$ is an abelian group, to be a function that is additive in each of the three variables (when we fix the other two),

$$f(ar,b,c) = f(a,rb,c), \quad \text{and} \quad f(a,bs,c) = f(a,b,sc)$$

for all $r \in R$ and $s \in S$. Consider the universal mapping problem described by the diagram

$$A \times B \times C \xrightarrow{\quad h \quad} T(A,B,C)$$
$$f \searrow \qquad \swarrow \widetilde{f}$$
$$G$$

where $G$ is an abelian group, $h$ and $f$ are triadditive, and $\widetilde{f}$ is a $\mathbb{Z}$-homomorphism. As for biadditive functions and tensor products of two modules, define $T(A,B,C) = F/N$, where $F$ is the free abelian group on all ordered triples $(a,b,c) \in A \times B \times C$, and $N$ is the obvious subgroup of relations. Define $h\colon A \times B \times C \to T(A,B,C)$ by

$$h\colon (a,b,c) \mapsto (a,b,c) + N$$

(denote $(a,b,c) + N$ by $a \otimes b \otimes c$). A routine check shows that this construction does give a solution to the universal mapping problem for triadditive functions.

We now show that $A \otimes_R (B \otimes_S C)$ is another solution to this universal problem. Define a triadditive function $\eta\colon A \times B \times C \to A \otimes_R (B \otimes_S C)$ by

$$\eta\colon (a,b,c) \mapsto a \otimes (b \otimes c);$$

we must find a homomorphism $\widetilde{f}\colon A \otimes_R (B \otimes_S C) \to G$ with $\widetilde{f}\eta = f$. For each $a \in A$, the $S$-biadditive function $f_a\colon B \times C \to G$, defined by $(b,c) \mapsto f(a,b,c)$, gives a unique homomorphism $\widetilde{f}_a\colon B \otimes_S C \to G$ taking $b \otimes c \mapsto f(a,b,c)$. If $a, a' \in A$, then $\widetilde{f}_{a+a'}(b \otimes c) = f(a+a',b,c) = f(a,b,c) + f(a',b,c) = \widetilde{f}_a(b \otimes c) + \widetilde{f}_{a'}(b \otimes c)$. It follows that the function $\varphi\colon A \times (B \otimes_S C) \to G$, defined by $\varphi(a, b \otimes c) = \widetilde{f}_a(b \otimes c)$, is additive in both variables. It is $R$-biadditive, for if $r \in R$, then $\varphi(ar, b \otimes c) = \widetilde{f}_{ar}(b \otimes c) = f(ar,b,c) = f(a,rb,c) = \widetilde{f}_a(rb \otimes c) = \varphi(a, r(b \otimes c))$. Therefore, there is a unique homomorphism $\widetilde{f}\colon A \otimes_R (B \otimes_S C) \to G$ with $a \otimes (b \otimes c) \mapsto \varphi(a, b \otimes c) = f(a,b,c)$; that is, $\widetilde{f}\eta = f$. Uniqueness of solutions to universal mapping problems shows that there is an isomorphism $T(A,B,C) \to A \otimes_R (B \otimes_S C)$ with $a \otimes b \otimes c \mapsto a \otimes (b \otimes c)$. Similarly, $T(A,B,C) \cong (A \otimes_R B) \otimes_S C$ via $a \otimes b \otimes c \mapsto (a \otimes b) \otimes c$, and so $A \otimes_R (B \otimes_S C) \cong (A \otimes_R B) \otimes_S C$ via $a \otimes (b \otimes c) \mapsto (a \otimes b) \otimes c$. $\bullet$

That the elements $a \otimes b \otimes c \in T(A,B,C)$ need no parentheses will be exploited in Chapter 8 when we construct tensor algebras.

We are tempted to invoke Corollary 1.25: Generalized Associativity holds in any semigroup $G$. In Proposition 6.121, we proved $A \otimes_R (B \otimes_S C) \cong (A \otimes_R B) \otimes_S C$, not $A \otimes_R (B \otimes_S C) = (A \otimes_R B) \otimes_S C$. But isomorphism is not equality, and an extra condition is needed to prove generalized associativity up to isomorphism (see the remark on page 707).

We need a special case of associativity for four factors in this chapter.

**Proposition 6.122 (4-Associativity).** *If $k$ is a commutative ring and $A, B, C, D$ are $k$-modules, then there is a $k$-isomorphism*

$$\theta\colon (A \otimes_k B) \otimes_k (C \otimes_k D) \to [A \otimes_k (B \otimes_k C)] \otimes_k D$$

*given by*

$$(a \otimes b) \otimes (c \otimes d) \mapsto [a \otimes (b \otimes c)] \otimes d.$$

**Proof.** The proof is a straightforward modification of the proof of Proposition 6.121, using 4-additive functions $A \times B \times C \times D \to M$, for a $k$-module $M$, in place of triadditive functions. We leave the details to the reader; note, however, that the proof is a bit less fussy because all modules here are $k$-modules. $\bullet$

**Definition.** If $k$ is a commutative ring, then a ring $R$ is a $k$-**algebra** if $R$ is a $k$-module and scalars in $k$ commute with everything:

$$a(rs) = (ar)s = r(as)$$

for all $a \in k$ and $r, s \in R$.

If $R$ and $S$ are $k$-algebras, then a ring homomorphism $f\colon R \to S$ is called a $k$-**algebra map** if

$$f(ar) = af(r)$$

for all $a \in k$ and $r \in R$; that is, $f$ is also a map of $k$-modules.

The reason that $k$ is assumed to be commutative in the definition of $k$-algebra can be seen in the important special case when $k$ is a subring of $R$; setting $s = 1$ and taking $r \in k$ gives $ar = ra$.

**Example 6.123.**

(i) Every ring $R$ is a $\mathbb{Z}$-algebra, and every ring homomorphism is a $\mathbb{Z}$-algebra map. This example shows why, in the definition of $R$-algebra, we do not demand that $k$ be isomorphic to a subring of $R$.

(ii) If $A = \mathbb{C}[x]$, then $A$ is a $\mathbb{C}$-algebra, and $\varphi \colon A \to A$, defined by $\varphi \colon \sum_j c_j x^j \mapsto \sum_j c_j(x-1)^j$ is a $\mathbb{C}$-algebra map. On the other hand, the function $\theta \colon A \to A$, defined by $\theta \colon \sum_j c_j x^j \mapsto \sum_j \bar{c}_j(x-1)^j$ (where $\bar{c}$ is the complex conjugate of $c$), is a ring map but it is not a $\mathbb{C}$-algebra map. For example, $\theta(ix) = -i(x-1)$ while $i\theta(x) = i(x-1)$. Now $\mathbb{C}[x]$ is also an $\mathbb{R}$-algebra, and $\theta$ is an $\mathbb{R}$-algebra map.

(iii) If $k$ is a subring contained in the center of $R$, then $R$ is a $k$-algebra.

(iv) If $k$ is a commutative ring, then $\mathrm{Mat}_n(k)$ is a $k$-algebra.

(v) If $k$ is a commutative ring and $G$ is a group, then the group ring $kG$ is a $k$-algebra.

(vi) If $k$ is a commutative ring and $A$ is a $k$-algebra, then $A \otimes_k B$ is a left $A$-module. This follows because a $k$-algebra $A$ is an $(A, k)$-bimodule. ◄

A consequence of the following construction is that we will be able to view bimodules as ordinary modules. Since rings are always $\mathbb{Z}$-algebras, the tensor product of any two rings always exists.

**Proposition 6.124.** *If $k$ is a commutative ring and $A$ and $B$ are $k$-algebras, then their tensor product $A \otimes_k B$ is a $k$-algebra if we define $(a \otimes b)(a' \otimes b') = aa' \otimes bb'$.*

**Proof.** First, $A \otimes_k B$ is a $k$-module, by Corollary 6.107. Let $\mu \colon A \times A \to A$ and $\nu \colon B \times B \to B$ be the given multiplications on the algebras $A$ and $B$, respectively. We must show that there is a multiplication on $A \otimes_k B$ as in the statement; that is, there is a well-defined $k$-bilinear function $\lambda \colon (A \otimes_k B) \times (A \otimes_k B) \to A \otimes_k B$ with $\lambda \colon (a \otimes b, a' \otimes b') \mapsto aa' \otimes bb'$. Indeed, $\lambda$ is the composite

$$(A \otimes B) \times (A \otimes B) \xrightarrow{h} (A \otimes B) \otimes (A \otimes B) \xrightarrow{\theta} [A \otimes (B \otimes A)] \otimes B$$

$$\xrightarrow{1 \otimes \tau \otimes 1} [A \otimes (A \otimes B)] \otimes B \xrightarrow{\theta^{-1}} (A \otimes A) \otimes (B \otimes B) \xrightarrow{\mu \otimes \nu} A \otimes B$$

(the map $\theta$ is 4-Associativity); on generators, these maps are

$$(a \otimes b, a' \otimes b') \mapsto (a \otimes b) \otimes (a' \otimes b') \mapsto [a \otimes (b \otimes a')] \otimes b'$$

$$\mapsto [a \otimes (a' \otimes b)] \otimes b' \mapsto (a \otimes a') \otimes (b \otimes b') \mapsto (aa') \otimes (bb').$$

It is now routine to check that the $k$-module $A \otimes_k B$ is a $k$-algebra. •

**Example 6.125.** Exercise 6.83 on page 485 shows that there is an isomorphism of abelian groups: $\mathbb{I}_m \otimes \mathbb{I}_n \cong \mathbb{I}_d$, where $d = (m, n)$. It follows that if $(m, n) = 1$, then $\mathbb{I}_m \otimes \mathbb{I}_n = \{0\}$. Of course, this tensor product is still $\{0\}$ if we regard $\mathbb{I}_m$ and $\mathbb{I}_n$ as $\mathbb{Z}$-algebras. Thus, in this case, the tensor product is the zero ring. Had we insisted, in the definition of ring, that $1 \neq 0$, then the tensor product of rings would not always be defined. ◄

We now show that the tensor product of algebras is an "honest" construction.

**Proposition 6.126.** *If $k$ is a commutative ring and $A$ and $B$ are commutative $k$-algebras, then $A \otimes_k B$ is the coproduct in the category of commutative $k$-algebras.*

**Proof.** Define $\rho \colon A \to A \otimes_k B$ by $\rho \colon a \mapsto a \otimes 1$, and define $\sigma \colon B \to A \otimes_k B$ by $\sigma \colon b \mapsto 1 \otimes b$. Let $R$ be a commutative $k$-algebra, and consider the diagram

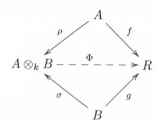

where $f$ and $g$ are $k$-algebra maps. The function $\varphi \colon A \times B \to R$, given by $(a, b) \mapsto f(a)g(b)$, is easily seen to be $k$-bilinear, and so there is a unique map of $k$-modules $\Phi \colon A \otimes_k B \to R$ with $\Phi(a \otimes b) = f(a)g(b)$. It remains to prove that $\Phi$ is a $k$-algebra map, for which it suffices to prove that $\Phi\big((a \otimes b)(a' \otimes b')\big) = \Phi(a \otimes b)\Phi(a' \otimes b')$. Now

$$\Phi\big((a \otimes b)(a' \otimes b')\big) = \Phi(aa' \otimes bb') = f(a)f(a')g(b)g(b').$$

On the other hand, $\Phi(a \otimes b)\Phi(a' \otimes b') = f(a)g(b)f(a')g(b')$. Since $R$ is commutative, $\Phi$ does preserve multiplication.  •

Bimodules can be viewed as left modules over a suitable ring.

**Proposition 6.127.** *Let $R$ and $S$ be $k$-algebras, where $k$ is a commutative ring. Every $(R, S)$-bimodule $M$ is a left $R \otimes_k S^{\mathrm{op}}$-module, where $S^{\mathrm{op}}$ is the opposite ring and $(r \otimes s)m = rms$.*

**Proof.** The function $R \times S^{\mathrm{op}} \times M \to M$, given by $(r, s, m) \mapsto rms$, is $k$-trilinear, and this can be used to prove that $(r \otimes s)m = rms$ is well-defined. Let us write $s * s'$ for the product in $S^{\mathrm{op}}$; that is, $s * s' = s's$. The only axiom that is not obvious is axiom (iii) in the definition of module: if $a, a' \in R \otimes_k S^{\mathrm{op}}$, then $(aa')m = a(a'm)$, and it is enough to check that this is true for generators $a = r \otimes s$ and $a' = r' \otimes s'$ of $R \otimes_k S^{\mathrm{op}}$. But

$$[(r \otimes s)(r' \otimes s')]m = [rr' \otimes s * s']m = (rr')m(s * s') = (rr')m(s's) = r(r'ms')s.$$

On the other hand,

$$(r \otimes s)[(r' \otimes s')m] = (r \otimes s)[r'(ms')] = r(r'ms')s.  \quad •$$

**Definition.** If $k$ is a commutative ring and $A$ is a $k$-algebra, then its **enveloping algebra** is

$$A^e = A \otimes_k A^{\mathrm{op}}.$$

**Corollary 6.128.** *If $k$ is a commutative ring and $A$ is a $k$-algebra, then $A$ is a left $A^e$-module whose submodules are the two-sided ideals.*

**Proof.** Since a $k$-algebra $A$ is an $(A, A)$-bimodule, it is a left $A^e$-module.  •

**Proposition 6.129.** *If $k$ is a commutative ring and $A$ is a $k$-algebra, then*

$$\operatorname{End}_{A^e}(A) \cong Z(A).$$

**Proof.** If $f: A \to A$ is an $A^e$-map, then it is a map of $A$ viewed only as a left $A$-module. Proposition 6.16 applies to say that $f$ is determined by $z = f(1)$, because $f(a) = f(a1) = af(1) = az$ for all $a \in A$. On the other hand, since $f$ is also a map of $A$ viewed as a right $A$-module, we have $f(a) = f(1a) = f(1)a = za$. Therefore, $z = f(1) \in Z(A)$; that is, the map $\varphi: f \mapsto f(1)$ is a map $\operatorname{End}_{A^e}(A) \to Z(A)$. The map $\varphi$ is surjective, for if $z \in Z(A)$, then $f(a) = za$ is an $A^e$-endomorphism with $\varphi(f) = z$; the map $\varphi$ is injective, for if $f \in \operatorname{End}_{A^e}(A)$ and $f(1) = 0$, then $f = 0$. •

# Exercises

**6.80.** Let $A$ and $B$ be abelian groups and let $\mu: A \to A$ be the multiplication map $a \mapsto na$.

   (i) Prove that the induced map $\mu_*: A \otimes_{\mathbb{Z}} B \to A \otimes_{\mathbb{Z}} B$ is also multiplication by $n$.

   (ii) Prove that $\mathbb{Q} \otimes_{\mathbb{Z}} A$ is a vector space over $\mathbb{Q}$.

**6.81.** Let $V$ and $W$ be finite-dimensional vector spaces over a field $k$, say, and let $v_1, \ldots, v_m$ and $w_1, \ldots, w_n$ be bases of $V$ and $W$, respectively. Let $S: V \to V$ be a linear transformation having matrix $A = [a_{ij}]$, and let $T: W \to W$ be a linear transformation having matrix $B = [b_{k\ell}]$. Show that the matrix of $S \otimes T: V \otimes_k W \to V \otimes_k W$, with respect to a suitable listing of the vectors $v_i \otimes w_j$, is their **Kronecker product**: the $nm \times nm$ matrix which we write in block form:

$$A \otimes B = \begin{bmatrix} a_{11}B & a_{12}B & \cdots & a_{1m}B \\ a_{21}B & a_{22}B & \cdots & a_{2m}B \\ \vdots & \vdots & \vdots & \vdots \\ a_{m1}B & a_{m2}B & \cdots & a_{mm}B \end{bmatrix}.$$

**6.82.** Let $R$ be a domain with $Q = \operatorname{Frac}(R)$. If $A$ is an $R$-module, prove that every element in $Q \otimes_R A$ has the form $q \otimes a$ for $q \in Q$ and $a \in A$ (instead of $\sum_i q_i \otimes a_i$). (Compare this result with Example 6.112.)

∗ **6.83.** Let $m$ and $n$ be positive integers, and let $d = (m, n)$.

   (i) Prove that there is an isomorphism of abelian groups

$$\mathbb{I}_m \otimes \mathbb{I}_n \cong \mathbb{I}_d.$$

   (ii) Prove that $\mathbb{I}_m \otimes_{\mathbb{Z}} \mathbb{I}_n \cong \mathbb{I}_d$ as commutative rings.

∗ **6.84.**   (i) Let $k$ be a commutative ring, and let $P$ and $Q$ be projective $k$-modules. Prove that $P \otimes_k Q$ is a projective $k$-module.

   (ii) Let $\varphi: R \to R'$ be a ring homomorphism. Prove that $R'$ is an $(R', R)$-bimodule if we define $r'r = r'\varphi(r)$ for all $r \in R$ and $r' \in R'$. Conclude that if $P$ is a left $R$-module, then $R' \otimes_R P$ is a left $R'$-module.

   (iii) Let $\varphi: R \to R'$ be a ring homomorphism. Prove that if $P$ is a projective left $R$-module, then $R' \otimes_R P$ is a projective left $R'$-module. Moreover, if $P$ is finitely generated, so is $R' \otimes_R P$.

* **6.85.** Assume that the following diagram commutes, and that the vertical arrows are isomorphisms:

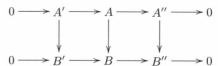

Prove that the bottom row is exact if and only if the top row is exact.

* **6.86. (Five Lemma)** Consider a commutative diagram with exact rows:

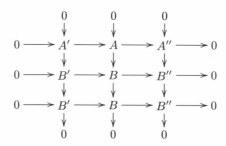

(i) If $h_2$ and $h_4$ are surjective and $h_5$ is injective, prove that $h_3$ is surjective.

(ii) If $h_2$ and $h_4$ are injective and $h_1$ is surjective, prove that $h_3$ is injective.

(iii) If $h_1$ is surjective, $h_2$, $h_4$ are isomorphisms, and $h_5$ is injective, prove that $h_3$ is an isomorphism.

(iv) Give an example of a commutative diagram in which no middle map $h_3$ exists.

* **6.87. ($3 \times 3$ Lemma)** Consider the commutative diagram of modules:

$$
\begin{array}{ccccc}
 & 0 & 0 & 0 & \\
 & \downarrow & \downarrow & \downarrow & \\
0 \longrightarrow & A' \longrightarrow & A \longrightarrow & A'' \longrightarrow & 0 \\
 & \downarrow & \downarrow & \downarrow & \\
0 \longrightarrow & B' \longrightarrow & B \longrightarrow & B'' \longrightarrow & 0 \\
 & \downarrow & \downarrow & \downarrow & \\
0 \longrightarrow & B' \longrightarrow & B \longrightarrow & B'' \longrightarrow & 0 \\
 & \downarrow & \downarrow & \downarrow & \\
 & 0 & 0 & 0 &
\end{array}
$$

with exact columns.

(i) If the bottom two rows are exact, prove that the top row is exact.

(ii) If the top two rows are exact, prove that the bottom row is exact.

* **6.88.** Let $u \colon A \to B$ be a map in $_R\mathbf{Mod}$.

(i) Prove that the inclusion $i \colon \ker u \to A$ solves the following universal mapping problem: $ui = 0$ and, for every $X$ and $g \colon X \to A$ with $ug = 0$, there exists a unique $\theta \colon X \to \ker u$ with $i\theta = g$.

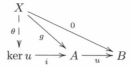

**Hint.** Use Proposition 6.117.

(ii) Prove that the natural map $\pi\colon B \to \operatorname{coker} u$ solves the following universal mapping problem: $\pi u = 0$ and, for every $Y$ and $h\colon B \to Y$ with $hu = 0$, there exists a unique $\theta\colon \operatorname{coker} u \to Y$ with $\theta\pi = h$.

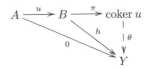

**Hint.** Use Proposition 6.116.

**6.89.** Let $\mathcal{A}, \mathcal{B}$, and $\mathcal{C}$ be categories. A **functor of two variables** (or **bifunctor**) is a function $T\colon \mathcal{A} \times \mathcal{B} \to \mathcal{C}$ that assigns to each ordered pair of objects $(A, B)$ an object $T(A, B) \in \operatorname{obj}(\mathcal{C})$, and to each ordered pair of morphisms $f\colon A \to A'$ and $g\colon B \to B'$ a morphism $T(f, g)\colon T(A, B) \to T(A', B')$, such that

(a) Fixing either variable is a functor; that is, for all $A \in \operatorname{obj}(\mathcal{A})$ and $B \in \operatorname{obj}(\mathcal{B})$,

$$T_A = T(A, \square)\colon \mathcal{B} \to \mathcal{C} \quad \text{and} \quad T_B = T(\square, B)\colon \mathcal{A} \to \mathcal{C}$$

are functors, where $T_A(B) = T(A, B)$ and $T_A(g) = T(1_A, g)$.

(b) The following diagram commutes:

$$
\begin{array}{ccc}
T(A, B) & \xrightarrow{T(1_A, g)} & T(A, B') \\
{\scriptstyle T(f, 1_B)}\downarrow & \searrow{\scriptstyle T(f,g)} & \downarrow{\scriptstyle T(f, 1_{B'})} \\
T(A', B) & \xrightarrow[T(1_{A'}, g)]{} & T(A', B')
\end{array}
$$

(i) Prove that $\square \otimes \square\colon \mathbf{Mod}_R \times {}_R\mathbf{Mod} \to \mathbf{Ab}$ is a functor of two variables.

(ii) Prove that direct sum, $\square \oplus \square\colon {}_R\mathbf{Mod} \times {}_R\mathbf{Mod} \to {}_R\mathbf{Mod}$, is a functor of two variables [if $f\colon A \to A'$ and $g\colon B \to B'$, then $f \oplus g\colon A \oplus B \to A' \oplus B'$ is defined by $(a, b) \mapsto (fa, gb)$].

(iii) Modify the definition of a functor of two variables to allow contravariance in a variable, and prove that $\operatorname{Hom}_R(\square, \square)\colon {}_R\mathbf{Mod} \times {}_R\mathbf{Mod} \to \mathbf{Ab}$ is a functor of two variables.

$*$ **6.90.** Let $\mathcal{A}$ be a category with finite products, let $A, B \in \operatorname{obj}(\mathcal{A})$, and let $i, j\colon A \to A \oplus A$ and $i', j'\colon B \to B \oplus B$ be injections. If $f, g\colon A \to B$, prove that $f \oplus g\colon A \oplus A \to B \oplus B$ is the unique map completing the coproduct diagram

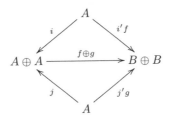

**6.91.** Let $0 \to A \to B \to C \to 0$ and $0 \to A' \to B' \to C' \to 0$ be, respectively, exact sequences of right $R$-modules and left $R$-modules. Prove that the following diagram is

commutative and all its rows and columns are exact:

$$
\begin{array}{ccccccc}
A \otimes_R A' & \longrightarrow & B \otimes_R A' & \longrightarrow & C \otimes_R A' & \longrightarrow & 0 \\
\downarrow & & \downarrow & & \downarrow & & \\
A \otimes_R B' & \longrightarrow & B \otimes_R B' & \longrightarrow & C \otimes_R B' & \longrightarrow & 0 \\
\downarrow & & \downarrow & & \downarrow & & \\
A \otimes_R C' & \longrightarrow & B \otimes_R C' & \longrightarrow & C \otimes_R C' & \longrightarrow & 0 \\
\downarrow & & \downarrow & & \downarrow & & \\
0 & & 0 & & 0 & &
\end{array}
$$

## Section 6.7. Adjoint Isomorphisms

There is a remarkable relationship between Hom and tensor: the Adjoint Isomorphisms. We begin by introducing the way to compare two functors.

**Definition.** Let $F, G \colon \mathcal{C} \to \mathcal{D}$ be covariant functors. A ***natural transformation*** is a family of morphisms $\tau = (\tau_C \colon FC \to GC)_{C \in \mathrm{obj}(\mathcal{C})}$, such that the following diagram commutes for all $f \colon C \to C'$ in $\mathcal{C}$:

$$
\begin{array}{ccc}
FC & \xrightarrow{\ Ff\ } & FC' \\
\tau_C \downarrow & & \downarrow \tau_{C'} \\
GC & \xrightarrow[\ Gf\ ]{} & GC'
\end{array}
$$

If each $\tau_C$ is an isomorphism, then $\tau$ is called a ***natural isomorphism*** and $F$ and $G$ are called ***naturally isomorphic***.

There is a similar definition of natural transformation between contravariant functors.

**Example 6.130.**

(i) If $P = \{p\}$ is a one-point set, we claim that $\mathrm{Hom}(P, \square) \colon \mathbf{Sets} \to \mathbf{Sets}$ is naturally isomorphic to the identity functor on $\mathbf{Sets}$. If $X$ is a set, define

$$\tau_X \colon \mathrm{Hom}(P, X) \to X \text{ by } f \mapsto f(p).$$

Each $\tau_X$ is a bijection, as is easily seen, and we now show that $\tau$ is a natural transformation. Let $X$ and $Y$ be sets, and let $h \colon X \to Y$; we must show that the following diagram commutes:

$$
\begin{array}{ccc}
\mathrm{Hom}(P, X) & \xrightarrow{\ h_*\ } & \mathrm{Hom}(P, Y) \\
\tau_X \downarrow & & \downarrow \tau_Y \\
X & \xrightarrow[\ h\ ]{} & Y
\end{array}
$$

where $h_* \colon f \mapsto hf$. Going clockwise, $f \mapsto hf \mapsto (hf)(p) = h(f(p))$, while going counterclockwise, $f \mapsto f(p) \mapsto h(f(p))$.

(ii) If $k$ is a field and $V$ is a vector space over $k$, then its dual space $V^*$ is the vector space $\operatorname{Hom}_k(V, k)$ of all linear functionals on $V$. The evaluation map $e_v \colon f \mapsto f(v)$ is a linear functional on $V^*$; that is, $e_v \in (V^*)^* = V^{**}$. Define $\tau_V \colon V \to V^{**}$ by

$$\tau_V \colon v \mapsto e_v.$$

The reader may check that $\tau$ is a natural transformation from the identity functor on ${}_k\mathbf{Mod}$ to the double dual functor. The restriction of $\tau$ to the subcategory of finite-dimensional vector spaces is a natural isomorphism. ◄

**Proposition 6.131.** *The isomorphisms $\varphi_M$ of* Corollary 6.64 *form a natural isomorphism* $\operatorname{Hom}_R(R, \square) \to 1_{{}_R\mathbf{Mod}}$, *the identity functor on* ${}_R\mathbf{Mod}$.

**Proof.**[20] The isomorphism $\varphi_M \colon \operatorname{Hom}_R(R, M) \to M$ is given by $f \mapsto f(1)$. To see that these isomorphisms $\varphi_M$ form a natural isomorphism, it suffices to show, for any module homomorphism $h \colon M \to N$, that the following diagram commutes:

$$
\begin{array}{ccc}
\operatorname{Hom}_R(R, M) & \xrightarrow{\;h_*\;} & \operatorname{Hom}_R(R, N) \\
{\scriptstyle\varphi_M}\downarrow & & \downarrow{\scriptstyle\varphi_N} \\
M & \xrightarrow{\;\;h\;\;} & N
\end{array}
$$

where $h_* \colon f \mapsto hf$. Let $f \colon R \to M$. Going clockwise, $f \mapsto hf \mapsto (hf)(1) = h(f(1))$, while going counterclockwise, $f \mapsto f(1) \mapsto h(f(1))$. •

**Proposition 6.132.** *The isomorphisms $\theta_M$ of* Corollary 6.108 *form a natural isomorphism* $R \otimes_R \square \to 1_{{}_R\mathbf{Mod}}$, *the identity functor on* ${}_R\mathbf{Mod}$.

**Proof.** The isomorphism $\theta_M \colon R \times_R M \to M$ is given by $r \otimes m \mapsto rm$. To see that these isomorphisms $\theta_M$ form a natural isomorphism, we must show, for any module homomorphism $h \colon M \to N$, that the following diagram commutes:

$$
\begin{array}{ccc}
R \otimes_R M & \xrightarrow{\;1 \otimes h\;} & R \otimes_R N \\
{\scriptstyle\theta_M}\downarrow & & \downarrow{\scriptstyle\theta_N} \\
M & \xrightarrow{\;\;h\;\;} & N
\end{array}
$$

It suffices to look at a generator $r \otimes m$ of $R \otimes_R M$. Going clockwise, $r \otimes m \mapsto r \otimes h(m) \mapsto rh(m)$, while going counterclockwise, $r \otimes m \mapsto rm \mapsto h(rm)$. These agree, for $h$ is an $R$-map, so that $h(rm) = rh(m)$. •

**Example 6.133.**

(i) We are now going to construct ***functor categories***. Given categories $\mathcal{A}$ and $\mathcal{C}$, we construct the category $\mathcal{C}^{\mathcal{A}}$ whose objects are (covariant) functors $F \colon \mathcal{A} \to \mathcal{C}$, whose morphisms are natural transformations $\tau \colon F \to G$, and whose composition is the only reasonable candidate: if $F \xrightarrow{\;\tau\;} G \xrightarrow{\;\sigma\;} H$ are natural transformations, define $\sigma\tau \colon F \to H$ by $(\sigma\tau)_A = \sigma_A \tau_A$ for every $A \in \operatorname{obj}(\mathcal{A})$.

---

[20]Note the similarity of this proof and the next with the argument in Example 6.130(i).

It would be routine to check that $\mathcal{C}^{\mathcal{A}}$ is a category if each $\mathrm{Hom}(F, G)$ were a set, where $\mathrm{Hom}(F, G) = \{$all natural transformations $F \to G\}$. If $\mathrm{obj}(\mathcal{A})$ is a proper class, then so is any natural transformation $\tau \colon F \to G$. In the usual Set Theory, however, a proper class is forbidden to be an element of a class: hence, $\tau \notin \mathrm{Hom}(F, G)$. A definition saves us.

**Definition.** A category $\mathcal{A}$ is a ***small category*** if $\mathrm{obj}(\mathcal{A})$ is a set.

The functor category $\mathcal{C}^{\mathcal{A}}$ actually is a category when $\mathcal{A}$ is a small category. If $F, G \colon \mathcal{A} \to \mathcal{C}$ are functors, then $\mathrm{Hom}_{\mathcal{C}^{\mathcal{A}}}(F, G)$ is a bona fide set; it is often denoted by $\mathrm{Nat}(F, G)$.

(ii) Let $\mathcal{C}$ be a category with objects $A, B$. In Exercise 6.42 on page 435, we constructed a category whose objects are sequences $A \xrightarrow{\alpha} X \xleftarrow{\beta} B$ and whose morphisms are triples $(1_A, \theta, 1_B)$ making the following diagram commute:

$$
\begin{array}{ccccc}
A & \xrightarrow{\alpha} & C & \xleftarrow{\beta} & B \\
{\scriptstyle 1_A}\downarrow & & \downarrow{\scriptstyle \theta} & & \downarrow{\scriptstyle 1_B} \\
A & \xrightarrow{\alpha'} & C' & \xleftarrow{\beta'} & B
\end{array}
$$

We saw that a coproduct of $A$ and $B$ in $\mathcal{C}$ is an initial object in this new category, and we used this fact to prove uniqueness of coproduct. If $\mathcal{A}$ is the (small) category with $\mathrm{obj}(\mathcal{A}) = \{1, 2, 3\}$ and $\mathrm{Hom}(1, 2) = \{i\}$ and $\mathrm{Hom}(3, 2) = \{j\}$, then a diagram

$$ 1 \xrightarrow{i} 2 \xleftarrow{j} 3 $$

is just a functor $\mathcal{A} \to \mathcal{C}$, and a commutative diagram is just a natural transformation. Hence, the category that arose in the exercise is just the functor category $\mathcal{C}^{\mathcal{A}}$.

(iii) Consider $\mathbb{Z}$ as a partially ordered set in which we reverse the usual inequalities. As in Example 6.38(vi), we consider the (small) category $\mathbf{PO}(\mathbb{Z})$ whose objects are integers and whose morphisms are identities $n \to n$ and composites of arrows $n \to n - 1$. Given a category $\mathcal{C}$, a functor $F \colon \mathbf{PO}(\mathbb{Z}) \to \mathcal{C}$ is a sequence

$$ \cdots \to F_{n+1} \to F_n \to F_{n-1} \to \cdots $$

and a natural transformation is just a sequence $(\tau_n)_{n \in \mathbb{Z}}$ making the following diagram commute:

$$
\begin{array}{ccccccccc}
\cdots & \longrightarrow & F_{n+1} & \xrightarrow{f} & F_n & \xrightarrow{g} & F_{n-1} & \longrightarrow & \cdots \\
& & {\scriptstyle \tau_{n+1}}\downarrow & & \downarrow{\scriptstyle \tau_n} & & \downarrow{\scriptstyle \tau_{n-1}} & & \\
\cdots & \longrightarrow & F'_{n+1} & \xrightarrow{f'} & F'_n & \xrightarrow{g'} & F'_{n-1} & \longrightarrow & \cdots
\end{array}
$$

Thus, the functor category $\mathcal{C}^{\mathbf{PO}(\mathbb{Z})}$ can be viewed as a category whose objects are sequences and whose morphisms are commutative diagrams.   ◀

The key idea behind the Adjoint Isomorphisms is that a function of two variables, say, $f: A \times B \to C$, can be viewed as a one-parameter family $(f_a)_{a \in A}$ of functions of one variable: fix $a \in A$ and define $f_a: B \to C$ by $f_a: b \mapsto f(a, b)$.

Recall Proposition 6.106: if $R$ and $S$ are rings, $A_R$ is a module, and $_R B_S$ is a bimodule, then $A \otimes_R B$ is a right $S$-module, where $(a \otimes b)s = a \otimes (bs)$. Furthermore, if $C_S$ is a module, then Proposition 6.62 shows that $\operatorname{Hom}_S(B, C)$ is a right $R$-module, where $(fr)(b) = f(rb)$. Thus, $\operatorname{Hom}_R(A, \operatorname{Hom}_S(B, C))$ makes sense, for it consists of $R$-maps between right $R$-modules. Finally, if $F \in \operatorname{Hom}_R(A, \operatorname{Hom}_S(B, C))$, view $F$ as a one-parameter family of functions $(F_a: B \to C)_{a \in A}$, where $F_a: b \mapsto F(a, b)$.

**Theorem 6.134 (Adjoint Isomorphism).** *Given modules $A_R$, $_R B_S$, and $C_S$, where $R$ and $S$ are rings, there is an isomorphism of abelian groups*

$$\tau_{A,B,C}: \operatorname{Hom}_S(A \otimes_R B, C) \to \operatorname{Hom}_R(A, \operatorname{Hom}_S(B, C));$$

*namely, for $f: A \otimes_R B \to C$, $a \in A$, and $b \in B$,*

$$\tau_{A,B,C}: f \mapsto f^* = (f_a^*: B \to C)_{a \in A}, \quad \text{where } f_a^*: b \mapsto f(a \otimes b).$$

*Indeed, fixing any two of $A, B, C$, the maps $\tau_{A,B,C}$ constitute natural isomorphisms*

$$\operatorname{Hom}_S(\square \otimes_R B, C) \to \operatorname{Hom}_R(\square, \operatorname{Hom}_S(B, C)),$$

$$\operatorname{Hom}_S(A \otimes_R \square, C) \to \operatorname{Hom}_R(A, \operatorname{Hom}_S(\square, C)),$$

*and*

$$\operatorname{Hom}_S(A \otimes_R B, \square) \to \operatorname{Hom}_R(A, \operatorname{Hom}_S(B, \square)).$$

**Proof.** To prove that $\tau = \tau_{A,B,C}$ is a $\mathbb{Z}$-homomorphism, let $f, g: A \otimes_R B \to C$. The definition of $f + g$ gives, for all $a \in A$,

$$\tau(f + g)_a: b \mapsto (f + g)(a \otimes b) = f(a \otimes b) + g(a \otimes b) = \tau(f)_a(b) + \tau(g)_a(b).$$

Therefore, $\tau(f + g) = \tau(f) + \tau(g)$.

Next, $\tau$ is injective. If $\tau(f)_a = 0$ for all $a \in A$, then $0 = \tau(f)_a(b) = f(a \otimes b)$ for all $a \in A$ and $b \in B$. Therefore, $f = 0$ because it vanishes on every generator of $A \otimes_R B$.

We now show that $\tau$ is surjective. If $F: A \to \operatorname{Hom}_S(B, C)$ is an $R$-map, define $\varphi: A \times B \to C$ by $\varphi(a, b) = F_a(b)$. Now consider the diagram

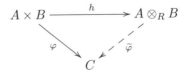

It is straightforward to check that $\varphi$ is $R$-biadditive, and so there exists a $\mathbb{Z}$-homomorphism $\widetilde{\varphi}: A \otimes_R B \to C$ with $\widetilde{\varphi}(a \otimes b) = \varphi(a, b) = F_a(b)$ for all $a \in A$ and $b \in B$. Therefore, $F = \tau(\widetilde{\varphi})$, so that $\tau$ is surjective.

We let the reader prove that the indicated maps form natural transformations by supplying diagrams and verifying that they commute. •

We merely state a second version of the Adjoint Isomorphism. The key idea now is to view a function $f\colon A \times B \to C$ of two variables as a one-parameter family $(f_b)_{b \in B}$ of functions of the second variable: fix $b \in B$ and define $f_b\colon A \to C$ by $f_b\colon a \mapsto f(a,b)$.

**Theorem 6.135 (Adjoint Isomorphism II).** *Given modules* $_RA$, $_SB_R$, *and* $_SC$, *where $R$ and $S$ are rings, there is an isomorphism of abelian groups*

$$\tau'_{A,B,C}\colon \operatorname{Hom}_S(B \otimes_R A, C) \to \operatorname{Hom}_R(A, \operatorname{Hom}_S(B,C));$$

*namely, for* $f\colon B \otimes_R A \to C$, $a \in A$, *and* $b \in B$,

$$\tau'_{A,B,C}\colon f \mapsto f^* = (f_a^*\colon B \to C)_{a \in A}, \quad \text{where } f_a^*\colon b \mapsto f(b \otimes a).$$

*Moreover,* $\tau'_{A,B,C}$ *is a natural isomorphism in each variable.*

As promised earlier, here is another proof of Theorem 6.113, the right exactness of tensor product.

**Proposition 6.136.** *If $A$ is a right $R$-module, then $A \otimes_R \square$ is a right exact functor; that is, if*

$$B' \xrightarrow{i} B \xrightarrow{p} B'' \to 0$$

*is an exact sequence of left $R$-modules, then*

$$A \otimes_R B' \xrightarrow{1_A \otimes i} A \otimes_R B \xrightarrow{1_A \otimes p} A \otimes_R B'' \to 0$$

*is an exact sequence of abelian groups.*

**Proof.** Regard a left $R$-module $B$ as an $(R, \mathbb{Z})$-bimodule, and note, for any abelian group $C$, that $\operatorname{Hom}_{\mathbb{Z}}(B,C)$ is a right $R$-module, by Proposition 6.62(iv). In light of Proposition 6.60, it suffices to prove that the top row of the following diagram is exact for every $C$:

$$
\begin{array}{ccccc}
0 \to \operatorname{Hom}_{\mathbb{Z}}(A \otimes_R B'', C) & \to & \operatorname{Hom}_{\mathbb{Z}}(A \otimes_R B, C) & \twoheadrightarrow & \operatorname{Hom}_{\mathbb{Z}}(A \otimes_R B', C) \\
\downarrow{\scriptstyle \tau''_{A,C}} & & \downarrow{\scriptstyle \tau_{A,C}} & & \downarrow{\scriptstyle \tau'_{A,C}} \\
0 \longrightarrow \operatorname{Hom}_R(A, H'') & \longrightarrow & \operatorname{Hom}_R(A, H) & \longrightarrow & \operatorname{Hom}_R(A, H')
\end{array}
$$

where $H'' = \operatorname{Hom}_{\mathbb{Z}}(B'', C)$, $H = \operatorname{Hom}_{\mathbb{Z}}(B,C)$, and $H' = \operatorname{Hom}_{\mathbb{Z}}(B',C)$. By the Adjoint Isomorphism, the vertical maps are isomorphisms and the diagram commutes. The bottom row is exact, for it arises from the given exact sequence $B' \to B \to B'' \to 0$ by first applying the left exact (contravariant) functor $\operatorname{Hom}_{\mathbb{Z}}(\square, C)$, and then applying the left exact (covariant) functor $\operatorname{Hom}_R(A, \square)$. Exactness of the top row now follows from Exercise 6.85 on page 486.  •

## Exercises

**6.92.** Let $F, G \colon {}_R\mathbf{Mod} \to \mathbf{Ab}$ be additive functors of the same variance. If $F$ and $G$ are naturally isomorphic, prove that the following properties of $F$ are also enjoyed by $G$: left exact; right exact; exact.

**6.93.** A functor $T \colon {}_R\mathbf{Mod} \to \mathbf{Ab}$ is called **representable** if it is naturally isomorphic to $\mathrm{Hom}_R(A, \square)$ for some $R$-module $A$. Prove that if $\mathrm{Hom}_R(A, \square) \cong \mathrm{Hom}_R(B, \square)$, then $A \cong B$. Conclude that if $T$ is naturally isomorphic to $\mathrm{Hom}_R(A, \square)$, then $T$ determines $A$ up to isomorphism.

**Hint.** Use Yoneda's Lemma (Rotman, *An Introduction to Homological Algebra*, p. 25). Let $\mathcal{C}$ be a category, let $A \in \mathrm{obj}(\mathcal{C})$, and let $G \colon \mathcal{C} \to \mathbf{Sets}$ be a covariant functor. Then there is a bijection $y \colon \mathrm{Nat}(\mathrm{Hom}_{\mathcal{C}}(A, \square), G) \to G(A)$ given by $y \colon \tau \mapsto \tau_A(1_A)$.

**6.94.** If ${}_k\mathbf{V}$ is the category of all finite-dimensional vector spaces over a field $k$, prove that the double dual, $V \mapsto V^{**}$, is naturally isomorphic to the identity functor.

**6.95.** Prove that there is a category, **Cat**, whose objects are small categories and whose morphisms are (covariant) functors.

## Section 6.8. Flat Modules

Flat modules arise from tensor products in the same way that projective and injective modules arise from Hom.

**Definition.** Let $R$ be a ring. A right $R$-module $A$ is **flat**[21] if $A \otimes_R \square$ is an exact functor. A left $R$-module $B$ is **flat** if $\square \otimes_R B$ is an exact functor.

Since $A \otimes_R \square$ is a right exact functor for every right $R$-module $A$, we see that $A$ is flat if and only if $1_A \otimes i \colon A \otimes_R B' \to A \otimes_R B$ is an injection whenever $i \colon B' \to B$ is an injection. Investigation of the kernel of $A \otimes_R B' \to A \otimes_R B$ is done in Homological Algebra; it is intimately related to a functor called $\mathrm{Tor}_1^R(A, \square)$.

We will see, in Exercise 6.98 on page 498, that *torsion-free abelian groups* (groups having no nonzero elements of finite order) are flat $\mathbb{Z}$-modules. The converse is proved in Corollary 8.7.

Here are some examples of flat modules over more general rings.

**Lemma 6.137.** *Let $R$ be an arbitrary ring.*

   (i) *The right $R$-module $R$ is a flat right $R$-module, and the left $R$-module $R$ is a flat left $R$-module.*

  (ii) *A direct sum $\bigoplus_j M_j$ of right $R$-modules is flat if and only each $M_j$ is flat.*

 (iii) *Every projective right $R$-module $F$ is flat.*

**Proof.**

---

[21] This term arose as the translation into algebra of a geometric property of varieties.

(i) Consider the commutative diagram

$$
\begin{array}{ccc}
A & \xrightarrow{\ i\ } & B \\
\sigma \downarrow & & \downarrow \tau \\
R \otimes_R A & \xrightarrow[1_R \otimes i]{} & R \otimes_R B
\end{array}
$$

where $i: A \to B$ is an injection, $\sigma: a \mapsto 1 \otimes a$, and $\tau: b \mapsto 1 \otimes b$. Now both $\sigma$ and $\tau$ are isomorphisms, by Proposition 6.108, and so $1_R \otimes i = \tau i \sigma^{-1}$ is an injection. Therefore, $R$ is a flat module over itself.

(ii) By Proposition 6.43, any family of $R$-maps $(f_j: U_j \to V_j)_{j \in J}$ can be assembled into an $R$-map $\varphi: \bigoplus_j U_j \to \bigoplus_j V_j$, where $\varphi: (u_j) \mapsto (f_j(u_j))$, and it is easy to check that $\varphi$ is an injection if and only if each $f_j$ is an injection.

Let $i: A \to B$ be an injection. There is a commutative diagram

$$
\begin{array}{ccc}
(\bigoplus_j M_j) \otimes_R A & \xrightarrow{\ 1 \otimes i\ } & (\bigoplus_j M_j) \otimes_R B \\
\downarrow & & \downarrow \\
\bigoplus_j (M_j \otimes_R A) & \xrightarrow[\varphi]{} & \bigoplus_j (M_j \otimes_R B)
\end{array}
$$

where $\varphi: (m_j \otimes a) \mapsto (m_j \otimes ia)$, where 1 is the identity map on $\bigoplus_j M_j$, and where the downward maps are the isomorphisms of Proposition 6.110.

By our initial observation, $1 \otimes i$ is an injection if and only if each $1_{M_j} \otimes i$ is an injection; this says that $\bigoplus_j M_j$ is flat if and only if each $M_j$ is flat.

(iii) A free right $R$-module, being a direct sum of copies of $R$, must be flat, by (i). But a module is projective if and only if it is a direct summand of a free module, so that (ii) shows that projective modules are flat.    •

This lemma cannot be improved without further assumptions on the ring (there exist rings $R$ for which every flat right $R$-module is projective).

We can now prove a result that we used earlier (in the proof of Theorem 6.96).

**Proposition 6.138.** *If $B$ is a flat right $R$-module and $D$ is a divisible abelian group, then $\operatorname{Hom}_{\mathbb{Z}}(B, D)$ is an injective left $R$-module. In particular, $\operatorname{Hom}_{\mathbb{Z}}(R, D)$ is an injective left $R$-module.*

**Proof.** Since $B$ is a $(\mathbb{Z}, R)$-bimodule, Proposition 6.62(i) shows that $\operatorname{Hom}_{\mathbb{Z}}(B, D)$ is a left $R$-module. It suffices to prove that $\operatorname{Hom}_R(\square, \operatorname{Hom}_{\mathbb{Z}}(B, D))$ is an exact functor. For any left $R$-module $A$, Adjoint Isomorphism II gives natural isomorphisms

$$\tau_A: \operatorname{Hom}_{\mathbb{Z}}(B \otimes_R A, D) \to \operatorname{Hom}_R(A, \operatorname{Hom}_{\mathbb{Z}}(B, D));$$

that is, the functors $\operatorname{Hom}_{\mathbb{Z}}(B \otimes_R \square, D)$ and $\operatorname{Hom}_R(\square, \operatorname{Hom}_{\mathbb{Z}}(B, D))$ are naturally isomorphic. Now $\operatorname{Hom}_{\mathbb{Z}}(B \otimes_R \square, D)$ is just the composite $A \mapsto B \otimes_R A \mapsto \operatorname{Hom}_{\mathbb{Z}}(B \otimes_R A, D)$. The first functor $B \otimes_R \square$ is exact because $B_R$ is flat, and the second functor $\operatorname{Hom}_{\mathbb{Z}}(\square, D)$ is exact because $D$ is divisible (hence $\mathbb{Z}$-injective). Since the composite of exact functors is exact, we have $\operatorname{Hom}_{\mathbb{Z}}(B, D)$ injective.    •

**Lemma 6.139.** *If every finitely generated submodule $N$ of a right $R$-module $M$ is contained in a flat submodule, then $M$ is flat. In particular, if every finitely generated submodule of $M$ is flat, then $M$ is flat.*

**Remark.** This result will be generalized in Corollary 6.161. ◀

**Proof.** Let $i\colon A \to B$ be an injective $R$-map between left $R$-modules. If $u \in M \otimes_R A$ lies in $\ker(1_M \otimes i)$, then $u = \sum_j m_j \otimes a_j$, where $m_j \in M$ and $a_j \in A$, and

$$(1_M \otimes i)(u) = \sum_{j=1}^{n} m_j \otimes i a_j = 0 \quad \text{in } M \otimes_R B.$$

As in the construction of the tensor product in Proposition 6.101, we have $M \otimes_R B \cong F/S$, where $F$ is the free abelian group with basis $M \times B$ and $S$ is the subgroup generated by all elements in $F$ of the form

$$(m, b + b') - (m, b) - (m, b'),$$
$$(m + m', b) - (m, b) - (m', b),$$
$$(mr, b) - (m, rb).$$

Let $N$ be the submodule of $M$ generated by $m_1, \ldots, m_n$ together with the (finite number of) first "coordinates" in $M$ exhibiting $\sum_j (m_j, i a_j)$ in $F$ as a linear combination of relators in $S$. Of course, $N$ is a finitely generated submodule of $M$. By hypothesis, there is a flat submodule $H$ of $M$ containing $N$. Consider the element $v = \sum m_j \otimes a_j \in H \otimes_R A$ (which is the version of $u$ lying in $H \otimes_R A$): if $\ell\colon H \to M$ is the inclusion, then $(\ell \otimes 1_A)(v) = u$. Now $v$ lies in $\ker(1_H \otimes i)$, for we have taken care that all the relations making $(1_M \otimes i)(u) = 0$ are still present in $H \otimes_R B$:

$$
\begin{array}{ccc}
M \otimes_R A & \xrightarrow{\ 1_M \otimes i\ } & M \otimes_R B \\
\uparrow{\scriptstyle \ell \otimes 1_A} & & \\
H \otimes_R A & \xrightarrow[\ 1_H \otimes i\ ]{} & H \otimes_R B
\end{array}
$$

Thus, $(1_H \otimes i)(v) = 0$; since $H$ is flat, we have $v = 0$ in $H \otimes_R A$. Finally, $u = (\ell \otimes 1_A)(v) = 0$. Therefore, $1_M \otimes i$ is injective, and $M$ is flat. •

**Corollary 6.140.** *If $R$ is a domain with $Q = \mathrm{Frac}(R)$, then $Q$ is a flat $R$-module.*

**Proof.** By Lemma 6.139, it suffices to prove that every finitely generated submodule $N = \langle x_1, \ldots, x_n \rangle \subseteq Q$ is contained in a flat submodule of $Q$. Now each $x_i = r_i/s_i$, where $r_i, s_i \in R$ and $s_i \neq 0$. If $s = s_1 \cdots s_n$, then $N \subseteq \langle 1/s \rangle \cong R$, which is flat. •

We are now going to give a connection between flat modules and injective modules (Proposition 6.142).

**Definition.** If $B$ is a right $R$-module, its **character group** $B^*$ is the left $R$-module

$$B^* = \mathrm{Hom}_{\mathbb{Z}}(B, \mathbb{Q}/\mathbb{Z}).$$

Recall that $B^*$ is a left $R$-module if we define $rf$ (for $r \in R$ and $f \colon B \to \mathbb{Q}/\mathbb{Z}$) by

$$rf \colon b \mapsto f(br).$$

The next lemma improves Proposition 6.60.

**Lemma 6.141.** *A sequence of right $R$-modules*

$$0 \to A \xrightarrow{\alpha} B \xrightarrow{\beta} C \to 0$$

*is exact if and only if the sequence of character groups*

$$0 \to C^* \xrightarrow{\beta^*} B^* \xrightarrow{\alpha^*} A^* \to 0$$

*is exact.*

**Proof.** Since divisible abelian groups are injective $\mathbb{Z}$-modules, by Corollary 6.93, $\mathbb{Q}/\mathbb{Z}$ is injective. Hence, $\mathrm{Hom}_{\mathbb{Z}}(\square, \mathbb{Q}/\mathbb{Z})$ is an exact contravariant functor, and exactness of the original sequence gives exactness of the sequence of character groups.

For the converse, it suffices to prove that $\ker \alpha = \operatorname{im} \beta$ without assuming that either $\alpha^*$ is surjective or $\beta^*$ is injective.

$\operatorname{im} \alpha \subseteq \ker \beta$: If $x \in A$ and $\alpha x \notin \ker \beta$, then $\beta \alpha(x) \neq 0$. By Exercise 6.79 on page 468, there is a map $f \colon C \to \mathbb{Q}/\mathbb{Z}$ with $f\beta\alpha(x) \neq 0$. Thus, $f \in C^*$ and $f\beta\alpha \neq 0$, which contradicts the hypothesis that $\alpha^* \beta^* = 0$.

$\ker \beta \subseteq \operatorname{im} \alpha$: If $y \in \ker \beta$ and $y \notin \operatorname{im} \alpha$, then $y + \operatorname{im} \alpha$ is a nonzero element of $B/\operatorname{im}\alpha$. Thus, there is a map $g \colon B/\operatorname{im}\alpha \to \mathbb{Q}/\mathbb{Z}$ with $g(y + \operatorname{im}\alpha) \neq 0$, by Exercise 6.79 on page 468. If $\nu \colon B \to B/\operatorname{im}\alpha$ is the natural map, define $g' = g\nu \in B^*$; note that $g'(y) \neq 0$, for $g'(y) = g\nu(y) = g(y + \operatorname{im}\alpha)$. Now $g'(\operatorname{im}\alpha) = \{0\}$, so that $0 = g'\alpha = \alpha^*(g')$ and $g' \in \ker \alpha^* = \operatorname{im}\beta^*$. Thus, $g' = \beta^*(h)$ for some $h \in C^*$; that is, $g' = h\beta$. Hence, $g'(y) = h\beta(y)$, which is a contradiction, for $g'(y) \neq 0$, while $h\beta(y) = 0$, because $y \in \ker\beta$. $\bullet$

**Proposition 6.142 (Lambek).** *A right $R$-module $B$ is flat if and only if its character group $B^*$ is an injective left $R$-module.*

**Proof.** If $B$ is flat, then Proposition 6.138 shows that the left $R$-module $B^* = \mathrm{Hom}_{\mathbb{Z}}(B, \mathbb{Q}/\mathbb{Z})$ is an injective left $R$-module (for $\mathbb{Q}/\mathbb{Z}$ is divisible).

Conversely, let $B^*$ be an injective left $R$-module and let $A' \to A$ be an injection between left $R$-modules $A'$ and $A$. Since $\mathrm{Hom}_R(A, B^*) = \mathrm{Hom}_R(A, \mathrm{Hom}_{\mathbb{Z}}(B, \mathbb{Q}/\mathbb{Z}))$, the Adjoint Isomorphism gives a commutative diagram in which the vertical maps are isomorphisms:

$$
\begin{array}{ccccc}
\mathrm{Hom}_R(A, B^*) & \longrightarrow & \mathrm{Hom}_R(A', B^*) & \longrightarrow & 0 \\
\downarrow & & \downarrow & & \\
\mathrm{Hom}_{\mathbb{Z}}(B \otimes_R A, \mathbb{Q}/\mathbb{Z}) & \longrightarrow & \mathrm{Hom}_{\mathbb{Z}}(B \otimes A', \mathbb{Q}/\mathbb{Z}) & \longrightarrow & 0 \\
= \downarrow & & \downarrow = & & \\
(B \otimes_R A)^* & \longrightarrow & (B \otimes_R A')^* & \longrightarrow & 0
\end{array}
$$

Exactness of the top row now gives exactness of the bottom row. By Lemma 6.141, the sequence $0 \to B \otimes_R A' \to B \otimes_R A$ is exact, and this gives $B$ flat. •

**Corollary 6.143.** *A right $R$-module $B$ is flat if and only if $0 \to B \otimes_R I \to B \otimes_R R$ is exact for every finitely generated left ideal $I$.*

**Proof.** If $B$ is flat, then the sequence $0 \to B \otimes_R I \to B \otimes_R R$ is exact for every left $R$-module $I$; in particular, this sequence is exact when $I$ is a finitely generated left ideal. Conversely, Lemma 6.139 shows that every (not necessarily finitely generated) left ideal $I$ is flat. There is an exact sequence $(B \otimes_R R)^* \to (B \otimes_R I)^* \to 0$ that, by the Adjoint Isomorphism, gives exactness of $\mathrm{Hom}_R(R, B^*) \to \mathrm{Hom}_R(I, B^*) \to 0$. This says that every map from a left ideal $I$ to $B^*$ extends to a map $R \to B^*$; thus, $B^*$ satisfies the Baer criterion, Theorem 6.89, and so $B^*$ is injective. By Proposition 6.142, $B$ is flat. •

We now seek further connections between flat modules and projectives.

**Lemma 6.144.** *Given modules $({}_R X, {}_R Y_S, Z_S)$, where $R$ and $S$ are rings, there is a natural transformation in $X, Y,$ and $Z$,*

$$\tau_{X,Y,Z} \colon \mathrm{Hom}_S(Y, Z) \otimes_R X \to \mathrm{Hom}_S(\mathrm{Hom}_R(X, Y), Z),$$

*given by*

$$\tau_{X,Y,Z}(f \otimes x) \colon g \mapsto f(g(x))$$

*(where $f \in \mathrm{Hom}_S(Y, Z)$ and $x \in X$), which is an isomorphism whenever $X$ is a finitely generated free left $R$-module.*

**Proof.** Note that both $\mathrm{Hom}_S(Y, Z)$ and $\mathrm{Hom}_R(X, Y)$ make sense, for $Y$ is a bimodule. It is straightforward to check that $\tau_{X,Y,Z}$ is a homomorphism natural in $X, Y, Z$, that $\tau_{R,Y,Z}$ is an isomorphism, and, by induction on the size of a basis, that $\tau_{X,Y,Z}$ is an isomorphism when $X$ is finitely generated and free. •

**Theorem 6.145.** *A finitely presented left $R$-module $B$ over any ring $R$ is flat if and only if it is projective.*

**Remark.** See Rotman, *An Introduction to Homological Algebra*, p. 142, for a different proof of this theorem. ◀

**Proof.** All projective modules are flat, by Lemma 6.137, and so only the converse is significant. Since $B$ is finitely presented, there is an exact sequence

$$F' \to F \to B \to 0,$$

where both $F'$ and $F$ are finitely generated free left $R$-modules. We begin by showing, for every left $R$-module $Y$ [which is necessarily an $(R, \mathbb{Z})$-bimodule], that the map $\tau_B = \tau_{B,Y,\mathbb{Q}/\mathbb{Z}} \colon Y^* \otimes_R B \to \mathrm{Hom}_R(B, Y)^*$ of Lemma 6.144 is an isomorphism.

Consider the following diagram:

$$
\begin{array}{ccccccc}
Y^* \otimes_R F' & \longrightarrow & Y^* \otimes_R F & \longrightarrow & Y^* \otimes_R B & \longrightarrow & 0 \\
\downarrow{\scriptstyle \tau_{F'}} & & \downarrow{\scriptstyle \tau_F} & & \downarrow{\scriptstyle \tau_B} & & \\
\mathrm{Hom}_R(F', Y)^* & \longrightarrow & \mathrm{Hom}_R(F, Y)^* & \longrightarrow & \mathrm{Hom}_R(B, Y)^* & \longrightarrow & 0
\end{array}
$$

By Lemma 6.144, this diagram commutes [for $Y^* \otimes_R F = \operatorname{Hom}_{\mathbb{Z}}(Y, \mathbb{Q}/\mathbb{Z}) \otimes_R F$ and $\operatorname{Hom}_R(F, Y)^* = \operatorname{Hom}_{\mathbb{Z}}(\operatorname{Hom}_R(F, Y), \mathbb{Q}/\mathbb{Z})$] and the first two vertical maps are isomorphisms. The top row is exact, because $Y^* \otimes_R \square$ is right exact. The bottom row is exact because $\operatorname{Hom}_R(\square, Y)^*$ is left exact: it is the composite of the contravariant left exact functor $\operatorname{Hom}_R(\square, Y)$ and the contravariant exact functor $\square^* = \operatorname{Hom}_{\mathbb{Z}}(\square, \mathbb{Q}/\mathbb{Z})$. Proposition 6.116 now shows that the third vertical arrow, $\tau_B \colon Y^* \otimes_R B \to \operatorname{Hom}_R(B, Y)^*$, is an isomorphism.

To prove that $B$ is projective, it suffices to prove that $\operatorname{Hom}(B, \square)$ preserves surjections: if $A \to A'' \to 0$ is exact, then $\operatorname{Hom}(B, A) \to \operatorname{Hom}(B, A'') \to 0$ is exact. By Lemma 6.141, it suffices to show that $0 \to \operatorname{Hom}(B, A'')^* \to \operatorname{Hom}(B, A)^*$ is exact. Consider the diagram

$$
\begin{array}{ccccc}
0 & \longrightarrow & A''^* \otimes_R B & \longrightarrow & A^* \otimes_R B \\
 & & \Big\downarrow{\scriptstyle \tau} & & \Big\downarrow{\scriptstyle \tau} \\
0 & \longrightarrow & \operatorname{Hom}(B, A'')^* & \longrightarrow & \operatorname{Hom}(B, A)^*
\end{array}
$$

Naturality of $\tau$ gives commutativity, and the vertical maps $\tau$ are isomorphisms, because $B$ is finitely presented. Since $A \to A'' \to 0$ is exact, $0 \to A''^* \to A^*$ is exact, and so the top row is exact, because $B$ is flat. It follows that the bottom row is also exact; that is, $0 \to \operatorname{Hom}(B, A'')^* \to \operatorname{Hom}(B, A'')^*$ is exact, which is what we were to show. Therefore, $B$ is projective. •

**Corollary 6.146.** *If $R$ is left noetherian, then a finitely generated left $R$-module $B$ is flat if and only if it is projective.*

**Proof.** This follows from the theorem once we recall Proposition 6.79: every finitely generated left module over a noetherian ring is finitely presented. •

## Exercises

* **6.96.** Let $k$ be a commutative ring, and let $P$ and $Q$ be flat $k$-modules. Prove that $P \otimes_k Q$ is a flat $k$-module.

**6.97.** For every positive integer $n$, prove that $\mathbb{I}_n$ is not a flat $\mathbb{Z}$-module.

* **6.98.** Prove that every torsion-free abelian group is a flat $\mathbb{Z}$-module. In particular, $\mathbb{R}$ and $\mathbb{C}$ are flat $\mathbb{Z}$-modules.

**Hint.** Imbed $A$ as a subgroup of a vector space $D$ over $\mathbb{Q}$, using Exercise 6.76 on page 468. Use character groups to show that $A$ is flat. (The converse is also true: see Corollary 8.7.)

**6.99.** Use the Basis Theorem to prove that if $A$ is a finite abelian group, then $A \cong A^* = \operatorname{Hom}_{\mathbb{Z}}(A, \mathbb{Q}/\mathbb{Z})$.

## Section 6.9. Limits

Many of the categorical constructions we have given, e.g., product, coproduct, pullback, pushout, are special cases of *inverse limits* or *direct limits*.

**Definition.** Let $I$ be a partially ordered set. An **inverse system of left $R$-modules** over $I$ is an ordered pair, denoted by $\{M_i, \psi_i^j\}$, of an indexed family of modules $(M_i)_{i \in I}$ and a family of morphisms $(\psi_i^j \colon M_j \to M_i)_{i \preceq j}$, with $\psi_i^i = 1_{M_i}$ for all $i$ and such that the following diagram commutes whenever $i \preceq j \preceq k$:

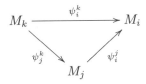

In Example 6.38(vi), we saw that a partially ordered set $I$ defines a category $\mathbf{PO}(I)$ whose objects are the elements of $I$ and whose morphisms are

$$\operatorname{Hom}(i, j) = \begin{cases} \{\kappa_j^i\} & \text{if } i \preceq j; \\ \varnothing & \text{otherwise.} \end{cases}$$

Define $F(i) = M_i$ and $F(\kappa_j^i) = \psi_i^j$; it is easy to see that $\{M_i, \psi_i^j\}$ is an inverse system if and only if $F \colon \mathbf{PO}(I) \to {}_R\mathbf{Mod}$ is a contravariant functor. We can now define inverse systems involving objects and morphisms in any category $\mathcal{C}$: every contravariant functor $F \colon \mathbf{PO}(I) \to \mathcal{C}$ yields one. For example, we can speak of inverse systems of commutative rings, groups, or topological spaces.

**Example 6.147.**

(i) If $I = \{1, 2, 3\}$ is the partially ordered set in which $1 \preceq 2$ and $1 \preceq 3$, then an inverse system over $I$ is a diagram of the form

$$\begin{array}{ccc} & & A \\ & & \downarrow{\scriptstyle g} \\ B & \xrightarrow{\;f\;} & C \end{array}$$

(ii) If $\mathcal{I}$ is a family of submodules of a module $A$, then it can be partially ordered under *reverse inclusion*; that is, $M \preceq M'$ in case $M \supseteq M'$. For $M \preceq M'$, the inclusion map $M' \to M$ is defined, and it is easy to see that the family of all $M \in \mathcal{I}$ with inclusion maps is an inverse system.

(iii) If $I$ is equipped with the **discrete partial order**, that is, $i \preceq j$ if and only if $i = j$, then an inverse system over $I$ is just an indexed family of modules.

(iv) If $\mathbb{N}$ is the natural numbers with the usual partial order, then an inverse system over $\mathbb{N}$ is a diagram

$$M_0 \leftarrow M_1 \leftarrow M_2 \leftarrow \cdots.$$

(v) If $J$ is an ideal in a commutative ring $R$, then its $n$th power is defined by

$$J^n = \Big\{ \sum a_1 \cdots a_n : a_i \in J \Big\}.$$

Each $J^n$ is an ideal and there is a decreasing sequence

$$R \supseteq J \supseteq J^2 \supseteq J^3 \supseteq \cdots.$$

If $A$ is an $R$-module, there is a sequence of submodules
$$A \supseteq JA \supseteq J^2A \supseteq J^3A \supseteq \cdots.$$
If $m \geq n$, define $\psi_n^m \colon A/J^mA \to A/J^nA$ by
$$\psi_n^m \colon a + J^mA \mapsto a + J^nA$$
(these maps are well-defined, for $m \geq n$ implies $J^mA \subseteq J^nA$). It is easy to see that
$$\{A/J^nA, \psi_n^m\}$$
is an inverse system over $\mathbb{N}$.

(vi) Let $G$ be a group and let $\mathcal{N}$ be the family of all the normal subgroups $N$ of $G$ having finite index partially ordered by reverse inclusion. If $N \preceq N'$ in $\mathcal{N}$, then $N' \leq N$; define $\psi_N^{N'} \colon G/N' \to G/N$ by $gN' \mapsto gN$. It is easy to see that the family of all such quotients together with the maps $\psi_N^{N'}$ form an inverse system over $\mathcal{N}$. ◄

**Definition.** Let $I$ be a partially ordered set, and let $\{M_i, \psi_i^j\}$ be an inverse system of left $R$-modules over $I$. The ***inverse limit*** (also called ***projective limit*** or ***limit***) is an $R$-module $\varprojlim M_i$ and a family of $R$-maps $(\alpha_i \colon \varprojlim M_i \to M_i)_{i \in I}$, such that

(i) $\psi_i^j \alpha_j = \alpha_i$ whenever $i \preceq j$;

(ii) for every module $X$ having maps $f_i \colon X \to M_i$ satisfying $\psi_i^j f_j = f_i$ for all $i \preceq j$, there exists a unique map $\theta \colon X \to \varprojlim M_i$ making the following diagram commute:

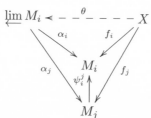

The notation $\varprojlim M_i$ for an inverse limit is deficient in that it does not display the morphisms of the corresponding inverse system (and $\varprojlim M_i$ does depend on them). However, this is standard practice.

As with any object defined as a solution to a universal mapping problem, the inverse limit of an inverse system is unique (up to isomorphism) if it exists.

**Proposition 6.148.** *The inverse limit of any inverse system $\{M_i, \psi_i^j\}$ of left $R$-modules over a partially ordered index set $I$ exists.*

**Proof.** Define
$$L = \left\{ (m_i) \in \prod M_i : m_i = \psi_i^j(m_j) \text{ whenever } i \preceq j \right\};^{22}$$

---

[22]An element $(m_i) \in \prod M_i$ is called a ***thread*** if $m_i = \psi_i^j$ for all $i \preceq j$. Thus, $L$ is the set of all threads.

it is easy to check that $L$ is a submodule of $\prod_i M_i$. If $p_i$ is the projection of the product to $M_i$, define $\alpha_i \colon L \to M_i$ to be the restriction $p_i|L$. It is clear that $\psi_i^j \alpha_j = \alpha_i$.

Assume that $X$ is a module having maps $f_i \colon X \to M_i$ satisfying $\psi_i^j f_j = f_i$ for all $i \preceq j$. Define $\theta \colon X \to \prod M_i$ by

$$\theta(x) = (f_i(x)).$$

That $\operatorname{im}\theta \subseteq L$ follows from the given equation $\psi_i^j f_j = f_i$ for all $i \preceq j$. Also, $\theta$ makes the diagram commute: $\alpha_i \theta \colon x \mapsto (f_i(x)) \mapsto f_i(x)$. Finally, $\theta$ is the unique map $X \to L$ making the diagram commute for all $i \preceq j$. If $\varphi \colon X \to L$, then $\varphi(x) = (m_i)$ and $\alpha_i \varphi(x) = m_i$. Thus, if $\varphi$ satisfies $\alpha_i \varphi(x) = f_i(x)$ for all $i$ and all $x$, then $m_i = f_i(x)$, and so $\varphi = \theta$. We conclude that $L \cong \varprojlim M_i$.   •

Inverse limits in categories other than module categories may exist; for example, inverse limits of commutative algebras exist, as do inverse limits of groups or of topological spaces.

The reader should verify the following assertions in which we describe the inverse limit of each of the inverse systems in Example 6.147.

**Example 6.149.**

(i) If $I$ is the partially ordered set $\{1, 2, 3\}$ with $1 \preceq 2$ and $1 \preceq 3$, then an inverse system is a diagram

and the inverse limit is the pullback.

(ii) We have seen that the intersection of two submodules of a module is a special case of pullback. Suppose now that $\mathcal{I}$ is a family of submodules of a module $A$, so that $\mathcal{I}$ and inclusion maps form an inverse system, as in Example 6.147(ii). The inverse limit of this inverse system is $\bigcap_{M \in \mathcal{I}} M$.

(iii) If $I$ is a discrete index set, then the inverse system $\{M_i, \varnothing\}$ has the product $\prod_i M_i$ as its inverse limit. Indeed, this is just the diagrammatic definition of product.

(iv) If $J$ is an ideal in a commutative ring $R$ and $M$ is an $R$-module, then the inverse limit of the inverse system $\{M/J^n M, \psi_n^m\}$ in Example 6.147(v) is usually called the *J-adic completion* of $M$; let us denote it by $\widehat{M}$.

Recall that a sequence $(x_n)$ in a metric space $X$ with metric $d$ **converges** to a *limit* $y \in X$ if, for every $\epsilon > 0$, there is an integer $N$ so that $d(x_n, y) < \epsilon$ whenever $n \geq N$; we denote $(x_n)$ converging to $y$ by

$$x_n \to y.$$

A difficulty with this definition is that we cannot tell if a sequence is convergent without knowing what its limit is. A sequence $(x_n)$ is a **Cauchy sequence** if, for every $\epsilon > 0$, there is $N$ so that $d(x_m, x_n) < \epsilon$ whenever

$m, n \geq N$. The virtue of this condition on a sequence is that it involves only the terms of the sequence and not its limit. If $X = \mathbb{R}$, then a sequence is convergent if and only if it is a Cauchy sequence. In general metric spaces, however, we can prove that convergent sequences are Cauchy sequences, but the converse may be false. For example, if $X$ is the set of all strictly positive real numbers with the usual metric $|x - y|$, then $(1/n)$ is a Cauchy sequence in $X$ which does not converge (because its limit 0 does not lie in $X$). A metric space $X$ is **complete** if every Cauchy sequence in $X$ converges to a limit in $X$.

**Definition.** A *completion* of a metric space $(X, d)$ is an ordered pair $(\widehat{X}, \varphi \colon X \to \widehat{X})$ such that:
(a) $(\widehat{X}, \widehat{d})$ is a **complete metric space**;
(b) $\varphi$ is an **isometry**; that is, $\widehat{d}(\varphi(x), \varphi(y)) = d(x, y)$ for all $x, y \in X$;
(c) $\varphi(X)$ is a *dense* subspace of $\widehat{X}$; that is, for every $\widehat{x} \in \widehat{X}$, there is a sequence $(x_n)$ in $X$ with $\varphi(x_n) \to \widehat{x}$.

It can be proved that completions exist (Kaplansky, *Set Theory and Metric Spaces*, p. 92) and that any two completions of a metric space $X$ are *isometric*: if $(\widehat{X}, \varphi)$ and $(Y, \psi)$ are completions of $X$, then there exists a unique bijective isometry $\theta \colon \widehat{X} \to Y$ with $\psi = \theta \varphi$. Indeed, a completion of $X$ is just a solution to the obvious universal mapping problem (density of im $\varphi$ gives the required uniqueness of $\theta$). One way to prove existence of a completion is to define its elements as equivalence classes of Cauchy sequences $(x_n)$ in $X$, where we define $(x_n) \equiv (y_n)$ if $d(x_n, y_n) \to 0$.

Let us return to the inverse system $\{M/J^n M, \psi_n^m\}$. A sequence

$$(a_1 + JM, a_2 + J^2 M, a_3 + J^3 M, \dots) \in \varprojlim(M/J^n M)$$

satisfies the condition $\psi_n^m(a_m + J^m M) = a_m + J^n M$ for all $m \geq n$, so that

$$a_m - a_n \in J^n M \quad \text{whenever} \quad m \geq n.$$

This suggests the following metric on $M$ in the (most important) special case when $\bigcap_{n=1}^{\infty} J^n M = \{0\}$. If $x \in M$ and $x \neq 0$, then there is $i$ with $x \in J^i M$ and $x \notin J^{i+1} M$; define $\|x\| = 2^{-i}$; define $\|0\| = 0$. It is a routine calculation to see that $d(x, y) = \|x - y\|$ is a metric on $M$ (without the intersection condition, $\|x\|$ would not be defined for a nonzero $x \in \bigcap_{n=1}^{\infty} J^n M$). Moreover, if a sequence $(a_n)$ in $M$ is a Cauchy sequence, then it is easy to construct an element $(b_n + JM) \in \varprojlim M/J^n M$ that is a limit of $(\varphi(a_n))$. In particular, when $M = \mathbb{Z}$ and $J = (p)$, where $p$ is prime, then the completion $\mathbb{Z}_p$ is called the ring of $p$-**adic integers**. It turns out that $\mathbb{Z}_p$ is a domain, and $\mathbb{Q}_p = \text{Frac}(\mathbb{Z}_p)$ is called the field of $p$-**adic numbers**.

(v) We have seen, in Example 6.147(vi), that the family $\mathcal{N}$ of all normal subgroups of finite index in a group $G$ forms an inverse system; the inverse limit of this system, $\varprojlim G/N$, denoted by $\widehat{G}$, is called the **profinite completion** of $G$. There is a map $G \to \widehat{G}$, namely, $g \mapsto (gN)$, and it is an injection if and only if $G$ is residually finite; that is, $\bigcap_{N \in \mathcal{N}} N = \{1\}$. In Corollary 4.81, we proved that every free group is residually finite.

There are some lovely results obtained making use of profinite completions. A group $G$ is said to have ***rank*** $r \geq 1$ if every subgroup of $G$ can be
generated by $r$ or fewer elements. If $G$ is a residually finite $p$-group (every
element in $G$ has order a power of $p$) of rank $r$, then $G$ is isomorphic to a
subgroup of $\mathrm{GL}(n, \mathbb{Z}_p)$ for some $n$ (not every residually finite group admits
such a linear imbedding). See Dixon–du Sautoy–Mann–Segal, *Analytic Pro-p
Groups*, p. 172. ◄

The next result, generalizing Theorem 6.45(i), says that $\mathrm{Hom}_R(A, \square)$ preserves
inverse limits.

**Proposition 6.150.** *If $\{M_i, \psi_i^j\}$ is an inverse system of left R-modules, then*

$$\mathrm{Hom}_R(A, \varprojlim M_i) \cong \varprojlim \mathrm{Hom}_R(A, M_i)$$

*for every left R-module A.*

**Proof.** Note that Exercise 6.101 on page 512 shows that $\{\mathrm{Hom}_R(A, M_i), (\varphi_j^i)_*\}$ is
an inverse system, so that $\varprojlim \mathrm{Hom}_R(A, M_i)$ makes sense.

This statement follows from inverse limit being the solution of a universal mapping problem. In more detail, consider the diagram

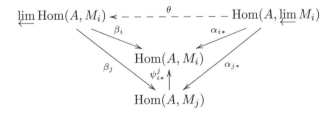

where the $\beta_i$ are the maps given in the definition of inverse limit.

To see that $\theta \colon \mathrm{Hom}(A, \varprojlim M_i) \to \varprojlim \mathrm{Hom}(A, M_i)$ is injective, suppose that
$f \colon A \to \varprojlim M_i$ and $\theta(f) = 0$. Then $0 = \beta_i \theta f = \alpha_i f$ for all $i$, and so the following
diagram commutes:

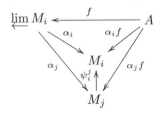

But the zero map in place of $f$ also makes the diagram commute, and so the
uniqueness of such a map gives $f = 0$; that is, $\theta$ is injective.

To see that $\theta$ is surjective, take $g \in \varprojlim \operatorname{Hom}(A, M_i)$. For each $i$, there is a map $\beta_i g \colon A \to M_i$ with $\psi_i^j \beta_i g = \beta_j g$.

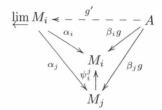

The definition of $\varprojlim M_i$ provides a map $g' \colon A \to \varprojlim M_i$ with $\alpha_i g' = \beta_i g$ for all $i$. It follows that $g = \theta(g')$; that is, $\theta$ is surjective. $\quad\bullet$

We now consider the dual construction.

**Definition.** Let $I$ be a partially ordered set. A **direct system of left $R$-modules** over $I$ is an ordered pair, denoted by $\{M_i, \varphi_j^i\}$, of an indexed family of modules $(M_i)_{i \in I}$ and a family of morphisms $(\varphi_j^i \colon M_i \to M_j)_{i \preceq j}$, with $\varphi_i^i = 1_{M_i}$ for all $i$ and such that the following diagram commutes whenever $i \preceq j \preceq k$:

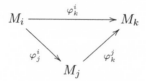

In Example 6.38(vi), we viewed $I$ as a category, $\mathbf{PO}(I)$. Define $F(i) = M_i$ and $F(\kappa_j^i) = \psi_j^i$; it is easy to see that $\{M_i, \psi_j^i\}$ is a direct system if and only if $F \colon \mathbf{PO}(I) \to {}_R\mathbf{Mod}$ is a covariant functor. We can now define direct systems involving objects and morphisms in any category $\mathcal{C}$: every covariant functor $F \colon \mathbf{PO}(I) \to \mathcal{C}$ yields one. For example, we can speak of direct systems of commutative algebras, groups, or topological spaces.

**Example 6.151.**

   (i) If $I = \{1, 2, 3\}$ is the partially ordered set in which $1 \preceq 2$ and $1 \preceq 3$, then a direct system over $I$ is a diagram of the form

$$
\begin{array}{ccc}
A & \overset{g}{\longrightarrow} & B \\
{\scriptstyle f}\downarrow & & \\
C & &
\end{array}
$$

   (ii) If $\mathcal{I}$ is a family of submodules of a module $A$, then it can be partially ordered under inclusion; that is, $M \preceq M'$ in case $M \subseteq M'$. For $M \preceq M'$, the inclusion map $M \to M'$ is defined, and it is easy to see that the family of all $M \in \mathcal{I}$ with inclusion maps is a direct system.

   (iii) If $I$ is equipped with the discrete partial order, then a direct system over $I$ is just a family of modules indexed by $I$. $\quad\blacktriangleleft$

**Definition.** Let $I$ be a partially ordered set, and let $\{M_i, \varphi_j^i\}$ be a direct system of left $R$-modules over $I$. The ***direct limit*** (also called ***inductive limit*** or ***colimit***) is a left $R$-module $\varinjlim M_i$ and a family of $R$-maps $(\alpha_i : M_i \to \varinjlim M_i)_{i \in I}$, such that

(i) $\alpha_j \varphi_j^i = \alpha_i$ whenever $i \preceq j$;

(ii) for every module $X$ having maps $f_i : M_i \to X$ satisfying $f_j \varphi_j^i = f_i$ for all $i \preceq j$, there exists a unique map $\theta : \varinjlim M_i \to X$ making the following diagram commute:

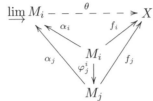

The notation $\varinjlim M_i$ for a direct limit is deficient in that it does not display the morphisms of the corresponding direct system (and $\varinjlim M_i$ does depend on them). However, this is standard practice.

As with any object defined as a solution to a universal mapping problem, the direct limit of a direct system is unique (to isomorphism) if it exists.

**Proposition 6.152.** *The direct limit of any direct system $\{M_i, \varphi_j^i\}$ of left $R$-modules over a partially ordered index set $I$ exists.*

**Proof.** For each $i \in I$, let $\lambda_i$ be the injection of $M_i$ into the sum $\bigoplus_i M_i$. Define

$$D = \left( \bigoplus_i M_i \right)/S,$$

where $S$ is the submodule of $\bigoplus M_i$ generated by all elements $\lambda_j \varphi_j^i m_i - \lambda_i m_i$ with $m_i \in M_i$ and $i \preceq j$. Now define $\alpha_i : M_i \to D$ by $\alpha_i : m_i \mapsto \lambda_i(m_i) + S$. It is routine to check that $D \cong \varinjlim M_i$. •

Thus, each element of $\varinjlim M_i$ has a representative of the form $\sum \lambda_i m_i + S$.

The argument in Proposition 6.152 can be modified to prove that direct limits in other categories exist; for example, direct limits of commutative rings, of groups, or of topological spaces exist.

The reader should verify the following assertions, in which we describe the direct limit of two of the direct systems in Example 6.151.

**Example 6.153.**

(i) If $I$ is the partially ordered set $\{1, 2, 3\}$ with $1 \preceq 2$ and $1 \preceq 3$, then a direct system is a diagram

$$\begin{array}{ccc} A & \xrightarrow{\ g\ } & B \\ {\scriptstyle f}\downarrow & & \\ C & & \end{array}$$

and the direct limit is the pushout.

(ii) If $I$ is a discrete index set, then the direct system is just the indexed family $\{M_i, \varnothing\}$, and the direct limit is the sum: $\varinjlim M_i \cong \bigoplus_i M_i$, for the submodule $S$ in the construction of $\varinjlim M_i$ is $\{0\}$. Alternatively, this is just the diagrammatic definition of a coproduct. ◄

The next result generalizes Theorem 6.45(ii).

**Proposition 6.154.** *If* $\{M_i, \varphi^i_j\}$ *is a direct system of left $R$-modules, then*

$$\mathrm{Hom}_R(\varinjlim M_i, B) \cong \varprojlim \mathrm{Hom}_R(M_i, B)$$

*for every left $R$-module $B$.*

**Proof.** This statement follows from direct limit being the solution of a universal mapping problem. The proof is dual to that of Proposition 6.150, and it is left to the reader. •

There is a special kind of partially ordered index set that is useful for direct limits.

**Definition.** A *directed set* is a partially ordered set $I$ such that, for every $i, j \in I$, there is $k \in I$ with $i \preceq k$ and $j \preceq k$.

**Example 6.155.**

(i) Let $\mathcal{I}$ be a chain of submodules of a module $A$; that is, if $M, M' \in \mathcal{I}$, then either $M \subseteq M'$ or $M' \subseteq M$. As in Example 6.151(ii), $\mathcal{I}$ is a partially ordered set; in fact, it is a directed set.

(ii) If $I$ is the partially ordered set $\{1, 2, 3\}$ with $1 \preceq 2$ and $1 \preceq 3$, then $I$ is *not* a directed set.

(iii) If $\{M_i : i \in I\}$ is some family of modules and $I$ is a discrete partially ordered index set, then $I$ is not directed. However, if we consider the family $\mathcal{F}$ of all *finite partial sums*

$$M_{i_1} \oplus \cdots \oplus M_{i_n},$$

where $n \geq 1$, then $\mathcal{F}$ is a directed set under inclusion.

(iv) If $A$ is a module, then the family $\mathrm{Fin}(A)$ of all the finitely generated submodules of $A$ is partially ordered by inclusion, as in Example 6.151(ii), and it is a directed set.

(v) If $R$ is a domain and $Q = \mathrm{Frac}(R)$, then the family of all cyclic $R$-submodules of $Q$ of the form $\langle 1/r \rangle$, where $r \in R$ and $r \neq 0$, is a partially ordered set, as in Example 6.151(ii); it is a directed set under inclusion, for given $\langle 1/r \rangle$ and $\langle 1/s \rangle$, then each is contained in $\langle 1/rs \rangle$.

(vi) Let $\mathcal{U}$ be the family of all the open intervals in $\mathbb{R}$ containing 0. Partially order $\mathcal{U}$ by reverse inclusion:

$$U \preceq V \quad \text{if} \quad V \subseteq U.$$

Notice that $\mathcal{U}$ is directed: given $U, V \in \mathcal{U}$, then $U \cap V \in \mathcal{U}$ and $U \preceq U \cap V$ and $V \preceq U \cap V$.

For each $U \in \mathcal{U}$, define

$$\mathcal{F}(U) = \{f \colon U \to \mathbb{R} : f \text{ is continuous}\},$$

and, if $U \preceq V$, that is, $V \subseteq U$, define $\rho_V^U \colon \mathcal{F}(U) \to \mathcal{F}(V)$ to be the restriction map $f \mapsto f|V$. Then $\{\mathcal{F}(U), \rho_V^U\}$ is a direct system. ◀

There are two reasons to consider direct systems with directed index sets. The first is that a simpler description of the elements in the direct limit can be given; the second is that $\varinjlim$ preserves short exact sequences.

**Proposition 6.156.** *Let $\{M_i, \varphi_j^i\}$ be a direct system of left $R$-modules over a directed index set $I$, and let $\lambda_i \colon M_i \to \bigoplus M_i$ be the $i$th injection, so that $\varinjlim M_i = (\bigoplus M_i)/S$, where*

$$S = \langle \lambda_j \varphi_j^i m_i - \lambda_i m_i : m_i \in M_i \text{ and } i \preceq j \rangle.$$

(i) *Each element of $\varinjlim M_i$ has a representative of the form $\lambda_i m_i + S$ (instead of $\sum_i \lambda_i m_i + S$).*

(ii) *$\lambda_i m_i + S = 0$ if and only if $\varphi_t^i(m_i) = 0$ for some $t \succeq i$.*

**Proof.**

(i) As in the proof of Proposition 6.152, the existence of direct limits, $\varinjlim M_i = (\bigoplus M_i)/S$, and so a typical element $x \in \varinjlim M_i$ has the form $x = \sum \lambda_i m_i + S$. Since $I$ is directed, there is an index $j$ with $j \succeq i$ for all $i$ occurring in the sum for $x$. For each such $i$, define $b^i = \varphi_j^i m_i \in M_j$, so that the element $b$, defined by $b = \sum_i b^i$, lies in $M_j$. It follows that

$$\sum \lambda_i m_i - \lambda_j b = \sum (\lambda_i m_i - \lambda_j b^i)$$
$$= \sum (\lambda_i m_i - \lambda_j \varphi_j^i m_i) \in S.$$

Therefore, $x = \sum \lambda_i m_i + S = \lambda_j b + S$, as desired.

(ii) If $\varphi_t^i m_i = 0$ for some $t \succeq i$, then

$$\lambda_i m_i + S = \lambda_i m_i + (\lambda_t \varphi_t^i m_i - \lambda_i m_i) + S = S.$$

Conversely, if $\lambda_i m_i + S = 0$, then $\lambda_i m_i \in S$, and there is an expression

$$\lambda_i m_i = \sum_j a_j (\lambda_k \varphi_k^j m_j - \lambda_j m_j) \in S,$$

where $a_j \in R$. We are going to normalize this expression. First, we introduce the following notation for relators: if $j \preceq k$, define

$$r(j, k, m_j) = \lambda_k \varphi_k^j m_j - \lambda_j m_j.$$

Since $a_j r(j, k, m_j) = r(j, k, a_j m_j)$, we may assume that the notation has been adjusted so that

$$\lambda_i m_i = \sum_j r(j, k, m_j).$$

As $I$ is directed, we may choose an index $t \in I$ larger than any of the indices $i, j, k$ occurring in the last equation. Now

$$\lambda_t \varphi_t^i m_i = (\lambda_t \varphi_t^i m_i - \lambda_i m_i) + \lambda_i m_i$$
$$= r(i, t, m_i) + \lambda_i m_i$$
$$= r(i, t, m_i) + \sum_j r(j, k, m_j).$$

Next,

$$r(j, k, m_j) = \lambda_k \varphi_k^j m_j - \lambda_j m_j$$
$$= (\lambda_t \varphi_t^j m_j - \lambda_j m_j) + \left[ \lambda_t \varphi_t^k(-\varphi_k^j m_j) - \lambda_k(-\varphi_k^j m_j) \right]$$
$$= r(j, t, m_j) + r(k, t, -\varphi_k^j m_j),$$

because $\varphi_t^k \varphi_k^i = \varphi_t^i$, by definition of direct system. Hence,

$$\lambda_t \varphi_t^i m_i = \sum_\ell r(\ell, t, x_\ell),$$

where $x_\ell \in M_\ell$. But it is easily checked, for $\ell \preceq t$, that

$$r(\ell, t, m_\ell) + r(\ell, t, m_\ell') = r(\ell, t, m_\ell + m_\ell').$$

Therefore, we may amalgamate all relators with the same smaller index $\ell$ and write

$$\lambda_t \varphi_t^i m_i = \sum_\ell r(\ell, t, x_\ell)$$
$$= \sum_\ell \lambda_t \varphi_t^\ell x_\ell - \lambda_\ell x_\ell$$
$$= \lambda_t \left( \sum_\ell \varphi_t^\ell x_\ell \right) - \sum_\ell \lambda_\ell x_\ell,$$

where $x_\ell \in M_\ell$ and all the indices $\ell$ are distinct. The unique expression of an element in a direct sum allows us to conclude, if $\ell \neq t$, that $\lambda_\ell x_\ell = 0$; it follows that $x_\ell = 0$, for $\lambda_\ell$ is an injection. The right side simplifies to $\lambda_t \varphi_t^t m_t - \lambda_t m_t = 0$, because $\varphi_t^t$ is the identity. Thus, the right side is 0 and $\lambda_t \varphi_t^i m_i = 0$. Since $\lambda_t$ is an injection, we have $\varphi_t^i m_i = 0$, as desired.   •

**Remark.** Our original construction of $\varinjlim M_i$ involved a quotient of $\bigoplus M_i$; that is, $\varinjlim M_i$ is a quotient of a coproduct. In the category **Sets**, coproduct is disjoint union $\bigsqcup_i M_i$. We may regard a "quotient" of a set $X$ as an *orbit space*, that is, as the family of equivalence classes of some equivalence relation on $X$. This categorical analogy suggests that we might be able to give a second construction of $\varinjlim M_i$ using an equivalence relation on $\bigsqcup_i M_i$. When the index set is directed, this can actually be done (Exercise 6.100 on page 512).   ◀

**Example 6.157.**

   (i) Let $\mathcal{I}$ be a chain of submodules of a module $A$; that is, if $M, M' \in \mathcal{I}$, then either $M \subseteq M'$ or $M' \subseteq M$. Then $\mathcal{I}$ is a directed set, and $\varinjlim M_i \cong \bigcup_i M_i$.

   (ii) If $\{M_i : i \in I\}$ is some family of modules, then $\mathcal{F}$, all **finite partial sums**, is a directed set under inclusion, and $\varinjlim M_i \cong \bigoplus_i M_i$.

(iii) If $A$ is a module, then the family $\text{Fin}(A)$ of all the finitely generated submodules of $A$ is a directed set and $\varinjlim M_i \cong A$.

(iv) If $R$ is a domain and $Q = \text{Frac}(R)$, then the family of all cyclic $R$-submodules of $Q$ of the form $\langle 1/r \rangle$, where $r \in R$ and $r \neq 0$, forms a directed set under inclusion, and $\varinjlim M_i \cong Q$; that is, $Q$ is a direct limit of cyclic modules. ◄

**Definition.** Let $\{A_i, \alpha_j^i\}$ and $\{B_i, \beta_j^i\}$ be direct systems over the same index set $I$. A ***transformation***[23] $r\colon \{A_i, \alpha_j^i\} \to \{B_i, \beta_j^i\}$ is an indexed family of homomorphisms

$$r = \{r_i \colon A_i \to B_i\}$$

that makes the following diagram commute for all $i \preceq j$:

$$\begin{array}{ccc} A_i & \xrightarrow{\ r_i\ } & B_i \\ {\scriptstyle \alpha_j^i}\downarrow & & \downarrow{\scriptstyle \beta_j^i} \\ A_j & \xrightarrow{\ r_j\ } & B_j \end{array}$$

A transformation $r\colon \{A_i, \alpha_j^i\} \to \{B_i, \beta_j^i\}$ determines a homomorphism

$$\vec{r}\colon \varinjlim A_i \to \varinjlim B_i$$

by

$$\vec{r}\colon \sum \lambda_i a_i + S \mapsto \sum \mu_i r_i a_i + T,$$

where $S \subseteq \bigoplus A_i$ and $T \subseteq \bigoplus B_i$ are the relation submodules in the construction of $\varinjlim A_i$ and $\varinjlim B_i$, respectively, and $\lambda_i$ and $\mu_i$ are the injections of $A_i$ and $B_i$ into the direct sums. The reader should check that $r$ being a transformation of direct systems implies that $\vec{r}$ is independent of the choice of coset representative, and hence it is a well-defined function.

**Proposition 6.158.** *Let $I$ be a directed set, and let $\{A_i, \alpha_j^i\}$, $\{B_i, \beta_j^i\}$, and $\{C_i, \gamma_j^i\}$ be direct systems over $I$. If $r\colon \{A_i, \alpha_j^i\} \to \{B_i, \beta_j^i\}$ and $s\colon \{B_i, \beta_j^i\} \to \{C_i, \gamma_j^i\}$ are transformations and*

$$0 \to A_i \xrightarrow{r_i} B_i \xrightarrow{s_i} C_i \to 0$$

*is exact for each $i \in I$, then there is an exact sequence*

$$0 \to \varinjlim A_i \xrightarrow{\vec{r}} \varinjlim B_i \xrightarrow{\vec{s}} \varinjlim C_i \to 0.$$

**Remark.** The hypothesis that $I$ be directed enters the proof only in showing that $\vec{r}$ is an injection. ◄

**Proof.** We prove only that $\vec{r}$ is an injection, for the proof of exactness of the rest is routine. Suppose that $\vec{r}(x) = 0$, where $x \in \varinjlim A_i$. Since $I$ is directed, Proposition 6.156(i) allows us to write $x = \lambda_i a_i + S$ (where $S \subseteq \bigoplus A_i$ is the relation submodule and $\lambda_i$ is the injection of $A_i$ into the direct sum). By definition, $\vec{r}(x + S) = \mu_i r_i a_i + T$ (where $T \subseteq \bigoplus B_i$ is the relation submodule and $\mu_i$ is

---

[23]If we recall that a direct system of $R$-modules over $I$ can be regarded as a functor $\mathbf{PO}(I) \to_R$ **Mod**, then transformations are natural transformations. Similarly, we can define transformations of inverse systems over an index set $I$.

the injection of $B_i$ into the direct sum). Now Proposition 6.156(ii) shows that $\mu_i r_i a_i + T = 0$ in $\varinjlim B_i$ implies that there is an index $k \succeq i$ with $\beta_k^i r_i a_i = 0$. Since $r$ is a transformation of direct systems, we have

$$0 = \beta_k^i r_i a_i = r_k \alpha_k^i a_i.$$

Finally, since $r_k$ is an injection, we have $\alpha_k^i a_i = 0$ and, hence, $x = \lambda_i a_i + S = 0$. Therefore, $\vec{r}$ is an injection.   •

An analysis of the proof of Proposition 6.150 shows that it can be generalized by replacing $\mathrm{Hom}(A, \square)$ by any (covariant) left exact functor $F \colon {}_R\mathbf{Mod} \to \mathbf{Ab}$ that preserves products. However, this added generality is only illusory, for it is a theorem of Watts, given such a functor $F$, that there exists a module $A$ with $F$ naturally isomorphic to $\mathrm{Hom}_R(A, \square)$; that is, these representable functors are characterized. Another theorem of Watts characterizes contravariant Hom functors: if $G \colon {}_R\mathbf{Mod} \to \mathbf{Ab}$ is a contravariant left exact functor that converts sums to products, then there exists a module $B$ with $G$ naturally isomorphic to $\mathrm{Hom}_R(\square, B)$. Watts also characterized tensor functors as right exact additive functors which preserve direct sums. Proofs of these theorems can be found in Rotman, *An Introduction to Homological Algebra*, pp. 261–266.

In Theorem 6.150, we proved that $\mathrm{Hom}(A, \square)$ preserves inverse limits; we now prove that $A \otimes \square$ preserves direct limits. Both of these results will follow from Theorem 6.164. However, we now give a proof based on the construction of direct limits.

**Theorem 6.159.** *If $A$ is a right $R$-module and $\{B_i, \varphi_j^i\}$ is a direct system of left $R$-modules (over any not necessarily directed index set $I$), then*

$$A \otimes_R \varinjlim B_i \cong \varinjlim (A \otimes_R B_i).$$

**Proof.** Note that Exercise 6.101 on page 512 shows that $\{A \otimes_R B_i, 1 \otimes \varphi_j^i\}$ is a direct system, so that $\varinjlim (A \otimes_R B_i)$ makes sense.

We begin by constructing $\varinjlim B_i$ as the cokernel of a certain map between sums. For each pair $i, j \in I$ with $i \preceq j$ in the partially ordered index set $I$, define $B_{ij}$ to be a module isomorphic to $B_i$ by a map $b_i \mapsto b_{ij}$, where $b_i \in B_i$, and define $\sigma \colon \bigoplus_{ij} B_{ij} \to \bigoplus_i B_i$ by

$$\sigma \colon b_i \mapsto \lambda_j \varphi_j^i b_i - \lambda_i b_i,$$

where $\lambda_i$ is the injection of $B_i$ into the sum. Note that $\mathrm{im}\,\sigma = S$, the submodule arising in the construction of $\varinjlim B_i$ in Proposition 6.152. Thus, coker $\sigma = (\bigoplus B_i)/S \cong \varinjlim B_i$, and there is an exact sequence

$$\bigoplus B_{ij} \overset{\sigma}{\to} \bigoplus B_i \to \varinjlim B_i \to 0.$$

Right exactness of $A \otimes_R \square$ gives exactness of

$$A \otimes_R \left( \bigoplus B_{ij} \right) \overset{1 \otimes \sigma}{\longrightarrow} A \otimes_R \left( \bigoplus B_i \right) \to A \otimes_R (\varinjlim B_i) \to 0.$$

By Theorem 6.110, the map $\tau \colon A \otimes_R \left( \bigoplus_i B_i \right) \to \bigoplus_i (A \otimes_R B_i)$, given by

$$\tau \colon a \otimes (b_i) \mapsto (a \otimes b_i),$$

is an isomorphism, and so there is a commutative diagram

$$
\begin{array}{ccccccc}
A \otimes \bigoplus B_{ij} & \xrightarrow{1 \otimes \sigma} & A \otimes \bigoplus B_i & \longrightarrow & A \otimes \varinjlim B_i & \longrightarrow & 0 \\
\downarrow{\scriptstyle \tau} & & \downarrow{\scriptstyle \tau'} & & \big\downarrow & & \\
\bigoplus (A \otimes B_{ij}) & \xrightarrow{\widetilde{\sigma}} & \bigoplus (A \otimes B_i) & \longrightarrow & \varinjlim (A \otimes B_i) & \longrightarrow & 0
\end{array}
$$

where $\tau'$ is another instance of the isomorphism of Theorem 6.110, and

$$\widetilde{\sigma} \colon a \otimes b_{ij} \mapsto (1 \otimes \lambda_j)(a \otimes \varphi_j^i b_i) - (1 \otimes \lambda_i)(a' \otimes b_i).$$

There is an isomorphism $A \otimes_R \varinjlim B_i \to \operatorname{coker} \widetilde{\sigma} \cong \varinjlim (A \otimes_R B_i)$, by Proposition 6.116. •

The reader has probably observed that we have actually proved a stronger result: any right exact functor that preserves direct sums must preserve all direct limits. Let us record this observation.

**Proposition 6.160.** *If* $T \colon {}_R\mathbf{Mod} \to \mathbf{Ab}$ *is a right exact functor that preserves all direct sums, then* $T$ *preserves all direct limits.*

**Proof.** This result is contained in the proof of Theorem 6.159. •

The dual result also holds, and it has a similar proof; every left exact functor that preserves products must preserve all inverse limits.

The next result generalizes Lemma 6.139.

**Corollary 6.161.** *If* $\{F_i, \varphi_j^i\}$ *is a direct system of flat right $R$-modules over a directed index set $I$, then* $\varinjlim F_i$ *is also flat.*

**Proof.** Let $0 \to A \xrightarrow{k} B$ be an exact sequence of left $R$-modules. Since each $F_i$ is flat, the sequence

$$0 \to F_i \otimes_R A \xrightarrow{1_i \otimes k} F_i \otimes_R B$$

is exact for every $i$, where $1_i$ abbreviates $1_{F_i}$. Consider the commutative diagram

$$
\begin{array}{ccccc}
0 & \longrightarrow & \varinjlim (F_i \otimes A) & \xrightarrow{\vec{k}} & \varinjlim (F_i \otimes B) \\
& & \downarrow{\scriptstyle \varphi} & & \downarrow{\scriptstyle \psi} \\
0 & \longrightarrow & (\varinjlim F_i) \otimes A & \xrightarrow[1 \otimes k]{} & (\varinjlim F_i) \otimes B
\end{array}
$$

where the vertical maps $\varphi$ and $\psi$ are the isomorphisms of Theorem 6.159, the map $\vec{k}$ is induced from the transformation of direct systems $\{1_i \otimes k\}$, and 1 is the identity map on $\varinjlim F_i$. Since each $F_i$ is flat, the maps $1_i \otimes k$ are injections; since the index set $I$ is directed, the top row is exact, by Proposition 6.158. Therefore, $1 \otimes k \colon (\varinjlim F_i) \otimes A \to (\varinjlim F_i) \otimes B$ is an injection, for it is the composite of injections $\psi \vec{k} \varphi^{-1}$. Therefore, $\varinjlim F_i$ is flat. •

Here are new proofs of Lemma 6.139 and Corollary 6.140.

**Corollary 6.162.**

 (i) *If every finitely generated submodule of a right R-module M is flat, then M is flat.*

 (ii) *If R is a domain with $Q = \mathrm{Frac}(R)$, then Q is a flat R-module.*

**Proof.**

 (i) In Example 6.157(iii), we saw that $M$ is a direct limit, over a directed index set, of its finitely generated submodules. Since every finitely generated submodule is flat, by hypothesis, the result follows from Corollary 6.161.

 (ii) In Example 6.155(v), we saw that $Q$ is a direct limit, over a directed index set, of cyclic submodules, each of which is isomorphic to $R$. Since $R$ is flat, the result follows from Corollary 6.161. •

A remarkable theorem of Lazard states that a left $R$-module over any ring $R$ is flat if and only if it is a direct limit (over a directed index set) of finitely generated free left $R$-modules (Rotman, *An Introduction to Homological Algebra*, p. 253).

## Exercises

∗ **6.100.** Let $\{M_i, \varphi_j^i\}$ be a direct system of left $R$-modules with index set $I$, and let $\bigsqcup_i M_i$ be the disjoint union. Define $m_i \sim m_j$ on $\bigsqcup_i M_i$, where $m_i \in M_i$ and $m_j \in M_j$, if there exists an index $k$ with $k \succeq i$ and $k \succeq j$ such that $\varphi_k^i m_i = \varphi_k^j m_j$.

 (i) Prove that $\sim$ is an equivalence relation on $\bigsqcup_i M_i$.

 (ii) Denote the equivalence class of $m_i$ by $[m_i]$, and let $L$ denote the family of all such equivalence classes. Prove that the following definitions give $L$ the structure of an $R$-module:
$$r[m_i] = [rm_i] \text{ if } r \in R;$$
$$[m_i] + [m_j'] = [\varphi_k^i m_i + \varphi_k^j m_j'], \text{ where } k \succeq i \text{ and } k \succeq j.$$

 (iii) Prove that $L \cong \varinjlim M_i$.
    **Hint.** Use Proposition 6.156.

∗ **6.101.** Let $\{M_i, \varphi_j^i\}$ be a direct system of left $R$-modules, and let $F: {}_R\mathbf{Mod} \to \mathcal{C}$ be a functor to some category $\mathcal{C}$. Prove that $\{FM_i, F\varphi_j^i\}$ is a direct system in $\mathcal{C}$ if $F$ is covariant, while it is an inverse system if $F$ is contravariant.

**6.102.** Give an example of a direct system of modules, $\{A_i, \alpha_j^i\}$, over some directed index set $I$, for which $A_i \neq \{0\}$ for all $i$ and $\varinjlim A_i = \{0\}$.

**6.103.** (i) Let $K$ be a cofinal subset of a directed index set $I$ (that is, for each $i \in I$, there is $k \in K$ with $i \preceq k$), let $\{M_i, \varphi_j^i\}$ be a direct system over $I$, and let $\{M_i, \varphi_j^i\}$ be the subdirect system whose indices lie in $K$. Prove that the direct limit over $I$ is isomorphic to the direct limit over $K$.

 (ii) A partially ordered set $I$ has a ***top element*** if there exists $\infty \in I$ with $i \preceq \infty$ for all $i \in I$. If $\{M_i, \varphi_j^i\}$ is a direct system over $I$, prove that
$$\varinjlim M_i \cong M_\infty.$$

(iii) Show that part (i) may not be true if the index set is not directed.
**Hint.** Pushout.

**6.104.** Prove that a ring $R$ is left noetherian if and only if every direct limit (with directed index set) of injective left $R$-modules is itself injective.

**Hint.** See Proposition 6.97.

**6.105.** Consider the ideal $(x)$ in $k[x]$, where $k$ is a commutative ring. Prove that the completion of the polynomial ring $k[x]$ in the $(x)$-adic topology [see Example 6.147(v)] is $k[[x]]$, the ring of formal power series.

**6.106.** Let $r\colon \{A_i, \alpha_j^i\} \to \{B_i, \beta_j^i\}$ and $s\colon \{B_i, \beta_j^i\} \to \{C_i, \gamma_j^i\}$ be transformations of inverse systems over an index set $I$. If

$$0 \to A_i \overset{r_i}{\to} B_i \overset{s_i}{\to} C_i$$

is exact for each $i \in I$, prove that there is an exact sequence

$$0 \to \varprojlim A_i \overset{\vec{r}}{\to} \varprojlim B_i \overset{\vec{s}}{\to} \varprojlim C_i.$$

* **6.107.** A commutative $k$-algebra $F$ is a ***free commutative $k$-algebra*** with **basis** $X$, where $X$ is a subset of $F$, if for every commutative $k$-algebra $A$ and every function $\varphi\colon X \to A$, there exists a unique $k$-algebra map $\widetilde{\varphi}$ with $\widetilde{\varphi}(x) = \varphi(x)$ for all $x \in X$.

(i) Let $\mathrm{Fin}(X)$ be the family of all finite subsets of a set $X$, partially ordered by inclusion. Prove that $\{k[Y], \varphi_Z^Y\}$, where the morphisms $\varphi_Z^Y\colon k[Y] \to k[Z]$ are the $k$-algebra maps induced by inclusions $Y \to Z$, is a direct system of commutative $k$-algebras over $\mathrm{Fin}(X)$.

(ii) Denote $\varinjlim k[Y]$ by $k[X]$, and prove that $k[X]$ is the free commutative $k$-algebra with basis $X$. (Another construction of $k[X]$ is given on page 713.)

**6.108.** If $I$ is a partially ordered set and $\mathcal{C}$ is a category, then a ***presheaf*** over $I$ in $\mathcal{C}$ is a contravariant functor $\mathcal{F}\colon \mathbf{PO}(I) \to \mathcal{C}$ [see Example 6.38(vi)].

(i) If $I$ is the family of all open intervals $U$ in $\mathbb{R}$ containing 0, show that $\mathcal{F}$ in Example 6.155(vi) is a presheaf of abelian groups.

(ii) Let $X$ be a topological space, and let $I$ be the partially ordered set whose elements are the open sets in $X$. Define a sequence of presheaves $\mathcal{F}' \to \mathcal{F} \to \mathcal{F}''$ over $I$ to **Ab** to be ***exact*** if

$$\mathcal{F}'(U) \to \mathcal{F}(U) \to \mathcal{F}''(U)$$

is an exact sequence for every $U \in I$. If $\mathcal{F}$ is a presheaf on $I$, define $\mathcal{F}_x$, the **stalk** at $x \in X$, by $\mathcal{F}_x = \varinjlim_{U \ni x} \mathcal{F}(U)$. If $\mathcal{F}' \to \mathcal{F} \to \mathcal{F}''$ is an exact sequence of presheaves, prove, for every $x \in X$, that there is an exact sequence of stalks

$$\mathcal{F}'_x \to \mathcal{F}_x \to \mathcal{F}''_x.$$

**6.109.** Prove that if $T\colon {}_R\mathbf{Mod} \to \mathbf{Ab}$ is an additive left exact functor preserving products, then $T$ preserves inverse limits.

**\* 6.110.** Generalize Proposition 4.4 to allow infinitely many summands. Let $(S_i)_{i \in I}$ be a family of submodules of an $R$-module $M$, where $R$ is a commutative ring. If $M = \langle \bigcup_{i \in I} S_i \rangle$, then the following conditions are equivalent.

 (i) $M = \bigoplus_{i \in I} S_i$.

 (ii) Every $a \in M$ has a unique expression of the form $a = s_{i_1} + \cdots + s_{i_n}$, where $s_{i_j} \in S_{i_j}$.

 (iii) For each $i \in I$,

$$S_i \cap \left\langle \bigcup_{j \neq i} S_j \right\rangle = \{0\}.$$

## Section 6.10. Adjoint Functors

The Adjoint Isomorphism, Theorem 6.134, gives a natural isomorphism

$$\tau \colon \operatorname{Hom}_S(A \otimes_R B, C) \to \operatorname{Hom}_R(A, \operatorname{Hom}_S(B, C)),$$

where $R$ and $S$ are rings and $A_R$, $_RB_S$, and $C_S$ are modules. Let us rewrite this by setting $F = \square \otimes_R B$ and $G = \operatorname{Hom}_S(B, \square)$:

$$\tau \colon \operatorname{Hom}_S(FA, C) \to \operatorname{Hom}_R(A, GC).$$

Thus, $F \colon \mathbf{Mod}_R \to \mathbf{Mod}_S$ and $G \colon \mathbf{Mod}_S \to \mathbf{Mod}_R$. If we pretend that $\operatorname{Hom}(\square, \square)$ is an inner product, then we are reminded of adjoints in Linear Algebra (we discuss them on page 695): if $T \colon V \to W$ is a linear transformation, then its *adjoint* is the linear transformation $T^* \colon W \to V$ such that

$$(Tv, w) = (v, T^*w)$$

for all $v \in V$ and $w \in W$.

**Definition.** Given categories $\mathcal{C}$ and $\mathcal{D}$, an ordered pair $(F, G)$ of functors,

$$F \colon \mathcal{C} \to \mathcal{D} \quad \text{and} \quad G \colon \mathcal{D} \to \mathcal{C}$$

is an **adjoint pair** if, for each pair of objects $C \in \mathcal{C}$ and $D \in \mathcal{D}$, there are bijections

$$\tau_{C,D} \colon \operatorname{Hom}_{\mathcal{D}}(FC, D) \to \operatorname{Hom}_{\mathcal{C}}(C, GD)$$

that are natural transformations in $C$ and in $D$.

In more detail, the following two diagrams commute for every $f \colon C' \to C$ in $\mathcal{C}$ and $g \colon D \to D'$ in $\mathcal{D}$:

$$
\begin{array}{ccc}
\operatorname{Hom}_{\mathcal{D}}(FC, D) \xrightarrow{(Ff)^*} \operatorname{Hom}_{\mathcal{D}}(FC', D) & & \operatorname{Hom}_{\mathcal{D}}(FC, D) \xrightarrow{g_*} \operatorname{Hom}_{\mathcal{D}}(FC, D') \\
\downarrow \tau_{C,D} \qquad\qquad \downarrow \tau_{C',D} & & \downarrow \tau_{C,D} \qquad\qquad \downarrow \tau_{C,D'} \\
\operatorname{Hom}_{\mathcal{C}}(C, GD) \xrightarrow[f^*]{} \operatorname{Hom}_{\mathcal{C}}(C', GD) & & \operatorname{Hom}_{\mathcal{C}}(C, GD) \xrightarrow[(Gg)_*]{} \operatorname{Hom}_{\mathcal{C}}(C, GD')
\end{array}
$$

**Example 6.163.**

(i) Recall Example 6.52(v): let $U \colon \mathbf{Groups} \to \mathbf{Sets}$ be the *forgetful functor* that assigns to each group $G$ its underlying set and views each homomorphism as a mere function, and let $F \colon \mathbf{Sets} \to \mathbf{Groups}$ be the ***free functor*** that assigns to each set $X$ the free group $FX$ having basis $X$. That $FX$ is free with basis $X$ says, for every group $H$, that every function $\varphi \colon X \to H$ corresponds to a unique homomorphism $\widetilde{\varphi} \colon FX \to H$. It follows that if $\varphi \colon X \to Y$ is any function, then $\widetilde{\varphi} \colon FX \to FY$; this is how $F$ is defined on morphisms: $F\varphi = \widetilde{\varphi}$. The reader should realize that the function $f \mapsto f|X$ is a bijection (whose inverse is $\varphi \mapsto \widetilde{\varphi}$)

$$\tau_{X,H} \colon \operatorname{Hom}_{\mathbf{Groups}}(FX, H) \to \operatorname{Hom}_{\mathbf{Sets}}(X, UH).$$

Indeed, $\tau_{X,H}$ is a natural bijection, showing that $(F, U)$ is an adjoint pair of functors.

This example can be generalized by replacing **Groups** with other categories having free objects; for example, $_R\mathbf{Mod}$ for any ring $R$.

(ii) Adjointness is a property of an *ordered pair* of functors. In (i), we saw that $(F, U)$ is an adjoint pair, where $F$ is a free functor and $U$ is the underlying functor. Were $(U, F)$ an adjoint pair, then there would be a natural bijection $\operatorname{Hom}_{\mathbf{Sets}}(UH, Y) \cong \operatorname{Hom}_{\mathbf{Groups}}(H, FY)$, where $H$ is a group and $Y$ is a set. This is false in general; if $H = \mathbb{I}_2$ and $Y$ is a set with more than one element, then $|\operatorname{Hom}_{\mathbf{Sets}}(UH, Y)| = |Y|^2$, while $|\operatorname{Hom}_{\mathbf{Groups}}(H, FY)| = 1$ (the free group $FY$ has no elements of order 2). Therefore, $(U, F)$ is not an adjoint pair.

(iii) Theorem 6.134 shows that if $R$ and $S$ are rings and $B$ is an $(R, S)$-bimodule, then

$$\left( \square \otimes_R B, \operatorname{Hom}_S(B, \square) \right)$$

is an adjoint pair of functors. ◀

For many more examples of adjoint pairs of functors, see Mac Lane, *Categories for the Working Mathematician*, Chapter 4, especially pp. 85–86, and Herrlich–Strecker, *Category Theory*, pp. 197–199.

Let $(F, G)$ be an adjoint pair of functors, where $F \colon \mathcal{C} \to \mathcal{D}$ and $G \colon \mathcal{D} \to \mathcal{C}$. If $C \in \operatorname{obj}(\mathcal{C})$, then setting $D = FC$ gives a bijection $\tau \colon \operatorname{Hom}_{\mathcal{D}}(FC, FC) \to \operatorname{Hom}_{\mathcal{C}}(C, GFC)$, so that $\eta_C$, defined by

$$\eta_C = \tau(1_{FC}),$$

is a morphism $C \to GFC$. Exercise 6.111 on page 518 shows that $\eta \colon 1_{\mathcal{C}} \to GF$ is a natural transformation; it is called the ***unit*** of the adjoint pair.

**Theorem 6.164.** *Let $(F, G)$ be an adjoint pair of functors, where $F \colon \mathcal{C} \to \mathcal{D}$ and $G \colon \mathcal{D} \to \mathcal{C}$. Then $F$ preserves all direct limits and $G$ preserves all inverse limits.*

**Remark.**

(i) There is no restriction on the index sets of the limits; in particular, they need not be directed.

(ii) A more precise statement is that if $\varinjlim C_i$ exists in $\mathcal{C}$, then $\varinjlim FC_i$ exists in $\mathcal{D}$, and $\varinjlim FC_i \cong F(\varinjlim C_i)$.  ◄

**Proof.** Let $I$ be a partially ordered set, and let $\{C_i, \varphi_j^i\}$ be a direct system in $\mathcal{C}$ over $I$. It is easy to see that $\{FC_i, F\varphi_j^i\}$ is a direct system in $\mathcal{D}$ over $I$. Consider the following diagram in $\mathcal{D}$:

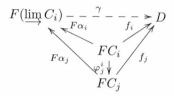

where $\alpha_i \colon C_i \to \varinjlim C_i$ are the maps in the definition of direct limit. We must show that there exists a unique morphism $\gamma \colon F(\varinjlim C_i) \to D$ making the diagram commute. The idea is to apply $G$ to this diagram, and to use the unit $\eta \colon 1_\mathcal{C} \to GF$ to replace $GF(\varinjlim C_i)$ and $GFC_i$ by $\varinjlim C_i$ and $C_i$, respectively. In more detail, there are morphisms $\eta$ and $\eta_i$, by Exercise 6.111 on page 518, making the following diagram commute:

$$
\begin{array}{ccc}
\varinjlim C_i & \xrightarrow{\ \eta\ } & GF(\varinjlim C_i) \\[4pt]
\alpha_i \uparrow & & \uparrow GF\alpha_i \\[4pt]
C_i & \xrightarrow{\ \eta_i\ } & GFC_i
\end{array}
$$

Combining this with $G$ applied to the original diagram gives commutativity of

$$
\varinjlim C_i \ \ \dashrightarrow^{\ \beta\ } \ \ GD
$$

By definition of direct limit, there exists a unique $\beta \colon \varinjlim C_i \to GD$ making the diagram commute; that is, $\beta \in \mathrm{Hom}_\mathcal{C}(\varinjlim C_i, GD)$. Since $(F, G)$ is an adjoint pair, there exists a natural bijection

$$
\tau \colon \mathrm{Hom}_\mathcal{D}(F(\varinjlim C_i), D) \to \mathrm{Hom}_\mathcal{C}(\varinjlim C_i, GD).
$$

Define

$$
\gamma = \tau^{-1}(\beta) \in \mathrm{Hom}_\mathcal{D}(F(\varinjlim C_i), D).
$$

We claim that $\gamma \colon F(\varinjlim C_i) \to D$ makes the first diagram commute. The first commutative square in the definition of adjointness gives commutativity of

$$
\begin{array}{ccc}
\mathrm{Hom}_\mathcal{C}(\varinjlim C_i, GD) & \xrightarrow{\ \alpha_i^*\ } & \mathrm{Hom}_\mathcal{C}(C_i, GD) \\[4pt]
\tau^{-1} \downarrow & & \downarrow \tau^{-1} \\[4pt]
\mathrm{Hom}_\mathcal{D}(F(\varinjlim C_i), D) & \xrightarrow{(F\alpha_i)^*} & \mathrm{Hom}_\mathcal{D}(FC_i, D)
\end{array}
$$

Hence, $\tau^{-1}\alpha_i^* = (F\alpha_i)^*\tau^{-1}$. Evaluating both functions on $\beta$, we have

$$(F\alpha_i)^*\tau^{-1}(\beta) = (F\alpha_i)^*\gamma = \gamma F\alpha_i.$$

On the other hand, since $\beta\alpha_i = (Gf_i)\eta_i$, we have

$$\tau^{-1}\alpha_i^*(\beta) = \tau^{-1}(\beta\alpha_i) = \tau^{-1}((Gf_i)\eta_i).$$

Therefore,

$$\gamma F\alpha_i = \tau^{-1}((Gf_i)\eta_i).$$

The second commutative square in the definition of adjointness gives commutativity of

$$
\begin{array}{ccc}
\mathrm{Hom}_{\mathcal{D}}(FC_i, FC_i) & \xrightarrow{\ (f_i)_*\ } & \mathrm{Hom}_{\mathcal{D}}(FC_i, D) \\
\Big\downarrow{\scriptstyle \tau} & & \Big\downarrow{\scriptstyle \tau} \\
\mathrm{Hom}_{\mathcal{C}}(C_i, GFC_i) & \xrightarrow[\ (Gf_i)_*\ ]{} & \mathrm{Hom}_{\mathcal{C}}(C_i, GD)
\end{array}
$$

that is,

$$\tau(f_i)_* = (Gf_i)_*\tau.$$

Evaluating at $1_{FC_i}$, the definition of $\eta_i$ gives $\tau(f_i)_*(1) = (Gf_i)_*\tau(1)$, and so $\tau f_i = (Gf_i)_*\eta_i$. Therefore,

$$\gamma F\alpha_i = \tau^{-1}((Gf_i)\eta_i) = \tau^{-1}\tau f_i = f_i,$$

so that $\gamma$ makes the original diagram commute.

We leave the proof of the uniqueness of $\gamma$ as an exercise for the reader, with the hint to use the uniqueness of $\beta$.

The dual proof shows that $G$ preserves inverse limits. $\quad\bullet$

There is a necessary and sufficient condition, called the **Adjoint Functor Theorem**, that a functor $F\colon \mathcal{C} \to \mathcal{D}$ be part of an adjoint pair; see Mac Lane, *Categories for the Working Mathematician*, p. 117. We state the special case of this theorem when $\mathcal{C}, \mathcal{D}$ are categories of modules.

**Theorem 6.165.** *If $F\colon \mathbf{Mod}_R \to \mathbf{Ab}$ is an additive functor, then the following statements are equivalent.*

(i) *$F$ preserves direct limits.*

(ii) *$F$ is right exact and preserves direct sums.*

(iii) *$F \cong \square \otimes_R B$ for some left $R$-module $B$.*

(iv) *$F$ has a right adjoint: there is a functor $G\colon \mathbf{Ab} \to \mathbf{Mod}_R$ so that $(F, G)$ is an adjoint pair.*

**Proof.** Rotman, *An Introduction to Homological Algebra*, p. 267. $\quad\bullet$

**Theorem 6.166.** *If $G\colon {}_R\mathbf{Mod} \to \mathbf{Ab}$ is an additive functor, then the following statements are equivalent.*

(i) *$G$ preserves inverse limits.*

(ii) *$G$ is left exact and preserves direct products.*

(iii) $G$ is **representable**; i.e., $G \cong \operatorname{Hom}_R(B, \square)$ *for some left R-module B*.

(iv) $G$ *has a left adjoint: there is a functor* $F \colon \mathbf{Ab} \to {}_R\mathbf{Mod}$ *so that* $(F, G)$ *is an adjoint pair.*

**Proof.** Ibid., p. 267.    •

## Exercises

* **6.111.** Let $(F, G)$ be an adjoint pair of functors, where $F \colon \mathcal{C} \to \mathcal{D}$ and $G \colon \mathcal{D} \to \mathcal{C}$, and let

(1)                               $\tau_{C,D} \colon \operatorname{Hom}(FC, D) \to \operatorname{Hom}(C, GD)$

be the natural bijection.

(i) If $D = FC$ in Equation (1), there is a natural bijection

$$\tau_{C,FC} \colon \operatorname{Hom}(FC, FC) \to \operatorname{Hom}(C, GFC)$$

with $\tau(1_{FC}) = \eta_C \in \operatorname{Hom}(C, GFC)$. Prove that $\eta \colon 1_{\mathcal{C}} \to GF$ is a natural transformation.

(ii) If $C = GD$ in Equation (1), there is a natural bijection

$$\tau_{GD,D}^{-1} \colon \operatorname{Hom}(GD, GD) \to \operatorname{Hom}(FGD, D)$$

with $\tau^{-1}(1_D) = \varepsilon_D \in \operatorname{Hom}(FGD, D)$. Prove that $\varepsilon \colon FG \to 1_{\mathcal{D}}$ is a natural transformation. (We call $\varepsilon$ the **counit** of the adjoint pair.)

**6.112.**  (i) Let $F \colon \mathbf{Groups} \to \mathbf{Ab}$ be the functor with $F(G) = G/G'$, where $G'$ is the commutator subgroup of a group $G$, and let $U \colon \mathbf{Ab} \to \mathbf{Groups}$ be the functor taking every abelian group $A$ into itself (that is, $UA$ regards $A$ as an object in **Groups**). Prove that $(F, U)$ is an adjoint pair of functors.

(ii) Prove that the unit of the adjoint pair $(F, U)$ is the natural map $G \to G/G'$.

**6.113.** Let $\varphi \colon k \to k^*$ be a ring homomorphism.

(i) Prove that if $F = \operatorname{Hom}_k(k^*, \square) \colon {}_k\mathbf{Mod} \to {}_{k^*}\mathbf{Mod}$, then both $({}_\varphi\square, F)$ and $(F, {}_\varphi\square)$ are adjoint pairs of functors, where ${}_\varphi\square$ is the change of rings functor (see Exercise 6.56 on page 449).

(ii) Using Theorem 6.164, conclude that both ${}_\varphi\square$ and $F$ preserve all direct limits and all inverse limits.

## Section 6.11.  Galois Theory for Infinite Extensions

We have investigated Galois Theory for finite extensions $E/k$, but there is also a theory of general extensions. In short, the Galois group $\operatorname{Gal}(E/k)$ will be made into a topological group, and there is a bijection between all the intermediate fields of $E/k$ and all the *closed* subgroups of $\operatorname{Gal}(E/k)$. We have waited until now to discuss this theory because it involves inverse limits.

We begin by reviewing some point-set topology. Given a set $X$, a family $(X_i)_{i \in I}$ of topological spaces, and a family $(\varphi_i \colon X \to X_i)_{i \in I}$ of functions, the **induced**

**topology** on $X$ is the smallest topology on $X$ making all $\varphi_i$ continuous. In particular, if $X$ is a subset of a topological space $Y$ and the family has only one member, the inclusion $\varphi \colon X \to Y$, then $X$ is called a **subspace** if it has the induced topology, and a subset $A$ is open in $X$ if and only if $A = \varphi^{-1}(U) = X \cap U$ for some open $U$ in $Y$. The **product topology** on a cartesian product $X = \prod_{i \in I} X_i$ of topological spaces is induced by the family of all projections $\varphi_i \colon X \to X_i$. If $U_j$ is an open subset of $X_j$, then $\varphi_j^{-1}(U_j) = \prod V_i$, where $V_j = U_j$ and $V_i = X_i$ for all $i \neq j$. A **cylinder** is a finite intersection of such sets; it is a subset of the form $\prod_{i \in I} V_i$, where $V_i$ is an open set in $X_i$ and almost all $V_i = X_i$. The family of all cylinders is a subbase of the product topology: every open set in $X$ is a union of cylinders. The **Tychonoff Theorem** says that products of compact spaces are compact.

**Lemma 6.167.** *Let $X = \prod_{i \in I} X_i$ be a product, and let $\varphi_i \colon X \to X_i$ be the $i$th projection.*

 (i) *If all $X_i$ are Hausdorff, then $X$ is Hausdorff.*

 (ii) *Let $Y$ be a topological space. Then a function $Y \to X$ is continuous if and only if $\varphi_i f \colon Y \to X_i$ is continuous for all $i$.*

 (iii) *If $(Y_i)_{i \in I}$ is a family of topological spaces and $(g_i \colon Y_i \to X_i)_{i \in I}$ is a family of continuous maps, then $g \colon \prod Y_i \to \prod X_i$, given by $(y_i) \mapsto (g_i(y_i))$, is continuous.*

 (iv) *If $\{G_i, \varphi_i^j\}$ is an inverse system in **Top**, where all the $G_i$ are discrete, then $\varprojlim_I G_i$ is a closed subset of $\prod_{i \in I} G_i$.*

**Proof.**

 (i) If $a = (a_i)$ and $b = (b_i)$ are distinct points in $X$, then $a_j \neq b_j$ for some $j$. Since $X_j$ is Hausdorff, there are disjoint open sets $U_j$ and $V_j$ in $X_j$ with $a_j \in U_j$ and $b_j \in V_j$. It follows that the cylinders $U_j \times \prod_{i \neq j} X_i$ and $V_j \times \prod_{i \neq j} X_i$ are disjoint neighborhoods of $a$ and $b$, respectively.

 (ii) If $f$ is continuous, then so are all the $\varphi_i f$, because the composite of continuous functions is continuous.

   Conversely, if $V \subseteq X$ is an open set, then $V = \bigcup_{i,j} \varphi_i^{-1}(U_i^j)$, where $U_i^j$ are open sets in $X_i$. Therefore,

$$f^{-1}(V) = f^{-1}\Big(\bigcup_{i,j} \varphi_i^{-1}(U_i^j)\Big) = \bigcup_{i,j} f^{-1}\varphi_i^{-1}(U_i^j) = \bigcup_{i,j} (\varphi_i f)^{-1}(U_i^j)$$

is open (for the $\varphi_i f$ are continuous), and so $f$ is continuous.

 (iii) If $\theta_j \colon \prod Y_i \to Y_j$ is the $j$th projection, then there is a commutative diagram

$$
\begin{array}{ccc}
\prod Y_i & \xrightarrow{\ g\ } & \prod X_i \\
\downarrow{\scriptstyle \theta_j} & & \downarrow{\scriptstyle \varphi_j} \\
Y_j & \xrightarrow[\ g_j\ ]{} & X_j
\end{array}
$$

Thus, $\varphi_j g = g_j \theta_j$ is continuous, being the composite of the continuous functions $g_j$ and $\theta_j$. It now follows from part (ii) that $g$ is continuous.

(iv) Let $L = \varprojlim_I G_i$; if $x = (x_i)$ is in the closure of $L$, then every open neighborhood $U$ of $x$ meets $L$. Choose $p \prec q$ in $I$, and let $U = \{x_p\} \times \{x_q\} \times \prod_{i \neq p,q} V_i$, where $V_i = G_i$ for all $i \neq p, q$. Note that $U$ is a cylinder: since $G_p$ and $G_q$ are discrete, $\{x_p\}$ and $\{x_q\}$ are open. There is $(g_i) \in L$ with $x_p = g_p$ and $x_q = g_q$; hence, $\varphi_p^q(x_q) = x_p$. This is true for all index pairs $p, q$ with $p \prec q$; hence, $x = (x_i) \in L$, and so $L$ is closed.  $\bullet$

**Definition.** A group $G$ is a ***topological group*** if it is a Hausdorff topological space such that inversion $\iota \colon G \to G$ (given by $\iota \colon g \mapsto g^{-1}$) and multiplication $\mu \colon G \times G \to G$ are continuous.

Of course, if a space $G$ is equipped with the discrete topology and $Y$ is any topological space, then every function $f \colon G \to Y$ is continuous: since every subset of $G$ is open, $f^{-1}(V)$ is open for every open $V \subseteq Y$. In particular, every discrete group is a topological group, for $G$ discrete implies that $G \times G$ is also discrete.

**Proposition 6.168.**

(i) *If $(G_i)_{i \in I}$ is a family of topological groups, then $\prod_{i \in I} G_i$ is a topological group.*

(ii) *If $\{G_i, \psi_i^j\}$ is an inverse system of topological groups, then $\varprojlim_I G_i$ is a topological group.*

**Proof.**

(i) By Lemma 6.167(i), the product $\prod_{i \in I} G_i$ is Hausdorff. Inversion $\iota \colon \prod G_i \to \prod G_i$ is given by $\iota \colon (x_i) \mapsto (x_i^{-1})$; since each $x_i \mapsto x_i^{-1}$ is continuous, so is $\iota$, by Lemma 6.167(iii). Finally, view $\prod_i G_i \times \prod_i G_i$ as $\prod_i (G_i \times G_i)$. Now multiplication $\mu \colon \prod_i G_i \times \prod_i G_i \to \prod_i G_i$ is continuous, by Lemma 6.167(iii), because each multiplication $G_i \times G_i \to G_i$ is continuous.

(ii) View $\varprojlim G_i$ as a subgroup of $\prod G_i$; every closed subgroup of a topological group is a topological group.  $\bullet$

Product spaces are related to function spaces. Given sets $X$ and $Y$, the function space $Y^X$ is the set of all $f \colon X \to Y$. Since elements of a product space $\prod_{i \in I} X_i$ are functions $f \colon I \to \bigcup_{i \in I} X_i$ with $f(i) \in X_i$ for all $i$, we can imbed $Y^X$ into $\prod_{x \in X} Z_x$ (where $Z_x = Y$ for all $x$) via $f \mapsto (f(x))$.

**Definition.** If $X$ and $Y$ are spaces, then the ***finite topology*** on the function space $Y^X$ has a subbase of open sets consisting of all sets

$$U(f; x_1, \ldots, x_n) = \{g \in Y^X : g(x_i) = f(x_i) \text{ for } 1 \leq i \leq n\},$$

where $f \colon X \to Y$, $n \geq 1$, and $x_1, \ldots, x_n \in X$.

**Proposition 6.169.** *If $Y$ is discrete, then the finite topology on $Y^X$ coincides with the topology induced by its being a subspace of $\prod_{x \in X} Z_x$ (where $Z_x = Y$ for all $x \in X$).*

**Proof.** When $Y$ is discrete, a cartesian product $\prod_{i \in I} V_i$, where $V_i = X$ for almost all $i$ and the other $V_i = \{x_i\}$ for some $x_i \in X$, is a cylinder. But these cylinders are precisely the subsets comprising the subbase of the finite topology.  $\bullet$

**Definition.** A ***profinite group*** is an inverse limit of finite groups.

If finite groups are given the discrete topology, then every profinite group is a topological group, by Proposition 6.168(ii). We can say more.

**Proposition 6.170.** *Every profinite group $G$ is a compact topological group.*

**Proof.** Each $G_i$ is compact, for it is finite, and the Tychonoff Theorem says that $\prod_{i \in I} G_i$ is compact. Now Lemma 6.167(iv) shows that $G$ is a closed subset of $\prod_{i \in I} G_i$, and so $G$ is compact. •

For example, the group $\mathbb{Z}_p$ of $p$-adic integers is compact as a topological space, for it is isomorphic to $\varprojlim_{\mathbb{N}} \mathbb{I}_{p^n}$ (in fact, $\mathbb{Z}_p$ is homeomorphic to the Cantor set).

Having developed the necessary topological tools, we now consider infinite extension fields.

**Definition.** A field extension $E/k$ is a ***Galois extension*** if it is algebraic and $E^G = k$, where $G = \mathrm{Gal}(E/k)$. If $E/k$ is a field extension, then its ***Galois group***, $\mathrm{Gal}(E/k)$, is the set of all those automorphisms of $E$ that fix $k$.

Theorem 3.36 shows that if $E/k$ is a finite extension, then this definition coincides with our earlier definition on page 196. Many properties of finite Galois extensions hold in the general case.

**Lemma 6.171.** *If $E/k$ is a Galois extension and $(K_i/k)_{i \in I}$ is the family of all finite Galois extensions $k \subseteq K_i \subseteq E$, then $E = \bigcup_{i \in I} K_i$.*

**Proof.** It suffices to prove that every $a \in E$ is contained in a finite Galois extension $K/k$. Now $\mathrm{irr}(a, k)$ is a separable polynomial in $k[x]$ having a root in $E$, by Theorem 3.36 (the finiteness hypothesis is not needed in proving this implication), and its splitting field $K$ over $k$ is a finite Galois extension contained in $E$. Therefore, $a \in K \subseteq E$. •

**Proposition 6.172.** *Let $k \subseteq B \subseteq E$ be a tower of fields, where $E/k$ and $B/k$ are Galois extensions.*

  (i) *If $\tau \in \mathrm{Gal}(E/k)$, then $\tau(B) = B$.*

  (ii) *If $\sigma \in \mathrm{Gal}(B/k)$, then there is $\widetilde{\sigma} \in \mathrm{Gal}(E/k)$ with $\widetilde{\sigma}|B = \sigma$.*

  (iii) *The map $\rho \colon \mathrm{Gal}(E/k) \to \mathrm{Gal}(B/k)$, given by $\sigma \mapsto \sigma|B$, is surjective; $\ker \rho = \mathrm{Gal}(E/B)$, and $\mathrm{Gal}(E/k)/\mathrm{Gal}(E/B) \cong \mathrm{Gal}(B/k)$.*

  (iv) *If $H \subseteq \mathrm{Gal}(E/k)$ and $E^H \subseteq B$, then $E^H = E^{\rho(H)}$.*

**Proof.**

  (i) By Lemma 6.171, we have $B = \bigcup_{i \in I} K_i$, where $(K_i/k)_{i \in I}$ is the family of all finite Galois extensions in $B$. But $\tau(K_i) = K_i$, by Theorem 3.17.

  (ii) Consider the family $X$ of all ordered pairs $(F, \varphi)$, where $B \subseteq F \subseteq E$ and $\varphi \colon F \to E$ is a field map extending $\sigma$. Partially order $X$ by $(F, \varphi) \preceq (F', \varphi')$ if $F \subseteq F'$ and $\varphi'|F = \varphi$. By Zorn's Lemma, there is a maximal element

$(F_0, \varphi_0)$ in $X$. The proof of Lemma 2.156, which proves this result for finite extensions, shows that $F_0 = E$.

(iii) The proof of Theorem 3.17 assumes that $E/k$ is a finite extension. However, parts (i) and (ii) show that this assumption is not necessary.

(iv) If $a \in E$, then $\sigma(a) = a$ for all $\sigma \in H$ if and only if $(\sigma|B)(a) = a$ for all $\sigma|B \in \rho(H)$.  •

At the moment, the Galois group $\mathrm{Gal}(E/k)$ of a Galois extension has no topology; we will topologize it using the next proposition.

**Proposition 6.173.** *If $E/k$ is a Galois extension and $(K_i/k)_{i \in I}$ is the family of all finite Galois extensions $k \subseteq K_i \subseteq E$, then there is an inverse system $\{\mathrm{Gal}(K_i/k), \varphi_i^j\}$ with*

$$\mathrm{Gal}(E/k) \cong \varprojlim_I \mathrm{Gal}(K_i/k).$$

**Proof.** Partially order $(K_i/k)_{i \in I}$ by inclusion. If $K_i \subseteq K_j$, then Theorem 3.17 shows that $\varphi_i^j \colon \mathrm{Gal}(K_j/k) \to \mathrm{Gal}(K_i/k)$, given by $\sigma \mapsto \sigma|K_i$, is well-defined, and $\{K_i, \varphi_i^j\}$ is an inverse system. By Theorem 3.17, the restriction $r_i \colon \sigma \mapsto \sigma|K_i$ is a homomorphism $\mathrm{Gal}(E/k) \to \mathrm{Gal}(K_i/k)$ making the following diagram commute:

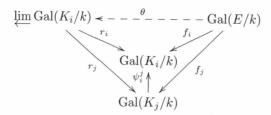

The universal property of inverse limit gives a map $\theta \colon \mathrm{Gal}(E/k) \to \varprojlim \mathrm{Gal}(K_i/k)$ which we claim is an isomorphism.

To see that $\theta$ is injective, take $\sigma \in \mathrm{Gal}(E/k)$ with $\sigma \neq 1$. There is $a \in E$ with $\sigma(a) \neq a$. By Lemma 6.171, there is a finite Galois extension $K_i$ with $a \in K_i$, and $\sigma|K_i \in \mathrm{Gal}(K_i/k)$. Now $(\sigma|K_i)(a) = \sigma(a) \neq a$, so that $\sigma|K_i \neq 1$. Thus, $r_i\theta(\sigma) \neq 1$, hence, $\theta(\sigma) \neq 1$, and so $\theta$ is injective.

To see that $\theta$ is surjective, take $\tau = (\tau_i) \in \varprojlim_I \mathrm{Gal}(K_i/k)$. If $a \in E$, then $a \in K_i$ for some $i$, by Lemma 6.171. Define $\sigma \colon E \to E$ by $\sigma(a) = \tau_i(a)$. This definition does not depend on $i$ because $(\tau_i)$ lies in the inverse limit: if $K_i \subseteq K_j$, then $\tau_i(a) = \tau_j(a)$. The reader may check that $\sigma$ lies in $\mathrm{Gal}(E/k)$ and that $\theta(\sigma) = \tau$.  •

**Corollary 6.174.** *If $E/k$ is a Galois extension, then $\mathrm{Gal}(E/k)$ is a profinite group and, hence, it is a compact topological group.*

**Proof.** The inverse limit $\varprojlim \mathrm{Gal}(K_i/k)$ is a profinite group, since each $\mathrm{Gal}(K_i/k)$ is finite and, hence, it is a compact topological group, by Proposition 6.170. Now use the isomorphism $\theta \colon \mathrm{Gal}(E/k) \to \varprojlim \mathrm{Gal}(K_i/k)$ in Proposition 6.173 to induce a topology on $\mathrm{Gal}(E/k)$.  •

The generalization to infinite Galois extensions of Theorem 3.45, the Fundamental Theorem of Galois Theory, is due to Krull. Let $E/k$ be a Galois extension, let

$$\mathrm{Sub}(\mathrm{Gal}(E/k))$$

denote the family of all *closed* subgroups of $\mathrm{Gal}(E/k)$, and let $\mathrm{Int}(E/k)$ denote the family of all intermediate fields $k \subseteq B \subseteq E$.

**Theorem 6.175 (Fundamental Theorem of Galois Theory II).** *Let $E/k$ be a Galois extension. The function $\gamma\colon \mathrm{Sub}(\mathrm{Gal}(E/k)) \to \mathrm{Int}(E/k)$, defined by*

$$\gamma\colon H \mapsto E^H,$$

*is an order-reversing bijection whose inverse, $\delta\colon \mathrm{Int}(E/k) \to \mathrm{Sub}(\mathrm{Gal}(E/k))$, is the order-reversing bijection*

$$\delta\colon B \mapsto \mathrm{Gal}(E/B).$$

*Moreover, an intermediate field $B/k$ is a Galois extension if and only if $\mathrm{Gal}(E/B)$ is a normal subgroup of $G$, in which case $\mathrm{Gal}(E/k)/\mathrm{Gal}(E/B) \cong \mathrm{Gal}(B/k)$.*

**Proof.** Proposition 3.31 proves that $\gamma$ is order-reversing: if $H \subseteq L$, then $E^L \subseteq E^H$. If $B$ is an intermediate field, then $\mathrm{Gal}(E/B)$ is a compact subgroup of $\mathrm{Gal}(E/k)$. Since $\mathrm{Gal}(E/k)$ is Hausdorff, every compact subset of it is closed; therefore, $\delta(B) = \mathrm{Gal}(E/B)$ is closed and, hence, it lies in $\mathrm{Sub}(\mathrm{Gal}(E/k))$. It is easy to prove that $\delta$ is order-reversing: if $B \subseteq C$, then $\mathrm{Gal}(E/C) \subseteq \mathrm{Gal}(E/B)$.

To see that $\gamma\delta = 1_{\mathrm{Int}(E/k)}$, we must show that if $B$ is an intermediate field, then $E^{\mathrm{Gal}(E/B)} = B$. Of course, $B \subseteq E^{\mathrm{Gal}(E/B)}$, for $\mathrm{Gal}(E/B)$ fixes $B$. For the reverse inclusion, let $a \in E$ with $a \notin B$. By Lemma 6.171, there is a finite Galois extension $K/B$ with $a \in K$. By finite Galois Theory, $B = K^{\mathrm{Gal}(K/B)}$, so there is $\sigma \in \mathrm{Gal}(K/B)$ with $\sigma(a) \neq a$. Now Proposition 6.172 says that $\sigma$ extends to $\widetilde{\sigma} \in \mathrm{Gal}(E/B)$; thus, $\widetilde{\sigma}(a) = \sigma(a) \neq a$, and so $a \notin E^{\mathrm{Gal}(E/B)}$.

To see that $\delta\gamma = 1_{\mathrm{Sub}(\mathrm{Gal}(E/k))}$, we must show that if $H$ is a closed subgroup of $\mathrm{Gal}(E/k)$, then $\mathrm{Gal}(E/E^H) = H$. Of course, $H \subseteq \mathrm{Gal}(E/E^H)$, for if $\sigma \in H$, then $\sigma \in \mathrm{Gal}(E/k)$ and $\sigma$ fixes $E^H$. For the reverse inclusion, let $\tau \in \mathrm{Gal}(E/E^H)$, and assume that $\tau \notin H$. Since $H$ is closed, its complement is open. Hence, there exists an open neighborhood $U$ of $\tau$ disjoint from $H$; we may assume that $U$ is a cylinder: $U = U(\tau; a_1, \ldots, a_n)$, where $a_1, \ldots, a_n \in E - E^H$. If $K/E^H(a_1, \ldots, a_n)$ is a finite Galois extension (where $E^H \subseteq K \subseteq E$), then Proposition 6.172(iii) says that restriction $\rho\colon \sigma \mapsto \sigma|K$ is a surjection $\mathrm{Gal}(E/E^H) \to \mathrm{Gal}(K/E^H)$. Now $\rho(\tau) = \tau|K \in \mathrm{Gal}(K/E^H)$, by Proposition 6.172(i); we claim that $\tau|K \notin \rho(H)$; that is, $\rho(H)$ is a proper subgroup of $\mathrm{Gal}(K/E^H)$. Otherwise, $\tau|K = \sigma|K$ for some $\sigma \in \mathrm{Gal}(E/E^H)$, contradicting $U(\tau; a_1, \ldots, a_n) \cap H = \varnothing$ [which says, for all $\sigma \in \mathrm{Gal}(E/E^H)$, that there is some $a_i$ with $\tau(a_i) \neq \sigma(a_i)$]. But finite Galois Theory says that $\rho(H) = \mathrm{Gal}(K/E^{\rho(H)}) = \mathrm{Gal}(K/E^H)$ [for $E^{\rho(H)} = E^H$, by Proposition 6.172(iv)], another contradiction. It follows that both $\gamma$ and $\delta$ are bijections. The last statement is just Proposition 6.172(iii). $\bullet$

The lattice-theoretic statements in the original Fundamental Theorem of Galois Theory, e.g., $\mathrm{Gal}(E/B) \cap \mathrm{Gal}(E/C) = \mathrm{Gal}(E/B \vee C)$, are valid in the general case as well, for their proof in Lemma 3.44 does not assume finiteness.

**Definition.** The ***absolute Galois group*** of a field $k$ is $\mathrm{Gal}(\overline{k}_s/k)$, where $\overline{k}_s$ is the ***separable algebraic closure*** of $k$; that is, $\overline{k}_s$ is the maximal separable extension of $k$ in $\overline{k}$.

Chapter IX of Neukirch–Schmidt–Wingberg, *Cohomology of Number Fields*, is entitled "The Absolute Galois Group of a Global Field." It begins by raising the question of "the determination of all extensions of a fixed base field $k$ (where the most important case is $k = \mathbb{Q}$), which means exploring how these extensions are built up over each other, how they are related, and how they can be classified. In other words, we want to study the structure of the absolute Galois group as a profinite group."

We mention that there is a Galois Theory of commutative ring extensions; see Chase–Harrison–Rosenberg, *Galois Theory and Cohomology of Commutative Rings*.

# Exercises

**6.114.** If $G$ is a group, $H$ is a discrete group, and $H^G$ has the product topology, prove that $\mathrm{Hom}(G, H) \subseteq H^G$ is a closed subset.

**6.115.** (i) If $G$ is a topological group and $a \in G$, prove that translation $g \mapsto ag$ is a homeomorphism $G \to G$. Conclude that if $U$ is an open subgroup of $G$, then the coset $aU$ is an open subset of $G$.

(ii) Prove that every open subgroup of a topological group is closed.

**6.116.** (i) Prove that a topological group $G$ is Hausdorff if and only if $\{1\}$ is closed.

(ii) Prove that if $N$ is a closed normal subgroup of a topological group $G$, then the quotient space $G/H$ is Hausdorff.

**6.117.** Give an example of a subgroup of the $p$-adic integers $\mathbb{Z}_p$ that is not closed.

**6.118.** (i) A topological space is ***totally disconnected*** if its components are its points. Prove that a topological group $G$ is totally disconnected if and only if $\bigcap_J V_j = \{1\}$, where $(V_j)_{j \in J}$ is the family of all the compact open neighborhoods of 1.

(ii) Prove that a topological group $G$ is profinite if and only if it is compact and totally disconnected.

**Hint.** See the article by Gruenberg, *Profinite Groups*, in Cassels–Fröhlich, *Algebraic Number Theory*.

**6.119.** Prove that every Galois extension $E/k$ is separable.

**Hint.** Use Proposition 3.41(iii).

**6.120.** Prove, for every prime $p$, that the absolute Galois group of $\mathbb{F}_p$ is an uncountable torsion-free group.

**6.121.** If $G$ is a profinite group, prove that $G \cong \varprojlim_I G/U_i$, where $(U_i)_{i \in I}$ is the family of all open normal subgroups of $G$.

# Chapter 7

# Representation Theory

Recall that a *representation* of a ring $R$ is a ring homomorphism $R \to \mathrm{End}_{\mathbb{Z}}(M)$, where $M$ is an abelian group. Now Proposition 6.14 shows that $R$-modules are just another way to view representations: every representation $R \to \mathrm{End}_{\mathbb{Z}}(M)$ equips $M$ with the structure of a left $R$-module, and conversely. In this chapter, we focus on $R = kG$, the group algebra of a finite group $G$ over a field $k$, and representations $kG \to \mathrm{End}_k(V)$, where $V$ is a finite-dimensional vector space over $k$. A key tool in this study is the notion of *character*, and we will use it to prove group-theoretic theorems of Burnside and of Frobenius. The chapter ends with a discussion of division rings as well as a theorem characterizing categories of modules which explains why matrix rings arise in the study of semisimple rings.

## Section 7.1. Chain Conditions

This chapter begins with rapid rephrasing of some earlier results about groups and rings into the language of modules. We have already proved the Jordan–Hölder Theorem for groups (Theorem 4.55); here is the version of this theorem for modules. [Both of these versions are special cases of a theorem about *operator groups* (Robinson, *A Course in the Theory of Groups*, p. 65).]

**Theorem 7.1 (Zassenhaus Lemma).** *Given submodules $A \subseteq A^*$ and $B \subseteq B^*$ of a left $R$-module $M$ over a ring $R$, there is an isomorphism*

$$\frac{A + (A^* \cap B^*)}{A + (A^* \cap B)} \cong \frac{B + (B^* \cap A^*)}{B + (B^* \cap A)}.$$

**Proof.** A straightforward adaptation of the proof of Lemma 4.52. •

**Definition.** A *series* (or *filtration*) of a left $R$-module $M$ over a ring $R$ is a sequence of submodules, $M = M_0, M_1, \ldots, M_n = \{0\}$, such that

$$M = M_0 \supseteq M_1 \supseteq \cdots \supseteq M_n = \{0\}.$$

The quotients $M_0/M_1, M_1/M_2, \ldots, M_{n-1}/M_n = M_{n-1}$ are called the **factor modules** of this series, and the number of strict inclusions is called the **length** of the series; equivalently, the length is the number of nonzero factor modules.

A **refinement** of a series is a series $M = M_0', M_1', \ldots, M_t' = \{0\}$ having the original series as a subsequence. Two series of a module $M$ are **equivalent** if there is a bijection between the lists of nonzero factor modules of each so that corresponding factor modules are isomorphic.

**Theorem 7.2 (Schreier Refinement Theorem).** *Any two series*

$$M = M_0 \supseteq M_1 \supseteq \cdots \supseteq M_n = \{0\} \quad and \quad M = N_0 \supseteq N_1 \supseteq \cdots \supseteq N_t = \{0\}$$

*of a left $R$-module $M$ have equivalent refinements.*

**Proof.** A straightforward adaptation, using the Zassenhaus Lemma, of the proof of Theorem 4.54.  •

Recall that a left $R$-module $M$ is *simple* (or *irreducible*) if $M \neq \{0\}$ and $M$ has no submodules other than $\{0\}$ and $M$ itself. The Correspondence Theorem shows that a submodule $N$ of a left $R$-module $M$ is a maximal submodule if and only if $M/N$ is simple; indeed, the proof of Corollary 6.25 (a left $R$-module $M$ is cyclic if and only if $M \cong R/I$ for some left ideal $I$) can be adapted to show that a left $R$-module is simple if and only if it is isomorphic to $R/I$ for some maximal left ideal $I$.

**Definition.** A **composition series** of a module is a series all of whose nonzero factor modules are simple.

A module need not have a composition series; for example, the abelian group $\mathbb{Z}$, considered as a $\mathbb{Z}$-module, has no composition series (Proposition 7.11). Notice that a composition series admits only insignificant refinements; we can only repeat terms (if $M_i/M_{i+1}$ is simple, then it has no proper nonzero submodules and, hence, there is no submodule $L$ with $M_i \supsetneq L \supsetneq M_{i+1}$). More precisely, any refinement of a composition series is equivalent to the original composition series.

**Theorem 7.3 (Jordan–Hölder Theorem).** *Any two composition series of a left $R$-module $M$ over a ring $R$ are equivalent. In particular, the length of a composition series, if one exists, is an invariant of $M$, called the **length** of $M$.*

**Proof.** As we have just remarked, any refinement of a composition series is equivalent to the original composition series. It now follows from the Schreier Refinement Theorem that any two composition series are equivalent; in particular, they have the same length.  •

**Corollary 7.4.** *If a left $R$-module $M$ has length $n$, then every ascending or descending chain of submodules of $M$ has length $\leq n$.*

**Proof.** There is a refinement of the given chain that is a composition series, and so the length of the given chain is at most $n$.  •

The Jordan–Hölder Theorem can be regarded as a kind of unique factorization theorem; for example, we used it in Corollary 4.56, to prove the Fundamental Theorem of Arithmetic. We now use it to prove Invariance of Dimension. If $V$ is an $n$-dimensional vector space over a field $k$, then $V$ has length $n$, for if $v_1, \ldots, v_n$ is a basis of $V$, then a composition series of $V$ is

$$V = \langle v_1, \ldots, v_n \rangle \supsetneq \langle v_2, \ldots, v_n \rangle \supsetneq \cdots \supsetneq \langle v_n \rangle \supsetneq \{0\}$$

(the factor modules are one-dimensional, hence they are simple $k$-modules).

In Chapter 5, we considered chain conditions on a ring and its ideals; we now consider chain conditions on modules and submodules.

**Definition.** A left $R$-module $M$ over a ring $R$ has **ACC**, the ***ascending chain condition***, if every ascending chain of submodules ***stops***; that is, if

$$S_1 \subseteq S_2 \subseteq S_3 \subseteq \cdots$$

is a chain of submodules, then there is some $t \geq 1$ with

$$S_t = S_{t+1} = S_{t+2} = \cdots .$$

A left $R$-module $M$ over a ring $R$ has **DCC**, the ***descending chain condition***, if every descending chain of submodules ***stops***; that is, if

$$S_1 \supseteq S_2 \supseteq S_3 \supseteq \cdots$$

is a chain of submodules, then there is some $t \geq 1$ with

$$S_t = S_{t+1} = S_{t+2} = \cdots .$$

We specialize these definitions to $R$ considered as a left $R$-module.

**Definition.** A ring $R$ is ***left noetherian*** if it has ACC on left ideals: every ascending chain of left ideals $I_1 \subseteq I_2 \subseteq I_3 \subseteq \cdots$ stops; that is, there is some $t \geq 1$ with $I_t = I_{t+1} = I_{t+2} = \cdots .$

If $k$ is a field, then every finite-dimensional $k$-algebra $A$ is both left and right noetherian, for if $\dim_k(A) = n$, then there are at most $n$ strict inclusions in any ascending chain of left ideals or of right ideals. In particular, if $G$ is a finite group, then $kG$ is finite-dimensional, and so it is left and right noetherian.

**Definition.** A ring $R$ is ***left artinian*** if it has DCC: every descending chain of left ideals $I_1 \supseteq I_2 \supseteq I_3 \supseteq \cdots$ stops; that is, there is some $t \geq 1$ with $I_t = I_{t+1} = I_{t+2} = \cdots .$

We define right artinian rings similarly, and there are examples of left artinian rings that are not right artinian (Exercise 7.10 on page 533). If $k$ is a field, then every finite-dimensional $k$-algebra $A$ is both left and right artinian, for if $\dim_k(A) = n$, then there are at most $n$ strict inclusions in any descending chain of left ideals or of right ideals. In particular, if $G$ is a finite group, then $kG$ is finite-dimensional, and so it is left and right artinian. We conclude that $kG$ has both chain conditions (on the left and on the right) when $k$ is a field and $G$ is a finite group.

The ring $\mathbb{Z}$ is (left) noetherian, but it is not (left) artinian, because the chain

$$\mathbb{Z} \supseteq (2) \supseteq (2^2) \supseteq (2^3) \supseteq \cdots$$

does not stop. In the next section, we will prove that left artinian implies left noetherian.

**Proposition 7.5.** *Let $R$ be a ring. The following conditions on a left $R$-module $M$ are equivalent.*

(i) *$M$ has ACC on submodules.*

(ii) *Every nonempty family of submodules of $M$ contains a maximal element.*

(iii) *Every submodule of $M$ is finitely generated.*

**Proof.** Adapt the proof of Proposition 5.33.   •

Proposition 5.33 is a special case of Theorem 7.5.

**Corollary 7.6 (= Proposition 5.33).** *The following conditions on a ring $R$ are equivalent.*

(i) *$R$ is left noetherian; that is, $R$ has ACC on left ideals.*

(ii) *Every nonempty family of left ideals of $R$ contains a maximal element.*

(iii) *Every left ideal is finitely generated.*

**Proof.** Consider $R$ as a left module over itself.   •

Here is the analog of Proposition 7.5.

**Proposition 7.7.** *Let $R$ be a ring. A left $R$-module $M$ has DCC on submodules if and only if every nonempty family of submodules of $M$ contains a minimal element.*

**Proof.** Adapt the proof of Proposition 5.33, replacing $\subseteq$ by $\supseteq$.   •

**Corollary 7.8.** *A ring $R$ has the DCC on left ideals if and only if every nonempty family of left ideals of $R$ contains a minimal element.*

**Proof.** Consider $R$ as a left module over itself.   •

**Definition.** A left ideal $L$ in a ring $R$ is a ***minimal left ideal*** if $L \neq (0)$ and there is no left ideal $J$ with $(0) \subsetneq J \subsetneq L$.

A ring need not contain minimal left ideals. For example, $\mathbb{Z}$ has no minimal ideals: every nonzero ideal $I$ in $\mathbb{Z}$ has the form $I = (n)$ for some nonzero integer $n$, and $I = (n) \supsetneq (2n) \neq (0)$.

**Example 7.9.** Let $R = \mathrm{Mat}_n(k)$, where $k$ is a field. For any $\ell$ between 1 and $n$, let $\mathrm{COL}(\ell)$ denote the $\ell$th columns; that is,

$$\mathrm{COL}(\ell) = \big\{ A = [a_{ij}] \in \mathrm{Mat}_n(k) : a_{ij} = 0 \text{ for all } j \neq \ell \big\}.$$

Let $e_1, \ldots, e_n$ be the standard basis of $k^n$. We identify $R = \mathrm{Mat}_n(k)$ with $\mathrm{End}_k(k^n)$, and so $\mathrm{COL}(\ell)$ is identified with

$$\mathrm{COL}(\ell) = \{T \colon k^n \to k^n : T(e_j) = 0 \text{ for } j \neq \ell\}.$$

We claim that $\mathrm{COL}(\ell)$ is a minimal left ideal in $R$. If $I$ is a nonzero left ideal with $I \subseteq \mathrm{COL}(\ell)$, choose a nonzero $F \in I$; now $F(e_\ell) = u \neq 0$; otherwise $F$ would kill every basis element and, hence, would be 0. To see that $\mathrm{COL}(\ell) \subseteq I$, take $T \in \mathrm{COL}(\ell)$, and write $T(e_\ell) = w$. Since $u \neq 0$, there is $S \in \mathrm{End}_k(k^n)$ with $S(u) = w$. Observe that

$$SF(e_j) = \begin{cases} 0 & \text{if } j \neq \ell; \\ S(u) = w & \text{if } j = \ell. \end{cases}$$

Therefore, $T = SF$, because they agree on a basis; since $I$ is a left ideal, $T \in I$. Therefore, $\mathrm{COL}(\ell) = I$, and $\mathrm{COL}(\ell)$ is a minimal left ideal.  ◀

**Proposition 7.10.**

(i) *Every minimal left ideal $L$ in a ring $R$ is a simple left $R$-module.*

(ii) *If $R$ is left artinian, then every nonzero left ideal $I$ contains a minimal left ideal.*

**Proof.**

(i) If $L$ contained a submodule $S$ with $\{0\} \subsetneq S \subsetneq L$, then $S$ would be a left ideal of $R$, contradicting the minimality of $L$.

(ii) If $\mathcal{F}$ is the family of all nonzero left ideals contained in $I$, then $\mathcal{F} \neq \varnothing$ because $I$ is nonzero. By Proposition 7.8, $\mathcal{F}$ has a minimal element, and any such is a minimal left ideal.  •

**Proposition 7.11.** *A left $R$-module $M$ over a ring $R$ has a composition series if and only if $M$ has both chain conditions on submodules.*

**Proof.** If $M$ has a composition series of length $n$, then no sequence of submodules can have length $> n$, lest we violate the Schreier Refinement Theorem (refining a series cannot shorten it). Therefore, $M$ has both chain conditions.

Conversely, let $\mathcal{F}_1$ be the family of all the proper submodules of $M$. By Proposition 7.8, the maximum condition gives a maximal submodule $M_1 \in \mathcal{F}_1$. Let $\mathcal{F}_2$ be the family of all proper submodules of $M_1$, and let $M_2$ be maximal such. Iterating, we have a descending sequence

$$M \supsetneq M_1 \supsetneq M_2 \supsetneq \cdots .$$

If $M_n$ occurs in this sequence, the only obstruction to constructing $M_{n+1}$ is if $M_n = \{0\}$. Since $M$ has both chain conditions, this chain must stop, and so $M_t = \{0\}$ for some $t$. This chain is a composition series of $M$, for each $M_i$ is a maximal submodule of its predecessor.  •

If $\Delta$ is a division ring, then a left $\Delta$-module $V$ is called a **left vector space** over $\Delta$. The following definition from Linear Algebra still makes sense here.

**Definition.** Let $V$ be a left vector space over a division ring $\Delta$. A list $X = x_1, \ldots, x_m$ in $V$ is *linearly dependent* if

$$x_i \in \langle x_1, \ldots, \widehat{x}_i, \ldots, x_m \rangle$$

for some $i$; otherwise, $X$ is called *linearly independent*.

As for vector spaces over fields, linear independence of $x_1, \ldots, x_m$ implies that

$$\langle x_1, \ldots, x_m \rangle = \langle x_1 \rangle \oplus \cdots \oplus \langle x_m \rangle.$$

The proper attitude is that theorems about vector spaces over fields have true analogs for left vector spaces over division rings, but the reader should not merely accept the word of a gentleman and scholar that this is so. Here is a proof of Invariance of Dimension similar to the proof (using the Jordan–Hölder Theorem) that we gave on page 527 for vector spaces over fields.

**Proposition 7.12.** *Let $V$ be a finitely generated[1] left vector space over a division ring $\Delta$.*

(i) *$V$ is a direct sum of copies of $\Delta$; that is, every finitely generated left vector space over $\Delta$ has a basis.*

(ii) *Any two bases of $V$ have the same number of elements.*

**Proof.**

(i) Let $V = \langle v_1, \ldots, v_n \rangle$, and consider the series

$$V = \langle v_1, \ldots, v_n \rangle \supseteq \langle v_2, \ldots, v_n \rangle \supseteq \langle v_3, \ldots, v_n \rangle \supseteq \cdots \supseteq \langle v_n \rangle \supseteq \{0\}.$$

Denote $\langle v_{i+1}, \ldots, v_n \rangle$ by $U_i$, so that $\langle v_i, \ldots, v_n \rangle = \langle v_i \rangle + U_i$. By the Second Isomorphism Theorem,

$$\langle v_i, \ldots, v_n \rangle / \langle v_{i+1}, \ldots, v_n \rangle = (\langle v_i \rangle + U_i)/U_i \cong \langle v_i \rangle / (\langle v_i \rangle \cap U_i).$$

Therefore, the $i$th factor module is isomorphic to a quotient of $\langle v_i \rangle \cong \Delta$ if $v_i \neq 0$. Since $\Delta$ is a division ring, its only quotients are $\Delta$ and $\{0\}$. After throwing away those $v_i$ corresponding to trivial factor modules $\{0\}$, we claim that the remaining $v$'s, denote them by $v_1, \ldots, v_m$, form a basis. For all $j$, we have $v_j \notin \langle v_{j+1}, \ldots, v_n \rangle$. The reader may now show, by induction on $m$, that $\langle v_1 \rangle, \ldots, \langle v_m \rangle$ generate their direct sum.

(ii) As in the proof of (i), a basis $v_1, v_2, \ldots, v_n$ of $V$ gives a series

$$V = \langle v_1, v_2, \ldots, v_n \rangle \supsetneq \langle v_2, \ldots, v_n \rangle \supsetneq \langle v_3, \ldots, v_n \rangle \supsetneq \cdots \supsetneq \langle v_n \rangle \supsetneq \{0\}.$$

This is a composition series, for every factor module is isomorphic to $\Delta$ and, hence, is simple, by Exercise 7.1 on page 532. By the Jordan–Hölder Theorem, the composition series arising from any other basis of $V$ must have the same length. ●

---

[1]This finiteness hypothesis will be removed on page 543.

Another proof of this proposition is sketched in Exercise 7.2 on page 532.

It now follows that every finitely generated left vector space $V$ over a division ring $\Delta$ has a left dimension; it will be denoted by $\dim(V)$.

If an abelian group $V$ is a left vector space and a right vector space over a division ring $\Delta$, must its left dimension equal its right dimension? There is an example (Jacobson, *Structure of Rings*, p. 158) of a division ring $\Delta$ and an abelian group $V$, which is a vector space over $\Delta$ on both sides, with left dimension 2 and right dimension 3.

We have just seen that dimension is well-defined for left vector spaces over division rings. On the other hand, recall our discussion of IBN: there are noncommutative rings $R$ with $R \cong R \oplus R$ as left $R$-modules; that is, bases of free modules may not have the same size.

Here is a surprising result.

**Theorem 7.13 (Wedderburn).** *Every finite division ring $\Delta$ is a field; that is, multiplication in $\Delta$ is commutative.*

**Proof. (Witt)**[2] If $Z$ denotes the center of $\Delta$, then $Z$ is a finite field, and so it has $q$ elements (where $q$ is a power of some prime). It follows that $\Delta$ is a vector space over $Z$, and so $|\Delta| = q^n$ for some $n \geq 1$; that is, if we define

$$[\Delta : Z] = \dim_Z(\Delta),$$

then $[\Delta : Z] = n$. The proof will be complete if we can show that $n > 1$ leads to a contradiction.

If $a \in \Delta$, define $C(a) = \{u \in \Delta : ua = au\}$. It is routine to check that $C(a)$ is a subdivision ring of $\Delta$ that contains $Z$: if $u, v \in \Delta$ commute with $a$, then so do $u + v, uv$, and $u^{-1}$ (when $u \neq 0$). Consequently, $|C(a)| = q^{d(a)}$ for some integer $d(a)$; that is, $[C(a) : Z] = d(a)$. We do not know whether $C(a)$ is commutative, but Exercise 7.5 on page 533 gives

$$[\Delta : Z] = [\Delta : C(a)][C(a) : Z],$$

where $[\Delta : C(a)]$ denotes the dimension of $\Delta$ as a left vector space over $C(a)$. That is, $n = [\Delta : C(a)]d(a)$, and so $d(a)$ is a divisor of $n$.

Since $\Delta$ is a division ring, its nonzero elements $\Delta^{\times}$ form a multiplicative group of order $q^n - 1$. By Exercise 7.4 on page 532, the center of the group $\Delta^{\times}$ is $Z^{\times}$ and, if $a \in \Delta^{\times}$, then its centralizer $C_{\Delta^{\times}}(a) = C(a)^{\times}$. Hence, $|Z(\Delta^{\times})| = q - 1$ and $|C_{\Delta^{\times}}(a)| = q^{d(a)} - 1$, where $d(a) \mid n$.

The class equation for $\Delta^{\times}$ is

$$|\Delta^{\times}| = |Z^{\times}| + \sum_i [\Delta^{\times} : C_{\Delta^{\times}}(a_i)],$$

where one $a_i$ is chosen from each noncentral conjugacy class. But

$$[\Delta^{\times} : C_{\Delta^{\times}}(a_i)] = |\Delta^{\times}|/|C_{\Delta^{\times}}(a_i)| = (q^n - 1)/(q^{d(a_i)} - 1),$$

---

[2]We shall give another proof of this in Theorem 7.127.

so that the class equation becomes

$$(1) \qquad\qquad q^n - 1 = q - 1 + \sum_i \frac{q^n - 1}{q^{d(a_i)} - 1}.$$

We have already noted that each $d(a_i)$ is a divisor of $n$, while the condition that $a_i$ is not central says that $d(a_i) < n$.

Recall that the $n$th cyclotomic polynomial is $\Phi_n(x) = \prod(x - \zeta)$, where $\zeta$ ranges over all the primitive $n$th roots of unity. In Corollary 2.72, we proved that $\Phi_n(q)$ is a common divisor of $q^n - 1$ and $(q^n - 1)/(q^{d(a_i)} - 1)$ for all $i$, and so Equation (1) gives

$$\Phi_n(q) \mid (q - 1).$$

If $n > 1$ and $\zeta$ is a primitive $n$th root of unity, then $\zeta \neq 1$, and hence $\zeta$ is a point on the unit circle to the left of the vertical line through $(1, 0)$. Since $q$ is a prime power, it is a point on the $x$-axis with $q \geq 2$, and so the distance $|q - \zeta| > q - 1$. Therefore,

$$|\Phi_n(q)| = \prod |q - \zeta| > q - 1,$$

and this contradicts $\Phi_n(q) \mid (q - 1)$. We conclude that $n = 1$; that is, $\Delta = Z$, and so $\Delta$ is commutative.   •

## Exercises

∗ **7.1.** Prove that a division ring $\Delta$ is a simple left $\Delta$-module.

∗ **7.2.** Let $\Delta$ be a division ring.

  (i) Generalize the proof of Lemma 5.68 to prove that $\alpha \preceq S$, defined by $\alpha \in \langle S \rangle$, is a dependency relation from $\Delta$ to $\mathcal{P}(\Delta)$ (*dependency relation* is defined on page 337).

  (ii) Use Theorem 5.70 to prove that every left vector space over $\Delta$ has a basis.

  (iii) Use Theorem 5.72 to prove that any two bases of a left vector space over $\Delta$ have the same cardinality.

∗ **7.3.** If $k$ is a field and $A$ is a finite-dimensional $k$-algebra, define

$$L = \{\lambda_a \in \mathrm{End}_k(A) : \lambda_a \colon x \mapsto ax\} \text{ and } R = \{\rho_a \in \mathrm{End}_k(A) : \rho_a \colon x \mapsto xa\}.$$

Prove that $L$ and $R$ are $k$-algebras, and that there are $k$-algebra isomorphisms

$$L \cong A \text{ and } R \cong A^{\mathrm{op}}.$$

**Hint.** Show that the function $A \to L$, defined by $a \mapsto \lambda_a$, is an injective $k$-algebra map which is surjective because $A$ is finite-dimensional.

∗ **7.4.** Let $\Delta$ be a division ring.

  (i) Prove that the center $Z(\Delta)$ is a field.

  (ii) If $\Delta^\times$ is the multiplicative group of nonzero elements of $\Delta$, prove that $Z(\Delta^\times) = Z(\Delta)^\times$; that is, the center of the multiplicative group $\Delta^\times$ consists of the nonzero elements of $Z(\Delta)$.

\* **7.5.** (i) Let $C$ be a subdivision ring of a division ring $\Delta$. Prove that $\Delta$ is a left vector space over $C$, and conclude that $[\Delta : C] = \dim_C(\Delta)$ is defined (if $\Delta$ is an infinite-dimensional vector space over $C$, we merely say $\dim_C(\Delta) = \infty$).

(ii) If $Z \subseteq C \subseteq D$ is a tower of division rings with $[\Delta : C]$ and $[C : Z]$ finite, prove that $[\Delta : Z]$ is finite and
$$[\Delta : Z] = [\Delta : C][C : Z].$$

**Hint.** If $u_1, \ldots, u_m$ is a basis of $\Delta$ as a left vector space over $C$ and $c_1, \ldots, c_d$ is a basis of $C$ as a left vector space over $Z$, show that the set of all $c_i u_j$ (in this order) is a basis of $\Delta$ over $Z$.

\* **7.6. (Modular Law)** Let $A$, $B$, and $A'$ be submodules of a module $M$. If $A' \subseteq A$, prove that $A \cap (B + A') = (A \cap B) + A'$.

\* **7.7.** Recall that a ring $R$ has *zero-divisors* if there exist nonzero $a, b \in R$ with $ab = 0$. More precisely, an element $a$ in a ring $R$ is called a **left zero-divisor** if $a \neq 0$ and there exists a nonzero $b \in R$ with $ab = 0$; the element $b$ is called a **right zero-divisor**. Prove that a left artinian ring $R$ having no left zero-divisors must be a division ring.

\* **7.8.** Let $0 \to A \to B \to C \to 0$ be an exact sequence of left $R$-modules over a ring $R$.

(i) Prove that if both $A$ and $C$ have DCC, then $B$ has DCC. Conclude, in this case, that $A \oplus B$ has DCC.

(ii) Prove that if both $A$ and $C$ have ACC, then $B$ has ACC. Conclude, in this case, that $A \oplus B$ has ACC.

(iii) Prove that every ring $R$ that is a direct sum of minimal left ideals is left artinian.

**7.9.** If $R$ is a (not necessarily commutative) ring, define the polynomial ring $R[x]$ as usual but with the indeterminate $x$ commuting with coefficients in $R$; thus, $x \in Z(R[x])$. Generalize the Hilbert Basis Theorem to such polynomial rings: if $R$ is left noetherian, then $R[x]$ is left noetherian.

\* **7.10.** Let $R$ be the ring of all $2 \times 2$ upper triangular matrices $\left[\begin{smallmatrix} a & b \\ 0 & c \end{smallmatrix}\right]$, where $a \in \mathbb{Q}$ and $b, c \in \mathbb{R}$. Prove that $R$ is right artinian but not left artinian.

**Hint.** The ring $R$ is not left artinian because, for every $V \subseteq \mathbb{R}$ that is a vector space over $\mathbb{Q}$,
$$\begin{bmatrix} 0 & V \\ 0 & 0 \end{bmatrix} = \left\{ \begin{bmatrix} 0 & v \\ 0 & 0 \end{bmatrix} : v \in V \right\}$$
is a left ideal.

\* **7.11.** If $k$ is a field of characteristic 0, then $\mathrm{End}_k(k[t])$ contains the operators
$$x \colon f(t) \mapsto \tfrac{d}{dt} f(t) \quad \text{and} \quad y \colon f(t) \mapsto t f(t).$$

(i) If $A_1(k)$ is the subalgebra of $\mathrm{End}_k(k[t])$ generated by $x$ and $y$, prove that
$$yx = xy + 1.$$

(ii) Prove that $A_1(k)$ is a left noetherian ring that satisfies the left and right cancellation laws (if $a \neq 0$, then either equation $ab = ac$ or $ba = ca$ implies $b = c$).

(iii) Prove that $A_1(k)$ has no proper nontrivial two-sided ideals.

**Remark.** This exercise can be generalized by replacing $k[t]$ with $k[t_1, \ldots, t_n]$ and the operators $x$ and $y$ with
$$x_i \colon f(t_1, \ldots, t_n) \mapsto \tfrac{d}{dt_i} f(t_1, \ldots, t_n) \quad \text{and} \quad y_i \colon f(t_1, \ldots, t_n) \mapsto t_i f(t_1, \ldots, t_n).$$

The subalgebra $A_n(k)$ of $\text{End}_k(k[t_1, \ldots, t_n])$ generated by $x_1, \ldots, x_n, y_1, \ldots, y_n$ is called the $n$th **Weyl algebra** over $k$. Weyl introduced this algebra to model momentum and position operators in Quantum Mechanics. It can be shown that $A_n(k)$ is a left noetherian simple domain for all $n \geq 1$ ( McConnell–Robson, *Noncommutative Noetherian Rings*, p. 19). ◄

**7.12.** Recall that an **idempotent** in a ring $A$ is an element $e \in A$ with $e \neq 0$ and $e^2 = e$. If $M$ is a left $R$-module over a ring $R$, prove that every direct summand $S \subseteq M$ determines an idempotent in $\text{End}_R(M)$.

**Hint.** See Corollary 6.28.

∗ **7.13.**  (i) **(Peirce Decomposition)** Prove that if $e$ is an idempotent in a ring $R$, then

$$R = Re \oplus R(1 - e).$$

  (ii) Let $R$ be a ring having left ideals $I$ and $J$ such that $R = I \oplus J$. Prove that there are idempotents $e \in I$ and $f \in J$ with $1 = e + f$; moreover, $I = Ie$ and $J = Jf$.
  **Hint.** Decompose $1 = e + f$, and show that $ef = 0 = fe$.

## Section 7.2. Jacobson Radical

The Jacobson radical, $J(R)$, of a ring $R$ is the analog of the Frattini subgroup in Group Theory; it is a two-sided ideal whose behavior has an impact on $R$. For example, *semisimple rings* (introduced in the next section) are rings generalizing the group algebra $\mathbb{C}G$ of a finite group $G$, and they will be characterized in terms of their Jacobson radical and chain conditions.

**Definition.** If $R$ is a ring, then its **Jacobson radical** $J(R)$ is defined to be the intersection of all the maximal left ideals in $R$. A ring $R$ is called **Jacobson semisimple** if $J(R) = (0)$.

Clearly, we can define another Jacobson radical: the intersection of all the maximal *right* ideals. In Proposition 7.20, we shall see that these coincide.

**Example 7.14.**

  (i) The ring $\mathbb{Z}$ is Jacobson semisimple. The maximal ideals in $\mathbb{Z}$ are the nonzero prime ideals $(p)$, and so $J(\mathbb{Z}) = \bigcap_{p \text{ prime}}(p) = (0)$.

  (ii) If $R$ is a local ring with unique maximal left ideal $P$, then $J(R) = P$. For example, $R = \{a/b \in \mathbb{Q} : b \text{ is odd}\}$ is such a ring; its unique maximal ideal is

$$R2 = (2) = \{2a/b : b \text{ is odd}\}.$$

  (iii) In Example 7.9, we saw that if $R = \text{Mat}_n(k)$, where $k$ is a field, then $\text{COL}(\ell)$ is a minimal left ideal, where $1 \leq \ell \leq n$ and

$$\text{COL}(\ell) = \big\{A = [a_{ij}] \in \text{Mat}_n(k) : a_{ij} = 0 \text{ for all } j \neq \ell\big\}.$$

We use these minimal left ideals to construct some maximal left ideals. The reader can show that that example generalizes to $R = \text{Mat}_n(\Delta)$, where $\Delta$ is a division ring. Define

$$\text{COL}^*(\ell) = \bigoplus_{j \neq \ell} \text{COL}(j);$$

$\text{COL}^*(\ell)$ is a left ideal with

$$R/\text{COL}^*(\ell) \cong \text{COL}(\ell)$$

as left $R$-modules. Since $\text{COL}(\ell)$ is a minimal left ideal, it is a simple left $R$-module, and hence $\text{COL}^*(\ell)$ is a maximal left ideal. Therefore,

$$J(R) \subseteq \bigcap_{\ell} \text{COL}^*(\ell) = (0),$$

so that $R = \text{Mat}_n(\Delta)$ is Jacobson semisimple. ◄

**Proposition 7.15.** *Given a ring $R$, the following conditions are equivalent for $x \in R$:*

(i) $x \in J(R)$;

(ii) *for every $r \in R$, the element $1 - rx$ has a left inverse; that is, there is $u \in R$ with $u(1 - rx) = 1$;*

(iii) $x(R/I) = (0)$ *for every maximal left ideal $I$ (equivalently, $xM = (0)$ for every simple left $R$-module $M$).*

**Proof.**

(i) $\Rightarrow$ (ii) If there is $r \in R$ with $1 - rx$ not having a left inverse, then $R(1 - rx)$ is a proper left ideal, for it does not contain 1. By Exercise on page 414, there is a maximal left ideal $I$ with $1 - rx \in R(1 - rx) \subseteq I$. Now $rx \in J(R) \subseteq I$, because $J(R)$ is a left ideal, and so $1 = (1 - rx) + rx \in I$, a contradiction.

(ii) $\Rightarrow$ (iii) As we mentioned on page 526, a left $R$-module $M$ is simple if and only if $M \cong R/I$, where $I$ is a maximal left ideal. Suppose there is a simple module $M$ for which $xM \neq (0)$; hence, there is $m \in M$ with $xm \neq 0$ (of course, $m \neq 0$). It follows that the submodule $Rxm \neq (0)$, for it contains $1xm$. Since $M$ is simple, it has only one nonzero submodule, namely, $M$ itself, and so $Rxm = M$. Therefore, there is $r \in R$ with $rxm = m$; that is, $(1 - rx)m = 0$. By hypothesis, $1 - rx$ has a left inverse, say, $u(1 - rx) = 1$. Hence, $0 = u(1 - rx)m = m$, a contradiction.

(iii) $\Rightarrow$ (i) If $x(R/I) = (0)$, then $x(1 + I) = x + I = I$; that is, $x \in I$. Therefore, if $x(R/I) = (0)$ for every maximal left ideal $I$, then $x \in \bigcap_I I = J(R)$. •

Notice that condition (ii) in Proposition 7.15 can be restated: $x \in J(R)$ if and only if $1 - z$ has a left inverse for every $z \in Rx$.

The following result is most frequently used in Commutative Algebra.

**Corollary 7.16 (Nakayama's Lemma).** *If $A$ is a finitely generated left $R$-module and $JA = A$, where $J = J(R)$ is the Jacobson radical, then $A = \{0\}$.*

*In particular, if $R$ is a commutative local ring with unique maximal ideal $P$ and $A$ is a finitely generated $R$-module with $PA = A$, then $A = \{0\}$.*

**Proof.** Let $a_1, \ldots, a_n$ be a generating set of $A$ that is minimal in the sense that no proper subset generates $A$. Since $JA = A$, we have $a_1 = \sum_{i=1}^{n} r_i a_i$, where $r_i \in J$. It follows that

$$(1 - r_1)a_1 = \sum_{i=2}^{n} r_i a_i.$$

Since $r_1 \in J$, Proposition 7.15 says that $1 - r_1$ has a left inverse, say, $u$, and so $a_1 = \sum_{i=2}^{n} u r_i a_i$. This is a contradiction, for now $A$ can be generated by the proper subset $\{a_2, \ldots, a_n\}$. The second statement follows at once because $J(R) = P$ when $R$ is a local ring with maximal ideal $P$.    •

**Remark.** The hypothesis in Nakayama's Lemma that the module $A$ be finitely generated is necessary. For example, it is easy to check that $R = \{a/b \in \mathbb{Q} : b \text{ is odd}\}$ is a local ring with maximal ideal $P = (2)$, while $\mathbb{Q}$ is an $R$-module with $P\mathbb{Q} = 2\mathbb{Q} = \mathbb{Q}$.  ◀

**Remark.** There are other characterizations of $J(R)$. One such will be given in Proposition 7.20, in terms of elements having two-sided inverses. Another characterization is in terms of *left quasi-regular* elements: those $x \in R$ for which there exist $y \in R$ with $y \circ x = 0$ (here, $y \circ x = x + y - yx$ is the *circle operation*); a left ideal is called **left quasi-regular** if each of its elements is left quasi-regular. It can be proved that $J(R)$ is the unique maximal left quasi-regular ideal in $R$ (Lam, *A First Course in Noncommutative Rings*, pp. 67–68).  ◀

Recall that an *element* $a$ in a ring $R$ is *nilpotent* if $a^m = 0$ for some $m \geq 1$.

**Definition.** A left ideal $A$ in a ring $R$ is **nilpotent** if there is some integer $m \geq 1$ with $A^m = (0)$.

The left ideal $A^m$ is the set of all sums of products of the form $a_1 \cdots a_m$, where $a_j \in A$ for all $j$; that is, $A^m = \{\sum_i a_{i1} \cdots a_{im} : a_{ij} \in A\}$. It follows that if $A$ is nilpotent, then every element $a \in A$ is nilpotent; that is, $a^m = 0$. On the other hand, if $a \in R$ is a nilpotent element, it does not follow that $Ra$, the left ideal generated by $a$, is a nilpotent ideal. For example, let $R = \mathrm{Mat}_2(k)$, for some commutative ring $k$, and let $a = \left[\begin{smallmatrix} 0 & 1 \\ 0 & 0 \end{smallmatrix}\right]$. Now $a^2 = \left[\begin{smallmatrix} 0 & 0 \\ 0 & 0 \end{smallmatrix}\right]$, but $Ra$ contains $e = \left[\begin{smallmatrix} 0 & 0 \\ 1 & 0 \end{smallmatrix}\right]\left[\begin{smallmatrix} 0 & 1 \\ 0 & 0 \end{smallmatrix}\right] = \left[\begin{smallmatrix} 0 & 0 \\ 0 & 1 \end{smallmatrix}\right]$, which is idempotent: $e^2 = e$. Hence, $e^m = e \neq 0$ for all $m$, and so $(Re)^m \neq (0)$.

**Corollary 7.17.** *If $R$ is a ring, then $I \subseteq J(R)$ for every nilpotent left ideal $I$ in $R$.*

**Proof.** Let $I^n = (0)$, and let $x \in I$. For every $r \in R$, we have $rx \in I$, and so $(rx)^n = 0$. The equation

$$(1 + rx + (rx)^2 + \cdots + (rx)^{n-1})(1 - rx) = 1$$

shows that $1 - rx$ is left invertible, and so $x \in J(R)$, by Proposition 7.15.    •

**Proposition 7.18.** *If $R$ is a left artinian ring, then $J(R)$ is a nilpotent ideal.*

**Proof.** Denote $J(R)$ by $J$ in this proof. The descending chain of left ideals,

$$J \supseteq J^2 \supseteq J^3 \supseteq \cdots,$$

stops, because $R$ is left artinian; say, $J^m = J^{m+1} = \cdots$. Define $I = J^m$; it follows that $I^2 = I$. We will assume that $I \neq (0)$ and reach a contradiction.

Let $\mathcal{F}$ be the family of all nonzero left ideals $B$ with $IB \neq (0)$; note that $\mathcal{F} \neq \varnothing$ because $I \in \mathcal{F}$. By Proposition 7.8, there is a minimal element $B_0 \in \mathcal{F}$. Choose $b \in B_0$ with $Ib \neq (0)$. Now

$$I(Ib) = I^2 b = Ib \neq (0),$$

so that $Ib \subseteq B_0 \in \mathcal{F}$, and minimality gives $B_0 = Ib$. Since $b \in B_0$, there is $x \in I \subseteq J = J(R)$ with $b = xb$. Hence, $0 = (1 - x)b$. But $1 - x$ has a left inverse, say, $u$, by Proposition 7.15, so that $0 = u(1 - x)b = b$, a contradiction. $\bullet$

The Jacobson radical is obviously a left ideal (for it is an intersection of left ideals), but it turns out to be a right ideal as well; that is, $J(R)$ is a two-sided ideal. We begin by giving another source of two-sided ideals (aside from kernels of ring maps).

**Definition.** If $R$ is a ring and $M$ is a left $R$-module, the ***annihilator*** of $M$ is

$$\operatorname{ann}(M) = \{a \in R : am = 0 \text{ for all } m \in M\}.$$

Let us show that $\operatorname{ann}(M)$ is a two-sided ideal in $R$. Now $\operatorname{ann}(M)$ is a left ideal, for if $am = 0$, then $(ra)m = r(am) = 0$; let us prove that it is a right ideal. Let $a \in \operatorname{ann}(M)$, $r \in R$, and $m \in M$. Since $M$ is a left $R$-module, we have $rm \in M$; since $a$ annihilates every element of $M$, we have $a(rm) = 0$. Finally, associativity gives $(ar)m = 0$ for all $m$, and so $ar \in \operatorname{ann}(M)$.

**Corollary 7.19.**

(i) $J(R) = \bigcap\limits_{\substack{I = \text{maximal} \\ \text{left ideal}}} \operatorname{ann}(R/I)$, *and so $J(R)$ is a two-sided ideal in $R$.*

(ii) $R/J(R)$ *is a Jacobson semisimple ring.*

**Proof.**

(i) If $x \in J(R)$, then $xM = \{0\}$ for every simple left $R$-module $M$, by Proposition 7.15(iii). But $M \cong R/I$ for some maximal left ideal $I$; that is, $x \in \operatorname{ann}(R/I)$. Thus, $x \in \bigcap\limits_{\substack{I = \text{maximal} \\ \text{left ideal}}} \operatorname{ann}(R/I)$.

For the reverse inclusion, if $x \in \bigcap\limits_{\substack{I = \text{maximal} \\ \text{left ideal}}} \operatorname{ann}(R/I)$, then $xM = \{0\}$ for every left $R$-module $M$ of the form $M \cong R/I$ for some maximal left ideal $I$. But every simple left $R$-module has this form. Therefore, $x \in J(R)$.

(ii) First, $R/J(R)$ is a ring, because $J(R)$ is a two-sided ideal. The Correspondence Theorem for rings shows that if $I$ is any two-sided ideal of $R$ contained in $J(R)$, then $J(R/I) = J(R)/I$; the result follows if $I = J(R)$. $\bullet$

Let us now show that we could have defined the Jacobson radical using right ideals instead of left ideals.

**Definition.** A **unit** in a ring $R$ is an element $u \in R$ having a two-sided inverse; that is, there is $v \in R$ with

$$uv = 1 = vu.$$

**Proposition 7.20.**

(i) *If $R$ is a ring, then*

$$J(R) = \{x \in R : 1 + rxs \text{ is a unit in } R \text{ for all } r, s \in R\}.$$

(ii) *If $R$ is a ring and $J'(R)$ is the intersection of all the maximal right ideals of $R$, then $J'(R) = J(R)$.*

**Proof.**

(i) Let $W$ be the set of all $x \in R$ such that $1 + rxs$ is a unit for all $r, s \in R$. If $x \in W$, then setting $s = -1$ gives $1 - rx$ a unit for all $r \in R$. Hence, $1 - rx$ has a left inverse, and so $x \in J(R)$, by Proposition 7.15. Therefore, $W \subseteq J(R)$. For the reverse inclusion, let $x \in J(R)$. Since $J(R)$ is a two-sided ideal, by Corollary 7.19, we have $xs \in J(R)$ for all $s \in R$. Proposition 7.15 says that $1 - rxs$ is left invertible for all $r \in R$; that is, there is $u \in R$ with $u(1 - rxs) = 1$. Thus, $u = 1 + urxs$. Now $(-ur)xs \in J(R)$, since $J(R)$ is a two-sided ideal, and so $u$ has a left inverse (Proposition 7.15 once again). On the other hand, $u$ also has a right inverse, namely, $1 - rxs$. By Exercise 6.13, $u$ is a unit in $R$. Therefore, $1 - rxs$ is a unit in $R$ for all $r, s \in R$. Finally, replacing $r$ by $-r$, we have $1 + rxs$ a unit, and so $J(R) \subseteq W$.

(ii) The description of $J(R)$ in part (i) is left-right symmetric. After proving right-sided versions of Proposition 7.15 and Corollary 7.19, one can see that $J'(R)$ is also described as in part (i). We conclude that $J'(R) = J(R)$.  •

## Exercises

**7.14.** (i) If $R$ is a commutative ring with $J(R) = (0)$, prove that $R$ has no nilpotent elements.

(ii) Give an example of a commutative ring $R$ having no nilpotent elements and for which $J(R) \neq (0)$.

**7.15.** Let $k$ be a field and $R = \text{Mat}_2(k)$. Prove that $a = \left[\begin{smallmatrix} 0 & 1 \\ 0 & 0 \end{smallmatrix}\right]$ is left quasi-regular, but that the principal left ideal $Ra$ is not a left quasi-regular ideal.

**7.16.** Prove that $R$ is Jacobson semisimple if and only if $R^{\text{op}}$ is.

**7.17.** Let $I$ be a two-sided ideal in a ring $R$. Prove that if $I \subseteq J(R)$, then

$$J(R/I) = J(R)/I.$$

## Section 7.3. Semisimple Rings

A group is an abstract object; it is a "cloud," a capital letter $G$. Of course, there are familiar concrete groups, such as the symmetric group $S_n$ and the general linear group $\mathrm{GL}(V)$ of all nonsingular linear transformations on a vector space $V$ over a field $k$. The idea underlying Representation Theory is that comparing an abstract group $G$ with familiar groups via homomorphisms can yield concrete information about $G$.

We begin by showing the connection between group representations and group algebras.

**Definition.** A *k-representation* of a group $G$ is a homomorphism

$$\sigma \colon G \to \mathrm{GL}(V),$$

where $V$ is a vector space over a field $k$.

Note that if $\dim(V) = n$, then $\mathrm{GL}(V)$ contains an isomorphic copy of $S_n$ [if $v_1, \ldots, v_n$ is a basis of $V$ and $\alpha \in S_n$, then there is a nonsingular linear transformation $T \colon V \to V$ with $T(v_i) = v_{\alpha(i)}$ for all $i$]; therefore, permutation representations (homomorphisms into $S_n$) are special cases of $k$-representations. Representations of groups can be translated into the language of $kG$-modules (compare the next proof with that of Proposition 6.14).

**Proposition 7.21.** *Every k-representation $\sigma \colon G \to \mathrm{GL}(V)$ equips $V$ with the structure of a left $kG$-module; denote this module by*

$$V^\sigma.$$

*Conversely, every left $kG$-module $V$ determines a k-representation $\sigma \colon G \to \mathrm{GL}(V)$.*

**Proof.** Given a homomorphism $\sigma \colon G \to \mathrm{GL}(V)$, denote $\sigma(g) \colon V \to V$ by $\sigma_g$, and define an action $kG \times V \to V$ by

$$\Big(\sum_{g \in G} a_g g\Big) v = \sum_{g \in G} a_g \sigma_g(v).$$

A routine calculation shows that $V$, equipped with this scalar multiplication, is a left $kG$-module.

Conversely, assume that $V$ is a left $kG$-module. If $g \in G$, then $v \mapsto gv$ defines a linear transformation $T_g \colon V \to V$; moreover, $T_g$ is nonsingular, for its inverse is $T_{g^{-1}}$. It is easily checked that the function $\sigma \colon G \to \mathrm{GL}(V)$, given by $\sigma \colon g \mapsto T_g$, is a $k$-representation.  •

If $\sigma, \tau \colon G \to \mathrm{GL}(V)$ are $k$-representations, and $V^\sigma, V^\tau$ are the $kG$-modules determined by $\sigma, \tau$ in Proposition 7.21, when is $V^\tau \cong V^\sigma$? Recall that if $T \colon V \to V$ is a linear transformation, then we made $V$ into a $k[x]$-module $V^T$ by defining $x^i v = T^i(v)$, and we saw, in Proposition 6.11, that if $S \colon V \to V$ is another linear transformation, then $V^S \cong V^T$ if and only if there is a nonsingular $\varphi \colon V \to V$ with $S = \varphi T \varphi^{-1}$.

**Proposition 7.22.** *Let $G$ be a group and let $\sigma, \tau \colon G \to \mathrm{GL}(V)$ be $k$-representations, where $k$ is a field. If $V^\sigma$ and $V^\tau$ are the corresponding $kG$-modules defined in* Proposition 7.21, *then $V^\sigma \cong V^\tau$ as $kG$-modules if and only if there exists a nonsingular $k$-linear transformation $\varphi \colon V \to V$ with*

$$\varphi \tau(g) = \sigma(g) \varphi$$

*for every $g \in G$.*

**Remark.** We often say that $\varphi$ **intertwines** $\sigma$ and $\tau$.  ◄

**Proof.** If $\varphi \colon V^\tau \to V^\sigma$ is a $kG$-isomorphism, then $\varphi \colon V \to V$ is an isomorphism of vector spaces with

$$\varphi \left( \sum a_g g v \right) = \left( \sum a_g g \right) \varphi(v)$$

for all $v \in V$ and all $g \in G$. But the definition of scalar multiplication in $V^\tau$ is $gv = \tau(g)(v)$, while the definition of scalar multiplication in $V^\sigma$ is $gv = \sigma(g)(v)$. Hence, for all $g \in G$ and $v \in V$, we have $\varphi(\tau(g)(v)) = \sigma(g)(\varphi(v))$. Therefore,

$$\varphi \tau(g) = \sigma(g) \varphi$$

for all $g \in G$.

Conversely, the hypothesis gives $\varphi \tau(g) = \sigma(g) \varphi$ for all $g \in G$, where $\varphi$ is a nonsingular $k$-linear transformation, and so $\varphi(\tau(g)v) = \sigma(g)\varphi(v)$ for all $g \in G$ and $v \in V$. It now follows easily that $\varphi$ is a $kG$-isomorphism; that is, $\varphi$ preserves scalar multiplication by $\sum_{g \in G} a_g g$.  •

We restate the last proposition in terms of matrices.

**Corollary 7.23.** *Let $G$ be a group and let $\sigma, \tau \colon G \to \mathrm{Mat}_n(k)$ be $k$-representations. Then $(k^n)^\sigma \cong (k^n)^\tau$ as $kG$-modules if and only if there is a nonsingular $n \times n$ matrix $P$ with*

$$P \tau(x) P^{-1} = \sigma(x)$$

*for every $x \in G$.*

**Example 7.24.** If $G$ is a finite group and $V$ is a vector space over a field $k$, then the **trivial homomorphism** $\sigma \colon G \to \mathrm{GL}(V)$ is defined by $\sigma(x) = 1_V$ for all $x \in G$. The corresponding $kG$-module $V^\sigma$ is called the **trivial** $kG$-module: if $v \in V$, then $xv = v$ for all $x \in G$. The trivial module $k$ (also called the **principal** $kG$-**module**) is denoted by

$$V_0(k). \quad ◄$$

We are going to study an important class of rings, *semisimple rings*, which contains most group algebras $kG$, but we first consider semisimple modules over any ring.

**Definition.** A left $R$-module is **semisimple** if it is a direct sum of (possibly infinitely many) simple modules.

We now characterize semisimple modules.

**Proposition 7.25.** *A left $R$-module $M$ over a ring $R$ is semisimple if and only if every submodule of $M$ is a direct summand.*

**Proof.** Suppose that $M$ is semisimple; hence, $M = \bigoplus_{j \in J} S_j$, where each $S_j$ is simple. For any subset $I \subseteq J$, define

$$S_I = \bigoplus_{j \in I} S_j.$$

If $B$ is a submodule of $M$, Zorn's Lemma provides a subset $K \subseteq J$ maximal with the property that $S_K \cap B = \{0\}$. We claim that $M = B \oplus S_K$. We must show that $M = B + S_K$, for their intersection is $\{0\}$ by hypothesis; it suffices to prove that $S_j \subseteq B + S_K$ for all $j \in J$. If $j \in K$, then $S_j \subseteq S_K \subseteq B + S_K$. If $j \notin K$, then maximality gives $(S_K + S_j) \cap B \neq \{0\}$. Thus,

$$s_K + s_j = b \neq 0,$$

where $s_K \in S_K$, $s_j \in S_j$, and $b \in B$. Note that $s_j \neq 0$, lest $s_K = b \in S_K \cap B = \{0\}$. Hence,

$$s_j = b - s_K \in S_j \cap (B + S_K),$$

and so $S_j \cap (B + S_K) \neq \{0\}$. But $S_j$ is simple, so that $S_j = S_j \cap (B + S_K)$, and so $S_j \subseteq B + S_K$, as desired. Therefore, $M = B \oplus S_K$.

Conversely, assume that every submodule of $M$ is a direct summand.

(i) Every nonzero submodule $B$ contains a simple summand.

Let $b \in B$ be nonzero. By Zorn's Lemma, there exists a submodule $C$ of $B$ maximal with $b \notin C$. By Corollary 6.29, $C$ is a direct summand of $B$: there is some submodule $D$ with $B = C \oplus D$. We claim that $D$ is simple. If $D$ is not simple, we may repeat the argument just given to show that $D = D' \oplus D''$ for nonzero submodules $D'$ and $D''$. Thus,

$$B = C \oplus D = C \oplus D' \oplus D''.$$

We claim that at least one of $C \oplus D'$ or $C \oplus D''$ does not contain the original element $b$. Otherwise, $b = c' + d' = c'' + d''$, where $c', c'' \in C$, $d' \in D'$, and $d'' \in D''$. But $c' - c'' = d'' - d' \in C \cap D = \{0\}$ gives $d' = d'' \in D' \cap D'' = \{0\}$. Hence, $d' = d'' = 0$, and so $b = c' \in C$, contradicting the definition of $C$. If, say, $b \notin C \oplus D'$, then this contradicts the maximality of $C$.

(ii) $M$ is semisimple.

By Zorn's Lemma, there is a family $(S_j)_{j \in I}$ of simple submodules of $M$ maximal such that the submodule $U$ they generate is their direct sum: $U = \bigoplus_{j \in I} S_j$. By hypothesis, $U$ is a direct summand: $M = U \oplus V$ for some submodule $V$ of $M$. If $V = \{0\}$, we are done. Otherwise, by part (i), there is some simple submodule $S$ contained in $V$ that is a summand: $V = S \oplus V'$ for some $V' \subseteq V$. The family $\{S\} \cup (S_j)_{j \in I}$ violates the maximality of the first family of simple submodules, for this larger family also generates its direct sum. Therefore, $V = \{0\}$ and $M$ is left semisimple.  •

**Corollary 7.26.** *Every submodule and every quotient module of a semisimple left module $M$ is itself semisimple.*

**Proof.** Let $B$ be a submodule of $M$. Every submodule $C$ of $B$ is, clearly, a submodule of $M$. Since $M$ is semisimple, $C$ is a direct summand of $M$ and so, by Corollary 6.29, $C$ is a direct summand of $B$. Hence, $B$ is semisimple, by Proposition 7.25.

Let $M/H$ be a quotient of $M$. Now $H$ is a direct summand of $M$, so that $M = H \oplus H'$ for some submodule $H'$ of $M$. But $H'$ is semisimple, by the first paragraph, and $M/H \cong H'$.    •

Recall that if a ring $R$ is viewed as a left $R$-module, then its submodules are its left ideals; moreover, a left ideal is minimal if and only if it is a simple left $R$-module.

**Definition.** A *ring* $R$ is **left semisimple**[3] if it is a direct sum of minimal left ideals.

The next proposition generalizes Example 7.14(iii).

**Proposition 7.27.** *Let $R$ be a left semisimple ring.*

(i) *$R$ is a direct sum of finitely many minimal left ideals.*

(ii) *$R$ has both chain conditions on left ideals.*

**Proof.**

(i) Since $R$ is left semisimple, it is a direct sum of minimal left ideals: $R = \bigoplus_i L_i$. Let $1 = \sum_i e_i$, where $e_i \in L_i$. If $r = \sum_i r_i \in \bigoplus_i L_i$, then $r = 1r$ and so $r_i = e_i r_i$. Hence, if $e_i = 0$, then $L_i = 0$. We conclude that there are only finitely many nonzero $L_i$; that is, $R = L_1 \oplus \cdots \oplus L_n$.

(ii) The series

$$R = L_1 \oplus \cdots \oplus L_n \supseteq L_2 \oplus \cdots \oplus L_n \supseteq \cdots \supseteq L_n \supseteq (0)$$

is a composition series, for the factor modules are $L_1, \ldots, L_n$, which are simple. It follows from Proposition 7.11 that $R$ (as a left $R$-module over itself) has both chain conditions.    •

**Corollary 7.28.** *The direct product $R = R_1 \times \cdots \times R_m$ of left semisimple rings $R_1, \ldots, R_m$ is also a left semisimple ring.*

**Proof.** Since each $R_i$ is left semisimple, it is a direct sum of minimal left ideals, say, $R_i = J_{i1} \oplus \cdots \oplus J_{i\,t(i)}$. Each $J_{ik}$ is a left ideal in $R$, not merely in $R_i$, as we saw in Example 6.5. It follows that $J_{ik}$ is a minimal left ideal in $R$. Hence, $R$ is a direct sum of minimal left ideals, and so it is a left semisimple ring.    •

**Corollary 7.29.**

(i) *If $R$ is a left semisimple ring, then every left $R$-module $M$ is a semisimple module.*

---

[3]We can define a ring to be **right semisimple** if it is a direct sum of minimal right ideals, but we shall see, in Corollary 7.45, that a ring is a left semisimple ring if and only if it is right semisimple.

(ii) *If $I$ is a two-sided ideal in a left semisimple ring $R$, then the quotient ring $R/I$ is also a semisimple ring.*

**Proof.**

(i) There is a free left $R$-module $F$ and a surjective $R$-map $\varphi \colon F \to M$. Now $R$ is a semisimple module over itself (this is the definition of semisimple ring), and so $F$ is a semisimple module (for $F$ is a direct sum of copies of $R$). Thus, $M$ is a quotient of the semisimple module $F$, and so it is itself semisimple, by Corollary 7.26.

(ii) First, $R/I$ is a ring, because $I$ is a two-sided ideal. The left $R$-module $R/I$ is semisimple, by (i), and so it is a direct sum $R/I \cong \bigoplus S_j$, where the $S_j$ are simple left $R$-modules. But each $S_j$ is also simple as a left $(R/I)$-module, for any $(R/I)$-submodule of $S_j$ is also an $R$-submodule of $S_j$. Therefore, $R/I$ is semisimple.  •

It follows that a finite direct product of fields is a commutative semisimple ring (we will prove the converse later). For example, if $n = p_1 \cdots p_t$ is a squarefree integer, then $\mathbb{I}_n \cong \mathbb{I}_{p_1} \times \cdots \times \mathbb{I}_{p_t}$ is a semisimple ring. Similarly, if $k$ is a field and $f(x) \in k[x]$ is a product of distinct irreducible polynomials, then $k[x]/(f(x))$ is a semisimple ring.

We can now generalize Proposition 7.12: every (not necessarily finitely generated) left vector space over a division ring $\Delta$ has a basis. Every division ring is a left semisimple ring, and $\Delta$ itself is the only minimal left ideal. Therefore, every left $\Delta$-module $M$ is a direct sum of copies of $\Delta$; say, $M = \bigoplus_{i \in I} \Delta_i$. If $x_i \in \Delta_i$ is nonzero, then $X = (x_i)_{i \in I}$ is a basis of $M$. This observation explains the presence of Zorn's Lemma in the proof of Proposition 7.25.

The next result shows that left semisimple rings can be characterized in terms of the Jacobson radical.

**Theorem 7.30.** *A ring $R$ is left semisimple if and only if it is left artinian and Jacobson semisimple; that is, $J(R) = (0)$.*

**Proof.** If $R$ is left semisimple, then there is a left ideal $I$ with $R = J(R) \oplus I$, by Proposition 7.25. It follows from Exercise 7.13 on page 534 that there are idempotents $e \in J(R)$ and $f \in I$ with $1 = e + f$. Since $e \in J(R)$, Proposition 7.15 says that $f = 1 - e$ has a left inverse; there is $u \in R$ with $uf = 1$. But $f$ is an idempotent, so that $f = f^2$. Hence, $1 = uf = uf^2 = (uf)f = f$, so that $e = 1 - f = 0$. Since $J(R)e = J(R)$, by Exercise 7.13 on page 534, we have $J(R) = (0)$. Finally, Proposition 7.27(ii) shows that $R$ is left artinian.

Conversely, assume that $R$ is left artinian and $J(R) = (0)$. We show first that if $I$ is a minimal left ideal of $R$, then $I$ is a direct summand of $R$. Now $I \neq (0)$, and so $I \nsubseteq J(R)$; therefore, there is a maximal left ideal $A$ not containing $I$. Since $I$ is minimal, it is simple, so that $I \cap A$ is either $I$ or $(0)$. But $I \cap A = I$ implies $I \subseteq A$, a contradiction, and so $I \cap A = (0)$. Maximality of $A$ gives $I + A = R$, and so $R = I \oplus A$.

Choose a minimal left ideal $I_1$, which exists because $R$ is left artinian. As we have just seen, $R = I_1 \oplus B_1$ for some left ideal $B_1$. Now $B_1$ contains a minimal left ideal, say, $I_2$, by Proposition 7.10, and so there is a left ideal $B_2$ with $B_1 = I_2 \oplus B_2$. This construction can be iterated to produce a strictly decreasing chain of left ideals $B_1 \supsetneq B_2 \supsetneq \cdots \supsetneq B_r$ as long as $B_r \neq (0)$. If $B_r \neq (0)$ for all $r$, then DCC is violated. Therefore, $B_r = (0)$ for some $r$, so that $R = I_1 \oplus \cdots \oplus I_r$ and $R$ is semisimple.    •

Note that the chain condition is needed. For example, $\mathbb{Z}$ is Jacobson semisimple, but $\mathbb{Z}$ is not a semisimple ring.

We can now prove the following remarkable result.

**Theorem 7.31 (Hopkins–Levitzki).** *If a ring $R$ is left artinian, then it is left noetherian.*

**Proof.** It suffices to prove that $R$, regarded as a left module over itself, has a composition series, for then Proposition 7.11 applies at once to show that $R$ has the ACC on left ideals (its submodules).

If $J = J(R)$ denotes the Jacobson radical, then $J^m = (0)$ for some $m \geq 1$, by Proposition 7.18, and so there is a descending chain

$$R = J^0 \supseteq J \supseteq J^2 \supseteq J^3 \supseteq \cdots \supseteq J^m = (0).$$

Since each $J^q$ is an ideal in $R$, it has DCC, as does its quotient $J^q/J^{q+1}$. Now $R/J$ is a semisimple ring, by Theorem 7.30 [it is left artinian, being a quotient of a left artinian ring, and Jacobson semisimple, by Corollary 7.19(ii)]. The factor module $J^q/J^{q+1}$ is an $(R/J)$-module; hence, by Corollary 7.26, $J^q/J^{q+1}$ is a semisimple module, and so it can be decomposed into a direct sum of (perhaps infinitely many) simple $(R/J)$-modules. But there can be only finitely many summands, for every $(R/J)$-submodule of $J^q/J^{q+1}$ is necessarily an $R$-submodule, and $J^q/J^{q+1}$ has DCC on $R$-submodules. Hence, there are simple $(R/J)$-modules $S_i$ with

$$J^q/J^{q+1} = S_1 \oplus S_2 \oplus \cdots \oplus S_p.$$

Throwing away one simple summand at a time yields a series of $J^q/J^{q+1}$ whose $i$th factor module is

$$(S_i \oplus S_{i+1} \oplus \cdots \oplus S_p)/(S_{i+1} \oplus \cdots \oplus S_p) \cong S_i.$$

Now the simple $(R/J)$-module $S_i$ is also a simple $R$-module, for it is an $R$-module annihilated by $J$, so that we have constructed a composition series for $J^q/J^{q+1}$ as a left $R$-module. Finally, refine the original series for $R$ in this way, for every $q$, to obtain a composition series for $R$.    •

The converse of Theorem 7.31 is false: $\mathbb{Z}$ is noetherian but not artinian.

The next result is fundamental.

**Theorem 7.32 (Maschke's Theorem).** *If $G$ is a finite group and $k$ is a field whose characteristic does not divide $|G|$, then $kG$ is a left semisimple ring.*

**Remark.** The hypothesis holds if $k$ has characteristic 0.   ◀

**Proof.** By Proposition 7.25, it suffices to prove that every left ideal $I$ of $kG$ is a direct summand. Since $k$ is a field, $kG$ is a vector space over $k$ and $I$ is a subspace. By Corollary 5.48, $I$ is a (vector space) direct summand: there is a subspace $V$ (which may not be a left ideal in $kG$) with $kG = I \oplus V$. There is a $k$-linear transformation $d: kG \to I$ with $d(b) = b$ for all $b \in I$ and with $\ker d = V$ [each $u \in kG$ has a unique expression of the form $u = b + v$, where $b \in I$ and $v \in V$, and $d(u) = b$]. Were $d$ a $kG$-map, not merely a $k$-map, then we would be done, by the criterion of Corollary 6.28 [$I$ is a summand of $kG$ if and only if it is a retract: there is a $kG$-map $D: kG \to I$ with $D(u) = u$ for all $u \in I$]. We now force $d$ to be a $kG$-map by an "averaging process."

Define $D : kG \to kG$ by

$$D(u) = \frac{1}{|G|} \sum_{x \in G} x d(x^{-1} u)$$

for all $u \in kG$. Note that $|G| \neq 0$ in $k$, by the hypothesis on the characteristic of $k$, and so $1/|G|$ is defined. It is obvious that $D$ is a $k$-map.

(i) $\operatorname{im} D \subseteq I$.

   If $u \in kG$ and $x \in G$, then $d(x^{-1}u) \in I$ (because $\operatorname{im} d \subseteq I$), and $x d(x^{-1}u) \in I$ because $I$ is a left ideal. Therefore, $D(u) \in I$, for each term in the sum defining $D(u)$ lies in $I$.

(ii) If $b \in I$, then $D(b) = b$.

   Since $b \in I$, so is $x^{-1}b$, and so $d(x^{-1}b) = x^{-1}b$. Hence, $x d(x^{-1}b) = x x^{-1}b = b$. Therefore, $\sum_{x \in G} x d(x^{-1}b) = |G|b$, and so $D(b) = b$.

(iii) $D$ is a $kG$-map.

   It suffices to prove that $D(gu) = gD(u)$ for all $g \in G$ and all $u \in kG$:

$$gD(u) = \frac{1}{|G|} \sum_{x \in G} gx d(x^{-1}u) = \frac{1}{|G|} \sum_{x \in G} gx d(x^{-1}g^{-1}gu)$$

$$= \frac{1}{|G|} \sum_{y = gx \in G} y d(y^{-1}gu) = D(gu)$$

(as $x$ ranges over all of $G$, so does $y = gx$).   •

The converse of Maschke's Theorem is true: if $G$ is a finite group and $k$ is a field whose characteristic $p$ divides $|G|$, then $kG$ is not left semisimple; a proof is outlined in Exercise 7.20 on page 549.

Before analyzing left semisimple rings further, let us give several characterizations of them.

**Proposition 7.33.** *The following conditions on a ring $R$ are equivalent.*

  (i) *$R$ is left semisimple.*

  (ii) *Every left $R$-module is a semisimple module.*

  (iii) *Every left $R$-module is injective.*

  (iv) *Every short exact sequence of left $R$-modules splits.*

  (v) *Every left $R$-module is projective.*

**Proof.**

(i) $\Rightarrow$ (ii). This follows at once from Corollary 7.26, which says that if $R$ is a semisimple ring, then every left $R$-module is a semisimple module.

(ii) $\Rightarrow$ (iii). Let $E$ be a left $R$-module; Proposition 6.84 says that $E$ is injective if every exact sequence $0 \to E \to B \to C \to 0$ splits. By hypothesis, $B$ is a semisimple module, and so Proposition 7.25 implies that the sequence splits; thus, $E$ is injective.

(iii) $\Rightarrow$ (iv). If $0 \to A \to B \to C \to 0$ is an exact sequence, then it must split because, as every module, $A$ is injective, by Proposition 6.84.

(iv) $\Rightarrow$ (v). Given a module $M$, there is an exact sequence

$$0 \to F' \to F \to M \to 0,$$

where $F$ is free. This sequence splits, by hypothesis, and so $F \cong M \oplus F'$. Therefore, $M$ is a direct summand of a free module, and hence it is projective, by Theorem 6.76.

(v) $\Rightarrow$ (i). If $I$ is a left ideal of $R$, then

$$0 \to I \to R \to R/I \to 0$$

is an exact sequence. By hypothesis, $R/I$ is projective, and so this sequence splits, by Proposition 6.73; thus, $I$ is a direct summand of $R$. By Proposition 7.25, $R$ is a semisimple left $R$-module; that is, $R$ is a left semisimple ring.  $\bullet$

Modules over semisimple rings are so nice that there is a notion of *global dimension* of a ring $R$ that measures how far removed $R$ is from being semisimple. We will discuss global dimension in Chapter 10.

In order to give more examples of left semisimple rings, we look at endomorphism rings of direct sums. Consider $\operatorname{Hom}_R(A, B)$, where both $A$ and $B$ are left $R$-modules that are finite direct sums: say, $A = \bigoplus_{i=1}^n A_i$ and $B = \bigoplus_{j=1}^m B_j$. Since a direct product of a finite family of modules is their direct sum, Theorem 6.45 gives

$$\operatorname{Hom}_R(A, B) \cong \bigoplus_{i,j} \operatorname{Hom}_R(A_i, B_j).$$

More precisely, if $\alpha_i \colon A_i \to A$ is the $i$th injection and $p_j \colon B \to B_j$ is the $j$th projection, then each $f \in \operatorname{Hom}_R(A, B)$ gives maps

$$f_{ji} = p_j f \alpha_i \in \operatorname{Hom}_R(A_i, B_j).$$

Thus, $f$ defines a **generalized $m \times n$ matrix** $[f_{ji}]$ (we call $[f_{ji}]$ a *generalized* matrix because entries in different positions need not lie in the same algebraic system). The map $f \mapsto [f_{ji}]$ is an isomorphism $\operatorname{Hom}_R(A, B) \to \bigoplus_{ij} \operatorname{Hom}_R(A_i, B_j)$. Similarly, if $g \colon B \to C$, where $C = \bigoplus_{k=1}^\ell C_k$, then $g$ defines a generalized $\ell \times m$ matrix $[g_{kj}]$, where $g_{kj} = q_k g \beta_j \colon B_j \to C_k$, $\beta_j \colon B_j \to B$ are the injections, and $q_k \colon C \to C_k$ are the projections.

The composite $gf \colon A \to C$ defines a generalized $\ell \times n$ matrix, and we claim that it is given by matrix multiplication: $(gf)_{ki} = \sum_j g_{kj} f_{ji}$:

$$\sum_j g_{kj} f_{ji} = \sum_j q_k g \beta_j p_j f \alpha_i = q_k g \Big( \sum_j \beta_j p_j \Big) f \alpha_i = q_k g f \alpha_i = (gf)_{ki},$$

because $\sum_j \beta_j p_j = 1_B$.

By adding some hypotheses, we can pass from generalized matrices to honest matrices.

**Proposition 7.34.** *Let $V = \bigoplus_{i=1}^n V_i$ be a left $R$-module. If there is a left $R$-module $L$ and, for each $i$, an isomorphism $\varphi_i \colon V_i \to L$, then there is a ring isomorphism*

$$\operatorname{End}_R(V) \cong \operatorname{Mat}_n(\operatorname{End}_R(L)).$$

**Proof.** Define $\theta \colon \operatorname{End}_R(V) \to \operatorname{Mat}_n(\operatorname{End}_R(L))$ by

$$\theta \colon f \mapsto [\varphi_j p_j f \alpha_i \varphi_i^{-1}],$$

where $\alpha_i \colon V_i \to V$ and $p_j \colon V \to V_j$ are injections and projections, respectively. That $\theta$ is an additive isomorphism is just the identity

$$\operatorname{Hom}\Big( \bigoplus_i V_i, \bigoplus_i V_i \Big) \cong \bigoplus_{i,j} \operatorname{Hom}(V_i, V_j),$$

which holds when the index sets are finite. In the paragraph above defining generalized matrices, the home of the $ij$ entry is $\operatorname{Hom}_R(V_i, V_j)$, whereas the present home of this entry is the isomorphic replica $\operatorname{Hom}_R(L, L) = \operatorname{End}_R(L)$.

We now show that $\theta$ preserves multiplication. If $g, f \in \operatorname{End}_R(V)$, then $\theta(gf) = [\varphi_j p_j gf \alpha_i \varphi_i^{-1}]$, while the matrix product is

$$\theta(g)\theta(f) = \Big[ \sum_k (\varphi_j p_j g \alpha_k \varphi_k^{-1})(\varphi_k p_k f \alpha_i \varphi_i^{-1}) \Big]$$

$$= \Big[ \sum_k \varphi_j p_j g \alpha_k p_k f \alpha_i \varphi_i^{-1} \Big]$$

$$= \Big[ \varphi_j p_j g \Big( \sum_k \alpha_k p_k \Big) f \alpha_i \varphi_i^{-1} \Big]$$

$$= \Big[ \varphi_j p_j gf \alpha_i \varphi_i^{-1} \Big]. \quad \bullet$$

**Corollary 7.35.** *If $V$ is an $n$-dimensional left vector space over a division ring $\Delta$, then there is an isomorphism of rings*

$$\operatorname{End}_\Delta(V) \cong \operatorname{Mat}_n(\Delta)^{\mathrm{op}}.$$

**Proof.** The isomorphism $\operatorname{End}_k(V) \cong \operatorname{Mat}_n(\Delta^{\mathrm{op}})$ is the special case of Proposition 7.34 for $V = V_1 \oplus \cdots \oplus V_n$, where each $V_i$ is one-dimensional, and hence is isomorphic to $\Delta$. Note that $\operatorname{End}_\Delta(\Delta) \cong \Delta^{\mathrm{op}}$, by Proposition 6.16. Now apply Proposition 6.17, which says that $\operatorname{Mat}_n(\Delta^{\mathrm{op}}) \cong \operatorname{Mat}_n(\Delta)^{\mathrm{op}}$. $\quad \bullet$

The next result involves a direct sum decomposition at the opposite extreme of that in Proposition 7.34.

**Corollary 7.36.** *Let an R-module $M$ be a direct sum $M = B_1 \oplus \cdots \oplus B_m$ in which* $\operatorname{Hom}_R(B_i, B_j) = \{0\}$ *for all $i \neq j$. Then there is a ring isomorphism*

$$\operatorname{End}_R(M) \cong \operatorname{End}_R(B_1) \times \cdots \times \operatorname{End}_R(B_m).$$

**Proof.** If $f, g \in \operatorname{End}_R(M)$, let $[f_{ij}]$ and $[g_{ij}]$ be their generalized matrices. It suffices to show that $[g_{ij}][f_{ij}]$ is the diagonal matrix

$$\operatorname{diag}(g_{11}f_{11}, \ldots, g_{mm}f_{mm}).$$

But if $i \neq j$, then $g_{ik}f_{kj} \in \operatorname{Hom}_R(B_i, B_j) = 0$; hence, $(gf)_{ij} = \sum_k g_{ik}f_{kj} = 0$.   •

We can now give more examples of semisimple rings. The Wedderburn–Artin Theorems will say that there are no others.

**Proposition 7.37.**

(i) *If $\Delta$ is a division ring and $V$ is a left vector space over $\Delta$ with $\dim(V) = n$, then $\operatorname{End}_\Delta(V) \cong \operatorname{Mat}_n(\Delta^{\operatorname{op}})$ is a left semisimple ring.*

(ii) *If $\Delta_1, \ldots, \Delta_m$ are division rings, then*

$$\operatorname{Mat}_{n_1}(\Delta_1) \times \cdots \times \operatorname{Mat}_{n_m}(\Delta_m)$$

*is a left semisimple ring.*

**Proof.**

(i) By Proposition 7.34, we have

$$\operatorname{End}_\Delta(V) \cong \operatorname{Mat}_n(\operatorname{End}_\Delta(\Delta));$$

by Proposition 6.16, $\operatorname{End}_\Delta(\Delta) \cong \Delta^{\operatorname{op}}$. Therefore, $\operatorname{End}_\Delta(V) \cong \operatorname{Mat}_n(\Delta^{\operatorname{op}})$.

Let us now show that $\operatorname{End}_\Delta(V)$ is semisimple. If $v_1, \ldots, v_n$ is a basis of $V$, define

$$\operatorname{COL}(j) = \{T \in \operatorname{End}_\Delta(V) : T(v_i) = 0 \text{ for all } i \neq j\}.$$

It is easy to see that $\operatorname{COL}(j)$ is a left ideal in $\operatorname{End}_\Delta(V)$: if $S \in \operatorname{End}_\Delta(V)$, then $S(Tv_i) = 0$ for all $i \neq j$. Recall Example 7.9: if we look in $\operatorname{Mat}_n(\Delta^{\operatorname{op}}) \cong \operatorname{End}_\Delta(V)$, then $\operatorname{COL}(j)$ corresponds to $\operatorname{COL}(j)$, all those matrices whose entries off the $j$th column are 0. It is obvious that

$$\operatorname{Mat}_n(\Delta^{\operatorname{op}}) = \operatorname{COL}(1) \oplus \cdots \oplus \operatorname{COL}(n).$$

Hence, $\operatorname{End}_\Delta(V)$ is also such a direct sum. We saw, in Example 7.14(iii), that each $\operatorname{COL}(\ell)$ is a minimal left ideal, and so $\operatorname{End}_\Delta(V)$ is a left semisimple ring.

(ii) This follows at once from part (i) and Corollary 7.28, for if $\Delta$ is a division ring, then so is $\Delta^{\operatorname{op}}$, by Exercise 6.12 on page 416.   •

**Corollary 7.38.** *If $V$ is an $n$-dimensional left vector space over a division ring $\Delta$, then the minimal left ideals $\operatorname{COL}(\ell)$, for $1 \leq \ell \leq n$, in $\operatorname{End}_\Delta(V)$ are all isomorphic.*

**Proof.** Let $v_1, \ldots, v_n$ be a basis of $V$. For each $\ell$, define $p_\ell \colon V \to V$ to be the linear transformation that interchanges $v_\ell$ and $v_1$ and that fixes all the other $v_j$. It is easy to see that $T \mapsto Tp_\ell$ is an isomorphism $\operatorname{COL}(1) \to \operatorname{COL}(\ell)$.   •

There may be minimal left ideals other than $\text{COL}(\ell)$ for some $\ell$. However, we will see [in Lemma 7.49(ii)] that all the minimal left ideals in $\text{End}_\Delta(V)$ are isomorphic to one of these.

**Definition.** A ring $R$ is *simple* if it is not the zero ring and it has no proper nonzero two-sided ideals.

Our language is a little deceptive. It is true that left artinian simple rings are semisimple (Proposition 7.47), but there are simple rings that are not semisimple.

**Proposition 7.39.** *If $\Delta$ is a division ring, then $R = \text{Mat}_n(\Delta)$ is a simple ring.*

**Proof.** A *matrix unit* $E_{pq}$ is the $n \times n$ matrix whose $p, q$ entry is 1 and all of whose other entries are 0. The matrix units form a basis for $\text{Mat}_n(\Delta)$ viewed as a left vector space over $\Delta$, for each matrix $A = [a_{ij}]$ has a unique expression

$$A = \sum_{i,j} a_{ij} E_{ij}.$$

[Of course, this says that $\dim_\Delta(\text{Mat}_n(\Delta)) = n^2$.] A routine calculation shows that matrix units multiply according to the following rule:

$$E_{ij} E_{k\ell} = \begin{cases} 0 & \text{if } j \neq k \\ E_{i\ell} & \text{if } j = k. \end{cases}$$

Suppose that $N$ is a nonzero two-sided ideal in $\text{Mat}_n(\Delta)$. If $A$ is a nonzero matrix in $N$, it has a nonzero entry; say, $a_{ij} \neq 0$. Since $N$ is a two-sided ideal, $N$ contains $E_{pi} A E_{jq}$ for all $p, q$. But

$$E_{pi} A E_{jq} = E_{pi} \sum_{k,\ell} a_{k\ell} E_{k\ell} E_{jq} = E_{pi} \sum_k a_{kj} E_{kq} = \sum_k a_{kj} E_{pi} E_{kq} = a_{ij} E_{pq}.$$

Since $a_{ij} \neq 0$ and $\Delta$ is a division ring, $a_{ij}^{-1} \in \Delta$, and so $E_{pq} \in N$ for all $p, q$. But the collection of all $E_{pq}$ span the left vector space $\text{Mat}_n(\Delta)$ over $\Delta$, and so $N = \text{Mat}_n(\Delta)$. •

## Exercises

**7.18.** Prove that a finitely generated left semisimple $R$-module $M$ over a ring $R$ is a direct sum of a finite number of simple left modules.

**7.19.** Let $A$ be an $n$-dimensional $k$-algebra over a field $k$. Prove that $A$ can be imbedded as a $k$-subalgebra of $\text{Mat}_n(k)$.

**Hint.** If $a \in A$, define $L_a : A \to A$ by $L_a : x \mapsto ax$.

∗ **7.20.** Let $G$ be a finite group, and let $k$ be a commutative ring. Define $\varepsilon : kG \to k$ by

$$\varepsilon\left(\sum_{g \in G} a_g g\right) = \sum_{g \in G} a_g$$

(this map is called the *augmentation*, and its kernel, denoted by $\mathcal{G}$, is called the *augmentation ideal*).

(i) Prove that $\varepsilon$ is a $kG$-map and that $kG/\mathcal{G} \cong k$ as $k$-algebras. Conclude that $\mathcal{G}$ is a two-sided ideal in $kG$.

(ii) Prove that $kG/\mathcal{G} \cong V_0(k)$, where $V_0(k)$ is $k$ viewed as a trivial $kG$-module.
    **Hint.** $\mathcal{G}$ is a two-sided ideal containing $xu - u = (x - 1)u$.

(iii) Use part (ii) to prove that if $kG = \mathcal{G} \oplus V$, then $V = \langle v \rangle$, where $v = \sum_{g \in G} g$.
    **Hint.** Argue as in Example 7.43.

(iv) Assume that $k$ is a field whose characteristic $p$ does divide $|G|$. Prove that $kG$ is not left semisimple.
    **Hint.** First show that $\varepsilon(v) = 0$, and then show that the short exact sequence
    $$0 \to \mathcal{G} \to kG \xrightarrow{\varepsilon} k \to 0$$
    does not split. (If $G$ is a finite $p$-group and $k$ is a field of characteristic $p$, then the Jacobson radical $J(kG)$ is the augmentation ideal; see Lam, *A First Course in Noncommutative Rings*, p. 131).

∗ **7.21.** Let $M$ be a left $R$-module over a semisimple ring $R$. Prove that $M$ is indecomposable if and only if $M$ is simple.

**7.22.** If $\Delta$ is a division ring, prove that every two minimal left ideals in $\mathrm{Mat}_n(\Delta)$ are isomorphic. (Compare Corollary 7.38.)

**7.23.** Let $T \colon V \to V$ be a linear transformation, where $V$ is a vector space over a field $k$, and let $k[T]$ be defined by
$$k[T] = k[x]/(m(x)),$$
where $m(x)$ is the minimum polynomial of $T$.

(i) If $m(x) = \prod_p p(x)^{e_p}$, where the $p(x) \in k[x]$ are distinct irreducible polynomials and $e_p \geq 1$, prove that $k[T] \cong \prod_p k[x]/(p(x)^{e_p})$.

(ii) Prove that $k[T]$ is a semisimple ring if and only if $m(x)$ is a product of distinct linear factors. (In Linear Algebra, we show that this last condition is equivalent to $T$ being *diagonalizable*; that is, any matrix of $T$ [arising from some choice of basis of $T$] is similar to a diagonal matrix.)

**7.24.** (i) If $\mathbb{H}$ is the division ring of real quaternions, prove that its multiplicative group $\mathbb{H}^\times$ has a finite subgroup that is not cyclic. Compare with Theorem 2.46.

(ii) If $\Delta$ is a division ring whose center is a field of characteristic $p > 0$, prove that every finite subgroup $G$ of $\Delta^\times$ is cyclic.
    **Hint.** Consider $\mathbb{F}_p G$, and use Theorem 7.13.

## Section 7.4. Wedderburn–Artin Theorems

We are now going to prove the converse of Proposition 7.37(ii): every left semisimple ring is isomorphic to a direct product of matrix rings over division rings. The first step shows how division rings arise.

**Theorem 7.40 (Schur's Lemma).** *Let $M$ and $M'$ be simple left $R$-modules over a ring $R$.*

(i) *Every nonzero $R$-map $f \colon M \to M'$ is an isomorphism.*

(ii) $\text{End}_R(M)$ *is a division ring. In particular, if $L$ is a minimal left ideal in a ring $R$, then $\text{End}_R(L)$ is a division ring.*

**Proof.**

(i) Since $M$ is simple, it has only two submodules: $M$ itself and $\{0\}$. Now the submodule $\ker f \neq M$ because $f \neq 0$, so that $\ker f = \{0\}$ and $f$ is an injection. Similarly, the submodule $\text{im} f \neq \{0\}$, so that $\text{im} f = M'$ and $f$ is a surjection.

(ii) If $f: M \to M$ and $f \neq 0$, then $f$ is an isomorphism, by part (i), and hence it has an inverse $f^{-1} \in \text{End}_R(M)$. Thus, $\text{End}_R(M)$ is a division ring.   •

The second step investigates minimal left ideals.

**Lemma 7.41.** *If $L$ and $L'$ are minimal left ideals in a ring $R$, then each of the following statements implies the one below it:*

(i) $LL' \neq (0)$.

(ii) $\text{Hom}_R(L, L') \neq \{0\}$ *and there exists $b' \in L'$ with $L' = Lb'$.*

(iii) $L \cong L'$ *as left $R$-modules.*

*If also $L^2 \neq (0)$, then (iii) implies (i) and the three statements are equivalent.*

**Proof.**

(i) $\Rightarrow$ (ii) If $LL' \neq (0)$, then there exists $b \in L$ and $b' \in L'$ with $bb' \neq 0$. Thus, the function $f: L \to L'$, defined by $x \mapsto xb'$, is a nonzero $R$-map, and so $\text{Hom}_R(L, L') \neq \{0\}$. Moreover, $Lb' = L'$, for it is a nonzero submodule of the minimal left ideal $L'$.

(ii) $\Rightarrow$ (iii) If $\text{Hom}_R(L, L') \neq \{0\}$, then there is a nonzero $f: L \to L'$, and $f$ is an isomorphism, by Schur's Lemma; that is, $L \cong L'$.

(iii) and $L^2 \neq (0) \Rightarrow$ (i) Assume now that $L^2 \neq (0)$, so there are $x, y \in L$ with $xy \neq 0$. If $g: L \to L'$ is an isomorphism, then $0 \neq g(xy) = xg(y) \in LL'$, and so $LL' \neq (0)$.   •

Note that if $J(R) = (0)$, then $L^2 \neq (0)$; otherwise, $L$ is a nilpotent left ideal and Corollary 7.17 gives $L \subseteq J(R) = (0)$, a contradiction.

**Proposition 7.42.** *If $R = \bigoplus_j L_j$ is a left semisimple ring, where the $L_j$ are minimal left ideals, then every simple $R$-module $S$ is isomorphic to some $L_j$.*

**Proof.** Now $S \cong \text{Hom}_R(R, S) \neq \{0\}$, by Corollary 6.64. But, if $\text{Hom}_R(L_j, S) = \{0\}$ for all $j$, then $\text{Hom}_R(R, S) = \{0\}$ (for $R = L_1 \oplus \cdots \oplus L_m$). Hence, $\text{Hom}_R(L_j, S) \neq \{0\}$ for some $j$. Since both $L_j$ and $S$ are simple, Theorem 7.40(i) gives $L_j \cong S$.   •

Here is a fancier proof of Proposition 7.42.

**Proof.** By Corollary 6.25, there is a left ideal $I$ with $S \cong R/I$, and so there is a series
$$R \supseteq I \supseteq (0).$$

In Proposition 7.27, we saw that

$$R = L_1 \oplus \cdots \oplus L_n \supseteq L_2 \oplus \cdots \oplus L_n \supseteq \cdots \supseteq L_n \supseteq (0)$$

is a composition series with factor modules $L_1, \ldots, L_n$. The Schreier Refinement Theorem (Theorem 7.2) now says that these two series have equivalent refinements. Since a composition series admits only refinements that repeat a term, the factor module $S$ occurring in the refinement of the first series must be isomorphic to one of the factor modules in the second series; that is, $S \cong L_i$ for some $i$.   •

**Example 7.43.** The trivial $kG$-module $V_0(k)$ (Example 7.24) is a simple $kG$-module (for it is one-dimensional, hence has no subspaces other than $\{0\}$ and itself). By Proposition 7.42, $V_0(k)$ is isomorphic to some minimal left ideal $L$ of $kG$. We shall find $L$ by searching for elements $u = \sum_{g \in G} a_g g$ in $kG$ with $hu = u$ for all $h \in G$. For such elements $u$,

$$hu = \sum_{g \in G} a_g hg = \sum_{g \in G} a_g g = u.$$

Since the elements in $G$ form a basis for the vector space $kG$, we may equate coefficients, and so $a_g = a_{hg}$ for all $g \in G$; in particular, $a_1 = a_h$. As this holds for every $h \in G$, all the coefficients $a_g$ are equal. Therefore, if we define $\gamma \in kG$ by

$$\gamma = \sum_{g \in G} g,$$

then $u$ is a scalar multiple of $\gamma$. It follows that $L = \langle \gamma \rangle$ is a left ideal isomorphic to the trivial module $V_0(k)$; moreover, $\langle \gamma \rangle$ is the unique such left ideal.   ◀

An abstract left semisimple ring $R$ is a direct sum of finitely many minimal left ideals: $R = \bigoplus_j L_j$, and we now know that $\mathrm{End}_R(L_j)$ is a division ring for every $j$. The next step is to find the direct summands of $R$ that will ultimately turn out to be matrix rings; they arise from a decomposition of $R$ into minimal left ideals by collecting isomorphic terms.

**Definition.** Let $R$ be a left semisimple ring, and let

$$R = L_1 \oplus \cdots \oplus L_n,$$

where the $L_j$ are minimal left ideals. Reindex the summands so that no two of the first $m$ ideals $L_1, \ldots, L_m$ are isomorphic, while every $L_j$ in the given decomposition is isomorphic to some $L_i$ for $1 \leq i \leq m$. The left ideals

$$B_i = \bigoplus_{L_j \cong L_i} L_j$$

are called the ***simple components*** of $R$ relative to the decomposition $R = \bigoplus_j L_j$.

We shall see, in Corollary 7.50, that the simple components do not depend on the particular decomposition of $R$ as a direct sum of minimal left ideals.

We divide Theorem 7.44, the Wedderburn–Artin[4] Theorem, into two parts: an existence theorem and a uniqueness theorem.

---

[4]Wedderburn proved Theorem 7.44 for semisimple $k$-algebras, where $k$ is a field; E. Artin generalized the theorem as it is stated here. This theorem is why *artinian* rings are so called.

**Theorem 7.44 (Wedderburn–Artin I).** *A ring $R$ is left semisimple if and only if $R$ is isomorphic to a direct product of matrix rings over division rings.*

**Proof.** Sufficiency is Proposition 7.37(ii).

For necessity, if $R$ is left semisimple, then it is the direct sum of its simple components:

$$R = B_1 \oplus \cdots \oplus B_m,$$

where each $B_i$ is a direct sum of isomorphic minimal left ideals. Proposition 6.16 says that there is a ring isomorphism

$$R^{\mathrm{op}} \cong \mathrm{End}_R(R),$$

where $R$ is regarded as a left module over itself. Now $\mathrm{Hom}_R(B_i, B_j) = \{0\}$ for all $i \neq j$, by Lemma 7.41, so that Corollary 7.36 applies to give a ring isomorphism

$$R^{\mathrm{op}} \cong \mathrm{End}_R(R) \cong \mathrm{End}_R(B_1) \times \cdots \times \mathrm{End}_R(B_m).$$

By Proposition 7.34, there are isomorphisms of rings

$$\mathrm{End}_R(B_i) \cong \mathrm{Mat}_{n_i}(\mathrm{End}_R(L_i)),$$

because $B_i$ is a direct sum of isomorphic copies of $L_i$. By Schur's Lemma, $\mathrm{End}_R(L_i)$ is a division ring, say, $\Delta_i$, and so

$$R^{\mathrm{op}} \cong \mathrm{Mat}_{n_1}(\Delta_1) \times \cdots \times \mathrm{Mat}_{n_m}(\Delta_m).$$

Hence,

$$R \cong [\mathrm{Mat}_{n_1}(\Delta_1)]^{\mathrm{op}} \times \cdots \times [\mathrm{Mat}_{n_m}(\Delta_m)]^{\mathrm{op}}.$$

Finally, Proposition 6.17 gives

$$R \cong \mathrm{Mat}_{n_1}(\Delta_1^{\mathrm{op}}) \times \cdots \times \mathrm{Mat}_{n_m}(\Delta_m^{\mathrm{op}}).$$

This completes the proof, for $\Delta_i^{\mathrm{op}}$ is also a division ring for all $i$, by Exercise 6.12 on page 416. •

**Corollary 7.45.** *A ring $R$ is left semisimple if and only if it is right semisimple.*

**Proof.** It is easy to see that a ring $R$ is right semisimple if and only if its opposite ring $R^{\mathrm{op}}$ is left semisimple. But we saw, in the middle of the proof of Theorem 7.44, that

$$R^{\mathrm{op}} \cong \mathrm{Mat}_{n_1}(\Delta_1) \times \cdots \times \mathrm{Mat}_{n_m}(\Delta_m),$$

where $\Delta_i = \mathrm{End}_R(L_i)$. •

As a consequence of this corollary, we say that a ring is **semisimple** without the adjectives left or right.

**Corollary 7.46.** *A commutative ring $R$ is semisimple if and only if it is isomorphic to a direct product of finitely many fields.*

**Proof.** A field is a semisimple ring, and so a direct product of finitely many fields is also semisimple, by Corollary 7.28. Conversely, if $R$ is semisimple, it is a direct product of matrix rings over division rings. Since $R$ is commutative, all the matrix rings must be of size $1 \times 1$ and all the division rings must be fields. •

Even though the name suggests it, it is not clear that simple rings are semisimple. Indeed, we must assume a chain condition: if $V$ is an infinite-dimensional vector space over a field $k$, then $R = \mathrm{End}_k(V)$ is a simple ring which is not semisimple (Lam, *A First Course in Noncommutative Rings*, pp. 43–44).

**Proposition 7.47.** *A simple left artinian ring $R$ is semisimple.*

**Proof. (Janusz)** Since $R$ is left artinian, it contains a minimal left ideal, say, $L$; of course, $L$ is a simple left $R$-module. For each $a \in R$, the function $f_a \colon L \to R$, defined by $f_a(x) = xa$, is a map of left $R$-modules: if $r \in R$, then

$$f_a(rx) = (rx)a = r(xa) = rf_a(x).$$

Now $\mathrm{im}\, f_a = La$, while $L$ being a simple module forces $\ker f_a = L$ or $\ker f_a = (0)$. In the first case, we have $La = (0)$; in the second case, we have $L \cong La$. Thus, $La$ is either $(0)$ or a minimal left ideal.

Consider the sum $I = \left\langle \bigcup_{a \in R} La \right\rangle \subseteq R$. Plainly, $I$ is a left ideal; it is a right ideal as well, for if $b \in R$ and $La \subseteq I$, then $(La)b = L(ab) \subseteq I$. Since $R$ is a simple ring, the nonzero two-sided ideal $I$ must equal $R$. We claim that $R$ is a sum of only finitely many $La$'s. As any element of $R$, the unit $1$ lies in some finite sum of $La$'s; say, $1 \in Le_1 + \cdots + Le_n$. If $b \in R$, then $b = b1 \in b(Le_1 + \cdots + Le_n) \subseteq Le_1 + \cdots + Le_n$ (because $Le_1 + \cdots + Le_n$ is a left ideal). Hence, $R = Le_1 + \cdots + Le_n$.

To prove that $R$ is semisimple, it remains to show that it is a *direct sum* of simple submodules. Choose $n$ minimal such that $R = Le_1 + \cdots + Le_n$; we claim that $R = Le_1 \oplus \cdots \oplus Le_n$. By Proposition 6.30, it suffices to show, for all $i$, that

$$Le_i \cap \left( \bigoplus_{j \neq i} Le_j \right) = (0).$$

If this intersection is not $(0)$, then simplicity of $Le_i$ says that $Le_i \cap (\bigoplus_{j \neq i} Le_j) = Le_i$; that is, $Le_i \subseteq \bigoplus_{j \neq i} Le_j$, and this contradicts the minimal choice of $n$. Therefore, $R$ is a semisimple ring.  •

The following corollary follows at once from Proposition 7.47 and Theorem 7.44, the Wedderburn–Artin Theorem.

**Corollary 7.48.** *If $A$ is a simple left artinian ring, then $A \cong \mathrm{Mat}_n(\Delta)$ for some $n \geq 1$ and some division ring $\Delta$.*

The next lemma gives some interesting properties enjoyed by semi-simple rings; it will be used to complete the Wedderburn–Artin Theorem by proving uniqueness of the constituent parts. In particular, it will say that the integer $n$ and the division ring $\Delta$ in Corollary 7.48 are uniquely determined by $A$.

**Lemma 7.49.** *Let $R$ be a semisimple ring, and let    ,*

$$R = L_1 \oplus \cdots \oplus L_n = B_1 \oplus \cdots \oplus B_m,$$

*where the $L_j$ are minimal left ideals and the $B_i$ are the corresponding simple components of $R$.*

(i) *Each $B_i$ is a ring that is also a two-sided ideal in $R$, and $B_i B_j = (0)$ if $j \neq i$.*

(ii) *If $L$ is any minimal left ideal in $R$, not necessarily occurring in the given decomposition of $R$, then $L \cong L_i$ for some $i$ and $L \subseteq B_i$.*

(iii) *Every two-sided ideal $D$ in $R$ is a direct sum of simple components.*

(iv) *Each $B_i$ is a simple ring.*

**Proof.**

(i) Each $B_i$ is a left ideal. To see that it is also a right ideal, consider
$$B_i R = B_i(B_1 \oplus \cdots \oplus B_m) \subseteq B_i B_1 + \cdots + B_i B_m.$$
Recall, for each $i$, that $B_i$ is a direct sum of left ideals $L$ isomorphic to $L_i$. If $L \cong L_i$ and $L' \cong L_j$, then the contrapositive, 'not (iii)' $\Rightarrow$ 'not (i)' in Lemma 7.41, applies to give $LL' = (0)$ if $j \neq i$. Hence, if $j \neq i$,
$$B_i B_j = \Big( \bigoplus_{L \cong L_i} L \Big) \Big( \bigoplus_{L' \cong L_j} L' \Big) \subseteq \bigoplus LL' = (0).$$
Thus, $B_i B_1 + \cdots + B_i B_m \subseteq B_i B_i$. Since $B_i$ is a left ideal, $B_i B_i \subseteq RB_i \subseteq B_i$. Therefore, $B_i R \subseteq B_i$, so that $B_i$ is a right ideal and, hence, is a two-sided ideal.

In the last step, proving that $B_i$ is a right ideal, we saw that $B_i B_i \subseteq B_i$; that is, $B_i$ is closed under multiplication. Therefore, to prove that $B_i$ is a ring, it now suffices to prove that it contains a unit element. If 1 is the unit element in $R$, then $1 = e_1 + \cdots + e_m$, where $e_i \in B_i$ for all $i$. If $b_i \in B_i$, then
$$b_i = 1b_i = (e_1 + \cdots + e_m)b_i = e_i b_i,$$
for $B_j B_i = (0)$ whenever $j \neq i$, by part (i). Similarly, the equation $b_i = b_i 1$ gives $b_i e_i = b_i$, and so $e_i$ is a unit in $B_i$. Thus, $B_i$ is a ring.[5]

(ii) By Proposition 7.42, a minimal left ideal $L$ is isomorphic to $L_i$ for some $i$. Now
$$L = RL = (B_1 \oplus \cdots \oplus B_m)L \subseteq B_1 L + \cdots + B_m L.$$
If $j \neq i$, then $B_j L = (0)$, by Lemma 7.41, so that
$$L \subseteq B_i L \subseteq B_i,$$
because $B_i$ is a right ideal.

(iii) A nonzero two-sided ideal $D$ in $R$ is a left ideal, and so it contains some minimal left ideal $L$, by Proposition 7.10. Now $L \cong L_i$ for some $i$, by Proposition 7.42; we claim that $B_i \subseteq D$. By Lemma 7.41, if $L'$ is any minimal left ideal in $B_i$, then $L' = Lb'$ for some $b' \in L'$. Since $L \subseteq D$ and $D$ is a right ideal, we have $L' = Lb' \subseteq LL' \subseteq DR \subseteq D$. We have shown that $D$ contains every left ideal isomorphic to $L_i$; as $B_i$ is generated by such ideals, $B_i \subseteq D$. Write $R = B_I \oplus B_J$, where $B_I = \bigoplus_i B_i$ with $B_i \subseteq D$ and $B_J = \bigoplus_j B_j$ with $B_j \nsubseteq D$. By Corollary 6.29 (which holds for modules over noncommutative rings), $D = B_I \oplus (D \cap B_J)$. But $D \cap B_J = (0)$; otherwise, it would contain a minimal left ideal $L \cong L_j$ for some $j \in J$ and, as above, this would force $B_j \subseteq D$. Therefore, $D = B_I$.

---

[5] $B_i$ is not a subring of $R$ because its unit $e_i$ is not the unit 1 in $R$.

(iv) A left ideal in $B_i$ is also a left ideal in $R$: if $a \in R$, then $a = \sum_j a_j$, where $a_j \in B_j$; if $b_i \in B_i$, then

$$ab_i = (a_1 + \cdots + a_m)b_i = a_i b_i \in B_i,$$

because $B_j B_i = (0)$ for $j \neq i$. Similarly, a right ideal in $B_i$ is a right ideal in $R$, and so a two-sided ideal $D$ in $B_i$ is a two-sided ideal in $R$. By part (iii), the only two-sided ideals in $R$ are direct sums of simple components, and so $D \subseteq B_i$ implies $D = (0)$ or $D = B_i$. Therefore, $B_i$ is a simple ring. •

**Corollary 7.50.** *If $R$ is a semisimple ring, then the simple component $B_i$ containing a minimal left ideal $L_i$ is the left ideal generated by all the minimal left ideals that are isomorphic to $L_i$. Therefore, the simple components $B_1, \ldots, B_m$ of a semisimple ring do not depend on a decomposition of $R$ as a direct sum of minimal left ideals.*

**Proof.** This follows from Lemma 7.49(ii). •

**Corollary 7.51.** *Let $A$ be a simple artinian ring.*

(i) *$A \cong \mathrm{Mat}_n(\Delta)$ for some division ring $\Delta$. If $L$ is a minimal left ideal in $A$, then every simple left $A$-module is isomorphic to $L$; moreover, $\Delta^{\mathrm{op}} \cong \mathrm{End}_A(L)$.*

(ii) *Two finitely generated left $A$-modules $M$ and $N$ are isomorphic if and only if $\dim_\Delta(M) = \dim_\Delta(N)$.*

**Proof.**

(i) Since $A$ is a semisimple ring, by Proposition 7.47, every left module $M$ is isomorphic to a direct sum of minimal left ideals. By Lemma 7.49(ii), all minimal left ideals are isomorphic, say, to $L$.

We now prove that $\Delta^{\mathrm{op}} \cong \mathrm{End}_A(L)$. We may assume that $A = \mathrm{Mat}_n(\Delta)$ and that $L = \mathrm{COL}(1)$, the minimal left ideal consisting of all the $n \times n$ matrices whose last $n-1$ columns are 0 (Proposition 7.37). Define $\varphi \colon \Delta \to \mathrm{End}_A(L)$ as follows: if $d \in \Delta$ and $\ell \in L$, then $\varphi_d \colon \ell \mapsto \ell d$. Note that $\varphi_d$ is an $A$-map: it is additive and, if $a \in A$ and $\ell \in L$, then $\varphi_d(a\ell) = (a\ell)d = a(\ell d) = a\varphi_d(\ell)$. Next, $\varphi$ is a ring antihomomorphism: $\varphi_1 = 1_L$, it is additive, and $\varphi_{dd'} = \varphi_{d'}\varphi_d$: if $\ell \in L$, then $\varphi_{d'}\varphi_d(\ell) = \varphi_d(\ell d') = \ell d'd = \varphi_{dd'}(\ell)$; that is, $\varphi$ is a ring homomorphism $\Delta^{\mathrm{op}} \to \mathrm{End}_A(L)$. To see that $\varphi$ is injective, note that each $\ell \in L \subseteq \mathrm{Mat}_n(\Delta)$ is a matrix with entries in $\Delta$; hence, $\ell d = 0$ implies $\ell = 0$. Finally, we show that $\varphi$ is surjective. Let $f \in \mathrm{End}_A(L)$. Now $L = AE_{11}$, where $E_{11}$ is the matrix unit (every simple module is generated by any nonzero element in it). If $u_i \in \Delta$, let $[u_1, \ldots, u_n]$ denote the $n \times n$ matrix in $L$ whose first column is $(u_1, \ldots, u_n)^{\top}$ and whose other entries are all 0. Write $f(E_{11}) = [d_1, \ldots, d_n]$. If $\ell \in L$, then $\ell$ has the form $[u_1, \ldots, u_n]$, and using only the definition of matrix multiplication, it is easy to see that

$[u_1, \ldots, u_n] = [u_1, \ldots, u_n]E_{11}$. Since $f$ is an $A$-map,

$$
\begin{aligned}
f([u_1, \ldots, u_n]) &= f([u_1, \ldots, u_n]E_{11}) \\
&= [u_1, \ldots, u_n]f(E_{11}) \\
&= [u_1, \ldots, u_n][d_1, \ldots, d_n] \\
&= [u_1, \ldots, u_n]d_1 = \varphi_{d_1}([u_1, \ldots, u_n]).
\end{aligned}
$$

Therefore, $f = \varphi_{d_1} \in \operatorname{im} \varphi$, as desired.

(ii) All minimal left ideals in $A$ are isomorphic to $L$, and so $M$ is a direct sum of $\dim_\Delta(M)/n$ copies of $L$. If $M \cong N$ as left $\operatorname{Mat}_n(\Delta)$-modules, then $M \cong N$ as left $\Delta$-modules, and so $\dim_\Delta(M) = \dim_\Delta(N)$. Conversely, if $\dim_\Delta(M) = nd = \dim_\Delta(N)$, then both $M$ and $N$ are direct sums of $d$ copies of $L$, and hence $M \cong N$ as left $A$-modules. $\bullet$

The number $m$ of simple components of $R$ is an invariant, for it is the number of nonisomorphic simple left $R$-modules (even better, we will see, in Theorem 7.58, that if $R = \mathbb{C}G$, then $m$ is the number of conjugacy classes in $G$). However, there is a much stronger uniqueness result.

**Theorem 7.52 (Wedderburn–Artin II).** *Every semisimple ring $R$ is a direct product,*

$$
R \cong \operatorname{Mat}_{n_1}(\Delta_1) \times \cdots \times \operatorname{Mat}_{n_m}(\Delta_m),
$$

*where $n_i \geq 1$ and $\Delta_i$ is a division ring, and the numbers $m$ and $n_i$, as well as the division rings $\Delta_i$, are uniquely determined by $R$.*

**Proof.** Let $R$ be a semisimple ring, and let $R = B_1 \oplus \cdots \oplus B_m$ be a decomposition into simple components arising from some decomposition of $R$ as a direct sum of minimal left ideals. Suppose that $R = B'_1 \times \cdots \times B'_t$, where each $B'_\ell$ is a two-sided ideal that is also a simple ring. By Lemma 7.49, each two-sided ideal $B'_\ell$ is a direct sum of $B_i$'s. But $B'_\ell$ cannot have more than one summand $B_i$, lest the simple ring $B'_\ell$ contain a proper nonzero two-sided ideal. Therefore, $t = m$ and, after reindexing, $B'_i = B_i$ for all $i$.

Dropping subscripts, it remains to prove that if $B = \operatorname{Mat}_n(\Delta) \cong \operatorname{Mat}_{n'}(\Delta') = B'$, then $n = n'$ and $\Delta \cong \Delta'$. In Proposition 7.37, we proved that $\operatorname{COL}(\ell)$, consisting of the matrices with $j$th columns 0 for all $j \neq \ell$, is a minimal left ideal in $B$, so that $\operatorname{COL}(\ell)$ is a simple $B$-module. Therefore,

$$
(0) \subseteq \operatorname{COL}(1) \subseteq [\operatorname{COL}(1) \oplus \operatorname{COL}(2)] \subseteq \cdots \subseteq [\operatorname{COL}(1) \oplus \cdots \oplus \operatorname{COL}(n)] = B
$$

is a composition series of $B$ as a module over itself. By the Jordan–Hölder Theorem (Theorem 7.3), $n$ and the factor modules $\operatorname{COL}(\ell)$ are invariants of $B$. Now $\operatorname{COL}(\ell) \cong \operatorname{COL}(1)$ for all $\ell$, by Corollary 7.51, and so it suffices to prove that $\Delta$ can be recaptured from $\operatorname{COL}(1)$. But this has been done in Corollary 7.51(i): $\Delta \cong \operatorname{End}_B(\operatorname{COL}(1))^{\operatorname{op}}$. $\bullet$

The description of the group algebra $kG$ simplifies when the field $k$ is algebraically closed. Here is the most useful version of Maschke's Theorem.

**Corollary 7.53 (Molien).** *If $G$ is a finite group and $k$ is an algebraically closed field whose characteristic does not divide $|G|$, then*

$$kG \cong \mathrm{Mat}_{n_1}(k) \times \cdots \times \mathrm{Mat}_{n_m}(k).$$

**Proof.** By Maschke's Theorem, $kG$ is a semisimple ring, and its simple components are isomorphic to matrix rings of the form $\mathrm{Mat}_n(\Delta)$, where $\Delta$ arises as $\mathrm{End}_{kG}(L)^{\mathrm{op}}$ for some minimal left ideal $L$ in $kG$. Therefore, it suffices to show that $\mathrm{End}_{kG}(L)^{\mathrm{op}} = \Delta = k$.

Now $\mathrm{End}_{kG}(L)^{\mathrm{op}} \subseteq \mathrm{End}_k(L)^{\mathrm{op}}$, which is finite-dimensional over $k$ because $L$ is; hence, $\Delta = \mathrm{End}_{kG}(L)^{\mathrm{op}}$ is finite-dimensional over $k$. Each $f \in \mathrm{End}_{kG}(L)$ is a $kG$-map, hence is a $k$-map; that is, $f(au) = af(u)$ for all $a \in k$ and $u \in L$. Therefore, the map $\varphi_a \colon L \to L$, given by $u \mapsto au$, commutes with $f$; that is, $k$ (identified with all $\varphi_a$) is contained in $Z(\Delta)$, the center of $\Delta$. If $\delta \in \Delta$, then $\delta$ commutes with every element in $k$, and so $k(\delta)$, the subdivision ring generated by $k$ and $\delta$, is a (commutative) field. As $\Delta$ is finite-dimensional over $k$, so is $k(\delta)$; that is, $k(\delta)$ is a finite extension of the field $k$, and so $\delta$ is algebraic over $k$, by Proposition 2.141. But $k$ is algebraically closed, so that $\delta \in k$ and $\Delta = k$.  •

The next corollary makes explicit a detail from the Wedderburn–Artin II, using Molien's simplification.

**Corollary 7.54.** *If $G$ is a finite group and $L_i$ is a minimal left ideal in $\mathbb{C}G$, then $\mathbb{C}G = B_1 \oplus \cdots \oplus B_m$, where $B_i$ is the ideal generated by all the minimal left ideals isomorphic to $L_i$. If $\dim_{\mathbb{C}}(L_i) = n_i$, then*

$$B_i = \mathrm{Mat}_{n_i}(\mathbb{C}).$$

**Example 7.55.** There are nonisomorphic finite groups $G$ and $H$ having isomorphic complex group algebras. If $G$ is an abelian group of order $d$, then $\mathbb{C}G$ is a direct product of matrix rings over $\mathbb{C}$, because $\mathbb{C}$ is algebraically closed. But $G$ abelian implies $\mathbb{C}G$ commutative. Hence, $\mathbb{C}G$ is the direct product of $d$ copies of $\mathbb{C}$ (for $\mathrm{Mat}_n(\mathbb{C})$ is commutative only when $n = 1$). It follows that if $H$ is any abelian group of order $d$, then $\mathbb{C}G \cong \mathbb{C}H$. In particular, $\mathbb{I}_4$ and $\mathbb{I}_2 \oplus \mathbb{I}_2$ are nonisomorphic groups having isomorphic complex group algebras. It follows from this example that certain properties of a group $G$ get lost in the group algebra $\mathbb{C}G$.  ◀

**Corollary 7.56.** *If $G$ is a finite group and $k$ is an algebraically closed field whose characteristic does not divide $|G|$, then $|G| = n_1^2 + n_2^2 + \cdots + n_m^2$, where the ith simple component $B_i$ of $kG$ consists of $n_i \times n_i$ matrices. Moreover, we may assume that $n_1 = 1$.[6]*

**Remark.** Theorem 7.98 says that all the $n_i$ are divisors of $|G|$.  ◀

**Proof.** As vector spaces over $k$, both $kG$ and $\mathrm{Mat}_{n_1}(k) \times \cdots \times \mathrm{Mat}_{n_m}(k)$ have the same dimension, for they are isomorphic, by Corollary 7.53. But $\dim(kG) = |G|$, and the dimension of the right side is $\sum_i \dim(\mathrm{Mat}_{n_i}(k)) = \sum_i n_i^2$.

---

[6] By Example 7.43, the group algebra $kG$ always has a unique minimal left ideal isomorphic to $V_0(k)$, even when $k$ is not algebraically closed.

Finally, Example 7.43 shows that there is a unique minimal left ideal isomorphic to the trivial module $V_0(k)$; the corresponding simple component, say, $B_1$, is one-dimensional, and so $n_1 = 1$.  •

The number $m$ of simple components in $\mathbb{C}G$ has a group-theoretic interpretation; we begin by finding the center of the group algebra.

**Definition.** Let $C_1, \ldots, C_r$ be the conjugacy classes in a finite group $G$. For each $C_j$, define the ***class sum*** to be the element $z_j \in \mathbb{C}G$ given by

$$z_j = \sum_{g \in C_j} g.$$

Here is a ring-theoretic interpretation of the number $r$ of conjugacy classes.

**Lemma 7.57.** *If $r$ is the number of conjugacy classes in a finite group $G$, then*

$$r = \dim_{\mathbb{C}}(Z(\mathbb{C}G)),$$

*where $Z(\mathbb{C}G))$ is the center of the group algebra. In fact, a basis of $Z(\mathbb{C}G)$ consists of all the class sums.*

**Proof.** If $z_j = \sum_{g \in C_j} g$ is a class sum, then we claim that $z_j \in Z(\mathbb{C}G)$. If $h \in G$, then $h z_j h^{-1} = z_j$, because conjugation by any element of $G$ merely permutes the elements in a conjugacy class. Note that if $j \neq \ell$, then $z_j$ and $z_\ell$ have no nonzero components in common, and so $z_1, \ldots, z_r$ is a linearly independent list. It remains to prove that the $z_j$ span the center.

Let $u = \sum_{g \in G} a_g g \in Z(\mathbb{C}G)$. If $h \in G$, then $huh^{-1} = u$, and so $a_{hgh^{-1}} = a_g$ for all $g \in G$. Thus, if $g$ and $g'$ lie in the same conjugacy class of $G$, then their coefficients in $u$ are the same. But this says that $u$ is a linear combination of the class sums $z_j$.  •

**Theorem 7.58.** *If $G$ is a finite group, then the number $m$ of simple components in $\mathbb{C}G$ is equal to the number $r$ of conjugacy classes in $G$.*

**Proof.** We have just seen, in Lemma 7.57, that $r = \dim_{\mathbb{C}}(Z(\mathbb{C}G))$. On the other hand, $Z(\mathrm{Mat}_{n_i}(\mathbb{C}))$, the center of a matrix ring, is the subspace of all scalar matrices, so that $m = \dim_{\mathbb{C}}(Z(\mathbb{C}G))$, by Lemma 7.57.  •

We began this section by seeing that $k$-representations of a group $G$ correspond to $kG$-modules. Let us now return to representations.

**Definition.** A $k$-representation of a group $G$ is ***irreducible*** if the corresponding $kG$-module is simple.

For example, a one-dimensional (necessarily irreducible) $k$-representation is a group homomorphism $\lambda \colon G \to k^{\times}$, where $k^{\times}$ is the multiplicative group of nonzero elements of $k$. The trivial $kG$-module $V_0(k)$ corresponds to the representation $\lambda_g = 1$ for all $g \in G$.

The next result is basic to the construction of the character table of a finite group.

**Theorem 7.59.** *If $G$ is a finite group, then the number of its irreducible complex representations is equal to the number $r$ of its conjugacy classes.*

**Proof.** By Proposition 7.42, every simple $\mathbb{C}G$-module is isomorphic to a minimal left ideal. Since the number of minimal left ideals is $m$ [the number of simple components of $\mathbb{C}G$], we see that $m$ is the number of irreducible $\mathbb{C}$-representations of $G$. But Theorem 7.58 equates $m$ with the number $r$ of conjugacy classes in $G$. •

**Example 7.60.**

(i) If $G = S_3$, then $\mathbb{C}G$ is six-dimensional. There are three simple components, for $S_3$ has three conjugacy classes (by Theorem 1.9, the number of conjugacy classes in $S_n$ is equal to the number of different cycle structures) having dimensions 1, 1, and 4, respectively. (We could have seen this without Theorem 7.58, for this is the only way to write 6 as a sum of squares aside from a sum of six 1's.) Therefore,

$$\mathbb{C}S_3 \cong \mathbb{C} \times \mathbb{C} \times \text{Mat}_2(\mathbb{C}).$$

One of the one-dimensional irreducible representations is the trivial one; the other is sgn (signum).

(ii) We now analyze $kG$ for $G = \mathbf{Q}$, the quaternion group of order 8. If $k = \mathbb{C}$, then Corollary 7.53 gives

$$\mathbb{C}\mathbf{Q} \cong \text{Mat}_{n_1}(\mathbb{C}) \times \cdots \times \text{Mat}_{n_r}(\mathbb{C}),$$

while Corollary 7.56 gives

$$|\mathbf{Q}| = 8 = 1 + n_2^2 + \cdots + n_r^2.$$

It follows that either all $n_i = 1$ or four $n_i = 1$ and one $n_i = 2$. The first case cannot occur, for it would imply that $\mathbb{C}\mathbf{Q}$ is a commutative ring, whereas the group $\mathbf{Q}$ of quaternions is not abelian. Therefore,

$$\mathbb{C}\mathbf{Q} \cong \mathbb{C} \times \mathbb{C} \times \mathbb{C} \times \mathbb{C} \times \text{Mat}_2(\mathbb{C}).$$

We could also have used Theorem 7.58, for $\mathbf{Q}$ has exactly five conjugacy classes, namely, $\{1\}$, $\{\overline{1}\}$, $\{i, \overline{i}\}$, $\{j, \overline{j}\}$, $\{k, \overline{k}\}$.

The group algebra $\mathbb{R}\mathbf{Q}$ is more complicated because $\mathbb{R}$ is not algebraically closed. Exercise 6.14 on page 416 shows that $\mathbb{H}$ is a quotient of $\mathbb{R}\mathbf{Q}$, hence $\mathbb{H}$ is isomorphic to a direct summand of $\mathbb{R}\mathbf{Q}$ because $\mathbb{R}\mathbf{Q}$ is semisimple. It turns out that

$$\mathbb{R}\mathbf{Q} \cong \mathbb{R} \times \mathbb{R} \times \mathbb{R} \times \mathbb{R} \times \mathbb{H}. \quad \blacktriangleleft$$

Here is an amusing application of the Wedderburn–Artin Theorems.

**Proposition 7.61.** *Let $R$ be a ring whose group of units $U = U(R)$ is finite and of odd order. Then $U$ is abelian and there are positive integers $m_i$ with*

$$|U| = \prod_{i=1}^{t} (2^{m_i} - 1).$$

**Proof.** First, we note that $1 = -1$ in $R$, lest $-1$ be a unit of even order. Consider the group algebra $kU$, where $k = \mathbb{F}_2$. Since $k$ has characteristic 2 and $|U|$ is odd, Maschke's Theorem says that $kU$ is semisimple. There is a ring map $\varphi \colon kU \to R$ carrying every $k$-linear combination of elements of $U$ to "itself." Now $R' = \operatorname{im} \varphi$ is a finite subring of $R$ containing $U$ (for $kU$ is finite); since dropping to a subring cannot create any new units, we have $U = U(R')$. By Corollary 7.29, the ring $R'$ is semisimple, so that Wedderburn–Artin Theorem I gives

$$R' \cong \prod_{i=1}^{t} \operatorname{Mat}_{n_i}(\Delta_i),$$

where each $\Delta_i$ is a division ring.

Now $\Delta_i$ is finite, because $R'$ is finite, and so $\Delta_i$ is a finite division ring. By the "other" theorem of Wedderburn, Theorem 7.13, each $\Delta_i$ is a field. But $-1 = 1$ in $R$ implies that $-1 = 1$ in $\Delta_i$ (for $\Delta_i \subseteq R'$), and so each field $\Delta_i$ has characteristic 2; hence,

$$|\Delta_i| = 2^{m_i}$$

for integers $m_i \geq 1$. All the matrix rings must be $1 \times 1$, for any matrix ring of larger size must contain an element of order 2, namely, $I + K$, where $K$ has entry 1 in the first position in the bottom row, and all other entries 0. For example,

$$\begin{bmatrix} 1 & 0 \\ 1 & 1 \end{bmatrix}^2 = \begin{bmatrix} 1 & 0 \\ 2 & 1 \end{bmatrix} = I.$$

Therefore, $R'$ is a direct product of finite fields of characteristic 2, and so $U = U(R')$ is an abelian group whose order is described in the statement. •

It follows, for example, that there is no ring having exactly five units.

The ***Jacobson–Chevalley Density Theorem***, an important generalization of the Wedderburn–Artin Theorems for certain nonartinian rings, was proved in the 1930s. Call a ring $R$ ***left primitive*** if there exists a faithful simple left $R$-module $S$; that is, $S$ is simple and, if $r \in R$ and $rS = \{0\}$, then $r = 0$. It can be proved that commutative primitive rings are fields, while left artinian left primitive rings are simple. Assume now that $R$ is a left primitive ring, that $S$ is a faithful simple left $R$-module, and that $\Delta$ denotes the division ring $\operatorname{End}_R(S)$. The Density Theorem says that if $R$ is left artinian, then $R \cong \operatorname{Mat}_n(\Delta)$, while if $R$ is not left artinian, then for every integer $n > 0$, there exists a subring $R_n$ of $R$ with $R_n \cong \operatorname{Mat}_n(\Delta)$. We refer the reader to Lam, *A First Course in Noncommutative Rings*, pp. 191–193.

The Wedderburn–Artin Theorems led to several areas of research, two of which are descriptions of division rings and of finite-dimensional algebras. Division rings will be considered in the context of central simple algebras in Chapter 8 and crossed product algebras in Chapter 9. Let us discuss finite-dimensional algebras now.

Thanks to the theorems of Maschke and Molien, the Wedderburn–Artin Theorems apply to ***ordinary*** representations of a finite group $G$; that is, to $kG$-modules, where $k$ is a field whose characteristic does not divide $|G|$. We know $kG$ is semisimple in this case. However, ***modular*** representations, that is, $kG$-modules for which the characteristic of $k$ does divide $|G|$, arise naturally. For example, if $G$ is a finite

$p$-group, for some prime $p$, then a minimal normal subgroup $N$ is a vector space over $\mathbb{F}_p$ (Rotman, *An Introduction to the Theory of Groups*, p. 106). Now $G$ acts on $N$ (by conjugation), and so $N$ is an $\mathbb{F}_p G$-module. Modular representations are used extensively in the classification of the finite simple groups. In his study of modular representations, Brauer observed that the important modules $M$ are *indecomposable* rather than irreducible. Recall that a module $M$ is indecomposable if $M \neq \{0\}$ and there are no nonzero modules $A$ and $B$ with $M = A \oplus B$ (in the ordinary case, a module is indecomposable if and only if it is irreducible [i.e., simple], but this is no longer true in the modular case). When $kG$ is semisimple, Proposition 7.42 says that there are only finitely many simple modules (corresponding to the minimal left ideals), which implies that there are only finitely many indecomposables. This is not true in the modular case, however. For example, if $k$ is an algebraically closed field of characteristic 2, $k\mathbf{V}$ and $kA_4$ have infinitely many nonisomorphic indecomposable modules.

A finite-dimensional $k$-algebra $R$ over a field $k$ is said to have ***finite representation type*** if there are only finitely many nonisomorphic finite-dimensional indecomposable $R$-modules. D. G. Higman proved, for a finite group $G$, that $kG$ has finite representation type for every field $k$ if and only if all its Sylow subgroups $G$ are cyclic (Curtis–Reiner, *Representation Theory of Finite Groups and Associative Algebras*, p. 431). In the 1950s, the following two problems, known as the ***Brauer–Thrall conjectures***, were posed. Let $R$ be a ring not of finite representation type.

(I) Are the dimensions of the indecomposable $R$-modules unbounded?

(II) Is there a strictly increasing sequence $n_1, n_2, \ldots$ with infinitely many nonisomorphic indecomposable $R$-modules of dimension $n_i$ for every $i$?

The positive solution of the first conjecture, by Roiter in 1968, had a great impact. Shortly thereafter, Gabriel introduced graph-theoretic methods, associating finite-dimensional algebras to certain oriented graphs, called *quivers*. He proved that a connected quiver has a finite number of nonisomorphic finite-dimensional representations if and only if the quiver is a Dynkin diagrams of type $A_n$, $D_n$, $E_6$, $E_7$, or $E_8$ (*Dynkin diagrams* are multigraphs that describe simple complex Lie algebras; see the discussion on page 748). Gabriel's result can be rephrased in terms of *left hereditary* $k$-algebras (all left ideals are projective modules). Dlab and Ringel extended Gabriel's result to Dynkin diagrams of any type, and they also extended the classification to certain left hereditary algebras.

A positive solution of Brauer–Thrall II for all finite-dimensional algebras over an algebraically closed field follows from results of Bautista, Gabriel, Roiter, and Salmerón. M. Auslander and Reiten created a theory involving *almost split sequences* and *Auslander–Reiten quivers*. As of this writing, Auslander–Reiten Theory is the most powerful tool in the study of representations of finite-dimensional algebras. For a discussion of these ideas, we refer the reader to Artin–Nesbitt–Thrall, *Rings with Minimum Condition*, Dlab–Ringel, *Indecomposable Representations of Graphs and Algebras*, Drozd–Kirichenko, *Finite-Dimensional Algebras*, Jacobson, *Structure of Rings*, and Rowen, *Ring Theory*.

## Exercises

**7.25.** Find $\mathbb{C}G$ if $G = D_8$, the dihedral group of order 8.

**7.26.** Find $\mathbb{C}G$ if $G = A_4$.

**Hint.** $A_4$ has four conjugacy classes.

**7.27.** (i) Let $k$ be a field, and view sgn: $S_n \to \{\pm 1\} \subseteq k$. Define Sig$(k)$ to be $k$ made into a $kS_n$-module (as in Proposition 7.21): if $\gamma \in S_n$ and $a \in k$, then $\gamma a = \text{sgn}(\gamma)a$. Prove that if Sig$(k)$ is an irreducible $kS_n$-module and $k$ does not have characteristic 2, then Sig$(k) \not\cong V_0(k)$.

(ii) Find $\mathbb{C}S_5$.

**Hint.** There are five conjugacy classes in $S_5$.

**7.28.** Let $G$ be a finite group, and let $k$ and $K$ be algebraically closed fields whose characteristics $p$ and $q$, respectively, do not divide $|G|$.

(i) Prove that $kG$ and $KG$ have the same number of simple components.

(ii) Prove that the degrees of the irreducible representations of $G$ over $k$ are the same as the degrees of the irreducible representations of $G$ over $K$.

## Section 7.5. Characters

*Characters* will enable us to use the preceding results to produce numerical invariants whose arithmetic properties help to prove theorems about finite groups. The first important instance of this technique is the following theorem.

**Theorem 7.62 (Burnside).** *Every group of order $p^m q^n$, where $p$ and $q$ are primes, is a solvable group.*

Notice that Burnside's Theorem cannot be improved to groups having orders with only three distinct prime factors, for $A_5$ is a simple group of order $60 = 2^2 \cdot 3 \cdot 5$.

Using representations, we will prove the following theorem.

**Theorem 7.63.** *If $G$ is a nonabelian finite simple group, then $\{1\}$ is the only conjugacy class whose size is a prime power.*

**Proposition 7.64.** Theorem 7.63 *implies Burnside's Theorem.*

**Proof.** Assume that Burnside's Theorem is false, and let $G$ be a "least criminal;" that is, $G$ is a counterexample of smallest order. If $G$ has a proper normal subgroup $H$ with $H \neq \{1\}$, then both $H$ and $G/H$ are solvable, for their orders are smaller than $|G|$ and are of the form $p^i q^j$. By Proposition 3.25, $G$ is solvable, and this is a contradiction. We may assume, therefore, that $G$ is a nonabelian simple group.

Let $Q$ be a Sylow $q$-subgroup of $G$. If $Q = \{1\}$, then $G$ is a $p$-group, contradicting $G$ being a nonabelian simple group; hence, $Q \neq \{1\}$. Since the center of $Q$

is nontrivial, by Theorem 1.113, there exists a nontrivial element $x \in Z(Q)$. Now $Q \subseteq C_G(x)$, for every element in $Q$ commutes with $x$, and so

$$[G : Q] = [G : C_G(x)][C_G(x) : Q];$$

that is, $[G : C_G(x)]$ is a divisor of $[G : Q] = p^m$. Of course, $[G : C_G(x)]$ is the number of elements in the conjugacy class $x^G$ of $x$ (Corollary 1.109), and so the hypothesis says that $|x^G| = 1$; hence, $x \in Z(G)$, contradicting $G$ being simple.  •

We now specialize the definition of $k$-representation on page 539 from arbitrary fields $k$ of scalars to the complex numbers $\mathbb{C}$.

**Definition.** A *representation* of a group $G$ is a homomorphism

$$\sigma : G \to \mathrm{GL}(V),$$

where $V$ is a vector space over $\mathbb{C}$. The *degree* of $\sigma$ is $\dim_{\mathbb{C}}(V)$.

*For the rest of this section, all groups are finite and all representations are finite-dimensional over $\mathbb{C}$.*

If $\sigma : G \to \mathrm{GL}(V)$ is a representation of degree $n$, then a choice of basis of $V$ allows each $\sigma(g)$ to be regarded as an $n \times n$ nonsingular complex matrix.

Representations can be translated into the language of modules. In Proposition 7.21, we proved that every representation $\sigma : G \to \mathrm{GL}(V)$ equips $V$ with the structure of a left $\mathbb{C}G$-module (and conversely): for each $g \in G$, we have $\sigma(g) : V \to V$ and, if $v \in V$, we define scalar multiplication $gv$ by

$$gv = \sigma(g)(v).$$

We denote $V$ made into a $\mathbb{C}G$-module in this way by $V^\sigma$, and we call it the *corresponding module*.

**Example 7.65.** Let $L$ be a left ideal in $\mathbb{C}G$ and let $\dim_{\mathbb{C}}(L) = d$. Then $\sigma : G \to \mathrm{GL}(L)$, defined by $\sigma(g) : u \mapsto gu$ for all $u \in L$, is a representation of degree $d$. The corresponding $\mathbb{C}G$-module $V^\sigma$ is just $L$ itself, for the original scalar multiplication on the left ideal $L$ coincides with the scalar multiplication given by $\sigma$.  ◄

**Example 7.66.** We now show that permutation representations, that is, $G$-sets,[7] give a special kind of representation. A $G$-set $X$ corresponds to a homomorphism $\pi : G \to S_X$, where $S_X$ is the symmetric group of all permutations of $X$. If $V$ is the complex vector space having $X$ as a basis, then we may regard $S_X \subseteq \mathrm{GL}(V)$ in the following way. Each permutation $\pi(g)$ of $X$, where $g \in G$, is now a permutation of a basis of $V$ and, hence, it determines a nonsingular linear transformation on $V$. With respect to the basis $X$, the matrix of $\pi(g)$ is a *permutation matrix*: it arises by permuting the columns of the identity matrix $I$ by $\pi(g)$; thus, it has exactly one entry equal to 1 in each row and column while all its other entries are 0.  ◄

One of the most important representations is the *regular representation*.

---

[7]Recall that if a group $G$ acts on a set $X$, then $X$ is called a *G-set*.

**Definition.** If $G$ is a group, then the representation $\rho\colon G \to \mathrm{GL}(\mathbb{C}G)$ defined, for all $g, h \in G$, by

$$\rho(g)\colon h \mapsto gh,$$

is called the **regular representation**.

By Example 7.65, the module corresponding to the regular representation is just $\mathbb{C}G$ considered as a left module over itself.

Two representations $\sigma\colon G \to \mathrm{GL}(V)$ and $\tau\colon G \to \mathrm{GL}(W)$ can be added.

**Definition.** If $\sigma\colon G \to \mathrm{GL}(V)$ and $\tau\colon G \to \mathrm{GL}(W)$ are representations, then their **sum** $\sigma + \tau\colon G \to \mathrm{GL}(V \oplus W)$ is defined by

$$(\sigma + \tau)(g)\colon (v, w) \mapsto (\sigma(g)v, \tau(g)w)$$

for all $g \in G$, $v \in V$, and $w \in W$.

In matrix terms, if $\sigma\colon G \to \mathrm{GL}(n, \mathbb{C})$ and $\tau\colon G \to \mathrm{GL}(m, \mathbb{C})$, then

$$\sigma + \tau\colon G \to \mathrm{GL}(n + m, \mathbb{C}),$$

and if $g \in G$, then $(\sigma + \tau)(g)$ is the direct sum of blocks $\sigma(g) \oplus \tau(g)$; that is,

$$(\sigma + \tau)(g) = \begin{bmatrix} \sigma(g) & 0 \\ 0 & \tau(g) \end{bmatrix}.$$

The following terminology is the common one used in Group Representations.

**Definition.** A representation $\sigma$ of a group $G$ is **irreducible** if the corresponding $\mathbb{C}G$-module is simple. A representation $\sigma$ is **completely reducible** if it is a direct sum of irreducible representations; that is, the corresponding $\mathbb{C}G$-module is semisimple.

**Example 7.67.** A representation $\sigma$ is **linear** if $\mathrm{degree}(\sigma) = 1$. The trivial representation of any group $G$ is linear, for the principal module $V_0(\mathbb{C})$ is one-dimensional. If $G = S_n$, then $\mathrm{sgn}\colon G \to \{\pm 1\}$ is also a linear representation.

Every linear representation is irreducible, for the corresponding $\mathbb{C}G$-module must be simple; after all, every submodule is a subspace, and $\{0\}$ and $V$ are the only subspaces of a one-dimensional vector space $V$. It follows that the trivial representation of any group $G$ is irreducible, as is the representation sgn of $S_n$. ◄

Recall the proof of the Wedderburn–Artin Theorem: there are pairwise non-isomorphic minimal left ideals $L_1, \ldots, L_r$ in $\mathbb{C}G$ and $\mathbb{C}G = B_1 \oplus \cdots \oplus B_r$, where $B_i$ is generated by all minimal left ideals isomorphic to $L_i$. By Corollary 7.53, $B_i \cong \mathrm{Mat}_{n_i}(\mathbb{C})$, where $n_i = \dim_{\mathbb{C}}(L_i)$. But all minimal left ideals in $\mathrm{Mat}_{n_i}(\mathbb{C})$ are isomorphic, by Lemma 7.49(ii), so that $L_i \cong \mathrm{COL}(1) \cong \mathbb{C}^{n_i}$ [see Example 7.14(iii)]. Therefore,

$$B_i \cong \mathrm{End}_{\mathbb{C}}(L_i).$$

**Proposition 7.68.**

   (i) *For each minimal left ideal $L_i$ in $\mathbb{C}G$, there is an irreducible representation $\lambda_i\colon G \to \mathrm{GL}(L_i)$, given by left multiplication:*

$$\lambda_i(g)\colon u_i \mapsto gu_i,$$

where $g \in G$ and $u_i \in L_i$; moreover, $\mathrm{degree}(\lambda_i) = n_i = \dim_{\mathbb{C}}(L_i)$.

(ii) *The representation $\lambda_i$ extends to a $\mathbb{C}$-algebra map $\widetilde{\lambda}_i \colon \mathbb{C}G \to \mathrm{End}_{\mathbb{C}}(L_i)$ if we define*

$$\widetilde{\lambda}_i(g)u_j = \begin{cases} gu_i & \text{if } j = i \\ 0 & \text{if } j \neq i \end{cases}$$

*for $g \in G$ and $u_j \in B_j$.*

**Proof.**

(i) Since $L_i$ is a left ideal in $\mathbb{C}G$, each $g \in G$ acts on $L_i$ by left multiplication, and so the representation $\lambda_i$ of $G$ is as stated. By Example 7.65, the corresponding module $L_i^\sigma$ is $L_i$, and so $\lambda_i$ is an irreducible representation because $L_i$, being a minimal left ideal, is a simple module.

(ii) If we regard $\mathbb{C}G$ and $\mathrm{End}_{\mathbb{C}}(L_i)$ as vector spaces over $\mathbb{C}$, then $\lambda_i$ extends to a linear transformation $\widetilde{\lambda}_i \colon \mathbb{C}G \to \mathrm{End}_{\mathbb{C}}(L_i)$ (because the elements of $G$ are a basis of $\mathbb{C}G$):

$$\widetilde{\lambda}_i \colon \sum_{g \in G} c_g g \mapsto \sum_{g \in G} c_g \lambda_i(g).$$

[remember that $\lambda_i(g) \in \mathrm{GL}(L_i) \subseteq \mathrm{End}_{\mathbb{C}}(L_i)$]. Let us show that $\widetilde{\lambda}_i \colon \mathbb{C}G \to \mathrm{End}_{\mathbb{C}}(L_i)$ is an algebra map. Now $\mathbb{C}G = B_1 \oplus \cdots \oplus B_r$, where the $B_j$ are two-sided ideals. To prove that $\widetilde{\lambda}_i$ is multiplicative, it suffices to check its values on products of basis elements. If $u_j \in B_j$ and $g, h \in G$, then

$$\widetilde{\lambda}_i(gh) \colon u_j \mapsto (gh)u_j,$$

while

$$\widetilde{\lambda}_i(g)\widetilde{\lambda}_i(h) \colon u_j \mapsto hu_j \mapsto g(hu_j);$$

these are the same, by associativity. Thus,

$$\widetilde{\lambda}_i(gh) = \widetilde{\lambda}_i(g)\widetilde{\lambda}_i.$$

Finally, $\widetilde{\lambda}_i(1) = \lambda_i(1) = 1_{L_i}$, and so $\widetilde{\lambda}_i$ is an algebra map. $\quad\bullet$

It is natural to call two representations *equivalent* if their corresponding modules are isomorphic. The following definition arises from Corollary 7.23, which gives a criterion that $\mathbb{C}G$-modules $(\mathbb{C}^n)^\sigma$ and $(\mathbb{C}^n)^\tau$ are isomorphic as $\mathbb{C}G$-modules.

**Definition.** Let $\sigma, \tau \colon G \to \mathrm{GL}(n, \mathbb{C})$ be representations of a group $G$. Then $\sigma$ and $\tau$ are **equivalent**, denoted by $\sigma \sim \tau$, if there is a nonsingular $n \times n$ matrix $P$ that intertwines them; that is, for every $g \in G$,

$$P\sigma(g)P^{-1} = \tau(g).$$

**Corollary 7.69.**

(i) *Every irreducible representation of a finite group $G$ is equivalent to one of the representations $\lambda_i$ given in* Proposition 7.68(i).

(ii) *Every irreducible representation of a finite abelian group is linear.*

(iii) *If $\sigma \colon G \to \mathrm{GL}(V)$ is a representation of a finite group $G$, then $\sigma(g)$ is similar to a diagonal matrix for each $g \in G$.*

**Proof.**

(i) If $\sigma \colon G \to \mathrm{GL}(V)$ is an irreducible representation $\sigma$, then the corresponding $\mathbb{C}G$-module $V^\sigma$ is a simple module. Therefore, $V^\sigma \cong L_i$, for some $i$, by Proposition 7.42. But $L_i \cong V^{\lambda_i}$, so that $V^\sigma \cong V^{\lambda_i}$ and $\sigma \sim \lambda_i$.

(ii) Since $G$ is abelian, $\mathbb{C}G = \bigoplus_i B_i$ is commutative, and so all $n_i = 1$ [we know that $B_i \cong \mathrm{Mat}_{n_i}(\mathbb{C})$, and the matrix ring is not commutative if $n_i \geq 2$]. But $\mathrm{degree}(\lambda_i) = n_i$, by Proposition 7.68(i).

(iii) If $\tau = \sigma|\langle g \rangle$, then $\tau(g) = \sigma(g)$. Now $\tau$ is a representation of the abelian group $\langle g \rangle$, and so part (ii) implies that the module $V^\tau$ is a direct sum of one-dimensional submodules. If $V^\tau = \langle v_1 \rangle \oplus \cdots \oplus \langle v_m \rangle$, then the matrix of $\sigma(g)$ with respect to the basis $v_1, \ldots, v_m$ is diagonal. $\bullet$

**Example 7.70.**

(i) The Wedderburn–Artin Theorems can be restated to say that every representation $\tau \colon G \to \mathrm{GL}(V)$ is completely reducible: $\tau = \sigma_1 + \cdots + \sigma_k$, where each $\sigma_j$ is irreducible; moreover, the multiplicity of each $\sigma_j$ is uniquely determined by $\tau$. Since each $\sigma_j$ is equivalent to the irreducible representation $\lambda_i$ arising from a minimal left ideal $L_i$, we usually collect terms and write $\tau \sim \sum_i m_i \lambda_i$, where the multiplicities $m_i$ are nonnegative integers.

(ii) The regular representation $\rho \colon G \to \mathrm{GL}(\mathbb{C}G)$ is important because every irreducible representation is a summand of it. Now $\rho$ is equivalent to the sum

$$\rho \sim n_1 \lambda_1 + \cdots + n_r \lambda_r,$$

where $n_i$ is the degree of $\lambda_i$ [recall that $\mathbb{C}G = \bigoplus_i B_i$, where $B_i \cong \mathrm{End}_{\mathbb{C}}(L_i) \cong \mathrm{Mat}_{n_i}(\mathbb{C})$; as a $\mathbb{C}G$-module, the simple module $L_i$ can be viewed as the first columns of $n_i \times n_i$ matrices, and so $B_i$ is a direct sum of $n_i$ copies of $L_i$]. ◄

Recall that the *trace* of an $n \times n$ matrix $A = [a_{ij}]$ with entries in a commutative ring $k$ is the sum of the diagonal entries: $\mathrm{tr}(A) = \sum_{i=1}^{n} a_{ii}$.

When $k$ is a field, then $\mathrm{tr}(A)$ turns out to be the sum of the eigenvalues of $A$ (we will assume this result now, but it is more convenient for us to prove it in the next chapter). Here are several other elementary facts about the trace that we will prove now.

**Proposition 7.71.**

(i) *If $I$ is the $n \times n$ identity matrix and $k$ is a field of characteristic $0$, then $\mathrm{tr}(I) = n$.*

(ii) *If $A = [a_{ij}]$ and $B = [b_{ij}]$ are $n \times n$ matrices with entries in a commutative ring $k$, then*

$$\mathrm{tr}(A + B) = \mathrm{tr}(A) + \mathrm{tr}(B) \quad and \quad \mathrm{tr}(AB) = \mathrm{tr}(BA).$$

(iii) *If $B = PAP^{-1}$, then $\mathrm{tr}(B) = \mathrm{tr}(A)$.*

**Proof.**

(i) The sum of the diagonal entries is $n$, which in not 0 because $k$ has characteristic 0.

(ii) The additivity of trace follows from the diagonal entries of $A + B$ being $a_{ii} + b_{ii}$. If $(AB)_{ii}$ denotes the $ii$ entry of $AB$, then

$$(AB)_{ii} = \sum_j a_{ij}b_{ji},$$

and so

$$\text{tr}(AB) = \sum_i (AB)_{ii} = \sum_{i,j} a_{ij}b_{ji}.$$

Similarly,

$$\text{tr}(BA) = \sum_{j,i} b_{ji}a_{ij}.$$

The entries commute because they lie in the commutative ring $k$, and so $a_{ij}b_{ji} = b_{ji}a_{ij}$ for all $i,j$. It follows that $\text{tr}(AB) = \text{tr}(BA)$, as desired.

(iii)
$$\text{tr}(B) = \text{tr}\big((PA)P^{-1}\big) = \text{tr}\big(P^{-1}(PA)\big) = \text{tr}(A). \quad \bullet$$

It follows from Proposition 7.71(iii) that we can define the trace of a linear transformation $T\colon V \to V$, where $V$ is a vector space over a field $k$, as the trace of any matrix arising from it: if $A$ and $B$ are matrices of $T$, determined by two choices of bases of $V$, then $B = PAP^{-1}$ for some nonsingular matrix $P$, and so $\text{tr}(B) = \text{tr}(A)$.

**Definition.** If $\sigma\colon G \to \text{GL}(V)$ is a representation, then its **character** is the function $\chi_\sigma\colon G \to \mathbb{C}$ defined by

$$\chi_\sigma(g) = \text{tr}(\sigma(g)).$$

We call $\chi_\sigma$ the character **afforded** by $\sigma$. An **irreducible character** is a character afforded by an irreducible representation. The **degree** of $\chi_\sigma$ is defined to be the degree of $\sigma$; that is,

$$\text{degree}(\chi_\sigma) = \text{degree}(\sigma) = \dim(V).$$

**Example 7.72.**

(i) The character $\theta$ afforded by a linear representation (Example 7.67) is called a **linear character**; that is, $\theta = \chi_\sigma$, where $\text{degree}(\sigma) = 1$. Since every linear representation is simple, every linear character is irreducible.

(ii) The representation $\lambda_i\colon G \to \text{GL}(L_i)$, given by $\lambda_i\colon u_i \mapsto gu_i$ if $u_i \in L_i$, is irreducible [see Proposition 7.68(i)]. Thus, the character $\chi_i$ afforded by $\lambda_i$, defined by

$$\chi_i = \chi_{\lambda_i},$$

is irreducible.

(iii) In light of Proposition 7.68(ii), it makes sense to speak of $\chi_i(u)$ for every $u \in \mathbb{C}G$. If we write $u = u_1 + \cdots + u_r \in B_1 \oplus \cdots \oplus B_r$, where $u_j \in B_j$, define $\chi_i(u) = \widetilde{\lambda}_i(u_i)$. In particular, $\chi_i(u_i) = \text{tr}(\widetilde{\lambda}_i(u_i))$ and $\chi_i(u_j) = 0$ if $j \neq i$.

(iv) If $\sigma\colon G \to \mathrm{GL}(V)$ is a representation, then $\sigma(1)$ is the identity matrix. Hence, Proposition 7.71(i) gives $\chi_\sigma(1) = n$, where $n$ is the degree of $\sigma$.

(v) Let $\sigma\colon G \to S_X$ be a homomorphism; as in Example 7.66, we may regard $\sigma$ as a representation on $V$, where $V$ is the vector space over $\mathbb{C}$ with basis $X$. For every $g \in G$, the matrix $\sigma(g)$ is a permutation matrix, and its $x$th diagonal entry is 1 if $\sigma(g)x = x$; otherwise, it is 0. Thus,

$$\chi_\sigma(g) = \mathrm{tr}(\sigma(g)) = \mathrm{Fix}(\sigma(g)),$$

the number of $x \in X$ fixed by $\sigma(g)$. In other words, if $X$ is a $G$-set, then each $g \in G$ acts on $X$, and the number of **fixed points** of the action of $g$ is a character value (see Example 7.92 for a related discussion).  ◀

Characters are compatible with addition of representations. If $\sigma\colon G \to \mathrm{GL}(V)$ and $\tau\colon G \to \mathrm{GL}(W)$, then $\sigma + \tau\colon G \to \mathrm{GL}(V \oplus W)$, and

$$\mathrm{tr}((\sigma + \tau)(g)) = \mathrm{tr}\left(\begin{bmatrix} \sigma(g) & 0 \\ 0 & \tau(g) \end{bmatrix}\right) = \mathrm{tr}(\sigma(g)) + \mathrm{tr}(\tau(g)).$$

Therefore,

$$\chi_{\sigma+\tau} = \chi_\sigma + \chi_\tau.$$

If $\sigma$ and $\tau$ are equivalent representations, then

$$\mathrm{tr}(\sigma(g)) = \mathrm{tr}(P\sigma(g)P^{-1}) = \mathrm{tr}(\tau(g))$$

for all $g \in G$; that is, they have the same characters: $\chi_\sigma = \chi_\tau$. It follows that if $\sigma\colon G \to \mathrm{GL}(V)$ is a representation, then its character $\chi_\sigma$ can be computed relative to any convenient basis of $V$.

### Proposition 7.73.

(i) *Every character $\chi_\sigma$ is a linear combination $\chi_\sigma = \sum_i m_i \chi_i$, where $m_i \geq 0$ are nonnegative integers and*

$$\chi_i = \chi_{\lambda_i}$$

*is the irreducible character afforded by the irreducible representation $\lambda_i$ arising from the minimal left ideal $L_i$.*

(ii) *Equivalent representations have the same character.*

(iii) *The only irreducible characters of $G$ are $\chi_1, \ldots, \chi_r$, the characters afforded by the irreducible representations $\lambda_i$.*

### Proof.

(i) The character $\chi_\sigma$ arises from a representation $\sigma$ of $G$, which, in turn, arises from a $\mathbb{C}G$-module $V$. But $V$ is a semisimple module (because $\mathbb{C}G$ is a semisimple ring), and so $V$ is a direct sum of simple modules: $V = \bigoplus_j S_j$. By Proposition 7.42, each $S_j \cong L_i$ for some minimal left ideal $L_i$. If, for each $i$, we let $m_i \geq 0$ be the number of $S_j$ isomorphic to $L_i$, then $\chi_\sigma = \sum_i m_i \chi_i$.

(ii) This follows from part (ii) of Proposition 7.71 and Corollary 7.69(i).

(iii) This follows from part (ii) and Corollary 7.69(i).  •

As a consequence of the proposition, we call $\chi_1, \ldots, \chi_r$ **the irreducible characters** of $G$.

**Example 7.74.**

(i) The (linear) character $\chi_1$ afforded by the trivial representation $\sigma\colon G \to \mathbb{C}$ with $\sigma(g) = 1$ for all $g \in G$ is called the **trivial character**. Thus, $\chi_1(g) = 1$ for all $g \in G$.

(ii) Let us compute the **regular character** $\psi = \chi_\rho$ afforded by the regular representation $\rho\colon G \to \mathrm{GL}(\mathbb{C}G)$, where $\rho(g)\colon u \mapsto gu$ for all $g \in G$ and $u \in \mathbb{C}G$. Any basis of $\mathbb{C}G$ can be used for this computation; we choose the usual basis comprised of the elements of $G$. If $g = 1$, then Example 7.72(iv) shows that $\psi(1) = \dim(\mathbb{C}G) = |G|$. On the other hand, if $g \neq 1$, then for all $h \in G$, we have $gh$ a basis element distinct from $h$. Therefore, the matrix of $\rho(g)$ has 0's on the diagonal, and so its trace is 0. Thus,

$$\psi(g) = \begin{cases} 0 & \text{if } g \neq 1 \\ |G| & \text{if } g = 1. \end{cases} \quad \blacktriangleleft$$

We have already proved that equivalent representations have the same character. The coming discussion will give the converse: if two representations have the same character, then they are equivalent.

**Definition.** A function $\varphi\colon G \to \mathbb{C}$ is a **class function** if it is constant on conjugacy classes; that is, if $h = xgx^{-1}$, then $\varphi(h) = \varphi(g)$.

Every character $\chi_\sigma$ afforded by a representation $\sigma$ is a class function: if $h = xgx^{-1}$, then

$$\sigma(h) = \sigma(xgx^{-1}) = \sigma(x)\sigma(g)\sigma(x)^{-1},$$

and so $\mathrm{tr}(\sigma(h)) = \mathrm{tr}(\sigma(g))$; that is,

$$\chi_\sigma(h) = \chi_\sigma(g).$$

Not every class function is a character. For example, if $\chi$ is a character, then $-\chi$ is a class function; it is not a character because $-\chi(1)$ is negative, and so it cannot be a degree.

**Definition.** We denote the set of all class functions $G \to \mathbb{C}$ by $\mathrm{cf}(G)$:

$$\mathrm{cf}(G) = \{\varphi\colon G \to \mathbb{C} : \varphi(g) = \varphi(xgx^{-1}) \text{ for all } x, g \in G\}.$$

It is easy to see that $\mathrm{cf}(G)$ is a vector space over $\mathbb{C}$.

An element $u = \sum_{g \in G} c_g g \in \mathbb{C}G$ is an $n$-tuple $(c_g)$ of complex numbers; that is, $u$ is a function $u\colon G \to \mathbb{C}$ with $u(g) = c_g$ for all $g \in G$. From this viewpoint, we see that $\mathrm{cf}(G)$ is a subring of $\mathbb{C}G$. Note that a class function is a scalar multiple of a class sum; therefore, Lemma 7.57 says that $\mathrm{cf}(G)$ is the center $Z(\mathbb{C}G)$, and so

$$\dim(\mathrm{cf}(G)) = r,$$

where $r$ is the number of conjugacy classes in $G$ (Theorem 7.58).

**Definition.** Write $\mathbb{C}G = B_1 \oplus \cdots \oplus B_r$, where $B_i \cong \operatorname{End}_\mathbb{C}(L_i)$, and let $e_i$ denote the identity element of $B_i$; hence,

$$1 = e_1 + \cdots + e_r,$$

where 1 is the identity element of $\mathbb{C}G$. The elements $e_i$ are called the ***idempotents*** in $\mathbb{C}G$.

Not only is each $e_i$ an idempotent, that is, $e_i^2 = e_i$, but it is easy to see that

$$e_i e_j = \delta_{ij} e_i,$$

where $\delta_{ij}$ is the Kronecker delta.

**Lemma 7.75.** *The irreducible characters $\chi_1, \ldots, \chi_r$ form a basis of $\operatorname{cf}(G)$.*

**Proof.** We have just seen that $\dim(\operatorname{cf}(G)) = r$, and so it suffices to prove that $\chi_1, \ldots, \chi_r$ is a linearly independent list, by Corollary 2.112(ii). We have already noted that $\chi_i(u_j) = 0$ for all $j \neq i$; in particular, $\chi_i(e_j) = 0$. On the other hand, $\chi_i(e_i) = n_i$, where $n_i$ is the degree of $\chi_i$, for it is the trace of the $n_i \times n_i$ identity matrix.

Suppose now that $\sum_i c_i \chi_i = 0$. It follows, for all $j$, that

$$0 = \Big( \sum_i c_i \chi_i \Big)(e_j) = c_j \chi_j(e_j) = c_j n_j.$$

Therefore, all $c_j = 0$, as desired. $\quad\bullet$

Since $\chi_i(1)$ is the trace of the $n_i \times n_i$ identity matrix, we have

$$(1) \qquad\qquad n_i = \chi_i(1) = \sum_j \chi_i(e_j) = \chi_i(e_i),$$

where $e_i$ is the identity element of $B_i$.

**Theorem 7.76.** *Two representations $\sigma, \tau$ of a finite group $G$ are equivalent if and only if they afford the same character: $\chi_\sigma = \chi_\tau$.*

**Proof.** We have already proved necessity, in Proposition 7.73(ii). For sufficiency, Proposition 7.73(i) says that every representation is completely reducible: there are nonnegative integers $m_i$ and $\ell_i$ with $\sigma \sim \sum_i m_i \lambda_i$ and $\tau \sim \sum_i \ell_i \lambda_i$. By hypothesis, the corresponding characters coincide:

$$\sum_i m_i \chi_i = \chi_\sigma = \chi_\tau = \sum_i \ell_i \chi_i.$$

As the irreducible characters $\chi_1, \ldots, \chi_r$ are a basis of $\operatorname{cf}(G)$, $m_i = \ell_i$ for all $i$, and so $\sigma \sim \tau$. $\quad\bullet$

There are relations between the irreducible characters that facilitate their calculation. We begin by finding the expression of the idempotents $e_i$ in terms of the basis $G$ of $\mathbb{C}G$. Observe, for all $y \in G$, that

$$(2) \qquad\qquad \chi_i(e_i y) = \chi_i(y),$$

for $y = \sum_j e_j y$, and so $\chi_i(y) = \sum_j \chi_i(e_j y) = \chi_i(e_i y)$, because $e_j y \in B_j$.

**Proposition 7.77.** *If $e_i = \sum_{g \in G} a_{ig} g$, where $a_{ig} \in \mathbb{C}$, then*

$$a_{ig} = \frac{n_i \chi_i(g^{-1})}{|G|}.$$

**Proof.** Let $\psi$ be the regular character; that is, $\psi$ is the character afforded by the regular representation. Now $e_i g^{-1} = \sum_h a_{ih} h g^{-1}$, so that

$$\psi(e_i g^{-1}) = \sum_{h \in G} a_{ih} \psi(hg^{-1}).$$

By Example 7.74(ii), $\psi(1) = |G|$ when $h = g$ and $\psi(hg^{-1}) = 0$ when $h \neq g$. Therefore,

$$a_{ig} = \frac{\psi(e_i g^{-1})}{|G|}.$$

On the other hand, since $\psi = \sum_j n_j \chi_j$, we have

$$\psi(e_i g^{-1}) = \sum_j n_j \chi_j(e_i g^{-1}) = n_i \chi_i(e_i g^{-1}),$$

by Proposition 7.68(ii). But $\chi_i(e_i g^{-1}) = \chi_i(g^{-1})$, by Equation (1). Therefore, $a_{ig} = n_i \chi_i(g^{-1})/|G|$. $\bullet$

It is now convenient to equip $\mathrm{cf}(G)$ with an inner product.

**Definition.** If $\alpha, \beta \in \mathrm{cf}(G)$, define

$$(\alpha, \beta) = \frac{1}{|G|} \sum_{g \in G} \alpha(g) \overline{\beta(g)},$$

where $\bar{c}$ denotes the complex conjugate of a complex number $c$.

It is easy to see that we have defined an inner product;[8] that is, for all $c_1, c_2 \in \mathbb{C}$,

(i) $(c_1 \alpha_1 + c_2 \alpha_2, \beta) = c_1(\alpha_1, \beta) + c_2(\alpha_2, \beta)$;

(ii) $(\beta, \alpha) = \overline{(\alpha, \beta)}$.

Note that $(\alpha, \alpha)$ is real, by (ii), and the inner product is *definite*; that is, $(\alpha, \alpha) > 0$ if $\alpha \neq 0$.

**Theorem 7.78.** *With respect to the inner product just defined, the irreducible characters $\chi_1, \ldots, \chi_r$ form an orthonormal basis; that is,*

$$(\chi_i, \chi_j) = \delta_{ij}.$$

**Proof.** By Proposition 7.77, we have

$$e_j = \frac{1}{|G|} \sum_g n_j \chi_j(g^{-1}) g.$$

---

[8]This inner product is *not* symmetric because we have $(\beta, \alpha) = \overline{(\alpha, \beta)}$, not $(\beta, \alpha) = (\alpha, \beta)$. Such a function is often called a *Hermitian form* or a *sesquilinear form* (*sesqui* means "one and a half").

Hence,

$$\chi_i(e_j)/n_j = \frac{1}{|G|} \sum_g \chi_j(g^{-1})\chi_i(g) = \frac{1}{|G|} \sum_g \chi_i(g)\overline{\chi_j(g)} = (\chi_i, \chi_j);$$

the next to last equation follows from Exercise 7.30 on page 588, for $\chi_j$ is a character (not merely a class function), and so $\chi_j(g^{-1}) = \overline{\chi_j(g)}$. The result now follows, for $\chi_i(e_j)/n_j = \delta_{ij}$, by Equations (1) and (2).  •

The inner product on $\mathrm{cf}(G)$ can be used to check irreducibility.

**Definition.** A **generalized character** $\varphi$ on a finite group $G$ is a $\mathbb{Z}$-linear combination

$$\varphi = \sum_i m_i \chi_i,$$

where $\chi_1, \dots, \chi_r$ are the irreducible characters of $G$ and all $m_i \in \mathbb{Z}$.

If $\theta$ is a character, then $\theta = \sum_i m_i \chi_i$, where all the coefficients are *nonnegative* integers, by Proposition 7.73.

**Corollary 7.79.** *A generalized character $\theta$ of a group $G$ is an irreducible character if and only if $\theta(1) > 0$ and $(\theta, \theta) = 1$.*

**Proof.** If $\theta$ is an irreducible character, then $\theta = \chi_i$ for some $i$, and so $(\theta, \theta) = (\chi_i, \chi_i) = 1$. Moreover, $\theta(1) = \deg(\chi_i) > 0$.

Conversely, let $\theta = \sum_j m_j \chi_j$, where $m_j \in \mathbb{Z}$, and suppose that $(\theta, \theta) = 1$. Then $1 = \sum_j m_j^2$; hence, some $m_i^2 = 1$ and all other $m_j = 0$. Therefore, $\theta = \pm\chi_i$, and so $\theta(1) = \pm\chi_i(1)$. Since $\chi_i(1) = \deg(\chi_i) > 0$, the hypothesis $\theta(1) > 0$ gives $m_i = 1$. Therefore, $\theta = \chi_i$, and so $\theta$ is an irreducible character.  •

Let us assemble the notation we will use from now on.

**Notation.** If $G$ is a finite group, we denote its conjugacy classes by

$$C_1, \dots, C_r,$$

a choice of elements, one from each conjugacy class, by

$$g_1 \in C_1, \dots, g_r \in C_r,$$

its irreducible characters by

$$\chi_1, \dots, \chi_r,$$

their degrees by

$$n_1 = \chi_1(1), \dots, n_r = \chi_r(1),$$

and the sizes of the conjugacy classes by

$$h_1 = |C_1|, \dots, h_r = |C_r|.$$

The matrix $[\chi_i(g_j)]$ is a useful way to display information.

**Definition.** The **character table** of $G$ is the $r \times r$ complex matrix whose $ij$ entry is $\chi_i(g_j)$.

We always assume that $C_1 = \{1\}$ and that $\chi_1$ is the trivial character. Thus, the first row consists of all 1's, while the first column consists of the degrees of the characters: $\chi_i(1) = n_i$ for all $i$, by Example 7.72(iv). The $i$th row of the character table consists of the values

$$\chi_i(1), \chi_i(g_2), \ldots, \chi_i(g_r).$$

There is no obvious way of labeling the other conjugacy classes (or the other irreducible characters), so that a finite group $G$ has many character tables. Nevertheless, we usually speak of "the" character table of $G$.

Since the inner product on $\mathrm{cf}(G)$ is summed over all $g \in G$, not just the chosen $g_i$ (one from each conjugacy class), it can be rewritten as a "weighted" inner product:

$$(\chi_i, \chi_j) = \frac{1}{|G|} \sum_{k=1}^{r} h_k \chi_i(g_k)\overline{\chi_j(g_k)}.$$

Theorem 7.78 says that the weighted inner product of distinct rows in the character table is 0, while the weighted inner product of any row with itself is 1.

**Example 7.80.** A character table can have complex entries. For example, it is easy to see that the character table for a cyclic group $G = \langle x \rangle$ of order 3 is given in Table 1, where $\omega = e^{2\pi i/3}$ is a primitive cube root of unity. ◄

| $g_i$ | 1 | $x$ | $x^2$ |
|---|---|---|---|
| $h_i$ | 1 | 1 | 1 |
| $\chi_1$ | 1 | 1 | 1 |
| $\chi_2$ | 1 | $\omega$ | $\omega^2$ |
| $\chi_3$ | 1 | $\omega^2$ | $\omega$ |

**Table 1.** Character table of $\mathbb{I}_3$.

**Example 7.81.** Write the four-group in additive notation:

$$\mathbf{V} = \{0, a, b, a+b\}.$$

As a vector space over $\mathbb{F}_2$, $\mathbf{V}$ has basis $a, b$, and the "coordinate functions" on $\mathbf{V}$, which take values in $\{1, -1\} \subseteq \mathbb{C}$, are linear; hence, they are irreducible representations. For example, the character $\chi_2$ arising from the function that is nontrivial on $a$ and trivial on $b$ is

$$\chi_2(v) = \begin{cases} -1 & \text{if } v = a \quad \text{or} \quad v = a+b \\ 1 & \text{if } v = 0 \quad \text{or} \quad v = b. \end{cases}$$

Table 2 is the character table for $\mathbf{V}$.

Table 3 is the character table for the symmetric group $G = S_3$. Since two permutations in $S_n$ are conjugate if and only if they have the same cycle structure, there are three conjugacy classes, and we choose elements 1, (1 2), and (1 2 3) from each. In Example 7.60(i), we saw that there are three irreducible representations: $\lambda_1 = $ the trivial representation, $\lambda_2 = \mathrm{sgn}$, and a third representation $\lambda_3$ of degree 2. We now give the character table, after which we discuss its entries.

| $g_i$ | 0 | $a$ | $b$ | $a+b$ |
|---|---|---|---|---|
| $h_i$ | 1 | 1 | 1 | 1 |
| $\chi_1$ | 1 | 1 | 1 | 1 |
| $\chi_2$ | 1 | $-1$ | 1 | $-1$ |
| $\chi_3$ | 1 | 1 | $-1$ | $-1$ |
| $\chi_4$ | 1 | $-1$ | $-1$ | 1 |

**Table 2.** Character table of **V**.

| $g_i$ | 1 | (1 2) | (1 2 3) |
|---|---|---|---|
| $h_i$ | 1 | 3 | 2 |
| $\chi_1$ | 1 | 1 | 1 |
| $\chi_2$ | 1 | $-1$ | 1 |
| $\chi_3$ | 2 | 0 | $-1$ |

**Table 3.** Character table of $S_3$.

We have already discussed the first row and column of any character table. Since $\chi_2 = $ sgn, the second row records the fact that 1 and (1 2 3) are even while (1 2) is odd. The third row has entries

$$2 \quad a \quad b,$$

where $a$ and $b$ are to be found. The weighted inner products of row 3 with the other two rows give the equations

$$2 + 3a + 2b = 0$$
$$2 - 3a + 2b = 0.$$

It follows easily that $a = 0$ and $b = -1$. ◀

The following lemma will be used to describe the inner products of the columns of the character table.

**Lemma 7.82.** *If $A$ is the character table of a finite group $G$, then $A$ is nonsingular and its inverse $A^{-1}$ has $ij$ entry*

$$(A^{-1})_{ij} = \frac{h_i \overline{\chi_j(g_i)}}{|G|}.$$

**Proof.** If $B$ is the matrix whose $ij$ entry is displayed in the statement, then

$$(AB)_{ij} = \frac{1}{|G|} \sum_k \chi_i(g_k) h_k \overline{\chi_j(g_k)} = \frac{1}{|G|} \sum_g \chi_i(g) \overline{\chi_j(g)} = (\chi_i, \chi_j) = \delta_{ij},$$

because $h_k \overline{\chi_j(g_k)} = \sum_{y \in C_k} \overline{\chi_j(y)}$. Therefore, $AB = I$. •

The next result is fundamental.

**Theorem 7.83 (Orthogonality Relations).** *Let $G$ be a finite group of order $n$ with conjugacy classes $C_1, \ldots, C_r$ of cardinalities $h_1, \ldots, h_r$, respectively, and choose elements $g_i \in C_i$. Let the irreducible characters of $G$ be $\chi_1, \ldots, \chi_r$, and let $\chi_i$ have degree $n_i$. Then the following relations hold:*

(i)

$$\sum_{k=1}^r h_k \chi_i(g_k) \overline{\chi_j(g_k)} = \begin{cases} 0 & \text{if } i \neq j; \\ |G| & \text{if } i = j. \end{cases}$$

(ii)

$$\sum_{i=1}^{r} \chi_i(g_k)\overline{\chi_i(g_\ell)} = \begin{cases} 0 & \text{if } k \neq \ell; \\ |G|/h_k & \text{if } k = \ell. \end{cases}$$

**Proof.**

(i) This is just a restatement of Theorem 7.78.

(ii) If $A$ is the character table of $G$ and $B = [h_i\overline{\chi_j(g_i)}/|G|]$, we proved, in Lemma 7.82, that $AB = I$. It follows that $BA = I$; that is, $(BA)_{k\ell} = \delta_{k\ell}$. Therefore,

$$\frac{1}{|G|} \sum_i h_k \overline{\chi_i(g_k)}\chi_i(g_\ell) = \delta_{k\ell},$$

and this is the second orthogonality relation.   •

In terms of the character table, the second orthogonality relation says that the usual (unweighted, but with complex conjugation) inner product of distinct columns is 0 while, for every $k$, the usual inner product of column $k$ with itself is $|G|/h_k$.

The orthogonality relations yield the following special cases.

**Corollary 7.84.**

(i) $|G| = \sum_{i=1}^{r} n_i^2$.

(ii) $\sum_{i=1}^{r} n_i\chi_i(g_k) = 0$   if $k > 1$.

(iii) $\sum_{k=1}^{r} h_k\chi_i(g_k) = 0$   if $i > 1$.

(iv) $\sum_{k=1}^{r} h_k|\chi_i(g_k)|^2 = |G|$.

**Proof.**

(i) This equation records the inner product of column 1 with itself: it is Theorem 7.83(ii) when $k = \ell = 1$.

(ii) This is the special case of Theorem 7.83(ii) with $\ell = 1$, for $\chi_i(1) = n_i$.

(iii) This is the special case of Theorem 7.83(i) in which $j = 1$.

(iv) This is the special case of Theorem 7.83(i) in which $j = i$.   •

We can now give another proof of Burnside's Lemma, Theorem 1.124, which counts the number of orbits of a $G$-set.

**Theorem 7.85 (Burnside's Lemma).** *Let $G$ be a finite group and let $X$ be a finite $G$-set. If $N$ is the number of orbits of $X$, then*

$$N = \frac{1}{|G|} \sum_{g \in G} \text{Fix}(g),$$

*where $\text{Fix}(g)$ is the number of $x \in X$ with $gx = x$.*

**Proof.** Let $V$ be the complex vector space having $X$ as a basis. As in Example 7.66, the $G$-set $X$ gives a representation $\sigma \colon G \to \mathrm{GL}(V)$ by $\sigma(g)(x) = gx$ for all $g \in G$ and $x \in X$; moreover, if $\chi_\sigma$ is the character afforded by $\sigma$, then Example 7.72(v) shows that $\chi_\sigma(g) = \mathrm{Fix}(g)$.

Let $\mathcal{O}_1, \ldots, \mathcal{O}_N$ be the orbits of $X$. We begin by showing that $N = \dim(V^G)$, where $V^G$ is the space of *fixed points*:

$$V^G = \{v \in V : gv = v \text{ for all } g \in G\}.$$

For each $i$, define $s_i$ to be the sum of all the $x$ in $\mathcal{O}_i$; it suffices to prove that these elements form a basis of $V^G$. It is plain that $s_1, \ldots, s_N$ is a linearly independent list in $V^G$, and it remains to prove that they span $V^G$. If $u \in V^G$, then $u = \sum_{x \in X} c_x x$, so that $gu = \sum_{x \in X} c_x(gx)$. Since $gu = u$, however, $c_x = c_{gx}$. Thus, given $x \in X$ with $x \in \mathcal{O}_j$, each coefficient of $gx$, where $g \in G$, is equal to $c_x$; that is, all the $x$ lying in the orbit $\mathcal{O}_j$ have the same coefficient, say, $c_j$, and so $u = \sum_j c_j s_j$. Therefore,

$$N = \dim(V^G).$$

Now define a linear transformation $T \colon V \to V$ by

$$T = \frac{1}{|G|} \sum_{g \in G} \sigma(g).$$

It is routine to check that $T$ is a $\mathbb{C}G$-map, that $T|(V^G) = $ identity, and that $\mathrm{im}\, T = V^G$. Since $\mathbb{C}G$ is semisimple, $V = V^G \oplus W$ for some submodule $W$. We claim that $T|W = 0$. If $w \in W$, then $\sigma(g)(w) \in W$ for all $g \in G$, because $W$ is a submodule, and so $T(w) \in W$. On the other hand, $T(w) \in \mathrm{im}\, T = V^G$, and so $T(w) \in V^G \cap W = \{0\}$, as claimed.

If $w_1, \ldots, w_\ell$ is a basis of $W$, then $s_1, \ldots, s_N, w_1, \ldots, w_\ell$ is a basis of $V = V^G \oplus W$. Note that $T$ fixes each $s_i$ and annihilates each $w_j$. Since trace preserves sums,

$$\mathrm{tr}(T) = \frac{1}{|G|} \sum_{g \in G} \mathrm{tr}(\sigma(g)) = \frac{1}{|G|} \sum_{g \in G} \chi_\sigma(g) = \frac{1}{|G|} \sum_{g \in G} \mathrm{Fix}(g).$$

It follows that

$$\mathrm{tr}(T) = \dim(V^G),$$

for the matrix of $T$ with respect to the chosen basis is the direct sum of an identity block and a zero block, and so $\mathrm{tr}(T)$ is the size of the identity block, namely, $\dim(V^G) = N$. Therefore,

$$N = \frac{1}{|G|} \sum_{g \in G} \mathrm{Fix}(g). \quad \bullet$$

Character tables can be used to detect normal subgroups.

**Definition.** If $\chi_\tau$ is the character afforded by a representation $\tau \colon G \to \mathrm{GL}(V)$, then

$$\ker \chi_\tau = \ker \tau.$$

**Proposition 7.86.** *Let $\theta = \chi_\tau$ be the character of a finite group $G$ afforded by a representation $\tau \colon G \to \mathrm{GL}(V)$.*

(i) *For each $g \in G$, we have*

$$|\theta(g)| \le \theta(1).$$

(ii)

$$\ker \theta = \{g \in G : \theta(g) = \theta(1)\}.$$

(iii) *If $\theta = \sum_j m_j \chi_j$, where $m_j$ are positive integers, then*

$$\ker \theta = \bigcap_j \ker \chi_j.$$

(iv) *If $N$ is a normal subgroup of $G$, there are irreducible characters $\chi_{i_1}, \dots, \chi_{i_s}$ with $N = \bigcap_{j=1}^s \ker \chi_{i_j}$.*

**Proof.**

(i) By Lagrange's Theorem, $g^{|G|} = 1$ for every $g \in G$; it follows that the eigen-values $\varepsilon_1, \dots, \varepsilon_d$ of $\tau(g)$, where $d = \theta(1)$, are $|G|$th roots of unity, and so $|\varepsilon_j| = 1$ for all $j$. By the triangle inequality in $\mathbb{C}$,

$$|\theta(g)| = \left| \sum_{j=1}^d \varepsilon_j \right| \le d = \theta(1).$$

(ii) If $g \in \ker \theta = \ker \tau$, then $\tau(g) = I$, the identity matrix, and $\theta(g) = \operatorname{tr}(I) = \theta(1)$. Conversely, suppose that $\theta(g) = \theta(1) = d$; that is, $\sum_{j=1}^d \varepsilon_j = d$. By Proposition 2.73, all the eigenvalues $\varepsilon_j$ are equal, say, $\varepsilon_j = \omega$ for all $j$. Therefore, $\tau(g) = \omega I$, by Corollary 7.69(iii), and so

$$\theta(g) = \theta(1)\omega.$$

But $\theta(g) = \theta(1)$, by hypothesis, and so $\omega = 1$; that is, $\tau(g) = I$ and $g \in \ker \tau$.

(iii) For all $g \in G$, we have

$$\theta(g) = \sum_j m_j \chi_j(g);$$

in particular,

$$\theta(1) = \sum_j m_j \chi_j(1).$$

If $g \in \ker \theta$, then $\theta(g) = \theta(1)$. Suppose that $\chi_{j'}(g) \ne \chi_{j'}(1)$ for some $j'$. Since $\chi_{j'}(g)$ is a sum of roots of unity, Proposition 2.73 applies to force $|\chi_{j'}(g)| < \chi_{j'}(1)$, and so $|\theta(g)| \le \sum_j m_j |\chi_j(g)| < \sum_j \chi_j(1) = 1$, which implies that $\theta(g) \ne \theta(1)$, a contradiction. Therefore, $g \in \bigcap_j \ker \chi_j$. For the reverse inclusion, if $g \in \ker \chi_j$, then $\chi_j(g) = \chi_j(1)$, and so

$$\theta(g) = \sum_j m_j \chi_j(g) = \sum_j m_j \chi_j(1) = \theta(1);$$

hence, $g \in \ker \theta$.

(iv) It suffices to find a representation of $G$ whose kernel is $N$. By part (ii) and Example 7.74(ii), the regular representation $\rho$ of $G/N$ is faithful (i.e., is an injection), and so its kernel is $\{1\}$. If $\pi\colon G \to G/N$ is the natural map, then $\rho\pi$ is a representation of $G$ having kernel $N$. If $\theta$ is the character afforded by $\rho\pi$, then $\theta = \sum_j m_j \chi_j$, where the $m_j$ are positive integers, by Lemma 7.75, and so part (iii) applies.  •

**Example 7.87.** We will construct the character table of $S_4$ in Example 7.97. We can see there that $\ker\chi_2 = A_4$ and $\ker\chi_3 = \mathbf{V}$ are the only two normal subgroups of $S_4$ (other than $\{1\}$ and $S_4$). Moreover, we can see that $\mathbf{V} \subseteq A_4$.

In Example 7.88, we can see that $\ker\chi_2 = \{1\} \cup z^G \cup y^G$ (where $z^G$ denotes the conjugacy class of $z$ in $G$) and $\ker\chi_3 = \{1\} \cup z^G \cup x^G$. Another normal subgroup occurs as $\ker\chi_2 \cap \ker\chi_3 = \{1\} \cup z^G$.  ◀

A normal subgroup described by characters is given as a union of conjugacy classes; this viewpoint can give another proof of the simplicity of $A_5$. In Exercise 1.97 on page 74, we saw that $A_5$ has five conjugacy classes, of sizes 1, 12, 12, 15, and 20. Since every subgroup contains the identity element, the order of a normal subgroup of $A_5$ is the sum of some of these numbers, including 1. But it is easy to see that 1 and 60 are the only such sums that are divisors of 60, and so the only normal subgroups are $\{1\}$ and $A_5$ itself.

There is a way to "lift" a representation of a quotient group to a representation of the group.

**Definition.** Let $H \lhd G$ and let $\sigma\colon G/H \to \mathrm{GL}(V)$ be a representation. If $\pi\colon G \to G/H$ is the natural map, then the representation $\sigma\pi\colon G \to \mathrm{GL}(V)$ is called a ***lifting*** of $\sigma$.

Scalar multiplication of $G$ on a $\mathbb{C}(G/H)$-module $V$ is given, for $v \in V$, by

$$gv = (gH)v.$$

Thus, every $\mathbb{C}(G/H)$-submodule of $V$ is also a $\mathbb{C}G$-submodule; hence, if $V$ is a simple $\mathbb{C}(G/H)$-module, then it is also a simple $\mathbb{C}G$-module. It follows that if $\sigma\colon G/H \to \mathrm{GL}(V)$ is an irreducible representation of $G/H$, then its lifting $\sigma\pi$ is also an irreducible representation of $G$.

**Example 7.88.** We know that $D_8$ and $\mathbf{Q}$ are nonisomorphic nonabelian groups of order 8; we now show that they have the same character tables.

If $G$ is a nonabelian group of order 8, then its center has order 2, say, $Z(G) = \langle z \rangle$. Now $G/Z(G)$ is not cyclic, by Exercise 1.77 on page 59, and so $G/Z(G) \cong \mathbf{V}$. Therefore, if $\sigma\colon \mathbf{V} \to \mathbb{C}$ is an irreducible representation of $\mathbf{V}$, then its lifting $\sigma\pi$ is an irreducible representation of $G$. This gives four (necessarily irreducible) linear characters of $G$, each of which takes value 1 on $z$. As $G$ is not abelian, there must be an irreducible character $\chi_5$ of degree $n_5 > 1$ (if all $n_i = 1$, then $\mathbb{C}G$ is commutative and $G$ is abelian). Since $\sum_i n_i^2 = 8$, we see that $n_5 = 2$. Thus, there are five irreducible representations and, hence, five conjugacy classes; choose representatives $g_i$ to be $1, z, x, y, w$. Table 4 on page 580 is the character table.

| $g_i$ | 1 | $z$ | $x$ | $y$ | $w$ |
|---|---|---|---|---|---|
| $h_i$ | 1 | 1 | 2 | 2 | 2 |
| $\chi_1$ | 1 | 1 | 1 | 1 | 1 |
| $\chi_2$ | 1 | 1 | $-1$ | 1 | $-1$ |
| $\chi_3$ | 1 | 1 | 1 | $-1$ | $-1$ |
| $\chi_4$ | 1 | 1 | $-1$ | $-1$ | 1 |
| $\chi_5$ | 2 | $-2$ | 0 | 0 | 0 |

**Table 4.** Character table of $D_8$ and of $\mathbf{Q}$.

The values for $\chi_5$ are computed from the orthogonality relations of the columns. For example, if the last row of the character table is

$$2 \quad a \quad b \quad c \quad d,$$

then the inner product of columns 1 and 2 gives the equation $4 + 2a = 0$, so that $a = -2$. The reader may verify that $0 = b = c = d$. ◄

The orthogonality relations help to complete a character table but, obviously, it would also be useful to have a supply of characters. One important class of characters consists of those afforded by *induced representations*; that is, representations of a group $G$ determined by representations of a subgroup $H$ of $G$.

The original construction of induced representations, due to Frobenius, is rather complicated. Tensor products make this construction more natural. The ring $\mathbb{C}G$ is a $(\mathbb{C}G, \mathbb{C}H)$-bimodule (for $\mathbb{C}H$ is a subring of $\mathbb{C}G$), so that if $V$ is a left $\mathbb{C}H$-module, then the tensor product $\mathbb{C}G \otimes_{\mathbb{C}H} V$ is defined; Proposition 6.106 says that this tensor product is, in fact, a left $\mathbb{C}G$-module.

**Definition.** Let $H$ be a subgroup of a group $G$. If $V$ is a left $\mathbb{C}H$-module, then the **induced module** is the left $\mathbb{C}G$-module

$$V\!\uparrow^G = \mathbb{C}G \otimes_{\mathbb{C}H} V.$$

The corresponding representation $\rho\!\uparrow^G \colon G \to V^G$ is called the **induced representation**. The character of $G$ afforded by $\rho\!\uparrow^G$ is called the **induced character**, and it is denoted by $\chi_\rho\!\uparrow^G$.

Let us recognize at the outset that the character of an induced representation need not restrict to the original representation of the subgroup. For example, we have seen that there is an irreducible character $\chi$ of $A_3 \cong \mathbb{I}_3$ having complex values, whereas every irreducible character of $S_3$ has (real) integer values. A related observation is that two elements may be conjugate in a group but not conjugate in a subgroup (for example, 3-cycles are conjugate in $S_3$, for they have the same cycle structure, but they are not conjugate in the abelian group $A_3$).

The next lemma will help us compute the character afforded by an induced representation.

**Lemma 7.89.**

(i) *If $H \subseteq G$, then $\mathbb{C}G$ is a free right $\mathbb{C}H$-module on $[G : H]$ generators.*

(ii) *If a left $\mathbb{C}H$-module $V$ has a (vector space) basis $e_1, \ldots, e_m$, then a (vector space) basis of the induced module $V\!\uparrow^G = \mathbb{C}G \otimes_{\mathbb{C}H} V$ is the family of all $t_i \otimes e_j$, where $t_1, \ldots, t_n$ is a transversal of $H$ in $G$.*

**Proof.**

(i) Since $t_1, \ldots, t_n$ is a transversal of $H$ in $G$ (of course, $n = [G : H]$), we see that $G$ is the disjoint union

$$G = \bigcup_i t_i H;$$

thus, for every $g \in G$, there is a unique $i$ and a unique $h \in H$ with $g = t_i h$. We claim that $t_1, \ldots, t_n$ is a basis of $\mathbb{C}G$ viewed as a right $\mathbb{C}H$-module.

If $u \in \mathbb{C}G$, then $u = \sum_g a_g g$, where $a_g \in \mathbb{C}$. Rewrite each term

$$a_g g = a_g t_i h = t_i a_g h$$

(scalars in $\mathbb{C}$ commute with everything), collect terms involving the same $t_i$, and obtain $u = \sum_i t_i \eta_i$, where $\eta_i \in \mathbb{C}H$.

To prove uniqueness of this expression, suppose that $0 = \sum_i t_i \eta_i$, where $\eta_i \in \mathbb{C}H$. Now $\eta_i = \sum_{h \in H} a_{ih} h$, where $a_{ih} \in \mathbb{C}$. Substituting,

$$0 = \sum_{i,h} a_{ih} t_i h.$$

But $t_i h = t_j h'$ if and only if $i = j$ and $h = h'$, so that $0 = \sum_{i,h} a_{ih} t_i h = \sum_{g \in G} a_{ih} g$, where $g = t_i h$. Since the elements of $G$ form a basis of $\mathbb{C}G$ (viewed as a vector space over $\mathbb{C}$), we have $a_{ih} = 0$ for all $i, h$, and so $\eta_i = 0$ for all $i$.

(ii) By Theorem 6.110,

$$\mathbb{C}G \otimes_{\mathbb{C}H} V \cong \bigoplus_i t_i \mathbb{C}H \otimes_{\mathbb{C}H} V.$$

It follows that every $u \in \mathbb{C}G \otimes_{\mathbb{C}H} V$ has a unique expression as a $\mathbb{C}$-linear combination of $t_i \otimes e_j$, and so these elements comprise a basis. •

**Notation.** If $H \subseteq G$ and $\chi \colon H \to \mathbb{C}$ is a function, then $\dot{\chi} \colon G \to \mathbb{C}$ is given by

$$\dot{\chi}(g) = \begin{cases} 0 & \text{if } g \notin H \\ \chi(g) & \text{if } g \in H. \end{cases}$$

**Theorem 7.90.** *If $\chi_\sigma$ is the character afforded by a representation $\sigma \colon H \to \mathrm{GL}(V)$ of a subgroup $H$ of a group $G$, then the induced character $\chi_\sigma\!\uparrow^G$ is given by*

$$\chi_\sigma\!\uparrow^G (g) = \frac{1}{|H|} \sum_{a \in G} \dot{\chi}_\sigma (a^{-1} g a).$$

**Proof.** Let $t_1, \ldots, t_n$ be a transversal of $H$ in $G$, so that $G$ is the disjoint union $G = \bigcup_i t_i H$, and let $e_1, \ldots, e_m$ be a (vector space) basis of $V$. By Lemma 7.89, a basis for the vector space $V^G = \mathbb{C}G \otimes_{\mathbb{C}H} V$ consists of all $t_i \otimes e_j$. If $g \in G$, we compute the matrix of left multiplication by $g$ relative to this basis. Note that

$$gt_i = t_{k(i)} h_i,$$

where $h_i \in H$, and so

$$g(t_i \otimes e_j) = (gt_i) \otimes e_j = t_{k(i)} h_i \otimes e_j = t_{k(i)} \otimes \sigma(h_i) e_j$$

(the last equation holds because we can slide any element of $H$ across the tensor sign). Now $g(t_i \otimes e_j)$ is written as a $\mathbb{C}$-linear combination of *all* the basis elements of $V\!\uparrow^G$, for the coefficients $t_p \otimes e_j$ for $p \neq k(i)$ are all 0. Hence, $\sigma\!\uparrow^G(g)$ gives the $nm \times nm$ matrix whose $m$ columns labeled by $t_i \otimes e_j$, for fixed $i$, are all zero except for an $m \times m$ block equal to

$$[a_{pq}(h_i)] = [a_{pq}(t_{k(i)}^{-1} gt_i)].$$

Thus, the big matrix is partitioned into $m \times m$ blocks, most of which are 0, and a nonzero block is on the diagonal of the big matrix if and only if $k(i) = i$; that is,

$$t_{k(i)}^{-1} gt_i = t_i^{-1} gt_i = h_i \in H.$$

The induced character is the trace of the big matrix, which is the sum of the traces of these blocks on the diagonal. Therefore,

$$\chi_\sigma\!\uparrow^G(g) = \sum_{t_i^{-1} gt_i \in H} \operatorname{tr}([a_{pq}(t_i^{-1} gt_i)]) = \sum_i \dot{\chi}_\sigma(t_i^{-1} gt_i)$$

(remember that $\dot{\chi}_\sigma$ is 0 outside of $H$). We now rewrite the summands (to get a formula that does not depend on the choice of the transversal): if $t_i^{-1} gt_i \in H$, then $(t_i h)^{-1} g(t_i h) = h^{-1}(t_i^{-1} gt_i)h$ in $H$, so that, for fixed $i$,

$$\sum_{h \in H} \dot{\chi}_\sigma\big((t_i h)^{-1} g(t_i h)\big) = |H| \dot{\chi}_\sigma(t_i^{-1} gt_i),$$

because $\dot{\chi}_\sigma$ is a class function on $H$. Therefore,

$$\chi_\sigma\!\uparrow^G(g) = \sum_i \dot{\chi}_\sigma(t_i^{-1} gt_i) = \frac{1}{|H|} \sum_{i,h} \dot{\chi}_\sigma\big((t_i h)^{-1} g(t_i h)\big) = \frac{1}{|H|} \sum_{a \in G} \dot{\chi}_\sigma(a^{-1} ga). \quad \bullet$$

**Remark.** We have been considering induced characters, but it is easy to generalize the discussion to **induced class functions**. If $H \subseteq G$, then a class function $\theta$ on $H$ has a unique expression as a $\mathbb{C}$-linear combination of irreducible characters of $H$, say, $\theta = \sum c_i \chi_i$, and so we can define

$$\theta\!\uparrow^G = \sum c_i \chi_i\!\uparrow^G.$$

It is plain that $\theta\!\uparrow^G$ is a class function on $G$, and that the formula in Theorem 7.90 extends to induced class functions. ◄

If, for $h \in H$, the matrix of $\sigma(h)$ (with respect to the basis $e_1, \ldots, e_m$ of $V$) is $B(h)$, then define $m \times m$ matrices $\dot{B}(g)$, for all $g \in G$, by

$$\dot{B}(g) = \begin{cases} 0 & \text{if } g \notin H; \\ B(g) & \text{if } g \in H. \end{cases}$$

The proof of Theorem 7.90 allows us to picture the matrix of the induced representation in block form

$$\sigma\!\uparrow^G(g) = \begin{bmatrix} \dot{B}(t_1^{-1}gt_1) & \dot{B}(t_1^{-1}gt_2) & \cdots & \dot{B}(t_1^{-1}gt_n) \\ \dot{B}(t_2^{-1}gt_1) & \dot{B}(t_2^{-1}gt_2) & \cdots & \dot{B}(t_2^{-1}gt_n) \\ \vdots & \vdots & \vdots & \vdots \\ \dot{B}(t_n^{-1}gt_1) & \dot{B}(t_n^{-1}gt_2) & \cdots & \dot{B}(t_n^{-1}gt_n) \end{bmatrix}.$$

**Corollary 7.91.** *Let $H$ be a subgroup of a finite group $G$ and let $\chi$ be a character on $H$.*

(i) $\chi\!\uparrow^G(1) = [G : H]\chi(1)$.

(ii) *If $H \lhd G$, then $\chi\!\uparrow^G(g) = 0$ for all $g \notin H$.*

**Proof.**

(i) For all $a \in G$, we have $a^{-1}1a = 1$, so that there are $|G|$ terms in the sum $\chi\!\uparrow^G(1) = \frac{1}{|H|} \sum_{a \in G} \dot{\chi}(a^{-1}ga)$ that are equal to $\chi(1)$; hence,

$$\chi\!\uparrow^G(1) = \frac{|G|}{|H|}\chi(1) = [G : H]\chi(1).$$

(ii) If $H \lhd G$, then $g \notin H$ implies that $a^{-1}ga \notin H$ for all $a \in G$. Therefore, $\dot{\chi}(a^{-1}ga) = 0$ for all $a \in G$, and so $\chi\!\uparrow^G(g) = 0$. •

**Example 7.92.** Let $H \subseteq G$ be a subgroup of index $n$, let $X = \{t_1H, \ldots, t_nH\}$ be the family of left cosets of $H$, and let $\varphi \colon G \to S_X$ be the (permutation) representation of $G$ on the cosets of $H$. As in Example 7.72(v), we may regard $\varphi \colon G \to \mathrm{GL}(V)$, where $V$ is the vector space over $\mathbb{C}$ having basis $X$; that is, $\varphi$ is a representation in the sense of this section.

We claim that if $\chi_\varphi$ is the character afforded by $\varphi$, then $\chi_\varphi = \epsilon\!\uparrow^G$, where $\epsilon$ is the trivial character on $H$. On the one hand, Example 7.72(v) shows that

$$\chi_\varphi(g) = \mathrm{Fix}(\varphi(g))$$

for every $g \in G$. On the other hand, suppose $\varphi(g)$ is the permutation (in two-rowed notation)

$$\varphi(g) = \begin{pmatrix} t_1H & \cdots & t_nH \\ gt_1H & \cdots & gt_nH \end{pmatrix}.$$

Now $gt_iH = t_iH$ if and only if $t_i^{-1}gt_i \in H$. Thus, $\dot{\epsilon}(t_i^{-1}gt_i) \neq 0$ if and only if $gt_iH = t_iH$, and so

$$\epsilon\!\uparrow^G(g) = \mathrm{Fix}(\varphi(g)). \quad \blacktriangleleft$$

Even though a character $\lambda$ of a subgroup $H$ is irreducible, its induced character need not be irreducible. For example, let $G = S_3$ and let $H$ be the cyclic subgroup generated by (1 2). The linear representation $\sigma = \mathrm{sgn}\colon H \to \mathbb{C}$ is irreducible, and it affords the character $\chi_\sigma$ with

$$\chi_\sigma(1) = 1 \quad \text{and} \quad \chi_\sigma((1\ 2)) = -1.$$

Using the formula for the induced character, we find that

$$\chi_\sigma\uparrow^{S_3}(1) = 3, \quad \chi_\sigma\uparrow^{S_3}((1\ 2)) = -1, \quad \text{and} \quad \chi_\sigma\uparrow^{S_3}((1\ 2\ 3)) = 0.$$

Corollary 7.79 shows that $\chi_\sigma\uparrow^{S_3}$ is not irreducible, for $(\chi_\sigma\uparrow^{S_3}, \chi_\sigma\uparrow^{S_3}) = 2$. It is easy to see that $\chi_\sigma\uparrow^{S_3} = \chi_2 + \chi_3$, the latter being the nontrivial irreducible characters of $S_3$.

Here is an important result of Brauer (Curtis–Reiner, *Representation Theory of Finite Groups and Associative Algebras*, p. 283). Call a subgroup $E$ of a finite group $G$ **elementary** if $E = Z \times P$, where $Z$ is cyclic and $P$ is a $p$-group for some prime $p$.

**Theorem 7.93 (Brauer).** *Every complex character $\theta$ on a finite group $G$ has the form*

$$\theta = \sum_i m_i \mu_i \uparrow^G,$$

*where $m_i \in \mathbb{Z}$ and the $\mu_i$ are linear characters on elementary subgroups of $G$.*

**Proof.** See Curtis–Reiner, *Representation Theory of Finite Groups and Associative Algebras*, p. 283. $\quad\bullet$

**Definition.** If $H$ is a subgroup of a group $G$, then every representation $\sigma\colon G \to \mathrm{GL}(V)$ gives, by restriction, a representation $\sigma|H\colon H \to \mathrm{GL}(V)$. (In terms of modules, every left $\mathbb{C}G$-module $V$ can be viewed as a left $\mathbb{C}H$-module.) We call $\sigma|H$ the **restriction** of $\sigma$, and we denote it by $\sigma\downarrow_H$. The character of $H$ afforded by $\sigma\downarrow_H$ is denoted by $\chi_\sigma\downarrow_H$.

The next result displays an interesting relation between characters on a group and characters on a subgroup. (Formally, it resembles the Adjoint Isomorphism.)

**Theorem 7.94 (Frobenius Reciprocity).** *Let $H$ be a subgroup of a group $G$, let $\chi$ be a class function on $G$, and let $\theta$ be a class function on $H$. Then*

$$(\theta\uparrow^G, \chi)_G = (\theta, \chi\downarrow_H)_H,$$

*where $(\square, \square)_G$ denotes the inner product on $\mathrm{cf}(G)$ and $(\square, \square)_H$ denotes the inner product on $\mathrm{cf}(H)$.*

**Proof.** We have

$$(\theta\!\uparrow^G, \chi)_G = \frac{1}{|G|} \sum_{g \in G} \theta\!\uparrow^G (g)\overline{\chi(g)}$$

$$= \frac{1}{|G|} \sum_{g \in G} \frac{1}{|H|} \sum_{a \in G} \dot\theta(a^{-1}ga)\overline{\chi(g)}$$

$$= \frac{1}{|G|}\frac{1}{|H|} \sum_{a,g \in G} \dot\theta(a^{-1}ga)\overline{\chi(a^{-1}ga)},$$

the last equation occurring because $\chi$ is a class function. For fixed $a \in G$, as $g$ ranges over $G$, then so does $a^{-1}ga$. Therefore, writing $x = a^{-1}ga$, the equations continue:

$$= \frac{1}{|G|}\frac{1}{|H|} \sum_{a,x \in G} \dot\theta(x)\overline{\chi(x)}$$

$$= \frac{1}{|G|}\frac{1}{|H|} \sum_{a \in G} \Big(\sum_{x \in G} \dot\theta(x)\overline{\chi(x)}\Big)$$

$$= \frac{1}{|G|}\frac{1}{|H|} |G| \sum_{x \in G} \dot\theta(x)\overline{\chi(x)}$$

$$= \frac{1}{|H|} \sum_{x \in G} \dot\theta(x)\overline{\chi(x)}$$

$$= (\theta, \chi\!\downarrow_H)_H,$$

the next to last equation holding because $\dot\theta(x)$ vanishes off the subgroup $H$.   •

The following elementary remark facilitates the computation of induced class functions.

**Lemma 7.95.** *Let $H$ be a subgroup of a finite group $G$, and let $\chi$ be a class function on $H$. Then*

$$\chi\!\uparrow^G(g) = \frac{1}{|H|} \sum_i |C_G(g_i)|\dot\chi(g_i^{-1}gg_i).$$

**Proof.** Let $|C_G(g_i)| = m_i$. If $a_0^{-1}g_ia_0 = g$, we claim that there are exactly $m_i$ elements $a \in G$ with $a^{-1}g_ia = g$. There are at least $m_i$ elements in $G$ conjugating $g_i$ to $g$; namely, all $aa_0$ for $a \in C_G(g_i)$. There are at most $m_i$ elements, for if $b^{-1}g_ib = g$, then $b^{-1}g_ib = a_0^{-1}g_ia_0$, and so $a_0b \in C_G(g_i)$. The result now follows by collecting terms involving $g_i$s in the formula for $\chi\!\uparrow^G(g)$.   •

**Example 7.96.** Table 5 is the character table of $A_4$, where $\omega = e^{2\pi i/3}$ is a primitive cube root of unity.

The group $A_4$ consists of the identity, eight 3-cycles, and three products of disjoint transpositions. In $S_4$, all the 3-cycles are conjugate; if $g = (1\ 2\ 3)$, then $[S_4 : C_{S_4}(g)] = 8$. It follows that $|C_{S_4}(g)| = 3$, and so $C_{S_4}(g) = \langle g \rangle$. Therefore, in $A_4$, the number of conjugates of $g$ is $[A_4 : C_{A_4}(g)] = 4$ [we know that $C_{A_4}(g) = A_4 \cap C_{S_4}(g) = \langle g \rangle$]. The reader may show that $g$ and $g^{-1}$ are not conjugate, and so we have verified the first two rows of the character table.

| $g_i$ | (1) | (1 2 3) | (1 3 2) | (1 2)(3 4) |
|-------|-----|---------|---------|------------|
| $h_i$ | 1   | 4       | 4       | 3          |
| $\chi_1$ | 1 | 1 | 1 | 1 |
| $\chi_2$ | 1 | $\omega$ | $\omega^2$ | 1 |
| $\chi_3$ | 1 | $\omega^2$ | $\omega$ | 1 |
| $\chi_4$ | 3 | 0 | 0 | $-1$ |

**Table 5.** Character table of $A_4$.

The rows for $\chi_2$ and $\chi_3$ are liftings of linear characters of $A_4/\mathbf{V} \cong \mathbb{I}_3$. Note that if $h = (1\ 2)(3\ 4)$, then $\chi_2(h) = \chi_2(1) = 1$, because $\mathbf{V}$ is the kernel of the lifted representation; similarly, $\chi_3(h) = 1$. Now $\chi_4(1) = 3$, because $3 + (n_4)^2 = 12$. The bottom row arises from orthogonality of the columns. (We can check, using Corollary 7.79, that the character of degree 3 is irreducible.)  ◀

**Example 7.97.** Table 6 is the character table of $S_4$.

| $g_i$ | (1) | (1 2) | (1 2 3) | (1 2 3 4) | (1 2)(3 4) |
|-------|-----|-------|---------|-----------|------------|
| $h_i$ | 1   | 6     | 8       | 6         | 3          |
| $\chi_1$ | 1 | 1 | 1 | 1 | 1 |
| $\chi_2$ | 1 | $-1$ | 1 | $-1$ | 1 |
| $\chi_3$ | 2 | 0 | $-1$ | 0 | 2 |
| $\chi_4$ | 3 | 1 | 0 | $-1$ | $-1$ |
| $\chi_5$ | 3 | $-1$ | 0 | 1 | $-1$ |

**Table 6.** Character table of $S_4$.

We know, for all $n$, that two permutations in $S_n$ are conjugate if and only if they have the same cycle structure; the sizes of the conjugacy classes in $S_4$ were computed in Table 1 on page 10.

The rows for $\chi_2$ and $\chi_3$ are liftings of irreducible characters of $S_4/\mathbf{V} \cong S_3$. The entries in the fourth column of these rows arise from $(1\ 2)\mathbf{V} = (1\ 2\ 3\ 4)\mathbf{V}$; the entries in the last column of these rows arise from $\mathbf{V}$ being the kernel (in either case), so that $\chi_j((1\ 2)(3\ 4)) = \chi_j(1)$ for $j = 2, 3$.

We complete the first column using $24 = 1 + 1 + 4 + n_4^2 + n_5^2$; thus, $n_4 = 3 = n_5$. Let us see whether $\chi_4$ is an induced character; if it is, then Corollary 7.91(i) shows that it arises from a linear character of a subgroup $H$ of index 3. Such a subgroup has order 8, and so it is a Sylow 2-subgroup; that is, $H \cong D_8$. Let us choose one such subgroup:

$$H = \langle \mathbf{V}, (1\ 3) \rangle = \mathbf{V} \cup \{(1\ 3), (2\ 4), (1\ 2\ 3\ 4), (1\ 4\ 3\ 2)\}.$$

The conjugacy classes are

$$
\begin{aligned}
C_1 &= \{1\}; \\
C_2 &= \{(1\ 3)(2\ 4)\}; \\
C_3 &= \{(1\ 2)(3\ 4), (1\ 4)(2\ 3)\}; \\
C_4 &= \{(1\ 3), (2\ 4)\}; \\
C_5 &= \{(1\ 2\ 3\ 4), (1\ 4\ 3\ 2)\}.
\end{aligned}
$$

Let $\theta$ be the character on $H$ defined by

| $C_1$ | $C_2$ | $C_3$ | $C_4$ | $C_5$ |
|-------|-------|-------|-------|-------|
| 1 | 1 | −1 | 1 | −1. |

Define $\chi_4 = \theta\!\uparrow^{S_4}$. Using the formula for induced characters, assisted by Lemma 7.95, we obtain the fourth row of the character table. However, before going on to row 5, we observe that Corollary 7.79 shows that $\chi_4$ is irreducible, for $(\chi_4, \chi_4) = 1$. Finally, the orthogonality relations allow us to compute row 5. ◀

At this point in the story, we must introduce algebraic integers. Since $G$ is a finite group, Lagrange's Theorem gives $g^{|G|} = 1$ for all $g \in G$. It follows that if $\sigma \colon G \to \mathrm{GL}(V)$ is a representation, then $\sigma(g)^{|G|} = I$ for all $g$; hence, all the eigenvalues of $\sigma(g)$ are $|G|$th roots of unity, and so all the eigenvalues are algebraic integers. By Proposition 2.70, the trace of $\sigma(g)$, being the sum of the eigenvalues, is also an algebraic integer.

We can now prove the following interesting result.

**Theorem 7.98.** *The degrees $n_i$ of the irreducible characters of a finite group $G$ are divisors of $|G|$.*

**Proof.** By Theorem 2.63, the rational number $\alpha = |G|/n_i$ is an integer if it is also an algebraic integer. Now Corollary 6.37(ii) says that $\alpha$ is an algebraic integer if there is a faithful $\mathbb{Z}[\alpha]$-module $M$ that is a finitely generated abelian group, where $\mathbb{Z}[\alpha]$ is the smallest subring of $\mathbb{C}$ containing $\alpha$.

By Proposition 7.77, we have

$$
e_i = \sum_{g \in G} \frac{n_i}{|G|} \chi_i(g^{-1})g = \sum_{g \in G} \frac{1}{\alpha} \chi_i(g^{-1})g.
$$

Hence, $\alpha e_i = \sum_{g \in G} \chi_i(g^{-1})g$. But $e_i$ is an idempotent: $e_i^2 = e_i$, and so

$$
\alpha e_i = \sum_{g \in G} \chi_i(g^{-1})g e_i.
$$

Define $M$ to be the abelian subgroup of $\mathbb{C}G$ generated by all elements of the form $\zeta g e_i$, where $\zeta$ is a $|G|$th root of unity and $g \in G$; of course, $M$ is a finitely generated abelian group.

To see that $M$ is a $\mathbb{Z}[\alpha]$-module, it suffices to show that $\alpha M \subseteq M$. But

$$\alpha \zeta g e_i = \zeta g \alpha e_i = \zeta g \sum_{h \in G} \chi_i(h^{-1}) h e_i = \sum_{h \in G} \chi_i(h^{-1}) \zeta g h e_i.$$

This last element lies in $M$, however, because $\chi_i(h^{-1})$ is a sum of $|G|$th roots of unity.

Finally, if $\beta \in \mathbb{C}$ and $u \in \mathbb{C}G$, then $\beta u = 0$ if and only if $\beta = 0$ or $u = 0$. Since $\mathbb{Z}[\alpha] \subseteq \mathbb{C}$ and $M \subseteq \mathbb{C}G$, however, it follows that $M$ is a faithful $\mathbb{Z}[\alpha]$-module, as desired. •

We will present two important applications of Character Theory in the next section; for other applications, as well as a more serious study of representations, the interested reader should look at the books by Curtis–Reiner, Feit, Huppert, and Isaacs.

Representation Theory is used throughout the proof of the Classification Theorem of Finite Simple Groups. An account of this theorem, describing the infinite families of such groups as well as the 26 sporadic simple groups, can be found in the ATLAS, by Conway et al. This book contains the character tables of every simple group of order under $10^{25}$ as well as the character tables of all the sporadic groups.

## Exercises

**7.29.** Prove that if $\theta$ is a generalized character of a finite group $G$, then there are characters $\chi$ and $\psi$ with $\theta = \chi - \psi$.

\* **7.30.** (i) Prove that if $z$ is a complex root of unity, then $z^{-1} = \bar{z}$.

(ii) Prove that if $G$ is a finite group and $\sigma \colon G \to \mathrm{GL}(V)$ is a representation, then

$$\chi_\sigma(g^{-1}) = \overline{\chi_\sigma(g)}$$

for all $g \in G$.

**Hint.** Use the fact that every eigenvalue of $\sigma(g)$ is a root of unity, as well as the fact that if $A$ is a nonsingular matrix over a field $k$ and if $u_1, \ldots, u_n$ are the eigenvalues of $A$ (with multiplicities), then the eigenvalues of $A^{-1}$ are $u_1^{-1}, \ldots, u_n^{-1}$; that is, $\overline{u_1}, \ldots, \overline{u_n}$.

**7.31.** If $\sigma \colon G \to \mathrm{GL}(n, \mathbb{C})$ is a representation, its ***contragredient*** $\sigma^* \colon G \to \mathrm{GL}(n, \mathbb{C})$ is the function given by

$$\sigma^*(g) = \sigma(g^{-1})^\top,$$

where $\square^\top$ denotes transpose.

(i) Prove that the contragredient of a representation $\sigma$ is a representation that is irreducible when $\sigma$ is irreducible.

(ii) Prove that the character $\chi_{\sigma^*}$ afforded by the contragredient $\sigma^*$ is

$$\chi_{\sigma^*}(g) = \overline{\chi_\sigma(g)},$$

where $\overline{\chi_\sigma(g)}$ is the complex conjugate. Conclude that if $\chi$ is a character of $G$, then $\overline{\chi}$ is also a character.

* **7.32.** Construct an irreducible representation of $S_3$ of degree 2.

**7.33.** (i) Let $g \in G$, where $G$ is a finite group. Prove that $g$ is conjugate to $g^{-1}$ if and only if $\chi(g)$ is real for every character $\chi$ of $G$.

(ii) Prove that every character of $S_n$ is real-valued. (It is a theorem of Frobenius that every character of $S_n$ is integer-valued.)

**7.34.** (i) Recall that the *character group* $G^*$ of a finite abelian group $G$ is

$$G^* = \operatorname{Hom}(G, \mathbb{C}^\times),$$

where $\mathbb{C}^\times$ is the multiplicative group of nonzero complex numbers ($\mathbb{C}^\times \cong \mathbb{R}/\mathbb{Z}$, by Corollary 8.30). Prove that $G^* \cong G$.

**Hint.** Use the Fundamental Theorem of Finite Abelian Groups.

(ii) Prove that $\operatorname{Hom}(G, \mathbb{C}^\times) \cong \operatorname{Hom}(G, \mathbb{Q}/\mathbb{Z})$ when $G$ is a finite abelian group.

* **7.35.** Prove that the only linear character of a simple group is the trivial character. Conclude that if $\chi_i$ is not the trivial character, then $n_i = \chi_i(1) > 1$.

* **7.36.** Let $\theta = \chi_\sigma$ be the character afforded by a representation $\sigma$ of a finite group $G$.

(i) If $g \in G$, prove that $|\theta(g)| = \theta(1)$ if and only if $\sigma(g)$ is a scalar matrix.
**Hint.** Use Proposition 2.73 on page 121.

(ii) If $\theta$ is an irreducible character, prove that

$$Z(G/\ker \theta) = \{g \in G : |\theta(g)| = \theta(1)\}.$$

* **7.37.** If $G$ is a finite group, prove that the number of its (necessarily irreducible) linear representations is $[G : G']$.

**7.38.** Let $G$ be a finite group.

(i) If $g \in G$, show that $|C_G(g)| = \sum_{i=1}^r |\chi_i(g)|^2$. Conclude that the character table of $G$ gives $|C_G(g)|$.

(ii) Show how to use the character table of $G$ to see whether $G$ is abelian.

(iii) Show how to use the character table of $G$ to find the lattice of normal subgroups of $G$ and their orders.

(iv) If $G$ is a finite group, show how to use its character table to find the commutator subgroup $G'$.
**Hint.** If $K \triangleleft G$, then the character table of $G/K$ is a submatrix of the character table of $G$, and so we can find the abelian quotient of $G$ having largest order.

(v) Show how to use the character table of a finite group $G$ to determine whether $G$ is solvable.

**7.39.** (i) Show how to use the character table of $G$ to find $|Z(G)|$.

(ii) Show how to use the character table of a finite group $G$ to determine whether $G$ is nilpotent.

**7.40.** Recall that the group $\mathbf{Q}$ of quaternions has the presentation

$$\mathbf{Q} = (A, B \mid A^4 = 1, A^2 = B^2, BAB^{-1} = A^{-1}).$$

(i) Show that there is a representation $\sigma \colon \mathbf{Q} \to \mathrm{GL}(2, \mathbb{C})$ with

$$A \mapsto \begin{bmatrix} i & 0 \\ 0 & -i \end{bmatrix} \text{ and } B \mapsto \begin{bmatrix} 0 & 1 \\ -1 & 0 \end{bmatrix}.$$

(ii) Prove that $\sigma$ is an irreducible representation.

**7.41.** (i) If $\sigma \colon G \to \mathrm{GL}(V)$ and $\tau \colon G \to \mathrm{GL}(W)$ are representations, prove that

$$\sigma \otimes \tau \colon G \to \mathrm{GL}(V \otimes W),$$

defined by

$$(\sigma \otimes \tau)(g) = \sigma(g) \otimes \tau(g)$$

is a representation.

(ii) Prove that the character afforded by $\sigma \otimes \tau$ is the pointwise product:

$$\chi_\sigma \chi_\tau \colon g \mapsto \mathrm{tr}(\sigma(g)) \, \mathrm{tr}(\tau(g)).$$

(iii) Prove that $\mathrm{cf}(G)$ is a commutative ring (usually called the **Burnside ring** of $G$).

## Section 7.6. Theorems of Burnside and of Frobenius

Character Theory will be used in this section to prove two important results in Group Theory: Burnside's $p^m q^n$ Theorem and a theorem of Frobenius. We begin with the following variation of Schur's Lemma.

**Proposition 7.99 (Schur's Lemma II).** *If $\sigma \colon G \to \mathrm{GL}(V)$ is an irreducible representation and a linear transformation $\varphi \colon V \to V$ satisfies*

$$\varphi \sigma(g) = \sigma(g) \varphi$$

*for all $g \in G$, then $\varphi$ is a scalar transformation: there exists $\omega \in \mathbb{C}$ with $\varphi = \omega 1_V$.*

**Proof.** The vector space $V$ is a $\mathbb{C}G$-module with scalar multiplication $gv = \sigma(g)(v)$ for all $v \in V$, and any linear transformation $\theta$ satisfying the equation $\theta \sigma(g) = \sigma(g)\theta$ for all $g \in G$ is a $\mathbb{C}G$-map $V^\sigma \to V^\sigma$. Since $\sigma$ is irreducible, the $\mathbb{C}G$-module $V^\sigma$ is simple. By Schur's Lemma, $\mathrm{End}(V^\sigma)$ is a division ring, and so every nonzero element in it is nonsingular. Now $\varphi - \omega 1_V \in \mathrm{End}(V^\sigma)$ for every $\omega \in \mathbb{C}$; in particular, this is so when $\omega$ is an eigenvalue of $\varphi$ (which lies in $\mathbb{C}$ because $\mathbb{C}$ is algebraically closed). The definition of eigenvalue says that $\varphi - \omega 1_V$ is singular, and so it must be 0; that is, $\varphi = \omega 1_V$, as desired.  •

Recall that if $L_i$ is a minimal left ideal in $\mathbb{C}G$ and $\lambda_i \colon G \to \mathrm{End}_{\mathbb{C}}(L_i)$ is the corresponding irreducible representation, then we extended $\lambda_i$ to a linear transformation $\widetilde{\lambda}_i \colon \mathbb{C}G \to \mathrm{End}_{\mathbb{C}}(L_i)$:

$$\widetilde{\lambda}_i(g) u_j = \begin{cases} g u_i & \text{if } j = i \\ 0 & \text{if } j \neq i \end{cases}$$

Thus, $\widetilde{\lambda}_i(g) = \lambda_i(g)$ for all $g \in G$. In Proposition 7.68(ii), we proved that $\widetilde{\lambda}_i$ is a $\mathbb{C}$-algebra map.

**Corollary 7.100.** *Let $L_i$ be a minimal left ideal in $\mathbb{C}G$, let $\lambda_i \colon G \to \operatorname{End}_{\mathbb{C}}(L_i)$ be the corresponding irreducible representation, and let $\widetilde{\lambda}_i \colon \mathbb{C}G \to \operatorname{End}_{\mathbb{C}}(L_i)$ be the algebra map of Proposition 7.68(ii).*

(i) *If $z \in Z(\mathbb{C}G)$, then there is $\omega_i(z) \in \mathbb{C}$ with*

$$\widetilde{\lambda}_i(z) = \omega_i(z)I.$$

(ii) *The function $\omega_i \colon Z(\mathbb{C}G) \to \mathbb{C}$, given by $z \mapsto \omega_i(z)$, is an algebra map.*

**Proof.**

(i) Let $z \in Z(\mathbb{C}G)$. We verify the hypothesis of Schur's Lemma II in the special case $V = L_i$, $\sigma = \lambda_i$, and $\varphi = \widetilde{\lambda}_i(z)$. For all $g \in G$, we have $\widetilde{\lambda}_i(z)\lambda_i(g) = \lambda_i(zg)$ (because $\widetilde{\lambda}_i$ is a multiplicative map extending $\lambda_i$), while $\lambda_i(g)\widetilde{\lambda}_i(z) = \lambda_i(gz)$. These are equal, for $zg = gz$ because $z \in Z(\mathbb{C}G)$. Proposition 7.99 now says that $\widetilde{\lambda}_i(z) = \omega_i(z)I$ for some $\omega_i(z) \in \mathbb{C}$.

(ii) This follows from the equation $\widetilde{\lambda}_i(z) = \omega_i(z)I$ and Proposition 7.68, which says that $\widetilde{\lambda}_i$ is an algebra map. $\quad\bullet$

Recall, from Lemma 7.57, that a basis for $Z(\mathbb{C}G)$ consists of the *class sums*

$$z_i = \sum_{g \in C_i} g,$$

where the conjugacy classes of $G$ are $C_1, \ldots, C_r$.

**Proposition 7.101.** *Let $z_1, \ldots, z_r$ be the class sums of a finite group $G$.*

(i) *For each $i, j$, we have*

$$\omega_i(z_j) = \frac{h_j \chi_i(g_j)}{n_i},$$

*where $g_j \in C_j$.*

(ii) *There are nonnegative integers $a_{ij\nu}$ with*

$$z_i z_j = \sum_\nu a_{ij\nu} z_\nu.$$

(iii) *The complex numbers $\omega_i(z_j)$ are algebraic integers.*

**Proof.**

(i) Computing the trace of $\widetilde{\lambda}_i(z_j) = \omega_i(z_j)I$ gives

$$n_i \omega_i(z_j) = \chi_i(z_j) = \sum_{g \in C_j} \chi_i(g) = h_j \chi_i(g_j),$$

for $\chi_i$ is constant on the conjugacy class $C_j$. Therefore, $\omega_i(z_j) = h_j \chi_i(g_j)/n_i$.

(ii) Choose $g_\nu \in C_\nu$. The definition of multiplication in the group algebra shows that the coefficient of $g_\nu$ in $z_i z_j$ is

$$|\{(g_i, g_j) \in C_i \times C_j : g_i g_j = g_\nu\}|,$$

the cardinality of a finite set, and hence it is a nonnegative integer. As all the coefficients of $z_\nu$ are equal [for we are in $Z(\mathbb{C}G)$], it follows that this number is $a_{ij\nu}$.

(iii) Let $M$ be the (additive) subgroup of $\mathbb{C}$ generated by all $\omega_i(z_j)$, for $j = 1, \ldots, r$. Since $\omega_i$ is an algebra map,

$$\omega_i(z_j)\omega_i(z_\ell) = \sum_\nu a_{j\ell\nu}\omega_i(z_\nu),$$

so that $M$ is a ring that is finitely generated as an abelian group (because $a_{ij\nu} \in \mathbb{Z}$). Hence, for each $j$, $M$ is a $\mathbb{Z}[\omega_i(z_j)]$-module that is a finitely generated abelian group. If $M$ is faithful, then Corollary 6.37(ii) will give $\omega_i(z_j)$ an algebraic integer. But $M \subseteq \mathbb{C}$, so that the product of nonzero elements is nonzero, and this implies that $M$ is a faithful $\mathbb{Z}[\omega_i(z_j)]$-module, as desired.   •

We are almost ready to complete the proof of Burnside's Theorem.

**Proposition 7.102.** *If $(n_i, h_j) = 1$ for some $i, j$, then either $|\chi_i(g_j)| = n_i$ or $\chi_i(g_j) = 0$.*

**Proof.** By hypothesis, there are integers $s$ and $t$ in $\mathbb{Z}$ with $sn_i + th_j = 1$, so that, for $g_j \in C_j$, we have

$$\frac{\chi_i(g_j)}{n_i} = s\chi_i(g_j) + \frac{th_j\chi_i(g_j)}{n_i}.$$

Hence, Proposition 7.101 gives $\chi_i(g_j)/n_i$ an algebraic integer, and so $|\chi_i(g_j)| \leq n_i$, by Proposition 7.86(i). Thus, it suffices to show that if $|\chi_i(g_j)/n_i| < 1$, then $\chi_i(g_j) = 0$.

Let $m(x) \in \mathbb{Z}[x]$ be the minimum polynomial of $\alpha = \chi_i(g_j)/n_i$; that is, $m(x)$ is the monic polynomial in $\mathbb{Z}[x]$ of least degree having $\alpha$ as a root. We proved, in Corollary 2.145, that $m(x)$ is irreducible in $\mathbb{Q}[x]$. If $\alpha'$ is a root of $m(x)$, then Proposition 3.14 shows that $\alpha' = \sigma(\alpha)$ for some $\sigma \in \mathrm{Gal}(E/\mathbb{Q})$, where $E/\mathbb{Q}$ is the splitting field of $m(x)(x^{|G|} - 1)$. But

$$\alpha = \frac{1}{n_i}\left(\varepsilon_1 + \cdots + \varepsilon_{n_i}\right),$$

where the $\varepsilon$'s are $|G|$th roots of unity, and so $\alpha' = \sigma(\alpha)$ is also such a sum. It follows that $|\alpha'| \leq 1$ [as in the proof of Proposition 7.86(i)]. Therefore, if $N(\alpha)$ is the norm of $\alpha$ [which is, by definition, the absolute value of the product of all the roots of $m(x)$], then $N(\alpha) < 1$ (for we are assuming that $|\alpha| < 1$). But $N(\alpha)$ is the absolute value of the constant term of $m(x)$, which is an integer. Therefore, $N(\alpha) = 0$, hence $\alpha = 0$, and so $\chi_i(g_j) = 0$, as claimed.   •

At last, we can prove Theorem 7.63.

**Theorem 7.103.** *If $G$ is a nonabelian finite simple group, then $\{1\}$ is the only conjugacy class whose size is a prime power. Therefore, Burnside's Theorem is true: every group of order $p^m q^n$, where $p$ and $q$ are primes, is solvable.*

**Proof.** Assume, on the contrary, that $h_j = p^e > 1$ for some $j$. By Exercise 7.36 on page 589, for all $i$, we have

$$Z(G/\ker\chi_i) = \{g \in G : |\chi_i(g)| = n_i\}.$$

Since $G$ is simple, $\ker\chi_i = \{1\}$ for all $i$, and so $Z(G/\ker\chi_i) = Z(G) = \{1\}$. By Proposition 7.102, if $(n_i, h_j) = 1$, then either $|\chi_i(g_j)| = n_i$ or $\chi_i(g_j) = 0$. Of course, $\chi_1(g_j) = 1$ for all $j$, where $\chi_1$ is the trivial character. If $\chi_i$ is not the trivial character, then we have just seen that the first possibility cannot occur, and so $\chi_i(g_j) = 0$. On the other hand, if $(n_i, h_j) \neq 1$, then $p \mid n_i$ (for $h_j = p^e$). Thus, for every $i \neq 1$, either $\chi_i(g_j) = 0$ or $p \mid n_i$.

Consider the orthogonality relation, Corollary 7.84(ii):

$$\sum_{i=1}^{r} n_i \chi_i(g_j) = 0.$$

Now $n_1 = 1 = \chi_1(g_j)$, while each of the other terms is either 0 or of the form $p\alpha_i$, where $\alpha_i$ is an algebraic integer. It follows that

$$0 = 1 + p\beta,$$

where $\beta$ is an algebraic integer. This implies that the rational number $-1/p$ is an algebraic integer, hence lies in $\mathbb{Z}$, and we have the contradiction that $-1/p$ is an integer. •

Another early application of characters is a theorem of Frobenius. We begin with a discussion of doubly transitive permutation groups. Let $G$ be a finite group and $X$ a finite $G$-set. Recall that if $x \in X$, then its *orbit* is $\mathcal{O}(x) = \{gx : g \in G\}$ and its *stabilizer* is $G_x = \{g \in G : gx = x\}$. Theorem 1.107 shows that $|\mathcal{O}(x)||G_x| = |G|$. A $G$-set $X$ is *transitive* if it has only one orbit: if $x, y \in X$, then there exists $g \in G$ with $y = gx$; in this case, $\mathcal{O}(x) = X$.

If $X$ is a $G$-set, then there is a homomorphism $\alpha\colon G \to S_X$, namely, $g \mapsto \alpha_g$, where $\alpha_g(x) = gx$. We say that $X$ is a **faithful** $G$-set if $\alpha$ is an injection; that is, if $gx = x$ for all $x \in X$, then $g = 1$. In this case, we may regard $G$ as a subgroup of $S_X$ acting as permutations of $X$.

Cayley's Theorem (Theorem 1.95) shows that every group $G$ can be regarded as a faithful transitive $G$-set.

**Definition.** A $G$-set $X$ is **doubly transitive** if, for every pair of 2-tuples $(x_1, x_2)$ and $(y_1, y_2)$ in $X \times X$ with $x_1 \neq x_2$ and $y_1 \neq y_2$, there exists $g \in G$ with $y_1 = gx_1$ and $y_2 = gx_2$.[9]

We often abuse language and call a *group* $G$ doubly transitive if there exists a doubly transitive $G$-set.

---

[9]More generally, we call a $G$-set $X$ $k$-transitive, where $1 \leq k \leq |X|$, if, for every pair of $k$-tuples $(x_1, \ldots, x_k)$ and $(y_1, \ldots, y_k)$ in $X \times \cdots \times X$ having distinct coordinates, there exists $g \in G$ with $y_i = gx_i$ for all $i \leq k$. Using the Classification Theorem of Finite Simple Groups, it can be proved that if $k > 5$, then the only faithful $k$-transitive groups are the symmetric groups and the alternating groups. The five **Mathieu groups** are interesting sporadic simple groups that are also highly transitive: $M_{22}$ is 3-transitive, $M_{11}$ and $M_{23}$ are 4-transitive, and $M_{12}$ and $M_{24}$ are 5-transitive.

Note that every doubly transitive $G$-set $X$ is transitive: if $x \neq y$, then $(x, y)$ and $(y, x)$ are 2-tuples as in the definition, and so there is $g \in G$ with $y = gx$ (and $x = gy$).

**Example 7.104.**

   (i) If $n \geq 2$, the symmetric group $S_n$ is doubly transitive; that is, $X = \{1, \dots, n\}$ is a doubly transitive $S_X$-set.

   (ii) The alternating group $A_n$ is doubly transitive if $n \geq 4$.

   (iii) Let $V$ be a finite-dimensional vector space over $\mathbb{F}_2$, and let $X = V - \{0\}$. Then $X$ is a doubly transitive $\mathrm{GL}(V)$-set, for every pair of distinct nonzero vectors $x_1, x_2$ in $V$ must be linearly independent (Exercise 2.74 on page 144). Since every linearly independent list can be extended to a basis, there is a basis $x_1, x_2, \dots, x_n$ of $V$. Similarly, if $y_1, y_2$ is another pair of distinct nonzero vectors, there is a basis $y_1, y_2, \dots, y_n$. But $\mathrm{GL}(V)$ acts transitively on the set of all bases of $V$, by Exercise 2.86 on page 155. Therefore, there is $g \in \mathrm{GL}(V)$ with $y_i = gx_i$ for all $i$, and so $X$ is a doubly transitive $\mathrm{GL}(V)$-set. ◄

**Proposition 7.105.** *A $G$-set $X$ is doubly transitive if and only if, for each $x \in X$, the $G_x$-set $X - \{x\}$ is transitive.*

**Proof.** Let $X$ be a doubly transitive $G$-set. If $y, z \in X - \{x\}$, then $(y, x)$ and $(z, x)$ are 2-tuples of distinct elements of $X$, and so there is $g \in G$ with $z = gy$ and $x = gx$. The latter equation shows that $g \in G_x$, and so $X - \{x\}$ is a transitive $G_x$-set.

To prove the converse, let $(x_1, x_2)$ and $(y_1, y_2)$ be 2-tuples of distinct elements of $X$. We must find $g \in G$ with $y_1 = gx_1$ and $y_2 = gy_2$. Let us denote $(gx_1, gx_2)$ by $g(x_1, x_2)$. There is $h \in G_{x_2}$ with $h(x_1, x_2) = (y_1, x_2)$: if $x_1 = y_1$, we may take $h = 1_X$; if $x_1 \neq y_1$, we use the hypothesis that $X - \{x_2\}$ is a transitive $G_{x_2}$-set. Similarly, there is $h' \in G_{y_1}$ with $h'(y_1, x_2) = (y_1, y_2)$. Therefore, $h'h(x_1, x_2) = (y_1, y_2)$, and $X$ is a doubly transitive $G$-set. •

**Example 7.106.** Let $k$ be a field, let $f(x) \in k[x]$ have no repeated roots, let $E/k$ be a splitting field, and let $G = \mathrm{Gal}(E/k)$ be the Galois group of $f(x)$. If $X = \{\alpha_1, \dots, \alpha_n\}$ is the set of all the roots of $f(x)$, then $X$ is a $G$-set (Theorem 3.3) that is transitive if and only if $f(x)$ is irreducible (Proposition 3.14). Now $f(x)$ factors in $k(\alpha_1)[x]$:

$$f(x) = (x - \alpha_1)f_1(x).$$

The reader may show that $G_1 = \mathrm{Gal}(E/k(\alpha_1)) \subseteq G$ is the stabilizer $G_{\alpha_1}$ and that $X - \{\alpha_1\}$ is a $G_1$-set. Thus, Proposition 7.105 shows that $X$ is a doubly transitive $G$-set if and only if both $f(x)$ and $f_1(x)$ are irreducible (over $k[x]$ and $k(\alpha_1)[x]$, respectively). ◄

Recall Example 1.100: if $H$ is a (not necessarily normal) subgroup of a group $G$ and $X = G/H$ is the family of all left cosets of $H$ in $G$, then $G$ acts on $G/H$ by $g \colon aH \mapsto gaH$. The $G$-set $X$ is transitive, and the stabilizer of $aH \in G/H$ is $aHa^{-1}$; that is, $gaH = aH$ if and only if $a^{-1}ga \in H$ if and only if $g \in aHa^{-1}$.

**Proposition 7.107.** *If $X$ is a doubly transitive $G$-set, then*

$$|G| = n(n-1)|G_{x,y}|,$$

*where $n = |X|$ and $G_{x,y} = \{g \in G : gx = x \text{ and } gy = y\}$. Moreover, if $X$ is a faithful $G$-set, then $|G_{x,y}|$ is a divisor of $(n-2)!$.*

**Proof.** First, Theorem 1.107 gives $|G| = n|G_x|$, because $X$ is a transitive $G$-set. Now $X - \{x\}$ is a transitive $G_x$-set, by Proposition 7.105, and so

$$|G_x| = |X - \{x\}||(G_x)_y| = (n-1)|G_{x,y}|,$$

because $(G_x)_y = G_{x,y}$. The last remark follows, in this case, from $G_{x,y}$ being a subgroup of $S_{X-\{x,y\}} \cong S_{n-2}$. •

It is now easy to give examples of groups that are not doubly transitive, for the orders of doubly transitive groups are constrained.

**Definition.** A transitive $G$-set $X$ is called **regular** if only the identity element of $G$ fixes any element of $X$; that is, $G_x = \{1\}$ for all $x \in X$.

For example, Cayley's Theorem shows that every group $G$ is isomorphic to a regular subgroup of $S_G$. The notion of regularity extends to doubly transitive groups.

**Definition.** A doubly transitive $G$-set $X$ is **sharply doubly transitive** if only the identity of $G$ fixes two elements of $X$; that is, $G_{x,y} = \{1\}$ for all distinct pairs $x, y \in X$.

**Proposition 7.108.** *The following conditions are equivalent for a faithful doubly transitive $G$-set $X$ with $|X| = n$.*

(i) *$X$ is sharply doubly transitive.*

(ii) *If $(x_1, x_2)$ and $(y_1, y_2)$ are 2-tuples in $X \times X$ with $x_1 \neq x_2$ and $y_1 \neq y_2$, then there is a unique $g \in G$ with $y_1 = gx_1$ and $y_2 = gy_2$.*

(iii) *$|G| = n(n-1)$.*

(iv) *$G_{x,y} = \{1\}$ for all distinct $x, y \in X$.*

(v) *For every $x \in X$, the $G_x$-set $X - \{x\}$ is regular.*

**Proof.** All the implications are routine. •

**Example 7.109.**

(i) $S_3$ and $A_4$ are sharply doubly transitive groups.

(ii) The affine group $\mathrm{Aff}(1, \mathbb{R})$, defined in Exercise 1.53 on page 45, is

$$\mathrm{Aff}(1, \mathbb{R}) = \{f \colon \mathbb{R} \to \mathbb{R} : f(x) = ax + b \text{ with } a \neq 0\}$$

under composition. It is isomorphic to the subgroup of $GL(2, \mathbb{R})$ consisting of all matrices of the form $\left[\begin{smallmatrix} a & b \\ 0 & 1 \end{smallmatrix}\right]$. It is plain that we can define $\mathrm{Aff}(1, k)$ for any field $k$ in a similar way. In particular, if $k$ is the finite field $\mathbb{F}_q$, then the affine group $\mathrm{Aff}(1, \mathbb{F}_q)$ is finite, and of order $q(q-1)$. The reader may check that $\mathbb{F}_q$ is a sharply doubly transitive $\mathrm{Aff}(1, \mathbb{F}_q)$-set. ◄

**Notation.** If $G$ is a group, then $G^{\#} = \{g \in G : g \neq 1\}$.

By Cayley's Theorem, every group acts regularly on itself. We now consider transitive groups $G$ such that each $g \in G^{\#}$ has at most one fixed point. In case every $g \in G^{\#}$ has no fixed points, we say that the action of $G$ is **fixed-point-free**. Thompson proved that if a finite group $H$ has a fixed-point-free automorphism $\alpha$ of prime order (that is, the action of the group $G = \langle \alpha \rangle$ on $H^{\#}$ is fixed-point-free), then $H$ is nilpotent (Robinson, *A Course in the Theory of Groups*, pp. 306–307). Thus, let us consider such actions in which there is some $g \in G^{\#}$ that has a fixed point; that is, the action of $G$ is not regular.

**Definition.** A finite group $G$ is a **Frobenius group** if there exists a transitive $G$-set $X$ such that

(i) every $g \in G^{\#}$ has at most one fixed point;

(ii) there is some $g \in G^{\#}$ that does have a fixed point.

If $x \in X$, we call $G_x$ a **Frobenius complement** of $G$.

Note that condition (i) implies that the $G$-set $X$ in the definition is necessarily faithful. Let us rephrase the two conditions: (i) if every $g \in G^{\#}$ has at most one fixed point, then $G_{x,y} = \{1\}$; (ii) if there is some $g \in G^{\#}$ that does have a fixed point, then $G_x \neq \{1\}$.

**Example 7.110.**

(i) The symmetric group $S_3$ is a Frobenius group: $X = \{1, 2, 3\}$ is a faithful transitive $S_3$-set; no $\alpha \in (S_3)^{\#}$ fixes two elements; each transposition $(i \ j)$ fixes one element. The cyclic subgroups $\langle (i \ j) \rangle$ are Frobenius complements (so Frobenius complements need not be unique). A permutation $\beta \in S_3$ has no fixed points if and only if $\beta$ is a 3-cycle. We are going to prove, in every Frobenius group, that 1 together with all those elements having no fixed points comprise a normal subgroup.

(ii) The example of $S_3$ in part (i) can be generalized. Let $X$ be a $G$-set, with at least three elements, which is a sharply doubly transitive $G$-set. Then $X$ is transitive, $G_{x,y} = \{1\}$, and $G_x \neq \{1\}$ (for if $x, y, z \in X$ are distinct, there exists $g \in G$ with $x = gx$ and $z = gy$). Therefore, every sharply doubly transitive group $G$ is a Frobenius group. ◄

**Proposition 7.111.** *A finite group $G$ is a Frobenius group if and only if it contains a proper nontrivial subgroup $H$ such that $H \cap gHg^{-1} = \{1\}$ for all $g \notin H$.*

**Proof.** Let $X$ be a $G$-set as in the definition of Frobenius group. Choose $x \in X$, and define $H = G_x$. Now $H$ is a proper subgroup of $G$, for transitivity does not permit $gx = x$ for all $g \in G$. To see that $H$ is nontrivial, choose $g \in G^{\#}$ having a fixed point; say, $gy = y$. If $y = x$, then $g \in G_x = H$. If $y \neq x$, then transitivity provides $h \in G$ with $hy = x$, and Exercise 1.109 on page 76 gives $H = G_x = hG_yh^{-1} \neq \{1\}$. If $g \notin H$, then $gx \neq x$. Now $g(G_x)g^{-1} = G_{gx}$. Hence, if $h \in H \cap gHg^{-1} = G_x \cap G_{gx}$, then $h$ fixes $x$ and $gx$; that is, $h \in G_{x,y} = \{1\}$.

For the converse, we take $X$ to be the $G$-set $G/H$ of all left cosets of $H$ in $G$, where $g: aH \mapsto gaH$ for all $g \in G$. We remarked earlier that $X$ is a transitive $G$-set and that the stabilizer of $aH \in G/H$ is the subgroup $aHa^{-1}$ of $G$. Since $H \neq \{1\}$, we see that $G_{aH} \neq \{1\}$. Finally, if $aH \neq bH$, then

$$G_{aH,bH} = G_{aH} \cap G_{bH} = aHa^{-1} \cap bHb^{-1} = a(H \cap a^{-1}bHb^{-1}a)a^{-1} = \{1\},$$

because $a^{-1}b \notin H$. Therefore, $G$ is a Frobenius group. •

The significance of this last proposition is that it translates the definition of Frobenius group from the language of $G$-sets into the language of abstract groups.

**Definition.** Let $X$ be a $G$-set. The ***Frobenius kernel*** of $G$ is the subset

$$N = \{1\} \cup \{g \in G : g \text{ has no fixed points}\}.$$

When $X$ is transitive, we can describe $N$ in terms of a stabilizer $G_x$. If $a \notin N^{\#}$, then there is some $y \in X$ with $ay = y$. Since $G$ acts transitively, there is $g \in G$ with $gx = y$, and $a \in G_y = gG_xg^{-1}$. Hence, $a \in \bigcup_{g \in G} gG_xg^{-1}$. For the reverse inclusion, if $a \in \bigcup_{g \in G} gG_xg^{-1}$, then $a \in gG_xg^{-1} = G_{gx}$ for some $g \in G$, and so $a$ has a fixed point; that is, $a \notin N$. We have proved that

$$N = \{1\} \cup \left(G - \left(\bigcup_{g \in G} gG_xg^{-1}\right)\right).$$

Exercise 1.96 on page 74 shows that if $G_x$ is a proper subgroup of $G$, then $G \neq \bigcup_{g \in G} gG_xg^{-1}$, and so $N \neq \{1\}$ in this case.

**Proposition 7.112.** *If $G$ is a Frobenius group with Frobenius complement $H$ and Frobenius kernel $N$, then $|N| = [G : H]$.*

**Proof.** By Proposition 7.111, there is a disjoint union

$$G = \{1\} \cup \left(\bigcup_{g \in G} gH^{\#}g^{-1}\right) \cup N^{\#}.$$

Note that $N_G(H) = H$: if $g \notin H$, then $H \cap gHg^{-1} = \{1\}$, and so $gHg^{-1} \neq H$. Hence, the number of conjugates of $H$ is $[G : N_G(H)] = [G : H]$ (Proposition 1.110). Therefore, $|\bigcup_{g \in G} gH^{\#}g^{-1}| = [G : H](|H| - 1)$, and so

$$|N| = |N^{\#}| + 1 = |G| - ([G : H](|H| - 1)) = [G : H]. \quad •$$

The Frobenius kernel may not be a subgroup of $G$. It is very easy to check that if $g \in N$, then $g^{-1} \in N$ and $aga^{-1} \in N$ for every $a \in G$; the difficulty is in proving that $N$ is closed under multiplication. For example, if $V = k^n$ is the vector space of all $n \times 1$ column vectors over a field $k$, then $V^{\#}$, the set of nonzero vectors in $V$, is a faithful transitive $\mathrm{GL}(V)$-set. Now $A \in \mathrm{GL}(V)$ has a fixed point if and only if there is some $v \in V^{\#}$ with $Av = v$; that is, $A$ has a fixed point if and only if 1 is an eigenvalue of $A$. Thus, the Frobenius kernel now consists of the identity matrix together with all linear transformations which do not have 1 as an eigenvalue. Let $|k| \geq 4$, and let $\alpha$ be a nonzero element of $k$ with $\alpha^2 \neq 1$. Then $A = \left[\begin{smallmatrix} \alpha & 0 \\ 0 & \alpha \end{smallmatrix}\right]$ and $B = \left[\begin{smallmatrix} \alpha^{-1} & 0 \\ 0 & \alpha \end{smallmatrix}\right]$ lie in $N$, but their product $AB = \left[\begin{smallmatrix} 1 & 0 \\ 0 & \alpha^2 \end{smallmatrix}\right]$ does not lie in $N$. However, if $G$ is a Frobenius group, then $N$ is a subgroup; the only known proof of this fact uses characters.

We have already remarked that if $\psi$ is a character on a subgroup $H$ of a group $G$, then the restriction $(\psi\vert^G)_H$ need not equal $\psi$. The next proof shows that irreducible characters of a Frobenius complement do extend to irreducible characters of $G$.

**Lemma 7.113.** *Let $G$ be a Frobenius group with Frobenius complement $H$ and Frobenius kernel $N$. For every irreducible character $\psi$ on $H$ other than the trivial character $\psi_1$, define the generalized character*

$$\varphi = \psi - d\psi_1,$$

*where $d = \psi(1)$. Then $\psi^* = \varphi\vert^G + d\chi_1$ is an irreducible character on $G$, and $\psi_H^* = \psi$; that is, $\psi^*(h) = \psi(h)$ for all $h \in H$.*

**Proof.** Note first that $\varphi(1) = 0$. We claim that the induced generalized character $\varphi\vert^G$ satisfies the equation

$$(\varphi\vert^G)_H = \varphi.$$

If $t_1 = 1, \ldots, t_n$ is a transversal of $H$ in $G$, then for $g \in G$, the matrix of $\varphi\vert^G(g)$ on page 583 has the blocks $\dot{B}(t_i^{-1}gt_i)$ on its diagonal, where $\dot{B}(t_i^{-1}gt_i) = 0$ if $t_i^{-1}gt_i \notin H$ (this is just the matrix version of Theorem 7.90). If $h \in H$, then $t_i^{-1}ht_i \notin H$ for all $i \neq 1$, and so $\dot{B}(t_i^{-1}ht_i) = 0$. Therefore, there is only one nonzero diagonal block, and

$$\operatorname{tr}(\varphi\vert^G(h)) = \operatorname{tr}(B(h));$$

that is,

$$\varphi\vert^G(h) = \varphi(h).$$

We have just seen that $\varphi\vert^G$ is a generalized character on $G$ such that $(\varphi\vert^G)_H = \varphi$. By Frobenius Reciprocity (Theorem 7.94),

$$(\varphi\vert^G, \varphi\vert^G)_G = (\varphi, (\varphi\vert^G)_H)_H = (\varphi, \varphi)_H.$$

But $\varphi = \psi - d\psi_1$, so that orthogonality of $\psi$ and $\psi_1$ gives

$$(\varphi, \varphi)_H = 1 + d^2.$$

Similarly,

$$(\varphi\vert^G, \chi_1)_G = (\varphi, \psi_1)_H = -d,$$

where $\chi_1$ is the trivial character on $G$. Define

$$\psi^* = \varphi\vert^G + d\chi_1.$$

Now $\psi^*$ is a generalized character on $G$, and

$$(\psi^*, \psi^*)_G = (\varphi\vert^G, \varphi\vert^G)_G + 2d(\varphi\vert^G, \chi_1)_G + d^2 = 1 + d^2 - 2d^2 + d^2 = 1.$$

We have

$$(\psi^*)_H = (\varphi\vert^G)_H + d(\chi_1)_H = \varphi + d\psi_1 = (\psi - d\psi_1) + d\psi_1 = \psi.$$

Since $\psi^*(1) = \psi(1) > 0$, Corollary 7.79 says that $\psi^*$ is an irreducible character on $G$. •

**Theorem 7.114 (Frobenius).** *Let $G$ be a Frobenius group with Frobenius complement $H$ and Frobenius kernel $N$. Then $N$ is a normal subgroup of $G$, $N \cap H = \{1\}$, and $NH = G$.*

**Remark.** A group $G$ having a subgroup $Q$ and a normal subgroup $K$ such that $K \cap Q = \{1\}$ and $KQ = G$ is called a *semidirect product*. We will discuss such groups in Chapter 9. ◀

**Proof.** For every irreducible character $\psi$ on $H$ other than the trivial character $\psi_1$, define the generalized character $\varphi = \psi - d\psi_1$, where $d = \psi(1)$. By Lemma 7.113, $\psi^* = \varphi \rvert^G + d\chi_1$ is an irreducible character on $G$. Define

$$N^* = \bigcap_{\psi \neq \psi_1} \ker \psi^*.$$

Of course, $N^*$ is a normal subgroup of $G$.

By Lemma 7.113, $\psi^*(h) = \psi(h)$ for all $h \in H$; in particular, if $h = 1$, we have

(1) $$\psi^*(1) = \psi(1) = d.$$

If $g \in N^{\#}$, then for all $a \in G$, we have $g \notin aHa^{-1}$ (for $g$ has no fixed points), and so $\dot{\varphi}(aga^{-1}) = 0$. The induced character formula, Theorem 7.90, now gives $\varphi \rvert^G(g) = 0$. Hence, if $g \in N^{\#}$, then Equation (5) gives

$$\psi^*(g) = \varphi \rvert^G(g) + d\chi_1(g) = d.$$

We conclude that if $g \in N$, then

$$\psi^*(g) = d = \psi^*(1);$$

that is, $g \in \ker \psi^*$. Therefore,

$$N \subseteq N^*.$$

The reverse inclusion will arise from a counting argument.

Let $h \in H \cap N^*$. Since $h \in H$, Lemma 7.113 gives $\psi^*(h) = \psi(h)$. On the other hand, since $h \in N^*$, we have $\psi^*(h) = \psi^*(1) = d$. Therefore, $\psi(h) = \psi^*(h) = d = \psi(1)$, so that $h \in \ker \psi$ for every irreducible character $\psi$ on $H$. Consider the regular character, afforded by the regular representation $\rho$ on $H$: $\chi_\rho = \sum_i n_i \psi_i$. Now $\chi_\rho(h) = \sum_i n_i \psi_i(h) \neq 0$, so that Example 7.74(ii) gives $h = 1$. Thus,

$$H \cap N^* = \{1\}.$$

Next, $|G| = |H|[G : H] = |H||N|$, by Proposition 7.112. Note that $HN^*$ is a subgroup of $G$, because $N^* \lhd G$. Now $|HN^*||H \cap N^*| = |H||N^*|$, by the Second Isomorphism Theorem; since $H \cap N^* = \{1\}$, we have $|H||N| = |G| \geq |HN^*| = |H||N^*|$. Hence, $|N| \geq |N^*|$. But $|N| \leq |N^*|$, because $N \subseteq N^*$, and so $N = N^*$. Therefore, $N \lhd G$, $H \cap N = \{1\}$, and $HN = G$. •

Much more can be said about the structure of Frobenius groups. Every Sylow subgroup of a Frobenius complement is either cyclic or generalized quaternion (Huppert, *Endliche Gruppen* I, p. 502), and it is a consequence of Thompson's Theorem on fixed-point-free automorphisms that every Frobenius kernel is nilpotent (Robinson, *A Course in the Theory of Groups*, p. 306); that is, $N$ is the direct product of its Sylow subgroups. The reader is referred to Curtis–Reiner, *Representation Theory of Finite Groups and Associative Algebras*, pp. 242–246, or Feit, *Characters of Finite Groups*, pp. 133–139.

## Exercises

**7.42.** Prove that the affine group $\mathrm{Aff}(1, \mathbb{F}_q)$ is sharply doubly transitive.

**7.43.** Assume that the family of left cosets $G/H$ of a subgroup $H \subseteq G$ is a $G$-set via the representation on cosets. Prove that $G/H$ is a faithful $G$-set if and only if $\bigcap_{a \in G} aHa^{-1} = \{1\}$. Give an example in which $G/H$ is not a faithful $G$-set.

**7.44.** Prove that every Sylow subgroup of $\mathrm{SL}(2, \mathbb{F}_5)$ is either cyclic or quaternion.

**7.45.** A subset $A$ of a group $G$ is a **T.I. set** (*trivial intersection set*) if $A \subseteq N_G(A)$ and $A \cap gAg^{-1} \subseteq \{1\}$ for all $g \notin N_G(A)$.

(i) Prove that a Frobenius complement $H$ in a Frobenius group $G$ is a T.I. set.

(ii) Let $A$ be a T.I. set in a finite group $G$, and let $N = N_G(A)$. If $\alpha$ is a class function vanishing on $N - A$ and $\beta$ is a class function on $N$ vanishing on $\left( \bigcup_{g \in G} (A^g \cap N) \right) - A$, prove, for all $g \in N^{\#}$, that $\alpha \uparrow^G (g) = \alpha(g)$ and $\beta \uparrow^G (g) = \beta(g)$.
**Hint.** See the proofs of Lemma 7.114 and Theorem 7.113.

(iii) If $\alpha(1) = 0$, prove that $(\alpha, \beta)_N = (\alpha \uparrow^G, \beta \uparrow^G)_G$.

(iv) Let $H$ be a *self-normalizing* subgroup of a finite group $G$; that is, $H = N_G(H)$. If $H$ is a T.I. set, prove that there is a normal subgroup $K$ of $G$ with $K \cap H = \{1\}$ and $KH = G$.
**Hint.** See Feit, *Characters of Finite Groups*, p. 124.

∗ **7.46.** Prove that there are no nonabelian simple groups of order $n$, where $60 < n \le 100$.
**Hint.** By Burnside's Theorem, the only candidates for $n$ in the given range are 66, 70, 78, 84, and 90; note that 90 was eliminated in Exercise 4.32 on page 251.

**7.47.** Prove that there are no nonabelian simple groups of order $n$, where $101 \le n < 168$. We remark that $\mathrm{PSL}(2, \mathbb{F}_7)$ is a simple group of order 168, and it is the unique such group up to isomorphism. In view of Proposition 4.44, Corollary 4.71, and Exercise 7.46, we see that $A_5$ is the only nonabelian simple group of order strictly less than 168.
**Hint.** By Burnside's Theorem, the only candidates for $n$ in the given range are 102, 105, 110, 120, 126, 130, 132, 138, 140, 150, 154,154, 156, and 165. Use Exercise 1.108 on page 76 and Exercise 4.33 on page 252.

## Section 7.7. Division Algebras

When applying the Wedderburn–Artin Theorems to group algebras $kG$, where $k$ is algebraically closed, we used Molien's Theorem (Corollary 7.53) to assume that the matrix rings have entries in $k$. If $k$ is not algebraically closed, then (noncommutative) division rings can occur. At the moment, the only example we know of such a ring is the quaternions $\mathbb{H}$, and it is not at all obvious how to construct other examples.

Linear representations of a finite group over a field $k$ are the simplest ones, for every finite subgroup of the multiplicative group $k^\times$ is cyclic (Theorem 2.46). Herstein proved that every finite subgroup of $D^\times$ is cyclic if $D$ is a division ring whose center is a field of characteristic $p > 0$, but it is false when $Z(D)$ has characteristic 0 (obviously, the group of quaternions is a subgroup of $\mathbb{H}^\times$). All finite subgroups of multiplicative groups of division rings were found by Amitsur, Finite subgroups of division rings, *Trans. Amer. Math. Soc.* 80 (1955), 361–386.

That the tensor product of algebras is, again, an algebra, is used in the study of division rings.

**Definition.** A *division algebra over a field* $k$ is a division ring regarded as an algebra over its center $k$.

Let us begin by considering the wider class of simple algebras.

**Definition.** A finite-dimensional[10] $k$-algebra $A$ over a field $k$ is *central simple* if it is simple (no two-sided ideals other than $A$ and $\{0\}$) and its center $Z(A) = k$.

**Notation.** If $A$ is an algebra over a field $k$, then we write

$$[A : k] = \dim_k(A).$$

**Example 7.115.**

(i) Every division algebra $\Delta$ that is finite-dimensional over its center $k$ is a central simple $k$-algebra. The ring $\mathbb{H}$ of quaternions is a central simple $\mathbb{R}$-algebra, and every field is a central simple algebra over itself.

(ii) If $k$ is a field, then $\mathrm{Mat}_n(k)$ is a central simple $k$-algebra (it is simple, by Proposition 7.39, and its center consists of all scalar matrices $\{aI : a \in k\}$, by Exercise 6.9 on page 415).

(iii) If $A$ is a central simple $k$-algebra, then its opposite algebra $A^{\mathrm{op}}$ is also a central simple $k$-algebra.  ◄

**Theorem 7.116.** *Let $A$ be a central simple $k$-algebra. If $B$ is a simple $k$-algebra, then $A \otimes_k B$ is a central simple $Z(B)$-algebra. In particular, if $B$ is a central simple $k$-algebra, then $A \otimes_k B$ is a central simple $k$-algebra.*

**Proof.** Each $x \in A \otimes_k B$ has an expression of the form

$$(1) \qquad\qquad x = a_1 \otimes b_1 + \cdots + a_n \otimes b_n,$$

where $a_i \in A$ and $b_i \in B$. For nonzero $x$, define the *length* of $x$ to be $n$ if there is no such expression having fewer than $n$ terms. We claim that if $x$ has length $n$, that is, if Equation (1) is a shortest such expression, then $b_1, \ldots, b_n$ is a linearly independent list in $B$ (viewed as a vector space over $k$). Otherwise, there is some $j$ and $u_i \in k$, not all zero, with $b_j = \sum_i u_i b_i$. Substituting and collecting terms gives

$$x = \sum_{i \neq j} (a_i + u_i a_j) \otimes b_i,$$

which is a shorter expression for $x$.

---

[10]We assume that central simple algebras are finite-dimensional, but some authors do not. Hilbert gave an example of an infinite-dimensional division algebra (Drozd–Kirichenko, *Finite-Dimensional Algebras*, p. 81).

Let $I \neq (0)$ be a two-sided ideal in $A \otimes_k B$. Choose $x$ to be a (nonzero) element in $I$ of smallest length, and assume that Equation (1) is a shortest expression for $x$. Now $a_1 \neq 0$. Since $Aa_1A$ is a two-sided ideal in $A$, simplicity gives $A = Aa_1A$. Hence, there are elements $a'_p$ and $a''_p$ in $A$ with $1 = \sum_p a'_p a_1 a''_p$. Since $I$ is a two-sided ideal,

$$(2) \qquad x' = \sum_p a'_p x a''_p = 1 \otimes b_1 + c_2 \otimes b_2 + \cdots + c_n \otimes b_n$$

lies in $I$, where, for $i \geq 2$, we have $c_i = \sum_p a'_p a_i a''_p$. At this stage, we do not know whether $x' \neq 0$, but we do know, for every $a \in A$, that $(a \otimes 1)x' - x'(a \otimes 1) \in I$. Now

$$(3) \qquad (a \otimes 1)x' - x'(a \otimes 1) = \sum_{i \geq 2}(ac_i - c_i a) \otimes b_i.$$

First, this element is 0, lest it be an element in $I$ of length smaller than the length of $x$. Since $b_1, \ldots, b_n$ is a linearly independent list, the $k$-subspace it generates is $\langle b_1, \ldots, b_n \rangle = \langle b_1 \rangle \oplus \cdots \oplus \langle b_n \rangle$, and so

$$A \otimes_k \langle b_1, \ldots, b_n \rangle = A \otimes_k \langle b_1 \rangle \oplus \cdots \oplus A \otimes_k \langle b_n \rangle.$$

It follows from Equation (3) that each term $(ac_i - c_i a) \otimes b_i$ must be 0. Hence, $ac_i = c_i a$ for all $a \in A$; that is, each $c_i \in Z(A) = k$. Equation (2) becomes

$$\begin{aligned} x' &= 1 \otimes b_1 + c_2 \otimes b_2 + \cdots + c_n \otimes b_n \\ &= 1 \otimes b_1 + 1 \otimes c_2 b_2 + \cdots + 1 \otimes c_n b_n \\ &= 1 \otimes (b_1 + c_2 b_2 + \cdots + c_n b_n). \end{aligned}$$

Now $b_1 + c_2 b_2 + \cdots + c_n b_n \neq 0$, because $b_1, \ldots, b_n$ is a linearly independent list, and so $x' \neq 0$. Therefore, $I$ contains a nonzero element of the form $1 \otimes b$. But simplicity of $B$ gives $BbB = B$, and so there are $b'_q, b''_q \in B$ with $\sum_q b'_q b b''_q = 1$. Hence, $I$ contains $\sum_q (1 \otimes b'_q)(1 \otimes b)(1 \otimes b''_q) = 1 \otimes 1$, which is the unit in $A \otimes_k B$. Therefore, $I = A \otimes_k B$ and $A \otimes_k B$ is simple.

We now seek the center of $A \otimes_k B$. Clearly, $k \otimes_k Z(B) \subseteq Z(A \otimes_k B)$. For the reverse inequality, let $z \in Z(A \otimes_k B)$ be nonzero, and let

$$z = a_1 \otimes b_1 + \cdots + a_n \otimes b_n$$

be a shortest such expression for $z$. As in the preceding argument, $b_1, \ldots, b_n$ is a linearly independent list over $k$. For each $a \in A$, we have

$$0 = (a \otimes 1)z - z(a \otimes 1) = \sum_i (aa_i - a_i a) \otimes b_i.$$

It follows, as above, that $(aa_i - a_i a) \otimes b_i = 0$ for each $i$. Hence, $aa_i - a_i a = 0$, so that $aa_i = a_i a$ for all $a \in A$ and each $a_i \in Z(A) = k$. Thus, $z = 1 \otimes x$, where $x = a_1 b_1 + \cdots + a_n b_n \in B$. But if $b \in B$, then

$$0 = z(1 \otimes b) - (1 \otimes b)z = (1 \otimes x)(1 \otimes b) - (1 \otimes b)(1 \otimes x) = 1 \otimes (xb - bx).$$

Therefore, $xb - bx = 0$ and $x \in Z(B)$. We conclude that $z \in k \otimes_k Z(B)$. $\quad \bullet$

It is not generally true that the tensor product of simple $k$-algebras is again simple; we must pay attention to the centers. In Exercise 7.52 on page 612, we saw that if $E/k$ is a field extension, then $E \otimes_k E$ need not be a field. The tensor product of division algebras need not be a division algebra, as we see in the next example.

**Example 7.117.** The algebra $\mathbb{C} \otimes_{\mathbb{R}} \mathbb{H}$ is an eight-dimensional $\mathbb{R}$-algebra, but it is also a four-dimensional $\mathbb{C}$-algebra: a basis is

$$1 = 1 \otimes 1, \ 1 \otimes i, \ 1 \otimes j, \ 1 \otimes k.$$

We let the reader prove that the vector space isomorphism $\mathbb{C} \otimes_{\mathbb{R}} \mathbb{H} \to \mathrm{Mat}_2(\mathbb{C})$ with

$$1 \otimes 1 \mapsto \begin{bmatrix} 1 & 0 \\ 0 & 1 \end{bmatrix}, \quad 1 \otimes i \mapsto \begin{bmatrix} i & 0 \\ 0 & -i \end{bmatrix}, \quad 1 \otimes j \mapsto \begin{bmatrix} 0 & 1 \\ -1 & 0 \end{bmatrix}, \quad 1 \otimes k \mapsto \begin{bmatrix} 0 & i \\ i & 0 \end{bmatrix},$$

is an isomorphism of $\mathbb{C}$-algebras. ◄

Another way to see that $\mathbb{C} \otimes_{\mathbb{R}} \mathbb{H} \cong \mathrm{Mat}_2(\mathbb{C})$ arises from Example 7.60(ii). We remarked then that

$$\mathbb{R}\mathbf{Q} \cong \mathbb{R} \times \mathbb{R} \times \mathbb{R} \times \mathbb{R} \times \mathbb{H};$$

tensoring by $\mathbb{C}$ gives

$$\mathbb{C}\mathbf{Q} \cong \mathbb{C} \otimes_{\mathbb{R}} \mathbb{R}\mathbf{Q} \cong \mathbb{C} \times \mathbb{C} \times \mathbb{C} \times \mathbb{C} \times (\mathbb{C} \otimes_{\mathbb{R}} \mathbb{H}).$$

It follows from the uniqueness in Wedderburn–Artin Theorem II that $\mathbb{C} \otimes_{\mathbb{R}} \mathbb{H} \cong \mathrm{Mat}_2(\mathbb{C})$ (we give yet another proof of this in the next theorem).

The next theorem puts the existence of the isomorphism in Example 7.117 into the context of central simple algebras.

**Theorem 7.118.** *Let $k$ be a field and let $A$ be a central simple $k$-algebra.*

(i) *If $\overline{k}$ is the algebraic closure of $k$, then there is an integer $n$ with*

$$\overline{k} \otimes_k A \cong \mathrm{Mat}_n(\overline{k}).$$

*In particular, $\mathbb{C} \otimes_{\mathbb{R}} \mathbb{H} \cong \mathrm{Mat}_2(\mathbb{C})$.*

(ii) *If $A$ is a central simple $k$-algebra, then there is an integer $n$ with*

$$[A : k] = n^2.$$

**Proof.**

(i) By Theorem 7.116, $\overline{k} \otimes_k A$ is a simple $\overline{k}$-algebra. Hence, Wedderburn's Theorem (actually, Corollary 7.51) gives $\overline{k} \otimes_k A \cong \mathrm{Mat}_n(D)$ for some $n \geq 1$ and some division ring $D$. Since $D$ is a finite-dimensional division algebra over $\overline{k}$, the argument in Molien's Corollary 7.53 shows that $D = \overline{k}$. In particular, since $\mathbb{C}$ is the algebraic closure of $\mathbb{R}$, we have $\mathbb{C} \otimes_{\mathbb{R}} \mathbb{H} \cong \mathrm{Mat}_n(\mathbb{C})$ for some $n$; as $\dim_{\mathbb{R}}(\mathbb{C} \otimes_{\mathbb{R}} \mathbb{H}) = 4$, we have $n = 2$.

(ii) We claim that $[A : k] = [\overline{k} \otimes_k A : \overline{k}]$, for if $a_1, \ldots, a_m$ is a basis of $A$ over $k$, then $1 \otimes a_1, \ldots, 1 \otimes a_m$ is a basis of $\overline{k} \otimes_k A$ over $\overline{k}$ (essentially because tensor product commutes with direct sum). Therefore,

$$[A : k] = [\overline{k} \otimes_k A : \overline{k}] = [\mathrm{Mat}_n(\overline{k}) : \overline{k}] = n^2. \quad \bullet$$

**Definition.** A ***splitting field*** for a central simple $k$-algebra $A$ is a field extension $E/k$ for which there exists an integer $n$ such that $E \otimes_k A \cong \mathrm{Mat}_n(E)$.

Theorem 7.118 says that the algebraic closure $\overline{k}$ of a field $k$ is a splitting field for every central simple $k$-algebra $A$. We are going to see that there always exists a splitting field that is a finite extension of $k$, but we first develop some tools in order to prove it.

**Definition.** If $A$ is a $k$-algebra and $X \subseteq A$ is a subset, then its ***centralizer***, $C_A(X)$, is defined by

$$C_A(X) = \{ a \in A : ax = xa \text{ for every } x \in X \}.$$

It is easy to check that centralizers are always subalgebras.

The key idea in the next proof is that a subalgebra $B$ of $A$ makes $A$ into a $(B, A)$-bimodule, and that the centralizer of $B$ can be described in terms of an endomorphism ring.

**Theorem 7.119 (Double Centralizer).** *Let $A$ be a central simple algebra over a field $k$ and let $B$ be a simple subalgebra of $A$.*

(i) *$C_A(B)$ is a simple $k$-algebra.*

(ii) *$B \otimes_k A^{\mathrm{op}} \cong \mathrm{Mat}_s(\Delta)$ and $C_A(B) \cong \mathrm{Mat}_r(\Delta)$ for some division algebra $\Delta$, where $r \mid s$.*

(iii) *$[B : k][C_A(B) : k] = [A : k]$.*

(iv) *$C_A(C_A(B)) = B$.*

**Proof.** Associativity of the multiplication in $A$ shows that $A$ can be viewed as a $(B, A)$-bimodule. As such, it is a left $(B \otimes_k A^{\mathrm{op}})$-module (Proposition 6.127), where $(b \otimes a)x = bxa$ for all $x \in A$; we denote this module by $A^*$. But $B \otimes_k A^{\mathrm{op}}$ is a simple $k$-algebra, by Theorem 7.116, so that Corollary 7.51 gives $B \otimes_k A^{\mathrm{op}} \cong \mathrm{Mat}_s(\Delta)$ for some integer $s$ and some division algebra $\Delta$ over $k$; in fact, $B \otimes_k A^{\mathrm{op}}$ has a unique (to isomorphism) minimal left ideal $L$, and $\Delta^{\mathrm{op}} \cong \mathrm{End}_{B \otimes_k A^{\mathrm{op}}}(L)$. Therefore, as $(B \otimes_k A^{\mathrm{op}})$-modules, Corollary 7.28 gives $A^* \cong L^r$, the direct sum of $r$ copies of $L$, and so $\mathrm{End}_{B \otimes_k A^{\mathrm{op}}}(A^*) \cong \mathrm{Mat}_r(\Delta)$.

We claim that

$$C_A(B) \cong \mathrm{End}_{B \otimes_k A^{\mathrm{op}}}(A^*) \cong \mathrm{Mat}_r(\Delta);$$

this will prove (i) and most of (ii). If $\varphi \in \mathrm{End}_{B \otimes_k A^{\mathrm{op}}}(A^*)$, then it is, in particular, an endomorphism of $A$ as a right $A$-module. Hence, for all $a \in A$, we have

$$\varphi(a) = \varphi(1a) = \varphi(1)a = ua,$$

where $u = \varphi(1)$. In particular, if $b \in B$, then $\varphi(b) = ub$. On the other hand, taking the left action of $B$ into account, we have $\varphi(b) = \varphi(b1) = b\varphi(1) = bu$. Therefore, $ub = bu$ for all $b \in B$, and so $u \in C_A(B)$. Thus, $\varphi \mapsto \varphi(1)$ is a function $\mathrm{End}_{B \otimes_k A^{\mathrm{op}}}(A^*) \to C_A(B)$. It is routine to check that this function is an injective $k$-algebra map; it is also surjective, for if $u \in C_A(B)$, then the map $A \to A$, defined by $a \mapsto ua$, is a $(B \otimes_k A^{\mathrm{op}})$-map.

We now compute dimensions. Define $d = [\Delta : k]$. Since $L$ is a minimal left ideal in $\text{Mat}_s(\Delta)$, we have $\text{Mat}_s(\Delta) \cong L^s$ [concretely, $L = \text{COL}(1)$, all the first columns of $s \times s$ matrices over $\Delta$]. Therefore, $[\text{Mat}_s(\Delta) : k] = s^2[\Delta : k]$ and $[L^s : k] = s[L : k]$, so that

$$[L : k] = sd.$$

Also,

$$[A : k] = [A^* : k] = [L^r : k] = rsd.$$

It follows that

$$[A : k][B : k] = [B \otimes_k A^{\text{op}} : k] = [\text{Mat}_s(\Delta) : k] = s^2 d.$$

Therefore, $[B : k] = s^2 d / rsd = s/r$, and so $r \mid s$. Hence,

$$[B : k][C_A(B) : k] = [B : k][\text{Mat}_r(\Delta) : k] = \frac{s}{r} \cdot r^2 d = rsd = [A : k],$$

because we have already proved that $C_A(B) \cong \text{Mat}_r(\Delta)$.

Finally, we prove (iv). It is easy to see that $B \subseteq C_A(C_A(B))$: after all, if $b \in B$ and $u \in C_A(B)$, then $bu = ub$, and so $b$ commutes with every such $u$. But $C_A(B)$ is a simple subalgebra, by (i), and so the equation in (iii) holds if we replace $B$ by $C_A(B)$:

$$[C_A(B) : k][C_A(C_A(B)) : k] = [A : k].$$

We conclude that $[B : k] = [C_A(C_A(B)) : k]$; together with $B \subseteq C_A(C_A(B))$, this equality gives $B = C_A(C_A(B))$.  •

Here is a minor variant of the theorem.

**Corollary 7.120.** *If $B$ is a simple subalgebra of a central simple $k$-algebra $A$, where $k$ is a field, then there is a division algebra $D$ with $B^{\text{op}} \otimes_k A \cong \text{Mat}_s(D)$.*

**Proof.** By Theorem 7.119(ii), we have $B \otimes_k A^{\text{op}} \cong \text{Mat}_s(\Delta)$ for some division algebra $\Delta$. Hence, $(B \otimes_k A^{\text{op}})^{\text{op}} \cong (\text{Mat}_s(\Delta))^{\text{op}}$. But $(\text{Mat}_s(\Delta))^{\text{op}} \cong \text{Mat}_s(\Delta^{\text{op}})$, by Proposition 6.17, while $(B \otimes_k A^{\text{op}})^{\text{op}} \cong B^{\text{op}} \otimes_k A$, by Exercise 7.50 on page 612. Setting $D = \Delta^{\text{op}}$ completes the proof.  •

If $\Delta$ is a division algebra over a field $k$ and $\delta \in \Delta$, then the subdivision algebra generated by $k$ and $\delta$ is a field, because elements in the center $k$ commute with $\delta$. We are interested in maximal subfields of $\Delta$.

**Lemma 7.121.** *If $\Delta$ is a division algebra over a field $k$, then a subfield $E$ of $\Delta$ is a maximal subfield if and only if $C_\Delta(E) = E$.*

**Proof.** If $E$ is a maximal subfield of $\Delta$, then $E \subseteq C_\Delta(E)$ because $E$ is commutative. For the reverse inclusion, it is easy to see that if $\delta \in C_\Delta(E)$, then the division algebra $E'$ generated by $E$ and $\delta$ is a field. Hence, if $\delta \notin E$, then $E \subsetneq E'$, and the maximality of $E$ is contradicted.

Conversely, suppose that $E$ is a subfield with $C_\Delta(E) = E$. If $E$ is not a maximal subfield of $\Delta$, then there exists a subfield $E'$ with $E \subsetneq E'$. Now $E' \subseteq C_\Delta(E)$, so that if there is some $a' \in E'$ with $a' \notin E$, then $E \neq C_\Delta(E)$. Therefore, $E$ is a maximal subfield.  •

After proving an elementary lemma about tensor products, we will extend the next result from division algebras to central simple algebras (Theorem 7.131).

**Theorem 7.122.** *If $D$ is a division algebra over a field $k$ and $E$ is a maximal subfield of $D$, then $E$ is a splitting field for $D$; that is, $E \otimes_k D \cong \mathrm{Mat}_s(E)$, where $s = [D : E] = [E : k]$.*

**Proof.** Let us specialize the algebras in Theorem 7.119. Here, $A = D$, $B = E$, and $C_A(E) = E$, by Lemma 7.121. Now the condition $C_A(B) \cong \mathrm{Mat}_r(\Delta)$ becomes $E \cong \mathrm{Mat}_r(\Delta)$; since $E$ is commutative, $r = 1$ and $\Delta = E$. Thus, Corollary 7.120 says that $E \otimes_k D = E^{\mathrm{op}} \otimes_k D \cong \mathrm{Mat}_s(E)$.

The equality in Theorem 7.119(iii) is now $[D : k] = [E : k][E : k] = [E : k]^2$. But $[E \otimes_k D : k] = [\mathrm{Mat}_s(E) : k] = s^2[E : k]$, so that $s^2 = [D : k] = [E : k]^2$ and $s = [E : k]$. $\bullet$

**Corollary 7.123.** *If $D$ is a division algebra over a field $k$, then all maximal subfields have the same degree over $k$.*

**Remark.** It is not true that maximal subfields in arbitrary division algebras are isomorphic; see Exercise 7.62 on page 613. ◄

**Proof.** For every maximal subfield $E$, we have $[E : k] = [D : E] = \sqrt{[D : k]}$. $\bullet$

This corollary can be illustrated by Example 7.117. The quaternions $\mathbb{H}$ is a four-dimensional $\mathbb{R}$-algebra, so that a maximal subfield must have degree 2 over $\mathbb{R}$; this is so, for $\mathbb{C}$ is a maximal subfield.

We now prove a technical theorem that will yield wonderful results. Recall that a *unit* in a noncommutative ring $A$ is an element having a two-sided inverse in $A$.

**Theorem 7.124.** *Let $k$ be a field, let $B$ be a simple $k$-algebra, and let $A$ be a central simple $k$-algebra. If there are algebra maps $f, g \colon B \to A$, then there exists a unit $u \in A$ with*
$$g(b) = uf(b)u^{-1}$$
*for all $b \in B$.*

**Proof.** The map $f$ makes $A$ into a left $B$-module if we define the action of $b \in B$ on an element $a \in A$ as $f(b)a$. This action makes $A$ into a $(B, A)$-bimodule, for the associative law in $A$ gives $\big(f(b)x\big)a = f(b)(xa)$ for all $x \in A$. As usual, this $(B, A)$-bimodule is a left $(B \otimes_k A^{\mathrm{op}})$-module, where $(b \otimes a')a = baa'$ for all $a \in A$; denote it by $_fA$. Similarly, $g$ can be used to make $A$ into a left $(B \otimes_k A^{\mathrm{op}})$-module we denote by $_gA$. By Theorem 7.116, $B \otimes_k A^{\mathrm{op}}$ is a simple $k$-algebra. Now
$$[_fA : \Delta] = [A : \Delta] = [_gA : \Delta],$$
so that $_fA \cong {_gA}$ as $(B \otimes_k A^{\mathrm{op}})$-modules, by Corollary 7.51. If $\varphi \colon {_fA} \to {_gA}$ is a $(B \otimes_k A^{\mathrm{op}})$-isomorphism, then

(4)                          $\varphi(f(b)aa') = g(b)\varphi(a)a'$

for all $b \in B$ and $a, a' \in A$. Since $\varphi$ is an automorphism of $A$ as a right module over itself, $\varphi(a) = \varphi(1a) = ua$, where $u = \varphi(1) \in A$. To see that $u$ is a unit, note that

$\varphi^{-1}(a) = u'a$ for all $a \in A$. Now $a = \varphi\varphi^{-1}(a) = \varphi(u'a) = uu'a$ for all $a \in A$; in particular, when $a = 1$, we have $1 = uu'$. The equation $\varphi^{-1}\varphi = 1_A$ gives $1 = u'u$, as desired. Substituting into Equation (4), we have

$$uf(b)a = \varphi(f(b)a) = g(b)\varphi(a) = g(b)ua$$

for all $a \in A$. In particular, if $a = 1$, then $uf(b) = g(b)u$ and $g(b) = uf(b)u^{-1}$. •

**Corollary 7.125 (Skolem–Noether).** *Let $A$ be a central simple $k$-algebra over a field $k$, and let $B$ and $B'$ be isomorphic simple $k$-subalgebras of $A$. If $\psi\colon B \to B'$ is an isomorphism, then there exists a unit $u \in A$ with $\psi(b) = ubu^{-1}$ for all $b \in B$.*

**Proof.** In the theorem, take $f\colon B \to A$ to be the inclusion, define $B' = \operatorname{im}\psi$, and define $g = i\psi$, where $i\colon B' \to A$ is the inclusion. •

There is an analog of the Skolem–Noether Theorem in Group Theory. A theorem of G. Higman, B. H. Neumann, and H. Neumann says that if $B$ and $B'$ are isomorphic subgroups of a group $G$, say, $\varphi\colon B \to B'$ is an isomorphism, then there exists a group $G^*$ containing $G$ and an element $u \in G^*$ with $\varphi(b) = ubu^{-1}$ for every $b \in B$. There is a proof in Rotman, *An Introduction to the Theory of Groups*, p. 404.

**Corollary 7.126.** *Let $k$ be a field. If $\psi$ is an automorphism of $\operatorname{Mat}_n(k)$, then there exists a nonsingular matrix $P \in \operatorname{Mat}_n(k)$ with*

$$\psi(T) = PTP^{-1}$$

*for every matrix $T$ in $\operatorname{Mat}_n(k)$.*

**Proof.** The matrix ring $A = \operatorname{Mat}_n(k)$ is a central simple $k$-algebra. Set $B = B' = A$ in the Skolem–Noether Theorem. •

Here is another proof of Wedderburn's Theorem 7.13 in the present spirit.

**Theorem 7.127 (Wedderburn).** *Every finite division ring $D$ is a field.*

**Proof. (van der Waerden)** Let $Z = Z(D)$, and let $E$ be a maximal subfield of $D$. If $d \in D$, then $Z(d)$ is a subfield of $D$, and hence there is a maximal subfield $E_d$ containing $Z(d)$. By Corollary 7.123, all maximal subfields have the same degree, hence have the same order. By Corollary 2.158, all maximal subfields here are isomorphic (this is not generally true; see Exercise 7.62). For every $d \in D$, the Skolem–Noether Theorem says that there is $x_d \in D$ with $E_d = x_d E x_d^{-1}$. Therefore, $D = \bigcup_x xEx^{-1}$, and so

$$D^\times = \bigcup_x xE^\times x^{-1}.$$

If $E$ is a proper subfield of $D$, then $E^\times$ is a proper subgroup of $D^\times$, and this equation contradicts Exercise 1.96 on page 74. Therefore, $D = E$ is commutative. •

**Theorem 7.128 (Frobenius).** *If $D$ is a noncommutative finite-dimensional real division algebra, then $D \cong \mathbb{H}$.*

**Proof.** If $E$ is a maximal subfield of $D$, then $[D : E] = [E : \mathbb{R}] \leq 2$. If $[E : \mathbb{R}] = 1$, then $[D : \mathbb{R}] = 1^2 = 1$ and $D = \mathbb{R}$. Hence, $[E : \mathbb{R}] = 2$ and $[D : \mathbb{R}] = 4$. Let us identify $E$ with $\mathbb{C}$ (we know they are isomorphic). Now complex conjugation is an automorphism of $E$, so that the Skolem–Noether Theorem gives $x \in D$ with $\bar{z} = xzx^{-1}$ for all $z \in E$. In particular, $-i = xix^{-1}$. Hence,

$$x^2 i x^{-2} = x(-i)x^{-1} = -xix^{-1} = i,$$

and so $x^2$ commutes with $i$. Therefore, $x^2 \in C_D(E) = E$, by Lemma 7.121, and so $x^2 = a + bi$ for $a, b \in \mathbb{R}$. But

$$a + bi = x^2 = xx^2 x^{-1} = x(a + bi)x^{-1} = a - bi,$$

so that $b = 0$ and $x^2 \in \mathbb{R}$. If $x^2 > 0$, then there is $t \in \mathbb{R}$ with $x^2 = t^2$. Now $(x + t)(x - t) = 0$ gives $x = \pm t \in \mathbb{R}$, contradicting $-i = xix^{-1}$. Therefore, $x^2 = -r^2$ for some real $r$. The element $j$, defined by $j = x/r$, satisfies $j^2 = -1$ and $ji = -ij$. The list $1, i, j, ij$ is linearly independent over $\mathbb{R}$: if $a+bi+cj+dij = 0$, then $(-di-c)j = a+ib \in \mathbb{C}$. Since $j \notin \mathbb{C}$ (lest $x \in \mathbb{C}$), we must have $-di-c = 0 = a+bi$. Hence, $a = b = 0 = c = d$. Since $[D : \mathbb{R}] = 4$, the list $1, i, j, ij$ is a basis of $D$. It is now routine to see that if we define $k = ij$, then $ki = j = -ik$, $jk = i = -kj$, and $k^2 = -1$, and so $D \cong \mathbb{H}$.   •

In 1929, Brauer introduced the Brauer group to study division rings. Since construction of division rings was notoriously difficult, he considered the wider class of central simple algebras. Brauer introduced the following relation on central simple $k$-algebras.

**Definition.** Two central simple $k$-algebras $A$ and $B$ are ***similar***, denoted by

$$A \sim B,$$

if there are integers $n$ and $m$ with $A \otimes_k \mathrm{Mat}_n(k) \cong B \otimes_k \mathrm{Mat}_m(k)$.

If $A$ is a (finite-dimensional) central simple $k$-algebra, then Corollary 7.48 and Wedderburn–Artin II show that $A \cong \mathrm{Mat}_n(\Delta)$ for a unique $k$-division algebra $\Delta$. We shall see that $A \sim B$ if and only if they determine the same division algebra.

**Lemma 7.129.** *Let $A$ be a finite-dimensional algebra over a field $k$. If $S$ and $T$ are $k$-subalgebras of $A$ such that*

(i) *$st = ts$ for all $s \in S$ and $t \in T$,*

(ii) *$A = ST$,*

(iii) *$[A : k] = [S : k][T : k]$,*

*then $A \cong S \otimes_k T$.*

**Proof.** There is a $k$-linear transformation $f : S \otimes_k T \to A$ with $s \otimes t \mapsto st$, because $(s, t) \mapsto st$ is a $k$-bilinear function $S \times T \to A$. Condition (i) implies that $f$ is an algebra map, for

$$f\big((s \otimes t)(s' \otimes t')\big) = f(ss' \otimes tt') = ss'tt' = sts't' = f(s \otimes t)f(s' \otimes t').$$

Since $A = ST$, by condition (ii), the $k$-linear transformation $f$ is a surjection; since $\dim_k(S \otimes_k T) = \dim_k(A)$, by condition (iii), $f$ is a $k$-algebra isomorphism.   •

**Lemma 7.130.** *Let $k$ be a field.*

(i) *If $A$ is a $k$-algebra, then*
$$A \otimes_k \operatorname{Mat}_n(k) \cong \operatorname{Mat}_n(A).$$

(ii) $\operatorname{Mat}_n(k) \otimes_k \operatorname{Mat}_m(k) \cong \operatorname{Mat}_{nm}(k).$

(iii) $A \sim B$ *is an equivalence relation.*

(iv) *If $A$ is a central simple algebra, then*
$$A \otimes_k A^{\operatorname{op}} \cong \operatorname{Mat}_n(k),$$

*where $n = [A : k]$.*

**Proof.**

(i) Define $k$-subalgebras of $\operatorname{Mat}_n(A)$ by
$$S = \operatorname{Mat}_n(k) \quad \text{and} \quad T = \{aI : a \in A\}.$$

If $s \in S$ and $t \in T$, then $st = ts$ (for the entries of matrices in $S$ commute with elements $a \in A$. Now $S$ contains every matrix unit $E_{ij}$ (whose $i,j$ entry is 1 and whose other entries are 0), so that $ST$ contains all matrices of the form $a_{ij}E_{ij}$ for all $i,j$, where $a_{ij} \in A$; hence, $ST = \operatorname{Mat}_n(A)$. Finally, $[S : k][T : k] = n^2[A : k] = [\operatorname{Mat}_n(A) : k]$. Therefore, Lemma 7.129 gives the desired isomorphism.

(ii) If $V$ and $W$ are vector spaces over $k$ of dimensions $n$ and $m$, respectively, it suffices to prove that $\operatorname{End}_k(V) \otimes_k \operatorname{End}_k(W) \cong \operatorname{End}_k(V \otimes_k W)$. Define $S$ to be all $f \otimes 1_W$, where $f \in \operatorname{End}_k(V)$, and define $T$ to be all $1_V \otimes g$, where $g \in \operatorname{End}_k(W)$. It is routine to check that the three conditions in Lemma 7.129 hold.

(iii) Since $k = \operatorname{Mat}_1(k)$, we have $A \cong A \otimes_k k \cong A \otimes_k \operatorname{Mat}_1(k)$, so that $\sim$ is reflexive. Symmetry is obvious; for transitivity, suppose that $A \sim B$ and $B \sim C$; that is,
$$A \otimes_k \operatorname{Mat}_n(k) \cong B \otimes_k \operatorname{Mat}_m(k) \quad \text{and} \quad B \otimes_k \operatorname{Mat}_r(k) \cong C \otimes_k \operatorname{Mat}_s(k).$$

Then $A \otimes_k \operatorname{Mat}_n(k) \otimes_k \operatorname{Mat}_r(k) \cong A \otimes_k \operatorname{Mat}_{nr}(k)$, by part (ii). On the other hand,
$$\begin{aligned}
A \otimes_k \operatorname{Mat}_n(k) \otimes_k \operatorname{Mat}_r(k) &\cong B \otimes_k \operatorname{Mat}_m(k) \otimes_k \operatorname{Mat}_r(k) \\
&\cong C \otimes_k \operatorname{Mat}_m(k) \otimes_k \operatorname{Mat}_s(k) \\
&\cong C \otimes_k \operatorname{Mat}_{ms}(k).
\end{aligned}$$

Therefore, $A \sim C$, and so $\sim$ is an equivalence relation.

(iv) Define $f \colon A \times A^{\operatorname{op}} \to \operatorname{End}_k(A)$ by $f(a, c) = \lambda_a \circ \rho_c$, where $\lambda_a \colon x \mapsto ax$ and $\rho_c \colon x \mapsto xc$; it is routine to check that $\lambda_a$ and $\rho_c$ are $k$-maps (so their composite is also a $k$-map), and that $f$ is $k$-biadditive. Hence, there is a $k$-map $\widehat{f} \colon A \otimes_k A^{\operatorname{op}} \to \operatorname{End}_k(A)$ with $\widehat{f}(a \otimes c) = \lambda_a \circ \rho_c$. Associativity $a(xc) = (ax)c$ in $A$ says that $\lambda_a \circ \rho_c = \rho_c \circ \lambda_a$, from which it easily follows that $\widehat{f}$ is a $k$-algebra map. As $A \otimes_k A^{\operatorname{op}}$ is a simple $k$-algebra and $\ker \widehat{f}$ is a proper two-sided ideal, we have $\widehat{f}$ injective. Now $\dim_k(\operatorname{End}_k(A)) = \dim_k(\operatorname{Hom}_k(A, A)) = n^2$, where $n = [A : k]$. Since $\dim_k(\operatorname{im} \widehat{f}) = \dim_k(A \otimes_k A^{\mathrm{op}}) = n^2$, it follows that $\widehat{f}$ is a $k$-algebra isomorphism: $A \otimes_k A^{\mathrm{op}} \cong \operatorname{End}_k(A)$.  •

We now extend Theorem 7.122 from division algebras to central simple algebras.

**Theorem 7.131.** *Let $A$ be a central simple $k$-algebra over a field $k$, so that $A \cong \operatorname{Mat}_r(\Delta)$, where $\Delta$ is a division algebra over $k$. If $E$ is a maximal subfield of $\Delta$, then $E$ splits $A$; that is, there is an integer $n$ and an isomorphism*

$$E \otimes_k A \cong \operatorname{Mat}_n(E).$$

*More precisely, if $[\Delta : E] = s$, then $n = rs$ and $[A : k] = (rs)^2$.*

**Proof.** By Theorem 7.122, $\Delta$ is split by a maximal subfield $E$ (which is, of course, a finite extension of $k$): $E \otimes_k \Delta \cong \operatorname{Mat}_s(E)$, where $s = [\Delta : E] = [E : k]$. Hence,

$$E \otimes_k A \cong E \otimes_k \operatorname{Mat}_r(\Delta) \cong E \otimes_k (\Delta \otimes_k \operatorname{Mat}_r(k))$$
$$\cong (E \otimes_k \Delta) \otimes_k \operatorname{Mat}_r(k) \cong \operatorname{Mat}_s(E) \otimes_k \operatorname{Mat}_r(k) \cong \operatorname{Mat}_{rs}(E).$$

Therefore, $A \cong \operatorname{Mat}_r(\Delta)$ gives $[A : k] = r^2[\Delta : k] = r^2 s^2$.  •

**Definition.** If $[A]$ denotes the equivalence class of a central simple $k$-algebra $A$ under similarity, define the ***Brauer group*** $\operatorname{Br}(k)$ to be the set

$$\operatorname{Br}(k) = \big\{ [A] : A \text{ is a central simple } k\text{-algebra} \big\}$$

with binary operation

$$[A][B] = [A \otimes_k B].$$

**Theorem 7.132.** $\operatorname{Br}(k)$ *is an abelian group for every field $k$. Moreover, if $A \cong \operatorname{Mat}_n(\Delta)$ for a division algebra $\Delta$, then $\Delta$ is a central simple $k$-algebra and $[A] = [\Delta]$ in $\operatorname{Br}(k)$.*

**Proof.** We show that the operation is well-defined: if $A, A', B, B'$ are $k$-algebras with $A \sim A'$ and $B \sim B'$, then $A \otimes_k B \sim A' \otimes_k B'$. The isomorphisms

$$A \otimes_k \operatorname{Mat}_n(k) \cong A' \otimes_k \operatorname{Mat}_m(k) \quad \text{and} \quad B \otimes_k \operatorname{Mat}_r(k) \cong B' \otimes_k \operatorname{Mat}_s(k)$$

give $A \otimes_k B \otimes_k \operatorname{Mat}_n(k) \otimes_k \operatorname{Mat}_r(k) \cong A' \otimes_k B' \otimes_k \operatorname{Mat}_m(k) \otimes_k \operatorname{Mat}_s(k)$ (we are using commutativity and associativity of tensor product), so that Lemma 7.130(ii) gives $A \otimes_k B \otimes_k \operatorname{Mat}_{nr}(k) \cong A' \otimes_k B' \otimes_k \operatorname{Mat}_{ms}(k)$. Therefore, $A \otimes_k B \sim A' \otimes_k B'$.

That $[k]$ is the identity follows from $k \otimes_k A \cong A$, associativity and commutativity follow from associativity and commutativity of tensor product, and Lemma 7.130(iv) shows that $[A]^{-1} = [A^{\mathrm{op}}]$. Therefore, $\operatorname{Br}(k)$ is an abelian group.

If $A$ is a central simple $k$-algebra, then $A \cong \operatorname{Mat}_r(\Delta)$ for some finite-dimensional division algebra $\Delta$ over $k$. Hence, $k = Z(A) \cong Z(\operatorname{Mat}_r(\Delta)) \cong Z(\Delta)$, by Theorem 7.116. Thus, $\Delta$ is a central simple $k$-algebra, $[\Delta] \in \operatorname{Br}(k)$, and $[\Delta] = [A]$ (because $\Delta \otimes_k \operatorname{Mat}_r(k) \cong \operatorname{Mat}_r(\Delta) \cong A \cong A \otimes_k k \cong A \otimes_k \operatorname{Mat}_1(k)$).  •

The next proposition shows the significance of the Brauer group.

**Proposition 7.133.** *If $k$ is a field, then there is a bijection from $\mathrm{Br}(k)$ to the family $\mathcal{D}$ of all isomorphism classes of finite-dimensional division algebras over $k$, and so $|\mathrm{Br}(k)| = |\mathcal{D}|$. Therefore, there exists a noncommutative division ring, finite-dimensional over its center $k$, if and only if $\mathrm{Br}(k) \neq \{0\}$.*

**Proof.** Define a function $\varphi \colon \mathrm{Br}(k) \to \mathcal{D}$ by setting $\varphi([A])$ to be the isomorphism class of $\Delta$ if $A \cong \mathrm{Mat}_n(\Delta)$. Note that Theorem 7.132 shows that $[A] = [\Delta]$ in $\mathrm{Br}(k)$. Let us see that $\varphi$ is well-defined. If $[\Delta] = [\Delta']$, then $\Delta \sim \Delta'$, so there are integers $n$ and $m$ with $\Delta \otimes_k \mathrm{Mat}_n(k) \cong \Delta' \otimes_k \mathrm{Mat}_m(k)$. Hence, $\mathrm{Mat}_n(\Delta) \cong \mathrm{Mat}_m(\Delta')$. By the uniqueness in the Wedderburn–Artin Theorems, $\Delta \cong \Delta'$ (and $n = m$). Therefore, $\varphi([\Delta]) = \varphi([\Delta'])$.

Clearly, $\varphi$ is surjective, for if $\Delta$ is a finite-dimensional division algebra over $k$, then the isomorphism class of $\Delta$ is equal to $\varphi([\Delta])$. To see that $\varphi$ is injective, suppose that $\varphi([\Delta]) = \varphi([\Delta'])$. Then, $\Delta \cong \Delta'$, which implies $\Delta \sim \Delta'$. •

**Example 7.134.**

(i) If $k$ is an algebraically closed field, then $\mathrm{Br}(k) = \{0\}$, by Theorem 7.118.

(ii) If $k$ is a finite field, then Wedderburn's Theorem 7.127 (= Theorem 7.13) shows that $\mathrm{Br}(k) = \{0\}$.

(iii) If $k = \mathbb{R}$, then Frobenius's Theorem 7.128 shows that $\mathrm{Br}(\mathbb{R}) \cong \mathbb{I}_2$.

(iv) It is proved, using Class Field Theory, that $\mathrm{Br}(\mathbb{Q}_p) \cong \mathbb{Q}/\mathbb{Z}$, where $\mathbb{Q}_p$ is the field of $p$-adic numbers. Moreover, there is an exact sequence

$$0 \to \mathrm{Br}(\mathbb{Q}) \to \mathrm{Br}(\mathbb{R}) \oplus \bigoplus_p \mathrm{Br}(\mathbb{Q}_p) \to \mathbb{Q}/\mathbb{Z} \to 0.$$

In a series of deep papers, $\mathrm{Br}(k)$ was computed for the most interesting fields $k$ arising in Algebraic Number Theory (*local fields*, one of which is $\mathbb{Q}_p$, and *global fields*) by Albert, Brauer, Hasse, and Noether. ◀

**Proposition 7.135.** *If $E/k$ is a field extension, then there is a homomorphism*

$$f_{E/k} \colon \mathrm{Br}(k) \to \mathrm{Br}(E)$$

*given by $[A] \mapsto [E \otimes_k A]$.*

**Proof.** If $A$ and $B$ are central simple $k$-algebras, then $E \otimes_k A$ and $E \otimes_k B$ are central simple $E$-algebras, by Theorem 7.116. If $A \sim B$, then $E \otimes_k A \sim E \otimes_k B$ as $E$-algebras, by Exercise 7.59 on page 613. It follows that the function $f_{E/k}$ is well-defined. Finally, $f_{E/k}$ is a homomorphism, because

$$(E \otimes_k A) \otimes_E (E \otimes_k B) \cong (E \otimes_E E) \otimes_k (A \otimes_k B) \cong E \otimes_k (A \otimes_k B),$$

by Proposition 6.121, associativity of tensor product. •

**Definition.** If $E/k$ is a field extension, then the ***relative Brauer group***, $\mathrm{Br}(E/k)$, is the kernel of the homomorphism $f_{E/k} \colon \mathrm{Br}(k) \to \mathrm{Br}(E)$:

$$\mathrm{Br}(E/k) = \ker f_{E/k} = \big\{[A] \in \mathrm{Br}(k) : A \text{ is split by } E\big\}.$$

**Corollary 7.136.** *For every field $k$, we have*

$$\mathrm{Br}(k) = \bigcup_{E/k \text{ finite Galois}} \mathrm{Br}(E/k).$$

**Proof.** This follows from Theorem 7.131 after showing that we may assume that $E/k$ is Galois.  •

In a word, the Brauer group arose as a way to study division rings. It is an interesting object, but we have not really used it seriously. For example, we have not yet seen any noncommutative division rings other than the real division algebra $\mathbb{H}$ (and its variants for subfields $k$ of $\mathbb{R}$). We will remedy this when we introduce *crossed product algebras* in Chapter 9. For example, we will see, in Corollary 9.143, that there exists a division ring whose center is a field of characteristic $p > 0$. For further developments, we refer the reader to Jacobson, *Finite-Dimensional Division Algebras over Fields*, and Reiner, *Maximal Orders*.

## Exercises

**7.48.**  (i) If $k$ is a subfield of a field $K$, prove that the ring $K \otimes_k k[x]$ is isomorphic to $K[x]$.

(ii) Suppose that $k$ is a field, $p(x) \in k[x]$ is irreducible, and $K = k(\alpha)$, where $\alpha$ is a root of $p(x)$. Prove that, as rings, $K \otimes_k K \cong K[x]/(p(x))$, where $(p(x))$ is the principal ideal in $K[x]$ generated by $p(x)$.

(iii) The polynomial $p(x)$, though irreducible in $k[x]$, may factor in $K[x]$. Give an example showing that the ring $K \otimes_k K$ need not be semisimple.

(iv) Prove that if $K/k$ is a finite separable extension, then $K \otimes_k K$ is semisimple. (The converse is also true.)

**7.49.** If $A \cong A'$ and $B \cong B'$ are $k$-algebras, where $k$ is a commutative ring, prove that $A \otimes_k B \cong A' \otimes_k B'$ as $k$-algebras.

* **7.50.** If $k$ is a commutative ring and $A$ and $B$ are $k$-algebras, prove that

$$(A \otimes_k B)^{\mathrm{op}} \cong A^{\mathrm{op}} \otimes_k B^{\mathrm{op}}.$$

**7.51.** If $R$ is a commutative $k$-algebra, where $k$ is a field and $G$ is a group, prove that $R \otimes_k kG \cong RG$.

* **7.52.**  (i) If $k$ is a subring of a commutative ring $R$, prove that $R \otimes_k k[x] \cong R[x]$ as $R$-algebras.

(ii) If $f(x) \in k[x]$ and $(f)$ is the principal ideal in $k[x]$ generated by $f(x)$, prove that $R \otimes_k (f)$ is the principal ideal in $R[x]$ generated by $f(x)$. More precisely, there is a commutative diagram

$$
\begin{array}{ccccc}
0 & \longrightarrow & R \otimes_k (f) & \longrightarrow & R \otimes_k k[x] \\
  &                 & \downarrow      &                 & \downarrow \\
0 & \longrightarrow & (f)             & \longrightarrow & R[x]
\end{array}
$$

(iii) Let $k$ be a field and $E \cong k[x]/(f)$, where $f(x) \in k[x]$ is irreducible. Prove that $E \otimes_k E \cong E[x]/(f)_E$, where $(f)_E$ is the principal ideal in $E[x]$ generated by $f(x)$.

(iv) Give an example of a field extension $E/k$ with $E \otimes_k E$ not a field.

   **Hint.** If $f(x) \in k[x]$ factors into $g(x)h(x)$ in $E[x]$, where $(g, h) = 1$, then the Chinese Remainder Theorem applies.

**7.53.** Let $k$ be a field and let $f(x) \in k[x]$ be irreducible. If $K/k$ is a field extension, then $f(x) = p_1(x)^{e_1} \cdots p_n(x)^{e_n} \in K[x]$, where the $p_i(x)$ are distinct irreducible polynomials in $K[x]$ and $e_i \geq 1$.

   (i) Prove that $f(x)$ is separable if and only if all $e_i = 1$.

   (ii) Prove that a finite field extension $K/k$ is separable if and only if $K \otimes_k K$ is a semisimple ring.

   **Hint.** First, observe that $K/k$ is a simple extension, so there is an exact sequence $0 \to (f) \to k[x] \to K \to 0$. Second, use the Chinese Remainder Theorem.

**7.54.** Prove that $\mathbb{H} \otimes_{\mathbb{R}} \mathbb{H} \cong \mathrm{Mat}_4(\mathbb{R})$ as $\mathbb{R}$-algebras.

**Hint.** Use Corollary 7.48 for the central simple $\mathbb{R}$-algebra $\mathbb{H} \otimes_{\mathbb{R}} \mathbb{H}$.

**7.55.** We have given one isomorphism $\mathbb{C} \otimes_{\mathbb{R}} \mathbb{H} \cong \mathrm{Mat}_2(\mathbb{C})$ in Example 7.117. Describe all possible isomorphisms between these two algebras.

**Hint.** Use the Skolem–Noether Theorem.

**7.56.** Prove that $\mathbb{C} \otimes_{\mathbb{R}} \mathbb{C} \cong \mathbb{C} \times \mathbb{C}$ as $\mathbb{R}$-algebras.

**7.57.** (i) Let $\mathbb{C}(x)$ and $\mathbb{C}(y)$ be function fields. Prove that $R = \mathbb{C}(x) \otimes_{\mathbb{C}} \mathbb{C}(y)$ is isomorphic to a subring of $\mathbb{C}(x, y)$. Conclude that $R$ has no zero-divisors.

   (ii) Prove that $\mathbb{C}(x) \otimes_{\mathbb{C}} \mathbb{C}(y)$ is not a field.

   **Hint.** Show that $R$ is isomorphic to the subring of $\mathbb{C}(x, y)$ consisting of polynomials of the form $f(x, y)/g(x)h(y)$.

   (iii) Use Exercise 7.7 on page 533 to prove that the tensor product of artinian algebras need not be artinian.

* **7.58.** Let $A$ be a central simple $k$-algebra. If $A$ is split by a field $E$, prove that $A$ is split by any field extension $E'$ of $E$.

* **7.59.** Let $E/k$ be a field extension. If $A$ and $B$ are central simple $k$-algebras with $A \sim B$, prove that $E \otimes_k A \sim E \otimes_k B$ as central simple $E$-algebras.

**7.60.** If $D$ is a finite-dimensional division algebra over $\mathbb{R}$, prove that $D$ is isomorphic to either $\mathbb{R}$, $\mathbb{C}$, or $\mathbb{H}$.

**7.61.** Prove that $\mathrm{Mat}_2(\mathbb{H}) \cong \mathbb{H} \otimes_{\mathbb{R}} \mathrm{Mat}_2(\mathbb{R})$ as $\mathbb{R}$-algebras.

* **7.62.** (i) Let $A$ be a four-dimensional vector space over $\mathbb{Q}$, and let $1, i, j, k$ be a basis. Show that $A$ is a division algebra if we define 1 to be the identity and

$$i^2 = -1, \qquad\qquad j^2 = -2, \qquad\qquad k^2 = -2,$$
$$ij = k, \qquad\qquad jk = 2i, \qquad\qquad ki = j,$$
$$ji = -k, \qquad\qquad kj = -2i, \qquad\qquad ik = -j.$$

   Prove that $A$ is a division algebra over $\mathbb{Q}$.

   (ii) Prove that $\mathbb{Q}(i)$ and $\mathbb{Q}(j)$ are nonisomorphic maximal subfields of $A$.

**7.63.** Let $D$ be the $\mathbb{Q}$-subalgebra of $\mathbb{H}$ having basis $1, i, j, k$.

(i) Prove that $D$ is a division algebra over $\mathbb{Q}$.
   **Hint.** Compute the center $Z(D)$.

(ii) For any pair of nonzero rationals $p$ and $q$, prove that $D$ has a maximal subfield isomorphic to $\mathbb{Q}(\sqrt{-p^2 - q^2})$.
   **Hint.** Compute $(pi + qj)^2$.

**7.64. (Dickson)** If $D$ is a division algebra over a field $k$, then each $d \in D$ is algebraic over $k$. Prove that $d, d' \in D$ are conjugate in $D$ if and only if $\mathrm{irr}(d, k) = \mathrm{irr}(d', k)$.

**Hint.** Use the Skolem–Noether Theorem.

**7.65.** Prove that if $A$ is a central simple $k$-algebra with $A \sim \mathrm{Mat}_n(k)$, then $A \cong \mathrm{Mat}_m(k)$ for some integer $m$.

**7.66.** Prove that if $A$ is a central simple $k$-algebra with $[A]$ of finite order $m$ in $\mathrm{Br}(k)$, then there is an integer $r$ with

$$A \otimes_k \cdots \otimes_k A \cong \mathrm{Mat}_r(k)$$

(there are $m$ factors equal to $A$). In Chapter 9, we shall see that every element in $\mathrm{Br}(k)$ has finite order.

## Section 7.8. Abelian Categories

Since representations of a group $G$ are just another way of considering $\mathbb{C}G$-modules, contemplating all the representations of $G$ is the same as contemplating $_{\mathbb{C}G}\mathbf{Mod}$. It is natural to ask, more generally, to what extent a category $_R\mathbf{Mod}$ determines a ring $R$. We now prepare the answer to this question; the answer itself is in the next section.

Recall that an object $A$ in a category $\mathcal{C}$ is an *initial object* if there is a unique morphism $A \to X$ for every $X \in \mathrm{obj}(\mathcal{C})$; an object $\Omega$ in $\mathcal{C}$ is a *terminal object* if there is a unique morphism $X \to \Omega$ for every $X \in \mathrm{obj}(\mathcal{C})$; and an object is a *zero object* if it is both initial and terminal (Exercise 6.43 on page 435).

**Definition.** A category $\mathcal{A}$ is **additive** if

(i) $\mathrm{Hom}_{\mathcal{A}}(A, B)$ is an (additive) abelian group for every $A, B \in \mathrm{obj}(\mathcal{A})$;

(ii) the distributive laws hold: given morphisms

$$X \xrightarrow{k} A \underset{g}{\overset{f}{\rightrightarrows}} B \xrightarrow{h} Y,$$

where $X$ and $Y \in \mathrm{obj}(\mathcal{A})$, then

$$h(f + g) = hf + hg \quad \text{and} \quad (f + g)k = fk + gk;$$

(iii) $\mathcal{A}$ has a zero object;

(iv) $\mathcal{A}$ has finite products and finite coproducts: for all objects $A, B$ in $\mathcal{A}$, both $A \sqcap B$ and $A \sqcup B$ exist in $\mathrm{obj}(\mathcal{A})$.

Let $\mathcal{A}$ and $\mathcal{C}$ be additive categories. A functor $T \colon \mathcal{A} \to \mathcal{C}$ (of either variance) is **additive** if, for all $A, B \in \mathrm{obj}(\mathcal{A})$ and all $f, g \in \mathrm{Hom}_{\mathcal{A}}(A, B)$, we have

$$T(f + g) = Tf + Tg;$$

that is, the function $\text{Hom}_{\mathcal{A}}(A, B) \to \text{Hom}_{\mathcal{C}}(TA, TB)$, given by $f \mapsto Tf$, is a homomorphism of abelian groups.

**Example 7.137.** It is easy to see that the Hom functors are additive. Let $\mathcal{A}$ be an additive category, let $X \in \text{obj}(\mathcal{A})$, and let $T = \text{Hom}_{\mathcal{A}}(X, \square) \colon \mathcal{A} \to \textbf{Ab}$. If $h \colon M \to N$, then $Th = h_* \colon \text{Hom}_{\mathcal{A}}(X, M) \to \text{Hom}_{\mathcal{A}}(X, N)$ is given by $h_* \colon f \mapsto h_*(f) = hf$. Hence, if $f, g \in \text{Hom}_{\mathcal{A}}(X, A)$, then

$$T(f + g) = h_*(f + g) = h(f + g) = hf + hg = h_*f + h_*g = Tf + Tg.$$

A similar argument shows that contravariant Hom functors are additive. ◀

Of course, if $T$ is an additive functor, then $T(0) = 0$, where $0$ is either a zero object or a zero morphism. Lemma 6.13 shows that $_R\textbf{Mod}$ and $\textbf{Mod}_R$ are additive categories, while Exercise 7.67 on page 625 shows that neither **Groups** nor **ComRings** is an additive category. We have just seen that if $\mathcal{A}$ is additive, then the Hom functors $\mathcal{A} \to \textbf{Ab}$ are additive, while Theorem 6.104 shows that the tensor functors $_R\textbf{Mod} \to \textbf{Ab}$ and $\textbf{Mod}_R \to \textbf{Ab}$ are additive.

That finite coproducts and products coincide for modules is a special case of a more general fact: finite products and finite coproducts coincide in all additive categories.

**Lemma 7.138.** *Let $\mathcal{C}$ be an additive category, and let $M, A, B \in \text{obj}(\mathcal{C})$. Then $M \cong A \sqcap B$ if and only if there are morphisms $i \colon A \to M$, $j \colon B \to M$, $p \colon M \to A$, and $q \colon M \to B$ such that*

$$pi = 1_A, \quad qj = 1_B, \quad pj = 0, \quad qi = 0, \quad and \quad ip + jq = 1_M.$$

*Moreover, $A \sqcap B$ is also a coproduct with injections $i$ and $j$, and so*

$$A \sqcap B \cong A \sqcup B.$$

**Proof.** The proof of the first statement, left to the reader, is a variation of the proof of Proposition 6.26. The proof of the second statement is a variation of the proof of Proposition 6.42, and it, too, is left to the reader. The last statement holds because two coproducts, here $A \sqcup B$ and $A \sqcap B$, must be isomorphic. •

If $A$ and $B$ are objects in an additive category, then $A \sqcap B \cong A \sqcup B$; their common value, denoted by $A \oplus B$, is called their **direct sum** (or **biproduct**).

Addition of homomorphisms in **Ab** can be described without elements. Define the *diagonal* $\Delta \colon A \to A \oplus A$ by $\Delta \colon a \mapsto (a, a)$; dually, the *codiagonal* $\nabla \colon B \oplus B \to B$ is defined by $\nabla \colon (b, b') \mapsto b + b'$. If $f, g \colon A \to B$, we claim that

$$\nabla(f \oplus g)\Delta = f + g.$$

If $a \in A$, then $\nabla(f \oplus g)\Delta \colon a \mapsto (a, a) \mapsto (fa, ga) \mapsto fa + ga = (f + g)(a)$. As usual, the advantage of definitions given in terms of maps (rather than in terms of elements) is that they can be recognized by functors. Diagonals and codiagonals can be defined and exist in additive categories.

**Definition.** Let $\mathcal{A}$ be an additive category. If $A \in \mathrm{obj}(\mathcal{A})$, then the **diagonal** $\Delta_A \colon A \to A \oplus A$ is the unique morphism with $p\Delta_A = 1_A$ and $q\Delta_A = 1_A$, where $p, q$ are projections:

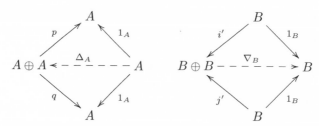

If $B \in \mathrm{obj}(\mathcal{A})$, then the **codiagonal** $\nabla_B \colon B \oplus B \to B$ is the unique morphism with $\nabla_B i' = 1_B$ and $\nabla_B j' = 1_B$, where $i', j'$ are injections.

The reader should check that these definitions, when specialized to **Ab**, give our original diagonal and codiagonal homomorphisms.

**Lemma 7.139.** *If $\mathcal{A}$ is an additive category and $f, g \in \mathrm{Hom}_{\mathcal{A}}(A, B)$, then*

$$\nabla_B(f \oplus g)\Delta_A = f + g.$$

**Proof.** Let $p, q \colon A \oplus A \to A$ be projections, and let $i, j \colon A \to A \oplus A$ and $i', j' \colon B \to B \oplus B$ be injections. We compute:

$$
\begin{aligned}
\nabla_B(f \oplus g)\Delta_A &= \nabla_B(f \oplus g)(ip + jq)\Delta_A \\
&= \nabla_B(f \oplus g)(ip\Delta_A + jq\Delta_A) \\
&= \nabla_B(f \oplus g)(i + j) \quad \text{(because } p\Delta_A = 1_A = q\Delta_A) \\
&= \nabla_B(f \oplus g)i + \nabla_B(f \oplus g)j \\
&= \nabla_B i' f + \nabla_B j' g \quad \text{(Exercise 6.90 on page 487)} \\
&= f + g. \qquad \text{(because } \nabla_B i' = 1_B = \nabla_B j') \quad \bullet
\end{aligned}
$$

**Definition.** A functor $T \colon \mathcal{A} \to \mathcal{B}$ between additive categories **preserves finite direct sums** if, for all $A, B \in \mathrm{obj}(\mathcal{A})$, whenever $A \oplus B$ is a direct sum with projections $p, q$ and injections $i, j$, then $TA \oplus TB$ is a direct sum with projections $Tp, Tq$ and injections $Ti, Tj$.

Recall Exercise 6.90 on page 487: if $i, j \colon A \to A \oplus A$ and $i', j' \colon B \to B \oplus B$ are injections and $f, g \colon A \to B$, then $f \oplus g \colon A \oplus A \to B \oplus B$ is the unique map completing the coproduct diagram

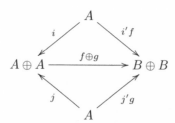

It follows that if $T$ preserves finite direct sums, then $T(f \oplus g) = Tf \oplus Tg$.

**Proposition 7.140.** *A functor* $T: \mathcal{A} \to \mathcal{B}$ *between additive categories is additive if and only if $T$ preserves finite direct sums.*

**Proof.** If $T$ is additive, then $T$ preserves finite direct sums, by Lemma 7.138.

Conversely, let $T$ preserve finite direct sums. If $f, g: A \to B$, then

$$
\begin{aligned}
T(f + g) &= T\big(\nabla_B(f \oplus g)\Delta_A\big) \qquad \text{(by Lemma 7.139)} \\
&= (T\nabla_B)T(f \oplus g)(T\Delta_A) \\
&= \nabla_{TB}T(f \oplus g)\Delta_{TA} \\
&= \nabla_{TB}(Tf \oplus Tg)\Delta_{TA} \qquad \text{(Exercise 6.90)} \\
&= Tf + Tg. \quad \bullet
\end{aligned}
$$

We have been reluctant to discuss injections and surjections in categories; after all, morphisms in a category need not be functions. On the other hand, it is often convenient to have them.

**Definition.** A morphism $u: B \to C$ in a category $\mathcal{C}$ is a **monomorphism**[11] (or is **monic**) if $u$ can be canceled from the left; that is, for all objects $A$ and all morphisms $f, g: A \to B$, we have $uf = ug$ implies $f = g$.

$$
A \underset{g}{\overset{f}{\rightrightarrows}} B \overset{u}{\longrightarrow} C.
$$

It is clear that $u: B \to C$ is monic if and only if, for all $A$, the induced map $u_*: \operatorname{Hom}(A, B) \to \operatorname{Hom}(A, C)$ is an injection. In an additive category, $\operatorname{Hom}(A, B)$ is an abelian group, and so $u$ is monic if and only if $ug = 0$ implies $g = 0$. Exercise 7.71 on page 625 shows that monomorphisms and injections coincide in **Sets**, $_R$**Mod**, and **Groups**. Even in a category whose morphisms are actually functions, however, monomorphisms need not be injections (see Exercise 7.72 on page 625).

Here is the dual definition.

**Definition.** A morphism $v: B \to C$ in a category $\mathcal{C}$ is an **epimorphism** (or is **epic**) if $v$ can be canceled from the right; that is, for all objects $D$ and all morphisms $h, k: C \to D$, we have $hv = kv$ implies $h = k$.

$$
B \overset{v}{\longrightarrow} C \underset{k}{\overset{h}{\rightrightarrows}} D.
$$

It is clear that $v: B \to C$ is epic if and only if, for all $D$, the induced map $v^*: \operatorname{Hom}(C, D) \to \operatorname{Hom}(B, D)$ is an injection. In an additive category, $\operatorname{Hom}(A, B)$ is an abelian group, and so $v$ is monic if and only if $gv = 0$ implies $g = 0$. Exercise 7.71 on page 625 shows that epimorphisms and surjections coincide in **Sets** and in $_R$**Mod**. Every surjective homomorphism in **Groups** is epic, but we must be clever to show this (Exercise 6.70 on page 460). Even in a category whose morphisms are actually functions, epimorphisms need not be surjections. For example, if $R$ is a domain, then the ring homomorphism $\varphi: R \to \operatorname{Frac}(R)$, given by $r \mapsto r/1$, is an epimorphism in **ComRings**. If $A$ is a commutative ring and

---

[11]A useful notation for a monomorphism $A \to B$ is $A \rightarrowtail B$, while a notation for an epimorphism $B \to C$ is $B \twoheadrightarrow C$.

$h, k \colon \mathrm{Frac}(R) \to A$ are ring homomorphisms agreeing on $R$, then $h = k$; thus, the inclusion $\mathbb{Z} \to \mathbb{Q}$ is epic.

Projectives and injectives can be defined in any category, but the definitions involve epimorphisms and monomorphisms. Recognizing these morphisms in general categories is too difficult, and so we usually consider projective and injective objects only in additive categories.

**Definition.** An object $P$ in an additive category $\mathcal{A}$ is ***projective*** if, for every epic $g \colon B \to C$ and every $f \colon P \to C$, there exists $h \colon P \to B$ with $f = gh$:

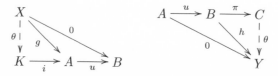

An object $E$ in an additve category $\mathcal{A}$ is ***injective*** if, for every monic $g \colon A \to B$ and every $f \colon A \to E$, there exists $h \colon B \to E$ with $f = hg$.

We can describe kernels and cokernels categorically. Recall that we have seen, in Example 6.46, that kernel is a pullback and, in Example 6.49, that cokernel is a pushout. Thus, kernel and cokernel are solutions to universal mapping problems. We promote this fact to a definition.

**Definition.** Let $\mathcal{A}$ be an additive category. If $u \colon A \to B$ and $i \colon K \to A$ in $\mathcal{A}$, then the ***kernel*** of $u$, denoted by $i = \ker u$, is a solution to the universal mapping problem illustrated in the diagram below on the left: if $ug = 0$, then there exists a unique $\theta$ with $i\theta = g$:

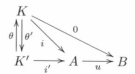

Similarly, the ***cokernel*** of $u$, denoted by $\pi \colon C \to Y$, is a solution to the universal mapping problem illustrated in the above diagram on the right: if $hu = 0$, then there exists a unique $\theta$ with $\theta\pi = h$.

Kernels and cokernels are now morphisms; this is not as strange as it first appears, for domains and targets are ingredients of functions and of morphisms (remember that Hom sets in a category are pairwise disjoint, so that each morphism has a unique domain and target).

Let us consider uniqueness of kernels and cokernels. Suppose that $i'$ is another kernel of $u \colon A \to B$; thus, $i' \colon K' \to A$ is another solution to the universal mapping problem:

The maps $\theta'$ and $\theta$ are inverses, hence are isomorphisms. Thus, the domain of $\ker u$ is unique up to isomorphism.

Let us compare these definitions with the usual notions in **Ab**. If $u \colon A \to B$ is a homomorphism, then $\ker u$ is a subgroup of $A$ and $\operatorname{coker} u = B/\operatorname{im} u$ is a quotient of $B$. Exercise 6.88 says that the inclusion $i \colon \ker u \to A$ is the kernel in **Ab** and the natural map $\pi \colon B \to B/\operatorname{im} u$ is the cokernel in **Ab**.

**Proposition 7.141.** *Let $\mathcal{A}$ be an additive category.*

(i) *A morphism $u$ is a monomorphism if and only if $\ker u = 0$, and $u$ is an epimorphism if and only if $\operatorname{coker} u = 0$.*

(ii) *If $i = \ker u$, then $i$ is a monomorphism and $\pi = \operatorname{coker} u$, then $\pi$ is an epimorphism.*

**Proof.**

(i) Let $u$ be monic. Since $i = \ker u$, we have $ui = 0$. But $u0 = 0$, so canceling $u$ gives $i = 0$. Conversely, if $i = 0$ and $ug = u0$, then $g = i\theta = i0 = 0$; hence, $u$ is monic. The proof for cokernels is dual.

(ii) Suppose that $X \xrightarrow{g} K \xrightarrow{i} A$ and $ig = 0$. Since $uig = 0$, there is a unique $\theta \colon X \to K$ with $i\theta = ig = 0$. Obviously, $\theta = 0$ satisfies this equation, and so uniqueness gives $g = \theta = 0$; therefore, $i$ is monic. A dual argument shows that cokernels are epimorphisms. $\bullet$

The converse of Proposition 7.141(ii) is true in *abelian categories*, which are additive categories in which a reasonable notion of exactness can be defined. They are so called because of their resemblance to **Ab**.

**Definition.** A category $\mathcal{A}$ is an ***abelian category*** if it is an additive category such that

(i) every morphism has a kernel and a cokernel;

(ii) every monomorphism is a kernel and every epimorphism is a cokernel.

In more detail, axiom (i) says that if $u$ is a morphism in $\mathcal{A}$, then $i = \ker u$ and $\pi = \operatorname{coker} u$ exist and lie in $\mathcal{A}$. Axiom (ii) says that if $i$ is monic, then there is a morphism $u$ in $\mathcal{A}$ with $i = \ker u$. Similarly, if $\pi$ is epic, then there is $v$ in $\mathcal{A}$ with $\pi = \operatorname{coker} v$.

We can define *image* and *exactness* in any abelian category. If $u \colon A \to B$ in **Ab**, we can choose $\operatorname{coker} u$ to be the natural map $\pi \colon B \to B/\operatorname{im} u$; therefore, $\ker \pi = \operatorname{im} u$. Thus, in **Ab**, we have $\ker(\operatorname{coker} u) = \ker \pi = \operatorname{im} u$. This motivates the definition of *image* in abelian categories.

**Definition.** Let $u \colon A \to B$ be a morphism in an abelian category, and let $\operatorname{coker} u$ be $\pi \colon B \to C$ for some object $C$. Then its ***image*** is

$$\operatorname{im} u = \ker(\operatorname{coker} u) = \ker \pi.$$

A sequence $A \xrightarrow{u} B \xrightarrow{v} C$ in $\mathcal{A}$ is ***exact*** if there is equality

$$\ker v = \operatorname{im} u.$$

**Remark.** We have been careless. If $B$ is an object in an additive category $\mathcal{A}$, consider all monomorphisms $j$ with target $B$. Call two such morphisms $j \colon A \to B$ and $j' \colon A' \to B$ *equivalent* if there exists an isomorphism $\theta \colon A' \to A$ with $j' = j\theta$. Define a *subobject* of $B$ to be an equivalence class $[j]$ of such monomorphisms:

Our discussion of uniqueness of kernels on page 618 shows that kernels may be viewed as subobjects. The reader can give a dual discussion describing *quotient objects*.

We need not be so fussy if an abelian category has "honest" subobjects, for we can then choose a "favorite" representative and call it the subobject. For example, if $u \colon B \to C$ in **Ab**, this formal definition says that the kernel of $u$ is an equivalence class of morphisms $i \colon K \to A$. We usually choose $K$ to be the submodule of $A$ and $i$ to be its inclusion.  ◀

**Remark.** Mac Lane (*Categories for the Working Mathematician*, Chapter VIII, Sections 1 and 3) views exactness in another way (which agrees with the definition above). If $u \colon A \to B$ is a morphism in an abelian category, then $u = me$, where $m = \ker(\operatorname{coker} u)$ is monic and $e = \operatorname{coker}(\ker u)$ is epic. Moreover, this factorization is unique in the following sense. If $u = m'e'$, where $m'$ is monic and $e'$ is epic, then there is equivalence of $m$ and $m'$ and of $e$ and $e'$. In light of this, he defines exactness of a sequence in an abelian category as follows: when $u = me$ and $v = m'e'$ (where $m, m'$ are monic and $e, e'$ are epic), then $A \xrightarrow{u} B \xrightarrow{v} C$ is exact if and only if $e$ and $m'$ are equivalent.  ◀

The next two propositions construct new abelian categories from old ones.

**Definition.** A category $\mathcal{S}$ is a *subcategory* of a category $\mathcal{C}$ if

   (i) $\operatorname{obj}(\mathcal{S}) \subseteq \operatorname{obj}(\mathcal{C})$;

   (ii) $\operatorname{Hom}_{\mathcal{S}}(A, B) \subseteq \operatorname{Hom}_{\mathcal{C}}(A, B)$ for all $A, B \in \operatorname{obj}(\mathcal{S})$;

   (iii) if $f \in \operatorname{Hom}_{\mathcal{S}}(A, B)$ and $g \in \operatorname{Hom}_{\mathcal{S}}(B, C)$, the composite $gf \in \operatorname{Hom}_{\mathcal{S}}(A, C)$ is equal to the composite $gf \in \operatorname{Hom}_{\mathcal{C}}(A, C)$;

   (iv) if $A \in \operatorname{obj}(\mathcal{S})$, then $1_A \in \operatorname{Hom}_{\mathcal{S}}(A, A)$ is equal to $1_A \in \operatorname{Hom}_{\mathcal{C}}(A, A)$.

A subcategory $\mathcal{S}$ of a category $\mathcal{C}$ is *full* if $\operatorname{Hom}_{\mathcal{S}}(A, B) = \operatorname{Hom}_{\mathcal{C}}(A, B)$ for all $A, B \in \operatorname{obj}(\mathcal{S})$.

It is easy to see that the inclusion of a subcategory is a functor. The subcategory **Ab** is a full subcategory of **Groups**, but if we regard **Top** as a subcategory of **Sets**, then it is not a full subcategory, for there are functions between spaces that are not continuous.

**Example 7.142.**

(i) For every ring $R$, both $_R\mathbf{Mod}$ and $\mathbf{Mod}_R$ are abelian categories. In particular, $_\mathbb{Z}\mathbf{Mod} = \mathbf{Ab}$ is abelian.

(ii) The full subcategory of $\mathbf{Ab}$ of all finitely generated abelian groups is an abelian category, as is the full subcategory of all torsion abelian groups.

(iii) The full subcategory of $\mathbf{Ab}$ of all torsion-free abelian groups is not an abelian category, for there are morphisms having no cokernel; for example, the inclusion $2\mathbb{Z} \to \mathbb{Z}$ has cokernel $\mathbb{I}_2$ which is not torsion-free.

(iv) The category $\mathbf{Groups}$ is not abelian (it is not even additive). If $S \subseteq G$ is a nonnormal subgroup of a group $G$, then the inclusion $i\colon S \to G$ has no cokernel. However, if $K$ is a normal subgroup of $G$ with inclusion $j\colon K \to G$, then $\mathrm{coker}\, j$ does exist. Thus, axiom (ii) in the definition of abelian category essentially says that every subobject in an abelian category is normal. ◄

**Remark.** Abelian categories are ***self-dual*** in the sense that the dual of every axiom in its definition is itself an axiom; it follows that if $\mathcal{A}$ is an abelian category, then so is its opposite $\mathcal{A}^{\mathrm{op}}$. A theorem using only these axioms in its proof is true in every abelian category; moreover, its dual is also a theorem in every abelian category, and its proof is dual to the original proof. The categories $_R\mathbf{Mod}$ and $\mathbf{Mod}_R$ are abelian categories having extra properties; for example, $R$ is a special type of object. Module categories are not self-dual, and this explains why a theorem and its dual, both of which are true in every module category, may have very different proofs. For example, the statements "every module is a quotient of a projective module" and "every module can be imbedded in an injective module" are dual and are always true. The proofs are not dual because these statements are not true in every abelian category. Exercise 7.79 on page 626 shows that the abelian category of all torsion abelian groups has no nonzero projectives, and Exercise 7.80 shows that the abelian category of all finitely generated abelian groups has no nonzero injectives. ◄

Here are more examples of abelian categories.

**Proposition 7.143.** *Let $\mathcal{S}$ be a full subcategory of an abelian category $\mathcal{A}$. If*

(i) *a zero object in $\mathcal{A}$ lies in $\mathcal{S}$,*

(ii) *for all $A, B \in \mathrm{obj}(\mathcal{S})$, the direct sum $A \oplus B$ in $\mathcal{A}$ lies in $\mathcal{S}$,*

(iii) *for all $A, B \in \mathrm{obj}(\mathcal{S})$ and all $f\colon A \to B$, both $\ker f$ and $\mathrm{coker}\, f$ lie in $\mathcal{S}$,*

*then $\mathcal{S}$ is an abelian category.*

**Remark.** When we say that a morphism $i\colon K \to A$ in $\mathcal{A}$ lies in a subcategory $\mathcal{S}$, we assume that its domain $K$ and target $A$ lie in $\mathrm{obj}(\mathcal{S})$. ◄

**Proof.** That $\mathcal{S}$ is a full subcategory of $\mathcal{A}$ satisfying (i) and (ii) gives $\mathcal{S}$ additive, by Exercise 7.70 on page 625.

If $f\colon A \to B$ is a morphism in $\mathcal{S} \subseteq \mathcal{A}$, then $\ker f$ and $\mathrm{coker}\, f$ lie in $\mathcal{A}$, and hence they lie in $\mathcal{S}$, by (iii). Thus, axiom (i) in the definition of abelian category holds; it remains to verify axiom (ii).

Let $u$ be a monomorphism in $\mathcal{S}$; we have just seen that $\ker u$ lies in $\mathcal{S}$. Since $u$ is monic, Proposition 7.141 gives $\ker u = 0$ in $\mathcal{S}$. By hypothesis, $\ker u$ is the same in $\mathcal{A}$ as in $\mathcal{S}$, so that $\ker u = 0$ in $\mathcal{A}$; hence, Proposition 7.141 says that $u$ is a monomorphism in $\mathcal{A}$. As $\mathcal{A}$ is abelian, we have $u = \ker v$ for some $v\colon A \to B$. By hypothesis, $\ker v$ lies in $\mathcal{S}$; that is, $u = \ker v$. The dual argument shows that epimorphisms in $\mathcal{S}$ are cokernels. Therefore, $\mathcal{A}$ is abelian.   •

**Proposition 7.144.** *If $\mathcal{A}$ is an abelian category and $\mathcal{C}$ is a small category, then the functor category $\mathcal{A}^{\mathcal{C}}$ is an abelian category.*

**Proof.** We assume that $\mathcal{C}$ is small to guarantee that $\mathcal{A}^{\mathcal{C}}$ is a category (Example 6.133). The zero object in $\mathcal{A}^{\mathcal{C}}$ is the constant functor with value 0, where 0 is a zero object in $\mathcal{A}$. If $\tau, \sigma \in \mathrm{Hom}(F, G) = \mathrm{Nat}(F, G)$, where $F, G\colon \mathcal{C} \to \mathcal{A}$ are functors, define $\tau + \sigma\colon F \to G$ by $(\tau + \sigma)_C = \tau_C + \sigma_C\colon FC \to GC$ for all $C \in \mathrm{obj}(\mathcal{C})$. Finally, define $F \oplus G$ by $(F \oplus G)C = FC \oplus GC$. It is straightforward to check that $\mathcal{A}^{\mathcal{C}}$, with these definitions, is an additive category.

If $\tau\colon F \to G$, define $K\colon \mathcal{C} \to \mathcal{A}$ on objects by

$$KC = \ker(\tau_C).$$

In the following commutative diagram with exact rows, where $f\colon C \to C'$ in $\mathcal{C}$, there is a unique $Kf\colon KC \to KC'$ making the augmented diagram commute:

$$
\begin{array}{ccccccc}
0 & \longrightarrow & KC & \xrightarrow{\iota_C} & FC & \longrightarrow & GC \\
 & & \downarrow{\scriptstyle Kf} & & \downarrow{\scriptstyle Ff} & & \downarrow{\scriptstyle Gf} \\
0 & \longrightarrow & KC' & \xrightarrow{\iota_{C'}} & FC' & \longrightarrow & GC'
\end{array}
$$

The reader may check that $K$ is a functor, $\iota\colon K \to F$ is a natural transformation, and $\ker \tau = \iota$; dually, cokernels exist in $\mathcal{A}^{\mathcal{C}}$. Verification of the axioms for an abelian category is routine.   •

The following construction is of fundamental importance in Algebraic Topology and in Homological Algebra.

**Definition.** A *complex*[12] in an abelian category $\mathcal{A}$ is a sequence of objects and morphisms in $\mathcal{A}$ (called *differentials*),

$$\to A_{n+1} \xrightarrow{d_{n+1}} A_n \xrightarrow{d_n} A_{n-1} \to,$$

such that the composite of adjacent morphisms is 0:

$$d_n d_{n+1} = 0 \qquad \text{for all } n \in \mathbb{Z}.$$

We usually denote this complex by $(\mathbf{C}, d)$ or by $\mathbf{C}$.

If $(\mathbf{C}, d)$ and $(\mathbf{C}', d')$ are complexes, then a *chain map*

$$f\colon (\mathbf{C}, d) \to (\mathbf{C}', d')$$

---

[12] These are also called *chain complexes* in the literature.

is a sequence of morphisms $(f_n \colon C_n \to C_n')_{n \in \mathbb{Z}}$ making the following diagram commute:

$$
\begin{array}{ccccccc}
\cdots \longrightarrow & C_{n+1} & \xrightarrow{d_{n+1}} & C_n & \xrightarrow{d_n} & C_{n-1} & \longrightarrow \cdots \\
& \downarrow{f_{n+1}} & & \downarrow{f_n} & & \downarrow{f_{n-1}} & \\
\cdots \longrightarrow & C_{n+1}' & \xrightarrow{d_{n+1}'} & C_n' & \xrightarrow{d_n'} & C_{n-1}' & \longrightarrow \cdots
\end{array}
$$

In Example 6.133(iii), we viewed a complex as a functor $\mathrm{PO}(\mathbb{Z}) \to \mathcal{A}$ and a chain map as a natural transformation.

**Definition.** If $\mathcal{A}$ is an abelian category, then $\mathbf{Comp}(\mathcal{A})$, the abelian category of *complexes over* $\mathcal{A}$, is the full subcategory of $\mathcal{A}^{\mathbf{PO}(\mathbb{Z})}$ generated by all complexes. If $\mathcal{A} = \mathbf{Ab}$, then we write $\mathbf{Comp}$ instead of $\mathbf{Comp}(\mathbf{Ab})$.

In Algebraic Topology, the simplicial homology groups $H_n(K)$ of a simplicial complex $K$, for $n \geq 0$, are constructed in two steps. The first step is geometric, constructing *simplicial chain groups* $C_n(K)$ and boundary maps, giving a sequence

$$
\mathbf{S}(K) = \to C_{n+1}(K) \xrightarrow{\partial_{n+1}} C_n(K) \xrightarrow{\partial_n} C_{n-1}(K) \to .
$$

The second step is algebraic: define $H_n(K) = \ker \partial_n / \operatorname{im} \partial_{n+1}$; the second step is the basic idea underlying Homological Algebra, as we shall see in Chapter 9.

**Theorem 7.145.** *If $\mathcal{A}$ is an abelian category, then $\mathbf{Comp}(\mathcal{A})$ is also an abelian category.*

**Proof.** Since $\mathbf{PO}(\mathbb{Z})$ is a small category, Proposition 7.144 says that the functor category $\mathcal{A}^{\mathbf{PO}(\mathbb{Z})}$ is abelian. Proposition 7.143 now says that the full subcategory $\mathbf{Comp}(\mathcal{A})$ is abelian if it satisfies several conditions.

(i) The *zero complex* is the complex each of whose terms is 0.

(ii) The *direct sum* $(\mathbf{C}, d) \oplus (\mathbf{C}', d')$ is the complex whose $n$th term is $C_n \oplus C_n'$ and whose $n$th differential is $d_n \oplus d_n'$.

(iii) If $f = (f_n) \colon (\mathbf{C}, d) \to (\mathbf{C}', d')$ is a chain map, define

$$
\mathbf{ker} f = \to \ker f_{n+1} \xrightarrow{\delta_{n+1}} \ker f_n \xrightarrow{\delta_n} \ker f_{n-1} \to,
$$

where $\delta_n = d_n | \ker f_n$, and

$$
\mathbf{im} f = \to \operatorname{im} f_{n+1} \xrightarrow{\Delta_{n+1}} \operatorname{im} f_n \xrightarrow{\Delta_n} \operatorname{im} f_{n-1} \to,
$$

where $\Delta_n = d_n' | \operatorname{im} f_n$.

Since these complexes lie in the full subcategory $\mathbf{Comp}(\mathcal{A})$, Proposition 7.143 applies to prove the theorem.   •

We end this section by describing the Full Imbedding Theorem, which says, for all intents and purposes, that working in abelian categories is the same as working in $\mathbf{Ab}$.

**Definition.** Let $\mathcal{C}, \mathcal{D}$ be categories, and let $F\colon \mathcal{C} \to \mathcal{D}$ be a functor. Then $F$ is **faithful** if, for all $A, B \in \mathrm{obj}(\mathcal{C})$, the functions $\mathrm{Hom}_{\mathcal{C}}(A, B) \to \mathrm{Hom}_{\mathcal{D}}(FA, FB)$, given by $f \mapsto Ff$, are injections; $F$ is **full** if these functions are surjective.

If $\mathcal{A}$ is an abelian category, then a functor $F\colon \mathcal{A} \to \mathbf{Ab}$ is **exact** if $A' \to A \to A''$ exact in $\mathcal{A}$ implies $FA' \to FA \to FA''$ exact in $\mathbf{Ab}$.

**Theorem 7.146 (Freyd–Heron–Lubkin).**[13]   *If $\mathcal{A}$ is a small abelian category, then there is a covariant faithful exact functor $F\colon \mathcal{A} \to \mathbf{Ab}$.*

**Proof.** See Mitchell, *Theory of Categories*, p. 101.   •

This imbedding theorem can be improved so that its image is a *full* subcategory of $\mathbf{Ab}$.

**Theorem 7.147 (Mitchell).** *If $\mathcal{A}$ is a small abelian category, then there is a covariant full faithful exact functor $F\colon \mathcal{A} \to \mathbf{Ab}$.*

**Proof.** Mitchell, *Theory of Categories*, p. 151.   •

In his *Theory of Categories*, Mitchell writes, "Let us say that a statement about a diagram in an abelian category is **categorical** if it states that certain parts of the diagram are or are not commutative, that certain sequences in the diagram are or are not exact, and that certain parts of the diagram are or are not (inverse) limits or (direct) limits. Then we have the following metatheorem."

**Metatheorem.**   *Let $\mathcal{A}$ be a (not necessarily small) abelian category.*

(i) *Let $\Sigma$ be a statement of the form p implies q, where p and q are categorical statements about a diagram in $\mathcal{A}$. If $\Sigma$ is true in $\mathbf{Ab}$, then $\Sigma$ is true in $\mathcal{A}$.*

(ii) *Let $\Sigma'$ be a statement of the form p implies q, where p is a categorical statement concerning a diagram in $\mathcal{A}$, while q states that additional morphisms exist between certain objects in the diagram and that some categorical statement is true of the extended diagram. If the statement can be proved in $\mathbf{Ab}$ by constructing the additional morphisms through diagram chasing, then the statement is true in $\mathcal{A}$.*

**Proof.** See Mitchell, *Theory of Categories*, p. 97.   •

Part (i) follows from the Freyd–Heron–Lubkin Imbedding Theorem. To illustrate, the Five Lemma is true in $\mathbf{Ab}$, as is the $3 \times 3$ Lemma (Exercise 6.87 on page 486), and so they are true in every abelian category.

---

[13]I have been unable to find any data about Heron other than that he was a student at Oxford around 1960.

Part (ii) follows from Mitchell's Full Imbedding Theorem. To illustrate, recall Proposition 6.116: given a commutative diagram of abelian groups with exact rows,

$$
\begin{array}{ccccccc}
A' & \xrightarrow{\ i\ } & A & \xrightarrow{\ p\ } & A'' & \longrightarrow & 0 \\
{\scriptstyle f}\downarrow & & {\scriptstyle g}\downarrow & & \downarrow{\scriptstyle h} & & \\
B' & \xrightarrow{\ j\ } & B & \xrightarrow{\ q\ } & B'' & \longrightarrow & 0
\end{array}
$$

there exists a unique map $h\colon A'' \to B''$ making the augmented diagram commute. Suppose now that the diagram lies in an abelian category $\mathcal{A}$. Applying the imbedding functor $F\colon \mathcal{A} \to \mathbf{Ab}$ of the Full Imbedding Theorem, we have a diagram in $\mathbf{Ab}$ as above, and so there is a homomorphism in $\mathbf{Ab}$, say, $h\colon FA'' \to FB''$, making the diagram commute: $F(q)F(g) = hF(p)$. Since $F$ is a full imbedding, there exists $\eta \in \operatorname{Hom}_{\mathcal{A}}(A'', B'')$ with $h = F(\eta)$; hence, $F(qg) = F(q)F(g) = hF(p) = F(\eta)F(p) = F(\eta p)$. But $F$ is faithful, so that $qg = \eta p$.

## Exercises

\* **7.67.** Prove that neither **Groups** nor **ComRings** is an additive category.

**Hint.** Use Lemma 7.138.

**7.68.** Let $\mathcal{A}$ be an additive category with zero object $0$. If $A \in \operatorname{obj}(\mathcal{A})$, prove that the unique morphism $A \to 0$ and the unique morphism $0 \to A$ are the identity elements of the abelian groups $\operatorname{Hom}_{\mathcal{A}}(A, 0)$ and $\operatorname{Hom}_{\mathcal{A}}(0, A)$.

\* **7.69.** If $\mathcal{A}$ is an additive category and $A \in \operatorname{obj}(\mathcal{A})$, prove that $\operatorname{End}_{\mathcal{A}}(A) = \operatorname{Hom}_{\mathcal{A}}(A, A)$ is a ring with composition as product.

\* **7.70.** Let $\mathcal{S}$ be a subcategory of an additive category $\mathcal{A}$. Prove that $\mathcal{S}$ is an additive category if $\mathcal{S}$ is full, $\mathcal{S}$ contains a zero object of $\mathcal{A}$, and $\mathcal{S}$ contains the direct sum $A \oplus B$ (in $\mathcal{A}$) of all $A, B \in \operatorname{obj}(\mathcal{S})$.

\* **7.71.**  (i) Prove that a function is epic in **Sets** if and only if it is surjective, and that a function is monic in **Sets** if and only if it is injective.

   (ii) Prove that an $R$-map is epic in ${}_R\mathbf{Mod}$ if and only if it is surjective, and that an $R$-map is monic in ${}_R\mathbf{Mod}$ if and only if it is injective.

\* **7.72.**  (i) Let $\mathcal{C}$ be the category of all divisible abelian groups. Prove that the natural map $\mathbb{Q} \to \mathbb{Q}/\mathbb{Z}$ is monic in $\mathcal{C}$. Conclude that $\mathcal{C}$ is a category whose morphisms are functions and in which monomorphisms and injections do not coincide.

   (ii) Let $\mathbf{Top}_2$ be the category of all Hausdorff spaces. If $D \subsetneq X$ is a dense subspace of a space $X$, prove that the inclusion $i\colon D \to X$ is an epimorphism. Conclude that $\mathbf{Top}_2$ is a category whose morphisms are functions and in which epimorphisms and surjections do not coincide.
   **Hint.** Two continuous functions agreeing on a dense subspace of a Hausdorff space must be equal.

**7.73.** Prove, in every abelian category, that the injections of a coproduct are monic and the projections of a product are epic.

\* **7.74.** (i) Prove that every isomorphism in an additive category is both monic and epic.

(ii) Prove that a morphism in an abelian category is an isomorphism if and only if it is both monic and epic.

(iii) Let $R$ be a domain that is not a field, and let $\varphi \colon R \to \mathrm{Frac}(R)$ be given by $r \mapsto r/1$. In **ComRings**, prove that $\varphi$ is both monic and epic, but that $\varphi$ is not an isomorphism.

\* **7.75.** Let $G$ be a (possibly nonabelian) group. If $A$ and $G$ are groups, prove that a homomorphism $\varphi \colon A \to G$ is surjective if and only if it is an epimorphism in **Groups**.

**Hint.** Use Exercise 6.70 on page 460.

**7.76.** State and prove the First Isomorphism Theorem in an abelian category $\mathcal{A}$.

\* **7.77.** (i) Let $\mathcal{S}$ be a full subcategory of an abelian category $\mathcal{A}$ which satisfies the hypotheses of Proposition 7.143. Prove that if $A \xrightarrow{f} B \xrightarrow{g} C$ is an exact sequence in $\mathcal{S}$, then it is an exact sequence in $\mathcal{A}$.

(ii) If $\mathcal{A}$ is an abelian category and $\mathcal{C}$ is a small category, prove that if $A \xrightarrow{f} B \xrightarrow{g} C$ is an exact sequence in $\mathcal{A}$, then it is an exact sequence in $\mathcal{A}^{\mathcal{C}}$.

**7.78.** Prove that every object in **Sets** is projective and injective (where we use injections and surjections in the definitions of projective and injective).

\* **7.79.** Let $\mathcal{T}$ be the category of all torsion abelian groups.

(i) Prove that $\mathcal{T}$ is an abelian category having infinite coproducts.

(ii) Prove that $\mathcal{T}$ has infinite products.
**Hint.** If $(A_i)_{i \in I}$ is a family of groups in $\mathcal{T}$, prove that $t\left(\prod_{i \in I} A_i\right)$ is a categorical product.

(iii) Prove that $\mathcal{T}$ has enough injectives.

(iv) Prove that $\mathcal{T}$ has no nonzero projective objects. Conclude that $\mathcal{T}$ is not isomorphic to a category of modules.

\* **7.80.** Prove that $\mathcal{T}'$, the abelian category of all finitely generated abelian groups, is an abelian category that has no nonzero injectives.

**7.81.** A direct limit $\varinjlim_I F$ or an inverse limit $\varprojlim_I F$ is called **finite** if the index set $I$ is finite.

Prove that if $\mathcal{A}$ is an additive category having kernels and cokernels, then $\mathcal{A}$ has all finite inverse limits and direct limits. Conclude that $\mathcal{A}$ has pullbacks and pushouts.

## Section 7.9. Module Categories

When is a category isomorphic to a module category $\mathbf{Mod}_R$?

**Definition.** A functor $F \colon \mathcal{C} \to \mathcal{D}$ is an **isomorphism** if there exists a functor $G \colon \mathcal{D} \to \mathcal{C}$ with both composites $GF$ and $FG$ being identity functors.

Every category is isomorphic to itself; Exercise 7.85 on page 634 shows that if $R$ is a ring with opposite ring $R^{\mathrm{op}}$, then $\mathbf{Mod}_R$ is isomorphic to $_{R^{\mathrm{op}}}\mathbf{Mod}$.

Empirically, isomorphism of functors turns out to be uninteresting, for it does not arise very often. The following example suggests another reason for modifying this notion of isomorphism. Consider the category $\mathcal{V}$ of all finite-dimensional vector spaces over a field $k$ and its full subcategory $\mathcal{W}$ generated by all vector spaces equal to $k^n$ for $n \in \mathbb{N}$. Since functors take identity morphisms to identity morphisms, an isomorphism $F \colon \mathcal{V} \to \mathcal{W}$ would give a bijection $\mathrm{obj}(\mathcal{V}) \to \mathrm{obj}(\mathcal{W})$. But $\mathcal{W}$ is a small category ($|\mathrm{obj}(\mathcal{W})| = \aleph_0$) while $\mathcal{V}$ is not small, and so these categories are not isomorphic. Carefully distinguishing between two such categories does not seem to be a worthy enterprise.

Here is a weaker but more useful definition.

**Definition.** A functor $F \colon \mathcal{C} \to \mathcal{D}$ is an **equivalence** if there is a functor $G \colon \mathcal{D} \to \mathcal{C}$ such that $GF$ and $FG$ are naturally isomorphic to the identity functors $1_{\mathcal{C}}$ and $1_{\mathcal{D}}$, respectively. When $\mathcal{C}$ and $\mathcal{D}$ are abelian categories, we will further assume that an equivalence $F \colon \mathcal{C} \to \mathcal{D}$ is an additive functor.

It is easy to see that the two nonisomorphic categories $\mathcal{V}$ and $\mathcal{W}$ of finite-dimensional vector spaces are equivalent.

**Proposition 7.148.** *A functor $F \colon \mathcal{C} \to \mathcal{D}$ is an equivalence if and only if*

(i) $F$ *is full and faithful: the function* $\mathrm{Hom}_{\mathcal{C}}(C, C') \to \mathrm{Hom}_{\mathcal{D}}(FC, FC')$, *given by* $f \mapsto Ff$, *is a bijection for all* $C, C' \in \mathrm{obj}(\mathcal{C})$;

(ii) *every* $D \in \mathrm{obj}(\mathcal{D})$ *is isomorphic to* $FC$ *for some* $C \in \mathrm{obj}(\mathcal{C})$.

**Proof.** If $F$ is an equivalence, then there exists a functor $G \colon \mathcal{D} \to \mathcal{C}$ with $GF \cong 1_{\mathcal{C}}$ and $FG \cong 1_{\mathcal{D}}$; let $\tau \colon GF \to 1_{\mathcal{C}}$ and $\sigma \colon FG \to 1_{\mathcal{D}}$ be natural isomorphisms. For each $D \in \mathrm{obj}(\mathcal{D})$, there is an isomorphism $\sigma_D \colon FGD \to D$. Thus, if we define $C = GD$, then $FC \cong D$, which proves (ii).

Given a morphism $f \colon C \to C'$ in $\mathcal{C}$, there is a commutative diagram

$$
\begin{array}{ccc}
GFC & \xrightarrow{\ \tau_C\ } & C \\
{\scriptstyle GFf}\downarrow & & \downarrow{\scriptstyle f} \\
GFC' & \xrightarrow[\ \tau_{C'}\ ]{} & C'
\end{array}
$$

Since $\tau$ is a natural isomorphism, each $\tau_C$ is an isomorphism; hence,

$$(1) \qquad\qquad f = \tau_{C'}(GFf)\tau_C^{-1}.$$

$F$ is faithful: if $f, f' \in \mathrm{Hom}_{\mathcal{C}}(C, C')$ and $Ff' = Ff$ in $\mathrm{Hom}_{\mathcal{D}}(FC, FC')$, then

$$f' = \tau_{C'}(GFf')\tau_C^{-1} = \tau_{C'}(GFf)\tau_C^{-1} = f.$$

Similarly, $FG \cong 1_{\mathcal{D}}$ implies that $G$ is faithful.

Finally, $F$ is full: if $g \colon FC \to FC'$, define a morphism $f = \tau_{C'}(Gg)\tau_C^{-1}$. Now $f = \tau_{C'}(GFf)\tau_C^{-1}$, by Equation (1), so that $GFf = Gg$. Since $G$ is faithful, we have $Ff = g$.

Conversely, assume that $F \colon \mathcal{C} \to \mathcal{D}$ satisfies (i) and (ii). For each $D \in \mathrm{obj}(\mathcal{D})$, (ii) gives a unique $C = C_D \in \mathrm{obj}(\mathcal{C})$ with $D \cong FC_D$; choose an isomorphism

$h_D \colon D \to FC_D$. Define a functor $G \colon \mathcal{D} \to \mathcal{C}$ on objects by $GD = C_D$. If $g \colon D \to D'$ is a morphism in $\mathcal{D}$, (i) gives a unique morphism $f \colon C_D \to C_{D'}$ with $Ff = h_{D'}gh_D^{-1}$. It is routine to check that $G$ is a functor, $GF \cong 1_{\mathcal{C}}$, and $FG \cong 1_{\mathcal{D}}$.   •

**Example 7.149.**

(i) If $R$ is a ring with opposite ring $R^{\mathrm{op}}$, then $\mathbf{Mod}_R$ is equivalent to $_{R^{\mathrm{op}}}\mathbf{Mod}$, for we have already observed that these categories are isomorphic.

(ii) If $R$ is a ring, then $\mathbf{Mod}_R$ and $_R\mathbf{Mod}$ need *not* be equivalent (Exercise 7.87 on page 634).

(iii) If $\mathcal{V}$ is the category of all finite-dimensional vector spaces over a field $k$, then *double dual* $F \colon \mathcal{V} \to \mathcal{V}$, sending $V \mapsto V^{**}$, is an equivalence ($V^*$ is the dual space), for $F$ satisfies the conditions in Proposition 7.148.

(iv) If $\mathcal{C}$ is a category, let $\mathcal{S} \subseteq \mathrm{obj}(\mathcal{C})$ consist of one object from each isomorphism class of objects. The full subcategory generated by $\mathcal{S}$ (also denoted by $\mathcal{S}$) is called a *skeletal subcategory* of $\mathcal{C}$. The inclusion functor $\mathcal{S} \to \mathcal{C}$ is an equivalence, by Proposition 7.148; thus, every category $\mathcal{C}$ is equivalent to a skeletal subcategory. For example, if $\mathcal{V}$ is the category of all finite-dimensional vector spaces over a field $k$, then the full category $\mathcal{W}$ of $\mathcal{V}$ generated by all $k^n$ for $n \in \mathbb{N}$ is a skeletal subcategory. Hence, $\mathcal{V}$ and $\mathcal{W}$ are equivalent.   ◄

We rephrase our original question. When is a category equivalent to a module category $\mathbf{Mod}_R$? The answer will place the Wedderburn–Artin Theorems in perspective.

We know that $\mathbf{Mod}_R$, for any ring $R$, is an abelian category; it also has arbitrary direct sums.

**Definition.** An abelian category $\mathcal{A}$ is *cocomplete* if it contains $\bigoplus_{i \in I} A_i$ for every family $(A_i)_{i \in I}$ of objects, where the index set $I$ may be infinite.

We now describe some categorical properties of the object $R$ in $\mathbf{Mod}_R$.

**Definition.** An object $P$ in a cocomplete abelian category $\mathcal{A}$ is *small* if the covariant Hom functor $\mathrm{Hom}_{\mathcal{A}}(P, \square) \colon \mathcal{A} \to \mathbf{Ab}$ preserves all coproducts; that is, if $(B_i)_{i \in I}$ is an indexed family of objects and $(j_i)_{i \in I}$, $(p_i)_{i \in I}$ are the injections, projections of $\bigoplus_{i \in I} B_i$, then there is an isomorphism $\theta \colon \mathrm{Hom}_{\mathcal{A}}(A, \bigoplus_{i \in I} B_i) \to \bigoplus_{i \in I} \mathrm{Hom}_{\mathcal{A}}(A, B_i)$ whose injections and projections are $(j_i \theta)_{i \in I}$ and $(\theta p_i)_{i \in I}$.

**Example 7.150.**

(i) Proposition 6.64 shows that the ring $R$ is a small $R$-module.

(ii) Every finite direct sum of small modules is small, and every direct summand of a small module is small.

(iii) Since a ring $R$ is a small $R$-module, it follows from (i) and (ii) that every finitely generated projective $R$-module is small.   ◄

The object $R$ in $\mathbf{Mod}_R$ is not only small; it is projective. If $P$ is a small projective object, then $\mathrm{Hom}_{\mathcal{A}}(P, \square)$ is a right exact functor (in fact, it is an exact functor).

**Definition.** An object $P$ in a cocomplete abelian category $\mathcal{A}$ is a **generator of** $\mathcal{A}$ if every $M \in \mathrm{obj}(\mathcal{A})$ is a quotient of a direct sum of copies of $P$.

It is clear that $R$ is a generator of $\mathbf{Mod}_R$, as is any free right $R$-module. However, a projective right $R$-module may not be a generator. For example, if $R = \mathbb{I}_6$, then $R = P \oplus Q$, where $P = \{[0], [2], [4]\} \cong \mathbb{I}_3$, and the (small) projective module $P$ is not a generator (for $Q \cong \mathbb{I}_2$ is not a quotient of a direct sum of copies of $P$). You will prove, in Exercise 7.86 on page 634, that small projective generators of $\mathbf{Mod}_R$ are finitely generated.

Recall that a functor $F \colon \mathcal{A} \to \mathcal{B}$ is *faithful* if, for all $A, A' \in \mathrm{obj}(\mathcal{A})$, the function $\mathrm{Hom}_{\mathcal{A}}(A, A') \to \mathrm{Hom}_{\mathcal{B}}(FA, FA')$, given by $\varphi \mapsto F\varphi$, is injective.

**Proposition 7.151.** *A right $R$-module $P$ is a generator of $\mathbf{Mod}_R$ if and only if $\mathrm{Hom}_R(P, \square)$ is a faithful functor.*

**Proof.** Assume that $\mathrm{Hom}_R(P, \square)$ is faithful. Given a right $R$-module $A$ and a map $f \colon P \to A$, let $P_f$ be an isomorphic copy of $P$, and let $Y = \bigoplus_{f \in \mathrm{Hom}_R(P, A)} P_f$. Define $\varphi \colon Y \to A$ by $(g_f) \mapsto \sum_f f(g_f)$. If $\varphi$ is not surjective, then the natural map $\nu \colon A \to A/\mathrm{im}\,\varphi$ is nonzero. Since $\mathrm{Hom}_R(P, \square)$ is faithful, $\nu_* \colon \mathrm{Hom}_R(P, A) \to \mathrm{Hom}_R(P, A/\mathrm{im}\,\varphi)$ is also nonzero. Thus, there is $f \in \mathrm{Hom}_R(P, A)$ such that $\nu_*(f) = \nu f \neq 0$. But $f(A) \subseteq \mathrm{im}\,\varphi$ so that $\nu f(A) = \{0\}$; that is, $\nu_* f = \nu f = 0$, a contradiction. Therefore, $P$ is a generator.

Conversely, assume that every module $A$ is a quotient of a direct sum $Y = \bigoplus_{i \in I} P_i$, where $P_i \cong P$ for all $i$; say, there is a surjective $\varphi \colon Y \to A$. Now $\varphi = \sum_i \varphi \lambda_i$, where $(\lambda_i \colon P_i \to Y)_{i \in I}$ are the injections. If $\alpha \colon A \to A'$ is nonzero, then $\alpha \varphi = \alpha \sum_i \varphi \lambda_i = \sum_i (\alpha \varphi \lambda_i)$. Now $\alpha \varphi \neq 0$, because $\varphi$ is surjective. Hence, $\alpha \varphi \lambda_i \neq 0$ for some $i$. But $P_i \cong P$, so that $0 \neq \alpha \varphi \lambda_i = \alpha_*(\varphi \lambda_i)$; that is, $\alpha_* \neq 0$, and so $\mathrm{Hom}_R(P, \square)$ is faithful. $\bullet$

Here is the characterization of module categories.

**Theorem 7.152 (Gabriel–Mitchell).** *A category $\mathcal{A}$ is equivalent to a module category $\mathbf{Mod}_R$ if and only if $\mathcal{A}$ is a cocomplete abelian category having a small projective generator*[14] *$P$. Moreover, $R \cong \mathrm{End}_{\mathcal{A}}(P)$ in this case.*

**Proof.** The proof of necessity is easy: $\mathbf{Mod}_R$ is a cocomplete abelian category and $R$ is a small projective generator.

For the converse, define $F = \mathrm{Hom}_{\mathcal{A}}(P, \square) \colon \mathcal{A} \to \mathbf{Ab}$. Note that $F$ is additive, by Example 7.137, and that $R = \mathrm{End}_{\mathcal{A}}(P)$ is a ring, by Exercise 7.85 on page 634. For each $A \in \mathrm{obj}(\mathcal{A})$, we claim that $FA$ is a right $R$-module. If $f \in FA = \mathrm{Hom}_{\mathcal{A}}(P, A)$ and $\varphi \colon P \to P$ lies in $R = \mathrm{End}(P)$, define scalar multiplication $f\varphi$ to be the composite $P \xrightarrow{\varphi} P \xrightarrow{f} A$. It is routine to check that $F$ actually takes values in $\mathbf{Mod}_R$.

Let us prove that $F$ is an equivalence. Now $F = \mathrm{Hom}_{\mathcal{A}}(P, \square) \colon \mathcal{A} \to \mathbf{Mod}_R$ [where $R = \mathrm{End}(P)$] is faithful, by Proposition 7.151. It remains to prove, by Proposition 7.148, that $F$ is full [the maps $\mathrm{Hom}_{\mathcal{A}}(Y, X) \to \mathrm{Hom}_R(FY, FX)$, given

---

[14] A small projective generator is often called a **progenerator**.

by $\varphi \mapsto F\varphi$, are all surjective] and that every $M \in \mathrm{obj}(\mathbf{Mod}_R)$ is isomorphic to $FA$ for some $A \in \mathrm{obj}(\mathcal{A})$.

For fixed $Y \in \mathrm{obj}(\mathcal{A})$, define the class

$$\mathcal{E} = \mathcal{E}_Y = \{X \in \mathrm{obj}(\mathcal{A}) : \mathrm{Hom}_{\mathcal{A}}(X, Y) \to \mathrm{Hom}_R(FX, FY) \text{ is surjective}\}.$$

We will prove three properties of $\mathcal{E}$.

(i)  $P \in \mathcal{E}$.

(ii) If $(X_i)_{i \in I}$ is a family of objects in $\mathcal{E}$, then $\bigoplus_{i \in I} X_i \in \mathcal{E}$.

(iii) If $X, Z \in \mathcal{E}$ and $f \colon X \to Z$ is any morphism, then $\mathrm{coker}\, f \in \mathcal{E}$.

These properties imply $\mathcal{E} = \mathrm{obj}(\mathcal{A})$. By (i) and (ii), every direct sum of copies of $P$ lies in $\mathcal{E}$; since $P$ is a generator of $\mathcal{A}$, every $Z \in \mathrm{obj}(\mathcal{A})$ is a cokernel of $\bigoplus_{i \in I} P_i \to \bigoplus_{j \in J} P_j$, where all $P_i, P_j$ are isomorphic to $P$. Thus, $Z \in \mathcal{E}_Y$, which says that $\mathrm{Hom}_{\mathcal{A}}(Z, Y) \to \mathrm{Hom}_{FP}(FZ, FY)$ is surjective; that is, $F$ is full. We now verify these three properties.

To see that $P \in \mathcal{E}_Y$, we must show that $\mathrm{Hom}_{\mathcal{A}}(P, Y) \to \mathrm{Hom}_R(FP, FY)$ is surjective. Since $P$ is a generator of $\mathcal{A}$, there is an exact sequence

$$(2) \qquad \bigoplus_{i \in I} P_i \to \bigoplus_{j \in J} P_j \to Y \to 0$$

which gives the commutative diagram (details below)

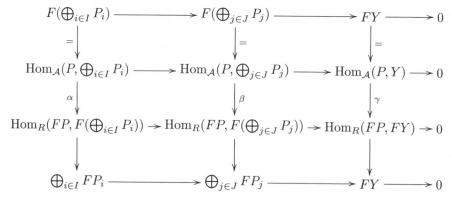

Now $F = \mathrm{Hom}_{\mathcal{A}}(P, \square)$ is an exact functor (because $P$ is projective), and so the top two rows arise by applying $F$ to (2); the vertical maps between these rows are identities. The third row arises from applying $\mathrm{Hom}_R(FP, \square)$ to the top row; the vertical maps are the maps $\mathrm{Hom}(X, Y) \to \mathrm{Hom}(FX, FY)$ given by $\varphi \mapsto F\varphi$. The bottom row arises from the third row, for $F$ preserves direct sums (because $P$ is small), and $\mathrm{Hom}_R(FP, FY) = \mathrm{Hom}_R(R, FY) = FY$. Finally, since $\alpha$ and $\beta$ are surjections, so is $\gamma$ (use the Five Lemma, by adding $\to 0$ at the ends of the middle two rows).

For (ii), let $(X_i)_{i \in I}$ be a family for which all $\mathrm{Hom}(X_i, Y) \to \mathrm{Hom}(FX_i, FY)$ are surjections. To see that $\mathrm{Hom}(\bigoplus X_i, Y) \to \mathrm{Hom}(F(\bigoplus X_i), FY)$ is surjective, use the facts that $\mathrm{Hom}(\bigoplus X_i, Y) \cong \prod \mathrm{Hom}(X_i, Y)$ and $\mathrm{Hom}(F(\bigoplus X_i), FY) \cong \mathrm{Hom}(\bigoplus FX_i, FY) \cong \prod \mathrm{Hom}(FX_i, FY)$ (because $F$ preserves direct sums).

For (iii), use the Five Lemma on a variant of the commutative diagram above. We conclude that $F$ is full.

Lastly, for every right $R$-module $M$, we show that there is $A \in \text{obj}(\mathcal{A})$ with $M \cong FA$. There is an exact sequence $\bigoplus_{i \in I} R_i \xrightarrow{f} \bigoplus_{j \in J} R_j \to M \to 0$ in $\textbf{Mod}_R$, where $R_i, R_j$ are isomorphic to $R$. If we view each $R_i, R_j$ as $FP_i, FP_j$, where all $P_i, P_j$ are isomorphic to $P$, then

$$f \in \text{Hom}(\bigoplus_{i \in I} R_i, \bigoplus_{j \in J} R_j) = \text{Hom}(\bigoplus_{i \in I} FP_i, \bigoplus_{j \in J} FP_j)$$

$$= \text{Hom}(F(\bigoplus_{i \in I} P_i), F(\bigoplus_{j \in J} P_j)).$$

Since $F$ is full, there is $\varphi \in \text{Hom}(\bigoplus_{i \in I} P_i, \bigoplus_{j \in J} P_j)$ with $F\varphi = f$. Using the Five Lemma again, the reader may show that $M \cong F(\text{coker}\,\varphi)$. •

**Corollary 7.153.** *If $R$ is a ring and $n \geq 1$, there is an equivalence of categories*

$$\textbf{Mod}_R \cong \textbf{Mod}_{\text{Mat}_n(R)}.$$

**Proof.** For any integer $n \geq 1$, the free module $P = \bigoplus_{i=1}^{n} R_i$, where $R_i \cong R$, is a small projective generator of $\textbf{Mod}_R$. Theorem 7.152 gives an equivalence $\textbf{Mod}_R \cong \textbf{Mod}_S$, where $S = \text{End}_R(P) \cong \text{Mat}_n(R)$. •

We can now understand Proposition 7.37: $\text{Mat}_n(\Delta)$ is semisimple when $\Delta$ is a division ring. By Proposition 7.33, a ring $R$ is semisimple if and only if every $R$-module is projective; that is, every object in $\textbf{Mod}_R$ is projective. But every $\Delta$-module is projective (even free), so that equivalence of the categories shows that every object in $\textbf{Mod}_{\text{Mat}_n(\Delta)}$ is also projective. Therefore, $\text{Mat}_n(\Delta)$ is semisimple. We have just seen why matrix rings arise in the study of semisimple rings.

Given rings $R$ and $S$, Corollary 7.153 raises the question when $\textbf{Mod}_R$ and $\textbf{Mod}_S$ are equivalent. The answer is provided by **Morita Theory**, which arose by analyzing the proof of Theorem 7.152. We merely report the main results; for details, see Jacobson, *Basic Algebra* II, pp. 177–184, Lam, *Lectures on Modules and Rings*, Chapters 18 and 19, McConnell–Robson, *Noncommutative Noetherian Rings*, Chapter 3, §5, Reiner, *Maximal Orders*, Chapter 4, and Rowen, *Ring Theory* I, Chapter 4.

**Definition.** Call rings $R$ and $S$ **Morita equivalent** if their module categories $\textbf{Mod}_R$ and $\textbf{Mod}_S$ are equivalent.

For example, Corollary 7.153 says that every ring $R$ is Morita equivalent to the matrix ring $\text{Mat}_n(R)$, where $n \geq 1$.

Given a ring $R$, every right $R$-module $P$ determines a **Morita context**

$$(P, R, Q, S, \alpha, \beta).$$

Here, $Q = \text{Hom}_R(P, R)$ and $S = \text{End}_R(P)$. Both $P$ and $Q$ turn out to be bimodules: $P = {}_S P_R$ and $Q = {}_R Q_S$, and there is an $(R, R)$-map $\alpha: Q \otimes_S P \to R$ and an $(R, R)$-map $\beta: P \otimes_R Q \to S$. When $P_R$ is a small projective generator, both $\alpha$ and $\beta$ are isomorphisms; in this case, $Q_S$ is also a small projective generator.

**Theorem 7.154 (Morita I).** *Let $P_R$ be a small projective generator with Morita context $(P, R, Q, S, \alpha, \beta)$.*

(i) $\mathrm{Hom}_R(P, \square)\colon \mathbf{Mod}_R \to \mathbf{Mod}_S$ *is an equivalence, and its inverse is* $\mathrm{Hom}_S(Q, \square)\colon \mathbf{Mod}_S \to \mathbf{Mod}_R$.

(ii) $\mathrm{Hom}_R(Q, \square)\colon {}_R\mathbf{Mod} \to {}_S\mathbf{Mod}$ *is an equivalence and its inverse is* $\mathrm{Hom}_S(P, \square)\colon {}_S\mathbf{Mod} \to {}_R\mathbf{Mod}$.

**Proof.** Lam, *Lectures on Modules and Rings*, states and proves this with tensor products, using $\mathrm{Hom}_R(P, \square) \cong \square \otimes_R Q$ and $\mathrm{Hom}_S(Q, \square) \cong \square \otimes_S P$ in part (i) and $\mathrm{Hom}_S(P, \square) \cong Q \otimes_S \square$ and $\mathrm{Hom}_R(Q, \square) \cong P \otimes_R \square$ in part (ii).   •

**Theorem 7.155 (Morita II).** *Let $R$ and $S$ be rings, and let $F\colon \mathbf{Mod}_R \to \mathbf{Mod}_S$ be an equivalence with inverse $G\colon \mathbf{Mod}_S \to \mathbf{Mod}_R$. Then $F$ and $G$ arise as in Morita I; that is, $F \cong \mathrm{Hom}_R(P, \square)$ and $G \cong \mathrm{Hom}_S(Q, \square)$, where $P = G(S)$ and $Q = F(R)$.*

**Corollary 7.156.** $\mathbf{Mod}_R$ *and* $\mathbf{Mod}_S$ *are equivalent if and only if* ${}_R\mathbf{Mod}$ *and* ${}_S\mathbf{Mod}$ *are equivalent.*

Exercise 7.85 on page 634 shows that $\mathbf{Mod}_R$ and ${}_{R^{op}}\mathbf{Mod}$ are equivalent, but Exercise 7.87 shows that $\mathbf{Mod}_R$ and ${}_R\mathbf{Mod}$ may not be equivalent.

**Corollary.** *Two rings $R$ and $S$ are Morita equivalent if and only if $S \cong \mathrm{End}_R(P)$ for some small projective generator $P$ of* $\mathbf{Mod}_R$.

If $\mathcal{A}$ is a category, then an **endomorphism** of the identity functor $1_\mathcal{A}$ is a natural transformation $\tau\colon 1_\mathcal{A} \to 1_\mathcal{A}$; that is, for every pair of objects $A$ and $B$ and every morphism $f\colon A \to B$, there is a commutative diagram

$$
\begin{array}{ccc}
A & \xrightarrow{\ \tau_A\ } & A \\
{\scriptstyle f}\downarrow & & \downarrow{\scriptstyle f} \\
B & \xrightarrow[\ \tau_B\ ]{} & B
\end{array}
$$

**Definition.** If $R$ is a ring, define

$$\mathrm{End}(\mathbf{Mod}_R) = \{\tau\colon 1_{\mathbf{Mod}_R} \to 1_{\mathbf{Mod}_R} \mid \tau \text{ is an endomorphism}\}.$$

It is easy to see that the pointwise sum of two endomorphisms is another such; that is, define $\tau + \sigma$ as the family $(\tau_A + \sigma_A)_{A \in \mathbf{Mod}_R}$. It should follow that $\mathrm{End}(\mathbf{Mod}_R)$ is a ring with composition as multiplication, but it is not obvious whether $\mathrm{End}(\mathbf{Mod}_R)$ is a set.

**Proposition 7.157.** *For any ring $R$, there is a ring isomorphism*

$$Z(R) \cong \mathrm{End}(\mathbf{Mod}_R).$$

**Proof.** If $c \in Z(R)$ and $A$ is a right $R$-module, define $\tau_A^c\colon A \to A$ to be multiplication by $c$:

$$\tau_A^c\colon a \mapsto ac.$$

Since $c \in Z(R)$, the function $\tau_A^c$ is an $R$-map. It is easily checked that $\tau^c = (\tau_A^c)_{A \in \mathbf{Mod}_R}$ is an endomorphism of $1_{\mathbf{Mod}_R}$. Define $\varphi \colon Z(R) \to \mathrm{End}(\mathbf{Mod}_R)$ by

$$\varphi \colon c \mapsto \tau^c = (\tau_A^c)_{A \in \mathbf{Mod}_R}.$$

We claim that $\varphi$ is a bijection [so that $\mathrm{End}(\mathbf{Mod}_R)$ is a set] and a ring isomorphism. The only point which is not obvious is whether $\varphi$ is surjective. Let $\sigma$ be an endomorphism of $1_A$ and let $A$ be a right $R$-module. If $a \in A$, define $f \colon R \to A$ by $f(1) = a$. There is a commutative diagram

$$\begin{array}{ccc} R & \xrightarrow{\;\sigma_R\;} & R \\ {\scriptstyle f}\downarrow & & \downarrow{\scriptstyle f} \\ A & \xrightarrow[\;\sigma_A\;]{} & A \end{array}$$

Define $c = \sigma_R(1)$. Now $f\sigma_R(1) = f(c) = f(1 \cdot c) = f(1)c = ac$. On the other hand, $\sigma_A f(1) = \sigma_A(a)$. Commutativity gives $\sigma_A(a) = ac$; that is, $\sigma_A = \tau_A^c$. •

**Corollary 7.158.** *Let $R$ and $S$ be rings.*

(i) *If $R$ and $S$ are Morita equivalent, then $Z(R) \cong Z(S)$.*

(ii) *If $R$ and $S$ are commutative, then $\mathbf{Mod}_R$ and $\mathbf{Mod}_S$ are equivalent if and only if $R \cong S$.*

**Proof.**

(i) If $\mathcal{A}$ and $\mathcal{B}$ are equivalent abelian categories, then $\mathrm{End}(1_{\mathcal{A}}) \cong \mathrm{End}(1_{\mathcal{B}})$. If $\mathcal{A} = \mathbf{Mod}_R$ and $\mathcal{B} = \mathbf{Mod}_S$, then $Z(R) \cong Z(S)$, by Proposition 7.157.

(ii) Sufficiency is obvious. For necessity, part (i) gives $Z(R) \cong Z(S)$. Since $R$ and $S$ are commutative, $R = Z(R) \cong Z(S) = S$. •

## Exercises

**7.82.** Let $F \colon \mathcal{C} \to \mathcal{D}$ be an *isomorphism* of categories with inverse $G \colon \mathcal{D} \to \mathcal{C}$; that is, $GF = 1_{\mathcal{C}}$ and $FG = 1_{\mathcal{D}}$. Prove that both $(F, G)$ and $(G, F)$ are adjoint pairs.

**7.83.** If $F \colon \mathcal{A} \to \mathcal{B}$ is an equivalence of abelian categories, prove the following statements.

(i) If $f$ is monic in $\mathcal{A}$, then $Ff$ is monic in $\mathcal{B}$.

(ii) If $f$ is epic in $\mathcal{A}$, then $Ff$ is epic in $\mathcal{B}$.

(iii) If $A \in \mathrm{obj}(\mathcal{A})$, then $f \mapsto Ff$ is a ring isomorphism $\mathrm{End}(A) \to \mathrm{End}(FA)$.

(iv) If $0 \to A' \to A \to A'' \to 0$ is exact in $\mathcal{A}$, then $0 \to FA' \to FA \to FA'' \to 0$ is exact in $\mathcal{B}$. Moreover, the first sequence is split if and only if the second sequence is split.

(v) If $(A_i)_{i \in I}$ is a family of objects in $\mathcal{A}$, then

$$F\Big(\bigoplus_{i \in I} A_i\Big) \cong \bigoplus_{i \in I} FA_i \quad \text{and} \quad F\Big(\prod_{i \in I} A_i\Big) \cong \prod_{i \in I} FA_i.$$

(vi) If $P$ is projective in $\mathcal{A}$, then $FP$ is projective in $\mathcal{B}$.

(vii) If $P$ is injective in $\mathcal{A}$, then $FP$ is injective in $\mathcal{B}$.

**7.84.** Let $F\colon \mathbf{Mod}_R \to \mathbf{Mod}_S$ be an equivalence. Prove that a right $R$-module $A$ has any of the following properties if and only if $FA$ does: simple; semisimple; ACC; DCC; indecomposable. Moreover, $A$ has a composition series if and only if $FA$ does; both have the same length, and $S_1, \ldots, S_n$ are composition factors of $A$ if and only if $FS_1, \ldots, FS_n$ are composition factors of $FA$.

∗ **7.85.** If $R$ is a ring with opposite ring $R^{\mathrm{op}}$, prove that $\mathbf{Mod}_R$ is equivalent to $_{R^{\mathrm{op}}}\mathbf{Mod}$.

**Hint.** For each right $R$-module $M$, show that there is an isomorphism $\tau$ with $\tau(M) = M'$, where $M'$ is $M$ made into a left $R$-module as in Proposition 6.15.

∗ **7.86.** Prove that every small projective generator $P$ of $\mathbf{Mod}_R$ is finitely generated.

∗ **7.87.**   (i) Let $R$ and $S$ be rings, and let $F\colon \mathbf{Mod}_R \to \mathbf{Mod}_S$ (or $F\colon \mathbf{Mod}_R \to {}_S\mathbf{Mod}$) be an equivalence. Prove that a right $R$-module $M$ is finitely generated if and only if $FM$ is a finitely generated right (or left) $S$-module.

   **Hint.** Use Exercise 6.17 on page 416.

   (ii) Call a category $\mathbf{Mod}_R$ (or $_R\mathbf{Mod}$) *noetherian* if every submodule of a finitely generated right (or left) $R$-module $M$ is finitely generated.

   Let $\mathcal{A}$ and $\mathcal{B}$ be equivalent categories of modules; that is, $\mathcal{A}$ is equivalent to $\mathbf{Mod}_R$ or $_R\mathbf{Mod}$ for some ring $R$, and $\mathcal{B}$ is equivalent to $\mathbf{Mod}_S$ or $_S\mathbf{Mod}$ for some ring $S$. Prove that $\mathcal{A}$ is noetherian if and only if $\mathcal{B}$ is noetherian.

   (iii) Prove that $\mathbf{Mod}_R$ is a noetherian category if and only if $R$ is a right noetherian ring, and that $_R\mathbf{Mod}$ is a noetherian category if and only if $R$ is a left noetherian ring.

   (iv) Give an example of a ring $R$ such that $_R\mathbf{Mod}$ and $\mathbf{Mod}_R$ are not equivalent.
   **Hint.** Let $R$ be the ring in Exercise 6.16 on page 416 which is left noetherian but not right noetherian.

# Chapter 8

# Advanced Linear Algebra

This chapter begins with the study of modules over PIDs, including characterizations of their projective, injective, and flat modules. Our emphasis, however, is on finitely generated modules, because the generalization of the Fundamental Theorem of Finite Abelian Groups, when applied to such $k[x]$-modules, yields the rational and Jordan canonical forms for matrices. The Smith normal form is also discussed, for it can be used to compute the invariants of a matrix. We then consider bilinear and quadratic forms on a vector space, which lead to symplectic and orthogonal groups. Multilinear algebra is the next step, leading to tensor algebras, exterior algebras, and determinants. We end with an introduction to Lie algebras, which can be viewed as a way of dealing with certain families of linear transformations instead of with individual ones.

## Section 8.1. Modules over PIDs

The structure theorems for finite abelian groups will now be generalized to finitely generated modules over PIDs. As we have just said, this is not mere generalization for its own sake; the module version will yield canonical forms for matrices. Not only do the theorems generalize, but the proofs of the theorems generalize as well, as we shall see.

In Chapter 7, we defined the annihilator ideal of an element of a module. This ideal has another name when $R$ is commutative.

**Definition.** Let $R$ be a commutative ring, and let $M$ be an $R$-module. If $m \in M$, then its **order ideal** (or **annihilator**) is

$$\mathrm{ann}(m) = \{r \in R : rm = 0\}.$$

We say that $m$ has **finite order** (or is a *torsion element*) if $\mathrm{ann}(m) \neq (0)$; otherwise, $m$ has **infinite order**.

When a commutative ring $R$ is regarded as a module over itself, its identity element 1 has infinite order, for $\operatorname{ann}(1) = (0)$.

Order ideals generalize the group-theoretic notion of the order of an element. Let $G$ be an additive abelian group, and let $g \in G$ have finite order $d$. We claim that the principal ideal $(d)$ in $\mathbb{Z}$ coincides with $\operatorname{ann}(g)$ when $G$ is viewed as a $\mathbb{Z}$-module. If $k \in \operatorname{ann}(g)$, then $kg = 0$; thus, $d \mid k$, by Proposition 1.26, and so $k \in (d)$. For the reverse inclusion, if $n \in (d)$, then $n = ad$ for some $a \in \mathbb{Z}$; hence, $ng = adg = 0$, and so $n \in \operatorname{ann}(g)$.

If $M = \langle m \rangle$ is a cyclic $R$-module, define $\varphi \colon R \to M$ by $r \mapsto rm$. Then $\varphi$ is surjective, $\ker \varphi = \operatorname{ann}(m)$, and the First Isomorphism Theorem gives

$$M = \langle m \rangle \cong R/\operatorname{ann}(m).$$

For the remainder of this section, $R$ will be a domain (indeed, it will soon be restricted even further).

**Definition.** If $M$ is an $R$-module, where $R$ is a domain,[1] then its **torsion submodule** $tM$ is defined by

$$tM = \{m \in M : m \text{ has finite order}\}.$$

**Proposition 8.1.** *If $R$ is a domain and $M$ is an $R$-module, then $tM$ is a submodule of $M$.*

**Proof.** If $m, m' \in tM$, then there are nonzero elements $r, r' \in R$ with $rm = 0$ and $r'm' = 0$. Clearly, $rr'(m + m') = 0$. Since $R$ is a domain, $rr' \neq 0$, and so $\operatorname{ann}(m + m') \neq (0)$; therefore, $m + m' \in tM$.

Let $m \in tM$ and $r \in \operatorname{ann}(m)$, where $r \neq 0$. If $s \in R$, then $sm \in tM$, for $r \in \operatorname{ann}(sm)$ because $r(sm) = s(rm) = 0$. •

This proposition may be false if $R$ is not a domain. For example, let $R = \mathbb{I}_6$. Viewing $\mathbb{I}_6$ as a module over itself, both $[3]$ and $[4]$ have finite order, for $[2] \in \operatorname{ann}([3])$ and $[3] \in \operatorname{ann}([4])$. But $[3] + [4] = [1]$ has infinite order because $\operatorname{ann}([1]) = ([0])$.

**Definition.** Let $R$ be a domain and let $M$ be an $R$-module. Then $M$ is **torsion** if $tM = M$, while $M$ is **torsion-free** if $tM = \{0\}$.

**Proposition 8.2.** *Let $M$ and $M'$ be $R$-modules, where $R$ is a domain.*

(i) *$M/tM$ is torsion-free.*

(ii) *If $M \cong M'$, then $tM \cong tM'$ and $M/tM \cong M'/tM'$.*

**Proof.**

(i) Assume that $m + tM \neq 0$ in $M/tM$; that is, $m \notin tM$ so that $m$ has infinite order. If $m + tM$ has finite order, then there is some $r \in R$ with $r \neq 0$ such that $0 = r(m + tM) = rm + tM$; that is, $rm \in tM$. Thus, there is $s \in R$

---

[1]There is a generalization of the torsion submodule, called the **singular submodule**, which is defined for left $R$-modules over any not necessarily commutative ring. See Dauns, *Modules and Rings*, pp. 231–238.

with $s \neq 0$ and with $0 = s(rm) = (sr)m$. But $sr \neq 0$, since $R$ is a domain, and so $\operatorname{ann}(m) \neq (0)$; this contradicts $m$ having infinite order.

(ii) If $\varphi \colon M \to M'$ is an isomorphism, then $\varphi(tM) \subseteq tM'$, for if $rm = 0$ with $r \neq 0$, then $r\varphi(m) = \varphi(rm) = 0$ (this is true for any $R$-homomorphism). Hence, $\varphi|tM \colon tM \to tM'$ is an isomorphism (with inverse $\varphi^{-1}|tM'$). For the second statement, the map $\overline{\varphi} \colon M/tM \to M'/tM'$, defined by $\overline{\varphi} \colon m + tM \mapsto \varphi(m) + tM'$, is easily seen to be an isomorphism. $\quad \bullet$

Here is a fancy proof of Proposition 8.2(ii). There is a functor $t \colon {}_R\mathbf{Mod} \to {}_R\mathbf{Mod}$ defined on modules by $M \mapsto tM$ and on morphisms by $\varphi \mapsto \varphi|tM$. That $tM \cong tM'$ follows from the fact that every functor preserves isomorphisms.

If a commutative ring $R$ is not noetherian, it has an ideal that is not finitely generated. Now $R$, viewed as a module over itself, is finitely generated; it is even cyclic (with generator 1). Thus, it is possible that a submodule of a finitely generated module not be finitely generated. This cannot happen when $R$ is a PID; in fact, we have proved, in Proposition 6.36, that $R$ is noetherian if and only if every submodule $S$ of a finitely generated $R$-module $M$ is finitely generated. If $R$ is a PID, then its modules have a stronger property: Proposition 2.69 says that if an $R$-module $M$ can be generated by $n$ elements, then every submodule $S \subseteq M$ can be generated by $n$ or fewer elements.

**Theorem 8.3.** *If $R$ is a PID, then every finitely generated torsion-free $R$-module $M$ is free.*

**Proof.** We prove the theorem by induction on $n$, where $M = \langle v_1, \dots, v_n \rangle$.

If $n = 1$, then $M$ is cyclic; hence, $M = \langle v_1 \rangle \cong R/\operatorname{ann}(v_1)$. Since $M$ is torsion-free, $\operatorname{ann}(v_1) = (0)$, so that $M \cong R$ and $M$ is free.

For the inductive step, let $M = \langle v_1, \dots, v_{n+1} \rangle$, and define

$$S = \{m \in M : \text{there is a nonzero } r \in R \text{ with } rm \in \langle v_{n+1} \rangle\};$$

it is easy to check that $S$ is a submodule of $M$. We show that $M/S$ is torsion-free. If $x \in M$, $x \notin S$, and $r(x + S) = 0$, then $rx \in S$; hence, there is $r' \in R$ with $r' \neq 0$ and $rr'x \in \langle v_{n+1} \rangle$. Since $rr' \neq 0$, we have $x \in S$, a contradiction.

Plainly, $M/S$ can be generated by $n$ elements, namely, $v_1 + S, \dots, v_n + S$, and so $M/S$ is free, by the inductive hypothesis. Since free modules are projective, Proposition 6.73 gives

$$M \cong S \oplus (M/S).$$

Thus, it suffices to prove that $S \cong R$ or $S = \{0\}$.

If $x \in S$, then there is some nonzero $r \in R$ with $rx \in \langle v_{n+1} \rangle$; that is, there is $a \in R$ with $rx = av_{n+1}$. Define $\varphi \colon S \to Q = \operatorname{Frac}(R)$, the fraction field of $R$, by $\varphi \colon x \mapsto a/r$. It is a straightforward calculation, left to the reader, that $\varphi$ is a (well-defined) injective $R$-map. Now $\operatorname{im} \varphi \cong S$ is finitely generated, for $S$ is a direct summand, hence an image, of $M$.

The proof will be complete if we prove that every finitely generated submodule $D$ of $Q$ (e.g., $D = \operatorname{im} \varphi$) is cyclic. Now

$$D = \langle b_1/c_1, \ldots, b_m/c_m \rangle,$$

where $b_i, c_i \in R$. Let $c = \prod_i c_i$, and define $f \colon D \to R$ by $f \colon d \mapsto cd$ for all $d \in D$ (it is plain that $f$ has values in $R$, for multiplication by $c$ clears all denominators). Since $D$ is torsion-free, $f$ is an injective $R$-map, and so $D$ is isomorphic to a submodule of $R$; that is, $D$ is isomorphic to an ideal of $R$. Since $R$ is a PID, every nonzero ideal in $R$ is isomorphic to $R$; hence, $S \cong \operatorname{im} \varphi = D \cong R$ or $S = \{0\}$. •

**Corollary 8.4.** *Let $R$ be a PID.*

(i) *Every submodule $S$ of a finitely generated free $R$-module $F$ is itself free, and $\operatorname{rank}(S) \leq \operatorname{rank}(F)$.*

(ii) *Every finitely generated projective $R$-module $P$ is free.*

**Proof.**

(i) Now $F$ is torsion-free and, hence, $S$ is torsion-free. By Proposition 2.69, the submodule $S$ is finitely generated; in fact, $S$ can be generated by $n$ or fewer elements, where $n = \operatorname{rank}(F)$. Theorem 8.3 now applies to give $S$ free, and $\operatorname{rank}(S) \leq n = \operatorname{rank}(F)$.

(ii) By Theorem 6.76, $P$ is a direct summand of a free $R$-module $F$. Since $P$ is finitely generated, we may choose $F$ to be finitely generated. Thus, $P$ is isomorphic to a submodule of $F$, which is free, by (i). •

**Remark.** Both statements in Corollary 8.4 are true without the finiteness hypothesis, and we shall soon prove them. On the other hand, both statements in the next corollary do require the finitely generated hypothesis. There exist abelian groups $G$ whose torsion subgroup $tG$ is not a direct summand of $G$ [see Exercise 8.3 on page 651]. ◄

**Corollary 8.5.** *Let $R$ be a PID.*

(i) *Every finitely generated $R$-module $M$ is a direct sum,*

$$M = tM \oplus F,$$

*where $F$ is a finitely generated free $R$-module.*

(ii) *If $M$ and $M'$ are finitely generated $R$-modules, then $M \cong M'$ if and only if $tM \cong tM'$ and $\operatorname{rank}(M/tM) = \operatorname{rank}(M'/tM')$.*

**Proof.**

(i) The quotient module $M/tM$ is finitely generated, because $M$ is finitely generated, and it is torsion-free, by Proposition 8.2(i). Therefore, $M/tM$ is free, by Theorem 8.3, and hence $M/tM$ is projective. Finally, $M \cong tM \oplus (M/tM)$, by Corollary 6.74.

(ii) By Proposition 8.2(ii), if $M \cong M'$, then $tM \cong tM'$ and $M/tM \cong M'/tM'$. Since $M/tM$ is finitely generated torsion-free, it is a free module, as is $M'/tM'$, and these are isomorphic if they have the same rank.

Conversely, since $M \cong tM \oplus (M/tM)$ and $M' \cong tM' \oplus (M'/tM')$, Proposition 6.43 assembles the isomorphisms on each summand into an isomorphism $M \to M'$. •

We now consider flat modules.

**Corollary 8.6.** *If $R$ is a domain, then every flat $R$-module $M$ is torsion-free.*

**Proof.** If $M$ is not torsion-free, then there is a nonzero $m \in M$ with $rm = 0$ for some nonzero $r \in R$. Let $0 \to R \xrightarrow{i} \text{Frac}(R)$ be the inclusion. Now $m \otimes 1 \neq 0$ in $M \otimes_R R$, for the map $m \otimes 1 \mapsto m$ is an isomorphism $M \otimes_R R \to M$, by Proposition 6.108. On the other hand,

$$(1_M \otimes i)(m \otimes 1) = m \otimes 1 = m \otimes \frac{r}{r} = rm \otimes \frac{1}{r} = 0 \quad \text{in } M \otimes_R \text{Frac}(R).$$

Therefore, $m \otimes 1$ is a nonzero element in $\ker(1_M \otimes i)$, so that $M$ is not flat. •

The converse is true for PIDs.

**Corollary 8.7.** *If $R$ is a PID, then an $R$-module $M$ is flat if and only if it is torsion-free.*

**Proof.** If $M$ is torsion-free, then every (finitely generated) submodule $S$ is torsion-free, hence, free (Theorem 8.3), hence, flat (Lemma 6.137). Therefore, $M$ itself is flat, by Lemma 6.139. The converse is Corollary 8.6. •

Before continuing the saga of finitely generated modules, we pause to prove an important result: the generalization of Corollary 8.4 in which we no longer assume that free modules $F$ are finitely generated. We begin with a second proof of the finitely generated case.

**Proposition 8.8.** *If $R$ is a PID, then every submodule $H$ of a finitely generated free $R$-module $F$ is itself free and $\text{rank}(H) \leq \text{rank}(F)$.*

**Proof.** The proof is by induction on $n = \text{rank}(F)$. If $n = 1$, then $F \cong R$. Thus, $H$ is isomorphic to an ideal in $R$; but all ideals are principal, and hence are isomorphic to $(0)$ or $R$. Therefore, $H$ is a free module of rank $\leq 1$.

Let us now prove the inductive step. If $\{x_1, \ldots, x_{n+1}\}$ is a basis of $F$, define $F' = \langle x_1, \ldots, x_n \rangle$, and let $H' = H \cap F'$. By induction, $H'$ is free of rank $\leq n$. Now

$$H/H' = H/(H \cap F') \cong (H + F')/F' \subseteq F/F' \cong R.$$

By the base step, either $H/H' = \{0\}$ or $H/H' \cong R$. In the first case, $H = H'$, and we are done. In the second case, Corollary 6.74 gives $H = H' \oplus \langle h \rangle$ for some $h \in H$, where $\langle h \rangle \cong R$, and so $H$ is free abelian of rank $\leq n + 1$. •

We now remove the finiteness hypothesis from Corollary 8.4.

**Theorem 8.9.** *Let $R$ be a PID.*

(i) *Every submodule $H$ of a free $R$-module $F$ is free and $\text{rank}(H) \leq \text{rank}(F)$.*

(ii) *Every projective $R$-module $H$ is free.*

**Proof.**

(i) We are going to use the statement, equivalent to the Axiom of Choice and to Zorn's Lemma, that every set can be well-ordered. In particular, we may assume that $\{x_k : k \in K\}$ is a basis of $F$ having a well-ordered index set $K$.

For each $k \in K$, define

$$F'_k = \langle x_j : j \prec k \rangle \quad \text{and} \quad F_k = \langle x_j : j \preceq k \rangle = F'_k \oplus \langle x_k \rangle;$$

note that $F = \bigcup_k F_k$. Define

$$H'_k = H \cap F'_k \quad \text{and} \quad H_k = H \cap F_k.$$

Now $H'_k = H \cap F'_k = H_k \cap F'_k$, so that

$$H_k / H'_k = H_k / (H_k \cap F'_k) \cong (H_k + F'_k)/F'_k \subseteq F_k/F'_k \cong R.$$

By Corollary 6.74, $H_k = H'_k$ or $H_k = H'_k \oplus \langle h_k \rangle$, where $h_k \in H_k \subseteq H$ and $\langle h_k \rangle \cong R$. We claim that $H$ is a free $R$-module with basis the set of all $h_k$. It will then follow that $\mathrm{rank}(H) \leq \mathrm{rank}(F)$ (of course, these ranks may be infinite cardinals).

Since $F = \bigcup F_k$, each $f \in F$ lies in some $F_k$. Since $K$ is well-ordered, there is a smallest index $k \in K$ with $f \in F_k$, and we denote this smallest index by $\mu(f)$. In particular, if $h \in H$, then

$$\mu(h) = \text{ smallest index } k \text{ with } h \in F_k.$$

Note that if $h \in H'_k \subseteq F'_k$, then $\mu(h) \prec k$. Let $H^*$ be the submodule of $H$ generated by all the $h_k$.

Suppose that $H^*$ is a proper submodule of $H$. Let $j$ be the smallest index in

$$\{\mu(h) : h \in H \text{ and } h \notin H^*\},$$

and choose $h' \in H$ to be such an element having index $j$; that is, $h' \notin H^*$ and $\mu(h') = j$. Now $h' \in H \cap F_j$, because $\mu(h') = j$, and so

$$h' = a + rh_j, \quad \text{where } a \in H'_j \text{ and } r \in R.$$

Thus, $a = h' - rh_j \in H'_j$ and $a \notin H^*$; otherwise $h' \in H^*$ (because $h_j \in H^*$). Since $\mu(a) \prec j$, we have contradicted $j$ being the smallest index of an element of $H$ not in $H^*$. We conclude that $H^* = H$; that is, every $h \in H$ is a linear combination of $h_k$'s.

It remains to prove that an expression of any $h \in H$ as a linear combination of $h_k$'s is unique. By subtracting two such expressions, it suffices to prove that if

$$0 = r_1 h_{k_1} + r_2 h_{k_2} + \cdots + r_n h_{k_n},$$

then all the coefficients $r_i = 0$. Arrange the terms so that $k_1 \prec k_2 \prec \cdots \prec k_n$. If $r_n \neq 0$, then $r_n h_{k_n} \in \langle h_{k_n} \rangle \cap H'_{k_n} = \{0\}$, a contradiction. Therefore, all $r_i = 0$, and so $H$ is a free module with basis $\{h_k : k \in K\}$.

(ii) By Theorem 6.76, $P$ is a direct summand of a free $R$-module $F$. Thus, $P$ is isomorphic to a submodule of $F$, which is free, by (i).   •

The special case of this theorem for $\mathbb{Z}$-modules follows from the Nielsen–Schreier Theorem, for if $F$ is a free group with basis $X$, then $F/F'$ is a free abelian group with basis $\{xF' : x \in X\}$, where $F'$ is the commutator subgroup of $F$.

We return to the discussion of finitely generated modules. In light of Proposition 8.2(ii), the problem of classifying finitely generated modules over a PID is reduced to classifying finitely generated torsion modules. Let us prove at once that these modules are precisely the generalization of finite abelian groups.

**Proposition 8.10.** *An abelian group $G$ is finite if and only if it is a finitely generated torsion $\mathbb{Z}$-module.*

**Proof.** If $G$ is finite, it is obviously finitely generated; moreover, Lagrange's Theorem says that $G$ is torsion.

Conversely, suppose that $G = \langle x_1, \ldots, x_n \rangle$ and there are nonzero integers $d_i$ with $d_i x_i = 0$ for all $i$. It follows that each $g \in G$ can be written as

$$g = m_1 x_1 + \cdots + m_n x_n,$$

where $0 \leq m_i < d_i$ for all $i$. Therefore, $|G| \leq \prod_i d_i$, and so $G$ is finite. $\quad \bullet$

**Definition.** Let $R$ be a PID and $M$ be an $R$-module. If $(p)$ is a nonzero prime ideal in $R$, then $M$ is $(p)$-**primary** if, for each $m \in M$, there is $n \geq 1$ with $p^n m = 0$.

If $M$ is any $R$-module, then its $(p)$-**primary component** is

$$M_{(p)} = \{m \in M : p^n m = 0 \text{ for some } n \geq 1\}.$$

It is clear that $(p)$-primary components are submodules. If we do not want to specify the prime $(p)$, we will say that a module is *primary* [instead of $(p)$-primary].

All of the coming theorems in this section were first proved for abelian groups and, later, generalized to modules over PIDs. The translation from abelian groups to modules is straightforward, but let us see this explicitly by generalizing the primary decomposition to modules over PIDs by adapting the proof given in Chapter 4 for abelian groups. For the reader's convenience, we reproduce this proof with the finiteness hypothesis eliminated.

**Theorem 8.11 (Primary Decomposition).**

(i) *Every torsion abelian group $G$ is a direct sum of its p-primary components:*

$$G = \bigoplus_{p \in P} G_p,$$

 *where $P$ is the set of all primes.*

(ii) *Every torsion $R$-module $M$ over a PID $R$ is a direct sum of its $(p)$-primary components:*

$$M = \bigoplus_{(p)} M_{(p)}.$$

**Proof.**

(i) Let $x$ be a nonzero element of $G$ having order $d$. By the Fundamental Theorem of Arithmetic, there are distinct primes $p_1, \ldots, p_n$ and positive exponents $e_1, \ldots, e_n$ with

$$d = p_1^{e_1} \cdots p_n^{e_n}.$$

Define $r_i = d/p_i^{e_i}$, so that $p_i^{e_i} r_i = d$. It follows that $r_i x \in G_{p_i}$ for each $i$. But the gcd of $r_1, \ldots, r_n$ is 1, and so there are integers $s_1, \ldots, s_n$ with $1 = \sum_i s_i r_i$. Therefore,

$$x = \sum_i s_i r_i x \in \left\langle \bigcup_p G_p \right\rangle.$$

For each prime $p$, write $H_p = \left\langle \bigcup_{q \neq p} G_q \right\rangle$. By Exercise 6.110 on page 514, it suffices to prove that if

$$x \in G_p \cap H_p,$$

then $x = 0$. Since $x \in G_p$, we have $p^\ell x = 0$ for some $\ell \geq 0$; since $x \in H_p$, we have $ux = 0$, where $u = q_1^{f_1} \cdots q_n^{f_n}$, $q_i \neq p$, and $f_i \geq 1$ for all $i$. But $p^\ell$ and $u$ are relatively prime, so there exist integers $s$ and $t$ with $1 = sp^\ell + tu$. Therefore,

$$x = (sp^\ell + tu)x = sp^\ell x + tux = 0.$$

(ii) We now translate the proof in (i) into the language of modules. If $m \in M$ is nonzero, its order ideal $\operatorname{ann}(m) = (d)$, for some $d \in R$. By unique factorization, there are irreducible elements $p_1, \ldots, p_n$, no two of which are associates, and positive exponents $e_1, \ldots, e_n$ with

$$d = p_1^{e_1} \cdots p_n^{e_n}.$$

By Proposition 5.16, $(p_i)$ is a prime ideal for each $i$. Define $r_i = d/p_i^{e_i}$, so that $p_i^{e_i} r_i = d$. It follows that $r_i m \in M_{(p_i)}$ for each $i$. But the gcd of the elements $r_1, \ldots, r_n$ is 1, and so there are elements $s_1, \ldots, s_n \in R$ with $1 = \sum_i s_i r_i$. Therefore,

$$m = \sum_i s_i r_i m \in \left\langle \bigcup_{(p)} M_{(p)} \right\rangle.$$

For each prime $(p)$, write $H_{(p)} = \left\langle \bigcup_{Q \neq (p)} G_Q \right\rangle$. By Exercise 6.110 on page 514, it suffices to prove that if

$$m \in M_{(p)} \cap H_{(p)},$$

then $m = 0$. Since $m \in M_{(p)}$, we have $p^\ell m = 0$ for some $\ell \geq 0$; since $m \in H_{(p)}$, we have $um = 0$, where $u = q_1^{f_1} \cdots q_n^{f_n}$, $Q_i = (q_i)$, and $f_i \geq 1$. But $p^\ell$ and $u$ are relatively prime, so there exist $s, t \in R$ with $1 = sp^\ell + tu$. Therefore,

$$m = (sp^\ell + tu)m = sp^\ell m + tum = 0. \quad \bullet$$

**Proposition 8.12.** *Two torsion modules $M$ and $M'$ over a PID are isomorphic if and only if $M_{(p)} \cong M'_{(p)}$ for every nonzero prime ideal $(p)$.*

**Proof.** If $f\colon M \to M'$ is an $R$-map, then $f(M_{(p)}) \subseteq M'_{(p)}$ for every prime ideal $(p)$, for if $p^\ell m = 0$, then $0 = f(p^\ell m) = p^\ell f(m)$. If $f$ is an isomorphism, then so is $f^{-1}\colon M' \to M$. It follows that each restriction $f|M_{(p)}\colon M_{(p)} \to M'_{(p)}$ is an isomorphism, with inverse $f^{-1}|M'_{(p)}$. Conversely, if there are isomorphisms $f_{(p)}\colon M_{(p)} \to M'_{(p)}$ for all $(p)$, then there is an isomorphism $\varphi\colon \bigoplus_{(p)} M_{(p)} \to \bigoplus_{(p)} M'_{(p)}$ given by $\sum_{(p)} m_{(p)} \mapsto \sum_{(p)} f_{(p)}(m_{(p)})$. •

There is a fancy proof here, just as there is one for Proposition 8.2. Define a "$(p)$-torsion functor" $t_{(p)}\colon {}_R\mathbf{Mod} \to {}_R\mathbf{Mod}$ on modules by $M \mapsto (tM)_{(p)}$ and on morphisms by $\varphi \mapsto \varphi|(tM)_{(p)}$. That $(tM)_{(p)} \cong (tM')_{(p)}$ follows from the fact that every functor preserves isomorphisms.

For the remainder of this section, we shall merely give definitions and statements of results; the reader should have no difficulty in adapting proofs of theorems about abelian groups in Chapter 4 to proofs of theorems about modules over PIDs.

**Theorem 8.13 (Basis Theorem).** *If $R$ is a PID, then every finitely generated $R$-module $M$ is a direct sum of cyclic modules in which each cyclic summand is either primary or is isomorphic to $R$.*

**Proof.** By Corollary 8.5, $M = tM \oplus F$, where $F$ is finitely generated free; see Theorem 4.20 for the abelian group version of the Basis Theorem. •

**Corollary 8.14.** *Every finitely generated abelian group is a direct sum of cyclic groups, each of prime power order or infinite.*

When are two finitely generated modules $M$ and $M'$ over a PID isomorphic?

Before stating the next lemma, recall that $M/pM$ is a vector space over $R/(p)$, and we define
$$d(M) = \dim(M/pM).$$
In particular, $d(pM) = \dim(pM/p^2M)$ and, more generally,
$$d(p^n M) = \dim(p^n M/p^{n+1}M).$$

**Definition.** If $M$ is a finitely generated $(p)$-primary $R$-module, where $R$ is a PID and $(p)$ is a nonzero prime ideal, then
$$U_{(p)}(n, M) = d(p^n M) - d(p^{n+1}M).$$

**Theorem 8.15.** *If $R$ is a PID and $(p)$ is a nonzero prime ideal in $R$, then any two decompositions of a finitely generated $(p)$-primary $R$-module $M$ into direct sums of cyclic modules have the same number of cyclic summands of each type. More precisely, for each $n \geq 0$, the number of cyclic summands having order ideal $(p^{n+1})$ is $U_{(p)}(n, M)$.*

**Proof.** See Theorem 4.26. •

**Corollary 8.16.** *Let $M$ and $M'$ be $(p)$-primary $R$-modules, where $R$ is a PID. Then $M \cong M'$ if and only if $U_{(p)}(n, M) = U_{(p)}(n, M')$ for all $n \geq 0$.*

**Proof.** See Corollary 4.27. •

**Definition.** If $M$ is a $(p)$-primary $R$-module over a PID $R$, then the *elementary divisors* of $M$ are the ideals $(p^{n+1})$, each repeated with multiplicity $U_{(p)}(n, M)$.

If $M$ is a (not necessarily primary) finitely generated torsion $R$-module, then its *elementary divisors* are the elementary divisors of all its primary components.

The next definition is motivated by Corollary 4.33: if $G$ is a finite abelian group with elementary divisors $\{p_i^{e_{ij}}\}$, then

$$|G| = \prod_{ij} p_i^{e_{ij}}.$$

**Definition.** If $M$ is a finitely generated torsion $R$-module, where $R$ is a PID, then the *order* of $M$ is the principal ideal generated by the product of its elementary divisors, namely, $\left(\prod_{ij} p_i^{e_{ij}}\right)$.

**Example 8.17.** How many $k[x]$-modules are there of order $(x-1)^3(x+1)^2$, where $k$ is a field? By the primary decomposition, every $k[x]$-module of order $(x-1)^3(x+1)^2$ is the direct sum of primary modules of order $(x-1)^3$ and $(x+1)^2$, respectively. There are three modules of order $(x-1)^3$, described by the elementary divisors

$$(x - 1, x - 1, x - 1), \quad (x - 1, (x - 1)^2), \quad \text{and} \quad (x - 1)^3;$$

there are two modules of order $(x+1)^2$, described by the elementary divisors

$$(x + 1, x + 1) \quad \text{and} \quad (x + 1)^2.$$

Therefore, up to isomorphism, there are six modules of order $(x - 1)^3(x + 1)^2$.

The reader has probably noticed that this argument is the same as that in Example 4.29 on page 238 classifying all abelian groups of order $72 = 2^3 3^2$. ◀

**Theorem 8.18 (Fundamental Theorem of Finitely Generated Modules).**
*If $R$ is a PID, then two finitely generated $R$-modules are isomorphic if and only if their torsion submodules have the same elementary divisors and their free parts have the same rank.*

**Proof.** By Theorem 8.11, $M \cong M'$ if and only if, for all nonzero primes $(p)$, the primary components $M_{(p)}$ and $M'_{(p)}$ are isomorphic. Corollary 8.16, Proposition 8.5(ii), and Proposition 8.12 now complete the proof. •

Here is a second type of decomposition of a finitely generated torsion module into a direct sum of cyclics that does not mention primary modules.

**Proposition 8.19.** *If $R$ is a PID, then every finitely generated torsion $R$-module $M$ is a direct sum of cyclic modules*

$$M = R/(c_1) \oplus R/(c_2) \oplus \cdots \oplus R/(c_t),$$

*where $t \geq 1$ and $c_1 \mid c_2 \mid \cdots \mid c_t$.*

**Proof.** See Proposition 4.30. •

**Definition.** If $M$ is a finitely generated torsion $R$-module, where $R$ is a PID, and

$$M = R/(c_1) \oplus R/(c_2) \oplus \cdots \oplus R/(c_t),$$

where $t \geq 1$ and $c_1 \mid c_2 \mid \cdots \mid c_t$, then the ideals $(c_1), (c_2), \ldots, (c_t)$ are called the *invariant factors* of $M$.

**Corollary 8.20.** *If $M$ is a finitely generated torsion module over a PID $R$, then*

$$(c_t) = \{r \in R : rM = \{0\}\},$$

*where $(c_t)$ is the last ideal occurring in the decomposition of $M$ in* Proposition 8.19.

*In particular, if $R = k[x]$, where $k$ is a field, then $c_t$ is the monic polynomial of least degree for which $c_t M = \{0\}$.*

**Proof.** For the first statement, see Corollary 4.31.

The second statement follows from the fact that every nonzero ideal in $k[x]$ is generated by the monic polynomial of least degree in it. •

**Definition.** If $M$ is an $R$-module, then its *exponent* (or *annihilator*) is the ideal

$$\mathrm{ann}(M) = \{r \in R : rM = \{0\}\}.$$

Corollary 8.20 computes the exponent of a finitely generated torsion module over a PID; it is the last invariant factor $(c_t)$.

**Corollary 8.21.** *If $M$ is a finitely generated torsion $R$-module, where $R$ is a PID, with invariant factors $c_1, \ldots, c_t$, then the order of $M$ is $\left(\prod_{i=1}^{t} c_i\right)$.*

**Proof.** See Corollary 4.33. The reader should check that the principal ideal generated by the product of the elementary divisors (which is the definition of the order of $M$) is equal to the principal ideal $\left(\prod_{i=1}^{t} c_i\right)$. •

**Example 8.22.** We displayed the elementary divisors of $k[x]$-modules of order $(x-1)^3(x+1)^2$ in Example 8.17; here are their invariant factors.

| Elementary divisors | $\leftrightarrow$ | Invariant factors |
|---|---|---|

$$(x-1, x-1, x-1, x+1, x+1) \leftrightarrow x-1 \mid (x-1)(x+1) \mid (x-1)(x+1)$$

$$(x-1, (x-1)^2, x+1, x+1) \leftrightarrow (x-1)(x+1) \mid (x-1)^2(x+1)$$

$$((x-1)^3, x+1, x+1) \leftrightarrow x+1 \mid (x-1)^3(x+1)$$

$$(x-1, x-1, x-1, (x+1)^2) \leftrightarrow x-1 \mid x-1 \mid (x-1)(x+1)^2$$

$$(x-1, (x-1)^2, (x+1)^2) \leftrightarrow x-1 \mid (x-1)^2(x+1)^2$$

$$((x-1)^3, (x+1)^2) \leftrightarrow (x-1)^3(x+1)^2 \quad \blacktriangleleft$$

**Theorem 8.23 (Invariant Factors).** *If $R$ is a PID, then two finitely generated $R$-modules are isomorphic if and only if their torsion submodules have the same invariant factors and their free parts have the same rank.*

**Proof.** By Corollary 8.5(i), every finitely generated $R$-module $M$ is a direct sum $M = tM \oplus F$, where $F$ is free, and $M \cong M'$ if and only if $tM \cong tM'$ and $F \cong F'$. Corollary 8.5(ii) shows that the free parts $F \cong M/tM$ and $F' \cong M'/tM'$ are isomorphic, and a straightforward generalization of Theorem 4.35 shows that the torsion submodules are isomorphic.   •

**8.1.1. Divisible Groups.** The reader should now be comfortable when we say that a theorem can easily be generalized from abelian groups to modules over PIDs. Consequently, we will state and prove theorems only for abelian groups, leaving the straightforward generalizations to modules to the reader.

Let us now consider modules that are not finitely generated. Recall that an abelian group $D$ is ***divisible*** if, for each $d \in D$ and each positive integer $n$, there exists $d' \in D$ with $d = nd'$. Every quotient of a divisible group is divisible, as is every direct sum of divisible groups. Now Corollary 6.93 states that an abelian group $D$ is an injective $\mathbb{Z}$-module if and only if it is divisible, so that classifying divisible abelian groups describes all injective abelian groups.

**Proposition 8.24.** *A torsion-free abelian group $D$ is divisible if and only if it is a vector space over $\mathbb{Q}$.*

**Proof.** If $D$ is a vector space over $\mathbb{Q}$, then it is a direct sum of copies of $\mathbb{Q}$, for every vector space has a basis. But $\mathbb{Q}$ is a divisible group, and any direct sum of divisible groups is itself divisible.

Let $D$ be torsion-free and divisible; we must show that $D$ admits scalar multiplication by rational numbers. Suppose that $d \in D$ and $n$ is a positive integer. Since $D$ is divisible, there exists $d' \in D$ with $nd' = d$ [of course, $d'$ is a candidate for $(1/n)d$]. Note, since $D$ is torsion-free, that $d'$ is the unique such element: if also $nd'' = d$, then $n(d' - d'') = 0$, so that $d' - d''$ has finite order, and hence is 0. If $m/n \in \mathbb{Q}$, define $(m/n)d = md'$, where $nd' = d$. The reader can prove that this scalar multiplication is well-defined [if $m/n = a/b$, then $(m/n)d = (a/b)d$] and that the various axioms in the definition of vector space hold.   •

**Definition.** If $G$ is an abelian group, then $dG$ is the subgroup generated by all the divisible subgroups of $G$.

**Proposition 8.25.**

  (i) *For any abelian group $G$, the subgroup $dG$ is the unique maximal divisible subgroup of $G$.*

  (ii) *Every abelian group $G$ is a direct sum*

$$G = dG \oplus R,$$

  *where $dR = \{0\}$. Hence, $R \cong G/dG$ has no nonzero divisible subgroups.*

**Proof.**

  (i) It suffices to prove that $dG$ is divisible, for then it is obviously the largest such. If $x \in dG$, then $x = x_1 + \cdots + x_t$, where $x_i \in D_i$ and the $D_i$ are divisible subgroups of $G$. If $n$ is a positive integer, then there are $y_i \in D_i$ with $x_i = ny_i$, because $D_i$ is divisible. Hence, $y = y_1 + \cdots + y_t \in dG$ and $x = ny$, so that $dG$ is divisible.

(ii) Since $dG$ is divisible, it is injective, and Proposition 6.84 on page 461 gives

$$G = dG \oplus R,$$

where $R$ is a subgroup of $G$. If $R$ has a nonzero divisible subgroup $D$, then $R = D \oplus S$ for some subgroup $S$, by Proposition 6.84. But $dG \oplus D$ is a divisible subgroup of $G$ properly containing $dG$, contradicting part (i).  •

**Definition.** An abelian group $G$ is **reduced** if $dG = \{0\}$; that is, $G$ has no nonzero divisible subgroups.

Exercise 8.18 on page 654 says that an abelian group $G$ is reduced if and only if $\mathrm{Hom}(\mathbb{Q}, G) = \{0\}$.

We have just shown that $G/dG$ is always reduced. The reader should compare the roles of the maximal divisible subgroup $dG$ of a group $G$ with that of $tG$, its torsion subgroup: $G$ is torsion if $tG = G$, and it is torsion-free if $tG = \{0\}$; $G$ is divisible if $dG = G$, and it is reduced if $dG = \{0\}$. There are exact sequences

$$0 \to dG \to G \to G/dG \to 0$$

and

$$0 \to tG \to G \to G/tG \to 0;$$

the first sequence always splits, but we will see, in Exercise 8.3 on page 651, that the second sequence may not split.

The following group has some remarkable properties. If $p$ is a prime and $n \geq 1$, let us denote the primitive $p^n$th root of unity by

$$z_n = e^{2\pi i/p^n}.$$

Of course, every complex $p^n$th root of unity is a power of $z_n$.

**Definition.** The **Prüfer group of type** $p^\infty$ (or **quasicyclic group**[2]) is the subgroup of the multiplicative group $\mathbb{C}^\times$:

$$\mathbb{Z}(p^\infty) = \langle z_n : n \geq 1 \rangle = \langle e^{2\pi i/p^n} : n \geq 1 \rangle.$$

Note, for every integer $n \geq 1$, that the subgroup $\langle z_n \rangle$ is the unique subgroup of $\mathbb{Z}(p^\infty)$ of order $p^n$, for the polynomial $x^{p^n} - 1 \in \mathbb{C}[x]$ has at most $p^n$ complex roots.

**Proposition 8.26.** *Let $p$ be a prime.*

(i) *$\mathbb{Z}(p^\infty)$ is isomorphic to the $p$-primary component of $\mathbb{Q}/\mathbb{Z}$.*

(ii) *$\mathbb{Z}(p^\infty)$ is a divisible $p$-primary abelian group.*

(iii) *The subgroups of $\mathbb{Z}(p^\infty)$ are*

$$\{1\} \subsetneq \langle z_1 \rangle \subsetneq \langle z_2 \rangle \subsetneq \cdots \subsetneq \langle z_n \rangle \subsetneq \langle z_{n+1} \rangle \subsetneq \cdots \subsetneq \mathbb{Z}(p^\infty),$$

*and so they are well-ordered by inclusion.*

---

[2]The group $\mathbb{Z}(p^\infty)$ is called *quasicyclic* because every proper subgroup of it is cyclic (Proposition 8.26(iii)).

(iv) $\mathbb{Z}(p^\infty)$ *has* DCC *on subgroups but not* ACC.[3]

**Proof.**

(i) Define $\varphi\colon \bigoplus_p \mathbb{Z}(p^\infty) \to \mathbb{Q}/\mathbb{Z}$ by $\varphi\colon (e^{2\pi i c_p/p^{n_p}}) \mapsto \sum_p c_p/p^{n_p} + \mathbb{Z}$, where $c_p \in \mathbb{Z}$. It is easy to see that $\varphi$ is an injective homomorphism. The proof that $\varphi$ is surjective is really contained in the proof of Theorem 4.15, but here it is again. Let $a/b \in \mathbb{Q}/\mathbb{Z}$, and write $b = \prod_p p^{n_p}$. Since the numbers $b/p^{n_p}$ are relatively prime, there are integers $m_p$ with $1 = \sum_p m_p(b/p^{n_p})$. Therefore, $a/b = \sum_p am_p/p^{n_p} = \varphi((e^{a2\pi i m_p/p^{n_p}}))$.

(ii) Since a direct summand is always a homomorphic image, $\mathbb{Z}(p^\infty)$ is a homomorphic image of the divisible group $\mathbb{Q}/\mathbb{Z}$; but every quotient of a divisible group is itself divisible.

(iii) Let $S$ be a proper subgroup of $\mathbb{Z}(p^\infty)$. Since $\{z_n : n \geq 1\}$ generates $\mathbb{Z}(p^\infty)$, we may assume that $z_m \notin S$ for some $m$. It follows that $z_\ell \notin S$ for all $\ell > m$; otherwise $z_m = z_\ell^{p^{\ell-m}} \in S$. If $S \neq \{0\}$, we claim that $S$ contains some $z_n$; indeed, we show that $S$ contains $z_1$. Now $S$ must contain some element $x$ of order $p$, and $x = z_1^c$, where $1 \leq c < p$ [for $\langle z_1 \rangle$ contains all the elements in $\mathbb{Z}(p^\infty)$ of order $p$]. Since $p$ is prime, $(c, p) = 1$, and there are integers $u, v$ with $1 = cu + pv$; hence, $z_1 = z_1^{cu+pv} = z_1^{cu} = x^u \in S$. Let $d$ be the largest integer with $z_d \in S$. Clearly, $\langle z_d \rangle \subseteq S$. For the reverse inclusion, let $s \in S$. If $s$ has order $p^n > p^d$, then $\langle s \rangle$ contains $z_n$, because $\langle z_n \rangle$ contains all the elements of order $p^n$ in $\mathbb{Z}(p^\infty)$. But this contradicts our observation that $z_\ell \notin S$ for all $\ell > d$. Hence, $s$ has order $\leq p^d$, and so $s \in \langle z_d \rangle$; therefore, $S = \langle z_d \rangle$.

As the only proper nonzero subgroups of $\mathbb{Z}(p^\infty)$ are the groups $\langle z_n \rangle$, it follows that the subgroups are well-ordered by inclusion.

(iv) First, $\mathbb{Z}(p^\infty)$ does not have ACC, as the chain of subgroups

$$\{1\} \subsetneq \langle z_1 \rangle \subsetneq \langle z_2 \rangle \subsetneq \cdots$$

illustrates. Now every strictly decreasing sequence in a well-ordered set is finite [if $x_1 \succ x_2 \succ x_3 \succ \cdots$ is infinite, the subset $(x_n)_{n \geq 1}$ has no smallest element]. It follows that $\mathbb{Z}(p^\infty)$ has DCC on subgroups. •

**Notation.** If $G$ is an abelian group and $n$ is a positive integer, then

$$G[n] = \{g \in G : ng = 0\}.$$

It is easy to see that $G[n]$ is a subgroup of $G$. Note that if $p$ is prime, then $G[p]$ is a vector space over $\mathbb{F}_p$.

**Lemma 8.27.** *If $G$ and $H$ are divisible $p$-primary abelian groups, then $G \cong H$ if and only if $G[p] \cong H[p]$.*

**Proof.** If there is an isomorphism $f\colon G \to H$, then it is easy to see that its restriction $f|G[p]$ is an isomorphism $G[p] \to H[p]$ (whose inverse is $f^{-1}|H[p]$).

---

[3]Recall Theorem 7.31, the Hopkins–Levitzki Theorem: a ring with DCC must also have ACC. This result shows that the analogous result for groups is false.

For sufficiency, assume that $f\colon G[p] \to H[p]$ is an isomorphism. Composing with the inclusion $H[p] \to H$, we may assume that $f\colon G[p] \to H$. Since $H$ is injective, $f$ extends to a homomorphism $F\colon G \to H$; we claim that any such $F$ is an isomorphism.

(i) $F$ is an injection.

If $g \in G$ has order $p$, then $F(g) = f(g) \neq 0$, by hypothesis. Suppose that $g$ has order $p^n$ for $n \geq 2$. If $F(g) = 0$, then $F(p^{n-1}g) = 0$, and this contradicts the hypothesis, because $p^{n-1}g$ has order $p$. Therefore, $F$ is an injection.

(ii) $F$ is a surjection.

We show, by induction on $n \geq 1$, that if $h \in H$ has order $p^n$, then $h \in \operatorname{im} F$. If $n = 1$, then $h \in H[p] = \operatorname{im} f \subseteq \operatorname{im} F$. For the inductive step, assume that $h \in H$ has order $p^{n+1}$. Now $p^n h \in H[p]$, so there exists $g \in G$ with $F(g) = f(g) = p^n h$. Since $G$ is divisible, there is $g' \in G$ with $p^n g' = g$; thus, $p^n(h - F(g')) = 0$. By induction, there is $x \in G$ with $F(x) = h - F(g')$. Therefore, $F(x + g') = h$, as desired. •

The next theorem classifies all divisible abelian groups.

**Definition.** If $D$ is a divisible abelian group, define[4]

$$\delta_\infty(D) = \dim_{\mathbb{Q}}(D/tD)$$

and, for all primes $p$, define

$$\delta_p(D) = \dim_{\mathbb{F}_p}(D[p]).$$

Of course, dimensions may be infinite cardinals.

**Theorem 8.28.**

(i) *An abelian group $D$ is an injective $\mathbb{Z}$-module if and only it is a divisible group.*

(ii) *Every divisible abelian group is isomorphic to a direct sum of copies of $\mathbb{Q}$ and of copies of $\mathbb{Z}(p^\infty)$ for various primes $p$.*

(iii) *Let $D$ and $D'$ be divisible abelian groups. Then $D \cong D'$ if and only if $\delta_\infty(D) = \delta_\infty(D')$ and $\delta_p(D) = \delta_p(D')$ for all primes $p$.*

**Proof.**

(i) This is proved in Corollary 6.93.

(ii) If $x \in D$ has finite order, $n$ is a positive integer, and $x = ny$, then $y$ has finite order. It follows that if $D$ is divisible, then its torsion subgroup $tD$ is also divisible, and hence

$$D = tD \oplus V,$$

where $V$ is torsion-free (by Proposition 6.84 on page 461). Since every quotient of a divisible group is divisible, $V$ is torsion-free and divisible, and hence it is a vector space over $\mathbb{Q}$, by Proposition 8.24.

---

[4]Recall Exercise 6.71 on page 467: every torsion-free divisible abelian group is a vector space over $\mathbb{Q}$.

Now $tD$ is the direct sum of its primary components: $tD = \bigoplus_p T_p$, each of which is $p$-primary and divisible, and so it suffices to prove that each $T_p$ is a direct sum of copies of $\mathbb{Z}(p^\infty)$. If $\dim(T_p[p]) = r$ ($r$ may be infinite), define $W$ to be a direct sum of $r$ copies of $\mathbb{Z}(p^\infty)$, so that $\dim(W[p]) = r$. Lemma 8.27 now shows that $T_p \cong W$.

(iii) By Proposition 8.2(ii), if $D \cong D'$, then $D/tD \cong D'/tD'$ and $tD \cong tD'$; hence, the $p$-primary components $(tD)_p \cong (tD')_p$ for all $p$. But $D/tD$ and $D'/tD'$ are isomorphic vector spaces over $\mathbb{Q}$, and hence have the same dimension; moreover, the vector spaces $(tD)_p[p]$ and $(tD')_p[p]$ are also isomorphic, so they, too, have the same dimension.

For the converse, write $D = V \oplus \bigoplus_p T_p$ and $D' = V' \oplus \bigoplus_p T'_p$, where $V$ and $V'$ are torsion-free divisible, and $T_p$ and $T'_p$ are $p$-primary divisible. By Lemma 8.27, $\delta_p(D) = \delta_p(D')$ implies $T_p \cong T'_p$, while $\delta_\infty(D) = \delta_\infty(D')$ implies that the vector spaces $V$ and $V'$ are isomorphic. Now use Proposition 6.43 to assemble these isomorphisms into an isomorphism $D \cong D'$.   •

We can now describe some familiar groups. The additive group of a field $K$ is easy to describe: it is a vector space over its prime field $k$, and so the only question is computing its degree $[K : k] = \dim_k(K)$. In particular, if $K = \overline{k}$ is the algebraic closure of $k = \mathbb{F}_p$ or of $k = \mathbb{Q}$, then $[\overline{k} : k] = \aleph_0$.

**Corollary 8.29.** *Let $K$ be an algebraically closed field, and let $K^\times$ be its multiplicative group.*

(i) *If $t(K^\times)$ is the torsion subgroup of its prime field, then $K^\times \cong t(K^\times) \oplus V$, where $V$ is a vector space over $\mathbb{Q}$.*

(ii) *$t(\overline{\mathbb{Q}}^\times) \cong \mathbb{Q}/\mathbb{Z} \cong \bigoplus_p \mathbb{Z}(p^\infty)$, where $\overline{\mathbb{Q}}$ is the algebraic closure of $\mathbb{Q}$.*

(iii) *$t(\overline{\mathbb{F}}_p^\times) \cong \bigoplus_{q \neq p} \mathbb{Z}(q^\infty)$, where $\overline{\mathbb{F}}_p$ is the algebraic closure of $\mathbb{F}_p$.*

**Proof.**

(i) Since $K$ is algebraically closed, the polynomials $x^n - a$ have roots in $K$ whenever $a \in K$; this says that every $a$ has an $n$th root in $K$, which is the multiplicative way of saying that $K^\times$ is a divisible abelian group. An element $a \in K$ has finite order if and only if $a^n = 1$ for some positive integer $n$; that is, $a$ is an $n$th root of unity. It is easy to see that the torsion subgroup $T$ of $K^\times$ is divisible and, hence, it is a direct summand: $K^\times = T \oplus V$, by Lemma 8.25. The complementary summand $V$ is a vector space over $\mathbb{Q}$, for $V$ is torsion-free divisible. Finally, we claim that $T = t(K^\times)$, for all roots of unity in $K^\times$ are already present in the algebraic closure of the prime field.

(ii) If $K = \overline{\mathbb{Q}}$ is the algebraic closure of $\mathbb{Q}$, there is no loss in generality in assuming that $K \subseteq \mathbb{C}$. Now the torsion subgroup $T$ of $K$ consists of all the roots of unity $e^{2\pi i r}$, where $r \in \mathbb{Q}$. It follows easily that the map $r \mapsto e^{2\pi i r}$ is a surjection $\mathbb{Q} \to T$ having kernel $\mathbb{Z}$, so that $T \cong \mathbb{Q}/\mathbb{Z}$.

(iii) Let us examine the primary components of $t(\overline{\mathbb{F}}_p^\times)$. If $q \neq p$ is a prime, then the polynomial $f(x) = x^q - 1$ has no repeated roots [for $\gcd(f(x), f'(x)) = 1$], and so there is some $q$th root of unity other than 1. Thus, the $q$-primary component is nontrivial, and there is at least one summand isomorphic to $\mathbb{Z}(q^\infty)$ (since $t(\overline{\mathbb{F}}_p^\times)$ is a torsion divisible abelian group, it is a direct sum of copies of Prüfer groups). Were there more than one such summand, there would be more than $q$ elements of order $q$, and this would provide too many roots for $x^q - 1$ in $\overline{\mathbb{F}}_p$. Finally, there is no summand isomorphic to $\mathbb{Z}(p^\infty)$, for $x^p - 1 = (x - 1)^p$ in $(\overline{\mathbb{F}}_p)[x]$, and so 1 is the only $p$th root of unity. $\quad\bullet$

**Corollary 8.30.** *The following abelian groups $G$ are isomorphic:*

$$\mathbb{C}^\times; \quad (\mathbb{Q}/\mathbb{Z}) \oplus \mathbb{R}; \quad \mathbb{R}/\mathbb{Z}; \quad \prod_p \mathbb{Z}(p^\infty); \quad S^1$$

*($S^1$ is the circle group; that is, the multiplicative group of all complex numbers $z$ with $|z| = 1$).*

**Proof.** All the groups $G$ on the list are divisible. Theorem 8.28(iii) shows they are isomorphic, since $\delta_p(G) = 1$ for all primes $p$ and $\delta_\infty(G) = c$ (the cardinal of the continuum). $\quad\bullet$

## Exercises

* **8.1.** If $M$ is an $R$-module, where $R$ is a domain, and $r \in R$, let $\mu_r\colon M \to M$ be multiplication by $r$; that is, $\mu_r\colon m \mapsto rm$ [see Example 6.10].

  (i) Prove that $\mu_r$ is an injection for every $r \neq 0$ if and only if $M$ is torsion-free.

  (ii) Prove that $\mu_r$ is a surjection for every $r \neq 0$ if and only if $M$ is divisible.

  (iii) Prove that $M$ is a vector space over $Q$ if and only if, for every $r \neq 0$, the map $\mu_r\colon M \to M$ is an isomorphism.

* **8.2.** Let $R$ be a domain with $Q = \mathrm{Frac}(R)$, and let $M$ be an $R$-module.

  (i) Prove that $M$ is a vector space over $Q$ if and only if it is torsion-free and divisible. (This generalizes Exercise 6.71 on page 467.)

  (ii) Let $\mu_r\colon M \to M$ be multiplication by $r$, where $r \in R$. For every $R$-module $A$, prove that the induced maps

  $$(\mu_r)_*\colon \mathrm{Hom}_R(A, M) \to \mathrm{Hom}_R(A, M) \text{ and } (\mu_r)^*\colon \mathrm{Hom}_R(M, A) \to \mathrm{Hom}_R(M, A)$$

  are also multiplication by $r$.

  (iii) Prove that both $\mathrm{Hom}_R(Q, M)$ and $\mathrm{Hom}_R(M, Q)$ are vector spaces over $Q$.

* **8.3.** Let $G = \prod_p \langle a_p \rangle$, where $p$ varies over all the primes, and $\langle a_p \rangle \cong \mathbb{I}_p$.

  (i) Prove that $tG = \bigoplus_p \langle a_p \rangle$.
  **Hint.** Use Exercise 4.7 on page 241.

  (ii) Prove that $G/tG$ is a divisible group.

  (iii) Prove that $tG$ is not a direct summand of $G$.
  **Hint.** Show that $\mathrm{Hom}(\mathbb{Q}, G) = \{0\}$ but that $\mathrm{Hom}(\mathbb{Q}, G/tG) \neq \{0\}$. Conclude that $G \ncong tG \oplus G/tG$.

**8.4.** Prove that if $R$ is a domain that is not a field, then an $R$-module $M$ that is both projective and injective must be $\{0\}$.

**Hint.** Use Exercise 6.63 on page 459.

**8.5.** If $M$ is a torsion $R$-module, where $R$ is a PID, prove that

$$\operatorname{Hom}_R(M, M) \cong \prod_{(p)} \operatorname{Hom}_R(M_{(p)}, M_{(p)}),$$

where $M_{(p)}$ is the $(p)$-primary component of $M$.

* **8.6.**  (i) If $G$ is a torsion abelian group with $p$-primary components $\{G_p : p \in P\}$, where $P$ is the set of all primes, prove that $G = t\left(\prod_{p \in P} G_p\right)$.

  (ii) Prove that $\left(\prod_{p \in P} G_p\right) / \left(\bigoplus_{p \in P} G_p\right)$ is torsion-free and divisible.
     **Hint.** Use Exercise 4.7 on page 241.

**8.7.**  Let $R$ be a PID.

  (i) If $M$ and $N$ are finitely generated torsion $R$-modules, prove, for all primes $(p)$ and all $n \geq 0$, that

$$U_{(p)}(n, M \oplus N) = U_{(p)}(n, M) + U_{(p)}(n, N).$$

  (ii) If $A$, $B$, and $C$ are finitely generated $R$-modules, prove that $A \oplus B \cong A \oplus C$ implies $B \cong C$.

  (iii) If $A$ and $B$ are finitely generated $R$-modules, prove that $A \oplus A \cong B \oplus B$ implies $A \cong B$.

**8.8.** Let $R$ be a PID, and let $M$ be an $R$-module, not necessarily primary. Define a submodule $S \subseteq M$ to be a ***pure submodule*** if $S \cap rM = rS$ for all $r \in R$.

  (i) Prove that if $M$ is a $(p)$-primary module, where $(p)$ is a nonzero prime ideal in $R$, then a submodule $S \subseteq M$ is pure as just defined if and only if $S \cap p^n M = p^n S$ for all $n \geq 0$.

  (ii) Prove that every direct summand of $M$ is a pure submodule.

  (iii) Prove that the torsion submodule $tM$ is a pure submodule of $M$.

  (iv) Prove that if $M/S$ is torsion-free, then $S$ is a pure submodule of $M$.

  (v) Prove that if $\mathcal{S}$ is a family of pure submodules of a module $M$ that is a chain under inclusion (that is, if $S, S' \in \mathcal{S}$, then either $S \subseteq S'$ or $S' \subseteq S$), then $\bigcup_{S \in \mathcal{S}} S$ is a pure submodule of $M$.

  (vi) Give an example of a pure submodule that is not a direct summand.

**8.9.**  (i) If $F$ is a finitely generated free $R$-module, where $R$ is a PID, prove that every pure submodule of $F$ is a direct summand.

  (ii) Let $R$ be a PID and let $M$ be a finitely generated $R$-module. Prove that a submodule $S \subseteq M$ is a pure submodule of $M$ if and only if $S$ is a direct summand of $M$.

* **8.10.** Call a subset $X$ of an abelian group $A$ ***independent*** if, whenever $\sum_i m_i x_i = 0$, where $m_i \in \mathbb{Z}$ and almost all $m_i = 0$, then $m_i = 0$ for all $i$. Define $\operatorname{rank}(A)$ to be the number of elements in a maximal independent subset of $A$.

  (i) If $X$ is independent, prove that $\langle X \rangle = \bigoplus_{x \in X} \langle x \rangle$, a direct sum of cyclic groups.

(ii) If $A$ is torsion, prove that rank$(A) = 0$.

(iii) If $A$ is free abelian, prove that the two notions of rank coincide [the earlier notion defined rank$(A)$ as the number of elements in a basis of $A$].

(iv) Prove that rank$(A) = \dim(\mathbb{Q} \otimes_{\mathbb{Z}} A)$, and conclude that every two maximal independent subsets of $A$ have the same number of elements; that is, rank$(A)$ is well-defined.

(v) If $0 \to A \to B \to C \to 0$ is an exact sequence of abelian groups, prove that rank$(B) = $ rank$(A) + $ rank$(C)$.

**8.11. (Kulikov)** Let $G$ be an abelian $p$-group. Call a subset $X \subseteq G$ *pure-independent* if $X$ is independent (Exercise 8.10) and $\langle X \rangle$ is a pure subgroup.

(i) Prove that $G$ has a maximal pure-independent subset.

(ii) If $X$ is a maximal pure-independent subset of $G$, the subgroup $B = \langle X \rangle$ is called a *basic subgroup* of $G$. Prove that if $B$ is a basic subgroup of $G$, then $G/B$ is divisible. (See Fuchs, *Infinite Abelian Groups* I, Chapter VI, for more about basic subgroups.)

**8.12.** Prove that if $G$ and $H$ are torsion abelian groups, then $G \otimes_{\mathbb{Z}} H$ is a direct sum of cyclic groups.

**Hint.** Use an exact sequence $0 \to B \to G \to G/B \to 0$, where $B$ is a basic subgroup, along with the following theorem: if $0 \to A' \xrightarrow{i} A \to A'' \to 0$ is an exact sequence of abelian groups and $i(A')$ is a pure subgroup of $A$, then

$$0 \to A' \otimes_{\mathbb{Z}} B \to A \otimes_{\mathbb{Z}} B \to A'' \otimes_{\mathbb{Z}} B \to 0$$

is exact for every abelian group $B$ (Rotman, *An Introduction to Homological Algebra*, p. 150).

\* **8.13.** Let $M$ be a $(p)$-primary $R$-module, where $R$ is a PID and $(p)$ is a prime ideal. Define, for all $n \geq 0$,

$$V_{(p)}(n, M) = \dim\left((p^n M \cap M[p])/(p^{n+1} M \cap M[p])\right),$$

where $M[p] = \{m \in M : pm = 0\}$.

(i) Prove that $V_{(p)}(n, M) = U_{(p)}(n, M)$ when $M$ is finitely generated. (The invariant $V_{(p)}(n, M)$ is introduced because we cannot subtract infinite cardinal numbers.)

(ii) Let $M = \bigoplus_{i \in I} C_i$ be a direct sum of cyclic modules $C_i$, where $I$ is any index set, possibly infinite. Prove that the number of summands $C_i$ having order ideal $(p^n)$ is $V_{(p)}(n, M)$, and hence it is an invariant of $M$.

(iii) Let $M$ and $M'$ be torsion modules that are direct sums of cyclic modules. Prove that $M \cong M'$ if and only if $V_{(p)}(n, M) = V_{(p)}(n, M')$ for all $n \geq 0$ and all prime ideals $(p)$.

**8.14.** (i) If $p$ is a prime and $G = t\left(\prod_{k \geq 1} \langle a_k \rangle\right)$, where $\langle a_k \rangle$ is a cyclic group of order $p^k$, prove that $G$ is an uncountable $p$-primary abelian group with $V_p(n, G) = 1$ for all $n \geq 0$.

(ii) Use Exercise 8.13 to prove that the primary group $G$ in part (i) is not a direct sum of cyclic groups.

\* **8.15.** Generalize Proposition 6.118 as follows: if $R$ is a domain, $D$ is a divisible $R$-module, and $T$ is a torsion $R$-module, then $D \otimes_R T = \{0\}$.

**8.16.** Prove that there is an additive functor $d \colon \mathbf{Ab} \to \mathbf{Ab}$ that assigns to each group $G$ its maximal divisible subgroup $dG$.

**8.17.**    (i) Prove that $\mathbb{Z}(p^\infty)$ has no maximal subgroups.

  (ii) Prove that $\mathbb{Z}(p^\infty) \cong \varinjlim \mathbb{I}_{p^n}$.

  (iii) Prove that a presentation of $\mathbb{Z}(p^\infty)$ is
$$(a_n, \ n \geq 1 \mid pa_1 = 0, pa_{n+1} = a_n \text{ for } n \geq 1).$$

∗ **8.18.** Prove that an abelian group $G$ is reduced if and only if $\operatorname{Hom}_\mathbb{Z}(\mathbb{Q}, G) = \{0\}$.

**8.19.** If $0 \to A \to B \to C \to 0$ is exact and both $A$ and $C$ are reduced, prove that $B$ is reduced.

**Hint.** Use left exactness of $\operatorname{Hom}_\mathbb{Z}(\mathbb{Q}, \square)$.

**8.20.** If $\{D_i : i \in I\}$ is a family of divisible abelian groups, prove that $\prod_{i \in I} D_i$ is isomorphic to a direct *sum* $\bigoplus_{j \in J} E_j$, where each $E_j$ is divisible.

**8.21.** Prove that $\mathbb{Q}^\times \cong \mathbb{I}_2 \oplus F$, where $F$ is a free abelian group of infinite rank.

**8.22.** Prove that $\mathbb{R}^\times \cong \mathbb{I}_2 \oplus \mathbb{R}$.

**Hint.** Use $e^x$.

**8.23.**    (i) Prove, for every group homomorphism $f \colon \mathbb{Q} \to \mathbb{Q}$, that there exists $r \in \mathbb{Q}$ with $f(x) = rx$ for all $x \in \mathbb{Q}$.

  (ii) Prove that $\operatorname{Hom}_\mathbb{Z}(\mathbb{Q}, \mathbb{Q}) \cong \mathbb{Q}$.

  (iii) Prove that $\operatorname{End}_\mathbb{Z}(\mathbb{Q}) \cong \mathbb{Q}$ as rings.

**8.24.** Prove that if $G$ is a nonzero abelian group, then $\operatorname{Hom}_\mathbb{Z}(G, \mathbb{Q}/\mathbb{Z}) \neq \{0\}$.

**8.25.** Prove that an abelian group $G$ is injective if and only if every nonzero quotient group is infinite.

**8.26.** Prove that if $G$ is an infinite abelian group all of whose proper subgroups are finite, then $G \cong \mathbb{Z}(p^\infty)$ for some prime $p$.[5]

**8.27.**    (i) Let $D = \bigoplus_{i=1}^n D_i$, where each $D_i \cong \mathbb{Z}(p_i^\infty)$ for some prime $p_i$. Prove that every subgroup of $D$ has DCC.

  (ii) Prove, conversely, that if an abelian group $G$ has DCC, then $G$ is isomorphic to a subgroup of a direct sum of a finite number of copies of $\mathbb{Z}(p_i^\infty)$.

**8.28.** If $G = \prod_{p \in P} \mathbb{Z}(p^\infty)$, where $P$ is the set of all primes, prove that $tG = \bigoplus_{p \in P} \mathbb{Z}(p^\infty)$ and $G/tG \cong \mathbb{R}$.

**8.29.** Let $R = k[x, y]$ be the polynomial ring in two variables over a field $k$, and let $I = (x, y)$.

  (i) Prove that $x \otimes y - y \otimes x \neq 0$ in $I \otimes_R I$.
  **Hint.** Show that this element has a nonzero image in $(I/I^2) \otimes_R (I/I^2)$.

  (ii) Prove that $x \otimes y - y \otimes x$ is a torsion element in $I \otimes_R I$, and conclude that the tensor product of torsion-free modules need not be torsion-free.

---

[5]There exist infinite *nonabelian* groups all of whose proper subgroups are finite. Indeed, Ol'shanskii proved that there exist infinite groups, called ***Tarski monsters***, all of whose proper subgroups have prime order.

## Section 8.2. Rational Canonical Forms

In Chapter 2, we saw that if $T\colon V \to V$ is a linear transformation and $X = x_1, \dots, x_n$ is a basis of $V$, then $T$ determines the matrix $A = {}_X[T]_X$ whose $i$th column consists of the coordinate list of $T(x_i)$ with respect to $X$. If $Y$ is another basis of $V$, then the matrix $B = {}_Y[T]_Y$ may be different from $A$, but Corollary 2.130 says that $A$ and $B$ are *similar*; that is, there exists a nonsingular matrix $P$ with $B = PAP^{-1}$.

**Corollary 2.130.** *Let $T\colon V \to V$ be a linear transformation on a vector space $V$ over a field $k$. If $X$ and $Y$ are bases of $V$, then there is a nonsingular matrix $P$ with entries in $k$, namely, $P = {}_Y[1_V]_X$, so that*

$$ {}_Y[T]_Y = P\big({}_X[T]_X\big)P^{-1}. $$

*Conversely, if $B = PAP^{-1}$, where $B, A$, and $P$ are $n \times n$ matrices with entries in $k$ and $P$ is nonsingular, then there is a linear transformation $T\colon k^n \to k^n$ and bases $X$ and $Y$ of $k^n$ such that $B = {}_Y[T]_Y$ and $A = {}_X[T]_X$.*

We now consider how to determine whether two given matrices are similar.

**Example 8.31.** Recall Example 6.8(iv): if $T\colon V \to V$ is a linear transformation, where $V$ is a vector space over a field $k$, then $V$ is a $k[x]$-module: it admits a scalar multiplication by polynomials $f(x) \in k[x]$:

$$ f(x)v = \Big(\sum_{i=0}^{m} c_i x^i\Big)v = \sum_{i=0}^{m} c_i T^i(v), $$

where $T^0$ is the identity map $1_V$, and $T^i$ is the composite of $T$ with itself $i$ times if $i \geq 1$. We denote this $k[x]$-module by $V^T$.

We now show that if $V$ is $n$-dimensional, then $V^T$ is a finitely generated torsion $k[x]$-module. To see that $V^T$ is finitely generated, note that if $X = v_1, \dots, v_n$ is a (vector space) basis of $V$, then $V^T = \langle v_1, \dots, v_n \rangle$.[6] To see that $V^T$ is torsion, note that Corollary 2.114 says, for each $v \in V$, that the list $v, T(v), T^2(v), \dots, T^n(v)$ must be linearly dependent (for it contains $n + 1$ vectors). Therefore, there are $c_i \in k$, not all 0, with $\sum_{i=0}^{n} c_i T^i(v) = 0$, and this says that $g(x) = \sum_{i=0}^{n} c_i x^i$ lies in the order ideal $\mathrm{ann}(v)$.

An important special case of the construction of the $k[x]$-module $V^T$ arises from an $n \times n$ matrix $A$ with entries in $k$. Define $T\colon k^n \to k^n$ by $T(v) = Av$ (the elements of $k^n$ are $n \times 1$ column vectors $v$, and $Av$ is matrix multiplication). This $k[x]$-module $(k^n)^T$ is denoted by $(k^n)^A$; explicitly, the action is given by

$$ f(x)v = \Big(\sum_{i=0}^{m} c_i x^i\Big)v = \sum_{i=0}^{m} c_i A^i v. \quad \blacktriangleleft $$

It is shown in Example 6.8(iv) that $V^T \cong (k^n)^A$ as $k[x]$-modules.

---

[6]Most likely, $V^T$ can be generated by a proper sublist of $X$, since to say that $X$ generates $V$ is to say, for each $v \in V$, that $v = \sum_i a_i v_i$ for $a_i \in k$, while $X$ generates $V^T$ says that $v = \sum_i f_i(x)_i v_i$ for $f_i(x) \in k[x]$.

We now interpret the results in the previous section (about modules over general PIDs) for the special $k[x]$-modules $V^T$ and $(k^n)^A$. If $T\colon V \to V$ is a linear transformation, then a submodule $W$ of $V^T$ is called an **invariant subspace**; that is, $W$ is a subspace of $V$ with $T(W) \subseteq W$, and so the restriction $T|W$ is a linear transformation on $W$; that is, $T|W\colon W \to W$.

**Definition.** If $A$ is an $r \times r$ matrix and $B$ is an $s \times s$ matrix, then their **direct sum** $A \oplus B$ is the $(r + s) \times (r + s)$ matrix

$$A \oplus B = \begin{bmatrix} A & 0 \\ 0 & B \end{bmatrix}.$$

**Lemma 8.32.** *If $V^T = W \oplus W'$, where $W$ and $W'$ are submodules, then*

$$_{B \cup B'}[T]_{B \cup B'} = {}_B[T|W]_B \oplus {}_{B'}[T|W']_{B'},$$

*where $B = w_1, \ldots, w_r$ is a basis of $W$ and $B' = w'_1, \ldots, w'_s$ is a basis of $W'$.*

**Proof.** Since $W$ and $W'$ are submodules, we have $T(W) \subseteq W$ and $T(W') \subseteq W'$; that is, the restrictions $T|W$ and $T|W'$ are linear transformations on $W$ and $W'$, respectively. Since $V = W \oplus W'$, the union $B \cup B'$ is a basis of $V$. Finally, the matrix $_{B \cup B'}[T]_{B \cup B'}$ is a direct sum: $T(w_i) \in W$, so that it is a linear combination of $w_1, \ldots, w_r$, and hence it requires no nonzero coordinates from the $w'_j$; similarly, $T(w'_j) \in W'$, and so it requires no nonzero coordinates from the $w_i$. $\bullet$

When we studied permutations, we saw that the cycle notation allowed us to recognize important properties that are masked by the conventional functional notation. We now ask whether there is an analogous way to denote matrices; for example, if $V^T$ is a cyclic $k[x]$-module, can we find a basis $B$ of $V$ so that the corresponding matrix $_B[T]_B$ displays the order ideal of $T$?

**Lemma 8.33.** *Let $T\colon V \to V$ be a linear transformation on a vector space $V$ over a field $k$, and let $W$ be a submodule of $V^T$. Then $W$ is cyclic with generator $v$ of finite order if and only if there is an integer $s \geq 1$ such that*

$$v,\ Tv,\ T^2v,\ \ldots,\ T^{s-1}v$$

*is a (vector space) basis of $W$. If $T^sv + \sum_{i=0}^{s-1} c_iT^iv = 0$, then $\operatorname{ann}(v) = (g)$, where $g(x) = x^s + c_{s-1}x^{s-1} + \cdots + c_1x + c_0$, and*

$$W \cong k[x]/(g).$$

**Proof.** Since the cyclic module $W = \langle v \rangle = \{f(x)v : f(x) \in k[x]\}$ has finite order, there is a nonzero polynomial $f(x) \in k[x]$ with $f(x)v = 0$. If $g(x)$ is the monic polynomial of least degree with $g(x)v = 0$, then $(g) = \operatorname{ann}(v)$ and $W \cong k[x]/(g)$; let $\deg(g) = s$. We claim that the list $v, Tv, T^2v, \ldots, T^{s-1}v$ is linearly independent: a nontrivial linear combination of them being zero would give a polynomial $h(x)$ with $h(x)v = 0$ and $\deg(h) < \deg(g)$. This list spans $W$: if $w \in W$, then $w = f(x)v$ for some $f(x) \in k[x]$. The Division Algorithm gives $q, r \in k[x]$ with $f = qg + r$ and $r = 0$ or $\deg(r) < s$, and $w = r(x)v$ lies in the subspace spanned by $v, Tv, T^2v, \ldots, T^{s-1}v$. Therefore, this list is a vector space basis of $W$.

To prove the converse, assume that there is a vector $v \in W$ and an integer $s \geq 1$ such that the list $v, Tv, T^2v, \ldots, T^{s-1}v$ is a (vector space) basis of $W$. It suffices to show that $W = \langle v \rangle$ and that $v$ has finite order. Now $\langle v \rangle \subseteq W$, for $W$ is a submodule of $V^T$ containing $v$. For the reverse inclusion, each $w \in W$ is a linear combination of the basis: there are $c_i \in k$ with $w = \sum_i c_i T^i v$. Hence, if $f(x) = \sum_i c_i x^i$, then $w = f(x)v \in \langle v \rangle$. Therefore, $W = \langle v \rangle$. Finally, $v$ has finite order. Adjoining the vector $T^s v \in W$ to the basis $v, Tv, T^2v, \ldots, T^{s-1}v$ gives a linearly dependent list, and a nontrivial $k$-linear combination gives a nonzero polynomial in $\operatorname{ann}(v)$.   •

**Definition.** If $g(x) = x + c_0$, then its **companion matrix** $C(g)$ is the $1 \times 1$ matrix $[-c_0]$; if $s \geq 2$ and $g(x) = x^s + c_{s-1}x^{s-1} + \cdots + c_1 x + c_0$, then its **companion matrix** $C(g)$ is the $s \times s$ matrix

$$C(g) = \begin{bmatrix} 0 & 0 & 0 & \cdots & 0 & -c_0 \\ 1 & 0 & 0 & \cdots & 0 & -c_1 \\ 0 & 1 & 0 & \cdots & 0 & -c_2 \\ 0 & 0 & 1 & \cdots & 0 & -c_3 \\ \vdots & \vdots & \vdots & \vdots & \vdots & \vdots \\ 0 & 0 & 0 & \cdots & 1 & -c_{s-1} \end{bmatrix}.$$

Obviously, we can recapture the polynomial $g(x)$ from the last column of the companion matrix $C(g)$.

**Lemma 8.34.** *Let $T \colon V \to V$ be a linear transformation on a vector space $V$ over a field $k$, and let $V^T$ be a cyclic $k[x]$-module with generator $v$. If $\operatorname{ann}(v) = (g)$, where $g(x) = x^s + c_{s-1}x^{s-1} + \cdots + c_1 x + c_0$, then $B = v, Tv, T^2v, \ldots, T^{s-1}v$ is a basis of $V$ and the matrix ${}_B[T]_B$ is the companion matrix $C(g)$.*

**Proof.** Let $A = {}_B[T]_B$. By definition, the first column of $A$ consists of the coordinate list of $T(v)$, the second column, the coordinate list of $T(Tv) = T^2v$, and, more generally, for $i < s - 1$, we have $T(T^i v) = T^{i+1}v$; that is, $T$ sends each basis vector into the next one. However, for the last basis vector, $T(T^{s-1}v) = T^s v = -\sum_{i=0}^{s-1} c_i T^i v$, where $g(x) = x^s + \sum_{i=0}^{s-1} c_i x^i$. Thus, ${}_B[T]_B$ is the companion matrix $C(g)$.   •

**Theorem 8.35.**

(i) *Let $A$ be an $n \times n$ matrix with entries in a field $k$. If*

$$(k^n)^A = W_1 \oplus \cdots \oplus W_t,$$

*where each $W_i$ is cyclic, say, with order ideal $(g_i)$, then $A$ is similar to a direct sum of companion matrices*

$$C(g_1) \oplus \cdots \oplus C(g_t).$$

(ii) *Every $n \times n$ matrix $A$ over a field $k$ is similar to a direct sum of companion matrices*

$$C(g_1) \oplus \cdots \oplus C(g_t)$$

*in which the $g_i(x)$ are monic polynomials and*

$$g_1(x) \mid g_2(x) \mid \cdots \mid g_t(x).$$

**Proof.** Define $V = k^n$ and define $T \colon V \to V$ by $T(y) = Ay$, where $y$ is a column vector.

(i) By Lemma 8.34, each $W_i$ has a basis $B_i$ such that the matrix of $T|W_i$ with respect to $B_i$ is $C(g_i)$, the companion matrix of $g_i(x)$. Now $B_1 \cup \cdots \cup B_t$ is a basis of $V$, and Proposition 8.32 shows that $T$ has the desired matrix with respect to this basis. By Corollary 2.130, $A$ is similar to $C(g_1) \oplus \cdots \oplus C(g_t)$.

(ii) By Example 8.31, the $k[x]$-module $V^T$ is a finitely generated torsion module, and so the consequence of the Basis Theorem, Proposition 8.19, gives

$$(k^n)^A = W_1 \oplus W_2 \oplus \cdots \oplus W_t,$$

where each $W_i$ is cyclic, say, with generator $v_i$ having order ideal $(g_i)$, and $g_1(x) \mid g_2(x) \mid \cdots \mid g_t(x)$. The statement now follows from part (i).  •

**Definition.** A **_rational canonical form_**[7] is a matrix $R$ that is a direct sum of companion matrices,

$$R = C(g_1) \oplus \cdots \oplus C(g_t),$$

where the $g_i(x)$ are monic polynomials with $g_1(x) \mid g_2(x) \mid \cdots \mid g_t(x)$.

If a matrix $A$ is similar to a rational canonical form $C(g_1) \oplus \cdots \oplus C(g_t)$, where $g_1(x) \mid g_2(x) \mid \cdots \mid g_t(x)$, then its **_invariant factors_** are $g_1(x), g_2(x), \ldots, g_t(x)$.

We have just proved that every $n \times n$ matrix over a field is similar to a rational canonical form, and so it has invariant factors. Can a matrix $A$ have more than one list of invariant factors?

**Theorem 8.36.** *Let $k$ be a field.*

(i) *Two $n \times n$ matrices $A$ and $B$ with entries in $k$ are similar if and only if they have the same invariant factors.*

(ii) *An $n \times n$ matrix $A$ over $k$ is similar to exactly one rational canonical form.*

**Proof.**

(i) By Corollary 2.130, $A$ and $B$ are similar if and only if $(k^n)^A \cong (k^n)^B$. By Theorem 8.23, $(k^n)^A \cong (k^n)^B$ if and only if their invariant factors are the same.

---

[7]The usage of the adjective *rational* in *rational canonical form* arises as follows. If $E/k$ is a field extension, then we call the elements of the ground field $k$ *rational* (so that every $e \in E$ not in $k$ is irrational; this generalizes our calling numbers in $\mathbb{R}$ not in $\mathbb{Q}$ irrational). Now all the entries of a rational canonical form lie in the field $k$ and not in some extension of it. In contrast, the Jordan canonical form, to be discussed in the next section, involves the eigenvalues of a matrix which may not lie in $k$.

The adjective *canonical* originally meant something dictated by ecclesiastical law, as *canonical hours* being those times devoted to prayers. The meaning broadened to mean things of excellence, leading to the mathematical meaning of something given by a general rule or formula.

(ii) If $C(g_1) \oplus \cdots \oplus C(g_t)$ and $C(h_1) \oplus \cdots \oplus C(h_r)$ are rational canonical forms of $A$, then part (i) says that the $k[x]$-modules $k[x]/(g_1) \oplus \cdots \oplus k[x]/(g_t)$ and $k[x]/(h_1) \oplus \cdots \oplus k[x]/(h_r)$ are isomorphic. Theorem 8.23 gives $t = r$ and $g_i = h_i$ for all $i$. •

Recall Corollary 2.58: if $k$ is a subfield of a field $K$ and $f(x), g(x) \in k[x]$, then their gcd in $k[x]$ is equal to their gcd in $K[x]$. Here is an analog of this result.

**Corollary 8.37.**

(i) *Let $k$ be a subfield of a field $K$, and let $A$ and $B$ be $n \times n$ matrices with entries in $k$. If $A$ and $B$ are similar over $K$, then they are similar over $k$ (i.e., if there is a nonsingular matrix $P$ having entries in $K$ with $B = PAP^{-1}$, then there is a nonsingular matrix $Q$ having entries in $k$ with $B = QAQ^{-1}$).*

(ii) *If $\overline{k}$ is the algebraic closure of a field $k$, then two $n \times n$ matrices $A$ and $B$ with entries in $k$ are similar over $k$ if and only if they are similar over $\overline{k}$.*

**Proof.**

(i) Suppose that $g_1(x), \ldots, g_t(x)$ are the invariant factors of $A$ regarded as a matrix over $k$, while $G_1(x), \ldots, G_r(x)$ are the invariant factors of $A$ regarded as a matrix over $K$. By Theorem 8.36(ii), the two lists of polynomials coincide, for both are invariant factors for $A$ as a matrix over $K$. Now $B$ has the same invariant factors as $A$, for they are similar over $K$; since these invariant factors lie in $k$, however, $A$ and $B$ are similar over $k$.

(ii) Immediate from part (i). •

For example, suppose that $A$ and $B$ are matrices with real entries that are similar over the complexes; that is, if there is a nonsingular complex matrix $P$ such that $B = PAP^{-1}$, then there exists a nonsingular real matrix $Q$ such that $B = QAQ^{-1}$.

Does a linear transformation $T$ on a finite-dimensional vector space $V$ over a field $k$ leave any one-dimensional subspaces of $V$ invariant; that is, is there a nonzero vector $v \in V$ with $T(v) = \alpha v$ for some $\alpha \in k$? If $T \colon \mathbb{R}^2 \to \mathbb{R}^2$ is rotation by $90°$, then its matrix with respect to the standard basis is $\left[\begin{smallmatrix} 0 & -1 \\ 1 & 0 \end{smallmatrix}\right]$. Now

$$T \colon \begin{bmatrix} x \\ y \end{bmatrix} \mapsto \begin{bmatrix} 0 & -1 \\ 1 & 0 \end{bmatrix} \begin{bmatrix} x \\ y \end{bmatrix} = \begin{bmatrix} -y \\ x \end{bmatrix}.$$

If $v = \left[\begin{smallmatrix} x \\ y \end{smallmatrix}\right]$ is a nonzero vector and $T(v) = \alpha v$ for some $\alpha \in \mathbb{R}$, then $\alpha x = -y$ and $\alpha y = x$; it follows that $(\alpha^2 + 1)x = x$ and $(\alpha^2 + 1)y = y$. Since $v \neq 0$, $\alpha^2 + 1 = 0$ and $\alpha \notin \mathbb{R}$. Thus, $T$ has no one-dimensional invariant subspaces. Note that $\left[\begin{smallmatrix} 0 & -1 \\ 1 & 0 \end{smallmatrix}\right]$ is the companion matrix of $x^2 + 1$.

**Definition.** Let $V$ be a vector space over a field $k$ and let $T \colon V \to V$ be a linear transformation. If $Tv = \alpha v$, where $\alpha \in k$ and $v \in V$ is nonzero, then $\alpha$ is called an **eigenvalue** of $T$ and $v$ is called an **eigenvector** of $T$ for $\alpha$.

Let $A$ be an $n \times n$ matrix over a field $k$. If $Av = \alpha v$, where $\alpha \in k$ and $v \in k^n$ is a nonzero column, then $\alpha$ is called an **eigenvalue** of $A$ and $v$ is called an **eigenvector** of $A$ for $\alpha$.

Rotation by $90°$ has no (real) eigenvalues. At the other extreme, can a linear transformation have infinitely many eigenvalues?

Recall a fundamental result of Linear Algebra.

**Theorem 2.123.** *If $T: k^n \to k^n$ is a linear transformation, then there exists a unique $n \times n$ matrix $A$ such that $T(v) = Av$ for all $v \in k^n$.*

The proof writes $v$ as a column vector with respect to the standard basis $E = e_1, \ldots, e_n$ and $A = {}_E[T]_E$, the matrix whose $i$th column is the coordinate list of $Te_i$. It is now clear that an eigenvalue $\alpha$ of $T$ is an eigenvalue of $A$, because $T(v) = Av$ for all $v$. To say that $Av = \alpha v$ for $v$ nonzero is to say that $v$ is a nontrivial solution of the homogeneous system $(A - \alpha I)v = 0$; that is, $A - \alpha I$ is a singular matrix. But a matrix with entries in a field is singular if and only if its determinant is 0.[8]

**Definition.** The **characteristic polynomial** of an $n \times n$ matrix $A$ is

$$\psi_A(x) = \det(xI - A) \in k[x].$$

Thus, the eigenvalues of an $n \times n$ matrix $A$ over a field $k$ are the roots of $\psi_A(x)$, a polynomial of degree $n$, and so $A$ has at most $n$ eigenvalues in $k$. The eigenvalues of $A$ may not lie in $k$, but they do lie in $\overline{k}$, the algebraic closure of $k$.

Recall that the *trace* of an $n \times n$ matrix $A = [a_{ij}]$ is $\operatorname{tr}(A) = \sum_{i=1}^{n} a_{ii}$.

**Proposition 8.38.** *If $A = [a_{ij}]$ is an $n \times n$ matrix over a field $k$ having eigenvalues (with multiplicity) $\alpha_1, \ldots, \alpha_n$, then*

$$\operatorname{tr}(A) = -\sum_i \alpha_i \quad and \quad \det(A) = \prod_i \alpha_i.$$

**Proof.** As we observed on page 173, for any polynomial $f(x) \in k[x]$, if $f(x) = x^n + c_{n-1}x^{n-1} + \cdots + c_1 x + c_0 = (x - \alpha_1) \cdots (x - \alpha_n)$, then $c_{n-1} = -\sum_i \alpha_i$ and $c_0 = (-1)^n \prod_i \alpha_i$. In particular, $\psi_A(x) = \prod_{i=1}^{n}(x - \alpha_i)$, so that $c_{n-1} = -\sum_i \alpha_i = -\operatorname{tr}(A)$. Now the constant term of any polynomial $f(x)$ is just $f(0)$; setting $x = 0$ in $\psi_A(x) = \det(xI - A)$ gives $\psi_A(0) = \det(-A) = (-1)^n \det(A)$. Hence, $\det(A) = \prod_i \alpha_i$.  •

Here are some elementary facts about eigenvalues.

**Corollary 8.39.** *Let $A$ be an $n \times n$ matrix with entries in a field $k$.*

(i) *$A$ is singular if and only if $0$ is an eigenvalue of $A$.*

(ii) *If $\alpha$ is an eigenvalue of $A$, then $\alpha^n$ is an eigenvalue of $A^n$.*

(iii) *If $A$ is nonsingular and $\alpha$ is an eigenvalue of $A$, then $\alpha \neq 0$ and $\alpha^{-1}$ is an eigenvalue of $A^{-1}$.*

---

[8]We continue using familiar properties of determinants even though they will not be proved until Section 8.7.

**Proof.**

(i) If $A$ is singular, then the homogeneous system $AX = 0$ has a nontrivial solution; that is, there is a nonzero $v$ with $Av = 0$. But this just says that $Av = 0x$ (here, $0$ is a scalar), and so $0$ is an eigenvalue.

    Conversely, if $0$ is an eigenvalue, then $0 = \det(0I - A) = \pm \det(A)$, so that $\det(A) = 0$ and $A$ is singular.

(ii) There is a nonzero vector $v$ with $Av = \alpha v$. It follows by induction on $n \geq 1$ that $A^n v = \alpha^n v$.

(iii) If $v$ is an eigenvector of $A$ for $\alpha$, then
$$v = A^{-1} A v = A^{-1} \alpha v = \alpha A^{-1} v.$$

Therefore, $\alpha \neq 0$ (because eigenvectors are nonzero) and $\alpha^{-1} v = A^{-1} v.$   •

Let us return to rational canonical forms.

**Lemma 8.40.** *If $C(g)$ is the companion matrix of $g(x) \in k[x]$, then*
$$\det\big(xI - C(g)\big) = g(x).$$

**Proof.** If $g(x) = x + c_0$, then $C(g)$ is the $1 \times 1$ matrix $[-c_0]$, and $\det(xI - C(g)) = x + c_0 = g(x)$. If $\deg(g) = s \geq 2$, then

$$xI - C(g) = \begin{bmatrix} x & 0 & 0 & \cdots & 0 & c_0 \\ -1 & x & 0 & \cdots & 0 & c_1 \\ 0 & -1 & x & \cdots & 0 & c_2 \\ \vdots & \vdots & \vdots & \vdots & \vdots & \vdots \\ 0 & 0 & 0 & \cdots & -1 & x + c_{s-1} \end{bmatrix},$$

and Laplace expansion across the first row gives
$$\det(xI - C(g)) = x \det(L) + (-1)^{1+s} c_0 \det(M),$$

where $L$ is the matrix obtained by erasing the top row and first column, and $M$ is the matrix obtained by erasing the top row and last column. Now $M$ is a triangular $(s-1) \times (s-1)$ matrix having $-1$'s on the diagonal, while $L = xI - C\big((g(x)-c_0)/x\big)$. By induction, $\det(L) = (g(x) - c_0)/x$, while $\det(M) = (-1)^{s-1}$. Therefore,
$$\det(xI - C(g)) = x[(g(x) - c_0)/x] + (-1)^{(1+s)+(s-1)} c_0 = g(x). \quad •$$

If $R = C(g_1) \oplus \cdots \oplus C(g_t)$ is a rational canonical form, then
$$xI - R = \big[xI - C(g_1)\big] \oplus \cdots \oplus \big[xI - C(g_t)\big].$$

Let us anticipate Proposition 8.142: given square matrices $B_1, \ldots, B_t$, we have $\det(B_1 \oplus \cdots \oplus B_t) = \prod_{i=1}^{t} \det(B_i)$. With Lemma 8.40, this gives

$$\psi_R(x) = \prod_{i=1}^{t} \psi_{C(g_i)}(x) = \prod_{i=1}^{t} g_i(x).$$

Thus, the characteristic polynomial is the product of the invariant factors; in light of Corollary 8.21, the characteristic polynomial of an $n \times n$ matrix $A$ over a field $k$ is the analog for $(k^n)^A$ of the order of a finite abelian group.

**Example 8.41.** We now show that similar matrices have the same characteristic polynomial. If $B = PAP^{-1}$, then since $xI$ commutes with every matrix, we have $PxIP^{-1} = (xI)PP^{-1} = xI$. Therefore,

$$\begin{aligned} \psi_B(x) = \det(xI - B) &= \det(PxIP^{-1} - PAP^{-1}) \\ &= \det(P[xI - A]P^{-1}) = \det(P)\det(xI - A)\det(P^{-1}) \\ &= \det(xI - A) = \psi_A(x). \end{aligned}$$

It follows that if $A$ is similar to $C(g_1) \oplus \cdots \oplus C(g_t)$, then

$$\psi_A(x) = \prod_{i=1}^{t} g_i(x).$$

Therefore, similar matrices have the same eigenvalues with multiplicities. Hence, Proposition 8.38 says that similar matrices have the same trace and the same determinant. ◀

**Theorem 8.42 (Cayley–Hamilton).** *If $A$ is an $n \times n$ matrix with characteristic polynomial $\psi_A(x) = x^n + b_{n-1}x^{n-1} + \cdots + b_1 x + b_0$, then $\psi_A(A) = 0$; that is,*

$$A^n + b_{n-1}A^{n-1} + \cdots + b_1 A + b_0 I = 0.$$

**Proof.** We may assume that $A = C(g_1) \oplus \cdots \oplus C(g_t)$ is a rational canonical form, by Example 8.41, where $\psi_A(x) = g_1(x) \cdots g_t(x)$. If we regard $k^n$ as the $k[x]$-module $(k^n)^A$, then Corollary 8.20 says that $g_t(A)y = 0$ for all $y \in k^n$. Thus, $g_t(A) = 0$. As $g_t(x) \mid \psi_A(x)$, however, we have $\psi_A(A) = 0$.  •

There are proofs of the Cayley–Hamilton Theorem without rational canonical forms; for example, see Birkhoff–Mac Lane, *A Survey of Modern Algebra*, p. 341.

The Cayley–Hamilton Theorem is the analog of Corollary 1.50 to Lagrange's Theorem: if $G$ is a finite group, then $a^{|G|} = 1$ for all $a \in G$; in additive notation, $|G|a = 0$ for all $a \in G$. If $M = (k^n)^A$ is the $k[x]$-module corresponding to an $n \times n$ matrix $A$, then, as we mentioned above, the characteristic polynomial corresponds to the order of $M$.

**Definition.** The *minimal polynomial* $m_A(x)$ of an $n \times n$ matrix $A$ is the monic polynomial $f(x)$ of least degree with the property that $f(A) = 0$.

Recall that if $M$ is an $R$-module, then

$$\mathrm{ann}(M) = \{r \in R : rm = 0 \text{ for all } m \in M\}.$$

In particular, given an $n \times n$ matrix $A$, let $M = (k^n)^A$ be its corresponding $k[x]$-module. Since $k[x]$ is a PID, the ideal $\mathrm{ann}(M)$ is principal, and $m_A(x)$ is its monic generator.

**Proposition 8.43.** *The minimal polynomial $m_A(x)$ is a divisor of the characteristic polynomial $\psi_A(x)$, and every eigenvalue of $A$ is a root of $m_A(x)$.*

**Proof.** By the Cayley–Hamilton Theorem, $\psi_A \in \mathrm{ann}((k^n)^A)$. But $\mathrm{ann}((k^n)^A) = (m_A)$, so that $m_A \mid \psi_A$.

Corollary 8.20 implies that $g_t(x)$ is the minimal polynomial of $A$, where $g_t(x)$ is the invariant factor of $A$ of highest degree. It follows from the fact that

$$\psi_A(x) = g_1(x) \cdots g_t(x),$$

where $g_1(x) \mid g_2(x) \mid \cdots \mid g_t(x)$, that $m_A(x) = g_t(x)$ is a polynomial having every eigenvalue of $A$ as a root [of course, the multiplicity of a root of $m_A(x)$ may be less than its multiplicity as a root of the characteristic polynomial $\psi_A(x)$].   •

**Corollary 8.44.** *If all the eigenvalues of an $n \times n$ matrix $A$ are distinct, then $m_A(x) = \psi_A(A)$; that is, the minimal polynomial coincides with the characteristic polynomial.*   .

**Proof.** This is true because every root of $\psi_A(x)$ is a root of $m_A(x)$.   •

**Corollary 8.45.**

(i) *A finite abelian group $G$ is cyclic if and only if its exponent equals its order.*

(ii) *An $n \times n$ matrix $A$ is similar to a companion matrix if and only if*

$$m_A(x) = \psi_A(x).$$

**Remark.** An $n \times n$ matrix $A$ whose minimum polynomial is equal to its characteristic polynomial is called **nonderogatory**.   ◄

**Proof.**

(i) A cyclic group of order $n$ has only one invariant factor, namely, $n$; but Corollary 8.20 identifies the exponent as the last invariant factor.

If the exponent of $G$ is equal to its order $|G|$, then $G$ has only one invariant factor, namely, $|G|$. Hence, $G$ and $\mathbb{I}_{|G|}$ have the same invariant factors, and so they are isomorphic.

(ii) A companion matrix $C(g)$ has only one invariant factor, namely, $g(x)$; but Corollary 8.20 identifies the minimal polynomial as the last invariant factor.

If $m_A(x) = \psi_A(x)$, then $A$ has only one invariant factor, namely, $\psi_A(x)$, by Corollary 8.21. Hence, $A$ and $C(\psi_A(x))$ have the same invariant factors, and so they are similar.   •

## Exercises

**8.30.**   (i) How many $10 \times 10$ matrices $A$ over $\mathbb{R}$ are there, up to similarity, with $A^2 = I$?

(ii) How many $10 \times 10$ matrices $A$ over $\mathbb{F}_p$ are there, up to similarity, with $A^2 = I$?
**Hint.** The answer depends on the parity of $p$.

**8.31.** Find the rational canonical forms of

$$A = \begin{bmatrix} 1 & 2 \\ 3 & 4 \end{bmatrix}, \quad B = \begin{bmatrix} 2 & 0 & 0 \\ 1 & 2 & 0 \\ 0 & 0 & 3 \end{bmatrix}, \quad \text{and} \quad C = \begin{bmatrix} 2 & 0 & 0 \\ 1 & 2 & 0 \\ 0 & 1 & 2 \end{bmatrix}.$$

∗ **8.32.** If $A$ is similar to $A'$ and $B$ is similar to $B'$, prove that $A \oplus B$ is similar to $A' \oplus B'$.

**8.33.** Let $k$ be a field, and let $f(x)$ and $g(x)$ lie in $k[x]$. If $g(x) \mid f(x)$ and every root of $f(x)$ is a root of $g(x)$, show that there exists a matrix $A$ having minimal polynomial $m_A(x) = g(x)$ and characteristic polynomial $\psi_A(x) = f(x)$.

**8.34.**   (i) Give an example of two nonisomorphic finite abelian groups having the same order and the same exponent.

   (ii) Give an example of two nonsimilar matrices having the same characteristic polynomial and the same minimal polynomial.

---

## Section 8.3. Jordan Canonical Forms

If $k$ is a finite field, then $\mathrm{GL}(n, k)$ is a finite group, and so every element in it has finite order. Consider the group-theoretic question: what is the order of $A = \begin{bmatrix} 0 & 0 & 1 \\ 1 & 0 & 4 \\ 0 & 1 & 3 \end{bmatrix}$ in $\mathrm{GL}(3, \mathbb{F}_7)$? Of course, we can compute the powers $A^2, A^3, \ldots$, and Lagrange's Theorem guarantees that there is some $m \geq 1$ with $A^m = I$; but this procedure for finding the order of $A$ is rather tedious. We recognize $A$ as the companion matrix of

$$g(x) = x^3 - 3x^2 - 4x - 1 = x^3 - 3x^2 + 3x - 1 = (x - 1)^3$$

(remember that $g(x) \in \mathbb{F}_7[x]$). Now $A$ and $PAP^{-1}$ are conjugates in the group $\mathrm{GL}(3, \mathbb{F}_7)$ and, hence, they have the same order. But the powers of a companion matrix are complicated (e.g., the square of a companion matrix is not a companion matrix). We now give a second canonical form whose powers are easily calculated, and we shall use it to compute the order of $A$ later in this section.

**Definition.** Let $k$ be a field and let $\alpha \in k$. A $1 \times 1$ **Jordan block** is a matrix $J(\alpha, 1) = [\alpha]$ and, if $s \geq 2$, an $s \times s$ **Jordan block** is a matrix $J(\alpha, s)$ of the form

$$J(\alpha, s) = \begin{bmatrix} \alpha & 0 & 0 & \cdots & 0 & 0 \\ 1 & \alpha & 0 & \cdots & 0 & 0 \\ 0 & 1 & \alpha & \cdots & 0 & 0 \\ \vdots & \vdots & \vdots & \vdots & \vdots & \vdots \\ 0 & 0 & 0 & \cdots & \alpha & 0 \\ 0 & 0 & 0 & \cdots & 1 & \alpha \end{bmatrix}.$$

Here is a more compact description of a Jordan block when $s \geq 2$. Let $L$ denote the $s \times s$ matrix having all entries 0 except for 1's on the subdiagonal just below the main diagonal. With this notation, a Jordan block $J(\alpha, s)$ can be written as

$$J(\alpha, s) = \alpha I + L.$$

Let us regard $L$ as a linear transformation on $k^s$. If $e_1, \ldots, e_s$ is the standard basis, then $Le_i = e_{i+1}$ if $i < s$ while $Le_s = 0$. It follows easily that the matrix $L^2$ is all 0's except for 1's on the second subdiagonal below the main diagonal; $L^3$ is all 0's except for 1's on the third subdiagonal; $L^{s-1}$ has 1 in the $s, 1$ position, with 0's everywhere else, and $L^s = 0$. Thus, $L$ is nilpotent.

**Lemma 8.46.** *If* $J = J(\alpha, s) = \alpha I + L$ *is an* $s \times s$ *Jordan block, then for all* $m \geq 1$,

$$J^m = \alpha^m I + \sum_{i=1}^{s-1} \binom{m}{i} \alpha^{m-i} L^i.$$

**Proof.** Since $L$ and $\alpha I$ commute (the scalar matrix $\alpha I$ commutes with every matrix), the ring generated by $\alpha I$ and $L$ is commutative, and the Binomial Theorem applies. Finally, note that all terms involving $L^i$ for $i \geq s$ are 0 because $L^s = 0$.  •

**Example 8.47.** Different powers of $L$ are "disjoint"; that is, if $m \neq n$ and the $i, j$ entry of $L^n$ is nonzero, then the $i, j$ entry of $L^m$ is zero. For example,

$$\begin{bmatrix} \alpha & 0 \\ 1 & \alpha \end{bmatrix}^m = \begin{bmatrix} \alpha^m & 0 \\ m\alpha^{m-1} & \alpha^m \end{bmatrix}$$

and

$$\begin{bmatrix} \alpha & 0 & 0 \\ 1 & \alpha & 0 \\ 0 & 1 & \alpha \end{bmatrix}^m = \begin{bmatrix} \alpha^m & 0 & 0 \\ m\alpha^{m-1} & \alpha^m & 0 \\ \binom{m}{2}\alpha^{m-2} & m\alpha^{m-1} & \alpha^m \end{bmatrix}. \quad \blacktriangleleft$$

**Lemma 8.48.** *If* $g(x) = (x - \alpha)^s$, *then the companion matrix* $C(g)$ *is similar to the* $s \times s$ *Jordan block* $J(\alpha, s)$.

**Proof.** If $T: k^s \to k^s$ is defined by $z \mapsto C(g)z$, then the proof of Lemma 8.34 gives a basis of $k^s$ of the form $v, Tv, T^2v, \ldots, T^{s-1}v$. We claim that the list $Y = y_0, \ldots, y_{s-1}$ is also a basis of $k^s$, where

$$y_0 = v, \ y_1 = (T - \alpha I)v, \ \ldots, \ y_{s-1} = (T - \alpha I)^{s-1}v.$$

It is easy to see that the list $Y$ spans $V$, because $T^i v \in \langle y_0, \ldots, y_i \rangle$ for all $0 \leq i \leq s - 1$. Since there are $s$ elements in $Y$, Proposition 2.111 shows that $Y$ is a basis.

We now compute $J = {}_Y[T]_Y$, the matrix of $T$ with respect to $Y$. If $j + 1 \leq s$, then

$$\begin{aligned} Ty_j &= T(T - \alpha I)^j v \\ &= (T - \alpha I)^j Tv \\ &= (T - \alpha I)^j [\alpha I + (T - \alpha I)]v \\ &= \alpha(T - \alpha I)^j v + (T - \alpha I)^{j+1}v. \end{aligned}$$

Thus, if $j + 1 < s$, then

$$Ty_j = \alpha y_j + y_{j+1}.$$

If $j + 1 = s$, then $(T - \alpha I)^{j+1}v = (T - \alpha I)^s v = 0$, by the Cayley–Hamilton Theorem [for $\psi_{C(g)}(x) = (x - \alpha)^s$ here]; hence,

$$Ty_{s-1} = \alpha y_{s-1}.$$

Therefore, $J$ is the Jordan block $J(\alpha, s)$. By Corollary 2.130, $C(g)$ and $J(\alpha, s)$ are similar.  •

It follows that Jordan blocks also correspond to polynomials (just as companion matrices do); in particular, the characteristic polynomial of $J(\alpha, s)$ is the same as that of $C((x - \alpha)^s)$:

$$\psi_{J(\alpha,s)}(x) = (x - \alpha)^s.$$

**Theorem 8.49.** *Let $A$ be an $n \times n$ matrix with entries in a field $k$. If $k$ contains all the eigenvalues of $A$ (in particular, if $k$ is algebraically closed), then $A$ is similar to a direct sum of Jordan blocks.*

**Proof.** Instead of using the invariant factors $g_1 \mid g_2 \mid \cdots \mid g_t$, we are now going to use the elementary divisors $f_i(x)$ occurring in the Basis Theorem itself; that is, each $f_i(x)$ is a power of an irreducible polynomial in $k[x]$. By Theorem 8.35(i), a decomposition of $(k^n)^A$ into a direct sum of cyclic $k[x]$-modules $W_i$ yields a direct sum of companion matrices

$$U = C(f_1) \oplus \cdots \oplus C(f_r)$$

[where $(f_i)$ is the order ideal of the $k[x]$-module $W_i$] and $U$ is similar to $A$. Since $\psi_A(x) = \prod_i f_i(x)$, however, the hypothesis on $k$ says that each $f_i(x)$ splits over $k$; that is, $f_i(x) = (x - \alpha_i)^{s_i}$ for some $s_i \geq 1$, where $\alpha_i$ is an eigenvalue of $A$. By Lemma 8.48, $C(f_i)$ is similar to a Jordan block and, by Exercise 8.32 on page 663, $A$ is similar to a direct sum of Jordan blocks. •

**Definition.** A **Jordan canonical form** is a direct sum of Jordan blocks.

If a matrix $A$ is similar to the Jordan canonical form

$$J(\alpha_1, s_1) \oplus \cdots \oplus J(\alpha_r, s_r),$$

then we say that $A$ has **elementary divisors** $(x - \alpha_1)^{s_1}, \ldots, (x - \alpha_r)^{s_r}$.

Theorem 8.49 says that every square matrix $A$ having entries in a field containing all the eigenvalues of $A$ is similar to a Jordan canonical form. Can a matrix be similar to several Jordan canonical forms? The answer is yes, but not really.

**Example 8.50.** Let $I_r$ be the $r \times r$ identity matrix, and let $I_s$ be the $s \times s$ identity matrix. Then interchanging blocks in a direct sum yields a similar matrix:

$$\begin{bmatrix} B & 0 \\ 0 & A \end{bmatrix} = \begin{bmatrix} 0 & I_r \\ I_s & 0 \end{bmatrix} \begin{bmatrix} A & 0 \\ 0 & B \end{bmatrix} \begin{bmatrix} 0 & I_s \\ I_r & 0 \end{bmatrix}.$$

Since every permutation is a product of transpositions, it follows that permuting the blocks of a matrix of the form $A_1 \oplus A_2 \oplus \cdots \oplus A_t$ yields a matrix similar to the original one. ◄

**Theorem 8.51.**

(i) *If $A$ and $B$ are $n \times n$ matrices over a field $k$ containing all their eigenvalues, then $A$ and $B$ are similar if and only if they have the same elementary divisors.*

(ii) *If a matrix $A$ is similar to two Jordan canonical forms, say, $H$ and $H'$, then $H$ and $H'$ have the same Jordan blocks (i.e., $H'$ arises from $H$ by permuting its Jordan blocks).*

**Remark.** The hypothesis that all the eigenvalues of $A$ and $B$ lie in $k$ is not a serious problem. Recall that Corollary 8.37(ii) says that if $K/k$ is a field extension and $A$ and $B$ are similar over $K$, then they are similar over $k$. Thus, if $A$ and $B$ are matrices over $k$, define $K = k(\alpha_1, \ldots, \alpha_t)$, where $\alpha_1, \ldots, \alpha_t$ are their eigenvalues. Use Jordan canonical forms to determine whether $A$ and $B$ are similar over $K$, and then invoke Corollary 8.37(ii) to conclude that they are similar over $k$. ◀

**Proof.**

(i) By Corollary 2.130, $A$ and $B$ are similar if and only if $(k^n)^A \cong (k^n)^B$. By Theorem 8.23, $(k^n)^A \cong (k^n)^B$ if and only if their elementary divisors are the same.

(ii) In contrast to the invariant factors, which are given in a specific order (each dividing the next), $A$ determines only a *set* of elementary divisors, hence only a set of Jordan blocks. By Example 8.50, the different Jordan canonical forms obtained from a given Jordan canonical form by permuting its Jordan blocks are all similar. •

Here are some applications of canonical forms.

**Proposition 8.52.** *If $A$ is an $n \times n$ matrix with entries in a field $k$, then $A$ is similar to its transpose $A^\top$.*

**Proof.** First, Corollary 8.37(ii) allows us to assume that $k$ contains all the eigenvalues of $A$. Now if $B = PAP^{-1}$, then $B^\top = (P^\top)^{-1}A^\top P^\top$; that is, if $B$ is similar to $A$, then $B^\top$ is similar to $A^\top$. Thus, it suffices to prove that $H$ is similar to $H^\top$ for a Jordan canonical form $H$; by Exercise 8.32 on page 663, it is enough to show that a Jordan block $J = J(\alpha, s)$ is similar to $J^\top$.

We illustrate the idea for $J(\alpha, 3)$. Let $Q$ be the matrix having 1's on the "wrong" diagonal and 0's everywhere else; notice that $Q = Q^{-1}$:

$$\begin{bmatrix} 0 & 0 & 1 \\ 0 & 1 & 0 \\ 1 & 0 & 0 \end{bmatrix} \begin{bmatrix} \alpha & 0 & 0 \\ 1 & \alpha & 0 \\ 0 & 1 & \alpha \end{bmatrix} \begin{bmatrix} 0 & 0 & 1 \\ 0 & 1 & 0 \\ 1 & 0 & 0 \end{bmatrix} = \begin{bmatrix} \alpha & 1 & 0 \\ 0 & \alpha & 1 \\ 0 & 0 & \alpha \end{bmatrix}.$$

Let $v_1, \ldots, v_s$ be a basis of a vector space $W$, define $Q \colon W \to W$ by $Q \colon v_i \mapsto v_{s-i+1}$, and define $J \colon W \to W$ by $J \colon v_i \mapsto \alpha v_i + v_{i+1}$ for $i < s$ and $J \colon v_s \mapsto \alpha v_s$. The reader can now prove that $Q = Q^{-1}$ and $QJ(\alpha, s)Q^{-1} = J(\alpha, s)^\top$. •

**Example 8.53.** At the beginning of this section, we asked for the order of the matrix

$$A = \begin{bmatrix} 0 & 0 & 1 \\ 1 & 0 & 4 \\ 0 & 1 & 3 \end{bmatrix}$$

in the group $\mathrm{GL}(3, \mathbb{F}_7)$. Now $A$ is the companion matrix of $(x-1)^3$; since $\psi_A(x)$ is a power of $x - 1$, the eigenvalues of $A$ are all equal to 1 and, hence, lie in $\mathbb{F}_7$. By

Lemma 8.48, $A$ is similar to the Jordan block

$$J = \begin{bmatrix} 1 & 0 & 0 \\ 1 & 1 & 0 \\ 0 & 1 & 1 \end{bmatrix}.$$

By Example 8.47,

$$J^m = \begin{bmatrix} 1 & 0 & 0 \\ m & 1 & 0 \\ \binom{m}{2} & m & 1 \end{bmatrix},$$

and it follows that $J^7 = I$ because, in $\mathbb{F}_7$, we have $[7] = [0]$ and $[\binom{7}{2}] = [21] = [0]$. Hence, $A$ has order 7 in $\mathrm{GL}(3, \mathbb{F}_7)$.  ◄

Exponentiating a matrix is used to find solutions to systems of linear differential equations; it is also very useful in setting up the relation between a Lie group and its corresponding Lie algebra. An $n \times n$ complex matrix $A$ consists of $n^2$ entries, and so $A$ may be regarded as a point in $\mathbb{C}^{n^2}$. This allows us to define convergence of a sequence of $n \times n$ complex matrices: $A_1, A_2, \ldots, A_k, \ldots$ **converges** to a matrix $M$ if, for each $i, j$, the sequence of $i, j$ entries converges. As in Calculus, convergence of a series means convergence of the sequence of its partial sums.

**Definition.** If $A$ is an $n \times n$ complex matrix, then

$$e^A = \sum_{k=0}^{\infty} \frac{1}{k!} A^k = I + A + \tfrac{1}{2}A^2 + \tfrac{1}{6}A^3 + \cdots.$$

This series converges for every matrix $A$ (see Exercise 8.41 on page 670), and the function $A \mapsto e^A$ is continuous; that is, if $\lim_{k\to\infty} A_k = M$, then

$$\lim_{k\to\infty} e^{A_k} = e^M$$

We are now going to show that the Jordan canonical form of $A$ can be used to compute $e^A$.

**Proposition 8.54.** *Let* $A = [a_{ij}]$ *be an* $n \times n$ *complex matrix.*

(i) *If* $P$ *is nonsingular, then* $Pe^A P^{-1} = e^{PAP^{-1}}$.

(ii) *If* $AB = BA$, *then* $e^A e^B = e^{A+B}$.

(iii) *For every matrix* $A$, *the matrix* $e^A$ *is nonsingular; indeed,*

$$(e^A)^{-1} = e^{-A}.$$

(iv) *If* $L$ *is the* $n \times n$ *matrix having 1's just below the main diagonal and 0's elsewhere, then* $e^L$ *is a lower triangular matrix with 1's on the diagonal.*

(v) *If* $D$ *is a diagonal matrix, say,* $D = \mathrm{diag}(\alpha_1, \alpha_2, \ldots, \alpha_n)$, *then*

$$e^D = \mathrm{diag}(e^{\alpha_1}, e^{\alpha_2}, \ldots, e^{\alpha_n}).$$

(vi) *If* $\alpha_1, \ldots, \alpha_n$ *are the eigenvalues of* $A$ *(with multiplicities), then* $e^{\alpha_1}, \ldots, e^{\alpha_n}$ *are the eigenvalues of* $e^A$ *(with multiplicities).*

(vii) *We can compute* $e^A$.

(viii) *If* $\mathrm{tr}(A) = 0$, *then* $\det(e^A) = 1$.

**Proof.**

(i) We use the continuity of matrix exponentiation:

$$Pe^A P^{-1} = P\left(\lim_{n\to\infty}\sum_{k=0}^{n}\frac{1}{k!}A^k\right)P^{-1}$$

$$= \lim_{n\to\infty}\sum_{k=0}^{n}\frac{1}{k!}\left(PA^k P^{-1}\right)$$

$$= \lim_{n\to\infty}\sum_{k=0}^{n}\frac{1}{k!}\left(PAP^{-1}\right)^k$$

$$= e^{PAP^{-1}}.$$

(ii) The coefficient of the $k$th term of the power series for $e^{A+B}$ is

$$\frac{1}{k!}(A+B)^k,$$

while the $k$th term of $e^A e^B$ is

$$\sum_{i+j=k}\frac{1}{i!}A^i\frac{1}{j!}B^j = \sum_{i=0}^{k}\frac{1}{i!(k-i)!}A^i B^{k-i} = \frac{1}{k!}\sum_{i=0}^{k}\binom{k}{i}A^i B^{k-i}.$$

Since $A$ and $B$ commute, the Binomial Theorem shows that both $k$th coefficients are equal. (See Exercise 8.43 on page 671 for an example where this is false if $A$ and $B$ do not commute.)

(iii) This follows immediately from part (ii), for $-A$ and $A$ commute and $e^0 = I$.

(iv) The equation

$$e^L = I + L + \frac{1}{2}L^2 + \cdots + \frac{1}{(s-1)!}L^{s-1}$$

holds because $L^s = 0$, and the result follows by Lemma 8.46. For example, when $s = 5$,

$$e^L = \begin{bmatrix} 1 & 0 & 0 & 0 & 0 \\ 1 & 1 & 0 & 0 & 0 \\ \frac{1}{2} & 1 & 1 & 0 & 0 \\ \frac{1}{6} & \frac{1}{2} & 1 & 1 & 0 \\ \frac{1}{24} & \frac{1}{6} & \frac{1}{2} & 1 & 1 \end{bmatrix}.$$

(v) This is clear from the definition:

$$e^D = I + D + \tfrac{1}{2}D^2 + \tfrac{1}{6}D^3 + \cdots,$$

for $D^k = \mathrm{diag}(\alpha_1^k, \alpha_2^k, \ldots, \alpha_n^k)$.

(vi) Since $\mathbb{C}$ is algebraically closed, $A$ is similar to its Jordan canonical form $J$: there is a nonsingular matrix $P$ with $PAP^{-1} = J$. Now $A$ and $J$ have the same characteristic polynomial and, hence, the same eigenvalues with multiplicities. But $J$ is a lower triangular matrix with the eigenvalues $\alpha_1, \ldots, \alpha_n$ of $A$ on the diagonal, and so the definition of matrix exponentiation gives

$e^J$ lower triangular with $e^{\alpha_1}, \ldots, e^{\alpha_n}$ on the diagonal. Since $e^A = e^{P^{-1}JP} = P^{-1}e^JP$, it follows that the eigenvalues of $e^A$ are as claimed.

(vii) By the *Jordan Decomposition* (Exercise 8.36 on page 670), there is a nonsingular matrix $P$ with $PAP^{-1} = \Delta + L$, where $\Delta$ is a diagonal matrix, $L^n = 0$, and $\Delta L = L\Delta$. Hence,

$$Pe^AP^{-1} = e^{PAP^{-1}} = e^{\Delta+L} = e^\Delta e^L.$$

But $e^\Delta$ is computed in part (v) and $e^L$ is computed in part (iv). Hence, $e^A = P^{-1}e^\Delta e^L P$ is computable.

(viii) By Proposition 8.38, $-\operatorname{tr}(A)$ is the sum of its eigenvalues, while $\det(A)$ is the product of the eigenvalues. Since the eigenvalues of $e^A$ are $e^{\alpha_1}, \ldots, e^{\alpha_n}$, we have

$$\det(e^A) = \prod_i e^{\alpha_i} = e^{\sum_i \alpha_i} = e^{-\operatorname{tr}(A)}.$$

Hence, $\operatorname{tr}(A) = 0$ implies $\det(e^A) = 1$.  •

## Exercises

**8.35.** Find all $n \times n$ matrices $A$ over a field $k$ for which $A$ and $A^2$ are similar.

∗ **8.36. (Jordan Decomposition)** Prove that every $n \times n$ matrix $A$ over an algebraically closed field $k$ can be written as

$$A = D + N,$$

where $D$ is **diagonalizable** (i.e., $D$ is similar to a diagonal matrix), $N$ is **nilpotent** (i.e., $N^m = 0$ for some $m \geq 1$), and $DN = ND$. We remark that the Jordan decomposition of a matrix is unique if $k$ is a perfect field.

**8.37.** Give an example of an $n \times n$ complex matrix that is not diagonalizable. [It is known that every *hermitian* matrix $A$ is diagonalizable ($A$ is **hermitian** if $A = A^*$, where the $i, j$ entry of $A^*$ is $\overline{a_{ji}}$).]

**Hint.** A rotation (not the identity) about the origin on $\mathbb{R}^2$ sends no line through the origin into itself.

**8.38.** (i) Prove that all the eigenvalues of a nilpotent matrix are 0.

(ii) Use the Jordan form to prove the converse: if all the eigenvalues of a matrix $A$ are 0, then $A$ is nilpotent. (This result also follows from the Cayley–Hamilton Theorem.)

**8.39.** How many similarity classes of $6 \times 6$ nilpotent matrices are there over a field $k$?

**8.40.** If $A$ and $B$ are similar and $A$ is nonsingular, prove that $B$ is nonsingular and that $A^{-1}$ is similar to $B^{-1}$.

∗ **8.41.** Let $A = [a_{ij}]$ be an $n \times n$ complex matrix.

(i) If $M = \max_{ij} |a_{ij}|$, prove that no entry of $A^s$ has absolute value greater than $(nM)^s$.

(ii) Prove that the series defining $e^A$ converges.

(iii) Prove that $A \mapsto e^A$ is a continuous function.

\* **8.42.** (i) Prove that every nilpotent matrix $N$ is similar to a strictly lower triangular matrix (i.e., all entries on and above the diagonal are 0).

(ii) If $N$ is a nilpotent matrix, prove that $I + N$ is nonsingular.

\* **8.43.** Let $A = \begin{bmatrix} 1 & 0 \\ 0 & 0 \end{bmatrix}$ and $B = \begin{bmatrix} 0 & 1 \\ 0 & 0 \end{bmatrix}$. Prove that $e^A e^B \neq e^B e^A$ and $e^A e^B \neq e^{A+B}$.

**8.44.** How many conjugacy classes are there in the group $\mathrm{GL}(3, \mathbb{F}_7)$?

**8.45.** We know that $\mathrm{PSL}(3, \mathbb{F}_4)$ is a simple group of order $20\,160 = \frac{1}{2}8!$. Now $A_8$ contains an element of order 15, namely, $(1\ 2\ 3\ 4\ 5)(6\ 7\ 8)$. Prove that $\mathrm{PSL}(3, \mathbb{F}_4)$ has no element of order 15, and conclude that $\mathrm{PSL}(3, \mathbb{F}_4) \not\cong A_8$. Conclude further that there exist nonisomorphic finite simple groups of the same order.

**Hint.** Use Corollary 8.37 to replace $\mathbb{F}_4$ by a larger field containing any needed eigenvalues of a matrix. Compute the order [in the group $\mathrm{PSL}(3, \mathbb{F}_4)$] of the possible Jordan canonical forms

$$A = \begin{bmatrix} a & 0 & 0 \\ 1 & a & 0 \\ 0 & 1 & a \end{bmatrix}, \ B = \begin{bmatrix} a & 0 & 0 \\ 0 & b & 0 \\ 0 & 1 & b \end{bmatrix}, \text{ and } C = \begin{bmatrix} a & 0 & 0 \\ 0 & b & 0 \\ 0 & 0 & c \end{bmatrix}.$$

## Section 8.4. Smith Normal Forms

There is a defect in our account of canonical forms: how do we find the invariant factors of a given matrix? The coming discussion will give an algorithm for computing them; in particular, it will enable us to compute minimal polynomials.

In Chapter 2, we showed that a linear transformation $T \colon V \to W$ between finite-dimensional vector spaces determines a matrix ${}_Z[T]_Y$ once bases $Y$ of $V$ and $Z$ of $W$ are chosen. We now generalize that calculation to $R$-maps between free $R$-modules, where $R$ is any commutative ring.

**Definition.** Let $R$ be a commutative ring and let $T \colon R^t \to R^n$ be an $R$-map, where $R^t$ and $R^n$ are free $R$-modules of ranks $t$ and $n$, respectively. If $Y = y_1, \ldots, y_t$ is a basis of $R^t$ and $Z = z_1, \ldots, z_n$ is a basis of $R^n$, then a **presentation matrix** for $\mathrm{coker}\, T$ is

$$_Z[T]_Y = [a_{ij}],$$

the $n \times t$ matrix over $R$ whose columns are the coordinate lists $T(y_i)$: for all $i$,

$$T(y_i) = \sum_{j=1}^{n} a_{ji} z_j.$$

There is no misprint; the double subscript $ji$ in $\sum_{j=1}^{n} a_{ji} z_j$ is correct. For example, $T(y_1) = a_{11} z_1 + a_{21} z_2 + a_{31} z_3 + \cdots$, and the coordinate list $(a_{11}, a_{21}, a_{31}, \ldots)$ of $T(y_1)$ is the first column of ${}_Z[T]_Y$.

Every finitely presented $R$-module $M$ has an $n \times t$ presentation matrix for some $n, t$. We are now going to compare two such matrices that arise from different choices of bases in $R^t$ and in $R^n$ (one could try to compare presentation matrices of different sizes, but we shall not). The next proposition generalizes Corollary 2.130 on page 152 from vector spaces to free modules over a commutative ring.

**Proposition 8.55.** *Let* $T\colon R^t \to R^n$ *be an R-map between free R-modules, where R is a commutative ring. Choose bases* $Y$ *and* $Y'$ *of* $R^t$ *and* $Z$ *and* $Z'$ *of* $R^n$. *There exist invertible*[9] *matrices* $P$ *and* $Q$ *(where* $P$ *is* $t \times t$ *and* $Q$ *is* $n \times n$*), with*

$$\Gamma' = Q\Gamma P^{-1},$$

*where* $\Gamma' = {}_{Z'}[T]_{Y'}$ *and* $\Gamma = {}_{Z}[T]_{Y}$ *are the corresponding presentation matrices.*

*Conversely, if* $\Gamma$ *and* $\Gamma'$ *are* $n \times t$ *matrices with* $\Gamma' = Q\Gamma P^{-1}$ *for some invertible matrices* $P$ *and* $Q$*, then there is an R-map* $T\colon R^t \to R^n$*, bases* $Y$ *and* $Y'$ *of* $R^t$*, and bases* $Z$ *and* $Z'$ *of* $R^n$*, respectively, such that* $\Gamma = {}_{Z}[T]_{Y}$ *and* $\Gamma' = {}_{Z'}[T]_{Y'}$.

**Proof.** This is the same calculation we did in Corollary 2.130 when we applied the formula

$$\big({}_{Z}[S]_{Y}\big)\big({}_{Y}[T]_{X}\big) = {}_{Z}[ST]_{X},$$

where $T\colon V \to V'$ and $S\colon V' \to V''$ and $X$, $Y$, and $Z$ are bases of $V$, $V'$, and $V''$, respectively. Note that the original proof never used the inverse of any matrix entry, so that the earlier hypothesis that the entries lie in a field can be relaxed to allow entries to lie in any commutative ring.  •

**Definition.** Two $n \times t$ matrices $\Gamma$ and $\Gamma'$ with entries in a commutative ring $R$ are **R-equivalent** if there are invertible matrices $P$ and $Q$ with entries in $R$ with

$$\Gamma' = Q\Gamma P^{-1}.$$

Of course, equivalence as just defined is an equivalence relation on the set of all (rectangular) $n \times t$ matrices over $R$. Thus, Proposition 8.55 says that any two $n \times t$ presentation matrices of a finitely presented $R$-module are $R$-equivalent.

**Corollary 8.56.** *Let* $M$ *and* $M'$ *be finitely presented R-modules over a commutative ring* $R$. *Assume that there are exact sequences*

$$R^t \xrightarrow{\lambda} R^n \xrightarrow{\pi} M \to 0 \quad and \quad R^t \xrightarrow{\lambda'} R^n \xrightarrow{\pi'} M' \to 0,$$

*and that bases* $Y, Y'$ *of* $R^t$ *and* $Z, Z'$ *of* $R^n$ *are chosen. If* $\Gamma = {}_{Z}[\lambda]_{Y}$ *and* $\Gamma' = {}_{Z'}[\lambda']_{Y'}$ *are R-equivalent, then* $M \cong M'$.

**Proof.** Since $\Gamma$ and $\Gamma'$ are $R$-equivalent, there are invertible matrices $P$ and $Q$ with $\Gamma' = Q\Gamma P^{-1}$. Now $P$ determines an $R$-isomorphism $\theta\colon R^n \to R^n$, and $Q$ determines an $R$-isomorphism $\varphi\colon R^t \to R^t$. The equation $\Gamma' = Q\Gamma P^{-1}$ gives commutativity of the diagram

$$
\begin{array}{ccccccc}
R^t & \xrightarrow{\lambda} & R^n & \xrightarrow{\pi} & M & \longrightarrow & 0 \\
\varphi \downarrow & & \theta \downarrow & & \downarrow \nu & & \\
R^t & \xrightarrow{\lambda'} & R^n & \xrightarrow{\pi'} & M' & \longrightarrow & 0
\end{array}
$$

---

[9]A matrix $P$ is *invertible* if it is square and there exists a matrix $P'$ with $PP' = I$ and $P'P = I$. In Proposition 8.139, we shall see that a matrix with entries in an arbitrary commutative ring $R$ is invertible if and only if $\det(P)$ is a unit in $R$.

Define an $R$-map $\nu\colon M \to M'$ as follows: if $m \in M$, then surjectivity of $\pi$ gives an element $u \in R^n$ with $\pi(u) = m$; set $\nu(m) = \pi'\theta(u)$ (see Proposition 6.116). Diagram chasing shows that $\nu$ is a well-defined isomorphism.   •

If $V$ is an $R$-module over a commutative ring $R$, then we saw, in Example 6.8(v), how to construct an $R[x]$-module $V^\varphi$ from an $R$-endomorphism $\varphi$ of $V$. For each $f(x) = \sum c_i x^i \in R[x]$ and $v \in V$, define

$$f(x)v = \sum_i c_i \varphi^i(v)$$

(this generalizes the construction of the $k[x]$-module $V^T$ when $V$ is a vector space over a field $k$ and $T\colon V \to V$ is a linear transformation).

If $R$ is a commutative ring, then the polynomial ring $R[x]$, as an $R$-module, is the direct sum $R[x] = \bigoplus_{i \geq 0}\langle x^i\rangle$. If $V$ is an $R$-module, then

$$R[x] \otimes_R V = \bigoplus_{i \geq 0}(\langle x^i\rangle \otimes_R V)$$

is an $R[x]$-module, where $x(x^i \otimes v) = x^{i+1} \otimes v$. If $u \in \langle x^i\rangle \otimes_R V$, then $u = cx^i \otimes v = x^i \otimes cv$ for $c \in R$ and $v \in V$. Thus, every element in $\langle x^i\rangle \otimes_R V$ has the form $x^i \otimes v$ and, since $\langle x^i\rangle \otimes_R V \cong R \otimes_R V \cong V$, we have $x^i \otimes v = 0$ implies $v = 0$.

We are now going to give a nice presentation of the $R[x]$-module $V^\varphi$.

**Theorem 8.57 (Characteristic Sequence).** *Let $V$ be an $R$-module over a commutative ring $R$, and let $\varphi\colon V \to V$ be an $R$-homomorphism.*

(i) *If $V$ is a free $R$-module with basis $X$, then $R[x] \otimes_R V$ is a free $R[x]$-module with basis $1 \otimes X = \{1 \otimes x : x \in X\}$.*

(ii) *There is an exact sequence of $R[x]$-modules*

$$0 \to R[x] \otimes_R V \xrightarrow{\lambda} R[x] \otimes_R V \xrightarrow{\pi} V^\varphi \to 0,$$

*where $\lambda\colon x^i \otimes v \mapsto x^{i+1} \otimes v - x^i \otimes \varphi(v)$ and $\pi\colon x^i \otimes v \mapsto \varphi^i(v)$.*

(iii) *Let $A$ be an $n \times n$ matrix over $R$, let $V = R^n$, and let $\varphi\colon V \to V$ be given by $\varphi(v) = Av$ for all $v \in V$. If $E$ is the standard basis of $R^n$, then the presentation matrix $_{1\otimes E}[\lambda]_{1\otimes E}$ arising from the presentation of $V^\varphi = (R^n)^A$ in part (ii) is $xI - A$.*

**Remark.** The formula for $\lambda$ can be rewritten as $\lambda\colon u \mapsto (x \otimes 1 - 1 \otimes x)u$, where $u \in k[x] \otimes_k V$. The element $x \otimes 1 - 1 \otimes x$ arises quite often.   ◀

**Proof.**

(i) If $B$ is a basis of $V$, then $V = \bigoplus_{b \in B}\langle b\rangle$, where $\langle b\rangle \cong R$. Since tensor product commutes with direct sum, $R[x] \otimes_R V \cong R[x] \otimes_R \bigoplus_{b \in B}\langle b\rangle = \bigoplus_{b \in B}\langle 1 \otimes b\rangle$ is a direct sum of copies of $R[x]$.

(ii) $\lambda$ is an $R[x]$-map:

$$\lambda(x(x^i \otimes v)) = \lambda(x^{i+1} \otimes v) = x^{i+2} \otimes v - x^{i+1} \otimes \varphi(v)$$
$$= x(x^{i+1} \otimes v - x^i \otimes \varphi(v)) = x\lambda(x^i v).$$

$\pi$ is an $R[x]$-map:

$$\pi(x(x^i \otimes v)) = \pi(x^{i+1} \otimes v) = \varphi^{i+1}(v) = x\varphi^i(v) = x\pi(x^i \otimes v).$$

$\pi$ is surjective:

If $v \in V$, then $v \in \operatorname{im} \pi$ because $\pi(1 \otimes v) = \pi(x^0 \otimes v) = \varphi^0(v) = v$.

$\operatorname{im} \lambda \subseteq \ker \pi$:

$$\pi\lambda(x^i \otimes v) = \pi(x^{i+1} \otimes v - x^i \otimes v) = \varphi^{i+1}(v) - \varphi^i(\varphi(v)) = 0.$$

$\ker \pi \subseteq \operatorname{im} \lambda$:

If $u = \sum_{i=0}^m x^i \otimes v_i \in \ker \pi$, then $\sum_{i=0}^m \varphi^i(v_i) = 0$. Hence,

$$u = \sum_{i=0}^m x^i \otimes v_i - \sum_{i=0}^m 1 \otimes \varphi^i(v_i) = \sum_{i=1}^m x^i \otimes v_i - \sum_{i=1}^m 1 \otimes \varphi^i(v_i),$$

because $x^0 \otimes v_i - 1 \otimes \varphi^0(v_i) = 1 \otimes v_i - 1 \otimes v_i = 0$. For each $i \geq 1$, we will write $x^i \otimes v_i - 1 \otimes \varphi^i(v_i)$ as a telescoping sum. If $0 \leq j < i$, then

$$\lambda(x^{i-j-1} \otimes \varphi^j(v_i)) = x^{i-j} \otimes \varphi^j(v_i) - x^{i-j-1} \otimes \varphi^{j+1}(v_i).$$

In the next sum, all cancels except the first and last terms, and

$$\lambda\Big(\sum_{j=0}^{i-1} x^{i-j-1} \otimes \varphi^j(v_i)\Big) = x^i \otimes v_i - 1 \otimes \varphi^i(v_i).$$

Hence, $x^i \otimes v_i - 1 \otimes \varphi^i(v_i) \in \operatorname{im} \lambda$, and $u = \sum_{i=1}^m [x^i \otimes v_i - 1 \otimes \varphi^i(v_i)] \in \operatorname{im} \lambda$.

$\lambda$ is injective:

Suppose that $u = \sum_{i=0}^q x^i \otimes v_i \in \ker \lambda$, where $x^q \otimes v_q \neq 0$; hence, $v_q \neq 0$ and $x^{q+1} \otimes v_q \neq 0$. Now

$$0 = \lambda(u) = \lambda\Big(\sum_{i=0}^q x^i \otimes v_i\Big) = \sum_{i=0}^q (x^{i+1} \otimes v_i - x^i \otimes \varphi(v_i)).$$

Therefore,

$$x^{q+1} \otimes v_q = -\sum_{i=0}^{q-1}(x^{i+1} \otimes v_i) + \sum_{i=0}^q x^i \otimes \varphi(v_i)$$

$$\in \big\langle x^{q+1} \otimes 1\big\rangle \cap \bigoplus_{i=0}^q \big\langle x^i \otimes 1\big\rangle = \{0\}.$$

This contradiction gives $\ker \lambda = \{0\}$.

(iii) The $i$th column of $_{1\otimes E}[\lambda]_{1\otimes E}$ arises from writing $\lambda(1 \otimes e_i)$ in terms of $1 \otimes E$, where $E = e_1, \dots, e_n$ is the standard basis of $R^n$. Now

$$\lambda(1 \otimes e_i) = x \otimes e_i - 1 \otimes \varphi(e_i)$$

$$= x \otimes e_i - 1 \otimes \sum_j a_{ji} e_j$$

$$= x(1 \otimes e_i) - \sum_j a_{ji}(1 \otimes e_j).$$

Now $I = [\delta_{ji}]$, where $\delta_{ji}$ is the Kronecker delta, so that $1 \otimes e_i = \sum_j \delta_{ji}(1 \otimes e_j)$. Hence,

$$x(1 \otimes e_i) - \sum_j a_{ji}(1 \otimes e_j) = \sum_j x\delta_{ji}(1 \otimes e_j) - \sum_j a_{ji}(1 \otimes e_j)$$
$$= \sum_j (x\delta_{ji} - a_{ji})(1 \otimes e_j).$$

Therefore, the matrix of $\lambda$ is $xI - A$. •

We specialize Theorem 8.57(iii) for $R$ a field.

**Corollary 8.58.** *Two $n \times n$ matrices $A$ and $B$ over a field $k$ are similar if and only if the matrices $\Gamma = xI - A$ and $\Gamma' = xI - B$ are $k[x]$-equivalent.*

**Proof.** If $A$ is similar to $B$, there is a nonsingular matrix $P$ with entries in $k$ such that $B = PAP^{-1}$. But

$$P(xI - A)P^{-1} = xI - PAP^{-1} = xI - B,$$

because the scalar matrix $xI$ commutes with $P$ (it commutes with every matrix). Thus, $xI - A$ and $xI - B$ are $k[x]$-equivalent.

Conversely, suppose that the matrices $xI - A$ and $xI - B$ are $k[x]$-equivalent. By Theorem 8.57(iii), $(k[x]^n)^A$ and $(k[x]^n)^B$ are finitely presented $k[x]$-modules having presentation matrices $xI - A$ and $xI - B$, respectively. Now Corollary 8.56 shows that $(k^n)^A \cong (k^n)^B$ as $k[x]$-modules, and so Corollary 6.12 gives $A$ and $B$ similar. •

Corollary 8.56 is a criterion that two finitely presented $R$-modules be isomorphic, but it is virtually useless; for most commutative rings $R$, there is no way to determine whether matrices $\Gamma$ and $\Gamma'$ with entries in $R$ are $R$-equivalent. Now Corollary 8.58 reduces the question of similarity of matrices over a field $k$ to a problem of equivalence of matrices over $k[x]$. Fortunately, *Gaussian elimination*, a method for solving systems of linear equations whose coefficients lie in a field, can be adapted when $R$ is a Euclidean ring, and we will be able to use it to find a computable normal form of a matrix. Let us generalize the ingredients of Gaussian elimination from matrices over fields to matrices over arbitrary commutative rings.

In what follows, we denote the $i$th row of a matrix $A$ by ROW($i$) and the $j$th column by COL($j$).

**Definition.** There are three *elementary row operations* on a matrix $A$ with entries in a commutative ring $R$:

**I.** Multiply ROW($j$) by a unit $u \in R$.

**II.** Replace ROW($i$) by ROW($i$) + $c$ROW($j$), where $j \neq i$ and $c \in R$; that is, add $c$ROW($j$) to ROW($i$).

**III.** Interchange ROW($i$) and ROW($j$).

There are analogous *elementary column operations*.

Notice that an operation of type III'(an interchange) can be accomplished by operations of the other two types. We indicate this schematically.

$$\begin{bmatrix} a & b \\ c & d \end{bmatrix} \rightarrow \begin{bmatrix} a-c & b-d \\ c & d \end{bmatrix} \rightarrow \begin{bmatrix} a-c & b-d \\ a & b \end{bmatrix} \rightarrow \begin{bmatrix} -c & -d \\ a & b \end{bmatrix} \rightarrow \begin{bmatrix} c & d \\ a & b \end{bmatrix}$$

**Definition.** An ***elementary matrix*** is a matrix obtained from the identity matrix $I$ by applying an elementary row[10] operation to it.

Thus, there are three types of elementary matrix. Performing an elementary operation is the same as multiplying on the left by an elementary matrix. For example,

$$L_{\mathrm{I}} = \begin{bmatrix} u & 0 & 0 \\ 0 & 1 & 0 \\ 0 & 0 & 1 \end{bmatrix}; \quad L_{\mathrm{II}} = \begin{bmatrix} 1 & 0 & 0 \\ 0 & 1 & 0 \\ c & 0 & 1 \end{bmatrix}; \quad L_{\mathrm{III}} = \begin{bmatrix} 0 & 1 & 0 \\ 1 & 0 & 0 \\ 0 & 0 & 1 \end{bmatrix}.$$

Given a $3 \times 3$ matrix $A$, the product $L_{\mathrm{I}}A$ is $A$ with its first row multiplied by a unit $u$; the product $L_{\mathrm{II}}A$ is $A$ after adding $c$ times its first row to its third row; the product $L_{\mathrm{III}}A$ is $A$ with its first and second rows interchanged. Applying the corresponding elementary column operation to $A$ gives the matrix $AL$. It is also easy to see that every elementary matrix is invertible, and its inverse is elementary of the same type. It follows that every product of elementary matrices is invertible.

**Definition.** Let $R$ be a commutative ring. Then a matrix $\Gamma'$ is ***Gaussian equivalent*** to a matrix $\Gamma$ if there is a sequence of elementary row and column operations

$$\Gamma = \Gamma_0 \rightarrow \Gamma_1 \rightarrow \cdots \rightarrow \Gamma_r = \Gamma'.$$

Gaussian equivalence is an equivalence relation on the family of all $n \times t$ matrices over $R$. It follows that if $\Gamma'$ is Gaussian equivalent to $\Gamma$, then there are matrices $P$ and $P'$, each a product of elementary matrices, with $\Gamma' = P\Gamma P'$. Now two matrices $\Gamma'$ and $\Gamma$ are *$R$-equivalent* if there are invertible matrices $P$ and $Q$ with $\Gamma' = P\Gamma Q^{-1}$. Hence, if $\Gamma'$ is Gaussian equivalent to $\Gamma$, then $\Gamma'$ and $\Gamma$ are $R$-equivalent, for the inverse of an elementary matrix is elementary. We shall see that the converse is true when $R$ is Euclidean.

**Theorem 8.59 (Smith Normal Form[11]).** *Every nonzero $n \times t$ matrix $\Gamma$ with entries in a Euclidean ring $R$ is Gaussian equivalent to a matrix of the form*

$$\begin{bmatrix} \Sigma & 0 \\ 0 & 0 \end{bmatrix},$$

*where $\Sigma = \mathrm{diag}(\sigma_1, \ldots, \sigma_q)$ and $\sigma_1 \mid \sigma_2 \mid \cdots \mid \sigma_q$ are nonzero (the lower blocks of 0's or the blocks of 0's on the right may not be present).*

**Proof.** If $\sigma \in R$ is nonzero, let $\partial(\sigma)$ denote its degree in the Euclidean ring $R$. Among all the nonzero entries of matrices Gaussian equivalent to $\Gamma$, let $\sigma_1$ have

---

[10]Applying elementary column operations to $I$ gives the same collection of elementary matrices.

[11]This theorem and the corresponding uniqueness result, soon to be proved, were found by Smith in 1861.

the smallest degree, and let $\Delta$ be a matrix Gaussian equivalent to $\Gamma$ that has $\sigma_1$ as an entry, say, in position $k, \ell$. We claim that $\sigma_1 \mid \eta_{kj}$ for all $\eta_{kj}$ in ROW$(k)$ of $\Delta$. Otherwise, there is $j \neq \ell$ and an equation $\eta_{kj} = \kappa \sigma_1 + \rho$, where $\partial(\rho) < \partial(\sigma_1)$. Adding $(-\kappa)$COL$(\ell)$ to COL$(j)$ gives a matrix $\Delta'$ having $\rho$ as an entry. But $\Delta'$ is Gaussian equivalent to $\Gamma$, and it has an entry $\rho$ whose degree is smaller than $\partial(\sigma_1)$, a contradiction. The same argument shows that $\sigma_1$ divides every entry in its column. Let us return to $\Delta$, a matrix Gaussian equivalent to $\Gamma$ that contains $\sigma_1$ as an entry. We claim that $\sigma_1$ divides every entry of $\Delta$, not merely those entries in $\sigma_1$'s row and column; let $a$ be such an entry. Schematically, we are focusing on a submatrix $\begin{bmatrix} a & b \\ c & \sigma_1 \end{bmatrix}$, where $b = u\sigma_1$ and $c = v\sigma_1$. Now replace ROW$(1)$ by ROW$(1) + (1-u)$ROW$(2) = [a + (1-u)c, \ \sigma_1]$. As above, $\sigma_1$ divides $a + (1-u)c$. Since $\sigma_1 \mid c$, we have $\sigma_1 \mid a$. We conclude that we may assume that $\sigma_1$ is an entry of $\Gamma$ which divides every entry of $\Gamma$.

Let us normalize $\Gamma$ further. By interchanges, there is a matrix that is Gaussian equivalent to $\Gamma$ and that has $\sigma_1$ in the $1, 1$ position. If $\eta_{1j}$ is another entry in the first row, then $\eta_{1j} = \kappa_j \sigma_1$, and adding $(-\kappa_j)$COL$(1)$ to COL$(j)$ gives a new matrix whose $1, j$ entry is $0$. Thus, we may also assume that $\Gamma$ has $\sigma_1$ in the $1, 1$ position and with $0$'s in the rest of the first row.

Having normalized $\Gamma$, we now complete the proof by induction on the number $n \geq 1$ of its rows. If $n = 1$, we have just seen that a nonzero $1 \times t$ matrix is Gaussian equivalent to $[\sigma_1 \ 0 \ \dots \ 0]$. For the inductive step, we may assume that $\sigma_1$ is in the $1, 1$ position and that all other entries in the first row are $0$. Since $\sigma_1$ divides all entries in the first column, $\Gamma$ is Gaussian equivalent to a matrix having all $0$'s in the rest of the first column as well. Thus, $\Gamma$ is Gaussian equivalent to a matrix of the form $\begin{bmatrix} \sigma_1 & 0 \\ 0 & \Omega \end{bmatrix}$. By induction, the matrix $\Omega$ is Gaussian equivalent to a matrix $\begin{bmatrix} \Sigma' & 0 \\ 0 & 0 \end{bmatrix}$, where $\Sigma' = \mathrm{diag}(\sigma_2, \dots, \sigma_q)$ and $\sigma_2 \mid \sigma_3 \mid \cdots \mid \sigma_q$. Hence, $\Gamma$ is Gaussian equivalent to $\begin{bmatrix} \sigma_1 & 0 & 0 \\ 0 & \Sigma' & 0 \\ 0 & 0 & 0 \end{bmatrix}$, and so $\sigma_1$ divides every entry of this matrix. In particular, $\sigma_1 \mid \sigma_2$. $\bullet$

**Definition.** The matrix $\begin{bmatrix} \Sigma & 0 \\ 0 & 0 \end{bmatrix}$ in the statement of the theorem is called a ***Smith normal form*** of $\Gamma$.

Thus, the theorem states that every nonzero (rectangular) matrix with entries in a Euclidean ring $R$ is Gaussian equivalent to a Smith normal form.

**Corollary 8.60.** *Let $R$ be a Euclidean ring.*

(i) *Every invertible $n \times n$ matrix $\Gamma$ with entries in $R$ is a product of elementary matrices.*

(ii) *Two matrices $\Gamma$ and $\Gamma'$ over $R$ are $R$-equivalent if and only if they are Gaussian equivalent.*

**Proof.**

(i) We now know that $\Gamma$ is Gaussian equivalent to a Smith normal form $\begin{bmatrix} \Sigma & 0 \\ 0 & 0 \end{bmatrix}$, where $\Sigma$ is diagonal. Since $\Gamma$ is a (square) invertible matrix, there can be no blocks of $0$'s, and so $\Gamma$ is Gaussian equivalent to $\Sigma$; that is, there are matrices

$P$ and $Q$ that are products of elementary matrices such that

$$P\Gamma Q = \Sigma = \mathrm{diag}(\sigma_1, \ldots, \sigma_n).$$

Hence, $\Gamma = P^{-1}\Sigma Q^{-1}$. Now the inverse of an elementary matrix is again elementary, so that $P^{-1}$ and $Q^{-1}$ are products of elementary matrices. Since $\Sigma$ is invertible, $\det(\Sigma) = \sigma_1 \cdots \sigma_n$ is a unit in $R$. It follows that each $\sigma_i$ is a unit, and so $\Sigma$ is a product of $n$ elementary matrices [arising from the elementary operations of multiplying ROW($i$) by the unit $\sigma_i$].

(ii) It is always true that if $\Gamma'$ and $\Gamma$ are Gaussian equivalent, then they are $R$-equivalent, for if $\Gamma' = P\Gamma Q$, where $P$ and $Q$ are products of elementary matrices, then $P$ and $Q$ are invertible. Conversely, if $\Gamma'$ is $R$-equivalent to $\Gamma$, then $\Gamma' = P\Gamma Q$, where $P$ and $Q$ are invertible, and part (i) shows that $\Gamma'$ and $\Gamma$ are Gaussian equivalent.  •

There are examples showing that this proposition may be false for PIDs that are not Euclidean.[12] Investigating this phenomenon was important in the beginnings of *Algebraic K-Theory* (see Milnor, *Introduction to Algebraic K-Theory*).

**Theorem 8.61 (Simultaneous Bases).** *Let $R$ be a Euclidean ring, let $F$ be a finitely generated free $R$-module, and let $S$ be a submodule of $F$. Then there exists a basis $z_1, \ldots, z_n$ of $F$ and nonzero $\sigma_1, \ldots, \sigma_q$ in $R$, where $0 \le q \le n$, such that $\sigma_1 \mid \cdots \mid \sigma_q$ and $\sigma_1 z_1, \ldots, \sigma_q z_q$ is a basis of $S$.*

**Proof.** If $M = F/S$, then Corollary 8.4 shows that $S$ is free of rank $\le n$, and so

$$0 \to S \overset{\lambda}{\to} F \to M \to 0$$

is a presentation of $M$, where $\lambda$ is the inclusion. Now any choice of bases of $S$ and $F$ associates a (possibly rectangular) presentation matrix $\Gamma$ to $\lambda$. According to Proposition 8.55, there are new bases $X$ of $S$ and $Y$ of $F$ relative to which $\Gamma = {}_Y[\lambda]_X$ is $R$-equivalent to a Smith normal form; these new bases are as described in the theorem.  •

**Corollary 8.62.** *Let $R$ be a Euclidean ring, let $\Gamma$ be the $n \times t$ presentation matrix associated to an $R$-map $\lambda\colon R^t \to R^n$ relative to some choice of bases, and let $M = \mathrm{coker}\,\lambda$.*

(i) *If $\Gamma$ is $R$-equivalent to a Smith normal form $\mathrm{diag}(\sigma_1, \ldots, \sigma_q) \oplus 0$, then those $\sigma_1, \ldots, \sigma_q$ that are not units are the invariant factors of $M$.*

(ii) *If $\mathrm{diag}(\eta_1, \ldots, \eta_s) \oplus 0$ is another Smith normal form of $\Gamma$, then $s = q$ and there are units $u_i$ with $\eta_i = u_i \sigma_i$ for all $i$; that is, the diagonal entries are associates.*

---

[12]There is a version for general PIDs obtained by augmenting the collection of elementary matrices by *secondary* matrices; see Exercise 8.49 on page 682.

**Proof.**

(i) If $\mathrm{diag}(\sigma_1, \ldots, \sigma_q) \oplus 0$ is a Smith normal form of $\Gamma$, then there are bases $y_1, \ldots, y_t$ of $R^t$ and $z_1, \ldots, z_n$ of $R^n$ with $\lambda(y_1) = \sigma_1 z_1, \ldots, \lambda(y_q) = \sigma_q z_q$ and $\lambda(y_j) = 0$ for all $y_j$ with $j > q$, if any. If $\sigma_s$ is the first $\sigma_i$ that is not a unit, then

$$M \cong R^{n-q} \oplus \frac{R}{(\sigma_s)} \oplus \cdots \oplus \frac{R}{(\sigma_q)},$$

a direct sum of cyclic modules for which $\sigma_s \mid \cdots \mid \sigma_q$. The Fundamental Theorem of Finitely Generated $R$-Modules identifies $\sigma_s, \ldots, \sigma_q$ as the invariant factors of $M$.

(ii) Part (i) proves the essential uniqueness of the Smith normal form, for the invariant factors, being generators of order ideals, are only determined up to associates. $\bullet$

With a slight abuse of language, we may now speak of *the* Smith normal form of a matrix $\Gamma$.

**Corollary 8.63.** *Two $n \times n$ matrices $A$ and $B$ over a field $k$ are similar if and only if $xI - A$ and $xI - B$ have the same Smith normal form over $k[x]$.*

**Proof.** By Theorem 8.58, $A$ and $B$ are similar if and only if $xI - A$ is $k[x]$-equivalent to $xI - B$, and Corollary 8.62 shows that $xI - A$ and $xI - B$ are $k[x]$-equivalent if and only if they have the same Smith normal form. $\bullet$

**Corollary 8.64.** *Let $F$ be a finitely generated free abelian group, and let $S$ be a subgroup of $F$ having finite index. Let $y_1, \ldots, y_n$ be a basis of $F$, let $z_1, \ldots, z_n$ be a basis of $S$, and let $A = [a_{ij}]$ be the $n \times n$ matrix with $z_i = \sum_j a_{ji} y_j$. Then*

$$[F : S] = |\det(A)|.$$

**Proof.** Changing bases of $S$ and of $F$ replaces $A$ by a matrix $B$ that is $\mathbb{Z}$-equivalent to it:

$$B = QAP,$$

where $Q$ and $P$ are invertible matrices with entries in $\mathbb{Z}$. Since the only units in $\mathbb{Z}$ are $1$ and $-1$, we have $|\det(B)| = |\det(A)|$. In particular, if we choose $B$ to be a Smith normal form, then $B = \mathrm{diag}(g_1, \ldots, g_n)$, and so $|\det(B)| = g_1 \cdots g_n$. But $g_1, \ldots, g_n$ are the invariant factors of $F/S$; by Corollary 4.33, their product is the order of $F/S$, which is the index $[F : S]$. $\bullet$

We have not yet kept our promise to give an algorithm computing the invariant factors of a matrix with entries in a field.

**Theorem 8.65.** *Let $\Sigma = \mathrm{diag}(\sigma_1, \ldots, \sigma_q)$ be the diagonal block in the Smith normal form of a matrix $\Gamma$ with entries in a Euclidean ring $R$. Define $d_i(\Gamma)$ inductively: $d_0(\Gamma) = 1$ and, for $i > 0$,*

$$d_i(\Gamma) = \gcd(\text{all } i \times i \text{ minors of } \Gamma).$$

*Then, for all $i \geq 1$,*

$$\sigma_i = d_i(\Gamma)/d_{i-1}(\Gamma).$$

**Proof.** We are going to show that if $\Gamma$ and $\Gamma'$ are $R$-equivalent, then

$$d_i(\Gamma) \sim d_i(\Gamma')$$

for all $i$, where we write $a \sim b$ to denote $a$ and $b$ being associates in $R$. This will suffice to prove the theorem, for if $\Gamma'$ is the Smith normal form of $\Gamma$ whose diagonal block is $\mathrm{diag}(\sigma_1, \ldots, \sigma_q)$, then $d_i(\Gamma') = \sigma_1 \sigma_2 \cdots \sigma_i$. Hence,

$$\sigma_i(x) = d_i(\Gamma')/d_{i-1}(\Gamma') \sim d_i(\Gamma)/d_{i-1}(\Gamma).$$

By Proposition 8.60, it suffices to prove that

$$d_i(\Gamma) \sim d_i(L\Gamma) \quad \text{and} \quad d_i(\Gamma) \sim d_i(\Gamma L)$$

for every elementary matrix $L$. Indeed, it suffices to prove that $d_i(\Gamma L) \sim d_i(\Gamma)$, because $d_i(\Gamma L) = d_i([\Gamma L]^\top) = d_i(L^\top \Gamma^\top)$ [the $i \times i$ submatrices of $\Gamma^\top$ are the transposes of the $i \times i$ submatrices of $\Gamma$; now use the facts that $L^\top$ is elementary and that, for every square matrix $M$, we have $\det(M^\top) = \det(M)$].

As a final simplification, it suffices to consider only elementary operations of types I and II, for we have seen on page 676 that an operation of type III, interchanging two rows, can be accomplished using the other two types.

$L$ is of type I:   If we multiply ROW($\ell$) of $\Gamma$ by a unit $u$, then an $i \times i$ submatrix either remains unchanged or one of its rows is multiplied by $u$. In the first case, the minor, namely, its determinant, is unchanged; in the second case, the minor is changed by a unit. Therefore, every $i \times i$ minor of $L\Gamma$ is an associate of the corresponding $i \times i$ minor of $\Gamma$, and so $d_i(L\Gamma) \sim d_i(\Gamma)$.

$L$ is of type II:   If $L$ replaces ROW($\ell$) by ROW($\ell$) + $r$ROW($j$), then only ROW($\ell$) of $\Gamma$ is changed. Thus, an $i \times i$ submatrix of $\Gamma$ either does not involve this row or it does. In the first case, the corresponding minor of $L\Gamma$ is unchanged; in the second case, it has the form $m + rm'$, where $m$ and $m'$ are $i \times i$ minors of $\Gamma$ (for det is a multilinear function of the rows of a matrix). It follows that $d_i(\Gamma) \mid d_i(L\Gamma)$, for $d_i(\Gamma) \mid m$ and $d_i(\Gamma) \mid m'$. Since $L^{-1}$ is also an elementary matrix of type II, this argument shows that $d_i(L^{-1}(L\Gamma)) \mid d_i(L\Gamma)$. Of course, $L^{-1}(L\Gamma) = \Gamma$, so that $d_i(\Gamma)$ and $d_i(L\Gamma)$ divide each other. As $R$ is a domain, we have $d_i(L\Gamma) \sim d_i(\Gamma)$.   •

**Theorem 8.66.** *There is an algorithm to compute the elementary divisors of any square matrix $A$ with entries in a field $k$.*

**Proof.** By Corollary 8.63, it suffices to find a Smith normal form for $\Gamma = xI - A$ over the ring $k[x]$; by Corollary 8.62, the invariant factors of $A$ are those nonzero diagonal entries that are not units.

Here are two algorithms: compute $d_i(xI - A)$ for all $i$ (of course, this is not a very efficient algorithm for large matrices); put $xI - A$ into Smith normal form using Gaussian elimination over $k[x]$. The reader should now have no difficulty in writing a program to compute the elementary divisors.   •

**Example 8.67.** Find the invariant factors, over $\mathbb{Q}$, of

$$A = \begin{bmatrix} 2 & 3 & 1 \\ 1 & 2 & 1 \\ 0 & 0 & -4 \end{bmatrix}.$$

We are going to use a combination of the two modes of attack: Gaussian elimination and gcd's of minors. Now

$$xI - A = \begin{bmatrix} x-2 & -3 & -1 \\ -1 & x-2 & -1 \\ 0 & 0 & x+4 \end{bmatrix}.$$

It is plain that $g_1 = 1$, for it is the gcd of all the entries of $A$, some of which are nonzero constants. Interchange ROW(1) and ROW(2), and then change sign in the top row to obtain

$$\begin{bmatrix} 1 & -x+2 & 1 \\ x-2 & -3 & -1 \\ 0 & 0 & x+4 \end{bmatrix}.$$

Add $-(x-2)$ROW(1) to ROW(2) to obtain

$$\begin{bmatrix} 1 & -x+2 & 1 \\ 0 & x^2-4x+1 & -x+1 \\ 0 & 0 & x+4 \end{bmatrix} \to \begin{bmatrix} 1 & 0 & 0 \\ 0 & x^2-4x+1 & -x+1 \\ 0 & 0 & x+4 \end{bmatrix}.$$

The gcd of the entries in the $2 \times 2$ submatrix '

$$\begin{bmatrix} x^2-4x+1 & -x+1 \\ 0 & x+4 \end{bmatrix}$$

is 1, for $-x+1$ and $x+4$ are distinct irreducibles, and so $g_2 = 1$. We have shown that there is only one invariant factor of $A$, namely, $(x^2 - 4x + 1)(x+4) = x^3 - 15x + 4$, and it must be the characteristic polynomial of $A$. It follows that the characteristic and minimal polynomials of $A$ coincide, and Corollary 8.45 shows that the rational canonical form of $A$ is

$$\begin{bmatrix} 0 & 0 & -4 \\ 1 & 0 & 15 \\ 0 & 1 & 0 \end{bmatrix}. \quad \blacktriangleleft$$

**Example 8.68.** Find the abelian group $G$ having generators $a, b, c$ and relations

$$\begin{aligned} 7a + 5b + 2c &= 0 \\ 3a + 3b &= 0 \\ 13a + 11b + 2c &= 0. \end{aligned}$$

Using elementary operations over $\mathbb{Z}$, we find the Smith normal form of the matrix of relations:

$$\begin{bmatrix} 7 & 5 & 2 \\ 3 & 3 & 0 \\ 13 & 11 & 2 \end{bmatrix} \to \begin{bmatrix} 1 & 0 & 0 \\ 0 & 6 & 0 \\ 0 & 0 & 0 \end{bmatrix}.$$

It follows that $G \cong (\mathbb{Z}/1\mathbb{Z}) \oplus (\mathbb{Z}/6\mathbb{Z}) \oplus (\mathbb{Z}/0\mathbb{Z})$. Simplifying, $G \cong \mathbb{I}_6 \oplus \mathbb{Z}$. $\quad \blacktriangleleft$

## Exercises

**8.46.** Find the invariant factors, over $\mathbb{Q}$, of the matrix

$$\begin{bmatrix} -4 & 6 & 3 \\ -3 & 5 & 4 \\ 4 & -5 & 3 \end{bmatrix}.$$

**8.47.** Find the invariant factors, over $\mathbb{Q}$, of the matrix

$$\begin{bmatrix} -6 & 2 & -5 & -19 \\ 2 & 0 & 1 & 5 \\ -2 & 1 & 0 & -5 \\ 3 & -1 & 2 & 9 \end{bmatrix}.$$

**8.48.** If $k$ is a field, prove that there is an additive exact functor $_k\mathbf{Mod} \to {}_{k[x]}\mathbf{Mod}$ taking any vector space $V$ to $V[x]$. [See Theorem 8.57(iii).]

∗ **8.49.** Let $R$ be a PID, and let $a, b \in R$.

(i) If $d$ is the gcd of $a$ and $b$, prove that there is a $2 \times 2$ matrix $Q = \begin{bmatrix} x & y \\ x' & y' \end{bmatrix}$ with $\det(Q) = 1$ so that

$$Q \begin{bmatrix} a & * \\ b & * \end{bmatrix} = \begin{bmatrix} d & * \\ d' & * \end{bmatrix},$$

where $d \mid d'$.
**Hint.** If $d = xa + yb$, define $x' = b/d$ and $y' = -a/d$.

(ii) Call an $n \times n$ matrix $U$ **secondary** if it can be partitioned

$$U = \begin{bmatrix} Q & 0 \\ 0 & I \end{bmatrix},$$

where $Q$ is a $2 \times 2$ matrix of determinant 1. Prove that every $n \times n$ matrix $A$ with entries in a PID can be transformed into a Smith canonical form by a sequence of elementary and secondary matrices.

## Section 8.5. Bilinear Forms

In this section, $V$ will be a vector space over a field $k$, usually finite-dimensional. We continue to use familiar properties of determinants even though we will not prove the basic theorems about them until Section 8.7.

### 8.5.1. Inner Product Spaces.

**Definition.** A **bilinear form** (or **inner product**) on a finite-dimensional vector space $V$ over a field $k$ is a bilinear function

$$f \colon V \times V \to k.$$

The ordered pair $(V, f)$ is called an **inner product space** over $k$.

Of course, $(k^n, f)$ is an inner product space if $f$ is the familiar **dot product**

$$f(u, v) = \sum_i u_i v_i,$$

where $u = (u_1, \ldots, u_n)^\top$ and $v = (v_1, \ldots, v_n)^\top$ (the superscript $\top$ denotes transpose; remember that the elements of $k^n$ are $n \times 1$ column vectors). In terms of matrix multiplication, we have

$$f(u, v) = u^\top v.$$

Two types of bilinear forms are of special interest.

**Definition.** A bilinear form $f \colon V \times V \to k$ is **symmetric** if

$$f(u, v) = f(v, u)$$

for all $u, v \in V$; we call an inner product space $(V, f)$ a **symmetric space** when $f$ is symmetric.

A bilinear form $f \colon V \times V \to k$ is **alternating** if

$$f(v, v) = 0$$

for all $v \in V$; we call an inner product space $(V, f)$ an **alternating space** when $f$ is alternating.

Dot product $k^n \times k^n \to k$ is an example of a symmetric bilinear form.

If we view the elements of $V = k^2$ as column vectors, then we may identify $\mathrm{Mat}_2(k)$ with $V \times V$. The function $f \colon V \times V \to k$, given by

$$f \colon \left( \begin{bmatrix} a \\ b \end{bmatrix}, \begin{bmatrix} c \\ d \end{bmatrix} \right) \mapsto \det \begin{bmatrix} a & c \\ b & d \end{bmatrix} = ad - bc,$$

is an alternating bilinear form: if two columns of $A$ are equal, then $\det(A) = 0$. This example will be generalized to determinants of $n \times n$ matrices $A$ in Section 8.7.

Every bilinear form over a field of characteristic not 2 can be expressed in terms of symmetric and alternating bilinear forms.

**Proposition 8.69.** *Let $k$ be a field of characteristic not 2, and let $f$ be a bilinear form defined on a vector space $V$ over $k$. Then there are unique bilinear forms $f_s$ and $f_a$, where $f_s$ is symmetric and $f_a$ is alternating, such that $f = f_s + f_a$.*

**Proof.** By hypothesis, $\frac{1}{2} \in k$, and so we may define

$$f_s(u, v) = \tfrac{1}{2}\big(f(u, v) + f(v, u)\big)$$

and

$$f_a(u, v) = \tfrac{1}{2}\big(f(u, v) - f(v, u)\big).$$

It is clear that $f = f_s + f_a$, that $f_s$ is symmetric, and that $f_a$ is alternating. Let us prove uniqueness. If $f = f_s' + f_a'$, where $f_s'$ is symmetric and $f_a'$ is alternating, then $f_s + f_a = f_s' + f_a'$, so that $f_s - f_s' = f_a' - f_a$. If we define $g$ to be the common value, $f_s - f_s' = g = f_a' - f_a$, then $g$ is both symmetric and alternating. By Exercise 8.53 on page 703, we have $g = 0$, and so $f_s = f_s'$ and $f_a = f_a'$. $\bullet$

**Definition.** A bilinear form $g$ on a vector space $V$ is called **skew** if

$$g(v, u) = -g(u, v)$$

for all $u, v \in V$.

**Proposition 8.70.** *If $k$ is a field of characteristic not 2, then $g$ is alternating if and only if $g$ is skew.*

**Proof.** If $g$ is any bilinear form, we have

$$g(u + v, u + v) = g(u, u) + g(u, v) + g(v, u) + g(v, v).$$

Therefore, if $g$ is alternating, then $0 = g(u, v) + g(v, u)$, so that $g$ is skew. (This implication does not assume that $k$ has characteristic not 2.)

Conversely, if $g$ is skew, then set $u = v$ in the equation $g(u, v) = -g(v, u)$ to get $g(u, u) = -g(u, u)$; that is, $2g(u, u) = 0$. Since $k$ does not have characteristic 2, $g(u, u) = 0$, and $g$ is alternating.  •

**Definition.** Let $(V, f)$ be an inner product space over $k$. If $E = e_1, \ldots, e_n$ is a basis of $V$, then the **inner product matrix** of $f$ **relative to** $E$ is

$$[f(e_i, e_j)].$$

Suppose that $(V, f)$ is an inner product space, $E = e_1, \ldots, e_n$ is a basis of $V$, and $A = [f(e_i, e_j)]$ is the inner product matrix of $f$ relative to $E$. If $b = \sum b_i e_i$ and $c = \sum c_i e_i$ are vectors in $V$, then

$$f(b, c) = f\left(\sum b_i e_i, \sum c_i e_i\right) = \sum_{i,j} b_i f(e_i, e_j) c_j.$$

If $b = (b_1, \ldots, b_n)^\top$ and $c = (c_1, \ldots, c_n)^\top$ are column vectors, then the displayed equation can be rewritten in matrix form:

(1) $$f(b, c) = b^\top A c.$$

Thus, an inner product matrix determines $f$ completely.

**Proposition 8.71.** *Let $V$ be an $n$-dimensional vector space over a field $k$.*

(i) *Every $n \times n$ matrix $A$ over a field $k$ is the inner product matrix of some bilinear form $f$ defined on $V$.*

(ii) *If $f$ is symmetric, then its inner product matrix $A$ relative to any basis of $V$ is a symmetric matrix (i.e., $A^\top = A$).*

(iii) *If $f$ is alternating and $k$ has characteristic not 2, then the inner product matrix of $f$ relative to any basis of $V$ is skew-symmetric (i.e., $A^\top = -A$). If $k$ has characteristic 2, then every inner product matrix is symmetric with 0's on the diagonal.*

(iv) *Given $n \times n$ matrices $A$ and $A'$, if $b^\top A c = b^\top A' c$ for all column vectors $b$ and $c$, then $A = A'$.*

(v) *Let $A$ and $A'$ be inner product matrices of bilinear forms $f$ and $f'$ on $V$ relative to bases $E$ and $E'$, respectively. Then $f = f'$ if and only if $A$ and $A'$ are* **congruent**; *that is, there exists a nonsingular matrix $P$ with*

$$A' = P^\top A P.$$

*In fact, $P$ is the transition matrix $_E 1_{E'}$.*

**Proof.**

(i) For any matrix $A$, the function $f : k^n \times k^n \to k$, defined by $f(b, c) = b^\top A c$, is easily seen to be a bilinear form, and $A$ is its inner product matrix relative to the standard basis $e_1, \dots, e_n$. The reader may easily transfer this construction to any vector space $V$ once a basis of $V$ is chosen.

(ii) If $f$ is symmetric, then so is its inner product matrix $A = [a_{ij}]$, for $a_{ij} = f(e_i, e_j) = f(e_j, e_i) = a_{ji}$.

(iii) Assume that $f$ is alternating. If $k$ does not have characteristic 2, then $f$ is skew: $a_{ij} = f(e_i, e_j) = -f(e_j, e_i) = -a_{ji}$, and so $A$ is skew-symmetric. If $k$ has characteristic 2, then $f(e_i, e_j) = -f(e_j, e_i) = f(e_j, e_i)$, while $f(e_i, e_i) = 0$ for all $i$; that is, $A$ is symmetric with 0's on the diagonal.

(iv) If $b = \sum_i b_i e_i$ and $c = \sum_i c_i e_i$, then we have seen that $f(b, c) = b^\top A c$, where $b$ and $c$ are the column vectors of the coordinate lists of $b$ and $c$ with respect to $E$. In particular, if $b = e_i$ and $c = e_j$, then $f(e_i, e_j) = a_{ij}$ is the $i, j$ entry of $A$.

(v) Let the coordinate lists of $b$ and $c$ with respect to the basis $E'$ be $b'$ and $c'$, respectively, so that $f'(b, c) = (b')^\top A' c'$, where $A' = [f(e_i', e_j')]$. If $P$ is the transition matrix $_E[1]_{E'}$, then $b = P b'$ and $c = P c'$. Hence, $f(b, c) = b^\top A c = (P b')^\top A (P c') = (b')^\top (P^\top A P) c'$. By part (iv), we must have $P^\top A P = A'$.

For the converse, the given matrix equation $A' = P^\top A P$ yields equations:

$$[f'(e_i', e_j')] = A' = P^\top A P = \left[ \sum_{\ell, q} p_{\ell i} f(e_\ell, e_q) p_{qj} \right]$$

$$= \left[ f\left( \sum_\ell p_{\ell i} e_\ell, \sum_q p_{qj} e_q \right) \right] = [f(e_i', e_j')].$$

Hence, $f'(e_i', e_j') = f(e_i', e_j')$ for all $i, j$, from which it follows that $f'(b, c) = f(b, c)$ for all $b, c \in V$. Therefore, $f = f'$. $\bullet$

**Corollary 8.72.** *If $(V, f)$ is an inner product space and $A$ and $A'$ are inner product matrices of $f$ relative to different bases of $V$, then there exists a nonzero $d \in k$ with*

$$\det(A') = d^2 \det(A).$$

*Consequently, $A'$ is nonsingular if and only if $A$ is nonsingular.*

**Proof.** This follows from familiar facts about determinants over fields (see Section 8.7): $\det(P^\top) = \det(P)$; $\det(AB) = \det(A) \det(B)$. Thus,

$$\det(A') = \det(P^\top A P) = \det(P)^2 \det(A). \quad \bullet$$

The most important bilinear forms are the *nondegenerate* ones.

**Definition.** A bilinear form $f$ is ***nondegenerate*** if it has a nonsingular inner product matrix.

For example, the dot product on $k^n$ is nondegenerate, for its inner product matrix relative to the standard basis is the identity matrix $I$.

The *discriminant* of a bilinear form is essentially the determinant of its inner product matrix. However, since the inner product matrix depends on a choice of basis, we must complicate the definition a bit.

**Definition.** If $k$ is a field, then its multiplicative group of nonzero elements is denoted by $k^\times$. Define $(k^\times)^2 = \{a^2 : a \in k^\times\}$. The ***discriminant*** of a bilinear form $f$ is either 0 or

$$\det(A)(k^\times)^2 \in k^\times/(k^\times)^2,$$

where $A$ is an inner product matrix of $f$.

It follows from Corollary 8.72 that the discriminant of $f$ is well-defined. Quite often, however, we are less careful and say that $\det(A)$ is the discriminant of $f$, where $A$ is some inner product matrix of $f$.

The next definition will be used in characterizing nondegeneracy.

**Definition.** If $(V, f)$ is an inner product space and $W \subseteq V$ is a subspace of $V$, then the ***left orthogonal complement*** of $W$ is

$$W^{\perp L} = \{b \in V : f(b, w) = 0 \text{ for all } w \in W\};$$

the ***right orthogonal complement*** of $W$ is

$$W^{\perp R} = \{c \in V : f(w, c) = 0 \text{ for all } w \in W\}.$$

It is easy to see that both $W^{\perp L}$ and $W^{\perp R}$ are subspaces of $V$. Moreover, $W^{\perp L} = W^{\perp R}$ if $f$ is either symmetric or alternating, in which case we write $W^\perp$.

Let $(V, f)$ be an inner product space, and let $A$ be the inner product matrix of $f$ relative to a basis $e_1, \ldots, e_n$ of $V$. We claim that $b \in V^{\perp L}$ if and only if $b$ is a solution of the homogeneous system $A^\top x = 0$. If $b \in V^{\perp L}$. then $f(b, e_j) = 0$ for all $j$. Writing $b = \sum_i b_i e_i$, we see that $0 = f(b, e_j) = f\left(\sum_i b_i e_i, e_j\right) = \sum_i b_i f(e_i, e_j)$. In matrix terms, $b = (b_1, \ldots, b_n)^\top$ and $b^\top A = 0$; transposing, $b$ is a solution of the homogeneous system $A^\top x = 0$. The proof of the converse is left to the reader. A similar argument shows that $c \in V^{\perp R}$ if and only if $c$ is a solution of the homogeneous system $Ax = 0$.

**Theorem 8.73.** *Let $(V, f)$ be an inner product space. Then $f$ is nondegenerate if and only if $V^{\perp L} = \{0\} = V^{\perp R}$; that is, if $f(b, c) = 0$ for all $c \in V$, then $b = 0$, and if $f(b, c) = 0$ for all $b \in V$, then $c = 0$.*

**Proof.** Our remarks above show that $b \in V^{\perp L}$ if and only if $b$ is a solution of the homogeneous system $A^\top x = 0$. Therefore, $V^{\perp L} \neq \{0\}$ if and only if there is a nontrivial solution $b$, and Exercise 2.75 on page 144 shows that this holds if and only if $\det(A^\top) = 0$. Since $\det(A^\top) = \det(A)$, we have $f$ degenerate. A similar argument shows that $V^{\perp R} \neq \{0\}$ if and only if there is a nontrivial solution to $Ax = 0$.  •

Here is another characterization of nondegeneracy, in terms of the dual space. This is quite natural, for if $f$ is a bilinear form on a vector space $V$ over a field $k$, then the function $f(\Box, u)\colon V \to k$ is a linear functional for any fixed $u \in V$.

**Theorem 8.74.** *Let $(V, f)$ be an inner product space, and let $e_1, \ldots, e_n$ be a basis of $V$. Then $f$ is nondegenerate if and only if the dual basis $f(\Box, e_1), \ldots, f(\Box, e_n)$ is a basis of the dual space $V^*$.*

**Proof.** Assume that $f$ is nondegenerate. Since $\dim(V^*) = n$, it suffices to prove linear independence. If there are scalars $c_1, \ldots, c_n$ with $\sum_i c_i f(\Box, e_i) = 0$, then

$$\sum_i c_i f(v, e_i) = 0 \quad \text{for all } v \in V.$$

If we define $u = \sum_i c_i e_i$, then $f(v, u) = 0$ for all $v$, so that nondegeneracy gives $u = 0$. But $e_1, \ldots, e_n$ is a linearly independent list, so that all $c_i = 0$; hence, $f(\Box, e_1), \ldots, f(\Box, e_n)$ is also linearly independent, and hence it is a basis of $V^*$.

Conversely, assume that the given linear functionals are a basis of $V^*$. If $f(v, u) = 0$ for all $v \in V$, where $u = \sum_i c_i e_i$, then $\sum_i c_i f(\Box, e_i) = 0$. Since these linear functionals are linearly independent, all $c_i = 0$, and so $u = 0$; that is, $f$ is nondegenerate. •

**Corollary 8.75.** *If $(V, f)$ is an inner product space with $f$ nondegenerate, then every linear functional $g \in V^*$ has the form*

$$g = f(\Box, u)$$

*for a unique $u \in V$.*

**Proof.** Let $e_1, \ldots, e_n$ be a basis of $V$, and let $f(\Box, e_1), \ldots, f(\Box, e_n)$ be its dual basis. Since $g \in V^*$, there are scalars $c_i$ with $g = \sum_i c_i f(\Box, e_i)$. If we define $u = \sum_i c_i e_i$, then $g(v) = f(v, u)$.

To prove uniqueness, suppose that $f(\Box, u) = f(\Box, u')$. Then $f(v, u - u') = 0$ for all $v \in V$, and so nondegeneracy of $f$ gives $u - u' = 0$. •

**Corollary 8.76.** *Let $(V, f)$ be an inner product space with $f$ nondegenerate. If $e_1, \ldots, e_n$ is a basis of $V$, then there exists a basis $b_1, \ldots, b_n$ of $V$ with*

$$f(e_i, b_j) = \delta_{ij}.$$

**Proof.** Since $f$ is nondegenerate, the function $V \to V^*$, given by $v \mapsto f(\Box, v)$, is an isomorphism. Hence, the following diagram commutes:

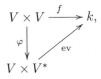

where ev is evaluation $(x, g) \mapsto g(x)$ and $\varphi\colon (x, y) \mapsto (x, f(\Box, y))$. For each $i$, let $g_i \in V^*$ be the $i$th coordinate function: if $v \in V$ and $v = \sum_j c_j e_j$, then

$g_i(v) = c_i$. By Corollary 8.75, there are $b_1, \ldots, b_n \in V$ with $g_i = f(\square, b_i)$ for all $i$. Commutativity of the diagram gives

$$f(e_i, b_j) = \mathrm{ev}(e_i, g_j) = \delta_{ij}. \quad \bullet$$

**Example 8.77.** Let $(V, f)$ be an inner product space, and let $W \subseteq V$ be a subspace. It is possible that $f$ is nondegenerate, while its restriction $f|(W \times W)$ is degenerate. For example, let $V = k^2$, and let $f$ have the inner product matrix $A = \left[\begin{smallmatrix} 0 & 1 \\ 1 & 0 \end{smallmatrix}\right]$ relative to the standard basis $e_1, e_2$. It is clear that $A$ is nonsingular, so that $f$ is nondegenerate. On the other hand, if $W = \langle e_1 \rangle$, then $f|(W \times W) = 0$, and hence it is degenerate. ◄

**Proposition 8.78.** *Let $(V, f)$ be an inner product space, and let $W$ be a subspace of $V$. If $f|(W \times W)$ is nondegenerate, then*

$$V = W \oplus W^{\perp R} = W \oplus W^{\perp L}.$$

**Remark.** We do not assume that $f$ itself is nondegenerate; even if we did, it would not force $f|(W \times W)$ to be nondegenerate, as we have seen in Example 8.77. ◄

**Proof.** If $u \in W \cap W^\perp$, then $f(w, u) = 0$ for all $w \in W$. Since $f|(W \times W)$ is nondegenerate and $u \in W$, we have $u = 0$; hence, $W \cap W^\perp = \{0\}$. If $v \in V$, then $f(\square, v)|W$ is a linear functional on $W$; that is, $f(\square, v)|W \in W^*$. By Corollary 8.75, there is $w_0 \in W$ with $f(w, v) = f(w, w_0)$ for all $w \in W$. Hence, $v = w_0 + (v - w_0)$, where $w_0 \in W$ and $v - w_0 \in W^\perp$. ◦

There is a name for direct sum decompositions as in the proposition.

**Definition.** Let $(V, f)$ be an inner product space. Then a direct sum

$$V = W_1 \oplus \cdots \oplus W_r$$

is an ***orthogonal direct sum*** if, for all $i \neq j$, we have $f(w_i, w_j) = 0$ for all $w_i \in W_i$ and $w_j \in W_j$. (Some authors denote orthogonal direct sum by $V = W_1 \perp \cdots \perp W_r$.)

We are now going to look more carefully at special bilinear forms; first we examine alternating forms, then symmetric ones.

We begin by constructing all alternating bilinear forms $f$ on a two-dimensional vector space $V$ over a field $k$. As always, $f = 0$ is an example. Otherwise, there exist two vectors $e_1, e_2 \in V$ with $f(e_1, e_2) \neq 0$; say, $f(e_1, e_2) = c$. If we replace $e_1$ by $e_1' = c^{-1} e_1$, then $f(e_1', e_2) = 1$. Since $f$ is alternating, the inner product matrix $A$ of $f$ relative to the basis $e_1', e_2$ is $A = \left[\begin{smallmatrix} 0 & 1 \\ -1 & 0 \end{smallmatrix}\right]$. This is even true when $k$ has characteristic 2; in this case, $A = \left[\begin{smallmatrix} 0 & 1 \\ 1 & 0 \end{smallmatrix}\right]$.

**Definition.** A ***hyperbolic plane*** over a field $k$ is a two-dimensional vector space over $k$ equipped with a nonzero alternating bilinear form.

We have just seen that every two-dimensional alternating space $(V, f)$ in which $f$ is not identically zero has an inner product matrix $A = \left[\begin{smallmatrix} 0 & 1 \\ -1 & 0 \end{smallmatrix}\right]$.

**Theorem 8.79.** *Let $(V, f)$ be an alternating space, where $V$ is a vector space over any field $k$. If $f$ is nondegenerate, then there is an orthogonal direct sum*

$$V = H_1 \oplus \cdots \oplus H_m,$$

*where each $H_i$ is a hyperbolic plane.*

**Proof.** The proof is by induction on $\dim(V) \geq 1$. For the base step, note that $\dim(V) \geq 2$, because an alternating form on a one-dimensional space must be 0, hence degenerate. If $\dim(V) = 2$, then we saw that $V$ is a hyperbolic plane. For the inductive step, note that there are vectors $e_1, e_2 \in V$ with $f(e_1, e_2) \neq 0$ (because $f$ is nondegenerate, hence, nonzero), and we may normalize so that $f(e_1, e_2) = 1$: if $f(e_1, e_2) = d$, replace $e_2$ by $d^{-1}e_2$. The subspace $H_1 = \langle e_1, e_2 \rangle$ is a hyperbolic plane, and the restriction $f|(H_1 \times H_1)$ is nondegenerate. Thus, Proposition 8.78 gives $V = H_1 \oplus H_1^{\perp}$. Since the restriction of $f$ to $H_1^{\perp}$ is nondegenerate, by Exercise 8.55 on page 703, the inductive hypothesis applies. •

**Corollary 8.80.** *Let $(V, f)$ be an alternating space, where $V$ is a vector space over a field $k$. If $f$ is nondegenerate, then $\dim(V)$ is even.*

**Proof.** By the theorem, $V$ is a direct sum of two-dimensional subspaces. •

**Definition.** Let $(V, f)$ be an alternating space with $f$ nondegenerate. A ***symplectic basis***[13] is a basis $x_1, y_1, \ldots, x_m, y_m$ such that $f(x_i, y_i) = 1$, $f(y_i, x_i) = -1$ for all $i$, and all other $f(x_i, x_j)$, $f(y_i, y_j)$, $f(x_i, y_j)$, and $f(y_j, x_i)$ are 0.

**Corollary 8.81.** *Let $(V, f)$ be an alternating space with $f$ nondegenerate, and let $A$ be an inner product matrix for $f$ (relative to some basis of $V$).*

(i) *There exists a symplectic basis $x_1, y_1, \ldots, x_m, y_m$ for $V$, and $A$ is a $2m \times 2m$ matrix for some $m \geq 1$.*

(ii) *If $k$ has characteristic not 2, then $A$ is congruent to a matrix direct sum of blocks of the form $\begin{bmatrix} 0 & 1 \\ -1 & 0 \end{bmatrix}$, and the latter is congruent to $\begin{bmatrix} 0 & I \\ -I & 0 \end{bmatrix}$, where $I$ is the $m \times m$ identity matrix.*[14] *If $k$ has characteristic 2, then remove the minus signs, for $-1 = 1$.*

(iii) *Every nonsingular skew-symmetric matrix $A$ over a field $k$ of characteristic not 2 is congruent to a direct sum of $2 \times 2$ blocks $\begin{bmatrix} 0 & 1 \\ -1 & 0 \end{bmatrix}$. If $k$ has characteristic 2, then remove the minus signs.*

**Proof.**

(i) A symplectic basis exists, by Theorem 8.79, and so $V$ is even-dimensional.

---

[13] The term *symplectic* was coined by H. Weyl. On p. 165 of his book, *The Classical Groups; Their Invariants and Representations*, he wrote, "The name 'complex group' formerly advocated by me in allusion to line complexes, as these are defined by the vanishing of antisymmetric bilinear forms, has become more and more embarrassing through collision with the word 'complex' in the connotation of complex number. I therefore propose to replace it by the corresponding Greek adjective 'symplectic.' Dickson calls the group the 'Abelian linear group' in homage to Abel who first studied it."

[14] If the form $f$ is degenerate, then $A$ is congruent to a direct sum of $2 \times 2$ blocks $\begin{bmatrix} 0 & 1 \\ -1 & 0 \end{bmatrix}$ and a block of 0's.

(ii) An inner product matrix $A$ is congruent to the inner product matrix relative to a symplectic basis arising from a symplectic basis $x_1, y_1, \ldots, x_m, y_m$. The second inner product matrix arises from a reordered symplectic basis $x_1, \ldots, x_m, y_1, \ldots, y_m$.

(iii) A routine calculation.   $\bullet$

We now consider symmetric bilinear forms.

**Definition.** Let $(V, f)$ be a symmetric space, and let $E = e_1, \ldots, e_n$ be a basis of $V$. Then $E$ is an ***orthogonal basis*** if $f(e_i, e_j) = 0$ for all $i \neq j$, and $E$ is an ***orthonormal basis*** if $f(e_i, e_j) = \delta_{ij}$, where $\delta_{ij}$ is the Kronecker delta.

If $e_1, \ldots, e_n$ is an orthogonal basis of a symmetric space $(V, f)$, then $V = \langle e_1 \rangle \oplus \cdots \oplus \langle e_n \rangle$ is an orthogonal direct sum. In Corollary 8.76, we saw that if $(V, f)$ is a symmetric space with $f$ nondegenerate and $e_1, \ldots, e_n$ is a basis of $V$, then there exists a basis $b_1, \ldots, b_n$ of $V$ with $f(e_i, b_j) = \delta_{ij}$. If $E$ is an orthonormal basis, then we can set $b_i = e_i$ for all $i$.

**Theorem 8.82.** *Let $(V, f)$ be a symmetric space, where $V$ is a vector space over a field $k$ of characteristic not 2.*

(i) *$V$ has an orthogonal basis, and so every symmetric matrix $A$ with entries in $k$ is congruent to a diagonal matrix.*

(ii) *If $C = \operatorname{diag}[c_1^2 d_1, \ldots, c_n^2 d_n]$, then $C$ is congruent to $D = \operatorname{diag}[d_1, \ldots, d_n]$.*

(iii) *If $f$ is nondegenerate and every element in $k$ has a square root in $k$, then $V$ has an orthonormal basis. Every nonsingular symmetric matrix $A$ with entries in $k$ is congruent to $I$.*

**Proof.**

(i) If $f = 0$, then every basis is an orthogonal basis. We may now assume that $f \neq 0$. By Exercise 8.53 on page 703, which applies because $k$ does not have characteristic 2, there is some $v \in V$ with $f(v, v) \neq 0$ (otherwise, $f$ is both symmetric and alternating). If $W = \langle v \rangle$, then $f|(W \times W)$ is nondegenerate, so that Proposition 8.78 gives $V = W \oplus W^\perp$. The proof is now completed by induction on $\dim(W)$.

If $A$ is a symmetric $n \times n$ matrix, then Proposition 8.71(i) shows that there is a symmetric bilinear form $f$ and a basis $U = u_1, \ldots, u_n$ so that $A$ is the inner product matrix of $f$ relative to $U$. We have just seen that there exists an orthogonal basis $v_1, \ldots, v_n$, so that Proposition 8.71(v) shows that $A$ is congruent to the diagonal matrix $\operatorname{diag}[f(v_1, v_1), \ldots, f(v_n, v_n)]$.

(ii) If an orthogonal basis consists of vectors $v_i$ with $f(v_i, v_i) = c_i^2 d_i$, then replacing each $v_i$ by $v_i' = c_i^{-1} v_i$ gives an orthogonal basis with $f(v_i', v_i') = d_i$. It follows that the inner product matrix of $f$ relative to the basis $v_1', \ldots, v_n'$ is $D = \operatorname{diag}[d_1, \ldots, d_n]$.

(iii) By (i), there exists an orthogonal basis $v_1, \ldots, v_n$; let $f(v_i, v_i) = c_i$ for each $i$. Since $f$ is nondegenerate, $c_i \neq 0$ for all $i$ (the determinant of the inner product matrix relative to this orthogonal basis is $c_1 c_2 \cdots c_n$); since each $c_i$ is a square, by hypothesis, we may replace each $v_i$ by $v_i' = (\sqrt{c_i})^{-1}v_i$, as in (ii); this new basis is orthonormal. The final statement follows because the inner product matrix relative to an orthonormal basis is the identity $I$. •

Notice that Theorem 8.82 does not say that any two diagonal matrices over a field $k$ of characteristic not 2 are congruent; this depends on $k$. For example, if $k = \mathbb{C}$, then all (nonsingular) diagonal matrices are congruent to $I$, but we now show that this is false if $k = \mathbb{R}$.

**Definition.** A symmetric bilinear form $f$ on a vector space $V$ over $\mathbb{R}$ is *positive definite* if $f(v,v) > 0$ for all nonzero $v \in V$, while $f$ is *negative definite* if $f(v,v) < 0$ for all nonzero $v \in V$.

The next result, and its matrix corollary, was proved by Sylvester. When $n = 2$, it classifies the conic sections, and when $n = 3$, it classifies the quadric surfaces.

**Lemma 8.83.** *If $f$ is a symmetric bilinear form on a vector space $V$ over $\mathbb{R}$ of dimension $m$, then there is an orthogonal direct sum*

$$V = W_+ \oplus W_- \oplus W_0,$$

*where $f|W_+$ is positive definite, $f|W_-$ is negative definite, and $f|W_0$ is identically $0$. Moreover, the dimensions of these three subspaces are uniquely determined by $f$.*

**Proof.** By Theorem 8.82, there is an orthogonal basis $v_1, \ldots, v_m$ of $V$. Denote $f(v_i, v_i)$ by $d_i$. As any real number, each $d_i$ is either positive, negative, or 0, and we rearrange the basis vectors so that $v_1, \ldots, v_p$ have positive $d_i$, $v_{p+1}, \ldots, v_{p+r}$ have negative $d_i$, and the last vectors have $d_i = 0$. It follows easily that $V$ is the orthogonal direct sum

$$V = \langle v_1, \ldots, v_p \rangle \oplus \langle v_{p+1}, \ldots, v_{p+r} \rangle \oplus \langle v_{p+r+1}, \ldots, v_m \rangle,$$

and that the restrictions of $f$ to each summand are positive definite, negative definite, and zero.

Now $W_0 = V^\perp$ depends only on $f$, and hence its dimension depends only on $f$ as well. To prove uniqueness of the other two dimensions, suppose that there is a second orthogonal direct sum $V = W_+' \oplus W_-' \oplus W_0$. If $T \colon V \to W_+$ is the projection, then $\ker T = W_- \oplus W_0$. It follows that if $\varphi = T|W_+'$, then

$$\ker \varphi = W_+' \cap \ker T = W_+' \cap (W_- \oplus W_0).$$

However, if $v \in W_+'$, then $f(v,v) \geq 0$, while if $v \in W_- \oplus W_0$, then $f(v,v) \leq 0$; hence, if $v \in \ker \varphi$, then $f(v,v) = 0$. But $f|W_+'$ is positive definite, for this is one of the defining properties of $W_+'$, so that $f(v,v) = 0$ implies $v = 0$. We conclude that $\ker \varphi = \{0\}$, and $\varphi \colon W_+' \to W_+$ is an injection; therefore, $\dim(W_+') \leq \dim(W_+)$. The reverse inequality is proved similarly, so that $\dim(W_+') = \dim(W_+)$. Finally, the formula $\dim(W_-) = \dim(V) - \dim(W_+) - \dim(W_0)$, and its primed version, give $\dim(W_-') = \dim(W_-)$. •

**Theorem 8.84 (Law of Inertia).** *Every symmetric $n \times n$ matrix $A$ over $\mathbb{R}$ is congruent to a matrix of the form*

$$\begin{bmatrix} I_p & 0 & 0 \\ 0 & -I_r & 0 \\ 0 & 0 & 0 \end{bmatrix}.$$

*Moreover, the **signature** $s$ of $f$, defined by $s = p - r$, is well-defined, and two symmetric real $n \times n$ matrices are congruent if and only if they have the same rank and the same signature.*

**Proof.** By Theorem 8.82, $A$ is congruent to a diagonal matrix $\mathrm{diag}[d_1, \ldots, d_n]$, where $d_1, \ldots, d_p$ are positive, $d_{p+1}, \ldots, d_{p+r}$ are negative, and $d_{p+r+1}, \ldots, d_n$ are 0. But every positive real is a square, while every negative real is the negative of a square; it now follows from Theorem 8.82(ii) that $A$ is congruent to a matrix as in the statement of the theorem.

It is clear that congruent $n \times n$ matrices have the same rank and the same signature. Conversely, let $A$ and $A'$ have the same rank and the same signature. Now $A$ is congruent to the matrix direct sum $I_p \oplus -I_r \oplus 0$ and $A'$ is congruent to $I_{p'} \oplus -I_{r'} \oplus 0$. Since $\mathrm{rank}(A) = \mathrm{rank}(A')$, we have $p' + r' = p + r$; since the signatures are the same, we have $p' - r' = p - r$. It follows that $p' = p$ and $r' = r$, so that both $A$ and $A'$ are congruent to the same diagonal matrix of 1's, $-1$'s, and 0's, and hence they are congruent to each other.  •

It would be simplest if a symmetric space $(V, f)$ with $f$ nondegenerate always had an orthonormal basis; that is, if every symmetric matrix were congruent to the identity matrix. This need not be so: the real $2 \times 2$ matrix $-I$ is not congruent to $I$ because their signatures are different ($I$ has signature 2 and $-I$ has signature $-2$).

Closely related to a bilinear form $f$ is a *quadratic form* $Q$, given by $Q(v) = f(v, v)$. Recall that the *length* of a vector $v = (x_1, \ldots, x_n) \in \mathbb{R}^n$ is $\sqrt{x_1^2 + \cdots + x_n^2}$. Thus, if $f$ is the dot product on $\mathbb{R}^n$, then

$$\|v\|^2 = \left( \sqrt{x_1^2 + \cdots + x_n^2} \right)^2 = f(v, v) = Q(v).$$

**Definition.** Let $V$ be a vector space over a field $k$. A **quadratic form** is a function $Q \colon V \to k$ such that

(i)  $Q(cv) = c^2 Q(v)$ for all $v \in V$ and $c \in k$;

(ii) the function $f \colon V \times V \to k$, defined by

$$f(u, v) = Q(u + v) - Q(u) - Q(v),$$

   is a bilinear form. We call $f$ the **associated bilinear form**.

If $Q$ is a quadratic form, it is clear that its associated bilinear form $f$ is symmetric: $f(u, v) = f(v, u)$.

**Example 8.85.**

(i) If $g$ is a bilinear form on a vector space $V$ over a field $k$, we claim that $Q$, defined by $Q(v) = g(v, v)$, is a quadratic form. Now $Q(cv) = g(cv, cv) = c^2 g(v, v) = c^2 Q(v)$, giving the first axiom in the definition. If $u, v \in V$, then

$$Q(u + v) = g(u + v, u + v)$$
$$= g(u, u) + g(u, v) + g(v, u) + g(v, v)$$
$$= Q(u) + Q(v) + f(u, v),$$

where

$$f(u, v) = g(u, v) + g(v, u).$$

It is easy to check that $f$ is a bilinear form.

(ii) We have just seen that every bilinear form $g$ determines a quadratic form $Q$; the converse is true if $g$ is symmetric and $k$ does not have characteristic 2. In this case, $Q$ determines $g$; in fact, the formula $2g(u, v) = f(u, v)$ gives

$$g(u, v) = \tfrac{1}{2} f(u, v).$$

(iii) If $f$ is the usual dot product defined on $\mathbb{R}^n$, then the corresponding quadratic form is $Q(v) = \|v\|^2$, where $\|v\|$ is the length of the vector $v$.

(iv) If $f$ is a bilinear form on a vector space $V$ with inner product matrix $A = [a_{ij}]$ relative to some basis $e_1, \ldots, e_n$, and $u = \sum c_i e_i$, then

$$Q(u) = \sum_{i,j} a_{ij} c_i c_j.$$

If $n = 2$, for example, we have

$$Q(u) = a_{11} c_1^2 + (a_{12} + a_{21}) c_1 c_2 + a_{22} c_2^2.$$

Thus, quadratic forms are really homogeneous quadratic polynomials.  ◀

We have just observed, in Example 8.85(ii), that if a field $k$ does not have characteristic 2, then symmetric bilinear forms and quadratic forms are merely two different ways of viewing the same thing, for each determines the other. Thus, we have classified quadratic forms $Q$ over $\mathbb{C}$ [Theorem 8.82(iii)] and over $\mathbb{R}$ (Theorem 8.84). The classification over the prime fields (even over $\mathbb{F}_2$) is also known, as is the classification over the finite fields. Given a quadratic form $Q$ defined on a finite-dimensional vector space $V$ over a field $k$, its associated bilinear form is

$$f(x, y) = Q(x + y) - Q(x) - Q(y).$$

Call two quadratic forms **equivalent** if their associated bilinear forms have congruent inner product matrices, and call a quadratic form **nondegenerate** if its bilinear form $f$ is nondegenerate. As we saw in Example 8.85, $f$ is a symmetric bilinear form (which is uniquely determined by $Q$ when $k$ does not have characteristic 2).

We now state (without proof) the results when $Q$ is nondegenerate. If $k$ is a finite field of odd characteristic, then two nondegenerate quadratic forms over $k$ are equivalent if and only if they have the same discriminant (Kaplansky, *Linear Algebra and Geometry; A Second Course*, pp. 14–15). If $k$ is a finite field of characteristic 2, the theory is a bit more complicated. In this case, the associated symmetric bilinear form

$$f(x, y) = Q(x + y) + Q(x) + Q(y)$$

must also be alternating, for $f(x, x) = Q(2x) + 2Q(x) = 0$. Therefore, $V$ has a symplectic basis $x_1, y_1, \ldots, x_m, y_m$. The **Arf invariant** of $Q$ is defined by

$$\text{Arf}(Q) = \sum_{i=1}^{m} Q(x_i)Q(y_i)$$

[it is not at all obvious that the Arf invariant is an invariant; i.e., that $\text{Arf}(Q)$ does not depend on the choice of symplectic basis; see R. L. Dye, On the Arf Invariant, *Journal of Algebra* 53 (1978), pp. 36–39, for an elegant proof]. If $k$ is a finite field of characteristic 2, then two nondegenerate quadratic forms over $k$ are equivalent if and only if they have the same discriminant and the same Arf invariant (Ibid., pp. 27–33). The classification of quadratic forms over $\mathbb{Q}$ is much deeper. Just as $\mathbb{R}$ can be obtained from $\mathbb{Q}$ by completing with respect to the usual metric $d(a, b) = |a - b|$, so, too, can we complete $\mathbb{Z}$, for every prime $p$, with respect to the $p$-adic metric [Example 6.149(iv)]. The completion $\mathbb{Z}_p$ is called the *$p$-adic integers*. The $p$-adic metric on $\mathbb{Z}$ can be extended to $\mathbb{Q}$, and its completion $\mathbb{Q}_p$ [which turns out to be $\text{Frac}(\mathbb{Z}_p)$] is called the *$p$-adic numbers*. The **Hasse–Minkowski Theorem** (Borevich–Shafarevich, *Number Theory*, pp. 61) says that two quadratic forms over $\mathbb{Q}$ are equivalent if and only if they are equivalent over $\mathbb{R}$ and over $\mathbb{Q}_p$ for all primes $p$.

**8.5.2. Isometries.** The first theorems of Linear Algebra consider the structure of vector spaces in order to pave the way for a discussion of linear transformations. Similarly, the first theorems of inner product spaces enable us to discuss appropriate linear transformations.

**Definition.** If $(V, f)$ is an inner product space with $f$ nondegenerate, then an *isometry* is a linear transformation $\varphi \colon V \to V$ such that, for all $u, v \in V$,

$$f(u, v) = f(\varphi u, \varphi v).$$

For example, if $f$ is the dot product on $\mathbb{R}^n$ and $v = (x_1, \ldots, x_n)$, then we saw in Example 8.85(iii) that $\|v\|^2 = f(v, v)$. If $\varphi \colon \mathbb{R}^n \to \mathbb{R}^n$ is an isometry, then

$$\|\varphi(v)\|^2 = f(\varphi v, \varphi v) = f(v, v) = \|v\|^2,$$

so that $\|\varphi(v)\| = \|v\|$. Since the distance between two points $u, v \in \mathbb{R}^n$ is $\|u - v\|$, every isometry $\varphi$ preserves distance; it follows that isometries here are continuous.

**Proposition 8.86.** *Let $(V, f)$ be an inner product space with $f$ nondegenerate, let $E = e_1, \ldots, e_n$ be a basis of $V$, and let $A$ be the inner product matrix relative to $E$. Then $\varphi \in \text{GL}(V)$ is an isometry if and only if its matrix $M = {}_E[\varphi]_E$ satisfies the equation $M^\top A M = A$.*

**Proof.** Recall Equation (1) on page 684:

$$f(b, c) = b^\top A c,$$

where $b, c \in V$ (elements of $k^n$ are $n \times 1$ column vectors). If $e_1, \ldots, e_n$ is the standard basis of $k^n$, then

$$\varphi(e_i) = M e_i$$

for all $i$, because $M e_i$ is the $i$th column of $M$ [which is the coordinate list of $\varphi(e_i)$]. Therefore,

$$f(\varphi e_i, \varphi e_j) = (M e_i)^\top A(M e_j) = e_i^\top (M^\top A M) e_j.$$

If $\varphi$ is an isometry, then

$$f(\varphi e_i, \varphi e_j) = f(e_i, e_j) = e_i^\top A e_j,$$

so that $f(e_i, e_j) = e_i^\top A e_j = e_i^\top (M^\top A M) e_j$. Hence, Proposition 8.71(iv) gives $M^\top A M = A$.

Conversely, if $M^\top A M = A$, then

$$f(\varphi e_i, \varphi e_j) = e_i^\top (M^\top A M) e_j = e_i^\top A e_j = f(e_i, e_j),$$

and $\varphi$ is an isometry. •

**Definition.** Let $(V, f)$ be an inner product space with $f$ nondegenerate. Then

$$\mathrm{Isom}(V, f) = \{\text{all isometries } V \to V\}.$$

**Proposition 8.87.** *If $(V, f)$ is an inner product space with $f$ nondegenerate, then* $\mathrm{Isom}(V, f)$ *is a subgroup of* $\mathrm{GL}(V)$.

**Proof.** Let us see that every isometry $\varphi \colon V \to V$ is nonsingular. If $u \in V$ and $\varphi u = 0$, then, for all $v \in V$, we have $0 = f(\varphi u, \varphi v) = f(u, v)$. Since $f$ is nondegenerate, $u = 0$ and so $\varphi$ is an injection. Hence, $\dim(\mathrm{im}\,\varphi) = \dim(V)$, so that $\mathrm{im}\,\varphi = V$, by Corollary 2.115(ii). Thus, $\varphi \in \mathrm{GL}(V)$, and $\mathrm{Isom}(V, f) \subseteq \mathrm{GL}(V)$.

We now show that $\mathrm{Isom}(V, f)$ is a subgroup. Of course, $1_V$ is an isometry. The inverse of an isometry $\varphi$ is also an isometry: for all $u, v \in V$,

$$f(\varphi^{-1} u, \varphi^{-1} v) = f(\varphi \varphi^{-1} u, \varphi \varphi^{-1} v) = f(u, v).$$

Finally, the composite of two isometries $\varphi$ and $\theta$ is also an isometry:

$$f(u, v) = f(\varphi u, \varphi v) = f(\theta \varphi u, \theta \varphi v). \quad •$$

Computing the inverse of a general nonsingular matrix is quite time-consuming, but it is easier for isometries. We introduce the *adjoint* of a linear transformation to aid us.

**Definition.** Let $(V, f)$ be an inner product space with $f$ nondegenerate. The **adjoint** of a linear transformation $T \colon V \to V$ is a linear transformation $T^* \colon V \to V$ such that, for all $u, v \in V$,

$$f(Tu, v) = f(u, T^* v).$$

Let us see that adjoints exist.

**Proposition 8.88.** *If $(V, f)$ is an inner product space with $f$ nondegenerate, then every linear transformation $T \colon V \to V$ has an adjoint.*

**Proof.** Let $e_1, \ldots, e_n$ be a basis of $V$. For each $j$, the function $\varphi_j \colon V \to k$, defined by

$$\varphi_j(v) = f(Tv, e_j),$$

is easily seen to be a linear functional. By Corollary 8.75, there exists $u_j \in V$ with $\varphi_j(v) = f(v, u_j)$ for all $v \in V$. Define $T^* \colon V \to V$ by $T^*(e_j) = u_j$, and note that

$$f(Te_i, e_j) = \varphi_j(e_i) = f(e_i, u_j) = f(e_i, T^*e_j). \quad \bullet$$

**Proposition 8.89.** *Let $(V, f)$ be an inner product space with $f$ nondegenerate. If $T \colon V \to V$ is a linear transformation with adjoint $T^*$, then $T$ is an isometry if and only if $T^*T = 1_V$, in which case $T^* = T^{-1}$.*

**Proof.** If $T^*T = 1_V$, then, for all $u, v \in V$, we have

$$f(Tu, Tv) = f(u, T^*Tv) = f(u, v),$$

so that $T$ is an isometry.

Conversely, assume that $T$ is an isometry. Choose $v \in V$; for all $u \in V$, we have

$$f(u, T^*Tv - v) = f(u, T^*Tv) - f(u, v) = f(Tu, Tv) - f(u, v) = 0.$$

Since $f$ is nondegenerate, $T^*Tv - v = 0$; that is, $T^*Tv = v$. As this is true for all $v \in V$, we have $T^*T = 1_V$. $\quad \bullet$

**Definition.** Let $(V, f)$ be an inner product space with $f$ nondegenerate.

(i) If $f$ is alternating, then $\mathrm{Isom}(V, f)$ is called the ***symplectic group***, and it is denoted by $\mathrm{Sp}(V, f)$.

(ii) If $f$ is symmetric, then $\mathrm{Isom}(V, f)$ is called the ***orthogonal group***, and it is denoted by $O(V, f)$.

As always, a choice of basis $E$ of an $n$-dimensional vector space $V$ over a field $k$ gives an isomorphism $\mu \colon \mathrm{GL}(V) \to \mathrm{GL}(n, k)$, the group of all nonsingular $n \times n$ matrices over $k$. In particular, let $(V, f)$ be an alternating space with $f$ nondegenerate, and let $E = x_1, y_1, \ldots, x_m, y_m$ be a symplectic basis of $V$ (which exists, by Corollary 8.81); recall that $n = \dim(V)$ is even; say, $n = 2m$. Denote the image of $\mathrm{Sp}(V, f)$ by $\mathrm{Sp}(2m, k)$. Similarly, if $(V, f)$ is a symmetric space with $f$ nondegenerate and $E$ is an orthogonal basis (which exists when $k$ does not have characteristic 2, by Theorem 8.82), denote the image of $O(V, f)$ by $O(n, f)$. The description of orthogonal groups when $k$ has characteristic 2 is more complicated; see our discussion on page 700.

Let $(V, f)$ be an inner product space with $f$ nondegenerate. We find adjoints, first when $f$ is symmetric, then when $f$ is alternating. This will enable us to recognize orthogonal matrices and symplectic matrices.

**Proposition 8.90.** *Let $(V, f)$ be a symmetric space with $f$ nondegenerate, let $T \colon V \to V$ be an isometry, let $E = e_1, \ldots, e_n$ be an orthogonal basis, and let $B = [b_{ij}] = {}_ET_E$.*

(i) *$B^*$ is the "weighted" transpose $B^* = [c_i^{-1} c_j b_{ji}]$, where $f(e_i, e_i) = c_i$ for all $i$.*

(ii) *If $E$ is an orthonormal basis, then $B^* = B^\top$ and $B$ is orthogonal if and only if $B^\top B = I$.*

**Proof.** We have

$$f(Be_i, e_j) = f\left(\sum_\ell b_{\ell i} e_\ell, e_j\right) = \sum_\ell b_{\ell i} f(e_\ell, e_j) = b_{ji} c_j.$$

If $B^* = [b_{ij}^*]$, then a similar calculation gives

$$f(e_i, B^* e_j) = \sum_\ell b_{\ell j}^* f(e_i, e_\ell) = c_i b_{ij}^*.$$

Since $f(Be_i, e_j) = f(e_i, B^* e_j)$, we have $b_{ji} c_j = c_i b_{ij}^*$ for all $i, j$. Since $f$ is nondegenerate, all $c_i \neq 0$, and so

$$b_{ij}^* = c_i^{-1} c_j b_{ji},$$

because $B$ is the matrix of the isometry $T$. Statement (ii) follows from Proposition 8.89, for $c_i = 1$ for all $i$ when $E$ is orthonormal. •

How can we recognize symplectic matrices?

**Proposition 8.91.** *Let $(V, f)$ be an alternating space with $f$ nondegenerate, where $V$ is a $2m$-dimensional vector space. If $B = \begin{bmatrix} P & Q \\ S & T \end{bmatrix}$ is a $2m \times 2m$ matrix partitioned into $m \times m$ blocks, then the adjoint of $B$ is*

$$B^* = \begin{bmatrix} T^\top & -Q^\top \\ -S^\top & P^\top \end{bmatrix},$$

*and $B$ is symplectic if and only if $B^* B = I$.*

**Proof.** Let $E$ be a symplectic basis ordered as $x_1, \ldots, x_m, y_1, \ldots, y_m$, and assume that the partition of $B$ respects $E$; that is,

$$f(Bx_i, x_j) = f\left(\sum_\ell p_{\ell i} x_\ell + s_{\ell i} y_\ell, x_j\right) = \sum_\ell p_{\ell i} f(x_\ell, x_j) + \sum_\ell s_{\ell i} f(y_\ell, x_j) = -s_{ji}$$

[the definition of symplectic basis says that $f(x_\ell, x_j) = 0$ and $f(y_\ell, x_j) = -\delta_{\ell j}$ for all $i, j$]. Partition the adjoint $B^*$ into $m \times m$ blocks:

$$B^* = \begin{bmatrix} \Pi & K \\ \Sigma & \Omega \end{bmatrix}.$$

Hence,

$$f(x_i, B^* x_j) = f\left(x_i, \sum_\ell \pi_{\ell j} x_\ell + \sigma_{\ell j} y_\ell\right) = \sum_\ell \pi_{\ell j} f(x_i, x_\ell) + \sum_\ell \sigma_{\ell j} f(x_i, y_\ell) = \sigma_{ij}$$

[for $f(x_i, x_\ell) = 0$ and $f(x_i, y_\ell) = \delta_{i\ell}$]. Since $f(Bx_i, x_j) = f(x_i, B^* x_j)$, we have $\sigma_{ij} = -s_{ji}$. Hence, $\Sigma = -S^\top$. Computation of the other blocks of $B^*$ is similar, and is left to the reader. The last statement follows from Proposition 8.89. •

The next question is whether $\text{Isom}(V, f)$ depends on the choice of nondegenerate bilinear form $f$. We shall see that it does not depend on $f$ when $f$ is alternating, and so there is only one symplectic group $\text{Sp}(V)$ (however, when $f$ is symmetric, then $\text{Isom}(V, f)$ does depend on $f$ and there are several types of orthogonal groups).

Observe that $\mathrm{GL}(V)$ acts on $k^{V \times V}$: if $\varphi \in \mathrm{GL}(V)$ and $f: V \times V \to k$, define $\varphi f = f^\varphi$, where

$$f^\varphi(b, c) = f(\varphi^{-1}b, \varphi^{-1}c).$$

This formula does yield an action: if $\theta \in \mathrm{GL}(V)$, then $(\varphi\theta)f = f^{\varphi\theta}$, where

$$(\varphi\theta)f(b, c) = f^{\varphi\theta}(b, c) = f((\varphi\theta)^{-1}b, (\varphi\theta)^{-1}c) = f(\theta^{-1}\varphi^{-1}b, \theta^{-1}\varphi^{-1}c).$$

On the other hand, $\varphi(\theta f)$ is defined by

$$(f^\theta)^\varphi(b, c) = f^\theta(\varphi^{-1}b, \varphi^{-1}c) = f(\theta^{-1}\varphi^{-1}b, \theta^{-1}\varphi^{-1}c),$$

so that $(\varphi\theta)f = \varphi(\theta f)$.

**Definition.** Let $V$ and $W$ be finite-dimensional vector spaces over a field $k$, and let $f: V \times V \to k$ and $g: W \times W \to k$ be bilinear forms. Then $f$ and $g$ are **equivalent** if there is an isometry $\varphi: V \to W$; that is, $f(u, v) = g(\varphi u, \varphi v)$ for all $u, v \in V$.

**Lemma 8.92.** *If $f, g$ are bilinear forms on a finite-dimensional vector space $V$, then the following statements are equivalent.*

(i) *$f$ and $g$ are equivalent.*

(ii) *If $E = e_1, \ldots, e_n$ is a basis of $V$, then the inner product matrices of $f$ and $g$ with respect to $E$ are congruent.*

(iii) *There is $\varphi \in \mathrm{GL}(V)$ with $g = f^\varphi$.*

**Proof.**

(i) $\Rightarrow$ (ii) If $\varphi: V \to V$ is an isometry, then $g(\varphi(b), \varphi(c)) = f(b, c)$ for all $b, c \in V$. If $E = e_1, \ldots, e_n$ is a basis of $V$, then $E' = \varphi(e_1), \ldots, \varphi(e_n)$ is also a basis, because isometries are isomorphisms. Now the inner product matrix $A'$ of $g$ with respect to the basis $E'$ is $A' = [g(\varphi e_i, \varphi e_j)]$, while the inner product matrix $A$ of $f$ with respect to the basis $E$ is $A = [f(e_i, e_j)]$. By Proposition 8.71(v), the inner product matrix of $g$ with respect to $E$ is congruent to $A$.

(ii) $\Rightarrow$ (iii) If $A = [f(e_i, e_j)]$ and $A' = [g(e_i, e_j)]$, then there exists a nonsingular matrix $Q = [q_{ij}]$ with $A' = Q^\top A Q$, by hypothesis. Define $\theta: V \to V$ to be the linear transformation with $\theta(e_j) = \sum_\nu q_{\nu j} e_\nu$. Finally, $g = f^{\theta^{-1}}$:

$$[g(e_i, e_j)] = A' = Q^\top A Q = \left[ f\left( \sum_\nu q_{\nu i} e_\nu, \sum_\lambda q_{\lambda j} e_\lambda \right) \right]$$

$$= [f(\theta(e_i), \theta(e_j))] = [f^{\theta^{-1}}(e_i, e_j)].$$

(iii) $\Rightarrow$ (i) It is obvious from the definition that $\varphi^{-1}: (V, g) \to (V, f)$ is an isometry:

$$g(b, c) = f^\varphi(b, c) = f(\varphi^{-1}b, \varphi^{-1}c).$$

Hence, $\varphi$ is an isometry, and $g$ is equivalent to $f$.   •

**Lemma 8.93.**

(i) *Let $(V, f)$ be an inner product space with $f$ nondegenerate. The stabilizer $\mathrm{GL}(V)_f$ of $f$ under the action on $k^{V \times V}$ is $\mathrm{Isom}(V, f)$.*

(ii) *If* $g \colon V \times V \to k$ *lies in the same orbit as* $f$, *then* $\mathrm{Isom}(V, f)$ *and* $\mathrm{Isom}(V, g)$ *are isomorphic; in fact, they are conjugate subgroups of* $\mathrm{GL}(V)$.

**Proof.**

(i) By definition of stabilizer, $\varphi \in \mathrm{GL}(V)_f$ if and only if $f^\varphi = f$; that is, for all $b, c \in V$, we have $f(\varphi^{-1}b, \varphi^{-1}c) = f(b, c)$. Thus, $\varphi^{-1}$, and hence $\varphi$, is an isometry.

(ii) By Exercise 1.109 on page 76, we have $\mathrm{GL}(V)_g = \tau (\mathrm{GL}(V)_f) \tau^{-1}$ for some $\tau \in \mathrm{GL}(V)$; that is, $\mathrm{Isom}(V, g) = \tau \mathrm{Isom}(V, f) \tau^{-1}$. •

It follows from Lemma 8.93 that equivalent bilinear forms have isomorphic isometry groups. We can now show that the symplectic group is, up to isomorphism, independent of the choice of nondegenerate alternating form.

**Theorem 8.94.** *If* $(V, f)$ *and* $(V, g)$ *are alternating spaces with* $f$ *and* $g$ *nondegenerate, then* $f$ *and* $g$ *are equivalent and*

$$\mathrm{Sp}(V, f) \cong \mathrm{Sp}(V, g).$$

**Proof.** By Corollary 8.81(ii), the inner product matrix of any nondegenerate alternating bilinear form is congruent to $\left[\begin{smallmatrix} 0 & I \\ -I & 0 \end{smallmatrix}\right]$, where $I$ is the identity matrix. The result now follows from Lemma 8.92. •

When $k$ is a finite field, say, $k = \mathbb{F}_q$ for some prime power $q$, the matrix group $\mathrm{GL}(n, k)$ is often denoted by $\mathrm{GL}(n, q)$. A similar notation is used for other groups arising from $\mathrm{GL}(n, k)$. For example, if $V$ is a $2m$-dimensional space over $\mathbb{F}_q$ equipped with a nondegenerate alternating form $g$, then $\mathrm{Sp}(V, f)$ may be denoted by $\mathrm{Sp}(2m, q)$ (we have just seen that this group does not depend on $f$).

Symplectic and orthogonal groups give rise to simple groups. We summarize the main facts below; a full discussion can be found in the following books: Artin, *Geometric Algebra*; Carter, *Simple Groups of Lie Type*, as well as the article by Carter in Kostrikin–Shafarevich, *Algebra* IX; Conway et al, *Atlas of Finite Groups*; Dieudonné, *La Géometrie des Groupes Classiques*; Suzuki, *Group Theory* I.

Symplectic groups yield the following simple groups. If $k$ is a field, define

$$\mathrm{PSp}(2m, k) = \mathrm{Sp}(2m, k)/Z(2m, k),$$

where $Z(2m, k)$ is the subgroup of all scalar matrices in $\mathrm{Sp}(2m, k)$. The groups $\mathrm{PSp}(2m, k)$ are simple for all $m \geq 1$ and all fields $k$ with only three exceptions: $\mathrm{PSp}(2, \mathbb{F}_2) \cong S_3$, $\mathrm{PSp}(2, \mathbb{F}_3) \cong A_4$, and $\mathrm{PSp}(4, \mathbb{F}_2) \cong S_6$.

The orthogonal groups, that is, isometry groups of a symmetric space $(V, f)$ when $f$ is nondegenerate, also give rise to simple groups. In contrast to symplectic groups, however, they depend on properties of the field $k$. We restrict our attention to finite fields $k$.

Assume that $k$ has odd characteristic $p$.

There is only one orthogonal group, $O(n, p^m)$, when $n$ is odd, but when $n$ is even, there are two groups, $O^+(n, p^m)$ and $O^-(n, p^m)$. Simple groups are defined from these groups as follows: first form $SO^\epsilon(n, p^m)$ (where $\epsilon = +$ or $\epsilon = -$) as all orthogonal matrices having determinant 1; next, form $PSO^{\epsilon}(n, p^m)$ by dividing by all scalar matrices in $SO^{\epsilon}(n, p^m)$. Finally, we define a subgroup $\Omega^{\epsilon}(n, p^m)$ of $PSO^{\epsilon}(n, p^m)$ (essentially the commutator subgroup), and these groups are simple with only a finite number of exceptions (which can be explicitly listed).

Assume that $k$ has characteristic 2.

We usually begin with a quadratic form instead of a symmetric bilinear form. In this case, there is also only one orthogonal group $O(n, 2^m)$ when $n$ is odd, but there are two, which are also denoted by $O^+(n, 2^m)$ and $O^-(n, 2^m)$, when $n$ is even. If $n$ is odd, say, $n = 2\ell + 1$, then $O(2\ell + 1, 2^m) \cong \mathrm{Sp}(2\ell, 2^m)$, so that we consider only orthogonal groups $O^{\epsilon}(2\ell, 2^m)$ arising from symmetric spaces of even dimension. Each of these groups gives rise to a simple group in a manner analogous to the odd characteristic case.

Quadratic forms are of great importance in Number Theory. For an introduction to this aspect of the subject, see Hahn, *Quadratic Algebras, Clifford Algebras, and Arithmetic Witt Groups*, Lam, *The Algebraic Theory of Quadratic Forms*, and O'Meara, *Introduction to Quadratic Forms*.

**Definition.** Let $(V, f)$ be an inner product space with $f$ nondegenerate. A linear transformation $T\colon V \to V$ is ***self-adjoint*** if $T = T^*$.

For example, if $f$ is symmetric, then Proposition 8.90(ii) shows that the matrix of a self-adjoint linear transformation $T$ relative to an orthonormal basis of $V$ is symmetric. We shall see that a matrix being self-adjoint influences its eigenvalues.

In Chapter 7, we defined an inner product on the complex vector space $\mathrm{cf}(G)$ of all class functions on a finite group $G$. We now generalize this inner product to arbitrary finite-dimensional complex vector spaces.

**Definition.** If $V$ is a finite-dimensional vector space over $\mathbb{C}$, define the ***complex inner product*** $h\colon V \times V \to \mathbb{C}$ by

$$h(u, v) = \sum_{j=1}^{n} u_j \bar{v}_j,$$

where $u = (u_1, \ldots, u_n), v = (v_1, \ldots, v_n) \in V$, and $\bar{c}$ denotes the complex conjugate of a complex number $c$..

Here are some elementary properties of $h$.

**Proposition 8.95.** *Let $V$ be a finite-dimensional vector space over $\mathbb{C}$.*

   (i) $h(u + u', v) = h(u, v) + h(u', v)$ *and* $h(u, v + v') = h(u, v) + h(u, v')$ *for all* $u, u', v, v' \in V$.

  (ii) $h(cu, v) = ch(u, v)$ *and* $h(u, cv) = \bar{c}h(u, v)$ *for all* $c \in \mathbb{C}$ *and* $u, v \in V$.

 (iii) $h(v, u) = \overline{h(u, v)}$ *for all* $u, v \in V$,

 (iv) $h(u, u) = 0$ *if and only if* $u = 0$.

  (v) *The standard basis* $e_1, \ldots, e_n$ *is an orthonormal basis; that is,* $h(e_i, e_j) = \delta_{ij}$.

 (vi) $Q(v) = h(v, v)$ *is a real-valued quadratic form.*

**Remark.** It follows from (ii) that $h$ is *not* bilinear, for it does not preserve scalar multiplication in the second variable. ◄

**Proof.** All verifications are routine; nevertheless, we check nondegeneracy. If $h(u, u) = 0$, then

$$0 = \sum_{j=1}^{n} u_j \overline{u_j} = \sum_{j=1}^{n} |u_j|^2.$$

Since $|u_j|^2$ is a nonnegative real, each $u_j = 0$ and $u = 0$. This last computation also shows that $Q$ is real-valued. •

**Definition.** Let $V$ be a finite-dimensional complex vector space. An isometry $T: V \to V$ [that is, $h(Tu, Tv) = h(u, v)$ for all $u, v \in V$] is called **unitary**.

The matrix $A$ of a unitary transformation relative to the standard basis is called a **unitary matrix**.

It is easy to see, as in the proof of Proposition 8.87, that all unitary matrices form a subgroup of $\mathrm{GL}(n, \mathbb{C})$.

**Definition.** The **unitary group** $U(n, \mathbb{C})$ is the set of all $n \times n$ unitary linear matrices. The **special unitary group** $SU(n, \mathbb{C})$ is the subgroup of $U(n, \mathbb{C})$ consisting of all unitary matrices having determinant 1.

Even though the complex inner product $h$ is not bilinear, its resemblance to "honest" inner products allows us to define the **adjoint** of a linear transformation $T: V \to V$ as a linear transformation $T^*: V \to V$ such that, for all $u, v \in V$,

$$h(Tu, v) = h(u, T^*v).$$

**Proposition 8.96.** *Let $V$ be a finite-dimensional complex vector space, and let $T: V \to V$ be a linear transformation.*

(i) *$T$ is a unitary transformation if and only if $T^*T = 1_V$.*

(ii) *If $A = [a_{ij}]$ is the matrix of $T$ relative to the standard basis $E$, then the matrix $A^* = [a_{ij}^*]$ of $T^*$ relative to $E$ is its **conjugate transpose**: for all $i, j$,*

$$a_{ij}^* = \overline{a_{ji}}.$$

**Proof.** Adapt the proofs of Propositions 8.89 and 8.90. •

We are now going to see that self-adjoint matrices are useful.

**Definition.** A complex $n \times n$ matrix $A$ is called **hermitian** if $A = A^*$.

Thus, $A = [a_{ij}]$ is hermitian if and only if $a_{ji} = \overline{a_{ij}}$ for all $i, j$ and its diagonal entries are real; a *real* matrix is hermitian if and only if it is symmetric.

What are the eigenvalues of a real symmetric $2 \times 2$ matrix $A$? If $A = \left[\begin{smallmatrix} p & q \\ q & r \end{smallmatrix}\right]$, then its characteristic polynomial is

$$\det(xI - A) = \det\left(\begin{bmatrix} x - p & -q \\ -q & x - r \end{bmatrix}\right) = (x - p)(x - r) - q^2 = x^2 - (p + r)x - q^2,$$

and its eigenvalues are given by the quadratic formula:

$$\tfrac{1}{2}\big(-(p+r) \pm \sqrt{(p+r)^2 + 4q^2}\big).$$

The eigenvalues are real because the discriminant $(p+r)^2 + 4q^2$, being a sum of squares, is nonnegative. Therefore, the eigenvalues of a real symmetric $2 \times 2$ matrix are real.

One needs great courage to extend this method to prove that the eigenvalues of a real symmetric $3 \times 3$ matrix are real, even if one assumes the characteristic polynomial is a reduced cubic and uses the cubic formula.

The next result is half of the Principal Axis Theorem. (The other half says that if $A$ is a hermitian matrix, then there is an unitary matrix $U$ with $UAU^{-1} = UAU^*$ diagonal; if $A$ is a real symmetric matrix, then there is a real orthogonal matrix $O$ with $OAO^{-1} = OAO^\top$ diagonal.)

**Theorem 8.97.** *The eigenvalues of a hermitian $n \times n$ matrix $A$ are real. In particular, the eigenvalues of a symmetric real $n \times n$ matrix are real.*

**Proof.** The second statement follows from the first, for real hermitian matrices are symmetric.

Since $\mathbb{C}$ is algebraically closed, all the eigenvalues of $A$ lie in $\mathbb{C}$. If $c$ is an eigenvalue, then $Au = cu$ for some nonzero vector $u$. Now $h(Au, u) = h(cu, u) = ch(u, u)$. On the other hand, since $A$ is hermitian, we have $A^* = A$ and $h(Au, u) = h(u, A^*u) = h(u, Au) = h(u, cu) = \bar{c}h(u, u)$. Therefore, $(c - \bar{c})h(u, u) = 0$. But $h(u, u) \neq 0$, and so $c = \bar{c}$; that is, $c$ is real.  •

The definition of the complex inner product $h$ can be extended to vector spaces over any field $k$ that has an automorphism $\sigma$ of order 2 (in place of complex conjugation on $\mathbb{C}$); for example, if $k$ is a finite field with $|k| = q^2 = p^{2n}$ elements, then $\sigma \colon a \mapsto a^\sigma = a^q$ is an automorphism of order 2. If $V$ is a finite-dimensional vector space over such a field $k$, call a function $g \colon V \times V \to k$ **hermitian** it satisfies the first four properties of $h$ in Proposition 8.95.

(i) $g(u + u', v) = g(u, v) + g(u', v)$ and $g(u, v + v') = g(u, v) + g(u, v')$ for all $u, u', v, v' \in V$.

(ii) $g(au, v) = ag(u, v)$ and $g(u, av) = a^\sigma g(u, v)$ for all $a \in k$ and $u, v \in V$.

(iii) $g(v, u) = g(u, v)^\sigma$ for all $u, v \in V$,

(iv) $g(u, u) = 0$ if and only if $u = 0$.

If $A = [a_{ij}] \in \mathrm{GL}(n, k)$, define $A^* = [a_{ji}^\sigma]$. Call $A$ **unitary** if $AA^* = I$, and define the **unitary group** $U(n, k)$ to be the family of all unitary $n \times n$ matrices over $k$; it is a subgroup of $\mathrm{GL}(n, k)$. The **special unitary group** $\mathrm{SU}(n, k)$ is the subgroup of $U(n, k)$ consisting of all unitary matrices having determinant 1. The **projective unitary group** $\mathrm{PSU}(n, k) = \mathrm{SU}(n, k)/Z(n, k)$, where $Z(n, k)$ is the center of $\mathrm{SU}(n, k)$ consisting of all scalar matrices $aI$ with $aa^\sigma = 1$. When $k$ is a finite field of order $q^2$, then every $\mathrm{PSU}(n, k)$ is a simple group except $\mathrm{PSU}(2, \mathbb{F}_4)$, $\mathrm{PSU}(2, \mathbb{F}_9)$, and $\mathrm{PSU}(3, \mathbb{F}_4)$.

# Exercises

**8.50.** It is shown in Analytic Geometry that if $\ell_1$ and $\ell_2$ are lines with slopes $m_1$ and $m_2$, respectively, then $\ell_1$ and $\ell_2$ are perpendicular if and only if $m_1 m_2 = -1$. If

$$\ell_i = \{\alpha v_i + u_i : \alpha \in \mathbb{R}\},$$

for $i = 1, 2$, prove that $m_1 m_2 = -1$ if and only if the dot product $v_1 \cdot v_2 = 0$. (Since both lines have slopes, neither of them is vertical.)

**Hint.** The slope of a vector $v = (a, b)$ is $m = b/a$.

**8.51.** (i) In Calculus, a line in space passing through a point $u$ is defined as

$$\{u + \alpha w : \alpha \in \mathbb{R}\} \subseteq \mathbb{R}^3,$$

where $w$ is a fixed nonzero vector. Show that every line through the origin is a one-dimensional subspace of $\mathbb{R}^3$.

(ii) In Calculus, a plane in space passing through a point $u$ is defined as the subset

$$\{v \in \mathbb{R}^3 : (v - u) \cdot n = 0\} \subseteq \mathbb{R}^3,$$

where $n \neq 0$ is a fixed *normal vector*. Prove that a plane through the origin is a two-dimensional subspace of $\mathbb{R}^3$.

**Hint.** To determine the dimension of a plane through the origin, find an orthogonal basis of $\mathbb{R}^3$ containing $n$.

**8.52.** If $k$ is a field of characteristic not 2, prove that for every $n \times n$ matrix $A$ with entries in $k$, there are unique matrices $B$ and $C$ with $B$ symmetric, $C$ skew-symmetric (i.e., $C^\top = -C$), and $A = B + C$.

**∗ 8.53.** Let $(V, f)$ be an inner product space, where $V$ is a vector space over a field $k$ of characteristic not 2. Prove that if $f$ is both symmetric and alternating, then $f = 0$.

**8.54.** If $(V, f)$ is an inner product space, define $u \perp v$ to mean $f(u, v) = 0$. Prove that $\perp$ is a symmetric relation if and only if $f$ is either symmetric or alternating.

**∗ 8.55.** Let $(V, f)$ be an inner product space with $f$ nondegenerate. If $W$ is a proper subspace and $V = W \oplus W^\perp$, prove that $f|(W^\perp \times W^\perp)$ is nondegenerate.

**8.56.** (i) Let $(V, f)$ be an inner product space, where $V$ is a vector space over a field $k$ of characteristic not 2. Prove that if $f$ is symmetric, then there is a basis $e_1, \ldots, e_n$ of $V$ and scalars $c_1, \ldots, c_n$ such that $f(x, y) = \sum_i c_i x_i y_i$, where $x = \sum x_i e_i$ and $y = \sum y_i e_i$. Moreover, if $f$ is nondegenerate and $k$ has square roots, then the basis $e_1, \ldots, e_n$ can be chosen so that $f(x, y) = \sum_i x_i y_i$.

(ii) If $k$ is a field of characteristic not 2, then every symmetric matrix $A$ with entries in $k$ is congruent to a diagonal matrix. Moreover, if $A$ is nonsingular and $k$ has square roots, then $A = P^\top P$ for some nonsingular matrix $P$.

**8.57.** Give an example of two real symmetric $m \times m$ matrices having the same rank and the same discriminant but that are not congruent.

**8.58.** For every field $k$, prove that $\text{Sp}(2, k) = \text{SL}(2, k)$.

**Hint.** By Corollary 8.81(ii), we know that if $P \in \text{Sp}(2m, k)$, then $\det(P) = \pm 1$. However, Proposition 8.90 shows that $\det(P) = 1$ for $P \in \text{Sp}(2, k)$ [it is true, for all $m \geq 1$, that $\text{Sp}(2m, k) \subseteq \text{SL}(2m, k)$].

**8.59.** If $A$ is an $m \times m$ matrix with $A^{\top}A = I$, prove that $\left[\begin{smallmatrix} A & 0 \\ 0 & A \end{smallmatrix}\right]$ is a symplectic matrix. Conclude, if $k$ is a finite field of odd characteristic, that $O(m, k) \subseteq \mathrm{Sp}(2m, k)$.

**8.60.** Let $(V, f)$ be an alternating space with $f$ nondegenerate. Prove that $T \in \mathrm{GL}(V)$ is an isometry [i.e., $T \in \mathrm{Sp}(V, f)$] if and only if, whenever $E = x_1, y_1, \ldots, x_m, y_m$ is a symplectic basis of $V$, then $T(E) = Tx_1, Ty_1, \ldots, Tx_m, Ty_m$ is also a symplectic basis of $V$.

**8.61.** Prove that the group $\mathbf{Q}$ of quaternions is isomorphic to a subgroup of the special unitary group $SU(2, \mathbb{C})$.

**Hint.** Recall that $\mathbf{Q} = \langle A, B \rangle \subseteq \mathrm{GL}(2, \mathbb{C})$, where $A = \left[\begin{smallmatrix} 0 & 1 \\ -1 & 0 \end{smallmatrix}\right]$ and $B = \left[\begin{smallmatrix} 0 & i \\ i & 0 \end{smallmatrix}\right]$.

## Section 8.6. Graded Algebras

We are now going to use tensor products of several modules in order to construct some useful rings. This topic is often called *Multilinear Algebra*.

> *Throughout this section, $k$ denotes a commutative ring.*

**Definition.** A $k$-algebra $A$ is a **graded $k$-algebra** if there are $k$-submodules $A^p$, for $p \geq 0$, such that

(i) $A = \bigoplus_{p \geq 0} A^p$;

(ii) for all $p, q \geq 0$, if $x \in A^p$ and $y \in A^q$, then $xy \in A^{p+q}$; that is,

$$A^p A^q \subseteq A^{p+q}.$$

An element $x \in A^p$ is called **homogeneous** of **degree** $p$.

Notice that 0 is homogeneous of any degree, but that most elements in a graded ring are not homogeneous and, hence, have no degree. Note also that (ii) implies that any product of homogeneous elements is itself homogeneous.

**Example 8.98.**

(i) The polynomial ring $A = k[x]$ is a graded $k$-algebra if we define

$$A^p = \{rx^p : r \in k\}.$$

The homogeneous elements are the monomials and, in contrast to ordinary usage, only monomials (including 0) have degrees. On the other hand, $x^p$ has degree $p$ in both usages of the term *degree*.

(ii) The polynomial ring $A = k[x_1, x_2, \ldots, x_n]$ is a graded $k$-algebra if we define

$$A^p = \left\{ rx_1^{e_1} x_2^{e_2} \cdots x_n^{e_n} : r \in k \text{ and } \sum e_i = p \right\};$$

that is, $A^p$ consists of all monomials of *total degree* $p$.

(iii) In Algebraic Topology, we assign a sequence of (abelian) *cohomology groups* $H^p(X, k)$ to a space $X$, where $k$ is a commutative ring and $p \geq 0$, and we define a multiplication on $\bigoplus_{p \geq 0} H^p(X, k)$, called *cup product*, making it a graded $k$-algebra (called the **cohomology ring**).

If $A$ is a graded $k$-algebra and $u \in A^r$, then multiplication by $u$ gives $k$-maps $A^p \to A^{p+r}$ for all $p$. This elementary observation arises in applications of the cohomology ring of a space. ◄

Just as the degree of a polynomial is often useful, so, too, is the degree of a homogeneous element in a graded algebra.

**Definition.** If $A$ and $B$ are graded $k$-algebras, then a ***graded map***[15] is a $k$-algebra map $f \colon A \to B$ with $f(A^p) \subseteq B^p$ for all $p \geq 0$.

It is easy to see that all graded $k$-algebras and graded maps form a category, which we denote by $\mathbf{Gr}_k\mathbf{Alg}$.

**Definition.** If $A$ is a graded $k$-algebra, then a ***graded ideal*** (or ***homogeneous ideal***) is a two-sided ideal $I$ in $A$ with $I = \bigoplus_{p \geq 0} I^p$, where $I^p = I \cap A^p$.

Here are some first properties of graded algebras.

**Proposition 8.99.** *Let $A$ and $B$ be graded $k$-algebras.*

(i) *If $f \colon A \to B$ is a graded map, then $\ker f$ is a graded ideal.*

(ii) *Let $I$ be a graded ideal in $A$. Then $A/I$ is a graded $k$-algebra if we define*
$$(A/I)^p = (A^p + I)/I.$$
*Moreover, $A/I = \bigoplus_p (A/I)^p \cong \bigoplus_p (A^p/I^p)$.*

(iii) *A two-sided ideal $I$ in $A$ is graded if and only if it is generated by homogeneous elements.*

(iv) *The identity element $1$ in $A$ is homogeneous of degree $0$.*

**Proof.**

(i) This is routine.

(ii) Since $I$ is a graded ideal, the Second Isomorphism Theorem gives
$$(A/I)^p = (A^p + I)/I \cong A^p/(I \cap A^p) = A^p/I^p.$$

(iii) If $I$ is graded, then $I = \bigoplus_p I^p$, so that $I$ is generated by $\bigcup_p I^p$. But $\bigcup_p I^p$ consists of homogeneous elements because $I^p = I \cap A^p \subseteq A^p$ for all $p$.

Conversely, suppose that $I$ is generated by a set $X$ of homogeneous elements. We must show that $I = \bigoplus_p (I \cap A^p)$, and it is only necessary to prove $I \subseteq \bigoplus_p (I \cap A^p)$, for the reverse inclusion always holds. Since $I$ is the two-sided ideal generated by $X$, a typical element in $I$ has the form $\sum_i a_i x_i b_i$, where $a_i, b_i \in A$ and $x_i \in X$. It suffices to show that each $a_i x_i b_i$ lies in $\bigoplus_p (I \cap A^p)$, and so we drop the subscript $i$. Since $a = \sum a_j$ and $b = \sum b_\ell$ (where each $a_j$ and $b_\ell$ is homogeneous), we have $axb = \sum_{j,\ell} a_j x b_\ell$. But each $a_j x b_\ell$ lies in $I$ (because $I$ is generated by $X$), and it is homogeneous, being the product of homogeneous elements.

---

[15]There is a more general definition of graded map $f \colon A \to B$. Given $d \in \mathbb{Z}$, then a $k$-algebra map $f$ is a ***graded map of degree $d$*** if $f(A^p) \subseteq B^{p+d}$ for all $p \geq 0$. Thus, graded maps defined above are graded maps of degree $0$.

(iv) Write $1 = e_0 + e_1 + \cdots + e_t$, where $e_i \in A^i$. If $a_p \in A^p$, then

$$a_p - e_0 a_p = e_1 a_p + \cdots + e_t a_p \in A^p \cap (A^{p+1} \oplus \cdots \oplus A^{p+t}) = \{0\}.$$

It follows that $a_p = e_0 a_p$ for all homogeneous elements $a_p$, and so $a = e_0 a$ for all $a \in A$. A similar argument, examining $a_p = a_p 1$ instead of $a_p = 1 a_p$, shows that $a = a e_0$ for all $a \in A$; that is, $e_0$ is also a right identity. Therefore, $1 = e_0$, by the uniqueness of the identity element in a ring.   •

**Example 8.100.** The quotient $k[x]/(x^{13})$ is a graded $k$-algebra. However, there is no obvious grading on the algebra $k[x]/(x^{13} + 1)$. After all, what degree should be assigned to the coset of $x^{13}$, which is the same as the coset of $-1$?   ◄

**8.6.1. Tensor Algebra.** For every commutative ring $k$, we are going to construct a functor $T: {}_k\mathbf{Mod} \to \mathbf{Gr}_k\mathbf{Alg}$ such that, if $V$ is a free $k$-module with basis $X$, then $T(V)$ is a free $k$-algebra with basis $X$; that is, $T(V)$ consists of polynomials in noncommuting variables $X$.

We begin by considering generalized associativity of tensor product.

**Definition.** Let $k$ be a commutative ring and let $M_1, \ldots, M_p$ be $k$-modules. A *k-multilinear function* $f: M_1 \times \cdots \times M_p \to N$, where $N$ is a $k$-module, is a function that is additive in each of the $p$ variables (when we fix the other $p - 1$ variables) and, if $1 \le i \le p$, then

$$f(m_1, \ldots, rm_i, \ldots, m_p) = rf(m_1, \ldots, m_i, \ldots, m_p),$$

where $r \in k$ and $m_\ell \in M_\ell$ for all $\ell$.

**Proposition 8.101.** *Let $k$ be a commutative ring and let $M_1, \ldots, M_p$ be $k$-modules.*

(i) *There exists a $k$-module $U[M_1, \ldots, M_p]$ that is a solution to the universal mapping problem posed by multilinearity:*

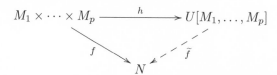

*that is, there is a $k$-multilinear $h$ such that, whenever $f$ is $k$-multilinear, there exists a unique $k$-homomorphism $\widetilde{f}$ making the diagram commute.*

(ii) *If $f_i: M_i \to M_i'$ are $k$-maps, then there is a unique $k$-map*

$$u[f_1, \ldots, f_p]: U[M_1, \ldots, M_p] \to U[M_1', \ldots, M_p']$$

*taking $h(m_1, \ldots, m_p) \mapsto h'(f_1(m_1), \ldots, f_p(m_p))$, where $h': M_1' \times \cdots \times M_p' \to U[M_1', \ldots, M_p']$.*

**Proof.**

(i) This is a straightforward generalization of Theorem 6.101, the existence of tensor products, using multilinear functions instead of bilinear ones. Let $F_p$

be the free $k$-module with basis $M_1 \times \cdots \times M_p$, and let $S$ be the submodule of $F_p$ generated by all elements of the following two types:

$$(A, m_i + m_i', B) - (A, m_i, B) - (A, m_i', B),$$

$$(A, rm_i, B) - r(A, m_i, B),$$

where $A = m_1, \ldots, m_{i-1}$, $B = m_{i+1}, \ldots, m_p$, $r \in k$, $m_i, m_i' \in M_i$, and $1 \le i \le p$ (of course, $A$ is empty if $i = 1$ and $B$ is empty if $i = p$).

Define $U[M_1, \ldots, M_p] = F_p/S$ and $h \colon M_1 \times \cdots \times M_p \to U[M_1, \ldots, M_p]$ by

$$h \colon (m_1, \ldots, m_p) \mapsto (m_1, \ldots, m_p) + S.$$

The reader should check that $h$ is $k$-multilinear. The remainder of the proof is merely an adaptation of the proof of Proposition 6.101, and it is also left to the reader.

(ii) The function $M_1 \times \cdots \times M_p \to U[M_1', \ldots, M_p']$, given by

$$(m_1, \ldots, m_p) \mapsto h'(f_1(m_1), \ldots, f_n(m_p)),$$

is easily seen to be $k$-multilinear; by universality, there exists a unique $k$-homomorphism as described in the statement. $\bullet$

Observe that there are no parentheses needed in the argument of the generator $h(m_1, \ldots, m_p)$; that is, $h(m_1, \ldots, m_p)$ depends only on the $p$-tuple $(m_1, \ldots, m_p)$ and not on any association of its coordinates. The next proposition relates this construction to iterated tensor products. Once this is done, we will change the notation $U[M_1, \ldots, M_p]$ (to $M_1 \otimes \cdots \otimes M_p$).

**Proposition 8.102 (Generalized Associativity).** *Let $k$ be a commutative ring and let $M_1, \ldots, M_p$ be $k$-modules. If $M_1 \otimes_k \cdots \otimes_k M_p$ is an iterated tensor product in some association, then there is a $k$-isomorphism*

$$U[M_1, \ldots, M_p] \to M_1 \otimes_k \cdots \otimes_k M_p$$

*taking $h(m_1, \ldots, m_p) \mapsto m_1 \otimes \cdots \otimes m_p$.*

**Remark.** We are tempted to quote Corollary 1.25: associativity for three factors implies associativity for many factors, for we have proved the associative law for three factors in Proposition 6.121. However, we did not prove $A \otimes_k (B \otimes_k C) = (A \otimes_k B) \otimes_k C$; we only constructed an isomorphism. There is an extra condition, due, independently, to Mac Lane and Stasheff: if the associative law holds up to isomorphism and a certain "pentagonal" diagram commutes, then generalized associativity holds up to isomorphism (Mac Lane, *Categories for the Working Mathematician*, pp. 157–161). ◀

**Proof.** The proof is by induction on $p \ge 2$. The base step is true, for $U[M_1, M_2] = M_1 \otimes_k M_2$. For the inductive step, let us assume that

$$M_1 \otimes_k \cdots \otimes_k M_p = U[M_1, \ldots, M_i] \otimes_k U[M_{i+1}, \ldots, M_p].$$

We have indicated the final factors in the given association; for example,

$$\big((M_1 \otimes_k M_2) \otimes_k M_3\big) \otimes_k (M_4 \otimes_k M_5) = U[M_1, M_2, M_3] \otimes_k U[M_4, M_5].$$

By induction, there are multilinear functions $h'\colon M_1 \times \cdots \times M_i \to M_1 \otimes_k \cdots \otimes_k M_i$ and $h''\colon M_{i+1} \times \cdots \times M_p \to M_{i+1} \otimes_k \cdots \otimes_k M_p$ with $h'(m_1,\ldots,m_i) = m_1 \otimes \cdots \otimes m_i$ associated as in $M_1 \otimes_k \cdots \otimes_k M_i$, and with $h''(m_{i+1},\ldots,m_p) = m_{i+1} \otimes \cdots \otimes m_p$ associated as in $M_{i+1} \otimes_k \cdots \otimes_k M_p$. Induction gives isomorphisms $\varphi'\colon U[M_1,\ldots,M_i] \to M_1 \otimes_k \cdots \otimes_k M_i$ and $\varphi''\colon U[M_{i+1},\ldots,M_p] \to M_{i+1} \otimes_k \cdots \otimes_k M_p$ with $\varphi'h' = h|(M_1 \times \cdots \times M_i)$ and $\varphi''h'' = h|(M_{i+1} \times \cdots \times M_p)$. By Corollary 6.105, $\varphi' \otimes \varphi''$ is an isomorphism $U[M_1,\ldots,M_i] \otimes_k U[M_{i+1},\ldots,M_p] \to M_1 \otimes_k \cdots \otimes_k M_p$.

We now show that $U[M_1,\ldots,M_i] \otimes_k U[M_{i+1},\ldots,M_p]$ is a solution to the universal problem for multilinear functions. Consider the diagram

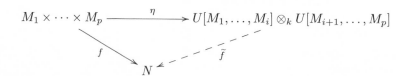

where $\eta(m_1,\ldots,m_p) = h'(m_1,\ldots,m_i) \otimes h''(m_{i+1},\ldots,m_p)$, $N$ is a $k$-module, and $f$ is multilinear. We must find a homomorphism $\widetilde{f}$ making the diagram commute.

If $(m_1,\ldots,m_i) \in M_1 \times \cdots \times M_i$, the function $f_{(m_1,\ldots,m_i)}\colon M_{i+1} \times \cdots \times M_p \to N$, defined by $(m_{i+1},\ldots,m_p) \mapsto f(m_1,\ldots,m_i,h''(m_{i+1},\ldots,m_p))$, is multilinear; hence, there is a unique homomorphism $\widetilde{f}_{(m_1,\ldots,m_i)}\colon U[M_{i+1},\ldots,M_p] \to N$ with

$$\widetilde{f}_{(m_1,\ldots,m_i)}\colon h''(m_{i+1},\ldots,m_p) \mapsto f(m_1,\ldots,m_p).$$

If $r \in k$ and $1 \leq j \leq i$, then

$$\widetilde{f}_{(m_1,\ldots,rm_j,\ldots,m_i)}(h''(m_{i+1},\ldots,m_p)) = f(m_1,\ldots,rm_j,\ldots,m_p)$$
$$= rf(m_1,\ldots,m_j,\ldots,m_i)$$
$$= r\widetilde{f}_{(m_1,\ldots,m_i)}(h''(m_{i+1},\ldots,m_p)).$$

Similarly, if $m_j, m_j' \in M_j$, where $1 \leq j \leq i$, then

$$\widetilde{f}_{(m_1,\ldots,m_j+m_j',\ldots,m_i)} = \widetilde{f}_{(m_1,\ldots,m_j,\ldots,m_i)} + \widetilde{f}_{(m_1,\ldots,m_j',\ldots,m_i)}.$$

The function of $i+1$ variables $M_1 \times \cdots \times M_i \times U[M_{i+1},\ldots,M_p] \to N$, defined by $(m_1,\ldots,m_i,u'') \mapsto \widetilde{f}_{(m_1,\ldots,m_i)}(u'')$, is multilinear, and so it gives a bilinear function $U[M_1,\ldots,M_i] \times U[M_{i+1},\ldots,M_p] \to N$, namely, $(u',u'') \mapsto (h'(u'),h''(u''))$. Thus, there is a unique homomorphism $\widetilde{f}\colon U[M_1,\ldots,M_i] \otimes_k U[M_{i+1},\ldots,M_p] \to N$ which takes $h'(m_1,\ldots,m_i) \otimes h''(m_{i+1},\ldots,m_p) \mapsto \widetilde{f}_{(m_1,\ldots,m_i)}(h''(m_{i+1},\ldots,m_p)) = f(m_1,\ldots,m_p)$; that is, $\widetilde{f}\eta = f$. Therefore, $U[M_1,\ldots,M_i] \otimes_k U[M_{i+1},\ldots,M_p]$ is a solution to the universal mapping problem. By uniqueness of such solutions, there is an isomorphism $\theta\colon U[M_1,\ldots,M_p] \to U[M_1,\ldots,M_i] \otimes_k U[M_{i+1},\ldots,M_p]$ with $\theta h(m_1,\ldots,m_p) = h'(m_1,\ldots,m_i) \otimes h''(m_{i+1},\ldots,m_p) = \eta(m_1,\ldots,m_p)$. Finally, $(\varphi' \otimes \varphi'')\theta$ is the desired isomorphism $U[M_1,\ldots,M_p] \cong M_1 \otimes_k \cdots \otimes_k M_p$.   •

We now abandon the notation in Proposition 8.101. From now on, we shall write

$$U[M_1, \ldots, M_p] = M_1 \otimes_k \cdots \otimes_k M_p,$$
$$h(m_1, \ldots, m_p) = m_1 \otimes \cdots \otimes m_p,$$
$$u[f_1, \ldots, f_p] = f_1 \otimes \cdots \otimes f_p.$$

This notation is modified when all $M_i = M$.

**Definition.** Let $k$ be a commutative ring, and let $M$ be a $k$-module. Define

$$\bigotimes\nolimits^0 M = k, \quad \bigotimes\nolimits^1 M = M, \quad \bigotimes\nolimits^p M = M \otimes_k \cdots \otimes_k M \ (p \text{ times}) \text{ if } p \geq 2.$$

Thus, when $p \geq 2$, the $k$-module $\bigotimes^p M$ is generated by symbols $m_1 \otimes \cdots \otimes m_p$ in which no parentheses occur.

We now construct the *tensor algebra* on a $k$-module $M$. Recall Proposition 6.124: if $k$ is a commutative ring and $A$ and $B$ are $k$-algebras, then $A \otimes_k B$ is a $k$-algebra with multiplication $(a \otimes b)(a' \otimes b') = aa' \otimes bb'$ (to prove this, we used Proposition 6.122, the special case of generalized associativity of a tensor product having four factors).

**Proposition 8.103.** *If $M$ is a $k$-module, then there is a graded $k$-algebra*

$$T(M) = \bigoplus_{p \geq 0} \left( \bigotimes\nolimits^p M \right)$$

*with action of $r \in k$ on $\bigotimes^p M$ given by*

$$r(y_1 \otimes \cdots \otimes y_p) = (ry_1) \otimes y_2 \otimes \cdots \otimes y_p,$$

*and multiplication $\bigotimes^p M \times \bigotimes^q M \to \bigotimes^{p+q} M$, for $p, q \geq 1$, given by*

$$(x_1 \otimes \cdots \otimes x_p, \ y_1 \otimes \cdots \otimes y_q) \mapsto x_1 \otimes \cdots \otimes x_p \otimes y_1 \otimes \cdots \otimes y_q.$$

**Proof.** First, note that

$$(ry_1) \otimes y_2 \otimes \cdots \otimes y_p = y_1 \otimes (ry_2) \otimes \cdots \otimes y_p = \cdots = y_1 \otimes \cdots \otimes (ry_p),$$

for all tensor products are over $k$. Now define the product of two homogeneous elements by the formula in the statement. Multiplication $\mu \colon T(M) \times T(M) \to T(M)$ must now be

$$\mu \colon \left( \sum_p x_p, \sum_q y_q \right) \mapsto \sum_{p,q} x_p \otimes y_q,$$

where $x_p \in \bigotimes^p M$ and $y_q \in \bigotimes^q M$. Multiplication is associative because no parentheses are needed in describing generators $m_1 \otimes \cdots \otimes m_p$ of $\bigotimes^p M$; the distributive laws hold because multiplication is $k$-bilinear. Finally, $1 \in k = \bigotimes^0 M$ is the identity, each element of $k$ commutes with every element of $T(M)$, and $[\bigotimes^p M][\bigotimes^q M] \subseteq \bigotimes^{p+q} M$, so that $T(M)$ is a graded $k$-algebra.    •

For example, the reader may check that if $M = k$, then $T(M) \cong k[x]$.

**Definition.** If $k$ is a commutative ring and $M$ is a $k$-module, then $T(M)$ is called the ***tensor algebra*** on $M$.

If $k$ is a commutative ring and $A$ and $B$ are $k$-modules, define a **word-module** on $A$ and $B$ of **length** $p \geq 0$ to be a $k$-module of the form

$$W(A, B)_p = \bigotimes^{e_1} A \otimes_k \bigotimes^{f_1} B \otimes_k \cdots \otimes_k \bigotimes^{e_r} A \otimes_k \bigotimes^{f_r} B,$$

where $\sum_i (e_i + f_i) = p$, all $e_i, f_i$ are integers, $e_1 \geq 0$, $f_r \geq 0$, and all the other exponents are positive.

**Proposition 8.104.** *If $A$ and $B$ are $k$-modules, then for all $p \geq 0$,*

$$\bigotimes^p (A \oplus B) \cong \bigoplus_{j=0}^{p} W(A, B)_j \otimes_k W'(A, B)_{p-j},$$

*where $W(A, B)_j$, $W'(A, B)_{p-j}$ range over all word-modules on $A$ and $B$ of length $j$ and $p - j$, respectively.*

**Proof.** The proof is by induction on $p \geq 0$. For the base step,

$$\bigotimes^0 (A \oplus B) = k \cong k \otimes_k k \cong \bigotimes^0 A \otimes_k \bigotimes^0 B.$$

For the inductive step,

$$\begin{aligned}
\bigotimes^{p+1} (A \oplus B) &= \bigotimes^p (A \oplus B) \otimes_k (A \oplus B) \\
&\cong \left( \bigotimes^p (A \oplus B) \otimes_k A \right) \oplus \left( \bigotimes^p (A \oplus B) \otimes_k B \right) \\
&\cong \bigoplus_{j=0}^{p} W(A, B)_j \otimes_k W'(A, B)_{p-j} \otimes_k X,
\end{aligned}$$

where $X \cong A$ or $X \cong B$. This completes the proof, for every word of length $p - j + 1$ has the form $W'(A, B) \otimes_k X$. •

**Proposition 8.105.** *Tensor algebra defines a functor $T \colon {}_k\mathbf{Mod} \to \mathbf{Gr}_k\mathbf{Alg}$ that preserves surjections.*

**Proof.** We have already defined $T$ on every $k$-module $M$: it is the tensor algebra $T(M)$. If $f \colon M \to N$ is a $k$-homomorphism, then Proposition 8.101 provides maps

$$f \otimes \cdots \otimes f \colon \bigotimes^p M \to \bigotimes^p N,$$

for each $p$, which give a $k$-algebra map $T(M) \to T(N)$. It is a simple matter to check that $T$ preserves identity maps and composites.

Assume that $f \colon M \to N$ is a surjective $k$-map. If $n_1 \otimes \cdots \otimes n_p \in \bigotimes^p N$, then surjectivity of $f$ provides $m_i \in M$, for all $i$, with $f(m_i) = n_i$, and so

$$T(f) \colon m_1 \otimes \cdots \otimes m_p \mapsto n_1 \otimes \cdots \otimes n_p. \quad •$$

We now generalize the notion of free module to free algebra.

**Definition.** Let $X$ be a subset of a $k$-algebra $F$. Then $F$ is a **free $k$-algebra** with **basis** $X$ if, for every $k$-algebra $A$ and every function $\varphi \colon X \to A$, there exists a unique $k$-algebra map $\widetilde{\varphi}$ with $\widetilde{\varphi}(x) = \varphi(x)$ for all $x \in X$. In other words, the following diagram commutes, where $i \colon X \to F$ is the inclusion:

$$
\begin{array}{ccc}
 & F & \\
i \uparrow & \diagdown \; \widetilde{\varphi} & \\
X & \xrightarrow[\varphi]{} & A
\end{array}
$$

If $k$ is a commutative ring and $V$ is a free $k$-module with basis $X$, then $T(V)$ is called the ring of polynomials over $k$ in **noncommuting variables** $X$, and it is denoted by

$$k\langle X \rangle.$$

If $V$ is the free $k$-module with basis $X$, then each element $u$ in $k\langle X \rangle = T(V)$ has a unique expression

$$u = \sum_{\substack{p \geq 0 \\ i_1, \ldots, i_p}} r_{i_1, \ldots, i_p} x_{i_1} \otimes \cdots \otimes x_{i_p},$$

where $r_{i_1, \ldots, i_p} \in k$ and $x_{i_j} \in X$. We obtain the usual notation for such a polynomial by erasing the tensor product symbols. For example, if $X = \{x, y\}$, then

$$u = r_0 + r_1 x + r_2 y + r_3 x^2 + r_4 y^2 + r_5 xy + r_6 yx + \cdots.$$

In the next proposition, we regard the graded $k$-algebra $k\langle X \rangle = T(V)$ merely as a $k$-algebra.

**Proposition 8.106.** *If $V$ is a free $k$-module with basis $X$, where $k$ is a commutative ring, then $k\langle X \rangle = T(V)$ is a free $k$-algebra with basis $X$.*

**Proof.** Consider the diagram

$$
\begin{array}{ccc}
T(V) & & \\
j \uparrow & \diagdown \; T(\widetilde{\varphi}) & \\
V & & T(A) \\
i \uparrow & \diagdown \; \widetilde{\varphi} & \downarrow \mu \\
X & \xrightarrow[\varphi]{} & A
\end{array}
$$

where $i \colon X \to V$ and $j \colon V \to T(V)$ are inclusions, and $A$ is a $k$-algebra. Viewing $A$ only as a $k$-module gives a $k$-module map $\widetilde{\varphi} \colon V \to A$, for $V$ is a free $k$-module with basis $X$. Applying the functor $T$ gives a $k$-algebra map $T(\widetilde{\varphi}) \colon T(V) \to T(A)$. For existence of a $k$-algebra map $T(V) \to A$, it suffices to define a $k$-algebra map $\mu \colon T(A) \to A$ such that the composite $\mu \circ T(\widetilde{\varphi})$ is a $k$-algebra map extending $\varphi$. For each $p$, consider the diagram

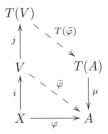

where $h_p\colon (a_1,\ldots,a_p) \mapsto a_1 \otimes \cdots \otimes a_p$ and $m_p\colon (a_1,\ldots,a_p) \mapsto a_1 \cdots a_p$, the latter being the product of the elements $a_1,\ldots,a_p$ in the $k$-algebra $A$. Of course, $m_p$ is $k$-multilinear, and so it induces a $k$-map $\mu_p$ making the diagram commute. Now define $\mu\colon T(A) \to A$ by $\mu = \sum_p \mu_p$. To see that $\mu$ is multiplicative, it suffices to show that

$$\mu_{p+q}\big((a_1 \otimes \cdots \otimes a_p) \otimes (a_1' \otimes \cdots \otimes a_q')\big) = \mu_p(a_1 \otimes \cdots \otimes a_p)\mu_q(a_1' \otimes \cdots \otimes a_q').$$

But this equation follows from the associative law in $A$:

$$(a_1 \cdots a_p)(a_1' \cdots a_q') = a_1 \cdots a_p a_1' \cdots a_q'.$$

Finally, uniqueness of this $k$-algebra map follows from $V$ generating $T(V)$ as a $k$-algebra [after all, every homogeneous element in $T(V)$ is a product of elements of degree 1]. $\bullet$

**Corollary 8.107.** *Let $k$ be a commutative ring.*

  (i) *If $A$ is a $k$-algebra, then there is a surjective $k$-algebra map $T(A) \to A$.*

 (ii) *Every $k$-algebra $A$ is a quotient of a free $k$-algebra.*

**Proof.**

  (i) Regard $A$ only as a $k$-module. If, for each $p \geq 2$, we denote the cartesian product of $A$ with itself $p$ times by

$$\times^p A,$$

  then multiplication $\times^p A \to A$ is $k$-multilinear, and there is a unique $k$-module map $\bigotimes^p A \to A$. These maps can be assembled to give a $k$-module map $T(A) = \bigoplus_p [\bigotimes^p A] \to A$. This map is surjective because $A$ has a unit 1, and it is easily seen to be a map of $k$-algebras; that is, it preserves multiplication.

 (ii) Let $V$ be a free $k$-module for which there exists a surjective $k$-map $\varphi\colon V \to A$. By Proposition 8.105, the induced map $T(\varphi)\colon T(V) \to T(A)$ is surjective. Now $T(V)$ is a free $k$-algebra, and if we compose $T(\varphi)$ with the surjection $T(A) \to A$, then $A$ is a quotient of $T(V)$. $\bullet$

**Example 8.108.** Just as for modules, we can now construct rings ($\mathbb{Z}$-algebras) by generators and relations. The first example of a ring that is left noetherian but not right noetherian was given by Dieudonné; it is the ring $R$ generated by elements $x$ and $y$ satisfying the relations $yx = 0$ and $y^2 = 0$. Proving that such a ring $R$ exists is now easy: let $V$ be the free abelian group with basis $u, v$, let $R = T(V)/I$, where $I$ is the two-sided ideal generated by $vu$ and $v^2$, and set $x = u + I$ and $y = v + I$. Note that since the ideal $I$ is generated by homogeneous elements of degree 2, we have $\bigotimes^1 V = V \cap I = \{0\}$, and so $x \neq 0$ and $y \neq 0$. $\blacktriangleleft$

We now construct polynomial rings in commuting variables.

**Definition.** Let $X$ be a subset of a commutative $k$-algebra $F$. Then $F$ is a ***free commutative $k$-algebra*** with ***basis*** $X$ if, for every $k$-algebra $A$ and every function

$\varphi \colon X \to A$, there exists a unique $k$-algebra map $\widetilde{\varphi}$ with $\widetilde{\varphi}(x) = \varphi(x)$ for all $x \in X$. In other words, the following diagram commutes, where $i \colon X \to F$ is the inclusion:

$$
\begin{array}{ccc}
F & & \\
\uparrow i & \searrow{\scriptstyle\widetilde{\varphi}} & \\
X & \xrightarrow[\varphi]{} & A
\end{array}
$$

**Proposition 8.109.** *Let $k$ be a commutative ring. Given any set $X$, there exists a free commutative $k$-algebra having $X$ as a basis.*

**Proof.** Let $V$ be the free $k$-module with basis $X$, and let $I$ be the two-sided ideal in the tensor algebra $T(V)$ generated by all $v \otimes v' - v' \otimes v$, where $v, v' \in V$. [16] The reader may show that $I$ is a graded ideal, so that $T(V)/I$ is a graded $k$-algebra. The argument that $T(V)/I$ is free is essentially the same as that of Lemma 4.78: if $F$ is a free group with basis $X = \{x_1, \ldots, x_n\}$, then $F/F'$ is a free abelian group with basis $X' = \{x_1 F', \ldots, x_n F'\}$, where $F'$ is the commutator subgroup of $F$.

Define $X' = \{x + I : x \in X\}$, and note that $\nu \colon x \mapsto x + I$ is a bijection $X \to X'$. It follows from $X$ generating $V$ that $X'$ generates $T(V)/I$. Consider the diagram

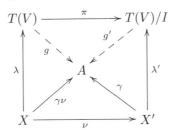

Here, $A$ is an arbitrary commutative $k$-algebra, $\lambda$ and $\lambda'$ are inclusions, $\pi$ is the natural map, $\nu \colon x \mapsto x + I$, and $\gamma \colon X' \to A$ is a function. Let $g \colon T(V) \to A$ be the unique homomorphism with $g\lambda = \gamma\nu$ [for $T(V)$ is a free $k$-algebra], and define $g' \colon T(V)/I' \to A$ by $w + I \mapsto g(w)$ ($g'$ is well-defined because $A$ commutative forces $I \subseteq \ker g$). Now $g'\lambda' = \gamma$, for

$$g'\lambda'\nu = g'\pi\lambda = g\lambda = \gamma\nu;$$

since $\nu$ is a surjection, it follows that $g'\lambda' = \gamma$. Finally, $g'$ is the unique such map, for if $g''$ satisfies $g''\lambda' = \gamma$, then $g'$ and $g''$ agree on the generating set $X'$, hence they are equal. $\bullet$

**Definition.** If $k$ is a commutative ring and $V$ is a free $k$-module with basis $X$, then $T(V)/I$ is called the ring of polynomials over $k$ in **commuting variables** $X$, and it is denoted by $k[X]$.

As usual, solutions to universal mapping problems are unique up to isomorphism. If $X = \{x_1, \ldots, x_n\}$ is finite, then Theorem 2.25 shows that the usual polynomial ring $k[x_1, \ldots, x_n]$ is the free commutative $k$-algebra on $X$. We remark

---

[16]The **symmetric algebra** $S(M)$ of a $k$-module $M$ is defined as $T(M)/I$, where $I$ is the two-sided ideal in $T(V)$ generated by all $m \otimes m' - m' \otimes m$, where $m, m' \in M$. The construction here is the special case when $M$ is a free $k$-module.

that we assumed the existence of big polynomial rings in Section 5.4 in order to construct the algebraic closure of a field. We have just constructed $k[X]$ as a quotient of a tensor algebra; earlier, in Exercise 6.107 on page 513, we constructed $k[X]$ as a direct limit.

Our earlier definition of $k[x, y]$ as $A[y]$, where $A = k[x]$, was careless. For example, it does not imply that $k[x, y] = k[y, x]$, although these two rings are isomorphic. However, if $V$ is the free $k$-module with basis $x, y$, then $y, x$ is also a basis of the $k$-module $V$, and so $k[x, y] = k[y, x]$.

We now mention a class of rings generalizing commutative rings. If $k$ is a commutative ring, then a **polynomial identity** on a $k$-algebra $A$ is an element $f(X) \in k\langle X \rangle$ (the ring of polynomials over $k$ in noncommuting variables $X$) all of whose substitutions in $A$ are 0. For example, when $f(x, y) = xy - yx \in k\langle x, y \rangle$, we have $f$ a polynomial identity on a $k$-algebra $A$ if $ab - ba = 0$ for all $a, b \in A$; that is, $A$ is a commutative $k$-algebra.

**Definition.** A $k$-algebra $A$ is a **PI-*algebra*** if $A$ satisfies some polynomial identity at least one of whose coefficients is 1.

Every $k$-algebra generated by $n$ elements satisfies the **standard identity**:

$$s_m(x_1, \ldots, x_m) = \sum_{\sigma \in S_m} \mathrm{sgn}(\sigma) x_{\sigma(1)} \cdots x_{\sigma(m)}$$

for all $m > n$.

We can prove that the matrix algebra $\mathrm{Mat}_m(k)$ satisfies the standard identity $s_{m^2+1}$, and Amitsur and Levitzki proved that $\mathrm{Mat}_m(k)$ satisfies $s_{2m}$; moreover, $2m$ is the lowest possible degree of such a polynomial identity. There is a short proof of this due to Rosset, "A New Proof of the Amitsur-Levitski Identity," *Israel Journal of Mathematics* 23, 1976, pp. 187–188.

**Definition.** A **central polynomial identity** on a $k$-algebra $A$ is a polynomial $f(X) \in k\langle X \rangle$ on $A$ all of whose values $f(a_1, a_2, \ldots)$ (as the $a_i$ vary over all elements of $A$) lie in $Z(A)$.

It was proved, independently, by Formanek and Razmyslov, that $\mathrm{Mat}_m(k)$ satisfies central polynomial identities.

There are theorems showing, in several respects, that PI-algebras behave like commutative algebras. For example, recall that a ring $R$ is *primitive* if it has a faithful simple left $R$-module; commutative primitive rings are fields. Kaplansky proved that every primitive quotient of a PI-algebra is simple and finite-dimensional over its center. The reader is referred to Procesi, *Rings with Polynomial Identities*.

Another interesting area of current research involves *Noncommutative Algebraic Geometry*. In essence, this involves the study of varieties now defined as zeros of ideals in $k\langle x_1, \ldots, x_n \rangle$ (the free $k$-algebra in $n$ noncommuting variables) instead of in $k[x_1, \ldots, x_n]$.

**8.6.2. Exterior Algebra.** In Calculus, the ***differential*** $df$ of a differentiable function $f(x, y)$ at a point $P = (x_0, y_0)$ is defined by

$$df|_P = \frac{\partial f}{\partial x}\bigg|_P (x - x_0) + \frac{\partial f}{\partial y}\bigg|_P (y - y_0).$$

If $(x, y)$ is a point near $P$, then $df|_P$ approximates the difference between the true value $f(x, y)$ and $f(x_0, y_0)$. The quantity $df$ is considered "small," and so its square, a second-order approximation, is regarded as negligible. For the moment, let us take being negligible seriously: suppose that

$$(df)^2 \approx 0$$

for all differentials $df$. There is a curious consequence: if $du$ and $dv$ are differentials, then so is $du + dv = d(u + v)$. But $(du + dv)^2 \approx 0$ gives

$$0 \approx (du + dv)^2 \approx (du)^2 + du\,dv + dv\,du + (dv)^2 \approx du\,dv + dv\,du,$$

and so $du$ and $dv$ anticommute:

$$dv\,du \approx -du\,dv.$$

Now consider a double integral $\iint_D f(x, y) dx\, dy$, where $D$ is some region in the plane. Equations

$$x = F(u, v)$$
$$y = G(u, v)$$

lead to the change of variables formula:

$$\iint_D f(x, y) dx\, dy = \iint_\Delta f(F(u, v), G(u, v)) J(u, v) du\, dv,$$

where $\Delta$ is some new region and $J(u, v)$ is the ***Jacobian***: $J(u, v) = \left| \det \left[ \begin{smallmatrix} F_u & F_v \\ G_u & G_v \end{smallmatrix} \right] \right|$. A key idea in proving this formula is that the graph of a differentiable function $f(x, y)$ looks, locally, like a real vector space—its tangent plane. Let us denote a basis of the tangent plane at a point by $dx, dy$. If $du, dv$ is another basis of this tangent plane, then the chain rule defines a linear transformation by the following system of linear equations:

$$dx = F_u du + F_v dv$$
$$dy = G_u du + G_v dv.$$

The Jacobian $J$ now arises in a natural way:

$$
\begin{aligned}
dx\, dy &= (F_u du + F_v dv)(G_u du + G_v dv) \\
&= F_u du G_u du + F_u du G_v dv + F_v dv G_u du + F_v dv G_v dv \\
&= F_u G_u (du)^2 + F_u G_v du\, dv + F_v G_u dv\, du + F_v G_v (dv)^2 \\
&\approx F_u G_v du\, dv + F_v G_u dv\, du \\
&\approx (F_u G_v - F_v G_u) du\, dv \\
&= \det \left[ \begin{smallmatrix} F_u & F_v \\ G_u & G_v \end{smallmatrix} \right] du\, dv.
\end{aligned}
$$

Analytic considerations, involving orientation, force us to use the absolute value of the determinant when proving the change of variables formula.

In the preceding equations, we used the distributive and associative laws, together with anticommutativity; that is, we assumed that the differentials form a ring in which all squares are 0. The following construction puts this kind of reasoning on a firm basis.

**Definition.** If $M$ is a $k$-module, where $k$ is a commutative ring, then its **exterior algebra**[17] is $\bigwedge M = T(M)/J$, pronounced "wedge $M$," where $J$ is the two-sided ideal in the tensor algebra $T(M)$ generated by all $m \otimes m$ with $m \in M$. The coset $m_1 \otimes \cdots \otimes m_p + J$ in $\bigwedge M$ is denoted by

$$m_1 \wedge \cdots \wedge m_p.$$

Notice that $J$ is generated by homogeneous elements (of degree 2). Moreover, Proposition 8.99 says that $J$ is a graded ideal in $T(M)$ and $\bigwedge M = T(M)/J$ is a graded $k$-algebra:

$$\bigwedge M = k \oplus M \oplus \bigwedge^2 M \oplus \bigwedge^3 M \oplus \cdots,$$

where, for $p \geq 2$, we have $\bigwedge^p M = (\bigotimes^p M)/J^p$ and $J^p = J \cap \bigotimes^p M$. Finally, $\bigwedge M$ is generated, as a $k$-algebra, by $\bigwedge^1 M = M$.

**Definition.** We call $\bigwedge^p M$ the **pth exterior power** of a $k$-module $M$.

**Lemma 8.110.** *Let $k$ be a commutative ring, and let $M$ be a $k$-module.*

(i) *If $m, m' \in M$, then $m \wedge m' = -m' \wedge m$ in $\bigwedge^2 M$.*

(ii) *If $p \geq 2$ and $m_i = m_j$ for some $i \neq j$, then $m_1 \wedge \cdots \wedge m_p = 0$ in $\bigwedge^p M$.*

**Proof.**

(i) Recall that $\bigwedge^2 M = (M \otimes_k M)/J^2$, where $J^2 = J \cap (M \otimes_k M)$. If $m, m' \in M$, then

$$(m + m') \otimes (m + m') = m \otimes m + m \otimes m' + m' \otimes m + m' \otimes m'.$$

Therefore, $m \otimes m' + J^2 = -m' \otimes m + J^2$, because $J^2$ contains the elements $(m + m') \otimes (m + m')$, $m \otimes m$, and $m' \otimes m'$. It follows, for all $m, m' \in M$, that

$$m \wedge m' = -m' \wedge m.$$

(ii) As we saw in the proof of Proposition 8.99, $\bigwedge^p M = (\bigotimes^p M)/J^p$, where $J^p = J \cap \bigotimes^p M$ consists of all elements of degree $p$ in the ideal $J$ generated by all elements in $\bigotimes^2 M$ of the form $m \otimes m$. In more detail, $J^p$ consists of all sums of homogeneous elements $\alpha \otimes m \otimes m \otimes \beta$, where $m \in M$, $\alpha \in \bigotimes^q M$, $\beta \in \bigotimes^r M$, and $q + r + 2 = p$; it follows that $m_1 \wedge \cdots \wedge m_p = 0$ if there are two equal *adjacent* factors, say, $m_i = m_{i+1}$. Since multiplication in $\bigwedge M$ is associative, however, we can (anti)commute a factor $m_i$ of $m_1 \wedge \cdots \wedge m_p$ several steps away at the possible cost of a change in sign, and so we can force any pair of factors to be adjacent.  •

---

[17] The original adjective in this context—the German *äußer*, meaning "outer"—was introduced by Grassmann in 1844. Grassmann used it in contrast to *inner product*. The first usage of the translation *exterior* can be found in work of Cartan in 1945, who wrote that he was using terminology of Kaehler. The wedge notation seems to have been introduced by Bourbaki.

One of our goals is to give a "basis-free" construction of determinants, and the idea is to focus on some properties that such a function has. If we regard an $n \times n$ matrix $A$ as consisting of its $n$ columns, then its determinant, $\det(A)$, is a function of $n$ variables (each ranging over $n$-tuples). One property of determinants is that $\det(A) = 0$ if two columns of $A$ are equal, and another property is that it is multilinear. Corollary 8.128 will show that these two properties almost characterize the determinant.

**Definition.** If $M$ and $N$ are $k$-modules, a $k$-multilinear function $f\colon \times^p M \to N$ (where $\times^p M$ is the cartesian product of $M$ with itself $p$ times) is ***alternating*** if

$$f(m_1, \ldots, m_p) = 0$$

whenever $m_i = m_j$ for some $i \neq j$.

An alternating $\mathbb{R}$-bilinear function arises naturally when considering (signed) areas in the plane $\mathbb{R}^2$. Informally, if $v_1, v_2 \in \mathbb{R}^2$, let $A(v_1, v_2)$ denote the area of the parallelogram having sides $v_1$ and $v_2$. It is clear that

$$A(rv_1, sv_2) = rsA(v_1, v_2)$$

for all $r, s \in \mathbb{R}$ (but we must say what this means when these numbers are negative), and a geometric argument can be given to show that

$$A(w_1 + v_1, v_2) = A(w_1, v_2) + A(v_1, v_2);$$

that is, $A$ is $\mathbb{R}$-bilinear. Now $A$ is alternating, for $A(v_1, v_1) = 0$ because the degenerate "parallelogram" having sides $v_1$ and $v_1$ has zero area. A similar argument shows that volume is an alternating $\mathbb{R}$-multilinear function on $\mathbb{R}^3$, as we see in vector calculus using the cross product.

**Theorem 8.111.** *For all $p \geq 0$ and all $k$-modules $M$, the pth exterior power $\bigwedge^p M$ solves the universal mapping problem posed by alternating multilinear functions:*

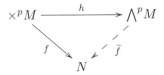

*If $h\colon \times^p M \to \bigwedge^p M$ is defined by $h(m_1, \ldots, m_p) = m_1 \wedge \cdots \wedge m_p$, then for every alternating multilinear function $f$, there exists a unique $k$-homomorphism $\tilde{f}$ making the diagram commute.*

**Proof.** Consider the diagram

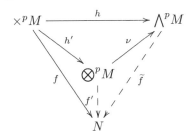

where $h'(m_1, \ldots, m_p) = m_1 \otimes \cdots \otimes m_p$ and $\nu(m_1 \otimes \cdots \otimes m_p) = m_1 \wedge \cdots \wedge m_p$. Since $f$ is multilinear, there is a $k$-map $f' \colon \bigotimes^p M \to N$ with $f'h' = f$; since $f$ is alternating, $J \cap \bigotimes^p M \subseteq \ker f'$, and so $f'$ induces a map

$$\tilde{f} \colon \bigotimes{}^p M / \left( J \cap \bigotimes{}^p M \right) \to N$$

with $\tilde{f}\nu = f'$. Hence,

$$\tilde{f}h = \tilde{f}\nu h' = f'h' = f.$$

But $\bigotimes^p M / (J \cap \bigotimes^p M) = \bigwedge^p M$, as desired. Finally, $\tilde{f}$ is the unique such map because $\operatorname{im} h$ generates $\bigwedge^p M$. •

**Proposition 8.112.** *For each $p \geq 0$, the pth exterior power is a functor*

$$\bigwedge{}^p \colon {}_k\mathbf{Mod} \to {}_k\mathbf{Mod}.$$

**Proof.** Now $\bigwedge^p M$ has been defined on modules; it remains to define it on morphisms. Suppose that $g \colon M \to M'$ is a $k$-homomorphism. Consider the diagram

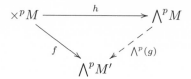

where $f(m_1, \ldots, m_p) = gm_1 \wedge \cdots \wedge gm_p$. It is easy to see that $f$ is an alternating multilinear function, and so universality yields a unique map

$$\bigwedge{}^p(g) \colon \bigwedge{}^p M \to \bigwedge{}^p M'$$

with $m_1 \wedge \cdots \wedge m_p \mapsto gm_1 \wedge \cdots \wedge gm_p$.

If $g$ is the identity map on a module $M$, then $\bigwedge^p(g)$ is also the identity map, for it fixes a set of generators. Finally, suppose that $g' \colon M' \to M''$ is a $k$-map. It is routine to check that both $\bigwedge^p(g'g)$ and $\bigwedge^p(g')\bigwedge^p(g)$ make the following diagram commute:

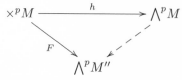

where $F(m_1, \ldots, m_p) = (g'gm_1) \wedge \cdots \wedge (g'gm_p)$. Uniqueness of such a dashed arrow gives $\bigwedge^p(g'g) = \bigwedge^p(g')\bigwedge^p(g)$, as desired. •

We will soon see that $\bigwedge^p$ is not as nice as Hom or tensor, for it is not an additive functor.

**Theorem 8.113 (Anticommutativity).** *If $M$ is a $k$-module, $x \in \bigwedge^p M$, and $y \in \bigwedge^q M$, then*

$$x \wedge y = (-1)^{pq} y \wedge x.$$

**Remark.** This identity holds only for products of homogeneous elements.   ◀

**Proof.** If $x \in \bigwedge^0 M = k$, then $\bigwedge M$ being a $k$-algebra implies that $x \wedge y = y \wedge x$ for all $y \in \bigwedge M$, and so the identity holds, in particular, for $y \in \bigwedge^q M$ for any $q$. A similar argument holds if $y$ is homogeneous of degree 0. Therefore, we may assume that $p, q \geq 1$; we do a double induction.

*Base Step*: $p = 1$ and $q = 1$. Suppose that $x, y \in \bigwedge^1 M = M$. Now

$$
\begin{aligned}
0 = (x + y) \wedge (x + y) \\
= x \wedge x + x \wedge y + y \wedge x + y \wedge y \\
= x \wedge y + y \wedge x.
\end{aligned}
$$

It follows that $x \wedge y = -y \wedge x$, as desired.

*Inductive Step*: $(p, 1) \Rightarrow (p + 1, 1)$. The inductive hypothesis gives

$$
(x_1 \wedge \cdots \wedge x_p) \wedge y = (-1)^p y \wedge (x_1 \wedge \cdots \wedge x_p).
$$

Using associativity, we have

$$
\begin{aligned}
(x_1 \wedge \cdots \wedge x_{p+1}) \wedge y &= x_1 \wedge [(x_2 \wedge \cdots \wedge x_{p+1}) \wedge y] \\
&= x_1 \wedge (-1)^p [y \wedge (x_2 \wedge \cdots \wedge x_{p+1})] \\
&= [x_1 \wedge (-1)^p y] \wedge (x_2 \wedge \cdots \wedge x_{p+1}) \\
&= (-1)^{p+1} (y \wedge x_1) \wedge (x_2 \wedge \cdots \wedge x_{p+1}).
\end{aligned}
$$

*Inductive Step*: $(p, q) \Rightarrow (p, q + 1)$. Assume that

$$
(x_1 \wedge \cdots \wedge x_p) \wedge (y_1 \wedge \cdots \wedge y_q) = (-1)^{pq} (y_1 \wedge \cdots \wedge y_q) \wedge (x_1 \wedge \cdots \wedge x_p).
$$

We let the reader prove, using associativity, that

$$
\begin{aligned}
(x_1 \wedge \cdots \wedge x_p) &\wedge (y_1 \wedge \cdots \wedge y_{q+1}) \\
&= (-1)^{p(q+1)} (y_1 \wedge \cdots \wedge y_{q+1}) \wedge (x_1 \wedge \cdots \wedge x_p). \quad \bullet
\end{aligned}
$$

**Definition.** Let $n$ be a positive integer and let $1 \leq p \leq n$. An **increasing $p \leq n$-list** is a list

$$
H = i_1, \ldots, i_p
$$

for which $1 \leq i_1 < i_2 < \cdots < i_p \leq n$.

If $H = i_1, \ldots, i_p$ is an increasing $p \leq n$ list, we write

$$
e_H = e_{i_1} \wedge e_{i_2} \wedge \cdots \wedge e_{i_p}.
$$

Of course, the number of increasing $p \leq n$ lists is the same as the number of $p$-subsets of a set with $n$ elements, namely, $\binom{n}{p}$.

**Proposition 8.114.** *Let $M$ be finitely generated, say, $M = \langle e_1, \ldots, e_n \rangle$. If $p \geq 1$, then $\bigwedge^p M$ is generated by all elements of the form $e_H$, where $H = i_1, \ldots, i_p$ is an increasing $p \leq n$ list.*

**Proof.** Every element of $M$ has some expression of the form $\sum a_i e_i$, where $a_i \in k$. We prove the proposition by induction on $p \geq 1$. Let $m_1 \wedge \cdots \wedge m_{p+1}$ be a typical generator of $\bigwedge^{p+1} M$. By induction,

$$m_1 \wedge \cdots \wedge m_p = \sum_H a_H e_H,$$

where $a_H \in k$ and $H$ is an increasing $p \leq n$ list. If $m_{p+1} = \sum b_j e_j$, then

$$m_1 \wedge \cdots \wedge m_{p+1} = \left( \sum_H a_H e_H \right) \wedge \left( \sum_j b_j e_j \right).$$

Each $e_j$ in $\sum b_j e_j$ can be moved to any position in $e_H = e_{i_1} \wedge \cdots \wedge e_{i_p}$ (with a possible change in sign) by (anti)commuting it from right to left. Of course, if $e_j = e_{i_\ell}$ for any $\ell$, then this term is 0, and so we can assume that all the factors in surviving wedges are distinct and are arranged with indices in ascending order. •

**Corollary 8.115.** *If $M$ can be generated by $n$ elements, then $\bigwedge^p M = \{0\}$ for all $p > n$.*

**Proof.** Any wedge of $p$ factors must be 0, for it must contain a repetition of one of the generators. •

**Definition.** If $V$ is a free $k$-module of rank $n$, then a **Grassmann algebra** on $V$ is a $k$-algebra $G(V)$ with identity element, denoted by $e_0$, such that

    (a)   $G(V)$ contains $\langle e_0 \rangle \oplus V$ as a submodule, where $\langle e_0 \rangle \cong k$;

    (b)   $G(V)$ is generated, as a $k$-algebra, by $\langle e_0 \rangle \oplus V$;

    (c)   $v^2 = 0$ for all $v \in V$;

    (d)   $G(V)$ is a free $k$-module of rank $2^n$.

The computation on page 715 shows that the condition $v^2 = 0$ for all $v \in V$ implies $vu = -uv$ for all $u, v \in V$. A candidate for $G(V)$ is $\bigwedge V$ but, at this stage, it is not clear how to show that $\bigwedge V$ is free and of the desired rank.

Grassmann algebras carry a generalization of complex conjugation, and this fact is the key to proving their existence. If $A$ is a $k$-algebra, then an **algebra automorphism** is a $k$-algebra isomorphism of $A$ with itself.

**Theorem 8.116.** *Let $V$ be a free $k$-module with basis $e_1, \ldots, e_n$, where $n \geq 1$.*

   (i)  *A Grassmann algebra $G(V)$ exists; moreover, it has a $k$-algebra automorphism $u \mapsto \overline{u}$, called **conjugation**, such that*

$$\overline{\overline{u}} = u;$$
$$\overline{e_0} = e_0;$$
$$\overline{v} = -v \quad \text{for all } v \in V.$$

  (ii)  *The Grassmann algebra $G(V)$ is a graded $k$-algebra*

$$G(V) = \bigoplus_p G^p(V),$$

*where*

$$G^p(V) = \langle e_H : H \text{ is an increasing } p \leq n \text{ list} \rangle$$

*[we have extended the notation* $e_H = e_{i_1} \wedge \cdots \wedge e_{i_p}$ *in* $\bigwedge^p V$ *to* $e_H = e_{i_1} \cdots e_{i_p}$
*in* $G^p(V)$*]. Moreover,* $G^p(V)$ *is a free* $k$-module with

$$\text{rank}(G^p(V)) = \binom{n}{p}.$$

**Proof.**

(i) The proof is by induction on $n \geq 1$. The base step is clear: if $V = \langle e_1 \rangle \cong k$,
set $G(V) = \langle e_0 \rangle \oplus \langle e_1 \rangle$; note that $G(V)$ is a free $k$-module of rank 2. Define
a multiplication on $G(V)$ by

$$e_0 e_0 = e_0; \quad e_0 e_1 = e_1 = e_1 e_0; \quad e_1 e_1 = 0.$$

It is routine to check that $G(V)$ is a $k$-algebra that satisfies the axioms of
a Grassmann algebra. There is no choice in defining the automorphism; we
must have

$$\overline{ae_0 + be_1} = ae_0 - be_1.$$

Finally, it is easy to see that $u \mapsto \overline{u}$ is the automorphism we seek.

For the inductive step, let $V$ be a free $k$-module of rank $n + 1$ and let
$e_1, \ldots, e_{n+1}$ be a basis of $V$. If $W = \langle e_1, \ldots, e_n \rangle$, then the inductive hypothesis provides a Grassmann algebra $G(W)$, free of rank $2^n$, and an automor-
phism $u \mapsto \overline{u}$ for all $u \in G(W)$. Define $G(V) = G(W) \oplus G(W)$, so that $G(V)$
is a free module of rank $2^n + 2^n = 2^{n+1}$. We make $G(V)$ into a $k$-algebra by
defining

$$(x_1, x_2)(y_1, y_2) = (x_1 y_1, x_2 \overline{y}_1 + x_1 y_2).$$

We now verify the four parts in the definition of Grassmann algebra.

(a) At the moment, $V$ is not a submodule of $G(V)$. Each $v \in V$ has a unique
expression of the form $v = w + ae_{n+1}$, where $w \in W$ and $a \in k$. The $k$-map
$V \to G(V)$, given by

$$v = w + ae_{n+1} \mapsto (w, ae_0),$$

is an isomorphism of $k$-modules, and we identify $V$ with its image in $G(V)$.
In particular, $e_{n+1}$ is identified with $(0, e_0)$. Note that the identity element
$e_0 \in G(W)$ in $G(W)$ has been identified with $(e_0, 0)$ in $G(V)$, and that the
definition of multiplication in $G(V)$ shows that $(e_0, 0)$ is the identity in $G(V)$.

(b) By induction, we know that the elements of $\langle e_0 \rangle \oplus W$ generate $G(W)$ as
a $k$-algebra; that is, all $(x_1, 0) \in G(W)$ arise from elements of $W$. Next, by
our identification, $e_{n+1} = (0, e_0)$,

$$(x_1, 0)e_{n+1} = (x_1, 0)(0, e_0) = (0, x_1),$$

and so the elements of $V$ generate all pairs of the form $(0, x_2)$. Since addition
is coordinatewise, all $(x_1, x_2) = (x_1, 0) + (0, x_2)$ arise from $V$ using algebra
operations.

(c) If $v \in V$, then $v = w + ae_{n+1}$, where $w \in W$, and $v$ is identified with $(w, ae_0)$ in $G(V)$. Hence,

$$v^2 = (w, ae_0)(w, ae_0) = (w^2, ae_0\overline{w} + ae_0 w).$$

Now $w^2 = 0$, and $\overline{w} = -w$, so that $v^2 = 0$.

(d) $\operatorname{rank} G(V) = 2^{n+1}$ because $G(V) = G(W) \oplus G(W)$.

We have shown that $G(V)$ is a Grassmann algebra. Finally, define conjugation by

$$\overline{(x_1, x_2)} = (\overline{x}_1, -\overline{x}_2).$$

The reader may check that this defines a function with the desired properties.

(ii) We prove, by induction on $n \geq 1$, that

$$G^p(V) = \left\langle e_H : H \text{ is an increasing } p \leq n \text{ list} \right\rangle$$

is a free $k$-module with the displayed products as a basis. The base step is obvious: if $\operatorname{rank}(V) = 1$, say, with basis $e_1$, then $G(V) = \langle e_0, e_1 \rangle$; moreover, both $G^0(V)$ and $G^1(V)$ are free of rank 1.

For the inductive step, assume that $V$ is free with basis $e_1, \ldots, e_{n+1}$. As in the proof of part (i), let $W = \langle e_1, \ldots, e_n \rangle$. By induction, $G^p(W)$ is a free $k$-module of rank $\binom{n}{p}$ with basis all $e_H$, where $H$ is an increasing $p \leq n$ list. Here are two types of elements of $G^p(V)$: elements $e_H \in G(W)$, where $H$ is an increasing $p \leq n$ list; elements $e_H = e_{i_1} \cdots e_{i_{p-1}} e_{n+1}$, where $H$ is an increasing $p \leq (n+1)$ list that involves $e_{n+1}$. We know that the elements of the first type comprise a basis of $G(W)$. The definition of multiplication in $G(V)$ gives $e_H e_{n+1} = (e_H, 0)(0, e_0) = (0, e_H)$. Thus, the number of such products is $\binom{n}{p-1}$. As $G(V) = G(W) \oplus G(W)$, we see that the union of these two types of products form a basis for $G^p(V)$, and so $\operatorname{rank}(G^p(V)) = \binom{n}{p} + \binom{n}{p-1} = \binom{n+1}{p}$.

It remains to prove that $G^p(V)G^q(V) \subseteq G^{p+q}(V)$. Consider a product $e_{i_1} \cdots e_{i_p} e_{j_1} \cdots e_{j_q}$. If some subscript $i_r$ equals a subscript $j_s$, then the product is 0, because it has a repeated factor; if all the subscripts are distinct, then the product lies in $G^{p+q}(V)$, as desired. Therefore, $G(V)$ is a graded $k$-algebra whose graded part of degree $p$ is a free $k$-module of rank $\binom{n}{p}$. $\quad \bullet$

**Theorem 8.117 (Binomial Theorem).** *If $V$ is a free $k$-module of rank $n$, then there is an isomorphism of graded $k$-algebras,*

$$\bigwedge V \cong G(V).$$

*Thus, $\bigwedge^p V$ is a free $k$-module, for all $p \geq 1$, with basis all increasing $p \leq n$ lists, and hence*

$$\operatorname{rank}\left(\bigwedge\nolimits^p V\right) = \binom{n}{p}.$$

**Proof.** For any $p \geq 2$, consider the diagram

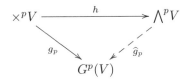

where $h(v_1, \ldots, v_p) = v_1 \wedge \cdots \wedge v_p$ and $g_p(v_1, \ldots, v_p) = v_1 \cdots v_p$. Since $v^2 = 0$ in $G^p(V)$ for all $v \in V$, the function $g_p$ is alternating multilinear. By the universal property of exterior power, there is a unique $k$-homomorphism $\widehat{g}_p \colon \bigwedge^p V \to G^p(V)$ making the diagram commute; that is,

$$\widehat{g}_p(v_1 \wedge \cdots \wedge v_p) = v_1 \cdots v_p.$$

If $e_1, \ldots, e_n$ is a basis of $V$, then we have just seen that $G^p(V)$ is a free $k$-module with basis all $e_{i_1} \cdots e_{i_p}$, and so $\widehat{g}_p$ is surjective. Now $\bigwedge^p V$ is generated by all $e_{i_1} \wedge \cdots \wedge e_{i_p}$, by Proposition 8.114. If some $k$-linear combination $\sum_H a_H e_H$ lies in $\ker \widehat{g}_p$, then $\sum a_H \widehat{g}_p(e_H) = 0$ in $G^p(V)$. But the list of images $\widehat{g}_p(e_H)$ forms a basis of the free $k$-module $G^p(V)$, so that all the coefficients $a_H = 0$. Therefore, $\ker \widehat{g}_p = \{0\}$, and so $\widehat{g}_p$ is a $k$-isomorphism.

Define $\gamma \colon \bigwedge V \to G(V)$ by $\gamma(\sum_{p=0}^n u_p) = \sum_{p=0}^n \widehat{g}_p(u_p)$, so that $\gamma(\bigwedge^p V) \subseteq G^p(V)$. We are done if we can show that $\gamma$ is an algebra map: $\gamma(u \wedge v) = \gamma(u)\gamma(v)$. But this is clear for homogeneous elements of $\bigwedge V$, and hence it is true for all elements.   •

**Corollary 8.118.** *If $V$ is a free $k$-module with basis $e_1, \ldots, e_n$, then*

$$\bigwedge^n V = \langle e_1 \wedge \cdots \wedge e_n \rangle \cong k.$$

**Proof.** By Proposition 8.114, we know that $\bigwedge^n V$ is a cyclic module generated by $e_1 \wedge \cdots \wedge e_n$, but we cannot conclude from this proposition whether or not this element is zero. However, the Binomial Theorem not only says that this element is nonzero; it also says that it generates a cyclic module isomorphic to $k$.   •

Proposition 6.55 says that if $T \colon {}_k\mathbf{Mod} \to {}_k\mathbf{Mod}$ is an additive functor, then $T(V \oplus V') \cong T(V) \oplus T(V')$. It follows, for $p \geq 2$, that $\bigwedge^p$ is *not* an additive functor: if $V$ is a free $k$-module of rank $n$, then $\bigwedge^p(V \oplus V)$ is free of rank $\binom{2n}{p}$, whereas $\bigwedge^p V \oplus \bigwedge^p V$ is free of rank $2\binom{n}{p}$.

An astute reader will have noticed that our construction of a Grassmann algebra $G(V)$ depends not only on the free $k$-module $V$ but also on a choice of basis of $V$. Had we chosen a second basis of $V$, would the second Grassmann algebra be isomorphic to the first one?

**Corollary 8.119.** *Let $V$ be a free $k$-module, and let $B$ and $B'$ be bases of $V$. If $G(V)$ is the Grassmann algebra defined using $B$ and $G'(V)$ is the Grassmann algebra defined using $B'$, then $G(V) \cong G'(V)$ as graded $k$-algebras.*

**Proof.** Both $G(V)$ and $G'(V)$ are isomorphic to $\bigwedge V$, and the latter has been defined without any choice of basis.   •

A second proof of the Binomial Theorem follows from the next result.

**Theorem 8.120.** *For all $p \geq 0$ and all $k$-modules $A$ and $B$, where $k$ is a commutative ring,*

$$\bigwedge\nolimits^{p}(A \oplus B) \cong \bigoplus_{i=0}^{p} \left( \bigwedge\nolimits^{i} A \otimes_k \bigwedge\nolimits^{p-i} B \right).$$

**Proof.** We sketch a proof. Let $\mathcal{A}$ be the category of all alternating anticommutative graded $k$-algebras $R = \bigoplus_{p \geq 0} R^p$ ($r^2 = 0$ for all $r \in R$ of odd degree and $rs = (-1)^{pq} sr$ if $r \in R^p$ and $s \in S^q$); by Theorem 8.113, the exterior algebra $\bigwedge A \in \mathrm{obj}(\mathcal{A})$ for every $k$-module $A$. If $R, S \in \mathrm{obj}(\mathcal{A})$, then one verifies that $R \otimes_k S = \bigoplus_{p \geq 0} \left( \bigoplus_{i=0}^{p} R^i \otimes_k S^{p-i} \right) \in \mathrm{obj}(\mathcal{A})$; using anticommutativity, a modest generalization of Proposition 6.126 shows that $\mathcal{A}$ has coproducts.

We claim that $(\bigwedge, D)$ is an adjoint pair of functors, where $\bigwedge \colon {}_k\mathbf{Mod} \to \mathcal{A}$ sends $A \mapsto \bigwedge A$, and $D \colon \mathcal{A} \to {}_k\mathbf{Mod}$ sends $\sum_{p \geq 0} R^p \mapsto R^1$, the terms of degree 1. If $R = \bigoplus_p R^p$, then there is a map $\pi_R \colon \bigwedge R^1 \to R$; define $\tau_{A,R} \colon \mathrm{Hom}_{\mathcal{A}}(\bigwedge A, R) \to \mathrm{Hom}_k(A, R^1)$ by $\varphi \mapsto \pi_R(\varphi|A)$. It follows from Theorem 6.164 that $\bigwedge$ preserves coproducts: $\bigwedge(A \oplus B) \cong \bigwedge A \otimes_k \bigwedge B$ and $\bigwedge^p (A \oplus B) \cong \bigoplus_{i=0}^{p} \left( \bigwedge^i A \otimes_k \bigwedge^{p-i} B \right)$ for all $p$.   •

Here is an explicit formula for an isomorphism. In $\bigwedge^3(A \oplus B)$, we have

$$(a_1 + b_1) \wedge (a_2 + b_2) \wedge (a_3 + b_3) = a_1 \wedge a_2 \wedge a_3 + a_1 \wedge b_2 \wedge a_3$$
$$+ b_1 \wedge a_2 \wedge a_3 + b_1 \wedge b_2 \wedge a_3 + a_1 \wedge a_2 \wedge b_3$$
$$+ a_1 \wedge b_2 \wedge b_3 + b_1 \wedge a_2 \wedge b_3 + b_1 \wedge b_2 \wedge b_3.$$

By anticommutativity, this can be rewritten so that each $a$ precedes all the $b$'s:

$$(a_1 + b_1) \wedge (a_2 + b_2) \wedge (a_3 + b_3) = a_1 \wedge a_2 \wedge a_3 - a_1 \wedge a_3 \wedge b_2$$
$$+ a_2 \wedge a_3 \wedge b_1 + a_3 \wedge b_1 \wedge b_2 + a_1 \wedge a_2 \wedge b_3$$
$$+ a_1 \wedge b_2 \wedge b_3 - a_2 \wedge b_1 \wedge b_3 + b_1 \wedge b_2 \wedge b_3.$$

An *i-**shuffle*** is a partition of $\{1, 2, \ldots, p\}$ into two disjoint subsets $\mu_1 < \cdots < \mu_i$ and $\nu_1 < \cdots < \nu_{p-i}$; it gives the permutation $\sigma \in S_p$ with $\sigma(j) = \mu_j$ for $j \leq i$ and $\sigma(i+\ell) = \nu_\ell$ for $j = i + \ell > i$. Each "mixed" term in $(a_1 + b_1) \wedge (a_2 + b_2) \wedge (a_3 + b_3)$ defines a shuffle, with the $a$'s giving the $\mu$ and the $b$'s giving the $\nu$; for example, $a_1 \wedge b_2 \wedge a_3$ is a 2-shuffle and $b_1 \wedge a_2 \wedge b_3$ is a 1-shuffle. Now $\mathrm{sgn}(\sigma)$ counts the total number of leftward moves of $a$'s so that they precede all the $b$'s, and the reader may check that the signs in the rewritten expansion are $\mathrm{sgn}(\sigma)$. Define $f \colon \bigwedge^p(A \oplus B) \to \bigoplus_{i=0}^{p} \left( \bigwedge^i A \otimes_k \bigwedge^{p-i} B \right)$ by

$$f(a_1 + b_1, \ldots, a_p + b_p) = \sum_{i=0}^{p} \left( \sum_{i\text{-shuffles } \sigma} \mathrm{sgn}(\sigma) a_{\mu_1} \wedge \cdots \wedge a_{\mu_i} \otimes b_{\nu_1} \wedge \cdots \wedge b_{\nu_{p-i}} \right).$$

**Corollary 8.121.** *If $k$ is a commutative ring and $V$ is a free $k$-module of rank $n$, then $\bigwedge^p V$ is a free $k$-module of rank $\binom{n}{p}$.*

**Proof.** Write $V = k \oplus B$ and use induction on $\mathrm{rank}(V)$.   •

We will use Exterior Algebra in the next section to prove theorems about determinants, but let us first note a nice result when $k$ is a field and, hence, $k$-modules are vector spaces.

**Proposition 8.122.** *Let $k$ be a field, let $V$ be a vector space over $k$, and let $v_1, \ldots, v_p$ be vectors in $V$. Then $v_1 \wedge \cdots \wedge v_p = 0$ in $\bigwedge V$ if and only if $v_1, \ldots, v_p$ is a linearly dependent list.*

**Proof.** Since $k$ is a field, a linearly independent list $v_1, \ldots, v_p$ can be extended to a basis $v_1, \ldots, v_p, \ldots, v_n$ of $V$. By Corollary 8.118, $v_1 \wedge \cdots \wedge v_n \neq 0$. But $v_1 \wedge \cdots \wedge v_p$ is a factor of $v_1 \wedge \cdots \wedge v_n$, so that $v_1 \wedge \cdots \wedge v_p \neq 0$.

Conversely, if $v_1, \ldots, v_p$ is linearly dependent, there is an $i$ with $v_i = \sum_{j \neq i} a_j v_j$, where $a_j \in k$. Hence,

$$v_1 \wedge \cdots \wedge v_i \wedge \cdots \wedge v_p = v_1 \wedge \cdots \wedge \sum_{j \neq i} a_j v_j \wedge \cdots \wedge v_p$$

$$= \sum_{j \neq i} a_j v_1 \wedge \cdots \wedge v_j \wedge \cdots \wedge v_p.$$

After expanding, each term has a repeated factor $v_j$, and so this is 0.  •

We introduced Exterior Algebra, at the beginning of this section, by looking at Jacobians; we now end this section by applying Exterior Algebra to differential forms. Let $X$ be a connected open[18] subset of Euclidean space $\mathbb{R}^n$. A function $f : X \to \mathbb{R}$ is called a $C^\infty$-***function*** if, for all $p \geq 1$, the $p$th partials $\partial^p f / \partial^p x_i$ exist for all $i = 1, \ldots, n$, as do all the mixed partials.

**Definition.** If $X$ is a connected open subset of $\mathbb{R}^n$, define

$$A(X) = \{f : X \to \mathbb{R} : f \text{ is a } C^\infty\text{-function}\}.$$

The condition that $X$ be a connected open subset of $\mathbb{R}^n$ is present so that $C^\infty$-functions are defined. It is easy to see that $A(X)$ is a commutative ring under pointwise operations:

$$f + g : x \mapsto f(x) + g(x); \qquad fg : x \mapsto f(x)g(x).$$

In the free $A(X)$-module $A(X)^n$ of all $n$-tuples, rename the standard basis

$$dx_1, \ldots, dx_n.$$

The Binomial Theorem says that a basis for $\bigwedge^p A(X)^n$ consists of all elements of the form $dx_{i_1} \wedge \cdots \wedge dx_{i_p}$, where $i_1, \ldots, i_p$ is an increasing $p \leq n$ list. But Proposition 8.103 says that if $M$ is a module over a commutative ring $k$, then $k$ acts on $\bigotimes^p M$ by $r(m_1 \otimes \cdots \otimes m_p) = (rm_1) \otimes \cdots \otimes m_p$. It follows that each $\omega \in \bigwedge^p A(X)^n$ has a unique expression

$$\omega = \sum_{i_1, \ldots, i_p} (f_{i_1, \ldots, i_p} dx_{i_1}) \wedge \cdots \wedge dx_{i_p},$$

---

[18]A subset $X$ is ***open*** if, for each $x \in X$, there is some $r > 0$ so that all points $y$ with distance $|y - x| < r$ also lie in $X$. An open subset $X$ of $\mathbb{R}^n$ is ***connected*** if we can join any pair of points in $X$ by a path lying wholly in $X$.

where $f_{i_1,\ldots,i_p} \in A(X)$ is a $C^\infty$-function on $X$ and $i_1, \ldots, i_p$ is an increasing $p \leq n$ list. We write

$$\Omega^p(X) = \bigwedge\nolimits^p A(X)^n,$$

and we call its elements **differential $p$-forms** on $X$.

**Definition.** The **exterior derivative** $d^p \colon \Omega^p(X) \to \Omega^{p+1}(X)$ is defined as follows:

(i) if $f \in \Omega^0(X) = A(X)$, then $d^0 f = \sum_{j=1}^n \frac{\partial f}{\partial x_j} dx_j$;

(ii) if $p \geq 1$ and $\omega \in \Omega^p(X)$, then $\omega = \sum_{i_1 \ldots i_p} f_{i_1 \ldots i_p} dx_{i_1} \wedge \cdots \wedge dx_{i_p}$, and

$$d^p \omega = \sum_{i_1 \ldots i_p} d^0(f_{i_1 \ldots i_p}) \wedge dx_{i_1} \wedge \cdots \wedge dx_{i_p}.$$

If $X$ is a connected open subset of $\mathbb{R}^n$, exterior derivatives give a sequence of $A(X)$-maps, called the **de Rham complex**:

$$0 \to \Omega^0(X) \xrightarrow{d^0} \Omega^1(X) \xrightarrow{d^1} \cdots \xrightarrow{d^{n-1}} \Omega^n(X) \to 0.$$

**Proposition 8.123.** *If $X$ is a connected open subset of $\mathbb{R}^n$, then*

$$d^{p+1} d^p \colon \Omega^p(X) \to \Omega^{p+2}(X)$$

*is the zero map for all $p \geq 0$.*

**Proof.** It suffices to prove that $dd\omega = 0$, where $\omega = f dx_I$ (we are using an earlier abbreviation: $dx_I = dx_{i_1} \wedge \cdots \wedge dx_{i_p}$, where $I = i_1, \ldots, i_p$ is an increasing $p \leq n$ list). Now

$$dd\omega = d(d^0 f \wedge x_I)$$

$$= d\left( \sum_i \frac{\partial f}{\partial x_i} dx_i \wedge dx_I \right)$$

$$= \sum_{i,j} \frac{\partial^2 f}{\partial x_i \partial x_j} dx_j \wedge dx_i \wedge dx_I.$$

Compare the $i, j$ and $j, i$ terms in this double sum: the first is

$$\frac{\partial^2 f}{\partial x_i \partial x_j} dx_j \wedge dx_i \wedge dx_I,$$

the second is

$$\frac{\partial^2 f}{\partial x_j \partial x_i} dx_i \wedge dx_j \wedge dx_I,$$

and these cancel each other because the mixed second partials are equal and $dx_i \wedge dx_j = -dx_j \wedge dx_i$.   •

**Example 8.124.** Consider the special case of the de Rham complex for $n = 3$:

$$0 \to \Omega^0(X) \xrightarrow{d^0} \Omega^1(X) \xrightarrow{d^1} \Omega^2(X) \xrightarrow{d^2} \Omega^3(X) \to 0.$$

If $\omega \in \Omega^0(X)$, then $\omega = f(x, y, z) \in A(X)$, and

$$d^0 f = \frac{\partial f}{\partial x} \, dx + \frac{\partial f}{\partial y} \, dy + \frac{\partial f}{\partial z} \, dz,$$

a 1-form resembling $\mathrm{grad}(f)$.

If $\omega \in \Omega^1(X)$, then $\omega = f dx + g dy + h dz$, and a simple calculation gives

$$d^1 \omega = \left( \frac{\partial g}{\partial x} - \frac{\partial f}{\partial y} \right) dx \wedge dy + \left( \frac{\partial h}{\partial y} - \frac{\partial g}{\partial z} \right) dy \wedge dz + \left( \frac{\partial f}{\partial z} - \frac{\partial h}{\partial x} \right) dz \wedge dx,$$

a 2-form resembling $\mathrm{curl}(\omega)$.

If $\omega \in \Omega^2(X)$, then $\omega = F dy \wedge dz + G dz \wedge dx + H dx \wedge dy$. Now

$$d^2 \omega = \frac{\partial F}{\partial x} + \frac{\partial G}{\partial y} + \frac{\partial H}{\partial z},$$

a 3-form resembling $\mathrm{div}(\omega)$.

These are not mere resemblances. Since $\Omega^1(X)$ is a free $A(X)$-module with basis $dx, dy, dz$, we see that $d^0 \omega$ is $\mathrm{grad}(\omega)$ when $\omega$ is a 0-form. Now $\Omega^2(X)$ is a free $A(X)$-module, but we choose a basis $dx \wedge dy, dy \wedge dz, dz \wedge dx$ instead of the usual basis $dx \wedge dy, dx \wedge dz, dy \wedge dz$; it follows that $d^1 \omega$ is $\mathrm{curl}(\omega)$ in this case. Finally, $\Omega^3(X)$ has a basis $dx \wedge dy \wedge dz$, and so $d^3 \omega$ is $\mathrm{div}(\omega)$ when $\omega$ is a 2-form. We have shown that the de Rham complex is

$$0 \to \Omega^0(X) \xrightarrow{\mathrm{grad}} \Omega^1(X) \xrightarrow{\mathrm{curl}} \Omega^2(X) \xrightarrow{\mathrm{div}} \Omega^3(X) \to 0.$$

Proposition 8.123 now gives the familiar identities from Advanced Calculus:

$$\mathrm{curl} \cdot \mathrm{grad} = 0 \quad \text{and} \quad \mathrm{div} \cdot \mathrm{curl} = 0.$$

We call a 1-form $\omega$ **closed** if $d\omega = 0$, and we call it **exact** if $\omega = \mathrm{grad} f$ for some $C^\infty$-function $f$. More generally, call a $p$-form $\omega$ **closed** if $d^p \omega = 0$, and call it **exact** if $\omega = d^{p-1} \omega'$ for some $(p-1)$-form $\omega'$. Thus, $\omega \in \Omega^p(X)$ is closed if and only if $\omega \in \ker d^p$, and $\omega$ is exact if and only if $\omega \in \mathrm{im}\, d^{p-1}$. Therefore, the de Rham complex is an exact sequence of $A(X)$-modules if and only if every closed form is exact; indeed, this is the etymology of the adjective *exact* in "exact sequence." It can be proved that the de Rham complex is an exact sequence whenever $X$ is a simply connected open subset of $\mathbb{R}^n$. For any (not necessarily simply connected) space $X$, we have $\mathrm{im}\, \mathrm{grad} \subseteq \ker \mathrm{curl}$ and $\mathrm{im}\, \mathrm{curl} \subseteq \ker \mathrm{div}$, and the $\mathbb{R}$-vector spaces $\ker \mathrm{curl} / \mathrm{im}\, \mathrm{grad}$ and $\ker \mathrm{div} / \mathrm{im}\, \mathrm{curl}$ are called the *cohomology groups* of $X$ (Bott–Tu, *Differential Forms in Algebraic Topology*, Chapter I). ◀

## Exercises

**8.62.** Prove that the ring $R$ in Example 8.108 is left noetherian but not right noetherian.

**Hint.** See Cartan–Eilenberg, *Homological Algebra*, p. 16.

**8.63.** Let $G$ be a group. Then a $k$-algebra $A$ is called $G$-**graded** if there are $k$-submodules $A^g$, for all $g \in G$, such that

(i) $A = \bigoplus_{g \in G} A^g$;

(ii) for all $g, h \in G$, $A^g A^h \subseteq A^{gh}$.

An $\mathbb{I}_2$-graded algebra is called a ***superalgebra***. If $A$ is a $G$-graded algebra and $e$ is the identity element of $G$, prove that $1 \in A^e$.

**8.64.** If $A$ is a $k$-algebra generated by $n$ elements, prove that $A$ satisfies the standard identity defined on page 714.

**8.65.** Let $G(V)$ be the Grassmann algebra of a free $k$-module $V$, and let $u = \sum_p u_p \in G(V)$, where $u_p \in G^p(V)$ is homogeneous of degree $p$. If $\bar{u}$ is the conjugate of $u$ in Theorem 8.116, prove that $\bar{u} = \sum_p (-1)^p u_p$.

**8.66.** (i) Let $p$ be a prime. Show that $\bigwedge^2(\mathbb{I}_p \oplus \mathbb{I}_p) \neq 0$, where $\mathbb{I}_p \oplus \mathbb{I}_p$ is viewed as a $\mathbb{Z}$-module (i.e., as an abelian group).

(ii) Let $D = \mathbb{Q}/\mathbb{Z} \oplus \mathbb{Q}/\mathbb{Z}$. Prove that $\bigwedge^2 D = 0$, and conclude that if $i \colon \mathbb{I}_p \oplus \mathbb{I}_p \to D$ is an inclusion, then $\bigwedge^2(i)$ is not an injection.

**8.67.** (i) If $k$ is a commutative ring and $N$ is a direct summand of a $k$-module $M$, prove that $\bigwedge^p N$ is a direct summand of $\bigwedge^p M$ for all $p \geq 0$.

**Hint.** Use Corollary 6.28 on page 409.

(ii) If $k$ is a field and $i \colon W \to V$ is an injection of vector spaces over $k$, prove that $\bigwedge^p(i)$ is an injection for all $p \geq 0$.

**8.68.** Prove, for all $p$, that the functor $\bigwedge^p$ preserves surjections.

**8.69.** If $P$ is a projective $k$-module, where $k$ is a commutative ring, prove that $\bigwedge^q P$ is a projective $k$-module for all $q$.

**8.70.** Let $k$ be a field, and let $V$ be a vector space over $k$. Prove that two linearly independent lists $u_1, \ldots, u_p$ and $v_1, \ldots, v_p$ span the same subspace of $V$ if and only if there is a nonzero $c \in k$ with $u_1 \wedge \cdots \wedge u_p = c v_1 \wedge \cdots \wedge v_p$.

$*$ **8.71.** If $U$ and $V$ are $k$-modules over a commutative ring $k$ and $U' \subseteq U$ and $V' \subseteq V$ are submodules, prove that

$$(U/U') \otimes_k (V/V') \cong (U \otimes_k V)/(U' \otimes_k V + U \otimes_k V').$$

**Hint.** Compute the kernel and image of $\varphi \colon U \otimes_k V \to (U/U') \otimes_k (V/V')$ defined by $\varphi \colon u \otimes v \mapsto (u + U') \otimes v + u \otimes (v + V')$.

**8.72.** Let $V$ be a finite-dimensional vector space over a field $k$, and let $q \colon V \to k$ be a quadratic form on $V$. Define the ***Clifford algebra*** $C(V, q)$ as the quotient $C(V, q) = T(V)/J$, where $J$ is the two-sided ideal generated by all elements of the form $v \otimes v - q(v)1$ (note that $J$ is not a graded ideal). For $v \in V$, denote the coset $v + J$ by $[v]$, and define $h \colon V \to C(V, q)$ by $h(v) = [v]$. Prove that $C(V, q)$ is a solution to the following universal problem:

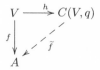

where $A$ is a $k$-algebra and $f \colon V \to A$ is a $k$-module map with $f(v)^2 = q(v)$ for all $v \in V$.

If $\dim(V) = n$ and $q$ is nondegenerate, then it can be proved that $\dim(C(V, q)) = 2^n$.

In particular, if $k = \mathbb{R}$ and $n = 2$, then the Clifford algebra has dimension 4 and $C(v, q) \cong \mathbb{H}$, the division ring of quaternions. Clifford algebras are used in the study of quadratic forms, hence of orthogonal groups; see Jacobson, *Basic Algebra* II, pp. 228–245.

## Section 8.7. Determinants

We have been using familiar properties of determinants, even though the reader may have seen their verifications only over fields and not over general commutative rings. Since determinants of matrices whose values lie in a commutative ring are of interest, the time has come to establish these properties in general, for Exterior Algebra is now available to help us.

If $k$ is a commutative ring, we claim that every $k$-module map $\gamma \colon k \to k$ is just multiplication by some $d \in k$: if $\gamma(1) = d$, then

$$\gamma(a) = \gamma(a1) = a\gamma(1) = ad = da$$

for all $a \in k$, because $\gamma$ is a $k$-module map. Here is a slight generalization: if $V = \langle v \rangle \cong k$, then every $k$-map $\gamma \colon V \to V$ has the form $\gamma \colon av \mapsto dav$, where $\gamma(v) = dv$. Suppose now that $V$ is a free $k$-module with basis $e_1, \ldots, e_n$; Corollary 8.118 shows that $\bigwedge^n V$ is free of rank 1 with generator $e_1 \wedge \cdots \wedge e_n$. It follows that every $k$-map $\gamma \colon \bigwedge^n V \to \bigwedge^n V$ has the form $\gamma(a(e_1 \wedge \cdots \wedge e_n)) = d(a(e_1 \wedge \cdots \wedge e_n))$.

**Definition.** If $V$ is a free $k$-module with basis $e_1, \ldots, e_n$ and $f \colon V \to V$ is a $k$-homomorphism, then the **determinant** of $f$, denoted by $\det(f)$, is the element $\det(f) \in k$ for which

$$\bigwedge^n(f) \colon e_1 \wedge \cdots \wedge e_n \mapsto f(e_1) \wedge \cdots \wedge f(e_n) = \det(f)(e_1 \wedge \cdots \wedge e_n).$$

If $A = [a_{ij}]$ is an $n \times n$ matrix with entries in $k$, then $A$ defines a $k$-map $f \colon k^n \to k^n$ by $f(x) = Ax$, where $x \in k^n$ is a column vector. If $e_1, \ldots, e_n$ is the standard basis of $k^n$, then $f(e_i) = \sum_j a_{ji} e_j$, and the matrix $A = [a_{ij}]$ associated to $f$ has as its $i$th column the coordinate-list of $f(e_i) = Ae_i$. We define $\det(A) = \det(f)$:

$$Ae_1 \wedge \cdots \wedge Ae_n = \det(A)(e_1 \wedge \cdots \wedge e_n).$$

Thus, the wedge of the columns of $A$ in $\bigwedge^n k^n$ is a constant multiple of $e_1 \wedge \cdots \wedge e_n$, and $\det(A)$ is that constant.

**Example 8.125.** If $A = \begin{bmatrix} a & c \\ b & d \end{bmatrix}$, then the wedge product of the columns of $A$ is

$$(ae_1 + be_2) \wedge (ce_1 + de_2) = ace_1 \wedge e_1 + ade_1 \wedge e_2 + bce_2 \wedge e_1 + bde_2 \wedge e_2$$
$$= ade_1 \wedge e_2 + bce_2 \wedge e_1$$
$$= ade_1 \wedge e_2 - bce_1 \wedge e_2$$
$$= (ad - bc)(e_1 \wedge e_2).$$

Therefore, $\det(A) = ad - bc$. ◀

The reader has probably noticed that this calculation is a repetition of the calculation on page 715 where we computed the Jacobian of a change of variables in a double integral. The next example considers triple integrals.

**Example 8.126.** Let us change variables in a triple integral $\iiint_D f(x, y, z)dxdydz$ using equations:

$$x = F(u, v, w);$$
$$y = G(u, v, w);$$
$$z = H(u, v, w).$$

Denote a basis of the tangent space $T$ of $f(x, y, z)$ at a point $P = (x_0, y_0, z_0)$ by $dx$, $dy$, $dz$. T If $du, dv, dw$ is another basis of $T$, then the chain rule defines a system of linear transformations on $T$ by the equations:

$$dx = F_u du + F_v dv + F_w dw$$
$$dy = G_u du + G_v dv + G_w dw$$
$$dz = H_u du + H_v dv + H_w dw.$$

If we write the differential $dxdydz$ in the integrand as $dx \wedge dy \wedge dz$, then the change of variables gives the new differential

$$dx \wedge dy \wedge dz = \det \left( \begin{bmatrix} F_u & F_v & F_w \\ G_u & G_v & G_w \\ H_u & H_v & H_w. \end{bmatrix} \right) du \wedge dv \wedge dw.$$

Expand

$$(F_u du + F_v dv + F_w dw) \wedge (G_u du + G_v dv + G_w dw) \wedge (H_u du + H_v dv + H_w dw)$$

to obtain nine terms, three of which involve $(du)^2$, $(dv)^2$, or $(dw)^2$, and hence are 0. Of the remaining six terms, three have a minus sign, and it is now easy to see that this sum is the determinant.  ◄

**Proposition 8.127.** *Let $k$ be a commutative ring.*

(i) *If $I$ is the identity matrix, then $\det(I) = 1$.*

(ii) *If $A$ and $B$ are $n \times n$ matrices with entries in $k$, then*

$$\det(AB) = \det(A)\det(B).$$

**Proof.** Both results follow from Proposition 8.112: $\bigwedge^n : {}_k\mathbf{Mod} \to {}_k\mathbf{Mod}$ is a functor.

(i) The linear transformation corresponding to the identity matrix is $1_{k^n}$, and every functor takes identities to identities.

(ii) If $f$ and $g$ are the linear transformations on $k^n$ arising from $A$ and $B$, respectively, then $fg$ is the linear transformation arising from $AB$. If we denote $e_1 \wedge \cdots \wedge e_n$ by $e_N$, then

$$\det(fg)e_N = \bigwedge^n (fg)(e_N)$$
$$= \bigwedge^n (f)\left( \bigwedge^n (g)(e_N) \right)$$
$$= \bigwedge^n (f)(\det(g)e_N)$$
$$= \det(g)\bigwedge^n (f)(e_N)$$
$$= \det(f)\det(g)e_N;$$

the next to last equation uses the fact that $\bigwedge^n(f)$ is a $k$-map. Therefore,

$$\det(AB) = \det(fg) = \det(f)\det(g) = \det(A)\det(B). \quad \bullet$$

**Corollary 8.128.** *If $k$ is a commutative ring, then* $\det\colon \operatorname{Mat}_n(k) \to k$ *is the unique alternating multilinear function with* $\det(I) = 1$.

**Proof.** The definition of determinant (as the wedge of the columns) shows that it is an alternating multilinear function $\det\colon \times^n V \to k$, where $V = k^n$, and Proposition 8.127 shows that $\det(I) = 1$. The uniqueness of such a function follows from the universal property of $\bigwedge^n$:

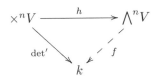

If $\det'$ is multilinear, then there exists a $k$-map $f\colon \bigwedge^n V \to k$ with $fh = \det'$; if $\det'(e_1,\ldots,e_n) = 1$, then $f(e_1 \wedge \cdots \wedge e_n) = 1$. Since $\bigwedge^n V \cong k$, every $k$-map $f\colon \bigwedge^n V \to k$ is determined by $f(e_1 \wedge \cdots \wedge e_n)$. Thus, the map $f$ is the same for $\det'$ as it is for $\det$, and so $\det' = fh = \det$. $\quad \bullet$

We now show that the determinant just defined coincides with the familiar determinant function.

**Lemma 8.129.** *Let $e_1,\ldots,e_n$ be a basis of a free $k$-module, where $k$ is a commutative ring. If $\sigma$ is a permutation of $1, 2, \ldots, n$, then*

$$e_{\sigma(1)} \wedge \cdots \wedge e_{\sigma(n)} = \operatorname{sgn}(\sigma)(e_1 \wedge \cdots \wedge e_n) = \operatorname{sgn}(\sigma)e_N.$$

**Proof.** Since $m \wedge m' = -m' \wedge m$, it follows that interchanging adjacent factors in the product $e_N = e_1 \wedge \cdots \wedge e_n$ gives

$$e_1 \wedge \cdots \wedge e_i \wedge e_{i+1} \wedge \cdots \wedge e_n = -e_1 \wedge \cdots \wedge e_{i+1} \wedge e_i \wedge \cdots \wedge e_n.$$

More generally, if $i < j$, then we can interchange $e_i$ and $e_j$ by a sequence of interchanges of adjacent factors, each of which causes a sign change. By Exercise 1.16 on page 15, this can be accomplished with an odd number of interchanges of adjacent factors. Hence, for any transposition $\tau \in S_n$, we have

$$\begin{aligned}
e_{\tau(1)} \wedge \cdots \wedge e_{\tau(n)} &= e_1 \wedge \cdots \wedge e_j \wedge \cdots \wedge e_i \wedge \cdots \wedge e_n \\
&= -[e_1 \wedge \cdots \wedge e_i \wedge \cdots \wedge e_j \wedge \cdots \wedge e_n] \\
&= \operatorname{sgn}(\tau)(e_1 \wedge \cdots \wedge e_n) = \operatorname{sgn}(\tau)e_N.
\end{aligned}$$

We prove the general statement by induction on $m$, where $\sigma$ is a product of $m$ transpositions. The base step having just been proven, we proceed to the inductive step. Write $\sigma = \tau_1\tau_2 \cdots \tau_{m+1}$, and denote $\tau_2 \cdots \tau_{m+1}$ by $\sigma'$. By the inductive hypothesis,

$$e_{\sigma'(1)} \wedge \cdots \wedge e_{\sigma'(n)} = \operatorname{sgn}(\sigma')e_N,$$

and so

$$e_{\sigma(1)} \wedge \cdots \wedge e_{\sigma(n)} = e_{\tau_1 \sigma'(1)} \wedge \cdots \wedge e_{\tau_1 \sigma'(n)}$$
$$= -e_{\sigma'(1)} \wedge \cdots \wedge e_{\sigma'(n)} \quad \text{(base step)}$$
$$= -\operatorname{sgn}(\sigma')e_N \quad \text{(inductive step)}$$
$$= \operatorname{sgn}(\tau_1)\operatorname{sgn}(\sigma')e_N$$
$$= \operatorname{sgn}(\sigma)e_N. \quad \bullet$$

**Remark.** There is a simpler proof of this lemma in the special case when $k$ is a field. If $k$ has characteristic 2, then Lemma 8.129 is obviously true, and so we may assume that the characteristic of $k$ is not 2. Let $e_1, \ldots, e_n$ be the standard basis of $k^n$. If $\sigma \in S_n$, define a linear transformation $\varphi_\sigma \colon k^n \to k^n$ by $\varphi_\sigma \colon e_i \mapsto e_{\sigma(i)}$. Since $\varphi_{\sigma\tau} = \varphi_\sigma\varphi_\tau$, as is easily verified, there is a group homomorphism $d \colon S_n \to k^\times$ given by $d \colon \sigma \mapsto \det(\varphi_\sigma)$. If $\sigma$ is a transposition, then $\sigma^2 = (1)$ and $d(\sigma)^2 = 1$ in $k^\times$. Since $k$ is a field, $d(\sigma) = \pm 1$. As every permutation is a product of transpositions, it follows that $d(\sigma) = \pm 1$ for every permutation $\sigma$, and so $\operatorname{im}(d) \subseteq \{\pm 1\}$. Now there are only two homomorphisms $S_n \to \{\pm 1\}$: the trivial homomorphism with kernel $S_n$ and sgn. To show that $d = \operatorname{sgn}$, it suffices to show that $d((1\ 2)) \neq 1$. But $d((1\ 2)) = \det(\varphi_{(1\,2)})$; that is,

$$\det(\varphi_{(1\,2)})e_N = \det(\varphi_{(1\,2)})(e_1 \wedge \cdots \wedge e_n)$$
$$= \varphi_{(1\,2)}(e_1) \wedge \cdots \wedge \varphi_{(1\,2)}(e_n)$$
$$= e_2 \wedge e_1 \wedge e_3 \wedge \cdots \wedge e_n$$
$$= -(e_1 \wedge \cdots \wedge e_n) = -e_N.$$

Therefore, $d((1\ 2)) = -1 \neq 1$, because $k$ does not have characteristic 2, and so, for all $\sigma \in S_n$, $d(\sigma) = \det(\varphi_\sigma) = \operatorname{sgn}(\sigma)$; that is, $e_{\sigma(1)} \wedge \cdots \wedge e_{\sigma(n)} = \operatorname{sgn}(\sigma)e_N$. ◀

**Proposition 8.130 (Complete Expansion).** *Let $e_1, \ldots, e_n$ be a basis of a free $k$-module, where $k$ is a commutative ring. If $A = [a_{ij}]$ is an $n \times n$ matrix with entries in $k$, then*

$$\det(A) = \sum_{\sigma \in S_n} \operatorname{sgn}(\sigma)a_{\sigma(1),1}a_{\sigma(2),2}\cdots a_{\sigma(n),n}.$$

**Proof.** The $j$th column $x_j = \sum_i a_{ij}e_i$. Since it is hazardous to use the same symbol to mean different things in a single equation, we denote the $j_q$th column by $x_{j_q} = \sum_{i_q} a_{i_q j_q}e_{i_q}$, where $1 \leq q \leq n$. Expand the wedge of the columns of $A$:

$$x_1 \wedge \cdots \wedge x_n = \sum_{i_1} a_{i_1 1}e_{i_1} \wedge \sum_{i_2} a_{i_2 2}e_{i_2} \wedge \cdots \wedge \sum_{i_n} a_{i_n n}e_{i_n}$$
$$= \sum_{i_1,i_2,\ldots,i_n} a_{i_1 1}e_{i_1} \wedge a_{i_2 2}e_{i_2} \wedge \cdots \wedge a_{i_n n}e_{i_n}.$$

Any summand in which $e_{i_p} = e_{i_q}$ must be 0, for it has a repeated factor, and so we may assume, in any surviving term, that $i_1, i_2, \ldots, i_n$ are all distinct; that is, for each summand, there is a permutation $\sigma \in S_n$ with $i_q = \sigma(q)$ for all $1 \leq q \leq n$.

The original product now has the form

$$\sum_{\sigma \in S_n} a_{\sigma(1)1} a_{\sigma(2)2} \cdots a_{\sigma(n)n} e_{\sigma(1)} \wedge e_{\sigma(2)} \wedge \cdots \wedge e_{\sigma(n)}.$$

By Lemma 8.129, $e_{\sigma(1)} \wedge e_{\sigma(2)} \wedge \cdots \wedge e_{\sigma(n)} = \operatorname{sgn}(\sigma) e_N$. Therefore, the wedge of the columns is equal to $\left( \sum_{\sigma \in S_n} \operatorname{sgn}(\sigma) a_{\sigma(1)1} a_{\sigma(2)2} \cdots a_{\sigma(n)n} \right) e_N$, and this completes the proof. •

Quite often, the complete expansion is taken as the definition of the determinant.

**Corollary 8.131.** *Let $k$ be a commutative ring, and let $A$ be an $n \times n$ matrix with entries in $k$. If $u \in k$, then the characteristic polynomial $\det(xI - A) = \psi_A(x) \in k[x]$, is a monic polynomial of degree $n$. Moreover, the coefficient of $x^{n-1}$ in $\psi_A(x)$ is $-\operatorname{tr}(A)$.*

**Proof.** Let $A = [a_{ij}]$ and let $B = [b_{ij}]$, where $b_{ij} = u\delta_{ij} - a_{ij}$ (where $\delta_{ij}$ is the Kronecker delta). By the Complete Expansion,

$$\det(B) = \sum_{\sigma \in S_n} \operatorname{sgn}(\sigma) b_{\sigma(1),1} b_{\sigma(2),2} \cdots b_{\sigma(n),n}.$$

If $\sigma = (1)$, then the corresponding term in the complete expansion is

$$b_{11} b_{22} \cdots b_{nn} = \prod_i (u - a_{ii}) = g(u),$$

where $g(x) = \prod_i (x - a_{ii})$ is a monic polynomial in $k[x]$ of degree $n$. If $\sigma \neq (1)$, then the $\sigma$th term in the complete expansion cannot have exactly $n - 1$ factors from the diagonal of $uI - A$, for if $\sigma$ fixes $n - 1$ indices, then $\sigma = (1)$. Therefore, the sum of the terms over all $\sigma \neq (1)$ is either 0 or a polynomial in $k[x]$ of degree at most $n - 2$. It follows that $\deg(\psi_A) = n$ and the coefficient of $x^{n-1}$ is $-\sum_i a_{ii} = -\operatorname{tr}(A)$. •

Let $f(x) \in k[x]$, where $k$ is a field. If $f(x) = (x - \alpha_1) \cdots (x - \alpha_n) = x^n + a_{n-1} x^{n-1} + \cdots + a_0$, then $a_{n-1} = -(\alpha_1 + \cdots + \alpha_n)$; that is, $-a_{n-1}$ is the sum of the roots of $f(x)$. In particular, since $-\operatorname{tr}(A)$ is the coefficient of $x^{n-1}$ in the characteristic polynomial of an $n \times n$ matrix $A$, we see that $\operatorname{tr}(A)$ is the sum (with multiplicities) of the eigenvalues of $A$.

**Proposition 8.132.** *If $A$ is an $n \times n$ matrix with entries in a commutative ring $k$, then*

$$\det(A^\top) = \det(A),$$

*where $A^\top$ is the transpose of $A$.*

**Proof.** If $A = [a_{ij}]$, write the complete expansion of $\det(A)$ more compactly:

$$\det(A) = \sum_{\sigma \in S_n} \operatorname{sgn}(\sigma) \prod_i a_{\sigma(i),i}.$$

For any permutation $\tau \in S_n$, we have $i = \tau(j)$ for all $i$, and so

$$\prod_i a_{\sigma(i),i} = \prod_j a_{\sigma(\tau(j)),\tau(j)},$$

for this merely rearranges the factors in the product. Choosing $\tau = \sigma^{-1}$ gives

$$\prod_j a_{\sigma(\tau(j)),\tau(j)} = \prod_j a_{j,\sigma^{-1}(j)}.$$

Therefore,

$$\det(A) = \sum_{\sigma \in S_n} \operatorname{sgn}(\sigma) \prod_j a_{j,\sigma^{-1}(j)}.$$

Now $\operatorname{sgn}(\sigma) = \operatorname{sgn}(\sigma^{-1})$ [if $\sigma = \tau_1 \cdots \tau_q$, where the $\tau$ are transpositions, then $\sigma^{-1} = \tau_q \cdots \tau_1$]; moreover, as $\sigma$ varies over $S_n$, so does $\sigma^{-1}$. Hence, writing $\sigma^{-1} = \rho$ gives

$$\det(A) = \sum_{\rho \in S_n} \operatorname{sgn}(\rho) \prod_j a_{j,\rho(j)}.$$

Now write $A^\top = [b_{ij}]$, where $b_{ij} = a_{ji}$. Then

$$\det(A^\top) = \sum_{\rho \in S_n} \operatorname{sgn}(\rho) \prod_j b_{\rho(j),j} = \sum_{\rho \in S_n} \operatorname{sgn}(\rho) \prod_j a_{j,\rho(j)} = \det(A). \quad \bullet$$

Recall that similar matrices have the same determinant and the same trace. The next corollary generalizes these facts, for all the coefficients of their characteristic polynomials coincide.

**Corollary 8.133.** *If $A$ and $B$ are similar $n \times n$ matrices with entries in a field $k$, then $A$ and $B$ have the same characteristic polynomial.*

**Proof.** There is a nonsingular matrix $P$ with $B = PAP^{-1}$, and

$$\begin{aligned}
\psi_B(x) &= \det(xI - B) \\
&= \det(xI - PAP^{-1}) \\
&= \det\left(P(xI - A)P^{-1}\right) \\
&= \det(P)\det(xI - A)\det(P)^{-1} \\
&= \det(xI - A) \\
&= \psi_A(x). \quad \bullet
\end{aligned}$$

We have already seen that the eigenvalues $\alpha_1, \ldots, \alpha_n$ of an $n \times n$ matrix $A$ with entries in a field $k$ are the roots of the characteristic polynomial

$$\psi_A(x) = \det(xI - A) \in k[x].$$

We remind the reader of Proposition 8.38: $\operatorname{tr}(A) = \alpha_1 + \alpha_2 + \cdots + \alpha_n$.

**Definition.** Let $A$ be an $n \times n$ matrix with entries in a commutative ring $k$. If $H = i_1, \ldots, i_p$ and $L = j_1, \ldots, j_p$ are increasing $p \leq n$ lists (that is, $1 \leq i_1 < i_2 < \cdots < i_p \leq n$ and $1 \leq j_1 < j_2 < \cdots < j_p \leq n$), then $A_{H,L}$ is the $p \times p$ **submatrix** $[a_{st}]$, where $(s,t) \in H \times L$. A **minor of order** $p$ is the determinant of a $p \times p$ submatrix.

The submatrix $A_{H,L}$ is obtained from $A$ by deleting all $i$th rows for $i$ not in $H$ and all $j$th columns for $j$ not in $L$. For example, every entry $a_{ij}$ is a minor of $A = [a_{ij}]$ (for it is the determinant of the $1 \times 1$ submatrix obtained from $A$ by deleting all rows except the $i$th and all columns except the $j$th). If

$$A = \begin{bmatrix} a_{11} & a_{12} & a_{13} \\ a_{21} & a_{22} & a_{23} \\ a_{31} & a_{32} & a_{33} \end{bmatrix},$$

then some minors of order 2 are

$$\det \begin{bmatrix} a_{11} & a_{12} \\ a_{21} & a_{22} \end{bmatrix} \text{ and } \det \begin{bmatrix} a_{12} & a_{13} \\ a_{32} & a_{33} \end{bmatrix}.$$

If $1 \leq i \leq n$, let $i'$ denote the increasing $n - 1 \leq n$ list in which $i$ is omitted; thus, an $(n-1) \times (n-1)$ submatrix has the form $A_{i',j'}$, and its determinant is a minor of order $n - 1$. Note that $A_{i',j'}$ is the submatrix obtained from $A$ by deleting its $i$th row and $j$th column.

**Lemma 8.134.** *Let $k$ be a commutative ring, let $e_1, \ldots, e_n$ be the standard basis of $k^n$, let $A = [a_{ij}]$ be an $n \times n$ matrix over $k$, and let $L = j_1, \ldots, j_p$ be an increasing $p \leq n$ list. If $x_{j_1}, \ldots, x_{j_p}$ are the corresponding columns of $A$, then*

$$x_{j_1} \wedge \cdots \wedge x_{j_p} = \sum_H \det(A_{H,L}) e_H,$$

*where $H$ varies over all increasing $p \leq n$ lists $i_1, \ldots, i_p$ and $e_H = e_{i_1} \wedge \cdots \wedge e_{i_p}$.*

**Proof.** The proof is quite similar to the proof of the Complete Expansion. For $q = 1, 2, \ldots, p$, write $x_{j_q} = \sum_{t_q} a_{t_q j_q} e_{t_q}$, so that

$$x_{j_1} \wedge \cdots \wedge x_{j_p} = \sum_{i_1} a_{i_1 j_1} e_{i_1} \wedge \cdots \wedge \sum_{i_p} a_{i_p j_p} e_{i_p} = \sum_{i_1, \ldots, i_p} a_{i_1 j_1} \cdots a_{i_p j_p} e_{i_1} \wedge \cdots \wedge e_{i_p}.$$

All terms involving a repeated index are 0, so that we may assume that the sum is over all $i_1, \ldots, i_p$ having no repetitions; that is, for each summand, there is a permutation $\sigma \in S_p$ with $i_1 = \sigma(1), \ldots, i_p = \sigma(p)$. With this notation,

$$a_{i_1 j_1} \cdots a_{i_p j_p} e_{i_1} \wedge \cdots \wedge e_{i_p} = a_{i_{\sigma(1)} j_1} \cdots a_{i_{\sigma(p)} j_p} e_{i_1} \wedge \cdots \wedge e_{i_p}$$
$$= \operatorname{sgn}(\sigma) a_{i_{\sigma(1)} j_1} \cdots a_{i_{\sigma(p)} j_p} e_L.$$

Summing over all $H$ gives the desired formula

$$\sum_H a_{i_1 j_1} \cdots a_{i_p j_p} e_{i_1} \wedge \cdots \wedge e_{i_p} = \sum_H \det(A_{H,L}) e_H. \quad \bullet$$

Multiplication in the algebra $\bigwedge V$ is determined by the products $e_H \wedge e_K$ of pairs of basis elements. Let us introduce the following notation: if $H = t_1, \ldots, t_p$ and $K = \ell_1, \ldots, \ell_q$ are disjoint increasing lists, then define

$$\tau_{H,K}$$

to be the permutation that rearranges the list $t_1, \ldots, t_p, \ell_1, \ldots, \ell_q$ into an increasing list, denoted by $H * K$. Define

$$\rho_{H,K} = \operatorname{sgn}(\tau_{H,K}).$$

With this notation, Lemma 8.129 says that

$$e_H \wedge e_K = \begin{cases} 0 & \text{if } H \cap K \neq \varnothing \\ \rho_{H,K} e_{H*K} & \text{if } H \cap K = \varnothing. \end{cases}$$

**Example 8.135.** The lists $H = 1, 3, 4$ and $K = 2, 6$ are increasing;

$$H * K = 1, 2, 3, 4, 6,$$

and

$$\tau_{H,K} = \begin{pmatrix} 1 & 3 & 4 & 2 & 6 \\ 1 & 2 & 3 & 4 & 6 \end{pmatrix} = (2\ 4\ 3).$$

Therefore,

$$\rho_{H,K} = \operatorname{sgn} \tau_{H,K} = +1,$$

and

$$e_H \wedge e_K = (e_1 \wedge e_3 \wedge e_4) \wedge (e_2 \wedge e_6) = e_1 \wedge e_2 \wedge e_3 \wedge e_4 \wedge e_6 = e_{H*K}. \quad \blacktriangleleft$$

**Proposition 8.136.** *Let $A = [a_{ij}]$ be an $n \times n$ matrix with entries in a commutative ring $k$.*

(i) *If $I = i_1, \ldots, i_p$ is an increasing $p \leq n$ list and $x_{i_1}, \ldots, x_{i_p}$ are the corresponding columns of $A$, then denote $x_{i_1} \wedge \cdots \wedge x_{i_p}$ by $x_I$. If $J = j_1, \ldots, j_q$ is an increasing $q \leq n$ list, then*

$$x_I \wedge x_J = \sum_{H,K} \rho_{H,K} \det(A_{H,I}) \det(A_{K,J}) e_{H*K},$$

*where $H * K$ is the increasing $(p + q) \leq n$ list formed from $H \cup K$ when $H \cap K = \varnothing$.*

(ii) ***Laplace expansion down the $j$th column***: *For each fixed $j$,*

$$\det(A) = (-1)^{1+j} a_{1j} \det(A_{1'j'}) + \cdots + (-1)^{n+j} a_{nj} \det(A_{n'j'}),$$

*where $A_{i',j'}$ is the $(n-1) \times (n-1)$ submatrix obtained from $A$ by deleting its $i$th row and $j$th column.*

(iii) ***Laplace expansion across the $i$th row***: *For each fixed $i$,*

$$\det(A) = (-1)^{i+1} a_{i1} \det(A_{i',1'}) + \cdots + (-1)^{i+n} a_{in} \det(A_{i',n'}).$$

**Proof.**

(i) By Lemma 8.134,

$$\begin{aligned}
x_I \wedge x_J &= \sum_H \det(A_{H,I}) e_H \wedge \sum_K \det(A_{K,J}) e_K \\
&= \sum_{H,K} \det(A_{H,I}) e_H \wedge \det(A_{K,J}) e_K \\
&= \sum_{H,K} \det(A_{H,I}) \det(A_{K,J}) e_H \wedge e_K \\
&= \sum_{H,K} \rho_{H,K} \det(A_{H,I}) \det(A_{K,J}) e_{H*K}.
\end{aligned}$$

(ii) If $I = j$ has only one element and $J = j' = 1, \ldots, \widehat{j}, \ldots, n$ is its complement, then

$$x_j \wedge x_{j'} = x_j \wedge x_1 \wedge \cdots \wedge \widehat{x_j} \wedge \cdots \wedge x_n$$
$$= (-1)^{j-1} x_1 \wedge \cdots \wedge x_n$$
$$= (-1)^{j-1} \det(A) e_1 \wedge \cdots \wedge e_n,$$

because $j, 1, \ldots, \widehat{j}, \ldots, n$ can be put in increasing order by $j-1$ transpositions. On the other hand, we can evaluate $x_j \wedge x_{j'}$ using part (i):

$$x_j \wedge x_{j'} = \sum_{H,K} \rho_{H,K} \det(A_{H,j}) \det(A_{K,j'}) e_{H*K}.$$

In this sum, $H$ has just one element, say, $H = i$, while $K$ has $n-1$ elements; thus, $K = \ell'$ for some element $\ell$. Since $e_h \wedge e_{\ell'} = 0$ if $\{i\} \cap \ell' \neq \varnothing$, we may assume that $i \notin \ell'$; that is, we may assume that $\ell' = i'$. Now, $\det(A_{i,j}) = a_{ij}$ (this is a $1 \times 1$ minor), while $\det(A_{K,j'}) = \det(A_{i',j'})$; that is, $A_{i',j'}$ is the submatrix obtained from $A$ by deleting its $j$th column and its $i$th row. Hence, if $e_N = e_1 \wedge \cdots \wedge e_n$,

$$x_j \wedge x_{j'} = \sum_{H,K} \rho_{H,K} \det(A_{H,j}) \det(A_{K,j'}) e_{H*K}$$
$$= \sum_i \rho_{i,i'} \det(A_{ij}) \det(A_{i',j'}) e_N$$
$$= \sum_i (-1)^{i-1} a_{ij} \det(A_{i',j'}) e_N.$$

Therefore, equating both values for $x_j \wedge x_{j'}$ gives

$$\det(A) = \sum_i (-1)^{i+j} a_{ij} \det(A_{i',j'}).$$

(iii) Laplace expansion across the $i$th row of $A$ is Laplace expansion down the $i$th column of $A^\top$, and so the result follows because $\det(A^\top) = \det(A)$. •

Determinant is independent of the row or column used in Laplace expansion.

**Corollary 8.137.** *Given any $n \times n$ matrix $A$ over a commutative ring $k$, Laplace expansion across any row or down any column always has the same value.*

**Proof.** All expansions equal $\det(A)$. •

The Laplace expansions resemble the sums arising in matrix multiplication, and the following matrix was invented to make this resemblance a reality.

**Definition.** If $A = [a_{ij}]$ is an $n \times n$ matrix with entries in a commutative ring $k$, then the **adjoint**[19] of $A$ is the matrix

$$\mathrm{adj}(A) = [C_{ij}],$$

---

[19]There is no connection between the adjoint of a matrix as just defined and the adjoint of a matrix with respect to an inner product defined on page 695.

where
$$C_{ij} = (-1)^{i+j} \det(A_{j'i'}).$$
The reversing of indices is deliberate. In words, $\mathrm{adj}(A)$ is the transpose of the matrix whose $i, j$ entry is $(-1)^{i+j} \det(A_{i'j'})$. We call $C_{ij}$ the $ij$-***cofactor*** of $A$.

**Corollary 8.138.** *If $A$ is an $n \times n$ matrix with entries in a commutative ring $k$, then*
$$A \, \mathrm{adj}(A) = \det(A)I = \mathrm{adj}(A)A.$$

**Proof.** Denote the $ij$ entry of $A \, \mathrm{adj}(A)$ by $b_{ij}$. The definition of matrix multiplication gives
$$b_{ij} = \sum_{p=1}^{n} a_{ip} C_{pj} = \sum_{p=1}^{n} a_{ip}(-1)^{j+p} \det(A_{j'p'}).$$
If $j = i$, Proposition 8.136 gives
$$b_{ii} = \det(A).$$
If $j \neq i$, consider the matrix $M$ obtained from $A$ by replacing row $j$ with row $i$. Of course, $\det(M) = 0$, for it has two identical rows. On the other hand, we may compute $\det(M)$ using Laplace expansion across its "new" row $j$. All the submatrices $M_{j'p'} = A_{j'p'}$, and so all the corresponding cofactors of $M$ and $A$ are equal. The matrix entries of the new row $j$ are $a_{ip}$, so that
$$0 = \det(M) = (-1)^{i+1} a_{i1} \det(A_{j'1'}) + \cdots + (-1)^{i+n} a_{in} \det(A_{j'n'}).$$
We have shown that $A \, \mathrm{adj}(A)$ is a diagonal matrix having each diagonal entry equal to $\det(A)$. The similar proof that $\det(A)I = \mathrm{adj}(A)A$ is left to the reader.  •

**Definition.** An $n \times n$ matrix $A$ with entries in a commutative ring $k$ is ***invertible over*** $k$ if there is a matrix $B$ with entries in $k$ such that
$$AB = I = BA.$$

If $k$ is a field, then invertible matrices are usually called *nonsingular*, and they are characterized by having a nonzero determinant. Consider the matrix with entries in $\mathbb{Z}$:
$$A = \begin{bmatrix} 3 & 1 \\ 1 & 1 \end{bmatrix}.$$
Now $\det(A) = 2 \neq 0$, but it is not invertible over $\mathbb{Z}$. Suppose
$$\begin{bmatrix} 3 & 1 \\ 1 & 1 \end{bmatrix} \begin{bmatrix} a & c \\ b & d \end{bmatrix} = \begin{bmatrix} 3a+b & 3c+d \\ a+b & c+d \end{bmatrix}.$$
If this product is $I$, then
$$3a + b = 1 = c + d$$
$$3c + d = 0 = a + b.$$
Hence, $b = -a$ and $1 = 3a + b = 2a$; as there is no solution to $1 = 2a$ in $\mathbb{Z}$, the matrix $A$ is not invertible over $\mathbb{Z}$. Of course, $A$ is invertible over $\mathbb{Q}$.

**Proposition 8.139.** *Let $k$ be a commutative ring and let $A \in \mathrm{Mat}_n(k)$. Then $A$ is invertible if and only if $\det(A)$ is a unit in $k$.*

**Proof.** If $A$ is invertible, then there is a matrix $B$ with $AB = I$. Hence,

$$1 = \det(I) = \det(AB) = \det(A)\det(B);$$

this says that $\det(A)$ is a unit in $k$.

Conversely, assume that $\det(A)$ is a unit in $k$, so there is an element $u \in k$ with $u\det(A) = 1$. Define

$$B = u\operatorname{adj}(A).$$

By Corollary 8.138,

$$AB = Au\operatorname{adj}(A) = u\det(A)I = I = u\operatorname{adj}(A)A = BA.$$

Thus, $A$ is invertible.  •

The next result generalizes Corollary 2.131 to matrices over commutative rings.

**Corollary 8.140.** *Let $A$ and $B$ be $n \times n$ matrices over a commutative ring $k$. If $AB = I$, then $BA = I$.*

**Proof.** If $AB = I$, then $\det(A)\det(B) = 1$; that is, $\det(A)$ is a unit in $k$. Therefore, $A$ is invertible, by Proposition 8.139; that is, $AB = I = BA$.  •

**Corollary 8.141 (Cramer's Rule).** *If $A$ is an invertible $n \times n$ matrix over a commutative ring $k$ and $B = [b_i]$ is an $n \times 1$ column matrix, then the solution of the linear system $AX = B$ is $X = (x_1, \ldots, x_n)^\top$, where $x_j = \det(M_j)\det(A)^{-1}$ and $M_j$ is obtained from $A$ by replacing its $j$th column by $B$.*

**Proof.** Multiply $AX = B$ by $\operatorname{adj}(A)$ to obtain

$$\det(A)X = \operatorname{adj}(A)B.$$

As in the proof of Corollary 8.138, Laplace expansion down the $j$th column gives

$$\det(M_j) = b_1 A_{j'1'} + \cdots + b_n A_{j'n'} = \operatorname{adj}(A)B.  •$$

Here is a proof by Exterior Algebra of the computation of the determinant of a matrix in block form.

**Proposition 8.142.** *Let $k$ be a commutative ring, and let*

$$X = \begin{bmatrix} A & C \\ 0 & B \end{bmatrix}$$

*be an $(m+n) \times (m+n)$ matrix with entries in $k$, where $A$ is an $m \times m$ submatrix, and $B$ is an $n \times n$ submatrix. Then*

$$\det(X) = \det(A)\det(B).$$

**Proof.** Let $e_1, \ldots, e_{m+n}$ be the standard basis of $k^{m+n}$, let $\alpha_1, \ldots, \alpha_m$ be the columns of $A$ (which are also the first $m$ columns of $X$), and let $\gamma_i + \beta_i$ be the $(m+i)$th column of $X$, where $\gamma_i$ is the $i$th column of $C$ and $\beta_i$ is the $i$th column of $B$.

Now $\gamma_i \in \langle e_1, \ldots, e_m \rangle$, so that $\gamma_i = \sum_{j=1}^{m} c_{ji} e_j$. Therefore, if $H = 1, 2, \ldots, n$, then

$$e_H \wedge \gamma_i = e_H \wedge \sum_{j=1}^{m} c_{ji} e_j = 0,$$

because each term has a repeated $e_j$. Using associativity, we see that

$$\begin{aligned}
e_H \wedge (\gamma_1 + \beta_1) &\wedge (\gamma_2 + \beta_2) \wedge \cdots \wedge (\gamma_n + \beta_n) \\
&= e_H \wedge \beta_1 \wedge (\gamma_2 + \beta_2) \wedge \cdots \wedge (\gamma_n + \beta_n) \\
&= e_H \wedge \beta_1 \wedge \beta_2 \wedge \cdots \wedge (\gamma_n + \beta_n) \\
&= e_H \wedge \beta_1 \wedge \beta_2 \wedge \cdots \wedge \beta_n.
\end{aligned}$$

Hence, if $J = m+1, m+2, \ldots, m+n$,

$$\begin{aligned}
\det(X) e_H \wedge e_J &= \alpha_1 \wedge \cdots \wedge \alpha_m \wedge (\gamma_1 + \beta_1) \wedge \cdots \wedge (\gamma_n + \beta_n) \\
&= \det(A) e_H \wedge (\gamma_1 + \beta_1) \wedge \cdots \wedge (\gamma_n + \beta_n) \\
&= \det(A) e_H \wedge \beta_1 \wedge \cdots \wedge \beta_n \\
&= \det(A) e_H \wedge \det(B) e_J \\
&= \det(A) \det(B) e_H \wedge e_J.
\end{aligned}$$

Therefore, $\det(X) = \det(A) \det(B)$.   •

**Corollary 8.143.** *If $A = [a_{ij}]$ is a triangular $n \times n$ matrix, that is, $a_{ij} = 0$ for all $i < j$ (lower triangular) or $a_{ij} = 0$ for all $i > j$ (upper triangular), then*

$$\det(A) = \prod_{i=1}^{n} a_{ii};$$

*that is, $\det(A)$ is the product of the diagonal entries.*

**Proof.** An easy induction on $n \geq 1$, using Laplace expansion down the first column (for upper triangular matrices) and the proposition for the inductive step.   •

Although the definition of determinant of a matrix $A$ in terms of the wedge of its columns gives an obvious algorithm for computing it, there is a more efficient means of calculating $\det(A)$ when its entries lie in a field. Using Gaussian elimination, there are elementary row operations changing $A$ into an upper triangular matrix $T$:

$$A \to A_1 \to \cdots \to A_r = T.$$

Keep a record of the operations used. For example, if $A \to A_1$ is an operation of Type I, which multiplies a row by a unit $c$, then $c \det(A) = \det(A_1)$ and so $\det(A) = c^{-1} \det(A_1)$; if $A \to A_1$ is an operation of Type II, which adds a multiple of some row to another one, then $\det(A) = \det(A_1)$; if $A \to A_1$ is an operation of Type III, which interchanges two rows, then $\det(A) = -\det(A_1)$. Thus, the record allows us, eventually, to write $\det(A)$ in terms of $\det(T)$. But since $T$ is upper triangular, $\det(T)$ is the product of its diagonal entries.

Another application of Exterior Algebra constructs the trace of a map.

**Definition.** Let $k$ be a commutative ring and let $A$ be a $k$-algebra. A ***derivation*** of $A$ is a homomorphism $d: A \to A$ of $k$-modules for which

$$d(ab) = (da)b + a(db).$$

In words, a derivation acts like ordinary differentiation in Calculus, for we are saying that the product rule, $(fg)' = f'g + fg'$, holds.

**Lemma 8.144.** *Let $k$ be a commutative ring, and let $M$ be a $k$-module.*

(i) *If $\varphi: M \to M$ is a $k$-map, then there exists a unique derivation $D_\varphi: T(M) \to T(M)$, where $T(M)$ is the tensor algebra on $M$, which is a graded map (of degree 0) with $D_\varphi | M = \varphi$; that is, for all $p \geq 0$,*

$$D_\varphi\left(\bigotimes^p M\right) \subseteq \bigotimes^p M.$$

(ii) *If $\varphi: M \to M$ is a $k$-map, then there exists a unique derivation $d_\varphi: \bigwedge M \to \bigwedge M$, which is a graded map (of degree 0) with $d_\varphi | M = \varphi$; that is, for all $p \geq 0$,*

$$d_\varphi\left(\bigwedge^p M\right) \subseteq \bigwedge^p M.$$

**Proof.**

(i) Define $D_\varphi | k = 1_k$ (recall that $\bigotimes^0 M = k$), and define $D_\varphi | \bigotimes^1 M = \varphi$ (recall that $\bigotimes^1 M = M$). If $p \geq 2$, define $D_\varphi^p: \bigotimes^p M \to \bigotimes^p M$ by

$$D_\varphi^p(m_1 \otimes \cdots \otimes m_p) = \sum_{i=1}^p m_1 \otimes \cdots \otimes \varphi(m_i) \otimes \cdots \otimes m_p.$$

For each $i$, the $i$th summand in the sum is well-defined, because it arises from the $k$-multilinear function $(m_1, \ldots, m_p) \mapsto m_1 \otimes \cdots \otimes \varphi(m_i) \otimes \cdots \otimes m_p$; it follows that $D_\varphi$ is well-defined.

It is clear that $D_\varphi$ is a map of $k$-modules. To check that $D_\varphi$ is a derivation, it suffices to consider its action on homogeneous elements $u = u_1 \otimes \cdots \otimes u_p$ and $v = v_1 \otimes \cdots \otimes v_q$:

$$
\begin{aligned}
D_\varphi(uv) &= D_\varphi(u_1 \otimes \cdots \otimes u_p \otimes v_1 \otimes \cdots \otimes v_q) \\
&= \sum_{i=1}^p u_1 \otimes \cdots \otimes \varphi(u_i) \otimes \cdots \otimes u_p \otimes v \\
&\quad + \sum_{j=1}^q u \otimes v_1 \otimes \cdots \otimes \varphi(v_j) \otimes \cdots \otimes v_q \\
&= D_\varphi(u)v + u D_\varphi(v).
\end{aligned}
$$

We leave the proof of uniqueness to the reader.

(ii) Define $d_\varphi: \bigwedge M \to \bigwedge M$ using the same formula as that for $D_\varphi$ after replacing $\otimes$ by $\wedge$. To see that this is well-defined, we must show that $D_\varphi(J) \subseteq J$, where $J$ is the two-sided ideal generated by all elements of the form $m \otimes m$. It suffices to prove, by induction on $p \geq 2$, that $D_\varphi(J^p) \subseteq J$, where $J^p = J \cap \bigotimes^p M$. The base step $p = 2$ follows from the identity, for $a, b \in M$,

$$a \otimes b + b \otimes a = (a+b) \otimes (a+b) - a \otimes a - b \otimes b \in J.$$

The inductive step follows from the identity, for $a, c \in M$ and $b \in J^{p-1}$,

$$a \otimes b \otimes c + J = -a \otimes c \otimes b + J = c \otimes a \otimes b + J = -c \otimes b \otimes a + J \quad \bullet$$

**Proposition 8.145.** *Let $k$ be a commutative ring, and let $M$ be a finitely generated free $k$-module with basis $e_1, \ldots, e_n$. If $\varphi \colon M \to M$ is a $k$-map and $d_\varphi \colon \bigwedge M \to \bigwedge M$ is the derivation it determines, then*

$$d_\varphi \big| \bigwedge^n M = \mathrm{tr}(\varphi) e_N,$$

*where $e_N = e_1 \wedge \cdots \wedge e_n$.*

**Proof.** By Lemma 8.144(ii), we have $d_\varphi \colon \bigwedge^n M \to \bigwedge^n M$. Since $M$ is a free $k$-module of rank $n$, the Binomial Theorem gives $\bigwedge^n M \cong k$. Hence, $d_\varphi(e_N) = c e_N$ for some $c \in k$; we show that $c = \mathrm{tr}(\varphi)$. Now $\varphi(e_i) = \sum a_{ji} e_j$, and

$$d_\varphi(e_N) = \sum_r e_1 \wedge \cdots \wedge \varphi(e_r) \wedge \cdots \wedge e_n$$

$$= \sum_r e_1 \wedge \cdots \wedge \sum a_{jr} e_j \wedge \cdots \wedge e_n$$

$$= \sum_r e_1 \wedge \cdots \wedge a_{rr} e_r \wedge \cdots \wedge e_n$$

$$= \sum_r a_{rr} e_N$$

$$= \mathrm{tr}(\varphi) e_N. \quad \bullet$$

# Exercises

**8.73.** Let $k$ be a commutative ring, and let $V$ and $W$ be free $k$-modules of ranks $m$ and $n$, respectively.

(i) Prove that if $f \colon V \to V$ is a $k$-map, then

$$\det(f \otimes 1_W) = [\det(f)]^n.$$

(ii) Prove that if $f \colon V \to V$ and $g \colon W \to W$ are $k$-maps, then

$$\det(f \otimes g) = [\det(f)]^n [\det(g)]^m.$$

$*$ **8.74.** (i) Let $z_1, \ldots, z_n$ be elements in a commutative ring $k$, and consider the **Vandermonde matrix**

$$V(z_1, \ldots, z_n) = \begin{bmatrix} 1 & 1 & \cdots & 1 \\ z_1 & z_2 & \cdots & z_n \\ z_1^2 & z_2^2 & \cdots & z_n^2 \\ \vdots & \vdots & \vdots & \vdots \\ z_1^{n-1} & z_2^{n-1} & \cdots & z_n^{n-1} \end{bmatrix}.$$

Prove that $\det(V(z_1, \ldots, z_n)) = \prod_{i < j}(z_j - z_i)$.

(ii) If $f(x) = \prod_i (x - z_i)$ has discriminant $D$, prove that $D = \det(V(z_1, \ldots, z_n))$.

(iii) Prove that if $z_1, \ldots, z_n$ are distinct elements of a field $k$, then $V(z_1, \ldots, z_n)$ is nonsingular.

**8.75.** Define a **tridiagonal matrix** to be an $n \times n$ matrix of the form

$$T[x_1, \ldots, x_n] = \begin{bmatrix} x_1 & 1 & 0 & 0 & \cdots & 0 & 0 & 0 & 0 \\ -1 & x_2 & 1 & 0 & \cdots & 0 & 0 & 0 & 0 \\ 0 & -1 & x_3 & 1 & \cdots & 0 & 0 & 0 & 0 \\ 0 & 0 & -1 & x_4 & \cdots & 0 & 0 & 0 & 0 \\ & & \vdots & & \ddots & & \vdots & & \\ 0 & 0 & 0 & 0 & \cdots & x_{n-3} & 1 & 0 & 0 \\ 0 & 0 & 0 & 0 & \cdots & -1 & x_{n-2} & 1 & 0 \\ 0 & 0 & 0 & 0 & \cdots & 0 & -1 & x_{n-1} & 1 \\ 0 & 0 & 0 & 0 & \cdots & 0 & 0 & -1 & x_n \end{bmatrix}.$$

(i) If $D_n = \det(T[x_1, \ldots, x_n])$, prove that $D_1 = x_1$, $D_2 = x_1 x_2 + 1$, and, for all $n > 2$,

$$D_n = x_n D_{n-1} + D_{n-2}.$$

(ii) Prove that if all $x_i = 1$, then $D_n = F_{n+1}$, the $n$th Fibonacci number. (Recall that $F_0 = 0, F_1 = 1$, and $F_n = F_{n-1} + F_{n-2}$ for all $n \geq 2$.)

**8.76.** If a matrix $A$ is a direct sum of square blocks,

$$A = B_1 \oplus \cdots \oplus B_t,$$

prove that $\det(A) = \prod_i \det(B_i)$.

**8.77.** If $A$ and $B$ are $n \times n$ matrices with entries in a commutative ring $k$, prove that $AB$ and $BA$ have the same characteristic polynomial.

**Hint. (Goodwillie)**

$$\begin{bmatrix} I & B \\ 0 & I \end{bmatrix} \begin{bmatrix} 0 & 0 \\ A & AB \end{bmatrix} \begin{bmatrix} I & -B \\ 0 & I \end{bmatrix} = \begin{bmatrix} BA & 0 \\ A & 0 \end{bmatrix}.$$

## Section 8.8. Lie Algebras

There are interesting examples of nonassociative algebras, the most important of which are Lie algebras. In the late nineteenth century, Lie (pronounced LEE) studied the solution space $S$ of a system of partial differential equations using a group $G$ of transformations of $S$. The underlying set of $G$ is a differentiable manifold and its group operation is a $C^\infty$-function; such groups are called **Lie groups**. The solution space is intimately related to its Lie group $G$; in turn, $G$ is studied using its *Lie algebra*, a considerably simpler object, which arises as the tangent space at the identity element of $G$. Moreover, the classification of the simple finite-dimensional complex Lie algebras, due to Killing and Cartan at the turn of the twentieth century, served as a model for the recent classification of all finite simple groups. Chevalley recognized that one could construct analogous families of finite simple groups by imitating the construction of simple Lie algebras, and Chevalley's groups are called **groups of Lie type**. Aside from this fundamental reason for their study, Lie algebras turn out to be the appropriate way to deal with families of linear transformations on a vector space (in contrast to the study of canonical forms of a single linear transformation given in the first sections of this chapter).

Before defining Lie algebras, we first generalize the notion of derivation.

**Definition.** A ***not-necessarily-associative*** $k$-***algebra*** $A$ over a commutative ring $k$ is a $k$-module equipped with a binary operation $A \times A \to A$, denoted by $(a, b) \mapsto ab$, such that

  (i)   $a(b + c) = ab + ac$ and $(b + c)a = ba + ca$ for all $a, b, c \in A$;

  (ii)   $ua = au$ for all $u \in k$ and $a \in A$;

  (iii)   $a(ub) = (au)b = u(ab)$ for all $u \in k$ and $a, b \in A$.

A ***derivation*** of $A$ is a $k$-map $d\colon A \to A$ such that, for all $a, b \in A$,

$$d(ab) = (da)b + a(db).$$

Aside from ordinary differentiation in Calculus, which is a derivation because the product rule $(fg)' = f'g + fg'$ holds, another example is provided by the $\mathbb{R}$-algebra $A$ of all real-valued functions $f(x_1, \dots, x_n)$ of several variables: the partial derivatives $\partial/\partial x_i$ are derivations, for $i = 1, \dots, n$.

The composite of two derivations need not be a derivation. For example, if $d\colon A \to A$ is a derivation, then $d^2 = d \circ d\colon A \to A$ satisfies the equation

$$d^2(fg) = d^2(f)g + 2d(f)d(g) + fd^2(g);$$

thus, the mixed term $2d(f)d(g)$ is the obstruction to $d^2$ being a derivation. The Leibniz formula for ordinary differentiation on the ring of all $C^\infty$-functions generalizes to derivations on any not-necessarily-associative algebra $A$. If $f, g \in A$, then

$$d^n(fg) = \sum_{i=0}^{n} \binom{n}{i} d^i f \cdot d^{n-i} g.$$

It is worthwhile to compute the composite of two derivations $d_1$ and $d_2$. If $A$ is a not-necessarily-associative algebra and $f, g \in A$, then

$$\begin{aligned} d_1 d_2(fg) &= d_1 \left[ (d_2 f)g + f(d_2 g) \right] \\ &= (d_1 d_2 f)g + (d_2 f)(d_1 g) + (d_1 f)(d_2 g) + f(d_1 d_2 g). \end{aligned}$$

Of course,

$$d_2 d_1(fg) = (d_2 d_1 f)g + (d_1 f)(d_2 g) + (d_2 f)(d_1 g) + f(d_2 d_1 g).$$

If we denote $d_1 d_2 - d_2 d_1$ by $[d_1, d_2]$, then

$$[d_1, d_2](fg) = ([d_1, d_2]f)g + f([d_1, d_2]g);$$

that is, $[d_1, d_2] = d_1 d_2 - d_2 d_1$ is a derivation.

**Example 8.146.** If $k$ is a commutative ring, equip $\mathrm{Mat}_n(k)$ with the ***bracket operation***:

$$[A, B] = AB - BA.$$

Of course, $A$ and $B$ commute if and only if $[A, B] = 0$. It is easy to find examples showing that the bracket operation is not associative. However, for any fixed $n \times n$ matrix $M$, the function

$$\text{ad}_M \colon \text{Mat}_n(k) \to \text{Mat}_n(k),$$

defined by

$$\text{ad}_M \colon A \mapsto [M, A],$$

is a derivation:

$$[M, [A, B]] = [[M, A], B] + [A, [M, B]].$$

The verification of this identity should be done once in one's life. ◄

The definition of Lie algebra involves a vector space with a binary operation generalizing the bracket operation.

**Definition.** A *Lie algebra* is a vector space $L$ over a field $k$ equipped with a bilinear operation $L \times L \to L$, denoted by $(a, b) \mapsto [a, b]$ (and called *bracket*), such that

(i) $[a, a] = 0$ for all $a \in L$;

(ii) for each $a \in L$, the function $\text{ad}_a \colon b \mapsto [a, b]$ is a derivation.

For all $u, v \in L$, bilinearity gives

$$[u + v, u + v] = [u, u] + [u, v] + [v, u] + [v, v],$$

which, when coupled with the first axiom $[a, a] = 0$, gives

$$[u, v] = -[v, u];$$

that is, bracket is *anticommutative*. The second axiom is often written out in more detail. If $b, c \in L$, then their product in $L$ is $[b, c]$; that $\text{ad}_a$ is a derivation is to say that

$$[a, [b, c]] = [[a, b], c] + [b, [a, c]];$$

rewriting,

$$[a, [b, c]] - [b, [a, c]] - [[a, b], c] = 0.$$

The anticommutativity from the first axiom now gives the *Jacobi identity*:

$$[a, [b, c]] + [b, [c, a]] + [c, [a, b]] = 0 \quad \text{for all } a, b, c \in L$$

(just cyclically permute $a, b, c$). Thus, a vector space $L$ is a Lie algebra if and only if $[a, a] = 0$ for all $a \in L$ and the Jacobi identity holds.

Here are some examples of Lie algebras.

**Example 8.147.**

(i) If $V$ is a vector space over a field $k$, define $[a, b] = 0$ for all $a, b \in V$. It is obvious that $V$ so equipped is a Lie algebra; it is called an *abelian* Lie algebra.

(ii) In $\mathbb{R}^3$, define $[u, v] = u \times v$, the vector product (or cross product) defined in Calculus. It is routine to check that $v \times v = 0$ and that the Jacobi identity holds, so that $\mathbb{R}^3$ is a Lie algebra. This example may be generalized: for every field $k$, cross product can be defined on the vector space $k^3$ making it a Lie algebra.

(iii) A **subalgebra** $S$ of a Lie algebra $L$ is a subspace that is closed under bracket: if $a, b \in S$, then $[a, b] \in S$. It is easy to see that every subalgebra of a Lie algebra is itself a Lie algebra.

(iv) If $k$ is a field, then $\text{Mat}_n(k)$ is a Lie algebra with bracket defined by

$$[A, B] = AB - BA.$$

We usually denote this Lie algebra by $\mathfrak{gl}(n, k)$. This example is quite general, for it is a theorem of Ado that every finite-dimensional Lie algebra over a field $k$ of characteristic 0 is isomorphic to a subalgebra of $\mathfrak{gl}(n, k)$ for some $n$ (Jacobson, *Lie Algebras*, p. 202).

(v) An interesting subalgebra of $\mathfrak{gl}(n, k)$ is $\mathfrak{sl}(n, k)$, which consists of all $n \times n$ matrices of trace 0. In fact, if $G$ is a Lie group whose associated Lie algebra is $\mathfrak{g}$, then there is an analog of exponentiation $\mathfrak{g} \to G$. In particular, if $\mathfrak{g} = \mathfrak{gl}(n, \mathbb{C})$, then this map is exponentiation $A \mapsto e^A$. Thus, Proposition 8.54(viii) shows that exponentiation sends $\mathfrak{sl}(n, \mathbb{C})$ into $\text{SL}(n, \mathbb{C})$.

(vi) If $A$ is any algebra over a field $k$, then

$$\mathfrak{Der}(A/k) = \{\text{all derivations } d \colon A \to A\},$$

with bracket $[d_1, d_2] = d_1 d_2 - d_2 d_1$, is a Lie algebra.

It follows from the Leibniz rule that if $k$ has characteristic $p > 0$, then $d^p$ is a derivation for every $d \in \mathfrak{Der}(A/k)$, since $\binom{p}{i} \equiv 0 \bmod p$ whenever $0 < i < p$. (This is an example of what is called a **restricted Lie algebra** of characteristic $p$.)

There is a Galois Theory for certain purely inseparable extensions (Jacobson, *Basic Algebra* II, pp. 533–536). If $k$ is a field of characteristic $p > 0$ and $E/k$ is a finite purely inseparable extension of *height* 1, that is, $\alpha^p \in k$ for all $\alpha \in E$, then there is a bijection between the family of all intermediate fields and the restricted Lie subalgebras of $\mathfrak{Der}(E/k)$, given by

$$B \mapsto \mathfrak{Der}(E/B);$$

the inverse of this function is given by

$$\mathfrak{L} \mapsto \{e \in E : D(e) = 0 \text{ for all } D \in \mathfrak{L}\}. \quad \blacktriangleleft$$

Not surprisingly, all Lie algebras over a field $k$ form a category.

**Definition.** Let $L$ and $L'$ be Lie algebras over a field $k$. Then a function $f \colon L \to L'$ is a **Lie homomorphism** if $f$ is a $k$-linear transformation that preserves bracket: for all $a, b \in L$,

$$f([a, b]) = [fa, fb].$$

**Definition.** An **ideal** of a Lie algebra $L$ is a subspace $I$ such that $[x, a] \in I$ for every $x \in L$ and $a \in I$.

Even though a Lie algebra need not be commutative, its anticommutativity shows that every left ideal (as just defined) is necessarily a right ideal; that is, every ideal is two-sided.

A Lie algebra $L$ is called *simple* if $L \neq \{0\}$ and $L$ has no nonzero proper ideals.

**Definition.** If $I$ is an ideal in $L$, then the *quotient* $L/I$ is the quotient space (considering $L$ as a vector space and $I$ as a subspace) with bracket defined by

$$[a + I, b + I] = [a, b] + I.$$

It is easy to check that this bracket on $L/I$ is well-defined. If $a' + I = a + I$ and $b' + I = b + I$, then $a - a' \in I$ and $b - b' \in I$, and so

$$[a', b'] - [a, b] = [a', b'] - [a', b] + [a', b] - [a, b]$$
$$= [a', b' - b] + [a' - a, b'] \in I.$$

**Example 8.148.**

(i) If $f \colon L \to L'$ is a Lie homomorphism, then its *kernel* is defined as usual:

$$\ker f = \{a \in L : f(a) = 0\}.$$

It is easy to see that $\ker f$ is an ideal in $L$.

Conversely, the *natural map* $\nu \colon L \to L/I$, defined by $a \mapsto a + I$, is a Lie homomorphism whose kernel is $I$. Thus, a subspace of $L$ is an ideal if and only if it is the kernel of some Lie homomorphism.

(ii) If $I$ and $J$ are ideals in a Lie algebra $L$, then

$$IJ = \Big\{ \sum_r [i_r, j_r] : i_r \in I \text{ and } j_r \in J \Big\}.$$

In particular, $L^2 = LL$ is the analog for Lie algebras of the commutator subgroup of a group: $L^2 = \{0\}$ if and only if $L$ is abelian.

(iii) There is an analog for Lie algebras of the derived series of a group. The *derived series* of a Lie algebra $L$ is defined inductively:

$$L^{(0)} = L; \quad L^{(n+1)} = (L^{(n)})^2.$$

A Lie algebra $L$ is called *solvable* if there is some $n \geq 0$ with $L^{(n)} = \{0\}$.

(iv) There is an analog for Lie algebras of the descending central series of a group. The *descending central series* is defined inductively:

$$L_1 = L; \quad L_{n+1} = LL_n.$$

A Lie algebra $L$ is called *nilpotent* if there is some $n \geq 0$ with $L_n = \{0\}$. ◀

We merely mention the first two theorems in the subject. If $L$ is a Lie algebra and $a \in L$, then $\mathrm{ad}_a \colon L \to L$, given by $\mathrm{ad}_a \colon x \mapsto [a, x]$, is a linear transformation on $L$ (viewed merely as a vector space). We say that $a$ is *ad-nilpotent* if $\mathrm{ad}_a$ is a nilpotent operator; that is, $(\mathrm{ad}_a)^m = 0$ for some $m \geq 1$.

**Theorem 8.149 (Engel's Theorem).**

(i) *Let $L$ be a finite-dimensional Lie algebra over any field $k$. Then $L$ is nilpotent if and only if every $a \in L$ is ad-nilpotent.*

(ii) *Let $L$ be a Lie subalgebra of $\mathfrak{gl}(n,k)$ all of whose elements $A$ are nilpotent matrices. Then $L$ can be put into strict upper triangular form (all diagonal entries are 0); that is, there is a nonsingular matrix $P$ so that $PAP^{-1}$ is strictly upper triangular for every $A \in L$.*

**Proof.** See Humphreys, *Introduction to Lie Algebras and Representation Theory*, p. 12. •

Compare Engel's Theorem with Exercise 8.42 on page 671, which is the much simpler version for a single nilpotent matrix. Nilpotent Lie algebras are so called because of Engel's Theorem; it is likely that nilpotent groups are so called by analogy. Corollary 4.51, which states that every finite $p$-group can be imbedded as a subgroup of unitriangular matrices over $\mathbb{F}_p$, may be viewed as a group-theoretic analog of Engel's Theorem.

**Theorem 8.150 (Lie's Theorem).** *Every solvable subalgebra $L$ of $\mathfrak{gl}(n,k)$, where $k$ is an algebraically closed field, can be put into (not necessarily strict) upper triangular form; that is, there is a nonsingular matrix $P$ so that $PAP^{-1}$ is upper triangular for every $A \in L$.*

**Proof.** Ibid., p. 16. •

Further study of Lie algebras leads to the classification of all finite-dimensional simple Lie algebras over an algebraically closed field of characteristic 0, due to Cartan and Killing. To each such algebra, they associated a certain geometric configuration called a *root system*, which is characterized by a *Cartan matrix*. Cartan matrices are, in turn, characterized by the **Dynkin diagrams** on the next page. Thus, Dynkin diagrams arise from simple Lie algebras over $\mathbb{C}$, and two such algebras are isomorphic if and only if they have the same Dynkin diagram. We refer the reader to Humphreys, *Introduction to Lie Algebras and Representation Theory*, Chapter IV, and Jacobson, *Lie Algebras*, Chapter IV. Recently, Block, Premet, Strade, and Wilson classified all finite-dimensional simple Lie algebras over algebraically closed fields of characteristic $p \geq 5$: there are two other types aside from analogs given by Dynkin diagrams. See Strade, *Simple Lie Algebras over Fields of Positive Characteristic* I.

Dynkin diagrams have also appeared in work of Gabriel in classifying finite-dimensional algebras over fields: Gabriel, Manuscripta Math. 6 (1972), pp. 71–103; correction, ibid. 6 (1972), p. 309.

There are other not-necessarily-associative algebras of interest. **Jordan algebras** are algebras $A$ which are commutative (instead of anti-commutative) and in which the Jacobi identity is replaced by

$$(x^2 y)x = x^2(yx)$$

for all $x, y \in A$. These algebras were introduced by P. Jordan to provide an algebraic setting for doing Quantum Mechanics. An example of a Jordan algebra is a subspace of all $n \times n$ matrices, over a field of characteristic not 2, equipped with the binary operation $A * B$, where

$$A * B = \tfrac{1}{2}(AB + BA).$$

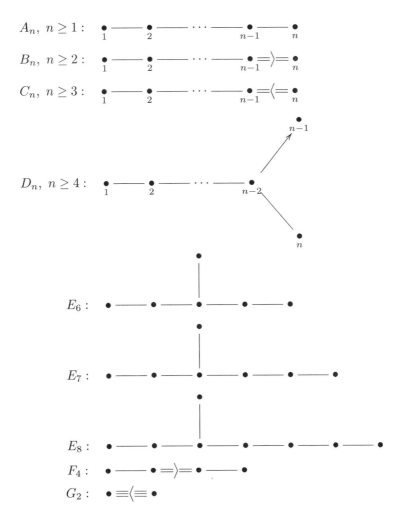

**Figure 8.1.** Dynkin diagrams.

Another source of not-necessarily-associative algebras comes from Combinatorics. The usual construction of a projective plane $P(k)$ over a field $k$, as the family of all lines in $k^3$ passing through the origin, leads to descriptions of its points by "homogeneous coordinates" $[x, y, z]$, where $x, y, z \in k$. Define an abstract ***projective plane*** to be an ordered pair $(X, \mathcal{L})$, where $X$ is a finite set and $\mathcal{L}$ is a family of subsets of $X$, called *lines*, subject to the following axioms:

(i) All lines have the same number of points.

(ii) Given any two points in $X$, there is a unique line containing them.

We want to introduce homogeneous coordinates to describe the points of such a projective plane, but there is no field $k$ given at the outset. Instead, we look at a collection $\mathcal{K}$ of functions on $X$, called *collineations*, and we equip $\mathcal{K}$ with two binary operations (called addition and multiplication). In general, $\mathcal{K}$ is a not-necessarily-associative algebra, but certain algebraic properties follow from the geometry of the plane. A theorem of Desargues holds if and only if multiplication is associative and $\mathcal{K}$ is a division ring; a theorem of Pappus also holds if and only if multiplication is commutative and $\mathcal{K}$ is a field (Reid–Szendrői, *Geometry and Topology*, p. 88). Thus, Wedderburn's Theorem 7.13 implies that if Desargues's Theorem holds in a finite projective plane, then so does the Pappus Theorem.

An interesting nonassociative algebra is the algebra of **Cayley numbers** (sometimes called *octonions*), which is an eight-dimensional real vector space containing the quaternions as a subalgebra (see the article by Curtis in Albert, *Studies in Modern Algebra*). Indeed, the Cayley numbers almost form a real division algebra ("almost," for even though every nonzero element has a multiplicative inverse, multiplication in the Cayley numbers is not associative). The algebra of Cayley numbers acquires added interest (as do other not-necessarily-associative algebras) because its automorphism group has interesting properties. For example, the exceptional simple Lie algebra $E_8$ is isomorphic to the Lie algebra of all the derivations of the Cayley numbers, while the Monster, the largest sporadic finite simple group, is the automorphism group of a certain nonassociative algebra constructed by Griess.

## Exercises

**8.78.** Consider the de Rham complex when $n = 2$:

$$0 \to \Omega^0(X) \xrightarrow{d^0} \Omega^1(X) \xrightarrow{d^1} \Omega^2(X) \to 0.$$

Prove that if $f(x, y) \in A(X) = \Omega^0(X)$, then

$$d^0 f = \frac{\partial f}{\partial x}\, dx + \frac{\partial f}{\partial y}\, dy,$$

and that if $P dx + Q dy$ is a 1-form, then

$$d^1(P dx + Q dy) = \left( \frac{\partial Q}{\partial x} - \frac{\partial P}{\partial y} \right) dx \wedge dy.$$

**8.79.** Prove that if $L$ and $L'$ are nonabelian two-dimensional Lie algebras, then $L \cong L'$.

**8.80.**  (i) Prove that the **center** of a Lie algebra $L$, defined by

$$Z(L) = \{a \in L : [a, x] = 0 \text{ for all } x \in L\},$$

is an abelian ideal in $L$.

  (ii) Give an example of a Lie algebra $L$ for which $Z(L) = \{0\}$.

  (iii) If $L$ is nilpotent and $L \neq \{0\}$, prove that $Z(L) \neq \{0\}$.

**8.81.** Prove that if $L$ is an $n$-dimensional Lie algebra, then $Z(L)$ cannot have dimension $n - 1$. (Compare Exercise 1.77 on page 59.)

**8.82.** Equip $\mathbb{C}^3$ with a cross product (using the same formula as the cross product on $\mathbb{R}^3$). Prove that

$$\mathbb{C}^3 \cong \mathfrak{sl}(2, \mathbb{C}).$$

# Chapter 9

# Homology

## Section 9.1. Simplicial Homology

When I was a graduate student, Homological Algebra was an unpopular subject. The general attitude was that it was a grotesque formalism, boring to learn, and not very useful once one had learned it. Perhaps an algebraic topologist was forced to know this stuff, but surely no one else should waste time on it. The few true believers were viewed as workers at the fringe of Mathematics who kept tinkering with their elaborate machine, smoothing out rough patches here and there.

This attitude changed dramatically when Serre characterized regular local rings using Homological Algebra (Theorem 10.198: they are the commutative noetherian local rings of "finite global dimension"), for this enabled him to prove that any localization of a regular local ring is itself regular (until then, only special cases of this were known). At the same time, Auslander and Buchsbaum completed work of Nagata by using global dimension to prove that every regular local ring is a UFD (Theorem 10.206). As more applications were found and as more homology and cohomology theories were invented to solve outstanding problems, resistance to Homological Algebra waned. Today, it is just another standard tool in a mathematician's kit.

Homology arose from Green's Theorem in Analysis. Let $X$ be an open subset of the plane, and fix points $a, b \in X$. Given a pair $P(x, y)$ and $Q(x, y)$ of real-valued continuously differentiable functions on $X$ and a path $\beta$ in $X$ from $a$ to $b$ [that is, if $\mathbf{I}$ denotes the closed unit interval $[0, 1]$ and $\beta \colon \mathbf{I} \to X$ is a continuous function with $\beta(0) = a$ and $\beta(1) = b$], is the line integral $\int_\beta P dx + Q dy$ independent of $\beta$? That is, if $\beta'$ is another path in $X$ from $a$ to $b$, do we have $\int_{\beta'} P\, dx + Q\, dy = \int_\beta P\, dx + Q\, dy$? Thus, if $\gamma = \beta - \beta'$ is the closed path at $a$, that is, $\gamma$ goes from $a$ to $b$ via $\beta$, and then goes back via $-\beta'$ from $b$ to $a$, where $-\beta' \colon t \mapsto \beta'(1 - t)$, then the integral is independent of the paths $\beta$ and $\beta'$ if and only if $\int_\gamma P\, dx + Q\, dy = 0$. Suppose there are "bad" points $z_1, \ldots, z_n$ deleted from $X$ (for example, if $P$ or $Q$ has a singularity at some $z_i$). The line integral along a *simple closed path* $\gamma$ (that is, $\gamma$ is a homeomorphism from the unit circle $S^1$ to im $\gamma \subseteq \mathbb{R}^2$) is affected by whether any of these bad points lies inside $\gamma$.[1] In Figure 9.1, each path $\gamma_i$ is a simple closed path in $X$ containing $z_i$ inside, while all the other $z_j$ are outside $\gamma_i$. If $\gamma$ is oriented counterclockwise and each $\gamma_i$ is oriented clockwise, then **Green's Theorem** states that

$$\int_\gamma P\,dx + Q\,dy + \int_{\gamma_1} P\,dx + Q\,dy + \cdots + \int_{\gamma_n} P\,dx + Q\,dy = \iint_R \left(\frac{\partial Q}{\partial x} - \frac{\partial P}{\partial y}\right) dx dy,$$

where $R$ is the shaded two-dimensional region in Figure 9.1. We are tempted to

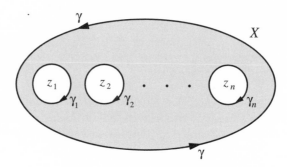

**Figure 9.1.** Green's Theorem.

write $\int_{\gamma_1} P\,dx + Q\,dy + \cdots + \int_{\gamma_n} P\,dx + Q\,dy$ more concisely, as $\int_{\sum_i \gamma_i} P\,dx + Q\,dy$. Instead of mentioning orientations explicitly, we may write sums and differences of paths, where a negative coefficient reverses direction; moreover, if we allow paths to wind around some $z_i$ several times, it makes sense to write formal $\mathbb{Z}$-linear combinations of paths; that is, we may allow integer coefficients other than $\pm 1$. Thus, Green's Theorem involves the free abelian group $C_1(X)$ with basis the (huge) set of all paths $\sigma\colon \mathbf{I} \to X$. Intuitively, elements of $C_1(X)$ are unions of finitely many (not necessarily closed) paths $\sigma$ in $X$.

Now consider those ordered pairs $(P, Q)$ of functions $X \to \mathbb{R}$ satisfying the Cauchy–Riemann equation: $\partial Q/\partial x = \partial P/\partial y$. The double integral in Green's Theorem vanishes, and so the left side $\int_{m\gamma + \sum_i m_i \gamma_i} P\,dx + Q\,dy = 0$. An equivalence relation on $C_1(X)$ suggests itself. Call $\beta, \beta' \in C_1(X)$ **equivalent** if, for all $(P, Q)$ with $\partial Q/\partial x = \partial P/\partial y$, the values of their line integrals agree:

$$\int_\beta P\,dx + Q\,dy = \int_{\beta'} P\,dx + Q\,dy.$$

The equivalence class of $\beta$ is called its **homology class**, from the Latin word *homologia* meaning *agreement*. Integration is independent of paths lying in the same homology class.

---

[1]The *Jordan Curve Theorem* says that if $\gamma$ is a simple closed path in the plane $\mathbb{R}^2$, then the complement $\mathbb{R}^2 - \text{im}\,\gamma$ has exactly two components: one, called its *inside*, is bounded (its closure is compact); the other, necessarily unbounded component, is called its *outside*.

Homology can be defined without integration. Poincaré recognized that whether a topological space $X$ has different types of holes is a kind of connectivity. To illustrate, suppose that $X$ can be *triangulated* into finitely many *n-simplexes* for $n \geq 0$, where 0-simplexes are points, say, $v_1, \ldots, v_q$, 1-simplexes are certain edges $[v_i, v_j]$ with endpoints $v_i$ and $v_j$, 2-simplexes are certain triangles $[v_i, v_j, v_k]$ (with vertices $v_i, v_j, v_k$), 3-simplexes are certain tetrahedra $[v_i, v_j, v_k, v_\ell]$, and there are $n$-dimensional analogs $[v_{i_0}, \ldots, v_{i_n}]$ for $n \geq 4$. The question to ask is whether a union of $n$-simplexes in $X$ that "ought" to be the boundary of some $(n+1)$-simplex actually is such a boundary. For example, when $n = 0$, two points $a$ and $b$ in $X$ ought to be the endpoints—the boundary—of a path in $X$; if there is no such path, then $X$ has a 0-dimensional hole. (A space $X$ is called **path connected** if, for each pair of points $a, b \in X$, there is a path in $X$ from $a$ to $b$.) For an example of a one-dimensional hole, let $X$ be the **punctured plane**; that is, $X$ is the plane with the origin deleted. The perimeter of a triangle $\Delta$ ought to be the boundary of a 2-simplex, but it is not if $\Delta$ contains the origin in its interior; thus, $X$ has a one-dimensional hole. If the interior of $X$ were missing a line segment containing the origin, or even a small disk containing the origin, this hole would still be one-dimensional. The size of the possible boundary is important, not the size of the hole. Keep your eye upon the doughnut and not upon the hole!

Here are precise definitions.

**Definition.** An **abstract simplicial complex** $K$ is a nonempty set $\text{Vert}(K)$, called **vertices**, together with a family of nonempty finite subsets $\sigma \subseteq \text{Vert}(K)$, called **simplexes**, such that

(i) $\{v\}$ is a simplex for every $v \in \text{Vert}(K)$;

(ii) every nonempty subset of a simplex is itself a simplex.

A simplex $\sigma$ with $|\sigma| = n + 1$ is called an *n***-simplex**. If $K$ and $L$ are abstract simplicial complexes, then a **simplicial map** is a function $\varphi \colon \text{Vert}(K) \to \text{Vert}(L)$ such that $\sigma$ a simplex in $K$ implies that $\varphi(\sigma)$ is a simplex in $L$. [We do not assume that $\varphi$ is injective; if $\sigma$ is an $n$-simplex, then $\varphi(\sigma)$ need not be an $n$-simplex. For example, a constant function $\text{Vert}(K) \to \text{Vert}(L)$ is a simplicial map.] All abstract simplicial complexes and simplicial maps form a category we denote by

$$\textbf{ASC}.$$

Every abstract simplicial complex $K$ has a *geometric realization* $|K|$ (see Munkres, *Elements of Algebraic Topology*, §1.3), and a *triangulated space* $X$ is a space homeomorphic to the geometric realization of some abstract simplicial complex $K$.

We are going to associate an algebraic construct to each abstract simplicial complex; here are its ingredients.

**Definition.** Let $K$ be an abstract simplicial complex. For each $n \geq 0$, define $C_n(K)$ to be the free abelian group with basis all $n$-simplexes $[v_{i_0}, \ldots, v_{i_n}]$. Define $C_{-1}(K) = \{0\}$.

**Definition.** Let $K$ be an abstract simplicial complex. If $n \geq 1$, define

$$\partial_n \colon C_n(K) \to C_{n-1}(K)$$

by

$$\partial_n[v_0, \dots, v_n] = \sum_{i=0}^{n} (-1)^i [v_0, \dots, \widehat{v}_i, \dots, v_n]$$

(as usual, $\widehat{v}_i$ means that $v_i$ is omitted). Define $\partial_0 \colon C_0(K) \to C_{-1}(K)$ to be identically zero [since $C_{-1}(K) = \{0\}$, this definition is forced on us]. As $C_n(X)$ is the free abelian group with basis the $n$-simplexes, the formula extends by linearity to a homomorphism $\partial_n$, called the $n$th **simplicial boundary map**.

The presence of signs gives the following fundamental result.

**Proposition 9.1.** *If $K$ is an abstract simplicial complex and $\partial_n \colon C_n(K) \to C_{n-1}(K)$ is its simplicial boundary map, then for all $n \geq 0$,*

$$\partial_{n-1} \partial_n = 0.$$

**Proof.** Each term of $\partial_n[x_0, \dots, x_n]$ has the form $(-1)^i[x_0, \dots, \widehat{x}_i, \dots, x_n]$. Hence, $\partial_n[x_0, \dots, x_n] = \sum_i (-1)^i [x_0, \dots, \widehat{x}_i, \dots, x_n]$, and

$$
\begin{aligned}
\partial_{n-1}[x_0, \dots, \widehat{x}_i, \dots, x_n] = &\; [\widehat{x}_0, \dots, \widehat{x}_i, \dots, x_n] + \cdots \\
&+ (-1)^{i-1}[x_0, \dots, \widehat{x}_{i-1}, \widehat{x}_i, \dots, x_n] \\
&+ (-1)^i [x_0, \dots, \widehat{x}_i, \widehat{x}_{i+1} \dots, x_n] + \cdots \\
&+ (-1)^{n-1}[x_0, \dots, \widehat{x}_i, \dots, \widehat{x}_n]
\end{aligned}
$$

(when $k \geq i+1$, the sign of $[x_0, \dots, \widehat{x}_i, \dots, \widehat{x}_k, \dots, x_n]$ is $(-1)^{k-1}$, for the vertex $x_k$ is the $(k-1)$st term in the list $x_0, \dots, \widehat{x}_i, \dots, x_k, \dots, x_n$). Thus,

$$
\begin{aligned}
\partial_{n-1}[x_0, \dots, \widehat{x}_i, \dots, x_n] = &\sum_{j=0}^{i-1} (-1)^j [x_0, \dots, \widehat{x}_j, \cdots, \widehat{x}_i, \dots, x_n] \\
&+ \sum_{k=i+1}^{n} (-1)^{k-1}[x_0, \dots, \widehat{x}_i, \dots, \widehat{x}_k, \dots, x_n].
\end{aligned}
$$

Now $[x_0, \dots, \widehat{x}_i, \dots, \widehat{x}_j, \dots, x_n]$ occurs twice in $\partial_{n-1}\partial_n[x_0, \dots, x_n]$: once from $\partial_{n-1}[x_0, \dots, \widehat{x}_i, \dots, x_n]$ and once from $\partial_{n-1}[x_0, \dots, \widehat{x}_j, \dots, x_n]$. The first occurrence has sign $(-1)^{i+j}$, while the second has sign $(-1)^{i+j-1}$. Thus, the $(n-2)$-tuples cancel in pairs, and $\partial_{n-1}\partial_n = 0$.  •

Recall that a *complex* in **Ab** is a sequence of abelian groups and homomorphisms

$$\mathbf{C} = \to C_2 \xrightarrow{d_2} C_1 \xrightarrow{d_1} C_0 \xrightarrow{d_0} C_{-1} \xrightarrow{d_{-1}} C_{-2} \to$$

in which all composites are zero: $d_{n-1}d_n = 0$ for all $n \in \mathbb{Z}$.

**Definition.** The **chain complex** of an abstract simplicial complex $K$ is the complex of abelian groups

$$\mathbf{C}(K) = \to C_2(K) \xrightarrow{\partial_2} C_1(K) \xrightarrow{\partial_1} C_0(K) \xrightarrow{\partial_0} 0.$$

Elements of the free abelian group $C_n(K)$ are called **simplicial $n$-chains**; they are $\mathbb{Z}$-linear combinations of $n$-simplexes. For each $n \in \mathbb{Z}$, the subgroup $\ker \partial_n \subseteq C_n(K)$ is denoted by $Z_n(K)$, and its elements are called **simplicial $n$-cycles**.

The subgroup $\operatorname{im}\partial_{n+1} \subseteq C_n$ is denoted by $B_n(K)$, and its elements are called *simplicial n-boundaries*.

**Corollary 9.2.** *Let* $\mathbf{C}(K)$ *be the chain complex of an abstract simplicial complex* $K$. *For all* $n \in \mathbb{Z}$,

$$B_n(K) \subseteq Z_n(K).$$

**Proof.** If $\alpha \in B_n$, then $\alpha = \partial_{n+1}(\beta)$ for some $(n+1)$-chain $\beta$. Hence, $\partial_n(\alpha) = \partial_n\partial_{n+1}(\beta) = 0$, so that $\alpha \in \ker\partial_n = Z_n$. •

Some simplicial $n$-chains, namely, simplicial $n$-cycles, ought to be boundaries of some sum (union) of simplicial $(n+1)$-simplexes. What survives in $Z_n(K)/B_n(K)$ are $n$-dimensional *holes*: those $n$-cycles that are not $n$-boundaries. For example, triangulate the plane $\mathbb{R}^2$ with a homeomorphism $\varphi\colon |K| \to \mathbb{R}^2$, where $K$ is an abstract simplicial complex having a 2-simplex $\Delta = [a, b, c]$ with the origin in the interior of $\varphi(\Delta)$. Removing $\Delta$ from $K$ gives an abstract simplicial complex which triangulates the punctured plane $X$, and $[b, c] - [a, c] + [a, b]$ is a 1-cycle which is not a 1-boundary.

The interesting groups are the quotients $Z_n(K)/B_n(K)$.

**Definition.** The $n$th *simplicial homology group* of an abstract simplicial complex $K$ is

$$H_n(K) = Z_n(K)/B_n(K) = \ker\partial_n/\operatorname{im}\partial_{n+1}.$$

If $z$ is a simplicial $n$-cycle, then the element $z + B_n(K) \in H_n(K)$ is called the *homology class* of $z$, and it is denoted by

$$\mathrm{cls}(z).$$

If $K$ is an abstract simplicial complex, then $H_n(K) = \{0\}$ means that $K$ has no $n$-dimensional holes; in particular, if $H_0(K) = \{0\}$, then the geometric realization of $K$ is path connected. On the other hand, $H_1(K) \neq \{0\}$ if $K$ triangulates the punctured plane.

Eventually, homology will be defined for mathematical objects $X$ other than abstract simplicial complexes. It is always a good idea to translate $H_n(X) = \{0\}$ into concrete terms, if possible, as some interesting property of $X$. One can then interpret the elements of $H_n(X)$ as being *obstructions* to whether $X$ enjoys this property.

We are now going to show that each $H_n$ is a functor $\mathbf{ASC} \to \mathbf{Ab}$. If $\varphi\colon K \to L$ is a simplicial map and $\sigma$ is a simplex in $K$, then $\varphi\sigma$ is a simplex in $L$. Define $\varphi_{\#}\colon C_n(K) \to C_n(L)$ by

$$\varphi_{\#}\colon \sum_{\sigma} m_\sigma \sigma \mapsto \sum_{\sigma} m_\sigma \varphi\sigma.$$

It is usually reckless to be careless with notation, but the next lemma shows that one can sometimes do so without causing harm (moreover, it is easier to follow an argument when notational clutter is absent). Strictly speaking, notation for $\varphi_{\#}$ should display $n$, $K$, and $L$, while boundary maps $\partial_n\colon C_n(K) \to C_{n-1}(K)$ obviously depend on $K$.

**Lemma 9.3.** *If $\varphi\colon K \to K'$ is a simplicial map, then there is a commutative diagram for each $n$ :*

$$
\begin{array}{ccc}
C_n(K) & \xrightarrow{\ \partial_n\ } & C_{n-1}(K) \\
{\scriptstyle \varphi_\#}\downarrow & & \downarrow{\scriptstyle \varphi_\#} \\
C_n(K') & \xrightarrow[\ \partial_n'\ ]{} & C_{n-1}(K')
\end{array}
$$

**Proof.** To see that $\partial_n'\varphi_\# = \varphi_\#\partial_n$ it suffices to evaluate each composite on a basis element $[v_0, \ldots, v_n]$ of $C_n(K)$. Going clockwise,

$$[v_0, \ldots, v_n] \mapsto \sum_i (-1)^i [v_0, \ldots, \widehat{v_i}, \ldots, v_n] \mapsto \sum_i (-1)^i [\varphi v_0, \ldots, \widehat{\varphi v_i}, \ldots, \varphi v_n];$$

going counterclockwise,

$$[v_0, \ldots, v_n] \mapsto [\varphi v_0, \ldots, \varphi v_n] \mapsto \sum_i (-1)^i [\varphi v_0, \ldots, \widehat{\varphi v_i}, \ldots, \varphi v_n]. \quad \bullet$$

**Theorem 9.4.** *Simplicial homology $H_n\colon \mathbf{ASC} \to \mathbf{Ab}$ is a functor for all $n \geq 0$.*

**Proof.** Having already defined $H_n$ on objects by $H_n(K) = Z_n(K)/B_n(K)$, we need only define it on morphisms. If $\varphi\colon K \to L$ is a simplicial map, define $H_n(\varphi)\colon H_n(K) \to H_n(L)$ by

$$H_n(\varphi)\colon \mathrm{cls}(z_n) \mapsto \mathrm{cls}(\varphi_\# z_n),$$

where $z_n$ is a simplicial $n$-cycle in $K$. In more detail, if $z_n = \sum m_\sigma \sigma$, then $H_n(\varphi)\colon \mathrm{cls}(z_n) \mapsto \mathrm{cls}(\sum m_\sigma \varphi\sigma)$. We claim that $\varphi_\# z_n$ is an $n$-cycle in $L$. Since $z_n$ is a cycle, Lemma 9.3 gives $\partial \varphi_\# z_n = \varphi_\# \partial z_n = 0$. Thus, $\varphi_\#(Z_n(K)) \subseteq Z_n(L)$. But we also have $\varphi_\#(B_n(K)) \subseteq B_n(L)$, for if $\partial\beta \in B_n(K)$, then Lemma 9.3 gives $\varphi_\# \partial\beta = \partial\varphi_\#\beta \in B_n(L)$. It follows that $H_n(\varphi)$ is a well-defined function. We let the reader prove that $H_n(\varphi)$ is a homomorphism, that $H_n(1_K) = 1_{H_n(K)}$, and that if $\psi\colon L \to L'$ is a simplicial map, then $H_n(\psi\varphi) = H_n(\psi)H_n(\varphi)$. $\quad \bullet$

Recall that if $\mathbf{C}$ and $\mathbf{C}'$ are complexes, then a *chain map* $f\colon \mathbf{C} \to \mathbf{C}'$ is a sequence of maps $f = (f_n)_{n\in\mathbb{Z}}$ making the following diagram commute:

$$
\begin{array}{ccccccc}
\longrightarrow & C_{n+1} & \xrightarrow{\ d_{n+1}\ } & C_n & \xrightarrow{\ d_n\ } & C_{n-1} & \longrightarrow \\
& {\scriptstyle f_{n+1}}\downarrow & & \downarrow{\scriptstyle f_n} & & \downarrow{\scriptstyle f_{n-1}} & \\
\longrightarrow & C_{n+1}' & \xrightarrow[\ d_{n+1}'\ ]{} & C_n' & \xrightarrow[\ d_n'\ ]{} & C_{n-1}' & \longrightarrow
\end{array}
$$

All complexes of abelian groups and chain maps form a category, **Comp**, and we proved, in Theorem 7.145, that **Comp** is an abelian category. Simplicial homology $H_n$ is really the composite

$$\mathbf{ASC} \to \mathbf{Comp} \to \mathbf{Ab},$$

where the first functor sends $K \mapsto \mathbf{C}(K)$, and the second (soon to be defined) sends $\mathbf{C}(K) \mapsto H_n(K)$. Now $K \mapsto \mathbf{C}(K)$ is geometrically inspired, with its origins in triangulating spaces, while $\mathbf{C}(K) \mapsto H_n(K)$ is completely algebraic. Homological Algebra is the study of this second, algebraic half.

## Section 9.2. Semidirect Products

We now investigate a basic problem in Group Theory whose solution leads to homology. A group $G$ having a normal subgroup $K$ can be "factored" into $K$ and $G/K$; the study of extensions involves the inverse question: how much of $G$ can be recovered from a normal subgroup $K$ and the quotient $Q = G/K$? For example, we know that $|G| = |K||Q|$ if $K$ and $Q$ are finite.

Exactness of a sequence of nonabelian groups,

$$\to G_{n+1} \xrightarrow{d_{n+1}} G_n \xrightarrow{d_n} G_{n-1} \to,$$

is defined just as it is for abelian groups: $\operatorname{im} d_{n+1} = \ker d_n$ for all $n$. Of course, each $\ker d_n$ is a normal subgroup of $G_n$.

**Definition.** If $K$ and $Q$ are groups, then an **extension** of $K$ by $Q$ is a short exact sequence

$$1 \to K \xrightarrow{i} G \xrightarrow{p} Q \to 1.$$

The notation $K$ is to remind us of kernel, and the notation $Q$ is to remind us of quotient. The elements of $K$ will be denoted by $a, b, c, \ldots$, and the elements of $Q$ will be denoted by $x, y, z, \ldots$.

There is an alternative usage of the term *extension*, which calls the (middle) group $G$ (not the short exact sequence) an *extension* if it contains a normal subgroup $K_1$ with $K_1 \cong K$ and $G/K_1 \cong Q$. As most people, we use the term in both senses.

**Example 9.5.**

(i) The direct product $K \times Q$ is an extension of $K$ by $Q$; it is also an extension of $Q$ by $K$.

(ii) Both $S_3$ and $\mathbb{I}_6$ are extensions of $\mathbb{I}_3$ by $\mathbb{I}_2$. On the other hand, $\mathbb{I}_6$ is an extension of $\mathbb{I}_2$ by $\mathbb{I}_3$, but $S_3$ is not, for $S_3$ contains no *normal* subgroup of order 2. ◄

We have just seen, for any given ordered pair of groups, that there always exists an extension of one by the other (their direct product), but there may be other extensions as well. The **extension problem** is to classify all possible extensions of a given pair of groups $K$ and $Q$.

Recall that we discussed the relation between the extension problem and the Jordan–Hölder Theorem in Chapter 4. If a group $G$ has a composition series

$$G = K_0 \supseteq K_1 \supseteq K_2 \supseteq \cdots \supseteq K_{n-1} \supseteq K_n = \{1\}$$

with simple factor groups $Q_i = K_{i-1}/K_i$ for all $i \geq 0$, then $G$ could be recaptured from $Q_{n-1}, \ldots, Q_1$ by solving the extension problem $n - 1$ times. Since all finite simple groups have been classified, we could survey all finite groups if we could solve the extension problem.

Let us begin by recalling the partition of a group into the cosets of a subgroup. We have already defined a *transversal* of a subgroup $K$ of a group $G$ as a subset $T$ of $G$ consisting of exactly one element from each coset[2] $Kt$ of $K$.

**Definition.** If $1 \to K \to G \xrightarrow{p} Q \to 1$ is an extension, then a ***lifting*** is a function $\ell \colon Q \to G$, not necessarily a homomorphism, with $p\ell = 1_Q$ and $\ell(1) = 1$.

Given a transversal, we can construct a lifting. For each $x \in Q$, surjectivity of $p$ provides $\ell x \in G$ with $p(\ell x) = x$; of course, we may choose $\ell 1 = 1$. Thus, the function $x \mapsto \ell x$ is a lifting. Conversely, given a lifting, we claim that $\operatorname{im} \ell$ is a transversal of $K$. If $Kg$ is a coset, then $p(g) \in Q$; say, $p(g) = x$. Then $p(g(\ell x)^{-1}) = 1$, so that $g(\ell x)^{-1} \in K$ and $Kg = K\ell x$. Thus, every coset has a representative in $\ell(Q)$. Finally, we must show that $\ell(Q)$ does not contain two elements in the same coset. If $K\ell x = K\ell y$, then there is $a \in K$ with $a\ell x = \ell y$. Apply $p$ to this equation; since $p(a) = 1$, we have $x = y$ and so $\ell x = \ell y$.

Of course, extensions are defined for arbitrary groups $K$, but we are going to restrict our attention to the special case when $K$ is abelian. If $G$ is an extension of $K$ by $Q$, it would be confusing to write $G$ multiplicatively and its subgroup $K$ additively. Hence, we shall use the following notational convention: even though $G$ may not be abelian, additive notation will be used for the operation in $G$. Corollary 9.8 gives the main reason for this decision.

**Proposition 9.6.** *Let* $0 \to K \xrightarrow{i} G \xrightarrow{p} Q \to 1$ *be an extension of an abelian group* $K$ *by a group* $Q$, *and let* $\ell \colon Q \to G$ *be a lifting.*

(i) *For every* $x \in Q$, *conjugation* $\theta_x \colon K \to K$, *defined by*

$$\theta_x \colon a \mapsto \ell x + a - \ell x,$$

*is independent of the choice of lifting* $\ell x$ *of* $x$. *[For convenience, we have assumed that* $i$ *is an inclusion; this allows us to write* $a$ *instead of* $i(a)$.*]*

(ii) *The function* $\theta \colon Q \to \operatorname{Aut}(K)$, *defined by* $x \mapsto \theta_x$, *is a homomorphism.*

**Proof.**

(i) Let us show that $\theta_x$ is independent of the choice of lifting $\ell x$ of $x$. If $\ell'$ is another lifting, there is $b \in K$ with $\ell' x = \ell x + b$, for $-\ell x + \ell' x \in \ker p = \operatorname{im} i = K$. Therefore,

$$\ell' x + a - \ell' x = \ell x + b + a - b - \ell x = \ell x + a - \ell x,$$

because $K$ is abelian.

(ii) Now $\theta_x(a) \in K$ because $K \lhd G$, so that each $\theta_x \colon K \to K$; also, $\theta_x$ is an automorphism of $K$, because conjugations are automorphisms.

It remains to prove that $\theta \colon Q \to \operatorname{Aut}(K)$ is a homomorphism. If $x, y \in Q$ and $a \in K$, then

$$\theta_x(\theta_y(a)) = \theta_x(\ell y + a - \ell y) = \ell x + \ell y + a - \ell y - \ell x,$$

---

[2]In this chapter, the subgroup $K$ will be a normal subgroup, in which case $tK = Kt$ for all $t \in G$. Thus, using right cosets or left cosets is only a matter of notational convenience.

while
$$\theta_{xy}(a) = \ell(xy) + a - \ell(xy).$$

The equality $\theta_x\theta_y = \theta_{xy}$ follows from (i), because both $\ell(xy)$ and $\ell x + \ell y$ are liftings of $xy$. •

Roughly speaking, the homomorphism $\theta$ tells "how" $K$ is normal in $G$, for isomorphic copies of a group can sit as normal subgroups of $G$ in different ways. For example, let $K$ be a cyclic group of order 3 and let $Q = \langle x \rangle$ be cyclic of order 2. If $G = K \times Q$, then $G$ is abelian and $K$ lies in the center of $G$. In this case, $\ell x + a - \ell x = a$ for all $a \in K$ and $\theta_x = 1_K$. On the other hand, if $G = S_3$, then $K = A_3$ which does not lie in the center; if $\ell x = (1\ 2)$, then $(1\ 2)(1\ 2\ 3)(1\ 2) = (1\ 3\ 2)$ and $\theta_x$ is not $1_K$.

The homomorphism $\theta$ makes a kernel $K$ into a left $\mathbb{Z}Q$-module.

**Proposition 9.7.** *Let $K$ and $Q$ be groups with $K$ abelian. Then a homomorphism $\theta\colon Q \to \mathrm{Aut}(K)$ makes $K$ into a left $\mathbb{Z}Q$-module if scalar multiplication is defined by*
$$xa = \theta_x(a)$$
*for all $a \in K$ and $x \in Q$. Conversely, if $K$ is a left $\mathbb{Z}Q$-module, then $x \mapsto \theta_x$ defines a homomorphism $\theta\colon Q \to \mathrm{Aut}(K)$, where $\theta_x\colon a \mapsto xa$.*

**Proof.** Define scalar multiplication as follows. Each $u \in \mathbb{Z}Q$ has a unique expression of the form $u = \sum_{x \in Q} m_x x$, where $m_x \in \mathbb{Z}$ and almost all $m_x = 0$; define
$$\Big(\sum_x m_x x\Big)a = \sum_x m_x\theta_x(a) = \sum_x m_x(xa).$$

We verify the module axioms. Since $\theta$ is a homomorphism, $\theta(1) = 1_K$, and so $1a = \theta_1(a)$ for all $a \in K$. That $\theta_x \in \mathrm{Aut}(K)$ implies $x(a + b) = xa + xb$, from which it follows that $u(a + b) = ua + ub$ for all $u \in \mathbb{Z}Q$. Similarly, we check easily that $(u + v)a = ua + va$ for $u, v \in \mathbb{Z}Q$. Finally, $(uv)a = u(va)$ will follow from $(xy)a = x(ya)$ for all $x, y \in Q$; but
$$(xy)a = \theta_{xy}(a) = \theta_x(\theta_y(a)) = \theta_x(ya) = x(ya).$$

The proof of the converse is also routine. •

**Corollary 9.8.** *Let*
$$0 \to K \xrightarrow{i} G \xrightarrow{p} Q \to 1$$
*be an extension of an abelian group $K$ by a group $Q$. Then $K$ is a left $\mathbb{Z}Q$-module if we define*
$$xa = \ell x + a - \ell x = \theta_x(a),$$
*where $\ell\colon Q \to G$ is a lifting, $x \in Q$, and $a \in K$; moreover, the scalar multiplication is independent of the choice of lifting $\ell$.*

**Proof.** Propositions 9.6 and 9.7. •

From now on, we will abbreviate the term "left $\mathbb{Z}Q$-module" to "$Q$-module."

Recall that a short exact sequence of left $R$-modules

$$0 \to A \xrightarrow{i} B \xrightarrow{p} C \to 0$$

is *split* if there exists a homomorphism $j \colon C \to B$ with $pj = 1_C$; in this case, the middle module is isomorphic to the direct sum $A \oplus C$. Here is the analogous definition for groups.

**Definition.** An extension of groups

$$0 \to K \xrightarrow{i} G \xrightarrow{p} Q \to 1$$

is ***split*** if there is a homomorphism $j \colon Q \to G$ with $pj = 1_Q$. The middle group $G$ of a split extension is called a ***semidirect product***[3] of $K$ by $Q$.

Thus, an extension is split if and only if there is a lifting, namely, $j$, that is also a homomorphism.

**Proposition 9.9.** *Let $G$ be an additive group having a normal subgroup $K$.*

(i) *If $0 \to K \xrightarrow{i} G \xrightarrow{p} Q \to 1$ is a split extension, where $j \colon Q \to G$ satisfies $pj = 1_Q$, then $i(K) \cap j(Q) = \{0\}$ and $i(K) + j(Q) = G$.*

(ii) *In this case, each $g \in G$ has a unique expression $g = i(a) + j(x)$, where $a \in K$ and $x \in Q$.*

(iii) *Let $K$ and $Q$ be subgroups of a group $G$ with $K \lhd G$. Then $G$ is a semidirect product of $K$ by $Q$ if and only if $K \cap Q = \{0\}$, $K + Q = G$, and each $g \in G$ has a unique expression $g = a + x$, where $a \in K$ and $x \in Q$.*

**Proof.**

(i) If $g \in i(K) \cap j(Q)$, then $g = i(a) = j(x)$ for $a \in K$ and $x \in Q$. Now $g = j(x)$ implies $p(g) = pj(x) = x$, while $g = i(a)$ implies $p(g) = pi(a) = 0$. Therefore, $x = 0$ and $g = j(x) = 0$.

   If $g \in G$, then $p(g) = pjp(g)$ (because $pj = 1_Q$), and so $g - (jp(g)) \in \ker p = \operatorname{im} i$; hence, there is $a \in K$ with $g - (jp(g)) = i(a)$, and so $g = i(a) + j(pg) \in i(K) + j(Q)$.

(ii) Every element $g \in G$ has a factorization $g = i(a) + j(pg)$ because $G = i(K) + j(Q)$. To prove uniqueness, suppose that $i(a) + j(x) = i(b) + j(y)$, where $b \in K$ and $y \in Q$. Then $-i(b) + i(a) = j(y) - j(x) \in i(K) \cap j(Q) = \{0\}$, so that $i(a) = i(b)$ and $j(x) = j(y)$.

(iii) Necessity is the special case of (ii) when both $i$ and $j$ are inclusions. Conversely, each $g \in G$ has a unique factorization $g = ax$ for $a \in K$ and $x \in Q$; define $p \colon G \to Q$ by $p(ax) = x$. It is easy to check that $p$ is a surjective homomorphism with $\ker p = K$. •

---

[3]This definition makes sense for any not necessarily abelian group $K$, and such semidirect products arise naturally. For example, the symmetric group $S_n$ is a semidirect product of the alternating group $A_n$ by $\mathbb{I}_2$. In order to keep hypotheses uniform, however, we assume in the text (except in some exercises) that $K$ is abelian, even though this assumption is not always needed.

A semidirect product is so called because a direct product $G$ of $K$ and $Q$ requires, in addition to $K + Q = G$ and $K \cap Q = \{1\}$, that both subgroups $K$ and $Q$ be normal (Proposition 1.86); here, only one subgroup must be normal.

**Definition.** If $K \subseteq G$ and $C \subseteq G$ satisfy $C \cap K = \{1\}$ and $K + C = G$, then $C$ is called a **complement** of $K$.

In a semidirect product $G$, the subgroup $K$ is normal; on the other hand, $j(Q)$, which Proposition 9.9 shows to be a complement of $K$, may not be normal. For example, if $G = S_3$ and $K = A_3 = \langle (1\ 2\ 3) \rangle$, we may take $C = \langle \tau \rangle$, where $\tau$ is any transposition in $S_3$; this example also shows that complements need not be unique. However, any two complements of $K$ are isomorphic, for each is isomorphic to $G/K$.

**Example 9.10.**

(i) A direct product $K \times Q$ is a semidirect product of $K$ by $Q$ (and of $Q$ by $K$).

(ii) An abelian group $G$ is a semidirect product if and only if it is a direct product (usually called a direct sum), for every subgroup of an abelian group is normal.

(iii) The dihedral group $D_{2n}$ is a semidirect product of $\mathbb{I}_n$ by $\mathbb{I}_2$. If $D_{2n} = \langle a, b \rangle$, where $a^n = 1$, $b^2 = 1$, and $bab = a^{-1}$, then $\langle a \rangle$ is a normal subgroup having $\langle b \rangle$ as a complement.

(iv) Every Frobenius group is a semidirect product of its Frobenius kernel by its Frobenius complement (Theorem 7.114).

(v) Let $G = \mathbb{H}^{\times}$, the multiplicative group of nonzero quaternions. It is easy to see that if $\mathbb{R}^+$ is the multiplicative group of positive reals, then the *norm* $N \colon G \to \mathbb{R}^+$, given by

$$N(a + bi + cj + dk) = a^2 + b^2 + c^2 + d^2,$$

is a homomorphism. There is a "polar decomposition" $h = rs$, where $r > 0$ and $s \in \ker N$, and $G$ is a semidirect product of $\ker N$ by $\mathbb{R}^+$. (The normal subgroup $\ker N$ is the 3-sphere.) In Exercise 9.4 on page 764, we will see that $\ker N \cong SU(2, \mathbb{C})$, the special unitary group.

(vi) Cyclic groups of prime power order are *not* semidirect products, for they cannot be a direct sum of two proper subgroups. ◄

**Definition.** Let $K$ be a $Q$-module. An extension $G$ of $K$ by $Q$ **realizes the operators** if, for all $x \in Q$ and $a \in K$, we have

$$xa = \ell x + a - \ell x;$$

that is, the given scalar multiplication of $\mathbb{Z}Q$ on $K$ coincides with the scalar multiplication of Corollary 9.8 arising from conjugation.

Here is the construction.

**Definition.** Let $Q$ be a group and let $K$ be a $Q$-module. Define

$$G = K \rtimes Q$$

to be the set of all ordered pairs $(a, x) \in K \times Q$ with binary operation[4]

$$(a, x) + (b, y) = (a + xb, xy).$$

Notice that $(a, 1) + (0, x) = (a, x)$ in $K \rtimes Q$.

**Proposition 9.11.** *Given a group $Q$ and a $Q$-module $K$, then $G = K \rtimes Q$ is a semidirect product of $K$ by $Q$ that realizes the operators.*

**Proof.** We begin by proving that $G$ is a group. For associativity,

$$[(a, x) + (b, y)] + (c, z) = (a + xb, xy) + (c, z) = (a + xb + (xy)c, (xy)z).$$

On the other hand,

$$(a, x) + [(b, y) + (c, z)] = (a, x) + (b + yc, yz) = (a + x(b + yc), x(yz)).$$

Of course, $(xy)z = x(yz)$, because of associativity in $Q$. The first coordinates are also equal: since $K$ is a $Q$-module, we have

$$x(b + yc) = xb + x(yc) = xb + (xy)c.$$

Thus, the operation is associative. The identity element of $G$ is $(0, 1)$, for

$$(0, 1) + (a, x) = (0 + 1a, 1x) = (a, x),$$

and the inverse of $(a, x)$ is $(-x^{-1}a, x^{-1})$, for

$$(-x^{-1}a, x^{-1}) + (a, x) = (-x^{-1}a + x^{-1}a, x^{-1}x) = (0, 1).$$

Therefore, $G$ is a group, by Exercise 1.31 on page 27.

Define a function $p \colon G \to Q$ by $p \colon (a, x) \mapsto x$. Since the only "twist" occurs in the first coordinate, $p$ is a surjective homomorphism with $\ker p = \{(a, 1) \colon a \in K\}$. If we define $i \colon K \to G$ by $i \colon a \mapsto (a, 1)$, then

$$0 \to K \xrightarrow{i} G \xrightarrow{p} Q \to 1$$

is an extension. Define $j \colon Q \to G$ by $j \colon x \mapsto (0, x)$. It is easy to see that $j$ is a homomorphism, for $(0, x) + (0, y) = (0, xy)$. Now $pjx = p(0, x) = x$, so that $pj = 1_Q$, and the extension splits; that is, $G$ is a semidirect product of $K$ by $Q$. Finally, $G$ realizes the operators: if $x \in Q$, then every lifting of $x$ has the form $\ell x = (b, x)$ for some $b \in K$, and

$$\begin{aligned}
(b, x) + (a, 1) - (b, x) &= (b + xa, x) + (-x^{-1}b, x^{-1}) \\
&= (b + xa + x(-x^{-1}b), xx^{-1}) \\
&= (b + xa - b, 1) \\
&= (xa, 1). \quad \bullet
\end{aligned}$$

**Theorem 9.12.** *Let $K$ be an abelian group. If a group $G$ is a semidirect product of $K$ by a group $Q$, then there is a $Q$-module structure on $K$ so that $G \cong K \rtimes Q$.*

---

[4]This binary operation is more natural in multiplicative notation:

$$(ax)(by) = a(xbx^{-1})xy.$$

**Proof.** Regard $G$ as a group with normal subgroup $K$ that has $Q$ as a complement. We continue writing $G$ additively (even though it may not be abelian), and so we write its subgroup $K$ additively as well. If $a \in K$ and $x \in Q$, define

$$xa = x + a - x;$$

that is, $xa$ is the conjugate of $a$ by $x$. By Proposition 9.9, each $g \in G$ has a unique expression as $g = a + x$, where $a \in K$ and $x \in Q$. It follows that $\varphi \colon G \to K \rtimes Q$, defined by $\varphi \colon a + x \mapsto (a, x)$, is a bijection. We show that $\varphi$ is an isomorphism:

$$\varphi((a + x) + (b + y)) = \varphi(a + x + b + (-x + x) + y)$$
$$= \varphi(a + (x + b - x) + x + y)$$
$$= (a + xb, x + y).$$

The definition of addition in $K \rtimes Q$ now gives

$$(a + xb, x + y) = (a, x) + (b, y) = \varphi(a + x) + \varphi(b + y). \quad \bullet$$

Let us use semidirect products to construct some groups.

**Example 9.13.** If $K = \langle a \rangle \cong \mathbb{I}_3$, then an automorphism of $K$ is completely determined by the image of the generator $a$; either $a \mapsto a$ and the automorphism is $1_K$, or $a \mapsto 2a$. Therefore, $\text{Aut}(K) \cong \mathbb{I}_2$; let us denote its generator by $\varphi$, so that $\varphi(a) = 2a$ and $\varphi(2a) = a$; that is, $\varphi$ multiplies by 2. Let $Q = \langle x \rangle \cong \mathbb{I}_4$, and define $\theta \colon Q \to \text{Aut}(K)$ by $\theta_x = \varphi$; hence

$$xa = 2a \quad \text{and} \quad x2a = a.$$

The group

$$T = \mathbb{I}_3 \rtimes \mathbb{I}_4$$

is a group of order 12. If we define $s = (2a, x^2)$ and $t = (0, x)$, then the reader may check that $6s = 0$ and $2t = 3s = 2(s + t)$.

The reader knows four other groups of order 12. The Fundamental Theorem of Finite Abelian Groups says that there are only two abelian groups of this order: $\mathbb{I}_{12} \cong \mathbb{I}_3 \times \mathbb{I}_4$ and $\mathbb{I}_2 \times \mathbb{I}_6 \cong \mathbf{V} \times \mathbb{I}_3$. Two nonabelian groups of order 12 are $A_4$ and $S_3 \times \mathbb{I}_2$ (Exercise 9.7 on page 765 asks the reader to prove that $A_4 \not\cong S_3 \times \mathbb{I}_2$). The group $T$ just constructed is a new example, and Exercise 9.17 on page 781 says that every group of order 12 is isomorphic to one of these five. [Note that Exercise 1.92 on page 74 states that $D_{12} \cong S_3 \times \mathbb{I}_2$.] ◀

**Example 9.14.** Let $p$ be a prime and let $K = \mathbb{I}_p \oplus \mathbb{I}_p$. Hence, $K$ is a vector space over $\mathbb{F}_p$, and so $\text{Aut}(K) \cong \text{GL}(K)$. We choose a basis $a, b$ of $K$, and this gives an isomorphism $\text{Aut}(K) \cong \text{GL}(2, p)$. Let $Q = \langle x \rangle$ be a cyclic group of order $p$.

Define $\theta \colon Q \to \text{GL}(2, p)$ by $\theta \colon x^n \mapsto \left[ \begin{smallmatrix} 1 & 0 \\ n & 1 \end{smallmatrix} \right]$ for all $n \in \mathbb{Z}$. Thus,

$$xa = a + b \quad \text{and} \quad xb = b.$$

It is easy to check that the commutator $x + a - x - a = xa - a = b$, and so $G = K \rtimes Q$ is a group of order $p^3$ with $G = \langle a, b, x \rangle$; these generators satisfy relations

$$pa = pb = px = 0, \quad b = [x, a], \quad \text{and} \quad [b, a] = 0 = [b, x].$$

If $p$ is odd, then we have the nonabelian group of order $p^3$ and exponent $p$ in Proposition 4.48. If $p = 2$, then $|G| = 8$, and the reader is asked to prove, in Exercise 9.8 on page 765, that $G \cong D_8$; that is, $D_8 \cong \mathbf{V} \rtimes \mathbb{I}_2$. In Example 9.10(iii), we saw that $D_8$ is a semidirect product of $\mathbb{I}_4$ by $\mathbb{I}_2$. Thus, $\mathbf{V} \rtimes \mathbb{I}_2 \cong \mathbb{I}_4 \rtimes \mathbb{I}_2$, and so a group can have different factorizations as a semidirect product.  ◀

**Example 9.15.** Let $k$ be a field and let $k^\times$ be its multiplicative group. Now $k^\times$ acts on $k$ by multiplication (if $a \in k$ and $a \neq 0$, then the additive homomorphism $x \mapsto ax$ is an automorphism whose inverse is $x \mapsto a^{-1}x$). Therefore, the semidirect product $k \rtimes k^\times$ is defined. In particular, if $(b, a), (d, c) \in k \rtimes k^\times$, then

$$(b, a) + (d, c) = (ad + b, ac).$$

Recall that an *affine map* is a function $f \colon k \to k$ of the form $f \colon x \mapsto ax + b$, where $a, b \in k$ and $a \neq 0$, and the collection of all affine maps under composition is the group $\mathrm{Aff}(1, k)$. Note that if $g(x) = cx + d$, then

$$(f \circ g)(x) = f(cx + d) = a(cx + d) + b = (ac)x + (ad + b).$$

It is now easy to see that the function $\varphi \colon (b, a) \mapsto f$, where $f(x) = ax + b$, is an isomomorphism $k \rtimes k^\times \to \mathrm{Aff}(1, k)$.  ◀

# Exercises

The group $K$ in the first three exercises need not be abelian, but it is assumed to be abelian in all the other exercises.

**9.1.** Kernels in this exercise may not be abelian groups.

(i) Prove that $\mathrm{SL}(2, \mathbb{F}_5)$ is an extension of $\mathbb{I}_2$ by $A_5$ that is not a semidirect product.

(ii) If $k$ is a field, prove that $\mathrm{GL}(n, k)$ is a semidirect product of $\mathrm{SL}(n, k)$ by $k^\times$.
  **Hint.** A complement consists of all matrices $\mathrm{diag}\{1, \ldots, 1, a\}$ with $a \in k^\times$.

∗ **9.2.** Let $G$ be a group of order $mn$, where $(m, n) = 1$. Prove that a normal subgroup $K$ of order $m$ has a complement in $G$ if and only if there exists a subgroup $C \subseteq G$ of order $n$. (Kernels in this exercise may not be abelian groups.)

∗ **9.3. (Baer)** Call a (not necessarily abelian) group $E$ *injective* if, whenever $S$ is a subgroup of some group $G$, then every homomorphism $f \colon S \to E$ can be extended to a homomorphism $G \to E$. Prove that a group $G$ is injective[5] if and only if $G = \{1\}$.

**Hint.** Let $A$ be free with basis $\{x, y\}$, and let $B$ be the semidirect product $B = A \rtimes \langle z \rangle$, where $z$ is an element of order 2 that acts on $A$ by $zxz = y$ and $zyz = x$.

∗ **9.4.** Let $SU(2, \mathbb{C})$ be the *special unitary group* consisting of all unitary matrices $A = \begin{bmatrix} a & b \\ c & d \end{bmatrix}$ of determinant 1. If $S$ is the subgroup of $\mathbb{H}^\times$ in Example 9.10(v), prove that $S \cong SU(2, \mathbb{C})$.

**Hint.** Use Exercise 6.6 on page 415.

---

[5]The term *injective* had not yet been coined when Baer (who introduced the notion of injective module) proved this result. After recognizing that injective groups are duals of free groups, he jokingly called such groups ***fascist***, and he was pleased to note that they are trivial.

**9.5.** Give an example of a split extension of groups $0 \to K \xrightarrow{i} G \xrightarrow{p} Q \to 1$ for which there does not exist a homomorphism $q \colon G \to K$ with $qi = 1_K$. Compare with Exercise 6.32.

**9.6.** Prove that the group $\mathbf{Q}$ of quaternions is not a semidirect product.

**Hint.** Recall that $\mathbf{Q}$ has a unique element of order 2.

∗ **9.7.** (i) Prove that $A_4 \not\cong S_3 \times \mathbb{I}_2$.

      **Hint.** Use Proposition 1.70 saying that $A_4$ has no subgroup of order 6.

  (ii) Prove that no two of the nonabelian groups of order 12: $A_4$, $S_3 \times \mathbb{I}_2$, and $T$ are isomorphic. (See Example 9.13.)

  (iii) The affine group $\mathrm{Aff}(1, \mathbb{F}_4)$ (Example 9.15) is a nonabelian group of order 12. Is it isomorphic to $A_4$, $S_3 \times \mathbb{I}_2$, or $T = \mathbb{I}_3 \rtimes \mathbb{I}_4$?

∗ **9.8.** Prove that the group $G$ of order 8 constructed in Example 9.14 is isomorphic to $D_8$.

**9.9.** If $K$ and $Q$ are solvable groups, prove that a semidirect product of $K$ by $Q$ is also solvable.

**9.10.** Let $K$ be an abelian group, let $Q$ be a group, and let $\theta \colon Q \to \mathrm{Aut}(K)$ be a homomorphism. Prove that $K \rtimes Q \cong K \times Q$ if and only if $\theta$ is the trivial map ($\theta_x = 1_K$ for all $x \in Q$).

∗ **9.11.** (i) If $K$ is cyclic of prime order $p$, prove that $\mathrm{Aut}(K)$ is cyclic of order $p - 1$.

  (ii) Let $G$ be a group of order $pq$, where $p > q$ are primes. If $q \nmid (p - 1)$, prove that $G$ is cyclic. Conclude, in particular, that every group of order 15 is cyclic.

∗ **9.12.** Let $G$ be an additive abelian $p$-group, where $p$ is prime.

  (i) If $(m, p) = 1$, prove that the function $a \mapsto ma$ is an automorphism of $G$.

  (ii) If $p$ is an odd prime and $G = \langle g \rangle$ is a cyclic group of order $p^2$, prove that $\varphi \colon G \to G$, given by $\varphi \colon a \mapsto 2g$, is the only automorphism of $G$ with $\varphi(pg) = 2pg$.

---

## Section 9.3. General Extensions and Cohomology

We now proceed to the study of the general extension problem: given a group $Q$ and an abelian group $K$, find all (not necessarily split) extensions $G$ of $K$ by $Q$. In light of our discussion of semidirect products, that is, of split extensions, it is reasonable to refine the problem by assuming that $K$ is a $Q$-module and then to seek all those extensions realizing the operators.

One way to describe a group $G$ is to give a multiplication table for it; that is, to list all its elements $a_1, a_2, \ldots$ and all products $a_i a_j$. Indeed, this is how we constructed semidirect products: the elements are all ordered pairs $(a, x)$ with $a \in K$ and $x \in Q$, and multiplication (really addition, because we have chosen to write $G$ additively) is

$$(a, x) + (b, y) = (a + xb, xy).$$

Schreier solved the extension problem in 1926 by finding all possible multiplication tables, and we present his solution in this section. The proof is not deep; rather, it involves manipulating and organizing a long series of elementary calculations.

We must point out that Schreier's solution does not allow us to determine the number of nonisomorphic middle groups $G$. A group $G$ of order $n$ has $n!$ multiplication tables, one for each ordering of its elements. Thus, if $G$ and $H$ are groups of order $n$, then the problem of determining whether $G \cong H$ is essentially that of comparing the two families of multiplication tables of each to see whether there is one for $G$ and one for $H$ which coincide. Even computers do not like so many comparisons.

**9.3.1. $H^2(Q, K)$ and Extensions.** Schreier's strategy is to extract enough properties of a given extension $G$ that suffice to reconstruct $G$. Thus, we may assume that $K$ is a $Q$-module, that $G$ is an extension of $K$ by $Q$ that realizes the operators, and that a transversal $\ell \colon Q \to G$ has been chosen. With these initial data, we see that each $g \in G$ has a unique expression of the form

$$g = a + \ell x, \text{ where } a \in K \text{ and } x \in Q;$$

this follows from $G$ being the disjoint union of the cosets $K + \ell x$. Furthermore, if $x, y \in Q$, then $\ell x + \ell y$ and $\ell(xy)$ are both representatives of the same coset (we do not say these representatives are the same!), and so there is an element $f(x, y) \in K$ such that

$$\ell x + \ell y = f(x, y) + \ell(xy).$$

**Definition.** Given a lifting $\ell \colon Q \to G$ of an extension $G$ of $K$ by $Q$,[6] a ***factor set***[7] (or *cocycle*) is a function $f \colon Q \times Q \to K$ such that

$$\ell x + \ell y = f(x, y) + \ell(xy)$$

for all $x, y \in Q$.

We remind the reader that our definition of *lifting* assumes $\ell 1 = 0$, and so we have incorporated this condition into the definition of factor set; our factor sets are often called ***normalized factor sets***.

Of course, a factor set depends on the choice of lifting $\ell$. When $G$ is a split extension, then there exists a lifting that is a homomorphism; the corresponding factor set is identically 0. Therefore, we can regard a factor set as the obstruction to a lifting being a homomorphism; that is, factor sets describe how an extension differs from being a split extension.

**Proposition 9.16.** *Let $Q$ be a group, $K$ a $Q$-module, and $0 \to K \to G \to Q \to 1$ an extension realizing the operators. If $\ell \colon Q \to G$ is a lifting and $f \colon Q \times Q \to K$ is the corresponding factor set, then*

(i) *for all $x, y \in Q$,*

$$f(1, y) = 0 = f(x, 1);$$

(ii) *the **cocycle identity** holds: for all $x, y, z \in Q$, we have*

$$f(x, y) + f(xy, z) = xf(y, z) + f(x, yz).$$

---

[6] By definition, liftings here satisfy $\ell 1 = 0$.

[7] If we switch to multiplicative notation, we see that a factor set occurs in the factorization $\ell x \ell y = f(x, y) \ell(xy)$. This is probably why factor sets are so called.

**Proof.** Set $x = 1$ in $\ell x + \ell y = f(x, y) + \ell(xy)$, the equation defining $f(x, y)$, to see that $\ell y = f(1, y) + \ell y$ [since $\ell 1 = 0$], and hence $f(1, y) = 0$. Setting $y = 1$ gives the other equation of (i).

The cocycle identity follows from associativity in $G$. For all $x, y, z \in Q$, we have

$$[\ell x + \ell y] + \ell z = f(x, y) + \ell(xy) + \ell z = f(x, y) + f(xy, z) + \ell(xyz).$$

On the other hand,

$$\begin{aligned} \ell x + [\ell y + \ell z] &= \ell x + f(y, z) + \ell(yz) \\ &= x f(y, z) + \ell x + \ell(yz) \\ &= x f(y, z) + f(x, yz) + \ell(xyz). \quad \bullet \end{aligned}$$

It is more interesting that the converse is true. The next result generalizes the construction of $K \rtimes Q$ in Proposition 9.11.

**Theorem 9.17.** *Given a group $Q$ and a $Q$-module $K$, a function $f \colon Q \times Q \to K$ is a factor set if and only if it satisfies the cocycle identity[8]*

$$x f(y, z) - f(xy, z) + f(x, yz) - f(x, y) = 0$$

*and $f(1, y) = 0 = f(x, 1)$ for all $x, y, z \in Q$.*

*More precisely, there is an extension $G$ of $K$ by $Q$ realizing the operators, namely*

$$G = G(K, Q, f),$$

*and there is a transversal $\ell \colon Q \to G$ whose corresponding factor set is $f$.*

**Proof.** Necessity is Proposition 9.16. For the converse, define $G = G(K, Q, f)$ to be the set of all ordered pairs $(a, x)$ in $K \times Q$ equipped with the operation

$$(a, x) + (b, y) = (a + xb + f(x, y), xy).$$

(Thus, if $f$ is identically 0, then $G = K \rtimes Q$.) The proof that $G$ is a group is similar to the proof of Proposition 9.11. The cocycle identity is used to prove associativity:

$$\begin{aligned} \big((a, x) + (b, y)\big) + (c, z) &= (a + xb + f(x, y), xy) + (c, z) \\ &= (a + xb + f(x, y) + xyc + f(xy, z), xyz) \end{aligned}$$

and

$$\begin{aligned} (a, x) + \big((b, y) + (c, z)\big) &= (a, x) + (b + yc + f(y, z), yz) \\ &= (a + xb + xyc + x f(y, z) + f(x, yz), xyz). \end{aligned}$$

The cocycle identity shows that these elements are equal.

We let the reader prove that the identity is $(0, 1)$ and the inverse of $(a, x)$ is

$$-(a, x) = (-x^{-1}a - x^{-1}f(x, x^{-1}), x^{-1}).$$

Define $p \colon G \to Q$ by $p \colon (a, x) \mapsto x$. Because the only "twist" occurs in the first

---

[8]The cocycle identity is reminiscent of the alternating sum formulas in Section 9.1 defining simplicial cycles.

coordinate, it is easy to see that $p$ is a surjective homomorphism with $\ker p = \{(a, 1) : a \in K\}$. If we define $i\colon K \to G$ by $i\colon a \mapsto (a, 1)$, then we have an extension $0 \to K \xrightarrow{i} G \xrightarrow{p} Q \to 1$.

To see that this extension realizes the operators, we must show, for every lifting $\ell$, that $xa = \ell x + a - \ell x$ for all $a \in K$ and $x \in Q$. Now $\ell x = (b, x)$ for some $b \in K$ and

$$\ell x + (a, 1) - \ell x = (b, x) + (a, 1) - (b, x)$$
$$= (b + xa, x) + (-x^{-1}b - x^{-1}f(x, x^{-1}), x^{-1})$$
$$= (b + xa + x[-x^{-1}b - x^{-1}f(x, x^{-1})] + f(x, x^{-1}), 1)$$
$$= (xa, 1).$$

Finally, we must show that $f$ is the factor set determined by $\ell$. Choose the lifting $\ell x = (0, x)$ for all $x \in Q$. The factor set $F$ determined by $\ell$ is defined by

$$F(x, y) = \ell x + \ell y - \ell(xy)$$
$$= (0, x) + (0, y) - (0, xy)$$
$$= (f(x, y), xy) + (-(xy)^{-1}f(xy, (xy)^{-1}), (xy)^{-1})$$
$$= (f(x, y) + xy[-(xy)^{-1}f(xy, (xy)^{-1})] + f(xy, (xy)^{-1}), xy(xy)^{-1})$$
$$= (f(x, y), 1). \quad \bullet$$

The next result shows that we have found all the extensions of a $Q$-module $K$ by a group $Q$.

**Theorem 9.18.** *Let $Q$ be a group, let $K$ be a $Q$-module, and let $G$ be an extension of $K$ by $Q$ realizing the operators. Then there exists a factor set $f\colon Q \times Q \to K$ with $G \cong G(K, Q, f)$.*

**Proof.** Let $\ell\colon Q \to G$ be a lifting, and let $f\colon Q \times Q \to K$ be the corresponding factor set: that is, for all $x, y \in Q$, we have

$$\ell x + \ell y = f(x, y) + \ell(xy).$$

Since $G$ is the disjoint union of the cosets, $G = \bigcup_{x \in Q} K + \ell x$, each $g \in G$ has a unique expression $g = a + \ell x$ for $a \in K$ and $x \in Q$. Uniqueness implies that the function $\varphi\colon G \to G(K, Q, f)$, given by

$$\varphi\colon g = a + \ell x \mapsto (a, x),$$

is a well-defined bijection. We now show that $\varphi$ is an isomorphism:

$$\varphi(a + \ell x + b + \ell y) = \varphi(a + \ell x + b - \ell x + \ell x + \ell y)$$
$$= \varphi(a + xb + \ell x + \ell y)$$
$$= \varphi(a + xb + f(x, y) + \ell(xy))$$
$$= (a + xb + f(x, y), xy)$$
$$= (a, x) + (b, y)$$
$$= \varphi(a + \ell x) + \varphi(b + \ell y). \quad \bullet$$

**Remark.** For later use, note that if $a \in K$, then $\varphi(a) = \varphi(a + \ell 1) = (a, 1)$, and if $x \in Q$, then $\varphi(\ell x) = (0, x)$. This would not be so had we not insisted that liftings $\ell$ satisfy $\ell 1 = 0$. ◄

We have now described all extensions in terms of factor sets. But factor sets are determined by liftings, and extensions have many different liftings; hence, our descriptions, which depend on choices of lifting, must have repetitions.

**Lemma 9.19.** *Given a group $Q$ and a $Q$-module $K$, let $G$ be an extension of $K$ by $Q$ realizing the operators, and let $\ell$, $\ell'$ be liftings giving rise to factor sets $f$, $f'$, respectively. Then there exists a function $h \colon Q \to K$ with $h(1) = 0$ and, for all $x, y \in Q$,*

$$f'(x, y) - f(x, y) = xh(y) - h(xy) + h(x).$$

**Proof.** For each $x \in Q$, both $\ell x$ and $\ell' x$ lie in the same coset of $K$ in $G$, and so there exists an element $h(x) \in K$ with

$$\ell'(x) = h(x) + \ell x.$$

Since $\ell 1 = 0 = \ell' 1$, we have $h(1) = 0$. The main formula is derived as follows:

$$\ell'(x) + \ell'(y) = [h(x) + \ell x] + [h(y) + \ell y] = h(x) + xh(y) + \ell x + \ell y,$$

because $G$ realizes the operators. The equations continue,

$$\begin{aligned}
\ell'(x) + \ell'(y) &= h(x) + xh(y) + f(x, y) + \ell(xy) \\
&= h(x) + xh(y) + f(x, y) - h(xy) + \ell'(xy).
\end{aligned}$$

By definition, $f'$ satisfies $\ell'(x) + \ell'(y) = f'(x, y) + \ell'(xy)$. Therefore,

$$f'(x, y) = h(x) + xh(y) + f(x, y) - h(xy),$$

and so

$$f'(x, y) - f(x, y) = xh(y) - h(xy) + h(x). \quad \bullet$$

**Definition.** Given a group $Q$ and a $Q$-module $K$, a function $g \colon Q \times Q \to K$ is called a **coboundary** if there exists a function $h \colon Q \to K$ with $h(1) = 0$ such that, for all $x, y \in Q$,

$$g(x, y) = xh(y) - h(xy) + h(x).$$

The term *coboundary* arises because its formula is analogous to the alternating sum formulas for simplicial boundaries in Section 9.1. We have just shown that if $f$ and $f'$ are factor sets of an extension $G$ that arise from different liftings, then $f' - f$ is a coboundary.

**Definition.** Given a group $Q$ and a $Q$-module $K$, define

$$Z^2(Q, K) = \{\text{all factor sets } f \colon Q \times Q \to K\}$$

and

$$B^2(Q, K) = \{\text{all coboundaries } g \colon Q \times Q \to K\}.$$

**Proposition 9.20.** *Given a group $Q$ and a $Q$-module $K$, then $Z^2(Q,K)$ is an abelian group with operation pointwise addition,*

$$f + f' : (x,y) \mapsto f(x,y) + f'(x,y),$$

*and $B^2(Q,K)$ is a subgroup of $Z^2(Q,K)$.*

**Proof.** To see that $Z^2$ is a group, it suffices to prove that $f + f'$, $0$, and $-f$ are factor sets. This is obvious, for they satisfy the two identities in Proposition 9.16.

To see that $B^2$ is a subgroup of $Z^2$, we must first show that every coboundary $g$ is a factor set; that is, $g$ satisfies the two identities in Proposition 9.16. This, too, is routine and is left to the reader. Next, we must show that $B^2$ is a nonempty subset; but the zero function, $g(x,y) = 0$ for all $x, y \in Q$, is clearly a coboundary. Finally, we show that $B^2$ is closed under subtraction. If $h, h' : Q \to K$ display $g$ and $g'$ as coboundaries, that is, $g(x,y) = xh(y) - h(xy) + h(x)$ and $g'(x,y) = xh'(y) - h'(xy) + h'(x)$, then

$$(g - g')(x,y) = x(h - h')(y) - (h - h')(xy) + (h - h')(x). \quad \bullet$$

A given extension has many liftings and, hence, many factor sets, but the difference of any two of these factor sets is a coboundary. Therefore, the following quotient group suggests itself.

**Definition.** Given a group $Q$ and a $Q$-module $K$, the ***second cohomology group*** is defined by

$$H^2(Q,K) = Z^2(Q,K)/B^2(Q,K).$$

**Definition.** Given a group $Q$ and a $Q$-module $K$, two extensions $G$ and $G'$ of $K$ by $Q$ that realize the operators are called ***equivalent*** if there is a factor set $f$ of $G$ and a factor set $f'$ of $G'$ so that $f' - f$ is a coboundary.

**Proposition 9.21.** *Given a group $Q$ and a $Q$-module $K$, two extensions $G$ and $G'$ of $K$ by $Q$ that realize the operators are equivalent if and only if there exists an isomorphism $\gamma : G \to G'$ making the following diagram commute:*

$$
\begin{array}{ccccccccc}
0 & \longrightarrow & K & \overset{i}{\longrightarrow} & G & \overset{p}{\longrightarrow} & Q & \longrightarrow & 1 \\
& & \downarrow{\scriptstyle 1_K} & & \vert{\scriptstyle \gamma} & & \downarrow{\scriptstyle 1_Q} & & \\
0 & \longrightarrow & K & \underset{i'}{\longrightarrow} & G' & \underset{p'}{\longrightarrow} & Q & \longrightarrow & 1
\end{array}
$$

**Remark.** A diagram chase shows that any homomorphism $\gamma$ making the diagram commute is necessarily an isomorphism.  ◄

**Proof.** Assume that the two extensions are equivalent. We begin by setting up notation. Let $\ell : Q \to G$ and $\ell' : Q \to G'$ be liftings, and let $f, f'$ be the corresponding factor sets; that is, for all $x, y \in Q$, we have

$$\ell x + \ell y = f(x,y) + \ell(xy),$$

with a similar equation for $f'$ and $\ell'$. Equivalence means that there is a function $h\colon Q \to K$ with $h(1) = 0$ and

$$f(x,y) - f'(x,y) = xh(y) - h(xy) + h(x)$$

for all $x, y \in Q$. Since $G = \bigcup_{x \in Q} K + \ell x$ is a disjoint union, each $g \in G$ has a unique expression $g = a + \ell x$ for $a \in K$ and $x \in Q$; similarly, each $g' \in G'$ has a unique expression $g' = a + \ell'(x)$.

This part of the proof generalizes that of Theorem 9.18. Define $\gamma\colon G \to G'$ by

$$\gamma(a + \ell x) = a + h(x) + \ell'(x).$$

This function makes the diagram commute. If $a \in K$, then

$$\gamma(a) = \gamma(a + \ell 1) = a + h(1) + \ell' 1 = a;$$

furthermore,

$$p'\gamma(a + \ell x) = p'(a + h(x) + \ell' x) = x = p(a + \ell x).$$

Finally, $\gamma$ is a homomorphism:

$$\begin{aligned}
\gamma\big([a + \ell x] + [b + \ell y]\big) &= \gamma(a + xb + f(x,y) + \ell(xy)) \\
&= a + xb + f(x,y) + h(xy) + \ell'(xy),
\end{aligned}$$

while

$$\begin{aligned}
\gamma(a + \ell x) + \gamma(b + \ell y) &= \big(a + h(x) + \ell'(x)\big) + \big(b + h(y) + \ell'(y)\big) \\
&= a + h(x) + xb + xh(y) + f'(x,y) + \ell'(xy) \\
&= a + xb + \big(h(x) + xh(y) + f'(x,y)\big) + \ell'(xy) \\
&= a + xb + f(x,y) + h(xy) + \ell'(xy).
\end{aligned}$$

We have used the given equation for $f - f'$ [remember that the terms other than $\ell'(xy)$ all lie in the abelian group $K$, and so they may be rearranged].

Conversely, assume that there exists an isomorphism $\gamma$ making the diagram commute, so that $\gamma(a) = a$ for all $a \in K$ and

$$x = p(\ell x) = p'\gamma(\ell x)$$

for all $x \in Q$. It follows that $\gamma\ell\colon Q \to G'$ is a lifting. Applying $\gamma$ to the equation $\ell x + \ell y = f(x,y) + \ell(xy)$ that defines the factor set $f$, we see that $\gamma f$ is the factor set determined by the lifting $\gamma\ell$. But $\gamma f(x,y) = f(x,y)$ for all $x, y \in Q$ because $f(x,y) \in K$. Therefore, $f$ is also a factor set of the second extension. On the other hand, if $f'$ is any other factor set for the second extension, then Lemma 9.19 shows that $f - f' \in B^2$; that is, the extensions are equivalent. $\bullet$

We say that the isomorphism $\gamma$ in Proposition 9.21 ***implements*** the equivalence. The remark after Theorem 9.18 shows that the isomorphism $\gamma\colon G \to G(K, Q, f)$ implements an equivalence of extensions.

**Example 9.22.** If two extensions of $K$ by $Q$ realizing the operators are equivalent, then their middle groups are isomorphic. However, the converse is false: we give an example of two inequivalent extensions with isomorphic middle groups. Let $p$ be an odd prime, and consider the following diagram:

$$
\begin{array}{ccccccccc}
0 & \longrightarrow & K & \overset{i}{\longrightarrow} & G & \overset{\pi}{\longrightarrow} & Q & \longrightarrow & 1 \\
& & \downarrow{\scriptstyle 1_K} & & \vdots & & \downarrow{\scriptstyle 1_Q} & & \\
0 & \longrightarrow & K & \underset{i'}{\longrightarrow} & G' & \underset{\pi'}{\longrightarrow} & Q & \longrightarrow & 1
\end{array}
$$

Define $K = \langle a \rangle$, a cyclic group of order $p$, $G = \langle g \rangle = G'$, a cyclic group of order $p^2$, and $Q = \langle x \rangle$, where $x = g + K$. In the top row, define $i(a) = pg$ and $\pi$ to be the natural map; in the bottom row define $i'(a) = 2pg$ and $\pi'$ to be the natural map. Note that $i'$ is injective because $p$ is odd.

Suppose there is an isomorphism $\gamma \colon G \to G'$ making the diagram commute. Commutativity of the first square implies $\gamma(pa) = 2pa$, and this forces $\gamma(g) = 2g$, by Exercise 9.12 on page 765; commutativity of the second square gives $g + K = 2g + K$; that is, $g \in K$. We conclude that the two extensions are not equivalent. ◄

The next theorem summarizes the calculations in this subsection.

**Theorem 9.23 (Schreier).** *Let $Q$ be a group, let $K$ be a $Q$-module, and let $e(Q, K)$ denote the family of all the equivalence classes of extensions of $K$ by $Q$ realizing the operators. There is a bijection*

$$\varphi \colon H^2(Q, K) \to e(Q, K)$$

*that takes $0$ to the class of the split extension.*

**Proof.** Denote the equivalence class of an extension

$$0 \to K \to G \to Q \to 1$$

by $[G]$. Define $\varphi \colon H^2(Q, K) \to e(Q, K)$ by

$$\varphi \colon f + B^2 \mapsto [G(K, Q, f)],$$

where $f$ is a factor set of the extension and the target extension is that constructed in Theorem 9.17.

First, $\varphi$ is a well-defined injection: $f$ and $g$ are factor sets with $f + B^2 = g + B^2$ if and only if $[G(K, Q, f)] = [G(K, Q, g)]$, by Proposition 9.21. To see that $\varphi$ is a surjection, let $[G] \in e(Q, K)$. By Theorem 9.18 and the remark following it, $[G] = [G(K, Q, f)]$ for some factor set $f$, and so $[G] = \varphi(f + B^2)$. Finally, the zero factor set corresponds to the semidirect product. •

If $H$ is a group and there is a bijection $\varphi \colon H \to X$, where $X$ is a set, then there is a unique operation defined on $X$ making $X$ a group and $\varphi$ an isomorphism: given $x, y \in X$, there are $g, h \in H$ with $x = \varphi(g)$ and $y = \varphi(h)$, and we define $xy = \varphi(gh)$. In particular, there is a way to add two equivalence classes of extensions; it is called **Baer sum** (see Section 9.6).

**Corollary 9.24.** *If $Q$ is a group, $K$ is a $Q$-module, and $H^2(Q, K) = \{0\}$, then every extension of $K$ by $Q$ realizing the operators is a semidirect product.*

**Proof.** By the theorem, $e(Q, K)$ has only one element; since the split extension always exists, this one element must be the equivalence class of the split extension. Therefore, every extension of $K$ by $Q$ realizing the operators is split, and so its middle group is a semidirect product.  •

We now apply Schreier's Theorem.

**Theorem 9.25.** *Let $G$ be a finite group of order $mn$, where $(m, n) = 1$. If $K$ is an abelian normal subgroup of order $m$, then $K$ has a complement and $G$ is a semidirect product.*

**Proof.** Define $Q = G/K$. By Corollary 9.24, it suffices to prove that every factor set $f: Q \times Q \to K$ is a coboundary. Define $\sigma: Q \to K$ by

$$\sigma(x) = \sum_{y \in Q} f(x, y);$$

$\sigma$ is well-defined because $Q$ is finite and $K$ is abelian. Now sum the cocycle identity

$$xf(y, z) - f(xy, z) + f(x, yz) - f(x, y) = 0$$

over all $z \in Q$ to obtain

$$x\sigma(y) - \sigma(xy) + \sigma(x) = nf(x, y)$$

(as $z$ varies over all of $Q$, so does $yz$). Since $(m, n) = 1$, there are integers $s$ and $t$ with $sm + tn = 1$. Define $h: Q \to K$ by

$$h(x) = t\sigma(x).$$

Note that $h(1) = 0$ and

$$xh(y) - h(xy) + h(x) = f(x, y) - msf(x, y).$$

But $sf(x, y) \in K$, and so $msf(x, y) = 0$. Therefore, $f$ is a coboundary.  •

**Remark.** P. Hall[9] proved that if $G$ is a finite solvable group of order $mn$, where $(m, n) = 1$, then $G$ has a subgroup of order $m$ (Rotman, *An Introduction to the Theory of Groups*, p. 108) and any two such are conjugate (Ibid, p. 110). In particular, in a solvable group, every (not necessarily normal) Sylow subgroup has a complement. Because of this theorem, a subgroup $H$ of a finite group $G$ is called a ***Hall subgroup*** if $(|H|, [G : H]) = 1$. Thus, Theorem 9.25 is often phrased as saying that every normal Hall subgroup of an arbitrary finite group has a complement.  ◄

We now use some Group Theory to remove the hypothesis that $K$ be abelian.

**Theorem 9.26 (Schur–Zassenhaus Lemma[10]).** *Let $G$ be a finite group of order $mn$, where $(m, n) = 1$. If $K$ is a normal subgroup of order $m$, then $K$ has a complement and $G$ is a semidirect product.*

---

[9]There were two contemporaneous group theorists named Hall: Philip Hall and Marshall Hall, Jr.

[10]Schur proved this theorem, in 1904, for the special case when $Q$ is cyclic. Zassenhaus, in 1938, proved the theorem for arbitrary finite $Q$.

**Proof.** By Exercise 9.2 on page 764, it suffices to prove that $G$ contains a subgroup of order $n$; we prove the existence of such a subgroup by induction on $m \geq 1$. Of course, the base step $m = 1$ is true.

Suppose that there is a proper subgroup $T$ of $K$ with $\{1\} \subsetneq T \lhd G$. Then $K/T \lhd G/T$ and $(G/T)/(K/T) \cong G/K$ has order $n$. Since $T \subsetneq K$, we have $|K/T| < |K| = m$, and so the inductive hypothesis provides a subgroup $N/T \subseteq G/T$ with $|N/T| = n$. Now $|N| = n|T|$, where $(|T|, n) = 1$ [because $|T|$ is a divisor of $|K| = m$], so that $T$ is a normal subgroup of $N$ whose order and index are relatively prime. Since $|T| < |K| = m$, the inductive hypothesis provides a subgroup of $N$ (which is obviously a subgroup of $G$) of order $n$.

We may now assume that $K$ is a minimal normal subgroup of $G$; that is, there is no normal subgroup $T$ of $G$ with $\{1\} \subsetneq T \subsetneq K$. Let $p$ be a prime divisor of $|K|$ and let $P$ be a Sylow $p$-subgroup of $K$. By the Frattini Argument (Exercise 4.24 on page 251), we have $G = KN_G(P)$. Therefore,

$$G/K = KN_G(P)/K \cong N_G(P)/(K \cap N_G(P)) = N_G(P)/N_K(P).$$

Hence, $|N_K(P)|n = |N_K(P)||G/K| = |N_G(P)|$. If $N_G(P)$ is a proper subgroup of $G$, then $|N_K(P)| < m$, and induction provides a subgroup of $N_G(P) \subseteq G$ of order $n$. Therefore, we may assume that $N_G(P) = G$; that is, $P \lhd G$.

Since $\{1\} \subsetneq P \subseteq K$ and $P$ is normal in $G$, we must have $P = K$, because $K$ is a minimal normal subgroup. But $P$ is a $p$-group, and so its center, $Z(P)$, is nontrivial. By Exercise 4.22(v) on page 250, we have $Z(P) \lhd G$, and so $Z(P) = P$, again because $P = K$ is a minimal normal subgroup of $G$. It follows that $P$ is abelian, and we have reduced the problem to Theorem 9.25.   •

**Corollary 9.27.** *If a finite group $G$ has a normal Sylow $p$-subgroup $P$, for some prime divisor $p$ of $|G|$, then $G$ is a semidirect product; more precisely, $P$ has a complement.*

**Proof.** The order and index of a Sylow subgroup are relatively prime.   •

**9.3.2. $H^1(Q, K)$ and Conjugacy.** There is another part of the Schur–Zassenhaus Lemma that we have not stated: if $K$ is a normal subgroup of $G$ whose order and index are relatively prime, then any two complements of $K$ are conjugate subgroups. We are now going to see that there is an analog of $H^2(K, Q)$ whose vanishing implies conjugacy of complements when $K$ is abelian. This group, $H^1(K, Q)$, arises, as did $H^2(K, Q)$, from a series of elementary calculations.

We begin with a computational result. Let $Q$ be a group, let $K$ be a $Q$-module, and let $0 \to K \to G \to Q \to 1$ be a split extension. Choose a lifting $\ell \colon Q \to G$, so that every element $g \in G$ has a unique expression of the form

$$g = a + \ell x,$$

where $a \in K$ and $x \in Q$.

**Definition.** An automorphism $\varphi$ of a group $G$ **stabilizes** an extension $0 \to K \to G \to Q \to 1$ if the following diagram commutes:

$$
\begin{array}{ccccccccc}
0 & \longrightarrow & K & \overset{i}{\longrightarrow} & G & \overset{p}{\longrightarrow} & Q & \longrightarrow & 1 \\
& & \Big\downarrow{\scriptstyle 1_K} & & \Big\downarrow{\scriptstyle \varphi} & & \Big\downarrow{\scriptstyle 1_Q} & & \\
0 & \longrightarrow & K & \overset{i}{\longrightarrow} & G & \overset{p}{\longrightarrow} & Q & \longrightarrow & 1
\end{array}
$$

The set of all stabilizing automorphisms of an extension of $K$ by $Q$, where $K$ is a $Q$-module, forms a group under composition, denoted by

$$\mathrm{Stab}(Q, K).$$

Note that a stabilizing automorphism is an isomorphism that implements an equivalence of an extension with itself. We shall see, in Proposition 9.30, that $\mathrm{Stab}(Q, K)$ does not depend on the extension.

**Proposition 9.28.** *Let $Q$ be a group, let $K$ be a $Q$-module, and let*

$$0 \to K \overset{i}{\to} G \overset{p}{\to} Q \to 1$$

*be a split extension. If $\ell \colon Q \to G$ is a lifting, then every stabilizing automorphism $\varphi \colon G \to G$ has the form*

$$\varphi(a + \ell x) = a + d(x) + \ell x,$$

*where $d(x) \in K$ is independent of the choice of lifting $\ell$. Moreover, this formula defines a stabilizing automorphism if and only if, for all $x, y \in Q$, the function $d \colon Q \to K$ satisfies*

$$d(xy) = d(x) + x d(y).$$

**Proof.** If $\varphi$ is stabilizing, then $\varphi i = i$, where $i \colon K \to G$, and $p\varphi = p$. Since we are assuming that $i$ is the inclusion [which is merely a convenience to allow us to write $a$ instead of $i(a)$], we have $\varphi(a) = a$ for all $a \in K$. To use the second constraint on $\varphi$, suppose that $\varphi(\ell x) = d(x) + \ell y$ for some $d(x) \in K$ and $y \in Q$. Then

$$x = p(\ell x) = p\varphi(\ell x) = p(d(x) + \ell y) = y;$$

that is, $x = y$. Therefore,

$$\varphi(a + \ell x) = \varphi(a) + \varphi(\ell x) = a + d(x) + \ell x.$$

To see that the formula for $d$ holds, we first show that $d$ is independent of the choice of lifting. Suppose that $\ell' \colon Q \to G$ is another lifting, so that $\varphi(\ell' x) = d'(x) + \ell' x$ for some $d'(x) \in K$. Now there is $k(x) \in K$ with $\ell' x = k(x) + \ell x$, for $p\ell' x = x = p\ell x$. Therefore,

$$d'(x) = \varphi(\ell' x) - \ell' x = \varphi(k(x) + \ell x) - \ell' x = k(x) + d(x) + \ell x - \ell' x = d(x),$$

because $k(x) + \ell x - \ell' x = 0$.

Since $d(x)$ is independent of the choice of lifting $\ell$, and since the extension splits, we may assume that $\ell$ is a homomorphism: $\ell x + \ell y = \ell(xy)$. We compute $\varphi(\ell x + \ell y)$ in two ways. On the one hand,

$$\varphi(\ell x + \ell y) = \varphi(\ell(xy)) = d(xy) + \ell(xy).$$

On the other hand,

$$\varphi(\ell x + \ell y) = \varphi(\ell x) + \varphi(\ell y) = d(x) + \ell x + d(y) + \ell y = d(x) + xd(y) + \ell(xy).$$

The proof of the converse, if $\varphi(a + \ell x) = a + d(x) + \ell x$, where $d$ satisfies the given identity, then $\varphi$ is a stabilizing isomorphism, is a routine argument and is left to the reader. •

We give a name to functions like $d$.

**Definition.** Let $Q$ be a group and let $K$ be a $Q$-module. A *derivation*[11] (or *crossed homomorphism*) is a function $d \colon Q \to K$ such that

$$d(xy) = xd(y) + d(x).$$

The set of all derivations, $\mathrm{Der}(Q, K)$, is an abelian group under pointwise addition [if $K$ is a trivial $Q$-module, then $\mathrm{Der}(Q, K) = \mathrm{Hom}(Q, K)$].

**Example 9.29.**

(i) If $d$ is a derivation, then $d(1 \cdot 1) = 1d(1) + d(1) \in K$, and so $d(1) = 0$.

(ii) If $Q$ is a group and $K$ is a $Q$-module, then a function $u \colon Q \to K$ of the form $u(x) = xa_0 - a_0$, where $a_0 \in K$, is a derivation:

$$\begin{aligned}
u(x) + xu(y) &= xa_0 - a_0 + x(ya_0 - a_0) \\
&= xa_0 - a_0 + xya_0 - xa_0 \\
&= xya_0 - a_0 \\
&= u(xy).
\end{aligned}$$

A derivation $u$ of the form $u(x) = xa_0 - a_0$ is called a *principal derivation*. If the action of $Q$ on $K$ is conjugation, $xa = x + a - x$, then

$$xa_0 - a_0 = x + a_0 - x - a_0;$$

that is, $xa_0 - a_0$ is the commutator of $x$ and $a_0$.

(iii) It is easy to check that the set $\mathrm{PDer}(Q, K)$ of all the principal derivations is a subgroup of $\mathrm{Der}(Q, K)$. ◄

Recall that $\mathrm{Stab}(Q, K)$ denotes the group of all the stabilizing automorphisms of an extension of $K$ by $Q$.

**Proposition 9.30.** *If $Q$ is a group, $K$ is a $Q$-module, and $0 \to K \to G \to Q \to 1$ is a split extension, then there is an isomorphism $\mathrm{Stab}(Q, K) \to \mathrm{Der}(Q, K)$.*

**Proof.** Let $\varphi$ be a stabilizing automorphism. If $\ell \colon Q \to G$ is a lifting, then Proposition 9.28 says that $\varphi(a + \ell x) = a + d(x) + \ell x$, where $d$ is a derivation. Since this proposition further states that $d$ is independent of the choice of lifting, $\varphi \mapsto d$ is a well-defined function $\mathrm{Stab}(Q, K) \to \mathrm{Der}(Q, K)$, which is easily seen to be a homomorphism.

---

[11]Earlier, we defined a derivation of a (not necessarily associative) ring $R$ as a function $d \colon R \to R$ with $d(xy) = d(x)y + xd(y)$. Derivations here are defined on modules, not on rings.

To see that this map is an isomorphism, we construct its inverse. If $d \in$ $\mathrm{Der}(Q, K)$, define $\varphi \colon G \to G$ by $\varphi(a + \ell x) = a + d(x) + \ell x$. Now $\varphi$ is stabilizing, by Proposition 9.28, and $d \mapsto \varphi$ is the desired inverse function. $\bullet$

It is not obvious from its definition that $\mathrm{Stab}(Q, K)$ is abelian, for its binary operation is composition. However, $\mathrm{Stab}(Q, K)$ is abelian, for $\mathrm{Der}(Q, K)$ is.

Recall that an automorphism $\varphi$ of a group $G$ is called an *inner automorphism* if it is a conjugation; that is, there is $c \in G$ with $\varphi(g) = c + g - c$ for all $g \in G$ (if $G$ is written additively).

**Lemma 9.31.** *Let $0 \to K \to G \to Q \to 1$ be a split extension, and let $\ell \colon Q \to G$ be a lifting. Then a function $\varphi \colon G \to G$ is an inner stabilizing automorphism by some $a_0 \in K$ if and only if*

$$\varphi(a + \ell x) = a + x a_0 - a_0 + \ell x.$$

**Proof.** If we write $d(x) = x a_0 - a_0$, then $\varphi(a + \ell x) = a + d(x) + \ell x$. But $d$ is a (principal) derivation, and so $\varphi$ is a stabilizing automorphism, by Proposition 9.28. Finally, $\varphi$ is conjugation by $-a_0$, for

$$-a_0 + (a + \ell x) + a_0 = -a_0 + a + x a_0 + \ell x = \varphi(a + \ell x).$$

Conversely, assume that $\varphi$ is a stabilizing conjugation. That $\varphi$ is stabilizing says that $\varphi(a + \ell x) = a + d(x) + \ell x$; that $\varphi$ is conjugation by $a_0 \in K$ says that $\varphi(a + \ell x) = a_0 + a + \ell x - a_0$. But $a_0 + a + \ell x - a_0 = a_0 + a - x a_0 + \ell x$, so that $d(x) = a_0 - x a_0$, as desired. $\bullet$

**Definition.** If $Q$ is a group and $K$ is a $Q$-module, define

$$H^1(Q, K) = \mathrm{Der}(Q, K) / \mathrm{PDer}(Q, K),$$

where $\mathrm{PDer}(Q, K)$ is the subgroup of $\mathrm{Der}(Q, K)$ consisting of all the principal derivations.

**Proposition 9.32.** *Let $0 \to K \to G \to Q \to 1$ be a split extension, and let $C$ and $C'$ be complements of $K$ in $G$. If $H^1(Q, K) = \{0\}$, then $C$ and $C'$ are conjugate.*

**Proof.** Since $G$ is a semidirect product, there are liftings $\ell \colon Q \to G$, with image $C$, and $\ell' \colon Q \to G$, with image $C'$, that are homomorphisms. Thus, the factor sets $f$ and $f'$ determined by each of these liftings are identically zero, and so $f' - f = 0$. But Lemma 9.19 says that there exists $h \colon Q \to K$, namely, $h(x) = \ell' x - \ell x$, with

$$0 = f'(x, y) - f(x, y) = x h(y) - h(xy) + h(x);$$

thus, $h$ is a derivation. Since $H^1(Q, K) = \{0\}$, $h$ is a principal derivation: there is $a_0 \in K$ with

$$\ell' x - \ell x = h(x) = x a_0 - a_0$$

for all $x \in Q$. Since addition in $G$ satisfies $\ell' x - a_0 = -x a_0 + \ell' x$, we have

$$\ell x = a_0 - x a_0 + \ell' x = a_0 + \ell' x - a_0.$$

But $\mathrm{im}\, \ell = C$ and $\mathrm{im}\, \ell' = C'$, and so $C$ and $C'$ are conjugate via $a_0$. $\bullet$

We can now supplement the Schur–Zassenhaus Lemma.

**Theorem 9.33.** *Let $G$ be a finite group of order $mn$, where $(m, n) = 1$. If $K$ is an abelian normal subgroup of order $m$, then $G$ is a semidirect product of $K$ by $G/K$, and any two complements of $K$ are conjugate.*

**Proof.** By Proposition 9.32, it suffices to prove that $H^1(Q, K) = \{0\}$, where $Q = G/K$. Note, first, that $|Q| = |G|/|K| = mn/m = n$.

Let $d \colon Q \to K$ be a derivation: for all $x, y \in Q$, we have

$$d(xy) = xd(y) + d(x).$$

Sum this equation over all $y \in Q$ to obtain

$$\Delta = x\Delta + nd(x),$$

where $\Delta = \sum_{y \in Q} d(y)$ (as $y$ varies over $Q$, so does $xy$). Since $(m, n) = 1$, there are integers $s$ and $t$ with $sn + tm = 1$. Hence,

$$d(x) = snd(x) + tmd(x) = snd(x),$$

because $d(x) \in K$ and so $md(x) = 0$. Therefore,

$$d(x) = s\Delta - xs\Delta.$$

Setting $a_0 = -s\Delta$, we see that $d$ is a principal derivation.    •

Removing the assumption in Theorem 9.33 that $K$ is abelian is much more difficult than removing this assumption in Theorem 9.25. One first proves that complements are conjugate if either $K$ or $Q$ is a solvable group (Robinson, *A Course in the Theory of Groups*, p. 248). Since $|Q|$ and $|K|$ are relatively prime, at least one of $K$ or $Q$ has odd order. The deep Feit–Thompson Theorem, which says that every group of odd order is solvable, now completes the proof.

There are other applications of homology in Group Theory besides the Schur–Zassenhaus Lemma. For example, if $G$ is a group, $a \in G$, and $\gamma_a \colon g \mapsto aga^{-1}$ is conjugation by $a$, then $\gamma_a^n \colon g \mapsto a^n g a^{-n}$ for all $n$. Hence, if $a$ has prime order $p$ and $a \notin Z(G)$, then $\gamma_a$ is an automorphism of order $p$. A theorem of Gaschütz (Rotman, *An Introduction to Homological Algebra*, p. 589) uses cohomology to prove that every finite nonabelian $p$-group has an automorphism of order $p$ that is not conjugation by an element of $G$.

Let us contemplate the formulas that have arisen:

$$\begin{aligned}
\text{factor set} : \quad & 0 = xf(y, z) - f(xy, z) + f(x, yz) - f(x, y), \\
\text{coboundary} : \quad & f(x, y) = xh(y) - h(xy) + h(x), \\
\text{derivation} : \quad & 0 = xd(y) - d(xy) + d(x), \\
\text{principal derivation} : \quad & d(x) = xa_0 - a_0.
\end{aligned}$$

All these formulas involve alternating sums; factor sets and derivations seem to be in kernels, and coboundaries and principal derivations seem to be in images. Let us make this more precise.

Denote the cartesian product of $n$ copies of $Q$ by $Q^n$; for clarity, we denote an element of $Q^n$ by $[x_1, \ldots, x_n]$ instead of $(x_1, \ldots, x_n)$. Factor sets and coboundaries are certain functions $Q^2 \to K$, and derivations are certain functions $Q^1 \to K$. Let $F_n$ be the free $Q$-module with basis $Q^n$. By the definition of basis, every function $f\colon Q^n \to K$ gives a unique $Q$-homomorphism $\widetilde{f}\colon F_n \to K$ extending $f$, for $K$ is a $Q$-module.

We now define maps that are suggested by the various formulas:

$$d_3\colon F_3 \to F_2: \quad d_3[x,y,z] = x[y,z] - [xy,z] + [x,yz] - [x,y];$$
$$d_2\colon F_2 \to F_1: \quad d_2[x,y] = x[y] - [xy] + [x].$$

In fact, we define one more map: let $Q^0 = \{1\}$ be a 1-point set, so that $F_0 = \mathbb{Z}Q$ is the free $Q$-module on the single generator, 1. Now define

$$d_1\colon F_1 \to F_0: \quad d_1[x] = x - 1.$$

We have defined each of $d_3$, $d_2$, and $d_1$ on bases of free modules, and so each extends to a $Q$-map.

**Proposition 9.34.** *The sequence*

$$F_3 \xrightarrow{d_3} F_2 \xrightarrow{d_2} F_1 \xrightarrow{d_1} F_0 \xrightarrow{\varepsilon} \mathbb{Z} \to 0$$

*is an exact sequence of $Q$-modules, where $\mathbb{Z}$ is a trivial $\mathbb{Z}Q$-module; that is, $xm = m$ for all $x \in G$ and $m \in \mathbb{Z}$.*

**Proof.** Define $\varepsilon\colon \mathbb{Z}Q \to \mathbb{Z}$ by

$$\varepsilon\colon \sum_{x \in Q} m_x x \mapsto \sum_{x \in Q} m_x.$$

Now $\varepsilon$ is a $Q$-map, for if $x \in Q$, then $\varepsilon(x) = 1$; on the other hand, $\varepsilon(x) = \varepsilon(x \cdot 1) = x\varepsilon(1) = 1$, because $\mathbb{Z}$ is a trivial $Q$-module. It is clear that $\varepsilon$ is a surjection and that $\operatorname{im} d_1 \subseteq \ker \varepsilon$ [because $\varepsilon(x-1) = 0$]. For the reverse inclusion, if $\sum_{x \in Q} m_x x \in \ker \varepsilon$, then $\sum_{x \in Q} m_x = 0$. Hence,

$$\sum_{x \in Q} m_x x = \sum_{x \in Q} m_x x - \Big(\sum_{x \in Q} m_x\Big)1 = \sum_{x \in Q} m_x(x-1) \in \operatorname{im} d_1.$$

Therefore, $\operatorname{im} d_1 = \ker \varepsilon$, and there is exactness at $F_0$.

We will only check that $d_1 d_2 = 0$ and $d_2 d_3 = 0$; that is, $\operatorname{im} d_2 \subseteq \ker d_1$ and $\operatorname{im} d_3 \subseteq \ker d_2$. The (trickier) reverse inclusions will be proved in Theorem 9.122 after we introduce some Homological Algebra:

$$\begin{aligned}
d_1 d_2[x,y] &= d_1(x[y] - [xy] + [x]) \\
&= x d_1[y] - d_1[xy] + d_1[x] \\
&= x(y-1) - (xy-1) + (x-1) = 0
\end{aligned}$$

(the equation $d_1 x[y] = x d_1[y]$ holds because $d_1$ is a $Q$-map). The reader should note that this is the same calculation as in Proposition 9.20:

$$
\begin{aligned}
d_2 d_3[x, y, z] &= d_2(x[y, z] - [xy, z] + [x, yz] - [x, y]) \\
&= x d_2[y, z] - d_2[xy, z] + d_2[x, yz] - d_2[x, y] \\
&= x(y[z] - [yz] + [y]) - (xy[z] - [xyz] + [xy]) \\
&\quad + (x[yz] - [xyz] + [x]) - (x[y] - [xy] + [x]) = 0. \quad \bullet
\end{aligned}
$$

Applying the contravariant functor $\mathrm{Hom}_{\mathbb{Z}Q}(\square, K)$ to the sequence in Proposition 9.34, we obtain a (not necessarily exact) sequence

$$
\mathrm{Hom}(F_3, K) \xleftarrow{d_3^*} \mathrm{Hom}(F_2, K) \xleftarrow{d_2^*} \mathrm{Hom}(F_1, K) \xleftarrow{d_1^*} \mathrm{Hom}(F_0, K).
$$

Let us regard a function $f \colon Q^n \to K$ as the restriction of the $Q$-map $\widetilde{f} \colon F_n \to K$ which extends it. Suppose that $f \colon Q^2 \to K$ lies in $\ker d_3^*$. Then $0 = d_3^*(f) = f d_3$. Hence, for all $x, y, z \in Q$, we have

$$
\begin{aligned}
0 &= f d_3[x, y, z] \\
&= f(x[y, z] - [xy, z] + [x, yz] - [x, y]) \\
&= x f[y, z] - f[xy, z] + f[x, yz] - f[x, y];
\end{aligned}
$$

the equation $f(x[y, z]) = x f[y, z]$ holds because $f$ is the restriction of a $Q$-map. Thus, $f$ is a factor set. If $f$ lies in $\mathrm{im}\, d_2^*$, then there is some $h \colon Q \to K$ with $f = d_2^*(h) = h d_2$. Thus,

$$
f[x, y] = h d_2[x, y] = h(x[y] - [xy] + [x]) = x h[y] - h[xy] + h[x];
$$

the equation $h(x[y]) = x h[y]$ holds because $h$ is the restriction of a $Q$-map. Thus, $f$ is a coboundary.

A similar analysis shows that if $g \colon Q \to K$ lies in $\ker d_2^*$, then $g$ is a derivation. Let us now compute $\mathrm{im}\, d_1^*$. If $k \colon \{1\} \to K$ is the trivial homomorphism, then

$$
d_1^*(k) = k d_1(x) = k((x - 1)1) = (x - 1)k(1),
$$

because $k$ is the restriction of a $Q$-map. Now $k(1)$ is merely an element of $K$; indeed, if we identify $k$ with its (1-point) image $k(1) = a_0$, then we see that $d_1^*(k)$ is a principal derivation.

Observe that $d_2 d_3 = 0$ implies $d_3^* d_2^* = 0$, which is equivalent to $\mathrm{im}\, d_2^* \subseteq \ker d_3^*$; that is, every coboundary is a factor set, which is Proposition 9.20. Similarly, $d_1 d_2 = 0$ implies $\mathrm{im}\, d_1^* \subseteq \ker d_2^*$; that is, every principal derivation is a derivation, which is Example 9.29.

We discuss Homological Algebra in the next section, for it is the proper context to understand these constructions.

As long as we are computing kernels and images, what is $\ker d_1^*$? If $k \colon \{1\} \to K$ and $k(1) = a_0$, then $k \in \ker d_1^*$ says that

$$
0 = d_1^*(k) = k d_1(x) = (x - 1)k(1) = (x - 1)a_0,
$$

so that $x a_0 = a_0$ for all $x \in Q$. We have been led to the following definition.

**Definition.** If $Q$ is a group and $K$ is a $Q$-module, then the submodule of **fixed points** is defined by

$$H^0(Q, K) = \{a \in K : xa = a \text{ for all } x \in Q\}.$$

The groups $H^2(Q, K)$, $H^1(Q, K)$, and $H^0(Q, K)$ were obtained by applying the functor $\operatorname{Hom}_{\mathbb{Z}Q}(\square, K)$ to the exact sequence $F_3 \to F_2 \to F_1 \to F_0$. In algebraic topology, we would also apply the functor $\square \otimes_{\mathbb{Z}Q} K$, obtaining **homology groups** [the tensor product is defined because we may view the free $Q$-modules $F_n$ as right $Q$-modules, as in Example 6.61(v)]:

$$H_i(Q, K) = \ker(d_i \otimes 1)/\operatorname{im}(d_{i+1} \otimes 1), \quad i = 0, 1, 2.$$

We can show that $H_0(Q, K)$ is the maximal $Q$-trivial quotient of $K$. In the special case of $K = \mathbb{Z}$ viewed as a trivial $Q$-module, we see that $H_1(Q, \mathbb{Z}) \cong Q/Q'$, where $Q'$ is the commutator subgroup of $Q$.

## Exercises

**9.13.** Let $Q$ be a group and let $K$ be a $Q$-module. Prove that any two split extensions of $K$ by $Q$ realizing the operators are equivalent.

**9.14.** Let $Q$ be a group and let $K$ be a $Q$-module.

(i) If $K$ and $Q$ are finite groups, prove that $H^2(Q, K)$ is also finite.

(ii) Let $\tau(K, Q)$ denote the number of nonisomorphic middle groups $G$ that occur in extensions of $K$ by $Q$ realizing the operators. Prove that

$$\tau(K, Q) \leq |H^2(Q, K)|.$$

(iii) Give an example showing that the inequality in part (ii) can be strict.
  **Hint.** Observe that $\tau(\mathbb{I}_p, \mathbb{I}_p) = 2$ (note that the kernel is the trivial module because every group of order $p^2$ is abelian).

**9.15.** Recall that a **generalized quaternion group** $\mathbf{Q}_n$ is a group of order $2^n$ (for $n \geq 2$) that is generated by two elements $a$ and $b$ such that

$$a^{2^{n-1}} = 1, \quad bab^{-1} = a^{-1}, \quad \text{and} \quad b^2 = a^{2^{n-2}}.$$

(i) Prove that $\mathbf{Q}_n$ has a unique element $z$ of order 2 and that $Z(\mathbf{Q}_n) = \langle z \rangle$. Conclude that $\mathbf{Q}_n$ is not a semidirect product.

(ii) Prove that $\mathbf{Q}_n$ is a **central extension** (i.e., $\theta$ is trivial) of $\mathbb{I}_2$ by $D_{2^{n-1}}$.

(iii) Using factor sets, give another proof of the existence of $\mathbf{Q}_n$.

**9.16.** Prove that every group $G$ of order $2p$, where $p \geq 2$ is a prime, is a semidirect product of $\mathbb{I}_p$ by $\mathbb{I}_2$, and conclude that either $G$ is cyclic or $G \cong D_{2p}$ (recall that $D_4 = \mathbf{V}$, the four-group). (See Exercise 4.64 on page 285.)

*\* **9.17.** Show that every group $G$ of order 12 is isomorphic to one of the following five groups:

$$\mathbb{I}_{12}, \quad \mathbf{V} \times \mathbb{I}_3, \quad A_4, \quad S_3 \times \mathbb{I}_2, \quad T,$$

where $T$ is the group in Example 9.13.

* **9.18.** If $Q$ is a group and $K$ is a $Q$-module, let $E$ be a semidirect product of $K$ by $Q$ and let $\ell\colon G \to E$ be a lifting. Prove that $\ell(x) = (d(x), x)$, where $d\colon Q \to K$, and $\ell$ is a homomorphism if and only if $d$ is a derivation.

**9.19.** Let $\Phi\colon \mathbf{Set} \to {}_{\mathbb{Z}Q}\mathbf{Mod}$ be the **free functor** that assigns to each set $X$ the free $Q$-module $\Phi(X)$ with basis $X$ (Exercise 6.54 on page 449). If $U\colon {}_{\mathbb{Z}Q}\mathbf{Mod} \to \mathbf{Sets}$ is the forgetful functor which assigns to each module its underlying set of elements, prove that the ordered pair $(\Phi, U)$ is an adjoint pair of functors.

## Section 9.4. Homology Functors

We are now going to study *resolutions* in order to place the foregoing discussion in its proper context. Given a module $M$, there is a free module $F_0$ and a surjection $\varepsilon\colon F_0 \to M$; thus, there is an exact sequence

$$0 \to \Omega_1 \xrightarrow{\ i\ } F_0 \xrightarrow{\ \varepsilon\ } M \to 0,$$

where $\Omega_1 = \ker \varepsilon$, $i\colon \Omega_1 \to F_0$ is the inclusion, and $M \cong \operatorname{coker} i$. This gives a presentation of $M$ by generators and relations. Thus, if $X$ is a basis of $F_0$, then we say that $M$ has generators $X$ [or $\varepsilon(X)$] and relations $\Omega_1$. The idea now is to find generators and relations of $\Omega_1$, getting "second-order" relations $\Omega_2$;[12] iterating this construction gives a *free resolution* of $M$, which should be regarded as a more detailed presentation of $M$ by generators and relations. We saw, in Section 9.1, that an abstract simplicial complex $K$ is also replaced by such a sequence, its chain complex of simplicial chains.

**Definition.** A **projective resolution** of a module $M$ is an exact sequence

$$\to P_n \to P_{n-1} \to \cdots \to P_1 \to P_0 \to M \to 0$$

in which each module $P_n$ is projective. A **free resolution** is a projective resolution in which each module $P_n$ is free.

Free resolutions exist.

**Proposition 9.35.** *Every module $M$ has a free resolution (and, hence, it has a projective resolution).*

**Proof.** By Proposition 6.70, there is a free module $F_0$ and an exact sequence

$$0 \to \Omega_1 \xrightarrow{\ i_1\ } F_0 \xrightarrow{\ \varepsilon\ } M \to 0.$$

Similarly, there is a free module $F_1$, a surjection $\varepsilon_1\colon F_1 \to \Omega_1$, and an exact sequence

$$0 \to \Omega_2 \xrightarrow{\ i_2\ } F_1 \xrightarrow{\ \varepsilon_1\ } \Omega_1 \to 0.$$

---

[12]Hilbert called higher order relations of a module its *syzygies*.

Define $d_1 \colon F_1 \to F_0$ to be the composite $i_1 \varepsilon_1$. It is plain that $\operatorname{im} d_1 = \Omega_1 = \ker \varepsilon$ and $\ker d_1 = \Omega_2$, so there is an exact sequence

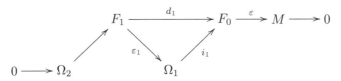

Clearly, this construction can be iterated for all $n \geq 0$ (the ultimate exact sequence is infinitely long). $\bullet$

At the end of our discussion of Schreier's investigation of the extension problem, Proposition 9.34 displayed an exact sequence of free $Q$-modules,

$$F_3 \xrightarrow{d_3} F_2 \xrightarrow{d_2} F_1 \xrightarrow{d_1} F_0 \xrightarrow{\varepsilon} \mathbb{Z} \to 0,$$

where $F_n$ is the free $Q$-module with basis $Q^n$ for $1 \leq n \leq 3$, and the module $F_0 = \mathbb{Z}Q$ is free on the generator 1 (actually, we did not prove exactness of this sequence; we will complete the proof in Theorem 9.122). Proposition 9.35 below shows that this sequence consists of the first few terms of a free resolution of the trivial $Q$-module $\mathbb{Z}$.

The definition of projective resolution can be dualized.

**Definition.** An ***injective resolution*** of a module $M$ is an exact sequence

$$0 \to M \to E^0 \to E^1 \to \cdots \to E^n \to E^{n+1} \to$$

in which each module $E^n$ is injective.

**Proposition 9.36.** *Every module $M$ has an injective resolution.*

**Proof.** We use Theorem 6.96, which states that every module can be imbedded as a submodule of an injective module. Thus, there is an injective module $E^0$, an injection $\eta \colon M \to E^0$, and an exact sequence

$$0 \to M \xrightarrow{\eta} E^0 \xrightarrow{p} \Sigma^1 \to 0,$$

where $\Sigma^1 = \operatorname{coker} \eta$ and $p$ is the natural map. Now repeat: there is an injective module $E^1$ and an imbedding $\eta^1 \colon \Sigma^1 \to E^1$, yielding an exact sequence

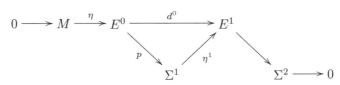

where $d^0$ is the composite $d^0 = \eta^1 p$. This construction can be iterated. $\bullet$

A *complex* $\mathbf{C} = (\mathbf{C}, d)$ in an abelian category $\mathcal{A}$ is a sequence of morphisms (called *differentials*)

$$\to C_{n+1} \xrightarrow{d_{n+1}} C_n \xrightarrow{d_n} C_{n-1} \to$$

in which $d_n d_{n+1} = 0$ for all $n \in \mathbb{Z}$. We introduced complexes in Chapter 7 and proved Theorem 7.145: $\mathbf{Comp}(\mathcal{A})$, the category of all complexes in $\mathcal{A}$ and chain maps, is an abelian category.

We simplify our exposition by writing $\mathbf{Comp}(\mathcal{A}) = \mathbf{Comp}$; indeed, the reader may assume that $\mathcal{A} = \mathbf{Ab}$.

**Example 9.37.**

(i) Every exact sequence is a complex, for the equalities $\operatorname{im} d_{n+1} = \ker d_n$ imply $\operatorname{im} d_{n+1} \subseteq \ker d_n$; that is, $d_n d_{n+1} = 0$.

(ii) The sequence of chain groups of an abstract simplicial complex $K$,

$$\to C_3(K) \xrightarrow{\partial_3} C_2(K) \xrightarrow{\partial_2} C_1(K) \xrightarrow{\partial_1} C_0(K),$$

is not yet a complex, for it does not have a module for every $n \in \mathbb{Z}$. We repair this by defining $C_n(K) = \{0\}$ for all negative $n$; there is no problem defining differentials $d_n \colon C_n(K) \to C_{n-1}(K)$ for $n \leq 0$, for only the zero map exists from any module into $\{0\}$.

(iii) In Chapter 8, we considered the de Rham complex of a connected open subset $X$ of $\mathbb{R}^n$:

$$0 \to \Omega^0(X) \xrightarrow{d^0} \Omega^1(X) \xrightarrow{d^1} \Omega^2(X) \to \cdots \to \Omega^{n-1}(X) \xrightarrow{d^{n-1}} \Omega^n(X) \to 0,$$

where the maps are the exterior derivatives.

(iv) The **zero complex** $\mathbf{0}$ is the complex $(\mathbf{C}, d)$ with all terms $C_n = \{0\}$ and, necessarily, with all differentials $d_n = 0$.

(v) If $(M_n)_{n \in \mathbb{Z}}$ is any sequence of modules, then $(\mathbf{M}, d)$ is a complex with $n$th term $M_n$ and with $d_n = 0$ for all $n$.

(vi) Every homomorphism is a differential in some complex. For example, if $f \colon A \to B$ is a homomorphism, define a complex $(\mathbf{C}, d)$ with $C_1 = A$, $C_0 = B$, $d_1 = f$, and all other terms and differentials zero.

(vii) Every projective resolution

$$\to P_1 \to P_0 \to M \to 0$$

is a complex if we add $\{0\}$s to the right.

(viii) Every injective resolution

$$0 \to M \to E^0 \to E^1 \to$$

is a complex if we add $\{0\}$s to the left.

We have used a convenient notation. According to the definition of complex, differentials lower the index: $d_n \colon C_n \to C_{n-1}$. The simplest way to satisfy the definition is to use negative indices: define $C_{-n} = E^n$, and

$$0 \to M \to C_0 \to C_{-1} \to C_{-2} \to$$

is a complex. Most people prefer avoiding negative indices when possible.

(ix) If $\mathbf{C} = \to C_n \xrightarrow{d_n} C_{n-1} \to$ is a complex and $F$ is an additive (covariant) functor, then $F\mathbf{C} = \to FC_n \xrightarrow{Fd_n} FC_{n-1} \to$ is also a complex, because

$$0 = F(0) = F(d_n d_{n+1}) = F(d_n)F(d_{n+1})$$

[the equation $0 = F(0)$ holds because $F$ is additive]. Note that even if the original complex is exact, the functored complex $F\mathbf{C}$ may not be exact.

(x) If $F$ is a contravariant additive functor, it is also true that $F\mathbf{C}$ is a complex, but we have to arrange notation so that differentials lower indices by 1. In more detail, after applying $F$, we have

$$F\mathbf{C} = \leftarrow FC_n \xleftarrow{Fd_n} FC_{n-1} \leftarrow;$$

thus, the differentials $Fd_n$ increase indices by 1. Introducing negative indices almost solves the problem. If we define $X_{-n} = F(C_n)$, then the sequence is

$$F\mathbf{C} = \to X_{-n+1} \xrightarrow{Fd_n} X_{-n} \to .$$

However, the index on the map should be $-n + 1$, and not $n$. Define

$$\delta_{-n+1} = F(d_n),$$

and now the relabeled sequence reads properly:

$$F\mathbf{C} = \to X_{-n+1} \xrightarrow{\delta_{-n+1}} X_{-n} \to .$$

As in (viii), negative indices are awkward. Again, we change the sign of the index by raising it to a superscript: write

$$\delta^n = \delta_{-n}.$$

The final version of the functored sequence now looks like this:

$$F\mathbf{C} = \to X^{n-1} \xrightarrow{\delta^{n-1}} X^n \to . \quad \blacktriangleleft$$

Let us generalize the following terms from the chain complex $\mathbf{C}(K)$ arising from an abstract simplicial complex $K$.

**Definition.** If $\mathbf{C} = \to C_n \xrightarrow{d_n} C_{n-1} \to \cdots$ is a complex and $n \in \mathbb{Z}$, then

$$
\begin{aligned}
\boldsymbol{n\text{-}chains}: \quad & C_n; \\
\boldsymbol{n\text{-}cycles}: \quad & Z_n(\mathbf{C}) = \ker d_n; \\
\boldsymbol{n\text{-}boundaries}: \quad & B_n(\mathbf{C}) = \operatorname{im} d_{n+1}; \\
\boldsymbol{n\text{th homology}}: \quad & H_n(\mathbf{C}) = Z_n(\mathbf{C})/B_n(\mathbf{C}).^{[13]}
\end{aligned}
$$

If $z$ is an $n$-cycle, its **homology class** is the coset

$$\operatorname{cls}(z) = z + B_n(\mathbf{C}).$$

---

[13] Every boundary $dc$ is a cycle, for $d(dc) = (dd)c = 0$; hence, $B_n \subseteq Z_n$ and $Z_n/B_n$ makes sense.

**Example 9.38.** A complex is an exact sequence if and only if $\ker d_n = \operatorname{im} d_{n+1}$ for all $n$; that is, all its homology groups $H_n(\mathbf{C}) = \{0\}$. Thus, the homology groups measure the deviation of a complex from being an exact sequence. An exact sequence is often called an **acyclic complex** (*acyclic* means "no cycles"; that is, no cycles that are not boundaries). ◀

**Example 9.39.** In Example 9.37(vi), we saw that every homomorphism $f\colon A \to B$ can be viewed as part of a complex $\mathbf{C}$ with $C_1 = A$, $C_0 = B$, $d_1 = f$, and having $\{0\}$s to the left and to the right. Now $d_2 = 0$ implies $\operatorname{im} d_2 = 0$, and $d_0 = 0$ implies $\ker d_0 = B$; it follows that

$$H_n(\mathbf{C}) = \begin{cases} \ker f & \text{if } n = 1; \\ \operatorname{coker} f & \text{if } n = 0; \\ \{0\} & \text{otherwise.} \end{cases} \quad \blacktriangleleft$$

**Proposition 9.40.** *For each $n \in \mathbb{Z}$, homology $H_n\colon \mathbf{Comp} \to \mathbf{Ab}$ is an additive functor.*

**Proof.** We have just defined $H_n$ on objects; it remains to define $H_n$ on morphisms. If $f\colon (\mathbf{C}, d) \to (\mathbf{C}', d')$ is a chain map, define $H_n(f)\colon H_n(\mathbf{C}) \to H_n(\mathbf{C}')$ by

$$H_n(f)\colon \operatorname{cls}(z_n) \mapsto \operatorname{cls}(f_n z_n).$$

We must show that $f_n z_n$ is a cycle and that $H_n(f)$ is independent of the choice of cycle $z_n$; both of these assertions follow from $f$ being a chain map; that is, from commutativity of the following diagram:

$$
\begin{array}{ccccc}
C_{n+1} & \xrightarrow{d_{n+1}} & C_n & \xrightarrow{d_n} & C_{n-1} \\
\downarrow{\scriptstyle f_{n+1}} & & \downarrow{\scriptstyle f_n} & & \downarrow{\scriptstyle f_{n-1}} \\
C'_{n+1} & \xrightarrow[d'_{n+1}]{} & C'_n & \xrightarrow[d'_n]{} & C'_{n-1}
\end{array}
$$

First, let $z$ be an $n$-cycle in $Z_n(\mathbf{C})$, so that $d_n z = 0$. Then commutativity of the diagram gives

$$d'_n f_n z = f_{n-1} d_n z = 0.$$

Therefore, $f_n z$ is an $n$-cycle.

Next, assume that $\operatorname{cls}(z) = \operatorname{cls}(y)$; that is $z - y \in B_n(\mathbf{C})$, so that $z - y = d_{n+1} c$ for some $c \in C_{n+1}$. Applying $f_n$ gives $f_n z - f_n y = f_n d_{n+1} c = d'_{n+1} f_{n+1} c \in B_n(\mathbf{C}')$. Thus, $\operatorname{cls}(f_n z) = \operatorname{cls}(f_n y)$. Therefore, $H_n(f)$ is well-defined.

Let us see that $H_n$ is a functor. It is obvious that $H_n(1_{\mathbf{C}})$ is the identity. If $f$ and $g$ are chain maps whose composite $gf$ is defined, then for every $n$-cycle $z$, we have

$$
\begin{aligned}
H_n(gf)\colon \operatorname{cls}(z) &\mapsto (gf)_n \operatorname{cls}(z) = g_n f_n \operatorname{cls}(z) \\
&= H_n(g) \operatorname{cls}(f_n z) = H_n(g) H_n(f) \operatorname{cls}(z).
\end{aligned}
$$

Finally, $H_n$ is additive: if $g \colon (\mathbf{C}, d) \to (\mathbf{C}', d')$ is another chain map, then

$$H_n(f + g) \colon \mathrm{cls}(z) \mapsto (f_n + g_n)\,\mathrm{cls}(z)$$
$$= f_n\,\mathrm{cls}(z) + g_n\,\mathrm{cls}(z)$$
$$= \big(H_n(f) + H_n(g)\big)\,\mathrm{cls}(z). \quad \bullet$$

**Definition.** We call $H_n(f)$ the **induced map**; we usually suppress $n$ and denote $H_n(f)$ by $f_*$.

We now introduce a notion that arises in Topology.

**Definition.** A chain map $f \colon (\mathbf{C}, d) \to (\mathbf{C}', d')$ is **nullhomotopic** if, for all $n$, there are maps $s_n \colon A_n \to A'_{n+1}$ with $f_n = d'_{n+1} s_n + s_{n-1} d_n$:

$$
\begin{array}{ccccccc}
\longrightarrow & A_{n+1} & \xrightarrow{\ d_{n+1}\ } & A_n & \xrightarrow{\ d_n\ } & A_{n-1} & \longrightarrow \\
& \Big\downarrow{\scriptstyle f_{n+1}} & {\scriptstyle s_n}\ \swarrow & \Big\downarrow{\scriptstyle f_n} & {\scriptstyle s_{n-1}}\ \swarrow & \Big\downarrow{\scriptstyle f_{n-1}} & \\
\longrightarrow & A'_{n+1} & \xrightarrow[\ d'_{n+1}\ ]{} & A'_n & \xrightarrow[\ d'_n\ ]{} & A'_{n-1} & \longrightarrow
\end{array}
$$

Let $f, g \colon \mathbf{C} \to \mathbf{C}'$ be chain maps. Then $f$ is **homotopic**[14] to $g$, denoted by $f \simeq g$, if $f - g$ is nullhomotopic.

**Proposition 9.41.** *Homotopic chain maps induce the same homomorphism between homology groups: if $f, g \colon (\mathbf{C}, d) \to (\mathbf{C}', d')$ are chain maps and $f \simeq g$, then*

$$f_* = g_* \colon H_n(\mathbf{C}) \to H_n(\mathbf{C}').$$

**Proof.** If $z$ is an $n$-cycle, then $d_n z = 0$ and

$$f_n z - g_n z = d'_{n+1} s_n z + s_{n-1} d_n z = d'_{n+1} s_n z.$$

Therefore, $f_n z - g_n z \in B_n(\mathbf{C}')$, and so $f_* = g_*$. $\quad \bullet$

**Definition.** A complex $(\mathbf{C}, d)$ has a **contracting homotopy**[15] if its identity $1_{\mathbf{C}}$ is nullhomotopic.

**Proposition 9.42.** *A complex $(\mathbf{C}, d)$ having a contracting homotopy is acyclic; that is, it is an exact sequence.*

**Proof.** We use Example 9.38. Now $1_{\mathbf{C}} \colon H_n(\mathbf{C}) \to H_n(\mathbf{C})$ is the identity map, while $0_* \colon H_n(\mathbf{C}) \to H_n(\mathbf{C})$ is the zero map. Since $1_{\mathbf{C}} \simeq 0$, however, these maps are the same. It follows that $H_n(\mathbf{C}) = \{0\}$ for all $n$; that is, $\ker d_n = \operatorname{im} d_{n+1}$ for all $n$, and this is the definition of exactness. $\quad \bullet$

---

[14]Two continuous functions $f, g \colon X \to Y$ are called **homotopic** if $f$ can be "deformed" into $g$; that is, there exists a continuous $F \colon X \times \mathbf{I} \to Y$, where $\mathbf{I} = [0, 1]$ is the closed unit interval, with $F(x, 0) = f(x)$ and $F(x, 1) = g(x)$ for all $x \in X$. Now every continuous $f \colon X \to Y$ induces homomorphisms $f_* \colon H_n(X) \to H_n(Y)$, and one proves that if $f$ and $g$ are homotopic, then $f_* = g_*$. The algebraic definition of homotopy given here has been distilled from the proof of this topological theorem.

[15]A topological space is **contractible** if its identity map is homotopic to a constant map.

Once we complete the free resolution of the trivial $Q$-module $\mathbb{Z}$ whose first few terms were given in Proposition 9.34, we will prove that it is an exact sequence by showing that it has a contracting homotopy as a complex of abelian groups.

In order to better study the homology functors, it is necessary to recall that their domain **Comp** is an abelian category. Here is a concrete description of some of its ingredients.

- The reader should check that a chain map $f \colon \mathbf{C} \to \mathbf{C}'$ is an isomorphism in **Comp** if and only if $f_n \colon C_n \to C_n'$ is an isomorphism for all $n \in \mathbb{Z}$ (it must be checked that the sequence of inverses $f_n^{-1}$ is a chain map; that is, that the appropriate diagram commutes).

- A complex $(\mathbf{A}, \delta)$ is a **subcomplex** of a complex $(\mathbf{C}, d)$ if, for every $n \in \mathbb{Z}$, we have $A_n$ a subgroup of $C_n$ and $\delta_n = d_n | A_n$.

  If $i_n \colon A_n \to C_n$ is the inclusion, then it is easy to see that $\mathbf{A}$ is a subcomplex of $\mathbf{C}$ if and only if $i \colon \mathbf{A} \to \mathbf{C}$ is a chain map.

- If $\mathbf{A}$ is a subcomplex of $\mathbf{C}$, then the **quotient complex** is

$$\mathbf{C}/\mathbf{A} = \to C_n/A_n \xrightarrow{d_n''} C_{n-1}/A_{n-1} \to,$$

  where $d_n'' \colon c_n + A_n \mapsto d_n c_n + A_{n-1}$ ($d_n''$ is well-defined: if $c_n + A_n = b_n + A_n$, then $d_n c_n + A_{n-1} = d_n b_n + A_{n-1}$). If $\pi_n \colon C_n \to C_n/A_n$ is the natural map, then $\pi \colon \mathbf{C} \to \mathbf{C}/\mathbf{A}$ is a chain map.

- If $f \colon (\mathbf{C}, d) \to (\mathbf{C}', d')$ is a chain map, define

$$\mathbf{ker} f = \to \ker f_{n+1} \xrightarrow{\delta_{n+1}} \ker f_n \xrightarrow{\delta_n} \ker f_{n-1} \to,$$

  where $\delta_n = d_n | \ker f_n$, and define

$$\mathbf{im} f = \to \operatorname{im} f_{n+1} \xrightarrow{\Delta_{n+1}} \operatorname{im} f_n \xrightarrow{\Delta_n} \operatorname{im} f_{n-1} \to,$$

  where $\Delta_n = d_n' | \operatorname{im} f_n$. It is easy to see that $\mathbf{ker} f$ is a subcomplex of $\mathbf{C}$, that $\mathbf{im} f$ is a subcomplex of $\mathbf{C}'$, and that the **First Isomorphism Theorem** holds:

$$\mathbf{C}/\mathbf{ker} f \cong \mathbf{im} f.$$

- A sequence of complexes and chain maps

$$\to \mathbf{C}^{n+1} \xrightarrow{f^{n+1}} \mathbf{C}^n \xrightarrow{f^n} \mathbf{C}^{n-1} \to$$

  is an **exact sequence** if, for all $n \in \mathbb{Z}$,

$$\mathbf{im} f^{n+1} = \mathbf{ker} f^n.$$

We may check that if $\mathbf{A}$ is a subcomplex of $\mathbf{C}$, then there is an exact sequence of complexes

$$0 \to \mathbf{A} \xrightarrow{i} \mathbf{C},$$

where $\mathbf{0}$ is the zero complex and $i$ is the chain map of inclusions. More generally, let $i \colon \mathbf{C} \to \mathbf{C}'$ be a chain map; each $i_n$ is injective if and only if there is an exact sequence $0 \to \mathbf{C} \xrightarrow{i} \mathbf{C}'$. Similarly, if $p \colon \mathbf{C} \to \mathbf{C}''$ is a chain map, then each $p_n$ is surjective if and only if there is an exact sequence

$$\mathbf{C} \xrightarrow{p} \mathbf{C}'' \to \mathbf{0}.$$

A **short exact sequence** of complexes is an exact sequence

$$0 \to \mathbf{C}' \xrightarrow{i} \mathbf{C} \xrightarrow{p} \mathbf{C}'' \to 0.$$

The reader should realize that this notation is very compact. For example, if we write a complex as a column, then a short exact sequence of complexes is really the infinite commutative diagram with exact rows:

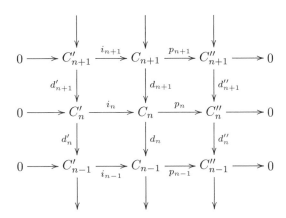

A sequence of complexes $\to \mathbf{C}^{n+1} \xrightarrow{f^{n+1}} \mathbf{C}^n \xrightarrow{f^n} \mathbf{C}^{n-1} \to$ is exact if and only if, for every $m \in \mathbb{Z}$, the row $\to C_m^{n+1} \to C_m^n \to C_m^{n-1} \to$ is an exact sequence of abelian groups.

- If $\left( (\mathbf{C}^i, d^i) \right)_{i \in I}$ is a family of complexes, then their **direct sum** is the complex

$$\bigoplus_{i \in I} \mathbf{C}^i = \to \bigoplus_{i \in I} C_{n+1}^i \xrightarrow{\oplus d_{n+1}^i} \bigoplus_{i \in I} C_n^i \xrightarrow{\oplus d_n^i} \bigoplus_{i \in I} C_{n-1}^i \to,$$

where $\bigoplus_i d_n^i$ acts coordinatewise; that is, $\bigoplus_i d_n^i \colon (c_n^i) \mapsto (d_n^i c_n^i)$.

To summarize, **Comp** is an abelian category; indeed, in light of Mitchell's Full Imbedding Theorem on page 624, we can view complexes as generalized abelian groups.

The next construction is fundamental; it gives a relation between different homology groups. The proof is a series of diagram chases. Ordinarily, we would just say that the proof is routine but, because of the importance of the result, we present (perhaps too many) details.

**Proposition 9.43 (Connecting Homomorphism).** *Given a short exact sequence of complexes* $0 \to \mathbf{C}' \xrightarrow{i} \mathbf{C} \xrightarrow{p} \mathbf{C}'' \to 0$, *then*

$$\partial_n \colon H_n(\mathbf{C}'') \to H_{n-1}(\mathbf{C}'),$$

*defined by* $\partial_n \colon \mathrm{cls}(z_n'') \mapsto \mathrm{cls}(i_{n-1}^{-1} d_n p_n^{-1} z_n'')$, *is a homomorphism for each* $n \in \mathbb{Z}$.

**Proof.** We will make many notational abbreviations in this proof. Consider the commutative diagram having exact rows:

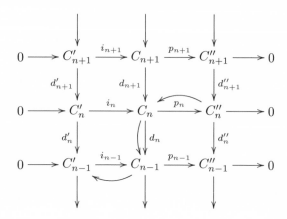

Suppose that $z'' \in C_n''$ and $d''z'' = 0$. Since $p_n$ is surjective, there is $c \in C_n$ with $pc = z''$. Now push $c$ down to $dc \in C_{n-1}$. By commutativity, $p_{n-1}dc = d''p_nc = d''z'' = 0$, so that $dc \in \ker p_{n-1} = \operatorname{im} i_{n-1}$. Therefore, there is a unique $c' \in C_{n-1}'$ with $i_{n-1}c' = dc$, for $i_{n-1}$ is an injection. Thus, $i_{n-1}^{-1}dp_n^{-1}z''$ makes sense; that is, the claim is that

$$\partial_n(\operatorname{cls}(z'')) = \operatorname{cls}(c')$$

is a well-defined homomorphism.

First, let us show independence of the choice of lifting. Suppose that $p_n\widehat{c} = z''$, where $\widehat{c} \in C_n$. Then $c - \widehat{c} \in \ker p_n = \operatorname{im} i_n$, so that there is $u' \in C_n'$ with $i_nu' = c - \widehat{c}$. By commutativity of the first square, we have

$$i_{n-1}d'u' = di_nu' = dc - d\widehat{c}.$$

Hence, $i^{-1}dc - i^{-1}d\widehat{c} = d'u' \in B_{n-1}'$; that is, $\operatorname{cls}(i^{-1}dc) = \operatorname{cls}(i^{-1}d\widehat{c})$. Thus, the formula gives a well-defined function

$$Z_n'' \to C_{n-1}'/B_{n-1}'.$$

Second, the function $Z_n'' \to C_{n-1}'/B_{n-1}'$ is a homomorphism. If $z'', z_1'' \in Z_n''$, let $pc = z''$ and $pc_1 = z_1''$. Since the definition of $\partial$ is independent of the choice of lifting, choose $c + c_1$ as a lifting of $z'' + z_1''$. This step may now be completed in a routine way.

Third, we show that if $i_{n-1}c' = dc$, then $c'$ is a cycle: $0 = ddc = dic' = idc'$, and so $d'c' = 0$ because $i$ is an injection. Hence, the formula gives a homomorphism

$$Z'' \to Z'/B' = H_{n-1}.$$

Finally, the subgroup $B_n''$ goes into $B_{n-1}'$. Suppose that $z'' = d''c''$, where $c'' \in C_{n+1}''$, and let $pu = c''$, where $u \in C_{n+1}$. Commutativity gives $pdu = d''pu = d''c'' = z''$. Since $\partial(z'')$ is independent of the choice of lifting, we choose $du$ with $pdu = z''$, and so $\partial(\operatorname{cls}(z'')) = \operatorname{cls}(i^{-1}ddu) = \operatorname{cls}(0)$. Therefore, the formula does give a homomorphism $\partial_n \colon H_n(\mathbf{C}'') \to H_{n-1}(\mathbf{C}')$.  •

The first question we ask is what homology functors do to short exact sequences of complexes. The next theorem is also proved by diagram chasing and, again, we give too many details because of the importance of the result. The reader should try to prove the theorem before looking at the proof we give.

**Theorem 9.44 (Long Exact Sequence).** *If*

$$0 \to \mathbf{C}' \xrightarrow{i} \mathbf{C} \xrightarrow{p} \mathbf{C}'' \to 0$$

*is an exact sequence of complexes, then there is an exact sequence of abelian groups*

$$\to H_{n+1}(\mathbf{C}'') \xrightarrow{\partial_{n+1}} H_n(\mathbf{C}') \xrightarrow{i_*} H_n(\mathbf{C}) \xrightarrow{p_*} H_n(\mathbf{C}'') \xrightarrow{\partial_n} H_{n-1}(\mathbf{C}') \to .$$

**Proof.** Our notation is abbreviated, and there are six inclusions to verify.

(i) $\operatorname{im} i_* \subseteq \ker p_*$.

$$p_* i_* = (pi)_* = 0_* = 0.$$

(ii) $\ker p_* \subseteq \operatorname{im} i_*$.

If $p_* \operatorname{cls}(z) = \operatorname{cls}(pz) = \operatorname{cls}(0)$, then $pz = d''c''$ for some $c'' \in C''_{n+1}$. But $p$ surjective gives $c'' = pc$ for some $c \in C_{n+1}$, so that $pz = d''pc = pdc$, because $p$ is a chain map, and so $p(z - dc) = 0$. By exactness, there is $c' \in C'_n$ with $ic' = z - dc$. Now $c'$ is a cycle, for $id'c' = dic' = dz - ddc = 0$, because $z$ is a cycle; since $i$ is injective, $d'c' = 0$. Therefore,

$$i_* \operatorname{cls}(c') = \operatorname{cls}(ic') = \operatorname{cls}(z - dc) = \operatorname{cls}(z).$$

(iii) $\operatorname{im} p_* \subseteq \ker \partial$.

If $p_*(\operatorname{cls}(c)) = \operatorname{cls}(pc) \in \operatorname{im} p_*$, then $\partial(\operatorname{cls}(pz)) = \operatorname{cls}(z')$, where $iz' = dp^{-1}pz$. Since this formula is independent of the choice of lifing of $pz$, let us choose $p^{-1}pz = z$. Now $dp^{-1}pz = dz = 0$, because $z$ is a cycle. Thus, $iz' = 0$, and hence $z' = 0$, because $i$ is injective.

(iv) $\ker \partial \subseteq \operatorname{im} p_*$.

If $\partial(\operatorname{cls}(z'')) = \operatorname{cls}(0)$, then $z' = i^{-1}dp^{-1}z'' \in B'$; that is, $z' = d'c'$ for some $c' \in C'$. But $iz' = id'c' = dic' = dp^{-1}z''$, so that $d(p^{-1}z'' - ic') = 0$; that is, $p^{-1}z'' - ic'$ is a cycle. Moreover, since $pi = 0$ because of exactness of the original sequence,

$$p_*(\operatorname{cls}(p^{-1}z'' - ic')) = pp^{-1}\operatorname{cls}(z'' - pic') = \operatorname{cls}(z'').$$

(v) $\operatorname{im} \partial \subseteq \ker i_*$.

We have $i_*\partial(\operatorname{cls}(z'')) = \operatorname{cls}(iz')$, where $iz' = dp^{-1}z'' \in B$; that is, $i_*\partial = 0$.

(vi) $\ker i_* \subseteq \operatorname{im} \partial$.

If $i_*(\operatorname{cls}(z')) = \operatorname{cls}(iz') = \operatorname{cls}(0)$, then $iz' = dc$ for some $c \in C$. Since $p$ is a chain map, $d''pc = pdc = piz' = 0$, by exactness of the original sequence, and so $pc$ is a cycle. But

$$\partial(\operatorname{cls}(pc)) = \operatorname{cls}(i^{-1}dp^{-1}pc) = \operatorname{cls}(i^{-1}dc) = \operatorname{cls}(i^{-1}iz') = \operatorname{cls}(z'). \quad \bullet$$

Theorem 9.44 is often called the ***exact triangle*** because of the diagram

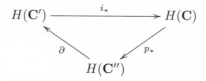

**Corollary 9.45 (Snake Lemma).** *Given a commutative diagram with exact rows,*

$$
\begin{array}{ccccccccc}
0 & \longrightarrow & A' & \longrightarrow & A & \longrightarrow & A'' & \longrightarrow & 0 \\
 & & \downarrow{\scriptstyle f} & & \downarrow{\scriptstyle g} & & \downarrow{\scriptstyle h} & & \\
0 & \longrightarrow & B' & \longrightarrow & B & \longrightarrow & B'' & \longrightarrow & 0
\end{array}
$$

*there is an exact sequence*

$$0 \to \ker f \to \ker g \to \ker h \to \operatorname{coker} f \to \operatorname{coker} g \to \operatorname{coker} h \to 0.$$

**Proof.** If we view each of the vertical maps $f$, $g$, and $h$ as a complex [as in Example 9.37(vi)], then the given commutative diagram can be viewed as a short exact sequence of complexes. The homology groups of each of these complexes has only two nonzero terms: thus, Example 9.39 shows that the homology groups of the first column are $H_1 = \ker f$, $H_0 = \operatorname{coker} f$, and all other $H_n = \{0\}$. The Snake Lemma now follows at once from the long exact sequence.   •

**Theorem 9.46 (Naturality of $\partial$).** *Given a commutative diagram of complexes with exact rows,*

$$
\begin{array}{ccccccccc}
\mathbf{0} & \longrightarrow & \mathbf{C}' & \overset{i}{\longrightarrow} & \mathbf{C} & \overset{p}{\longrightarrow} & \mathbf{C}'' & \longrightarrow & \mathbf{0} \\
 & & \downarrow{\scriptstyle f} & & \downarrow{\scriptstyle g} & & \downarrow{\scriptstyle h} & & \\
\mathbf{0} & \longrightarrow & \mathbf{A}' & \underset{j}{\longrightarrow} & \mathbf{A} & \underset{q}{\longrightarrow} & \mathbf{A}'' & \longrightarrow & \mathbf{0}
\end{array}
$$

*there is a commutative diagram of abelian groups with exact rows,*

$$
\begin{array}{ccccccccc}
\longrightarrow & H_n(\mathbf{C}') & \overset{i_*}{\longrightarrow} & H_n(\mathbf{C}) & \overset{p_*}{\longrightarrow} & H_n(\mathbf{C}'') & \overset{\partial}{\longrightarrow} & H_{n-1}(\mathbf{C}') & \longrightarrow \\
 & \downarrow{\scriptstyle f_*} & & \downarrow{\scriptstyle g_*} & & \downarrow{\scriptstyle h_*} & & \downarrow{\scriptstyle f_*} & \\
\longrightarrow & H_n(\mathbf{A}') & \underset{j_*}{\longrightarrow} & H_n(\mathbf{A}) & \underset{q_*}{\longrightarrow} & H_n(\mathbf{A}'') & \underset{\partial'}{\longrightarrow} & H_{n-1}(\mathbf{A}') & \longrightarrow
\end{array}
$$

**Proof.** Exactness of the rows is Theorem 9.44, while commutativity of the first two squares follows from $H_n$ being a functor. To prove commutativity of the square involving the connecting homomorphism, let us first display the chain maps and differentials in one (three-dimensional!) diagram:

If $\mathrm{cls}(z'') \in H_n(\mathbf{C}'')$, we must show that

$$f_*\partial(\mathrm{cls}(z'')) = \partial' h_*(\mathrm{cls}(z'')).$$

Let $c \in C_n$ be a lifting of $z''$; that is, $pc = z''$. Now $\partial(\mathrm{cls}(z'')) = \mathrm{cls}(z')$, where $iz' = dc$. Hence, $f_*\partial(\mathrm{cls}(z'')) = \mathrm{cls}(fz')$. On the other hand, since $h$ is a chain map, we have $qgc = hpc = hz''$. In computing $\partial'(\mathrm{cls}(hz''))$, we choose $gc$ as the lifting of $hz''$. Hence, $\partial'(\mathrm{cls}(hz'')) = \mathrm{cls}(u')$, where $ju' = \delta gc$. But

$$jfz' = giz' = gdc = \delta gc = ju',$$

and so $fz' = u'$, because $j$ is injective. $\quad\bullet$

We shall apply these general results in the next section.

## Exercises

$*$ **9.20.** If $\mathbf{C}$ is a complex with $C_n = \{0\}$ for some $n$, prove that $H_n(\mathbf{C}) = \{0\}$.

$*$ **9.21.** Prove that isomorphic complexes have the same homology: if $\mathbf{C}$ and $\mathbf{D}$ are isomorphic, then $H_n(\mathbf{C}) \cong H_n(\mathbf{D})$ for all $n$.

**9.22.** Regard the map $d\colon \mathbb{Z} \to \mathbb{Z}$, defined by $d\colon m \mapsto 2m$, as a complex, as in Example 9.37(vi). Prove that it is not a projective object in the category **Comp** even though each of its terms is a projective $\mathbb{Z}$-module. (Projectives in $\mathbf{Comp}(\mathcal{A})$ for arbitrary abelian categories $\mathcal{A}$ are characterized in Rotman, *An Introduction to Homological Algebra*, p. 652.)

$*$ **9.23.** In this exercise, we prove that the Snake Lemma implies the Long Exact Sequence (the converse is Corollary 9.45). Consider a commutative diagram with exact rows (note that two zeros are "missing" from this diagram):

  (i) Prove that $\Delta\colon \ker\gamma \to \mathrm{coker}\,\alpha$, defined by $\Delta\colon z \mapsto i^{-1}\beta p^{-1}z + \mathrm{im}\,\alpha$, is a well-defined homomorphism.

  (ii) Prove that there is an exact sequence

$$\ker\alpha \to \ker\beta \to \ker\gamma \xrightarrow{\Delta} \mathrm{coker}\,\alpha \to \mathrm{coker}\,\beta \to \mathrm{coker}\,\gamma.$$

(iii) Given a commutative diagram with exact rows,

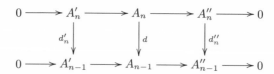

prove that the following diagram is commutative and has exact rows:

$$A_n'/\operatorname{im} d_{n+1}' \longrightarrow A_n/\operatorname{im} d_{n+1} \longrightarrow A_n''/\operatorname{im} d_{n+1}'' \longrightarrow 0$$

$$0 \longrightarrow \ker d_{n-1}' \longrightarrow \ker d_{n-1} \longrightarrow \ker d_{n-1}''$$

(iv) Use part (ii) and this last diagram to give another proof of the Long Exact Sequence.

* **9.24.** If $T$ is an exact additive functor, prove that $T$ commutes with homology; that is, for every complex $\mathbf{C}$ and for every $n \in \mathbb{Z}$, there is an isomorphism $H_n(TC) \cong TH_n(\mathbf{C})$.

* **9.25.** Let $f, g \colon \mathbf{C} \to \mathbf{C}'$ be chain maps, and let $F \colon \mathbf{C} \to \mathbf{C}'$ be an additive functor. If $f \simeq g$, prove that $Ff \simeq Fg$; that is, if $f$ and $g$ are homotopic, then $Ff$ and $Fg$ are homotopic.

**9.26.** Let $0 \to \mathbf{C}' \xrightarrow{i} \mathbf{C} \xrightarrow{p} \mathbf{C}'' \to 0$ be an exact sequence of complexes in which $\mathbf{C}'$ and $\mathbf{C}''$ are acyclic; prove that $\mathbf{C}$ is also acyclic.

**9.27.** Let $(\mathbf{C}, d)$ be a complex each of whose differentials $d_n$ is the zero map. Prove that $H_n(\mathbf{C}) \cong C_n$ for all $n$.

**9.28.** Prove that homology commutes with direct sums: for all $n$, there are natural isomorphisms for all $n$,

$$H_n\Big(\bigoplus_\alpha \mathbf{C}^\alpha\Big) \cong \bigoplus_\alpha H_n(\mathbf{C}^\alpha).$$

**9.29.** Let $\mathcal{A}$ be an abelian category having direct limits.

(i) Define a direct system of complexes $\{\mathbf{C}^i, \varphi_j^i\}$ in $\mathbf{Comp}(\mathcal{A})$, and prove that $\varinjlim \mathbf{C}^i$ exists in $\mathbf{Comp}(\mathcal{A})$.

(ii) If $\{\mathbf{C}^i, \varphi_j^i\}$ is a direct system of complexes in $\mathbf{Comp}(\mathcal{A})$, prove that there are natural isomorphisms for all $n$,

$$H_n\Big(\varinjlim \mathbf{C}^i\Big) \cong \varinjlim H_n(\mathbf{C}^i).$$

* **9.30.** Suppose that a complex $(\mathbf{C}, d)$ of left $R$-modules has a contracting homotopy in which the maps $s_n \colon C_n \to C_{n+1}$ satisfying

$$1_{C_n} = d_{n+1}s_n + s_{n-1}d_n$$

are only $\mathbb{Z}$-maps. Prove that $(\mathbf{C}, d)$ is an exact sequence.

**9.31.** (i) Let $0 \to F_n \to F_{n-1} \to \cdots \to F_0 \to 0$ be an exact sequence of finitely generated free $k$-modules, where $k$ is a commutative ring. Prove that

$$\sum_{i=0}^n (-1)^i \operatorname{rank}(F_i) = 0.$$

(ii) Let

$$0 \to F_n \to F_{n-1} \to \cdots \to F_0 \to M \to 0$$

and

$$0 \to F'_m \to F'_{m-1} \to \cdots \to F'_0 \to M \to 0$$

be free resolutions of a $k$-module $M$ in which all $F_i$ and $F'_j$ are finitely generated free $k$-modules. Prove that

$$\sum_{i=0}^{n} (-1)^i \operatorname{rank}(F_i) = \sum_{j=0}^{m} (-1)^j \operatorname{rank}(F'_j).$$

The common value, denoted by $\chi(M)$, is called the ***Euler–Poincaré characteristic*** of $M$.

**Hint.** Use Schanuel's Lemma.

**9.32.** (i) Let $\mathbf{C} : 0 \to C_n \to C_{n-1} \to \cdots \to C_0 \to 0$ be a complex of finitely generated free $k$-modules over a commutative ring $k$. Prove that

$$\sum_{i=0}^{n} (-1)^i \operatorname{rank}(C_i) = \sum_{i=0}^{n} (-1)^i \operatorname{rank}(H_i(\mathbf{C})).$$

(ii) Let $0 \to M' \to M \to M'' \to 0$ be an exact sequence of $k$-modules. If two of the modules have an Euler-Poincaré characteristic, prove that the third module does, too, and that

$$\chi(M) = \chi(M') + \chi(M'').$$

**9.33.** (i) **(Barratt–Whitehead)** Consider the commutative diagram with exact rows:

$$
\begin{array}{ccccccccccc}
A_n & \xrightarrow{i_n} & B_n & \xrightarrow{p_n} & C_n & \xrightarrow{\partial_n} & A_{n-1} & \longrightarrow & B_{n-1} & \longrightarrow & C_{n-1} \\
\downarrow{f_n} & & \downarrow{g_n} & & \downarrow{h_n} & & \downarrow{f_{n-1}} & & \downarrow{g_{n-1}} & & \downarrow{h_{n-1}} \\
A'_n & \xrightarrow{j_n} & B'_n & \xrightarrow{q_n} & C'_n & \longrightarrow & A'_{n-1} & \longrightarrow & B'_{n-1} & \longrightarrow & C'_{n-1}
\end{array}
$$

If each $h_n$ is an isomorphism, prove that there is an exact sequence

$$A_n \xrightarrow{(f_n, i_n)} A'_n \oplus B_n \xrightarrow{j_n - q_n} B'_n \xrightarrow{\partial_n h_n^{-1} q_n} A_{n-1} \to A'_{n-1} \oplus B_{n-1} \to B'_{n-1},$$

where $(f_n, i_n): a_n \mapsto (f_n a_n, i_n a_n)$ and $j_n - g_n: (a'_n, b_n) \mapsto j_n a'_n - g_n b_n$.

(ii) **(Mayer–Vietoris)** Assume, in the diagram of Theorem 9.46, that every third vertical map $h_*$ is an isomorphism. Prove that there is an exact sequence

$$\to H_n(\mathbf{C}') \to H_n(\mathbf{A}') \oplus H_n(\mathbf{C}) \to H_n(\mathbf{A}) \to H_{n-1}(\mathbf{C}') \to .$$

**9.34.** (i) Define a direct system of complexes $\{\mathbf{C}^i, \varphi^i_j\}$, and prove that $\varinjlim \mathbf{C}^i$ exists.

(ii) If $\{\mathbf{C}^i, \varphi^i_j\}$ is a direct system of complexes over a directed index set, prove, for all $n \geq 0$, that

$$H_n(\varinjlim \mathbf{C}^i) \cong \varinjlim H_n(\mathbf{C}^i).$$

**Remark.** The ***Eilenberg–Steenrod axioms*** characterize homology functors on the category $\mathbf{Top}^2$ of all pairs of topological spaces and continuous maps [an object in $\mathbf{Top}^2$ is an ordered pair of topological spaces $(X, A)$, where $A \subseteq X$, and a morphism $f: (X, A) \to (Y, B)$ is a continuous map $f: X \to Y$ with $f(A) \subseteq B$]. If $h_n: \mathbf{Top}^2 \to \mathbf{Ab}$ is a sequence of functors, for all $n \geq 0$, satisfying the long exact sequence, naturality of connecting homomorphisms, $h_n(f) = h_n(g)$ whenever $f$ and $g$ are homotopic, $h_0(X, \varnothing) = \mathbb{Z}$ and $h_n(X) = \{0\}$ for all $n > 0$ when $X$ is a 1-point space, and *Excision*, then there are natural isomorphisms $h_n \to H_n$ for all $n$. In the presence of the other axioms, Excision can be replaced by exactness of the Mayer–Vietoris sequence. ◄

## Section 9.5. Derived Functors

In order to apply general results about homology, we need a source of short exact sequences of complexes, as well as of commutative diagrams in which they sit. The idea is to replace modules by resolutions (which are complexes but, more importantly, are glorified presentations of the modules); replacing each term in a short exact sequence of modules by a resolution will give a short exact sequence of complexes. We then apply either Hom or $\otimes$ to these complexes, and the resulting homology groups are called Ext or Tor.

This section is fairly dry, but it is necessary to establish the existence of homology functors. The most useful theorems in this section are Theorem 9.48 (Comparison Theorem), Proposition 9.55 (Horseshoe Lemma), Corollary 9.59 (Long Exact Sequence), and Proposition 9.60 (Naturality of the Connecting Homomorphism).

For those readers who are interested in using Tor (the left derived functors of tensor) and Ext (the right derived functors of Hom) immediately, and who are willing to defer looking at mazes of arrows, the next theorem gives a set of axioms characterizing the contravariant functors $\text{Ext}^n(\square, M)$. There are similar axiomatic descriptions of the covariant Ext functors $\text{Ext}^n(M, \square)$ and of the Tor functors $\text{Tor}_n(\square, M)$ and $\text{Tor}_n(M, \square)$ (see Exercises 9.44 and 9.45 on page 815).

**Theorem 9.47 (Axioms for Ext($\square, M$)).** *Let* $\text{EXT}^n \colon {}_R\mathbf{Mod} \to \mathbf{Ab}$ *be a sequence of contravariant functors, for $n \geq 0$, such that*

(i) *for every short exact sequence $0 \to A \to B \to C \to 0$, there is a long exact sequence and natural connecting homomorphisms*

$$\to \text{EXT}^n(C) \to \text{EXT}^n(B) \to \text{EXT}^n(A) \xrightarrow{\Delta_n} \text{EXT}^{n+1}(C) \to;$$

(ii) *there is a left $R$-module $M$ with $\text{EXT}^0$ and $\text{Hom}_R(\square, M)$ naturally isomorphic;*

(iii) $\text{EXT}^n(P) = \{0\}$ *for all projective modules $P$ and all $n \geq 1$.*

*If $\text{Ext}^n(\square, M)$ is another sequence of contravariant functors satisfying these same axioms, then $\text{EXT}^n$ is naturally isomorphic to $\text{Ext}^n(\square, M)$ for all $n \geq 0$.*

**Proof.** We proceed by induction on $n \geq 0$. The base step is axiom (ii).

For the inductive step, given a module $A$, choose a short exact sequence

$$0 \to L \to P \to A \to 0,$$

where $P$ is projective. By axiom (i), there is a diagram with exact rows:

$$
\begin{array}{ccccccc}
\text{EXT}^0(P) & \longrightarrow & \text{EXT}^0(L) & \xrightarrow{\Delta_0} & \text{EXT}^1(A) & \longrightarrow & \text{EXT}^1(P) \\
\big\downarrow{\scriptstyle\tau_P} & & \big\downarrow{\scriptstyle\tau_L} & & \big\downarrow & & \\
\text{Hom}(P,M) & \longrightarrow & \text{Hom}(L,M) & \xrightarrow[\partial_0]{} & \text{Ext}^1(A,M) & \longrightarrow & \text{Ext}^1(P,M)
\end{array}
$$

where the maps $\tau_P$ and $\tau_L$ are the isomorphisms given by axiom (ii). This diagram commutes because of the naturality of the isomorphism $\text{EXT}^0 \to \text{Hom}(\square, M)$. By axiom (iii), $\text{Ext}^1(P,M) = \{0\}$ and $\text{EXT}^1(P) = \{0\}$. It follows that the maps $\Delta_0$ and $\partial_0$ are surjective. This is a diagram as in Proposition 6.116, and so there exists an isomorphism $\text{EXT}^1(A) \to \text{Ext}^1(A,M)$ making the augmented diagram commute.

We may now assume that $n \geq 1$, and we look further out in the long exact sequence. By axiom (i), there is a diagram with exact rows

$$
\begin{array}{ccccccc}
\text{EXT}^n(P) & \longrightarrow & \text{EXT}^n(L) & \xrightarrow{\Delta_n} & \text{EXT}^{n+1}(A) & \longrightarrow & \text{EXT}^{n+1}(P) \\
& & \big\downarrow{\scriptstyle\sigma} & & \big\downarrow & & \\
\text{Ext}^n(P,M) & \longrightarrow & \text{Ext}^n(L,M) & \xrightarrow[\partial_n]{} & \text{Ext}^{n+1}(A,M) & \longrightarrow & \text{Ext}^{n+1}(P,M)
\end{array}
$$

where $\sigma \colon \text{EXT}^n(L) \to \text{Ext}^n(L,M)$ is an isomorphism given by the inductive hypothesis. Since $n \geq 1$, all four terms involving the projective $P$ are $\{0\}$; it follows from exactness of the rows that both $\Delta_n$ and $\partial_n$ are isomorphisms. Finally, the composite $\partial_n \sigma \Delta_n^{-1} \colon \text{EXT}^{n+1}(A) \to \text{Ext}^{n+1}(A,M)$ is an isomorphism.

It remains to prove that the isomorphisms $\text{EXT}^n(A) \to \text{Ext}^n(A,M)$ constitute a natural transformation. It is here that the assumed naturality in axiom (i) of the connecting homomorphism is used, and this is left for the reader to do. $\quad\bullet$

Such slow-starting induction proofs, using long exact sequences, vanishing for special (e.g., projective) arguments, and two base steps $n = 0$ and $n = 1$ before proving the inductive step, are called *dimension shifting*.

The rest of this section consists of constructing functors that satisfy these axioms. We prove the existence of Ext and Tor using derived functors (there are other proofs as well). As these functors are characterized by short lists of axioms, analogous to the list in Theorem 9.47, we can usually work with Ext and Tor without being constantly aware of the details of their construction.

**9.5.1. Left Derived Functors.** Let us say at the outset that we state and prove all theorems for complexes of abelian groups. However, all we do holds true, in particular, for complexes of modules. Indeed, the category **Ab** of abelian groups can be replaced by any abelian category, using Theorem 7.147 and the subsequent discussion.

We begin with a technical definition.

**Definition.** If $\to P_2 \to P_1 \xrightarrow{d_1} P_0 \to A \to 0$ is a projective resolution of a module $A$, then its **deleted projective resolution** is the complex

$$\mathbf{P}_A = \to P_2 \to P_1 \xrightarrow{d^0} P_0 \to 0.$$

Similarly, if $0 \to A \to E^0 \xrightarrow{d^0} E^1 \to E^2 \to$ is an injective resolution of a module $A$, then a **deleted injective resolution** is the complex

$$\mathbf{E}^A = 0 \to E^0 \xrightarrow{d^0} E^1 \to E^2 \to .$$

In either case, deleting $A$ loses no information, for $A$ can be recaptured: $A \cong \operatorname{coker} d_1$ in the first case, and $A \cong \ker d^0$ in the second case. Of course, a deleted resolution is no longer exact; indeed, $H_0(\mathbf{P}_A) \neq \{0\}$ if $A \neq \{0\}$:

$$H_0(\mathbf{P}_A) = \ker(P_0 \to \{0\})/\operatorname{im} d_1 = P_0/\operatorname{im} d_1 \cong A.$$

Similarly, $H_0(\mathbf{E}^A) \cong A$.

We know that a module has many resolutions, and so the next result is fundamental.

**Theorem 9.48 (Comparison Theorem).** *Given a map $f\colon A \to A'$, consider the diagram*

$$
\begin{array}{ccccccccc}
\longrightarrow & P_2 & \xrightarrow{d_2} & P_1 & \xrightarrow{d_1} & P_0 & \xrightarrow{\varepsilon} & A & \longrightarrow 0 \\
& \downarrow{\widehat{f}_2} & & \downarrow{\widehat{f}_1} & & \downarrow{\widehat{f}_0} & & \downarrow{f} & \\
\longrightarrow & P_2' & \xrightarrow{d_2'} & P_1' & \xrightarrow{d_1'} & P_0' & \xrightarrow{\varepsilon'} & A' & \longrightarrow 0
\end{array}
$$

*where the rows are complexes. If each $P_n$ in the top row is projective and the bottom row is exact, then there exists a chain map $\widehat{f}\colon \mathbf{P} \to \mathbf{P}'$ making the completed diagram commute. Moreover, any two such chain maps are homotopic.*

**Remark.** The dual of the Comparison Theorem is also true. Now the complexes go off to the right, the top row is assumed exact, and every term in the bottom row other than $A'$ is injective. ◀

**Proof.**

(i) We prove the existence of $\widehat{f}_n$ by induction on $n \geq 0$. For the base step $n = 0$, consider the diagram

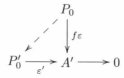

Since $\varepsilon'$ is surjective and $P_0$ is projective, there exists a map $\widehat{f}_0\colon P_0 \to P_0'$ with $\varepsilon'\widehat{f}_0 = f\varepsilon$.

For the inductive step, consider the diagram

$$
\begin{array}{ccccc}
P_{n+1} & \xrightarrow{d_{n+1}} & P_n & \xrightarrow{d_n} & P_{n-1} \\
 & & \downarrow{\widehat{f}_n} & & \downarrow{\widehat{f}_{n-1}} \\
P'_{n+1} & \xrightarrow[d'_{n+1}]{} & P'_n & \xrightarrow[d'_n]{} & P'_{n-1}
\end{array}
$$

If we can show that $\operatorname{im} \widehat{f}_n d_{n+1} \subseteq \operatorname{im} d'_{n+1}$, then we will have the diagram

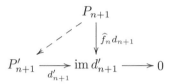

Projectivity of $P_{n+1}$ will provide a map $\widehat{f}_{n+1} \colon P_{n+1} \to P'_{n+1}$ with $d'_{n+1}\widehat{f}_{n+1} = \widehat{f}_n d_{n+1}$. To check that the inclusion does hold, note that exactness at $P'_n$ of the bottom row of the original diagram gives $\operatorname{im} d'_{n+1} = \ker d'_n$, and so it suffices to prove that $d'_n \widehat{f}_n d_{n+1} = 0$. But $d'_n \widehat{f}_n d_{n+1} = \widehat{f}_{n-1} d_n d_{n+1} = 0$.

(ii) We now prove uniqueness of $\widehat{f}$ up to homotopy. If $h \colon \mathbf{P} \to \mathbf{P}'$ is a chain map also satisfying $\varepsilon' h_0 = f\varepsilon$, then we construct the terms $s_n \colon P_n \to P'_{n+1}$ of a homotopy $s$ by induction on $n \geq -1$; that is, we want

$$
h_n - \widehat{f}_n = d'_{n+1} s_n + s_{n-1} d_n.
$$

Let us now begin the induction. First, define $\widehat{f}_{-1} = f = h_{-1}$. If we define $s_{-1} = 0 = s_{-2}$, then

$$
h_{-1} - \widehat{f}_{-1} = f - f = 0 = d'_0 s_{-1} + s_{-2} d_{-1}
$$

for any choice of $d'_0$ and $d_{-1}$; define $d'_0 = \varepsilon'$ and $d_{-1} = 0$.

For the inductive step, it suffices to prove, for all $n \geq -1$, that

$$
\operatorname{im}(h_{n+1} - \widehat{f}_{n+1} - s_n d_{n+1}) \subseteq \operatorname{im} d'_{n+2},
$$

for we would then have a diagram with exact row

$$
\begin{array}{c}
 & & P_{n+1} \\
 & \swarrow & \downarrow{h_{n+1}-\widehat{f}_{n+1}-s_n d_{n+1}} \\
P'_{n+2} & \xrightarrow[d'_{n+2}]{} \operatorname{im} d'_{n+2} & \xrightarrow{\phantom{xx}} 0
\end{array}
$$

and projectivity of $P_{n+1}$ would give a map $s_{n+1} \colon P_{n+1} \to P'_{n+2}$ satisfying the desired equation. As in the proof of part (i), exactness of the bottom row of the original diagram gives $\operatorname{im} d'_{n+2} = \ker d'_{n+1}$, and so it suffices to prove

$$
d'_{n+1}(h_{n+1} - \widehat{f}_{n+1} - s_n d_{n+1}) = 0.
$$

But

$$d'_{n+1}(h_{n+1} - \widehat{f}_{n+1} - s_n d_{n+1}) = d'_{n+1}(h_{n+1} - \widehat{f}_{n+1}) - d'_{n+1} s_n d_{n+1}$$
$$= d'_{n+1}(h_{n+1} - \widehat{f}_{n+1}) - (h_n - \widehat{f}_n - s_{n-1} d_n) d_{n+1}$$
$$= d'_{n+1}(h_{n+1} - \widehat{f}_{n+1}) - (h_n - \widehat{f}_n) d_{n+1},$$

and the last term is 0 because $h$ and $\widehat{f}$ are chain maps.  •

We say that the chain map $\widehat{f}$ just constructed is *over* $f$.

**Definition.** If $f: A \to A'$ is a map of abelian groups and $\mathbf{P}_A$ and $\mathbf{P}'_{A'}$ are deleted projective resolutions of $A$ and $A'$, respectively, then a chain map $\widehat{f}: \mathbf{P}_A \to \mathbf{P}'_{A'}$,

$$
\begin{array}{ccccccccc}
\longrightarrow & P_2 & \xrightarrow{d_2} & P_1 & \xrightarrow{d_1} & P_0 & \xrightarrow{\varepsilon} & A & \longrightarrow & 0 \\
& \downarrow{\widehat{f}_2} & & \downarrow{\widehat{f}_1} & & \downarrow{\widehat{f}_0} & & \downarrow{f} & & \\
\longrightarrow & P'_2 & \xrightarrow{d'_2} & P'_1 & \xrightarrow{d'_1} & P'_0 & \xrightarrow{\varepsilon'} & A' & \longrightarrow & 0
\end{array}
$$

is said to be **over** $f$ if the square containing $f: A \to A'$ commutes:

$$f\varepsilon = \varepsilon' \widehat{f}_0.$$

Thus, the Comparison Theorem implies, given a homomorphism $f: A \to A'$, that a chain map over $f$ always exists between deleted projective resolutions of $A$ and $A'$; moreover, such a chain map is unique to homotopy.

Given a covariant additive functor $T$, we are now going to construct, for all $n \in \mathbb{Z}$, its **left derived functors** $L_n T$. The definition is suggested by two examples. First, in Algebraic Topology, we apply $T = \square \otimes G$ to the chain complex of an abstract simplicial complex $K$ to get homology groups $H_n(K; G)$ of $K$ with **coefficients** in an abelian group $G$. Second, we saw that the formulas that arise in Schreier's investigation of group extensions suggest constructing a free resolution of the trivial $\mathbb{Z}Q$-module $\mathbb{Z}$, where $Q$ is a group; we shall see, for any left $\mathbb{Z}Q$-module $M$, that applying $T = \square \otimes M$ to this resolution yields *group homology*.

The definition of left derived functors is in two parts: first on objects; then on morphisms. Choose, once for all, a deleted projective resolution $\mathbf{P}_A$ of every module $A$. As in Example 9.37(ix), form the complex $T\mathbf{P}_A$, and take homology:

$$(L_n T)A = H_n(T\mathbf{P}_A).$$

We now define $(L_n T)f$, where $f: A \to A'$ is a homomorphism. By the Comparison Theorem, there is a chain map $\widehat{f}: \mathbf{P}_A \to \mathbf{P}'_{A'}$ over $f$. It follows that $T\widehat{f}: T\mathbf{P}_A \to T\mathbf{P}'_{A'}$ is also a chain map, and we define $(L_n T)f: (L_n T)A \to (L_n T)A'$ by

$$(L_n T)f = H_n(T\widehat{f}) = (T\widehat{f})_*.$$

In more detail, if $z \in \ker T d_n$, then

$$(L_n T)f: z + \mathrm{im}(T d_{n+1}) \mapsto (T\widehat{f}_n)z + \mathrm{im}(T d'_{n+1});$$

that is, $(L_n T)f: \mathrm{cls}(z) \mapsto \mathrm{cls}((T\widehat{f}_n)z)$.

In pictures, look at the chosen projective resolutions:

$$\longrightarrow P_1 \longrightarrow P_0 \longrightarrow A \longrightarrow 0$$

$$\longrightarrow P_1' \longrightarrow P_0' \longrightarrow A' \longrightarrow 0$$

Fill in a chain map $\widehat{f}$ over $f$, delete $A$ and $A'$, and apply $T$ to this diagram:

$$\longrightarrow TP_1 \longrightarrow TP_0 \longrightarrow 0$$

$$\longrightarrow TP_1' \longrightarrow TP_0' \longrightarrow 0$$

Now take the map induced by $T\widehat{f}$ in homology: $(L_nT)f\colon \mathrm{cls}(z) \mapsto \mathrm{cls}((T\widehat{f}_n)z)$.

**Proposition 9.49.** *If* $T\colon {}_R\mathbf{Mod} \to \mathbf{Ab}$ *is a covariant additive functor, for some ring $R$, then*

$$L_nT\colon {}_R\mathbf{Mod} \to \mathbf{Ab}$$

*is a covariant additive functor for every $n \in \mathbb{Z}$.*

**Proof.** We will prove that $L_nT$ is well-defined on morphisms; it is then routine to check that it is a covariant additive functor (remember that $H_n$ is a covariant additive functor $\mathbf{Comp} \to \mathbf{Ab}$).

If $h\colon \mathbf{P}_A \to \mathbf{P}_{A'}'$ is another chain map over $f$, then the Comparison Theorem says that $h \simeq \widehat{f}$; therefore, $Th \simeq T\widehat{f}$, by Exercise 9.25 on page 794, and so $H_n(Th) = H_n(T\widehat{f})$, by Proposition 9.41.   •

Recall Example 6.10(iii): if $r \in Z(R)$ is a central element in a ring $R$ and $A$ is a left $R$-module, then *multiplication by $r$*, defined by $\mu_r^A\colon a \mapsto ra$, is an $R$-map $\mu_r^A\colon A \to A$. We usually suppress $A$ in this notation.

**Definition.** A functor $T\colon {}_R\mathbf{Mod} \to {}_R\mathbf{Mod}$, of either variance, ***preserves multiplications*** if $T(\mu_r)\colon TA \to TA$ is multiplication by $r$ for all $r \in Z(R)$.

Tensor and Hom preserve multiplications in both variables.

**Example 9.50.** We show that if $T$ preserves multiplications, then so does $L_nT$:

$$(L_nT)\mu_r^A = \text{ multiplication by } r = \mu_r^{(L_nT)A}.$$

If $\mathbf{P}$ is a projective resolution of $A$, consider the diagram

$$\longrightarrow P_2 \xrightarrow{d_2} P_1 \xrightarrow{d_1} P_0 \xrightarrow{\varepsilon} A \longrightarrow 0$$

$$\longrightarrow P_2 \xrightarrow{d_2} P_1 \xrightarrow{d_1} P_0 \xrightarrow{\varepsilon} A \longrightarrow 0$$

where $\widehat{\mu}_n\colon P_n \to P_n$ is also multiplication by $r$ for all $n \geq 0$. Then $\widehat{\mu} = (\widehat{\mu}_n)$ is a chain map over $\mu_r$. Since $T$ preserves multiplications, the terms of the chain map $T\widehat{\mu}$ are multiplication by $r$, and so the induced maps in homology are also multiplication by $r$: if $z_n \in \ker(Td_n)$, then

$$(T\widehat{\mu})_* \colon \operatorname{cls}(z_n) \mapsto \operatorname{cls}((T\widehat{\mu}_n)z_n) = \operatorname{cls}(rz_n) = r\operatorname{cls}(z_n). \quad \blacktriangleleft$$

**Proposition 9.51.** *If $T \colon {}_R\mathbf{Mod} \to \mathbf{Ab}$ is a covariant additive functor, then $(L_nT)A = \{0\}$ for all negative $n$ and for all $A$.*

**Proof.** Since the $n$th term of $\mathbf{P}_A$ is $\{0\}$ when $n$ is negative, Exercise 9.20 on page 793 gives $(L_nT)A = \{0\}$ for all $n < 0$. $\quad \bullet$

**Definition.** If $B$ is a left $R$-module and $T = \square \otimes_R B$, define

$$\operatorname{Tor}_n^R(\square, B) = L_nT.$$

Thus, if $\mathbf{P}_A = \to P_2 \xrightarrow{d_2} P_1 \xrightarrow{d_1} P_0 \to 0$ is the chosen deleted projective resolution of a right $R$-module $A$, then

$$\operatorname{Tor}_n^R(A, B) = H_n(\mathbf{P}_A \otimes_R B) = \frac{\ker(d_n \otimes 1_B)}{\operatorname{im}(d_{n+1} \otimes 1_B)}.$$

The domain of $\operatorname{Tor}_n^R(\square, B)$ is $\mathbf{Mod}_R$, the category of right $R$-modules; its target is $\mathbf{Ab}$, the category of abelian groups. However, if $R$ is commutative, then $A \otimes_R B$ is an $R$-module and induced homomorphisms are $R$-maps, and so the values of their quotient $\operatorname{Tor}_n^R(\square, B)$ lie in ${}_R\mathbf{Mod}$ [for $d_n \otimes 1_B$ is an $R$-map, hence, $\ker(d_n \otimes 1_B)$ and $\operatorname{im}(d_n \otimes 1_B)$ are $R$-modules, and so their quotient $\operatorname{Tor}_n^R(A, B)$ is an $R$-module].

**Definition.** If $A$ is a right $R$-module and $T = A \otimes_R \square$, define $\operatorname{tor}_n^R(A, \square) = L_nT$. Thus, if $\mathbf{Q}_B = \to Q_2 \xrightarrow{d_2} Q_1 \xrightarrow{d_1} Q_0 \to 0$ is the chosen deleted projective resolution of a left $R$-module $B$, then

$$\operatorname{tor}_n^R(A, B) = H_n(A \otimes_R \mathbf{Q}_B) = \frac{\ker(1_A \otimes d_n)}{\operatorname{im}(1_A \otimes d_{n+1})}.$$

The domain of $\operatorname{tor}_n^R(A, \square)$ is ${}_R\mathbf{Mod}$, the category of left $R$-modules; its target is $\mathbf{Ab}$, the category of abelian groups and, as above, if $R$ is commutative, then the values of $\operatorname{tor}_n^R(\square, B)$ lie in ${}_R\mathbf{Mod}$.

One of the nice theorems of Homological Algebra is, for all $A$ and $B$ (and for all $R$ and $n$), that

$$\operatorname{Tor}_n^R(A, B) \cong \operatorname{tor}_n^R(A, B).$$

There is a proof using spectral sequences (a sketch is given on page 857); there is also an elementary (but tricky) proof due to Zaks (Rotman, *An Introduction to Homological Algebra*, p. 355). We assume this result and identify Tor and tor.

The definition of $L_nT$ uses a choice of deleted projection resolution of each module. Does $L_nT$ depend on this choice?

**Proposition 9.52.** *Denote the left derived functors arising from new choices $\widetilde{\mathbf{P}}$ of projective resolutions by $\widetilde{L}_nT$. Given a ring $R$ and a covariant additive functor $T \colon {}_R\mathbf{Mod} \to \mathbf{Ab}$, then, for each $n$, the functors $L_nT$ and $\widetilde{L}_nT$ are naturally isomorphic. In particular, for all $A$,*

$$(L_nT)A \cong (\widetilde{L}_nT)A,$$

*and so these groups are independent of the choice of projective resolution of $A$.*

**Proof.** Consider the diagram

$$\longrightarrow P_2 \longrightarrow P_1 \longrightarrow P_0 \longrightarrow A \longrightarrow 0$$
$$\downarrow^{1_A}$$
$$\longrightarrow \widetilde{P}_2 \longrightarrow \widetilde{P}_1 \longrightarrow \widetilde{P}_0 \longrightarrow A \longrightarrow 0$$

where the top row is the chosen projective resolution of $A$ used to define $L_n T$ and the bottom is that used to define $\widetilde{L}_n T$. By the Comparison Theorem, there is a chain map $\iota\colon \mathbf{P}_A \to \widetilde{\mathbf{P}}_A$ over $1_A$. Applying $T$ gives a chain map $T\iota\colon T\mathbf{P}_A \to T\widetilde{\mathbf{P}}_A$ over $T1_A = 1_{TA}$. This last chain map induces homomorphisms, one for each $n$,

$$\tau_A = (T\iota)_*\colon (L_n T)A \to (\widetilde{L}_n T)A.$$

We now prove that each $\tau_A$ is an isomorphism (thereby proving the last statement in the theorem) by constructing its inverse. Turn the preceding diagram upside down, so that the chosen projective resolution $\mathbf{P}_A \to A \to 0$ is now the bottom row. Again, the Comparison Theorem gives a chain map, say, $\kappa\colon \widetilde{\mathbf{P}}_A \to \mathbf{P}_A$. Now the composite $\kappa\iota$ is a chain map from $\mathbf{P}_A$ to itself over $1_{\mathbf{P}_A}$. By the uniqueness statement in the Comparison Theorem, $\kappa\iota \simeq 1_{\mathbf{P}_A}$; similarly, $\iota\kappa \simeq 1_{\widetilde{\mathbf{P}}_A}$. It follows that $T(\iota\kappa) \simeq 1_{T\widetilde{\mathbf{P}}_A}$ and $T(\kappa\iota) \simeq 1_{T\mathbf{P}_A}$. Hence, $1 = (T\iota\kappa)_* = (T\iota)_*(T\kappa)_*$ and $1 = (T\kappa\iota)_* = (T\kappa)_*(T\iota)_*$. Therefore, $\tau_A = (T\iota)_*$ is an isomorphism.

We now prove that the isomorphisms $\tau_A$ constitute a natural isomorphism; that is, if $f\colon A \to B$ is a homomorphism, then the following diagram commutes:

$$
\begin{array}{ccc}
(L_n T)A & \xrightarrow{\ \tau_A\ } & (\widetilde{L}_n T)A \\
{\scriptstyle (L_n T)f}\downarrow & & \downarrow{\scriptstyle (\widetilde{L}_n T)f} \\
(L_n T)B & \xrightarrow[\ \tau_B\ ]{} & (\widetilde{L}_n T)B
\end{array}
$$

To evaluate in the clockwise direction, consider

$$\longrightarrow P_1 \longrightarrow P_0 \longrightarrow A \longrightarrow 0$$
$$\downarrow^{1_A}$$
$$\longrightarrow \widetilde{P}_1 \longrightarrow \widetilde{P}_0 \longrightarrow A \longrightarrow 0$$
$$\downarrow^{f}$$
$$\longrightarrow \widetilde{Q}_1 \longrightarrow \widetilde{Q}_0 \longrightarrow B \longrightarrow 0$$

where the bottom row is some projective resolution of $B$. The Comparison Theorem gives a chain map $\mathbf{P}_A \to \widetilde{\mathbf{Q}}_B$ over $f1_A = f$. Going counterclockwise, the picture will now have the chosen projective resolution of $B$ as its middle row, and we get a chain map $\mathbf{P}_A \to \widetilde{\mathbf{Q}}_B$ over $1_B f = f$. The uniqueness statement in the Comparison Theorem tells us that these two chain maps are homotopic, so that they give the same homomorphism in homology. Thus, the appropriate diagram commutes, showing that $\tau\colon L_n T \to \widetilde{L}_n T$ is a natural isomorphism. $\bullet$

**Corollary 9.53.** $\operatorname{Tor}_n^R(A, B)$ *is independent of the choices of projective resolutions of $A$ and of $B$.*

**Proof.** Proposition 9.52 applies at once to the left derived functors of $\square \otimes_R B$, namely, $\operatorname{Tor}_n^R(\square, B)$, and to the left derived functors of $A \otimes_R \square$, namely $\operatorname{tor}_n^R(A, \square)$. But we have already cited the fact that $\operatorname{Tor}_n^R(A, B) \cong \operatorname{tor}_n^R(A, B)$.   •

**Corollary 9.54.** *Let $T \colon {}_R\mathbf{Mod} \to \mathbf{Ab}$ be a covariant additive functor. If $P$ is a projective left $R$-module, then $(L_n T)P = \{0\}$ for all $n \geq 1$.*

*In particular, if either $A$ is a projective right $R$-module or $B$ is a projective left $R$-module, then $\operatorname{Tor}_n^R(A, B) = \{0\}$ for all $n \geq 1$.*

**Proof.** Since $P$ is projective, a projective resolution of it is

$$\mathbf{C} = \to 0 \to 0 \to P \xrightarrow{1_P} P \to 0,$$

where $C_0 = P$ and $C_n = \{0\}$ for $n \geq 1$, and so the corresponding deleted projective resolution $\mathbf{C}_P$ has only one nonzero term. It follows that $T\mathbf{C}_P$ is a complex having $n$th term $\{0\}$ for all $n \geq 1$, and so $(L_n T)P = H_n(T\mathbf{C}_P) = \{0\}$ for all $n \geq 1$, by Exercise 9.20 on page 793. The second statement follows from our recognizing that Tor = tor; that is, Tor does not depend on which variable is resolved.   •

We are now going to show that there is a long exact sequence of left derived functors. We begin with a useful lemma; it says that if we are given an exact sequence of modules as well as a projective resolution of its first and third terms, then we can "fill in the horseshoe"; that is, there is a projective resolution of the middle term that fits in the middle.

**Proposition 9.55 (Horseshoe Lemma).** *Given a diagram*

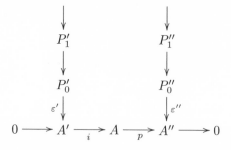

*where the columns are projective resolutions and the row is exact, then there exists a projective resolution of $A$ and chain maps so that the three columns form an exact sequence of complexes.*

**Remark.** The dual theorem, in which projective resolutions are replaced by injective resolutions and all arrows are reversed, is also true.   ◄

**Proof.** We show first that there is a projective $P_0$ and a commutative $3 \times 3$ diagram with exact columns and rows:

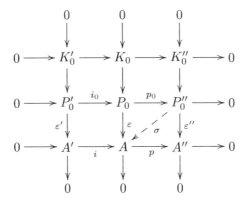

Define $P_0 = P_0' \oplus P_0''$; it is projective because both $P_0'$ and $P_0''$ are projective. Define $i_0 \colon P_0' \to P_0' \oplus P_0''$ by $x' \mapsto (x', 0)$, and define $p_0 \colon P_0' \oplus P_0'' \to P_0''$ by $(x', x'') \mapsto x''$. It is clear that

$$0 \to P_0' \xrightarrow{i_0} P_0 \xrightarrow{p_0} P_0'' \to 0$$

is exact. Since $P_0''$ is projective, there exists a map $\sigma \colon P_0'' \to A$ with $p\sigma = \varepsilon''$. Now define $\varepsilon \colon P_0 \to A$ by $\varepsilon \colon (x', x'') \mapsto i\varepsilon'x' + \sigma x''$. It is left as a routine exercise that if $K_0 = \ker \varepsilon$, then there are maps $K_0' \to K_0$ and $K_0 \to K_0''$ (where $K_0' = \ker \varepsilon'$ and $K_0'' = \ker \varepsilon''$), so that the resulting $3 \times 3$ diagram commutes. Exactness of the top row is Exercise 6.87 on page 486, and surjectivity of $\varepsilon$ follows from the Five Lemma (Exercise 6.86 on page 486).

We now prove, by induction on $n \geq 0$, that the bottom $n$ rows of the desired diagram can be constructed. For the inductive step, assume that the first $n$ steps have been filled in, and let $K_n = \ker(P_n \to P_{n-1})$, etc. Now construct the $3 \times 3$ diagram whose bottom row is $0 \to K_n' \to K_n \to K_n'' \to 0$, and splice it to the $n$th diagram, as illustrated below (note that the map $P_{n+1} \to P_n$ is defined as the composite $P_{n+1} \to K_n \to P_n$):

$$
\begin{array}{ccccccccc}
0 & \longrightarrow & K_{n+1}' & \longrightarrow & K_{n+1} & \longrightarrow & K_{n+1}'' & \longrightarrow & 0 \\
& & \searrow & & \searrow & & \searrow & & \\
& 0 \longrightarrow & P_{n+1}' & \longrightarrow & P_{n+1} & \longrightarrow & P_{n+1}'' & \longrightarrow & 0 \\
0 & \longrightarrow & K_n' & \dashrightarrow & K_n & \dashrightarrow & K_n'' & \dashrightarrow & 0 \\
& & \searrow \downarrow & & \searrow \downarrow & & \searrow \downarrow & & \\
& 0 \longrightarrow & P_n' & \longrightarrow & P_n & \longrightarrow & P_n'' & \longrightarrow & 0 \\
& 0 \longrightarrow & P_{n-1}' & \longrightarrow & P_{n-1} & \longrightarrow & P_{n-1}'' & \longrightarrow & 0
\end{array}
$$

The columns of the new diagram are exact because, for example, $\mathrm{im}(P_{n+1} \to P_n) = K_n = \ker(P_n \to P_{n-1})$. $\quad\bullet$

**Theorem 9.56.** *If $0 \to A' \xrightarrow{i} A \xrightarrow{p} A'' \to 0$ is an exact sequence of modules and $T \colon {}_R\mathbf{Mod} \to \mathbf{Ab}$ is a covariant additive functor, then there is a long exact* *sequence:*

$$\to (L_nT)A' \xrightarrow{(L_nT)i} (L_nT)A \xrightarrow{(L_nT)p} (L_nT)A'' \xrightarrow{\partial_n}$$

$$(L_{n-1}T)A' \xrightarrow{(L_{n-1}T)i} (L_{n-1}T)A \xrightarrow{(L_{n-1}T)p} (L_{n-1}T)A'' \xrightarrow{\partial_{n-1}}$$

*that ends with*

$$\to (L_0T)A' \to (L_0T)A \to (L_0T)A'' \to 0.$$

**Proof.** Let $\mathbf{P}'_{A'}$ and $\mathbf{P}''_{A''}$ be the chosen deleted projective resolutions of $A'$ and of $A''$, respectively. By Proposition 9.55, there is a deleted projective resolution $\widetilde{\mathbf{P}}_A$ of $A$ with

$$0 \to \mathbf{P}'_{A'} \xrightarrow{j} \widetilde{\mathbf{P}}_A \xrightarrow{q} \mathbf{P}''_{A''} \to 0$$

(in the notation of the Comparison Theorem, $j = \widehat{i}$ is a chain map over $i$, and $q = \widehat{p}$ is a chain map over $p$). Applying $T$ gives the sequence of complexes

$$0 \to T\mathbf{P}'_{A'} \xrightarrow{Tj} T\widetilde{\mathbf{P}}_A \xrightarrow{Tq} T\mathbf{P}''_{A''} \to 0.$$

To see that this sequence is exact, note that each row $0 \to P'_n \xrightarrow{j_n} \widetilde{P}_n \xrightarrow{q_n} P''_n \to 0$ is a split exact sequence (because $P''_n$ is projective), and additive functors preserve split short exact sequences.[16] There is thus a long exact sequence

$$\to H_n(T\mathbf{P}'_{A'}) \xrightarrow{(Tj)_*} H_n(T\widetilde{\mathbf{P}}_A) \xrightarrow{(Tq)_*} H_n(T\mathbf{P}''_{A''}) \xrightarrow{\partial_n} H_{n-1}(T\mathbf{P}'_{A'}) \to;$$

that is, there is an exact sequence

$$\to (L_nT)A' \xrightarrow{(Tj)_*} (\widetilde{L}_nT)A \xrightarrow{(Tq)_*} (L_nT)A'' \xrightarrow{\partial_n} (L_{n-1}T)A' \to .$$

We do not know whether the projective resolution of $A$ given by the Horseshoe Lemma is the resolution originally chosen, and this is why we have $(\widetilde{L}_nT)A$ instead of $(L_nT)A$. But there is a natural isomorphism $\tau \colon L_nT \to \widetilde{L}_nT$, and so there is an exact sequence

$$\to (L_nT)A' \xrightarrow{\tau_A^{-1}(Tj)_*} (L_nT)A \xrightarrow{(Tq)_*\tau_A} (L_nT)A'' \xrightarrow{\partial_n} (L_{n-1}T)A' \to .$$

This sequence does terminate with $\{0\}$, for $L_{-n}T = 0$ for all negative $n$, by Proposition 9.51.

It remains to show that $\tau_A^{-1}(Tj)_* = (L_nT)i = T(j)_*$ (remember that $j = \widehat{i}$, a chain map over $i$) and $(Tq)_*\tau_A = (L_nT)p$. Now $\tau_A^{-1} = (T\kappa)_*$, where $\kappa \colon \widetilde{\mathbf{P}}_A \to \mathbf{P}_A$ is a chain map over $1_A$, and so

$$\tau_A^{-1}(Tj)_* = (T\kappa)_*(Tj)_* = (T\kappa Tj)_* = \big(T(\kappa j)\big)_*.$$

Both $\kappa j$ and $j$ are chain maps $\widetilde{\mathbf{P}}_A \to \mathbf{P}_A$ over $1_A$, so they are homotopic, by the Comparison Theorem. Hence, $T(\kappa j)$ and $Tj$ are homotopic, and so they induce the same map in homology: $(T(\kappa j))_* = (Tj)_* = (L_nT)i$. Therefore, $\tau_A^{-1}(Tj)_* = (L_nT)i$. That $(Tq)_*\tau_A = (L_nT)p$ is proved in the same way.  •

**Corollary 9.57.** *If $T \colon {}_R\mathbf{Mod} \to \mathbf{Ab}$ is a covariant additive functor, then the functor $L_0T$ is right exact.*

---

[16]The exact sequence of complexes need *not* be split because the sequence of splitting maps need not constitute a chain map $\mathbf{P}''_{A''} \to \widetilde{\mathbf{P}}_A$.

**Proof.** If $A \to B \to C \to 0$ is exact, then $(L_0 T)A \to (L_0 T)B \to (L_0 T)C \to 0$ is exact. $\bullet$

**Theorem 9.58.**

(i) *If a covariant additive functor $T \colon {}_R\mathbf{Mod} \to \mathbf{Ab}$ is right exact, then $T$ is naturally isomorphic to $L_0 T$.*

(ii) *The functor $\square \otimes_R B$ is naturally isomorphic to $\operatorname{Tor}_0^R(\square, B)$. Hence, for all right $R$-modules $A$, there is a natural isomorphism*

$$A \otimes_R B \cong \operatorname{Tor}_0^R(A, B).$$

**Proof.**

(i) Let $\mathbf{P}_A$ be the chosen deleted projective resolution of $A$, and let

$$\to P_1 \xrightarrow{d_1} P_0 \xrightarrow{\varepsilon} A \to 0$$

be the chosen projective resolution. By definition, $(L_0 T)A = \operatorname{coker} T(d_1)$. But right exactness of $T$ gives an exact sequence

$$TP_1 \xrightarrow{Td_1} TP_0 \xrightarrow{T\varepsilon} TA \to 0.$$

Now $T\varepsilon$ induces an isomorphism $\sigma_A \colon \operatorname{coker} T(d_1) \to TA$, by the First Isomorphism Theorem; that is, $\sigma_A \colon (L_0 T)A \to TA$. It is left as a routine exercise that $\sigma \colon L_0 T \to T$ is a natural isomorphism.

(ii) $\square \otimes_R B$ is a covariant right exact additive functor. $\bullet$

The next corollary specializes Theorem 9.56; it shows that Tor repairs the loss of exactness that may occur after tensoring a short exact sequence.

**Corollary 9.59.** *If $0 \to A' \to A \to A'' \to 0$ is a short exact sequence of right $R$-modules, then there is a long exact sequence*

$$\to \operatorname{Tor}_2^R(A', B) \to \operatorname{Tor}_2^R(A, B) \to \operatorname{Tor}_2^R(A'', B)$$
$$\to \operatorname{Tor}_1^R(A', B) \to \operatorname{Tor}_1^R(A, B) \to \operatorname{Tor}_1^R(A'', B)$$
$$\to A' \otimes_R B \to A \otimes_R B \to A'' \otimes_R B \to 0.$$

The next proposition shows that the functors $\operatorname{Tor}_n(\square, B)$ satisfy the covariant version of Theorem 9.47 (see Exercise 9.45 on page 815).

**Proposition 9.60.** *Given a commutative diagram of right $R$-modules having exact rows,*

$$
\begin{array}{ccccccccc}
0 & \longrightarrow & A' & \xrightarrow{\ i\ } & A & \xrightarrow{\ p\ } & A'' & \longrightarrow & 0 \\
& & \Big\downarrow{\scriptstyle f} & & \Big\downarrow{\scriptstyle g} & & \Big\downarrow{\scriptstyle h} & & \\
0 & \longrightarrow & C' & \xrightarrow[\ j\ ]{} & C & \xrightarrow[\ q\ ]{} & C'' & \longrightarrow & 0
\end{array}
$$

*there is, for all n, a commutative diagram with exact rows*

$$\begin{array}{ccccccc}
\operatorname{Tor}_n^R(A',B) & \xrightarrow{i_*} & \operatorname{Tor}_n^R(A,B) & \xrightarrow{p_*} & \operatorname{Tor}_n^R(A'',B) & \xrightarrow{\partial_n} & \operatorname{Tor}_{n-1}^R(A',B) \\
\downarrow{\scriptstyle f_*} & & \downarrow{\scriptstyle g_*} & & \downarrow{\scriptstyle h_*} & & \downarrow{\scriptstyle f_*} \\
\operatorname{Tor}_n^R(C',B) & \xrightarrow{j_*} & \operatorname{Tor}_n^R(C,B) & \xrightarrow{q_*} & \operatorname{Tor}_n^R(C'',B) & \xrightarrow{\partial_n'} & \operatorname{Tor}_{n-1}^R(C',B)
\end{array}$$

*There is a similar diagram if the first variable is fixed.*

**Proof.** Exactness of $0 \to A' \to A \to A'' \to 0$ gives exactness of the sequence of deleted complexes $0 \to \mathbf{P}_{A'} \to \mathbf{P}_A \to \mathbf{P}_{A''} \to 0$. If $T = \square \otimes_R B$, then $0 \to T\mathbf{P}_{A'} \to T\mathbf{P}_A \to T\mathbf{P}_{A''} \to 0$ is still exact, for every row splits because each term of $\mathbf{P}_{A''}$ is projective. Therefore, the naturality of the connecting homomorphism, Theorem 9.46, applies at once. $\quad\bullet$

In the next section, we will show how Tor can be computed and used. But, before leaving this section, let us give the same treatment to Hom that we have just given to tensor.

**9.5.2. Right Derived Covariant Functors.** *Left* derived functors of a functor $T$ are defined so that $T\mathbf{P}_A$ is a complex with all its nonzero terms on the *left* side; that is, all terms of negative degree are $\{0\}$. The reason for this is that if $T$ is right exact (as are $\square \otimes B$ and $A \otimes \square$), then $L_0 T \cong T$. Since the Hom functors are left exact, however, we define derived functors $R^n T$ in terms of resolutions $\mathbf{C}$ for which $T\mathbf{C}$ is on the right, for this implies that $R^0 T$ is naturally isomorphic to $T$ when $T$ is left exact.

Given a covariant additive functor $T$ [for example, $T = \operatorname{Hom}_R(B, \square)$], we are now going to construct, for all $n \in \mathbb{Z}$, its **right derived functors** $R^n T$.

Choose, once for all, a deleted injective resolution $\mathbf{E}^A$ of every module $A$, form the complex $T\mathbf{E}^A$, and take homology:

$$(R^n T)A = H^n(T\mathbf{E}^A) = \frac{\ker T d^n}{\operatorname{im} T d^{n-1}}.$$

The reader should reread Example 9.37(x) to recall the index raising convention: with indices lowered, the definition is

$$(R^n T)A = H_{-n}(T\mathbf{E}^A) = \frac{\ker T d_{-n}}{\operatorname{im} T d_{-n+1}}.$$

Notice that we have raised the index on homology groups as well; we write $H^n$ instead of $H_{-n}$.

The definition of $(R^n T)f$, where $f \colon A \to A'$ is a homomorphism, is similar to that for left derived functors. By the dual of the Comparison Theorem, there is a chain map $\widehat{f} \colon \mathbf{E}^A \to \mathbf{E}'^{A'}$ over $f$, unique to homotopy, and a unique map $(R^n T)f \colon H^n(T\mathbf{E}^A) \to H^n(T\mathbf{E}^{A'})$, namely, $(T\widehat{f}_n)_*$, induced in homology.

In pictures, look at the chosen injective resolutions:

$$0 \longrightarrow A' \longrightarrow E'^0 \longrightarrow E'^1 \longrightarrow$$

$$f \Big\uparrow$$

$$0 \longrightarrow A \longrightarrow E^0 \longrightarrow E^1 \longrightarrow$$

Fill in a chain map $\widehat{f}$ over $f$, delete $A$ and $A'$, then apply $T$ to this diagram, and then take the map induced by $T\widehat{f}$ in homology.

**Proposition 9.61.** *Given a ring $R$ and a covariant additive functor $T \colon {}_R\mathbf{Mod} \to \mathbf{Ab}$, then*

$$R^n T \colon {}_R\mathbf{Mod} \to \mathbf{Ab}$$

*is a covariant additive functor for every $n$.*

The proof of this proposition, as well as the proofs of other propositions about right derived functors soon to be stated, are duals of the proofs we have already given, and so they will be omitted.

**Proposition 9.62.** *If $T \colon {}_R\mathbf{Mod} \to \mathbf{Ab}$ is a covariant additive functor, then $(R^n T)A = \{0\}$ for all negative $n$ and for all $A$.*

**Definition.** If $T = \operatorname{Hom}_R(B, \square)$, define $\operatorname{Ext}_R^n(B, \square) = R^n T$. Thus, if

$$\mathbf{E}^A = 0 \to E^0 \xrightarrow{d^0} E^1 \xrightarrow{d^1} E^2 \to$$

is the chosen deleted injective resolution of a module $A$, then

$$\operatorname{Ext}_R^n(B, A) = H^n(\operatorname{Hom}_R(B, \mathbf{E}^A)) = \frac{\ker(d^n)_*}{\operatorname{im}(d^{n-1})_*},$$

where $(d^n)_* \colon \operatorname{Hom}_R(B, E^n) \to \operatorname{Hom}_R(B, E^{n+1})$ is defined, as usual, by

$$(d^n)_* \colon f \mapsto d^n f.$$

The domain of $R^n T$, in particular, the domain of $\operatorname{Ext}_R^n(B, \square)$, is ${}_R\mathbf{Mod}$, the category of all left $R$-modules, and its target is $\mathbf{Ab}$, the category of abelian groups. The target may be larger; for example, it is ${}_R\mathbf{Mod}$ if $R$ is commutative.

**Example 9.63.** If $T$ is a covariant additive functor that preserves multiplications and $\mu_r \colon A \to A$ is multiplication by $r$, where $r \in Z(R)$ is a central element, then $R^n T$ also preserves multiplications (Example 9.50). ◄

**Proposition 9.64.** *Denote the right derived functors arising from new choices $\widetilde{\mathbf{E}}$ of injective resolutions by $\widetilde{R}^n T$. Given a ring $R$ and a covariant additive functor $T \colon {}_R\mathbf{Mod} \to \mathbf{Ab}$, then, for each $n$, the functors $R^n T$ and $\widetilde{R}^n T$ are naturally isomorphic. In particular, for all $A$,*

$$(R^n T)A \cong (\widetilde{R}^n T)A,$$

*and so these groups are independent of the choice of injective resolution of $A$.*

**Corollary 9.65.** $\operatorname{Ext}_R^n(B, A)$ *is independent of the choice of injective resolution of $A$.*

**Corollary 9.66.** *Let* $T\colon {}_R\mathbf{Mod} \to \mathbf{Ab}$ *be a covariant additive functor. If* $E$ *is an injective left* $R$-*module, then* $(R^nT)E = \{0\}$ *for all* $n \geq 1$.

*In particular, if* $E$ *is an injective* $R$-*module, then* $\mathrm{Ext}_R^n(B, E) = \{0\}$ *for all* $n \geq 1$ *and all left* $R$-*modules* $B$.

**Theorem 9.67.** *If* $0 \to A' \xrightarrow{i} A \xrightarrow{p} A'' \to 0$ *is an exact sequence of left* $R$-*modules and* $T\colon {}_R\mathbf{Mod} \to \mathbf{Ab}$ *is a covariant additive functor, then there is a long exact sequence:*

$$\to (R^nT)A' \xrightarrow{(R^nT)i} (R^nT)A \xrightarrow{(R^nT)p} (R^nT)A'' \xrightarrow{\partial^n}$$
$$(R^{n+1}T)A' \xrightarrow{(R^{n+1}T)i} (R^{n+1}T)A \xrightarrow{(R^{n+1}T)p} (R^{n+1}T)A'' \xrightarrow{\partial^{n+1}}$$

*that begins with*

$$0 \to (R^0T)A' \to (R^0T)A \to (R^0T)A'' \to .$$

**Corollary 9.68.** *If* $T\colon {}_R\mathbf{Mod} \to \mathbf{Ab}$ *is a covariant additive functor, then the functor* $R^0T$ *is left exact.*

**Theorem 9.69.**

(i) *If a covariant additive functor* $T\colon {}_R\mathbf{Mod} \to \mathbf{Ab}$ *is left exact, then* $T$ *is naturally isomorphic to* $R^0T$.

(ii) *If* $B$ *is a left* $R$-*module, the functor* $\mathrm{Hom}_R(B, \square)$ *is naturally isomorphic to* $\mathrm{Ext}_R^0(B, \square)$. *Hence, for all left* $R$-*modules* $A$, *there is a natural isomorphism*

$$\mathrm{Hom}_R(B, A) \cong \mathrm{Ext}_R^0(B, A).$$

The next corollary shows that $\mathrm{Ext}(B, \square)$ repairs the loss of exactness that may occur after applying $\mathrm{Hom}(B, \square)$ to a short exact sequence.

**Corollary 9.70.** *If* $0 \to A' \to A \to A'' \to 0$ *is a short exact sequence of left* $R$-*modules, then there is a long exact sequence*

$$0 \to \mathrm{Hom}_R(B, A') \to \mathrm{Hom}_R(B, A) \to \mathrm{Hom}_R(B, A'')$$
$$\to \mathrm{Ext}_R^1(B, A') \to \mathrm{Ext}_R^1(B, A) \to \mathrm{Ext}_R^1(B, A'')$$
$$\to \mathrm{Ext}_R^2(B, A') \to \mathrm{Ext}_R^2(B, A) \to \mathrm{Ext}_R^2(B, A'') \to .$$

**Proposition 9.71.** *Given a commutative diagram of left* $R$-*modules having exact rows,*

$$
\begin{array}{ccccccccc}
0 & \longrightarrow & A' & \xrightarrow{\ i\ } & A & \xrightarrow{\ p\ } & A'' & \longrightarrow & 0 \\
& & \downarrow{\scriptstyle f} & & \downarrow{\scriptstyle g} & & \downarrow{\scriptstyle h} & & \\
0 & \longrightarrow & C' & \xrightarrow{\ j\ } & C & \xrightarrow{\ q\ } & C'' & \longrightarrow & 0
\end{array}
$$

*there is, for all n, a commutative diagram with exact rows*

$$\begin{array}{ccccccc}
\operatorname{Ext}_R^n(B,A') & \xrightarrow{i_*} & \operatorname{Ext}_R^n(B,A) & \xrightarrow{p_*} & \operatorname{Ext}_R^n(B,A'') & \xrightarrow{\partial^n} & \operatorname{Ext}_R^{n+1}(B,A') \\
\downarrow{f_*} & & \downarrow{g_*} & & \downarrow{h_*} & & \downarrow{f_*} \\
\operatorname{Ext}_R^n(B,C') & \xrightarrow[j_*]{} & \operatorname{Ext}_R^n(B,C) & \xrightarrow[q_*]{} & \operatorname{Ext}_R^n(B,C'') & \xrightarrow[\partial'^n]{} & \operatorname{Ext}_R^{n+1}(B,C')
\end{array}$$

**9.5.3. Right Derived Contravariant Functors.** Finally, we discuss derived functors of contravariant functors $T$. If we define right derived functors $R^n T$ in terms of deleted resolutions $\mathbf{C}$ for which $T\mathbf{C}$ is on the right, then we must start with a deleted *projective* resolution $\mathbf{P}_A$, for then the contravariance of $T$ puts $T\mathbf{P}_A$ on the right; it will follow that if $T$ is left exact, then $R^0 T \cong T$.[17]

Given a contravariant additive functor $T$, we are now going to construct, for all $n \in \mathbb{Z}$, its **right derived functors** $R^n T$. Choose, once for all, a projective resolution $\mathbf{P}$ of every module $A$, delete $A$, form the complex $T\mathbf{P}_A$, and take homology:

$$(R^n T)A = H^n(T\mathbf{P}_A) = \frac{\ker T d_{n+1}}{\operatorname{im} T d_n}.$$

If $f: A \to A'$, define $(R^n T)f: (R^n T)A' \to (R^n T)A$ as we did for left derived functors. By the Comparison Theorem, there is a chain map $\widehat{f}: \mathbf{P}_A \to \mathbf{P}'_{A'}$ over $f$, unique to homotopy, which induces a map $(R^n T)f: H^n(T\mathbf{P}'_{A'}) \to H^n(T\mathbf{P}_A)$ in homology, namely, $(T\widehat{f}_n)_*$.

**Example 9.72.** If $T$ is a contravariant additive functor that preserves multiplications and $\mu_r: A \to A$ is multiplication by $r$, where $r \in Z(R)$ is a central element, then $R^n T$ also preserves multiplications (see Example 9.50). ◄

**Proposition 9.73.** *Given a ring $R$ and a contravariant additive functor $T: {}_R\mathbf{Mod} \to \mathbf{Ab}$, then*

$$R^n T: {}_R\mathbf{Mod} \to \mathbf{Ab}$$

*is a contravariant additive functor for every $n$.*

**Proposition 9.74.** *If $T: {}_R\mathbf{Mod} \to \mathbf{Ab}$ is a contravariant additive functor, then $(R^n T)A = \{0\}$ for all negative $n$ and all $A$.*

**Definition.** If $T = \operatorname{Hom}_R(\Box, C)$, define $\operatorname{ext}_R^n(\Box, C) = R^n T$. Thus, if

$$\to P_2 \xrightarrow{d_2} P_1 \xrightarrow{d_1} P_0 \to 0$$

is the chosen deleted projective resolution of a module $A$, then

$$\operatorname{ext}_R^n(A, C) = H^n(\operatorname{Hom}_R(\mathbf{P}_A, C)) = \frac{\ker(d_{n+1})^*}{\operatorname{im}(d_n)^*},$$

where $(d^n)^*: \operatorname{Hom}_R(P_{n-1}, C) \to \operatorname{Hom}_R(P_n, C)$ is defined, as usual, by

$$(d_n)^*: f \mapsto f d_n.$$

---

[17]If we were interested in left derived functors of a contravariant $T$, but we are not, then we would use injective resolutions.

The same phenomenon that holds for Tor holds for Ext: for all $A$ and $C$ (and all $R$ and $n$),

$$\mathrm{Ext}_R^n(A, C) \cong \mathrm{ext}_R^n(A, C).$$

The proof that shows that Tor is independent of the variable resolved also works for Ext. In light of this theorem, we will dispense with the two notations for Ext.

**Proposition 9.75.** *Denote the right derived functors arising from new choices $\widetilde{\mathbf{P}}$ of projective resolutions by $\widetilde{R}^n T$. Given a ring $R$ and a contravariant additive functor $T \colon {}_R\mathbf{Mod} \to \mathbf{Ab}$, then, for each $n$, the functors $R^n T$ and $\widetilde{R}^n T$ are naturally isomorphic. In particular, for all $A$,*

$$(R^n T)A \cong (\widetilde{R}^n T)A,$$

*and so these groups are independent of the choice of projective resolution of $A$.*

**Corollary 9.76.** $\mathrm{Ext}_R^n(A, C)$ *is independent of the choice of projective resolution of $A$.*

**Corollary 9.77.** *Let $T \colon {}_R\mathbf{Mod} \to \mathbf{Ab}$ be a contravariant additive functor. If $P$ is a projective module, then $(R^n T)P = \{0\}$ for all $n \geq 1$.*

*In particular, if $P$ is a projective left $R$-module, then $\mathrm{Ext}_R^n(P, B) = \{0\}$ for all $n \geq 1$ and all left $R$-modules $B$.*

**Theorem 9.78.** *If $0 \to A' \xrightarrow{i} A \xrightarrow{p} A'' \to 0$ is an exact sequence of left $R$-modules and $T \colon {}_R\mathbf{Mod} \to \mathbf{Ab}$ is a contravariant additive functor, then there is a long exact sequence*

$$\to (R^n T)A'' \xrightarrow{(R^n T)p} (R^n T)A \xrightarrow{(R^n T)i} (R^n T)A' \xrightarrow{\partial^n}$$

$$(R^{n+1} T)A'' \xrightarrow{(R^{n+1} T)p} (R^{n+1} T)A \xrightarrow{(R^{n+1} T)i} (R^{n+1} T)A' \xrightarrow{\partial^{n+1}}$$

*that begins with*

$$0 \to (R^0 T)A'' \to (R^0 T)A \to (R^0 T)A' \to .$$

**Corollary 9.79.** *If $T \colon {}_R\mathbf{Mod} \to \mathbf{Ab}$ is a contravariant additive functor, then the functor $R^0 T$ is left exact.*

**Theorem 9.80.**

(i) *If a contravariant additive functor $T \colon {}_R\mathbf{Mod} \to \mathbf{Ab}$ is left exact, then $T$ is naturally isomorphic to $R^0 T$.*

(ii) *If $C$ is a left $R$-module, the functor $\mathrm{Hom}_R(\square, C)$ is naturally isomorphic to $\mathrm{Ext}_R^0(\square, C)$. Hence, for all left $R$-modules $A$, there is a natural isomorphism*

$$\mathrm{Hom}_R(A, C) \cong \mathrm{Ext}_R^0(A, C).$$

The next corollary shows that Ext repairs the loss of exactness that may occur after applying Hom to a short exact sequence.

**Corollary 9.81.** *If* $0 \to A' \to A \to A'' \to 0$ *is a short exact sequence of left R-modules, then there is a long exact sequence*

$$0 \to \operatorname{Hom}_R(A'', C) \to \operatorname{Hom}_R(A, C) \to \operatorname{Hom}_R(A', C)$$
$$\to \operatorname{Ext}_R^1(A'', C) \to \operatorname{Ext}_R^1(A, C) \to \operatorname{Ext}_R^1(A', C)$$
$$\to \operatorname{Ext}_R^2(A'', C) \to \operatorname{Ext}_R^2(A, C) \to \operatorname{Ext}_R^2(A', C) \to .$$

**Proposition 9.82.** *Given a commutative diagram of left R-modules having exact rows,*

$$
\begin{array}{ccccccccc}
0 & \longrightarrow & A' & \overset{i}{\longrightarrow} & A & \overset{p}{\longrightarrow} & A'' & \longrightarrow & 0 \\
 & & \downarrow{f} & & \downarrow{g} & & \downarrow{h} & & \\
0 & \longrightarrow & C' & \underset{j}{\longrightarrow} & C & \underset{q}{\longrightarrow} & C'' & \longrightarrow & 0
\end{array}
$$

*there is, for all n, a commutative diagram with exact rows*

$$
\begin{array}{ccccccc}
\operatorname{Ext}_R^n(A'', B) & \overset{p^*}{\longrightarrow} & \operatorname{Ext}_R^n(A, B) & \overset{i^*}{\longrightarrow} & \operatorname{Ext}_R^n(A', B) & \overset{\partial^n}{\longrightarrow} & \operatorname{Ext}_R^{n+1}(A'', B) \\
\uparrow{h^*} & & \uparrow{g^*} & & \uparrow{f^*} & & \uparrow{h^*} \\
\operatorname{Ext}_R^n(C'', B) & \underset{q^*}{\longrightarrow} & \operatorname{Ext}_R^n(C, B) & \underset{j^*}{\longrightarrow} & \operatorname{Ext}_R^n(C', B) & \underset{\partial^{n\prime}}{\longrightarrow} & \operatorname{Ext}_R^{n+1}(C'', B)
\end{array}
$$

**Remark.** When $T$ is a covariant functor, then we call the ingredients of $L_n T$ chains, cycles, boundaries, and homology. When $T$ is contravariant, we often add the prefix "co," and the ingredients of $R^n T$ are usually called *cochains* $C^n$, *cocycles* $Z^n$, *coboundaries* $B^n$, and *cohomology* $H^n$. Unfortunately, this clear distinction is blurred because the Hom functor is contravariant in one variable but covariant in the other. In spite of this, we usually use the "co" prefix for both derived functors $\operatorname{Ext}^n$ of Hom. ◄

We repeat what we said earlier. Theorem 9.47 characterizes the sequence of contravariant Ext functors (Exercises 9.44 and 9.45 on page 815 are similar axioms characterizing the sequence of covariant Ext functors and the sequence of Tor functors). Therefore, the reader can use these functors without having to be constantly aware of the intricacies of derived functors. Derived functors are an important way of proving the existence of such sequences of functors, for some of the results needed in the proof, such as the Comparison Theorem, will be used later. (In the next section, along with more properties of Ext and Tor, we shall describe some other constructions of Ext and Tor.)

## Exercises

**9.35.** If $\tau \colon F \to G$ is a natural transformation between additive functors, prove that $\tau$ gives chain maps $\tau_\mathbf{C} \colon F\mathbf{C} \to G\mathbf{C}$ for every complex $\mathbf{C}$. If $\tau$ is a natural isomorphism, prove that $F\mathbf{C} \cong G\mathbf{C}$.

**9.36.** (i) Let $T\colon {}_R\mathbf{Mod} \to \mathbf{Ab}$ be an exact additive functor, where $R$ is a ring, and suppose that $P$ projective implies $TP$ projective. If $B$ is a left $R$-module and $\mathbf{P}_B$ is a deleted projective resolution of $B$, prove that $T\mathbf{P}_{TB}$ is a deleted projective resolution of $TB$.

(ii) Let $A$ be a $k$-algebra, where $k$ is a commutative ring, that is flat as a $k$-module. Prove that if $B$ is an $A$-module (and hence a $k$-module), then

$$A \otimes_k \operatorname{Tor}_n^k(B, C) \cong \operatorname{Tor}_n^A(B, A \otimes_k C)$$

for all $k$-modules $C$ and all $n \geq 0$.

**9.37.** Let $R$ be a semisimple ring. Prove that $\operatorname{Tor}_n^R(A, B) = \{0\} = \operatorname{Ext}_R^n(A, B)$ for all $n \geq 1$, all right $R$-modules $A$, and all left $R$-modules $B$.

**Hint.** If $R$ is semisimple, then every (left or right) $R$-module is projective.

* **9.38.** Let $R$ be a PID.

(i) Prove that $\operatorname{Tor}_n^R(A, B) = \{0\} = \operatorname{Ext}_R^n(A, B)$ for all $n \geq 2$ and all $R$-modules $A$ and $B$.
**Hint.** Use Theorem 8.9.

(ii) Prove that $\operatorname{Tor}_1^R(A, \square)\colon {}_R\mathbf{Mod} \to {}_R\mathbf{Mod}$ is a left exact additive functor and that $\operatorname{Ext}_R^1(A, \square)\colon {}_R\mathbf{Mod} \to {}_R\mathbf{Mod}$ is a right exact additive functor.

**9.39.** Consider the covariant additive functor $T = \operatorname{Ext}_{\mathbb{Z}}^1(A, \square)\colon \mathbf{Ab} \to \mathbf{Ab}$, where $A$ is an abelian group. The previous exercise says that $T$ is right exact.

(i) Prove that $L_1 T$ is a left exact functor.

(ii) Compute $L_n T$ for all $n \geq 0$. Is $L_1 T \cong \operatorname{Hom}_{\mathbb{Z}}(A, \square)$?

**9.40.** Let $R$ be a domain, let $Q = \operatorname{Frac}(R)$, and let $A$ be an $R$-module.

(i) Prove that if the multiplication $\mu_r\colon A \to A$ is an injection for all $r \neq 0$, then $A$ is torsion-free.

(ii) Prove that if the multiplication $\mu_r\colon A \to A$ is a surjection for all $r \neq 0$, then $A$ is divisible.

(iii) Prove that if the multiplication $\mu_r\colon A \to A$ is an isomorphism for all $r \neq 0$, then $A$ is a vector space over $Q$.
**Hint.** A module $A$ is a vector space over $Q$ if and only if it is torsion-free and divisible.

(iv) If either $C$ or $A$ is a vector space over $Q$, prove that $\operatorname{Tor}_n^R(C, A)$ and $\operatorname{Ext}_R^n(C, A)$ are also vector spaces over $Q$.

* **9.41.** Let $R$ be a domain and let $Q = \operatorname{Frac}(R)$.

(i) Let $r \in R$ be nonzero and $A$ be an $R$-module for which $rA = \{0\}$; that is, $ra = 0$ for all $a \in A$. Prove that $\operatorname{Ext}_R^n(Q, A) = \{0\} = \operatorname{Tor}_n^R(Q, A)$ for all $n \geq 0$.
**Hint.** If $V$ is a vector space over $Q$ for which $rV = \{0\}$, then $V = \{0\}$.

(ii) Prove that $\operatorname{Ext}_R^n(V, A) = \{0\} = \operatorname{Tor}_n^R(V, A)$ for all $n \geq 0$ whenever $V$ is a vector space over $Q$ and $A$ is an $R$-module for which $rA = \{0\}$ for some nonzero $r \in R$.

* **9.42.** Let $A$ and $B$ be $R$-modules. For $f\colon A' \to B$, where $A'$ is a submodule of $A$, define its ***obstruction*** to be $\partial(f)$, where $\partial\colon \operatorname{Hom}_R(A', B) \to \operatorname{Ext}_R^1(A/A', B)$ is the connecting homomorphism. Prove that $f$ can be extended to a homomorphism $A \to B$ if and only if its obstruction is 0.

**9.43.** If $T\colon \mathbf{Ab} \to \mathbf{Ab}$ is a left exact functor, prove that $L_0T$ is an exact functor. Conclude, for any abelian group $B$, that $(L_0 \operatorname{Hom})(B, \square)$ is not naturally isomorphic to $\operatorname{Hom}(B, \square)$.

* **9.44.** Prove the following analog of Theorem 9.47. Let $\mathcal{E}^n\colon {}_R\mathbf{Mod} \to \mathbf{Ab}$ be a sequence of covariant functors, for $n \geq 0$, such that

    (i) for every short exact sequence $0 \to A \to B \to C \to 0$, there is a long exact sequence and natural connecting homomorphisms

$$\to \mathcal{E}^n(A) \to \mathcal{E}^n(B) \to \mathcal{E}^n(C) \xrightarrow{\Delta_n} \mathcal{E}^{n+1}(A) \to;$$

    (ii) there is a left $R$-module $M$ such that $\mathcal{E}^0$ and $\operatorname{Hom}_R(M, \square)$ are naturally isomorphic;

    (iii) $\mathcal{E}^n(E) = \{0\}$ for all injective modules $E$ and all $n \geq 1$.

    Prove that $\mathcal{E}^n$ is naturally isomorphic to $\operatorname{Ext}^n(M, \square)$ for all $n \geq 0$.

* **9.45.** Let $\operatorname{TOR}^n\colon {}_R\mathbf{Mod} \to \mathbf{Ab}$ be a sequence of covariant functors, for $n \geq 0$, such that

    (i) for every short exact sequence $0 \to A \to B \to C \to 0$, there is a long exact sequence and natural connecting homomorphisms

$$\to \operatorname{TOR}_n(A) \to \operatorname{TOR}_n(B) \to \operatorname{TOR}_n(C) \xrightarrow{\Delta_n} \operatorname{TOR}_{n-1}(A) \to;$$

    (ii) there is a left $R$-module $M$ such that $\operatorname{TOR}_0$ and $\square \otimes_R M$ are naturally isomorphic;

    (iii) $\operatorname{TOR}_n(P) = \{0\}$ for all projective modules $P$ and all $n \geq 1$.

    Prove that $\operatorname{TOR}_n$ is naturally isomorphic to $\operatorname{Tor}_n^R(\square, M)$ for all $n \geq 0$. [There is a similar result for $\operatorname{Tor}_n^R(M, \square)$.]

---

## Section 9.6. Ext

We now examine Ext more closely (we will examine Tor in the next section). As we said in the last section, all properties of these functors must follow from the axioms characterizing them (Theorem 9.47 on page 796 and Exercises 9.44 and 9.45 above). In particular, their construction as derived functors need not be used.

We begin by showing that Ext behaves like Hom with respect to sums and products.

**Proposition 9.83.** *If $(A_k)_{k \in K}$ is a family of left $R$-modules, then there are natural isomorphisms, for all $n$,*

$$\operatorname{Ext}_R^n\Big(\bigoplus_{k \in K} A_k, B\Big) \cong \prod_{k \in K} \operatorname{Ext}_R^n(A_k, B).$$

**Proof.** The proof is by dimension shifting; that is, by induction on $n \geq 0$. The base step is Theorem 6.45(iii), for $\operatorname{Ext}_R^0(\square, B)$ is naturally isomorphic to the contravariant functor $\operatorname{Hom}_R(\square, B)$.

For the inductive step, choose, for each $k \in K$, a short exact sequence

$$0 \to L_k \to P_k \to A_k \to 0,$$

where $P_k$ is projective. There is an exact sequence

$$0 \to \bigoplus_k L_k \to \bigoplus_k P_k \to \bigoplus_k A_k \to 0,$$

and $\bigoplus_k P_k$ is projective, for every direct sum of projectives is projective. There is a commutative diagram with exact rows:

$$\begin{array}{ccccccc}
\mathrm{Hom}(\bigoplus P_k, B) & \twoheadrightarrow & \mathrm{Hom}(\bigoplus L_k, B) & \overset{\partial}{\twoheadrightarrow} & \mathrm{Ext}^1(\bigoplus A_k, B) & \twoheadrightarrow & \mathrm{Ext}^1(\bigoplus P_k, B) \\
{\scriptstyle \tau}\downarrow & & {\scriptstyle \sigma}\downarrow & & \downarrow & & \downarrow \\
\prod \mathrm{Hom}(P_k, B) & \twoheadrightarrow & \prod \mathrm{Hom}(L_k, B) & \underset{d}{\twoheadrightarrow} & \prod \mathrm{Ext}^1(A_k, B) & \twoheadrightarrow & \prod \mathrm{Ext}^1(P_k, B)
\end{array}$$

where the maps in the bottom row are just the usual induced maps in each coordinate, and the maps $\tau$ and $\sigma$ are the isomorphisms given by Theorem 6.45(iii). Now $\mathrm{Ext}^1(\bigoplus P_k, B) = \{0\} = \prod \mathrm{Ext}^1(P_k, B)$, because $\bigoplus P_k$ and each $P_k$ are projective, so that the maps $\partial$ and $d$ are surjective. This is precisely the sort of diagram in Proposition 6.116, and so there exists an isomorphism $\mathrm{Ext}^1(\bigoplus A_k, B) \to \prod \mathrm{Ext}^1(A_k, B)$ making the augmented diagram commute.

We may now assume that $n \geq 1$, and we look further out in the long exact sequence. There is a commutative diagram

$$\begin{array}{ccccccc}
\mathrm{Ext}^n(\bigoplus P_k, B) & \twoheadrightarrow & \mathrm{Ext}^n(\bigoplus L_k, B) & \overset{\partial}{\twoheadrightarrow} & \mathrm{Ext}^{n+1}(\bigoplus A_k, B) & \twoheadrightarrow & \mathrm{Ext}^{n+1}(\bigoplus P_k, B) \\
& & {\scriptstyle \sigma}\downarrow & & \downarrow & & \\
\prod \mathrm{Ext}^n(P_k, B) & \twoheadrightarrow & \prod \mathrm{Ext}^n(L_k, B) & \underset{d}{\twoheadrightarrow} & \prod \mathrm{Ext}^{n+1}(A_k, B) & \twoheadrightarrow & \prod \mathrm{Ext}^{n+1}(P_k, B)
\end{array}$$

where $\sigma \colon \mathrm{Ext}^n(\bigoplus L_k, B) \to \prod \mathrm{Ext}^n(L_k, B)$ is an isomorphism that exists by the inductive hypothesis. Since $n \geq 1$, all four Ext's whose first variable is projective are $\{0\}$; it follows from exactness of the rows that both $\partial$ and $d$ are isomorphisms. Finally, the composite $d\sigma\partial^{-1} \colon \mathrm{Ext}^{n+1}(\bigoplus A_k, B) \to \prod \mathrm{Ext}^{n+1}(A_k, B)$ is an isomorphism, as desired.   $\bullet$

There is a dual result in the second variable.

**Proposition 9.84.** *If* $(B_k)_{k \in K}$ *is a family of left R-modules, then there are natural isomorphisms, for all $n$,*

$$\mathrm{Ext}_R^n\Big(A, \prod_{k \in K} B_k\Big) \cong \prod_{k \in K} \mathrm{Ext}_R^n(A, B_k).$$

**Proof.** The proof is also by dimension shifting. The base step is Theorem 6.45(i), for $\mathrm{Ext}_R^0(A, \square)$ is naturally isomorphic to the covariant functor $\mathrm{Hom}_R(A, \square)$.

For the inductive step, choose, for each $k \in K$, a short exact sequence

$$0 \to B_k \to E_k \to N_k \to 0,$$

where $E_k$ is injective. There is an exact sequence

$$0 \to \prod_k B_k \to \prod_k E_k \to \prod_k N_k \to 0$$

and $\prod_k E_k$ is injective, for every direct product of injectives is injective, by Proposition 6.87. Hence, there is a commutative diagram with exact rows:

$$\operatorname{Hom}(A, \textstyle\prod E_k) \to \operatorname{Hom}(A, \textstyle\prod N_k) \xrightarrow{\partial} \operatorname{Ext}^1(A, \textstyle\prod B_k) \to \operatorname{Ext}^1(A, \textstyle\prod E_k)$$

$$\prod \operatorname{Hom}(A, E_k) \to \prod \operatorname{Hom}(A, N_k) \xrightarrow{d} \prod \operatorname{Ext}^1(A, B_k) \to \prod \operatorname{Ext}^1(A, E_k)$$

where the maps in the bottom row are just the usual induced maps in each coordinate, and the maps $\tau$ and $\sigma$ are the isomorphisms given by Theorem 6.45(i). The proof now finishes as that of Proposition 9.83. $\quad\bullet$

It follows that $\operatorname{Ext}^n$ commutes with finite direct sums in either variable.

**Remark.** These last two propositions cannot be generalized by replacing direct sums with direct limits or direct products with inverse limits. $\quad\blacktriangleleft$

When the ring $R$ is noncommutative, the abelian group $\operatorname{Hom}_R(A, B)$ need not be an $R$-module, but it is a $Z(R)$-module. And so it is with Ext.

**Proposition 9.85.**

(i) $\operatorname{Ext}_R^n(A, B)$ *is a $Z(R)$-module for all $n \geq 0$. In particular, $\operatorname{Ext}_R^n(A, B)$ is an $R$-module if $R$ is commutative.*

(ii) *If $A$ and $B$ are left $R$-modules and $r \in Z(R)$ is a central element, then the induced map $\mu_r^*\colon \operatorname{Ext}_R^n(A, B) \to \operatorname{Ext}_R^n(A, B)$, where $\mu_r\colon B \to B$ is multiplication by $r$, is also multiplication by $r$. A similar statement is true in the other variable.*

**Proof.**

(i) By Example 9.50, $\mu_r$ is an $R$-map, and so it induces a homomorphism on $\operatorname{Ext}_R^n(A, B)$. It is straightforward to check that $x \mapsto \mu_r^*(x)$ defines a scalar multiplication $Z(R) \times \operatorname{Ext}_R^n(A, B) \to \operatorname{Ext}_R^n(A, B)$.

(ii) This follows from (i), for we define scalar multiplication by $r$ to be $\mu_r^*$. $\quad\bullet$

**Example 9.86.**

(i) We show, for every abelian group $B$, that

$$\operatorname{Ext}_{\mathbb{Z}}^1(\mathbb{I}_n, B) \cong B/nB.$$

There is an exact sequence

$$0 \to \mathbb{Z} \xrightarrow{\mu_n} \mathbb{Z} \to \mathbb{I}_n \to 0,$$

where $\mu_n$ is multiplication by $n$. Applying $\operatorname{Hom}_{\mathbb{Z}}(\square, B)$ gives exactness of $\operatorname{Hom}(\mathbb{Z}, B) \xrightarrow{\mu_n^*} \operatorname{Hom}(\mathbb{Z}, B) \to \operatorname{Ext}^1(\mathbb{I}_n, B) \to \operatorname{Ext}^1(\mathbb{Z}, B)$. Now the last term $\operatorname{Ext}^1(\mathbb{Z}, B)$ is $\{0\}$, because $\mathbb{Z}$ is projective. Moreover, $\mu_n^*$ is also multiplication by $n$, while $\operatorname{Hom}(\mathbb{Z}, B) = B$. More precisely, $\operatorname{Hom}(\mathbb{Z}, \square)$ is naturally isomorphic to the identity functor on **Ab**, and so there is a commutative diagram with exact rows

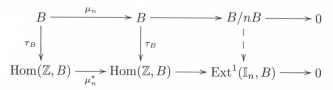

By Proposition 6.116, there is an isomorphism $B/nB \cong \operatorname{Ext}^1(\mathbb{I}_n, B)$.

(ii) We can now compute $\operatorname{Ext}^1_{\mathbb{Z}}(A, B)$ whenever $A$ and $B$ are finitely generated abelian groups. By the Fundamental Theorem, both $A$ and $B$ are direct sums of cyclic groups. Since Ext commutes with finite direct sums, $\operatorname{Ext}^1_{\mathbb{Z}}(A, B)$ is the direct sum of groups $\operatorname{Ext}^1_{\mathbb{Z}}(C, D)$, where $C$ and $D$ are cyclic. We may assume that $C$ is finite; otherwise, $C \cong \mathbb{Z}$ and $\operatorname{Ext}^1(C, D) = \{0\}$. This calculation can be completed using part (i) and Exercise 4.8 on page 242, which says that if $D$ is a cyclic group of finite order $m$, then $D/nD$ is a cyclic group of order $d$, where $d = (m, n)$ is their gcd. ◀

We define an *extension* of modules by analogy with that of groups.

**Definition.** Given $R$-modules $C$ and $A$, an ***extension*** of $A$ by $C$ is a short exact sequence
$$0 \to A \xrightarrow{i} B \xrightarrow{p} C \to 0.$$
An extension is ***split*** if there exists an $R$-map $s \colon C \to B$ with $ps = 1_C$.

Of course, if $0 \to A \to B \to C \to 0$ is a split extension, then $B \cong A \oplus C$.

Whenever meeting a homology group, we must ask what it means for it to be zero, for its elements can then be construed as being obstructions. For example, factor sets explain why a group extension may not be split. Exercise 9.42 on page 814 says that if $A' \subseteq A$, then elements of $\operatorname{Ext}^1_R(A/A', B)$ can be interpreted as obstructions to extending a homomorphism $A' \to B$ to $A \to B$. In this section, we will show that $\operatorname{Ext}^1_R(C, A) = \{0\}$ if and only if every extension of $A$ by $C$ splits. Thus, nonzero elements of any $\operatorname{Ext}^1_R(C', A')$ describe nonsplit extensions (indeed, this result is why Ext is so called).

We begin with a definition motivated by Proposition 9.21.

**Definition.** Given modules $C$ and $A$, two extensions $\xi : 0 \to A \to B \to C \to 0$ and $\xi' : 0 \to A \to B' \to C \to 0$ of $A$ by $C$ are ***equivalent*** if there exists a map $\varphi \colon B \to B'$ making the following diagram commute:

$$
\begin{array}{ccccccccc}
\xi : 0 & \longrightarrow & A & \xrightarrow{i} & B & \xrightarrow{p} & C & \longrightarrow & 0 \\
 & & \downarrow{1_A} & & \downarrow{\varphi} & & \downarrow{1_C} & & \\
\xi' : 0 & \longrightarrow & A & \xrightarrow{i'} & B' & \xrightarrow{p'} & C & \longrightarrow & 0
\end{array}
$$

If two extensions are equivalent, then the Five Lemma shows that the map $\varphi$ must be an isomorphism; it follows that equivalence is, indeed, an equivalence relation (for we can now prove symmetry). We denote the equivalence class of an extension $\xi$ by $[\xi]$, and we define

$$e(C, A) = \big\{ [\xi] : \xi \text{ is an extension of } A \text{ by } C \big\}.$$

Note, however, that there are inequivalent extensions having isomorphic middle terms, as we saw in Example 9.22 (all groups in this example are abelian, and so we may view it as an example of $\mathbb{Z}$-modules).

**Proposition 9.87.** *If* $\mathrm{Ext}^1_R(C, A) = \{0\}$, *then every extension* $0 \to A \xrightarrow{i} B \xrightarrow{p} C \to 0$ *is split.*

**Proof.** Apply the functor $\mathrm{Hom}_R(C, \square)$ to the extension to obtain an exact sequence

$$\mathrm{Hom}(C, B) \xrightarrow{p_*} \mathrm{Hom}(C, C) \xrightarrow{\partial} \mathrm{Ext}^1(C, A).$$

By hypothesis, $\mathrm{Ext}^1(C, A) = \{0\}$, so that $p_*$ is surjective. Hence, there exists $s \in \mathrm{Hom}(C, B)$ with $1_C = p_*(s)$; that is, $1_C = ps$, and this says that the extension splits. $\bullet$

**Corollary 9.88.** *A left* $R$-*module* $P$ *is projective if and only if* $\mathrm{Ext}^1_R(P, B) = \{0\}$ *for every left* $R$-*module* $B$.

**Proof.** If $P$ is projective, then Corollary 9.77 gives $\mathrm{Ext}^1_R(P, B) = \{0\}$ for all $B$. Conversely, if $\mathrm{Ext}^1_R(P, B) = \{0\}$ for all $B$, then Proposition 9.87 says that every exact sequence $0 \to B \to X \to P \to 0$ splits, and so $P$ is projective, by Proposition 6.73. $\bullet$

We are going to prove the converse of Proposition 9.87 by showing that there is a bijection $\psi \colon e(C, A) \to \mathrm{Ext}^1(C, A)$.

Given an extension $\xi : 0 \to A \to B \to C \to 0$ and a projective resolution of $C$, form the diagram

$$
\begin{array}{ccccccccc}
P_2 & \xrightarrow{d_2} & P_1 & \xrightarrow{d_1} & P_0 & \longrightarrow & C & \longrightarrow & 0 \\
\big\downarrow & & \alpha\big\downarrow & & \big\downarrow & & \big\downarrow{\scriptstyle 1_C} & & \\
0 & \longrightarrow & A & \longrightarrow & B & \longrightarrow & C & \longrightarrow & 0
\end{array}
$$

By the Comparison Theorem (Theorem 9.48), we may fill in dashed arrows to obtain a commutative diagram. In particular, there is a map $\alpha \colon P_1 \to A$ with $\alpha d_2 = 0$; that is, $d_2^*(\alpha) = 0$, so that $\alpha \in \ker d_2^*$ is a 1-cocycle. The Comparison Theorem also says that any two fillings in of the diagram are homotopic; thus, if $\alpha' \colon P_1 \to A$ is part of a second filling in, there are maps $s_0$ and $s_1$ with $\alpha' - \alpha = 0 \cdot s_1 + s_0 d_1 = s_0 d_1$:

$$
\begin{array}{ccccc}
P_2 & \xrightarrow{d_2} & P_1 & \xrightarrow{d_1} & P_0 \\
 & {\scriptstyle s_1}\swarrow & {\scriptstyle \alpha'}\big\Vert{\scriptstyle \alpha} & \swarrow{\scriptstyle s_0} & \\
0 & \longrightarrow & A & \longrightarrow & B
\end{array}
$$

Thus, $\alpha' - \alpha \in \operatorname{im} d_1^*$, and so $\operatorname{cls}(\alpha) \in \operatorname{Ext}^1(C, A)$ is well-defined. We leave as an exercise for the reader that equivalent extensions $\xi$ and $\xi'$ determine the same element of Ext. Thus,

$$\psi\colon e(C, A) \to \operatorname{Ext}^1(C, A),$$

given by

$$\psi\colon [\xi] \mapsto \operatorname{cls}(\alpha),$$

is a well-defined function. Note that if $\xi$ is a split extension, then $\psi([\xi]) = 0$. In order to prove that $\psi$ is a bijection, we first analyze the diagram containing the map $\alpha$.

**Lemma 9.89.** *Let* $\Xi\colon\ 0 \to X_1 \xrightarrow{j} X_0 \xrightarrow{\varepsilon} C \to 0$ *be an extension of a module* $X_1$ *by a module* $C$. *Given a module* $A$, *consider the diagram*

$$
\begin{array}{ccccccccc}
0 & \longrightarrow & X_1 & \xrightarrow{\ j\ } & X_0 & \xrightarrow{\ \varepsilon\ } & C & \longrightarrow & 0 \\
  &        & \ \downarrow{\alpha} &        & &        & \ \downarrow{1_C} & & \\
  &        & A &        & &        & C & & \\
\end{array}
$$

(i) *There exists a commutative diagram with exact rows completing the given diagram:*

$$
\begin{array}{ccccccccc}
0 & \longrightarrow & X_1 & \xrightarrow{\ j\ } & X_0 & \xrightarrow{\ \varepsilon\ } & C & \longrightarrow & 0 \\
  &        & \ \downarrow{\alpha} &        & \ \downarrow{\beta} &        & \ \downarrow{1_C} & & \\
0 & \longrightarrow & A & \xrightarrow{\ i\ } & B & \xrightarrow{\ \eta\ } & C & \longrightarrow & 0 \\
\end{array}
$$

(ii) *Any two bottom rows of completed diagrams are equivalent extensions.*

**Proof.**

(i) We define $B$ as the pushout of $j$ and $\alpha$. Thus,

$$B = (A \oplus X_0)/S,\ , \text{ where } S = \big\{(\alpha x_1, -j x_1) \in A \oplus X_0 : x_1 \in X_1\big\};$$

define $i\colon a \mapsto (a, 0) + S$, $\beta\colon x_0 \mapsto (0, x_0) + S$, and $\eta\colon (a, x_0) + S \mapsto \varepsilon x_0$. That $\eta$ is well-defined, that the diagram commutes, and that the bottom row is exact are left for the reader to check.

(ii) Let

$$
\begin{array}{ccccccccc}
0 & \longrightarrow & X_1 & \xrightarrow{\ j\ } & X_0 & \xrightarrow{\ \varepsilon\ } & C & \longrightarrow & 0 \\
  &        & \ \downarrow{\alpha} &        & \ \downarrow{\beta'} &        & \ \downarrow{1_C} & & \\
0 & \longrightarrow & A & \xrightarrow{\ i'\ } & B' & \xrightarrow{\ \eta'\ } & C & \longrightarrow & 0 \\
\end{array}
$$

be a second completion of the diagram. Define $f\colon A \oplus X_0 \to B'$ by

$$f\colon (a, x_0) \mapsto i'a + \beta' x_0.$$

We claim that $f$ is surjective. If $b' \in B'$, then $\eta' b' \in C$, and so there is $x_0 \in X_0$ with $\varepsilon x_0 = \eta' b'$. Commutativity gives $\eta' \beta' x_0 = \varepsilon x_0 = \eta' b'$. Hence, $b' - \beta' x_0 \in \ker \eta' = \operatorname{im} i'$, and so there is $a \in A$ with $i'a = b' - \beta' x_0$. Therefore, $b' = i'a + \beta' x_0 \in \operatorname{im} f$, as desired.

We now show that $\ker f = S$. If $(\alpha x_1, -jx_1) \in S$, then $f(\alpha x_1, -jx_1) = i'\alpha x_1 - \beta' jx_1 = 0$, by commutativity of the first square of the diagram, and so $S \subseteq \ker f$. For the reverse inclusion, let $(a, x_0) \in \ker f$, so that $i'a + \beta' x_0 = 0$. Commutativity of the second square gives $\varepsilon x_0 = \eta' \beta' x_0 = -\eta' i'a = 0$. Hence, $x_0 \in \ker \varepsilon = \operatorname{im} j$, so there is $x_1 \in X_1$ with $jx_1 = x_0$. Thus, $i'a = -\beta' x_0 = -\beta' jx_1 = -i'\alpha x_1$. Since $i'$ is injective, we have $a = -\alpha x_1$. Replacing $x_1$ by $y_1 = -x_1$, we have $(a, x_0) = (\alpha y_1, -jy_1) \in S$, as desired.

Finally, define $\varphi \colon B \to B'$ by

$$\varphi \colon (a, x_0) + S \mapsto f(a, x_0) = i'a + \beta' x_0$$

[$\varphi$ is well-defined because $B = (A \oplus X_0)/S$ and $S = \ker f$]. To show commutativity of the diagram

$$
\begin{array}{ccccccccc}
0 & \longrightarrow & A & \stackrel{i}{\longrightarrow} & B & \stackrel{\eta}{\longrightarrow} & C & \longrightarrow & 0 \\
 & & {\scriptstyle 1_A}\big\downarrow & & {\scriptstyle \varphi}\big\downarrow & & {\scriptstyle 1_C}\big\downarrow & & \\
0 & \longrightarrow & A & \stackrel{i'}{\longrightarrow} & B' & \stackrel{\eta'}{\longrightarrow} & C & \longrightarrow & 0
\end{array}
$$

we use the definitions of the maps $i$ and $\eta$ in part (i). For the first square, if $a \in A$, then $\varphi i a = \varphi((a, 0) + S) = i'a$. For the second square,

$$\eta' \varphi \colon (a, x_0) + S \mapsto \eta'(i'a + \beta' x_0) = \eta' \beta' x_0 = \varepsilon x_0 = \eta((a, x_0) + S).$$

Therefore, the two bottom rows are equivalent extensions.   •

**Notation.** Denote the extension of $A$ by $C$ just constructed by

$$\alpha \Xi.$$

The dual result is true; it is related to the construction of Ext using injective resolutions of the second variable $A$.

**Lemma 9.90.** *Let $A$ and $Y_0$ be modules, and let $\Xi' : 0 \to A \to Y_0 \to Y_1 \to 0$ be an extension of $A$ by $Y_1$. Given a module $C$, consider the diagram*

$$
\begin{array}{ccccccccc}
 & & A & & & & C & & \\
 & & {\scriptstyle 1_A}\big\downarrow & & & & {\scriptstyle \gamma}\big\downarrow & & \\
0 & \longrightarrow & A & \longrightarrow & Y_0 & \stackrel{p}{\longrightarrow} & Y_1 & \longrightarrow & 0
\end{array}
$$

(i) *There exists a commutative diagram with exact rows completing the given diagram:*

$$
\begin{array}{ccccccccc}
0 & \longrightarrow & A & \longrightarrow & B & \longrightarrow & C & \longrightarrow & 0 \\
 & & {\scriptstyle 1_A}\big\downarrow & & \big\downarrow & & {\scriptstyle \gamma}\big\downarrow & & \\
0 & \longrightarrow & A & \longrightarrow & Y_0 & \stackrel{p}{\longrightarrow} & Y_1 & \longrightarrow & 0
\end{array}
$$

(ii) *Any two top rows of completed diagrams are equivalent extensions.*

**Proof.** Dual to that of Lemma 9.89; in particular, construct the top row using the pullback of $\gamma$ and $p$.   •

**Notation.** Denote the extension of $A$ by $C$ just constructed by

$$\Xi'\gamma.$$

**Theorem 9.91.** *The function* $\psi\colon e(C,A) \to \mathrm{Ext}^1_R(C,A)$ *is a bijection.*

**Proof.** We construct the inverse $\theta\colon \mathrm{Ext}^1(C,A) \to e(C,A)$ of $\psi$. Choose a projective resolution of $C$, so there is an exact sequence

$$\to P_2 \xrightarrow{d_2} P_1 \xrightarrow{d_1} P_0 \to C \to 0,$$

and choose a 1-cocycle $\alpha\colon P_1 \to A$. Since $\alpha$ is a cocycle, we have $0 = d_2^*(\alpha) = \alpha d_2$, so that $\alpha$ induces a homomorphism $\alpha'\colon P_1/\operatorname{im} d_2 \to A$ [if $x_1 \in P_1$, then $\alpha' : x_1 + \operatorname{im} d_2 \mapsto \alpha(x_1)$]. Let $\Xi$ denote the extension

$$\Xi : 0 \to P_1/\operatorname{im} d_2 \to P_0 \to C \to 0.$$

As in Lemma 9.90, there is a commutative diagram with exact rows:

$$
\begin{array}{ccccccccc}
0 & \longrightarrow & P_1/\operatorname{im} d_2 & \longrightarrow & P_0 & \longrightarrow & C & \longrightarrow & 0 \\
 & & \downarrow{\scriptstyle \alpha'} & & \downarrow{\scriptstyle \beta} & & \downarrow{\scriptstyle 1_C} & & \\
0 & \longrightarrow & A & \underset{i}{\longrightarrow} & B & \longrightarrow & C & \longrightarrow & 0
\end{array}
$$

Define $\theta\colon \mathrm{Ext}^1(C,A) \to e(C,A)$ using the construction in the lemma:

$$\theta(\mathrm{cls}(\alpha)) = [\alpha'\Xi].$$

We begin by showing that $\theta$ is independent of the choice of cocycle $\alpha$. If $\zeta$ is another representative of the coset $\alpha + \operatorname{im} d_1^* = \mathrm{cls}(\alpha)$, then there is a map $s\colon P_0 \to A$ with $\zeta = \alpha + sd_1$. But it is easy to see that the following diagram commutes:

$$
\begin{array}{ccccccccc}
P_2 & \xrightarrow{d_2} & P_1 & \xrightarrow{d_1} & P_0 & \longrightarrow & C & \longrightarrow & 0 \\
\downarrow & & \downarrow{\scriptstyle \alpha+sd_1} & & \downarrow{\scriptstyle \beta+is} & & \downarrow{\scriptstyle 1_C} & & \\
0 & \longrightarrow & A & \underset{i}{\longrightarrow} & B & \longrightarrow & C & \longrightarrow & 0
\end{array}
$$

As the bottom row has not changed, we have $[\alpha'\Xi] = [\zeta'\Xi]$.

It remains to show that the composites $\psi\theta$ and $\theta\psi$ are identities. If $\mathrm{cls}(\alpha) \in \mathrm{Ext}^1(C,A)$, then $\theta(\mathrm{cls}(\alpha)) = \theta(\alpha + \operatorname{im} d_1^*)$ is the bottom row of the diagram

$$
\begin{array}{ccccccccc}
0 & \longrightarrow & P_1/\operatorname{im} d_2 & \longrightarrow & P_0 & \longrightarrow & C & \longrightarrow & 0 \\
 & & \downarrow{\scriptstyle \alpha'} & & \downarrow{\scriptstyle \beta} & & \downarrow{\scriptstyle 1_C} & & \\
0 & \longrightarrow & A & \underset{i}{\longrightarrow} & B & \longrightarrow & C & \longrightarrow & 0
\end{array}
$$

and $\psi\theta(\mathrm{cls}(\alpha))$ is the homology class of a cocycle fitting this diagram. Clearly, $\alpha$ is such a cocycle; and so $\psi\theta$ is the identity. For the other composite, start with an extension $\xi$, and then imbed it as the bottom row of a diagram

$$
\begin{array}{ccccccc}
P_2 & \xrightarrow{d_2} & P_1 & \xrightarrow{d_1} & P_0 & \longrightarrow & C & \longrightarrow & 0 \\
\big\downarrow & & \alpha\big\downarrow & & \big\downarrow & & \big\downarrow{\scriptstyle 1_C} & & \\
0 & \longrightarrow & A & \xrightarrow{\;i\;} & B & \longrightarrow & C & \longrightarrow & 0
\end{array}
$$

Both $\xi$ and $\alpha'\Xi$ are bottom rows of such a diagram, and so Lemma 9.89(ii) shows that $[\xi] = [\alpha'\Xi]$. $\bullet$

We can now prove the converse of Proposition 9.87.

**Corollary 9.92.** *For all left $R$-modules $C$ and $A$, every extension of $A$ by $C$ splits if and only if $\mathrm{Ext}^1_R(C, A) = \{0\}$.*

**Proof.** If every extension is split, then $|e(C, A)| = 1$, by Exercise 9.46 on page 825, and so $|\mathrm{Ext}^1_R(C, A)| = 1$, by Theorem 9.91; hence, $\mathrm{Ext}^1_R(C, A) = \{0\}$. Conversely, if $\mathrm{Ext}^1_R(C, A) = \{0\}$, then Proposition 9.87 says that every extension is split. $\bullet$

**Example 9.93.** If $p$ is a prime, then $\mathrm{Ext}^1_{\mathbb{Z}}(\mathbb{I}_p, \mathbb{I}_p) \cong \mathbb{I}_p$, as we saw in Example 9.86. On the other hand, it follows from Theorem 9.91 that there are $p$ equivalence classes of extensions $0 \to \mathbb{I}_p \to B \to \mathbb{I}_p \to 0$. But $|B| = p^2$, so there are only two choices for $B$ to isomorphism: $B \cong \mathbb{I}_{p^2}$ or $B \cong \mathbb{I}_p \oplus \mathbb{I}_p$. Of course, this is consistent with Example 9.22. Thus, if $p$ is an odd prime, then $|\mathrm{Ext}^1_{\mathbb{Z}}(C, A)| < g(C, A)$, where $g(C, A)$ is the number of nonisomorphic middle abelian groups $B$ occurring in extensions $0 \to A \to B \to C \to 0$. $\blacktriangleleft$

Here is a minor application of Ext.

**Proposition 9.94.**

  (i) *If $F$ is a torsion-free abelian group and $T$ is an abelian group of **bounded order** (that is, $nT = \{0\}$ for some positive integer $n$), then $\mathrm{Ext}^1_{\mathbb{Z}}(F, T) = \{0\}$.*

  (ii) *Let $G$ be an abelian group. If the torsion subgroup $tG$ of $G$ is of bounded order, then $tG$ is a direct summand of $G$.*

**Proof.**

  (i) Since $F$ is torsion-free, it is a flat $\mathbb{Z}$-module, by Corollary 8.7, so that exactness of $0 \to \mathbb{Z} \to \mathbb{Q}$ gives exactness of $0 \to \mathbb{Z} \otimes F \to \mathbb{Q} \otimes F$. Thus, $F \cong \mathbb{Z} \otimes F$ can be imbedded in a vector space $V$ over $\mathbb{Q}$, namely, $V = \mathbb{Q} \otimes F$. Applying the contravariant functor $\mathrm{Hom}(\square, T)$ to $0 \to F \to V \to V/F \to 0$ gives an exact sequence

$$
\mathrm{Ext}^1(V, T) \to \mathrm{Ext}^1(F, T) \to \mathrm{Ext}^2(V/F, T).
$$

Now the last term is $\{0\}$, by Exercise 9.38 on page 814, and $\mathrm{Ext}^1(V, T)$ is (torsion-free) divisible, by Example 9.72, so that $\mathrm{Ext}^1(F, T)$ is divisible. Since $T$ has bounded order, Exercise 9.41 on page 814 gives $\mathrm{Ext}^1(F, T) = \{0\}$.

(ii) To prove that the extension $0 \to tG \to G \to G/tG \to 0$ splits, it suffices to prove that $\text{Ext}^1(G/tG, tG) = \{0\}$. Since $G/tG$ is torsion-free, this follows from part (i) and Corollary 9.92. •

The torsion subgroup of a group may not be a direct summand; the following proof using homology is quite different from that of Exercise 8.3 on page 651.

**Proposition 9.95.** *There exists an abelian group $G$ whose torsion subgroup is not a direct summand of $G$; in fact, we may choose $tG = \bigoplus_p \mathbb{I}_p$, where the index set is the family of all primes $p$.*

**Proof.** It suffices to prove that $\text{Ext}^1(\mathbb{Q}, \bigoplus_p \mathbb{I}_p) \neq 0$, for this will give a nonsplit extension $0 \to \bigoplus_p \mathbb{I}_p \to G \to \mathbb{Q} \to 0$, by Corollary 9.92; moreover, since $\mathbb{Q}$ is torsion-free, it follows that $\bigoplus_p \mathbb{I}_p = tG$.

Consider the exact sequence $0 \to \bigoplus_p \mathbb{I}_p \to \prod_p \mathbb{I}_p \to D \to 0$. By Exercise 8.6 on page 652, we know that $D$ is divisible.[18] There is an exact sequence

$$\text{Hom}\left(\mathbb{Q}, \prod_p \mathbb{I}_p\right) \to \text{Hom}(\mathbb{Q}, D) \xrightarrow{\partial} \text{Ext}^1\left(\mathbb{Q}, \bigoplus_p \mathbb{I}_p\right) \to \text{Ext}^1\left(\mathbb{Q}, \prod_p \mathbb{I}_p\right).$$

But $\partial$ is an isomorphism: $\text{Ext}^1(\mathbb{Q}, \prod_p \mathbb{I}_p) \cong \prod \text{Ext}^1(\mathbb{Q}, \mathbb{I}_p)$, by Proposition 9.83, $\text{Ext}^1(\mathbb{Q}, \mathbb{I}_p) = \{0\}$, by Proposition 9.94, and $\text{Hom}(\mathbb{Q}, \prod_p \mathbb{I}_p) \cong \prod \text{Hom}(\mathbb{Q}, \mathbb{I}_p) = \{0\}$, by Theorem 6.45(iii). Since $\text{Hom}(\mathbb{Q}, D) \neq \{0\}$, we have $\text{Ext}^1(\mathbb{Q}, \bigoplus_p \mathbb{I}_p) \neq \{0\}$. •

**Remark.** A theorem of Baer says that a torsion abelian group $T$ is a direct summand of any group containing it as its torsion subgroup if and only if $T \cong B \oplus D$, where $B$ has bounded order and $D$ is divisible. ◀

We now describe another construction of Ext. If $X$ is a set and $\psi \colon X \to G$ is a bijection to a group $G$, then there is a unique group structure on $X$ making it a group and $\psi$ an isomorphism [if $x, x' \in X$, then $x = \psi^{-1}(g)$ and $x' = \psi^{-1}(g')$; define $xx' = \psi^{-1}(gg')$]. In particular, Theorem 9.91 implies that there is a group structure on $e(C, A)$; here are the necessary definitions.

Recall that we constructed the addition of maps on page 615. The *diagonal map* $\Delta_C \colon C \to C \oplus C$ is defined by $\Delta_C \colon c \mapsto (c, c)$, and the *codiagonal map* $\nabla_A \colon A \oplus A \to A$ is defined by $\nabla_A \colon (a_1, a_2) \mapsto a_1 + a_2$. If $f, f' \colon C \to A$ is a homomorphism, then the composite $\nabla_A(f \oplus f')\Delta$ maps $C \to C \oplus C \to A \oplus A \to A$. Evaluating,

$$\nabla_A(f \oplus f')\Delta \colon a \mapsto (a, a) \mapsto (fa, f'a) \mapsto fa + f'a = (f + f')a,$$

so that this formula describes addition in $\text{Hom}(C, A)$. Now Ext is a generalized Hom, and so we mimic this definition to define addition in $e(C, A)$.

If $\xi \colon 0 \to A \to B \to C \to 0$ and $\xi' \colon 0 \to A' \to B' \to C' \to 0$ are extensions, then their **direct sum** is the extension

$$\xi \oplus \xi' \colon 0 \to A \oplus A' \to B \oplus B' \to C \oplus C' \to 0.$$

---

[18] In truth, $D \cong \mathbb{R}$: it is a torsion-free divisible group, hence it is a vector space over $\mathbb{Q}$, by Proposition 8.24, and we check that $\dim(D) = $ continuum, which is the dimension of $\mathbb{R}$ as a vector space over $\mathbb{Q}$.

***Baer sum*** is the binary operation on $e(C, A)$ given by $([\xi], [\xi']) \mapsto [\xi] + [\xi'] = [\nabla_A(\xi \oplus \xi')\Delta_C]$ (we have already defined $\alpha\Xi$ and $\Xi'\gamma$). To show that Baer sum is well-defined, first prove that $\alpha(\Xi'\gamma)$ is equivalent to $(\alpha\Xi')\gamma$; then prove that $e(C, A)$ is a group under this operation by showing that $\psi([\xi] + [\xi']) = \psi([\nabla_A(\xi \oplus \xi')\Delta_C])$ (for details, see Rotman, *An Introduction to Homological Algebra*, p. 431). The identity element is the class of the split extension, and the inverse of $[\xi]$ is $[(-1_A)\xi]$.

This description of $\text{Ext}^1$ has been generalized by Yoneda to a description of $\text{Ext}^n$ for all $n \geq 1$. Elements of Yoneda's $\text{Ext}^n(C, A)$ are certain equivalence classes of exact sequences

$$0 \to A \to B_1 \to \cdots \to B_n \to C \to 0,$$

and we add them by a generalized Baer sum (Mac Lane, *Homology*, pp. 82–87). Thus, there is a construction of Ext that does not use derived functors. Indeed, we can construct $\text{Ext}^n$ without using projectives or injectives.

## Exercises

\* **9.46.** Prove that any two split extensions of modules $A$ by $C$ are equivalent.

**9.47.** Prove that if $A$ is an abelian group with $nA = A$ for some positive integer $n$, then every extension $0 \to A \to E \to \mathbb{I}_n \to 0$ splits.

**9.48.** If $A$ is a torsion abelian group, prove that $\text{Ext}^1_{\mathbb{Z}}(A, \mathbb{Z}) \cong \text{Hom}_{\mathbb{Z}}(A, S^1)$, where $S^1$ is the circle group.

\* **9.49.** Prove that a left $R$-module $E$ is injective if and only if $\text{Ext}^1_R(A, E) = \{0\}$ for every left $R$-module $A$.

**9.50.** For any ring $R$, prove that a left $R$-module $B$ is injective if and only if $\text{Ext}^1_R(R/I, B) = \{0\}$ for every left ideal $I$.

**Hint.** Use the Baer criterion.

**9.51.** Prove that an abelian group $G$ is injective if and only if $\text{Ext}^1_{\mathbb{Z}}(\mathbb{Q}/\mathbb{Z}, G) = \{0\}$.

**9.52.** Prove that an abelian group $G$ is free abelian if and only if $\text{Ext}^1_{\mathbb{Z}}(G, F) = \{0\}$ for every free abelian group $F$.[19]

## Section 9.7. Tor

It is now Tor's turn. Here is a result that has no analog for Ext [there is rarely an isomorphism $\text{Hom}_R(A, B) \to \text{Hom}_R(B, A)$].

**Theorem 9.96.** *If $R$ is a ring, $A$ is a right $R$-module, and $B$ is a left $R$-module, then*

$$\text{Tor}^R_n(A, B) \cong \text{Tor}^{R^{\text{op}}}_n(B, A)$$

*for all $n \geq 0$, where $R^{\text{op}}$ is the opposite ring of $R$.*

---

[19]The question whether $\text{Ext}^1(G, \mathbb{Z}) = \{0\}$ implies that $G$ is free abelian is known as ***Whitehead's problem***. It turns out that if $G$ is countable, then it must be free abelian, but Shelah proved that it is undecideable whether uncountable such $G$ must be free abelian.

**Proof.** Recall Proposition 6.15: every left $R$-module is a right $R^{\mathrm{op}}$-module, and every right $R$-module is a left $R^{\mathrm{op}}$-module. Choose a deleted projective resolution $\mathbf{P}_A$ of $A$. It is easy to see that $t \colon \mathbf{P}_A \otimes_R B \to B \otimes_{R^{\mathrm{op}}} \mathbf{P}_A$ is a chain map of $\mathbb{Z}$-complexes, where $t_n$ is given by

$$t_n \colon x_n \otimes b \mapsto b \otimes x_n.$$

Since each $t_n$ is an isomorphism of abelian groups (its inverse is $b \otimes x_n \mapsto x_n \otimes b$), the chain map $t$ is an isomorphism of complexes. By Exercise 9.21 on page 793,

$$\mathrm{Tor}_n^R(A, B) = H_n(\mathbf{P}_A \otimes_R B) \cong H_n(B \otimes_{R^{\mathrm{op}}} \mathbf{P}_A)$$

for all $n$. But $\mathbf{P}_A$, viewed as a complex of left $R^{\mathrm{op}}$-modules, is a deleted projective resolution of $A$ qua left $R^{\mathrm{op}}$-module, and so $H_n(B \otimes_{R^{\mathrm{op}}} \mathbf{P}_A) \cong \mathrm{Tor}_n^{R^{\mathrm{op}}}(B, A)$. $\quad\bullet$

In light of this result, we will no longer have to say "similarly in the other variable;" theorems about $\mathrm{Tor}(A, \square)$ will yield results about $\mathrm{Tor}(\square, B)$.

**Corollary 9.97.** *If $R$ is a commutative ring and $A$ and $B$ are $R$-modules, then for all $n \geq 0$,*

$$\mathrm{Tor}_n^R(A, B) \cong \mathrm{Tor}_n^R(B, A).$$

We know that $\mathrm{Tor}_n$ vanishes on projectives; we now show that it vanishes on flat modules.

**Proposition 9.98.** *If a right $R$-module $F$ is flat, then $\mathrm{Tor}_n^R(F, M) = \{0\}$ for all $n \geq 1$ and every left $R$-module $M$. Conversely, if $\mathrm{Tor}_1^R(F, M) = \{0\}$ for every left $R$-module $M$, then $F$ is flat.*

**Proof.** Let $0 \to N \xrightarrow{i} P \to M \to 0$ be exact, where $P$ is projective. There is an exact sequence

$$\mathrm{Tor}_1(F, P) \to \mathrm{Tor}_1(F, M) \to F \otimes N \xrightarrow{1 \otimes i} F \otimes P.$$

Now $\mathrm{Tor}_1(F, P) = \{0\}$, because $P$ is projective, so that $\mathrm{Tor}_1(F, M) = \ker(1 \otimes i)$. Since $F$ is flat, however, $\ker(1 \otimes i) = \{0\}$, and so $\mathrm{Tor}_1(F, M) = \{0\}$. The result for all $n \geq 1$ follows by dimension shifting.

For the converse, $0 \to A \xrightarrow{i} B$ exact implies exactness of

$$0 = \mathrm{Tor}_1(F, B/A) \to F \otimes A \xrightarrow{1 \otimes i} F \otimes B.$$

Hence, $1 \otimes i$ is an injection, and so $F$ is flat. $\quad\bullet$

**Proposition 9.99.** *If $(B_k)_{k \in K}$ is a family of left $R$-modules, then there are natural isomorphisms, for all $n \geq 0$,*

$$\mathrm{Tor}_n^R\Big(A, \bigoplus_{k \in K} B_k\Big) \cong \bigoplus_{k \in K} \mathrm{Tor}_n^R(A, B_k).$$

*There is also an isomorphism if the direct sum is in the first variable.*

**Proof.** The proof is by induction on $n \geq 0$. The base step is Theorem 6.110, for $\mathrm{Tor}_0(A, \square)$ is naturally isomorphic to $A \otimes \square$.

For the inductive step, choose, for each $k \in K$, a short exact sequence

$$0 \to N_k \to P_k \to B_k \to 0,$$

where $P_k$ is projective. There is an exact sequence

$$0 \to \bigoplus_k N_k \to \bigoplus_k P_k \to \bigoplus_k B_k \to 0,$$

and $\bigoplus_k P_k$ is projective, for every direct sum of projectives is projective. There is a commutative diagram with exact rows:

$$
\begin{array}{ccccccccc}
\mathrm{Tor}_1(A, \bigoplus P_k) & \longrightarrow & \mathrm{Tor}_1(A, \bigoplus B_k) & \overset{\partial}{\longrightarrow} & A \otimes \bigoplus N_k & \longrightarrow & A \otimes \bigoplus P_k \\
& & \downarrow & & \downarrow{\scriptstyle \tau} & & \downarrow{\scriptstyle \sigma} \\
\bigoplus \mathrm{Tor}_1(A, P_k) & \longrightarrow & \bigoplus \mathrm{Tor}_1(A, B_k) & \underset{\partial'}{\longrightarrow} & \bigoplus A \otimes N_k & \longrightarrow & \bigoplus A \otimes P_k
\end{array}
$$

where the maps in the bottom row are just the usual induced maps in each coordinate, and the maps $\tau$ and $\sigma$ are the isomorphisms given by Theorem 6.110. The proof is completed by dimension shifting.    •

**Example 9.100.**

(i) We show, for every abelian group $B$, that

$$\mathrm{Tor}_1^{\mathbb{Z}}(\mathbb{I}_n, B) \cong B[n] = \{b \in B : nb = 0\}.$$

There is an exact sequence

$$0 \to \mathbb{Z} \overset{\mu_n}{\longrightarrow} \mathbb{Z} \to \mathbb{I}_n \to 0,$$

where $\mu_n$ is multiplication by $n$. Applying $\square \otimes B$ gives exactness of

$$\mathrm{Tor}_1(\mathbb{Z}, B) \to \mathrm{Tor}_1(\mathbb{I}_n, B) \to \mathbb{Z} \otimes B \overset{1 \otimes \mu_n}{\longrightarrow} \mathbb{Z} \otimes B.$$

Now $\mathrm{Tor}_1(\mathbb{Z}, B) = \{0\}$, because $\mathbb{Z}$ is projective. Moreover, $1 \otimes \mu_n$ is also multiplication by $n$, while $\mathbb{Z} \otimes B = B$. More precisely, $\mathbb{Z} \otimes \square$ is naturally isomorphic to the identity functor on **Ab**, and so there is a commutative diagram with exact rows

$$
\begin{array}{ccccccc}
0 & \longrightarrow & B[n] & \longrightarrow & B & \overset{\mu_n}{\longrightarrow} & B \\
& & \downarrow & & \downarrow{\scriptstyle \tau_B} & & \downarrow{\scriptstyle \tau_B} \\
0 & \longrightarrow & \mathrm{Tor}_1(\mathbb{I}_n, B) & \longrightarrow & \mathbb{Z} \otimes B & \underset{1 \otimes \mu_n}{\longrightarrow} & \mathbb{Z} \otimes B
\end{array}
$$

By Proposition 6.117, there is an isomorphism $B[n] \cong \mathrm{Tor}_1^{\mathbb{Z}}(\mathbb{I}_n, B)$.

(ii) We can now compute $\mathrm{Tor}_1^{\mathbb{Z}}(A, B)$ whenever $A$ and $B$ are finitely generated abelian groups. By the Fundamental Theorem, both $A$ and $B$ are direct sums of cyclic groups. Since Tor commutes with direct sum, $\mathrm{Tor}_1^{\mathbb{Z}}(A, B)$ is the direct sum of groups $\mathrm{Tor}_1^{\mathbb{Z}}(C, D)$, where $C$ and $D$ are cyclic. We may assume that $C$ and $D$ are finite; otherwise they are projective and $\mathrm{Tor}_1 = \{0\}$. This calculation can be completed using part (i) and Exercise 4.9 on page 242, which says that if $D$ is a cyclic group of finite order $m$, then $D[n]$ is a cyclic group of order $d$, where $d = (m, n)$ is their gcd. ◄

In contrast to Ext, Proposition 9.99 can be generalized by replacing direct sum with direct limit.

**Proposition 9.101.** *If $\{B_i, \varphi_j^i\}$ is a direct system of left $R$-modules over a directed index set $I$, then there is an isomorphism, for all right $R$-modules $A$ and for all $n \geq 0$,*

$$\operatorname{Tor}_n^R\big(A, \varinjlim B_i\big) \cong \varinjlim \operatorname{Tor}_n^R(A, B_i).$$

**Proof.** The proof is by dimension shifting. The base step is Theorem 6.159, for $\operatorname{Tor}_0(A, \square)$ is naturally isomorphic to $A \otimes \square$.

For the inductive step, choose, for each $i \in I$, a short exact sequence

$$0 \to N_i \to P_i \to B_i \to 0,$$

where $P_i$ is projective. Since the index set is directed, Proposition 6.158 says that there is an exact sequence

$$0 \to \varinjlim N_i \to \varinjlim P_i \to \varinjlim B_i \to 0.$$

Now $\varinjlim P_i$ is flat: every $P_i$ is flat, for it is projective, and Corollary 6.161 says that a direct limit (over a directed index set) of flat modules is flat. There is a commutative diagram with exact rows:

$$
\begin{array}{ccccccc}
\operatorname{Tor}_1(A, \varinjlim P_i) & \longrightarrow & \operatorname{Tor}_1(A, \varinjlim B_i) & \overset{\partial}{\longrightarrow} & A \otimes \varinjlim N_i & \longrightarrow & A \otimes \varinjlim P_i \\
\downarrow & & \downarrow & & \downarrow{\scriptstyle \tau} & & \downarrow{\scriptstyle \sigma} \\
\varinjlim \operatorname{Tor}_1(A, P_i) & \longrightarrow & \varinjlim \operatorname{Tor}_1(A, B_i) & \underset{\vec{\partial}}{\longrightarrow} & \varinjlim A \otimes N_i & \longrightarrow & \varinjlim A \otimes P_i
\end{array}
$$

where the maps in the bottom row are just the usual induced maps between direct limits, and the maps $\tau$ and $\sigma$ are the isomorphisms given by Theorem 6.159. This proves the step $n = 1$; the step $n \geq 2$ is routine. •

The last proposition generalizes Lemma 6.139, which says that if every finitely generated submodule of a module $M$ is flat, then $M$ itself is flat. After all, by Example 6.155(iv), $M$ is a direct limit, over a directed index set, of its finitely generated submodules.

When the ring $R$ is noncommutative, the abelian group $A \otimes_R B$ need not be an $R$-module, but it is a $Z(R)$-module. And so it is with Tor.

**Proposition 9.102.**

(i) *Let $r \in Z(R)$ be a central element, let $A$ be a right $R$-module, and let $B$ be a left $R$-modules. If $\mu_r \colon B \to B$ is multiplication by $r$, then the induced map*

$$\mu_{r*} \colon \operatorname{Tor}_n^R(A, B) \to \operatorname{Tor}_n^R(A, B)$$

*is also multiplication by $r$.*

(ii) *If $R$ is a commutative ring, then $\operatorname{Tor}_n^R(A, B)$ is an $R$-module for all $R$-modules $A$ and $B$ and all $n \geq 0$.*

**Proof.**

(i) This follows at once from Example 9.50.

(ii) This follows from part (i) if we define scalar multiplication by $r$ to be $\mu_{r*}$.  •

We are now going to assume that $R$ is a domain, so that the notion of torsion submodule is defined. This discussion will also show why Tor is so called.

**Lemma 9.103.** *Let $R$ be a domain, let $Q = \operatorname{Frac}(R)$, and let $K = Q/R$.*

(i) *There is a natural isomorphism $\operatorname{Tor}_1^R(K, A) \to A$ if $A$ is a torsion $R$-module.*

(ii) *For every $R$-module $A$, we have $\operatorname{Tor}_n^R(K, A) = \{0\}$ for all $n \geq 2$.*

(iii) *If $A$ is a torsion-free $R$-module, then $\operatorname{Tor}_1^R(K, A) = \{0\}$.*

**Proof.**

(i) Exactness of $0 \to R \to Q \to K \to 0$ gives exactness of

$$\operatorname{Tor}_1(Q, A) \to \operatorname{Tor}_1(K, A) \to R \otimes A \to Q \otimes A.$$

Now $Q$ is flat, by Corollary 6.140, and so $\operatorname{Tor}_1(Q, A) = \{0\}$, by Proposition 9.98. The last term $Q \otimes A = \{0\}$ because $Q$ is divisible and $A$ is torsion, by Exercise 8.15 on page 653, and so the middle map $\operatorname{Tor}_1(K, A) \to R \otimes A$ is an isomorphism. Of course, $R \otimes_R A \cong A$; it is natural because it is the connecting homomorphism.

(ii) There is an exact sequence

$$\operatorname{Tor}_n(Q, A) \to \operatorname{Tor}_n(K, A) \to \operatorname{Tor}_{n-1}(R, A).$$

Since $n \geq 2$, we have $n - 1 \geq 1$, and so both the first and third Tor are $\{0\}$, because $Q$ and $R$ are flat. Therefore, exactness gives $\operatorname{Tor}_n(K, A) = \{0\}$.

(iii) By Theorem 6.96, there is an injective $R$-module $E$ containing $A$ as a submodule. Since $A$ is torsion-free, however, $A \cap tE = \{0\}$, and so $A$ is imbedded in $E/tE$. By Lemma 6.92, injective modules are divisible, and so $E$ is divisible, as is its quotient $E/tE$. Now $E/tE$ is a vector space over $Q$, for it is a torsion-free divisible $R$-module (Exercise 8.1 on page 651). Let us denote $E/tE$ by $V$. Since every vector space has a basis, $V$ is a direct sum of copies of $Q$. Corollary 6.140 says that $Q$ is flat, and Lemma 6.137 says that a direct sum of flat modules is flat. We conclude that $V$ is flat.[20]

Exactness of $0 \to A \to V \to V/A \to 0$ gives exactness of

$$\operatorname{Tor}_2(K, V/A) \to \operatorname{Tor}_1(K, A) \to \operatorname{Tor}_1(K, V).$$

Now $\operatorname{Tor}_2(K, V/A) = \{0\}$, by part (ii), and $\operatorname{Tor}_1(K, V) = \{0\}$, because $V$ is flat. We conclude from exactness that $\operatorname{Tor}_1(K, A) = \{0\}$.  •

---

[20]Torsion-free $\mathbb{Z}$-modules are flat, but there exist domains $R$ having torsion-free modules that are not flat. In fact, domains for which every torsion-free module is flat, called *Prüfer rings*, are characterized as those such that every finitely generated ideal is a projective module.

The next result shows why Tor is so called.

**Theorem 9.104.**

(i) *If $R$ is a domain, $Q = \mathrm{Frac}(R)$, and $K = Q/R$, then the functor $\mathrm{Tor}_1^R(K, \square)$ is naturally isomorphic to the torsion functor.*

(ii) $\mathrm{Tor}_1^R(K, A) \cong tA$ *for all $R$-modules $A$.*

**Proof.** Exactness of $0 \to tA \xrightarrow{i_A} A \to A/tA \to 0$ gives exactness of

$$\mathrm{Tor}_2(K, A/tA) \to \mathrm{Tor}_1(K, tA) \xrightarrow{\iota_{A*}} \mathrm{Tor}_1(K, A) \to \mathrm{Tor}_1(K, A/tA).$$

The first term is $\{0\}$, by Lemma 9.103(ii); the last term is $\{0\}$, by Lemma 9.103(iii). Therefore, the map $\iota_{A*}\colon \mathrm{Tor}_1(K, tA) \to \mathrm{Tor}_1(K, A)$ is an isomorphism. Now the connecting homomorphism is a natural isomorphism $\partial\colon \mathrm{Tor}_1(K, A) \to A$ when $A$ is torsion, by Lemma 9.103(i). It follows that $\sigma_A\colon \mathrm{Tor}_1(K, tA) \to tA$, defined by $\sigma_A = \partial\iota_{A*}$, is an isomorphism. We let the reader prove that $\sigma\colon \mathrm{Tor}_1(K, \square) \to t\square$ is natural; that is, if $f\colon A \to B$, then the following diagram commutes [recall that $tf\colon tA \to tB$ is just $f|tA$].

$$
\begin{array}{ccc}
\mathrm{Tor}_1(K, A) & \xrightarrow{\sigma_A} & tA \\
{\scriptstyle f_*}\downarrow & & \downarrow{\scriptstyle tf} \\
\mathrm{Tor}_1(K, B) & \xrightarrow[\sigma_B]{} & tB
\end{array}
$$

Part (ii) follows at once from part (i).  •

We describe another construction of $\mathrm{Tor}_1^{\mathbb{Z}}(A, B)$ which uses generators and relations. Consider all triples $(a, n, b)$, where $a \in A$, $b \in B$, $na = 0$, and $nb = 0$. Then $\mathrm{Tor}_1^{\mathbb{Z}}(A, B)$ is generated by all such triples subject to the relations (whenever both sides are defined):

$$
\begin{aligned}
(a + a', n, b) &= (a, n, b) + (a', n, b) \\
(a, n, b + b') &= (a, n, b) + (a, n, b') \\
(ma, n, b) &= (a, mn, b) = (a, m, nb).
\end{aligned}
$$

For a proof of this result, and its generalization to $\mathrm{Tor}_n^R(A, B)$ for arbitrary rings $R$, see Mac Lane, *Homology*, pp. 150–159.

The Tor functors are very useful in Algebraic Topology. The *Universal Coefficient Theorem* gives a formula for the homology groups $H_n(X; G)$ with coefficients in an abelian group $G$.

**Theorem 9.105 (Universal Coefficients).** *For every topological space $X$ and every abelian group $G$, there are isomorphisms for all $n \geq 0$,*

$$H_n(X; G) \cong H_n(X) \otimes_{\mathbb{Z}} G \ \oplus \ \mathrm{Tor}_1^{\mathbb{Z}}(H_{n-1}(X), G).$$

**Proof.** See Rotman, *An Introduction to Algebraic Topology*, p. 261.  •

If we know the homology groups of spaces $X$ and $Y$, then the *Künneth Formula* gives a formula for the homology groups of the product space $X \times Y$, and this, too, involves Tor in an essential way.

**Theorem 9.106 (Künneth Formula).** *For every pair of topological spaces $X$ and $Y$, there are isomorphisms for every $n \geq 0$,*

$$H_n(X \times Y) \cong \bigoplus_i H_i(X) \otimes_{\mathbb{Z}} H_{n-i}(Y) \ \oplus \ \bigoplus_p \operatorname{Tor}_1^{\mathbb{Z}}(H_p(X), H_{n-1-p}(Y)).$$

**Proof.** Ibid., p. 269.   •

## Exercises

**9.53.** If $0 \to A \to B \to C \to 0$ is an exact sequence of right $R$-modules with both $A$ and $C$ flat, prove that $B$ is flat.

**9.54.** If $A$ and $B$ are finite abelian groups, prove that $\operatorname{Tor}_1^{\mathbb{Z}}(A, B) \cong A \otimes_{\mathbb{Z}} B$.

**9.55.** Let $R$ be a domain, $Q = \operatorname{Frac}(R)$, and $K = Q/R$.

  (i) Prove, for every $R$-module $A$, that there is an exact sequence

$$0 \to tA \to A \to Q \otimes A \to K \otimes A \to 0.$$

  (ii) Prove that a module $A$ is torsion if and only if $Q \otimes A = \{0\}$.

**9.56.** Let $R$ be a domain.

  (i) If $B$ is a torsion $R$-module, prove that $\operatorname{Tor}_n^R(A, B)$ is a torsion $R$-module for all $R$-modules $A$ and for all $n \geq 0$.

  (ii) For all $R$-modules $A$ and $B$, prove that $\operatorname{Tor}_n^R(A, B)$ is a torsion $R$-module for all $n \geq 1$.

**9.57.** Let $k$ be a field, let $R = k[x, y]$, and let $I$ be the ideal $(x, y)$.

  (i) Prove that $x \otimes y - y \otimes x \in I \otimes_R I$ is nonzero.
  **Hint.** Consider $(I/I^2) \otimes (I/I^2)$.

  (ii) Prove that $x(x \otimes y - y \otimes x) = 0$, and conclude that $I \otimes_R I$ is not torsion-free.

## Section 9.8. Cohomology of Groups

Recall that Proposition 9.34 (only partially proved) says that there is an exact sequence

$$F_3 \xrightarrow{d_3} F_2 \xrightarrow{d_2} F_1 \xrightarrow{d_1} F_0 \xrightarrow{\epsilon} \mathbb{Z} \to 0,$$

where $F_0$, $F_1$, $F_2$, and $F_3$ are free $Q$-modules and $\mathbb{Z}$ is viewed as a trivial $Q$-module. In light of the discussion in Section 9.3, the following definition should now seem reasonable.

**Definition.** Let $G$ be a group, let $A$ be a $G$-module (i.e., a left $\mathbb{Z}G$-module), and let $\mathbb{Z}$ be the integers viewed as a trivial $G$-module (i.e., $gm = m$ for all $g \in G$ and $m \in \mathbb{Z}$). The ***cohomology groups*** of $G$ are

$$H^n(G, A) = \operatorname{Ext}^n_{\mathbb{Z}G}(\mathbb{Z}, A);$$

the ***homology groups*** of $G$ are

$$H_n(G, A) = \operatorname{Tor}^{\mathbb{Z}G}_n(\mathbb{Z}, A).$$

The history of cohomology of groups is quite interesting. The subject began with the discovery, by the topologist Hurewicz in the 1930s, that if $X$ is a connected *aspherical space* (in modern language, if the higher homotopy groups of $X$ are all trivial), then all the homology and cohomology groups of $X$ are determined by the fundamental group $\pi = \pi_1(X)$. This led to the question of whether $H_n(X)$ could be described algebraically in terms of $\pi$. For example, Hurewicz proved that $H_1(X) \cong \pi/\pi'$, where $\pi'$ is the commutator subgroup. In 1942, Hopf proved that if $\pi$ has a presentation $F/R$, where $F$ is free, then $H_2(X) \cong (R \cap F')/[F, R]$, where $[F, R]$ is the subgroup generated by all commutators of the form $frf^{-1}r^{-1}$ for $f \in F$ and $r \in R$. These results led to the creation of cohomology of groups.

In what follows, we will write $\operatorname{Hom}_G$, instead of $\operatorname{Hom}_{\mathbb{Z}G}$, and $\otimes_G$, instead of $\otimes_{\mathbb{Z}G}$. Recall that the ***augmentation***

$$\varepsilon \colon \mathbb{Z}G \to \mathbb{Z}$$

is defined by

$$\varepsilon \colon \sum_{x \in G} m_x x \mapsto \sum_{x \in G} m_x.$$

We have seen, in Exercise 7.20 on page 549, that $\varepsilon$ is a surjective ring homomorphism, and so its kernel, the ***augmentation ideal*** $\mathcal{G}$, is a two-sided ideal in $\mathbb{Z}G$. Thus, there is an exact sequence

$$0 \to \mathcal{G} \to \mathbb{Z}G \xrightarrow{\varepsilon} \mathbb{Z} \to 0.$$

The augmentation and the augmentation ideal play an important role in cohomology of groups because of the distinguished role played by the trivial $G$-module $\mathbb{Z}$.

**Proposition 9.107.** *Let $G$ be a group with augmentation ideal $\mathcal{G}$. As an abelian group, $\mathcal{G}$ is free abelian with basis $G - 1 = \{g - 1 : g \in G \text{ and } g \neq 1\}$.*

**Proof.** An element $u = \sum_g m_g g \in \mathbb{Z}G$ lies in $\ker \varepsilon = \mathcal{G}$ if and only if $\sum_g m_g = 0$. Therefore, if $u \in \mathcal{G}$, then

$$u = u - \left( \sum_g m_g \right) 1 = \sum_g m_g (g - 1).$$

Thus, $\mathcal{G}$ is generated by the nonzero $g - 1$ for $g \in G$.

Suppose that $\sum_{g \neq 1} n_g (g - 1) = 0$. Then $\sum_{g \neq 1} n_g g - \left( \sum_{g \neq 1} n_g \right) 1 = 0$ in $\mathbb{Z}G$, which, as an abelian group, is free abelian with basis the elements of $G$. Hence, $n_g = 0$ for all $g \neq 1$. Therefore, the nonzero $g - 1$ comprise a basis of $\mathcal{G}$. $\bullet$

We begin by examining homology groups.

**Proposition 9.108.** *If $A$ is a $G$-module, then*

$$H_0(G, A) = \mathbb{Z} \otimes_G A \cong A/\mathcal{G}A.$$

**Proof.** By definition, $H_0(G, A) = \operatorname{Tor}_0^{\mathbb{Z}G}(\mathbb{Z}, A) = \mathbb{Z} \otimes_G A$. Applying the right exact functor $\square \otimes_G A$ to the exact sequence

$$0 \to \mathcal{G} \to \mathbb{Z}G \to \mathbb{Z} \to 0$$

gives exactness of the top row of the commutative diagram

$$
\begin{array}{ccccccc}
\mathcal{G} \otimes_G A & \longrightarrow & \mathbb{Z}G \otimes_G A & \longrightarrow & \mathbb{Z} \otimes_G A & \longrightarrow & 0 \\
\downarrow & & \downarrow & & \vdots & & \\
\mathcal{G}A & \longrightarrow & A & \longrightarrow & A/\mathcal{G}A & \longrightarrow & 0
\end{array}
$$

The two solid vertical arrows are given by $u \otimes a \mapsto ua$; the first is a surjection and the second is an isomorphism. By Proposition 6.116, there is an isomorphism $\mathbb{Z} \otimes_G A \to A/\mathcal{G}A$ making the diagram commute. •

It is easy to see that $A/\mathcal{G}A$ is $G$-trivial; indeed, it is the largest $G$-trivial quotient of $A$.

**Example 9.109.** Suppose that $E$ is a semidirect product of an abelian group $A$ by a group $G$. Recall that $[G, A]$ is the subgroup generated by all commutators of the form $[x, a] = xax^{-1}a^{-1}$, where $x \in G$ and $a \in A$. If we write commutators additively, as we did at the beginning of this chapter, then

$$[x, a] = x + a - x - a = xa - a = (x - 1)a$$

(recall that $G$ acts on $A$ by conjugation). Therefore, $A/\mathcal{G}A = A/[G, A]$ here. ◀

We are now going to use independence of the choice of projective resolution to compute the homology groups of a finite cyclic group $G$.

**Lemma 9.110.** *Let $G = \langle x \rangle$ be a cyclic group of finite order $k$. Define elements $D$ and $N$ in $\mathbb{Z}G$ by*

$$D = x - 1 \quad and \quad N = 1 + x + x^2 + \cdots + x^{k-1}.$$

*Then the following sequence is a $G$-free resolution of $Z$:*

$$\to \mathbb{Z}G \xrightarrow{N} \mathbb{Z}G \xrightarrow{D} \mathbb{Z}G \xrightarrow{N} \mathbb{Z}G \xrightarrow{D} \mathbb{Z}G \xrightarrow{\varepsilon} \mathbb{Z} \to 0,$$

*where $\varepsilon$ is the augmentation and the other maps are multiplication by $N$ and $D$, respectively.*

**Proof.** Obviously, every term $\mathbb{Z}G$ is free; moreover, since $\mathbb{Z}G$ is commutative, the maps are $G$-maps. Now $DN = ND = x^k - 1 = 0$, while if $u \in \mathbb{Z}G$, then

$$\varepsilon D(u) = \varepsilon\big((x - 1)u\big) = \varepsilon(x - 1)\varepsilon(u) = 0,$$

because $\varepsilon$ is a ring map. Thus, we have a complex, and it only remains to prove exactness.

We have already noted that $\varepsilon$ is surjective. Now $\ker \varepsilon = \mathcal{G} = \operatorname{im} D$, by Proposition 9.107, and so we have exactness at the zeroth step.

Suppose $u = \sum_{i=0}^{k-1} m_i x^i \in \ker D$; that is, $(x-1)u = 0$. Expanding, and using the fact that $\mathbb{Z}G$ has basis $\{1, x, x^2, \ldots, x^{k-1}\}$, we have

$$m_0 = m_1 = \cdots = m_{k-1},$$

so that $u = m_0 N \in \operatorname{im} N$, as desired.

Finally, if $u = \sum_{i=0}^{k-1} m_i x^i \in \ker N$, then $0 = \varepsilon(Nu) = \varepsilon(N)\varepsilon(u) = k\varepsilon(u)$, so that $\varepsilon(u) = \sum_{i=0}^{k-1} m_i = 0$. Therefore,

$$u = -D\big(m_0 1 + (m_0 + m_1)x + \cdots + (m_0 + \cdots + m_{k-1})x^{k-1}\big) \in \operatorname{im} D. \quad \bullet$$

**Definition.** If $A$ is a $G$-module, define submodules

$$A[N] = \{a \in A : Na = 0\}$$

and

$$A^G = \{a \in A : ga = a \text{ for all } g \in G\}.$$

**Theorem 9.111.** *If $G$ is a cyclic group of finite order $k$ and $A$ is a $G$-module, then*

$$H_0(G, A) = A/\mathcal{G}A;$$
$$H_{2n-1}(G, A) = A^G/NA \quad \text{for all } n \geq 1;$$
$$H_{2n}(G, A) = A[N]/\mathcal{G}A \quad \text{for all } n \geq 1.$$

**Proof.** Apply $\square \otimes_G A$ to the resolution of $\mathbb{Z}$ in Lemma 9.110 after noting that $\mathbb{Z}G \otimes_G A \cong A$. The calculation of $\ker / \operatorname{im}$ is now simple, using $\operatorname{im} D = \mathcal{G}A$, which follows from Proposition 9.107 and the fact that $(x-1) \mid (x^i - 1)$. $\quad \bullet$

**Corollary 9.112.** *If $G$ is a finite cyclic group of order $k$ and $A$ is a trivial $G$-module, then*

$$H_0(G, A) = A;$$
$$H_{2n-1}(G, A) = A/kA \quad \text{for all } n \geq 1;$$
$$H_{2n}(G, A) = A[k] \quad \text{for all } n \geq 1.$$

*In particular,*

$$H_0(G, \mathbb{Z}) = \mathbb{Z};$$
$$H_{2n-1}(G, \mathbb{Z}) = \mathbb{Z}/k\mathbb{Z} \quad \text{for all } n \geq 1;$$
$$H_{2n}(G, \mathbb{Z}) = \{0\} \quad \text{for all } n \geq 1.$$

**Proof.** Since $A$ is $G$-trivial, we have $A^G = A$ and $\mathcal{G}A = \{0\}$ (for $Da = (x-1)a = 0$ because $xa = a$). $\quad \bullet$

We now compute low-dimensional homology groups of not necessarily cyclic groups.

**Lemma 9.113.** *For any group $G$, we have*

$$H_1(G, \mathbb{Z}) \cong \mathcal{G}/\mathcal{G}^2.$$

**Proof.** The long exact sequence arising from $0 \to \mathcal{G} \to \mathbb{Z}G \xrightarrow{\varepsilon} \mathbb{Z} \to 0$ ends with

$$H_1(G, \mathbb{Z}G) \to H_1(G, \mathbb{Z}) \xrightarrow{\partial} H_0(G, \mathcal{G}) \to H_0(G, \mathbb{Z}G) \xrightarrow{\varepsilon_*} H_0(G, \mathbb{Z}) \to 0.$$

Now $H_1(G, \mathbb{Z}G) = \{0\}$, because $\mathbb{Z}G$ is projective, so that $\partial$ is an injection. Also,

$$H_0(G, \mathbb{Z}G) \cong \mathbb{Z},$$

by Proposition 9.108. Since $\varepsilon_*$ is surjective, it must be injective as well (if $\ker \varepsilon_* \neq \{0\}$, then $\mathbb{Z}/\ker \varepsilon_*$ is finite; on the other hand, $\mathbb{Z}/\ker \varepsilon_* \cong \operatorname{im} \varepsilon_* = \mathbb{Z}$, which is torsion-free). Exactness of the sequence of homology groups ($\partial$ is surjective if and only if $\varepsilon_*$ is injective) now gives $\partial$ a surjection. We conclude that

$$\partial \colon H_1(G, \mathbb{Z}) \cong H_0(G, \mathcal{G}) \cong \mathcal{G}/\mathcal{G}^2,$$

by Proposition 9.108.  •

**Proposition 9.114.** *For any group $G$, we have*

$$H_1(G, \mathbb{Z}) \cong G/G',$$

*where $G'$ is the commutator subgroup of $G$.*

**Proof.** It suffices to prove that $G/G' \cong \mathcal{G}/\mathcal{G}^2$. Define $\theta \colon G \to \mathcal{G}/\mathcal{G}^2$ by

$$\theta \colon x \mapsto (x - 1) + \mathcal{G}^2.$$

To see that $\theta$ is a homomorphism, note that

$$xy - 1 - (x - 1) - (y - 1) = (x - 1)(y - 1) \in \mathcal{G}^2,$$

so that

$$\begin{aligned}
\theta(xy) &= xy - 1 + \mathcal{G}^2 \\
&= (x - 1) + (y - 1) + \mathcal{G}^2 \\
&= x - 1 + \mathcal{G}^2 + y - 1 + \mathcal{G}^2 \\
&= \theta(x) + \theta(y).
\end{aligned}$$

Since $\mathcal{G}/\mathcal{G}^2$ is abelian, $G' \subseteq \ker \theta$, and so $\theta$ induces a homomorphism $\theta' \colon G/G' \to \mathcal{G}/\mathcal{G}^2$, namely, $xG' \mapsto x - 1 + \mathcal{G}^2$.

We now construct the inverse of $\theta'$. By Proposition 9.107, $\mathcal{G}$ is a free abelian group with basis all $x - 1$, where $x \in G$ and $x \neq 1$. It follows that there is a (well-defined) homomorphism $\varphi \colon \mathcal{G} \to G/G'$, given by

$$\varphi \colon x - 1 \mapsto xG'.$$

If $\mathcal{G}^2 \subseteq \ker \varphi$, then $\varphi$ induces a homomorphism $\mathcal{G}/\mathcal{G}^2 \to G/G'$ that, obviously, is the inverse of $\theta'$, and this will complete the proof.

If $u \in \mathcal{G}^2$, then

$$u = \left(\sum_{x \neq 1} m_x(x-1)\right)\left(\sum_{y \neq 1} n_y(y-1)\right)$$

$$= \sum_{x,y} m_x n_y(x-1)(y-1)$$

$$= \sum_{x,y} m_x n_y\big((xy-1) - (x-1) - (y-1)\big).$$

Therefore, $\varphi(u) = \prod_{x,y}(xyx^{-1}y^{-1})^{m_x n_y}G' = G'$, and so $u \in \ker \varphi$, as desired. •

The following question is not easily answered. If groups $G$ and $H$ have isomorphic group rings, that is, if $\mathbb{Z}G \cong \mathbb{Z}H$, is $G \cong H$? The answer is 'yes' if $G$ and $H$ are abelian: Proposition 9.114 gives

$$G \cong G/G' \cong H_1(G, \mathbb{Z}) \cong H_1(H, \mathbb{Z}) \cong H/H' \cong H.$$

The general question was only recently answered: there are nonisomorphic finite groups with $\mathbb{Z}G \cong \mathbb{Z}H$ (Hertwick, A counterexample to the isomorphism problem for integral group rings, *Annals Math.* 154 (2001), 115–138).

The group $H_2(G, \mathbb{Z})$ is useful; it is called the **Schur multiplier** of $G$. For example, suppose that $G = F/R$, where $F$ is a free group; that is, we have a presentation of a group $G$. Then **Hopf's formula** is

$$H_2(G, \mathbb{Z}) \cong (R \cap F)/[F, R]$$

(Rotman, *An Introduction to Homological Algebra*, p. 547). It follows that the group $(R \cap F)/[F, R]$ depends only on $G$ and not upon the choice of presentation of $G$.

The Schur multiplier also enters into the construction of *universal central extensions*; these are used in constructing "covers" of simple groups.

**Definition.** An exact sequence $0 \to A \to E \to G \to 1$ is a **central extension** of a group $G$ if $A \subseteq Z(E)$. A **universal central extension** of $G$ is a central extension $0 \to M \to U \to G \to 1$ for which there always exists a commutative diagram

$$
\begin{array}{ccccccccc}
0 & \longrightarrow & A & \longrightarrow & E & \longrightarrow & G & \longrightarrow & 1 \\
  &   & \vdots &   & \vdots &   & \downarrow 1_G &   &   \\
0 & \longrightarrow & M & \longrightarrow & U & \longrightarrow & G & \longrightarrow & 1
\end{array}
$$

**Theorem 9.115.** *Let $G$ be a finite group. Then $G$ has a universal central extension if and only if $G = G'$, in which case $M \cong H_2(G, \mathbb{Z})$. In particular, every finite simple group has a universal central extension.*

**Proof.** Milnor, *Introduction to Algebraic K-Theory*, pp. 43–46. •

We now consider low-dimensional cohomology groups.

**Proposition 9.116.** *Let $G$ be a group, let $A$ be a $G$-module, and let $\mathbb{Z}$ be viewed as a trivial $G$-module. Then*

$$H^0(G, A) = \text{Hom}_G(\mathbb{Z}, A) \cong A^G.$$

**Proof.** By definition,

$$H^0(G, A) = \text{Ext}^0_{\mathbb{Z}G}(\mathbb{Z}, A) = \text{Hom}_G(\mathbb{Z}, A).$$

Define $\tau_A \colon \text{Hom}_G(\mathbb{Z}, A) \to A^G$ by $f \mapsto f(1)$. Note that $f(1) \in A^G$: if $g \in G$, then $gf(1) = f(g \cdot 1)$ (because $f$ is a $G$-map), and $g \cdot 1 = 1$ (because $\mathbb{Z}$ is $G$-trivial); therefore, $gf(1) = f(1)$, and $f(1) \in A^G$. That $\tau_A$ is an isomorphism is a routine calculation. $\bullet$

It follows that $H^0(G, A)$ is the largest $G$-trivial submodule of $A$.

**Theorem 9.117.** *Let $G = \langle \sigma \rangle$ be a cyclic group of finite order $k$, and let $A$ be a $G$-module. If $N = \sum_{i=0}^{k-1} \sigma^i$ and $D = \sigma - 1$, then*

$$H^0(G, A) = A^G;$$
$$H^{2n-1}(G, A) = \ker N/(\sigma - 1)A \quad \text{for all } n \geq 1;$$
$$H^{2n}(G, A) = A^G/NA \quad \text{for all } n \geq 1.$$

**Proof.** Apply the contravariant $\text{Hom}_G(\square, A)$ to the resolution of $\mathbb{Z}$ in Lemma 9.110, noting that $\text{Hom}_G(\mathbb{Z}G, A) \cong A$. The calculation of $\ker / \text{im}$ is now as given in the statement. $\bullet$

Note that Proposition 9.107 gives $\text{im } D = \mathcal{G}A$.

**Corollary 9.118.** *If $G$ is a cyclic group of finite order $k$ and $A$ is a trivial $G$-module, then*

$$H^0(G, A) = A;$$
$$H^{2n-1}(G, A) = A[k] \quad \text{for all } n \geq 1;$$
$$H^{2n}(G, A) = A/kA \quad \text{for all } n \geq 1.$$

*In particular,*

$$H^0(G, \mathbb{Z}) = \mathbb{Z};$$
$$H^{2n-1}(G, \mathbb{Z}) = \{0\} \quad \text{for all } n \geq 1;$$
$$H^{2n}(G, \mathbb{Z}) = \mathbb{Z}/k\mathbb{Z} \quad \text{for all } n \geq 1.$$

**Remark.** A finite group $G$ for which there exists a nonzero integer $d$ such that

$$H^n(G, A) \cong H^{n+d}(G, A),$$

for all $n \geq 1$ and all $G$-modules $A$, is said to have ***periodic cohomology***. It can be proved that a group $G$ has periodic cohomology if and only if its Sylow $p$-subgroups are cyclic for all odd primes $p$, while its Sylow 2-subgroups are either cyclic or generalized quaternion (Adem–Milgram, *Cohomology of Finite Groups*, p. 148). For example, $G = \text{SL}(2, 5)$ has periodic cohomology: it is a group of order

$120 = 8 \cdot 3 \cdot 5$, so its Sylow 3-subgroups and its Sylow 5-subgroups are cyclic, of prime order, while its Sylow 2-subgroups are isomorphic to the quaternions. ◄

We can interpret $H^1(G, A)$ and $H^2(G, A)$, where $G$ is any not necessarily cyclic group, in terms of derivations and extensions if we can show that the formulas in Section 9.3 do, in fact, arise from a projective resolution of $\mathbb{Z}$.

**Definition.** If $G$ is a group, define $B_0(G)$ to be the free $G$-module on the single generator $[\ ]$ (hence, $B_0(G) \cong \mathbb{Z}G$) and, for $n \geq 1$, define $B_n(G)$ to be the free $G$-module with basis all symbols $[x_1 \mid x_2 \mid \cdots \mid x_n]$, where $x_i \in G$. Define $\varepsilon \colon B_0(G) \to \mathbb{Z}$ by $\varepsilon([\ ]) = 1$ and, for $n \geq 1$, define $d_n \colon B_n(G) \to B_{n-1}(G)$ by

$$d_n \colon [x_1 \mid \cdots \mid x_n] \mapsto x_1[x_2 \mid \cdots \mid x_n]$$

$$+ \sum_{i=1}^{n-1} (-1)^i [x_1 \mid \cdots \mid x_i x_{i+1} \mid \cdots \mid x_n]$$

$$+ (-1)^n [x_1 \mid \cdots \mid x_{n-1}].$$

The **bar resolution**[21] is the sequence

$$\mathbf{B}(G) = \to B_2(G) \xrightarrow{d_2} B_1(G) \xrightarrow{d_1} B_0(G) \xrightarrow{\varepsilon} \mathbb{Z} \to 0.$$

Let us look at the low-dimensional part of the bar resolution:

$$d_1 \colon [x] \mapsto x[\ ] - [\ ];$$
$$d_2 \colon [x \mid y] \mapsto x[y] - [xy] + [x];$$
$$d_3 \colon [x \mid y \mid z] \mapsto x[y \mid z] - [xy \mid z] + [x \mid yz] - [x \mid y].$$

These are the formulas that arose in the earlier sections, but without the added conditions $[x \mid 1] = 0 = [1 \mid y]$ and $[1] = 0$. In fact, there are two bar resolutions; the bar resolution just defined, and another we shall soon see, called the *normalized bar resolution*.

The bar resolution $\mathbf{B}(G)$ is a free resolution of $\mathbb{Z}$, but it is not a routine calculation to see this; we prove that $\mathbf{B}(G)$ is a resolution by comparing it to a resolution familiar to topologists.

**Definition.** If $G$ is a group, let $P_n(G)$ be the free abelian group with basis all $(n + 1)$-tuples of elements of $G$; make $P_n(G)$ into a $G$-module by defining

$$x(x_0, x_1, \ldots, x_n) = (xx_0, xx_1, \ldots, xx_n).$$

Define $\partial_n \colon P_n(G) \to P_{n-1}(G)$, whenever $n \geq 1$, by

$$\partial_n \colon (x_0, x_1, \ldots, x_n) \mapsto \sum_{i=0}^{n} (-1)^i (x_0, \ldots, \widehat{x}_i, \ldots, x_n),$$

where $\widehat{x}_i$ means that $x_i$ has been deleted. We call $\mathbf{P}(G)$ the **homogeneous resolution** of $\mathbb{Z}$.

---

[21]The reason for the name is unexciting; to differentiate the symbols $[x_1 \mid \cdots \mid x_n]$ from other tuples, Eilenberg and Mac Lane used bars instead of commas.

Note that $P_0(G)$ is the free abelian group with basis all $(y)$, for $y \in G$, made into a $G$-module by $x(y) = (xy)$. In other words, $P_0(G) = \mathbb{Z}G$.

**Proposition 9.119.** *The sequence*

$$\mathbf{P}(G): \to P_2(G) \xrightarrow{\partial_2} P_1(G) \xrightarrow{\partial_1} P_0(G) \xrightarrow{\varepsilon} \mathbb{Z} \to 0,$$

*where $\varepsilon$ is the augmentation, is a $G$-free resolution of $\mathbb{Z}$.*

**Proof.** That $\mathbf{P}(G)$ is a complex, that is, that $\partial_n \partial_{n+1} = 0$, follows from Proposition 9.1. The reader can prove that $P_n(G)$ is a free $G$-module with basis all symbols of the form $(1, x_1, \ldots, x_n)$.

In order to prove exactness of $\mathbf{P}(G)$, it suffices, by Exercise 9.30 on page 794, to construct a contracting homotopy: a sequence $(s_n)_{n \geq -1}$ of $\mathbb{Z}$-maps

$$\leftarrow P_2(G) \xleftarrow{s_1} P_1(G) \xleftarrow{s_0} P_0(G) \xleftarrow{s_{-1}} \mathbb{Z}$$

with $\varepsilon s_{-1} = 1_{\mathbb{Z}}$ and

$$\partial_{n+1} s_n + s_{n-1} \partial_n = 1_{P_n(G)}, \quad \text{for all } n \geq 0.$$

Define $s_{-1} \colon \mathbb{Z} \to P_0(G)$ by $m \mapsto m(1)$, where the 1 in the parentheses is the identity element of the group $G$, and, for $n \geq 0$, define $s_n \colon P_n(G) \to P_{n+1}(G)$ by

$$s_n \colon (x_0, x_1, \ldots, x_n) \mapsto (1, x_0, x_1, \ldots, x_n).$$

Here are the computations:

$$\varepsilon s_{-1}(1) = \varepsilon\big((1)\big) = 1.$$

If $n \geq 0$, then

$$\partial_{n+1} s_n(x_0, \ldots, x_n) = \partial_{n+1}(1, x_0, \ldots, x_n)$$

$$= (x_0, \ldots, x_n) + \sum_{i=0}^{n} (-1)^{i+1}(1, x_0, \ldots, \widehat{x}_i, \ldots, x_n)$$

[the range of summation has been rewritten because $x_i$ sits in the $(i+1)$st position in $(1, x_0, \ldots, x_n)$]. On the other hand,

$$s_{n-1} \partial_n(x_0, \ldots, x_n) = s_{n-1} \sum_{j=0}^{n} (-1)^j (x_0, \ldots, \widehat{x}_j, \ldots, x_n)$$

$$= \sum_{j=0}^{n} (-1)^j (1, x_0, \ldots, \widehat{x}_j, \ldots, x_n).$$

It follows that $\big(\partial_{n+1} s_n + s_{n-1} \partial_n\big)(x_0, \ldots, x_n) = (x_0, \ldots, x_n)$. $\quad \bullet$

**Proposition 9.120.** *The bar resolution $\mathbf{B}(\mathbf{G})$ is a $G$-free resolution of $\mathbb{Z}$.*

**Proof.** For each $n \geq 0$, define $\tau_n \colon P_n(G) \to B_n(G)$ by

$$\tau_n \colon (x_0, \ldots, x_n) \mapsto x_0[x_0^{-1}x_1 \mid x_1^{-1}x_2 \mid \cdots \mid x_{n-1}^{-1}x_n],$$

and define $\sigma_n \colon B_n(G) \to P_n(G)$ by

$$\sigma_n \colon [x_1 \mid \cdots \mid x_n] \mapsto (1, x_1, x_1 x_2, x_1 x_2 x_3, \ldots, x_1 x_2 \cdots x_n).$$

It is routine to check that $\tau_n$ and $\sigma_n$ are inverse, and so each $\tau_n$ is an isomorphism. The reader can also check that $\tau\colon \mathbf{P}(\mathbf{G}) \to \mathbf{B}(\mathbf{G})$ is a chain map; that is, the following diagram commutes:

$$
\begin{array}{ccc}
P_n(G) & \xrightarrow{\ \tau_n\ } & B_n(G) \\
\downarrow{\scriptstyle \partial_n} & & \downarrow{\scriptstyle d_n} \\
P_{n-1}(G) & \xrightarrow[\ \tau_{n-1}\ ]{} & B_{n-1}(G)
\end{array}
$$

Finally, Exercise 9.21 on page 793 shows that both complexes have the same homology groups. By Proposition 9.119, the complex $\mathbf{P}(G)$ is an exact sequence, so that all its homology groups are $\{0\}$. It follows that all the homology groups of $\mathbf{B}(G)$ are $\{0\}$, and so it, too, is an exact sequence. •

**Corollary 9.121.** $\mathrm{Ext}^1_{\mathbb{Z}G}(\mathbb{Z}, K) \cong \mathrm{Der}(G, K)/\mathrm{PDer}(G, K)$; *that is, the* $\mathrm{Ext}$ *version of* $H^1(G, K)$ *coincides with Schreier's group* $H^1(G, K)$.

**Proof.** Compute $\mathrm{Ext}^1_{\mathbb{Z}G}(\mathbb{Z}, K)$ by applying the contravariant functor $\mathrm{Hom}_G(\Box, K)$ to the bar resolution:

$$\mathrm{Hom}_G(B_2, K) \xleftarrow{\ d_2^*\ } \mathrm{Hom}_G(B_1, K) \xleftarrow{\ d_1^*\ } \mathrm{Hom}_G(B_0, K);$$

by definition, $\mathrm{Ext}^1(\mathbb{Z}, K) = \ker d_2^* / \operatorname{im} d_1^*$.

There is no loss in generality in identifying a $G$-map $g \in \mathrm{Hom}_G(B_1, K)$ with its restriction to the basis $G^2$; hence, $g(x[y]) = xg[y]$ for all $x, y \in G$. Moreover, we may extend any function $\delta\colon G \to K$ to a $G$-map $B_1 \to K$ by $[y] \mapsto \delta(y)$; in particular, we may regard $\mathrm{Der}(G, K)$ as a subset of $\mathrm{Hom}_G(B_1, K)$.

If $g\colon G \to K$ lies in $\ker d_2^*$, then

$$0 = (d_2^* g)[x \mid y] = g d_2[x \mid y] = g(x[y] - [xy] + [x]) = xg[y] - g[xy] + g[x].$$

Thus, $g[xy] = xg[y] + g[x]$, $g$ is a derivation, and $\ker d_2^* \subseteq \mathrm{Der}(G, K)$. For the reverse inclusion, take $\delta \in \mathrm{Der}(G, K)$. Then

$$(d_2^* \delta)[x \mid y] = \delta d_2[x \mid y] = \delta\big(x[y] - [xy] + [x]\big) = x\delta[y] - \delta[xy] + \delta[x] = 0,$$

because $\delta$ is (the restriction of) a $G$-map. Hence, $\ker d_2^* = \mathrm{Der}(G, K)$.

Let us compute $\operatorname{im} d_1^*$. If $t \in \mathrm{Hom}_G(B_0, K)$, then $t[\ ] = a_0 \in K$. Now

$$(d_1^* t)[x] = t d_1[x] = t\big(x[\ ] - [\ ]\big) = xt[\ ] - t[\ ] = xa_0 - a_0,$$

because $t$ is (the restriction of) a $G$-map. Thus, $d_1^* t$ is a principal derivation; that is, $\operatorname{im} d_1^* \subseteq \mathrm{PDer}(G, K)$. For the reverse inclusion, let $p\colon G \to K$ be a principal derivation: $p(x) = xb_0 - b_0$ for some $b_0 \in K$. Now define $u\colon B_0 \to K$ by $u[\ ] = b_0$, so that $p = d_1^* u$ and $\mathrm{PDer}(G, K) \subseteq \operatorname{im} d_1^*$.

Therefore, $\mathrm{Ext}^1(\mathbb{Z}, K) = \mathrm{Der}(G, K)/\mathrm{PDer}(G, K)$, which is the definition of Schreier's $H^1(G, K)$. •

Let us now consider $\operatorname{Ext}^2(G, K) = \ker d_3^* / \operatorname{im} d_2^*$. If $f \colon G^2 \to K$ lies in $\ker d_3^*$, then $0 = d_3^* f = f d_3$. Hence, for all $x, y, z \in G$, we have

$$0 = f d_3[x \mid y \mid z] = f(x[y \mid z] - [xy \mid z] + [x \mid yz] - [x \mid y])$$
$$= x f[y \mid z] - f[xy \mid z] + f[x \mid yz] - f[x \mid y],$$

the equation $f(x[y \mid z]) = x f[y \mid z]$ holding because $f$ is (the restriction of) a $G$-map. Thus, $f$ satisfies the cocycle identity; it would be a factor set if $f[1 \mid y] = 0 = f[x \mid 1]$ for all $x, y \in G$. Alas, we do not know this. If $f$ lies in $\operatorname{im} d_2^*$, then there is some $h \colon G \to K$ with $f = d_2^* h = h d_2$. Thus,

$$f[x \mid y] = h d_2[x \mid y] = h(x[y] - [xy] + [x]) = x h[y] - h[xy] + h[x];$$

the equation $h(x[y]) = x h[y]$ holding because $h$ is (the restriction of) a $G$-map. Now $f$ is almost a coboundary, but it may not be one because we cannot guarantee $h(1) = 0$. Thus, the bar resolution is inadequate to prove that Schreier's second cohomology group is $\operatorname{Ext}^2(\mathbb{Z}, \square)$. That these two groups are isomorphic is shown in Theorem 9.123. We now introduce another projective resolution of $\mathbb{Z}$ by modifying the bar resolution $\mathbf{B}(G)$.

Define $U_n \subseteq B_n$ to be the $G$-submodule generated by all $[x_1 \mid \cdots \mid x_n]$ having at least one $x_i = 1$. It is easy to check that $d_n(U_n) \subseteq U_{n-1}$, for if $x_i = 1$, then every term in the expression for $d_n[x_1 \mid \cdots \mid x_n]$ involves $x_i$ except two (those involving $x_{i-1} x_i$ and $x_i x_{i+1}$) that cancel. Hence, $\mathbf{U}(G) = (U_n)$ is a subcomplex of $\mathbf{B}(G)$.

**Definition.** The ***normalized bar resolution*** $\mathbf{B}^\star(G)$ is the quotient complex

$$\mathbf{B}^\star(G) = \mathbf{B}(G)/\mathbf{U}(G) = \to B_2^\star \xrightarrow{d_2^\star} B_1^\star \xrightarrow{d_1^\star} B_0^\star \xrightarrow{\epsilon} \mathbb{Z} \to 0.$$

Note that $B_0^\star = B_0$ (for $U_0 = \{0\}$) and, when $n \geq 1$, $B_n^\star = B_n/U_n$ is the free $G$-module with basis all cosets $[x_1 \mid \cdots \mid x_n]^\star = [x_1 \mid \cdots \mid x_n] + U_n$ in which all $x_i \neq 1$. Moreover, each differential $d_n^\star$ has the same formula as the map $d_n$ in the bar resolution except that all symbols $[x_1 \mid \cdots \mid x_n]$ now occur with stars; in particular, $[x_1 \mid \cdots \mid x_n]^\star = 0$ if some $x_i = 1$.

**Theorem 9.122.** *The normalized bar resolution $\mathbf{B}^\star(G)$ is a $G$-free resolution of $\mathbb{Z}$.*

**Proof.** To prove exactness of $\mathbf{B}^\star(G)$, it suffices, by Proposition 9.42 and Exercise 9.30 on page 794, to construct a contracting homotopy

$$\leftarrow B_2^\star \xleftarrow{t_1} B_1^\star \xleftarrow{t_0} B_0^\star \xleftarrow{t_{-1}} \mathbb{Z},$$

where each $t_n$ is merely a $\mathbb{Z}$-map. Define $t_{-1} \colon \mathbb{Z} \to B_0^\star$ by $t_{-1} \colon m \mapsto m[\ ]$. Since the additive group of $\mathbb{Z}G$ is free abelian, $B_n^\star$ is also a free abelian group; it is easy to see that a basis is $\{x[x_1 \mid \cdots \mid x_n]^\star : x, x_i \in G \text{ and } x_i \neq 1\}$ ($x = 1$ is allowed). To define $t_n$ for $n \geq 0$, we take advantage of the fact that $t_n$ need only be a $\mathbb{Z}$-map. It suffices to define $t_n$ on this basis; moreover, freeness allows us to choose the values without restriction. Thus, for $n \geq 0$, we define $t_n \colon B_n^\star \to B_{n+1}^\star$ by

$$t_n \colon x[x_1 \mid \cdots \mid x_n]^\star \mapsto [x \mid x_1 \mid \cdots \mid x_n]^\star.$$

That we have constructed a contracting homotopy is routine; the reader may check that $\epsilon t_{-1} = 1_{\mathbb{Z}}$ and, for $n \geq 0$, that $d_{n+1}^{\star} t_n + t_{n-1} d_n^{\star} = 1_{B_n^{\star}}$. •

**Theorem 9.123.** $\operatorname{Ext}^2_{\mathbb{Z}G}(\mathbb{Z}, K)$ *coincides with Schreier's group* $H^2(G, K)$.

**Proof.** When we use the normalized bar resolution $\mathbf{B}^{\star}(G)$, the identities $f(1, y) = 0 = f(x, 1)$ and $h(1) = 0$ that were lacking in the calculation on page 841 are now present. •

There is a group-theoretic interpretation of $H^3(Q, K)$. Given an extension with a nonabelian kernel $N$, say, $1 \to N \to G \to Q \to 1$, the normality of $N$ need not give a homomorphism $Q \to \operatorname{Aut}(N)$; instead, there is a homomorphism $\theta \colon Q \to \operatorname{Out}(N)$, called a ***coupling***, where $\operatorname{Out}(N) = \operatorname{Aut}(N)/\operatorname{Inn}(N)$. Not every homomorphism $\theta \colon Q \to \operatorname{Out}(N)$ actually corresponds to a coupling arising from some extension, and the group $H^3(Q, Z(N)) = \{0\}$ if and only if every such $\theta$ can be so realized (Robinson, *A Course in the Theory of Groups*, Section 11.4). Robinson says, "While group-theoretic interpretations of the cohomology groups in dimensions greater than 3 are known, no really convincing applications to Group Theory have been made." On the other hand, higher cohomology groups will be used on page 844 to define the *cohomological dimension* $\operatorname{cd}(G)$ of a group $G$.

**Proposition 9.124.** *If* $G$ *is a finite group of order* $m$, *then* $mH^n(G, A) = \{0\}$ *for all* $n \geq 1$ *and all* $G$-modules $A$.

**Proof.** For $f \in \operatorname{Hom}_G(B_n, A)$, define $g \colon B_{n-1} \to A$ by

$$g(x_1 \mid \cdots \mid x_{n-1}) = \sum_{y \in G} f(x_1 \mid \cdots \mid x_{n-1} \mid y).$$

As in the proof of Theorem 9.25, sum the cocycle formula to obtain, for all $x_{n+1} = y$ in $G$,

$$(df)(x_1 \mid \cdots \mid x_n \mid x_{n+1}) = x_1 f(x_2 \mid \cdots \mid x_{n+1})$$

$$+ \sum_{i=1}^{n-2} (-1)^i f(x_1 \mid \cdots \mid x_i x_{i+1} \mid \cdots \mid x_{n+1})$$

$$+ (-1)^{n-1} f(x_1 \mid \cdots \mid x_n x_{n+1}) + (-1)^n f(x_1 \mid \cdots \mid x_n).$$

In the next to last term, as $x_{n+1}$ varies over $G$, so does $x_n x_{n+1}$. Therefore, if $f$ is a cocycle, then $df = 0$ and

$$0 = x_1 g(x_2 \mid \cdots \mid x_{n-1}) + \sum_{i=1}^{n-2} (-1)^i g(x_1 \mid \cdots \mid x_i x_{i+1} \mid \cdots \mid x_n)$$

$$+ (-1)^{n-1} g(x_1 \mid \cdots \mid x_{n-1}) + m(-1)^n f(x_1 \mid \cdots \mid x_n)$$

(the last term is independent of $x_{n+1}$). Hence

$$0 = dg + (-1)^n mf,$$

and $mf$ is a coboundary. •

**Corollary 9.125.** *If* $G$ *is a finite group of order* $m$ *and* $A$ *is a finitely generated* $G$-module, *then* $H^n(G, A)$ *is finite for all* $n \geq 0$.

**Proof.** $H^n(G, A)$ is a finitely generated abelian group (because $A$ is finitely generated) of finite exponent (if $u \in H^n = \langle h_1, \ldots, h_k \rangle$, then $u = \sum_i a_i h_i$, and we may assume that $|a_i| < m$ for all $i$). •

Both Proposition 9.124 and its corollary are true for homology groups as well.

We now consider the cohomology of free groups.

**Lemma 9.126.** *If $G$ is a free group with basis $X$, then its augmentation ideal $\mathcal{G}$ is a free $G$-module with basis*

$$X - 1 = \{x - 1 : x \in X \text{ and } x \neq 1\}.$$

**Remark.** If $G$ is any, not necessarily free, group, we proved, in Proposition 9.107, that $\mathcal{G}$ is a free abelian group with basis $G - 1 = \{g - 1 : g \in G \text{ and } g \neq 1\}$. ◀

**Proof.** We show first that $\mathcal{G}$ is generated by all $X - 1$. The identities

$$xy - 1 = (x - 1) + x(y - 1)$$

and

$$x^{-1} - 1 = -x^{-1}(x - 1)$$

show that if $w$ is any word in $X$, then $w - 1$ can be written as a $G$-linear combination of $X - 1$.

To show that $\mathcal{G}$ is a free $G$-module with basis $X - 1$, it suffices, by Proposition 6.67, to show that the following diagram can be completed:

$$
\begin{array}{ccc}
& \mathcal{G} & \\
& \uparrow \diagdown {\scriptstyle \Phi} & \\
X - 1 & \xrightarrow[\varphi]{} & A
\end{array}
$$

where $A$ is any $G$-module and $\varphi$ is any function (uniqueness of such a map $\Phi$ follows from $X - 1$ generating $\mathcal{G}$). Thus, we are seeking $\Phi \in \mathrm{Hom}_G(\mathcal{G}, A)$. By Exercise 9.59 on page 847, we have $\mathrm{Hom}_G(\mathcal{G}, A) \cong \mathrm{Der}(G, A)$ via $f \colon x \mapsto f(x - 1)$, where $f \in \mathcal{G} \to A$, and so we seek a derivation.

Consider the (necessarily split) extension $0 \to A \to E \to G \to 1$, so that $E$ consists of all ordered pairs $(a, g) \in A \times G$. The given function $\varphi \colon X - 1 \to A$ defines a lifting $\ell$ of the generating set $X$ of $G$, namely,

$$\ell(x) = (\varphi(x - 1), x).$$

Since $G$ is free with basis $X$, the function $\ell \colon X \to E$ extends to a homomorphism $L \colon G \to E$. We claim, for every $g \in G$, that $L(g) = (d(g), g)$, where $d \colon G \to A$. Each $g \in G$ has a unique expression as a reduced word $g = x_1^{e_1} \cdots x_n^{e_n}$, where $x_i \in X$ and $e_i = \pm 1$. We prove the claim by induction on $n \geq 1$. The base step is clear, while

$$
\begin{aligned}
L(g) = L(x_1^{e_1} \cdots x_n^{e_n}) &= L(x_1^{e_1}) \cdots L(x_n^{e_n}) \\
&= (\varphi(x_1 - 1), x_1)^{e_1} \cdots (\varphi(x_n - 1), x_n)^{e_n} = (d(g), g),
\end{aligned}
$$

and so the first coordinate $d(g)$ lies in $A$. Finally, $d$ is a derivation because $L$ is a homomorphism.

Exercise 9.59 now says that there is a homomorphism $\Phi\colon \mathcal{G} \to A$ defined by $\Phi(g-1) = d(g)$ for all $g \in G$. In particular, $\Phi(x-1) = d(x) = \varphi(x-1)$, so that $\Phi$ does extend $\varphi$.   •

**Theorem 9.127.** *If $G$ is a free group, then $H^n(G, A) = \{0\}$ for all $n > 1$ and all $G$-modules $A$.*

**Proof.** The sequence $0 \to \mathcal{G} \to \mathbb{Z}G \to \mathbb{Z} \to 0$ is a free resolution of $\mathbb{Z}$ because $\mathcal{G}$ is now a free $G$-module. Thus, the only nonzero terms in the deleted resolution occur in positions 0 and 1, and so all cohomology groups vanish for $n > 1$.   •

We are going to state an interesting result, the *Stallings–Swan Theorem*, which was discovered using homological methods.

If $G$ is a group and $S \subseteq G$ is a subgroup, then every $G$-module $A$ can be viewed as an $S$-module, for $\mathbb{Z}S$ is a subring of $\mathbb{Z}G$.

**Definition.** A group $G$ has **cohomological dimension** $\leq n$, denoted by

$$\mathrm{cd}(G) \leq n,$$

if $H^{n+1}(S, A) = \{0\}$ for all $G$-modules $A$ and every subgroup $S$ of $G$. We write $\mathrm{cd}(G) = \infty$ if no such integer $n$ exists, and we say that $\mathrm{cd}(G) = n$ if $\mathrm{cd}(G) \leq n$ but it is not true that $\mathrm{cd}(G) \leq n - 1$.

**Example 9.128.**

(i) If $G = \{1\}$, then $\mathrm{cd}(G) = 0$; this follows from Theorem 9.117 because $G$ is a cyclic group of order 1.

(ii) If $G$ is a finite cyclic group of order $k > 1$, then $\mathrm{cd}(G) = \infty$, as we see from Corollary 9.118 with $A = \mathbb{Z}$.

(iii) Suppose that $G$ is an infinite cyclic group. Since every subgroup $S \subseteq G$ is cyclic, Theorem 9.127 gives $\mathrm{cd}(G) \leq 1$. If $\mathrm{cd}(G) = 0$, then $H^1(S, A) = \{0\}$ for all subgroups $S$ and all modules $A$. In particular, if $S \cong \mathbb{Z}$ and $A \neq \{0\}$ is a trivial module, then $H^1(S, A) = \mathrm{Der}(S, A)/\mathrm{PDer}(S, A) \cong \mathrm{Hom}(\mathbb{Z}, A) \neq \{0\}$. Hence, $\mathrm{cd}(G) = 1$.

(iv) If $G \neq \{1\}$ is a free group, then Theorem 9.127 shows that $\mathrm{cd}(G) = 1$, for every subgroup of a free group is free.

(v) A free abelian group $G$ of finite rank $n$ has $\mathrm{cd}(G) = n$ (Rotman, *An Introduction to Homological Algebra*, p. 666).   ◄

If $S$ is a subgroup of a group $G$ and $A$ is an $S$-module, then $A^* = \mathrm{Hom}_S(\mathbb{Z}G, A)$ is a $G$-module: if $y \in S$, $f\colon \mathbb{Z}G \to A$ is an $S$-map, and $x \in G$, define $yg\colon \mathbb{Z}G \to A$ by

$$yg\colon x \mapsto f(y^{-1}x).$$

The following proposition is usually called Shapiro's Lemma, but it was discovered, independently, by Eckmann at the same time.

**Proposition 9.129 (Eckmann–Shapiro Lemma).** *Let $G$ be a group and let $S \subseteq G$ be a subgroup. If $A$ is a $\mathbb{Z}S$-module, then for all $n \geq 0$,*

$$H^n(S, A) \cong H^n(G, \operatorname{Hom}_{\mathbb{Z}S}(\mathbb{Z}G, A)).$$

**Proof.** Let $\to P_1 \to P_0 \to \mathbb{Z} \to 0$ be a $\mathbb{Z}G$-free resolution. Denote $\operatorname{Hom}_{\mathbb{Z}S}(\mathbb{Z}G, A)$ by $A^*$, so that

$$H^n(G, A^*) = H^n(\operatorname{Hom}_{\mathbb{Z}G}(\mathbf{P}, A^*)).$$

By the Adjoint Isomorphism,

$$\operatorname{Hom}_{\mathbb{Z}G}(P_i, A^*) = \operatorname{Hom}_{\mathbb{Z}G}(P_i, \operatorname{Hom}_{\mathbb{Z}S}(\mathbb{Z}G, A))$$
$$\cong \operatorname{Hom}_{\mathbb{Z}S}(P_i \otimes_{\mathbb{Z}G} \mathbb{Z}G, A)$$
$$\cong \operatorname{Hom}_{\mathbb{Z}S}(P_i, A).$$

But (the proof of) Lemma 7.89(i) shows that $\mathbb{Z}G$ is a free $S$-module, and so the free $G$-modules $P_i$ are also free $S$-modules. It follows that we may regard $\mathbf{P}$ as an $S$-free resolution of $\mathbb{Z}$, and there is an isomorphism of complexes:

$$\operatorname{Hom}_{\mathbb{Z}S}(\mathbf{P}, A) \cong \operatorname{Hom}_{\mathbb{Z}G}(\mathbf{P}, A^*).$$

Hence, their homology groups are isomorphic; that is, $H^n(S, A) \cong H^n(G, A^*)$.   •

**Corollary 9.130.** *If $G$ is a group and $S \subseteq G$ is a subgroup, then $\operatorname{cd}(S) \leq \operatorname{cd}(G)$.*

**Proof.** We may assume that $\operatorname{cd}(G) = n < \infty$. If $m > n$ and there is an $S$-module $A$ with $H^m(S, A) \neq \{0\}$, then $H^m(G, \operatorname{Hom}_{\mathbb{Z}S}(\mathbb{Z}G, A)) \cong H^m(S, A) \neq \{0\}$, by the Eckmann–Shapiro Lemma, and this contradicts $\operatorname{cd}(G) = n$.   •

**Corollary 9.131.** *A group $G$ of finite cohomological dimension is torsion-free.*

**Proof.** If $G$ is not torsion-free, it has a subgroup $S$ that is cyclic of finite order $k > 1$. Hence, $\operatorname{cd}(S) = \infty$, by Example 9.128(ii), contradicting Corollary 9.130.   •

**Corollary 9.132.** *A group $G = \{1\}$ if and only if $\operatorname{cd}(G) = 0$.*

**Proof.** If $G = \{1\}$, then $\operatorname{cd}(G) = 0$, by Example 9.128(i). Conversely, if $\operatorname{cd}(G) = 0$, then Corollary 9.130 gives $\operatorname{cd}(S) = 0$ for every cyclic subgroup $S = \langle g \rangle \subseteq G$. By Example 9.128(ii), we have $\langle g \rangle = \{1\}$ for all $g \in G$, and so $G = \{1\}$.   •

Are there groups $G$ with $\operatorname{cd}(G) = 1$ that are not free? In 1970, Stallings proved the following lovely theorem ($\mathbb{F}_2 G$ denotes the group algebra over $\mathbb{F}_2$).

**Theorem 9.133.** *If $G$ is a finitely presented group for which $H^1(G, \mathbb{F}_2 G)$ has more than two elements, then $G$ is a free product.*[22]

As a consequence, he proved the following results.

**Corollary 9.134.** *If $G$ is a finitely generated group with $\operatorname{cd}(G) = 1$, then $G$ is free.*

**Corollary 9.135.** *If $G$ is a torsion-free finitely generated group having a free subgroup of finite index, then $G$ is free.*

---

[22]Free product is the coproduct in **Groups**, and free groups are free products of copies of $\mathbb{Z}$.

Swan showed that both corollaries remain true if we remove the hypothesis that $G$ be finitely generated.

**Theorem 9.136 (Stallings–Swan).** [23] *A torsion-free group that has a free subgroup of finite index must be free.*

**Proof.** See Cohen, *Groups of Cohomological Dimension* 1.      •

There are important relations between the cohomology of a group and the cohomology of its subgroups and its quotient groups. For example, there are two standard maps, which we will define in terms of the bar resolution. The first is **restriction**. If $S$ is a subgroup of a group $G$, then every function $f: B_n(G) \to A$ is defined on all $[x_1 \mid \cdots \mid x_n]$ with $x_i \in G$; of course, $f$ is defined on all $n$-tuples of the form $[s_1 \mid \cdots \mid s_n]$ with $s_i \in S \subseteq G$, so that its restriction, which we denote by $f|S$, maps $B_n(S) \to A$. Then

$$\mathrm{Res}: H^n(G, A) \to H^n(S, A)$$

is defined by $\mathrm{Res}(\mathrm{cls}\, f) = \mathrm{cls}(f|S)$. One result is that if $G$ is a finite group, $S_p$ is a Sylow $p$-subgroup, and $n \geq 1$, then $\mathrm{Res}: H^n(G, A) \to H^n(S_p, A)$ is injective on the $p$-primary component of $H^n(G, A)$; thus, the cohomology of $G$ is strongly influenced by the cohomology of its Sylow subgroups.

The second standard map is called **inflation**. Suppose that $N$ is a normal subgroup of a group $G$. If $A$ is a $G$-module, then $A^N$ is a $G/N$-module if we define $(gN)a = ga$ for $a \in A^N$ [if $gN = hN$, then $h = gx$ for some $x \in N$, and so $ha = (gx)a = g(xa) = ga$, because $xa = a$]. Define

$$\mathrm{Inf}: H^n(G/N, A^N) \to H^n(G, A)$$

by $\mathrm{cls}(f) \mapsto \mathrm{cls}(f^{\#})$, where

$$f^{\#}: [g_1 \mid \cdots \mid g_n] \mapsto f[g_1 N \mid \cdots \mid g_n N].$$

A useful result here is the **Five Term Exact Sequence**: if $N \lhd G$ and $A$ is a $G$-module, then $H^1(N, A)$ is a $G/N$-module and there is an exact sequence of abelian groups:

$$0 \to H^1(G/N, A^N) \xrightarrow{\mathrm{Inf}} H^1(G, A) \xrightarrow{\mathrm{Res}} H^1(N, A)^{G/N}$$
$$\longrightarrow H^2(G/N, A^N) \longrightarrow H^2(G, A)$$

(there is a version of this sequence in homology as well). For an excellent discussion of these ideas, we refer the reader to Serre, *Corps Locaux*, pp. 135–138.

If $S \subseteq G$, there is a map

$$\mathrm{Tr}: H^n(S, A) \to H^n(G, A)$$

in the reverse direction, called **transfer**, which is defined when $S$ has finite index in $G$. We first define $\mathrm{Tr}$ in dimension 0; that is, $\mathrm{Tr}^0: A^S \to A^G$, by $a \mapsto \sum_{t \in T} ta$, where $T$ is a left transversal of $S$ in $G$ (of course, we must check that $\mathrm{Tr}^0$ is a

---

[23]If a group $G$ has a subgroup of finite index with a certain property $P$, then we say that $G$ is **virtually** $P$. Thus, the Stallings–Swan Theorem says that virtually free torsion-free groups are free.

homomorphism that is independent of the choice of transversal); this map is often called the *norm*. There is a standard way of extending a map in dimension 0 to maps in higher dimensions (essentially by dimension shifting), and if $[G : S] = m$, then

$$\operatorname{Tr}^n \circ \operatorname{Res}^n \colon H^n(G, A) \to H^n(G, A) = m;$$

that is, the composite is multiplication by $m$. Similarly, in homology, the map

$$\operatorname{Tr}_0 \colon A/\mathcal{G}A \to A/\mathcal{S}A,$$

defined by $a + \mathcal{G}A \mapsto \sum_{t \in T} t^{-1}a + \mathcal{S}A$, extends to maps in higher dimensions. When $n = 1$ and $A = \mathbb{Z}$, we have $\operatorname{Tr}_1 \colon H_1(G, \mathbb{Z}) \to H_1(S, \mathbb{Z})$; that is, $\operatorname{Tr}_1 \colon G/G' \to S/S'$. There is such a homomorphism well-known to group theorists, also called the **transfer** $V_{G \to S}$, and it turns out that $\operatorname{Tr}_1 = V_{G \to S}$ (Rotman, *An Introduction to Homological Algebra*, p. 578).

We can force cohomology groups to be a graded ring by defining **cup product** on $H(G, R) = \bigoplus_{n \geq 0} H^n(G, R)$, where $R$ is any commutative ring (Evens, *The Cohomology of Groups*), and this added structure has important applications.

# Exercises

**9.58.** (i) Prove that the isomorphisms in Proposition 9.108 constitute a natural isomorphism $\mathbb{Z} \otimes_G \square$ to the functor $A \mapsto A/\mathcal{G}A$.

(ii) Prove that the isomorphisms in Proposition 9.116 constitute a natural isomorphism $\operatorname{Hom}_G(\mathbb{Z}, \square)$ to $A \mapsto A^G$.

∗ **9.59.** For a fixed group $G$, prove that the functors $\operatorname{Hom}_G(\mathcal{G}, \square)$ and $\operatorname{Der}(G, \square)$ are naturally isomorphic.

**Hint.** If $f \colon \mathcal{G} \to A$ is a homomorphism, then $x \mapsto f(x - 1)$ is a derivation.

**9.60.** (i) If $G$ is a finite cyclic group and $0 \to A \to B \to C \to 0$ is an exact sequence of $G$-modules, prove that there is an **exact hexagon**; that is, kernel = image at each vertex of the diagram

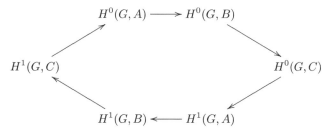

This exercise is a key lemma in Class Field Theory.

(ii) If $G$ is a finite cyclic group and $A$ is a $G$-module for which both $H^0(G, A)$ and $H^1(G, A)$ are finite, define the **Herbrand quotient** by

$$h(A) = |H^0(G, A)|/|H^1(G, A)|.$$

Let $0 \to A \to B \to C \to 0$ be an exact sequence of $G$-modules. Prove that if the Herbrand quotient is defined for two of the modules $A, B, C$, then it is defined

for the third one, and
$$h(B) = h(A)h(C).$$

**9.61.** If $G$ is a group, define $\bigotimes_1^n \mathbb{Z}G = \mathbb{Z}G \otimes_\mathbb{Z} \mathbb{Z}G \otimes_\mathbb{Z} \cdots \otimes_\mathbb{Z} \mathbb{Z}G$, the tensor product over $\mathbb{Z}$ of $\mathbb{Z}G$ with itself $n$ times. Prove that
$$P_n(G) \cong \bigotimes_1^{n+1} \mathbb{Z}G,$$
where $P_n(G)$ is the $n$th term in the homogeneous resolution $\mathbf{P}(G)$.

**9.62.** If $G$ is a finite cyclic group, prove, for all $G$-modules $A$ and for all $n \geq 1$, that $H^n(G, A) \cong H_{n+1}(G, A)$.

**9.63.** Let $G$ be a group.

(i) Show, for any abelian group $A$, that $A^* = \text{Hom}_\mathbb{Z}(\mathbb{Z}G, A)$ is a left $\mathbb{Z}G$-module ($A^*$ is called a **coinduced module**).
   **Hint.** If $\varphi \colon \mathbb{Z}G \to A$ and $g \in G$, define $g\varphi$ by $x \mapsto g\varphi(g^{-1}x)$.

(ii) For any left $\mathbb{Z}G$-module $B$, prove that $\text{Hom}_{\mathbb{Z}G}(B, A^*) \cong \text{Hom}_\mathbb{Z}(B, A)$.
   **Hint.** Use the Adjoint Isomorphism, Theorem 6.134.

(iii) If $A^*$ is a coinduced module, prove that $H^n(G, A^*) = \{0\}$ for all $n \geq 1$.

**9.64.** (i) Prove that $\mathbb{Z}G \otimes_\mathbb{Z} A$ is a $G$-module, where $A$ is an abelian group.

(ii) Prove that $H_n(G, \mathbb{Z}G \otimes_\mathbb{Z} A) = \{0\}$ for all $n \geq 1$ and all abelian groups $A$ ($\mathbb{Z}G \otimes_\mathbb{Z} A$ is called an **induced module**).

## Section 9.9. Crossed Products

This section is essentially descriptive, showing how cohomology groups are used to study division rings. Let us begin with a return to Galois Theory.

**Theorem 9.137.** *Let $E/k$ be a finite Galois extension with Galois group $G = \text{Gal}(E/k)$. The multiplicative group $E^\times$ is a $kG$-module, and*
$$H^1(G, E^\times) = \{0\}.$$

**Proof.** If $c \colon G \to E^\times$ is a 1-cocycle, denote $c(\sigma)$ by $c_\sigma$. In multiplicative notation, the cocycle condition is the identity $\sigma(c_\tau)c_{\sigma\tau}^{-1}c_\sigma = 1$ for all $\sigma, \tau \in G$; that is,

(1)
$$\sigma(c_\tau) = c_{\sigma\tau}c_\sigma^{-1}.$$

For $e \in E^\times$, consider
$$b = \sum_{\tau \in G} c_\tau \tau(e).$$

By Independence of Characters, Proposition 3.32, there is some $e \in E^\times$ with $b \neq 0$. For such an element $e$, we have, using Equation (1),
$$\sigma(b) = \sum_{\tau \in G} \sigma(c_\tau)\sigma\tau(e) = \sum_{\tau \in G} c_{\sigma\tau}c_\sigma^{-1}\sigma\tau(e)$$
$$= c_\sigma^{-1} \sum_{\tau \in G} c_{\sigma\tau}\sigma\tau(e) = c_\sigma^{-1} \sum_{\omega \in G} c_\omega\omega(e) = c_\sigma^{-1}b.$$

Hence, $c_\sigma = b\sigma(b)^{-1}$, and $c$ is a coboundary. Therefore, $H^1(G, E^\times) = \{0\}$. •

Theorem 9.137 implies Theorem 3.53, which describes the elements of norm 1 in a cyclic extension. We now give a homological proof of Theorem 3.53.

**Corollary 9.138 (Hilbert's Theorem 90).** *Let $E/k$ be a Galois extension whose Galois group $G = \mathrm{Gal}(E/k)$ is cyclic, say, with generator $\sigma$. If $u \in E^\times$, then $Nu = 1$ if and only if there is $v \in E^\times$ with*

$$u = \sigma(v)v^{-1}.$$

**Proof.** By Theorem 9.117, we have $H^1(G, E^\times) = \ker N / \operatorname{im} D$, where $N$ is the norm (remember that $E^\times$ is a multiplicative group) and $De = \sigma(e)e^{-1}$. Theorem 9.137 gives $H^1(G, E^\times) = \{0\}$, so that $\ker N = \operatorname{im} D$. Hence, if $u \in E^\times$, then $Nu = 1$ if and only if there is $v \in E^\times$ with $u = \sigma(v)v^{-1}$. •

Theorem 9.137 is one of the first results in what is called *Galois cohomology*. Another early result is that $H^n(G, E) = \{0\}$ for all $n \geq 1$, where $E$ (in contrast to $E^\times$) is the additive group of the Galois extension (this result follows easily from Theorem 10.100, the Normal Basis Theorem). We are now going to see that $H^2(G, E^\times)$ is useful in studying division rings.

Essentially, only one example of a noncommutative division ring has been given in the text so far: the quaternions $\mathbb{H}$ (this is an $\mathbb{R}$-algebra) and its $k$-algebra analogs for every subfield $k \subseteq \mathbb{R}$ (actually, another example is given in Exercise 7.62 on page 613). Hamilton discovered the quaternions in 1843, and Frobenius, in 1880, proved that the only division $\mathbb{R}$-algebras are $\mathbb{R}$, $\mathbb{C}$, and $\mathbb{H}$ (Theorem 7.128). No other examples of noncommutative division rings were known until *cyclic algebras* were found in the early 1900s, by Wedderburn and by Dickson. In 1932, Albert found an example of a *crossed product algebra* that is not a cyclic algebra, and in 1972, Amitsur found an example of a noncommutative division ring that is not a crossed product algebra. Simpler examples were found by Brussel, Noncrossed products and nonabelian crossed products over $\mathbb{Q}(t)$ and $\mathbb{Q}((t))$, *Amer. J. Math.* 117 (1995), no. 2, 377–393.

Wedderburn proved that every finite division ring is a field (Theorem 7.13). Are there any noncommutative division rings of prime characteristic?

We begin with an elementary calculation. Suppose that $V$ is a vector space over a field $E$ having basis $\{u_\sigma : \sigma \in G\}$ for some set $G$; thus, each $v \in V$ has a unique expression as a finite linear combination $v = \sum_\sigma a_\sigma u_\sigma$ for $a_\sigma \in E$. Given a function $\mu \colon V \times V \to V$, denote $\mu(u_\sigma, u_\tau)$ by $u_\sigma u_\tau$, and define scalars $c_\alpha^{\sigma,\tau} \in E$, called **structure constants**, by

$$u_\sigma u_\tau = \sum_{\alpha \in G} c_\alpha^{\sigma,\tau} u_\alpha.$$

To have the associative law, we must have $u_\sigma(u_\tau u_\omega) = (u_\sigma u_\tau)u_\omega$; expanding this equation, the coefficient of each $u_\beta$ is

$$\sum_\alpha c_\alpha^{\sigma,\tau} c_\beta^{\alpha,\omega} = \sum_\gamma c_\gamma^{\tau,\omega} c_\beta^{\sigma,\gamma}.$$

We can simplify these equations when $G$ is a finite group if we assume that $u_\sigma u_\tau = f(\sigma, \tau) u_{\sigma\tau}$, where $f(\sigma, \tau) = c_{\sigma\tau}^{\sigma,\tau}$; that is, $c_\alpha^{\sigma,\tau} = 0$ unless $\alpha = \sigma\tau$. The function $f \colon G \times G \to E^\times$, given by $f(\sigma, \tau) = c_{\sigma\tau}^{\sigma,\tau}$, satisfies the following equation for all $\sigma, \tau, \omega \in G$:

$$f(\sigma, \tau) f(\sigma\tau, \omega) = f(\tau, \omega) f(\sigma, \tau\omega),$$

an equation reminiscent of the cocycle identity written in multiplicative notation. This is why factor sets enter into the next definition.

Let $E/k$ be a Galois extension with $\mathrm{Gal}(E/k) = G$, and let $f \colon G \times G \to E^\times$ be a factor set: in multiplicative notation

$$f(\sigma, 1) = 1 = f(1, \tau) \quad \text{for all } \sigma, \tau \in G$$

and, if we denote the action of $\sigma \in G$ on $a \in E^\times$ by $a^\sigma$, then

$$f(\sigma, \tau) f(\sigma\tau, \omega) = f(\tau, \omega)^\sigma f(\sigma, \tau\omega).$$

**Definition.** Given a Galois extension $E/k$ with Galois group $G = \mathrm{Gal}(E/k)$ and a factor set $f \colon G \times G \to E^\times$, define the **crossed product algebra**

$$(E, G, f)$$

as the vector space over $E$ with basis all symbols $\{u_\sigma : \sigma \in G\}$ and multiplication

$$(a u_\sigma)(b u_\tau) = ab^\sigma f(\sigma, \tau) u_{\sigma\tau}$$

for all $a, b \in E$. If $G$ is a cyclic group, then the crossed product algebra $(E, G, f)$ is called a **cyclic algebra**.

Since every element in $(E, G, f)$ has a unique expression of the form $\sum a_\sigma u_\sigma$, the definition of multiplication extends by linearity to all of $(E, G, f)$. We note two special cases:

$$u_\sigma b = b^\sigma u_\sigma;$$
$$u_\sigma u_\tau = f(\sigma, \tau) u_{\sigma\tau}.$$

**Proposition 9.139.** *If $E/k$ is a Galois extension with Galois group $G = \mathrm{Gal}(E/k)$ and $f \colon G \times G \to E^\times$ is a factor set, then $(E, G, f)$ is a central simple $k$-algebra that is split by $E$.*

**Proof.** Denote $(E, G, f)$ by $A$. First, we show that $A$ is a $k$-algebra. To prove that $A$ is associative, it suffices to prove that

$$a u_\sigma (b u_\tau c u_\omega) = (a u_\sigma b u_\tau) c u_\omega,$$

where $a, b, c \in E$. Using the definition of multiplication,

$$\begin{aligned}
a u_\sigma (b u_\tau c u_\omega) &= a u_\sigma (bc^\tau f(\tau, \omega) u_{\tau\omega}) \\
&= a \big(bc^\tau f(\tau, \omega)\big)^\sigma f(\sigma, \tau\omega) u_{\sigma\tau\omega} \\
&= a b^\sigma c^{\sigma\tau} f(\tau, \omega)^\sigma f(\sigma, \tau\omega) u_{\sigma\tau\omega}.
\end{aligned}$$

Similarly,

$$(au_\sigma bu_\tau)cu_\omega = ab^\sigma f(\sigma, \tau)u_{\sigma\tau}cu_\omega$$
$$= ab^\sigma f(\sigma, \tau)c^{\sigma\tau}f(\sigma\tau, \omega)u_{\sigma\tau\omega}$$
$$= ab^\sigma c^{\sigma\tau}f(\sigma, \tau)f(\sigma\tau, \omega)u_{\sigma\tau\omega}.$$

The cocycle identity shows that multiplication in $A$ is associative.

That $u_1$ is the unit in $A$ follows from our assuming that factor sets are normalized:

$$u_1 u_\tau = f(1, \tau)u_{1\tau} = u_\tau \quad \text{and} \quad u_\sigma u_1 = f(\sigma, 1)u_{\sigma 1} = u_\sigma.$$

We have shown that $A$ is a ring. We claim that $ku_1 = \{au_1 : a \in k\}$ is the center $Z(A)$. If $a \in E$, then $u_\sigma au_1 = a^\sigma u_\sigma$. If $a \in k = E^G$, then $a^\sigma = a$ for all $\sigma \in G$, and so $k \subseteq Z(A)$. For the reverse inclusion, suppose that $z = \sum_\sigma a_\sigma u_\sigma \in Z(A)$. For any $b \in E$, we have $zbu_1 = bu_1 z$. But

$$zbu_1 = \sum a_\sigma u_\sigma bu_1 = \sum a_\sigma b^\sigma u_\sigma.$$

On the other hand,

$$bu_1 z = \sum ba_\sigma u_\sigma.$$

For every $\sigma \in G$, we have $a_\sigma b^\sigma = ba_\sigma$, so that if $a_\sigma \neq 0$, then $b^\sigma = b$. If $\sigma \neq 1$ and $H = \langle\sigma\rangle$, then $E^H \neq E^{\{1\}} = E$, by Theorem 3.35, and so there exists $b \in E$ with $b^\sigma \neq b$. We conclude that $z = a_1 u_1$. For every $\sigma \in G$, the equation $(a_1 u_1)u_\sigma = u_\sigma(a_1 u_1)$ gives $a_1^\sigma = a_1$, and so $a_1 \in E^G = k$. Therefore, $Z(A) = ku_1$.

We now show that $A$ is simple. Observe first that each $u_\sigma$ is invertible, for its inverse is $f(\sigma^{-1}, \sigma)^{-1}u_{\sigma^{-1}}$ (remember that $\operatorname{im} f \subseteq E^\times$, so that its values are nonzero). Let $I$ be a nonzero two-sided ideal in $A$, and choose a nonzero $y = \sum_\sigma c_\sigma u_\sigma \in I$ of shortest length; that is, $y$ has the smallest number of nonzero coefficients. Multiplying by $(c_\sigma u_\sigma)^{-1}$ if necessary, we may assume that $y = u_1 + c_\tau u_\tau + \cdots$. Suppose that $c_\tau \neq 0$. Since $\tau \neq 1_E$, there is $a \in E$ with $a^\tau \neq a$. Now $I$ contains $ay - ya = b_\tau u_\tau + \cdots$, where $b_\tau = c_\tau(a - a^\tau) \neq 0$. Hence, $I$ contains $y - c_\tau b_\tau^{-1}(ay - ya)$, which is shorter than $y$ (it involves $u_1$ but not $u_\tau$). We conclude that $y$ must have length 1; that is, $y = c_\sigma u_\sigma$. But $y$ is invertible, and so $I = A$ and $A$ is simple.

Finally, Theorem 7.131 says that $A$ is split by $K$, where $K$ is any maximal subfield of $A$. The reader may show, using Lemma 7.121, that $Eu_1 \cong E$ is a maximal subfield. •

In light of Proposition 9.139, it is natural to expect a connection between relative Brauer groups and cohomology.

**Theorem 9.140.** *Let $E/k$ be a Galois extension with $G = \operatorname{Gal}(E/k)$. There is an isomorphism $H^2(G, E^\times) \to \operatorname{Br}(E/k)$ with $\operatorname{cls}(f) \mapsto [(G, E, f)]$.*

**Proof.** The usual proofs of this theorem are rather long. Each of the following items must be checked, and the proofs are computational: the isomorphism is a well-defined function; it is a homomorphism; it is injective; it is surjective. For example, see Herstein, *Noncommutative Rings*, pp. 110–116, or Rosenberg, *Algebraic*

*K-Theory and Its Applications*, pp. 231–233. There is a less computational proof in Serre, *Corps Locaux*, pp. 164–167, using the method of *descent*.  •

What is the advantage of this isomorphism? In Corollary 7.136, we saw that

$$\mathrm{Br}(k) = \bigcup_{E/k \text{ finite Galois}} \mathrm{Br}(E/k).$$

**Corollary 9.141.** *Let $k$ be a field.*

(i) *The Brauer group $\mathrm{Br}(k)$ is a torsion group.*

(ii) *Let $A$ be a central simple $k$-algebra. If $r$ is the order of $[A]$ in $\mathrm{Br}(k)$, then there is an integer $n$ so that the tensor product of $A$ with itself $r$ times is a matrix algebra:*

$$A \otimes_k A \otimes_k \cdots \otimes_k A \cong \mathrm{Mat}_n(k).$$

**Proof.**

(i) $\mathrm{Br}(k)$ is the union of the relative Brauer groups $\mathrm{Br}(E/k)$, where $E/k$ is finite Galois. Proposition 9.124, which says that $|G| \, H^2(G, E^\times) = \{0\}$, may now be invoked.

(ii) Tensor product is the binary operation in the Brauer group.  •

Recall Proposition 7.133: there exists a noncommutative division $k$-algebra over a field $k$ if and only if $\mathrm{Br}(k) \neq \{0\}$.

**Corollary 9.142.** *Let $k$ be a field. If there is a cyclic Galois extension $E/k$ such that the norm $N \colon E^\times \to k^\times$ is not surjective, then there exists a noncommutative $k$-division algebra.*

**Proof.** If $G$ is a finite cyclic group, then Theorem 9.117 gives

$$H^2(G, E^\times) = (E^\times)^G / \operatorname{im} N = k^\times / \operatorname{im} N.$$

Hence, $\mathrm{Br}(E/k) \neq \{0\}$ if $N$ is not surjective, and this implies that $\mathrm{Br}(k) \neq \{0\}$.  •

If $k$ is a finite field and $E/k$ is a finite extension, then it follows from Wedderburn's Theorem on finite division rings (Theorem 7.13) that the norm $N \colon E^\times \to k^\times$ is surjective.

**Corollary 9.143.** *If $p$ is a prime, then there exists a noncommutative division algebra of characteristic $p$.*

**Proof.** If $k$ is a field of characteristic $p$, it suffices to find a cyclic extension $E/k$ for which the norm $N \colon E^\times \to k^\times$ is not surjective; that is, we must find some $z \in k^\times$ which is not a norm.

If $p$ is an odd prime, let $k = \mathbb{F}_p(x)$. Since $p$ is odd, $t^2 - x$ is a separable irreducible polynomial, and so $E = k(\sqrt{x})$ is a Galois extension of degree 2. If $u \in E$, then there are polynomials $a, b, c \in \mathbb{F}_p[x]$ with $u = (a + b\sqrt{x})/c$. Moreover,

$$N(u) = (a^2 - b^2 x)/c^2.$$

We claim that $x^2 + x$ is not a norm. Otherwise,

$$a^2 - b^2 x = c^2(x^2 + x).$$

Since $c \neq 0$, the polynomial $c^2(x^2 + x) \neq 0$, and it has even degree. On the other hand, if $b \neq 0$, then $a^2 - b^2 x$ has odd degree, and this is a contradiction. If $b = 0$, then $u = a/c$; since $a^2 = c^2(x^2 + x)$, we have $c^2 \mid a^2$, hence $c \mid a$, and so $u \in \mathbb{F}_p[x]$ is a polynomial. But it is easy to see that $x^2 + x$ is not the square of a polynomial. We conclude that $N \colon E^\times \to k^\times$ is not surjective.

Here is an example in characteristic 2. Let $k = \mathbb{F}_2(x)$, and let $E = k(\alpha)$, where $\alpha$ is a root of $f(t) = t^2 + t + x + 1$ [$f(t)$ is irreducible and separable; its other root is $\alpha + 1$]. As before, each $u \in E$ can be written in the form $u = (a + b\alpha)/c$, where $a, b, c \in \mathbb{F}_2[x]$. Of course, we may assume that $x$ is not a common divisor of the three polynomials $a, b$, and $c$. Moreover,

$$N(u) = \big((a + b\alpha)(a + b\alpha + b)\big)/c^2 = \big(a^2 + ab + b^2(x+1)\big)/c^2.$$

We claim that $x$ is not a norm. Otherwise,

(2) $$a^2 + ab + b^2(x+1) = c^2 x.$$

Now $a(0)$, the constant term of $a$, is either 0 or 1. Consider the four cases arising from the constant terms of $a$ and $b$; that is, evaluate Equation (2) at $x = 0$. We see that $a(0) = 0 = b(0)$; that is, $x \mid a$ and $x \mid b$. Hence, $x^2 \mid a^2$ and $x^2 \mid b^2$, so that Equation (2) has the form $x^2 d = c^2 x$, where $d \in \mathbb{F}_2[x]$. Dividing by $x$ gives $xd = c^2$, which forces $c(0) = 0$; that is, $x \mid c$, and this is a contradiction. •

For further discussion of the Brauer group, see Jacobson, *Basic Algebra* II, pp. 471–481, Neukirch–Schmidt–Wingberg, *Cohomology of Number Fields*, Reiner, *Maximal Orders*, Chapters 5, 7, and 8, the article by Platonov–Yanchevskii, *Finite-Dimensional Division Algebras*, in Kostrikin–Shafarevich, *Encyclopaedia of Mathematical Sciences, Algebra* IX, and the article by Serre in Cassels–Fröhlich, *Algebraic Number Theory*. Here are some highlights. A **global field** is a field which is either an arithmetic number field [i.e., a finite extension of $\mathbb{Q}$] or a *function field* [a finite extension of $k(x)$, where $k$ is a finite field]. To each global field, we assign a family of *local fields*. These fields are best defined in terms of discrete valuations. A **discrete valuation**[24] on a field $L$ is a function $v \colon L \to \Gamma \cup \{0\}$, where $\Gamma$ is a multiplicative infinite cyclic group, such that, for all $a, b \in L$,

$$v(a) = 0 \quad \text{if and only if } a = 0;$$
$$v(ab) = v(a)v(b);$$
$$v(a + b) = \max\{v(a), v(b)\}.$$

Now $R = \{a \in L : v(a) \leq 1\}$ is a domain and $P = \{a \in L : v(a) < 1\}$ is a maximal ideal in $R$. We call $R/P$ the **residue field** of $L$ with respect to the discrete valuation $v$. A **local field** is a field which is complete with respect to the metric arising from a discrete valuation on it, and whose residue field is finite. It turns out that every local field is either a finite extension of the $p$-adic numbers $\mathbb{Q}_p$ (the fraction field of the $p$-adic integers $\mathbb{Z}_p$) or is isomorphic to $\mathbb{F}_q[[x]]$, the ring of

---

[24]See Exercise 10.18 for an additive version of this definition.

formal power series in one variable over a finite field $\mathbb{F}_q$. If $k$ is a local field, then $\mathrm{Br}(k) \cong H^2(k_s, k^\times)$, where $k_s/k$ is the maximal separable extension of $k$ in the algebraic closure $\overline{k}$. If $A$ is a central simple $K$-algebra, where $K$ is a global field and $K_v$ is a local field of $K$, then $K_v \otimes_K A$ is a central simple $K_v$-algebra. The **Hasse–Brauer–Noether–Albert Theorem** states that if $A$ is a central simple algebra over a global field $K$, then $[A] = [K]$ in $\mathrm{Br}(K)$ if and only if $[K_v \otimes_K A] = [K_v]$ in $\mathrm{Br}(K_v)$ for all associated local fields $K_v$. We merely mention that these results were used by Chevalley to develop *Class Field Theory* [the branch of Algebraic Number Theory involving (possibly infinite) Galois extensions having abelian Galois groups].

For generalizations of the Brauer group $\mathrm{Br}(k)$ for $k$ a commutative ring and ties to Morita Theory, see Orzech–Small, *The Brauer Group of Commutative Rings*, and Caenepeel, *Brauer Groups, Hopf Algebras, and Galois Theory*.

## Exercises

**9.65.** Show that the structure constants in the crossed product $(E, G, f)$ are

$$
g_\alpha^{\sigma, \tau} = \begin{cases} f(\sigma, \tau) & \text{if } \alpha = \sigma\tau; \\ 0 & \text{otherwise.} \end{cases}
$$

**9.66.** Prove that $\mathbb{H} \otimes_{\mathbb{R}} \mathbb{H} \cong \mathrm{Mat}_4(\mathbb{R})$.

## Section 9.10. Introduction to Spectral Sequences

Spectral sequences are used in computing homology groups, comparing homology groups of composites of functors, and giving certain five term exact sequences. This brief section merely describes the setting for spectral sequences in the hope that it will ease the reader's first serious encounter with them. For a more complete account, we refer the reader to Mac Lane, *Homology*, Chapter XI, McCleary, *User's Guide to Spectral Sequences*, or Rotman, *An Introduction to Homological Algebra*, Chapter 10.

Call a series of submodules of a module $K$,

$$(1) \qquad\qquad K = K_0 \supseteq K_1 \supseteq K_2 \supseteq \cdots \supseteq K_\ell = \{0\},$$

a **filtration** (instead of a normal series), and call the quotients $K_i/K_{i+1}$ the **factor modules** of the filtration. We know that a module $K$ may not be determined by the factor modules of a filtration; on the other hand, knowledge of the factor modules does give some information about $K$. For example, if all the factor modules are zero, then $K = \{0\}$; if all the factor modules are finite, then $K$ is finite (and $|K|$ is the product of the orders of the factor modules); or, if all the factor modules are finitely generated, then $K$ is finitely generated.

**Definition.** If $K$ is a module, then a **subquotient** of $K$ is a module isomorphic to $S/T$, where $T \subseteq S \subseteq K$ are submodules.

Thus, a subquotient of $K$ is a quotient of a submodule. It is also easy to see that a subquotient of $K$ is also a submodule of a quotient (because $S/T \subseteq K/T$).

**Example 9.144.**

(i) The factor modules of a filtration of a module $K$ are subquotients of $K$.

(ii) The $n$th homology group of a complex $\mathbf{C}$ is a subquotient of $C_n$. ◄

A spectral sequence computes a homology group $H_n$ in the sense that it computes the factor modules of some filtration of $H_n$. In general, this gives only partial information about $H_n$; but, if the factor modules are heavily constrained, they can give much more information and, indeed, might even determine $H_n$ completely. For example, suppose that only one of the factor modules of $K$ in Equation (1) is nonzero, say, $K_j/K_{j+1} = \{0\}$ for all $j \neq i$; we claim that $K = K_i \cong K_i/K_{i+1}$. The beginning of the filtration is

$$K = K_0 \supseteq K_1 \supseteq \cdots \supseteq K_i.$$

Since $K_0/K_1 = \{0\}$, we have $K = K_0 = K_1$. More generally, $K_j/K_{j+1} = \{0\}$ for all $j < i$ gives $K = K_0 = K_1 = \cdots = K_i$. Similar reasoning computes the end of the filtration: $\{0\} = K_\ell = K_{\ell-1} = \cdots = K_{i+1}$. Thus, Equation (1) is now

$$K = K_0 = \cdots = K_i \supseteq K_{i+1} = \cdots = K_\ell = \{0\},$$

and so $K = K_i \cong K_i/\{0\} = K_i/K_{i+1}$.

In order to appreciate spectral sequences, we must recognize an obvious fact: very general statements can become useful if extra simplifying hypotheses can be imposed.

Spectral sequences usually arise in the following context. A **bigraded module** $M$ is a doubly indexed family of modules $M_{p,q}$, where $p, q \in \mathbb{Z}$; we picture a bigraded module as a collection of modules, one sitting on each lattice point $(p, q)$ in the plane. Thus, there are **first quadrant** bigraded modules, for example, with $M_{p,q} = \{0\}$ if either $p$ or $q$ is negative; similarly, there are **third quadrant** bigraded modules. A **bicomplex** is a bigraded module that has vertical arrows $d''_{p,q} \colon M_{p,q} \to M_{p,q-1}$ making the columns complexes, horizontal arrows $d'_{p,q} \colon M_{p,q} \to M_{p-1,q}$ making the rows complexes, and whose squares anticommute: that is, $d'd'' + d''d' = 0$. The reason for anticommutativity is to allow us to define the **total complex**, $\mathrm{Tot}(M)$, of a bicomplex $M$: its term in degree $n$ is:

$$\mathrm{Tot}(M)_n = \bigoplus_{p+q=n} M_{p,q};$$

its differential $d_n \colon \mathrm{Tot}(M)_n \to \mathrm{Tot}(M)_{n-1}$ is given by

$$d_n = \sum_{p+q=n} d'_{p,q} + d''_{p,q}.$$

Anticommutativity forces $d_{n-1}d_n = 0$:

$$dd = (d' + d'')(d' + d'') = d'd' + (d'd'' + d''d') + d''d'' = 0;$$

thus, $\mathrm{Tot}(M)$ is a complex.

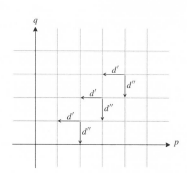

**Figure 9.2.** Bicomplex $M$.

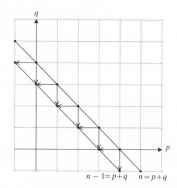

**Figure 9.3.** Tot($M$).

All bigraded modules form a category. Given an ordered pair of integers $(a, b)$, a family of maps $f_{p,q} \colon M_{p,q} \to L_{p+a,q+b}$ is called a **map** $f \colon M \to L$ of **bidegree** $(a, b)$. For example, the maps $d' \colon M \to M$ and $d'' \colon M \to M$ above have respective bidegrees $(-1, 0)$ and $(0, -1)$. It is easy to check that all bigraded modules and all maps having some bidegree form a category. One nice feature of composition is that bidegrees add: if $f$ has bidegree $(a, b)$ and $f'$ has bidegree $(a', b')$, then their composite $f'f$ has bidegree $(a + a', b + b')$. Maps of bigraded modules are used in establishing certain exact sequences. For example, one proof of the five term exact sequence on page 846 uses these maps.

A **spectral sequence** is a sequence of bicomplexes, $(E_{p,q}^r)_{r \geq 2}$, where each $E_{p,q}^{r+1}$ is a subquotient of $E_{p,q}^r$ whose maps $d'$ and $d''$ arise from those of $E_{p,q}^r$. Most spectral sequences arise from filtrations of $\mathrm{Tot}(M)$, where $M$ is a bicomplex. In particular, there are two "usual" filtrations (if $M$ is either first quadrant or third quadrant); the spectral sequences they determine are denoted by $({}^{\mathrm{I}}E_{p,q}^r)_r$ and $({}^{\mathrm{II}}E_{p,q}^r)_r$.

We say that a spectral sequence $(E_{p,q}^r)$ **converges** to a (singly graded) module $H$, denoted by $E_{p,q}^2 \Rightarrow H_n$, if each $H_n$ has a filtration $(\Phi^p H_n)_{s=s(n) \leq p \leq t = t(n)}$,

$$H_n = \Phi^t H_n \supseteq \Phi^{t-1} H_n \supseteq \cdots \supseteq \Phi^{s-1} H_n \supseteq \Phi^s H_n = \{0\},$$

with each factor module $\Phi^p H_n / \Phi^{p-1} H_n$ a subquotient of $E_{p,q}^r$ for all $r$, where $p + q = n$. Here are two theorems facilitating the use of spectral sequences.

**Theorem I.** *If $M$ is a first quadrant or third quadrant bicomplex, then there are two convergent spectral sequences:*

$${}^{\mathrm{I}}E_{p,q}^2 \Rightarrow H_n(\mathrm{Tot}(M)) \quad \text{and} \quad {}^{\mathrm{II}}E_{p,q}^2 \Rightarrow H_n(\mathrm{Tot}(M)).$$

Thus, the two usual filtrations of $\mathrm{Tot}(M)$ yield spectral sequences $({}^{\mathrm{I}}E_{p,q}^r)_r$ and $({}^{\mathrm{II}}E_{p,q}^r)_r$, both of which converge to the same thing; that is, each $H_n$ has one filtration whose factor modules are subquotients of ${}^{\mathrm{I}}E_{p,q}^2$ (where $p + q = n$), as well as a second filtration whose factor modules are subquotients of ${}^{\mathrm{II}}E_{p,q}^2$.

**Theorem II.** *If $M$ is a first quadrant or third quadrant bicomplex, then there are formulas for ${}^{\mathrm{I}}E_{p,q}^2$ and ${}^{\mathrm{II}}E_{p,q}^2$ for every $p, q$.*

Theorem II offers the possibility that subquotients of $^{\text{I}}E_{p,q}^2$ and $^{\text{II}}E_{p,q}^2$ can be computed.

We can now sketch a proof that $\text{Tor}_n(A, B)$ does not depend on the variable resolved: the value of $\text{Tor}_n(A, B)$, defined as $H_n(\mathbf{P}_A \otimes B)$, where $\mathbf{P}_A$ is a deleted projective resolution of $A$, coincides with $\text{tor}_n(A, B)$, defined as $H_n(A \otimes \mathbf{Q}_B)$, where $\mathbf{Q}_B$ is a deleted projective resolution of $B$. The idea is to resolve both variables simultaneously, using projective resolutions of each. Define a first quadrant bigraded module $M = \mathbf{P}_A \otimes \mathbf{Q}_B$ whose $p, q$ term is $P_p \otimes Q_q$; make this into a bicomplex by defining vertical arrows $d_{p,q}'' = (-1)^p 1 \otimes \partial_q \colon P_p \otimes Q_q \to P_p \otimes Q_{q-1}$ and horizontal arrows $d_{p,q}' = \Delta_p \otimes 1 \colon P_p \otimes Q_q \to P_{p-1} \otimes Q_q$, where the $\partial_n$ are the differentials in $\mathbf{Q}_B$ and the $\Delta_n$ are the differentials in $\mathbf{P}_A$ (the signs force anticommutativity). The formula whose existence is stated in Theorem II for the first spectral sequence $^{\text{I}}E_{p,q}^2$ gives, in this case,

$$^{\text{I}}E_{p,q}^2 = \begin{cases} \{0\} & \text{if } q > 0; \\ H_p(\mathbf{P}_A \otimes B) & \text{if } q = 0. \end{cases}$$

Since a subquotient of $\{0\}$ must be $\{0\}$, all but one of the factor modules of a filtration of $H_n(\text{Tot}(M))$ are zero, and so

$$H_n(\text{Tot}(M)) \cong H_n(\mathbf{P}_A \otimes B).$$

Similarly, the formula alluded to in Theorem II for the second spectral sequence gives

$$^{\text{II}}E_{p,q}^2 = \begin{cases} \{0\} & \text{if } p > 0; \\ H_q(A \otimes \mathbf{Q}_B) & \text{if } p = 0. \end{cases}$$

Again, there is a filtration of $H_n(\text{Tot}(M))$ with only one possible nonzero factor module, and so

$$H_n(\text{Tot}(M)) \cong H_n(A \otimes \mathbf{Q}_B).$$

Therefore,

$$H_n(\mathbf{P}_A \otimes B) \cong H_n(\text{Tot}(M)) \cong H_n(\mathbf{P}_A \otimes B).$$

We have shown that Tor is independent of the variable resolved.

Here is a cohomology result illustrating how spectral sequences can be used to compute composite functors. The index raising convention extends here, so that one denotes the modules in a third quadrant bicomplex by $M^{p,q}$ instead of by $M_{-p,-q}$.

**Theorem 9.145 (Grothendieck).** *Let $F \colon \mathcal{B} \to \mathcal{C}$ and $G \colon \mathcal{A} \to \mathcal{B}$ be additive functors, where $\mathcal{A}, \mathcal{B},$ and $\mathcal{C}$ are module categories. If $F$ is left exact and $E$ injective in $\mathcal{A}$ implies $(R^m F)(GE) = \{0\}$ for all $m > 0$ (where $R^m F$ are the right derived functors of $F$), then for every module $A \in \mathcal{A}$, there is a third quadrant spectral sequence*

$$E_2^{p,q} = (R^p F)(R^q G(A)) \Rightarrow R^n (FG)(A).$$

**Proof.** Rotman, *An Introduction to Homological Algebra*, pp. 656–660. •

The next result shows that if $N$ is a normal subgroup of a group $\pi$, then the cohomology groups of $N$ and of $\pi/N$ can be used to compute the cohomology groups of $\pi$.

**Theorem 9.146 (Lyndon–Hochschild–Serre).** *Let $\pi$ be a group with normal subgroup $N$. For each $\pi$-module $A$, there is a third quadrant spectral sequence with*

$$E_2^{p,q} = H^p(\pi/N, H^q(N, A)) \Rightarrow H^n(\pi, A).$$

**Proof.** Define functors $G\colon {}_{\mathbb{Z}\pi}\mathbf{Mod} \to {}_{\mathbb{Z}(\pi/N)}\mathbf{Mod}$ and $F\colon {}_{\mathbb{Z}(\pi/N)}\mathbf{Mod} \to \mathbf{Ab}$ by $G = \operatorname{Hom}_N(\mathbb{Z}, \square)$ and $F = \operatorname{Hom}_{\pi/N}(\mathbb{Z}, \square)$. Of course, $F$ is left exact, and it is easy to see that $FG = \operatorname{Hom}_\pi(\mathbb{Z}, \square)$. It is true that $H^m(\pi/N, E) = \{0\}$ whenever $E$ is an injective $\pi$-module and $m > 0$ (Ibid., p. 661). The result now follows from Theorem 9.145. •

This theorem was found by Lyndon, in his dissertation in 1948, in order to compute the cohomology groups of finitely generated abelian groups. Hochschild and Serre put the result into its present form in 1953.

There is a five term exact sequence arising from the Lyndon–Hochschild–Serre spectral sequence. We can prove that if $N \lhd \pi$, then the abelian group $H^1(\pi, A)$ is a $(\pi/n)$-module, and so $H^1(\pi, A)^{\pi/N}$ makes sense.

**Theorem 9.147 (Lyndon–Hochschild–Serre).** *Let $\pi$ be a group with normal subgroup $N$. For each $\pi$-module $A$, there is an exact sequence*

$$0 \to H^1(\pi/N, A) \to H^1(\pi, A) \to H^1(\pi, A)^{\pi/N} \to H^2(\pi/N, A) \to H^2(\pi, A).$$

**Proof.** There is a general result giving a five term exact sequence from a spectral sequence (Ibid., p. 645). •

## Section 9.11. Grothendieck Groups

This section may be regarded as an introduction to Algebraic $K$-Theory.[25] In contrast to Homological Algebra, this topic is often called *Homotopical Algebra*, for there are topological interpretations involving homotopy groups as well as homology groups. There are two different constructions of *Grothendieck groups*, $K_0$ and $G_0$. Although we will define $K_0(X)$ and $G_0(X)$ for various types of $X$ (semigroups, categories, rings), the most important versions have $X$ a ring. We shall see that $K_0(R) \cong G_0(R)$ for every ring $R$.

Exercise 9.67 on page 871 shows that $K_0(R) \cong K_0(R^{\mathrm{op}})$ for every ring $R$, so that "$R$-module" in this section will always mean "left $R$-module."

**9.11.1. The Functor $K_0$.** We begin by generalizing the construction of the additive group of the integers $\mathbb{Z}$ from the additive semigroup of the natural numbers $\mathbb{N}$.

---

[25] It would have been appropriate for this section to appear in the previous chapter, Advanced Linear Algebra, but I believe that a reader who has seen homology is more receptive to these ideas.

**Definition.** If $S$ is a commutative semigroup, then $K_0(S)$ is an abelian group solving the following universal problem in the category **CS** of commutative semigroups: there is a semigroup map $\nu\colon S \to K_0(S)$ such that, for every abelian group $G$ and every semigroup map $\varphi\colon S \to G$, there exists a unique group homomorphism $\Phi\colon K_0(S) \to G$ with $\Phi\nu = \varphi$,

$$
\begin{array}{ccc}
S & \xrightarrow{\ \nu\ } & K_0(S) \\
 & \searrow{\scriptstyle\varphi} & \big\downarrow{\scriptstyle\Phi} \\
 & & G
\end{array}
$$

We call $K_0(S)$ the **Grothendieck group** of the semigroup $S$.

**Proposition 9.148.** *For every commutative semigroup $S$, the group $K_0(S)$ exists.*

**Proof.** Let $F(S)$ be the free abelian group with basis $S$, and let

$$H = \langle x + y - z : x + y = z \text{ in } S\rangle.$$

Define $K_0(S) = F(S)/H$, and define $\nu\colon S \to K_0(S)$ by $\nu\colon x \mapsto x + H$. A semigroup map $\varphi\colon S \to G$, being a function defined on a basis, has a unique extension to a group homomorphism $\widetilde{\varphi}\colon F(S) \to G$; note that $H \subseteq \ker\widetilde{\varphi}$, because $\varphi(x + y) = \varphi(x) + \varphi(y)$. Define $\Phi\colon K_0(S) \to G$ by $x + H \mapsto \widetilde{\varphi}(x)$. $\quad\bullet$

If $x \in S$, we will denote $\nu(x) = x + H$ by $[x]$.

As always, solutions to a universal mapping problem are unique to isomorphism.

**Corollary 9.149.** *If $\mathbb{N}$ is viewed as an additive commutative semigroup, then*

$$K_0(\mathbb{N}) = \mathbb{Z}.$$

**Proof.** In the diagram below, let $\nu\colon \mathbb{N} \to \mathbb{Z}$ be the inclusion. Define $\Phi(n) = \varphi(n)$ if $n \geq 0$, and define $\Phi(n) = -\varphi(-n)$ if $n < 0$.

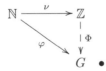

$\bullet$

**Example 9.150.** The last corollary is not really a construction of $\mathbb{Z}$ from $\mathbb{N}$, because the construction of $K_0(\mathbb{N})$ in Proposition 9.148 uses free abelian groups. We now sketch a second construction of $K_0(S)$, where $S$ is a commutative semigroup, which remedies this.

Define a relation on $S \times S$ by

$$(x, y) \sim (a, b) \text{ if there exists } s \in S \text{ with } x + b + s = a + y + s.$$

It is easily verified that this is an equivalence relation [if $R$ is a domain, then $R-\{0\}$ is a multiplicative semigroup, and this relation is essentially the cross product relation arising in the usual construction of $\mathrm{Frac}(R)$]. Denote the equivalence class of $(x, y)$ by $[x, y]$, and note that $[x, x] = [y, y]$ for all $x, y \in S$. Define

$$[x, y] + [x', y'] = [x + x', y + y'].$$

A routine check shows that the orbit space $(S \times S)/\sim$ is an abelian group: addition is well-defined and associative; the identity element is $[x, x]$; the inverse of $[x, y]$ is $[y, x]$. Finally, define $\nu \colon S \to (S \times S)/\sim$ by $x \mapsto [x + x, x]$. The group $K_0(S) = (S \times S)/\sim$ is a solution of the universal problem.

For example, if $S = \mathbb{N}$, then $(2, 5) \sim (5, 8)$ because $2 + 8 + 0 = 5 + 5 + 0$. We usually write $-3 = [2, 5]$; that is, we view $[x, y]$ as "$x - y$" (this explains the definition of $\nu(x) = [x + x, x]$). The definition of $\sim$ for $\mathbb{N}$ is simpler than that for an arbitrary commutative semigroup because $\mathbb{N}$ is a monoid which satisfies the cancellation law ($u + s = v + s$ implies $u = v$). ◄

**Proposition 9.151.** *The Grothendieck group defines a functor $K_0 \colon \mathbf{CS} \to \mathbf{Ab}$, where $\mathbf{CS}$ is the category of all commutative semigroups.*

**Proof.** If $f \colon S \to T$ is a semigroup map, define $K_0(f) \colon K_0(S) \to K_0(T)$ by $[s] \mapsto [f(s)]$. The proof that $K_0$ is a functor is straightforward. •

Semigroups arising from categories provide interesting examples of Grothendieck groups. Since most categories are not small, we first deal with a set-theoretic problem this might cause. If $\mathcal{C}$ is a category, let

$$\mathrm{Iso}(\mathcal{C})$$

denote the family of all isomorphism classes of objects in $\mathcal{C}$. We will abuse notation by allowing an object $A$ to denote its own isomorphism class.

**Definition.** A category $\mathcal{C}$ is ***virtually small*** if its family $\mathrm{Iso}(\mathcal{C})$ of isomorphism classes is a *set*; that is, $\mathrm{Iso}(\mathcal{C})$ has a cardinal number.

**Example 9.152.**

(i) The category $\mathcal{C}$ of all countable abelian groups is not a small category [indeed, its (full) subcategory of all groups of order 1 is not small]. However, $\mathcal{C}$ is virtually small.

(ii) Any *full* subcategory $\mathcal{U}$ of a virtually small category $\mathcal{C}$ is virtually small. [We must assume that $\mathcal{U}$ is full; otherwise, objects in $\mathcal{U}$ which are isomorphic in $\mathcal{C}$ might not be isomorphic in $\mathcal{U}$. For example, the discrete subcategory (having the same objects but whose only morphisms are identities) of the category of all countable abelian groups is not virtually small.]

(iii) For any ring $R$, the full subcategory of $_R\mathbf{Mod}$ consisting of all finitely generated $R$-modules is virtually small. Using part (ii), we see that the full subcategory

$$R\text{-}\mathbf{Proj}$$

generated by all finitely generated projective $R$-modules is virtually small. ◄

**Lemma 9.153.** *Let $\mathcal{C}$ be a full additive subcategory of $_R\mathbf{Mod}$ for some ring $R$. If $\mathcal{C}$ is virtually small, then the family $\mathrm{Iso}(\mathcal{C})$ of all its isomorphism classes is a semigroup in which*

$$A + A' = A \oplus A', \text{ where } A, A' \in \mathrm{Iso}(\mathcal{C}).$$

**Proof.** First, $\mathrm{Iso}(\mathcal{C})$ is a set, for $\mathcal{C}$ is virtually small. Since $\mathcal{C}$ is additive, $A, A' \in \mathrm{Iso}(\mathcal{C})$ implies $A \oplus A' \in \mathrm{Iso}(\mathcal{C})$ (we are using our notational convention of allowing an object to denote its own isomorphism class). Thus, direct sum gives a well-defined binary operation on $\mathrm{Iso}(\mathcal{C})$. This operation is associative, for $A \oplus (A' \oplus A'') \cong (A \oplus A') \oplus A''$ gives $A \oplus (A' \oplus A'') = (A \oplus A') \oplus A''$ in $\mathrm{Iso}(\mathcal{C})$. $\bullet$

**Definition.** If $R$ is a ring, the **Grothendieck group** $K_0(R)$ is defined by

$$K_0(R) = K_0(\mathrm{Iso}(R\text{-}\mathbf{Proj})),$$

where $R$-**Proj** is the full subcategory of all finitely generated projective $R$-modules.

We have defined $K_0(\mathbb{Z})$ in two different ways. On the one hand, it was defined by viewing $\mathbb{Z}$ as a semigroup (and $K_0(\mathbb{Z}) \cong \mathbb{Z}$). On the other hand, we have just defined $K_0(\mathbb{Z})$ as $K_0(\mathrm{Iso}(\mathbb{Z}\text{-}\mathbf{Proj}))$. But $\mathbb{Z}$-**Proj** is just the semigroup of isomorphism classes of finitely generated projective $\mathbb{Z}$-modules (that is, of finitely generated free abelian groups). We shall see, in Proposition 9.155 below, that this second $K_0(\mathbb{Z})$ is also isomorphic to $\mathbb{Z}$.

**Remark.**

(i) It is not interesting to remove the finitely generated hypothesis in the definition of $K_0(R)$. If $\mathcal{C}$ is closed under countable direct sums, then $A \in \mathrm{obj}(\mathcal{C})$ implies $A' \in \mathrm{obj}(\mathcal{C})$, where $A'$ is the direct sum of countably many copies of $A$. But the isomorphism $A \oplus A' \cong A'$ gives the equation $[A] + [A'] = [A']$ in the group $K_0$, which forces $[A] = 0$. Thus, this Grothendieck group is trivial.

(ii) If $R$ is commutative, then tensor product makes $K_0(R)$ into a commutative ring. If $A, B \in R$-**Proj**, then $A \otimes_R B \in R$-**Proj** (Exercise 6.84 on page 485). Moreover, if $A \cong A'$ and $B \cong B'$, then $A \otimes_R B \cong A' \otimes_R B'$, so that $(A, B) \mapsto A \otimes_R B$ gives a well-defined multiplication on $K_0(R)$. The reader may check that $K_0(R)$ is a commutative ring with unit $[R]$.

(iii) Forming a Grothendieck group on $R$-**Proj** when $R$ is commutative, with binary operation tensor product instead of direct sum, leads to the notion of *Picard group* in Algebraic Geometry. $\blacktriangleleft$

We have seen that the Grothendieck group defines a functor on the category of commutative semigroups; we now show that it also defines a functor on **Rings**, the category of rings.

**Proposition 9.154.** *There is a functor $K_0 \colon \mathbf{Rings} \to \mathbf{Ab}$ with $R \mapsto K_0(R)$. Moreover, if $\varphi \colon R \to S$ is a ring homomorphism, then $K_0(\varphi)$ takes free $R$-modules to free $S$-modules; that is, $K_0(\varphi) \colon [R^n] \mapsto [S^n]$ for all $n \geq 0$.*

**Proof.** If $\varphi \colon R \to S$ is a ring homomorphism, then we may view $S$ as an $(S, R)$-bimodule by defining $sr = s\varphi(r)$ for all $r \in R$ and $s \in S$. For any left $R$-module $P$, the tensor product $S \otimes_R P$ now makes sense, and it is a left $S$-module, by Corollary 6.107. The restriction of the functor $S \otimes_R \square \colon {}_R\mathbf{Mod} \to {}_S\mathbf{Mod}$ takes $R$-**Proj** $\to S$-**Proj**, for if $P$ is a finitely generated projective $R$-module, then $S \otimes_R P$ is a finitely generated projective $S$-module [Exercise 6.84(iii) on page 485]. Now define $K_0(\varphi) \colon K_0(R) \to K_0(S)$ by $[P] \mapsto [S \otimes_R P]$. In particular, if $P = R^n$ is a finitely generated free $R$-module, then $K_0(\varphi) \colon [R^n] \mapsto [S \otimes_R R^n] = [S^n]$. •

For every ring $R$, there is a ring homomorphism $\iota \colon \mathbb{Z} \to R$ which takes the integer 1 to the unit element of $R$. Since $K_0$ is a functor on the category of rings, there is a homomorphism $K_0(\iota) \colon K_0(\mathbb{Z}) \to K_0(R)$. By Proposition 9.154, im $K_0(\iota)$ is the subgroup of $K_0(R)$ generated by all finitely generated free $R$-modules. If $V$ is an infinite-dimensional vector space over a field $k$, then $R = \operatorname{End}_k(V)$ has the property that $R \cong R \oplus R$ as $R$-modules (Rotman, *An Introduction to Homological Algebra*, p. 58). It follows that im $K_0(\iota) = \{0\}$ for this ring. In fact, $K_0(R) = \{0\}$ (Rosenberg, *Algebraic K-Theory and Its Applications*, p. 10).

**Definition.** For any ring $R$, the ***reduced Grothendieck group*** is defined by

$$\widetilde{K}_0(R) = \operatorname{coker} K_0(\iota) = K_0(R)/\operatorname{im} K_0(\iota).$$

In Theorem 10.131, we will show that $\widetilde{K}_0(R)$ is the class group of the ring of integers $R$ in an algebraic number field.

Since im $K_0(\iota)$ is the subgroup of $K_0(R)$ generated by all finitely generated free $R$-modules, the reduced Grothendieck group is the important part of $K_0(R)$, for it describes nonfree projective $R$-modules. For example, if $R$ is a ring for which every finitely generated projective $R$-module is free, then $K_0(R) = \operatorname{im} K_0(\iota)$ and $\widetilde{K}_0(R) = \{0\}$. We shall see, on page 866, that the converse of this is false.

Let us compute $K_0(R)$ for some nice rings $R$.

**Proposition 9.155.** *Let $R$ be a ring all of whose finitely generated projective $R$-modules are free and which has a rank function; that is, $R^m \cong R^n$ implies $m = n$. Then $K_0(R) \cong \mathbb{Z}$.[26] In particular, $K_0(R) \cong \mathbb{Z}$ if $R$ is a division ring, a PID, a local ring, or $k[x_1, \ldots, x_n]$ when $k$ is a field.*

**Proof.** The function $\rho \colon \operatorname{Iso}(R\text{-}\mathbf{Proj}) \to \mathbb{N}$ induced by rank, namely, $\rho(R^n) = n$, is well-defined; it is a semigroup map, for $\rho(V \oplus W) = \rho(V) + \rho(W)$, and it is an isomorphism because we are assuming that two free modules $P$ and $Q$ are isomorphic if and only if $\rho(P) = \rho(Q)$. Since $K_0 \colon \mathbf{CS} \to \mathbf{Ab}$ is a functor, $K_0(\rho)$ is an isomorphism $K_0(R) \to K_0(\mathbb{N}) = \mathbb{Z}$.

Each of the stated rings satisfies the hypotheses: see Proposition 7.12 for division rings, Corollary 8.4(ii) for PIDs, Theorem 10.26 for local rings, and the Quillen–Suslin Theorem (Rotman, *An Introduction to Homological Algebra*, p. 209) for polynomial rings in several variables over a field. •

**9.11.2. The Functor $G_0$.** As we said earlier, we are now going to define a new Grothendieck group, $G_0(\mathcal{C})$, which coincides with $K_0(R)$ when $\mathcal{C} = R\text{-}\mathbf{Proj}$.

**Definition.** A *G-category* $\mathcal{C}$ is a virtually small full subcategory of $_R\mathbf{Mod}$ for some ring $R$.

---

[26]The ring $R = \operatorname{End}_k(V)$ mentioned on page 862 does not have a rank function.

**Example 9.156.**

  (i) $R$-**Proj** is a $G$-category.

 (ii) All finitely generated $R$-modules form a $G$-category.

(iii) The category of all countable torsion-free abelian groups is a $G$-category.  ◀

We continue to denote the family of all isomorphism classes of objects in $\mathcal{C}$ by $\text{Iso}(\mathcal{C})$, and we continue letting an object $A$ denote its own isomorphism class.

**Definition.** If $\mathcal{C}$ is a $G$-category, let $\mathcal{F}(\mathcal{C})$ be the free abelian group with basis $\text{Iso}(\mathcal{C})$,[27] and let $\mathcal{E}$ be the subgroup of $\mathcal{F}(\mathcal{C})$ generated by all elements of the form

$$A + C - B \quad \text{if} \quad 0 \to A \to B \to C \to 0 \text{ is exact.}$$

The **Grothendieck group** $G_0(\mathcal{C})$ is the abelian group

$$G_0(\mathcal{C}) = \mathcal{F}(\mathcal{C})/\mathcal{E}.$$

For any object $A$ in $\mathcal{C}$, the coset $A + \mathcal{E}$ in $G_0(\mathcal{C})$ is denoted by $[A]$.

In particular, if $R$ is a ring, define

$$G_0(R) = G_0(R\text{-}\mathbf{Proj}).$$

**Proposition 9.157.** *For every ring $R$, we have $G_0(R) \cong K_0(R)$.*

**Proof.** Recall Proposition 9.148; if $S$ is a commutative semigroup, then $K_0(S) = F(S)/H$, where $H = \langle x + y - z \colon x + y = z \text{ in } S \rangle$. By definition, $K_0(R) = K_0(S)$, where $S$ is the semigroup $\text{Iso}(R\text{-}\mathbf{Proj})$ with operation $A + B = A \oplus B$; hence, $K_0(R) = F(\text{Iso}(R\text{-}\mathbf{Proj}))/H$, where $H$ is generated by all $A + B - A \oplus B$. Now $G_0(R) = F(R\text{-}\mathbf{Proj})/\mathcal{E}$, where $\mathcal{E}$ is generated by all $A + B - X$ with $0 \to A \to X \to B \to 0$ exact. But every exact sequence $0 \to A \to X \to B \to 0$ splits, because every object in $R$-**Proj** is projective, and so $X \cong A \oplus B$.  •

When are two elements in $G_0(\mathcal{C})$ equal?

**Proposition 9.158.** *Let $\mathcal{C}$ be a $G$-category.*

  (i) *If $x \in G_0(\mathcal{C})$, then $x = [A] - [B]$ for $A, B \in \text{obj}(\mathcal{C})$.*

 (ii) *Let $A, B \in \text{obj}(\mathcal{C})$. Then $[A] = [B]$ in $G_0(\mathcal{C})$ if and only if there are $C, U, V \in \text{obj}(\mathcal{C})$ and exact sequences*

$$0 \to U \to A \oplus C \to V \to 0 \quad \text{and} \quad 0 \to U \to B \oplus C \to V \to 0.$$

---

[27]Since a basis of a free abelian group must be a set and not a proper class, we are using the hypothesis that $\mathcal{C}$ is virtually small.

**Proof.**

(i) Since $G_0(\mathcal{C})$ is generated by $\mathrm{Iso}(\mathcal{C})$, each element $x \in G_0(\mathcal{C})$ is a $\mathbb{Z}$-linear combination of objects: $x = \sum_i m_i C_i$ (we allow $C_i$ to be repeated; that is, we assume each $m_i = \pm 1$). If we denote those $C$ with positive coefficient $m_i$ by $A_1, \ldots, A_r$ and those with negative $m_i$ by $B_1, \ldots, B_t$, then

$$x = \sum_{i=1}^{r} [A_i] - \sum_{j=1}^{s} [B_j].$$

Define $A = A_1 \oplus \cdots \oplus A_r$. That $[A] = [A_1 \oplus \cdots \oplus A_r]$ is proved by induction on $r \geq 2$. If $r = 2$, then exactness of $0 \to A_1 \to A_1 \oplus A_2 \to A_2 \to 0$ gives $[A_1] + [A_2] = [A_1 \oplus A_2]$; the inductive step is also easy. Similarly, if $B = B_1 \oplus \cdots \oplus B_s$, then $[B] = [B_1 \oplus \cdots \oplus B_s]$. Therefore, $x = [A] - [B]$.

(ii) If there exist modules $C, U,$ and $V$ as in the statement, then

$$[A \oplus C] = [U] + [V] = [B \oplus C].$$

But exactness of $0 \to A \to A \oplus C \to C \to 0$ gives $[A \oplus C] = [A] + [C]$. Similarly, $[B \oplus C] = [B] + [C]$, so that $[A] + [C] = [B] + [C]$ and $[A] = [B]$.

Conversely, if $[A] = [B]$ in $G_0(\mathcal{C})$, then $A - B \in \mathcal{E}$, and there is an equation in $F(\mathcal{C})$ [as in part (i), we allow repetitions, so that all coefficients are $\pm 1$],

$$A - B = \sum_i (X_i' + X_i'' - X_i) - \sum_j (Y_j' + Y_j'' - Y_j),$$

where there are exact sequences $0 \to X_i' \to X_i \to X_i'' \to 0$ and $0 \to Y_j' \to Y_j \to Y_j'' \to 0$. Transposing to eliminate negative coefficients,

$$A + \sum_i (X_i' + X_i'') + \sum_j Y_j = B + \sum_i X_i + \sum_j (Y_j' + Y_j'').$$

This is an equation in a free abelian group, so that expressions as linear combinations of a basis are unique. Therefore, the set $\{A, X_i', X_i'', Y_j\}$ of objects on the left-hand side, with multiplicities, coincides with $\{B, X_i, Y_j', Y_j''\}$, the set of objects on the right-hand side, with multiplicities. Therefore, the direct sum of the objects on the left is isomorphic to the direct sum of the objects on the right:

$$A \oplus X' \oplus X'' \oplus Y \cong B \oplus X \oplus Y' \oplus Y'',$$

where $X' = \bigoplus_i X_i'$, $X = \bigoplus_i X_i$, and so forth. Let $C$ denote their common value:

$$A \oplus X' \oplus X'' \oplus Y = C = B \oplus X \oplus Y' \oplus Y''.$$

By Exercise 6.25 on page 417, there are exact sequences

$$0 \to X' \oplus Y'' \to X \oplus Y'' \to X'' \to 0,$$

$$0 \to X' \oplus Y'' \to (X \oplus Y'') \oplus Y' \to X'' \oplus Y' \to 0,$$

and

$$0 \to X' \oplus Y'' \to A \oplus (X \oplus Y' \oplus Y'') \to A \oplus (X'' \oplus Y') \to 0.$$

The middle object is $C$. Applying Exercise 6.25 once again, there is an exact sequence

$$0 \to X' \oplus Y' \to B \oplus C \to B \oplus (A \oplus X'' \oplus Y'') \to 0.$$

Define $U = X' \oplus Y'$ and $V = B \oplus A \oplus X'' \oplus Y''$; with this notation, the last exact sequence is

$$0 \to U \to B \oplus C \to V \to 0.$$

Similar manipulation yields an exact sequence $0 \to U \to A \oplus C \to V \to 0$.   •

**Corollary 9.159.** *Let $R$ be a ring.*

(i) *If $x \in G_0(R)$, then there are projective $R$-modules $P$ and $Q$ with*

$$x = [P] - [Q].$$

(ii) *If $A$ and $B$ are projective $R$-modules, then $[A] = [B]$ in $G_0(R)$ if and only if there is a projective $R$-module $C$ with*

$$A \oplus C \cong B \oplus C.$$

**Proof.** Recall that $G_0(R) = G_0(R\text{-}\mathbf{Proj})$, so that objects are projective $R$-modules. Thus, (i) follows at once from Proposition 9.158. The second statement in the Proposition says that there are projectives $C, U, V$ (for every object in $R$-**Proj** is projective) and exact sequences $0 \to U \to A \oplus C \to V \to 0$ and $0 \to U \to B \oplus C \to V \to 0$. These sequences split, because $V$ is projective; that is, $A \oplus C \cong U \oplus V$ and $B \oplus C \cong U \oplus V$. Hence, $A \oplus C \cong B \oplus C$.   •

**Corollary 9.160.** *Let $A$ and $B$ be projective $R$-modules. Then $[A] = [B]$ in $K_0(R)$ if and only if there is a projective $R$-module $P$ with*

$$A \oplus P \cong B \oplus P.$$

**Proof.** This follows from Corollary 9.159 because $K_0(R) \cong G_0(R)$.   •

**Definition.** Let $R$ be a commutative ring, and let $\mathcal{C}$ be a subcategory of $_R\mathbf{Mod}$. Two $R$-modules $A$ and $B$ are called ***stably isomorphic in*** $\mathcal{C}$ if there exists a module $C \in \mathrm{obj}(\mathcal{C})$ with $A \oplus C \cong B \oplus C$.

With this terminology, Corollary 9.160 says that two modules determine the same element of $K_0(R)$ if and only if they are stably isomorphic. It is clear that isomorphic modules are stably isomorphic; the next example shows that the converse need not hold.

**Example 9.161.**

(i) If $\mathcal{F}$ is the category of all finite abelian groups, then Exercise 4.13 on page 242 shows that two finite abelian groups are stably isomorphic in $\mathcal{F}$ if and only if they are isomorphic.

(ii) Stably isomorphic finitely generated projective modules need not be isomorphic; here is an example of Kaplansky as described in Làm–Siu, $K_0$ and $K_1$—An Introduction to Algebraic $K$-Theory, *Amer. Math. Monthly* 82 (1975), 329–364. "Let $A$ be the coordinate ring of the real 2-sphere; i.e.,

$A = \mathbb{R}[x, y, z]$ with the relation $x^2 + y^2 + z^2 = 1$. Let $\varepsilon \colon A^3 \to A$ be the $A$-linear functional given by $\varepsilon(\alpha, \beta, \gamma) = \alpha x + \beta y + \gamma z$. Since $\varepsilon(x, y, z) = 1$, we have a splitting $A^3 \cong \ker \varepsilon \oplus A$, so $P = \ker \varepsilon$ is stably free. Assume, for the moment, that $P$ is actually free, so (necessarily) $P \cong A^2$. Then, $A^3$ will have a new basis consisting of $(x, y, z)$ and two other triples $(f, g, h), (f', g', h')$. The matrix

$$\begin{bmatrix} x & f & f' \\ y & g & g' \\ z & h & h' \end{bmatrix}$$

is therefore invertible over $A$, and has determinant equal to a unit $e \in A$. We think of $f, g, h, e, e^{-1}, \cdots$ as functions on $S^2$ (they are polynomial expressions in the 'coordinate functions' $x, y, z$). Consider the continuous vector field on $S^2$ given by $v \in S^2 \mapsto (f(v), g(v), h(v)) \in \mathbb{R}^3$. Since $e = f' \cdot (yh - zg) - g' \cdot (xh - zf) + h' \cdot (xg - yf)$ is clearly nowhere zero on $S^2$, the vector $(f(v), g(v), h(v))$ is nowhere collinear with the vector $v$. Taking the orthogonal projections onto the tangent planes of $S^2$, we obtain a continuous vector field of nowhere vanishing tangents. This is well-known to be impossible by elementary topology. It follows that $P$ cannot be free over $A$." Here is the heart of this construction.

**Theorem (Swan).** *Let $X$ be a compact Hausdorff space and let $R$ be the ring of continuous real-valued functions on $X$. If $E \xrightarrow{p} X$ is a vector bundle with global sections*

$$\Gamma(X) = \{\text{continuous } s \colon X \to E \mid ps = 1_X\},$$

*then $\Gamma(X)$ is a finitely generated projective $R$-module, and every such $R$-module arises in this way.*

See Swan, Vector bundles and projective modules, *Trans. AMS* 105 (1962), 264–277, Bass, *Algebraic K-Theory*, p. 735, or Rosenberg, *Algebraic K-Theory and Its Applications*, pp. 32–36.  ◄

Grothendieck proved that if $R$ is left noetherian, then $K_0(R) \cong K_0(R[x])$ (Rosenberg, *Algebraic K-Theory and Its Applications*, p. 141). It follows, for $k$ a field, that $K_0(k[x_1, \ldots, x_n]) \cong K_0(k) = \mathbb{Z}$, and so the reduced Grothendieck group $\widetilde{K}_0(R) = \{0\}$ [recall that $\widetilde{K}_0(R) = K_0(R)/\operatorname{im} K_0(\iota)$, where $\operatorname{im} K_0(\iota)$ is the subgroup generated by free $R$-modules]. Serre wondered whether the converse holds for $R = k[x_1, \ldots, x_n]$: are finitely generated projective $k[x_1, \ldots, x_n]$-modules free? The Quillen–Suslin Theorem (Rotman, *An Introduction to Homological Algebra*, p. 209) proves that this is so.

A finitely generated projective $R$-module $P$ is **stably free** if it is stably isomorphic to a free module. Free modules are, obviously, stably free, but the coordinate ring of the 2-sphere in Example 9.161(ii) shows there are stably free modules which are not free. A ring $R$ is called a **Hermite ring** if every finitely generated projective $R$-module is stably free. In light of Corollary 9.160, a ring $R$ is Hermite if and only if $\widetilde{K}_0(R) = \{0\}$. If $R$ is a ring for which every finitely generated projective $R$-module is free, then $R$ is a Hermite ring. The converse is not true. Ojanguren–Sridharan, Cancellation of Azumaya algebras, *J. Algebra* 18 (1971), 501–505, showed that if $\Delta$

is a noncommutative division ring, then $R = \Delta[x, y]$ (where coefficients commute with indeterminates) is a Hermite ring having projectives which are not free.

We now present computations of $G_0(\mathcal{C})$ for some familiar $G$-categories $\mathcal{C}$. The next result shows that filtrations of modules give rise to sums in $G_0$.

**Proposition 9.162.** *Let $R$ be a commutative ring and let $\mathcal{C}$ be the category of all finitely generated $R$-modules. If $M \in \mathrm{obj}(\mathcal{C})$ and*

$$M = M_0 \supseteq M_1 \supseteq M_2 \supseteq \cdots \supseteq M_n = \{0\}$$

*has factor modules $Q_i = M_{i-1}/M_i$, then*

$$[M] = [Q_1] + \cdots + [Q_n] \quad in \ G_0(\mathcal{C}).$$

**Proof.** Since $Q_i = M_{i-1}/M_i$, there is a short exact sequence

$$0 \to M_i \to M_{i-1} \to Q_i \to 0,$$

so that $[Q_i] = [M_{i-1}] - [M_i]$ in $G_0(\mathcal{C})$. We now have a telescoping sum:

$$\sum_{i=1}^{n} [Q_i] = \sum_{i=1}^{n} \left( [M_{i-1}] - [M_i] \right) = [M_0] - [M_n] = [M]. \quad \bullet$$

**Definition.** Let $\mathcal{C}$ be a $G$-category. Call an object $S$ **simple in $\mathcal{C}$** if, for every object $X$ and every monic $u \colon X \to S$ in $\mathcal{C}$, either $u = 0$ or $u$ is an isomorphism. A filtration

$$M = M_0 \supseteq M_1 \supseteq M_2 \supseteq \cdots \supseteq M_n = \{0\}$$

in a category $\mathcal{C}$ of modules is called a **composition series** of $M$ if each of its factors $Q_i = M_{i-1}/M_i$ is simple in $\mathcal{C}$.

A **Jordan–Hölder category** is a $G$-category $\mathcal{C}$ such that:

(i) each object $M$ has a composition series;

(ii) for every two composition series

$$M = M_0 \supseteq M_1 \supseteq M_2 \supseteq \cdots \supseteq M_n = \{0\}$$

and

$$M = M_0' \supseteq M_1' \supseteq M_2' \supseteq \cdots \supseteq M_m' = \{0\},$$

we have $m = n$ and a permutation $\sigma \in S_n$ such that $Q_j' \cong Q_{\sigma j}$ for all $j$, where $Q_i = M_{i-1}/M_i$ and $Q_j' = M_{j-1}'/M_j'$.

Define the **length** $\ell(M)$ of a module $M$ in a Jordan–Hölder category to be the number $n$ of terms in a composition series. If the simple factors of a composition series are $Q_1, \ldots, Q_n$, we define

$$\mathrm{jh}(M) = Q_1 \oplus \cdots \oplus Q_n.$$

A composition series may have several isomorphic factors, and $\mathrm{jh}(M)$ records their multiplicity. Since $Q_1 \oplus Q_2 \cong Q_2 \oplus Q_1$, $\mathrm{jh}(M)$ does not depend on the ordering $Q_1, \ldots, Q_n$.

**Lemma 9.163.** *Let $\mathcal{C}$ be a Jordan–Hölder category, and let $Q_1, \ldots, Q_n, Q_1', \ldots, Q_m'$ be simple objects in $\mathcal{C}$.*

(i) If $Q_1 \oplus \cdots \oplus Q_n \cong Q_1' \oplus \cdots \oplus Q_m'$, then $m = n$ and there is a permutation $\sigma \in S_n$ such that $Q_j' \cong Q_{\sigma j}$ for all $j$, where $Q_i = M_{i-1}/M_i$ and $Q_j' = M_{j-1}'/M_j'$.

(ii) If $M$ and $M'$ are modules in $\mathrm{obj}(\mathcal{C})$ and there is a simple object $S$ in $\mathcal{C}$ with

$$S \oplus \mathrm{jh}(M) \cong S \oplus \mathrm{jh}(M'),$$

then $\mathrm{jh}(M) \cong \mathrm{jh}(M')$.

**Proof.**

(i) Since $Q_1, \ldots, Q_n$ are simple, the filtration

$$Q_1 \oplus \cdots \oplus Q_n \supseteq Q_2 \oplus \cdots \oplus Q_n \supseteq Q_3 \oplus \cdots \oplus Q_n \supseteq \cdots$$

is a composition series of $Q_1 \oplus \cdots \oplus Q_n$ with factors $Q_1, \ldots, Q_n$; similarly, $Q_1' \oplus \cdots \oplus Q_m'$ has a composition series with factors $Q_1', \ldots, Q_m'$. The result follows from $\mathcal{C}$ being a Jordan–Hölder category.

(ii) This result follows from part (i) because $S$ is simple.   •

**Lemma 9.164.** *If $0 \to A \to B \to C \to 0$ is an exact sequence in a Jordan–Hölder category, then*

$$\mathrm{jh}(B) \cong \mathrm{jh}(A) \oplus \mathrm{jh}(C).$$

**Proof.** The proof is by induction on the length $\ell(C)$. Let $A = A_0 \supseteq A_1 \supseteq \cdots \supseteq A_n = \{0\}$ be a composition series for $A$ with factors $Q_1, \ldots, Q_n$. If $\ell(C) = 1$, then $C$ is simple, and so

$$B \supseteq A \supseteq A_1 \supseteq \cdots \supseteq A_n = \{0\}$$

is a composition series for $B$ with factors $C, Q_1, \ldots, Q_n$. Therefore,

$$\mathrm{jh}(B) = C \oplus Q_1 \oplus \cdots \oplus Q_n = \mathrm{jh}(C) \oplus \mathrm{jh}(A).$$

For the inductive step, let $\ell(C) > 1$. Choose a maximal submodule $C_1$ of $C$ (which exists because $C$ has a composition series). If $\nu \colon B \to C$ is the given surjection, define $B_1 = \nu^{-1}(C_1)$. There is a commutative diagram (with vertical arrows inclusions)

$$
\begin{array}{ccccccccc}
0 & \longrightarrow & A & \longrightarrow & B & \overset{\nu}{\longrightarrow} & C & \longrightarrow & 0 \\
 & & \big\uparrow & & \big\uparrow & & \big\uparrow & & \\
0 & \longrightarrow & A & \longrightarrow & B_1 & \longrightarrow & C_1 & \longrightarrow & 0
\end{array}
$$

Since $C_1$ is a maximal submodule of $C$, the quotient module

$$C'' = C/C_1$$

is simple. Note that $B/B_1 \cong (B/A)/(B_1/A) \cong C/C_1 = C''$. By the base step, we have

$$\mathrm{jh}(C) = C'' \oplus \mathrm{jh}(C_1) \quad \text{and} \quad \mathrm{jh}(B) = C'' \oplus \mathrm{jh}(B_1).$$

By the inductive hypothesis,

$$\mathrm{jh}(B_1) = \mathrm{jh}(A) \oplus \mathrm{jh}(C_1).$$

Therefore,

$$\mathrm{jh}(B) = C'' \oplus \mathrm{jh}(B_1)$$
$$\cong C'' \oplus \mathrm{jh}(A) \oplus \mathrm{jh}(C_1)$$
$$\cong \mathrm{jh}(A) \oplus C'' \oplus \mathrm{jh}(C_1)$$
$$\cong \mathrm{jh}(A) \oplus \mathrm{jh}(C). \quad \bullet$$

**Theorem 9.165.** *Let $\mathcal{C}$ be a $G$-category in which every $M$ in $\mathcal{C}$ has a composition series. Then $\mathcal{C}$ is a Jordan–Hölder category if and only if $G_0(\mathcal{C})$ is a free abelian group with basis the set $\mathcal{B}$ of all nonisomorphic simple $S$ in $\mathcal{C}$.*

**Proof.** Assume that $G_0(\mathcal{C})$ is free abelian with basis $\mathcal{B}$. Since $[0]$ is not a member of a basis, we have $[S] \neq [0]$ for every simple object $S$; moreover, if $S \not\cong S'$, then $[S] \neq [S']$, for a basis repeats no elements. Let $M \in \mathrm{obj}(\mathcal{C})$, and let $Q_1, \ldots, Q_n$ and $Q'_1, \ldots, Q'_m$ be simple quotients arising, respectively, as factors of two composition series of $M$. By Proposition 9.162, we have

$$[Q_1] + \cdots + [Q_n] = [M] = [Q'_1] + \cdots + [Q'_m].$$

Uniqueness of expression in terms of the basis $\mathcal{B}$ says, for each $Q'_j$, that there exists $Q_i$ with $[Q_i] = [Q'_j]$; in fact, the number of any $[Q_i]$ on the left-hand side is equal to the number of copies of $[Q'_j]$ on the right-hand side. Therefore, $\mathcal{C}$ is a Jordan–Hölder category.

Conversely, assume that the Jordan–Hölder Theorem holds for $\mathcal{C}$. Since every $M \in \mathrm{obj}(\mathcal{C})$ has a composition series, Proposition 9.162 shows that $\mathcal{B}$ generates $G_0(\mathcal{C})$. Let $S$ in $\mathrm{obj}(\mathcal{C})$ be simple. If $[S] = [T]$, then Proposition 9.158 says there are $C, U, V \in \mathrm{obj}(\mathcal{C})$ and exact sequences $0 \to U \to S \oplus C \to V \to 0$ and $0 \to U \to T \oplus C \to V \to 0$. Lemma 9.164 gives

$$\mathrm{jh}(S) \oplus \mathrm{jh}(C) \cong \mathrm{jh}(U) \oplus \mathrm{jh}(V) \cong \mathrm{jh}(T) \oplus \mathrm{jh}(C).$$

By Lemma 9.163, we may cancel the simple summands one by one until we are left with $S \cong T$. A similar argument shows that if $S$ is simple, then $[S] \neq [0]$. Finally, let us show that every element in $G_0(\mathcal{C})$ has a unique expression as a linear combination of elements in $\mathcal{B}$. Suppose there are positive integers $m_i$ and $n_j$ so that

$$(1) \qquad \sum_i m_i[S_i] - \sum_j n_j[T_j] = [0],$$

where the $S_i$ and $T_j$ are simple and $S_i \not\cong T_j$ for all $i, j$. If we denote the direct sum of $m_i$ copies of $S_i$ by $m_i S_i$, then Equation (1) gives

$$\left[ \bigoplus_i m_i S_i \right] = \left[ \bigoplus_j n_j T_j \right].$$

By Proposition 9.158, there are modules $C, U, V$ and exact sequences

$$0 \to U \to C \oplus \bigoplus_i m_i S_i \to V \to 0 \quad \text{and} \quad 0 \to U \to C \oplus \bigoplus_j n_j T_j \to V \to 0,$$

and Lemma 9.164 gives

$$\mathrm{jh}\Big(\bigoplus_i m_i S_i\Big) \cong \mathrm{jh}\Big(\bigoplus_j n_j T_j\Big).$$

By Lemma 9.163, some $S_i$ occurs on the right-hand side, contradicting $S_i \ncong T_j$ for all $i, j$. Therefore, Equation (1) cannot occur.   •

The reader is probably curious why the Grothendieck group $K_0(R)$ has a subscript (he or she may also be curious about why the letter K is used). Grothendieck constructed $K_0(R)$ to generalize the Riemann–Roch Theorem in Algebraic Geometry; he used the letter K to abbreviate the German *Klasse*. Atiyah–Hirzebruch applied this idea to the category of vector bundles over compact Hausdorff spaces $X$ obtaining *Topological K-Theory*: a sequence of functors $K^{-n}(X)$ for $n \geq 0$. This sequence satisfies all the Eilenberg–Steenrod Axioms characterizing the cohomology groups of topological spaces [actually, of pairs $(X, Y)$, where $X$ is a compact Hausdorff space and $Y \subseteq X$ is a closed subspace] except for the *Dimension Axiom*.[28] In particular, there is a long exact sequence

$$\to K^{-1}(X, Y) \to K^{-1}(X) \to K^{-1}(Y) \to K^0(X, Y) \to K^0(X) \to K^0(Y).$$

At that time, the Grothendieck group $K_0(R)$ had no need for a subscript, for there was no analog of the higher groups $K^{-i}$. This was remedied by Bass–Schanuel who defined $K_1$ as well as relative groups $K_0(R, I)$ and $K_1(R, I)$, where $I$ is a two-sided ideal in $R$. Moreover, they exhibited a six-term exact sequence involving three $K_0$ (now subscripted) and three $K_1$. In his review (MR0249491(40#2736)) of Bass's book *Algebraic K-Theory*, 1968, Heller wrote,

> Algebraic $K$-theory flows from two sources. The (J. H. C.) Whitehead torsion, introduced in order to study the topological notion of simple homotopy type, leads to the groups $K_1$. The Grothendieck group of projective modules over a ring leads to $K_0$. The latter notion was applied by Atiyah and Hirzebruch in order to construct a new cohomology theory which has been enormously fruitful in topology.
>
> The observation that these two ideas could be unified in a beautiful and powerful theory with widespread applications in algebra is due to Bass, who is also responsible for a major portion of those applications.

We have seen that $K_0$ can be regarded as a functor on the category of rings, so that a ring homomorphism $\varphi \colon R \to S$ gives a homomorphism $K_0(R) \to K_0(S)$; studying its kernel leads to relative $K_0$. Milnor defined a functor $K_2$ extending the exact sequence of lower $K$s (Milnor, *Introduction to Algebraic K-Theory*) which has applications to Field Theory and Group Theory. Quillen then constructed an infinite sequence $(K_n(\mathcal{C}))_{n \geq 0}$ of abelian groups by associating a topological space $X(\mathcal{C})$ to certain categories $\mathcal{C}$. He then defined $K_n(\mathcal{C}) = \pi_n(X(\mathcal{C}))$ for all $n \geq 0$,

---

[28]The Dimension Axiom says that if $X$ is a one-point space and $G$ is an abelian group, then $H^0(X, G) \cong G$ and $H^i(X, G) \cong \{0\}$ for all $i > 0$.

the homotopy groups of this space, and proved that his $K_n$ coincide with those for $n = 0, 1, 2$ (Rosenberg, *Algebraic K-Theory and Its Applications*).

Here are two interesting applications. The Merkurjev–Suslin Theorem relates $K$-Theory to Brauer groups. Let $n$ be a positive integer and let $F$ be a field containing a primitive $n$th root of unity. If the characteristic of $F$ is either 0 or does not divide $n$, then $K_2(F) \otimes_{\mathbb{Z}} \mathbb{I}_n \cong \mathrm{Br}(F)[n]$, the subgroup of the Brauer group $\mathrm{Br}(F)$ consisting of all elements whose order divides $n$. The second application involves what are called the ***Lichtenbaum Conjectures***, which relate the orders of certain $K$-groups with values of zeta functions.

## Exercises

∗ **9.67.** For any ring $R$, prove that $K_0(R) \cong K_0(R^{\mathrm{op}})$.

**Hint.** Prove that $\mathrm{Hom}_R(\square, R)\colon R\text{-}\mathbf{Proj} \to R^{\mathrm{op}}\text{-}\mathbf{Proj}$ is an equivalence of categories.

**9.68.** Prove that $K_0(R) \cong K_0(\mathrm{Mat}_n(R))$ for every ring $R$.

**Hint.** The functor $\mathrm{Hom}_R(R^n, \square)$ preserves finitely generated projectives.

**9.69.** If a ring $R$ is a direct product, $R = R_1 \times R_2$, prove that
$$K_0(R) \cong K_0(R_1) \oplus K_0(R_2).$$

**9.70.** If a commutative semigroup $S$ is an abelian group, prove that $K_0(S) \cong S$.

**9.71.** Let $R$ be a domain, and let $a \in R$ be neither 0 nor a unit. If $\mathcal{C}$ is the category of all finitely generated $R$-modules, prove that $[R/Ra] = 0$ in $G_0(\mathcal{C})$.

**Hint.** Use the exact sequence $0 \to R \xrightarrow{\mu_a} R \to R/Ra \to 0$, where $\mu_a\colon r \mapsto ar$.

**9.72.** If $\mathcal{C}$ and $\mathcal{A}$ are $G$-categories, prove that every exact functor $F\colon \mathcal{C} \to \mathcal{A}$ defines a homomorphism $K_0(\mathcal{C}) \to K_0(\mathcal{A})$ with $[C] \mapsto [FC]$ for all $C \in \mathrm{obj}(\mathcal{C})$.

**9.73.** Let $\mathcal{C}$ be the category of all finitely generated abelian groups.

  (i) Prove that $K_0(\mathcal{C})$ is free abelian of countably infinite rank.

  (ii) Prove that $G_0(\mathcal{C}) \cong \mathbb{Z}$.

  (iii) When are two finitely generated abelian groups stably isomorphic?

**9.74.** If $\mathcal{C}$ is the category of all finitely generated $R$-modules over a PID $R$, compute $K_0(\mathcal{C})$ and $G_0(\mathcal{C})$.

**9.75.** If $\mathcal{C}$ is the category of all finitely generated $\mathbb{I}_6$-modules, prove that $K_0(\mathcal{C}) \cong \mathbb{Z} \oplus \mathbb{Z}$.

**9.76.** If $R = k[x_1, \ldots, x_n]$, where $k$ is a field, use Hilbert's Syzygy Theorem (Corollary 10.167) to prove that $G_0(R) \cong \mathbb{Z}$. (Of course, this also follows from the Quillen–Suslin Theorem that finitely generated projective $R$-modules are free.)

∗ **9.77. (Eilenberg)** Prove that if $P$ is a projective $R$-module (over some commutative ring $R$), then there exists a free $R$-module $Q$ with $P \oplus Q$ a free $R$-module. Conclude that $K_0(\mathcal{C}) = \{0\}$ for $\mathcal{C}$ the category of countably generated projective $R$-modules.

**Hint.** $Q$ need not be finitely generated.

# Chapter 10

# Commutative Rings III

### Section 10.1. Local and Global

It is often easier to examine algebraic structures "one prime at a time." For example, let $G$ and $H$ be finite groups. If $G \cong H$, then their Sylow $p$-subgroups are isomorphic for all primes $p$; studying $G$ and $H$ *locally* means studying their $p$-subgroups. This local information is not enough to determine whether $G \cong H$; for example, the symmetric group $S_3$ and the cyclic group $\mathbb{I}_6$ are nonisomorphic groups having isomorphic Sylow 2-subgroups and isomorphic Sylow 3-subgroups. The *global problem* assumes that the Sylow $p$-subgroups of groups $G$ and $H$ are isomorphic, for all primes $p$, and asks what else is necessary to conclude that $G \cong H$. For general groups, this global problem is intractible (although there are partial results). However, this strategy does succeed for finite *abelian* groups. The local problem involves primary components (Sylow subgroups), which are direct sums of cyclic groups, and the global problem is solved by Primary Decomposition: every finite abelian group is the direct sum of its primary components. In this case, the local information is sufficient to solve the global problem. Even though local problems are simpler than global ones, their solutions can give very useful information.

**10.1.1. Subgroups of $\mathbb{Q}$.** We begin with another group-theoretic illustration of local and global investigation, after which we will consider localization of commutative rings.

**Definition.** Let $R$ be a domain with $Q = \mathrm{Frac}(R)$. If $M$ is an $R$-module, define

$$\mathrm{rank}(M) = \dim_Q(Q \otimes_R M).$$

For example, the rank of an abelian group $G$ is defined as $\dim_{\mathbb{Q}}(\mathbb{Q} \otimes_{\mathbb{Z}} G)$. Recall, when $R$ is a domain, that an $R$-module $M$ is *torsion-free* if it has no nonzero elements of finite order; that is, if $r \in R$ and $m \in M$ are nonzero, then $rm$ is nonzero.

**Lemma 10.1.** *Let $R$ be a domain with $Q = \mathrm{Frac}(R)$, and let $M$ be an $R$-module.*

(i) *If $0 \to M' \to M \to M'' \to 0$ is a short exact sequence of $R$-modules, then*
$$\mathrm{rank}(M) = \mathrm{rank}(M') + \mathrm{rank}(M'').$$

(ii) *An $R$-module $M$ is torsion if and only if $\mathrm{rank}(M) = 0$.*

(iii) *Let $M$ be torsion-free. Then $M$ has rank $1$ if and only if it is isomorphic to a nonzero $R$-submodule of $Q$.*

**Proof.**

(i) By Corollary 6.140, the fraction field $Q$ is a flat $R$-module. Therefore, $0 \to Q \otimes_R M' \to Q \otimes_R M \to Q \otimes_R M'' \to 0$ is a short exact sequence of vector spaces over $Q$, and the result is a standard result of Linear Algebra, Exercise 2.80 on page 145.

(ii) If $M$ is torsion, then $Q \otimes_R M = \{0\}$, by a routine generalization of Proposition 6.118 [divisible $\otimes$ torsion $= \{0\}$]. Hence, $\mathrm{rank}(M) = 0$.

If $M$ is not torsion, then there is an exact sequence $0 \to R \to M$ (if $m \in M$ has infinite order, then $r \mapsto rm$ is an injection $R \to M$). Since $Q$ is flat, the sequence $0 \to Q \otimes_R R \to Q \otimes_R M$ is exact. By Proposition 6.108, $Q \otimes_R R \cong Q \neq \{0\}$, so that $Q \otimes_R M \neq \{0\}$. Therefore, $\mathrm{rank}(M) > 0$.

(iii) If $\mathrm{rank}(M) = 1$, then $M \neq \{0\}$, and exactness of $0 \to R \to Q$ gives exactness of $\mathrm{Tor}_1^R(Q/R, M) \to R \otimes_R M \to Q \otimes_R M$. Since $M$ is torsion-free, Lemma 9.103(iii) gives $\mathrm{Tor}_1^R(Q/R, M) = \{0\}$. As always, $R \otimes_R M \cong M$, while $Q \otimes_R M \cong Q$ because $M$ has rank $1$. Therefore, $M$ is isomorphic to an $R$-submodule of $Q$.

Conversely, if $M$ is isomorphic to an $R$-submodule of $Q$, there is an exact sequence $0 \to M \to Q$. Since $Q$ is flat, by Corollary 6.140, we have exactness of $0 \to Q \otimes_R M \to Q \otimes_R Q$. Now this last sequence is an exact sequence of vector spaces over $Q$. Since $Q \otimes_R Q \cong Q$ is one-dimensional, its nonzero subspace $Q \otimes_R M$ is also one-dimensional; that is, $\mathrm{rank}(M) = 1$.  $\bullet$

**Example 10.2.** The following abelian groups are torsion-free of rank 1.

(i) The group $\mathbb{Z}$ of integers.

(ii) The additive group $\mathbb{Q}$.

(iii) The **2-adic fractions** $\mathbb{Z}_{(2)} = \{a/b \in \mathbb{Q} : b \text{ is odd}\}$.

(iv) The set of all rationals having a finite decimal expansion.

(v) The set of all rationals having squarefree denominator.  $\blacktriangleleft$

**Proposition 10.3.** *Let $R$ be a domain with $Q = \mathrm{Frac}(R)$. Two submodules $A$ and $B$ of $Q$ are isomorphic if and only if there is $c \in Q$ with $B = cA$.*

**Proof.** If $B = cA$, then $A \cong B$ via $a \mapsto ca$. Conversely, suppose that $f : A \to B$ is an isomorphism. Regard $f$ as an $R$-injection $A \to Q$ by composing it with the inclusion $B \to Q$. Since $Q$ is an injective $R$-module, by Proposition 6.90(ii), there is an $R$-map $g : Q \to Q$ extending $f$. But Proposition 6.90(iii) says that there is $c \in Q$ with $g(x) = cx$ for all $x \in Q$. Thus, $B = f(A) = g(A) = cA$.  $\bullet$

Recall, for each prime $p$, that the ring of *$p$-adic fractions* is the subring of $\mathbb{Q}$:

$$\mathbb{Z}_{(p)} = \{a/b \in \mathbb{Q} : (b, p) = 1\}.$$

For example, if $p = 2$, then $\mathbb{Z}_{(2)}$ is the group of 2-adic fractions.

**Proposition 10.4.**

(i) *For each prime $p$, the ring $\mathbb{Z}_{(p)}$ has a unique maximal ideal; that is, it is a local[1] PID.*

(ii) *If $G$ is a torsion-free abelian group of rank 1, then $\mathbb{Z}_{(p)} \otimes_{\mathbb{Z}} G$ is a torsion-free $\mathbb{Z}_{(p)}$-module of rank 1.*

(iii) *If $M$ is a torsion-free $\mathbb{Z}_{(p)}$-module of rank 1, then $M \cong \mathbb{Z}_{(p)}$ or $M \cong \mathbb{Q}$.*

**Proof.**

(i) We show that the only nonzero ideals $I$ in $\mathbb{Z}_{(p)}$ are $(p^n)$, for $n \geq 0$; it will then follow that $\mathbb{Z}_{(p)}$ is a PID and that $(p)$ is its unique maximal ideal. Each nonzero $x \in \mathbb{Z}_{(p)}$ has the form $a/b$ for integers $a$ and $b$, where $(b, p) = 1$. But $a = p^n a'$, where $n \geq 0$ and $(a', p) = 1$; that is, there is a unit $u \in \mathbb{Z}_{(p)}$, namely, $u = a'/b$, with $x = up^n$. Let $I \neq (0)$ be an ideal. Of all the nonzero elements in $I$, choose $x = up^n \in I$, where $u$ is a unit, with $n$ minimal. Then $I = (x) = (p^n)$, for if $y \in I$, then $y = vp^m$, where $v$ is a unit and $n \leq m$. Hence, $p^n \mid y$ and $y \in (p^n)$.

(ii) Since $\mathbb{Z}_{(p)} \subseteq \mathbb{Q}$, it is an additive torsion-free abelian group of rank 1, and so it is flat (Corollary 8.7). Hence, exactness of $0 \to G \to \mathbb{Q}$ gives exactness of

$$0 \to \mathbb{Z}_{(p)} \otimes_{\mathbb{Z}} G \to \mathbb{Z}_{(p)} \otimes_{\mathbb{Z}} \mathbb{Q}.$$

By Exercise 10.4 on page 880, $\mathbb{Z}_{(p)} \otimes_{\mathbb{Z}} \mathbb{Q} \cong \mathbb{Q} = \mathrm{Frac}(\mathbb{Z}_{(p)})$, so that $\mathbb{Z}_{(p)} \otimes_{\mathbb{Z}} G$ is a torsion-free $\mathbb{Z}_{(p)}$-module of rank 1.

(iii) There is no loss in generality in assuming that $M \subseteq \mathbb{Q}$ and that $1 \in M$. Consider the equations $p^n y_n = 1$ for $n \geq 0$. We claim that if all these equations are solvable for $y_n \in M$, then $M = \mathbb{Q}$. If $a/b \in \mathbb{Q}$, then $a/b = a/p^n b'$, where $(b', p) = 1$, and so $a/b = (a/b')y_n$; as $a/b' \in \mathbb{Z}_{(p)}$, we have $a/b \in M$. We may now assume that there is a largest $n \geq 0$ for which the equation $p^n y_n = 1$ is solvable for $y_n \in M$. We claim that $M = \langle y_n \rangle$, the cyclic submodule generated by $y_n$, which will show that $M \cong \mathbb{Z}_{(p)}$. If $m \in M$, then $m = c/d = p^r c'/p^s d' = (c'/d')(1/p^{s-r})$, where $(c', p) = 1 = (d', p)$. Since $c'/d'$ is a unit in $\mathbb{Z}_{(p)}$, we have $1/p^{s-r} \in M$, and so $s - r \leq n$; that is, $s - r = n - \ell$ for some $\ell \geq 0$. Hence, $1/p^{s-r} = 1/p^{n-\ell} = p^\ell/p^n = p^\ell y_n$, and so $m = (c'p^\ell/d')y_n \in \langle y_n \rangle$. $\bullet$

**Definition.** A *discrete valuation ring*, abbreviated DVR, is a local PID that is not a field.

---

[1] In this section, *local rings* are commutative rings having a unique maximal ideal. Often, local rings are also assumed to be noetherian. On the other hand, the term sometimes means a noncommutative ring having a unique maximal left or right ideal.

Some examples of DVRs are $\mathbb{Z}_{(p)}$, the $p$-adic integers $\mathbb{Z}_p$, and formal power series $k[[x]]$, where $k$ is a field.

**Definition.** Two abelian groups $G$ and $H$ are ***locally isomorphic*** if $\mathbb{Z}_{(p)} \otimes_{\mathbb{Z}} G \cong \mathbb{Z}_{(p)} \otimes_{\mathbb{Z}} H$ for all primes $p$.

We have solved the local problem for torsion-free abelian groups $G$ of rank 1; associate to $G$ the family $\mathbb{Z}_{(p)} \otimes_{\mathbb{Z}} G$ of $\mathbb{Z}_{(p)}$-modules, one for each prime $p$.

**Example 10.5.** Let $G$ be the subgroup of $\mathbb{Q}$ consisting of those rationals having squarefree denominator. Then $G$ and $\mathbb{Z}$ are locally isomorphic, but they are not isomorphic, because $\mathbb{Z}$ is finitely generated and $G$ is not.  ◄

We are now going to solve the global problem for torsion-free abelian groups of rank 1.

**Definition.** Let $G$ be an abelian group, let $x \in G$, and let $p$ be a prime. Then $x$ is ***divisible by $p^n$ in $G$*** if there exists $y_n \in G$ with $p^n y_n = x$. Define the $p$-***height*** of $x$, denoted by $h_p(x)$, by

$$h_p(x) = \begin{cases} \infty & \text{if } x \text{ is divisible by } p^n \text{ in } G \text{ for all } n \geq 0, \\ k & \text{if } x \text{ is divisible by } p^k \text{ in } G \text{ but not by } p^{k+1}. \end{cases}$$

The ***height sequence*** (or ***characteristic***) of $x$ in $G$, where $x$ is nonzero, is the sequence

$$\chi(x) = \chi_G(x) = (h_2(x), h_3(x), h_5(x), \ldots, h_p(x), \ldots).$$

Thus, $\chi(x)$ is a sequence $(h_p)$, where $h_p = \infty$ or $h_p \in \mathbb{N}$. Let $G \subseteq \mathbb{Q}$ and let $x \in G$ be nonzero. If $\chi(x) = (h_p)$ and $a = p_1^{f_1} \cdots p_n^{f_n}$, then $\frac{1}{a}x \in G$ if and only if $f_{p_i} \leq h_{p_i}$ for $i = 1, \ldots, n$.

**Example 10.6.** Each of the groups in Example 10.2 contains $x = 1$.

(i) In $\mathbb{Z}$,
$$\chi_{\mathbb{Z}}(1) = (0, 0, 0, \ldots).$$

(ii) In $\mathbb{Q}$,
$$\chi_{\mathbb{Q}}(1) = (\infty, \infty, \infty, \ldots).$$

(iii) In $\mathbb{Z}_{(2)}$,
$$\chi_{\mathbb{Z}_{(2)}}(1) = (0, \infty, \infty, \ldots).$$

(iv) If $G$ is the group of all rationals having a finite decimal expansion, then
$$\chi_G(1) = (\infty, 0, \infty, 0, 0, \ldots).$$

(v) If $H$ is the group of all rationals having squarefree denominator, then
$$\chi_H(1) = (1, 1, 1, \ldots).  \quad ◄$$

**Lemma 10.7.** *Let $(k_2, k_3, \ldots, k_p, \ldots)$ be a sequence, where $k_p \in \mathbb{N} \cup \{\infty\}$ for all $p$. There exists a unique subgroup $G \subseteq \mathbb{Q}$ containing 1 such that*
$$\chi_G(1) = (k_2, k_3, \ldots, k_p, \ldots).$$

**Proof.** Define

$$G = \{a/c \in \mathbb{Q} : c = p_1^{e_1} \cdots p_n^{e_n} \text{ and } e_p \leq k_p \text{ whenever } k_p < \infty\}.$$

That $G$ is a subgroup follows from the elementary fact that if $c = p_1^{e_1} \cdots p_n^{e_n}$ and $d = p_1^{f_1} \cdots p_n^{f_n}$, then $1/c - 1/d = m/p_1^{\ell_1} \cdots p_n^{\ell_n}$, where $m \in \mathbb{Z}$ and $\ell_p \leq \max\{e_p, f_p\} \leq k_p$ for all $p$. It is clear that $\chi_G(1) = (k_2, k_3, \ldots, k_p, \ldots)$. To prove uniqueness of $G$, let $H$ be another such subgroup. Suppose $k_p < \infty$, and let $u/v \in \mathbb{Q}$, where $u/v$ is in lowest terms. If $v = p_1^{r_1} \cdots p_n^{r_n}$ and $r_p > k_p$, then $u/v \notin G$. But if $u/v \in H$, then $h_p(1) \geq r_p > k_p$, contradicting $\chi_H(1) = (k_2, k_3, \ldots, k_p, \ldots)$. Hence, $u/v \notin H$ if and only if $u/v \notin G$; that is, $G = H$. $\bullet$

Different elements in a torsion-free abelian group of rank 1 may have different height sequences. For example, if $G$ is the group of rationals having finite decimal expansions, then 1 and $\frac{63}{8}$ lie in $G$, and

$$\chi(1) = (\infty, 0, \infty, 0, 0, 0, \ldots) \quad \text{and} \quad \chi(\tfrac{63}{8}) = (\infty, 2, \infty, 1, 0, 0, \ldots).$$

Thus, these height sequences agree when $h_p = \infty$, but they disagree for some finite $p$-heights: $h_3(1) \neq h_3(\frac{63}{8})$ and $h_7(1) \neq h_7(\frac{63}{8})$.

**Definition.** Two height sequences $(h_2, h_3, \ldots, h_p, \ldots)$ and $(k_2, k_3, \ldots, k_p, \ldots)$ are **equivalent**, denoted by

$$(h_2, h_3, \ldots, h_p, \ldots) \sim (k_2, k_3, \ldots, k_p, \ldots),$$

if there are only finitely many $p$ for which $h_p \neq k_p$ and, for such primes $p$, neither $h_p$ nor $k_p$ is $\infty$.

It is easy to see that equivalence just defined is, in fact, an equivalence relation.

**Lemma 10.8.** *If $G$ is a torsion-free abelian group of rank 1 and $x, y \in G$ are nonzero, then their height sequences $\chi(x)$ and $\chi(y)$ are equivalent.*

**Proof.** By Lemma 10.1(iii), we may assume that $G \subseteq \mathbb{Q}$. If $b = p_1^{e_1} \cdots p_n^{e_n}$, then it is easy to see that $h_p(bx) = h_p(x)$ for all $p \notin \{p_1, \ldots, p_n\}$, while

$$h_{p_i}(bx) = e_i + h_{p_i}(x)$$

for $i = 1, \ldots, n$ (we agree that $e_i + \infty = \infty$). Hence, $\chi(x) \sim \chi(bx)$. Since $x, y \in G \subseteq \mathbb{Q}$, we have $x/y = a/b$ for integers $a, b$, so that $bx = ay$. Therefore, $\chi(x) \sim \chi(bx) = \chi(ay) \sim \chi(y)$. $\bullet$

**Definition.** The equivalence class of a height sequence is called a **type**. If $G$ is a torsion-free abelian group of rank 1, then its **type**, denoted by $\tau(G)$, is the type of a height sequence $\chi(x)$, where $x$ is a nonzero element of $G$.

Lemma 10.8 shows that $\tau(G)$ depends only on $G$ and not on the choice of nonzero element $x \in G$. We now solve the global problem for subgroups of $\mathbb{Q}$.

**Theorem 10.9.** *If $G$ and $H$ are torsion-free abelian groups of rank 1, then $G \cong H$ if and only if $\tau(G) = \tau(H)$.*

**Proof.** Let $\varphi\colon G \to H$ be an isomorphism. If $x \in G$ is nonzero, it is easy to see that $\chi(x) = \chi(\varphi(x))$, and so $\tau(G) = \tau(H)$.

For the converse, choose nonzero $x \in G$ and $y \in H$. By the definition of equivalence, there are primes $p_1, \ldots, p_n, q_1, \ldots, q_m$ with $h_{p_i}(x) < h_{p_i}(y) < \infty$, with $\infty > h_{q_j}(x) > h_{q_j}(y)$, and with $h_p(x) = h_p(y)$ for all other primes $p$. If we define $b = \prod p_i^{h_{p_i}(y) - h_{p_i}(x)}$, then $bx \in G$ and $h_{p_i}(bx) = \big(h_{p_i}(y) - h_{p_i}(x)\big) + h_{p_i}(x) = h_{p_i}(y)$. A similar construction, using $a = \prod q_j^{h_{q_j}(x) - h_{q_j}(y)}$, gives $\chi_G(bx) = \chi_H(ay)$. We have found elements $x' = bx \in G$ and $y' = ay \in H$ having the same height sequence.

Since $G$ has rank 1, there is an injection $f\colon G \to \mathbb{Q}$; write $A = \operatorname{im} f$, so that $A \cong G$. If $u = f(x')$, then $\chi_A(u) = \chi_G(x')$. If $g\colon \mathbb{Q} \to \mathbb{Q}$ is defined by $q \mapsto q/u$, then $g(A)$ is a subgroup of $\mathbb{Q}$ containing 1 with $\chi_{g(A)}(1) = \chi_A(u)$. Of course, $G \cong g(A)$. Similarly, there is a subgroup $C \subseteq \mathbb{Q}$ containing 1 with $\chi_C(1) = \chi_H(y') = \chi_G(x')$. By Lemma 10.7, $G \cong g(A) = C \cong H$.  •

We can also solve the global problem for subrings of $\mathbb{Q}$.

**Corollary 10.10.** *The following are equivalent for subrings $R$ and $S$ of $\mathbb{Q}$.*

(i) $R = S$.

(ii) $R \cong S$.

(iii) *$R$ and $S$ are locally isomorphic: $R_{(p)} \cong S_{(p)}$ for all primes $p$.*

**Proof.**

(i) $\Rightarrow$ (ii) This is obvious.

(ii) $\Rightarrow$ (iii) This, too, is obvious.

(iii) $\Rightarrow$ (i) If $1/b \in R$, where $b \in \mathbb{Z}$, then its prime factorization is $b = \pm p_1^{e_1} \cdots p_t^{e_t}$, where $e_i > 0$. Thus, $h_{p_i}(1) > 0$, and so $h_{p_i}(1) = \infty$. Every ring contains 1; here, its height is a sequence of 0's and $\infty$'s. Since $R_{(p)} \cong S_{(p)}$, the height of 1 in $R$ is the same as its height in $S$: $\chi_R(1) = \chi_S(1)$. Therefore $R = S$, by the uniqueness in Lemma 10.7.  •

The uniqueness theorems for groups and rings just proved are complemented by existence theorems.

**Corollary 10.11.** *Given any type $\tau$, there exists a torsion-free abelian group $G$ of rank 1, unique up to isomorphism, with $\tau(G) = \tau$. Hence, there are uncountably many nonisomorphic subgroups of $\mathbb{Q}$.*

**Proof.** The existence of $G$ follows at once from Lemma 10.7, while uniqueness up to isomorphism follows from Theorem 10.9. But there are uncountably many types; for example, two height sequences of 0s and $\infty$s are equivalent if and only if they are equal.  •

**Corollary 10.12.**

(i) *If $R$ is a subring of $\mathbb{Q}$, then the height sequence of 1 consists of 0's and $\infty$'s.*

(ii) *There are uncountably many nonisomorphic subrings of $\mathbb{Q}$. In fact, the additive groups of distinct subrings of $\mathbb{Q}$ are not isomorphic.*

**Proof.**

(i) If $h_p(1) > 0$, then $\frac{1}{p} \in R$. Since $R$ is a ring, $\left(\frac{1}{p}\right)^n = \frac{1}{p^n} \in R$ for all $n \geq 1$, and so $h_p(1) = \infty$.

(ii) If $R$ and $S$ are distinct subrings of $\mathbb{Q}$, then the height sequences of 1 are distinct, by part (i). Both statements follow from the observation that two height sequences whose only terms are 0 and $\infty$ are equivalent if and only if they are equal.   •

Kurosh classified torsion-free $\mathbb{Z}_{(p)}$-modules $G$ of finite rank with invariants $\mathrm{rank}(G) = n$, $p\text{-}\boldsymbol{rank}(G) = \dim_{\mathbb{F}_p}(\mathbb{F}_p \otimes G)$, and an equivalence class of $n \times n$ nonsingular matrices $M_p$ over the $p$-adic numbers $\mathbb{Q}_p$ (Fuchs, *Infinite Abelian Groups* II, pp. 154–158). This theorem was globalized by Derry; the invariants are rank, $p$-ranks for all primes $p$, and an equivalence class of matrix sequences $(M_p)$, where $M_p$ is an $n \times n$ nonsingular matrix over $\mathbb{Q}_p$. Alas, there is no normal form for the matrix sequences (nor for the matrices in the local case), and so this theorem does not have many applications.

An abelian group is ***decomposable*** if it is a direct sum of two proper subgroups; otherwise, it is ***indecomposable***. For example, every torsion-free abelian group of rank 1 is indecomposable. A torsion-free abelian group $G$ is called ***completely decomposable*** if it a direct sum of (possibly infinitely many) groups of rank 1. Baer proved (Ibid., p. 114) that every direct summand of a completely decomposable group is itself completely decomposable. Moreover, if $G = \bigoplus_{i \in I} A_i = \bigoplus_{i \in I} B_i$, where all $A_i$ and $B_i$ have rank 1, then, for each type $\tau$, the number of $A_i$ of type $\tau$ is equal to the number of $B_i$ of type $\tau$ (Ibid., p. 112). This nice behavior of decompositions does not hold more generally. Every torsion-free abelian group $G$ of finite rank is a direct sum of indecomposable groups (this is not true for infinite rank), but there is virtually no uniqueness for such a decomposition. For example, there exists a group $G$ of rank 6 with

$$G = A_1 \oplus A_2 = B_1 \oplus B_2 \oplus B_3,$$

with all the direct summands indecomposable, with $\mathrm{rank}(A_1) = 1$, $\mathrm{rank}(A_2) = 5$, and $\mathrm{rank}(B_j) = 2$ for $j = 1, 2, 3$ (Ibid., p. 135). Thus, the number of indecomposable direct summands in a decomposition is not uniquely determined by $G$, nor is the isomorphism class of any of its indecomposable direct summands. Here is an interesting theorem of Corner that can be used to produce bad examples of torsion-free groups such as the group $G$ of rank 6 just mentioned. Let $R$ be a ring whose additive group is countable, torsion-free, and reduced (it has no nonzero divisible subgroups). Then there exists an abelian group $G$, also countable, torsion-free, and reduced, with $\mathrm{End}(G) \cong R$. Moreover, if the additive group of $R$ has finite rank $n$, then $G$ can be chosen to have rank $2n$ (Ibid, p. 231). Jónsson introduced the notion of *quasi-isomorphism*: two torsion-free abelian groups $G$ and $H$ of finite rank are ***quasi-isomorphic***, denoted by $G \doteq H$, if each is isomorphic to a subgroup of finite index in the other, and he defined a group to be ***strongly indecomposable*** if it is not quasi-isomorphic to a decomposable group. He then proved that every torsion-free abelian group $G$ of finite rank is quasi-isomorphic to a direct sum of strongly indecomposable groups, and this decomposition is unique in the following sense: if $G \doteq A_1 \oplus \cdots \oplus A_n \doteq B_1 \oplus \cdots \oplus B_m$, then $n = m$ and, after possible reindexing, $A_i \doteq B_i$ for all $i$ (Ibid., p. 150). Torsion-free abelian groups still hide many secrets.

# Exercises

**10.1.** Prove that $\mathbb{Z}_{(p)} \not\cong \mathbb{Q}$ as $\mathbb{Z}_{(p)}$-modules.

**10.2.** Prove that the following statements are equivalent for a torsion-free abelian group $G$ of rank 1.

   (i) $G$ is finitely generated.

   (ii) $G$ is cyclic.

   (iii) If $x \in G$ is nonzero, then $h_p(x) \neq \infty$ for all primes $p$ and $h_p(x) = 0$ for almost all $p$.

   (iv) $\tau(G) = \tau(\mathbb{Z})$.

**10.3.**   (i) If $G$ is a torsion-free abelian group of rank 1, prove that the additive group of $\text{End}(G)$ is torsion-free of rank 1. Hint: Use Proposition 6.90(iii).

   (ii) Let $x \in G$ be nonzero with $\chi(x) = (h_2(x), h_3(x), \ldots, h_p(x), \ldots)$, and let $R$ be the subring of $\mathbb{Q}$ in which $\chi(1) = (k_2, k_3, \ldots, k_p, \ldots)$, where

$$k_p = \begin{cases} \infty & \text{if } h_p(x) = \infty, \\ 0 & \text{if } h_p(x) \text{ is finite.} \end{cases}$$

   Prove that $\text{End}(G) \cong R$. Prove that there are infinitely many $G$ with $\text{Aut}(G) \cong \mathbb{I}_2$.

   (iii) Prove that $G$ and $\text{End}(G)$ are locally isomorphic abelian groups.

* **10.4.**   (i) Prove that if $G$ and $H$ are torsion-free abelian groups of finite rank, then

$$\text{rank}(G \otimes_{\mathbb{Z}} H) = \text{rank}(G)\,\text{rank}(H).$$

   (ii) If $G$ and $H$ are torsion-free abelian groups of rank 1, then $G \otimes_{\mathbb{Z}} H$ is torsion-free of rank 1, by part (i). If $(h_p)$ is the height sequence of a nonzero element $x \in G$ and $(k_p)$ is the height sequence of a nonzero element $y \in H$, prove that the height sequence of $x \otimes y$ is $(m_p)$, where $m_p = h_p + k_p$ (we agree that $\infty + k_p = \infty$).

**10.5.** Let $G$ and $G'$ be nonzero subgroups of $\mathbb{Q}$, and let $\chi_G(g) = (k_p)$ and $\chi_{G'}(g') = (k'_p)$ for $g \in G$ and $g' \in G'$.

   (i) Let $T$ be the set of all types. Define $\tau \leq \tau'$, where $\tau, \tau' \in T$, if there are height sequences $(k_p) \in \tau$ and $(k'_p) \in \tau'$ with $k_p \leq k'_p$ for all primes $p$. Prove that $\leq$ is a well-defined relation which makes $T$ a partially ordered set.

   (ii) Prove that $G$ is isomorphic to a subgroup of $G'$ if and only if $\tau(G) \leq \tau(G')$.

   (iii) Note that $G \cap G'$ and $G + G'$ are subgroups of $\mathbb{Q}$. Prove that there are elements $x \in G \cap G'$ and $y \in G + G'$ with $\chi_{G \cap G'}(x) = (\min\{k_p, k'_p\})$ and $\chi_{G+G'}(y) = (\max\{k_p, k'_p\})$. Define $\tau(G) \wedge \tau(G')$ to be the type of $(\min\{k_p, k'_p\})$ and $\tau(G) \vee \tau(G')$ to be the type of $(\max\{k_p, k'_p\})$. Prove that these are well-defined operations which make $T$ a lattice.

(iv) Prove that there is an exact sequence $0 \to A \to G \oplus G' \to B \to 0$ with $A \cong G \cap G'$ and $B \cong G + G'$.

(v) Prove that $H = \operatorname{Hom}(G, G')$ is a torsion-free abelian group with $\operatorname{rank}(H) \leq 1$, and that $H \neq \{0\}$ if and only if $\tau(G) \leq \tau(G')$. Prove that $\operatorname{End}(G) = \operatorname{Hom}(G, G)$ has height sequence $(h_p)$, where $h_p = \infty$ if $k_p = \infty$ and $h_p = 0$ otherwise.

(vi) Prove that $G \otimes_{\mathbb{Z}} G'$ has rank 1, and that $\chi_{G \otimes_{\mathbb{Z}} G'}(g \otimes g') = (k_p + k'_p)$.

**10.6.** Let $A$ be a torsion-free abelian group. If $a \in A$ is nonzero, define $\tau(a) = \tau(\chi_A(a))$, and define
$$A(\tau) = \{a \in A : \tau(a) \geq \tau\} \cup \{0\}.$$
Prove that $A(\tau)$ is a fully invariant subgroup of $A$ and that $A/A(\tau)$ is torsion-free.

**10.7.** If $G$ is a $p$-primary abelian group, prove that $G$ is a $\mathbb{Z}_{(p)}$-module.

**10.8.** Let $\mathcal{C}$ be the category of all torsion-free abelian groups of finite rank.

(i) Prove that the Grothendieck group $G_0(\mathcal{C})$ is generated by all $[A]$ with $\operatorname{rank}(A) = 1$.

(ii) Prove that the Grothendieck group $G_0(\mathcal{C})$ is a commutative ring if we define multiplication by
$$[A][B] = [A \otimes_{\mathbb{Z}} B],$$
and prove that rank induces a ring homomorphism $G_0(\mathcal{C}) \to \mathbb{Z}$.
**Hint.** See the remark on page 861.

**Remark.** Rotman, The Grothendieck group of torsion-free abelian groups of finite rank, *Proc. London Math. Soc.* 13 (1963), 724–732, showed that $G_0(\mathcal{C})$ is a commutative ring isomorphic to the subring of $\mathbb{Z}^{\mathbb{N}}$ consisting of all functions having finite image.

Consider the Grothendieck group $G'_0(\mathcal{C}) = \mathcal{F}(\mathcal{C})/\mathcal{E}'$, where $\mathcal{F}(\mathcal{C})$ is the free abelian group with basis $\operatorname{Iso}(\mathcal{C})$ and $\mathcal{E}'$ is the subgroup of $\mathcal{F}(\mathcal{C})$ generated by all *split* short exact sequences (thus, the relations are $[A \oplus C] = [A] + [C]$). This Grothendieck group was studied by Lady, Nearly isomorphic torsion-free abelian groups, *J. Algebra* 35 (1975), 235–238: $G'_0(\mathcal{C})/tG'_0(\mathcal{C})$ is a free abelian group of uncountable rank. He also characterized $tG'_0(\mathcal{C})$ as all $[A] - [B]$ with $A$ and $B$ **nearly isomorphic**: for each $n > 0$, $B$ contains a subgroup $A_n$ with $A_n \cong A$ and finite index $[B : A_n]$ relatively prime to $n$. ◄

## Section 10.2. Localization

### All rings in this section are commutative.

The ring $\mathbb{Z}$ has infinitely many prime ideals, but $\mathbb{Z}_{(2)}$ has only one nonzero prime ideal, namely, $(2)$ [all odd primes in $\mathbb{Z}$ are invertible in $\mathbb{Z}_{(2)}$]. Now $\mathbb{Z}_{(2)}$-modules are much simpler than $\mathbb{Z}$-modules. For example, Proposition 10.4(iii) says that there are only two $\mathbb{Z}_{(2)}$-submodules of $\mathbb{Q}$ (up to isomorphism): $\mathbb{Z}_{(2)}$ and $\mathbb{Q}$. On the other hand, Corollary 10.11 says that there are uncountably many nonisomorphic $\mathbb{Z}$-submodules (= subgroups) of $\mathbb{Q}$. Similar observations lead to a localization-globalization strategy to attack algebraic and number-theoretic problems. The fundamental assumption underlying this strategy is that the local case is simpler than the global. Given a prime ideal $\mathfrak{p}$ in a commutative ring $R$, we will construct local rings $R_{\mathfrak{p}}$; localization looks at problems involving the rings $R_{\mathfrak{p}}$, while globalization uses all such local information to answer questions about $R$. We confess that this section is rather dry and formal.

**Definition.** Every ring $R$ is a monoid under multiplication. A subset $S \subseteq R$ of a commutative ring $R$ is ***multiplicative*** if $S$ is a submonoid not containing 0; that is, $0 \notin S$, $1 \in S$, and $s, s' \in S$ implies $ss' \in S$.

**Example 10.13.**

(i) If $\mathfrak{p}$ is a prime ideal in $R$, then its set-theoretic complement $S = R - \mathfrak{p}$ is multiplicative (if $a \notin \mathfrak{p}$ and $b \notin \mathfrak{p}$, then $ab \notin \mathfrak{p}$).

(ii) If $R$ is a domain, then the set $S = R^{\times}$ of all its nonzero elements is multiplicative [this is a special case of part (i), for (0) is a prime ideal in a domain].

(iii) If $a \in R$ is not nilpotent, then the set of its powers $S = \{a^n : n \geq 0\}$ is multiplicative. ◀

**Definition.** If $S \subseteq R$ is multiplicative, a ***localization*** of $R$ is an ordered pair $(S^{-1}R, h)$ which is a solution to the following universal mapping problem: given a commutative $R$-algebra $A$ and an $R$-algebra map $\varphi \colon R \to A$ with $\varphi(s)$ invertible in $A$ for all $s \in S$, there exists a unique $R$-algebra map $\widetilde{\varphi} \colon S^{-1}R \to A$ with $\widetilde{\varphi}h = \varphi$:

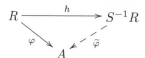

The map $h \colon R \to S^{-1}R$ is called the ***localization map***.

As is any solution to a universal mapping problem, a localization $S^{-1}R$ is unique up to isomorphism if it exists, and so we call $S^{-1}R$ *the* localization at $S$. The reason for excluding 0 from a multiplicative set is now apparent, for 0 is invertible only in the zero ring.

Given a multiplicative subset $S \subseteq R$, most authors construct the localization $S^{-1}R$ by generalizing the (tedious) construction of the fraction field of a domain $R$. They define a relation on $R \times S$ by $(r, s) \equiv (r', s')$ if there exists $s'' \in S$ with $s''(rs' - r's) = 0$ (when $R$ is a domain and $S = R^{\times}$ is the subset of its nonzero elements, this definition reduces to the usual definition of equality of fractions involving cross multiplication). After proving that $\equiv$ is an equivalence relation, $S^{-1}R$ is defined to be the set of all equivalence classes, addition and multiplication are defined and proved to be well-defined, all the $R$-algebra axioms are verified, and the elements of $S$ are shown to be invertible. We prefer to develop the existence and first properties of $S^{-1}R$ in another manner,[2] which is less tedious and which will show that the equivalence relation generalizing cross multiplication arises naturally.

**Theorem 10.14.** *If $S \subseteq R$ is multiplicative, then the localization $(S^{-1}R, h)$ exists.*

---

[2]I first saw this expounded in M.I.T. lecture notes of M. Artin.

**Proof.** Let $X = (x_s)_{s \in S}$ be an indexed set with $x_s \mapsto s$ a bijection $X \to S$, and let $R[X]$ be the polynomial ring over $R$ with indeterminates $X$. Define

$$S^{-1}R = R[X]/J,$$

where $J$ is the ideal generated by $\{sx_s - 1 : s \in S\}$, and define $h: R \to S^{-1}R$ by $h: r \mapsto r + J$, where $r$ is a constant polynomial. It is clear that $S^{-1}R$ is an $R$-algebra, that $h$ is an $R$-algebra map, and that each $h(s)$ is invertible. Assume now that $A$ is an $R$-algebra, and that $\varphi: R \to A$ is an $R$-algebra map with $\varphi(s)$ invertible for all $s \in S$. Consider the diagram in which the top arrow $\iota: R \to R[X]$ sends each $r \in R$ to the constant polynomial $r$, and $\nu: R[X] \to R[X]/J = S^{-1}R$ is the natural map:

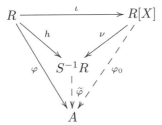

The top triangle commutes because both $h$ and $\nu\iota$ send $r \in R$ to $r + J$. Define an $R$-algebra map $\varphi_0: R[X] \to A$ by $\varphi_0(x_s) = \varphi(s)^{-1}$ for all $x_s \in X$. Clearly, $J \subseteq \ker \varphi_0$, for $\varphi_0(sx_s - 1) = 0$, and so there is an $R$-algebra map $\widetilde{\varphi}: S^{-1}R = R[X]/J \to A$ making the diagram commute. The map $\widetilde{\varphi}$ is the unique such map because $S^{-1}R$ is generated by $\operatorname{im} h \cup \{h(s)^{-1} : s \in S\}$ as an $R$-algebra. $\bullet$

We now describe the elements in $S^{-1}R$.

**Proposition 10.15.** *If $S \subseteq R$ is multiplicative, then each $y \in S^{-1}R$ has a (not necessarily unique) factorization $y = h(r)h(s)^{-1}$, where $h: R \to S^{-1}R$ is the localization map, $r \in R$, and $s \in S$.*

**Proof.** Define $A = \{y \in S^{-1}R : y = h(r)h(s)^{-1}, r \in R, s \in S\}$. It is easy to check that $A$ is an $R$-subalgebra of $S^{-1}R$ containing $\operatorname{im} h$. Since $\operatorname{im} h \subseteq A$, there is an $R$-algebra map $h': R \to A$ that is obtained from $h$ by changing its target. Consider the diagram

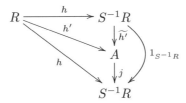

where $j: A \to S^{-1}R$ is the inclusion and $\widetilde{h'}: S^{-1}R \to A$ is given by universality (so the top triangle commutes). The lower triangle commutes, because $h(r) = h'(r)$ for all $r \in R$, and so the large triangle commutes: $(j\widetilde{h'})h = h$. But $1_{S^{-1}R}$ also makes this diagram commute, so that uniqueness gives $j\widetilde{h'} = 1_{S^{-1}R}$. By Set Theory, $j$ is surjective; that is, $S^{-1}R = A$. $\bullet$

In light of this proposition, the elements of $S^{-1}R$ can be regarded as "fractions" $h(r)h(s)^{-1}$, where $r \in R$ and $s \in S$.

**Notation.** Let $h\colon R \to S^{-1}R$ be the localization map. If $r \in R$ and $s \in S$, define

$$r/s = h(r)h(s)^{-1}.$$

In particular, $r/1 = h(r)$.

Is the localization map $h\colon r \mapsto r/1$ an injection?

**Proposition 10.16.** *If $S \subseteq R$ is multiplicative and $h\colon R \to S^{-1}R$ is the localization map, then*

$$\ker h = \{r \in R : sr = 0 \text{ for some } s \in S\}.$$

**Proof.** If $sr = 0$, then $0 = h(s)h(r)$ in $S^{-1}R$, so that $0 = h(s)^{-1}h(s)h(r) = h(r)$ $[h(s)$ is a unit]. Hence, $r \in \ker h$, and $\{r \in R : sr = 0 \text{ for some } s \in S\} \subseteq \ker h$.

For the reverse inclusion, suppose that $h(r) = 0$ in $S^{-1}R$. Since $S^{-1}R = R[X]/J$, where $J = (sx_s - 1 : s \in S)$, there is an equation $r = \sum_{i=1}^{n} f_i(X)(s_i x_{s_i} - 1)$ in $R[X]$ which involves only finitely many elements $\{s_1, \dots, s_n\} \subseteq S$; let $S_0$ be the submonoid of $S$ they generate (so $S_0$ is multiplicative). If $h_0\colon R \to S_0^{-1}R$ is the localization map, then $r \in \ker h_0$. In fact, if $s = s_1 \cdots s_n$ and $h'\colon R \to \langle s \rangle^{-1}R$ is the localization map [where $\langle s \rangle$ is the multiplicative set $\{s^n : n \geq 0\}$], then every $h'(s_i)$ is invertible, for $s_i^{-1} = s^{-1}s_1 \cdots \widehat{s_i} \cdots s_n$ (omit the factor $s_i$). Now $\langle s \rangle^{-1}R = R[x]/(sx - 1)$, so that $r \in \ker h'$ says that there is $f(x) = \sum_{i=0}^{m} a_i x^i \in R[x]$ with

$$r = f(x)(sx - 1) = \left(\sum_{i=0}^{m} a_i x^i\right)(sx - 1) = \sum_{i=0}^{m}(sa_i x^{i+1} - a_i x^i) \text{ in } R[x].$$

Expanding and equating coefficients of like powers of $x$ gives

$$r = -a_0, \quad sa_0 = a_1, \quad \dots, \quad sa_{m-1} = a_m, \quad sa_m = 0.$$

Hence, $sr = -sa_0 = -a_1$, and, by induction, $s^i r = -a_i$ for all $i$. In particular, $s^m r = -a_m$, and so $s^{m+1}r = -sa_m = 0$, as desired. $\quad \bullet$

$+$

When are two "fractions" $r/s$ and $r'/s'$ equal? The next corollary shows why the equivalence relation in the usual account of localization arises.

**Corollary 10.17.** *Let $S \subseteq R$ be multiplicative. If both $r/s, r'/s' \in S^{-1}R$, where $s, s' \in S$, then $r/s = r'/s'$ if and only if there exists $s'' \in S$ with $s''(rs' - r's) = 0$ in $R$.*

**Remark.** If $S$ contains no zero-divisors, then $s''(rs' - r's) = 0$ if and only if $rs' - r's = 0$, because $s''$ is a unit, and so $rs' = r's$. $\quad \blacktriangleleft$

**Proof.** If $r/s = r'/s'$, then multiplying by $ss'$ gives $(rs' - r's)/1 = 0$ in $S^{-1}R$. Hence, $rs' - r's \in \ker h$, and Proposition 10.16 gives $s'' \in S$ with $s''(rs' - r's) = 0$ in $R$.

Conversely, if $s''(rs' - r's) = 0$ in $R$ for some $s'' \in S$, then $h(s'')h(rs' - r's) = 0$ in $S^{-1}R$. As $h(s'')$ is a unit, we have $h(r)h(s') = h(r')h(s)$; as $h(s)$ and $h(s')$ are units, $h(r)h(s)^{-1} = h(r')h(s')^{-1}$; that is, $r/s = r'/s'$. $\bullet$

**Corollary 10.18.** *Let $S \subseteq R$ be multiplicative.*

(i) *If $S$ contains no zero-divisors, then the localization map $h \colon R \to S^{-1}R$ is an injection.*

(ii) *If $R$ is a domain with $Q = \mathrm{Frac}(R)$, then $S^{-1}R \subseteq Q$. Moreover, if $S = R^{\times}$, then $S^{-1}R = Q$.*

**Proof.**

(i) This follows easily from Proposition 10.16.

(ii) The localization map $h \colon R \to S^{-1}R$ is an injection, by Proposition 10.16. The result now follows from Proposition 10.15. $\bullet$

If $R$ is a domain and $S \subseteq R$ is multiplicative, then Corollary 10.18 says that $S^{-1}R$ consists of all elements $a/s \in \mathrm{Frac}(R)$ with $a \in R$ and $s \in S$.

Let us now investigate the ideals in $S^{-1}R$.

**Definition.** If $S \subseteq R$ is multiplicative and $I$ is an ideal in $R$, then we denote the ideal in $S^{-1}R$ generated by $h(I)$ by $S^{-1}I$.

**Example 10.19.**

(i) If $S \subseteq R$ is multiplicative and $I$ is an ideal in $R$ containing an element $s \in S$ (that is, $I \cap S \neq \varnothing$), then $S^{-1}I$ contains $s/s = 1$, and so $S^{-1}I = S^{-1}R$.

(ii) Let $S$ consist of all the odd integers [that is, $S$ is the complement of the prime ideal $(2)$], let $I = (3)$, and let $I' = (5)$. Then $S^{-1}I = S^{-1}\mathbb{Z} = S^{-1}I'$. Therefore, the function from the ideals in $\mathbb{Z}$ to the ideals in $S^{-1}\mathbb{Z} = \mathbb{Z}_{(2)} = \{a/b \in \mathbb{Q} : b \text{ is odd}\}$, given by $I \mapsto S^{-1}I$, is not injective. $\blacktriangleleft$

**Corollary 10.20.** *Let $S \subseteq R$ be multiplicative.*

(i) *Every ideal $J$ in $S^{-1}R$ is of the form $S^{-1}I$ for some ideal $I$ in $R$. In fact, if $R$ is a domain and $I = J \cap R$, then $J = S^{-1}I$; in the general case, if $I = h^{-1}(h(R) \cap J)$, then $J = S^{-1}I$.*

(ii) *Let $I$ be an ideal in $R$. Then $S^{-1}I = S^{-1}R$ if and only if $I \cap S \neq \varnothing$.*

(iii) *If $\mathfrak{q}$ is a prime ideal in $R$ with $\mathfrak{q} \cap S = \varnothing$, then $S^{-1}\mathfrak{q}$ is a prime ideal in $S^{-1}R$.*

(iv) *The function $f \colon \mathfrak{q} \mapsto S^{-1}\mathfrak{q}$ is a bijection from the family of all prime ideals in $R$ disjoint from $S$ to the family of all prime ideals in $S^{-1}R$.*

(v) *If $R$ is noetherian, then $S^{-1}R$ is also noetherian.*

**Proof.**

(i) Let $J = (j_\lambda : \lambda \in \Lambda)$. By Proposition 10.15, we have $j_\lambda = h(r_\lambda)h(s_\lambda)^{-1}$, where $r_\lambda \in R$ and $s_\lambda \in S$. Define $I$ to be the ideal in $R$ generated by

$\{r_\lambda : \lambda \in \Lambda\}$; that is, $I = h^{-1}(h(R) \cap J)$. It is clear that $S^{-1}I = J$; in fact, since all $s_\lambda$ are units in $S^{-1}R$, we have $J = (h(r_\lambda) : \lambda \in \Lambda)$.

(ii) If $s \in I \cap S$, then $s/1 \in S^{-1}I$. But $s/1$ is a unit in $S^{-1}R$, and so $S^{-1}I = S^{-1}R$. Conversely, if $S^{-1}I = S^{-1}R$, then $h(a)h(s)^{-1} = 1$ for some $a \in I$ and $s \in S$. Therefore, $s - a \in \ker h$, and so there is $s'' \in S$ with $s''(s - a) = 0$. Therefore, $s''s = s''a \in I$. Since $S$ is multiplicative, $s''s \in I \cap S$.

(iii) Suppose that $\mathfrak{q}$ is a prime ideal in $R$. First, $S^{-1}\mathfrak{q}$ is a proper ideal, for $\mathfrak{q} \cap S = \varnothing$. If $(a/s)(b/t) = c/u$, where $a, b \in R$, $c \in \mathfrak{q}$, and $s, t, u \in S$, then there is $s'' \in S$ with $s''(uab - stc) = 0$. Hence, $s''uab \in \mathfrak{q}$. Now $s''u \notin \mathfrak{q}$ (because $s''u \in S$ and $S \cap \mathfrak{q} = \varnothing$); hence, $ab \in \mathfrak{q}$ (because $\mathfrak{q}$ is prime). Thus, either $a$ or $b$ lies in $\mathfrak{q}$, and either $a/s$ or $b/t$ lies in $S^{-1}\mathfrak{q}$. Therefore, $S^{-1}\mathfrak{q}$ is a prime ideal.

(iv) Suppose that $\mathfrak{p}$ and $\mathfrak{q}$ are prime ideals in $R$ with $f(\mathfrak{p}) = S^{-1}\mathfrak{p} = S^{-1}\mathfrak{q} = f(\mathfrak{q})$; we may assume that $\mathfrak{p} \cap S = \varnothing = \mathfrak{q} \cap S$. If $a \in \mathfrak{p}$, then there is $b \in \mathfrak{q}$ and $s \in S$ with $a/1 = b/s$. Hence, $sa - b \in \ker h$, where $h$ is the localization map, and so there is $s' \in S$ with $s'sa = s'b \in \mathfrak{q}$. But $s's \in S$, so that $s's \notin \mathfrak{q}$. Since $\mathfrak{q}$ is prime, we have $a \in \mathfrak{q}$; that is, $\mathfrak{p} \subseteq \mathfrak{q}$. The reverse inclusion is proved similarly. Thus, $f$ is injective.

Let $\mathfrak{P}$ be a prime ideal in $S^{-1}R$. By (i), there is some ideal $I$ in $R$ with $\mathfrak{P} = S^{-1}I$. We must show that $I$ can be chosen to be a prime ideal in $R$. Now $h(R) \cap \mathfrak{P}$ is a prime ideal in $h(R)$, and so $\mathfrak{p} = h^{-1}(h(R) \cap \mathfrak{P})$ is a prime ideal in $R$. By (i), $\mathfrak{P} = S^{-1}\mathfrak{p}$, and so $f$ is surjective.

(v) If $J$ is an ideal in $S^{-1}R$, then (i) shows that $J = S^{-1}I$ for some ideal $I$ in $R$. Since $R$ is noetherian, we have $I = (r_1, \ldots, r_n)$, and so $J = (r_1/1, \ldots, r_n/1)$. Hence, every ideal in $S^{-1}R$ is finitely generated, and so $S^{-1}R$ is noetherian. $\bullet$

**Notation.** If $\mathfrak{p}$ is a prime ideal in a commutative ring $R$ and $S = R - \mathfrak{p}$, then $S^{-1}R$ is denoted by

$$R_\mathfrak{p}.$$

**Example 10.21.** If $p$ is a nonzero prime in $\mathbb{Z}$, then $\mathfrak{p} = (p)$ is a prime ideal, and $\mathbb{Z}_\mathfrak{p} = \mathbb{Z}_{(p)}$. ◄

**Proposition 10.22.** *If $R$ is a domain, then $\bigcap_\mathfrak{m} R_\mathfrak{m} = R$, where the intersection is over all the maximal ideals $\mathfrak{m}$ in $R$.*

**Proof.** Since $R$ is a domain, $R_\mathfrak{m} \subseteq \operatorname{Frac}(R)$ for all $\mathfrak{m}$, and so the intersection in the statement is defined. Moreover, it is plain that $R \subseteq R_\mathfrak{m}$ for all $\mathfrak{m}$, so that $R \subseteq \bigcap_m R_\mathfrak{m}$. For the reverse inclusion, let $a \in \bigcap R_\mathfrak{m}$. Consider the colon ideal

$$I = (R : a) = \{r \in R : ra \in R\}.$$

If $I = R$, then $1 \in I$, and $a = 1a \in R$, as desired. If $I$ is a proper ideal, then there exists a maximal ideal $\mathfrak{m}$ with $I \subseteq \mathfrak{m}$. Now $a/1 \in R_\mathfrak{m}$, so there is $r \in R$ and $\sigma \notin \mathfrak{m}$ with $a/1 = r/\sigma$; that is, $\sigma a = r \in R$. Hence, $\sigma \in I \subseteq \mathfrak{m}$, contradicting $\sigma \notin \mathfrak{m}$. Therefore, $R = \bigcap_\mathfrak{m} R_\mathfrak{m}$. $\bullet$

The next proposition explains why $S^{-1}R$ is called localization.

**Theorem 10.23.** *If $\mathfrak{p}$ is a prime ideal in a commutative ring $R$, then $R_\mathfrak{p}$ is a local ring with unique maximal ideal $\mathfrak{p}R_\mathfrak{p} = \{r/s : r \in \mathfrak{p} \text{ and } s \notin \mathfrak{p}\}$.*

**Proof.** If $x \in R_\mathfrak{p}$, then $x = r/s$, where $r \in R$ and $s \notin \mathfrak{p}$. If $r \notin \mathfrak{p}$, then $r/s$ is a unit in $R_\mathfrak{p}$; that is, all nonunits lie in $\mathfrak{p}R_\mathfrak{p}$. Hence, if $I$ is any ideal in $R_\mathfrak{p}$ that contains an element $r/s$ with $r \notin \mathfrak{p}$, then $I = R_\mathfrak{p}$. It follows that every proper ideal in $R_\mathfrak{p}$ is contained in $\mathfrak{p}R_\mathfrak{p}$, and so $R_\mathfrak{p}$ is a local ring with unique maximal ideal $\mathfrak{p}R_\mathfrak{p}$. •

Here is an application of localization. Recall that a prime ideal $\mathfrak{p}$ in a commutative ring $R$ is a ***minimal prime ideal*** if there is no prime ideal strictly contained in it. In a domain, $(0)$ is a minimal prime ideal, and it is the unique such.

**Proposition 10.24.** *If $\mathfrak{p}$ is a minimal prime ideal in a commutative ring $R$, then every $x \in \mathfrak{p}$ is nilpotent; that is, $x^n = 0$ for some $n = n(x) \geq 1$.*

**Proof.** Let $x \in \mathfrak{p}$ be nonzero. By Corollary 10.20(iv), there is only one prime ideal in $R_\mathfrak{p}$, namely, $\mathfrak{p}R\mathfrak{p}$, and $x/1$ is a nonzero element in it. Indeed, $x$ is nilpotent if and only if $x/1$ is nilpotent, by Proposition 10.16. Thus, we have normalized the problem; we may now assume that $x \in \mathfrak{p}$ and that $\mathfrak{p}$ is the only prime ideal in $R$. If $x$ is not nilpotent, then $S = \{1, x, x^2, \dots\}$ is multiplicative. By Exercise 5.39 on page 321, there exists an ideal $I$ maximal with $I \cap S = \varnothing$; by Exercise 5.7 on page 301, $I$ is a prime ideal; that is, $\mathfrak{p} = I$. But $x \in S \cap \mathfrak{p} = S \cap I = \varnothing$, a contradiction. Therefore, $x$ is nilpotent. •

The structure of projective modules over a general ring can be quite complicated, but the next proposition shows that projective modules over local rings are free.

**Lemma 10.25.** *Let $R$ be a local ring with maximal ideal $\mathfrak{m}$. An element $r \in R$ is a unit if and only if $r \notin \mathfrak{m}$.*

**Proof.** It is clear that if $r$ is a unit, then $r \notin \mathfrak{m}$, for $\mathfrak{m}$ is a proper ideal. Conversely, assume that $r$ is not a unit. By Zorn's Lemma, there is a maximal ideal $\mathfrak{m}'$ containing the principal ideal $(r)$. Since $R$ is local, $\mathfrak{m}$ is the only maximal ideal; hence, $\mathfrak{m}' = \mathfrak{m}$ and $r \in \mathfrak{m}$. •

**Proposition 10.26.** *If $R$ is a local ring, then every finitely generated[3] projective $R$-module $B$ is free.*

**Proof.** Let $R$ be a local ring with maximal ideal $\mathfrak{m}$, and let $\{b_1, \dots, b_n\}$ be a *smallest set of generators* of $B$ in the sense that $B$ cannot be generated by fewer than $n$ elements. Let $F$ be the free $R$-module with basis $x_1, \dots, x_n$, and define $\varphi \colon F \to B$ by $\varphi(x_i) = b_i$ for all $i$. Thus, there is an exact sequence

$$(1) \qquad\qquad 0 \to K \to F \xrightarrow{\varphi} B \to 0,$$

where $K = \ker \varphi$.

---

[3]It is a theorem of Kaplansky [Projective modules, *Annals Math.* 68 (1958), 372–377] that the finiteness hypothesis can be omitted: every projective module over a local ring is free. He even proves freeness when $R$ is a noncommutative local ring.

We claim that $K \subseteq \mathfrak{m}F$. If, on the contrary, $K \subsetneq \mathfrak{m}F$, there is an element $y = \sum_{i=1}^n r_i x_i \in K$ which is not in $\mathfrak{m}F$; that is, some coefficient, say, $r_1 \notin \mathfrak{m}$. Lemma 10.25 says that $r_1$ is a unit. Now $y \in K = \ker \varphi$ gives $\sum r_i b_i = 0$. Hence, $b_1 = -r_1^{-1}\left(\sum_{i=2}^n r_i b_i\right)$, which implies that $B = \langle b_2, \ldots, b_n \rangle$, contradicting the original generating set being smallest.

Returning to the exact sequence (1), projectivity of $B$ gives $F = K \oplus B'$, where $B'$ is a submodule of $F$ with $B' \cong B$. Hence, $\mathfrak{m}F = \mathfrak{m}K \oplus \mathfrak{m}B'$. Since $\mathfrak{m}K \subseteq K \subseteq \mathfrak{m}F$, Corollary 6.29 gives

$$K = \mathfrak{m}K \oplus (K \cap \mathfrak{m}B').$$

But $K \cap \mathfrak{m}B' \subseteq K \cap B' = \{0\}$, so that $K = \mathfrak{m}K$. The submodule $K$ is finitely generated, being a direct summand (and hence a homomorphic image) of the finitely generated module $F$, so that Nakayama's Lemma (Corollary 7.16) gives $K = \{0\}$. Therefore, $\varphi$ is an isomorphism and $B$ is free.  •

Having localized a commutative ring, we now localize its modules. Recall a general construction we introduced in Exercise 6.56 on page 449. Given a ring homomorphism $\varphi \colon R \to R^*$, every $R^*$-module $A^*$ can be viewed as an $R$-module: if $a^* \in A^*$ and $r \in R$, define

$$ra^* = \varphi(r)a^*$$

and denote $A^*$ viewed as an $R$-module in this way by $_\varphi A^*$. Indeed, $\varphi$ induces a *change of rings functor*; namely, $_\varphi \square \colon {}_{R^*}\mathbf{Mod} \to {}_R\mathbf{Mod}$, which is additive and exact. In particular, the localization map $h \colon R \to S^{-1}R$ allows us to view an $S^{-1}R$-module $M$ as an *induced $R$-module*

$$_h M,$$

where $rm = h(r)m = (r/1)m$ for all $r \in R$ and $m \in M$.

If $M$ is an $R$-module and $s \in R$, let $\mu_s$ denote the multiplication map $M \to M$ defined by $m \mapsto sm$. Given a subset $S \subseteq R$, note that the map $\mu_s \colon M \to M$ is invertible for every $s \in S$ (that is, every $\mu_s$ is an automorphism) if and only if $M$ is an $S^{-1}R$-module.

**Definition.** Let $M$ be an $R$-module and let $S \subseteq R$ be multiplicative. A *localization* of $M$ is an ordered pair $(S^{-1}M, h_M)$, where $S^{-1}M$ is an $S^{-1}R$-module and $h_M \colon M \to S^{-1}M$ is an $R$-map (called the *localization map*), which is a solution to the universal mapping problem: if $M'$ is an $S^{-1}R$-module and $\varphi \colon M \to M'$ is an $R$-map, then there exists a unique $S^{-1}R$-map $\widetilde{\varphi} \colon S^{-1}M \to M'$ with $\widetilde{\varphi}h_M = \varphi$.

The obvious candidate for $(S^{-1}M, h_M)$, namely, $(S^{-1}R \otimes_R M, h \otimes 1_M)$, where $h \colon R \to S^{-1}R$ is the localization map, actually is the localization.

**Proposition 10.27.** *Let $R$ be a commutative ring, and let $S \subseteq R$ be multiplicative.*

(i) *If $M$ is an $S^{-1}R$-module, then $M$ is naturally isomorphic to $S^{-1}R \otimes_{R\ h}M$ via $m \mapsto 1 \otimes m$, where $_hM$ is the $R$-module induced from $M$.*

(ii) *If $M$ is an $R$-module, then $(S^{-1}R \otimes_R M, h_M)$ is a localization of $M$, where $h_M \colon M \to S^{-1}R \otimes_R M$ is given by $m \mapsto 1 \otimes m$.*

**Proof.**

(i) As above, there is an $R$-module $_hM$ with $rm = (r/1)m$ for $r \in R$ and $m \in M$. Define $g \colon M \to S^{-1}R \otimes_{R\ h}M$ by $m \mapsto 1 \otimes m$. Now $g$ is an $S^{-1}R$-map: if $s \in S$ and $m \in M$, then

$$g(s^{-1}m) = 1 \otimes s^{-1}m = s^{-1}s \otimes s^{-1}m = s^{-1} \otimes m = s^{-1}g(m).$$

To see that $g$ is an isomorphism, we construct its inverse. Since $M$ is an $S^{-1}R$-module, the function $S^{-1}R \times {}_hM \to M$, defined by $(rs^{-1}, m) \mapsto (rs^{-1})m$, is an $R$-bilinear function, and it induces an $R$-map $S^{-1}R \otimes_R ({}_hM) \to M$, which is obviously inverse to $g$. Proof of naturality of $g$ is left to the reader.

(ii) Consider the diagram

$$
\begin{array}{ccc}
M & \xrightarrow{\ h_M\ } & S^{-1}R \otimes_{R\ h}M \\
{\scriptstyle\varphi}\searrow & & \nearrow{\scriptstyle\tilde{\varphi}} \\
& M' &
\end{array}
$$

where $M$ is an $R$-module, $h_M \colon m \mapsto 1 \otimes m$, $M'$ is an $S^{-1}R$-module, and $\varphi \colon M \to M'$ is an $R$-map. Since $\varphi$ is merely an $R$-map, we may regard it as a map $\varphi \colon M \to {}_hM'$. By (i), there is an isomorphism $g \colon M' \to S^{-1}R \otimes_{R\ h}M'$. Define an $R$-map $\tilde{\varphi} \colon S^{-1}R \otimes_R M \to M'$ by $g^{-1}(1 \otimes \varphi)$. $\quad \bullet$

One of the most important properties of $S^{-1}R$ is that it is flat as an $R$-module. To prove this, we first generalize the argument in Proposition 10.16.

**Proposition 10.28.** *Let $S \subseteq R$ be multiplicative. If $M$ is an $R$-module and $h_M \colon M \to S^{-1}M$ is the localization map, then*

$$\ker h_M = \{m \in M : sm = 0 \text{ for some } s \in S\}.$$

**Proof.** Denote $\{m \in M : sm = 0 \text{ for some } s \in S\}$ by $K$. If $sm = 0$, for $m \in M$ and $s \in S$, then $h_M(m) = (1/s)h_M(sm) = 0$, and so $K \subseteq \ker h_M$. For the reverse inclusion, proceed as in Proposition 10.16: if $m \in K$, there is $s \in S$ with $sm = 0$. Reduce to $S = \langle s \rangle$ for some $s \in S$, where $\langle s \rangle = \{s^n : n \geq 0\}$, so that $S^{-1}R = R[x]/(sx - 1)$. Now $R[x] \otimes_R M \cong \bigoplus_i Rx^i \otimes_R M$, because $R[x]$ is the free $R$-module with basis $\{1, x, x^2, \dots\}$. Hence, each element in $R[x] \otimes_R M$ has a unique expression of the form $\sum_i x^i \otimes m_i$, where $m_i \in M$. Hence,

$$\ker h_M = \Big\{m \in M : 1 \otimes m = (sx - 1)\sum_{i=0}^{n} x^i \otimes m_i\Big\}.$$

The proof now finishes as the proof of Proposition 10.16. Expanding and equating coefficients gives equations

$$1 \otimes m = -1 \otimes m_0, \quad x \otimes sm_0 = x \otimes m_1, \ \ldots,$$

$$x^n \otimes sm_{n-1} = x^n \otimes m_n, \quad x^{n+1} \otimes sm_n = 0.$$

It follows that

$$m = -m_0, \quad sm_0 = m_1, \quad \ldots, \quad sm_{n-1} = m_n, \quad sm_n = 0.$$

Hence, $sm = -sm_0 = -m_1$, and, by induction, $s^i m = -m_i$ for all $i$. In particular, $s^n m = -m_n$ and so $s^{n+1} m = -sm_n = 0$ in $M$. Therefore, $\ker h_M \subseteq K$. •

We now generalize Proposition 10.16 from rings to modules.

**Corollary 10.29.** *Let $S \subseteq R$ be multiplicative and let $M$ be an $R$-module.*

(i) *Every element $u \in S^{-1}M = S^{-1} \otimes_R M$ has the form $u = s^{-1} \otimes m$ for some $s \in S$ and some $m \in M$.*

(ii) *$s_1^{-1} \otimes m_1 = s_2^{-1} \otimes m_2$ in $S^{-1} \otimes_R M$ if and only if $s(s_2 m_1 - s_1 m_2) = 0$ in $M$ for some $s \in S$.*

**Proof.**

(i) If $u \in S^{-1} \otimes_R M$, then $u = \sum_i (r_i/s_i) \otimes m_i$, where $r_i \in R$, $s_i \in S$, and $m_i \in M$. If we define $s = \prod s_i$ and $\hat{s}_i = \prod_{j \neq i} s_j$, then

$$u = \sum (1/s_i) r_i \otimes m_i = \sum (\hat{s}_i/s) r_i \otimes m_i$$

$$= (1/s) \sum \hat{s}_i r_i \otimes m_i = (1/s) \otimes \sum \hat{s}_i r_i m_i = (1/s) \otimes m,$$

where $m = \sum \hat{s}_i r_i m_i \in M$.

(ii) If $s \in S$ with $s(s_2 m_1 - s_1 m_2) = 0$ in $M$, then $(s/1)(s_2 \otimes m_1 - s_1 \otimes m_2) = 0$ in $S^{-1}R \otimes_R M$. As $s/1$ is a unit, $s_2 \otimes m_1 - s_1 \otimes m_2 = 0$, and so $s_1^{-1} \otimes m_1 = s_2^{-1} \otimes m_2$.

Conversely, suppose that $s_1^{-1} \otimes m_1 = s_2^{-1} \otimes m_2$ in $S^{-1} \otimes_R M$; then we have $(1/s_1 s_2)(s_2 \otimes m_1 - s_1 \otimes m_2) = 0$. Since $1/s_1 s_2$ is a unit, we have $(s_2 \otimes m_1 - s_1 \otimes m_2) = 0$ and $s_2 m_1 - s_1 m_2 \in \ker h_M$. By Proposition 10.28, there exists $s \in S$ with $s(s_2 m_1 - s_1 m_2) = 0$ in $M$. •

**Theorem 10.30.** *If $S \subseteq R$ is multiplicative, then $S^{-1}R$ is a flat $R$-module.*

**Proof.** We must show that if $0 \to A \xrightarrow{f} B$ is exact, then so is

$$0 \to S^{-1}R \otimes_R A \xrightarrow{1 \otimes f} S^{-1}R \otimes_R B.$$

By Corollary 10.29, every $u \in S^{-1}A$ has the form $u = s^{-1} \otimes a$ for some $s \in S$ and $a \in A$. In particular, if $u \in \ker(1 \otimes f)$, then $(1 \otimes f)(u) = s^{-1} \otimes f(a) = 0$. Multiplying by $s$ gives $1 \otimes f(a) = 0$ in $S^{-1}B$; that is, $f(a) \in \ker h_B$. By Proposition 10.28, there is $t \in S$ with $0 = tf(a) = f(ta)$. Since $f$ is an injection, $ta \in \ker f = \{0\}$. Hence, $0 = 1 \otimes ta = t(1 \otimes a)$. But $t$ is a unit in $S^{-1}R$, so that $1 \otimes a = 0$ in $S^{-1}A$. Therefore, $1 \otimes f$ is an injection, and $S^{-1}R$ is a flat $R$-module. •

**Corollary 10.31.** *If $S \subseteq R$ is multiplicative, then localization $M \mapsto S^{-1}M = S^{-1}R \otimes_R M$ defines an exact functor $_R\mathbf{Mod} \to {}_{S^{-1}R}\mathbf{Mod}$.*

**Proof.** Localization is the functor $S^{-1}R \otimes_R \square$, and it is exact because $S^{-1}R$ is a flat $R$-module. $\bullet$

Since tensor product commutes with direct sum, it is clear that if $M$ is a free (or projective) $R$-module, then $S^{-1}M$ is a free (or projective) $S^{-1}R$-module.

**Proposition 10.32.** *Let $S \subseteq R$ be multiplicative. If $A$ and $B$ are $R$-modules, then there is a natural isomorphism*

$$\varphi \colon S^{-1}(B \otimes_R A) \to S^{-1}B \otimes_{S^{-1}R} S^{-1}A.$$

**Proof.** Every element $u \in S^{-1}(B \otimes_R A)$ has the form $u = s^{-1}m$ for $s \in S$ and $m \in B \otimes_R A$, by Corollary 10.29; hence, $u = s^{-1} \sum a_i \otimes b_i$:

$$
\begin{array}{ccccc}
A \times B & \longrightarrow & A \otimes_R B & \longrightarrow & S^{-1}(A \otimes_R B) \\
& \searrow & \downarrow & \swarrow & \\
& & S^{-1}B \otimes_{S^{-1}R} S^{-1}A & &
\end{array}
$$

The idea is to define $\varphi(u) = \sum s^{-1}a_i \otimes b_i$ but, as usual with tensor product, the problem is whether obvious maps are well-defined. We suggest that the reader use the universal property of localization to complete the proof. $\bullet$

**Proposition 10.33.** *Let $S \subseteq R$ be multiplicative. If $B$ is a flat $R$-module, then $S^{-1}B$ is a flat $S^{-1}R$-module.*

**Proof.** Recall Proposition 10.27(i): if $A$ is an $S^{-1}R$-module, then $A$ is naturally isomorphic to $S^{-1}R \otimes_R {}_hA$ . The isomorphism of Proposition 10.32,

$$S^{-1}B \otimes_{S^{-1}R} A = (S^{-1}R \otimes_R B) \otimes_{S^{-1}R} A \to S^{-1}R \otimes_R (B \otimes_R A),$$

can now be used to give a natural isomorphism

$$S^{-1}B \otimes_{S^{-1}R} \square \to (S^{-1}B \otimes_R \square)(B \otimes_R \square)_h\square,$$

where $_h\square \colon {}_{S^{-1}R}\mathbf{Mod} \to {}_R\mathbf{Mod}$ is the change of rings functor induced by the localization map $h \colon R \to S^{-1}R$. As each factor is an exact functor, by Exercise 6.56 on page 449, so is the composite $S^{-1}B \otimes_{S^{-1}R} \square$; that is, $S^{-1}B$ is flat. $\bullet$

We now investigate localization of injective modules. Preserving injectivity is more subtle than preserving projectives and flats. We will need an analog for Hom of Proposition 10.32, but the next example shows that the obvious analog is not true without some extra hypothesis.

**Example 10.34.** If the analog of Proposition 10.32 for Hom were true, then we would have $S^{-1}\operatorname{Hom}_R(B, A) \cong \operatorname{Hom}_{S^{-1}R}(S^{-1}B, S^{-1}A)$, but such an isomorphism may not exist. If $R = \mathbb{Z}$ and $S^{-1}R = \mathbb{Q}$, then

$$\mathbb{Q} \otimes_{\mathbb{Z}} \operatorname{Hom}_{\mathbb{Z}}(\mathbb{Q}, \mathbb{Z}) \ncong \operatorname{Hom}_{\mathbb{Q}}(\mathbb{Q} \otimes_{\mathbb{Z}} \mathbb{Q}, \mathbb{Q} \otimes_{\mathbb{Z}} \mathbb{Z}).$$

The left-hand side is $\{0\}$, because $\operatorname{Hom}_{\mathbb{Z}}(\mathbb{Q}, \mathbb{Z}) = \{0\}$. On the other hand, the right-hand side is $\operatorname{Hom}_{\mathbb{Q}}(\mathbb{Q}, \mathbb{Q}) \cong \mathbb{Q}$. $\blacktriangleleft$

**Lemma 10.35.** *Let $S \subseteq R$ be multiplicative, and let $M$ and $A$ be $R$-modules with $A$ finitely presented. Then there is a natural isomorphism*

$$\tau_A \colon S^{-1} \operatorname{Hom}_R(A, M) \to \operatorname{Hom}_{S^{-1}R}(S^{-1}A, S^{-1}M).$$

**Proof.** It suffices to construct natural isomorphisms

$$\varphi_A \colon S^{-1} \operatorname{Hom}_R(A, M) \to \operatorname{Hom}_R(A, S^{-1}M)$$

and

$$\theta_A \colon \operatorname{Hom}_R(A, S^{-1}M) \to \operatorname{Hom}_{S^{-1}R}(S^{-1}A, S^{-1}M),$$

for then we can define $\tau_A = \theta_A \varphi_A$.

If, now, $A$ is a finitely presented $R$-module, then there is an exact sequence

$$(2) \qquad\qquad\qquad R^t \to R^n \to A \to 0.$$

Applying the contravariant functors $\operatorname{Hom}_R(\square, M')$ and $\operatorname{Hom}_{S^{-1}R}(\square, M')$, where $M' = S^{-1}M$ is first viewed as an $R$-module, gives a commutative diagram with exact rows

$$
\begin{array}{ccccccc}
0 \longrightarrow & \operatorname{Hom}_R(A, M') & \longrightarrow & \operatorname{Hom}_R(R^n, M') & \longrightarrow & \operatorname{Hom}_R(R^t, M') \\
& \theta_A \Big\downarrow & & \theta_{R^n} \Big\downarrow & & \Big\downarrow \theta_{R^t} \\
0 \to & \operatorname{Hom}_{S^{-1}R}(S^{-1}A, M') & \to & \operatorname{Hom}_{S^{-1}R}((S^{-1}R)^n, M') & \to & \operatorname{Hom}_{S^{-1}R}((S^{-1}R)^t, M')
\end{array}
$$

Since the vertical maps $\theta_{R^n}$ and $\theta_{R^t}$ are isomorphisms, there is a dotted arrow $\theta_A$ which must be an isomorphism, by Proposition 6.117. If $\beta \in \operatorname{Hom}_R(A, M)$, then the reader may check that

$$\theta_A(\beta) = \widetilde{\beta} \colon a/s \mapsto \beta(a)/s,$$

from which it follows that the isomorphisms $\theta_A$ are natural.

Construct $\varphi_A \colon S^{-1} \operatorname{Hom}_R(A, M) \to \operatorname{Hom}_R(A, S^{-1}M)$ by defining $\varphi_A \colon g/s \mapsto g_s$, where $g_s(a) = g(a)/s$. Note that $\varphi_A$ is well-defined, for it arises from the $R$-bilinear function $S^{-1}R \times \operatorname{Hom}_R(A, M) \to \operatorname{Hom}_R(A, S^{-1}M)$ given by $(r/s, g) \mapsto rg_s$ [remember that $S^{-1} \operatorname{Hom}_R(A, M) = S^{-1}R \otimes_R \operatorname{Hom}_R(A, M)$, by Proposition 10.27]. Observe that $\varphi_A$ is an isomorphism when $A$ is finitely generated free, and consider the commutative diagram

$$
\begin{array}{ccccccc}
0 \longrightarrow & S^{-1} \operatorname{Hom}_R(A, M) & \longrightarrow & S^{-1} \operatorname{Hom}_R(R^n, M) & \longrightarrow & S^{-1} \operatorname{Hom}_R(R^t, M) \\
& \varphi_A \Big\downarrow & & \varphi_{R^n} \Big\downarrow & & \Big\downarrow \varphi_{R^t} \\
0 \longrightarrow & \operatorname{Hom}_R(A, S^{-1}M) & \longrightarrow & \operatorname{Hom}_R(R^n, S^{-1}M) & \longrightarrow & \operatorname{Hom}_R(R^t, S^{-1}M)
\end{array}
$$

The top row is exact, for it arises from Equation (2) by first applying the left exact contravariant functor $\operatorname{Hom}_R(\square, M)$, and then applying the exact localization functor. The bottom row is exact, for it arises from Equation (2) by applying the left exact contravariant functor $\operatorname{Hom}_R(\square, S^{-1}M)$. The Five Lemma shows that $\varphi_A$ is an isomorphism.

Assume first that $A = R^n$ is a finitely generated free $R$-module. If $a_1, \ldots, a_n$ is a basis of $A$, then $a_1/1, \ldots, a_n/1$ is a basis of $S^{-1}A = S^{-1}R \otimes_R R^n$. The map

$$\theta_{R^n} \colon \operatorname{Hom}_R(A, S^{-1}M) \to \operatorname{Hom}_{S^{-1}R}(S^{-1}A, S^{-1}M),$$

given by $f \mapsto \widetilde{f}$, where $\widetilde{f}(a_i/s) = f(a_i)/s$, is easily seen to be a well-defined $R$-isomorphism. •

**Theorem 10.36.** *Let $R$ be noetherian and let $S \subseteq R$ be multiplicative. If $E$ is an injective $R$-module, then $S^{-1}E$ is an injective $S^{-1}R$-module.*

**Proof.** By Baer's Criterion, it suffices to extend any map $I \to S^{-1}E$ to a map $S^{-1}R \to S^{-1}E$, where $I$ is an ideal in $S^{-1}R$; that is, if $i \colon I \to S^{-1}R$ is the inclusion, then the induced map

$$i^* \colon \operatorname{Hom}_{S^{-1}R}(S^{-1}R, S^{-1}E) \to \operatorname{Hom}_{S^{-1}R}(I, S^{-1}E)$$

is a surjection. Now $S^{-1}R$ is noetherian because $R$ is, and so $I$ is finitely generated; say, $I = (r_1/s_1, \ldots, r_n/s_n)$, where $r_i \in R$ and $s_i \in S$. There is an ideal $J$ in $R$, namely, $J = (r_1, \ldots, r_n)$, with $I = S^{-1}J$. Naturality of the isomorphism in Lemma 10.35 gives a commutative diagram

$$\begin{array}{ccc} S^{-1}\operatorname{Hom}_R(R, E) & \longrightarrow & S^{-1}\operatorname{Hom}_R(J, E) \\ \downarrow & & \downarrow \\ \operatorname{Hom}_{S^{-1}R}(S^{-1}R, S^{-1}E) & \longrightarrow & \operatorname{Hom}_{S^{-1}R}(S^{-1}J, S^{-1}E) \end{array}$$

Now $\operatorname{Hom}_R(R, E) \to \operatorname{Hom}_R(J, E)$ is a surjection, because $E$ is an injective $R$-module, and so $S^{-1} = S^{-1}R \otimes_R \square$ being right exact implies that the top arrow is also a surjection. But the vertical maps are isomorphisms, and so the bottom arrow is a surjection; that is, $S^{-1}E$ is an injective $S^{-1}R$-module. •

**Remark.** Theorem 10.36 may be false if $R$ is not noetherian. Dade, Localization of injective modules, *J. Algebra* 69 (1981), 415–425, showed, for every commutative ring $k$, that if $R = k[X]$, where $X$ is an uncountable set of indeterminates, then there is a multiplicative $S \subseteq R$ and an injective $R$-module $E$ such that $S^{-1}E$ is not an injective $S^{-1}R$-module. If, however, $R = k[X]$, where $k$ is noetherian and $X$ is countable, then $S^{-1}E$ is an injective $S^{-1}R$-module for every injective $R$-module $E$ and every multiplicative $S \subseteq R$. ◄

Here are some globalization tools.

**Notation.** In the special case $S = R - \mathfrak{p}$, where $\mathfrak{p}$ is a prime ideal in $R$, we write

$$S^{-1}M = S^{-1}R \otimes_R M = R_{\mathfrak{p}} \otimes_R M = M_{\mathfrak{p}}.$$

If $f \colon M \to N$ is an $R$-map, write $f_{\mathfrak{p}} \colon M_{\mathfrak{p}} \to N_{\mathfrak{p}}$, where $f_{\mathfrak{p}} = 1_{R_{\mathfrak{p}}} \otimes f$.

We restate Corollary 10.20(iv) in this notation. The function $f \colon \mathfrak{q} \mapsto \mathfrak{q}_{\mathfrak{p}}$ is a bijection from the family of all prime ideals in $R$ that are contained in $\mathfrak{p}$ to the family of prime ideals in $R_{\mathfrak{p}}$.

**Proposition 10.37.** *Let $I$ and $J$ be ideals in a domain $R$. If $I_{\mathfrak{m}} = J_{\mathfrak{m}}$ for every maximal ideal $\mathfrak{m}$, then $I = J$.*

**Proof.** Take $b \in J$, and define

$$(I : b) = \{r \in R : rb \in I\}.$$

Let $\mathfrak{m}$ be a maximal ideal in $R$. Since $I_{\mathfrak{m}} = J_{\mathfrak{m}}$, there are $a \in I$ and $s \notin \mathfrak{m}$ with $b/1 = a/s$. As $R$ is a domain, $sb = a \in I$, so that $s \in (I : b)$; but $s \notin \mathfrak{m}$, so that $(I : b) \subsetneq \mathfrak{m}$. Thus, $(I : b)$ cannot be a proper ideal, for it is not contained in any maximal ideal. Therefore, $(I : b) = R$; hence, $1 \in (I : b)$ and $b = 1b \in I$. We have proved that $J \subseteq I$, and the reverse inclusion is proved similarly. •

**Proposition 10.38.** *Let $R$ be a commutative ring, and let $M, N$ be $R$-modules.*

(i) *If $M_{\mathfrak{m}} = \{0\}$ for every maximal ideal $\mathfrak{m}$, then $M = \{0\}$.*

(ii) *If $f \colon M \to N$ is an $R$-map and $f_{\mathfrak{m}} \colon M_{\mathfrak{m}} \to N_{\mathfrak{m}}$ is an injection for every maximal ideal $\mathfrak{m}$, then $f$ is an injection.*

(iii) *If $f \colon M \to N$ is an $R$-map and $f_{\mathfrak{m}} \colon M_{\mathfrak{m}} \to N_{\mathfrak{m}}$ is a surjection for every maximal ideal $\mathfrak{m}$, then $f$ is a surjection.*

(iv) *If $f \colon M \to N$ is an $R$-map and $f_{\mathfrak{m}} \colon M_{\mathfrak{m}} \to N_{\mathfrak{m}}$ is an isomorphism for every maximal ideal $\mathfrak{m}$, then $f$ is an isomorphism.*

**Proof.**

(i) If $M \neq \{0\}$, then there is $m \in M$ with $m \neq 0$. It follows that the annihilator $I = \{r \in R : rm = 0\}$ is a proper ideal in $R$, for $1 \notin I$, and so there is some maximal ideal $\mathfrak{m}$ containing $I$. Now $1 \otimes m = 0$ in $M_{\mathfrak{m}}$, so that $m \in \ker h_M$. Proposition 10.28 gives $s \notin \mathfrak{m}$ with $sm = 0$ in $M$. Hence, $s \in I \subseteq \mathfrak{m}$, and this is a contradiction. Therefore, $M = \{0\}$.

(ii) There is an exact sequence $0 \to K \to M \xrightarrow{f} N$, where $K = \ker f$. Since localization is an exact functor, there is an exact sequence

$$0 \to K_{\mathfrak{m}} \to M_{\mathfrak{m}} \xrightarrow{f_{\mathfrak{m}}} N_{\mathfrak{m}}$$

for every maximal ideal $\mathfrak{m}$. By hypothesis, each $f_{\mathfrak{m}}$ is an injection, so that $K_{\mathfrak{m}} = \{0\}$ for all maximal ideals $\mathfrak{m}$. Part (i) now shows that $K = \{0\}$, and so $f$ is an injection.

(iii) There is an exact sequence $M \xrightarrow{f} N \to C \to 0$, where $C = \operatorname{coker} f = N/\operatorname{im} f$. Since tensor product is right exact, $C_{\mathfrak{m}} = \{0\}$ for all maximal ideals $\mathfrak{m}$, and so $C = \{0\}$. But $f$ is surjective if and only if $C = \operatorname{coker} f = \{0\}$.

(iv) This follows at once from parts (ii) and (iii). •

We cannot weaken the hypothesis of Proposition 10.38(iv) to $M_{\mathfrak{m}} \cong N_{\mathfrak{m}}$ for all maximal ideals $\mathfrak{m}$; we must assume that all the local isomorphisms arise from a given map $f \colon M \to N$. If $G$ is the subgroup of $\mathbb{Q}$ consisting of all $a/b$ with $b$ squarefree, then we saw, in Example 10.5, that $G_{(p)} \cong \mathbb{Z}_{(p)}$ for all primes $p$, but $G \not\cong \mathbb{Z}$.

Localization commutes with Tor, essentially because $S^{-1}R$ is a flat $R$-module.

**Proposition 10.39.** *If $S$ is a subset of a commutative ring $R$, then there are isomorphisms*

$$S^{-1}\operatorname{Tor}_n^R(A, B) \cong \operatorname{Tor}_n^{S^{-1}R}(S^{-1}A, S^{-1}B)$$

*for all $n \geq 0$ and for all $R$-modules $A$ and $B$.*

**Proof.** First consider the case $n = 0$. For any $R$-module $A$, there is a natural isomorphism

$$\tau_B \colon S^{-1}(A \otimes_R B) \to S^{-1}A \otimes_{S^{-1}R} S^{-1}B,$$

for either side is a solution $U$ of the universal mapping problem

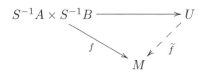

where $M$ is an $(S^{-1}R)$-module, $f$ is $(S^{-1}R)$-bilinear, and $\widetilde{f}$ is an $(S^{-1}R)$-map.

Let $\mathbf{P}_B$ be a deleted projective resolution of $B$. Since the localization functor is exact and preserves projectives, $S^{-1}(\mathbf{P}_B)$ is a deleted projective resolution of $S^{-1}B$. Naturality of the isomorphisms $\tau_A$ gives an isomorphism of complexes

$$S^{-1}(A \otimes_R \mathbf{P}_B) \cong S^{-1}A \otimes_{S^{-1}R} S^{-1}(\mathbf{P}_B),$$

so that their homology groups are isomorphic. Since localization is an exact functor, Exercise 9.24 on page 794 applies, and

$$H_n(S^{-1}(A \otimes_R \mathbf{P}_B)) \cong S^{-1}H_n(A \otimes_R \mathbf{P}_B) \cong S^{-1}\operatorname{Tor}_n^R(A, B).$$

On the other hand, since $S^{-1}(\mathbf{P}_B)$ is a deleted projective resolution of $S^{-1}B$, the definition of Tor gives

$$H_n(S^{-1}A \otimes_{S^{-1}R} S^{-1}(\mathbf{P}_B)) \cong \operatorname{Tor}_n^{S^{-1}R}(S^{-1}A, S^{-1}B). \quad \bullet$$

**Corollary 10.40.** *Let $A$ be an $R$-module over a commutative ring $R$. If $A_{\mathfrak{m}}$ is a flat $R_{\mathfrak{m}}$-module for every maximal ideal $\mathfrak{m}$, then $A$ is a flat $R$-module.*

**Proof.** The hypothesis, together with Proposition 9.98, gives $\operatorname{Tor}_n^{R_{\mathfrak{m}}}(A_{\mathfrak{m}}, B_{\mathfrak{m}}) = \{0\}$ for all $n \geq 1$, for every $R$-module $B$, and for every maximal ideal $\mathfrak{m}$. But Proposition 10.39 gives $\operatorname{Tor}_n^R(A, B)_{\mathfrak{m}} = \{0\}$ for all maximal ideals $\mathfrak{m}$ and all $n \geq 1$. Finally, Proposition 10.38 shows that $\operatorname{Tor}_n^R(A, B) = \{0\}$ for all $n \geq 1$. Since this is true for all $R$-modules $B$, we have $A$ flat. $\quad \bullet$

We must add some hypotheses to get a similar result for Ext (Exercise 10.24 on page 898).

**Lemma 10.41.** *If $R$ is a left noetherian ring and $A$ is a finitely generated left $R$-module, then there is a projective resolution $\mathbf{P}$ of $A$ in which each $P_n$ is finitely generated.*

**Proof.** Since $A$ is finitely generated, there exists a finitely generated free left $R$-module $P_0$ and a surjective $R$-map $\varepsilon \colon P_0 \to A$. Since $R$ is left noetherian, $\ker \varepsilon$ is finitely generated, and so there exists a finitely generated free left $R$-module $P_1$

and a surjective $R$-map $d_1 \colon P_1 \to \ker \varepsilon$. If we define $D_1 \colon P_1 \to P_0$ as the composite $id_1$, where $i \colon \ker \varepsilon \to P_0$ is the inclusion, then there is an exact sequence

$$0 \to \ker D_1 \to P_1 \xrightarrow{D_1} P_0 \xrightarrow{\varepsilon} A \to 0.$$

This construction can be iterated, for $\ker D_1$ is finitely generated, and the proof can be completed by induction. (We remark that we have, in fact, constructed a free resolution of $A$.)  •

**Proposition 10.42.** *Let $S \subseteq R$ be a subset of a commutative noetherian ring $R$. If $A$ is a finitely generated $R$-module, then there are isomorphisms*

$$S^{-1} \operatorname{Ext}_R^n(A, B) \cong \operatorname{Ext}_{S^{-1}R}^n(S^{-1}A, S^{-1}B)$$

*for all $n \geq 0$ and for all $R$-modules $B$.*

**Proof.** Since $R$ is noetherian and $A$ is finitely generated, Lemma 10.41 says there is a projective resolution $\mathbf{P}$ of $A$ each of whose terms is finitely generated. By Lemma 10.35, there is a natural isomorphism

$$\tau_A \colon S^{-1} \operatorname{Hom}_R(A, B) \to \operatorname{Hom}_{S^{-1}R}(S^{-1}A, S^{-1}B)$$

for every $R$-module $B$ (a finitely generated module over a noetherian ring must be finitely presented). Now $\tau_A$ gives an isomorphism of complexes

$$S^{-1}(\operatorname{Hom}_R(\mathbf{P}_A, B)) \cong \operatorname{Hom}_{S^{-1}R}(S^{-1}(\mathbf{P}_A), S^{-1}B).$$

Taking homology of the left hand side gives

$$H_n(S^{-1}(\operatorname{Hom}_R(\mathbf{P}_A, B))) \cong S^{-1} H_n(\operatorname{Hom}_R(\mathbf{P}_A, B)) \cong S^{-1} \operatorname{Ext}_R^n(A, B),$$

because localization is an exact functor (Exercise 9.24 on page 794). On the other hand, homology of the right hand side is

$$H_n(\operatorname{Hom}_{S^{-1}R}(S^{-1}(\mathbf{P}_A), S^{-1}B)) = \operatorname{Ext}_{S^{-1}R}^n(S^{-1}A, S^{-1}B),$$

because $S^{-1}(\mathbf{P}_A)$ is an $(S^{-1}R)$-projective resolution of $S^{-1}A$.  •

**Remark.** An alternative proof of the Proposition 10.42 can be given using a deleted injective resolution $\mathbf{E}_B$ in the second variable. We must still assume that $A$ is finitely generated, in order to use Lemma 10.35, but we can now use the fact that localization preserves injectives when $R$ is noetherian.  ◀

**Corollary 10.43.** *Let $A$ be a finitely generated $R$-module over a commutative noetherian ring $R$. Then $A_{\mathfrak{m}}$ is a projective $R_{\mathfrak{m}}$-module for every maximal ideal $\mathfrak{m}$ if and only if $A$ is a projective $R$-module.*

**Proof.** Sufficiency follows from localization preserving direct sum, and necessity follows from Proposition 10.42: for every $R$-module $B$ and maximal ideal $\mathfrak{m}$, we have

$$\operatorname{Ext}_R^1(A, B)_{\mathfrak{m}} \cong \operatorname{Ext}_{R_{\mathfrak{m}}}^1(A_{\mathfrak{m}}, B_{\mathfrak{m}}) = \{0\},$$

because $A_{\mathfrak{m}}$ is projective. By Proposition 10.38, $\operatorname{Ext}_R^1(A, B) = \{0\}$, which says that $A$ is projective.  •

**Example 10.44.** Let $R$ be a Dedekind ring which is not a PID (Dedekind rings are defined in the next section), and let $\mathfrak{p}$ be a nonzero prime ideal in $R$. Then we shall see, in Proposition 10.103, that $R_{\mathfrak{p}}$ is a local PID. Hence, if $P$ is a projective $R$-module, then $P_{\mathfrak{p}}$ is a projective $R_{\mathfrak{p}}$-module, and so it is free, by Proposition 10.26. In particular, if $\mathfrak{b}$ is a non-principal ideal in $R$, then $\mathfrak{b}$ is not free even though all its localizations are free. ◀

# Exercises

**10.9.** If $R$ is a domain with $Q = \text{Frac}(R)$, determine whether every $R$-subalgebra $A$ of $Q$ is a localization of $R$. If true, prove it; if false, give a counterexample.

**10.10.** Prove that every localization of a PID is a PID. Conclude that if $\mathfrak{p}$ is a nonzero prime ideal in a PID $R$, then $R_{\mathfrak{p}}$ is a DVR.

**10.11.** If $R$ is a Boolean ring and $\mathfrak{m}$ is a maximal ideal in $R$, prove that $R_{\mathfrak{m}}$ is a field.

**10.12.** Let $A$ be an $R$-algebra, and let $N$ be a finitely presented $R$-module. For every $A$-module $M$, prove that

$$\theta\colon \text{Hom}_R(N, M) \to \text{Hom}_A(N \otimes_R A, M)$$

given by $\theta\colon f \mapsto \widetilde{f}$, is a natural isomorphism, where $\widetilde{f}(n \otimes 1) = f(n)$ for all $n \in N$.

**10.13.** Let $B$ be a flat $R$-module, and let $N$ be a finitely presented $R$-module. For every $R$-module $M$, prove that

$$\psi\colon B \otimes_R \text{Hom}_R(N, M) \to \text{Hom}_R(N, M \otimes_R B),$$

given by $b \otimes g \mapsto g_b$, is a natural isomorphism, where $g_b(n) = g(n) \otimes b$ for all $n \in N$.

∗ **10.14.** Let $S \subseteq R$ be multiplicative, and let $I$ and $J$ be ideals in $R$.

(i) Prove that $S^{-1}(IJ) = (S^{-1}I)(S^{-1}J)$.

(ii) Prove that $S^{-1}(I : J) = (S^{-1}I : S^{-1}J)$.

**10.15.** Recall that if $k$ is a commutative ring, then the ring of *Laurent polynomials* over $k$ is the subring of rational functions

$$k[x, x^{-1}] = \Big\{ \sum_{i=m}^{n} a_i x^i \in k(x) : m \leq n \text{ and } m, n \in \mathbb{Z} \Big\}.$$

Prove that if $R$ is noetherian, then $k[x, x^{-1}]$ is noetherian.

**Hint.** See Exercise 6.7 on page 415, and use Corollary 10.20.

**10.16.** A *valuation ring* is a domain $R$ such that, for all $a, b \in R$, either $a \mid b$ or $b \mid a$.

(i) Prove that every DVR is a valuation ring.

(ii) Let $R$ be a domain with $F = \text{Frac}(R)$. Prove that $R$ is a valuation ring if and only if $a \in R$ or $a^{-1} \in R$ for each nonzero $a \in F$.

**10.17.** (i) Prove that every finitely generated ideal in a valuation ring is principal.

(ii) Prove that every finitely generated ideal in a valuation ring is projective.

∗ **10.18.** An abelian group $\Gamma$ is ***ordered*** if it is a partially ordered set in which $a+b \leq a'+b'$ whenever $a \leq a'$ and $b \leq b'$; call $\Gamma$ a ***totally ordered abelian group*** if the partial order is a chain. A ***valuation*** on a field $k$ is a function $v\colon k^\times \to \Gamma$, where $\Gamma$ is a totally ordered abelian group, such that

$$v(ab) = v(a) + v(b);$$
$$v(a + b) \geq \min\{v(a), v(b)\}.$$

  (i) If $a/b \in \mathbb{Q}$ is nonzero, write $a = p^m a'$ and $b = p^n b'$, where $m, n \geq 0$ and $(a', p) = 1 = (b', p)$. Prove that $v\colon \mathbb{Q}^\times \to \mathbb{Z}$, defined by $v(a/b) = m - n$, is a valuation.

  (ii) If $v\colon k^\times \to \Gamma$ is a valuation on a field $k$, define $R = \{0\} \cup \{a \in k^\times : v(a) \geq 0\}$. Prove that $R$ is a valuation ring. (Every valuation ring arises in this way from a suitable valuation on its fraction field. Moreover, the valuation ring is discrete when the totally ordered abelian group $\Gamma$ is isomorphic to $\mathbb{Z}$.)

  (iii) Prove that $a \in R$ is a unit if and only if $v(a) = 0$.

  (iv) Prove that every valuation ring is a (not necessarily noetherian) local ring.
  **Hint.** Show that $\mathfrak{m} = \{a \in R : v(a) > 0\}$ is the unique maximal ideal in $R$.

**10.19.** Let $\Gamma$ be a totally ordered abelian group and let $k$ be a field. Define $k[\Gamma]$ to be the group ring (consisting of all functions $f\colon \Gamma \to k$ almost all of whose values are 0). As usual, if $f(\gamma) = r_\gamma$, we denote $f$ by $\sum_{\gamma \in \Gamma} r_\gamma \gamma$.

  (i) Define the ***degree*** of $f = \sum_{\gamma \in \Gamma} r_\gamma \gamma$ to be $\alpha$ if $\alpha$ is the largest index $\gamma$ with $r_\gamma \neq 0$. Prove that $k[\Gamma]$ is a valuation ring, where $v(f)$ is the degree of $f$.

  (ii) Give an example of a non-noetherian valuation ring.

∗ **10.20.** A multiplicative subset $S$ of a commutative ring $R$ is ***saturated*** if $ab \in S$ implies $a \in S$ and $b \in S$.

  (i) Prove that $U(R)$, the set of all units in $R$, is a saturated subset of $R$.

  (ii) An element $r \in R$ is a ***zero-divisor*** on an $R$-module $A$ if there is some nonzero $a \in A$ with $ra = 0$. Prove that $\mathrm{Zer}(A)$, the set of all zero-divisors on an $R$-module $A$, is a saturated subset of $R$.

  (iii) If $S \subseteq R$ is multiplicative, prove that there exists a unique smallest saturated subset $S'$ containing $S$ (called the ***saturation*** of $S$), and that $(S')^{-1}R \cong S^{-1}R$.

  (iv) Prove that a multiplicative subset $S$ is saturated if and only if its complement $R - S$ is a union of prime ideals.

**10.21.** Let $S$ be a multiplicative subset of a commutative ring $R$, and let $M$ be a finitely generated $R$-module. Prove that $S^{-1}M = \{0\}$ if and only if there is $s \in S$ with $sM = \{0\}$.

**10.22.** Let $S$ be a subset of a commutative ring $R$, and let $A$ be an $R$-module.

  (i) If $A$ is finitely generated, prove that $S^{-1}A$ is a finitely generated $(S^{-1}R)$-module.

  (ii) If $A$ is finitely presented, prove that $S^{-1}A$ is a finitely presented $(S^{-1}R)$-module.

**10.23.** If $\mathfrak{p}$ is a nonzero prime ideal in a commutative ring $R$ and $A$ is a projective $R$-module, prove that $A_\mathfrak{p}$ is a free $R_\mathfrak{p}$-module.

∗ **10.24.**  (i) Give an example of an abelian group $B$ for which $\mathrm{Ext}^1_\mathbb{Z}(\mathbb{Q}, B) \neq \{0\}$.

  (ii) Prove that $\mathbb{Q} \otimes_\mathbb{Z} \mathrm{Ext}^1_\mathbb{Z}(\mathbb{Q}, B) \neq \{0\}$ for the abelian group $B$ in part (i).

(iii) Prove that Proposition 10.42 may be false if $R$ is noetherian but $A$ is not finitely generated.

∗ **10.25.** Let $R$ be a commutative $k$-algebra, where $k$ is a commutative ring, and let $M$ be a $k$-module. Prove, for all $n \geq 0$, that

$$R \otimes_k \bigwedge^n (M) \cong \bigwedge^n (R \otimes_k M)$$

[of course, $\bigwedge^n (R \otimes_k M)$ means the $n$th exterior power of the $R$-module $R \otimes_k M$]. Conclude, for all maximal ideals $\mathfrak{m}$ in $k$, that

$$\left( \bigwedge^n (M) \right)_{\mathfrak{m}} \cong \bigwedge^n (M_{\mathfrak{m}}).$$

**Hint.** Show that $R \otimes_k \bigwedge^n (M)$ is a solution to the universal mapping problem for alternating $n$-multilinear $R$-functions.

**10.26.** Let $R$ be a commutative noetherian ring. If $A$ and $B$ are finitely generated $R$-modules, prove that $\mathrm{Tor}_n^R(A, B)$ and $\mathrm{Ext}_R^n(A, B)$ are finitely generated $R$-modules for all $n \geq 0$.

## Section 10.3. Dedekind Rings

A **_Pythagorean triple_** is a triple $(a, b, c)$ of positive integers such that $a^2 + b^2 = c^2$; examples are $(3, 4, 5)$, $(5, 12, 13)$, and $(7, 24, 25)$. All Pythagorean triples were classified by Diophantus, ca 250 AD. Fermat proved that there do not exist positive integers $(a, b, c)$ with $a^4 + b^4 = c^4$ and, in 1637, he wrote in the margin of his copy of a book by Diophantus that he had a wonderful proof that there are no positive integers $(a, b, c)$ with $a^n + b^n = c^n$ for any $n > 2$. Fermat's proof was never found, and his remark (that was merely a note to himself) became known only several years after his death, when Fermat's son published a new edition of Diophantus in 1670 containing his father's notes. There were other notes of Fermat, many of them true, some of them false, and this statement, the only one unresolved by 1800, was called **_Fermat's Last Theorem_**, perhaps in jest. It remained one of the outstanding challenges in Number Theory until 1995, when Wiles proved Fermat's Last Theorem.

Every positive integer $n > 2$ is a multiple of 4 or of some odd prime $p$. Thus, if there do not exist positive integers $(a, b, c)$ with $a^p + b^p = c^p$ for every odd prime $p$, then Fermat's Last Theorem is true [if $n = pm$, then $a^n + b^n = c^n$ implies $(a^m)^p + (b^m)^p = (c^m)^p$]. Over the centuries, there were many attempts to prove it. For example, Euler published a proof (with gaps, later corrected) for the case $n = 3$, Dirichlet published a proof (with gaps, later corrected) for the case $n = 5$, and Lamé published a correct proof for the case $n = 7$.

The first major progress (not dealing only with particular primes $p$) was due to Kummer, in the middle of the 19th century. If $a^p + b^p = c^p$, where $p$ is an odd prime, then a natural starting point of investigation is the identity

$$c^p = a^p + b^p = (a + b)(a + \zeta b)(a + \zeta^2 b) \cdots (a + \zeta^{p-1} b),$$

where $\zeta = \zeta_p$ is a primitive $p$th root of unit (perhaps this idea occurred to Fermat). Kummer proved that if $\mathbb{Z}[\zeta_p]$ is a UFD, where $\mathbb{Z}[\zeta_p] = \{ f(\zeta_p) : f(x) \in \mathbb{Z}[x] \}$, then there do not exist positive integers $a, b, c$ with $a^p + b^p = c^p$. On the other hand, he also showed that there do exist primes $p$ for which $\mathbb{Z}[\zeta_p]$ is not a UFD. To restore unique factorization, he invented "ideal numbers" that he adjoined to $\mathbb{Z}[\zeta_p]$. Later, Dedekind recast Kummer's ideal numbers into our present notion of ideal. Thus, Fermat's Last Theorem has served as a catalyst in the development of both modern Algebra and Algebraic Number Theory. *Dedekind rings* are the appropriate generalization of rings like $\mathbb{Z}[\zeta_p]$, and we will study them in this section.

**10.3.1. Integrality.** The notion of algebraic integer is a special case of the notion of integral element.

**Definition.** A *ring extension*[4] $R^*/R$ is a commutative ring $R^*$ containing $R$ as a subring. Let $R^*/R$ be a ring extension. Then an element $a \in R^*$ is *integral* over $R$ if it is a root of a monic polynomial in $R[x]$. A ring extension $R^*/R$ is an *integral extension* if every $a \in R^*$ is integral over $R$.

**Example 10.45.** The *Noether Normalization Theorem* (Matsumura, *Commutative Ring Theory*, p. 262) is often used to prove the Nullstellensatz. It says that if $A$ is a finitely generated $k$-algebra over a field $k$, then there exist algebraically independent elements $a_1, \ldots, a_n$ in $A$ so that $A$ is integral over $k[a_1, \ldots, a_n]$. ◄

Recall that a complex number is an algebraic integer if it is a root of a monic polynomial in $\mathbb{Z}[x]$, so that algebraic integers are integral over $\mathbb{Z}$. The reader should compare the next lemma with Proposition 2.70.

**Lemma 10.46.** *If $R^*/R$ is a ring extension, then the following conditions on a nonzero element $u \in R^*$ are equivalent.*

(i) *$u$ is integral over $R$.*

(ii) *There is a finitely generated $R$-submodule $B$ of $R^*$ with $uB \subseteq B$.*

(iii) *There is a finitely generated faithful $R$-submodule $B$ of $R^*$ with $uB \subseteq B$; that is, if $dB = \{0\}$ for some $d \in R$, then $d = 0$.*

**Proof.**

(i) $\Rightarrow$ (ii). If $u$ is integral over $R$, there is a monic polynomial $f(x) \in R[x]$ with $f(u) = 0$; that is, there are $r_i \in R$ with $u^n = \sum_{i=0}^{n-1} r_i u^i$. Define $B = \langle 1, u, u^2, \ldots, u^{n-1} \rangle$. It is clear that $uB \subseteq B$.

(ii) $\Rightarrow$ (iii). If $B = \langle b_1, \ldots, b_m \rangle$ is a finitely generated $R$-submodule of $R^*$ with $uB \subseteq B$, define $B' = \langle 1, b_1, \ldots, b_m \rangle$. Now $B'$ is finitely generated, faithful (because $1 \in B'$), and $uB' \subseteq B'$.

(iii) $\Rightarrow$ (i). Suppose there is a faithful $R$-submodule of $R^*$, say, $B = \langle b_i, \ldots, b_n \rangle$, with $uB \subseteq B$. There is a system of $n$ equations $ub_i = \sum_{j=1}^{n} p_{ij} b_j$ with $p_{ij} \in R$. If $P = [p_{ij}]$ and if $X = (b_1, \ldots, b_n)^\top$ is an $n \times 1$ column vector, then the $n \times n$ system can be rewritten in matrix notation: $(uI - P)X = 0$. Now $0 = \big(\mathrm{adj}(uI - P)\big)(uI - P)X = dX$, where $d = \det(uI - P)$, by Corollary 8.138. Since $dX = 0$, we have $db_i = 0$ for all $i$, and so $dB = \{0\}$. Therefore, $d = 0$,

---

[4]We use the same notation for ring extensions as we do for field extensions.

because $B$ is faithful. On the other hand, Corollary 8.131 gives $d = f(u)$, where $f(x) \in R[x]$ is a monic polynomial of degree $n$; hence, $u$ is integral over $R$. •

Being an integral extension is transitive.

**Proposition 10.47.** *If $T \subseteq S \subseteq R$ are commutative rings with $S$ integral over $T$ and $R$ integral over $S$, then $R$ is integral over $T$.*

**Proof.** If $r \in R$, there is an equation $r^n + s_{n-1}r^{n-1} + \cdots + r_0 = 0$, where $s_i \in S$ for all $i$. By Lemma 10.46, the subring $S' = T[s_{n-1}, \ldots, s_0]$ is a finitely generated $T$-module. But $r$ is integral over $S'$, so that the ring $S'[r]$ is a finitely generated $S'$-module. Therefore, $S'[r]$ is a finitely generated $T$-module, and so $r$ is integral over $T$. •

**Proposition 10.48.** *Let $E/R$ be a ring extension.*

(i) *If $u, v \in E$ are integral over $R$, then both $uv$ and $u + v$ are integral over $R$.*

(ii) *The commutative ring $\mathcal{O}_{E/R}$, defined by*

$$\mathcal{O}_{E/R} = \{u \in E : u \text{ is integral over } R\},$$

*is an $R$-subalgebra of $E$.*

**Proof.**

(i) If $u$ and $v$ are integral over $R$, then Lemma 10.46(ii) says that there are $R$-submodules $B = \langle b_1, \ldots, b_n \rangle$ and $C = \langle c_1, \ldots, c_m \rangle$ of $E$ with $uB \subseteq B$ and $vC \subseteq C$; that is, $ub_i \in B$ for all $i$ and $vc_j \in C$ for all $j$. Define $BC$ to be the $R$-submodule of $E$ generated by all $b_i c_j$; of course, $BC$ is finitely generated. Now $uvBC \subseteq BC$, for $uvb_i c_j = (ub_i)(vc_j)$ is an $R$-linear combination of $b_k c_\ell$'s, and so $uv$ is integral over $R$. Similarly, $u + v$ is integral over $R$, for $(u + v)b_i c_j = (ub_i)c_j + (vc_j)b_i \in BC$.

(ii) Part (i) shows that $\mathcal{O}_{E/R}$ is closed under multiplication and addition. Now $R \subseteq \mathcal{O}_E$, for if $r \in R$, then $r$ is a root of $x - r$. It follows that $1 \in \mathcal{O}_{E/R}$ and that $\mathcal{O}_{E/R}$ is an $R$-subalgebra of $E$. •

Here is a second proof of Proposition 10.48(i) for a *domain* $E$ using tensor products and Linear Algebra. Let $f(x) \in R[x]$ be the minimal polynomial of $u$, let $A$ be the companion matrix of $f(x)$, and let $y$ be an eigenvector [over the algebraic closure of $\text{Frac}(E)$]: $Ay = uy$. Let $g(x)$ be the minimal polynomial of $v$, let $B$ be the companion matrix of $g(x)$, and let $Bz = vz$. Now

$$(A \otimes B)(y \otimes z) = Ay \otimes Bz = uy \otimes vz = uv(y \otimes z).$$

Therefore, $uv$ is an eigenvalue of $A \otimes B$; that is, $uv$ is a root of the monic polynomial $\det(xI - A \otimes B)$, which lies in $R[x]$ because both $A$ and $B$ have all their entries in $R$. Therefore, $uv$ is integral over $R$. Similarly, the equation

$$(A \otimes I + I \otimes B)(y \otimes z) = Ay \otimes z + y \otimes Bz = (u + v)y \otimes z$$

shows that $u + v$ is integral over $R$.

**Definition.** Let $E/R$ be a ring extension. The $R$-subalgebra $\mathcal{O}_{E/R}$ of $E$, consisting of all those elements integral over $R$, is called the ***integral closure*** of $R$ in $E$. If $\mathcal{O}_{E/R} = R$, then $R$ is called ***integrally closed in*** $E$. If $R$ is a domain and $R$ is integrally closed in $F = \operatorname{Frac}(R)$, that is, $\mathcal{O}_{F/R} = R$, then $R$ is called ***integrally closed***.

Thus, $R$ is integrally closed if, whenever $\alpha \in \operatorname{Frac}(R)$ is integral over $R$, we have $\alpha \in R$.

**Example 10.49.** The ring $\mathcal{O}_{\mathbb{Q}/\mathbb{Z}} = \mathbb{Z}$, for if a rational number $a$ is a root of a monic polynomial in $\mathbb{Z}[x]$, then Theorem 2.63 shows that $a \in \mathbb{Z}$. Hence, $\mathbb{Z}$ is integrally closed. ◄

**Proposition 10.50.** *Every* UFD $R$ *is integrally closed. In particular, every* PID *is integrally closed.*

**Proof.** Let $F = \operatorname{Frac}(R)$, and suppose that $u \in F$ is integral over $R$. Thus, there is an equation

$$u^n + r_{n-1}u^{n-1} + \cdots + r_1 u + r_0 = 0,$$

where $r_i \in R$. We may write $u = b/c$, where $b, c \in R$ and $(b, c) = 1$ (gcd's exist because $R$ is a UFD, and so every fraction can be put in lowest terms). Substituting and clearing denominators, we obtain

$$b^n + r_{n-1}b^{n-1}c + \cdots + r_1 bc^{n-1} + r_0 c^n = 0.$$

Hence, $b^n = -c\big(r_{n-1}b^{n-1} + \cdots + r_1 bc^{n-2} + r_0 c^{n-1}\big)$, so that $c \mid b^n$ in $R$. But $(b, c) = 1$ implies $(b^n, c) = 1$, so that $c$ must be a unit in $R$; that is, $c^{-1} \in R$. Therefore, $u = b/c = bc^{-1} \in R$, and so $R$ is integrally closed. •

We now understand Example 5.21. If $k$ is a field, the subring $R$ of $k[x]$, consisting of all polynomials $f(x) \in k[x]$ having no linear term, is not integrally closed. It is easy to check that $\operatorname{Frac}(R) = k(x)$, since $x = x^3/x^2 \in \operatorname{Frac}(R)$. But $x \in k(x)$ is a root of the monic polynomial $t^2 - x^2 \in R[t]$, and $x \notin R$. Hence, $R$ is not a UFD.

**Definition.** An ***algebraic number field*** is a finite field extension of $\mathbb{Q}$. If $E$ is an algebraic number field, then $\mathcal{O}_{E/\mathbb{Z}}$ is usually denoted by $\mathcal{O}_E$ instead of by $\mathcal{O}_{E/\mathbb{Z}}$, and it is called the ***ring of integers*** in $E$.

Because of this new use of the word *integers*, algebraic number theorists often speak of the ring of ***rational integers*** when referring to $\mathbb{Z}$.

**Proposition 10.51.** *Let $E$ be an algebraic number field and let $\mathcal{O}_E$ be its ring of integers.*

   (i) *If $\alpha \in E$, then there is a nonzero integer $m$ with $m\alpha \in \mathcal{O}_E$.*

   (ii) $\operatorname{Frac}(\mathcal{O}_E) = E$.

   (iii) $\mathcal{O}_E$ *is integrally closed.*

**Proof.**

(i) If $\alpha \in E$, then there is a monic polynomial $f(x) \in \mathbb{Q}[x]$ with $f(\alpha) = 0$. Clearing denominators gives an integer $m$ with

$$m\alpha^n + c_{n-1}\alpha^{n-1} + c_{n-2}\alpha^{n-2} + \cdots + c_1\alpha + c_0 = 0,$$

where all $c_i \in \mathbb{Z}$. Multiplying by $m^{n-1}$ gives

$$(m\alpha)^n + c_{n-1}(m\alpha)^{n-1} + mc_{n-2}(m\alpha)^{n-2} + \cdots + c_1 m^{n-2}(m\alpha) + m^{n-1}c_0 = 0.$$

Thus, $m\alpha \in \mathcal{O}_E$.

(ii) It suffices to show that if $\alpha \in E$, then there are $a, b \in \mathcal{O}_E$ with $\alpha = a/b$. But $m\alpha \in \mathcal{O}_E$ [by part (i)], $m \in \mathbb{Z} \subseteq \mathcal{O}_E$, and $\alpha = (m\alpha)/m$.

(iii) Suppose that $\alpha \in \operatorname{Frac}(\mathcal{O}_E) = E$ is integral over $\mathcal{O}_E$. By transitivity of integral extensions, Proposition 10.47, we have $\alpha$ integral over $\mathbb{Z}$. But this means that $\alpha \in \mathcal{O}_E$, which is, by definition, the set of all those elements in $E$ that are integral over $\mathbb{Z}$. Therefore, $\mathcal{O}_E$ is integrally closed.   •

**Example 10.52.** We shall see, in Proposition 10.82, that if $E = \mathbb{Q}(i)$, then $\mathcal{O}_E = \mathbb{Z}[i]$, the Gaussian integers. Now $\mathbb{Z}[i]$ is a PID, because it is a Euclidean ring, and hence it is a UFD. The generalization of this example which replaces $\mathbb{Q}(i)$ by an algebraic number field $E$ is more subtle. It is true that $\mathcal{O}_E$ is integrally closed, but it may not be true that the elements of $\mathcal{O}_E$ are $\mathbb{Z}$-linear combinations of $\alpha$. Moreover, the rings $\mathcal{O}_E$ may not be UFDs. We will investigate rings of integers at the end of this section.   ◄

Given a ring extension $R^*/R$, what is the relation between ideals in $R^*$ and ideals in $R$?

**Definition.** Let $R^*/R$ be a ring extension. If $I$ is an ideal in $R$, define its ***extension*** $I^e$ to be $R^*I$, the ideal in $R^*$ generated by $I$. If $I^*$ is an ideal in $R^*$, define its ***contraction*** $I^{*c} = R \cap I^*$.

**Remark.** The definition can be generalized. Let $h \colon R \to R^*$ be a ring homomorphism, where $R$ and $R^*$ are any two commutative rings. Define the extension of an ideal $I$ in $R$ to be the ideal in $R^*$ generated by $h(I)$; define the contraction of an ideal $I^*$ in $R^*$ to be $h^{-1}(I^*)$. If $R^*/R$ is a ring extension, then taking $h \colon R \to R^*$ to be the inclusion gives the definition above. Another interesting instance is the localization map $h \colon R \to S^{-1}R$.   ◄

**Example 10.53.**

(i) It is easy to see that if $R^*/R$ is a ring extension and $\mathfrak{p}^*$ is a prime ideal in $R^*$, then its contraction $\mathfrak{p}^* \cap R$ is also a prime ideal: if $a, b \in R$ and $ab \in \mathfrak{p}^* \cap R \subseteq \mathfrak{p}^*$, then $\mathfrak{p}^*$ prime gives $a \in \mathfrak{p}^*$ or $b \in \mathfrak{p}^*$; as $a, b \in R$, either $a \in \mathfrak{p}^* \cap R$ or $b \in \mathfrak{p}^* \cap R$. Thus, contraction defines a function $c \colon \operatorname{Spec}(R^*) \to \operatorname{Spec}(R)$.

(ii) By (i), contraction $\mathfrak{p}^* \mapsto \mathfrak{p}^* \cap R$ induces a function $c \colon \operatorname{Spec}(R^*) \to \operatorname{Spec}(R)$; in general, this contraction function is neither an injection nor a surjection. For example, $c \colon \operatorname{Spec}(\mathbb{Q}) \to \operatorname{Spec}(\mathbb{Z})$ is not surjective, while $c \colon \operatorname{Spec}(\mathbb{Q}[x]) \to \operatorname{Spec}(\mathbb{Q})$ is not injective.

(iii) The contraction of a maximal ideal, though necessarily prime, need not be maximal. For example, if $R^*$ is a field, then $(0)^*$ is a maximal ideal in $R^*$, but if $R$ is not a field, then the contraction of $(0)^*$, namely, $(0)$, is not a maximal ideal in $R$.  ◀

**Example 10.54.**

(i) Let $\mathcal{I}(R)$ denote the family of all the ideals in a commutative ring $R$. Extension defines a function $e\colon \mathcal{I}(R) \to \mathcal{I}(R^*)$; in general, it is neither injective nor surjective. If $R^*$ is a field and $R$ is not a field, then $e\colon \mathcal{I}(R) \to \mathcal{I}(R^*)$ is not injective. If $R$ is a field and $R^*$ is not a field, then $e\colon \mathcal{I}(R) \to \mathcal{I}(R^*)$ is not surjective.

(ii) If $R^*/R$ is a ring extension and $\mathfrak{p}$ is a prime ideal in $R$, then its extension $R^*\mathfrak{p}$ need not be a prime ideal. Observe first that if $(a) = Ra$ is a principal ideal in $R$, then its extension is the principal ideal $R^*a$ in $R^*$ generated by $a$. Now let $R = \mathbb{R}[x]$ and $R^* = \mathbb{C}[x]$. The ideal $(x^2 + 1)$ is prime, because $x^2 + 1$ is irreducible in $\mathbb{R}[x]$, but its extension is not a prime ideal because $x^2 + 1$ factors in $\mathbb{C}[x]$.  ◀

There are various elementary properties of extension and contraction, such as $I^{*\,ce} \subseteq I^*$ and $I^{ec} \supseteq I$; they are collected in Exercise 10.29 on page 907.

Is there a reasonable condition on a ring extension $R^*/R$ that will give a good relationship between prime ideals in $R$ and prime ideals in $R^*$? This question was posed and answered by Cohen and Seidenberg. We say that a ring extension $R^*/R$ satisfies **Lying Over** if, for every prime ideal $\mathfrak{p}$ in $R$, there exists a prime ideal $\mathfrak{p}^*$ in $R^*$ which contracts to $\mathfrak{p}$; that is, $\mathfrak{p}^* \cap R = \mathfrak{p}$. We say that a ring extension $R^*/R$ satisfies **Going Up** if $\mathfrak{p} \subseteq \mathfrak{q}$ are prime ideals in $R$ and $\mathfrak{p}^*$ lies over $\mathfrak{p}$, then there exists a prime ideal $\mathfrak{q}^* \supseteq \mathfrak{p}^*$ which lies over $\mathfrak{q}$; that is, $\mathfrak{q}^* \cap R = \mathfrak{q}$.

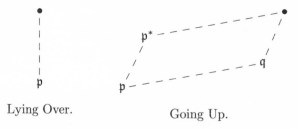

Lying Over.                               Going Up.

We are going to see that extension and contraction are well-behaved in the presence of integral extensions.

**Lemma 10.55.** *Let $R^*$ be an integral extension of $R$.*

(i) *If $\mathfrak{p}$ is a prime ideal in $R$ and $\mathfrak{p}^*$ lies over $\mathfrak{p}$, then $R^*/\mathfrak{p}^*$ is integral over $R/\mathfrak{p}$.*

(ii) *If $S$ is a multiplicative subset of $R$, then $S^{-1}R^*$ is integral over $S^{-1}R$.*

**Proof.**

(i) We view $R/\mathfrak{p}$ as a subring of $R^*/\mathfrak{p}^*$, by the Second Isomorphism Theorem:

$$R/\mathfrak{p} = R/(\mathfrak{p}^* \cap R) \cong (R + \mathfrak{p}^*)/\mathfrak{p}^* \subseteq R^*/\mathfrak{p}^*.$$

Each element in $R^*/\mathfrak{p}^*$ has the form $\alpha + \mathfrak{p}^*$, where $\alpha \in R^*$. Since $R^*$ is integral over $R$, there is an equation

$$\alpha^n + r_{n-1}\alpha^{n-1} + \cdots + r_0 = 0,$$

where $r_i \in R$. Now view this equation mod $\mathfrak{p}^*$ to see that $\alpha + \mathfrak{p}^*$ is integral over $R/\mathfrak{p}$.

(ii) If $\alpha^* \in S^{-1}R^*$, then $\alpha^* = \alpha/s$, where $\alpha \in R^*$ and $s \in S$. Since $R^*$ is integral over $R$, there is an equation $\alpha^n + r_{n-1}\alpha^{n-1} + \cdots + r_0 = 0$ with $r_i \in R$. Multiplying by $1/s^n$ in $S^{-1}R^*$ gives

$$(\alpha/s)^n + (r_{n-1}/s)(\alpha/s)^{n-1} + \cdots + r_0/s^n = 0,$$

which shows that $\alpha/s$ is integral over $S^{-1}R$. $\quad\bullet$

When $R^*/R$ is a ring extension and $R$ is a field, every proper ideal in $R^*$ contracts to $(0)$ in $R$. The following proposition eliminates this collapse when $R^*$ is an integral extension of $R$.

**Proposition 10.56.** *Let $R^*/R$ be a ring extension of domains with $R^*$ integral over $R$. Then $R^*$ is a field if and only if $R$ is a field.*

**Proof.** Assume that $R^*$ is a field. If $u \in R$ is nonzero, then $u^{-1} \in R^*$, and so $u^{-1}$ is integral over $R$. Therefore, there is an equation $(u^{-1})^n + r_{n-1}(u^{-1})^{n-1} + \cdots + r_0 = 0$, where the $r_i \in R$. Multiplying by $u^{n-1}$ gives $u^{-1} = -u(r_{n-1} + \cdots + r_0 u^{n-1})$. Therefore, $u^{-1} \in R$ and $R$ is a field.

Conversely, assume that $R$ is a field. If $\alpha \in R^*$ is nonzero, then there is a monic $f(x) \in R[x]$ with $f(\alpha) = 0$. Thus, $\alpha$ is algebraic over $R$, and so we may assume that $f(x) = \mathrm{irr}(\alpha, R)$; that is, $f(x)$ is irreducible. If $f(x) = \sum_{i=0}^{n} r_i x^i$, then

$$\alpha(\alpha^{n-1} + r_{n-1}\alpha^{n-1} + \cdots + r_1) = -r_0.$$

Irreducibility of $f(x)$ gives $r_0 \neq 0$, so that $\alpha^{-1}$ lies in $R^*$. Thus, $R^*$ is a field. $\quad\bullet$

**Corollary 10.57.** *Let $R^*/R$ be an integral extension. If $\mathfrak{p}$ is a prime ideal in $R$ and $\mathfrak{p}^*$ is a prime ideal lying over $\mathfrak{p}$, then $\mathfrak{p}$ is a maximal ideal if and only if $\mathfrak{p}^*$ is a maximal ideal.*

**Proof.** By Lemma 10.55(i), the domain $R^*/\mathfrak{p}^*$ is integral over the domain $R/\mathfrak{p}$. But now Proposition 10.56 says that $R^*/\mathfrak{p}^*$ is a field if and only if $R/\mathfrak{p}$ is a field; that is, $\mathfrak{p}^*$ is a maximal ideal in $R^*$ if and only if $\mathfrak{p}$ is a maximal ideal in $R$. $\quad\bullet$

The next corollary gives an important property of rings of integers $\mathcal{O}_E$.

**Corollary 10.58.** *If $E$ is an algebraic number field, then every nonzero prime ideal in $\mathcal{O}_E$ is a maximal ideal.*

**Proof.** Let $\mathfrak{p}$ be a nonzero prime ideal in $\mathcal{O}_E$. If $\mathfrak{p} \cap \mathbb{Z} \neq (0)$, then there is a prime $p$ with $\mathfrak{p} \cap \mathbb{Z} = (p)$, by Example 10.53(i). But $(p)$ is a maximal ideal in $\mathbb{Z}$, so that $\mathfrak{p}$ is a maximal ideal, by Corollary 10.57. It remains to show that $\mathfrak{p} \cap \mathbb{Z} \neq (0)$. Let $\alpha \in \mathfrak{p}$ be nonzero. Since $\alpha$ is integral over $\mathbb{Z}$, there is an equation

$$\alpha^n + c_{n-1}\alpha^{n-1} + \cdots + c_1\alpha + c_0 = 0,$$

where $c_i \in \mathbb{Z}$ for all $i$. If we choose such an equation with $n$ minimal, then $c_0 \neq 0$. Since $\alpha \in \mathfrak{p}$, we have $c_0 = -\alpha(\alpha_{n-1} + c_{n-1}\alpha^{n-2} + \cdots + c_1) \in \mathfrak{p} \cap \mathbb{Z}$, so that $\mathfrak{p} \cap \mathbb{Z}$ is nonzero. •

**Corollary 10.59.** *Let $R^*$ be integral over $R$, let $\mathfrak{p}$ be a prime ideal in $R$, and let $\mathfrak{p}^*$ and $\mathfrak{q}^*$ be prime ideals in $R^*$ lying over $\mathfrak{p}$. If $\mathfrak{p}^* \subseteq \mathfrak{q}^*$, then $\mathfrak{p}^* = \mathfrak{q}^*$.*

**Proof.** Lemma 10.55(ii) and Corollary 10.20(iii) show that the hypotheses are preserved by localizing at $\mathfrak{p}$; that is, $R_{\mathfrak{p}}^*$ is integral over $R_{\mathfrak{p}}$ and $\mathfrak{p}^* R_{\mathfrak{p}}^* \subseteq \mathfrak{q}^* R_{\mathfrak{p}}^*$ are prime ideals. Hence, replacing $R^*$ and $R$ by their localizations, we may assume that $R^*$ and $R$ are local rings and that $\mathfrak{p}$ is a maximal ideal in $R$ (by Proposition 10.23). But Corollary 10.57 says that maximality of $\mathfrak{p}$ forces maximality of $\mathfrak{p}^*$. Since $\mathfrak{p}^* \subseteq \mathfrak{q}^*$, we have $\mathfrak{p}^* = \mathfrak{q}^*$. •

Here are the Cohen–Seidenberg theorems.

**Theorem 10.60 (Lying Over).** *Let $R^*/R$ be a ring extension with $R^*$ integral over $R$. If $\mathfrak{p}$ is a prime ideal in $R$, then there is a prime ideal $\mathfrak{p}^*$ in $R^*$ lying over $\mathfrak{p}$; that is, $\mathfrak{p}^* \cap R = \mathfrak{p}$.*

**Proof.** There is a commutative diagram

$$\begin{array}{ccc} R & \xrightarrow{\ i\ } & R^* \\ {\scriptstyle h}\downarrow & & \downarrow{\scriptstyle h^*} \\ R_{\mathfrak{p}} & \xrightarrow[\ j\ ]{} & S^{-1}R^* \end{array}$$

where $h$ and $h^*$ are localization maps and $i$ and $j$ are inclusions. Let $S = R - \mathfrak{p}$; then $S^{-1}R^*$ is an extension of $R_{\mathfrak{p}}$ (since localization is an exact functor, $R$ contained in $R^*$ implies that $R_{\mathfrak{p}}$ is contained in $S^{-1}R^*$); by Lemma 10.55, $S^{-1}R^*$ is integral over $R_{\mathfrak{p}}$. Choose a maximal ideal $\mathfrak{m}^*$ in $S^{-1}R^*$. By Corollary 10.57, $\mathfrak{m}^* \cap R_{\mathfrak{p}}$ is a maximal ideal in $R_{\mathfrak{p}}$. But $R_{\mathfrak{p}}$ is a local ring with unique maximal ideal $\mathfrak{p}R_{\mathfrak{p}}$, so that $\mathfrak{m}^* \cap R_{\mathfrak{p}} = \mathfrak{p}R_{\mathfrak{p}}$. Since the inverse image of a prime ideal (under any ring map) is always prime, the ideal $\mathfrak{p}^* = (h^*)^{-1}(\mathfrak{m}^*)$ is a prime ideal in $R^*$. Now

$$(h^*i)^{-1}(\mathfrak{m}^*) = i^{-1}(h^*)^{-1}(\mathfrak{m}^*) = i^{-1}(\mathfrak{p}^*) = \mathfrak{p}^* \cap R,$$

while

$$(jh)^{-1}(\mathfrak{m}^*) = h^{-1}j^{-1}(\mathfrak{m}^*) = h^{-1}(\mathfrak{m}^* \cap R_{\mathfrak{p}}) = h^{-1}(\mathfrak{p}R_{\mathfrak{p}}) = \mathfrak{p}.$$

Therefore, $\mathfrak{p}^*$ is a prime ideal lying over $\mathfrak{p}$. •

**Theorem 10.61 (Going Up).** *Let $R^*/R$ be a ring extension with $R^*$ integral over $R$. If $\mathfrak{p} \subseteq \mathfrak{q}$ are prime ideals in $R$ and $\mathfrak{p}^*$ is a prime ideal in $R^*$ lying over $\mathfrak{p}$, then there exists a prime ideal $\mathfrak{q}^*$ lying over $\mathfrak{q}$ with $\mathfrak{p}^* \subseteq \mathfrak{q}^*$.*

**Proof.** Lemma 10.55 says that $(R^*/\mathfrak{p}^*)/(R/\mathfrak{p})$ is an integral ring extension, where $R/\mathfrak{p}$ is imbedded in $R^*/\mathfrak{p}^*$ as $(R + \mathfrak{p}^*)/\mathfrak{p}^*$. Replacing $R^*$ and $R$ by these quotient rings, we may assume that both $\mathfrak{p}^*$ and $\mathfrak{p}$ are $(0)$. The theorem now follows at once from the Lying Over Theorem. •

There is also a *Going Down Theorem*, but it requires an additional hypothesis.

**Theorem 10.62 (Going Down).** *Let $R^*/R$ be an integral extension and assume that $R$ is integrally closed. If $\mathfrak{p}_1 \supseteq \mathfrak{p}_2 \supseteq \cdots \supseteq \mathfrak{p}_n$ is a chain of prime ideals in $R$ and, for $m < n$, $\mathfrak{p}_1^* \supseteq \mathfrak{p}_2^* \supseteq \cdots \supseteq \mathfrak{p}_m^*$ is a chain of prime ideals in $R^*$ with each $\mathfrak{p}_i^*$ lying over $\mathfrak{p}_i$, then the chain in $R^*$ can be extended to $\mathfrak{p}_1^* \supseteq \mathfrak{p}_2^* \supseteq \cdots \supseteq \mathfrak{p}_n^*$ with $\mathfrak{p}_i^*$ lying over $\mathfrak{p}_i$ for all $i \leq n$.*

**Proof.** Atiyah–Macdonald, *Introduction to Commutative Algebra*, p. 64. •

# Exercises

∗ **10.27.** If $R$ is an integrally closed domain and $S \subseteq R$ is multiplicative, prove that $S^{-1}R$ is also integrally closed.

**10.28.** Prove that every valuation ring is integrally closed.

∗ **10.29.** Let $R^*/R$ be a ring extension. If $I$ is an ideal in $R$, denote its extension by $I^e$; if $I^*$ is an ideal in $R^*$, denote its contraction by $I^{*c}$. Prove each of the follow assertions.

(i) Both $e$ and $c$ preserve inclusion: if $I \subseteq J$, then $I^e \subseteq J^e$; if $I^* \subseteq J^*$, then $I^{*c} \subseteq J^{*c}$.

(ii) $I^{*ce} \subseteq I^*$ and $I^{ec} \supseteq I$.

(iii) $I^{*cec} = I^{*c}$ and $I^{ece} = I^e$.

(iv) $(I^* + J^*)^c \supseteq I^{*c} + J^{*c}$ and $(I + J)^e = I^e + J^e$.

(v) $(I^* \cap J^*)^c = I^{*c} \cap J^{*c}$ and $(I \cap J)^e \subseteq I^e \cap J^e$.

(vi) $(I^*J^*)^c \supseteq I^{*c}J^{*c}$ and $(IJ)^e = I^eJ^e$.

(vii) $\left(\sqrt{I^*}\right)^c = \sqrt{I^{*c}}$ and $\left(\sqrt{I}\right)^e \subseteq \sqrt{I^e}$.

(viii) $(J^* : I^*)^c \subseteq (J^{*c} : I^{*c})$ and $(I : J)^e \subseteq (I^e : J^e)$.

**10.30.** If $\mathbb{A}$ is the field of all algebraic numbers, denote the ring of all algebraic integers by $\mathcal{O}_{\mathbb{A}}$. Prove that
$$\mathcal{O}_{\mathbb{A}} \cap \mathbb{Q} = \mathbb{Z}.$$
Conclude, for every algebraic number field $E$, that $\mathcal{O}_E \cap \mathbb{Q} = \mathbb{Z}$.

**10.31.** Let $R^*/R$ be an integral ring extension.

(i) If $a \in R$ is a unit in $R^*$, prove that $a$ is a unit in $R$.

(ii) Prove that $J(R) = R \cap J(R^*)$, where $J(R)$ is the Jacobson radical.

**10.32.** Generalize Theorem 10.61 as follows. Let $R$ be integrally closed and let $R^*/R$ be an integral extension. If $\mathfrak{p}_1 \subseteq \mathfrak{p}_2 \subseteq \cdots \subseteq \mathfrak{p}_n$ is a chain of prime ideals in $R$ and, for $m < n$, $\mathfrak{p}_1^* \subseteq \mathfrak{p}_2^* \subseteq \cdots \subseteq \mathfrak{p}_m^*$ is a chain of prime ideals in $R^*$ with each $\mathfrak{p}_i^*$ lying over $\mathfrak{p}_i$, prove that the chain in $R^*$ can be extended to $\mathfrak{p}_1^* \subseteq \mathfrak{p}_2^* \subseteq \cdots \subseteq \mathfrak{p}_n^*$ with $\mathfrak{p}_i^*$ lying over $\mathfrak{p}_i$ for all $i \leq n$.

∗ **10.33.** Let $R^*/R$ be an integral extension. If every nonzero prime ideal in $R$ is a maximal ideal, prove that every nonzero prime ideal in $R^*$ is also a maximal ideal.

**Hint.** See the proof of Corollary 10.58.

∗ **10.34.** Let $\alpha$ be algebraic over $\mathbb{Q}$, let $E/\mathbb{Q}$ be a splitting field, and let $G = \text{Gal}(E/\mathbb{Q})$ be its Galois group.

(i) Prove that if $\alpha$ is integral over $\mathbb{Z}$, then, for all $\sigma \in G$, $\sigma(\alpha)$ is also integral over $\mathbb{Z}$.

(ii) Prove that $\alpha$ is an algebraic integer if and only if $\mathrm{irr}(\alpha, \mathbb{Q}) \in \mathbb{Z}[x]$. Compare this proof with that of Corollary 2.145.

(iii) Let $E$ be an algebraic number field and let $R \subseteq E$ be integrally closed. If $\alpha \in E$, prove that $\mathrm{irr}(\alpha, \mathrm{Frac}(R)) \in R[x]$.
   **Hint.** If $\widehat{E}$ is a Galois extension of $\mathrm{Frac}(R)$ containing $\alpha$, then $G = \mathrm{Gal}(\widehat{E}/\mathrm{Frac}(R))$ acts transitively on the roots of $\alpha$.

---

**10.3.2. Nullstellensatz Redux.** We now prove the Nullstellensatz for arbitrary algebraically closed fields (the proof we gave in Chapter 5 for $\mathbb{C}$ actually works for all uncountable fields, but it does not apply, for example, to the algebraic closures of prime fields, which are all countable). There are several different proofs of this result, and we present the proof of Goldman, as expounded by Kaplansky in his book *Commutative Rings*.

**Definition.** A ring extension $A/R$ is ***finitely generated*** if there is a surjective $R$-algebra map $\varphi \colon R[x_1, \ldots, x_n] \to A$. If $\varphi(x_i) = a_i$, then we write

$$A = R[a_1, \ldots, a_n].$$

If $I$ is an ideal in a commutative ring $R$, then the nilpotent elements in $R/I$ arise from elements of the radical $\sqrt{I}$. We now begin working toward a theorem of Krull that characterizes the nilpotent elements in a commutative ring, for this will give us information about radicals of ideals.

**Lemma 10.63.** *Let $R$ be a domain with $F = \mathrm{Frac}(R)$. Then $F/R$ is a finitely generated ring extension if and only if $F/R$ is a simply generated ring extension; that is, there is $u \in R$ with $F = R[u^{-1}]$.*

**Proof.** Sufficiency is obvious; we prove necessity. If $F = R[a_1/b_1, \ldots, a_n/b_n]$, define $u = \prod_i b_i$. We claim that $F = R[u^{-1}]$. Clearly, $F \supseteq R[u^{-1}]$. For the reverse inclusion, note that $a_i/b_i = a_i \widehat{u}_i/u \in R[u^{-1}]$, where $\widehat{u}_i = b_1 \cdots \widehat{b}_i \cdots b_n$.   •

**Definition.** If $R$ is a domain with $F = \mathrm{Frac}(R)$, then $R$ is a ***G-domain*** if $F/R$ is a finitely generated ring extension. An ideal $I$ in a commutative ring $R$ is a ***G-ideal*** [5] if it is prime and $R/I$ is a $G$-domain.

Every field is a $G$-domain, and so every maximal ideal in a commutative ring is a $G$-ideal. Corollary 10.67 below will say that $\mathbb{Z}$ is not a $G$-domain; it will then follow that the prime ideal $(x)$ in $\mathbb{Z}[x]$ is not a $G$-ideal.

**Proposition 10.64.** *Let $E/R$ be a ring extension in which both $E$ and $R$ are domains. If $E$ is a finitely generated $R$-algebra and each $\alpha \in E$ is algebraic over $R$ (that is, $\alpha$ is a root of a nonzero polynomial in $R[x]$), then $R$ is a $G$-domain if and only if $E$ is a $G$-domain.*

---

[5] $G$-domains and $G$-ideals are named after Goldman.

**Proof.** Let $R$ be a $G$-domain, so that $F = \text{Frac}(R) = R[u^{-1}]$ for some nonzero $u \in R$, by Lemma 10.63. Now $E[u^{-1}] \subseteq \text{Frac}(E)$, because $u \in R \subseteq E$,. But $E[u^{-1}]$ is a domain algebraic over the field $F = R[u^{-1}]$, so that $E[u^{-1}]$ is a field, by Exercise 10.36 on page 914. Since $\text{Frac}(E)$ is the smallest field containing $E$, we have $E[u^{-1}] = \text{Frac}(E)$, and so $E$ is a $G$-domain.

If $E$ is a $G$-domain, then there is $v \in E$ with $\text{Frac}(E) = E[v^{-1}]$. By hypothesis, $E = R[b_1, \ldots, b_n]$, where $b_i$ is algebraic over $F = \text{Frac}(R)$ for all $i$. Now $v \in E$, so that $v$ algebraic over $R$ implies $v^{-1}$ is algebraic over $F$, by Proposition 5.53(iii). Thus, there are monic polynomials $f_0(x), f_i(x) \in F[x]$ with $f_0(v^{-1}) = 0$ and $f_i(b_i) = 0$ for all $i \geq 1$. Clearing denominators, we obtain equations $\beta_i f_i(b_i) = 0$, for $i \geq 0$, with coefficients in $R$:

$$\beta_0 (v^{-1})^{d_0} + \cdots = 0,$$
$$\beta_i b_i^{d_i} + \cdots = 0.$$

Define $R^* = R[\beta_0^{-1}, \beta_1^{-1}, \ldots, \beta_n^{-1}]$. Each $b_i$ is integral over $R^*$, for each $\beta_i$ is a unit in $R^*$. Clearly, $E[v^{-1}] = R^*[v^{-1}, b_1, \ldots, b_n]$. Thus, the field $E[v^{-1}]$ is integral over $R^*$, by Proposition 10.48 (since $E[v^{-1}] = R^*[v^{-1}, b_1, \ldots, b_n]$ and each of the displayed generators is integral over $R^*$), and this forces $R^*$ to be a field, by Proposition 10.56. But $R^* = R[\beta_0^{-1}, \beta_1^{-1}, \ldots, \beta_n^{-1}] \subseteq F$, because $\beta_i \in R$ for all $i$, so that $R^* = F$. Therefore, $F = R[\beta_0^{-1}, \beta_1^{-1}, \ldots, \beta_n^{-1}]$ is a finitely generated ring extension of $R$; that is, $R$ is a $G$-domain. $\bullet$

The next lemma leads to Corollary 10.66, an "internal" characterization of $G$-domains, phrased solely in terms of $R$, with no mention of $\text{Frac}(R)$.

**Lemma 10.65.** *Let $R$ be a domain with $F = \text{Frac}(R)$. The following conditions are equivalent for a nonzero element $u \in R$.*

(i) *$u$ lies in every nonzero prime ideal of $R$.*

(ii) *for every nonzero ideal $I$ in $R$, there is an integer $n = n(I)$ with $u^n \in I$.*

(iii) *$R$ is a $G$-domain; that is, $F = R[u^{-1}]$.*

**Proof.**

(i) $\Rightarrow$ (ii). Suppose there is a nonzero ideal $I$ for which $u^n \notin I$ for all $n \geq 0$. If $S = \{u^n : n \geq 0\}$, then $I \cap S = \varnothing$. By Zorn's Lemma, there is an ideal $\mathfrak{p}$ maximal with $I \subseteq \mathfrak{p}$ and $\mathfrak{p} \cap S = \varnothing$. Now $\mathfrak{p}$ is a prime ideal, by Exercise 5.7 on page 301, and this contradicts $u$ lying in every prime ideal.

(ii) $\Rightarrow$ (iii). If $b \in R$ and $b \neq 0$, then $u^n \in (b)$ for some $n \geq 1$, by hypothesis. Hence, $u^n = rb$ for some $r \in R$, and so $b^{-1} = ru^{-n} \in R[u^{-1}]$. Therefore, $F = R[u^{-1}]$.

(iii) $\Rightarrow$ (i). Let $\mathfrak{p}$ be a nonzero prime ideal. If $b \in \mathfrak{p}$ is nonzero, then $b^{-1} = \sum_{i=0}^n r_i u^{-i}$, where $r_i \in R$, because $F = R[u^{-1}]$. Hence $u^n = b\left(\sum_i r_i u^{n-i}\right)$ lies in $\mathfrak{p}$, because $b \in \mathfrak{p}$ and $\sum_i r_i u^{n-i} \in R$. Since $\mathfrak{p}$ is a prime ideal, $u \in \mathfrak{p}$. $\bullet$

**Corollary 10.66.** *A domain $R$ is a $G$-domain if and only if $\bigcap_{\substack{\mathfrak{p} \text{ prime} \\ \mathfrak{p} \neq 0}} \mathfrak{p} \neq (0)$.*

**Proof.** By Lemma 10.65, $R$ is a $G$-domain if and only if it has a nonzero element $u$ lying in every nonzero prime ideal.   •

**Corollary 10.67.** *If $R$ is a PID, then $R$ is a $G$-domain if and only if $R$ has only finitely many prime ideals.*

**Proof.** If $R$ is a $G$-domain, then $I = \bigcap \mathfrak{p} \neq (0)$, where $\mathfrak{p}$ ranges over all nonzero prime ideals. Suppose that $R$ has infinitely many prime ideals, say, $(p_1), (p_2), \dots$. If $a \in I$, then $p_i \mid a$ for all $i$. But $a = q_1^{e_1} \cdots q_n^{e_n}$, where the $q_j$ are distinct prime elements, contradicting unique factorization in the PID $R$.

Conversely, if $R$ has only finitely many nonzero prime ideals, say, $(p_1), \dots, (p_m)$, then the product $p_1 \cdots p_m$ is a nonzero element lying in $\bigcap_i (p_i)$. Therefore, $R$ is a $G$-domain.   •

It follows, for example, that every DVR is a $G$-domain.

**Definition.** If $R$ is a commutative ring, then its ***nilradical*** is

$$\mathrm{nil}(R) = \{r \in R : r \text{ is nilpotent}\}.$$

We note that $\mathrm{nil}(R)$ is an ideal. If $r, s \in R$ are nilpotent, then $r^n = 0 = s^m$, for positive integers $m$ and $n$. Hence,

$$(r + s)^{m+n-1} = \sum_{i=0}^{m+n-1} \binom{m+n-1}{i} r^i s^{m+n-1-i}.$$

If $i \geq n$, then $r^i = 0$ and the $i$th term in the sum is $0$; if $i < n$, then $m+n-i-1 \geq m$, $s^{m+n-1-i} = 0$, and the $i$th term in the sum is $0$ in this case as well. Thus, $(r + s)^{m+n-1} = 0$ and $r + s$ is nilpotent. Finally, $rs$ is nilpotent, for $(rs)^{mn} = r^{mn} s^{ms} = 0$.

The next theorem is a modest improvement of a theorem of Krull which characterizes the nilradical as the intersection of all the prime ideals.

**Theorem 10.68 (Krull).** *If $R$ is a commutative ring, then*

$$\mathrm{nil}(R) = \bigcap_{\substack{\mathfrak{p}=\text{prime} \\ \text{ideal}}} \mathfrak{p} = \bigcap_{\mathfrak{p}=G\text{-ideal}} \mathfrak{p}.$$

**Remark.** If $R$ is a domain, then $(0)$ is a prime ideal, and so $\mathrm{nil}(R) = (0)$ (alternatively, there are no nonzero nilpotent elements in a domain). The intersection of all the nonzero prime ideals in a commutative ring $R$ may be larger than $\mathrm{nil}(R)$; this happens, for example, when $R$ is a DVR.   ◀

**Proof.** There are inclusions $\mathrm{nil}(R) \subseteq \bigcap_{\mathfrak{p}=\text{prime ideal}} \mathfrak{p} \subseteq \bigcap_{\mathfrak{p}=G\text{-ideal}} \mathfrak{p}$: nilpotent elements lie in every prime ideal; every $G$-ideal is a prime ideal. Thus, it suffices to prove that $\bigcap_{\mathfrak{p}=G\text{-ideal}} \mathfrak{p} \subseteq \mathrm{nil}(R)$. Suppose that $u \notin \bigcap_{\mathfrak{p}=G\text{-ideal}} \mathfrak{p}$. It follows that $u^n \neq 0$ for all $n \geq 1$, for if $u^n = 0$, then $u^n \in \mathfrak{p}$ for every $G$-ideal $\mathfrak{p}$, and so $u \in \mathfrak{p}$, because $G$-ideals are prime. Therefore, the subset $S = \{u^n : n \geq 1\}$ is multiplicative, By Exercise 5.39 on page 321, there exists an ideal $\mathfrak{q}$ maximal with $\mathfrak{q} \cap S = \varnothing$; now $\mathfrak{q}$ is a prime ideal, by Exercise 5.7 on page 301, and so $R/\mathfrak{q}$ is a domain. We claim that $\mathfrak{q}$ is a $G$-ideal, which will give $u \notin \bigcap_{\mathfrak{p}=G\text{-ideal}} \mathfrak{p}$. If there is a nonzero prime ideal $\mathfrak{p}^*$ in $R/\mathfrak{q}$ not containing $u + \mathfrak{q}$, then there is an ideal $\mathfrak{p} \supsetneq \mathfrak{q}$ in $R$ with $\mathfrak{p}^* = \mathfrak{p}/\mathfrak{q}$ [for $\mathfrak{p}^* \neq (0)$], contradicting the maximality of $\mathfrak{q}$. Therefore, $u + \mathfrak{q}$ lies in every nonzero prime ideal in $R/\mathfrak{q}$. By Corollary 10.66, $R/\mathfrak{q}$ is a $G$-domain, and so $\mathfrak{q}$ is a $G$-ideal. •

The next corollary follows easily from Krull's Theorem.

**Corollary 10.69.** *If $I$ is an ideal in a commutative ring $R$, then $\sqrt{I}$ is the intersection of all the $G$-ideals containing $I$.*

**Proof.** By definition, $\sqrt{I} = \{r \in R : r^n \in I \text{ for some } n \geq 1\}$. Therefore, $\sqrt{I}/I = \operatorname{nil}(R/I) = \bigcap_{\mathfrak{p}^*=G\text{-ideal}} \mathfrak{p}^*$. For each $\mathfrak{p}^*$, there is an ideal $\mathfrak{p}$ containing $I$ with $\mathfrak{p}^* = \mathfrak{p}/I$, and $\sqrt{I} = \bigcap_{\mathfrak{p}/I=G\text{-ideal}} \mathfrak{p}$. Finally, every $\mathfrak{p}$ involved in the intersection is a $G$-ideal, because $(R/I)/\mathfrak{p}^*$ is a $G$-domain, and $R/\mathfrak{p} \cong (R/I)/(\mathfrak{p}/I) = (R/I)/\mathfrak{p}^*$. •

We now focus on the relation between ideals in $R[x]$ and those in $R$.

**Proposition 10.70.** *An ideal $I$ in a commutative ring $R$ is a $G$-ideal if and only if $I$ is the contraction of a maximal ideal in $R[x]$.*

**Proof.** If $I$ is a $G$-ideal in $R$, then $R/I$ is a $G$-domain. Hence, there is $u \in \operatorname{Frac}(R/I)$ with $\operatorname{Frac}(R/I) = (R/I)[u^{-1}]$. Let $\varphi \colon (R/I)[x] \to (R/I)[u^{-1}]$ be the $R$-algebra map taking $x \mapsto u^{-1}$. Since $\varphi$ is a surjection onto the field $(R/I)[u^{-1}] = \operatorname{Frac}(R/I)$, its kernel $\mathfrak{m}$ is a maximal ideal in $(R/I)[x]$. Since $\varphi|(R/I)$ is an injection, we have $\mathfrak{m} \cap (R/I) = (0)$. By Exercise 5.2 on page 301, there is an ideal, necessarily maximal, $\mathfrak{m}'$ in $R[x]$ with $\mathfrak{m}'/I = \mathfrak{m}$, and $\mathfrak{m}' \cap R = I$.

Conversely, assume that $\mathfrak{m}$ is a maximal ideal in $R[x]$ with $\mathfrak{m} \cap R = I$. If $\nu \colon R[x] \to R[x]/\mathfrak{m}$ is the natural map and $u = \nu(x)$, then $\operatorname{im} \nu = (R/I)[u]$ is a field. Hence, $R/I$ is a $G$-domain, by Proposition 10.64, and so $I$ is a $G$-ideal. •

**Notation.** If $I$ is an ideal in a commutative ring $R$ and $f(x) \in R[x]$, then $\overline{f}(x)$ denotes the polynomial in $(R/I)[x]$ obtained from $f(x)$ by reducing its coefficients mod $I$; that is, if $f(x) = \sum_i a_i x^i$, where $a_i \in R$, then

$$\overline{f}(x) = \sum_i (a_i + I)x^i.$$

**Corollary 10.71.** *Let $R$ be a commutative ring, and let $\mathfrak{m}$ be a maximal ideal in $R[x]$. If the contraction $\mathfrak{m}' = \mathfrak{m} \cap R$ is a maximal ideal in $R$, then $\mathfrak{m} = (\mathfrak{m}', f(x))$, where $f(x) \in R[x]$ and $\overline{f}(x) \in (R/\mathfrak{m}')[x]$ is irreducible. If $R/\mathfrak{m}'$ is algebraically closed, then $\mathfrak{m} = (\mathfrak{m}', x - a)$ for some $a \in R$.*

**Proof.** First, Proposition 10.70 says that $\mathfrak{m}' = \mathfrak{m} \cap R$ is a $G$-ideal in $R$. Consider the map $\varphi \colon R[x] \to (R/\mathfrak{m}')[x]$ which reduces coefficients mod $\mathfrak{m}'$. Since $\varphi$ is a surjection, the ideal $\varphi(\mathfrak{m})$ is a maximal ideal; that is, $\varphi(\mathfrak{m}) = (g(x))$, where $g(x) \in (R/\mathfrak{m}')[x]$ is irreducible. Therefore, $\mathfrak{m} = (\mathfrak{m}', f(x))$, where $\varphi(f) = g$; that is, $\overline{f}(x) = g(x)$. •

Maximal ideals are always $G$-ideals, and $G$-ideals are always prime ideals. The next definition imposes the condition that $G$-ideals be maximal. In light of Proposition 10.70, this will force the contraction of maximal ideals in $R[x]$ to be maximal ideals in $R$.

**Definition.** A commutative ring $R$ is a ***Jacobson ring***[6] if every $G$-ideal is a maximal ideal.

**Example 10.72.**

(i) Every field is a Jacobson ring.

(ii) By Corollary 10.67, a PID $R$ is a $G$-domain if and only if it has only finitely many prime ideals. Such a ring cannot be a Jacobson ring, for $(0)$ is a $G$-ideal which is not maximal $[R/(0) \cong R$ is a $G$-domain$]$. On the other hand, if $R$ has infinitely many prime ideals, then $R$ is not a $G$-domain and $(0)$ is not a $G$-ideal. The $G$-ideals, which are now nonzero prime ideals, must be maximal. Therefore, a PID is a Jacobson ring if and only if it has infinitely many prime ideals.

(iii) We note that if $R$ is a Jacobson ring, then so is any quotient $R^* = R/I$. If $\mathfrak{p}^*$ is a $G$-ideal in $R^*$, then $R^*/\mathfrak{p}^*$ is a $G$-domain. Now $\mathfrak{p}^* = \mathfrak{p}/I$ for some ideal $\mathfrak{p}$ in $R$, and $R/\mathfrak{p} \cong (R/I)/(\mathfrak{p}/I) = R^*/\mathfrak{p}^*$. Thus, $\mathfrak{p}$ is a $G$-ideal in $R$. Since $R$ is a Jacobson ring, $\mathfrak{p}$ is a maximal ideal, and $R/\mathfrak{p} \cong R^*/\mathfrak{p}^*$ is a field. Therefore, $\mathfrak{p}^*$ is a maximal ideal, and so $R^*$ is also a Jacobson ring.

(iv) By Corollary 10.69, every radical ideal in a commutative ring $R$ is the intersection of all the $G$-ideals containing it. Therefore, if $R$ is a Jacobson ring, then every radical ideal is an intersection of maximal ideals.  ◄

Example 10.72(iv) suggests the following result.

**Proposition 10.73.** *A commutative ring $R$ is a Jacobson ring if and only if every prime ideal in $R$ is an intersection of maximal ideals.*

**Proof.** By Corollary 10.69, every radical ideal, hence, every prime ideal, is the intersection of all the $G$-ideals containing $I$. But in a Jacobson ring, every $G$-ideal is maximal.

Conversely, assume that every prime ideal in $R$ is an intersection of maximal ideals. We let the reader check that this property is inherited by quotient rings. Let $\mathfrak{p}$ be a $G$-ideal in $R$, so that $R/\mathfrak{p}$ is a $G$-domain. Thus, there is $u \neq 0$ in $R/\mathfrak{p}$ with $\mathrm{Frac}(R/\mathfrak{p}) = (R/\mathfrak{p})[u^{-1}]$. By Lemma 10.65, $u$ lies in every nonzero prime ideal of $R/\mathfrak{p}$, and so $u$ lies in every nonzero maximal ideal. Now every prime ideal in $R/\mathfrak{p}$ is an intersection of maximal ideals; in particular, since $R/\mathfrak{p}$ is a domain, there are maximal ideals $\mathfrak{m}_\alpha$ with $(0) = \bigcap_\alpha \mathfrak{m}_\alpha$. If all these $\mathfrak{m}_\alpha$ are nonzero, then $u \in \bigcap_\alpha \mathfrak{m}_\alpha = (0)$, a contradiction. We conclude that $(0)$ is a maximal ideal. Therefore, $R/\mathfrak{p}$ is a field, the $G$-ideal $\mathfrak{p}$ is maximal, and $R$ is a Jacobson ring.  •

---

[6]These rings are called ***Hilbert rings*** by some authors. In 1951, Krull and Goldman, independently, published proofs of the Nullstellensatz using the techniques in this subsection. Krull introduced the term *Jacobson ring* in his paper.

**Corollary 10.74.** *A commutative ring $R$ is a Jacobson ring if and only if $J(R/I) =$ $\mathrm{nil}(R/I)$ for every ideal $I$. In particular, $J(R) = \mathrm{nil}(R)$.*

**Proof.** Let $R$ be a Jacobson ring. If $I$ is an ideal in $R$, then $\sqrt{I} = \bigcap \mathfrak{m}$, where $\mathfrak{m}$ is a maximal ideal containing $I$. Now $J(R/I)$ is the intersection of all the maximal ideals in $R/I$; that is, $J(R) = \bigcap(\mathfrak{m}/I) = (\bigcap \mathfrak{m})/I = \sqrt{I}/I$. On the other hand, $\mathrm{nil}(R/I)$ consists of all the nilpotent elements in $R/I$. But $0 = (f + I)^n = f^n + I$ holds if and only if $f^n \in I$; that is, $f \in \sqrt{I}$. To prove the converse, note that hypothesis says that every radical ideal in $R$ is an intersection of maximal ideals. In particular, every prime ideal is such an intersection, and so $R$ is a Jacobson ring. •

The next result will give many examples of Jacobson rings.

**Theorem 10.75.** *A commutative ring $R$ is a Jacobson ring if and only if $R[x]$ is a Jacobson ring.*

**Proof.** We have seen that every quotient of a Jacobson ring is a Jacobson ring. Hence, if $R[x]$ is a Jacobson ring, then $R \cong R[x]/(x)$ is also a Jacobson ring.

Conversely, suppose that $R$ is a Jacobson ring. If $\mathfrak{q}$ is a $G$-ideal in $R[x]$, then we may assume that $\mathfrak{q} \cap R = (0)$, by Exercise 10.37 on page 914. If $\nu \colon R[x] \to R[x]/\mathfrak{q}$ is the natural map, then $R[x]/\mathfrak{q} = R[u]$, where $u = \nu(x)$. Now $R[u]$ is a $G$-domain, because $\mathfrak{q}$ is a $G$-ideal; hence, if $K = \mathrm{Frac}(R[u])$, then there is $v \in K$ with $K = R[u][v^{-1}]$. If $\mathrm{Frac}(R) = F$, then

$$K = R[u][v^{-1}] \subseteq F[u][v^{-1}] \subseteq K,$$

so that $F[u][v^{-1}] = K$; that is, $F[u]$ is a $G$-domain. But $F[u]$ is not a $G$-domain if $u$ is transcendental over $F$, by Corollary 10.67, for $F[x] \cong F[u]$ has infinitely many prime ideals. Thus, $u$ is algebraic over $F$, and hence $u$ is algebraic over $R$. Since $R[u]$ is a $G$-domain, Proposition 10.64 says that $R$ is a $G$-domain. Now $R$ is a Jacobson ring, and so $R$ is a field, by Exercise 10.35 on page 914. But if $R$ is a field, so is $R[u]$, for $u$ is algebraic over $R$. Therefore, $R[u] = R[x]/\mathfrak{q}$ is a field, so that $\mathfrak{q}$ is a maximal ideal, and $R[x]$ is a Jacobson ring. •

**Corollary 10.76.** *If $k$ is a field, then $k[x_1, \dots, x_n]$ is a Jacobson ring.*

**Proof.** The proof is by induction on $n \geq 1$. For the base step, $k[x]$ is a PID having infinitely many prime ideals, by Exercise 10.41 on page 915, and so it is a Jacobson ring, by Example 10.72(ii). For the inductive step, the inductive hypothesis gives $R = k[x_1, \dots, x_{n-1}]$ a Jacobson ring, and Theorem 10.75 applies. •

**Theorem 10.77.** *If $\mathfrak{m}$ is a maximal ideal in $k[x_1, \dots, x_n]$, where $k$ is an algebraically closed field, then there are $a_1, \dots, a_n \in k$ such that*

$$\mathfrak{m} = (x_1 - a_1, \dots, x_n - a_n).$$

**Proof.** The proof is by induction on $n \geq 1$. If $n = 1$, then $\mathfrak{m} = (p(x))$, where $p(x) \in k[x]$ is irreducible. Since $k$ is algebraically closed, $p(x)$ is linear. For the inductive step, let $R = k[x_1, \dots, x_{n-1}]$. Corollary 10.76 says that $R$ is a Jacobson ring, and so $\mathfrak{m} \cap R$ is a $G$-ideal in $R$, by Proposition 10.70. Since $R$ is a Jacobson ring,

$\mathfrak{m}' = \mathfrak{m} \cap R$ is a maximal ideal. Corollary 10.71 now applies to give $\mathfrak{m} = (\mathfrak{m}', f(x_n))$, where $f(x_n) \in R[x_n]$ and $\overline{f}(x_n) \in (R/\mathfrak{m}')[x_n]$ is irreducible. As $k$ is algebraically closed and $R/\mathfrak{m}'$ is a finitely generated $k$-algebra, $R/\mathfrak{m}' \cong k$, and we may assume that $f(x_n)$ is linear; there is $a_n \in k$ with $f_n(x) = x_n - a_n$. By the inductive hypothesis, $\mathfrak{m}' = (x_1 - a_1, \ldots, x_{n-1} - a_{n-1})$ for $a_1, \ldots, a_{n-1} \in k$, and this completes the proof.   •

We use Theorem 10.77 to prove the Weak Nullstellensatz, Theorem 5.98. Recall that only the special case of the Nullstellensatz for $k = \mathbb{C}$ was proved in Chapter 5.

**Theorem 10.78 (Weak Nullstellensatz).** *Let $f_1(X), \ldots, f_t(X) \in k[X]$, where $k$ is an algebraically closed field. Then $I = (f_1, \ldots, f_t)$ is a proper ideal in $k[X]$ if and only if $\mathrm{Var}(f_1, \ldots, f_t) \neq \varnothing$.*

**Proof.** If $I$ is a proper ideal, then there is a maximal ideal $\mathfrak{m}$ containing it. By Theorem 5.98, there is $a = (a_1, \ldots, a_n) \in k^n$ with $\mathfrak{m} = (x_1 - a_1, \ldots, x_n - a_n)$. Now $I \subseteq \mathfrak{m}$ implies $\mathrm{Var}(\mathfrak{m}) \subseteq \mathrm{Var}(I)$. But $a \in \mathrm{Var}(\mathfrak{m})$, and so $\mathrm{Var}(I) \neq \varnothing$.   •

We could now repeat the proof of the Nullstellensatz over $\mathbb{C}$, Theorem 5.99, to obtain the Nullstellensatz over any algebraically closed field. However, the following proof is easier.

**Theorem 10.79 (Nullstellensatz).** *Let $k$ be an algebraically closed field. If $I$ is an ideal in $k[x_1, \ldots, x_n]$, then $\mathrm{Id}(\mathrm{Var}(I)) = \sqrt{I}$.*

**Proof.** The inclusion $\mathrm{Id}(\mathrm{Var}(I)) \supseteq \sqrt{I}$ is easy to see. If $f^n(a) = 0$ for all $a \in \mathrm{Var}(I)$, then $f(a) = 0$ for all $a \in \mathrm{Var}(I)$, because the values of $f$ lie in the field $k$. Hence, $f \in \mathrm{Id}(\mathrm{Var}(I))$. For the reverse inclusion, note first that $k[x_1, \ldots, x_n]$ is a Jacobson ring, by Corollary 10.76; hence, Example 10.72(iv) shows that $\sqrt{I}$ is an intersection of maximal ideals. Let $g \in \mathrm{Id}(\mathrm{Var}(I))$. If $\mathfrak{m}$ is a maximal ideal containing $I$, then $\mathrm{Var}(\mathfrak{m}) \subseteq \mathrm{Var}(I)$, and so $\mathrm{Id}(\mathrm{Var}(I)) \subseteq \mathrm{Id}(\mathrm{Var}(\mathfrak{m}))$. But $\mathrm{Id}(\mathrm{Var}(\mathfrak{m})) = \mathfrak{m}$: $\mathrm{Id}(\mathrm{Var}(I)) \supseteq \sqrt{\mathfrak{m}} = \mathfrak{m}$, because $\mathfrak{m}$ is a maximal, hence prime ideal. Therefore, $g \in \bigcap \mathfrak{m} = \sqrt{I}$, as desired.   •

## Exercises

*   **10.35.** Prove that a commutative ring $R$ is a field if and only if $R$ is both a Jacobson ring and a $G$-domain.

*   **10.36.** Let $E/R$ be a ring extension in which $R$ is a field and $E$ is a domain.

    (i) Let $b \in E$ be algebraic over $R$. Prove that there exists an equation
    $$b^n + r_{n-1}b^{n-1} + \cdots + r_1 b + r_0 = 0,$$
    where $r_i \in R$ for all $i$ and $r_0 \neq 0$.

    (ii) If $E = R[b_1, \ldots, b_m]$, where each $b_j$ is algebraic over $R$, prove that $E$ is a field.

*   **10.37.** Let $R$ be a Jacobson ring, and assume that $(R/\mathfrak{q}')[x]$ is a Jacobson ring for every $G$-ideal $\mathfrak{q}$ in $R[x]$, where $\mathfrak{q}' = \mathfrak{q} \cap R$. Prove that $R[x]$ is a Jacobson ring.

**10.38.** (i) Prove that $\mathfrak{m} = (x^2 - y, y^2 - 2)$ is a maximal ideal in $\mathbb{Q}[x, y]$.

(ii) Prove that there do not exist $f(x) \in \mathbb{Q}[x]$ and $g(y) \in \mathbb{Q}[y]$ with $\mathfrak{m} = \big(f(x), g(y)\big)$.

**10.39.** Let $k$ be a field and let $\mathfrak{m}$ be a maximal ideal in $k[x_1, \ldots, x_n]$. Prove that
$$\mathfrak{m} = \big(f_1(x_1), f_2(x_1, x_2), \ldots, f_{n-1}(x_1, \ldots, x_{n-1}), f_n(x_1, \ldots, x_n)\big).$$

**Hint.** Use Corollary 10.71.

\* **10.40.** Prove that if $R$ is noetherian, then $\mathrm{nil}(R)$ is a nilpotent ideal.

\* **10.41.** If $k$ is a field, prove that $k[x]$ has infinitely many prime ideals.

---

**10.3.3. Algebraic Integers.** We have mentioned that Kummer investigated the ring $\mathbb{Z}[\zeta_p]$, where $p$ is an odd prime and $\zeta_p$ is a primitive $p$th root of unity. We now study rings of integers in algebraic number fields $E$ further. Recall the definition:
$$\mathcal{O}_E = \{\alpha \in E : \alpha \text{ is integral over } \mathbb{Z}\}.$$

We begin with a consequence of Gauss's Lemma.

**Lemma 10.80.** *Let $E$ be an algebraic number field with $[E : \mathbb{Q}] = n$, and let $\alpha \in E$ be an algebraic integer. Then $\mathrm{irr}(\alpha, \mathbb{Q}) \in \mathbb{Z}[x]$ and $\deg(\mathrm{irr}(\alpha, \mathbb{Q})) \mid n$.*

**Proof.** By Corollary 2.145, $\mathrm{irr}(\alpha, \mathbb{Q}) \in \mathbb{Z}[x]$, and so the result follows from Proposition 2.141(v). $\bullet$

**Definition.** A *quadratic field* is an algebraic number field $E$ with $[E : \mathbb{Q}] = 2$.

**Proposition 10.81.** *Every quadratic field $E$ has the form $E = \mathbb{Q}(\sqrt{d})$, where $d$ is a squarefree integer.*

**Proof.** We know that $E = \mathbb{Q}(\alpha)$, where $\alpha$ is a root of a quadratic polynomial; say, $\alpha^2 + b\alpha + c = 0$, where $b, c \in \mathbb{Q}$. If $D = b^2 - 4c$, then the quadratic formula gives $\alpha = -\frac{1}{2}b \pm \frac{1}{2}\sqrt{D}$, and so $E = \mathbb{Q}(\alpha) = \mathbb{Q}(\sqrt{D})$. Write $D$ in lowest terms: $D = U/V$, where $U, V \in \mathbb{Z}$ and $(U, V) = 1$. Now $U = ur^2$ and $V = vs^2$, where $u, v$ are squarefree; hence, $uv$ is squarefree, because $(u, v) = 1$. Therefore, $\mathbb{Q}(\sqrt{D}) = \mathbb{Q}(\sqrt{u/v}) = \mathbb{Q}(\sqrt{uv})$, for $\sqrt{u/v} = \sqrt{uv/v^2} = \sqrt{uv}/v$. $\bullet$

We now describe the integers in quadratic fields.

**Proposition 10.82.** *Let $E = \mathbb{Q}(\sqrt{d})$, where $d$ is a squarefree integer (which implies that $d \not\equiv 0 \bmod 4$).*

(i) *If $d \equiv 2 \bmod 4$ or $d \equiv 3 \bmod 4$, then $\mathcal{O}_E = \mathbb{Z}[\sqrt{d}]$.*

(ii) *If $d \equiv 1 \bmod 4$, then $\mathcal{O}_E$ consists of all $\frac{1}{2}\big(u + v\sqrt{d}\big)$ with $u$ and $v$ rational integers having the same parity.*

**Proof.** If $\alpha \in E = \mathbb{Q}(\sqrt{d})$, then there are $a, b \in \mathbb{Q}$ with $\alpha = a + b\sqrt{d}$. We first show that $\alpha \in \mathcal{O}_E$ if and only if

(1)
$$2a \in \mathbb{Z} \quad \text{and} \quad a^2 - db^2 \in \mathbb{Z}.$$

If $\alpha \in \mathcal{O}_E$, then Lemma 10.80 says that $p(x) = \mathrm{irr}(\alpha, \mathbb{Q}) \in \mathbb{Z}[x]$ is quadratic. Now $\mathrm{Gal}(E/\mathbb{Q}) = \langle \sigma \rangle$, where $\sigma \colon E \to E$ carries $\sqrt{d} \mapsto -\sqrt{d}$; that is,

$$\sigma(\alpha) = a - b\sqrt{d}.$$

Since $\sigma$ permutes the roots of $p(x)$, the other root of $p(x)$ is $\sigma(\alpha)$; that is,

$$p(x) = (x - \alpha)(x - \sigma(\alpha)) = x^2 - 2ax + (a^2 - db^2).$$

Hence, the formulas in (1) hold, because $p(x) \in \mathbb{Z}[x]$.

Conversely, if (1) holds, then $\alpha \in \mathcal{O}_E$, because $\alpha$ is a root of a monic polynomial in $\mathbb{Z}[x]$, namely, $x^2 - 2ax + (a^2 - db^2)$.

We now show that $2b \in \mathbb{Z}$. Multiplying the second formula in (1) by 4 gives $(2a)^2 - d(2b)^2 \in \mathbb{Z}$. Since $2a \in \mathbb{Z}$, we have $d(2b)^2 \in \mathbb{Z}$. Write $2b$ in lowest terms: $2b = m/n$, where $(m, n) = 1$. Now $dm^2/n^2 \in \mathbb{Z}$, so that $n^2 \mid dm^2$. But $(n^2, m^2) = 1$ forces $n^2 \mid d$; as $d$ is squarefree, $n = 1$ and $2b = m/n \in \mathbb{Z}$.

We have shown that $a = \frac{1}{2}u$ and $b = \frac{1}{2}v$, where $u, v \in \mathbb{Z}$. Substituting these values into the second formula in (1) gives

$$(2) \qquad\qquad\qquad\qquad u^2 \equiv dv^2 \bmod 4.$$

Note that squares are congruent mod 4, either to 0 or to 1. If $d \equiv 2 \bmod 4$, then the only way to satisfy (2) is $u^2 \equiv 0 \bmod 4$ and $v^2 \equiv 0 \bmod 4$. Thus, both $u$ and $v$ must be even, and so $\alpha = \frac{1}{2}u + \frac{1}{2}v\sqrt{d} \in \mathbb{Z}[\sqrt{d}]$. Therefore, $\mathcal{O}_E = \mathbb{Z}[\sqrt{d}]$ in this case, for $\mathbb{Z}[\sqrt{d}] \subseteq \mathcal{O}_E$ is easily seen to be true. A similar argument works when $d \equiv 3 \bmod 4$. However, if $d \equiv 1 \bmod 4$, then $u^2 \equiv v^2 \bmod 4$. Hence, $v$ is even if and only if $u$ is even; that is, $u$ and $v$ have the same parity. If $u$ and $v$ are both even, then $a, b \in \mathbb{Z}$ and $\alpha \in \mathcal{O}_E$. If $u$ and $v$ are both odd, then $u^2 \equiv 1 \equiv v^2 \bmod 4$, and so $u^2 \equiv dv^2 \bmod 4$, because $d \equiv 1 \bmod 4$. Therefore, (1) holds, and so $\alpha$ lies in $\mathcal{O}_E$. $\bullet$

If $E = \mathbb{Q}(\sqrt{d})$, where $d \in \mathbb{Z}$, then $\mathbb{Z}[\sqrt{d}] \subseteq \mathcal{O}_E$, but we now see that this inclusion may be strict. For example, $\frac{1}{2}(1 + \sqrt{5})$ is an algebraic integer (it is a root of $x^2 - x - 6$). Therefore, $\mathbb{Z}[\sqrt{5}] \subsetneqq \mathcal{O}_E$, where $E = \mathbb{Q}(\sqrt{5})$.

The coming brief digression into Linear Algebra will enable us to prove that rings of integers $\mathcal{O}_E$ are noetherian.

**Definition.** Let $E/k$ be a field extension in which $E$ is finite-dimensional. If $u \in E$, then multiplication $\Gamma_u \colon E \to E$, given by $\Gamma_u \colon y \mapsto uy$, is a $k$-map. If $e_1, \ldots, e_n$ is a basis of $E$, then $\Gamma_u$ is represented by a matrix $A = [a_{ij}]$ with entries in $k$; that is,

$$\Gamma_u(e_i) = ue_i = \sum a_{ij} e_j.$$

Define the **trace** and **norm**:

$$\mathrm{tr}(u) = \mathrm{tr}(\Gamma_u) \quad \text{and} \quad N(u) = \det(\Gamma_u).$$

The characteristic polynomial of a linear transformation, and hence, any of its coefficients, is independent of any choice of basis of $E/k$, and so the definitions of trace and norm do not depend on the choice of basis. It is easy to see that $\mathrm{tr} \colon E \to k$ is a linear functional and that $N \colon E^\times \to k^\times$ is a (multiplicative) homomorphism.

If $u \in k$, then the matrix of $\Gamma_u$, with respect to any basis of $E/k$, is the scalar matrix $uI$. Hence, if $u \in k$, then

$$\operatorname{tr}(u) = [E : k]u \quad \text{and} \quad N(u) = u^{[E:k]}.$$

**Definition.** The **trace form** is the function $t \colon E \times E \to k$ given by

$$t(u, v) = \operatorname{tr}(uv) = \operatorname{tr}(\Gamma_{uv}).$$

It is a routine exercise, left to the reader, to check that the trace form is a symmetric bilinear form.

**Example 10.83.** If $E = \mathbb{Q}(\sqrt{d})$ is a quadratic field, then a basis for $E/\mathbb{Q}$ is $1, \sqrt{d}$. If $u = a + b\sqrt{d}$, then the matrix of $\Gamma_u$ is

$$\begin{bmatrix} a & bd \\ b & a \end{bmatrix},$$

so that

$$\operatorname{tr}(u) = 2a \quad \text{and} \quad N(u) = a^2 - db^2 = u\overline{u}.$$

Thus, trace and norm arose in the description of the integers in quadratic fields, in relations (1).

We now show that $u = a + b\sqrt{d}$ is a unit in $\mathcal{O}_E$ if and only if $N(u) = \pm 1$. If $u$ is a unit, then there is $v \in \mathcal{O}_E$ with $1 = uv$. Hence, $1 = N(1) = N(uv) = N(u)N(v)$, so that $N(u)$ is a unit in $\mathbb{Z}$; that is, $N(u) = \pm 1$. Conversely, if $N(u) = \pm 1$, then $N(\overline{u}) = N(u) = \pm 1$, where $\overline{u} = a - b\sqrt{d}$. Therefore, $N(u\overline{u}) = 1$. But $u\overline{u} \in \mathbb{Q}$, so that $1 = N(u\overline{u}) = (u\overline{u})^2$. Therefore $u\overline{u} = \pm 1$, and so $u$ is a unit. ◄

**Lemma 10.84.** *Let $E/k$ be a field extension of finite degree $n$, and let $u \in E$. If $u = u_1, \dots, u_s$ are the roots (with multiplicity) of $\operatorname{irr}(u, k)$ in some extension field of $E$; that is, $\operatorname{irr}(u, k) = \prod_{i=1}^{s}(x - u_i)$, then*

$$\operatorname{tr}(u) = [E : k(u)] \sum_{i=1}^{s} u_i \quad \text{and} \quad N(u) = \left( \prod_{i=1}^{s} u_i \right)^{[E:k(u)]}.$$

**Remark.** Of course, if $u$ is separable over $k$, then $\operatorname{irr}(u, k)$ has no repeated roots and each $u_i$ occurs exactly once in the formulas. ◄

**Proof.** We sketch the proof. A basis of $k(u)$ over $k$ is $1, u, u^2, \dots, u^{s-1}$, and the matrix $C_1$ of $\Gamma_u | k(u)$ with respect to this basis is the companion matrix of $\operatorname{irr}(u, k)$. If $1, v_2, \dots, v_r$ is a basis of $E$ over $k(u)$, then the list

$$1, u, \dots, u^{s-1}, v_1, v_1 u, \dots, v_1 u^{s-1}, \dots, v_r, v_r u, \dots, v_r u^{s-1}$$

is a basis of $E$ over $k$. Each of the subspaces $k(u)$ and $\langle v_j, v_j u, \dots, v_j u^{s-1} \rangle$ for $j \geq 2$ is $\Gamma_u$-invariant, and so the matrix of $\Gamma_u$ relative to the displayed basis of $E$ over $k$ is a direct sum of blocks $C_1 \oplus \cdots \oplus C_r$. In fact, the reader may check that each $C_j$ is the companion matrix of $\operatorname{irr}(u, k)$. The trace and norm formulas now follow from $\operatorname{tr}(C_1 \oplus \cdots \oplus C_r) = \sum_j \operatorname{tr}(C_j)$ and $\det(C_1 \oplus \cdots \oplus C_r) = \prod_j \det(C_j)$. •

If $E/k$ is a field extension and $u \in E$, then a more precise notation for the trace and norm is

$$\operatorname{tr}_{E/k}(u) \quad \text{and} \quad N_{E/k}(u).$$

Indeed, the formulas in Lemma 10.84 display the dependence on the larger field $E$.

**Proposition 10.85.** *Let $R$ be a domain with $Q = \mathrm{Frac}(R)$, let $E/Q$ be a field extension of finite degree $[E : Q] = n$, and let $u \in E$ be integral over $R$. If $R$ is integrally closed, then*
$$\mathrm{tr}(u) \in R \quad and \quad N(u) \in R.$$

**Proof.** The formulas for $\mathrm{tr}(u)$ and $N(u)$ in Lemma 10.84 express each as an elementary symmetric function of the roots $u = u_1, \dots, u_s$ of $\mathrm{irr}(u, Q)$. Since $u$ is integral over $R$, Exercise 10.34 on page 907 says that $\mathrm{irr}(u, Q) \in R[x]$. Therefore, $\sum_i u_i$ and $\prod_i u_i$ lie in $R$, and hence $\mathrm{tr}(u)$ and $N(u)$ lie in $R$. •

In Example 3.37, we saw that if $E/k$ is a finite separable extension, then its normal closure $\widehat{E}$ is a Galois extension of $k$. Recall from the Fundamental Theorem of Galois Theory, Theorem 3.45, that if $G = \mathrm{Gal}(\widehat{E}/k)$ and $H = \mathrm{Gal}(\widehat{E}/E) \subseteq G$, then $[G : H] = [E : k]$.

**Lemma 10.86.** *Let $E/k$ be a separable field extension of finite degree $n = [E : k]$ and let $\widehat{E}$ be a normal closure of $E$. Write $G = \mathrm{Gal}(\widehat{E}/k)$ and $H = \mathrm{Gal}(\widehat{E}/E)$, and let $T$ be a transversal of $H$ in $G$; that is, $G$ is the disjoint union $G = \bigcup_{\sigma \in T} \sigma H$.*

(i) *For all $u \in E$,*
$$\prod_{\sigma \in T} (x - \sigma(u)) = \mathrm{irr}(u, k)^{[E:k(u)]}.$$

(ii) *For all $u \in E$,*
$$\mathrm{tr}(u) = \sum_{\sigma \in T} \sigma(u) \quad and \quad N(u) = \prod_{\sigma \in T} \sigma(u).$$

**Proof.**

(i) Denote $\prod_{\sigma \in T}(x - \sigma(u))$ by $h(x)$; of course, $h(x) \in \widehat{E}[x]$.

We claim that the set $X$, defined by $X = \{\sigma(u) : \sigma \in T\}$, satisfies $\tau(X) = X$ for every $\tau \in G$. If $\sigma \in T$, then $\tau\sigma \in \sigma'H$ for some $\sigma' \in T$, because $T$ is a left transversal; hence, $\tau\sigma = \sigma'\eta$ for some $\eta \in H$. But $\tau\sigma(u) = \sigma'\eta(u) = \sigma'(u)$, because $\eta \in H$, and every element of $H$ fixes $E$. Therefore, $\tau\sigma(u) = \sigma'(u) \in X$. Thus, the function $\varphi_\tau$, defined by $\sigma(u) \mapsto \tau\sigma(u)$, is a function $X \to X$. In fact, $\varphi_\tau$ is a permutation, because $\tau$ is an isomorphism, hence $\varphi_\tau$ is an injection, and hence it is a bijection, by the Pigeonhole Principle. It follows that every elementary symmetric function on $X = \{\sigma(u) : \sigma \in T\}$ is fixed by every $\tau \in G$. Since $\widehat{E}/k$ is a Galois extension, each value of these elementary symmetric functions lies in $k$. We have shown that all the coefficients of $h(x)$ lie in $k$, and so $h(x) \in k[x]$. Now compare $h(x)$ and $\mathrm{irr}(u, k)$. If $\sigma \in G$, then $\sigma$ permutes the roots of $\mathrm{irr}(u, k)$, so that every root $\sigma(u)$ of $h(x)$ is also a root of $\mathrm{irr}(u, k)$. By Exercise 2.94 on page 170, we have
$$h(x) = \mathrm{irr}(u, k)^m$$
for some $m \geq 1$, and so it only remains to compute $m$. Now
$$\deg(h) = m \deg(\mathrm{irr}(u, k)) = m[k(u) : k].$$

But $\deg(h) = [G : H] = [E : k]$, and so $m = [E : k]/[k(u) : k] = [E : k(u)]$.

(ii) Recall our earlier notation: $\mathrm{irr}(u, k) = \prod_{i=1}^{s}(x - u_i)$. Since

$$\prod_{\sigma \in T}(x - \sigma(u)) = \mathrm{irr}(u, k)^{[E:k(u)]} = \left(\prod_{i=1}^{s}(x - u_i)\right)^{[E:k(u)]},$$

their constant terms are the same,

$$\pm \prod_{\sigma \in T} \sigma(u) = \pm\left(\prod_{i=1}^{s} u_i\right)^{[E:k(u)]},$$

and their penultimate coefficients are the same,

$$-\sum_{\sigma \in T} \sigma(u) = -[E : k(u)]\sum_{i=1}^{s} u_i.$$

By Lemma 10.84, $\mathrm{tr}(u) = [E : k(u)]\sum_{i=1}^{s} u_i$ and $N(u) = \left(\prod_{i=1}^{s} u_i\right)^{[E:k(u)]}$. It follows that

$$\mathrm{tr}(u) = [E : k(u)]\sum_{i=1}^{s} u_i = \sum_{\sigma \in T} \sigma(u)$$

and

$$N(u) = \left(\prod_{i=1}^{s} u_i\right)^{[E:k(u)]} = \prod_{\sigma \in T} \sigma(u). \quad \bullet$$

**Definition.** Let $E/k$ be a finite field extension, let $\widehat{E}$ be a normal closure of $E$, and let $T$ be a left transversal of $\mathrm{Gal}(\widehat{E}/E)$ in $\mathrm{Gal}(\widehat{E}/k)$. If $u \in E$, then the elements $\sigma(u)$, where $\sigma \in T$, are called the **conjugates** of $u$.

If $E = \mathbb{Q}(i)$, then $\mathrm{Gal}(E/\mathbb{Q}) = \langle\sigma\rangle$, where $\sigma$ is complex conjugation. If $u = a + ib \in E$, then $\sigma(u) = a - ib$, so that we have just generalized the usual notion of conjugate.

If $E/k$ is a separable extension, then the conjugates of $u$ are the roots of $\mathrm{irr}(u, k)$; in the inseparable case, the conjugates may occur with multiplicities.

**Corollary 10.87.** *If $E/k$ is a finite Galois extension with $G = \mathrm{Gal}(E/k)$, then*

$$\mathrm{tr}(u) = \sum_{\sigma \in G} \sigma(u) \quad and \quad N(u) = \prod_{\sigma \in G} \sigma(u).$$

*Moreover, $\mathrm{tr}$ is nonzero.*

**Proof.** Since $E/k$ is a Galois extension, $E$ is its own normal closure, and so a transversal $T$ of $G$ in itself is just $G$. If $\mathrm{tr} = 0$, then

$$\mathrm{tr}(u) = \sum_{\sigma \in G} \sigma(u) = 0$$

for all $u \in E$, contradicting Independence of Characters, Proposition 3.32. $\quad \bullet$

This last corollary shows that the norm here coincides with the norm occurring in Chapter 3 in the proof of Hilbert's Theorem 90.

Let $V$ be a vector space over a field $k$, and let $f : V \times V \to k$ be a bilinear form. If $e_1, \ldots, e_n$ is a basis of $V$, recall from Chapter 8 that the *discriminant* is defined by

$$D(e_1, \ldots, e_n) = \det([f(e_i, e_j)]),$$

and that $f$ is *nondegenerate* if there is a basis whose discriminant is nonzero (it then follows that the discriminant of $f$ with respect to any other basis of $V$ is also nonzero).

**Lemma 10.88.** *If $E/k$ is a finite separable field extension, then the trace form is nondegenerate.*[7]

**Proof.** We compute the discriminant using Lemma 10.86 (which uses separability). Let $T = \{\sigma_1, \ldots, \sigma_n\}$ be a transversal of $\mathrm{Gal}(\widehat{E}/E)$ in $\mathrm{Gal}(\widehat{E}/k)$, where $\widehat{E}$ is a normal closure of $E$. We have

$$
\begin{aligned}
D(e_1, \ldots, e_n) &= \det([t(e_i, e_j)]) \\
&= \det([\mathrm{tr}(e_i e_j)]) \\
&= \det\left[\sum_\ell \sigma_\ell(e_i e_j)\right] \qquad \text{(Lemma 10.86)} \\
&= \det\left(\left[\sum_\ell \sigma_\ell(e_i)\sigma_\ell(e_j)\right]\right) \\
&= \det([\sigma_\ell(e_i)]) \det([\sigma_\ell(e_j)]) \\
&= \det([\sigma_\ell(e_i)])^2.
\end{aligned}
$$

Suppose that $\det([\sigma_\ell(e_i)]) = 0$. If $[\sigma_\ell(e_i)]$ is singular, there is a column matrix $C = [c_1, \ldots, c_n]^\top \in \widehat{E}^n$ with $[\sigma_\ell(e_i)]C = 0$. Hence,

$$c_1 \sigma_1(e_j) + \cdots + c_n \sigma_n(e_j) = 0$$

for $j = 1, \ldots, n$. It follows that

$$c_1 \sigma_1(v) + \cdots + c_n \sigma_n(v) = 0$$

for every linear combination $v$ of the $e_i$. But this contradicts Independence of Characters.   •

**Proposition 10.89.** *Let $R$ be integrally closed, and let $Q = \mathrm{Frac}(R)$. If $E/Q$ is a finite separable field extension of degree $n$ and $\mathcal{O} = \mathcal{O}_{E/R}$ is the integral closure of $R$ in $E$, then $\mathcal{O}$ can be imbedded as a submodule of a free $R$-module of rank $n$.*

**Proof.** Let $e_1, \ldots, e_n$ be a basis of $E/Q$. By Proposition 10.51, for each $i$ there is $r_i \in R$ with $r_i e_i \in \mathcal{O}$; changing notation if necessary, we may assume that each $e_i \in \mathcal{O}$. Now Corollary 8.76 (which assumes nondegeneracy of bilinear forms) says that there is a basis $f_1, \ldots, f_n$ of $E$ with $t(e_i, f_j) = \mathrm{tr}(e_i f_j) = \delta_{ij}$.

---

[7]If $E/k$ is inseparable, then the trace form is identically 0. See Zariski–Samuel, *Commutative Algebra* I, p. 95.

Let $\alpha \in \mathcal{O}$. Since $f_1, \ldots, f_n$ is a basis, there are $c_j \in Q$ with $\alpha = \sum c_j f_j$. For each $i$, where $1 \leq i \leq n$, we have $e_i \alpha \in \mathcal{O}$ (because $e_i \in \mathcal{O}$). Therefore, $\mathrm{tr}(e_i \alpha) \in R$, by Proposition 10.85. But

$$\mathrm{tr}(e_i \alpha) = \mathrm{tr}\left(\sum_j c_j e_i f_j\right) = \sum_j c_j \,\mathrm{tr}(e_i f_j) = c_j \delta_{ij} = c_i.$$

Therefore, $c_i \in R$ for all $i$, and so $\alpha = \sum_i c_i f_i$ lies in the free $R$-module with basis $f_1, \ldots, f_n$. •

**Definition.** If $E$ is an algebraic number field, then an ***integral basis*** for $\mathcal{O}_E$ is a list $\beta_1, \ldots, \beta_n$ in $\mathcal{O}_E$ such that every $\alpha \in \mathcal{O}_E$ has a unique expression

$$\alpha = c_1 \beta_1 + \cdots + c_n \beta_n,$$

where $c_i \in \mathbb{Z}$ for all $i$.

If $B = \beta_1, \ldots, \beta_n$ is an integral basis of $\mathcal{O}_E$, then $\mathcal{O}_E$ is a free abelian group with basis $B$. We now prove that integral bases always exist.

**Proposition 10.90.** *Let $E$ be an algebraic number field.*

(i) *The ring of integers $\mathcal{O}_E$ has an integral basis, and hence it is a free abelian group of finite rank under addition.*

(ii) *$\mathcal{O}_E$ is a noetherian domain.*

**Proof.**

(i) Since $\mathbb{Q}$ has characteristic 0, the field extension $E/\mathbb{Q}$ is separable. Hence, Proposition 10.89 applies to show that $\mathcal{O}_E$ is a submodule of a free $\mathbb{Z}$-module of finite rank; that is, $\mathcal{O}_E$ is a subgroup of a finitely generated free abelian group. By Corollary 8.4, $\mathcal{O}_E$ is itself a free abelian group. But a basis of $\mathcal{O}_E$ as a free abelian group is an integral basis.

(ii) Any ideal $I$ in $\mathcal{O}_E$ is a subgroup of a finitely generated free abelian group, and hence $I$ is itself a finitely generated abelian group, by Proposition 8.8. A fortiori, $I$ is a finitely generated $\mathcal{O}_E$-module; that is, $I$ is a finitely generated ideal. •

**Example 10.91.** We show that $\mathcal{O}_E$ need not be a UFD and, hence, it need not be a PID. Let $E = \mathbb{Q}(\sqrt{-5})$. Since $-5 \equiv 3 \bmod 4$, Proposition 10.82 gives $\mathcal{O}_E = \mathbb{Z}[\sqrt{-5}]$. By Example 10.83, the only units in $\mathcal{O}_E$ are elements $u$ with $N(u) = \pm 1$. If $a^2 + 5b^2 = \pm 1$, where $a, b \in \mathbb{Z}$, then $b = 0$ and $a = \pm 1$, and so the only units in $\mathcal{O}_E$ are $\pm 1$. Consider the factorizations in $\mathcal{O}_E$:

$$2 \cdot 3 = 6 = \left(1 + \sqrt{-5}\right)\left(1 - \sqrt{-5}\right).$$

Note that no two of these factors are associates (the only units are $\pm 1$), and we now show that each of them is irreducible. If $v \in \mathcal{O}_E$ divides any of these four factors (but is not an associate of it), then $N(v)$ is a proper divisor in $\mathbb{Z}$ of $4, 9$, or $6$, for these are the norms of the four factors $[N(1 + \sqrt{-5}) = 6 = N(1 - \sqrt{-5})]$. It is quickly checked, however, that there are no such divisors in $\mathbb{Z}$ of the form $a^2 + 5b^2$ other than $\pm 1$. Therefore, $\mathcal{O}_E = \mathbb{Z}[\sqrt{-5}]$ is not a UFD. ◄

Trace and norm can be used to find other rings of integers.

**Definition.** If $n \geq 2$, then a **cyclotomic field** is $E = \mathbb{Q}(\zeta_n)$, where $\zeta_n$ is a primitive $n$th root of unity.

Recall that if $p$ is prime, then the cyclotomic polynomial

$$\Phi_p(x) = x^{p-1} + x^{p-2} + \cdots + x + 1 \in \mathbb{Z}[x]$$

is irreducible, so that $\mathrm{irr}(\zeta_p, \mathbb{Q}) = \Phi_p(x)$ and $[\mathbb{Q}(\zeta_p)/\mathbb{Q}] = p - 1$ (recall our remark on page 122: $\Phi_d(x)$ is irreducible for all, not necessarily prime, natural numbers $d$). Moreover,

$$\mathrm{Gal}(\mathbb{Q}(\zeta_p)/\mathbb{Q}) = \{\sigma_1, \ldots, \sigma_{p-1}\},$$

where $\sigma_i \colon \zeta_p \mapsto \zeta_p^i$ for $i = 1, \ldots, p-1$.

We do some elementary calculations in $E = \mathbb{Q}(\zeta_p)$ to enable us to describe $\mathcal{O}_E$.

**Lemma 10.92.** *Let $p$ be an odd prime, and let $E = \mathbb{Q}(\zeta)$, where $\zeta = \zeta_p$ is a primitive $p$th root of unity.*

(i) $\mathrm{tr}(\zeta^i) = -1$ *for* $1 \leq i \leq p-1$.

(ii) $\mathrm{tr}(1 - \zeta^i) = p$ *for* $1 \leq i \leq p-1$.

(iii) $p = \prod_{i=1}^{p-1} (1 - \zeta^i) = N(1 - \zeta)$.

(iv) $\mathcal{O}_E(1 - \zeta) \cap \mathbb{Z} = p\mathbb{Z}$.

(v) $\mathrm{tr}(u(1 - \zeta)) \in p\mathbb{Z}$ *for every* $u \in \mathcal{O}_E$.

**Proof.**

(i) We have $\mathrm{tr}(\zeta) = \sum_{i=1}^{p-1} \zeta^i = \Phi_p(\zeta) - 1$, which is also true for every primitive $p$th root of unity $\zeta^i$. The result follows from $\Phi_p(\zeta) = 0$.

(ii) Since $\mathrm{tr}(1) = [E : \mathbb{Q}] = p - 1$ and $\mathrm{tr}$ is a linear functional,

$$\mathrm{tr}(1 - \zeta^i) = \mathrm{tr}(1) - \mathrm{tr}(\zeta^i) = (p - 1) - (-1) = p.$$

(iii) Since $\Phi(x) = x^{p-1} + x^{p-2} + \cdots + x + 1$, we have $\Phi_p(1) = p$. On the other hand, the primitive $p$th roots of unity are the roots of $\Phi_p(x)$, so that

$$\Phi_p(x) = \prod_{i=1}^{p-1}(x - \zeta^i).$$

Evaluating at $x = 1$ gives the first equation. The second equation holds because the $1 - \zeta^i$ are the conjugates of $1 - \zeta$.

(iv) The first equation in (iii) shows that $p \in \mathcal{O}_E(1-\zeta) \cap \mathbb{Z}$, so that $\mathcal{O}_E(1-\zeta) \cap \mathbb{Z} \supseteq p\mathbb{Z}$. If this inclusion is strict, then $\mathcal{O}_E(1-\zeta) \cap \mathbb{Z} = \mathbb{Z}$, because $p\mathbb{Z}$ is a maximal ideal in $\mathbb{Z}$. In this case, $\mathcal{O}_E(1 - \zeta) \cap \mathbb{Z} = \mathbb{Z}$, hence $\mathbb{Z} \subseteq \mathcal{O}_E(1 - \zeta)$, and so $1 \in \mathcal{O}_E(1 - \zeta)$. Thus, there is $v \in \mathcal{O}_E$ with $v(1 - \zeta) = 1$; that is, $1 - \zeta$ is a unit in $\mathcal{O}_E$. But if $1 - \zeta$ is a unit, then $N(1 - \zeta) = \pm 1$, contradicting the second equation in (iii).

(v) Each conjugate $\sigma_i(u(1 - \zeta)) = \sigma_i(u)(1 - \zeta^i)$ is, obviously, divisible by $1 - \zeta^i$ in $\mathcal{O}_E$. But $1 - \zeta^i$ is divisible by $1 - \zeta$ in $\mathcal{O}_E$, because

$$1 - \zeta^i = (1 - \zeta)(1 + \zeta + \zeta^2 + \cdots + \zeta^{i-1}).$$

Hence, $\sigma_i(1 - \zeta^i) \in \mathcal{O}_E(1 - \zeta)$ for all $i$, and so $\sum_i(u(1 - \zeta^i)) \in \mathcal{O}_E(1 - \zeta)$. By Corollary 10.87, $\sum_i(u(1 - \zeta)) = \mathrm{tr}(u(1 - \zeta))$. Therefore, $\mathrm{tr}(u(1 - \zeta)) \in \mathcal{O}_E(1 - \zeta) \cap \mathbb{Z} = p\mathbb{Z}$, by (iv), for $\mathrm{tr}(u(1 - \zeta)) \in \mathbb{Z}$, by Proposition 10.85. $\bullet$

**Proposition 10.93.** *If $p$ is an odd prime and $E = \mathbb{Q}(\zeta_p)$ is a cyclotomic field, then*

$$\mathcal{O}_E = \mathbb{Z}[\zeta_p].$$

**Proof.** Let us abbreviate $\zeta_p$ to $\zeta$. It is always true that $\mathbb{Z}[\zeta] \subseteq \mathcal{O}_E$, and we now prove the reverse inclusion. By Lemma 10.80, each element $u \in \mathcal{O}_E$ has an expression

$$u = c_0 + c_1\zeta + c_2\zeta^2 + \cdots + c_{p-2}\zeta^{p-2},$$

where $c_i \in \mathbb{Q}$ (remember that $[E : \mathbb{Q}] = p - 1$). We must show that $c_i \in \mathbb{Z}$ for all $i$. Multiplying by $1 - \zeta$ gives

$$u(1 - \zeta) = c_0(1 - \zeta) + c_1(\zeta - \zeta^2) + \cdots + c_{p-2}(\zeta^{p-2} - \zeta^{p-1}).$$

By Lemma 10.92(i), $\mathrm{tr}(\zeta^i - \zeta^{i+1}) = \mathrm{tr}(\zeta^i) - \mathrm{tr}(\zeta^{i+1}) = 0$ for $1 \le i \le p - 2$, so that $\mathrm{tr}(u(1 - \zeta)) = c_0 \mathrm{tr}(1 - \zeta)$; hence, $\mathrm{tr}(u(1 - \zeta)) = pc_0$, because $\mathrm{tr}(1 - \zeta) = p$, by Lemma 10.92(ii). On the other hand, $\mathrm{tr}(u(1 - \zeta)) \in p\mathbb{Z}$, by Lemma 10.92(iv). Hence, $pc_0 = mp$ for some $m \in \mathbb{Z}$, and so $c_0 \in \mathbb{Z}$. Now $\zeta^{-1} = \zeta^{p-1} \in \mathcal{O}_E$, so that

$$(u - c_0)\zeta^{-1} = c_1 + c_2\zeta + \cdots + c_{p-2}\zeta^{p-3} \in \mathcal{O}_E.$$

The argument just given shows that $c_1 \in \mathbb{Z}$. Indeed, repetition of this argument shows that all $c_i \in \mathbb{Z}$, and so $u \in \mathbb{Z}[\zeta]$. $\bullet$

There are other interesting bases of finite Galois extensions.

**Definition.** Let $E/k$ be a finite Galois extension with $\mathrm{Gal}(E/k) = \{\sigma_1, \ldots, \sigma_n\}$. A **normal basis** of $E/k$ is a vector space basis $e_1, \ldots, e_n$ of $E$ for which there exists $u \in E$ with $e_i = \sigma_i(u)$ for all $i$.

If $E/k$ is an extension with $\mathrm{Gal}(E/k) = G$, then $E$ is a $kG$-module, where $\sigma e = \sigma(e)$ for all $e \in E$ and $\sigma \in G$.

**Proposition 10.94.** *Let $E/k$ be a finite Galois extension with $\mathrm{Gal}(E/k) = G$. Then $E \cong kG$ as $kG$-modules if and only if $E/k$ has a normal basis.*

**Proof.** Suppose $\varphi \colon kG \to E$ is a $kG$-isomorphism, where $G = \{1 = \sigma_1, \ldots, \sigma_n\}$. Now $kG$ is a vector space over $k$ with basis $1 = \sigma_1, \ldots, \sigma_n$. If $u = \varphi(1)$, then $\varphi(\sigma_i) = \sigma_i\varphi(1) = \sigma_i u = \sigma_i(u)$. As $\varphi$ is a $k$-linear isomorphism, $\{\sigma_i(u) : 1 \le i \le n\}$ is a basis of $E$; that is, $E/k$ has a normal basis.

Conversely, assume that there is $u \in E$ with $\{\sigma_i(u) : 1 \le i \le n\}$ a basis of $E/k$. It is easily checked that $\varphi \colon kG \to E$, given by $\sigma_i \mapsto \sigma_i(u)$, is a $kG$-map; $\varphi$ is an isomorphism because it carries a basis of $kG$ onto a basis of $E$. $\bullet$

**Proposition 10.95.** *Let $E/k$ be a Galois extension with $\mathrm{Gal}(E/k) = \{\sigma_1, \ldots, \sigma_n\}$, and let $e_1, \ldots, e_n \in E$. Then $e_1, \ldots, e_n$ is a basis of $E$ over $k$ if and only if the matrix $[\sigma_i(e_j)]$ is nonsingular.*

**Proof.** Assume that $M = [\sigma_i(e_j)]$ is nonsingular. Since $\dim_k(E) = n$, it suffices to prove that $e_1, \ldots, e_n$ is linearly independent. Otherwise, there are $a_i \in k$ with

$$a_1 e_1 + \cdots + a_n e_n = 0.$$

For each $i$, we have

$$a_1 \sigma_i(e_1) + \cdots + a_n \sigma_i(e_n) = 0.$$

Thus, $MA = 0$, where $A$ is the column vector $[a_1, \ldots, a_n]^\top$ in $k^n \subseteq E^n$. Since $M$ is nonsingular over $E$, we have $A = 0$, and so $e_1, \ldots, e_n$ is a basis.

Conversely, if $M = [\sigma_i(e_j)]$, a matrix with entries in $E$, is singular, there is a nontrivial solution $X = [\alpha_1, \ldots, \alpha_n]^\top$ of $MX = 0$; that is, for all $i$, there are $\alpha_1, \ldots, \alpha_n \in E$ with

$$\alpha_1 \sigma_i(e_1) + \alpha_2 \sigma_i(e_2) + \cdots + \alpha_n \sigma_i(e_n) = 0.$$

For notational convenience, we may assume that $\alpha_1 \neq 0$. By Corollary 10.87, there is $u \in E$ with $\mathrm{tr}(u) \neq 0$. Multiply each equation above by $u\alpha_1^{-1}$ to obtain equations, for all $i$,

$$u\sigma_i(e_1) + \beta_2 \sigma_i(e_2) + \cdots + \beta_n \sigma_i(e_n) = 0,$$

where $\beta_j = u\alpha_1^{-1}\alpha_j$ for $j \geq 2$. For each $i$, apply $\sigma_i^{-1}$ to the $i$th equation:

$$\sigma_1^{-1}(u)e_1 + \sigma_1^{-1}(\beta_2)e_2 + \cdots + \sigma_1^{-1}(\beta_n)e_n = 0$$
$$\sigma_2^{-1}(u)e_1 + \sigma_2^{-1}(\beta_2)e_2 + \cdots + \sigma_2^{-1}(\beta_n)e_n = 0$$
$$\vdots \qquad\qquad\qquad\qquad\qquad \vdots$$
$$\sigma_n^{-1}(u)e_1 + \sigma_n^{-1}(\beta_2)e_2 + \cdots + \sigma_n^{-1}(\beta_n)e_n = 0.$$

Note that as $\sigma_i$ varies over all of $G$, so does $\sigma_i^{-1}$. Adding these equations gives

$$\mathrm{tr}(u)e_1 + \mathrm{tr}(\beta_2)e_2 + \cdots + \mathrm{tr}(\beta_n)e_n = 0.$$

This is a nontrivial linear combination of $e_1, \ldots, e_n$, because $\mathrm{tr}(u) \neq 0$, and this contradicts linear independence.  •

**Corollary 10.96.** *Let $E/k$ be a Galois extension with $\mathrm{Gal}(E/k) = \{\sigma_1, \ldots, \sigma_n\}$, and let $u \in E$. Then $\sigma_1(u), \ldots, \sigma_n(u)$ is a normal basis of $E$ over $k$ if and only if the matrix $[\sigma_i \sigma_j(u)]$ is nonsingular.*

**Proof.** Let $e_1 = \sigma_1(u), \ldots, e_n = \sigma_n(u)$ in Proposition 10.95.  •

**Remark.** There is another way to view the matrix $[\sigma_i \sigma_j]$. If $\alpha_1, \ldots, \alpha_n \in E$, define

$$M(\alpha_1, \ldots, \alpha_n) = [\sigma_j(\alpha_i)].$$

If $\beta_1, \ldots, \beta_n \in E$, then $M(\alpha_1, \ldots, \alpha_n)M(\beta_1, \ldots, \beta_n)^\top = [\mathrm{tr}(\alpha_i\beta_j)]$. In particular,

$$(3) \qquad\qquad M(\alpha_1, \ldots, \alpha_n)M(\alpha_1, \ldots, \alpha_n)^\top = [\mathrm{tr}(\alpha_i\alpha_j)].$$

The latter matrix is just the inner product matrix of the trace form defined on page 917. In fact, in light of (3), Lemma 10.88 (which says that $[\mathrm{tr}(\alpha_1, \ldots, \alpha_n)]$ is nonsingular) gives another proof of the nonsingularity of $[M(\alpha_1, \ldots, \alpha_n)]$. ◀

**Proposition 10.97.**

(i) *If $E/k$ is a finite Galois extension with cyclic Galois group $G$, then $E$ has a normal basis.*

(ii) *If $E$ is a finite field and $k$ is a subfield, then $E/k$ has a normal basis.*

**Proof.**

(i) Let $G = \langle \sigma \rangle$, where $\sigma$ has order $n = [E : k]$, and view $\sigma \colon E \to E$ as a $k$-linear transformation. Now $\sigma^n - 1 = 0$; on the other hand, Independence of Characters says that whenever $c_1, \ldots, c_n \in k$ and $\sum_i c_i \sigma^i(e) = 0$ for all $e \in E$, all the $c_i = 0$; thus, if $0 \neq f(x) = \sum_{i=0}^{n-1} c_i x^i \in k[x]$, then $f(\sigma) \neq 0$. It follows that the characteristic and minimal polynomials of $\sigma$ coincide, and so Corollary 8.45 applies: $E$ is a cyclic $k[x]$-module, say, with generator $u \in E$. Thus, every element of $E$ is a $k$-linear combination of $u, \sigma(u), \ldots, \sigma^{n-1}(u)$ [for if $m \geq n$, then $\sigma^m(u) = \sigma^i(u)$, where $m \equiv i \bmod n$]. Therefore, $E/k$ has a normal basis.

(ii) Exercise 3.11 on page 192 says that $\mathrm{Gal}(E/k)$ is cyclic. •

**Lemma 10.98.** *If $k$ is an infinite field and $E/k$ is a finite Galois extension with $G = \mathrm{Gal}(E/k)$, then the elements $\{\sigma_1, \ldots, \sigma_n\}$ of $G$ are algebraically independent over $E$: if $f(\sigma_1(v), \ldots, \sigma_n(v)) = 0$ for all $v \in E$, then $f$ is the zero polynomial.*

**Proof.** Suppose that $f(x_1, \ldots, x_n) \in E[x_1, \ldots, x_n]$ and $f(\sigma_1(v), \ldots, \sigma_n(v)) = 0$ for all $v \in E$. If $e_1, \ldots, e_n$ is a basis of $E/k$, then $v = \sum_q a_q e_q$ for $a_q \in k$, and so

$$0 = f\Big(\sigma_1\Big(\sum_q a_q e_q\Big), \ldots, \sigma_n\Big(\sum_q a_q e_q\Big)\Big)$$
$$= f\Big(\sum_q a_q \sigma_1(e_q), \ldots, \sum_q a_q \sigma_n(e_q)\Big).$$

Define

$$g(x_1, \ldots, x_n) = f\Big(\sum_q \sigma_1(e_q) x_q, \ldots, \sum_q \sigma_n(e_q) x_q\Big).$$

Thus, $g(a_1, \ldots, a_n) = 0$ for all $a_1, \ldots, a_n \in k$, and so Proposition 2.45 (whose hypothesis has $k$ infinite) gives $g(x_1, \ldots, x_n) = 0$. Since $e_1, \ldots, e_n$ is a basis of $E/k$, Proposition 10.95 says that the matrix $[\sigma_p(e_q)]$ over $E$ is nonsingular. Let $[w_{iq}]$ be its inverse, so that $\sum_p w_{ip} \sigma_p(e_q) x_q = \delta_{iq}$ (Kronecker delta). For every $i$, we have

$$\sum_{p,q} w_{ip} \sigma_p(e_q) x_q = \sum_q \Big(\sum_p w_{ip} \sigma_p(e_q)\Big) x_q = \sum_q \delta_{iq} x_q = x_i.$$

Thus, $0 = g(\sum_{p,q} w_{1p} \sigma_p(e_q) x_q, \ldots, \sum_{p,q} w_{np} \sigma_p(e_q) x_q) = f(x_1, \ldots, x_n)$, and so $\sigma_1, \ldots, \sigma_n$ are algebraically independent over $E$. •

If $X$ is a set with $n$ elements, then a ***Latin square*** based on $X$ is an $n \times n$ matrix $L$ whose rows and columns are permutations of $X$.

**Lemma 10.99.** *Let $E$ be a field. If $\{x_1, \ldots, x_n\} \subseteq E[x_1, \ldots, x_n]$ and $L$ is a Latin square based on $\{x_1, \ldots, x_n\}$, then $\det(L)$ is a nonzero polynomial in $E[x_1, \ldots, x_n]$.*

**Proof.** If $R = E[x_1, \ldots, x_n]$, then a ring map $\mu \colon R \to E$ induces a ring map $\mu_* \colon \mathrm{Mat}_n(R) \to \mathrm{Mat}_n(E)$; moreover, $\det(\mu_*(L)) = \mu(\det(L))$. In particular, if $\mu$ is given by $x_1 \mapsto 1$ and $x_i \mapsto 0$ for $i \geq 2$, then $\mu_*(L) = P$, where $P$ is the (permutation) matrix obtained from $L$ by setting $x_1 = 1$ and all other $x_i = 0$. Since $\det(P) = \pm 1 \neq 0$, we have $\det(L) \neq 0$. Proposition 8.130, the complete expansion of $\det(L)$, shows that $\det(L) \in E[x_1, \ldots, x_n]$. •

**Theorem 10.100 (Normal Basis Theorem).** *Every finite Galois extension $E/k$ has a normal basis.*

**Proof.** By Proposition 10.97, we may assume that $k$ is an infinite field.

The matrix $L = [x_i x_j]$ over the polynomial ring $E[x_1, \ldots, x_n]$ is a Latin square based on $\{x_1, \ldots, x_n\}$, and so $f(x_1, \ldots, x_n) = \det([x_{i(j)}])$ is a nonzero polynomial, by Lemma 10.99. Since $G = \mathrm{Gal}(E/k) = \{\sigma_1, \ldots, \sigma_n\}$ is a group, we have $\sigma_i \sigma_j = \sigma_{i(j)} \in G$. Thus, for each $v \in E$, the entries in the $n$-tuple $(\sigma_{i(1)}(v), \ldots, \sigma_{i(n)}(v))$ form a permutation of the entries in $(\sigma_1(v), \ldots, \sigma_n(v))$, and so $f(\sigma_{i(1)}(v), \ldots, \sigma_{i(n)}(v)) = \det([\sigma_i(\sigma_j(v))])$. Lemma 10.98 gives $u \in E$ with $f(\sigma_{i(1)}(u), \ldots, \sigma_{i(n)}(u)) \neq 0$. Hence, $f(\sigma_{i(1)}(u), \ldots, \sigma_{i(n)}(u)) = \det[\sigma_{i(j)}(u)] = \det[\sigma_i \sigma_j(u)] \neq 0$. Corollary 10.96 now applies: $\sigma_1(u), \ldots, \sigma_n(u)$ is a normal basis of $E/k$. •

Before we leave this discussion of algebraic number fields, we must mention a beautiful theorem of Dirichlet. In order to state the theorem, we cite two results whose proofs can be found in Samuel, *Algebraic Theory of Numbers*, Chapter 4. An algebraic number field $E$ of degree $n$ has exactly $n$ imbeddings into $\mathbb{C}$. If $r_1$ is the number of such imbeddings with image in $\mathbb{R}$, then $n - r_1$ is even; say, $n - r_1 = 2r_2$.

**Theorem (Dirichlet Unit Theorem).** *Let $E$ be an algebraic number field of degree $n$. Then $n = r_1 + 2r_2$ (where $r_1$ is the number of imbeddings of $E$ into $\mathbb{R}$), and the multiplicative group $U(\mathcal{O}_E)$ of units in $\mathcal{O}_E$ is a finitely generated abelian group. More precisely,*

$$U(\mathcal{O}_E) \cong \mathbb{Z}^{r_1 + r_2 - 1} \times T,$$

*where $T$ is a finite cyclic group consisting of the roots of unity in $E$.*

**Proof.** Borevich–Shafarevich, *Number Theory*, p. 112. •

---

## Exercises

**10.42.** (i) If $E = \mathbb{Q}(\sqrt{-3})$, prove that the only units in $\mathcal{O}_E$ are

$$\pm 1, \quad \tfrac{1}{2}(1 \pm \sqrt{-3}), \quad \tfrac{1}{2}(-1 \pm \sqrt{-3}).$$

(ii) Let $d$ be a negative squarefree integer with $d \neq -1$ and $d \neq -3$. If $E = \mathbb{Q}(\sqrt{d})$, prove that the only units in $\mathcal{O}_E$ are $\pm 1$.

**10.43.** (i) Prove that if $E = \mathbb{Q}(\sqrt{2}) \subseteq \mathbb{R}$, then there are no units $u \in \mathcal{O}_E$ with $1 < u < 1 + \sqrt{2}$.

(ii) If $E = \mathbb{Q}(\sqrt{2})$, prove that $\mathcal{O}_E$ has infinitely many units.
**Hint.** Use (i) to prove that all powers of $1 + \sqrt{2}$ are distinct.

**Definition.** If $\mathcal{O}_E$ is the ring of integers in an algebraic number field $E$, then a **discriminant** of $\mathcal{O}_E$ is
$$\Delta(\mathcal{O}_E) = \det[\mathrm{tr}(\alpha_i \alpha_j)],$$
where $\alpha_1, \ldots, \alpha_n$ is an integral basis of $\mathcal{O}_E$.

**10.44.** Let $d$ be a squarefree integer, and let $E = \mathbb{Q}(\sqrt{d})$.

(i) If $d \equiv 2 \bmod 4$ or $d \equiv 3 \bmod 4$, prove that $1, \sqrt{d}$ is an integral basis of $\mathcal{O}_E$, and prove that a discriminant of $\mathcal{O}_E$ is $4d$.

(ii) If $d \equiv 1 \bmod 4$, prove that $1, \frac{1}{2}(1 + \sqrt{d})$ is an integral basis of $\mathcal{O}_E$, and prove that a discriminant of $\mathcal{O}_E$ is $d$.

**10.45.** Let $p$ be an odd prime, and let $E = \mathbb{Q}(\zeta_p)$ be the cyclotomic field.

(i) Show that $1, 1 - \zeta_p, (1 - \zeta_p)^2, \ldots, (1 - \zeta_p)^{p-2}$ is an integral basis for $\mathcal{O}_E$.

(ii) Prove that a discriminant of $\mathcal{O}_E$ is $(-1)^{\frac{1}{2}(p-1)} p^{p-2}$.
**Hint.** See Pollard, *The Theory of Algebraic Numbers*, p. 67.

\* **10.46.** (i) If $\mathbb{A}$ is the field of all algebraic numbers, prove that $\mathcal{O}_{\mathbb{A}}$ is not noetherian.

(ii) Prove that every nonzero prime ideal in $\mathcal{O}_{\mathbb{A}}$ is a maximal ideal.
**Hint.** Use the proof of Corollary 10.58.

---

**10.3.4. Characterizations of Dedekind Rings.** The following definition involves some of the ring-theoretic properties enjoyed by the ring of integers $\mathcal{O}_E$ in an algebraic number field $E$.

**Definition.** A domain $R$ is a **Dedekind ring** if it is integrally closed, noetherian, and all its nonzero prime ideals are maximal ideals.

**Example 10.101.**

(i) The ring $\mathcal{O}_E$ in an algebraic number field $E$ is a Dedekind ring, by Proposition 10.51, Proposition 10.90, and Corollary 10.58.

(ii) Every PID $R$ is a Dedekind ring (Proposition 10.50 says that $R$ is integrally closed). ◄

It is shown, in Example 10.91, that $R = \mathbb{Z}[\sqrt{-5}]$ is a Dedekind ring that is not a UFD and, hence, it is not a PID. We now characterize discrete valuation rings, and then show that localizations of Dedekind rings are well-behaved.

**Lemma 10.102.** *A domain $R$ is a DVR if and only if it is noetherian, integrally closed, and has a unique nonzero prime ideal.*

**Proof.** If $R$ is a DVR, then it does have the required properties (recall that $R$ is a PID, hence it is integrally closed).

The converse, which requires us to show that $R$ is a PID, is not as simple as we might expect. Let $\mathfrak{p}$ be the nonzero prime ideal, and choose a nonzero $a \in \mathfrak{p}$. Define $M = R/Ra$, and consider the family $\mathcal{A}$ of all the annihilators $\operatorname{ann}(m)$ as $m$ varies over all the nonzero elements of $M$. Since $R$ is noetherian, it satisfies the maximum condition, and so there is a nonzero element $b + Ra \in M$ whose annihilator $\mathfrak{q} = \operatorname{ann}(b + Ra)$ is maximal in $\mathcal{A}$. We claim that $\mathfrak{q}$ is a prime ideal. Suppose that $x, y \in R$, $xy \in \mathfrak{q}$, and $x, y \notin \mathfrak{q}$. Then $y(b + Ra) = yb + Ra$ is a nonzero element of $M$, because $y \notin \mathfrak{q}$. But $\operatorname{ann}(yb + Ra) \supsetneq \operatorname{ann}(b + Ra)$, because $x \notin \operatorname{ann}(b + Ra)$, contradicting the maximality property of $\mathfrak{q}$. Therefore, $\mathfrak{q}$ is a prime ideal. Since $R$ has a unique nonzero prime ideal $\mathfrak{p}$, we have

$$\mathfrak{q} = \operatorname{ann}(b + Ra) = \mathfrak{p}.$$

Note that

$$b/a \notin R:$$

otherwise, $b + Ra = 0 + Ra$, contradicting $b + Ra$ being a nonzero element of $M = R/Ra$.

We now show that $\mathfrak{p}$ is principal, with generator $a/b$ (we do not yet know whether $a/b \in \operatorname{Frac}(R)$ lies in $R$). First, we have $\mathfrak{p}b = \mathfrak{q}b \subseteq Ra$, so that $\mathfrak{p}(b/a) \subseteq R$; that is, $\mathfrak{p}(b/a)$ is an ideal in $R$. If $\mathfrak{p}(b/a) \subseteq \mathfrak{p}$, then $b/a$ is integral over $R$, for $\mathfrak{p}$ is a finitely generated $R$-submodule of $\operatorname{Frac}(R)$, as required in Lemma 10.46. As $R$ is integrally closed, this puts $b/a \in R$, contradicting what we noted at the end of the previous paragraph. Therefore, $\mathfrak{p}(b/a)$ is not a proper ideal, so that $\mathfrak{p}(b/a) = R$ and $\mathfrak{p} = R(a/b)$. It follows that $a/b \in R$ and $\mathfrak{p}$ is a principal ideal.

Denote $a/b$ by $t$. The proof is completed by showing that the only nonzero ideals in $R$ are the principal ideals generated by $t^n$, for $n \geq 0$. Let $I$ be a nonzero ideal in $R$, and consider the chain of submodules of $\operatorname{Frac}(R)$:

$$I \subseteq It^{-1} \subseteq It^{-2} \subseteq \cdots.$$

We claim that this chain is strictly increasing. If $It^{-n} = It^{-n-1}$, then the finitely generated $R$-module $It^{-1}$ satisfies $t^{-1}(It^{-n}) \subseteq It^{-n}$, so that $t^{-1} = b/a$ is integral over $R$. As above, $R$ integrally closed forces $b/a \in R$, a contradiction. Since $R$ is noetherian, this chain can contain only finitely many ideals in $R$. Thus, there is $n$ with $It^{-n} \subseteq R$ and $It^{-n-1} \nsubseteq R$. If $It^{-n} \subseteq \mathfrak{p} = Rt$, then $It^{-n-1} \subseteq R$, a contradiction. Therefore, $It^{-n} = R$ and $I = Rt^n$, as desired.  •

**Proposition 10.103.** *Let $R$ be a noetherian domain. Then $R$ is a Dedekind ring if and only if the localizations $R_{\mathfrak{p}}$ are DVRs for every nonzero prime ideal $\mathfrak{p}$.*

**Remark.** Exercise 10.46 on page 927 shows that it is necessary to assume that $R$ is noetherian.  ◀

**Proof.** If $R$ is a Dedekind ring and $\mathfrak{p}$ is a maximal ideal, Corollary 10.20(iv) shows that $R_{\mathfrak{p}}$ has a unique nonzero prime ideal. Moreover, $R_{\mathfrak{p}}$ is noetherian [Corollary 10.20(v)], a domain (Corollary 10.18), and integrally closed (Exercise 10.27 on page 907). By Lemma 10.102, $R_{\mathfrak{p}}$ is a DVR.

For the converse, we must show that $R$ is integrally closed and that its nonzero prime ideals are maximal. Let $u/v \in \operatorname{Frac}(R)$ be integral over $R$. For every nonzero prime ideal $\mathfrak{p}$, the element $u/v$ is integral over $R_\mathfrak{p}$ [note that $\operatorname{Frac}(R_\mathfrak{p}) = \operatorname{Frac}(R)$]. But $R_\mathfrak{p}$ is a PID, hence is integrally closed, and so $u/v \in R_\mathfrak{p}$. We conclude that $u/v \in \bigcap_\mathfrak{p} R_\mathfrak{p} = R$, by Proposition 10.22. Therefore, $R$ is integrally closed.

Suppose there are nonzero prime ideals $\mathfrak{p} \subsetneq \mathfrak{q}$ in $R$. By Corollary 10.20(iv), $\mathfrak{p}_\mathfrak{q} \subsetneq \mathfrak{q}_\mathfrak{q}$ in $R_\mathfrak{q}$, contradicting the fact that DVRs have unique nonzero prime ideals. Therefore, nonzero prime ideals are maximal, and $R$ is a Dedekind ring. $\quad\bullet$

Let $R$ be a domain with $Q = \operatorname{Frac}(R)$, and let $I = Ra$ be a nonzero principal ideal in $R$. If we define $J = Ra^{-1} \subseteq Q$, the cyclic $R$-submodule generated by $a^{-1}$, then it is easy to see that

$$IJ = \{uv : u \in I \text{ and } v \in J\} = R.$$

**Definition.** If $R$ is a domain with $Q = \operatorname{Frac}(R)$, then a ***fractional ideal*** is a finitely generated nonzero $R$-submodule $I$ of $Q$. If $I$ is a fractional ideal, define

$$I^{-1} = \{v \in Q : vI \subseteq R\}.$$

It is always true that $I^{-1}I \subseteq R$; a fractional ideal $I$ is ***invertible*** if $I^{-1}I = R$.

Every nonzero finitely generated ideal in $R$ is also a fractional ideal. In this context, we often call such ideals (which are the usual ideals!) ***integral ideals*** when we want to contrast them with more general fractional ideals.

We claim that if $I = Ra$ is a nonzero principal ideal in $R$, then $I^{-1} = Ra^{-1}$. Clearly, $(ra^{-1})(r'a) = rr' \in R$ for all $r' \in R$, so that $Ra^{-1} \subseteq I^{-1}$. For the reverse inclusion, suppose that $(u/v)a \in R$, where $u, v \in R$. Then $v \mid ua$ in $R$, so there is $r \in R$ with $rv = ua$. Hence, in $Q$, we have $u = rva^{-1}$, so that $u/v = (rva^{-1})/v = ra^{-1}$. Therefore, every nonzero principal ideal in $R$ is invertible.

**Lemma 10.104.** *Let $R$ be a domain with $Q = \operatorname{Frac}(R)$. Then a fractional ideal $I$ is invertible if and only if there exist $a_1, \ldots, a_n \in I$ and $q_1, \ldots, q_n \in Q$ with*

(i) *$q_i I \subseteq R$ for $i = 1, \ldots, n$;*

(ii) *$1 = \sum_{i=1}^n q_i a_i$.*

**Proof.** If $I$ is invertible, then $I^{-1}I = R$. Since $1 \in I^{-1}I$, there are $a_1, \ldots, a_n \in R$ and $q_1, \ldots, q_n \in I^{-1}$ with $1 = \sum_i q_i a_i$. Since $q_i \in I^{-1}$, we have $q_i I \subseteq R$.

To prove the converse, note that the $R$-submodule $J$ of $Q$ generated by $q_1, \ldots, q_n$ is a fractional ideal. Since $1 = \sum_{i=1}^n q_i a_i \in JI$, $JI$ is an $R$-submodule of $R$ containing 1; that is, $JI = R$. To see that $I$ is invertible, it remains to prove that $J = I^{-1}$. Clearly, each $q_i \in I^{-1}$, so that $J \subseteq I^{-1}$. For the reverse inclusion, assume that $u \in Q$ and $uI \subseteq R$. Since $1 = \sum_i q_i a_i$, we have $u = \sum_i (ua_i)q_i \in J$ because $ua_i \in R$ for all $i$. $\quad\bullet$

**Remark.** This lemma leads to a homological characterization of Dedekind rings; see Theorem 10.111(ii). $\quad\blacktriangleleft$

**Corollary 10.105.** *Every invertible ideal $I$ in a domain $R$ is finitely generated.*

**Proof.** Since $I$ is invertible, there exist $a_1, \ldots, a_n \in I$ and $q_1, \ldots, q_n \in Q$ as in the lemma. If $b \in I$, then $b = b1 = \sum_i bq_i a_i \in I$, because $bq_i \in R$. Therefore, $I$ is generated by $a_1, \ldots, a_n \in I$. •

**Proposition 10.106.** *The following conditions are equivalent for a domain $R$.*

(i) *$R$ is a Dedekind ring.*

(ii) *Every fractional ideal is invertible.*

(iii) *The set of all the fractional ideals $\mathcal{F}(R)$ forms an abelian group under multiplication of ideals.*

**Proof.** The structure of the proof is unusual, for we will need the equivalence of (ii) and (iii) in order to prove (iii) $\Rightarrow$ (i).

(i) $\Rightarrow$ (ii). Let $J$ be a fractional ideal in $R$. Since $R$ is a Dedekind ring, its localization $R_\mathfrak{p}$ is a PID, and so $J_\mathfrak{p}$, as every nonzero principal ideal, is invertible (in Theorem 8.3, in the course of proving that finitely generated torsion-free abelian groups are free abelian, we really proved that fractional ideals of PIDs are cyclic modules). Now Exercise 10.51 on page 937 gives

$$(J^{-1}J)_\mathfrak{p} = (J^{-1})_\mathfrak{p} J_\mathfrak{p} = (J_\mathfrak{p})^{-1} J_\mathfrak{p} = R_\mathfrak{p}.$$

Proposition 10.37 gives $J^{-1}J = R$, and so $J$ is invertible.

(ii) $\Leftrightarrow$ (iii). If $I, J \in \mathcal{F}(R)$, then they are finitely generated, by Corollary 10.105, and

$$IJ = \left\{ \sum a_\ell b_\ell : a_\ell \in I \text{ and } b_\ell \in J \right\}$$

is a finitely generated $R$-submodule of $\mathrm{Frac}(R)$. If $I = (a_1, \ldots, a_n)$ and $J = (b_1, \ldots, b_m)$, then $IJ$ is generated by all $a_i b_j$. Hence, $IJ$ is finitely generated and $IJ \in \mathcal{F}(R)$. Associativity does hold, the identity is $R$, and the inverse of a fractional ideal $J$ is $J^{-1}$, because $J$ is invertible. It follows that $\mathcal{F}(R)$ is an abelian group.

Conversely, if $\mathcal{F}(R)$ is an abelian group and $I \in \mathcal{F}(R)$, then there is $J \in \mathcal{F}(R)$ with $JI = R$. We must show that $J = I^{-1}$. But

$$R = JI \subseteq I^{-1}I \subseteq R,$$

so that $JI = I^{-1}I$. Canceling $I$ in the group $\mathcal{F}(R)$ gives $J = I^{-1}$, as desired.

(iii) $\Rightarrow$ (i). First, $R$ is noetherian, for (iii) $\Rightarrow$ (ii) shows that every nonzero ideal $I$ is invertible, and Corollary 10.105 shows that $I$ is finitely generated.

Second, we show that every nonzero prime ideal $\mathfrak{p}$ is a maximal ideal. Let $I$ be an ideal with $\mathfrak{p} \subsetneq I$ (we allow $I = R$). Then $\mathfrak{p}I^{-1} \subseteq II^{-1} = R$, so that $\mathfrak{p}I^{-1}$ is an (integral) ideal in $R$. Now $(\mathfrak{p}I^{-1})I = \mathfrak{p}$, because multiplication is associative in $\mathcal{F}(R)$. Since $\mathfrak{p}$ is a prime ideal, Proposition 5.5 says that either $\mathfrak{p}I^{-1} \subseteq \mathfrak{p}$ or $I \subseteq \mathfrak{p}$. The second option does not hold, so that $\mathfrak{p}I^{-1} \subseteq \mathfrak{p}$. Multiplying by $\mathfrak{p}^{-1}I$ gives $R \subseteq I$. Therefore, $I = R$, and so $\mathfrak{p}$ is a maximal ideal.

Third, if $a \in \mathrm{Frac}(R)$ is integral over $R$, then Lemma 10.46 gives a finitely generated $R$-submodule $J$ of $\mathrm{Frac}(R)$, i.e., a fractional ideal, with $aJ \subseteq J$. Since $J$ is invertible, there are $q_1, \ldots, q_n \in \mathrm{Frac}(R)$ and $a_1, \ldots, a_n \in J$ with

$q_i J \subseteq R$ for all $i$ and $1 = \sum q_i a_i$. Hence, $a = \sum_i q_i a_i a$. But $a_i a \in J$ and $q_i J \subseteq R$ gives $a = \sum_i q_i (a_i a) \in R$. Therefore, $R$ is integrally closed, and hence it is a Dedekind ring. $\bullet$

**Proposition 10.107.**

(i) *If $R$ is a UFD, then a nonzero ideal $I$ in $R$ is invertible if and only if it is principal.*

(ii) *A Dedekind ring $R$ is a UFD if and only if it is a PID.*

**Proof.**

(i) We have already seen that every nonzero principal ideal is invertible. Conversely, if $I$ is invertible, there are elements $a_1, \ldots, a_n \in I$ and $q_1, \ldots, q_n \in$ Frac$(R)$ with $1 = \sum_i q_i a_i$ and $q_i I \subseteq R$ for all $i$. Let $q_i = b_i / c_i$, where $b_i, c_i \in R$. Since $R$ is a UFD, we may assume that $q_i$ is in lowest terms; that is, $(b_i, c_i) = 1$. But $(b_i / c_i) a_j \in R$ says that $c_i \mid b_i a_j$, so that $c_i \mid a_j$ for all $i, j$, by Exercise 5.15(i) on page 311. We claim that $I = Rc$, where $c = \text{lcm}\{c_1, \ldots, c_n\}$. First, $c \in I$, for $cb_i / c_i \in R$ and $c = c1 = \sum_i (cb_i / c_i) a_i$. Hence, $Rc \subseteq I$. For the reverse inclusion, Exercise 5.15(ii) shows that $c \mid a_j$ for all $j$, so that $a_j \in Rc$, for all $j$, and so $I \subseteq Rc$.

(ii) Since every nonzero ideal in a Dedekind ring is invertible, it follows from (i) that if $R$ is a UFD, then every ideal in $R$ is principal. $\bullet$

**Definition.** If $R$ is a Dedekind ring, then its **class group** $C(R)$ is defined by

$$C(R) = \mathcal{F}(R)/\mathcal{P}(R),$$

where $\mathcal{P}(R)$ is the subgroup of all nonzero principal ideals.

Dirichlet proved, for every algebraic number field $E$, that the class group of $C(\mathcal{O}_E)$ is finite; the order $|C(R)|$ is called the **class number** of $\mathcal{O}_E$. The usual proof of finiteness of the class number uses a geometric theorem of Minkowski which says that sufficiently large parallelepipeds in Euclidean space must contain lattice points (see Samuel, *Algebraic Theory of Numbers*, pp. 57–58).

Claborn proved, for every (not necessarily finite) abelian group $G$, that there is a Dedekind ring $R$ with $C(R) \cong G$ (Every Abelian group is a class group, *Pacific J. Math* 18 (1966), 219–222).

We can now prove the result linking Kummer and Dedekind.

**Theorem 10.108.** *If $R$ is a Dedekind ring, then every proper nonzero ideal has a unique factorization as a product of prime ideals.*

**Remark.** In *Commutative Algebra* I, Zariski and Samuel define a Dedekind ring to be a domain in which every nonzero ideal is a product of prime ideals (and they then prove that such a factorization must be unique). They prove the converse of this theorem on p. 275. ◀

**Proof.** Let $\mathcal{S}$ be the family of all proper nonzero ideals in $R$ that are not products of prime ideals. If $\mathcal{S} = \varnothing$, then every nonzero ideal in $R$ is a product of prime ideals. If $\mathcal{S} \neq \varnothing$, then $\mathcal{S}$ has a maximal element $I$, because noetherian rings satisfy the maximum condition (Proposition 5.33). Now $I$ cannot be a maximal ideal in $R$, for a "product of prime ideals" is allowed to have only one factor. Let $\mathfrak{m}$ be a maximal ideal containing $I$. Since $I \subsetneq \mathfrak{m}$, we have $\mathfrak{m}^{-1}I \subsetneq \mathfrak{m}^{-1}\mathfrak{m} = R$; that is, $\mathfrak{m}^{-1}I$ is a proper ideal properly containing $I$. Neither $\mathfrak{m}$ nor $\mathfrak{m}^{-1}I$ lies in $\mathcal{S}$, for each is strictly larger than a maximal element, namely, $I$, and so each of them is a product of prime ideals. Therefore, $I = \mathfrak{m}(\mathfrak{m}^{-1}I)$ (equality holding because $R$ is a Dedekind ring) is a product of prime ideals, contradicting $I$ being in $\mathcal{S}$. Therefore, $\mathcal{S} = \varnothing$, and every proper nonzero ideal in $R$ is a product of prime ideals.

Suppose that $\mathfrak{p}_1 \cdots \mathfrak{p}_r = \mathfrak{q}_1 \cdots \mathfrak{q}_s$, where the $\mathfrak{p}_i$ and $\mathfrak{q}_j$ are prime ideals. We prove unique factorization by induction on $\max\{r, s\}$. The base step $r = 1 = s$ is obviously true. For the inductive step, note that $\mathfrak{p}_1 \supseteq \mathfrak{q}_1 \cdots \mathfrak{q}_s$, so that Proposition 5.5 gives $\mathfrak{q}_j$ with $\mathfrak{p}_1 \supseteq \mathfrak{q}_j$. Hence, $\mathfrak{p}_1 = \mathfrak{q}_j$, because prime ideals are maximal. Now multiply the original equation by $\mathfrak{p}_1^{-1}$ and use the inductive hypothesis. •

**Corollary 10.109.** *If $R$ is a Dedekind ring, then $\mathcal{F}(R)$ is a free abelian group with basis all the nonzero prime ideals.*

**Proof.** Of course, $\mathcal{F}(R)$ is written multiplicatively. That every fractional ideal is a product of primes shows that the set of primes generates $\mathcal{F}(R)$; uniqueness of the factorization says the set of primes is a basis. •

In light of Theorem 10.108, many of the usual formulas of arithmetic extend to ideals in Dedekind rings. Observe that in $\mathbb{Z}$, the ideal $(3)$ contains $(9)$. In fact, $\mathbb{Z}m \supseteq \mathbb{Z}n$ if and only if $m \mid n$. We will now see that the relation "contains" for ideals is the same as "divides," and that the usual formulas for gcd's and lcm's [Proposition 2.61(ii)] generalize to Dedekind rings.

**Proposition 10.110.** *Let $I$ and $J$ be nonzero ideals in a Dedekind ring $R$, and let their prime factorizations be*

$$I = \mathfrak{p}_1^{e_1} \cdots \mathfrak{p}_n^{e_n} \quad and \quad J = \mathfrak{p}_1^{f_1} \cdots \mathfrak{p}_n^{f_n},$$

*where $e_i \geq 0$ and $f_i \geq 0$ for all $i$.*

(i) *$J \supseteq I$ if and only if $I = JL$ for some ideal $L$.*

(ii) *$J \supseteq I$ if and only if $f_i \leq e_i$ for all $i$.*

(iii) *If $m_i = \min\{e_i, f_i\}$ and $M_i = \max\{e_i, f_i\}$, then*

$$I \cap J = \mathfrak{p}_1^{M_1} \cdots \mathfrak{p}_n^{M_n} \quad and \quad I + J = \mathfrak{p}_1^{m_1} \cdots \mathfrak{p}_n^{m_n}.$$

*In particular, $I + J = R$ if and only if $\min\{e_i, f_i\} = 0$ for all $i$.*

(iv) *Let $R$ be a Dedekind ring, and let $I = \mathfrak{p}_1^{e_1} \cdots \mathfrak{p}_n^{e_n}$ be a nonzero ideal in $R$. Then*

$$R/I = R/\mathfrak{p}_1^{e_1} \cdots \mathfrak{p}_n^{e_n} \cong \left(R/\mathfrak{p}_1^{e_1}\right) \times \cdots \times \left(R/\mathfrak{p}_n^{e_n}\right).$$

**Proof.**

(i) If $I \subseteq J$, then $J^{-1}I \subseteq R$, and

$$J(J^{-1}I) = I.$$

Conversely, if $I = JL$, then $I \subseteq J$ because $JL \subseteq JR = J$.

(ii) This follows from (i) and the unique factorization of nonzero ideals as products of prime ideals.

(iii) We prove the formula for $I+J$. Let $I+J = \mathfrak{p}_1^{r_1} \cdots \mathfrak{p}_n^{r_n}$ and let $A = \mathfrak{p}_1^{m_1} \cdots \mathfrak{p}_n^{m_n}$. Since $I \subseteq I + J$ and $J \subseteq I + J$, we have $r_i \leq e_i$ and $r_i \leq f_i$, so that $r_i \leq \min\{e_i, f_i\} = m_i$. Hence, $A \subseteq I + J$. For the reverse inclusion, $A \subseteq I$ and $A \subseteq J$, so that $A = II'$ and $A = JJ'$ for ideals $I'$ and $J'$, by (i). Therefore, $I + J = AI' + AJ' = A(I' + J')$, and so $I + J \subseteq A$. The proof of the formula for $IJ$ is left to the reader.

(iv) This is just the Chinese Remainder Theorem, Exercise 5.9(iii) on page 301, so that it suffices to verify the hypothesis that $\mathfrak{p}_i^{e_i}$ and $\mathfrak{p}_j^{e_j}$ are coprime when $i \neq j$; that is, $\mathfrak{p}_i^{e_i} + \mathfrak{p}_j^{e_j} = R$. But this follows from (iii).   •

Recall Proposition 6.78: an $R$-module $A$ is projective if and only if it has a *projective basis*: there exist elements $\{a_j : j \in J\} \subseteq A$ and $R$-maps $\{\varphi_j \colon A \to R : j \in J\}$ such that

(i) for each $x \in A$, almost all $\varphi_j(x) = 0$;

(ii) for each $x \in A$, we have $x = \sum_{j \in J}(\varphi_j x)a_j$.

The coming characterizations of Dedekind rings, Theorems 10.111, 10.114, and 10.123, are useful in Homological Algebra.

**Theorem 10.111.**

(i) *A nonzero ideal $I$ in a domain $R$ is invertible if and only if $I$ is a projective $R$-module.*

(ii) *A domain $R$ is a Dedekind ring if and only if every ideal in $R$ is projective.*

**Proof.**

(i) If $I$ is invertible, there are elements $a_1, \ldots, a_n \in I$ and $q_1, \ldots, q_n \in \mathrm{Frac}(R)$ with $1 = \sum_i q_i a_i$ and $q_i I \subset R$ for all $i$. Define $\varphi_i \colon I \to R$ by $\varphi_i \colon a \mapsto q_i a$ (note that $\mathrm{im}\, \varphi_i \subseteq I$ because $q_i I \subseteq R$). If $a \in I$, then

$$\sum_i \varphi_i(a)a_i = \sum_i q_i a a_i = a \sum_i q_i a_i = a.$$

Therefore, $I$ has a projective basis, and so $I$ is a projective $R$-module.

Conversely, if an ideal $I$ is a projective $R$-module, it has a projective basis $\{\varphi_j : j \in J\}, \{a_j : j \in J\}$. If $b \in I$ is nonzero, define $q_j \in \mathrm{Frac}(R)$ by

$$q_j = \varphi_j(b)/b.$$

This element does not depend on the choice of nonzero $b$: if $b' \in I$ is nonzero, then $b'\varphi_j(b) = \varphi_j(b'b) = b\varphi_j(b')$, so that $\varphi_j(b)/b = \varphi_j(b')/b'$. To see that $q_j I \subseteq R$, note that if $b \in I$ is nonzero, then $q_j b = (\varphi_j(b)/b)b = \varphi_j(b) \in R$. By item (i) in the definition of projective basis, almost all $\varphi_j(b) = 0$, and so there are only finitely many nonzero $q_j = \varphi_j(b)/b$ (remember that $q_j$ does not depend on the choice of nonzero $b \in I$). Item (ii) in the definition of projective basis gives, for $b \in I$,

$$b = \sum_j \varphi_j(b)a_j = \sum_j (q_j b)a_j = b\Big(\sum_j q_j a_j\Big).$$

Canceling $b$ gives $1 = \sum_j q_j a_j$. Finally, the set of those $a_j$ with indices $j$ for which $q_j \neq 0$ completes the data necessary to show that $I$ is an invertible ideal.

(ii) This follows at once from (i) and Proposition 10.106.   •

**Example 10.112.** We have seen that $R = \mathbb{Z}\big[\sqrt{-5}\,\big]$ is a Dedekind ring that is not a PID. Any non-principal ideal gives an example of a projective $R$-module that is not free.   ◀

**Remark.** Noncommutative analogs of Dedekind rings are called *left hereditary*. A ring $R$ is **left hereditary** if every left ideal is a projective $R$-module. [Small's example of a right noetherian ring that is not left noetherian (Exercise 6.16 on page 416) is right hereditary but not left hereditary.] Aside from Dedekind rings, some examples of left hereditary rings are semisimple rings, noncommutative principal ideal rings, and FIRs (*free ideal rings*—all left ideals are free $R$-modules). Cohn, *Free Rings and Their Relations*, proved that if $k$ is a field, then free $k$-algebras $k\langle X \rangle$ (polynomials in noncommuting variables $X$) are FIRs, and so there exist left hereditary rings that are not left noetherian.   ◀

Projective and injective modules over a Dedekind ring are well-behaved.

**Lemma 10.113.** *A left $R$-module $P$ (over any ring $R$) is projective if and only if every diagram below with $E$ injective can be completed to a commutative diagram. The dual characterization of injective modules is also true.*

$$\begin{array}{ccc} & & P \\ & \swarrow & \downarrow \\ E & \longrightarrow E'' & \longrightarrow 0 \end{array}$$

**Proof.** If $P$ is projective, then the diagram can be completed for every not necessarily injective module $E$. Conversely, we must show that the diagram

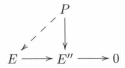

can be completed for any module $A$ and any surjection $g \colon A \to A''$. By Theorem 6.96, there is an injective $R$-module $E$ and an injection $\sigma \colon A \to E$. Define

$E'' = \operatorname{coker}\sigma i = E/\operatorname{im}\sigma i$, and consider the commutative diagram with exact rows

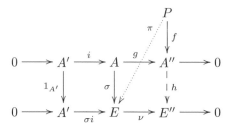

where $\nu\colon E \to E'' = \operatorname{coker}\sigma i$ is the natural map and $h\colon A'' \to E''$ exists, by Proposition 6.116. By hypothesis, there exists a map $\pi\colon P \to E$ with $\nu\pi = hf$. We claim that $\operatorname{im}\pi \subseteq \operatorname{im}\sigma$. For $x \in P$, surjectivity of $g$ gives $a \in A$ with $ga = fx$. Then $\nu\pi x = hfx = hga = \nu\sigma a$, and so $\pi x - \sigma a \in \ker\nu = \operatorname{im}\sigma i$; hence, $\pi x - \sigma a = \sigma i a'$ for some $a' \in A'$, and so $\pi x = \sigma(a + ia') \in \operatorname{im}\sigma$. Therefore, if $x \in P$, there is a unique $a \in A$ with $\sigma a = \pi x$ ($a$ is unique because $\sigma$ is an injection). Thus, there is a well-defined function $\pi'\colon P \to A$, given by $\pi' x = a$, where $\sigma a = \pi x$. The reader may check that $\pi'$ is an $R$-map and that $g\pi' = f$. $\bullet$

**Theorem 10.114 (Cartan-Eilenberg).** *The following conditions are equivalent for a domain $R$.*

(i) *$R$ is a Dedekind ring.*

(ii) *Every submodule of a projective $R$-module is projective.*

(iii) *Every quotient of an injective $R$-module is injective.*

**Proof.**

(i) $\Leftrightarrow$ (ii). If $R$ is Dedekind, then we can adapt the proof of Theorem 8.9 (which proves that every subgroup of a free abelian group is free abelian) to prove that every submodule of a free $R$-module is projective (Exercise 10.48 on page 936); in particular, every submodule of a projective $R$-module is projective. Conversely, since $R$ itself is a projective $R$-module, its submodules are also projective, by hypothesis; that is, the ideals of $R$ are projective. Theorem 10.111 now shows that $R$ is a Dedekind ring.

(ii) $\Leftrightarrow$ (iii). Assume (iii), and consider the diagram with exact rows

$$
\begin{array}{ccccc}
P & \longleftarrow & P' & \longleftarrow & 0 \\
\downarrow & \searrow & \downarrow f & & \\
E & \longrightarrow & E'' & \longrightarrow & 0
\end{array}
$$

where $P$ is projective and $E$ is injective; note that the hypothesis gives $E''$ injective. To prove projectivity of $P'$, it suffices, by Lemma 10.113, to find a map $P' \to E$ making the diagram commute. Since $E''$ is injective, there exists a map $P \to E''$ giving commutativity. Since $P$ is projective, there is a map $P \to E$ also giving commutativity. The composite $P' \to P \to E$ is the desired map.

The converse is the dual of this, using the dual of Lemma 10.113. $\bullet$

The next result generalizes Exercise 9.38 on page 814 from PIDs to Dedekind rings.

**Corollary 10.115.** *Let $R$ be a Dedekind ring.*

(i) *For all $R$-modules $C$ and $A$ and all $n \geq 2$,*
$$\operatorname{Ext}_R^n(C, A) = \{0\} \quad and \quad \operatorname{Tor}_n^R(C, A) = \{0\}.$$

(ii) *Let $0 \to A' \to A \to A'' \to 0$ be a short exact sequence. For every module $C$, there are exact sequences*

$$0 \to \operatorname{Hom}(C, A') \to \operatorname{Hom}(C, A) \to \operatorname{Hom}(C, A'')$$
$$\to \operatorname{Ext}^1(C, A') \to \operatorname{Ext}^1(C, A) \to \operatorname{Ext}^1(C, A'') \to 0$$

*and*

$$0 \to \operatorname{Tor}_1^R(C, A') \to \operatorname{Tor}_i^R(C, A) \to \operatorname{Tor}_1^R(C, A'')$$
$$\to C \otimes_R A' \to C \otimes_R A \to C \otimes_R A'' \to 0.$$

**Proof.**

(i) By definition, if $\cdots \to P_2 \xrightarrow{d_2} P_1 \xrightarrow{d_1} P_0 \to C \to 0$ is a projective resolution of $C$, then $\operatorname{Ext}^n(C, A) = \ker d_{n+1}^* / \operatorname{im} d_n^*$; moreover, $\operatorname{Ext}^n(C, A)$ does not depend on the choice of projective resolution, by Corollary 9.76. Now $C$ is a quotient of a free module $F$, and so there is an exact sequence

(4)                        $$0 \to K \to F \xrightarrow{\varepsilon} C \to 0,$$

where $K = \ker \varepsilon$. Since $R$ is Dedekind, the submodule $K$ of the free $R$-module $F$ is projective, so that (3) defines a projective resolution of $C$ with $P_0 = F$, $P_1 = K$, and $P_n = \{0\}$ for all $n \geq 2$. Hence, $\ker d_{n+1}^* \subseteq \operatorname{Hom}(P_n, A) = \{0\}$ for all $n \geq 2$, and so $\operatorname{Ext}_R^n(C, A) = \{0\}$ for all $A$ and for all $n \geq 2$. A similar argument works for Tor.

(ii) This follows from Corollary 9.70 and Corollary 9.59, the long exact sequence for Ext and for Tor, respectively.  •

We will use this result in the next section to generalize Proposition 9.94.

---

## Exercises

*  **10.47.** Let $R$ be a commutative ring and let $M$ be a finitely generated $R$-module. Prove that if $IM = M$ for some ideal $I$ of $R$, then there exists $a \in I$ with $(1 - a)M = \{0\}$.

**Hint.** If $M = \langle m_1 \ldots, m_n \rangle$, then each $m_i = \sum_j a_{ij} m_j$, where $a_{ij} \in I$. Use the adjoint matrix (the matrix of cofactors) as in the proof of Lemma 10.46.

*  **10.48.** Generalize the proof of Theorem 8.9 to prove that if $R$ is a left hereditary ring, then every submodule of a free left $R$-module $F$ is isomorphic to a direct sum of ideals, and hence is projective.

*  **10.49.** Let $R$ be a Dedekind ring, and let $\mathfrak{p}$ be a nonzero prime ideal in $R$.

(i) If $a \in \mathfrak{p}$, prove that $\mathfrak{p}$ occurs in the prime factorization of $Ra$.

(ii) If $a \in \mathfrak{p}^e$ and $a \notin \mathfrak{p}^{e+1}$, prove that $\mathfrak{p}^e$ occurs in the prime factorization of $Ra$, but that $\mathfrak{p}^{e+1}$ does not occur in the prime factorization of $Ra$.

**10.50.** Let $I$ be a nonzero ideal in a Dedekind ring $R$. Prove that if $\mathfrak{p}$ is a prime ideal, then $I \subseteq \mathfrak{p}$ if and only if $\mathfrak{p}$ occurs in the prime factorization of $I$.

\* **10.51.** If $J$ is a fractional ideal of a Dedekind ring $R$, prove that $(J^{-1})_\mathfrak{p} = (J_\mathfrak{p})^{-1}$ for every maximal ideal $\mathfrak{p}$.

\* **10.52.** Let $I_1, \ldots, I_n$ be ideals in a Dedekind ring $R$. If there is no nonzero prime ideal $\mathfrak{p}$ with $I_i = \mathfrak{p}L_i$ for all $i$ for ideals $L_i$, then

$$I_1 + \cdots + I_n = R.$$

**10.53.** Give an example of a projective $\mathbb{Z}[\sqrt{-5}]$-module that is not free.

**Hint.** See Example 10.112.

\* **10.54.** (i) A commutative ring $R$ is called a ***principal ideal ring*** if every ideal in $R$ is a principal ideal ($R$ would be a PID if it were a domain). For example, $\mathbb{I}_n$ is a principal ideal ring. Prove that $\mathbb{Z} \times \mathbb{Z}$ is a principal ideal ring.

(ii) Let $I_1, \ldots, I_n$ be pairwise coprime ideals in a commutative ring $R$. If $R/I_i$ is a principal ideal ring for each $i$, prove that $R/(I_1 \cdots I_n)$ is a principal ideal ring.
**Hint.** Use the Chinese Remainder Theorem, Exercise 5.9 on page 301.

**10.55.** Let $a$ be a nonzero element in a Dedekind ring $R$. Prove that there are only finitely many ideals $I$ in $R$ containing $a$.

**Hint.** If $a \in I$, then $Ra = IL$ for some ideal $L \subseteq R$.

---

**10.3.5. Finitely Generated Modules over Dedekind Rings.** We saw, in Chapter 8, that theorems about abelian groups generalize to theorems about modules over PIDs. We are now going to see that such theorems can be further generalized to modules over Dedekind rings.

**Proposition 10.116.** *Let $R$ be a Dedekind ring.*

(i) *If $I \subseteq R$ is a nonzero ideal, then every ideal in $R/I$ is principal.*

(ii) *Every fractional ideal $J$ can be generated by two elements. More precisely, for any nonzero $a \in J$, there exists $b \in J$ with $J = (a, b)$.*

**Proof.**

(i) Let $I = \mathfrak{p}_1^{e_1} \cdots \mathfrak{p}_n^{e_n}$ be the prime factorization of $I$. Since the ideals $\mathfrak{p}_i^{e_i}$ are pairwise coprime, it suffices, by Exercise 10.54 on page 937, to prove that $R/\mathfrak{p}_i^{e_i}$ is a principal ideal ring for each $i$. Now right exactness of $R_{\mathfrak{p}_i} \otimes_R \square$ shows that $(R/\mathfrak{p}_i^{e_i})_{\mathfrak{p}_i} \cong R_{\mathfrak{p}_i}/(\mathfrak{p}_i^{e_i})_{\mathfrak{p}_i}$. Since $R_{\mathfrak{p}_i}$ is a PID (it is even a DVR, by Proposition 10.103), any quotient ring of it is a principal ideal ring.

(ii) Assume first that $J$ is an integral ideal. Choose a nonzero $a \in J$. By (i), the ideal $J/Ra$ in $R/Ra$ is principal; say, $J/Ra$ is generated by $b + Ra$, where $b \in J$. It follows that $J = (a, b)$.

For the general case, there is a nonzero $c \in R$ with $cJ \subseteq R$ (if $J$ is generated by $u_1/v_1, \ldots, u_m/v_m$, take $c = \prod_J v_j$). Since $cJ$ is an integral ideal, given any nonzero $a \in J$, there is $cb \in cJ$ with $cJ = (ca, cb)$. It follows that $J = (a, b)$. •

The next corollary says that we can force nonzero ideals to be coprime.

**Corollary 10.117.** *If $I$ and $J$ are fractional ideals over a Dedekind ring $R$, then there are $a, b \in \mathrm{Frac}(R)$ with*

$$aI + bJ = R.$$

**Proof.** Choose a nonzero $a \in I^{-1}$. Now $aI \subseteq I^{-1}I = R$, so that $aIJ^{-1} \subseteq J^{-1}$. By Proposition 10.116(ii), there is $b \in J^{-1}$ with

$$J^{-1} = aIJ^{-1} + Rb.$$

Since $b \in J^{-1}$, we have $bJ \subseteq R$, and so

$$R = JJ^{-1} = J(aIJ^{-1} + Rb) = aI + RbJ = aI + bJ. \quad •$$

Let us now investigate the structure of $R$-modules. Recall that if $R$ is a domain and $M$ is an $R$-module, then $\mathrm{rank}(M) = \dim_Q(Q \otimes_R M)$, where $Q = \mathrm{Frac}(R)$. We can restate Lemma 10.1(iii) in our present language. If $R$ is a domain and $M$ is a nonzero finitely generated torsion-free $R$-module, then $\mathrm{rank}(M) = 1$ if and only if $M$ is isomorphic to a nonzero integral ideal.

**Proposition 10.118.** *If $R$ is a Dedekind ring and $M$ is a finitely generated torsion-free $R$-module, then*

$$M \cong I_1 \oplus \cdots \oplus I_n,$$

*where $I_i$ is an ideal in $R$.*

**Proof.** The proof is by induction on $\mathrm{rank}(M) \geq 0$. If $\mathrm{rank}(M) = 0$, then $M$ is torsion, by Lemma 10.1(ii). Since $M$ is torsion-free, $M = \{0\}$. Assume now that $\mathrm{rank}(M) = n + 1$. Choose a nonzero $m \in M$, so that $\mathrm{rank}(Rm) = 1$. The sequence

$$0 \to Rm \to M \xrightarrow{\nu} M'' \to 0$$

is exact, where $M'' = R/Rm$ and $\nu$ is the natural map. Note that $\mathrm{rank}(M'') = n$, by Lemma 10.1(i). Now $M$ finitely generated implies that $M''$ is also finitely generated. If $T = t(M'')$ is the torsion submodule of $M''$, then $M''/T$ is a finitely generated torsion-free $R$-module with $\mathrm{rank}(M''/T) = \mathrm{rank}(M'') = n$, because $\mathrm{rank}(T) = 0$. By induction, $M''/T$ is a direct sum of ideals, hence is projective, by Theorem 10.111. Define

$$M' = \nu^{-1}(T) = \{m \in M : rm \in Rm \text{ for some } r \neq 0\} \subseteq M.$$

There is an exact sequence $0 \to M' \to M \to M''/T \to 0$; this sequence splits, by Corollary 6.74, because $M''/T$ is projective; that is, $M \cong M' \oplus (M''/T)$. Hence

$$\mathrm{rank}(M') = \mathrm{rank}(M) - \mathrm{rank}(M''/T) = 1.$$

Since $R$ is noetherian, every submodule of a finitely generated $R$-module is itself finitely generated; hence, $M'$ is finitely generated. Therefore, $M'$ is isomorphic to an ideal, by Lemma 10.1(ii), and this completes the proof. •

**Corollary 10.119.** *If $R$ is a Dedekind ring and $M$ is a finitely generated torsion-free $R$-module, then $M$ is projective.*

**Proof.** By Theorem 10.111, every ideal in a Dedekind ring is projective. It now follows from Proposition 10.118 that $M$ is a direct sum of ideals, and hence it is projective. •

Corollary 10.119 can also be proved by localization. For every maximal ideal $\mathfrak{m}$, the $R_{\mathfrak{m}}$-module $M_{\mathfrak{m}}$ is finitely generated torsion-free. Since $R_{\mathfrak{m}}$ is a PID (even a DVR), $M_{\mathfrak{m}}$ is a free module, and hence it is projective. The result now follows from Corollary 10.43.

**Corollary 10.120.** *If $M$ is a finitely generated $R$-module, where $R$ is a Dedekind ring, then the torsion submodule $tM$ is a direct summand of $M$.*

**Proof.** The quotient module $M/tM$ is a finitely generated torsion-free $R$-module, so that it is projective, by Corollary 10.119. Therefore, $tM$ is a direct summand of $M$, by Corollary 6.74. •

**Corollary 10.121.** *Let $R$ be a Dedekind ring. Then an $R$-module $A$ is flat if and only if it is torsion-free.*

**Proof.** By Lemma 6.139, it suffices to prove that every finitely generated submodule of $A$ is flat. But such submodules are torsion-free, hence projective, and projective modules are always flat, by Lemma 6.137. The converse is Corollary 8.6: every flat $R$-module over a domain $R$ is torsion-free. •

**Corollary 10.122.** *Let $R$ be a Dedekind ring with $Q = \mathrm{Frac}(R)$.*

(i) *If $C$ is a torsion-free $R$-module and $T$ is a torsion module with $\mathrm{ann}(T) \neq (0)$ (that is, there is a nonzero $r \in R$ with $rT = \{0\}$), then $\mathrm{Ext}_R^1(C, T) = \{0\}$.*

(ii) *Let $M$ be an $R$-module. If $\mathrm{ann}(tM) \neq (0)$, where $tM$ is the torsion submodule of $M$, then $tM$ is a direct summand of $M$.*

**Proof.**

(i) We generalize the proof of Proposition 9.94. Since $C$ is torsion-free, it is a flat $R$-module, by Corollary 10.121, so that exactness of $0 \to R \to Q$ gives exactness of $0 \to R \otimes_R C \to Q \otimes_R C$. Thus, $C \cong R \otimes_R C$ can be imbedded in a vector space $V$ over $Q$, namely, $V = Q \otimes_R C$. Applying the contravariant functor $\mathrm{Hom}_R(\square, T)$ to $0 \to C \to V \to V/C \to 0$ gives an exact sequence

$$\mathrm{Ext}_R^1(V, T) \to \mathrm{Ext}_R^1(C, T) \to \mathrm{Ext}^2(V/C, T).$$

Now the last term is $\{0\}$, by Corollary 10.115, and $\mathrm{Ext}_R^1(V, T)$ is (torsion-free) divisible, by (a straightforward generalization of) Example 9.72; hence, $\mathrm{Ext}_R^1(C, T)$ is divisible. Since $\mathrm{ann}(T) \neq (0)$, Exercise 9.41 on page 814 gives $\mathrm{Ext}_R^1(C, T) = \{0\}$.

(ii) To prove that the extension $0 \to tM \to M \to M/tM \to 0$ splits, it suffices to prove that $\mathrm{Ext}_R^1(M/tM, tM) = \{0\}$. Since $M/tM$ is torsion-free, this follows from part (i) and Corollary 9.92. •

The next result generalizes Proposition 6.93.

**Theorem 10.123.** *Let $R$ be a domain. Then $R$ is Dedekind if and only if divisible $R$-modules are injective.*

**Proof.** Let $R$ be a Dedekind ring and let $E$ be a divisible $R$-module. By the Baer criterion, Theorem 6.89, it suffices to complete the diagram

$$
\begin{array}{ccc}
 & E & \\
 & \uparrow^{f} \ \ \diagdown^{g} & \\
0 \longrightarrow I & \underset{i}{\longrightarrow} & R
\end{array}
$$

where $I$ is an ideal and $i\colon I \to R$ is the inclusion. Of course, we may assume that $I$ is nonzero, and so $I$ is invertible: there are elements $a_1, \dots, a_n \in I$ and elements $q_1, \dots, q_n \in Q$ with $q_i I \subseteq R$ and $1 = \sum_i q_i a_i$. Since $E$ is divisible, there are elements $e_i \in E$ with $f(a_i) = a_i e_i$. Note, for every $b \in I$, that

$$
f(b) = f\Big(\sum_i q_i a_i b\Big) = \sum_i (q_i b) f(a_i) = \sum_i (q_i b) a_i e_i = b \sum_i (q_i a_i) e_i.
$$

Hence, if we define $e = \sum_i (q_i a_i) e_i$, then $e \in E$ and $f(b) = be$ for all $b \in I$. Defining $g\colon R \to E$ by $g(r) = re$ shows that the diagram can be completed, and so $E$ is injective.

Conversely, if $E$ is an injective $R$-module, then $E$ is divisible [that every injective $R$-module is divisible was proved (for arbitrary domains $R$) in Lemma 6.92]. If $E'$ is a quotient of $E$, then $E'$ is divisible and, hence, is injective, by hypothesis. Therefore, every quotient of an injective module is injective, and so $R$ is Dedekind, by Theorem 10.114.  •

Having examined torsion-free modules, let us now look at torsion modules.

**Proposition 10.124.** *Let $\mathfrak{p}$ be a nonzero prime ideal in a Dedekind ring $R$. If $M$ is an $R$-module with $\operatorname{ann}(M) = \mathfrak{p}^e$ for some $e > 0$, then the localization map $M \to M_{\mathfrak{p}}$ is an isomorphism (and hence $M$ may be regarded as an $R_{\mathfrak{p}}$-module).*

**Proof.** It suffices to prove that $M \cong R_{\mathfrak{p}} \otimes_R M$. If $m \in M$ is nonzero and $s \in R$ with $s \notin \mathfrak{p}$, then

$$\mathfrak{p}^e + Rs = R,$$

by Proposition 10.110. Hence, there exist $u \in \mathfrak{p}^e$ and $r \in R$ with $1 = u + rs$, and so

$$m = um + rsm = rsm.$$

If $1 = u' + r's$, where $u' \in \mathfrak{p}$ and $r' \in R$, then $s(r - r')m = 0$, so that

$$s(r - r') \in \operatorname{ann}(m) = \mathfrak{p}^e.$$

Since $s \notin \mathfrak{p}^e$, it follows that $r - r' \in \mathfrak{p}^e$ [see Exercise 10.49 on page 936; the prime factorization of $Rs$ does not contain $\mathfrak{p}^e$; if the prime factorization of $R(r - r')$ does not contain $\mathfrak{p}^e$, then neither does the prime factorization of $Rs(r - r')$]. Hence, $rm = r'm$. Define $s^{-1}m = rm$, and define $f\colon R_{\mathfrak{p}} \times M \to M$ by $f(r/s, m) = s^{-1}(rm)$. It is straightforward to check that $f$ is $R$-bilinear, and so there is an

$R$-map $\widetilde{f}\colon R_{\mathfrak{p}} \otimes M \to M$ with $\widetilde{f}\colon (r/s \otimes m) = s^{-1}rm$. In particular, $\widetilde{f}(1 \otimes m) = m$, so that $\widetilde{f}$ is surjective. On the other hand, the localization map $h_M\colon M \to R_{\mathfrak{p}} \otimes_R M$, defined by $h_M(m) = 1 \otimes m$, is easily seen to be the inverse of $\widetilde{f}$. $\quad\bullet$

**Definition.** Let $\mathfrak{p}$ be a nonzero prime ideal in a Dedekind ring $R$. An $R$-module $M$ is called $\mathfrak{p}$-***primary*** if, for each $m \in M$, there is $e > 0$ with $\mathrm{ann}(m) = \mathfrak{p}^e$.

**Theorem 10.125 (Primary Decomposition).** *Let $R$ be a Dedekind ring, and let $T$ be a finitely generated torsion $R$-module. If $I = \mathrm{ann}(T) = \mathfrak{p}_1^{e_1} \cdots \mathfrak{p}_n^{e_n}$, then*

$$T = T[\mathfrak{p}_1] \oplus \cdots \oplus T[\mathfrak{p}_n],$$

*where*

$$T[\mathfrak{p}_i] = \{m \in T : \mathrm{ann}(m) \text{ is a power of } \mathfrak{p}_i\}.$$

$T[\mathfrak{p}_i]$ *is called the* $\mathfrak{p}_i$-***primary component*** *of $T$.*

**Proof.** It is easy to see that the $\mathfrak{p}_i$-primary components $T[\mathfrak{p}_i]$ are submodules of $T$. We now check the conditions in Proposition 6.30. If $W_i$ is the submodule of $T$ generated by all $T[\mathfrak{p}_j]$ with $j \neq i$, we must show that $T[\mathfrak{p}_i] \cap W_i = \{0\}$. Let $x \in T[\mathfrak{p}_i] \cap W_i$. If

$$I_i = \mathfrak{p}_1^{e_1} \cdots \widehat{\mathfrak{p}_i^{e_i}} \cdots \mathfrak{p}_n^{e_n},$$

then $\mathfrak{p}_i$ and $I_i$ are coprime: $\mathfrak{p}_i + I_i = R$. Hence, there are $a_i \in \mathfrak{p}_i$ and $r_i \in I_i$ with $1 = a_i + r_i$, and so $x = a_i x + r_i x$. But $a_i x = 0$, because $x \in T[\mathfrak{p}_i]$, and $r_i x = 0$, because $x \in W_i$ and $I_i = \mathrm{ann}(W_i)$. Therefore, $x = 0$.

By Exercise 10.52 on page 937, we have

$$I_1 + \cdots + I_n = R.$$

Thus, there are $b_i \in I_i$ with $b_1 + \cdots + b_n = 1$. If $t \in T$, then $t = b_1 t + \cdots + b_n t$. But if $c_i \in \mathfrak{p}_i^{e_i}$, then $c_i b_i \in \mathfrak{p}_i^{e_i} I_i = I = \mathrm{ann}(T)$ and so $c_i(b_i t) = 0$. Hence, $\mathfrak{p}_i^{e_i} \subseteq \mathrm{ann}(b_i t)$, so that $\mathrm{ann}(b_i t) = \mathfrak{p}_i^e$ for some $e > 0$. Therefore, $b_i t \in T[\mathfrak{p}_i]$, and so

$$T = T[\mathfrak{p}_1] + \cdots + T[\mathfrak{p}_n].$$

The result now follows from Proposition 6.30. $\quad\bullet$

**Theorem 10.126.** *Let $R$ be a Dedekind ring.*

(i) *Two finitely generated torsion $R$-modules $T$ and $T'$ are isomorphic if and only if $T[\mathfrak{p}_i] \cong T'[\mathfrak{p}_i]$ for all $i$.*

(ii) *Every finitely generated $\mathfrak{p}$-primary $R$-module $T$ is a direct sum of cyclic $R$-modules, and the number of summands of each type is an invariant of $T$.*

**Proof.**

(i) The result follows easily from the observation that if $f\colon T \to T'$ is an isomorphism, then $\mathrm{ann}(t) = \mathrm{ann}(f(t))$ for all $t \in T$.

(ii) The Primary Decomposition shows that $T$ is the direct sum of its primary components $T[\mathfrak{p}_i]$. By Proposition 10.124, $T[\mathfrak{p}_i]$ is an $R_{\mathfrak{p}_i}$-module. But $R_{\mathfrak{p}_i}$ is a PID, and so the Basis Theorem and the Fundamental Theorem hold: each $T[\mathfrak{p}_i]$ is a direct sum of cyclic modules, and the numbers and isomorphism types of the cyclic summands are uniquely determined. $\quad\bullet$

We now know, when $R$ is Dedekind, that every finitely generated $R$-module $M$ is a direct sum of cyclic modules and ideals. What uniqueness is there in such a decomposition? Since the torsion submodule is a fully invariant direct summand, we may focus on torsion-free modules.

Recall Proposition 10.3: two ideals $J$ and $J'$ in a domain $R$ are isomorphic if and only if there is $a \in \mathrm{Frac}(R)$ with $J' = aJ$.

**Lemma 10.127.** *Let $M$ be a finitely generated torsion-free $R$-module, where $R$ is a Dedekind ring, so that $M \cong I_1 \oplus \cdots \oplus I_n$, where the $I_i$ are ideals. Then*

$$M \cong R^{n-1} \oplus J,$$

*where $J = I_1 \cdots I_n$.*

**Remark.** We call $R^{n-1} \oplus J$ a ***Steinitz normal form*** for $M$. We will prove, in Theorem 10.129, that $J$ is unique up to isomorphism. ◀

**Proof.** It suffices to prove that $I \oplus J \cong R \oplus IJ$, for the result then follows easily by induction on $n \geq 2$. By Corollary 10.117, there are nonzero $a, b \in \mathrm{Frac}(R)$ with $aI + bJ = R$. Since $aI \cong I$ and $bJ \cong J$, we may assume that $I$ and $J$ are coprime integral ideals. There is an exact sequence

$$0 \to I \cap J \xrightarrow{\delta} I \oplus J \xrightarrow{\alpha} I + J \to 0,$$

where $\delta \colon x \mapsto (x, x)$ and $\alpha \colon (u, v) \mapsto u - v$. Since $I$ and $J$ are coprime, however, we have $I \cap J = IJ$ and $I + J = R$, by Exercise 5.9 on page 301. As $R$ is projective, this sequence splits; that is, $I \oplus J \cong R \oplus IJ$. •

The following cancellation lemma, while true for Dedekind rings, can be false for some other rings. In Example 9.161(ii), we described an example of Kaplansky showing that if $R = \mathbb{R}[x, y, z]/(x^2 + y^2 + z^2 - 1)$ is the real coordinate ring of the 2-sphere, then there is a finitely generated stably free $R$-module $M$ that is not free. Hence, there are free $R$-modules $F$ and $F'$ with $M \oplus F \cong F' \oplus F$ but $M \not\cong F'$.

**Lemma 10.128.** *Let $R$ be a Dedekind ring. If $R \oplus G \cong R \oplus H$, where $G$ and $H$ are $R$-modules, then $G \cong H$.*

**Proof.** We may assume that there is a module $E = A \oplus G = B \oplus H$, where $A \cong R \cong B$. Let $p \colon E \to B$ be the projection $p \colon (b, h) \mapsto b$, and let $p' = p|G$. Now

$$\ker p' = G \cap H \quad \text{and} \quad \mathrm{im}\, p' \subseteq B \cong R.$$

Thus, $\mathrm{im}\, p' \cong L$, where $L$ is an ideal in $R$.

If $\mathrm{im}\, p' = \{0\}$, then $G \subseteq \ker p = H$. Since $E = A \oplus G$, Corollary 6.29 gives $H = G \oplus (H \cap A)$. On the one hand, $E/G = (A \oplus G)/G \cong A \cong R$; on the other hand, $E/G = (B \oplus H)/G \cong B \oplus (H/G) \cong R \oplus (H/G)$. Thus, $R \cong R \oplus (H/G)$. Since $R$ is a domain, this forces $H/G = \{0\}$: if $R = X \oplus Y$, then $X$ and $Y$ are ideals; if $x \in X$ and $y \in Y$ are both nonzero, then $xy \in X \cap Y = (0)$, giving zero-divisors in $R$. It follows that $H/G = \{0\}$ and $G = H$.

We may now assume that $L = \operatorname{im} p'$ is a nonzero ideal. The First Isomorphism Theorem gives $G/(G \cap H) \cong L$. Since $R$ is a Dedekind ring, $L$ is a projective module, and so

$$G = I \oplus (G \cap H),$$

where $I \cong L$. Similarly,

$$H = J \oplus (G \cap H),$$

where $J$ is isomorphic to an ideal. Therefore,

$$E = A \oplus G = A \oplus I \oplus (G \cap H);$$
$$E = B \oplus H = B \oplus J \oplus (G \cap H).$$

It follows that

$$A \oplus I \cong E/(G \cap H) \cong B \oplus J.$$

If we can prove that $I \cong J$, then

$$G = I \oplus (G \cap H) \cong J \oplus (G \cap H) = H.$$

Therefore, we have reduced the theorem to the special case where $G$ and $H$ are nonzero ideals.

We will prove that if $\alpha \colon R \oplus I \to R \oplus J$ is an isomorphism, then $I \cong J$. As in our discussion of generalized matrices on page 546, $\alpha$ determines a $2 \times 2$ matrix

$$A = \begin{bmatrix} a_{11} & a_{12} \\ a_{21} & a_{22} \end{bmatrix},$$

where $a_{11} \colon R \to R$, $a_{21} \colon R \to J$, $a_{12} \colon I \to R$, and $a_{22} \colon I \to J$. Indeed, as maps between ideals are just multiplications by elements of $Q = \operatorname{Frac}(R)$, we may regard $A$ as a matrix in $\mathrm{GL}(2, Q)$. Now $a_{21} \in J$ and $a_{22} I \subseteq J$, so that if $d = \det(A)$, then

$$dI = (a_{11} a_{22} - a_{12} a_{12}) I \subseteq J.$$

Similarly, $\beta = \alpha^{-1}$ determines a $2 \times 2$ matrix $B = A^{-1}$. But

$$d^{-1} J = \det(B) J \subseteq I,$$

so that $J \subseteq dI$. We conclude that $J = dI$, and so $J \cong I$. $\quad \bullet$

Let us sketch another proof of Lemma 10.128 using exterior algebra. Let $R$ be a Dedekind ring, and let $I$ and $J$ be fractional ideals; we show that $R \oplus I \cong R \oplus J$ implies $I \cong J$. The fact that $2 \times 2$ determinants are used in the original proof suggests that second exterior powers may be useful. By Theorem 8.120,

$$\bigwedge\nolimits^2 (R \oplus I) \cong \left( R \otimes \bigwedge\nolimits^2 I \right) \oplus \left( \bigwedge\nolimits^1 R \otimes_R \bigwedge\nolimits^1 I \right) \oplus \left( \bigwedge\nolimits^2 R \otimes_R I \right).$$

Now $\bigwedge^2 R = \{0\}$, by Corollary 8.115, and $\bigwedge^1 R \otimes_R \bigwedge^1 I \cong R \otimes_R I \cong I$. We now show, for every maximal ideal $\mathfrak{m}$, that $\left( \bigwedge^2 I \right)_{\mathfrak{m}} = \{0\}$. By Exercise 10.25 on page 899,

$$\left( \bigwedge\nolimits^n I \right)_{\mathfrak{m}} \cong \bigwedge\nolimits^n I_{\mathfrak{m}}.$$

But $R_{\mathfrak{m}}$ is a PID, so that $I_{\mathfrak{m}}$ is a principal ideal, and hence $\bigwedge^2 I_{\mathfrak{m}} = \{0\}$, by Corollary 8.115. It now follows from Proposition 10.38(i) that $\bigwedge^2 I = \{0\}$. Therefore, $\bigwedge^2 (R \oplus I) \cong I$. Similarly, $\bigwedge^2 (R \oplus J) \cong J$, and so $I \cong J$.

**Theorem 10.129 (Steinitz).** *Let $R$ be a Dedekind ring, and let $M \cong I_1 \oplus \cdots \oplus I_n$ and $M' \cong I_1' \oplus \cdots \oplus I_\ell'$ be finitely generated torsion-free $R$-modules. Then $M \cong M'$ if and only if $n = \ell$ and $I_1 \cdots I_n \cong I_1' \cdots I_n'$.*

**Proof.** Lemma 10.1(iii) shows that $\operatorname{rank}(I_i) = 1$ for all $i$, and Lemma 10.1(i) shows that $\operatorname{rank}(M) = n$; similarly, $\operatorname{rank}(M') = \ell$. Since $M \cong M'$, we have $Q \otimes_R M \cong Q \otimes_R M'$, so that $\operatorname{rank}(M) = \operatorname{rank}(M')$ and $n = \ell$. By Lemma 10.127, it suffices to prove that if $R^n \oplus I \cong R^n \oplus J$, then $I \cong J$. But this follows at once from repeated use of Lemma 10.128. •

Recall that two $R$-modules $A$ and $B$ over a commutative ring $R$ are *stably isomorphic* if there exists an $R$-module $C$ with $A \oplus C \cong B \oplus C$.

**Corollary 10.130.** *Let $R$ be a Dedekind ring, and let $M$ and $M'$ be finitely generated torsion-free $R$-modules. Then $M$ and $M'$ are stably isomorphic if and only if they are isomorphic.*

**Proof.** Isomorphic modules are always stably isomorphic. To prove the converse, assume that there is a finitely generated torsion-free $R$-module $X$ with

$$M \oplus X \cong M' \oplus X.$$

By Lemma 10.127, there are ideals $I, J, L$ with $M \cong R^{n-1} \oplus I$, $M' \cong R^{n-1} \oplus J$, and $X \cong R^{m-1} \oplus L$, where $n = \operatorname{rank}(M) = \operatorname{rank}(M')$. Hence,

$$M \oplus X \cong R^{n-1} \oplus I \oplus R^{m-1} \oplus L \cong R^{n+m-1} \oplus IL.$$

Similarly,

$$M' \oplus X \cong R^{n+m-1} \oplus JL.$$

By Theorem 10.129, $IL \cong JL$, and so there is a nonzero $a \in \operatorname{Frac}(R)$ with $aIL = JL$, by Lemma 10.3. Multiplying by $L^{-1}$ gives $aI = J$, and so $I \cong J$. Therefore,

$$M \cong R^{n-1} \oplus I \cong R^{n-1} \oplus J \cong M'. \quad •$$

If $R$ is a ring, then $R$-**Proj** is the full subcategory of $_R$**Mod** generated by all finitely generated projective left $R$-modules. Recall that the *Grothendieck group* $K_0(R)$ has generators all $[P]$ and relations $[P] + [P'] = [P \oplus P']$, where $[P]$ is the isomorphism class of $P$. There is a relation between the class group $C(R)$ of a Dedekind ring $R$ and its Grothendieck group $K_0(R)$. If $I$ is a nonzero ideal in a Dedekind ring $R$, denote the corresponding element in $C(R)$ by $\operatorname{cls}(I)$. Corollary 9.160 says that two $R$-modules determine the same element of $K_0(R)$ if and only if they are stably isomorphic. Thus, if $R$ is Dedekind, then Corollary 10.130 says that two finitely generated projective $R$-modules are isomorphic if and only if they determine the same element of $K_0(R)$.

**Theorem 10.131.** *If $R$ is a Dedekind ring, then*

$$K_0(R) \cong C(R) \oplus \mathbb{Z},$$

*where $C(R)$ is the class group of $R$.*

**Proof.** If $P$ is a nonzero finitely generated projective $R$-module, then $P$ has a Steinitz normal form: by Lemma 10.127, $P \cong R^{n-1} \oplus J$, where $\text{rank}(P) = n$ and $J$ is a nonzero finitely generated ideal. Moreover, the isomorphism class of $J$ is uniquely determined by $[P]$, by Theorem 10.129. Therefore, the function $\varphi \colon K_0(R) \to C(R) \oplus \mathbb{Z}$, given by $\varphi \colon [P] \mapsto (\text{cls}(J), n)$, is a well-defined homomorphism. Clearly, $\varphi$ is surjective, and it is injective by Corollary 10.130.  •

**Corollary 10.132.** *If $R$ is a Dedekind ring, then $\widetilde{K}_0(R) \cong C(R)$.*

**Proof.** By definition, the reduced Grothendieck group $\widetilde{K}_0(R) = \ker \rho$, where $\rho \colon K_0(R) \to \mathbb{Z}$ is defined by $[P] \mapsto \text{rank}(P)$. Here, $\ker \rho = C(R)$.  •

# Exercises

**10.56.** Let $R$ be a Dedekind ring, and let $I \subseteq R$ be a nonzero ideal. Prove that there exists an ideal $J \subseteq R$ with $I + J = R$ and $IJ$ principal.

**Hint.** Let $I = \mathfrak{p}_1^{e_1} \cdots \mathfrak{p}_n^{e_n}$, and choose $r_i \in \mathfrak{p}_i^{e_i} - \mathfrak{p}_i^{e_i+1}$. Use the Chinese Remainder Theorem to find an element $a \in R$ with $a \in \mathfrak{p}_i^{e_i}$ and $a \notin \mathfrak{p}_i^{e_i+1}$, and consider the prime factorization of $Ra$.

**10.57.** (i) If $R$ is a nonzero commutative ring, prove that $R^n \cong R^m$ implies $n = m$. Conclude that the rank of a free $R$-module is well-defined.

　　　**Hint.** Corollary 6.69.

　　(ii) If $R$ is any commutative ring, prove that $\mathbb{Z}$ is a direct summand of $K_0(R)$.

**10.58.** If $I$ is a fractional ideal in a Dedekind ring $R$, prove that $I \otimes_R I^{-1} \cong R$.

**Hint.** Use invertibility of $I$.

# Section 10.4. Homological Dimensions

### Rings in this section need not be commutative.

Semisimple and Dedekind rings can be characterized in terms of their projective modules: Proposition 7.33 says that a ring $R$ is semisimple if and only if every $R$-module is projective; Theorem 10.111(ii) says that a domain $R$ is Dedekind if and only if every ideal is projective. The notion of global dimension allows us to classify arbitrary rings in this spirit.

**Definition.** Let $R$ be a ring and let $A$ be a left $R$-module. If there is a finite projective resolution

$$0 \to P_n \to \cdots \to P_1 \to P_0 \to A \to 0,$$

then we write $\text{pd}(A) \leq n$.[8] If $n \geq 0$ is the smallest integer such that $\text{pd}(A) \leq n$, then we say that $A$ has ***projective dimension*** $n$; if there is no finite projective resolution of $A$, then $\text{pd}(A) = \infty$.

---

[8]Sometimes it is convenient to display the ring $R$, and we will write $\text{pd}_R(M)$ in this case.

**Example 10.133.**

(i) A module $A$ is projective if and only if $\mathrm{pd}(A) = 0$. We may thus regard $\mathrm{pd}(A)$ as a measure of how far away $A$ is from being projective.

(ii) If $R$ is a Dedekind ring, we claim that $\mathrm{pd}(A) \leq 1$ for every $R$-module $A$. By Theorem 10.114, every submodule of a free $R$-module is projective. Hence, if $F$ is a free $R$-module and $\varepsilon \colon F \to A$ is a surjection, then $\ker \varepsilon$ is projective, and

$$0 \to \ker \varepsilon \to F \xrightarrow{\varepsilon} A \to 0$$

is a projective resolution of $A$.  ◄

**Definition.** Let $\mathbf{P} = \to P_2 \xrightarrow{d_2} P_1 \xrightarrow{d_1} P_0 \xrightarrow{\varepsilon} A \to 0$ be a projective resolution of a module $A$. If $n \geq 0$, then the $n$th **syzygy**[9] is

$$\Omega_n(A, \mathbf{P}) = \begin{cases} \ker \varepsilon & \text{if } n = 0 \\ \ker d_n & \text{if } n \geq 1. \end{cases}$$

**Proposition 10.134.** *For every $n \geq 1$, for all left $R$-modules $A$ and $B$, and for every projective resolution $\mathbf{P}$ of $B$, there is an isomorphism*

$$\mathrm{Ext}_R^{n+1}(A, B) \cong \mathrm{Ext}_R^1(\Omega_{n-1}(A, \mathbf{P}), B).$$

**Proof.** The proof is by induction on $n \geq 1$. Exactness of the projective resolution

$$\mathbf{P} = \to P_2 \xrightarrow{d_2} P_1 \xrightarrow{d_1} P_0 \xrightarrow{\varepsilon} A \to 0$$

gives exactness of

$$\to P_2 \xrightarrow{d_2} P_1 \to \Omega_0(A, \mathbf{P}) \to 0,$$

which is a projective resolution $\mathbf{P}^+$ of $\Omega_0(A, \mathbf{P})$. In more detail, define

$$P_n^+ = P_{n+1} \qquad \text{and} \qquad d_n^+ = d_{n+1}.$$

Since $\mathrm{Ext}^1$ is independent of the choice of projective resolution of the first variable,

$$\mathrm{Ext}_R^1(\Omega_0(A, \mathbf{P}), B) = \frac{\ker(d_2^+)^*}{\mathrm{im}(d_1^+)^*} = \frac{\ker(d_3)^*}{\mathrm{im}(d_2)^*} = \mathrm{Ext}_R^2(A, B).$$

The inductive step is proved in the same way, noting that

$$\to P_{n+2} \xrightarrow{d_{n+2}} P_{n+1} \to \Omega_n(A, \mathbf{P}) \to 0$$

is a projective resolution of $\Omega_n(A, \mathbf{P})$.  •

**Corollary 10.135.** *For all left $R$-modules $A$ and $B$, for all $n \geq 0$, and for any projective resolutions $\mathbf{P}$ and $\mathbf{P}'$ of $A$, there is an isomorphism*

$$\mathrm{Ext}_R^1(\Omega_n(A, \mathbf{P}), B) \cong \mathrm{Ext}_R^1(\Omega_n(A, \mathbf{P}'), B).$$

**Proof.** By Proposition 10.134, both sides are isomorphic to $\mathrm{Ext}_R^{n+1}(A, B)$.  •

---

[9]The term *syzygy* was introduced into mathematics by Sylvester in 1850. It is a Greek word meaning *yoke*, and it was used in astronomy to mean an alignment of three celestial figures.

Two left $R$-modules $\Omega$ and $\Omega'$ are called ***projectively equivalent*** if there exist projective left $R$-modules $P$ and $P'$ with $\Omega \oplus P \cong \Omega' \oplus P'$. Exercise 10.59 on page 954 shows that any two $n$th syzygies of a left $R$-module $A$ are projectively equivalent. We often abuse notation and speak of *the* $n$th syzygy of a module, writing $\Omega_n(A)$ instead of $\Omega_n(A, \mathbf{P})$.

Syzygies help compute projective dimension.

**Lemma 10.136.** *The following conditions are equivalent for a left $R$-module $A$.*

(i) $\mathrm{pd}(A) \leq n$.

(ii) $\mathrm{Ext}_R^k(A, B) = \{0\}$ *for all left $R$-modules $B$ and all $k \geq n + 1$.*

(iii) $\mathrm{Ext}_R^{n+1}(A, B) = \{0\}$ *for all left $R$-modules $B$.*

(iv) *For every projective resolution $\mathbf{P}$ of $A$, the $(n-1)$st syzygy $\Omega_{n-1}(A, \mathbf{P})$ is projective.*

(v) *There exists a projective resolution $\mathbf{P}$ of $A$ with $\Omega_{n-1}(A, \mathbf{P})$ projective.*

**Proof.**

(i) $\Rightarrow$ (ii). By hypothesis, there is a projective resolution $\mathbf{P}$ of $A$ with $P_k = \{0\}$ for all $k \geq n + 1$. Necessarily, all the maps $d_k \colon P_k \to P_{k-1}$ are zero for $k \geq n + 1$, and so

$$\mathrm{Ext}_R^k(A, B) = \frac{\ker(d_{k+1})^*}{\mathrm{im}(d_k)^*} = \{0\}.$$

(ii) $\Rightarrow$ (iii). Obvious.

(iii) $\Rightarrow$ (iv). If $\mathbf{P} = \to P_n \to P_{n-1} \to \cdots \to P_1 \to P_0 \to A \to 0$ is a projective resolution of $A$, then $\mathrm{Ext}_R^{n+1}(A, B) \cong \mathrm{Ext}_R^1(\Omega_{n-1}(A, \mathbf{P}), B)$, by Proposition 10.134. But the last group is $\{0\}$, by hypothesis, so that $\Omega_{n-1}(A)$ is projective, by Corollary 9.88.

(iv) $\Rightarrow$ (v). Obvious.

(v) $\Rightarrow$ (i). If

$$\to P_n \to P_{n-1} \to \cdots \to P_1 \to P_0 \to A \to 0$$

is a projective resolution of $A$, then

$$0 \to \Omega_{n-1}(A) \to P_{n-1} \to \cdots \to P_1 \to P_0 \to A \to 0$$

is an exact sequence. Since $\Omega_{n-1}(A)$ is projective, the last sequence is a projective resolution of $A$, and so $\mathrm{pd}(A) \leq n$. $\bullet$

**Example 10.137.** Let $G$ be a finite cyclic group with $|G| > 1$. If $\mathbb{Z}$ is viewed as a trivial $\mathbb{Z}G$-module, then $\mathrm{pd}(\mathbb{Z}) = \infty$, because Corollary 9.112 gives $H^n(G, \mathbb{Z}) = \mathrm{Ext}_{\mathbb{Z}G}^n(\mathbb{Z}, \mathbb{Z}) \neq \{0\}$ for all odd $n$. ◄

The following definition will soon be simplified.

**Definition.** If $R$ is a ring, then its ***left projective global dimension*** is defined by

$$\ell\mathrm{pD}(R) = \sup\{\mathrm{pd}(A) : A \in \mathrm{obj}(_R\mathbf{Mod})\}.$$

Of course, if $R$ is commutative, we write $\mathrm{pD}(R)$ instead of $\ell\mathrm{pD}(R)$.

**Proposition 10.138.** *For any ring $R$,*

$$\ell\mathrm{pD}(R) \le n \ \text{ if and only if } \ \mathrm{Ext}_R^{n+1}(A, B) = \{0\}$$

*for all left $R$-modules $A$ and $B$.*

**Proof.** This follows from the equivalence of (i) and (iii) in Lemma 10.136.  •

A ring $R$ is semisimple if and only if $\ell\mathrm{pD}(R) = 0$. Thus, global dimension is a measure of how far a ring is from being semisimple.

A similar discussion can be given using injective resolutions.

**Definition.** Let $R$ be a ring and let $B$ be a left $R$-module. If there is an injective resolution

$$0 \to B \to E^0 \to E^1 \to \cdots \to E^n \to 0,$$

then we write $\mathrm{id}(B) \le n$. If $n \ge 0$ is the smallest integer such that $\mathrm{id}(B) \le n$, then we say that $B$ has ***injective dimension*** $n$; if there is no finite injective resolution of $B$, then $\mathrm{id}(B) = \infty$.

**Example 10.139.**

(i) A left $R$-module $B$ is injective if and only if $\mathrm{id}(B) = 0$. We may thus regard $\mathrm{id}(B)$ as a measure of how far away $B$ is from being injective.

(ii) The injective and projective dimensions of a left $R$-module $A$ can be distinct. For example, the abelian group $A = \mathbb{Z}$ has $\mathrm{pd}(A) = 0$ and $\mathrm{id}(A) = 1$.

(iii) If $R$ is a Dedekind ring, then Theorem 10.114 says that every quotient module of an injective $R$-module is injective. Hence, if $\eta \colon B \to E$ is an imbedding of an $R$-module $B$ into an injective $R$-module $E$, then $\mathrm{coker}\,\eta$ is injective, and

$$0 \to B \xrightarrow{\ \eta\ } E \to \mathrm{coker}\,\eta \to 0$$

is an injective resolution of $B$. It follows that $\mathrm{id}(B) \le 1$.  ◄

**Definition.** Let $\mathbf{E} = 0 \to B \xrightarrow{\ \eta\ } E^0 \xrightarrow{\ d^0\ } E^1 \xrightarrow{\ d^1\ } E^2 \to \cdots$ be an injective resolution of a left $R$-module $B$. If $n \ge 0$, then the $n$th ***cosyzygy*** is

$$\mho^n(B, \mathbf{E}) = \begin{cases} \mathrm{coker}\,\eta & \text{if } n = 0 \\ \mathrm{coker}\,d^{n-1} & \text{if } n \ge 1. \end{cases}$$

**Proposition 10.140.** *For every $n \ge 1$, for all left $R$-modules $A$ and $B$, and for every injective resolution $\mathbf{E}$ of $A$, there is an isomorphism*

$$\mathrm{Ext}_R^{n+1}(A, B) \cong \mathrm{Ext}_R^1(A, \mho^{n-1}(B, \mathbf{E})).$$

**Proof.** Dual to the proof of Proposition 10.134.  •

**Corollary 10.141.** *For all left $R$-modules $A$ and $B$, for all $n \ge 0$, and for any injective resolutions $\mathbf{E}$ and $\mathbf{E}'$ of $B$, there is an isomorphism*

$$\mathrm{Ext}_R^1(A, \mho^n(B, \mathbf{E})) \cong \mathrm{Ext}_R^1(A, \mho^n(B, \mathbf{E}')).$$

**Proof.** Dual to the proof of Corollary 10.135.  •

Two left $R$-modules $\mho$ and $\mho'$ are called ***injectively equivalent*** if there exist injective left $R$-modules $E$ and $E'$ with $\mho \oplus E \cong \mho' \oplus E'$. Exercise 10.60 on page 954 shows that any two $n$th cosyzygies of a left $R$-module $B$ are injectively equivalent. We often abuse notation and speak of *the* $n$th cosyzygy of a module, writing $\mho^n(B)$ instead of $\mho^n(B, \mathbf{E})$.

Cosyzygies help compute injective dimension.

**Lemma 10.142.** *The following conditions are equivalent for a left $R$-module $B$.*

(i) $\mathrm{id}(B) \leq n$.

(ii) $\mathrm{Ext}_R^k(A, B) = \{0\}$ *for all left $R$-modules $A$ and all $k \geq n+1$.*

(iii) $\mathrm{Ext}_R^{n+1}(A, B) = \{0\}$ *for all left $R$-modules $A$.*

(iv) *For every injective resolution $\mathbf{E}$ of $B$, the $(n-1)$st cosyzygy $\mho^{n-1}(B, \mathbf{E})$ is injective.*

(v) *There exists an injective resolution $\mathbf{E}$ of $B$ with $\mho^{n-1}(B, \mathbf{E})$ injective.*

**Proof.** Dual to that of Lemma 10.136, using Exercise 9.49 on page 825  •

**Definition.** If $R$ is a ring, then its ***left injective global dimension*** is defined by
$$\ell\mathrm{iD}(R) = \sup\{\mathrm{id}(B) : B \in \mathrm{obj}(_R\mathbf{Mod})\}.$$

**Proposition 10.143.** *For any ring $R$,*
$$\ell\mathrm{iD}(R) \leq n \ \text{ if and only if } \ \mathrm{Ext}_R^{n+1}(A, B) = \{0\}$$
*for all left $R$-modules $A$ and $B$.*

**Proof.** This follows from the equivalence of (i) and (iii) in Lemma 10.142.  •

**Theorem 10.144.** *For every ring $R$,*
$$\ell\mathrm{pD}(R) = \ell\mathrm{iD}(R).$$

**Proof.** This follows at once from Propositions 10.138 and 10.143, for each number is equal to the smallest $n$ for which $\mathrm{Ext}_R^{n+1}(A, B) = \{0\}$ for all left $R$-modules $A$ and $B$.  •

**Definition.** The ***left global dimension*** of a ring $R$ is the common value of the left projective global dimension and the left injective global dimension:
$$\ell\mathrm{D}(R) = \ell\mathrm{pD}(R) = \ell\mathrm{iD}(R).$$

If $R$ is commutative, then we denote its global dimension by $\mathrm{D}(R)$.

There is also a ***right global dimension*** $r\mathrm{D}(R) = \ell\mathrm{D}(R^{\mathrm{op}})$ of a ring $R$; thus, we consider projective and injective dimensions of right $R$-modules. If $R$ is commutative, then $\ell\mathrm{D}(R) = r\mathrm{D}(R)$ and we write $\mathrm{D}(R)$. Since left semisimple rings are also right semisimple, by Corollary 7.45, we have $\ell\mathrm{D}(R) = 0$ if and only if $r\mathrm{D}(R) = 0$. On the other hand, these two dimensions can differ. Jategaonkar, A counterexample in Ring Theory and Homological Algebra, *J. Algebra* 12 (1966), 97–105, proved that if $1 \leq m \leq n \leq \infty$, then there exists a ring $R$ with $\ell\mathrm{D}(R) = m$ and $r\mathrm{D}(R) = n$.

**Theorem 10.145.** *A domain $R$ is a Dedekind ring if and only if $D(R) \leq 1$.*

**Proof.** Let $R$ be a Dedekind ring. If $A$ is an $R$-module, there is a free $R$-module $F$ and an exact sequence $0 \to K \to F \to A \to 0$. But $K$ is projective, by Theorem 10.114 (which says that a domain $R$ is Dedekind if and only if every submodule of a projective $R$-module is projective). Thus, $\mathrm{pd}(A) \leq 1$ and $D(R) \leq 1$.

Conversely, assume that $D(R) \leq 1$. If $S$ is a submodule of a projective $R$-module $F$, then there is an exact sequence $0 \to S \to F \to F/S \to 0$. But $\mathrm{pd}(F/S) \leq 1$, so that $S$ is projective, by the equivalence of parts (i) and (iv) of Lemma 10.136. Theorem 10.114 shows that $R$ is Dedekind.  •

The next theorem says that $\ell D(R)$ can be computed from $\mathrm{pd}(M)$ for finitely generated $R$-modules $M$; in fact, $\ell D(R)$ can even be computed from $\mathrm{pd}(M)$ for $M$ cyclic.

**Lemma 10.146.** *A left $R$-module $B$ is injective if and only if $\mathrm{Ext}^1_R(R/I, B) = \{0\}$ for every left ideal $I$.*

**Proof.** If $B$ is injective, then $\mathrm{Ext}^1_R(A, B)$ vanishes for every right $R$-module $A$. Conversely, suppose that $\mathrm{Ext}^1_R(R/I, B) = \{0\}$ for every left ideal $I$. Applying $\mathrm{Hom}_R(\square, B)$ to the exact sequence $0 \to I \to R \to R/I \to 0$ gives exactness of

$$\mathrm{Hom}_R(R, B) \to \mathrm{Hom}_R(I, B) \to \mathrm{Ext}^1_R(R/I, B) = 0.$$

That is, every $R$-map $f \colon I \to B$ can be extended to an $R$-map $R \to B$ (Proposition 6.83). But this is precisely the Baer criterion, Theorem 6.89, and so $B$ is injective.  •

**Theorem 10.147 (Auslander).** *For any ring $R$,*

$$\ell D(R) = \sup\{\mathrm{pd}(R/I) : I \text{ is a left ideal}\}.$$

**Proof.** (Matlis) If $\sup\{\mathrm{pd}(R/I)\} = \infty$, we are done. Therefore, we may assume there is an integer $n \geq 0$ with $\mathrm{pd}(R/I) \leq n$ for every left ideal $I$. By Lemma 10.142, $\mathrm{Ext}^{n+1}_R(R/I, B) = \{0\}$ for every left $R$-module $B$. But $\ell \mathrm{pD}(R) = \ell \mathrm{iD}(R)$, by Theorem 10.144, so that it suffices to prove that $\mathrm{id}(B) \leq n$ for every $B$. Let $\mathbf{E}$ be an injective resolution of $B$, with $(n-1)$st cosyzygy $\mho^{n-1}(B)$. By Corollary 10.140, $\{0\} = \mathrm{Ext}^{n+1}_R(R/I, B) \cong \mathrm{Ext}^1_R(R/I, \mho^{n-1}(B))$. Now Lemma 10.146 gives $\mho^{n-1}(B)$ injective, and so Lemma 10.142 gives $\mathrm{id}(B) \leq n$, as desired.  •

This theorem explains why every ideal in a Dedekind ring being projective is such a strong condition.

Just as Ext defines the global dimension of a ring $R$, we can use Tor to define the *weak dimension* (or *Tor-dimension*) of a ring $R$.

**Definition.** Let $R$ be a ring and let $A$ be a right $R$-module. A **flat resolution** of $A$ is an exact sequence

$$\to F_n \to \cdots \to F_1 \to F_0 \to A \to 0$$

in which each $F_n$ is a flat right $R$-module.

If there is a finite flat resolution

$$0 \to F_n \to \cdots \to F_1 \to F_0 \to A \to 0,$$

then we write $\mathrm{fd}(A) \leq n$. If $n \geq 0$ is the smallest integer such that $\mathrm{fd}(A) \leq n$, then we say that $A$ has **flat dimension** $n$; if there is no finite flat resolution of $A$, then $\mathrm{fd}(A) = \infty$.

**Example 10.148.**

(i) A right $R$-module $A$ is flat if and only if $\mathrm{fd}(A) = 0$. We may thus regard $\mathrm{fd}(A)$ as a measure of how far away $A$ is from being flat.

(ii) Since projective right $R$-modules are flat, every projective resolution of $A$ is a flat resolution. It follows that if $R$ is any ring, then $\mathrm{fd}(A) \leq \mathrm{pd}(A)$ for every right $R$-module $A$. The same argument applies to left $R$-modules.

(iii) If $R$ is a Dedekind ring and $A$ is an $R$-module, then $\mathrm{fd}(A) \leq \mathrm{pd}(A) \leq 1$, by (ii). Corollary 10.121 says that $A$ is flat if and only if $A$ is torsion-free. Hence, $\mathrm{fd}(A) = 1$ if and only if $A$ is not torsion-free.

(iv) Schanuel's Lemma (Proposition 6.80) does not hold for flat resolutions. Recall that a $\mathbb{Z}$-module if flat if and only if it is torsion-free. Consider the flat resolutions

$$0 \to \mathbb{Z} \to \mathbb{Q} \to \mathbb{Q}/\mathbb{Z} \to 0,$$
$$0 \to K \to F \to \mathbb{Q}/\mathbb{Z} \to 0,$$

where $F$ (and its subgroup $K$) is a free abelian group. There is no isomorphism $K \oplus \mathbb{Q} \cong \mathbb{Z} \oplus F$, for $\mathbb{Q}$ is not projective. ◀

**Definition.** Let $\mathbf{F} = \to F_2 \xrightarrow{d_2} F_1 \xrightarrow{d_1} F_0 \xrightarrow{\varepsilon} B \to 0$ be a flat resolution of a left $R$-module $B$. If $n \geq 0$, then the $n$th **yoke** is

$$Y_n(B, \mathbf{F}) = \begin{cases} \ker \varepsilon & \text{if } n = 0 \\ \ker d_n & \text{if } n \geq 1. \end{cases}$$

The term *yoke* is not standard; it is a translation of the Greek word *syzygy*.

**Proposition 10.149.** *For every $n \geq 1$, for all right $R$-modules $A$ and left $R$-modules $B$, and for every flat resolution $\mathbf{F}$ of $B$, there is an isomorphism*

$$\mathrm{Tor}_{n+1}^R(A, B) \cong \mathrm{Tor}_1^R(A, Y_{n-1}(B, \mathbf{F})).$$

**Proof.** The same as the proof of Proposition 10.134. •

**Corollary 10.150.** *For every right $R$-module $A$ and left $R$-module $B$, for all $n \geq 0$, and for any flat resolutions $\mathbf{F}$ and $\mathbf{F}'$ of $B$, there is an isomorphism*

$$\mathrm{Tor}_1^R(A, Y_n(B, \mathbf{F})) \cong \mathrm{Tor}_1^R(A, Y_n(B, \mathbf{F})).$$

**Proof.** The same as the proof of Corollary 10.135. •

**Lemma 10.151.** *The following conditions are equivalent for a left $R$-module $B$.*

(i) $\mathrm{fd}(B) \leq n$.

  (ii) $\mathrm{Tor}_k^R(A, B) = \{0\}$ *for all $k \geq n + 1$ and all right $R$-modules $A$.*

  (iii) $\mathrm{Tor}_{n+1}^R(A, B) = \{0\}$ *for all right $R$-modules $A$.*

  (iv) *For every flat resolution $\mathbf{F}$ of $B$, the $(n-1)$st yoke $Y_{n-1}(B, \mathbf{F})$ is flat.*

  (v) *There exists a flat resolution $\mathbf{F}$ of $B$ with flat $(n-1)$st yoke $Y_{n-1}(B, \mathbf{F})$.*

**Proof.** The same as the proof of Lemma 10.136.  •

**Definition.** The ***right weak dimension*** of a ring $R$ is defined by

$$\mathrm{rwD}(R) = \sup\{\mathrm{fd}(A) : A \in \mathrm{obj}(\mathbf{Mod}_R)\}.$$

**Proposition 10.152.** *For any ring $R$, $\mathrm{rwD}(R) \leq n$ if and only if $\mathrm{Tor}_{n+1}^R(A, B) = \{0\}$ for every left $R$-module $B$.*

**Proof.** This follows at once from Lemma 10.151.  •

    We define the flat dimension of left $R$-modules in the obvious way, in terms of its flat resolutions, and taking the supremum over all left $R$-modules gives the left weak dimension.

**Definition.** The ***left weak dimension*** of a ring $R$ is

$$\ell\mathrm{wD}(R) = \sup\{\mathrm{fd}(B) : B \in \mathrm{obj}(_R\mathbf{Mod})\}.$$

**Theorem 10.153.** *For any ring $R$,*

$$\mathrm{rwD}(R) = \ell\mathrm{wD}(R).$$

**Proof.** If either dimension is finite, then the left or right weak dimension is the smallest $n \geq 0$ with $\mathrm{Tor}_{n+1}^R(A, B) = \{0\}$ for all right $R$-modules $A$ and all left $R$-modules $B$.  •

**Definition.** The ***weak dimension*** of a ring $R$, denoted by $\mathrm{wD}(R)$, is the common value of $\mathrm{rwD}(R)$ and $\ell\mathrm{wD}(R)$.

    As we have remarked earlier, there are (noncommutative) rings whose left global dimension and right global dimension can be distinct. In contrast, weak dimension has no left/right distinction, because tensor and Tor involve both left and right modules simultaneously.

**Example 10.154.** A ring $R$ has $\mathrm{wD}(R) = 0$ if and only if every left or right module is flat. These rings turn out to be ***von Neumann regular***: for each $a \in R$, there exists $a' \in R$ with $aa'a = a$. Examples of such rings are Boolean rings (rings $R$ in which $r^2 = r$ for all $r \in R$), and $\mathrm{End}_k(V)$, where $V$ is a (possibly infinite-dimensional) vector space over a field $k$ (Rotman, *An Introduction to Homological Algebra*, pp. 159–160). For these rings $R$, we have $\mathrm{wD}(R) = 0 < \ell\mathrm{D}(R)$.  ◀

    The next proposition explains why weak dimension is so called.

**Proposition 10.155.** *For any ring $R$,*

$$\mathrm{wD}(R) \leq \min\{\ell\mathrm{D}(R), r\mathrm{D}(R)\}.$$

**Proof.** It suffices to prove that $\text{fd}(A) \leq \text{pd}(A)$ for any right or left $R$-module $A$; but we saw this in Example 10.148(ii). •

**Corollary 10.156.** *Suppose that* $\text{Ext}_R^n(A, B) = \{0\}$ *for all left $R$-modules $A$ and $B$. Then* $\text{Tor}_n^R(C, D) = \{0\}$ *for all right $R$-modules $C$ and all left $R$-modules $D$.*

**Proof.** If $\text{Ext}_R^n(A, B) = \{0\}$ for all $A$, $B$, then $\ell\text{D}(R) \leq n-1$; if $\text{Tor}_n^R(C, D) \neq \{0\}$ for some $C$, $D$, then $n \leq \text{wD}(DR)$. Hence,

$$\ell\text{D}(R) \leq n - 1 < n \leq \text{wD}(R),$$

contradicting Proposition 10.155. •

**Lemma 10.157.** *A left $R$-module $B$ is flat if and only if* $\text{Tor}_1^R(R/I, B) = \{0\}$ *for every right ideal $I$.*

**Proof.** Exactness of $0 \to I \xrightarrow{i} R \to R/I \to 0$ gives exactness of

$$0 = \text{Tor}_1^R(R, B) \to \text{Tor}_1^R(R/I, B) \to I \otimes_R B \xrightarrow{i \otimes 1} R \otimes_R B.$$

Therefore, $i \otimes 1$ is an injection if and only if $\text{Tor}_1^R(R/I, B) = \{0\}$.

On the other hand, $B$ is flat if and only if $i \otimes 1$ is an injection for every right ideal $I$, by Corollary 6.143. •

As global dimension, weak dimension can be computed from cyclic modules.

**Corollary 10.158.** *For any ring $R$,*

$$\text{wD}(R) = \sup\{\text{fd}(R/I) : I \text{ is a right ideal of } R\}$$
$$= \sup\{\text{fd}(R/J) : J \text{ is a left ideal of } R\}.$$

**Proof.** This proof is similar to the proof of Theorem 10.147, using Lemma 10.157 instead of Lemma 10.146. •

**Proposition 10.159.** *If $R$ is a commutative ring and $S \subseteq R$ is multiplicative, then*

$$\text{wD}(S^{-1}R) \leq \text{wD}(R).$$

**Proof.** We may assume that $\text{wD}(R) = n < \infty$. If $A$ is an $S^{-1}R$-module, we must prove that $\text{fd}(A) \leq n$. By Proposition 10.27(i), $A \cong S^{-1}M$ for some $R$-module $M$ (indeed, $M = {}_hA$, the $S^{-1}R$-module $A$ viewed as an $R$-module). By hypothesis, there is an $R$-flat resolution

$$0 \to F_n \to \cdots \to F_0 \to M \to 0.$$

But localization is an exact functor, so that $0 \to S^{-1}F_n \to \cdots \to S^{-1}F_0 \to S^{-1}M \to 0$ is an exact sequence of $S^{-1}R$-modules. Finally, each $S^{-1}F_i$ is flat, by Exercise 6.96 on page 498, so that $\text{fd}(S^{-1}M) = \text{fd}(A) \leq n$. •

**Theorem 10.160.** *Let $R$ be a left noetherian ring.*

(i) *If $A$ is a finitely generated left $R$-module, then*

$$\text{pd}(A) = \text{fd}(A).$$

(ii) $\text{wD}(R) = \ell\text{D}(R).$

(iii) *If $R$ is a commutative noetherian ring, then*

$$\mathrm{wD}(R) = \mathrm{D}(R).$$

**Proof.**

(i) It is always true that $\mathrm{fd}(A) \leq \mathrm{pd}(A)$, since every projective resolution is a flat resolution [Example 10.148(ii)]. For the reverse inequality, it is enough to prove that if $\mathrm{fd}(A) \leq n$, then $\mathrm{pd}(A) \leq n$. By Lemma 10.41, there is a projective resolution of $A$,

$$\rightarrow P_n \rightarrow \cdots \rightarrow P_0 \rightarrow A \rightarrow 0,$$

in which every $P_i$ is finitely generated. Now this is also a flat resolution, so that, by Lemma 10.151, $\mathrm{fd}(A) \leq n$ implies that $Y_n = \ker(P_{n-1} \rightarrow P_{n-2})$ is flat. But every finitely generated flat left $R$-module is projective, by Corollary 6.146 (because $R$ is left noetherian), and so

$$0 \rightarrow Y_{n-1} \rightarrow P_{n-1} \rightarrow \cdots \rightarrow P_0 \rightarrow A \rightarrow 0$$

is a projective resolution. Therefore, $\mathrm{pd}(A) \leq n$.

(ii) By Theorem 10.147, we have that $\ell\mathrm{D}(R)$ is the supremum of projective dimensions of cyclic left $R$-modules, and by Corollary 10.158, we have that $\mathrm{wd}(R)$ is the supremum of flat dimensions of cyclic left $R$-modules. But part (i) gives $\mathrm{fd}(A) = \mathrm{pd}(A)$ for every finitely generated right $R$-module $A$, and this suffices to prove the result.

(iii) This is a special case of (ii).  $\bullet$

## Exercises

* **10.59.** (i) If $A \rightarrow B \xrightarrow{f} C \rightarrow D$ is an exact sequence and $X$ is any module, prove that there is an exact sequence

$$A \rightarrow B \oplus X \xrightarrow{f \oplus 1_X} C \oplus X \rightarrow D.$$

(ii) Let

$$\mathbf{P} = \rightarrow P_2 \xrightarrow{d_2} P_1 \xrightarrow{d_1} P_0 \xrightarrow{\varepsilon} A \rightarrow 0$$

and

$$\mathbf{P}' = \rightarrow P_2' \xrightarrow{d_2'} P_1' \xrightarrow{d_1'} P_0' \xrightarrow{\varepsilon'} A \rightarrow 0$$

be projective resolutions of a left $R$-module $A$. For all $n \geq 0$, prove that there are projective modules $Q_n$ and $Q_n'$ with

$$\Omega_n(A, \mathbf{P}) \oplus Q_n \cong \Omega_n(A, \mathbf{P}') \oplus Q_n'.$$

**Hint.** Proceed by induction on $n \geq 0$, using Schanuel's Lemma, Proposition 6.80.

* **10.60.** Let

$$0 \rightarrow B \rightarrow E^0 \rightarrow E^1 \rightarrow E^2 \rightarrow$$

and

$$0 \rightarrow B \rightarrow E'^0 \rightarrow E'^1 \rightarrow E'^2 \rightarrow$$

be injective resolutions of a left $R$-module $B$. For all $n \geq 0$, prove that there are injective modules $I_n$ and $I'_n$ with

$$\mho^n(B, \mathbf{E}) \oplus I_n \cong \mho^n(B, \mathbf{E}') \oplus I'_n.$$

**Hint.** The proof is dual to that of Exercise 10.59 on page 954.

$*$ **10.61.** Let $0 \to M' \to M \to M'' \to 0$ be an exact sequence of left $R$-modules (over some ring $R$). Use the long exact Ext sequence to prove the following statements.

(i) If $\mathrm{pd}(M') < \mathrm{pd}(M)$, prove that $\mathrm{pd}(M'') = \mathrm{pd}(M)$.

(ii) If $\mathrm{pd}(M') > \mathrm{pd}(M)$, prove that $\mathrm{pd}(M'') = \mathrm{pd}(M') + 1$.

(iii) If $\mathrm{pd}(M') = \mathrm{pd}(M)$, prove that $\mathrm{pd}(M'') \leq \mathrm{pd}(M') + 1$.

**10.62.** (i) If $0 \to A' \to A \to A'' \to 0$ is an exact sequence, prove that

$$\mathrm{pd}(A) \leq \max\{\mathrm{pd}(A'), \mathrm{pd}(A'')\}.$$

(ii) If the sequence in part (i) is not split and $\mathrm{pd}(A') = \mathrm{pd}(A'') + 1$, prove that

$$\mathrm{pd}(A) = \max\{\mathrm{pd}(A'), \mathrm{pd}(A'')\}.$$

$*$ **10.63.** If $A$ is an $R$-module with $\mathrm{pd}(A) = n$, prove that there exists a free $R$-module $F$ with $\mathrm{Ext}_R^n(A, F) \neq \{0\}$.

**Hint.** Every module is a quotient of a free module.

**10.64.** If $G$ is a finite cyclic group of order not 1, prove that

$$\ell\mathrm{D}(\mathbb{Z}G) = \infty = r\mathrm{D}(\mathbb{Z}G).$$

**Hint.** Use Theorem 9.111.

**10.65. (Auslander)** If $R$ is both left noetherian and right noetherian, prove that

$$\ell\mathrm{D}(R) = r\mathrm{D}(R).$$

**Hint.** Use weak dimension.

**10.66.** Prove that a noetherian von Neumann regular ring is semisimple.

**Hint.** See Example 10.154.

$*$ **10.67.** If $(M_\alpha)_{\alpha \in A}$ is a family of left $R$-modules, prove that

$$\mathrm{pd}\Big(\bigoplus_{\alpha \in A} M_\alpha\Big) = \sup_{\alpha \in A}\{\mathrm{pd}(M_\alpha)\}.$$

$*$ **10.68.** Let $R$ be a commutative ring, let $c \in R$, let $R^* = R/cR$, and let $\nu\colon R \to R^*$ be the natural map.

(i) If $M$ is an $R$-module, define $M^* = M/cM$. Prove that $M^* \cong R^* \otimes_R M$.

(ii) If $A^*$ is an $R^*$-module, define $_\nu A^*$ to be $A^*$ viewed as an $R$-module, as on page 888. Prove that $_\nu A^* \cong \mathrm{Hom}_R(R^*, A^*)$. Conclude that $M \mapsto M^*$ and $A^* \mapsto {}_\nu A^*$ form an adjoint pair of functors.

$*$ **10.69.** If $\nu\colon R \to R^*$ is a ring homomorphism and $A^*$ and $B^*$ are $R^*$-modules, prove that there is a natural isomorphism

$$\mathrm{Hom}_{R^*}(A^*, B^*) \to \mathrm{Hom}_R(_\nu A^*, {}_\nu B^*),$$

where $_\nu A^*$ is $A^*$ viewed as an $R$-module via change of rings.

∗ **10.70.**   (i) If $I$ is a left ideal in a ring $R$, prove that either $R/I$ is projective or $\mathrm{pd}(R/I) = 1 + \mathrm{pd}(I)$ (we agree that $1 + \infty = \infty$).

(ii) If $0 \to M' \to M \to M'' \to 0$ is exact and two of the modules have finite projective (or injective) dimension, prove that the third module has finite projective (or injective) dimension as well.

## Section 10.5.   Hilbert's Theorem on Syzygies

We are going to compute the global dimension $\mathrm{D}(k[x_1, \ldots, x_n])$ of polynomial rings. The key result is that $\mathrm{D}(R[x]) = \mathrm{D}(R) + 1$ for any, possibly noncommutative, ring $R$ (where $R[x]$ is the polynomial ring in which the indeterminate $x$ commutes with all the constants in $R$). However, we will assume that $R$ is commutative for clarity of exposition.

**Henceforth, all rings are assumed commutative unless we say otherwise.**

**Lemma 10.161.** *If* $0 \to A' \to A \to A'' \to 0$ *is a short exact sequence, then*

$$\mathrm{pd}(A'') \leq 1 + \max\{\mathrm{pd}(A), \mathrm{pd}(A')\}.$$

**Proof.** We may assume that the right side is finite, or there is nothing to prove; let $\mathrm{pd}(A) \leq n$ and $\mathrm{pd}(A') \leq n$. Applying $\mathrm{Hom}(\square, B)$, where $B$ is any module, to the short exact sequence gives the long exact sequence

$$\to \mathrm{Ext}^{n+1}(A', B) \to \mathrm{Ext}^{n+2}(A'', B) \to \mathrm{Ext}^{n+2}(A, B) \to .$$

The two outside terms are $\{0\}$, by Lemma 10.136(ii), so that exactness forces $\mathrm{Ext}^{n+2}(A'', B) = \{0\}$ for all $B$. The same lemma gives $\mathrm{pd}(A'') \leq n + 1$.   •

We wish to compare global dimension of $R$ and $R[x]$, and so we consider the $R[x]$-module $R[x] \otimes_R M$ arising from an $R$-module $M$.

**Definition.** If $M$ is an $R$-module over a commutative ring $R$, define

$$M[x] = \bigoplus_{i \geq 0} M_i,$$

where $M_i \cong M$ for all $i$. The $R$-module $M[x]$ is an $R[x]$-module if we define

$$x\left(\sum_i x^i m_i\right) = \sum_i x^{i+1} m_i.$$

Note that $M[x] \cong R[x] \otimes_R M$ via $(x^n m_n) \mapsto 1 \otimes (\sum m_n)$. Thus, if $M$ is a free $R$-module over a commutative ring $R$, then $M[x]$ is a free $R[x]$-module, for tensor product commutes with direct sum. The next result generalizes this from $\mathrm{pd}_R(M) = 0$ to higher dimensions.

**Lemma 10.162.** *For every $R$-module $M$, where $R$ is a commutative ring,*

$$\mathrm{pd}_R(M) = \mathrm{pd}_{R[x]}(M[x]).$$

**Proof.** It suffices to prove that if one of the dimensions is finite and at most $n$, then so is the other.

If $\mathrm{pd}_R(M) \le n$, then there is an $R$-projective resolution

$$0 \to P_n \to \cdots \to P_0 \to M \to 0.$$

Since $R[x]$ is a free $R$-module, it is flat, and there is an exact sequence of $R[x]$-modules

$$0 \to R[x] \otimes_R P_n \to \cdots \to R[x] \otimes_R P_0 \to R[x] \otimes_R M \to 0.$$

But $R[x] \otimes_R M \cong M[x]$ and $R[x] \otimes_R P_n$ is a projective $R[x]$-module (for a projective is a direct summand of a free module). Therefore, $\mathrm{pd}_{R[x]}(M[x]) \le n$.

If $\mathrm{pd}_{R[x]}(M[x]) \le n$, then there is an $R[x]$-projective resolution

$$0 \to Q_n \to \cdots \to Q_0 \to M[x] \to 0.$$

As an $R$-module, $M[x] \cong \bigoplus_{n \ge 1} M_n$, where $M_n \cong M$ for all $n$. By Exercise 10.67 on page 955, $\mathrm{pd}_R(M[x]) = \mathrm{pd}_R(M)$. Each projective $R[x]$-module $Q_i$ is an $R[x]$-summand of a free $R[x]$-module $F_i$; a fortiori, $Q_i$ is an $R$-direct summand of $F_i$. But $R[x]$ is a free $R$-module, so that $F_i$ is also a free $R$-module. Therefore, $Q_i$ is projective as an $R$-module, and so $\mathrm{pd}_R(M) \le n = \mathrm{pd}_{R[x]}(M[x])$. $\bullet$

**Corollary 10.163.** *If $R$ is a commutative ring and $\mathrm{D}(R) = \infty$, then $\mathrm{D}(R[x]) = \infty$.*

**Proof.** If $\mathrm{D}(R) = \infty$, then for every integer $n$, there exists an $R$-module $M_n$ with $n < \mathrm{pd}(M_n)$. By Lemma 10.162, $n < \mathrm{pd}_{R[x]}(M_n[x])$ for all $n$. Therefore, $\mathrm{D}(R[x]) = \infty$. $\bullet$

**Proposition 10.164.** *For every commutative ring $R$,*

$$D(R[x]) \le \mathrm{D}(R) + 1.$$

**Proof.** Recall the characteristic sequence, Theorem 8.57: if $M$ is an $R$-module and $S: M \to M$ is an $R$-map, then there is an exact sequence of $R[x]$-modules

$$0 \to M[x] \to M[x] \to M^S \to 0,$$

where $M^S$ is the $R[x]$-module $M$ with scalar multiplication given by $ax^i m = aS^i(m)$. If $M$ is already an $R[x]$-module and $S: M \to M$ is the $R$-map $m \mapsto xm$, then $M^S = M$. By Lemma 10.161,

$$\mathrm{pd}_{R[x]}(M) \le 1 + \mathrm{pd}_{R[x]}(M[x]) = 1 + \mathrm{pd}_R(M) \le 1 + D(R). \quad \bullet$$

We proceed to prove the reverse inequality.

**Definition.** If $M$ is an $R$-module, where $R$ is a commutative ring, then a nonzero element $c \in R$ is **regular** on $M$ (or is $M$-**regular**) if the $R$-map $M \to M$, given by $m \mapsto cm$, is injective. Otherwise, $c$ is a **zero-divisor** on $M$; that is, there is some nonzero $m \in M$ with $cm = 0$.

Before stating the next theorem, let us explain the notation. Suppose that $R$ is a commutative ring, $c \in R$, and $R^* = R/Rc$. If $M$ is an $R$-module, then $M/cM$ is an $(R/Rc)$-module; that is, $M/cM$ is an $R^*$-module. On the other hand, there is the notion of *Change of Rings* in Exercise 6.56 on page 449. If $\nu\colon R \to R/Rc$ is the natural map, $r \in R$, and $a^* \in A^*$, define

$$ra^* = \nu(r)a^*.$$

This makes $A^*$ into an $R$-module, denoted by $_\nu A^*$. Exercise 10.68 on page 955 asks you to prove that $_\nu A^* \cong \operatorname{Hom}_R(R/cR, A^*)$, so that these constructions involve an adjoint pair of functors, namely, $(\Box^*, {}_\nu\Box)$.

**Proposition 10.165 (Rees Lemma).** *Let $R$ be a commutative ring, let $c \in R$ be neither a unit nor a zero-divisor, and let $R^* = R/Rc$. If $c$ is regular on an $R$-module $M$, then there are natural isomorphisms, for every $R^*$-module $A^*$ and all $n \geq 0$,*

$$\operatorname{Ext}_{R^*}^n(A^*, M/cM) \cong \operatorname{Ext}_R^{n+1}({}_\nu A^*, M),$$

*where $_\nu A^*$ is the $R^*$-module $A^*$ viewed as an $R$-module.*

**Proof.** Recall Theorem 9.47, the axioms characterizing Ext functors. Given a sequence of contravariant functors $G^n\colon {}_{R^*}\mathbf{Mod} \to \mathbf{Ab}$, for $n \geq 0$, such that:

(i) for every short exact sequence $0 \to A^* \to B^* \to C^* \to 0$ of $R^*$-modules, there is a long exact sequence with natural connecting homomorphisms

$$\to G^n(C^*) \to G^n(B^*) \to G^n(A^*) \to G^{n+1}(C^*) \to,$$

(ii) $G^0$ and $\operatorname{Hom}_{R^*}(\Box, L^*)$ are naturally equivalent, for some $R^*$-module $L^*$,

(iii) $G^n(P^*) = 0$ for all projective $R^*$-modules $P^*$ and all $n \geq 1$,

then $G^n$ is naturally equivalent to $\operatorname{Ext}_{R^*}^n(\Box, L^*)$ for all $n \geq 0$.

Define contravariant functors $G^n\colon {}_{R^*}\mathbf{Mod} \to \mathbf{Ab}$ by $G^n = \operatorname{Ext}_R^{n+1}({}_\nu\Box, M)$; that is, for all $R^*$-modules $A^*$,

$$G^n(A^*) = \operatorname{Ext}_R^{n+1}({}_\nu A^*, M).$$

Since axiom (i) holds for the functors $\operatorname{Ext}^n$, it also holds for the functors $G^n$. Let us prove axiom (ii). The map $\mu_c\colon M \to M$, defined by $m \mapsto cm$, is an injection, because $c$ is $M$-regular, and so the sequence $0 \to M \xrightarrow{\mu_c} M \to M/cM \to 0$ is exact. Consider the portion of the long exact sequence, where $A^*$ is an $R^*$-module:

$$\operatorname{Hom}_R({}_\nu A^*, M) \to \operatorname{Hom}_R({}_\nu A^*, M/cM) \xrightarrow{\partial} \operatorname{Ext}_R^1({}_\nu A^*, M) \xrightarrow{\mu_{c*}} \operatorname{Ext}_R^1({}_\nu A^*, M).$$

We claim that $\partial$ is an isomorphism. If $a \in {}_\nu A^*$, then $ca = 0$, because $A^*$ is an $R^*$-module (remember that $R^* = R/cR$). Hence, if $f \in \operatorname{Hom}_R({}_\nu A^*, M)$, then $cf(a) = f(ca) = f(0) = 0$. Since $\mu_c\colon M \to M$ is an injection, $f(a) = 0$ for all $a \in A^*$. Thus, $f = 0$, $\ker \partial = \operatorname{Hom}_R({}_\nu A^*, M) = \{0\}$, and $\partial$ is an injection. The map $\mu_{c*}\colon \operatorname{Ext}_R^1({}_\nu A^*, M) \to \operatorname{Ext}_R^1({}_\nu A^*, M)$ is multiplication by $c$, by Example 9.63. On the other hand, Example 9.72 shows that if $\mu_c'\colon {}_\nu A^* \to {}_\nu A^*$ is multiplication by $c$, then the induced map $\mu_c'^*$ on Ext is also multiplication by $c$. But $\mu_c' = 0$, because $A^*$ is a $(R/cR)$-module, and so $\mu_c'^* = 0$. Hence, $\mu_{c*} = \mu_c'^* = 0$. Therefore, $\operatorname{im} \partial = \ker(\mu_{c*}) = \operatorname{Ext}_R^1({}_\nu A^*, M)$, and so $\partial$ is a surjection. It follows that

$$\partial\colon \operatorname{Hom}_R({}_\nu A^*, M/cM) \to \operatorname{Ext}_R^1({}_\nu A^*, M)$$

is an isomorphism, natural because it is the connecting homomorphism. By Exercise 10.69 on page 955, there is a natural isomorphism

$$\mathrm{Hom}_{R^*}(A^*, M/cM) \to \mathrm{Hom}_R({}_\nu A^*, M/cM).$$

The composite

$$\mathrm{Hom}_{R^*}(A^*, M/cM) \to \mathrm{Hom}_R({}_\nu A^*, M/cM) \to \mathrm{Ext}^1_R({}_\nu A^*, M) = G^0(A^*)$$

is a natural isomorphism; hence, its inverse defines a natural isomorphism

$$G^0 \to \mathrm{Hom}_{R^*}(\Box, M/cM).$$

Setting $L^* = M/cM$ completes the verification of axiom (ii).

It remains to verify axiom (iii): $G^n(P^*) = \{0\}$ whenever $P^*$ is a projective $R^*$-module and $n \geq 1$. In fact, since $G^n$ is an additive functor and every projective is a direct summand of a free module, we may assume that $P^*$ is a free $R^*$-module with basis, say, $E$. If $Q = \bigoplus_{e \in E} Re$ is the free $R$-module with basis $E$, then there is an exact sequence of $R$-modules, where $\lambda_c$ is multiplication by $c$,

$$(1) \qquad\qquad 0 \to Q \xrightarrow{\lambda_c} Q \to P^* \to 0.$$

The first arrow is an injection because $c$ is not a zero-divisor on $R$; the last arrow is a surjection because

$$Q/cQ = \left(\bigoplus_{e \in E} Re\right) \Big/ \left(\bigoplus_{e \in E} Rce\right) \cong \bigoplus_{e \in E}(R/Rc)e = \bigoplus_{e \in E} R^*e = P^*.$$

The long exact sequence arising from (1) contains

$$\to \mathrm{Ext}^n_R(Q, M) \to \mathrm{Ext}^{n+1}_R({}_\nu P^*, M) \to \mathrm{Ext}^{n+1}_R(Q, M) \to .$$

Since $Q$ is $R$-free and $n \geq 1$, the outside terms are $\{0\}$, and exactness gives $G^n(P^*) = \mathrm{Ext}^{n+1}_R({}_\nu P^*, M) = \{0\}$. Therefore,

$$\mathrm{Ext}^{n+1}_R({}_\nu A^*, M) = G^n(A^*) \cong \mathrm{Ext}^n_{R^*}(A^*, M/cM). \quad \bullet$$

**Theorem 10.166.** *For every commutative ring $k$,*

$$\mathrm{D}(k[x]) = \mathrm{D}(k) + 1.$$

**Proof.** We have proved that $\mathrm{D}(k[x]) \leq \mathrm{D}(k) + 1$ in Proposition 10.164, and so it suffices to prove the reverse inequality. We may assume that $\mathrm{D}(k) = n < \infty$, by Corollary 10.163.

In the notation of the Rees Lemma, Proposition 10.165, let us write $R = k[x]$, $c = x$, and $R^* = k$. Let $A$ be a $k$-module with $\mathrm{pd}_k(A) = n$. By Exercise 10.63 on page 955, there is a free $k$-module $F$ with $\mathrm{Ext}^n_k(A, F) \neq \{0\}$; of course, multiplication by $x$ is an injection $F \to F$. As in the proof of the Rees Lemma, there is a free $k[x]$-module $Q = k[x] \otimes_k F$ with $Q/xQ \cong F$. The Rees Lemma gives

$$\mathrm{Ext}^{n+1}_{k[x]}(A, Q) \cong \mathrm{Ext}^n_k({}_\nu A, Q/xQ) \cong \mathrm{Ext}^n_k({}_\nu A, F) \neq \{0\}$$

($A$ is viewed as a $k[x]$-module ${}_\nu A$ via $k[x] \to k$). Therefore, $\mathrm{pd}_{k[x]}(A) \geq n+1$, and so $\mathrm{D}(k[x]) \geq n + 1 = \mathrm{D}(k) + 1$. $\quad \bullet$

**Corollary 10.167 (Hilbert's Theorem on Syzygies).** *If $k$ is a field, then*

$$\mathrm{D}(k[x_1,\ldots,x_n]) = n.$$

**Proof.** Since $\mathrm{D}(k) = 0$ and $\mathrm{D}(k[x]) = 1$ for every field $k$, the result follows from Theorem 10.166 by induction on $n \geq 0$.   •

Hilbert's Theorem on Syzygies implies that if $R = k[x_1,\ldots,x_n]$, where $k$ is a field, then every finitely generated $R$-module $M$ has has a resolution

$$0 \to P_n \to P_{n-1} \to \cdots \to P_0 \to M \to 0,$$

where $P_i$ is free for all $i < n$ and $P_n$ is projective.

**Definition.** We say that a (necessarily finitely generated) $R$-module $M$, over an arbitrary commutative ring $R$, has **FFR** (***finite free resolution***) if it has a resolution in which every $P_i$, including the last $P_n$, is a finitely generated free module.

Hilbert's Theorem on Syzygies can be improved to the theorem that if $k$ is a field, then every finitely generated $k[x_1,\ldots,x_n]$-module has FFR (Rotman, *An Introduction to Homological Algebra*, p. 480). [Of course, this result also follows from the more difficult Quillen–Suslin Theorem, which says that every projective $k[x_1,\ldots,x_n]$-module, where $k$ is a field, is free (Ibid., p. 209).]

The next corollary is another application of the Rees Lemma.

**Corollary 10.168.** *Let $R$ be a commutative ring, let $x \in R$ not be a zero-divisor, and let $x$ be regular on an $R$-module $M$. If $\mathrm{pd}_R(M) = n < \infty$, then*

$$\mathrm{pd}_{R/xR}(M/xM) \leq n - 1.$$

**Proof.** Since $\mathrm{pd}_R(M) = n < \infty$, we have $\mathrm{Ext}_R^{n+1}(L, M) = \{0\}$ for all left $R$-modules $L$; in particular, $\mathrm{Ext}_R^{n+1}(L^*, M) = \{0\}$ for all left $R^*$-modules $L^*$ (by Exercise 10.68 on page 955, $L^*$ can be viewed as a left $R$-module). By the Rees Lemma with $R^* = R/xR$, we have $\mathrm{Ext}_{R^*}^n(L^*, M/xM) \cong \mathrm{Ext}_R^{n+1}(L^*, M)$. Therefore, $\mathrm{Ext}_{R^*}^n(L^*, M/xM) = \{0\}$ for all $R^*$-modules $L^*$, and so $\mathrm{pd}_{R^*}(M/xM) \leq n-1$; that is, $\mathrm{pd}_{R/xR}(M/xM) \leq n-1$.   •

Here is a change of rings theorem, simpler than the theorem of Rees, which gives another proof of the inequality $\mathrm{D}(k[x]) \geq \mathrm{D}(k) + 1$ in the proof of Theorem 10.166: let $R = k[x]$, so that $R^* = R/(x) = k[x]/(x) = k$.

**Proposition 10.169 (Kaplansky).** *Let $R$ be a* (*not necessarily commutative*) *ring, let $x \in Z(R)$ be neither a unit nor a zero-divisor, let $R^* = R/(x)$, and let $M^*$ be a left $R^*$-module. If $\mathrm{pd}_{R^*}(M^*) = n < \infty$, then $\mathrm{pd}_R(M^*) = n + 1$.*

**Proof.** We note that a left $R^*$-module is merely a left $R$-module $M$ with $xM = \{0\}$.

The proof is by induction on $n \geq 0$. If $n = 0$, then $M^*$ is a projective left $R^*$-module. Since $x$ is not a zero-divisor, there is an exact sequence of left $R$-modules

$$0 \to R \xrightarrow{x} R \to R^* \to 0,$$

so that $\operatorname{pd}_R(R^*) \leq 1$. Now $M^*$, being a projective left $R^*$-module, is a direct summand of a free left $R^*$-module $F^*$. But $\operatorname{pd}_R(F^*) \leq 1$, for it is a direct sum of copies of $R^*$, and so $\operatorname{pd}_R(M^*) \leq 1$. Finally, if $\operatorname{pd}_R(M^*) = 0$, then $M^*$ would be a projective left $R$-module; but this contradicts Exercise 6.58 on page 459, for $xM = \{0\}$. Therefore, $\operatorname{pd}_R(M^*) = 1$.

If $n \geq 1$, then there is an exact sequence of left $R^*$-modules

$$(2) \qquad\qquad 0 \to K^* \to F^* \to M^* \to 0$$

with $F^*$ free. Now $\operatorname{pd}_{R^*}(K^*) = n - 1$, so induction gives $\operatorname{pd}_R(K^*) = n$.

If $n = 1$, then $\operatorname{pd}_R(K^*) = 1$ and $\operatorname{pd}_R(K^*) \leq 2$, by Exercise 10.61 on page 955. There is an exact sequence of left $R$-modules

$$(3) \qquad\qquad 0 \to L \to F \to M^* \to 0$$

with $F$ free. Since $xM^* = \{0\}$, we have $xF \subseteq \ker(F \to M^*) = L$, and this gives an exact sequence of $R^*$-modules (each term is annihilated by $x$)

$$0 \to L/xF \to F/xF \to M^* \to 0.$$

Thus, $\operatorname{pd}_{R^*}(L/xF) = \operatorname{pd}_{R^*}(M^*) - 1 = 0$, because $F/xF$ is a free $R^*$-module, so the exact sequence of $R^*$-modules

$$0 \to xF/xL \to L/xL \to L/xF \to 0$$

splits. Since $M^* \cong F/L \cong xF/xL$, we see that $M^*$ is a direct summand of $L/xL$. Were $L$ a projective left $R$-module, then $L/xL$ and, hence, $M^*$ would be projective left $R^*$-modules, contradicting $\operatorname{pd}_{R^*}(M^*) = 1$. Exact sequence (3) shows that $\operatorname{pd}_R(M^*) = 1 + \operatorname{pd}_R(L) \geq 2$, and so $\operatorname{pd}_R(M^*) = 2$.

Finally, assume that $n \geq 2$. Exact sequence (2) gives $\operatorname{pd}_{R^*}(K^*) = n - 1 > 1 \geq \operatorname{pd}_R(F^*)$; hence, Exercise 10.61 gives

$$\operatorname{pd}_R(M^*) = \operatorname{pd}_R(K^*) + 1 = n + 1. \quad \bullet$$

## Section 10.6. Commutative Noetherian Rings

We now investigate relations between ideals and prime ideals. These results will be used in the next section to prove the theorems of Auslander–Buchsbaum and Serre about regular local rings. Unless we say otherwise,

all rings in this section are commutative and noetherian.

Recall that a prime ideal $\mathfrak{p}$ is *minimal* if there is no prime ideal $\mathfrak{q}$ with $\mathfrak{q} \subsetneq \mathfrak{p}$. In a domain, $(0)$ is the only minimal prime ideal.

**Definition.** A ***prime chain*** of ***length*** $n$ in a commutative ring $R$ is a strictly decreasing chain of prime ideals

$$\mathfrak{p}_0 \supsetneq \mathfrak{p}_1 \supsetneq \cdots \supsetneq \mathfrak{p}_n.$$

The ***height*** $\operatorname{ht}(\mathfrak{p})$ of a prime ideal $\mathfrak{p}$ is the length of the longest prime chain with $\mathfrak{p} = \mathfrak{p}_0$. Thus, $\operatorname{ht}(\mathfrak{p}) \leq \infty$.

**Example 10.170.**

(i) If $\mathfrak{p}$ is a prime ideal, then $\mathrm{ht}(\mathfrak{p}) = 0$ if and only if $\mathfrak{p}$ is a minimal prime ideal. If $R$ is a domain, then $\mathrm{ht}(\mathfrak{p}) = 0$ if and only if $\mathfrak{p} = (0)$.

(ii) If $R$ is a Dedekind ring and $\mathfrak{p}$ is a nonzero prime ideal in $R$, then $\mathrm{ht}(\mathfrak{p}) = 1$.

(iii) Let $k$ be a field and let $R = k[X]$ be the polynomial ring in infinitely many variables $X = \{x_1, x_2, \dots\}$. If $\mathfrak{p}_i = (x_i, x_{i+1}, \dots)$, then $\mathfrak{p}_i$ is a prime ideal (because $R/\mathfrak{p}_i \cong k[x_1, \dots, x_{i-1}]$ is a domain) and, for every $n \geq 1$,

$$\mathfrak{p}_1 \supsetneq \mathfrak{p}_2 \supsetneq \cdots \supsetneq \mathfrak{p}_{n+1}$$

is a prime chain of length $n$. It follows that $\mathrm{ht}(\mathfrak{p}_1) = \infty$. ◀

**Definition.** The *Krull dimension* of a ring $R$ is

$$\dim(R) = \sup\{\mathrm{ht}(\mathfrak{p}) : \mathfrak{p} \in \mathrm{Spec}(R)\};$$

that is, $\dim(R)$ is the length of a longest prime chain in $R$.

If $R$ is a Dedekind ring, then $\dim(R) = 1$, for every nonzero prime is a maximal ideal; if $R$ is a domain, then $\dim(R) = 0$ if and only if $R$ is a field. If $R = k[X]$ is a polynomial ring in infinitely many variables, then $\dim(R) = \infty$. The next proposition characterizes the noetherian rings of Krull dimension 0.

**Proposition 10.171.** *Let $R$ be a commutative ring. Then $R$ is noetherian with $\dim(R) = 0$ if and only if every finitely generated $R$-module $M$ has a composition series.*

**Proof.** Assume that $R$ is noetherian with Krull dimension 0. Since $R$ is noetherian, Corollary 5.118(iii) says that there are only finitely many minimal prime ideals. Since $\dim(R) = 0$, every prime ideal is a minimal prime ideal (as well as a maximal ideal). We conclude that $R$ has only finitely many prime ideals, say, $\mathfrak{p}_1, \dots, \mathfrak{p}_n$. Now $\mathrm{nil}(R) = \bigcap_{i=1}^{n} \mathfrak{p}_i$ is nilpotent, by Exercise 10.40 on page 915; say, $\mathrm{nil}(R)^m = (0)$. Define

$$N = \mathfrak{p}_1 \cdots \mathfrak{p}_n \subseteq \mathfrak{p}_1 \cap \cdots \cap \mathfrak{p}_n = \mathrm{nil}(R),$$

so that

$$N^m = (\mathfrak{p}_1 \cdots \mathfrak{p}_n)^m \subseteq \mathrm{nil}(R)^m = (0).$$

Let $M$ be a finitely generated $R$-module, and consider the chain

$$M \supseteq \mathfrak{p}_1 M \supseteq \mathfrak{p}_1 \mathfrak{p}_2 M \supseteq \cdots \supseteq NM.$$

The factor module $\mathfrak{p}_1 \cdots \mathfrak{p}_{i-1} M / \mathfrak{p}_1 \cdots \mathfrak{p}_i M$ is an $(R/\mathfrak{p}_i)$-module; that is, it is a vector space over the field $R/\mathfrak{p}_i$ (for $\mathfrak{p}_i$ is a maximal ideal). Since $M$ is finitely generated, the factor module is finite-dimensional, and so the chain can be refined so that all the factor modules be simple. Finally, repeat this argument for the chains

$$N^j M \supseteq \mathfrak{p}_1 N^j M \supseteq \mathfrak{p}_1 \mathfrak{p}_2 N^j M \supseteq \cdots \supseteq N^{j+1} M.$$

Since $N^m = \{0\}$, we have constructed a composition series for $M$.

Conversely, if every finitely generated $R$-module has a composition series, then the cyclic $R$-module $R$ has a composition series; say, of length $\ell$. It follows that any ascending chain of ideals has length at most $\ell$, and so $R$ is noetherian. To prove that $\dim(R) = 0$, we must show that $R$ does not contain any pair of prime ideals with $\mathfrak{p} \supsetneq \mathfrak{q}$. Passing to the quotient ring $R/\mathfrak{q}$, we may restate the hypotheses: $R$ is a domain having a nonzero prime ideal as well as a composition series $R \supseteq I_1 \supseteq \cdots \supseteq I_d \neq (0)$. The last ideal $I_d$ is a minimal ideal; choose a nonzero element $x \in I_d$. Of course, $xI_d \subseteq I_d$; since $R$ is a domain, $xI_d \neq (0)$, so that minimality of $I_d$ gives $xI_d = I_d$. Hence, there is $y \in I_d$ with $xy = x$; that is, $1 = y \in I_d$, and so $I_d = R$. We conclude that $R$ is a field, contradicting its having a nonzero prime ideal. $\bullet$

Low-dimensional cases become more interesting once we recognize that they can be used in tandem with formation of quotient rings and localization, for these methods often reduce a more general case to these ones. We shall see this in the proof of the next theorem.

We are going to prove a theorem of Krull, the *Principal Ideal Theorem*, which implies that every prime ideal (in a noetherian ring) has finite height. Our proof is Kaplansky's adaptation of a proof of Rees.

**Lemma 10.172.** *Let $a$ and $b$ be nonzero elements in a domain $R$. If there exists $c \in R$ such that $ca^2 \in (b)$ implies $ca \in (b)$, then the series $(a, b) \supseteq (a) \supseteq (a^2)$ and $(a^2, b) \supseteq (a^2, ab) \supseteq (a^2)$ have isomorphic factor modules.*[10]

**Proof.** Now $(a,b)/(a) \cong (a^2, ab)/(a^2)$, for multiplication by $a$ sends $(a,b)$ onto $(a^2, ab)$ and $(a)$ onto $(a^2)$.

The module $(a)/(a^2)$ is cyclic with annihilator $(a)$; that is, $(a)/(a^2) \cong R/(a)$. The module $(a^2, b)/(a^2, ab)$ is also cyclic, for the generator $a^2$ lies in $(a^2, ab)$. Now $A = \mathrm{ann}\big((a^2, b)/(a^2, ab)\big)$ contains $(a)$, and so it suffices to prove that $A = (a)$; that is, if $cb = ua^2 + vab$, then $c \in (a)$. This equation gives $ua^2 \in (b)$, and so the hypothesis gives $ua = rb$ for some $r \in R$. Substituting, $cb = rab + vab$, and canceling $b$ gives $c = ra + va \in (a)$. Therefore, $(a)/(a^2) \cong (a^2, b)/(a^2, ab)$. $\bullet$

**Definition.** A prime ideal $\mathfrak{p}$ is **minimal over an ideal** $I$ if $I \subseteq \mathfrak{p}$ and there is no prime ideal $\mathfrak{q}$ with $I \subseteq \mathfrak{q} \subsetneq \mathfrak{p}$; equivalently, $\mathfrak{p}/I$ is a minimal prime ideal in $R/I$.

**Theorem 10.173 (Principal Ideal Theorem).** *Let $(r)$ be a principal ideal in a commutative noetherian ring $R$. If $(r)$ is proper (that is, if $r$ is not a unit) and $\mathfrak{p}$ is a prime ideal minimal over $(r)$, then $\mathrm{ht}(\mathfrak{p}) \leq 1$.*

**Proof.** If, on the contrary, $\mathrm{ht}(\mathfrak{p}) \geq 2$, then there is a prime chain

$$\mathfrak{p} \supsetneq \mathfrak{p}_1 \supsetneq \mathfrak{p}_2.$$

We normalize the problem in two ways. First, replace $R$ by $R/\mathfrak{p}_2$; second, localize at $\mathfrak{p}/\mathfrak{p}_2$. The hypotheses are modified accordingly: $R$ is now a local domain whose maximal ideal $\mathfrak{m}$ is minimal over a proper principal ideal $(x)$, and there is a prime ideal $\mathfrak{q}$ with

$$R \supsetneq \mathfrak{m} \supsetneq \mathfrak{q} \supsetneq (0).$$

---

[10]Our notation for ideals is not consistent. The principal ideal generated by an element $a \in R$ is sometimes denoted by $(a)$ and sometimes denoted by $Ra$.

Choose a nonzero element $b \in \mathfrak{q}$, and define

$$I_i = ((b) : x^i) = \{c \in R : cx^i \in (b)\}.$$

The ascending chain $I_1 \subseteq I_2 \subseteq \cdots$ must stop, because $R$ is noetherian: say, $I_n = I_{n+1} = \cdots$. It follows that if $c \in I_{2n}$, then $c \in I_n$; that is, if $cx^{2n} \in (b)$, then $cx^n \in (b)$. If we set $a = x^n$, then $ca^2 \in (b)$ implies $ca \in (b)$.

If $R^* = R/(a^2)$, then $\dim(R^*) = 0$, for it has exactly one prime ideal. By Proposition 10.171, the $R^*$-module $(a, b)/(a^2)$ (as every finitely generated $R^*$-module) has finite length $\ell$ (the length of its composition series). But Lemma 10.172 implies that both $(a, b)$ and its submodule $(a^2, b)$ have length $\ell$. The Jordan–Hölder Theorem says that this can happen only if $(a^2, b) = (a, b)$, which forces $a \in (a^2, b)$: thus, there are $s, t \in R$ with $a = sa^2 + tb$. Since $sa \in \mathfrak{m}$, the element $1 - sa$ is a unit [for $R$ is a local ring with maximal ideal $\mathfrak{m}$]. Hence, $-a(1 - sa) = tb \in (b)$ gives $a \in (b) \subseteq \mathfrak{q}$. But $a = x^n$ gives $x \in \mathfrak{q}$, contradicting $\mathfrak{m}$ being a prime ideal minimal over $(x)$.  $\bullet$

We now generalize the Principal Ideal Theorem to finitely generated ideals.

**Theorem 10.174 (Generalized Principal Ideal Theorem).** *If $I = (a_1, \ldots, a_n)$ is a proper ideal in a ring $R$ and $\mathfrak{p}$ is a prime ideal minimal over $I$, then $\mathrm{ht}(\mathfrak{p}) \leq n$.*

**Proof.** The hypotheses still hold after localizing at $\mathfrak{p}$, so we may now assume that $R$ is a local ring with $\mathfrak{p}$ as its maximal ideal.

The proof is by induction on $n \geq 1$, the base step being the Principal Ideal Theorem. Let $I = (a_1, \ldots, a_{n+1})$, and assume, by way of contradiction, that $\mathrm{ht}(\mathfrak{p}) > n + 1$: thus, there is a prime chain

$$\mathfrak{p} = \mathfrak{p}_0 \supsetneq \mathfrak{p}_1 \supsetneq \cdots \supsetneq \mathfrak{p}_{n+1}.$$

We may assume that there are no prime ideals strictly between $\mathfrak{p}$ and $\mathfrak{p}_1$ (if there were such a prime ideal, insert it, thereby lengthening the chain; such insertions can be done only finitely many times, for the module $\mathfrak{p}/\mathfrak{p}_1$ has ACC). Now $I \not\subseteq \mathfrak{p}_1$, because $\mathfrak{p}$ is a prime ideal minimal over $I$. Reindexing the generators of $I$ if necessary, $a_1 \notin \mathfrak{p}_1$. Hence, $(a_1, \mathfrak{p}_1) \supsetneq \mathfrak{p}_1$. We claim that $\mathfrak{p}$ is the only prime ideal containing $(a_1, \mathfrak{p}_1)$; there can be no prime ideal $\mathfrak{p}'$ with $(a_1, \mathfrak{p}_1) \subseteq \mathfrak{p}' \subseteq \mathfrak{p}$ (the second inclusion holds because $\mathfrak{p}$ is the only maximal ideal in $R$) because there are no prime ideals strictly between $\mathfrak{p}$ and $\mathfrak{p}_1$. Therefore, in the ring $R/(a_1, \mathfrak{p}_1)$, the image of $\mathfrak{p}$ is the unique nonzero prime ideal. As such, it must be the nilradical, and hence it is nilpotent, by Exercise 10.40 on page 915. There is an integer $m$ with $\mathfrak{p}^m \subseteq (a_1, \mathfrak{p}_1)$, and so there are equations

$$(1) \qquad a_i^m = r_i a_1 + b_i, \qquad r_i \in R, \quad b_i \in \mathfrak{p}_1, \quad \text{and} \quad i \geq 2.$$

Define $J = (b_2, \ldots, b_{n+1})$. Now $J \subseteq \mathfrak{p}_1$, while $\mathrm{ht}(\mathfrak{p}_1) > n$. By induction, $\mathfrak{p}_1$ cannot be a prime ideal minimal over $J$, and so there exists a prime ideal $\mathfrak{q}$ minimal over $J$:

$$J \subseteq \mathfrak{q} \subsetneq \mathfrak{p}_1.$$

Now $a_i^m \in (a_1, \mathfrak{q})$ for all $i$, by (1). Thus, any prime ideal $\mathfrak{p}'$ containing $(a_1, \mathfrak{q})$ must contain all $a_i^m$, hence all $a_i$, and hence $I$. As $\mathfrak{p}$ is the unique maximal ideal, $I \subseteq \mathfrak{p}' \subseteq \mathfrak{p}$. But $\mathfrak{p}$ is a prime ideal minimal over $I$, and so $\mathfrak{p}' = \mathfrak{p}$. Therefore, $\mathfrak{p}$ is the unique prime ideal containing $(a_1, \mathfrak{q})$. If $R^* = R/\mathfrak{q}$, then $\mathfrak{p}^* = \mathfrak{p}/\mathfrak{q}$ is a prime ideal minimal over the principal ideal $(a_1 + \mathfrak{q})$. On the other hand, $\mathrm{ht}(\mathfrak{p}^*) \geq 2$, for $\mathfrak{p}^* \supsetneqq \mathfrak{p}_1^* \supsetneqq (0)$ is a prime chain, where $\mathfrak{p}_1^* = \mathfrak{p}_1/\mathfrak{q}$. This contradiction to the Principal Ideal Theorem completes the proof. •

**Corollary 10.175.** *Every prime ideal in a (noetherian) ring $R$ has finite height, and so $\mathrm{Spec}(R)$ has DCC.*

**Proof.** Every prime ideal $\mathfrak{p}$ is finitely generated, because $R$ is noetherian; say, $\mathfrak{p} = (a_1, \ldots, a_n)$. But $\mathfrak{p}$ is a minimal prime ideal over itself, so that Theorem 10.174 gives $\mathrm{ht}(\mathfrak{p}) \leq n$. •

A noetherian ring may have infinite Krull dimension, for there may be no uniform bound on the length of prime chains. We will see that this cannot happen for local rings.

The Generalized Principal Ideal Theorem bounds the height of a prime ideal that is minimal over an ideal; more generally, the next result bounds the height of a prime ideal that merely contains an ideal, but which may not be minimal over it.

**Corollary 10.176.** *Let $I = (a_1, \ldots, a_n)$ be an ideal in $R$, and let $\mathfrak{p}$ be a prime ideal in $R$ with $\mathfrak{p} \supseteq I$. If $\mathrm{ht}(\mathfrak{p}/I)$ denotes the height of $\mathfrak{p}/I$ in $R/I$, then*

$$\mathrm{ht}(\mathfrak{p}) \leq n + \mathrm{ht}(\mathfrak{p}/I).$$

**Proof.** The proof is by induction on $h = \mathrm{ht}(\mathfrak{p}/I) \geq 0$. If $h = 0$, then Exercise 10.76 on page 969 says that $\mathfrak{p}$ is minimal over $I$, and so the base step is the Generalized Principal Ideal Theorem. For the inductive step $h > 0$, $\mathfrak{p}$ is not minimal over $I$. By Corollary 5.118(iii), there are only finitely many minimal primes in $R/I$, and so Exercise 10.76 on page 969 says that there are only finitely many prime ideals minimal over $I$; say, $\mathfrak{q}_1, \ldots, \mathfrak{q}_s$. Since $\mathfrak{p}$ is not minimal over $I$, $\mathfrak{p} \not\subseteq \mathfrak{q}_i$ for any $i$; hence, Proposition 5.15 says that $\mathfrak{p} \not\subseteq \mathfrak{q}_1 \cup \cdots \cup \mathfrak{q}_s$, and so there is $y \in \mathfrak{p}$ with $y \notin \mathfrak{q}_i$ for any $i$. Define $J = (I, y)$.

We now show, in $R/J$, that $\mathrm{ht}(\mathfrak{p}/J) \leq h - 1$. Let

$$\mathfrak{p}/J \supsetneqq \mathfrak{p}_1/J \supsetneqq \cdots \supsetneqq \mathfrak{p}_r/J$$

be a prime chain in $R/J$. Since $I \subsetneqq J$, there is a surjective ring map $R/I \to R/J$. The prime chain lifts to a prime chain in $R/I$:

$$\mathfrak{p}/I \supsetneqq \mathfrak{p}_1/I \supsetneqq \cdots \supsetneqq \mathfrak{p}_r/I.$$

Now $\mathfrak{p}_r \supseteq J \supsetneqq I$, and $J = (I, y)$ does not contain any $\mathfrak{q}_i$. But the ideals $\mathfrak{q}_i/I$ are the minimal prime ideals in $R/I$, by Exercise 10.76 on page 969, so that $\mathfrak{p}_r$ is not a minimal prime ideal in $R$. Therefore, there is a prime chain starting at $\mathfrak{p}$ of length $r + 1$. We conclude that $r + 1 \leq h$, and so $\mathrm{ht}(\mathfrak{p}/J) \leq h - 1$.

Since $J = (I, y) = (a_1, \ldots, a_n, y)$ is generated by $n + 1$ elements, the inductive hypothesis gives

$$\mathrm{ht}(\mathfrak{p}) \leq n + 1 + \mathrm{ht}(\mathfrak{p}/J) = (n+1) + (h-1) = n + h = n + \mathrm{ht}(\mathfrak{p}/I). \quad •$$

Recall that a nonzero element $c \in R$ is *regular* on $R$ if the multiplication map $\mu_c \colon R \to R$, given by $r \mapsto cr$, is an injection; that is, $c$ is not a zero-divisor.

**Definition.** A sequence $x_1, \ldots, x_n$ in a ring $R$ is an **$R$-*sequence*** if $x_1$ is regular on $R$, $x_2$ is regular on $R/(x_1)$, $x_3$ is regular on $R/(x_1, x_2), \ldots, x_n$ is regular on $R/(x_1, \ldots, x_{n-1})$.

For example, if $R = k[x_1, \ldots, x_n]$ is a polynomial ring over a field $k$, then it is easy to see that $x_1, \ldots, x_n$ is an $R$-sequence. Exercise 10.75 on page 969 gives an example of a permutation of an $R$-sequence that is not an $R$-sequence. However, if $R$ is local, then every permutation of an $R$-sequence is also an $R$-sequence (Kaplansky, *Commutative Rings*, p. 86).

The Generalized Principal Ideal Theorem gives an upper bound on the height of a prime ideal; the next lemma gives a lower bound.

**Lemma 10.177.**

(i) *If $x$ is not a zero-divisor in a ring $R$, then $x$ lies in no minimal prime ideal.*

(ii) *If $\mathfrak{p}$ is a prime ideal in $R$ and $x \in \mathfrak{p}$ is not a zero-divisor, then*

$$1 + \operatorname{ht}(\mathfrak{p}/(x)) \le \operatorname{ht}(\mathfrak{p}).$$

(iii) *If a prime ideal $\mathfrak{p}$ in $R$ contains an $R$-sequence $x_1, \ldots, x_d$, then*

$$d \le \operatorname{ht}(\mathfrak{p}).$$

**Proof.**

(i) Suppose, on the contrary, that $\mathfrak{p}$ is a nonzero minimal prime ideal containing $x$. Now $R_\mathfrak{p}$ is a ring with only one nonzero prime ideal, namely, $\mathfrak{p}_\mathfrak{p}$, which must be the nilradical. Thus, $x/1$, as every element in $\mathfrak{p}_\mathfrak{p}$, is nilpotent. If $x^m/1 = 0$ in $R_\mathfrak{p}$, then there is $s \notin \mathfrak{p}$ (so that $s \ne 0$) with $sx = 0$, contradicting $x$ not being a zero-divisor.

(ii) If $h = \operatorname{ht}(\mathfrak{p}/(x))$, then there is a prime chain in $R/(x)$:

$$\mathfrak{p}/(x) \supsetneq \mathfrak{p}_1/(x) \supsetneq \cdots \supsetneq \mathfrak{p}_h/(x).$$

Lifting back to $R$, there is a prime chain $\mathfrak{p} \supsetneq \mathfrak{p}_1 \supsetneq \cdots \supsetneq \mathfrak{p}_h$ with $\mathfrak{p}_h \supseteq (x)$. Since $x$ is not a zero-divisor, part (i) says that $\mathfrak{p}_h$ is not a minimal prime. Therefore, there exists a prime ideal $\mathfrak{p}_{h+1}$ properly contained in $\mathfrak{p}_h$, which shows that $\operatorname{ht}(\mathfrak{p}) \ge 1 + h$.

(iii) The proof is by induction on $d \ge 1$. For the base step $d = 1$, suppose, on the contrary, that $\operatorname{ht}(\mathfrak{p}) = 0$; then $\mathfrak{p}$ is a minimal prime ideal, and this contradicts part (i). For the inductive step, part (ii) gives $\operatorname{ht}(\mathfrak{p}/(x_1)) + 1 \le \operatorname{ht}(\mathfrak{p})$. Now $\mathfrak{p}/(x_1)$ contains an $(R/(x_1))$-sequence $x_2 + (x_1), \ldots, x_d + (x_1)$, by Exercise 10.80(ii) on page 983, so that the inductive hypothesis gives $d - 1 \le \operatorname{ht}(\mathfrak{p}/(x_1))$. Therefore, part (ii) gives $d \le \operatorname{ht}(\mathfrak{p})$.   •

**Definition.** A subset $X$ of an $R$-module $M$ is **scalar-closed** if $x \in X$ implies that $rx \in X$ for all $r \in R$.

Every submodule of a module $M$ is scalar-closed, An example of a scalar-closed subset (of $R$) that is not a submodule is

$$\text{Zer}(R) = \{r \in R : r = 0 \text{ or } r \text{ is a zero-divisor}\}.$$

**Definition.** Let $X \subseteq M$ be a scalar-closed subset. The **annihilator** of $x \in X$ is

$$\text{ann}(x) = \{r \in R : rx = 0\},$$

the **annihilator** of $X$ is

$$\text{ann}(X) = \{r \in R : rx = 0 \text{ for all } x \in X\},$$

and

$$\mathcal{A}(X) = \{\text{ann}(x) : x \in X \text{ and } x \neq 0\}.$$

Note that $\text{ann}(x)$ and $\text{ann}(X)$ are ideals [$\text{ann}(0) = R$, and so $\mathcal{A}(X)$ is a family of proper ideals if $M \neq \{0\}$].

**Lemma 10.178.** *Let $X$ be a nonempty scalar-closed subset of a nonzero finitely generated $R$-module $M$.*

(i) *An ideal $I$ maximal among the ideals in $\mathcal{A}(X)$ is a prime ideal.*

(ii) *There is a descending chain*

$$M = M_0 \supseteq M_1 \supseteq M_2 \supseteq \cdots \supseteq M_n = \{0\}$$

*whose factor modules $M_i/M_{i+1} \cong R/\mathfrak{p}_i$ for prime ideals $\mathfrak{p}_i$.*

**Proof.**

(i) Since $R$ is noetherian, the nonempty family $\mathcal{A}(X)$ contains a maximal element, by Proposition 5.33; call it $I = \text{ann}(x)$. Suppose that $a, b$ are elements in $R$ with $ab \in I$ and $b \notin I$; that is, $abx = 0$ but $bx \neq 0$. Thus, $\text{ann}(bx) \supseteq I + Ra \supseteq I$. If $a \notin I$, then $\text{ann}(bx) \supseteq I + Ra \supsetneq I$. But $bx \in X$ because $X$ is scalar-closed, so that $\text{ann}(bx) \in \mathcal{A}(X)$, which contradicts the maximality of $I = \text{ann}(x)$. Therefore, $a \in I$ and $I$ is a prime ideal.

(ii) Since $R$ is noetherian and $M$ is finitely generated, $M$ has the maximum condition on ideals. Thus, the nonempty family $\mathcal{A}(M)$ has a maximal element, say, $\mathfrak{p}_1 = \text{ann}(x_1)$, which is prime, by part (i). Define $M_1 = \langle x_1 \rangle$, and note that $M_1 \cong R/\text{ann}(x_1) = R/\mathfrak{p}_1$. Now repeat this procedure. Let $\mathfrak{p}_2 = \text{ann}(x_2 + M_1)$ be a maximal element of $\mathcal{A}(M/M_1)$, so that $\mathfrak{p}_2$ is prime, and define $M_2 = \langle x_2, x_1 \rangle$. Note that $\{0\} \subseteq M_1 \subseteq M_2$ and that $M_2/M_1 \cong R/\text{ann}(x_2 + M_1) = R/\mathfrak{p}_2$. By Exercise 6.18 on page 416, the module $M$ has ACC, and so this process terminates, say, with $M^* \subseteq M$. We must have $M^* = M$, however, lest the process continue for another step. Now reindex the subscripts to get the desired statement. $\quad\bullet$

**Lemma 10.179 (Prime Avoidance).** *Let $\mathfrak{p}_1, \ldots, \mathfrak{p}_n$ be prime ideals in a ring $R$. If $J$ is an ideal with $J \subseteq \mathfrak{p}_1 \cup \cdots \cup \mathfrak{p}_n$, then $J$ is contained in some $\mathfrak{p}_i$.*

**Proof.** The proof is by induction on $n \geq 1$, and the base step is trivially true. For the inductive step, let $J \subseteq \mathfrak{p}_1 \cup \cdots \cup \mathfrak{p}_{n+1}$, and define

$$D_i = \mathfrak{p}_1 \cup \cdots \cup \widehat{\mathfrak{p}_i} \cup \cdots \cup \mathfrak{p}_{n+1}.$$

We may assume that $J \subsetneq D_i$ for all $i$, since otherwise the inductive hypothesis can be invoked to complete the proof. Hence, for each $i$, there exists $a_i \in J$ with $a_i \notin D_i$; since $J \subseteq D_i \cup \mathfrak{p}_i$, we must have $a_i \in \mathfrak{p}_i$. Consider the element

$$b = a_1 + a_2 \cdots a_{n+1}.$$

Now $b \in J$ because all the $a_i$ are. We claim that $b \notin \mathfrak{p}_1$. Otherwise, $a_2 \cdots a_{n+1} = b - a_1 \in \mathfrak{p}_1$; but $\mathfrak{p}_1$ is a prime ideal, and so $a_i \in \mathfrak{p}_1$ for some $i \geq 2$. This is a contradiction, for $a_i \in \mathfrak{p}_1 \subseteq D_i$ and $a_i \notin D_i$. Therefore, $b \notin \mathfrak{p}_i$ for any $i$, contradicting $J \subseteq \mathfrak{p}_1 \cup \cdots \cup \mathfrak{p}_n$.   •

**Proposition 10.180.** *There are finitely many prime ideals* $\mathfrak{p}_1, \ldots, \mathfrak{p}_n$ *with*

$$\mathrm{Zer}(R) = \{r \in R : r = 0 \text{ or } r \text{ is a zero-divisor}\} \subseteq \mathfrak{p}_1 \cup \cdots \cup \mathfrak{p}_n.$$

**Proof.** Since $\mathrm{Zer}(R)$ is scalar-closed, Lemma 10.178(i) shows that any ideal $I$ that is a maximal member of the family $\mathcal{A}(X) = (\mathrm{ann}(x))_{x \in \mathrm{Zer}(R)}$ is prime (maximal members exist because $R$ is noetherian). Let $(\mathfrak{p}_\alpha)_{\alpha \in A}$ be the family of all such maximal members. If $x$ is a zero-divisor, then there is a nonzero $r \in R$ with $rx = 0$; that is, $x \in \mathrm{ann}(r)$ [of course, $r \in \mathrm{Zer}(R)$]. It follows that every zero-divisor $x$ lies in some $\mathfrak{p}_\alpha$, and so $\mathrm{Zer}(R) \subseteq \bigcup_{\alpha \in A} \mathfrak{p}_\alpha$. It remains to prove that we may choose the index set $A$ to be finite.

Each $\mathfrak{p}_\alpha = \mathrm{ann}(x_\alpha)$ for some $x_\alpha \in X$; let $S$ be the submodule of $M$ generated by all the $x_\alpha$. Since $R$ is noetherian and $M$ is finitely generated, the submodule $S$ is generated by finitely many of the $x_\alpha$; say, $S = \langle x_1, \ldots, x_n \rangle$. We claim that $\mathrm{ann}(X) \subseteq \mathfrak{p}_1 \cup \cdots \cup \mathfrak{p}_n$, and it suffices to prove that $\mathfrak{p}_\alpha \subseteq \mathfrak{p}_1 \cup \cdots \cup \mathfrak{p}_n$ for all $\alpha$. Now $\mathfrak{p}_\alpha = \mathrm{ann}(x_\alpha)$, and $x_\alpha \in S$; hence,

$$x_\alpha = r_1 x_1 + \cdots + r_n x_n$$

for $r_i \in R$. If $a \in \mathfrak{p}_1 \cap \cdots \cap \mathfrak{p}_n = \mathrm{ann}(x_1) \cap \cdots \cap \mathrm{ann}(x_n)$, then $a x_i = 0$ for all $i$ and so $a x_\alpha = 0$. Therefore,

$$\mathfrak{p}_1 \cap \cdots \cap \mathfrak{p}_n \subseteq \mathrm{ann}(x_\alpha) = \mathfrak{p}_\alpha.$$

But $\mathfrak{p}_\alpha$ is prime, so that $\mathfrak{p}_i \subseteq \mathfrak{p}_\alpha$ for some $i$,[11] and this contradicts the maximality of $\mathfrak{p}_i$.   •

---

[11]Assume that $I_1 \cap \cdots \cap I_n \subseteq \mathfrak{p}$, where $\mathfrak{p}$ is prime. If $I_i \not\subseteq \mathfrak{p}$ for all $i$, then there are $u_i \in I_i$ with $u_i \notin \mathfrak{p}$. But $u_1 \cdots u_n \in I_1 \cap \cdots \cap I_n \subseteq \mathfrak{p}$; since $\mathfrak{p}$ is prime, some $u_i \in \mathfrak{p}$, a contradiction.

# Exercises

**10.71.** If $R = k[x_1, \ldots, x_n]$, where $k$ is a field, and $\mathfrak{p} = (x_1, \ldots, x_n)$, prove that $\operatorname{ht}(\mathfrak{p}) = n$.

**10.72.** Let $R$ be a commutative ring with FFR. Prove that every finitely generated projective $R$-module $P$ has a finitely generated free complement; that is, there is a finitely generated free $R$-module $F$ such that $P \oplus F$ is a free $R$-module. Compare with Exercise 9.77 on page 871.

**10.73.** If $x_1, x_2, \ldots, x_n$ is an $R$-sequence on an $R$-module $M$, prove that $(x_1) \subseteq (x_1, x_2) \subseteq \cdots \subseteq (x_1, x_2, \ldots, x_n)$ is a strictly ascending chain.

**10.74.** Let $R = k[x_1, \ldots, x_n]$, where $k$ is a field, and let $\mathfrak{m} = (x_1, \ldots, x_n)$. Prove that $\operatorname{ht}(\mathfrak{m}) = n$.

∗ **10.75.** Let $R = k[x, y, z]$, where $k$ is a field.

  (i) Prove that $x, y(1 - x), z(1 - x)$ is an $R$-sequence.

  (ii) Prove that $y(1 - x), z(1 - x), x$ is not an $R$-sequence.

∗ **10.76.** Let $R$ be a commutative ring. Prove that a prime ideal $\mathfrak{p}$ in $R$ is minimal over an ideal $I$ if and only if $\operatorname{ht}(\mathfrak{p}/I) = 0$ in $R/I$.

∗ **10.77.** If $k$ is a field, prove that $k[[x_1, \ldots, x_n]]$ is noetherian.

**Hint.** Define the *order* $o(f)$ of a nonzero formal power series $f = (f_0, f_1, f_2, \ldots)$ to be the smallest $n$ with $f_n \neq 0$. Find a proof similar to that of the Hilbert Basis Theorem (Zariski–Samuel, *Commutative Algebra* II, p. 138).

## Section 10.7. Regular Local Rings

We are going to prove that local rings have finite global dimension if and only if they are *regular local* rings (such rings arise quite naturally in Algebraic Geometry), in which case they are UFDs. Let us begin with a localization result.

**All rings in this section are commutative and noetherian.**

**Proposition 10.181.**

  (i) *If $A$ is a finitely generated $R$-module, then*

$$\operatorname{pd}(A) = \sup_{\mathfrak{m}} \operatorname{pd}(A_{\mathfrak{m}}),$$

  *where $\mathfrak{m}$ ranges over all the maximal ideals of $R$.*

  (ii) *For every commutative ring $R$, we have*

$$\mathrm{D}(R) = \sup_{\mathfrak{m}} \mathrm{D}(R_{\mathfrak{m}}),$$

  *where $\mathfrak{m}$ ranges over all the maximal ideals of $R$.*

**Proof.**

(i) We first prove that $\text{pd}(A) \geq \text{pd}(A_{\mathfrak{m}})$ for every maximal ideal $\mathfrak{m}$. There is nothing to prove if $\text{pd}(A) = \infty$, and so we may assume that $\text{pd}(A) = n < \infty$. Thus, there is a projective resolution

$$0 \to P_n \to P_{n-1} \to \cdots \to P_0 \to A \to 0.$$

Since $R_{\mathfrak{m}}$ is a flat $R$-module, by Theorem 10.30,

$$0 \to R_{\mathfrak{m}} \otimes P_n \to R_{\mathfrak{m}} \otimes P_{n-1} \to \cdots \to R_{\mathfrak{m}} \otimes P_0 \to A_{\mathfrak{m}} \to 0$$

is exact; it is a projective resolution of $A_{\mathfrak{m}}$ because all $R_{\mathfrak{m}} \otimes P_i$ are projective $R_{\mathfrak{m}}$-modules. Therefore, $\text{pd}(A_{\mathfrak{m}}) \leq n$. (Neither hypothesis $R$ noetherian nor $A$ finitely generated is needed for this implication.)

For the reverse inequality, we may assume that $\sup_{\mathfrak{m}} \text{pd}(A_{\mathfrak{m}}) = n < \infty$. Since $R$ is noetherian, Theorem 10.160(i) says that $\text{pd}(A) = \text{fd}(A)$. Now $\text{pd}(A_{\mathfrak{m}}) \leq n$ if and only if $\text{Tor}_{n+1}^{R_{\mathfrak{m}}}(A_{\mathfrak{m}}, B_{\mathfrak{m}}) = \{0\}$ for all $R_{\mathfrak{m}}$-modules $B_{\mathfrak{m}}$, by Lemma 10.151. However, Proposition 10.39 gives $\text{Tor}_{n+1}^{R_{\mathfrak{m}}}(A_{\mathfrak{m}}, B_{\mathfrak{m}}) \cong \left( \text{Tor}_{n+1}^{R}(A, B) \right)_{\mathfrak{m}}$. Therefore, $\text{Tor}_{n+1}^{R}(A, B) = \{0\}$, by Proposition 10.38(i). We conclude that $n \geq \text{pd}(A)$.

(ii) This follows at once from part (i), for $D(R) = \sup_A \{\text{pd}(A)\}$, where $A$ ranges over all finitely generated (even cyclic) $R$-modules [Theorem 10.147].  •

**Definition.** If $R$ is a local ring with maximal ideal $\mathfrak{m}$, then $k = R/\mathfrak{m}$ is called its *residue field*. We shall say that

$$(R, \mathfrak{m}, k)$$

is a local ring.

Theorem 10.147 allows us to compute global dimension as the supremum of projective dimensions of cyclic modules. When $(R, \mathfrak{m}, k)$ is a local ring, there is a dramatic improvement: global dimension is determined by the projective dimension of one cyclic module—its residue field $k$.

**Lemma 10.182.** *Let $(R, \mathfrak{m}, k)$ be a local ring. If $A$ is a finitely generated $R$-module, then*

$$\text{pd}(A) \leq n \quad \textit{if and only if} \quad \text{Tor}_{n+1}^{R}(A, k) = \{0\}.$$

**Proof.** Suppose $\text{pd}(A) \leq n$. By Example 10.148, we have $\text{fd}(A) \leq \text{pd}(A)$, so that $\text{Tor}_{n+1}^{R}(A, B) = \{0\}$ for every $R$-module $B$. In particular, $\text{Tor}_{n+1}^{R}(A, k) = \{0\}$.

We prove the converse by induction on $n \geq 0$. For the base step $n = 0$, we must prove that $\text{Tor}_1^{R}(A, k) = \{0\}$ implies $\text{pd}(A) = 0$; that is, $A$ is projective (hence free, since $R$ is local, by Proposition 10.26). Let $\{a_1, \ldots, a_r\}$ be a *minimal set of generators* of $A$ (that is, no proper subset generates $A$),[12] let $F$ be the free $R$-module with basis $\{e_1, \ldots, e_r\}$, and let $\varphi \colon F \to A$ be the $R$-map with $\varphi(e_i) = a_i$. There is an exact sequence $0 \to N \xrightarrow{i} F \xrightarrow{\varphi} A \to 0$, where $N = \ker \varphi$ and $i$ is the inclusion. Since $R$ is noetherian, the submodule $N$ is finitely generated. For later

---

[12]A minimal generating set for $A = (2)$ in $\mathbb{Z}$ is $\{4, 6\}$; of course, there is a generating set of smaller cardinality.

use, we observe, as in the proof of Proposition 10.26, that

$$N \subseteq \mathfrak{m}F.$$

Now the sequence $0 \to N \otimes_R k \xrightarrow{i \otimes 1} F \otimes_R k \xrightarrow{\varphi \otimes 1} A \otimes_R k \to 0$ is exact, because $\mathrm{Tor}_1^R(A, k) = \{0\}$. Hence, $i \otimes 1 \colon N \otimes_R k \to F \otimes_R k$ is an injection.

Since $\square \otimes_R N$ is right exact, exactness of $0 \to \mathfrak{m} \xrightarrow{j} R \to k \to 0$ gives exactness of $\mathfrak{m} \otimes_R N \xrightarrow{j \otimes 1} R \otimes_R N \to k \otimes_R N \to 0$, so that $k \otimes_R N \cong R \otimes_R N / \mathrm{im}(j \otimes 1)$. Under the isomorphism $\lambda \colon R \otimes_R N \to N$, given by $r \otimes n \mapsto rn$, we have $\lambda(\mathrm{im}(j \otimes 1)) = \mathfrak{m}N$. Indeed, the map $\tau_N \colon N \otimes_R k \to N/\mathfrak{m}N$, given by $n \otimes b \mapsto n + \mathfrak{m}N$ (where $n \in N$ and $b \in k$), is a natural isomorphism. Thus, there is a commutative diagram

$$
\begin{array}{ccc}
0 \longrightarrow N \otimes_R k & \xrightarrow{i \otimes 1} & F \otimes_R k \\
{\scriptstyle \tau_N} \downarrow & & \downarrow {\scriptstyle \tau_F} \\
N/\mathfrak{m}N & \xrightarrow[\overline{i}]{} & F/\mathfrak{m}F
\end{array}
$$

where $\overline{i} \colon n + \mathfrak{m}N \mapsto n + \mathfrak{m}F$. Since $i \otimes 1$ is an injection, so is $\overline{i}$. But $N \subseteq \mathfrak{m}F$ implies that the map $\overline{i}$ is the zero map. Therefore, $N/\mathfrak{m}N = \{0\}$ and $N = \mathfrak{m}N$. By Corollary 7.16, Nakayama's Lemma, $N = \{0\}$, and so $\varphi \colon F \to A$ is an isomorphism; that is, $A$ is free.

For the inductive step, we must prove that if $\mathrm{Tor}_{n+2}^R(A, k) = \{0\}$, then $\mathrm{pd}(A) \leq n + 1$. Take a projective resolution $\mathbf{P}$ of $A$, and let $\Omega_n(A, \mathbf{P})$ be its $n$th syzygy. Since $\mathbf{P}$ is also a flat resolution of $A$, we have $Y_n(A, \mathbf{P}) = \Omega_n(A, \mathbf{P})$. By Proposition 10.149, $\mathrm{Tor}_{n+2}^R(A, k) \cong \mathrm{Tor}_1^R(Y_n(A, \mathbf{P}), k)$. The base step shows that $Y_n(A, \mathbf{P}) = \Omega_n(A, \mathbf{P})$ is free, and this gives $\mathrm{pd}(A) \leq n + 1$, by Lemma 10.136. $\bullet$

**Corollary 10.183.** *Let $(R, \mathfrak{m}, k)$ be a local ring. If $A$ is a finitely generated $R$-module, then*

$$\mathrm{pd}(A) = \sup\{i : \mathrm{Tor}_i^R(A, k) \neq \{0\}\}.$$

**Proof.** If $n = \sup\{i : \mathrm{Tor}_i^R(A, k) \neq \{0\}\}$, then $\mathrm{pd}(A) \leq n$. But $\mathrm{pd}(A) \not< n$; that is, $\mathrm{pd}(A) = n$. $\bullet$

**Theorem 10.184.** *Let $(R, \mathfrak{m}, k)$ be a local ring.*

(i) *$\mathrm{D}(R) \leq n$ if and only if $\mathrm{Tor}_{n+1}^R(k, k) = \{0\}$.*

(ii) *$\mathrm{D}(R) = \mathrm{pd}(k)$.*

**Proof.**

(i) If $\mathrm{D}(R) \leq n$, then Lemma 10.182 gives $\mathrm{Tor}_{n+1}^R(k, k) = \{0\}$.

Conversely, if $\mathrm{Tor}_{n+1}^R(k, k) = \{0\}$, the same lemma gives $\mathrm{pd}(k) \leq n$. By Lemma 10.151, we have $\mathrm{Tor}_{n+1}^R(A, k) = \{0\}$ for every $R$-module $A$. In particular, if $A$ is finitely generated, then Lemma 10.182 gives $\mathrm{pd}(A) \leq n$. Finally, Theorem 10.147 shows that $\mathrm{D}(R) = \sup_A\{\mathrm{pd}(A)\}$, where $A$ ranges over all finitely generated (even cyclic) $R$-modules. Therefore, $\mathrm{D}(R) \leq n$.

(ii) Immediate from part (i). $\bullet$

If $(R, \mathfrak{m}, k)$ is a local ring, then $\mathfrak{m}/\mathfrak{m}^2$ is an $(R/\mathfrak{m})$-module; that is, it is a vector space over $k$. Recall that a generating set $X$ of a module $M$ (over any ring) is *minimal* if no proper subset of $X$ generates $M$.

**Proposition 10.185.** *Let* $(R, \mathfrak{m}, k)$ *be a local ring.*

(i) *Elements* $x_1, \ldots, x_d$ *form a minimal generating set for* $\mathfrak{m}$ *if and only if the cosets* $x_i^* = x_i + \mathfrak{m}^2$ *form a basis of* $\mathfrak{m}/\mathfrak{m}^2$.

(ii) *Any two minimal generating sets of* $\mathfrak{m}$ *have the same number of elements.*

**Proof.**

(i) If $x_1, \ldots, x_d$ is a minimal generating set for $\mathfrak{m}$, then $X^* = x_1^*, \ldots, x_d^*$ spans the vector space $\mathfrak{m}/\mathfrak{m}^2$. If $X^*$ is linearly dependent, then there is some $x_i^* = \sum_{j \neq i} r_j' x_j^*$, where $r_j' \in k$. Lifting this equation to $\mathfrak{m}$, we have $x_i \in \sum_{j \neq i} r_j x_j + \mathfrak{m}^2$. Thus, if $B = \langle x_j : j \neq i \rangle$, then $B + \mathfrak{m}^2 = \mathfrak{m}$. Hence,

$$\mathfrak{m}\left(\mathfrak{m}/B\right) = (B + \mathfrak{m}^2)/B = \mathfrak{m}/B.$$

By Nakayama's Lemma, $\mathfrak{m}/B = \{0\}$, and so $\mathfrak{m} = B$. This contradicts $x_1, \ldots, x_d$ being a minimal generating set. Therefore, $X^*$ is linearly independent, and hence it is a basis of $\mathfrak{m}/\mathfrak{m}^2$.

Conversely, assume that $x_1^*, \ldots, x_d^*$ is a basis of $\mathfrak{m}/\mathfrak{m}^2$, where $x_i^* = x_i + \mathfrak{m}^2$. If we define $A = \langle x_1, \ldots, x_d \rangle$, then $A \subseteq \mathfrak{m}$. If $y \in \mathfrak{m}$, then $y^* = \sum r_i' x_i^*$, where $r_i' \in k$, so that $y \in A + \mathfrak{m}^2$. Hence, $\mathfrak{m} = A + \mathfrak{m}^2$, and, as in the previous paragraph, Nakayama's Lemma gives $\mathfrak{m} = A$; that is, $x_1, \ldots, x_d$ generate $\mathfrak{m}$. If a proper subset of $x_1, \ldots, x_d$ generates $\mathfrak{m}$, then the vector space $\mathfrak{m}/\mathfrak{m}^2$ could be generated by fewer than $d$ elements, contradicting $\dim_k(\mathfrak{m}/\mathfrak{m}^2) = d$.

(ii) The number of elements in any minimal generating set is $\dim_k(\mathfrak{m}/\mathfrak{m}^2)$.  •

**Definition.** If $(R, \mathfrak{m}, k)$ is a local ring, then $\mathfrak{m}/\mathfrak{m}^2$ is a finite-dimensional vector space over $k$. Write

$$V(R) = \dim_k(\mathfrak{m}/\mathfrak{m}^2).$$

Proposition 10.185 shows that all minimal generating sets of $\mathfrak{m}$ have the same number of elements, namely, $V(R)$.

**Proposition 10.186.** *Let* $(R, \mathfrak{m}, k)$ *be a local ring. If* $x \in \mathfrak{m} - \mathfrak{m}^2$, *define* $R^* = R/(x)$ *and* $\mathfrak{m}^* = \mathfrak{m}/(x)$. *Then* $(R^*, \mathfrak{m}^*, k)$ *is a local ring, and*

$$V(R) = V(R^*) + 1.$$

**Proof.** The reader may show that $\mathfrak{m}^*$ is the unique maximal ideal in $R^*$. Note that $R^*/\mathfrak{m}^* = [R/(x)]/[\mathfrak{m}/(x)] \cong R/\mathfrak{m} = k$.

Let $\{y_1^*, \ldots, y_t^*\}$ be a minimal generating set of $\mathfrak{m}^*$, and let $y_i^* = y_i + \mathfrak{m}$. It is clear that $\{x, y_1, \ldots, y_t\}$ generates $\mathfrak{m}$, and we now show that it is a minimal generating set; that is, their cosets mod $\mathfrak{m}$ form a basis of $\mathfrak{m}/\mathfrak{m}^2$.

If $rx + \sum_i r_i y_i \in \mathfrak{m}^2$, where $r, r_i \in R$, then we must show that each term lies in $\mathfrak{m}^2$; that is, all $r, r_i \in \mathfrak{m}$. Passing to $R^*$, we have $\sum_i r_i^* y_i^* \in \mathfrak{m}^2$, because $r^* x^* = 0$ [where $*$ denotes coset mod $(x)$]. But $\{y_1^*, \ldots, y_t^*\}$ is a basis of $\mathfrak{m}^*/(\mathfrak{m}^*)^2$, so that

$r_i^* \in \mathfrak{m}^*$ and $r_i \in \mathfrak{m}$ for all $i$. Therefore, $rx \in \mathfrak{m}^2$. But $x \notin \mathfrak{m}^2$, and so $r \in \mathfrak{m}$, as desired. $\bullet$

**Corollary 10.187.** *If $(R, \mathfrak{m}, k)$ is a local ring, then $\mathrm{ht}(\mathfrak{m}) \leq V(R)$, and*

$$\dim(R) \leq V(R).$$

**Proof.** If $V(R) = d$, then $\mathfrak{m} = (x_1, \ldots, x_d)$. Since $\mathfrak{m}$ is obviously a minimal prime over itself, Theorem 10.174, the Generalized Principal Ideal Theorem, gives $\mathrm{ht}(\mathfrak{m}) \leq d = V(R)$.

If $\mathfrak{p} \neq \mathfrak{m}$ is a prime ideal in $R$, then any prime chain, $\mathfrak{p} = \mathfrak{p}_0 \supsetneq \mathfrak{p}_1 \supsetneq \cdots \supsetneq \mathfrak{p}_h$, can be lengthened to a prime chain $\mathfrak{m} \supsetneq \mathfrak{p}_0 \supsetneq \mathfrak{p}_1 \supsetneq \cdots \supsetneq \mathfrak{p}_h$ of length $h + 1$. Therefore, $h < V(R)$, and so $\dim(R) = \mathrm{ht}(\mathfrak{m}) \leq V(R)$. $\bullet$

**Definition.** A ***regular local ring*** is a local ring $(R, \mathfrak{m}, k)$ such that

$$\dim(R) = V(R).$$

It is clear that every field is a regular local ring of dimension 0, and every DVR is a regular local ring of dimension 1.

**Example 10.188.** Regular local rings arise in Algebraic Geometry: varieties are viewed as algebraic analogs of differentiable manifolds, and geometric definitions are adapted to an algebraic setting. Recall some definitions from §5.5. Let $I$ be an ideal in $k[X] = k[x_1, \ldots, x_n]$, where $k$ is an algebraically closed field; we regard each $f(X) \in k[X]$ as a $k$-valued function. Define the ***affine variety*** of $I$ by

$$V = \mathrm{Var}(I) = \{a \in k^n : f(a) = 0 \text{ for all } f \in I\},$$

and define its *coordinate ring* $k[V]$ by

$$k[V] = \{f|V : f(X) \in k[X]\}.$$

Now $k[V] \cong k[X]/\mathrm{Id}(V)$, where $\mathrm{Id}(V) = \{f(X) \in k[X] : f(v) = 0 \text{ for all } v \in V\}$ is an ideal in $k[X]$. Affine varieties $V$ are usually assumed to be *irreducible*; that is, $\mathrm{Id}(V)$ is a prime ideal (see Proposition 5.105), so that the coordinate ring $k[V]$ is a domain.

If $a$ is a point in an affine variety $V$, then $\mathfrak{m}_a = \{f + \mathrm{Id}(V) \in k[V] : f(a) = 0\}$ is a maximal ideal in $k[V]$. Define $\mathcal{O}_a$ to be the localization

$$\mathcal{O}_a = k[V]_{\mathfrak{m}_a}.$$

Identify the maximal ideal $\mathfrak{m}_a$ with its localization in $k[V]_{\mathfrak{m}_a}$, and call $(\mathcal{O}_a, \mathfrak{m}_a, k)$ the ***local ring of $V$ at $a$***. Now $V = \mathrm{Var}(I)$, where $I = (f_1(X), \ldots, f_t(X))$. Each $f_i(X)$ has partial derivatives $\partial f_i / \partial x_j$ (which are defined formally, as derivatives of polynomials of a single variable are defined). The ***tangent space*** $T_a$ of $V$ at $a$ can be defined, and there is an isomorphism of the dual space $T_a^* \cong \mathfrak{m}_a / \mathfrak{m}_a^2$ (as vector spaces over $k$); thus, $\dim_k(\mathfrak{m}_a / \mathfrak{m}_a^2) = \dim_k(T_a)$. The ***Jacobian*** of $f_1(X), \ldots, f_t(X)$ is the $t \times n$ matrix over $k[X]$,

$$\mathrm{Jac}(f_1, \ldots, f_t) = [\partial f_i / \partial x_j].$$

If $a \in V$, evaluating each entry in $\mathrm{Jac}(f_1, \ldots, f_t)$ at $a$ gives a matrix $\mathrm{Jac}(f_1, \ldots, f_t)_a$ over $k$. We say that $a$ is ***nonsingular*** if $\mathrm{rank}(\mathrm{Jac}(f_1, \ldots, f_n)_a) = n - \dim(V)$.

There are several equivalent ways of expressing nonsingularity of $a$, and all of them say that $a$ is nonsingular if and only if $(\mathcal{O}_a, \mathfrak{m}_a, k)$ is a regular local ring (Hartshorne, *Algebraic Geometry*, §1.5). (There is a notion of *projective variety*, which is essentially a compactification of an affine variety, and this discussion can be adapted to them as well.)  ◄

**Proposition 10.189.** *Let $(R, \mathfrak{m}, k)$ be a local ring. If $\mathfrak{m}$ can be generated by an $R$-sequence $x_1, \ldots, x_d$, then $R$ is a regular local ring and*

$$d = \dim(R) = V(R).$$

**Remark.** We will soon prove the converse: in a regular local ring, the maximal ideal can be generated by an $R$-sequence.  ◄

**Proof.** Consider the inequalities

$$d \leq \mathrm{ht}(\mathfrak{m}) \leq V(R) \leq d.$$

The first inequality holds by Lemma 10.177; the second by Corollary 10.187, and the third by Proposition 10.185. It follows that all the inequalities are, in fact, equalities, and the proposition follows because $\dim(R) = \mathrm{ht}(\mathfrak{m})$.  •

**Example 10.190.** Let $k$ be a field, and let $R = k[[x_1, \ldots, x_r]]$ be the ring of *formal power series* in $r$ variables $x_1, \ldots, x_r$. Recall that an element $f \in R$ is a sequence

$$f = (f_0, f_1, f_2, \ldots, f_n, \ldots),$$

where $f_n$ is a homogeneous polynomial of total degree $n$ in $k[x_1, \ldots, x_r]$, and that multiplication is defined by

$$(f_0, f_1, f_2, \ldots)(g_0, g_1, g_2, \ldots) = (h_0, h_1, h_2, \ldots),$$

where $h_n = \sum_{i+j=n} f_i g_j$. We claim that $R$ is a local ring with maximal ideal $\mathfrak{m} = (x_1, \ldots, x_r)$ and residue field $k$. First, $R/\mathfrak{m} \cong k$, so that $\mathfrak{m}$ is a maximal ideal. To see that $\mathfrak{m}$ is the unique maximal ideal, it suffices to prove that if $f \in R$ and $f \notin \mathfrak{m}$, then $f$ is a unit. Now $f \notin \mathfrak{m}$ if and only if $f_0 \neq 0$, and we now show that $f$ is a unit if and only if $f_0 \neq 0$. If $fg = 1$, then $f_0 g_0 = 1$, and $f_0 \neq 0$; conversely, if $f_0 \neq 0$, we can solve $(f_0, f_1, f_2, \ldots)(g_0, g_1, g_2, \ldots) = 1$ recursively for $g_n$, and $fg = 1$, where $g = (g_0, g_1, g_2, \ldots)$.

Exercise 10.77 on page 969 shows that the ring $R$ is noetherian. But the quotient ring $R/(x_1, \ldots, x_{i-1})$ is a domain, because it is isomorphic to $k[[x_i, \ldots, x_r]]$, and so $x_i$ is a regular element on it. Hence, Proposition 10.189 shows that $R = k[[x_1, \ldots, x_r]]$ is a regular local ring, for $x_1, \ldots, x_r$ is an $R$-sequence.  ◄

The next lemma prepares us for induction.

**Lemma 10.191.** *Let $(R, \mathfrak{m}, k)$ be a regular local ring. If $x \in \mathfrak{m} - \mathfrak{m}^2$, then $R/(x)$ is regular and $\dim(R/(x)) = \dim(R) - 1$.*

**Proof.** Since $R$ is regular, we have $\dim(R) = V(R)$. Let us note at the outset that $\dim(R) = \mathrm{ht}(\mathfrak{m})$. We must show that $\mathrm{ht}(\mathfrak{m}^*) = V(R/(x)) = \dim_k(\mathfrak{m}^*/\mathfrak{m}^{*2})$, where

$\mathfrak{m}^* = \mathfrak{m}/(x)$. By Corollary 10.176, $\operatorname{ht}(\mathfrak{m}) \leq \operatorname{ht}(\mathfrak{m}^*) + 1$. Hence,

$$\operatorname{ht}(\mathfrak{m}) - 1 \leq \operatorname{ht}(\mathfrak{m}^*) \leq V(R/(x)) = V(R) - 1 = \operatorname{ht}(\mathfrak{m}) - 1.$$

The next to last equation is Proposition 10.186; the last equation holds because $R$ is regular. Therefore, $\dim(R/(x)) = \operatorname{ht}(\mathfrak{m}^*) = V(\mathfrak{m}^*)$, and so $R/(x)$ is regular with $\dim(R/(x)) = \dim(R) - 1$. $\bullet$

We are now going to prove that regular local rings are domains, and we will then use this to prove the converse of Proposition 10.189.

**Proposition 10.192.** *Every regular local ring* $(R, \mathfrak{m}, k)$ *is a domain.*

**Proof.** The proof is by induction on $d = \dim(R)$. If $d = 0$, then $R$ is a field, by Exercise 10.79 on page 983. If $d > 0$, let $\mathfrak{p}_1, \ldots, \mathfrak{p}_s$ be the minimal prime ideals in $R$ (there are only finitely many such, by Corollary 5.118). If $\mathfrak{m} - \mathfrak{m}^2 \subseteq \mathfrak{p}_1 \cup \cdots \cup \mathfrak{p}_s$, then Lemma 10.179 would give $\mathfrak{m} \subseteq \mathfrak{p}_i$, which cannot occur because $d = \operatorname{ht}(\mathfrak{m}) > 0$. Therefore, there is $x \in \mathfrak{m} - \mathfrak{m}^2$ with $x \notin \mathfrak{p}_i$ for all $i$. By Lemma 10.191, $R/(x)$ is regular of dimension $d - 1$. Now $R/(x)$ is a domain, by the inductive hypothesis, and so $(x)$ is a prime ideal. It follows that $(x)$ contains a minimal prime ideal; say, $\mathfrak{p}_i \subseteq (x)$.

If $\mathfrak{p}_i = (0)$, then $(0)$ is a prime ideal and $R$ is a domain. Hence, we may assume that $\mathfrak{p}_i \neq (0)$. For each nonzero $y \in \mathfrak{p}_i$, there exists $r \in R$ with $y = rx$. Since $x \notin \mathfrak{p}_i$, we have $r \in \mathfrak{p}_i$, so that $y \in x\mathfrak{p}_i$. Thus, $\mathfrak{p}_i \subseteq x\mathfrak{p}_i \subseteq \mathfrak{m}\mathfrak{p}_i$. As the reverse inclusion $\mathfrak{m}\mathfrak{p}_i \subseteq \mathfrak{p}_i$ is always true, we have $\mathfrak{p}_i = \mathfrak{m}\mathfrak{p}_i$. Nakayama's Lemma now applies, giving $\mathfrak{p}_i = (0)$, a contradiction. $\bullet$

**Proposition 10.193.** *A local ring* $(R, \mathfrak{m}, k)$ *is regular if and only if* $\mathfrak{m}$ *is generated by an* $R$-*sequence* $x_1, \ldots, x_d$. *Moreover, in this case,*

$$d = V(R).$$

**Proof.** We have already proved sufficiency, in Proposition 10.189. If $R$ is regular, we prove the result by induction on $d \geq 1$, where $d = \dim(R)$. The base step holds, for $R$ is a domain and so $x$ is a regular element; that is, $x$ is not a zero-divisor. For the inductive step, the ring $R/(x)$ is regular of dimension $d - 1$, by Lemma 10.191. Therefore, its maximal ideal is generated by an $(R/(x))$-sequence $x_1^*, \ldots, x_{d-1}^*$. In view of Proposition 10.186, a minimal generating set for $\mathfrak{m}$ is $x, x_1, \ldots, x_{d-1}$. Finally, this is an $R$-sequence, by Exercise 10.80(i) on page 983, because $x$ is not a zero-divisor. $\bullet$

We are now going to characterize regular local rings by their global dimension.

**Lemma 10.194.** *Let* $(R, \mathfrak{m}, k)$ *be a local ring, and let* $A$ *be an* $R$-*module with* $\operatorname{pd}(A) = n$. *If* $x \in \mathfrak{m}$ *and multiplication* $\mu_x \colon A \to A$, *given by* $a \mapsto xa$, *is injective, then* $\operatorname{pd}(A/xA) = n + 1$.

**Proof.** By hypothesis, there is an exact sequence

$$0 \to A \xrightarrow{\mu_x} A \to A/xA \to 0,$$

where $\mu_x \colon a \mapsto xa$. By Lemma 10.161, we have $\operatorname{pd}(A/xA) \leq n + 1$.

Consider the portion of the long exact sequence arising from tensoring by $k$:

$$0 = \text{Tor}^R_{n+1}(A, k) \to \text{Tor}^R_{n+1}(A/xA, k) \xrightarrow{\partial} \text{Tor}^R_n(A, k) \xrightarrow{(\mu_x)_*} \text{Tor}^R_n(A, k).$$

The first term is $\{0\}$, for $\text{pd}(A) \leq n$ if and only if $\text{Tor}^R_{n+1}(A, k) = \{0\}$, by Lemma 10.182. The induced map $(\mu_x)_*$ is multiplication by $x$. But if $\mu'_x : k \to k$ is multiplication by $x$, then $x \in \mathfrak{m}$ implies $\mu'_x = 0$; therefore, $(\mu_x)_* = (\mu'_x)_* = 0$. Exactness now implies that $\partial : \text{Tor}^R_{n+1}(A/xA, k) \to \text{Tor}^R_n(A, k)$ is an isomorphism. Since $\text{pd}(A) = n$, we have $\text{Tor}^R_n(A, k) \neq \{0\}$, so that $\text{Tor}^R_{n+1}(A/xA, k) \neq \{0\}$. Therefore, $\text{pd}(A/xA) \geq n + 1$, as desired.   •

**Proposition 10.195.** *If $(R, \mathfrak{m}, k)$ is a regular local ring, then*

$$\text{D}(R) = V(R) = \dim(R).$$

**Proof.** Since $R$ is regular, Proposition 10.193 says that $\mathfrak{m}$ can be generated by an $R$-sequence $x_1, \ldots, x_d$. Applying Lemma 10.194 to the modules

$$R, \ R/(x_1), \ R/(x_1, x_2), \ldots, R/(x_1, \ldots, x_d) = R/\mathfrak{m} = k,$$

we see that $\text{pd}(k) = d$. By Proposition 10.189, $d = V(R) = \dim(R)$. On the other hand, Theorem 10.184(ii) gives $d = \text{pd}(k) = \text{D}(R)$.   •

The converse of Proposition 10.195: a noetherian local ring of finite global dimension is regular, is more difficult to prove. The following proof is essentially that in Lam, *Lectures on Modules and Rings*, Chapter 2, § 5F.

**Lemma 10.196.** *Let $(R, \mathfrak{m}, k)$ be a local ring of finite global dimension. If $V(R) \leq \text{D}(R)$ and $\text{D}(R) \leq d$, where $d$ is the length of a longest $R$-sequence in $\mathfrak{m}$, then $R$ is a regular local ring.*

**Proof.** By Corollary 10.187, $\dim(R) \leq V(R)$. By hypothesis, $V(R) \leq \text{D}(R) \leq d$, while Lemma 10.177 gives $d \leq \text{ht}(\mathfrak{m}) = \dim(R)$. Therefore, $\dim(R) = V(R)$, and so $R$ is a regular local ring.   •

In proving that $\text{D}(R)$ finite implies $R$ regular, we cannot assume that $R$ is a domain (though this will turn out to be true); hence, we must deal with zero-divisors. Recall that $\text{Zer}(R)$ is the set of all zero-divisors in $R$.

**Proposition 10.197.** *Let $(R, \mathfrak{m}, k)$ be a local ring.*

(i) *If $\mathfrak{m} - \mathfrak{m}^2$ consists of zero-divisors, then there is a nonzero $a \in R$ with $a\mathfrak{m} = (0)$.*

(ii) *If $0 < \text{D}(R) = n < \infty$, then there exists a nonzero-divisor $x \in \mathfrak{m} - \mathfrak{m}^2$.*

**Proof.**

(i) By Proposition 10.180, there are prime ideals $\mathfrak{p}_1, \ldots, \mathfrak{p}_n$ with

$$\mathfrak{m} - \mathfrak{m}^2 \subseteq \mathrm{Zer}(R) \subseteq \mathfrak{p}_1 \cup \cdots \cup \mathfrak{p}_n.$$

If we can show that

(1) $$\mathfrak{m} \subseteq \mathfrak{p}_1 \cup \cdots \cup \mathfrak{p}_n,$$

then Prime Avoidance, Lemma 10.179, gives $\mathfrak{m} \subseteq \mathfrak{p}_i$ for some $i$. But $\mathfrak{p}_i = \mathrm{ann}(a)$ for some $a \in \mathfrak{m}$, so that $a\mathfrak{m} = (0)$, as desired.

To verify (1), it suffices to prove $\mathfrak{m}^2 \subseteq \mathfrak{p}_1 \cup \cdots \cup \mathfrak{p}_n$. Now $\mathfrak{m} \neq \mathfrak{m}^2$, by Nakayama's Lemma (we may assume that $\mathfrak{m} \neq (0)$, for the result is trivially true otherwise), and so there exists $x \in \mathfrak{m} - \mathfrak{m}^2 \subseteq \mathfrak{p}_1 \cup \cdots \cup \mathfrak{p}_n$. Let $y \in \mathfrak{m}^2$. For every integer $s \geq 1$, we have $x + y^s \in \mathfrak{m} - \mathfrak{m}^2 \subseteq \mathfrak{p}_1 \cup \cdots \cup \mathfrak{p}_n$; that is, $x + y^s \in \mathfrak{p}_j$ for some $j = j(s)$. By the Pigeonhole Principle, there is an integer $j$ and integers $s < t$ with $x + y^s, x + y^t \in \mathfrak{p}_j$. Subtracting, $y^s(1 - y^{t-s}) \in \mathfrak{p}_j$. But $1 - y^{t-s}$ is a unit [if $u \in \mathfrak{m}$, then $1 - u$ is a unit; otherwise, $(1 - u)$ is a proper ideal, $(1 - u) \subseteq \mathfrak{m}$, $1 - u \in \mathfrak{m}$, and $1 \in \mathfrak{m}$]. Since $\mathfrak{p}_j$ is a prime ideal, $y \in \mathfrak{p}_j$.

(ii) **(Griffith)** In light of (i), it suffices to show that there is no nonzero $a \in \mathfrak{m}$ with $a\mathfrak{m} = (0)$. If, on the contrary, such an $a$ exists and $\mu \colon R \to R$ is given by $\mu \colon r \mapsto ar$, then $\mathfrak{m} \subseteq \ker \mu$. This inclusion cannot be strict; if $b \in \ker \mu$, then $ab = 0$, but if $b \notin \mathfrak{m}$, then $b$ is a unit (for $Rb$ is not contained in the maximal ideal), and so $ab \neq 0$. Hence, $\mathfrak{m} = \ker \mu$. Thus, $k = R/\mathfrak{m} = R/\ker \mu \cong \mathrm{im}\,\mu = Ra$; that is, $k \cong Ra$. Consider the exact sequence $0 \to Ra \to R \to R/Ra \to 0$; by Exercise 10.70 on page 956, either $\mathrm{pd}(R/Ra)\,\mathrm{pd}(Ra) + 1 = \mathrm{pd}(k) + 1$ or $\mathrm{pd}(R/Ra) = 0$. In the first case, $\mathrm{pd}(R/Ra) = \mathrm{pd}(k) + 1 > \mathrm{pd}(k)$, contradicting Theorem 10.184(ii) [which says that $\mathrm{pd}(k) = \mathrm{D}(R)$]. In the second case, $0 = \mathrm{pd}(Ra) = \mathrm{pd}(k)$, contradicting $\mathrm{pd}(k) = \mathrm{D}(R) > 0$. $\bullet$

**Theorem 10.198 (Serre–Auslander–Buchsbaum).** *A local ring* $(R, \mathfrak{m}, k)$ *is regular if and only if* $\mathrm{D}(R)$ *is finite; in fact,*

$$\mathrm{D}(R) = V(R) = \dim(R).$$

**Proof.** Necessity is Proposition 10.195. We prove the converse by induction on $\mathrm{D}(R) = n \geq 0$. If $n = 0$, then $R$ is semisimple. Since $R$ is commutative, it is the direct product of finitely many fields; since $R$ is local, it is a field, and hence it is regular.

If $n \geq 1$, then Proposition 10.197(ii) says that $\mathfrak{m} - \mathfrak{m}^2$ contains a nonzero-divisor $x$. Now $(R^*, \mathfrak{m}^*, k)$ is a local ring, where $R^* = R/(x)$ and $\mathfrak{m}^* = \mathfrak{m}/(x)$. Since $x$ is not a zero-divisor, it is regular on $\mathfrak{m}$; since $\mathrm{pd}_R(\mathfrak{m}) < \infty$, Corollary 10.168 gives $\mathrm{pd}_{R^*}(\mathfrak{m}/x\mathfrak{m}) < \infty$.

Consider a short exact sequence of $R$-modules

(2) $$0 \to k \xrightarrow{\alpha} B \to C \to 0$$

in which $\alpha(1) = e$, where $e \in B - \mathfrak{m}B$. Now the coset $e + \mathfrak{m}B$ is part of a basis of the $k$-vector space $B/\mathfrak{m}B$, and so there is a $k$-map $\beta \colon B/\mathfrak{m}B \to k$ with $\beta(e + \mathfrak{m}B) = 1$.

We can use the composite $\pi\colon B \xrightarrow{\text{nat}} B/\mathfrak{m}B \xrightarrow{\beta} k$ to show that the exact sequence
(2) splits, for $\pi\alpha = 1_k$. In particular, this applies when $B = \mathfrak{m}/x\mathfrak{m}$ and $k$ is the
cyclic submodule generated by $e + x\mathfrak{m}$. Thus, $k$ is a direct summand of $\mathfrak{m}/x\mathfrak{m}$. It
follows that $\mathrm{pd}_{R^*}(k) < \infty$, so that Proposition 10.169 gives $\mathrm{pd}_{R^*}(k) = \mathrm{pd}_R(k) - 1$.
Theorem 10.184 now gives

$$\mathrm{D}(R^*) = \mathrm{pd}_{R^*}(k) = n - 1.$$

By induction, $R^*$ is a regular local ring and $\dim(R^*) = n - 1$. Hence, there is a
prime chain of length $n - 1$ in $R^*$,

$$\mathfrak{p}_0^* \supsetneq \mathfrak{p}_1^* \supsetneq \cdots \supsetneq \mathfrak{p}_{n-1}^* = (0)$$

[we have $\mathfrak{p}_{n-1}^* = (0)$ because $R^*$ is a domain (being regular) so that $(0)$ is a prime
ideal]. Taking inverse images gives a prime chain in $R$:

$$\mathfrak{p}_0 \supsetneq \mathfrak{p}_1 \supsetneq \cdots \supsetneq \mathfrak{p}_{n-1} = (x).$$

Were $(x)$ a minimal prime ideal, then every element in it would be nilpotent, by
Proposition 10.24. Since $x$ is not a zero-divisor, it is not nilpotent, and so there is
a prime ideal $\mathfrak{q} \subsetneq (x)$. Hence, $\dim(R) \geq n$.

Since $x \in \mathfrak{m} - \mathfrak{m}^2$, Proposition 10.186 says that $V(R) = 1 + V(R^*) = 1 +
\dim(R^*) = n$. Therefore,

$$n = V(R) \geq \dim(R) \geq n$$

[the inequality $V(R) \geq \dim(R)$ always being true], and $\dim(R) = V(R)$; that is, $R$
is a regular local ring of dimension $n$.   •

**Corollary 10.199.** *If $S \subseteq R$ is multiplicative, where $(R, \mathfrak{m}, k)$ is a regular local
ring, then $S^{-1}R$ is also a regular local ring. In particular, if $\mathfrak{p}$ is a prime ideal in
$R$, then $R_{\mathfrak{p}}$ is regular.*

**Proof.** Theorem 10.160(ii) says that $\mathrm{D}(R) = \mathrm{wD}(R)$ in this case. But Proposition 10.159 says that $\mathrm{wD}(S^{-1}R) \leq \mathrm{wD}(R)$. Therefore,

$$\mathrm{D}(S^{-1}R) = \mathrm{wD}(S^{-1}R) \leq \mathrm{wD}(R) = \mathrm{D}(R) < \infty.$$

It follows from Corollary 10.20 that $S^{-1}R$ is a local ring; therefore, $S^{-1}R$ is regular,
by Theorem 10.198. The second statement follows: if $S = R - \mathfrak{p}$, then $R_{\mathfrak{p}}$ is a local
ring, and so the Serre–Auslander–Buchsbaum Theorem says that $R_{\mathfrak{p}}$ is regular.   •

There are several proofs that regular local rings are unique factorization domains; most use the notion of *depth* that was used in the original proof of Auslander
and Buchsbaum. If $M$ is a finitely generated $R$-module, we generalize the notion
of $R$-sequence: the **depth** of $M$ is the maximal length of a sequence $r_1, \ldots, r_n$
in $R$ such that $x_1$ is regular on $M$, $x_2$ is regular on $M/(x_1)M$, $\ldots$, $x_n$ is regular on
$M/(x_1, \ldots, x_{n-1})M$. The depth of $M$ was originally called its *codimension* because
of the following result of Auslander and Buchsbaum.

**Theorem 10.200 (Codimension Theorem).** *Let $(R, \mathfrak{m}, k)$ be a local ring, and
let $M$ be a finitely generated $R$-module with $\mathrm{pd}(M) < \infty$. Then*

$$\mathrm{pd}(M) + \mathrm{depth}(M) = \mathrm{depth}(R).$$

*In particular, if $R$ is regular, then* $\mathrm{depth}(R) = \mathrm{D}(R)$ *and*

$$\mathrm{pd}(M) + \mathrm{depth}(M) = \mathrm{D}(R).$$

**Proof.** Eisenbud, *Commutative Algebra with a View Toward Algebraic Geometry*, p. 479. •

Nagata [A General Theory of Algebraic Geometry over Dedekind Rings II, *Amer. J. Math.* 80 (1958), 382–420] proved, using the Serre–Auslander–Buchsbaum Theorem, that if one knew that every regular local ring $R$ with $\mathrm{D}(R) = 3$ is a UFD, then it would follow that every regular local ring is a UFD.

Here is a sketch of the original proof of Auslander–Buchsbaum that every regular local ring $R$ is a unique factorization domain. They began with a standard result of Commutative Algebra.

**Lemma 10.201.** *If $R$ is a noetherian domain, then $R$ is a UFD if and only if every prime ideal of height 1 is principal.*

**Proof.** Let $R$ be a UFD, and let $\mathfrak{p}$ be a prime ideal of height 1. If $a \in \mathfrak{p}$ is nonzero, then $a = \pi_1^{e_1} \cdots \pi_n^{e_n}$, where the $\pi_i$ are irreducible and $e_i \geq 1$. Since $\mathfrak{p}$ is prime, one of the factors, say, $\pi_j \in \mathfrak{p}$. Of course, $R\pi_j \subseteq \mathfrak{p}$. But $R\pi_j$ is a prime ideal, by Proposition 5.16, so that $R\pi_j = \mathfrak{p}$, because $\mathrm{ht}(\mathfrak{p}) = 1$.

Conversely, since $R$ is noetherian, Lemma 5.17 shows that every nonzero nonunit in $R$ is a product of irreducibles, and so Proposition 5.16 says that it suffices to prove, for every irreducible $\pi \in R$, that $R\pi$ is a prime ideal. Choose a prime ideal $\mathfrak{p}$ that is minimal over $R\pi$. By the Principal Ideal Theorem, Theorem 10.173, we have $\mathrm{ht}(\mathfrak{p}) = 1$, and so the hypothesis gives $\mathfrak{p} = Ra$ for some $a \in R$. Therefore, $\pi = ua$ for some $u \in R$. Since $\pi$ is irreducible, we must have $u$ a unit, and so $R\pi = Ra = \mathfrak{p}$, as desired. •

In light of Nagata's result, Auslander and Buchsbaum could focus on the three-dimensional case.

**Lemma 10.202.** *If $(R, \mathfrak{m}, k)$ is a local ring with $\mathrm{D}(R) = 3$ and $\mathfrak{p} \neq \mathfrak{m}$ is a prime ideal in $R$, then*

$$\mathrm{pd}(\mathfrak{p}) \leq 1.$$

**Proof.** By hypothesis, there exists $x \in \mathfrak{m} - \mathfrak{p}$. Now $x$ is regular on $R/\mathfrak{p}$: if $x(r + \mathfrak{p}) = \mathfrak{p}$, then $xr \in \mathfrak{p}$ and $r \in \mathfrak{p}$, for $x \notin \mathfrak{p}$ and $\mathfrak{p}$ is prime. Therefore, $\mathrm{depth}(R/\mathfrak{p}) \geq 1$ and so $\mathrm{pd}(R/\mathfrak{p}) \leq 2$, by the Codimension Theorem. But there is an exact sequence $0 \to \mathfrak{p} \to R \to R/\mathfrak{p} \to 0$, which shows that $\mathrm{pd}(\mathfrak{p}) \leq 1$. •

Auslander and Buchsbaum showed that $\mathrm{pd}(\mathfrak{p}) = 1$ gives a contradiction, so that $\mathrm{pd}(\mathfrak{p}) = 0$; that is, $\mathfrak{p}$ is a projective, hence free, $R$-module. But any ideal in a domain $R$ that is free as an $R$-module must be principal. This completed the proof.

A second proof, not using Nagata's difficult proof, is based on the following criterion.

**Proposition 10.203.** *If $R$ is a noetherian domain for which every finitely gener-ated $R$-module has FFR, then $R$ is a unique factorization domain.*

**Proof.** This is Theorem 184 in Kaplansky, *Commutative Rings*. The proof uses a criterion for a domain to be a unique factorization domain (his Theorem 179), which involves showing that if a commutative ring $R$ has the property that every finitely generated $R$-module has FFR, then so does $R[x]$.  •

Unique factorization for regular local rings follows easily. If $D(R) = n$, then every finitely generated $R$-module $M$ has a projective resolution $0 \to P_n \to \cdots \to P_0 \to M \to 0$ in which every $P_i$ is finitely generated (Lemma 10.41). But (finitely generated) projective $R$-modules are free [Proposition 10.26], and so every finitely generated $R$-module has FFR.

We now give a complete (third) proof that every regular local ring is a UFD; we begin with two elementary lemmas.

**Lemma 10.204.** *Let $R$ be a noetherian domain, let $x \in R$ be a nonzero element with $Rx$ a prime ideal, and denote $S^{-1}R$ by $R_x$, where $S = \{x^n : n \geq 0\}$. Then $R$ is a UFD if and only if $R_x$ is a UFD for all such $x$.*

**Proof.** We leave necessity as a routine exercise for the reader. For sufficiency, assume that $R_x$ is a UFD. Let $\mathfrak{p}$ be a prime ideal in $R$ of height 1. If $x \in \mathfrak{p}$, then $Rx \subseteq \mathfrak{p}$ and, since $\mathrm{ht}(\mathfrak{p}) = 1$, we have $Rx = \mathfrak{p}$ (for $Rx$ is prime), and so $\mathfrak{p}$ is principal in this case. We may now assume that $x \notin \mathfrak{p}$; that is, $S \cap \mathfrak{p} = \varnothing$. It follows that $\mathfrak{p}R_x$ is a prime ideal in $R_x$ of height 1, and so it is principal, by hypothesis. There is some $a \in \mathfrak{p}$ and $n \geq 0$ with $\mathfrak{p}R_x = R_x(a/x^n) = R_x a$, for $x$ is a unit in $R_x$. We may assume that $a \notin Rx$. If $a = a_1 x$ and $a_1 \notin Rx$, then replace $a$ by $a_1$, for $R_x a = R_x a_1$. If $a_1 = a_2 x$ and $a_2 \notin Rx$, then replace $a_1$ by $a_2$, for $R_x a_1 = R_x a_2$. If this process does not stop, there are equations $a_m = a_{m+1} x$ for all $m \geq 1$, which give rise to an ascending sequence $Ra_1 \subseteq Ra_2 \subseteq \cdots$. Since $R$ is noetherian, $Ra_m = Ra_{m+1}$ for some $m$. Hence, $a_{m+1} = ra_m$ for some $r \in R$, and $a_m = a_{m+1}x = ra_m x$. Since $R$ is a domain, $1 = rx$; thus, $x$ is a unit, contradicting $Rx$ being a prime (hence, proper) ideal. Clearly, $Ra \subseteq \mathfrak{p}$; we claim that $Ra = \mathfrak{p}$. If $b \in \mathfrak{p}$, then $b = (r/x^m)a$ in $R_x$, where $r \in R$ and $m \geq 0$. Hence, $x^m b = ra$ in $R$. Choose $m$ minimal. If $m > 0$, then $ra = x^m b \in Rx$; since $Rx$ is prime, either $r \in Rx$ or $a \in Rx$. But $a \notin Rx$ since $S \cap \mathfrak{p} = \varnothing$, so that $r = xr'$. As $R$ is a domain, this gives $r'a = x^{m-1}b$, contradicting the minimality of $m$. We conclude that $m = 0$, and so $\mathfrak{p} = Ra$ is principal. Lemma 10.201 now shows that $R$ is a UFD.  •

The following lemma is true when the localizing ideal is prime; however, we will use it only in the case the ideal is maximal.

**Lemma 10.205.** *Let $R$ be a domain, and let $I$ be a nonzero projective ideal in $R$. If $\mathfrak{m}$ is a maximal ideal in $R$, then*

$$I_{\mathfrak{m}} \cong R_{\mathfrak{m}}.$$

**Proof.** Since $I$ is a projective $R$-module, $I_{\mathfrak{m}}$ is a projective $R_{\mathfrak{m}}$-module. As $R_{\mathfrak{m}}$ is a local ring, however, $I_{\mathfrak{m}}$ is a free $R_{\mathfrak{m}}$-module. But $I_{\mathfrak{m}}$ is an ideal in a domain $R_{\mathfrak{m}}$, and so it must be principal; that is, $I_{\mathfrak{m}} \cong R_{\mathfrak{m}}$. •

**Theorem 10.206 (Auslander–Buchsbaum).** *Every regular local ring* $(R, \mathfrak{m}, k)$ *is a unique factorization domain.*

**Proof. (Kaplansky)** The proof is by induction on the Krull dimension $\dim(R)$, the cases $n = 0$ ($R$ is a field) and $n = 1$ ($R$ is a DVR) being obvious (Exercise 10.79 on page 983). For the inductive step, choose $x \in \mathfrak{m} - \mathfrak{m}^2$. By Lemma 10.191, $R/Rx$ is a regular local ring with $\dim(R/Rx) < \dim(R)$; by Proposition 10.192, $R/Rx$ is a domain, and so $Rx$ is a prime ideal. It suffices, by Lemma 10.204, to prove that $R_x$ is a UFD (where $R_x = S^{-1}R$ for $S = \{x^n : n \geq 0\}$). Let $\mathfrak{P}$ be a prime ideal of height 1 in $R_x$; we must show that $\mathfrak{P}$ is principal. Define $\mathfrak{p} = \mathfrak{P} \cap R$ (since $R$ is a domain, $R_x \subseteq \mathrm{Frac}(R)$, so that the intersection makes sense). Since $R$ is a regular local ring, $\mathrm{D}(R) < \infty$, and so the $R$-module $\mathfrak{p}$ has a free resolution of finite length:

$$0 \to F_n \to F_{n-1} \to \cdots \to F_0 \to \mathfrak{p} \to 0.$$

Tensoring by $R_x$, which is a flat $R$-module (Theorem 10.30), gives a free $R_x$-resolution of $\mathfrak{P}$ (for $\mathfrak{P} = R_x\mathfrak{p}$):

$$(3) \qquad\qquad 0 \to F'_n \to F'_{n-1} \to \cdots \to F'_0 \to \mathfrak{P} \to 0,$$

where $F'_i = R_x \otimes_R F_i$.

We claim that $\mathfrak{P}$ is projective. By Proposition 10.181, it suffices to show that every localization $\mathfrak{P}_{\mathfrak{M}}$ is projective, where $\mathfrak{M}$ is a maximal ideal in $R_x$. Now $(R_x)_{\mathfrak{M}}$ is a localization of $R$, and so it is a regular local ring, by Corollary 10.202; its dimension is smaller than $D(R)$, and so it is a UFD, by induction. Now $\mathfrak{P}_{\mathfrak{M}}$, being a height 1 prime ideal in the UFD $(R_x)_{\mathfrak{M}}$, is principal. But principal ideals in a domain are free, hence projective, and so $\mathfrak{P}_{\mathfrak{M}}$ is projective. Therefore, $\mathfrak{P}$ is projective.

The exact sequence (3) "factors" into split short exact sequences. Since $\mathfrak{P}$ is projective, we have $F'_0 \cong \mathfrak{P} \oplus \Omega_0$, where $\Omega_0 = \ker(F'_0 \to \mathfrak{P})$. Thus, $\Omega_0$ is projective, being a summand of a free module, and so $F'_1 \cong \Omega_1 \oplus \Omega_0$, where $\Omega_1 = \ker(F'_1 \to F'_0)$. More generally, $F'_i \cong \Omega_i \oplus \Omega_{i-1}$ for all $i \geq 1$. Hence,

$$F'_0 \oplus F'_1 \oplus \cdots \oplus F'_n \cong (\mathfrak{P} \oplus \Omega_0) \oplus (\Omega_1 \oplus \Omega_0) \oplus \cdots .$$

Since projective modules over a local ring are free, we see that there are finitely generated free $R_x$-modules $Q$ and $Q'$ with

$$Q \cong \mathfrak{P} \oplus Q'.$$

Recall that $\mathrm{rank}(Q) = \dim_K(K \otimes_{R_x} Q)$, where $K = \mathrm{Frac}(R_x)$; now $\mathrm{rank}(\mathfrak{P}) = 1$ and $\mathrm{rank}(Q') = r$, say, so that $\mathrm{rank}(Q) = r + 1$.

We must still show that $\mathfrak{P}$ is principal. Now

$$\bigwedge^{r+1} Q \cong \bigwedge^{r+1}(\mathfrak{P} \oplus Q').$$

Since $Q$ is free of rank $r+1$, the Binomial Theorem (Theorem 8.117) gives $\bigwedge^{r+1} Q \cong R_x$. On the other hand, Theorem 8.120 gives

$$(4) \qquad \bigwedge^{r+1}(\mathfrak{P} \oplus Q') \cong \sum_{i=0}^{r+1}\left(\bigwedge^i \mathfrak{P} \otimes_{R_x} \bigwedge^{r+1-i} Q'\right).$$

We claim that $\bigwedge^i \mathfrak{P} = \{0\}$ for all $i > 1$. By Lemma 10.205, we have $\mathfrak{P}_{\mathfrak{M}} \cong (R_x)_{\mathfrak{M}}$ for every maximal ideal $\mathfrak{M}$ in $R_x$. Now Exercise 10.25 on page 899 gives

$$\left(\bigwedge^i \mathfrak{P}\right)_{\mathfrak{M}} \cong \bigwedge^i (\mathfrak{P}_{\mathfrak{M}}) \cong \bigwedge^i (R_x)_{\mathfrak{M}}$$

for all maximal ideals $\mathfrak{M}$ and all $i$. But $\bigwedge^i (R_x)_{\mathfrak{M}} = \{0\}$ for all $i > 1$ (by the Binomial Theorem or by the simpler Corollary 8.115), so that Proposition 10.38 gives $\bigwedge^i \mathfrak{P} = \{0\}$ for all $i > 1$.

We have just seen that most of the terms in (4) are $\{0\}$; what survives is:

$$\bigwedge^{r+1}(\mathfrak{P} \oplus Q') \cong \left(\bigwedge^0 \mathfrak{P} \otimes_{R_x} \bigwedge^{r+1} Q'\right) \oplus \left(\bigwedge^1 \mathfrak{P} \otimes_{R_x} \bigwedge^r Q'\right).$$

But $\bigwedge^{r+1} Q' = \{0\}$ and $\bigwedge^r Q' \cong R_x$, because $Q'$ is free of rank $r$. Therefore, $\bigwedge^{r+1}(\mathfrak{P} \oplus Q') \cong \mathfrak{P}$. Since $\mathfrak{P} \cong \bigwedge^{r+1}(\mathfrak{P} \oplus Q') \cong \bigwedge^{r+1} Q \cong R_x$, we have $\mathfrak{P} \cong R_x$ principal. Thus, $R_x$, and hence $R$, is a UFD. •

Having studied localization, we turn, briefly, to globalization, merely describing its setting. To a commutative noetherian ring $R$, we have associated a family of local rings $R_{\mathfrak{p}}$, one for each prime ideal $\mathfrak{p}$, and an $R$-module $M$ has localizations $M_{\mathfrak{p}} = R_{\mathfrak{p}} \otimes_R M$. These data are assembled into a **sheaf** over $\mathrm{Spec}(R)$, where $\mathrm{Spec}(R)$ is equipped with the Zariski topology. There are two ways to view a sheaf. One way resembles covering spaces in Algebraic Topology. Equip the disjoint union

$$E(M) = \bigcup_{\mathfrak{p} \in \mathrm{Spec}(R)} M_{\mathfrak{p}}$$

with a topology so that the **projection** $\pi \colon E(M) \to \mathrm{Spec}(R)$, defined by $\pi(e) = \mathfrak{p}$ if $e \in M_{\mathfrak{p}}$, is a local homeomorphism. All such sheaves form a category $\mathbf{Sh}(R)$, where a morphism $\varphi \colon E(M) \to E(M')$ is a continuous map with $\varphi|M_{\mathfrak{p}} \colon M_{\mathfrak{p}} \to M'_{\mathfrak{p}}$ an $R_{\mathfrak{p}}$-map for all $\mathfrak{p}$. Now $\mathbf{Sh}(R)$ is an abelian category (Rotman, *An Introduction to Homological Algebra*, p. 309), which has enough injectives (Ibid, p. 314). A **section** of $E(M)$ over an open set $U \subseteq \mathrm{Spec}(R)$ is a continuous function $s \colon U \to E(M)$ with $\pi s = 1_U$. The family $\Gamma(U, E(M))$ of all sections over $U$ is an abelian group. Sections give rise to the second way of viewing a sheaf, for $\Gamma(\square, E(M))$ is a presheaf of abelian groups. We call $\Gamma(\mathrm{Spec}(R), E(M))$ the **global sections** of $E(M)$. Now global sections $\Gamma(\mathrm{Spec}(R), \square) \colon \mathbf{Sh}(R) \to \mathbf{Ab}$ is a left exact additive functor (Ibid, p. 378), and its derived functors $H^n(\mathrm{Spec}(R), \square)$ are called **sheaf cohomology**. These cohomology groups are the most important tool in globalizing. We strongly recommend the article by Serre, Faisceaux algèbriques cohérents, *Annals of Math.* 61 (1955), pp. 97–278, for a lucid discussion.

# Exercises

**10.78.** If $(R, \mathfrak{m}, k)$ is a noetherian local ring and $B$ is a finitely generated $R$-module, prove that
$$\operatorname{depth}(B) = \min\{i : \operatorname{Ext}_R^i(k, B) \neq \{0\}\}.$$

* **10.79.** Let $R$ be a regular local ring.

(i) Prove that $R$ is a field if and only if $\dim(R) = 0$.

(ii) Prove that $R$ is a DVR if and only if $\dim(R) = 1$.

* **10.80.** (i) Let $(R, \mathfrak{m}, k)$ be a noetherian local ring, and let $x \in R$ be a regular element; i.e., $x$ is not a zero-divisor. If $x_1 + (x), \ldots, x_s + (x)$ is an $(R/(x))$-sequence, prove that $x, x_1, \ldots, x_s$ is an $R$-sequence.

(ii) Let $R$ be a commutative ring. If $x_1, \ldots, x_d$ is an $R$-sequence, prove that the cosets $x_2 + (x_1), \ldots, x_d + (x_1)$ form an $(R/(x_1))$-sequence.

**10.81.** Let $R$ be a noetherian (commutative) ring with Jacobson radical $J = J(R)$. If $B$ is a finitely generated $R$-module, prove that
$$\bigcap_{n \geq 1} J^n B = \{0\}.$$
Conclude that if $(R, \mathfrak{m}, k)$ is a noetherian local ring, then $\bigcap_{n \geq 1} \mathfrak{m}^n B = \{0\}$.

**Hint.** Let $D = \bigcap_{n \geq 1} J^n B$, observe that $JD = D$, and use Nakayama's Lemma.

**10.82.** Use the Rees Lemma to prove a weaker version of Proposition 10.195: if $(R, \mathfrak{m}, k)$ is a regular local ring, then $D(R) \geq V(R) = \dim(R)$.

**Hint.** If $\operatorname{Ext}_R^d(k, R) \neq \{0\}$, then $D(R) > d - 1$; that is, $D(R) \geq d$. Let $\mathfrak{m} = (x_1, \ldots, x_d)$, where $x_1, \ldots, x_d$ is an $R$-sequence. Then
$$\operatorname{Ext}_R^d(k, R) \cong \operatorname{Ext}_{R/(x_1)}^{d-1}(k, R/(x_1))$$
$$\cong \operatorname{Ext}_{R/(x_1, x_2)}^{d-2}(k, R/(x_1, x_2)) \cong \cdots$$
$$\cong \operatorname{Ext}_k^0(k, k) \cong \operatorname{Hom}_k(k, k) \cong k \neq \{0\}.$$

**10.83.** If $k$ is a field, prove that the ring of formal power series $k[[x_1, \ldots, x_n]]$ is a UFD.

# Bibliography

Adem, A., and Milgram, R. J., *Cohomology of Finite Groups*, Springer-Verlag, Berlin, 1994.

Albert, A. A., editor, *Studies in Modern Algebra*, MAA Studies in Mathematics, Vol. 2, Mathematical Association of America, Washington, 1963.

Anderson, F. W., and Fuller, K. R., *Rings and Categories of Modules*, 2nd ed., Springer-Verlag, Berlin, 1992.

Artin, E., *Galois Theory*, 2nd ed., Notre Dame, 1955; Dover reprint, Mineola, 1998.

————, *Geometric Algebra*, Interscience Publishers, New York, 1957.

Artin, E., Nesbitt, C. J., and Thrall, R. M., *Rings with Minimum Condition*, University of Michigan Press, Ann Arbor, 1968.

Aschbacher, M., *Finite Group Theory*, Cambridge University Press, Cambridge, 1986.

Atiyah, M., and Macdonald, I. G., *Introduction to Commutative Algebra*, Addison–Wesley, Reading, 1969.

Bass, H., *Algebraic K-Theory*, W. A. Benjamin, New York, 1968.

Becker, T., and Weispfenning, V., *Gröbner Bases: A Computational Approach to Commutative Algebra*, Springer-Verlag, New York, 1993.

Biggs, N. L., *Discrete Mathematics*, 2nd ed., Oxford University Press, 2002.

Birkhoff, G., and Mac Lane, S., *A Survey of Modern Algebra*, 4th ed., Macmillan, New York, 1977.

Blyth, T. S., *Module Theory; An Approach to Linear Algebra*, Oxford University Press, 1990.

Borevich, Z. I., and Shafarevich, I. R., *Number Theory*, Academic Press, Orlando, 1966.

Bott, R., and Tu, L. W., *Differential Forms in Algebraic Topology*, Springer-Verlag, New York, 1982.

Bourbaki, N., *Elements of Mathematics; Algebra* I; *Chapters* 1–3, Springer-Verlag, New York, 1989.

————, *Elements of Mathematics; Commutative Algebra*, Addison–Wesley, Reading, 1972.

Brown, K. S., *Cohomology of Groups*, Springer-Verlag, Berlin, 1982.

Bruns, W., and Herzog, J., *Cohen–Macaulay Rings*, Cambridge University Press, 1993.

Buchberger, B., and Winkler, F., editors, *Gröbner Bases and Applications*, LMS Lecture Note Series 251, Cambridge University Press, 1998.

Burnside, W., *The Theory of Groups of Finite Order*, 2nd ed., Cambridge University Press, 1911; Dover reprint, Mineola, 1955.

Caenepeel, S., *Brauer Groups, Hopf Algebras, and Galois Theory*, Kluwer, Dordrecht, 1998.

Cajori, F., *A History of Mathematical Notation*, Open Court, 1928; Dover reprint, Mineola, 1993.

Carmichael, R., *An Introduction to the Theory of Groups*, Ginn, New York, 1937.

Carter, R., *Simple Groups of Lie Type*, Cambridge University Press, Cambridge, 1972.

Cassels, J. W. S., and Fröhlich, A., *Algebraic Number Theory*, Thompson Book Co., Washington, D.C., 1967.

Chase, S. U., Harrison, D. K., and Rosenberg, A., *Galois Theory and Cohomology of Commutative Rings*, Mem. Amer. Math. Soc. No. 52, Providence, 1965, pp. 15–33.

Cohen, D. E., *Groups of Cohomological Dimension* 1, Lecture Notes in Mathematics, Vol. 245, Springer-Verlag, New York, 1972.

Cohn, P. M., *Free Rings and Their Relations*, Academic Press, New York, 1971.

Collins, D. J., Grigorchuk, R. I., Kurchanov, P. F., and Zieschang, H., *Combinatorial Group Theory and Applications to Geometry*, 2nd ed., Springer-Verlag, New York, 1998.

Conway, J. H., Curtis, R. T., Norton, S. P., Parker, R. A., and Wilson, R. A., *ATLAS of Finite Groups*, Oxford University Press, 1985.

Cox, D., Little, J., and O'Shea, D., *Ideals, Varieties, and Algorithms*, 2nd ed., Springer-Verlag, New York, 1997.

Coxeter, H. S. M., and Moser, W. O. J., *Generators and Relations for Discrete Groups*, Springer-Verlag, New York, 1972.

Curtis, C. W., and Reiner, I., *Representation Theory of Finite Groups and Associative Algebras*, Interscience, New York, 1962.

Dauns, J., *Modules and Rings*, Cambridge University Press, 1994.

Dieudonné, J., *La Géometrie des Groupes Classiques*, Springer-Verlag, Berlin, 1971.

Dixon, J. D., du Sautoy, M. P. F., Mann, A., and Segal, D., *Analytic Pro-p Groups*, 2nd ed., Cambridge University Press, 2003.

Dlab, V., and Ringel, C. M., *Indecomposable representations of graphs and algebras*, Mem. Amer. Math.Soc. 6 (1976), no. 173.

Dornhoff, L., *Group Representation Theory, Part* A, *Ordinary Representation Theory*, Marcel Dekker, New York, 1971.

Drozd, Yu. A., and Kirichenko, V. V., *Finite-Dimensional Algebras*, Springer-Verlag, New York, 1994.

Dummit, D. S., and Foote, R. M., *Abstract Algebra*, 2nd ed., Prentice Hall, Upper Saddle River, 1999.

Eisenbud, D., *Commutative Algebra with a View Toward Algebraic Geometry*, Springer-Verlag, New York, 1995.

Evens, L., *The Cohomology of Groups*, Oxford Mathematical Monographs, Oxford University Press, 1991.

Farb, B., and Dennis, R. K., *Noncommutative Algebra*, Springer-Verlag, New York, 1993.

Feit, W., *Characters of Finite Groups*, W. A. Benjamin, New York, 1967.

Fröhlich, A., and Taylor, M. J., *Algebraic Number Theory*, Cambridge Studies in Advanced Mathematics 27, Cambridge University Press, 1991.

Fuchs, L., *Infinite Abelian Groups* I, Academic Press, Orlando, 1970.

———, *Infinite Abelian Groups* II, Academic Press, Orlando, 1973.

Fulton, W., *Algebraic Curves*, Benjamin, New York, 1969.

———, *Algebraic Topology; A First Course*, Springer-Verlag, New York, 1995.

Gaal, L., *Classical Galois Theory with Examples*, 4th ed., Chelsea, American Mathematical Society, Providence, 1998.

Gelfand, S. I., and Manin, Y. I., *Methods of Homological Algebra*, 2nd ed., Springer-Verlag, New York, 2003.

Gorenstein, D., Lyons, R. and Solomon, R., *The Classification of the Finite Simple Groups*, Math. Surveys and Monographs, Vol. 40, American Mathematical Society, Providence, 1994.

Greub, W. H., *Multilinear Algebra*, Springer-Verlag, New York, 1967.

Hadlock, C., *Field Theory and Its Classical Problems*, Carus Mathematical Monographs, No. 19, Mathematical Association of America, Washington, 1978.

Hahn, A. J., *Quadratic Algebras, Clifford Algebras, and Arithmetic Witt Groups*, Universitext, Springer-Verlag, New York, 1994.

Hall, M., Jr., *The Theory of Groups*, Macmillan, New York, 1959.

Hardy, G. H., and Wright, E. M., *An Introduction to the Theory of Numbers*, 4th ed., Oxford University Press, 1960.

Harris, J., *Algebraic Geometry*, Springer-Verlag, New York, 1992.

Hartshorne, R., *Algebraic Geometry*, Springer-Verlag, New York, 1977.

Herrlich, H., and Strecker, G. E., *Category Theory. An Introduction*, Allyn & Bacon, Boston, 1973.

Herstein, I. N., *Topics in Algebra*, 2nd ed., Wiley, New York, 1975.

———, *Noncommutative Rings*, Carus Mathematical Monographs, No. 15, Mathematical Association of America, Washington, 1968.

Higgins, P. J., *Notes on Categories and Groupoids*, van Nostrand–Reinhold, London, 1971.

Humphreys, J. E., *Introduction to Lie Algebras and Representation Theory*, Springer-Verlag, New York, 1972.

Huppert, B., *Character Theory of Finite Groups*, de Gruyter, Berlin, 1998.

———, *Endliche Gruppen* I, Springer-Verlag, New York, 1967.

Hurewicz, W., and Wallman, H., *Dimension Theory*, Princeton University Press, Princeton, 1948.

Isaacs, I. M., *Character Theory of Finite Groups*, Academic Press, San Diego, 1976.

———, *Algebra, A Graduate Course*, Brooks/Cole Publishing, Pacific Grove, 1994.

Jacobson, N., *Basic Algebra* I, Freeman, San Francisco, 1974.

———, *Basic Algebra* II, Freeman, San Francisco, 1980.

———, *Finite-Dimensional Division Algebras over Fields*, Springer-Verlag, New York, 1996.

———, *Lie Algebras*, Interscience Tracts, Number 10, Wiley, New York, 1962.

———, *Structure of Rings*, Colloquium Publications, 37, American Mathematical Society, Providence, 1956.

Johnson, D. L., *Topics in the Theory of Group Presentations*, Cambridge University Press, 1980.

Kaplansky, I., *Commutative Rings*, University of Chicago Press, 1974.

———, *Fields and Rings*, 2nd ed., University of Chicago Press, 1972.

———, *Infinite Abelian Groups*, 2nd ed., University of Michigan Press, Ann Arbor, 1969.

———, *Linear Algebra and Geometry; a Second Course*, Allyn & Bacon, Boston, 1969.

———, *Set Theory and Metric Spaces*, Chelsea, American Mathematical Society, Providence, 1977.

King, R. B., *Beyond the Quartic Equation*, Birkhäuser, Boston, 1996.

Kostrikin, A. I., and Shafarevich, I. R. (editors), *Algebra IX. Finite Groups of Lie Type; Finite-Dimensional Division Algebras*, Encyclopaedia of Mathematical Sciences, 77, Springer-Verlag, New York, 1996.

Kurosh, A. G., *The Theory of Groups*, Chelsea, New York, 1956.

Lam, T. Y., *The Algebraic Theory of Quadratic Forms*, Benjamin, Reading, 1973, 2nd revised printing, 1980.

———, *A First Course in Noncommutative Rings*, Springer-Verlag, New York, 1991.

———, *Lectures on Modules and Rings*, Springer-Verlag, New York, 1999.

Lang, S., *Algebra*, Addison–Wesley, Reading, 1965.

Lyndon, R. C., and Schupp, P. E., *Combinatorial Group Theory*, Springer-Verlag, New York, 1977.

Macdonald, I. G., *Algebraic Geometry; Introduction to Schemes*, Benjamin, New York, 1968.

Mac Lane, S., *Categories for the Working Mathematician*, Springer-Verlag, New York, 1971.

———, *Homology*, Springer-Verlag, New York, 3rd corrected printing, 1975.

Mac Lane, S., and Birkhoff, G., *Algebra*, MacMillan, New York, 1967.

Malle, G., and Matzat, B., *Inverse Galois Theory*, Springer-Verlag, New York, 1999.

Matsumura, H., *Commutative Ring Theory*, Cambridge University Press, 1986.

McCleary, J., *User's Guide to Spectral Sequences*, Publish or Perish, Wilmington, 1985.

McConnell, J. C., and Robson, J. C., *Noncommutative Noetherian Rings*, Wiley, New York, 1987.

McCoy, N. H., and Janusz, G. J., *Introduction to Modern Algebra*, 5th ed., W. C. Brown Publishers, Dubuque, Iowa, 1992.

Milnor, J., *Introduction to Algebraic K-Theory*, Annals of Mathematical Studies, No. 72, Princeton University Press, 1971.

Mitchell, B., *Theory of Categories*, Academic Press, New York, 1965.

Montgomery, S., and Ralston, E. W., *Selected Papers on Algebra*, Raymond W. Brink Selected Mathematical Papers, Vol. 3, Mathematical Association of America, Washington, 1977.

Mumford, D., *The Red Book of Varieties and Schemes*, Lecture Notes in Mathematics, 1358, Springer-Verlag, New York, 1988.

Munkres, J. R., *Topology, A First Course*, Prentice Hall, Upper Saddle River, 1975.

———, *Elements of Algebraic Topology*, Addison–Wesley, Reading, 1984.

Neukirch, J., Schmidt, A., and Wingberg, K., *Cohomology of Number Fields*, Grundlehren der mathematischen Wissenschaften, Vol. 323, Springer-Verlag, New York, 2000.

Niven, I., and Zuckerman, H. S., *An Introduction to the Theory of Numbers*, Wiley, New York, 1972.

Northcott, D. G., *Ideal Theory*, Cambridge University Press, 1953.

Ol'shanskii, A. Y., *Geometry of Defining Relations in Groups*, Kluwer Academic Publishers, Dordrecht, 1991.

O'Meara, O. T., *Introduction to Quadratic Forms*, Springer-Verlag, New York, 1971.

Orzech, M., and Small, C., *The Brauer Group of Commutative Rings*, Lecture Notes in Pure and Applied Mathematics, Marcel Dekker, New York, 1975.

Osborne, M. S., *Basic Homological Algebra*, Springer-Verlag, New York, 2000.

Pollard, H., *The Theory of Algebraic Numbers*, Carus Mathematical Monographs No. 9, Mathematical Association of America, Washington, 1950.

Procesi, C., *Rings with Polynomial Identities*, Marcel Dekker, New York, 1973.

Reid, M., and Szendrői, B., *Geometry and Topology*, Cambridge University Press, 2005.

Reiner, I., *Maximal Orders*, Academic Press, London, 1975; Oxford University Press, 2003.

Robinson, D. J. S., *A Course in the Theory of Groups*, 2nd ed., Springer-Verlag, New York, 1996.

Roman, S., *Field Theory*, 2nd ed., Graduate Texts in Mathematics, 158, Springer, New York, 2006.

Rosenberg, J., *Algebraic K-Theory and Its Applications*, Springer-Verlag, New York, 1994.

Rotman, J. J., *A First Course in Abstract Algebra*, 3rd ed., Prentice Hall, Upper Saddle River, NJ, 2006.

———, *Galois Theory*, 2nd ed., Springer-Verlag, New York, 1998.

———, *An Introduction to Homological Algebra*, 2nd ed., Springer, New York, 2009.

———, *An Introduction to the Theory of Groups*, 4th ed., Springer-Verlag, New York, 1995.

———, *Journey into Mathematics*, Prentice Hall, Upper Saddle River, 1998, Dover reprint, Mineola, 2007.

Rowen, L. H., *Polynomial Identities in Ring Theory*, Academic Press, New York, 1980.

———, *Ring Theory*, Vols. I, II, Academic Press, Boston, 1988.

Samuel, P. *Algebraic Theory of Numbers*, Houghton Mifflin, Boston, 1970.

Serre, J.-P., *Algèbre Locale: Multiplicités*, Lecture Notes in Mathematics 11, 3rd ed., Springer-Verlag, New York, 1975; English transl., *Local Algebra*, Springer Monographs in Mathematics, Springer–Verlag, 2000.

———, *Corps Locaux*, Hermann, Paris, 1968; English transl., *Local Fields*, Graduate Texts in Mathematics, 67, Springer–Verlag, 1979.

———, *Topics in Galois Theory*, Jones and Bartlett, Boston, 1992.

———, *Trees*, Springer-Verlag, New York, 1980.

Shafarevich, I. R., *Algebra I. Basic Notions of Algebra*, Encyclopaedia of Mathematical Sciences, 11, Springer-Verlag, Berlin, 1990.

Sims, C. C., *Computation with Finitely Presented Groups*, Cambridge University Press, 1994.

Small, C., *Arithmetic of Finite Fields*, Monographs and Textbooks in Pure and Applied Mathematics 148, Marcel Dekker, Inc., New York, 1991.

Stewart, I., *Galois Theory*, 3rd ed., Chapman & Hall/CRC, Boca Raton, 2004.

Stillwell, J., *Classical Topology and Combinatorial Group Theory*, 2nd ed., Springer-Verlag, New York, 1993.

———, J., *Mathematics and Its History*, Springer-Verlag, New York, 1989.

Strade, H., *Simple Lie Algebras over Fields of Positive Characteristic* I, de Gruyter, Berlin, 2004.

Suzuki, M., *Group Theory* I, Springer-Verlag, New York, 1982.

Tignol, J.-P., *Galois' Theory of Algebraic Equations*, World Scientific Publishing Co., Inc., River Edge, 2001.

van der Waerden, B. L., *Geometry and Algebra in Ancient Civilizations*, Springer-Verlag, New York, 1983.

———, *A History of Algebra*, Springer-Verlag, New York, 1985.

———, *Modern Algebra*, Vols. I, II, 4th ed., Ungar, New York, 1966.

———, *Science Awakening*, Wiley, New York, 1963.

Weibel, C., *An Introduction to Homological Algebra*, Cambridge University Press, 1994.

Weyl, H., *The Classical Groups; Their Invariants and Representations*, Princeton, 1946.

Weiss, E., *Cohomology of Groups*, Academic Press, Orlando, 1969.

Zariski, O., and Samuel, P., *Commutative Algebra* I, van Nostrand, Princeton, 1958.

———, *Commutative Algebra* II, van Nostrand, Princeton, 1960.

# Index